Fuzzy Techniques for Decision Making 2018

Fuzzy Techniques for Decision Making 2018

Special Issue Editor

José Carlos R. Alcantud

MDPI • Basel • Beijing • Wuhan • Barcelona • Belgrade • Manchester • Tokyo • Cluj • Tianjin

Special Issue Editor
José Carlos R. Alcantud
University of Salamanca
Spain

Editorial Office
MDPI
St. Alban-Anlage 66
4052 Basel, Switzerland

This is a reprint of articles from the Special Issue published online in the open access journal *Symmetry* (ISSN 2073-8994) (available at: https://www.mdpi.com/journal/symmetry/special_issues/Fuzzy_Techniques_Decision_Making_2018).

For citation purposes, cite each article independently as indicated on the article page online and as indicated below:

LastName, A.A.; LastName, B.B.; LastName, C.C. Article Title. *Journal Name* **Year**, *Article Number*, Page Range.

ISBN 978-3-03936-415-2 (Hbk)
ISBN 978-3-03936-416-9 (PDF)

Contents

About the Special Issue Editor

José Carlos R. Alcantud, Ph.D., completed his MSc in Mathematics in 1991 at the University of Valencia, Spain, and his Ph.D. in Mathematics in 1996 at the University of Santiago de Compostela, Spain. He is currently a professor at, and the Director of, the Department of Economics and Economic History at the University of Salamanca, Spain. He has authored over 100 papers on topics including mathematical economics, social choice, fuzzy mathematics and soft computing. He serves on the editorial board of Symmetry and Global & Local Economic Review, and the advisory board of Sci. He has been a reviewer for more than 50 SCI journals, including *IEEE Transactions on Fuzzy Systems, Information Sciences, Information Fusion, Expert Systems with Applications, Fuzzy Sets and Systems, IEEE Access*, etc. As project leader, he has directed seven research projects at the national or provincial level. He has also been a member of an EU-funded research project.

Preface to "Fuzzy Techniques for Decision Making 2018"

This book contains the successful invited submissions for the Special Issue of Symmetry on the subject of "Fuzzy Techniques for Decision Making 2018". We invited contributions addressing novel techniques and tools for decision making (e.g., group or multi-criteria decision making), with notions that overcome the problem of finding the membership degree of each element in Zadeh's original model. We were able to attract interesting articles in a variety of setups as well as applications. As a result, this Special Issue includes some novel techniques and tools for making decisions, such as—instrumental tools for analysis, like the approximation of ordered fuzzy numbers or distance measures and aggregation operators in various settings, novel contributions to methodologies, like programming-based algorithms or TOPSIS, new methodologies for hybrid models which are inclusive of new models (picture hesitant fuzzy sets) and theoretical novelties (intertemporal choices in the context of fuzzy soft sets), applications to assessment of enterprise performance, (green) supplier selection, risk level evaluation, arctic route planning, measurement of customer loyalty, contractors' bidding decisions, automatic parking methods, emergency decisions, selection of mine ventilation systems, or medical diagnosis. The response to our call for papers had the following statistics: submissions (63); publications (38); rejections (25); article types: research article (38). The published submissions are related to various settings, such as linguistic term sets, fuzzy soft sets, hesitant fuzzy sets, (fuzzy) soft rough sets, and neutrosophic sets, as well as other hybrid models. I have found the edition and selections of papers for this book very inspiring and rewarding. I also thank the editorial staff and reviewers for their efforts and help during the process.

José Carlos R. Alcantud
Special Issue Editor

symmetry

MDPI

Article

A Programming-Based Algorithm for Probabilistic Uncertain Linguistic Intuitionistic Fuzzy Group Decision-Making

Kaixin Gongand Chunfang Chen *

Department of Mathematics, Nanchang University, Nanchang 330031, China; gkxflw@163.com
* Correspondence:ccfygd@sina.com

Received: 7 January 2019; Accepted: 11 February 2019; Published: 15 February 2019

Abstract: As an effective tool to express the subjective preferences of decision makers, the linguistic term sets (LTS) have been widely used in group decision-making (GDM) problems, such as hesitant fuzzy LTS, linguistic hesitant fuzzy sets, probabilistic LTS, etc. However, due to the increasing complexity of practical decision-making (DM) problems, LTS still has a lot of room to expand in fuzzy theory. Qualitative uncertainty information in the application of GDM is yet to be improved. Therefore, in order to improve the applicability of linguistic terms in DM problems, a probabilistic uncertain linguistic intuitionistic fuzzy set (PULIFS) that can fully express the decision-maker's (DM's) evaluation information is first proposed. To improve the rationality of DM results, we give a method for determining individual weights in the probabilistic uncertain linguistic intuitionistic fuzzy preference relation (PULIFPR) environment. In addition, we present two consistency definitions of PULIFPR to reflect both the assessment information and risk attitudes of decision makers. Subsequently, a series of goal programming models (GPMs) are established, which effectively avoid the consistency check and correction process of existing methods. Finally, the developed method is applied to an empirical example concerning the selection of a virtual reality (VR) project. The advantages of the proposed method are demonstrated by comparative analysis.

Keywords: group decision-making (GDM); probabilistic uncertain linguistic intuitionistic fuzzy preference relation (PULIFPR); consistency; goal programming model (GPM); risk preference

MSC: 34A08; 34B10

---

1. Introduction

Because of the inherent subjective ambiguity of human thinking and the complexity of practical decision-making (DM) problems, the use of qualitative information is almost an indispensable link in DM. As the most commonly used qualitative information expression tool, linguistic terms (LT) have been extensively studied by scholars. Since Zadeh [1] proposed linguistic variables in 1975, various extended forms of LT have been proposed to model qualitative information and improve its calculation. In order to have a general understanding of these extended LT, we will present the general development process of LT in the form of Table 1.

It is easy to see from the table 1 that the development of LT can be mainly divided into two stages. The first stage is some traditional linguistic models, whose main research object is single LT. The second stage is the complex linguistic expression stage, whose linguistic information expression form is generally more than one LT or implied multiple linguistic information. In addition, it is easy to find that the LT of the later stage mostly introduces probability to comprehensively reflect the subjective uncertainty of decision makers (DMs) and the randomness of objective existence. All of these proposed sets give expression methods of qualitative information from different perspectives, and all of them have been applied reasonably in the DM problem. However, the qualitative information expressed by these sets has different degrees of defects. For example, although probabilistic uncertain linguistic term sets (PULTS) expresses the DM's preference information and its probability distribution, it fails to consider the DM's non-preference information, while linguistic intuitionistic fuzzy sets (LIFS) only expressed the subjective hesitation of DMs from the perspective of preference and non-preference, failing to consider the probability distribution of its information. Therefore, in order to improve the expression of qualitative information and promote the use of LT in DM problems, this paper further proposes a probabilistic uncertain linguistic intuitionistic fuzzy set (PULIFS) based on the above research, which integrates the advantages of LIFS and PULTS.

Table 1. A brief history of the types of linguistic terms.

	Year	Event
Traditional linguistic models	1975	Zadeh proposes the linguistic variable and introduced the fuzzy linguistic approach [1–3].
	1981	Yager presents an ordered structure model [4].
	1988	Degani and Bortolan present the semantic model [5].
	1993	Delgado and Verdegay propose the symbolic model [6].
	2000	Herrera and Martinez introduce the two-tuple linguistic model [7].
	2004	Xu defines the virtual linguistic model [8].
Complex linguistic expression	2004	Xu introduces the uncertain linguistic term (ULT) [9].
	2012	Rodriguez et al. present the concept of hesitant fuzzy linguistic term sets (HFLTS) [10].
	2014	Meng et al. propose the linguistic hesitant fuzzy sets (LHFS) [11].
	2014	Zhang gives the concept of linguistic intuitionistic fuzzy sets (LIFS) [12].
	2015	Ye presents the single-valued neutrosophic linguistic sets (SVNLS) [13].
	2016	Pang et al. present the probabilistic linguistic term sets (PLTS) [14].
	2107	Lin et al. define the probabilistic uncertain linguistic term sets (PULTS) [15].
	2018	Bai et al. present the interval-valued probabilistic linguistic term sets (IVPLTS) [16].
	2018	Zhang et al. propose the dual hesitant fuzzy linguistic term sets (DHFLTS) [17].

For example, when a decision team needs to evaluate and compare some alternatives, due to the complexity of the actual decision-making environment, the decision-makers can only provide qualitative preference and non-preference information based on the linguistic term set (LTS) $S = \{s_0$: extremely poor, s_1: very poor, s_2: poor, s_3: slightly poor, s_4: fair, s_5: slightly good, s_6: good, s_7: very good, s_8: extremely good$\}$. Among them, 40% of the DMs gave preference information between the very poor and the slightly poor, and the non-preference information was between the fair and the slightly good. While 60% of DMs gave preference information between fair and good, and non-preference information was between very poor and poor. Then the preference information given by this decision team can be represented by PULIFS as

$$P = \{\langle([s_1, s_3], [s_4, s_5]), 0.4\rangle, \langle([s_4, s_6], [s_1, s_2]), 0.6\rangle\}$$

From the above example, it can be seen that PULIFSs not only expresses the qualitative preference and non-preference information of DMs, but also provides flexible linguistic selection space for DMs and gives the probability distribution information of an uncertain linguistic. Moreover, the above example is only one application case of PULIFS proposed, besides, PULIFS can also be used to express individual preference information. So it is natural that we want to apply PULIFS to group decision-making (GDM) problems to compensate for the application limitations of existing sets, thus improving the application of qualitative information in fuzzy theory. This is the first focus of this paper.

Considering the cognitive uncertainty and fuzziness of DMs in complex decision-making environment, the application of uncertainty theory in decision-making has been widely studied. For example, Pamucar et al. [18] combined with linguistic neutrosophic numbers presented the selection method of power generation technology, and Liu et al. [19] established the selection model of transportation service provider with single valued neutrosophic number. In addition, preference relation (PR) has been widely used in GDM as an effective tool to express DMs' preferences over alternatives. Its main types include fuzzy PR [20], multiplicative PR [21] and linguistic PR (LPR) [22]. On this basis, many forms of preference relations have been proposed, such as interval fuzzy PR [23], interval multiplicative PR [24], intuitionistic PR [25], intuitionistic multiplicative PR [26], linguistic intuitionistic PR [27], etc. These PRs all express the DM's preference information in different forms from different perspectives. However, the application of existing PRs in GDM have the following defects:

(1) Most of the PRs fail to reflect the distribution of information given by DMs.
(2) Most studies on PRs ignore the information that cannot be grasped by DMs or fail to take into account information loss caused by certain objective factors.
(3) In the process of solving the priority weights, most of the GPMs only consider the principle of minimum consistency deviation and ignore the risk attitude of decision makers, which may result in the loss of original information and reduce the rationality of the ranking results.
(4) Almost all methods, none can guarantee the consistency of PRs in the process of solving priority weights. They all need to test and improve the consistency of PRs, which greatly reduces the accuracy of the results.

Therefore, in order to make up for the above defects of the existing methods, this paper further proposes probabilistic uncertain linguistic intuitionistic fuzzy preference relation (PULIFPR) based on the excellent nature of PULIFS proposed. To ensure the reasonable application of PR in GDM, we divide the uncertain information represented by PULIFPR into vagueness uncertain information and non-vagueness uncertain information, and its consistency is studied from two spatial dimensions respectively. Among them, non-vagueness uncertain information refers to some relevant information held by the decision maker for the alternatives to be compared. While vagueness uncertain information refers to the decision information that cannot be given by decision makers due to lack of relevant experience and knowledge, or the information loss caused by some objective factors. The non-vagueness uncertainty information in PULIFPR is mainly presented in the form of qualitative preference and non-preference information. For ease of understanding, the relationship between the uncertain information of each dimension expressed by PULIFPR is shown in Figure 1.

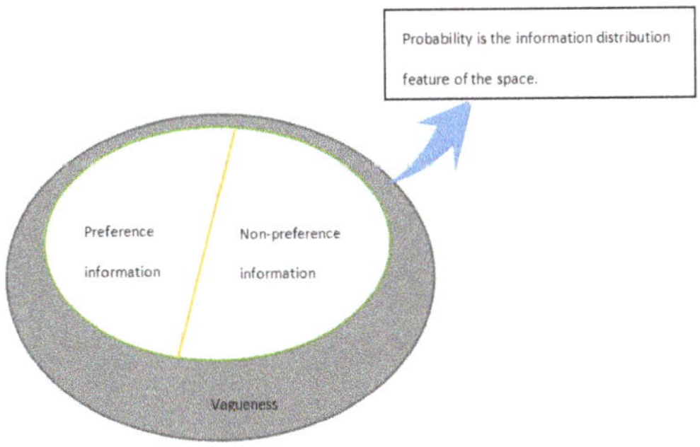

Figure 1. The uncertainty space of probabilistic uncertain linguistic intuitionistic fuzzy preference relation (PULIFPR).

From Figure 1, we can see intuitively that the uncertain space of PULIFPR is divided into vagueness subspace and non-vagueness subspace, and non-vagueness subspace can be further divided into preference information and non-preference information. Therefore, this paper will discuss the consistency of PULIFPR from the perspectives of preference, non-preference and vagueness, so as to guarantee the rationality and accuracy of the final results to the greatest extent. We will consider the DM's risk preference comprehensively based on the consistency proposed, so as to establish a reasonable GPM and get a reasonable ranking result.

Based on the above analysis, the main contributions of this paper is organized as follows:

(1) We put forward PULIFS, which is of great significance for improving the application of LT in fuzzy theory and effectively promoting the application of qualitative information in GDM.
(2) We extracted fuzzy and non-fuzzy uncertain information from PULIFPR, and then used it to define the distance measure of PULIFPRs, thus solving the problem of determining the individual weight in GDM.
(3) We built a series of GPMs by taking into account the DMs' qualitative preference, non-preference and fuzzy information, and then give a reasonable ranking results of PULIFPR.
(4) We avoided the consistency test and correction of preference relation in GDM, thus simplifying the process of GDM and improving the accuracy of decision result.
(5) The proposed method is applied to the industrial docking of virtual reality (VR) industry conference, which solves the problem of project selection before industrial docking.

To sum up, compared with the existing group decision-making methods, the main advantages of the proposed method are as follows:

(1) Most of the decision-making methods directly use the information provided by the decision-maker to model and make judgments, but ignore the information that the decision-maker fails to grasp or the information loss caused by some objective factors. In this paper, uncertain information is divided into fuzzy uncertain information and non-fuzzy uncertain information for comprehensive discussion, which improves the utilization of information and ensures the rationality of decision-making results.
(2) Most of the existing decision-making models fail to consider the risk attitude of DMs and fail to guarantee the consistency of preference information given by DMs. In this paper, two extreme attitudes of DMs under uncertain conditions are considered to establish programming models, which ensures the consistency of preference relations, simplifies GDM process and improves the accuracy of decision-making results.

The remainder of this paper is organized as follows: Section 2 recalls some basic concepts, including LIFS, PLTS, PULTS. Section 3 introduces the concepts of PULIFS and PULIFPR, and gives the definition of the distance measure of PULIFSs. Section 4 discusses the consistency of PULIFPR and establishes the corresponding GPM to obtain its comprehensive priority ranking weight. Then a specific algorithm is developed for GDM with PULIFPRs. In Sections 5, a practical example about VR industry and comparative analysis are given to demonstrate the proposed method. Finally, Section 6 is concluding remarks.

2. Preliminaries

In this part, we review some basic concepts of LIFS, PLTS and PULTS, and point out the main disadvantage of these fuzzy sets.

2.1. PLTS and PULTS

For convenience, all the LST mentioned in this article are represented by $S = \{s_\alpha | \alpha \in [0, 2\tau]\}$ except for special explanations. In order to present the probability distribution information of the HFLTS, Pang et al. [14] proposed PLTS.

Definition 1 ([14]). *Let $S = \{s_\alpha | \alpha \in [0, 2\tau]\}$ be a continuous LTS, then a PLTS is defined as*

$$L(p) = \{L^{(k)}(p^{(k)}) | L^{(k)} \in S, p^{(k)} \geq 0, k = 1, 2, \cdots, \#L(p), \sum_{k=1}^{\#L(p)} p^{(k)} \leq 1\}, \quad (1)$$

where $L^{(k)}(p^{(k)})$ *is the linguistic term* $L^{(k)}$ *associated with the probability* $p^{(k)}$, *and* $\#L(p)$ *is the number of all different linguistic terms in* $L(p)$. *To further reflect the hesitation of DMs, Lin et al.* [15] *expanded PLTS into PULTS:*

$$S(p) = \{\langle [L^k, U^k], p^k \rangle | p^k \geq 0, k = 1, 2 \cdots, \#S(p), \sum_{k=1}^{\#S(p)} p^k \leq 1\}, \tag{2}$$

where $\langle [L^k, U^k], p^k \rangle$ *represents the uncertain linguistic variable* $[L^k, U^k]$ *associated with its probability* p^k. $L^k, U^k \in S$ *are the linguistic terms,* $L^k \leq U^k$, *and* $\#S(p)$ *is the cardinality of* $S(p)$.

2.2. LIFS

To reflect the DM's qualitative non-preference information, Zhang [12] proposes LIFS.

Definition 2 ([12]). *Let X be a finite universal set and* $S = \{s_\alpha | \alpha \in [0, 2\tau]\}$ *be a continuous linguistic term set. Then a LIFS L in X is given as*

$$L = \{(x, s_\theta(x), s_\sigma(x)) | x \in X\} \tag{3}$$

where $s_\theta(x), s_\sigma(x) \in S$ *stand for the linguistic membership degree and linguistic nonmembership of the element x to L, respectively, and* $0 \leq \theta + \sigma \leq 2\tau$ *for all* $x \in X$.

PULTS only takes into account the DM's qualitative preference information and its probability distribution, however, in actual DM, DMs may need to give preference and non-preference information from both sides due to various uncertainties. Although LIFS takes into account the DMs' non-preference information, it requires the DMs to give only single linguistic terms as decision information, which cannot reflect the decision makers' hesitation in a complex environment. Therefore, in order to avoid the limitations mentioned above in actual DM, this paper proposes PULIFS in combination with the advantages of PULTS and LIFS.

3. PULIFS and PULIFPR

3.1. PULIFS

Definition 3. *Let* $S = \{s_\alpha | \alpha \in [0, 2\tau]\}$ *be a continuous LTS, then a PULIFS on S is expressed by a mathematical symbol:*

$$U(p) = \{\langle ([s_{\underline{u}^k}, s_{\overline{u}^k}], [s_{\underline{v}^k}, s_{\overline{v}^k}]), p^k \rangle | p^k \geq 0, k = 1, 2 \cdots, \#U(p), \sum_{k=1}^{\#U(p)} p^k \leq 1\}, \tag{4}$$

where $\langle ([s_{\underline{u}^k}, s_{\overline{u}^k}], [s_{\underline{v}^k}, s_{\overline{v}^k}]), p^k \rangle$ *is a PULIF element (PULIFE), which denotes the k-th uncertain linguistic intuitionistic variable (ULIV)* $([s_{\underline{u}^k}, s_{\overline{u}^k}], [s_{\underline{v}^k}, s_{\overline{v}^k}])$ *associated with its probability* p^k *in* $U(p)$, *and* $[s_{\underline{u}^k}, s_{\overline{u}^k}] \subseteq [s_0, s_{2\tau}]$, $[s_{\underline{v}^k}, s_{\overline{v}^k}] \subseteq [s_0, s_{2\tau}]$ *represent non-vagueness qualitative uncertain preference and non-preference information respectively.* $s_{\underline{u}^k}, s_{\overline{u}^k}, s_{\underline{v}^k}, s_{\overline{v}^k} \in S$, *are the linguistic terms,* $s_{\underline{u}^k} \leq s_{\overline{u}^k}, s_{\underline{v}^k} \leq s_{\overline{v}^k}, \overline{u}^k + \overline{v}^k \leq 2\tau$, *and* $\#U(p)$ *is the cardinality of* $U(p)$. *Similarly, the uncertain linguistic variable* $s_{\pi^k} = [s_{\underline{\pi}^k}, s_{\overline{\pi}^k}]$ *represent vagueness uncertain information, where* $\underline{\pi}^k = 2\tau - \overline{u}^k - \overline{v}^k, \overline{\pi}^k = 2\tau - \underline{u}^k - \underline{v}^k$.

In actual DM, DMs tend to compare two alternatives and give preference information instead of directly giving evaluation information to one alternative. Therefore, we further give the concept of PULIFPR based on PULIFS. For convenience, we use $u(p) = \{([s_{\underline{u}^k}, s_{\overline{u}^k}], [s_{\underline{v}^k}, s_{\overline{v}^k}]), p^k\}$ to represent the PULIFS, where $k = 1, 2, \cdots, \#u(p)$, and $\#u(p)$ is the number of PULIFE in $u(p)$.

3.2. PULIFPR

Definition 4. *Let* $U = (u(p)_{ij})_{n\times n}$ *be a matrix on the object set* $X = \{x_1, x_2, \cdots, x_n\}$ *for the LTS* $S = \{s_\alpha | \alpha \in [0, 2\tau]\}$, *where* $u(p)_{ij} = \{(s_{u_{ij}^k}, s_{v_{ij}^k}), p_{ij}^k\}$ *is a PULIFS,* $s_{u_{ij}^k} = [s_{\underline{u}_{ij}^k}, s_{\overline{u}_{ij}^k}]$ *represents the preference of DMs for* x_i *over* x_j, $s_{v_{ij}^k} = [s_{\underline{v}_{ij}^k}, s_{\overline{v}_{ij}^k}]$ *represents the non-preference of DMs for* x_i *over* x_j, *and* $s_{\pi_{ij}^k} = [s_{\underline{\pi}_{ij}^k}, s_{\overline{\pi}_{ij}^k}]$ *indicates the hesitancy (vagueness) degree to the preference of DMs for* x_i *over* x_j. $\underline{\pi}_{ij}^k = 2\tau - \overline{u}_{ij}^k - \overline{v}_{ij}^k, \overline{\pi}_{ij}^k = 2\tau - \underline{u}_{ij}^k - \underline{v}_{ij}^k$, $k = 1, 2, \cdots, \#u(p)_{ij}$, *and* $\#u(p)_{ij}$ *is the number of PULIFE in* $u(p)_{ij}$. *U is called a PULIFPR, if it satisfies the following conditions:*

(1) $p_{ij}^k = p_{ji}^k, p_{ii}^k = 1$;
(2) $s_{u_{ij}^k} = s_{v_{ji}^k}, s_{v_{ij}^k} = s_{u_{ji}^k}$;
(3) $u(p)_{ii} = \{([s_\tau, s_\tau], [s_\tau, s_\tau]), 1\} = s_\tau$;
(4) $\#u(p)_{ij} = \#u(p)_{ji}$;

for all $i, j = 1, 2, \cdots, n$ *with* $i \neq j$, *and* $\sum_{k=1}^{\#u(p)_{ij}} p_{ij}^k \leq 1, 0 \leq \overline{u}_{ij}^k + \overline{v}_{ij}^k \leq 2\tau, s_{u_{ij}^k} \subseteq [s_0, s_{2\tau}], s_{v_{ij}^k} \subseteq [s_0, s_{2\tau}]$. *In particular, when* $\#u(p)_{ij} = 1$ *and* $(s_{u_{ij}^k}, s_{v_{ij}^k}) \in \{([s_0, s_0], [s_{2\tau}, s_{2\tau}]), ([s_{2\tau}, s_{2\tau}], [s_0, s_0])\}$, *it means that the preference information given by the decision maker is certain and extreme for* x_i *over* x_j. *However, in a complex decision-making environment, the decision maker often does not give such a judgment with extreme certainty, so this paper only considers the case of* $(s_{u_{ij}^k}, s_{v_{ij}^k}) \notin \{([s_0, s_0], [s_{2\tau}, s_{2\tau}]), ([s_{2\tau}, s_{2\tau}], [s_0, s_0])\}$. *In addition,* $\sum_{k=1}^{\#u(p)_{ij}} p_{ij}^k = 0$ *means that the decision maker cannot give preference information for* x_i *over* x_j. *Therefore, in order to ensure the completeness of information, we assume that* $\sum_{k=1}^{\#u(p)_{ij}} p_{ij}^k > 0$.

In GDM, it is often necessary to aggregate individual preference information into group preference information. However, due to the knowledge and experience gaps between individuals, the determination of individual weight in the aggregation process is particularly important. Therefore, in order to determine a reasonable individual weight, we first introduce the definition of the distance measure of PULIFPRs

3.3. The Distance Measure of PULIFSs

Considering that different PULIFS may have different numbers of PULIFE, it may be too complicated to give the distance measurement directly. Therefore, before giving the distance measure of PULIFS, we need to convert PULIFS. According to the partition of uncertain space of PULIFS in Figure 1, we transform the information expressed by PULIFS into two parts: non-fuzzy uncertain information and fuzzy uncertain information. Inspired by the conversion method of probabilistic interval-valued intuitionistic hesitant fuzzy set (PIVIHFS) proposed by Zhai et al. [28], we present the conversion function as follows

Definition 5. *Let* $u(p) = \{([s_{\underline{u}^k}, s_{\overline{u}^k}], [s_{\underline{v}^k}, s_{\overline{v}^k}]), p^k\}$ *be a PULIFS associated with S, then its non-fuzzy uncertain information transformation function f is defined as*

$$f(u(p)) = \sum_{k=1}^{\#u(p)} p^k \times \frac{I(s_{\underline{u}^k}) - I(s_{\overline{v}^k}) + I(s_{\overline{u}^k}) - I(s_{\underline{v}^k}) + 4\tau}{8\tau} \tag{5}$$

and its fuzzy uncertain information transformation function g is defined as

$$g(u(p)) = \sum_{k=1}^{\#u(p)} p^k \times \frac{I(s_{\underline{\pi}^k}) + I(s_{\overline{\pi}^k})}{4\tau} \tag{6}$$

where $\#u(p)$ *is the number of PULIFE in* $u(p)$, $I(\cdot)$ *is the subscript function of the linguistic term, that is* $I(s_t) = t$. *Moreover,* $f(u(p))$ *represents the non-fuzzy information part of PULIFS,* $I(s_{\underline{u}^k}) - I(s_{\overline{v}^k})$ *and*

$I(s_{\overline{u}^k}) - I(s_{\underline{v}^k})$ in Equation (5) can be respectively interpreted as the pessimistic and optimistic attitude values of DMs. On the contrary, $g(u(p))$ denotes the fuzzy information part of PULIFS, which can be interpreted as the average of information that DMs fail to grasp or ignore.

Remark 1. *On the premise that the original meaning expressed by PULIFS is not lost, we used Equations (5) and (6) to transform the qualitative non-fuzzy and fuzzy information into the specific values in [0,1], so as to simplify the calculation of distance measure. For convenience, we used $v = (f, g)$ to represent the converted PULIFS and call it the conversion set (CS). Thus, for each PULIFPR $U = (u(p)_{ij})_{n \times n}$, there is a transformation matrix $V = (v_{ij})_{n \times n}$, where $v_{ij} = (f_{ij}, g_{ij})$. Now, we give the definition of the distance measure of PULIFSs.*

Definition 6. *Let $u_1(p)$ and $u_2(p)$ be two PULIFSs associated with S, $v_1 = (f_1, g_1)$ and $v_2 = (f_2, g_2)$ be the corresponding CSs of $u_1(p)$ and $u_2(p)$, then the Hamming distance between $u_1(p)$ and $u_2(p)$ is:*

$$d(u_1(p), u_2(p)) = d(v_1, v_2) = \frac{1}{2}(|f_1 - f_2| + |g_1 - g_2|) \tag{7}$$

the Euclidean distance between $u_1(p)$ and $u_2(p)$ is:

$$d(u_1(p), u_2(p)) = d(v_1, v_2) = [\frac{(f_1 - f_2)^2 + (g_1 - g_2)^2}{2}]^{\frac{1}{2}} \tag{8}$$

It is obvious that the given distance measure satisfies the following properties:

(1) $0 \leq d(v_1, v_2) \leq 1$;
(2) $d(v_1, v_2) = 0$ *if and only if* $v_1 = v_2$;
(3) $d(v_1, v_2) = d(v_2, v_1)$.

For convenience, this paper only takes hamming distance for discussion, and based on the relationship between distance measure and similarity degree, we further give the similarity degree of PULIFSs.

$$s(u_1(p), u_2(p)) = 1 - d(u_1(p), u_2(p)) = 1 - \frac{1}{2}(|f_1 - f_2| + |g_1 - g_2|) \tag{9}$$

Lets give a simple example to show the distance calculation between PULIFSs $u_1(p)$ and $u_2(p)$.

Example 1. *Let LTS $S = \{s_\alpha | \alpha \in [0, 8]\}$, and the two PULIFSs are shown below:*
$u_1(p) = \{\langle([s_1, s_2], [s_4, s_5]), 0.2\rangle, \langle([s_0, s_2], [s_3, s_5]), 0.3\rangle, \langle([s_2, s_3], [s_4, s_5]), 0.5\rangle\}$
$u_2(p) = \{\langle([s_4, s_6], [s_0, s_1]), 0.45\rangle, \langle([s_3, s_5], [s_1, s_2]), 0.5\rangle\}$
the values of non-fuzzy function and fuzzy function corresponding to $u_1(p)$ and $u_2(p)$ can be easily obtained from Equations (5) and (6) are as follows
$f_1 = 0.2 \times \frac{1-5+2-4+16}{32} + 0.3 \times \frac{0-5+2-3+16}{32} + 0.5 \times \frac{2-5+3-4+16}{32} = 0.3438,$
$g_1 = 0.2 \times \frac{8-2-5+8-1-4}{16} + 0.3 \times \frac{8-2-5+8-0-3}{16} + 0.5 \times \frac{8-3-5+8-2-4}{16} = 0.2250,$
$f_2 = 0.45 \times \frac{4-1+6-0+16}{32} + 0.5 \times \frac{3-2+5-1+16}{32} = 0.6797,$
$g_2 = 0.45 \times \frac{8-6-1+8-4-0}{16} + 0.5 \times \frac{8-5-2+8-3-1}{16} = 0.2969,$
then the distance between $u_1(p)$ and $u_2(p)$ is:
$d(u_1(p), u_2(p)) = \frac{1}{2}(|f_1 - f_2| + |g_1 - g_2|) = \frac{1}{2}(|0.3438 - 0.6797| + |0.225 - 0.2969|) = 0.2039,$
and the corresponding similarity degree is $s(u_1(p), u_2(p)) = 1 - d(u_1(p), u_2(p)) = 0.7961$.

Remark 2. *From the above example, it is not difficult to find that compared with the distance measure defined in general literature, the distance measure proposed in this paper does not need to normalize the initial set, which allows different sets to have different elements and allows for the absence of probability information ($0 < \sum_{k=1}^{\#u(p)} p^k \leq 1$). In addition, under the premise that the original information is not lost, multiple elements in PULIFS are integrated into two parts of fuzzy information and non-fuzzy information, which greatly simplifies the calculation between sets.*

Based on the distance between PULIFSs, we give the distance measure of PULIFPRs.

Definition 7. *Let $U^l = (u(p)^l_{ij})_{n\times n}$ and $U^m = (u(p)^m_{ij})_{n\times n}$ be two PULIFPRs, and their corresponding transformation matrices are $V^l = (v^l_{ij})_{n\times n}$ and $V^m = (v^m_{ij})_{n\times n}$, where $v^l_{ij} = (f^l_{ij}, g^l_{ij}), v^m_{ij} = (f^m_{ij}, g^m_{ij})$. Similar to Equation (7), the hamming distance between individual PULIFPRs U^l and U^m is defined as:*

$$d(U^l, U^m) = d(V^l, V^m) = \frac{1}{n \times (n-1)} \sum_{i<j}^{n} (|f^l_{ij} - f^m_{ij}| + |g^l_{ij} - g^m_{ij}|) \tag{10}$$

where $f^l_{ij} = \sum_{k=1}^{\#u(p)^l_{ij}} p^k_{ij} \times \frac{I(s_{\underline{u}^k_{ij}}) - I(s_{\overline{v}^k_{ij}}) + I(s_{\overline{u}^k_{ij}}) - I(s_{\underline{v}^k_{ij}}) + 4\tau}{8\tau}$ *and* $g^l_{ij} = \sum_{k=1}^{\#u(p)^l_{ij}} p^k_{ij} \times \frac{I(s_{\underline{\pi}^k_{ij}}) + I(s_{\overline{\pi}^k_{ij}})}{4\tau}$.

Then the similarity degree between U^l and U^m is defined as:

$$s(U^l, U^m) = 1 - d(U^l, U^m) \tag{11}$$

Next, we have used the distance measure and similarity degree between individual PULIFPRs to present the aggregation process of GDM.

3.4. Deriving Individual Weights and Aggregating Individual PULIFPRs

For GDM problems, without loss of generality, we supposed there are q DMs $D = \{d_1, d_2, \cdots, d_q\}$ who are invited to compare n alternatives $X = \{x_1, x_2, \cdots, x_n\}$, and $U^l = (u(p)^l_{ij})_{n\times n}$ be the individual PULIFPR provided by the DMs $d_l, (l = 1, \cdots, q)$. Then, based on the similarity degree of PULIFPRs given by the DMs, we defined the confidence degree of the l-th decision maker d_l as:

$$cs_l = \sum_{m=1, m\neq l}^{q} s(U^l, U^m) (l = 1, 2, \cdots, q) \tag{12}$$

Obviously, the higher the confidence degree of a decision maker, the higher the overall similarity between the decision maker and other DMs, and the greater the importance of the decision maker in GDM. Therefore, we regarded the normalized confidence degree cs^N_l as the weight of individual in GDM, where $cs^N_l = \frac{cs_l}{\sum_{l=1}^{q} cs_l}$. Let the weight of the l-th decision maker be $w^l = cs^N_l$, then $\sum_{l=1}^{q} w^l = 1$ and $0 \leq w^l \leq 1, (l = 1, 2, \cdots, q)$.

In order to aggregate individual PULIFPRs into a collective one, the basic operational laws between PULIFSs $u_1(p) = \{([s_{\underline{u}^k_1}, s_{\overline{u}^k_1}], [s_{\underline{v}^k_1}, s_{\overline{v}^k_1}]), p^k_1\}$ and $u_2(p) = \{([s_{\underline{u}^k_2}, s_{\overline{u}^k_2}], [s_{\underline{v}^k_2}, s_{\overline{v}^k_2}]), p^k_2\}$ is given as follows:

$$u_1(p) \bigoplus u_2(p) = \bigcup_{k\in(1,\cdots,\#u_1(p))} \{\langle([s_{\underline{u}^k_1+\underline{u}^k_2}, s_{\overline{u}^k_1+\overline{u}^k_2}], [s_{\underline{v}^k_1+\underline{v}^k_2}, s_{\overline{v}^k_1+\overline{v}^k_2}]), \frac{p^k_1 + p^k_2}{2}\rangle\} \tag{13}$$

$$\lambda u_1(p) = \bigcup_{k\in(1,\cdots,\#u_1(p))} \{([s_{\lambda\underline{u}^k_1}, s_{\lambda\overline{u}^k_1}], [s_{\lambda\underline{v}^k_1}, s_{\lambda\overline{v}^k_1}]), p^k_1\} \tag{14}$$

where $\#u_1(p) = \#u_2(p)$, when $\#u_1(p) \neq \#u_2(p)$, we normalized it by the following method:

If $\#u_1(p) \neq \#u_2(p)$, $\#u_1(p) > \#u_2(p)$, then we have added $\#u_1(p) - \#u_2(p)$ PULIFEs to $u_2(p)$ so that the PULIFSs $u_1(p)$ and $u_2(p)$ have the same number of elements. The added uncertain linguistic intuitionistic variables (ULIVs) are the smallest one(s) in $u_2(p)$, and the probabilities of the added ULIVs are zero. In addition, the comparison method of two PULIFE $e_k = \langle([s_{\underline{u}^k}, s_{\overline{u}^k}], [s_{\underline{v}^k}, s_{\overline{v}^k}]), p^k\rangle$ and

$e_l = \langle(([s_{\underline{u}^l}, s_{\overline{u}^l}], [s_{\underline{v}^l}, s_{\overline{v}^l}]), p^l\rangle$ in PULIFS is as follows
Let

$$\begin{aligned} f_i &= p^i \times [I(s_{\underline{u}^i}) - I(s_{\overline{v}^i}) + I(s_{\overline{u}^i}) - I(s_{\underline{v}^i})], \\ g_i &= p^i \times [I(s_{\underline{\pi}^i}) + I(s_{\overline{\pi}^i})], (i = k, l). \end{aligned} \tag{15}$$

(1) If $g_k > g_l$, then $e_k < e_l$;
(2) If $g_k = g_l$, then
  (a) If $f_k > f_l$, then $e_k > e_l$;
  (b) If $f_k = f_l$, then $e_k = e_l$.

The larger the PULIFE, the larger its corresponding ULIV. Based on this, we give the definition of probabilistic uncertain linguistic intuitionistic weighted average (PULIWA) operator.

Definition 8. *Given q PULIFSs* $u_i(p) = \{(([s_{\underline{u}_i^k}, s_{\overline{u}_i^k}], [s_{\underline{v}_i^k}, s_{\overline{v}_i^k}]), p_i^k\}, (i = 1, 2, \cdots, q), k = 1, 2 \cdots, \#u_i(p)$, *the weight vector* $W = (w_1, w_2, \cdots, w_q), w_i \in [0,1], \sum_{i=1}^q w_i = 1$, *then we called*

$$\begin{aligned} PWA(u_1(p), \cdots, u_q(p)) &= \bigoplus_{i=1}^{q} w_i u_i(p) \\ &= \bigcup_{k=1,2\cdots,\#u_i(p)} \{(([s_{\sum_{i=1}^q w_i \underline{u}_i^k}, s_{\sum_{i=1}^q w_i \overline{u}_i^k}], [s_{\sum_{i=1}^q w_i \underline{v}_i^k}, s_{\sum_{i=1}^q w_i \overline{v}_i^k}]), \frac{\sum_{i=1}^q p_i^k}{q}\} \end{aligned} \tag{16}$$

the PULIWA operator.

Example 2. *Continuing with Example 1, assuming that the weight values of both PULIFS* $u_1(p)$ *and* $u_2(p)$ *are 0.5. Since* $\#u_2(p) = 2 < \#u_1(p) = 3$, *we can easily know from Equation (15) that* $g_{21} = 0.45 \cdot [8 - 1 - 6 + 8 - 4 - 0] = 2.25$ *and* $g_{22} = 0.5 \cdot [8 - 5 - 2 + 8 - 3 - 1] = 2.5$ *in* $u_2(p)$. *Therefore,* $g_{21} < g_{22}, e_{21} > e_{22}$, *the normalized* $u_2(p) = \{\langle([s_4, s_6], [s_0, s_1]), 0.45\rangle, \langle([s_3, s_5], [s_1, s_2]), 0.5\rangle, \langle([s_3, s_5], [s_1, s_2]), 0\rangle\}$, *and the PULIWA operator* $PWA(u_1(p), u_2(p)) = \{\langle([s_{2.5}, s_4], [s_2, s_3]), 0.325\rangle, \langle([s_{1.5}, s_{3.5}], [s_2, s_{3.5}]), 0.4\rangle, \langle([s_{2.5}, s_4], [s_{2.5}, s_{3.5}]), 0.5\rangle\}$ *of* $u_1(p)$ *and* $u_2(p)$ *can be obtained by using Equation (16).*

Obviously, it is easy to aggregate individual PR into collective PR by using Equation (16). Therefore, the process of obtaining priority weights is given in the following discussion based on the consistency of collective PULIFPR $\tilde{U}$.

4. Consistency Analysis of PULIFPR and Acquisition of Its Priority Weight

4.1. Consistency Analysis of PULIFPR

At present, the research on the consistency of PR is mainly divided into two categories: multiplicative consistency and additive consistency. Without loss of generality, this paper discusses PULIFPR consistency based on multiplicative consistency. Therefore, before giving the definition of PULIFPR consistency, lets review the multiplicative consistency of fuzzy preference relations (FPRs).

Definition 9. *[29]. For the FPR* $R = (r_{ij})_{n \times n}, (i, j = 1, 2, \cdots, n), r_{ij} \in [0, 1]$, *if we have*

$$r_{ij} = \frac{w_i}{w_i + w_j}. \tag{17}$$

for all $i, j = 1, 2, \cdots, n$, *and which satisfies: 1)* $r_{ii} = 0.5$; *2)* $r_{ij} + r_{ji} = 1$; *3)* $\sum_{i=1}^n w_i = 1$. *then we called the FPR R is multiplicative consistent, where* r_{ij} *is the preference degree of the objectives* x_i *over* x_j, *and* $w_i \in [0, 1]$, $w = (w_1, w_2, \cdots, w_n)$ *is the priority vector of R.*

Inspired by this, we presented the following definition of consistency by combining the preferences, non-preferences and vagueness information expressed by PULIFPR.

Definition 10. *Let $\tilde{U} = (u(p)_{ij})_{n\times n}$ be a PULIFPR on the object set $X = \{x_1, x_2, \cdots, x_n\}$ for the LTS $S = \{s_\alpha | \alpha \in [0, 2\tau]\}$, its corresponding transformation matrix is $V = (v_{ij})_{n\times n}$, where $v_{ij} = (f_{ij}, g_{ij})$ is a CS. Based on this, we can extract the FPR $H = (h_{ij})_{n\times n}$ from the PULIFPR $\tilde{U}$, where*

$$h_{ij} = \begin{cases} \theta_{ij} f_{ij} + (1-\theta_{ij}) g_{ij}, & If \ \ i < j \\ 0.5, & If \ \ i = j \\ 1 - h_{ji}, & If \ \ i > j \end{cases} \tag{18}$$

If we have

$$h_{ij} = \frac{w_i}{w_i + w_j} \tag{19}$$

for all $i, j = 1, 2, \cdots, n$, then we called PULIFPR $\tilde{U}$ multiplicative consistent, where $\theta_{ij} \in [0,1]$ represents the importance of non-fuzzy information f_{ij} extracted from $u(p)_{ij}$, and $w = (w_1, w_2, \cdots, w_n)$ is the priority vector of $\tilde{U}$, satisfying $w_i \in [0,1]$ and $\sum_{i=1}^{n} w_i = 1$.

Remark 3. *From Equations (5) and (6), it is not difficult to see that the values of non-fuzzy information f_{ij} and fuzzy information g_{ij} extracted from PULIFPR are all located in [0,1]. Furthermore, it is easy to know that $f_{ii} = 0.5$ and $g_{ii} = 0$ by the nature of PULIFPR. Therefore, the FPR $H = (h_{ij})_{n\times n}$ is generated by combining the non-fuzzy and fuzzy information of the DMs, and the consistency of PULIFPR is transformed into the consistency of the FPR. However, this definition of consistency only considers the fuzzy and non-fuzzy space of PULIFPR in general. In order to make full use of the decision-making information expressed by PULIFPR, we have considered the decision maker's risk attitude and further discuss its consistency with the preference and non-preference information in the non-fuzzy space.*

Definition 11. *Based on Definitions 4 and 5, we set $a_{ij} = \sum_{k=1}^{\#u(p)} p^k \times \frac{I(s_{\overline{u}^k_{ij}}) - I(s_{\underline{v}^k_{ij}}) + 2\tau}{4\tau}$ as the maximum preference (the most optimistic judgment) of DMs for x_i over x_j, while $b_{ij} = \sum_{k=1}^{\#u(p)} p^k \times \frac{I(s_{\underline{u}^k_{ij}}) - I(s_{\overline{v}^k_{ij}}) + 2\tau}{4\tau}$ as the minimum preference (the most pessimistic judgment) of DMs for x_i over x_j. So similarly, we can extract a FPR $D = (d_{ij})_{n\times n}$ from the PULIFPR $\tilde{U}$, where*

$$d_{ij} = \begin{cases} t_{ij} a_{ij} + (1-t_{ij}) b_{ij}, & If \ \ i < j \\ 0.5, & If \ \ i = j \\ 1 - d_{ji}, & If \ \ i > j \end{cases} \tag{20}$$

If we have

$$d_{ij} = \frac{w'_i}{w'_i + w'_j} \tag{21}$$

for all $i, j = 1, 2, \cdots, n$, then we also called PULIFPR $\tilde{U}$ multiplicative consistent, where $t_{ij} \in [0,1]$ indicates the degree of optimism of the DMs, the bigger the values of t_{ij}, the higher DM's optimistic degree, and $w = (w'_1, w'_2, \cdots, w'_n)$ is the priority vector of $\tilde{U}$, satisfying $w'_i \in [0,1]$ and $\sum_{i=1}^{n} w'_i = 1$.

Based on the two consistency definitions given above, we give the method to obtain the priority weight of collective PULIFPR.

4.2. Determine the Priority Weights of PULIFPR through the GPM

The consistency definition given by Equation (19) integrates the fuzzy uncertainty and non-fuzzy uncertainty of the information given by the decision-maker. However, in actual decision-making, we hope that the fuzzy uncertainty degree of information expressed by PULIFPR is as small as possible, so as to make the ranking result as reasonable and accurate as possible. Therefore, the higher the value of parameter θ_{ij}, which indicates the importance of non-fuzzy uncertainty information, the more reasonable the result will be. Based on this principle, we establish the following GPM to obtain the priority weight of the PULIFPR.

$$\max \theta = \textstyle\sum_{i<j} \theta_{ij}$$

$$s.t. \begin{cases} h_{ij} = \frac{w_i}{w_i + w_j}, & (1) \\ \sum_{i \neq j} w_i > w_j - 0.5, & (2) \\ \sum_{i=1}^{n} w_i = 1, w_i \geq 0, \\ 0 \leq \theta_{ij} \leq 1, (i,j = 1,2,\cdots,n; i<j). \end{cases} \quad (22)$$

In Equation (22), the constraint condition (1) guarantees the consistency of the FPR H extracted from PULIFPR $\tilde{U}$, and the constraint condition (2) avoids the occurrence of extreme judgment caused by individual subjective preference, thus guaranteeing the objectivity of decision-making process. In addition, the literature [30] shows that the consistency of the FPR only needs to discuss the upper triangular part of it. So to simplify the calculation, we have $i < j$.

If the feasible region of Equation (22) is nonempty, the optimal solutions θ_{ij} and priority weight vector $w_i, (i = 1,2,\cdots,n)$ can be obtained by solving it. However, it does not guarantee that there will always be nonempty feasible regions. Therefore, when the feasible region is empty, we expand the feasible region of the model by appropriately increasing the fuzzy uncertain information value g_{ij} and reducing the non-fuzzy uncertain information value f_{ij}, and the expanded model is as follows

$$\max \theta = \textstyle\sum_{i<j} \theta_{ij} - \sum_{i<j} (\phi_{ij} + \psi_{ij})$$

$$s.t. \begin{cases} \theta_{ij}(f_{ij} - \phi_{ij}) + (1-\theta_{ij})(g_{ij} + \psi_{ij}) = \frac{w_i}{w_i + w_j}, \\ \sum_{i \neq j} w_i > w_j - 0.5, \\ \sum_{i=1}^{n} w_i = 1, w_i \geq 0, \quad \phi_{ij}, \psi_{ij} \geq 0, \\ 0 \leq \theta_{ij} \leq 1, (i,j = 1,2,\cdots,n; i<j). \end{cases} \quad (23)$$

where ϕ_{ij} and ψ_{ij} are the deviation variables, satisfying $\phi_{ij} \geq 0, \psi_{ij} \geq 0$. Then, by solving (23), the optimal solutions θ_{ij} and priority weight vector $w_i, (i = 1,2,\cdots,n)$ can be obtained.

Similarly, it is easy to know from Definition 11 that the bigger the value of t_{ij}, the higher the degree of optimism of the decision maker. Therefore, we combine the two extreme attitudes of the decision maker, the most optimistic and the most pessimistic, and respectively present the following GPM.

$$\max t = \textstyle\sum_{i<j} t_{ij}$$

$$s.t. \begin{cases} d_{ij} = \frac{w_i^+}{w_i^+ + w_j^+}, \\ \sum_{i \neq j} w_i^+ > w_j^+ - 0.5, \\ \sum_{i<j} t_{ij} < \frac{n \times (n-1)}{2} - 1, \\ \sum_{i=1}^{n} w_i^+ = 1, w_i^+ \geq 0, \\ 0 \leq t_{ij} \leq 1, (i,j = 1,2,\cdots,n; i<j) \end{cases} \quad (24)$$

$$\min t = \textstyle\sum_{i<j} t_{ij}$$

$$s.t.\begin{cases} d_{ij}=\frac{w_i^-}{w_i^-+w_j^-}, \\ \sum_{i\neq j}w_i^- > w_j^- - 0.5, \\ \sum_{i<j}t_{ij} > 1, \\ \sum_{i=1}^{n}w_i^- = 1, w_i^- \geq 0, \\ 0\leq t_{ij}\leq 1, (i,j=1,2,\cdots,n; i<j) \end{cases} \quad (25)$$

where $w^+ = (w_1^+, w_2^+, \cdots, w_n^+)$ represents the most optimistic weight vector and $w^- = (w_1^-, w_2^-, \cdots, w_n^-)$ represents the most pessimistic weight vector. It is noted that different from Model (22), Equations (24) and (25) have added restriction condition $\sum_{i<j}t_{ij} < \frac{n\times(n-1)}{2} - 1$ and $\sum_{i<j}t_{ij} > 1$ respectively, which ensures that the decision maker does not show overly optimistic or pessimistic judgment information when in a rational state. Similarly, for Model (23), when the feasible regions of Equations (24) and (25) are empty, we give the expansion model as follows

$$\max t = \sum_{i<j}t_{ij} - \sum_{i<j}(\alpha_{ij}+\beta_{ij})$$

$$s.t.\begin{cases} t_{ij}(a_{ij}-\alpha_{ij})+(1-t_{ij})(b_{ij}+\beta_{ij}) = \frac{w_i^+}{w_i^++w_j^+}, \\ \sum_{i\neq j}w_i^+ > w_j^+ - 0.5, \\ \sum_{i<j}t_{ij} < \frac{n\times(n-1)}{2} - 1, \\ \sum_{i=1}^{n}w_i^+ = 1, w_i^+ \geq 0, \ \alpha_{ij}, \beta_{ij} \geq 0. \\ 0\leq t_{ij}\leq 1, (i,j=1,2,\cdots,n; i<j) \end{cases} \quad (26)$$

$$\min t = \sum_{i<j}t_{ij} + \sum_{i<j}(\alpha_{ij}+\beta_{ij})$$

$$s.t.\begin{cases} t_{ij}(a_{ij}+\alpha_{ij})+(1-t_{ij})(b_{ij}-\beta_{ij}) = \frac{w_i^-}{w_i^-+w_j^-}, \\ \sum_{i\neq j}w_i^- > w_j^- - 0.5, \\ \sum_{i<j}t_{ij} > 1, \\ \sum_{i=1}^{n}w_i^- = 1, w_i^- \geq 0, \ \alpha_{ij}, \beta_{ij} \geq 0. \\ 0\leq t_{ij}\leq 1, (i,j=1,2,\cdots,n; i<j) \end{cases} \quad (27)$$

where α_{ij} and β_{ij} are the deviation variables, satisfying $\alpha_{ij} \geq 0, \beta_{ij} \geq 0$.

By solving Equations (26) and (27), the optimal weight vectors w_i^+ and $w_i^- (i = 1,2,\cdots,n)$ can be obtained respectively. Combining w_i^+ with w_i^- , the compromise weight vector can be obtained as follows:

$$w_i' = \lambda w_i^+ + (1-\lambda)w_i^-, i = 1,2,\cdots,n \quad (28)$$

where $\lambda \in [0,1]$ represents the risk attitude of DMs. If $0 \leq \lambda < 0.5$, DMs are risk averse; If $\lambda = 0.5$, DMs are risk neutral; If $0.5 < \lambda \leq 1$, DMs are risk taking.

Considering that the decision maker pays more attention to the final result in the actual decision, we take the average value of priority weight obtained under the two consistency definitions as the final ranking weight, namely $\bar{w}_i = \frac{w_i+w_i'}{2}, i = 1,2,\cdots,n$.

Remark 4. *Compared with general programming models for solving priority weights, the main advantages of the programming models presented in this paper are as follows:*

(1) *At present, most of the programming models proposed in many literatures only consider the principle of minimum consistency deviation, such as literatures [27,31–34]. In this paper, the consistency of the newly proposed PULIFPR is considered comprehensively from the three aspects of fuzzy and non-fuzzy uncertain information and DM's risk attitude. Therefore, the rationality of decision result is greatly improved.*

(2) *Currently, most of the research on PR needs to test its consistency, and some literatures that needs to test the consistency of acceptable PR fails to provide a reasonable test method, such as the research on triangular FPR by Wang [35], and the research on interval-valued intuitionistic FPR by Wan et al. [36]. In this paper, the priority weight of consistent PULIFPR can be obtained directly through the proposed programming models without considering the consistency test, which greatly simplifies the DM process.*

4.3. A New Algorithm for Solving GDM with PULIFPR

Summarizing above analyses, a new method for GDM with PULIFPR is developed as follows:

Step 1. Calculate the distance measure $d(U^l, U^m)$ and similar measure $s(U^l, U^m)(l, m = 1, 2, \cdots, q, l \neq m)$ between individual PULIFPRs by Equations (10) and (11).

Step 2. Use Equation (12) to calculate the confidence degree cs_l and determine the individual weight $w^l(l = 1, 2, \cdots, q)$.

Step 3. Aggregating individual PULIFPR U into collective PULIFPR $\tilde{U}$ by Equation (16).

Step 4. When feasible regions of Models (22), (24) and (25) are nonempty, priority weights w_i, w_i^+ and w_i^- can be solved respectively. Otherwise, priority weights w_i, w_i^+ and w_i^- shall be obtained by solving Equations (23), (26) and (27).

Step 5. Determining the risk parameter value λ, and then the compromise weight w_i' is obtained by Equation (28).

Step 6. Combining w_i and w_i', the comprehensive weight $\overline{w}_i = \frac{w_i + w_i'}{2}$ is obtained.

Step 7. According to the comprehensive weight value to compare the alternatives, the best alternative has the bigger value.

The graphical process of solving GDM using PULIFPR is shown in Figure 2.

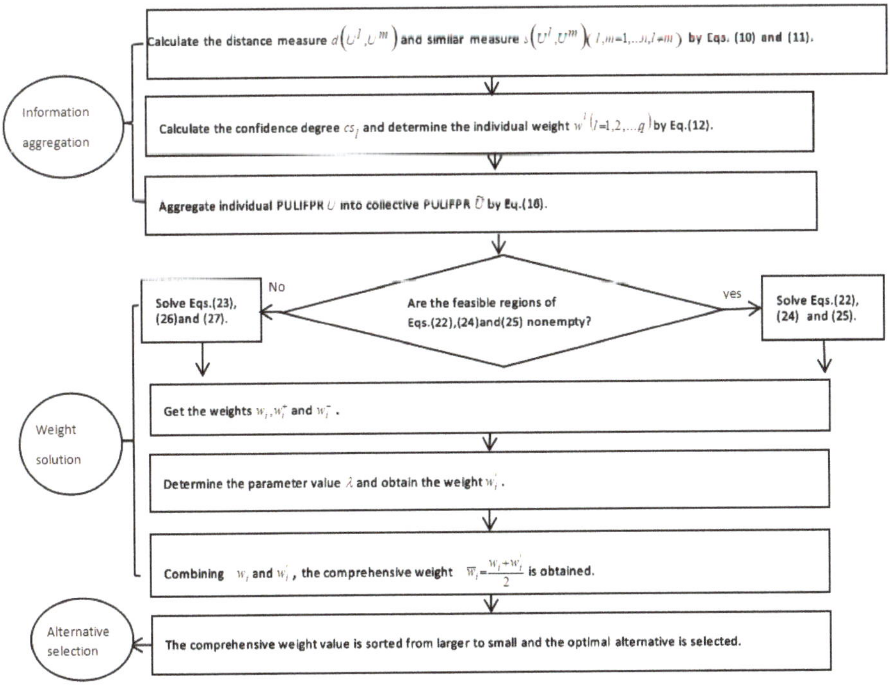

Figure 2. Process of group decision-making (GDM) with PULIFPRs.

5. Case Application and Comparative Analysis

In order to demonstrate the effectiveness and practicability of the proposed method, this section is mainly divided into two parts. The first part discusses the application of the proposed method in the world VR industry conference 2018. The second part gives the comparative analysis between the proposed method and other methods.

5.1. Application in VR Project Selection

In recent years, virtual reality (VR) technology has received unprecedented attention from all sectors of society, and it is regarded as the portal of the next generation general computing platform and Internet along with augmented reality (AR) and mixed reality (MR). In addition, as an important force leading a new round of industrial reform in the world, it plays an important role in promoting new economic development. Therefore, in order to explore the key and common problems in the development of VR, as well as the industrial development trend and solutions, the 2018 world VR industry conference was successfully held in nanchang, jiangxi province on 19 October. As one of the important activities of the conference, the industrial counterpart conference was successfully held in nanchang on 20 October.

However, in order to ensure the successful holding of the industrial docking conference, it is particularly important for the organizers to have extensive and in-depth communication with the investors in the early stage of the conference. On the one hand, it can enable investors to have a deep and sufficient understanding of each VR project in our province so that investors can select the best cooperation project. On the other hand, it is convenient for every VR industry company in our province to select the best partner or investor. Finally, the cooperation agreements reached at the industry conference are guaranteed. Therefore, the communication and mutual selection process is an important preparation work in the early stage of the conference.

Due to the complexity of VR technology, VR project selection is a very challenging task for investors. It requires investors to make a comprehensive analysis and judgment on the competitive advantage, profitability, viability and development potential of VR project from the perspectives of simulation technology and computer graphics, man-machine interface technology, sensor technology and network technology,etc. Therefore, the project selection process is often a GDM. Without loss of generality, in order to demonstrate the GDM process using the proposed method, we take the four important projects selected by Microsoft as an example. The four projects are Touch display integration project x_1, Optoelectronic project x_2, Network security industry center project x_3 and Intelligent VR visual equipment project x_4 respectively.

In view of the complexity of VR project, Microsoft sent two investment teams (e_1, e_2) to inspect the project content and one technical team (e_3) to inspect the company's technical equipment. Due to the wide range of knowledge involved in VR project and the complexity of factors to be considered by DMs, the decision team can only give judgment information from positive and negative aspects based on the LTS $S = \{s_0$: extremely poor, s_1: very poor, s_2: poor, s_3: slightly poor, s_4: fair, s_5: slightly good, s_6: good, s_7: very good, s_8: extremely good$\}$.

For example, by analyzing and comparing projects x_1 and x_2, the decision team e_1 gave the following judgment information:

$$u(p)_{12} = \{\langle([s_4, s_6], [s_1, s_1]), 0.45\rangle, \langle([s_3, s_5], [s_1, s_2]), 0.5\rangle\}$$

where $[s_4, s_6]$ indicates that the DM's preference degree for x_1 over x_2 is between fair and good, $[s_1, s_1]$ expresses that the DM's non-preference degree for x_1 over x_2 is very poor. The probability 0.45 indicates that 45% of the people in investment teams e_1 give interval intuitionistic judgment information $([s_4, s_6], [s_1, s_1])$. Similarly, PULIFE $\langle([s_3, s_5], [s_1, s_2]), 0.5\rangle$ indicated that 50% of the people in investment team e_1 gave the interval intuitionistic judgment information as $([s_3, s_5], [s_1, s_2])$. In addition, 5% of the people failed to give any judgment information.

Now, we regard the three teams e_1, e_2 and e_3 sent by Microsoft as three individuals, and take the four projects of jiangxi province, x_1, x_2, x_3, and x_4 as the alternatives. The preference information given by the three teams in the form of PULIFPR is as follows

$U_1 = (u_1(p)_{ij})_{4\times 4}, (i,j = 1,2,3,4)$

where

$$u_1(p)_{11} = u_1(p)_{22} = u_1(p)_{33} = u_1(p)_{44} = s_4;$$
$$u_1(p)_{12} = \{\langle([s_1,s_3],[s_4,s_5]),0.3\rangle, \langle([s_1,s_2],[s_5,s_6]),0.6\rangle\}$$
$$u_1(p)_{13} = \{\langle([s_4,s_6],[s_0,s_1]),0.8\rangle, \langle([s_5,s_6],[s_1,s_2]),0.2\rangle\}$$
$$u_1(p)_{14} = \{\langle([s_3,s_4],[s_4,s_4]),0.6\rangle, \langle([s_2,s_3],[s_4,s_5]),0.2\rangle\}$$
$$u_1(p)_{23} = \{\langle([s_0,s_2],[s_6,s_6]),0.7\rangle, \langle([s_1,s_3],[s_5,s_5]),0.3\rangle\}$$
$$u_1(p)_{24} = \{\langle([s_5,s_6],[s_0,s_1]),0.2\rangle, \langle([s_6,s_7],[s_0,s_1]),0.8\rangle\}$$
$$u_1(p)_{34} = \{\langle([s_0,s_1],[s_5,s_6]),0.9\rangle, \langle([s_0,s_1],[s_7,s_7]),0.1\rangle\}$$

$U_2 = (u_2(p)_{ij})_{4\times 4}, (i,j = 1,2,3,4)$

where

$$u_2(p)_{11} = u_2(p)_{22} = u_2(p)_{33} = u_2(p)_{44} = s_4;$$
$$u_2(p)_{12} = \{\langle([s_2,s_3],[s_4,s_4]),0.4\rangle, \langle([s_3,s_4],[s_4,s_5]),0.6\rangle\}$$
$$u_2(p)_{13} = \{\langle([s_1,s_2],[s_5,s_5]),0.3\rangle, \langle([s_2,s_3],[s_4,s_5]),0.5\rangle\}$$
$$u_2(p)_{14} = \{\langle([s_5,s_6],[s_1,s_1]),0.3\rangle, \langle([s_6,s_6],[s_1,s_2]),0.6\rangle\}$$
$$u_2(p)_{23} = \{\langle([s_6,s_7],[s_1,s_1]),0.4\rangle, \langle([s_5,s_6],[s_1,s_2]),0.5\rangle\}$$
$$u_2(p)_{24} = \{\langle([s_0,s_2],[s_5,s_6]),0.7\rangle, \langle([s_1,s_2],[s_6,s_6]),0.3\rangle\}$$
$$u_2(p)_{34} = \{\langle([s_3,s_5],[s_2,s_2]),0.4\rangle, \langle([s_2,s_5],[s_1,s_2]),0.4\rangle\}$$

$U_3 = (u_3(p)_{ij})_{4\times 4}, (i,j = 1,2,3,4)$

where

$$u_3(p)_{11} = u_1(p)_{22} = u_1(p)_{33} = u_1(p)_{44} = s_4;$$
$$u_3(p)_{12} = \{\langle([s_2,s_4],[s_3,s_4]),0.2\rangle, \langle([s_4,s_5],[s_3,s_3]),0.5\rangle\}$$
$$u_3(p)_{13} = \{\langle([s_1,s_2],[s_5,s_6]),0.8\rangle, \langle([s_2,s_3],[s_5,s_5]),0.2\rangle\}$$
$$u_3(p)_{14} = \{\langle([s_7,s_8],[s_0,s_0]),0.3\rangle, \langle([s_6,s_7],[s_1,s_1]),0.6\rangle\}$$
$$u_3(p)_{23} = \{\langle([s_0,s_1],[s_6,s_6]),0.6\rangle, \langle([s_1,s_1],[s_6,s_7]),0.3\rangle\}$$
$$u_3(p)_{24} = \{\langle([s_4,s_5],[s_1,s_2]),0.7\rangle, \langle([s_5,s_5],[s_2,s_3]),0.3\rangle\}$$
$$u_3(p)_{34} = \{\langle([s_7,s_7],[s_0,s_1]),0.3\rangle, \langle([s_5,s_6],[s_1,s_2]),0.6\rangle\}$$

Since the upper triangle of PULIFPR has a one-to-one correspondence with the lower triangle, we only give the upper triangle of the preference relation. According to Section 4.3, we can solve the GDM problem about project selection as follows:

Step 1: According to Equation (10), the distance measure between U_1 and U_2 is $d(U_1,U_2) = \frac{1}{4\times 3}(|f_{12}^1 - f_{12}^2| + |g_{12}^1 - g_{12}^2| + |f_{13}^1 - f_{13}^2| + |g_{13}^1 - g_{13}^2| + |f_{14}^1 - f_{14}^2| + |g_{14}^1 - g_{14}^2| + |f_{23}^1 - f_{23}^2| + |g_{23}^1 - g_{23}^2| + |f_{24}^1 - f_{24}^2| + |g_{24}^1 - g_{24}^2| + |f_{34}^1 - f_{34}^2| + |g_{34}^1 - g_{34}^2|) = \frac{1}{12}(|0.2531 - 0.425| + |0.775 - 0.2719| + |0.3563 - 0.7031| + |0.225 - 0.7125| + |0.8625 - 0.2188| + |0.1781 - 0.5| + |0.1313 - 0.075| + |0.275 - 0.1188| + |0.0625 - 0.0938| + |0.125 - 0.0875| + |0.15 - 0.15| + |0.2313 - 0.25|) = 0.2313$, Similarly, we can calculate $d(U_1,U_3) = 0.1930$ and $d(U_2,U_3) = 0.1398$ respectively, so the corresponding similarity degree is $s(U_1,U_2) = 0.7687, s(U_1,U_3) = 0.8070$ and $s(U_2,U_3) = 0.8602$ respectively.

Step 2: According to Equation (12), the confidence degree of the three teams can be calculated as $cs_1 = s(U_1,U_2) + s(U_1,U_3) = 1.5758, cs_2 = 1.6289$ and $cs_3 = 1.6672$, so the weight of each team can be further determined as $w^1 = \frac{cs_1}{cs_1+cs_2+cs_3} = 0.3234, w^2 = 0.3344$, and $w^3 = 0.3422$.

Step 3: By using Equation (16), the collective PULIFPR $\tilde{U}$ can be obtained as follows

$\tilde{U} = (\tilde{u}(p)_{ij})_{4\times 4}, (i,j = 1,2,3,4)$

where

$\tilde{u}(p)_{11} = \tilde{u}(p)_{22} = \tilde{u}(p)_{33} = \tilde{u}(p)_{44} = s_4;$
$\tilde{u}(p)_{12} = \{\langle([s_{1.6766}, s_{3.3442}], [s_{3.6578}, s_{4.3234}]), 0.3\rangle, \langle([s_{2.6953}, s_{3.6953}], [s_{3.0717}, s_{4.6390}]), 0.5667\rangle\};$
$\tilde{u}(p)_{13} = \{\langle([s_{1.9703}, s_{3.2938}], [s_{3.3828}, s_{4.0484}]), 0.6333\rangle, \langle([s_{2.9703}, s_{3.9703}], [s_{3.3719}, s_{4.0297}]), 0.3\rangle\};$
$\tilde{u}(p)_{14} = \{\langle([s_{5.0375}, s_{6.0375}], [s_{1.6281}, s_{1.6281}]), 0.4\rangle, \langle([s_{4.7062}, s_{5.3719}], [s_{1.9703}, s_{2.6281}]), 0.4667\rangle\};$
$\tilde{u}(p)_{23} = \{\langle([s_{2.0061}, s_{3.3295}], [s_{4.3283}, s_{4.3283}]), 0.5667\rangle, \langle([s_{2.3374}, s_{3.3186}], [s_{4.0048}, s_{4.6814}]), 0.3667\rangle\};$
$\tilde{u}(p)_{24} = \{\langle([s_{2.9860}, s_{4.3204}], [s_{2.0140}, s_{3.0140}]), 0.5333\rangle, \langle([s_{3.9860}, s_{4.6438}], [s_{2.6905}, s_{3.3562}]), 0.4667\rangle\};$
$\tilde{u}(p)_{34} = \{\langle([s_{3.3985}, s_{4.3906}], [s_{2.2859}, s_{2.9516}]), 0.5333\rangle, \langle([s_{2.3797}, s_{4.0484}], [s_{2.9407}, s_{3.6172}]), 0.3667\rangle\}.$

Step 4: The non-fuzzy uncertain information values $f_{ij}(i,j = 1,2,3,4, i < j)$ and the fuzzy uncertain information values $g_{ij}(i,j = 1,2,3,4, i < j)$ of PULIFPR $\tilde{U}$ are calculated respectively. By substituting them into Equation (22), the following model can be obtained

$$\max \theta = \theta_{12} + \theta_{13} + \theta_{14} + \theta_{23} + \theta_{24} + \theta_{34}$$

$$s.t. \begin{cases} [0.3822\theta_{12} + 0.1235(1-\theta_{12})](w_1 + w_2) - w_1 = 0, \\ [0.4195\theta_{13} + 0.1619(1-\theta_{13})](w_1 + w_3) - w_1 = 0, \\ [0.6110\theta_{14} + 0.0803(1-\theta_{14})](w_1 + w_4) - w_1 = 0, \\ [0.3731\theta_{23} + 0.1091(1-\theta_{23})](w_2 + w_3) - w_2 = 0, \\ [0.5756\theta_{24} + 0.1608(1-\theta_{24})](w_2 + w_4) - w_2 = 0, \\ [0.4910\theta_{34} + 0.1682(1-\theta_{34})](w_3 + w_4) - w_3 = 0, \\ w_1 + w_2 + w_3 > w_4 - 0.5,\ w_1 + w_2 + w_4 > w_3 - 0.5, \\ w_1 + w_3 + w_4 > w_2 - 0.5,\ w_2 + w_3 + w_4 > w_1 - 0.5, \\ w_1 + w_2 + w_3 + w_4 = 1,\ w_1, w_2, w_3, w_4 \geq 0, \\ 0 \leq \theta_{12}, \theta_{13}, \theta_{14}, \theta_{23}, \theta_{24}, \theta_{34} \leq 1. \end{cases} \tag{29}$$

By solving this model, priority weights and parameter values can be obtained as $w_1 = 0.1227, w_2 = 0.1984, w_3 = 0.3333, w_4 = 0.3456, \theta_{13} = 0.4162, \theta_{14} = 0.3425, \theta_{24} = 0.4916, \theta_{12} = \theta_{23} = \theta_{34} = 1$.

Step 5: By calculating the optimistic judgment values $a_{ij} = \sum_{k=1}^{\#u(p)} p^k \times \frac{I(s_{\overline{u}_{ij}^k}) - I(s_{\underline{v}_{ij}^k}) + 2\tau}{4\tau}$ and pessimistic judgment values $b_{ij} = \sum_{k=1}^{\#u(p)} p^k \times \frac{I(s_{\underline{u}_{ij}^k}) - I(s_{\overline{v}_{ij}^k}) + 2\tau}{4\tau}$, $(i,j = 1,2,3,4, i < j)$ and substituting them into Equations (24) and (25) respectively to solve the weight. But their feasible regions are all empty. Therefore, substitute the values of a_{ij} and b_{ij} into Equations (26) and (27) respectively, then the priority weights can be obtained as follows
$w_1^+ = 0.2149, w_2^+ = 0.2631, w_3^+ = 0.3700, w_4^+ = 0.1520,$
$w_1^- = 0.1395, w_2^- = 0.3034, w_3^- = 0.2432, w_4^- = 0.3139.$

Without loss of generality, assume that the value of risk parameter λ determined by Microsoft is 0.5. Then the priority weights can be obtained as $w_1' = 0.5w_1^+ + (1-0.5)w_1^- = 0.1783, w_2' = 0.2823, w_3' = 0.3058, w_4' = 0.2336$.

Step 6: Combining the results obtained in steps 4 and 5, the comprehensive ranking weight of $\tilde{U}$ can be obtained as $\varpi_1 = 0.1505, \varpi_2 = 0.2404, \varpi_3 = 0.3195, \varpi_4 = 0.2896$. Therefore, the final ranking result is $\varpi_3 > \varpi_4 > \varpi_2 > \varpi_1$, namely, x_3 is the best candidate partner of Microsoft.

In addition, the sorting results for different risk parameter values λ are shown in Table 2.

Table 2. Ranking orders of alternatives with different parameter values λ.

λ	ϖ_1	ϖ_2	ϖ_3	ϖ_4	Ranking Order
0.1	0.1350	0.2488	0.2945	0.3217	$x_4 > x_3 > x_2 > x_1$
0.2	0.1388	0.2467	0.3008	0.3137	$x_4 > x_3 > x_2 > x_1$
0.3	0.1427	0.2446	0.3070	0.3057	$x_3 > x_4 > x_2 > x_1$
0.4	0.1466	0.2425	0.3133	0.2976	$x_3 > x_4 > x_2 > x_1$
0.5	0.1505	0.2404	0.3195	0.2896	$x_3 > x_4 > x_2 > x_1$
0.6	0.1544	0.2382	0.3258	0.2816	$x_3 > x_4 > x_2 > x_1$
0.7	0.1583	0.2361	0.3321	0.2735	$x_3 > x_4 > x_2 > x_1$
0.8	0.1622	0.2340	0.3383	0.2655	$x_3 > x_4 > x_2 > x_1$
0.9	0.1660	0.2319	0.3446	0.2575	$x_3 > x_4 > x_2 > x_1$

It can be seen from Table 2 that different sorting results may occur for different risk parameter values λ. When $0 < \lambda < 0.3$, the sorting result is $x_4 > x_3 > x_2 > x_1$, and when $0.3 \leq \lambda \leq 0.9$, the sorting result is $x_3 > x_4 > x_2 > x_1$. This fully demonstrates the importance of DM's risk attitude in GDM and the rationality of the method proposed in this paper. In addition, to further reflect the impact of risk parameter value λ on GDM. We give the variation trend diagram of the compromise weight w_i' and the comprehensive weight ϖ (see Figure 3).

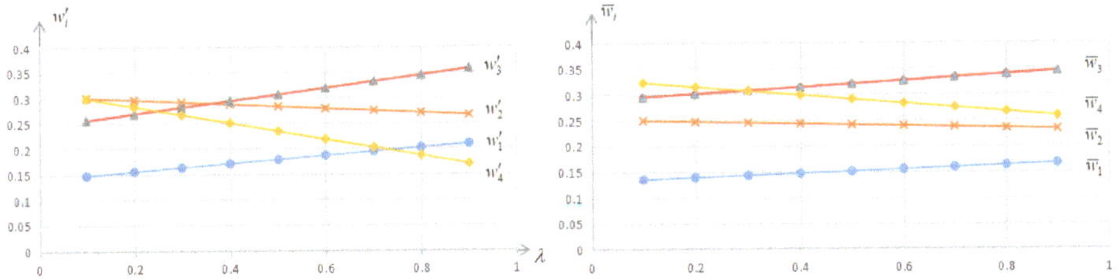

Figure 3. The variation trend of weights w_i' and ϖ_i based on different parameter values λ.

It can be seen intuitively from Figure 3 that the variation trend of the compromise weight w_i' and the comprehensive weight ϖ_i with the increase of λ. Furthermore, by comparing w_i' and w_i , it is easy to see that project x_4 is greatly influenced by λ when only taking into account DM's risk attitude (as the value of λ increases, the value of w_4' decreases from the maximum to the minimum), while its comprehensive weight ϖ_4 is less affected by λ. This further illustrates the necessity and rationality of comprehensive consideration of risk attitude, fuzzy and non-fuzzy uncertain information in GDM problems.

5.2. Comparison Analyses

As a new preference relation, PULIFPR expands the application scope of qualitative information in fuzzy theory and improves the applicability of linguistic terms in GDM. Moreover, as an extension form of LPR, it can be transformed into various LPRs through corresponding changes. Therefore, the method proposed in this paper is also applicable to other types of preference relations, and its specific advantages compared with existing series methods are shown in Table 3.

Table 3. Comparison of different methods.

Methods	Preference Relations	Considering the Non-Preference Information	Considering the Probability Distribution	Considering the Fuzzy Uncertainty (Ignorance Information)	Determining the Individual Weight	Avoiding Consistency Checks and Corrections	Considering the Risk Attitudes of DMs
The proposed method	PULIFPRs	Yes	Yes	Yes	Yes	Yes	Yes
Meng et al.'s method [27]	LIFPRs	Yes	No	No	No	No	No
Xie et al.'s method [37]	PULPRs	No	Yes	No	No	No	No
Zhang et al.'s method [38]	PLPRs	No	Yes	No	No	No	No
Wan et al.'s method [36]	IVIFPRs	Yes	No	Yes	Yes	No	Yes
Liao et al.'s method [39]	IFPRs	Yes	No	No	No	No	No
Wan et al.'s method [32]	IVFPRs	No	No	No	Yes	No	Yes
Zhao et al.'s method [40]	LPRs	No	No	Yes	Yes	No	No
Meng et al.'s method [31]	IVIFPRs	Yes	No	No	Yes	No	No

It is easy to see from Table 3 that compared with other methods, the methods proposed in this paper have many advantages, which not only make up for the deficiencies of current methods, but also avoid the detection and correction of consistency in GDM problems. Specifically speaking, compared with the model proposed by the existing methods, the specific advantages of the model proposed in this paper are as follows

(1) Compared with models M-1 and M-3 in method [27], the model proposed in this paper can directly obtain the priority weight of preference relation without consistency test and correction.
(2) Compared with Algorithms 1 and 2 in method [37], the algorithm proposed in this paper provides a method to determine the individual weight, and the consensus collective preference relation can be obtained directly without iterative calculation.
(3) Compared with the model proposed by wan et al. [36], the model proposed in this paper considers the probability distribution of uncertain information, which is more suitable for large-scale GDM problems in complex environments and can ensure the consistency of collective preference relations.
(4) Compared with the GPM proposed by liao et al. [39], the model proposed in this paper considers both the risk attitude of DMs and the information that they fail to grasp, which improves the rationality and accuracy of decision-making results.

In addition, PULIFS proposed in this paper is a comprehensive extension of the LTS, which can be converted into other sets according to the practical needs of decision problems. Therefore, PULIFS is more general and representative than many existing fuzzy sets, and it is more flexible in the application of decision problem. Furthermore, we classify the information expressed by PULIFPR as fuzzy and non-fuzzy uncertain information to fully consider the preferences, non-preferences and unknown information of the decision-maker. Thus, the method proposed in this paper comprehensively reflects the subjective hesitation, uncertainty and objective randomness existing in actual decision-making problems, and thus ensures the rationality of the DM results.

To sum up, the advantages of the proposed method in practical application can be summarized as follows

(1) Compared with the general preference relation, the PULIFPR proposed in this paper can express both individual preference and group preference, which is more suitable for the increasingly complex decision-making environment. Therefore, the decision-making method proposed in this paper has a broad application prospect. Such as the selection of investment projects, the formulation of enterprise marketing plans, the introduction of talents in institutions, etc.
(2) The method proposed in this paper comprehensively considers the risk attitude and fuzzy uncertain information of DMs, which is more in line with the actual decision-making situation and is easily accepted and adopted by DMs.
(3) The proposed model can guarantee the consistency of the collective preference relation without checking and revising, so it is more simple and accurate in practical application.

However, although the method proposed in this paper has many advantages, it also has some limitations. On the one hand, this paper considers the risk attitude of decision makers, but fails to give a method to determine the value of risk parameters; on the other hand, this paper does not consider the group decision-making problem in the context of incomplete information. Therefore, the method of determining the risk parameter value and extending the proposed method to an incomplete environment will be the future research direction.

6. Conclusions

This paper first briefly summarizes the development history of LTS and puts forward PULIFS, which extends the application of LT in fuzzy theory and promotes the application of qualitative information in GDM. Secondly, the definition of PULIFPR is proposed, which can fully express the subjective hesitation of the decision-maker in the DM problem as well as the uncertainty and randomness of the objective existence. We then defined the distance measure between PULIFSs and used it to determine the individual objective weight, thus increasing the accuracy of information aggregation. Subsequently, a series of GPMs for solving priority weights were established, which not only fully considered the fuzziness of information and DM's risk attitude, but also avoided the test and correction of consistency in GDM. Moreover, we take the project selection of world VR industry conference 2018 as an example demonstrates the effectiveness and practicality of the proposed method.

It is worth noting that this paper only discusses the application of qualitative information in GDM, so it will be an interesting research direction to apply the method proposed in this paper to the quantitative decision-making field, such as IVIFS [41] or PIVIHFS [28], and the decision problem in heterogeneous environment (both qualitative and quantitative information should be considered). In addition, since this paper studies the uncertain problem in a complex environment, it will be a worthy research direction to combine the proposed method with the complex network with fuzzy logic units [42].

Author Contributions: All authors contributed equally.

Funding: This work was supported by National Natural Science Foundation of China (Grant No. 11661053, 11771198) and the Provincial Natural Science Foundation of Jiangxi, China (Grant No. 20181BAB201003).

Acknowledgments: The authors would like to thank the editors and anonymous reviewers for their insightful and constructive commendations that have lead to an improved version of this paper.

Conflicts of Interest: The authors declare no conflict of interest.

References

1. Zadeh, L.A. The concept of a linguistic variable and its application to approximate reasoning—I. *Inf. Sci.* **1975**, *8*, 199–249. [CrossRef]
2. Zadeh, L.A. The concept of a linguistic variable and its application to approximate reasoning—II. *Inf. Sci.* **1975**, *8*, 301–357. [CrossRef]
3. Zadeh, L.A. The concept of a linguistic variable and its application to approximate reasoning—III. *Inf. Sci.* **1975**, *9*, 43–80. [CrossRef]
4. Yager, R.R. A new methodology for ordinal multiobjective decisions based on fuzzy sets. *Decis. Sci.* **1981**, *12*, 589–600. [CrossRef]
5. Degadni, R.; Bortolan, G. The problem of linguistic approximation in clinical decision making. *Int. J. Approx. Reason.* **1988**, *2*, 143–162. [CrossRef]
6. Delgado, M.; Verdegay, J.L.; Vila, M.A. On aggregation operations of linguistic labels. *Int. J. Intell. Syst.* **1993**, *8*, 351-370. [CrossRef]
7. Herrera, F.; Martinez, L. A 2-tuple fuzzy linguistic representation model for computing with words. *IEEE Trans. Fuzzy Syst.* **2000**, *8*, 746–752.
8. Xu, Z.S. A method based on linguistic aggregation operators for group decision making with linguistic preference relations. *Inf. Sci.* **2004**, *166*, 19–30. [CrossRef]
9. Xu, Z.S. Uncertain linguistic aggregation operators based approach to multiple attribute group decision making under uncertain linguistic environment. *Inf. Sci.* **2004**, *168*, 171–184. [CrossRef]
10. Rodriguez, R.M.; Martinez, L.; Herrera, F. Hesitant Fuzzy Linguistic Term Sets for Decision Making. *IEEE Trans. Fuzzy Syst.* **2012**, *20*, 109–119. [CrossRef]
11. Meng, F.; Chen, X.; Zhang, Q. Multi-attribute decision analysis under a linguistic hesitant fuzzy environment. *Inf. Sci.* **2014**, *267*, 287–305. [CrossRef]
12. Zhang, H.M. Linguistic Intuitionistic Fuzzy Sets and Application in MAGDM. *J. Appl. Math.* **2014**, 1–11. [CrossRef]
13. Ye, J. An extended TOPSIS method for multiple attribute group decision making based on single valued neutrosophic linguistic numbers. *J. Intell. Fuzzy Syst.* **2015**, *28*, 247–255.
14. Pang, Q.; Wang, H.; Xu, Z.S. Probabilistic linguistic term sets in multi-attribute group decision making. *Inf. Sci.* **2016**, *369*, 128–143. [CrossRef]

15. Lin, M.W.; Xu, Z.S.; Zhai, Y.L.; Yao, Z.Q. Multi-attribute group decision-making under probabilistic uncertain linguistic environment. *J. Oper. Res. Soc.* **2017**, *22*, 1–15. [CrossRef]
16. Bai, C.; Zhang, R.; Shen, S.; Huang, C.; Fan, X. Interval valued probabilistic linguistic term sets in multi-criteria group decision making. *Int. J. Intell. Syst.* **2018**, *33*, 1301–1321. [CrossRef]
17. Zhang, R.C.; Li, Z.M.; Liao, H.C. Multiple-attribute decision-making method based on the correlation coefficient between dual hesitant fuzzy linguistic term sets. *Knowl.-Based Syst.* **2018**, *159*, 186–192. [CrossRef]
18. Pamucar, D.; Badi, I.; Sanja, K.; Obradovic, R. A Novel Approach for the Selection of Power-Generation Technology Using a Linguistic Neutrosophic CODAS Method: A Case Study in Libya. *Energies* **2018**, *11*, 2489. [CrossRef]
19. Liu, F.; Aiwu, G.; Lukovac, V.; Vukic, M. A multicriteria model for the selection of the transport service provider: A single valued neutrosophic DEMATEL multicriteria model. *Decision Mak. Appl. Manag. Eng.* **2018**, *1*, 121–130. [CrossRef]
20. Orlorski, S.A. Decision making with a fuzzy preference relation. *Fuzzy Sets Syst.* **1978**, *3*, 155–167.
21. Saaty, T.L. *The Analytical Hierarchy Process*; McGraw-Hill: New York, NY, USA, 1980.
22. Herrera, F.; Herrera-Viedma, E.; Verdegay, J.L. A model of consensus in group decision making under linguistic assessments. *Fuzzy Sets Syst.* **1996**, *78*, 73–87. [CrossRef]
23. Xu, Z.S. On compatibility of interval fuzzy preference relations. *Fuzzy Optim. Decis. Mak.* **2004**, *3*, 217–225. [CrossRef]
24. Saaty, T.L.; Vargas, L.G. Uncertainty and rank order in the analytic hierarchy process. *Eur. J. Oper. Res.* **1987**, *32*, 107–117. [CrossRef]
25. Xu, Z.S. Intuitionistic preference relations and their apllication in group decision making. *Inf. Sci.* **2007**, *177*, 2363–2379. [CrossRef]
26. Xia, M.M.; Xu, Z.S.; Liao, H.C. Preference relations based on intuitionistic multiplicative information. *IEEE Trans. Fuzzy Syst.* **2013**, *21*, 113–133.
27. Meng, F.Y.; Tang, J.; Fujita, H. Linguistic intuitionistic fuzzy preference relations and their application to multi-criteria decision making. *Inf. Fusion* **2019**, *46*, 77–90. [CrossRef]
28. Zhai, Y.L.; Xu, Z.S.; Liao, H.C. Measures of Probabilistic Interval-Valued Intuitionistic Hesitant Fuzzy Sets and the Application in Reducing Excessive Medical Examinations. *IEEE Trans. Fuzzy Syst.* **2018**, *26*, 1651–1670. [CrossRef]
29. Tanino, T. Fuzzy preference orderings in group decision making. *Fuzzy Sets Syst.* **1984**, *12*, 117–131. [CrossRef]
30. Chiclana, F.; Herrera-Viedma, E.; Alonso, S.; Herrera, F. Cardinal consistency of reciprocal preference relations: A characterization of mulitiplicative transitivity. *IEEE Trans. Fuzzy Syst.* **2009**, *17*, 14–23. [CrossRef]
31. Meng, F.Y.; Tang, J.; Wang, P.; Chen, X.H. A programming-based algorithm for interval-valued intuitionistic fuzzy group decision making. *Knowl.-Based Syst.* **2018**, *144*, 122–143. [CrossRef]
32. Wan, S.P.; Wang, F.; Dong, J.Y. A group decision making method with interval valued fuzzy preference relations based on the geometric consistency. *Inf. Fusion* **2018**, *40*, 87–100. [CrossRef]
33. Wu, J.; Chiclana, F.; Liao, H.C. Isomorphic Multiplicative Transitivity for Intuitionistic and Interval-Valued Fuzzy Preference Relations and Its Application in Deriving Their Priority Vectors. *IEEE Trans. Fuzzy Syst.* **2018**, *26*, 193–202. [CrossRef]
34. Zhou, W.; Xu, Z.S. Probability Calculation and Element Optimization of Probabilistic Hesitant Fuzzy Preference Relations Based on Expected Consistency. *IEEE Trans. Fuzzy Syst.* **2018**, *26*, 1367–1378. [CrossRef]
35. Wang, Z.J. A Goal Programming Based Heuristic Approach to Deriving Fuzzy Weights in Analytic Form from Triangular Fuzzy Preference Relations. *IEEE Trans. Fuzzy Syst.* **2018**. [CrossRef]
36. Wan, S.P.; Wang, F.; Dong, J.Y. Three-Phase Method for Group Decision Making With Interval-Valued Intuitionistic Fuzzy Preference Relations. *IEEE Trans. Fuzzy Syst.* **2018**, *26*, 998–1010. [CrossRef]
37. Xie, W.Y.; Ren, Z.L.; Xu, Z.S.; Wang, H. The consensus of probabilistic uncertain linguistic preference relations and the application on the virtual reality industry. *Knowl.-Based Syst.* **2018**, *162*, 14–28. [CrossRef]
38. Zhang, Y.X.; Xu, Z.S.; Liao, H.C. A consensus process for group decision making with probabilistic linguistic preference relations. *Inf. Sci.* **2017**, *414*, 260–275. [CrossRef]
39. Liao, H.C.; Xu, Z.; Zeng, X.J.; Merigó, J.M. Framework of group decision making with intuitionistic fuzzy preference information. *IEEE Trans. Fuzzy Syst.* **2015**, *23*, 1211–1227. [CrossRef]

40. Zhao, M.; Ma, X.Y.; Wei, D.W. A method considering and adjusting individual consistency and group consensus for group decision making with incomplete linguistic preference relations. *Appl. Soft Comput.* **2017**, *54*, 322–346. [CrossRef]
41. Atanassov, K.; Gargov, G. Interval valued intuitionistic fuzzy sets. *Fuzzy Sets Syst.* **1989**, *31*, 343–349. [CrossRef]
42. Bucolo, M.; Fortuna, L.; La Rosa, M. Complex dynamics through fuzzy chains. *IEEE Trans. Fuzzy Syst.* **2004**, *12*, 289–295. [CrossRef]

symmetry MDPI

Article

Assessment of Enterprise Performance Based on Picture Fuzzy Hamacher Aggregation Operators

Chiranjibe Jana * and Madhumangal Pal

Department of Applied Mathematics with Oceanology and Computer Programming, Vidyasagar University, Midnapore 721102, India; mmpalvu@gmail.com

* Correspondence: jana.chiranjibe7@gmail.com; Tel.: +91-96-4713-5650

Received: 28 November 2018; Accepted: 25 December 2018; Published: 11 January 2019

Abstract: In the age of the knowledge-based economy and the rapid development of information technology, enterprise management is facing great challenges and has entered an era of prudent management. Traditional enterprise performance evaluation focuses on the interests of shareholders. Investors take financial data as their base and pay attention to the study of material attraction and the results; if they do not, they cannot adjust to a new economy period. Therefore, enterprise performance reflects the interests of shareholders and business strategists for the needs of stakeholders, which is important for the future of lively competition. With that in mind, aggregation of information is an important research tool that has recently drawn the attention of researchers for information analysis. In this paper, we have developed multiple-attribute decision-making methods for enterprise performance evaluation with picture fuzzy information. We have applied Hamacher aggregation operators such as the picture fuzzy Hamacher weighted averaging (PFHWA) operator and picture fuzzy Hamacher weighted geometric (PFHWG) operator in picture fuzzy environment for the assessment of the best enterprise selection. Finally, we justified the proposed approach with the existing methods for feasibility and effectiveness.

Keywords: multiple attribute decision making (MADM); picture fuzzy numbers; Hamacher operations; picture fuzzy Hamacher aggregation operators; evaluation of enterprise performance

1. Introduction

Financial management is an important part of strategic management research, which investigates how enterprises exploit proper strategies to create and maintain competitive advantages. Presently, research into competition has grown exponentially. Mintzberg et al. [1] criticized overly analytical orientation, upper management slant, lack of attention to action and learning, and neglect of the elements that lead to the creation of strategies. Shrivastava [2] focused his research on organizational learning processes; this has the potential to offer insights into these identified drawbacks. Brockman and Morgan [3] showed that organizational knowledge is the basis for gaining and defending competitive advantage and a key variable in the amplification of firm performance. Furthermore, some studies showed evidence of a useful relationship between organizational learning and inflexible performance. For example, Baker and Sinkula [4] proved that learning intention has a direct effect on firm performance. Ussahawanitchakit [5] practiced an advance measure of knowledge and obtained similar issues. Enterprise performance estimation is not only development of market economy within a certain time, but also a scientific method and constructive tool to supervise enterprises for a nation with current market economy. Work for the evaluation of enterprise performance of our country, how to change our economy, social environment, and the trend of internationalization and how to show a system of performance appraisal that fits our economic development plays an especially important operational significance for enhancing the health of our enterprises, improving the management level, sharpening the enterprise competitiveness, and further improving the economic growth quality. Many MCDM or multiple attribute decision making (MADM) problems (such as business and strategic financial management, medical diagnosis etc.) have been developed with the use of aggregation [6–8] operators under probabilistic environment. Merigo et al. [9] introduced order weighted averaging operator, induce aggregation operators, weighted averaging operator and then used these operators to develop strategic decision making theory. Li [10] have studied decision and game theory in management in the environment of intuitionistic fuzzy information. The study of aggregation functions under different fuzzy environment is an important research instrument in decision science. In next paragraph, we briefly overview of some fuzzy aggregation functions and their corresponding decision making problems.

It is very difficult to take real attribute values, because of complexity presented in serious level in the field of decision environment. In 1965, Zadeh [11] introduced theory of fuzzy sets (FS), a new mathematical notion to handle easily multi-criteria decision making MCDM [12,13] problems and multi-criteria group decision making MCGDM [14,15] problems. It is known to all that intuitionistic fuzzy sets (IFSs) and interval-valued intuitionistic fuzzy sets (IVIFSs) [16] are the generalization of IFS. All these traditional theories are very interesting research topics which have engaged attention of the researchers, because these theories have successful applications in different directions such as decision-making, medical diagnosis, pattern recognition, cluster analysis etc. Then many scholars expressed intuitionistic fuzzy information in their excellent study as follows, Chen and Chou [17] used particle swarm optimization method to develop a MADM problem under the information IVIF numbers. Du and Liu [18] proposed to study a MADM problem based on VIKOR method using intuitionistic trapezoidal fuzzy information. Garg [19] studied an MCDM problem with unknown attribute values by defining a new score and accuracy function under IVIF environment. Kumar and Garg [20] contributed an MADM problem for INVIF environment based on SPA. Li [21] proposed a MADM to develop a closeness coefficient of nonlinear programming method based on IVIF values in which preference values of the attribute are unknown. Lourenzutti and Krohling [22] used IFVs to develop MCDM method using fuzzy-TODIM technique. Wan and Li [23,24] used to develop heterogeneous MAGDM problems with A-IF (A-IVIF) information and then applicability of the proposed method is justified with an example of real supplier selection. Ye [25] motivated to analyze MCDM based on novel accuracy function under IVIF environment. Recently, researchers have drawn attention to model covering-based IF rough sets with their applications to MADM problems [26,27]. Additionally, intuitionistic fuzzy rough graphs are helpful to understand these models and can solved decision-making problems, which is explained in [28].

All the execution of the criteria for alternatives, weighted and order weighted aggregation operators [29,30], have a major role in the course of the document aggregation. In that view, Xu [31,32] introduced some weighted aggregation operators to solve MAGDM problems. Recently, aggregation of information for operators is an interesting research subject, receiving great attention of the researchers in the light of Hamacher operations. The operation introduced by Hamacher [33], known as "Hamacher operation (HO)", is a combination of algebraic TN and TCN, and Einstein TN and TCN [34]. Huang [35] introduced intuitionistic fuzzy Hamacher aggregation operators and gave application of these operators in MADM. Liu [36] presented some interval-valued intuitionistic fuzzy Hamacher aggregation operators and applied it to MAGDM problems. Xiao [37] proposed order weighted Hamacher geometric operator under IVIF environment. Li [38] gave the attention to the study of Hamacher correlated averaging (HCA) operator with the help of Hamacher sum and the Hamacher product operations based on IVIF information. The readers can get more information about Hamacher and other aggregation functions and developed decision-making methods from the following references [39–51].

Although IFS and IVIFS have been successfully applied to solve real world problems, in reality there are some conditions which cannot be handled by IFSs. Suppose, that in case of voting, human outlook involved more responses such as yes, abstain, no, refusal, which cannot be perfectly presented by traditional FS and IFS. To overcome this situation, the notion of picture fuzzy (PFS) set was originated by Cuong [52–54] as a new mathematical tool for computational intelligence problems. A PFS is identified by functions called (positive, neutral, refusal)-membership function during the information analysis. With this point of view, a PFS can be considered as a generalization of IF and the fuzzy sets. Recently, many researchers have studied PFSs and its applications: Sing [55] suggested correlation coefficient of PFS, and gave an application in clustering analysis. Jana et al. [56] used weighted Dombi aggregation operators in picture fuzzy environment, and using these operators developed software selection method. Based on some new fuzzy algorithms on the basis of PFSs environment, Son and others [57,58] provided weather forecasting and time series forecasting. Thong [59] studied a novel fuzzy hybrid model for PF-clustering, and IF-systems for Medical diagnosis and health care support systems. Son [60] investigated generalized PF-distance measure, and applied it to solve clustering analysis problems under PFSs environment. Wei and others [61,62] used PF-information aggregation to find ranking of EPR systems, picture 2-tuple linguistic Bonferroni mean-based model, picture 2-tuple linguistic model for the solution of multi attribute decision making problems. To see more information about PFS applied to risk management, picture preference relation and picture 2-tuple linguistic [63–65] information used to find their corresponding decision making. Recently, Wei [66] introduced Hamacher aggregation operator on picture fuzzy set. He studied different kinds of Hamacher aggregation operators under picture fuzzy environment such as (PFHA) operators, (PFHG) operators, (PFHCA) operators, (IPFHA) operators, (IPFHCA) operators, (PFHPA) operators, (PFHPA) operators, and then provided a MADM problem for the utility and flexibility of proposed method. Therefore, based on Hamacher operation, how to aggregate these PFS is a very useful topic. In this paper, we shall define some PF Hamacher aggregation operators on the basis of traditional arithmetic [29,30,32], geometric operations [31], and Hamacher operations [33,66].

The PFS has a powerful ability to model the ambiguous and imprecise information of the real world. In literature, there are many different works related to applications of fuzzy aggregation method in decision making problem based on Hamacher operations. With this motivation, we used picture fuzzy Hamacher weighted averaging (PFHWA) operator, picture fuzzy Hamacher weighted geometric (PFHWGA) operator to access the best enterprise on the basis of performance evaluation of enterprises.

2. Preliminaries

In this section, some basic definitions and operations related intuitionistic fuzzy sets and picture fuzzy sets are recalled briefly.

2.1. Intuitionistic Fuzzy Sets

Let X be universe of discourse. Then, an intuitionistic fuzzy set (IFS) [16] is defined as follows:

$$A = \{\langle \hat{\mu}_A(x), \hat{\nu}_A(x)\rangle | x \in X\}, \tag{1}$$

where $\hat{\mu}_A : X \to [0,1]$ and $\hat{\nu}_A : X \to [0,1]$ are called membership function and non-membership function of IFS A, respectively. Here $0 \leq \hat{\mu}_A + \hat{\nu}_A \leq 1$ for all $x \in X$ and $\pi = 1 - (\hat{\mu}_A(x) + \hat{\nu}_A(x)$ is called degree of indeterminacy of $x \in X$ in IFS A. The pair $\langle \hat{\mu}_A, \hat{\nu}_A \rangle$ is called intuitionistic fuzzy value (IFV) or intuitionistic fuzzy number (IFN) by Xu [31].

2.2. Picture Fuzzy Sets

Let X be a universe of discourse objects. A picture fuzzy set over X, denoted by $\hat{P}$, is defined in [52,53] as follows:

$$\hat{P} = \{\langle \hat{\mu}_{\hat{P}}(x), \hat{\eta}_{\hat{P}}(x), \hat{\nu}_{\hat{P}}(x)\rangle : x \in X\}, \tag{2}$$

where $\hat{\mu}_{\hat{P}} : X \to [0,1]$, $\hat{\eta}_{\hat{P}} : X \to [0,1]$ and $\hat{\nu}_{\hat{P}} : X \to [0,1]$ are called positive (neutral, negative)-degree of membership of picture fuzzy set $\hat{P}$, respectively. Here $0 \leq \hat{\mu}_{\hat{P}}(x) + \hat{\eta}_{\hat{P}}(x) + \hat{\nu}_{\hat{P}}(x) \leq 1$ for all $x \in X$. Besides, $\pi_{\hat{P}}(x)$ denotes degree of refusal of $x \in X$, and is defined as $\pi_{\hat{P}}(x) = 1 - (\hat{\mu}_{\hat{P}}(x) + \hat{\eta}_{\hat{P}}(x) + \hat{\nu}_{\hat{P}}(x))$. The pair $(\hat{\mu}_{\hat{P}}, \hat{\eta}_{\hat{P}}, \hat{\nu}_{\hat{P}})$ is called picture fuzzy value (PFV) or picture fuzzy element (PFE).

Definition 1 ([67]). *Let $\hat{P} = (\hat{\mu}_{\hat{P}}, \hat{\eta}_{\hat{P}}, \hat{\nu}_{\hat{P}})$ be a PFN. Then, the score function $\hat{S}$ of PFN $\hat{P}$, denoted by $\hat{S}(\hat{P})$, is defined as follows:*

$$\hat{S}(\hat{P}) = \hat{\mu}_{\hat{P}} - \hat{\nu}_{\hat{P}}, \ \ \hat{S}(\hat{P}) \in [-1,1]. \tag{3}$$

Definition 2 ([67]). *Let $\hat{P} = (\hat{\mu}_{\hat{P}}, \hat{\eta}_{\hat{P}}, \hat{\nu}_{\hat{P}})$ be a PFN. Then, the accuracy function $\hat{H}$ of PFN $\hat{P}$, denoted by $\hat{H}(\hat{P})$, is defined as follows:*

$$\hat{H}(\hat{P}) = \hat{\mu}_{\hat{P}} + \hat{\eta}_{\hat{P}} + \hat{\nu}_{\hat{P}}, \hat{H}(\hat{P}) \in [0,1]. \tag{4}$$

Here, the larger value of $\hat{H}(\hat{P})$ implies a greater degree of accuracy of the PFE $\hat{P} = (\hat{\mu}_{\hat{P}}, \hat{\eta}_{\hat{P}}, \hat{\nu}_{\hat{P}})$.

3. Hamacher Operations (HOs) on the Picture Fuzzy Set

3.1. Hamacher Operations

The TN and TCN are useful notions in fuzzy set theory, that are used to define general union and intersection of fuzzy sets [68]. The definitions and conditions of TN and TCN are proposed by Roychowdhury and Wang [69]. The generalized union and generalized intersection of intuitionistic fuzzy sets based on TN and TCN were provided by Deschrijver and Kerre [70]. In 1978, Hamacher [33] introduced HOs known as Hamacher product ($\otimes$) and Hamacher sum ($\oplus$), which are examples of TN and TCN, respectively. Hamacher TN and Hamacher TCN are provided in the following definition

$$T_H(u,v) = u \bigotimes v = \frac{uv}{\xi + (1-\xi)(u+v-uv)} \tag{5}$$

$$T_H^*(u,v) = u \bigoplus v = \frac{u+v-uv-(1-\xi)uv}{1-(1-\xi)uv}. \tag{6}$$

Usually, when $\xi = 1$, then Hamacher TN and TCN will reduce to the form

$$T_H(u,v) = u \bigotimes v = uv \tag{7}$$

$$T_H^*(u,v) = u \bigoplus v = u + v - uv \tag{8}$$

which represent algebraic TN and TCN. When $\xi = 2$, then Hamacher TN and Hamacher TCN will conclude to the form

$$T_H(u,v) = u \bigotimes v = \frac{uv}{1+(1-u)(1-v)} \tag{9}$$

$$T_H^*(u,v) = u \bigoplus v = \frac{u+v}{1+uv} \tag{10}$$

which are called Einstein TN and TCN respectively.

3.2. Hamacher Operations(HOs) of Picture Fuzzy Set

Here, given some Hamacher operations on PFNs which are provided by Wei [66]. Let A and B be two PFSs and $\kappa > 0$. Then, Hamacher product and Hamacher sum of the two PFSs A and B are denoted by $(\hat{p}_1 \otimes \hat{p}_2)$ and $(\hat{p}_1 \oplus \hat{p}_2)$, respectively, and defined by

- $\hat{p}_1 \oplus \hat{p}_2 = \left(\frac{\hat{\mu}_1+\hat{\mu}_2-\hat{\mu}_1\hat{\mu}_2-(1-\xi)\hat{\mu}_1\hat{\mu}_2}{1-(1-\xi)\hat{\mu}_1\hat{\mu}_2}, \frac{\hat{\eta}_1\hat{\eta}_2}{\xi+(1-\xi)(\hat{\eta}_1+\hat{\eta}_2-\hat{\eta}_1\hat{\eta}_2)}, \frac{\hat{v}_1\hat{v}_2}{\xi+(1-\xi)(\hat{v}_1+\hat{v}_2-\hat{v}_1\hat{v}_2)} \right)$
- $\hat{p}_1 \otimes \hat{p}_2 = \left(\frac{\hat{\mu}_1\hat{\mu}_2}{\xi+(1-\xi)(\hat{\mu}_1+\hat{\mu}_2-\hat{\mu}_1\hat{\mu}_2)}, \frac{\hat{\eta}_1+\hat{\eta}_2-\hat{\eta}_1\hat{\eta}_2-(1-\xi)\hat{\eta}_1\hat{\eta}_2}{1-(1-\xi)\hat{\eta}_1\hat{\eta}_2}, \frac{\hat{v}_1+\hat{v}_2-\hat{v}_1\hat{v}_2-(1-\xi)\hat{v}_1\hat{v}_2}{1-(1-\xi)\hat{v}_1\hat{v}_2} \right)$
- $\kappa\hat{p}_1 = \left(\frac{(1+(\xi-1)\hat{\mu}_1)^\kappa-(1-\hat{\mu}_1)^\kappa}{(1+(\xi-1)\hat{\mu}_1)^\kappa+(\xi-1)(1-\hat{\mu}_1)^\kappa}, \frac{\xi(\hat{\eta})^\kappa}{(1+(\xi-1)(1-\hat{\eta}_1))^\kappa+(\xi-1)(\hat{\eta})^\kappa}, \frac{\xi(\hat{v}_1)^\kappa}{(1+(\xi-1)(1-\hat{v}_1))^\kappa+(\xi-1)(\hat{v}_1)^\kappa} \right), \kappa > 0$
- $\hat{p}_1^\kappa = \left(\frac{\xi(\hat{\mu})^\kappa}{(1+(\xi-1)(1-\hat{\mu}_1))^\kappa+(\xi-1)(\hat{\mu})^\kappa}, \frac{(1+(\xi-1)\hat{\eta}_1)^\kappa-(1-\hat{\eta}_1)^\kappa}{(1+(\xi-1)\hat{\eta}_1)^\kappa+(\xi-1)(1-\hat{\eta}_1)^\kappa}, \frac{(1+(\xi-1)\hat{v}_1)^\kappa-(1-\hat{v}_1)^\kappa}{(1+(\xi-1)\hat{v}_1)^\kappa+(\xi-1)(1-\hat{v}_1)^\kappa} \right), \kappa > 0.$

Now, we have drawn the attention on picture fuzzy Hamacher weighted averaing operator (PFHWA) and picture fuzzy Hamacher weighted geometric (PFHWG) operator introduced by Wei [66] that are as follows.

Definition 3 ([66]). *Let $\hat{p}_q = (\hat{\mu}_q, \hat{\eta}_q, \hat{v}_q)$ $(q = 1, 2, \ldots, s)$ be several picture fuzzy number (PFNs).*

A picture fuzzy Hamacher weighted average (PFHWA) operator is defined as a mapping from $\tilde{P}^s$ to $\tilde{P}$ as follows:

$$PFHWA_\Psi(\hat{p}_1, \hat{p}_2, \ldots, \hat{p}_s) = \bigoplus_{q-1}^{s} (\Psi_q \hat{p}_q) \tag{11}$$

where $\Psi = (\Psi_1, \Psi_2, \ldots, \Psi_s)^T$ is the weight vector of $\hat{p}_q$ $(q = 1, 2, \ldots, s)$ with $\Psi_q > 0$ and $\sum_{q=1}^{s} \Psi_q = 1$.

Now, we considered two special cases subsequently for the PFHWA operator when the parameter ξ takes the values 1 or 2.

Case 1. If $\xi = 1$, then PFHWA operator will reduce to PFWA operator (Wei, 2017):

$$\begin{aligned} PFWA_\Psi(\hat{p}_1, \hat{p}_2, \ldots, \hat{p}_s) &= \bigoplus_{q=1}^{s} (\Psi_q \hat{p}_q) \\ &= \left(1 - \prod_{q=1}^{s} (1-\hat{\mu}_q)^{\Psi_q}, \prod_{q=1}^{s} (\hat{\eta}_q)^{\Psi_q}, \prod_{q=1}^{s} (\hat{v}_q)^{\Psi_q} \right). \end{aligned}$$

Case 2. If $\xi = 2$, then PFHWA operator will reduce to picture fuzzy Einstein weighted averaging (PFEWA) operator:

$$PFEWA_{\Psi}(\hat{p}_1, \hat{p}_2, \ldots, \hat{p}_s) = \bigoplus_{q=1}^{s}(\Psi_q \hat{p}_q)$$
$$= \left(\frac{\prod_{q=1}^{s}(1+\hat{\mu}_q)^{\Psi_q} - \prod_{q=1}^{s}(1-\hat{\mu}_q)^{\Psi}}{\prod_{q=1}^{s}(1+\hat{\mu}_q)^{\Psi_q} + \prod_{q=1}^{s}(1-\hat{\mu}_q)^{\Psi_q}}, \frac{2\prod_{q=1}^{s}(\hat{\eta}_q)^{\Psi_q}}{\prod_{q=1}^{s}(2-\hat{\eta}_q)^{\Psi_q} + \prod_{q=1}^{s}(\hat{\eta}_q)^{\Psi_q}}, \frac{2\prod_{q=1}^{s}(\hat{v}_q)^{\Psi_q}}{\prod_{q=1}^{s}(2-\hat{v}_q)^{\Psi_q} + \prod_{q=1}^{s}(\hat{v}_q)^{\Psi_q}} \right). \quad (12)$$

Definition 4 ([66]). *Let $\hat{p}_q = (\hat{\mu}_q, \hat{\eta}_q, \hat{v}_q)$ $(q = 1, 2, \ldots, s)$ be several PFNs. A picture fuzzy Hamacher weighted geometric (PFHWG) operator is defined as a mapping $PFHWG : \hat{P}^s \to \hat{P}$ by*

$$PFHWG_{\Psi}(\hat{p}_1, \hat{p}_2, \ldots, \hat{p}_s) = \bigotimes_{q=1}^{s}(\hat{p}_q)^{\Psi_q} \quad (13)$$

where $\Psi = (\Psi_1, \Psi_2, \ldots, \Psi_s)^T$ is the weight vector of $\hat{p}_q$ $(q = 1, 2, \ldots, s)$ such that $\Psi_q > 0$ and $\sum_{q=1}^{s} \Psi_q = 1$.

Case 1. If $\xi = 1$, PFHWG operator reduces to picture fuzzy weighted geometric (PFWG) operator:

$$PFWG_{\Psi}(\hat{p}_1, \hat{p}_2, \ldots, \hat{p}_s) = \bigotimes_{q=1}^{s}(\hat{p}_q)^{\Psi_q}$$
$$= \left(\prod_{q=1}^{s}(\hat{\mu}_q)^{\Psi_q}, 1 - \prod_{q=1}^{s}(1-\hat{\eta}_q)^{\Psi_q}, 1 - \prod_{q=1}^{s}(1-\hat{\eta}_q)^{\Psi_q} \right). \quad (14)$$

Case 2. If $\xi = 2$, then PFHWG operator reduces to a picture fuzzy Einstein weighted geometric (PFEWG) operator:

$$PFEWG_{\Psi}(\hat{p}_1, \hat{p}_2, \ldots, \hat{p}_s) = \bigotimes_{q=1}^{s}(\hat{p}_q)^{\Psi_q}$$

$$= \left(\frac{2\prod_{q=1}^{s}(\hat{\mu}_q)^{\Psi_q}}{\prod_{q=1}^{s}(2-\hat{\mu}_q)^{\Psi_q} + \prod_{q=1}^{s}(\hat{\mu}_q)^{\Psi_q}}, \frac{\prod_{q=1}^{s}(1+\hat{\eta}_q)^{\Psi_q} - \prod_{q=1}^{s}(1-\hat{\eta}_q)^{\Psi}}{\prod_{q=1}^{s}(1+\hat{\eta}_q)^{\Psi_q} + \prod_{q=1}^{s}(1-\hat{\eta}_q)^{\Psi}}, \frac{\prod_{q=1}^{s}(1+\hat{\eta}_q)^{\Psi_q} - \prod_{q=1}^{s}(1-\hat{\eta}_q)^{\Psi}}{\prod_{q=1}^{s}(1+\hat{\eta}_q)^{\Psi_q} + \prod_{q=1}^{s}(1-\hat{\eta}_q)^{\Psi}} \right).$$

4. Model for MADM Using Picture Fuzzy Information

To this part, multiple attribute decision making (MADM) method is proposed based on PFHA operators of which weights of attributes are real numbers and values of attributes are PFNs. To illustrate effectiveness of the proposed MADM method, an application in evaluation of enterprises performance under picture fuzzy information is given. Let $Q = \{Q_1, Q_2, \ldots, Q_r\}$ be the discrete set of alternatives and $G = \{G_1, G_2, \ldots, G_s\}$ be the set of attributes.

Let $\Psi = (\Psi_1, \Psi_2, \ldots, \Psi_s)$ be the weight vector of the attribute such that $\Psi_b > 0$ $(b = 1, 2, \ldots, s)$ and $\sum_{b=1}^{s} \Psi_b = 1$, and $R = (\hat{\mu}_{ab}, \hat{\eta}_{ab}, \hat{v}_{ab})_{r \times s}$ be a picture fuzzy decision matrix. Here, $\hat{\mu}_{ab}$ is the degree of the positive membership for which alternative Q_a satisfies the attribute G_b given by the decision makers, $\hat{\eta}_{ab}$ denote the degree of neutral membership such that alternative Q_a does not satisfy the attribute G_b, and $\hat{v}_{ab}$ provides the degree that the alternative Q_a does not satisfy the attribute G_b given by the decision maker, where $\hat{\mu}_{ab} \subset [0,1]$, $\hat{\eta}_{ab} \subset [0,1]$ and $\hat{\upsilon}_{ab} \subset [0,1]$ such that $0 \leq \hat{\mu}_{ab} + \hat{\eta}_{ab} + \hat{\upsilon}_{ab} \leq 1$, $(a = 1,2,\ldots,r)$ and $(b = 1,2,\ldots,s)$.

In the following algorithm, a MADM method using PFHWA and PFHWG operators is proposed to solve problems involving picture fuzzy information.

Step 1. Construction of decision matrix R by decision makers under PF-information:

$$R = \begin{pmatrix} \hat{\beta}_{11} & \hat{\beta}_{12} & \cdots & \hat{\beta}_{1r} \\ \hat{\beta}_{21} & \hat{\beta}_{22} & \cdots & \hat{\beta}_{2r} \\ \vdots & \vdots & \ddots & \vdots \\ \hat{\beta}_{a1} & \hat{\beta}_{s2} & \cdots & \hat{\beta}_{ab} \end{pmatrix}$$

Step 2. Finding of values of $\hat{\beta}_a$ $(a = 1,2,\ldots r)$ based on decision matrix R: These values are found by using PFHWA (or PFHWG) given as follow:

$$\begin{aligned} \hat{\beta}_a &= PFHWA(\hat{\beta}_{a1}, \hat{\beta}_{a2}, \ldots, \hat{\beta}_{ab}) = \bigoplus_{b=1}^{s} (\Psi_b \hat{\beta}_{ab}) \\ &= \left(\frac{\prod_{b=1}^{s}(1+(\xi-1)\hat{\mu}_b)^{\Psi_b} - \prod_{b=1}^{s}(1-\hat{\mu}_b)^{\Psi_b}}{\prod_{b=1}^{s}(1+(\xi-1)\hat{\mu}_b)^{\Psi_b} + (\xi-1)\prod_{b=1}^{s}(1-\hat{\mu}_b)^{\Psi_b}}, \frac{\xi \prod_{b=1}^{s}(\hat{\eta}_b)^{\Psi_b}}{\prod_{b=1}^{s}(1+(\xi-1)(1-\hat{\eta}_b))^{\Psi_b} + (\xi-1)\prod_{b=1}^{s}(\hat{\eta}_b)^{\Psi_b}}, \right. \\ &\qquad \left. \frac{\xi \prod_{b=1}^{s}(\hat{\eta}_b)^{\Psi_b}}{\prod_{b=1}^{s}(1+(\xi-1)(1-\hat{\eta}_b))^{\Psi_b} + (\xi-1)\prod_{b=1}^{s}(\hat{\eta}_b)^{\Psi_b}} \right), \end{aligned} \tag{15}$$

$(a = 1,2,\ldots,r)$ or $\hat{\beta}_a = PFHWG(\hat{\beta}_{a1}, \hat{\beta}_{a2}, \ldots, \hat{\beta}_{ab}) = \bigotimes_{b=1}^{s} (\hat{\beta}_{ab})^{\Psi_b}$

$$\begin{aligned} &= \left(\frac{\xi \prod_{b=1}^{s}(\hat{\mu}_b)^{\Psi_b}}{\prod_{b=1}^{s}(1+(\xi-1)(1-\hat{\mu}_b))^{\Psi_b} + (\xi-1)\prod_{b=1}^{s}(\hat{\mu}_b)^{\Psi_b}}, \frac{\prod_{b=1}^{s}(1+(\varsigma-1)\hat{\eta}_b)^{\Psi_b} - \prod_{b=1}^{s}(1-\hat{\eta}_b)^{\Psi_b}}{\prod_{b=1}^{s}(1+(\xi-1)\hat{\eta}_b)^{\Psi_b} + (\xi-1)\prod_{b=1}^{s}(1-\hat{\eta}_b)^{\Psi_b}}, \right. \\ &\qquad \left. \frac{\prod_{q=1}^{s}(1+(\xi-1)\hat{\eta}_b)^{\Psi_b} - \prod_{b=1}^{s}(1-\hat{\eta}_b)^{\Psi_b}}{\prod_{b=1}^{s}(1+(\xi-1)\hat{\eta}_b)^{\Psi_b} + (\xi-1)\prod_{b=1}^{s}(1-\hat{\eta}_b)^{\Psi_b}} \right), \end{aligned} \tag{16}$$

$(a = 1,2,\ldots,r)$ to obtain the overall preference values $\hat{\beta}_a$ $(a = 1,2,\ldots,r)$ of the alternative Q_r.

Step 3. Calculate the score $\hat{S}(\hat{\beta}_a)$ $(a = 1,2,\ldots,r)$ by using Equation (3) based on overall PF-information $\hat{\beta}_a$ $(a = 1,2,\ldots,r)$ in order to rank all the alternative Q_u $(a - 1,2,\ldots,r)$ to choose the best choice Q_a. If score values of $\hat{S}(\hat{\beta}_a)$ and $\hat{S}(\hat{\beta}_c)$ are equal, accuracy degrees of $\hat{H}(\hat{\beta}_a)$ and $\hat{H}(\hat{\beta}_c)$ based on overall picture fuzzy information of $\hat{\beta}_a$ and $\hat{\beta}_c$ are calculated, and rank the alternative Q_a depending with the accuracy of $\hat{H}(\hat{\beta}_a)$ and $\hat{H}(\hat{\beta}_c)$.

Step 4. To rank the alternatives Q_a $(a = 1,2,\ldots,r)$, choose the best one(s) in accordance with $\hat{S}(\hat{\beta}_a)$ $(a = 1,2,\ldots,r)$.

Step 5. Select the best alternative.

Step 6. Stop.

5. Numerical Example and Comparative Analysis

5.1. Numerical Example

The long-term stable development of enterprise hampered due to these issues: Development of production, environmental pollution, poor quality production, waste of resources, and lack of protection of the interests of the employees, as a result shareholders lose interest to invest their wealth, and they urge to special-purpose investment to the company and bear the investment risk. Thus, an enterprise's growth and survival depends on its ability to effectively deal with the relationship among various shareholders. The strategic management experts gradually realized that it is a small-minded behavior for enterprises if they want to achieve the goal of shareholder value in the production of process, regardless of the interest of other stakeholders requirements. From the standpoint of stakeholders, as a supervision and management system, the enterprise's financial performance is not only an enterprise's important self-monitoring, self-restraint, self-evaluation, but also have a vital instrumentation to effectively communicate with stakeholders, coordinating each stakeholder's interest, and finally achieving the strategic management goal of enterprise. In this part, we shall present a project for the selection of best enterprise alternative(s) on the basis of the present trend of enterprise financial performances in order to investigate our proposed method. Here, we have evaluated the enterprise overall performance of five possible enterprises Q_t $(t = 1, 2, 3, 4, 5)$. A company invests its money to an enterprize with the enterprise performances, and seeks to maximize the expected profit. In that view, it is required to calculate the enterprise performance of five possible enterprises as to select the desirable one. The whole decision-making process is presented by a flow-chart in Figure 1. The investment company take a decision depending on the following four attributes:

G_1 Financial performance
G_2 Customer performance
G_3 Internal processes of performance
G_4 Staff performance..

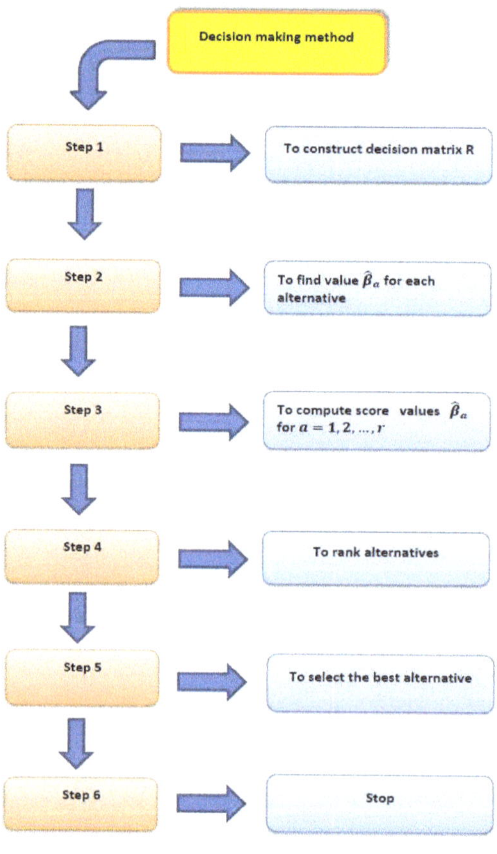

Figure 1. A flow chart of PFNs based on multiple attribute decision making (MADM) problem.

To keep away from dominating each other, decision makers are required to exempted the five possible enterprises Q_a $(q = 1,2,3,4,5)$ under the considered attributes whose weight vector $(0.2, 0.1, 0.3, 0.4)$ determined by decision makers. According to opinions of decision makers, decision matrix $\widetilde{R} = (\hat{\beta}_{ab})_{5\times 4}$ is constructed under picture fuzzy information as in Table 1.

- **Step 1.** Decision matrix R is constructed by decision maker or expert under PF information as follows:

Table 1. Decision matrix R under picture fuzzy (PF)-information.

	G_1	G_2	G_3	G_4
Q_1	(0.56, 0.34, 0.10)	(0.90, 0.06, 0.04)	(0.40, 0.33, 0.19)	(0.09, 0.79, 0.03)
Q_2	(0.70, 0.10, 0.09)	(0.10, 0.66, 0.20)	(0.06, 0.81, 0.12)	(0.72, 0.14, 0.09)
Q_3	(0.88, 0.09, 0.03)	(0.08, 0.10, 0.06)	(0.05, 0.83, 0.09)	(0.65, 0.25, 0.07)
Q_4	(0.80, 0.07, 0.04)	(0.70, 0.15, 0.11)	(0.03, 0.88, 0.05)	(0.07, 0.82, 0.05)
Q_5	(0.85, 0.06, 0.03)	(0.64, 0.07, 0.22)	(0.06, 0.88, 0.05)	(0.13, 0.77, 0.09)

- **Step 2.** Let $\xi = 3$. By using the PFHWA operator of the overall performance values $\hat{\beta}_a$ of enterprises, Q_a $(a = 1,2,3,4,5)$ are obtained as follows:

$$\hat{\beta}_1 = \left(\frac{[(1+2\times 0.56)^{0.2}\times(1+2\times 0.90)^{0.1}\times(1+2\times 0.40)^{0.3}\times(1+2\times 0.09)^{0.4}]-[(1-0.56)^{0.2}\times(1-0.90)^{0.1}\times(1-0.40)^{0.3}\times(1-0.09)^{0.4}]}{[(1+2\times 0.56)^{0.2}\times(1+2\times 0.90)^{0.1}\times(1+2\times 0.40)^{0.3}\times(1+2\times 0.09)^{0.4}]+2\times[(1-0.56)^{0.2}\times(1-0.90)^{0.1}\times(1-0.40)^{0.3}\times(1-0.09)^{0.4}]}, \right.$$

$$\frac{3\times[(0.34)^{0.2}\times(0.06)^{0.1}\times(0.33)^{0.3}\times(0.79)^{0.4}]}{[(1+2\times(1-0.34))^{0.2}\times(1+2\times(1-0.06))^{0.1}\times(1+2\times(1-0.33))^{0.3}\times(1+2\times(1-0.79))^{0.4}]-2\times[(0.34)^{0.2}\times(0.06)^{0.1}\times(0.33)^{0.3}\times(0.79)^{0.4}]},$$

$$\left. \frac{3\times[(0.10)^{0.2}\times(0.04)^{0.1}\times(0.19)^{0.3}\times(0.03)^{0.4}]}{[(1+2\times(1-0.10))^{0.2}\times(1+2\times(1-0.04))^{0.1}\times(1+2\times(1-0.19))^{0.3}\times(1+2\times(1-0.03))^{0.4}]-2\times[(0.10)^{0.2}\times(0.04)^{0.1}\times(0.19)^{0.3}\times(0.03)^{0.4}]} \right)$$

$= (0.394, 0.434, 0.070)$
by a similar way, $\hat{\beta}_2, \hat{\beta}_3$, $\hat{\beta}_4$, and $\hat{\beta}_5$ are obtained as follows: $\hat{\beta}_2 = (0.492, 0.294, 0.107)$, $\hat{\beta}_3 = (0.520, 0.298, 0.063)$, $\hat{\beta}_4 = (0.301, 0.517, 0.052)$, $\hat{\beta}_5 = (0.351, 0.462, 0.067)$.

- **Step 3.** By using Equation (3) the score values $\hat{S}(\hat{\beta}_a)$ $(a = 1,2,3,4,5)$ of the overall PFNs $\hat{\beta}_a$ $(a = 1,2,3,4,5)$ are obtained as follows:

  $\hat{S}(\hat{\beta}_1) = 0.394 - 0.070 = 0.324$. By a similar way, $\hat{S}(\hat{\beta}_2) = 0.386$, $\hat{S}(\hat{\beta}_3) = 0.457$, $\hat{S}(\hat{\beta}_4) = 0.249$, $\hat{S}(\hat{\beta}_5) = 0.284$.

- **Step 4.** The ranking order in the performance of enterprises Q_a $(a = 1,2,3,4,5)$ in accordance with the value of the score functions $\hat{S}(\hat{\beta}_a)$ $(s = 1,2,\ldots,5)$ of the overall PFNs is as follows: $Q_3 \succ Q_2 \succ Q_1 \succ Q_5 \succ Q_4$.

- **Step 5.** Q_3 is selected as the most desirable enterprises.

- **Step 6.** Stop.

Figures 2 and 3 show the graph of score values of $\hat{\beta}_a$ obtained by two different operators.

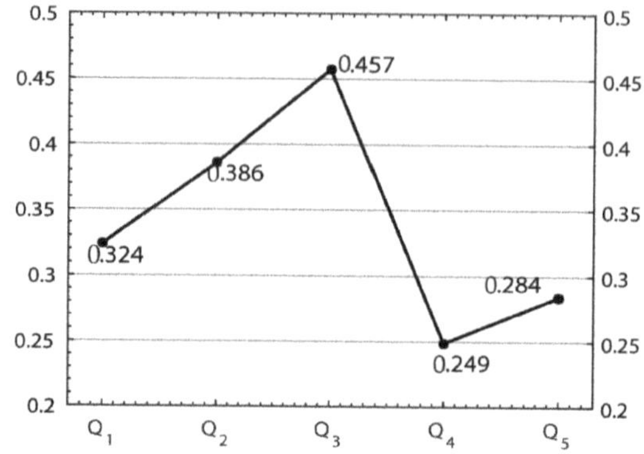

Figure 2. Graph of score values of $\hat{\beta}_a$ obtained by picture fuzzy Hamacher weighted averaging (PFHWA) operator.

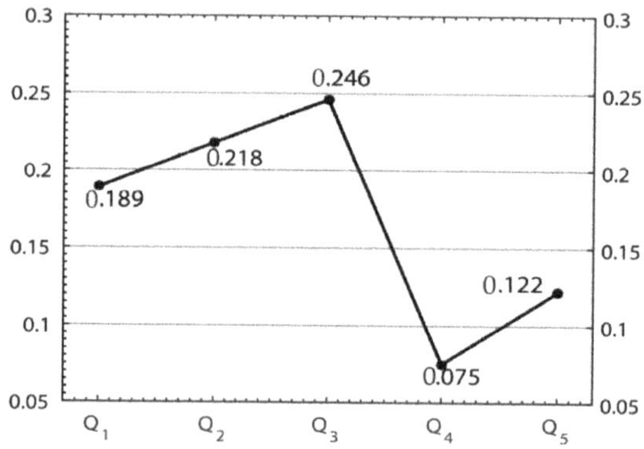

Figure 3. Graph of score values of $\hat{\beta}_a$ obtained by picture fuzzy Hamacher weighted geometric (PFHWG) operator.

If PFHWG operator is implemented instead, then the problem can be solved similarly as above.

- **Step 1.** Let us consider Table 1.
- **Step 2.** Let $\xi = 3$, using the PFHWG operator to evaluate the overall performance values $\hat{\beta}_a$ of enterprises Q_a $(a = 1,2,3,4,5)$

$$\hat{\beta}_1 = \Bigg(\frac{3\times[(0.56)^{0.2}\times(0.90)^{0.1}\times(0.40)^{0.3}\times(0.09)^{0.4}]}{[(1+2\times(1-0.56))^{0.2}\times(1+2\times(1-0.90))^{0.1}\times(1+2\times(1-0.40))^{0.3}\times(1+2\times(1-0.09))^{0.4}]-2\times[(0.56)^{0.2}\times(0.90)^{0.1}\times(0.40)^{0.3}\times(0.09)^{0.4}]},$$

$$\frac{[(1+2\times0.34)^{0.2}\times(1+2\times0.06)^{0.1}\times(1+2\times0.33)^{0.3}\times(1+2\times0.79)^{0.4}]-[(1-0.34)^{0.2}\times(1-0.06)^{0.1}\times(1-0.33)^{0.3}\times(1-0.79)^{0.4}]}{[(1+2\times0.34)^{0.2}\times(1+2\times0.06)^{0.1}\times(1+2\times0.33)^{0.3}\times(1+2\times0.79)^{0.4}]+2\times[(1-0.34)^{0.2}\times(1-0.06)^{0.1}\times(1-0.33)^{0.3}\times(1-0.79)^{0.4}]},$$

$$\frac{[(1+2\times0.10)^{0.2}\times(1+2\times0.04)^{0.1}\times(1+2\times0.19)^{0.3}\times(1+2\times0.03)^{0.4}]-[(1-0.10)^{0.2}\times(1-0.04)^{0.1}\times(1-0.19)^{0.3}\times(1-0.03)^{0.4}]}{[(1+2\times0.10)^{0.2}\times(1+2\times0.04)^{0.1}\times(1+2\times0.19)^{0.3}\times(1+2\times0.03)^{0.4}]+2\times[(1-0.10)^{0.2}\times(1-0.04)^{0.1}\times(1-0.19)^{0.3}\times(1-0.03)^{0.4}]} \Bigg)$$

$= (0.281, 0.531, 0.092)$

by a similar way, $\hat{\beta}_2, \hat{\beta}_3, \hat{\beta}_4$, and $\hat{\beta}_5$ are obtained as $\hat{\beta}_2 = (0.327, 0.435, 0.110)$, $\hat{\beta}_3 = (0.312, 0.430, 0.067)$, $\hat{\beta}_4 = (0.129, 0.704, 0.054)$, $\hat{\beta}_5 = (0.200, 0.669, 0.078)$.

- **Step 3.** Calculate the values of the score functions $\hat{S}(\hat{\beta}_a)$ $(a = 1,2,3,4,5)$ of the overall picture fuzzy numbers $\hat{\beta}_a$ $(a = 1,2,3,4,5)$ as follows:
  $\hat{S}(\hat{\beta}_1) = 0.281 - 0.092 = 0.189$, by a similar way, the other score values are obtained as follows $\hat{S}(\hat{\beta}_2) = 0.218$, $\hat{S}(\hat{\beta}_3) = 0.246$, $\hat{S}(\hat{\beta}_4) = 0.075$, $\hat{S}(\hat{\beta}_5) = 0.122$.

- **Step 4.** Rank all of the enterprises Q_a $(s = 1, 2, \ldots, 5)$ according to score values of the overall PFNs $\hat{\beta}_a$ $(a = 1, 2, 3, 4, 5)$ as $Q_3 \succ Q_2 \succ Q_1 \succ Q_5 \succ Q_4$.
- **Step 5.** Return Q_3 is selected as the most desirable enterprise.
- **Step 6.** Stop.

From the analysis, it is clear that although overall rating values of the alternatives are different for these two operators, graphically presented in Figures 2 and 3, the ranking orders of the alternatives are similar, and the most desirable enterprise is Q_3.

5.2. Comparison Analysis

In ordered to compare our proposed method more effective to the existing method [56,71], we use PFDWA (PFDWGA) and PFWA (PFWGA) operators to aggregate picture fuzzy input arguments (Table 1) for the given decision matrix in Table 2 and their corresponding score values are given in Table 3 as follows:

Table 2. Aggregated values of the alternatives using PFWA (PFWGA) and PFDWA (PFDWGA) operators.

Alternative(Q_s)	*PFWA*	*PFWGA*	*PFDWA*	*PFDWG*
Q_1	(0.4431, 0.3969, 0.0683)	(0.2555, 0.5656, 0.0957)	(0.5825, 0.2914, 0.0515)	(0.1766, 0.6381, 0.0984)
Q_2	(0.5412, 0.2588, 0.1063)	(0.2789, 0.4972, 0.1106)	(0.6040, 0.1859, 0.1045)	(0.1462, 0.6094, 0.1113)
Q_3	(0.5801, 0.2665, 0.0627)	(0.2595, 0.4914, 0.0672)	(0.6908, 0.1929, 0.0575)	(0.1236, 0.6196, 0.0675)
Q_4	(0.3815, 0.4320, 0.0517)	(0.1113, 0.7415, 0.0542)	(0.5175, 0.2298, 0.0502)	(0.0621, 0.8022, 0.0544)
Q_5	(0.4264, 0.3785, 0.0662)	(0.1760, 0.7117, 0.0806)	(0.5816, 0.1779, 0.0569)	(0.1181, 0.7801, 0.0824)

Table 3. Score values of alternatives using PFWA (PFWGA) and PFDWA (PFDWG) operators.

Alternative(Q_s)	*PFWA*	*PFWGA*	*PFDWA*	*PFDWG*
Q_1	0.3748	0.1598	0.7655	0.5391
Q_2	0.4349	0.1683	0.7498	0.5175
Q_3	0.5174	0.1923	0.8167	0.5281
Q_4	0.3298	0.0571	0.7337	0.5039
Q_5	0.3602	0.0954	0.7624	0.5179

It follows from Table 4 that although overall rating values of the alternatives are different for these two operators, the most desirable alternative is Q_3. In comparison with the other existing method [56,71], the ranking order of alternatives is slightly different but the optimum alternative is almost same. Thus, our proposed method is stable and can be applicable to handle different uncertain environments. It is also notified in order to compare the effectiveness of the proposed technique for MADM problems using PF-Hamacher aggregation, operators with other existing methods for MADM problems based on IF Hamacher aggregation operators [35] and bipolar fuzzy Hamacher aggregation operators [50] have some restraints and are not provided overall information about the situation. Picture fuzzy set is a more generalization of IFS. Therefore, picture fuzzy Hamacher set has provided more information (positive, neutral, negative, and refusal)-membership degrees to analyze systems of information, whereas IF-Hamacher set provides (membership, non-membership)-degree and BF-Hamacher set gives (positive, negative)-membership degree only. Therefore, the developed models PF-Hamacher set can be regarded as a further generalization of IF-Hamacher set [35]. Thus, our developed models are careful about the degrees of (positive, neutral, negative)-membership, and the soundness of the information of refusal degree of membership. Thus, existing models for IF-Hamacher set are particular cases of the proposed models of PF-Hamacher set. Hence, the developed models and algorithms in this paper not only solve MADM technique under PF-Hamacher environment, but also the MADM method with IF-Hamacher information, although the method given in [35] is only suitable for MADM problems for IF-Hamacher information.

Table 4. Ranking order of the alternatives.

Aggregation Operator	*Ranking Ordered*
Wei [71] *PFWA* operator	$Q_3 \succ Q_2 \succ Q_1 \succ Q_5 \succ Q_4$
Wei [71] *PFWGA* operator	$Q_3 \succ Q_2 \succ Q_1 \succ Q_5 \succ Q_4$
Jana et al. [56] *PFDWA* operator	$Q_3 \succ Q_1 \succ Q_5 \succ Q_2 \succ Q_4$
Jana et al. [56] *PFDWGA* operator	$Q_1 \succ Q_3 \succ Q_5 \succ Q_2 \succ Q_4$
Proposed *PFHWA* operator	$Q_3 \succ Q_1 \succ Q_5 \succ Q_2 \succ Q_4$
Proposed *PFHWGA* operator	$Q_3 \succ Q_2 \succ Q_1 \succ Q_5 \succ Q_4$

6. Conclusions

Enterprises are an important factor of stockholders, employees, creditor, customer, government, and other stakeholders. In the performance of enterprises, two characteristics should be considered: Economic and society, hence we should consider all stakeholders' benefit in performance of enterprise evaluating time. We set up a performance evaluating system on the basis of stakeholder benefits. In this article, we have studied a multi-attribute decision-making problem for emerging technology enterprise performance evaluation with picture fuzzy information. We used a picture fuzzy Hamacher weighted averaging (PFHWA) operator and a picture fuzzy Hamacher weighted geometric (PFHWGA) operator to assess the best enterprise on the basis of performance evaluation of enterprises. In the future, the application of our proposed model can be applied in decision-making theory, risk evaluation, and other domains under ambiguous environments.

Author Contributions: Conceptualization, C.J.; formal analysis, M.P.; writing—original draft preparation, C.J.; writing—review and editing, C.J. and M.P.; supervision, M.P.

Funding: This research received no external funding.

Conflicts of Interest: The authors declare no conflict of interest.

References

1. Mintzberg, H.; Ahlstrand, B.W.; Lampel, J. Strategy safari: A guided tour through the wilds of strategic management. *Manag. Serv.* **2009**, *48*, 152.
2. Shrivastava, P. *Advances in Strategic Management: Organizational Learning and Strategic Management*; Emerald Group: Greenwicht, UK, 1997.
3. Brockman, B.K.; Morgan, R.M. The Role of Existing Knowledge in New Product Innovativeness and Performance. *Dec. Sci.* **2003**, *34*, 385–419. [CrossRef]
4. Baker, W.E.; Sinkula, J.M. The synergistic effect of market orientation and learning orientation on organizational performance. *J. Acad. Mark. Sci.* **1999**, *27*, 411–427. [CrossRef]
5. Ussahawanitchakit, P. Impacts of organizational learning on innovation orientation and firm efficiency: An empirical assessment of accounting firms in Thailand. *Int. J. Bus. Res.* **2008**, *8*, 1–12.
6. Merigó, J.M. Probabilities in the OWA operator. *Expert Syst. Appl.* **2012**, *39*, 11456–11467. [CrossRef]
7. Merigó, J.M.; Casanovas, M.; Yang, J.B. Group decision making with expertons and uncertain generalized probabilistic weighted aggregation operators. *Eur. J. Oper. Res.* **2014**, *235*, 215–224. [CrossRef]
8. Merigó, J.M. Decision-making under risk and uncertainty and its application in strategic management. *J. Bus. Econ. Manag.* **2015**, *16*, 1–24. [CrossRef]
9. Merigó, J.M.; Gil-Lafuente, A.M.; Aluja, J.G. A new aggregation method for strategic decision making and its application in assignment theory. *Afr. J. Bus. Manag.* **2011**, *5*, 4033–4043.
10. Li, D.F. *Decision and Game Theory in Management with Intuitionistic Fuzzy Sets*; Springer: Berlin, Germany, 2014; Volume 308, pp. 1–441, ISBN 978-3-642-40711-6.
11. Zadeh, L.A. Fuzzy sets. *Inform. Control* **1965**, *8*, 338–353. [CrossRef]
12. Adrian, C.; Abdullah, R.; Atan, R.; Jusoh, Y.Y. Conceptual Model Development of Big Data Analytics Implementation Assessment Effect on Decision-Making. *Int. J. Interact. Multimed. Artif. Intell.* **2017**, *4*, 28–33. [CrossRef]

13. Idrus, A.; Mahmoud, M.A.; Ahmad, M.S.; Yahya, A.; Husen, H. A Solution Generator Algorithm for Decision Making-based Automated Negotiation in the Construction Domain. *Int. J. Artif. Intell. Interact. Multimed.* **2018**, in press. [CrossRef]
14. Morente-Molinera, J.A.; Kou, G.; González-Crespo, R.; Corchado, J.M.; Herrera-Viedma, E. Solving multi-criteria group decision making problems under environments with a high number of alternatives using fuzzy ontologies and multi-granular linguistic modelling methods. *Knowl.-Based Syst.* **2017**, *137*, 54–64. [CrossRef]
15. Morente-Molinera, J.A.; Kou, G.; González-Crespo, R.; Corchado, J.M.; Herrera-Viedma, E. Managing multi-criteria group decision making environments with high number of alternatives using fuzzy ontologies. *SOMET* **2018**, *303*, 493–506.
16. Atanassov, K.T. *Intuitionistic Fuzzy Sets: Theory and Applications, Studies in Fuzziness and Soft Computing*; Physica-Verlag: Heidelberg, Germany; New York, NY, USA, 1999; Volume 35.
17. Chen, S.M.; Chiou, C.H. Multiattribute decision making based on interval-valued intuitionistic fuzzy sets, PSO techniques, and evidential reasoning methodology. *IEEE Trans. Fuzzy Syst.* **2015**, *23*, 1905–1916. [CrossRef]
18. Du, Y.; Liu, P. Extended fuzzy VIKOR method with intuitionistic trapezoidal fuzzy numbers. *Inf. Int. Interdiscip. J.* **2011**, *14*, 2575–2583.
19. Garg, H. A new generalized improved score function of interval-valued intuitionistic fuzzy sets and applications in expert systems. *Appl. Soft Comput.* **2016**, *38*, 988–999. [CrossRef]
20. Kumar, K.; Garg, H. TOPSIS method based on the connection number of set pair analysis under interval-valued intuitionistic fuzzy set environment. *Comput. Appl. Math.* **2016**. [CrossRef]
21. Li, D.F. Closeness coefficient-based nonlinear programming method for interval-valued intuitionistic fuzzy multiattribute decision making with incomplete preference information. *Appl. Soft Comput.* **2011**, *11*, 3402–3418. [CrossRef]
22. Lourenzutti, R.; Krohling, R.A. A study of TODIM in a intuitionistic fuzzy and random environment. *Expert Syst. Appl.* **2013**, *40*, 6459–6468. [CrossRef]
23. Wan, S.P.; Li, D.F. Atanassovs intuitionistic fuzzy programming method for heterogeneous multiattribute group decision making with atanassovs intuitionistic fuzzy truth degrees. *IEEE Trans. Fuzzy Syst.* **2014**, *22*, 300–312. [CrossRef]
24. Wan, S.P.; Li, D.F. Fuzzy mathematical programming approach to heterogeneous multiattribute decision-making with interval-valued intuitionistic fuzzy truth degrees. *Inform. Sci.* **2015**, *325*, 484–503. [CrossRef]
25. Ye, J. Multicriteria fuzzy decision-making method based on a novel accuracy function under interval-valued intuitionistic fuzzy environment. *Expert Syst. Appl.* **2009**, *36*, 899–6902. [CrossRef]
26. Zhang, L.; Zhan, J.; Xu, Z.S. Covering-based generalized IF rough sets with applications to multi-attribute decision-making. *Inform. Sci.* **2019**, *478*, 275–302. [CrossRef]
27. Zhan, J.; Sun, B. Covering-based intuitionistic fuzzy rough sets and applications in multi-attribute decision-making. *Artif. Intell. Rev.* **2018**. [CrossRef]
28. Zhan, J.; Malik, H.M.; Akram, M. Novel decision-making algorithms based on intuitionistic fuzzy rough environment. *Int. J. Mach. Learn. Cybern.* **2018**. 0827-4. [CrossRef]
29. Yager, R.R. On ordered weighted avergaing aggregation operators in multi-criteria decision making. *IEEE Trans. Syst. Man Cybern.* **1988**, *18*, 183–190. [CrossRef]
30. Yager, R.R.; Kacprzyk, J. *The Ordered Weighted Averaging Operators: Theory and Applications*; Kluwer: Boston, MA, USA, 1997.
31. Xu, Z.S.; Yager, R.R. Some geometric aggregation operators-based on intuitionistic fuzzy sets. *Int. J. Gen. Syst.* **2006**, *35*, 417–433. [CrossRef]
32. Xu, Z.S. Intuitionistic fuzzy aggregation operators. *IEEE Trans. Fuzzy Syst.* **2007**, *15*, 1179–1187.
33. Hamacher, H. Uber logische verknunpfungenn unssharfer Aussagen undderen Zugenhorige Bewertungsfunktione. In *Progress in Cybernatics and Systems Research*; Trappl, K.R., Ed.; Hemisphere: Washington, DC, USA, 1978; Volume 3, pp. 276–288.
34. Beliakov, G.; Pradera, A.; Calvo, T. *Aggregation Functions: A Guide For Practitioner*; Springer: Berlin/Heidelberg, Germany, 2007.

35. Huang, J.Y. Intuitionistic fuzzy Hamacher aggregation operators and their application to multiple attribute decision making. *J. Intell. Fuzzy Syst.* **2014**, *27*, 505–513.
36. Liu, P.D. Some Hamacher aggregation operators based on the interval-valued intuitionistic fuzzy numbers and their application to group decision making. *IEEE Trans. Fuzzy Syst.* **2014**, *22*, 83–97. [CrossRef]
37. Xiao, S. Induced interval-valued intuitionistic fuzzy Hamacher ordered weighted geometric operator and their application to multiple attribute decision making. *J. Intell. Fuzzy Syst.* **2014**, *27*, 527–534.
38. Li, W. Approaches to decision making with interval-valued intuitionistic fuzzy information and their application to enterprise financial performance assessment. *J. Intell. Fuzzy Syst.* **2014**, *27*, 1–8.
39. He, Y.D.; Chen, H.Y.; He, Z.; Zhou, L.G. Multi-attribute decision making based on neutral averaging operators for intuitionistic fuzzy information. *Appl. Soft Comput.* **2015**, *27*, 64–76. [CrossRef]
40. Jana, C.; Pal, M.; Wang, J.-Q. Bipolar fuzzy Dombi aggregation operators and its application in multiple attribute decision making process. *J. Ambient Intell. Humaniz. Comput.* **2018**. [CrossRef]
41. Jana, C.; Pal, M.; Karaaslan, F.; Wang, J.-Q. Trapezoidal neutrosophic aggregation operators and its application in multiple attribute decision -making process. *Sci. Iran. E* **2019**, in press.
42. Karaaslan, F. Correlation coefficients of single-valued neutrosophic refined soft sets and their applications in clustering analysis. *Neural Comput. Appl.* **2017**, *28*, 2781–2793. [CrossRef]
43. Karaaslan, F. Possibility neutrosophic soft sets and PNS-decision making method. *Appl. Soft Comput.* **2017**, *54*, 403–414. [CrossRef]
44. Khameneh, A.Z.; Kilicman, A. m-polar fuzzy soft weighted aggregation operators and their applications in group decision-making. *Symmetry* **2018**, *10*, 636. [CrossRef]
45. Lu, M.; Wei, G.W.; Alsaadi, F.E.; Hayat, T.; Alsaedi, A. Hesitant pythagorean fuzzy hamacher aggregation operators and their application to multiple attribute decision making. *J. Intell. Fuzzy Syst.* **2017**, *33*, 1105–1117. [CrossRef]
46. Tan, C.Q.; Yi, W.T.; Chen, X.H. Hesitant fuzzy Hamacher aggregation operators for multicriteria decision making. *Appl. Soft Comput.* **2015**, *26*, 325–349. [CrossRef]
47. Wang, L.; Wang, Y.; Liu, X.D. Prioritized aggregation operators and correlated aggregation operators for hesitant 2-tuple linguistic variables. *Symmetry* **2018**, *10*. [CrossRef]
48. Wei, G.W. Some geometric aggregation functions and their application to dynamic multiple attribute decision making in intuitionistic fuzzy setting. *Int. J. Uncertain. Fuzz. Knowl.-Based Syst.* **2009**, *17*, 179–196. [CrossRef]
49. Wei, G.W.; Lu, M. Dual hesitant pythagorean fuzzy Hamacher aggregation operators in multiple attribute decision making. *Arch. Control Sci.* **2017**, *27*, 365–395. [CrossRef]
50. Wei, G.W.; Alsaadi, F.E.; Hayat, T.; Alsaedi, A. Bipolar fuzzy Hamacher aggregation operators in multiple attribute decision Making. *Int. J. Fuzzy Syst.* **2018**, *20*, 1–12. [CrossRef]
51. Zhou, L.Y.; Zhao, X.F.; Wei, G.W. Hesitant fuzzy Hamacher aggregation operators and their application to multiple attribute decision making. *J. Intell. Fuzzy Syst.* **2014**, *26*, 2689–2699.
52. Cuong, B.C. *Picture Fuzzy Sets-First Results, Part 1, Seminar Neuro-Fuzzy Systemswith Applications*; Tech. Rep.; Instiute of Mathematics: Hanoi, Vietnam, 2013.
53. Cuong, B.C. *Picture Fuzzy Sets-First Results, Part 2, Seminar Neuro-Fuzzy Systemswith Applications*; Tech. Rep.; Instiute of Mathematics: Hanoi, Vietnam, 2013.
54. Cuong, B.C. Picture fuzzy sets. *J. Comput. Sci. Cybern.* **2014**, *30*, 409–420.
55. Singh, P. Correlation coefficients for picture fuzzy sets. *J. Intell. Fuzzy Syst.* **2014**, *27*, 2857–2868.
56. Jana, C.; Senapati, T.; Pal, M.; Yager, R.R. Picture fuzzy Dombi aggregation operators: Application to MADM process. *Appl. Soft Comput.* **2019**, *74*, 99–109. [CrossRef]
57. Son, L. DPFCM: A novel distributed picture fuzzy clustering method on picture fuzzy sets. *Expert Syst. Appl.* **2015**, *2*, 51–66. [CrossRef]
58. Thong, P.H.; Son, L.H. A new approach to multi-variables fuzzy forecasting using picture fuzzy clustering and picture fuzzy rules interpolation method. In Proceedings of the 6th International Conference on Knowledge and Systems Engineering, Hanoi, Vietnam, 9–11 October 2014; pp. 679–690.
59. Thong, N.T. HIFCF An effective hybrid model between picture fuzzy clustering and intuitionistic fuzzy recommender systems for medical diagnosis. *Expert Syst. Appl.* **2015**, *42*, 3682–3701. [CrossRef]
60. Son, L.H. Generalized picture distance measure and applications to picture fuzzy clustering. *Appl. Soft Comput.* **2016**, *46*, 284–295. [CrossRef]

61. Wei, G.W. Picture fuzzy cross-entropy for multiple attribute decision making problems. *J. Bus. Econ. Manag.* **2016**, *17*, 491–502. [CrossRef]
62. Wei, G.W. Picture 2-tuple linguistic Bonferroni mean operators and their application to multiple attribute decision making. *Int. J. Fuzzy Syst.* **2017**, *19*, 997–1010. [CrossRef]
63. Nie, R.X.; Wang, J.Q.; Li, L. A shareholder voting method for proxy advisory firm selection based on 2-tuple linguistic picture preference relation. *Appl. Soft Comput.* **2017**, *60*, 520–539. [CrossRef]
64. Wang, L.; Zhang, H.Y.; Wang, J.Q.; Li, L. Picture fuzzy normalized projection-based VIKOR method for the risk evaluation of construction project. *J. Intell. Fuzzy Syst.* **2018**, *64*, 216–226. [CrossRef]
65. Zhang, X.Y.; Wang, J.Q.; Hu, J.H. On novel operational laws and aggregation operators of picture 2-tuple linguistic information for MCDM problems. *Int. J. Fuzzy Syst.* **2017**. [CrossRef]
66. Wei, G.W. Picture fuzzy Hamacher aggregation operators and their application to multiple attribute decision making. *Fundam. Inform.* **2018**, *157*, 271–320. [CrossRef]
67. Chen, S.M.; Tan, J.M. Handling multicriteria fuzzy decision-making problems based on vague set theory. *Fuzzy Sets Syst.* **1994**, *67*, 163–172. [CrossRef]
68. Deschrijver, G.; Cornelis, C.; Kerre, E.E. On the representation of intuitionistic fuzzy t-norms and t-conorms. *IEEE Trans. Fuzzy Syst.* **2004**, *12*, 45–61. [CrossRef]
69. Roychowdhury, S.; Wang, B.H. On generalized Hamacher families of triangular operators. *Int. J. Approx. Reason.* **1998**, *19*, 419–439. [CrossRef]
70. Deschrijver, G.; Kerre, E.E. Ageneralization of operators on intuitionistic fuzzy sets using triangular norms and conorms. *Notes Intuitionistic Fuzzy Sets* **2002**, *8*, 19–27.
71. Wei, G.W. Picture fuzzy aggregation operators and their application to multiple attribute decision making. *J. Intell. Fuzzy Syst.* **2017**, *33*, 713–724. [CrossRef]

symmetry

MDPI

Article

A Multi-Criteria Group Decision Making Model for Green Supplier Selection under the Ordered Weighted Hesitant Fuzzy Environment

Yumin Liu, Linlin Jin * and Feng Zhu

Business School, Zhengzhou University, Zhengzhou 450001, China; yuminliu@zzu.edu.cn (Y.L.); zhufeng1994@163.com (F.Z.)

* Correspondence: nclzzu@163.com

Received: 24 November 2018; Accepted: 20 December 2018; Published: 25 December 2018

Abstract: The green supplier selection (GSS) problem is one of the most pressing issues that can directly affect manufacturer performance. GSS has been studied in previous literature, which is considered to be a typical multiple criteria group decision making (MCGDM) problem. The ordered weighted hesitant fuzzy MCGDM method can present the importance of each possible value, and the priority relationship among criteria has rarely been studied. In this study, we first extend the prioritized average (PA) operator to the ordered weighted hesitant fuzzy set (OWHFS) for solving the both problems. The generalized ordered weighted hesitant fuzzy prioritized weighted average operator (GOWHFPWA) is recommended, and some desirable properties are discussed. Based on this operator, a novel MCGDM method for GSS is developed. A numerical example of GSS is then given to prove the robustness of the proposed approach, and a sensitivity analysis is used to identify the robustness of the proposed method. Finally, a comparative analysis based on the MCGDM approach with the hesitant fuzzy prioritized weighted average (HFPWA) operator is illustrated to indicate the validity and advantages of the proposed approach.

Keywords: green supplier selection; ordered weighted hesitant fuzzy set; GOWHFPWA operator; multi-criteria group decision making

---

1. Introduction

Nowadays, with the increasingly global awareness of environmental responsibility, green production has already become the development orientation of industrial production for most manufacturing firms. Growing environmental concerns mean that it is necessary for manufacturing companies to be more concerned about green supply chain management (GSCM) to reduce environmental pollution from industrial sectors [1]. The green supplier selection (GSS) is a critical link of GSCM, which can directly affect the sustainable development and performance of manufacturing enterprises [2]. GSS can be regarded as a multiple criteria group decision making (MCGDM) problem that involves many conflicting assessment criteria [3], such as cost, materials, recycling capacity, green competencies, green technology, and green certification. Essentially, the act of decision making is more complicated than in the traditional supplier selection since some environmental criteria need to be considered, and these criteria are qualitative in nature and the weights cannot be provided in advance [3,4]. Therefore, how to choose a suitable green supplier in GSCM has become a key strategic consideration.

Researchers have come up with and applied a range of multi-criteria decision making (MCDM) approaches for green supplier decision making problems [5,6]. To synthesize multiple qualitative or quantitative environmental criteria and obtain a clear evaluation result, some MCDM approaches based on precise information are used in green supplier decision making. Handfield et al. [7] evaluated the environmental standards of green suppliers by using the analytic hierarchy process(AHP). Likewise, Lu et al. [8] used AHP to evaluate and coordinate green suppliers. Hsu and Hu [9] applied the the analytic network process(ANP) for GSS. Kuo et al. [10] integrated artificial neural network and MCDM approaches to GSS. Bai and Sarkis [11] came up with an analytical evaluation on the basis of rough set theory. Yeh and Chuang [12] introduced an optimal mathematical planning approach for selecting a green supplier. The ANP and radial basis function neural network approaches of choosing green suppliers for China chemical industries was proposed by Zhou et al [13]. Kuo et al. [14] integrated ANP with the data envelopment analysis (DEA) to evaluate green suppliers. A mathematical model based on DEA for choosing green suppliers was proposed by Jauhar et al. [15]. Dobos and Vörösmarty [16] used a DEA approach towards environmental issues. Freeman and Chen [17] designed an approach for GSS by combining technique for order preferenceby similarity to ideal solution (TOPSIS), the AHP model, and entropy approach. Hashemi et al. [18] combined the GSS approach with the ANP method. Yazdani et al. [19] recommended a novel integrated MCDM basis of selecting most suitable green suppliers. Liu et al. [20] expanded a linguistic group decision-making method in assessing big projects.

The major issue obstructing the ability to determine the right mathematical method for choosing a green supplier is the absence of the ability to handle uncertain and inadequate information which mostly happens in real-life conditions. In the practical problems of GSS, a great number of assessment detailed information is unknown, and additionally, several criteria are affected by uncertainty. Meanwhile, decision makers (DMs) usually cannot make completely reasonable judgements due to uncertain and ambiguous information. DMs judgments are usually uncertain and difficult to measure by exact numerical values, so a fuzzy set theory proposed by Zadeh [21] has become essential for solving the complications characterized by vagueness and imprecision. Recently, several studies have applied the typical MCDM methods to a range of fuzzy environments [22–25]. Chiou et al. [26] applied a fuzzy AHP for GSS in China electronic industries. Lee et al. [27] extended a fuzzy AHP decision model to identify GSS for high-tech industries. Tsai and Huang [28] came up with a fuzzy goal programming technique for GSS. Tuzkaya et al. [29] developed a hybrid fuzzy MCDM model, and Büyüközkan and Cifci [30] recommended a unique hybrid MCDM method to evaluate green suppliers base on reference [29]. Datta et al. [31] presented a VlseKriterijuska Optimizacija I Komoromisno Resenje (VIKOR) method together with the interval valued fuzzy set to choose the best green supplier. Shen et al. [32] presented a fuzzy MCDM as basis for selecting green supplier with linguistic preference. Wang and Chan [33] proposed the hierarchical fuzzy TOPSIS model to choose the green supplier. Cao et al. [34] presented a unique intuitionistic fuzzy judgment matrix integrated with the TOPSIS approach to define the subjective and objective weights in green supplier assessment and selection. Kannan et al. [35] utilized a fuzzy axiomatic design method to choose the most suitable green supplier. Hamdan and Cheaitou [36] proposed fuzzy TOPSIS and AHP methods to define preference weights of respective supplier and criterion. Guo et al. [37] developed a fuzzy MCDM method to solve the GSS in apparel manufacturing.

GSS is known as a MCGDM problem that involves both several interrelated evaluation criteria and several DMs behavior characters. Moreover, the complexity of MCGDM problems is increased when several DMs might be considered in assessment of the problems [38]. Tsui et al. [39] came up with a hybrid MCGDM method based on entropy and AHP to assess GSS problems in manufacturing enterprise. Based on group decision analysis, Darabi and Heydari [40] presented an interval-valued hesitant fuzzy ranking method for selecting green suppliers. Gitinavard et al. [41] developed a unique interval-valued hesitant fuzzy group outranking method for choosing green suppliers. Qin et al. [42] recommended a comprehensive MCGDM approach for GSS in interval type-2 fuzzy sets. Tang [43] employed the hesitant fuzzy Hamacher power weighted average operator to solve the GSS complexities with hesitant fuzzy information.

As evidently shown in the above reviewed literature, various MCDM methods for GSS have been extended to intuitionistic fuzzy sets [44], linguistic fuzzy sets [32], interval-valued fuzzy sets, and type-2 fuzzy sets [31,42]. However, little study has been done on GSS by using a hesitant fuzzy set (HFS), which was first introduced by Torra [45,46]. As a generalization of fuzzy sets, HFS can describe the situations that permit the expert's preference judgment for a particular criterion that have few different values, which is a very suitable method for tackling uncertain information and for expressing DMs' hesitancy in real group decision making [47–49]. Nowadays, to be able to solve the MCGDM problems, varieties of extensions of the HFS have been proposed by scholars, such as generalized HFS, dual hesitant fuzzy sets, hesitant fuzzy linguistic term sets, the higher order HFS, and NaP-HFS [50–56].

However, the HFS method has its own shortcomings, because it only expresses the expert's judgment as several probable values lack considerations of their importance. In several applied MCGDM problems, especially in GSS, experts usually come from the same field, and might often make the same judgments on a given criterion. Thus, the possible value repeated many times is more significant than that displayed only one time. For this reason, Zhang and Wu [57] developed the model of weighted hesitant fuzzy set (WHFS), in which the importance of possible values provided by DMs has been considered. Farhadinia and Xu [58] modified the definition of WHFS and proposed a new extension of HFS as the ordered weighted hesitant fuzzy set (OWHFS), in which the importance of DMs' judgments is defined as the repetition rate of the possible values. Therefore, OWHFS can not only express the experts' judgments as several possible values but also give the importance of each possible value.

Besides the importance of DMs' judgments, the priority relationship among criteria of GSS selection for OWHFS is one of the most critical research topics at present. To be able to aggregate the evaluation values of criteria for an alternative, Yager [59] first presented a prioritized scoring operator and prioritized average (PA) operator. Recently, several studies have concentrated on aggregation operators for HFS and their application in MCDM. Xia and Xu [60] investigated a series of aggregation operators for hesitant fuzzy information. Wei [61] developed hesitant fuzzy prioritized operators. Qua et al. [62] examined induced generalized dual hesitant fuzzy Shapley hybrid operators. Wei et al. [63] utilized Pythagorean hesitant fuzzy Hamacher aggregation operators. Farhadinia and Xu [58] first presented several aggregation operators for OWHFS and used them for MCDM. However, as far as we know, the priority relationship among criteria for OWHFS has rarely been investigated.

Moreover, by reviewing the existing literature, the criteria of GSS can usually be classified into two categories: General and green criteria [64,65]. Generally, organizations consider criteria such as cost, quality, and delivery performance when evaluating supplier performance. However, due to enterprises facing double pressures of environmental laws and regulations and the increasing demands of environmental protection, environmental performance is considered by many enterprises in selecting suppliers. To solve the complexity of GSS problems in practice, the criteria of green supplier evaluation were studied by scholars. For instance, Lee et al. [27] mentioned that quality, technology capability, environment management, and green competencies are the most commonly referred criteria in green supplier evaluation literature. Yeh and Chuang [12] developed assessment criteria for GSS such as green image, product recycling, green design, green supply chain management, pollution treatment cost, and environment performance assessment criteria. A summary of the most critical standards for GSS are shown in Table 1.

In summary, the concept of GSS is a typical MCGDM problem, of which there are two critical issues of concern. The first issue depicts the importance of DMs' judgments. Another is mathematically expressing the priority relationship among criteria. The focus of this study is to develop a novel group decision making approach with ordered weighted hesitant fuzzy information for GSS that addresses both of the above problems.

The remainder of this study is established as follows: Section 2 briefly introduces the basic principles of OWHFS and the PA operator. Section 3 develops the generalized ordered weighted hesitant fuzzy prioritized weighted average (GOWHFPWA) operator and investigates its desirable properties. Section 4 proposes a novel MCGDM method for GSS with a GOWHFPWA operator. Section 5 presents a numerical example of GSS to demonstrate the superiority and effectiveness of the proposed approach. Section 6 provides performance analysis and comparison, including sensitivity and validity analysis of the proposed approach. Finally, conclusions and recommendations are discussed in Section 7.

Table 1. Key criteria for green supplier selection.

Variable	Criterion	Definition	Authors
c_1	Cost	Total cost of product and service	Yeh and Chuang [12], Govindan et al. [6], Mousakhani et al. [65]
c_2	Quality	The quality of product and service	Omurca [66], Govindan et al. [6], Mousakhani et al. [65]
c_3	Service	Performance in terms of product service and social service	Omurca [66], Kannan et al. [35], Govindan et al. [6]
c_4	Environment	Environmental protection; certification and materials recycling capacity	Govindan et al. [6], Mousakhani et al. [65], Lee et al. [27]
c_5	Technology	Ability to facilitate the development of green products	Lee et al. [27], Govindan et al. [6], Mousakhani et al. [65]
c_6	Management	Capcity for environmental management	Kuo et al. [12], Tseng et al. [24], Mousakhani et al. [65]
c_7	Responsibility	Including safety production, social morality and public interest	Galankashi, et al. [6], Mousakhani et al. [65]

2. Preliminaries

In this section, some basic concepts related to ordered weighted hesitant fuzzy set (OWHFS) and PA operator are reviewed, which will be useful for later analysis.

Definition 1. [58] *Let X be the universe of discourse. An ordered weighted hesitant fuzzy set (OWHFS) on X is defined as:*

$$^{\omega}H = \{< x, {}^{\omega}h(x) > | x \in X\}$$

where, $^{\omega}h(x) = \underset{1 \le j \le L_x}{\cup} \{< h^{\delta(j)}(x), w^{\delta(j)}(x) >\}$, *referred to as the ordered weighted hesitant fuzzy element (OWHFE), is a set of some different values in [0,1]. It denotes all possible membership degrees of the element* $x \in X$ *to the set* $^{\omega}H$ *, and* $w^{\delta(j)}(x) \in [0,1]$ *is the weight of* $h^{\delta(j)}(x)$ *such that* $\sum_{j=1}^{L_x} w^{\delta(j)}(x) = 1$ *for any* $x \in X$.

It is worth noting that when $w^{\delta(1)}(x) = w^{\delta(2)}(x) = \ldots = w^{\delta(L_x)}(x) = 1/L_x$ for any $x \in X$, then the OWHFS $^{\omega}H$ will become a typical HFS. For the convenience of representation, OWHFE can be denoted by $^{\omega}h = {}^{\omega}h(x) = \underset{1 \le j \le L_x}{\cup} \{< h^{\delta(j)}, w^{\delta(j)} >\}$

Suppose that the membership degrees provided by k experts, of the element x in the set $^{\omega}H$, where $h^{\delta(i)}(x)$ is given by k_i experts, $i = 1, 2, \ldots, L$, $\sum_{i=1}^{L} k_i = k$. It should be noted that every expert cannot persuade other experts to change their opinions. In such a situation, the membership degree of the element x in the set $^{\omega}H$ has L possible values $h^{\delta(1)}(x), h^{\delta(2)}(x), \ldots$, and $h^{\delta(L)}(x)$ associated with weights $w^{\delta(1)}(x) = \frac{k_1}{k}$, $w^{\delta(2)}(x) = \frac{k_2}{k}, \ldots$, and $w^{\delta(L)}(x) = \frac{k_L}{k}$ respectively.

Definition 2. [58] *Let* $^{\omega}h = \underset{1 \le j \le L}{\cup} \left\{< h^{\delta(j)}, w^{\delta(j)} >\right\}$, $^{\omega}h_1 = \underset{1 \le j \le L}{\cup} \{< h_1^{\delta(j)}, w_1^{\delta(j)} >\}$ *and* $^{\omega}h_2 = \underset{1 \le j \le L}{\cup} \{< h_2^{\delta(j)}, w_2^{\delta(j)} >\}$ *be three OWHFEs. Then, some operations on the OWHFEs* $^{\omega}h$, $^{\omega}h_1$ *and* $^{\omega}h_2$ *are defined as follows:*

(1) $\quad {}^{\omega}h^{\lambda} = \underset{1 \le j \le L}{\cup} \left\{\left\langle \left(h^{\delta(j)}\right)^{\lambda}, w^{\delta(j)} \right\rangle\right\};$

(2) $\lambda^{\omega}h = \underset{1\leq j\leq L}{\cup}\left\{\left\langle 1-(1-h^{\delta(j)})^{\lambda}, w^{\delta(j)}\right\rangle\right\}$;

(3) ${}^{\omega}h_1 \oplus {}^{\omega}h_2 = \underset{1\leq j\leq L}{\cup}\left\{\left\langle h_1^{\delta(j)} + h_2^{\delta(j)} - h_1^{\delta(j)}h_2^{\delta(j)}, \overline{(w_1^{\delta(j)} + w_2^{\delta(j)})}\right\rangle\right\}$;

where $\lambda > 0$ *and* $\overline{(w_1^{\delta(j)} + w_2^{\delta(j)})} = \frac{w_1^{\delta(j)}+w_2^{\delta(j)}}{\sum\limits_{j=1}^{L}(w_1^{\delta(j)}+w_2^{\delta(j)})}(j = 1,2,\ldots,L)$.

Definition 3. [59] *Let* ${}^{\omega}h = \underset{1\leq j\leq L}{\cup}\left\{< h^{\delta(j)}, w^{\delta(j)} >\right\}$, ${}^{\omega}h_1 = \underset{1\leq j\leq L}{\cup}\left\{< h_1^{\delta(j)}, w_1^{\delta(j)} >\right\}$ *and* ${}^{\omega}h_2 = \underset{1\leq j\leq L}{\cup}\left\{< h_2^{\delta(j)}, w_2^{\delta(j)} >\right\}$ *be three OWHFEs.* $\Delta({}^{\omega}h) = \sum\limits_{j=1}^{L} h^{\delta(j)}w^{\delta(j)}$ *is called the score function of* ${}^{\omega}h$, *and* $\nabla({}^{\omega}h) = \sum\limits_{j=1}^{L}(\Delta({}^{\omega}h) - h^{\delta(j)})^2 w^{\delta(j)}$ *is called the deviation function of* ${}^{\omega}h$.

(1) *If* $\Delta({}^{\omega}h_1) > \Delta({}^{\omega}h_2)$, *then* ${}^{\omega}h_1 >^{\omega} h_2$
(2) *If* $\Delta({}^{\omega}h_1) < \Delta({}^{\omega}h_2)$, *then* ${}^{\omega}h_1 <^{\omega} h_2$
(3) *If* $\Delta({}^{\omega}h_1) = \Delta({}^{\omega}h_2)$, *then* $\begin{cases} \nabla({}^{\omega}h_1) > \nabla({}^{\omega}h_2) \Rightarrow {}_{\omega}h_1 <^{\omega} h_2 \\ \nabla({}^{\omega}h_1) = \nabla({}^{\omega}h_2) \Rightarrow {}_{\omega}h_1 =^{\omega} h_2 \\ \nabla({}^{\omega}h_1) < \nabla({}^{\omega}h_2) \Rightarrow {}_{\omega}h_1 >^{\omega} h_2 \end{cases}$

Definition 4. [59] *Let* $C = \{C_1, C_2, \ldots, C_n\}$ *be a set of criteria, and there is a prioritization among the criteria expressed by the linear ordering* $C_1 \succ C_2 \succ \ldots \succ C_n$, *which indicates that criterion* C_j *has a higher priority than* C_i, *if* $j < i$. *The value* $C_j(x)$ *is the performance of any alternative* x *under criterion* C_j, *and satisfies* $C_j(x) \in [0,1]$. *If*

$$PA(C(x)) = \sum_{j=1}^{n} w_j C_j(x) \tag{1}$$

where $w_j = \frac{T_j}{\sum\limits_{i=1}^{n} T_i}$, $T_j = \prod\limits_{l=1}^{j-1} C_l(x)(j = 1,2,\ldots,n)$, $T_1 = 1$. *Then PA is called the prioritized average operator.*

3. GOWHFPWA Operator and Its Properties

In this section, the GOWHFPWA operator is proposed to aggregate the OWHFEs, and some properties are studied.

The PA operator has been commonly used in situations where the DMs' judgments are the exact values [59]. In this part, we shall extend the PA operator to ordered weighted hesitant fuzzy environments and define the GOWHFPWA operator.

Definition 5. *Let* ${}^{\omega}h_1, {}^{\omega}h_2, \ldots, {}^{\omega}h_n$ *be a set of OWHFEs, then the GOWHFPWA operator is defined as follows:*

$$GOWHFPWA({}^{\omega}h_1, {}^{\omega}h_2, \cdots, {}^{\omega}h_n) = \left(\frac{T_1}{\sum\limits_{i=1}^{n} T_i}({}^{\omega}h_1)^{\alpha} \oplus \frac{T_2}{\sum\limits_{i=1}^{n} T_i}({}^{\omega}h_2)^{\alpha} \oplus \cdots \oplus \frac{T_n}{\sum\limits_{i=1}^{n} T_i}({}^{\omega}h_n)^{\alpha}\right)^{1/\alpha} = \left(\oplus \frac{T_i({}^{\omega}h_i)^{\alpha}}{\sum\limits_{i=1}^{n} T_i}\right)^{1/\alpha} \tag{2}$$

where, $\alpha > 0$ *is a parameter of GOWHFPWA operator,* $T_i = \prod\limits_{l=1}^{i-1} \Delta({}^{\omega}h_l)(i = 1,2,\cdots,n)$, $T_1 = 1$ *and* $\Delta({}^{\omega}h_k)$ *is the score function of* ${}^{\omega}h_k$.

Theorem 1. *Let $^{\omega}h_1, ^{\omega}h_2, \cdots, ^{\omega}h_n$ be a set of OWHFEs, then their aggregated value by using the GOWHFPWA operator is also an OWHFE, and*

$$GOWHFPWA(^{\omega}h_1, ^{\omega}h_2, \cdots, ^{\omega}h_n) = \underset{1 \leq j \leq L}{\cup} \left\{ \left\langle \left(1 - \prod_{i=1}^{n} (1 - (h_i^{\delta(j)})^{\alpha})^{\frac{T_i}{\sum_{i=1}^{n} T_i}} \right)^{1/\alpha}, \overline{(\sum_{i=1}^{n} w_i^{\delta(j)})} \right\rangle \right\} \tag{3}$$

where, $T_i = \prod_{l=1}^{i-1} \Delta(^{\omega}h_l)(i = 1, 2, \cdots, n)$, $T_1 = 1$, $\Delta(^{\omega}h_k)$ *is the score function of* $^{\omega}h_k$, *and L is the number of basic units in* $^{\omega}h_i(i = 1, 2, \cdots, n)$.

Proof. For $n = 1$, the result can be obtained easily by Definition 5. In the following, we prove the equation

$$GOWHFPWA(^{\omega}h_1, ^{\omega}h_2, \cdots, ^{\omega}h_n) = \underset{1 \leq j \leq L}{\cup} \left\{ \left\langle \left(1 - \prod_{i=1}^{n} (1 - (h_i^{\delta(j)})^{\alpha})^{\frac{T_i}{\sum_{i=1}^{n} T_i}} \right)^{1/\alpha}, \overline{(\sum_{i=1}^{n} w_i^{\delta(j)})} \right\rangle \right\}$$

by using mathematical induction for $n(n \geq 2)$.

For $n = 2$, since

$$\frac{T_1}{\sum_{i=1}^{2} T_i} {}^{\omega}h_1^{\alpha} = \underset{1 \leq j \leq L}{\cup} \left\{ \left\langle 1 - (1 - (h_1^{\delta(j)})^{\alpha})^{\frac{T_1}{\sum_{i=1}^{2} T_i}}, w_1^{\delta(j)} \right\rangle \right\}$$

$$\frac{T_2}{\sum_{i=1}^{2} T_i} {}^{\omega}h_2^{\alpha} = \underset{1 \leq j \leq L}{\cup} \left\{ \left\langle 1 - (1 - (h_2^{\delta(j)})^{\alpha})^{\frac{T_1}{\sum_{i=1}^{2} T_i}}, w_2^{\delta(j)} \right\rangle \right\}$$

then

$$\frac{T_1}{\sum_{i=1}^{2} T_i} {}^{\omega}h_1^{\alpha} \oplus \frac{T_2}{\sum_{i=1}^{2} T_i} {}^{\omega}h_2^{\alpha} =$$

$$= \underset{1 \leq j \leq L}{\cup} \left\{ \left\langle 1 - (1 - (h_1^{\delta(j)})^{\alpha})^{\frac{T_1}{\sum_{i=1}^{2} T_i}} + 1 - (1 - (h_2^{\delta(j)})^{\alpha})^{\frac{T_1}{\sum_{i=1}^{2} T_i}} - (1 - (1 - (h_1^{\delta(j)})^{\alpha})^{\frac{T_1}{\sum_{i=1}^{2} T_i}}) \times (1 - (1 - (h_2^{\delta(j)})^{\alpha})^{\frac{T_1}{\sum_{i=1}^{2} T_i}}), \overline{(w_1^{\delta(j)} + w_2^{\delta(j)})} \right\rangle \right\}$$

$$= \underset{1 \leq j \leq L}{\cup} \left\{ \left\langle 1 - \prod_{i=1}^{2} (1 - (h_i^{\delta(j)})^{\alpha})^{\frac{T_i}{\sum_{i=1}^{2} T_i}}, \overline{(w_1^{\delta(j)} + w_2^{\delta(j)})} \right\rangle \right\}$$

That is, Equation (7) holds when $n = 2$.

Suppose that Equation (3) also holds when for $n = l$,

$$GOWHFPWA(^{\omega}h_1, ^{\omega}h_2, \cdots, ^{\omega}h_l) = \underset{1 \leq j \leq L}{\cup} \left\{ \left\langle \left(1 - \prod_{i=1}^{l} (1 - (h_i^{\delta(j)})^{\alpha})^{\frac{T_i}{\sum_{i=1}^{n} T_i}} \right)^{1/\alpha}, \overline{(\sum_{i=1}^{l} w_i^{\delta(j)})} \right\rangle \right\}$$

when $n = l+1$, the operational laws described in Definition 2 state that

$$GOWHFPWA({}^{\omega}h_1, {}^{\omega}h_2, \cdots, {}^{\omega}h_l, {}^{\omega}h_{l+1}) = (\frac{1}{\alpha}\bigoplus_{i=1}^{l}(\frac{T_i {}^{\omega}h_i^{\alpha}}{\sum_{i=1}^{l+1}T_i})) \oplus \frac{1}{\alpha}(\frac{T_{l+1}{}^{\omega}h_{l+1}^{\alpha}}{\sum_{i=1}^{l+1}T_i})$$

$$= \bigcup_{1\le j\le L}\left\{\left\langle\left(1-\prod_{i=1}^{l}(1-(h_i^{\delta(j)})^{\alpha})^{\frac{T_i}{\sum_{i=1}^{l+1}T_i}}\right)^{1/\alpha}, \overline{(\sum_{i=1}^{l} w_i^{\delta(j)})}\right\rangle\right\} + \bigcup_{1\le j\le L}\left\{\left\langle\left(1-(1-(h_{l+1}^{\delta(j)})^{\alpha})^{\frac{T_{l+1}}{\sum_{i=1}^{l+1}T_i}}\right)^{1/\alpha}, w_{l+1}^{\delta(j)}\right\rangle\right\}$$

$$= \bigcup_{1\le j\le L}\left\{\left\langle\left(1-\prod_{i=1}^{l+1}(1-(h_i^{\delta(j)})^{\alpha})^{\frac{T_i}{\sum_{i=1}^{l+1}T_i}}\right)^{1/\alpha}, \overline{(\sum_{i=1}^{l+1} w_i^{\delta(j)})}\right\rangle\right\}$$

That is, Equation (3) holds for $n = l+1$.
Thus, Equation (3) holds for all n.
Then,

$$GOWHFPWA({}^{\omega}h_1, {}^{\omega}h_2, \cdots, {}^{\omega}h_n) = \bigcup_{1\le j\le L}\left\{\left\langle\left(1-\prod_{i=1}^{n}(1-(h_i^{\delta(j)})^{\alpha})^{\frac{T_i}{\sum_{i=1}^{n}T_i}}\right)^{1/\alpha}, \overline{(\sum_{i=1}^{n} w_i^{\delta(j)})}\right\rangle\right\}$$

Now, consider some desirable properties of the GOWHFPWA operator.

Theorem 2. *(Idempotency). Let ${}^{\omega}h_1, {}^{\omega}h_2, \cdots, {}^{\omega}h_n$ be a set of OWHFs, where $T_i = \prod_{l=1}^{i-1}\Delta({}^{\omega}h_l)(i = 1, 2, \cdots, n)$, $T_1 = 1$ and $\Delta({}^{\omega}h_l)$ is the score function of ${}^{\omega}h_l$. If ${}^{\omega}h_1 = {}^{\omega}h_2 = \cdots = {}^{\omega}h_n = {}^{\omega}h$, then*

$$GOWHFPWA({}^{\omega}h_1, {}^{\omega}h_2, \cdots, {}^{\omega}h_n) = \bigcup_{1\le j\le L}\left\{\left\langle 1-\prod_{i=1}^{n}(1-h_i^{\delta(j)})^{\frac{T_i}{\sum_{i=1}^{n}T_i}}, \overline{(\sum_{i=1}^{n} w_i^{\delta(j)})}\right\rangle\right\} = {}^{\omega}h \quad (4)$$

Proof. If ${}^{\omega}h_1 = {}^{\omega}h_2 = \cdots = {}^{\omega}h_n = {}^{\omega}h = \bigcup_{1\le j\le N}\{< h^{\delta(j)}, w^{\delta(j)} >\}$, then $\overline{\sum_{i=1}^{n} w_i^{\delta(j)}} = w^{\delta(j)}$,

$$GOWHFPWA({}^{\omega}h_1, {}^{\omega}h_2, \cdots, {}^{\omega}h_n) = \bigcup_{1\le j\le L}\left\{\left\langle\left(1-\prod_{i=1}^{n}(1-(h_i^{\delta(j)})^{\alpha})^{\frac{T_i}{\sum_{i=1}^{n}T_i}}\right)^{1/\alpha}, \overline{(\sum_{i=1}^{n} w_i^{\delta(j)})}\right\rangle\right\}$$

$$= \bigcup_{1\le j\le L}\left\{\left\langle\left(1-\prod_{i=1}^{n}(1-(h_i^{\delta(j)})^{\alpha})^{\frac{T_i}{\sum_{i=1}^{n}T_i}}\right)^{1/\alpha}, w^{\delta(j)}\right\rangle\right\} = \bigcup_{1\le j\le L}\left\{\left\langle 1-(1-h^{\delta(j)}), w^{\delta(j)}\right\rangle\right\} = \bigcup_{1\le j\le N}\left\{\left\langle h^{\delta(j)}, w^{\delta(j)}\right\rangle\right\}. \square$$

Theorem 3. *(Boundedness). Let ${}^{\omega}h_1, {}^{\omega}h_2, \cdots, {}^{\omega}h_n$ be a collection of OWHFEs, where $T_i = \prod_{l=1}^{i-1}\Delta({}^{\omega}h_l)(i = 1, 2, \cdots, n)$, $T_1 = 1$, $\Delta({}^{\omega}h_l)$ is the score function of ${}^{\omega}h_l$. Let ${}^{\omega}h^- = \{< h^-, 1 >\}$ ${}^{\omega}h^+ = \{< h^+, 1 >\}$, $h^- = \min(\min_{h_1^{\delta(j)}\in{}^{\omega}h_1}(h_1^{\delta(j)}), \min_{h_2^{\delta(j)}\in{}^{\omega}h_2}(h_2^{\delta(j)}), \cdots, \min_{h_n^{\delta(j)}\in{}^{\omega}h_n}(h_n^{\delta(j)}))$ and $h^+ = \max(\max_{h_1^{\delta(j)}\in{}^{\omega}h_1}(h_1^{\delta(j)}), \max_{h_2^{\delta(j)}\in{}^{\omega}h_2}(h_2^{\delta(j)}), \cdots, \max_{h_n^{\delta(j)}\in{}^{\omega}h_n}(h_n^{\delta(j)}))$. Then*

$${}^{\omega}h^- \le GOWHFPWA({}^{\omega}h_1, {}^{\omega}h_2, \cdots, {}^{\omega}h_n) \le {}^{\omega}h^+ \quad (5)$$

Proof. Since $f(x) = (1-x)^a (a \in (0,1))$ is a decreasing function about $x \in [0,1]$, then, $h^- = \left(1-\prod_{i=1}^{n}(1-(h^-)^\alpha)^{\frac{T_i}{\sum_{i=1}^{n} T_i}}\right)^{1/\alpha} \leq \left(1-\prod_{i=1}^{n}(1-\min_{h_i^{\delta(j)}\in {}^\omega h_i}(h_i^{\delta(j)})^\alpha)^{\frac{T_i}{\sum_{i=1}^{n} T_i}}\right)^{1/\alpha} \leq \left(1-\prod_{i=1}^{n}(1-(h_i^{\delta(j)})^\alpha)^{\frac{T_i}{\sum_{i=1}^{n} T_i}}\right)^{1/\alpha} \leq \left(1-\prod_{i=1}^{n}(1-\max_{h_i^{\delta(j)}\in {}^\omega h_i}(h_i^{\delta(j)})^\alpha)^{\frac{T_i}{\sum_{i=1}^{n} T_i}}\right)^{1/\alpha} \leq \left(1-\prod_{i=1}^{n}(1-(h^+)^\alpha)^{\frac{T_i}{\sum_{i=1}^{n} T_i}}\right)^{1/\alpha} = h^+$, thus $\Delta({}^\omega h^-) \leq \Delta({}^\omega h_i) \leq \Delta({}^\omega h^+)$ and ${}^\omega h^- \leq GOWHFPWA({}^\omega h_1, {}^\omega h_2, \cdots, {}^\omega h_n) \leq {}^\omega h^+$. □

Theorem 4. *(Monotonicity). Let* ${}^\omega h_1, {}^\omega h_2, \cdots, {}^\omega h_n$ *and* ${}^\omega h_1', {}^\omega h_2', \cdots, {}^\omega h_n'$ *be two sets of OWHFEs, where* $T_i = \prod_{l=1}^{i-1}\Delta({}^\omega h_l)(i = 1,2,\cdots,n)$, $T_i' = \prod_{l=1}^{i-1}\Delta({}^\omega h_l')(i = 1,2,\cdots,n)$, $T_1 = T_1' = 1$, $\Delta({}^\omega h_l)$ *is the score function of* ${}^\omega h_l$ *and* $\Delta({}^\omega h_l')$ *is the score function of* ${}^\omega h_l'$, *if* $h_i^{\delta(j)} \leq h_i'^{\delta(j)} (i = 1,2,\cdots,n, j = 1,2,\cdots,L)$ *and* $w_i^{\delta(j)} = w_i'^{\delta(j)} (i = 1,2,\cdots,n, j = 1,2,\cdots,L)$, *then*

$$GOWHFPWA({}^\omega h_1, {}^\omega h_2, \cdots, {}^\omega h_n) \leq GOWHFPWA({}^\omega h_1', {}^\omega h_2', \cdots, {}^\omega h_n') \tag{6}$$

Proof. According to the proof of Theorem 3, it is easy to prove that the GOWHFPWA operator satisfies the above monotonicity, thus the proof process is omitted. □

Theorem 5. *Let* ${}^\omega h_1, {}^\omega h_2, \cdots, {}^\omega h_n$ *be a set of OWHFEs, where* $T_i = \prod_{l=1}^{i-1}\Delta({}^\omega h_l)(i = 1,2,\cdots,n)$, $T_1 = 1$ *and* $\Delta({}^\omega h_l)$ *is the score function of* ${}^\omega h_l$. *If* ${}^\omega g$ *is an OWHFE. Then*

$$GOWHFPWA({}^\omega h_1 \oplus {}^\omega g, {}^\omega h_2 \oplus {}^\omega g, \cdots, {}^\omega h_n \oplus {}^\omega g) = GOWHFPWA({}^\omega h_1, {}^\omega h_2, \cdots, {}^\omega h_n) \oplus {}^\omega g \tag{7}$$

Theorem 6. *Let* ${}^\omega h_1, {}^\omega h_2, \cdots, {}^\omega h_n$ *be a set of OWHFEs, where* $T_i = \prod_{l=1}^{i-1}\Delta({}^\omega h_l)(i = 1,2,\cdots,n)$, $T_1 = 1$ *and* $\Delta({}^\omega h_l)$ *is the score function of* ${}^\omega h_l$. *Then*

$$GOWHFPWA(r{}^\omega h_1, r{}^\omega h_2, \cdots, r{}^\omega h_n) = rGOWHFPWA({}^\omega h_1, {}^\omega h_2, \cdots, {}^\omega h_n) \tag{8}$$

where r is an arbitrary number greater than 0.

Theorem 7. *Let* ${}^\omega h_1, {}^\omega h_2, \cdots, {}^\omega h_n$ *be a set of OWHFEs, where* $T_i = \prod_{l=1}^{i-1}\Delta({}^\omega h_l)(i = 1,2,\cdots,n)$, $T = 1$ *and* $\Delta({}^\omega h_l)$ *and is the score function of* ${}^\omega h_l$. *If* ${}^\omega g$ *is an OWHFE. Then*

$$GOWHFPWA(r{}^\omega h_1, r{}^\omega h_2, \cdots, r{}^\omega h_n) \oplus {}^\omega g = rGOWHFPWA({}^\omega h_1, {}^\omega h_2, \cdots, {}^\omega h_n) \oplus {}^\omega g \tag{9}$$

where r is an arbitrary number greater than 0.

Theorem 8. *Let* ${}^\omega h_1, {}^\omega h_2, \cdots, {}^\omega h_n$ *and* ${}^\omega g_1, {}^\omega g_2, \cdots, {}^\omega g_n$ *be two set of OWHFEs, where* $T_i = \prod_{l=1}^{i-1}\Delta({}^\omega h_l)(i = 1,2,\cdots,n)$, $T_1 = 1$ *and* $\Delta({}^\omega h_l)$ *is the score function of* ${}^\omega h_l$. *Then*

$$GOWHFPWA({}^\omega h_1, {}^\omega h_2, \cdots, {}^\omega h_n) \oplus GOWHFPWA({}^\omega g_1, {}^\omega g_2, \cdots, {}^\omega g_n) = GOWHFPWA({}^\omega h_1 \oplus {}^\omega g_1, {}^\omega h_2 \oplus {}^\omega g_2, \cdots, {}^\omega h_n \oplus {}^\omega g_n) \tag{10}$$

Proof. According to Definition 2, it is easy to prove that the GOWHFPWA operator satisfies Theorem 5, 6, 7, and 8, so the proof process is omitted. □

4. The MCGDM Approach with Order Weighted Hesitant Fuzzy Information

In this section, we present a novel MCGDM method based on ordered weighted hesitant fuzzy information, which utilizes the above GOWHFPWA operator to rank the alternatives of GSS. Consider a MCGDM for GSS problem, let $X = \{x_1, x_2, \ldots x_m\}$ be a set of suppliers, $C = \{c_1, c_2, \ldots c_n\}$ be a set of criteria, and $E = \{e_1, e_2, \ldots e_k\}$ be a set of DMs. In practice, there is a priority relationship among the GSS evaluation criteria. For example, if DMs believe that environmental protection is the most important criterion, they should take precedence over price, quality, and other criteria. Secondly, if price is more important than quality and other criteria, the priority of price is higher than quality, and so on. Such a prioritization among the criteria can be expressed by the ordering $c_1 \succ c_2 \succ \ldots \succ c_n$, in which criterion c_j has a higher priority than c_i if $j < i$.

For an alternative under a criterion, all the DMs provide their evaluated values anonymously. The evaluation values of alternative x_p under criteria c_q are provided by DM $e_u(u = 1,2,\ldots,k)$, which can be represented by an OWHFE ${}^{\omega}h_{pq}$. The ordered weighted hesitant fuzzy group decision matrix $M = ({}^{\omega}h_{pq})_{m\times n}$ is constructed from all of these OWHFEs.

In view of the above analysis, the procedure of the proposed approach is described under the following steps:

Step 1. Calculate the values of $T_{pq}(p = 1,2,\ldots m; q = 1,2,\ldots,n)$ based on Equation (11).

$$T_{pq} = \prod_{q=1}^{n-1} \Delta({}^{\omega}h_{pq})(p = 1,2,\ldots,m, q = 1,2,\ldots,n) \quad (11)$$

where $T_{p1} = 1$.

Step 2. Aggregate the OWHFEs ${}^{\omega}h_{pq}$ for each supplier $x_p(p = 1,2,\ldots,m)$ by the GOWHFPWA operator, then we can get the overall OWHFE ${}^{\omega}h_p(p = 1,2,\ldots,m)$ for the supplier $x_p(p = 1,2,\ldots,m)$ as follows:

$${}^{\omega}h_p = GOWHFPWA({}^{\omega}h_{p1}, {}^{\omega}h_{p2}, \ldots, {}^{\omega}h_{pn}) = \bigcup_{1\le j\le L_p} \left\{ \left\langle \left(1 - \prod_{q=1}^{n} \left(1 - \left(h_{pq}^{\delta(j)}\right)^{\alpha}\right)^{\frac{T_{pq}}{\sum_{q=1}^{n} T_{pq}}}\right)^{1/\alpha}, \left(\overline{\sum_{q=1}^{n} w_{pq}^{\delta(j)}}\right) \right\rangle \right\} = \bigcup_{1\le j\le L_p} \left\{ < h_p^{\delta(j)}, w_p^{\delta(j)} > \right\}. \quad (12)$$

Step 3. Calculate the score functions $\Delta({}^{\omega}h_p)(p = 1,2,\ldots,m)$ of the OWHFE ${}^{\omega}h_p(p = 1,2,\ldots,m)$ for the supplier $x_p(p = 1,2,\ldots,m)$, that is,

$$\Delta({}^{\omega}h_p) = \sum_{j=1}^{L_p} h_p^{\delta(j)} w_p^{\delta(j)} \quad (13)$$

Step 4. Rank the score functions $\Delta({}^{\omega}h_p)$ in ascending order. Then, the supplier with the highest priority is the most desirable green supplier.

5. Numerical Example

In light of the above discussion, we will further illustrate the procedure of the proposed method by an example of GSS. The GSCM of manufacturing enterprises is affected by its green suppliers' performance, and GSCM is considered as a strategic decision for manufacturing enterprises to maintain a competitive advantage in the international market. Inspired by the advantages of GSCM, there is a bus manufacturing enterprise who wants to choose the most appropriate green supplier for purchasing the key components of its new bus equipment. After initial screening, five potential suppliers x_i (i = 1, 2, 3, 4, 5) have been determined for further assessment. In order to choose the most suitable supplier, the company established a team of six DMs $e_u(u = 1,2,\ldots,6)$ from the department of purchasing, quality, and production who have abundant knowledge and experience in GSCM. Finally, four criteria are chosen from the Table 1 criteria list by experts to evaluate possible green suppliers. The four selected criteria are quality (c_1), technology(c_2), environment(c_3), cost(c_4), and the priority relationship among the criteria is $c_1 \succ c_2 \succ c_3 \succ c_4$ in the evaluation process. For a supplier under a criterion, six DMs need to give their evaluation values. As an instance, for the supplier x_1 under the criterion c_1, the evaluation values 0.3, 0.5, and 0.8 are provided by two, one and three DMs, respectively, and then an OWHFE ${}^{\omega}h_{11}$ can be represented by {<0.3,2/6>,<0.5,1/6>,<0.8,3/6>}.

In the same manner, all of OWHFEs ${}^{\omega}h_{pq}(p = 1,2,\ldots,5, q = 1,2,3,4)$ can be obtained, as shown in Table 2.

Table 2. Ordered weighted hesitant fuzzy decision matrix.

	c_1	c_2	c_3	c_4
x_1	{<0.3,2/6>,<0.5, 1/6>,<0.8,3/6>}	{<0.3,2/6>,<0.6, 1/6>,<0.7,3/6>}	{<0.3,2/6>,<0.6, 1/6>,<0.7,3/6>}	{<0.4,2/6>,<0.5, 1/6>,<0.6,3/6>}
x_2	{<0.1,2/6>,<0.4, 1/6>,<0.5,3/6>}	{<0.2,2/6>,<0.3, 1/6>,<0.5,3/6>}	{<0.1,2/6>,<0.4, 1/6>,<0.5,3/6>}	{<0.2,2/6>,<0.3, 1/6>,<0.4,3/6>}
x_3	{<0.1,2/6>,<0.2, 1/6>,<0.3,3/6>}	{<0.1,2/6>,<0.2, 1/6>,<0.4,3/6>}	{<0.1,2/6>,<0.2, 1/6>,<0.3,3/6>}	{<0.1,2/6>,<0.2, 1/6>,<0.4,3/6>}
x_4	{<0.3,2/6>,<0.4, 1/6>,<0.7,3/6>}	{<0.2,2/6>,<0.3, 1/6>,<0.6,3/6>}	{<0.1,2/6>,<0.5, 1/6>,<0.7,3/6>}	{<0.3,2/6>,<0.4, 1/6>,<0.5,3/6>}
x_5	{<0.7,2/6>,<0.8, 1/6>,<0.9,3/6>}	{<0.5,2/6>,<0.7, 1/6>,<0.8,3/6>}	{<0.4,2/6>,<0.6, 1/6>,<0.7,3/6>}	{<0.5,2/6>,<0.6, 1/6>,<0.7,3/6>}

Step 1. According to Equation (11), $T_{pq}(p = 1,2,\ldots,5, q = 1,2,3,4)$ are calculated as follows:

$$T_{5\times 4} = \begin{pmatrix} 1.0000 & 0.5833 & 0.3208 & 0.1764 \\ 1.0000 & 0.3500 & 0.1283 & 0.0449 \\ 1.0000 & 0.2167 & 0.0578 & 0.0125 \\ 1.0000 & 0.5167 & 0.2153 & 0.1005 \\ 1.0000 & 0.8167 & 0.5581 & 0.3255 \end{pmatrix}$$

Step 2. Aggregate ${}^{\omega}h_{pq}(p = 1,2,\ldots,5, q = 1,2,3,4)$ by using a GOWHFPWA ($\alpha = 1$) operator to derive the overall OWHFEs ${}^{\omega}h_p(p = 1,2,\ldots,5)$ for the supplier $x_p(p = 1,2,\ldots,5)$.

$$\begin{aligned} {}^{\omega}h_1 &= \{< 0.3091, 2/6 >, < 0.5462, 1/6 >, < 0.7470, 3/6 >\} \\ {}^{\omega}h_2 &= \{< 0.1305, 2/6 >, < 0.3755, 1/6 >, < 0.4973, 3/6 >\} \\ {}^{\omega}h_3 &= \{< 0.1000, 2/6 >, < 0.2000, 1/6 >, < 0.3190, 3/6 >\} \\ {}^{\omega}h_4 &= \{< 0.2514, 2/6 >, < 0.3866, 1/6 >, < 0.6654, 3/6 >\} \\ {}^{\omega}h_5 &= \{< 0.5703, 2/6 >, < 0.7163, 1/6 >, < 0.8233, 3/6 >\} \end{aligned}$$

Step 3. Calculate the score functions $\Delta({}^{\omega}h_p)(p = 1,2,\ldots,5)$ of the OWHFEs ${}^{\omega}h_p(p = 1,2,\ldots,5)$ for the supplier $x_p(p = 1,2,\ldots,5)$, that is,

$$\Delta({}^{\omega}h_2) = 0.5676, \Delta({}^{\omega}h_2) = 0.3547, \ \Delta({}^{\omega}h_3) = 0.2262, \Delta({}^{\omega}h_4) = 0.4809, \Delta({}^{\omega}h_5) = 0.7211$$

Step 4. Rank all the suppliers $x_p(p = 1,2,\ldots,5)$ in accordance with the score functions $\Delta({}^{\omega}h_p)(p = 1,2,\ldots,5)$ and the priority relationship of five suppliers can be obtained, that is,

$$x_5 \succ x_1 \succ x_4 \succ x_2 \succ x_3$$

Thus, the most desirable green supplier is x_5.

6. Performance Analysis and Comparation Analysis

In this section, performance analysis is provided based on the numerical example above to prove the validation and verification of the proposed method, including sensitivity analysis and effectiveness analysis. Additionally, the proposed GOWHFPWA operator is further compared with the hesitant fuzzy prioritized weighted average (HFPWA) operator suggested by Wei [61].

The sensitivity analysis is used to identify and determine the robustness of the proposed method. In Equation (2), the parameter α may affect the final ranking result, so the sensitivity analysis can be carried out by taking different α. The score functions $\Delta({}^{\omega}h_p)$ with different α can be calculated, and all of the results are presented in Table 3 and Figure 1.

Table 3. The results of the generalized ordered weighted hesitant fuzzy prioritized weighted average operator (GOWHFPWA) operator with different α.

α	x_1	x_2	x_3	x_4	x_5	Rankings
0.1	0.5666	0.3527	0.2255	0.4777	0.7176	$x_5 \succ x_1 \succ x_4 \succ x_2 \succ x_3$
0.2	0.5667	0.3529	0.2256	0.4780	0.7180	$x_5 \succ x_1 \succ x_4 \succ x_2 \succ x_3$
0.5	0.5670	0.3535	0.2258	0.4791	0.7192	$x_5 \succ x_1 \succ x_4 \succ x_2 \succ x_3$
1	0.5676	0.3547	0.2262	0.4809	0.7211	$x_5 \succ x_1 \succ x_4 \succ x_2 \succ x_3$
2	0.5689	0.3581	0.2270	0.4847	0.7254	$x_5 \succ x_1 \succ x_4 \succ x_2 \succ x_3$
5	0.5737	0.3703	0.2309	0.4941	0.7389	$x_5 \succ x_1 \succ x_4 \succ x_2 \succ x_3$
10	0.5836	0.3854	0.2389	0.5045	0.7581	$x_5 \succ x_1 \succ x_4 \succ x_2 \succ x_3$

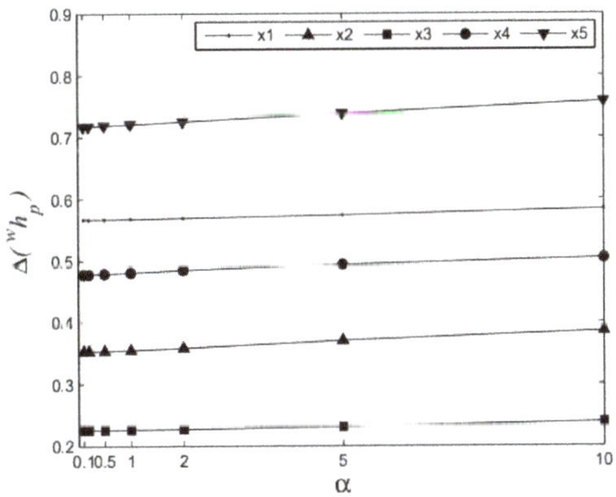

Figure 1. The curve of the score function with different α.

It can be seen from Table 3 that as parameter α takes different values, the priority relationships of five suppliers are unchanged and the most desirable supplier is still x_5. Therefore, the parameter α is insensitive to the proposed method and the obtained result ranking is robustness.

Meanwhile, it can be observed from Figure 1 that the values of the score function for each alternative will increase as α increases. From this point of view, the parameter α can be regarded as a DM's risk attitude. As the DMs can select different α in accordance with their own risk preferences, the proposed GOWHFPWA operator can offer more choice opportunities for the DMs in the actual GSS problems.

Additionally, since we proposed the GOWHFPWA operator based on the HFPWA operator [61], a comparative analysis was conducted in order to illustrate the effectiveness of the proposed GOWHFPWA operator. For convenience of comparison, we apply the HFPWA operator to the above numerical example in this paper. The hesitant fuzzy decision matrix is shown in Table 4.

Table 4. Hesitant fuzzy decision matrix.

	c_1	c_2	c_3	c_4
x_1	{0.3,0.5,0.8}	{0.3,0.6,0.7}	{0.3,0.6,0.7}	{0.4,0.5,0.6}
x_2	{0.1,0.4,0.5}	{0.2,0.3,0.5}	{0.1,0.4,0.5}	{0.2,0.3,0.4}
x_3	{0.1,0.2,0.3}	{0.1,0.2,0.4}	{0.1,0.2,0.3}	{0.1,0.2,0.4}
x_4	{0.3,0.4,0.7}	{0.2,0.3,0.6}	{0.1,0.5,0.7}	{0.3,0.4,0.5}
x_5	{0.7,0.8,0.9}	{0.5,0.7,0.8}	{0,4,0.6,0.7}	{0.5,0.6,0.7}

Then $t_{pq}(p = 1, 2, \ldots, 5, q = 1, 2, 3, 4)$ are calculated as follows:

$$t_{5\times4} = \begin{pmatrix} 1.0000 & 0.5333 & 0.2844 & 0.1517 \\ 1.0000 & 0.3333 & 0.1111 & 0.0370 \\ 1.0000 & 0.2000 & 0.0467 & 0.0093 \\ 1.0000 & 0.4667 & 0.1711 & 0.0741 \\ 1.0000 & 0.8000 & 0.5333 & 0.3022 \end{pmatrix}$$

We aggregate all hesitant fuzzy elements $h_{pq}(p = 1, 2, \ldots, 5, q = 1, 2, 3, 4)$ by using the HFPWA operator to derive the overall hesitant fuzzy elements $h_p(p = 1, 2, \ldots, 5)$ of the suppliers $x_p(p = 1, 2, \ldots, 5)$. Taking supplier x_1 as an example, we have $h_1 = HFPWA(h_{11}, h_{12}, h_{13}, h_{14}) =$ {0.3083,0.3179,0.3295,0.3620,0.3709,0.3816,0.3879,0.3965,0.4067,0.4055,0.4138,0.4238,0.4517,0.4593,0.4685, 0.4740,0.4813,0.4902,0.4501,0.4578,0.4670,0.4928,0.4998,0.5084,0.5134,0.5202,0.5284,0.4169,0.4250,0.4348, 0.4622,0.4697,0.4787,0.4841,0.4912,0.4999,0.4989,0.5059,0.5143,0.5378,0.5442,0.5520,0.5566,0.5628,0.5702, 0.5365,0.5429,0.5507,0.5724,0.5784,0.5856,0.5898,0.5956,0.6024,0.6338,0.6389,0.6451,0.6623,0.6670,0.6727, 0.6760,0.6805,0.6860,0.6853,0.6897,0.6950,0.7098,0.7138,0.7187,0.7216,0.7255,0.7301,0.7089,0.7130,0.7179, 0.7315,0.7353,0.7398,0.7424,0.7460,0.7504}.

The scores $s(h_p)(p = 1, 2, \ldots, 5)$ of the suppliers $x_p(p = 1, 2, \ldots, 5)$ are obtained as the following: $s(h_1) = 0.5539$, $s(h_2) = 0.3408$, $s(h_3) = 0.2080$, $s(h_4) = 0.4524$, $s(h_5) = 0.7174$. Finally, ranking all the suppliers $x_p(p = 1, 2, \ldots, 5)$ according to the scores $s(h_p)(p = 1, 2, \ldots, 5)$, we can get the priority relationship of six suppliers, that is,

$$x_5 \succ x_1 \succ x_4 \succ x_2 \succ x_3$$

Thus, the most desirable supplier by using the HFPWA operator proposed by Wei [61] is also x_5. The comparative results can be shown in Table 5.

Table 5. The result of different approaches.

Methods	x_1	x_2	x_3	x_4	x_5	Ranking Order
GOWHFPWA	0.5676	0.3547	0.2262	0.4809	0.7211	$x_5 \succ x_1 \succ x_4 \succ x_2 \succ x_3$
HFPWA	0.5689	0.3581	0.2270	0.4847	0.7254	$x_5 \succ x_1 \succ x_4 \succ x_2 \succ x_3$

From Table 5, despite the evaluation result obtained by using the HFPWA operator being the same as that of the GOWHFPWA operator, the proposed method has some advantages over the previous method. Firstly, the proposed method in this paper extends a prioritized weighted average operator from HFS to OWHFS which can solve the problem of the importance of the experts' evaluation results that the previous method cannot solve. Secondly, the computational complexity of the proposed approach is much lower than that of the previous method. Therefore, the introduced model for GSS in practice is more objective and reasonable than that obtained by using the HFPWA operator proposed by Wei [61].

7. Conclusions and Further Directions

In this paper, in order to overcome the limitation of MCGDM problems with GSS in practice, we have focused on a novel MCGDM approach with a priority relationship under the ordered weighted hesitant fuzzy environment to evaluate green suppliers, which can present the importance of each DM's judgment. Firstly, based on the ideal of the PA operator and HFPWA operator, the OWHFPWA operator was introduced and the prominent characteristics of the propose operator were studied. Secondly, we have utilized the OWHFPWA operator to develop MCGDM approaches to solve the GSS problem. Finally, a practical example of GSS in bus manufacturing enterprise was given to verify the practicality of the proposed method, meanwhile, its feasibility and effectiveness in dealing with MCGDM problems was carried out by the performance analysis and comparative analysis.

In future research, we will develop another hesitant fuzzy prioritized aggregation operator to solve the ordered weighted hesitant fuzzy MCGDM for GSS problems, namely, the generalized ordered weighted hesitant fuzzy prioritized weighted geometric (GOWHFPWG) operator. Moreover, we will combine the expanded hesitant fuzzy set (EHFS) [67] with the PA operator to deal with the MCDGM for GSS problems for future research, which take into account that a single DM gives several hesitant fuzzy elements in MCDGM problems.

Author Contributions: Authors Y.L. and L.J. conceived and designed the model for research, pre-processed and analyzed the data and the obtained inference. Anthors F.Z. and L.J. processed the data collected and wrote the paper. The final manuscript has been read and approved by all authors.

Funding: The work is supported by the National Natural Science Foundation of China (NSFC) under Projects 71672182, 71711540309, U1604262 and 71272207.

Acknowledgments: The authors are grateful to the editors and the anonymous reviewers for providing us with insightful comments and suggestions throughout the revision process.

Conflicts of Interest: The authors declare no conflict of interest.

References

1. Guo, S.; Shen, B.; Choi, T.M.; Jung, S. A review on supply chain contracts in reverse logistics: Supply chain structures and channel leaderships. *J. Clean. Prod.* **2017**, *144*, 387–402. [CrossRef]
2. Blome, C.; Hollos, D.; Paulraj, A. Green procurement and green supplier development: Antecedents and effects on supplier performance. *Int. J. Prod. Econ.* **2014**, *124*, 252–264. [CrossRef]
3. Ho, W.; Xu, X.; Dey, P.K. Multi-criteria decision making approaches for supplier evaluation and selection: A literature review. *Eur. J. Oper. Res.* **2010**, *202*, 16–24. [CrossRef]
4. Chai, J.; Liu, J.N.; Ngai, E.W. Application of decision-making techniques in supplier selection: A systematic review of literature. *Expert. Syst. Appl.* **2013**, *40*, 3872–3885. [CrossRef]
5. Deng, X.; Hu, Y.; Deng, Y.; Mahadevan, S. Supplier selection using AHP methodology extended by D numbers. *Expert. Syst. Appl.* **2014**, *41*, 156–167. [CrossRef]
6. Govindan, K.; Rajendran, S.; Sarkis, J.; Murugesan, P. Multi criteria decision making approaches for green supplier evaluation and selection: A literature review. *J. Clean. Prod.* **2015**, *98*, 66–83. [CrossRef]
7. Handfield, R.; Walton, S.V.; Sroufe, R.; Melnyk, S.A. Applying environmental criteria to supplier assessment: A study in the application of the analytical hierarchy process. *Eur. J. Oper. Res.* **2002**, *141*, 70–87. [CrossRef]
8. Lu, L.Y.Y.; Wu, C.H.; Kuo, T.C. Environmental principles applicable to green supplier evaluation by using multi-objective decision analysis. *Int. J. Prod. Res.* **2007**, *45*, 4317–4331. [CrossRef]
9. Hsu, C.W.; Hu, A.H. Applying hazardous substance management to supplier selection using analytic network process. *J. Clean. Prod.* **2009**, *17*, 255–264. [CrossRef]
10. Kuo, R.J.; Wang, Y.C.; Tien, F.C. Integration of artificial neural network and MAMD methods for green supplier selection. *J. Clean. Prod.* **2010**, *18*, 1161–1170. [CrossRef]
11. Bai, C.; Sarkis, J. Green supplier development: Analytical evaluation using rough set theory. *J. Clean. Prod.* **2010**, *18*, 1200–1210. [CrossRef]
12. Yeh, W.C.; Chuang, M.C. Using multi-objective genetic algorithm for partner selection in green supply chain problems. *Expert. Syst. Appl.* **2011**, *38*, 4244–4253. [CrossRef]

13. Zhou, R.; Ma, X.; Li, S.; Li, J. The green supplier selection method for chemical industry with analytic network process and radial basis function neural network. *Adv. Inf. Sci. Serv. Sci.* **2012**, *4*, 147–158.
14. Kuo, R.J.; Lin, Y.J. Supplier selection using analytic network process and data envelopment analysis. *Int. J. Prod. Res.* **2012**, *50*, 2852–2863. [CrossRef]
15. Jauhar, S.K.; Pant, M.; Deep, A. An approach to solve multi-criteria supplier selection while considering environmental aspects using differential evolution. In *Proceedings of Swarm, Evolutionary, and Memetic Computing*; Springer International Publishing: Switzerland, 2013; Volume 8297, pp. 199–208.
16. Dobos, I.; Vörösmarty, G. Green supplier selection and evaluation using DEA-type composite indicators. *Int. J. Prod. Econ.* **2014**, *157*, 273–278. [CrossRef]
17. Freeman, J.; Chen, T. Green supplier selection using an AHP-Entropy-TOPSIS framework. *Supply Chain Manag. An Int. J.* **2015**, *20*, 327–340. [CrossRef]
18. Hashemi, S.H.; Karimi, A.; Tavana, M. An integrated green supplier selection approach with analytic network process and improved Grey relational analysis. *Int. J. Prod. Econ.* **2015**, *159*, 178–191. [CrossRef]
19. Yazdani, M.; Chatterjee, P.; Zavadskas, E.K.; Zolfani, S.H. Integrated QFD-MCDM framework for green supplier selection. *J. Clean. Prod.* **2017**, *142*, 3728–3740. [CrossRef]
20. Liu, B.; Yang, X.; Huo, T.; Shen, G.Q.; Wang, X. A linguistic group decision making framework for bid evaluation in mega public projects considering carbon dioxide emissions reduction. *J. Clean. Prod.* **2017**, *148*, 811–825. [CrossRef]
21. Zadeh, L.A. Fuzzy sets. *Inf. Control.* **1965**, *8*, 338–356. [CrossRef]
22. Akman, G. Evaluating suppliers to include green supplier development programs via Fuzzy c-means and VIKOR methods. *Comput. Ind. Eng.* **2015**, *86*, 69–82. [CrossRef]
23. Lourenzutti, R.; Krohling, R.A. The hellinger distance in multicriteria decision making: An illustration to the TOPSIS and TODIM methods. *Expert. Syst. Appl.* **2014**, *41*, 4414–4421. [CrossRef]
24. Tseng, M.L.; Lin, Y.H.; Tan, K.; Chen, R.H.; Chen, Y.H. Using TODIM to evaluate green supply chain practices under uncertainty. *Appl. Math. Model.* **2014**, *38*, 2983–2995. [CrossRef]
25. Govindan, K.; Khodaverdi, R.; Jafarian, A. A fuzzy multi criteria approach for measuring sustainability performance of a supplier based on triple bottom line approach. *J. Clean. Prod.* **2013**, *47*, 345–354. [CrossRef]
26. Chiou, C.Y.; Hsu, C.W.; Hwang, W.Y. Comparative investigation on green supplier selection of the American, Japanese and Taiwanese electronics industry in China. In Proceedings of the IEEE International Conference on IE&EM, Singapore, 8–11 December 2008; pp. 1909–1914.
27. Lee, A.H.I.; Kang, H.Y.; Hsu, C.F.; Hung, H.C. A green supplier selection model for high-tech industry. *Expert. Syst. Appl.* **2009**, *36*, 7917–7927. [CrossRef]
28. Tsai, W.H.; Hung, S.J. A fuzzy goal programming approach for green supply chain optimization under activity-based costing and performance evaluation with a value-chain structure. *Int. J. Prod. Res.* **2009**, *47*, 4991–5017. [CrossRef]
29. Tuzkaya, G.; Ozgen, A.; Ozgen, D.; Tuzkaya, U.R. Environmental performance evaluation of suppliers: A hybrid fuzzy multi-criteria decision approach. *Int. J. Environ. Sci. Technol.* **2009**, *6*, 477–490. [CrossRef]
30. Büyüközkan, G.; Çifçi, G. A novel hybrid MCDM approach based on fuzzy DEMATEL, fuzzy ANP and fuzzy TOPSIS to evaluate green suppliers. *Expert. Syst. Appl.* **2012**, *39*, 3000–3011. [CrossRef]
31. Datta, S.; Samantra, C.; Mahapatra, S.S.; Banerjee, S.; Bandyopadhya, A. Green supplier evaluation and selection using VIKOR method embedded in fuzzy expert system with interval-valued fuzzy numbers. *Int. J. Procure. Manag.* **2012**, *5*, 647–678. [CrossRef]
32. Shen, L.; Olfat, L.; Govindan, K.; Khodaverdi, R.; Diabat, A. A fuzzy multi criteria approach for evaluating green supplier's performance in green supply chain with linguistic preferences. *Resour. Conserv. Recycl.* **2013**, *74*, 170–179. [CrossRef]
33. Wang, X.; Chan, H.K. A hierarchical fuzzy TOPSIS approach to assess improvement areas when implementing green supply chain initiatives. *Int. J. Prod. Res.* **2013**, *51*, 3117–3130. [CrossRef]
34. Cao, Q.W.; Wu, J.; Liang, C.Q. An intuitionsitic fuzzy judgment matrix and TOPSIS integrated multi-criteria decision making method for green supplier selection. *J. Intell. Fuzzy Syst.* **2014**, *28*, 117–126.
35. Kannan, D.; Govindan, K.; Rajendran, S. Fuzzy axiomatic design approach based green supplier selection: A case study from Singapore. *J. Clean. Prod.* **2015**, *96*, 194–208. [CrossRef]
36. Hamdan, S.; Cheaitou, A. Supplier selection and order allocation with green criteria: An MCDM and multi-objective optimization approach. *Comput. Oper. Res.* **2016**, *81*, 282–304. [CrossRef]

37. Guo, Z.X.; Liu, H.T.; Zhang, D.Q.; Yang, J. Green supplier evaluation and selection in apparel manufacturing using a fuzzy multi-criteria decision-making approach. *Sustainability* **2017**, *9*, 650–663.
38. Xu, Y.; Cabrerizo, F.J.; Herrera-Viedma, E. A consensus model for hesitant fuzzy preference relations and its application in water allocation management. *Appl. Soft Comput.* **2017**, *58*, 265–284. [CrossRef]
39. Tsui, C.W.; Tzeng, G.H.; Wen, P.U. A hybrid MCDM approach for improving the performance of green suppliers in the TFT-LCD Industry. *Int. J. Prod. Res.* **2015**, *53*, 6436–6454. [CrossRef]
40. Darabi, S.; Heydari, J. An interval-valued hesitant fuzzy ranking method based on group decision analysis for green supplier selection. *IFAC PapersOnLine* **2016**, *49*, 12–17. [CrossRef]
41. Gitinavard, H.; Ghaderi, H.; Pishvaeeet, S.M. Green supplier evaluation in manufacturing systems: A novel interval-valued hesitant fuzzy group outranking approach. *Soft Comput.* **2017**, *3*, 1–20. [CrossRef]
42. Qin, J.; Liu, X.; Pedrycz, W. An extended todim multi-criteria group decision making method for green supplier selection in interval type-2 fuzzy environment. *Eur. J. Oper. Res.* **2017**, *258*, 626–638. [CrossRef]
43. Tang, S.L. Green supplier selection model with hesitant fuzzy information. *J. Intell Fuzzy. Syst.* **2017**, *32*, 189–195. [CrossRef]
44. Krohling, R.A.; Pacheco, A.G.C.; Siviero, A.L.T. IF-TODIM: An intuitionistic fuzzy TODIM to multi-criteria decision making. *Knowl. Base Syst.* **2013**, *53*, 142–146. [CrossRef]
45. Torra, V.; Narukawa, Y. On hesitant fuzzy sets and decision. In Proceedings of the IEEE International Conference on Fuzzy Systems Fuzzy Systems, Jeju Island, South Korea, 20–24 August 2009; pp. 1378–1382.
46. Torra, V. Hesitant fuzzy sets. *Int. J. Intell. Syst.* **2010**, *25*, 529–539. [CrossRef]
47. Rodríguez, R.M.; Martínez, L.; Torra, V.; Xu, Z.S.; Herrera, F. Hesitant fuzzy sets: State of the art and future directions. *Int. J. Intell. Syst.* **2014**, *29*, 495–524. [CrossRef]
48. Chen, N.; Xu, Z.S.; Xia, M.M. Interval-valued hesitant preference relations and their applications to group decision making. *Knowl. Based. Syst.* **2013**, *37*, 528–540. [CrossRef]
49. Yu, D.J.; Zhang, W.Y.; Xu, Y.J. Group decision making under hesitant fuzzy environment with application to personnel evaluation. *Knowl. Based Syst.* **2013**, *52*, 1–10. [CrossRef]
50. Peng, D.H.; Gao, C.Y.; Gao, Z.F. Generalized hesitant fuzzy synergetic weighted distance measures and their application to multiple criteria decision making. *Appl. Math. Model.* **2013**, *37*, 5837–5850. [CrossRef]
51. Farhadinia, B. Correlation for dual hesitant fuzzy sets and dual interval-valued hesitant fuzzy sets. *Int. J. Intell. Syst.* **2014**, *29*, 184–205. [CrossRef]
52. Rodríguez, R.M.; Martínez, L.; Herrera, F. Hesitant fuzzy linguistic term sets for decision making. *IEEE Trans. Syst.* **2012**, *20*, 109–119. [CrossRef]
53. Dong, Y.C.; Chen, X.; Herrera, F. Minimizing adjusted simple terms in the consensus reaching process with hesitant linguistic assessments in group decision making. *Inf. Sci.* **2015**, *297*, 95–117. [CrossRef]
54. Farhadinia, B. Distance and similarity measures for higher order hesitant fuzzy sets. *Knowl. Based Syst.* **2014**, *55*, 43–48. [CrossRef]
55. Morente-Molinera, J.A.; Kou, G.; González-Crespo, R.; Corchado, J.M.; Herrera-Viedma, E. Solving multi-criteria group decision making problems under environments with a high number of alternatives using fuzzy ontologies and multi-granular linguistic modelling methods. *Knowl. Syst.* **2017**, *137*, 54–64. [CrossRef]
56. Alcantud, J.C.R.; Giarlotta, A. Necessary and possible hesitant fuzzy sets: A novel model for group Decision making. *Inf. Fusion* **2018**, *46*, 63–76. [CrossRef]
57. Zhang, Z.; Wu, C. Weighted hesitant fuzzy sets and their application to multi-criteria decision making. *British J. Math. Comput. Sci.* **2014**, *4*, 1091–1123. [CrossRef]
58. Farhadinia, B.; Xu, Z.S. Distance and aggregation-based methodologies for hesitant fuzzy decision making. *Cogn. Comput.* **2017**, *9*, 81–94. [CrossRef]
59. Yager, R.R. Prioritized aggregation operators. *Int. J. Approx. Reason.* **2008**, *48*, 263–274. [CrossRef]
60. Xia, M.; Xu, Z.S. Hesitant fuzzy information aggregation in decision making. *Int. J. Approx. Reason.* **2011**, *52*, 395–407. [CrossRef]
61. Wei, G.W. Hesitant fuzzy prioritized operators and their application to multiple attribute decision making. *Knowl. Based Syst.* **2012**, *31*, 176–182. [CrossRef]
62. Qua, G.H.; Zhang, H.P.; Qua, W.H.; Zhang, Z.H. Induced generalized dual hesitant fuzzy Shapley hybrid operators and their application in multi-attributes decision making. *J. Intell. Fuzzy Syst.* **2016**, *31*, 633–650. [CrossRef]

63. Wei, G.W.; Lu, M.; Tang, X.Y.; Wei, Y. Pythagorean hesitant fuzzy Hamacher aggregation operators and their application to multiple attribute decision making. *Int. J. Intell. Syst.* **2018**, *33*, 1197–1233. [CrossRef]
64. Galankashi, M.R.; Chegeni, A.; Soleimanynanadegany, A.; Memari, A.; Anjomshoae, A.; Helmi, S.A.; Dargi, A. Prioritizing green supplier selection criteria using fuzzy analytical network process. *Procedia CIRP* **2015**, *26*, 689–694. [CrossRef]
65. Mousakhani, S.; Nazari-Shirkouhi, S.; Bozorgi-Amiri, A. A novel interval type-2 fuzzy evaluation model based group decision analysis for green supplier selection problems: A case study of battery industry. *J. Clean. Prod.* **2017**, *168*, 205–218. [CrossRef]
66. Omurca, S.I. An intelligent supplier evaluation, selection and development system. *Appl. Soft Comput.* **2013**, *13*, 690–697. [CrossRef]
67. Alcantud, J.C.R.; Santos-García, G. Expanded hesitant fuzzy sets and group decision making. In Proceedings of the IEEE International Conference on Fuzzy Systems, Naples, Italy, 9–12 July 2017.

symmetry

MDPI

Article

Risk Level Evaluation on Construction Project Lifecycle Using Fuzzy Comprehensive Evaluation and TOPSIS

Tsegay Gebrehiwet * and Hanbin Luo

Institute of Construction Management, School of Civil Engineering and Mechanics, Huazhong University of Science and Technology, Wuhan 1037, China; luohbcem@hust.edu.cn

* Correspondence: tseggt77@hust.edu.cn; Tel.: +86-131-250-96204

Received: 17 November 2018; Accepted: 17 December 2018; Published: 24 December 2018

Abstract: A risk is a predominant problem in the developing countries construction projects. Although numerous studies have been concerned on risk, there is a limited study on a mechanism to identify the typical risks and effects level. This paper presents an approach for evaluating the risks in case of schedule delays at the various lifecycles of construction projects. The methodology applied is an integrated model of the technique for order preference by similarity to ideal solution (TOPSIS) and fuzzy comprehensive evaluation (FCE). In this study, common criteria, sub-criteria, and attributes are constructed to make a decision concerning the influence level of risk of delay at the construction project lifecycle. The results showed that the construction stage (44%) is highly influenced, the second highly influenced stage is post construction (37%), and the least risked stage is pre-construction (35%). The construction projects in Ethiopia have faced an average delay risk of 38% at a high and very high-risk level. This work is expected to serve as a tool to assist managers in the management and control of schedule delays to mitigate their risks.

Keywords: FCE; Construction project; Evaluation of schedule delay risk; TOPSIS

1. Introduction

Construction projects have a complex nature owing to the site work difficulty, labour change for every site, adverse weather effect, and the higher exposure to error [1]. Consequently, numerous types of risks are occurred in different phases of projects and have an impact on time overrun, cost overrun, quality, and safety [2]. Schedule delay risk has many effects such as increased cost, late completion, disruption, third-party claims, loss of productivity and quality, disputes, and termination of contracts [3]. Therefore, controlling risks in a construction project has been a fundamental part of management in construction projects for decades [4].

A schedule is the main concern for construction projects, because of it influenced by several risks, such as environmental condition, equipment efficiency, productivity, material delivery, and soil types [5]. These factors are caused to delay and cost overruns that often jeopardize safety and quality performance [6]. For example, 13% of the Australian construction projects have faced 40% time overrun [7]. In the United Arab Emirates (UAE), the construction projects encountered 50% schedule delay [8]. In Malaysia, the government projects faced approximately 20% of delay [9]. Construction projects in Saudi Arabia experienced 70% time overrun, among 56% of consultants and 76% of contractors experienced 10–30% of delay, which causes to approximately 50% of cost overruns [10]. In Ethiopia, 40–60% of the construction projects occurred delay [11]. The studies indicate that the risk of schedule delay in developed and developing countries is critical and requires further investigation.

Risk of Schedule delay has a major negative effect on participating parties and projects [12]. Schedule delay often occurs because of many reasons attributable to owners or contractors. The significant contribution of owners often related to issues such as the settlement of claims, slow decision-making, early planning and design, change in the scope, schedule delay in payment, and excessive bureaucracy. Conversely, the contribution of contractors includes cash flow problems, difficulty in obtaining permits, and ineffective planning and scheduling [13]. The main groups of schedule delay factors were classified as owner, contractor, design, consultant, labour, equipment, material, external, and project-related [14]. The other important schedule delay factors mentioned are a slow decision by authorities, lack of funding, errors in work, improper planning, and lack of need identification [15].

Different scholars have conducted different studies on risk. However, the problem of risk is still prevalent. Moreover, the problem requires a pioneering decision-making mechanism to evaluate risk [16]. The evaluation of risk helps to quantify the risks level to mitigate their effects. Therefore, managers need to emphasize on a decision-making mechanism for schedule risk evaluation [6]. The evaluation mechanism should consider the risk throughout the construction lifecycle. Evaluation of risks on construction projects improves quality, safety, reduces cost, and increases the satisfaction of stakeholders.

The objective of this study is to evaluate the delay risk level at the construction lifecycle and comparing the lifecycle to each other. The study evaluates the risk impact level for criteria and sub-criteria based on the different attributes. Finally, the study proposes a method for delay risk evaluation using an integrated decision model of the Technique for Order Preference by Similarity to Ideal Solution (TOPSIS) and fuzzy comprehensive evaluation (FCE).

2. Literature Review

Improper risk management is found to be the main cause for the time overrun, cost overrun, and the problem of quality and safety [4]. Risk management is a crucial and important part of the decision-making process for construction projects to reduce risk [17]. Risk management has been investigated from a different viewpoint among certain countries because of the fact that the risk effect varies in different instances [18]. Various researchers have studied the causes and effects of risk for decades to identify the risk factors. Most of the previous studies focused more on the identification of general risk factors and effects than identifying the typical level of impact at the different construction lifecycles.

The studies in different countries identified different risk problems using different methods. However, the method has a certain limitation. One of the most vital methods is the application of the integrated methods of a relative importance index and fuzzy analytical hierarchy process, have been used to reduce risk [19]. This assessment method has limitations in imprecision and involves subjective judgment [14]. The integration of interpretive structural modelling with fuzzy logic has an important role in risk assessment. The model used a computer program to form a relationship between factors in a system [18]. However, this method overlooked the weight of the relationship between factors that makes difficult to use in a complex system [20]. Bayesian belief network developed to quantify the probability of delay risk in construction project considering the cause-effect relations among factors [21,22]. In addition, System Dynamic (SD) plays a respective advantage for schedule delay risk effects investigation based on the causal loop diagrams considering the simulation-based scheduling [23]. Although the System Dynamics and Bayesian Belief Network are important for risk assessment, the methods had restrictions on the structural relationship. Different studies may have different structures of relations and assumptions between factors, this makes confined the models because it may vary from one expert to another [22,24]. The structural equation modelling considers the complex nature of factors and used a causal relationships modelling technique for analyzing the relationship between the factors [25]. However, the method has a certain curb that the risk paths in the design and the dynamic behaviour of the risk factors were not considered [18]. The different methods may have different contribution with certain limitations, especially for risk assessment [26]. In addition, the different studies in the different methods have still lacked a systematic approach to consider the complex and dynamic nature of construction projects risk for better understanding and effective risk management [27]. The aforementioned methods contributed for better assessment of schedule delay risk but the studies still have some confined. First, the evaluation of risk impact level throughout the construction lifecycle has limitations. Second, the assessment methods have been identifying the relations and ranking of factors, but most of the studies were confined to evaluate the level of impact of factors on different construction activities.

To fill this gap, this study aims to propose a hybrid model for delay risk assessment in construction projects with the aid of FCE and TOPSIS. The hybrid model helps managers to have appropriate decision-making mechanism to minimize schedule risk impact. The integrated model of FCE and TOPSIS is an important mechanism to detect typical delay risk impact in construction projects, which is beyond the identification of factors and relationships between factors.

3. Research Methodology

FCE and TOPSIS are essential for selecting and ranking a set of factors with usually incommensurate and independent attributes [28]. FCE and TOPSIS are supporting tools for decision-making in a finite number of alternatives [29]. The integrated model plays the respective advantage of flexibility in order to achieve a better result and make the right decision by identifying the typical level delay risk on construction activities. The layout of the evaluation mechanism of schedule delay risk using integrated models of FCE and TOPSIS is described in Figure 1.

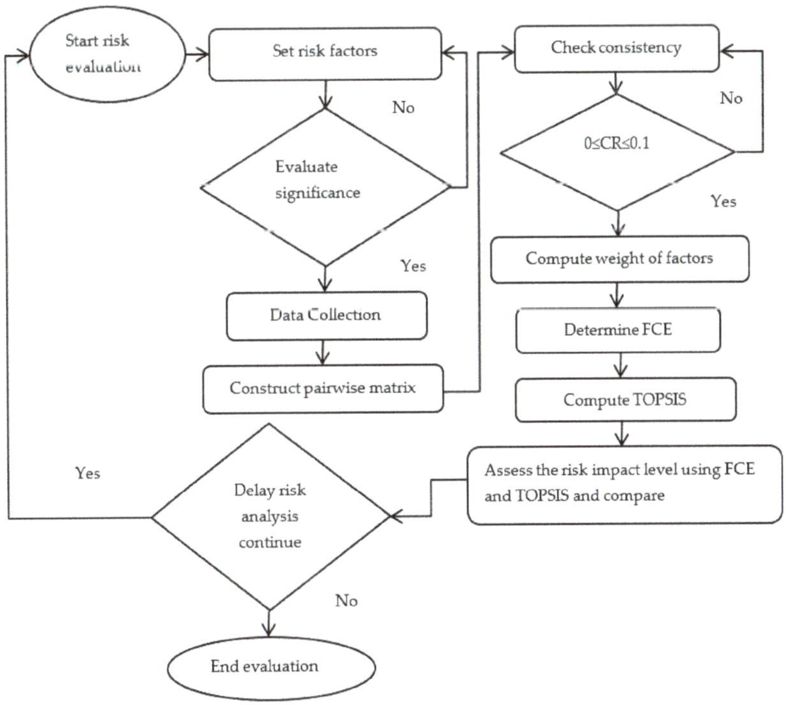

Figure 1. The layout of delay risk evaluation using integrated models of fuzzy comprehensive evaluation (FCE) and the technique for order preference by similarity to ideal solution (TOPSIS).

3.1. Set Significant Delay Risk Factors

Set significant delay risk factors: it involves the identification and classification of delay risk factors and their consequences [19]. The procedure of selecting factors expressed as, based on the extensive literature review, a questionnaire for pilot study was developed, then the pilot study was assessed in fifteen different projects and participant were interviewed and responded to the questionnaire. Next based on the literature review and pilot study final questionnaire was developed. Finally, the developed questionnaire was validated to confirm clarity, completeness, and applicability in the case study.

In this study, the construction project lifecycle considers the three basic stages of pre-construction stage, construction stage, and post-construction stage according to the pilot to fit for the empirical assessment. The risk factors listed in Table 1 were selected based on the impact that directly or indirectly associated with the three-construction lifecycle. However, the influence level of risk is completely different among the lifecycles, which is essential for decision-making in risk. The final questionnaire was consisted of four criteria, eight sub-criteria, and fifty-two attributes, as summarized in Table 1. Then based on this questionnaire data were collected.

Table 1. Schedule delay risk factors evaluation indexing system and reference.

Criteria Related	Sub-Criteria Related	Attribute Related	Reference
(A) Responsibility	A1 Client	A11 Lack of on-time payments and finance	[4,19,30,31]
		A12 Client interference	[4,31,32]
		A13 Leisureliness in decision-making	[15,19]
		A14 Late at delivery of site for design and construction	[4,17,30]
		A15 Inadequate feasibility study	[4,19,31]
		A16 Poor coordination and communication with other	[4,19,31]
	A2 Contractor	A21 Problems related to subcontractor	[4,19,31]
		A22 Poor site performance and management	[15,33]
		A23 Ineffective scheduling and planning	[4,19,30,32]
		A24 Improper construction methods	[19,22,33]
		A25 Poor coordination and communication with other	[4,30,31]
		A26 Inadequate experience of the contractor	[19,22,25]
		A27 Rework for unsatisfactory work	[17,19,22]
	A3 Consultant	A31 Inadequate consultant experience	[15,19]
		A32 Late in receiving and approving of work	[4,19,22,30]
		A33 Late in performing inspection and testing and poor supervision	[34]
		A34 Poor coordination and communication with other	[19]

Table 1. *Cont.*

Criteria Related	Sub-Criteria Related	Attribute Related	Reference
	A4 Designer	A41 In adequate details and unclear specification	[19,25,30,33]
		A42 Late design documents and design	[19]
		A43 Design errors and mistakes	[4,19,30,31]
		A44 Misunderstanding requirements of client's	[15]
(B) Resource	B1 Material	B11 Absence of quality materials	[4,30,31]
		B12 Leisurely delivery of material	[17,19,25,30]
		B13 Changes in specifications and material types	[25,31]
		B14 Damage of materials	[17,25,32]
		B15 Inflation of price for materials	[4,31–33]
	B2 Finance	B21 Problem for a financial claim	[25]
		B22 Problem of funding processes from government	[2,15]
		B23 Late release budget	[2,35]
		B24 Global financial crisis	[17,19,25,33]
	B3 Labor	B31 Less productivity	[4,17,25]
		B32 Low morale and motivation	[18]
		B33 Unqualified workers	[17,19,23,30]
		B34 Discipline problem (conflicts)	[15]
		B35 Labour injuries and accidents	[17,32]
	B4 Equipment	B41 Insufficient or equipment shortage	[4,22,33]
		B42 Low efficiency and productivity of equipment	[34]
		B42 Lack of spare parts and failures of equipment	[4,17,30,31]
		B43 Problem of mobilization and allocation	[16]
		B45 Equipment outdated	[16,34]
(C) Contrac condition		C1 Absence of alternative dispute resolution (ADR)	[4,30,33]
		C2 Discrepancies and mistakes in contract	[33]
		C3 Unrealistic cost and duration contract	[4,17,30]
		C4 Poor incentives and inadequate penalties in contract	[4]
		C5 Insufficient details of the contract documents	[25,33]
		C6 Absence of clear understanding for contract	[15,25]
(D) External		D1 Adverse weather condition	[4,22,30,32]
		D2 Force of majeure	[4,19,30,31]
		D3 Corruption	[35]
		D4 Effect of cultural and social factors	[19,33]
		D5 Government commitment and policy	[4,19,30,31]
		D6 Unavailability of utilities in a site	[25,32]

3.2. Fuzzy Comprehensive Evaluation (FCE)

FCE model is used to obtain reliable results in evaluating alternatives widely in an uncertain environment [35]. The FCE approach has strong evaluation abilities that lead to appropriate decision-making. The FCE generates the weight of factors and priorities factors based on expert judgment [36]. FCE determines the risk of criteria and sub-criteria of delay factors on the construction project lifecycle. To get priorities in an organized way, the FCE technique requires to decompose the decision through the following steps, based on the different studies aforementioned.

Firstly, set criteria, sub-criteria, and attributes. This step was segmented the complex problem of delay risk factors into a structured hierarchy of three levels, which are level one (criteria), level two (sub-criteria), and level three (attributes), as shown in Table 1.

Secondly, develop a pairwise comparison matrix. This matrix was constructed using the rate of scale that denotes how much one element dominates over another with respect to a given attribute. The scale value measures the degree of relative importance in a pairwise comparison matrix using linguistic variables for delay risk factors, as described in Table 2.

Table 2. The Linguistic Variables for Schedule Delay Risk Factor.

Linguistic Terms	Fuzzy Number
Very small risk level	1
Small risk level	3
Medium risk level	5
High risk level	7
Very high risk level	9

The matrix for pairwise comparison was developed based on the judgment of respondents, as described in the matrix (1):

$$A = \begin{bmatrix} 1 & \frac{w_1}{w_2} & \cdots & \frac{w_1}{w_n} \\ \frac{w_2}{w_1} & 1 & \cdots & \frac{w_2}{w_n} \\ \vdots & \vdots & \ddots & \vdots \\ \frac{w_n}{w_1} & \frac{w_n}{w_2} & \cdots & 1 \end{bmatrix} = \begin{bmatrix} a_{11} & a_{12} & \cdots & a_{1n} \\ a_{21} & a_{22} & \cdots & a_{2n} \\ \vdots & \vdots & \ddots & \vdots \\ a_{n1} & a_{n2} & \cdots & a_{nn} \end{bmatrix}. \quad (1)$$

where A is the matrix with a_{ij} element of the pairwise matrix in i column and j row and the rated weight of factors $W = (w_1, w_2, \ldots, w_n)$, where $(i, j = 1, 2, \ldots, n)$ and n is the number of factors.

Thirdly, estimate normalized weight. It computed from the pairwise matrix using the arithmetic mean. The normalized weight computed by Equation (2):

$$w_i = \frac{\overline{w}_i}{\sum_{i=1}^{n} \overline{w}_i} \text{ but } \overline{w}_i = \sum_{j=1}^{n} a_{ij} \; (i = 1, 2, \ldots, n) \text{ and } (j = 1, 2, 3, \ldots, n). \quad (2)$$

where w_i is normalized weight from the cumulative weight $\overline{w} = (\overline{w}_1, \overline{w}_2, \ldots, \overline{w}_n)$.

Determining consistency ratio aids to identify consistency of decision. First, compute random indexing (RI) and consistency index (CI) then determine the consistency ratio (CR). Finally, if CR is greater than or equal to zero and less than or equal to 0.1, the judgment is consistent [37]. The relations for the checking consistency can be expressed as Equations (3):

$$CR = \frac{CI}{RI} \text{ but } CI = \frac{\lambda_{max} - n}{n - 1} \quad \text{Where; } \lambda_{max} = \sum_{i}^{n} a_{ij} \times w_i \,. \quad (3)$$

Finally, compute multi-comprehensive evaluation. In a multi-comprehensive evaluation, the first step is to establish a grade factor and element set. According to the delay risk factors in Table 1, the main element set is $U = \{A, B, C, D\}$, which represents responsibility, resource, contract condition, and external related factors, respectively. The five grade factors determined by $V = \{v_1\ v_2\ v_3\ v_4\ v_5\}$, which represents very small risk level (1), small risk level (2), medium risk level (3), high-risk level (4), and very high-risk level (5), respectively. The second step, establish a single fuzzy matrix. The single fuzzy matrix uses the respondent's rate to evaluate the risk level of attributes. The single fuzzy matrix computed using Equation (4):

$$r_{ik} = \frac{r_i}{N} \text{ then } R = \begin{bmatrix} r_{11} & r_{12} & r_{13} & r_{14} & r_{15} \\ r_{21} & r_{22} & r_{23} & r_{24} & r_{25} \\ \vdots & \vdots & \vdots & \vdots & \vdots \\ r_{n1} & r_{n2} & r_{n3} & r_{n4} & r_{n5} \end{bmatrix}. \tag{4}$$

where n is attributes or element set, grade factor $k = \{1,2,3,4,5\}$, the rate of respondents for grade factor r_i, the total population of respondents N and single fuzzy matrix R.

Lastly, evaluate the comprehensive and multi comprehensive. Comprehensive evaluation compiles a single fuzzy matrix of attributes with a normalized weight of sub-criteria from the pairwise matrix to priorities influence of sub-criteria. This can be expressed using Equation (5):

$$B_i = W_i \times R_i = \begin{bmatrix} W_1 \\ W_2 \\ \cdots \\ W_n \end{bmatrix} \times \begin{bmatrix} r_{11} & r_{12} & r_{13} & r_{14} & r_{15} \\ r_{21} & r_{22} & r_{23} & r_{24} & r_{25} \\ \vdots & \vdots & \vdots & \vdots & \vdots \\ r_{n1} & r_{n2} & r_{n3} & r_{n4} & r_{n5} \end{bmatrix} = \begin{bmatrix} b_{i1} & b_{i2} & b_{i3} & b_{i4} & b_{i5} \end{bmatrix}. \tag{5}$$

where normalized weight of sub-criteria W_i, single fuzzy matrix R_i, and B_i is prioritized result of sub-criteria, namely $\begin{bmatrix} b_{i1} & b_{i2} & b_{i3} & b_{i4} & b_{i5} \end{bmatrix}$.

The multi comprehensive evaluation used to priorities the influence of criteria on alternatives. It can be computed based on Equation (6):

$$B_i = W_i \times R_i = \begin{bmatrix} W_1 \\ W_2 \\ \cdots \\ W_n \end{bmatrix} \times \begin{bmatrix} b_{11} & b_{12} & b_{13} & b_{14} & b_{15} \\ b_{21} & b_{22} & b_{23} & b_{24} & b_{25} \\ \vdots & \vdots & \vdots & \vdots & \vdots \\ b_{n1} & b_{n2} & b_{n3} & b_{n4} & b_{n5} \end{bmatrix} = \begin{bmatrix} b_{i1} & b_{i2} & b_{i3} & b_{i4} & b_{i5} \end{bmatrix}. \tag{6}$$

where B_i is prioritized result of criteria, R_i is a matrix of prioritized result from Equation (5) and W_i is a normalized weight of criteria from the pairwise matrix

3.3. *Technique For Order Preference by Similarity to Ideal Solution (TOPSIS)*

TOPSIS is important to reasonable and rational decision-making under uncertainty, subjectivity, ambiguity, and biases [29]. TOPSIS suggests both the best and worst alternatives. The best alternative has the short distance from the positive ideal solution and the farthest distance from the negative ideal solution and vice versa [28].

In this study, precise scores of alternatives were determined by formulating a decision matrix from the linguistic variables and then forming a normalized decision matrix. Multiplying the normalized decision matrix with the normalized weight of fuzzy comprehensive evaluation has a result of weighted normalized decision matrix of attributes. Based on the weighted normalized decision matrix, the negative and positive ideal solutions, Euclidean distance, and closeness coefficient were computed. Comparing the closeness coefficient, the typical delay risk on construction project lifecycle was detected. The procedure for implementing TOPSIS presented as:

Firstly, compute normalized decision matrix. This matrix is important because each factor has a different rated value and this rate values should convert into a single unit to enable direct comparison. The normalized decision matrix computed using Equation (7):

$$r_{ij} = \frac{a_{ij}}{\sqrt{\sum_{i=1}^{n} a_{ij}^2}}, j = 1,2,3,\ldots,n\ J,\ i = 1,2,3,\ldots,n \tag{7}$$

where r_{ij} is normalized decision matrix, a_{ij} is rated value of ith factor and j^{th} alternative and n is the total number of alternatives in the construction lifecycle.

Secondly, estimate the weighted normalized decision matrix. In this phase, the normalized weight from the pairwise matrix of fuzzy comprehensive evaluation multiplied with the normalized decision matrix. The weighted normalized decision matrix v_{ij} calculated based on expression as (8):

$$v_{ij} = w_i \times r_{ij}, j = 1,2,3,\ldots,n\ J,\ i = 1,2,3,\ldots,n. \tag{8}$$

where w_i is the normalized weight of ith factors and r_{ij} is normalized decision matrix.

Thirdly, identify the negative ideal solution (NIS) and positive ideal solution (PIS). The other important thing in TOPSIS is to identify the positive ideal solution and negative ideal solution. These can be determined from weighted normalized decision matrix using Equations (9) and (10):

$$A^* = \{v_1^*, v_2^*, v_3^*, \ldots, v_n^*\} \text{ maximum values, where } v_i^* \{max(v_{ij})\ if\ j \in J\} \tag{9}$$

$$A^- = \{v_1^-, v_2^-, v_3^-, \ldots, v_n^-\} \text{ maximum values, where } v_i^- \{min(v_{ij})\ if\ j \in J\}. \tag{10}$$

where negative ideal solution (NIS) and positive ideal solution (PIS) denoted by A^- and A^*, respectively.

Fourthly, determine the Euclidean distance of alternative. The Euclidean distance from the positive ideal solution (PIS) is represented by d_i^* and the Euclidean distance from the negative ideal solution (NIS) is represented by d_i^- can be determined by Equations (11) and (12):

$$d_i^* = \sqrt{\sum_{j=1}^{n} (v_{ij} - v_j^*)^2}, j = 1,2,3,\ldots,J \tag{11}$$

$$d_i^- = \sqrt{\sum_{j=1}^{n} (v_{ij} - v_j^-)^2}, j = 1,2,3,\ldots,J. \tag{12}$$

Finally, estimate closeness coefficient (CC). The risk level of delay determines based on closeness coefficient (CC) that deems to measure the influence level of the construction project lifecycle. The closeness coefficient (CC) expressed by Equation (13):

$$CC_i = \frac{d_i^-}{d_i^* + d_i^-}, i = 1,2,3,\ldots J. \tag{13}$$

The closeness coefficient (CC_i) ideal solution estimates one and the closeness coefficient (CC_i) non-ideal solution estimates zero. Therefore, the risk level of schedule delay risk on the construction project lifecycle estimates based on the value of closeness coefficient (CC_i) ideal solution.

4. Descriptive Case Study

The study demonstrated through an empirical analysis in the construction projects in Ethiopia. To gather data, participants were nominated based on the purposive sampling based on their experience in the different parts of the country. The participants were experienced and they serve as clients, contractors, consultants, managers, engineers, designers, surveyors, and other concerned positions. To minimize the subjectivity data were collected in discussion form, from 77 different groups in the participating construction projects and each group consists of at least five experts. For the questionnaire survey, the five-point Likert scale was employed, where each point corresponded to very small level risk (1), small level risk (2), medium level risk (3), high-level risk (4), and very high-level risk (5).

Based on the pairwise matrix the normalized weight of factors are computed, this used as a weight for comprehensive and multi-comprehensive evaluation in FCE and normalized weight for TOPSIS. The product of the normalized weight with a single fuzzy matrix can give the risk impact level of sub-criteria. The impact level of sub-criteria multiplied with the weight of criteria will give risk impact level of criteria. Then, the risk level of criteria multiplied with weight will give the level of risk for the construction lifecycle, as described from Formula 3 to 6. Whereas for TOPSIS, from the normalized weight of attributes and the normalized decision matrix, the weighted normalized decision matrix can be determined so based on the weighted normalized decision matrix the negative ideal solution (NIS), positive ideal solution (PIS), and Euclidian distance is estimated, then, finally, the closeness coefficient (CC) can be computed, as shown in formula 7 to 13.

The comprehensive evaluation of the sub-criteria such as client, contractor, consultant, and designer represented by {A1, A2, A3, A4}, respectively, under responsibility related criteria and the evaluation of sub-criteria, finance, material, labour, and equipment represented by {B1, B2, B3, B4} respectively under resource related criteria, computed in Table 3.

Table 3. Comprehensive Evaluation of Sub-Criteria Risk on Construction Process.

Sub-Criteria	Pre-Construction					Construction					Post-Construction				
	1	2	3	4	5	1	2	3	4	5	1	2	3	4	5
A1	0.11	0.24	0.31	0.21	0.13	0.09	0.24	0.30	0.26	0.11	0.09	0.21	0.35	0.17	0.18
A2	0.18	0.29	0.23	0.19	0.11	0.09	0.22	0.24	0.28	0.17	0.08	0.22	0.31	0.24	0.15
A3	0.10	0.28	0.29	0.19	0.14	0.09	0.23	0.24	0.27	0.17	0.14	0.30	0.30	0.15	0.11
A4	0.09	0.22	0.27	0.20	0.22	0.07	0.24	0.25	0.22	0.22	0.14	0.27	0.28	0.21	0.10
B1	0.08	0.22	0.30	0.20	0.20	0.05	0.22	0.28	0.23	0.22	0.13	0.24	0.28	0.21	0.14
B2	0.12	0.23	0.31	0.19	0.15	0.10	0.18	0.29	0.24	0.19	0.09	0.22	0.32	0.21	0.16
B3	0.13	0.25	0.27	0.19	0.16	0.10	0.22	0.29	0.21	0.15	0.09	0.31	0.30	0.17	0.13
B4	0.17	0.26	0.29	0.18	0.10	0.08	0.21	0.27	0.25	0.19	0.08	0.22	0.25	0.20	0.25

The multi-comprehensive evaluation results of delay risk on alternatives based on the criteria of {A, B, C, D} that represent responsibility, resource, contract condition, and external related, described in Table 4.

Table 4. Multi-Comprehensive Evaluation of Criteria Risk on Construction Process.

Criteria	Pre-Construction					Construction					Post-Construction				
	1	2	3	4	5	1	2	3	4	5	1	2	3	4	5
A	0.12	0.26	0.28	0.20	0.14	0.09	0.23	0.26	0.27	0.15	0.13	0.24	0.28	0.21	0.14
B	0.11	0.23	0.30	0.19	0.17	0.08	0.21	0.28	0.24	0.19	0.09	0.23	0.32	0.20	0.16
C	0.10	0.32	0.28	0.18	0.12	0.09	0.26	0.26	0.23	0.16	0.09	0.30	0.30	0.18	0.13
D	0.10	0.20	0.28	0.18	0.24	0.07	0.15	0.23	0.27	0.28	0.08	0.22	0.25	0.20	0.25

4.1. Evaluation of Schedule Delay Risk on Pre-Construction Stage

The pre-construction stage has the least risk alternative with 35% of a high and very high-risk level of schedule delay as described in Table 5. The most influential criteria responsibility related is embraced by the substantial sub-criteria sequenced as designer, client, consultant, and contractor related (Tables 3 and 4). The second important criteria external related include influential factors such as corruption and unavailability of utilities at a site (Table 4). The third critical criteria resource related has momentous sub-criteria ranked as material, finance, labour, and equipment related (Tables 3 and 4).

The least one is a contract condition related that has top contributing attributes of unrealistic cost and duration in the contract and insufficient details of the contract documents (Table 4).

4.2. Evaluation of Schedule Delay Risk on Construction Stage

The construction stage has the highest risk, which exposes 44% of the high and very high-risk level of schedule delay as denoted in Table 5. In the construction stage, the most influential criterion is responsibility related, with dominant sub-criteria sequenced as consultant, designer, contractor, and client related (Tables 3 and 4). The second influential criterion is resource that has important sub-criteria ranked as finance, construction material, labour, and equipment related (Tables 3 and 4). The third contract condition, which considerate influential attributes of unrealistic cost and duration in contract, and poor incentives and inadequate penalties in contract (Table 4). The external related is least one, counting the most contributing factors of corruption and unavailability of utilities at a site (Table 4).

4.3. Evaluation of Schedule Delay Risk on Post-Construction Stage

The second influenced by risk is the post-construction stage that covered 37% of the high and very high-risk level of schedule delay (Table 5). In this stage, the most persuasive criterion is responsibility related, with the extensive risk of sub-criteria sequenced as consultant, designer, contractor, and client related (Tables 3 and 4). The second influential criteria external related include the dominant attributes of corruption and unavailability of utilities at a site (Table 4). Contract condition is the third noticeable criterion, with top contributing attributes of unrealistic cost and duration in contract, and poor incentives and inadequate penalties in contract (Table 4). The least contributing criteria resource related, containing influential sub-criteria ranked as finance, construction material, equipment, and labour (Tables 3 and 4).

4.4. The Schedule Delay Risk Comparison at Each Construction Lifecycle

From multi-comprehensive evaluation matrix result of Table 5, the construction stage (44%) is high-risk, the second high-risk stage is post construction (37%), and the least risk stage is pre-construction (35%).

Table 5. Risk of Schedule Delay on Construction Process.

Alternatives	1	2	3	4	5
Pre-construction	0.11	0.25	0.29	0.19	0.16
Construction	0.08	0.22	0.26	0.25	0.19
Post-construction	0.10	0.25	0.28	0.20	0.17
Average schedule delay risk of Ethiopia construction projects	0.10	0.24	0.28	0.21	0.17

From the weighted normalized decision matrix, negative ideal solution, positive ideal solution, and Euclidean distance (Table 6) then the closeness coefficients (CC) is computed in Table 7. The higher value of closeness coefficient for a specific alternative denotes higher risk influence.

Table 6. Weighted Normalized Decision Matrix, positive ideal solution (PIS), negative ideal solution (NIS), and Euclidian Distance of Alternatives.

Factors	Pre-Construction	Construction	Post-Construction	PIS(A*)	NIS(A⁻)
A11	0.120	0.168	0.120	0.168	0.120
A12	0.149	0.106	0.064	0.149	0.064
A13	0.062	0.062	0.021	0.062	0.021
A14	0.162	0.116	0.116	0.162	0.116
A15	0.168	0.056	0.056	0.168	0.056
A16	0.056	0.019	0.019	0.056	0.019
A21	0.041	0.095	0.123	0.123	0.041
A22	0.109	0.140	0.140	0.140	0.109
A23	0.147	0.082	0.115	0.147	0.082
A24	0.043	0.099	0.043	0.099	0.043
A25	0.014	0.041	0.040	0.041	0.014
A26	0.042	0.099	0.099	0.099	0.042
A27	0.012	0.035	0.081	0.081	0.012
A31	0.122	0.073	0.073	0.122	0.073
A32	0.171	0.171	0.220	0.220	0.171
A33	0.121	0.121	0.170	0.170	0.121
A34	0.122	0.171	0.171	0.171	0.122
A41	0.197	0.154	0.110	0.197	0.110
A42	0.206	0.206	0.160	0.206	0.160
A43	0.197	0.154	0.110	0.197	0.110
A44	0.070	0.070	0.070	0.070	0.070
B11	0.136	0.136	0.136	0.136	0.136
B12	0.136	0.136	0.136	0.136	0.136
B13	0.104	0.104	0.074	0.104	0.074
B14	0.099	0.043	0.071	0.099	0.043
B15	0.136	0.136	0.136	0.136	0.136
B21	0.142	0.142	0.142	0.142	0.142
B22	0.137	0.098	0.176	0.176	0.098
B23	0.140	0.180	0.180	0.180	0.140
B24	0.099	0.139	0.139	0.139	0.099
B31	0.174	0.104	0.104	0.174	0.104
B32	0.102	0.170	0.102	0.170	0.102
B33	0.245	0.245	0.175	0.245	0.175
B34	0.036	0.036	0.036	0.036	0.036
B35	0.030	0.089	0.030	0.089	0.030
B41	0.181	0.325	0.253	0.325	0.181
B42	0.097	0.289	0.161	0.289	0.097
B43	0.040	0.040	0.040	0.040	0.040
B44	0.040	0.040	0.040	0.040	0.040
B45	0.040	0.040	0.040	0.040	0.040
C1	0.018	0.018	0.088	0.088	0.018
C2	0.025	0.02	0.025	0.025	0.025
C3	0.123	0.172	0.172	0.172	0.123
C4	0.069	0.115	0.161	0.161	0.069
C5	0.126	0.126	0.126	0.126	0.126
C6	0.125	0.075	0.075	0.125	0.075
D1	0.052	0.017	0.017	0.052	0.017
D2	0.055	0.018	0.055	0.055	0.018
D3	0.182	0.182	0.182	0.182	0.182
D4	0.020	0.020	0.020	0.020	0.020
D5	0.099	0.099	0.139	0.139	0.099
D6	0.182	0.182	0.182	0.182	0.182
d*	0.343	0.253	0.311		
d⁻	0.264	0.325	0.250		

On the other hand, based on TOPSIS ideal solution of closeness coefficient rank in Table 7, construction stage (0.563) is high-risk, post construction stage (0.446) the second, and pre-construction stage (0.435) the least.

Table 7. Ranking of Risk Level of Schedule Delay on Construction Process.

Rank	Alternatives	Closeness Coefficient
1	Construction	0.563
2	Post construction	0.446
3	Pre-construction	0.435

As a result, the risk of schedule delay at the different lifecycle of construction projects has a difference or the risk impact different either in level or in type. These differences are important to mitigation delay risk in the construction projects of Ethiopia by recognizing the exact influence level of risk at each activity for a better decision. Overall, Figure 2 shows the comparison of the risk of the major schedule delay risk factors, external, resource, responsibility, and contract condition related.

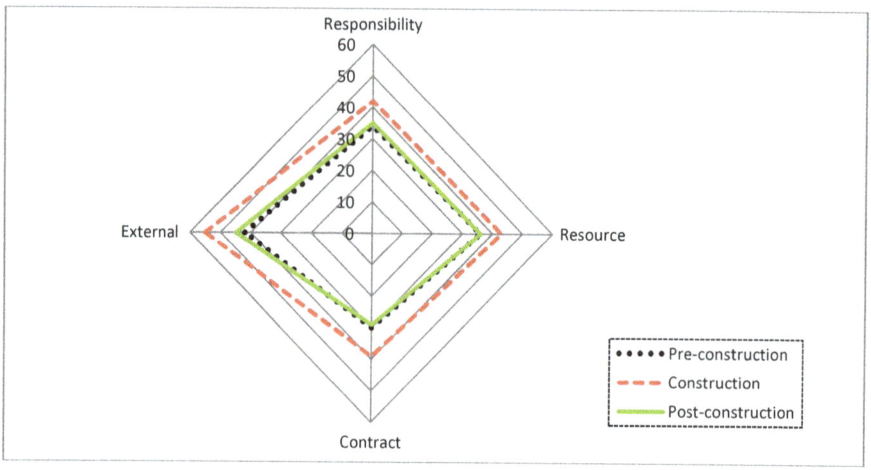

Figure 2. Risk of criteria on construction stage.

The study detects the influence of schedule delay risk in the construction projects of Ethiopia. The construction projects in Ethiopia have faced an average of 38% schedule delay risk at high and very high-risk level (Table 6). The rank, based on TOPSIS Closeness Coefficient and multi-comprehensive evaluations of risk at high and very high-risk level of alternatives are denoted in Figure 3.

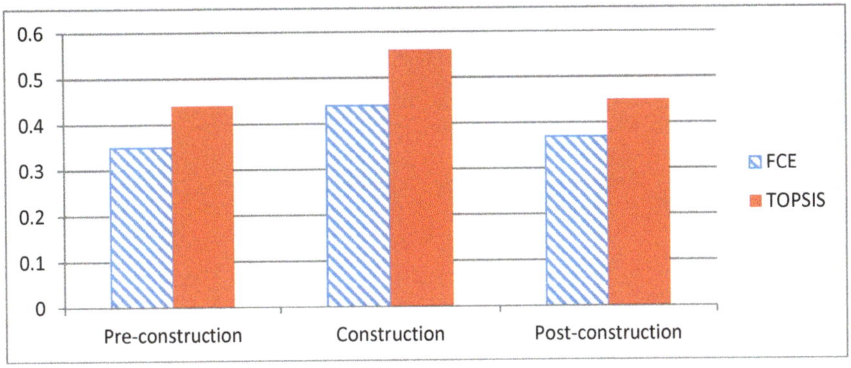

Figure 3. Comparison of risk of schedule delay by FCE and TOPSIS on the construction lifecycle.

5. Conclusions

Schedule delay risk was analyzed to provide a decision regarding the typical risk of delay in construction lifecycle to mitigate risk. This was achieved through the application of the hybrid model of FCE and TOPSIS. The application of this combined model can evaluate not only the typical risk of the schedule delay but also reflects the general risk of schedule delay. Based on the empirical study, the results identified that the risk level of the schedule delay varies at the construction projects lifecycle. The comparison on the empirical study shown that the construction stage (44%) is high risk, the second high-risk stage is post construction (37%), and the least risk is pre-construction (35%). The construction projects in Ethiopian have faced an average of 38% of schedule delay risk at the high and very high-risk level. This helps the construction managers to identify which schedule delay risk factor is highly influential and which construction lifecycle is highly risky and to make the right decision on how to mitigate the risk. The application of the integrated model of FCE and TOPSIS is a viable tool in delay risk management of construction project. The model can be applied at any time, with different project types by adapting the factors and evaluated weight.

Author Contributions: The author's responsibility and contribution during the study expressed as T.G. had the responsibility and contribution to propose the research and develop a methodology, collected and analyzed data, and obtained results; H.L. assisted and organized the study throughout the research process and provided extensive advice. The authors have carefully read and approved the manuscript.

Funding: "Huazhong University of Science and Technology under the National 12th Five-Year Plan Major Scientific and Technological Issues through grant 2015BAK33B04" funded this research.

Conflicts of Interest: We have no conflict of interest to disclose.

References

1. Mohammadsoroush, T. Dynamic Risk Analysis of Construction Delays Using Fuzzy-Failure Mode Effects Analysis. Master's Dissertation, University of Nevada, Las Vegas, NV, USA, 2017. Available online: https://digitalscholarship.unlv.edu/thesesdissertations/3043 (accessed on 23 December 2018).
2. Zou, P.X.W.; Zhang, G.; Wang, J. Understanding the key risks in construction projects in China. *Int. J. Proj. Manag.* **2007**, *25*, 601–614. [CrossRef]
3. Mahamid, I.; Bruland, A.; Dmaidi, N. Causes of Delay in Road Construction Projects. *J. Manag. Eng.* **2012**, *28*, 300–310. [CrossRef]
4. Choudhry, R.M.; Aslam, M.A.; Hinze, J.W.; Arain, F.M. Cost and Schedule Risk Analysis of Bridge Construction in Pakistan: Establishing Risk Guidelines. *J. Constr. Eng. Manag.* **2014**, *140*. [CrossRef]
5. Szymański, P. Risk management in construction projects. *Procedia Eng.* **2017**, *208*, 174–182. [CrossRef]
6. Gonzalez, P.; Gonzalez, V.; Molenaar, K.; Orozco, F. Analysis of Causes of Delay and Time Performance in Construction Projects. *J. Constr. Eng. Manag.* **2014**, *140*. [CrossRef]

7. Han, S.; Love, P.; Peña-Mora, F. A system dynamics model for assessing the impacts of design errors in construction projects. *Math. Comput. Model.* **2013**, *57*, 2044–2053. [CrossRef]
8. Faridi, A.S.; El-Sayegh, S.M. Significant factors causing schedule delay in the UAE construction industry. *Constr. Manag. Econ.* **2006**, *24*, 1167–1176. [CrossRef]
9. Sambasivan, M.; Soon, Y.W. Causes and effects of delays in Malaysian construction industry. *Int. J. Proj. Manag.* **2007**, *25*, 517–526. [CrossRef]
10. Assaf, S.A.; Al-Hejji, S. Causes of schedule delay in large construction projects. *Int. J. Proj. Manag.* **2006**, *24*, 349–357. [CrossRef]
11. Ayalew, T.; Dakhli, Z.; Lafhaj, Z. Assessment on Performance and Challenges of Ethiopian Construction Industry. *J. Arch. Civ. Eng.* **2016**, *2*, 1–11.
12. Abd El-Razek, M.E.; Bassioni, H.A.; Mobarak, A.M. Causes of Schedule delay in Building Construction Projects in Egypt. *J. Constr. Eng. Manag.* **2008**, *134*, 831–841. [CrossRef]
13. Frimpong, Y.; Oluwoye, J.; Crawford, L. Causes of schedule delay and cost overruns in construction of groundwater projects in a developing countries; Ghana as a case study. *Int. J. Proj. Manag.* **2003**, *21*, 321–326. [CrossRef]
14. Gunduz, M.; Nielsen, Y.; Ozdemir, M. Quantification of Delay Factors Using the Relative Importance Index Method for Construction Projects in Turkey. *J. Manag. Eng.* **2013**, *29*, 133–139. [CrossRef]
15. Larsen, J.K.; Shen, G.Q.; Lindhard, S.M.; Brunoe, T.D. Factors Affecting Schedule Schedule delay, Cost Overrun, and Quality Level in Public Construction Projects. *J. Manag. Eng.* **2016**, *32*. [CrossRef]
16. Xu, X.; Wang, J.; Li, C.Z.; Huang, W.; Xia, N. Schedule risk analysis of infrastructure projects: A hybrid dynamic approach. *Autom. Constr.* **2018**, *95*, 20–34. [CrossRef]
17. Abd El-Karim, M.S.B.; Mosa El Nawawy, O.A.; Abdel-Alim, A.M. Identification and assessment of risk factors affecting construction projects. *HBRC J.* **2015**, *13*, 202–216. [CrossRef]
18. Tavakolan, M.; Etemadinia, H. Fuzzy Weighted Interpretive Structural Modeling: Improved Method for Identification of Risk Interactions in Construction Projects. *J. Constr. Eng. Manag.* **2017**, *143*. [CrossRef]
19. Hossen, M.M.; Kang, S.; Kim, J. Construction schedule delay risk assessment by using combined AHP-RII methodology for an international NPP project. *Nucl. Eng. Technol.* **2015**, *47*, 362–379. [CrossRef]
20. Attri, R.; Grover, S.; Dev, N.; Kumar, D. Analysis of barriers total productive maintenance (TPM). *Int. J. Syst. Assur. Eng. Manag.* **2013**, *4*, 313–326. [CrossRef]
21. Kim, D.Y.; Han, S.H.; Kim, H.; Park, H. Structuring the prediction model of project performance for international construction projects: A comparative analysis. *Expert Syst. Appl.* **2009**, *36 Pt 1*, 1961–1971. [CrossRef]
22. Luu, V.T.; Kim, S.-Y.; Tuan, N.V.; Ogunlana, S.O. Quantifying schedule risk in construction projects using Bayesian belief networks. *Int. J. Proj. Manag.* **2009**, *27*, 39–50. [CrossRef]
23. Wang, J.; Yuan, H. System Dynamics Approach for Investigating the Risk Effects on Schedule Delay in Infrastructure Projects. *J. Manag. Eng.* **2017**, *33*. [CrossRef]
24. Adams, F.K. Risk perception and Bayesian analysis of international construction contract risks: The case of payment delays in a developing economy. *Int. J. Proj. Manag.* **2008**, *26*, 138–148. [CrossRef]
25. Eybpoosh, M.; Dikmen, I.; Birgonul, M.T. Identification of Risk Paths in International Construction Projects Using Structural Equation Modeling. *J. Constr. Eng. Manag.* **2011**, *137*, 1164–1175. [CrossRef]
26. Wang, W.-C.; Lin, C.-L.; Wang, S.-H.; Liu, J.-J.; Lee, M.-T. Application of importance-satisfaction analysis and influence-relations map to evaluate design delay factors. *J. Civ. Eng. Manag.* **2014**, *20*, 497–510. [CrossRef]
27. Boateng, P.; Chen, Z.; Ogunlana, S. A dynamic framework for managing the complexities of risks in megaprojects. *Int. J. Technol. Manag. Res.* **2016**, *5*, 1–13.
28. Karim, R.; Karmaker, C. Machine Selection by AHP and TOPSIS Methods. *Am. J. Ind. Eng.* **2016**, *4*, 7–13. [CrossRef]
29. Jang, W.; Hong, H.; Baek, S. Optimal Supply Vendor Selection Model for LNG Plant Projects Using Fuzzy-TOPSIS Theory. *J. Constr. Eng. Manag.* **2017**, *33*. [CrossRef]
30. El-Sayegh, S.M. Risk assessment and allocation in the UAE construction industry. *Int. J. Proj. Manag.* **2008**, *26*, 431–438. [CrossRef]
31. Yoon, Y.; Tamer, Z.; Hastak, M. Protocol to Enhance Profitability by Managing Risks in Construction Projects. *J. Manag. Eng.* **2015**, *31*. [CrossRef]

32. Kuo, Y.-C.; Lu, S.-T. Using fuzzy multiple criteria decision making approach to enhance risk assessment for metropolitan construction projects. *Int. J. Proj. Manag.* **2013**, *31*, 602–614. [CrossRef]
33. Dikmen, I.; Birgonul, M.T.; Anac, C.; Tah, J.H.M.; Aouad, G. Learning from risks: A tool for post-project risk assessment. *Autom. Constr.* **2008**, *18*, 42–50. [CrossRef]
34. Rao, R.; Zhang, X.; Shi, Z.; Luo, K.; Tan, Z.; Feng, Y. A Systematical Framework of Schedule Risk Management for Power Grid Engineering Projects' Sustainable Development. *Sustainability* **2014**, *6*. [CrossRef]
35. Chatterjee, K.; Zavadskas, K.E.; Tamošaitienė, J.; Adhikary, K.; Kar, S. A Hybrid MCDM Technique for Risk Management in Construction Projects. *Symmetry* **2018**, *10*. [CrossRef]
36. Nguyen, A.T.; Nguyen, L.D.; Long, L.-H.; Dang, C.N. Quantifying the complexity of transportation projects using the fuzzy analytic hierarchy process. *Int. J. Proj. Manag.* **2015**, *33*, 1364–1376. [CrossRef]
37. Lee, M.C. Information security risk analysis method and research trends: AHP and fuzzy comprehensive method. *Int. J. Comput. Sci. Inf. Technol.* **2014**, *6*, 29.

symmetry

MDPI

Article

An Improved A* Algorithm Based on Hesitant Fuzzy Set Theory for Multi-Criteria Arctic Route Planning

Yangjun Wang [1], Ren Zhang [1,*] and Longxia Qian [2]

1 College of Meteorology and Oceanography, National University of Defense Technology, Nanjing 211101, China; wyjscholar@sina.com

2 Institute of Navigation, Jiangsu Maritime Institute, Nanjing 211170, China; qianlongxiawmx@163.com

* Correspondence: zrpaper@163.com; Tel: +86-137-7052-7306

Received: 17 October 2018; Accepted: 13 December 2018; Published: 17 December 2018

Abstract: This paper presents a new route planning system for the purpose of evaluating the strategic prospects for future Arctic routes. The route planning problem can be regarded as a multi criteria decision making problem with large uncertainties originating from multi-climate models and experts' knowledge and can be solved by a modified A* algorithm where the hesitant fuzzy set theory is incorporated. Compared to the traditional A* algorithm, the navigability of the Arctic route is firstly analyzed as a measure to determine the obstacle nodes and three key factors to the vessel navigation including sailing time, economic cost and risk are overall considered in the HFS-A* algorithm. A numerical experiment is presented to test the performance of the proposed algorithm.

Keywords: hesitant fuzzy sets; multi criteria decision making; A* algorithm; route optimization; navigability

1. Introduction

The dramatic variation of sea ice in the Arctic region, due to global warming, has attracted many researchers in science and engineering, where shipping in the Arctic water is one of the hottest issues. Compared to the traditional shipping routes, the Arctic routes have shorter distances linking Asia and North America, as well as linking Asia and Europe, and are a more open navigation environment, more access to the abundant oil and gas resources and lower piracy risk [1]. Therefore, navigation through Arctic routes is considered to be a money-making opportunity for shipping and oil gas companies.

Compared to the traditional ship path planning problem [2–6], sea ice condition becomes a key factor to the route planning in the Arctic region due to its impact on travel time and fuel consumption, as well as the risk of being stuck in the ice [7].

Reference [7] introduced a system for route optimization in ice-covered water, which consisted of an ice model, a ship transit model, and an end-user system. The system was operated on commercial vessels in the Baltic Sea, and its performance was tested. Reference [8] developed an ice navigation system combined with a sea ice model, transit model and a model for route planning which simulated the whole Arctic area. The system employed a modified transit model devised by [9], which integrated various parameters such as ice-breaking fee, port charge, capital cost, etc. Reference [10] developed another ice navigation system where the uncertainty of sea ice prediction and the extremely severe conditions were taken into consideration. The route optimization problem in ice-covered water was regarded as a dynamic stochastic path planning problem, and a heuristic route optimization model was proposed to solve it. Reference [11] devised an automatic ice navigation support system to find the safest- and- shortest routes in the Arctic area for different types of vessels with a geographic information system.

However, unlike the real-time path planning system, the model in this paper is mainly devised to evaluate the strategic prospects for future Arctic routes. Therefore, most data are incorporated for future prediction, where large uncertainties arise from the bias of current multi-climate models and the inconsistency of experts' cognition. Additionally, for commercial navigating in the Arctic routes, sailing safety, as well as economic benefits, should be guaranteed according to the harsh weather conditions in the Arctic area. Therefore, sailing time, economic cost, and navigation safety are all key factors to influence the route planning, which makes the problem a multi-criteria decision making (MCDM) problem.

Owing to the MCDM problem with large uncertainties in the route optimization model, information on each grid has variation so that a new path planning method is required for the model to handle this uncertain decision problem. Many studies have examined that the hesitant fuzzy sets theory is a powerful tool to solve the mentioned kind of problem [12–17]. Therefore, this paper develops a new ice navigation system with a modified A* path planning algorithm called HFS-A* algorithm, where the hesitant fuzzy set theory is incorporated to improve the traditional A* algorithm. Three key factors, including sailing time, economic benefits, and navigation safety, are considered to the final decision-making in this system where multi-models of sea ice prediction and multiple experts' knowledge are used as input. More details related to hesitant fuzzy set theory and A* algorithm can be seen in Section 2. Section 3 introduces the establishment of the HFS-A* algorithm. A numerical experiment has been used to examine the proposed model in Section 4, and the conclusion can be seen in Section 5.

2. Preliminaries

2.1. Traditional A* Algorithm

2.1.1. Basic Concepts

A* algorithm is a heuristic algorithm widely used for finding an optimal path in static road network presented by [18], which is derived from the Dijkstra algorithm [19] and the Greedy algorithm [20]. The Dijkstra algorithm can find the shortest path, but has to traverse the entire network with low efficiency, and the Greedy algorithm has fast search speed but cannot guarantee to find the best path. The A* algorithm can balance both search speed and global optimality by using the specific utility function $f(n)$, which consists of a kind of cost function $g(n)$ and a kind of cost function $q(n)$:

$$f(n) = g(n) + q(n) \tag{1}$$

where $g(n)$ represents the actual cost from initial node to the current node, and $q(n)$ is the estimated cost from the current node to the end node. When $q(n) = 0$, only $q(n)$ works, then the A* algorithm degenerates to the Dijkstra algorithm, which can only guarantee finding the optimal route. When $h(n) \leq q(n)$, then the A* algorithm can maintain the search speed and the global optimality, and the search speed will be slower when the value of $q(n)$ becomes smaller. When $h(n) \gg q(n)$, then the A* algorithm degenerates to the Greedy algorithm, which can run faster but may fall into local optimum.

2.1.2. Work Flow

The flow of the algorithm can be seen in Figure 1.

Step 1 Initiate two ordered lists called "OPEN" list and "CLOSE" list and generate two nodes called "START" node and "END" node.

Step 2 The utility function $f(n)$ is calculated by Equation (1) at "START" node and put the "START" node into "OPEN" list. Where, $f(n)$ is the estimated value from the "START" node to the "END" node through the current node n; $g(n)$ is the actual value from the "START" node to the current node n; $g(n)$ is the estimated value from the current node n to the "END" node.

Step 3 Take out the node of minimum utility from "OPEN" list and mark it as the current node n. This node will be saved in "CLOSE" list.

Step 4 If and only if the node n is not the "END" node, continue the algorithm.

Step 5 Evaluate each adjacent node of node n and skip the one which has already existed in "CLOSE" list. Then, compute the utility of this node if it is not in "OPEN" list and save it in "OPEN" list. If the node has already existed in "OPEN" list, recalculate the utility of this node and choose the smaller value by comparing the utility with the previous one. Finally, node n is assigned as the parent node of the node.

Step 6 If "OPEN" list is not empty, back to **Step 3**. Otherwise, exit and report the failure of route search.

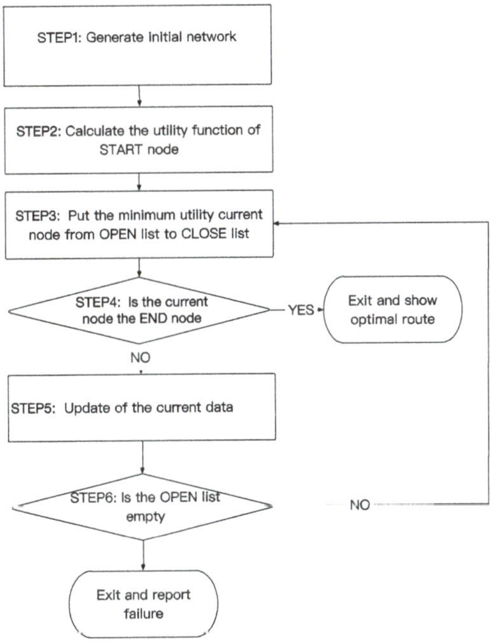

Figure 1. Work flow of the traditional A* algorithm.

2.2. Basic Concepts of Hesitant Fuzzy Set

2.2.1. Hesitant Fuzzy Set

Hesitant fuzzy set, proposed by [21], is a more general fuzzy set. An HFS is defined in terms of a function that returns a set of membership values for each element in the domain [21].

Definition 1 ([21]). *A hesitant fuzzy set A on X is a function h^A that when applied to X returns a finite subset of [0,1], which can be represented as the following mathematical symbol:*

$$A = \{\langle x, h_A(x)\rangle | x \in X\}, \tag{2}$$

where $h_A(x)$ is a set of some values in [0,1], denoting the possible membership degrees of the element $x \in X$ to the set A. For convenience, $h_A(x)$ is named a hesitant fuzzy element (HFE) [22].

Definition 2 ([21]). *For a hesitant fuzzy set represented by its membership function $h = h_A(x)$, we define its complement as follows:*

$$h^c = \cup_{\gamma \in h}\{1 - \gamma\}. \tag{3}$$

Definition 3 ([22]). *For an HFE h, $Sc(h) = \frac{1}{l_h}\sum_{\gamma \in h}\gamma$, is called the score function of h, where l_h is the number of elements in h and $Sc(h) \in [0,1]$. For two HFEs h_1 and h_2, if $Sc(h_1) > Sc(h_2)$, then $h_2 \prec h_1$, if $Sc(h_1) = Sc(h_2)$, $h_2 \approx h_1$.*

Some operations on the HFEs (h, h_1 and h_2) and the scalar number λ are defined by [22]:

$$h_1 \oplus h_2 = \cup_{\gamma_1 \in h_1, \gamma_2 \in h_2}\{\gamma_1 + \gamma_2\gamma_1\gamma_2\}, \tag{4}$$

$$h_1 \otimes h_2 = \cup_{\gamma_1 \in h_1, \gamma_2 \in h_2}\{\gamma_1\gamma_2\}, \tag{5}$$

$$h^\lambda = \cup_{\gamma \in h}\left\{\gamma^\lambda\right\}, \tag{6}$$

$$h^\lambda = \cup_{\gamma \in h}\left\{\gamma^\lambda\right\}. \tag{7}$$

2.2.2. The Aggregation Operators for Hesitant Fuzzy Information

Reference [23] proposed an aggregation principle for HFEs:

Definition 4 ([23]). *Let $A = \{h_1, h_2, \ldots, h_n\}$ be a set of n HFEs, Θ be a function on A, $\Theta : [0,1]^N \to [0,1]$, then*

$$\Theta_A = \cup_{\gamma \in \{h_1 \times h_2 \times \ldots \times h_n\}}\{\Theta(\gamma)\}. \tag{8}$$

Based on Definition 4, some new aggregation operators for HFEs were given by [22]:

Definition 5 ([22]). *Let h_i $(i = 1, 2, \ldots, n)$ be a collection of HFEs, $w = (w_1, w_2, \ldots, w_n)^T$ be the weight vector of them, such that $w_i \in [0,1]$ and $\sum_{i=1}^n w_i = 1$. A generalized hesitant fuzzy weighted averaging (GHFWA) operator is a mapping $H^n \to H$, and*

$$GHFWA_\lambda(h_1, h_2, \ldots, h_n) = \left(\oplus_{i=1}^n (w_i h_i^\lambda)\right)^{\frac{1}{\lambda}} = \cup_{\gamma_1 \in h_1, \gamma_2 \in h_2, \ldots, \gamma_n \in h_n}\left\{\left(1 - \prod_{i=1}^n (1 - \gamma_i^\lambda)^{w_i}\right)^{\frac{1}{\lambda}}\right\}. \tag{9}$$

Definition 6 ([22]). *Let h_i $(i = 1, 2, \ldots, n)$ be a collection of HFEs, $w = (w_1, w_2, \ldots, w_n)^T$ be the weight vector of them, such that $w_i \in [0,1]$ and $\sum_{i=1}^n w_i = 1$. A generalized hesitant fuzzy weighted geometric (GHFWG) operator is a mapping $H^n \to H$, such that*

$$\begin{aligned} GHFWG_\lambda(h_1, h_2, \ldots, h_n) &= \tfrac{1}{\lambda}\left(\otimes_{i=1}^n (\lambda h_i)^{w_i}\right) \\ &= \cup_{\gamma_1 \in h_1, \gamma_2 \in h_2, \ldots, \gamma_n \in h_n}\left\{1 - \left(1 - \prod_{i=1}^n \left(1 - (1 - \gamma_i)^\lambda\right)^{w_i}\right)^{\frac{1}{\lambda}}\right\}. \end{aligned} \tag{10}$$

2.2.3. Decision Making Based on Hesitant Fuzzy Information

The decision method based on the above definitions can be derived as follows:

Step 1 The possible alternative X_i of the attribute A_i provided by decision makers or other sources are denoted by the hesitant fuzzy elements $h_i (i = 1, 2, \ldots, n)$.

Step 2 The aggregation operators mentioned above are utilized to obtain the hesitant fuzzy elements $h_i(i = 1,2,\ldots,m)$ for the possible alternative $X_i(i = 1,2,\ldots,m)$.

Step 3 The score values $Sc(h_i)$ of $h_i(j = 1,2,\ldots,m)$ are calculated by Definition 3.

Step 4 Choose the optimal alternative X^* by the comparison of $Sc(h_i)$.

Example 1. *The vessel, which includes four directions $X_i(i = 1,2,3,4)$ to navigate, is planed to determine the optimal Arctic route for the following year. Suppose there are three factors $A_i(i = 1,2,3)$ that affect the decision making—A_1: navigation time; A_2: economic cost; A_3: navigation risk. It should be noted that all of them are of the minimization type. The weight vector of the attributes is $w = (0.3, 0.4, 0.3)^T$.*

Then, the optimal route can be determined by using the mentioned method.

Step 1 The decision matrix $H = (h_{ij})_{n\times n}$ is presented in Table 1, where $h_{ij}(i = 1,2,3,4; j = 1,2,3)$ are in the form of HFEs.

Step 2 Two operators, GHFWA and GHFWG, are used to obtain the HFE $h_i(i = 1,2,3,4)$ for the directions $X_i(i = 1,2,3,4)$. Take direction X_1 as an example and let $\lambda = 2$; we have

$$h_1^A = GHFWA_2(h_{11},\ h_{12},\ h_{13}) = \left(\oplus_{j=1}^{3}\left(w_j h_{1j}^2\right)\right)^{\frac{1}{2}}$$

$$= \cup_{\gamma_{11}\in h_{11},\gamma_{12}\in h_{12},\gamma_{13}\in h_{13}}\left\{\left(1-\prod_{j=1}^{3}\left(1-\gamma_{1j}^2\right)^{w_j}\right)^{\frac{1}{2}}\right\}$$

$$= \{0.2157, 0.2652, 0.3503, 0.4351, 0.4577, 0.5040, 0.5966, 0.6099, 0.6383, 0.2630, 0.3039, 0.3789, 0.4566, 0.4788, 0.5213, 0.6093, 0.6220, 0.6492, 0.3166, 0.3503, 0.4149, 0.4847, 0.5040, 0.5442, 0.6263, 0.6383, 0.6639\}.$$

$$h_1^G = GHFWG_2(h_{11},\ h_{12},\ h_{13}) = \frac{1}{2}\left(\otimes_{j=1}^{3}(2h_{1j})^{w_j}\right)$$

$$= \cup_{\gamma_{11}\in h_{11},\gamma_{12}\in h_{12},\gamma_{13}\in h_{13}}\left\{1-\left(1-\prod_{j=1}^{3}\left(1-(1-\gamma_{1j})^2\right)^{w_j}\right)^{\frac{1}{2}}\right\}$$

$$= \{0.1814, 0.2541, 0.2915, 0.2672, 0.3857, 0.4514, 0.2848, 0.4142, 0.4875, 0.1958, 0.2754, 0.3168, 0.2898, 0.4226, 0.4983, 0.3092, 0.4551, 0.5411, 0.2065, 0.2915, 0.3361, 0.3071, 0.4514, 0.5361, 0.3279, 0.4875, 0.5851\}.$$

Step 3 The score values $Sc(h_i)$, $i = 1,2,3,4$ are calculated by **Definition 3,** which can be seen in Table 2.

Step 4 From Table 2, X_4 will be chosen as the optimal direction based on both the GHFWA operator and GHFWG operator where λ is set as 2.

Table 1. Hesitant fuzzy decision matrix.

	A_1	A_2	A_3
X_1	(0.1, 0.3, 0.5)	(0.2, 0.6, 0.8)	(0.3, 0.4, 0.5)
X_2	(0.2, 0.4, 0.7)	(0.1, 0.2, 0.4)	(0.3, 0.4, 0.6)
X_3	(0.3, 0.5, 0.6)	(0.2, 0.4, 0.5)	(0.3, 0.5, 0.7)
X_4	(0.3, 0.5, 0.6)	(0.4, 0.5, 0.6)	(0.2, 0.5, 0.6)

Table 2. Score values obtained by the GHFWA operator and GHFWG operator and the rankings of alternatives ($\lambda = 2$).

	X_1	X_2	X_3	X_4	**Ranking**
$GHFWA_2$	0.4777	0.3993	0.4644	0.4887	$X_4 > X_1 > X_3 > X_2$
$GHFWG_2$	0.3649	0.3052	0.3052	0.3052	$X_4 > X_3 > X_1 > X_2$

3. An Improved A* Algorithm (HFS-A*)

In light of the harsh weather conditions in the Arctic region, the primary task for route planning is to identify the obstacles (e.g., sea ice). The Coupled Model Intercomparison Project, phase 5 (CMIP5) provided 39 Global Climate Models (GCMs) to predict sea ice data, from history to 21st century, under different representative concentration pathways (RCPs) [23,24]. Unlike the route planning in other regions, for the current sea ice forecasts in the Arctic region, there exists large uncertainty among these GCMs [25,26], which leads to the uncertainty of the length of the navigation season and the economic risk of exploiting the Arctic routes [27]. Therefore, only treating the shortest distance as the optimal route in the Arctic region is not reasonable; more factors, including the navigation risk, the navigation time, and the economic cost during navigation should be considered. Compared to the traditional A* algorithm, the HFS-A* algorithm is used to tackle the multi-criteria decision-making (MCDM) problem with large uncertainty derived from multi-model outputs and expert knowledge. The improved parts mainly focus on t, the identification of obstacles, and the construction of utility function.

3.1. Navigability of the Arctic Routes

With the impact of global warming, the extent of Arctic sea ice continues to decline [27]. Human's enthusiasm to explore and develop the Arctic routes are aroused by shorter sailing distance, longer navigation season and increased access to natural resources. There are three criteria related to sea ice conditions for evaluating the navigability in the Arctic area.

Criterion 1 (navigation uncertainty). *Sea ice concentration is considered only for no ice-breaking or ice-strengthening ships, and it is navigable when sea ice concentration is less than 15%* [28–31].

Criterion 2 (navigation time). *Sea ice thickness derived from an empirical regression model is considered and for no ice-breaking or ice-strengthening ships, it is navigable when sea ice thickness is no more than 1.2 m* [32,33].

Criterion 3 (navigation economic cost). *Both sea ice concentration and thickness are considered by computing the Ice Numeral (IN) index from the Arctic Ice Regime Shipping System (AIRSS) provided by the Canada Transport* [34–37]. *The Ice Numeral is given by*

$$IN = C_a IM_a + C_b IM_b + \ldots + C_n IM_n \tag{11}$$

where C_n is the concentration in tenths of ice type n, and IM_n is the Ice Multiplier for ice type n. Ice type describes the specific stage of development of ice, which is closely related to the ice age. Ice Multipliers, determined by ship class and ice type, are a series of integers, which are used to reflect the impact of sea ice type to the specific vessel. A negative IM represents the obstacle effect of vessel sailing. Ice types are determined by [34,38], *which are presented in Appendix A. For no ice-breaking or ice-strengthening ships, it is navigable when the IN index is larger than zero. Details about vessel type and IM can also be seen in Appendix A.*

Additionally, geographical environment, including water depth and channel width, is also a key factor for ships to navigate, which is related to the vessel type and dimension.

Overall, various evaluation criteria will add to the uncertainty of the sea ice navigability projection. In this paper, we take all three criteria into consideration with geographical restriction to make sure the navigability of the Arctic region, in another words, for one region, can be defined as navigable if and only if the criteria mentioned above are all reached.

3.2. Route Planning Criterion 1: Uncertainty of Sea Ice Condition

In this paper, a series of GCMs have been chosen based on their reasonable projections for future sea ice conditions evaluated by literature studies [39–42] (see Appendix B). In order to obtain the uncertainty of each model, these model outputs were compared with the Pan-Arctic Ice Ocean Modeling and Assimilation System (PIOMAS) estimate data set, which is a reanalysis with good spatial and temporal consistency constrained by the quality of the assimilated observations [43–46].

Suppose we have M GCMs, let each model data-set be X_i, $i = 1, 2, \ldots, M$, the PIOMAS data set be $\widetilde{X}$, then the uncertainty of each model can be obtained as follows:

$$h_{i1} = 0.5 - 0.5 \times \frac{X_i \cdot \widetilde{X}}{\|X_i\| \times \|\widetilde{X}\|}, \quad i = 1, 2, \ldots, M \tag{12}$$

where, $\frac{X_i \cdot \widetilde{X}}{\|X_i\| \times \|\widetilde{X}\|}$ is called cosine similarity, which is a measure of similarity between two non-zero vectors introduced by [47]. h_{i1}, $i = 1, 2, \ldots, M$ represents the bias from the model to the "real state", which can also be regarded as model uncertainty. When h_{i1} is equal to zero, it represents that the model data sets can well reflect the "real state", while when h_{i1} is equal to one, the model uncertainty reaches its maximum.

3.3. Route Planning Criterion 2: Time for Navigation on the Arctic Routes

The navigation time for each grid can be depicted as follows:

$$T_i = \frac{S_i}{V_i}, \; i = 1, 2, \ldots, M, \tag{13}$$

where, S_i is the distance of each grid, and v_i is the velocity of the vessel on each grid. h_{i2} is the normalization of T_i.

On the Northern Sea Route (NSR), the vessel speed is mainly impacted by sea ice conditions. A h-v curve presented by [48] can reflect, well, the relationship between sea ice thickness and vessel speed. In this h-v curve, the ice resistance and the net thrust of the engine to overcome the ice resistance should both be considered.

Step 1 The ice resistance can be presented as follows:

$$\begin{aligned} R_{ch} = {} & 0.5\mu_B \rho_\Delta g H_F^2 K_P \left(\tfrac{1}{2} + \tfrac{H_M}{2H_F}\right)^2 [B + 2H_F(cos\delta - 1/tan\psi)](\mu_h cos\phi + sin\psi sin\alpha) \\ & + \mu_B \rho_\Delta g K_0 \mu_h L_{par} H_F^2 + \rho_\Delta g \left(\tfrac{LT}{B^2}\right)^3 H_M A_{WF} F_n^2, \end{aligned} \tag{14}$$

where $\mu_B = 0.5$ represents the ice porosity factor, $\rho_\Delta = 0.8$ is the density difference between ice and water. In this model, these two factors are considered to be constant for simple situations, e.g., when the temperature changes. Variables include the waterline area of the foreship A_{WF}, the Froude number F_n, the length L, the parallel midbody length at waterline L_{par}, the width B, the friction coefficient μ_h and the vessel draft T. K_P and K_0 are mechanical factors of ice found by [49]. The thickness of the brash ice H_F can be determined as follows:

$$H_F = H_M + \frac{B}{2} tan\gamma + (tan\gamma + tan\delta)\sqrt{B\left(\frac{H_M + \frac{B}{4} tan\gamma}{tan\gamma + tan\delta}\right)}. \tag{15}$$

Both γ and δ represent the slope angles of the brash ice.

The formula can be simplified by an approximation when $B > 10$ m and $H_M > 4$ m [39]:

$$H_F = 0.26 + (BH_M)^{0.5}. \tag{16}$$

The flare angle ψ mentioned above can be obtained with the bow angles ϕ and α:

$$\psi = \arctan\left(\frac{tan\phi}{sin\alpha}\right). \tag{17}$$

Step 2 The net thrust can be calculated by the following formula:

$$T_{tot}(v)(1-t) = R_{ow}(v) + R_i(v), \tag{18}$$

$$T_{tot}(v)(1-t) = R_{ow}(v) + R_i(v), \tag{19}$$

where T_{tot} is the total thrust, $(1-t)$ represents the thrust deduction factor, and $R_i(v)$ and $R_{ow}(v)$ are the resistance in ice and in open water respectively.

The effect of vessel speed is approximated by a quadratic factor called bollard pull T_B [50], with the maximum open water speed v_{ow}, and the net thrust can be rewritten as

$$T_{net}(v) = \left(1 - \frac{\frac{1}{3}v}{v_{ow}} - \frac{2}{3}\left(\frac{v}{v_{ow}}\right)^2\right) T_B, \tag{20}$$

$$T_B = K_e(P_D D_P)^{\frac{2}{3}}, \tag{21}$$

where v is the vessel speed in the ice, K_e is the bollard pull quality factor, D_P is the propeller diameter and P_D is the actual power delivered.

3.4. Route Planning Criterion 3: Economic Cost for Navigation on the Arctic Routes

The economic cost of navigation on the Arctic routes are consist of four parts, which are capital cost, fuel cost, operation cost, and transit cost:

$$\text{Cost}_{\text{economic}} = Cost_{capital} + Cost_{fuel} + Cost_{operation} + Cost_{transit} \tag{22}$$

3.4.1. Model for Capital Cost

Capital cost is related to the price of new ship building or the cost of a ship with loans and depreciations [51]. Generally, the annual capital cost can be computed as follows [52]:

$$Cost_{capital} = \left(P \times (1-eq) \times \frac{(1+r)^y \times r}{(1+r)^y - 1}\right) + (eq \times C)/y \tag{23}$$

where P is a new ship price, eq is the equity, r is the interest rate, and y is the term of loan. The former item is called annual interests, while the latter item is called cash price.

3.4.2. Model for Fuel Cost

Fuel cost is often depicted as the major cost on the marine transportation, which is the largest single cost factor to most related simulations, ranging from 36.7% to 61% [53].

Fuel cost is impacted by the rate of fuel consumption and the price of fuel, which can be descripted as follows:

$$Cost_{fuel} = Fuel_{consumption} \times Fuel_{price} \tag{24}$$

The fuel consumption of a vessel can be influenced by ship dimensions (e.g., the ship size, hull design, engine profile, speed) and external factors (e.g., sea ice, wind, wave, current, foggy) [51]. The fuel consumption for a specific type of vessel is basically computed by multiplying SFOC (specific fuel oil consumption) (g/kWh), engine power (kw) and sailing hours (h) [54].

Most authors consider speed as the key factor impacting the fuel consumption when the type of a vessel is determined and a simple exponential law based on empirical data are derived: the fuel consumption per unit distance is proportional to the square of the speed [55–57], which can be presented as

$$\frac{V_1^2}{F_1} = \frac{V_2^2}{F_2}, \tag{25}$$

where F_1 and F_2 are the fuel consumption rate under the velocity V_1 and V_2, respectively. Fuel price is affected by the fluctuation of the global economic market. Low fuel price indicates the depression of global economy, while high fuel price reflects the booming global market.

3.4.3. Model for Operation Cost

Operation cost mainly includes crew cost, insurance cost and maintenance fee which is presented as follows:

$$Cost_{operation} = Cost_{crew} + Cost_{maintenance} + Cost_{insurance} \tag{26}$$

1. Crew cost

Crew cost is determined by the vessel type, automation level and numerous other factors [51]. Compared to the open water, the crews in the Arctic region require additional ice navigation experience and the ability to cope with hash weather conditions, which may increase the crew cost [53]. The increased crew cost may come from the higher wage [58] for each member or the larger size of members [59].

2. Maintenance cost

For the purpose of preventing the occurrence of breakdowns and following the scheduled maintenance program, the cost of regular maintenance is needed for vessels.

3. Insurance cost

In face of the risk of Arctic navigation (e.g., collision, engine damage, propeller damage, local hull damage, grounding, etc.) analyzed IN some studies [60,61], maritime insurance is a good tool to mitigate the associated risks, which can be approximately separated into three major components: protection and indemnity (P&I), hull and machinery (H&M), and cargo insurance. The third-party liabilities encountered during the commercial operation of a ship are charged by P&I. H&M covers the cost of damage done to the ship or its equipment. Cargo insurance provides the payment for the damage to the cargo itself [62].

3.4.4. Model for Transit Cost

On the NSR, the transit fee based on vessel type and ice conditions mainly includes ice pilot fee and ice breaking fee and can be given as

$$Cost_{transit} = Cost_{icebreaking} \times Load \times Pay_{load} + Cost_{icepilot} \tag{27}$$

These services are mainly provided by the Russian icebreaking service provider, Atomflot, and are compulsively charged subject to the law of the Russian Federation, which is dependent on the vessel type (e.g., the size and the ice-class of a vessel), and the navigation length and pilotage distance [62]. In general, higher ice-classed vessels are charged with lower icebreaking fees.

3.5. *Work Flow of HFS-A* Algorithm*

In this HFS-A* algorithm, the work flow can be seen in Figure 2. The method of obstacle identification has been discussed in Section 3.1 while the modified utility function can be described as follows:

$$f^*(i) = g^*(i) + q^*(i) \tag{28}$$

When the current node i is determined, the actual cost $g^*(i)$ is equal to the score value $\mathrm{Sc}(h_i)$, which can be computed by Definition 3.

More specifically, we assume each selected node has three criteria $C_j (j = 1, 2, 3)$ that affect the decision making—C_1: uncertainty of sea ice condition; C_2: navigation time; C_3: navigation economic cost. It should be noted that all of them are of the minimization type. The weight vector of the attributes is $w = (w_1, \ldots, w_j)^T$.

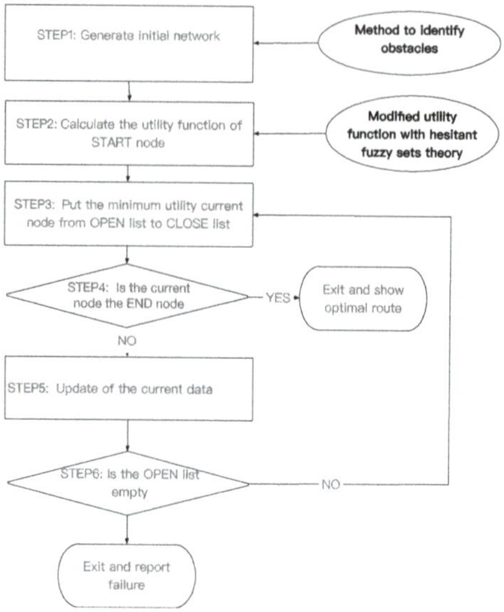

Figure 2. Work flow of the HFS-A* algorithm.

The heuristic estimated cost function can be approximately evaluated:

$$q^*(i) = Sc(h_i) \times D \tag{29}$$

where, D is the heuristic distance (Manhattan, Euclidean or Chebyshev) from the evaluated node to the END node [63].

Step 1 Map initialization

- Initialize map grid and interpolate the mentioned data into grid.
- Set the "START" node, "END" node, "OPEN" list and "CLOSE" list.
- Find the obstacle nodes in terms of the constrain conditions mentioned in Section 3.1.

Step 2 The construction of utility function

- Each time, compare all the adjacent nodes i of the current node n by

$$f^*(i) = Sc(h_i) \times (D + 1) \tag{30}$$

where, $Sc(h_i) = GHFWG_\lambda(h_{i1}, h_{i2}, h_{i3})$ or $Sc(h_i) = GHFWA_\lambda(h_{i1}, h_{i2}, h_{i3})$, h_{i1} is the HFEs of navigation uncertainty (see Section 3.2), h_{i2} is the HFEs of navigation time (see Section 3.3), and h_{i3} is the HFEs of navigation economic cost (see Appendix ??).

Step 3–6 The same as the traditional A* algorithm.

4. Case Study and Conclusion

4.1. Study Area and Data Description

This experiment is to find the optimal route on the Northern Sea Route (NSR) based on the proposed method from Shanghai to Bergen port for an IB-classed 3800TEU container vessel.

Data related to water depth is derived from a product called ETOPO1 provided by the National Geophysical Data Center (NGDC), with a resolution of 1 arc-minute [64]. Data related to AIRSS system can be seen in Appendix A. Data related to sea ice conditions (both sea ice thickness and sea ice concentration) can be seen in Appendix B. Data related to vessel information can be seen in Appendix C. Data related to economic cost can be seen in Appendix D. All the climate model outputs and data related to water depth are interpolated to the grid size of 360 × 120 for the comparison with PIOMAS estimate data set, which the spatial coverage is 45° N to 90° N and the temporal resolution is monthly.

4.2. Route Planning by HFS-A Algorithm*

4.2.1. Navigability of the NSR

The numerical simulation firstly examines the navigability of the IB-classed 3800 TEU container vessels on the NSR for each month in the year of 2050 (see Figure 3). According to the ensemble model predictions, the open time of the NSR for that vessel to access may last for 3 to 5 months in the year of 2050. Most model outputs show the navigable time starts from August to the October, while merely 2 to 4 models extend the navigable time (from July to November).

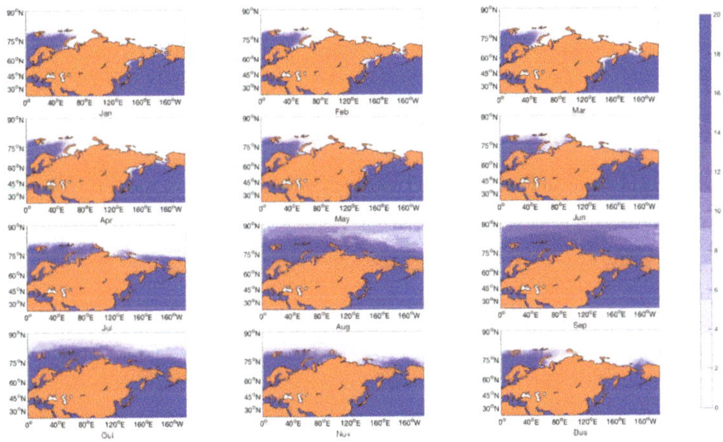

Figure 3. The navigability of IB-classed 3800 TEU container vessels on the NSR for each month in the year of 2050 based on multi-models. (The color of orange in the map reflects the geographic information, the white color represents the area that cannot access during that month, different blue colors reflect different amount of the models that give the navigable prediction for each grid during that month.).

4.2.2. Selection of the Aggregation Operators and Route Optimization

Secondly, route planning criteria (uncertainty, time and economic cost) have been calculated based on the models mentioned in Sections 3.2–3.4, and the results can be seen in Appendix D. These factors can be normalized respectively between 0 and 1, which can be described as:

$$\frac{h_i - \min(h_i)}{max(h_i) - \min(h_i)},\ i = 1, 2, \ldots,\ M \tag{31}$$

Thirdly, the optimal routes for IB-classed 3800 TEU container vessels from Shanghai to Bergen port on the NSR in September of 2050 can be found by two kinds of the aggregation operators ($GHFHA_\lambda$, $\lambda = 1,\ 2,\ 6$ and $GHFHG_\lambda$, $\lambda = 1,\ 2,\ 6$), which are used to aggregate the normalized results mentioned above. The detailed results can be seen in Table 3. Compared these two kinds of aggregation operators, the performance of the $GHFHG_\lambda$, $\lambda = 1,\ 2,\ 6$ operators in the route optimization is better than the $GHFHA_\lambda$, $\lambda = 1, 2, 6$ operators from the view of total sailing distance, sailing time, economic cost and average uncertainty. In the light of the comparison of different λ for each operator, the performance of the $GHFHG_\lambda$ becomes better with the λ increase, while the performance of the $GHFHG_\lambda$ becomes better when λ decreases. Therefore, the $GHFHG_1$ operator has been examined as the best aggregation operator in this numerical study (The weights vectors for this experiment are assigned as 0.4, 0.3, and 0.3 for three criteria).

Table 3. Results of route optimization for IB-classed 3800 TEU container vessels on the NSR from Shanghai to Bergen port in September of 2050 with different aggregation operators based on HFS-A* algorithm.

	Distance (Nautical Miles)	Sailing Time (day)	Economic Cost (Million USD)	Uncertainty (Mean)
GHFHA1	9900	25.99	1.20	0.5105
GHFHA2	10,110	26.53	1.21	0.5104
GHFHA6	10,142	26.98	1.23	0.5102
GHFHG1	8451	22.61	1.08	0.4700
GHFHG2	8513	22.78	1.09	0.4704
GHFHG6	8341	22.33	1.07	0.4901

Finally, the optimal route determined by HFS-A* algorithm with the $GHFHG_1$ operator can be seen in Figure 4. Three other routes according to simple single criterion (uncertainty, time, or economic cost) are also drawn in Figure 4. The detailed results can be seen in Table 4, where it can be found that path planning based on a single factor shows a slight advantage in its related aspect, but shows significant disadvantage in any other aspect compared with the optimal route. In other words, the optimal route can better balance these three key factors and show more realistic performance of the proposed route planning algorithm than the other three single factor route planning.

Table 4. Results of route optimization for IB-classed 3800 TEU container vessels on the NSR from Shanghai to Bergen port in September of 2050 based on HFS-A* algorithm and simple A* algorithm (The percentages in brackets are compared with the values in GHFHG1).

	Distance (Nautical Miles)	Sailing Time (Day)	Economic Cost (Million USD)	Uncertainty (Mean)
GHFHG1	8451	22.61	1.08	0.4700
Uncertainty	10,674 (+26.0)	27.87 (+23.3%)	1.26 (+16.7%)	0.4404 (-6.3%)
Time	8603 (+1.7%)	21.97 (−2.8%)	1.07 (-0.9%)	0.5148 (+9.5%)
Economic cost	8167 (−3.4%)	21.07 (−6.8%)	1.04 (-3.7%)	0.5516 (+17.4%)

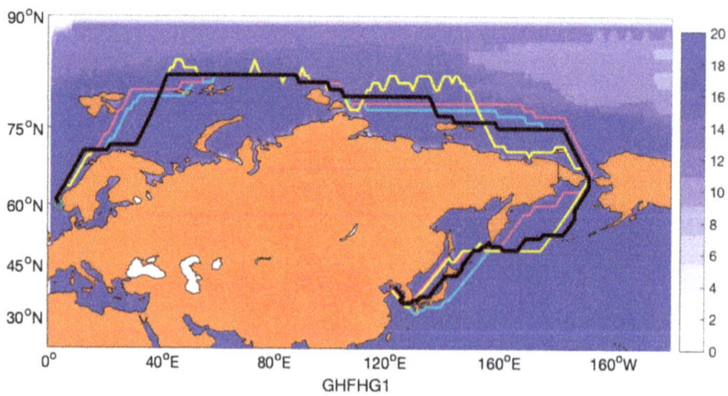

Figure 4. Route optimization for IB-classed 3800 TEU container vessels on the NSR from Shanghai to Bergen port in September of 2050 by HFS-A* algorithm. (The red line represents the route planning based on navigation uncertainty, the blue line based on the navigation time, the yellow line based on navigation economic cost, the dark line represents the optimal route integrated of these three criteria by the $GHFHA_1$ operator.)

5. Conclusions

The opening of Arctic routes will be no longer a dream in the coming future with climate change; route planning is necessary for vessels to navigation on the Arctic region from different points of view (safe, economic cost, time etc.). This paper presents a modified A* algorithm where the hesitant fuzzy set theory is incorporated for the purpose of solving the MCDM problem in Arctic route planning with large uncertainties originating from multi-climate models and experts' knowledge. Compared to the traditional A* algorithm, the navigability of the Arctic route is firstly analyzed as a measure to determine the obstacle nodes, and three key factors to vessel navigation, including sailing time, economic cost and risk are overall considered in the HFS-A* algorithm.

A numerical experiment, which is to find the optimal route between Bergen port and Shanghai port on the NSR, is presented to test the performance of the proposed algorithm. Multi-model ensemble forecast displays that the IB-class 3800 TEU container vessels can navigate on the NSR lasting for 3 to 5 months in the year of 2050. Most model outputs show the navigable time starts from August to October, while merely 2 to 4 models extend the navigable time (from July to November). The sensitivity analysis for the aggregation operators examines that the $GHFHG_1$ operator has an advantage over other aggregation operators in route optimization, and its performance of integrating the three key factors in route planning is better than the performance of any other single factor.

In this paper, the improvement effects for this new approach have been evaluated theoretically and practically. Theoretically speaking, the simple A* algorithm cannot handle the Arctic path planning problem which has multi-criteria attribution with large uncertainties. Even if we can synthesize the time, economic and uncertainty factors by addition and multiplication, the uncertainties existing in climate model prediction and expert knowledge cannot be portrayed by a simple A* algorithm. Practically speaking, we compared the route planning result of HFS-A* algorithm and single factor route planning result (see Figure 4). It can be found that there is a more realistic performance of the HFS-A* route planning algorithm than compared with the simple A* route planning algorithm. Overall, this new HFS-A* algorithm can be well-applied to the Arctic region and to evaluate the strategic prospects for future Arctic routes.

Author Contributions: Conceptualization, Y.W..; methodology, Y.W.; software, Y.W.; validation, Y.W.; formal analysis, Y.W.; investigation, Y.W.; resources, L.Q.; data curation, L.Q.; writing—original draft preparation, Y.W.; writing—review and editing, R.Z.; supervision, R.Z.; funding acquisition, R.Z., L.Q.

Funding: This work has received funding from the National Natural Science Foundation of China under grant agreement number 51609254 and number 41375002.

Conflicts of Interest: The authors declare no conflicts of interest.

Appendix A

Table A1. Ice Type [28,29].

Ice Type	Characteristic
New (Open Water)	Newly formed ice, include ice crystal, grease like ice, crushed ice clusters, etc. These types of ice are just loosely frozen together, can only been seen in the process of floating. The ice thickness is less than 10 cm.
Grey	Young ice, has a thickness of 10–15 cm which is lower than that of nilas and is easy to expand and break
Grey-white	Young ice has a thickness of 15–30 cm
Thin first year 1st stage	One year ice, which the formation time does not exceed one winter, has a thickness of 30–50 cm.
Thin first year 2 nd stage	One year ice, which the formation time does not exceed one winter, has a thickness of 50–70 cm.
Medium first year	One year ice has a thickness of 70–120 cm

Table A1. *Cont.*

Ice Type	Characteristic
Thick first year	One year ice has a thickness of 120–220 cm
Second year	Adult ice, which has gone through at least one summer's melting, has a thickness of 220–250 cm
Multiyear	Multiyear ice, which has gone through at least two summers' melting, has a thickness beyond 250 cm

Table A2. Ice Multiplier for ice type [28].

	Open Water	Grey Ice	Grey White Ice	Thin First Year 1st Stage	Thin FIRST Year 2nd Stage	Medium First Year	Thick First Year	Second Year	Multi Year
CAC 3	2	2	2	2	2	2	2	1	−1
CAC 4	2	2	2	2	2	2	1	−2	−3
Type A	2	2	2	2	2	1	−1	−3	−4
Type B	2	2	1	1	1	−1	−2	−4	−4
Type C	2	2	1	1	−1	−2	−3	−4	−4
Type D	2	2	1	−1	−1	−2	−3	−4	−4
Type E	2	1	−1	−1	−1	−2	−3	−4	−4

Table A3. Vessel type [28].

Vessel Type	Description
CAC3	Commercial cargo ship, which can navigate in the area on all kinds of ice types but will escape the area of multiyear ice
CAC4	Commercial cargo ship, which is able to navigate on the area of arbitrary one-year-old ice, while the speed on the area of multiyear ice will be extremely reduced.
Type A (IAS, PC6)	Vessel, which can navigate on the area of thick first year ice.
Type B (IA, PC7)	Vessel, which can navigate on the area of medium first year ice.
Type C (IB)	Vessel, which can navigate on the area of thin first year ice
Type D (IC)	Vessel, which can navigate on the area of grey-white ice.
Type E (ID)	Vessel, which can navigate on the area of grey-ice.

Appendix B

Table A4. Key sources of Global Cliamte Models (GCMs) used in this paper.

Model	Country	Oceanic Resolution
ACCESS1.0	Australia	1° × 1° L50
ACCESS1.3	Australia	1° × 1° L50
BNU-ESM	China	360(lon) × 200(lat) L50
CCSM4	United States	Nominal 1° (1.125° in longitude, 0.27–0.64° in latitude) L50
CESM1-BGC	United States	Nominal 1° (1.125° in longitude, 0.27–0.64° in latitude) L60
CESM1-CAM5	United States	Nominal 1° (1.125° in longitude, 0.27–0.64° in latitude) L60
CNRM-CM5	France	Average 0.7° L42
HADGEM2-CC	United Kingdom	1.875° × 1.25°
MIROC5	Japan	0.5–1.4° × 1.4° L50
MPI-ESM-MR	Japan	Approx. 1° × 1° L40

Appendix C

Table A5. Ship dimensions [57].

Vessel Type		IB	
Deadweight	50,000 ton	SFOC	145.8 g/kWh
Payload	3800 TEU	engine power	35,000 kWh
Load Factor	0.65	Engine load	0.8
L	250 m	v_{OW}	24 knots
L_{PAR}	130 m	α	23°
B	32.2 m	ϕ	90°
T	12 m	μ_h	0.02
P_D	19,600 kW	A_{WF}	806.5 m^2
K_e	0.78	D_P	7.5 m
K_p	6.5		

Appendix D

Appendix D.1 Data Related to Economic Cost

An ordinary 3800 TEU container ship, which usually navigates on the open water, should cost 60 million, the equity is 30%, and the interest rate is 3% with a 20-year loan [65]. Therefore, we can obtain the ordinary capital cost by Equation (23): 10,200 USD/day.

On the NSR, an IB-classed 3800 TEU container ship may have extra building cost due to the principal structure and strength, propulsion system, hull form, etc. [59]. Literature related to extra building cost can be seen in Table A6. From Table A6, the range of the extra building cost for IB-classed vessel is from 5% to 35%.

Suppose C is a new ship price, *eq* is the equity, *r* is the interest rate and *i* is the term of loan.

Table A6. Extra building cost for a commercial ice-classed vessel.

Ice Class Category Considered	Extra Building Cost	Resource
CAC3	+30%	[58]
IAS	+20%	[53]
IB	+7.5%	[66]
PC7 to PC4 [1]	+20%	[56]
Ice class	+30%	Expert suggestions
IA-IAS	+5%–7%	[67]
IB	+20%–30%	[59]
Ice class	+10%–35%	[68]

[1] According to approximate equivalence of ice class classification systems made by [44], PC6 is equal to 1AS and PC7 is equal to 1A.

The fuel consumption of an ordinary 3800 TEU vessel under the designed speed can be derived based on Table A5: 97.98 ton/day. The current fuel price can be obtained by Bunker Index (www.bunkerindex.com): 450 USD/ton.

It is relatively well-documented in the literature that the fuel consumption of an ice-classed vessel will be more than an ordinary vessel with the same size due to extra weight, bow shape, hull appendages, which increases the frictional resistance [66]. Data related to extra fuel consumption can be seen in Table A7.

Table A7. Extra fuel consumption rate for a commercial ice-class ship.

	Extra Fuel Consumption Rate	Resource
IB	+67%	[59]
IAS	+20%	[53]
Ice class	+30%	[52]
Ice class	+10%	[54]
IB	+3%	[65]

The crew cost for a 3800 TEU ordinary vessel is 2740 USD/day [57]. Data related to extra crew cost can be found in Table A8.

Table A8. Extra crew cost for a 3800 TEU IB-classed vessel.

Extra Crew Cost	Resource
+0%	Japan Ship owners Association (JSA), 2012
+0%	[65]
+10%	[59]
+10%	[53]
+11%–14%	[58]

The Maintenance cost of a 3800 TEU ordinary vessel is 1644 USD/day [57]. On the NSR, the IB-classed vessel needs to undergo more challenging operation conditions, which will put extra cost on vessel and equipment (see Table A9).

Table A9. Extra maintenance cost for a 3800 TEU IB-classed vessel.

Extra Maintenance Cost	Resource
+20%	[53]
+20%	[56]
+23%	[68]
+100%	[59]
+100%	[69]
+150%	[58]

The maintenance cost of a 3800 TEU ordinary vessel is 1644 USD/day [57]. Data related to the extra insurance fee of the IB-classed vessel can be seen in Table A10.

Table A10. Extra insurance cost for a 3800 TEU IB-classed vessel.

Extra Insurance Cost			Resource
P&I	H&M	Cargo	
+0%	+0%	+0%	[66]
+25%	+100%	+0%	[59]
+43%	+100%	+0%	[70]
+50%	+0%	+0%	[53]
+100%	+30%	+0%	[67]

Ice breaking and ice pilot remain necessary for safe navigation for an IB-classed vessel [71]. The ice-breaking cost of an IB-classed vessel is 356250 USD/trip [62].

Appendix D.2 The Calculation of Economic Cost Based on HFS

The economic cost of an IB-classed 3800 TEU vessel can be derived as follows:

$$\begin{aligned} Cost_{EC\_NSR} = & \left(1 + E_{capital}\right) \times Cost_{capital} + \left(1 + E_{fuel}\right) \times Cost_{fuel} \\ & +(1 + E_{crew}) \times Cost_{crew} + (1 + E_{maintenance}) \times Cost_{maintenance} \\ & +(1 + E_{insurance}) \times Cost_{insurance} + Cost_{icebreaking} \end{aligned} \quad (A1)$$

where, the value of each kind of extra cost on the NSR can be determined by 1000 times random sampling from the information mentioned above by Monte Carlo methods using the software of Matlab (Table A11).

Table A11. Parameters related to the economic cost of an IB-classed 3800 TEU vessel.

Parameters	Value
$E_{capital}$	0.196
E_{fuel}	0.260
E_{crew}	0.065
$E_{maintenance}$	0.688
$E_{insurance}$	0.299

From Equations (22)–(27), the economic cost of an IB-classed 3800 TEU vessel for each grid can be obtained as follows:

$$\mathrm{EC}_i = \left(20,028 + 55,555 \times \sqrt[3]{\frac{v_i}{24}}\right) \times \frac{S_i}{24v_i} \quad (A2)$$

Specially, the icebreaking fee for each route on the NSR is the same (356,250 USD), so it can be omitted when used as a criterion of route optimization. h_{i3} is the normalization of EC_i.

References

1. Arctic Council. *Arctic Council Arctic Marine Shipping Assessment 2009 Report*; Arctic Council: Tromsø, Norway, 2009.
2. Benjamin, M.R.; Curcio, J.A.; Leonard, J.J.; Newman, P.M. Navigation of Unmanned Marine Vehicles in Accordance with the Rules of the Road. In Proceedings of the 2006 IEEE International Conference on Robotics and Automation, Orlando, FL, USA, 15–19 May 2006.
3. Szlapczynski, R. A New Method of Ship Routing on Raster Grids, with Turn Penalties and Collision Avoidance. *J. Navig.* **2006**, *59*, 27–42. [CrossRef]
4. Walther, L.; Rizvanolli, A.; Wendebourg, M.; Jahn, C. Modeling and Optimization Algorithms in Ship Weather Routing. *Int. J. E-Navig. Marit. Econ.* **2016**, *4*, 31–45. [CrossRef]
5. Wang, S.; Meng, Q.; Liu, Z. Bunker Consumption Optimization Methods in Shipping: A Critical Review and Extensions. *Transp. Res. Part E* **2013**, *53*, 49–62. [CrossRef]
6. Witt, J.; Dunbabin, M. Go with the Flow: Optimal Path Planning in Coastal Environments. In Proceedings of the Australasian Conference on Robotics & Automation (ACRA), Canberra, Australia, 3–5 December 2008.
7. Kotovirta, V.; Jalonen, R.; Axell, L.; Riska, K.; Berglund, R. A System for Route Optimization in Ice-Covered Waters. *Cold Reg. Sci. Technol.* **2009**, *55*, 52–62. [CrossRef]
8. Nam, J.H.; Park, I.; Lee, H.J.; Kwon, M.O.; Choi, K.; Seo, Y.K. Simulation of Optimal Arctic Routes Using a Numerical Sea Ice Model Based on an Ice-Coupled Ocean Circulation Method. *Int. J. Nav. Archit. Ocean Eng.* **2013**, *5*, 210–226. [CrossRef]
9. Mulherin, N.D.; Eppler, D.T.; Proshutinsky, T.O.; Farmer, L.D.; Smith, O.P. *Development and Results of a Northern Sea Route Transit Model*; Cold Regions Research and Engineering Lab: Hanover, NH, USA, 1996.
10. Choi, Mi.; Chung, H.; Yamaguchi, H.; Nagakawa, K. Arctic Sea Route Path Planning Based on an Uncertain Ice Prediction Model. *Cold Reg. Sci. Technol.* **2015**, *109*, 61–69. [CrossRef]
11. Liu, X.; Sattar, S.; Li, S. Towards an Automatic Ice Navigation Support System in the Arctic Sea. *ISPRS Int. J. Geo-Inf.* **2016**, *5*, 36. [CrossRef]

12. Xia, M.; Xu, Z. Hesitant Fuzzy Information Aggregation in Decision Making. *Int. J. Approx. Reason.* **2011**, *52*, 395–407. [CrossRef]
13. Faizi, S.; Rashid, T.; Sałabun, W.; Zafar, S.; Wątróbski, J. Decision Making with Uncertainty Using Hesitant Fuzzy Sets. *Int. J. Fuzzy Syst.* **2017**, *20*, 93–103. [CrossRef]
14. Zhang, X.; Xu, Z. Novel Distance and Similarity Measures on Hesitant Fuzzy Sets with Applications to Clustering Analysis. *J. Intell. Fuzzy Syst.* **2015**, *28*, 2279–2296.
15. Zhang, Z. Hesitant Fuzzy Power Aggregation Operators and Their Application to Multiple Attribute Group Decision Making. *Inf. Sci.* **2013**, *234*, 150–181. [CrossRef]
16. Rodriguez, R.M.; Martinez, L.; Herrera, F. Hesitant Fuzzy Linguistic Term Sets for Decision Making. *IEEE Trans. Fuzzy Syst.* **2012**, *20*, 109–119. [CrossRef]
17. Song, C.; Xu, Z.; Zhao, H. A Novel Comparison of Probabilistic Hesitant Fuzzy Elements in Multi-Criteria Decision Making. *Symmetry* **2018**, *10*, 177. [CrossRef]
18. Hart, P.E.; Nilsson, N.J.; Raphael, B. A Formal Basis for the Heurist Dennination of Minimum Cost Paths. *IEE Trans. Syst. Sci. Cyberetics* **1968**, *4*, 61.
19. Dijkstra, E.W. A Note on Two Problems in Connexion with Graphs. *Numer. Math.* **1959**, *1*, 269–271. [CrossRef]
20. Zhang, Z.; Schwartz, S.; Wagner, L.; Miller, W. A Greedy Algorithm for Aligning DNA Sequences. *J. Comput. Biol.* **2000**, *7*, 203–214. [CrossRef] [PubMed]
21. Torra, V. Hesitant Fuzzy Sets. *Int. J. Intell. Syst.* **2010**, *25*, 529–539. [CrossRef]
22. Torra, V.; Narukawa, Y. On Hesitant Fuzzy Sets and Decision. In Proceedings of the IEEE International Conference on Fuzzy Systems, Jeju Island, Korea, 20–24 August 2009.
23. Taylor, K.E.; Stouffer, R.J.; Meehl, G.A. An overview of cmip5 and the experiment design. *Bull. Amer. Meteor. Soc.* **2012**, *93*, 485–498. [CrossRef]
24. Vuuren, D.P.V.; Edmonds, J.A.; Kainuma, M.; Riahi, K.; Weyant, J. A special issue on the rcps. *Clim. Chang.* **2011**, *109*. [CrossRef]
25. Blanchardwrigglesworth, E.; Bitz, C.M. Characteristics of arctic sea-ice thickness variability in gcms. *J. Clim.* **2014**, *27*, 8244–8258. [CrossRef]
26. Melia, N.; Haines, K.; Hawkins, E. Improved arctic sea ice thickness projections using bias-corrected cmip5 simulations. *The Cryosphere* **2015**, *9*, 3821–3857. [CrossRef]
27. Aksenov, Y.; Popova, E.E.; Yool, A.; Nurser, A.G.; Williams, T.D.; Bertino, L.; Bergh, J. On the Future of Arctic Sea Routes: High-Resolution Projections of the Arctic Ocean and Sea Ice. *Mar. Policy* **2017**, *75*, 300–317. [CrossRef]
28. Khon, V.; Mokhov, I. Arctic Climate Changes and Possible Conditions of Arctic Navigation in the 21st Century. *Izvestiya Atmos. Ocean. Phys.* **2010**, *46*, 14–20. [CrossRef]
29. Khon, V.C.; Mokhov, I.I.; Latif, M.; Semenov, V.A.; Park, W. Perspectives of Northern Sea Route and Northwest Passage in the Twenty-First Century. *Climat. Chang.* **2010**, *100*, 757–768. [CrossRef]
30. Khon, V.C.; Mokhov, I.I.; Semenov, V.A. Transit Navigation through Northern Sea Route from Satellite Data and CMIP5 Simulations. *Environ. Res. Lett.* **2017**, *12*, 024010. [CrossRef]
31. Mokhov, I.I.; Khon, V.C.; Prokof'eva, M.A. New Model Estimates of Changes in the Duration of the Navigation Period for the Northern Sea Route in the 21st Century. *Dokl. Earth Sci.* **2016**, *468*, 641–645. [CrossRef]
32. Melia, N.; Haines, K.; Hawkins, E. Sea Ice Decline and 21st Century Trans-Arctic Shipping Routes. *Geophys. Res. Lett.* **2016**, *43*, 9720–9728. [CrossRef]
33. Stephenson, S.R.; Smith, L.C. Influence of Climate Model Variability on Projected Arctic Shipping Futures. *Earth's Future* **2015**, *3*, 331–343. [CrossRef]
34. CanadaTransport. *Arctic Ice Regime Shipping System (AIRSS) Standards*; CanadaTransport: Ottawa, ON, Canada, 1998.
35. Howell, S.E.L.; Yackel, J.J. A Vessel Transit Assessment of Sea Ice Variability in the Western Arctic, 1969–2002: Implications for Ship Navigation. *Can. J. Remote Sens.* **2004**, *30*, 205–215. [CrossRef]
36. Smith, L.C.; Stephenson, S.R. New Trans-Arctic Shipping Routes Navigable by Midcentury. *Proc. Natl. Acad. Sci. USA* **2013**, *110*, E1191–E1195. [CrossRef]
37. Stephenson, S.R.; Smith, L.C.; Brigham, L.W.; Agnew, J.A. Projected 21st-Century Changes to Arctic Marine Access. *Clim. Chang.* **2013**, *118*, 885–899. [CrossRef]
38. Johnston, M.E. Cold Reg. Sci. Technol. Seasonal Changes in the Properties of Fi Rst-Year, Second-Year and Multi-Year Ice. *Cold Reg. Sci. Technol.* **2017**, *141*, 36–53. [CrossRef]

39. Jahn, A.; Sterling, K.; Holland, M.M.; Kay, J.E.; Maslanik, J.A.; Bitz, C.M.; Bailey, D.A.; Stroeve, J.; Hunke, E.C.; Lipscomb, W.H. Late-Twentieth-Century Simulation of Arctic Sea Ice and Ocean Properties in the CCSM4. *J. Clim.* **2012**, *25*, 1431–1452. [CrossRef]
40. Liu, J.; Song, M.; Horton, R.M.; Hu, Y. Reducing Spread in Climate Model Projections of a September Ice-Free Arctic. *Proc. Natl. Acad. Sci. USA* **2013**, *110*, 12571–12576. [CrossRef]
41. Massonnet, F.; Fichefet, T.; Goosse, H.; Vancoppenolle, M.; Mathiot, P.; König Beatty, C. On the Influence of Model Physics on Simulations of Arctic and Antarctic Sea Ice. *Cryosphere* **2011**, *5*, 687–699. [CrossRef]
42. Vavrus, S.J.; Holland, M.M.; Jahn, A.; Bailey, D.A.; Blazey, B.A. Twenty-First-Century Arctic Climate Change in CCSM4. *J. Clim.* **2012**, *25*, 2696–2710. [CrossRef]
43. Laxon, S.W.; Giles, K.A.; Ridout, A.L.; Wingham, D.J.; Willatt, R.; Cullen, R.; Kwok, R.; Schweiger, A.; Zhang, J.; Haas, C.; et al. CryoSat-2 Estimates of Arctic Sea Ice Thickness and Volume. *Geophys. Res. Lett.* **2013**, *40*, 732–737. [CrossRef]
44. Lindsay, R. New Unified Sea Ice Thickness Climate Data Record. *Eos* **2010**, *91*, 405–406. [CrossRef]
45. Schweiger, A.; Lindsay, R.; Zhang, J.; Steele, M.; Stern, H.; Kwok, R. Uncertainty in Modeled Arctic Sea Ice Volume. *J. Geophys. Res. Oceans* **2011**. [CrossRef]
46. Stroeve, J.; Barrett, A.; Serreze, M.; Schweiger, A. Using Records from Submarine, Aircraft and Satellites to Evaluate Climate Model Simulations of Arctic Sea Ice Thickness. *Cryosphere* **2014**, *8*, 1839–1854. [CrossRef]
47. Salton, G.; McGill, M.J. *Introduction to Modern Information Retrieval*; McGraw-Hill: New York, NY, USA.
48. Juva, M.; Riska, K. *On the Power Requirement in the Finnish-Swedish Ice Class Rules*; Research Report No.47; Helsinki University of Technology, Ship Laboratory: Espoo, Finland, 2002.
49. Kujala, P.; Sundell, T. *Performance of Ice-Strengthened Ships in the Northern Baltic Sea in Winter 1991*; Helsinki University of Technology, Laboratory of Naval Architecture and Marine Engineering: Espoo, Finland, 1992.
50. Riska, K.; Wilhelmson, M.; Englund, K.; Leiviskä, T. *Performance of Merchant Vessels in the Baltic*; Research report No. 52; Helsinki University of Technology, Ship Laboratory, Winter Navigation Research Board: Espoo, Finland, 1997.
51. Wijnolst, N.; Wergeland, T. *Shipping Innovation*; IOS Press: Amsterdam, The Netherlands, 2009.
52. Zhang, Y.; Meng, Q.; Ng, S.H. Shipping Efficiency Comparison between Northern Sea Route and the Conventional Asia-Europe Shipping Route via Suez Canal. *J. Transp. Geogr.* **2016**, *57*, 241–249. [CrossRef]
53. Lasserre, F. Case Studies of Shipping along Arctic Routes. Analysis and Profitability Perspectives for the Container Sector. *Transp. Res. Part A Policy Pract.* **2014**, *66*, 144–161. [CrossRef]
54. Furuichi, M. Economic Feasibility of Finished Vehicle and Container Transport by NSR/SCR-Combined Shipping between East Asia and Northwest Europe. In Proceedings of the IAME 2014 Conference, Norfolk, VA, USA, 15–18 July 2014.
55. Furuichi, M. Cost Analysis of the Northern Sea Route (NSR) and the Conventional Route Shipping. In Proceedings of the IAME 2013 Conference 2013, Marseille, France, 3–5 July 2013; Volume 13.
56. Schøyen, H.; Bråthen, S. The Northern Sea Route versus the Suez Canal: Cases from Bulk Shipping. *J. Transp. Geogr.* **2011**, *19*, 977–983. [CrossRef]
57. Xu, H.; Yin, Z.; Jia, D.; Jin, F.; Ouyang, H. The Potential Seasonal Alternative of Asia-Europe Container Service via Northern Sea Route under the Arctic Sea Ice Retreat. *Marit. Policy Manag.* **2011**, *38*, 541–560. [CrossRef]
58. Somanathan, S.; Flynn, P.; Szymanski, J. The Northwest Passage: A Simulation. *Transp. Res. Part A Policy Pract.* **2009**, *43*, 127–135. [CrossRef]
59. Liu, M.; Kronbak, J. The Potential Economic Viability of Using the Northern Sea Route (NSR) as an Alternative Route between Asia and Europe. *J. Transp. Geogr.* **2010**, *18*, 434–444. [CrossRef]
60. Goerlandt, F.; Montewka, J. A Framework for Risk Analysis of Maritime Transportation Systems: A Case Study for Oil Spill from Tankers in a Ship–Ship Collision. *Saf. Sci.* **2015**, *76*, 42–66. [CrossRef]
61. Valdez Banda, O.A.; Goerlandt, F.; Montewka, J.; Kujala, P. A Risk Analysis of Winter Navigation in Finnish Sea Areas. *Accid. Anal. Prev.* **2015**, *79*, 100–116. [CrossRef]
62. Kiiski, T. *Feasibility of Commercial Cargo Shipping Along the Northern Sea Route*; University of Turku: Turku, Finland, 2017.
63. Duchoň, F.; Babinec, A.; Kajan, M.; Beňo, P.; Florek, M.; Fico, T.; Jurišica, L. Path Planning with Modified A Star Algorithm for a Mobile Robot. *Procedia Eng.* **2014**, *96*, 59–69. [CrossRef]

64. Amante, C.; Eakins, B.W. *ETOPO1 1 Arc-Minute Global Relief Model: Procedures, Data Sources and Analysis*; NOAA Technical Memorandum NESDIS NGDC-24; National Oceanic and Atmospheric Administration: Boulder, CO, USA, 2009.
65. Omre, A. An Economic Transport System of the next Generation Integrating the Northern and Southern Passages. Master's Thesis, Institutt for marin teknikk, Trondheim, Norway, 2012.
66. Erikstad, S.O.; Ehlers, S. Decision Support Framework for Exploiting Northern Sea Route Transport Opportunities. *Ship Technol. Res.* **2012**, *59*, 34–42. [CrossRef]
67. Pruyn, J.F.J. Will the Northern Sea Route Ever Be a Viable Alternative? *Marit. Policy Manag.* **2016**, *43*, 661–675. [CrossRef]
68. Ostreng, W.; Eger, K.M.; Fløistad, B.; Jørgensen-Dahl, A.; Lothe, L.; Mejlænder-Larsen, M.; Wergeland, T. *Shipping in Arctic Waters: A Comparison of the Northeast, Northwest and Trans Polar Passages*; Springer: Chichester, UK, 2013.
69. Verny, J.; Grigentin, C. Container Shipping on the Northern Sea Route. *Int. J. Prod. Econ.* **2009**, *122*, 107–117. [CrossRef]
70. Srinath, B.N. Arctic Shipping—Commercial Viability of the Arctic Sea Routes. Master's Thesis, Delft University of Technology, Delft, The Netherlands, 2010.
71. Gritsenko, D.; Kiiski, T. A Review of Russian Ice-Breaking Tariff Policy on the Northern Sea Route 1991–2014. *Polar Rec.* **2016**, *52*, 144–158. [CrossRef]

symmetry

MDPI

Article

A Fuzzy Logic Based Intelligent System for Measuring Customer Loyalty and Decision Making

Usman Ghani *, Imran Sarwar Bajwa and Aimen Ashfaq

Department of Computer Science & IT, The Islamia University Bahawalpur, Bahawalpur 63100, Pakistan; imran.sarwar@iub.edu.pk (I.S.B.); aimenashfaq62@gmail.com (A.A.)
* Correspondence: usmanrazi@hotmail.com

Received: 10 November 2018; Accepted: 12 December 2018; Published: 17 December 2018

Abstract: In this paper, an intelligent approach is presented to measure customers' loyalty to a specific product and assist new customers regarding a product's key features. Our approach uses an aggregated sentiment score of a set of reviews in a dataset and then uses a fuzzy logic model to measure customer's loyalty to a product. Our approach uses a novel idea of measuring customer's loyalty to a product and can assist a new customer to take a decision about a particular product considering its various features and reviews of previous customers. In this study, we use a large sized data set of online reviews of customers from Amazon.com to test the performance of the customer's reviews. The proposed approach pre-processes the input text via tokenization, Lemmatization and removal of stop words and then applies fuzzy logic approach to take decisions. To find similarity and relevance to a topic, various libraries and API are used in this work such as SentiWordNet, Stanford Core NLP, etc. The approach utilized focuses on identifying polarity of the reviews that may be positive, negative and neutral. To find customer's loyalty and help in decision making, the fuzzy logic approach is applied using a set of membership functions and rule-based system of fuzzy sets that classify data in various types of loyalty. The implementation of the approach provides high accuracy of 94% of correct loyalty to the e-commerce products that outperforms the previous approaches.

Keywords: fuzzy logic; decision making; customer loyalty; customer reviews

---

1. Introduction

The World Wide Web (WWW) has revolutionized our lives with many different services to facilitate its users such as online shopping, online study courses, online banking and many more. For the last decade, e-commerce (the act of buying and selling of products through internet) is growing day by day and has emerged into the future of shopping. The trend setters in modern e-commerce are Amazon, E-Bay, Ali Baba Express, olx, Daraz.com, and many others. One of the largest retailing e-commerce website is AMAZON.com. Recently, there are approximately 244 million Active buyer accounts, 200 million active products on Amazon and 2.2 billion sales in the past 12 months (average 6 million sales a day [1]. Shopping by e-commerce creates much ease for the customers and businesses as well. However, a challenge faced by the e-commerce users is the need for a better and improved platform to compare and select products and its prices for best choice selection [2]. If such a platform is available, it can save a customer's time, money and energy and can help in buying better products that fulfill their requirements. A big source of knowledge is customers' reviews and feedbacks of a product at social media and e-commerce websites that can effectively guide new customers about previous customers' opinions, interests, past experience and brand loyalty [3–7]. Such information can be very helpful for new customers to buy online with satisfaction and select a right product.

To know about a customer's loyalty to a product, the easiest and widely used technique for measuring customer satisfaction is to understand their sentiments or opinions, which they expressed in the form of comments [8–10]. The most important way to understand their feelings, mood and sentiments or what they are trying to say is to judge their reviews and comments about the product and services [11]. After collecting the information about the consumer's opinion, we can distinguish what is necessary and what is not. The tracking of opinions, feeling, responses and mood of the customers is known as opinion mining and sentiment analysis [12]. The recent type of text analysis that targets to conclude the opinion and polarity of reviews is referred to as Sentiment Analysis. It is a kind of text analysis that deals with a wide aspect of natural language processing, computational semantics and text mining [13].

The current web is a huge repository of valuable information in the scattered form such as micro-blogging websites, such as Twitter or Facebook, have billions of comments and opinions uploaded on daily basis. Sentiments, such as opinions, attitudes, views and emotions, are personal experiences of individuals that are not open to impartial observation. They are stated in language that uses subjective opinions which express sentiment analysis. Most of the organizations carried opinion mining and sentiment analysis of the reviews of online posts [14–17]. The opinions expressed on social networking sites are very effective for the decision making process of business organizations. Organizations use these posts to extract the opinions of the people and to perform sentiment analysis. Sentiment analysis provides a part of text to be positive, neutral or negative in sense.

Previously, general purpose sentiment analysis of tweets and posts have been carried out [3–12], however a task-oriented sentiment analysis of users' reviews of a product to find key features liked by the users and measuring their confidence level is a new idea. A challenge in performing a task-oriented sentiment analysis is measuring a customer's loyalty to a specific product on the basis of customers' views about a product. In this paper, we propose a novel idea of using sentiment score of each customer review of a product and then take the aggerate of the sentiment score and then use such a score to measure customers' loyalty with a product. In this paper, a fuzzy logic method is used for measuring customers' loyalty to a product with the help of sentiment analysis score as shown in Figure 1.

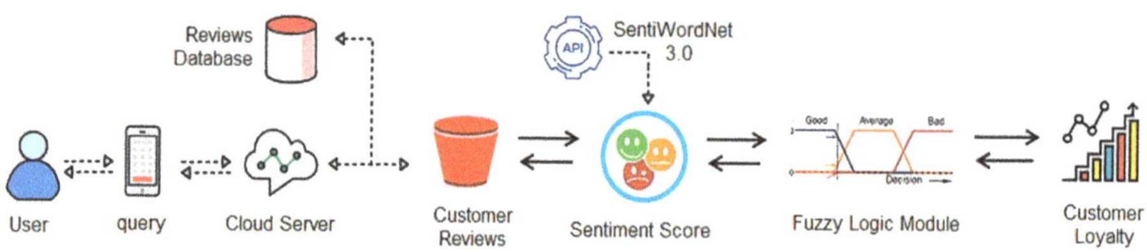

Figure 1. A sketch of proposed approach for Customer Loyalty Measurement.

In our approach, we identify sentiments of users by reading their comments of social network users and by analyzing this we can view them as positive, neutral or negative. We measure the "PN-polarity" of subjective terms, i.e., recognizes whether a text can be positive or negative in which opinions and emotions are expressed. Stanford core NLP is a set of tools and techniques that provides sense to the computer to understand the speech of a human. Stanford Core NLP is transcribed in Java and requires Java 1.8+. Java is required to be connected to execute Core NLP. However, other languages for code writing, e.g., Python or JS (Java Script), can be used and some other languages [18]. With the help of Core NLP, our approach easily understands what people are trying to express through their words. To retrieve sentiments and polarity of input text apply, SentiWordNet library [19] is used to measure the customer involvement level towards a product. Here, we apply the sentiment analysis on the products reviews and also performs P-N polarity on this set of data that tells the positivity, neutral and negativity of reviews and tries to provides accurate results.

Finally, to measure customers' loyalty on the basis of sentiment score calculated from the reviews, the fuzzy logic method is applied. Fuzzy Logic is a process of reasoning that looks a lot like human reasoning [20]. This approach replicates the way of decision making in a human being that includes all the possibilities between digital values YES and NO. The standard logic that a computer can easily understand is to takes specific input and produces a certain output as TRUE or FALSE, or1 or 0, which is equal to the human YES or NO. The fuzzy logic works on the levels of possibilities of input to achieve the definite output and is also called many valued logic which only deals with the truth values [4]. It is also known as many valued logic and deals with truth values only. The values of truth varies from all the values in between 0 and 1. These truth values can encompasses all the numbers between 0 and 1. It does not hold only with both true and false values such as Boolean algebra. The membership functions organized these truth values. It basically provides approximate reasoning.

The rest of the paper is structured into a set of sections. Section 2 discusses the related work of sentiment analysis. Section 3 presents an architecture of the designed approach based on fuzzy logic for measuring customer loyalty using sentiment analysis. Section 4 presents the results of the experiments and the paper is concluded with the future work in Section 5.

2. Literature Survey

In the recent years, sentiment analysis has gained much attention in the field of research. It has many paybacks and useful applications in the field of business, most probably in e-commerce [3–7]. It can give business many profitable gains and visions into how customers think and feel about products and services [8]. It also provides people a better option when they are trying to buy anything online. They know everything which they want to know just by clicking the button and reading the previous reviews about the product [9,10]. Sentiment analysis is a vast area of research because it is a very valuable action for businesses running online. Many people performed research on the sentiment analysis from the previous years and it always provides a remarkable gain in the business. Many researchers show much interest towards it and nowadays it gains major attention [11]. It takes a wide range of importance in industry as well as from a study point of view. Sentiment analysis provides measurable study for mining out the knowledge coming from a consumer's opinion, moods, emotions and feelings towards the product and their characteristics [12]. Today the world has become a global village and the use of internet is excessively growing day by day. So, the demand of the internet is also increased and people prefer online shopping rather than going to malls. So, the review (sentiments) from online customers becomes a need for businesses, other consumers and producers as well [13].

Fuzzy logic is a method which calculates value based on degrees of truth other than the typical 1 or 0. The modern computer is based on Boolean logic (True or False). A lot of work has done on sentiment analysis by using fuzzy logic approach. A method for feature mining from the online reviews of the product was suggested by Indhuja et al. [14]. The feature-based sentiment extraction method categorized into positive, negative and neutral features. Research has been done on it to eliminate noises and for feature mining. It was prolonged to include the result of linguistic borders and fuzzy roles to copy the product of concentrators, transformers and also dilators. The technique was evaluated on SFU (Simon Fraser University) review corpus and the conclusions indicated that fuzzy logic executed flawlessly in sentiment analysis. A theory based on fuzzy logic approach in which sentiment sorting of Chinese sentence-level was projected [15]. This theory of fuzzy set provides the direct way to allocate the core fuzziness between the polarity modules of sentiments [20,21]. For a further procedure of fuzzy sentiment extraction, at the beginning it mentions a technique for measuring the intensity of sentiment sentences. After this it describes fuzzy set which determines the sentiment polarity score. It provides three fuzzy sets which are positive, negative and neutral sentiments. It builds a membership functions on the basis of sentiment intensities which designate the sentiment text measure in many fuzzy sets. The conclusion gives polarity of sentiment sentence level by the use of the maximum membership value.

A technique used for the collection of reviews, blogs and comments from the social networking sites, it differentiates subjective and objective reviews. We take a subjective type review in order to extract sentiment scores from the dictionary of SentiWordNet. Here the polarity of relative sentence structure is obtained from the SentiWordNet dictionary which are positive, negative and neutral scores.

This technique of research performs machine learning and word-level approaches [17]. This proposed technique attains a precision of 97.8% at the view andfeedback level and 86.6% at the sentence level.

In a paper addressing sentiment analyzing techniques using movie reviews using sentiment sorting methods [18], the text at document level yields the polarity scores of the person discussed in reviews. It uses a dictionary of SentiWordNet to analyze every word scores involved in the reviews or comments. There are three types of scores of sentiment words which are positive, negative and neutral as well. It also uses a fuzzy logic technique and its rule base method for carrying out the output. It also uses precision, Recall and accuracy method in order to determine the efficiency of the project. In a similar research, a fuzzy logic approach was used to solve the cloudiness in natural languages. This paper proposed an aspect oriented sentiment classification. They use fuzzy logic for extracting the polarity scores of opinions such as positive, strongly positive, negative and strongly negative [20,21]. It includes objective and subjective types of sentences. It also involves non-opinionated reviews by using the IMS (Imputation of Missing Sentiment) technique. IMS is used for extracting accurate results. Researchers used fuzzy logic for the sentiment modules of reviews. The results explore that for mining of the effective conclusions, this framework is feasible [22].

A model [23] was proposed which provides broadcasting of the fuzzy logic for conception polarities. The researchers describe the ambiguity created by the fuzzy logic useful to diverse areas. This technique joined two linguistic properties, which are named as SenticNet and WordNet. After that a graph is plotted by the propagation algorithm of consequent data. It was broadcasted sentiment of characterized (labeled and un-labeled) datasets. The proposed work was implemented and performed on the dataset. The conclusions show the achievability in problems. Applications of Sentiment analysis took a very vital role in the social networking sites [24]. Nowadays social media becomes a place where mostly people express their emotion, feelings and also comment about their current shopping from any social networking. A particular attention should be given also to the application of sentiment analysis in social networks. The social network environment explores new tasks because many different behaviors and people show their opinions, as defined in this paper, which discuss "noisy data", which is actually the main obstacle in the analysis of the text extracted from social networks [25,26].

Negation recognition and polarity enhancer influence the polarity score in a very unusual way. So, the polarity of a specific word is not sufficient and dependable for overall results. This paper describes all the probable techniques which are used to sense problems for the exact polarity of sentences and for the accuracy of sentiment analysis [27–29]. Some other works in sentiment analysis and opinion mining are addressing the problem in general [30–32]. None of these works target task-oriented sentiment analysis.

3. Materials and Methods

An approach is presented for measuring sentiments of users regarding their comments of a particular product. In our approach, we have attributed polarity analysis and then used a fuzzy logic approach to attribute the loyalty of a customer to a product. The used approach also involves a set of libraries such as core NLP, SentiWordNet library, etc. The users' comments, or reviews are collected from social media and famous e-shopping website AMAZON.com. The sentiment analysis is performed on the products' reviews to measure P-N polarity. Afterwards, to measure customer loyalty on the basis of sentiment score calculated from the reviews, a fuzzy logic method [20] is applied. This approach replicates the way of decision making in human being that includes all the possibilities between digital values YES and NO. The standard logic that a computer can easily understand takes specific input and produces a certain output as TRUE or FALSE or 1 or 0, which is equal to the human YES or NO. The fuzzy logic works on the levels of possibilities of input to achieve the definite output and is also called many valued logic, which only deals with the truth values [4]. It is also known as many valued logic and deals with truth values only. The values of truth varies from all the values in between 0 and 1. These truth values can encompasses all the numbers between 0 and 1. It does not hold only with both true and false values such as Boolean algebra. The membership functions organized these truth values. It basically provides approximate reasoning.

Figure 2 shows the basic structure of sentiment analysis architecture. Sentiment analysis has many different structures based on a phrase, sentence and documents level. The process of collection of data and recognition is the calculating the data obtained from different means.

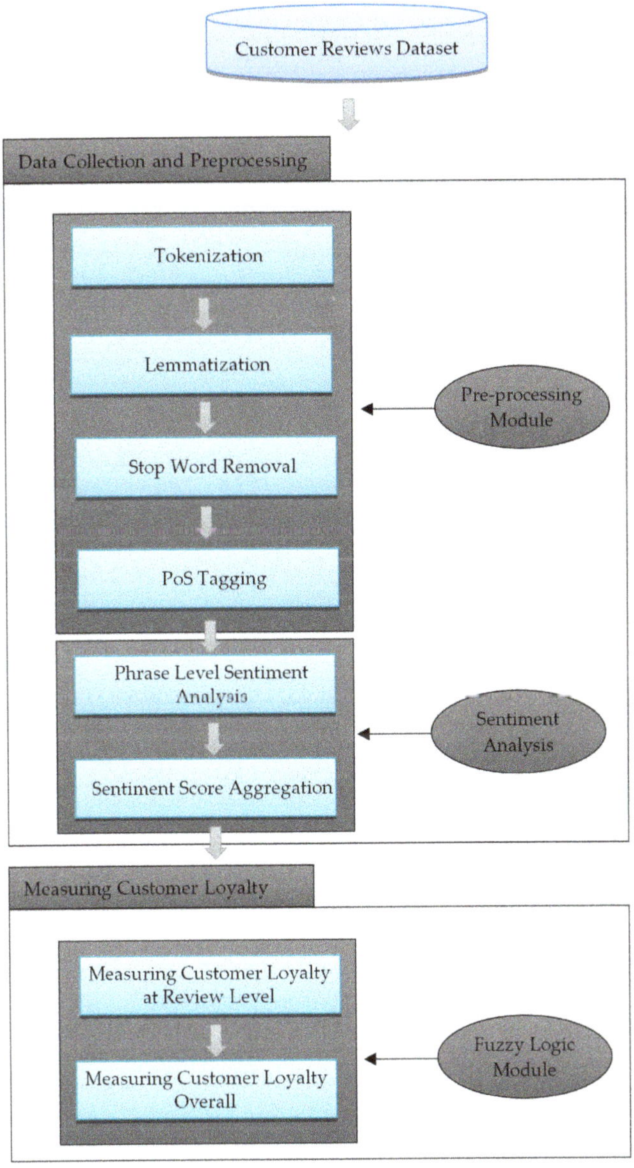

Figure 2. Research Architecture of proposed methodology.

After the lemmatization process, we tagged text by PoS (Parts-of-Speech) tagger. We take POS tagger of Stanford Core NLP (natural language processing). A PoS tagger is very beneficial for sentiment analysis because a POS tagger can differentiate words that can be used in different parts of speech and it is capable of filtering out the words which are not necessary, i.e., we do not need nouns or pronouns because they do not contain any type of sentiments and at the same time adjectives express the sentiments. After this step, we do the most important thing which is sentiment analysis on the text reviews which are being parsed by Stanford POS tagger. We use SentiWordNet 3.0.0 (ISTI, CNR, Rome, Italy) for the analysis. We use a technique for calculating in which a review is positive, negative or neutral and calculate the polarity of reviews by focusing upon adjectives because an adjective names an attribute or quality from which one canit easily discern the positivity, negativity and neutrality scores of the reviews. Then we find out the polarity scores using SentiWordNet database dictionary.

3.1. Data Collection of Customer Reviews

The processing of the used approach starts with the collection of users' reviews, comments, posts and tweets regarding a particular product from various sources such as social media, shopping websites, etc. In our approach, we have collected the dataset from Facebook and AMAZON.com website. The data is collected for a particular product suggested by the user. In this study, the customers' views and reviews of Apple products (such as Apple iPhone 6 and iPhone 7) are collected. The user gives reviews dependent of their feelings, experience or like and dislike of the product. In this study 3500 reviews were collected from social media and Amazon's website.

3.2. Tokenization

Each review in the data set is individually processed. The preprocessing of the reviews starts by the tokenization phase that splits a piece of review into small units such tokens. A typical tokenization process can confiscate punctuation marks from the given text and create tokens of the text. A token can be anything, a word or a symbol, etc. Here, we use Core NLP PTB Tokenizer which is actually PENN TREEBANK way of tokenization of English writing and it splits the reviews into sentences in order to make a simple review file.

3.3. Stop Words Removal

A set of meaningless or irrelevant words in a piece of text can seriously affect the accuracy of the output. Hence, removal of such stop words from the input text is an important phase in sentiment analysis of the text. In the collected user reviews, a stop word can be a number, a preposition or a person's name, a product's name, etc. Each review after tokenization goes through the stop words removal phase. The used approach uses Core NLP library [33], which helps in identifying a list of stop words.

3.4. Lemmatization

Lemmatization is a process that extracts core form of a word to a common base. The used approach banks on lemmatization phase to extract core form of a token or a word to achieve more accurate results in sentiment analysis phase [34]. It can drive linked forms of words to a mutual base. Many textual documents use dissimilar forms of a word, e.g., mobile, mobiles, mobile's are all attributed to 'mobile'.

3.5. Parts-of-Speech Tagging

After the lemmatization phase, the review's text is Parts-of-Speech tagged to identify the lexical position and significance of that word in the sentence. Such lexical position and significance helps in identifying the impact of the word in the sentence. The used approach performs PoS tagging with the help of the Stanford POS tagger that is part of the Stanford CoreNLP library [33]. In this PoS tagging phase, each word in a review text, gives a list of its parts of speech, e.g., Noun, Verb, Adjective, etc. The used PoS tagger "Penn Treebank Tag set" is used for PoS tagging. Besides its three English models, here we use a POS tagger which is also an English tagger and it is known as the "Penn Treebank Tag" set. It can also tokenize the sentence which means it splits the sentences for the quick understanding. It can break down the text into pieces, e.g.,

- Input: This phone has best features e.g., screen, sound system, etc.
- Output: [This/DT] [phone/NN] [has/VBZ] [best/JJS] [features/NNS] [e.g.,/VBG] [screen/NN] [,/,] [sound/JJ] [system/NN] [,/,] [etc./FW] [./.]

3.6. Polarity Analysis of Reviews

Measuring polarity of a customer's review is a key phase in the used approach. In the used approach, the SentiWordNet 3.0.0 library [35] is used to identify the polarity score of each word in a user's review. The polarity score of each word is further accumulated to find the accumulative polarity score of each review. It can formed by examining an automated classifier Φ to coordinate to each synsets of WordNet. It produces numerical scores of three types, $\Phi(s, p)$ (for p P = {Positive, Negative, Objective}) telling the powerfulness of the words in s, which consists of each of these three score values. The hypothesis shows change terms to synsets is that dissimilar nature of the same term with unlike opinion properties sometimes. Each of the three $\Phi(s, p)$ scores ranges from 0.0 to 1.0, and their sum is 1.0 for each synsets.

The Figure 3 shows the graphical representation used by SentiWordNet which represents the properties of opinion of a synset [13]. This shows that for all of the three classes, synset may have non-zero scores that specify the similar terms have, in the sense for the synset. Therefore, it shows that SentiWordNet is used for the identifying and extracting polarity for subjectivity sentences. Table 1 shows output of PoS tag process and Table 2 shows the processed example of a review statement.

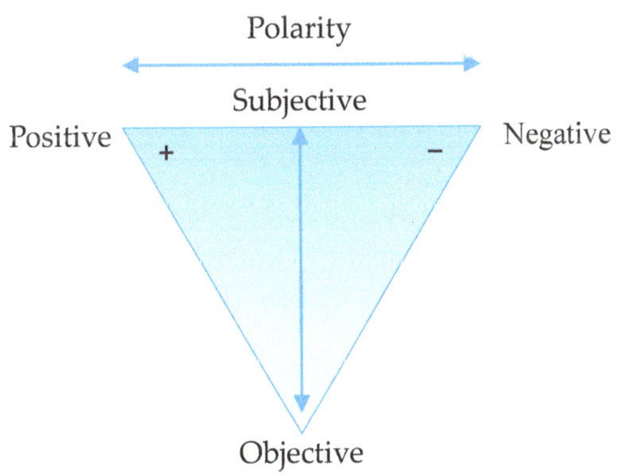

Figure 3. The graphical representation of sentiment analysis.

Table 1. PoS Type output of user reviews.

Pos_ID	Pos_Name	Pos_Abbreviation	SentiWordNet_Abr
1	Noun	NN	N
2	Adjective	JJ	A
3	Verb	VB	V
4	Adverb	RB	R
5	Noun plural	NNS	N
6	Adjective Superlative	JJS	A
7	Verbs	VBZ	V

Table 2. Methodology applied for sentiment analysis.

Type	Values
Original sentence	iPhone 6 is one of the good models of Apple phone.
Sentence After Drop Stop-Words	iPhone 6 + one + good + models + Apple phone.
Tagged Stanford POS tagger To Sentence	iPhone/NNP 6/CD is/VBZ one/CD of/IN the/DT good/JJ models/NNS of/IN Apple/NNP phone/NN ./.
After Lemmatized Sentence	iPhone 6 + one + good + model + Apple phone
Tagged SentiWordNet POS tagger To Sentence	iPhone#n 6#n one#n good#a model#n Apple#n phone#n
Sentence token score per word:	iPhone#n ==> SentiWordNet Score: 0.0 one#v ==> SentiWordNet Score: 0.0 good#a ==> SentiWordNet Score: 0.634 model#n ==> SentiWordNet Score: 0.0 Apple#n ==> SentiWordNet Score: 0.0 phone#n ==> SentiWordNet Score: 0.0 review#n ==> SentiWordNet Score: 0.053
scoreSum:	0.343
Sentence Score:	Positive
Positive:	34.35%
Negative:	0.0%
Neutral:	5.0%

By applying all the methods and techniques of sentiment analysis process, we reach our results. The first line explains that we enter a simple review in a sentence form, then we remove stop words from a review in the second step. In the third step, we apply lemmatization on that review. In the fourth step, we use the Stanford Parts-of-Speech (POS) tagger which is used specify the important and useful parts of speech in the context. After applying POS tagging, we use another tagger of SentiWordNet POS tagger in the fifth step, which is almost same as that of the POS tagger but it calculates the score of that POS words by its weights. Here, we apply some constraints on it that it only calculates the score of adjectives in the given reviews. We only focus on the adjective based reviews because adjective is a quality word or the word that describes a noun, which is clearly represents the sentiment behind the reviews.

In the sixth step, we calculate Sentence token score per word using Equation (1) but we only use the score of adjective words in the text. In the seventh step, the score sum is used to identify the sum of all sentiment words in the given sentence. Equation (1) shows how the score sum is calculated by adding score of all words in a review:

$$Sum\_Score = \sum_{k=0}^{n} \binom{n}{k} W^k \tag{1}$$

After that the eighth step shows the most important feature of sentiment analysis, which is the sentence type of the review. The sentence type of the review shows that whether the review is considered positive, neutral or negative. The sentence type of this review is positive obtained by using SentiWordNet dictionary. In the last three lines, the code executes that how much a review is positive, neutral or negative and the final result shows that it is positive because it has the highest positive score percentage.

3.7. Used Fuzzy Logic System

For finding the customer loyalty to a product, a fuzzy logic system is used. This system is based on the fuzzy set theory [36]. The fuzzy sets and rule-based approach provides high performance and working for the sentiment analysis purpose. It provides a degree of truth and human reasoning. It is also used in decision making techniques. The used fuzzy logic system is based on following principles of fuzzy logic [37]:

(1) In fuzzy logic, accurate reasoning is experimented as a case of limit for approximate reasoning.
(2) All relation used are the relation of degree in fuzzy logic.

(3) It also provides that each logical method can be fuzzified.
(4) Fuzzy logic restricts on the choice of on a collecting the variables and knowledge is understood as a flexible collection.
(5) The result of inference system is broadcasting of flexible limitation.

The used fuzzy logic system introduces fractional truth values, between YES and NO.

$$A = \{(x,\ u_A(x))|\ x\ \in X\} \tag{2}$$

Here, Equation (2) shows that μ_A (X) is called the membership function or grade of membership, it is also a degree of truth, of x in A that plots X to the membership position M. While M contains only the two points 0 and 1, A is non-fuzzy and μ_A (X) is alike to the distinctive function of a non-fuzzy set. Zero degree elements of membership are usually not taken. It can show the fractional membership to that set. It shows that the element from the set has particular degree and some particular membership functions are used that provides the degree of membership of fuzzy logic. These membership functions are the trapezoidal membership function, triangular membership function, Bell membership function and Gaussian membership function. In the proposed research, we apply a triangular membership function which is completely discussed in fuzzy membership functions approach. The core of a membership function for some fuzzy set A is defined asthat area of the universe that is specified by the whole membership in the set A. It shows that the core consists of those elements x of the universe such that $\mu_A(x) = 1$. The membership function's support for some fuzzy set A is defined as the area of the universe that is indicated by nonzero membership in the set A. Figure 4 shows, the support contains by the elements x of the universe such that $\mu_A(x) > 0$.

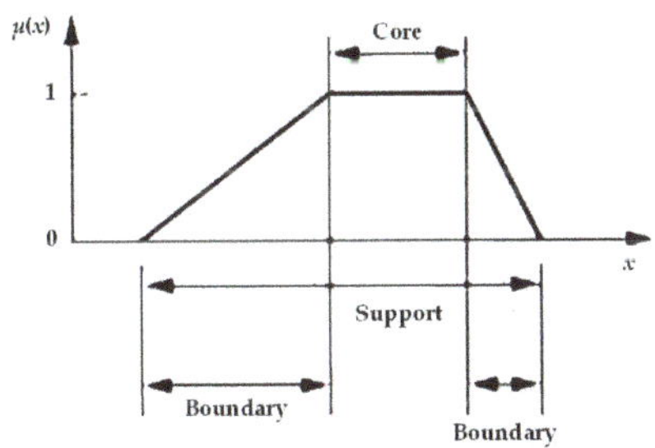

Figure 4. Support of element x in a membership function.

3.7.1. Fuzzification

The first step in the used fuzzy logic systems is to recognize the input and output variables. In this process, the crisp input data is converted into a fuzzy set with the membership functions [38]. Input variables of the fuzzy logic system are represented on the fuzzy sets by use of linguistic terms, membership functions and linguistic variables. The linguistic terms and variables are frequently the terms or the complete sentences of used natural language. When we are setting the linguistic variables, we are confident enough that no numerical values are used in the linguistic variables. The two vital points Fuzzy sets and fuzzy membership functions which are needed to be used to obtain the fuzzified values. The conversion of crisp input values into fuzzy values are performed by use of Membership Functions and this method of transformation is known as fuzzification. Every membership function signifies a feature of the linguistic variable being fuzzified. As we take this the membership function approach of linguistic variables in our research, we take "Sentiment Score" and "Customer Loyalty" as an input variables which may "Pos" "Neu" "Neg", and the membership function of linguistic variable "Customer Loyalty" is "Pesudo" and "Latent", "True". We described the fuzzified set by following relation:

$$A = \mu_1 K(x_1) + \mu_2 K(x_2) + \ldots + \mu_n K(x_n) \tag{3}$$

In equation (3), the fuzzy set $K(x_i)$ is called as kernel of fuzzification. To apply this technique, μ_A is constant and x_A is being converted to a fuzzy set $K(x_i)$. This equation is used in the fuzzification process in which Universe of Discourse and membership function are being applied.

In our paper, we take the sentiment analysis score as an input linguistic variable and customer loyalty as an output linguistic variable as shown in Table 3.

Table 3. Input and output Linguistic Variables for proposed method.

Type	Linguistic Variables
Input Linguistic Variable	Sentiment analysis score (SA)
Output Linguistic Variable	Customer loyalty (LO)

We again define linguistic terms for each input and output linguistic variables. The input linguistic variable is sentiment scores and we assigned mainly three linguistic terms. These linguistic terms are Positive, neutral, negative as shown in Table 4.

Table 4. Input Linguistic Variable and Terms.

Type	Linguistic Variable	Linguistic Terms
Input	Sentiment analysis score (SA)	{Positive, neutral, negative}

The output linguistic variable we taken Customer loyalty (LO) also have three linguistic terms, these linguistic terms are True loyalty, pseudo loyalty and latent loyalty, shown in Table 5.

Table 5. Output Linguistic Variable and Terms.

Type	Linguistic Variable	Linguistic Terms
Output	Customer loyalty (LO)	{True loyalty, pseudo loyalty, latent loyalty}

3.7.2. Membership Function

For taking decision on the input crisp values, a triangular membership function is used in our approach. The function of fuzzy sets that are achieved by crisp values of linguistic variables and show the relationships of these crisp values to the set are divided as a membership function. It is actually degree of truth that occurs between 0 and 1. There are many different kinds of membership functions, i.e., triangular MF, Trapezoidal MF, Gaussian MF, etc. It is used to plot the values of non-fuzzy sets to linguistic fuzzy sets.

Triangular Membership Function: In our research we use a triangular membership function which describes in fuzzy membership functions approach as shown in Figure 5. Fuzzy logic involves precise logical operations and these are little bit unlike those used in logic of approximate degree of truth, they are conjunction, disjunction and negation. In order to get the smallest values from the all available fuzzy variables, we use a minimum function known as conjunction.

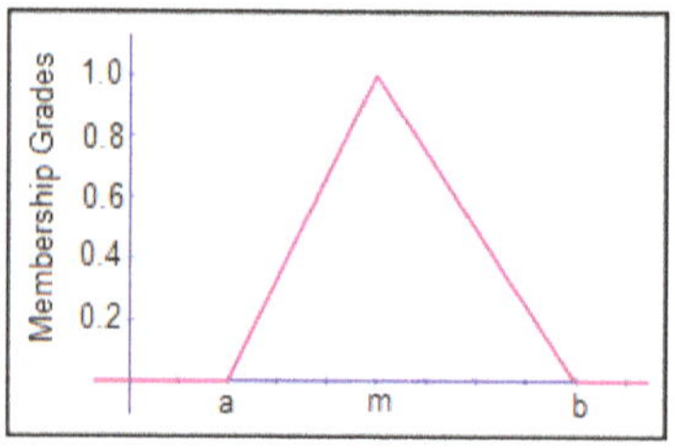

Figure 5. Graphical representation of Triangular Membership Function.

Figure 5 shows a triangular membership function. For example, we take three fuzzy variables a, b and m and also with their truth values of 0.3, 0.6 and 0.9, correspondingly; as shown in Equation (4):

$$a \wedge b \wedge m = \min(a; b; m) = 0{:}3. \quad (4)$$

Just like the last solved example, we take the max function now just as we find the min function. Here, disjunction involves the maximum function as shown in Equation (5):

$$a \_ b \_ m = \max(a; b; m) = 0{:}9. \quad (5)$$

In our research, we use a triangular membership function. We use this membership function because it maintains three variables and creates a relation between them. Here we categorize sentiments analysis score into three linguistic terms that identifies the sentiment scoring of reviews. These linguistic variables are used for evaluating customer loyalty. These terms are Positive (a), Negative (b), Neutral (m). Here we take only the subjective reviews for sentiment analysis because subjective reviews can easily state the opinion of the consumer. Here we prefer triangular membership function also known as trimf because we take three linguistic variables, i.e., a, b and x, where trimf describe by a lower limit a, an upper limit b, and a value c, where $a < c < b$ as shown in Equation (6).

$$\text{Triangular}(x; a, b, m) = \begin{cases} x < a & 0 \\ a \leq x \leq m & \frac{(x-a)}{(m-a)} \\ m \leq x \leq b & \frac{(b-x)}{(b-m)} \\ m \leq x & 0 \end{cases} \quad (6)$$

where a, b and m represent the x-coordinates for triangle, x represents the crisp value from the isolated variable fuzzy universe of discourse. We classified the sentiment score into three parts:

- positive score;
- neutral score;
- negative score.

We use these sentiment scores in order to evaluate loyalty of customers towards online products. In our proposed method, we use three types of loyalty which distinguish how much the consumer is loyal towards the product and services and their values lies according the fuzzy logic triangular membership function.

- In Pseudo Loyalty, the value in trimf lies between $0.0 < x < 0.30$ because the consumer is not long-lasting whether they are buying from you in the future or choose any other opportunity. It is referred as low loyalty.
- In Latent Loyalty, the value in trimf lies between $0.30 \leq x < 0.70$ because the consumer prefers not to purchase anything from any brand but if they are going to purchase they will always buy from one brand. It is referred as medium loyalty.
- In True loyalty, the value in trimf lies between $0.70 \leq x \leq 1.0$ because the consumer is only loyal to a product. They are trustworthy and always refer the product to their family, friends and relatives. They will never switch from the brand. It is also known as High Loyalty.

Here we take three different types of customer loyalty which are denoted by triangular membership functions as shown in Equation (7):

$$LO(x) = \begin{cases} \text{if } 0.0 \leq x < 0.3 & \text{PseudoLoyalty} \\ \text{if } 0.3 \leq x < 0.7 & \text{LatentLoyalty} \\ \text{if } 0.7 \leq x \leq 1.0 & \text{TrueLoyalty} \end{cases} \quad (7)$$

Fuzzy Rules Based System: Three most common types of fuzzy rule based systems, which are named as Mamdani, Sugeno, Tsukamoto, etc. These first two kinds of fuzzy rule-based systems are used to executed on regression problems and the output of these systems is a real value, and the third type is used to implement to problems which relates to categorization. We use Mamdani inference system in our research.

Mamdani Fuzzy Inference System: Mamdani fuzzy inference system is proposed by Ebrahim Mamdani it in 1975. It is most general and highly useful approach used in the research methods. It was the first control system constructed by the use of fuzzy set theory. It has six basic stages:

(1) Building of fuzzy rules.
(2) By using membership function, find fuzzification of input.
(3) The fuzzified inputs are shared by following the fuzzy set theory.
(4) The allocation of rule strength and output membership function to find results of the rules. MISO (Multiple Input Single Output) and MIMO (Multiple Input Multiple Output) systems is used in Mamdani FIS.
(5) In order to obtain an allocation of output just by sum up the outcomes.
(6) Output membership function can be Defuzzified.

In this study, we use Mamdani fuzzy inference: Mamdani systems are instinctive that means it is usually based on what a person feels about something to be true even without knowing a reasonable answer. It is fully appropriate to human input.

We used Mamdani rule based systems which is being implemented on MATLAB. When we plot a graph for membership functions, the curve of membership function is built in MATLAB and it is used to plot membership values between 0 and 1 where 0 shows the starting point and 1 is the peak point. The values occurs between 0 and 1 represents that how input is plotted to membership function value. The membership function for sentiment analysis is given in Equation (8):

$$A = \{x, p_A(x), o_A(x), n_A(x) \mid x \in X\} \quad (8)$$

where "x" is the review taken from the file, "$p_A(x)$" is the membership of positive reviews, $o_A(x)$ is the membership of neutral reviews and $n_A(x)$ is the membership of negative reviews.

We take sentiment analysis as an input, which shows that the value lies between

- 0.0 to 0.3 is taken as negative.
- 0.3 to 0.7 is taken as neutral.
- 0.7 to 1.0 is taken as positive.

We plot a graph showing these sentiment values and terms by using MATLAB which is given below:

By taking the input linguistic variables- Sentiment analysis score. The Figure 6 shows indicates that Sentiment analysis score greater 0.7 is positive. Hence, all scores lie between 0.7 and 1 will always be positive.

Fuzzy Rules: The backbone of any fuzzy logic system is its fuzzy rules. By using these rules, we can easily describe the controlled output and the conclusion is taken. These are simple IF-ELSE rules. Suppose we have a variable x included in the problem (which is our sentiment score), so the loyalty output has its own membership function which is low, medium and high e.g., when we apply rules (shown in Table 6), it will give:

- If x is low THEN loyalty is low.
- If x is medium THEN loyalty is medium.
- If x is high THEN loyalty is high.

Table 6. The fuzzy rules for calculating Customer Loyalty.

S#	RULE
1	if ("SENTIMENT SCORE IS NEGATIVE") then CUSTOMER LOYALTY is "PSEUDO"
2	if ("SENTIMENT SCORE IS NEUTRAL") then CUSTOMER LOYALTY is "LATENT"
3	If ("SENTIMENT SCORE IS POSITIVE") then CUSTOMER LOYALTY is "TRUE"

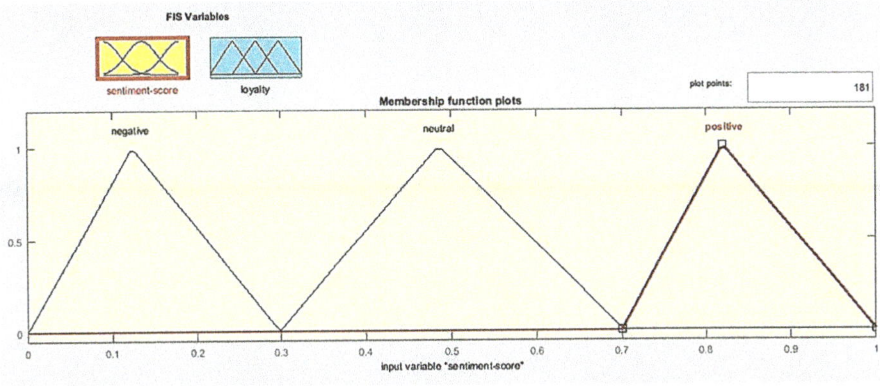

Figure 6. The triangular fuzzy membership function plot for sentiment analysis as inputs.

Table 6 defines rules which are written in the form of given technique in MATLAB. The given figure shows that if our sentiment score is positive, i.e., it lies in the values between 0.7 to 1, then our loyalty is true.

By following these rules, suppose the degree of membership for x is 0.45 to the MF medium, then the loyalty will be also 0.45 medium.

3.7.3. Defuzzification

Defuzzification is the method which generates quantifiable results in crisp logic which is achieved from fuzzy sets and membership functions with consistent degrees. It is the method that plots a fuzzy set to a crisp set. It uses a set of rules that change a number of variables into a fuzzy result. It produces computable results which contains fuzzy sets and membership functions. It performs mapping output of fuzzy sets into crisp values. Here, we take triangular MF which defines the exact conclusions. If the degree of the membership function is not equal to 1, we have to use a trapezoidal shape instead of a triangle shape. Here are some rules which tell us relation between sentiment score and type of loyalty [14].

The last step in the fuzzy logic system is the defuzzification. After the implementation is complete in the inference step, we achieve an output value and the output value obtained from it is known as fuzzy value. In order to signify this fuzzy value in a proper way, we required to convert it into Crisp Output Value. The process of converting the fuzzy value into Crisp Output Value is known as Defuzzification

Output Membership Function for "Customer Loyalty": We convert the fuzzy output to the crisp output which is formed by the steps of fuzzy inference system, the Customer Loyalty Membership function is taken as output MF. It consists of the different types of Customer Loyalty which are calculated by firm value of sentiment score. Such as, if sentiment score is in between 0.75 then loyalty will also be increase at the almost same level of 0.75 and this type of loyalty is known as "True Loyalty". We apply defuzzification rules to clarify the relation process between sentiment score and customer loyalty.

Defuzzification rules: Here are some rules of defuzzification where 'x' denotes the sentiment score while 'y' denotes the type of loyalty:

$$\text{if } (0.0 \leq x < 0.30), \text{ then } y = \text{'Pseudo Loyalty'}$$

$$\text{if } (0.30 \leq x < 0.70), \text{ then } y = \text{'Latent Loyalty'}$$

$$\text{if } (0.70 \leq x \leq 1.0), \text{ then } y = \text{'True Loyalty'}$$

Figure 7 shows a graphical representation of triangular membership functions in which we consider sentiment score on the x-axis while membership on the y-axis. These score of sentiments analysis shows how much loyalty we achieved from the reviews of the online products. We noticed that the most of the sentiment values occurs between 0 and 1; this shows our graph gives almost positive results. Here is algorithm which presents the functionality and working of triangular membership function graph. This graph shows x as a vector and three points a, b and c are the scalar, where "a" is the lower limit from where our sentiment score starts increasing, it is also known as "min" function while "b" is the peak limit or level from the sentiment score are stop increasing and it is also known as "max" function and "c" is the middle point or the value where our sentiment score achieve its highest point, such as in the following chart, the curve of the graph is at the highest level of 0.7, which means that most of our reviews lies at this point between positive and neutral. Here x-axis denotes the sentiment score lie between 0.0–1.0 and the y-axis denotes the values of membership.

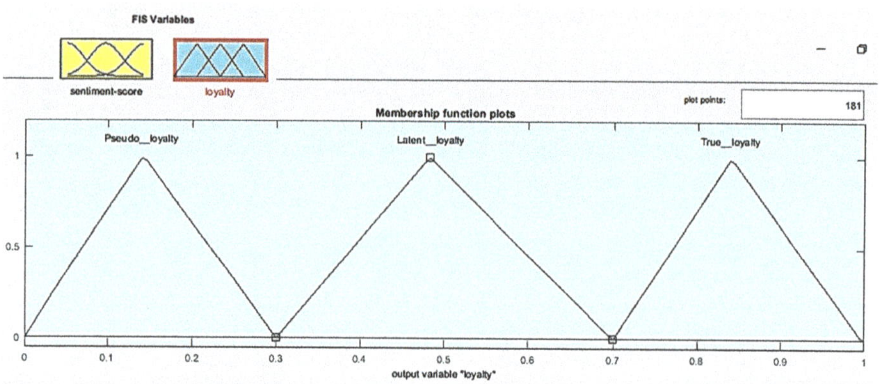

Figure 7. The triangular fuzzy membership function plot for loyalty as an outputs.

Algorithm:

x = 0:0.1:1;
y = trimf(x, [0.30 0.70 1.0]);
plot(x,y)
$x$label('trimf, P = [0.3 0.7 1.0]')
$y$lim([−0.05 1.05]

4. Experiments and Results

The results can be obtained by the use of algorithm of SentiWordNet and fuzzy logic. We collect reviews as an input. These are opinion sentences collection which is collected from the website www.Amazon.com and these are the reviews or comments expressed by the customers. We collect and store 1000 comments for two different Apple products. Once the reviews are taken out, then we applied pre-processing on it, parsed, tokenize and lemmatize these reviews. These sentences show positive, negative and neutral type sentiments scores. So the results of sentiments scores are measured by using SentiWordNet 3.0 software. Here we show sentiment analysis process of a single review and we applied these above mentioned techniques in Java by using the platform "Eclipse". We choose this software because it takes less effort and made our task easier. We take simple review as an input and achieve our desired results.

We take the complete results of the sentiments analysis achieved by the reviews. We take a sentence level approach in which we take a single review and apply sentiment analysis on it. In order to calculate the percentage of total number of positive reviews, total number of neutral reviews and total number of negative reviews, we collect the total results obtained from sentiment analysis and take their percentage. From the above mentioned results shows that we have 320 reviews are positive, 105 are neutral reviews while 75 are negative reviews from the collection of 500 reviews. In order to calculate the percentage of positive, neutral and negative, we use a following formula.

- Positive Sentiment Percentage:
  Positive (%) = (Number of Neutral Sentiments/Total Number of Reviews) × 100 = (320/500) × 100
  Positive (%) = 64%
- Neutral Sentiment Percentage:
  Neutral (%) = (Number of Neutral Sentiments/Total Number of Reviews) × 100 = (105/500) × 100
  Neutral (%) = 21%
- Negative Sentiment Percentage:
  Negative (%) = (Number of Negative Sentiments/Total Number of Reviews) × 100 = (75/500) × 100
  Negative (%) = 15%

Table 7 shows the accuracy of different types of reviews. We can more explain the number of Positive, neutral and negative reviews in the form of bar graph and their corresponding percentages in the form of pie graph of Apple iPhone 6s plus as shown in Figures 8 and 9.

Table 7. Overall percentage of Sentiment Analysis of Samsung Galaxy S8.

Sentiment Positioning	Sentence Level Accuracy
Positive	94%
Negative	91%
Neutral	85%

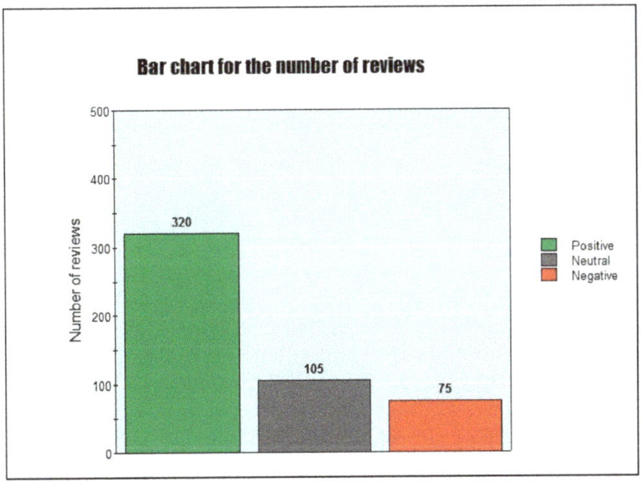

Figure 8. Bar chart for the number of reviews of Apple iPhone 6s plus.

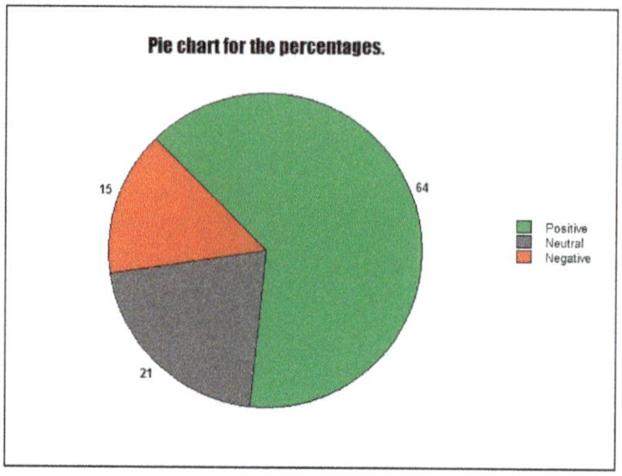

Figure 9. Pie chart for the sentiment score Apple iPhone 6s plus (In percentage).

We also use fuzzy logic for the evaluating the loyalty with sentiment scores. We apply the rules which are simulated in MATLAB, these rules show the relation between sentiment score with types of loyalty, e.g., we see in the following figure that if we have sentiment score say 0.5 than our loyalty is also 0.5. Sentiment score are directly proportional to the type of customer loyalty, we can also say that if our sentiment score is 0.5, it is considered as neutral, the loyalty also lies at 0.468, very close to the value of sentiment score and this type of loyalty is considered as Latent loyalty (see Figure 10).

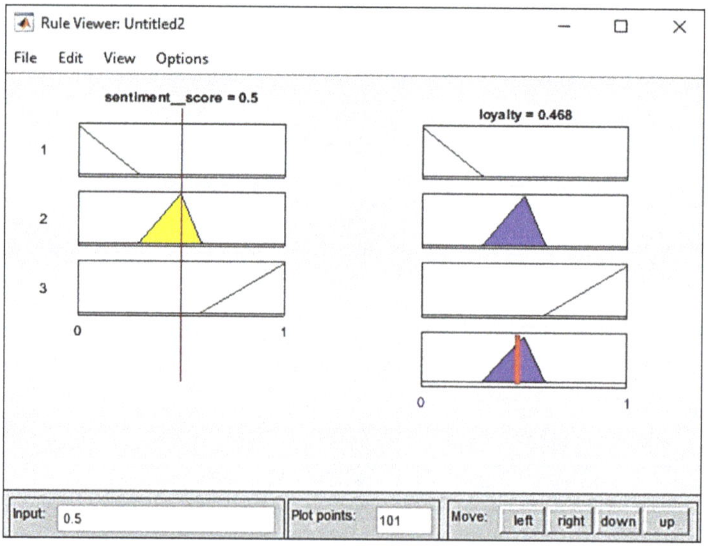

Figure 10. Rule Inference System in MATLAB.

We reached the conclusion that the online customers of Apple iPhone 6s plus mobile are very loyal as compared to other mobiles. It achieves the loyalty score of 64%. The features of Apple iPhone 6s plus are more reliable and their most of the online customers are satisfied with this product and services as well. It supports all the new versions and feature such as camera, memory and battery timings, etc. Table 8 shows a comparison of the results of our approach with the results of previous approaches.

The results shown in Table 8 represent that previous approaches mainly performed in precision that varies from 58.2 to 87.5% whereas recall for the previous approaches is quite low and ranges from 52.0 to 79.44%. Similarly, the F-Score of previous approaches is also lower 53.0 to 77.98%. Our approach performs better as precision of our approach is 89.32%, recall is 80.36% and F-Score is 83.69%. The improvement in precision is minor however, the major improvement is in recall and F-measure.

Table 8. Comparison of results with other approaches.

Sr. No.	Work	Application	Precision %	Recall %	F-Score %
1	Grabner, et al. [2]	Sentiment Analysis of customer reviews	83.0	40.0	53.0
2	Bagheri, et al. [3]	Sentiment Analysis of customer reviews	87.5	65.0	70.3
3	Guzman, et al. [39]	Sentiment Detection on Twitter	58.2	52.0	54.9
4	Thet et al. [40]	Sentiment Analysis of Movie Reviews	76.5	79.44	77.98
5	This Approach	Sentiment Analysis for Customer Loyalty	89.32	80.36	83.69

The results of the presented approach for measuring customer loyalty to a product using sentiment analysis are shown in Figure 11 and the results are also compared with the previous approaches. The results show that our approach performs better than the previous approaches available in literature.

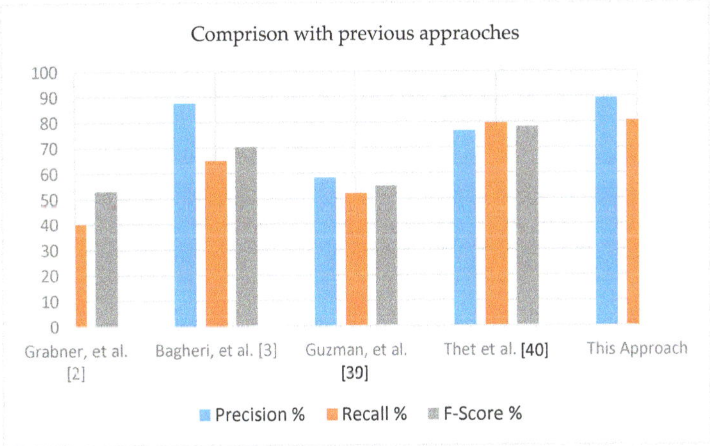

Figure 11. Rule Inference System in MATLAB.

A limitation of the presented implementation is that it processes only English language text that is grammatically correct and has no spelling mistakes in text.

5. Conclusions and Future Work

This paper addresses an important problem of measuring customer's loyalty to a specific product. Previously, general purpose sentiment analysis of tweets and posts are carried out however a task-oriented sentiment analysis of users' reviews of a product to find key features liked by the users and their confidence level is a new idea. In this paper, we presented a novel idea of using a fuzzy logic approach for measuring customer's loyalty to a product with the help of a sentiment analysis score. We use a Fuzzy logic approach which used membership functions and rule-based system of fuzzy sets which is used classifies the types of loyalty. It attained the average accuracy of 94% of positive which shows the number of customers which are loyal to the e-commerce products.

In this study we have experimented with the small sized reviews that are processing in separate sentences. In future, we aim to extend the ability of the implementation to process and handle large sized text. In the future, this work can be extended by considering both sentence types, i.e., subjective as well as objective. It aims to achieve more accuracy by these techniques. It also improves the speed when dealing with a large amount of data. Additionally, every organization or e-commerce site can use sentiment analysis because it is a very beneficial technique and, by using this, organizations take their business at the peak and will grow rapidly.

Author Contributions: U.G. is the main author of this paper and he has contributed in investigation of the problem, research design, experiments design and writing the original draft. I.S.B. has supervised this research work and contributed in design of the approach, writing, review and editing this paper. A.A. has contributed in implementation and coding of this research.

Funding: No funding is involved in this research.

Conflicts of Interest: The authors declare no conflict of interest.

References

1. Inc.com. Amazon Just Eclipsed Records, Selling over 600 Items Per Second. Available online: https://www.inc.com/tom-popomaronis/amazon-just-eclipsed-records-selling-over-600-items-per-second.html (accessed on 31 May 2018).
2. Gräbner, D.; Zanker, M.; Fliedl, G.; Fuchs, M. Classification of customer reviews based on sentiment analysis. In Proceedings of the 19th Conference on Information and Communication Technologies in Tourism (ENTER), Helsingborg, Sweden, 24–27 January 2012; pp. 460–470.
3. Bagheri, A.; Saraee, M.; De Jong, F. Care more about customers: Unsupervised domain-independent aspect detection for sentiment analysis of customer reviews. *Knowl.-Based Syst.* **2013**, *52*, 201–213. [CrossRef]
4. Wilson, T.; Wiebe, J.; Hoffmann, P. Recognizing contextual polarity in phrase-level sentiment analysis. In Proceedings of the Conference on Human Language Technology and Empirical Methods in Natural Language Processing, Vancouver, BC, Canada, 6–8 October 2005; pp. 347–354.
5. Hu, M.; Liu, B. Mining and summarizing customer reviews. In Proceedings of the Tenth ACM SIGKDD International Conference on Knowledge Discovery and Data Mining, Seattle, WA, USA, 22–25 August 2014; pp. 168–177.
6. Neri, F.; Aliprandi, C.; Capeci, F.; Cuadros, M.; By, T. Sentiment analysis on social media. In Proceedings of the IEEE/ACM International Conference on Advances in Social Networks Analysis and Mining (ASONAM), Istanbul, Turkey, 26–29 August 2012; pp. 919–926.
7. He, W.; Zha, S.; Li, L. Social media competitive analysis and text mining: A case study in the pizza industry. *Int. J. Inf. Manag.* **2013**, *33*, 464–472. [CrossRef]
8. Chandrashekaran, M.; Rotte, K.; Tax, S.S.; Grewal, R. Satisfaction strength and customer loyalty. *J. Mark. Res.* **2007**, *44*, 153–163. [CrossRef]
9. Kang, D.; Park, Y. Review-based measurement of customer satisfaction in mobile service: Sentiment analysis and VIKOR approach. *Expert Syst. Appl.* **2014**, *41*, 1041–1050. [CrossRef]
10. Chamlertwat, W.; Bhattarakosol, P.; Rungkasiri, T.; Haruechaiyasak, C. Discovering Consumer Insight from Twitter via Sentiment Analysis. *J. UCS* **2012**, *18*, 973–992.
11. Bauman, K.; Liu, B.; Tuzhilin, A. Recommending Items with Conditions Enhancing User Experiences Based on Sentiment Analysis of Reviews. In Proceedings of the CBRecSys@ RecSys; CEUR Workshop 2016, Boston, MA, USA, 16 September 2016; Volume 1677, pp. 19–22.
12. Dubey, A.; Gupta, A.; Raturi, N.; Saxena, P. Item-Based Collaborative Filtering Using Sentiment Analysis of User Reviews. In Proceedings of the International Conference on Application of Computing and Communication Technologies, Delhi, India, 9 March 2018; pp. 77–87.
13. Artificial Intelligence Fuzzy Logic Systems. Available online: https://www.tutorialspoint.com/artificial_intelligence/artificial_intelligence_fuzzy_logic_systems.htm (accessed on 31 May 2018).
14. Indhuja, K.; Reghu, P.C. Fuzzy Logic Based Sentiment Analysis of Product Review Documents. In Proceedings of the 1st International Conference on Computational Systems and Communications, Trivandrum, India, 17–18 December 2014; pp. 18–22.
15. Guohong, F.; Wang, X. *Chinese Sentence-Level Sentiment Classification Based on Fuzzy Sets*; COLING: Beijing, China, 2010; pp. 312–319.
16. Tumsare, P.; Sambare, A.S.; Jain, S.R. Opinion Mining in Natural Language Processing using SentiWordNet and Fuzzy. *Int. J. Emerg. Trends Technol. Comput. Sci.* **2014**, 3, 154–158.

17. Khan, A.; Baharudin, B.; Khan, K. Sentiment Classification Using Sentence-level Lexical Based Semantic Orientation of Online Reviews. *Trends Appl. Sci. Res.* **2011**, *6*, 1141–1157. [CrossRef]
18. Stanfordnlp.github.io. Stanford CoreNLP—Natural Language Software | Stanford CoreNLP, 2018. Available online: https://stanfordnlp.github.io/CoreNLP/ (accessed on 31 May 2018).
19. SentiWordNet. n.d. Available online: https://sentiwordnet.isti.cnr.it/ (accessed on 12 February 2018).
20. Zadeh, L.A. Toward a theory of fuzzy information granulation and its centrality in human reasoning and fuzzy logic. *Fuzzy Sets Syst.* **1997**, *90*, 111–127. [CrossRef]
21. Mamdani, E.H.; Assilian, S. An experiment in linguistic synthesis with a fuzzy logic controller. *Int. J. Man-Mach. Stud.* **1975**, *7*, 1–13. [CrossRef]
22. Pang, B.; Lee, L. Opinion mining and sentiment analysis. *Found. Trends Inf. Retrieval.* **2008**, *2*, 1–135. [CrossRef]
23. Yuan, H.; Xu, W.; Li, Q.; Lau, R. Topic sentiment mining for sales performance prediction in e-commerce. *Ann. Oper. Res.* **2018**, *270*, 553–576. [CrossRef]
24. Al-Smadi, M.; Qawasmeh, O.; Al-Ayyoub, M.; Jararweh, Y.; Gupta, B. Deep Recurrent neural network vs. support vector machine for aspect-based sentiment analysis of Arabic hotels' reviews. *J. Comput. Sci.* **2018**, *27*, 386–393. [CrossRef]
25. Pham, D.H.; Le, A.C. Learning multiple layers of knowledge representation for aspect based sentiment analysis. *Data Knowl. Eng.* **2018**, *114*, 26–39. [CrossRef]
26. Mubarok, M.S.; Adiwijaya; Aldhi, M.D. Aspect-based sentiment analysis to review products using Naïve Bayes. In *AIP Conference Proceedings*; AIP Publishing: Melville, NY, USA, 2017; Volume 1867, p. 020060.
27. Liu, Y.; Bi, J.W.; Fan, Z.P. Ranking products through online reviews: A method based on sentiment analysis technique and intuitionistic fuzzy set theory. *Inf. Fusion* **2017**, *36*, 149–161. [CrossRef]
28. Giatsoglou, M.; Vozalis, M.G.; Diamantaras, K.; Vakali, A.; Sarigiannidis, G.; Chatzisavvas, K.C. Sentiment analysis leveraging emotions and word embeddings. *Exp. Syst. Appl.* **2017**, *69*, 214–224. [CrossRef]
29. Phillips, P.; Barnes, S.; Zigan, K.; Schegg, R. Understanding the impact of online reviews on hotel performance: An empirical analysis. *J. Travel Res.* **2017**, *56*, 235–249. [CrossRef]
30. Salehan, M.; Kim, D.J. Predicting the performance of online consumer reviews: A sentiment mining approach to big data analytics. *Decis. Support Syst.* **2016**, *81*, 30–40. [CrossRef]
31. Jin, J.; Ji, P.; Gu, R. Identifying comparative customer requirements from product online reviews for competitor analysis. *Eng. Appl. Artif. Intell.* **2016**, *49*, 61–73. [CrossRef]
32. Jothi, A.J.; Santiago, M.S.; Arockiam, L. A Methodological Framework to Identify the Students' Opinion using Aspect based Sentiment Analysis. *Int. J. Eng. Res.* **2016**, *5*. [CrossRef]
33. Manning, C.; Surdeanu, M.; Bauer, J.; Finkel, J.; Bethard, S.; McClosky, D. The Stanford CoreNLP natural language processing toolkit. In Proceedings of the 52nd Annual Meeting of the Association for Computational Linguistics: System Demonstrations, Baltimore, MD, USA, 23–24 June 2014; pp. 55–60.
34. Barbosa, L.; Feng, J. Robust sentiment detection on twitter from biased and noisy data. In Proceedings of the 23rd International Conference on Computational Linguistics, Beijing, China, 36–44 August 2010.
35. Baccianella, S.; Esuli, A.; Sebastiani, F. Sentiwordnet 3.0: An enhanced lexical resource for sentiment analysis and opinion mining. *Lrec* **2010**, *10*, 2200–2204.
36. Zadeh, L.A. Fuzzy sets. *Inf. Control* **1965**, *8*, 338–353. [CrossRef]
37. Zadeh, L.A.; Fu, K.S.; Tanaka, K. Fuzzy sets and their applications to cognitive and decision processes. In Proceedings of the US-Japan Seminar on Fuzzy Sets and Their Applications, University of California, Berkeley, CA, USA, 1–4 July 1974.
38. Herrera-Viedma, E. Fuzzy sets and fuzzy logic in multi-criteria decision making. The 50th anniversary of Prof. Lotfi Zadeh's theory: Introduction. *Technol. Econ. Dev. Econ.* **2015**, *21*, 677–683. [CrossRef]
39. Guzman, E.; Maalej, W. How do users like this feature? A fine grained sentiment analysis of app reviews. In Proceedings of the 2014 IEEE 22nd International on Requirements Engineering Conference (RE), Karlskrona, Sweden, 25–29 August 2014; pp. 153–162.
40. Thet, T.T.; Na, J.C.; Khoo, C.S. Aspect-based sentiment analysis of movie reviews on discussion boards. *J. Inf. Sci.* **2010**, *36*, 823–848. [CrossRef]

symmetry

MDPI

Article

Fuzzy Attribute Expansion Method for Multiple Attribute Decision-Making with Partial Attribute Values and Weights Unknown and Its Applications

Jinbao Zhuo, Weifeng Shi * and Ying Lan

Logistics Engineering College, Shanghai Maritime University, Shanghai 201306, China; 201640211007@stushmtu.edu.cn (J.Z.); ylan@shmtu.edu.cn (Y.L.)
* Correspondence: wfshi@shmtu.edu.cn; Tel.: +86-136-1192-3783

Received: 22 November 2018; Accepted: 30 November 2018; Published: 4 December 2018

Abstract: In the real world, there commonly exists types of multiple attribute decision-making (MADM) problems with partial attribute values and weights totally unknown. Symmetry among some attribute information that is already known and unknown, and symmetry between the pure attribute set and fuzzy attribute membership set can be a considerable way to solve this type of MADM problem. In this paper, a fuzzy attribute expansion method is proposed to solve this type of problem based on two key techniques: the spline interpolation technique and the attribute weight reconfiguration technique, which are respectively used for the determination of attribute values and the reconfiguration of attribute weights. The spline interpolation technique to expand attribute values can enhance the performance of some regression methods and clustering methods by the comparisons between the results of these methods dealing with practical cases with and without the application of the technique, which further illustrates the effectiveness of this technique. For MADM problems with partial attribute values and weights totally unknown, compared with traditional fuzzy comprehensive evaluation (FCE), FCE with the application of fuzzy attribute expansion method can obtain results more similar with the ones when all attribute values and weights are known, which is proved by the practical power quality evaluation example.

Keywords: fuzzy set; fuzzy attribute expansion; MADM

---

1. Introduction

Since fuzzy set was proposed by Zadeh [1] in 1965, fuzzy theory is used to quantitatively depict the fuzziness of processes or attributes of things, especially for multiple attribute decision-making (MADM) in some real-life problems.

The applications of fuzzy theory and related techniques in the MADM area can be basically classified into two categories: (1) Applications based on modifications of fuzzy set theory and its extension theories. Fuzzy comprehensive evaluation (FCE) is one of the most classic applications of fuzzy set theory to MADM. One-level FCE can effectively deal with an evaluation problem with a small number of evaluation index parameters (also called attributes). Zhao [2] proposed an electrocardiogram signal quality evaluation method based on one-level FCE and simple heuristic fusion. Yang [3] proposed a method to evaluate exposure, sensitivity, and adaptive capacity based on one-level FCE for better flood vulnerability assessment. Multi-level FCE is preferred as the number of attributes is large. To map porphyry-copper prospectivity in the Gangdese district, Tibet, western China, Zuo [4] established a two-level binary geoscience FCE including favorable rocks, intrusive rocks, faults, and geochemical anomalies. Besides, FCE also has been successfully applied to other MADM problems, e.g., image analysis [5,6], risk assessment [7,8], energy management [9,10], personnel selection [11], etc. As a successful extension of fuzzy set, intuitionistic fuzzy set (IFS) was initiated by Atanassov [12]. IFS uses membership degree, non-membership degree, and hesitancy degree to deal with fuzziness and uncertainty information, which is very useful for the resolution of MADM problems with incomplete attribute weights and uncertain attribute information [13–17]. Xu [13] proposed the models for MADM with intuitionistic fuzzy information based on IFS theory. Wan [14] proposed a new risk attitudinal method for IFS and applied it to the MADM of the teacher selection problem. Furthermore, with the initiation of interval-valued intuitionistic fuzzy set (IVIFS) by Atanassov [18], MADM based on IVIFS becomes a hot topic for researchers [19–27]. The applications of type-2 fuzzy sets [28–30], hesitant fuzzy sets [31,32], and dual hesitant fuzzy linguistic term sets [33] were also reported recently. From these applications, we can find that the extension theories provided researchers with more and more profound theoretical models of fuzzy theory to depict and solve the complicated real-life MADM problems. (2) Applications based on the combination between fuzzy theory and other MADM methods. There are about 20 MADM methods in the literature [34]. The analytic hierarchy process (AHP) [35,36] and technique for order of preference by similarity to ideal solution (TOPSIS) [37–39] are two of the most popular methods combined with fuzzy theory in the area of supplier evaluation [40]. These applications are based on the methodology that one MADM method can be modified by the combination of the other MADM methods. It could be useful in most cases, but the disadvantage is that an increase in computational complexity is also obvious, which is rarely discussed by researchers.

The applications mentioned above focus on the representation and calculation of fuzziness and uncertainty of attributes in MADM problems, the attribute information of which is completely given. As the deepening of understanding of MADM problems grows, researchers began to study MADM problems with incomplete attribute information in recent years. Here, we call such property of attribute information as the incompleteness of attribute, which has two forms, i.e., incomplete weights or values of partial attributes. For the type of problems with incomplete attribute weights information, Park [41] provided mathematical tools for interactive MADM from the perspective of pairwise dominance. Xu [42] determined the attribute weights by the optimization model based on the maximizing deviation method. Wei [43] proposed a gray relationship analysis method to calculate the weights for IFS. Bao [44] proposed an intuitionistic fuzzy decision method based on prospect theory and the evidential reasoning approach. For the type of problems with incomplete attribute values and weights, Eum [45] established dominance and potential optimality to evaluate whether the alternative outperforms for a fixed feasible region denoted by the constraints. There also exists the third type of MADM problems with some partial attribute values and weights totally unknown, i.e., no constraints of incomplete attributes, which is the extreme type of the above two types of problem. Actually, this type of problem is common in the real world, e.g., decision maker could not provide some attribute values and weights because of discreet principles or cognitive impairment, or some attribute values could not be obtained because of the failure of the data acquisition system, or there exists some unclear or undefined attributes of new things or processes. Thus, it is necessary to study and find out a solution to these type of MADM problems.

In the real world, attributes are partially correlated and continuous in the attribute space. Hence, if the fuzzy mapping is linear, the fuzzy values and weights of these attributes are also continuous, which means that for the correlated attributes of things or processes, fuzzy values and weights of some unknown attributes can be approximately estimated by the ones of some known attributes. Thus, based on the above new cognition of the mapping relationship between the objective world and fuzzy attribute set, we propose a fuzzy attribute expansion method, which consisted of the spline interpolation technique and attribute weight reconfiguration technique, to deal with the MADM problems with some partial attribute values and weights totally unknown.

The rest of the paper is organized as follows: Section 2 provides some basic definitions about the fuzzy set and some related sets. In Section 3, the formulaic expression of the third type of MADM problem is given based on the definitions in Section 2. The geometric analysis of the pure attribute set (PAS), the measurable attribute set (MAS), and the fuzzy attribute membership set (FAMS) of the problem is conducted in Section 4, which is the theoretic basis of the fuzzy attribute expansion method proposed in Section 5. Applications in regression, clustering, and power quality evaluation are presented in Section 6. In Section 7, the conclusions of the paper are given.

2. Basic Definitions

In this section, some basic concepts related to the method proposed in this paper are introduced and defined.

Definition 1. ([1]) *A fuzzy set A in the universe of discourse* $X = \{x_1, x_2, \cdots, x_n\}$ is defined as follows:

$$A = \{\langle x, \mu_A(x)\rangle | x \in X\}, \tag{1}$$

where $\mu_A(x) : X \to [0,1]$ *is the membership function.*

Definition 2. *Assume that x has m kind of striking attributes (denoted as* $\boldsymbol{a}_j^x (j = 1, \cdots, m)$ *), if a set* U_p^x *of x* satisfying the following conditions:

(a) $\forall \boldsymbol{a}_j^x (j = 1, \cdots, m) \in U_p^x$

(b) $\boldsymbol{a}_i^x \propto \boldsymbol{a}_j^x (i \neq j)$,

(c) $\boldsymbol{a}_i^x \cong h_i(\boldsymbol{a}_1^x, \cdots, \boldsymbol{a}_{i-1}^x, \boldsymbol{a}_{i+1}^x \cdots, \boldsymbol{a}_m^x)$,

where h_i *is relationship function, then* U_p^x *is defined as the pure attribute set (PAS) of x.*

If t $(t \ll m)$ *kind of attributes of x are only known, then define these t kind of attributes as knowable fuzzy attributes (KFA) and the other* $(m - t)$ *kind of attributes as unknowable fuzzy attributes (UFA).*

Definition 3. *If a set* U_m^x *of x satisfying the following conditions:*

(a) $\forall \boldsymbol{a}_j^x (j = 1, \cdots, m) \in U_p^x$,

(b) $\forall a_j^x =_j \left(\boldsymbol{a}_j^x\right) (j = 1, \cdots, m) \in U_m^x$, $_j\left(\boldsymbol{a}_j^x\right) : \boldsymbol{a}_j^x \to R^{1\times 1}$,

where $_j$ *is an unknown function to measure the attribute j in the real world,* a_j^x *is the projection of attribute vector* $\boldsymbol{a}_j^x$ *in some kind of space and expressed in numerical form, then* U_m^x *is defined as the measurable attribute set (MAS) of x.*

Definition 4. *If a set* U_f^x *of x satisfying the following conditions:*

(a) $\forall \boldsymbol{a}_j^x (j = 1, \cdots, m) \in U_p^x$,

(b) $\forall a_j^x =_j \left(\boldsymbol{a}_j^x\right) (j = 1, \cdots, m) \in U_m^x$,

(c) $\forall \widetilde{a}_j^x = v_j\left(a_j^x\right) (j = 1, \cdots, m) \in U_f^x$, $v_j\left(a_j^x\right) : a_j^x \to [0,1]$,

where $v_j\left(a_j^x\right) : a_j^x \to [0,1]$ *is the attribute membership function,* $\widetilde{a}_j^x$ *is fuzzy membership grade for attribute j of x, then* U_f^x *is defined as the fuzzy attribute membership set (FAMS) of x.*

For example, if $\boldsymbol{a}_j^x$ *is the volume attribute of a box,* a_j^x *could be the length of the box and the length unit is meter,* v_j *is the length attribute membership of satisfaction for customer, then* $\widetilde{a}_j^x$ *is the fuzzy membership grade of length of the box.*

3. Problem

In this section, some further definitions are defined based on the following theorems. Then, the main problem this paper focuses on is raised and expressed by these definitions.

Theorem 1. *PAS and FAMS of x are equivalent:* PAS $\sim$ FAMS.

Proof. $\because\ \forall \widetilde{a}_j^x = v_j\left(a_j^x\right)(j = 1, \cdots, m)$, $\forall a_j^x =_j \left(\boldsymbol{a}_j^x\right)(j = 1, \cdots, m)$. $\therefore\ \forall \widetilde{a}_j^x = v_j\left(o_j\left(\boldsymbol{a}_j^x\right)\right)(j = 1, \cdots, m)$, which means there exists one-to-one correspondence (bijection) from PAS to FAMS. Besides, $|\text{PAS}| = |\text{FAMS}| = m$. Thus, PAS $\sim$ FAMS is proved. □

Theorem 2. *PAS and FAMS of x are countable sets.*

Proof. Choose random two attributes from FAMS of x and denote them as $\widetilde{a}_1^x$ and $\widetilde{a}_m^x$, the other attributes are ranked by the similarity with $\widetilde{a}_1^x$ and $\widetilde{a}_m^x$ from large to small and inserted between $\widetilde{a}_1^x$ and $\widetilde{a}_m^x$, which comes out as the sequence

$$\widetilde{S}^x = \langle \widetilde{a}_1^x, \widetilde{a}_2^x, \cdots, \widetilde{a}_m^x \rangle.$$

By this method to rank PAS of x, it also comes out as the sequence

$$\boldsymbol{S}^x = \langle \boldsymbol{a}_1^x, \boldsymbol{a}_2^x, \cdots, \boldsymbol{a}_m^x \rangle.$$

Thus, PAS and FAMS of x are both countable sets. □

Definition 5. *Based on Theorem 2, the attribute vector sequence of x is defined as follows:*

$$\boldsymbol{S}^x = \langle \boldsymbol{a}_1^x, \boldsymbol{a}_2^x, \cdots, \boldsymbol{a}_m^x \rangle. \tag{2}$$

The measurable attribute sequence is defined as:

$$\begin{aligned} S^x &= \left\langle a_1^x, a_2^x, \cdots, a_m^x \right\rangle \\ &= \underbrace{\left\langle a_{i-1}^x, \cdots, a_{j-1}^x, \cdots, a_{k-1}^x \right\rangle}_{S_{\text{KFA}}^x} + \underbrace{\left\langle a_i^x, \cdots, a_j^x, \cdots, a_k^x \right\rangle}_{S_{\text{UFA}}^x}. \end{aligned} \tag{3}$$

The fuzzy measurable attribute sequence of x has the following form:

$$\begin{aligned} \widetilde{S}^x &= \left\langle \widetilde{a}_1^x, \widetilde{a}_2^x, \cdots, \widetilde{a}_m^x \right\rangle \\ &= \underbrace{\left\langle \widetilde{a}_{i-1}^x, \cdots, \widetilde{a}_{j-1}^x, \cdots, \widetilde{a}_{k-1}^x \right\rangle}_{\widetilde{S}_{\text{KFA}}^x} + \underbrace{\left\langle \widetilde{a}_i^x, \cdots, \widetilde{a}_j^x, \cdots, \widetilde{a}_k^x \right\rangle}_{\widetilde{S}_{\text{UFA}}^x}. \end{aligned} \tag{4}$$

Definition 6. *It is affirmative that KFA and UFA are correlated:*

$$\widetilde{S}_{\text{UFA}}^x \propto \widetilde{S}_{\text{KFA}}^x,$$

consider the function relations among these attributes is undefined:

$$\widetilde{a}_{j-1}^x = f\left(\widetilde{a}_j^x\right),$$

where function f has no exact analytic expression, so UFA can only be depicted by approximate estimation:

$$\hat{\widetilde{S}}_{\text{UFA}}^x \cong g\left(\widetilde{S}_{\text{KFA}}^x\right),$$

where function g is the function for approximately estimating $\widetilde{S}_{\text{UFA}}^x$ with $\widetilde{S}_{\text{KFA}}^x$ as the independent variable.

Based on the above definitions, the main problem this paper focused on can be described as follows:

Problem description: Given a set of fuzzy membership grades of KFA:

$$\left\{\left\langle i, \widetilde{S}_{\mathrm{KFA}}^{x_i}\right\rangle | x_i \in X, i = 1, \cdots, n\right\},$$

evaluate x_i under the following conditions:

(a) The length of $\widetilde{S}_{\mathrm{KFA}}^{x_i}$ is t, $t \ll m$;
(b) The sequence of attribute weights is $\lambda = \langle \lambda_1, \lambda_2, \cdots, \lambda_t \rangle$;
(c) Evaluation value of x_i falls into the interval $[0, 1]$, the greater the evaluation value is, the higher the evaluation is.
(d) The evaluation values are as similar to those obtained under the condition that some UFAs are given as possible.

Research about the problem with all conditions as mentioned above is rarely conducted. In fact, the condition (d) is an important index to evaluation. To satisfy condition (d), we suggest a fuzzy attribute expansion method to evaluate x_i: before the final evaluation, use KFAs to approximately estimate UFAs.

4. Geometric Analysis of PAS, MAS, and FAMS

In this section, the geometric analysis of PAS, MAS, and FAMS of the problem is conducted. Firstly, the generalized geometric structures (GGS) of PAS, MAS, and FAMS are modeled and represented in the form of a diagrammatic sketch. Secondly, the geometric relationship between GGS of PAS and GGS of FAMS is analyzed. Thirdly, GGS of x in FAMS can be approximately estimated by the $\widetilde{S}_{\mathrm{KFA}}^{x}$ of x based on interpolation technique is discussed.

These three kinds of sets depict attributes, the relationships between attributes, and attribute membership degrees from different spatial cognition. PAS is an abstract set to characterize nonlinear relationships between attributes, especially the vague attributes. Each vector represents different attributes, all of the vectors represent x. MAS and FAMS are the numerical mapping of PAS. In fact, they are the projection of the PAS in the distance space. Both of them fail to show the attribute correlation because of the loss of vector directivity.

To understand these three kinds of sets intuitively, the generalized geometric structures (GGS) in different sets are modeled based on Theorem 2 as follows:

GGS in PAS: Use different dotted lines with direction to represent different attributes. These dotted lines are straight lines or curves, which depends on the linear relationship between each of attributes. To reduce complexity, choose one attribute as the unified reference attribute, then compare other attributes with it; if the relationship is linear (or nonlinear), the dotted lines of these attributes are straight lines (or curves). PAS of x is depicted by the combination of the m vectors whose ends are located on the curves, as Figure 1 illustrates. The surface consisting of all vector ends is defined as the GGS of x in PAS. The GGS in PAS is a smooth and continuous surface.

GGS in MAS: Use different line segments to represent the projections of different attributes in distance space (such as Euclidean space) which are arranged in sequence, such as the integer sequence. S_{KFA}^{x} and S_{UFA}^{x} are respectively indicated with solid line segments and dotted line segments. MAS of x is depicted by the combination of m line segments, as Figure 2 illustrates. The dotted curve consisting of m line segment ends is defined as the GGS of x in MAS.

GGS in FAMS: Use different line segments to represent the projections of different measurable attributes in $[0, 1]$, which are arranged in the same sequence with MAS. $\widetilde{S}_{\mathrm{KFA}}^{x}$ and $\widetilde{S}_{\mathrm{UFA}}^{x}$ are respectively indicated with solid line segments and dotted line segments. If the fuzzy membership function is linear, FAMS of x is depicted by the combination of m line segments, as Figure 3 illustrates. The dotted curve consisting of m line segment ends is defined as the GGS of x in FAMS. The GGS in FAMS is a smooth and continuous curve.

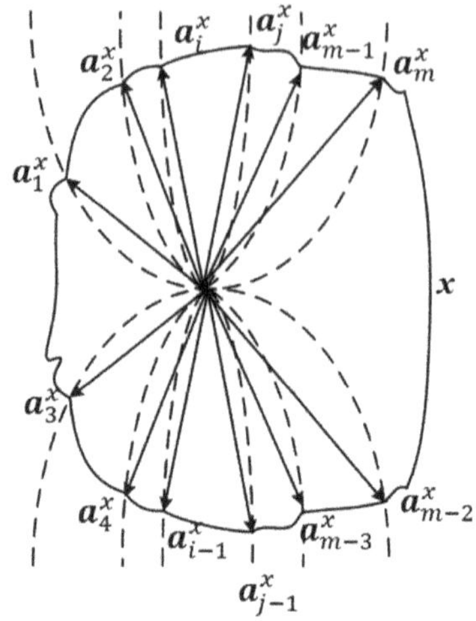

Figure 1. Illustration of generalized geometric structures (GGS) in a pure attribute set (PAS).

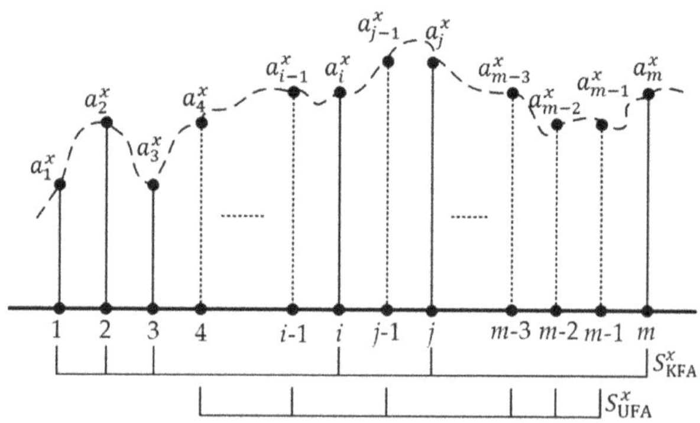

Figure 2. Illustration of GGS in a measurable attribute set (MAS).

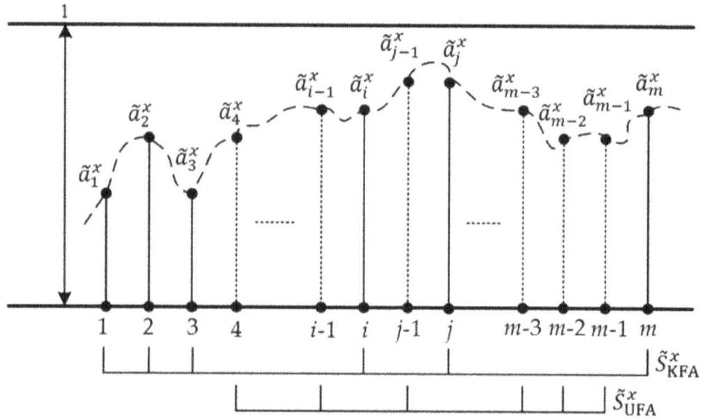

Figure 3. Illustration of GGS in a fuzzy attribute membership set (FAMS) with a linear fuzzy membership function.

From the above set models, it is easy to intuitively understand that MAS and FAMS are the low-dimensional embedding of PAS. The value ranges of set elements are different for MAS and FAMS. FAMS is linear mapping of MAS, which is determined by the fuzzy membership function of each attribute. In FAMS, some fuzzy attribute membership grades may approach 1 while the value of it in MAS is quite small or even negative. Furthermore, after defining and studying the GGS of x in these three sets, we find that GGS of x in PAS is a surface, while in MAS it is a curve and in FAMS it is a curve. The curves intercept the surface.

Thus, if we want to depict the GGS of x in PAS (the surface) more precisely to solve the problem raised at the beginning of the paper, the GGS of FAMS (the curve) should be calculated more precisely by the $\widetilde{S}^x_{\text{KFA}}$ of x, which is a feasible and considerable way. Interpolation technique is a commonly used and effective technique to approximate estimation. From all of the interpolation techniques, the spline interpolation has the best smoothing ability which is the most important for the estimation for the curve. Since the GGSs in PAS and FAMS are smooth and continuous, the spline interpolation technique can be used as the interpolation technique to approximately estimate the GGS in FAMS.

5. The Fuzzy Attribute Expansion Method

In this section, the new fuzzy attribute expansion method to solve the problem is proposed. The method is basically consisted of two sub methods: (1) the method to approximately estimate UFA and (2) the method to generate the final evaluation. The detailed descriptions of these methods are given as follows.

5.1. The Technique to Approximate Estimate UFAs Based on Interpolation

The basic idea of the method is: UFAs can be approximately estimated by inputting specified attribute sequence numbers into the interpolation function, which is the result of applying the curve interpolation technique to KFAs. Notably, the UFAs and KFAs here are correlated. Otherwise, the technique will not work. Basically, this technique can be divided into five steps.

Step 1: Rearrange $\widetilde{S}^x_{\text{KFA}}$. For new samples, rearrangement of $\widetilde{S}^x_{\text{KFA}}$ based on Theorem 2 is needed.

Step 2: Determine the attribute sequence number. There are no special restrictions on the selection of sequence number form. Normally, we can simply number the attribute sequence with the form of the positive integer sequence:

$$N_{\text{KFA}} = \langle 1, 2, \cdots, t \rangle. \tag{5}$$

Step 3: Generate the interpolation function of KFAs. Choose a suitable interpolation. Taking N_{KFA} as the independent variable $\boldsymbol{E} = \begin{bmatrix} e_1 & e_2 & \cdots & e_t \end{bmatrix}^{\text{T}}$ and $\widetilde{S}^x_{\text{KFA}}$ as the dependent variable $\boldsymbol{Y} = \begin{bmatrix} y_1 & y_2 & \cdots & y_t \end{bmatrix}^{\text{T}}$, minimizes the objective:

$$J = p\sum_i (y_i - s(e_i))^2 + (1-p)\int \left(\frac{d^2 s}{de^2}\right)^2 de, \tag{6}$$

where s is the smoothing spline, p is smoothing parameter which is defined between 0 and 1 ($p = 0.95$ in this paper). When the optimum solution is found, the finalist s is the interpolation function.

Step 4: Generate UFA number sequence.

$$N_{\text{UFA}} = \underbrace{\langle 1+\alpha_1, 1+\alpha_2, \cdots, 1+\alpha_i, \cdots, 1+\alpha_r \rangle}_{r}, \tag{7}$$

whose length is r with $1+\alpha_1$ as the start and $1+\alpha_r$ as the end, $1+\alpha_1 \geq 1, 1+\alpha_r \leq t, r \geq m-t$. For different problems, r is different and determined by the heuristic knowledge, which can be shown in the examples in Section 6. α_r is determined by two adjacent attributes of the attribute sequence. The sequence is distributed evenly and linear, step size α_i should be fixed as:

$$\alpha_i = i \times \left[\frac{t-1}{r} \right], \tag{8}$$

and the UFA number sequence becomes:

$$N_{\text{UFA}} = \langle 1 + \left[\frac{t-1}{r} \right], \cdots, 1 + r \times \left[\frac{t-1}{r} \right] \rangle. \tag{9}$$

Step 5: Approximately estimate UFAs finally. Input N_{UFA} into s, so the UFA membership grades sequence can be estimated as:

$$\hat{\tilde{S}}^x_{\text{UFA}} = s(N_{\text{UFA}}), \tag{10}$$

and the total estimated membership grades sequence is:

$$\begin{aligned} \hat{\tilde{S}}^x &= \left\langle \tilde{a}^x_1, \hat{\tilde{a}}^x_{1+\alpha_1}, \cdots, \hat{\tilde{a}}^x_{1+\alpha_j}, \cdots, \tilde{a}^x_i, \cdots, \hat{\tilde{a}}^x_{1+\alpha_r}, \tilde{a}^x_t \right\rangle \\ &= \underbrace{\langle \tilde{a}^x_1, \cdots, \tilde{a}^x_i, \cdots, \tilde{a}^x_t \rangle}_{\tilde{S}^x_{\text{KFA}}} + \underbrace{\left\langle \hat{\tilde{a}}^x_{1+\alpha_1}, \cdots, \hat{\tilde{a}}^x_{1+\alpha_j}, \cdots, \hat{\tilde{a}}^x_{1+\alpha_r} \right\rangle}_{\hat{\tilde{S}}^x_{\text{UFA}}} . \end{aligned} \tag{11}$$

For this technique, membership grades of UFA and KFA are considered as vertical coordinate values of the attribute vector sequence curve. The interpolation technique is used to depict the curve. Once the function is interpolated, the estimation result could be calculated after inputting the customized horizontal ordinate values. The result $\hat{\tilde{S}}^x_{\text{UFA}}$ is an estimation of $\tilde{S}^x_{\text{UFA}}$ to some degree. Its accuracy depends on the quality of the interpolation technique, which means the interpolation function s is the key to the method. For spline interpolation, s is the optimal result of all. Meanwhile, the attribute number sequence is another key. The cognition of the KFA determines the generation of the sequence. For instance, if some attributes of KFA are more important for x, which is usually judged artificially, then steps between each of them should be smaller than others.

5.2. The Technique to Generate the Final Evaluation Based on Attribute Weight Reconfiguration

Since the UFA membership grades sequence has been approximately estimated, we propose a new technique to generate the final evaluation based on attribute weight reconfiguration.

Step 1: Regenerate a new sequence of attribute weights. Let every element of λ be divided into certain parts, the number of which is equal to how many estimated UFAs locate between two KFAs, then the new sequence of attribute weights can be written as:

$$\hat{\lambda} = \left\langle \hat{\lambda}^1_1, \cdots, \hat{\lambda}^{i-1}_{i-1}, \cdots, \underline{\hat{\lambda}^{i-1\_i}_{1+\alpha_j}}, \cdots, \hat{\lambda}^i_i, \cdots, \hat{\lambda}^t_t \right\rangle, \tag{12}$$

$$\hat{\lambda}^i_i = \frac{\lambda_i}{d_i}, \tag{13}$$

$$\hat{\lambda}^{i-1\_i}_{1+\alpha_j} = \frac{\lambda_{i-1}}{d_{i-1}} + \frac{\lambda_i}{d_i}, \tag{14}$$

where if $i \in (1, t)$, then d_i is the size of N_{UFA} within the $[i-1, i+1]$. If $i = 1$, then d_1 is the size of N_{UFA} within the $[1, 2]$. If $i = t$, then d_t is the size of N_{UFA} within the $[t-1, t]$.

Step 2: Calculate the multiplication of corresponding elements from $\hat{\lambda}$ and $\hat{\tilde{S}}^x$:

$$\left\langle \hat{\lambda}^1_1 \times \tilde{a}^x_1, \hat{\lambda}^{1\_2}_{1+\alpha_1} \times \hat{\tilde{a}}^x_{1+\alpha_1}, \cdots, \hat{\lambda}^{i-1\_i}_{1+\alpha_j} \times \hat{\tilde{a}}^x_{1+\alpha_j}, \cdots, \hat{\lambda}^i_i \times \tilde{a}^x_i, \cdots, \hat{\lambda}^{t-1\_t}_{1+\alpha_r} \times \hat{\tilde{a}}^x_{1+\alpha_r}, \hat{\lambda}^t_t \times \tilde{a}^x_t \right\rangle. \tag{15}$$

Step 3: Sum and obtain final evaluation of x:

$$\hat{E} = \hat{\lambda}^1_1 \times \tilde{a}^x_1 + \hat{\lambda}^{1\_2}_{1+\alpha_1} \times \hat{\tilde{a}}^x_{1+\alpha_1} + \cdots + \hat{\lambda}^{i-1\_i}_{1+\alpha_j} \times \hat{\tilde{a}}^x_{1+\alpha_j} + \cdots + \hat{\lambda}^i_i \times \tilde{a}^x_i + \cdots + \hat{\lambda}^{t-1\_t}_{1+\alpha_r} \times \hat{\tilde{a}}^x_{1+\alpha_r} + \hat{\lambda}^t_t \times \tilde{a}^x_t. \tag{16}$$

6. Applications

In this section, the proposed methods are respectively used to enhance the performance of some research methods related to fuzzy attributes membership: regression, clustering, and fuzzy evaluation. Four samples are presented with the detailed description and comparisons of the results from different methods.

6.1. Applications for Regression

For some regression problems, the size of KFA is small for some reason, which definitely affects the regression performance. These KFAs are partially correlated, which can be figured out by a known mechanism or experience for researchers. To deal with this type of problem, the technique to approximately estimate UFA in Section 4 is applied to enhance the performance of regression. We firstly use the estimation method to approximately estimate the UFA of samples and combine the estimated UFAs and KFA as the attributes of predictor before regressing. The support vector machine regression model (SVMR) and Gaussian kernel regression model using random feature expansion (GKR) are chosen as the regression models. For GKR, there are two kinds of learners: SVM learner and linear regression via ordinary least squares.

Example 1. *Let* $X = \{x_1, x_2, \cdots, x_n\}$, $n = 392$ *be a set of car samples, the measurable attribute sequence of node* x_i *is defined as:*

$$S^{x_i} = \langle a_1^{x_i}(weight), a_2^{x_i}(cylinders), a_3^{x_i}(horsepower), a_4^{x_i}(model\ year), a_5^{x_i}(MPG)\rangle. \tag{17}$$

Let fuzzy membership grade for attribute j of x be calculated by:

$$\widetilde{a}_j^{x_i} = v_j\left(a_j^{x_i}\right), \tag{18}$$

$$v_j\left(a_j^{x_i}\right) = \frac{a_j^{x_i} - \min(a^x)}{\max(a^x) - \min(a^x)}, \tag{19}$$

where $v_j\left(a_j^x\right) : a_j^x \to [0,1]$, $j = 1,2,3,4$, *is the attribute membership function.*
The fuzzy measurable attribute sequence of x_i *is:*

$$\widetilde{S}^{x_i} = \langle \widetilde{a}_1^{x_i}, \widetilde{a}_2^{x_i}, \cdots, \widetilde{a}_9^{x_i}\rangle, \tag{20}$$

Choose some attributes from $\widetilde{a}_1^x$, $\widetilde{a}_2^x$, $\widetilde{a}_3^x$, $\widetilde{a}_4^x$ *as predictor variables,* $\widetilde{a}_5^x$ *as the response variable, calculate the regression model between predictor variables and the response variable by SVMR and GKR, respectively, with and without the application of the proposed method to approximately estimate UFA in Section 4. To compare these regression results, calculate loss indexes: Huber loss (HL), mean squared error (MSE), and epsilon-insensitive function (EI) of results. The smaller the values of these three indexes are, the better performances of the regressions are. EI index is appropriate for SVM learners only.*

If $\widetilde{a}_1^x$, $\widetilde{a}_2^x$, $\widetilde{a}_3^x$ are predictor variables, $\widetilde{a}_5^x$ is the response variable, then $\widetilde{a}_1^x$, $\widetilde{a}_2^x$, $\widetilde{a}_3^x$ are KFAs, so the KFA attribute sequence number is $N_{\mathrm{KFA}} = \langle 1,2,3\rangle$ in Step 2 of the technique to approximately estimate UFAs. Calculate the regressions respectively by SVMR, GKR-SVM learner (denoted as GKR-1), GKR-linear regression (denoted as GKR-2) with and without the application of the proposed method. Take 20% of samples out as test samples. The comparisons of loss indexes of results are shown in Table 1. Notably, for the SVMR method, the more estimated UFAs to learn, the better performance of learning is. Meanwhile, the GKR method can use the small size of UFAs to expand features. If the size of UFAs is too large, it can result in the over learning for the GKR methods. Thus, the size of UFAs for the SVMR method should be much larger than the ones for GKR methods. After lots of tests, the best N_{UFA} in Step 4 for different methods is shown in Table 1.

Table 1. The comparisons of loss indexes of results if $\widetilde{a}_1^x$, $\widetilde{a}_2^x$, $\widetilde{a}_3^x$ are knowable fuzzy attributes (KFAs).

Regression Model	A [1]	HL (Train/Test)	MSE (Train/Test)	EI (Train/Test)	N_{UFA} [2]
SVMR	0	2.6426/2.5052	18.4283/18.3963	2.3938/2.3395	/
	1	2.6427/2.4991	18.3377/18.3808	2.3949/2.3379	<1.08:0.08:2.92>
GKR-1	0	2.1724/2.2637	15.5750/17.2569	1.9958/2.0781	/
	1	2.1362/2.1400	15.4048/16.1185	1.9793/1.9573	<1.08:0.8:2.68>
GKR-2	0	2.0406/2.2523	12.8566/17.0757	/	/
	1	1.9960/2.0937	12.5828/15.6837	/	<1.08:0.8:2.68>

[1] A = 1: with the application of proposed method, A = 0: without the application of proposed method. [2] $\langle \alpha : \beta : \gamma \rangle$: the sequence (the starting point is α, step is β, the endpoint is γ). HL: Huber loss; MSE: mean squared error; EI: epsilon-insensitive function; SVMR: support vector machine regression; GKR: Gaussian kernel regression.

If $\widetilde{a}_1^x$, $\widetilde{a}_2^x$, $\widetilde{a}_3^x$, $\widetilde{a}_4^x$ are predictor variables, $\widetilde{a}_5^x$ is the response variable, then $\widetilde{a}_1^x$, $\widetilde{a}_2^x$, $\widetilde{a}_3^x$ $\widetilde{a}_4^x$ are KFAs, so the KFA attribute sequence number is $N_{KFA} = \langle 1, 2, 3, 4 \rangle$ in Step 2 of the technique to approximately estimate UFAs. Calculate the regressions, respectively, by SVMR, GKR-1, and GKR-2 with and without the application of the proposed method. Take 20% of samples out as test samples. The UFA number sequences of each regression and the comparisons of loss indexes of results are shown in Table 2.

Table 2. The comparisons of loss indexes of results if $\widetilde{a}_1^x$, $\widetilde{a}_2^x$, $\widetilde{a}_3^x$,$\widetilde{a}_4^x$ are KFAs.

Regression Model	A [1]	HL (Train/Test)	MSE (Train/Test)	EI (Train/Test)	N_{UFA} [2]
SVMR	0	2.0199/1.8340	12.5788/10.6975	1.8338/1.6461	/
	1	2.0220/1.8295	12.4889/10.5811	1.8323/1.6459	<1.08:0.08:3.96>
GKR-1	0	1.7104/1.7270	11.0732/13.1882	1.5813/1.6434	/
	1	1.5242/1.5924	10.5474/10.5426	1.4230/1.4778	<1.8:1:3.8>
GKR-2	0	1.4146/1.7246	7.1309/11.8654	/	/
	1	1.1562/1.5241	5.9034/9.0240	/	<1.8:1:3.8>

[1] A = 1: with the application of proposed method, A = 0: without the application of proposed method. [2] $\langle \alpha : \beta : \gamma \rangle$: the sequence (the starting point is α, step is β, the endpoint is γ).

From Tables 1 and 2, we can find that the index values of the regression results (both training results and test results of the regression) with the application of the proposed method are all smaller than ones without the application, especially in Table 2. Thus, it is concluded that the application of the proposed method in the regression problem, where the size of KFAs is quite small and known attributes are correlated, is effective.

6.2. Applications for Clustering

For some clustering problems, only some attributes of given samples are given, which definitely lower the accurate rates of clustering results. In this section, the proposed methods are respectively used to enhance the performance of fuzzy c-means clustering (FCM), K-means clustering (K-means), and K-medoids clustering (K-medoids): the estimation method proposed is applied to approximately estimate the UFA of the sample and the estimated UFAs and KFAs are combined as the attributes of samples before clustering. These three clustering methods are partition-based clustering methods. To analyze and compare the clustering results, accurate rate (AR), Rand index (RI), normalized mutual information (NMI) of clustering results are calculated. The larger the values of these indexes are, the better the performances of clustering are.

Example 2. *Let* $X = \{x_1, x_2, \cdots, x_n\}$, $n = 150$ *be a set of iris samples, the measurable attribute sequence of node* x_i *is defined as:*

$$S^{x_i} = \langle a_1^{x_i}(sepal\ length), a_2^{x_i}(sepal\ width), a_3^{x_i}(petal\ length), a_4^{x_i}(petal\ width) \rangle. \tag{21}$$

Let fuzzy membership grade for attribute j of x be calculated by Equations (18) and (19).

The fuzzy measurable attribute sequence of x_i is:

$$\widetilde{S}^{x_i} = \langle \widetilde{a}_1^{x_i}, \widetilde{a}_2^{x_i}, \cdots, \widetilde{a}_4^{x_i} \rangle. \tag{22}$$

For all 150 iris samples, they can be divided into three clusters, each cluster contains 50 samples.

$\widetilde{a}_1^x, \widetilde{a}_2^x, \widetilde{a}_3^x, \widetilde{a}_4^x$ are KFAs, so the KFA attribute sequence number is $N_{KFA} = \langle 1, 2, 3, 4 \rangle$ in Step 2 of the technique to approximately estimate UFAs. For the iris, the four attributes of the sample set are the depiction of its geometry. Thus, the size of UFAs can be large. Calculate the clustering respectively by FCM, K-means, and K-medoids with and without the application of the proposed method. Every method is calculated 100 times. All clustering calculations choose Euclidean distance. The UFA number sequences of each clustering and the comparisons of clustering indexes of results are shown in Table 3.

Table 3. The comparisons of clustering indexes of results.

Clustering Method	A [1]	AR (Worst/Mean/Best)	RI (Worst/Mean/Best)	NMI (Worst/Mean/Best)	N_{UFA}
FCM	0	0.8800/0.8879/0.8933	0.8679/0.8748/0.8797	0.7225/0.7328/0.743	/
	1	0.9200/0.9200/0.9200	0.9055/0.9055/0.9055	0.7855/0.7855/0.785	<1.8:0.5:3.8>
K-means	0	0.5800/0.7822/0.8867	0.7214/0.8210/0.8737	0.5927/0.6830/0.741	/
	1	0.5067/0.8373/0.9533	0.7204/0.8648/0.9417	0.6011/0.7653/0.846	<1.8:0.5:3.8>
K-medoids	0	0.9000/0.9040/0.9067	0.8859/0.8897/0.8923	0.7578/0.7596/0.761	/
	1	0.9333/0.9333/0.9333	0.9195/0.9195/0.9135	0.8038/0.8038/0.804	<1.8:0.5:3.8>

[1] A = 1: with the application of proposed method, A = 0: without the application of proposed method. AR: accurate rate; RI: Rand index; NMI: normalized mutual information; FCM: fuzzy c-means.

From Table 3, we can find that the index values of clustering results (worst, mean, and best values of AR, RI, and NMI) with the application of the proposed method are all larger than ones without the application. It is concluded that the application of the proposed method in the clustering problem, where the size of KFAs is quite small and known attributes are correlated, is effective.

6.3. Applications for Power Quality Evaluation

Let $X = \{x_1(node1), x_2(node2), x_3(node3), x_4(node4), x_5(node5)\}$ be a set of power net nodes, the measurable attribute sequence of node x_i is defined as:

$$\begin{aligned} S^{x_i} = \; & \langle a_1^{x_i}(frequency\ deviation), a_2^{x_i}(voltage\ deviation), a_3^{x_i}(voltage\ sag), \\ & a_4^{x_i}(three\ phase\ imbalance), a_5^{x_i}(voltage\ fluctuation), a_6^{x_i}(voltage\ flicker), \\ & a_7^{x_i}(voltage\ harmonics), a_8^{x_i}(reliability\ index), a_9^{x_i}(service\ index) \rangle \end{aligned} \tag{23}$$

which detailed values are shown in Table 4, the smaller the value of the measurable attribute is, the better the power quality is.

Table 4. Power quality values.

Measurable Attributes	Node1	Node2	Node3	Node4	Node5
a_1^x (Frequency deviation)	0.0922	0.1562	0.1180	0.1787	0.1892
a_2^x (Voltage deviation)	3.2120	6.6800	4.3500	5.3300	4.2200
a_3^x (Voltage sag)	79.6300	15.8900	51.5600	58.5600	48.6300
a_4^x (Three phase imbalance)	0.8300	1.3600	1.3500	1.7400	1.8300
a_5^x (Voltage fluctuation)	1.3300	1.5300	1.9500	1.3700	1.5800
a_6^x (Voltage flicker)	0.4730	0.8470	0.6340	0.8260	0.8280
a_7^x (Voltage harmonics)	1.7200	4.3800	2.6700	3.3600	4.5700
a_8^x (Unreliability index)	0.1670	0.2380	0.2040	0.2600	0.2360
a_9^x (Unserviceable index)	0.1680	0.2870	0.1360	0.3160	0.2170

Let the fuzzy membership grade for attribute j of x be calculated by Equations (18) and (19). The fuzzy measurable attribute sequence of x_i is:

$$\widetilde{S}^{x_i} = \langle \widetilde{a}_1^{x_i}, \widetilde{a}_2^{x_i}, \cdots, \widetilde{a}_9^{x_i} \rangle, \tag{24}$$

the detailed values of $\widetilde{S}^{x_i}$ are shown as follows:

$$\begin{aligned}
\widetilde{S}^{x_1} = & \Big\langle \widetilde{a}_1^{x_1}(0.0000), \ \widetilde{a}_2^{x_1}(0.0000), \ \widetilde{a}_3^{x_1}(1.0000), \\
& \widetilde{a}_4^{x_1}(0.0000), \ \widetilde{a}_5^{x_1}(0.0000), \ \widetilde{a}_6^{x_1}(0.0000), \\
& \widetilde{a}_7^{x_1}(0.0000), \ \widetilde{a}_8^{x_1}(0.0000), \ \widetilde{a}_9^{x_1}(0.1778) \Big\rangle, \\
\widetilde{S}^{x_2} = & \Big\langle \widetilde{a}_1^{x_2}(0.6598), \ \widetilde{a}_2^{x_2}(1.0000), \ \widetilde{a}_3^{x_2}(0.0000), \\
& \widetilde{a}_4^{x_2}(0.5300), \ \widetilde{a}_5^{x_2}(0.3226), \ \widetilde{a}_6^{x_2}(1.0000), \\
& \widetilde{a}_7^{x_2}(0.9333), \ \widetilde{a}_8^{x_2}(0.7634), \ \widetilde{a}_9^{x_2}(0.8389) \Big\rangle, \\
\widetilde{S}^{x_3} = & \Big\langle \widetilde{a}_1^{x_3}(0.2660), \ \widetilde{a}_2^{x_3}(0.3281), \ \widetilde{a}_3^{x_3}(0.5570), \\
& \widetilde{a}_4^{x_3}(0.5200), \ \widetilde{a}_5^{x_3}(1.0000), \ \widetilde{a}_6^{x_3}(0.4304) \\
& \widetilde{a}_7^{x_3}(0.3333), \ \widetilde{a}_8^{x_3}(0.3978), \ \widetilde{a}_9^{x_3}(0.0000), \Big\rangle, \\
\widetilde{S}^{x_4} = & \Big\langle \widetilde{a}_1^{x_4}(0.8918), \ \widetilde{a}_2^{x_4}(0.6107), \ \widetilde{a}_3^{x_4}(0.6694), \\
& \widetilde{a}_4^{x_4}(0.9100), \ \widetilde{a}_5^{x_4}(0.0645), \ \widetilde{a}_6^{x_4}(0.9439), \\
& \widetilde{a}_7^{x_4}(0.5754), \ \widetilde{a}_8^{x_4}(1.0000), \ \widetilde{a}_9^{x_4}(1.0000) \Big\rangle, \\
\widetilde{S}^{x_5} = & \Big\langle \widetilde{a}_1^{x_5}(1.0000), \ \widetilde{a}_2^{x_5}(0.2907), \ \widetilde{a}_3^{x_5}(0.5136), \\
& \widetilde{a}_4^{x_5}(1.0000), \ \widetilde{a}_5^{x_5}(0.4032), \ \widetilde{a}_6^{x_5}(0.9492), \\
& \widetilde{a}_7^{x_5}(1.0000), \ \widetilde{a}_8^{x_5}(0.7419), \ \widetilde{a}_9^{x_5}(0.4500) \Big\rangle.
\end{aligned}$$

If the sequences of attribute weights of each node are same and the attribute weights are equal:

$$\lambda = \Big\langle \ \lambda_1\left(\tfrac{1}{9}\right) \ \lambda_2\left(\tfrac{1}{9}\right) \ \lambda_3\left(\tfrac{1}{9}\right) \ \lambda_4\left(\tfrac{1}{9}\right), \ \lambda_5\left(\tfrac{1}{9}\right), \ \lambda_6\left(\tfrac{1}{9}\right), \ \lambda_7\left(\tfrac{1}{9}\right), \ \lambda_8\left(\tfrac{1}{9}\right), \ \lambda_9\left(\tfrac{1}{9}\right) \ \Big\rangle, \tag{25}$$

then the evaluation of the power quality of nodes by the traditional fuzzy evaluation method is calculated by:

$$E^{x_i} = \widetilde{S}^{x_i} \cdot \lambda^{x_i} = \sum_{j=1}^{9} \widetilde{a}_j^{x_i} \times \lambda_j. \tag{26}$$

The evaluation results are shown in Table 5. The smaller the evaluation value is, the better the power quality of the node is. From Table 5, we can conclude that: $x_1 \succ x_3 \succ x_2 \succ x_5 \succ x_4$, where $\succ$ means better than.

Table 5. Evaluation results by the traditional method with all attribute information known.

Node1	Node2	Node3	Node4	Node5
0.1309	0.6720	0.4262	0.7405	0.7054

Furthermore, if the conditions mentioned in Section 2 are taken into consideration, evaluate the following examples based on the proposed method mentioned in Section 5, and compare the result with other methods.

Example 3. *Assume that only measurable attributes $\widetilde{a}_1^x, \widetilde{a}_2^x, \widetilde{a}_3^x, \widetilde{a}_4^x, \widetilde{a}_5^x$ are known, the other four attributes are unknown in this example. According to the power quality theory, these unknown attributes are correlated with those five attributes, which means the fuzzy measurable attribute sequence of x_i is redefined as follows:*

$$\begin{aligned} \widetilde{S}^{x_i} &= \langle \widetilde{a}_1^{x_i}, \widetilde{a}_2^{x_i}, \cdots, \widetilde{a}_9^{x_i} \rangle \\ &= \underbrace{\langle \widetilde{a}_1^{x_i}, \widetilde{a}_2^{x_i}, \widetilde{a}_3^{x_i}, \widetilde{a}_4^{x_i}, \widetilde{a}_5^{x_i} \rangle}_{\widetilde{S}_{KFA}^{x_i}} + \underbrace{\langle \widetilde{a}_6^{x_i}, \widetilde{a}_7^{x_i}, \widetilde{a}_8^{x_i}, \widetilde{a}_9^{x_i} \rangle}_{\widetilde{S}_{UFA}^{x_i}} . \end{aligned} \tag{27}$$

Additionally, the sequence of attribute weights is redefined as follows:

$$\lambda = \left\langle \frac{1}{5}, \frac{1}{5}, \frac{1}{5}, \frac{1}{5}, \frac{1}{5} \right\rangle. \tag{28}$$

Apply the proposed evaluation method to solve this example, and the process is:

Firstly, approximately estimate $\widetilde{S}_{\text{UFA}}^{x_i}$ based on the proposed method. Taking $N_{\text{KFA}} = \langle 1, 2, \cdots, 5 \rangle$ as the independent variable $\boldsymbol{E}$ in Step 2 of the technique to approximately estimate UFAs and $\widetilde{S}_{\text{KFA}}^{x_i}$ as the dependent variable $\boldsymbol{Y}$, interpolate the function curve of each node. There is less heuristic knowledge for us to determine the size of UFA r. Thus, $r = 4$ is the best choice conservatively. We can choose the median values of adjacent pairs of N_{KFA} as UFA. Generate the UFA number sequence $N_{\text{UFA}} = \langle 1.5, 2.5, 3.5, 4.5 \rangle$ in Step 4 of the technique to approximately estimate UFAs. Calculate the estimation based on the fit function with N_{UFA} as its input and the results are:

$$\begin{aligned} \hat{\tilde{S}}_{\text{UFA}}^{x_1} &= \left\langle \ \hat{\tilde{a}}_{1.5}^{x_1}(-0.4063), \ \ \hat{\tilde{a}}_{2.5}^{x_1}(0.6563), \ \ \hat{\tilde{a}}_{3.5}^{x_1}(0.6564), \ \ \hat{\tilde{a}}_{4.5}^{x_1}(-0.4062) \ \right\rangle, \\ \hat{\tilde{S}}_{\text{UFA}}^{x_2} &= \left\langle \ \hat{\tilde{a}}_{1.5}^{x_2}(1.2571), \ \ \hat{\tilde{a}}_{2.5}^{x_2}(0.4079), \ \ \hat{\tilde{a}}_{3.5}^{x_2}(0.1352), \ \ \hat{\tilde{a}}_{4.5}^{x_2}(0.7405) \ \right\rangle, \\ \hat{\tilde{S}}_{\text{UFA}}^{x_3} &= \left\langle \ \hat{\tilde{a}}_{1.5}^{x_3}(0.2291), \ \ \hat{\tilde{a}}_{2.5}^{x_3}(0.4695), \ \ \hat{\tilde{a}}_{3.5}^{x_3}(0.5435), \ \ \hat{\tilde{a}}_{4.5}^{x_3}(0.6264) \ \right\rangle, \\ \hat{\tilde{S}}_{\text{UFA}}^{x_4} &= \left\langle \ \hat{\tilde{a}}_{1.5}^{x_4}(0.7162), \ \ \hat{\tilde{a}}_{2.5}^{x_4}(0.5901), \ \ \hat{\tilde{a}}_{3.5}^{x_4}(0.8289), \ \ \hat{\tilde{a}}_{4.5}^{x_4}(0.7196) \ \right\rangle, \\ \hat{\tilde{S}}_{\text{UFA}}^{x_5} &= \left\langle \ \hat{\tilde{a}}_{1.5}^{x_5}(0.4976), \ \ \hat{\tilde{a}}_{2.5}^{x_5}(0.3168), \ \ \hat{\tilde{a}}_{3.5}^{x_5}(0.7975), \ \ \hat{\tilde{a}}_{4.5}^{x_5}(0.9317) \ \right\rangle. \end{aligned}$$

Secondly, generate the final evaluation based on the attribute weight reconfiguration. Because four attributes are estimated from five attributes, the new sequence of attribute weights is calculated by Step 1 of the technique to generate the final evaluation:

$$\begin{aligned} \hat{\lambda}^x &= \left\langle \hat{\lambda}_1^1, \ \hat{\lambda}_{\underline{1.5}}^{\underline{1\ 2}}, \ \hat{\lambda}_2^2, \ \hat{\lambda}_{\underline{2.5}}^{\underline{2\ 3}}, \ \hat{\lambda}_3^3, \ \hat{\lambda}_{\underline{3.5}}^{\underline{3\ 4}}, \ \hat{\lambda}_4^4, \ \hat{\lambda}_{\underline{4.5}}^{\underline{4\ 5}}, \ \hat{\lambda}_5^5 \right\rangle \\ &= \left\langle \frac{1}{10}, \frac{1}{6}, \frac{1}{15}, \frac{2}{15}, \frac{1}{15}, \frac{2}{15}, \frac{1}{15}, \frac{1}{6}, \frac{1}{10} \right\rangle \end{aligned} , \tag{29}$$

and the evaluation of power quality of nodes is calculated by Step 3 of the technique to generate the final evaluation:

$$\hat{E}^{x_i} = \tilde{a}_1^{x_i} \times \hat{\lambda}_1^1 + \hat{\tilde{a}}_{1.5}^{x_i} \times \hat{\lambda}_{\underline{1.5}}^{\underline{1\ 2}} + \tilde{a}_2^{x_i} \times \hat{\lambda}_2^2 + \hat{\tilde{a}}_{2.5}^{x_i} \times \hat{\lambda}_{\underline{2.5}}^{\underline{2\ 3}} + \tilde{a}_3^{x_i} \times \hat{\lambda}_3^3 + \hat{\tilde{a}}_{3.5}^{x_i} \times \hat{\lambda}_{\underline{3.5}}^{\underline{3\ 4}} + \tilde{a}_4^{x_i} \times \hat{\lambda}_4^4 + \hat{\tilde{a}}_{4.5}^{x_i} \times \hat{\lambda}_{\underline{4.5}}^{\underline{4\ 5}} + \tilde{a}_5^{x_i} \times \hat{\lambda}_5^5. \tag{30}$$

The evaluation results are shown in Table 3. The smaller the evaluation value is, the better the power quality of the node is. From Table 6, we can find that: the evaluation result by the traditional method is $x_1 \succ x_2 \succ x_3 \succ x_4 \succ x_5$, while by the proposed method is $x_1 \succ x_3 \succ x_2 \succ x_5 \succ x_4$, which is the same as the result by the traditional method with all attributes known. To compare the results quantitatively, hamming distance can be used to calculate the similarity between two evaluation results. The Hamming distance between two equal-length sequences is defined as the ratio of the minimum number of substitutions required to change one of them into another to the length of the ranking. Hamming distance between the decision result by the traditional method with all attributes known and the decision result by the traditional method with KFAs only known is 0.8, while the proposed method is 0, i.e., the proposed method to deal with the decision-making problem with incomplete attribute information can obtain the same decision result of the problem with complete attribute information calculated by the traditional method.

Table 6. Evaluation results with KFAs known.

Methods	Node1	Node2	Node3	Node4	Node5	Hamming Distance
Traditional	0.2000	0.5025	0.5347	0.6293	0.6415	0.8
Proposed	0.1063	0.6281	0.4757	0.6884	0.6436	0

Example 4. *Assume that only measurable attributes* $\tilde{a}_1^x$, $\tilde{a}_2^x$, $\tilde{a}_4^x$, $\tilde{a}_5^x$ *are known, the other five attributes are unknown in this example. According to the power quality theory, these unknown attributes are correlated with those four attributes, which means the fuzzy measurable attribute sequence of* x_i *is redefined as follows:*

$$\begin{aligned}\tilde{S}^{x_i} &= \langle \tilde{a}_1^{x_i}, \tilde{a}_2^{x_i}, \cdots, \tilde{a}_9^{x_i} \rangle \\ &= \underbrace{\langle \tilde{a}_1^{x_i}, \tilde{a}_2^{x_i}, \tilde{a}_4^{x_i}, \tilde{a}_5^{x_i} \rangle}_{\tilde{S}^{x_i}_{KFA}} + \underbrace{\langle \tilde{a}_3^{x_i}, \tilde{a}_6^{x_i}, \tilde{a}_7^{x_i}, \tilde{a}_8^{x_i}, \tilde{a}_9^{x_i} \rangle}_{\tilde{S}^{x_i}_{UFA}} .\end{aligned} \tag{31}$$

Additionally, the sequence of attribute weights is redefined as follows:

$$\lambda = \left\langle \frac{1}{4}, \frac{1}{4}, \frac{1}{4}, \frac{1}{4} \right\rangle. \tag{32}$$

Apply the proposed evaluation method to solve this example, and the process is:

Firstly, approximately estimate $\tilde{S}^{x_i}_{\text{UFA}}$ based on the proposed method. Taking $N_{\text{KFA}} = \langle 1, 2, 3, 4 \rangle$ as the independent variable $\boldsymbol{E}$ in Step 2 of the technique to approximately estimate UFAs and $\tilde{S}^{x_i}_{\text{KFA}}$ as the dependent variable $\boldsymbol{Y}$, interpolate the function curve of each node. Here, we can also let $r = 5$ from a conservative perspective. To avoid the excessive effect of $\tilde{a}_1^{x_i}$, we chose 1.5 as the starting point of UFA. Generate the UFA number sequence $N_{\text{UFA}} = \langle 1.5, 2.1, 2.7, 3.3, 3.9 \rangle$ in Step 4 of the technique to approximately estimate UFAs. Calculate the estimation based on the fit function with N_{UFA} as its input and the results are:

$$\begin{aligned}
\hat{\tilde{S}}^{x_1}_{\text{UFA}} &= \left\langle \hat{\tilde{a}}^{x_1}_{1.5}(0),\ \hat{\tilde{a}}^{x_1}_{2.1}(0),\ \hat{\tilde{a}}^{x_1}_{2.7}(0),\ \hat{\tilde{a}}^{x_1}_{3.3}(0),\ \hat{\tilde{a}}^{x_1}_{3.9}(0) \right\rangle, \\
\hat{\tilde{S}}^{x_2}_{\text{UFA}} &= \left\langle \hat{\tilde{a}}^{x_2}_{1.5}(0.9982),\ \hat{\tilde{a}}^{x_2}_{2.1}(0.9718),\ \hat{\tilde{a}}^{x_2}_{2.7}(0.6922),\ \hat{\tilde{a}}^{x_2}_{3.3}(0.3914),\ \hat{\tilde{a}}^{x_2}_{3.9}(0.3009) \right\rangle, \\
\hat{\tilde{S}}^{x_3}_{\text{UFA}} &= \left\langle \hat{\tilde{a}}^{x_3}_{1.5}(0.2908),\ \hat{\tilde{a}}^{x_3}_{2.1}(0.3389),\ \hat{\tilde{a}}^{x_3}_{2.7}(0.4394),\ \hat{\tilde{a}}^{x_3}_{3.3}(0.6265),\ \hat{\tilde{a}}^{x_3}_{3.9}(0.9345) \right\rangle, \\
\hat{\tilde{S}}^{x_4}_{\text{UFA}} &= \left\langle \hat{\tilde{a}}^{x_4}_{1.5}(0.5709),\ \hat{\tilde{a}}^{x_4}_{2.1}(0.6430),\ \hat{\tilde{a}}^{x_4}_{2.7}(0.8619),\ \hat{\tilde{a}}^{x_4}_{3.3}(0.8550),\ \hat{\tilde{a}}^{x_4}_{3.9}(0.2497) \right\rangle, \\
\hat{\tilde{S}}^{x_5}_{\text{UFA}} &= \left\langle \hat{\tilde{a}}^{x_5}_{1.5}(0.2977),\ \hat{\tilde{a}}^{x_5}_{2.1}(0.3427),\ \hat{\tilde{a}}^{x_5}_{2.7}(0.8004),\ \hat{\tilde{a}}^{x_5}_{3.3}(1.0821),\ \hat{\tilde{a}}^{x_5}_{3.9}(0.5993) \right\rangle.
\end{aligned}$$

Secondly, generate the final evaluation based on the attribute weight reconfiguration. Because four attributes are estimated from five attributes, the new sequence of attribute weights is calculated by Step 1 of the technique to generate the final evaluation:

$$\begin{aligned}\hat{\lambda}^x &= \left\langle \hat{\lambda}^1_1, \underline{\hat{\lambda}^{1\_2}_{1.5}}, \hat{\lambda}^2_2, \underline{\hat{\lambda}^{2\_3}_{2.1}}, \underline{\hat{\lambda}^{2\_3}_{2.7}}, \hat{\lambda}^3_3, \underline{\hat{\lambda}^{3\_4}_{3.3}}, \underline{\hat{\lambda}^{3\_4}_{3.9}}, \hat{\lambda}^4_4 \right\rangle \\ &= \left\langle \frac{1}{8}, \frac{1}{6}, \frac{1}{16}, \frac{9}{80}, \frac{9}{80}, \frac{1}{20}, \frac{2}{15}, \frac{2}{15}, \frac{1}{12} \right\rangle\end{aligned}, \tag{33}$$

and the evaluation of the power quality of nodes is calculated by Step 3 of the technique to generate the final evaluation:

$$\hat{E}^{x_i} = \tilde{a}_1^{x_i} \times \hat{\lambda}^1_1 + \hat{\tilde{a}}^{x_i}_{1.5} \times \hat{\lambda}^{1\_2}_{1.5} + \tilde{a}_2^{x_i} \times \hat{\lambda}^2_2 + \hat{\tilde{a}}^{x_i}_{2.1} \times \hat{\lambda}^{2\_3}_{2.1} + \hat{\tilde{a}}^{x_i}_{2.7} \times \hat{\lambda}^{2\_3}_{2.7} + \tilde{a}_3^{x_i} \times \hat{\lambda}^3_3 + \hat{\tilde{a}}^{x_i}_{3.3} \times \hat{\lambda}^{3\_4}_{3.3} + \hat{\tilde{a}}^{x_i}_{3.9} \times \hat{\lambda}^{3\_4}_{3.9} + \tilde{a}_4^{x_i} \times \hat{\lambda}^4_4. \tag{34}$$

The evaluation results are shown in Table 4. The smaller the evaluation value is, the better the power quality of the node is. From Table 7, we can find that: the evaluation result by the traditional method is $x_1 \succ x_3 \succ x_4 \succ x_2 \succ x_5$, while by the proposed method is $x_1 \succ x_3 \succ x_4 \succ x_5 \succ x_2$.

Hamming distance between the result by the traditional method with KFAs only known and the result by the traditional method with all attributes known is 0.6, while the proposed method is 0.4.

Table 7. Evaluation results with KFAs known.

Methods	Node1	Node2	Node3	Node4	Node5	Hamming Distance
Traditional	0	0.6281	0.5285	0.6192	0.6735	0.6
Proposed	0	0.6650	0.5133	0.6242	0.6354	0.4

Thus, we can conclude that the proposed method is more effective than the traditional method to deal with this type of power quality evaluation problem.

7. Conclusions

In this paper, the GGS of PAS, MAS, and PAMS was modeled and analyzed, which gives out a new idea to enhance the existing methods involving fuzzy membership to deal with MADM problems with partial attribute values and weights unknown. A fuzzy attributes expansion method was proposed. The proposed method can be applied to some research fields, such as regression, clustering, and fuzzy evaluation, which is proven to be effective in four examples when only part KFAs were given. By application of this method, the results of FCE were more consistent with the actual situation than traditional FCE.

For this method, it was necessary and critical to find out the appropriate size of the UFA number sequence for different practical problems by experiments. We are now considering applying this method to evaluate the power system under massive attack, where partial attribute information of some nodes could be incomplete or even totally unknown for researchers. Meanwhile, extending the proposed method from fuzzy set theory to other related fuzzy theories is also worth considering.

Author Contributions: Conceptualization, J.Z.; methodology, J.Z and W.S.; software, J.Z.; validation, J.Z., W.S., and Y.L.; formal analysis, J.Z.; investigation, J.Z.; resources, J.Z.; data curation, J.Z.; writing—original draft preparation, J.Z.; writing—review and editing, J.Z.; visualization, J.Z.; supervision, W.S. and Y.L.; project administration, W.S.; funding acquisition, Y.L.

Funding: This research was funded by the National Natural Science Foundation of China (grant number 61503240) and the Shanghai Maritime University Graduate Student Innovation Fund Project (grant number 2016ycx078).

Acknowledgments: The authors would like to thank classmates and teachers in the Shanghai Maritime University—Schneider Electric Joint Laboratory. The authors would like to thank all the reviewers for their valuable comments.

Conflicts of Interest: The authors declare no conflict of interest.

References

1. Zadeh, L.A. Fuzzy sets. *Inf. Control* **1965**, *8*, 338–353. [CrossRef]
2. Zhao, Z.D.; Zhang, Y.F. SQI quality evaluation mechanism of single-lead ECG signal based on simple heuristic fusion and fuzzy comprehensive evaluation. *Front. Physiol.* **2018**, *9*, 727. [CrossRef] [PubMed]
3. Yang, W.C.; Xu, K.; Lian, J.J.; Bin, L.L.; Ma, C. Multiple flood vulnerability assessment approach based on fuzzy comprehensive evaluation method and coordinated development degree model. *J. Environ. Manag.* **2018**, *213*, 440–450. [CrossRef] [PubMed]
4. Zuo, R.G.; Cheng, Q.M.; Agterberg, F.P. Application of a hybrid method combining multilevel fuzzy comprehensive evaluation with asymmetric fuzzy relation analysis to mapping prospectivity. *Ore Geol. Rev.* **2009**, *35*, 101–108. [CrossRef]
5. Bo, C.X.; Zhang, X.H. New operations of picture fuzzy relations and fuzzy comprehensive evaluation. *Symmetry* **2017**, *9*, 268. [CrossRef]
6. Jin, C. Adaptive robust image watermark approach based on fuzzy comprehensive evaluation and analytic hierarchy process. *Signal Image Video Process.* **2012**, *6*, 317–324. [CrossRef]

7. Zeng, D.Z.; He, Q.Y.; Yu, Z.M.; Jia, W.Y.; Zhang, S.; Liu, Q.P. Risk assessment of sustained casing pressure in gas wells based on the fuzzy comprehensive evaluation method. *J. Nat. Gas Sci. Eng.* **2017**, *46*, 756–763. [CrossRef]
8. Liu, Y.L.; Huang, X.L.; Duan, J.; Zhang, H.M. The assessment of traffic accident risk based on grey relational analysis and fuzzy comprehensive evaluation method. *Nat. Hazards* **2017**, *88*, 1409–1422. [CrossRef]
9. Li, Y.; Sun, Z.D.; Han, L.; Mei, N. Fuzzy comprehensive evaluation method for energy management systems based on an internet of things. *IEEE Access* **2017**, *5*, 21312–21322. [CrossRef]
10. Wei, B.; Wang, S.; Li, L. Fuzzy comprehensive evaluation of district heating systems. *Energy Policy* **2010**, *38*, 5947–5955. [CrossRef]
11. Medina, J.; Ojeda-Aciego, M. Multi-adjoint t-concept lattices. *Inf. Sci.* **2010**, *180*, 712–725. [CrossRef]
12. Atanassov, K.T. Intuitionistic fuzzy sets. *Fuzzy Sets Syst.* **1986**, *20*, 87–96. [CrossRef]
13. Xu, Z.S. Models for multiple attribute decision making with intuitionistic fuzzy information. *Int. J. Uncertain. Fuzziness Knowl. Based Syst.* **2007**, *15*, 285–297. [CrossRef]
14. Wan, S.P.; Wan, F.; Dong, J.Y. A novel risk attitudinal ranking method for intuitionistic fuzzy values and application to MADM. *Appl. Soft Comput.* **2016**, *40*, 98–112. [CrossRef]
15. Qin, J.D.; Liu, X.W. An approach to intuitionistic fuzzy multiple attribute decision making based on Maclaurin symmetric mean operators. *J. Intell. Fuzzy Syst.* **2014**, *27*, 2177–2190. [CrossRef]
16. Liu, P.D.; Mahmood, T.; Khan, Q. Multi-attribute decision-making based on prioritized aggregation operator under hesitant intuitionistic fuzzy linguistic environment. *Symmetry* **2017**, *9*, 270. [CrossRef]
17. Abdullah, L.; Najib, L. A new preference scale of intuitionistic fuzzy analytic hierarchy process in multi-criteria decision making problems. *J. Intell. Fuzzy Syst.* **2014**, *26*, 1039–1049. [CrossRef]
18. Atanassov, K.T. Interval-valued intuitionistic fuzzy sets. *Fuzzy Sets Syst.* **1989**, *31*, 343–349. [CrossRef]
19. Liu, P.D. Some hamacher aggregation operators based on the interval-valued intuitionistic fuzzy numbers and their application to group decision making. *IEEE Trans. Fuzzy Syst.* **2014**, *22*, 83–97. [CrossRef]
20. Wei, G.W. Interval valued hesitant fuzzy uncertain linguistic aggregation operators in multiple attribute decision making. *Int. J. Mach. Learn. Cybern.* **2016**, *7*, 1093–1114. [CrossRef]
21. Gupta, P.; Lin, C.T.; Mehlawat, M.K.; Grover, N. A new method for intuitionistic fuzzy multiattribute decision making. *IEEE Trans. Syst. Man. Cybern. Syst.* **2016**, *46*, 1167–1179. [CrossRef]
22. Wu, J.; Cao, Q.W.; Li, H. An approach for MADM problems with interval-valued intuitionistic fuzzy sets based on nonlinear functions. *Technol. Econ. Dev. Econ.* **2016**, *22*, 336–356. [CrossRef]
23. Tang, X.A.; Fu, C.; Xu, D.L.; Yang, S.L. Analysis of fuzzy Hamacher aggregation functions for uncertain multiple attribute decision making. *Inf. Sci.* **2017**, *387*, 19–33. [CrossRef]
24. Hussain, S.A.I.; Mandal, U.K.; Mondal, S.P. Decision maker priority index and degree of vagueness coupled decision making method: A synergistic approach. *Int. J. Fuzzy Syst.* **2018**, *20*, 1551–1566. [CrossRef]
25. Chen, S.M.; Han, W.H. A new multiattribute decision making method based on multiplication operations of interval-valued intuitionistic fuzzy values and linear programming methodology. *Inf. Sci.* **2018**, *429*, 421–432. [CrossRef]
26. Dahooie, J.H.; Zavadskas, E.K.; Abolhasani, M.; Vanaki, A.; Turskis, Z. A novel approach for evaluation of projects using an interval-valued fuzzy additive ratio assessment (ARAS) method: A case study of oil and gas well drilling projects. *Symmetry* **2018**, *10*, 45. [CrossRef]
27. Chen, B.; Guo, Y.Y.; Gao, X.E.; Wang, Y.M. A novel multi-attribute decision making approach: Addressing the complexity of time dependent and interdependent data. *IEEE Access* **2018**, *6*, 55838–55849. [CrossRef]
28. Yazici, I.; Kahraman, C. VIKOR method using interval type two fuzzy sets. *J. Intell. Fuzzy Syst.* **2015**, *29*, 411–421. [CrossRef]
29. Yao, D.B.; Liu, X.X.; Zhang, X.; Wang, C.C. Type-2 fuzzy cross-entropy and entropy measures and their applications. *J. Intell. Fuzzy Syst.* **2016**, *30*, 2169–2180. [CrossRef]
30. Keshavarz-Ghorabaee, M.; Amiri, M.; Zavadskas, E.K.; Turskis, Z.; Antucheviciene, J. An extended step-wise weight assessment ratio analysis with symmetric interval type-2 fuzzy sets for determining the subjective weights of criteria in multi-criteria decision-making problems. *Symmetry* **2018**, *10*, 91. [CrossRef]
31. Yang, Y.; Lang, L.; Lu, L.L.; Sun, Y.M. A new method of multiattribute decision-making based on interval-valued hesitant fuzzy soft sets and its application. *Math. Probl. Eng.* **2017**, *2017*, 9376531. [CrossRef]
32. Tong, X.; Yu, L.Y. MADM based on distance and correlation coefficient measures with decision-maker preferences under a hesitant fuzzy environment. *Soft Comput.* **2016**, *20*, 4449–4461. [CrossRef]

33. Zhang, R.C.; Li, Z.M.; Liao, H.C. Multiple-attribute decision-making method based on the correlation coefficient between dual hesitant fuzzy linguistic term sets. *Knowl. Based Syst.* **2018**, *159*, 186–192. [CrossRef]
34. Kahraman, C.; Onar, S.Ç.; Öztaysi, B. Fuzzy multicriteria decision-making: A literature review. *Int. J. Comput. Intell. Syst.* **2015**, *8*, 637–666. [CrossRef]
35. Li, M.Z.; Du, Y.N.; Wang, Q.Y.; Sun, C.M.; Ling, X.; Yu, B.Y.; Tu, J.S.; Xiong, Y.R. Risk assessment of supply chain for pharmaceutical excipients with AHP-fuzzy comprehensive evaluation. *Drug Dev. Ind. Pharm.* **2016**, *42*, 676–684. [CrossRef]
36. Goyal, R.K.; Kaushal, S.; Sangaiah, A.K. The utility based non-linear fuzzy AHP optimization model for network selection in heterogeneous wireless networks. *Appl. Soft Comput.* **2018**, *67*, 800–811. [CrossRef]
37. Selvachandran, G.; Quek, S.G.; Smarandache, F.; Broumi, S. An extended technique for order preference by similarity to an ideal solution (TOPSIS) with maximizing deviation method based on integrated weight measure for single-valued neutrosophic sets. *Symmetry* **2018**, *10*, 236. [CrossRef]
38. Sun, G.D.; Guan, X.; Yi, X.; Zhou, Z. An innovative TOPSIS approach based on hesitant fuzzy correlation coefficient and its applications. *Appl. Soft Comput.* **2018**, *68*, 249–267. [CrossRef]
39. Baykasoglu, A.; Golcuk, I. Development of an interval type-2 fuzzy sets based hierarchical MADM model by combining DEMATEL and TOPSIS. *Expert Syst. Appl.* **2017**, *70*, 37–51. [CrossRef]
40. Keshavarz Ghorabaee, M.; Amiri, M.; Zavadskas, E.K.; Antucheviciene, J. Supplier evaluation and selection in fuzzy environments: A review of MADM approaches. *Econ. Res. Ekon. Istraž.* **2017**, *30*, 1073–1118. [CrossRef]
41. Park, K.S.; Kim, S.H. Tools for interactive multiattribute decisionmaking with incompletely identified information. *Eur. J. Oper. Res.* **1997**, *98*, 111–123. [CrossRef]
42. Xu, Z.S.; Zhang, X.L. Hesitant fuzzy multi-attribute decision making based on TOPSIS with incomplete weight information. *Knowl. Based Syst.* **2013**, *52*, 53–64. [CrossRef]
43. Wei, G.W. GRA method for multiple attribute decision making with incomplete weight information in intuitionistic fuzzy setting. *Knowl. Based Syst.* **2010**, *23*, 243–247. [CrossRef]
44. Bao, T.T.; Xie, X.L.; Long, P.Y.; Wei, Z.K. MADM method based on prospect theory and evidential reasoning approach with unknown attribute weights under intuitionistic fuzzy environment. *Expert Syst. Appl.* **2017**, *88*, 305–317. [CrossRef]
45. Eum, Y.S.; Park, K.S.; Kim, S.H. Establishing dominance and potential optimality in multi-criteria analysis with imprecise weight and value. *Comput. Oper. Res.* **2001**, *28*, 397–409. [CrossRef]

symmetry

MDPI

Article

A Decision-Making Approach Based on a Multi Q-Hesitant Fuzzy Soft Multi-Granulation Rough Model

Kholood Mohammad Alsager [1], Noura Omair Alshehri [2] and Muhammad Akram [3,*]

[1] Department of Mathematics, Faculty of Sciences, King Abdulaziz University, Jeddah 21589, Saudi Arabia; kolood1987@hotmail.com

[2] Department of Mathematics, Faculty of Sciences, University of Jeddah, Jeddah 21589, Saudi Arabia; noal-shehri@uj.edu.sa

[3] Department of Mathematics, University of the Punjab, New Campus, Lahore 4590, Pakistan

* Correspondence: m.akram@pucit.edu.pk

Received: 20 October 2018; Accepted: 14 November 2018; Published: 4 December 2018

Abstract: In this paper, we propose a new hybrid model, multi Q-hesitant fuzzy soft multi-granulation rough set model, by combining a multi Q-hesitant fuzzy soft set and multi-granulation rough set. We demonstrate some useful properties of these multi Q-hesitant fuzzy soft multi-granulation rough sets. Furthermore, we define multi Q-hesitant fuzzy soft (M^kQHFS) rough approximation operators in terms of M^kQHFS relations and M^kQHFS multi-granulation rough approximation operators in terms of M^kQHFS relations. We study the main properties of lower and upper M^kQHFS rough approximation operators and lower and upper M^kQHFS multi-granulation rough approximation operators. Moreover, we develop a general framework for dealing with uncertainty in decision-making by using the multi Q-hesitant fuzzy soft multi-granulation rough sets. We analyze the photovoltaic systems fault detection to show the proposed decision methodology.

Keywords: Q-hesitant fuzzy soft set; multi Q-hesitant fuzzy soft rough set; photovoltaic systems fault detection approach; decision-making method

1. Introduction

The notion of rough set theory was introduced by Pawlak in 1982 [1]. It is a mathematical approach concerning uncertainty that comes from noisy, inexact or incomplete information. In rough set theory, the equivalence relation plays a significant role in creating the upper and lower approximations of the set. Currently, rough set approximations [2] have been constructed into fuzzy sets [3], intuitionistic fuzzy sets [4], hesitant fuzzy sets [5] and covering sets [6]. The soft set theory, originally initiated by Molodtsov [7], is a general tool for dealing with uncertainty. Different from some traditional tools for dealing with uncertainties, such as the theory of fuzzy sets [3], the theory of probability and the theory of rough sets [1], the advantage of soft set theory is that it is free from the inadequacy of the parametrization tools of those theories. According to Molodtsov [7], the soft set theory applied successfully to many fields such as functions' smoothness, game theory, theory of measurement and so on. Maji and Roy [8] introduced the soft set into the decision-making problems with the help of the rough theory. Necessary and possible hesitant fuzzy sets, and probabilistic soft sets and dual probabilistic soft sets in decision-making have discussed in [9,10]. Moreover, many new rough set models have been established by combining the Pawlak rough set with other uncertainty theories such as soft set theory. Feng [11] provided a framework to combine fuzzy set, rough set, and soft set all together, which gave rise to several interesting new concepts such as rough soft set, soft rough set and soft rough fuzzy set [12]. Zhang et al. [13] proposed the notion of soft rough intuitionistic fuzzy sets and intuitionistic fuzzy soft rough sets, which are generalized soft rough set models. Akram et al. [14] presented a new hybrid model, a hesitant N-soft set model for group decision-making. Several research works have been done to solve different real life decision-making problems (see [15–19]). All of these models have always been described by the expression of a one-dimensional membership function that can not be able to deal with the information that appears in a two-dimensional universal set. From this point of view, the idea of Q-fuzzy sets was came out. Afterwards, the concept of multi Q-fuzzy soft sets [20–24] was established to combine the key feature of soft sets and Q-fuzzy sets with multi membership values. The notion of multi Q-hesitant fuzzy soft sets is the generalization of multi Q-fuzzy soft sets. This extension can easily handle the difficulty more objectively than other developed Q-fuzzy set approaches. The combination of multi Q-hesitant fuzzy soft sets and rough sets will be an improved model of hesitant fuzzy rough approaches that concern both areas theoretical and practical applications. Qian et al. [25] proposed the model of multi-granulation rough sets. The main idea of this model is based on defined multiple equivalence relations in a given universe that eliminated the restrictions that may occur through the single equivalence relations in classical rough sets [1] perfectly. The notions of multi-granulation fuzzy rough sets and multi-granulation hesitant fuzzy rough sets are presented by Sun et al. [26] and Zhang et al. [27], respectively, to solve decision-making problems. For other notations and terminologies not mentioned in this paper, the readers are referred to [28–33].

In the field of electrical engineering, photovoltaic systems fault detection is one of the challenging tasks that electrical experts have faced in recent years dealing with a substantial amount of uncertain information. Different experts would give their different judgments towards the systems fault detection data. Hence, by combining multi Q-hesitant fuzzy soft sets with multi-granulation rough sets, we constructed the concept of a multi Q-hesitant fuzzy soft multi-granulation rough set model and its application in photovoltaic systems fault detection through developing a new data analysis model in fault detection procedures under the framework of Q-hesitant fuzzy soft information. In this paper, we propose a new hybrid model, multi Q-hesitant fuzzy soft multi-granulation rough set model, by combining a multi Q-hesitant fuzzy soft set and a multi-granulation rough set. We present some of its fundamental properties. We develop a general framework for dealing with uncertainty decision-making by using the multi Q-hesitant fuzzy soft multi-granulation rough sets. We use the photovoltaic systems fault detection to indicate the principle steps of the decision methodology.

The presentation of the article is organized as follows: In Section 2, we recalled some basic concepts of rough sets, soft sets and hesitant fuzzy soft sets. In Section 3, we have presented multi Q-hesitant fuzzy soft sets and discussed some properties. In Section 4, we have introduced a rough set model based on multi Q-hesitant fuzzy soft relation and have examined some properties of this model. In Section 5, we have generalized the notion of multi Q-hesitant fuzzy soft rough sets into multi Q-hesitant fuzzy soft multi-granulation rough set model. In Section 6, we have established a general approach to decision-making based on multi Q-hesitant fuzzy soft multi-granulation rough sets and illustrated the principal steps of the proposed decision method by a numerical example. Finally, in Section 7, we have concluded the paper with a summary and outlook for further research.

2. Preliminaries

In this section, we recall some basic notions and definitions which will be used in this paper.

Definition 1 ([1])**.** *Let U be a non-empty finite universe and R be an equivalence relation on U. We use U/R to denote the family of all equivalence classes of R (or classifications of U), and $[x]_R$ to denote an equivalence class of R containing the element $x \in U$. The pair (U, R) is called an approximation space. For any $X \subseteq U$, we can define the lower and upper approximations of X as follows:*

$$\underline{R}(X) = \{x \in U : [x]_R \subseteq X\},$$

$$\overline{R}(X) = \{x \in U : [x]_R \cap X \neq \phi\}.$$

The pair $(\underline{R}(X), \overline{R}(X))$ is referred to as the rough set of X. The rough set $(\underline{R}(X), \overline{R}(X))$ gives rise to a description of X under the present knowledge, i.e., the classification of U.

Furthermore, the positive region, negative region, and boundary region of X about the approximation space (U, R) are defined as follows, respectively:

$$pos(X) = \underline{R}(X), neg(X) = \sim \overline{R}(X), bn(X) = \overline{R}(X) - \underline{R}(X),$$

where $\sim X$ stands for complementation of the set X.

Definition 2 ([7]). *Let E be the set of parameters with the connection to the objects in U. A pair (F, E) is called a soft set over U, where F is a mapping given by $F : E \longrightarrow P(U)$, $P(U)$ is a set of all subsets of U.*

This definition shows that a soft set over U is a parameterized family of subsets of the universe U. For $e \in E, F(e)$ is regarded as the set of e-approximate elements of the soft set (F, E).

Definition 3 ([5]). *Given a non-empty subset A of X, a hesitant fuzzy set $H_X = \{(x, h_X(x) : x \in X)\}$ on X satisfying the following condition:*

$$h_X(x) = \phi \text{ for all } x \notin A$$

is called a hesitant fuzzy set related to A (briefly, A-hesitant fuzzy set) on X and is represented by $H_A = \{(x, h_A(x) : x \in X)\}$, where h_A is a mapping from X to $p([0, 1])$ with $h_A(x) = \phi$ for all $x \notin A$.

Definition 4 ([34]). *Let $\tilde{H}(U)$ be the set of all hesitant fuzzy sets in U. A pair $(\tilde{F}, \tilde{A})$ is called a hesitant fuzzy soft set over U, where $\tilde{F}$ is a mapping given by*

$$\tilde{F} : A \longrightarrow \tilde{H}(U).$$

A hesitant fuzzy soft set is a mapping from parameters to $\tilde{H}(U)$. It is a parameterized family of hesitant fuzzy subsets of U. For $e \in A$, $\tilde{F}(e)$ may be considered as the set of e-approximate elements of the hesitant fuzzy soft set $(\tilde{F}, A)$.

3. Multi Q-Hesitant Fuzzy Soft Sets

We first introduce the notion of Q-hesitant fuzzy soft sets as a generalization of Q-fuzzy soft sets.

Definition 5. *Let U be a universal set and Q be non-empty set. A Q-hesitant fuzzy set A_Q is a set given by*

$$A_Q = \{\langle (uq), h_{A_Q}(uq)\rangle : u \in U, q \in Q\},$$

where $h_{A_Q} : U \times Q \longrightarrow [0, 1]$. The function $h_{A_Q}(uq)$ is called the membership function of Q-hesitant fuzzy set, and the set of all Q-hesitant fuzzy sets over $U \times Q$ will be denoted by $QHF(U \times Q)$.

Definition 6. *Let U be a non-empty finite universe and Q be a non-empty set. For any $A_Q, B_Q \in QHF(U \times Q)$, then, for all $u \in U, q \in Q$, we have*

1. $h_{A_Q^c}(uq) = \sim h_{A_Q}(uq) = \bigcup_{\gamma \in h_{A_Q^c}(uq)} \{1 - \gamma\}$.
2. $A_Q \cup B_Q = \{\langle (uq), h_{A_Q}(uq) \vee h_{B_Q}(uq)\rangle, u \in U, q \in Q\}$.
3. $A_Q \cap B_Q = \{\langle (uq), h_{A_Q}(uq) \wedge h_{B_Q}(uq)\rangle, u \in U, q \in Q\}$.
4. $A_Q \oplus B_Q = \bigcup_{\gamma_1 \in h_{A_Q}(uq), \gamma_2 \in h_{B_Q}(uq)} \{\gamma_1 + \gamma_2 - \gamma_1\gamma_2\}$.
5. $A_Q \otimes B_Q = \bigcup_{\gamma_1 \in h_{A_Q}(uq), \gamma_2 \in h_{B_Q}(uq)} \{\gamma_1\gamma_2\}$.

Definition 7. *Let U be a universal set and Q be non-empty set, I be a unit interval $[0, 1]$ and k be a positive integer. A multi Q-hesitant fuzzy set $\tilde{H}_Q$ in $U \times Q$ is a set defining by*

$$\tilde{H}_Q = \{\langle (uq), h^i_{\tilde{H}_Q}(uq)\rangle : u \in U, q \in Q \text{ for all } i = 1, 2, \cdots, k\},$$

where $h^i_{\tilde{H}_Q} : U \times Q \longrightarrow I^k$ *for all* $i = 1, 2, \cdots, k$*. The function* $h^1_{\tilde{H}_Q}(uq), h^2_{\tilde{H}_Q}(uq), \cdots, h^k_{\tilde{H}_Q}(uq)$ *is called the membership function of multi Q-hesitant fuzzy set and k is called the dimension of* $h^i_{\tilde{H}_Q}$*. The set of all multi Q-hesitant fuzzy set of dimension k in* $U \times Q$ *is denoted by* $M^kQHFS(U \times Q)$*.*

Definition 8. *Let* A_Q*,* B_Q *be a multi Q-hesitant fuzzy sets over* $U \times Q$*. Then,* A_Q *is said to be a multi Q-hesitant fuzzy subset of* B_Q *if*

$$h^i_{A_Q}(uq) \leq h^i_{B_Q}(uq)$$

holds for any $u \in U, q \in Q, i = i, 2, \cdots, k$ *and it is denoted by* $A_Q \subseteq B_Q$*.*

Definition 9. *Let U be a universal set and be non-empty set, E be the set of parameters and* $M^kQHF(U \times Q)$ *be the set of all multi Q-hesitant fuzzy sets on* $U \times Q$ *with the dimension k. Let* $A \subseteq E$ *the pair* (H_Q, A) *is called a multi Q-hesitant fuzzy soft set* (M^kQHFSS) *over U, where* (H_Q, A) *is given by the form*

$$(H_Q, A) = \{(e, h^i_Q(e)) : e \in A, h^i_Q(e) \in M^kQHFS(U \times Q)\},$$

where $h^i_Q : A \longrightarrow M^kQHF(U \times Q)$ *such that* $h^i_Q(e) \neq \phi$ *if* $e \notin A$*. The set of all multi Q-hesitant fuzzy soft sets over* $U \times Q$ *will be denoted by* $M^kQHFSS(U \times Q)$*.*

Example 1. *Suppose that a company wants to buy three types of products from two brands and wants to take the opinion of two specialists about these products (k= 2). Let* $U = \{u_1, u_2, u_3\}$ *be a set of products,* $Q = \{p, q\}$ *be a set of brands, and* $E = \{e_1 = easy\ to\ use, e_2 = quality, e_3 = price\}$ *is the set of decision parameters. Then we can define the multi Q-hesitant fuzzy soft sets*(H_Q, A) *as follows:*

$$(H_Q, A) = \{\langle e_1, (\frac{u_1p}{(0.2,0.3)(0.1)}), (\frac{u_1q}{(0.1,0.3)(0.4,0.8)}), (\frac{u_3p}{(0.6,0.5)(0.2,0.2)}), (\frac{u_3q}{(0.2,0.4)(0.1)})\rangle,$$

$$\langle e_2, (\frac{u_2p}{(0.3,0.1)(0.2,0.3,0,6)}), (\frac{u_2q}{(0.5,0.3)(0.5,0.5,0.2)}), (\frac{u_3p}{(0.2,0.2)(0.4)}), (\frac{u_3q}{(0.7,0.3)(0.2,0.9)})\rangle,$$

$$\langle e_3, (\frac{u_1p}{(0.1,0.1)(0.4,0.4)}), (\frac{u_1q}{(0.1,0.3)(0.7,0.6)}), (\frac{u_2p}{(0.4,0.3)(0.4,0.1)}), (\frac{u_2q}{(0.2,0.6)(0.7,0.3)})\rangle\}.$$

Definition 10. *Let* (H_Q, A) *and* (F_Q, B) *be two multi Q-hesitant fuzzy soft sets,* (H_Q, A) *is said to be multi Q-hesitant fuzzy soft subset of* (F_Q, B) *if* $A \subseteq B$ *and* $H_Q(e) \subseteq F_Q(e)$ *for all* $e \in E$ *and denoted by* $(H_Q, A) \subseteq (F_Q, B)$*.*

Proposition 1. *Let* (H_Q, A)*,* (F_Q, B) *and* (G_Q, C) *be three multi Q-hesitant fuzzy soft sets. Then,*

1. $(H_Q, A) \subseteq (U, E)$,
2. $(\phi, A) \subseteq (H_Q, B)$,
3. If $(H_Q, A) \subseteq (F_Q, B)$ and $(F_Q, B) \subseteq (G_Q, C)$, then $(H_Q, A) \subseteq (G_Q, C)$.

Definition 11. *A multi Q-hesitant fuzzy soft set* (H_Q, A) *of dimension k over* $U \times Q$ *is called the null multi Q-hesitant fuzzy soft set if* $H_Q(e) = \phi_k$ *for all* $e \in A$ *and it is denoted by* ϕ^k_A*.*

Definition 12. *A multi Q-hesitant fuzzy soft set* (H_Q, A) *of dimension k over* $U \times Q$ *is called the absolute multi Q-hesitant fuzzy soft set if* $H_Q(e) = 1_k$ *for all* $e \in A$ *and it is denoted by* U^k_A*.*

Definition 13. *Let* (H_Q, A) *be a multi Q-hesitant fuzzy soft set of dimension k over* $U \times Q$*. Then, the complement of* (H_Q, A) *is denoted by* $(H_Q, A)^c$ *and defined by* $(H_Q, A)^c = (H^c_Q, A)$*, where* $H^c_Q : A \longrightarrow M^kQHFS(U \times Q)$ *is mapping given by* $H^c_Q(e) = (H_Q(e))^c$ *for all* $e \in A$*.*

Remark 1. *Clearly,* $((H_Q, A)^c)^c = (H_Q, A)$ *and* $(\phi_A^k)^c = U_A^k$, $(U_A^k)^c = \phi_A^k$.

Definition 14. *The union of two multi Q-hesitant fuzzy soft sets of dimension k over U,* (H_Q, A) *and* (F_Q, B) *is the multi Q-hesitant fuzzy soft set* (G_Q, C), *where* $C = A \cup B$, *and for all* $e \in C$, $G_Q(e) = H_Q(e) \cup F_Q(e)$. *We write* $(H_Q, A) \cup (F_Q, B) = (G_Q, C)$.

Definition 15. *The intersection of of two multi Q-hesitant fuzzy soft sets of dimension k over U,* (H_Q, A) *and* (F_Q, B) *with* $A \cap B \neq \phi$ *is the multi Q-hesitant fuzzy soft set* (G_Q, C), *where* $C = A \cap B$, *and for all* $e \in C$,

$$G_Q(e) = \begin{cases} H_Q(e) & \text{for} \quad e \in A - B, \\ F_Q(e) & \text{for} \quad e \in B - A, \\ H_Q(e) \cup F_Q(e) & \text{for} \quad e \in A \cap B. \end{cases}$$

In this case, we write $(H_Q, A) \cap (F_Q, B) = (G_Q, C)$.

Theorem 1. *Let* (H_Q, A) *and* (F_Q, B) *be two multi Q-hesitant fuzzy soft sets of dimension k over* $U \times Q$. *Then,*

1. $(H_Q, A) \cup (H_Q, A) = (H_Q, A)$,
2. $(H_Q, A) \cap (H_Q, A) = (H_Q, A)$,
3. $(H_Q, A) \cup \phi_A^k = (H_Q, A)$,
4. $(H_Q, A) \cap \phi_A^k = \phi_A^k$,
5. $(H_Q, A) \cup U_A^k = U_A^k$,
6. $(H_Q, A) \cap U_A^k = (H_Q, A)$,
7. $(H_Q, A) \cup (F_Q, B) = (F_Q, B) \cup (H_Q, A)$,
8. $(H_Q, A) \cap (F_Q, B) = (F_Q, B) \cap (H_Q, A)$.

4. Multi Q-Hesitant Fuzzy Soft Rough Set

Definition 16. *Let* (H_Q, A) *be a multi Q-hesitant fuzzy soft set over* $U \times Q$. *A multi Q-hesitant fuzzy subset of* $(U \times Q) \times (E \times Q)$ *is called a multi Q-hesitant fuzzy soft relation* (M^kQHFSR) *from* $(U \times Q)$ *to* $(E \times Q)$ *given by*

$$R_Q = \{\langle (uq, eq), h^i_{R_Q}(uq, eq) \rangle, uq \in U \times Q, eq \in E \times Q, i = 1, 2, \cdots, k\},$$

where $h^i_{R_Q} : (U \times Q) \times (E \times Q) \longrightarrow [0,1]^k$.

Definition 17. *Let U be nonempty universe, Q be a nonempty set and E be the set of parameters.* R_Q *is a multi Q-hesitant fuzzy soft relation* $R_Q \in M^kQHFSR((U \times Q) \times (E \times Q))$ *and the triple* $((U, Q), (E, Q), R_Q)$ *is multi Q-hesitant fuzzy soft approximation space. For any* $A_Q \in M^kQHFS(E)$, *the lower and upper approximations of* A_Q *with respect to* (U, E, Q, R_Q) *denoted by* $\underline{R_Q}(A_Q)$ *and* $\overline{R_Q}(A_Q)$, *are two multi Q-hesitant fuzzy soft sets, respectively, defined as follows:*

$$\underline{R_Q}(A_Q) = \{\langle (uq), h_{\underline{R_Q}(A_Q)}(uq) \rangle : (uq) \in U \times Q\},$$
$$\overline{R_Q}(A_Q) = \{\langle (uq), h_{\overline{R_Q}(A_Q)}(uq) \rangle : (uq) \in U \times Q\},$$

where

$$h_{\underline{R_Q}(A_Q)}(uq) = \{\langle \bigwedge_{e \in E} \{(1 - h^i_{R_Q}(uq, eq)) \vee h^i_{A_Q}(eq)\} \rangle : (uq) \in U \times Q, i = 1, 2, ..., k\},$$
$$h_{\overline{R_Q}(A_Q)}(uq) = \{\langle \bigvee_{e \in E} \{h^i_{R_Q}(uq, eq) \wedge h^i_{A_Q}(eq)\} \rangle : (uq) \in U \times Q, i = 1, 2, ..., k\}.$$

$\underline{R_Q}(A_Q)$ and $\overline{R_Q}(A_Q)$ are, respectively, called the lower and upper Q-hesitant fuzzy soft rough approximations' operators. The pair $(\underline{R_Q}(A_Q), \overline{R_Q}(A_Q))$ is called the multi Q-hesitant fuzzy soft rough set of A_Q with respect to (U, E, Q, R_Q). Moreover, if $\underline{R_Q}(A_Q) = \overline{R_Q}(A_Q)$, then A_Q is called definable.

Example 2. *Suppose that $U = \{u_1, u_2, u_3\}$ is the set of cars that Mr X wants to buy and $Q = \{q_1, q_2\}$ represents the companies of the different cars. They form the universe (U,Q) and let $E = \{e_1 =$ size$, e_2 =$ price$, e_3 =$ colour$\}$ be the set of parameters. Consider a multi Q-hesitant fuzzy soft relation $R_Q : U \times Q \longrightarrow E \times Q$ with dimension k = 2 is given by Table 1.*

Table 1. Multi Q-hesitant fuzzysoft relation R_Q.

R_Q	e_1q_1	e_1q_2	e_2q_1	e_2q_2
(u_1q_1)	{(0.2)(0.6,0.4)}	{(0.3,0.7)(0.6)}	{(0.5,0.4,0.6)(0.6,0.5)}	{(0.4,0.2)(0.1,0.3)}
(u_1q_2)	{(0.8,0.5)(0.2)}	{(0.6,0.9)(0.2,0.9)}	{(0.3)(0.2,0.7)}	{(0.5,0.2,0.1)(0.1,0.5)}
(u_2q_1)	{(0.1,0.3)(0.9,0.7,0.2)}	{(0.5,0.1,)(0.6,0.2)}	{(0.4)(0.5)}	{(0.2,0.4)(0.2,0.8)}
(u_2q_2)	{(0.5)(0.6)}	{(0.9,0.5)(0.6,0.7,0.4)}	{(0.6)(0.3,0.1)}	{(0.2)(0.6,0.1)}

Now, if Mr X gives the optimum decision object $A_Q \in M^kQHF(E)$, which is a Q-hesitant fuzzy subset defined as follows:

$$A_Q = \{\langle((e_1q_1), \{(0.1, 0.3)(0.4, 0.5)\}), ((e_1q_2), \{(0.2, 0.4)(0.5, 0.6)\})\rangle, \langle((e_2q_1), \{(0.3, 0.6)(0.6, 0.7)\}), ((e_2q_2), \{(0.2, 0.5), (0.2, 0.8)\})\rangle\}.$$

Then, by Definition 17, we have

$$h_{\underline{R_Q}}(u_1q_1) = \bigwedge_{e\in E}\{(1 - h^2_{R_Q})(u_1q_1, eq) \vee h^2_{A_Q}(eq)\}$$
$$= (\{(0.8), (0.4, 0.6)\} \vee \{(0.1, 0.3)(0.4, 0.5)\}) \wedge (\{(0.7, 0.3), (0.4)\} \vee \{(0.2, 0.4)(0.5, 0.6)\})$$
$$\wedge (\{(0.5, 0.6, 0.4), (0.4, 0.5)\} \vee \{(0.3, 0.6)(0.6, 0.7)\}) \wedge (\{(0.6, 0.8), (0.9, 0.7)\} \vee \{(0.2, 0.5), (0.2, 0.8)\})$$
$$= \{(0.8, 0.8), (0.4, 0.6)\} \wedge \{(0.7, 0.4), (0.5, 0.6)\} \wedge \{(0.5, 0.6, 0.6), (0.6, 0.7)\} \wedge \{(0.6, 0.8), (0.9, 0.8)\}$$
$$= \{(0.5, 0.4, 0.4), (0.4, 0.6)\}.$$

Similarly, we have

$$h_{\underline{R_Q}}(u_1q_2) = \{(0.2, 0.4, 0.4)(0.8, 0.6)\},$$
$$h_{\underline{R_Q}}(u_2q_1) = \{(0.5, 0.6)(0.4, 0.5, 0.7)\},$$
$$h_{\underline{R_Q}}(u_2q_2) = \{(0.2, 0.5)(0.4, 0.5, 0.5)\},$$
$$h_{\overline{R_Q}}(u_1q_1) = \{(0.3, 0.4, 0.6)(0.6, 0.6)\},$$
$$h_{\overline{R_Q}}(u_1q_2) = \{(0.3, 0.4, 0.4)(0.2, 0.7)\},$$
$$h_{\overline{R_Q}}(u_2q_1) = \{(0.3, 0.4)(0.5, 0.8, 0.8)\},$$
$$h_{\overline{R_Q}}(u_2q_2) = \{(0.3, 0.6)(0.5, 0.6, 0.5)\}.$$

Thus, we conclude that:

$$\underline{R_Q}(A_Q) = \{\langle(u_1q_1), \{(0.5, 0.4, 0.4), (0.4, 0.6)\}\rangle, \langle(u_1q_2), \{(0.2, 0.4, 0.4)(0.8, 0.6)\}\rangle, \langle(u_2q_1), \{(0.5, 0.6)(0.4, 0.5, 0.7)\}\rangle, \langle(u_2q_2), \{(0.2, 0.5)(0.4, 0.5, 0.5)\}\rangle\},$$

$$\overline{R_Q}(A_Q) = \{\langle(u_1q_1), \{(0.3, 0.4, 0.6)(0.6, 0.6)\rangle, \langle(u_1q_2), \{(0.3, 0.4, 0.4)(0.2, 0.7)\}\rangle, \langle(u_2q_1), \{(0.3, 0.4)(0.5, 0.8, 0.8)\}\rangle, \langle(u_2q_2), \{0.3, 0.6)(0.5, 0.6, 0.5)\}\rangle\}.$$

The pair $(\underline{R_Q}(A_Q), \overline{R_Q}(A_Q))$ is called a multi Q-hesitant fuzzy soft rough set with dimension 2.

Theorem 2. *Let* (U, E, Q, R_Q) *be multi Q-hesitant fuzzy soft approximation space. The lower and upper Q-hesitant fuzzy soft rough approximations operators* $\underline{R_Q}(A_Q)$ *and* $\overline{R_Q}(A_Q)$*, respectively, for any* $A_Q, B_Q \in M^kQHF(E)$ *satisfy the following properties:*

1. $\underline{R_Q}(A^c_Q) = (\overline{R_Q}(A_Q))^c, \overline{R_Q}(A^c_Q) = (\underline{R_Q}(A_Q))^c,$
2. $A_Q \subseteq B_Q \Rightarrow \underline{R_Q}(A_Q) \subseteq (\underline{R_Q}(B_Q))\, , A_Q \subseteq B_Q \Rightarrow \overline{R_Q}(A_Q) \subseteq (\overline{R_Q}(A_Q)),$
3. $\underline{R_Q}(A_Q \cap B_Q) = \underline{R_Q}(A_Q) \cap (\underline{R_Q}(B_Q)), \overline{R_Q}(A_Q \cup B_Q) = \overline{R_Q}(A_Q) \cup (\overline{R_Q}(B_Q)),$
4. $\underline{R_Q}(A_Q \cup B_Q) \supseteq \underline{R_Q}(A_Q) \cup (\underline{R_Q}(B_Q)), \overline{R_Q}(A_Q \cap B_Q) \subseteq \overline{R_Q}(A_Q) \cap (\overline{R_Q}(B_Q)).$

Proof. 1. By Definition 17, we have

$$\underline{R_Q}(A^c_Q) = \{\langle (uq), h^i_{\underline{R_Q}(\sim A_Q)}(uq)\rangle : uq \in U \times Q, i = 1,2,...,k\}$$
$$= \{\langle (uq), \wedge_{e\in E}\{h^i_{\sim R_Q}(uq, eq) \vee h^i_{\sim A_Q}(eq)\rangle : uq \in U \times Q, i = 1,2,...,k\}$$
$$= \{\langle (uq), \sim (\vee_{e\in E}\{h^i_{R_Q}(uq, eq) \wedge h^i_{A_Q}(eq)\rangle : uq \in U \times Q, i = 1,2,...,k\}$$
$$= \{\langle (uq), \sim h^i_{\overline{R_Q}(A_Q)}(uq)\rangle : (uq) \in U \times Q, i = 1,2,...,k\}$$
$$= (\overline{R_Q}(A_Q))^c.$$

Similarly, we can obtain that $\overline{R_Q}(A^c_Q) = (\underline{R_Q}(A_Q))^c$.

2. If $A_Q \subseteq B_Q$, by Definition 8, $h^i_{A_Q}(uq) \leq h^i_{B_Q}(uq)$ for all $u \in U, q \in Q$. Therefore, $\wedge_{e\in E}\{(1 - h^i_{R_Q})(uq, eq) \vee h^i_{A_Q}(eq)\} \leq \wedge_{e\in E}\{(1 - h^i_{R_Q})(uq, eq) \vee h^i_{B_Q}(eq)\}$, thus $h^i_{\underline{R_Q}(A_Q)}(uq) \leq h^i_{\underline{R_Q}(B_Q)}(uq)$. It follows that $\underline{R_Q}(A_Q) \subseteq \underline{R_Q}(B_Q)$.

3. $$\underline{R_Q}(A_Q \cap B_Q) = \{\langle (uq), h^i_{\underline{R_Q}(A_Q\cap B_Q)}(uq)\rangle : uq \in U \times Q, i = 1,2,...,k\}$$
$$= \{\langle (uq), \wedge_{e\in E}(1 - h^i_{R_Q})(uq, eq) \vee h^i_{A_Q\cap B_Q}(eq)\rangle : uq \in U \times Q, i = 1,2,...,k\}$$
$$= \{\langle (uq), \wedge_{e\in E}(1 - h^i_{R_Q})(uq, eq) \vee (h^i_{A_Q}(eq) \wedge h^i_{B_Q}(eq))\rangle : uq \in U \times Q, i = 1,2,...,k\}$$
$$= \{\langle (uq), \Big(\wedge_{e\in E}((1 - h^i_{R_Q})(uq, eq) \vee h^i_{A_Q}(eq))\Big) \wedge \Big(\wedge_{e\in E}((1 - h^i_{R_Q})(uq, eq) \vee h^i_{B_Q}(eq))\Big)\rangle : uq \in U \times Q, i = 1,2,...,k\}$$
$$= \{\langle (uq), h^i_{\underline{R_Q}(A_Q)}(uq) \wedge h^i_{\underline{R_Q}(B_Q)}(uq)\rangle : uq \in U \times Q\}$$
$$= \underline{R_Q}(A_Q) \cap \underline{R_Q}(B_Q).$$

Hence, $\underline{R_Q}(A_Q \cap B_Q) = \underline{R_Q}(A_Q) \cap \underline{R_Q}(B_Q)$.

Similarly, we can prove that $\overline{R_Q}(A_Q \cap B_Q) = \overline{R_Q}(A_Q) \cap \overline{R_Q}(B_Q)$.

4. $$\underline{R_Q}(A_Q \cup B_Q) = \{\langle (uq), h^i_{\underline{R_Q}(A_Q\cup B_Q)}(uq)\rangle : uq \in U \times Q, i = 1,2,...,k\}$$
$$= \{\langle (uq), \wedge_{e\in E}(1 - h^i_{R_Q})(uq, eq) \vee h^i_{A_Q\cup B_Q}(eq)\rangle : uq \in U \times Q, i = 1,2,...,k\}$$
$$= \{\langle (uq), \wedge_{e\in E}(1 - h^i_{R_Q})(uq, eq) \vee (h^i_{A_Q}(eq) \vee h^i_{B_Q}(eq))\rangle : uq \in U \times Q, i = 1,2,...,k\}$$
$$= \{\langle (uq), \Big(\wedge_{e\in E}((1 - h^i_{R_Q})(uq, eq) \vee h^i_{A_Q}(eq))\Big) \vee \Big(\wedge_{e\in E}((1 - h^i_{R_Q})(uq, eq) \vee h^i_{B_Q}(eq))\Big)\rangle : uq \in U \times Q, i = 1,2,...,k\}$$
$$= \{\langle (uq), h^i_{\underline{R_Q}(A_Q)}(uq) \vee h^i_{\underline{R_Q}(B_Q)}(uq)\rangle : uq \in U \times Q, i = 1,2,...,k\}$$
$$= \underline{R_Q}(A_Q) \cup (\underline{R_Q}(B_Q)).$$

Hence, $\underline{R_Q}(A_Q \cup B_Q) = \underline{R_Q}(A_Q) \cup \underline{R_Q}(B_Q)$.

Similarly, we can prove that $\overline{R_Q}(A_Q \cup B_Q) = \overline{R_Q}(A_Q) \cup \overline{R_Q}(B_Q)$.

□

Theorem 3. *Let R_Q, S_Q be multi Q-hesitant fuzzy soft relations from $(U \times Q)$ to $(E \times Q)$, if $R_Q \subseteq S_Q$, for any $A \in QHF(E)$, then:*

1. $\underline{R_Q}(A_Q) \supseteq \underline{S_Q}(A_Q)$,
2. $\overline{R_Q}(A_Q) \subseteq \overline{S_Q}(A_Q)$.

Proof. 1. If $R_Q \subseteq S_Q$, then, by Definition 8, we have $h^i_{R_Q}(uq, eq) \leq h^i_{S_Q}(uq, eq)$ for all $uq \in U \times Q$, $eq \in E \times Q$, then

$$\begin{aligned} \underline{R_Q}(A_Q) &= \{\langle (uq), h^i_{\underline{R_Q(A_Q)}}(uq)\rangle : uq \in U \times Q, i = 1, 2, ..., k\} \\ &= \{\langle (uq), \wedge_{e \in E}\{(1 - h^i_{R_Q})(uq, eq) \vee h^i_{A_Q}(eq)\rangle : uq \in U \times Q, i = 1, 2, ..., k\} \\ &\geq \{\langle (uq), \wedge_{e \in E}\{(1 - h^i_{S_Q})(uq, eq) \vee h^i_{A_Q}(eq)\rangle : uq \in U \times Q, i = 1, 2, ..., k\} \\ &= \{\langle (uq), h^i_{\underline{S_Q(A_Q)}}(uq)\rangle : uq \in U \times Q, i = 1, 2, ..., k\} \\ &= \underline{S_Q}(A_Q). \end{aligned}$$

2. Similarly, it can be proved.

□

5. Multi Q-Hesitant Fuzzy Soft Multi-Granulation Rough Set

Definition 18. *Let U be a universal set and Q be non-empty set, and E be the set of parameters and R_{Q_j},(j=1,2,...,m) be multi Q_m-hesitant fuzzy soft relations over $(U \times Q) \times (E \times Q)$, and (U, E, Q, R_{Q_j}) be called multi Q-hesitant fuzzy soft multi-granulation approximation space, for any $A_Q \in M^kQHF(E)$, the optimistic lower and upper approximation of A_Q with respect to (U, E, Q, R_{Q_j}) are defined as follows:*

$$\underline{\sum_{j=1}^{m} R_{Q_j}}^{o}(A_Q) = \{\langle (uq), h^i_{\underline{\sum_{j=1}^{m} R_{Q_j}}^{o}(A_Q)}(u, q)\rangle : uq \in U \times Q\},$$

$$\overline{\sum_{j=1}^{m} R_{Q_j}}^{o}(A_Q) = \{\langle (uq), h^i_{\overline{\sum_{j=1}^{m} R_{Q_j}}^{o}(A_Q)}(u, q)\rangle : uq \in U \times Q\},$$

where

$$h_{\underline{\sum_{j=1}^{m} R_{Q_j}}^{o}(A_Q)}(uq) = \{\langle \bigvee_{j=1}^{m} \bigwedge_{i=1}^{k} \{(1 - h^i_{R_{Q_j}})(uq, eq) \vee h^i_{A_Q}(eq)\}\rangle : uq \in U \times Q\},$$

$$h_{\overline{\sum_{j=1}^{m} R_{Q_j}}^{o}(A_Q)}(uq) = \{\langle \bigwedge_{j=1}^{m} \bigvee_{i=1}^{k} \{h^i_{R_{Q_i}}(uq, eq) \wedge h^i_{A_Q}(eq)\}\rangle : uq \in U \times Q\}.$$

The pair $(\underline{\sum_{j=1}^{m} R_{Q_j}}^{o}(A_Q), \overline{\sum_{j=1}^{m} R_{Q_j}}^{o}(A_Q))$ is called an optimistic multi Q-hesitant fuzzy soft multi-granulation rough set of A_Q with respect to (U, E, Q, R_{Q_j}).

Theorem 4. *Let (U, E, Q, R_{Q_j}) be multi Q-hesitant fuzzy soft multi-granulation approximation space and $R_{Q_j} \in M^kQHFSR((U \times Q) \times (E \times Q))$,(j =1,2,...,m) be multi Q_m hesitant fuzzy soft relations over $(U \times Q) \times (E \times Q)$, for any $A_Q, B_Q \in M^kQHF(E)$, the optimistic lower and upper approximation satisfy the following properties:*

1. $\underline{\sum_{j=1}^{m} R_{Q_j}}^{o}(A^c_Q) = (\overline{\sum_{j=1}^{m} R_{Q_j}}^{o}(A_Q))^c$,
   $\overline{\sum_{j=1}^{m} R_{Q_j}}^{o}(A^c_Q) = (\underline{\sum_{j=1}^{m} R_{Q_j}}^{o}(A_Q))^c$.
2. $A_Q \subseteq B_Q \Rightarrow \underline{\sum_{j=1}^{m} R_{Q_j}}^{o}(A_Q) \subseteq \underline{\sum_{j=1}^{m} R_{Q_j}}^{o}(B_Q)$,
   $A_Q \subseteq B_Q \Rightarrow \overline{\sum_{j=1}^{m} R_{Q_j}}^{o}(A_Q) \subseteq \overline{\sum_{j=1}^{m} R_{Q_j}}^{o}(B_Q)$.

3. $\underline{\sum_{j=1}^{m} R_{Q_j}}^{o}(A_Q \cap B_Q) = \underline{\sum_{j=1}^{m} R_{Q_j}}^{o}(A_Q) \cap \underline{\sum_{j=1}^{m} R_{Q_j}}^{o}(B_Q)$,
$\overline{\sum_{j=1}^{m} R_{Q_j}}^{o}(A_Q \cup B_Q) = \overline{\sum_{j=1}^{m} R_{Q_j}}^{o}(A_Q) \cup \overline{\sum_{j=1}^{m} R_{Q_j}}^{o}(B_Q)$.
4. $\underline{\sum_{j=1}^{m} R_{Q_j}}^{o}(A_Q \cup B_Q) \supseteq \underline{\sum_{j=1}^{m} R_{Q_j}}^{o}(A_Q) \cup \underline{\sum_{j=1}^{m} R_{Q_j}}^{o}(B_Q)$,
$\overline{\sum_{j=1}^{m} R_{Q_j}}^{o}(A_Q \cap B_Q) \subseteq \overline{\sum_{j=1}^{m} R_{Q_j}}^{o}(A_Q) \cap \overline{\sum_{j=1}^{m} R_{Q_j}}^{o}(B_Q)$.

Proof. 1. By Definition 18, we have,

$$\begin{aligned}
\underline{\sum_{j=1}^{m} R_{Q_j}}^{o}(A_Q^c) &= \{\langle (uq), h^i_{\underline{\sum_{j=1}^{m} R_{Q_j}}^{o}(\sim A_Q)}(uq)\rangle : uq \in U \times Q, i = 1,2,...,k\} \\
&= \{\langle (uq), \vee_{j=1}^{m} \wedge_{i=1}^{k} \{\sim h^i_{R_{Q_j}}(uq,eq) \vee h^i_{\sim A_Q}(eq)\rangle : uq \in U \times Q\} \\
&= \{\langle (uq), \sim (\wedge_{j=1}^{m} \vee_{i=1}^{k} \{h^i_{R_{Q_j}}(uq,eq) \wedge h^i_{A_Q}(eq))\rangle : uq \in U \times Q\} \\
&= \{\langle (uq), \sim h^i_{\overline{\sum_{j=1}^{m} R_{Q_j}}^{o}(A_Q)}(uq)\rangle : uq \in U \times Q, i = 1,2,...,k\} \\
&= (\overline{\sum_{j=1}^{m} R_{Q_j}}^{o}(A_Q))^c.
\end{aligned}$$

Similarly, we can obtain that $\overline{\sum_{j=1}^{m} R_{Q_j}}^{o}(A_Q^c) = (\underline{\sum_{j=1}^{m} R_{Q_j}}^{o}(A_Q))^c$.

2. If $A_Q \subseteq B_Q$, by Definition 8, $h^i_{A_Q}(u,q) \leq h^i_{B_Q}(uq)$ for all $u \in U, q \in Q$, therefore, $\vee_{j=1}^{m} \wedge_{i=1}^{k} \{(1 - h^i_{R_{Q_j}})(uq,eq) \vee h_{A_Q}(e,q)\} \leq \vee_{j=1}^{m} \wedge_{i=1}^{k} \{(1 - h^i_{R_{Q_j}})(uq,eq) \vee h^i_{B_Q}(eq)\}$, thus $h^i_{\underline{\sum_{j=1}^{m} R_{Q_j}}^{o}(A_Q)}(uq) \leq h^i_{\underline{\sum_{j=1}^{m} R_{Q_j}}^{o}(B_Q)}(uq)$ it follows that $\underline{\sum_{j=1}^{m} R_{Q_j}}^{o}(A_Q) \subseteq \underline{\sum_{j=1}^{m} R_{Q_j}}^{o}(B_Q)$.

3. $$\begin{aligned}
\underline{\sum_{j=1}^{m} R_{Q_j}}^{o}(A_Q \cap B_Q) &= \{\langle (uq), h^i_{\underline{\sum_{j-1}^{m} R_{Q_j}}^{o}(A_Q \cap B_Q)}(uq)\rangle : uq \in U \times Q, i = 1,2,...,k\} \\
&= \{\langle (uq), \vee_{j=1}^{m} \wedge_{i=1}^{k} (1 - h^i_{R_{Q_j}})(uq,eq) \vee h^i_{A_Q \cap B_Q}(eq)\rangle : uq \in U \times Q\} \\
&= \{\langle (uq), \vee_{j=1}^{m} \wedge_{i=1}^{k} (1 - h^i_{R_{Q_j}})(uq,eq) \vee (h^i_{A_Q}(eq) \wedge h^i_{B_Q}(eq))\rangle : uq \in U \times Q\} \\
&= \{\langle (uq), \left(\vee_{i=1}^{m} \wedge_{i=1}^{k} ((1 - h^i_{R_{Q_j}})(uq,eq) \vee h^i_{A_Q}(eq))\right) \wedge \\
&\quad \left(\vee_{j=1}^{m} \wedge_{i=1}^{k} ((1 - h^i_{R_{Q_j}})(uq,eq) \vee h^i_{B_Q}(eq))\right)\rangle : uq \in U \times Q\} \\
&= \{\langle (uq), h^i_{\underline{\sum_{i=1}^{m} R_{Q_i}}^{o}(A_Q)}(uq) \wedge h_{\underline{\sum_{i=1}^{m} R_{Q_i}}^{o}(B_Q)}(uq)\rangle : uq \in U \times Q\} \\
&= \underline{\sum_{j=1}^{m} R_{Q_j}}^{o}(A_Q) \cap (\underline{\sum_{j=1}^{m} R_{Q_j}}^{o}(B_Q)).
\end{aligned}$$

Hence, $\underline{\sum_{j=1}^{m} R_{Q_j}}^{o}(A_Q \cap B_Q) = \underline{\sum_{j=1}^{m} R_{Q_j}}^{o}(A_Q) \cap \underline{\sum_{j=1}^{m} R_{Q_j}}^{o}(B_Q)$.

Similarly, we can prove that $\overline{\sum_{j=1}^{m} R_{Q_j}}^{o}(A_Q \cap B_Q) = \overline{\sum_{j=1}^{m} R_{Q_j}}^{o}(A_Q) \cap \overline{\sum_{j=1}^{m} R_{Q_j}}^{o}(B_Q)$.

4. $$\begin{aligned}
\underline{\sum_{j=1}^{m} R_{Q_j}}^{o}(A_Q \cup B_Q) &= \{\langle (uq), h^i_{\underline{\sum_{j=1}^{m} R_{Q_j}}^{o}(A_Q \cup B_Q)}(uq)\rangle : uq \in U \times Q, i = 1,2,...,k\} \\
&= \{\langle (uq), \vee_{j=1}^{m} \wedge_{i=1}^{k} (1 - h^i_{R_{Q_j}})(uq,eq) \vee h^i_{A_Q \cup B_Q}(eq)\rangle : uq \in U \times Q\} \\
&= \{\langle (uq), \vee_{j=1}^{m} \wedge_{i=1}^{k} (1 - h^i_{R_{Q_j}})(uq,eq) \vee (h^i_{A_Q}(eq) \vee h^i_{B_Q}(eq))\rangle : uq \in U \times Q\} \\
&= \{\langle (uq), \left(\vee_{j=1}^{m} \wedge_{i=1}^{k} ((1 - h^i_{R_{Q_j}})(uq,eq) \vee h^i_{A_Q}(eq))\right) \vee \\
&\quad \left(\vee_{i=1}^{m} \wedge_{i=1}^{k} ((1 - h^i_{R_{Q_j}})(uq,eq) \vee h^i_{B_Q}(eq))\right)\rangle : (uq) \in U \times Q\} \\
&= \{\langle (uq), h^i_{\underline{\sum_{j=1}^{m} R_{Q_j}}^{o}(A_Q)}(uq) \vee h^i_{\underline{\sum_{j=1}^{m} R_{Q_j}}^{o}(B_Q)}(uq)\rangle : uq \in U \times Q, i = 1,2,...,k\} \\
&= \underline{\sum_{j=1}^{m} R_{Q_j}}^{o}(A_Q) \cup (\underline{\sum_{j=1}^{m} R_{Q_j}}^{o}(B_Q)).
\end{aligned}$$

Hence, $\underline{\sum_{j=1}^{m} R_{Q_j}{}^{o}}(A_Q \cup B_Q) = \underline{\sum_{j=1}^{m} R_{Q_j}{}^{o}}(A_Q) \cup \underline{R_Q}(B_Q)$.

Similarly, we can prove that $\overline{\sum_{j=1}^{m} R_{Q_j}{}^{o}}(A_Q \cup B_Q) = \overline{\sum_{j=1}^{m} R_{Q_j}{}^{o}}(A_Q) \cup \overline{\sum_{j=1}^{m} R_{Q_j}{}^{o}}(B_Q)$.
□

Theorem 5. *Let $R_{Q_j}, S_{Q_j} \in M^kQHFSR((U \times Q) \times (E \times Q))$ $(j = 1, 2, ..., m)$ be multi Q_m hesitant fuzzy soft relations over $(U \times Q) \times (E \times Q)$, if $R_{Q_j} \subseteq S_{Q_j}$, for any $A_Q \in M^kQHF(E)$, the following properties are true:*

1. $\underline{\sum_{j=1}^{m} R_{Q_j}{}^{o}}(A_Q) \supseteq \underline{\sum_{j=1}^{m} S_{Q_j}{}^{o}}(A_Q)$,
2. $\overline{\sum_{j=1}^{m} R_{Q_j}{}^{o}}(A_Q) \subseteq \overline{\sum_{j=1}^{m} S_{Q_j}{}^{o}}(A_Q)$.

Proof. 1. If $R_{Q_j} \subseteq S_{Q_j}$, then, by Definition 8, we have $h^i_{R_{Q_j}}(uq, eq) \leq h^i_{S_{Q_j}}(uq, eq)$ for all $(u, q) \in U \times Q$, $eq \in E \times Q$, then

$\underline{\sum_{j=1}^{m} R_{Q_j}{}^{o}}(A_Q) = \{\langle (uq), h^i_{\underline{\sum_{j=1}^{m} R_{Q_j}{}^{o}(A_Q)}}(uq)\rangle : uq \in U \times Q, i = 1, 2, ..., k\}$

$= \{\langle (uq), \bigvee_{j=1}^{m} \bigwedge_{i=1}^{k} \{(1 - h^i_{R_{Q_j}})(uq, eq) \vee h^i_{A_Q}(eq)\rangle : uq \in U \times Q\}$

$\geq \{\langle (uq), \bigvee_{j=1}^{m} \bigwedge_{i=1}^{k} \{(1 - h^i_{S_{Q_j}})(uq, eq) \vee h^i_{A_Q}(eq)\rangle : uq \in U \times Q\}$

$= \{\langle (uq), h^i_{\underline{\sum_{j=1}^{m} S_{Q_j}{}^{o}(A_Q)}}(uq)\rangle : uq \in U \times Q\}$

$= \underline{\sum_{j=1}^{m} S_{Q_j}{}^{o}}(A_Q)$.

2. It can be proved similarly to 1.
□

Definition 19. *Let U be a universal set and Q be a non-empty set, and E be the set of parameters and R_{Q_j},(j=1,2,...,m) are multi Q_m-hesitant fuzzy soft relations over $(U \times Q) \times (E \times Q)$, the triple$(U, E, Q, R_{Q_j})$ is called multi Q-hesitant fuzzy soft multi-granulation approximation space, for any $A_Q \in M^kQHF(E)$, and the pessimistic lower and upper approximation of A_Q with respect to (U, E, Q, R_{Q_j}) are defined as follows:*

$$\underline{\sum_{j=1}^{m} R^p_{Q_j}}(A_Q) = \{\langle (uq), h^i_{\underline{\sum_{j=1}^{m} R^p_{Q_j}(A_Q)}}(u, q)\rangle : uq \in U \times Q\},$$

$$\overline{\sum_{j=1}^{m} R^p_{Q_j}}(A_Q) = \{\langle (uq), h^i_{\overline{\sum_{j=1}^{m} R_{Q_j}{}^{p}(A_Q)}}(u, q)\rangle : uq \in U \times Q\},$$

where

$$h_{\underline{\sum_{j=1}^{m} R_{Q_j}{}^{p}(A_Q)}}(uq) = \{\langle \bigwedge_{j=1}^{m} \bigwedge_{i=1}^{k} \{(1 - h^i_{R_{Q_j}})(uq, eq) \vee h^i_{A_Q}(eq)\}\rangle : uq \in U \times Q\},$$

$$h_{\overline{\sum_{j=1}^{m} R_{Q_j}{}^{p}(A_Q)}}(uq) = \{\langle \bigvee_{j=1}^{m} \bigvee_{i=1}^{k} \{h^i_{R_{Q_i}}(uq, eq) \wedge h^i_{A_Q}(eq)\}\rangle : uq \in U \times Q\}.$$

The pair $(\underline{\sum_{j=1}^{m} R_{Q_j}{}^{p}}(A_Q), \overline{\sum_{j=1}^{m} R_{Q_j}{}^{p}}(A_Q))$ is called an pessimistic multi Q-hesitant fuzzy soft multi-granulation rough set of A_Q with respect to (U, E, Q, R_{Q_j}).

Theorem 6. *Let (U, E, Q, R_{Q_j}) be multi Q-hesitant fuzzy soft multi-granulation approximation space and $R_{Q_j} \in M^kQHFSR((U \times Q) \times (E \times Q)$,(i=1,2,...,m) be multi Q_m hesitant fuzzy soft relations over $(U \times Q) \times (E \times Q)$, for any $A_Q, B_Q \in M^kQHF(E)$, the pessimistic lower and upper approximation satisfy the following properties:*

1. $\underline{\sum_{j=1}^{m} R_{Q_j}}^{p}(A_Q^c) = (\overline{\sum_{j=1}^{m} R_{Q_j}^{p}}(A_Q))^c, \overline{\sum_{j=1}^{m} R_{Q_j}^{p}}(A_Q^c) = (\underline{\sum_{j=1}^{m} R_{Q_j}}^{p}(A_Q))^c.$
2. $A_Q \subseteq B_Q \Rightarrow \underline{\sum_{j=1}^{m} R_{Q_j}}^{p}(A_Q) \subseteq \underline{\sum_{j=1}^{m} R_{Q_j}}^{p}(B_Q)$,
   $A_Q \subseteq B_Q \Rightarrow \overline{\sum_{j=1}^{m} R_{Q_j}^{p}}(A_Q) \subseteq \overline{\sum_{j=1}^{m} R_{Q_j}^{p}}(B_Q).$
3. $\underline{\sum_{j=1}^{m} R_{Q_j}}^{o}(A_Q \cap B_Q) = \underline{\sum_{j=1}^{m} R_{Q_j}}^{p}(A_Q) \cap \underline{\sum_{j=1}^{m} R_{Q_j}}^{p}(B_Q)$,
   $\overline{\sum_{j=1}^{m} R_{Q_j}^{p}}(A_Q \cup B_Q) = \overline{\sum_{j=1}^{m} R_{Q_j}^{p}}(A_Q) \cup \overline{\sum_{j=1}^{m} R_{Q_j}^{p}}(B_Q).$
4. $\underline{\sum_{i=1}^{m} R_{Q_i}}^{p}(A_Q \cup B_Q) \supseteq \underline{\sum_{j=1}^{m} R_{Q_j}}^{p}(A_Q) \cup \underline{\sum_{j=1}^{m} R_{Q_j}}^{p}(B_Q)$,
   $\overline{\sum_{j=1}^{m} R_{Q_j}^{p}}(A_Q \cap B_Q) \subseteq \overline{\sum_{j=1}^{m} R_{Q_j}^{p}}(A_Q) \cap \overline{\sum_{j=1}^{m} R_{Q_j}^{p}}(B_Q).$

Proof. It can easily be proved by using Theorem 4 and Definition 19. □

Theorem 7. *Let (U, E, Q, R_{Q_j}) be multi Q-hesitant fuzzy soft multi-granulation approximation space and $R_{Q_j}, S_{Q_j} \in M^kQHFSR((U \times Q) \times (E \times Q)$,(i=1,2,...,m) be multi Q_m hesitant fuzzy soft relations over $(U \times Q) \times (E \times Q)$, if $R_{Q_j} \subseteq S_{Q_j}$, for any $A_Q \in M^kQHF(E)$, the following properties are true :*

1. $\underline{\sum_{j=1}^{m} R_{Q_j}}^{p}(A_Q) \supseteq \underline{\sum_{j=1}^{m} S_{Q_j}}^{p}(A_Q)$,
2. $\overline{\sum_{j=1}^{m} R_{Q_j}^{p}}(A_Q) \subseteq \overline{\sum_{j=1}^{m} S_{Q_j}^{p}}(A_Q).$

Proof. It can be easily proved by Theorem 5 and Definition 19. □

6. Photovoltaic Systems Fault Detection Approach

Fuzzy sets and rough sets are both mathematical tools to handle uncertainties, they have a wide applications in many practical problems, especially in the area of decision-making. In many instances, we can not successfully utilize these classical methods to deal with decision-making problems since various types of uncertainties involved in these problems which require that second dimension must be added to the expression of the membership value.

Inspired by this, we construct a new model to the decision-making problem of photovoltaic system fault detection depending on the notion of M^kQHFS multi-granulation rough set.

6.1. The Application Model

Photovoltaic systems (solar panel) can be explained as a piece of equipment converting sunlight (photons) to electric energy. Loss of power in photovoltaic systems can occur suddenly any time. Therefore, it is necessary to detect faults as early as possible. Unexpected power loss is usually detected by comparing the output to a reference figure.

By employing the model of multi Q-hesitant fuzzy soft multi-granulation rough sets, we can indicate the loss of power in photovoltaic systems expressed as multi Q-hesitant fuzzy soft elements.

Let $U = \{u_1, u_2, ..., u_v\}$ be the fault type set, $Q = \{q_1, q_2\}$ represents the set of condition degrees and $E = \{e_1, e_2, ..., e_s\}$ be the set of power measurement. Let $R_{Q_j} \in M^kQHFSR((U \times Q) \times (E \times Q))$ $(j = 1, 2, ..., m)$, which was employed to indicate the electrical information given by m experts via the membership degrees between the fault detected with condition degrees and the power measurement with condition degrees. In addition, $A_Q \in M^kQHF(E)$ represents the power measurements with the condition degree of each measurement. Then, we construct a multi Q-hesitant fuzzy soft decision information system (U, E, Q, R_{Q_j}) of the electrical detection procedure.

First, based on the the score function definition given by Xia and Xu [29], we define the score function of M^kQHFS element as follows:

Definition 20. *Let $h^i_Q(uq)$ be M^kQHFS element, then the score function can be fined as follows:*

$$S(h^i_Q(uq)) = \{\frac{1}{l(h^i_Q)} \sum_{\gamma \in h^i_Q} \gamma, i = 1, 2, ..., k\},$$

where $l(h^i_Q)$ is the number of values in $(h^i_Q(uq))$.

By Definition 20, we can define the sum of A_Q and B_Q as follows:

Definition 21. *Letting A_Q and B_Q be two M^kQHFSS in $U \times Q$, we define the sum of $h^i_{A_Q}(uq)$ and $h^i_{B_Q}(uq)$ such that $i = 1, 2, ..., k$ by*

$$h^i_{A_Q}(uq) \oplus h^i_{B_Q}(uq) = \{\langle h^1_{A_Q}(uq) + h^1_{B_Q}(uq) - h^1_{A_Q}(uq) h^1_{B_Q}(uq), h^2_{A_Q}(uq) h^2_{B_Q}(uq), ..., h^k_{A_Q}(uq) h^k_{B_Q}(uq) \rangle\}.$$

Based on the decision-making strategy developed in [14], we introduce the following three measurement indices which are denoted by:

$T_1 = \left\{(S, T) | \max_{u_s q_t} S\left(\underline{\sum_{j=1}^{m} R_{Q_j}}^{o}(A_Q)(u_s q_t) \oplus \overline{\sum_{j=1}^{m} R_{Q_j}}^{o}(A_Q)(u_s q_t)\right)\right\}$,

$T_2 = \left\{(X, Y) | \max_{u_x q_y} S\left(\underline{\sum_{j=1}^{m} R_{Q_j}}^{p}(A_Q)(u_x q_y) \oplus \overline{\sum_{j=1}^{m} R_{Q_j}}^{p}(A_Q)(u_x q_y)\right)\right\}$,

$T_3 = \left\{(V, N) | \max_{u_v q_n} S\left((\underline{\sum_{j=1}^{m} R_{Q_j}}^{o}(A_Q) \oplus \overline{\sum_{i=1}^{m} R_{Q_j}}^{o}(A_Q)) \oplus (\underline{\sum_{i=1}^{m} R_{Q_j}}^{p}(A_Q) \oplus \overline{\sum_{j=1}^{m} R_{Q_j}}^{p}(A_Q))\right)\right\}$.

Now, the decision rules for photovoltaic systems fault detection by using a multi Q-hesitant fuzzy soft multi-granulation rough set are given as follows :

1. If $T_1 \cap T_2 \cap T_3 \neq \phi$, then the decision maker will choose (m, n) as the optimal object, where (m, n) $\in T_1 \cap T_2 \cap T_3$.
2. If $T_1 \cap T_2 \cap T_3 = \phi$ and $T_1 \cap T_2 \neq \phi$, then the decision maker will choose (m, n) as the optimal object, where $(m, n) \in T_1 \cap T_2$.
3. If $T_1 \cap T_2 \cap T_3 = \phi$ and $T_1 \cap T_2 = \phi$, then $(m, n) \in T_3$ is the determined fault type in level.

In the following, we present our method in an Algorithm 1 for the photovoltaic systems fault detection model by using a multi Q-hesitant fuzzy soft multi-granulation rough set.

Algorithm 1. Photovoltaic systems fault detection

1. Input the universal set (U,Q).
2. Input the set (E,Q).
3. Construct multi Q-hesitant fuzzy soft relation according to m experts.
4. Give the testing set $A_Q \in M^kQHF(E)$.
5. Compute the M^kQHFS operators $\underline{\sum_{j=1}^{m} R_{Q_j}}^{o}(A_Q)$, $\overline{\sum_{j=1}^{m} R_{Q_j}}^{o}(A_Q)$,$\underline{\sum_{j=1}^{m} R_{Q_j}}^{p}(A_Q)$, $\overline{\sum_{j=1}^{m} R_{Q_j}}^{p}(A_Q)$.
6. Calculate $\underline{\sum_{j=1}^{m} R_{Q_j}}^{o}(A_Q) \oplus \overline{\sum_{j=1}^{m} R_{Q_j}}^{o}(A_Q)$,$\underline{\sum_{j=1}^{m} R_{Q_j}}^{p}(A_Q) \oplus \overline{\sum_{j=1}^{m} R_{Q_j}}^{p}(A_Q)$ and $\left((\underline{\sum_{j=1}^{m} R_{Q_j}}^{o}(A_Q) \oplus \overline{\sum_{j=1}^{m} R_{Q_j}}^{o}(A_Q)) \oplus (\underline{\sum_{j=1}^{m} R_{Q_j}}^{p}(A_Q) \oplus \overline{\sum_{j=1}^{m} R_{Q_j}}^{p}(A_Q))\right)$.
7. Determine the score function values of $\underline{\sum_{j=1}^{m} R_{Q_j}}^{o}(A_Q) \oplus \overline{\sum_{j=1}^{m} R_{Q_j}}^{o}(A_Q)$,$\underline{\sum_{j=1}^{m} R_{Q_j}}^{p}(A_Q) \oplus \overline{\sum_{j=1}^{m} R_{Q_j}}^{p}(A_Q)$ and $\left((\underline{\sum_{j=1}^{m} R_{Q_j}}^{o}(A_Q) \oplus \overline{\sum_{j=1}^{m} R_{Q_j}}^{o}(A_Q)) \oplus (\underline{\sum_{j=1}^{m} R_{Q_j}}^{p}(A_Q) \oplus \overline{\sum_{j=1}^{m} R_{Q_j}}^{p}(A_Q))\right)$.
8. Compute $T_1 \cap T_2 \cap T_3$ and $T_1 \cap T_2$, and confirm the determined fault type and its degree.

6.2. Example

For illustrating the efficiency of the proposed algorithm, we use a photovoltaic system fault diagnose problem with multi Q-hesitant fuzzy soft decision information.
Suppose that $U = \{u_1, u_2, u_3\}$ be the set of fault type where u_v stands for, partial shading, delamination, cracks in cells, respectively. $Q = \{q_1 = low, q_2 = high\}$ represent the set of status levels and $E = \{e_1, e_2, e_3\}$ be the set of power measurement where e_s stands for current, voltage, and series resistance, respectively. The photovoltaic system fault detection knowledge base with M^kQHFS information with dimension k = 1 is presented in Tables 2–4.

In photovoltaic system fault detection, assume that we take a fault testing sample, which is presented by the following multi Q-hesitant fuzzy soft information:

$$A_Q = \{\langle((e_1, q_1), 0.9, 0.4), ((e_1, q_2), 0.6, 0.8, 0.4)\rangle, \langle((e_2, q_1), 0.1, 0.9), ((e_2, q_2), 0.2, 0.5)\rangle, \langle((e_3, q_1), 0.2, 0.4, 0.1), ((e_3, q_2), 0.3, 0.7)\rangle\}.$$

Table 2. Knowledge given by expert 1.

R_{Q_1}	e_1q_1	e_1q_2	e_2q_1	e_2q_2	e_3q_1	e_3q_2
(u_1q_1)	{0.1,0.4,0.8}	{0.9,0.3}	{0.5,0.7,0.1}	{0.2,0.6,}	{0.9,0.3}	{0.1,0.2,0.4}
(u_1q_2)	{0.5,0.2}	{0.2,0.6}	{0.3,0.1,0.4}	{0.1,0.8}	{0.1,0.3}	{0.4,0.9,0.3}
(u_2q_1)	{0.8,0.1}	{0.1,0.8,0.7 }	{0.8,0.3}	{0.2,0.6,0.3}	{0.2,0.4,0.9}	{0.6,0.3}
(u_2q_2)	{0.3,0.4,0.5}	{0.8,0.3}	{0.5,0.4,0.3}	{0.1,0.6,0.7}	{0.2,0.9}	{0.3,0.1,0.6}
(u_3q_1)	{0.2,0.1}	{0.4,0.7,0.8}	{0.6,0.9,0.4}	{0.7,0.1}	{0.8,0.7,0.2}	{(0.4,0.5)}
(u_3q_2)	{0.1,0.2,0.4}	{0.7,0.2,0.5}	{0.5,0.6}	{0.1,0.2,0.8}	{0.4,0.2}	{0.7,0.3,0.1}

Table 3. Knowledge given by expert 2.

R_{Q_2}	e_1q_1	e_1q_2	e_2q_1	e_2q_2	e_3q_1	e_3q_2
(u_1q_1)	{0.6,0.2,0.7}	{0.3}	{0.4,0.8,0.2}	{0.1,0.4}	{0.2,0.7,0.3}	{0.5,0.9}
(u_1q_2)	{0.2,0.6}	{0.3,0.4}	{0.2,0.3}	{0.6,0.2}	{0.3,0.9}	{0.1,0.6,0.3}
(u_2q_1)	{0.4,0.2,0.6}	{0.1,0.2}	{0.7,0.5,0.7 }	{0.8,0.3,0.9}	{0.9,0.8,0.4}	{0.4,0.3}
(u_2q_2)	{0.9,0.6}	{0.4,0.8}	{0.3,0.1,0.9}	{0.6,0.5}	{0.7,0.3,0.6}	{0.1,0.7}
(u_3q_1)	{0.2,0.1,0.2}	{0.7,0.4}	{0.1,0.5,0.6}	{0.7,0.1,0.3}	{0.2,0.1}	{0.5,0.9,0.6}
(u_3q_2)	{0.7,0.8}	{0.3}	{0.4,0.8}	{0.1,0.2,0.4}	{0.2,0.7,0.3}	{0.4,0.5}

Table 4. Knowledge given by expert 3.

R_{Q_3}	e_1q_1	e_1q_2	e_2q_1	e_2q_2	e_3q_1	e_3q_2
(u_1q_1)	{0.6,0.2,0.1}	{0.2,0.3}	{0.1,0.2,0.9}	{0.2,0.8}	{0.8,0.5,0.6}	{0.7,0.3,0.6}
(u_1q_2)	{0.5,0.3}	{0.3,0.1,0.4}	{0.2,0.3}	{0.9,0.1,0.6}	{0.5,0.4}	{0.2,0.7,0.1}
(u_2q_1)	{0.4,0.6,0.5}	{0.5,0.1}	{0.2,0.8,0.7 }	{0.8,0.7}	{0.5,0.2,0.1}	{0.4,0.3}
(u_2q_2)	{0.3,0.4}	{0.8,0.2,0.5}	{0.4,0.9}	{0.1,0.2}	{0.8,0.5,0.3}	{0.5,0.3}
(u_3q_1)	{0.4,0.3,0.6}	{0.5,0.4}	{0.4,0.7,0.5}	{0.4,0.6}	{0.7,0.6,0.2}	{0.8,0.9,0.2}
(u_3q_2)	{0.8,0.2}	{0.3,0.1,0.3}	{0.9,0.1}	{0.4,0.6,0.7}	{0.3,0.8}	{0.6,0.4,0.7}

Now, by applying the steps of algorithm that we mentioned above, we first calculate the lower and upper approximation of optimistic and pessimistic multi Q-hesitant fuzzy soft multi-granulation rough sets of A_Q with respect to (U, E, Q, R_{Q_j}), respectively:

$$\underline{\sum_{j=1}^{3} R_{Q_j}}^{o}(A_Q) = \{\langle(u_1q_1), \{0.5, 0.5, 0.4\}\rangle, \langle(u_1q_2), \{0.6, 0.6, 0.5\}\rangle, \langle(u_2q_1), \{0.2, 0.5, 0.5\}\rangle, \langle(u_2q_2), \{0.5, 0.5, 0.5\}\rangle, \langle(u_3q_1), \{0.3, 0.7, 0.6\}\rangle, \langle(u_3q_2), \{0.6, 0.7, 0.5\}\rangle\},$$

$$\overline{\sum_{j=1}^{3} R_{Q_j}}^{o}(A_Q) = \{\langle(u_1q_1), \{0.6, 0.5, 0.5\}\rangle, \langle(u_1q_2), \{0.3, 0.6, 0.4\}\rangle, \langle(u_2q_1), \{0.4, 0.5, 0.3\}\rangle, \langle(u_2q_2), \{0.6, 0.5, 0.6\}\rangle, \langle(u_3q_1), \{0.4, 0.7, 0.5\}\rangle, \langle(u_3q_2), \{0.6, 0.5, 0.6\}\rangle\},$$

and

$\underline{\sum_{j=1}^{3} R_{Q_j}}^{p}(A_Q) = \{\langle(u_1q_1), \{0.2, 0.4, 0.4\}\rangle, \langle(u_1q_2), \{0.2, 0.4, 0.1\}\rangle,$
$\langle(u_2q_1), \{0.2, 0.4, 0.1\}\rangle, \langle(u_2q_2), \{0.2, 0.4, 0.1\}\rangle, \langle(u_3q_1), \{0.2, 0.4, 0.4\}\rangle, \langle(u_3q_2), \{0.1, 0.4, 0.2\}\rangle\},$

$\overline{\sum_{j=1}^{3} R_{Q_j}}^{p}(A_Q) = \{\langle(u_1q_1), \{0.6, 0.8, 0.9\}\rangle, \langle(u_1q_2), \{0.5, 0.7, 0.5\}\rangle,$
$\langle(u_2q_1), \{0.8, 0.8, 0.7\}\rangle, \langle(u_2q_2), \{0.9, 0.9, 0.9\}\rangle, \langle(u_3q_1), \{0.6, 0.9, 0.6\}\rangle, \langle(u_3q_2), \{0.8, 0.8, 0.8\}\rangle\}.$

Then, by Definition 21, we have:

$\underline{\sum_{j=1}^{3} R_{Q_j}}^{o}(A_Q) \oplus \overline{\sum_{j=1}^{3} R_{Q_j}}^{o}(A_Q) = \{\langle(u_1q_1), \{0.8, 0.75, 0.7\}\rangle, \langle(u_1q_2), \{0.72, 0.84, 0.7\}\rangle,$
$\langle(u_2q_1), \{0.52, 0.75, 0.65\}\rangle, \langle(u_2q_2), \{0.8, 0.75, 0.8\}\rangle,$
$\langle(u_3q_1), \{0.58, 0.91, 0.8\}\rangle, \langle(u_3q_2), \{0.84, 0.85, 0.8\}\rangle\},$

$\underline{\sum_{j=1}^{3} R_{Q_j}}^{p}(A_Q) \oplus \overline{\sum_{j=1}^{3} R_{Q_j}}^{p}(A_Q) = \{\langle(u_1q_1), \{0.68, 0.88, 0.94\}\rangle, \langle(u_1q_2), \{0.6, 0.82, 0.55\}\rangle,$
$\langle(u_2q_1), \{0.84, 0.88, 0.73\}\rangle, \langle(u_2q_2), \{0.92, 0.94, 0.91\}\rangle,$
$\langle(u_3q_1), \{0.86, 0.94, 0.76\}\rangle, \langle(u_3q_2), \{0.82, 0.88, 0.84\}\rangle\},$

$\left(\left(\underline{\sum_{j=1}^{3} R_{Q_j}}^{o}(A_Q) \oplus \overline{\sum_{j=1}^{3} R_{Q_j}}^{o}(A_Q)\right) \oplus \left(\underline{\sum_{j=1}^{3} R_{Q_j}}^{p}(A_Q) \oplus \overline{\sum_{j=1}^{3} R_{Q_j}}^{p}(A_Q)\right)\right)$
$= \{\langle(u_1q_1), \{0.936, 0.97, 0.982\}\rangle, \langle(u_1q_2), \{0.888, 0.9712, 0.865\}\rangle,$
$\langle(u_2q_1), \{0.9232, 0.97, 0.905\}\rangle, \langle(u_2q_2), \{0.984, 0.985, 0.982\}\rangle,$
$\langle(u_3q_1), \{0.8656, 0.9946, 0.952\}\rangle, \langle(u_3q_2), \{0.9712, 0.982, 0.968\}\rangle\}.$

In what follows, according to Definition 20, we calculate the score function values of multi Q-hesitant fuzzy soft elements

$S\left(\underline{\sum_{j=1}^{3} R_{Q_j}}^{o}(A_Q) \oplus \overline{\sum_{j=1}^{3} R_{Q_j}}^{o}(A_Q)\right) = \{\langle(u_1q_1), \{0.75\}\rangle, \langle(u_1q_2), \{0.753\}\rangle,$
$\langle(u_2q_1), \{0.64\}\rangle, \langle(u_2q_2), \{0.78\}\rangle, \langle(u_3q_1), \{0.76\}\rangle, \langle(u_3q_2), \{0.83\}\rangle\}.$

$S\left(\underline{\sum_{j=1}^{3} R_{Q_j}}^{p}(A_Q) \oplus \overline{\sum_{j=1}^{3} R_{Q_j}}^{p}(A_Q)\right) = \{\langle(u_1q_1), \{0.83\}\rangle, \langle(u_1q_2), \{0.65\}\rangle,$
$\langle(u_2q_1), \{(0.81\}\rangle, \langle(u_2q_2), \{0.92\}\rangle, \langle(u_3q_1), \{0.79\}\rangle, \langle(u_3q_2), \{0.84\}\rangle\}.$

$S\left(\left(\underline{\sum_{j=1}^{3} R_{Q_j}}^{o}(A_Q) \oplus \overline{\sum_{j=1}^{3} R_{Q_j}}^{o}(A_Q)\right) \oplus \left(\underline{\sum_{j=1}^{3} R_{Q_j}}^{p}(A_Q) \oplus \overline{\sum_{j=1}^{3} R_{Q_j}}^{p}(A_Q)\right)\right)$
$= \{\langle(u_1q_1), \{0.96\}\rangle, \langle(u_1q_2), \{0.90\}\rangle, \langle(u_2q_1), \{0.93\}\rangle, \langle(u_2q_2), \{0.98\}\rangle,$
$\langle(u_3q_1), \{0.94\}\rangle, \langle(u_3q_2), \{0.97\}\rangle\}.$

Then, we obtain that

$$T_1 = \left\{(S, T) | \max_{u_sq_t} S\left(\underline{\sum_{j=1}^{m} R_{Q_j}}^{o}(A_Q)(u_sq_t) \oplus \overline{\sum_{j=1}^{m} R_{Q_j}}^{o}(A_Q)(u_sq_t)\right)\right\} = (3, 2),$$

$$T_2 = \left\{(X, Y) | \max_{u_xq_y} S\left(\underline{\sum_{j=1}^{m} R_{Q_j}}^{p}(A_Q)(u_xq_y) \oplus \overline{\sum_{j=1}^{m} R_{Q_j}}^{p}(A_Q)(u_xq_y)\right)\right\} = (2, 2),$$

$$T_3 = \left\{(V, N) | \max_{u_vq_n} S\left((\underline{\sum_{j=1}^{m} R_{Q_j}}^{o}(A_Q) \oplus \overline{\sum_{j=1}^{m} R_{Q_j}}^{o}(A_Q)) \oplus (\underline{\sum_{j=1}^{m} R_{Q_j}}^{p}(A_Q) \oplus \overline{\sum_{j=1}^{m} R_{Q_j}}^{p}(A_Q))\right)\right\} = (2, 2).$$

According to the above results, the decision maker will choose the type of fault u_2 and condition degree q_2. Thus, we find that the photovoltaic systems fault is initiated by a high degree of delamination.

6.3. Comparative Analysis and Discussion

To explore the effectiveness of the proposed model based on multi-Q hesitant fuzzy soft multi-granulation rough sets, we compare it with the method proposed in [27]. The method given in [27] deals with the decision-making problems of one-dimensional universal sets U and V with hesitant fuzzy information, while the model proposed in the present paper can handle the decision-making problems of two-dimensional universal sets $U \times Q$ and $E \times Q$ with multi hesitant fuzzy soft information that contains much more information to deal with uncertainties in data related to an object with parameter value and the information expressed more precisely and objectively during the decision-making process. Thus, the proposed method is more general and its application domain is wider than that of the method in [27]. Reference [27] proposed a decision-making method based on the TODIM approach, and the basic parts of the previous method compute the dominance degree $\zeta(p_i, p_k) = \sum_{j=1}^{n} \Phi_j(p_i, p_k)$ of each alternative p_i over each alternative p_k and the overall prospect values $\zeta(p_i)$ for alternative p_i according to the following expression, respectively:

$$\Phi_j(p_i, p_k) = \begin{cases} \sqrt{w_{jr}(h_{ij} - h_{kj})/(\sum_{j=1}^{n} w_{jr})} & \text{if} \quad h_{ij} - h_{kj} > 0, \\ 0 & \text{if} \quad h_{ij} - h_{kj} = 0, \\ -\frac{\sqrt{(\sum_{j=1}^{n} w_{jr})(h_{ij} - h_{kj})/w_{jr}}}{\theta} & \text{if} \quad h_{ij} - h_{kj} < 0, \end{cases}$$

and

$$\zeta(p_i) = \frac{\sum_{j=1}^{n} \Phi_j(p_i, p_k) - min_i\{\sum_{j=1}^{n} \Phi_j(p_i, p_k)\}}{max_i\{\sum_{j=1}^{n} \Phi_j(p_i, p_k)\} - min_i\{\sum_{j=1}^{n} \Phi_j(p_i, p_k)\}}.$$

As presented in [27], the optimistic decision criterion $\underline{\sum_{j=1}^{m} R_{Q_j}}^{o}(A_Q) \oplus \overline{\sum_{j=1}^{m} R_{Q_j}}^{o}(A_Q)$, pessimistic decision criterion $\underline{\sum_{j=1}^{m} R_{Q_j}}^{p}(A_Q) \oplus \overline{\sum_{j=1}^{m} R_{Q_j}}^{p}(A_Q)$ and the weighted decision criterion

$$\frac{1}{2}\left(\underline{\sum_{j=1}^{m} R_{Q_j}}^{o}(A_Q) \oplus \overline{\sum_{j=1}^{m} R_{Q_j}}^{o}(A_Q)\right) \oplus \frac{1}{2}\left(\underline{\sum_{j=1}^{m} R_{Q_j}}^{p}(A_Q) \oplus \overline{\sum_{j=1}^{m} R_{Q_j}}^{p}(A_Q)\right)$$

are three alternatives, the fault types with condition degrees are the criteria, and the obtained evaluation values of the alternative with respect to the criterion are the elements in the decision matrix. The alternative with the largest overall prospect value is the optimal alternative. Then, in the optimal alternative, the fault type and condition degree with the largest score value are the determined fault type with its degree. Through utilizing the above procedure, we could obtain that $\zeta(p_1) = 0.22$, $\zeta(p_2) = 0.35$ and $\zeta(p_3) = 0.36$. Since the greater $\zeta(p_i)$ is, the better alternative p_i will be, the weighted decision criterion can be considered as the best alternative.

Then, we compute the score value of the fault types with condition degrees in the weighted decision criterion, which means the type of fault u_2 and condition degree q_2. Thus, we find that the photovoltaic systems fault is initiated by a high degree of delamination.

Discussion: Based on the above analysis, the results obtained by the proposed method in this paper are consistent with the one obtained using the compared method in [27], which further demonstrate the effectiveness and feasibility of the proposed model. There are two advantages of a multi Q-hesitant fuzzy soft multi-granulation rough set model in photovoltaic systems fault detection procedure. One advantage is that the hesitancy membership function in multi Q-hesitant fuzzy soft sets provides the electrical engineers with much more access to convey their understanding about the electrical knowledge base and another advantage is that the decision makers can control the size of the loss of information by adding another dimension to the universal sets. In light of the above, the greatness of the multi Q-hesitant fuzzy soft multi-granulation rough set model could decline the uncertainty to a great extent and enhance the accuracy and reliability of electrical detection effectively.

7. Conclusions

A multi Q-hesitant fuzzy soft multi-granulation rough set is a new hybrid model, which is a combination of powerful topics: multi Q-hesitant fuzzy soft sets and multi-granulation rough sets. We have defined M^kQHFS rough approximation operators in terms of M^kQHFS relations and M^kQHFS multi-granulation rough approximation operators in terms of M^kQHFS relations. We have investigated the properties of lower and upper M^kQHFS rough approximation operators and lower and upper M^kQHFS multi-granulation rough approximation operators. Finally, we have developed a general framework for dealing with uncertainty decision-making by using the multi Q-hesitant fuzzy soft multi-granulation rough sets. We have used the photovoltaic systems fault detection to indicate the principle steps of the decision methodology. In the future, we will mainly focus on investigating uncertain measures and knowledge reductions of the M^kQHFS rough sets.

Author Contributions: K.M.A., N.O.A. and M.A. conceived of the presented idea. K.M.A. and N.O.A. developed the theory and performed the computations. K.M.A. and M.A. wrote the paper.

Funding: The authors received no specific funding for this work.

Conflicts of Interest: The authors declare that there is no conflict of interests regarding the publication of this paper.

References

1. Pawlak, Z. Rough sets. *Int. J. Comput. Inf. Sci.* **1982**, *11*, 341–356. [CrossRef]
2. Huang, J.; Guan, Y.; Shen, J.; Wang, H. Rough approximations in tolerance rough set models. In Proceedings of the 11th International Conference on Fuzzy Systems and Knowledge Discovery, Xiamen, China, 19–21 August 2014; pp. 61–66.
3. Zadeh, L.A. Fuzzy sets. *Inf. Control* **1965**, *8*, 338–353. [CrossRef]
4. Atanassov, K.T. Intuitionistic fuzzy sets. *Fuzzy Sets Syst.* **1968**, *20*, 87–96. [CrossRef]
5. Torra, V. Hesitant fuzzy sets. *Int J. Intell. Syst.* **2010**, *25*, 529–539. [CrossRef]
6. Zhan, J.; Sun, B.; Alcantud, J.C.R. Covering based multigranulation (I, T)-Fuzzy rough set models and applications in multi-attribute group decision-making. *Inf. Sci.* **2019**, *476*, 290–318. [CrossRef]
7. Molodtsov, D. Soft set theory. *Comput. Math. Appl.* **1999**, *37*, 19–31. [CrossRef]
8. Maji, P.K.; Roy, A.R. An application of soft sets in a decision-making problem. *Comput. Math. Appl.* **2002**, *44*, 1077–1083. [CrossRef]
9. Fatimah, F.; Rosadi, D.; Hakim, R.F.; Alcantud, J.C.R. Probabilistic soft sets and dual probabilistic soft sets in decision-making. *Neural Comput. Appl.* **2017**, 1–11. [CrossRef]
10. Alcantud, J.C.R.; Giarlotta, A. Necessary and possible hesitant fuzzy sets: A novel model for group decision making. *Inf. Fusion* **2019**, *46*, 63–76. [CrossRef]
11. Feng, F. Soft rough sets applied to multi-criteria group decision-making. *Ann. Fuzzy Math. Inf.* **2011**, *2*, 69–80.
12. Meng, D.; Zhang, H.; Qin, K. Soft rough fuzzy sets and soft fuzzy rough sets. *Comput. Math. Appl.* **2011**, *12*, 4635–4645. [CrossRef]
13. Zhang, H.; Shu, L.; Liao, S. Intuitionistic fuzzy soft rough set and its application in decision-making. In *Abstract and Applied Analysis*; Articale ID 287314; Hindawi Publishing Corporation: Cairo, Egypt, 2014; 13p.
14. Akram, M.; Adeel, A.; Alcantud, J.C.R. Group decision-making methods based on hesitant N-soft sets. *Expert Syst. Appl.* **2019**, *115*, 95–105. [CrossRef]
15. Akram, M.; Adeel, A.; Alcantud, J.C.R. Fuzzy N-soft sets: A novel model with applications. *J. Intell. Fuzzy Syst.* **2018**, *35*, 4757–4771. [CrossRef]
16. Akram, M.; Adeel, A. TOPSIS Approach for MAGDM Based on Interval-Valued Hesitant Fuzzy N-Soft Environment. *J. Intell. Fuzzy Syst.* **2018**, in press.
17. Akram, M.; Ali, G.; Waseem, N.; Davvaz, B. Decision-making methods based on hybrid mF models. *J. Intell. Fuzzy Syst.* **2018**, *35*, 3387–3403. [CrossRef]
18. Akram, M.; Ali, G.; Alshehri, N.O. A new multi-attribute decision-making method based on m-polar fuzzy soft rough sets. *Symmetry* **2017**, *9*, 271. [CrossRef]

19. Akram, M.; Ali, G. Hybrid models for decision-making based on rough Pythagorean fuzzy bipolar soft information. *Granul. Comput.* **2018**, 1–15. [CrossRef]
20. Adam, F.; Hassan, N. *Q*-fuzzy soft set. *Appl. Math. Sci.* **2014**, *8*, 8689–8695. [CrossRef]
21. Adam, F.; Hassan, N. Multi *Q*-fuzzy parameterized soft set and its application. *J. Intell. Fuzzy Syst.* **2014**, *27*, 419–424.
22. Adam, F.; Hassan, N. Properties on the multi *Q*-fuzzy soft matrix. *AIP Conf. Proc.* **2014**, *1614*, 834–839.
23. Adam, F.; Hassan, N. *Q*-fuzzy soft matrix and its application. *AIP Conf. Proc.* **2014**, *1602*, 772–778.
24. Adam, F.; Hassan, N. Operations on *Q*-fuzzy soft sets. *Appl. Math. Sci.* **2014**, *8*, 8697–8701. [CrossRef]
25. Qian, Y.; Liang, J.; Yao, Y.; Dang, C. A multi-granulation rough set. *Inf. Sci.* **2010**, *180*, 949–970. [CrossRef]
26. Sun, B.; Ma, W.; Qian, Y. multi-granulation fuzzy rough set over two universes and its application to decision-making. *Knowl.-Based Syst.* **2017**, *123*, 61–74. [CrossRef]
27. Zhang, C.; Li, D.; Yang, Y. Multi-granulation rough set in hesitant fuzzy information systems and its application in person-job fit. *Int. J. Mach. Learn. Cybern.* **2017**. [CrossRef]
28. Sun, B.; Ma, W.; Zhao, H. A fuzzy rough set approach to emergency material demand prediction over two universes. *Appl. Math. Model.* **2013**, *37*, 7062–7070. [CrossRef]
29. Xia, M.; Xu, Z. Hesitant fuzzy information aggregation in decision-making. *Int. J. Approx Reason.* **2011**, *52*, 395–407. [CrossRef]
30. Alghamd, M.A.; Alshehri, N.O.; Akram, M. Multi-criteria decision-making methods in bipolar fuzzy environment. *Int. J. Fuzzy Syst.* **2018**, *20*, 2057–2064. [CrossRef]
31. Akram, M.; Shumaiza; Smarandache, F. Decision-Making with Bipolar Neutrosophic TOPSIS and Bipolar Neutrosophic ELECTRE-I. *Axioms* **2018**, *7*, 33. [CrossRef]
32. Naz, S.; Ashraf, S.; Akram, M. A novel approach to decision-making with Pythagorean fuzzy information. *Mathematics* **2018**, *6*, 95. [CrossRef]
33. Zafar, F.; Akram, M. A novel decision-making method based on rough fuzzy information. *Int. J. Fuzzy Syst.* **2018**, *20*, 1000–1014. [CrossRef]
34. Wang, F.; Xihua, L.; Xiaohong, C. Hesitant fuzzy soft set and its application in multicriteria decision-making. *J. Appl. Math.* **2014**. [CrossRef]

symmetry

MDPI

Article

Two Types of Single Valued Neutrosophic Covering Rough Sets and an Application to Decision Making

Jingqian Wang [1] and Xiaohong Zhang [2,*]

1 College of Electrical and Information Engineering, Shaanxi University of Science and Technology, Xi'an 710021, China; wangjingqianw@163.com or 81157@sust.edu.cn

2 School of Arts and Sciences, Shaanxi University of Science and Technology, Xi'an 710021, China

* Correspondence: zxhonghz@263.net or zhangxiaohong@sust.edu.cn

Received: 8 November 2018; Accepted: 27 November 2018; Published: 3 December 2018

Abstract: In this paper, to combine single valued neutrosophic sets (SVNSs) with covering-based rough sets, we propose two types of single valued neutrosophic (SVN) covering rough set models. Furthermore, a corresponding application to the problem of decision making is presented. Firstly, the notion of SVN β-covering approximation space is proposed, and some concepts and properties in it are investigated. Secondly, based on SVN β-covering approximation spaces, two types of SVN covering rough set models are proposed. Then, some properties and the matrix representations of the newly defined SVN covering approximation operators are investigated. Finally, we propose a novel method to decision making (DM) problems based on one of the SVN covering rough set models. Moreover, the proposed DM method is compared with other methods in an example.

Keywords: covering; single valued neutrosophic; matrix representation; decision making

1. Introduction

Rough set theory, as a a tool to deal with various types of data in data mining, was proposed by Pawlak [1,2] in 1982. Since then, rough set theory has been extended to generalized rough sets based on other notions such as binary relations, neighborhood systems and coverings.

Covering-based rough sets [3–5] were proposed to deal with the type of covering data. In application, they have been applied to knowledge reduction [6,7], decision rule synthesis [8,9], and other fields [10–12]. In theory, covering-based rough set theory has been connected with matroid theory [13–16], lattice theory [17,18] and fuzzy set theory [19–22].

Zadeh's fuzzy set theory [23] addresses the problem of how to understand and manipulate imperfect knowledge. It has been used in various applications [24–27]. Recent investigations have attracted more attention on combining covering-based rough set and fuzzy set theories. There are many fuzzy covering rough set models proposed by researchers, such as Ma [28] and Yang et al. [20].

Wang et al. [29] presented single valued neutrosophic sets (SVNSs) which can be regarded as an extension of IFSs [30]. Neutrosophic sets and rough sets both can deal with partial and uncertain information. Therefore, it is necessary to combine them. Recently, Mondal and Pramanik [31] presented the concept of rough neutrosophic set. Yang et al. [32] presented a SVN rough set model based on SVN relations. However, SVNSs and covering-based rough sets have not been combined up to now. In this paper, we present two types of SVN covering rough set models. This new combination is a bridge, linking SVNSs and covering-based rough sets.

As we know, the multiple criteria decision making (MCDM) is an important tool to deal with more complicated problems in our real world [33,34]. There are many MCDM methods presented based on different problems or theories. For example, Liu et al. [35] dealt with the challenges of many criteria in the MCDM problem and decision makers with heterogeneous risk preferences. Watróbski et al. [36] proposed a framework for selecting suitable MCDA methods for a particular decision situation.

Faizi et al. [37,38] presented an extension of the MCDM method based on hesitant fuzzy theory. Recently, many researchers have studied decision making (DM) problems by rough set models [39–42]. For example, Zhan et al. [39] applied a type of soft rough model to DM problems. Yang et al. [32] presented a method for DM problems under a type of SVN rough set model. By investigation, we have observed that no one has applied SVN covering rough set models to DM problems. Therefore, we construct the covering SVN decision information systems according to the characterizations of DM problems. Then, we present a novel method to DM problems under one of the SVN covering rough set models. Moreover, the proposed decision making method is compared with other methods, which were presented by Yang et al. [32], Liu [43] and Ye [44].

The rest of this paper is organized as follows. Section 2 reviews some fundamental definitions about covering-based rough sets and SVNSs. In Section 3, some notions and properties in SVN β-covering approximation space are studied. In Section 4, we present two types of SVN covering rough set models, based on the SVN β-neighborhoods and the β-neighborhoods. In Section 5, some new matrices and matrix operations are presented. Based on this, the matrix representations of the SVN approximation operators are shown. In Section 6, a novel method to decision making (DM) problems under one of the SVN covering rough set models is proposed. Moreover, the proposed DM method is compared with other methods. This paper is concluded and further work is indicated in Section 7.

2. Basic Definitions

Suppose U is a nonempty and finite set called universe.

Definition 1 (Covering [45,46]). *Let U be a universe and $\mathbf{C}$ a family of subsets of U. If none of subsets in $\mathbf{C}$ is empty and $\bigcup \mathbf{C} = U$, then $\mathbf{C}$ is called a covering of U.*

The pair $(U, \mathbf{C})$ is called a covering approximation space.

Definition 2 (Single valued neutrosophic set [29]). *Let U be a nonempty fixed set. A single valued neutrosophic set (SVNS) A in U is defined as an object of the following form:*

$$A = \{\langle x, T_A(x), I_A(x), F_A(x)\rangle : x \in U\},$$

where $T_A(x) : U \to [0,1]$ is a truth-membership function, $I_A(x) : U \to [0,1]$ is an indeterminacy-membership function and $F_A(x) : U \to [0,1]$ is a falsity-membership function for any $x \in U$. They satisfy $0 \leq T_A(x) + I_A(x) + F_A(x) \leq 3$ for all $x \in U$. The family of all single valued neutrosophic sets in U is denoted by $SVN(U)$. For convenience, a SVN number is represented by $\alpha = \langle a, b, c\rangle$, where $a, b, c \in [0,1]$ and $a + b + c \leq 3$.

Specially, for two SVN numbers $\alpha = \langle a, b, c\rangle$ and $\beta = \langle d, e, f\rangle$, $\alpha \leq \beta \Leftrightarrow a \leq d, b \geq e$ and $c \geq f$. Some operations on $SVN(U)$ are listed as follows [29,32]: for any $A, B \in SVN(U)$,

(1) $A \subseteq B$ iff $T_A(x) \leq T_B(x)$, $I_B(x) \leq I_A(x)$ and $F_B(x) \leq F_A(x)$ for all $x \in U$.
(2) $A = B$ iff $A \subseteq B$ and $B \subseteq A$.
(3) $A \cap B = \{\langle x, T_A(x) \wedge T_B(x), I_A(x) \vee I_B(x), F_A(x) \vee F_B(x)\rangle : x \in U\}$.
(4) $A \cup B = \{\langle x, T_A(x) \vee T_B(x), I_A(x) \wedge I_B(x), F_A(x) \wedge F_B(x)\rangle : x \in U\}$.
(5) $A' = \{\langle x, F_A(x), 1 - I_A(x), T_A(x)\rangle : x \in U\}$.
(6) $A \oplus B = \{\langle x, T_A(x) + T_B(x) - T_A(x) \cdot T_B(x), I_A(x) \cdot I_B(x), F_A(x) \cdot F_B(x)\rangle : x \in U\}$.

3. Single Valued Neutrosophic β-Covering Approximation Space

In this section, we present the notion of SVN β-covering approximation space. There are two basic concepts in this new approximation space: SVN β-covering and SVN β-neighborhood. Then, some of their properties are studied.

Definition 3. *Let U be a universe and $SVN(U)$ be the SVN power set of U. For a SVN number $\beta = \langle a,b,c \rangle$, we call $\widehat{C} = \{C_1, C_2, \cdots, C_m\}$, with $C_i \in SVN(U)(i = 1,2,...,m)$, a SVN β-covering of U, if for all $x \in U$, $C_i \in \widehat{C}$ exists such that $C_i(x) \geq \beta$. We also call $(U, \widehat{C})$ a SVN β-covering approximation space.*

Definition 4. *Let $\widehat{C}$ be a SVN β-covering of U and $\widehat{C} = \{C_1, C_2, \ldots, C_m\}$. For any $x \in U$, the SVN β-neighborhood $\widetilde{\mathbb{N}}_x^\beta$ of x induced by $\widehat{C}$ can be defined as:*

$$\widetilde{\mathbb{N}}_x^\beta = \cap\{C_i \in \widehat{C} : C_i(x) \geq \beta\}. \tag{1}$$

Note that $C_i(x)$ is a SVN number $\langle T_{C_i}(x), I_{C_i}(x), F_{C_i}(x) \rangle$ in Definitions 3 and 4. Hence, $C_i(x) \geq \beta$ means $T_{C_i}(x) \geq a$, $I_{C_i}(x) \leq b$ and $F_{C_i}(x) \leq c$ where SVN number $\beta = \langle a,b,c \rangle$.

Remark 1. *Let $\widehat{C}$ be a SVN β-covering of U, $\beta = \langle a,b,c \rangle$ and $\widehat{C} = \{C_1, C_2, \ldots, C_m\}$. For any $x \in U$,*

$$\widetilde{\mathbb{N}}_x^\beta = \cap\{C_i \in \widehat{C} : T_{C_i}(x) \geq a,\ I_{C_i}(x) \leq b,\ F_{C_i}(x) \leq c\}. \tag{2}$$

Example 1. *Let $U = \{x_1, x_2, x_3, x_4, x_5\}$, $\widehat{C} = \{C_1, C_2, C_3, C_4\}$ and $\beta = \langle 0.5, 0.3, 0.8 \rangle$. We can see that $\widehat{C}$ is a SVN β-covering of U in Table 1.*

Table 1. The tabular representation of single valued neutrosophic (SVN) β-covering $\widehat{C}$.

U	C_1	C_2	C_3	C_4
x_1	$\langle 0.7,0.2,0.5 \rangle$	$\langle 0.6,0.2,0.4 \rangle$	$\langle 0.4,0.1,0.5 \rangle$	$\langle 0.1,0.5,0.6 \rangle$
x_2	$\langle 0.5,0.3,0.2 \rangle$	$\langle 0.5,0.2,0.8 \rangle$	$\langle 0.4,0.5,0.4 \rangle$	$\langle 0.6,0.1,0.7 \rangle$
x_3	$\langle 0.4,0.5,0.2 \rangle$	$\langle 0.2,0.3,0.6 \rangle$	$\langle 0.5,0.2,0.4 \rangle$	$\langle 0.6,0.3,0.4 \rangle$
x_4	$\langle 0.6,0.1,0.7 \rangle$	$\langle 0.4,0.5,0.7 \rangle$	$\langle 0.3,0.6,0.5 \rangle$	$\langle 0.5,0.3,0.2 \rangle$
x_5	$\langle 0.3,0.2,0.6 \rangle$	$\langle 0.7,0.3,0.5 \rangle$	$\langle 0.6,0.3,0.5 \rangle$	$\langle 0.8,0.1,0.2 \rangle$

Then,

$$\widetilde{\mathbb{N}}_{x_1}^\beta = C_1 \cap C_2, \widetilde{\mathbb{N}}_{x_2}^\beta = C_1 \cap C_2 \cap C_4, \widetilde{\mathbb{N}}_{x_3}^\beta = C_3 \cap C_4, \widetilde{\mathbb{N}}_{x_4}^\beta = C_1 \cap C_4, \widetilde{\mathbb{N}}_{x_5}^\beta = C_2 \cap C_3 \cap C_4.$$

Hence, all SVN β-neighborhoods are shown in Table 2.

Table 2. The tabular representation of $\widetilde{\mathbb{N}}_{x_k}^\beta$ $(k = 1,2,3,4,5)$.

$\widetilde{\mathbb{N}}_{x_k}^\beta$	x_1	x_2	x_3	x_4	x_5
$\widetilde{\mathbb{N}}_{x_1}^\beta$	$\langle 0.6,0.2,0.5 \rangle$	$\langle 0.5,0.3,0.8 \rangle$	$\langle 0.2,0.5,0.6 \rangle$	$\langle 0.4,0.5,0.7 \rangle$	$\langle 0.3,0.3,0.6 \rangle$
$\widetilde{\mathbb{N}}_{x_2}^\beta$	$\langle 0.1,0.5,0.6 \rangle$	$\langle 0.5,0.3,0.8 \rangle$	$\langle 0.2,0.5,0.6 \rangle$	$\langle 0.4,0.5,0.7 \rangle$	$\langle 0.3,0.3,0.6 \rangle$
$\widetilde{\mathbb{N}}_{x_3}^\beta$	$\langle 0.1,0.5,0.6 \rangle$	$\langle 0.4,0.5,0.7 \rangle$	$\langle 0.5,0.3,0.4 \rangle$	$\langle 0.3,0.6,0.5 \rangle$	$\langle 0.6,0.3,0.5 \rangle$
$\widetilde{\mathbb{N}}_{x_4}^\beta$	$\langle 0.1,0.5,0.6 \rangle$	$\langle 0.5,0.3,0.7 \rangle$	$\langle 0.4,0.5,0.4 \rangle$	$\langle 0.5,0.3,0.7 \rangle$	$\langle 0.3,0.2,0.6 \rangle$
$\widetilde{\mathbb{N}}_{x_5}^\beta$	$\langle 0.1,0.5,0.6 \rangle$	$\langle 0.4,0.5,0.8 \rangle$	$\langle 0.2,0.3,0.6 \rangle$	$\langle 0.3,0.6,0.7 \rangle$	$\langle 0.6,0.3,0.5 \rangle$

In a SVN β-covering approximation space $(U, \widehat{C})$, we present the following properties of the SVN β-neighborhood.

Theorem 1. *Let $\widehat{C}$ be a SVN β-covering of U and $\widehat{C} = \{C_1, C_2, \ldots, C_m\}$. Then, the following statements hold:*

(1) $\widetilde{\mathbb{N}}_x^\beta(x) \geq \beta$ *for each* $x \in U$.
(2) $\forall x, y, z \in U$, *if* $\widetilde{\mathbb{N}}_x^\beta(y) \geq \beta$, $\widetilde{\mathbb{N}}_y^\beta(z) \geq \beta$, *then* $\widetilde{\mathbb{N}}_x^\beta(z) \geq \beta$.
(3) *For two SVN numbers* β_1, β_2, *if* $\beta_1 \leq \beta_2 \leq \beta$, *then* $\widetilde{\mathbb{N}}_x^{\beta_1} \subseteq \widetilde{\mathbb{N}}_x^{\beta_2}$ *for all* $x \in U$.

Proof.

(1) For any $x \in U$, $\widetilde{\mathrm{N}}_x^{\beta}(x) = (\bigcap\limits_{C_i(x) \geq \beta} C_i)(x) = \bigwedge\limits_{C_i(x) \geq \beta} C_i(x) \geq \beta$.

(2) Let $I = \{1, 2, \cdots, m\}$. Since $\widetilde{\mathrm{N}}_x^{\beta}(y) \geq \beta$, for any $i \in I$, if $C_i(x) \geq \beta$, then $C_i(y) \geq \beta$. Since $\widetilde{\mathrm{N}}_y^{\beta}(z) \geq \beta$, for any $i \in I$, $C_i(z) \geq \beta$ when $C_i(y) \geq \beta$. Then, for any $i \in I$, $C_i(x) \geq \beta$ implies $C_i(z) \geq \beta$. Therefore, $\widetilde{\mathrm{N}}_x^{\beta}(z) \geq \beta$.

(3) For all $x \in U$, since $\beta_1 \leq \beta_2 \leq \beta$, $\{C_i \in \widehat{\mathbf{C}} : C_i(x) \geq \beta_1\} \supseteq \{C_i \in \widehat{\mathbf{C}} : C_i(x) \geq \beta_2\}$. Hence, $\widetilde{\mathrm{N}}_x^{\beta_1} = \cap\{C_i \in \widehat{\mathbf{C}} : C_i(x) \geq \beta_1\} \subseteq \cap\{C_i \in \widehat{\mathbf{C}} : C_i(x) \geq \beta_2\} = \widetilde{\mathrm{N}}_x^{\beta_2}$ for all $x \in U$.

□

Proposition 1. *Let $\widehat{\mathbf{C}}$ be a SVN β-covering of U. For any $x, y \in U$, $\widetilde{\mathrm{N}}_x^{\beta}(y) \geq \beta$ if and only if $\widetilde{\mathrm{N}}_y^{\beta} \subseteq \widetilde{\mathrm{N}}_x^{\beta}$.*

Proof. Suppose the SVN number $\beta = \langle a, b, c \rangle$.
($\Rightarrow$): Since $\widetilde{\mathrm{N}}_x^{\beta}(y) \geq \beta$,

$$T_{\widetilde{\mathrm{N}}_x^{\beta}}(y) = T_{\bigcap\limits_{\substack{T_{C_i}(x) \geq a \\ I_{C_i}(x) \leq b \\ F_{C_i}(x) \leq c}} C_i}(y) = \bigwedge_{\substack{T_{C_i}(x) \geq a \\ I_{C_i}(x) \leq b \\ F_{C_i}(x) \leq c}} T_{C_i}(y) \geq a,\ I_{\widetilde{\mathrm{N}}_x^{\beta}}(y) = I_{\bigcap\limits_{\substack{T_{C_i}(x) \geq a \\ I_{C_i}(x) \leq c \\ F_{C_i}(x) \leq b}} C_i}(y) = \bigvee_{\substack{T_{C_i}(x) \geq a \\ I_{C_i}(x) \leq b \\ F_{C_i}(x) \leq c}} I_{C_i}(y) \leq b,$$

and

$$F_{\widetilde{\mathrm{N}}_x^{\beta}}(y) = F_{\bigcap\limits_{\substack{T_{C_i}(x) \geq a \\ I_{C_i}(x) \leq c \\ F_{C_i}(x) \leq b}} C_i}(y) = \bigvee_{\substack{T_{C_i}(x) \geq a \\ I_{C_i}(x) \leq b \\ F_{C_i}(x) \leq c}} F_{C_i}(y) \leq c.$$

Then,

$\{C_i \in \widehat{\mathbf{C}} : T_{C_i}(x) \geq a, I_{C_i}(x) \leq b, F_{C_i}(x) \leq c\} \subseteq \{C_i \in \widehat{\mathbf{C}} : T_{C_i}(y) \geq a, I_{C_i}(y) \leq b, F_{C_i}(y) \leq c\}$.

Therefore, for each $z \in U$,

$$T_{\widetilde{\mathrm{N}}_x^{\beta}}(z) = \bigwedge_{\substack{T_{C_i}(x) \geq a \\ I_{C_i}(x) \leq b \\ F_{C_i}(x) \leq c}} T_{C_i}(z) \geq \bigwedge_{\substack{T_{C_i}(y) \geq a \\ I_{C_i}(y) \leq b \\ F_{C_i}(y) \leq c}} T_{C_i}(z) = T_{\widetilde{\mathrm{N}}_y^{\beta}}(z),$$

$$I_{\widetilde{\mathrm{N}}_x^{\beta}}(z) = \bigvee_{\substack{T_{C_i}(x) \geq a \\ I_{C_i}(x) \leq b \\ F_{C_i}(x) \leq c}} I_{C_i}(z) \leq \bigvee_{\substack{T_{C_i}(y) \geq a \\ I_{C_i}(y) \leq b \\ F_{C_i}(y) \leq c}} I_{C_i}(z) = I_{\widetilde{\mathrm{N}}_y^{\beta}}(z),$$

$$F_{\widetilde{\mathrm{N}}_x^{\beta}}(z) = \bigvee_{\substack{T_{C_i}(x) \geq a \\ I_{C_i}(x) \leq b \\ F_{C_i}(x) \leq c}} F_{C_i}(z) \leq \bigvee_{\substack{T_{C_i}(y) \geq a \\ I_{C_i}(y) \leq b \\ F_{C_i}(y) \leq c}} F_{C_i}(z) = F_{\widetilde{\mathrm{N}}_y^{\beta}}(z).$$

Hence, $\widetilde{\mathrm{N}}_y^{\beta} \subseteq \widetilde{\mathrm{N}}_x^{\beta}$.
($\Leftarrow$): For any $x, y \in U$, since $\widetilde{\mathrm{N}}_y^{\beta} \subseteq \widetilde{\mathrm{N}}_x^{\beta}$,

$$T_{\widetilde{\mathrm{N}}_x^{\beta}}(y) \geq T_{\widetilde{\mathrm{N}}_y^{\beta}}(y) \geq a,\ I_{\widetilde{\mathrm{N}}_x^{\beta}}(y) \leq I_{\widetilde{\mathrm{N}}_y^{\beta}}(y) \leq b \text{ and } F_{\widetilde{\mathrm{N}}_x^{\beta}}(y) \leq F_{\widetilde{\mathrm{N}}_y^{\beta}}(y) \leq c.$$

Therefore, $\widetilde{\mathrm{N}}_x^{\beta}(y) \geq \beta$. □

The notion of SVN β-neighborhood in the SVN β-covering approximation space in the following definition.

Definition 5. *Let $(U, \widehat{C})$ be a SVN β-covering approximation space and $\widehat{C} = \{C_1, C_2, \ldots, C_m\}$. For each $x \in U$, we define the β-neighborhood $\overline{\mathbb{N}}_x^\beta$ of x as:*

$$\overline{\mathbb{N}}_x^\beta = \{y \in U : \widetilde{\mathbb{N}}_x^\beta(y) \geq \beta\}. \tag{3}$$

Note that $\widetilde{\mathbb{N}}_x^\beta(y)$ is a SVN number $\langle T_{\widetilde{\mathbb{N}}_x^\beta}(y), I_{\widetilde{\mathbb{N}}_x^\beta}(y), F_{\widetilde{\mathbb{N}}_x^\beta}(y)\rangle$ in Definition 5.

Remark 2. *Let $\widehat{C}$ be a SVN β-covering of U, $\beta = \langle a, b, c\rangle$ and $\widehat{C} = \{C_1, C_2, \ldots, C_m\}$. For each $x \in U$,*

$$\overline{\mathbb{N}}_x^\beta = \{y \in U : T_{\widetilde{\mathbb{N}}_x^\beta}(y) \geq a, I_{\widetilde{\mathbb{N}}_x^\beta}(y) \leq b, F_{\widetilde{\mathbb{N}}_x^\beta}(y) \leq c\}. \tag{4}$$

Example 2 (Continued from Example 1). *Let $\beta = \langle 0.5, 0.3, 0.8\rangle$, then we have*

$$\overline{\mathbb{N}}_{x_1}^\beta = \{x_1, x_2\},\ \overline{\mathbb{N}}_{x_2}^\beta = \{x_2\},\ \overline{\mathbb{N}}_{x_3}^\beta = \{x_3, x_5\}, \overline{\mathbb{N}}_{x_4}^\beta = \{x_2, x_4\},\ \overline{\mathbb{N}}_{x_5}^\beta = \{x_5\}.$$

Some properties of the β-neighborhood in a SVN β-covering of U are presented in Theorem 2 and Proposition 2.

Theorem 2. *Let $\widehat{C}$ be a SVN β-covering of U and $\widehat{C} = \{C_1, C_2, \ldots, C_m\}$. Then, the following statements hold:*

(1) *$x \in \overline{\mathbb{N}}_x^\beta$ for each $x \in U$.*
(2) *$\forall x, y, z \in U$, if $x \in \overline{\mathbb{N}}_y^\beta$, $y \in \overline{\mathbb{N}}_z^\beta$, then $x \in \overline{\mathbb{N}}_z^\beta$.*

Proof.

(1) According to Theorem 1 and Definition 5, it is straightforward.
(2) For any $x, y, z \in U$, $x \in \overline{\mathbb{N}}_y^\beta \Leftrightarrow \widetilde{\mathbb{N}}_y^\beta(x) \geq \beta \Leftrightarrow \widetilde{\mathbb{N}}_x^\beta \subseteq \widetilde{\mathbb{N}}_y^\beta$, and $y \in \overline{\mathbb{N}}_z^\beta \Leftrightarrow \widetilde{\mathbb{N}}_z^\beta(y) \geq \beta \Leftrightarrow \widetilde{\mathbb{N}}_y^\beta \subseteq \widetilde{\mathbb{N}}_z^\beta$. Hence, $\widetilde{\mathbb{N}}_x^\beta \subset \widetilde{\mathbb{N}}_z^\beta$. By Proposition 1, we have $\widetilde{\mathbb{N}}_z^\beta(x) \geq \beta$, i.e., $x \in \overline{\mathbb{N}}_z^\beta$.

□

Proposition 2. *Let $\widehat{C}$ be a SVN β-covering of U and $\widehat{C} = \{C_1, C_2, \ldots, C_m\}$. Then, for all $x \in U$, $x \in \overline{\mathbb{N}}_y^\beta$ if and only if $\overline{\mathbb{N}}_x^\beta \subseteq \overline{\mathbb{N}}_y^\beta$.*

Proof. (⇒): For any $z \in \overline{\mathbb{N}}_x^\beta$, we know $\widetilde{\mathbb{N}}_x^\beta(z) \geq \beta$. Since $x \in \overline{\mathbb{N}}_y^\beta$, $\widetilde{\mathbb{N}}_y^\beta(x) \geq \beta$. According to (2) in Theorem 1, we have $\widetilde{\mathbb{N}}_y^\beta(z) \geq \beta$. Hence, $z \in \overline{\mathbb{N}}_y^\beta$. Therefore, $\overline{\mathbb{N}}_x^\beta \subseteq \overline{\mathbb{N}}_y^\beta$.
(⇐): According to (1) in Theorem 2, $x \in \overline{\mathbb{N}}_x^\beta$ for all $x \in U$. Since $\overline{\mathbb{N}}_x^\beta \subseteq \overline{\mathbb{N}}_y^\beta$, $x \in \overline{\mathbb{N}}_y^\beta$. □

The relationship between SVN β-neighborhoods and β-neighborhoods is presented in the following proposition.

Proposition 3. *Let $\widehat{C}$ be a SVN β-covering of U. For any $x, y \in U$, $\widetilde{\mathbb{N}}_x^\beta \subseteq \widetilde{\mathbb{N}}_y^\beta$ if and only if $\overline{\mathbb{N}}_x^\beta \subseteq \overline{\mathbb{N}}_y^\beta$.*

Proof. According to Propositions 1 and 2, it is straightforward. □

4. Two Types of Single Valued Neutrosophic Covering Rough Set Models

In this section, we propose two types of SVN covering rough set models on basis of the SVN β-neighborhoods and the β-neighborhoods, respectively. Then, we investigate the properties of the defined lower and upper approximation operators.

Definition 6. *Let $(U, \widehat{\mathbb{C}})$ be a SVN β-covering approximation space. For each $A \in SVN(U)$ where $A = \{\langle x, T_A(x), I_A(x), F_A(x)\rangle : x \in U\}$, we define the single valued neutrosophic (SVN) covering upper approximation $\widetilde{\mathbb{C}}(A)$ and lower approximation $\underset{\sim}{\mathbb{C}}(A)$ of A as:*

$$\begin{aligned}
\widetilde{\mathbb{C}}(A) &= \{\langle x, \vee_{y\in U}[T_{\widetilde{\mathbb{N}}_x^\beta}(y) \wedge T_A(y)], \vee_{y\in U}[I_{\widetilde{\mathbb{N}}_x^\beta}(y) \wedge I_A(y)], \wedge_{y\in U}[F_{\widetilde{\mathbb{N}}_x^\beta}(y) \vee F_A(y)]\rangle : x \in U\}, \\
\underset{\sim}{\mathbb{C}}(A) &= \{\langle x, \wedge_{y\in U}[F_{\widetilde{\mathbb{N}}_x^\beta}(y) \vee T_A(y)], \wedge_{y\in U}[(1 - I_{\widetilde{\mathbb{N}}_x^\beta}(y)) \vee I_A(y)], \vee_{y\in U}[T_{\widetilde{\mathbb{N}}_x^\beta}(y) \wedge F_A(y)]\rangle : x \in U\}.
\end{aligned} \tag{5}$$

If $\widetilde{\mathbb{C}}(A) \neq \underset{\sim}{\mathbb{C}}(A)$, then A is called the first type of SVN covering rough set.

Example 3 (Continued from Example 1). *Let $\beta = \langle 0.5, 0.3, 0.8\rangle$, $A = \frac{\langle 0.6,0.3,0.5\rangle}{x_1} + \frac{\langle 0.4,0.5,0.1\rangle}{x_2} + \frac{\langle 0.3,0.2,0.6\rangle}{x_3} + \frac{\langle 0.5,0.3,0.4\rangle}{x_4} + \frac{\langle 0.7,0.2,0.3\rangle}{x_5}$. Then,*

$$\begin{aligned}
\widetilde{\mathbb{C}}(A) &= \{\langle x_1, 0.6, 0.3, 0.5\rangle, \langle x_2, 0.4, 0.3, 0.6\rangle, \langle x_3, 0.6, 0.5, 0.5\rangle, \langle x_4, 0.5, 0.3, 0.6\rangle, \langle x_5, 0.6, 0.5, 0.5\rangle\}, \\
\underset{\sim}{\mathbb{C}}(A) &= \{\langle x_1, 0.6, 0.5, 0.5\rangle, \langle x_2, 0.6, 0.5, 0.4\rangle, \langle x_3, 0.4, 0.4, 0.5\rangle, \langle x_4, 0.4, 0.5, 0.4\rangle, \langle x_5, 0.6, 0.4, 0.3\rangle\}.
\end{aligned}$$

Some basic properties of the SVN covering upper and lower approximation operators are proposed in the following proposition.

Proposition 4. *Let $\widehat{\mathbb{C}}$ be a SVN β-covering of U. Then, the SVN covering upper and lower approximation operators in Definition 6 satisfy the following properties: for all $A, B \in SVN(U)$,*

(1) $\widetilde{\mathbb{C}}(A') = (\underset{\sim}{\mathbb{C}}(A))'$, $\underset{\sim}{\mathbb{C}}(A') = (\widetilde{\mathbb{C}}(A))'$.

(2) *If $A \subseteq B$, then $\underset{\sim}{\mathbb{C}}(A) \subseteq \underset{\sim}{\mathbb{C}}(B)$, $\widetilde{\mathbb{C}}(A) \subseteq \widetilde{\mathbb{C}}(B)$.*

(3) $\underset{\sim}{\mathbb{C}}(A \cap B) = \underset{\sim}{\mathbb{C}}(A) \cap \underset{\sim}{\mathbb{C}}(B)$, $\widetilde{\mathbb{C}}(A \cup B) = \widetilde{\mathbb{C}}(A) \cup \widetilde{\mathbb{C}}(B)$.

(4) $\underset{\sim}{\mathbb{C}}(A \cup B) \supseteq \underset{\sim}{\mathbb{C}}(A) \cup \underset{\sim}{\mathbb{C}}(B)$, $\widetilde{\mathbb{C}}(A \cap B) \subseteq \widetilde{\mathbb{C}}(A) \cap \widetilde{\mathbb{C}}(B)$.

Proof.
(1)

$$\begin{aligned}
\widetilde{\mathbb{C}}(A') &= \{\langle x, \vee_{y\in U}[T_{\widetilde{\mathbb{N}}_x^\beta}(y) \wedge T_{A'}(y)], \vee_{y\in U}[I_{\widetilde{\mathbb{N}}_x^\beta}(y) \wedge I_{A'}(y)], \wedge_{y\in U}[F_{\widetilde{\mathbb{N}}_x^\beta}(y) \vee F_{A'}(y)]\rangle : x \in U\} \\
&= \{\langle x, \vee_{y\in U}[T_{\widetilde{\mathbb{N}}_x^\beta}(y) \wedge F_A(y)], \vee_{y\in U}[I_{\widetilde{\mathbb{N}}_x^\beta}(y) \wedge (1 - I_A(y))], \wedge_{y\in U}[F_{\widetilde{\mathbb{N}}_x^\beta}(y) \vee T_A(y)]\rangle : x \in U\} \\
&= (\underset{\sim}{\mathbb{C}}(A))'.
\end{aligned}$$

If we replace A by A' in this proof, we can also prove $\underset{\sim}{\mathbb{C}}(A') = (\widetilde{\mathbb{C}}(A))'$.
(2) Since $A \subseteq B$, so $T_A(x) \leq T_B(x)$, $I_B(x) \leq I_A(x)$ and $F_B(x) \leq F_A(x)$ for all $x \in U$. Therefore,

$$\begin{aligned}
T_{\underset{\sim}{\mathbb{C}}(A)}(x) &= \wedge_{y\in U}[F_{\widetilde{\mathbb{N}}_x^\beta}(y) \vee T_A(y)] \leq \wedge_{y\in U}[F_{\widetilde{\mathbb{N}}_x^\beta}(y) \vee T_B(y)] = T_{\underset{\sim}{\mathbb{C}}(B)}(x), \\
I_{\underset{\sim}{\mathbb{C}}(A)}(x) &= \wedge_{y\in U}[(1 - I_{\widetilde{\mathbb{N}}_x^\beta}(y)) \vee I_A(y)] \geq \wedge_{y\in U}[(1 - I_{\widetilde{\mathbb{N}}_x^\beta}(y)) \vee I_B(y)] = I_{\underset{\sim}{\mathbb{C}}(B)}(x), \\
F_{\underset{\sim}{\mathbb{C}}(A)}(x) &= \vee_{y\in U}[T_{\widetilde{\mathbb{N}}_x^\beta}(y) \wedge F_A(y)] \geq \vee_{y\in U}[T_{\widetilde{\mathbb{N}}_x^\beta}(y) \wedge F_B(y)] = F_{\underset{\sim}{\mathbb{C}}(B)}(x).
\end{aligned}$$

Hence, $\underset{\sim}{\mathbb{C}}(A) \subseteq \underset{\sim}{\mathbb{C}}(B)$. In the same way, there is $\widetilde{\mathbb{C}}(A) \subseteq \widetilde{\mathbb{C}}(B)$.

(3)

$$\begin{aligned}
& \underset{\sim}{\mathbb{C}}(A \bigcap B) \\
= \; & \{\langle x, \wedge_{y \in U}[F_{\widetilde{\mathbb{N}}_x^\beta}(y) \vee T_{A \bigcap B}(y)], \wedge_{y \in U}[(1 - I_{\widetilde{\mathbb{N}}_x^\beta}(y)) \vee I_{A \bigcap B}(y)], \vee_{y \in U}[T_{\widetilde{\mathbb{N}}_x^\beta}(y) \wedge F_{A \bigcap B}(y)]\rangle : x \in U\} \\
= \; & \{\langle x, \wedge_{y \in U}[F_{\widetilde{\mathbb{N}}_x^\beta}(y) \vee (T_A(y) \wedge T_B(y))], \wedge_{y \in U}[(1 - I_{\widetilde{\mathbb{N}}_x^\beta}(y)) \vee (I_A(y) \vee I_B(y))], \vee_{y \in U}[T_{\widetilde{\mathbb{N}}_x^\beta}(y) \wedge (F_A(y) \\
& \vee F_B(y))]\rangle : x \in U\} \\
= \; & \{\langle x, \wedge_{y \in U}[(F_{\widetilde{\mathbb{N}}_x^\beta}(y) \vee T_A(y)) \wedge (F_{\widetilde{\mathbb{N}}_x^\beta}(y) \vee T_B(y))], \wedge_{y \in U}[((1 - I_{\widetilde{\mathbb{N}}_x^\beta}(y)) \vee I_A(y)) \vee (1 - I_{\widetilde{\mathbb{N}}_x^\beta}(y)) \vee \\
& I_B(y))], \vee_{y \in U}[(T_{\widetilde{\mathbb{N}}_x^\beta}(y) \wedge F_A(y)) \vee (T_{\widetilde{\mathbb{N}}_x^\beta}(y) \wedge F_B(y))]\rangle : x \in U\} \\
= \; & \underset{\sim}{\mathbb{C}}(A) \bigcap \underset{\sim}{\mathbb{C}}(B).
\end{aligned}$$

Similarly, we can obtain $\widetilde{\mathbb{C}}(A \bigcup B) = \widetilde{\mathbb{C}}(A) \bigcup \widetilde{\mathbb{C}}(B)$.

(4) Since $A \subseteq A \cup B$, $B \subseteq A \cup B$, $A \cap B \subseteq A$ and $A \cap B \subseteq B$,

$$\underset{\sim}{\mathbb{C}}(A) \subseteq \underset{\sim}{\mathbb{C}}(A \cup B), \underset{\sim}{\mathbb{C}}(B) \subseteq \underset{\sim}{\mathbb{C}}(A \cup B), \widetilde{\mathbb{C}}(A \cap B) \subseteq \widetilde{\mathbb{C}}(A) \text{ and } \widetilde{\mathbb{C}}(A \cap B) \subseteq \widetilde{\mathbb{C}}(B).$$

Hence, $\underset{\sim}{\mathbb{C}}(A \bigcup B) \supseteq \underset{\sim}{\mathbb{C}}(A) \bigcup \underset{\sim}{\mathbb{C}}(B), \widetilde{\mathbb{C}}(A \bigcap B) \subseteq \widetilde{\mathbb{C}}(A) \bigcap \widetilde{\mathbb{C}}(B)$. □

We propose the other SVN covering rough set model, which concerns the crisp lower and upper approximations of each crisp set in the SVN environment.

Definition 7. *Let $(U, \widehat{\mathbb{C}})$ be a SVN β-covering approximation space. For each crisp subset $X \in P(U)$ ($P(U)$ is the power set of U), we define the SVN covering upper approximation $\overline{\mathbb{C}}(X)$ and lower approximation $\underline{\mathbb{C}}(X)$ of X as:*

$$\begin{aligned}
\overline{\mathbb{C}}(X) &= \{x \in U : \overline{\mathbb{N}}_x^\beta \cap X \neq \emptyset\}, \\
\underline{\mathbb{C}}(X) &= \{x \in U : \overline{\mathbb{N}}_x^\beta \subseteq X\}.
\end{aligned} \tag{6}$$

If $\overline{\mathbb{C}}(X) \neq \underline{\mathbb{C}}(X)$, then X is called the second type of SVN covering rough set.

Example 4 (Continued from Example 2). *Let $\beta = \langle 0.5, 0.3, 0.8 \rangle$, $X = \{x_1, x_2\}$, $Y = \{x_2, x_4, x_5\}$. Then,*

$$\begin{aligned}
&\overline{\mathbb{C}}(X) = \{x_1, x_2, x_4\}, \underline{\mathbb{C}}(X) = \{x_1, x_2\}, \\
&\overline{\mathbb{C}}(Y) = \{x_1, x_2, x_3, x_4, x_5\}, \underline{\mathbb{C}}(Y) = \{x_2, x_4, x_5\}, \\
&\overline{\mathbb{C}}(U) = U, \underline{\mathbb{C}}(U) = U, \overline{\mathbb{C}}(\emptyset) = \emptyset, \underline{\mathbb{C}}(\emptyset) = \emptyset.
\end{aligned}$$

Proposition 5. *Let $\widehat{\mathbb{C}}$ be a SVN β-covering of U. Then, the SVN covering upper and lower approximation operators in Definition 7 satisfy the following properties: for all $X, Y \in P(U)$,*

(1) $\underline{\mathbb{C}}(\emptyset) = \emptyset, \overline{\mathbb{C}}(U) = U$.
(2) $\underline{\mathbb{C}}(U) = U, \overline{\mathbb{C}}(\emptyset) = \emptyset$.
(3) $\underline{\mathbb{C}}(X') = (\overline{\mathbb{C}}(X))', \overline{\mathbb{C}}(X') = (\underline{\mathbb{C}}(X))'$.
(4) *If $X \subseteq Y$, then* $\underline{\mathbb{C}}(X) \subseteq \underline{\mathbb{C}}(Y), \overline{\mathbb{C}}(X) \subseteq \overline{\mathbb{C}}(Y)$.
(5) $\underline{\mathbb{C}}(X \bigcap Y) = \underline{\mathbb{C}}(X) \bigcap \underline{\mathbb{C}}(Y), \overline{\mathbb{C}}(X \bigcup Y) = \overline{\mathbb{C}}(X) \bigcup \overline{\mathbb{C}}(Y)$.
(6) $\underline{\mathbb{C}}(X \bigcup Y) \supseteq \underline{\mathbb{C}}(X) \bigcup \underline{\mathbb{C}}(Y), \overline{\mathbb{C}}(X \bigcap Y) \subseteq \overline{\mathbb{C}}(X) \bigcap \overline{\mathbb{C}}(Y)$.
(7) $\underline{\mathbb{C}}(\underline{\mathbb{C}}(X)) \subseteq \underline{\mathbb{C}}(X), \overline{\mathbb{C}}(\overline{\mathbb{C}}(X)) \supseteq \overline{\mathbb{C}}(X)$.
(8) $\underline{\mathbb{C}}(X) \subseteq X \subseteq \overline{\mathbb{C}}(X)$.
(9) $X \subseteq Y$ *or* $Y \subseteq X \Leftrightarrow \underline{\mathbb{C}}(X \cap Y) = \underline{\mathbb{C}}(X) \cap \underline{\mathbb{C}}(Y), \overline{\mathbb{C}}(X \cup Y) = \overline{\mathbb{C}}(X) \cup \overline{\mathbb{C}}(Y)$.

Proof. It can be directly followed from Definitions 5 and 7. □

5. Matrix Representations of These Single Valued Neutrosophic Covering Rough Set Models

In this section, matrix representations of the proposed SVN covering rough set models are investigated. Firstly, some new matrices and matrix operations are presented. Then, we show the matrix representations of these SVN approximation operators defined in Definitions 6 and 7. The order of elements in U is given.

Definition 8. *Let $\widehat{\mathbf{C}}$ be a SVN β-covering of U with $U = \{x_1, x_2, \cdots, x_n\}$ and $\widehat{\mathbf{C}} = \{C_1, C_2, \cdots, C_m\}$. Then, $M_{\widehat{\mathbf{C}}} = (C_j(x_i))_{n\times m}$ is named a matrix representation of $\widehat{\mathbf{C}}$, and $M^{\beta}_{\widehat{\mathbf{C}}} = (s_{ij})_{n\times m}$ is called a β-matrix representation of $\widehat{\mathbf{C}}$, where*

$$s_{ij} = \begin{cases} 1, & C_j(x_i) \geq \beta; \\ 0, & otherwise. \end{cases}$$

Example 5 (Continued from Example 1). *Let $\beta = \langle 0.5, 0.3, 0.8\rangle$.*

$$M_{\widehat{\mathbf{C}}} = \begin{pmatrix} \langle 0.7,0.2,0.5\rangle & \langle 0.6,0.2,0.4\rangle & \langle 0.4,0.1,0.5\rangle & \langle 0.1,0.5,0.6\rangle \\ \langle 0.5,0.3,0.2\rangle & \langle 0.5,0.2,0.8\rangle & \langle 0.4,0.5,0.4\rangle & \langle 0.6,0.1,0.7\rangle \\ \langle 0.4,0.5,0.2\rangle & \langle 0.2,0.3,0.6\rangle & \langle 0.5,0.2,0.4\rangle & \langle 0.6,0.3,0.4\rangle \\ \langle 0.6,0.1,0.7\rangle & \langle 0.4,0.5,0.7\rangle & \langle 0.3,0.6,0.5\rangle & \langle 0.5,0.3,0.2\rangle \\ \langle 0.3,0.2,0.6\rangle & \langle 0.7,0.3,0.5\rangle & \langle 0.6,0.3,0.5\rangle & \langle 0.8,0.1,0.2\rangle \end{pmatrix}, \quad M^{\beta}_{\widehat{\mathbf{C}}} = \begin{pmatrix} 1 & 1 & 0 & 0 \\ 1 & 1 & 0 & 1 \\ 0 & 0 & 1 & 1 \\ 1 & 0 & 0 & 1 \\ 0 & 1 & 1 & 1 \end{pmatrix}.$$

Definition 9. *Let $A = (a_{ik})_{n\times m}$ and $B = (\langle b^+_{kj}, b_{kj}, b^-_{kj}\rangle)_{1\leq k\leq m, 1\leq j\leq l}$ be two matrices. We define $D = A * B = (\langle d^+_{ij}, d_{ij}, d^-_{ij}\rangle)_{1\leq i\leq n, 1\leq j\leq l}$, where*

$$\langle d^+_{ij}, d_{ij}, d^-_{ij}\rangle = \langle \wedge_{k=1}^{m}[(1-a_{ik}) \vee b^+_{kj}], 1 - \wedge_{k=1}^{m}[(1-a_{ik}) \vee (1-b_{kj})], 1 - \wedge_{k=1}^{m}[(1-a_{ik}) \vee (1-b^-_{kj})]\rangle. \quad (7)$$

Based on Definitions 8 and 9, all $\widetilde{\mathbb{N}}^{\beta}_{x}$ for any $x \in U$ can be obtained by matrix operations.

Proposition 6. *Let $\widehat{\mathbf{C}}$ be a SVN β-covering of U with $U = \{x_1, x_2, \cdots, x_n\}$ and $\widehat{\mathbf{C}} = \{C_1, C_2, \cdots, C_m\}$. Then*

$$M^{\beta}_{\widehat{\mathbf{C}}} * M^{T}_{\widehat{\mathbf{C}}} = (\widetilde{\mathbb{N}}^{\beta}_{x_i}(x_j))_{1\leq i\leq n, 1\leq j\leq n}, \quad (8)$$

where $M^{T}_{\widehat{\mathbf{C}}}$ is the transpose of $M_{\widehat{\mathbf{C}}}$.

Proof. Suppose $M^{T}_{\widehat{\mathbf{C}}} = (C_k(x_j))_{m\times n}$, $M^{\beta}_{\widehat{\mathbf{C}}} = (s_{ik})_{n\times m}$ and $M^{\beta}_{\widehat{\mathbf{C}}} * M^{T}_{\widehat{\mathbf{C}}} = (\langle d^+_{ij}, d_{ij}, d^-_{ij}\rangle)_{1\leq i\leq n, 1\leq j\leq n}$. Since $\widehat{\mathbf{C}}$ is a SVN β-covering of U, for each i $(1 \leq i \leq n)$, there exists k $(1 \leq k \leq m)$ such that $s_{ik} = 1$. Then,

$$\begin{aligned}
& \langle d^+_{ij}, d_{ij}, d^-_{ij}\rangle \\
= {} & \langle \wedge_{k=1}^{m}[(1-s_{ik}) \vee T_{C_k}(x_j)], 1 - \wedge_{k=1}^{m}[(1-s_{ik}) \vee (1-I_{C_k}(x_j))], 1 - \wedge_{k=1}^{m}[(1-s_{ik}) \vee (1-F_{C_k}(x_j))]\rangle \\
= {} & \langle \wedge_{s_{ik}=1}[(1-s_{ik}) \vee T_{C_k}(x_j)], 1 - \wedge_{s_{ik}=1}[(1-s_{ik}) \vee (1-I_{C_k}(x_j))], 1 - \wedge_{s_{ik}=1}[(1-s_{ik}) \vee (1-F_{C_k}(x_j))]\rangle \\
= {} & \langle \wedge_{s_{ik}=1} T_{C_k}(x_j), 1 - \wedge_{s_{ik}=1}(1-I_{C_k}(x_j)), 1 - \wedge_{s_{ik}=1}(1-F_{C_k}(x_j))\rangle \\
= {} & \langle \wedge_{C_k(x_i)\geq\beta} T_{C_k}(x_j), 1 - \wedge_{C_k(x_i)\geq\beta}(1-I_{C_k}(x_j)), 1 - \wedge_{C_k(x_i)\geq\beta}(1-F_{C_k}(x_j))\rangle \\
= {} & (\cap_{C_k(x_i)\geq\beta} C_k)(x_j) \\
= {} & \widetilde{\mathbb{N}}^{\beta}_{x_i}(x_j), 1 \leq i, j \leq n.
\end{aligned}$$

Hence, $M^{\beta}_{\widehat{\mathbf{C}}} * M^{T}_{\widehat{\mathbf{C}}} = (\widetilde{\mathbb{N}}^{\beta}_{x_i}(x_j))_{1\leq i\leq n, 1\leq j\leq n}$. □

Example 6 (Continued from Example 1).

$$
\begin{aligned}
& M_{\widehat{C}}^{\beta} * M_{\widehat{C}}^{T} \\
= & \begin{pmatrix} 1 & 1 & 0 & 0 \\ 1 & 1 & 0 & 1 \\ 0 & 0 & 1 & 1 \\ 1 & 0 & 0 & 1 \\ 0 & 1 & 1 & 1 \end{pmatrix} * \begin{pmatrix} \langle 0.7,0.2,0.5\rangle & \langle 0.6,0.2,0.4\rangle & \langle 0.4,0.1,0.5\rangle & \langle 0.1,0.5,0.6\rangle \\ \langle 0.5,0.3,0.2\rangle & \langle 0.5,0.2,0.8\rangle & \langle 0.4,0.5,0.4\rangle & \langle 0.6,0.1,0.7\rangle \\ \langle 0.4,0.5,0.2\rangle & \langle 0.2,0.3,0.6\rangle & \langle 0.5,0.2,0.4\rangle & \langle 0.6,0.3,0.4\rangle \\ \langle 0.6,0.1,0.7\rangle & \langle 0.4,0.5,0.7\rangle & \langle 0.3,0.6,0.5\rangle & \langle 0.5,0.3,0.2\rangle \\ \langle 0.3,0.2,0.6\rangle & \langle 0.7,0.3,0.5\rangle & \langle 0.6,0.3,0.5\rangle & \langle 0.8,0.1,0.2\rangle \end{pmatrix}^{T} \\
= & \begin{pmatrix} 1 & 1 & 0 & 0 \\ 1 & 1 & 0 & 1 \\ 0 & 0 & 1 & 1 \\ 1 & 0 & 0 & 1 \\ 0 & 1 & 1 & 1 \end{pmatrix} * \begin{pmatrix} \langle 0.7,0.2,0.5\rangle & \langle 0.5,0.3,0.2\rangle & \langle 0.4,0.5,0.2\rangle & \langle 0.6,0.1,0.7\rangle & \langle 0.3,0.2,0.6\rangle \\ \langle 0.6,0.2,0.4\rangle & \langle 0.5,0.2,0.8\rangle & \langle 0.2,0.3,0.6\rangle & \langle 0.4,0.5,0.7\rangle & \langle 0.7,0.3,0.5\rangle \\ \langle 0.4,0.1,0.5\rangle & \langle 0.4,0.5,0.4\rangle & \langle 0.5,0.2,0.4\rangle & \langle 0.3,0.6,0.5\rangle & \langle 0.6,0.3,0.5\rangle \\ \langle 0.1,0.5,0.6\rangle & \langle 0.6,0.1,0.7\rangle & \langle 0.6,0.3,0.4\rangle & \langle 0.5,0.3,0.2\rangle & \langle 0.8,0.1,0.2\rangle \end{pmatrix} \\
= & \begin{pmatrix} \langle 0.6,0.2,0.5\rangle & \langle 0.5,0.3,0.8\rangle & \langle 0.2,0.5,0.6\rangle & \langle 0.4,0.5,0.7\rangle & \langle 0.3,0.3,0.6\rangle \\ \langle 0.1,0.5,0.6\rangle & \langle 0.5,0.3,0.8\rangle & \langle 0.2,0.5,0.6\rangle & \langle 0.4,0.5,0.7\rangle & \langle 0.3,0.3,0.6\rangle \\ \langle 0.1,0.5,0.6\rangle & \langle 0.4,0.5,0.7\rangle & \langle 0.5,0.3,0.4\rangle & \langle 0.3,0.6,0.5\rangle & \langle 0.6,0.3,0.5\rangle \\ \langle 0.1,0.5,0.6\rangle & \langle 0.5,0.3,0.7\rangle & \langle 0.4,0.5,0.4\rangle & \langle 0.5,0.3,0.7\rangle & \langle 0.3,0.2,0.6\rangle \\ \langle 0.1,0.5,0.6\rangle & \langle 0.4,0.5,0.8\rangle & \langle 0.2,0.3,0.6\rangle & \langle 0.3,0.6,0.7\rangle & \langle 0.6,0.3,0.5\rangle \end{pmatrix} \\
= & (N_{x_i}^{\beta}(x_j))_{1\le i\le 5, 1\le j\le 5}.
\end{aligned}
$$

Definition 10. *Let $A = (\langle c_{ij}^{+}, c_{ij}, c_{ij}^{-}\rangle)_{m\times n}$ and $B = (\langle d_j^{+}, d_j, d_j^{-}\rangle)_{n\times 1}$ be two matrices. We define $C = A\circ B = (\langle e_i^{+}, e_i, e_i^{-}\rangle)_{m\times 1}$ and $D = A\diamond B = (\langle f_i^{+}, f_i, f_i^{-}\rangle)_{m\times 1}$, where*

$$
\begin{aligned}
\langle e_i^{+}, e_i, e_i^{-}\rangle &= \langle \vee_{j=1}^{n}(c_{ij}^{+}\wedge d_j^{+}), \vee_{j=1}^{n}(c_{ij}\wedge d_j), \wedge_{j=1}^{n}(c_{ij}^{-}\vee d_j^{-})\rangle, \\
\langle f_i^{+}, f_i, f_i^{-}\rangle &= \langle \wedge_{j=1}^{n}(c_{ij}^{-}\vee d_j^{+}), \wedge_{j=1}^{n}[(1-c_{ij})\vee d_j], \vee_{j=1}^{n}(c_{ij}^{+}\wedge d_j^{-})\rangle.
\end{aligned} \tag{9}
$$

According to Proposition 6 and Definition 10, the set representations of $\widetilde{\mathbb{C}}(A)$ and $\underset{\sim}{\mathbb{C}}(A)$ (for any $A\in SVN(U)$) can be converted to matrix representations.

Theorem 3. *Let $\widehat{C}$ be a SVN β-covering of U with $U = \{x_1, x_2, \cdots, x_n\}$ and $\widehat{C} = \{C_1, C_2, \cdots, C_m\}$. Then, for any $A\in SVN(U)$,*

$$
\begin{aligned}
\widetilde{\mathbb{C}}(A) &= (M_{\widehat{C}}^{\beta} * M_{\widehat{C}}^{T})\circ A, \\
\underset{\sim}{\mathbb{C}}(A) &= (M_{\widehat{C}}^{\beta} * M_{\widehat{C}}^{T})\diamond A,
\end{aligned} \tag{10}
$$

where $A = (a_i)_{n\times 1}$ with $a_i = \langle T_A(x_i), I_A(x_i), F_A(x_i)\rangle$ is the vector representation of the SVNS A. $\widetilde{\mathbb{C}}(A)$ and $\underset{\sim}{\mathbb{C}}(A)$ are also vector representations.

Proof. According to Proposition 6 and Definitions 6 and 10, for any x_i $(i = 1, 2, \cdots, n)$,

$$
\begin{aligned}
((M_{\widehat{C}}^{\beta} * M_{\widehat{C}}^{T})\circ A)(x_i) &= \langle \vee_{j=1}^{n}(T_{\widetilde{\mathbb{N}}_{x_i}^{\beta}}(x_j)\wedge T_A(x_j)), \vee_{j=1}^{n}(I_{\widetilde{\mathbb{N}}_{x_i}^{\beta}}(x_j)\wedge I_A(x_j)), \wedge_{j=1}^{n}(F_{\widetilde{\mathbb{N}}_{x_i}^{\beta}}(x_j)\vee F_A(x_j))\rangle \\
&= (\widetilde{\mathbb{C}}(A))(x_i),
\end{aligned}
$$

and

$$
\begin{aligned}
((M_{\widehat{C}}^{\beta} * M_{\widehat{C}}^{T})\diamond A)(x_i) &= \langle \wedge_{j=1}^{n}(F_{\widetilde{\mathbb{N}}_{x_i}^{\beta}}(x_j)\vee T_A(x_j)), \wedge_{j=1}^{n}[(1-I_{\widetilde{\mathbb{N}}_{x_i}^{\beta}}(x_j))\vee I_A(x_j)], \vee_{j=1}^{n}(T_{\widetilde{\mathbb{N}}_{x_i}^{\beta}}(x_j)\wedge F_A(x_j))\rangle \\
&= (\underset{\sim}{\mathbb{C}}(A))(x_i).
\end{aligned}
$$

Hence, $\widetilde{\mathbb{C}}(A) = (M_{\widehat{C}}^{\beta} * M_{\widehat{C}}^{T})\circ A$, $\underset{\sim}{\mathbb{C}}(A) = (M_{\widehat{C}}^{\beta} * M_{\widehat{C}}^{T})\diamond A$. □

Example 7 (Continued from Example 3). *Let* $\beta = \langle 0.5, 0.3, 0.8\rangle$, $A = \frac{\langle 0.6,0.3,0.5\rangle}{x_1} + \frac{\langle 0.4,0.5,0.1\rangle}{x_2} + \frac{\langle 0.3,0.2,0.6\rangle}{x_3} + \frac{\langle 0.5,0.3,0.4\rangle}{x_4} + \frac{\langle 0.7,0.2,0.3\rangle}{x_5}$. *Then,*

$$
\begin{aligned}
&\widetilde{\overline{\mathbb{C}}}(A)\\
=\;& (M^{\beta}_{\widehat{C}} * M^{T}_{\widehat{C}}) \circ A\\
=\;& \begin{pmatrix}
\langle 0.6,0.2,0.5\rangle & \langle 0.5,0.3,0.8\rangle & \langle 0.2,0.5,0.6\rangle & \langle 0.4,0.5,0.7\rangle & \langle 0.3,0.3,0.6\rangle\\
\langle 0.1,0.5,0.6\rangle & \langle 0.5,0.3,0.8\rangle & \langle 0.2,0.5,0.6\rangle & \langle 0.4,0.5,0.7\rangle & \langle 0.3,0.3,0.6\rangle\\
\langle 0.1,0.5,0.6\rangle & \langle 0.4,0.5,0.7\rangle & \langle 0.5,0.3,0.4\rangle & \langle 0.3,0.6,0.5\rangle & \langle 0.6,0.3,0.5\rangle\\
\langle 0.1,0.5,0.6\rangle & \langle 0.5,0.3,0.7\rangle & \langle 0.4,0.5,0.4\rangle & \langle 0.5,0.3,0.7\rangle & \langle 0.3,0.2,0.6\rangle\\
\langle 0.1,0.5,0.6\rangle & \langle 0.4,0.5,0.8\rangle & \langle 0.2,0.3,0.6\rangle & \langle 0.3,0.6,0.7\rangle & \langle 0.6,0.3,0.5\rangle
\end{pmatrix} \circ \begin{pmatrix}
\langle 0.6,0.3,0.5\rangle\\ \langle 0.4,0.5,0.1\rangle\\ \langle 0.3,0.2,0.6\rangle\\ \langle 0.5,0.3,0.4\rangle\\ \langle 0.7,0.2,0.3\rangle
\end{pmatrix}\\
=\;& \begin{pmatrix}
\langle 0.6,0.3,0.5\rangle\\ \langle 0.4,0.3,0.6\rangle\\ \langle 0.6,0.5,0.5\rangle\\ \langle 0.5,0.3,0.6\rangle\\ \langle 0.6,0.5,0.5\rangle
\end{pmatrix},
\end{aligned}
$$

and

$$
\begin{aligned}
&\underset{\sim}{\underline{\mathbb{C}}}(A)\\
=\;& (M^{\beta}_{\widehat{C}} * M^{T}_{\widehat{C}}) \diamond A\\
=\;& \begin{pmatrix}
\langle 0.6,0.2,0.5\rangle & \langle 0.5,0.3,0.8\rangle & \langle 0.2,0.5,0.6\rangle & \langle 0.4,0.5,0.7\rangle & \langle 0.3,0.3,0.6\rangle\\
\langle 0.1,0.5,0.6\rangle & \langle 0.5,0.3,0.8\rangle & \langle 0.2,0.5,0.6\rangle & \langle 0.4,0.5,0.7\rangle & \langle 0.3,0.3,0.6\rangle\\
\langle 0.1,0.5,0.6\rangle & \langle 0.4,0.5,0.7\rangle & \langle 0.5,0.3,0.4\rangle & \langle 0.3,0.6,0.5\rangle & \langle 0.6,0.3,0.5\rangle\\
\langle 0.1,0.5,0.6\rangle & \langle 0.5,0.3,0.7\rangle & \langle 0.4,0.5,0.4\rangle & \langle 0.5,0.3,0.7\rangle & \langle 0.3,0.2,0.6\rangle\\
\langle 0.1,0.5,0.6\rangle & \langle 0.4,0.5,0.8\rangle & \langle 0.2,0.3,0.6\rangle & \langle 0.3,0.6,0.7\rangle & \langle 0.6,0.3,0.5\rangle
\end{pmatrix} \diamond \begin{pmatrix}
\langle 0.6,0.3,0.5\rangle\\ \langle 0.4,0.5,0.1\rangle\\ \langle 0.3,0.2,0.6\rangle\\ \langle 0.5,0.3,0.4\rangle\\ \langle 0.7,0.2,0.3\rangle
\end{pmatrix}\\
=\;& \begin{pmatrix}
\langle 0.6,0.5,0.5\rangle\\ \langle 0.6,0.5,0.4\rangle\\ \langle 0.4,0.4,0.5\rangle\\ \langle 0.4,0.5,0.6\rangle\\ \langle 0.6,0.4,0.3\rangle
\end{pmatrix}.
\end{aligned}
$$

Two operations of matrices are defined in [28]. We can use them to study the matrix representations of $\underline{\mathbb{C}}(X)$ and $\overline{\mathbb{C}}(X)$ of every crisp subset $X \in P(U)$.

Definition 11 ([28]). *Let $A = (a_{ik})_{n\times m}$ and $B = (b_{kj})_{m\times l}$ be two matrices. We define $C = A \cdot B = (c_{ij})_{n\times l}$ and $D = A \odot B = (d_{ij})_{n\times l}$ as follows:*

$$
\begin{aligned}
c_{ij} &= \vee_{k=1}^{m}(a_{ik} \wedge b_{kj}),\\
d_{ij} &= \wedge_{k=1}^{m}[(1-a_{ik}) \vee b_{kj}],\ \textit{for any } i = 1,2,\cdots,n, \textit{ and } j = 1,2,\cdots,l.
\end{aligned} \tag{11}
$$

Let $U = \{x_1, \cdots, x_n\}$ and $X \in P(U)$. Then, the characteristic function of the crisp subset X is defined as χ_X, where

$$
\chi_X(x_i) = \begin{cases} 1, & x_i \in X;\\ 0, & otherwise. \end{cases}
$$

Proposition 7. *Let $\widehat{C}$ be a SVN β-covering of U with $U = \{x_1, x_2, \cdots, x_n\}$ and $\widehat{C} = \{C_1, C_2, \cdots, C_m\}$. Then,*

$$
M^{\beta}_{\widehat{C}} \odot (M^{\beta}_{\widehat{C}})^{T} = (\chi_{\overline{\mathbb{N}}^{\beta}_{x_i}}(x_j))_{1\le i\le n, 1\le j\le n}, \tag{12}
$$

Proof. Suppose $M^{\beta}_{\widehat{\mathbb{C}}} = (s_{ik})_{n\times m}$ and $M^{\beta}_{\widehat{\mathbb{C}}} \odot (M^{\beta}_{\widehat{\mathbb{C}}})^T = (t_{ij})_{n\times n}$. Since $\widehat{\mathbb{C}}$ is a SVN β-covering of U, for each i $(1 \le i \le n)$ there exists k $(1 \le k \le m)$ such that $s_{ik} = 1$. If $t_{ij} = 1$, then $\wedge_{k=1}^{m}[(1-s_{ik}) \vee s_{jk}] = 1$.

It implies that if $s_{ik} = 1$, then $s_{jk} = 1$. Hence, $C_k(x_i) \ge \beta$ implies $C_k(x_j) \ge \beta$. Therefore, $x_j \in \overline{\mathbb{N}}^{\beta}_{x_i}$, i.e., $\chi_{\overline{\mathbb{N}}^{\beta}_{x_i}}(x_j) = 1 = t_{ij}$.

If $t_{ij} = 0$, then $\wedge_{k=1}^{m}[(1-s_{ik}) \vee s_{jk}] = 0$. This implies that if $s_{ik} = 1$, then $s_{jk} = 0$. Hence, $C_k(x_i) \ge \beta$ implies $C_k(x_j) < \beta$. Thus, we have $x_j \notin \overline{\mathbb{N}}^{\beta}_{x_i}$, i.e., $\chi_{\overline{\mathbb{N}}^{\beta}_{x_i}}(x_j) = 1 = t_{ij}$. □

Example 8 (Continued from Example 2). *According to $M^{\beta}_{\widehat{\mathbb{C}}}$ in Example 5, we have the following result.*

$$M^{\beta}_{\widehat{\mathbb{C}}} \odot (M^{\beta}_{\widehat{\mathbb{C}}})^T = \begin{pmatrix} 1 & 1 & 0 & 0 & 0 \\ 0 & 1 & 0 & 0 & 0 \\ 0 & 0 & 1 & 0 & 1 \\ 0 & 1 & 0 & 1 & 0 \\ 0 & 0 & 0 & 0 & 1 \end{pmatrix} = (\chi_{\overline{\mathbb{N}}^{\beta}_{x_i}}(x_j))_{1\le i\le 5, 1\le j\le 5}.$$

For any $X \in P(U)$, we also denote $\chi_X = (a_i)_{n\times 1}$ with $a_i = 1$ iff $x_i \in X$; otherwise, $a_i = 0$.

Then, the set representations of $\overline{\mathbb{C}}(X)$ and $\underline{\mathbb{C}}(X)$ (for any $X \in P(U)$) can be converted to matrix representations.

Theorem 4. *Let $\widehat{\mathbb{C}}$ be a SVN β-covering of U with $U = \{x_1, x_2, \cdots, x_n\}$ and $\widehat{\mathbb{C}} = \{C_1, C_2, \cdots, C_m\}$. Then, for any $X \in P(U)$,*

$$\begin{aligned} \chi_{\overline{\mathbb{C}}(X)} &= (M^{\beta}_{\widehat{\mathbb{C}}} \odot (M^{\beta}_{\widehat{\mathbb{C}}})^T) \cdot \chi_X, \\ \chi_{\underline{\mathbb{C}}(X)} &= (M^{\beta}_{\widehat{\mathbb{C}}} \odot (M^{\beta}_{\widehat{\mathbb{C}}})^T) \odot \chi_X. \end{aligned} \tag{13}$$

Proof. Suppose $(M^{\beta}_{\widehat{\mathbb{C}}} \odot (M^{\beta}_{\widehat{\mathbb{C}}})^T) \cdot \chi_X = (a_i)_{n\times 1}$ and $(M^{\beta}_{\widehat{\mathbb{C}}} \odot (M^{\beta}_{\widehat{\mathbb{C}}})^T) \odot \chi_X = (b_i)_{n\times 1}$. For any $x_i \in U$ $(i = 1, 2, \cdots, n)$,

$$\begin{aligned} x_i \in \overline{\mathbb{C}}(X) &\Leftrightarrow \chi_{\overline{\mathbb{C}}(X)}(x_i) = 1 \\ &\Leftrightarrow a_i = 1 \\ &\Leftrightarrow \vee_{k=1}^{n}[\chi_{\overline{\mathbb{N}}^{\beta}_{x_i}}(x_k) \wedge \chi_X(x_k)] = 1 \\ &\Leftrightarrow \exists k \in \{1, 2, \cdots, n\}, \text{ s.t., } \chi_{\overline{\mathbb{N}}^{\beta}_{x_i}}(x_k) = \chi_X(x_k) = 1 \\ &\Leftrightarrow \exists k \in \{1, 2, \cdots, n\}, \text{ s.t., } x_k \in \overline{\mathbb{N}}^{\beta}_{x_i} \cap X \\ &\Leftrightarrow \overline{\mathbb{N}}^{\beta}_{x_i} \cap X \neq \emptyset, \end{aligned}$$

and

$$\begin{aligned} x_i \in \underline{\mathbb{C}}(X) &\Leftrightarrow \chi_{\underline{\mathbb{C}}(X)}(x_i) = 1 \\ &\Leftrightarrow b_i = 1 \\ &\Leftrightarrow \wedge_{k=1}^{n}[(1-\chi_{\overline{\mathbb{N}}^{\beta}_{x_i}}(x_k)) \vee \chi_X(x_k)] = 1 \\ &\Leftrightarrow \chi_{\overline{\mathbb{N}}^{\beta}_{x_i}}(x_k) = 1 \to \chi_X(x_k) = 1,\ k = 1, 2, \cdots, n \\ &\Leftrightarrow x_k \in \overline{\mathbb{N}}^{\beta}_{x_i} \to x_k \in X,\ k = 1, 2, \cdots, n \\ &\Leftrightarrow \overline{\mathbb{N}}^{\beta}_{x_i} \subseteq X. \end{aligned}$$

□

Example 9 (Continued from Example 4). *Let $X = \{x_1, x_2\}$. By $M_{\widehat{C}}^{\beta} \odot (M_{\widehat{C}}^{\beta})^T$ in Example 8, we have*

$$(M_{\widehat{C}}^{\beta} \odot (M_{\widehat{C}}^{\beta})^T) \cdot \chi_X = \begin{pmatrix} 1 & 1 & 0 & 0 & 0 \\ 0 & 1 & 0 & 0 & 0 \\ 0 & 0 & 1 & 0 & 1 \\ 0 & 1 & 0 & 1 & 0 \\ 0 & 0 & 0 & 0 & 1 \end{pmatrix} \cdot \begin{pmatrix} 1 \\ 1 \\ 0 \\ 0 \\ 0 \end{pmatrix} = \begin{pmatrix} 1 \\ 1 \\ 0 \\ 1 \\ 0 \end{pmatrix} = \chi_{\overline{\mathbb{C}}(X)},$$

$$(M_{\widehat{C}}^{\beta} \odot (M_{\widehat{C}}^{\beta})^T) \odot \chi_X = \begin{pmatrix} 1 & 1 & 0 & 0 & 0 \\ 0 & 1 & 0 & 0 & 0 \\ 0 & 0 & 1 & 0 & 1 \\ 0 & 1 & 0 & 1 & 0 \\ 0 & 0 & 0 & 0 & 1 \end{pmatrix} \odot \begin{pmatrix} 1 \\ 1 \\ 0 \\ 0 \\ 0 \end{pmatrix} = \begin{pmatrix} 1 \\ 1 \\ 0 \\ 0 \\ 0 \end{pmatrix} = \chi_{\underline{\mathbb{C}}(X)}.$$

6. An Application to Decision Making Problems

In this section, we present a novel approach to DM problems based on the SVN covering rough set model. Then, a comparative study with other methods is shown.

6.1. The Problem of Decision Making

Let $U = \{x_k : k = 1, 2, \cdots, l\}$ be the set of patients and $V = \{y_i | i = 1, 2, \cdots, m\}$ be the m main symptoms (for example, cough, fever, and so on) for a Disease B. Assume that Doctor R evaluates every Patient x_k $(k = 1, 2, \cdots, l)$.

Assume that Doctor R believes each Patient $x_k \in U$ $(k = 1, 2, \cdots, l)$ has a symptom value C_i $(i = 1, 2, \cdots, m)$, denoted by $C_i(x_k) = \langle T_{C_i}(x_k), I_{C_i}(x_k, F_{C_i}(x_k)) \rangle$, where $T_{C_i}(x_k) \in [0, 1]$ is the degree that Doctor R confirms Patient x_k has symptom y_i, $I_{C_i}(x_k) \in [0, 1]$ is the degree that Doctor R is not sure Patient x_k has symptom y_i, $F_{C_i}(x_k) \in [0, 1]$ is the degree that Doctor R confirms Patient x_k does not have symptom y_i, and $T_{C_i}(x_k) + I_{C_i}(x_k) + F_{C_i}(x_k) \leq 3$.

Let $\beta = \langle a, b, c \rangle$ be the critical value. If any Patient $x_k \in U$, there is at least one symptom $y_i \in V$ such that the symptom value C_i for Patient x_k is not less than β, respectively, then $\widehat{\mathbf{C}} = \{C_1, C_2, \cdots, C_m\}$ is a SVN β-covering of U for some SVN number β.

If d is a possible degree, e is an indeterminacy degree and f is an impossible degree of Disease B of every Patient $x_k \in U$ that is diagnosed by Doctor R, denoted by $A(x_k) = \langle d, e, f \rangle$, then the decision maker (Doctor R) for the decision making problem needs to know how to evaluate whether Patients $x_k \in U$ have Disease B.

6.2. The Decision Making Algorithm

In this subsection, we give an approach for the problem of DM with the above characterizations by means of the first type of SVN covering rough set model. According to the characterizations of the DM problem in Section 6.1, we construct the SVN decision information system and present the Algorithm 1 of DM under the framework of the first type of SVN covering rough set model.

Algorithm 1 The decision making algorithm based on the SVN covering rough set model.

Input: SVN decision information system $(U, \widehat{\mathbb{C}}, \beta, A)$.
Output: The score ordering for all alternatives.

1: Compute the SVN β-neighborhood $\widetilde{\mathbb{N}}_x^\beta$ of x induced by $\widehat{\mathbb{C}}$, for all $x \in U$ according to Definition 4;
2: Compute the SVN covering upper approximation $\widetilde{\mathbb{C}}(A)$ and lower approximation $\underset{\sim}{\mathbb{C}}(A)$ of A, according to Definition 6;
3: Compute $\widetilde{R}_A = \widetilde{\mathbb{C}}(A) \oplus \underset{\sim}{\mathbb{C}}(A)$ according to (6) in the basic operations on $SVN(U)$;
4: Compute

$$s(x) = \frac{T_{\widetilde{R}_A}(x)}{\sqrt{(T_{\widetilde{R}_A}(x))^2+(I_{\widetilde{R}_A}(x))^2+(F_{\widetilde{R}_A}(x))^2}};$$

5: Rank all the alternatives $s(x)$ by using the principle of numerical size and select the most possible patient.

According to the above process, we can get the decision making according to the ranking. In Step 4, $S(x)$ is the cosine similarity measure between $\widetilde{R}_A(x)$ and the ideal solution $(1,0,0)$, which was proposed by Ye [44].

6.3. An Applied Example

Example 10. *Assume that $U = \{x_1, x_2, x_3, x_4, x_5\}$ is a set of patients. According to the patients' symptoms, we write $V = \{y_1, y_2, y_3, y_4\}$ to be four main symptoms (cough, fever, sore and headache) for Disease B. Assume that Doctor R evaluates every Patient x_k $(k = 1, 2, \cdots, 5)$ as shown in Table 1.*

Let $\beta = \langle 0.5, 0.3, 0.8\rangle$ be the critical value. Then, $\widehat{\mathbb{C}} = \{C_1, C_2, C_3, C_4\}$ is a SVN β-coverings of U. $\widetilde{\mathbb{N}}_{x_k}^\beta$ $(k = 1, 2, 3, 4, 5)$ are shown in Table 2.

Assume that Doctor R diagnoses the value $A = \frac{(0.6,0.3,0.5)}{x_1} + \frac{(0.4,0.5,0.1)}{x_2} + \frac{(0.3,0.2,0.6)}{x_3} + \frac{(0.5,0.3,0.4)}{x_4} + \frac{(0.7,0.2,0.3)}{x_5}$ of Disease B of every patient. Then,

$$\widetilde{\mathbb{C}}(A) = \{\langle x_1, 0.6, 0.3, 0.5\rangle, \langle x_2, 0.4, 0.3, 0.6\rangle, \langle x_3, 0.6, 0.5, 0.5\rangle, \langle x_4, 0.5, 0.3, 0.6\rangle, \langle x_5, 0.6, 0.5, 0.5\rangle\},$$
$$\underset{\sim}{\mathbb{C}}(A) = \{\langle x_1, 0.6, 0.5, 0.5\rangle, \langle x_2, 0.6, 0.5, 0.4\rangle, \langle x_3, 0.4, 0.4, 0.5\rangle, \langle x_4, 0.4, 0.5, 0.4\rangle, \langle x_5, 0.6, 0.4, 0.3\rangle\}.$$

Then,

$$\begin{aligned} & \widetilde{R}_A \\ = & \widetilde{\mathbb{C}}(A) \oplus \underset{\sim}{\mathbb{C}}(A) \\ = & \{\langle x_1, 0.84, 0.15, 0.25\rangle, \langle x_2, 0.76, 0.15, 0.24\rangle, \langle x_3, 0.76, 0.2, 0.25\rangle, \langle x_4, 0.7, 0.15, 0.24\rangle, \langle x_5, 0.84, 0.2, 0.15\rangle\}. \end{aligned}$$

Hence, we can obtain $s(x_k)$ $(k = 1, 2, \cdots, 5)$ in Table 3.

Table 3. $s(x_k)$ $(k = 1, 2, \cdots, 5)$.

U	x_1	x_2	x_3	x_4	x_5
$s(x_k)$	0.945	0.937	0.922	0.909	0.958

According to the principle of numerical size, we have:

$$s(x_4) < s(x_3) < s(x_2) < s(x_1) < s(x_5).$$

Therefore, Doctor R diagnoses Patient x_5 as more likely to be sick with Disease B.

6.4. A Comparison Analysis

To validate the feasibility of the proposed decision making method, a comparative study was conducted with other methods. These methods, which were introduced by Liu [43], Yang et al. [32] and Ye [44], are compared with the proposed approach using SVN information system.

6.4.1. The Results of Liu's Method

Liu's method is shown in Algorithm 2.

Algorithm 2 The decision making algorithm [43].

Input: A SVN decision matrix D, a weight vector $\mathbf{w}$ and γ.
Output: The score ordering for all alternatives.

1: Compute

$$\begin{aligned} n_k &= \langle T_{n_k}, I_{n_k}, F_{n_k} \rangle \\ &= HSVNNWA(n_{k1}, n_{k2}, \cdots, n_{km}) \\ &= \langle \frac{\prod_{i=1}^{m}(1+(\gamma-1)T_{ki})^{w_i} - \prod_{i=1}^{m}(1-T_{ki})^{w_i}}{\prod_{i=1}^{m}(1+(\gamma-1)T_{ki})^{w_i} + (\gamma-1)\prod_{i=1}^{m}(1-T_{ki})^{w_i}}, \\ &\quad \frac{\gamma \prod_{i=1}^{m} I_{ki}^{w_i}}{\prod_{i=1}^{m}(1+(\gamma-1)(1-I_{ki}))^{w_i} + (\gamma-1)\prod_{i=1}^{m} I_{ki}^{w_i}}, \\ &\quad \frac{\gamma \prod_{i=1}^{m} F_{ki}^{w_i}}{\prod_{i=1}^{m}(1+(\gamma-1)(1-F_{ki}))^{w_i} + (\gamma-1)\prod_{i=1}^{m} F_{ki}^{w_i}} \rangle \; (k = 1, 2, \cdots, l); \end{aligned}$$

2: Calculate $s(n_k) = \frac{T_{n_k}}{\sqrt{T_{n_k}^2 + I_{n_k}^2 + F_{n_k}^2}}$;

3: Obtain the ranking for all $s(n_k)$ by using the principle of numerical size and select the most possible patient.

Then, Algorithm 2 can be used for Example 10. Let $n_{ki} = \langle T_{ki}, I_{ki}, F_{ki} \rangle$ be the evaluation information of x_k on C_i in Table 1. That is to say, Table 1 is the SVN decision matrix D. We suppose the weight vector of the criteria is $\mathbf{w} = (0.35, 0, 25, 0.3, 0.1)$ and $\gamma = 1$.

Step 1: Based on HSVNNWA operator, we get

$$\begin{aligned} &n_1 = \langle 0.557, 0.178, 0.482 \rangle, \; n_2 = \langle 0.484, 0.283, 0.395 \rangle, \\ &n_3 = \langle 0.414, 0.318, 0.347 \rangle, \; n_4 = \langle 0.465, 0.286, 0.558 \rangle, \\ &n_5 = \langle 0.578, 0.233, 0.486 \rangle. \end{aligned}$$

Step 2: We get

$$s(n_1) = 0.735, s(n_2) = 0.706, s(n_3) = 0.660, s(n_4) = 0.596, s(n_5) = 0.734.$$

Step 3: According to the cosine similarity degrees $s(n_k)$ $(k = 1, 2, \cdots, 5)$, we obtain $x_4 < x_3 < x_2 < x_5 < x_1$.

Therefore, Patient x_1 is more likely to be sick with Disease B.

6.4.2. The Results of Yang's Method

Yang's method is shown in Algorithm 3.

Algorithm 3 The decision making algorithm [32].

Input: A generalized SVN approximation space $(U, V, \widetilde{R})$, $B \in SVN(V)$.
Output: The score ordering for all alternatives.

1: Calculate the lower and upper approximations $\overline{\widetilde{R}}(B)$ and $\underline{\widetilde{R}}(B)$;
2: Compute $n_{x_k} = (\overline{\widetilde{R}}(B) \oplus \underline{\widetilde{R}}(B))(x_k)$ $(k = 1, 2, \cdots, l)$;
3: Compute

$$s(n_{x_k}, n^*) = \frac{T_{n_{x_k}} \cdot T_{n^*} + I_{n_{x_k}} \cdot I_{n^*} + F_{n_{x_k}} \cdot F_{n^*}}{\sqrt{T^2_{n_{x_k}} + I^2_{n_{x_k}} + F^2_{n_{x_k}}} \cdot \sqrt{(T_{n^*})^2 + (I_{n^*})^2 + (F_{n^*})^2}} \ (k = 1, 2, \cdots, l),$$

where $n^* = \langle T_{n^*}, I_{n^*}, F_{n^*} \rangle = \langle 1, 0, 0 \rangle$;
4: Obtain the ranking for all $s(n_{x_k}, n^*)$ by using the principle of numerical size and select the most possible patient.

For Example 10, we suppose Disease $B \in SVN(V)$ and $B = \frac{(0.3,0.6,0.5)}{y_1} + \frac{(0.7,0.2,0.1)}{y_2} + \frac{(0.6,0.4,0.3)}{y_3} + \frac{(0.8,0.4,0.5)}{y_4}$. According to Table 1, the generalized SVN approximation space $(U, V, \widetilde{R})$ can be obtained in Table 4, where $U = \{x_1, x_2, x_3, x_4, x_5\}$ and $V = \{y_1, y_2, y_3, y_4\}$.

Table 4. The generalized SVN approximation space $(U, V, \widetilde{R})$.

$\widetilde{R}$	x_1	x_2	x_3	x_4	x_5
y_1	$\langle 0.7, 0.2, 0.5 \rangle$	$\langle 0.5, 0.3, 0.2 \rangle$	$\langle 0.4, 0.5, 0.2 \rangle$	$\langle 0.6, 0.1, 0.7 \rangle$	$\langle 0.3, 0.2, 0.6 \rangle$
y_2	$\langle 0.6, 0.2, 0.4 \rangle$	$\langle 0.5, 0.2, 0.8 \rangle$	$\langle 0.2, 0.3, 0.6 \rangle$	$\langle 0.4, 0.5, 0.7 \rangle$	$\langle 0.7, 0.3, 0.5 \rangle$
y_3	$\langle 0.4, 0.1, 0.5 \rangle$	$\langle 0.4, 0.5, 0.4 \rangle$	$\langle 0.5, 0.2, 0.4 \rangle$	$\langle 0.3, 0.6, 0.5 \rangle$	$\langle 0.6, 0.3, 0.5 \rangle$
y_4	$\langle 0.1, 0.5, 0.6 \rangle$	$\langle 0.6, 0.1, 0.7 \rangle$	$\langle 0.6, 0.3, 0.4 \rangle$	$\langle 0.5, 0.3, 0.2 \rangle$	$\langle 0.8, 0.1, 0.2 \rangle$

Step 1: We get

$$\overline{\widetilde{R}}(B) = \{\langle x_1, 0.6, 0.2, 0.4 \rangle, \langle x_2, 0.6, 0.2, 0.4 \rangle, \langle x_3, 0.6, 0.3, 0.4 \rangle, \langle x_4, 0.5, 0.4, 0.5 \rangle, \langle x_5, 0.8, 0.3, 0.5 \rangle\},$$

$$\underline{\widetilde{R}}(B) = \{\langle x_1, 0.5, 0.6, 0.5 \rangle, \langle x_2, 0.3, 0.6, 0.5 \rangle, \langle x_3, 0.3, 0.5, 0.5 \rangle, \langle x_4, 0.6, 0.6, 0.5 \rangle, \langle x_5, 0.6, 0.6, 0.5 \rangle\}.$$

Step 2:

$$\overline{\widetilde{R}}(B) \oplus \underline{\widetilde{R}}(B) = \{\langle x_1, 0.80, 0.12, 0.20 \rangle, \langle x_2, 0.72, 0.12, 0.20 \rangle, \langle x_3, 0.72, 0.15, 0.20 \rangle, \langle x_4, 0.80, 0.24, 0.25 \rangle, \langle x_5, 0.92, 0.18, 0.25 \rangle\}.$$

Step 3: Let $n^* = \langle 1, 0, 0 \rangle$. Then,

$$s(n_{x_1}, n^*) = 0.960,\ s(n_{x_2}, n^*) = 0.951,\ s(n_{x_3}, n^*) = 0.945,\ s(n_{x_4}, n^*) = 0.918,\ s(n_{x_5}, n^*) = 0.948.$$

Step 4:

$$s(n_{x_4}, n^*) < s(n_{x_3}, n^*) < s(n_{x_5}, n^*) < s(n_{x_2}, n^*) < s(n_{x_1}, n^*).$$

Therefore, Patient x_1 is more likely to be sick with Disease B.

6.4.3. The Results of Ye's Methods

Ye presented two methods [44]. Thus, Algorithms 4 and 5 are presented for Example 10.

Algorithm 4 The decision making algorithm [44].

Input: A SVN decision matrix D and a weight vector $\mathbf{w}$.
Output: The score ordering for all alternatives.

1: Compute

$$W_k(x_k, A^*) = \frac{\sum_{i=1}^{m} w_i[a_{ki} \cdot a_i^* + b_{ki} \cdot b_i^* + c_{ki} \cdot c_i^*]}{\sqrt{\sum_{i=1}^{m} w_i[a_{ki}^2 + b_{ki}^2 + c_{ki}^2]} \cdot \sqrt{\sum_{i=1}^{m} w_i[(a_i^*)^2 + (b_i^*)^2 + (c_i^*)^2]}} \quad (k = 1, 2, \cdots, l),$$

where $\alpha_i^* = \langle a_i^*, b_i^*, c_i^* \rangle = \langle 1, 0, 0 \rangle$ $(i = 1, 2, \cdots, m)$;

2: Obtain the ranking for all $W_k(x_k, A^*)$ by using the principle of numerical size and select the most possible patient.

For Example 10, Table 1 is the SVN decision matrix D. We suppose the weight vector of the criteria is $\mathbf{w} = (0.35, 0, 25, 0.3, 0.1)$.

Step 1:

$$W_1(x_1, A^*) = 0.677, W_2(x_2, A^*) = 0.608, W_3(x_3, A^*) = 0.580, W_4(x_4, A^*) = 0.511, W_5(x_5, A^*) = 0.666.$$

Step 2: The ranking order of $\{x_1, x_2, \cdots, x_5\}$ is $x_4 < x_3 < x_2 < x_5 < x_1$. Therefore, Patient x_1 is more likely to be sick with Disease B.

Algorithm 5 The other decision making algorithm [44].

Input: A SVN decision matrix D and a weight vector $\mathbf{w}$.
Output: The score ordering for all alternatives.

1: Compute

$$M_k(x_k, A^*) = \sum_{i=1}^{m} w_i \frac{a_{ki} \cdot a_i^* + b_{ki} \cdot b_i^* + c_{ki} \cdot c_i^*}{\sqrt{a_{ki}^2 + b_{ki}^2 + c_{ki}^2} \cdot \sqrt{(a_i^*)^2 + (b_i^*)^2 + (c_i^*)^2}} \quad (k = 1, 2, \cdots, l),$$

where $\alpha_i^* = \langle a_i^*, b_i^*, c_i^* \rangle = \langle 1, 0, 0 \rangle$ $(i = 1, 2, \cdots, m)$;

2: Obtain the ranking for all $M_k(x_k, A^*)$ by using the principle of numerical size and select the most possible patient.

By Algorithms 5, we have:

Step 1:

$$M_1(x_1, A^*) = 0.676, M_2(x_2, A^*) = 0.637, M_3(x_3, A^*) = 0.581,$$
$$M_4(x_4, A^*) = 0.521, M_5(x_5, A^*) = 0.654.$$

Step 2: The ranking order of $\{x_1, x_2, \cdots, x_5\}$ is $x_4 < x_3 < x_2 < x_5 < x_1$. Therefore, Patient x_1 is more likely to be sick with Disease B.

All results are shown in Table 5, Figures 1 and 2.

Table 5. The results utilizing the different methods of Example 10.

Methods	The Final Ranking	The Patient Is Most Sick With the Disease B
Algorithm 2 in Liu [43]	x_4, x_3, x_2, x_5, x_1	x_1
Algorithm 3 in Yang et al. [32]	x_4, x_3, x_5, x_2, x_1	x_1
Algorithm 4 in Ye [44]	x_4, x_3, x_2, x_5, x_1	x_1
Algorithm 5 in Ye [44]	x_4, x_3, x_2, x_5, x_1	x_1
Algorithm 1 in this paper	x_4, x_3, x_2, x_1, x_5	x_5

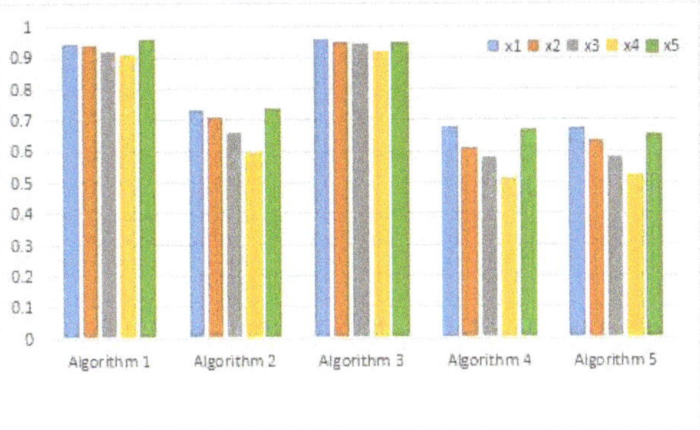

Figure 1. The first chat of different values of patient in utilizing different methods in Example 10.

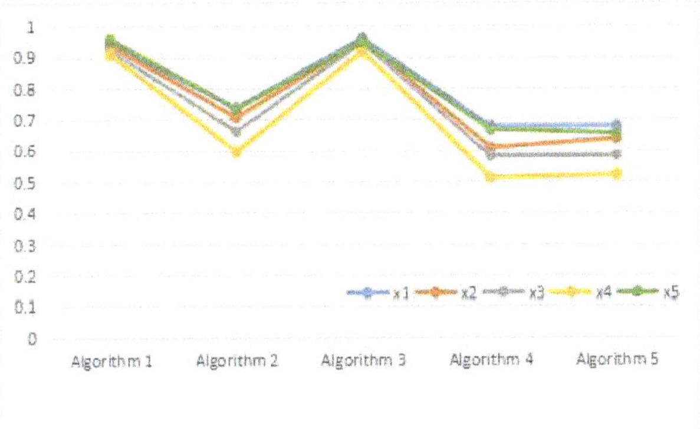

Figure 2. The second chat of different values of patient in utilizing different methods in Example 10.

Liu [43] and Ye [44] presented the methods by SVN theory. In their methods, the ranking order would be changed by different **w** and γ. We as well as Yang et al. [32] used different rough set models to make the decision. Yang et al. present a SVN rough set model based on SVN relations, while we present a new SVN rough set model based on coverings. The results are different by Yang's and our methods, although the methods are both based on an operator presented by Ye [44].

In any method, if there are more than one most possible patient, then each patient will be the optimal decision. In this case, we need other methods to make a further decision. By means of different methods, the obtained results may be different. To achieve the most accurate results, further diagnosis is necessary in combination with other hybrid methods.

7. Conclusions

This paper is a bridge, linking SVNSs and covering-based rough sets. By introducing some definitions and properties in SVN β-covering approximation spaces, we present two types of SVN covering rough set models. Then, their characterizations and matrix representations are investigated. Moreover, an application to the problem of DM is proposed. The main conclusions in this paper and the further work to do are listed as follows.

1. Two types of SVN covering rough set models are first presented, which combine SVNSs with covering-based rough sets. Some definitions and properties in covering-based rough set model, such as coverings and neighborhoods, are generalized to SVN covering rough set models.

Neutrosophic sets and related algebraic structures [47–49] will be connected with the research content of this paper in further research.

2. It would be tedious and complicated to use set representations to calculate SVN covering approximation operators. Therefore, the matrix representations of these SVN covering approximation operators make it possible to calculate them through the new matrices and matrix operations. By these matrix representations, calculations will become algorithmic and can be easily implemented by computers.
3. We propose a method to DM problems under one of the SVN covering rough set models. It is a novel method based on approximation operators specific to SVN covering rough sets firstly. The comparison analysis is very interesting to show the difference between the proposed method and other methods.

Author Contributions: All authors contributed equally to this paper. The idea of this whole thesis was put forward by X.Z., and he also completed the preparatory work of the paper. J.W. analyzed the existing work of rough sets and SVNSs, and wrote the paper.

Funding: This work was supported by the National Natural Science Foundation of China under Grant Nos. 61573240 and 61473239.

Conflicts of Interest: The authors declare no conflict of interest.

References

1. Pawlak, Z. Rough sets. *Int. J. Comput. Inf. Sci.* **1982**, *11*, 341–356.
2. Pawlak, Z. *Rough Sets: Theoretical Aspects of Reasoning about Data*; Kluwer Academic Publishers: Boston, MA, USA, 1991.
3. Bartol, W.; Miro, J.; Pioro, K.; Rossello, F. On the coverings by tolerance classes. *Inf. Sci.* **2004**, *166*, 193–211.
4. Bianucci, D.; Cattaneo, G.; Ciucci, D. Entropies and co-entropies of coverings with application to incomplete information systems. *Fundam. Inform.* **2007**, *75*, 77–105.
5. Zhu, W. Relationship among basic concepts in covering-based rough sets. *Inf. Sci.* **2009**, *179*, 2478–2486.
6. Yao, Y.; Zhao, Y. Attribute reduction in decision-theoretic rough set models. *Inf. Sci.* **2008**, *178*, 3356–3373.
7. Wang, J.; Zhang, X. Matrix approaches for some issues about minimal and maximal descriptions in covering-based rough sets. *Int. J. Approx. Reason.* **2019**, *104*, 126–143.
8. Li, F.; Yin, Y. Approaches to knowledge reduction of covering decision systems based on information theory. *Inf. Sci.* **2009**, *179*, 1694–1704.
9. Wu, W. Attribute reduction based on evidence theory in incomplete decision systems. *Inf. Sci.* **2008**, *178*, 1355–1371.
10. Wang, J.; Zhu, W. Applications of bipartite graphs and their adjacency matrices to covering-based rough sets. *Fundam. Inform.* **2017**, *156*, 237–254.
11. Dai, J.; Wang, W.; Xu, Q.; Tian, H. Uncertainty measurement for interval-valued decision systems based on extended conditional entropy. *Knowl.-Based Syst.* **2012**, *27*, 443–450.
12. Wang, C.; Chen, D.; Wu, C.; Hu, Q. Data compression with homomorphism in covering information systems. *Int. J. Approx. Reason.* **2011**, *52*, 519–525.
13. Li, X.; Yi, H.; Liu, S. Rough sets and matroids from a lattice-theoretic viewpoint. *Inf. Sci.* **2016**, *342*, 37–52.
14. Wang, J.; Zhang, X. Four operators of rough sets generalized to matroids and a matroidal method for attribute reduction. *Symmetry* **2018**, *10*, 418.
15. Wang, J.; Zhu, W. Applications of matrices to a matroidal structure of rough sets. *J. Appl. Math.* **2013**, *2013*, 493201.
16. Wang, J.; Zhu, W.; Wang, F.; Liu, G. Conditions for coverings to induce matroids. *Int. J. Mach. Learn. Cybern.* **2014**, *5*, 947–954.
17. Chen, J.; Li, J.; Lin, Y.; Lin, G.; Ma, Z. Relations of reduction between covering generalized rough sets and concept lattices. *Inf. Sci.* **2015**, *304*, 16–27.
18. Zhang, X.; Dai, J.; Yu, Y. On the union and intersection operations of rough sets based on various approximation spaces. *Inf. Sci.* **2015**, *292*, 214–229.

19. D'eer, L.; Cornelis, C.; Godo, L. Fuzzy neighborhood operators based on fuzzy coverings. *Fuzzy Sets Syst.* **2017**, *312*, 17–35.
20. Yang, B.; Hu, B. On some types of fuzzy covering-based rough sets. *Fuzzy Sets Syst.* **2017**, *312*, 36–65.
21. Zhang, X.; Miao, D.; Liu, C.; Le, M. Constructive methods of rough approximation operators and multigranulation rough sets. *Knowl.-Based Syst.* **2016**, *91*, 114–125.
22. Wang, J.; Zhang, X. Two types of intuitionistic fuzzy covering rough sets and an application to multiple criteria group decision making. *Symmetry* **2018**, *10*, 462.
23. Zadeh, L.A. Fuzzy sets. *Inf. Control* **1965**, *8*, 338–353.
24. Medina, J.; Ojeda-Aciego, M. Multi-adjoint t-concept lattices. *Inf. Sci.* **2010**, *180*, 712–725.
25. Pozna, C.; Minculete, N.; Precup, R.E.; Kóczy, L.T.; Ballagi, Á. Signatures: Definitions, operators and applications to fuzzy modeling. *Fuzzy Sets Syst.* **2012**, *201*, 86–104.
26. Jankowski, J.; Kazienko, P.; Watróbski, J.; Lewandowska, A.; Ziemba, P.; Zioło, M. Fuzzy multi-objective modeling of effectiveness and user experience in online advertising. *Expert Syst. Appl.* **2016**, *65*, 315–331.
27. Vrkalovic, S.; Lunca, E.C.; Borlea, I.D. Model-free sliding mode and fuzzy controllers for reverse osmosis desalination plants. *Int. J. Artif. Intell.* **2018**, *16*, 208–222.
28. Ma, L. Two fuzzy covering rough set models and their generalizations over fuzzy lattices. *Fuzzy Sets Syst.* **2016**, *294*, 1–17.
29. Wang, H.; Smarandache, F.; Zhang, Y.; Sunderraman, R. Single valued neutrosophic sets. *Multispace Multistruct.* **2010**, *4*, 410–413.
30. Atanassov, K. Intuitionistic fuzzy sets. *Fuzzy Sets Syst.* **1986**, *20*, 87–96.
31. Mondal, K.; Pramanik, S. Rough neutrosophic multi-attribute decision-making based on grey relational analysis. *Neutrosophic Sets Syst.* **2015**, *7*, 8–17.
32. Yang, H.; Zhang, C.; Guo, Z.; Liu, Y.; Liao, X. A hybrid model of single valued neutrosophic sets and rough sets: Single valued neutrosophic rough set model. *Soft Comput.* **2017**, *21*, 6253–6267.
33. Zhang, X.; Xu, Z. The extended TOPSIS method for multi-criteria decision making based on hesitant heterogeneous information. In Proceedings of the 2014 2nd International Conference on Software Engineering, Knowledge Engineering and Information Engineering (SEKEIE 2014), Singapore, 5–6 August 2014.
34. Cheng, J.; Zhang, Y.; Feng, Y.; Liu, Z.; Tan, J. Structural optimization of a high-speed press considering multi-source uncertainties based on a new heterogeneous TOPSIS. *Appl. Sci.* **2018**, *8*, 126.
35. Liu, J.; Zhao, H.; Li, J.; Liu, S. Decision process in MCDM with large number of criteria and heterogeneous risk preferences. *Oper. Res. Perspect.* **2017**, *4*, 106–112.
36. Watróbski, J.; Jankowski, J.; Ziemba, P.; Karczmarczyk, A.; Zioło, M. Generalised framework for multi-criteria method selection. *Omega* **2018**. [CrossRef]
37. Faizi, S.; Sałabun, W.; Rashid, T.; Wątróbski, J.; Zafar, S. Group decision-making for hesitant fuzzy sets based on characteristic objects method. *Symmetry* **2017**, *9*, 136.
38. Faizi, S.; Rashid, T.; Sałabun, W.; Zafar, S.; Wątróbski, J. Decision making with uncertainty using hesitant fuzzy sets. *Int. J. Fuzzy Syst.* **2018**, *20*, 93–103.
39. Zhan, J.; Ali, M.I.; Mehmood, N. On a novel uncertain soft set model: Z-soft fuzzy rough set model and corresponding decision making methods. *Appl. Soft Comput.* **2017**, *56*, 446–457.
40. Zhan, J.; Alcantud, J.C.R. A novel type of soft rough covering and its application to multicriteria group decision making. *Artif. Intell. Rev.* **2018**, *4*, 1–30.
41. Zhang, Z. An approach to decision making based on intuitionistic fuzzy rough sets over two universes. *J. Oper. Res. Soc.* **2013**, *64*, 1079–1089.
42. Akram, M.; Ali, G.; Alshehri, N.O. A new multi-attribute decision-making method based on m-polar fuzzy soft rough sets. *Symmetry* **2017**, *9*, 271.
43. Liu, P. The aggregation operators based on archimedean t-conorm and t-norm for single-valued neutrosophic numbers and their application to decision making. *Int. J. Fuzzy Syst.* **2016**, *18*, 849–863.
44. Ye, J. Multicriteria decision-making method using the correlation coefficient under single-valued neutrosophic environment. *Int. J. Gen. Syst.* **2013**, *42*, 386–394.
45. Bonikowski, Z.; Bryniarski, E.; Wybraniec-Skardowska, U. Extensions and intentions in the rough set theory. *Inf. Sci.* **1998**, *107*, 149–167.
46. Pomykala, J.A. Approximation operations in approximation space. *Bull. Pol. Acad. Sci.* **1987**, *35*, 653–662.

47. Zhang, X.; Bo, C.; Smarandache, F.; Dai, J. New inclusion relation of neutrosophic sets with applications and related lattice structure. *Int. J. Mach. Learn. Cybern.* **2018**, *9*, 1753–1763.
48. Zhang, X. Fuzzy anti-grouped filters and fuzzy normal filters in pseudo-BCI algebras. *J. Intell. Fuzzy Syst.* **2017**, *33*, 1767–1774.
49. Zhang, X.; Park, C.; Wu, S. Soft set theoretical approach to pseudo-BCI algebras. *J. Intell. Fuzzy Syst.* **2018**, *34*, 559–568.

symmetry

MDPI

Article

Shadowed Sets-Based Linguistic Term Modeling and Its Application in Multi-Attribute Decision-Making

Huidong Wang [1,*], Shifan He [1], Xiaohong Pan [1] and Chengdong Li [2]

[1] School of Management Science and Engineering, Shandong University of Finance and Economics, Jinan 250014, China; shifanhe0828@163.com (S.H.); xiaohongpan0303@126.com (X.P.)

[2] School of Information and Electrical Engineering, Shandong Jianzhu University, Jinan 250101, China; chengdong.li@foxmail.com

* Correspondence: huidong.wang@ia.ac.cn

Received: 20 October 2018; Accepted: 16 November 2018; Published: 2 December 2018

Abstract: For many multi-attribute decision-making (MADM) problems, linguistic variables are more convenient for people to express the attribute values. In this paper, a novel shadowed set-based method is proposed to deal with linguistic terms, where the linguistic term sets are symmetrical both in meaning and form. Firstly, to effectively express the linguistic variables, we develop a data-driven method to construct the shadowed set model for the linguistic terms. Secondly, the Pythagorean shadowed set is defined, and some theorems are subsequently explored. Thirdly, we propose the score function of the Pythagorean shadowed number and develop a new MADM method on the basis of the Pythagorean shadowed set. Finally, a case study of the supplier selection problem is provided to illustrate the effectiveness of the proposed method, and the superiority of our method is demonstrated by comparison analysis.

Keywords: Pythagorean fuzzy linguistic set; shadowed set; Pythagorean shadowed set; multi-attribute decision-making

---

1. Introduction

Multi-attribute decision-making (MADM) aims to select the best alternative solution(s) from multiple alternatives and has been widely used in various fields [1–3]. In MADM problems, linguistic terms are a convenient and natural way to describe evaluation information. For example, the decision-makers (DMs) can use linguistic terms such as 'Extremely low', 'Very low', 'Low', 'Fair', 'High', 'Very high', and 'Extremely high' to estimate service quality, product performance, and so forth. Therefore, MADM problems based on linguistic terms have received increasing attention. In [4], Aggarwal proposed a new aggregation operator for linguistic terms, and the effectiveness of the operator was illustrated by a case study on the supplier selection problem. Jin [5] developed two group decision-making methods to handle MADM problems under linguistic set environment, and comparative analysis with other methods was performed to demonstrate the validity and merits of the two methods. Yu [6] proposed an extended TODIM method with unbalanced hesitant fuzzy linguistic term sets for MADM problems. For linguistic decision-making problems, Pei [7] developed a new decision-making method by integrating the fuzzy linguistic multiset and TOPSIS methods, and two practical examples were utilized to verify the feasibility of the proposed approach. For the venture capital problem under a linguistic environment, Cheng [8] proposed an interaction approach. However, the methods mentioned above directly replace the linguistic variables with linguistic subscript in the decision-making process, which may cause distortion of information. To better express linguistic variables, the linguistic 2-tuple [9,10] and linguistic scale function [11,12] were introduced to deal with linguistic sets. Nevertheless, the linguistic 2-tuple and linguistic scale function methods still use linguistic subscript to express language variables in nature. Besides, it is difficult to explain the rationality in theory by simply replacing linguistic word with its linguistic subscript. Furthermore, people may have diverse opinions on identical words, but linguistic subscript can only depict a single meaning for one person, which may lead to information distortion.

As distinct from the linguistic 2-tuple and scale function, shadowed sets [13,14] can effectively construct linguistic terms using a data-driven method, and have recently been attracting more and more attention [15,16]. The membership value of the shadowed set is not a precise number, and its distribution is composed of three different zones: the core zone, shadowed zone, and exclusion zone. The core zone and the exclusion zone take the values of 1 and 0, which means all the elements of both zones are fully compatible with or completely excluded from the linguistic word described by a shadowed set. The shadowed zone is an entire unit interval perceived as a zone of uncertainty, which means we are not sure whether the shadowed zone elements represent the linguistic word described by a shadowed set.

In addition, in many situations, experts may hesitate as to what attribute values should be given by them, due to the increasing complexity. Consequently, Atanassov [17] proposed the intuitionistic fuzzy set (IFS) to express uncertainty, which involves not only membership degree but also non-membership degree. However, the limitation of IFS is that the sum of membership degree and non-membership degree must be no more than 1, which makes it difficult to sufficiently express the ideas of the DMs. Therefore, Yager [18] defined the Pythagorean fuzzy set (PFS), which can effectively express the certainty and uncertainty of experts. Recently, PFS has been introduced to deal with MADM problems [19–22]. Zhang and Xu [19] proposed the operation rules of PFS, and extended the TOPSIS method to PFS. By combining PFS with the hesitant fuzzy set (HFS), a new fuzzy set was defined by Liang and Xu [20], named the hesitant Pythagorean fuzzy set (HPFS), an extended TOPSIS method with HPFS was subsequently proposed. Zhang [21] extended PFS to the interval-valued case, and explored the basic operation rules of the Pythagorean fuzzy set (IVPFS). In addition, a Pythagorean fuzzy QUALIFLEX method was developed by integrating closeness index, and its effectiveness was demonstrated through a hierarchical MADM problem. Combining PFSs with linguistic variables, the definition of Pythagorean fuzzy linguistic set (PFLS) was proposed by Peng and Yang [22] and the operation rules of PFLS was defined, subsequently.

Inspired by the idea of the shadowed set and PFS, we propose a new approach to solve MADM problems under linguistic set environment. Firstly, we define Pythagorean shadowed set and explore some theorems of the shadowed set. Secondly, a score function of the Pythagorean shadowed number is defined and the detailed decision-making procedures-based upon the score function is proposed. Finally, a case study of supplier selection is adopted to verify the feasibility of the proposed approach.

The organization of this paper is as follows. Section 2 presents the preliminaries of the Pythagorean fuzzy set and shadowed set. In Section 3, the shadowed set model of seven-level language term is obtained by a data-driven method. A new score function of Pythagorean shadowed number is introduced in Section 4. Section 5 mainly addresses a new MADM method based on Pythagorean shadowed set. The effectiveness of the proposed approach is demonstrated through a supplier selection problem in Section 6, and comparative analysis is made with the other existing methods. Finally, some conclusions are drawn in Section 7.

2. Preliminaries

2.1. Pythagorean Fuzzy Set (PFS)

Definition 1 ([18,23]). *Suppose X is a fixed set. A PFS takes the form of:*

$$P = \{\langle\ x,\ P(u_P(x), v_P(x))\rangle | x \in X\}$$

where $v_p(x)$*:* $X \to [0,1]$ *and* $u_p(x)$*:* $X \to [0,1]$ *represent the non-membership function and membership function of* $x \in X$*, respectively,* $u_p^2(x) + v_p^2(x) \le 1$*. In addition,* $\pi_{p(x)} = \sqrt{1 - u_p^2(x) - v_p^2(x)}$ *denotes the hesitation degree of* $x \in X$*.*

For the sake of simplicity, Zhang and Xu [19] named $P(u_P(x), v_P(x))$ the Pythagorean fuzzy number (PFN), expressed by $\beta = P(u_\beta, v_\beta)$, where $u_p(x), v_p(x) \in [0,1]$, $\pi_{p(x)} = \sqrt{1 - u_p^2(x) - v_p^2(x)}$ and $u_p^2(x) + v_p^2(x) \le 1$.

Definition 2 ([9]). *Assume* $S = \{s_i | i = 0, \cdots, t, t \in R\}$ *is a linguistic term set,* s_i *is the linguistic evaluation value, t is the granularity of S. Take the seven-level linguistic term as an example: S =* {s_0 *= Extremely low,* s_1 *= Very low,* s_2 *= Low,* s_3 *= Fair,* s_4 *= High,* s_5 *= Very high,* s_6 *= Extremely high*}.

S must satisfy the following two properties:

(1) There is a negation operator: $neg(s_i) = s_{t-i}$;
(2) If $i < j$ then $S_i < S_j$;

Definition 3 ([22]). *Based on the definition of linguistic term set and PFS, the Pythagorean fuzzy linguistic set (PFLS) takes the form of* $D = \left\{ \left\langle s_{\tau(x)}, u_p(x), v_p(x) \right\rangle \middle| x \in X \right\}$*, and the Pythagorean fuzzy linguistic number (PFLN) is denoted as* $\left\langle s_{\tau(x)}, u_p(x), v_p(x) \right\rangle$*, where* $s_{\tau(x)}$ *is the linguistic evaluation value.*

When the attribute values are represented in the form of linguistic terms in MADM problems, the linguistic variable cannot be directly calculated. Therefore, Xu used the subscript of the linguistic term [24] for computation, Wang put forward a linguistic scale function [11] to convert linguistic terms into crisp numbers, and Herrera converted linguistic terms into fuzzy numbers [25]. However, all those methods still use linguistic subscript to express language variables in nature. To express the fuzziness and uncertainty of linguistic terms, we introduce shadowed set method to cope with linguistic term, and further put forward a new Pythagorean shadowed set.

2.2. Shadowed Set

Definition 4 ([13,14]). *A shadowed set S is a set-valued mapping as follows:*

$$S : U \to \{0, [0,1], 1\}$$

where U is a given universe of discourse.

The core of the shadowed set *S* is the area where the mapping values of the elements are equal to 1.

$$core(S) = \{x \in U | S(x) = 1\}$$

The elements of *U* whose mapping values are unit intervals in *S* compose the shadowed zone of the shadowed set and are expressed as follows,

$$CU(S) = \{x \in U | S(x) = [0,1]\}$$

The elements of *U* whose mapping values are equal to 0 will be excluded from the shadowed set *S*.

Definition 5. *$A = [a,b,c,d]$ is called a shadowed number (SN), where a,b are the lower and upper bound of the left-shoulder shadowed part, and c,d are the lower and upper bound of the right-shoulder shadowed part. Figure 1 shows an illustration of shadowed number.*

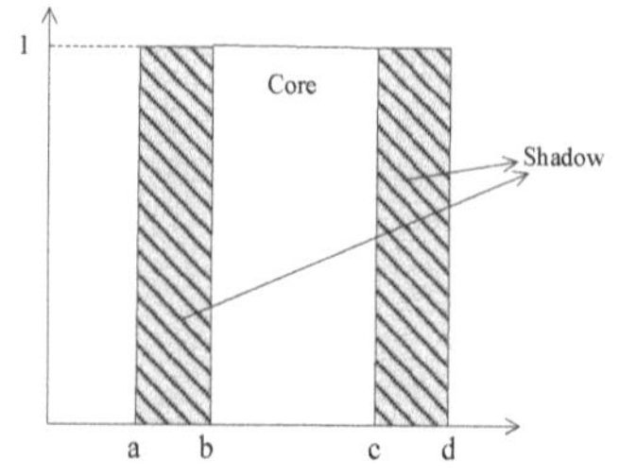

Figure 1. Shadowed number.

2.3. Pythagorean Shadowed Set (PSS)

In this section, we will define the Pythagorean shadowed set and give some properties for it.

Definition 6. *Suppose X is a fixed set, a Pythagorean shadowed set T over X takes the form of*

$$T = \{\langle A,\ P(u_P(x), v_P(x))\rangle | x \in X\}$$

where $A = [a,b,c,d]$, a,b are the lower and upper bound of the left-shoulder shadowed part, respectively, and c,d are the lower and upper bound of the right-shoulder shadowed part, respectively. Function $u_p(x)$: $X \to [0,1]$ and $v_p(x)$: $X \to [0,1]$ denote the membership function and non-membership function, respectively. ${u^2}_p(x) + {v^2}_p(x) \le 1$, and $\pi_{p(x)} = \sqrt{1 - u_p^2(x) - v_p^2(x)}$ denotes the hesitate degree of $x \in X$.

Definition 7. *A Pythagorean shadowed number (PSN) takes the form of:*

$$V = \langle A,\ P(u_P(x), v_P(x))\rangle$$

where a,b are the lower and upper bound of the left-shoulder shadowed part, c,d are the lower and upper bound of the right-shoulder shadowed part, and $u_p(x)$: $X \to [0,1]$ and $v_p(x)$: $X \to [0,1]$ represent membership function and non-membership function, respectively.

Let $V_1 = \langle A_1,\ P(u_P(x_1), v_P(x_1))\rangle$ and $V_2 = \langle A_2,\ P(u_P(x_2), v_P(x_2))\rangle$ be two PSNs, where $A_1 = [a_1, b_1, c_1, d_1]$ and $A_2 = [a_2, b_2, c_2, d_2]$, then the operation rules are as follows:

(1) $V_1 + V_2 = \left\langle [a_1 + a_2, b_1 + b_2, c_1 + c_2, d_1 + d_2],\ P\left(\sqrt{(u_p(x_1))^2 + (u_p(x_2))^2 - (u_p(x_1))^2 (u_p(x_2))^2}, v_p(x_1) v_p(x_2)\right)\right\rangle$

(2) $V_1 \times V_2 = \left\langle [a_1 \times a_2, b_1 \times b_2, c_1 \times c_2, d_1 \times d_2],\ P\left(u_p(x_1) u_p(x_2), \sqrt{(v_p(x_1))^2 + (v_p(x_2))^2 - (v_p(x_1))^2 (v_p(x_2))^2}\right)\right\rangle$

(3) $\lambda V_1 = \left\langle [\lambda a_1, \lambda b_1, \lambda c_1, \lambda d_1],\ P\left(\sqrt{1 - \left(1 - (u_p(x_1))^2\right)^{\lambda}}, (v_p(x_1))^{\lambda}\right)\right\rangle, \lambda \ge 0$

(4) $V_1^{\lambda} = \left\langle [a_1^{\lambda}, b_1^{\lambda}, c_1^{\lambda}, d_1^{\lambda}],\ P\left((u_p(x_1))^{\lambda}, \sqrt{1 - \left(1 - \left(v_p(x_1)\right)^2\right)^{\lambda}}\right)\right\rangle, \lambda \ge 0$

Theorem 1. *For any two PSNs $V_1 = \langle A_1,\ P(u_P(x_1), v_P(x_1))\rangle$ and $V_2 = \langle A_2,\ P(u_P(x_2), v_P(x_2))\rangle$, where $A_1 = [a_1, b_1, c_1, d_1]$ and $A_2 = [a_2, b_2, c_2, d_2]$, the calculation rules satisfy the following properties:*

(1) $V_1 + V_2 = V_2 + V_1$
(2) $V_1 \times V_2 = V_2 \times V_1$

(3) $\lambda(V_1 + V_2) = \lambda V_1 + \lambda V_2, \lambda \geq 0$

(4) $V_1^{\lambda_1+\lambda_2} = V_1^{\lambda_1} + V_2^{\lambda_2}, \lambda_1, \lambda_2 \geq 0$

(5) $\lambda_1 V_1 + \lambda_2 V_1 = (\lambda_1 + \lambda_2)V_1, \lambda_1, \lambda_2 \geq 0$

(6) $V_1^{\lambda} \times V_2^{\lambda} = (V_1 \times V_2)^{\lambda}, \lambda \geq 0$

3. Shadowed Set Model of Linguistic Terms

We collect the interval data for each language word in the form of the seven-level linguistic term listed in Definition 2 and use the collected interval data to construct the shadowed set models for the seven-level linguistic term. The interval data are obtained by means of questionnaire survey. The main framework of our questionnaire is designed to get a proper interval value for each word of the seven-level linguistic term from those respondents according to their experience, habits and common sense. It is necessary for the filled numbers to be accurate to the first decimal place.

We handed out our questionnaires via leaflets, emails and online survey websites to people in different fields, especially to those with a bachelor degree or above. In the end, we got 1205 valid questionnaires, and the questionnaire data were processed by the following interval data preprocessing method to obtain the shadowed number of the seven-level linguistic term.

3.1. Interval Data Preprocessing

Wu and Liu [26,27] proposed an efficient method to preprocess interval data, and we preprocessed the n interval endpoint data $[a_k, b_k](k = 1, 2, \ldots, n)$ based on this method, as follows:

Step 1: Bad data processing. This aims to remove unreasonable results from the surveyed people, whose answers were beyond the range of the universe of discourse U. If the interval endpoints satisfy the following conditions, the interval data are acceptable. Otherwise, they will be rejected.

$$\begin{cases} 0 \leq a_k \leq 10 \\ 0 \leq b_k \leq 10 \quad ,k = 1, 2, \ldots, n \\ b_k \geq a_k \end{cases}$$

By this step, some data will be abandoned, and $n^* < n$ interval data will be preserved.

Step 2: Outlier Processing. By using the Box and Whisker test [28], the data that are extremely large or small, i.e., outliers, can be eliminated. Outlier tests can be applied to process the endpoints of interval data and the lengths of interval data $L_k = b_k - a_k$, respectively. Consequently, only the interval endpoints and lengths satisfying the following conditions are kept:

$$\begin{cases} a_k \in [Q_a(0.25) - 1.5IQR_a, Q_a(0.75) + 1.5IQR_a] \\ b_k \in [Q_b(0.25) - 1.5IQR_b, Q_b(0.75) + 1.5IQR_b] \quad ,k = 1, 2, \ldots, n^* \\ L_k \in [Q_L(0.25) - 1.5IQR_L, Q_L(0.75) + 1.5IQR_L] \end{cases}$$

where Q_a and IQR_a are respectively the quartile and interquartile ranges of the left endpoints, Q_b and IQR_b are respectively the quartile and interquartile ranges of the right endpoints, Q_L and IQR_L are respectively the quartile and interquartile ranges of the interval data's length. $Q(0.25)$ and $Q(0.75)$ are the first and third quartiles, which include 25% and 75% of the data, respectively. In addition, the interquartile range IQR is the difference between $Q(0.25)$ and $Q(0.75)$; that is to say, IQR contains 50% of the data between $Q(0.25)$ and $Q(0.75)$. The points that are more than $1.5IQR$ below the first quartile or more than $1.5IQR$ above the third quartile are regarded as outliers.

After this step, $m^* < n^*$ interval data will remain.

Then, the following statistics of the m^* interval data are calculated: m_l and σ_l are mean values and standard deviations of the m^* left endpoints, respectively. Similarly, m_r and σ_r represent the mean values and standard deviations of the m^* right endpoints. m_L and σ_L denote the mean values and standard deviations of the lengths of the m^* interval data.

Step 3: Tolerance limit processing. If the remaining intervals satisfy the following conditions, then they will be accepted; otherwise, they will be rejected.

$$\begin{cases} a_k \in [m_l - \eta\sigma_l, m_l + \eta\sigma_l] \\ b_k \in [m_r - \eta\sigma_r, m_r + \eta\sigma_r] \\ L_k \in [m_L - \eta\sigma_L, m_L + \eta\sigma_L] \end{cases} , k = 1, 2, \ldots, m^*$$

where η is the tolerance factor, which represents that we can assure the given limits at least include the proportion $1 - \alpha$ of the measurements with $100 \cdot (1-\gamma)\%$ confidence level. The value of tolerance factor can be obtained from Table 1 [29].

Table 1. Tolerance factor η for several collected data.

m^*	$1-\gamma = 0.95$		$1-\gamma = 0.99$	
	$1-\alpha$		$1-\alpha$	
	0.90	**0.95**	**0.90**	**0.95**
10	2.839	3.379	3.582	4.265
15	2.480	2.954	2.945	3.507
20	2.310	2.752	2.659	3.168
30	2.140	2.549	2.358	2.841
50	1.996	2.379	2.162	2.576
100	1.874	2.233	1.977	2.355
1000	1.709	2.036	1.736	2.718
∞	1.645	1.960	1.645	1.960

After the processing of Step 3, $m^{**} < m^*(1 \leq m^{**} \leq n)$ interval data will be left, and the following statistical characteristics of the m^{**} data will be computed: m_l, σ_l, m_r, σ_r, m_L and σ_L of the left (right) endpoints the m^{**} interval data.

Step 4: Reasonable-interval processing. If the intervals satisfy the following conditions, they will be kept; otherwise, they will be rejected.

$$2m_l - \phi^* \leq a_k < \phi^* < b_k \leq 2m_r - \phi^*$$

where

$$\phi^* = \frac{(m_r\sigma_l^2 - m_l\sigma_r^2) \pm \sigma_l\sigma_r\left[(m_l - m_r)^2 + 2(\sigma_l^2 - \sigma_r^2)\ln(\sigma_l/\sigma_r)\right]^{1/2}}{\sigma_l^2 - \sigma_r^2}$$

After this step, there will be m interval data.

In a word, there will be m interval data after the four processing steps above, which is not greater than the n interval data at the beginning, as shown in Figure 2.

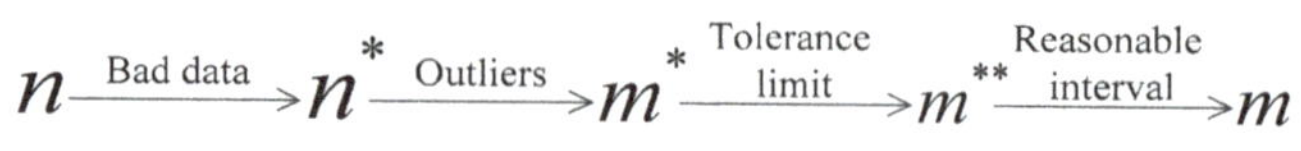

Figure 2. The process of data preprocessing.

3.2. Shadowed Set Model of Seven-Level Language Terms

After data preprocessing, the distribution of the remaining interval data is obtained as shown in Figure 3. The intervals of the left-end points and right-end points can reflect the linguistic word's uncertainties from different surveyed persons. Therefore, it is necessary to determine the representative intervals for the left-end points and right-end points to express the uncertainties. As shown in Figure 3, the core area can be determined even if the surveyed people cannot give accurate representative intervals. As a result, the core of the shadowed set is the core area and the uncertain bound of the shadowed set is the representative intervals.

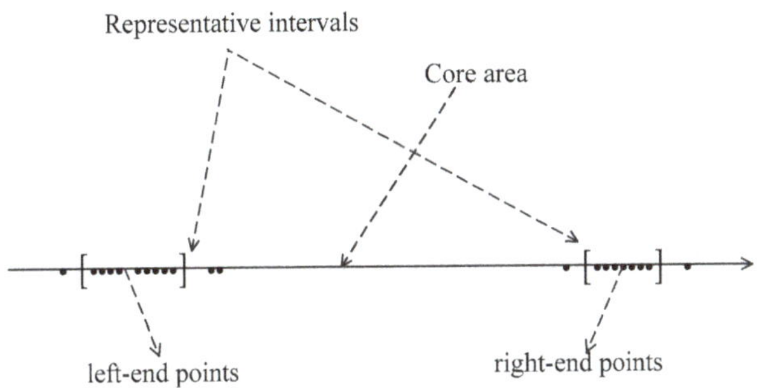

Figure 3. Distribution of the remaining interval data.

Next, we will estimate the representative intervals by the tolerance limit method via the following steps.

Step 1: Calculate the mean m_l and standard deviation σ_l of the remaining left-end points

$$m_l = \frac{\sum_{k=1}^{m} \hat{l}_k}{m} \tag{1}$$

$$\sigma_l = \sqrt{\frac{\sum_{k=1}^{m} \left(\hat{l}_k - m_l\right)^2}{m}} \tag{2}$$

where $\hat{l}_k$ denotes the left-end point of each remaining interval, m is the number of remaining intervals.

Step 2: Determine the representative interval. Let $[L_l, L_r]$ and $[R_l, R_r]$ be the representative intervals of the left-end points and right-end points, respectively.

$$L_l = m_l - \eta * \sigma_l \tag{3}$$

$$L_r = m_l + \eta * \sigma_l \tag{4}$$

where η is the tolerance factor in Table 1.

Then, the representative interval for the right-end points is calculated in the same way.

The parameters γ and α are set to 0.05 and 0.1 in this paper, respectively, and we can obtain a tolerance factor η of 1.709 from Table 1. Take the seven-level language terms as an example: based on the results above, the shadowed set models for seven-level language terms can be constructed as shown in Figure 4.

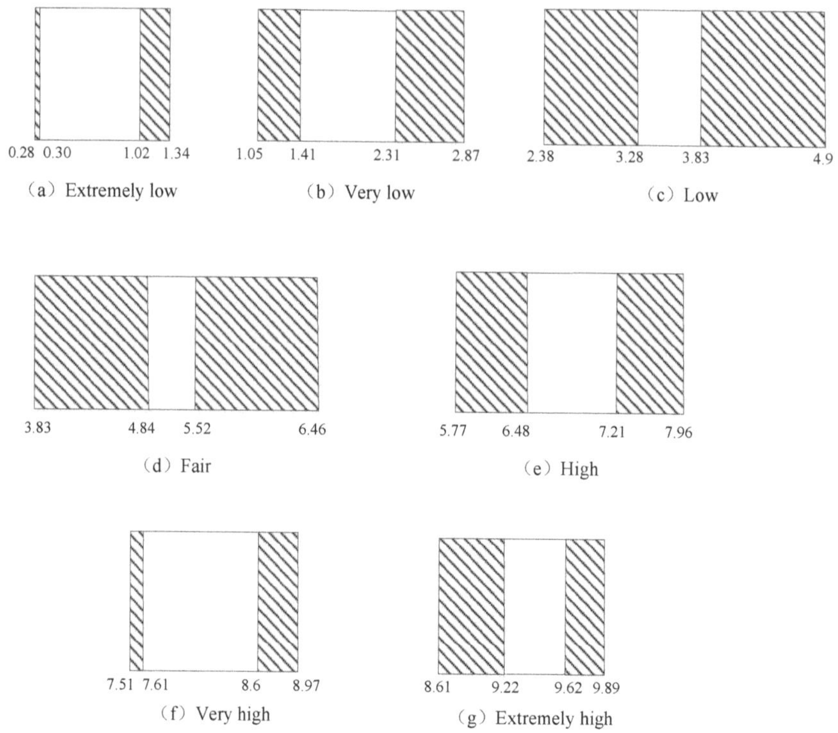

Figure 4. The shadowed set models for seven-level language terms.

4. The Score Function of Pythagorean Shadowed Number

Based on the concepts of shadowed number and Pythagorean shadowed number in Section 2, we will further present the score functions of shadowed number and Pythagorean shadowed number, respectively. Numerical examples will also be given to illustrate the specific calculation process of the two score functions.

According to the central limit theorem, the attribute value r_{ij} given by the decision-maker is stable and tends to be the most likely attribute value at a certain point, so it is believed that r_{ij} obeys the normal distribution within the fuzzy interval. From the tolerance limit method in Section 3.2, we can obtain the distribution of attribute value in the shadowed set $S = \{A_i|U\}$, $A_i = [a_i, b_i, c_i, d_i]$, as shown in Figure 5.

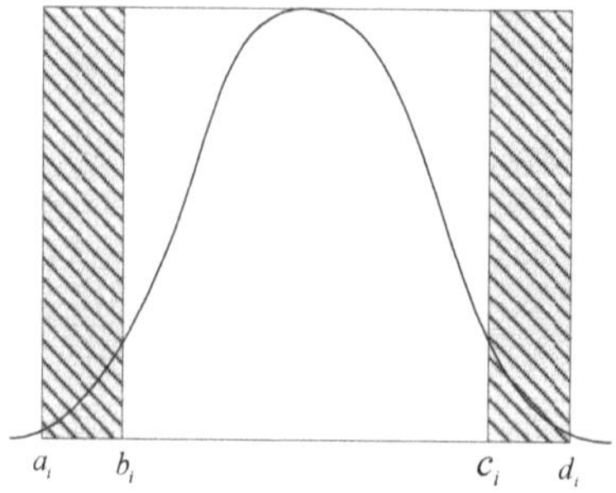

Figure 5. Normal distribution of attribute value.

Definition 8. *The score function of shadowed number A is defined as follows:*

$$\text{score}(A) = a + \int_a^b f(x)\,dx + c - b + \int_c^d f(x)\,dx + d \quad (5)$$

where $f(x) = \frac{1}{\sqrt{2\pi}\sigma} \cdot e^{-(x-u)^2/(2\cdot\sigma^2)}$.

According to the 3σ principle of normal distribution:

$$p(r \in [u - 3\sigma, u + 3\sigma]) = 0.9974,\ p(r \in [a, d]) = 0.9974$$

Then $u = \frac{a+d}{2}, \sigma = \frac{d-a}{6}$.

Example 1. *The score function value of shadowed set A_0 for 'High' in Figure 3 can be calculated as follows:*

$$u = a + d = 5.77 + 7.96 = 13.73, \sigma = \frac{d-a}{6} = \frac{7.96 - 5.77}{6} = 0.37$$

$f(x) = 1.08e^{-(x-6.87)^2/0.27}$

Then, we can gain the figure of shadowed number 'High' as shown in Figure 6.

$$\text{score}(A_0) = 5.77 + \int_{5.77}^{6.48} f(x)\,dx + 7.21 - 6.48 + \int_{7.21}^{7.96} f(x)\,dx + 7.96 = 17.9$$

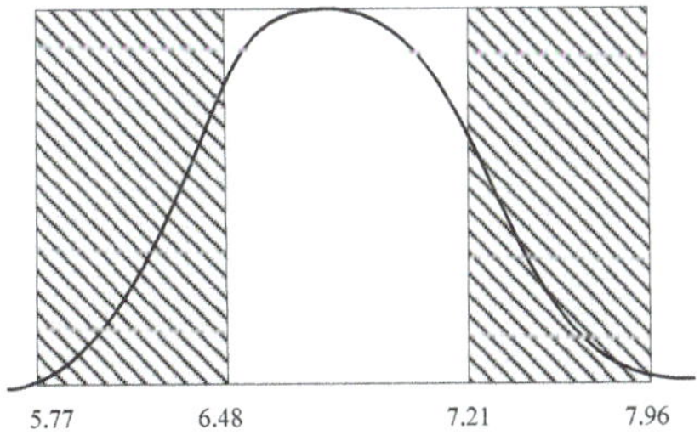

Figure 6. The shadowed number of 'High' and its normal distribution.

In the same way, we can get the score function of shadowed sets for the other six language terms in Figure 3.

Definition 9. *The score function of a Pythagorean shadowed number V is denoted as:*

$$\text{score}(V) = \left(a + \int_a^b f(x)\,dx + c - b + \int_c^d f(x)\,dx + d\right) \cdot (u/v) \quad (6)$$

Example 2. *For a Pythagorean shadowed number $V_1 = \{[5.77, 6.48, 7.21, 7.96], P(0.7, 0.4)\}$, the score function value is:*

$$score(V_1) = \left(5.77 + \int_{5.77}^{6.48} f(x)dx + 7.21 - 6.48 + \int_{7.21}^{7.96} f(x)dx + 7.96\right) \cdot (0.7/0.4) = 31.33$$

where $f(x) = 1.08e^{-(x-6.87)^2/0.27}$.

5. MADM Method Based on the Pythagorean Shadowed Set

With the concept of PSS in mind, we can put forward a novel MADM approach under Pythagorean fuzzy linguistic term circumstances. The diagram of the proposed method is shown in Figure 7. Firstly, present a description of the MADM problem under the Pythagorean linguistic fuzzy circumstances. Secondly, transform the PFLS into PSS through a data-driven method. Thirdly, determine the ranking order of all alternatives so as to obtain the best choice(s) by means of the score function of PSNs and OWA operator. The whole decision-making process is carried out in the following steps.

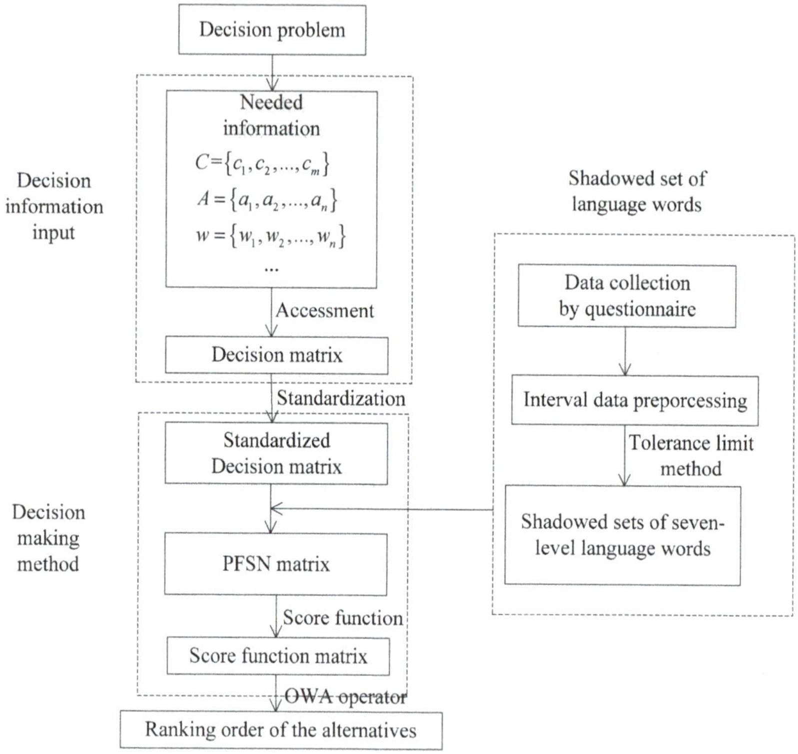

Figure 7. Diagram of the proposed method.

Step 1: Standardized decision matrix. For PFLVs $P_{ij} = \left\langle s_{\tau_{ij}}, P(u_p(x_{ij}), v_p(x_{ij}))\right\rangle$

For beneficial attributes, $\overline{P_{ij}} = P_{ij} = \left\langle s_{\tau_{ij}}, P(u_p(x_{ij}), v_p(x_{ij}))\right\rangle$

For cost attributes, $\overline{P_{ij}} = (P_{ij})^{-1} = \left\langle s_{(\tau_{ij})^{-1}}, P(v_p(x_{ij}), u_p(x_{ij}))\right\rangle$ where $(\tau_{ij})^{-1} = l + 1 - \tau_{ij}$ and l is the number of language term.

Step 2: Collect the data by questionnaire and get the shadowed set of language terms by processing the data. Transform Pythagorean fuzzy linguistic numbers into PSNs using Figure 4.

Step 3: Transform the PFSN decision matrix into score function matrix based on Equation (6).

Step 4: By OWA operator, the attribute values r_{ij} of each alternative a_i are aggregated to obtain the comprehensive attribute values z_i.

$$z_i = OWA_w(r_{i1}, r_{i2}, \ldots, r_{im}) = \sum_{j=1}^{m} w_i r_{ij},\ i = 1, 2, \ldots, n$$

where $w = (w_1, w_2, \ldots, w_m)$ is the criterion weight vector, n is the number of alternatives, m is the number of attribute.

Step 5: Determine the order of all the alternatives in the light of the comprehensive attribute values z_i.

6. Numerical Study

The proposed algorithm will be demonstrated by solving the problem of how to select the most suitable supplier for a company under various evaluation factors. At the same time, comparisons with the linguistic term subscript method and the linguistic scale function method are performed to show the advantages of our approach.

6.1. Supplier Selection Problem

A car company needs to choose appropriate supplier of spare parts. A total of five alternative suppliers are denoted as a_1, a_2, a_3, a_4, a_5. After synthetical consideration, four main factors are taken into account: c_1 Supply capacity, c_2 Delivery timeliness, c_3 Service quality, c_4 Scientific research ability. The criterion weight vector is $w = (0.3, 0.2, 0.4, 0.1)$. Language evaluation of the four attributes adopts the form of seven-level linguistic term, $S = \{s_0, s_1, s_2, s_3, s_4, s_5, s_6\}$ = {Extremely low, Very low, Low, Fair, High, Very high, and Extremely high}. The decision matrix given by experts is shown in Table 2:

Table 2. Decision matrix.

Alternatives	Attributes			
	c_1	c_2	c_3	c_4
a_1	$\langle s_4, P(0.7, 0.4)\rangle$	$\langle s_5, P(0.5, 0.6)\rangle$	$\langle s_2, P(0.7, 0.3)\rangle$	$\langle s_3, P(0.8, 0.4)\rangle$
a_2	$\langle s_5, P(0.6, 0.4)\rangle$	$\langle s_3, P(0.7, 0.4)\rangle$	$\langle s_3, P(0.6, 0.4)\rangle$	$\langle s_4, P(0.7, 0.5)\rangle$
a_3	$\langle s_6, P(0.6, 0.5)\rangle$	$\langle s_3, P(0.8, 0.3)\rangle$	$\langle s_5, P(0.6, 0.5)\rangle$	$\langle s_2, P(0.6, 0.4)\rangle$
a_4	$\langle s_3, P(0.7, 0.3)\rangle$	$\langle s_4, P(0.6, 0.5)\rangle$	$\langle s_3, P(0.7, 0.4)\rangle$	$\langle s_6, P(0.7, 0.6)\rangle$
a_5	$\langle s_4, P(0.7, 0.4)\rangle$	$\langle s_5, P(0.6, 0.5)\rangle$	$\langle s_4, P(0.7, 0.4)\rangle$	$\langle s_3, P(0.8, 0.4)\rangle$

Step 1: c_1, c_2, c_3, c_4 are beneficial attributes. Therefore, the standardized decision matrix is the same with Table 2.

Step 2: Transform PFLNs into Pythagorean shadowed numbers using Figure 4, and the result is shown in Table 3.

Table 3. Decision matrix with PFSN.

Alternatives	Attributes	
	c_1	c_2
a_1	$\langle [5.77, 6.48, 7.21, 7.96], P(0.7, 0.4)\rangle$	$\langle [7.51, 7.61, 8.60, 8.97], P(0.5, 0.6)\rangle$
a_2	$\langle [7.51, 7.61, 8.60, 8.97], P(0.6, 0.4)\rangle$	$\langle [3.83, 4.84, 5.52, 6.46], P(0.7, 0.4)\rangle$
a_3	$\langle [8.61, 9.22, 9.62, 9.89], P(0.6, 0.5)\rangle$	$\langle [3.83, 4.84, 5.52, 6.46], P(0.8, 0.3)\rangle$
a_4	$\langle [3.83, 4.84, 5.52, 6.46], P(0.7, 0.3)\rangle$	$\langle [5.77, 6.48, 7.21, 7.96], P(0.6, 0.5)\rangle$
a_5	$\langle [5.77, 6.48, 7.21, 7.96], P(0.7, 0.4)\rangle$	$\langle [7.51, 7.61, 8.60, 8.97], P(0.6, 0.5)\rangle$
Alternatives	**Attributes**	
	c_3	c_4
a_1	$\langle [2.38, 3.28, 3.83, 4.90], P(0.7, 0.3)\rangle$	$\langle [3.83, 4.84, 5.52, 6.46], P(0.8, 0.4)\rangle$
a_2	$\langle [3.83, 4.84, 5.52, 6.46], P(0.6, 0.4)\rangle$	$\langle [5.77, 6.48, 7.21, 7.96], P(0.7, 0.5)\rangle$
a_3	$\langle [7.51, 7.61, 8.60, 8.97], P(0.6, 0.5)\rangle$	$\langle [2.38, 3.28, 3.83, 4.90], P(0.6, 0.4)\rangle$
a_4	$\langle [3.83, 4.84, 5.52, 6.46], P(0.7, 0.4)\rangle$	$\langle [8.61, 9.22, 9.62, 9.89], P(0.7, 0.6)\rangle$
a_5	$\langle [5.77, 6.48, 7.21, 7.96], P(0.7, 0.4)\rangle$	$\langle [3.83, 4.84, 5.52, 6.46], P(0.8, 0.4)\rangle$

Step 3: Transform the PFSNs decision matrix into score function matrix (shown in Table 4) based on Equation (6).

Table 4. Score function matrix.

Alternatives	Attributes			
	c_1	c_2	c_3	c_4
a_1	31.33	18.39	25.29	26.76
a_2	33.11	23.42	20.07	25.06
a_3	40.97	35.68	26.49	16.26
a_4	31.22	21.48	23.42	39.83
a_5	31.33	26.49	31.33	26.76

Step 4: By OWA operator, the attribute values r_{ij} of each alternative a_i are aggregated to obtain the comprehensive attribute values z_i.

$z_1 = 25.87, z_2 = 25.15, z_3 = 31.65, z_4 = 27.01, z_5 = 29.91$**Step 5:** Rank the alternatives and obtain the best alternative(s) according to the comprehensive attribute values z_i in the Step 4.

$z_3 > z_5 > z_4 > z_1 > z_2$, that means, $a_3 \succ a_5 \succ a_4 \succ a_1 \succ a_2$.

And the alternative a_3 is the best choice of the supplier option problem.

6.2. Comparison Analysis

To verify the superiority of our method, comparations will be made between our approach and the other two approaches, i.e., the linguistic term subscript method [22] and the linguistic scale function method [11,19].

In [22], the score function of $p = \left\langle s_{\tau(x)}, u_A(x), v_A(x) \right\rangle$ is:

$$score(p) = \frac{\tau(x)}{t+1} * \left(\mu_\beta^2 - v_\beta^2\right) \tag{7}$$

where $\tau(x)$ is the subscript of the linguistic term, and t is the number of linguistic terms.

We can obtain the comprehensive attribute values z_i based on Equation (7) and the OWA operator.

$z_1 = 0.094, z_2 = 0.104, z_3 = 0.098, z_4 = 0.115, z_5 = 0.147$, and $z_5 > z_4 > z_2 > z_3 > z_1$.

Therefore, the alternative a_5 is the best choice.

In [19], the score function of PFN $\beta = P(u_\beta, v_\beta)$ is:

$$score(\beta) = \mu_\beta^2 - v_\beta^2 \tag{8}$$

In [11], the improved linguistic scale function is calculated as follows:

$$f(s_i) = \theta_i = \begin{cases} \frac{m^\alpha - (m-i)^\alpha}{2m^\alpha} & (i = 0, 1, 2, \ldots, m) \\ \frac{m^\beta + (i-m)^\beta}{2m^\beta} & (i = m+1, m+2m\ldots, t) \end{cases} \tag{9}$$

where $\alpha, \beta \in (0, 1]$, $m = \frac{t}{2}$, and t is the number of linguistic terms.

According to the improved linguistic scale Function (8) and score Function (9), we can obtain the score function of $p = \left\langle s_{\tau(x)}, u_A(x), v_A(x) \right\rangle$ as:

$$score(p) = f(s_i) * \left(\mu_\beta^2 - v_\beta^2\right) \tag{10}$$

Let $\alpha = \beta = 0.5$. We can obtain the comprehensive attribute values z_i based on Equation (10) and the OWA operator.

$z_1 = 0.07, z_2 = 0.15, z_3 = 0.2, z_4 = 0.14, z_5 = 0.16$, and $z_3 > z_5 > z_2 > z_4 > z_1$.

Therefore, the alternative a_3 is the best choice.

From Table 5, it can be observed that the ranking result obtained via our algorithm is different from the other two methods. By using the linguistic term subscript method, the ranking order is $a_5 \succ a_4 \succ a_2 \succ a_3 \succ a_1$, which is totally different form the results of our method and the language scale function method. The reason is that replacing linguistic words simply with linguistic subscript leads to distortion of information. In fact, the linguistic subscript cannot effectively reflect original decision information. Compared with the linguistic term subscript approach, the linguistic scale function method seems more reasonable for describing the linguistic term information with a so-called language scale function. However, the language scale function still replaces linguistic words with numbers in nature, and information loss or information distortion is still inevitable. On the other hand, different people may have different viewpoints on the same word, but the linguistic subscript and linguistic scale function can only express a single meaning for a word. Compared with the other two methods, we utilize a data-driven method to construct the shadowed set models for the linguistic terms, which cannot only maintain the original decision information as far as possible, but also take different views into account for a single word.

Table 5. Comparison analysis results.

Method	Order of Alternatives
Our method	$a_3 \succ a_5 \succ a_4 \succ a_1 \succ a_2$
Linguistic term subscript method [22]	$a_5 \succ a_4 \succ a_2 \succ a_3 \succ a_1$
Language scale function method [11,19]	$a_3 \succ a_5 \succ a_2 \succ a_4 \succ a_1$

7. Conclusions

A novel method for MADM problems under a linguistic term environment was proposed, combining shadowed sets and Pythagorean fuzzy sets. We defined Pythagorean shadowed numbers and subsequently described their operation rules and basic properties. Based on the operation rules, the score function of Pythagorean shadowed numbers was deduced, and a numerical example was provided to illustrate the computing process. Bearing the above results in mind, we proposed a new MADM approach to deal with linguistic terms. A supplier selection example was used to demonstrate the feasibility of our method. Compared with the linguistic term subscript method and the linguistic scale function method, a data-driven method was adopted to construct the shadowed set models for linguistic terms, which can avoid information loss or information distortion to a great extent. The comparative analysis shows that our method can provide more reasonable and accurate decision-making results by depicting linguistic terms in a more precise manner.

In future research, the proposed method can be extended to other types of shadowed sets, for example, left-shoulder, right-shoulder, non-cored, etc. Additionally, applications in other fields are also worth exploring with our approach.

Author Contributions: Methodology, H.W.; Writing—Original Draft Preparation, S.H.; Formal Analysis, C.L.; Investigation, X.P.; Writing—Review & Editing, H.W.

Funding: NSFC (No. 61402260, 61473176), and partly by the Taishan Scholar Project of Shandong Province.

Acknowledgments: The authors would like to thank the National Natural Science Foundation of China (No. 61402260, 61473176) and the Taishan Scholar Project of Shandong Province for the financial support.

Conflicts of Interest: The authors declare no conflict of interest.

References

1. Gao, S.; Zhao, H.R. Fuzzy best-worst multi-criteria decision-making method and its applications. *Knowl.-Based Syst.* **2017**, *121*, 23–31. [CrossRef]
2. Wu, T.; Liu, X.W.; Liu, F. An interval type-2 fuzzy TOPSIS model for large scale group decision making problems with social network information. *Inf. Sci.* **2018**, *432*, 392–410.

3. Li, C.D.; Ding, Z.X.; Yi, J.Q.; Lv, Y.; Zhang, G. Deep belief network based hybrid model for building energy consumption prediction. *Energies* **2018**, *11*, 242. [CrossRef]
4. Manish, A. Adaptive linguistic weighted aggregation operators for multi-criteria decision making. *Appl. Soft Comput.* **2017**, *58*, 690–699.
5. Jin, F.F.; Ni, Z.W.; Pei, L.D.; Chen, H.; Tao, Z.; Zhu, X.; Ni, L. Approaches to group decision making with linguistic preference relations based on multiplicative consistency. *Comput. Ind. Eng.* **2017**, *114*, 69–79. [CrossRef]
6. Yu, W.Y.; Zhang, Z.; Zhong, Q.Y.; Sun, L.L. Extended TODIM for multi-criteria group decision making based on unbalanced hesitant fuzzy linguistic term sets. *Comput. Ind. Eng.* **2017**, *114*, 316–328. [CrossRef]
7. Pei, Z.; Liu, J.; Hao, F.; Zhou, B. FLM-TOPSIS: The fuzzy linguistic multiset TOPSIS method and its application in linguistic decision making. *Inf. Fusion* **2019**, *45*, 266–281.
8. Cheng, X.; Gu, J.; Xu, Z.S. Venture capital group decision-making with interaction under probabilistic linguistic environment. *Knowl.-Based Syst.* **2018**, *140*, 82–91. [CrossRef]
9. Liu, H.B.; Jiang, L.; Martínez, L. A dynamic multi-criteria decision making model with bipolar linguistic term sets. *Exp. Syst. Appl.* **2018**, *95*, 104–112.
10. Wang, H.; Xu, Z.S.; Zeng, X.J. Hesitant fuzzy linguistic term sets for linguistic decision making: Current developments, issues and challenges. *Inf. Fusion* **2018**, *43*, 1–12.
11. Wang, J.Q.; Wu, J.T.; Wang, J.; Zhang, H.-Y.; Chen, X.-H. Interval-valued hesitant fuzzy linguistic sets and their applications in multi-criteria decision-making problems. *Inf. Sci.* **2014**, *288*, 55–72. [CrossRef]
12. Wang, H.D.; Pan, X.H.; He, S.F. A new interval type-2 fuzzy VIKOR method for multi-attribute decision making. *Int. J. Fuzzy Syst.* **2018**. [CrossRef]
13. Pedrycz, W. Shadowed sets: Representing and processing fuzzy sets. *IEEE Trans. Syst. Man Cybern. Part B Cybern.* **1998**, *28*, 103–109. [CrossRef] [PubMed]
14. Pedrycz, W. From fuzzy sets to shadowed sets: Interpretation and computing. *Int. J. Intell. Syst.* **2009**, *24*, 48–61. [CrossRef]
15. Yao, Y.Y.; Wang, S.; Deng, X.F. Construction shadowed sets and three-way approximations of fuzzy sets. *Inf. Sci.* **2017**, *412–413*, 132–153. [CrossRef]
16. Li, C.D.; Yi, J.Q.; Wang, H.K.; Zhang, G.; Li, J. Interval data driven construction of shadowed sets with application to linguistic words modelling. *Inf. Sci.* **2018**. [CrossRef]
17. Atanassov, K.T. Intuitionistic fuzzy sets. *Fuzzy Sets Syst.* **1986**, *20*, 87–96. [CrossRef]
18. Yager, L.L. Pythagorean membership grades in multicriteria decision making. *IEEE Trans. Fuzzy Syst.* **2014**, *22*, 958–965. [CrossRef]
19. Zhang, X.L.; Xu, Z.S. Extension of TOPSIS to Multiple Criteria Decision Making with Pythagorean Fuzzy Sets. *Int. J. Intell. Syst.* **2014**, *29*, 1061–1078. [CrossRef]
20. Liang, D.; Xu, Z.S. The new extension of TOPSIS method for multiple criteria decision making with hesitant Pythagorean fuzzy sets. *Appl. Soft Comput.* **2017**, *60*, 167–179. [CrossRef]
21. Zhang, X.L. Multicriteria Pythagorean fuzzy decision analysis: A hierarchical QUALIFLEX approach with the closeness index-based ranking methods. *Inf. Sci.* **2016**, *330*, 104–124. [CrossRef]
22. Peng, J.D.; Yang, Y. multi-attribute group decision making method based on Pythagorean fuzzy linguistic set. *Comput. Eng. Appl.* **2016**, *52*, 50–54. (In Chinese)
23. Yager, L.L. Pythagorean Membership Grades, Complex Numbers, and Decision Making. *Int. J. Intell. Syst.* **2013**, *28*, 436–452. [CrossRef]
24. Xu, Z.S. A method based on linguistic aggregation operators for group decision making with linguistic preference relations. *Inf. Sci.* **2004**, *166*, 19–30. [CrossRef]
25. Herrera, F.; Herrera-Viedma, E. Linguistic decision analysis: Steps for solving decision problems under linguistic information. *Fuzzy Sets Syst.* **2000**, *115*, 67–82. [CrossRef]
26. Liu, F.L.; Mendel, J.M. Encoding words into interval type-2 fuzzy sets using an interval approach. *IEEE Trans. Fuzzy Syst.* **2008**, *16*, 1503–1521.
27. Wu, D.R.; Mendel, J.M.; Coupland, S. Enhanced interval approach for encoding words into interval type-2 fuzzy sets and its convergence analysis. *IEEE Trans. Fuzzy* **2012**, *20*, 499–513.

28. Walpole, R.W.; Myers, R.H.; Myers, A.L.; Ye, K. *Probability & Statistics for Engineers and Scientists*, 8th ed.; Prentice-Hall: Upper Saddleback River, NJ, USA, 2007.
29. Walpole, R.E.; Myers, R.H.; Ye, K. *Probability and Statistics for Engineers and Scientists*; Pearson: London, UK, 2014.

symmetry

MDPI

Article

Algorithm for T-Spherical Fuzzy Multi-Attribute Decision Making Based on Improved Interactive Aggregation Operators

Harish Garg [1,*], Muhammad Munir [2], Kifayat Ullah [2], Tahir Mahmood [2] and Naeem Jan [2]

1 School of Mathematics, Thapar Institute of Engineering & Technology, Deemed University, Patiala 147004, Punjab, India

2 Department of Mathematics and Statistics, International Islamic University, Islamabad 44000, Pakistan; munir.phdma78@iiu.edu.pk (M.M.); kifayat.phdma72@iiu.edu.pk (K.U.); tahirbakhat@iiu.edu.pk (T.M.); naeem.phdma73@iiu.edu.pk (N.J.)

* Correspondence: harishg58iitr@gmail.com; Tel.: +91-8699031147

Received: 6 November 2018; Accepted: 23 November 2018; Published: 27 November 2018

Abstract: The objective of this manuscript is to present some new, improved aggregation operators for the T-spherical fuzzy sets, which is an extension of the several existing sets, such as intuitionistic fuzzy sets, picture fuzzy sets, neutrosophic sets, and Pythagorean fuzzy sets. In it, some new, improved operational laws and their corresponding properties are studied. Further, based on these laws, we propose some geometric aggregation operators and study their various relationships. Desirable properties, as well as some special cases of the proposed operators, are studied. Then, based on these proposed operators, we present a decision-making approach to solve the multi-attribute decision-making problems. The reliability of the presented decision-making method is explored with the help of a numerical example and the proposed results are compared with several prevailing studies' results. Finally, the superiority of the proposed approach is explained with a counter example to show the advantages of the proposed work.

Keywords: multi-attribute decision making; aggregation operators; spherical fuzzy sets; interactive geometric operators

1. Introduction

The term fuzzy set (FS) was developed by Zadeh [1] based on a characteristic function that described the degree of membership of an element. Atanassov [2] established the theory of intuitionistic fuzzy set (IFS) as a generalization of FS with the help of two characteristic functions, known as membership and non-membership functions, describing the positive and negative aspects of an element or object. In the framework of IFSs, there was a constraint on two characteristic functions, in that their sum must not exceed the unit interval, which restricted the selection of membership and non-membership grades. Accordingly, Atanassov and Gargov [3] extended the IFS to the interval-valued intuitionistic fuzzy sets (IVIFSs), which contain the degrees of agreeing and disagreeing as interval values instead of single digits. Keeping in mind the constraint on IFSs, Yager [4,5] introduced a new generalization of IFSs, known as Pythagorean fuzzy set (PyFS), with a condition that the sum of squares of membership and non-membership grades must not exceed the unit interval.

The frameworks of IFSs and PyFSs have importance in situations where the structure of FSs fails to be applied. But these structures have their own limitations, as in the circumstances of voting where opinion cannot be restricted to "yes" or "no" but some refusal degree and abstinence is also involved. Therefore, Cuong [6,7] developed a novel concept of picture fuzzy sets (PFSs), which is based on four characteristic functions known as membership, non-membership, abstinence, and refusal grades. Cuong's structure of PFSs is diverse in nature but, similar to IFSs, there is also a restriction in PFSs that the sum of all three membership grades must not exceed the unit interval. In the above-stated environments, various researchers have constructed their methodologies for solving the multi-attribute decision-making (MADM) problems, focusing on information measures, aggregation operators, etc. For instance, Xu [8] presented some weighted averaging aggregation operators (AOs) for intuitionistic fuzzy numbers (IFNs). Garg [9,10] presented some improved interactive AOs for IFNs. Wang and Liu [11] gave interval-valued intuitionistic fuzzy hybrid weighted AOs based on Einstein operations. Wang and Liu [12] presented some hybrid weighted AOs using Einstein norm operations for IFNs. Wang et al. [13] presented some AOs to aggregate various interval-valued intuitionistic fuzzy (IVIF) numbers (IVIFNs). Garg [14,15] presented generalized AOs using Einstein norm operations for Pythagorean fuzzy sets. Xu and Xia [16] proposed induced generalized aggregation tools and applied them in MADM. Garg and Kumar [17] presented some new similarity measures for IVIFNs based on the connection number of the set pair analysis theory. However, apart from these, a comprehensive overview of the different approaches under the IFSs, IVIFSs, PyFSs, etc., to solve MCDM problems are summarized in [18–33].

Apart from the above theories, the concept of the spherical fuzzy sets (SFSs) has been introduced by Mahmood et al. [34], which consists of three membership degrees with a condition that the sum of squares of all degrees must not exceeds one. Further, the concepts of SFSs are extended to T-spherical fuzzy sets T-SFSs, where there are no restrictions on their constants and, hence, T-SFSs can handle all the situations where the frameworks of FSs, IFSs, PyFSs, PFSs, and SFSs failed. For this environment, Mahmood et al. [34] presented some aggregation operators for T-SFSs. Later on, Ullah et al. [35] presented the concept of the symmetry measures for handling the uncertainties under the T-SFSs environment, and applied it to solve the decision-making problems. However, from the existing work, it is noticeable that the existing AOs under the IFSs, PFSs, etc., have failed to handle the situations under some certain cases. For instance, under the IFS environment, if we consider the two IFNs, $A = (0, n_A)$ and $B = (m_B, n_B)$ where $m_B, n_A,\ n_B$ represented degrees of membership grades that lies between zero and one, by applying the geometric AOs, as defined in [36], to such numbers we then get the aggregated numbers as $(0, n_A + n_B - n_A n_B)$. Thus, the final aggregated value of membership degree is zero, irrespective of value of m_B. Similarly, for T-SFSs, if we assume $A = (m_A, 0, n_A)$ and $B = (m_B, i_B, n_B)$ then, using the geometric aggregation operators of PFSs [37,38] and T-SFSs [34], we obtain the result of type $(some\ value, 0, some\ value)$. This shows that the abstinence value of B is not accounted for in aggregation. Further, by taking $A = (0, 0, n_A)$ and $B = (m_B, i_B, n_B)$, then using the operators defined in [34,37,38], we get $(some\ value, 0, some\ value)$. This shows that the membership and abstinence value is not accounted for in aggregation. These examples clearly point out the shortcomings that exist in the aggregation operators of PFSs and T-SFSs.

In order to overcome such shortcomings, and by utilizing the advantages of the T-SFSs over the several other existing theories, in this manuscript we have presented some new, improved geometric interactive aggregation operators. For it, firstly, we define some new operational laws by adding the degree of the hesitation into the operations. To do this, the concept of probability membership, non-membership, and heterogeneous are introduced and then some of their desirable properties are studied. Then, based on these proposed operational laws, some weighted, ordered weighted, and hybrid geometric aggregation operators, namely, T-spherical fuzzy weighted geometric interaction averaging (T-SFWGIA), T-spherical fuzzy ordered weighted geometric interaction averaging (T-SFOWGIA), and T-spherical fuzzy hybrid geometric interaction averaging (T-SFHGIA) operators are introduced in the paper. The desirable properties of such operators are investigated in detail. Then, based on such operators, we developed an algorithm for solving the decision-making problem under the T-SFS environment. The practical utility of the proposed approach is demonstrated through a numerical example, and comparative studies investigate the superiority of the approach. Finally, a counter example is provided to show the supremacy of the proposed operators with respect to the existing operators. Therefore, motivated from it, the objectives of the paper are summarized as follows:

(1) To propose some new operational laws based on the probability membership, non-membership, and heterogeneous laws.
(2) To define some new, improved weighted geometric aggregation operators under the T-SFSs environment.
(3) To develop an algorithm for solving the multi-attribute decision-making problems based on the proposed operators.
(4) To check numerical applicability of the approach to a real-life case, and to compare the outcomes with prevailing approaches.

To do so, the organization of this manuscript is summarized as follows: Section 2 gives a basic overview of the basic concepts of IFSs, PFSs, SFSs, and T-SFSs; Section 3 deals with some new multiplication operations laws and their corresponding weighed geometric AOs; inn Section 4, we present a MADM approach for solving the decision-making problem by using the proposed AOs (here the preferences related to each alternative are summarized in the form of T-SFS information); Section 5 presents a numerical example to illustrate the proposed approach and the comparative analysis; and finally, Section 6 concludes the paper with some concluding remarks.

2. Preliminaries

In this section, we present some basic concepts related to IFS, PyFS, PFS, SFS, and T-SFS over the universal set X.

Definition 1. [2] *An IFS on X consists of membership and non-membership functions defined as*

$$P = \{\langle x, m(x), n(x)\rangle \mid x \in X\}$$

such that $m, n : X \to [0,1]$ with a condition $0 \le m(x) + n(x) \le 1 \, \forall \, x \in X$ Further, the degree of refusal of x in P is $r(x) = 1 - (m(x) + n(x))$ and the pair (m, n) is regarded as an IFN.

Definition 2. [4] *A Pythagorean fuzzy set (PyFS) on X consists of membership and non-membership functions defined as*

$$P = \{\langle x, m(x), n(x)\rangle \mid x \in X\}$$

such that $m, n : X \to [0,1]$ with a condition that $0 \le m^2(x) + n^2(x) \le 1 \, \forall \, x \in X$. Further, the degree of refusal of x in P is $r(x) = \sqrt{1 - (m^2(x) + n^2(x))}$ and the pair (m, n) is regarded as a Pythagorean fuzzy number (PyFN).

Definition 3. [6] *A picture fuzzy set (PFS) on X consists of membership, abstinence, and non-membership functions defined as*

$$P = \{\langle x, m(x), i(x), n(x)\rangle \mid x \in X\}$$

such that $m, i, n : X \to [0,1]$ with a condition that $0 \le m(x) + i(x) + n(x) \le 1 \, \forall \, x \in X$ Further, the degree of refusal of x in P is $r(x) = 1 - (m(x) + i(x) + n(x))$ and (m, i, n) is regarded as a picture fuzzy number (PFN).

Definition 4. [34] *A spherical fuzzy set (SFS) on X consists of membership, abstinence, and non-membership functions defined as*

$$P = \{\langle x, m(x), i(x), n(x)\rangle \mid x \in X\}$$

such that $m, i, n : X \to [0,1]$ with a condition that $0 \le m^2(x) + i^2(x) + n^2(x) \le 1 \, \forall \, x \in X$ Further, the degree of refusal of x in P is $r(x) = \sqrt{1 - (m^2(x) + i^2(x) + n^2(x))}$ and (m, i, n) is regarded as a spherical fuzzy number (SFN).

Definition 5. [34] *A T-SFS on X consists of membership, abstinence, and non-membership functions defined as*

$$P = \{\langle x, m(x), i(x), n(x)\rangle \mid x \in X\}$$

such that $m, i, n : X \to [0,1]$ *with a condition that* $0 \leq m^t(x) + i^t(x) + n^t(x) \leq 1 \; \forall \; x \in X \; t = 1, 2, \ldots k$. *Further, the degree of refusal of x in P is* $r(x) = \sqrt[t]{1 - (m^t(x) + i^t(x) + n^t(x))}$ *and* (m, i, n) *is regarded as a T-spherical fuzzy number (T-SFN).*

Definition 6. [34] *Let* $P = (m, i, n)$ *be a T-SFS. Then the score value of P is defined as*

$$SC(P) = m^t - n^t$$

and accuracy function is defined as

$$AC(P) = m^t + i^t + n^t$$

The one which has a greater score is the superior value. If the score of two T-SFNs is equal, then we rank them using the accuracy value, and a number is called superior if it has greater accuracy. If again accuracy values of two T-SFNs become equal, then both numbers are considered as similar.

Definition 7. [39] *Let* $P = (m_P, n_P)$ *and* $P' = (m_{P'}, n_{P'})$ *be two IFNs. Then the existing operational laws between them are defined as*

(1) $P \otimes P' = ((1 - n_P)(1 - n_{P'}) - (1 - m_P - n_P)(1 - m_{P'} - n_{P'})), 1 - (1 - n_P)(1 - n_{P'}))$

(2) $P^\lambda = ((1 - n_P)^\lambda - (1 - m_P - n_P)^\lambda, 1 - (1 - n_P)^\lambda)$.

Definition 8. *For any collection of T-SFNs* $P_j = \langle m_j, i_j, n_j\rangle$ $(j = 1, 2, 3, \ldots\ldots, k)$, [34] *defined the T-spherical fuzzy weighted geometric aggregation operator (T-SFWGA) as*

$$\mathrm{T-SFWGA_w}(\mathrm{P_1, P_2, \ldots\ldots, P_k}) = \begin{pmatrix} \sqrt[t]{\prod_{j=1}^{k} (\mathrm{m}_j^t + \mathrm{i}_j^t)^{\mathrm{w_j}} - \prod_{j=1}^{k} (\mathrm{i}_j^t)^{\mathrm{w_j}}}, \prod_{j=1}^{k} (\mathrm{i_j})^{\mathrm{w_j}}, \\ \sqrt[t]{1 - \prod_{j=1}^{k} (1 - \mathrm{n}_j^t)^{\mathrm{w_j}}} \end{pmatrix} \tag{1}$$

where $w = (w_1, w_2, \ldots\ldots w_k)^T$ *be the weighting vector of T-SFNs* P_j *with* $w_j \in (0,1]$ *and* $\sum_{j=1}^{k} w_j = 1$ *and* $t = 1, 2, \ldots\ldots k$.

3. Proposed Operational Laws and Aggregation Operators

This section is divided into two subsections. One presents the improved operations laws for the T-SFSs, while other presents some improved geometric AOs under the T-SFS environment.

3.1. Improved Operational Laws

In this section, we present some new, improved operations laws by incorporating the features of the degree of refusal into the analysis.

Definition 9. *Let* $P_1 = (m_{P_1}, i_{P_1}, n_{P_1})$ *and* $P_2 = (m_{P_2}, i_{P_2}, n_{P_2})$ *be two T-SFNs. Then, the proposed operational laws are defined as*

(1) $$P_1 \otimes P_2 = \begin{pmatrix} \sqrt[t]{(1 - n_{P_1}^t)(1 - n_{P_2}^t) - (1 - m_{P_1}^t - i_{P_1}^t - n_{P_1}^t)(1 - m_{P_2}^t - i_{P_2}^t - n_{P_2}^t) - i_{P_1}^t i_{P_2}^t}, \\ \sqrt[t]{1 - (1 - i_{P_1}^t)(1 - i_{P_2}^t)}, \sqrt[t]{1 - (1 - n_{P_1}^t)(1 - n_{P_2}^t)} \end{pmatrix}$$

(2) $$P^\lambda = \left(\sqrt[t]{(1 - n_P^t)^\lambda - (1 - m_P^t - i_P^t - n_P^t)^\lambda - i_P^{t\lambda}}, \sqrt[t]{1 - (1 - i_P^t)^\lambda}, \sqrt[t]{1 - (1 - n_P^t)^\lambda}\right)$$

For two T-SFNs, $P_1 = (m_{P_1}, i_{P_1}, n_{P_1})$ and $P_2 = (m_{P_2}, i_{P_2}, n_{P_2})$, new operations of multiplication can be construed from four aspects, such as between:

(1) Two non-membership functions of different T-SFNs.
(2) Two membership functions of different T-SFNs.
(3) Membership and non-membership functions of different T-SFNs.
(4) Two neutral functions of different T-SFNs.

These multiplication rules are of the form:

1. $E(n_{P_1}, n_{P_2}) = n_{P_1}.n_{P_2}$. Therefore, $n_{P_1 \otimes P_2} = \sqrt[t]{(n_{P_1}^t + n_{P_2}^t - n_{P_1}^t n_{P_2}^t)}$ is considered as a probability non-membership (PN) function operator, that is,

$$PN(n_{P_1},\ n_{P_2}) = \sqrt[t]{n_{P_1}^t + n_{P_2}^t - n_{P_1}^t n_{P_2}^t}$$

2. $E(m_{P_1}, m_{P_2}) = (m_{P_1} + i_{P_1}).(m_{P_1} + i_{P_1})$. Therefore, $m_{P_1 \otimes P_2} = \sqrt[t]{1 - \left(1 - \left(m_{P_1}^t + i_{P_1}^t\right)\right)\left(1 - \left(m_{P_2}^t + i_{P_2}^t\right)\right)}$ is considered as a probability membership (PM) function operator, that is,

$$PM(m_{P_1}, m_{P_2}) = \sqrt[t]{1 - (1 - m_{P_1}^t - i_{P_1}^t)(1 - m_{P_1}^t - i_{P_1}^t)}$$

3. $I(n_{P_1}, m_{P_2}) = \sqrt[t]{(m_{P_2}^t + i_{P_2}^t) n_{P_1}^t . I(n_{P_1}, m_{P_2})}$ is considered as a probability heterogeneous (PH) function operator, that is,

$$PH(n_{P_1}, m_{P_2}) = \sqrt[t]{m_{P_2}^t n_{P_1}^t + i_{P_2}^t n_{P_1}^t}$$

4. $I(i_{P_1}, i_{P_2}) = i_{P_1}.i_{P_2}$. Therefore, $i_{P_1 \otimes P_2} = \sqrt[t]{(i_{P_1}^t + i_{P_2}^t - i_{P_1}^t i_{P_2}^t)} . i_{P_1 \otimes P_2}$ is considered as a probability neutral (PNe) function operator, that is,

$$PNe(i_{P_1},\ i_{P_2}) = \sqrt[t]{i_{P_1}^t + i_{P_2}^t - i_{P_1}^t i_{P_2}^t}$$

From the proposed laws, it is observed that the several existing laws can be considered as a special case of it. For instance,

(i) For $t = 2$, above operations become valid for SFNs.
(ii) For $t = 1$, above operations become valid for PFNs.
(iii) For $t = 2$ and $i = 0$, above operations become valid for PyFNs.
(iv) For $t = 1$ and $i = 0$, above operations become valid for IFNs.

Further, it is observed that for the above defined PN, PH satisfies the following properties:

Theorem 1. *Let $P = \langle m_P, i_P, n_P \rangle, Q = \langle m_Q, i_Q, n_Q \rangle, R = \langle m_R, i_R, n_R \rangle$ and $D = \langle m_D, i_D, n_D \rangle$ be four T-SFNs. Then, we have:*

(1) *Boundedness: $PN(1,1) = 1, PN(0,0) = 0, 0 \leq PN(n_P, n_Q) \leq 1$.*
(2) *Monotonicity: If $n_P \leq n_R$ and $n_Q \leq n_D$. Then $PN(n_P, n_Q) \leq PN(n_R, n_D)$.*
(3) *Commutativity: $PN(n_P, n_Q) = PN(n_Q, n_P)$.*

Proof.
(1) For two T-SFNs, P and Q, and by definition of PN, we have $PN(n_P,\ n_Q) = \sqrt[t]{n_P^t + n_Q^t - n_P^t n_Q^t}$. Thus, we have PN(1,1) = 1 and PN(0,0) = 0. Further, since $n_P, n_Q \in [0,1]$ and $t \in Z$, which implies that $n_P^t + n_Q^t - n_P^t n_Q^t = 1 - (1 - n_P^t)(1 - n_Q^t) \leq 1$. Also, $PN(n_P, n_Q) \geq 0$. Therefore, $0 \leq PN(n_P, n_Q) \leq 1$.

(2) Since $n_P \leq n_R$ and $n_Q \leq n_D$. Thus, for any $t \in Z$, we get $1 - n_P^t \geq 1 - n_R^t$ and $1 - n_Q^t \geq 1 - n_D^t$, and hence $1 - (1 - n_P^t)(1 - n_Q^t) \leq 1 - (1 - n_R^t)(1 - n_D^t)$. Thus, $PN(n_P, n_Q) \leq PN(n_R, n_D)$ holds.

(3) Holds trivial. □

Theorem 2. *Let $P = \langle m_P, i_P, n_P \rangle, Q = \langle m_Q, i_Q, n_Q \rangle, R = \langle m_R, i_R, n_R \rangle$ and $S = \langle m_S, i_S, n_S \rangle$ be four T-SFN. Then:*

(1) *Boundedness: $PH(1,0,1) = 1, PH(0,0,0) = 0, 0 \leq PH(m_P, i_P, n_P) \leq 1$.*
(2) *Monotonicity: If $m_P \leq m_R$, $i_P \leq i_R$ and $n_Q \leq n_S$. Then $PH(m_P, i_P, n_Q) \leq PH(m_R, i_R, n_S)$ and if $n_P \leq n_R$, $i_Q \leq i_S$ and $m_Q \leq m_S$. Then $PH(n_P, i_Q, n_Q) \leq PH(n_R, i_S, m_S)$*
(3) *Commutativity: $PH(m_P, i_P, n_P) = PH(n_P, i_P, m_P)$.*

Proof. Similar to Theorem 1, so we omit here.

Theorem 3. *If P and Q are two T-SFNs and $\lambda > 0$ is a real number, then $P \otimes Q$ and P^λ are also T-SFNs.*

Proof. Follows from the definition easily, so we omit here.

Theorem 4. *Let $P_1 = \langle m, i, n \rangle$, $P_2 = \langle m', i', n' \rangle$ be a T-SFNs, $\lambda, \lambda_1, \lambda_2 > 0$ be real numbers. Then we have*

(1) $P_1 \otimes P_2 = P_2 \otimes P_1$
(2) $(P_1 \otimes P_2)^\lambda = P_1^\lambda \otimes P_2^\lambda$
(3) $P_1^{\lambda_1} \otimes P_1^{\lambda_2} = P_1^{\lambda_1 + \lambda_2}$.

Proof. Follows from the definition easily, so we omit here.

3.2. Aggregation Operators

In this section, based on the above proposed operational laws, we have proposed some series of geometric interactive improved AOs, namely, T-SFWGIA, T-SFOWGIA, and T-SFHGIA, under the T-SFS environment.

Definition 10. *For any collection, $P_j = \langle m_j, i_j, n_j \rangle$ $(j = 1, 2, 3, \ldots\ldots, k)$ of T-SFNs. If the mapping*

$$T - SFWGIA_w(P_1, P_2, \ldots\ldots, P_k) = \otimes_{j=1}^{k} P_j^{w_j} \tag{2}$$

then $T - SFWGIA_w$ is called a T-Spherical fuzzy weighted geometric interactive averaging (T-SFWGIA) operator, where $w = (w_1, w_2, \ldots\ldots w_k)^T$ is the weighting vector of P_j with $w_j \in (0, 1]$ and $\sum_{j=1}^{k} w_j = 1$.

Theorem 5. *For any collection of T-SFNs, $P_j = \langle m_j, i_j, n_j \rangle$ $(j = 1, 2, 3, \ldots\ldots, k)$, the aggregated values obtained by using Definition 10 is still T-SFNs and is given by:*

$$T - SFWGIA_w(P_1, P_2, \ldots\ldots, P_k) = \left(\begin{array}{c} \sqrt[t]{\prod_{j=1}^{k} (1 - n_j^t)^{w_j} - \prod_{j=1}^{k} (1 - m_j^t - i_j^t - n_j^t)^{w_j} - \prod_{j=1}^{k} (i_j^t)^{w_j}}, \\ \sqrt[t]{1 - \prod_{j=1}^{k} (1 - i_j^t)^{w_j}}, \sqrt[t]{1 - \prod_{j=1}^{k} (1 - n_j^t)^{w_j}} \end{array} \right)$$

Proof. For any collection of T-SFNs, $P_j = \langle m_j, i_j, n_j \rangle$ $(j = 1, 2, 3, \ldots\ldots, k)$, we shall proof the result by induction on k.

For $k = 1$, we have:

$$T - SFWGIA_w(P_1) = P_1^{w_1} = (m_1, i_1, n_1)$$
$$= \left(\sqrt[t]{(1-n_1^t)^1 - (1-(m_1^t+i_1^t+n_1^t))^1 - (i_1^t)^1}, \sqrt[t]{1-1+(i_1^t)^1}, \sqrt[t]{1-1+(n_1^t)^1}\right)$$

Thus, hold for $k = 1$. Now, the result holds for $n = m$:

$$T - SFWGIA_w\left(P^1, P^2, \ldots\ldots, P_m\right) = \left(\begin{array}{c} \sqrt[t]{\prod_{j=1}^{m}\left(1-n_j^t\right)^{w_j} - \prod_{j=1}^{m}\left(1-\mathrm{m}_j^t-\mathrm{i}_j^t-n_j^t\right)^{\mathrm{w}_j} - \prod_{j=1}^{m}(i_j^t)^{w_j}}, \\ \sqrt[t]{1-\prod_{j=1}^{m}\left(1-i_j^t\right)^{w_j}}, \sqrt[t]{1-\prod_{j=1}^{m}\left(1-n_j^t\right)^{w_j}} \end{array}\right)$$

Then for $k = m + 1$, we have:

$$T - SFWGIA_w(P_1, P_2, \ldots\ldots\ldots, P_{m+1}) = \otimes_{j=1}^{m+1} P_j^{w_j}$$
$$= T - SFWGIA_w(P_1, P_2, \ldots\ldots\ldots, P_m) \otimes P_{m+1}^{w_{m+1}}$$
$$= \left(\begin{array}{c} \sqrt[t]{\prod_{j=1}^{m}(1-n_j^t)^{w_j} - \prod_{j=1}^{m}(1-\mathrm{m}_j^t-\mathrm{i}_j^t-n_j^t)^{\mathrm{w}_j} - \prod_{j=1}^{m}(i_j^t)^{w_j}}, \\ \sqrt[t]{1-\prod_{j=1}^{m}(1-i_j^t)^{w_j}}, \sqrt[t]{1-\prod_{j=1}^{m}(1-n_j^t)^{w_j}} \end{array}\right)$$
$$\otimes \left(\begin{array}{c} \sqrt[t]{(1-n_j^t)^{w_j} - (1-\mathrm{m}_j^t-\mathrm{i}_j^t-n_j^t)^{\mathrm{w}_j} - (i_j^t)^{w_j}}, \\ \sqrt[t]{1-(1-i_j^t)^{w_j}}, \sqrt[t]{1-(1-n_j^t)^{w_j}} \end{array}\right)$$
$$= \left(\begin{array}{c} \sqrt[t]{\prod_{j=1}^{m+1}(1-n_j^t)^{w_j} - \prod_{j=1}^{m+1}(1-\mathrm{m}_j^t-\mathrm{i}_j^t-n_j^t)^{\mathrm{w}_j} - \prod_{j=1}^{m+1}(i_j^t)^{w_j}}, \\ \sqrt[t]{1-\prod_{j=1}^{m+1}(1-i_j^t)^{w_j}}, \sqrt[t]{1-\prod_{j=1}^{m+1}(1-n_j^t)^{w_j}} \end{array}\right)$$

So, the result holds for $k = m + 1$. Therefore, by the principle of mathematical induction, the result holds for all $k \in Z^+$. □

Theorem 6. *If $P_j = (m_j, i_j, n_j), j = 1, \ldots, k$ are T-SFNs. Then the aggregated value using the T SFWGIA operator is also T-SFN.*

Proof. Since $P_j = (m_j, i_j, n_j)$ is a T-SFN, $j = 1, \ldots, k$, we have $0 \leq m_j, i_j, n_j \leq 1$. So $0 \leq m_j^t, i_j^t, n_j^t \leq 1$ and $0 \leq m_j^t + i_j^t + n_j^t \leq 1$. Then:

$$\leq \prod_{j=1}^{k}(1-n_j^t)^{w_j} - \prod_{j=1}^{k}(1-\mathrm{m}_j^t-\mathrm{i}_j^t-n_j^t)^{\mathrm{w}_j} - \prod_{j=1}^{k}(i_j^t)^{w_j} \leq 1$$
$$0 \leq 1 - \prod_{j=1}^{k}(1-i_j^t)^{w_j} \leq 1$$
$$0 \leq 1 - \prod_{j=1}^{k}(1-n_j^t)^{w_j} \leq 1$$

Now:

$$\sqrt[t]{\begin{array}{c} \prod_{j=1}^{k}(1-n_j^t)^{w_j} - \prod_{j=1}^{k}(1-(\mathrm{m}_j^t+\mathrm{i}_j^t+n_j^t))^{\mathrm{w}_j} - \prod_{j=1}^{k}(i_j^t)^{w_j} + \\ 1-\prod_{j=1}^{k}(1-i_j^t)^{w_j} + 1 - \prod_{j=1}^{k}(1-n_j^t)^{w_j} \end{array}}$$
$$= \sqrt[t]{2 - \prod_{j=1}^{k}(1-(\mathrm{m}_j^t+\mathrm{i}_j^t+n_j^t))^{\mathrm{w}_j} - \prod_{j=1}^{k}(i_j^t)^{w_j} - \prod_{j=1}^{k}(1-i_j^t)^{w_j}} \in [0,1]$$

Thus, $T - SFWGIA_w(P_1, \ldots\ldots\ldots, P_k)$ is T-SFN.

Further, it is observed that the proposed operator satisfies certain properties, which are listed as follows: □

Theorem 7. *If all T-SFNs, $P_j (j = 1, 2, \ldots, k)$, are equal to P_0, where P_0 is another T-SFN, then*

$$T - SFWGIA_w(P_1, \ldots\ldots\ldots, P_k) = P_0$$

Proof. Assume that $P_j = P_0 = (m_0, i_0, n_0)$ is a T-SFN $\forall j$. Then, by definition of T-SFWGIA operator, we have:

$$T - SFWGIA_w(P_1, P_2, \ldots, P_k) = \begin{pmatrix} \sqrt[t]{\prod_{j=1}^{k} (1 - n_j^t)^{w_j} - \prod_{j=1}^{k} (1 - (\mathrm{m}_j^t + \mathrm{i}_j^t + n_j^t))^{\mathrm{w}_j} - \prod_{j=1}^{k} (i_j^t)^{w_j}}, \\ \sqrt[t]{1 - \prod_{j=1}^{k} (1 - i_j^t)^{w_j}}, \sqrt[t]{1 - \prod_{j=1}^{k} (1 - n_j^t)^{w_j}} \end{pmatrix}$$

$$= \begin{pmatrix} \sqrt[t]{(1 - n_j^t)^{\sum_{j=1}^{k} w_j} - (1 - (\mathrm{m}_j^t + \mathrm{i}_j^t + n_j^t))^{\sum_{j=1}^{k} w_j} - (i_j^t)^{\sum_{j=1}^{k} w_j}}, \\ \sqrt[t]{1 - (1 - i_j^t)^{\sum_{j=1}^{k} w_j}}, \sqrt[t]{1 - (1 - n_j^t)^{\sum_{j=1}^{k} w_j}} \end{pmatrix}$$

$$= (m_0, i_0, n_0)$$

$$= P_0$$

Theorem 8. *If $P_j = (m_j, i_j, n_j)$ is a T-SFN and*

$$P^L = (max\{0, (min(m_j + i_j + n_j) - min\ i_j - max\ n_j)\}, min\ i_j, max\ n_j),$$

$$P^U = (max(m_j + i_j + n_j) - max\ i_j - min\ n_j, max\ i_j, min\ n_j). \text{ Then, we have}$$

$$P^L \leq T - SFWGIA_w(P_1, \ldots\ldots, P_k) \leq P^U$$

Proof is straightforward.

Theorem 9. *For a collection of two different T-SFNs, $A_j = (m_{A_j}, i_{A_j}, n_{A_j})$, $(j = 1, 2, \ldots, k)$ and $B_j = (m_{B_j}, i_{B_j}, n_{B_j})$, $(j = 1, 2, \ldots, k)$, which satisfy the following inequalities if $n_{A_j} \geq n_{B_j}$, $i_{A_j} \geq i_{B_j}$ and $m_{A_j}^t + i_{A_j}^t + n_{A_j}^t \leq m_{B_j}^t + i_{B_j}^t + n_{B_j}^t$ $\forall j$, then we have*

$$T - SFWGIA_w\ (A_1, A_2, \ldots, A_k) \leq T - SFWGIA_w\ (B_1, B_2, \ldots, B_k)$$

Proof. Since $n_{A_j} \geq n_{B_j}$, we have:

$$\sqrt[t]{1 - \prod_{j=1}^{k} \left(1 - n_{A_j}^t\right)^{w_j}} \geq \sqrt[t]{1 - \prod_{j=1}^{k} \left(1 - n_{B_j}^t\right)^{w_j}}$$

and $i_{A_j} \geq i_{B_j}$

$$\sqrt[t]{1 - \prod_{j=1}^{k} \left(1 - i_{A_j}^t\right)^{w_j}} \geq \sqrt[t]{1 - \prod_{j=1}^{k} \left(1 - i_{B_j}^t\right)^{w_j}}$$

As, $n_{A_j} \geq n_{B_j}$, $m^t_{A_j} + i^t_{A_j} + n^t_{A_j} \leq m^t_{B_j} + i^t_{B_j} + n^t_{B_j}$ $\forall j$ we have:

$$\left(\begin{array}{c} \sqrt[t]{\prod_{j=1}^{k}\left(1-n^t_{A_j}\right)^{w_j} - \prod_{j=1}^{k}\left(1-\left(\mathrm{m}^t_{\mathrm{A_j}} + \mathrm{i}^t_{\mathrm{A_j}} + n^t_{A_j}\right)\right)^{\mathrm{w_j}} - \prod_{j=1}^{k}\left(i^t_{A_j}\right)^{w_j}}, \\ \sqrt[t]{1-\prod_{j=1}^{k}\left(1-i^t_{A_j}\right)^{w_j}}, \sqrt[t]{1-\prod_{j=1}^{k}\left(1-n^t_{A_j}\right)^{w_j}} \end{array} \right) \leq \left(\begin{array}{c} \sqrt[t]{\prod_{j=1}^{k}\left(1-n^t_{B_j}\right)^{w_j} - \prod_{j=1}^{k}\left(1-\left(\mathrm{m}^t_{\mathrm{B_j}} + \mathrm{i}^t_{\mathrm{B_j}} + n^t_{B_j}\right)\right)^{\mathrm{w_j}} - \prod_{j=1}^{k}\left(i^t_{B_j}\right)^{w_j}}, \\ \sqrt[t]{1-\prod_{j=1}^{k}\left(1-i^t_{B_j}\right)^{w_j}}, \sqrt[t]{1-\prod_{j=1}^{k}\left(1-n^t_{B_j}\right)^{w_j}} \end{array} \right)$$

Therefore, we have:

$$T-SFWGIA_w(A_1, A_2, \ldots, A_k) \leq T-SFWGIA_w(B_1, B_2, \ldots, B_k)$$

Definition 11. [34] *For any collection,* $P_j = (m_j, i_j, n_j)$ $(j = 1, 2, \ldots, k)$ *of T-SFNs. The* $T-SFOWGA_w : \Omega^n \to \Omega$ *is a mapping defined as*

$$T-SFOWGA_w(P_1, P_2, \ldots\ldots, P_k) = \left(\begin{array}{c} \sqrt[t]{\prod_{j=1}^{k}\left(\mathrm{m}^t_{\sigma(j)} + \mathrm{i}^t_{\sigma(j)}\right)^{\mathrm{w_j}} - \prod_{j=1}^{k}\left(i^t_{\sigma(j)}\right)^{w_j}}, \prod_{j=1}^{k}\left(i_{\sigma(j)}\right)^{w_j}, \\ \sqrt[t]{1-\prod_{j=1}^{k}\left(1-n^t_{\sigma(j)}\right)^{w_j}} \end{array} \right) \quad (3)$$

where Ω *is the collection of all T-SFNs, then* $T-SFOWGA_w$ *is called a T-SFOWGA operator with weighting vector* $w = (w_1, w_2, \ldots\ldots w_k)^T$ *of* P_j *with* $w_j \in (0,1]$ *and* $\sum_{j=1}^{k} w_j = 1$.

Definition 12. *For any collection,* $P_j = (m_j, i_j, n_j)$, $(j = 1, 2, \ldots, k)$ *of T-SFNs. The* $T-SFOWGIA_w : \Omega^n \to \Omega$ *is a mapping defined as:*

$$T-SFOWGIA_w(P_1, P_2, \ldots\ldots, P_k) = \otimes_{j=1}^{k} P_{\sigma(j)}^{w_j} \quad (4)$$

then $T-SFOWGIA_w$ *is called T-SFOWGIA operator, where* $w = (w_1, w_2, \ldots\ldots\ldots w_k)^T$ *is the weighting vector of* P_j *with* $w_j \in (0,1]$ *and* $\sum_{j=1}^{k} w_j = 1$ *and* σ *is the permutation of* $\{1, 2, \ldots, k\}$*, such that* $\sigma(j-1) \geq \sigma(j)$.

Theorem 10. *For any collection* $P_j = (m_j, i_j, n_j)$, $(j = 1, 2, \ldots, k)$ *of T-SFNs. Then*

$$T-SFOWGIA_w(P_1, P_2, \ldots, P_k) = \left(\begin{array}{c} \sqrt[t]{\prod_{j=1}^{k}\left(1-n^t_{\sigma(j)}\right)^{w_j} - \prod_{j=1}^{k}\left(1-\left(m^t_{\sigma(j)} + i^t_{\sigma(j)} + n^t_{\sigma(j)}\right)\right)^{w_j} - \prod_{j=1}^{k}\left(i^t_{\sigma(j)}\right)^{w_j}}, \\ \sqrt[t]{1-\prod_{j=1}^{k}\left(1-i^t_{\sigma(j)}\right)^{w_j}}, \sqrt[t]{1-\prod_{j=1}^{k}\left(1-n^t_{\sigma(j)}\right)^{w_j}} \end{array} \right)$$

Proof is similar to Theorem 5.

Theorem 11. *If* $P_j = (m_j, i_j, n_j)$ *is a T-SFN,* $j = 1, \ldots, k$*. Then the aggregated value using the T-SFOWGIA operator is also T-SFN.*

Proof. Since $P_{\sigma(j)} = \left(m_{\sigma(j)}, i_{\sigma(j)}, n_{\sigma(j)}\right)$ is a T-SFN, $j = 1, \ldots, k$, we have $0 \leq m_{\sigma(j)}, i_{\sigma(j)}, n_{\sigma(j)} \leq 1$. So $0 \leq m^t_{\sigma(j)}, i^t_{\sigma(j)}, n^t_{\sigma(j)} \leq 1$ and $0 \leq m^t_{\sigma(j)} + i^t_{\sigma(j)} + n^t_{\sigma(j)} \leq 1$. Then:

$$0 \leq \prod_{j=1}^{k}\left(1-n^t_{\sigma(j)}\right)^{w_j} - \prod_{j=1}^{k}\left(1-\left(\mathrm{m}^t_{\sigma(j)}+\mathrm{i}^t_{\sigma(j)}+n^t_{\sigma(j)}\right)\right)^{\mathrm{w}_j} - \prod_{j=1}^{k}\left(i^t_{\sigma(j)}\right)^{w_j} \leq 1$$
$$0 \leq 1-\prod_{j=1}^{k}\left(1-i^t_{\sigma(j)}\right)^{w_j} \leq 1$$
$$0 \leq 1-\prod_{j=1}^{k}\left(1-n^t_{\sigma(j)}\right)^{w_j} \leq 1$$

Now:

$$\sqrt[t]{\begin{array}{c}\prod_{j=1}^{k}\left(1-n^t_{\sigma(j)}\right)^{w_j} - \prod_{j=1}^{k}\left(1-\left(\mathrm{m}^t_{\sigma(j)}+\mathrm{i}^t_{\sigma(j)}+n^t_{\sigma(j)}\right)\right)^{\mathrm{w}_j} - \prod_{j=1}^{k}\left(i^t_{\sigma(j)}\right)^{w_j} + 1 - \prod_{j=1}^{k}\left(1-i^t_{\sigma(j)}\right)^{w_j} \\ +1-\prod_{j=1}^{k}\left(1-n^t_{\sigma(j)}\right)^{w_j}\end{array}}$$
$$= \sqrt[t]{2-\prod_{j=1}^{k}\left(1-\left(\mathrm{m}^t_{\sigma(j)}+\mathrm{i}^t_{\sigma(j)}+n^t_{\sigma(j)}\right)\right)^{\mathrm{w}_j} - \prod_{j=1}^{k}\left(i^t_{\sigma(j)}\right)^{w_j} - \prod_{j=1}^{k}\left(1-i^t_{\sigma(j)}\right)^{w_j}} \in [0,1]$$

Thus, $T - SFOWGIA_w(P_1, \ldots\ldots\ldots, P_k)$ is T-SFN. □

Theorem 12. $T - SFOWGIA_w(P_1, \ldots\ldots\ldots, P_k) = P_0$ if $P_j = P_0 = \left(m_j, i_j, n_j\right)$ is a T-SFN $\forall j$.

Proof. We have:

$$T - SFOWGIA_w(P_1, \ldots\ldots\ldots P_k) =$$
$$\left(\begin{array}{c}\sqrt[t]{\prod_{j=1}^{k}\left(1-n^t_{\sigma(j)}\right)^{w_j} - \prod_{j=1}^{k}\left(1-\left(\mathrm{m}^t_{\sigma(j)}+\mathrm{i}^t_{\sigma(j)}+n^t_{\sigma(j)}\right)\right)^{\mathrm{w}_j} - \prod_{j=1}^{k}\left(i^t_{\sigma(j)}\right)^{w_j}}, \\ \sqrt[t]{1-\prod_{j=1}^{k}\left(1-i^t_{\sigma(j)}\right)^{w_j}}, \sqrt[t]{1-\prod_{j=1}^{k}\left(1-n^t_{\sigma(j)}\right)^{w_j}}\end{array}\right)$$
$$= \left(\begin{array}{c}\sqrt[t]{\left(1-n^t_{\sigma(j)}\right)^{\sum_{j=1}^{k} w_j} - \left(1-\left(\mathrm{m}^t_{\sigma(j)}+\mathrm{i}^t_{\sigma(j)}+n^t_{\sigma(j)}\right)\right)^{\sum_{j=1}^{k} w_j} - \left(i^t_{\sigma(j)}\right)^{\sum_{j=1}^{k} w_j}}, \\ \sqrt[t]{1-\left(1-i^t_{\sigma(j)}\right)^{\sum_{j=1}^{k} w_j}}, \sqrt[t]{1-\left(1-n^t_{\sigma(j)}\right)^{\sum_{j=1}^{k} w_j}}\end{array}\right)$$
$$= \left(m_{\sigma(0)}, i_{\sigma(0)}, n_{\sigma(0)}\right)$$
$$= P_0$$

□

Theorem 13. *If* $P_j = \left(m_j, i_j, n_j\right)$ *is a T-SFN and*

$$P^L = \left(max\left\{0, \left(min\left(m_j+i_j+n_j\right) - min\ i_j - max\ n_j\right)\right\}, min\ i_j, max\ n_j\right),$$
$$P^U = \left(max\left(m_j+i_j+n_j\right) - max\ i_j - min\ n_j\right)\}, max\ i_j, min\ n_j). \text{ Then}$$
$$P^L \leq T - SFOWGIA(P_1, \ldots\ldots, P_k) \leq P^U$$

Proof is straightforward.

Theorem 14. $T - SFOWGIA_w(B_1 B_2, \ldots\ldots, B_k) = T - SFOWGIA_w(A_1, \ldots\ldots, A_k)$ *if* $B_j = \left(m_{B_j}, i_{B_j}, n_{B_j}\right)$ *is any permutation of* $A_j = \left(m_{A_j}, i_{A_j}, n_{A_j}\right)$ *where* $j = 1, \ldots\ldots, k$.

Proof.

$$T-SFOWGIA_w(B_1,B_2\ldots\ldots B_k)=$$

$$\left(\begin{array}{c}\sqrt[t]{\prod_{j=1}^{k}\left(1-n^t_{B_{\sigma(j)}}\right)^{w_j}-\prod_{j=1}^{k}\left(1-\left(\mathrm{m}^t_{B_{\sigma(j)}}+\mathrm{i}^t_{B_{\sigma(j)}}+n^t_{B_{\sigma(j)}}\right)\right)^{\mathrm{w}_j}-\prod_{j=1}^{k}\left(i^t_{B_{\sigma(j)}}\right)^{w_j}},\\ \sqrt[t]{1-\prod_{j=1}^{k}\left(1-i^t_{B_{\sigma(j)}}\right)^{w_j}},\ \sqrt[t]{1-\prod_{j=1}^{k}\left(1-n^t_{B_{\sigma(j)}}\right)^{w_j}}\end{array}\right)$$

$$T-SFOWGIA_w(A_1,A_2\ldots\ldots A_k)=$$

$$\left(\begin{array}{c}\sqrt[t]{\prod_{j=1}^{k}\left(1-n^t_{A_{\sigma(j)}}\right)^{w_j}-\prod_{j=1}^{k}\left(1-\left(\mathrm{m}^t_{A_{\sigma(j)}}+\mathrm{i}^t_{A_{\sigma(j)}}+n^t_{A_{\sigma(j)}}\right)\right)^{\mathrm{w}_j}-\prod_{j=1}^{k}\left(i^t_{A_{\sigma(j)}}\right)^{w_j}},\\ \sqrt[t]{1-\prod_{j=1}^{k}\left(1-i^t_{A_{\sigma(j)}}\right)^{w_j}},\ \sqrt[t]{1-\prod_{j=1}^{k}\left(1-n^t_{A_{\sigma(j)}}\right)^{w_j}}\end{array}\right)$$

If $B_j=\left(m_{B_j},i_{B_j},n_{B_j}\right)$ is any permutation of $A_j=\left(m_{A_j},i_{A_j},n_{A_j}\right)$ then we have $B_{\sigma(j)}=A_{\sigma(j)}$. Thus, $T-SFOWGIA_w(B_1,\ldots\ldots,B_k)=T-SFOWGIA_w(A^1,\ldots\ldots,A_k)$. □

Definition 13. *For any collection,* $P_j=\langle m_j,i_j,n_j\rangle$ *of T-SFNs* $(j=1,2,3,\ldots\ldots,k)$. *If the mapping*

$$T-SFHGA_{\omega,w}(P_1,P_2,\ldots\ldots,P_k)=\otimes_{j=1}^{k}\left(\widetilde{P}_{\sigma(j)}\right)^{w_i} \quad (5)$$

then $T-SFHGA_{\omega,w}$ *is called a T-SFHGA operator, where* $\widetilde{P}_j=(P_j)^{n\omega_j}$ *and* $\omega=(\omega_1,\ldots\ldots\omega_k)^T$ *is the weighting vector of* P_j *with* $\omega_j\in(0,1]$ *and* $\sum_{j=1}^{k}\omega_j=1$.

Theorem 15. [34] *For any collection,* $P_j=\langle m_j,i_j,n_j\rangle$ $(j=1,2,3,\ldots\ldots,k)$ *of T-SFNs. If*

$$T-SFHGA_{\omega,w}(P_1,P_2,\ldots\ldots,P_k)=\left(\begin{array}{c}\sqrt[t]{\prod_{j=1}^{k}\left(\mathrm{m}^t_{\widetilde{P}_{\sigma(j)}}+\mathrm{i}^t_{\widetilde{P}_{\sigma(j)}}\right)^{\mathrm{w}_j}-\prod_{j=1}^{k}\left(i^t_{\widetilde{P}_{\sigma(j)}}\right)^{w_j}},\\ \prod_{j=1}^{k}\left(i_{\widetilde{P}_{\sigma(j)}}\right)^{w_j},\ \sqrt[t]{1-\prod_{j=1}^{k}\left(1-n^t_{\widetilde{P}_{\sigma(j)}}\right)^{w_j}}\end{array}\right)$$

then $T-SFHGA_{\omega,w}$ *is called a T-SFHGA operator with weighting vector* $\omega=(\omega_1,\ \omega_2,\ldots\ldots\omega_k)^T$ *of* P_j *with* $\omega_j\in(0,1]$ *and* $\sum_{j=1}^{k}\omega_j=1$.

Definition 14. *For any collection,* $P_j=\langle m_j,i_j,n_j\rangle$ $(j=1,2,3,\ldots\ldots,k)$ *of T-SFNs. If the mapping*

$$T-SFHGIA_{\omega,\ w}(P_1,P_2,\ldots\ldots,P_k)=\otimes_{j=1}^{k}\widetilde{P}_{\sigma(j)}^{w_j} \quad (6)$$

then $T-SFHGIA_{\omega,w}$ *is called a T-SFHGIA operator, where* $\omega=(\omega_1,\omega_2,\ldots\ldots\omega_k)^T$ *is the weighting vector of* P_j *with* $\omega_j\in[0,1]$ *and* $\sum_{j=1}^{k}w_j=1$.

Theorem 16. *For any collection,* $P_j=\langle m_j,i_j,n_j\rangle$ $(j=1,2,3,\ldots\ldots,k)$ *of T-SFNs. Then*

$$T-SFHGIA_{\omega,\ w}(P_1,P_2,\ldots\ldots,P_k)$$

$$=\left(\begin{array}{c}\sqrt[t]{\prod_{j=1}^{k}\left(1-n^t_{\widetilde{P}_{\sigma(j)}}\right)^{w_j}-\prod_{j=1}^{k}\left(1-\left(m^t_{\widetilde{P}_{\sigma(j)}}+i^t_{\widetilde{P}_{\sigma(j)}}+n^t_{\widetilde{P}_{\sigma(j)}}\right)\right)^{w_j}-\prod_{j=1}^{k}\left(i^t_{\widetilde{P}_{\sigma(j)}}\right)^{w_j}},\\ \sqrt[t]{1-\prod_{j=1}^{k}\left(1-i^t_{\widetilde{P}_{\sigma(j)}}\right)^{w_j}},\ \sqrt[t]{1-\prod_{j=1}^{k}\left(1-n^t_{\widetilde{P}_{\sigma(j)}}\right)^{w_j}}\end{array}\right)$$

The following example demonstrates these aggregation operators:

Example 1. *Let* $P_1 = (0.3, 0.8, 0.1)$, $P_2 = (0.4, 0.3, 0.6)$, $P_3 = (0.7, 0.1, 0.5)$, $P_4 = (0.9, 0.4, 0.1)$ *and* $P_5 = (0.2, 0.6, 0.7)$ *are T-SFN. The weight vector for* P_i $(i = 1, 2, \ldots, 5)$ *is* $\omega = (0.18, 0.22, 0.16, 0.21, 0.23)^T$. *With loss of generality, we use* $t = 2$ *for all calculations.*

Firstly, we utilized T-SFHGIA operators on this data to aggregate it.

$$P_1 = \left(\begin{array}{c} \sqrt{(1-0.1^2)^{5\times0.18} - (1-(0.3^2+0.8^2+0.1^2))^{5\times0.18} - (0.8^2)^{5\times0.18}}, \\ \sqrt{1-(1-0.8^2)^{5\times0.18}}, \sqrt{1-(1-0.1^2)^{5\times0.18}} \end{array} \right)$$
$$= (0.1559,\ 0.7754,\ 0.0949)$$

$$P_2 = \left(\begin{array}{c} \sqrt{(1-0.6^2)^{5\times0.22} - (1-(0.4^2+0.3^2+0.6^2))^{5\times0.22} - (0.3^2)^{5\times0.22}}, \\ \sqrt{1-(1-0.3^2)^{5\times0.22}}, \sqrt{1-(1-0.6^2)^{5\times0.22}} \end{array} \right)$$
$$= (0.4317,\ 0.3139,\ 0.6228)$$

$$P_3 = \left(\begin{array}{c} \sqrt{(1-0.5^2)^{5\times0.16} - (1-(0.7^2+0.1^2+0.5^2))^{5\times0.16} - (0.1^2)^{5\times0.16}}, \\ \sqrt{1-(1-0.1^2)^{5\times0.16}}, \sqrt{1-(1-0.5^2)^{5\times0.16}} \end{array} \right)$$
$$= (0.6629,\ 0.0895,\ 0.4534)$$

$$P_4 = \left(\begin{array}{c} \sqrt{(1-0.1^2)^{5\times0.21} - (1-(0.9^2+0.4^2+0.1^2))^{5\times0.21} - (0.4^2)^{5\times0.21}}, \\ \sqrt{1-(1-0.4^2)^{5\times0.21}}, \sqrt{1-(1-0.1^2)^{5\times0.21}} \end{array} \right)$$
$$= (0.9094, 0.4090, 0.1024)$$

$$P_5 = \left(\begin{array}{c} \sqrt{(1-0.7^2)^{5\times0.23} - (1-(0.2^2+0.6^2+0.7^2))^{5\times0.23} - (0.6^2)^{5\times0.23}}, \\ \sqrt{1-(1-0.6^2)^{5\times0.23}}, \sqrt{1-(1-0.7^2)^{5\times0.23}} \end{array} \right)$$
$$= (0.2705, 0.6336, 0.7342)$$

The score values corresponding to these aggregated numbers were obtained as $SC(P_1) = 0.0153$, $SC(P_2) = -0.2016$, $SC(P_3) = 0.2338$, $SC(P_4) = 0.8166$, $SC(P_5) = -0.4658$. Based on the score values, we had the following arrangement of data:

$P_{\sigma(1)} = (0.9094, 0.4090, 0.1024)$, $P_{\sigma(2)} = (0.6629,\ 0.0895,\ 0.4534)$, $P_{\sigma(3)} = (0.1559,\ 0.7754,\ 0.0949)$, $P_{\sigma(4)} = (0.4317,\ 0.3139,\ 0.6228)$, $P_{\sigma(5)} = (0.2705, 0.6336, 0.7342)$

By using the normal distribution-based method, we found $w = (0.1117, 0.2365, 0.3036, 0.2365, 0.1117)^T$ and by the definition of T-SFHGIA operator we had

$$T - SFHGIA_{\omega,w}(P_1, P_2, P_3, P_4, P_5) = (0.4688, 0.5643, 0.4792)$$

Theorem 17. *If* $P_j = (m_j, i_j, n_j)$ *is a T-SFN* , $j = 1, \ldots, k$, *then the aggregated value using the T-SFHGIA operator is also T-SFN.*

Proof is similar as in Theorem 11.

Theorem 18. $T - SFHGIA_{\omega,\, w}(P_1, P_2, \ldots\ldots, P_k) = P_0$ *if* $P_j = P_0 = (m_j, i_j, n_j)$ *is a T-SFN* $\forall j$.
Proof is similar as in Theorem 12.

Theorem 19. *If* $P_j = (m_j, i_j, n_j)$ *is a T-SFN and*

$$P^L = (max\{0, (min(m_j + i_j + n_j) - min\ i_j - max\ n_j)\}, min\ i_j, max\ n_j),$$
$$P^U = (max(m_j + i_j + n_j) - max\ i_j - min\ n_j, max\ i_j, min\ n_j). \text{ Then}$$
$$P^L \le T - SFHGIA_{\omega,w}(P_1, \ldots\ldots, P_k) \le P^U$$

Proof is straightforward.

Theorem 20. *$T - SFHGIA_{\omega,w}(B_1,\ldots\ldots,B_k) = T - SFHGIA_{\omega,w}(A_1,\ldots\ldots,A_k)$ if $B_j = \left(m_{B_j}, i_{B_j}, n_{B_j}\right)$ is any permutation of $A_j = \left(m_{A_j}, i_{A_j}, n_{A_j}\right)$ where $j = 1,\ldots\ldots,k$.*
Proof is similar as Theorem 14.

Whenever membership and neutral number of one T-SFN become zero then the membership and abstinence value is not accounted for in the aggregation [34]. However, the geometric interaction averaging operators that are developed in our manuscript overcome this problem. The example below will describe this more clearly.

Example 2. *Let $P_1 = (0.7, 0.5, 0.6), P_2 = (0.9, 0.5, 0.4), P_3 = (0, 0, 0.1), P_4 = (0.5, 0.3, 0.4)$ and $P_5 = (0.6, 0.4, 0.5)$ are T-SFN. The weight vector for P_i $(i = 1, 2, \ldots, 5)$ is $\omega = (0.18, 0.22, 0.16, 0.21, 0.23)^T$.*

For the solution, first we will find the T-SFHGA operator.
As, $0.7 + 0.5 + 0.6 = 1.8 \notin [0,1]$, $0.7^2 + 0.5^2 + 0.6^2 = 1.1 \notin [0,1]$ but $0.7^3 + 0.5^3 + 0.6^3 = 0.684 \in [0,1]$
Similarly, P_2 and P_4 satisfy the condition for $t = 3$.

$$\begin{aligned}
\widetilde{P}_1 &= \left(\sqrt[3]{(0.7^3+0.5^3)^{5\times0.18} - (0.5^3)^{5\times0.18}}, 0.5^{5\times0.18}, \sqrt[3]{1-(1-0.6^3)^{5\times0.18}}\right) \\
&= (0.7054, 0.5359, 0.5816) \\
\widetilde{P}_2 &= \left(\sqrt[3]{(0.9^3+0.5^3)^{5\times0.22} - (0.5^3)^{5\times0.22}}, 0.5^{5\times0.22}, \sqrt[3]{1-(1-0.4^3)^{5\times0.22}}\right) \\
&= (0.9041, 0.4665, 0.4125) \\
\widetilde{P}_3 &= \left(\sqrt[3]{(0^3+0^3)^{5\times0.16} - (0^3)^{5\times0.16}}, 0^{5\times0.16}, \sqrt[3]{1-(1-0.1^3)^{5\times0.16}}\right) \\
&= (0, 0, 0.0928) \\
\widetilde{P}_4 &= \left(\sqrt[3]{(0.5^3+0.3^3)^{5\times0.21} - (0.3^3)^{5\times0.21}}, 0.3^{5\times0.21}, \sqrt[3]{1-(1-0.4^3)^{5\times0.21}}\right) \\
&= (0.4874, 0.2885, 0.4063) \\
\widetilde{P}_5 &= \left(\sqrt[3]{(0.6^3+0.4^3)^{5\times0.23} - (0.4^3)^{5\times0.23}}, 0.4^{5\times0.23}, \sqrt[3]{1-(1-0.5^3)^{5\times0.23}}\right) \\
&= (0.5738, 0.3486, 0.5221)
\end{aligned}$$

Scores values for these aggregated numbers were obtained as $SC(\widetilde{P}_1) = 0.1543$, $SC(\widetilde{P}_2) = 0.6689$, $SC(\widetilde{P}_3) = -0.0008$, $SC(\widetilde{P}_4) = 0.0487$, $SC(\widetilde{P}_5) = 0.0466$, and, based on these score values, we had
$\widetilde{P}_{\sigma(1)} = (0.9041, 0.4665, 0.4125)$, $\widetilde{P}_{\sigma(2)} = (0.7054, 0.5359, 0.5816)$, $\widetilde{P}_{\sigma(3)} = (0.4874, 0.2885, 0.4063)$, $\widetilde{P}_{\sigma(4)} = (0.5738, 0.3486, 0.5221)$, $\widetilde{P}_{\sigma(5)} = (0, 0, 0.0928)$
By using the normal distribution-based method, we found $w = (0.1117, 0.2365, 0.3036, 0.2365, 0.1117)^T$, and, by the definition of T-SFHGA operator, we found

$$\mathrm{T - SFHGA_{\omega,w}(P_1, P_2, P_3, P_4, P_5)} = (0, 0, 0.4803) \tag{7}$$

This type of aggregated value seems meaningless, as whenever the membership and abstinence value is zero in any one of the T-SFN it will make the value of the membership and non-membership as zero in the whole aggregated value. This shows that the geometric aggregation operator of T-SFSs [34] does not possess the ability to aggregate such types of information effectively.

On the other hand, the proposed new geometric interactive aggregation operators can process any type of information effectively. Now, the Example 2 was solved using the proposed new aggregation operators in order to justify its effectiveness. For it, we aggregated the data using the T-SFHGIA operator:

$$P_1 = \left(\begin{array}{c} \sqrt[3]{(1-0.6^3)^{5\times0.18} - (1-(0.7^3+0.5^3+0.6^3))^{5\times0.18} - (0.5^3)^{5\times0.18}}, \\ \sqrt[3]{1-(1-0.5^3)^{5\times0.18}}, \sqrt[3]{1-(1-0.6^3)^{5\times0.18}} \end{array} \right)$$
$$= (0.6656,\ 0.5359,\ 0.5816)$$

$$P_2 = \left(\begin{array}{c} \sqrt[3]{(1-0.4^3)^{5\times0.22} - (1-(0.9^3+0.5^3+0.4^3))^{5\times0.22} - (0.5^3)^{5\times0.22}}, \\ \sqrt[3]{1-(1-0.5^3)^{5\times0.22}}, \sqrt[3]{1-(1-0.4^3)^{5\times0.22}} \end{array} \right)$$
$$= (0.9144,\ 0.4665,\ 0.4125)$$

$$P_3 = \left(\begin{array}{c} \sqrt[3]{(1-0.1^3)^{5\times0.16} - (1-(0^3+0^3+0.1^3))^{5\times0.16} - (0^3)^{5\times0.16}}, \\ \sqrt[3]{1-(1-0^3)^{5\times0.16}}, \sqrt[3]{1-(1-0.1^3)^{5\times0.16}} \end{array} \right)$$
$$= (0,\ 0,\ 0.0928)$$

$$P_4 = \left(\begin{array}{c} \sqrt[3]{(1-0.4^3)^{5\times0.21} - (1-(0.5^3+0.3^3+0.4^3))^{5\times0.21} - (0.3^3)^{5\times0.21}}, \\ \sqrt[3]{1-(1-0.3^3)^{5\times0.21}}, \sqrt[3]{1-(1-0.4^3)^{5\times0.21}} \end{array} \right)$$
$$= (0.5141, 0.2885, 0.4063)$$

$$P_5 = \left(\begin{array}{c} \sqrt[3]{(1-0.5^3)^{5\times0.23} - (1-(0.6^3+0.4^3+0.5^3))^{5\times0.23} - (0.4^3)^{5\times0.23}}, \\ \sqrt[3]{1-(1-0.4^3)^{5\times0.23}}, \sqrt[3]{1-(1-0.5^3)^{5\times0.23}} \end{array} \right)$$
$$(0.6422, 0.3486, 0.5221)$$

The score values of these numbers were obtained as $SC(P_1) = 0.0981$, $SC(P_2) = 0.6943$, $SC(P_3) = -0.0008$, $SC(P_4) = 0.0688$, $SC(P_5) = 0.1225$, and, based on score values, we had the following arrangement:

$$P_{\sigma(1)} = (0.9144, 0.4665, 0.4125),$$
$$P_{\sigma(2)} = (0.6422, 0.3486, 0.5221),\ P_{\sigma(3)} = (0.6656, 0.5359, 0.5816),$$
$$P_{\sigma(4)} = (0.5141, 0.2885, 0.4063), P_{\sigma(5)} = (0, 0, 0.0928)$$

Now, by using the definition of the T-SFHGIA operator, we found

$$\mathrm{T-SFHGIA}_{\omega,\mathrm{w}}(\mathrm{P}_1, \mathrm{P}_2, \mathrm{P}_3, \mathrm{P}_4, \mathrm{P}_5) = (0.8375, 0.4223, 0.4928) \tag{8}$$

Clearly, the aggregated value obtained in Equation (8) was an improvement of the one obtained in Equation (7), as it incorporated the zero values occurring in the membership and abstinence of T-SFNs efficiently. The analysis of Equations (7) and (8) proved the significance of proposed aggregation operators.

4. MADM Approach Based on Proposed Operators

Consider a decision-making problem which consists of a set of alternatives ($Y = \{y_1, y_2, \ldots\ldots, y_l\}$) and set of attributes ($Z = \{z_1, z_2, \ldots\ldots, z_q\}$) associated with weighted vector ($w = (w_1, w_2, \ldots\ldots, w_q)^T$), where $w_k \in (0,1]$ and $\Sigma_{k=1}^{q} w_k = 1$. Suppose every alternative (y_j) is represented by T-SFNs ($P_{jk} = \langle m_{jk}, i_{jk}, n_{jk} \rangle$), which show by which degree alternatives satisfy, neutral, and not satisfy the given attribute. Then, the following steps of the MADM approach, based on the proposed operators, are summarized as follows:

Step 1 Find the value of t for which the information of the decision matrix lies in the T-spherical fuzzy environment.

Step 2 Assume the weighting vector $\omega = (\omega_1, \ldots\ldots, \omega_q)^T$ of $P_{j1}, P_{j2}, \ldots\ldots, P_{jq}$. where $\omega_k \in (0,1]$ and $\Sigma_{k=1}^{q} \omega_k = 1$ we get $P_{jk} = P_{jk}^{l\omega_k}$.

Step 3 By calculating the scores of each attribute of all alternatives, we find:

$$P_{\sigma(j1)}, P_{\sigma(j2)}, \ldots\ldots, P_{\sigma(jk)}$$

Step 4 By using the normal-distribution based method we find w and then aggregate the data using the T-SFHGIA operator.

Step 5 Find the scores of all alternatives.

Step 6 With the help of score values, we find the best option.

5. Numerical Example

The above-mentioned approach has been illustrated with a real-life decision-making problem under the T-SFS environment, and obtained results have been compared with the other existing results.

5.1. Case Study

Jharkhand is the eastern state of India, which has 40 percent of the mineral resources of the country, and is the second leading state in terms of mineral wealth, after Chhattisgarh state. It is also known for its vast forest resources. Jamshedpur, Bokaro, and Dhanbad, cities in Jharkhand, are famous for industries from all over the world. After that, it is known as being the state in India that has widespread poverty state, because it is primarily a rural state, as 76 percent of the population lives in villages that depend on agriculture and wages from agriculture. Only 30 percent of the villages are connected by roads, and only 55 percent of the villages have access to electricity and other facilities. But in the today's life, many are looking for ways to make changes in order to better their lives, and, accordingly, many move to the urban cities for better jobs. To stop this emigration, the Jharkhand government wants to set up agricultural-based industries in the rural areas. For this, the government organized the "Momentum Jharkhand" global investor summit 2017, in Ranchi, to invite companies to invest in the rural areas. The government announced the various facilities that were available to be set up as five food processing plants in the rural areas, and the five attributes required for selection of the companies to set them up, namely, project cost (Q_1), technical capability (Q_2), financial status (Q_3), company background (Q_4), and other factors (Q_5). The three companies that were interested in this projects, Surya Food and Agro Pvt. Ltd. (s_1), Mother Dairy Fruit and Vegetable Pvt. Ltd. (s_2), and Parle Products Ltd. (s_3), were taken as in the form of the alternatives. Then, the main object of the government was to choose the best company among them for the task. In order to fulfill this, a decision maker evaluated these and gave their preferences in the term of T-SFS, and their preference values were summarized in the form of a decision-matrix, shown in Table 1 as follows.

Table 1. Input information related to each alternative.

	Q_1	Q_2	Q_3	Q_4	Q_5
s_1	$(0.7, 0.5, 0.6)$	$(0.9, 0.5, 0.4)$	$(0.4, 0.2, 0.1)$	$(0.5, 0.3, 0.4)$	$(0.6, 0.4, 0.5)$
s_2	$(0.5, 0.4, 0.6)$	$(0.7, 0.2, 0.3)$	$(0.5, 0.3, 0.6)$	$(0.4, 0.1, 0.6)$	$(0.5, 0.2, 0.4)$
s_3	$(0.4, 0.1, 0.2)$	$(0.5, 0.4, 0.1)$	$(0, 0, 0.5)$	$(0.6, 0.2, 0.2)$	$(0.6, 0.1, 0.5)$

The given problem was solved using two approaches. First it was solved using new interactive operators showing their applicability. Then it was solved using geometric aggregation operators proposed in [34], showing their failure.

Solution using proposed operators:

Step 1 With some calculations, it was found that all the values in Table 1 were T-SFNs for $t = 3$.

Step 2 By taking $\omega = (0.18, 0.22, 0.16, 0.21, 0.23)^T$ we found P_{jk} and their values were summarized as below.

$$\begin{array}{cccccc} & k=1 & k=2 & k=3 & k=4 & k=5 \\ j=1 & \begin{pmatrix} 0.6656, \\ 0.5359, \\ 0.5816 \end{pmatrix} & \begin{pmatrix} 0.9144, \\ 0.4665, \\ 0.4125 \end{pmatrix} & \begin{pmatrix} 0.3333, \\ 0.2759, \\ 0.0928 \end{pmatrix} & \begin{pmatrix} 0.5141, \\ 0.2825, \\ 0.4063 \end{pmatrix} & \begin{pmatrix} 0.6422, \\ 0.3486, \\ 0.5221 \end{pmatrix} \\ j=2 & \begin{pmatrix} 0.4520, \\ 0.4384, \\ 0.5816 \end{pmatrix} & \begin{pmatrix} 0.7194, \\ 0.1703, \\ 0.3095 \end{pmatrix} & \begin{pmatrix} 0.4212, \\ 0.3817, \\ 0.5614 \end{pmatrix} & \begin{pmatrix} 0.4053, \\ 0.0891, \\ 0.6086 \end{pmatrix} & \begin{pmatrix} 0.5264, \\ 0.1571, \\ 0.4184 \end{pmatrix} \\ j=3 & \begin{pmatrix} 0.3843, \\ 0.1259, \\ 0.1931 \end{pmatrix} & \begin{pmatrix} 0.5397, \\ 0.3650, \\ 0.1032 \end{pmatrix} & \begin{pmatrix} 0, \\ 0, \\ 0.4662 \end{pmatrix} & \begin{pmatrix} 0.6104, \\ 0.1845, \\ 0.2033 \end{pmatrix} & \begin{pmatrix} 0.6209, \\ 0.0708, \\ 0.5221 \end{pmatrix} \end{array}$$

Step 3 Now we had to find the score of each attribute of all alternatives, and their computed values were given as below

	k = 1	k = 2	k = 3	k = 4	k = 5
j = 1	0.0981	0.6943	0.0362	0.0688	0.1225
j = 2	−0.1043	0.3426	−0.1021	−0.1589	0.0726
j = 3	0.0495	0.1561	−0.1013	0.2190	0.0970

By comparing the score values, we had

$$\begin{aligned} SC(P_{12}) > SC(P_{15}) > SC(P_{11}) > SC(P_{14}) > SC(P_{13}) \\ SC(P_{22}) > SC(P_{25}) > SC(P_{23}) > SC(P_{21}) > SC(P_{24}) \\ SC(P_{34}) > SC(P_{32}) > SC(P_{35}) > SC(P_{31}) > SC(P_{33}) \end{aligned}$$

Based on above score analysis, we found $P_{\sigma(jk)}$ and summarized them as

$$\begin{array}{cccccc} & k=1 & k=2 & k=3 & k=4 & k=5 \\ j=1 & \begin{pmatrix} 0.9144, \\ 0.5150, \\ 0.4125 \end{pmatrix} & \begin{pmatrix} 0.6422, \\ 0.9857, \\ 0.5221 \end{pmatrix} & \begin{pmatrix} 0.6656, \\ 0.4838, \\ 0.5816 \end{pmatrix} & \begin{pmatrix} 0.5141, \\ 0.3048, \\ 0.4063 \end{pmatrix} & \begin{pmatrix} 0.3333, \\ 0.1857, \\ 0.0928 \end{pmatrix} \\ j=2 & \begin{pmatrix} 0.7194, \\ 0.2064, \\ 0.3095 \end{pmatrix} & \begin{pmatrix} 0.5264, \\ 0.9987, \\ 0.4184 \end{pmatrix} & \begin{pmatrix} 0.4212, \\ 0.2787, \\ 0.5614 \end{pmatrix} & \begin{pmatrix} 0.4520, \\ 0.9804, \\ 0.5816 \end{pmatrix} & \begin{pmatrix} 0.4053, \\ 0.1016, \\ 0.6086 \end{pmatrix} \\ j=3 & \begin{pmatrix} 0.6104, \\ 0.2033, \\ 0.2033 \end{pmatrix} & \begin{pmatrix} 0.5397, \\ 0.4125, \\ 0.1032 \end{pmatrix} & \begin{pmatrix} 0.6209, \\ 0.9999, \\ 0.5221 \end{pmatrix} & \begin{pmatrix} 0.3843, \\ 0.0966, \\ 0.1931 \end{pmatrix} & \begin{pmatrix} 0, \\ 0, \\ 0.4662 \end{pmatrix} \end{array}$$

Step 4 By using the normal distribution-based method, we got $w = (0.1117, 0.2365, 0.3036, 0.2365, 0.1117)^T$, and by using the defined aggregation operators, we had

$$P_1 = T - SFHGIA_{\omega,w}(P_{11}, P_{12}, P_{13}, P_{14}, P_{15})$$
$$= \left(\begin{array}{c} \sqrt[3]{\prod_{j=1}^{5}\left(1 - n^3_{\widetilde{P}_{\sigma(1k)}}\right)^{w_j} - \prod_{j=1}^{5}\left(1 - \left(m^3_{\widetilde{P}_{\sigma(1k)}} + i^3_{\widetilde{P}_{\sigma(1k)}} + n^3_{\widetilde{P}_{\sigma(1k)}}\right)\right)^{w_j}} - \prod_{j=1}^{5}\left(i^3_{\widetilde{P}_{\sigma(1k)}}\right)^{w_j}, \\ \sqrt[3]{1 - \prod_{j=1}^{5}\left(1 - i^3_{\widetilde{P}_{\sigma(1k)}}\right)^{w_j}}, \sqrt[3]{1 - \prod_{j=1}^{5}\left(1 - n^3_{\widetilde{P}_{\sigma(1k)}}\right)^{w_j}} \end{array} \right)$$
$$= (0.9380, 0.4264, 0.4928)$$

$$P_2 = T - SFHGIA_{\omega,w}(P_{21}, P_{22}, P_{23}, P_{24}, P_{25})$$
$$= \left(\begin{array}{c} \sqrt[3]{\prod_{j=1}^{5}\left(1 - n^3_{\widetilde{P}_{\sigma(2k)}}\right)^{w_j} - \prod_{j=1}^{5}\left(1 - \left(m^3_{\widetilde{P}_{\sigma(2k)}} + i^3_{\widetilde{P}_{\sigma(2k)}} + n^3_{\widetilde{P}_{\sigma(2k)}}\right)\right)^{w_j}} - \prod_{j=1}^{5}\left(i^3_{\widetilde{P}_{\sigma(2k)}}\right)^{w_j}, \\ \sqrt[3]{1 - \prod_{j=1}^{5}\left(1 - i^3_{\widetilde{P}_{\sigma(2k)}}\right)^{w_j}}, \sqrt[3]{1 - \prod_{j=1}^{5}\left(1 - n^3_{\widetilde{P}_{\sigma(2k)}}\right)^{w_j}} \end{array} \right)$$
$$= (0.9420, 0.3390, 0.5296)$$

$$P_3 = T - SFHGIA_{\omega,w}(P_{31}, P_{32}, P_{33}, P_{34}, P_{35})$$

$$= \left(\begin{array}{c} \sqrt[3]{\prod_{j=1}^{5}\left(1 - n^3_{\widetilde{P}_{\sigma(3k)}}\right)^{w_j} - \prod_{j=1}^{5}\left(1 - \left(m^3_{\widetilde{P}_{\sigma(3k)}} + i^3_{\widetilde{P}_{\sigma(3k)}} + n^3_{\widetilde{P}_{\sigma(3k)}}\right)\right)^{w_j} - \prod_{j=1}^{5}\left(i^3_{\widetilde{P}_{\sigma(3k)}}\right)^{w_j}} \\ \sqrt[3]{1 - \prod_{j=1}^{5}\left(1 - i^3_{\widetilde{P}_{\sigma(3k)}}\right)^{w_j}}, \sqrt[3]{1 - \prod_{j=1}^{5}\left(1 - n^3_{\widetilde{P}_{\sigma(3k)}}\right)^{w_j}}, \end{array} \right)$$

$$= (0.9779, 0.9713, 0.3906)$$

Step 5 The score values of three alternatives based on their aggregated values were computed as $SC(P_1) = 0.7056$, $SC(P^2) = 0.6874$, and $SC(P_3) = 0.8813$.

Step 6 By comparing score values, we got

$$SC(P_3) > SC(P_1) > SC(P_2)$$

The comparison of score values indicated that P_3 had a greater score value. So, the third company was the best option. Thus, by using the new geometric interaction averaging operators a MADM problem was successfully solved.

Solution using aggregation operators proposed in [34]:

Step 1 The input preferences related to each alternative was summarized in Table 1 for $t = 3$.

Step 2 By using weight vector $\omega = (0.18, 0.22, 0.16, 0.21, 0.23)^T$ we found P'_{jk} as follows

	k = 1	k = 2	k = 3	k = 4	k = 5
j = 1	(0.7054, 0.5359, 0.5816)	(0.9041, 0.4665, 0.4125)	(0.4655, 0.2759, 0.0928)	(0.4874, 0.2825 0.4063)	(0.5738, 0.3486, 0.5221)
j = 2	(0.5180, 0.4384, 0.5816)	(0.6776, 0.1703, 0.3095)	(0.7330, 0.3817, 0.5614)	(0.3826, 0.0891, 0.6086)	(0.4553, 0.1517, 0.4184)
j = 3	(0.4370, 0.1259, 0.1931)	(0.4811, 0.3650, 0.1032)	(0, 0, 0.4662)	(0.5863, 0.1845, 0.2033)	(0.5563, 0.0708, 0.5221)

Step 3 Now, we had to find the score of each attribute of all alternatives.

	k = 1	k = 2	k = 3	k = 4	k = 5
j = 1	0.1543	0.6689	0.1000	0.0487	0.0466
j = 2	−0.0577	0.2815	0.2169	−0.1695	0.0212
j = 3	0.0762	0.1103	−0.1013	0.1932	0.0298

By comparing the score values, we had

$$SC(P'_{12}) > SC(P'_{11}) > SC(P'_{13}) > SC(P'_{14}) > SC(P'_{15})$$
$$SC(P'_{22}) > SC(P'_{23}) > SC(P'_{25}) > SC(P'_{21}) > SC(P'_{24})$$
$$SC(P'_{34}) > SC(P'_{32}) > SC(P'_{31}) > SC(P'_{35}) > SC(P'_{33})$$

Based on above score analysis, we found $P'_{\sigma(jk)}$

$$
\begin{array}{cccccc}
 & k=1 & k=2 & k=3 & k=4 & k=5 \\
j=1 & \begin{pmatrix} 0.9041, \\ 0.4665, \\ 0.4125 \end{pmatrix} & \begin{pmatrix} 0.7054, \\ 0.5359, \\ 0.5816 \end{pmatrix} & \begin{pmatrix} 0.4655, \\ 0.2759, \\ 0.0928 \end{pmatrix} & \begin{pmatrix} 0.4874, \\ 0.2825, \\ 0.4063 \end{pmatrix} & \begin{pmatrix} 0.5738, \\ 0.3486, \\ 0.5221 \end{pmatrix} \\
j=2 & \begin{pmatrix} 0.6776, \\ 0.1703, \\ 0.3095 \end{pmatrix} & \begin{pmatrix} 0.7330, \\ 0.3817, \\ 0.5614 \end{pmatrix} & \begin{pmatrix} 0.4553, \\ 0.1571, \\ 0.4184 \end{pmatrix} & \begin{pmatrix} 0.5180, \\ 0.4384, \\ 0.5816 \end{pmatrix} & \begin{pmatrix} 0.3826, \\ 0.0891, \\ 0.6086 \end{pmatrix} \\
j=3 & \begin{pmatrix} 0.5863, \\ 0.1845, \\ 0.2033 \end{pmatrix} & \begin{pmatrix} 0.4811, \\ 0.3650, \\ 0.1032 \end{pmatrix} & \begin{pmatrix} 0.4370, \\ 0.1259, \\ 0.1931 \end{pmatrix} & \begin{pmatrix} 0.5563, \\ 0.0708, \\ 0.5221 \end{pmatrix} & \begin{pmatrix} 0, \\ 0, \\ 0.4662 \end{pmatrix}
\end{array}
$$

Step 4 By using the normal distribution-based method, we got $w = (0.1117, 0.2365, 0.3036, 0.2365, 0.1117)^T$, and by using the defined aggregation operators, we had

$$P'_1 = T - SFHGIA_{\omega,w}(P'_{11}, P'_{12}, P'_{13}, P'_{14}, P'_{15})$$

$$= \left(\sqrt[3]{\prod_{j=1}^{5}\left(m^3_{\widetilde{P}_{\sigma(1k)}} + i^3_{\widetilde{P}_{\sigma(1k)}}\right)^{w_j} - \prod_{j=1}^{5}(i^3_{\widetilde{P}_{\sigma(1k)}})^{w_j}},\ \prod_{j=1}^{5}\left(i_{\widetilde{P}_{\sigma(1k)}}\right)^{w_j},\ \sqrt[3]{1 - \prod_{j=1}^{5}\left(1 - n^3_{\widetilde{P}_{\sigma(1k)}}\right)^{w_j}} \right)$$

$$= (0.5750, 0.3533, 0.4473)$$

$$P'_2 = T - SFHGIA_{\omega,w}(P'_{21}, P'_{22}, P'_{23}, P'_{24}, P'_{25})$$

$$= \left(\sqrt[3]{\prod_{j=1}^{5}\left(m^3_{\widetilde{P}_{\sigma(2k)}} + i^3_{\widetilde{P}_{\sigma(2k)}}\right)^{w_j} - \prod_{j=1}^{5}(i^3_{\widetilde{P}_{\sigma(2k)}})^{w_j}},\ \prod_{j=1}^{5}\left(i_{\widetilde{P}_{\sigma(2k)}}\right)^{w_j},\ \sqrt[3]{1 - \prod_{j=1}^{5}\left(1 - n^3_{\widetilde{P}_{\sigma(2k)}}\right)^{w_j}} \right)$$

$$= (0.5384, 0.1970, 0.5721)$$

$$P'_3 = T - SFHGIA_{\omega,w}(P'_{31}, P'_{32}, P'_{33}, P'_{34}, P'_{35})$$

$$= \left(\sqrt[3]{\prod_{j=1}^{5}\left(m^3_{\widetilde{P}_{\sigma(3k)}} + i^3_{\widetilde{P}_{\sigma(3k)}}\right)^{w_j} - \prod_{j=1}^{5}(i^3_{\widetilde{P}_{\sigma(3k)}})^{w_j}},\ \prod_{j=1}^{5}\left(i_{\widetilde{P}_{\sigma(3k)}}\right)^{w_j},\ \sqrt[3]{1 - \prod_{j=1}^{5}\left(1 - n^3_{\widetilde{P}_{\sigma(3k)}}\right)^{w_j}} \right)$$

$$= (0,\ 0,\ 0.3692)$$

This seems meaningless because membership and abstinence of only one T-SFN is zero, but existing operators make a whole aggregated value zero.

Step 5 This step involved the computation of score values:

$$SC(P_1) = 0.1006$$
$$SC(P_2) = -0.0312$$
$$SC(P_3) = -0.0503$$

Step 6 By comparing score values, we got

$$SC(P_1) > SC(P_2) > SC(P_3)$$

From the above example, the applicability of the proposed operators could easily be checked by comparing the results obtained using new and existing geometric aggregation operators. It was noticed that whenever membership and abstinence of one TSFN became zero, then the aggregated value using existing aggregation operators seemed impractical. However, the aggregated value using new geometric interactive aggregation operators seemed significant and consistent.

5.2. Advantages of the Proposed Work

In this section, we prove the generalization of proposed work over the existing literature. Here we observed that under some certain conditions the proposed aggregation operators became the existing aggregation operators under different environment, which shows the superiority of our proposed work.

Consider the T-SFWGIA operator defined as

$$T-SFWGIA_w(P_1,P_2,\ldots\ldots,P_k)=\left(\begin{array}{c}\sqrt[t]{\prod_{j=1}^{k}\left(1-n_j^t\right)^{w_j}-\prod_{j=1}^{k}\left(1-\left(m_j^t+i_j^t+n_j^t\right)\right)^{w_j}-\prod_{j=1}^{k}(i_j^t)^{w_j}},\\ \sqrt[t]{1-\prod_{j=1}^{k}\left(1-i_j^t\right)^{w_j}},\sqrt[t]{1-\prod_{j=1}^{k}\left(1-n_j^t\right)^{w_j}}\end{array}\right) \quad (9)$$

(1) If we take $t=2$, the Equation (9) becomes spherical fuzzy weighted geometric interaction averaging operator (SFWGIA operator) and we have

$$SFWGIA_w(P_1,P_2,\ldots\ldots,P_k)=\left(\begin{array}{c}\sqrt{\prod_{j=1}^{k}\left(1-n_j^2\right)^{w_j}-\prod_{j=1}^{k}\left(1-\left(m_j^2+i_j^2+n_j^2\right)\right)^{w_j}-\prod_{j=1}^{k}(i_j^2)^{w_j}},\\ \sqrt{1-\prod_{j=1}^{k}\left(1-i_j^2\right)^{w_j}},\sqrt{1-\prod_{j=1}^{k}\left(1-n_j^2\right)^{w_j}}\end{array}\right)$$

(2) If we take $t=1$, the Equation (9) becomes picture fuzzy weighted geometric interaction averaging operator (PFWGIA operator) and we have

$$PFWGIA_w(P_1,P_2,\ldots\ldots,P_k)=\left(\begin{array}{c}\prod_{j=1}^{k}(1-n_j)^{w_j}-\prod_{j=1}^{k}(1-(m_j+i_j+n_j))^{w_j}-\prod_{j=1}^{k}(i_j)^{w_j},\\ 1-\prod_{j=1}^{k}(1-i_j)^{w_j},1-\prod_{j=1}^{k}(1-n_j)^{w_j}\end{array}\right)$$

(3) If we take $t=2$ and $i=0$, the Equation (9) becomes Pythagorean fuzzy weighted geometric interaction averaging operator (PyFWGIA operator) and we have

$$PyFWGIA_w(P_1,P_2,\ldots\ldots,P_k)=\left(\begin{array}{c}\sqrt{\prod_{j=1}^{k}\left(1-n_j^2\right)^{w_j}-\prod_{j=1}^{k}\left(1-\left(m_j^2+n_j^2\right)\right)^{w_j}},\\ \sqrt{1-\prod_{j=1}^{k}\left(1-n_j^2\right)^{w_j}}\end{array}\right)$$

(4) If we take $t=1$ and $i=0$, the Equation (9) becomes intuitionistic fuzzy weighted geometric interaction averaging operator (IFWGIA operator) and we have

$$IFWGIA_w(P_1,P_2,\ldots\ldots,P_k)=\left(\begin{array}{c}\prod_{j=1}^{k}(1-n_j)^{w_j}-\prod_{j=1}^{k}(1-(m_j+n_j))^{w_j},\\ 1-\prod_{j=1}^{k}(1-n_j)^{w_j}\end{array}\right)$$

Similarly, T-SFOWGIA and T-SFHGIA operators can be converted to the existing operators. All of this clearly indicated that our proposed work could be used in the problems described in existing literature, but the operators of existing literature are unable to deal with problems of T-spherical fuzzy information. For example, if we look at Example 2, it can be seen that none of the existing operators can be applied to such problems where information is in the form of T-SFNs.

5.3. Comparative Analysis

The significance of the proposed new geometric operators lies in the fact that the result obtained by using these operations were more justifiable than those developed earlier (i.e., [34,37,38]). Such operators could not deal with situations where if membership and abstinence value of any number becomes zero then the membership and abstinence value of their aggregated value is also zero. Hence the existing operators of PFSs and T-SFSs did possess the capability of dealing with any kinds of information. But, on the other hand, the new geometric operators of T-SFSs can deal with any type of data justifiably. This point is demonstrated in the case study described in Section 5.1.

The second main advantage of our proposed work is that it has the ability to aggregate the data available in the form of IFSs, PyFSs, PFSs, and SFSs. But, conversely, the existing operators could not handle the data provided in the T-spherical fuzzy environment. For example, if we look at Example 2, its data is purely in the form of T-SFNs based on four grades, being membership, abstinence, non-membership, and refusal degree with $t = 3$, which shows that the aggregation operators of IFSs, PyFSs, PFSs, and SFSs could not aggregate this data. But if we look at Example 3, its data is in the form of IFNs, and our proposed operators easily aggregated this type of data with $t = 1$ and $i = 0$.

Hence, by all means, the proposed work had superiority over the existing work.

Example 3. *Let* $P_1 = (0, 0.5), P_2 = (0.5, 0.4), P_3 = (0.4, 0.2), P_4 = (0.3, 0.3)$ *and* $P_5 = (0.7, 0.1) \in IFN$. *The weight vector for* $P_i (i = 1, 2, \ldots, 5)$ *is* $\omega = (0.18, 0.22, 0.16, 0.21, 0.23)^T$.

$$\begin{aligned}
P_1 &= \left((1-0.5)^{5\times 0.18} - (1-(0+0.5))^{5\times 0.18}, 1-(1-0.5)^{5\times 0.18}\right) \\
&= (0, 0.5796) \\
P_2 &= \left((1-0.4)^{5\times 0.22} - (1-(0.5+0.4))^{5\times 0.22}, 1-(1-0.4)^{5\times 0.22}\right) \\
&= (0.5039, 0.3183) \\
P_3 &= \left((1-0.2)^{5\times 0.16} - (1-(0.4+0.2))^{5\times 0.16}, 1-(1-0.2)^{5\times 0.16}\right) \\
&= (0.4000, 0.2000) \\
P_4 &= \left((1-0.3)^{5\times 0.21} - (1-(0.3+0.3))^{5\times 0.21}, 1-(1-0.3)^{5\times 0.21}\right) \\
&= (0.2870, 0.2746) \\
P_5 &= \left((1-0.1)^{5\times 0.23} - (1-(0.7+0.1))^{5\times 0.23}, 1-(1-0.1)^{5\times 0.23}\right) \\
&= (0.7203, 0.1094)
\end{aligned}$$

Scores values were

$$SC(P_1) = -0.5796,\ SC(P_2) = 0.1856,\ SC(P_3) = 0.2000,\ SC(P_4) = 0.0125,\ SC(P_5) = 0.6109.$$

Thus, $SC(P_5) > SC(P_3) > SC(P_2) > SC(P_4) > SC(P_1)$ and we had

$$\begin{aligned}
P_{\sigma(1)} &= (0.7203, 0.1094) \\
P_{\sigma(2)} &= (0.4000, 0.2000) \\
P_{\sigma(3)} &= (0.5039, 0.3183) \\
P_{\sigma(4)} &= (0.2870, 0.2746) \\
P_{\sigma(5)} &= (0, 0.5796)
\end{aligned}$$

By using the normal distribution-based method, we found $w = (0.1117, 0.2365, 0.3036, 0.2365, 0.1117)^T$.

Now, by using the definition of the T-SFHGIA operator, we found

$$T - SFHGIA_{\omega,w}(P_1, P_2, P_3, P_4, P_5) = (0.4093, 0.2919)$$

Here we got the same result as in [9,10,39]. Thus, the proposed new operators had the capability to solve the problems that lie in the existing structures.

6. Conclusions

In this manuscript, we utilized the concept of T-SFS to handle the uncertainty in the data, so as to capture the information with some more degree of freedom. For it, we defined some new, improved interactive aggregation operations by adding the degree of refusal into the analysis. Then, we studied some basic properties of them. Based on these operational laws, we defined some new weighted geometric aggregation operators and studied their desirable properties. Some of the counter examples were also provided, which showed that the proposed operators worked well in all cases where the existing ones failed to classify the objects. In addition to this, in a comprehensive scrutiny of T-SFSs and the decision-maker preferences, a MADM approach was presented, based on the proposed operator, to select the best alternatives among the feasible ones. Finally, the presented decision-making approach was explained with the help of a numerical example, and an extensive comparative analysis was conducted in relation to the existing decision-making theories. Additionally, the advantages as well as the superiority of the approach was tested with some examples. The advantages of the proposed operators were that a decision maker could choose the required operator in order to optimize their desired goals with more confidence level as compared the existing operators. Furthermore, it was concluded that the several existing operators could be deduced from the proposed one and, hence, the presented operators and algorithm were more generalized. In the future, there is the scope to extend the proposed method to some different environments, and to extend its application in various fields related to decision-theory [40–47].

Author Contributions: Conceptualization, H.G., K.U. and N.J.; Methodology, M.M., K.U., T.M. and N.J.; Validation, H.G.; Formal analysis, M.M., K.U., T.M. and N.J.; Investigation, M.M., K.U., T.M. and N.J.; Writing-original draft preparation, M.M., K.U., T.M. and N.J.; Writing-review and editing, H.G.; Visualization, H.G.

Funding: This research received no external funding.

Conflicts of Interest: The authors declare no conflict of interest.

References

1. Zadeh, L.A. Fuzzy sets. *Inf. Control.* **1965**, *8*, 338–353. [CrossRef]
2. Atanassov, K.T. Intuitionistic fuzzy sets. *Fuzzy Sets Syst.* **1986**, *20*, 87–96. [CrossRef]
3. Atanassov, K.; Gargov, G. Interval-valued intuitionistic fuzzy sets. *Fuzzy Sets Syst.* **1989**, *31*, 343–349. [CrossRef]
4. Yager, R.R. Pythagorean fuzzy subsets. In Proceedings of the Joint IFSA World Congress and NAFIPS Annual Meeting, Edmonton, AB, Canada, 24–28 June 2013; pp. 57–61.
5. Yager, R.R. Pythagorean membership grades in multicriteria decision making. *IEEE Trans. Fuzzy Syst.* **2014**, *22*, 958–965. [CrossRef]
6. Cuong, B.C. *Picture Fuzzy Sets-First Results. Part 1, Seminar "Neuro-Fuzzy Systems with Applications*; Institute of Mathematics: Hanoi, Vietnam, 2013.
7. Cuong, B.C. *Picture Fuzzy Sets-First Results. Part 2, Seminar "Neuro-Fuzzy Systems with Applications*; Institute of Mathematics: Hanoi, Vietnam, 2013.
8. Xu, Z.S. Intuitionistic fuzzy aggregation operators. *IEEE Trans. Fuzzy Syst.* **2007**, *15*, 1179–1187.
9. Garg, H. Generalized intuitionistic fuzzy interactive geometric interaction operators using Einstein t-norm and t-conorm and their application to decision making. *Comput. Ind. Eng.* **2016**, *101*, 53–69. [CrossRef]
10. Garg, H. Novel intuitionistic fuzzy decision making method based on an improved operation laws and its application. *Eng. Appl. Artif. Intell.* **2017**, *60*, 164–174. [CrossRef]
11. Wang, W.; Liu, X. Interval-valued intuitionistic fuzzy hybrid weighted averaging operator based on Einstein operation and its application to decision making. *J. Intell. Fuzzy Syst.* **2013**, *25*, 279–290.
12. Wang, W.; Liu, X. The multi-attribute decision making method based on interval-valued intuitionistic fuzzy Einstein hybrid weighted geometric operator. *Comput. Math. Appl.* **2013**, *66*, 1845–1856. [CrossRef]
13. Wang, W.; Liu, X.; Qin, Y. Interval-valued intuitionistic fuzzy aggregation operators. *J. Syst. Eng. Electron.* **2012**, *23*, 574–580. [CrossRef]

14. Garg, H. A new generalized Pythagorean fuzzy information aggregation using Einstein operations and its application to decision making. *Int. J. Intell. Syst.* **2016**, *31*, 886–920. [CrossRef]
15. Garg, H. Generalized Pythagorean fuzzy geometric interactive aggregation operators using Einstein operations and their application to decision making. *J. Exp. Theor. Artif. Intell.* **2018**, 1–32. [CrossRef]
16. Xu, Z.S.; Xia, M. Induced generalized intuitionistic fuzzy operators. *Knowl. Based Syst.* **2003**, *24*, 197–209. [CrossRef]
17. Garg, H.; Kumar, K. An advanced study on the similarity measures of intuitionistic fuzzy sets based on the set pair analysis theory and their application in decision making. *Soft Comput.* **2018**, *22*, 4959–4970. [CrossRef]
18. Wei, G.-W.; Xu, X.-R.; Deng, D.-X. Interval-valued dual hesitant fuzzy linguistic geometric aggregation operators in multiple attribute decision making. *Int. J. Knowl.-Based Intell. Eng. Syst.* **2016**, *20*, 189–196. [CrossRef]
19. Peng, X.; Dai, J.; Garg, H. Exponential operation and aggregation operator for q-rung orthopair fuzzy set and their decision-making method with a new score function. *Int. J. Intell. Syst.* **2018**, *33*, 2255–2282. [CrossRef]
20. Garg, H. Some robust improved geometric aggregation operators under interval-valued intuitionistic fuzzy environment for multi-criteria decision-making process. *J. Ind. Manag. Optim.* **2018**, *14*, 283–308. [CrossRef]
21. Liu, P. Some frank aggregation operators for interval-valued intuitionistic fuzzy numbers and their application to group decision making. *J. Mult.-Valued Logic Soft Comput.* **2017**, *29*, 183–223.
22. Chen, S.M.; Cheng, S.H.; Tsai, W.H. Multiple attribute group decision making based on interval-valued intuitionistic fuzzy aggregation operators and transformation techniques of interval—Valued intuitionistic fuzzy values. *Inf. Sci.* **2016**, *367–368*, 418–442. [CrossRef]
23. Garg, H. Some arithmetic operations on the generalized sigmoidal fuzzy numbers and its application. *Granul. Comput.* **2018**, *3*, 9–25. [CrossRef]
24. Chen, S.M.; Cheng, S.H.; Chiou, C.H. Fuzzy multiattribute group decision making based on intuitionistic fuzzy sets and evidential reasoning methodology. *Inf. Fusion* **2016**, *27*, 215–227. [CrossRef]
25. Garg, H. A new generalized improved score function of interval-valued intuitionistic fuzzy sets and applications in expert systems. *Appl. Soft Comput.* **2016**, *38*, 988–999. [CrossRef]
26. Wei, G.W. Interval valued hesitant fuzzy uncertain linguistic aggregation operators in multiple attribute decision making. *Int. J. Mach. Learn. Cybern.* **2016**, *7*, 1093–1114. [CrossRef]
27. Kumar, K.; Garg, H. Connection number of set pair analysis based TOPSIS method on intuitionistic fuzzy sets and their application to decision making. *Appl. Intell.* **2018**, *48*, 2112–2119. [CrossRef]
28. Kumar, K.; Garg, H. Prioritized linguistic interval-valued aggregation operators and their applications in group decision-making problems. *Mathematics* **2018**, *6*, 209. [CrossRef]
29. Adrian, C.; Abdullah, R.; Atan, R.; Jusoh, Y.Y. Conceptual model development of big data analytics implementation assessment effect on decision-making. *Int. J. Interact. Multimed. Artif. Intell.* **2017**, *4*, 28–33. [CrossRef]
30. Morente-Molinera, J.A.; Kou, G.; González-Crespo, R.; Corchado, J.M.; Herrera-Viedma, E. Solving multi-criteria group decision making problems under environments with a high number of alternatives using fuzzy ontologies and multi-granular linguistic modelling methods. *Knowl.-Based Syst.* **2017**, *137*, 54–64. [CrossRef]
31. Khan, M.U.; Mahmood, T.; Ullah, K.; Jan, N.; Deli, I.; Khan, Q. Some aggregation operators for bipolar-valued hesitant fuzzy information based on Einstein operational laws. *J. Eng. Appl. Sci.* **2017**, *36*, 63–72.
32. Garg, H. An improved cosine similarity measure for intuitionistic fuzzy sets and their applications to decision-making process. *Hacet. J. Math. Stat.* **2018**, *47*, 1585–1601. [CrossRef]
33. Garg, H. Some methods for strategic decision-making problems with immediate probabilities in Pythagorean fuzzy environment. *Int. J. Intell. Syst.* **2018**, *33*, 687–712. [CrossRef]
34. Mahmood, T.; Ullah, K.; Khan, Q.; Jan, N. An approach toward decision-making and medical diagnosis problems using the concept of spherical fuzzy sets. *Neural Comput. Appl.* **2018**, 1–13. [CrossRef]
35. Ullah, K.; Mahmood, T.; Jan, N. Similarity measures for t-spherical fuzzy sets with applications in pattern recognition. *Symmetry* **2018**, *10*, 193. [CrossRef]
36. Xu, Z.S.; Yager, R.R. Some geometric aggregation operators based on intuitionistic fuzzy sets. *Int. J. Gen. Syst.* **2006**, *35*, 417–433. [CrossRef]

37. Wei, G. Picture fuzzy aggregation operators and their application to multiple attribute decision making. *J. Intell. Fuzzy Syst.* **2017**, *33*, 713–724. [CrossRef]
38. Garg, H. Some picture fuzzy aggregation operators and their applications to multicriteria decision-making. *Arab. J. Sci. Eng.* **2017**, *42*, 5275–5290. [CrossRef]
39. He, Y.; Chen, H.; Zhau, L.; Liu, J.; Tao, Z. Intuitionistic fuzzy geometric interaction averaging operators and their application to multi-criteria decision making. *Inf. Sci.* **2014**, *259*, 142–159. [CrossRef]
40. Garg, H. Linguistic Pythagorean fuzzy sets and its applications in multiattribute decision-making process. *Int. J. Intell. Syst.* **2018**, *33*, 1234–1263. [CrossRef]
41. Garg, H. New logarithmic operational laws and their aggregation operators for Pythagorean fuzzy set and their applications. *Int. J. Intell. Syst.* **2019**, *34*, 82–106. [CrossRef]
42. Garg, H.; Arora, R. Generalized and group-based generalized intuitionistic fuzzy soft sets with applications in decision-making. *Appl. Intell.* **2018**, *48*, 343–356. [CrossRef]
43. Garg, H. New logarithmic operational laws and their applications to multiattribute decision making for single-valued neutrosophic numbers. *Cogn. Syst. Res.* **2018**, *52*, 931–946. [CrossRef]
44. Rani, D.; Garg, H. Complex intuitionistic fuzzy power aggregation operators and their applications in multi-criteria decision-making. *Expert Syst.* **2018**, e12325. [CrossRef]
45. Garg, H.; Rani, D. Complex interval-valued intuitionistic fuzzy sets and their aggregation operators. *Fundam. Inform.* **2019**, *164*, 61–101. [CrossRef]
46. Garg, H. Linguistic single-valued neutrosophic prioritized aggregation operators and their applications to multiple-attribute group decision-making. *J. Ambient. Intell. Humaniz. Comput.* **2018**, *9*, 1975–1997. [CrossRef]
47. Ullah, K.; Mahmood, T.; Jan, N.; Broumi, S.; Khan, Q. On bipolar-valued hesitant fuzzy sets and their applications in multi-attribute decision making. *Nucleus* **2018**, *55*, 93–101.

symmetry

MDPI

Article

Fuzzy AHP Application for Supporting Contractors' Bidding Decision

Agnieszka Leśniak [1,*], Daniel Kubek [1], Edyta Plebankiewicz [1], Krzysztof Zima [1] and Stanisław Belniak [1]

Cracow University of Technology, Faculty of Civil Engineering, 31-155 Kraków, Poland; dkubek@pk.edu.pl (D.K.); eplebank@L3.pk.edu.pl (E.P.); kzima@L3.pk.edu.pl (K.Z.); belniaks@ae.krakow.pl (S.B.)

* Correspondence: alesniak@L3.pk.edu.pl; Tel.: +48-12-628-23-30

Received: 14 October 2018; Accepted: 13 November 2018; Published: 16 November 2018

Abstract: This paper proposes the author's model based on the Fuzzy Analytic Hierarchy Process (FAHP) to improve the efficiency of contractor bidding decisions. The essence of the AHP method is to make pairwise comparisons of available options against all evaluation criteria. The results of these comparisons are recorded in a square matrix in which symmetrical elements are reciprocal. In the expert opinion, a 9-step, bipolar verbal scale was used so that the symmetry of the response was maintained. For contractors from countries where the tendering system is commonly used, the choice of the right tender in which to participate influences their image, financial condition, and their aspiration to succeed. The bid/no bid decision depends on numerous factors associated with the company itself, the environment, and the project concerning the tender. When facing tough competition, contractors search for a solution which increases their chances of winning the tender. The proposed model was based on factors selected by Polish contractors. The original element of the model involves 4 original criteria and 15 sub-criteria for the assessment of investment decision projects to the selection of the most advantageous contract, i.e., the contractor's participation in the bid. For verbal evaluations describing the criteria, symmetric triangular fuzzy numbers were assigned. The authors performed an extended analysis method combined with FAHP in the model. Fuzzy evaluations underwent elaborate analysis, the aim of which was to specify the synthetic priority weights for each criterion. As a result of the application of the method, to prove that the model works, an example from the Polish construction market was presented in which a bid/no bid decision about four possible tenders was to be taken. Despite the considered example applying to Polish conditions, the proposed model can be used also in other countries. The authors' rationale is to produce new and more flexible methodologies in order to realistically model a variety of concrete decision problems.

Keywords: bidding decision; Fuzzy Analytic Hierarchy Process (FAHP); contractors; construction industry; decision making

---

1. Introduction

Efforts to gain a construction project include making two vital decisions by the contractor. The first involves deciding whether to bid or not; the second concerns estimating the offer price as accurately as possible, especially the mark-up value that needs to be specified in the bid. An appropriate selection of tenders in which the company wishes to participate plays an important part in establishing its position in the market, and contributes to the contractor's success. Participation in tenders involving projects that do not fit the company's abilities may cause losses. On the other hand, the cancellation of a bid means losing the opportunity to profit, to establish relationships with new customers, and to expand the company's business. Yet, bidding and losing the bid causes financial loss and damages reputations.

The result of the tender (a win or loss) depends on the value of the bid, especially on the mark-up it includes.

Both of the decisions are complex, dynamic, and involve many factors [1]. A bidding decision, despite its being vital for the contractor, often needs to be made quickly and within a limited timeframe. The contractor typically relies on experience, intuition, and subjective information. To facilitate the contractor's reasoning, increase the efficiency of decision making, and limit mistakes and randomness, decision support models are frequently applied.

The article presents the authors' own proposition of a model supporting bidding decisions. The model is based on the Fuzzy Analytic Hierarchy Process (FAHP), facilitating the selection of the most appropriate projects on which to bid. Unlike the classic Analytic Hierarchy Process (AHP), FAHP uses fuzzy logic, which allows a more accurate evaluation of linguistic criteria. A verbal assessment model is presented in the form of a triangular membership function. The model is constructed on the basis of factors influencing a bidding decision identified in Poland. Limit values for the triangular membership function were adopted in such a way that the adopted values corresponded to the values in accordance with a scale from 1–9, as proposed by Saaty, defining the decision-maker's preferences by means of relative assessments of the validity of sub-criteria and variants. The authors of the article distinguished four main evaluation criteria: Company capabilities, Investment characteristics, Financial conditions, and Tender characteristics; these were divided into sub-criteria. In total, 15 project evaluation sub-criteria were obtained. The main aim of this article is to create a decision support model to join the bidding and selection of the best tender from the point of view of the contractor. In order to explain the procedure in the model, a simple calculation example is also shown.

2. Decision Making Processes in Construction Management

Decisions taken during the planning and preparation of a development project have a crucial impact on its profitability [2]. The models proposed so far proved to be helpful for the participants of the construction process, i.e., the investors [3,4] and contractors [5–7]. Decisions concerning such problems in most cases belong to multi-criteria issues; therefore, solutions to these issues typically involve various multi-criteria decision making methods. Table 1 summarizes the multi-criteria methods previously used to support decisions concerning the management of a construction project at the pre-investment stage.

Table 1. A summary of example methods used to support decision-making in construction management.

Method Name	Aim of Analysis	Number of Criterion Used	Source
Analytic Hierarchy Process (AHP) + PROMETHEE	subcontractor selection for main contractor	13	[8]
Data Envelopment Analysis (DEA)	subcontractor selection at short-listing stage	5–6 selected depending on the specific tender	[9]
Fuzzy AHP; The method of entropy; Method of criterion impact loss (CILOS); Integrated Determination of Objective CRIteria Weights (IDOCRIW) method; The SAW method; The TOPSIS method; The COPRAS method	comparing quality assurance in different contractor contracts	7	[10]
The EDAS method	comparing quality assurance in different contractor contracts	7	[11]
hybrid MCDM model of discrete zero-sum two-person matrix games with grey numbers	delays in Design-Bid-Build projects	8	[12]

Table 1. *Cont.*

Method Name	Aim of Analysis	Number of Criterion Used	Source
Integration of intuitionistic fuzzy sets I(FS) theory, ELECTRE and VIKOR along with Grey Relational Analysis (GRA)	contractor selection problem	20	[13]
Weighted Aggregated Sum Product Assessment with Grey Values (WASPAS-G)	evaluating and selecting contractors	6	[14]

Table 1 reveals that the use of multi-criteria methods in the decision-making processes of business management most frequently concerns the issue of the comparison of quality assurance in various contractor contracts. They include the following methods: fuzzy AHP, entropy, criterion impact loss, SAW, and TOPSIS. Frequently, a multi-criteria analysis is used to support the process of selecting contractors or subcontractors for construction works. The methods used for this purpose encompass AHP, PROMETHEE, Data Envelopment Analysis, ELECTRE, VIKOR, Grey Relational Analysis, and Weighted Aggregated Sum Product Assessment with Grey Values. The study in [15] proposes a hybrid method based on the Weighted Sum Model (WSM) and the Weighted Product Model (WPM). Most of the methods mentioned here are well-known; moreover, as shown by researchers, they may be applied, rendering very good results. Their advantage is that the criteria applied are easy to assess and are understandable for the potential decision maker. What is more, if complex calculations are necessary, their automation is possible. It is worth paying attention to the number of criteria, ranging from 5 to 20 (Table 1), used in analyses. The models mentioned are also applied to bid/no bid decision making processes. There were several early endeavors to develop a model facilitating bidding decisions, one of which was performed by Ahmad [16], who employed the weight model. In the following years, a number of models based on various mathematical devices were created.

One of the more recent models is found in the study by El-Mashaleh [17] presenting a data envelopment analysis (DEA), namely, an efficient non-parametric linear programming method which is applied to benchmarking procedures and selection decision making. DEA uses the contractor's database containing information about previous bidding decisions to create a "best-practice frontier" determined by favorable bidding opportunities. Consequently, the frontier allows us to evaluate new bidding options more efficiently, and to reach a more advantageous bid/no bid decision.

On the other hand, [18] presents an ANFIS model based on a MATLAB software program for processing a set of input data in the way that the human reasoning operates, namely, through neural network learning and fuzzy logic. The results of the analysis proved to be statistically significant.

The study in [19] describes improvements on the existing bid decision-making methods by means of the application of support vector machines and backward elimination regression. In particular, the method helps to attain a parsimonious support vector machine classifier facilitating bid/no bid decision making in offshore oil and gas platform fabrication projects. Then, the output of the support vector machine classifier is compared with other classifiers: the worth evaluation model, linear regression, and neural networks. What the study reveals is the significantly more efficient performance of the support vector machine classifier in comparison with the other methods, thus proving its great explanatory predictive power for bid/no bid decision making. What is more, once the insignificant input variables are removed, the generalization performance of the support vector machines increases. Other attempts to develop an efficient model of bid/no bid decision making include, for example, the fuzzy set theory [20,21], analytic hierarchy process [22,23], game theory [24], multi-criteria analysis methods [25], and artificial neural network [26,27].

In many models, the selection and evaluation of factors influencing the bid/no bid decision occurs prior to the decision making itself, as presented in numerous studies performed in various countries and on a number of markets. Early research concerned the American market in 1988 [28], in which 31 factors were specified. While some of them were found to be very important at the mark-up decision stage but not at the bid/no bid decision stage, other factors proved to be significant at both stages. Another study conducted in Saudi Arabia [29] enumerates 37 factors influencing bid/no bid decisions, as identified by the contractors operating in this market. Research conducted in Great Britain [1] helped to establish 55 potential factors influencing contractors' bidding decisions. Considerably fewer were found by Wanous et al. [27], whose formal questionnaire revealed 38 factors that affect the bid/no bid decision, ranked in accordance of their importance to Syrian contractors. On the other hand, the study performed in Saudi Arabia [30] established as many as 87 potential factors, ranked on the basis of 91 responses to the questionnaire. Similar studies, closely resembling those presented in the aforementioned articles, were performed in Palestine [31] and Australia [32]. One of the most recent studies was completed in Poland [33]. The authors proposed 15 factors, being a selection of the factors proposed in literature, and asked 61 contractors to evaluate them. In this way, a ranking list of these factors in order of their importance and frequency of their appearance in the Polish market was created.

Bageis and Fortune [30] found considerable correlations among a number of studies, since the top positions in various rankings were occupied by similar factors. They were not identical, though, as the specifics of the construction markets differ from country to country, so certain factors appeared to be significant in some regions but not in others. The implication is, therefore, that factors influencing bid/no bid decisions are conditioned by the particulars of the environment and the market in which the contractor works.

The authors proposed the Fuzzy AHP method to solve the problem of supporting contractors' bidding decisions. Although (as shown above) the classical AHP method, like other multicriteria methods, has already been used by other researchers and presented in the literature, it is difficult to find examples of using Fuzzy AHP to solve the problem of bidding decisions. The authors of the present paper introduced 15 criteria identified by Polish contractors, the number of which does not substantially exceed the size of the set of criteria proposed in multicriteria methods (Table 1). The proposed Fuzzy AHP method provides the expert with the possibility of independently performing evaluations of the opinions of other experts. It allows him/her to freely shape opinions without setting any sharp values. In the decision-making process, objective or subjective opinions or information, both quantitative and qualitative, play a very important role, and with the help of AHP, they can be easily assessed. Any amount of information characterizing the main purpose can be mentioned or even structured in this method. The use of the Fuzzy AHP method improves the way experts deliver opinions, without limiting them to one specific wording or parameter. It therefore increases the possibilities of the application of this method and the flexibility of the solutions obtained thereby.

3. Bidding Decision Support Systems Based on Fuzzy AHP—Methodology

The Analytic Hierarchy Process (AHP), developed by Saaty [34], is a method supporting the decision making process. Its aim is to quantify the relative priorities for a particular set of alternatives on a ratio scale on the basis of the decision maker's judgement. The model emphasizes the importance of the decision maker's intuitive judgements, and the consistency with which alternatives are compared in the decision making process [34]. The AHP is combined with other methods or techniques, such as mathematical programming, data envelopment analysis, fuzzy theory, and meta-heuristics [35].

According to bid decisions, most decision makers tend to rely on their knowledge and personal experience, which lead to highly unstructured and uncertain decisions. Although the aim of the AHP is to capture the decision maker's knowledge, the traditional AHP cannot fully reflect the human way of thinking. In the literature, the fuzzy AHP approach is widely used to deal with this inconvenience. The Fuzzy AHP (namely, FAHP) model is based on fuzzy sets theory, in which the membership of the given element is determined by the membership function. Fuzzy decision variable values are described by a membership function which is between zero and one. The membership function defines the degree of truth, that is, the fuzzy decision variable may range between completely true and completely false. This approach is more appropriate when the linguistic variables used are common in the decision process, such as expert judgment. Membership functions may assume various forms:

trapezoid, Gaussian, or triangular. The method described below involves triangular membership functions, as described in Chang's study [36].

Due to the mentioned characteristics of AHP and fuzzy AHP, these methods are widely used in issues related to decision making regarding various aspects of construction management. AHP and fuzzy AHP were applied, for example, to the ranking and selection of alternatives in construction project management [37], construction projects selection and risk assessment [38], performance evaluation of territorial units [39], and the development of an integrated discounting strategy based on vendors' expectations [40].

In the FAHP method, objects (that is, criteria and alternatives) are evaluated by triangular fuzzy values (TFN). The values of the TFN membership function are $\mu_M(x) : R \to [0,\ 1]$, so they generalize the classic Boolean logic. Each triangular fuzzy set is defined unambiguously by three parameters, namely, by triangular fuzzy values (*l*, *m*, *u*) which denote the beginning, middle, and end of the fuzzy triangle, respectively. The value of the membership function μ_M of triangular fuzzy values *M* in the set *R* can be specified by the following dependency:

$$\mu_M = \begin{cases} \frac{x}{m-l} - \frac{l}{m-l}, & x \in [l, m] \\ \frac{x}{m-u} - \frac{u}{m-u}, & x \in [m, u] \\ 0, & \text{in other cases} \end{cases} \quad (1)$$

The function is depicted in Figure 1.

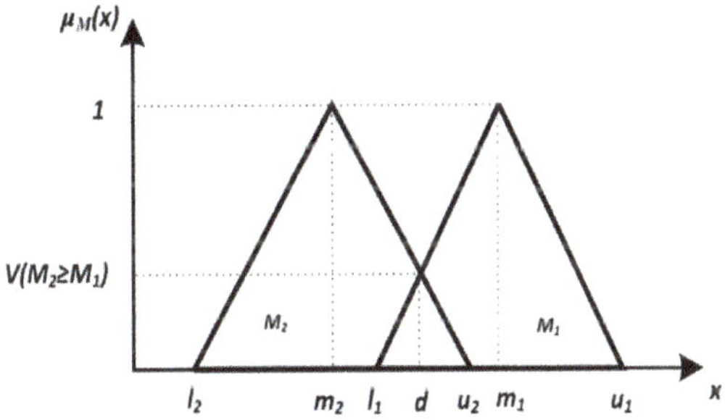

Figure 1. Triangular membership function and intersection between M_1 and M_2.

If $l = m = u$, then it is a conventional crisp value, as in the classic AHP.

Table 2 presents example TFN values. In the following part of the article, this scheme was applied to case study calculations.

Table 2. A fuzzy scheme of preference evaluation [41].

Qualitative Evaluation	Fuzzy Evaluation	AHP Equivalent
Extreme preference	(2; 5/2; 3)	9
Very strong preference	(3/2; 2; 5/2)	7
Strong preference	(1; 3/2; 2)	5
Moderate preference	(1; 1; 3/2)	3
Equal preference	(1; 1; 1)	1
Moderate inferiority	(2/3; 1; 1)	1/3
Strong inferiority	(1/2; 2/3; 1)	1/5
Very strong inferiority	(2/5; $\frac{1}{2}$; 2/3)	1/7
Extreme inferiority	(1/3; 2/5; 1/2)	1/9

As in the classic AHP method, the analysis should begin by designating a criteria priority matrix and alternatives preference matrix for each criterion. In the fuzzy AHP, this step involves TFN.

Fuzzy evaluations undergo an extent analysis, the aim of which is to specify the synthetic priority weights. The analysis consists of the following four steps.

Let $X = \{x_1, x_2, \ldots, x_n\}$ be a set of objects and $U = \{u_1, u_2, \ldots, u_m\}$ a set of aims. According to this method, each objects has to undergo an extent analysis for each aim of the problem. As a result of such extent analysis, m values for each object will be obtained, which will be represented as follows:

$$M_{g_i}^1,\ M_{g_i}^2,\ \ldots,\ M_{g_i}^m,\quad i = 1,\ 2, \ldots, n \tag{2}$$

where all $M_{g_i}^j$ for $j = 1, 2, \ldots, m$ are triangular fuzzy values.

Step 1: Computation of synthetic fuzzy values for each object of the analysis.

If $M_{g_i}^1,\ M_{g_i}^2,\ \ldots,\ M_{g_i}^m$ are the extent analysis values of the i-th object for an m-th aim, then the synthetic fuzzy value can be defined as:

$$S_i = \sum_{j=1}^{m} M_{g_i}^j \odot \left[\sum_{i=1}^{n}\sum_{j=1}^{m} M_{g_i}^j\right]^{-1} \tag{3}$$

Since the FAHP uses three values for the evaluation of a particular criterion, it is necessary to define the arithmetic operations involving these values.

If one assumes two TFNs, $M_1 = (l_1, m_1, u_1)$ and $M_2 = (l_2, m_2, u_2)$, then the operations are as follows:

$$(l_1, m_1, u_1) \oplus (l_2, m_2, u_2) = (l_1 + l_2, m_1 + m_2, u_1 + u_2) \tag{4}$$

$$(l_1, m_1, u_1) \odot (l_2, m_2, u_2) = (l_1 l_2,\ m_1 m_2,\ u_1 u_2) \tag{5}$$

$$(\lambda, \lambda, \lambda) \odot (l_2, m_2, u_2) = (\lambda l_2, \lambda m_2, \lambda u_2),\ \text{for } \lambda > 0,\ \lambda \epsilon (l_1, m_1, u_1)^{-1} = \left(\frac{1}{u_1}, \frac{1}{m_1}, \frac{1}{l_1}\right) \tag{6}$$

Step 2: Comparison of the degree of possibility that $M_2 \geq M_1$.

Another step in the FAHP analysis following the specification of synthetic fuzzy values involves computing the priority vector. To do so, each fuzzy set represented by a synthetic fuzzy value has to be compared with each other. The comparison of two TFNs, $M_1 = (l_1, m_1, u_1)$ and $M_2 = (l_2, m_2, u_2)$, allows to compute the degree of possibility that $M_1 \geq M_2$ and the degree of possibility that $M_2 \geq M_1$. The degree of possibility $V(M_2 \geq M_1)$ is expressed by:

$$V(M_2 \geq M_1) = \mu(d) = \begin{cases} 1, & \text{if } m_2 \geq m_1 \\ 0, & \text{if } l_1 \geq u_2 \\ \frac{l_1 - u_2}{(m_2 - u_2) - (m_1 - l_1)}, & \text{otherwise} \end{cases} \tag{7}$$

where d is the ordinate of the highest intersection point D between two convex membership functions μ_{M_1} and μ_{M_2}.

Step 3: Computation of the smallest degree of possibility $M_2 \geq M_1$.

To compare all the possible fuzzy values $M_i = (1, 2, \ldots, k)$, one needs to specify the minimum:

$$V(M \geq M_1, M_2, \ldots, M_k) = \min V(M \geq M_i), i = 1,\ 2, \ldots, k \tag{8}$$

Step four involves calculating the priority weight vector for variants. Let us assume that:

$$d'(A_i) = \min V(S_i \geq S_k) \text{ for } k\ =\ 1,\ 2,\ \ldots,\ n,\ i\, k \neq I \tag{9}$$

The weight vector for variants is represented as:

$$W' = (d'(A_1), d'(A_2), \ldots, d'(A_n))^T$$

To calculate the priority weight vector for individual variants, one needs to normalize vector W', which gives vector W:

$$W = (d(A_1), d(A_2), \ldots, d(A_n))^T \tag{10}$$

where W is a vector of crisp numbers.

This procedure of computing normalized priority weights should be applied to the evaluation of particular alternatives of each criterion (alternatives preference matrices). The particular steps of the FAHP and the steps of the procedure of calculating weight vectors are presented in Figure 2 (based on [37]). The final ranking can be obtained by a sum of products of particular criteria weights and weights of particular alternatives, as in the AHP method.

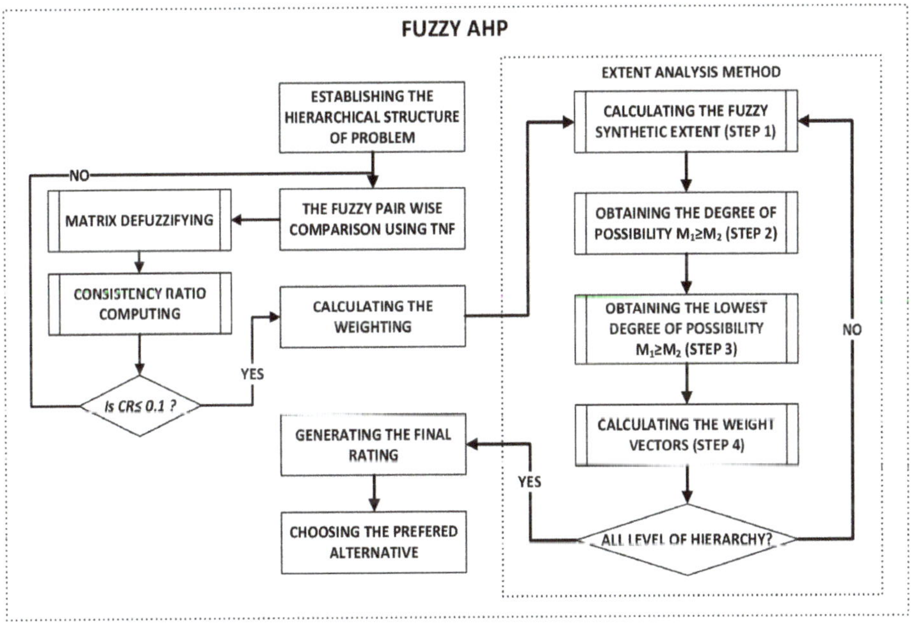

Figure 2. Fuzzy analytic hierarchy process flow chart.

An essential step in both the AHP and FAHP methodologies is checking the consistency ratio of the pairwise comparison matrix. In the classical approach, the consistency ratio (CR) is estimated on a consistency index (CI) of the comparison matrix and a random-like matrix (RI), which simulates highly inconsistent judgments in the comparison stage. Saaty [34] has shown that a CR of 0.10 or less is acceptable to continue the AHP analysis. In FAHP, the consistency ratio procedure is preceded by defuzzifying the pairwise comparison matrix. In the presented approach, the graded mean integration approach is utilized. It is assumed that TFN $M = (l, m, u)$ is transformed into a crisp number by formula:

$$P(M) = \frac{l + 4m + u}{6} \tag{11}$$

After the defuzzification of each pairwise comparison matrix, the Saaty's consistency is applied.

4. Project Selection for Bidding—Model Application

The FAHP method presented above was applied to the selection of the most advantageous contract, which indicates the contractor's participation in the bid. As previous research have observed, the evaluation of any complex object by human beings grows in complexity as they try to describe the object precisely, to the point where the evaluation becomes imprecise. Moreover, in [42], the authors revealed that a one-stage decision problem structure with multiple criteria may lead to the elimination of less significant criteria when the FAHP is used. Therefore, the decision problem was constructed as a two-stage one; criteria were classified into main criteria with sets of sub-criteria assigned to them. Due to such division, the evaluation of particular ventures was easier for the experts.

The FAHP model was proposed only for contractors because they decide about participating in the tender and performing actions aimed at the preparation of the offer without being sure of winning. The authors do not consider the result of the tender in the model, nor the way it is organized, but only participation in it. The model serves to support solely the contractor's decisions about choosing the most suitable tender to join from a number of various projects. Opinions of other participants do not affect the decision.

The example considered concerns Polish conditions, where the tender procedure is the most popular system for awarding construction works contracts. In 2014–2017, the tender was used in over 80% of all contracts awarded for construction work in the Polish public sector [43]. The interest of contractors in tenders for construction works is considerable, which is proven by the fact that, according to the Public Procurement Bulletin [43], for 45% of tenders announced in 2017, three or more offers were submitted. Polish contractors are often faced with the choice of which tender to participate and engage resources in before preparing the offer. Therefore, research was undertaken among Polish contractors regarding the factors influencing the decision to participate in the tender, which were presented in [44].

Polish contractors (61 out of 160) responded to the questionnaire, which made up 38%. Among the respondents, 38% of companies signed more than 75% of contracts resulting from bidding, while 19% of respondents signed no more than 25%. They were asked to specify the degree of importance of the 15 proposed factors, marking them on a 1–7 scale, where 1 was the factor with no influence on the decision and 7 was the one with the greatest significance in decision making. For each factor, an average score was established. On the basis of these data, 15 factors were selected which were then grouped into four main criteria, i.e., C1 (Company's capabilities), C2 (Investment characteristics), C3 (Financial conditions), and C4 (Tender characteristics) influencing the bidding decision. Subsequently, sets with sub-criteria were assigned to the main ones. Table 3 presents factor groups (criteria) deciding about the selection of a project and their average evaluations.

The contractor is considering a participation in one of the four potential tenders. Each of them concerns a different project: P1, P2, P3 and P4. Figure 3 presents the hierarchical structure of the model in which "Project selection for bidding" serves as the target hierarchy at the highest level, the influence factors function as criteria hierarchy at the intermediate levels, and alternatives constitute an alternatives hierarchy at the lowest level.

The projects P1, P2, P3 and P4 were evaluated by two experts—employees of one of the Polish construction companies invited to participate in these studies. The experts were the manager and the deputy head of the tender preparation department, whose professional experience amounted to more than 15 years. They were obliged by the company to make a decision about participating in one of 4 tenders (respectively for project: P1, P2, P3 and P4). The evaluation of the projects was made during the meeting of experts and the co-authors of the paper. The evaluation of the four projects in accordance with the sub-criteria adopted (on a 1–7 scale) is presented in Table 4.

Table 3. Average evaluation of the criteria involved in the decision process of selecting a project.

Criterion/Sub-Criterion	Name of the Criterion/Factor	Average Evaluation of Criterion/Factor *
C1	Company's capabilities	5.14
C1_1	Need of work	5.21
C1_2	Past experience with similar projects	5.95
C1_3	Location of the project	4.25
C2	Investment characteristics	4.48
C2_1	Size of the project (e.g., cubic measure)	4.95
C2_2	Time of project duration	4.49
C2_3	Type of works	5.98
C2_4	Degree of works complexity	3.25
C2_5	Necessity for specialized equipment	3.51
C2_6	Possible subcontractors	3.87
C2_7	Owner's reputation	5.31
C3	Financial conditions	5.35
C3_1	Value of the project	5.30
C3_2	Contract conditions	5.89
C3_3	Profits from similar past projects	4.87
C4	Tender characteristics	4.14
C4_1	Time for the preparation of the bid	3.89
C4_2	Criteria of bid selection	4.38

* According to research by [21].

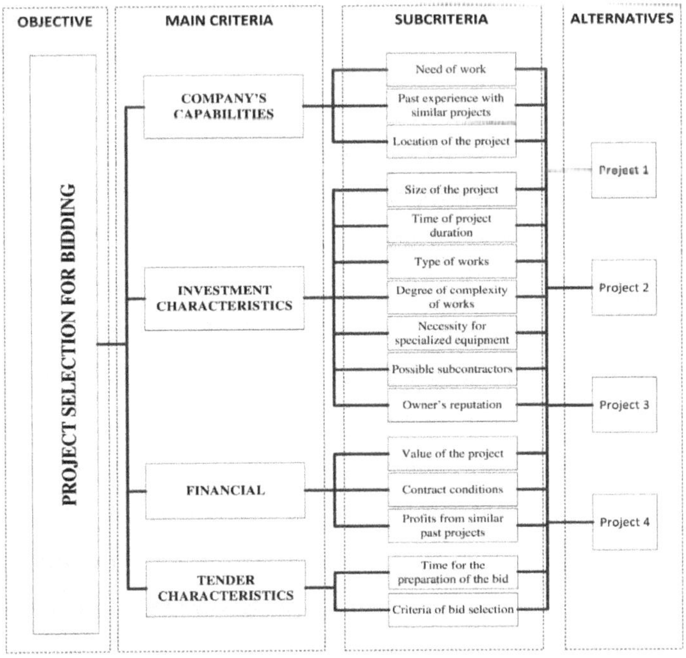

Figure 3. The hierarchy of the AHP model for tender selection.

Table 4. Evaluation of the four ventures under discussion.

Sub-Criterion/Factor	Project			
	P1	P2	P3	P4
C1_1	7	5	7	5
C1_2	4	7	7	7
C1_3	4	5	6	6
C2_1	3	3	5	2
C2_2	3	4	6	5
C2_3	5	7	7	6
C2_4	4	6	7	7
C2_5	4	6	6	6
C2_6	6	6	6	3
C2_7	6	4	7	4
C3_1	4	3	4	2
C3_2	5	4	6	4
C3_3	4	4	5	4
C4_1	6	5	7	6
C4_2	4	5	4	4

Since the judgment scale in the FAHP involves nine relative ranks, each evaluation presented in the research was transformed into the respective FAHP ranks. For this purpose, a simple method was utilized. To determine the relative rank of *i*-th and *j*-th objects, the difference between the obtained evaluation was counted. Afterward, a new threshold is computed, based on the range between absolute judgment obtained in the research and on a number of FAHP ranks. In the next step, in comparing object *i*-th and object *j*-th, the absolute value of the difference is used to directly search for the appropriate Saaty rank (the absolute value of the difference is compared with the newly-evaluated threshold).

While computing all the steps of the FAHP, one can obtain synthetic values of fuzzy triangles for the evaluation of the decision problem objects. Figure 4 illustrates an example of such an evaluation with synthetic TFN values for the main criteria (the consistency ratio for given example equals CR = 0.0043, which proves the consistency of comparison judgments).

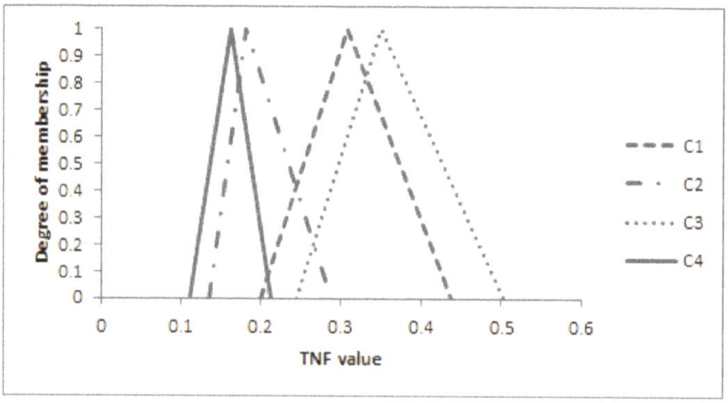

Figure 4. Synthetic TFN values for main criteria.

The characteristic feature of the FAHP analysis is the capability to obtain sets of fuzzy values of individual objects which are then analysed by TFN. This feature is similar to human reasoning.

Criterion C3: Financial condition received the experts' highest rating (5,35), so the synthetic TFN are the highest. The triangle in Figure 4 is the widest and the most right-oriented. This result is highly reliable. In addition, it shares a part with TFN for criterion C1 (company's capabilities) and C2

(Investment characteristics), which indicates a degree of superiority of one criterion over the other. A greater common set indicates a greater criteria equivalence. The lack of a set in common means large dominance of one over the other, as in the case of criterion C3 (financial conditions) and C4 (tender characteristics).

Table 5 presents the results obtained by the FAHP method—normalized priority weight vectors for each individual project.

Table 5. Normalized priority weight vectors for projects.

Names	Priority Weight Vector for Each Individual Project					
	Criteria	Sub-Criteria	P1	P2	P3	P4
C1_1	0.4045	0.3381	0.5000	0.0000	0.5000	0.0000
C1_2		0.6619	0.0000	0.3333	0.3333	0.3333
C1_3		0.0000	0.0309	0.2253	0.3719	0.3719
C2_1	0.1026	0.1990	0.0264	0.0264	0.9472	0.0000
C2_2		0.1287	0.0000	0.1086	0.5586	0.3329
C2_3		0.3544	0.0309	0.3719	0.3719	0.2253
C2_4		0.0000	0.0000	0.1870	0.4065	0.4065
C2_5		0.0028	0.0000	0.3333	0.3333	0.3333
C2_6		0.0547	0.3333	0.3333	0.3333	0.0000
C2_7		0.2603	0.3119	0.0000	0.6881	0.0000
C3_1	0.4928	0.2692	0.3719	0.2253	0.3719	0.0309
C3_2		0.7308	0.3694	0.0333	0.5640	0.0333
C3_3		0.0000	0.1688	0.1688	0.4937	0.1688
C4_1	0.0000	0.0000	0.2474	0.0809	0.4244	0.2474
C4_2		1.0000	0.1688	0.4937	0.1688	0.1688
		Solution	0.2627	0.1486	0.4707	0.1180

The most advantageous venture, as the method proved, is P3, which received the highest priority weight, i.e., 0.4707. Interestingly enough, the fourth criterion C4 (tender characteristics), which involved such factors as bid preparation time and the client's selection criteria, was not considered to be important by the decision makers. It is likely that the contractor's experience makes it possible to prepare a bid on time, as required by the client, or the preparation time is sufficient in practice.

The basic tool in the selection of the most advantageous offer for construction work in Poland is the accepted criteria for the evaluation of offers. Usually, the procuring entities do not apply more than 3 criteria, of which the most important weight is always assigned to the price criterion when evaluating the best offer [43]. Therefore, the type of the criteria used are not as vital from the point of view of the contractor. Additionally, FAHP eliminated some sub-criteria, namely: C1_3: location of the project, C2_4: degree of works complexity, C3_3: profits from similar past projects and C4_1: time for the preparation of the bid. The normalized priority weights for these sub-criteria equal zero. Each of the eliminated sub-criteria belongs to a different main criteria group. The location of the project (C1_3) and the degree of work's complexity (C2_4) are vital for the calculation of the investment costs, yet, they are irrelevant for the analysed bid/no bid decision case. Profits from similar past projects (C3_3) is the factor which, in surveys mentioned above [14], was placed 9th out of 15. Thus, it was not rated as particularly vital, and in the case under consideration, as rather immaterial too, which proves how individual and unique in character each construction project is.

5. Conclusions

Tendering is an obligatory and basic process (though not the only one) on the construction market in the Polish public sector. In 2017, the majority of construction orders resulted from bids, (according to the Public Procurement Bulletin [42], 86% of orders). This way of acquiring contractors is also used by clients representing the private sector, who themselves decide on the form of awarding contracts.

A tender has many advantages, as it is the most competitive process, while its procedure is not complicated. Polish contractors identified 15 factors influencing their bidding decisions. These factors were grouped into main criteria, with each group consisting of sets with sub-criteria. On this basis, a model was proposed, the aim of which was to facilitate the choice of a tender appropriate for the company. To construct the model, the Fuzzy AHP method was used. To prove that the model works, an example was presented in which a bid/no bid decision about four possible tenders was to be taken. As a result of the application of the method, the P3 project was shown to be the one with the highest priority weight (0.4707). It is noteworthy that, according to experts, the modelling performance proved that the most significant element of evaluation was the criterion concerning the financial conditions of the project under tender, while the least vital one was the specifications of the tender itself, namely, the time of bid preparation and the criteria of bid selection. The model proposed here is assumed to be universal, and may be applied to facilitate contractor bidding decisions not only on the Polish market. However, the type and influence of the factors on bidding decisions should be related to the environment in which the contractor works. In further studies, the authors plan to focus on further objectivization of the selected criteria for assessing the problem under investigation, as well as their validity. Attempts will also be made to analyze the flexibility of the final decision, by using sensitivity analyses, and measuring the consistency of assessments made by the decision-maker.

Author Contributions: The individual contribution and responsibilities of the authors were as follows: A.L. and E.P. made a review of the literature concerning models supporting bidding decisions and distinguished in the study factors influencing bid/no bid decisions. A.L designed the research main idea and collected the data. D.K. and K.Z. together analyzed the data and the obtained results. S.B. provided extensive advice throughout the study results and methodology. All the authors have read and approved the final manuscript.

Funding: This research received no external funding.

Conflicts of Interest: The authors declare no conflict of interest.

References

1. Shash, A.A. Factors considered in tendering decisions by top UK contractors. *Constr. Manag. Econ.* **1993**, *11*, 111–118. [CrossRef]
2. Gajzler, M.; Zima, K. Evaluation of Planned Construction Projects Using Fuzzy Logic. *Int. J. Civ. Eng.* **2017**, *15*, 641–652. [CrossRef]
3. Plebankiewicz, E. Modelling decision-making processes in bidding procedures with the use of the fuzzy sets theory. *Int. J. Strateg. Prop. Manag.* **2014**, *18*, 307–316. [CrossRef]
4. Zima, K. The concept of investment decision support model using fuzzy set theory. In Proceedings of the 11th International Conference of Numerical Analysis and Applied Mathematics 2013 (ICNAAM 2013), Rhodes, Greece, 21–27 September 2013; pp. 1307–1311.
5. Plebankiewicz, E.; Leśniak, A. Overhead costs and profit calculation by polish contractors. *Technol. Econ. Dev. Econ.* **2013**, *19*, 141–161. [CrossRef]
6. Zavadskas, E.K.; Antuchevičienė, J.; Kapliński, O. Multi-criteria decision making in civil engineering: Part I—A state-of-the-art survey. *Eng. Struct. Technol.* **2015**, *7*, 103–113. [CrossRef]
7. Hoła, B. Identification and evaluation of processes in a construction enterprise. *Arch. Civ. Mech. Eng.* **2015**, *15*, 419–426. [CrossRef]
8. Gul, P. Subcontractor selection using the integration of the AHP and PROMETHEE methods. *J. Civ. Eng. Manag.* **2016**, *22*, 1042–1054. [CrossRef]
9. Yang, J.-B.; Wang, H.-H.; Wang, W.-C.; Ma, S.-M. Using data envelopment analysis to support best-value contractor selection. *J. Civ. Eng. Manag.* **2016**, *22*, 199–209. [CrossRef]
10. Trinkūnienė, E.; Podvezko, V.; Zavadskas, E.K.; Jokšienė, I.; Vinogradova, I.; Trinkūnas, V. Evaluation of quality assurance in contractor contracts by multi-attribute decision-making methods. *Econ. Res.-Ekon. Istraž.* **2017**, *30*, 1152–1180. [CrossRef]
11. Keshavarz-Ghorabaee, M.; Amiri, M.; Zavadskas, E.K.; Turskis, Z.; Antucheviciene, J. A Dynamic Fuzzy Approach Based on the EDAS Method for Multi-Criteria Subcontractor Evaluation. *Information* **2018**, *9*, 68. [CrossRef]

12. Khanzadi, M.; Turskis, Z.; Amiri, G.G.; Chalekaee, A. A model of discrete zero-sum two-person matrix games with grey numbers to solve dispute resolution problems in construction. *J. Civ. Eng. Manag.* **2017**, *23*, 824–835. [CrossRef]
13. Hashemi, H.; Mousavi, S.M.; Zavadskas, E.K.; Chalekaee, A.; Turskis, Z. A New Group Decision Model Based on Grey-Intuitionistic Fuzzy-ELECTRE and VIKOR for Contractor Assessment Problem. *Sustainability* **2018**, *10*, 1635. [CrossRef]
14. Zavadskas, E.K.; Turskis, Z.; Antucheviciene, J. Selecting a Contractor by Using a Novel Method for Multiple Attribute Analysis: Weighted Aggregated Sum Product Assessment with Grey Values (WASPAS-G). *Stud. Inform. Control* **2015**, *24*, 141–150. [CrossRef]
15. Zavadskas, E.K.; Turskis, Z.; Antucheviciene, J.; Zakarevicius, A. Optimization of Weighted Aggregated Sum Product Assessment. *Electron. Electr. Eng.* **2012**, *122*, 3–6. [CrossRef]
16. Ahmad, I. Decision-support system for modeling bid/no-bid decision problem. *J. Constr. Eng. Manag.* **1990**, *116*, 595–608. [CrossRef]
17. El-Mashaleh, M.S. Empirical framework for making the bid/no-bid decision. *J. Manag. Eng.* **2013**, *29*, 200–205. [CrossRef]
18. Polat, G.; Bingol, B.N.; Uysalol, E. Modeling Bid/No Bid Decision Using Adaptive Neuro Fuzzy Inference System (ANFIS): A Case Study. In Proceedings of the 2014 Construction Research Congress: Construction in a Global Network, Atlanta, GA, USA, 19–21 May 2014; pp. 1083–1092.
19. Sonmez, R.; Sözgen, B. A support vector machine method for bid/no bid decision making. *J. Civ. Eng. Manag.* **2017**, *23*, 641–649. [CrossRef]
20. Cheng, M.-Y.; Hsiang, C.-C.; Tsai, H.-C.; Do, H.-L. Bidding decision making for construction company using a multi-criteria prospect model. *J. Civ. Eng. Manag.* **2011**, *17*, 424–436. [CrossRef]
21. Leśniak, A.; Plebankiewicz, E. Modeling the Decision-Making Process Concerning Participation in Construction Bidding. *J. Manag. Eng.* **2015**, *31*. [CrossRef]
22. Chou, J.S.; Pham, A.D.; Wang, H. Bidding strategy to support decision-making by integrating fuzzy AHP and regression-based simulation. *Autom. Constr.* **2013**, *35*, 517–527. [CrossRef]
23. Leśniak, A. The simplified bidding decision model based on AHP method in ordering of investor's supervision service. In Proceedings of the 14th GeoConference on Ecology, Economics, Education and Legislation, Multilateral Relations and Funding Opportunities, Albena, Bulgaria, 17–26 June 2014; Volume 3, pp. 501–508.
24. Liu, J.B.; Hong, R.; Guo, X.J. Study on the Game Mechanism of Construction Project Bidding. In Proceedings of the International Joint Conference on Artificial Intelligence, Pasadena, CA, USA, 11–17 July 2009; pp. 814–817.
25. Leśniak, A.; Radziejowska, A. Supporting bidding decision using multi-criteria analysis methods. *Procedia Eng.* **2017**, *208*, 76–81. [CrossRef]
26. Leśniak, A. Supporting contractors' bidding decision: RBF neural network application. In Proceedings of the International Conference on Numerical Analysis and Applied Mathematics (ICNAAM-2015), Rhodes, Greece, 23–29 September 2015; Volume 1738.
27. Wanous, M.; Boussabaine, A.H.; Lewis, J. A neural network bid/no bid model, the case for contractors in Syria. *Constr. Manag. Econ.* **2003**, *21*, 737–744. [CrossRef]
28. Ahmad, I.; Minkarah, I. Questionnaire survey on bidding in construction. *J. Manag. Eng. Div.* **1988**, *4*, 229–243. [CrossRef]
29. Shash, A.A.; Abdul-Hadi, N.H. Factors affecting a contractor's mark-up size decision in Saudi Arabia. *Constr. Manag. Econ.* **1992**, *10*, 415–429. [CrossRef]
30. Bageis, A.S.; Fortune, C. Factors affecting the bid/no bid decision in the Saudi Arabian construction contractors. *Constr. Manag. Econ.* **2009**, *27*, 53–71. [CrossRef]
31. Enshassi, A.; Mohamed, S.; El Karriri, A. Factors affecting the bid/no bid decision in the Palestinian construction industry. *J. Financ. Manag. Prop. Constr.* **2010**, *15*, 118–142. [CrossRef]
32. Shokri-Ghasabeh, M.; Chileshe, N. Critical factors influencing the bid/no bid decision in the Australian construction industry. *Constr. Innov.* **2016**, *16*, 127–157. [CrossRef]
33. Leśniak, A. Classification of the Bid/No Bid Criteria—Factor Analysis. *Arch. Civ. Eng.* **2015**, *61*, 79–90. [CrossRef]
34. Saaty, T.L. *The Analytic Hierarchy Process*; McGraw-Hill: New York, NY, USA, 1980.

35. Ho, W. Integrated analytic hierarchy process and its applications—A literature review. *Eur. J. Oper. Res.* **2008**, *186*, 211–228. [CrossRef]
36. Chang, D. Application of extent analysis method on fuzzy AHP. *Eur. J. Oper. Res.* **1996**, *95*, 649–655. [CrossRef]
37. Prascevic, N.; Prascevic, Z. Application of fuzzy AHP for ranking and selection of alternatives in construction project management. *J. Civ. Eng. Manag.* **2017**, *23*, 1123–1135. [CrossRef]
38. Taylana, O.; Bafail, A.O.; Abdulaala, R.M.S. Construction projects selection and risk assessment by fuzzy AHP and fuzzy TOPSIS methodologies. *Appl. Soft Comput.* **2014**, *17*, 105–116. [CrossRef]
39. Çalik, A.; Yapici Pehlivan, N.; Kahraman, C. An integrated fuzzy AHP/DEA approach for performance evaluation of territorial units in Turkey. *Technol. Econ. Dev. Econ.* **2018**, *24*, 1280–1302. [CrossRef]
40. Ucal Sari, I. Development of an integrated discounting strategy based on vendors' expectations using FAHP and FUZZY goal programming. *Technol. Econ. Dev. Econ.* **2018**, *24*, 635–652. [CrossRef]
41. Kutlu, A. Fuzzy failure modes and effects analysis by fuzzy TOPSIS-based fuzzy AHP. *Experts Syst. Appl.* **2012**, *39*, 61–67. [CrossRef]
42. Plebankiewicz, E.; Kubek, D. Multicriteria selection of the building material supplier using AHP and fuzzy AHP. *J. Constr. Eng. Manag.* **2015**, *142*, 40–57. [CrossRef]
43. Report of the Public Procurement Office on the Functioning of the Public Procurement System in 2017. Available online: https://www.uzp.gov.pl/baza-wiedzy/analizy-systemowe/sprawozdania-o-funkcjonowaniu-systemu-zamowien-publicznych (accessed on 26 October 2018).
44. Plebankiewicz, E.; Kozik, R. The transformation of the tender evaluation process in public procurement in Poland. *IOP Conf. Ser. Mater. Sci. Eng.* **2017**, *251*, 012042. [CrossRef]

symmetry

MDPI

Article

m-Polar Fuzzy Soft Weighted Aggregation Operators and Their Applications in Group Decision-Making

Azadeh Zahedi Khameneh [1] and Adem Kiliçman [1,2,*]

1 Institute for Mathematical Research, Universiti Putra Malaysia, UPM Serdang 43400, Selangor, Malaysia; azadeh503@gmail.com

2 Department of Mathematics, Faculty of Science, Universiti Putra Malaysia, UPM Serdang 43400, Selangor, Malaysia

* Correspondence: akilic@upm.edu.my; Tel.: +60-38946-6813

Received: 26 October 2018; Accepted: 6 November 2018; Published: 13 November 2018

Abstract: Aggregation operators are important tools for solving multi-attribute group decision-making (MAGDM) problems. The main challenging issue for aggregating data in a MAGDM problem is how to develop a symmetric aggregation operator expressing the decision makers' behavior. In the literature, there are some methods dealing with this difficulty; however, they lack an effective approach for multi-polar inputs. In this study, a new aggregation operator for m-polar fuzzy soft sets (M-pFSMWM) reflecting different agreement scenarios within a group is presented to proceed MAGDM problems in which both attributes and experts have different weights. Moreover, some desirable properties of M-pFSMWM operator, such as idempotency, monotonicity, and commutativity (symmetric), that means being invariant under any permutation of the input arguments, are studied. Further, m-polar fuzzy soft induced ordered weighted average (M-pFSIOWA) operator and m-polar fuzzy soft induced ordered weighted geometric (M-pFSIOWG) operator, which are extensions of IOWA and IOWG operators, respectively, are developed. Two algorithms are also designed based on the proposed operators to find the final solution in MAGDM problems with weighted multi-polar fuzzy soft information. Finally, the efficiency of the proposed methods is illustrated by some numerical examples. The characteristic comparison of the proposed aggregation operators shows the M-pFSMWM operator is more adaptable for solving MAGDM problems in which different cases of agreement affect the final outcome.

Keywords: m-polar fuzzy soft set; m-polar fuzzy soft preference relationship; aggregation operator; multi-attribute group decision-making

---

1. Introduction

A group decision-making process, which is also called multi-person decision-making, is the problem of finding the best option accepted by the majority of decision makers among a list of possible alternatives. The basic act in the group decision-making is the process of consensus between different decision makers. Although a unanimous consensus is the ideal case, in real situations, full agreement rarely happens. Decision makers usually share and discuss their opinions about the alternatives to obtain a consensus or partial agreement for making the final decision. However, sometimes, they provide their priorities about the alternatives individually and then try to reach a consensus on them. Regardless which approach is applied, using aggregation operators is one of the most often used techniques to reach the process of consensus in group decision-making problems.

The aim of an aggregation operator of dimension K is to aggregate the K-tuple of objects into a single object by a bounded non-decreasing function $A : \bigcup_{K \in \mathbb{N}}[0,1]^K \to [0,1]$. The mean value, the median, the minimum, the maximum, the t-norms, and the t-conorms are commonly used to reach the process of consensus in group decision-making problems. To aggregate a sequence of inputs with different importance degrees, the concept of weighting vector in aggregation operators has been studied, such as the weighted minimum and the weighted maximum [1–3]; ordered weighted average (OWA) [4], which is calculated based on the arithmetic mean; and ordered weighted geometric (OWG) [5], which is formulated based on the geometric mean. The main disadvantage of OWA and OWG operators, i.e., ignoring the importance of given arguments $x_1, \cdots, x_K$ for calculating the aggregated value, leads to definition of their extensions into the induced OWA (IOWA) operator [6], and induced OWG (IOWG) operator [7], respectively. However, these extensions have the inherent limitations from OWA operator and OWG operator, concerning the determination of associated weighting vector w for IOWA and IOWG operators. In practice, there is no unique strategy to find the associated weighting vector w. Usually, a quantifier function $Q : [0,1] \to [0,1]$ whose definition may change from one case to another one is applied to compute the associated weighting vector $w = (w_1, \cdots, w_K)^T$ based on the formulation $w_i = Q(\frac{i}{K}) - Q(\frac{i-1}{K})$ for all i [4,5,8,9].

Soft set theory (SS) [10], characterized by a set-valued function $f : P \to 2^U$, is defined by a parameterization family of the universe U. Thus, in comparison with fuzzy set theory [11], it allows us to have a more comprehensive description of U based on any type and number of parameters $p \in P$. The concepts of fuzzy soft set (FSS) [12] and intuitionistic fuzzy soft set (IFSS) [13] were also studied to handle more complicated problems. Although soft set theory was originally developed to cope with the lack of parameterization tool in fuzzy sets, its flexibility to deal with set-valued functions makes it a powerful tool for providing a new methodology in decision-making problems. The first adaptivity of soft sets in decision-making was conducted by Maji et al. [14]. However, they did not discuss the aggregation methods. Roy and Maji [15] gave an algorithm to solve a group decision-making problem based on fuzzy soft sets. The FS minimum operator was applied for finding the consensus of multi-source parameter sets. Later, Alcantud [16] overcame the disadvantages of Roy and Maji's algorithm by applying FS product operator rather than FS minimum operator. Cagman et al. [17,18] applied FS minimum operator, FS maximum operator, and FS product operator to produce different fuzzy soft aggregation operators in a MAGDM problem. Guan et al. [19] applied the FS intersection operator for aggregating data. Then, a new ranking system of objects, in which the rate of objects is computed based on the full unanimity of experts with respect to all parameters rather than the number of parameters that are owned by each object, was constructed to rank alternatives in a group decision problem. Zhang et al. [20] used the IFS maximum operator to check the process of consensus in MAGDM problem. Mao et al. [21] extended IFSOWA and IFSOWG operators for intuitionistic fuzzy soft sets under three different cases of experts' weights called completely known, partly known, and completely unknown. Das and Kar [22] used the IFS product and the IFS sum operators of intuitionistic fuzzy soft matrices to obtain a collective opinion among different decision makers. Zhang and Zhang [23] utilized the FS union operation to reach the process of consensus in MAGDM problems based on trapezoidal interval type-2 fuzzy soft sets (TIT2FSSs). The TOPSIS approach was then applied to select the best option. However, recently, Pandey and Kumar [24] modified this procedure so that the complement operation that is used by Zhang and Zhang can only be applied for TIT2FSSs in which left and right heights are equal. Tao et al. [25] proposed four aggregation functions, including the soft maximum operator, soft minimum operator, soft average operator, and soft weighted average operator, to aggregate data in MAGDM problems. Two selection tools based on the SAW method and comparison matrix were also developed to obtain the optimal decision. Zhu and Zhan [26] developed the new concept fuzzy parameterized fuzzy soft sets (FPFSS), where the parameters are considered as fuzzy sets and then the consensus stage proceeds by using the t-norm and t-conorm products of FPFSSs. A new choice value function is also introduced to handle the process of selection. On the other hand, some researchers attempted to develop novel aggregation methods for FSS and IFSS. For example, Zahedi et al. [27,28] extended the concept of fuzzy soft topology to reach the consensus based on a collective preorder relation.

Conventionally, soft-set-based aggregating tools have been developed for unipolar input data where the truth value belongs to $[0,1]$. Some situations require multi-polar arguments to be aggregated.

To handle the problem of multi-polarity in consensus process, some studies extended the aggregation operators from unipolar scale into the bipolar scale (i.e., the interval $[-1,1]$) [29] and multi-polar scale (i.e., the space $K \times [0,1]$ where K is a set of m different categories) [30,31]. These extensions allow dealing with inputs from different categories where the output is presented by a pair (k,x) that k shows the category of aggregated value x. However, in many decision-making problems, the attribute set contains multi-feature or multi-polar decision parameters where the final decision should reflect the best option based on all multi-polar attributes beyond their category. For example, in the hotel booking problem, "Location" is one of the most important parameters for finding a good hotel to stay, which is a multi-feature parameter depending on how close it is to the main road, city center, tourist attractions, etc. A hotel is selected if it has the best location in terms of all features from different categories not only one. In fact, there is no ideal category k in final solution. This issue becomes more complex when a group of people want to choose a hotel. In this case, different cases of agreement within a group including unanimous consensus and partial agreement, e.g., "almost all", "most", and "more than 50%", are considered to obtain the collective view based on individuals' opinions. Thus, an alternative aggregation operator is required to deal with multi-polar fuzzy attributes in group decision-making.

In a group decision process with multi-polar inputs, the performance of alternatives are judged by each decision maker with respect to each criterion. The main problem is to compare these judgements and reach a consensus among them. The existing aggregation methods usually consider weighted or unweighted cases under a unanimous agreement [15–23]. However, besides the importance degrees of experts, the consensus degree for a fuzzy majority of the experts and different choices of experts' judgments at a consensus level should be taken into account in the proposed alternative aggregation operator to reach more reliable methodology. Moreover, due to the extreme applicability of FSSs in MAGDM problems with multi-polar fuzzy soft input information, adaptability of the proposed aggregation operator for m-polar fuzzy soft sets should be studied. To do this, an extension of fuzzy soft sets into the m-polar fuzzy soft sets, where the values of membership functions f_p are extended from the unit interval $[0,1]$ into the cubic $[0,1]^m$, needs to be developed.

Until now, weighted aggregation operators for multi-polar fuzzy soft arguments have not been considered. Thus, this study was carried out to develop some weighted m-polar aggregation operators which cover different scenarios at the consensus degree for a fuzzy majority expressed by linguistic variables, such as "most", "much more than 70%", and "more than half". The main goal is to design FS-set-based algorithms for finding the best solution in group decision-making with weighted multi-polar input information according to the proposed aggregation methods. To achieve this goal, the following problems are addressed in this study: (i) how to express the multi-polarity of input data under fuzzy soft environment; (ii) how to generate an aggregation method based on the fuzzy majority concept for weighted multi-polar inputs; (iii) how to apply the proposed aggregation method for finding the solution in group decision-making problems; and (iv) how to analyze the final result obtained by the proposed algorithm. Accordingly, there are four main contributions of this research as follows: (i) to define a new concept m-polar fuzzy soft set; (ii) to introduce m-polar fuzzy soft weighted aggregation operators based on the fuzzy majority concept; (iii) to design m-polar FS-based algorithms for finding the solution in group decision-making; and (iv) to give some illustrative examples for validating and comparing the results.

The rest of this paper is organized as the following. Section 2 represents some basic definitions and concepts from the related works. Section 3 gives a new aggregation operator, called M-pFSMWM, for weighted m-polar fuzzy soft data. This new procedure aggregates the experts' judgments based on their importance degrees, a linguistic or numerical consensus level between the experts, and different choices of experts' judgments at the consensus level. Some of its desirable properties as well as special families of M-pFSMWM operator according to different values of consensus degree α and weighting vector ω are studied. In Section 4, a new score value function is developed to design an algorithm for ranking alternatives in MAGDM problems based on the M-pFSMWM operator and a new m-polar fuzzy soft preference relation. To compare the proposed M-pFSMWM operator with some existing aggregation methodologies, in Section 5, the m-polar fuzzy soft induced ordered weighted average (M-pFSIOWA) operator and the m-polar fuzzy soft induced ordered weighted geometric (M-pFSIOWG) operator, which are the extensions of IOWA and IOWG operators, respectively, are developed and their properties are considered. We also present an algorithm to solve MAGDM problems based on M-pFSIOWA and M-pFSIOWG operators. Section 6 focuses on the efficiency of proposed techniques by some numerical examples. Finally, in Section 7, we discuss the advantages and limitations of our approach.

2. Preliminaries

This section recalls some definitions about the weighted aggregation operators, m-polar fuzzy sets, and fuzzy soft sets to achieve our main aim, proposing new algorithms for solving group decision-making based on alternative m-polar fuzzy soft aggregation operators, in the next sections.

2.1. Weighted Aggregation Operator

The weighted minimum and the weighted maximum are two important aggregation operators dealing with objects having non-negative weights $\omega_1, \cdots, \omega_K$ such that $\sum_{i=1}^{K} \omega_i = 1$. However, there is no unique solution to formulate them. For instance, Fagin and Wimmers [2] introduced the below formula to obtain the weighted minimum and the weighted maximum, respectively.

$$\min^{\otimes}_{\omega_1,\cdots,\omega_K}(x_1, \cdots, x_K) = \sum_{i=1}^{K} \left[i \cdot (\omega_{\sigma(i)} - \omega_{\sigma(i+1)}) \cdot \min(x_{\sigma(1)}, \cdots, x_{\sigma(i)}) \right] \tag{1}$$

$$\max^{\otimes}_{\omega_1,\cdots,\omega_K}(x_1, \cdots, x_K) = \sum_{i=1}^{K} \left[i \cdot (\omega_{\sigma(i)} - \omega_{\sigma(i+1)}) \cdot \max(x_{\sigma(1)}, \cdots, x_{\sigma(i)}) \right] \tag{2}$$

where σ is a permutation that orders the weights as follows: $\omega_{\sigma(1)} \geq \omega_{\sigma(2)} \geq \cdots \geq \omega_{\sigma(K)}$ and $\omega_{\sigma(k+1)} = 0$.

The IOWA operator, introduced by Yager and Filev [6], and the IOWG operator, given by Xu and Da [7], are also some commonly used tools for aggregating weighted objects. The IOWA operator is defined by

$$IOWA(\langle \omega_1, x_1 \rangle, \cdots, \langle \omega_K, x_K \rangle) = \sum_{j=1}^{K} w_j \cdot y_j \tag{3}$$

where y_j is the value of x_i that has the jth largest ω_i and ω_i in $\langle \omega_i, x_i \rangle$ is referred to as the order inducing variable and x_i as the argument variable. The weights $w_1, \cdots, w_K$ such that $\sum_{i=1}^{K} \omega_i = 1$ are the associated weights to the IOWA operator that can be defined by a quantifier function $Q : [0,1] \to [0,1]$. Here, the re-ordering step of x_is is carried out by the variable ω_i rather than the value of x_i, that is used to handle the re-ordering step in OWA operator, i.e., the collection $x_1, \cdots, x_K$ is re-ordered as $\langle \max\{\omega_i\}, y_1 \rangle \geq \cdots \geq \langle \min\{\omega_i\}, y_K \rangle$.

An IOWG operator is defined by

$$IOWG(\langle \omega_1, x_1 \rangle, \cdots, \langle \omega_K, x_K \rangle) = \prod_{j=1}^{K} {y_j}^{w_j} \tag{4}$$

where y_j is the value of x_i that has the jth largest ω_i and ω_i in $\langle \omega_i, x_i \rangle$ is referred to as the order inducing variable and x_i as the argument variable. Note that here also the re-ordering step is based on the inducing variable ω_i and weights $w_1, \cdots, w_K$ such that $\sum_{i=1}^{K} \omega_i = 1$ are the associated weights to the IOWG operator.

2.2. Fuzzy Sets

Today, fuzzy sets are known as an effective tool for modeling vague data [32–35]. If U is a non-empty set of elements, then a fuzzy subset X of U is a set of ordered pairs $(u, \mu_X(u))$ such that $u \in U$ and $\mu_X : U \to [0,1]$ is a membership function where $\mu_X(u)$ shows the membership degree of element u in X. Any fuzzy set R in $U \times U$ is called a fuzzy relation on U. If R represents a fuzzy preference relation on U, then, for each pair $(u,v) \in U \times U$, the value $\mu_R(u,v)$ shows the preference degree of u over v. Here, $\mu_R(u,v) = 0.5$ indicates indifference between u and v $(u \sim v)$, while $\mu_R(u,v) \in (0.5, 1]$ shows u is preferred to v $(u \succ v)$. Moreover, generally, we have $\mu_R(u,v) + \mu_R(v,u) = 1$.

Theorem 1 (Multiplicative transitivity). *[36] If $u(.)$ is a utility function on the set $X = \{x_1, \cdots, x_n\}$ such that the value $u(x_i) = u_i$ shows the utility of alternative $x_i \in X$, then the fuzzy preference relation R defined by $\mu_R(x_i, x_j) = r_{ij} = \frac{u_i}{u_i + u_j}$, which is a fuzzy preference relation satisfying multiplicative transitivity condition, i.e., $\frac{r_{ji}}{r_{ij}} \frac{r_{kj}}{r_{jk}} = \frac{r_{ki}}{r_{ik}}$ for all $i, j, k \in \{1, \cdots, n\}$.*

To extend the traditional fuzzy sets dealing with unipolar data into the multi-polar information, the concept of m-polar fuzzy set is defined as below.

Definition 1 (m-polar fuzzy set). *[37] An m-polar fuzzy set (M-pFS) X on U is a mapping $\mu : U \to [0,1]^m$, where $[0,1]^m$ refers to as the multiplication of $[0,1] \times \cdots \times [0,1]$ m-times, such that $\mu(u) = ((\mu^1(u), \cdots, \mu^m(u)), \mathbf{0} = (0,0,\cdots,0)$ and $\mathbf{1} = (1,1,\cdots,1)$ are the least and greatest elements, respectively, and $\mu^c(u) = (1-\mu^1(u), \cdots, 1-\mu^m(u))$ shows its complement. The set of all m-polar fuzzy sets over U is represented by $m(U)$.*

If $\{\mu_k\}_k$ is a family of M-pFSs over the universe U, then for any $u \in U$:

1. If $\mu_i^s(u) \le \mu_j^s(u)$ for all $s = 1, \cdots, m$, then $\mu_i \le \mu_j$.
2. $(\vee_k \mu_k)(u) = \sup_k\{\mu_k(u)\} = (\sup_k\{\mu_k^1(u)\}, \cdots, \sup_k\{\mu_k^m(u)\})$.
3. $(\wedge_k \mu_k)(u) = \inf_k\{\mu_k(u)\} = (\inf_k\{\mu_k^1(u)\}, \cdots, \inf_k\{\mu_k^m(u)\})$.

2.3. Fuzzy Soft Sets

Theory of soft sets is presented based on the approximate descriptions of the set U. A soft set is characterized by a set-valued mapping $f : P \to 2^U$ where P is a set of parameters and 2^U shows the power set of U. By combining the definitions of fuzzy sets and soft sets a new concept called fuzzy soft set is proposed.

Definition 2 (Fuzzy soft set). *[12] A fuzzy soft set (FSS), denoted by f_P or (f, P), is a mapping $f : P \to [0,1]^U$ where for every $p \in P$, $f(p)$ is a fuzzy subset of U with membership function $f_p : U \to [0,1]$ where $\tilde{0}$ and $\tilde{1}$, defined by $\tilde{0}(p)(u) = 0$ and $\tilde{1}(p)(u) = 1$ $\forall u \in U$ and $p \in P$, is called the null fuzzy soft set and the absolute fuzzy soft set, respectively. Moreover, the complement of (f, P), denoted by (f^c, P), is defined by $f^c : P \to [0,1]^U$ where $\forall p \in P$, $f_p^c(u) = 1 - f_p(u)$ for all $u \in U$.*

If $\{(f_k, P_k)\}_k$ is a family of FSSs over the universe U, then, for any $u \in U$:

1. If $P_i \subset P_j$ and $f_i(p)(u) \le f_j(p)(u)$ for all $p \in P_i$, then $(f_i, P_i) \tilde{\le} (f_j, P_j)$.
2. $(\tilde{\vee}_k f_k)(p)(u) = \sup_k\{f_k(p)(u)\}$ for all $p \in \cup_k P_k$.
3. $(\tilde{\wedge}_k f_k)(p)(u) = \inf_k\{f_k(p)(u)\}$ for all $p \in \cap_k P_i$.

3. A New Weighted Aggregation Operator for M-pFSSs

In this section, we introduce a new weighted aggregation operator, called M-pFSMWM operator, to improve the aggregating tools for multi-polar inputs with non-negative weights under fuzzy soft environment. The advantages of this new operator is also demonstrated by some theorems and properties. To this end, we first develop the new concept of m-polar fuzzy soft sets (M-pFSS) and then introduce the M-pFSMWM operator in the domain of m-polar fuzzy soft sets.

3.1. m-Polar Fuzzy Soft Sets

Motivated by m-polar fuzzy sets given in Definition 1, the notion of m-polar fuzzy soft set is developed to model data dealing with multi-polar or multi-feature attributes. Basic operations of m-polar fuzzy soft sets are also discussed in this section.

Definition 3 (m-polar fuzzy soft set). *Let U and P be two non-empty sets of alternatives and parameters, respectively. The pair (f, P) where f is the mapping $f : P \to m(U)$ such that for any $p \in P$ the $f(p)$ is an m-polar fuzzy subset of U can be defined as an m-polar fuzzy soft set (M-pFSS) over U. It means, for each $p \in P$ and any $u \in U$, $f(p)(u)$ is an m-tuple $f_p(u) = (f_p^1(u), f_p^2(u), \cdots, f_p^m(u))$ such that the $f_p^s(u)$, for $s = 1, 2, \cdots, m$, represents the relation between object $u \in U$ and feature s of parameter p.*

The set of all m-polar fuzzy soft sets is shown by $mfs(U)$. Furthermore, an m-polar fuzzy soft set (f, P) is called a null M-pFSS, shown by $\tilde{\mathbf{0}}$, or an absolute M-pFSS, shown by $\tilde{\mathbf{1}}$, if for any $p \in P$, $f^s(p)(u) = 0$ and $f^s(p)(u) = 1$, respectively, for all $u \in U$ and $1 \leq s \leq m$. The complement of M-pFSS (f, P) is also an M-pFSS, shown by (f^c, P), where for any $p \in P$ and $u \in U$: $f^{sc}(p)(u) = 1 - f^s(p)(u)$ for all $s = 1, 2, \cdots, m$.

Example 1. *Let us suppose a person wants to rate four restaurants $\{x_1, x_2, x_3, x_4\}$ according to the parameters $\{p_1 = Location, p_2 = Meal, p_3 = Services\}$. Let he/she considers the different aspects of these parameters as follows: The location of the restaurants includes close to the main road, in green surroundings, and in shopping mall. The meal of the restaurants includes main course, appetizer (starter), and dessert. The services of the restaurants include parking lot, live music, and free Wi-Fi connectivity.*

Assume that the person uses the linguistic variables "No" (0), "Yes" (1), "Very Poor" (0), "Poor" (0.1), "Medium Poor" (0.3), "Medium" (0.5), "Medium Good"(0.7), "Good" (0.9), "Very Good" (1) "Very Far" (0), "Far" (0.1), "Medium Far" (0.3), "Medium Close"(0.7), "Close" (0.9), and "Very Close" (1) shown in Table 1 for describing the performance of each alternative with respect to these parameters.

Table 1. Linguistic variables for describing each alternative with respect to the parameters.

U		x_1	x_2	x_3	x_4
p_1:					
	p_1^1	Close	Far	Very Far	Very Close
	p_1^2	No	No	Yes	Yes
	p_1^3	Yes	Yes	No	No
p_2:					
	p_2^1	Very Good	Medium	Medium Poor	Good
	p_2^2	Very Poor	Good	Medium	Good
	p_2^3	Medium	Very Good	Poor	Medium
p_3:					
	p_3^1	Yes	Yes	Yes	No
	p_3^2	No	Yes	Yes	Yes
	p_3^3	Medium Good	Medium	Poor	Good

Thus, a three-polar fuzzy soft set (that is shown in Table 2) can help he/she to explain his/her opinion about these four restaurants.

Table 2. Tabular representation of M-pFSS (f, P).

U	p_1	p_2	p_3
x_1	(0.9,0,1)	(1,0,0.5)	(1,0,0.7)
x_2	(0.1,0,1)	(0.5,0.9,1)	(1,1,0.5)
x_3	(0,1,0)	(0.3,0.5,0.1)	(1,1,0.1)
x_4	(1,1,0)	(0.9,0.9,0.5)	(0,1,0.9)

For example, if the person considers the meal of the restaurant x_1, then the 3-tuple (1,0,0.5) means that the main course of the restaurant x_1 is very good while the starter and the dessert are very poor and medium, respectively.

Definition 4. *Let $\{(f_k, P_k)\}_k$ be a family of M-pFSSs over the common universe U and parameter sets P_k. Then, for any $u \in U$:*

1. $(f_i, P_j) \tilde{\leq} (f_j, P_j)$ *if* $P_i \subseteq P_j$ *and* $f_i^s(a)(u) \leq f_j^s(a)(u)$ *for all* $a \in P_i$ *and* $s = 1, 2, \cdots, m$.
2. $(\tilde{\bigvee}_k f_k)(a)(u) = \sup_k\{f_k(a)(u)\} = (\sup_k\{f_k^1(a)(u)\}, \cdots, \sup_k\{f_k^m(a)(u)\})$, *for all* $a \in \bigcup_{k \in K} P_k$.
3. $(\tilde{\bigwedge}_k f_k)(a)(u) = \inf_k\{f_k(a)(u)\} = (\inf_k\{f_k^1(a)(u)\}, \cdots, \inf_k\{f_k^m(a)(u)\})$, *for all* $a \in \bigcap_{k \in K} P_k$.

Proposition 1. *Let U and F be the universal sets of objects and parameters, respectively, and P, Q, and E are some subsets of F. Assume that $(f, P), (g, Q)$, and (h, E) are some m-polar fuzzy soft sets over U where $f_p, g_q, h_e \in m(U)$ for all $p \in P, q \in Q$, and $e \in E$. Then:*

1. $(f, P)\tilde{\vee}\tilde{0} = (f, P)$, $(f, P)\tilde{\wedge}\tilde{0} = \tilde{0}$ *and* $(f, P)\tilde{\vee}\tilde{1} = \tilde{1}$ *and* $(f, P)\tilde{\wedge}\tilde{1} = (f, P)$.
2. *(Idempotent)* $(f, P)\tilde{\vee}(f, P) = (f, P)$ *and* $(f, P)\tilde{\wedge}(f, P) = (f, P)$.
3. *(Commutative)* $(f, P)\tilde{\vee}(g, Q) = (g, Q)\tilde{\vee}(f, P)$ *and* $(f, P)\tilde{\wedge}(g, Q) = (g, Q)\tilde{\wedge}(f, P)$.
4. *(Associative)* $(f, P)\tilde{\vee}[(g, Q)\tilde{\vee}(h, E)] = [(f, P)\tilde{\vee}(g, Q)]\tilde{\vee}(h, E)$ *and* $(f, P)\tilde{\wedge}[(g, Q)\tilde{\wedge}(h, E)] = [(f, P)\tilde{\wedge}(g, Q)]\tilde{\wedge}(h, E)$.
5. *(Distributive)* $(f, P)\tilde{\vee}[(g, Q)\tilde{\wedge}(h, E)] = [(f, P)\tilde{\vee}(g, Q)]\tilde{\wedge}[(f, P)\tilde{\vee}(h, E)]$ *and* $(f, P)\tilde{\wedge}[(g, Q)\tilde{\vee}(h, E)] = [(f, P)\tilde{\wedge}(g, Q)]\tilde{\vee}[(f, P)\tilde{\wedge}(h, E)]$.

Proof. Trivial by Definitions 3 and 4. □

Proposition 2 (De Morgan Law). *Let U and F be the universal sets of objects and parameters, respectively. Assume that (f, P) and (g, Q) are two m-polar fuzzy soft sets over U where P and Q are the subsets of F and $f_p, g_q \in m(U)$ for all $p \in P$ and $q \in Q$. Then:*

1. $[(f, P)\tilde{\vee}(g, Q)]^c = (f^c, P)\tilde{\wedge}(g^c, Q)$.
2. $[(f, P)\tilde{\wedge}(g, Q)]^c = (f^c, P)\tilde{\vee}(g^c, Q)$.

Proof. It is proved easily by Definitions 3 and 4. □

3.2. The M-pFSMWM Operator

In this subsection, we develop the M-pFSMWM operator in the domain of M-pFSSs. This new aggregation operator is used to reach the process of consensus in group decision-making problems with weighted m-polar fuzzy soft inputs. Additionally, we show the M-pFSMWM is a well-defined operator having the behavioral properties.

Let $D_K = \{(f_k, P) | f_k : P \to m(u), f_k \in mfs(U), k = 1, 2, \cdots, K\}$ be a collection of m-polar fuzzy soft sets over U and P, such that for all k: $f_k(p)(u) = (f_k^1(p)(u), \cdots, f_k^m(p)(u)) \in [0, 1]^m$ for $p \in P$ and $u \in U$ where $[0, 1]^m$ refers to as the multiplication of $[0, 1] \times \cdots \times [0, 1]$ m-times, with non-negative weights $\omega_1, \cdots, \omega_K \in [0, 1]$ where $\sum_{k=1}^{K} \omega_k = 1$. In the following, we develop a new weighted aggregation operator for m-polar fuzzy soft sets (M-pFSMWM operator) based on the weighted minimum operator given in Equation (1) and M-pFS maximum defined in Definition 4.

Definition 5 (M-pFSMWM Operator). *Let* $D_K = \{(f_k, P) \in mfs(U) | k = 1, 2, \cdots, K\}$ *be a collection of m-polar fuzzy soft sets over U and P with non-negative weights* $\omega_1, \cdots, \omega_K \in [0,1]$ *such that* $\sum_{k=1}^{K} \omega_k = 1$. *Let value* α *where* $\alpha \in \{1, 2, \cdots, K\}$ *be the required consensus degree. An M-pFSMWM operator of dimension K and at consensus degree* α *is a mapping* $M - pFSMWM^{(K,\alpha,m)} : \bigcup_{K \in \mathbb{N}} (mfs(U))^K \to mfs(U)$ *that is defined by*

$$M - pFSMWM^{(K,\alpha,m)}\langle f_1, \cdots, f_K \rangle (p)(u) = \Bigg\langle \max_{l=1}^{C_{K,\alpha}} \Big\{ \sum_{k=1}^{\alpha} k \cdot (\omega_{\sigma(\delta_k(l))} - \omega_{\sigma(\delta_{k+1}(l))}) \cdot \min\{f^1_{\sigma(\delta_1(l))}(p)(u), \cdots, f^1_{\sigma(\delta_k(l))}(p)(u)\} \Big\}_{\delta_k(l) \in \Delta_{K,\alpha}(l)}, \cdots, \max_{l=1}^{C_{K,\alpha}} \Big\{ \sum_{k=1}^{\alpha} k \cdot (\omega_{\sigma(\delta_k(l))} - \omega_{\sigma(\delta_{k+1}(l))}) \cdot \min\{f^m_{\sigma(\delta_1(l))}(p)(u), \cdots, f^m_{\sigma(\delta_k(l))}(p)(u)\} \Big\}_{\delta_k(l) \in \Delta_{K,\alpha}(l)} \Bigg\rangle \tag{5}$$

for $u \in U$ *and* $p \in P$ *where the sum* $\sum_{k=1}^{\alpha} [...]$ *refers to as the weighted minimum over different choices* α *of K,* σ *is the permutation operator,* $C_{K,\alpha} = \frac{K!}{\alpha!(K-\alpha)!}$ *is the binomial coefficient, and* $\Delta_{K,\alpha}(l)$ *is an indexing set, where* $card(\Delta_{K,\alpha}(l)) = \alpha$, *including lth* α*-combination from a set of K elements. Thus,* $\{\delta_1(l) \cdots, \delta_\alpha(l)\}_{l=1}^{C_{K,\alpha}}$ *traverses all the* α*-combinations of the set* $\{1, 2, \cdots, K\}$ *and* $f^s_{\delta_k(l)}(p)(u)$ *represents the* δ_k*th element in lth* α*-combination of K for feature s;* $s = 1, 2, \cdots, m$.

In the following, the various properties of M-pFSMWM operator including idempotency, boundedness, monotonicity, and commutativity (symmetry) are discussed.

Theorem 2. *Let* $\{(f_k, P)\}_{k=1}^{K}$ *and* $\{(g_k, P)\}_{k=1}^{K}$, *for* $k = 1, 2, \cdots, K$, *be two collections of some m-polar fuzzy soft sets over U and P with non-negative weights* ω_k *that for all k:* $\omega_k \in [0,1]$ *and* $\sum_{k=1}^{K} \omega_k = 1$. *Let the required consensus degree* α *is given. Then, the M-pFSMWM operator has the following properties.*

1. *(Idempotency) If* $(f_k, P) = (f, P)$ *for all k, then*

$$M - pFSMWM^{(K,\alpha,m)}\langle (f_1, P), \cdots, (f_K, P) \rangle = (f, P)$$

2. *(Boundary Conditions)*

$$M - pFSMWM^{(K,\alpha,m)}\langle \tilde{\mathbf{0}}, \cdots, \tilde{\mathbf{0}} \rangle = \tilde{\mathbf{0}}$$

and

$$M - pFSMWM^{(K,\alpha,m)}\langle \tilde{\mathbf{1}}, \cdots, \tilde{\mathbf{1}} \rangle = \tilde{\mathbf{1}}$$

3. *(Monotonicity) If* $(f_k, P) \tilde{\leq} (g_k, P)$ *for all k, then*

$$M - pFSMWM^{(K,\alpha,m)}\langle (f_1, P), \cdots, (f_K, P) \rangle \tilde{\leq} M - pFSMWM^{(K,\alpha,m)}\langle (g_1, P), \cdots, (g_K, P) \rangle$$

4. *(Boundedness)*

$$\min_k \{(f_k, P)\}_{k=1}^{K} \tilde{\leq} M - pFSMWM^{(K,\alpha,m)}\langle (f_1, P), \cdots, (f_K, P) \rangle \tilde{\leq} \max_k \{(f_k, P)\}_{k=1}^{K}$$

5. *(Commutativity or Symmetry)*

$$M - pFSMWM^{(K,\alpha,m)}\langle (f_1, P), \cdots, (f_K, P) \rangle = M - pFSMWM^{(K,\alpha,m)}\langle (f_{\sigma(1)}, P), \cdots, (f_{\sigma(K)}, P) \rangle$$

where σ is any permutation of $\{1,2,\cdots,k\}$;

Proof. 1. Let for all k: $(f_k,P)=(f,P)$. Thus, it is clear that the distinct α-combinations of K objects is reduced to the trivial case K-combination of K with $C_{K,K}=1$ and $\omega_k=\frac{1}{K}$ for all k, i.e., the unweighted case. Thus,

$$\begin{aligned} M-pFSMWM^{(K,\alpha,m)}\langle f_1,\cdots,f_K\rangle(p)(u) &= M-pFSMWM^{(K,K,m)}\langle f,\cdots,f\rangle(p)(u) \\ &= (f^1(p)(u),\cdots,f^m(p)(u))=f(p)(u) \end{aligned}$$

since $\sum_{k=1}^{K} k.(\omega_k-\omega_{k+1}).\min\{f^1(p)(u),\cdots,f^1(p)(u)\} = \sum_{k=1}^{K-1} k.(\frac{1}{K}-\frac{1}{K}).f^1(p)(u)+K.\frac{1}{K}.f^1(p)(u)$.

2. First, assume for all k: $(f_k,P)=\tilde{\mathbf{0}}$. Then, by Property 1 of Theorem 2, we have $M-pFSMWM^{(K,\alpha,m)}\langle\tilde{\mathbf{0}},\cdots,\tilde{\mathbf{0}}\rangle=\tilde{\mathbf{0}}$. The property $M-pFSMWM^{(K,\alpha,m)}\langle\tilde{\mathbf{1}},\cdots,\tilde{\mathbf{1}}\rangle=\tilde{\mathbf{1}}$ follows the similar way for $(f_k,P)=\tilde{\mathbf{1}},\forall k$.
3. Let $(f_k,P)\tilde{\leq}(g_k,P)$ for all k. Then, for each $s=1,2,\cdots,m$: $f_k^s(p)(u)\leq g_k^s(p)(u)$. Thus, the condition is hold since $\min\{f^s_{\sigma(\delta_1(l))}(p)(u),\cdots,f^s_{\sigma(\delta_k(l))}(p)(u)\}_{\delta_k(l)\in\Delta_{K,\alpha}(l)}\leq\min\{g^s_{\sigma(\delta_1(l))}(p)(u),\cdots,g^s_{\sigma(\delta_k(l))}(p(u))\}_{\delta_k(l)\in\Delta_{K,\alpha}(l)}$: $s=1,2,\cdots,m$, for any l of the $C_{K,\alpha}$ possible choices of K.
4. Let for $p\in P$ and $u\in U$: $\min_k f_k^s(p)(u)=B_*{}^s$ and $\max_k f_k^s(p)(u)=B^{*s}$ for each $s=1,2,\cdots,m$. Then, for all k: $\boldsymbol{B}_*\leq f_k(p)(u)\leq\boldsymbol{B}^*$ where $\boldsymbol{B}_*=(B_*^1,\cdots,B_*^m)$ and $\boldsymbol{B}^*=(B^{*1},\cdots,B^{*m})$. Hence, by Properties 1 and 3 of Theorem 2, the inequality holds.
5. It is trivial from Definition 5.

□

Theorem 3. *Let $(f_1,P),\cdots,(f_K,P)$, where $K\geq2$, be some m-polar fuzzy soft sets over U and P such that for all k: $f_k(p)(u)=(f_k^1(p)(u),\cdots,f_k^m(p)(u))\in[0,1]^m$ for $p\in P$ and $u\in U$, with non-negative weights $\omega_1,\cdots,\omega_K\in[0,1]$ where $\sum_{k=1}^{K}\omega_k=1$. Then, the aggregated value $M-pFSMWM^{(K,\alpha,m)}\langle(f_1,P),\cdots,(f_K,P)\rangle$ is still an m-polar fuzzy soft set over U.*

Proof. Let $D_K=\{(f_k,P)|f_k(p)(u)=(f_k^1(p)(u),\cdots,f_k^m(p)(u))\in[0,1]^m;k=1,2,\cdots,K,p\in P\}$ be a set of m-polar fuzzy soft arguments. Since, for each k, $\mathbf{0}\leq f_k(p)(u)\leq\mathbf{1}$, then clearly by Theorem 2 we have $\mathbf{0}\leq M-pFSMWM^{(K,\alpha,m)}\langle f_1,\cdots,f_K\rangle(p)(u)\leq\mathbf{1}$. This means that $M-pFSMWM^{(K,\alpha,m)}\langle f_1,\cdots,f_K\rangle(p)(u)\in[0,1]^m$. Now, define the function $F:P\to m(U)$ such that for any $p\in P$ and $u\in U$, $F(p)(u)=M-pFSMWM^{(K,\alpha,m)}\langle f_1,\cdots,f_K\rangle(p)(u)$, given by Equation (5), which is an m-tuple of real numbers in the unit interval $[0,1]$. This shows the $M-pFSMWM$ operator of m-polar fuzzy soft sets is still an m-polar fuzzy soft set. □

Remark 1. *According to Definition 5 and Theorem 3, weights in M-pFSMWM operator first re-order the position of arguments, means in the re-ordered list the first object has the biggest weight. Then, the aggregated value is computed based on the weighting vector $\boldsymbol{\omega}=(\omega_1,\cdots,\omega_K)^T$ related to the m-polar fuzzy soft sets $(f_1,P),\cdots,(f_K,P)$. Thus, the weights reflect positions and importance degrees of input arguments in the aggregated value in comparison with IOWA and IOWG operators where weights only show the position of arguments. Moreover, if α shows the consensus degree, then the α-combinations of set $\{1,2,\cdots,K\}$ express different scenarios of agreement among K decision makers where the decision of first, second, ... , and last α individuals are checked one by one. Further, by choosing $l=1,2,\cdots,C_{K,\alpha}$, all possible choices of agreement between K experts at consensus degree α are considered. Thus, the concept of fuzzy majority, expressed by linguistic variables such as "half plus one", "more than 75%", "most", and "almost all" can be taken into account by choosing $\frac{K}{2}+1\leq\alpha\leq K$ if K is an even number and $\frac{K+1}{2}\leq\alpha\leq K$ if K is an odd number.*

From Theorems 2 and 3, the $M-pFSMWM$ operator degenerates to some special aggregation operators as follows.

Theorem 4. *Let $\{(f_k, P)\}_{k=1}^{K}$ be a set of m-polar fuzzy soft sets over U with non-negative weights ω_k that, for all k, $\omega_k \in [0,1]$ and $\sum_{k=1}^{K} \omega_k = 1$. Then, the $M-pFSMWM$ operator degenerates to some special aggregation operators as follows.*

1. *If $\omega = (0, \cdots, \underbrace{1}_{j-th}, \cdots, 0)^T$ i.e., $\omega_j = 1$ for $k = j$ and $\omega_k = 0$ for $k \neq j$, then*

$$M - pFSMWM^{(K,\alpha,m)}\langle f_1, \cdots, f_K\rangle(p)(u) = f_j(p)(u)$$

2. *When $K = \alpha$, we have*

$$M - pFSMWM^{(K,\alpha,m)}\langle f_1 \cdots, f_K\rangle(p)(u) = \langle \overset{\otimes}{\underset{\omega_1,\cdots,\omega_K}{\min}} \{f_k^1(p)(u)\}_{k=1}^{K}, \cdots, \overset{\otimes}{\underset{\omega_1,\cdots,\omega_K}{\min}} \{f_k^m(p)(u)\}_{k=1}^{K}\rangle$$

which is called the M-pFS weighted minimum operator.

3. *When $K = \alpha$: if $\omega = (\frac{1}{K}, \frac{1}{K}, \cdots, \frac{1}{K})^T$, then*

$$M - pFSMWM^{(K,\alpha,m)}\langle f_1 \cdots, f_K\rangle(p)(u) = \langle \min_k \{f_k^1(p)(u)\}_{k=1}^{K}, \cdots, \min_k \{f_k^m(p)(u)\}_{k=1}^{K}\rangle$$

which is the M-pFS minimum operator.

4. *When $K = \alpha$: if $f_{\sigma(1)} \tilde{\geq} \cdots \tilde{\geq} f_{\sigma(K)}$; and $\omega_{\sigma(1)} = 1$ and $\omega_{\sigma(k)} = 0$ for all $k \neq 1$, then*

$$M - pFSMWM^{(K,\alpha,m)}\langle f_1 \cdots, f_K\rangle(p)(u) = \langle \max_k \{f_k^1(p)(u)\}_{k=1}^{K}, \cdots, \max_k \{f_k^m(p)(u)\}_{k=1}^{K}\rangle$$

which is the M-pFS maximum operator.

5. *When $K = \alpha$: If $f_{\sigma(1)} \tilde{\geq} \cdots \tilde{\geq} f_{\sigma(K)}$; and $\omega_{\sigma(K)} = 1$ and $\omega_{\sigma(k)} = 0$ for all $k \neq K$, then*

$$M - pFSMWM^{(K,\alpha,m)}\langle f_1 \cdots, f_K\rangle(p)(u) = \langle \min_k \{f_k^1(p)(u)\}_{k=1}^{K}, \cdots, \min_k \{f_k^m(p)(u)\}_{k=1}^{K}\rangle$$

Proof. 1. Let $\omega = (\omega_1, \cdots, 1, \cdots, \omega_K)^T$, then in any lth α-combination of K objects involving jth element, the value of $f_j(p)(u)$ for $p \in P$ and $u \in U$ is interpreted as the first object, i.e., $f_{\sigma(\delta_1(l))}(p)(u)$, where $\omega_{\sigma(\delta_1(l))} = \omega_j = 1$ and for rest $\omega_k = 0$. Thus, by using Equation (5): $M - pFSMWM^{(K,\alpha,m)}\langle f_1, \cdots, f_K\rangle(p)(u) = (f_j^1(p)(u), \cdots, f_j^m(p)(u)) = f_j(p)(u)$.

2. Let $K = \alpha$. Then, we have $C_{K,K} = 1$ (only one trivial combination) and thus $\Delta_{K,K}(1) = \{1, \cdots, K\}$. Hence, by Equation (5):

$$\begin{aligned} M - pFSMWM^{(K,\alpha,m)}\langle f_1, \cdots, f_K\rangle(p)(u) &= M - pFSMWM^{(K,K,m)}\langle f_1, \cdots, f_K\rangle(p)(u) \\ &= \Big\langle \overset{\otimes}{\underset{\omega_1,\cdots,\omega_K}{\min}} \{f_1^1(p)(u), \cdots, f_K^1(p)(u)\}, \cdots, \\ &\quad \overset{\otimes}{\underset{\omega_1,\cdots,\omega_K}{\min}} \{f_1^m(p)(u), \cdots, f_K^m(p)(u)\}\Big\rangle \end{aligned}$$

3. When $\omega = (\frac{1}{K}, \frac{1}{K}, \cdots, \frac{1}{K})$, then the resultant weighted minimum in Part 2 of Theorem 4 acts as the standard (unweighted) minimum operator. Thus, the $M - pFMWM$ operator is derived by the minimum operator, easily.

4. When $\omega_{\sigma(1)} = 1$ and $\omega_{\sigma(k)} = 0$ for all $k \neq 1$, then by Part 1 of Theorem 4 we have $M - pFSMWM^{(K,\alpha,m)}\langle f_1, \cdots, f_K\rangle(p)(u) = f_{\sigma(1)}(p)(u)$ that is the largest argument since $f_{\sigma(1)} \tilde{\geq} \cdots \tilde{\geq} f_{\sigma(K)}$, i.e., the $M - pFMWM$ operator is derived by the maximum operator.

5. When $\omega_{\sigma(K)} = 1$ and $\omega_{\sigma(k)} = 0$ for all $k \neq K$, then by Part 1 of Theorem 4 we have $M - pFSMWM^{(K,\alpha,m)}\langle f_1, \cdots, f_K\rangle(p)(u) = f_{\sigma(K)}(p)(u)$ that is the lowest argument since $f_{\sigma(1)} \tilde{\geq} \cdots \tilde{\geq} f_{\sigma(K)}$, i.e., the $M - pFMWM$ operator is derived by the minimum operator.

□

4. Application of M-pFSMWM Operator in Group Decision-Making

In this section, the M-pFSMWM operator is applied to handle group decision-making problems with weighted m-polar fuzzy soft inputs.

In a group decision-making problem with m-polar fuzzy soft information, let $U = \{u_1, u_2, \cdots, u_N\}$ and $P = \{p_1, p_2, \cdots, p_M\}$ be the finite sets of alternatives and parameters, respectively, where $\boldsymbol{\lambda} = (\lambda_{p_1}, \lambda_{p_2}, \cdots, \lambda_{p_M})^T$ is the weighting vector for the parameter set P such that $\forall y$: $\lambda_{p_y} \in [0,1]$ and $\sum_{y=1}^{M} \lambda_{p_y} = 1$. Additionally, let each p_y be a multi-polar parameter with m different aspects or features such that $\boldsymbol{\lambda}_{p_y} = (\lambda_{p_y}^1, \lambda_{p_y}^2, \cdots, \lambda_{p_y}^m)^T$ is the weighting vector for the parameter $p_y \in P$ where $\forall s$: $\lambda_{p_y}^s \in [0,1]$ and $\sum_{s=1}^{m} \lambda_{p_y}^s = 1$. Suppose that $D_K = \{f_1, f_2, \cdots, f_K\}$ is the set of decision makers and $\boldsymbol{\omega} = (\omega_1, \omega_2, \cdots, \omega_K)^T$ is the weighting vector of f_k where, for all k: $\omega_k \in [0,1]$ and $\sum_{k=1}^{K} \omega_k = 1$. Assume that each decision maker f_k applies an m-polar fuzzy soft set to present the linguistic evaluation about alternatives such that $f_k(p_y)(u_i) = (f_k^1(p_y)(u_i), \cdots, f_k^m(p_y)(u_i)) \in [0,1]^m$ and each $f_k^s(p_y)(u_i)$ shows the satisfaction degree of alternative u_i about feature s of attribute p_y. Moreover, let the required consensus degree α mean an alternative may be selected if it is acceptable for at least α individuals.

After each expert prepares a linguistic or numerical judgment of alternatives based on the parameters p_y, the first stage is to reach consensus among a fuzzy majority or a partial agreement of them. This step is handled through the proposed aggregation operator M-pFSMWM by Equation (5) of the previous Section 3. The second stage of a MAGDM problem aims to find the best option with respect to the collective view. Thus, a ranking procedure is needed to derive the optimum choice. In the following subsections, we first define a fuzzy soft preference relationship over the universe U based on the collective view obtained by M-pFSMWM operator and then, propose a new score value function for ranking the preference order of objects.

4.1. A Fuzzy Soft Preference Relationship

The aim of this section is to define an square matrix $N \times N$ based on a fuzzy soft preference relationship over the set of alternatives U.

Let the m-tuple $M - pFSMWM^{(K,\alpha,m)}\langle f_1, \cdots, f_K\rangle(p_y)(u_i) = (u_{iy}^1, \cdots, u_{iy}^m)$ present the performance of alternative $u_i \in U$ based on parameter $p_y \in P$ and collective view obtained by M-pFSMWM operator. We define a fuzzy soft preference relationship on U by the mapping $\tilde{R} : P \to [0,1]^{U \times U}$ where for each $p_y \in P$, $\tilde{R}(p_y)$ is a fuzzy preference relationship on U which is characterized by the membership function $\tilde{R}(p_y) : U \times U \to [0,1]$. For any $(u_i, u_j) \in U \times U$ we define $\tilde{R}(p_y)(u_i, u_j) = \tilde{r}_{ij}(p_y)$ such that

$$\tilde{r}_{ij}(p_y) = \frac{\sum_{s=1}^{m} \lambda_{p_y}^s u_{iy}^s}{\sum_{s=1}^{m} \lambda_{p_y}^s u_{iy}^s + \sum_{s=1}^{m} \lambda_{p_y}^s u_{jy}^s} \tag{6}$$

Definition 6. *Suppose that $U = \{u_1, u_2, \cdots, u_N\}$ is the set of alternatives and $p_y \in P = \{p_1, p_2, \cdots, p_M\}$ is a parameter including m different aspects $\{p_y^1, p_y^2, \cdots, p_y^m\}$. Let $\boldsymbol{\lambda}_{p_y} = (\lambda_{p_y}^1, \cdots, \lambda_{p_y}^m)^T \in [0,1]^m$ show the weighting vector for parameter $p_y \in P$ where $\sum_{s=1}^{m} \lambda_{p_y}^s = 1$. If $\tilde{R}(p_y)$ is the fuzzy preference relationship on U defined by (6), then the $N \times N$ matrix $\tilde{R}(p_y) = [\tilde{r}_{ij}(p_y)]_{N \times N}$ defined by*

$$\tilde{R}(p_y) = [\tilde{r}_{ij}(p_y)]_{N \times N} = \begin{bmatrix} \tilde{r}_{11}(p_y) & \tilde{r}_{12}(p_y) & \cdots & \tilde{r}_{1N}(p_y) \\ \tilde{r}_{21}(p_y) & \tilde{r}_{22}(p_y) & \cdots & \tilde{r}_{2N}(p_y) \\ \vdots & \vdots & & \vdots \\ \tilde{r}_{N1}(p_y) & \tilde{r}_{N2}(p_y) & \cdots & \tilde{r}_{NN}(p_y) \end{bmatrix} \tag{7}$$

where $\tilde{r}_{ij}(p_y) \in [0,1]$ interprets the degree of preference of the alternative u_i over the alternative u_j with respect to the parameter $p_y \in P$. Moreover, $\tilde{r}_{ii}(p_y) = 0.5$ for all $1 \le i \le N$.

In Definition 6, $\tilde{r}_{ij}(p_y) = 0.5$ shows indifference between u_i and u_j based on the parameter $p_y \in P$, which is represented by $u_i \sim_{p_y} u_j$, while $\tilde{r}_{ij}(p_y) \in (0.5, 1]$ shows u_i is preferred to u_j based on the parameter $p_y \in P$ at degree $\tilde{r}_{ij}(p_y)$, i.e., $u_i \succ_{p_y} u_j$. Moreover, the fuzzy soft preference relationship $\tilde{R} : P \to [0,1]^{U \times U}$ can be represented by matrix $\tilde{R} = [\tilde{\boldsymbol{r}}_{ijy}]_{N \times N}$ where each entry $\tilde{\boldsymbol{r}}_{ijy} = [\tilde{r}_{ij}(p_y)]_{N \times N}$ is in fact the $N \times N$ matrix $\tilde{R}(p_y)$. Hence, we have

$$\tilde{R} = [\tilde{\boldsymbol{r}}_{ij}]_{N \times N} = \Big[[\tilde{r}_{ij}(p_1)]_{N \times N} \quad [\tilde{r}_{ij}(p_2)]_{N \times N} \quad \cdots \quad [\tilde{r}_{ij}(p_M)]_{N \times N} \Big]$$

Proposition 3. *The fuzzy preference relationship $\tilde{R}(p_y)$ clearly satisfies the following statements:*

1. $\tilde{r}_{ij}(p_y) + \tilde{r}_{ji}(p_y) = 1$
2. $(\frac{\tilde{r}_{ji}(p_y)}{\tilde{r}_{ij}(p_y)})(\frac{\tilde{r}_{kj}(p_y)}{\tilde{r}_{jk}(p_y)}) = \frac{\tilde{r}_{ki}(p_y)}{\tilde{r}_{ik}(p_y)}$
3. *If $\tilde{r}_{ij}(p_y) \geq 0.5$ and $\tilde{r}_{jk}(p_y) \geq 0.5$, then $\tilde{r}_{ik}(p_y) \geq \max\{\tilde{r}_{ij}(p_y), \tilde{r}_{jk}(p_y)\}$.*

for all $i, j, k = 1, 2, \cdots, N$ and $y = 1, 2, \cdots, M$.

Proof. Item 1 is easily checked by Equation (6). Parts 2 and 3 are obtained by using $\tilde{r}_{ij}(p_y) = \frac{\sum_{s=1}^{m} \lambda_{p_y}^s u_{iy}^s}{\sum_{s=1}^{m} \lambda_{p_y}^s u_{iy}^s + \sum_{s=1}^{m} \lambda_{p_y}^s u_{jy}^s} = \frac{1}{1 + \frac{\sum_{s=1}^{m} \lambda_{p_y}^s u_{jy}^s}{\sum_{s=1}^{m} \lambda_{p_y}^s u_{iy}^s}}$ and $\frac{\tilde{r}_{ji}(p_y)}{\tilde{r}_{ij}(p_y)} = \frac{1}{\tilde{r}_{ij}(p_y)} - 1$ (see also Theorem 1). □

Proposition 3 says that the fuzzy preference relationship $\tilde{R}(p_y)$ satisfies the reciprocity and the restricted max-max transitivity. This means that, if $u_i \succ_{p_y} u_j$ and $u_j \succ_{p_y} u_k$, then $u_i \succ_{p_y} u_k$.

Definition 7. *Let $\tilde{R}(p_y) = [\tilde{r}_{ij}(p_y)]_{N \times N}$ be the $N \times N$ matrix defined in Definition 6. Then, the $N \times N$ matrix $\tilde{A} = [\tilde{\boldsymbol{a}}_{ij}]_{N \times N}$ defined by*

$$\tilde{A} = [\tilde{\boldsymbol{a}}_{ij}]_{N \times N} = \begin{bmatrix} \tilde{\boldsymbol{a}}_{11} & \tilde{\boldsymbol{a}}_{12} & \cdots & \tilde{\boldsymbol{a}}_{1N} \\ \tilde{\boldsymbol{a}}_{21} & \tilde{\boldsymbol{a}}_{22} & \cdots & \tilde{\boldsymbol{a}}_{2N} \\ \vdots & \vdots & & \vdots \\ \tilde{\boldsymbol{a}}_{N1} & \tilde{\boldsymbol{a}}_{N2} & \cdots & \tilde{\boldsymbol{a}}_{NN} \end{bmatrix} \tag{8}$$

where is an M-polar fuzzy preference relationship on U characterized by the membership function $\tilde{A} : U \times U \to [0,1]^M$ and, for $1 \leq i, j \leq N$, $\tilde{\boldsymbol{a}}_{ij} = (\tilde{a}_{ij}^1, \tilde{a}_{ij}^2, \cdots, \tilde{a}_{ij}^M) \in [0,1]^M$ shows the degrees of preference of the alternative u_i over the alternative u_j with respect to the parameter set P. For each y: $\tilde{a}_{ij}^y = 1$ if $\tilde{r}_{ij}(p_y) > 0.5$, $\tilde{a}_{ij}^y = 0.5$ if $\tilde{r}_{ij}(p_y) = 0.5$, and $\tilde{a}_{ij}^y = 0$ if $\tilde{r}_{ij}(p_y) < 0.5$. Clearly, for all i, each entry $\tilde{\boldsymbol{a}}_{ii} = \underbrace{(0.5, 0.5, \cdots, 0.5)}_{M-times}$.

For all i, j: $\tilde{A}(u_i, u_j) = \tilde{\boldsymbol{a}}_{ij} = \underbrace{(0.5, 0.5, \cdots, 0.5)}_{M-times}$ shows indifference between u_i and u_j based on parameter set P, which is represented by $u_i \sim u_j$, while $\tilde{A}(u_i, u_j) = \tilde{\boldsymbol{a}}_{ij} = \underbrace{(1, 1, \cdots, 1)}_{M-times}$ shows u_i is preferred to u_j ($u_i \succ u_j$) for all parameters.

Proposition 4. *The following statements hold for M-polar fuzzy preference relation $\tilde{A}$.*

1. $\tilde{\boldsymbol{a}}_{ij} + \tilde{\boldsymbol{a}}_{ji} = \underbrace{(1, 1, \cdots, 1)}_{M-times}$.
2. *If $\tilde{\boldsymbol{a}}_{ij} \cdot \tilde{\boldsymbol{a}}_{jk} = (\tilde{a}_{ij}^1, \tilde{a}_{ij}^2, \cdots, \tilde{a}_{ij}^M) \cdot (\tilde{a}_{jk}^1, \tilde{a}_{jk}^2, \cdots, \tilde{a}_{jk}^M) = (\tilde{a}_{ij}^1 \cdot \tilde{a}_{jk}^1, \tilde{a}_{ij}^2 \cdot \tilde{a}_{jk}^2, \cdots, \tilde{a}_{ij}^M \cdot \tilde{a}_{jk}^M)$, then $\tilde{\boldsymbol{a}}_{ij} \cdot \tilde{\boldsymbol{a}}_{jk} = \tilde{\boldsymbol{a}}_{ik}$.*
3. *If $\tilde{\boldsymbol{a}}_{ij} = \underbrace{(1, 1, \cdots, 1)}_{M-times}$ and $\tilde{\boldsymbol{a}}_{jk} = \underbrace{(1, 1, \cdots, 1)}_{M-times}$, then $\tilde{\boldsymbol{a}}_{ik} = \underbrace{(1, 1, \cdots, 1)}_{M-times}$.*

for all $i, j, k \in \{1, 2, \cdots, N\}$.

Proof. It is obtained easily by Proposition 3. □

4.2. An Approach to Group Decision-Making Based on M-pFSMWM Operator

The second stage of a group decision-making is to reach the process of selection based on an overall performance of alternatives in terms of the crisp or partial agreement among the experts. In this section, we introduce a new procedure for solving group decision-making problems based on the M-pFSMWM operator and the pairwise comparisons of alternatives that are obtained by the $N \times N$ matrix $\tilde{A}$ defined in Equation (8).

By using Definitions 6 and 7, a new overall score value function $S : U \to R$ over the universe U is defined as below.

Definition 8. *The mapping* $S : U \to R$ *defined by*

$$S_i = \sum_{j=1, j \neq i}^{N} \sum_{y=1}^{M} \lambda_{p_y} \cdot \tilde{a}_{ij}^{y} \tag{9}$$

for $i = 1, \cdots, N$ *is called the score value function over* U *where* λ_{p_y} *shows the importance degree of parameter* $p_y \in P$ *and* $\tilde{\mathbf{a}}_{ij} = (\tilde{a}_{ij}^1, \tilde{a}_{ij}^2, \cdots, \tilde{a}_{ij}^M) \in [0, 1]$ *is the entry in ith row and jth column of matrix* $\tilde{A}$.

In the following, we apply the M-pFSMWM operator for solving MAGDM problems based on m-polar fuzzy soft information (Algorithm 1).The proposed procedure uses Equation (9) for ranking the preference order of objects. We also clarify the idea of the proposed method in Algorithm 1 by the given flowchart in Figure 1.

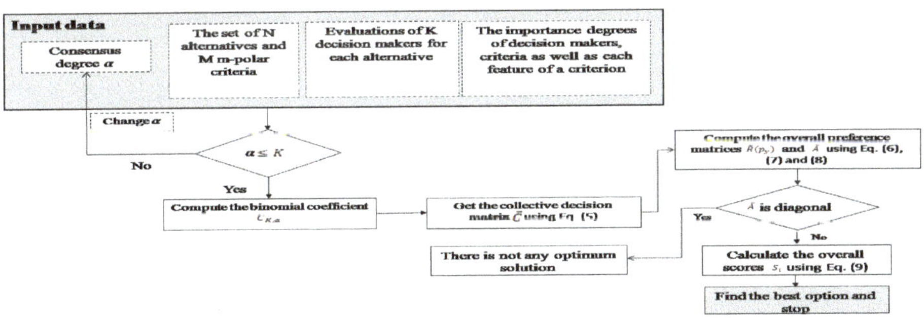

Figure 1. Flowchart of the proposed Algorithm 1.

Algorithm 1. Finding the optimum solution in MAGDM problems based on M-pFSMWM operator for M-pFSSs.

Input :K different m-polar fuzzy soft sets $(f_1, P), \cdots, (f_K, P)$ over the set U such that $|U| = N$ and $|P| = M$ where m is the number of different aspects of each parameter and K shows the number of decision makers. Weighting vectors $\boldsymbol{\omega} = (\omega_1, \omega_2, \cdots, \omega_K)^T$, $\boldsymbol{\lambda} = (\lambda_{p_1}, \lambda_{p_2}, \cdots, \lambda_{p_M})^T$, and $\lambda_{p_y} = (\lambda^1_{p_y}, \lambda^2_{p_y}, \cdots, \lambda^m_{p_y})^T$ for $y = 1, \cdots, M$. Consensus degree α where $\alpha \leq K$.

Output: Optimum solution.

```
begin
  Step 1. Calculate L = C_{K,α} = K! / (α!(K−α)!).
  Step 2. for l = 1,2,...,L do
      Find the lth α-combination of the set {1,2,··· ,K} presented by Δ_{K,α}(l) where |Δ_{K,α}(l)| = α.
  end
  Step 3. for i = 1,2,...,N do
      for y = 1,2,...,M do
          for s = 1,2,...,m do
              Compute the m-tuple M − pFSMWM^{(K,α,m)}(f_1,··· ,f_K)(p_y)(u_i) = (u^1_{iy},··· ,u^m_{iy}) by
               using Equation (5) to derive matrix C̄ = [c̄_{iy}]_{N×M} such that c̄_{iy} = (u^1_{iy}, u^2_{iy},··· ,u^m_{iy}).
          end
      end
  end
  Step 4. for y = 1,2,...,M do
      for i = 1,2,...,N do
          for j = 1,2,...,N do
              Utilize Equations (6) and (7) to compute matrix R̃(p_y) = [r̃_{ij}(p_y)]_{N×N}.
          end
      end
  end
  Step 5. Regarding fuzzy relation matrices R̃(p_y) and by Equation (8), construct the collective overall
   preference matrix Ã = [ã_{ij}]_{N×N}.
  if Ã is a diagonal matrix then
      There is no optimal option over U;
  else
      Go to the Step 6.
  end
  Step 6. for i = 1,2,...,N do
      Using Equation (9) to calculate the overall score value S_i.
  end
  Step 7. Rank the alternatives u_i based on S_i and then select the best one(s).
end
```

Remark 2 (Analysing Algorithm 1). *Let K decision makers evaluate N number of alternatives based on M number of parameters where the m-polar fuzzy soft sets are applied to present their linguistic evaluations of the alternatives. According to Algorithm 1, we first utilize the M-pFSMWM operator to obtain a collective view of decision makers. The M-pFSMWM operator allows us to have not only partial agreement within a group, such as "almost all", "most", "more than half" etc., but also different choices for a partial agreement at the consensus degree* α*.*

To this end, Algorithm 1 starts with finding the subsets $\Delta_{K,\alpha}(l) \subseteq \{1, 2, \cdots, K\}$ *where* $l = 1, 2, \cdots, C_{K,\alpha}$*. This helps us to check all possible cases of agreement between K decision makers at consensus level* α*. In fact, the value of* α *shows the number of possible iterations of Algorithm 1 (*$1 \leq \alpha \leq K$*). By repeating Steps 1 and 2 for different value* α *until* $\alpha \leq K$*, the aggregated value moves from the minimum value to the maximum value. This guaranties Algorithm 1 is convergent (please also see Theorems 2 and 4). Then, at Step 3, matrix* $\bar{C}$ *is driven. Each entry of* $\bar{C}$ *shows the performance of alternative* u_i *based on parameter* p_y *and the collective view of experts at degree* α*. In Step 4, the fuzzy preference relations* $\tilde{R}(p_y)$ *(for* $p_y \in P$*) give a comparison of objects based on the collective view of decision makers and each parameter* p_y*. The information of matrices* $\tilde{R}(p_y)$ *are then converted to the M-polar fuzzy soft preference relation* $\tilde{A} = [\tilde{\mathbf{a}}_{ij}]_{N \times N}$*, in Step 5, for providing comparison results where* $\tilde{\mathbf{a}}_{ij} = (\tilde{a}^1_{ij}, \tilde{a}^2_{ij}, \cdots, \tilde{a}^M_{ij}) \in [0,1]^M$ *defined by* $\tilde{a}^y_{ij} = 0.5$ *if* $\tilde{r}_{ij}(p_y) = 0.5$*,* $\tilde{a}^y_{ij} = 1$ *if* $\tilde{r}_{ij}(p_y) > 0.5$*, otherwise* $\tilde{a}^y_{ij} = 0$*. Moreover, if* $\tilde{A}$ *is an upper triangle matrix such that* $\tilde{\mathbf{a}}_{ii} = (0.5, 0.5, \cdots, 0.5)$*;* $\tilde{\mathbf{a}}_{ij} = (1, 1, \cdots, 1)$ *for* $i < j$*; and* $\tilde{\mathbf{a}}_{ij} = (0, 0, \cdots, 0)$ *for* $i > j$*, then we have the following descending chain* $u_1 \succ u_2 \succ \cdots \succ u_N$ *on U. If* $\tilde{A}$ *is a lower triangle matrix such that* $\tilde{\mathbf{a}}_{ii} = (0.5, 0.5, \cdots, 0.5)$*;* $\tilde{\mathbf{a}}_{ij} = (1, 1, \cdots, 1)$ *for* $i > j$*; and* $\tilde{\mathbf{a}}_{ij} = (0, 0, \cdots, 0)$ *for* $i < j$*, then we have the ascending chain* $u_N \succ u_{N-1} \succ \cdots \succ u_1$ *on U. However, if* $\tilde{A}$ *is a diagonal matrix, then there is no optimal option on U. In the last step of Algorithm 1, the best option is* *selected based on its rank in the resultant preference order. For MAGDM problems with benefit criteria, means more is better, the alternative with the highest score can be selected as the best option. However, for the problems dealing with cost criteria the counter condition should be considered.*

5. The M-pFSIOWA and M-pFSIOWG Operators

To compare different m-polar fuzzy soft weighted aggregation operators with the proposed operator in Section 3.2, in this section, we develop the m-polar fuzzy soft induced ordered weighted average (M-pFSIOWA) operator and m-polar fuzzy soft induced ordered weighted geometric (M-pFSIOWG) operator, which are the extensions of IOWA and IOWG operators, respectively. The re-ordering step of M-pFSIOWA and M-pFSIOWG operators are defined based on the weights of arguments $\omega = (\omega_1, \cdots \omega_K)^T$. Since the M-pFSIOWA and M-pFSIOWG operators are defined in the domain of M-pFSSs, these new families of IOWA and IOWG operators give more general methods for aggregating data than traditional IOWA and IOWG operators.

Motivated by development of OWA operator and OWG operator for FSs [38,39] and IFSSs [21], the extensions of these two aggregation operators for M-pFSSs are defined as below.

Definition 9. *1. The M-pFSIOWA operator of dimension k is the mapping $M - pFSIOWA : \bigcup_{K\in\mathbb{N}}(mfs(U))^K \to mfs(U)$ such that for an associated weighting vector $w = (w_1, w_2, \cdots, w_K)^T$, where $w_j \in [0,1]$ and $\sum_{j=1}^{K} w_j = 1$, is defined as below:*

$$M - pFSIOWA\langle f_1, \cdots, f_K\rangle(p_y)(u_i) = \left\langle \sum_{j=1}^{K} w_j \cdot F_{jyi}^1, \cdots, \sum_{j=1}^{K} w_j \cdot F_{jyi}^m \right\rangle \tag{10}$$

2. The M-pFSIOWG operator of dimension k is the mapping $M - pFSIOWG : \bigcup_{K\in\mathbb{N}}(mfs(U))^K \to mfs(U)$ such that for an associated weighting vector $w = (w_1, w_2, \cdots, w_K)^T$, where $w_j \in [0,1]$ and $\sum_{j=1}^{K} w_j = 1$, can be defined by

$$M - pFSIOWG\langle f_1, \cdots, f_K\rangle(p_y)(u_i) = \left\langle \prod_{j=1}^{K} (F_{jyi}^1)^{w_j}, \cdots, \prod_{j=1}^{K} (F_{jyi}^m)^{w_j} \right\rangle \tag{11}$$

where F_{jyi}^s is the kth value $f_k^s(p_y)(u_i)$ having the jth largest ω_j of the weighting vector $\omega = (\omega_1, \omega_2, \cdots, \omega_K)^T$ for M-pFSSs $(f_1, P), \cdots, (f_K, P)$.

The main steps of these operations are the re-ordering step according to the weighting vector $\omega = (\omega_1, \omega_2, \cdots, \omega_K)^T$ and then determining the associated weighting vector $w = (w_1, w_2, \cdots, w_K)^T$ to the aggregation operators M-pFSIOWA and M-pFSIOWG. Here, for each $1 \le s \le m$, $1 \le y \le M$, and $1 \le i \le N$, the collection: $f_1^s(p_y)(u_i), \cdots, f_K^s(p_y)(u_i)$ is re-ordered as $\langle \max\{\omega_k\}, F_{1yi}^s\rangle \ge \cdots \ge \langle \min\{\omega_k\}, F_{Kyi}^s\rangle$ where the weighting vector $\omega = (\omega_1, \omega_2, \cdots, \omega_K)^T$ shows the weights of different decision makers.

Theorem 5. *Let $(f_1, P), \cdots, (f_K, P)$, where $K \ge 2$, is some m-polar fuzzy soft set over U and P such that for all $1 \le k \le K$: $f_k(p_y)(u_i) = (f_k^1(p_y)(u_i), \cdots, f_k^m(p_y)(u_i)) \in [0,1]^m$ for $p_y \in P$ and $u_i \in U$, with non-negative weights $\omega_1, \cdots, \omega_K \in [0,1]$ where $\sum_{k=1}^{K} \omega_k = 1$. Then, the aggregated value $M - pFSIOWA\langle (f_1, P), \cdots, (f_K, P)\rangle$ and $M - pFSIOWG\langle (f_1, P), \cdots, (f_K, P)\rangle$ are still an m-polar fuzzy soft set over U.*

Proof. Define the function $F : P \to m(U)$ such that for any $p_y \in P$ and $u_i \in U$, $F(p_y)(u_i) = M - pFSIOWA\langle f_1, \cdots, f_K\rangle(p_y)(u_i)$ or $F(p_y)(u_i) = M - pFSIOWG\langle f_1, \cdots, f_K\rangle(p_y)(u_i)$. Then, the assertion is trivial from $f_k^s(p_y)(u_i) \in [0,1]$, $w_k \in [0,1]$, $\sum_{k=1}^{K} w_k = 1$, and convexity of $[0,1]$. □

The following properties are inherited to M-pFSIOWA and M-pFSIOWG operators from IOWA operator and IOWG operator, respectively.

Theorem 6. *Let $(f_1, P), \cdots, (f_K, P)$ be some m-polar fuzzy soft sets over U and P with non-negative weights $\omega_1, \cdots, \omega_K \in [0,1]$ where $\sum_{k=1}^{K} \omega_k = 1$. Let $w = (w_1, \cdots, w_K)^T$ be the associated weighting vector to the M-pFSIOWA and M-pFSIOWG operators. Then,*

1. *(Idempotency) If $(f_k, P) = (f, P)$ $\forall k$, then*

$$M - pFSIOWA\langle (f_1, P), \cdots, (f_K, P)\rangle = (f, P)$$

and

$$M - pFSIOWG\langle (f_1, P), \cdots, (f_K, P)\rangle = (f, P)$$

2. *(Monotonicity) If $(f_k, P)\tilde{\leq}(g_k, P)$ $\forall k$, then*

$$M - pFSIOWA\langle (f_1, P), \cdots, (f_K, P)\rangle \tilde{\leq} M - pFSIOWA\langle (g_1, P), \cdots, (g_K, P)\rangle$$

and

$$M - pFSIOWG\langle (f_1, P), \cdots, (f_K, P)\rangle \tilde{\leq} M - pFSIOWG\langle (g_1, P), \cdots, (g_K, P)\rangle$$

3. *(Boundedness)*

$$\min_k\{(f_k, P)\} \tilde{\leq} M - pFSIOWA\langle (f_1, P), \cdots, (f_K, P)\rangle \tilde{\leq} \max_k\{(f_k, P)\}$$

and

$$\min_k\{(f_k, P)\} \tilde{\leq} M - pFSIOWG\langle (f_1, P), \cdots, (f_K, P)\rangle \tilde{\leq} \max_k\{(f_k, P)\}$$

4. *(Commutativity or Symmetry)*

$$M - pFSIOWA\langle (f_1, P), \cdots, (f_K, P)\rangle = M - pFSIOWA\langle (f_{\sigma(1)}, P), \cdots, (f_{\sigma(K)}, P)\rangle$$

and

$$M - pFSIOWG\langle (f_1, P), \cdots, (f_K, P)\rangle = M - pFSIOWG\langle (f_{\sigma(1)}, P), \cdots, (f_{\sigma(K)}, P)\rangle$$

where σ is any permutation of $\{1, 2, \cdots, k\}$.

We can also obtain some spacial cases of M-pFSIOWA and M-pFSIOWG operators by using different choices for w.

Theorem 7. *Let $(f_1, P), \cdots, (f_K, P)$ be some m-polar fuzzy soft sets over U and P with non-negative weights $\omega_1, \cdots, \omega_K \in [0,1]$ where $\sum_{k=1}^{K} \omega_k = 1$. Let $w = (w_1, \cdots, w_K)^T$ be the associated weighting vector to the M-pFSIOWA and M-pFSIOWG operators. Then, the M-pFSIOWA operator and M-pFSIOWG operator degenerate to some special aggregation operators as follows.*

1. *If $w = (\frac{1}{K}, \cdots, \frac{1}{K})^T$, then*

$$M - pFSIOWA\langle f_1, \cdots, f_K\rangle(p_y)(u_i) = \langle \frac{1}{K}\sum_{j=1}^{K} F_{jyi}^1, \cdots, \frac{1}{K}\sum_{j=1}^{K} F_{jyi}^m\rangle$$

which we call the m-polar fuzzy soft arithmetic average operator, and

$$M - pFSIOWG\langle f_1, \cdots, f_K\rangle(p_y)(u_i) = \langle \prod_{j=1}^{K}(F_{jyi}^1)^{\frac{1}{K}}, \cdots, \prod_{j=1}^{K}(F_{jyi}^m)^{\frac{1}{K}}\rangle$$

which we call the m-polar fuzzy soft geometric average operator.

2. *If* $w = (1, 0, \cdots, 0)^T$, *then*

$$M - pFSIOWA\langle f_1, \cdots, f_K\rangle(p_y)(u_i) = \langle \max_j \{F^1_{jyi}\}^K_{j=1}, \cdots, \max_j \{F^m_{jyi}\}^K_{j=1}\rangle$$

and

$$M - pFSIOWG\langle f_1, \cdots, f_K\rangle(p_y)(u_i) = \langle \max_j \{F^1_{jyi}\}^K_{j=1}, \cdots, \max_j \{F^m_{jyi}\}^K_{j=1}\rangle$$

3. *If* $w = (0, \cdots, 0, 1)^T$, *then*

$$M - pFSIOWA\langle f_1, \cdots, f_K\rangle(p_y)(u_i) = \langle \min_j \{F^1_{jyi}\}^K_{j=1}, \cdots, \min_j \{F^m_{jyi}\}^K_{j=1}\rangle$$

and

$$M - pFSIOWG\langle f_1, \cdots, f_K\rangle(p_y)(u_i) = \langle \min_j \{F^1_{jyi}\}^K_{j=1}, \cdots, \min_j \{F^m_{jyi}\}^K_{j=1}\rangle$$

Application of M-pFSIOWA and M-pFSIOWG Operators in Group Decision-Making

In this section, similar to Algorithm 1, we apply the M-pFSIOWA operator and M-pFSIOWG operator to propose a procedure for solving MAGDM problems with M-pFS inputs as the following Algorithm 2.

Algorithm 2. Finding the optimum solution in MAGDM problems based on M-pFSIOWA or M-pFSIOWG operators for M-pFSSs.

Input :K different m-polar fuzzy soft sets $(f_1, P), \cdots, (f_K, P)$ over the set U such that $|U| = N$ and $|P| = M$ where m is the number of different aspects of each parameter and K shows the number of decision makers. Weighting vectors $\omega = (\omega_1, \cdots, \omega_K)^T$ related to the m-polar fuzzy soft sets $(f_1, P), \cdots, (f_K, P)$. Weighting vectors $w = (w_1, \cdots, w_K)^T$ related to the M-pFSIOWA or M-pFSIOWG operators. $\lambda = (\lambda_{p_1}, \lambda_{p_2}, \cdots, \lambda_{p_M})^T$, and $\lambda_{p_y} = (\lambda^1_{p_y}, \lambda^2_{p_y}, \cdots, \lambda^m_{p_y})^T$ for $y = 1, \cdots, M$.

Output:Optimum solution.

begin
- Step 1. **for** $i = 1, 2, \ldots, N$ **do**
  - **for** $y = 1, 2, \ldots, M$ **do**
    - **for** $s = 1, 2, \ldots, m$ **do**
      - Using the associated eighting vectors $w = (w_1, \cdots, w_K)^T$ to compute the m-tuple $M - pFSIOWA\langle f_1, \cdots, f_K\rangle(p_y)(u_i) = (u^1_{iy}, \cdots, u^m_{iy})$ or $M - pFSIOWG\langle f_1, \cdots, f_K\rangle(p_y)(u_i) = (u^1_{iy}, \cdots, u^m_{iy})$ by Equation (10) or Equation (11) and derive matrix $\tilde{C} = [\tilde{c}_{iy}]_{N\times M}$ such that $\tilde{c}_{iy} = (u^1_{iy}, u^2_{iy}, \cdots, u^m_{iy})$.
    - **end**
  - **end**
- **end**
- Step 2. **for** $y = 1, 2, \ldots, M$ **do**
  - **for** $i = 1, 2, \ldots, N$ **do**
    - **for** $j = 1, 2, \ldots, N$ **do**
      - Utilize Equation (6) and Equation (7) to compute matrix $\tilde{R}(p_y) = [\tilde{r}_{ij}(p_y)]_{N\times N}$ based on the matrix $\tilde{C} = [\tilde{c}_{iy}]_{N\times M}$ obtained in Step 1.
    - **end**
  - **end**
- **end**
- Step 3. Using Equation (8) to construct the collective overall preference matrix $\tilde{A} = [\tilde{a}_{ij}]_{N\times N}$ according to the matrices $\tilde{R}(p_y) = [\tilde{r}_{ij}(p_y)]_{N\times N}$ computed in Step 2.
- **if** $\tilde{A}$ *is a diagonal matrix* **then**
  - There is no optimal option over U;
- **else**
  - Go to the Step 4.
- **end**
- Step 4. **for** $i = 1, 2, \ldots, N$ **do**
  - Calculate the overall score value S_i based on the resultant matrix $\tilde{A}$ from Step 3 and Equation (9).
- **end**
- Step 5. Rank the alternatives u_i based on S_i and then select the best one(s).

end

Remark 3. *Note that Algorithm 2 starts with computing matrix $\bar{C}$ based on M-pFSIOWA or M-pFSIOWG operators rather than the M-pFSMWM operator used in Algorithm 1. In Steps 2 and 3, the entries of the resultant matrix $\bar{C}$ are used to compute the matrices $\tilde{R}(p_y)$ and $\tilde{A}$. Then, Algorithm 2 is followed similarly with Algorithm 1.*

Here, by repeating Step 1 for different iteration value α until $\alpha \leq K$, the aggregated value computed by M-pFSIOWA operator or M-pFSIOWG operator moves between the minimum value and the maximum value (please see Theorems 6 and 7). This guarantees Algorithm 2 is also convergent.

6. Illustrative Example

The general method for solving the MAGDM problems involves two main phases: (i) aggregation or consensus stage; and (ii) selection stage. The proposed methods in Equations (5), (10), and (11) can help us to aggregate different decision makers' judgments about the alternatives to obtain a collective decision matrix $\bar{C}$. Further, Equations (8) and (9) provide an overall preference matrix $\tilde{A}$ and the overall score values S_i for the alternatives, respectively, to cover the selection phase. In the following, we compare the proposed procedures in Algorithms 1 and 2 for MAGDM problems with m-polar fuzzy soft information by some numerical examples.

Hotel Booking Problem

In any trip, the problem of accommodation is one of the most important issues. The best option is always selected after comparing different residences based on some parameters, such as facilities and location of hotels and the budget. In this section, we discuss the problem of hotel booking, which is about selecting the best hotel to stay regarding a list of criteria, to provide a real-life example which shows the application of our method in decision-making problems. The input data are obtained from the "www.agoda.com" website, an online hotel booking service. This website provides some online questioners that passengers (guests) can fill them up to share their experiences about the hotel that they have staid. Guests are classified into the five main groups: families with young children, families with elder children, couples, solo travelers, and group of friends. Each hotel is characterized based on several criteria including "Comfort", "Services", "Location", and "Food" according to the guests' idea by numbers between zero and ten. Note that, here, all collected data are divided to ten to be in the unit interval [0, 1].

Example 2. *Let us to suppose that a travel agency in Iran wants to offer a luxury group tour for Kuala Lumpur, Malaysia, to their customers. The list of ten four-star hotels in Kuala Lumpur $H = \{h_1, \cdots, h_{10}\}$, which are compared with each other based on the following four parameters $P = \{p_1$=Comfort, p_2=Services, p_3=Location, p_4=Food$\}$, is chosen by this agency from the "www.agoda.com" website. The comments of five passengers, who filled on-line questionnaires, for five different categories "Families with Young Children", "Families with Elder Children", "Couples", "Solo Travelers", and "Group of Friends", whose weighting vector is $\omega = (0.25, 0.25, 0.2, 0.15, 0.15)^T$, are selected by the travel agency as an input data of five experts. A hotel may be selected as the best accommodation place if at least three individuals of five people are satisfied with it. Since, in general, the importance degree of all criteria for different decision makers are not the same, this company defined the weighting vector $\lambda = (0.3, 0.2, 0.35, 0.15)^T$ for the parameters based on the frequency of these parameters in the comments of the passengers. Let this travel agency also consider two different aspects for each parameters as follows: The parameter "Comfort" includes "Cleanliness" and "Staff Performance" with the weighting vector $\lambda_{p_1} = (0.7, 0.3)^T$. The parameter "Services" includes "Facilities" and "Free Wi-Fi connectivity" with the weighting vector $\lambda_{p_2} = (0.8, 0.2)^T$. The parameter "Location" includes "Close to Tourist Attractions" and "In the Green Surroundings" with the weighting vector $\lambda_{p_3} = (0.7, 0.3)^T$. The parameter "Food" includes "Breakfast" and "Lunch and Dinner" with the weighting vector $\lambda_{p_4} = (0.5, 0.5)^T$. Tables 3–7 show the evaluation of the hotels based on these five passengers' comments by using m-polar fuzzy soft sets.*

Table 3. Tabular representation of $MFSS$ (f_1, P).

H	p_1	p_2	p_3	p_4
h_1	(0.87,0.86)	(0.7,0.86)	(0.86,0.67)	(0.58,0.81)
h_2	(0.7,0.7)	(0.55,0.7)	(0.74,0.73)	(0.71,0.7)
h_3	(0.82,0.86)	(0.6,0.86)	(0.79,0.58)	(0.6,0.8)
h_4	(0.83,0.81)	(0.88,0.81)	(0.84,0.78)	(0.55,0.81)
h_5	(0.89,0.81)	(0.64,0.81)	(0.82,0.71)	(0.69,0.84)
h_6	(0.68,0.69)	(0.66,0.69)	(0.82,0.77)	(0.67,0.74)
h_7	(0.82,0.78)	(0.73,0.78)	(0.77,0.83)	(0.66,0.8)
h_8	(0.78,0.8)	(0.73,0.8)	(0.74,0.8)	(0.68,0.79)
h_9	(0.82,0.71)	(0.8,0.71)	(0.69,0.84)	(0.64,0.76)
h_{10}	(0.88,0.89)	(0.86,0.89)	(0.7,0.8)	(0.83,0.85)

Table 4. Tabular representation of $MFSS$ (f_2, P).

H	p_1	p_2	p_3	p_4
h_1	(0.89,0.79)	(0.72,0.84)	(0.84,0.7)	(0.59,0.8)
h_2	(0.74,0.76)	(0.6,0.76)	(0.77,0.75)	(0.69,0.71)
h_3	(0.77,0.68)	(0.73,0.77)	(0.71,0.6)	(0.57,0.69)
h_4	(0.78,0.72)	(0.87,0.79)	(0.85,0.65)	(0.54,0.78)
h_5	(0.91,0.83)	(0.8,0.91)	(0.81,0.71)	(0.66,0.88)
h_6	(0.69,0.7)	(0.68,0.71)	(0.82,0.75)	(0.69,0.75)
h_7	(0.86,0.78)	(0.8,0.86)	(0.74,0.8)	(0.66,0.8)
h_8	(0.78,0.78)	(0.75,0.8)	(0.74,0.77)	(0.64,0.8)
h_9	(0.71,0.75)	(0.85,0.69)	(0.8,0.84)	(0.49,0.72)
h_{10}	(0.89,0.94)	(0.86,0.89)	(0.82,0.86)	(0.85,0.92)

Table 5. Tabular representation of $MFSS$ (f_3, P).

H	p_1	p_2	p_3	p_4
h_1	(0.85,0.79)	(0.74,0.85)	(0.84,0.8)	(0.64,0.81)
h_2	(0.71,0.69)	(0.7,0.71)	(0.74,0.66)	(0.66,0.69)
h_3	(0.77,0.78)	(0.73,0.8)	(0.73,0.6)	(0.57,0.73)
h_4	(0.83,0.76)	(0.73,0.84)	(0.82,0.7)	(0.61,0.8)
h_5	(0.83,0.75)	(0.69,0.82)	(0.7,0.69)	(0.65,0.79)
h_6	(0.69,0.7)	(0.65,0.7)	(0.8,0.76)	(0.69,0.73)
h_7	(0.81,0.74)	(0.8,0.82)	(0.75,0.59)	(0.65,0.78)
h_8	(0.76,0.73)	(0.8,0.76)	(0.71,0.6)	(0.67,0.75)
h_9	(0.78,0.67)	(0.75,0.8)	(0.77,0.56)	(0.57,0.73)
h_{10}	(0.88,0.88)	(0.8,0.86)	(0.68,0.77)	(0.77,0.84)

Table 6. Tabular representation of $MFSS$ (f_4, P).

H	p_1	p_2	p_3	p_4
h_1	(0.83,0.82)	(0.6,0.83)	(0.75,0.7)	(0.68,0.83)
h_2	(0.72,0.72)	(0.4,0.71)	(0.69,0.71)	(0.64,0.7)
h_3	(0.74,0.76)	(0.8,0.76)	(0.75,0.59)	(0.62,0.68)
h_4	(0.88,0.84)	(0.74,0.89)	(0.83,0.77)	(0.69,0.84)
h_5	(0.79,0.79)	(0.64,0.83)	(0.75,0.76)	(0.61,0.82)
h_6	(0.69,0.69)	(0.76,0.68)	(0.79,0.81)	(0.66,0.72)
h_7	(0.81,0.76)	(0.53,0.81)	(0.73,0.84)	(0.66,0.78)
h_8	(0.73,0.72)	(0.7,0.73)	(0.68,0.79)	(0.64,0.72)
h_9	(0.82,0.81)	(0.6,0.89)	(0.79,0.87)	(0.71,0.79)
h_{10}	(0.87,0.87)	(1,0.87)	(0.7,0.76)	(0.75,0.83)

Table 7. Tabular representation of $MFSS\ (f_5, P)$.

H	p_1	p_2	p_3	p_4
h_1	(0.87,0.6)	(0.84,0.87)	(0.93,0.84)	(0.6,0.9)
h_2	(0.74,0.8)	(0.74,0.74)	(0.8,0.74)	(0.64,0.74)
h_3	(0.85,0.8)	(0.8,0.8)	(0.81,0.8)	(0.62,0.8)
h_4	(0.81,0.84)	(0.77,0.76)	(0.77,0.76)	(0.59,0.77)
h_5	(0.8,0.8)	(0.76,0.77)	(0.84,0.77)	(0.61,0.76)
h_6	(0.67,0.8)	(0.72,0.78)	(0.67,0.78)	(0.66,0.72)
h_7	(0.78,0.69)	(0.76,0.74)	(0.78,0.74)	(0.64,0.76)
h_8	(0.77,0.72)	(0.76,0.72)	(0.79,0.72)	(0.67,0.76)
h_9	(0.65,0.6)	(0.69,0.76)	(0.78,0.76)	(0.44,0.69)
h_{10}	(0.88,0.86)	(0.82,0.66)	(0.88,0.66)	(0.77,0.82)

Since the required consensus degree is $\alpha = 3$, *then* $L = \frac{5!}{3!(5-3)!} = 10$. *All 10 different combinations of three of five objects can be listed as follows:* $\Delta_{5,3}(1) = \{1,2,3\}$, $\Delta_{5,3}(2) = \{1,2,4\}$, $\Delta_{5,3}(3) = \{1,2,5\}$, $\Delta_{5,3}(4) = \{1,3,4\}$, $\Delta_{5,3}(5) = \{1,3,5\}$, $\Delta_{5,3}(6) = \{1,4,5\}$, $\Delta_{5,3}(7) = \{2,3,4\}$, $\Delta_{5,3}(8) = \{2,3,5\}$, $\Delta_{5,3}(9) = \{2,4,5\}$, *and* $\Delta_{5,3}(10) = \{3,4,5\}$. *(Steps 1,2).*

Step 3. Utilize data given in Tables 3–7 with the weighting vector $\omega = (0.25, 0.25, 0.2, 0.15, 0.15)^T$ *and M-pFSMWM operator proposed in Theorem 3 to get the collective matrix* $\bar{C} = [\bar{c}_{iy}]_{10\times 4}$ *as below:*

$$\bar{C} = \begin{bmatrix} (0.597, 0.553) & (0.49, 0.588) & (0.588, 0.469) & (0.406, 0.56) \\ (0.49, 0.484) & (0.385, 0.49) & (0.518, 0.4745) & (0.465, 0.484) \\ (0.539, 0.476) & (0.438, 0.539) & (0.497, 0.406) & (0.399, 0.483) \\ (0.546, 0.504) & (0.525, 0.553) & (0.576, 0.455) & (0.378, 0.546) \\ (0.587, 0.531) & (0.448, 0.567) & (0.537, 0.485) & (0.456, 0.558) \\ (0.476, 0.483) & (0.456, 0.483) & (0.562, 0.525) & (0.469, 0.512) \\ (0.568, 0.522) & (0.511, 0.546) & (0.518, 0.52) & (0.456, 0.548) \\ (0.534, 0.516) & (0.511, 0.536) & (0.5, 0.5005) & (0.448, 0.529) \\ (0.497, 0.473) & (0.53, 0.483) & (0.483, 0.546) & (0.3455, 0.504) \\ (0.616, 0.617) & (0.566, 0.605) & (0.478, 0.542) & (0.545, 0.589) \end{bmatrix}$$

Step 4. Utilize Equation (6) to compute matrices $\tilde{R}(p_y) = [\tilde{r}_{ij}(p_y)]_{10\times 10}$, *where* $y = 1, 2, 3, 4$, *as follows:*

$$\tilde{R}(p_1) = \begin{bmatrix} 0.5 & 0.54 & 0.53 & 0.52 & 0.506 & 0.55 & 0.51 & 0.52 & 0.54 & 0.49 \\ 0.45 & 0.5 & 0.48 & 0.48 & 0.46 & 0.505 & 0.47 & 0.48 & 0.49 & 0.44 \\ 0.47 & 0.51 & 0.5 & 0.49 & 0.47 & 0.52 & 0.48 & 0.49 & 0.51 & 0.45 \\ 0.47 & 0.52 & 0.506 & 0.5 & 0.48 & 0.52 & 0.49 & 0.502 & 0.52 & 0.46 \\ 0.49 & 0.53 & 0.52 & 0.51 & 0.5 & 0.54 & 0.507 & 0.51 & 0.53 & 0.48 \\ 0.45 & 0.49 & 0.47 & 0.47 & 0.45 & 0.5 & 0.46 & 0.47 & 0.49 & 0.43 \\ 0.48 & 0.53 & 0.51 & 0.509 & 0.49 & 0.53 & 0.5 & 0.51 & 0.53 & 0.47 \\ 0.47 & 0.51 & 0.504 & 0.49 & 0.48 & 0.52 & 0.48 & 0.5 & 0.51 & 0.46 \\ 0.45 & 0.5008 & 0.48 & 0.47 & 0.46 & 0.506 & 0.46 & 0.48 & 0.5 & 0.44 \\ 0.51 & 0.55 & 0.54 & 0.53 & 0.51 & 0.56 & 0.52 & 0.53 & 0.55 & 0.5 \end{bmatrix}$$

$$\tilde{R}(p_2) = \begin{bmatrix} 0.5 & 0.55 & 0.52 & 0.48 & 0.51 & 0.52 & 0.49 & 0.49 & 0.49 & 0.47 \\ 0.44 & 0.5 & 0.46 & 0.43 & 0.46 & 0.46 & 0.43 & 0.44 & 0.43 & 0.41 \\ 0.47 & 0.53 & 0.5 & 0.46 & 0.49 & 0.49 & 0.46 & 0.47 & 0.46 & 0.44 \\ 0.51 & 0.56 & 0.53 & 0.5 & 0.52 & 0.53 & 0.506 & 0.506 & 0.504 & 0.48 \\ 0.48 & 0.53 & 0.507 & 0.47 & 0.5 & 0.505 & 0.47 & 0.47 & 0.47 & 0.45 \\ 0.47 & 0.53 & 0.501 & 0.46 & 0.49 & 0.5 & 0.47 & 0.47 & 0.46 & 0.44 \\ 0.504 & 0.56 & 0.53 & 0.49 & 0.52 & 0.52 & 0.5 & 0.501 & 0.49 & 0.47 \\ 0.503 & 0.56 & 0.52 & 0.49 & 0.52 & 0.52 & 0.49 & 0.5 & 0.49 & 0.47 \\ 0.505 & 0.56 & 0.53 & 0.49 & 0.52 & 0.53 & 0.501 & 0.502 & 0.5 & 0.47 \\ 0.53 & 0.58 & 0.55 & 0.51 & 0.54 & 0.55 & 0.52 & 0.52 & 0.52 & 0.5 \end{bmatrix}$$

$$\tilde{R}(p_3) = \begin{bmatrix} 0.5 & 0.52 & 0.54 & 0.505 & 0.51 & 0.5006 & 0.51 & 0.52 & 0.52 & 0.52 \\ 0.47 & 0.5 & 0.51 & 0.48 & 0.49 & 0.47 & 0.49 & 0.502 & 0.501 & 0.503 \\ 0.45 & 0.48 & 0.5 & 0.46 & 0.47 & 0.46 & 0.47 & 0.48 & 0.48 & 0.48 \\ 0.49 & 0.51 & 0.53 & 0.5 & 0.508 & 0.49 & 0.509 & 0.51 & 0.51 & 0.52 \\ 0.48 & 0.508 & 0.52 & 0.49 & 0.5 & 0.48 & 0.501 & 0.51 & 0.509 & 0.51 \\ 0.49 & 0.52 & 0.53 & 0.505 & 0.51 & 0.5 & 0.51 & 0.52 & 0.52 & 0.52 \\ 0.48 & 0.506 & 0.52 & 0.49 & 0.49 & 0.48 & 0.5 & 0.509 & 0.508 & 0.51 \\ 0.47 & 0.49 & 0.51 & 0.48 & 0.48 & 0.47 & 0.49 & 0.5 & 0.49 & 0.501 \\ 0.47 & 0.49 & 0.51 & 0.48 & 0.49 & 0.47 & 0.49 & 0.5008 & 0.5 & 0.502 \\ 0.47 & 0.49 & 0.51 & 0.47 & 0.48 & 0.47 & 0.48 & 0.49 & 0.49 & 0.5 \end{bmatrix}$$

$$\tilde{R}(p_4) = \begin{bmatrix} 0.5 & 0.504 & 0.52 & 0.51 & 0.48 & 0.49 & 0.49 & 0.49 & 0.53 & 0.46 \\ 0.49 & 0.5 & 0.51 & 0.506 & 0.48 & 0.49 & 0.48 & 0.49 & 0.52 & 0.45 \\ 0.47 & 0.48 & 0.5 & 0.48 & 0.46 & 0.47 & 0.46 & 0.47 & 0.509 & 0.43 \\ 0.48 & 0.49 & 0.51 & 0.5 & 0.47 & 0.48 & 0.47 & 0.48 & 0.52 & 0.44 \\ 0.51 & 0.51 & 0.53 & 0.52 & 0.5 & 0.508 & 0.502 & 0.509 & 0.54 & 0.47 \\ 0.503 & 0.508 & 0.52 & 0.51 & 0.49 & 0.5 & 0.49 & 0.501 & 0.53 & 0.46 \\ 0.509 & 0.51 & 0.53 & 0.52 & 0.49 & 0.505 & 0.5 & 0.506 & 0.54 & 0.46 \\ 0.502 & 0.507 & 0.52 & 0.51 & 0.49 & 0.49 & 0.49 & 0.5 & 0.53 & 0.46 \\ 0.46 & 0.47 & 0.49 & 0.47 & 0.45 & 0.46 & 0.45 & 0.46 & 0.5 & 0.42 \\ 0.54 & 0.54 & 0.56 & 0.55 & 0.52 & 0.53 & 0.53 & 0.53 & 0.57 & 0.5 \end{bmatrix}$$

Step 5. *Now, the matrix* $\tilde{A} = [\tilde{\mathbf{a}}_{ij}]_{10\times 10}$ *is computed, using information given in matrices* $\tilde{R}(p_y)$ *(*$y = 1,2,3,4$*), to obtain a collective four-polar fuzzy soft preference matrix as the following:*

$$
\tilde{A} = \left[\begin{array}{ccccc}
(0.5\ 0.5\ 0.5\ 0.5) & (1\ 1\ 1\ 1) & (1\ 1\ 1\ 1) & (1\ 0\ 1\ 1) & (1\ 1\ 1\ 0) \\
(0\ 0\ 0\ 0) & (0.5\ 0.5\ 0.5\ 0.5) & (0\ 0\ 1\ 1) & (0\ 0\ 0\ 1) & (0\ 0\ 0\ 0) \\
(0\ 0\ 0\ 0) & (1\ 1\ 0\ 0) & (0.5\ 0.5\ 0.5\ 0.5) & (0\ 0\ 0\ 0) & (0\ 0\ 0\ 0) \\
(0\ 1\ 0\ 0) & (1\ 1\ 1\ 0) & (1\ 1\ 1\ 1) & (0.5\ 0.5\ 0.5\ 0.5) & (0\ 1\ 1\ 0) \\
(0\ 0\ 0\ 1) & (1\ 1\ 1\ 1) & (1\ 1\ 1\ 1) & (1\ 0\ 0\ 1) & (0.5\ 0.5\ 0.5\ 0.5) \\
(0\ 0\ 0\ 1) & (0\ 1\ 1\ 1) & (0\ 1\ 1\ 1) & (0\ 0\ 1\ 1) & (0\ 0\ 1\ 0) \\
(0\ 1\ 0\ 1) & (1\ 1\ 1\ 1) & (1\ 1\ 1\ 1) & (1\ 0\ 0\ 1) & (0\ 1\ 0\ 0) \\
(0\ 1\ 0\ 1) & (1\ 1\ 0\ 1) & (1\ 1\ 1\ 1) & (0\ 0\ 0\ 1) & (0\ 1\ 0\ 0) \\
(0\ 1\ 0\ 0) & (1\ 1\ 0\ 0) & (0\ 1\ 1\ 0) & (0\ 0\ 0\ 0) & (0\ 1\ 0\ 0) \\
(1\ 1\ 0\ 1) & (1\ 1\ 0\ 1) & (1\ 1\ 1\ 1) & (1\ 1\ 0\ 1) & (1\ 1\ 0\ 1)
\end{array} \right.
$$

$$
\left. \begin{array}{ccccc}
(1\ 1\ 1\ 0) & (1\ 0\ 1\ 0) & (1\ 0\ 1\ 0) & (1\ 0\ 1\ 1) & (0\ 0\ 1\ 0) \\
(1\ 0\ 0\ 0) & (0\ 0\ 0\ 0) & (0\ 0\ 1\ 0) & (0\ 0\ 1\ 1) & (0\ 0\ 1\ 0) \\
(1\ 0\ 0\ 0) & (0\ 0\ 0\ 0) & (0\ 0\ 0\ 0) & (1\ 0\ 0\ 1) & (0\ 0\ 0\ 0) \\
(1\ 1\ 0\ 0) & (0\ 1\ 1\ 0) & (1\ 1\ 1\ 0) & (1\ 1\ 1\ 1) & (0\ 0\ 1\ 0) \\
(1\ 1\ 0\ 1) & (1\ 0\ 1\ 1) & (1\ 0\ 1\ 1) & (1\ 0\ 1\ 1) & (0\ 0\ 1\ 0) \\
(0.5\ 0.5\ 0.5\ 0.5) & (0\ 0\ 1\ 0) & (0\ 0\ 1\ 1) & (0\ 0\ 1\ 1) & (0\ 0\ 1\ 0) \\
(1\ 1\ 0\ 1) & (0.5\ 0.5\ 0.5\ 0.5) & (1\ 1\ 1\ 1) & (1\ 0\ 1\ 1) & (0\ 0\ 1\ 0) \\
(1\ 1\ 0\ 0) & (0\ 0\ 0\ 0) & (0.5\ 0.5\ 0.5\ 0.5) & (1\ 0\ 0\ 1) & (0\ 0\ 1\ 0) \\
(1\ 1\ 0\ 0) & (0\ 1\ 0\ 0) & (0\ 1\ 1\ 0) & (0.5\ 0.5\ 0.5\ 0.5) & (0\ 0\ 1\ 0) \\
(1\ 1\ 0\ 1) & (1\ 1\ 0\ 1) & (1\ 1\ 0\ 1) & (1\ 1\ 0\ 1) & (0.5\ 0.5\ 0.5\ 0.5)
\end{array} \right]
$$

Step 6. Calculate the score value of each alternatives according to the collective 4-polar fuzzy soft preference matrix $\tilde{A}$ and Equation (9) as the following: $S_1 = 6.95$, $S_2 = 2.15$, $S_3 = 1.25$, $S_4 = 5.85$, $S_5 = 6$, $S_6 = 4.1$, $S_7 = 5.8$, $S_8 = 3.65$, $S_9 = 3.05$, and $S_{10} = 6.2$.

Step 7. Thus, we have: $h_1 \succ h_{10} \succ h_5 \succ h_4 \succ h_7 \succ h_6 \succ h_8 \succ h_9 \succ h_2 \succ h_3$. Thus, the first hotel, called h_1, is the best option to stay while h_3 should not be selected for accommodation.

In Example 2, the desirable alternative is accepted by most of the decision makers where "most" is interpreted as acceptable by 60% of decision makers , i.e., at least three individuals of total five decision makers are satisfied. To check the impact of consensus degree on the final solution, we compare the obtained results in Example 2 with the result of full agreement, given by the following example in which $\alpha = 5$ is used as the consensus degree.

Example 3. *(Example 2 continued) Let us reconsider the "Hotel Booking Problem" involving M-pFSSs $(f_1, P), (f_2, P), (f_3, P), (f_4, P)$, and (f_5, P) which is discussed earlier in Example 2 where $H = \{h_1, \cdots, h_{10}\}$ is the set of alternatives and $P = \{p_1, p_2, p_3, p_4\}$ is the set of parameters.*

Here, the proposed Algorithm 1 is applied for case $\alpha = 5$, where $L = C_{5,5} = 1$ and $\Delta_{5,5}(1) = \{1, 2, 3, 4, 5\}$, to get the most desirable alternative based on a unanimous consensus. After applying Algorithm 1, the scores of alternatives are computed as follows: $S_1 = 6.65$, $S_2 = 2.45$, $S_3 = 2.15$, $S_4 = 6.55$, $S_5 = 7.1$, $S_6 = 4.1$, $S_7 = 4.95$, $S_8 = 3.55$, $S_9 = 0.6$, and $S_{10} = 6.9$. Therefore, we get: $h_5 \succ h_{10} \succ h_1 \succ h_4 \succ h_7 \succ h_6 \succ h_8 \succ h_2 \succ h_3 \succ h_9$ which shows h_5 is the alternative accepted by all decision makers. Hotel h_1, which is the best option for accommodation according to the most decision makers' views, is the third most desirable alternative. Figure 2 shows the effect of different consensus degrees $\alpha = 3$ and $\alpha = 5$ on scores of alternatives.

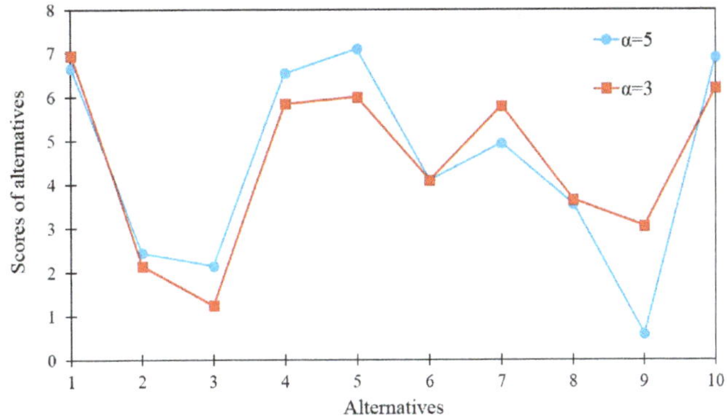

Figure 2. The effect of consensus degree on scores of alternatives based on the M-pFSMWM aggregation method.

Now, we reconsider Examples 2 and 3 according to Algorithm 2 which is based on M-pFSIOWA operator and M-pFSIOWG operator.

Example 4. *(Examples 2 and 3 continued) To apply Algorithm 2 for solving MAGDM problem given in Examples 2 and 3, we first need to re-order the M-pFSSs* $(f_1, P), \cdots, (f_5, P)$ *based on weighting vector* $\omega = (0.25, 0.25, 0.2, 0.15, 0.15)^T$ *for the decision makers* $f_1, \cdots, f_5$*. Subsequently, we get:* $F^s_{1yi} = f^s_1(p_y)(u_i)$, $F^s_{2yi} = f^s_2(p_y)(u_i)$, $F^s_{3yi} = f^s_3(p_y)(u_i)$, $F^s_{4yi} = f^s_4(p_y)(u_i)$, *and* $F^s_{5yi} = f^s_5(p_y)(u_i)$ *for* $s = 1, 2$; $i = 1, 2, \cdots, 10$; *and* $y = 1, 2, 3, 4$*. The associated weighting vector* $w = (w_1, w_2, \cdots, w_5)$ *is then generated by* $w_k = Q(\frac{k}{5}) - Q(\frac{k-1}{5})$ *for* $k = 1, \cdots, 5$ *where* $Q(.)$ *is a quantifier function* $Q : [0, 1] \to [0, 1]$.

First, for case "most of the decision makers" (i.e., $\alpha = 3$*), where "most" is interpreted as 60% or* $\frac{3}{5}$ *of all data, the quantifier function may be defined by*

$$Q_{most}(z) = \begin{cases} 0 & \text{if } z \le \frac{1}{5} \\ \frac{z-1/5}{2/5} & \text{if } \frac{1}{5} < z < \frac{3}{5} \\ 1 & \text{if } z \ge \frac{3}{5} \end{cases} \tag{12}$$

Therefore, we get $w_1 = w_4 = w_5 = 0$, $w_2 = w_3 = 0.5$ *and subsequently* $w = (0, 0.5, 0.5, 0, 0)^T$*. Thus, by using Equations (10) and (11), we have:*

$$[M - pFSIOWA\langle f_1, f_2, f_3, f_4, f_5\rangle]^s(p_y)(u_i) = \frac{F^s_{2yi} + F^s_{3yi}}{2}$$

and

$$[M - pFSIOWG\langle f_1, f_2, f_3, f_4, f_5\rangle]^s(p_y)(u_i) = \sqrt{F^s_{2yi} \cdot F^s_{3yi}}$$

for $s = 1, 2$; $i = 1, 2, \cdots, 10$; *and* $y = 1, 2, 3, 4$.

For the unanimous consensus (i.e., $\alpha = 5$*), the quantifier function may be defined by*

$$Q_{all}(z) = \begin{cases} 0 & \text{if } z < 1 \\ 1 & \text{if } z = \frac{5}{5} = 1 \end{cases} \tag{13}$$

Thus, the weighting vector is $w = (0, 0, 0, 0, 1)^T$*. Subsequently, by using Equations (10) and (11), we have*

$$[M - pFSIOWA\langle f_1, f_2, f_3, f_4, f_5\rangle]^s(p_y)(u_i) = F^s_{5yi}$$

and

$$[M - pFSIOWG\langle f_1, f_2, f_3, f_4, f_5\rangle]^s(p_y)(u_i) = F^s_{5yi}$$

for $s = 1, 2$; $i = 1, 2, \cdots, 10$; *and* $y = 1, 2, 3, 4$.

Thus, based on the M-pFSIOWA operator and M-pFSIOWG operator the resultant collective m-polar fuzzy soft matrix $\bar{C} = [\bar{c}_{iy}]_{10\times 4}$ *where*

$$\bar{c}_{iy} = (\frac{f^1_2(p_y)(u_i) + f^1_3(p_y)(u_i)}{2}, \frac{f^2_2(p_y)(u_i) + f^2_3(p_y)(u_i)}{2});$$

$$\bar{c}_{iy} = (\sqrt{f^1_2(p_y)(u_i).f^1_3(p_y)(u_i)}, \sqrt{f^2_2(p_y)(u_i).f^2_3(p_y)(u_i)});$$

and

$$\bar{c}_{iy} = (f^1_5(p_y)(u_i), f^2_5(p_y)(u_i))$$

is derived for different cases $\alpha = 3$ *and* $\alpha = 5$, *respectively. Then, Steps 4–7 of Algorithm 1 are used to compare the scores of alternatives for both cases* $\alpha = 3$ *and* $\alpha = 5$.

The obtained results from M-pFSMWM, M-pFSIOWA, and M-pFSIOWG operators are reported in Table 8.

Table 8. Comparison results of Examples 2–4 for different M-pFS-based aggregation methods.

Aggregation Methods	α	Number of Computational Steps	Scores of Alternatives (S_i)									
			h_1	h_2	h_3	h_4	h_5	h_6	h_7	h_8	h_9	h_{10}
M-pFSMWM	$\alpha = 3$	18	6.95	2.15	1.25	5.85	6	4.1	5.8	3.65	3.05	6.2
M-pFSMWM	$\alpha = 5$	9	6.65	2.45	2.15	6.55	7.1	4.1	4.95	3.55	0.6	6.9
M-pFSIOWA	$\alpha = 3$	9	6.6	1.8	1.45	5.65	5.825	3.825	5.15	3.375	3.375	7.95
M-pFSIOWA	$\alpha = 5$	9	7.65	3.875	7	3.8	5.9	1.025	2.85	3.6	1.4	7.9
M-pFSIOWG	$\alpha = 3$	9	6.75	1.8	1.45	5.65	5.85	3.9	5.15	3.3	3.2	7.95
M-pFSIOWG	$\alpha = 5$	9	7.65	3.875	7	3.8	5.9	1.025	2.85	3.6	1.4	7.9

α refers to consensus degree.

It can be seen that, for case $\alpha = 5$, h_{10} is the best option selected by all methods. For $\alpha = 3$, the M-pFSMWM method selects h_1, however the tenth hotel, h_{10}, is chosen by the two other methods. Accordingly, the scores of alternatives for different consensus degrees $\alpha = 3$ and $\alpha = 5$ are compared in Figure 3a–c.

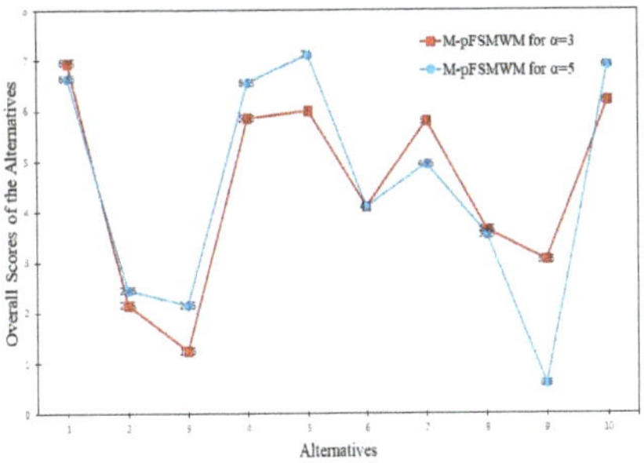

(**a**) M-pFSMWM Method

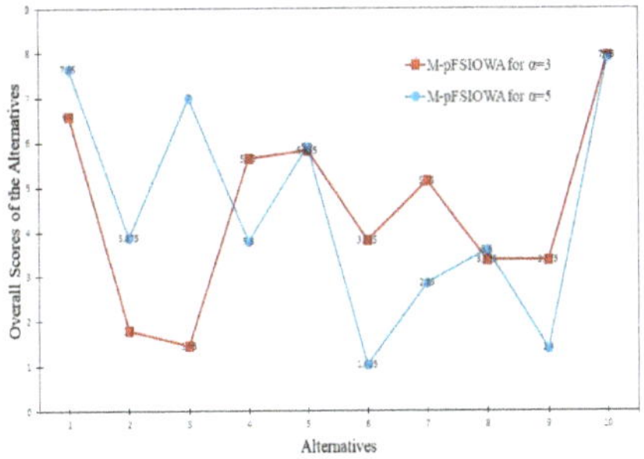

(**b**) M-pFSIOWA Method

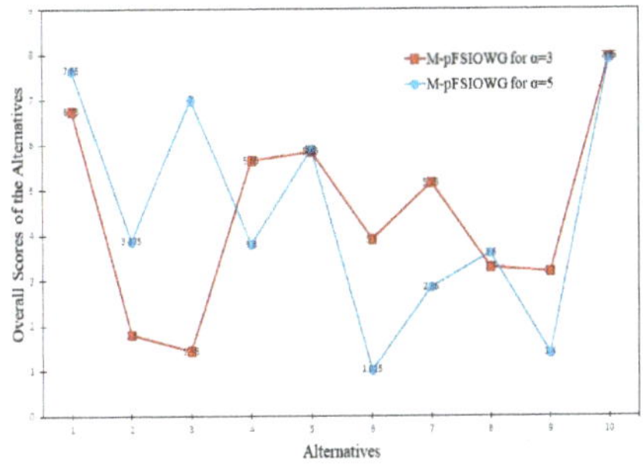

(**c**) M-pFSIOWG Method

Figure 3. The effect of consensus degree on scores of alternatives based on different M-pFS-based aggregation methods.

7. Discussion

To date, various soft set-based techniques have been applied to solve decision-making problems. Some of them have proposed novel methodology to find the solution [18,22,27,28], while some authors have made effort to adapt the well-known decision-making methods, such as SAW, TOPSIS, entropy, OWA, and OWG to the soft set theory [15,16,21,40,41]. However, a technique to solve decision-making problems based on m-polar fuzzy soft information has not been studied yet. Thus, new methodologies are proposed to handle the consensus stage and selection stage of MAGDM problem with M-pFS inputs.

For aggregating input data which take their values from $[0,1]^m$, some new m-polar aggregating methods, called M-pFSMWM operator, M-pFSIOWA operator, and M-pFSIOWG operator, are developed in Sections 3 and 5. The properties comparison of these operators are summarized in Table 9. It can be seen that the most interesting property of M-pFSMWM operator is it is sensitive for different scenarios of a partial agreement at the consensus degree α. This characteristic makes the M-pFSIOWA operator more adaptable for MAGDM problem in which not only the number of individuals satisfying an alternative is important but also the weight of decision makers who agree with this decision affect the final output. Moreover, by changing the value of consensus degree α different cases of agreement are obtained. In particular, when $\alpha \to K$, the partial agreement becomes a full agreement.

Table 9. Properties comparison of different aggregation operators.

Operators	Idempotency	Boundedness	Monotonicity	Symmetry	Flexible by Consensus Degree α	Flexible for Different Choices of K
M-pFSMWM	Yes	Yes	Yes	Yes	Yes	Yes
M-pFSIOWA	Yes	Yes	Yes	Yes	No	No
M-pFSIOWG	Yes	Yes	Yes	Yes	No	No

To reach the process of selection, we propose two procedures in Section 4.2, Algorithm 1 where the consensus stage is reached based on the new M-pFSMWM operator, and Section 5, Algorithm 2 in which the consensus stage is obtained by M-pFSIOWA or M-pFSIOWG operators are extended based on IOWA and IOWG, respectively. To reach the selection stage a new score value function, described by Equation (9), is applied. The main advantage of the proposed formulation is to rank and compare objects based on a collective m-polar fuzzy soft preference relationship. This allows us to have a ranking system of alternatives, from the most preferred element to the least preferred element, which may include some incomparable objects because of preference relationships nature.

Illustrative examples, given in Examples 2 and 3, show the application of Algorithm 1 to analyze MAGDM problems with multi-polar fuzzy soft information. The obtained results are then compared with the m-polar fuzzy soft extensions of two well-known aggregation operators IOWA and IOWG, i.e., M-pFSIOWA and M-pFSIOWG, in Example 4. Table 10 makes a comparison of the preference orders of the alternatives for methods using different aggregation operators including the new proposed M-pFSMWM method, the m-polar fuzzy soft induced ordered weighted average (M-pFSIOWA) method, and the m-polar fuzzy soft induced ordered weighted geometric (M-pFSIOWG) method. As can be seen, Hotel h_{10} is the best option for staying based on the all discussed methods except the M-pFSMWM-based method, where h_{10} is considered as the best second option for accommodation. According to final preference order obtained based on the M-pFSMWM operator, Hotel h_1, which has the second place based on the other methods, is the best option accepted by the majority, i.e., 60%, of decision makers. Hotel h_5 is the best place to stay in terms of all decision makers. The analysis derived in Table 10 shows a good agreement among thees methods, however the number of computational steps in M-pFSMWM-based algorithm is $C_{K,\alpha}+1$ in comparison with $K+2$ stages in Algorithm 2. On the other hand, the main disadvantage of M-pFSIOWA and M-pFSIOWG methods is that there is no unique approach to determine the associated weighting vector w related to the aggregation operators M-pFSIOWA and M-pFSIOWG. Finally, in Figure 4a, the overall scores of alternatives based on the different methods for case $\alpha = 3$ are shown. Figure 4b shows the scores of alternatives obtained by using different methods for case $\alpha = 5$. Note that, using some relative preference matrices to find the scores of alternatives in all methods, leads to record a similar trend in Figure 4b.

Table 10. Comparison of the alternatives' preference orders for different methods.

Methods	Consensus Degree α	Preference Order
M-pFSMWM	3	$h_1 \succ h_{10} \succ h_5 \succ h_4 \succ h_7 \succ h_6 \succ h_8 \succ h_9 \succ h_2 \succ h_3$
M-pFSMWM	5	$h_5 \succ h_{10} \succ h_1 \succ h_4 \succ h_7 \succ h_6 \succ h_8 \succ h_2 \succ h_3 \succ h_9$
M-pFSIOWA	3	$h_{10} \succ h_1 \succ h_5 \succ h_4 \succ h_7 \succ h_6 \succ h_8 \succ h_9 \succ h_2 \succ h_3$
M-pFSIOWA	5	$h_{10} \succ h_1 \succ h_3 \succ h_5 \succ h_2 \succ h_4 \succ h_8 \succ h_7 \succ h_9 \succ h_6$
M-pFSIOWG	3	$h_{10} \succ h_1 \succ h_5 \succ h_4 \succ h_7 \succ h_6 \succ h_8 \succ h_9 \succ h_2 \succ h_3$
M-pFSIOWG	5	$h_{10} \succ h_1 \succ h_3 \succ h_5 \succ h_2 \succ h_4 \succ h_8 \succ h_7 \succ h_9 \succ h_6$

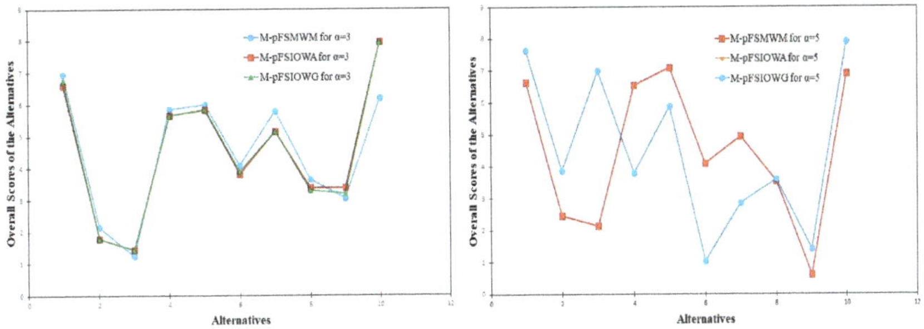

(**a**) The overall scores of alternatives for different methods when $\alpha = 3$

(**b**) The overall scores of alternatives for different methods when $\alpha = 5$

Figure 4. The effect of different methods on scores of alternatives for different consensus degrees.

8. Conclusions

Traditional aggregation operators, which usually deal with uni-polar information, fail to aggregate m-polar fuzzy soft information taking their values from $[0, 1]^m$. This study proposes a new aggregation method for processing the MAGDM problems with m-polar fuzzy soft information in which both attributes and experts have different weights. For this purpose, firstly the concept of m-polar fuzzy soft sets is introduced. Then, the new aggregation operator M-pFSMWM in the domain of m-polar fuzzy soft sets is defined. The advantage of proposed M-pFSMWM operator is to be sensitive for different partial agreement scenarios at a consensus degree α. Further, the m-polar fuzzy soft induced ordered weighted average (M-pFSIOWA) operator and the m-polar fuzzy soft induced ordered weighted geometric (M-pFSIOWG) operator, which are the extensions of IOWA and IOWG operators, respectively, are developed. Some desirable properties of M-pFSMWM, M-pFSIOWA, and M-pFSIOWG operators, such as idempotency, monotonicity, and commutativity are also studied. The characteristics of the proposed M-pFSMWM operator shows it is more adaptable for a wider range of MAGDM problems in comparison with M-pFSIOWA and M-pFSIOWG operators. In addition, a procedure for ranking m-polar fuzzy soft data based on a new score value function is proposed. Then, two algorithms are designed to MAGDM problems based on M-pFSMWM, M-pFSIOWA, and M-pFSIOWG operators Finally, to show the efficiency of proposed methods, some numerical examples are discussed.

Author Contributions: Conceptualization, A.Z.K.; Investigation, A.Z.K.; Methodology, A.Z.K. and A.K.; Validation, A.Z.K. and A.K.; Writing—original draft, A.Z.K.; and Writing—review and editing, A.Z.K. and A.K.

Acknowledgments: This research was supported by the Ministry of Education Malaysia (KPM) under Grant FRGS/1/2018/STG06/UPM/01/3.

Conflicts of Interest: The authors declare no conflict of interest.

References

1. Dubois, D.; Prade, H. Weighted minimum and maximum operations in fuzzy set theory. *Inf. Sci.* **1986**, *39*, 205–210. [CrossRef]
2. Fagin, R.; Wimmers, E.L. A formula for incorporating weights into scoring rules. *Theor. Comput. Sci.* **2000**, *239*, 309–338. [CrossRef]
3. Yager, R.R. A new methodology for ordinal multiobjective decisions based on fuzzy sets. In *Readings in Fuzzy Sets for Intelligent Systems*; Elsevier: San Francisco, CA, USA, 1993; pp. 751–756.
4. Yager, R.R. On ordered weighted averaging aggregation operators in multicriteria decisionmaking. In *Readings in Fuzzy Sets for Intelligent Systems*; Elsevier: San Francisco, CA, USA, 1993; pp. 80–87.
5. Chiclana, F.; Herrera, F.; Herrera-Viedma, E. The ordered weighted geometric operator: Properties and application in MCDM Problems. In *Technologies for Constructing Intelligent Systems 2* ; Physica: Heidelberg, Germany, 2002; pp. 173–183.

6. Yager, R.R.; Filev, D.P. Induced ordered weighted averaging operators. *IEEE Trans. Syst. Man Cybern. B Cybern.* **1999**, *29*, 141–150. [CrossRef] [PubMed]
7. Xu, Z.S.; Da, Q.L. An overview of operators for aggregating information. *Int. J. Intell. Inf. Technol.* **2003**, *18*, 953–969. [CrossRef]
8. Kacprzyk, J. Group decision making with a fuzzy linguistic majority. *Fuzzy Sets Syst.* **1986**, *18*, 105–118. [CrossRef]
9. Kacprzyk, J.; Fedrizzi, M.; Nurmi, H. Group decision making and consensus under fuzzy preferences and fuzzy majority. *Fuzzy Sets Syst.* **1992**, *49*, 21–31. [CrossRef]
10. Molodtsov, D. Soft set theory—First results. *Comput. Math. Appl.* **1999**, *37*, 19–31. [CrossRef]
11. Zadeh, L.A. Fuzzy sets. *Inf. Control* **1965**, *8*, 338–353. [CrossRef]
12. Maji, P.K.; Biswas, R.; Roy, A.R. Fuzzy soft sets. *J. Fuzzy Math.* **2001**, *9*, 589–602.
13. Maji, P.K.; Biswas, R.; Roy, A.R. Intuitionistic fuzzy soft sets. *J. Fuzzy Math.* **2001**, *9*, 677–692.
14. Maji, P.K.; Roy, A.R.; Biswas, R. An application of soft sets in a decision making problem. *Comput. Math. Appl.* **2002**, *44*, 1077–1083. [CrossRef]
15. Roy, A.R.; Maji, P.K. A fuzzy soft set theoretic approach to decision making problems. *J. Comput. Appl. Math.* **2007**, *203*, 412–418. [CrossRef]
16. Alcantud, J.C.R. A novel algorithm for fuzzy soft set based decision making from multiobserver input parameter data set. *Inf. Fusion* **2016**, *29*, 142–148. [CrossRef]
17. Cagman, N.; Enginoglu, S.; Citak, F. Fuzzy soft set theory and its applications. *Iran. J. Fuzzy Syst.* **2011**, *8*, 137–147.
18. Cagman, N.; Enginoglu, S. Fuzzy soft matrix theory and its application in decision making. *Iran. J. Fuzzy Syst.* **2012**, *9*, 109–119.
19. Guan, X.; Li, Y.; Feng, F. A new order relation on fuzzy soft sets and its application. *Soft Comput.* **2013**, *17*, 63–70. [CrossRef]
20. Zhang, Z. A rough set approach to intuitionistic fuzzy soft set based decision making. *Appl. Math. Model.* **2012**, *36*, 4605–4633. [CrossRef]
21. Mao, J.; Yao, D.; Wang, C. Group decision making methods based on intuitionistic fuzzy soft matrices. *Appl. Math. Model.* **2013**, *37*, 6425–6436. [CrossRef]
22. Das, S.; Kar, S. Group decision making in medical system: An intuitionistic fuzzy soft set approach. *Appl. Soft Comput.* **2014**, *24*, 196–211. [CrossRef]
23. Zhang, Z.; Zhang, S. A novel approach to multi attribute group decision making based on trapezoidal interval type-2 fuzzy soft sets. *Appl. Math. Model.* **2013**, *37*, 4948–4971. [CrossRef]
24. Pandey, A.; Kumar, A. A note on "A novel approach to multi attribute group decision making based on trapezoidal interval type-2 fuzzy soft sets". *Appl. Math. Model.* **2017**, *41*, 691–693. [CrossRef]
25. Tao, Z.; Chen, H.; Zhou, L.; Liu, J. 2-Tuple linguistic soft set and its application to group decision making. *Soft Comput.* **2015**, *19*, 1201–1213. [CrossRef]
26. Zhu, K.; Zhan, J. Fuzzy parameterized fuzzy soft sets and decision making. *Int. J. Mach. Learn. Cybern.* **2016**, *7*, 1207–1212. [CrossRef]
27. Zahedi-Khameneh, A.; Kılıçman, A.; Salleh, A.R. An adjustable approach to multi-criteria group decision-making based on a preference relationship under fuzzy soft information. *Int. J. Fuzzy Syst.* **2017**, *19*, 1840–1865. [CrossRef]
28. Zahedi-Khameneh, A.; Kılıçman, A.; Salleh, A.R. Application of a preference relationship in decision-making based on intuitionistic fuzzy soft sets. *J. Intell. Fuzzy Syst.* **2018**, *34*, 123–139. [CrossRef]
29. Mesiarova, A.; Lazaro, J. Bipolar aggregation operators. *Proc. AGOP* **2003**, *2003*, 119–123.
30. Mesiarova-Zemankova, A.; Ahmad, K. Extended multi-polarity and multi-polar-valued fuzzy sets. *Fuzzy Sets Syst.* **2014**, *234*, 61–78. [CrossRef]
31. Mesiarova-Zemankova, A. Multi-polar t-conorms and uninorms. *Inf. Sci.* **2015**, *301*, 227–240. [CrossRef]
32. Medina, J.; Ojeda-Aciego, M. Multi-adjoint t-concept lattices. *Inf. Sci.* **2010**, *180*, 712–725. [CrossRef]
33. Pozna, C.; Minculete, N.; Precup, R.E.; Koczy, L.; Ballagi, A. Signatures: Definitions, operators and applications to fuzzy modelling. *Fuzzy Sets Syst.* **2012**, *201*, 86–104. [CrossRef]
34. Jankowski, J.; Kazienko, P.; Watrobski, J.; Lewandowska, A. Fuzzy multi-objective modeling of effectiveness and user experience in online advertising. *Expert Syst. Appl.* **2016**, *65*, 315–331. [CrossRef]

35. Kumar, A.; Kumar, D.; Jarial, S.K. A hybrid clustering method based on improved artificial bee colony and fuzzy c-means algorithm. *Int. J. Artif. Intell.* **2017**, *15*, 40–60.
36. Tanino, T. Fuzzy preference relations in group decision making. In *Non-Conventional Preference Relations in Decision Making. Lecture Notes in Economics and Mathematical Systems*; Kacprzyk, J., Roubens, M., Eds.; Springer: Berlin/Heidelberg, Germany, 1988; pp. 54–71.
37. Chen, J.; Li, S.; Ma, S.; Wang, X. m-polar fuzzy sets: An extension of bipolar fuzzy sets. *Sci. World J.* **2014**, *2014*, 416530. [CrossRef] [PubMed]
38. Wei, G. Some induced geometric aggregation operators with intuitionistic fuzzy information and their application to group decision making. *Appl. Soft Comput.* **2010**, *10*, 423–431. [CrossRef]
39. Xu, Y.; Wang, H. The induced generalized aggregation operators for intuitionistic fuzzy sets and their application in group decision making. *Appl. Soft Comput.* **2012**, *12*, 1168–1179. [CrossRef]
40. Xiao, Z.; Xia, S.; Gong, K.; Li, D. The trapezoidal fuzzy soft set and its application in MCDM. *Appl. Math. Model.* **2012**, *36*, 5844–5855. [CrossRef]
41. Jiang, Y.; Tang, Y.; Liu, H.; Chen, Z. Entropy in intuitionistic fuzzy soft sets and on interval-valued fuzzy soft sets. *Inf. Sci.* **2013**, *240*, 95–114. [CrossRef]

symmetry

MDPI

Article

Novel Parameterized Distance Measures on Hesitant Fuzzy Sets with Credibility Degree and Their Application in Decision-Making

Jiaru Li [1], Fangwei Zhang [1,*], Qiang Li [2], Jing Sun [3,*], Janney Yee [4], Shuhong Wang [5] and Shujun Xiao [6]

1 College of Transport and Communications, Shanghai Maritime University, Shanghai 201306, China; cherishljr@163.com
2 Faculty of Public Security, Shanghai Police College, Shanghai 200137, China; zeeliqiang@aliyun.com
3 Department of Mathematics, University of Macau, Macau 999078, China
4 College of Social Sciences, University of Birmingham, Birmingham 1610681, UK; yeejanney@gmail.com
5 Shool of Computing, National University of Singapore, Singapore 118414, Singapore; wangshuhong001@gmail.com
6 Logistics Engineering College, Shanghai Maritime University, Shanghai 201306, China; meggiexsj@sina.com
* Correspondence: fangweizhang@aliyun.com (F.Z.); mb75406@umac.mo (J.S.)

Received: 5 September 2018; Accepted: 16 October 2018; Published: 1 November 2018

Abstract: The subject of this study is to explore the role of cardinality of hesitant fuzzy element (HFE) in distance measures on hesitant fuzzy sets (HFSs). Firstly, three parameters, i.e., credibility factor, conservative factor, and a risk factor are introduced, thereafter, a series of novel distance measures on HFSs are proposed using these three parameters. These newly proposed distance measures handle the relationship between the cardinal number and the element values of hesitant fuzzy set well, and are suitable to combine subjective and objective decision-making information. When using these functions, decision makers with different risk preferences are allowed to give different values for these three parameters. In particular, this study transfers the hesitance degree index to a credibility of the values in HFEs, which is consistent with people's intuition. Finally, the practicability of the newly proposed distance measures is verified by two examples.

Keywords: hesitant fuzzy sets; hesitant degree; credibility; distance measure; similarity measure

1. Introduction

To handle the uncertainty in real life problems effectively, Zadeh proposed the concept "fuzzy set" [1]. Thereafter, some extensions of fuzzy sets were proposed, for example, interval-valued fuzzy sets proposed by Zadeh [2–4], intuitionistic fuzzy sets proposed by Atanassov [5], and interval-valued intuitionistic fuzzy sets proposed by Atanassov and Gargov [6]. Most recently, Torra and Narukawa introduced hesitant fuzzy sets (HFSs) to deal with hesitant situations, which were not well managed by the previous tools [7,8]. In HFSs, the membership is a union of several memberships of fuzzy sets. Practices show that HFS is a useful mathematical tool for dealing with this kind of uncertainty. Nowadays, lots of branches of HFSs have been studied, such as intuitionistic hesitant fuzzy sets (see reference [9]), dual hesitant fuzzy sets (see reference [10]), etc.

Distance and similarity measures are two important research objects in fuzzy set theory and they have attracted the attention of many scholars. Zwick, Carlstein and Budescu [11], Pappis and Karacapilidis [12] proposed a comparative analysis on similarity measures on fuzzy sets, respectively. Wang introduced two influential similarity measures on fuzzy sets [13]. As for HFSs, Xu and Xia proposed a series of classical distance measures on HFSs [14,15]. Thereafter, Peng et al. proposed a novel hesitant fuzzy weighted distance measure [16]. Then, Rodríguez et al. gave a clear perspective of HFSs [17]. Li et al. pointed out that the existing distance and similarity measures fail to consider the cardinal numbers of HFEs [18]. Thereafter, Li et al. proposed a concept of hesitance degree of HFEs and HFSs to introduce a decision maker's hesitance situation. They also proposed a series of distance and similarity measures on HFSs, which take both the values and the cardinal numbers of HFEs into consideration. In addition, Tang et al. introduced some continuous hesitant fuzzy distance measures which also consider the element number of HFEs [19]. It is noteworthy that the distance measures on HFSs are important in decision-making. As for this application, Alcantud et al. summarized the latest related studies in their work [20].

The distance measures proposed by Li et al. are innovative [18]. In particular, they introduced the concept of hesitance degree on HFSs. This is a new beginning, where the proposed distance and similarity measures should be explored further with consideration of hesitant degree. The aims of this study are to proceed towards the direction where the distance and similarity measures should develop according to reference [18]. Specifically, this study proposes a series of novel distance measures on HFSs. The main characteristic of the proposed distance measures is that they contain three parameters, i.e., credibility factor, conservative factor, and a risk factor. These newly proposed distance measures handle the relationship between the cardinal number and the element values of hesitant fuzzy set well. When using these functions, decision makers with different risk preferences are allowed to give different values for the three parameters.

The remaining part of this study is arranged as follows: Section 1 reviewed some basic notions on HFSs and introduced some classical distance measures. Section 2 proposes a series of novel distance measures on HFSs. Section 3 provides two examples to show the validity of the novel distance measures. Finally, innovations of this study are concluded in Section 4.

2. Preliminaries

This section introduces some basic notions on HFSs. Throughout this paper, $X = \{x_1, x_2, \cdots, x_n\}$ is denoted as discourse set. In addition, denote h as HFS, denote $h(x)$ as HFE, and denote H as the set of all HFSs on X.

Definition 1. *[8,21] Let X be a fixed set, a HFS on X is a function such that for any element in X, there is a subset of* $[0,1]$ *corresponding to it. Symbolically, the function is represented as* $E = \{\langle x, h_E(x)\rangle \mid x \in X\}$*, where* $h_E(x)$ *is a value set in* $[0,1]$*, representing the possible membership degrees of* $x \in X$ *to the set E. For convenience's sake,* $h_E(x)$ *is called an HFE.*

Definition 2. *[14] Let* h_1 *and* h_2 *be two HFSs on* $X = \{x_1, x_2, \cdots, x_n\}$*; then, the distance measure between* h_1 *and* h_2 *is defined as* $d(h_1, h_2)$*, which satisfies: (i)* $0 \le d(h_1, h_2) \le 1$*; (ii)* $d(h_1, h_2) = 0$*, if and only if* $h_1 = h_2$*; (iii)* $d(h_1, h_2) = d(h_2, h_1)$*. The similarity measure between* h_1 *and* h_2 *is defined as* $s(h_1, h_2)$*, which satisfies the following properties: (i)* $0 \le s(h_1, h_2) \le 1$*; (ii)* $s(h_1, h_2) = 1$*, if and only if* $h_1 = h_2$*; (iii)* $s(h_1, h_2) = s(h_2, h_1)$*.*

To introduce HFSs clearly, Xu and Xia proposed two properties on HFSs as follows [14].

Property 1. *Assume that d is a distance measure between HFSs* h_1 *and* h_2*, then,* $s(h_1, h_2) = 1 - d(h_1, h_2)$ *is a similarity measure between HFSs* h_1 *and* h_2*. If s is a similarity measure between HFSs* h_1 *and* h_2*, then,* $d(h_1, h_2) = 1 - s(h_1, h_2)$ *is a distance measure between HFSs* h_1 *and* h_2*.*

Thereafter, Xu and Xia introduced the classical hesitant normalized Hamming distance, classical Euclidean distance and classical generalized hesitant normalized distance [14]. Limited to the layout, they are not introduced in this study. Reference [18] noticed that the divergence of HFSs h_1 and h_2 includes two parts, i.e., the difference of their cardinal numbers and the difference of their values. Following this idea, reference [18] officially introduced the concept of hesitance degree of HFEs as follows.

Definition 3. *[18] Let h be a HFS on $X = \{x_1, x_2, \cdots, x_n\}$. For any $x_i \in X$, denote $l(h(x_i))$ as the cardinal number of $h(x_i)$. Then, denote $u(h(x_i)) = 1 - \frac{1}{l(h(x_i))}$, and denote $u(h) = \frac{1}{n}\sum_{i=1}^{n} u(h(x_i))$. Understandably, $u(h(x_i))$ represents the hesitant degree of $h(x_i)$, and $u(h)$ represents the hesitant degree of h.*

Based on Definition 3, reference [18] proposed a series of novel distance and similarity measures on HFSs as follows.

Definition 4. *[18] Let h_1 and h_2 be two HFSs on $X = \{x_1, x_2, \cdots, x_n\}$; then, a normalized Hamming distance including hesitance degree between $h_1(x_i)$ and $h_2(x_i)$ is defined as*

$$d_{hh}(h_1, h_2) = \frac{1}{2n} \cdot \sum_{i=1}^{n} \left[|u(h_1(x_i)) - u(h_2(x_i))| + \frac{1}{l_{x_i}} \sum_{j=1}^{l_{x_i}} |h_1^j(x_i) - h_2^j(x_i)| \right]. \tag{1}$$

A normalized Euclidean distance is defined as

$$d_{he}(h_1, h_2) = \left[\frac{1}{2n} \cdot \sum_{i=1}^{n} \left(|u(h_1(x_i)) - u(h_2(x_i))|^2 + \frac{1}{l_{x_i}} \sum_{j=1}^{l_{x_i}} |h_1^j(x_i) - h_2^j(x_i)|^2 \right) \right]^{\frac{1}{2}}. \tag{2}$$

A normalized generalized distance is defined as

$$d_{hg}(h_1, h_2) = \left[\frac{1}{2n} \cdot \sum_{i=1}^{n} \left(|u(h_1(x_i)) - u(h_2(x_i))|^\lambda + \frac{1}{l_{x_i}} \sum_{j=1}^{l_{x_i}} |h_1^j(x_i) - h_2^j(x_i)|^\lambda \right) \right]^{\frac{1}{\lambda}}, \tag{3}$$

where $\lambda > 0$, $h_1^j(x_i)$ and $h_2^j(x_i)$ denote the jth ordinal values in $h_1(x_i)$ and $h_2(x_i)$ respectively.

When the different preference between the hesitance degrees and the membership values is considered, the distance measures with preference are proposed as

$$d_{phh}(h_1, h_2) = \frac{1}{n} \cdot \sum_{i=1}^{n} \left[\alpha|u(h_1(x_i)) - u(h_2(x_i))| + \frac{\beta}{l_{x_i}} \sum_{j=1}^{l_{x_i}} |h_1^j(x_i) - h_2^j(x_i)| \right], \tag{4}$$

$$d_{phe}(h_1, h_2) = \left[\frac{1}{n} \cdot \sum_{i=1}^{n} \left(\alpha|u(h_1(x_i)) - u(h_2(x_i))|^2 + \frac{\beta}{l_{x_i}} \sum_{j=1}^{l_{x_i}} |h_1^j(x_i) - h_2^j(x_i)|^2 \right) \right]^{\frac{1}{2}}, \tag{5}$$

and

$$d_{phg}(h_1, h_2) = \left[\frac{1}{n} \cdot \sum_{i=1}^{n} \left(\alpha|u(h_1(x_i)) - u(h_2(x_i))|^\lambda + \frac{\beta}{l_{x_i}} \sum_{j=1}^{l_{x_i}} |h_1^j(x_i) - h_2^j(x_i)|^\lambda \right) \right]^{\frac{1}{\lambda}}, \tag{6}$$

where $\lambda > 0$, $0 \le \alpha, \beta \le 1$, and $\alpha + \beta = 1$.

When $\alpha = 0$, it means that the influence of the hesitant degree of HFE is ignored; then, d_{phh}, d_{phe}, and d_{phg} are degenerated into the distance measure d_h, d_e, and d_g proposed in reference [14], respectively. When the weight of the element $x \in X$ is considered, the following weighted distance measures are proposed. Denote the *weight of* $x_i \in X$ *is* $w_i(i = \{1, 2, \cdots, n\}$, *where* $0 \leq w_i \leq 1$ *and* $\sum_{i=1}^{n} w_i = 1$; *then, reference [18] proposed the following weighted distance:*

$$d_{whh}(h_1, h_2) = \frac{1}{2} \cdot \sum_{i=1}^{n} w_i \left[|u(h_1(x_i)) - u(h_2(x_i))| + \frac{1}{l_{x_i}} \sum_{j=1}^{l_{x_i}} |h_1^j(x_i) - h_2^j(x_i)| \right], \tag{7}$$

$$d_{whe}(h_1, h_2) = \left[\frac{1}{2} \cdot \sum_{i=1}^{n} w_i \left(|u(h_1(x_i)) - u(h_2(x_i))|^2 + \frac{1}{l_{x_i}} \sum_{j=1}^{l_{x_i}} |h_1^j(x_i) - h_2^j(x_i)|^2 \right) \right]^{\frac{1}{2}}, \tag{8}$$

$$d_{whg}(h_1, h_2) = \left[\frac{1}{2} \cdot \sum_{i=1}^{n} w_i \left(|u(h_1(x_i)) - u(h_2(x_i))|^\lambda + \frac{1}{l_{x_i}} \sum_{j=1}^{l_{x_i}} |h_1^j(x_i) - h_2^j(x_i)|^\lambda \right) \right]^{\frac{1}{\lambda}}, \tag{9}$$

where $\lambda > 0$.

When the weight of each element $x \in X$, *and the different preference between the influence of hesitance degrees and membership values are all taken into account, a series of weighted distance measures with preference can also be proposed. For details, please refer to reference [18].*

3. Main Results

3.1. Analysis on Hesitance Degree

Reference [18] noticed that the cardinality of HFEs is very important in proposing distance and similarity measures on HFSs, and then reference[18] proposed the concept of hesitant degree. We think this work is a pioneer contribution to the theory of HFSs. Further analysis shows that the index hesitant degree only reflects the hesitance degree when decision makers consider the membership for an HFE, and it has no direct relationship with the distance between HFEs. To explain this issue further, considering that Equation (1) is the basis of the series of distance measures proposed by reference [18], one counter-intuitive case of Equation (1) is provided here.

Assume that there is a set $X = \{x\}$, and assume that there are two patterns which are described in HFSs setting, i.e., $h_1 = \{0.97, 0.95, 0.88, 0.86, 0.82\}$ and $h_2 = \{0.45\}$. Assume that there is a sample that is described by an HFE $h = \{0.43, 0.44, 0.45, 0.46, 0.47\}$. Then, which pattern does h belong to? To answer this question, a principle is considered when $d(h_{i0}, h) = \min\{d(h_1, h), d(h_2, h)\}$, one can get that the sample h belongs to pattern h_{i0}.

First, this study extends h_2 as $h_2 = \{0.45, 0.45, 0.45, 0.45, 0.45\}$. Then, it finds that the difference of the membership values between h and h_1 are much larger than that of the membership values between h and h_2. Though the hesitant degrees of h and h_1 are the same, it is very obvious that h belongs to the pattern h_2. Meanwhile, by Equation (1), it gets that $d_{hh}(h, h_1) = 0.2230$, $d_{hh}(h, h_1) = 0.406$. Thus, h belongs to the pattern h_1, which is counter-intuitive.

The introduced case illustrates that it is necessary to further consider the distance measures on HFSs. By borrowing concepts from statistics, the hesitant degree of the HFE can be transferred as credibility factor of the membership values of the HFE, where the bigger the hesitant degree, the lower the credibility of the membership values of the HFE. From this viewpoint, some novel distance measures are proposed in the coming subsection.

3.2. Novel Distance Measures with Three Factors

Before introducing the novel distance measures, a basic concept is introduced as follows.

Definition 5. *Denote h as a HFS on $X = \{x_1, x_2, \cdots, x_n\}$ and for any $x_i \in X$, denote $l(h(x_i))$ as the cardinal number of $h(x_i)$, denote $c(h(x_i)) = l(h(x_i))^{-1}$ as the credibility factor of $h(x_i)$.*

Thereafter, a series of novel distance measures are proposed as follows.

Definition 6. *Denote h_1 and h_2 as two HFSs on $X = \{x_1, x_2, \cdots, x_n\}$. Then, the normalized Hamming distance between $h_1(x_i)$ and $h_2(x_i)$ is defined as*

$$d_h(h_1(x_i), h_2(x_i)) = \frac{1}{l_{x_i}} \sum_{j=1}^{l_{x_i}} |h_1^j(x_i) - h_2^j(x_i)| \tag{10}$$

with a credibility factor $c(h_1(x_i), h_2(x_i)) = [c(h_1(x_i))c(h_2(x_i))]^{\frac{1}{2}}$. Denote

$$c^*(h_1(x_i), h_2(x_i)) = \frac{[c(h_1(x_i))c(h_2(x_i))]^{\frac{1}{2}}}{\sum_{i=1}^{n} [c(h_1(x_i))c(h_2(x_i))]^{\frac{1}{2}}} \tag{11}$$

as the normalized credibility factor. Then, a series of novel Hamming, Euclidean, and generalized distances between h_1 and h_2 are proposed as

$$d_{chh}(h_1, h_2) = \sum_{i=1}^{n} c^*(h_1(x_i), h_2(x_i)) \cdot d_h(h_1(x_i), h_2(x_i)), \tag{12}$$

$$d_{che}(h_1, h_2) = \left[\sum_{i=1}^{n} [c^*(h_1(x_i), h_2(x_i)) \cdot d_h(h_1(x_i), h_2(x_i))]^2\right]^{\frac{1}{2}}, \tag{13}$$

$$d_{chg}(h_1, h_2) = \left[\sum_{i=1}^{n} [c^*(h_1(x_i), h_2(x_i)) \cdot d_h(h_1(x_i), h_2(x_i))]^\lambda\right]^{\frac{1}{\lambda}}, \tag{14}$$

where $\lambda > 0$, $h_1^j(x_i)$ and $h_2^j(x_i)$ are the jth ordinal values in $h_1(x_i)$ and $h_2(x_i)$, respectively.

In the situation that the weight of the element $x \in X$ is considered, some weighted distance measures for HFSs are obtained. Denote the weight of $x_i \in X$ as $w_i (i = \{1, 2, \cdots, n\}$, where $0 \le w_i \le 1$ and $\sum_{i=1}^{n} w_i = 1$; then, a series of weighted distance measures are structured as

$$d_{wchh}(h_1, h_2) = c \cdot \sum_{i=1}^{n} w_i c^*(h_1(x_i), h_2(x_i)) \cdot d_h(h_1(x_i), h_2(x_i)), \tag{15}$$

$$d_{wche}(h_1, h_2) = \left[\sum_{i=1}^{n} w_i \cdot [c^*(h_1(x_i), h_2(x_i)) \cdot d_h(h_1(x_i), h_2(x_i))]^2\right]^{\frac{1}{2}}, \tag{16}$$

$$d_{wchg}(h_1, h_2) = \left[\sum_{i=1}^{n} w_i \cdot [c^*(h_1(x_i), h_2(x_i)) \cdot d_h(h_1(x_i), h_2(x_i))]^\lambda\right]^{\frac{1}{\lambda}}, \tag{17}$$

where $\lambda > 0$, $h_1^j(x_i)$ and $h_2^j(x_i)$ are the jth ordinal values in $h_1(x_i)$ and $h_2(x_i)$, respectively.

In order to deeply understand the relationship between the cardinalities and the values of HFEs, a conservative factor α and a risk factor β are considered, and a series of novel distance measures are proposed as

$$d_{wchh}(h_1,h_2) = \sum_{i=1}^{n} w_i c^*(h_1(x_i),h_2(x_i))^{\alpha} \cdot d_h(h_1(x_i),h_2(x_i))^{\beta}, \tag{18}$$

$$d_{wche}(h_1,h_2) = \left[\sum_{i=1}^{n} w_i \cdot \left[c^*(h_1(x_i),h_2(x_i))^{\alpha} \cdot d_h(h_1(x_i),h_2(x_i))^{\beta}\right]^2\right]^{\frac{1}{2}}, \tag{19}$$

$$d_{wchg}(h_1,h_2) = \left[\sum_{i=1}^{n} w_i \cdot \left[c^*(h_1(x_i),h_2(x_i))^{\alpha} \cdot d_h(h_1(x_i),h_2(x_i))^{\beta}\right]^{\lambda}\right]^{\frac{1}{\lambda}}, \tag{20}$$

where $\alpha, \beta \in [0,1]$.

In the following section, the usefulness of the proposed distance measures is illustrated by two numerical examples.

4. Numerical Examples

Example 1. *[18] Let Y be the set of all equilateral triangles, where* $Y = \{(\alpha,\beta,\gamma)|\alpha = \beta = \gamma = 60°\}$. *Then, every triangle could be considered as a fuzzy set in Y. For instance, for a triangle A with three angles as (60°, 85°, 35°), some people may be thought as an equilateral triangle, and take 0.7 as the membership value of fuzzy set; however, some other people may not think that it can be dealt with as an equilateral triangle, and take 0.3 as the membership value of fuzzy set. This means that triangle A can be dealt with by using an HFS concept. Suppose that there are two kinds of triangles which are denoted using HFEs as* $A_1 = \{0.7, 0.35\}$ *and* $A_2 = \{0.4\}$, *and a triangle* $A_0 = \{0.6\}$ *to be recognized.*

By using Equation (1), it gets that $d_{hh}(A_1,A_0) = 0.3375, d_{hh}(A_2,A_0) = 0.1$. By Equation (2), it gets that $d_{he}(A_1,A_0) = 0.3783, d_{he}(A_2,A_0) = 0.1414$. By comparing the two distances, it gets that A_0 belongs to A_2. Meanwhile, by using Equation (10), it gets that $d_h(A_1,A_0) = 0.175$ with a credibility factor $d_c(A_1,A_0) = 0.707$, and it gets that $d_h(A_2,A_0) = 0.2$ with a credibility factor $d_c(A_2,A_0) = 1$. Therefore, for decision makers who are willing to take risks, it is obtained that A_0 belongs to the class of A_1; for decision makers who are conservative, it is obtained that A_0 belongs to the class of A_2. The essence of the difference is that the element numbers of the HFEs are dealt with in different ways. This also illustrates the importance of the three parameters.

Example 2. *[14,18] Energy plays a very important role in socio-economic development in different countries. Suppose that there are five energy projects to be invested, which are defined as* A_i $(i = 1,2,\cdots,5)$. *Meanwhile, suppose that there are four attributes to be considered, which are technological* (P_1)*; environmental* (P_2)*; socio-political* (P_3)*; and economic* (P_4)*. The attribute weight is obtained as* $W = (0.15, 0.3, 0.2, 0.35)$. *Thereafter, a group of experts are invited to evaluate the performance of the five alternatives with respect to the four attributes on the concept "excellence". By using HFSs, the evaluation results are obtained as Table 1.*

Table 1. Hesitant fuzzy decision matrix.

Alternative	P_1	P_2	P_3	P_4
A_1	$\{0.5, 0.4, 0.3\}$	$\{0.9, 0.8, 0.7, 0.1\}$	$\{0.5, 0.4, 0.2\}$	$\{0.9, 0.6, 0.5, 0.3\}$
A_2	$\{0.5, 0.3\}$	$\{0.9, 0.7, 0.6, 0.5, 0.2\}$	$\{0.8, 0.6, 0.5, 0.1\}$	$\{0.7, 0.4, 0.3\}$
A_3	$\{0.7, 0.6\}$	$\{0.9, 0.6\}$	$\{0.7, 0.5, 0.3\}$	$\{0.6, 0.4\}$
A_4	$\{0.8, 0.7, 0.4, 0.3\}$	$\{0.7, 0.4, 0.2\}$	$\{0.8, 0.1\}$	$\{0.9, 0.8, 0.6\}$
A_5	$\{0.9, 0.7, 0.6, 0.3, 0.1\}$	$\{0.8, 0.7, 0.6, 0.4\}$	$\{0.9, 0.8, 0.7\}$	$\{0.9, 0.7, 0.6, 0.3\}$

Denote the "ideal alternative" as $A^* = \{1\}$. By using the technique for order preference by similarity to an ideal solution (see references [22,23]), and the newly proposed distance measures, the five energy projects (alternatives) are ranked, and the optimal one is obtained. Firstly, we extend the HFEs provided in Table 1, so that all the HFEs have the same cardinal number. Secondly, by using Equations (15)–(17), and taking $\alpha = 1$, $\beta = 1$, and $\lambda = 1, 2, 6, 10$, respectively, the deviations between each alternative and the ideal alternative are obtained, which are shown in Figure 1. Obviously, it gets that $A_5 \succ A_3 \succ A_4 \succ A_1 \succ A_2$, and the optimal alternative is A_5. This ranking results and the optimal alternative are consistent with the results proposed by reference [14].

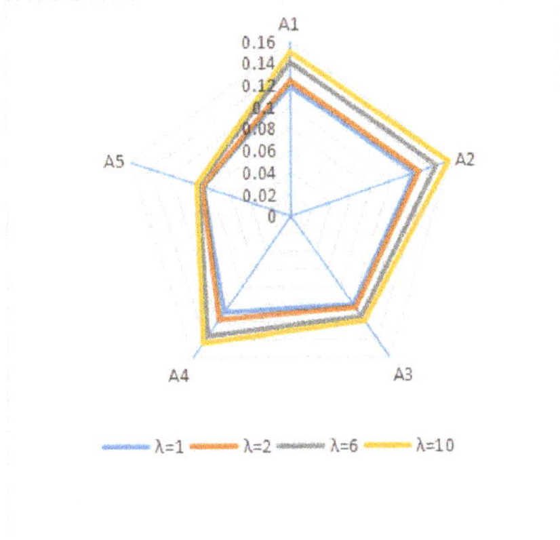

Figure 1. Results obtained by novel methods.

Thereafter, take (α, β) as $(0.9, 0.1)$, $(0.7, 0.3)$, $(0.5, 0.5)$, $(0.3, 0.7)$, $(0.1, 0.9)$, and take $\lambda = 1, 2, 6, 10$, respectively. By using Equation (20), the corresponding comprehensive deviations between each alternative and the ideal alternative are obtained, which are shown as Tables 2–6.

Table 2. Deviations between each alternative and the ideal alternative where $(\alpha, \beta) = (0.9, 0.1)$.

λ	A_1	A_2	A_3	A_4	A_5	**Ranking**
$\lambda = 1$	0.2612926	0.2612927	0.2631	0.2628	0.2582	$A_5 \succ A_1 \succ A_2 \succ A_4 \succ A_3$
$\lambda = 2$	0.2622	0.2645	0.2636	0.2646	0.2585	$A_5 \succ A_1 \succ A_2 \succ A_3 \succ A_4$
$\lambda = 6$	0.2661	0.2774	0.2656	0.2723	0.2593	$A_5 \succ A_3 \succ A_1 \succ A_4 \succ A_2$
$\lambda = 10$	0.2699	0.2891	0.2672	0.2801	0.2602	$A_5 \succ A_3 \succ A_1 \succ A_4 \succ A_2$

Table 3. Deviations between each alternative and the ideal alternative where $(\alpha, \beta) = (0.7, 0.3)$.

λ	A_1	A_2	A_3	A_4	A_5	**Ranking**
$\lambda = 1$	0.2983	0.3021	0.2869	0.2904	0.2751	$A_5 \succ A_3 \succ A_4 \succ A_1 \succ A_2$
$\lambda = 2$	0.3003	0.3054	0.2880	0.2940	0.2752	$A_5 \succ A_3 \succ A_4 \succ A_1 \succ A_2$
$\lambda = 6$	0.3083	0.3186	0.2921	0.3073	0.2758	$A_5 \succ A_3 \succ A_4 \succ A_1 \succ A_2$
$\lambda = 10$	0.3156	0.3304	0.2960	0.3178	0.2763	$A_5 \succ A_3 \succ A_1 \succ A_4 \succ A_2$

Table 4. Deviations between each alternative and the ideal alternative where $(\alpha, \beta) = (0.5, 0.5)$.

λ	A_1	A_2	A_3	A_4	A_5	**Ranking**
$\lambda = 1$	0.3409	0.3494	0.3143	0.3227	0.2944	$A_5 \succ A_3 \succ A_4 \succ A_1 \succ A_2$
$\lambda = 2$	0.3445	0.3528	0.3173	0.3298	0.2958	$A_5 \succ A_3 \succ A_4 \succ A_1 \succ A_2$
$\lambda = 6$	0.3588	0.3662	0.3274	0.3515	0.3002	$A_5 \succ A_3 \succ A_4 \succ A_1 \succ A_2$
$\lambda = 10$	0.3707	0.3780	0.3347	0.3641	0.3032	$A_5 \succ A_3 \succ A_4 \succ A_1 \succ A_2$

Table 5. Deviations between each alternative and the ideal alternative where $(\alpha, \beta) = (0.3, 0.7)$.

λ	A_1	A_2	A_3	A_4	A_5	**Ranking**
$\lambda = 1$	0.3900	0.4041	0.3459	0.3604	0.3166	$A_5 \succ A_3 \succ A_4 \succ A_1 \succ A_2$
$\lambda = 2$	0.3959	0.4076	0.3524	0.3729	0.3205	$A_5 \succ A_3 \succ A_4 \succ A_1 \succ A_2$
$\lambda = 6$	0.4192	0.4211	0.3715	0.4051	0.3316	$A_5 \succ A_3 \succ A_4 \succ A_1 \succ A_2$
$\lambda = 10$	0.4368	0.4329	0.3822	0.4197	0.3385	$A_5 \succ A_3 \succ A_4 \succ A_2 \succ A_1$

Table 6. Deviations between each alternative and the ideal alternative where $(\alpha, \beta) = (0.1, 0.9)$.

λ	A_1	A_2	A_3	A_4	A_5	**Ranking**
$\lambda = 1$	0.4465	0.4674	0.3823	0.4044	0.3418	$A_5 \succ A_3 \succ A_4 \succ A_1 \succ A_2$
$\lambda = 2$	0.4559	0.4711	0.3941	0.4247	0.3497	$A_5 \succ A_3 \succ A_4 \succ A_1 \succ A_2$
$\lambda = 6$	0.4914	0.4848	0.4250	0.4694	0.3702	$A_5 \succ A_3 \succ A_4 \succ A_2 \succ A_1$
$\lambda = 10$	0.5157	0.4963	0.4390	0.4864	0.3833	$A_5 \succ A_3 \succ A_4 \succ A_2 \succ A_1$

Tables 2–6 show that the alternative ranking order varies when the parameters are valued differently. Therefore, decision makers with different subjective preferences can choose specific parameters according to their experiences and attitudes. It means that the proposed parameterized distance measures are beneficial for the combination of subjective and objective decision-making information.

Moreover, the above results are not consistent with reference [18]. By using distance measures proposed in reference [18], the distances between each alternative and optimal alternative are obtained as Figure 2. In particular, the alternative ranking results are obtained as: (1) when $\lambda = 1$, it gets $A_3 \succ A_4 \succ A_5 \succ A_1 \succ A_2$; (2) when $\lambda = 2$, it gets $A_3 \succ A_4 \succ A_5 \succ A_2 \succ A_1$; (3) when $\lambda = 6$, it gets $A_3 \succ A_4 \succ A_5 \succ A_2 \succ A_1$; and (4) when $\lambda = 10$, it gets $A_3 \succ A_4 \succ A_5 \succ A_2 \succ A_1$.

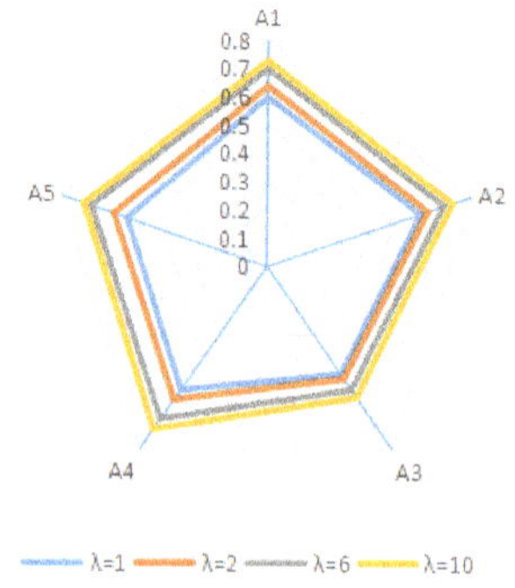

Figure 2. Results obtained by classical methods.

Contrastive analysis shows that the distance from alternative A_5 and the ideal alternative varies greatly. By investigation, the reasons for this results are concluded as follows: (*i*) The element numbers of HFE A_5 is bigger than those of the other four HFEs. When distance measures proposed in reference [18] are used, the element numbers of HFEs are viewed as a part of the distance between them; therefore, the distance between A_5 and the ideal alternative is larger. (*ii*) In the newly proposed distance measures, the cardinality of HFE is transferred to credibility factor; therefore, the corresponding distance between A_5 and the ideal alternative is smaller. (*iii*) The distance measures proposed in reference [14] is suitable to weight the values in HFEs. When calculating the distance between A_5 and the optimal alternative, unduly large or small deviations on the aggregation results are assigned low weights. Therefore, the calculation results obtained by reference [14] and this study are consistent with each other.

In essence, the characteristic of the distance measures proposed in this study is that they can combine the subjective and objective information well. They are good complements to decision-making theory. This case also illustrates that the decision-making process is not a pure mathematical calculation, and decision makers should choose the most suitable distance measure according to the specific decision-making environment. This is also the reason why decision-making is fascinating.

5. Conclusions

In this study, the role of cardinality of HFE in distance measures on HFSs is analysed. Moreover, a series of parameterized distance measures on HFSs are proposed. The main innovation points of this study are as follows:

1. In classical distance measures, the hesitance degree index of HFE is often calculated in addition to operations with the values of HFEs in classical distance measures. In contrast, the distance measures proposed in this study transfer the hesitance degree index to a credibility factor. Specifically, the credibility factor of HFE is calculated in multiplication operations with the values of HFEs in newly proposed distance measures, which handles the relationship between the cardinal number and the element values of hesitant fuzzy set well.
2. In the newly proposed distance measures, there are three parameters. These parameters can be adjusted by decision makers according to the specific decision-making environment, which is beneficial for combining subjective and objective decision-making information, making the decision-making results more objective.

However, every method does have its limitations, and it is hoped that these novel distance measures could become perfect step-by-step in practice.

Author Contributions: Conceptualization, J.R.L.; F.W.Z. and Q.L.; Methodology, F.W.Z. and J.S.; Software, J.Y.; Validation, S.H.W. and S.J.X.; Formal Analysis, S.J.X. and J.S.; Writing—Review and Editing, S.J.X.; Supervision, F.W.Z.; Project Administration, F.W.Z.; Funding Acquisition, F.W.Z.

Funding: This research was funded by National Natural Science Foundation of China (51508319, 61374195, 51409157), the research program of the National Special Authorized Social Science Fund of China (07@ZH005), and the Nature and Science Fund from the Zhejiang Province Ministry of Education (Y201327642).

Acknowledgments: The authors are thankful to the editor Ms. Jocelyn He, and the anonymous reviewers for their insightful and constructive comments for improving this paper.

Conflicts of Interest: The authors declare no conflict of interest.

Abbreviations

The following abbreviations are used in this manuscript:

HFS	Hesitant fuzzy set
HFE	Hesitant fuzzy element

References

1. Zadeh, L.A. Fuzzy sets. *Inf. Control* **1965**, *8*, 338–356. [CrossRef]
2. Zadeh, L.A. The concept of a linguistic variable and its application to approximate reasoning (I). *Inf. Sci.* **1975**, *8*, 199–249. [CrossRef]
3. Zadeh, L.A. The concept of a linguistic variable and its application to approximate reasoning (II). *Inf. Sci.* **1975**, *8*, 301–357. [CrossRef]
4. Zadeh, L.A. The concept of a linguistic variable and its application to approximate reasoning (III). *Inf. Sci.* **1975**, *9*, 43–80. [CrossRef]

5. Atanassov, K. Intuitionistic fuzzy sets. *Fuzzy Sets Syst.* **1986**, *1*, 87–96. [CrossRef]
6. Atanassov, K.; Gargov, G. Interval valued intuitionistic fuzzy sets. *Fuzzy Sets Syst.* **1989**, *3*, 343–349. [CrossRef]
7. Torra, V.; Narukawa, Y. On hesitant fuzzy sets and decision. In Proceedings of the 2009 IEEE International Conference on Fuzzy Systems, Jeju Island, Korea, 20–24 August 2009; pp. 1378–1382.
8. Torra, V. Hesitant fuzzy sets. *Int. J. Intell. Syst.* **2010**, *25*, 529–539. [CrossRef]
9. Beg, I.; Rashid, T. Group decision making using intuitionistic hesitant fuzzy sets. *Int. J. Fuzzy Logic Intell. Syst.* **2014**, *3*, 181–187. [CrossRef]
10. Wei, G.W. Interval-valued dual hesitant fuzzy uncertain linguistic aggregation operators in multiple attribute decision making. *J. Intell. Fuzzy Syst.* **2017**, *33*, 1881–1893. [CrossRef]
11. Zwick, R.; Carlstein, E.; Budescu, D.V. Measures of similarity among fuzzy concepts: A comparative analysis. *Int. J. Approx. Reason.* **1987**, *1*, 221–242. [CrossRef]
12. Pappis, C.P.; Karacapilidis, N.I. A comparative assessment of measures of similarity of fuzzy values. *Fuzzy Sets Syst.* **1993**, *56*, 171–174. [CrossRef]
13. Wang, W.J. New similarity measures on fuzzy sets and on elements. *Fuzzy Sets Syst.* **1997**, *85*, 305–309. [CrossRef]
14. Xu, Z.S.; Xia, M.M. Distance and similarity measures for hesitant fuzzy sets. *Inf. Sci.* **2011**, *181*, 2128–2138. [CrossRef]
15. Xu, Z.S.; Xia, M.M. On distance and correlation measures of hesitant fuzzy information. *Int. J. Intell. Syst.* **2011**, *26*, 410–425. [CrossRef]
16. Peng, D.H.; Gao, C.Y.; Gao, Z.F. Generalized hesitant fuzzy synergetic weighted distance measures and their application to multiple criteria decision-making. *Appl. Math. Model.* **2013**, *37*, 5837–5850. [CrossRef]
17. Rodríguez, R.M.; Martínez, L.; Torra, V.; Xu, Z.S.; Herrera, F. Hesitant fuzzy sets: State of the art and future directions. *Int. J. Intell. Syst.* **2014**, *29*, 495–524. [CrossRef]
18. Li, D.Q.; Zeng, W.Y.; Li, J.H. New distance and similarity measures on hesitant fuzzy sets and their applications in multiple criteria decision making. *Eng. Appl. Artif. Intell.* **2015**, *40*, 11–16. [CrossRef]
19. Tang, X.A.; Peng, Z.L.; Ding, H.N.; Cheng, M.L.; Yang, S.L. Novel distance and similarity measures for hesitant fuzzy sets and their applications to multiple attribute decision making. *J. Intell. Fuzzy Sys.* **2018**, *34*, 3903–3916. [CrossRef]
20. Alcantud, J.C.R.; de Andrés Calle, R.; Torrecillas, M.J.M. Hesitant fuzzy worth: An innovative ranking methodology for hesitant fuzzy subsets. *Appl. Soft Comput.* **2016**, *38*, 232–243. [CrossRef]
21. Wei, G.W. Hesitant fuzzy prioritized operators and their application to multiple attribute decision making. *Knowl.-Based Syst.* **2012**, *31*, 176–182. [CrossRef]
22. Behzadian, M.; Otaghsara, S.K.; Yazdani, M.; Ignatius, J. A state-of the-art survey of TOPSIS applications. *Expert Syst. Appl.* **2012**, *17*, 13051–13069. [CrossRef]
23. Muscat, J. *Functional Analysis: An Introduction to Metric Spaces, Hilbert Spaces, and Banach Algebras*; Springer International Publishing: Basel, Switzerland, 2014; pp. 13–26.

symmetry

MDPI

Article

On Approximation of Any Ordered Fuzzy Number by A Trapezoidal Ordered Fuzzy Number

Krzysztof Piasecki [1,*] and Anna Łyczkowska-Hanćkowiak [2]

[1] Department of Investment and Real Estate, Poznan University of Economics and Business, al. Niepodleglosci 10, 61-875 Poznań, Poland

[2] Department of Finance, WSB University in Poznań, ul. Powstańców Wielkopolskich 5, 61-895 Poznań, Poland; anna.lyczkowska-hanckowiak@wsb.poznan.pl

* Correspondence: krzysztof.piasecki@ue.poznan.pl; Tel.: +48-61-854-3531

Received: 5 September 2018; Accepted: 18 October 2018; Published: 21 October 2018

Abstract: In this paper, the model of imprecise quantity information is an ordered fuzzy number. The purpose of our study is to propose some methods of approximating any ordered fuzzy number using a trapezoidal ordered fuzzy number. The information ambiguity is evaluated by means of an energy measure. The information indistinctness is evaluated by Kosko's entropy measure. We discuss the problem of approximation of an arbitrary ordered fuzzy number by the nearest trapezoidal ordered fuzzy number. This way, we can simplify arithmetical operations on the linear space of ordered fuzzy numbers. The set of feasible trapezoidal ordered numbers is limited by the combination of the following conditions: invariance of energy measure, invariance of entropy measure, and invariance of information support. Evaluating the influence of individual limits combinations on the utility of given approximations, two combinations of those restraints, recommended for use, were chosen. It was also indicated that one of the recommended approximation problems can be used only for ordered fuzzy numbers characterized by a low level of entropy. The obtained results are currently used in such multi-criterial decision making models as financial portfolio management, evaluation of negotiations offers, the fuzzy TOPSIS model, and the fuzzy SAW model.

Keywords: ordered fuzzy numbers; energy measure; entropy measure; information ambiguity; information indistinctness

1. Introduction

Ordered fuzzy numbers (OFN) are defined in the perfect intuitive way by Kosiński et al. [1–4] as an extension of the concept of fuzzy number (FN). In this way, they were going to introduce a FN supplemented by a negative or positive orientation. The negative orientation means the order from bigger numbers to smaller ones. The positive orientation means the order from smaller numbers to bigger ones. The FN orientation is interpreted as a prediction of future FN changes. The Kosiński' theory has significant drawbacks. Kosiński [4] has shown that there exist improper OFNs that cannot be represented b. This means that we cannot apply any knowledge of fuzzy sets to solve practical problems described by improper OFN. Therefore, considerations that use improper OFN may not be fruitful. Kosiński [1–3] has determined OFNs' arithmetic as an extension of results obtained by Goetschel and Voxman [5] for FNs. Moreover, Kosiński [4] has shown that there exist such proper OFNs that their Kosiński's sum is equal to improper OFN. This means that the family of all proper OFNs is not closed under the Kosiński's addition. On the other hand, most mathematical applications require that the considered objects' family be closed under used arithmetic operations. Then again, the intuitive Kosiński's approach to the notion of OFN is very useful. For this reason, the original Kosiński's theory was revised in [6]. This paper is fully based on the revised OFNs' theory.

OFNs have already begun to find their use in operations research applied in decision making, economics, and finance [7–31]. All discussed applications are associated with the linear space of OFNs. However, a big inconvenience appears when using that space. The addition of an arbitrary OFN is a very complicated operation of significant formal complexity. An example of this addition is presented in Section 2. On the other hand, the addition of trapezoidal OFNs (TrOFN) is reduced to a simple addition of four-dimensional vectors of real numbers. Moreover, TrOFN are already applied for finance [20–22,25] and decision making [14,26,29,30]. These are the main premises to propose a substitution of OFNs by TrOFNs. It is obvious that any arbitrary OFN should be substituted by the nearest TrOFN. Therefore, we will apply the approximation task as a tool of substitution of OFN by TrOFN. When processing imprecise values, we use OFN only to follow the influence of the initial values' imprecision on the resulting information imprecision. Due to our approximation problem, we can impose the requirements of approximating by such a value that retains the estimates of the imprecision of an approximated value. The approximation problem of OFNs by TrOFNs has not, so far, been discussed in the literature [32].

The main aim of this paper is to propose the approximation methods of any OFN by trapezoidal OFNs. As a result, authors intend to indicate the recommended approximation methods. The paper is organised as follows. Section 2 presents the concept of OFNs and the linear space of OFNs. Section 3 briefly discusses the idea of imprecision [33]. The same chapter describes energy measure [34] as a measure tool of information ambiguity and entropy measure [35,36], as well as a measure tool of information indistinctness. In Section 4, the authors propose six approximation methods of any OFN by a TrOFN. Section 5 recommends chosen approximation methods and shows the possibility of implementing obtained results in a case of existing models of real objects. Finally, Section 6 concludes the article, summarizes the main findings of this research, and proposes some future research directions.

2. Ordered Fuzzy Numbers—Basic Facts

By $\mathcal{F}(\mathbb{R})$, we denote the family of all fuzzy subsets of a real line $\mathbb{R}$. An imprecise number is a family of values in which each considered value belongs to it in a varying degree. A commonly accepted model of an imprecise number is the fuzzy number (FN), defined as a fuzzy subset of the real line $\mathbb{R}$. The most general definition of FN was given by Dubois and Prade [37]. Thanks to the results obtained in [5], a concept of a fuzzy number can be equivalently denoted as follows:

Theorem 1. *For any FN $\mathcal{L}$ there exists such a nondecreasing sequence $(a,b,c,d) \subset \mathbb{R}$ that $\mathcal{L} \in \mathcal{F}(\mathbb{R})$ is represented by its membership function $\mu_{\mathcal{L}}(\cdot|a,b,c,d,L_L,R_L) \in [0;1]^{\mathbb{R}}$ given by the identity*

$$\mu_{\mathcal{L}}(x|a,b,c,d,L_L,R_L) = \begin{cases} 0, & x \notin [a,d], \\ L_L(x), & x \in [a,b], \\ 1, & x \in [b,c], \\ R_L(x), & x \in [c,d], \end{cases} \tag{1}$$

in which the left reference function $L_L \in [0;1]^{[a,b]}$ and the right reference function $R_L \in [0;1]^{[c,d]}$ are upper semi-continuous monotonic functions satisfying the following conditions:

$$L_L(b) = R_L(c) = 1, \tag{2}$$

$$\forall_{x \in]a,d[} : \mu_{\mathcal{L}}(x|a,b,c,d,L_L,R_L) > 0. \tag{3}$$

The family of all FN is denoted as $\mathbb{F}$. Moreover, Dubois and Prade [38] have introduced such arithmetic operations on FN that are coherent with the Zadeh Extension Principle. □

The concept of ordered fuzzy numbers (OFN) was intuitively introduced by Kosiński and his co-writers in the series of papers [1–4] as an extension of the concept of FN. A significant drawback of

Kosiński's theory is that there exist such OFNs that, in fact, are not FN [4]. What is more, the intuitive Kosiński's approach to the notion of OFN is very useful. The OFNs' usefulness results from the fact that an OFN is defined as FN supplemented by a negative or positive orientation. The negative orientation means the order from bigger numbers to smaller ones. The positive orientation means the order from smaller numbers to bigger ones. The FN orientation is interpreted as prediction of future FN changes. For this reason, the Kosiński's theory was revised in [6], in which OFNs are generally defined in a following way:

Definition 1. *Let any monotonic sequence* $(a, b, c, d) \subset \mathbb{R}$ *of ordered fuzzy number (OFN)* $\overleftrightarrow{\mathcal{L}}(a, b, c, d, S_L, E_L)$ *be defined as the pair of fuzzy number* $\mathcal{L} \in \mathbb{F}$ *and orientation* $a \rightarrowtail d = (a, d)$. *The fuzzy number* $\mathcal{L} \in \mathbb{F}$ *is determined by its membership function* $\mu_{\mathcal{L}}(\cdot|a, b, c, d, S_L, E_L) \in [0; 1]^{\mathbb{R}}$ *given by the identity*

$$\mu_{\mathcal{L}}(x|a, b, c, d, S_L, E_L) = \begin{cases} 0, & x \notin [a, d] = [d, a], \\ S_L(x), & x \in [a, b] = [b, a], \\ 1, & x \in [b, c] = [c, b], \\ E_L(x), & x \in [c, d] = [d, c] \end{cases} \tag{4}$$

in which the starting-function $S_L \in [0; 1]^{[a,b]}$ *and the ending-function* $E_L \in [0; 1]^{[c,d]}$ *are upper semi-continuous monotonic functions satisfying the conditions*

$$S_L(b) = E_L(c) = 1. \tag{5}$$

$$\forall_{x \in]a,d[} \; \mu_{\mathcal{L}}(x|a, b, c, d, S_L, E_L) > 0. \square \tag{6}$$

Remark: *Let us note that Equation (4) identity describes additionally extended notation of numerical intervals, which is used in this work.*

The space of all OFN is denoted by the symbol $\mathbb{K}$. The condition $a < d$ fulfilment determines the positive orientation $a \rightarrowtail d$ of OFN $\overleftrightarrow{\mathcal{L}}(a, b, c, d, S_L, E_L)$. In this case, the starting-function S_L is non-decreasing and the ending-function E_L is non-increasing. Any positively oriented OFN is interpreted as imprecise number, which may increase. The space of all positively oriented OFN is denoted by the symbol $\mathbb{K}^+$. The condition $a > d$ fulfilment determines the negative orientation of OFN $\overleftrightarrow{\mathcal{L}}(a, b, c, d, S_L, E_L)$. In this case, the starting-function S_L is non-increasing and the ending-function E_L is non-decreasing. Negatively oriented OFN is interpreted as an imprecise number, which may decrease. The space of all negatively oriented OFN is denoted by the symbol $\mathbb{K}^-$. For the case $a = d$, OFN $\overleftrightarrow{\mathcal{L}}(a, a, a, a, S_L, E_L)$ represents a crisp number $a \in \mathbb{R}$, which is not oriented. In summary, we can write

$$\mathbb{K} = \mathbb{K}^+ \cup \mathbb{R} \cup \mathbb{K}^-. \tag{7}$$

For the case $a \geq d$, any FN $\mathcal{L}(a, b, c, d, S_L, E_L)$ is equal to OFN $\overleftrightarrow{\mathcal{L}}(a, b, c, d, S_L, E_L)$. On the other hand, for this case any OFN $\overleftrightarrow{\mathcal{L}}(a, b, c, d, S_L, E_L)$ is equal to FN $\mathcal{L}(a, b, c, d, S_L, E_L)$. These facts imply that we have

$$\mathbb{F} = \mathbb{K}^+ \cup \mathbb{R}, \tag{8}$$

$$\mathbb{K} = \mathbb{F} \cup \mathbb{K}^-. \tag{9}$$

Also, the arithmetic proposed by Kosiński has a significant disadvantage. The space of ordered fuzzy numbers is not closed under Kosiński's addition. Therefore, Kosinski's theory is modified in this way so that the space of ordered fuzzy numbers is closed under revised arithmetic operations. The necessary arithmetic operators will generally be defined for OFNs using the following concepts.

Definition 2. *Cut-function* $L^{\star} \in [u,v]^{[0;1]}$ *of any upper semi-continuous nondecreasing function* $L \in [0;1]^{[u;v]}$ *is given by the identity*

$$L^{\star}(\alpha) = min\{x \in [u,v] : L(x) \geq \alpha\}. \square \tag{10}$$

Definition 3. *Cut-function* $R^{\star} \in [u,v]^{[0;1]}$ *of any upper semi-continuous nonincreasing function* $R \in [0;1]^{[u;v]}$ *is given by the identity*

$$R^{\star}(\alpha) = max\{x \in [u,v] : R(x) \geq \alpha\}. \square \tag{11}$$

Definition 4. *Pseudo inverse function* $l^{\triangleleft} \in [0;1]^{[l(0),l(1)]}$ *of any bounded continuous and nondecreasing function* $l \in [l(0),l(1)]^{[0;1]}$ *is given by the identity*

$$l^{\triangleleft}(x) = max\{\alpha \in [0;1] : l(\alpha) = x\}. \square \tag{12}$$

Definition 5. *Pseudo inverse function* $r^{\triangleleft} \in [0;1]^{[r(1),r(0)]}$ *of any bounded continuous and nonincreasing function* $r \in [r(0),r(1)]^{[0;1]}$ *is given by the identity*

$$r^{\triangleleft}(x) = min\{\alpha \in [0;1] : r(\alpha) = x\}. \square \tag{13}$$

The vast majority of practical implementations of OFNs is associated with the linear space $\langle \mathbb{K}, \boxed{+}, \odot \rangle$, in which

- the symbol $\boxed{+}$ denotes such addition operator on $\mathbb{K}$, which is an extension of the addition operator " $+$ " on $\mathbb{R}$;
- the symbol $\odot$ denotes such dot product operator on $\mathbb{K}$, which is an extension of the dot product operator "$\cdot$" on $\mathbb{R}$.

Therefore, the following considerations will be limited to the case of linear space $\langle \mathbb{K}, \boxed{+}, \odot \rangle$. Let us consider any pair of OFNs $\overset{\leftrightarrow}{\mathcal{K}}, \overset{\leftrightarrow}{\mathcal{M}} \in \mathbb{K}$ described as follows

$$\overset{\leftrightarrow}{\mathcal{K}} = \overset{\leftrightarrow}{\mathcal{L}}(a_K, b_K, c_K, d_K, S_K, E_K),\ \overset{\leftrightarrow}{\mathcal{M}} = \overset{\leftrightarrow}{\mathcal{L}}(a_M, b_M, c_M, d_M, S_M, E_M). \tag{14}$$

and the convention $\frac{0}{0} = 1$ is used. For any $\beta \in \mathbb{R}$, the dot product operation on $\mathbb{K}$ is defined by the identity

$$\beta \odot \overset{\leftrightarrow}{\mathcal{M}} = \overset{\leftrightarrow}{\mathcal{J}} = \overset{\leftrightarrow}{\mathcal{L}}(\beta \cdot a_M, \beta \cdot b_M, \beta \cdot c_M, \beta \cdot d_M, S_J, E_J), \tag{15}$$

in which we have

$$S_J(x) = S_M(x/\beta), \tag{16}$$

$$E_J(x) = S_M(x/\beta). \tag{17}$$

Example 1. *Let us take into account the OFN* $\overset{\leftrightarrow}{\mathcal{U}} = \overset{\leftrightarrow}{\mathcal{L}}(2,4,8,10,S_U,E_U)$, *in which*

$$\mu_U(x) = \left\{ \begin{array}{cc} 0, & x \notin [2,10], \\ S_U(x), & x \in [2,4], \\ 1, & x \in [4,8], \\ E_U(x), & x \in [8,10], \end{array} \right\} = \begin{cases} 0, & x \notin [2,10], \\ \frac{2x-4}{x}, & x \in [2,4], \\ 1, & x \in [4,8], \\ \frac{3x-30}{x-14}, & x \in [8,10]. \end{cases} \tag{18}$$

Then, for example, we have

$$\overset{\leftrightarrow}{\mathcal{W}} = 3\odot\overset{\leftrightarrow}{\mathcal{U}} = \overset{\leftrightarrow}{\mathcal{L}}(3\cdot 2, 3\cdot 4, 3\cdot 8, 3\cdot 10, S_W, E_W) = \overset{\leftrightarrow}{\mathcal{L}}(6, 12, 24, 30, S_W, E_W), \tag{19}$$

in which the conditions (16) and (17) imply

$$\mu_W(x) = \left\{\begin{array}{cc} 0, & x \notin [6,30], \\ S_W(x), & x \in [6,12], \\ 1, & x \in [12,24], \\ E_W(x), & x \in [24,30], \end{array}\right\} = \left\{\begin{array}{cc} 0, & x \notin [6,30], \\ S_U(\frac{x}{3}), & x \in [6,12], \\ 1, & x \in [12,24], \\ E_U(\frac{x}{3}), & x \in [24,30], \end{array}\right\} = \left\{\begin{array}{cc} 0, & x \notin [6,30], \\ \frac{2\cdot x-12}{x}, & x \in [6,12], \\ 1, & x \in [12,24], \\ \frac{3\cdot x-90}{x-42}, & x \in [24,30]. \end{array}\right. \square \tag{20}$$

The addition operation on $\mathbb{K}$ is defined by the identity [6]

$$\overset{\leftrightarrow}{\mathcal{K}} \boxed{+} \overset{\leftrightarrow}{\mathcal{M}} = \overset{\leftrightarrow}{\mathcal{J}} = \overset{\leftrightarrow}{\mathcal{L}}(a_J, b_J, c_J, d_J, S_J, E_J), \tag{21}$$

in which we have

$$\check{a}_J = a_K + a_M, \tag{22}$$

$$b_J = b_K + b_M, \tag{23}$$

$$c_J = c_K + c_M, \tag{24}$$

$$\check{d}_J = d_K + d_M, \tag{25}$$

$$a_J = \left\{\begin{array}{ll} min\{\check{a}_J, b_J\}, & (b_J < c_J) \vee (b_J = c_J \wedge \check{a}_J \le \check{d}_J), \\ max\{\check{a}_J, b_J\}, & (b_J > c_J) \vee (b_J = c_J \wedge \check{a}_J > \check{d}_J), \end{array}\right. \tag{26}$$

$$d_J = \left\{\begin{array}{ll} max\left\{\check{d}_J, c_J\right\}, & (b_J < c_J) \vee (b_J = c_J \wedge \check{a}_J \le \check{d}_J), \\ min\left\{\check{d}_J, c_J\right\}, & (b_J > c_J) \vee (b_J = c_J \wedge \check{a}_J > \check{d}_J), \end{array}\right. \tag{27}$$

$$\forall_{\alpha\in[0;1]} \quad s_J(\alpha) = \left\{\begin{array}{ll} S_K^{\star}(\alpha) + S_M^{\star}(\alpha), & a_J \ne b_J, \\ b_J, & a_J = b_J, \end{array}\right. \tag{28}$$

$$\forall_{\alpha\in[0;1]} \quad e_J(\alpha) = \left\{\begin{array}{ll} E_K^{\star}(\alpha) + E_M^{\star}(\alpha), & c_J \ne d_J, \\ c_J, & c_J = d_J, \end{array}\right. \tag{29}$$

$$\forall_{x\in[a_J,b_J]} \quad S_J(x) = s_J^{\lhd}(x), \tag{30}$$

$$\forall_{x\in[c_J,d_J]} \quad E_J(x) = e_J^{\lhd}(x) \tag{31}$$

The above described procedure of OFN's addition is widely justified in [6].

Example 2. *Let us take into account the OFNs* $\overset{\leftrightarrow}{\mathcal{U}}$ *given by (18) and* $\overset{\leftrightarrow}{\mathcal{V}} = \overset{\leftrightarrow}{\mathcal{L}}(15, 11, 6, 5, S_V, E_V)$*, in which*

$$\mu_V(x) = \left\{\begin{array}{cc} 0, & x \notin [15,5], \\ S_V(x), & x \in [15,11], \\ 1, & x \in [11,6], \\ E_V(x), & x \in [6,5], \end{array}\right\} = \left\{\begin{array}{ll} 0, & x \notin [15,5], \\ \frac{3x-45}{x-23}, & x \in [15,11], \\ 1, & x \in [11,6], \\ \frac{3x-15}{x-3}, & x \in [6,5]. \end{array}\right. \tag{32}$$

We determine the sum

$$\overleftrightarrow{\mathcal{Z}} = \overleftrightarrow{\mathcal{U}} \boxed{+} \overleftrightarrow{\mathcal{V}} = \overleftrightarrow{\mathcal{L}}(a_Z, b_Z, c_Z, d_Z, S_Z, E_Z). \tag{33}$$

From relations (22)–(25) we obtain

$$\begin{aligned} \check{a}_Z = a_U + a_V = 2 + 15 = 17, b_Z = b_U + b_V = 4 + 11 = 15, \\ c_Z = c_U + c_V = 8 + 6 = 14, \check{d}_Z = d_U + d_V = 10 + 5 = 15. \end{aligned} \tag{34}$$

Because we have $15 = b_Z > c_z = 14$, *from (26) we get* $a_Z = 17$, *and from (27) we get* $d_Z = 14$. *Consequently, we get*

$$\overleftrightarrow{\mathcal{Z}} = \overleftrightarrow{\mathcal{U}} \boxed{+} \overleftrightarrow{\mathcal{V}} = \overleftrightarrow{\mathcal{L}}(17, 15, 14, 14, S_Z, E_Z). \tag{35}$$

In the next step, we determine the variability of the ending-function $E_Z \in [0;1]^{[14,14]}$. *Due to satisfaction of the condition* $c_Z = 14 = d_Z$, *the Equation (29) implies*

$$\forall_{\alpha \in [0;1]} \quad e_Z(\alpha) = c_Z = 14. \tag{36}$$

Then, from the relations (13) and (31), we obtain

$$\forall_{x \in [14,14]} \quad E_Z(x) = E_Z(14) = e_M^{\triangleleft}(14) = max\{\alpha \in [0;1] : e_Z(\alpha) = 14\} = 1. \tag{37}$$

Next, we establish the variability of the starting-function $S_Z \in [0;1]^{[17,15]}$. *Here, we have* $a_Z = 17 \neq 15 = b_Z$. *Starting-function* $S_U \in [0;1]^{[2,4]}$ *is a non-decreasing one. It is very easy to check that this function is increasing. Therefore, from (10) for any* $\alpha \in [0;1]$ *we have*

$$S_U^{\star}(\alpha) = min\{x \in [2,4] : S_U(x) \geq \alpha\} = S_U^{-1}(\alpha) = \frac{4}{2 - \alpha}. \tag{38}$$

Starting-function $S_V \in [0;1]^{[15,11]}$ *is a non-increasing function. It is very easy to check that this function is decreasing. Therefore, from (10) for any* $\alpha \in [0;1]$ *we have*

$$S_V^{\star}(\alpha) = max\{x \in [15,11] : S_V(x) \geq \alpha\} = S_V^{-1}(\alpha) = \frac{23\alpha - 45}{\alpha - 3}. \tag{39}$$

Next, from (28) we obtain

$$s_Z(\alpha) = S_U^{\star}(\alpha) + S_V^{\star}(\alpha) = \frac{4}{2 - \alpha} + \frac{23\alpha - 45}{\alpha - 3} = \frac{23\alpha^2 - 95\alpha + 102}{\alpha^2 - 5\alpha + 6}. \tag{40}$$

It is very easy to check whether the function $s_Z \in [17,15]^{[0,1]}$ *is decreasing. Therefore, from the identities (13) and (30), we obtain*

$$\begin{aligned} S_Z(x) = s_Z^{\triangleleft}(x) = min\{\alpha \in [0;1] : s_Z(\alpha) = x\} = min\left\{\alpha \in [0;1] : \tfrac{23\alpha^2 - 95\alpha + 102}{\alpha^2 - 5\alpha + 6} = x\right\} \\ = min\left\{\tfrac{5(x-19) - \sqrt{x^2 + 10x - 359}}{2(x-23)}, \tfrac{5(x-19) + \sqrt{x^2 + 10x - 359}}{2(x-23)}\right\} = \tfrac{5(x-19) + \sqrt{x^2 + 10x - 359}}{2(x-23)}. \end{aligned} \tag{41}$$

Eventually, what we obtain is that the sum $\overleftrightarrow{\mathcal{K}} \boxed{+} \overleftrightarrow{\mathcal{M}} = \overleftrightarrow{\mathcal{Z}} = \overleftrightarrow{\mathcal{L}}(17, 15, 14, 14, S_Z, E_Z)$ *is unambiguously determined by its membership function*

$$\mu_Z(x) = \left\{ \begin{array}{cc} 0, & x \notin [17,14], \\ S_Z(x), & x \in [17,15], \\ 1, & x \in [15,14], \\ E_Z(x), & x \in [14,14], \end{array} \right\} = \left\{ \begin{array}{cc} 0, & x \notin [17,4], \\ \frac{5(x-19) + \sqrt{x^2 + 10x - 359}}{2(x-23)}, & x \in [17,15], \\ 1, & x \in [15,14], \\ 1, & x \in [14,14]. \end{array} \right. \quad \square \tag{42}$$

The above example sufficiently shows a high level of formal complexity upon the addition of any OFN. Therefore, in many practical applications researchers limit the use of OFN only to a form presented below.

Definition 2. *[6] For any monotonic sequence $(a,b,c,d) \subset \mathbb{R}$ the trapezoidal ordered fuzzy number (TrOFN) $\overleftrightarrow{Tr}(a,b,c,d)$ is defined as the pair of FN $\mathcal{T}\nabla(a,b,c,d) \in \mathbb{F}$ and the orientation $a \rightarrowtail d$. The fuzzy number $\mathcal{T}\nabla(a,b,c,d) \in \mathbb{F}$ is determined by its membership function $\mu_{\overleftrightarrow{Tr}}(\cdot|a,b,c,d) \in [0;1]^{\mathbb{R}}$ given by the identity*

$$\mu_{\overleftrightarrow{Tr}}(x|a,b,c,d) = \begin{cases} 0, & x \notin [a,d] = [d,a], \\ \frac{x-a}{b-a}, & x \in [a,b[=]b,a], \\ 1, & x \in [b,c] = [c,b], \\ \frac{x-d}{c-d} & x \in]c,d] = [d,c[, \end{cases} . \square \tag{43}$$

The space of all TrOFN is denoted by the symbol $\mathbb{K}_{Tr}$. The following numbers are a particular case of TrOFN.

Definition 3. *For any monotonic sequence $(a,b,c) \subset \mathbb{R}$, the triangle ordered fuzzy number (TOFN) $\overleftrightarrow{T}(a,b,c)$ is defined by the identity*

$$\overleftrightarrow{T}(a,b,c) = \overleftrightarrow{Tr}(a,b,b,c). \square \tag{44}$$

For the case of any real number $\beta \in \mathbb{R}$ and any TrOFN $\overleftrightarrow{Tr}(a,b,c,d)$, their dot product can be calculated as follows:

$$\beta \odot \overleftrightarrow{Tr}(a,b,c,d) = \overleftrightarrow{Tr}(\beta \cdot a, \beta \cdot b, \beta \cdot c, \beta \cdot d). \tag{45}$$

Example 3. *For example, we have*

$$\overleftrightarrow{\mathcal{W}} = 3 \odot \overleftrightarrow{Tr}(2,4,8,10) = \overleftrightarrow{Tr}(3 \cdot 2, 3 \cdot 4, 3 \cdot 8, 3 \cdot 10) = \overleftrightarrow{Tr}(6,12,24,30). \square \tag{46}$$

In agreement with arithmetic introduced in [6], for the case of any TrOFNs $\overleftrightarrow{Tr}(a,b,c,d)$ and $\overleftrightarrow{Tr}(p-a, q-b, r-c, s-d)$, their sum is determined as follows:

$$\begin{gathered} \overleftrightarrow{Tr}(a,b,c,d) \boxplus \overleftrightarrow{Tr}(p-a, q-b, r-c, s-d) \\ = \begin{cases} \overleftrightarrow{Tr}(min\{p,q\}, q, r, max\{r,s\}), & (q < r) \vee (q = r \wedge p \leq s), \\ \overleftrightarrow{Tr}(max\{p,q\}, q, r, min\{r,s\}), & (q > r) \vee (q = r \wedge p > s). \end{cases} \end{gathered} \tag{47}$$

Example 4. *We determine the sum*

$$\overleftrightarrow{\mathcal{W}} = \overleftrightarrow{Tr}(2,\ 4,\ 8,\ 10) \boxed{+} \overleftrightarrow{Tr}(15,\ 11,\ 6,\ 5\ ). \tag{48}$$

We have $p = 2 + 15 = 17$, $q = 4 + 11 = 15 > 14 = 8 + 6 = r$, and $s = 10 + 5 = 15$. Then, from (47) we get

$$\overleftrightarrow{\mathcal{W}} = \overleftrightarrow{Tr}(17,\ 15,\ 14,\ 14). \square \tag{49}$$

Comparing examples 1, 2, 3, and 4 we can easily notice the benefits resulting from using TrOFN when implementing OFN in real object models. The postulate of being constrained to TrOFN use when constructing models of real objects is not always possible to satisfy. A behavioural present value [17,18,39,40], described by such OFN that cannot be TrOFN, is an example. The attempts to use this concept in portfolio analysis produce many problems.

All these reservations lead to a postulate to approximate all OFNs by TrOFNs while modelling all real objects. Such an approximation task will be formulated in Section 4 of the paper.

3. Evaluation of Imprecision

After [33], we understand imprecision as a superposition of ambiguity and indistinctness of information. Ambiguity can be interpreted as a lack of a clear recommendation between one alternative and various others. Indistinctness is understood as a lack of explicit distinction between recommended and unrecommended alternatives.

Fuzzy subsets [41] are widely used as a model of imprecision information [42,43]. Any OFN is a particular kind of imprecision information. For this reason, in this chapter—if necessary—any OFN $\overleftrightarrow{\mathcal{L}}(a,b,c,d,S_L,E_L)$ will be considered as an ordered pair $(\mathcal{L},(a,d)) \in \mathcal{F}(\mathbb{R}) \times \mathbb{R}^2$ in which the fuzzy subset $\mathcal{L}$ is determined by its membership function $\mu_{\mathcal{L}} \in [0,1]^{\mathbb{R}}$ given by the identity (4).

An increase in information imprecision reduces the suitability of this information. Therefore, it is logical to consider the problem of imprecision assessment. A basic tool used in this chapter to measure the imprecision of fuzzy sets is Khalili's measure [44] determined by the identity

$$m(\mathcal{A}) = \int_{-\infty}^{+\infty} \mu_{\mathcal{A}}(x)dx, \tag{50}$$

in which the fuzzy subset $\mathcal{A} \in \mathcal{F}(\mathbb{R})$ has bounded support, i.e., its membership function $\mu_{\mathcal{L}} \in [0,1]^{\mathbb{R}}$ fulfils the condition

$$\exists_{(p,q)\in\mathbb{R}^2}\{x \in \mathbb{R} : \mu_{\mathcal{A}}(x) > 0\} \subset [p,q]. \tag{51}$$

The increase in the ambiguity of an OFN suggests a higher number of alternative recommendations to choose from. This leads to an increase in the risk of choosing an incorrect assessment from recommended alternative ones. This may result in making a decision, which will be ex post associated with a loss of chance. Therefore, an increase in the ambiguity of OFN implies the decrease in the utility of information described by OFN. The proper tool for measuring the ambiguity of FN is an energy measure $d \in \left[\mathbb{R}_0^+\right]^{\mathcal{F}(\mathbb{R})}$ proposed by de Luca and Termini [35]. In this article, for an arbitrary OFN $\overleftrightarrow{\mathcal{L}}(a,b,c,d,S_L,E_L) \in \mathbb{K}$ represented by $\mathcal{L} \in \mathcal{F}(\mathbb{R})$ we have

$$d(\overleftrightarrow{\mathcal{L}}(a,b,c,d,S_L,E_L)) = m(\mathcal{L}). \tag{52}$$

A new tool for OFN ambiguity evaluation is introduced in [19]. There, it is proposed that the oriented energy index $a \in \mathbb{R}^{\mathbb{K}}$, which assesses the ambiguity of any OFN $\overleftrightarrow{\mathcal{L}}(a,b,\ c,d,S_L,E_L)$ by integral, was as follows

$$a(\overleftrightarrow{\mathcal{L}}(a,b,c,d,S_L,E_L)) = \int_a^d \mu_{\mathcal{L}}(x)dx, \tag{53}$$

in which $\mu_{\mathcal{L}} \in [0,1]^{\mathbb{R}}$ is the membership function of the fuzzy subset $\mathcal{L} \in \mathcal{F}(\mathbb{R})$ representing evaluated OFN. Quite a new fact for evaluation is that for any negatively oriented OFN, its oriented energy index is negative and for any positively oriented OFN, its oriented energy index is positive. Moreover, it is obvious that for any OFN we have

$$d(\overleftrightarrow{\mathcal{L}}(a,b,c,d,S_L,E_L)) = |a(\overleftrightarrow{\mathcal{L}}(a,b,c,d,S_L,E_L))|. \tag{54}$$

This means that oriented energy index stores the information on energy index and the orientation of assessed OFN. This gives new perspectives for imprecision management.

Example 5. *Let us calculate oriented energy index: for OFNs $\overset{\leftrightarrow}{\mathcal{U}}$ determined by (18), for $\overset{\leftrightarrow}{\mathcal{V}}$ determined by (32), and for $\overset{\leftrightarrow}{\mathcal{Z}} = \overset{\leftrightarrow}{\mathcal{U}} \boxed{+} \overset{\leftrightarrow}{\mathcal{V}}$ determined by (42). Implementing (53), we get the following:*

$$a(\overset{\leftrightarrow}{\mathcal{U}}) = \int_2^4 \frac{2x-4}{x} dx + \int_4^8 dx + \int_8^{10} \frac{3x-30}{x-14} dx \approx 6.3614, \tag{55}$$

$$a(\overset{\leftrightarrow}{\mathcal{V}}) = \int_{15}^{11} \frac{3x-45}{x-23} dx + \int_{11}^{6} dx + \int_{6}^{5} \frac{3x-15}{x-3} dx \approx 7.8362, \tag{56}$$

$$a(\overset{\leftrightarrow}{\mathcal{Z}}) = \int_{17}^{15} \frac{5(x-19) + \sqrt{x^2+10x-359}}{2(x-23)} dx + \int_{15}^{14} dx \approx -2.0414. \square \tag{57}$$

An increase in the indistinctness of an OFN suggests that the differences between recommended and unrecommended decision alternatives are harder to differentiate. This leads to an increase in a risk of choosing a not recommended option. Therefore, increase in the indistinctness of OFN implies the decrease in the utility of information described by OFN. The right tool for measuring the indistinctness is the entropy measure, proposed also by de Luca and Termini [35] and modified by Piasecki [36]. In this article, the entropy measure $e \in [0;1]^{\mathcal{F}(\mathbb{R})}$ will be described like in Kosko [45]. For an arbitrary OFN $\overset{\leftrightarrow}{\mathcal{L}}(a,b,c,d,S_L,E_L) \notin \mathbb{R}$, we have

$$e(\overset{\leftrightarrow}{\mathcal{L}}(a,b,c,d,S_L,E_L)) = \frac{m(\mathcal{L} \cap \mathcal{L}^C)}{m((\mathcal{L} \cup \mathcal{L}^C) \cap [a,d])}, \tag{58}$$

in which the symbol $\mathcal{L}^C$ denotes the complement of the fuzzy subset $\mathcal{L} \in \mathcal{F}(\mathbb{R})$ describing evaluated OFN. Due to a good synthetic substantiation and universalism of the above-mentioned formula, the entropy measure proposed by Kosko [45] is now widely used.

Example 6. *Let us calculate entropy measure: for OFNs $\overset{\leftrightarrow}{\mathcal{U}}$ determined by (18), for $\overset{\leftrightarrow}{\mathcal{V}}$ determined by (32), and for $\overset{\leftrightarrow}{\mathcal{Z}} = \overset{\leftrightarrow}{\mathcal{U}} \boxed{+} \overset{\leftrightarrow}{\mathcal{V}}$ determined by (42). Using (58), we get the following:*

$$\begin{gathered} m(\mathcal{U} \cap \mathcal{U}^C) = \int_2^{10} min\left\{\mu_{\overset{\leftrightarrow}{\mathcal{U}}}(x), 1-\mu_{\overset{\leftrightarrow}{\mathcal{U}}}(x)\right\}dx \\ = \int_2^{\frac{8}{3}} \tfrac{2x-4}{x} dx + \int_{\frac{8}{3}}^{4} (1 - \tfrac{2x-4}{x}) dx + \int_4^8 0 dx + \int_8^{\frac{46}{5}} (1 - \tfrac{3x-30}{x-14}) dx + \int_{\frac{46}{5}}^{10} \tfrac{3x-30}{x-14} dx \approx 0.9609, \end{gathered} \tag{59}$$

$$\begin{gathered} m(\mathcal{U} \cup \mathcal{U}^C) = \int_2^{10} max\left\{\mu_{\overset{\leftrightarrow}{\mathcal{U}}}(x), 1-\mu_{\overset{\leftrightarrow}{\mathcal{U}}}(x)\right\}dx \\ = \int_2^{\frac{8}{3}} (1 - \tfrac{2x-4}{x}) dx + \int_{\frac{8}{3}}^{4} \tfrac{2x-4}{x} dx + \int_4^8 dx + \int_8^{\frac{46}{5}} \tfrac{3x-30}{x-14} dx + \int_{\frac{46}{5}}^{10} (1 - \tfrac{3x-30}{x-14}) dx \approx 7.0391, \end{gathered} \tag{60}$$

$$e(\overset{\leftrightarrow}{\mathcal{U}}) = \frac{m(\mathcal{U} \cap \mathcal{U}^C)}{m(\mathcal{U} \cup \mathcal{U}^C)} \approx 0.1365. \tag{61}$$

In analogous way we obtain

$$e(\overset{\leftrightarrow}{\mathcal{V}}) \approx 0.1396, e(\overset{\leftrightarrow}{\mathcal{Z}}) \approx 0.2041. \square \tag{62}$$

The values of oriented energy index entropy measure obtained in Examples 5 and 6 will be used in approximation tasks presented in a following chapter. As it can be easily proved, in case of

TrOFN $\overleftrightarrow{Tr}(a,b,c,d), \notin \mathbb{R}$-oriented energy index and entropy measure are determined basing on the following relation

$$a(\overleftrightarrow{Tr}(a,b,c,d)) = \frac{1}{2}(d+c-b-a), \tag{63}$$

$$e(\overleftrightarrow{Tr}(a,b,c,d)) = \frac{d-c+b-a}{3{\cdot}d+c-b-3{\cdot}a} \tag{64}$$

Hence, for TOFN $\overleftrightarrow{Tr}(a,b,c) \notin \mathbb{R}$, we have

$$a(\overleftrightarrow{T}(a,b,c)) = \frac{1}{2}(d-c), \tag{65}$$

$$e(\overleftrightarrow{T}(a,b,c)) = \frac{1}{3}. \tag{66}$$

From the point of view of real objects modelling, steadiness of indistinctness is a drawback of TOFN. This disadvantage is the result of a lack of any possibility to track the influence of occurring changes on the imprecision element. It is yet another premise to prefer TrOFN as a tool to describe imprecise numbers.

Finally, let's notice the following, simple characteristics of TrOFN.

Lemma 1. *For an arbitrary TrOFN* $\overleftrightarrow{Tr}(a,b,c,d) \notin \mathbb{R}$, *we have:*

$$e(\overleftrightarrow{Tr}(a,b,c,d)) \le \frac{1}{3}, \tag{67}$$

$$e(\overleftrightarrow{Tr}(a,b,c,d)) = \frac{1}{3} \Leftrightarrow \overleftrightarrow{Tr}(a,b,c,d) = \overleftrightarrow{T}(a,b,d) = \overleftrightarrow{T}(a,c,d). \ \square \tag{68}$$

Proof. For positively oriented TrOFN $\overleftrightarrow{Tr}(a,b,c,d)$, we have $c \ge b$, which implies

$$e(\overleftrightarrow{Tr}(a,b,c,d)) = \tfrac{d-c+b-a}{3{\cdot}d+c-b-3{\cdot}a} = \tfrac{-d+c-b+a}{-3{\cdot}d-c+b+3{\cdot}a} = \tfrac{(a-d)+(c-b)}{3{\cdot}(a-d)-(c-b)} \le \tfrac{(a-d)}{3{\cdot}(a-d)} = \tfrac{1}{3}. \tag{69}$$

For negatively oriented TrOFN $\overleftrightarrow{Tr}(a,b,c,d)$, we have $b \ge c$, which implies

$$e(\overleftrightarrow{Tr}(a,b,c,d)) = \frac{d-c+b-a}{3{\cdot}d+c-b-3{\cdot}a} = \frac{(d-a)+(b-c)}{3{\cdot}(d-a)-(b-c)} \le \frac{(d-a)}{3{\cdot}(d-a)} = \frac{1}{3} \tag{70}$$

The condition (58) follows immediately from the identities (64) and (66). □

This conclusion will be used in the next chapter when examining the characteristics of a suggested approximation model of any OFN utilizing TrOFN.

4. Approximation Problem

After reference [33], we understand imprecision as a superposition of ambiguity and indistinctness of information. Ambiguity can be interpreted as a lack of a clear recommendation between one alternative among various others. Indistinctness is understood as a lack of explicit distinction between recommended and unrecommended alternatives.

To estimate the distance between any pair of OFNs, we introduce a pseudo-metrics $\delta : \mathbb{K}^2 \to \mathbb{R}_0^+$ determined by a following identity:

$$\begin{aligned} &\delta(\overleftrightarrow{\mathcal{L}}(a_1,b_1,c_1,d_1,S_L^{(1)},E_L^{(1)}),\overleftrightarrow{\mathcal{L}}(a_2,b_2,c_2,d_2,S_L^{(2)},E_L^{(2)})) \\ &= \sqrt{(a_1-a_2)^2+(b_1-b_2)^2+(c_1-c_2)^2+(d_1-d_2)^2}. \end{aligned} \tag{71}$$

In this section, we will consider the approximation problem of an arbitrary OFN $\overset{\leftrightarrow}{\mathcal{L}}(a,b,c,d,S_L,E_L)$ by the nearest TrOFN $\overset{\leftrightarrow}{Tr}(p,q,r,s)$. Hence, the main criterion of approximation is to determine such TrOFN $\overset{\leftrightarrow}{Tr}(p_0,q_0,r_0,s_0)$, which will satisfy the following condition:

$$\begin{gathered}\delta(\overset{\leftrightarrow}{Tr}(p_0,q_0,r_0,s_0),\overset{\leftrightarrow}{\mathcal{L}}(a,b,c,d,S_L,E_L))\\ = min\left\{\delta(\overset{\leftrightarrow}{Tr}(p,q,r,s),\overset{\leftrightarrow}{\mathcal{L}}(a,b,c,d,S_L,E_L)) : \overset{\leftrightarrow}{Tr}(p,q,r,s)\in\mathbb{K}_{Tr}\right\}.\end{gathered} \tag{72}$$

Function $\delta : (\mathbb{K}_{Tr})^2 \to \mathbb{R}_0^+$ is a metric in $\mathbb{K}_{Tr}$. It implies that if any $\overset{\leftrightarrow}{\mathcal{L}}(a,b,c,d,S_L,E_L)\in\mathbb{K}_{Tr}$ then it is the unique solution of the minimization problem (72). Moreover, any crisp number $a\in\mathbb{R}$ is represented by TrOFN $\overset{\leftrightarrow}{Tr}(a,a,a,a)$. Therefore, we must consider a case of

$$\overset{\leftrightarrow}{\mathcal{L}}(a,b,c,d,S_L,E_L)\notin\mathbb{K}_{Tr}\supset\mathbb{R}. \tag{73}$$

The condition (72) will be equivalent to a problem

$$\Phi(p_0,q_0,r_0,s_0|a,b,c,d) = min\left\{\Phi(p,q,r,s|a,b,c,d) : (p,q,r,s)\in\mathbb{R}^4\right\} \tag{74}$$

of minimization of the objective function $\Phi(\cdot|a,b,c,d):\mathbb{R}^4\to\mathbb{R}_0^+$ given by an identity

$$\Phi(p,q,r,s|a,b,c,d) = (p-a)^2+(q-b)^2+(r-c)^2+(s-d)^2. \tag{75}$$

When processing imprecise values, we use FN or OFN only to follow the influence of initial values imprecision on the imprecision of obtained information. Due to our approximation problem, we can impose the requirements of approximating by such value that retains the measures of the imprecision of an approximated value. Using (53) and (58) in an uncomplicated way, we can determine the estimates of ambiguity and indistinctness of the approximated value

$$A = a(\overset{\leftrightarrow}{\mathcal{L}}(a,b,c,d,S_L,E_L)), \tag{76}$$

$$E = e(\overset{\leftrightarrow}{\mathcal{L}}(a,b,c,d,S_L,E_L)). \tag{77}$$

In this case, the conditions of imprecision we denote as

$$a(\overset{\leftrightarrow}{Tr}(p,q,r,s)) = A, \tag{78}$$

$$e(\overset{\leftrightarrow}{Tr}(p,q,r,s)) = E. \tag{79}$$

Let us also notice that due to the condition (78), OFNs $\overset{\leftrightarrow}{\mathcal{L}}(a,b,c,d,S_L,E_L)$ i $\overset{\leftrightarrow}{Tr}(p,q,r,s)$ are identically oriented. Juxtaposing the identity (63) and the condition (78) implies a linear equation describing invariance of ambiguity

$$s+r-q-p = 2\cdot A. \tag{80}$$

Pairing the identity (50) and the condition (79) implies a linear equation describing invariance of indistinctness

$$(1-3\cdot E)\cdot s-(1+E)\cdot r+(1+E)\cdot q-(1-3\cdot E)\cdot p = 0. \tag{81}$$

According to Lemma 1, the condition (81) can be used only in case of OFN $\overset{\leftrightarrow}{\mathcal{L}}(a,b,c,d,S_L,E_L)$ satisfying the condition

$$e(\overset{\leftrightarrow}{\mathcal{L}}(a,b,c,d,S_L,E_L)) = E \le \frac{1}{3} \tag{82}$$

For any OFN $\overleftrightarrow{\mathcal{L}}(a,b,c,d,S_L,E_L)$ represented, inter alia, by its membership function $\mu_{\mathcal{L}} \in [0,1]^{\mathbb{R}}$, we determine its support closure $[\overleftrightarrow{\mathcal{L}}(a,b,c,d,S_L,E_L)]_{0^+}$ in the following way

$$[\overleftrightarrow{\mathcal{L}}(a,b,c,d,S_L,E_L)]_{0^+} = \lim_{\alpha \to 0^+} \{x \in \mathbb{R} : \mu_{\mathcal{L}}(x) \geq \alpha\} = [a,d]. \tag{83}$$

The support closure $[\overleftrightarrow{\mathcal{L}}(a,b,c,d,S_L,E_L)]_{0^+}$ is interpreted as the smallest closed subset containing all possible values represented by OFN $\overleftrightarrow{\mathcal{L}}(a,b,c,d,S_L,E_L)$. Hence, Invariant Criterion

$$[a,d] = [\overleftrightarrow{\mathcal{L}}(a,b,c,d,S_L,E_L)]_{0^+} = [\overleftrightarrow{Tr}(p,q,r,s)]_{0^+} = [p,s] \tag{84}$$

can be treated as yet another constraint imposed on approximation method. This Criterion is represented by the equation system:

$$\begin{cases} p = a, \\ s = d. \end{cases} \tag{85}$$

Each TrOFN feasible in the approximation task will be called a feasible TrOFN. Conditions (80), (81), and (85) are restrictions that limit a set of feasible TrOFN. The solution of OFN approximation is the nearest feasible TrOFN, denoted by the symbol $\overleftrightarrow{Tr}_0$. In following subsections, we will analyse chosen problems of approximation OFN by TrOFN. Each of those problems will be distinguished by a chosen combination of restrictions limiting the set of feasible TrOFN. Each of approximation problems is called XYZ-approximation, in which the prefix XYZ is an acronym identifying the approximation problem. The solution of XYZ-approximation of any OFN $\overleftrightarrow{\mathcal{L}}(a,b,c,d,S_L,E_L)$ is denoted by a symbol $\overleftrightarrow{Tr}_{XYZ}(\overleftrightarrow{\mathcal{L}}(a,b,c,d,S_L,E_L))$.

4.1. Unconditional Approximation (UC-Approximation)

Initially, we will consider the approximation problem with no constraints imposed on the set of feasible TrOFN. This approximation problem is called UC-approximation. UC-approximation is determined just by the objective function (74). In the face of a lack of any restricting equations, the solution of UC-approximation problem of OFN $\overleftrightarrow{\mathcal{L}}(a,b,c,d,S_L,E_L)$ is TrOFN

$$\overleftrightarrow{Tr}(a,b,c,d) = \overleftrightarrow{Tr}_{UC}(\overleftrightarrow{\mathcal{L}}(a,b,c,d,S_L,E_L)). \tag{86}$$

It is obvious that a UC-approximation problem always has a solution.

Example 7. *For OFNs* $\overleftrightarrow{\mathcal{U}} = \overleftrightarrow{\mathcal{L}}(2,4,8,10,S_U,E_U)$ *determined by (18),* $\overleftrightarrow{\mathcal{V}} = \overleftrightarrow{\mathcal{L}}(15,11,6,5,S_V,E_V)$ *determined by (32), and* $\overleftrightarrow{\mathcal{Z}} = \overleftrightarrow{\mathcal{U}} \boxed{+} \overleftrightarrow{\mathcal{V}} = \overleftrightarrow{\mathcal{L}}(17,15,14,14,S_Z,E_Z)$ *determined by (42), we have*

$$\overleftrightarrow{Tr}(2,4,8,10) = \overleftrightarrow{Tr}_{UC}(\overleftrightarrow{\mathcal{U}}), \overleftrightarrow{Tr}(15,11,6,5) = \overleftrightarrow{Tr}_{UC}(\overleftrightarrow{\mathcal{V}}), \overleftrightarrow{Tr}(17,15,14,14) = \overleftrightarrow{Tr}_{UC}(\overleftrightarrow{\mathcal{Z}}). \tag{87}$$

Then, we evaluate an example influence that UC-approximation had on the imprecision measure. For OFNs $\overleftrightarrow{\mathcal{U}}, \overleftrightarrow{\mathcal{V}}$, *and* $\overleftrightarrow{\mathcal{Z}}$, *the results gathered in Examples 5 and 6 were used. In case of TrOFNs* $\overleftrightarrow{Tr}_{UC}(\overleftrightarrow{\mathcal{U}})$, $\overleftrightarrow{Tr}_{UC}(\overleftrightarrow{\mathcal{V}})$, *and* $\overleftrightarrow{Tr}_{UC}(\overleftrightarrow{\mathcal{Z}})$, *the imprecision characteristics were determined by formulas (63) and (64). All those values are presented in Table 1.*

Table 1. UC-approximation impact on imprecision evaluation.

$\overleftrightarrow{\mathcal{L}}$	$a(\overleftrightarrow{\mathcal{L}})$	$a(\overleftrightarrow{Tr}_{UC}(\overleftrightarrow{\mathcal{L}}))$	$e(\overleftrightarrow{\mathcal{L}})$	$e(\overleftrightarrow{Tr}_{UC}(\overleftrightarrow{\mathcal{L}}))$
$\overleftrightarrow{\mathcal{U}}$	6.3614	6.0000	0.1365	0.1429
$\overleftrightarrow{\mathcal{V}}$	−7.8362	−7.500	0.1396	0.1429
$\overleftrightarrow{\mathcal{Z}}$	−2.0414	−2.000	0.2041	0.2000

Source: Own elaboration. □

The results presented in this example show that UC-approximation to a certain extent distorts the imprecision evaluations. However, these evaluations are only the values representing the points on conventional scale. Therefore, the observed deviations of less than 5% of relative error can be considered insignificant.

On the other hand, the comparison of identities (15) and (45) lets us conclude that for any pair $(\beta, \overleftrightarrow{\mathcal{M}}) \in \mathbb{R} \times \mathbb{K}$ we have

$$\beta \odot \overleftrightarrow{Tr}_{UC}(\overleftrightarrow{\mathcal{M}}) = \overleftrightarrow{Tr}_{UC}(\beta \odot \overleftrightarrow{\mathcal{M}}). \tag{88}$$

Moreover, from the comparison of identities (21)–(27) and (47), it follows that for any pair $(\overleftrightarrow{\mathcal{K}}, \overleftrightarrow{\mathcal{M}}) \in \mathbb{K}^2$ we have

$$\overleftrightarrow{Tr}_{UC}(\overleftrightarrow{\mathcal{K}}) \boxed{+} \overleftrightarrow{Tr}_{UC}(\overleftrightarrow{\mathcal{M}}) = \overleftrightarrow{Tr}_{UC}(\overleftrightarrow{\mathcal{K}} \boxed{+} \overleftrightarrow{\mathcal{M}}). \tag{89}$$

It means that the function $\overleftrightarrow{Tr}_{UC} \in (\mathbb{K}_{Tr})^{\mathbb{K}}$ is a linear operator on $\mathbb{K}$. Due to this property, the errors of the estimations of oriented energy index and entropy measure did not succumb to propagation. This is a vital advantage of UC-approximation method.

4.2. *Approximation under Criterion of Constant Ambiguity (CA-Approximation)*

Let us consider the approximation problem in which each feasible TrOFN satisfies the condition (78). Such an approximation task we call CA-approximation problem. Then, the coordinates of all the feasible TrOFN $\overleftrightarrow{Tr}(p, q, r, s)$ can be presented as a general solution of the Equation (80). This solution is given as follows

$$\begin{cases} p = x, \\ q = y, \\ r = z, \\ s = 2 \cdot A + x + y - z, \end{cases} \quad x, y, z \in \mathbb{R}, \tag{90}$$

in which the sequence (x, y, z) is a monotonic sequence. Then—to solve the OFN $\overleftrightarrow{\mathcal{L}}(a, b, c, d, S_L, E_L)$ CA-approximation problem—the first step is to find the minimum of a function

$$\varphi(x, y, z) = \Phi(x, y, z, 2 \cdot A + x + y - z | a, b, c, d). \tag{91}$$

The minimum is reached in a point (x_0, y_0, z_0) of following coordinates

$$\begin{cases} x_0 = \frac{3 \cdot a - b + c + d - 2 \cdot A}{4}, \\ y_0 = \frac{-a + 3 \cdot b + c + d - 2 \cdot A}{4}, \\ z_0 = \frac{a + b + 3 \cdot c - d + 2 \cdot A}{4}. \end{cases} \tag{92}$$

When inserting those coordinates to (90), we conclude that the solution of OFN $\overleftrightarrow{\mathcal{L}}(a, b, c, d, S_L, E_L)$ CA-approximation problem is TrOFN

$$\overleftrightarrow{Tr}\left(\frac{3 \cdot a - b + c + d - 2 \cdot A}{4}, \frac{-a + 3 \cdot b + c + d - 2 \cdot A}{4}, \frac{a + b + 3 \cdot c - d + 2 \cdot A}{4}, \frac{a + b - c + 3 \cdot d + 2 \cdot A}{4}\right) = \overleftrightarrow{Tr}_{CA}(\overleftrightarrow{\mathcal{L}}(a, b, c, d, S_L, E_L)). \tag{93}$$

Example 8. *For OFNs* $\overleftrightarrow{\mathcal{U}} = \overleftrightarrow{\mathcal{L}}(2,4,8,10,S_U,E_U)$ *determined by (18),* $\overleftrightarrow{\mathcal{V}} = \overleftrightarrow{\mathcal{L}}(15,11,6,5,S_V,E_V)$ *determined by (32), and* $\overleftrightarrow{\mathcal{Z}} = \overleftrightarrow{\mathcal{U}} \boxplus \overleftrightarrow{\mathcal{V}} = \overleftrightarrow{\mathcal{L}}(17,15,14,14,S_Z,E_Z)$ *determined by (42), we have*

$$\overleftrightarrow{Tr}(1.8193, 3.8193, 8.1807, 10.1807) = \overleftrightarrow{Tr}_{CA}(\overleftrightarrow{\mathcal{U}}), \quad (94)$$

$$\overleftrightarrow{Tr}(15.1681, 11.1681, 5.8319, 4.8319) = \overleftrightarrow{Tr}_{CA}(\overleftrightarrow{\mathcal{V}}), \quad (95)$$

$$\overleftrightarrow{Tr}(17.0207, 15.0207, 13.9793, 13.9793) = \overleftrightarrow{Tr}_{CA}(\overleftrightarrow{\mathcal{Z}}). \quad (96)$$

Next, we assess an example influence that CA-approximation had on indistinctness evaluation. For OFNs $\overleftrightarrow{\mathcal{U}}$, $\overleftrightarrow{\mathcal{V}}$, $\overleftrightarrow{\mathcal{Z}}$, $\overleftrightarrow{Tr}_{UC}(\overleftrightarrow{\mathcal{U}})$, $\overleftrightarrow{Tr}_{UC}(\overleftrightarrow{\mathcal{V}})$, *and* $\overleftrightarrow{Tr}_{UC}(\overleftrightarrow{\mathcal{Z}})$, *the results from Example 7 were used. In case of TrOFNs* $\overleftrightarrow{Tr}_{CA}(\overleftrightarrow{\mathcal{U}})$, $\overleftrightarrow{Tr}_{CA}(\overleftrightarrow{\mathcal{V}})$, *and* $\overleftrightarrow{Tr}_{CA}(\overleftrightarrow{\mathcal{Z}})$, *their entropy measure was determined by the formula (64). All these values are presented in Table 2.*

Table 2. CA-approximation impact on indistinctness evaluation.

$\overleftrightarrow{\mathcal{L}}$	$e(\overleftrightarrow{\mathcal{L}})$	$e(\overleftrightarrow{Tr}_{CA}(\overleftrightarrow{\mathcal{L}}))$	$e(\overleftrightarrow{Tr}_{UC}(\overleftrightarrow{\mathcal{L}}))$
$\overleftrightarrow{\mathcal{U}}$	0.1365	0.1358	0.1429
$\overleftrightarrow{\mathcal{V}}$	0.1396	0.1375	0.1429
$\overleftrightarrow{\mathcal{Z}}$	0.2041	0.1967	0.2000

Source: Own elaboration.

Moreover, we have here

$$\begin{aligned}&\overleftrightarrow{Tr}_{CA}(\overleftrightarrow{\mathcal{U}}) \boxplus \overleftrightarrow{Tr}_{CA}(\overleftrightarrow{\mathcal{V}}) = \overleftrightarrow{Tr}(1.8193, 3.8193, 8.1807, 10.1807) \boxplus \overleftrightarrow{Tr}(15.1681, 11.1681, 5.8319, 4.8319)\\ &= \overleftrightarrow{Tr}(16.9874, 14.9874, 14.0126, 14.0126) \neq \overleftrightarrow{Tr}(17.0207, 15.0207, 13.9793, 13.9793) = \overleftrightarrow{Tr}_{CA}(\overleftrightarrow{\mathcal{Z}}).\end{aligned} \quad (97)$$

Hence, the function $\overleftrightarrow{Tr}_{AC} \in (\mathbb{K}_{Tr})^{\mathbb{K}}$ *is not an additive operator on* $\mathbb{K}$. *Furthermore, the two equations below show the existing effect of estimation errors propagation of oriented energy index and entropy measure.*

$$a(\overleftrightarrow{Tr}_{CA}(\overleftrightarrow{\mathcal{U}}) \boxplus \overleftrightarrow{Tr}_{CA}(\overleftrightarrow{\mathcal{V}})) = -1.9784 \neq -2.0414 = a(\overleftrightarrow{Tr}_{CA}(\overleftrightarrow{\mathcal{Z}})), \quad (98)$$

$$e(\overleftrightarrow{Tr}_{CA}(\overleftrightarrow{\mathcal{U}}) \boxplus \overleftrightarrow{Tr}_{CA}(\overleftrightarrow{\mathcal{V}})) = 0.2020 \neq 0.1967 = e(\overleftrightarrow{Tr}_{CA}(\overleftrightarrow{\mathcal{Z}})). \ \square \quad (99)$$

The above example shows that the influence of CA-approximation on the entropy measure evaluation is moderate, and when it comes to its scale it is similar to the influence of UC-approximation method. This example proves that the function $\overleftrightarrow{Tr}_{CA} \in (\mathbb{K}_{Tr})^{\mathbb{K}}$ is not a linear operator on $\mathbb{K}$. Calculations run in this example indicate the possibility of the occurrence of the error estimation propagation of oriented energy index and entropy measure. An extremely interesting phenomenon is revealed by Equation (98). IT appears that during the addition of OFNs approximations with precisely imitated ambiguity index, the error propagation of index estimation occurs. This phenomenon will be further called a zero-error propagation phenomenon.

4.3. Approximation under Criterion of Constant Imprecision (CI-Approximation)

Let us consider an approximation problem in which each feasible TrOFN satisfies conditions (78) and (79). Such a task will be called CI-approximation. Then, coordinates of all feasible TrOFN

$\overleftrightarrow{Tr}(p,q,r,s)$ can be presented as a general solution of equation system (80) and (81). This solution is given as

$$\begin{cases} p = x, \\ q = y, \\ r = y + \frac{1-3\cdot E}{1-E}\cdot A, \\ s = x + \frac{1+E}{1-E}\cdot A, \end{cases} \quad x, y \in \mathbb{R}. \tag{100}$$

Then, to solve the OFN $\overleftrightarrow{\mathcal{L}}(a,b,c,d,S_L,E_L)$ CI-approximation problem, the first step is to find the minimum of a function

$$\varphi(x,y) = \Phi(x,y,y + \frac{1-3\cdot E}{1-E}\cdot A, x + \frac{1+E}{1-E}\cdot A|a,b,c,d) \tag{101}$$

This minimum is reached in a point (x_0,y_0) of following coordinates

$$\begin{cases} x_0 = \frac{a+d}{2} - \frac{1+E}{2\cdot(1-E)}\cdot A, \\ y_0 = \frac{b+c}{2} - \frac{1-3\cdot E}{2\cdot(1-E)}\cdot A. \end{cases} \tag{102}$$

After inserting these coordinates to (98), we can conclude that the solution of CI-approximation problem of OFN $\overleftrightarrow{\mathcal{L}}(a,b,c,d,S_L,E_L)$ fulfilling condition (82) is TrOFN

$$\overleftrightarrow{Tr}(\tfrac{a+d}{2} - \tfrac{1+E}{2\cdot(1-E)}\cdot A, \tfrac{b+c}{2} - \tfrac{1-3\cdot E}{2\cdot(1-E)}\cdot A, \tfrac{b+c}{2} + \tfrac{1-3\cdot E}{2\cdot(1-E)}\cdot A, \tfrac{a+d}{2} + \tfrac{1+E}{2\cdot(1-E)}\cdot A) = \overleftrightarrow{Tr}_{CI}(\overleftrightarrow{\mathcal{L}}(a,b,c,d,S_L,E_L)). \tag{103}$$

Example 9. *For OFNs $\overleftrightarrow{\mathcal{U}} = \overleftrightarrow{\mathcal{L}}(2,4,8,10,S_U,E_U)$ determined by (18), $\overleftrightarrow{\mathcal{V}} = \overleftrightarrow{\mathcal{L}}(15,11,6,5,S_V,E_V)$ determined by (32), and $\overleftrightarrow{\mathcal{Z}} = \overleftrightarrow{\mathcal{U}} \boxplus \overleftrightarrow{\mathcal{V}} = \overleftrightarrow{\mathcal{L}}(17,15,14,14,S_Z,E_Z)$ determined by (42), we have*

$$\overleftrightarrow{Tr}(1.8137, 3.8248, 8.1751, 10.1863) = \overleftrightarrow{Tr}_{CI}(\overleftrightarrow{\mathcal{U}}), \tag{104}$$

$$\overleftrightarrow{Tr}(15.1610, 11.1752, 5.8248, 4.8390) = \overleftrightarrow{Tr}_{CI}(\overleftrightarrow{\mathcal{V}}), \tag{105}$$

$$\overleftrightarrow{Tr}(17.0442, 14.9972, 14.0038, 13.9558) = \overleftrightarrow{Tr}_{CI}(\overleftrightarrow{\mathcal{Z}}). \tag{106}$$

Furthermore, we have here

$$\begin{aligned} \overleftrightarrow{Tr}_{CI}(\overleftrightarrow{\mathcal{U}}) \boxplus \overleftrightarrow{Tr}_{CI}(\overleftrightarrow{\mathcal{V}}) &= \overleftrightarrow{Tr}(1.8137, 3.8248, 8.1751, 10.1863) \boxplus \overleftrightarrow{Tr}(15.1610, 11.1752, 5.8248, 4.8390) \\ &= \overleftrightarrow{Tr}(16.9747, 15.0000, 13.9999, 13.9999) \neq \overleftrightarrow{Tr}(17.0442, 14.9972, 14.0038, 13.9558) = \overleftrightarrow{Tr}_{CI}(\overleftrightarrow{\mathcal{Z}}). \end{aligned} \tag{107}$$

Therefore, the function $\overleftrightarrow{Tr}_{CI} \in (\mathbb{K}_{Tr})^{\mathbb{K}}$ is not an additive operator on $\mathbb{K}$. Moreover, the two equations presented below are examples of an occurrence of zero-error propagation phenomenon.

$$a(\overleftrightarrow{Tr}_{CI}(\overleftrightarrow{\mathcal{U}}) \boxplus \overleftrightarrow{Tr}_{CI}(\overleftrightarrow{\mathcal{V}})) = -1.9875 \neq -2.0414 = a(\overleftrightarrow{Tr}_{CI}(\overleftrightarrow{\mathcal{Z}})), \tag{108}$$

$$e(\overleftrightarrow{Tr}_{CI}(\overleftrightarrow{\mathcal{U}}) \boxplus \overleftrightarrow{Tr}_{CI}(\overleftrightarrow{\mathcal{V}})) = 0.1990 \neq 0.2041 = e(\overleftrightarrow{Tr}_{CI}(\overleftrightarrow{\mathcal{Z}})). \; \square \tag{109}$$

The above example proves that the function $\overleftrightarrow{Tr}_{CI} \in (\mathbb{K}_{Tr})^{\mathbb{K}}$ is not a linear operator on $\mathbb{K}$. The calculations in this example confirm that there is a possibility of an occurrence of zero-error propagation phenomenon of oriented energy index and entropy measure estimations.

4.4. Approximation under Invariant Criterion (IC-Approximation)

Let us consider an approximation problem in which each feasible TrOFN satisfies conditions (84). Such an approximation task will be called IC-approximation. Then, coordinates of all feasible TrOFN $\overleftrightarrow{Tr}(p,q,r,s)$ can be presented as a general solution of the Equation (85). This solution is given as

$$\begin{cases} p = a, \\ q = x, \\ r = y, \\ s = d, \end{cases} \quad x, y \in \mathbb{R}. \tag{110}$$

Then, to solve the OFN $\overleftrightarrow{\mathcal{L}}(a,b,c,d,S_L,E_L)$ IC-approximation problem, the first step is to find the minimum of a function

$$\varphi(x,y) = \Phi(a,x,y,d|a,b,c,d). \tag{111}$$

This minimum is reached in a point (x_0, y_0) of following coordinates

$$\begin{cases} x_0 = b, \\ y_0 = c. \end{cases} \tag{112}$$

In this instance, we can conclude that the solution of IC-approximation problem of OFN $\overleftrightarrow{\mathcal{L}}(a,b,c,d,S_L,E_L)$ is TrOFN

$$\overleftrightarrow{Tr}(a,b,c,d) = \overleftrightarrow{Tr}_{IC}(\overleftrightarrow{\mathcal{L}}(a,b,c,d,S_L,E_L)) = \overleftrightarrow{Tr}_{UC}(\overleftrightarrow{\mathcal{L}}(a,b,c,d,S_L,E_L)). \tag{113}$$

This means that UC-approximation and IC-approximation problems always have an identical solution. Therefore, there is no need to consider IC-approximation problem.

4.5. Approximation under Invariant Criterion with Constant Ambiguity (ICCA-Approximation)

Let us consider an approximation problem in which each feasible TrOFN satisfies conditions (78) and (84). Such an approximation task will be called ICCA-approximation problem. Then, coordinates of all feasible TrOFN $\overleftrightarrow{Tr}(p,q,r,s)$ can be presented as a general solution of equation system (80) and (85). This solution is given as

$$\begin{cases} p = a, \\ q = x, \\ r = x + 2 \cdot A + a - d, \\ s = d, \end{cases} \quad x \in \mathbb{R}. \tag{114}$$

Then, to solve the OFN $\overleftrightarrow{\mathcal{L}}(a,b,c,d,S_L,E_L)$ ICCA-approximation problem, the first step is to find the minimum of a function

$$\varphi(x) = \Phi(a,x,x+2 \cdot A + a - d,d|a,b,c,d). \tag{115}$$

This minimum is reached in a point

$$x_0 = \frac{-a+b+c+d}{2} - A. \tag{116}$$

After inserting that value to (114), we can conclude that the solution of ICCA-approximation problem of OFN $\overleftrightarrow{\mathcal{L}}(a,b,c,d,S_L,E_L)$ is TrOFN

$$\overleftrightarrow{Tr}(a, \frac{-a+b+c+d}{2} - A, \frac{a+b+c-d}{2} + A, d) = \overleftrightarrow{Tr}_{ICCA}(\overleftrightarrow{\mathcal{L}}(a,b,c,d,S_L,E_L)). \tag{117}$$

Example 10. *For OFNs* $\overleftrightarrow{\mathcal{U}} = \overleftrightarrow{\mathcal{L}}(2,4,8,10,S_U,E_U)$ *determined by (18),* $\overleftrightarrow{\mathcal{V}} = \overleftrightarrow{\mathcal{L}}(15,11,6,5,S_V,E_V)$ *determined by (32), and* $\overleftrightarrow{\mathcal{Z}} = \overleftrightarrow{\mathcal{U}} \boxed{+} \overleftrightarrow{\mathcal{V}} = \overleftrightarrow{\mathcal{L}}(17,15,14,14,S_Z,E_Z)$ *determined by (42), we have*

$$\overleftrightarrow{Tr}(2,3.6386,8.3614,10) = \overleftrightarrow{Tr}_{ICCA}(\overleftrightarrow{\mathcal{U}}), \tag{118}$$

$$\overleftrightarrow{Tr}(15,11.3362,5.6638,5) = \overleftrightarrow{Tr}_{ICCA}(\overleftrightarrow{\mathcal{V}}), \tag{119}$$

$$\overleftrightarrow{Tr}(17,15.0414,13.9586,14) = \overleftrightarrow{Tr}_{ICCA}(\overleftrightarrow{\mathcal{Z}}).(!) \tag{120}$$

Let us notice that the sequence $(17,15.0414,13.9586,14)$ *is not monotonic. Therefore, the object* $\overleftrightarrow{Tr}(17,15.0414,13.9586,14)$ *is not an OFN. It is described by its membership relation* $\mu_{\overleftrightarrow{Tr}}(\cdot|a,b,c,d) \in \mathbb{R} \times [0,1]$ *given by the identity (43). This object is an example of an improper OFN described in [6]. Summing up, the solution of an ICCA-approximation problem OFN* $\overleftrightarrow{\mathcal{Z}}$ *does not exist. In addition, this fact also results in conclusion that for the triple* $(\overleftrightarrow{\mathcal{U}}, \overleftrightarrow{\mathcal{V}}, \overleftrightarrow{\mathcal{Z}})$*, the function* $\overleftrightarrow{Tr}_{ICCI} \in (\mathbb{K}_{Tr})^{\mathbb{K}}$ *does not fulfil the condition (89).*

Next, we evaluate an example influence that ICCA-approximation had on indistinctness evaluation. For OFNs $\overleftrightarrow{\mathcal{U}}$, $\overleftrightarrow{\mathcal{V}}$, $\overleftrightarrow{\mathcal{Z}}$, $\overleftrightarrow{Tr}_{CA}(\overleftrightarrow{\mathcal{U}})$, $\overleftrightarrow{Tr}_{CA}(\overleftrightarrow{\mathcal{V}})$, $\overleftrightarrow{Tr}_{CA}(\overleftrightarrow{\mathcal{Z}})$, $\overleftrightarrow{Tr}_{UC}(\overleftrightarrow{\mathcal{U}})$, $\overleftrightarrow{Tr}_{UC}(\overleftrightarrow{\mathcal{V}})$, *and* $\overleftrightarrow{Tr}_{UC}(\overleftrightarrow{\mathcal{Z}})$, *the results from Example 8 were used. In case of TrOFNs* $\overleftrightarrow{Tr}_{ICCA}(\overleftrightarrow{\mathcal{U}})$ *and* $\overleftrightarrow{Tr}_{ICCA}(\overleftrightarrow{\mathcal{V}})$*, their entropy measures were determined based on formula (64). All these values are given in Table 3.*

Table 3. ICCA-approximation impact on indistinctness evaluation.

$\overleftrightarrow{\mathcal{L}}$	$e(\overleftrightarrow{\mathcal{L}})$	$e(\overleftrightarrow{Tr}_{ICCA}(\overleftrightarrow{\mathcal{L}}))$	$e(\overleftrightarrow{Tr}_{CA}(\overleftrightarrow{\mathcal{L}}))$	$e(\overleftrightarrow{Tr}_{UC}(\overleftrightarrow{\mathcal{L}}))$
$\overleftrightarrow{\mathcal{U}}$	0.1365	0.1141	0.1358	0.1429
$\overleftrightarrow{\mathcal{V}}$	0.1396	0.1213	0.1375	0.1429
$\overleftrightarrow{\mathcal{Z}}$	0.2041	-	0.1967	0.2000

Source: Own elaboration. □

The above example proves that the function $\overleftrightarrow{Tr}_{ICCA} \in (\mathbb{K}_{Tr})^{\mathbb{K}}$ is not a linear operator on $\mathbb{K}$. The lack of linearity of that function lets us also expect that the zero-error propagation phenomenon will occur. The calculations in this example also prove the thesis that for chosen OFNs, the solution of ICCA-approximation problem might not exist. Similar conclusion can be drawn in case of CA-approximation and CI-approximation.

4.6. Approximation under Invariant Criterion with Constant Imprecision (ICCI-Approximation)

Let us consider an approximation problem in which each feasible TrOFN satisfies conditions (78), (79), and (84). Such an approximation task will be called ICCI-approximation problem. Using the substitutions (85), the equation system (80), (81), and (85) can be transformed into the following equivalent equation system:

$$\begin{cases} r - q = 2 \cdot A - d + a, \\ (1+E) \cdot r - (1+E) \cdot q = (1 - 3 \cdot E) \cdot (-d + a). \end{cases} \tag{121}$$

According to Cramer's Rule [46], system of Equation (121) is consistent if it satisfies the following condition:

$$(2 \cdot A - d + a) \cdot (1 + E) = (1 - 3 \cdot E) \cdot (-d + a), \tag{122}$$

From the condition (122), we have

$$E = \frac{d - a}{2 \cdot A - (d - a)} \tag{123}$$

because the condition (73) implies $a \neq d$. Therefore, there is no need to consider ICCI-approximation problem. On the other hand, we can say that if OFN $\overleftrightarrow{\mathcal{L}}(a, b, c, d, S_L, E_L)$ satisfies the conditions (82) and (123) then the ICCA-approximation problem solution $\overleftrightarrow{Tr}_{ICCA}(\overleftrightarrow{\mathcal{L}}(a, b, c, d, S_L, E_L))$ additionally satisfies the condition (84).

5. Recommendations

OFNs are a form of a record of imprecise information. The purpose of approximation of OFNs by TrOFNs was justified in the paper. Following that premise, six OFN approximation problems by the nearest TrOFN were proposed and discussed in the paper. It appeared that the two remaining approximation problems UC-approximation ad IC-approximation always deliver identical solutions. Additionally, the ICCI-approximation and ICCA-approximation problem always have identical solutions. Therefore, they should not be differentiated. Each of remaining four approximation methods has its advantages and disadvantages.

The main advantage of UC-approximation method is the fact that it is the only discussed method that delivers a solution of any OFN. IT is also the only one of the discussed approximation methods free of the occurrence of error propagation phenomenon of imprecision characteristics, that is, ambiguity index and entropy measure. Thanks to using the UC-approximation for processing complex sets of information, it is always safe. The drawbacks of UC-approximation method are the discrepancies between the characteristics of imprecision of approximated OFN and approximating TrOFN. This distorts the usefulness evaluation of processed information. From the point of view of algebra, UC-approximation is a linear operator of all OFN space.

All other approximation methods might lack the solution of incidental OFN. Such cases can be easily recognised utilising the fact that every TrOFN is characterised by a monotonic sequence of parameters. If, during implementation of a chosen approximation method of a given OFN we will obtain a non-monotonic sequence of parameters, then such an approximation problem has no solution. Attempts should be made to approximate a given OFN using another approximation method. There always exists at least one approximation method with a solution for a given OFN.

CA-approximation method approximates a given OFN by such TrOFN that its oriented energy index is equal to oriented energy index of OFN. This lets us monitor the usefulness of processed information more credibly. A more precise approximation method is ICCS-approximation, which approximates a given OFN by TrOFN of the identical support closure and equally oriented energy index. It allows one to monitor the usefulness of processed information more credibly. This means that CA-approximation should be used only in cases in which ICCA-approximation method does not deliver the solution. For both these approximation methods, there are no universal conditions restraining the set of approximated OFNs.

On the contrary, in case of CI-approximation method, the approximated OFNs must be characterised by a low level of entropy that satisfies condition (82). The CA-approximation method approximated a given OFN by TrOFN with an equally oriented energy index and entropy measure. It lets us monitor the usefulness of processed information more credibly than in case of the approximation obtained by the CA-approximation method.

A significant, common disadvantage of CA-, CI-, and ICCA-approximation methods is the occurrence of zero-error propagation phenomenon, described in Section 4. This usually leads to the conclusion that using such methods of approximation in cases of big, complex sets of information creates a threat for the credible monitoring of processed data usefulness.

In this case, the procedure of approximation method choice depends on individual preferences of the researcher who wants to approximate OFNs by TrOFNs. Nevertheless, it can be clarified by the following procedure:

- If the researcher accepts the phenomenon of zero-error propagation, then one should choose CI- or ICCA-approximation method if applicable. Otherwise, one should choose UC-approximation.
- If, in a considered case, the approximation method provides no solution, then the researcher chooses the second of CI- or ICCA-approximation method,
- If, in the considered case, CI- and ICCA-approximation methods do not have solutions, then the researcher should choose the CA-approximation method,
- If, in the considered case, the CA-approximation method does not have a solution, then the researcher should choose UC-approximation.

The above-described procedure of the approximation method choice can be used separately for any number. It is acceptable to use different methods for different numbers in the same empirical implementation.

The results obtained in the paper let us, for instance, use the Oriented Behavioural Present Value [17,18] in portfolio models described in [19,21,22,24].

6. Conclusions

The results obtained in this article will facilitate the use of OFN in real object modelling.

In the face of conditions (7) and (8), the above-described approximation methods can be used for approximate FN by trapezoidal FN. The oriented energy index is then reduced to the energy measure.

Among other things, FN are applied in social choice and political sciences. In reference [47], FN describe imprecise linguistic evaluation labels used by voters. Each assessment can be irrationally undervalued or overvalued. Then, we can use OFN to describe the irrational deviations of the assessment. In [48], FN describe voters' opinion profiles. This is a static description of the population of voters. On the other side, voters' opinions are very dynamic. Replacement of FN by OFN would allow one to describe mentioned dynamics. The problems discussed in this paper can be a starting point for an attempt to create such approximation methods of OFN by TrOFN that such a constraint that a "vector of solution parameters is a monotonic sequence" will be included. Moreover, further research on the considered above approximation method should be devoted to the following problems:

- the impact of approximation on the ordering of OFNs,
- the impact of the approximation on a solution to equations determined by OFNs,
- the impact of approximation on formal modelling of real objects.

Solutions obtained to the last three problems will allow estimation of the approximation error.

Author Contributions: Conceptualization, K.P.; Methodology, K.P. and A.Ł.H.; Validation, A.Ł.H.; Formal Analysis, A.Ł.H.; Writing—Original Draft Preparation, K.P.; Writing—Review & Editing, K.P. and A.Ł.H.

Funding: This research received no external funding.

Acknowledgments: The authors are very grateful to the anonymous reviewers for their insightful and constructive comments and suggestions. We would like to thank to Professor José Carlos R. Alcantud for his valuable final remarks. Using these comments allowed us to improve this article.

Conflicts of Interest: The authors declare no conflict of interest.

References

1. Kosiński, W.; Słysz, P. Fuzzy numbers and their quotient space with algebraic operations. *Bull. Pol. Acad. Sci.* **1993**, *41*, 285–295.

2. Kosiński, W.; Prokopowicz, P.; Ślęzak, D. Fuzzy reals with algebraic operations: Algorithmic approach. In *Advances in Soft Computing, Proceedings of the Intelligent Information Systems 2002 Symposium, Sopot, Poland, 3–6 June 2002*; Klopotek, M., Wierzchoń, S.T., Michalewicz, M., Eds.; Springer: Heidelberg, Germany, 2002; pp. 311–320.
3. Kosiński, W.; Prokopowicz, P.; Ślęzak, D. Ordered fuzzy numbers. *Bull. Pol. Acad. Sci.* **2003**, *51*, 327–339.
4. Kosiński, W. On fuzzy number calculus. *Int. J. Appl. Math. Comput. Sci.* **2006**, *16*, 51–57.
5. Goetschel, R.; Voxman, W. Elementary fuzzy calculus. *Fuzzy Sets Syst.* **1986**, *18*, 31–43. [CrossRef]
6. Piasecki, K. Revision of the Kosiński's Theory of Ordered Fuzzy Numbers. *Axioms* **2018**, *7*, 16. [CrossRef]
7. Apiecionek, Ł. Fuzzy Observation of DDoS Attack. In *Theory and Applications of Ordered Fuzzy Number*; Prokopowicz, P., Czerniak, J., Mikołajewski, D., Apiecionek, Ł., Ślęzak, D., Eds.; Springer: Berlin, Germany, 2017; pp. 239–252.
8. Apiecionek, Ł. Fuzzy Control for Secure TCP Transfer. In *Theory and Applications of Ordered Fuzzy Number*; Prokopowicz, P., Czerniak, J., Mikołajewski, D., Apiecionek, Ł., Ślęzak, D., Eds.; Springer: Berlin, Germany, 2017; pp. 253–268.
9. Chwastyk, A.; Pisz, I. OFN Capital Budgeting Under Uncertainty and Risk. In *Theory and Applications of Ordered Fuzzy Number*; Prokopowicz, P., Czerniak, J., Mikołajewski, D., Apiecionek, Ł., Ślęzak, D., Eds.; Springer: Berlin, Germany, 2017; pp. 157–170.
10. Czerniak, J.M. OFN Ant Method Based on TSP Ant Colony Optimization. In *Theory and Applications of Ordered Fuzzy Number*; Prokopowicz, P., Czerniak, J., Mikołajewski, D., Apiecionek, Ł., Ślęzak, D., Eds.; Springer: Berlin, Germany, 2017; pp. 207–222.
11. Dobrosielski, W.T.; Czerniak, J.M.; Zarzycki, H.; Szczepański, J. Fuzzy Numbers Applied to a Heat Furnace Control. In *Theory and Applications of Ordered Fuzzy Number*; Prokopowicz, P., Czerniak, J., Mikołajewski, D., Apiecionek, Ł., Ślęzak, D., Eds.; Springer: Berlin, Germany, 2017; pp. 269–288.
12. Ewald, D.; Czerniak, J.M.; Paprzycki, M. A New OFN Bee Method as an Example of Fuzzy Observance Applied for ABC Optimization. In *Theory and Applications of Ordered Fuzzy Number*; Prokopowicz, P., Czerniak, J., Mikołajewski, D., Apiecionek, Ł., Ślęzak, D., Eds.; Springer: Berlin, Germany, 2017; pp. 223–238.
13. Kacprzak, D. Input-Output Model Based on Ordered Fuzzy Numbers. In *Theory and Applications of Ordered Fuzzy Number*; Prokopowicz, P., Czerniak, J., Mikołajewski, D., Apiecionek, Ł., Ślęzak, D., Eds.; Springer: Berlin, Germany, 2017; pp. 171–182.
14. Kacprzak, D. Objective weights based on ordered fuzzy numbers for fuzzy multiple criteria decision-making methods. *Entropy* **2017**, *19*, 373. [CrossRef]
15. Kacprzak, D.; Kosiński, W. Optimizing Firm Inventory Costs as a Fuzzy Problem. *Stud. Logic Grammar Rhetor.* **2014**, *37*, 89–105. [CrossRef]
16. Kacprzak, D.; Kosiński, W.; Kosiński, W.K. Financial Stock Data and Ordered Fuzzy Numbers. In Proceedings of the 12th International Conference on Artificial Intelligence and Soft Computing: (ICAISC 2013), Zakopane, Poland, 9–13 June 2013; pp. 259–270.
17. Łyczkowska-Hanćkowiak, A. Behavioural present value determined by ordered fuzzy number. *SSRN Electr. J.* **2017**. [CrossRef]
18. Łyczkowska-Hanćkowiak, A. Oriented behavioural present value—Revised approach. In *Methods and Models of Regional Development, Proceedings of the 9th International Scientific Conference: Analysis of International Relations 2018 (Summer Edition Conference), Katowice, Poland, 19–20 June 2018*; Szkutnik, W., Sączewska-Piotrowska, A., Hadaś-Dyduch, M., Acedański, J., Eds.; Publishing House of the University of Economics in Katowice: Katowice, Poland, 2018; pp. 14–21.
19. Łyczkowska-Hanćkowiak, A.; Piasecki, K. The expected discount factor determined for present value given as ordered fuzzy number. In *Methods and Models of Regional Development, Proceedings of the 9th International Scientific Conference Analysis of International Relations 2018 (Winter Edition Conference), Katowice, Poland, 12 January 2018*; Szkutnik, W., Sączewska-Piotrowska, A., Hadaś-Dyduch, M., Acedański, J., Eds.; Publishing House of the University of Economics in Katowice: Katowice, Poland, 2018; pp. 69–75.
20. Łyczkowska-Hanćkowiak, A.; Piasecki, K. On representation of Japanese candlesticks by ordered fuzzy numbers. In *Methods and Models of Regional Development, Proceedings of the 9th International Scientific Conference Analysis of International Relations 2018 (Winter Edition Conference), Katowice, Poland, 12 January 2018*; Szkutnik, W., Sączewska-Piotrowska, A., Hadaś-Dyduch, M., Acedański, J., Eds.; Publishing House of the University of Economics in Katowice: Katowice, Poland, 2018; pp. 61–68.

21. Łyczkowska-Hanćkowiak, A.; Piasecki, K. Two-assets portfolio with trapezoidal oriented fuzzy present values. In Proceedings of the 36th International Conference Mathematical Methods in Economics Conference, Jindřichův Hradec, Czech Republic, 12–14 September 2018; pp. 306–311.
22. Łyczkowska-Hanćkowiak, A.; Piasecki, K. Present value of portfolio of assets with present values determined by trapezoidal ordered fuzzy numbers. *Oper. Res. Decis.* **2018**, *28*, 41–56.
23. Marszałek, A.; Burczyński, T. Ordered fuzzy candlesticks. In *Theory and Applications of Ordered Fuzzy Number*; Prokopowicz, P., Czerniak, J., Mikołajewski, D., Apiecionek, Ł., Ślęzak, D., Eds.; Springer: Berlin, Germany, 2017; pp. 183–194.
24. Piasecki, K. Expected return rate determined as oriented fuzzy number. In Proceedings of the 35th International Conference Mathematical Methods in Economics Conference, Hradec Králové, Czech Republic, 13–15 September 2017; pp. 561–565.
25. Piasecki, K.; Łyczkowska-Hanćkowiak, A. Japanese candle for financial portfolio. In Proceedings of the 36th International Conference Mathematical Methods in Economics Conference, Jindřichův Hradec, Czech Republic, 12–14 September 2018; pp. 444–449.
26. Piasecki, K.; Roszkowska, E. On application of ordered fuzzy numbers in ranking linguistically evaluated negotiation offers. *Adv. Fuzzy Syst.* **2018**, in press.
27. Prokopowicz, P.; Mikołajewska, E.; Mikołajewski, D.; Kotlarz, P. Analysis of Temporospatial Gait Parameters. In *Theory and Applications of Ordered Fuzzy Number*; Prokopowicz, P., Czerniak, J., Mikołajewski, D., Apiecionek, Ł., Ślęzak, D., Eds.; Springer: Berlin, Germany, 2017; pp. 289–302.
28. Prokopowicz, P.; Mikołajewski, D. OFN-Based Brain Function Modeling. In *Theory and Applications of Ordered Fuzzy Number*; Prokopowicz, P., Czerniak, J., Mikołajewski, D., Apiecionek, Ł., Ślęzak, D., Eds.; Springer: Berlin, Germany, 2017; pp. 303–322.
29. Roszkowska, E.; Kacprzak, D. The fuzzy SAW and fuzzy TOPSIS procedures based on ordered fuzzy numbers. *Inf. Sci.* **2016**, *369*, 564–584. [CrossRef]
30. Rudnik, K.; Kacprzak, D. Fuzzy TOPSIS method with ordered fuzzy numbers for flow control in a manufacturing system. *Appl. Soft Comput.* **2017**, *52*, 1020–1041. [CrossRef]
31. Zarzycki, H.; Czerniak, J.M.; Dobrosielski, W.T. Detecting Nasdaq Composite Index Trends with OFNs. In *Theory and Applications of Ordered Fuzzy Number*; Prokopowicz, P., Czerniak, J., Mikołajewski, D., Apiecionek, Ł., Ślęzak, D., Eds.; Springer: Berlin, Germany, 2017; pp. 195–206.
32. Prokopowicz, P.; Czerniak, J.; Mikołajewski, D.; Apiecionek, Ł.; Ślęzak, D. *Theory and Applications of Ordered Fuzzy Number*; Springer: Berlin, Germany, 2017.
33. Klir, G.J. Developments in uncertainty-based information. *Adv. Comput.* **1993**, *36*, 255–332.
34. De Luca, A.; Termini, S. Entropy and energy measures of fuzzy sets. In *Advances in Fuzzy Set Theory and Applications*; Gupta, M.M., Ragade, R.K., Yager, R.R., Eds.; Elsevier: Amsterdam, The Netherlands, 1979; pp. 321–338.
35. De Luca, A.; Termini, S. A definition of a non-probabilistic entropy in the settings of fuzzy set theory. *Inf. Control* **1972**, *20*, 301–312. [CrossRef]
36. Piasecki, K. Some remarks on axiomatic definition of entropy measure. *J. Intell. Fuzzy Syst.* **2017**, *33*, 1945–1952. [CrossRef]
37. Dubois, D.; Prade, H. Operations on fuzzy numbers. *Int. J. Syst. Sci.* **1978**, *9*, 613–626. [CrossRef]
38. Dubois, D.; Prade, H. Fuzzy real algebra: some results. *Fuzzy Sets Syst.* **1979**, *2*, 327–348. [CrossRef]
39. Piasecki, K. Behavioural Present Value. *SSRN Electr. J.* **2011**. [CrossRef]
40. Piasecki, K.; Siwek, J. Behavioural Present Value Defined as Fuzzy Number a new Approach. *Folia Oecon. Stetin.* **2015**, *15*, 27–41. [CrossRef]
41. Zadeh, L.A. Fuzzy sets. *Inf. Control* **1965**, *8*, 338–353. [CrossRef]
42. Zadeh, L.A. *Computing with Words in Information/Intelligent Systems 1: Foundations*; Physica-Verlag: Heidelberg, Germany, 1999.
43. Zadeh, L.A. *Computing with Words in Information/Intelligent Systems 2: Applications*; Physica-Verlag: Heidelberg, Germany, 1999.
44. Khalili, S. Fuzzy measures and mappings. *J. Math. Anal. Appl.* **1979**, *68*, 92–99. [CrossRef]
45. Kosko, B. Fuzzy entropy and conditioning. *Inf. Sci.* **1986**, *40*, 165–174. [CrossRef]
46. Shores, T.S. *Applied Linear Algebra and Matrix Analysis*; Springer: Berlin, Germany, 2007.

47. Garcia-Lapresta, J.L.; Piggins, A. Voting on how to vote. In *Soft Computing in Humanities and Social Sciences*; Seising, R., Sanz, V., Eds.; Springer: Berlin, Germany, 2012; Volume 273, pp. 301–321.
48. Alcantud, J.C.R.; Biondo, A.E.; Giarlotta, A. Fuzzy politics I: The genesis of parties. *Fuzzy Sets Syst.* **2018**, *349*, 71–98. [CrossRef]

symmetry

MDPI

Article

Study on an Automatic Parking Method Based on the Sliding Mode Variable Structure and Fuzzy Logical Control

Ying Xu *, Zefeng Lu *, Xin Shan, Wenhao Jia, Bo Wei and Yingqing Wang

College of Mechatronics and Control Engineering, Shenzhen University, Shenzhen 518060, China; shanxin6812@126.com (X.S.); jiawenhao2018@163.com (W.J.); vincent305@126.com (B.W.); m13590202511_1@163.com (Y.W.)

* Correspondence: yxu@szu.edu.cn (Y.X.); Luzefeng_work@163.com (Z.L.); Tel.: +86-755-2653-2036 (Y.X.)

Received: 29 September 2018; Accepted: 16 October 2018; Published: 19 October 2018

Abstract: This paper discusses an automatic parking control method based on the combination of the sliding mode variable structure control (SMVSC) and fuzzy logical control. SMVSC is applied to drive the vehicle from a random initial position and pose, to the designated parking position and pose. Then, the vehicle is driven from the designated parking position to the target parking slot using the method of fuzzy logical control, whose rules are limited to the range of the effective initial position. To combine SMVSC with the fuzzy logical control, the experimental results demonstrate that effective parking can be guaranteed, even if the initial position is out of the effective parking area of the fuzzy logical control.

Keywords: automatic parking; sliding mode variable structure control (SMVSC); fuzzy logical control

1. Introduction

As a key part of autonomous driving technologies, automatic parking technology can release the human driver from complicated parking procedures and can park more efficiently. Accordingly, automatic parking technology has gained a lot of attention, and the correlative research is increasing [1].

According to the procedures of parking, the research of automatic parking technology has been divided into two major aspects, which include parking slots detection, and parking path planning and tracking. In the aspect of parking slots detection, Huang, C.C. and Wang, S.J. proposed a three-layer Bayesian hierarchical detection framework to detect parking slots [2]. Suhr, J.K. proposed a method based on estimating parallel line pairs so as to detect a parking slot [3], and used a hierarchical tree structure method to recognize various parking slot markings [4]. Yu Cheng proposed an approach for parking slots detection based on video images [5]. Jung, H.G. and Yun, H.L. proposed a method based on target position-designation to mark the parking slot [6]. In another aspect, research on parking path planning and tracking have been developed, for example, Vorobieva, H. proposed a path-planning method based on corresponding geometry and tracking the path with a controller, based on traveled distance [7]. Li, B. and Wang, K. used a simultaneous dynamic optimization method to optimize the maneuver planning [8]. Xu, J. proposed an automatic parking method based on computer vision [9]. Sugeno, M. and Murakami, K. designed fuzzy logical controller rules, based on the experience of human drivers, to park the vehicle [10]. On this foundation, Zhan, Y.N. and Collins, E.G. optimized the membership functions of the fuzzy logical controller by using a genetic algorithm to park the vehicle more efficiently [11]. Yin, Y.A. also improved the fuzzy controller based on images using a genetic algorithm [12].

The fuzzy logical control applied in automatic parking has gained a lot of attention, since Kong, S.G. applied an adaptive fuzzy logical control algorithm to back up a truck-and-trailer in 1992 [13].

Liang, Z. designed an automatic parking path tracking controller based on self-organizing fuzzy control [14]. Xiong, Z.B. proposed an automatic parking algorithm based on the preview fuzzy control [15]. Grzegorzewski, P. proposed an efficient algorithm for checking separability, which can be easily applied in practice [16]. The fuzzy logical control applied in automatic parking becomes the mainstream method, with the advantage of improving robustness against uncertainties and of simulating the nonlinear control of the human driver [17].

However, fuzzy logical control in automatic parking still has some limitations. In reality, the uncertainty of a vehicle's initial position results in the fuzzy logical controller not being able to park the vehicle successfully, for example, if the initial parking position is outside of the range of the effective parking position where the fuzzy logical controller can park the vehicle successfully. The effective parking position depends on the rules of the fuzzy logical controller. In other words, if the fuzzy logical controller is confirmed, the range of the effective parking position will be confirmed. Hence, the confirmed fuzzy logical controller cannot park successfully if the vehicle is outside the range of the effective position.

Our objective is solving this problem (the confirmed fuzzy logical controller cannot park successfully if the vehicle is outside of the range of the effective position) by sliding the mode variable structure control (SMVSC). The SMVSC is insensitive to the disturbance and responds quickly [18], so it is often used to track trajectory [19]. Recently, Yue, M. proposed a method based on a model predictive control (MPC) and SMVSC in order to track the coordinated trajectory of vehicles [20]. The solution can be divided into two steps, with the first step using the SMVSC method drive the vehicle from an initial position to the range of an effective parking position, and the second step parking the vehicle from the effective parking position to the target slot.

In this paper, a method based on the sliding mode variable structure control (SMVSC) and fuzzy logical control is proposed in order to park a car, from an initial position where it is outside the range of an effective position. The procedures can be divided as follows. SMVSC drives the vehicle to the range of the effective position to prepare for parking. Afterwards, the fuzzy logical controller parks the vehicle into the target slot. The results, based on MATLAB, show that the control method combining the SMVSC with fuzzy logical control can realize parking from a random initial position, where it is outside the range of an effective parking position.

In this paper, Section 2 presents an algorithm of SMVSC, to drive the vehicle from a random initial position to the effective position. Section 3 discusses the fuzzy logical controller that is used to park the vehicle from the designated position. Section 4 describes the results of the experiment, based on MATLAB Simulink. The conclusion remarks are presented in Section 5.

2. Algorithm of the Sliding Mode Variable Structure Control

The vehicle's position and pose in the 2D plane can be defined as (x, y, θ). As shown in Figure 1, (x, y) are the coordinates of the center point of a vehicle's rear axle, θ is the course of the vehicle, φ is the front-wheel corner, L is the wheelbase, and V is the speed of the vehicle. The dynamical equations of the vehicle can be expressed in Equation (1), as follows:

$$\begin{cases} \dot{x} = v\cos\theta\cos\varphi \\ \dot{y} = v\sin\theta\cos\varphi \\ \dot{\theta} = \frac{v}{L}\sin\varphi \end{cases}, \tag{1}$$

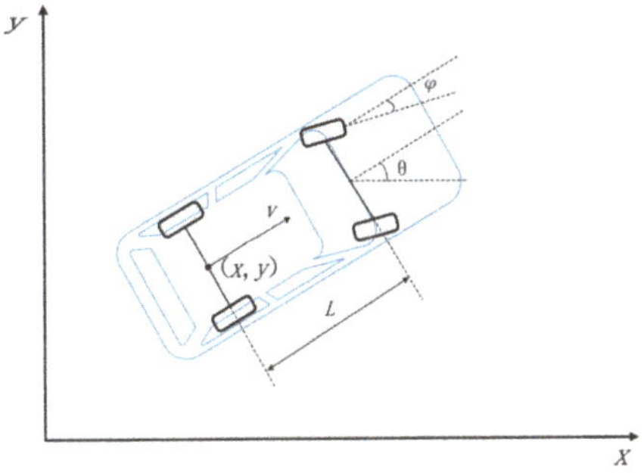

Figure 1. Mode of vehicle.

If φ is small enough, then $\cos\varphi \approx 1$ and $\sin\varphi \approx \tan\varphi$; thus, the dynamical equations of the vehicle are shown as follows:

$$\begin{cases} \dot{x} = v\cos\theta \\ \dot{y} = v\sin\theta \\ \dot{\theta} = \frac{v}{L}\tan\varphi = \omega \end{cases}, \tag{2}$$

where ω is the yaw rate.

The dynamical equations in matrix form are governed by the following equations:

$$\dot{P} = \begin{bmatrix} \dot{x} \\ \dot{y} \\ \dot{\theta} \end{bmatrix} = \begin{bmatrix} \cos\theta & 0 \\ \sin\theta & 0 \\ 0 & 1 \end{bmatrix} \begin{bmatrix} v \\ \omega \end{bmatrix} = Jq, \tag{3}$$

where $J = \begin{bmatrix} \cos\theta & 0 \\ \sin\theta & 0 \\ 0 & 1 \end{bmatrix}$ and $q = \begin{bmatrix} v \\ \omega \end{bmatrix}$.

As shown in Figure 2, (x_c, y_c, θ_c) is the vehicle's initial position and pose, and (x_r, y_r, θ_r) is the vehicle's ideal position and pose. According to the geometric relationship between (x_c, y_c, θ_c) and (x_r, y_r, θ_r), the error (x_e, y_e, θ_e) can be defined as Equation (4), as follows:

$$P_e = \begin{bmatrix} x_e \\ y_e \\ \theta_e \end{bmatrix} = \begin{bmatrix} \cos\theta_c & \sin\theta_c & 0 \\ -\sin\theta_c & \cos\theta_c & 0 \\ 0 & 0 & 1 \end{bmatrix} [P_r - P_c] \tag{4}$$

where P_r and P_c are $\begin{bmatrix} x_r \\ y_r \\ \theta_r \end{bmatrix}$ and $\begin{bmatrix} x_c \\ y_c \\ \theta_c \end{bmatrix}$, respectively.

According to the derivative of error, Equation (4), and the dynamical equation, Equation (2), the differential equation of error is proposed as Equation (5), as follows:

$$\dot{P}_e = \begin{bmatrix} \dot{x}_e \\ \dot{y}_e \\ \dot{\theta}_e \end{bmatrix} = \begin{bmatrix} y_e\omega_c - v_c + v_r\cos\theta_e \\ -x_e\omega_c + v_r\sin\theta_e \\ \omega_r - \omega_c \end{bmatrix} \tag{5}$$

where v_r and ω_r are the ideal speed and ideal yaw rate, respectively. v_c and ω_c are the current speed and current yaw rate, respectively.

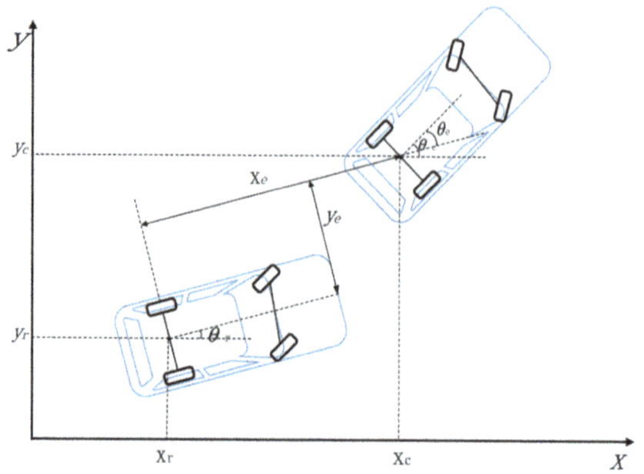

Figure 2. Diagram of the error between (x_c, y_c, θ_c) and (x_r, y_r, θ_r).

It can be pointed out from the above analysis that, according to the numerical values of P_e, v_r, and ω_r, SMVSC aims to output v_c and ω_c to make P_e converge to zero.

Lemma 1. [21] *If any* $x \in R$ *and* $|x| < \infty$, *then* $\varphi(x) = x\sin(\tan^{-1}x) \geq 0$, *if and only if for* $x = 0$, then the equality holds.

If $x_e = 0$, then the Lyapunov function can be described using Equation (6), as follows:

$$V_y = \frac{1}{2}y_e^2 \tag{6}$$

with the hypothesis of $\theta_e = -\tan^{-1}(v_r y_e)$, $\dot{V}_y = y_e\dot{y}_e = y_e(-x_e\omega_c + v_r\sin\theta_e) = -y_e x_e\omega_c - v_r y_e\sin(\tan^{-1}(v_r y_e))$, according to the Lemma 1, $v_r y_e\sin(\tan^{-1}(v_r y_e)) \geq 0$ (if and only if $v_r y_e = 0$, then the equality holds), hence $\dot{V}_y \leq 0$.

It can be seen that if x_e converges to zero and θ_e converges to $-\tan^{-1}(v_r y_e)$, then y_e converges to zero, thus the switching function is designed using Equation (7), as follows:

$$s = \begin{bmatrix} s_1 \\ s_2 \end{bmatrix} = \begin{bmatrix} x_e \\ \theta_e + \tan^{-1}(v_r y_e) \end{bmatrix} \tag{7}$$

We designed a sliding mode controller to let s_1 and s_2 converge to zero, which means that x_e converges to zero and θ_e converges to $-\tan^{-1}(v_r y_e)$ to make y_e and θ_e converge to zero.

The constant rate reaching law can be expressed as follows:

$$\dot{s} = -ksgn(s) \tag{8}$$

It is unavoidable for the sliding mode control to generate chattering effect, but the chattering effect can be decreased by replacing Equation (8) with Equation (9), as follows:

$$\dot{s} = -k\frac{s}{|s| + \delta} \tag{9}$$

where δ is a positive number.

Thus, the constant rate reaching law for Equation (7) can be described using Equation (10), as follows:

$$\dot{s} = \begin{bmatrix} \dot{s}_1 \\ \dot{s}_2 \end{bmatrix} = \begin{bmatrix} -k_1\frac{s_1}{|s_1|+\delta_1} \\ -k_2\frac{s_2}{|s_2|+\delta_2} \end{bmatrix} \tag{10}$$

Using Equation (5), as well as the derivative of Equations (7) and (10), Equation (11) is obtained, as follows:

$$\dot{s} = \begin{bmatrix} \dot{s}_1 \\ \dot{s}_2 \end{bmatrix} = \begin{bmatrix} -k_1 \frac{s_1}{|s_1|+\delta_1} \\ -k_2 \frac{s_2}{|s_2|+\delta_2} \end{bmatrix} = \begin{bmatrix} y_e\omega_c - v_c + v_r\cos\theta_e \\ \omega_r - \omega_c + \frac{\partial\alpha}{\partial v_r}\dot{v}_r + \frac{\partial\alpha}{\partial y_e}(-x_e\omega_c + v_r\sin\theta_e) \end{bmatrix} \tag{11}$$

where $\alpha = \tan^{-1}(v_r y_e)$, $\frac{\partial\alpha}{\partial v_r} = \frac{y_e}{1+(v_r y_e)^2}$, $\frac{\partial\alpha}{\partial y_e} = \frac{v_r}{1+(v_r y_e)^2}$.

Changing the form of Equation (11), the control law is designed and shown, as follows:

$$q_c = \begin{bmatrix} v_c \\ \omega_c \end{bmatrix} = \begin{bmatrix} y_e\omega_c + v_r\cos\theta_e + k_1\frac{s_1}{|s_1|+\delta_1} \\ \frac{\omega_r + \frac{\partial\alpha}{\partial v_r}\dot{v}_r + \frac{\partial\alpha}{\partial y_e}(v_r\sin\theta_e) + k_2\frac{s_2}{|s_2|+\delta_2}}{1+\frac{\partial\alpha}{\partial y_e}x_e} \end{bmatrix} \tag{12}$$

The function relationship between φ and q_c is given by Equation (2), and we changed the form of Equation (2), resulting in Equation (13), as follows:

$$\varphi_c = \sin x^{-1}\left(\frac{w_c L}{v_c}\right) \tag{13}$$

In this section, the purpose of SMVSC is for driving a vehicle from a random initial position and pose, to the ideal position and pose [22,23]. As shown in Figure 3, (x_r, y_r, θ_r) represents the ideal position and pose. ω_r and v_r are the ideal yaw rate and ideal speed of the vehicle, respectively. According to Equation (4), by combining the ideal position and pose (x_r, y_r, θ_r) with the current position and pose (x_c, y_c, θ_c), we defined the error of position and pose (x_e, y_e, θ_e), as shown in Figure 2, which is the input of SMVSC. The output of SMVSC, ω_c and v_c, which depend on (x_e, y_e, θ_e) and ω_r, v_r, decide the corner of front wheel, φ_c. Lastly, according to v_c and φ_c, SMVSC controls the vehicle.

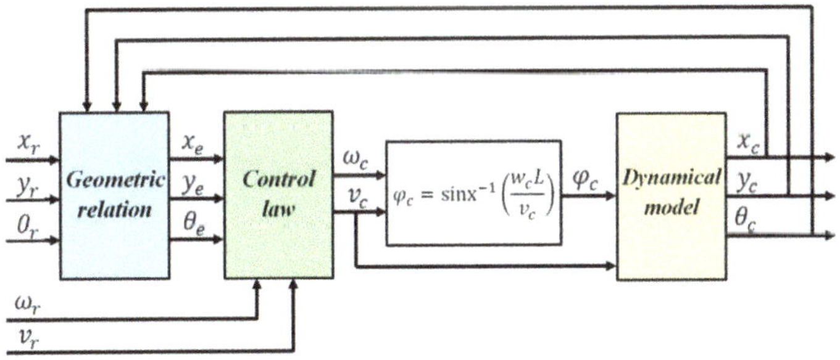

Figure 3. Relationship among variables of the sliding mode variable structure control (SMVSC).

3. Fuzzy Logical Controller

As shown in Figure 4, w is the weight of the parking slot and h is the height of the parking slot. (x, y) are the coordinates of the center point of the vehicle's rear axle, and θ is the course of the vehicle. In order to adapt to the different sizes of the parking slot, we replaced (x, y) with (x_a, y_a) [24], which is defined by Equation (14), as follows:

$$\begin{cases} x_a = \frac{x}{w} \\ y_a = \frac{y}{h} \end{cases} \tag{14}$$

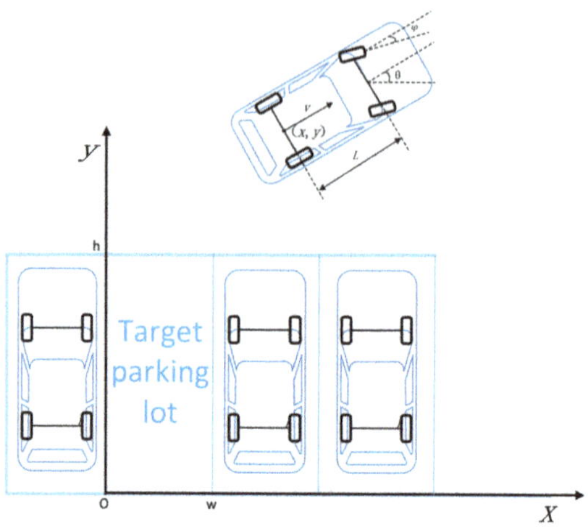

Figure 4. Coordinate system and target parking slot.

There are three inputs for the fuzzy logical controller, x_a, y_a, and θ. The output is the front-wheel corner, φ. The parking speed is the constant. The diagram of the fuzzy logical controller is shown in Figure 5, as follows:

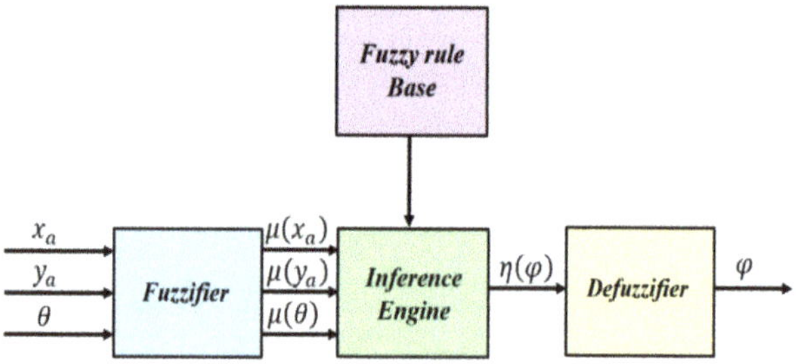

Figure 5. Diagram of the fuzzy logical controller.

The algorithm of the fuzzy logical controller is the Mamdani algorithm [25], and the membership functions are shown in Figures 6–9, the abbreviations in Figures 6–9 are the name of membership function, and the parameters of the membership functions are shown in Table 1.

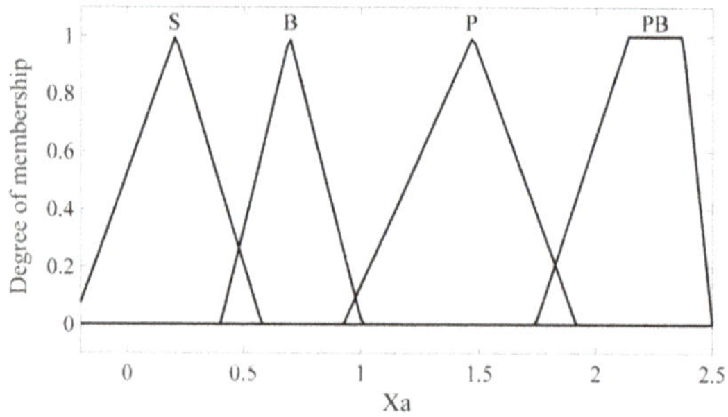

Figure 6. Generated membership function for x_a.

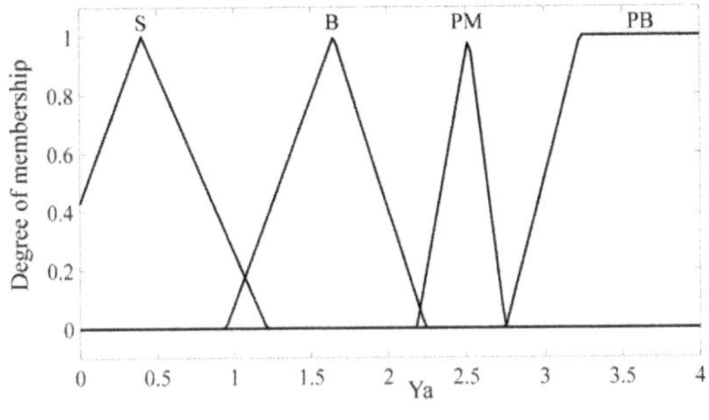

Figure 7. Generated membership function for y_a.

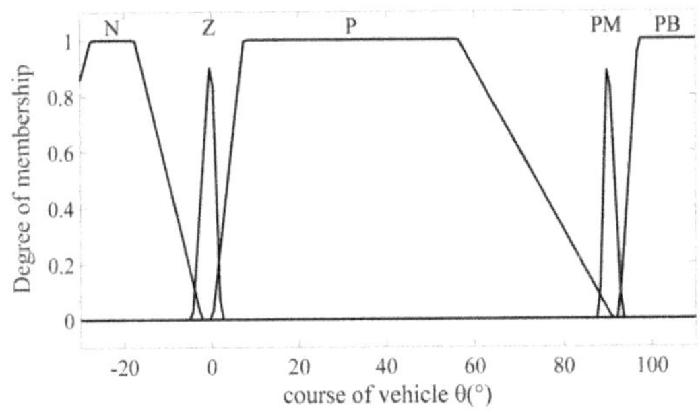

Figure 8. Generated membership function for θ.

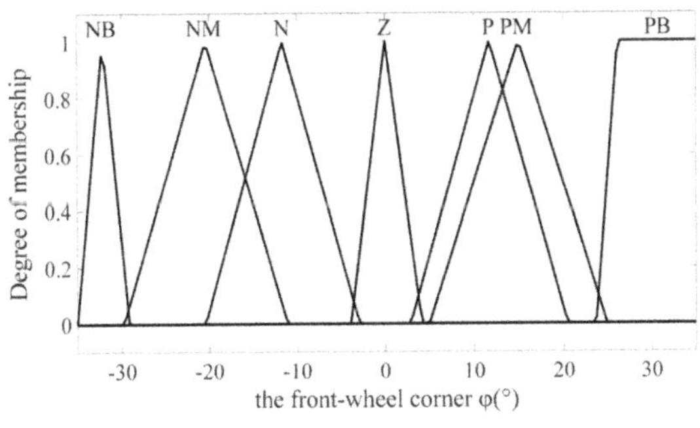

Figure 9. Generated membership function for φ.

Table 1. Parameter of the membership function.

	Input						Output	
		x_a		y_a		θ		φ
Fuzzy Subset	S	Triangular [−0.23,0.20,0.57]	S	Triangular [−0.30,0.40,1.21]	N	Trapezoidal [−44.6,−27.6,−17.6,−2.30]	NB	Triangular [−35,−32.14,−29.15]
	B	Triangular [0.40,0.70,1.00]	B	Triangular [0.94,1.65,2.24]	Z	Triangular [−4.46,0,2.03]	NM	Triangular [−29.77,−20.43,−11.09]
	P	Triangular [0.93,1.47,1.92]	PM	Triangular [2.18,2.52,2.75]	P	Trapezoidal [0.11,7.37,56.3,91]	N	Triangular [−20.43,−11.71,−2.87]
	PB	Trapezoidal [1.74,2.14,2.37,2.50]	PB	Trapezoidal [2.75,3.23,4.40,5.40]	PM	Triangular [88,90,93.2]	Z	Triangular [−3.85,0,4.12]
					PB	Trapezoidal [92.45,97,120,120]	P	Triangular [2.87,11.71,20.43]
							PM	Triangular [4.98,14.95,24.91]
							PB	Trapezoidal [23.67,26.16,37.37,37.37]

The parking strategy is described in Figure 10. Leave 2 m between the vehicle and the parking slot. Keep the vehicle in reverse, with the speed of v (m/s), until the distance between the extended line of the parking slot line and the tail of the vehicle reach 1 m. Then, let the front-wheel corner, φ, turn 35°. When the course of the vehicle, θ, reaches 90°, it returns to φ.

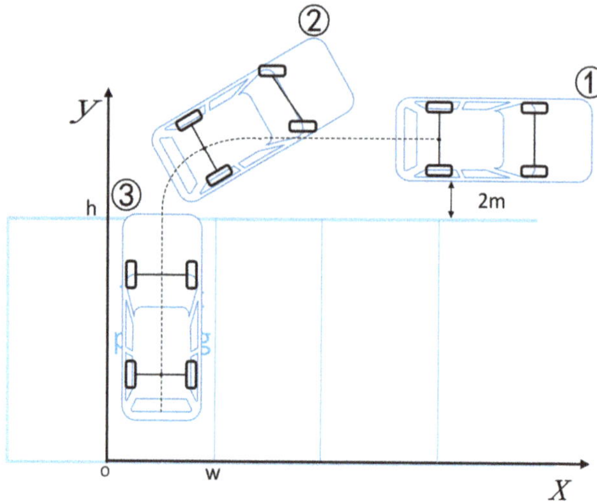

Figure 10. Diagram of the parking strategy.

According to the parking strategy, the fuzzy rules are designed and are shown in Tables 2–5.

Table 2. Fuzzy rules when x_a is S.

θ \ y_a	S	B	PM	PB
N	–	–	–	–
Z	–	–	–	–
P	NB	NB	–	–
PM	Z	–	–	–
PB	–	–	–	–

Table 3. Fuzzy rules when x_a is B.

θ \ y_a	S	B	PM	PB
N	–	–	–	–
Z	–	–	–	–
P	–	NB	–	–
PM	–	–	–	–
PB	–	–	–	–

Table 4. Fuzzy rules when x_a is P.

θ \ y_a	S	B	PM	PB
N	–	–	–	–
Z	–	NB	–	–
P	–	NB	–	–
PM	–	–	–	–
PB	–	–	–	–

Table 5. Fuzzy rules when x_a is P.

θ \ y_a	S	B	PM	PB
N	–	NB	–	–
Z	–	Z	–	–
P	–	PB	–	–
PM	–	–	–	–
PB	–	–	–	–

There are nine rules in the fuzzy rule base.

Finally, the center average is used in the defuzzifier to calculate φ.

4. Results

In this section, we show the effect of SMVSC and the fuzzy logical controller, respectively. The simulating parameters of SMVSC are shown as follows: $(x_r, y_r, \theta_r) = (0,0,0)$, $\omega_r = 0$, $v_r = 0$, $(x_c, y_c, \theta_c) = (2,2,0)$, $\omega_c = 0$, and $v_c = 0$. The results are shown in Figures 11–15.

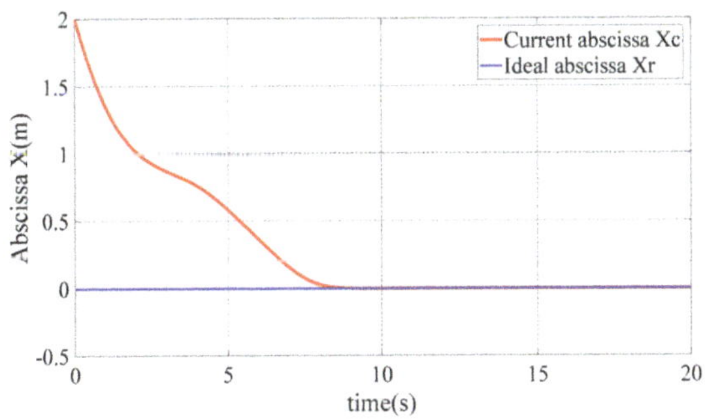

Figure 11. Abscissa X time response diagram.

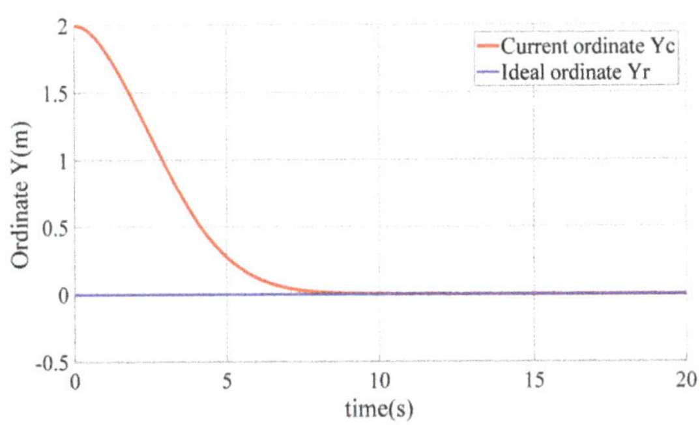

Figure 12. Ordinate Y time response diagram.

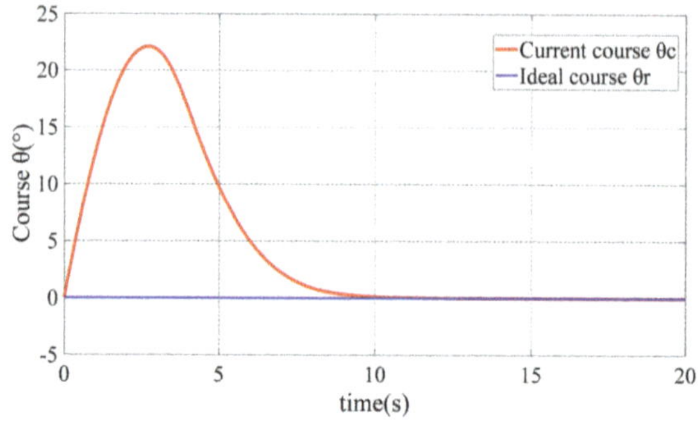

Figure 13. Course θ time response diagram.

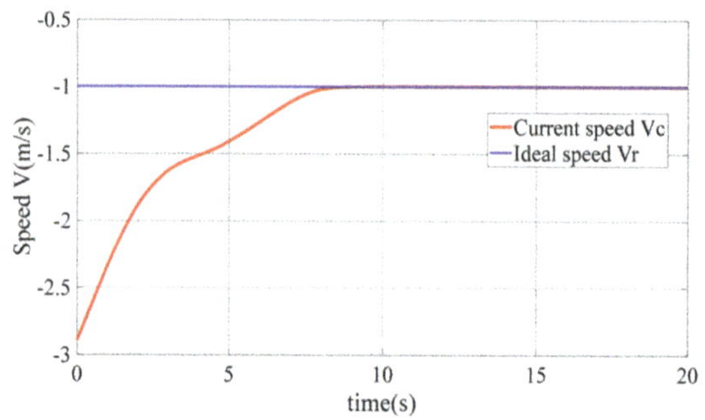

Figure 14. Speed V time response diagram.

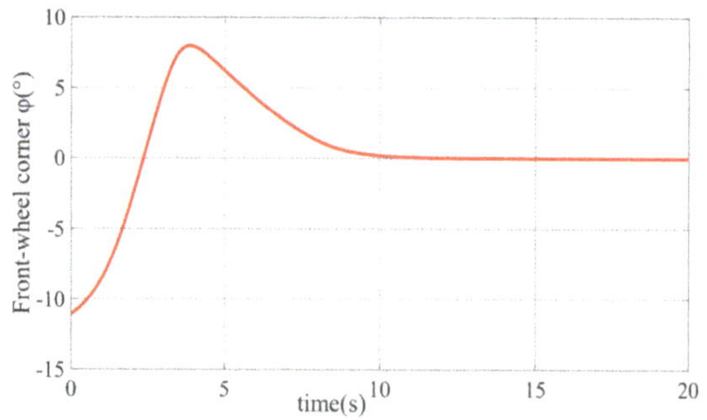

Figure 15. φ time response diagram.

As we can see from Figures 11–13, the ideal position and pose (x_r, y_r, θ_r) and the initial position and pose (x_c, y_c, θ_c) are changed from the initial error of $(-2, -2, 0)$ to $(0, 0, 0)$, by the control laws v_c and φ_c, which are shown in Figures 14 and 15. In other words, according to control laws v_c and φ_c, the vehicle is driven from the initial position (x_c, y_c, θ_c) and pose to the ideal position and pose (x_r, y_r, θ_r). It is noteworthy that the convergence time of (x_e, y_e, θ_e) in ten seconds. In Figures 14 and 15, the curvilinear trend is smooth, so that it can be implemented in reality [26].

The result of the fuzzy logical controller is given in Figure 16, which is the trajectory of the vehicle's center point of the rear axle.

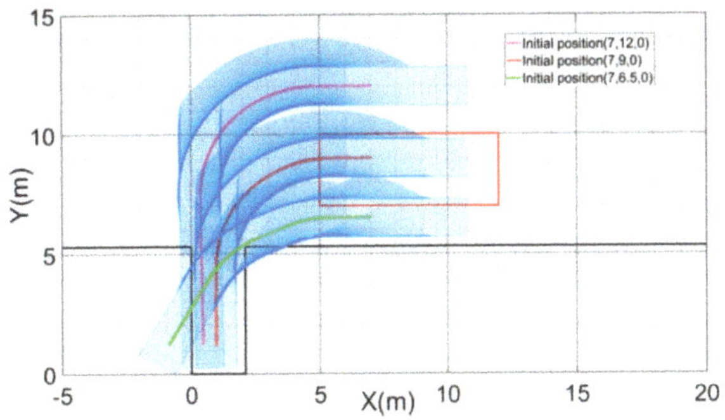

Figure 16. The trajectory of the fuzzy control parking to a different initial position.

As shown in Figure 16, the simulating parameters are shown as follows: $w = 2.5$, $h = 5.3$, speed $v = -1$ m/s, the initial position and pose of magenta trajectory is $(x, y, \theta) = (7, 12, 0)$, the initial position and pose of green trajectory is $(x, y, \theta) = (7, 6.5, 0)$, and both of the trajectories are outside the area of the effective parking position, so they park unsuccessfully. The initial position and pose of the red trajectory is $(x, y, \theta) = (7, 9, 0)$, which is inside the red box, which allows for parking successfully. According to the experiments, if the initial position is within the area of the red box, shown in Figure 16, the fuzzy logical controller described in Section 3 can park the vehicle successfully.

As discussed in the Section 1 (Introduction), this paper aims to park the vehicle from an initial position, which is out of the black box, by combining SMVSC with the fuzzy logical control. Figure 17 shows the effect of this method.

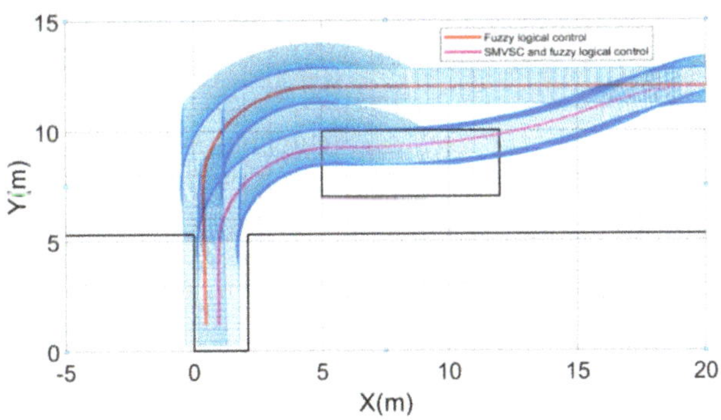

Figure 17. Trajectory of two different control methods.

The magenta trajectory is controlled by the hybrid method. The red trajectory is controlled by the fuzzy logical controller, which is the same as the hybrid method. The simulating parameters of Figure 17 are shown as follows: the ideal position and pose (x_r, y_r, θ_r) are the same as that of Figure 16 $(x, y, \theta) = (7, 9, 0)$. The initial position and pose are $(x_c, y_c, \theta_c) = (20, 12, 0)$, which is out of black box. The result of the simulating shows that the hybrid method can park the vehicle into the target slot successfully, but the fuzzy logical control cannot park successfully (the left side of the vehicle is not in the slot). The result shows that the hybrid method can expand the range of the effective parking position.

5. Conclusions

A control parking method, which combines SMVSC with the fuzzy logical control, is proposed in this paper. This method aims to expand the range of the effective parking position, which is confirmed by the fuzzy logical controller. The disadvantage of the fuzzy logical control is that the range of the effective initial position and pose are limited to its rules. In other words, the fuzzy logical controller cannot park the vehicle when the initial position is out of the effective initial position. The aim of this paper to solve this problem, of driving the vehicle from a random initial position to the effective parking position using the mothed of SMVSC, and parking the vehicle from the effective parking position using the method of fuzzy logical control.

SMVSC can control a vehicle from a random initial position to an effective parking position in limited amount of time; furthermore, the curve of the control laws is smooth enough to implement in reality, and the chattering effect is decreased. The results also show that the fuzzy logical controller has nine rules, which is according to the strategy of parking, and can park the vehicle from the effective parking position (area of red box). By comparing the fuzzy logical control and the hybrid control method, the experimental results verified that the fuzzy logical control cannot park from an initial position outside the range of an effective parking position, but the hybrid method can. In short, parking from a random initial position to parking outside of the effective parking position is realized by the hybrid method of combining the SMVSC and fuzzy logical control.

Author Contributions: Y.X. designed the main body of the study and revised the manuscript; Z.L. and X.S. wrote the main body of the paper; W.J. and B.W. guided the system design and analyzed the data; and Y.W. helped to revise the paper.

Funding: This research was funded by the National Natural Science Foundation of China (no. 61403259) and the Science and Technology Research and Development Foundation of Shenzhen (no. JCYJ20170302142107025).

Conflicts of Interest: The authors declare no conflict of interest.

References

1. Wang, W.; Song, Y.; Zhang, J.; Deng, H. Automatic parking of vehicles: A review of literatures. *Int. J. Autom. Technol.* **2014**, *15*, 967–978. [CrossRef]
2. Huang, C.C.; Wang, S.J. A Hierarchical Bayesian Generation Framework for Vacant Parking Space Detection. *IEEE Trans. Circ. Syst. Video Technol.* **2010**, 1770–1785. [CrossRef]
3. Suhr, J.K.; Jung, H.G. Automatic Parking Space Detection and Tracking for Underground and Indoor Environments. *IEEE Trans. Ind. Electron.* **2016**, *63*, 5687–5698. [CrossRef]
4. Suhr, J.K.; Jung, H.G. Full-automatic recognition of various parking slot markings using a hierarchical tree structure. *Opt. Eng.* **2013**, *52*, 7203. [CrossRef]
5. Li, Y.C.; Yan, J.L.; Wang, M.S.; Han, D.T. Approach for parking spaces detection based on video images. *Comput. Eng. Des.* **2012**, *33*, 281–285.
6. Jung, H.G.; Yun, H.L.; Kim, J. Uniform User Interface for Semiautomatic Parking Slot Marking Recognition. *IEEE Trans. Veh. Technol.* **2010**, *59*, 616–626. [CrossRef]
7. Vorobieva, H.; Glaser, S.; Minoiu-Enache, N.; Mammar, S. Automatic Parallel Parking in Tiny Spots: Path Planning and Control. *IEEE Trans. Intell. Transp. Syst.* **2015**, *16*, 396–410. [CrossRef]
8. Li, B.; Wang, K.; Shao, Z. Time-Optimal Maneuver Planning in Automatic Parallel Parking Using a Simultaneous Dynamic Optimization Approach. *IEEE Trans. Intell. Transp. Syst.* **2016**, *17*, 3263–3274. [CrossRef]
9. Xu, J.; Chen, G.; Xie, M. Vision-guided automatic parking for smart car. In Proceedings of the IEEE Intelligent Vehicles Symposium 2000, Dearborn, MI, USA, 5 October 2002; pp. 725–730.
10. Sugeno, M.; Murakami, K. Fuzzy parking control of model car. In Proceedings of the 23rd IEEE Conference on Decision and Control, Las Vegas, Nevada, USA, 12–14 December 1984; pp. 902–903.
11. Zhao, Y.; Emmanuel, G.C., Jr. Robust automatic parallel parking in tight spaces via fuzzy logic. *Robot. Auton. Syst.* **2005**, *51*, 111–127. [CrossRef]
12. Aye, Y.Y.; Watanabe, K.; Maeyama, S.; Nagai, I. An automatic parking system using an optimized image-based fuzzy controller by genetic algorithms. *Artif. Life Robot.* **2016**, *22*, 1–6. [CrossRef]
13. Kong, S.G.; Kosko, B. Adaptive fuzzy systems for backing up a truck-and-trailer. *IEEE Trans. Neural Netw.* **1992**, *3*, 211–223. [CrossRef] [PubMed]

14. Liang, Z.; Zheng, G.Q.; Li, J.-S. Tracking Controller Based on Self-Organizing Fuzzy Control. *Fire Control Commun. Control* **2013**, *38*, 50–54.
15. Xiong, Z.B.; Yang, W.; Ding, K.; Liang, F.-H.; Zheng, L.; Li, Y.-S.; Chang'an Auto Global R&D Center; Chang'an Automobile Co., Ltd.; The State Key Laboratory of Mechanical Transmissions, Chongqing University. Research on the Auto Parking Algorithm Based on the Preview Fuzzy Control. *J. Chongqing Univ. Technol.* **2017**, *17*, 323–325.
16. Grzegorzewski, P. On Separability of Fuzzy Relations. *Int. J. Fuzzy Logic Intell. Syst.* **2017**, *17*, 137–144. [CrossRef]
17. Zhao, Y.; Collins, E.G. Fuzzy parallel parking control of autonomous ground vehicles in tight spaces. In Proceedings of the IEEE International Symposium on Intelligent Control, Houston, TX, USA, 8 October 2003; pp. 811–816.
18. Edwards, C.; Spurgeon, S.K. *Sliding Mode Control: Theory and Applications*; Taylor & Francis: Boca Raton, FL, USA, 1988.
19. Schindele, D.; Aschemann, H. Trajectory tracking of a pneumatically driven parallel robot using higher-order SMC. *IEEE Int. Conf. Methods Models Autom. Robot.* **2010**, 387–392. [CrossRef]
20. Yue, M.; Hou, X.; Fan, M.; Jia, R. Coordinated trajectory tracking control for an underactuated tractor-trailer vehicle via MPC and SMC approaches. In Proceedings of the 2017 2nd International Conference on Advanced Robotics and Mechatronics (ICARM), Hefei, China, 27–31 August 2017; pp. 82–87.
21. Tahara, J.I.; Tsuboi, K.; Sawano, T.; Nagata, Y. An adaptive VSS control method with integral type switching gain. In Proceedings of the IASTED International Conference Robotics and Applications, Clearwater, FL, USA, 19–22 November 2001; pp. 106–111.
22. Jiang, L.; Zhongwei, W.U. Sliding Mode Control for Intelligent Vehicle Trajectory Tracking Based on Reaching Law. *Trans. Chin. Soc. Agric. Mach.* **2018**, *49*, 381–386.
23. He, H.; Peng, J.; Xiong, R.; Fan, H. An Acceleration Slip Regulation Strategy for Four-Wheel Drive Electric Vehicles Based on Sliding Mode Control. *Energies* **2014**, *7*, 3748–3763. [CrossRef]
24. Lanza, J.; Sánchez, L.; Gutiérrez, V.; Galache, J.A.; Santana, J.R.; Sotres, P.; Muñoz, L. Smart City Services over a Future Internet Platform Based on Internet of Things and Cloud: The Smart Parking Case. *Energies* **2016**, *9*, 719. [CrossRef]
25. Mamdani, E.H. Application of fuzzy algorithms for control of simple dynamic plant. *Proc. Inst. Electr. Eng.* **1974**, *121*, 1585–1588. [CrossRef]
26. Hua, Y.; Jiang, H.; Cai, Y.; Zhang, X.; Ma, S.; Wu, D. Path Tracking Control of Automatic Parking Cloud Model Considering the Influence of Time-delay. *Math. Probl. Eng.* **2017**, *2017*, 6590383. [CrossRef]

symmetry

MDPI

Article

An Emergency Decision Making Method for Different Situation Response Based on Game Theory and Prospect Theory

Zi-Xin Zhang [1,2], Liang Wang [1,2] and Ying-Ming Wang [1,*]

1 Decision Sciences Institute, Fuzhou University, Fuzhou 350116, China; zhangzixin768@outlook.com (Z.-X.Z.); wangliangg322@hotmail.com (L.W.)
2 Department of Computer Science, University of Jaén, 23071 Jaén, Spain
* Correspondence: msymwang@hotmail.com; Tel.: +86-591-2286-6677

Received: 20 September 2018; Accepted: 6 October 2018; Published: 11 October 2018

Abstract: Because of the continuous burst of emergency events (EEs) recently, emergency decision making (EDM) has become an active research topic due to its crucial role in relieving and reducing various losses and damages (property, lives, environment, etc.) caused by EEs. Current EDM studies based on prospect theory (PT) have considered decision maker's (DM's) psychological behavior, which is very important in the EDM process because it affects DM's decision behavior directly, particularly under the uncertainty decision environment. However, those studies neglected an important fact that different emergency situations should be handled by different measures to show the pertinence and effectiveness of the emergency response in the real world, which has been taken into consideration in EDM studies based on game theory (GT). Different behavior experiments show that DMs usually have limited rationality when involved in risk and an uncertain decision environment, in which their psychological behavior has distinct impacts on their decision choice and behavior. Nevertheless, the existing studies of EDM based on GT build on an assumption that DMs are totally rational; however, it is obvious that such an assumption is unreasonable and far from the real-world situation. Motivated by these limitations pointed out previously, this study proposes a novel EDM method combining GT and PT that considers not only the DM's psychological behavior, but also takes different situations' handling for EEs into account, which is closer to the EDM problems in reality. An example and comparison with other methods are provided to demonstrate the validity and rationality of the proposed method for coping with real-world EDM problems.

Keywords: emergency response; prospect theory; game theory; situation-response

1. Introduction

The definition of emergency event (EE) is [1] "events which suddenly take place causing or having the possibility to cause intense death and injury, property loss, ecological damage and social hazards", such as landslides, earthquakes, terrorist attacks, etc. In the World Disaster Report 2016, there were 6090 disasters that took place between 2006 and 2015 in the world. In these disasters, 771,911 people had been killed, 1,917,557 thousands people had been affected and the economic damage had reached 1,424,814 million dollars [2]. From such ghastly statistics, it is necessary to take some strategies to reduce such kinds of losses and impacts on mankind's daily life and socio-economic development. Fortunately, emergency decision making (EDM) is one such kind of strategy, which is defined as a process in which a decision maker (DM) selects the optimal alternative to respond to or control the EE in order that life and property protection and political and social stability can be achieved [3]. Because of the important role in reducing the losses and impacts caused by EEs, EDM has become an active research field in recent years [4–8].

The EDM problem is usually complex and dynamic because the EDM environment is full of risk and uncertainty [9]. Different behavior studies prove that DMs have limited rationality under an environment with risk and uncertainty, and the psychological behavior of DM is an important factor in the EDM process due to its direct influence on decision behavior and outcomes. Hence, some researchers pay close attention to DM's psychological behavior by means of prospect theory (PT), proposed by Kahneman and Tversky in 1979 [10], in the EDM process because of its greatest influence among different behavior theories (such as regret theory [11], disappointment theory [12], third-generation PT [13], etc.) and having achieved fruitful results [1,14–18].

All the research with and without PT has made important contributions to EDM; however, both of them have limitations that they do not take into consideration about the different emergency situations, which are caused by the dynamic evolution and uncertainty of EEs, nor do they consider DM's psychological behavior. Each emergency situation should be considered and be handled by proper measures because of the limited resources in the real world and the importance of DM's psychological behavior in the decision process.

Game theory (GT) is a useful tool for providing a mathematical process to select the optimal strategy for one player with respect to all possible strategies of the other ones throughout the game [19]. Thus, theoretically speaking, GT can help DM select proper measures to deal with different situations that may occur in real-world EEs. The EDM problem is a typical noncooperation game if we regard the EE and DM as the game players [20], in which the emergency situations and the measures are regarded as the strategies of EE and DM, respectively. Therefore, the EDM problems can be solved from the perspective of game theory.

In recent years, some EDM methods based on GT have been studied, which have taken into account different emergency situations dealt with by different measures [20–24]. However, it is necessary to point out that existing EDM methods based on GT build on an assumption that the player (decision maker) has total rationality [24,25]. Nevertheless, different studies [19,26,27] have shown that DMs have limited rationality under an environment with risk and uncertainty, and the DM's psychological behavior is very important to the decision process in EDM problems and must be considered.

To manage the limitations mentioned above, this study proposes a novel EDM method based on GT and PT that takes into account DM's psychological behavior by means of PT and different situations handled by using different measures based on GT.

The outline of this paper is as follows: Section 2 provides a brief introduction of PT and GT that will be utilized in our proposal together with a brief review of related works highlighting the importance of this study. A novel EDM method will be presented in Section 3 that considers both DM's psychological behavior and coping with different emergency situations. Section 4 offers a case study on a typhoon emergency and a comparison with existing studies. Section 5 provides the conclusions and future works of this paper.

2. Preliminaries

In this section, GT and PT will be briefly reviewed so that unfamiliar readers can understand our proposed method easily. In addition, some related works to illustrate the importance and necessity of this research are reviewed.

2.1. *Game Theory in Emergency Decision Making*

GT is a useful tool to solve decision making problems in which the situations either have conflict or cooperation and sometimes both [23]. These situations may happen when there are two or more players (DMs) involved in a same system and they attempt to achieve their own objectives using the same resources [28]. As a branch of mathematical analysis, GT provides a scientific process to choose the best strategies for each possible situation throughout the game [19]. Such a characterization of GT is suitable for EDM problems, in which the DM usually needs to have a corresponding response with respect to different emergency situations.

Generally, if a game has n players, it will be denoted as $G = \{(S_i; P_i), i = 1, 2, \ldots, n\}$, where S_i and P_i denote the strategies and payoffs of the i-th player, respectively. In the game process of EDM problems, there are usually two players, i.e., the EE and DM, in which the EE is a special player because it is unconscious about the benefits or costs. Thus, the game between EE and DM can be denoted as $G = \{(S_i; P_i)\}$, where $i = 1, 2$.

The game can be classified according to the relationship among the players [29]: if the relationship among the players is competitive, the game is a noncooperation game; otherwise, if the players are cooperative, it is a cooperation game. Obviously, the relationship between EE and DM in the game is noncooperation, so the game in EDM problems can be assumed as a typical noncooperation game, the zero-sum game, i.e., $P_1 + P_2 = 0$, which means if the DM gains $\triangle_i$, the EE loses $\triangle_i$, otherwise, the EE gains $\triangle_i$, while the DM loses $\triangle_i$.

Three basic notions of GT for the EDM problem are briefly introduced as follows:

1. Players: Players are always denoted by $i = 1, 2, \ldots, n$ and at least $i \geq 2$; this means that there are at least two players in one game. In EDM, there are two players, who are the decision maker (DM) and the EE. Thus, in the emergency game $G = \{(S_i, P_i)\}, i = 1, 2$, where 1 denotes the DM and 2 refers to EE.
2. Strategies: Let $S_i = \{S_{ik_i}\}$ be the set of action strategies of the i-th player who has k_i strategies. In EDM, $S_1 = \{S_{1\delta}\}$ refers to the set of different alternatives of DM, in which $S_{1\delta}$ denotes the δ-th alternatives, $\delta = 1, 2, \cdots k_1$. $S_2 = \{S_{2\theta}\}$ refers to the set of different situations of EE, where $S_{2\theta}$ denotes the θ-th possible situation of EE, $\theta = 1, 2, \cdots k_2$.
3. Payoffs: Let $P_i(S_i)$ be the payoffs of the i-th player, where $P_1(S_1) + P_2(S_2) = 0$.

The game can also be classified according to the action sequence among players [29]: if the players take the action simultaneously or the players do not know the exact information of the other player's action, the game is a static game; if not, the game is a dynamic one. The dynamic one is also called the extensive from game (EFG) [29]. Obviously, in the EDM problems, the player EE always takes the action firstly, so the game between DM and EE is an EFG problem. However, in the real world, because of the imprecise and incomplete information of the EE, which strategy the EE will take the DM does not know. Thus, when this situation occurs, the EFG problem can be regarded as the static one, and its game tree is shown in Figure 1.

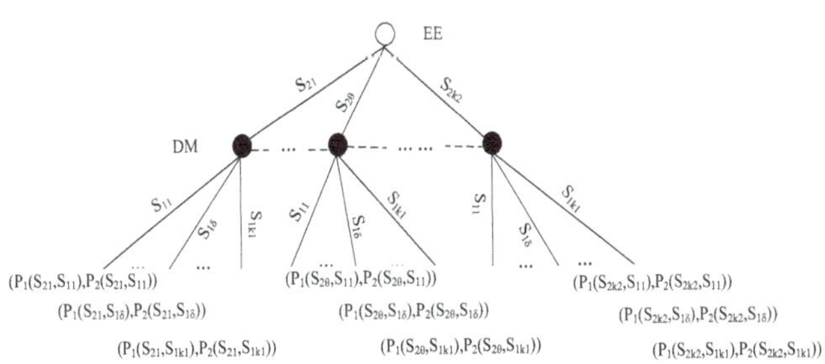

Figure 1. The game tree between an emergency event (EE) and decision maker (DM).

Based on the presentation mentioned above, since the EDM problem is a static game, therefore the payoff matrix of EE and DM can be simplified into Table 1 according to Figure 1.

Table 1. The payoff matrix of emergency event (EE) and decision maker (DM).

		EE				
		S_{21}	$\cdots$	$S_{2\theta}$	$\cdots$	S_{2k_2}
DM	S_{11}	$(P_1(S_{21},S_{11}),P_2(S_{21},S_{11}))$	$\cdots$	$(P_1(S_{2\theta},S_{11}),P_2(S_{2\theta},S_{11}))$	$\cdots$	$(P_1(S_{2k_2},S_{11}),P_2(S_{2k_2},S_{11}))$
	$\vdots$	$\vdots$	$\cdots$	$\vdots$	$\cdots$	$\vdots$
	$S_{1\delta}$	$(P_1(S_{21},S_{1\delta}),P_2(S_{21},S_{1\delta}))$	$\cdots$	$(P_1(S_{2\theta},S_{1\delta}),P_2(S_{2\theta},S_{1\delta}))$	$\cdots$	$(P_1(S_{2k_2},S_{1\delta}),P_2(S_{2k_2},S_{1\delta}))$
	$\vdots$	$\vdots$	$\cdots$	$\vdots$	$\cdots$	$\vdots$
	S_{1k_1}	$(P_1(S_{21},S_{1k_1}),P_2(S_{21},S_{1k_1}))$	$\cdots\cdots$	$(P_1(S_{2\theta},S_{1k_1}),P_2(S_{2\theta},S_{1k_1}))$	$\cdots\cdots$	$(P_1(S_{2k_2},S_{1k_1}),P_2(S_{2k_2},S_{1k_1}))$

Based on Figure 1 and Table 1, the game process between EE and DM can be described as shown in Figure 2. In our proposal, we assume that the EE chooses its strategy randomly.

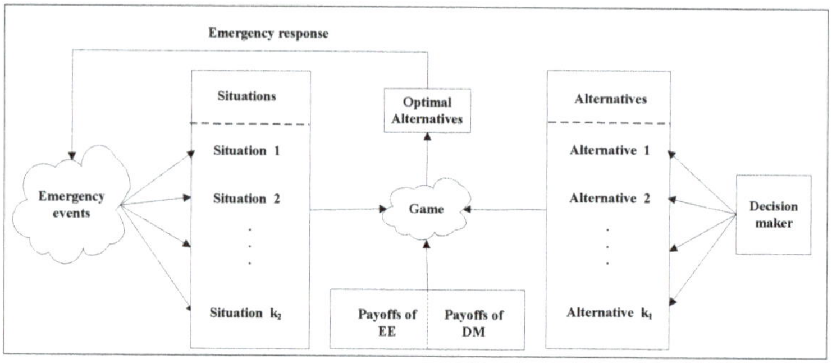

Figure 2. The game process between EE and DM.

The assumption presented in current EDM studies based on GT [20–24] in which the DM is completely rational is not fully reasonable. Due to the importance of the psychological behavior of DM, it will be taken into account in the phase of determining payoffs and will be introduced in detail in the third section of this proposal.

2.2. *Prospect Theory in Emergency Decision Making*

As was mentioned in the Introduction, DM's psychological behavior is a key and important factor in the EDM process especially, when DM is under pressure. However, such an important issue is neglected in the current EDM approaches based on GT; thus, it will be taken into account in this proposal by using PT.

PT is a useful tool to consider human being's psychological behavior issues, which was firstly presented in Kahneman and Tversky's study in 1979 [10] and was developed by them in 1992 [30] as an economic behavior theory. In the proposal of Kahneman and Tversky, they provided a simple and clear computation process to describe the psychological behavior using reference points (RPs), losses, gains and overall prospect values, which are important concepts in PT. Since PT has a simple calculation process and a clear logic, it has been widely applied in the field of decision making to solve the problems considering human being's psychological behavior [13,15,30–32]. Therefore, the PT will be utilized to address the DM's psychological behavior in our proposal.

Generally, in the process of decision making, PT was distinguished as three phases [30]:

1. An editing phase, in which the gains and losses can be calculated according to the RPs provided by DM.
2. An evaluation phase: in this phase, the prospect values can be obtained by a value function, then the overall prospect values will be calculated on the foundation of prospect values and the weighting vector.

3. A selection phase, in which the alternative with the highest overall prospect value will be selected as the best one to deal with the given decision problem.

According to PT, human beings are usually more sensitive to losses than the same gains, and their psychological behavior shows risk-seeking for losses and risk-aversion for gains [26]. Thus, PT can be depicted by means of an S-shaped value function that shows a concave shape in the loss domain and a convex shape in the gain domain, respectively (see Figure 3). The value function of PT is related to RPs and expressed by a power law presented as below [10]:

$$v(x) = \begin{cases} x^{\alpha}, & x \geq 0 \\ -\lambda(-x)^{\beta}, & x < 0 \end{cases} \quad (1)$$

where α is the parameter with respect to gains, while β is the parameter associated with losses, $0 \leq \alpha$, $\beta \leq 1$. x means gains with $x \geq 0$, and losses with $x < 0$. λ denote the parameter of risk aversion, $\lambda > 1$. The values of parameters α, β and λ are determined through experiments [26,33–35].

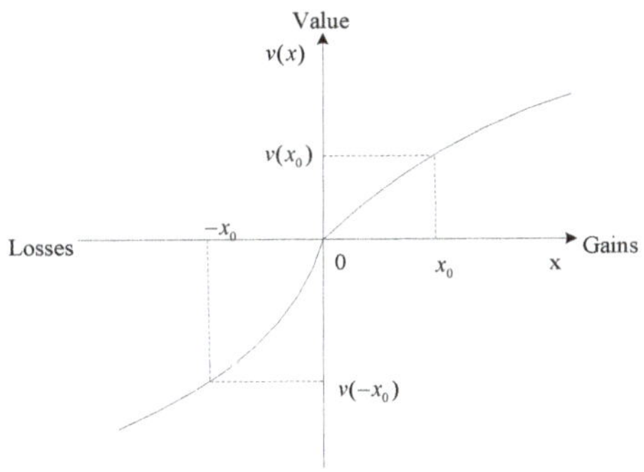

Figure 3. S-shaped value function of prospect theory (PT).

2.3. Related Works

In order to demonstrate the importance and necessity of this study, several important studies in the literature are briefly reviewed that are close to our research.

The DM's psychological behavior has been addressed in existing EDM studies by different researchers. For example, Fan et al. [14] proposed a risk decision analysis method for emergency response that addressed DM's psychological behavior in the decision process by employing PT. Wang et al. [16] developed an EDM method that considered not only DM's psychological behavior in the decision process by using PT, but also the dynamic evolution feature of EE. Due to the uncertainty information about EEs in real-world situations, it is usually a big challenge for DM to estimate possible losses by using crisp values that are employed in existing EDM studies [14,16,36]. Wang et al. [18] presented an EDM method based on PT considering DM's psychological behavior with interval values, which not only extended the scope of PT for dealing with interval values, but also made the EDM method close to the real world. With the increasing complexity of EEs in the real world, one DM alone [14,16,18,36] cannot make comprehensive judgments and proper decisions; therefore, Wang et al. [17] proposed a group EDM method for emergency situations by using group wisdom to support DM making a decision that takes into account experts' psychological behavior in the decision process by using PT. Due to the fact that there are various types of information about EEs in the real world, such as crisp values [14,16,36], interval values [18], linguistic information [37], and so on, none of the proposals considers various types of information at the same time; to do that, Wang et al. [38] proposed a group EDM method for not only considering various types of information at the same time, but also together with experts' psychological behavior and hesitation in qualitative contexts. Motivated by [38], Zhang et al. [39] presented an EDM method based on PT and hesitant fuzzy sets considering not only experts' psychological behavior, but also experts' hesitation in quantitative contexts.

Despite existing EDM studies based on PT having achieved fruitful results [14,16–18,36–40], they neglect an important fact that different emergency situations should be handled by using different measures because of the limited resources and dynamic evolution of EEs.

Nevertheless, to address such an important issue in the real world, GT has been employed in existing EDM studies. For example, Yang and Xu [20] proposed an engineering model based on sequential games considering different situations coping with a flood eruption EDM problem. Chen et al. [41] provided a game theory-based approach for evaluating possible terrorist attacks and corresponding deployment of emergency responses. Gupta et al. [23] proposed a game-theoretic EDM method for considering the optimal allocation solutions of resources to different situations of the EEs, particularly when the available resources are limited. Cheng and Zheng [42] proposed a game-theoretical analysis method considering possible solutions of emergency evacuation for different emergency cases. Rezazadeh et al. [43] presented a security risk assessment method based on game theory for considering the possible terrorist attacks on oil and gas pipelines. Gao et al. [44] proposed an approach for considering different scenarios coping with corporate environment risk based on game theory. Wu [45] presented two game theoretic models for search-and-rescue resource allocation and selection of an acceptable plan for different districts after devastating tsunamis.

Although the existing EDM studies based on GT have obtained remarkable results regarding the different situations coping with the problems of EEs, they build on an assumption that DM is totally rational in the decision process. However, different behavior studies [19,26,27] have proven that DM has limited rationality and his/her psychological behavior can affect the decision behavior directly, especially under a risk and uncertainty environment, and must be considered because of its importance in the decision process.

To overcome the limitations pointed out above and highlight the significance and importance of our research, this study combines the merits of PT and GT to propose a novel EDM method based on GT and PT that considers not only the different situations of coping with problems, but also DM's psychological behavior in the EDM process, which is introduced in detail in Section 3.

3. Emergency Decision Making Method Based on Game Theory and Prospect Theory

As previously mentioned, the proposed EDM method based on GT and PT is introduced in this section. The general framework of our proposal is illustrated in Figure 4, and it consists of three main phases:

1. Definition framework: this part introduces the basic notations and related terminology that are employed in this proposal.
2. Computation of overall prospect values: in this part, the value function will be used to compute the overall prospect values according to gains and losses.
3. Selecting the optimal alternative based on payoffs: the payoffs of DM including his/her psychological behavior and the payoffs of EE will be determined. Based on the payoffs, the optimal alternative will be selected to respond to corresponding emergency situation.

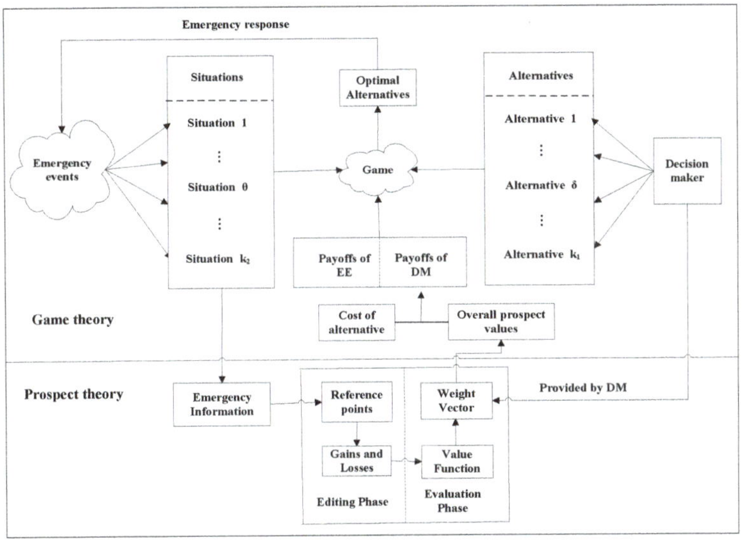

Figure 4. The general framework of the proposed method.

3.1. Definition Framework

Due to the information about EE usually being inadequate or incomplete, especially in the early stage in a real-world situation, and related emergency situations become more and more complicated with the dynamic evolution of EE across time, it is hard for DM to describe the EE using just one type of information; thus, for convenience, different types of information will be used to describe the situation of EE and emergency response alternatives [15,16]. Thus, in our proposal, both interval and numerical values are employed, in which the interval values are used to estimate the damages or losses caused by EE and numerical values are used to describe the cost of alternatives.

The following notations that will be used in our proposal are defined below:

- $S_1 = \{S_{1\delta}\}$: refers to the set of different alternatives, in which $S_{1\delta}$ denotes the δ-th alternative, $\delta = 1, 2, \ldots, k_1$.
- $S_2 = \{S_{2\theta}\}$: refers to the set of different situations, in which $S_{2\theta}$ denotes the θ th situations, $\theta = 1, 2, \ldots, k_2$.
- $X = \{X_m\}$: refers to the set of criteria, in which X_m represents the m-th criterion, $m = 1, 2, \ldots, M$.
- $W_{X_m} = (w_{X_1}, \ldots, w_{X_M})$: refers to the weighting vector, in which w_{X_m} represents the weight of the m-th criterion. The weighting vector is usually provided by the DM satisfying $\sum_{m=1}^{M} w_{X_m} = 1$, $w_{X_m} \in [0, 1]$, $m = 1, 2, \ldots, M$.
- C_δ: refers to the cost of the δ-th available emergency alternative, $\delta = 1, 2, \ldots, k_1$.
- $R_{\theta m} = [R^L_{\theta m}, R^H_{\theta m}], R^H_{\theta m} > R^L_{\theta m}$: refers to the values of RPs, in which $R^L_{\theta m}$ and $R^H_{\theta m}$ represent the lower and upper limits of RP provided by DM for the m-th criterion in the θ-th situation, respectively, $m = 1, 2, \ldots, M, \theta = 1, 2, \ldots, k_2$.
- $E_{\delta m} = [E^L_{\delta m}, E^H_{\delta m}], E^H_{\delta m} > E^L_{\delta m}$: refers to the value of the pre-defined effective control scope [18], in which $E^L_{\delta m}$ and $E^H_{\delta m}$ represent the lower and upper limits of losses' protection scope from EE with respect to the δ-th alternative concerning the m-th criteria, respectively. $E_{\delta m}$ is usually determined by the local government, $\delta = 1, 2, \ldots, k_1, m = 1, 2, \ldots, M$.

3.2. Calculation of Gains and Losses

When an EE occurs, it may have different possible emergency situations. The DM needs to collect related information about possible situations and losses to make a decision. According to the collected information, DM forms the corresponding RP, $R_{\theta m}$, of the m-th criterion X_m in the θ-th situation $S_{2\theta}$. Gains and losses can be determined on the basis of the RPs $R_{\theta m}$ and the pre-defined effective control scope $E_{\delta m}$ of different alternatives.

Because both the RPs and the pre-defined effective control scopes are expressed in the form of interval values, the relationship between the interval values $R_{\theta m}$ and $E_{\delta m}$ should be analyzed before determining the gains and losses. To simplify, the relationship between $R_{\theta m}$ and $E_{\delta m}$ and the computation formulas for obtaining gains and losses taken from Wang et al. [17] will be utilized in our proposal.

The positional relationship between $R_{\theta m}$ and $E_{\delta m}$ is summarized in Table 2. Tables 3 and 4 provide the computation formulas of gains and losses for all possible relationships between $R_{\theta m}$ and $E_{\delta m}$, in which Tables 3 and 4 are for cost criteria and benefit criteria, respectively.

Based on the computation formulas of gain and loss provided in Tables 3 and 4, the gain matrix GM_θ and the loss matrix LM_θ can then be formed. Afterwards, the overall prospect values can be calculated by the value function on the basis of the gain and loss matrix GM_θ, LM_θ.

Table 2. Positional relationship between interval values $R_{\theta m}$ and $E_{\delta m}$ [17].

Cases		Positional Relationship
Case 1	$E^H_{\delta m} < R^L_{\theta m}$	
Case 2	$R^H_{\theta m} < E^L_{\delta m}$	
Case 3	$E^L_{\delta m} < R^L_{\theta m} < E^H_{\delta m} < R^H_{\theta m}$	
Case 4	$R^L_{\theta m} < E^L_{\delta m} < R^H_{\theta m} < E^H_{\delta m}$	
Case 5	$E^L_{\delta m} < R^L_{\theta m} < R^H_{\theta m} < E^H_{\delta m}$	
Case 6	$R^L_{\theta m} < E^L_{\delta m} < E^H_{\delta m} < R^H_{\theta m}$	

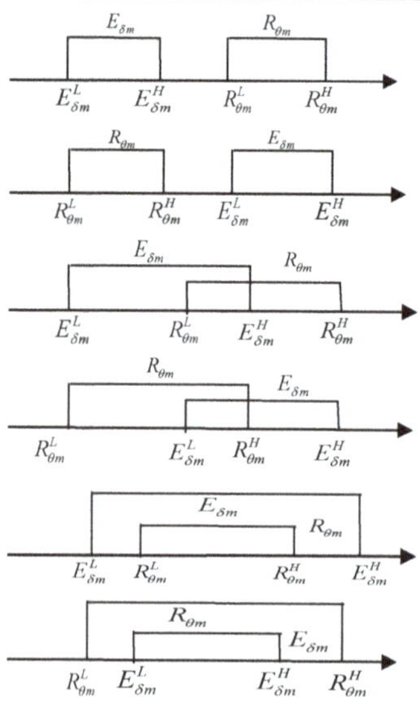

Table 3. Computation formulas of gain and loss for cost criteria [17].

Cases		Gain $G_{\delta m}$	Loss $L_{\delta m}$
Case 1	$E^H_{\delta m} < R^L_{\theta m}$	$R^L_{\theta m} - 0.5(E^L_{\delta m} + E^H_{\delta m})$	0
Case 2	$R^H_{\theta m} < E^L_{\delta m}$	0	$R^H_{\theta m} - 0.5(E^L_{\delta m} + E^H_{\delta m})$
Case 3	$E^L_{\delta m} < R^L_{\theta m} < E^H_{\delta m} < R^H_{\theta m}$	$0.5(R^L_{\theta m} - E^L_{\delta m})$	0
Case 4	$R^L_{\theta m} < E^L_{\delta m} < R^H_{\theta m} < E^H_{\delta m}$	0	$0.5(R^H_{\theta m} - E^H_{\delta m})$
Case 5	$E^L_{\delta m} < R^L_{\theta m} < R^H_{\theta m} < E^H_{\delta m}$	$0.5(R^L_{\theta m} - E^L_{\delta m})$	$0.5(R^H_{\theta m} - E^H_{\delta m})$
Case 6	$R^L_{\theta m} < E^L_{\delta m} < E^H_{\delta m} < R^H_{\theta m}$	0	0

Table 4. Computation formulas of gain and loss for benefit criteria [17].

Cases		Gain $G_{\delta m}$	Loss $L_{\delta m}$
Case 1	$E_{\delta m}^{H} < R_{\theta m}^{L}$	0	$0.5(E_{\delta m}^{L} + E_{\delta m}^{H}) - R_{\theta m}^{L}$
Case 2	$R_{\theta m}^{H} < E_{\delta m}^{L}$	$0.5(E_{\delta m}^{L} + E_{\delta m}^{H}) - R_{\theta m}^{H}$	0
Case 3	$E_{\delta m}^{L} < R_{\theta m}^{L} < E_{\delta m}^{H} < R_{\theta m}^{H}$	0	$0.5(E_{\delta m}^{L} - R_{\theta m}^{L})$
Case 4	$R_{\theta m}^{L} < E_{\delta m}^{L} < R_{\theta m}^{H} < E_{\delta m}^{H}$	$0.5(E_{\delta m}^{H} - R_{\theta m}^{H})$	0
Case 5	$E_{\delta m}^{L} < R_{\theta m}^{L} < R_{\theta m}^{H} < E_{\delta m}^{H}$	$0.5(E_{\delta m}^{H} - R_{\theta m}^{H})$	$0.5(E_{\delta m}^{L} - R_{\theta m}^{L})$
Case 6	$R_{\theta m}^{L} < E_{\delta m}^{L} < E_{\delta m}^{H} < R_{\theta m}^{H}$	0	0

3.3. *Computation of Overall Prospect Values*

Assume that the gain matrix of the θ-th situation is denoted by $GM_\theta = (G_{\theta\delta m})_{\delta\times m}$, and similarly, the loss matrix and value matrix of the θ-th situation are denoted by $LM_\theta = (L_{\theta\delta m})_{\delta\times m}$ and $VM_\theta = (v_{\theta\delta m})_{\delta\times m}$, respectively.

$$v_{\theta\delta m} = G_{\theta\delta m}{}^{\alpha} + [-\lambda(-L_{\theta\delta m})^{\beta}],\ \delta = 1,2,\ldots,k_1;\ \theta = 1,2,\ldots,k_2;\ m = 1,2,\ldots,M \quad (2)$$

where $v_{\theta\delta m}$ means the value with respect to the alternative $S_{1\delta}$, concerning criterion X_m, in the situation $S_{2\theta}$. According to [30], the parameters α, β and λ can employ different values. In this proposal, the following ones will be employed, i.e., $\alpha = \beta = 0.88, \lambda = 2.25$. According to PT, Equation (2) is usually utilized to measure the degree of gains and losses, in which different feelings of DM towards gains and losses are reflected by using prospect values; the greater $v_{\theta\delta m}$, the more DM satisfies, which denotes that the DM satisfies his/her decisions; otherwise, he/she regrets or feels depressed about his/her decisions. In this way, the DM's psychological behavior can be described clearly and comprehensively.

Due to $v_{\theta\delta m}$ not usually having the same units, a normalization process for removing the effect of units is needed. The normalized value matrix $\overline{VM_\theta} = (\overline{v}_{\theta\delta m})_{\delta\times m}$ can be obtained by using:

$$\overline{v}_{\theta\delta m} = \frac{v_{\theta\delta m}}{v_{\theta\delta}^{*}},\ \delta = 1,2,\ldots,k_1;\ \theta = 1,2,\ldots,k_2;\ m = 1,2,\ldots,M \quad (3)$$

where $v_{\theta\delta}{}^{*} = \max_{m\in M} |v_{\theta\delta m}|$.

On the basis of the normalized value matrix $\overline{VM_\theta}$ and the weighting vector W_{X_m} provided by DM, the overall prospect values of alternative $S_{1\delta}$ can be calculated by using the following equation,

$$O_{\theta\delta} = \sum_{m=1}^{M} \overline{v}_{\theta\delta m} w_{X_m},\ \delta = 1,2,\ldots,k_1;\ \theta = 1,2,\ldots,k_2;\ m = 1,2,\ldots,M \quad (4)$$

3.4. *Selecting Optimal Alternative Based on Payoffs*

In this section, the payoffs of EE and DM will be determined on the basis of the overall prospect values, $O_{\theta\delta}$, obtained above. Then, according to the payoffs of EE and DM, the optimal alternative can be selected as the proper response regarding different emergency situations.

3.4.1. Determining the Payoffs of the Players

Due to the fact that the game between EE and DM is a zero-sum game and EE is unconscious of the benefits or costs that it will get or lose, just determining the payoffs of the DM is adequate for the emergency response.

Because $O_{\theta\delta}$ is a comprehensive value that reflects the DM's psychological behavior, it is regarded as the part of the payoffs of DM. Since each alternative has its own cost, it is more reasonable to consider the prospect values of per unit cost rather than the overall prospect values. The payoffs of DM are determined as follows:

$$P_1(S_1) = f(O_{\theta\delta}, C_\delta) = \frac{O_{\theta\delta}}{C_\delta},\ \delta = 1, 2, \cdots k_1;\ \theta = 1, 2, \cdots k_2 \tag{5}$$

Then, the payoffs of EE can be obtained as:

$$P_2(S_2) = -P_1(S_1) \tag{6}$$

From Equations (5) and (6), the selection process of the optimal alternative can be determined in the coming subsection.

3.4.2. Selection of the Optimal Alternative with Respect to Each Emergency Situation

As mentioned previously, the game between EE and DM is a zero-sum game, and EE is a special player, which has no consciousness about the real world, so it is adequate to determine the optimal strategy of the DM.

The equation for selecting the optimal strategy of DM with respect to each possible emergency situation goes as follows:

$$P_1(S_{2\theta}, S_{1\delta}^*) = \max_{\delta \in k_1} P_1(S_{2\theta}, S_{1\delta}),\ \theta = 1, 2, \cdots k_2 \tag{7}$$

The vector strategy $(S_{2\theta}, S_{1\delta}^*)$ means if the EE has taken $S_{2\theta}$ as its strategy, the best response for the DM is the strategy $S_{1\delta}^*$. In other words, the strategy $S_{1\delta}^*$ will be the optimal strategy of DM to deal with the emergency situation $S_{2\theta}$.

For a clear understanding, the procedures of the new proposed method are summarized as the following steps:

1. Based on the information of $R_{\theta m}$ and $E_{\delta m}$, gains and losses can be calculated by using the equations provided in Tables 3 and 4, respectively.
2. The gain and loss matrix GM_θ, LM_θ can be formed on the basis of the obtained gains and losses, respectively. Then, the value matrix VM_θ and its normalized form $\overline{VM_\theta}$ can be obtained by using Equations (2) and (3), respectively. Afterwards, the overall prospect value $O_{\theta\delta}$ can be calculated by Equation (4).
3. Based on the overall prospect value $O_{\theta\delta}$ and the cost of each alternative, the payoffs of DM and EE can be determined by Equations (5) and (6), respectively.
4. Based on the obtained payoffs of DM and EE, the DM can select the optimal strategies for dealing with all possible emergency situations according to Equation (7).

4. Case Study and Comparison

4.1. Case Study

This part will provide a case study on a typhoon emergency event to demonstrate the validity and rationality of the proposed method.

In summer, it is quite common for coastal cities to suffer from different kinds of losses (lives, property, environment, etc.) caused by typhoons. In order to take effective measures to reduce the losses caused by typhoon as much as possible in the real world, this section takes typhoon landfall as an application background to demonstrate the validity and rationality of our proposal. Suppose that a typhoon is approaching and will possibly make landfall at one city located on the southeast coast of China. When it makes landfall, it might cause various losses, such as lives, properties, environment damages, etc. Thus, the following criteria are concerned in this case study:

c_1: The number of casualties.

c_2: Property losses (in 1000$).

c_3: The negative effects on the environment on a scale of 0–100 (0: no negative effect; 100: serious negative effect).

The emergency alternatives are described as follows:

Regarding the coming typhoon, the following alternatives can be carried out:

S_{11}: Broadcast and send short messages to remind citizens regarding the coming typhoon and suggest that citizens prepare food, water, medicine and other daily necessities in advance; furthermore, local government organizes related departments to check the evacuation solutions and paths to ensure the citizens' safety as much as possible;

S_{12}: Based on S_{11}, inform schools and plants to check the safety issues; classes and work can be stopped if necessary. Meanwhile, employees in ocean transport, fishermen and mariculture are required to come back to or go closer to harbors to take shelter from the typhoon. In addition, check the stability of high-altitude facilities and dangerous buildings.

S_{13}: Based on S_{12}, telecom operators and power supply departments strengthen their checking and maintenance to ensure all different lines of communication and power supply are open. Meanwhile, check the urban drainage pipelines to avoid urban waterlogging.

S_{14}: Based on S_{13}, vindicate public security in preventing criminal issues from occurring; meanwhile, hospitals prepare enough ambulances and staff to ensure that injured citizens can be rescued and treated immediately. Furthermore, the reservoirs and hydropower stations near the city should make reasonable schedules to avoid floods.

C_δ is the cost of the δ-th alternative (in 1000$). The criteria weights of each criterion are provided by DM in this case study. The pre-defined effective control scope $E_{\delta m}$, the cost C_δ and related weights w_{X_m} are given in Table 5.

Table 5. The $E_{\delta m}$, C_δ and w_{X_m} of the typhoon emergency.

	Criteria			
Alternatives	c_1**(0.5)**	c_2**(0.25)**	c_3**(0.25)**	C_δ
	$E_{\delta 1}$	$E_{\delta 2}$	$E_{\delta 3}$	C_δ
S_{11}	[3,5]	[200,400]	[40,50]	10
S_{12}	[6,14]	[800,1200]	[50,60]	30
S_{13}	[14,20]	[1200,1500]	[60,70]	70
S_{14}	[18,25]	[1500,1800]	[70,80]	130

Analyzing by the weather forecast and historical data, there are four possible situations of a typhoon in the coming 72 h, as follows:

S_{21}: The typhoon will not make landfall at the city, and it just brings light rain and wind;

S_{22}: The typhoon will make landfall at part of the area of the city and bring moderate rain and gales;

S_{23}: The typhoon will make landfall over the entire city and bring rainstorms and strong wind;

S_{24}: The typhoon will have a front landfall over the entire city and bring downpours and blustery weather;

The reference points $R_{\theta m}$ regarding the four possible emergency situations provided by DM are shown in Table 6.

Table 6. Reference points (RPs) regarding the four emergency situations.

Situations	Criteria		
	c_1	c_2	c_3
	$R_{\theta 1}$	$R_{\theta 2}$	$R_{\theta 3}$
S_{21}	[5,8]	[100,300]	[20,35]
S_{22}	[5,12]	[300,500]	[35,45]
S_{23}	[12,18]	[600,800]	[45,55]
S_{24}	[18,20]	[800,100]	[55,65]

According to the information shown in Tables 5 and 6, the positional relationship between $R_{\theta m}$ and $E_{\delta m}$ in Table 2 and the equations provided in Tables 3 and 4, the gain and loss matrix GM_θ, LM_θ can be obtained as follows,

$$GM_1 = \begin{bmatrix} 0 & 0 & 10 \\ 3 & 700 & 20 \\ 9 & 1050 & 30 \\ 13.5 & 1350 & 40 \end{bmatrix}, GM_2 = \begin{bmatrix} 0 & 0 & 2.5 \\ 1 & 500 & 10 \\ 5 & 850 & 20 \\ 9.5 & 1150 & 30 \end{bmatrix},$$

$$GM_3 = \begin{bmatrix} 0 & 0 & 0 \\ 0 & 200 & 2.5 \\ 1 & 550 & 10 \\ 3.5 & 850 & 20 \end{bmatrix}, GM_4 = \begin{bmatrix} 0 & 0 & 0 \\ 0 & 100 & 0 \\ 0 & 350 & 2.5 \\ 0 & 650 & 10 \end{bmatrix};$$

$$LM_1 = \begin{bmatrix} -1 & -50 & 0 \\ 0 & 0 & 0 \\ 0 & 0 & 0 \\ 0 & 0 & 0 \end{bmatrix}, LM_2 = \begin{bmatrix} -1 & -100 & 0 \\ 0 & 0 & 0 \\ 0 & 0 & 0 \\ 0 & 0 & 0 \end{bmatrix},$$

$$LM_3 = \begin{bmatrix} -8 & -300 & -2.5 \\ -3 & 0 & 0 \\ 0 & 0 & 0 \\ 0 & 0 & 0 \end{bmatrix}, LM_4 = \begin{bmatrix} -14 & -500 & -10 \\ -8 & 0 & -2.5 \\ -2 & 0 & 0 \\ 0 & 0 & 0 \end{bmatrix}.$$

Based on GM_θ and LM_θ, the value matrix VM_θ and its normalized form $\overline{VM_\theta}$ can be obtained according to Equations (2) and (3), respectively, i.e.,

$$VM_1 = \begin{bmatrix} -2.25 & -70.35 & 7.59 \\ 2.63 & 318.92 & 13.96 \\ 6.91 & 455.67 & 19.95 \\ 9.88 & 568.45 & 25.69 \end{bmatrix}, VM_2 = \begin{bmatrix} -2.25 & -129.47 & 2.24 \\ 1 & 237.19 & 7.59 \\ 4.12 & 378.35 & 13.96 \\ 7.25 & 493.64 & 19.95 \end{bmatrix},$$

$$VM_3 = \begin{bmatrix} -14.02 & -340.45 & -5.04 \\ -5.92 & 105.90 & 2.24 \\ 1 & 257.94 & 7.59 \\ 3.01 & 378.35 & 13.96 \end{bmatrix}, VM_4 = \begin{bmatrix} -22.95 & -533.67 & -17.07 \\ -14.02 & 57.54 & -5.04 \\ -4.14 & 173.29 & 2.24 \\ 0 & 298.79 & 7.59 \end{bmatrix}; \text{and}$$

$$\overline{VM_1} = \begin{bmatrix} -0.2278 & -0.1238 & 0.2952 \\ 0.2662 & 0.5610 & 0.5434 \\ 0.6999 & 0.8160 & 0.7763 \\ 1 & 1 & 1 \end{bmatrix}, \overline{VM_2} = \begin{bmatrix} -0.3103 & -0.2623 & 0.1123 \\ 0.1379 & 0.4805 & 0.3803 \\ 0.5685 & 0.7664 & 0.6999 \\ 1 & 1 & 1 \end{bmatrix},$$

$$\overline{VM_3} = \begin{bmatrix} -1 & -0.8998 & -0.3610 \\ -0.4218 & 0.2799 & 0.1604 \\ 0.0713 & 0.6818 & 0.5434 \\ 0.2147 & 1 & 1 \end{bmatrix}, \overline{VM_4} = \begin{bmatrix} -1 & -1 & -1 \\ -0.6111 & 0.1078 & -0.2952 \\ -0.1804 & 0.3247 & 0.1312 \\ 0 & 0.5599 & 0.4444 \end{bmatrix}.$$

According to Equation (4), the overall prospect values $O_{\theta\delta}$ of the θ-th alternatives in δ-th emergency situation are calculated and shown in Table 7.

Table 7. The overall prospect values $O_{\theta\delta}$ of the θ-th alternatives in the δ-th emergency situation.

$O_{\theta\delta}$		Situations			
		S_{21}	S_{22}	S_{23}	S_{24}
Alternative	S_{11}	−0.0710	−0.1927	−0.8152	−1.000
	S_{12}	0.4092	0.2841	−0.1008	−0.3524
	S_{13}	0.7444	0.6508	0.3419	0.0238
	S_{14}	1.000	1.000	0.6074	0.2511

Based on Equations (5) and (6) and the results of $O_{\theta\delta}$ shown in Table 7, the payoff matrix of EE and DM is provided in Table 8.

Table 8. The payoff matrix of EE and DM.

		EE			
		S_{21}	S_{22}	S_{23}	S_{24}
DM	S_{11}	$(-0.0071, 0.0071)$	$(-0.0193, 0.0193)$	$(-0.0815, 0.0815)$	$(-0.0100, 0.0100)$
	S_{12}	$(0.0136, -0.0136)$	$(0.0095, -0.0095)$	$(-0.0034, 0.0034)$	$(-0.0117, 0.0117)$
	S_{13}	$(0.0106, -0.0106)$	$(0.0093, -0.0093)$	$(0.0049, -0.0049)$	$(0.0003, -0.0003)$
	S_{14}	$(0.0077, -0.0077)$	$(0.0077, -0.0077)$	$(0.0047, -0.0047)$	$(0.0019, -0.0019)$

Then, based on Table 8 and Equation (7), the best strategy of DM for different possible situations can be obtained as follows:

If the EE has selected the strategy S_{21}, the best strategy of DM is the one with the biggest payoff value, which can be obtained by using Equation (7):

$$\begin{aligned} P_1(S_{21}, S_{1\delta}^*) &= \max_{\delta \in k_1} P_1(S_{21}, S_{1\delta}) \\ &= \max \ \{P_1(S_{21}, S_{11}), P_1(S_{21}, S_{12}), P_1(S_{21}, S_{13}), P_1(S_{21}, S_{14})\} \\ &= \max \ \{-0.0071, 0.0136, 0.0106, 0.0077\} \\ &= 0.0136 \end{aligned}$$

That is $P_1(S_{21}, S_{1\delta}^*) = P_1(S_{21}, S_{12})$, which means if the EE has selected the strategy S_{21}, the best strategy for DM is S_{12}.

Similarly, the best strategies of DM regarding different possible situations are the ones with the biggest payoffs, which are underlined and bolded in Table 9.

Table 9. Best strategies of DM.

		EE			
		S_{21}	S_{22}	S_{23}	S_{24}
DM	S_{11}	$(-0.0071, 0.0071)$	$(-0.0193, 0.0193)$	$(-0.0815, 0.0815)$	$(-0.0100, 0.0100)$
	S_{12}	$(\underline{\mathbf{0.0136}}, -0.0136)$	$(\underline{\mathbf{0.0095}}, -0.0095)$	$(-0.0034, 0.0034)$	$(-0.0117, 0.0117)$
	S_{13}	$(0.0106, -0.0106)$	$(0.0093, -0.0093)$	$(\underline{\mathbf{0.0049}}, -0.0049)$	$(0.0003, -0.0003)$
	S_{14}	$(0.0077, -0.0077)$	$(0.0077, -0.0077)$	$(0.0047, -0.0047)$	$(\underline{\mathbf{0.0019}}, -0.0019)$

The four optimal solutions with respect to each emergency situation are (S_{21}, S_{12}), (S_{22}, S_{12}), (S_{23}, S_{13}) and (S_{24}, S_{14}), which means if the EE has selected S_{21}, the best strategy of DM is to select S_{12}; if the EE has selected S_{22}, the best strategy of DM is to select S_{12}; if the EE has selected S_{23}, the best strategy of DM is to select S_{13}; the EE has selected S_{24}, the best strategy of DM is to select S_{14}.

4.2. Comparison with Other Methods

In order to demonstrate the superiority and novelty of our proposal, a comparison with other methods will be conducted. Because there are no existing approaches that are based on PT and GT simultaneously, thus, some characteristics have been studied to highlight the superiority of our proposal; see Table 10.

Table 10. Comparison with other emergency decision making (EDM) methods.

Literature	Considering DM's Psychological Behaviors	Considering Different Emergency Situations
[4–8,40,46]	No	No
[1,15–18]	Yes	No
[20–24]	No	Yes
Our proposal	Yes	Yes

According to Table 10, it can be seen clearly that our proposal considers not only the DM's psychological behavior, but also the coping with the different emergency situations. The proposed EDM method is closer to the real-world situations than other EDM methods.

5. Conclusions and Future Works

A new EDM method based on GT and PT is proposed in this paper aiming at overcoming the limitations in previous EDM approaches. Due to the inadequate and incomplete information about EEs, interval values are employed in our proposal to estimate the possible losses caused by different situations. DM's psychological behavior and coping with different emergency situations have been considered simultaneously, which is the significant difference between our proposal and the existing EDM approaches. An example about a typhoon and related comparison with existing EDM approaches have been conducted to demonstrate the novelty and rationality of our proposal. It is hoped that our proposed method can be applied to solve real-word problems in the near future.

The research in the near future should consider the different types of information in the game process, such as linguistic information, hesitant fuzzy linguistic information, and so on, which are common information types in the real world when DM hesitates in his/her assessments.

Author Contributions: All authors have contributed equally to this paper.

Funding: This work was partly supported by the National Natural Science Foundation of China (Project Nos. 61773123, 71701050, 71801050), the Spanish National Research Project (Project No. TIN2015-66524-P) and the Spanish Ministry of Economy and Finance Postdoctoral Fellow (IJCI-2015-23715)

Acknowledgments: The authors would like to thank the anonymous reviewers for their valuable time in reviewing our paper and for their helpful suggestions to improve our paper.

Conflicts of Interest: The authors declare no conflict of interest.

References

1. Liu, B.S.; Zhao, X.; Li, Y. Review and Prospect of Studies on Emergency Management. *Procedia Eng.* **2016**, *145*, 1501–1508. [CrossRef]
2. IFRC. Word Disaster Reprot 2016. Available online: http://media.ifrc.org/ifrc/publications/world-disasters-report-2016/ (accessed on 14 October 2016).
3. Xia, L.; Xiao, Y. Generation Mechanism of Emergency Decision-making: Links, Elements and Serial Processing. *J. Shanghai Adm. Inst.* **2011**, *4*, 005.
4. Domeneghetti, B.; Benamrane, Y.; Wybo, J.L. Analyzing nuclear expertise support to population protection decision making process during nuclear emergencies. *Saf. Sci.* **2018**, *101*, 155–163. [CrossRef]
5. Levy, J.K.; Taji, K. Group decision support for hazards planning and emergency management: A Group Analytic Network Process (GANP) approach. *Math. Comput. Model.* **2007**, *46*, 906–917. [CrossRef]
6. Peng, Y.; Zhang, Y.; Tang, Y.; Li, S.M. An incident information management framework based on data integration, data mining, and multi-criteria decision making. *Decis. Support Syst.* **2011**, *51*, 316–327. [CrossRef]
7. Qu, J.H.; Meng, X.L.; You, H. Multi-stage ranking of emergency technology alternatives for water source pollution accidents using a fuzzy group decision making tool. *J. Hazard. Mater.* **2016**, *310*, 68–81. [CrossRef] [PubMed]
8. Xu, X.H.; Du, Z.J.; Chen, X.H. Consensus model for multi-criteria large-group emergency decision making considering non-cooperative behaviors and minority opinions. *Decis. Support Syst.* **2015**, *79*, 150–160. [CrossRef]
9. Zhou, L.; Wu, X.H.; Xu, Z.S.; Fujita, H. Emergency Decision Making for Natural Disasters: An Overview. *Int. J. Disaster Risk Reduction* **2017**, *27*, 567–576. [CrossRef]
10. Kahneman, D.; Tversky, A. Prospect theory: An analysis of decision under risk. *Econometrica J. Econom. Soc.* **1979**, *47*, 263–291. [CrossRef]
11. Bell, D.E. Regret in decision making under uncertainty. *Oper. Res.* **1982**, *30*, 961–981. [CrossRef]
12. Bell, D.E. Disappointment in decision making under uncertainty. *Oper. Res.* **1985**, *33*, 1–27. [CrossRef]

13. Schmidt, U.; Starmer, C.; Sugden, R. Third-generation prospect theory. *J. Risk Uncertain.* **2008**, *36*, 203. [CrossRef]
14. Fan, Z.P.; Liu, Y.; Shen, R.J. Risk decision analysis method for emergency response based on prospect theory. *Syst. Eng.-Theory Pract.* **2012**, *5*, 011.
15. Fan, Z.P.; Zhang, X.; Chen, F.D.; Liu, Y. Multiple attribute decision making considering aspiration-levels: A method based on prospect theory. *Comput. Ind. Eng.* **2013**, *65*, 341–350. [CrossRef]
16. Wang, L.; Wang, Y.M.; Hu, B.X. Dynamic adjusting method of emergency alternatives based on prospect theory. *Control Decis.* **2016**, *1*, 99–104.
17. Wang, L.; Wang, Y.M.; Martínez, L. A group decision method based on prospect theory for emergency situations. *Inf. Sci.* **2017**, *418*, 119–135. [CrossRef]
18. Wang, L.; Zhang, Z.X.; Wang, Y.M. A prospect theory-based interval dynamic reference point method for emergency decision making. *Expert Syst. Appl.* **2015**, *42*, 9379–9388. [CrossRef]
19. Camerer, C. *Behavioral Game Theory: Experiments in Strategic Interaction*; Princeton University Press: Princeton, NJ, USA, 2003.
20. Yang, J.J.; Xu, C.H. Emergency decision engineering model based on sequential games. *Syst. Eng. Procedia* **2012**, *5*, 276–282. [CrossRef]
21. Chen, Y.M.; Wu, D.; Wu, C.K. A game theory approach for an emergency management security advisory system. In Proceedings of the 2008 IEEE International Conference on Intelligence and Security Informatics, Taipei, Taiwan, 17–20 June 2008; pp. 116–121.
22. Coles, J.; Zhuang, J. Decisions in disaster recovery operations: a game theoretic perspective on organization cooperation. *J. Homeland Secur. Emerg. Manag.* **2011**, *8*. [CrossRef]
23. Gupta, U.; Ranganathan, N. Multievent crisis management using noncooperative multistep games. *IEEE Trans. Comput.* **2007**, *56*, 577–589. [CrossRef]
24. Yao, J.; Ji, L.; Chi, H. Dynamic games analysis of emergency management. *Manag. Rev.* **2005**, *17*, 46–50.
25. Zhong, Q.; Li, Z.; Zhang, L. Analysis of the emergency procurement based on evolutionary game theory. In Proceedings of the KAM'09 Second International Symposium on Knowledge Acquisition and Modeling, Wuhan, China, 30 November–1 December 2009; Volume 2, pp. 261–264.
26. Abdellaoui, M.; Bleichrodt, H.; Paraschiv, C. Loss aversion under prospect theory: A parameter-free measurement. *Manag. Sci.* **2007**, *53*, 1659–1674. [CrossRef]
27. Camerer, C. Bounded rationality in individual decision making. *Exp. Econ.* **1998**, *1*, 163–183. [CrossRef]
28. Limaei, S.M. Mixed strategy game theory, application in forest industry. *For. Policy Econ.* **2010**, *12*, 527–531. [CrossRef]
29. Gibbons, R. *Game Theory for Applied Economists*; Princeton University Press: Princeton, NJ, USA, 1992.
30. Tversky, A.; Kahneman, D. Advances in prospect theory: Cumulative representation of uncertainty. *J. Risk Uncertain.* **1992**, *5*, 297–323. [CrossRef]
31. Schmidt, U.; Zank, H. A genuine foundation for prospect theory. *J. Risk Uncertain.* **2012**, *45*, 97–113. [CrossRef]
32. Tversky, A.; Kahneman, D. Loss aversion in riskless choice: A reference-dependent model. *Q. J. Econ.* **1991**, *106*, 1039–1061. [CrossRef]
33. Bleichrodt, H.; Schmidt, U.; Zank, H. Additive utility in prospect theory. *Manag. Sci.* **2009**, *55*, 863–873. [CrossRef]
34. Booij, A.S.; Van de Kuilen, G. A parameter-free analysis of the utility of money for the general population under prospect theory. *J. Econ. Psychol.* **2009**, *30*, 651–666. [CrossRef]
35. Wakker, P.P. *Prospect Theory: For Risk and Ambiguity*; Cambridge University Press: Cambridge, UK, 2010.
36. Liu, Y.; Fan, Z.P.; Zhang, Y. Risk decision analysis in emergency response: A method based on cumulative prospect theory. *Comput. Oper. Res.* **2014**, *42*, 75–82. [CrossRef]
37. Ju, Y.B.; Wang, A.H.; Liu, X.Y. Evaluating emergency response capacity by fuzzy AHP and 2-tuple fuzzy linguistic approach. *Expert Syst. Appl.* **2012**, *39*, 6972–6981. [CrossRef]
38. Wang, L.; Labella, Á.; Rodríguez, R.M.; Wang, Y.M.; Martínez, L. Managing non-homogeneous information and experts' psychological behavior in group emergency decision making. *Symmetry* **2017**, *9*, 234. [CrossRef]
39. Zhang, Z.X.; Wang, L.; Rodríguez, R.M.; Wang, Y.M.; Martínez, L. A hesitant group emergency decision making method based on prospect theory. *Complex Intell. Syst.* **2017**, *3*, 177–187. [CrossRef]

40. Wang, L.; Rodríguez, R.M.; Wang, Y.M. A dynamic multi-attribute group emergency decision making method considering experts' hesitation. *Int. J. Comput. Intell. Syst.* **2018**, *11*, 163–182. [CrossRef]
41. Chen, Y.M.; Wu, D.; Wu, C.K. A game theory approach for evaluating terrorist threats and deploying response agents in urban environments. *J. Homeland Secur. Emerg. Manag.* **2009**, *6*, 31. [CrossRef]
42. Cheng, Y.; Zheng, X.P. Emergence of cooperation during an emergency evacuation. *Appl. Math. Comput.* **2018**, *320*, 485–494. [CrossRef]
43. Rezazadeh, A.; Talarico, L.; Reniers, G.; Cozzani, V.; Zhang, L. Applying game theory for securing oil and gas pipelines against terrorism. *Reliab. Eng. Syst. Saf.* **2018**, doi:10.1016/j.ress.2018.04.021. [CrossRef]
44. Gao, Y.; Li, Z.W.; Wang, F.M.; Wang, F.; Tan, R.R.; Bi, J.; Jia, X.P. A game theory approach for corporate environmental risk mitigation. *Resour. Conserv. Recycl.* **2018**, *130*, 240–247. [CrossRef]
45. Wu, C.K. A game theory approach for assessing risk value and deploying search-and-rescue resources after devastating tsunamis. *Environ. Res.* **2018**, *162*, 18–26. [CrossRef] [PubMed]
46. Liu, X. Energy network dispatch optimization under emergency of local energy shortage with web tool for automatic large group decision-making. *Energy* **2017**, *120*, 740–750.

symmetry

MDPI

Article

Some q-Rung Dual Hesitant Fuzzy Heronian Mean Operators with Their Application to Multiple Attribute Group Decision-Making

Yuan Xu [1], Xiaopu Shang [1,*], Jun Wang [1], Wen Wu [1] and Huiqun Huang [2]

1 School of Economics and Management, Beijing Jiaotong University, Beijing 100044, China; 17120627@bjtu.edu.cn (Y.X.); 14113149@bjtu.edu.cn (J.W.); wuwen@bjtu.edu.cn (W.W.)
2 School of Software Engineering, Beijing Jiaotong University, Beijing 100044, China; 15301005@bjtu.edu.cn
* Correspondence: sxp@bjtu.edu.cn; Tel.: +86-10-5168-3854

Received: 21 September 2018; Accepted: 7 October 2018; Published: 10 October 2018

Abstract: The q-rung orthopair fuzzy sets (q-ROFSs), originated by Yager, are good tools to describe fuzziness in human cognitive processes. The basic elements of q-ROFSs are q-rung orthopair fuzzy numbers (q-ROFNs), which are constructed by membership and nonmembership degrees. As realistic decision-making is very complicated, decision makers (DMs) may be hesitant among several values when determining membership and nonmembership degrees. By incorporating dual hesitant fuzzy sets (DHFSs) into q-ROFSs, we propose a new technique to deal with uncertainty, called q-rung dual hesitant fuzzy sets (q-RDHFSs). Subsequently, we propose a family of q-rung dual hesitant fuzzy Heronian mean operators for q-RDHFSs. Further, the newly developed aggregation operators are utilized in multiple attribute group decision-making (MAGDM). We used the proposed method to solve a most suitable supplier selection problem to demonstrate its effectiveness and usefulness. The merits and advantages of the proposed method are highlighted via comparison with existing MAGDM methods. The main contribution of this paper is that a new method for MAGDM is proposed.

Keywords: q-rung orthopair fuzzy set; q-rung dual hesitant fuzzy; q-rung dual hesitant fuzzy Heronian mean; multiple attribute group decision-making

---

1. Introduction

With the rapid economic and technological development, competition among enterprises has become increasingly fierce. For manufacturing companies, choosing an appropriate supplier is of high importance. Generally speaking, companies need to collect relevant information for all suppliers and use some technologies to determine the most suitable one. In essence, supplier selection is a multiple attribute decision-making problem. Due to the complexity of modern decision-making problems, it is impossible for a single decision maker (DM) to grasp all the information of all decision objectives. Thus, many real decision-making problems often require group decision-making, i.e., multiple attribute group decision-making (MAGDM). Decision-making problems are constrained by a variety of internal and external factors. For example, as decision-making problems become increasingly complex, it is almost impossible to describe attribute values using crisp values. Decision-making problems often have enormous complexity and uncertainty. So, many scholars focus on how to deal with and describe uncertain phenomena. In 1986, Atanassov [1] proposed the concept of an intuitionistic fuzzy set (IFS) for coping with fuzziness and uncertainty. IFS is more powerful and useful than Zadeh's fuzzy set (FS) [2], as FS only has a membership degree, which makes it impossible to comprehensively describe imprecision. Since the appearance of IFS, it has been widely applied to medical diagnoses [3,4], pattern recognition [5,6], cluster analysis [7,8], and especially, MAGDM [9–12].

However, there are quite a few circumstances with which IFSs cannot cope. For instance, in some cases, the sum of the membership and nonmembership degrees provided by DMs is greater than that of their square sum being less than or equal to one. To effectively address these cases, the concept of the Pythagorean fuzzy set (PFS) was introduced by Yager [13]. Obviously, the PFS is a generalized form of an IFS and can describe a wider information range. Owing to the effectiveness and powerfulness of PFSs, MAGDM with Pythagorean fuzzy information have become a research topic of great interest. Studies on PFSs can be roughly divided into three categories. The first category includes extensions of classical decision-making methods to MAGDM with Pythagorean fuzzy information, the most representative of which are the Pythagorean fuzzy decision-making methods proposed by Zhang and Xu [14] and Khan et al. [15] based on TOPSIS (Technique for Order Preference by Similarity to an Ideal Solution), and the one developed by Ren et al. [16] on the basis on TODIM (an acronym in Portuguese for Interactive Multi-Criteria Decision Making). The second category contains MAGDM methods with Pythagorean fuzzy information based on aggregation operators. Aggregation operators play a significantly important role in MAGDM. Solving MAGDM in different scenarios requires different aggregation operators. For example, to fairly treat membership and nonmembership degrees of PFSs, Ma et al. [17] raised symmetry operations of PFSs and proposed a battery of Pythagorean fuzzy symmetric aggregation operators. Xing et al. [18] put forward Pythagorean fuzzy Choquet integral aggregation operators based on Frank t-norm and t-conorm. To capture the interrelationship between aggregated Pythagorean fuzzy numbers (PFNs), Wei and Lu [19] put forward Pythagorean fuzzy Maclaurin symmetric mean operators. To fully absorb the advantages of Bonferroni mean and generalized Bonferroni mean in capturing the relationship among variables, Liang et al. [20] and Zhang et al. [21] introduced the Pythagorean fuzzy Bonferroni mean and generalized Pythagorean fuzzy Bonferroni mean operators, respectively. Due to the complexity of decision-making issues and the lack of sufficient experience, DMs often make unreasonable assessments. These unreasonable evaluation values have a serious negative impact on the final decision results. Thus, Li et al. [22] proposed the Pythagorean fuzzy power Muirhead mean operators to eliminate such bad impacts. Analogously, to fully utilize the advantages of Pythagorean fuzzy interaction operational rules in dealing with the interaction between membership and nonmembership degrees, Xu et al. [23] proposed the Pythagorean fuzzy interaction Muirhead mean operators. The third category is the investigation of combining PFSs with linguistic term sets. In actual MAGDM problems, the evaluations made by DMs need to be expressed from both qualitative and quantitative perspectives. Thus, Teng et al. [24], Du et al. [25], and Xian et al. [26] investigated MAGDM with Pythagorean fuzzy linguistic sets and interval-valued Pythagorean fuzzy linguistic sets, respectively. Considering uncertain linguistic terms provides DMs with a more convenient method to express their assessments. Geng et al. [27], Liu et al. [28], and Liu et al. [29] proposed the concept of Pythagorean fuzzy uncertain linguistic sets and studied their applications in MAGDM. In addition, some scientists also investigated MAGDM issues with Pythagorean 2-tuple linguistic information [30–32].

Although in the majority of cases IFSs and PFSs can successfully describe the attribute values in MAGDM, there are quite a few situations in which IFSs and PFSs are insufficient. According to the constraints of IFSs and PFSs, when the square sum of membership and nonmembership degrees exceed one, then the attribute value cannot be represented by both IFSs and PFSs. To deal with such a case, more recently, Yager [33] introduced the concept of the q-rung orthopair fuzzy set (q-ROFS), which can be viewed as an extension of IFS and PFS. From the definition of q-ROFSs, it is not difficult to see that q-ROFSs give DMs great freedom and a wider space within which to evaluate alternatives. Therefore, the decision-making opinions of DMs are greatly preserved, resulting in less information distortion. Analogous to PFSs, quite a few aggregation operators for q-ROFSs have been proposed [34–37]. To deal with both DMs' quantitative and qualitative evaluations in MAGDM, Li et al. [38] proposed q-rung orthopair linguistic sets as well as their aggregation operators. Moreover, Li et al. [39] introduced q-rung picture fuzzy linguistic sets by taking DMs' neutrality degree into consideration.

Due to the extreme complexity of realistic decision-making problems, the abovementioned decision-making methods with q-ROFSs are still insufficient. In reality, it is very common to encounter the following issues: (1) The complexity of decision-making problems causes DMs to be highly hesitant. In quite a few real-life decision-making scenarios, DMs may feel hesitant among a group of values when determining the attribute values in q-ROFSs. By taking such hesitancy into consideration, Torra [40] originated the concept of the hesitant fuzzy set (HFS), in which the membership degree is denoted by several discrete values instead of a single value. Afterwards, Zhu et al. [41] pointed out the drawback of HFS is that it only contains membership degrees. Subsequently, they proposed the concept of the dual hesitant fuzzy set (DHFS), which has both membership and nonmembership degrees. Recently, Wei and Lu [42] extended DHFS to PFS and proposed the concept of dual hesitant Pythagorean fuzzy set (DHPFS) (It is noted that Khan et al. [43] and Liang and Xu [44] also proposed the so-called hesitant Pythagorean fuzzy set, however, their definitions are the same as Wei and Lu's [42] DHPFS). Analogously, DMs may feel that it is difficult to determine membership and nonmembership degrees by single values, as they prefer to use several values to represent them in q-ROFSs. Therefore, this paper proposes the concept of the q-rung dual hesitant fuzzy set (q-RDHFS), which is constructed by a set of q-rung membership degrees and q-rung nonmembership degrees. Compared with DHFS and DHPFS, the proposed q-RDHFS allows the sum and square sum of membership and nonmembership degrees to be greater than one, providing decision makers more freedom to express their assessments. Compared with q-ROFS, the proposed q-RDHFS can effectively deal with DMs' hesitancy when determining membership and nonmembership degrees, consequently resulting in less information loss. Thus, the q-RDHFS exhibits more usefulness, power, and flexibility over DHFS, DHPFS, and q-ROFS. In Section 2, we introduce the concept of q-RDHFS in detail. (2) In most MAGDM, there is a strong correlation between attributes. Thus, in the process of information integration, it is not only necessary to aggregate the attribute values themselves but also to collect the correlation between them. Heronian mean (HM) [45] is the most common information aggregation method that can reflect the correlation between variables. Thus, we extended HM to q-RDHFSs to integrate q-rung dual hesitant fuzzy information. Then, we applied the proposed operators to solve MAGDM problems.

The main significance of this paper is that it expands the theory of q-ROFSs and DHFSs and proposes a new, powerful tool for describing uncertain phenomena, called q-RDHFSs. Compared with many existing fuzzy set theories, the newly proposed q-RDHFSs show great flexibility and effectiveness and can very effectively express the decision-making opinions of DMs in a very hesitant state. We also investigated their applications in MAGDM. The remainder of the paper is organized as follows. Section 2 briefly recalls some basic concepts. Section 3 presents some q rung dual hesitant fuzzy Heronian mean operators. Section 4 introduces a novel approach to MAGDM. Section 5 provides a numerical example to demonstrate the validity and superiority of the proposed method. Finally, Section 6 summarizes the paper.

2. Basic Concepts

2.1. q-Rung Orthopair Fuzzy Set

Definition 1 [33]. *Let X be an ordinary fixed set. A q-ROFS A defined on X is given by*

$$A = \{\langle x, u_A(x), v_A(x)\rangle \,|x \in X\} \tag{1}$$

where $u_A(x)$ and $v_A(x)$ represent the membership and nonmembership degrees, respectively, satisfying $u_A(x) \in [0,1]$, $v_A(x) \in [0,1]$ and $0 \leq u_A(x)^q + v_A(x)^q \leq 1$, $(q \geq 1)$. The indeterminacy degree is defined as $\pi_A(x) = \left(u_A(x)^q + v_A(x)^q - u_A(x)^q v_A(x)^q\right)^{1/q}$. For convenience, $(u_A(x), v_A(x))$ is called a q-rung orthopair fuzzy number (q-ROFN) by Liu and Wang [34], which can be denoted by $A = (u_A, v_A)$.

From Definition 1, it is not difficult to find out that q-ROFS can describe a wider information range than IFSs and PFSs. To illustrate the difference among intuitionistic fuzzy numbers (IFNs), PFNs, and q-ROFNs, we present their space of acceptable membership degrees in Figure 1.

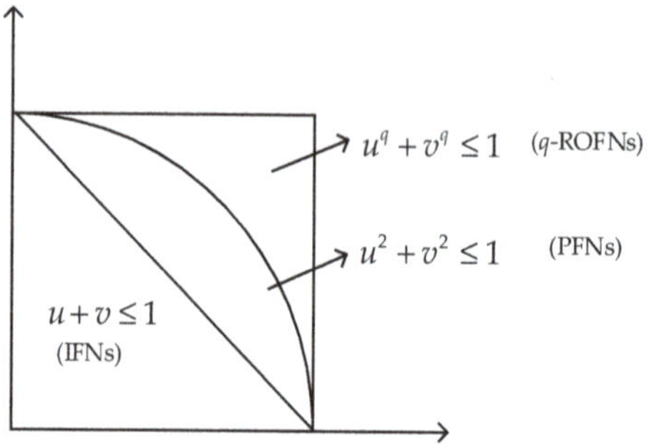

Figure 1. Comparison of grades of IFNs, PFNs, and q-ROFNs.

Figure 1 clearly shows that as the index of u and v increases, the range of information that the fuzzy numbers can describe also grows. Therefore, the q-ROFNs can expand the information that the attributes can describe and widen the space for experts to evaluate alternatives.

Definition 2 [34]. *Let $\widetilde{a}_1 = (u_1, v_1)$, $\widetilde{a}_2 = (u_2, v_2)$ be two q-ROFNs and λ be a positive real number. Then,*

1. $\widetilde{a}_1 \oplus \widetilde{a}_2 = \left(\left(u_1^q + u_2^q - u_1^q u_2^q\right)^{1/q}, v_1 v_2\right).$
2. $\widetilde{a}_1 \otimes \widetilde{a}_2 = \left(u_1 u_2, \left(v_1^q + v_2^q - v_1^q v_2^q\right)^{1/q}\right).$
3. $\lambda \widetilde{a}_1 = \left(\left(1 - \left(1 - u_1^q\right)^{\lambda}\right)^{1/q}, v_1^{\lambda}\right).$
4. $\widetilde{a}_1^{\lambda} = \left(u_1^{\lambda}, \left(1 - \left(1 - v_1^q\right)^{\lambda}\right)^{1/q}\right).$

Definition 3 [34]. *Let $\widetilde{a} = (u_a, v_a)$ be a q-ROFN. Then, the score of $\widetilde{a}$ is defined as $S(\widetilde{a}) = u_a^q - v_a^q$ and the accuracy of $\widetilde{a}$ is defined as $H(\widetilde{a}) = u_a^q + v_a^q$. For any two q-ROFNs, $\widetilde{a}_1 = (u_1, v_1)$ and $\widetilde{a}_2 = (u_2, v_2)$.* Then,

1. If $S(\widetilde{a}_1) > S(\widetilde{a}_2)$, then $\widetilde{a}_1 > \widetilde{a}_2$;
2. If $S(\widetilde{a}_1) = S(\widetilde{a}_2)$, then
   if $H(\widetilde{a}_1) > H(\widetilde{a}_2)$, then $\widetilde{a}_1 > \widetilde{a}_2$;
   if $H(\widetilde{a}_1) = H(\widetilde{a}_2)$, then $\widetilde{a}_1 = \widetilde{a}_2$.

2.2. *q-Rung Dual Hesitant Fuzzy Set*

In this subsection, we introduce q-RDHFS, which is a new extension of q-ROFS and DHFS. Clearly, the proposed q-RDHFS is constructed of a set of membership degrees and several nonmembership degrees.

Definition 4. *Let X be an ordinary fixed set. A q-RDHFS A defined on X is given by*

$$A = \{\langle x, h_A(x), g_A(x) \rangle | x \in X\}, \tag{2}$$

in which $h_A(x)$ and $g_A(x)$ are two sets of values in [0, 1] denoting the possible membership and nonmembership degrees of the element $x \in X$ to the set A, respectively, with the conditions

$$\gamma^q + \eta^q \leq 1 \ (q \geq 1),$$

where $\gamma \in h_A(x)$, $\eta \in g_A(x)$ for all $x \in X$. For convenience, the pair $d(x) = (h_A(x), g_A(x))$ is called a q-RDHFE denoted by $d = (h, g)$ with the conditions $\gamma \in h, \eta \in g$, $0 \leq \gamma, \eta \leq 1$, $0 \leq \gamma^q + \eta^q \leq 1$. Evidently, when q = 2, then q-RDHFS is reduced to Wei and Lu's [42] DHPFS, and when q = 1, then q-RDHFS is reduced to Zhu et al.'s [41] DHFS.

To compare any two q-RDHFEs, in the following, we propose a comparison law for q-RDHFEs.

Definition 5. *Let $d = (h, g)$ be a q-RDHFE, $S(d) = \left(\frac{1}{\#h}\sum\limits_{\gamma \in h}\gamma\right)^q - \left(\frac{1}{\#g}\sum\limits_{\eta \in g}\eta\right)^q$ be the score function of d, and $H(d) = \left(\frac{1}{\#h}\sum\limits_{\gamma \in h}\gamma\right)^q + \left(\frac{1}{\#g}\sum\limits_{\eta \in g}\eta^q\right)^q$ the accuracy function of d, where #h and #g are the numbers of the elements in h and g, respectively. Then, let $d_i = (h_i, g_i)(i = 1, 2)$ be any two q-RDHFEs. Thus, we have the following comparison laws:*

1. *If $S(d_1) > S(d_2)$, then d_1 is superior to d_2, denoted by $d_1 > d_2$;*
2. *If $S(d_1) > S(d_2)$, then*
   if $H(d_1) = H(d_2)$, then d_1 is equivalent to d_2, denoted by $d_1 = d_2$;
   if $H(d_1) > H(d_2)$, then d_1 is superior to d_2, denoted by $d_1 > d_2$.
   In the following, we define some operations of the q-RDHFEs.

Definition 6. *Let $d = (h, g)$, $d_1 = (h_1, g_1)$, and $d_2 = (h_2, g_2)$ be any three of q-RDHFEs and λ be a positive real number. Then,*

1. $d_1 \oplus d_2 = \cup_{\gamma_1 \in h_1, \gamma_2 \in h_2, \eta_1 \in g_1, \eta_2 \in g_2}\left\{\left\{\left(\gamma_1^q + \gamma_2^q - \gamma_1^q\gamma_2^q\right)^{\frac{1}{q}}\right\}, \{\eta_1\eta_2\}\right\}$;
2. $d_1 \otimes d_2 = \cup_{\gamma_1 \in h_1, \gamma_2 \in h_2, \eta_1 \in g_1, \eta_2 \in g_2}\left\{\{\gamma_1\gamma_2\}, \left\{\left(\eta_1^q + \eta_2^q - \eta_1^q\eta_2^q\right)^{\frac{1}{q}}\right\}\right\}$;
3. $\lambda d = \cup_{\gamma \in h, \eta \in g}\left\{\left\{\left(1 - (1 - \gamma^q)^\lambda\right)^{\frac{1}{q}}\right\}, \{\eta^\lambda\}\right\}, \lambda > 0$;
4. $d^\lambda = \cup_{\gamma \in h, \eta \in g}\left\{\{\gamma^\lambda\}, \left\{\left(1 - (1 - \eta^q)^\lambda\right)^{\frac{1}{q}}\right\}\right\}, \lambda > 0$.

2.3. Heronian Mean

The HM was first proposed by Sykora [45] for crisp numbers. It can process the interrelationship between arguments.

Definition 7 [45]. *Let $x_i (i = 1, 2, \ldots, n)$ be a group of real numbers, and $s, t > 0$. Then, the HM is defined as*

$$HM^{s,t}(x_1, x_2, \ldots, x_n) = \left(\frac{2}{n(n+1)}\sum_{i=1}^{n}\sum_{j=i}^{n}x_i^s x_j^t\right)^{\frac{1}{s+t}}. \quad (3)$$

Recently, Yu [46] introduced the concept of geometric Heronian mean (GHM).

Definition 8 [46]. *Let $x_i (i = 1, 2, \ldots, n)$ be a group of numbers, and $s, t > 0$. Then, the GHM is defined as*

$$GHM^{p,q}(x_1, x_2, \ldots, x_n) = \frac{1}{p+q}\left(\prod_{i=1}^{n}\prod_{j=i}^{n}(px_i + qx_j)^{\frac{2}{n(n+1)}}\right). \quad (4)$$

3. The q-Rung Dual Hesitant Fuzzy Heronian Mean Operators

In this subsection, we extend the HM and GHM to q-RDHFSs and propose some new q-rung dual hesitant fuzzy Heronian mean aggregation operators.

3.1. The q-Rung Dual Hesitant Fuzzy Heronian Mean Operator

Definition 9. *Let $s,t \geq 0$ and $d_j = (h_j, g_j)$ $(j = 1,2,...,n)$ be a q-RDHFE. If*

$$q-RDHFHM^{s,t}(d_1, d_2, ..., d_n) = \left(\frac{2}{n(n+1)}\sum_{i=1}^{n}\sum_{j=i}^{n} d_i^s d_j^t\right)^{\frac{1}{s+t}}, \tag{5}$$

then $q - RDHFHM^{s,t}$ is called the q-rung dual hesitant fuzzy Heronian mean (q-RDHFHM) operator.

Based on the operational laws of the q-RDHFEs shown in Definition 6, we can get Theorem 1.

Theorem 1. *Let $s,t \geq 0$ and $d_j = (h_j, g_j)$ $(j = 1,2,...,n)$ be a q-RDHFE. Then, the aggregated value by the q-RDHFHM is also a q-RDHFE, and*

$$\begin{aligned} q-RDHFHM^{s,t}(d_1, d_2, ..., d_n) = \cup_{\gamma_i \in h_i, \gamma_j \in h_j, \eta_i \in g_i, \eta_j \in g_j} \Bigg\{ \Bigg\{ \left(1 - \prod_{i=1}^{n}\prod_{j=i}^{n}\left(1 - \left(\gamma_i^s \gamma_j^t\right)^q\right)^{\frac{2}{n(n+1)}}\right)^{\frac{1}{q(s+t)}} \Bigg\}, \\ \Bigg\{ \left(1 - \left(1 - \prod_{i=1}^{n}\prod_{j=i}^{n}\left(1 - \left(1 - \eta_i^q\right)^s \left(1 - \eta_j^q\right)^t\right)^{\frac{2}{n(n+1)}}\right)^{\frac{1}{s+t}}\right)^{\frac{1}{q}} \Bigg\} \Bigg\} \end{aligned} \tag{6}$$

Proof. From Definition 6, we have

$$d_i^s = \cup_{\gamma_i \in h_i, \eta_i \in g_i} \left\{ \{\gamma_i^s\}, \left\{ \left(1 - \left(1 - \eta_i^q\right)^s\right)^{\frac{1}{q}} \right\} \right\},\ d_j^t = \cup_{\gamma_j \in h_j, \eta_j \in g_j} \left\{ \{\gamma_j^t\}, \left\{ \left(1 - \left(1 - \eta_j^q\right)^t\right)^{\frac{1}{q}} \right\} \right\}.$$

Therefore,

$$d_i^s d_j^t = \cup_{\gamma_i \in h_i, \gamma_j \in h_j, \eta_i \in g_i, \eta_j \in g_j} \left\{ \{\gamma_i^s \gamma_j^t\}, \left\{ \left(1 - \left(1 - \eta_i^q\right)^s \left(1 - \eta_j^q\right)^t\right)^{\frac{1}{q}} \right\} \right\},$$

$$\sum_{j=i}^{n} d_i^s d_j^t = \cup_{\gamma_i \in h_i, \gamma_j \in h_j, \eta_i \in g_i, \eta_j \in g_j} \left\{ \left\{ \left(1 - \prod_{j=i}^{n}\left(1 - \left(\gamma_i^s \gamma_j^t\right)^q\right)\right)^{\frac{1}{q}} \right\}, \left\{ \prod_{j=i}^{n}\left(1 - \left(1 - \eta_i^q\right)^s \left(1 - \eta_j^q\right)^t\right)^{\frac{1}{q}} \right\} \right\},$$

and

$$\sum_{i=1}^{n}\sum_{j=i}^{n} d_i^s d_j^t = \cup_{\gamma_i \in h_i, \gamma_j \in h_j, \eta_i \in g_i, \eta_j \in g_j} \left\{ \left\{ \left(1 - \prod_{i=1}^{n}\prod_{j=i}^{n}\left(1 - \left(\gamma_i^s \gamma_j^t\right)^q\right)\right)^{\frac{1}{q}} \right\}, \left\{ \prod_{i=1}^{n}\prod_{j=i}^{n}\left(1 - \left(1 - \eta_i^q\right)^s \left(1 - \eta_j^q\right)^t\right)^{\frac{1}{q}} \right\} \right\},$$

$$\begin{aligned} \frac{2}{n(n+1)}\sum_{i=1}^{n}\sum_{j=i}^{n} d_i^s d_j^t = \cup_{\gamma_i \in h_i, \gamma_j \in h_j, \eta_i \in g_i, \eta_j \in g_j} \Bigg\{ \Bigg\{ \left(1 - \prod_{i=1}^{n}\prod_{j=i}^{n}\left(1 - \left(\gamma_i^s \gamma_j^t\right)^q\right)^{\frac{2}{n(n+1)}}\right)^{\frac{1}{q}} \Bigg\}, \\ \Bigg\{ \prod_{i=1}^{n}\prod_{j=i}^{n}\left(1 - \left(1 - \eta_i^q\right)^s \left(1 - \eta_j^q\right)^t\right)^{\frac{2}{qn(n+1)}} \Bigg\} \Bigg\}. \end{aligned}$$

Thus,

$$\left(\frac{2}{n(n+1)}\sum_{i=1}^{n}\sum_{j=i}^{n} d_i^s d_j^t\right)^{\frac{1}{s+t}} = \cup_{\gamma_i\in h_i,\gamma_j\in h_j,\eta_i\in g_i,\eta_j\in g_j}\left\{\left\{\left(1-\prod_{i=1}^{n}\prod_{j=i}^{n}\left(1-\left(\gamma_i^s\gamma_j^t\right)^q\right)^{\frac{2}{n(n+1)}}\right)^{\frac{1}{q(s+t)}}\right\},\right.$$

$$\left.\left\{\left(1-\left(1-\prod_{i=1}^{n}\prod_{j=i}^{n}\left(1-\left(1-\eta_i^q\right)^s\left(1-\eta_j^q\right)^t\right)^{\frac{2}{n(n+1)}}\right)^{\frac{1}{s+t}}\right)^{\frac{1}{q}}\right\}\right\}.$$

The q-RDHFHM operator has the following properties. □

Theorem 2. *(Monotonicity) Let d_j and d'_j be two collections of q-RDHFEs. If $d_j \geq d'_j$ for all $j = 1, 2, \ldots, n$, then*

$$q - RDHFHM^{s,t}(d_1, d_2, \ldots, d_n) \geq q - RDHFHM^{s,t}(d'_1, d'_2, \ldots, d'_n) \tag{7}$$

Proof. Since $d_i \geq d'_i$ and $d_j \geq d'_j$ for $i = 1, 2, \ldots, n$ and $j = i, i+1, \ldots, n$, we have

$$d_i^s d_j^t \geq d'^s_i d'^t_j.$$

Then,

$$\sum_{i=1}^{n}\sum_{j=i}^{n} d_i^s d_j^t \geq \sum_{i=1}^{n}\sum_{j=i}^{n} d'^s_i d'^t_j,$$

and

$$\frac{2}{n(n+1)}\sum_{i=1}^{n}\sum_{j=i}^{n} d_i^s d_j^t \geq \frac{2}{n(n+1)}\sum_{i=1}^{n}\sum_{j=i}^{n} d'^s_i d'^t_j.$$

So,

$$\left(\frac{2}{n(n+1)}\sum_{i=1}^{n}\sum_{j=i}^{n} d_i^s d_j^t\right)^{\frac{1}{s+t}} \geq \left(\frac{2}{n(n+1)}\sum_{i=1}^{n}\sum_{j=t}^{n} d'^s_i d'^t_j\right)^{\frac{1}{s+t}}.$$

i.e.,

$$q - RDHFHM^{s,t}(d_1, d_2, \ldots, d_n) \geq q - RDHFHM^{s,t}(d'_1, d'_2, \ldots, d'_n).$$

□

Theorem 3. *(Idempotency) Let $d_j = (h_j, g_j)$, $j = 1, 2, \ldots, n$ be a collection of q-RDHFEs. If all the q-RDHFEs are equal, i.e., $d_j = d = (h, g)$, then*

$$q - RDHFHM^{s,t}(d_1, d_2, \ldots, d_n) = d. \tag{8}$$

Proof. Since $d_j = d$ for all i, we have

$$q - RDHFHM^{s,t}(d_1, d_2, \ldots, d_n) = \left(\frac{2}{n(n+1)}\sum_{i=1}^{n}\sum_{j=i}^{n} d_i^s d_j^t\right)^{\frac{1}{s+t}} = \left(\frac{2}{n(n+1)}\sum_{i=1}^{n}\sum_{j=i}^{n} d^{s+t}\right)^{\frac{1}{s+t}} = \left(d^{s+t}\right)^{\frac{1}{s+t}} = d.$$

□

Theorem 4. *(Boundedness) Let $d_j = (h_j, g_j)$, $j = 1, 2, \ldots, n$ be a collection of q-RDHFEs. If $d^+ = \max_j d_j$ and $d^- = \min_j d_j$, then*

$$d^+ \geq q - RDHFHM^{s,t}(d_1, d_2, \ldots, d_n) \geq d^-. \tag{9}$$

Proof. According to the Theorems 2 and 3, we can get

$$q-RDHFHM^{s,t}(d_1,d_2,\ldots,d_n) \le q-RDHFHM^{s,t}(d^+,d^+,\ldots,d^+),$$

and

$$q-RDHFHM^{s,t}(d_1,d_2,\ldots,d_n) \ge q-RDHFHM^{s,t}(d^+,d^+,\ldots,d^+).$$

Thus, we can get

$$d^+ \ge q-RDHFHM^{s,t}(d_1,d_2,\ldots,d_n) \ge d^-.$$

The advantages of q-RDHFHM are that it not only reflects the hesitation of DMs in the decision-making process and captures the correlation between attribute values, but it also shows great generality and flexibility. In the following, we can discuss some special cases of the q-RDHFHM operator. □

1. If $t \to 0$, then the q-RDHFHM reduces to a q-rung dual hesitant fuzzy generalized linear descending weighted mean operator, and we can obtain

$$\begin{aligned} q-RDHFHM^{s,0}(d_1,d_2,\ldots,d_n) &= \lim_{t\to 0}\left\{ \cup_{\gamma_i\in h_i,\gamma_j\in h_j,\eta_i\in g_i,\eta_j\in g_j}\left\{\left\{\left(1-\prod_{i=1}^{n}\prod_{j=i}^{n}\left(1-\left(\gamma_i^s\gamma_j^t\right)^q\right)^{\frac{2}{n(n+1)}}\right)^{\frac{1}{q(s+t)}}\right\},\right.\right. \\ &\left.\left.\left\{\left(1-\left(1-\prod_{i=1}^{n}\prod_{j=i}^{n}\left(1-\left(1-\eta_i^q\right)^s\left(1-\eta_j^q\right)^t\right)^{\frac{2}{n(n+1)}}\right)^{\frac{1}{s+t}}\right)^{\frac{1}{q}}\right\}\right\}\right\} \\ &= \cup_{\gamma_i\in h_i,\eta_i\in g_i}\left\{\left\{\left(1-\left(\prod_{i=1}^{n}\left(1-(\gamma_i^s)^q\right)^{n+1-i}\right)^{\frac{2}{n(n+1)}}\right)^{\frac{1}{qs}}\right\},\left\{\left(1-\left(1-\left(\prod_{i=1}^{n}\left(1-\left(1-\eta_i^q\right)^s\right)^{n+1-i}\right)^{\frac{2}{n(n+1)}}\right)^{\frac{1}{s}}\right)^{\frac{1}{q}}\right\}\right\} \end{aligned} \tag{10}$$

Evidently, it is equivalent to weight the information $(d_1,d_2,\ldots,d_n)$ with the weight values $(n,\ n-1,\ \ldots,\ 1)$.

2. If $s \to 0$, then the q-RDHFHM reduces to a q-rung dual hesitant fuzzy generalized liner ascending weighted mean operator, and we can obtain

$$\begin{aligned} q-RDHFHM^{0,t}(d_1,d_2,\ldots,d_n) &= \lim_{s\to 0}\left\{ \cup_{\gamma_i\in h_i,\gamma_j\in h_j,\eta_i\in g_i,\eta_j\in g_j}\left\{\left\{\left(1-\prod_{i=1}^{n}\prod_{j=i}^{n}\left(1-\left(\gamma_i^s\gamma_j^t\right)^q\right)^{\frac{2}{n(n+1)}}\right)^{\frac{1}{q(s+t)}}\right\},\right.\right. \\ &\left.\left.\left\{\left(1-\left(1-\prod_{i=1}^{n}\prod_{j=i}^{n}\left(1-\left(1-\eta_i^q\right)^s\left(1-\eta_j^q\right)^t\right)^{\frac{2}{n(n+1)}}\right)^{\frac{1}{s+t}}\right)^{\frac{1}{q}}\right\}\right\}\right\} \\ &= \cup_{\gamma_j\in h_j,\eta_j\in g_j}\left\{\left\{\left(1-\left(\prod_{i=1}^{n}\left(1-\left(\gamma_j^t\right)^q\right)^{i}\right)^{\frac{2}{n(n+1)}}\right)^{\frac{1}{qt}}\right\},\left\{\left(1-\left(1-\left(\prod_{i=1}^{n}\left(1-\left(1-\eta_j^q\right)^t\right)^{i}\right)^{\frac{2}{n(n+1)}}\right)^{\frac{1}{t}}\right)^{\frac{1}{q}}\right\}\right\} \end{aligned} \tag{11}$$

Obviously, it is equivalent to weight the information $(d_1,d_2,\ldots,d_n)$ with weight values $(1,\ 2,\ \ldots,\ n)$.

3. If $s=t=\frac{1}{2}$, then the q-RDHFHM reduces to a q-rung dual hesitant fuzzy basic Heronian mean operator, and we can obtain

$$\begin{aligned} &q-RDHFHM^{\frac{1}{2},\frac{1}{2}}(d_1,d_2,\ldots,d_n) = \\ &\cup_{\gamma_i\in h_i,\gamma_j\in h_j,\eta_i\in g_i,\eta_j\in g_j}\left\{\left\{\left(1-\prod_{i=1}^{n}\prod_{j=i}^{n}\left(1-\left(\sqrt{\gamma_i\gamma_j}\right)^q\right)^{\frac{2}{n(n+1)}}\right)^{\frac{1}{q}}\right\},\left\{\prod_{i=1}^{n}\prod_{j=i}^{n}\left(1-\sqrt{\left(1-\eta_i^q\right)\left(1-\eta_j^q\right)}\right)^{\frac{2}{nq(n+1)}}\right\}\right\} \end{aligned} \tag{12}$$

4. If $s = t = 1$, then the q-RDHFHM reduces to a q-rung dual hesitant fuzzy line Heronian mean operator. It follows that

$$q-RDHFHM^{1,1}(d_1, d_2, \ldots, d_n) = \cup_{\gamma_i \in h_i, \gamma_j \in h_j, \eta_i \in g_i, \eta_j \in g_j} \left\{ \begin{array}{l} \left\{ \left(1 - \prod\limits_{i=1}^{n} \prod\limits_{j=i}^{n} \left(1 - (\gamma_i \gamma_j)^q \right)^{\frac{2}{n(n+1)}} \right)^{\frac{1}{2q}} \right\}, \\ \left\{ \left(1 - \left(1 - \prod\limits_{i=1}^{n} \prod\limits_{j=i}^{n} \left(1 - \left(1 - \eta_i^q \right) \left(1 - \eta_j^q \right) \right)^{\frac{2}{n(n+1)}} \right)^{\frac{1}{2}} \right)^{\frac{1}{q}} \right\} \end{array} \right\} \tag{13}$$

5. If $q = 2$, then the q-RDHFHM reduces to a dual hesitant Pythagorean fuzzy Heronian mean operator. So, we can obtain

$$q-RDHFHM^{s,t}(d_1, d_2, \ldots, d_n) = \cup_{\gamma_i \in h_i, \gamma_j \in h_j, \eta_i \in g_i, \eta_j \in g_j} \left\{ \begin{array}{l} \left\{ \left(1 - \prod\limits_{i=1}^{n} \prod\limits_{j=i}^{n} \left(1 - \left(\gamma_i^s \gamma_j^t \right)^2 \right)^{\frac{2}{n(n+1)}} \right)^{\frac{1}{2(s+t)}} \right\}, \\ \left\{ \left(1 - \left(1 - \prod\limits_{i=1}^{n} \prod\limits_{j=i}^{n} \left(1 - \left(1 - \eta_i^2 \right)^s \left(1 - \eta_j^2 \right)^t \right)^{\frac{2}{n(n+1)}} \right)^{\frac{1}{s+t}} \right)^{\frac{1}{2}} \right\} \end{array} \right\} \tag{14}$$

6. If $q = 1$, then the q-RDHFHM reduces to the dual hesitant fuzzy Heronian mean operator proposed by Yu et al. [47]. It follows that

$$q-RDHFHM^{s,t}(d_1, d_2, \ldots, d_n) = \cup_{\gamma_i \in h_i, \gamma_j \in h_j, \eta_i \in g_i, \eta_j \in g_j} \left\{ \begin{array}{l} \left\{ \left(1 - \prod\limits_{i=1}^{n} \prod\limits_{j=i}^{n} \left(1 - \left(\gamma_i^s \gamma_j^t \right) \right)^{\frac{2}{n(n+1)}} \right)^{\frac{1}{s+t}} \right\}, \\ \left\{ \left(1 - \left(1 - \prod\limits_{i=1}^{n} \prod\limits_{j=i}^{n} \left(1 - (1 - \eta_i)^s (1 - \eta_j)^t \right)^{\frac{2}{n(n+1)}} \right)^{\frac{1}{s+t}} \right) \right\} \end{array} \right\} \tag{15}$$

3.2. The q-Rung Dual Hesitant Fuzzy Weighted Heronian Mean (q-RDHFWHM) Operator

Definition 10. *Let $s, t \geq 0$ and $d_j = (h_j, g_j)$ $(j = 1, 2, \ldots, n)$ be a collection q-RDHFEs. The q-RDHFWHM operator is defined as*

$$q - RDHFWHM^{s,t}(d_1, d_2, \ldots, d_n) = \left(\frac{2}{n(n+1)} \sum_{i=1}^{n} \sum_{j=i}^{n} (nw_i d_i)^s (nw_j d_j)^t \right)^{\frac{1}{s+t}}, \tag{16}$$

where $w = (w_1, w_2, \ldots, w_n)^T$ is the weight vector of $(d_1, d_2, \ldots, d_n)$, satisfying $w_j \in [0, 1]$, $\sum\limits_{j=1}^{n} w_j = 1$.

According to the operations for q-RDHFEs, the following theorem can be obtained.

Theorem 5. *Let $s, t \geq 0$ and $d_j = (h_j, g_j)$ $(j = 1, 2, \ldots, n)$ be a collection q-RDHFEs. The aggregated value by the q-RDHFWHM is also a q-RDHFE and*

$$q - RDHFWHM^{s,t}(d_1, d_2, \ldots, d_n) = \cup_{\gamma_i \in h_i, \gamma_j \in h_j, \eta_i \in g_i, \eta_j \in g_j} \left\{ \begin{array}{l} \left\{ \left(1 - \prod\limits_{i=1}^{n} \prod\limits_{j=i}^{n} \left(1 - \left(\left(1 - \left(1 - \gamma_i^q \right)^{nw_i} \right)^{\frac{1}{s}} \right) \left(\left(1 - \left(1 - \gamma_j^q \right)^{nw_j} \right)^{\frac{1}{t}} \right) \right)^{\frac{2}{n(n+1)}} \right)^{\frac{1}{q(s+t)}} \right\}, \\ \left\{ \left(1 - \left(1 - \prod\limits_{i=1}^{n} \prod\limits_{j=i}^{n} \left(1 - \left(1 - \eta_i^{qnw_i} \right)^s \left(1 - \eta_j^{qnw_j} \right)^t \right)^{\frac{2}{n(n+1)}} \right)^{\frac{1}{s+t}} \right)^{\frac{1}{q}} \right\} \end{array} \right\} \tag{17}$$

The proof of Theorem 5 is similar to that of Theorem 1.

Theorem 6. *Suppose* $w = (1/n, 1/n, ..., 1/n)^T$. *Then,*

$$q - RDHFWHM^{s,t}(d_1, d_2, ..., d_n) = q - RDHFHM^{s,t}(d_1, d_2, ..., d_n). \tag{18}$$

Proof. Since $w = (1/n, 1/n, ..., 1/n)^T$, then according to Equation (20),

$$\begin{aligned} q - RDHFWHM^{s,t}(d_1, d_2, ..., d_n) = & \left(\frac{2}{n(n+1)} \sum_{i=1}^{n} \sum_{j=i}^{n} (nw_i d_i)^s (nw_j d_j)^t \right)^{\frac{1}{s+t}} \\ &= \left(\frac{2}{n(n+1)} \sum_{i=1}^{n} \sum_{j=i}^{n} \left(n\frac{1}{n}d_i\right)^s \left(n\frac{1}{n}d_j\right)^t \right)^{\frac{1}{s+t}} \\ &= \left(\frac{2}{n(n+2)} \sum_{i=1}^{n} \sum_{j=i}^{n} d_i^s d_j^t \right)^{\frac{1}{s+t}} = q - RDHFWHM^{s,t}(d_1, d_2, ..., d_n) \end{aligned}$$

Moreover, it is easy to prove that the q-RDHFWHM operator has the properties of monotonicity and boundedness. □

3.3. The q-Rung Dual Hesitant Fuzzy Geometric Heronian Mean Operator

In this subsection, we shall extend the GHM to aggregate q-rung dual hesitant fuzzy information.

Definition 11. *Let* $s, t \geq 0$ *and* $d_j = (h_j, g_j)$ $(j = 1, 2, ..., n)$ *be a q-RDHFE. Then, the q-RDHFGHM operator is defined as*

$$q - RDHFGHM^{s,t}(d_1, d_2, ..., d_n) = \frac{1}{s+t} \left(\prod_{i=1}^{n} \prod_{j=i}^{n} (sd_i + td_j)^{\frac{2}{n(n+1)}} \right). \tag{19}$$

Based on the operational laws of q-RDHFEs, the following theorem can be obtained.

Theorem 7. *Let* $s, t \geq 0$ *and* $d_j = (h_j, g_j)$ $(j = 1, 2, ..., n)$ *be a q-RDHFE. The aggregated value by the q-RDHFGHM is also q-RDHFE and*

$$\begin{aligned} & q - RDHFGHM^{s,t}(d_1, d_2, ..., d_n) = \\ & \cup_{\gamma_i \in h_i, \gamma_j \in h_j, \eta_i \in g_i, \eta_j \in g_j} \left\{ \begin{array}{l} \left\{ \left(1 - \left(1 - \prod_{i=1}^{n} \prod_{j=i}^{n} \left(1 - \left(1 - \gamma_i^q\right)^s \left(1 - \gamma_j^q\right)^t \right)^{\frac{2}{n(n+1)}} \right)^{\frac{1}{s+t}} \right)^{\frac{1}{q}} \right\}, \\ \left\{ \left(1 - \prod_{i=1}^{n} \prod_{j=i}^{n} \left(1 - \left(\eta_i^s \eta_j^t\right)^q \right)^{\frac{2}{n(n+1)}} \right)^{\frac{1}{q(s+t)}} \right\} \end{array} \right\} \end{aligned} \tag{20}$$

Proof. According to Definition 6, we can get

$$sd_i = \cup_{\gamma_i \in h_i, \eta_i \in g_i} \left\{ \left\{ \left(1 - \left(1 - \gamma_i^q\right)^s \right)^{\frac{1}{q}} \right\}, \left\{ \eta_i^s \right\} \right\} \text{ and}$$
$$td_j = \cup_{\gamma_j \in h_j, \eta_j \in g_j} \left\{ \left\{ \left(1 - \left(1 - \gamma_j^q\right)^t \right)^{\frac{1}{q}} \right\}, \left\{ \eta_j^t \right\} \right\}.$$

Then,

$$sd_i + td_j = \cup_{\gamma_i \in h_i, \gamma_j \in h_j, \eta_i \in g_i, \eta_j \in g_j} \left\{ \left\{ \left(1 - \left(1 - \gamma_i^q\right)^s \left(1 - \gamma_j^q\right)^t \right)^{\frac{1}{q}} \right\}, \left\{ \eta_i^s \eta_j^t \right\} \right\},$$

and

$$(sd_i + td_j)^{\frac{2}{n(n+1)}} = \cup_{\gamma_i \in h_i, \gamma_j \in h_j, \eta_i \in g_i, \eta_j \in g_j} \left\{ \left\{ \left(1 - \left(1 - \gamma_i^q\right)^s \left(1 - \gamma_j^q\right)^t \right)^{\frac{2}{n(n+1)}} \right\}, \left\{ \left(1 - \left(1 - \left(\eta_i^s \eta_j^t\right)^q\right)^{\frac{2}{n(n+1)}} \right)^{\frac{1}{q}} \right\} \right\}.$$

Thus,

$$\prod_{j=i}^{n} (sd_i + td_j)^{\frac{2}{n(n+1)}} = \cup_{\gamma_i \in h_i, \gamma_j \in h_j, \eta_i \in g_i, \eta_j \in g_j} \left\{ \left\{ \prod_{j=i}^{n} \left(1 - \left(1 - \gamma_i^q\right)^s \left(1 - \gamma_j^q\right)^t \right)^{\frac{2}{n(n+1)}} \right\}, \left\{ \left(1 - \prod_{j=i}^{n} \left(1 - \left(\eta_i^s \eta_j^t\right)^q\right)^{\frac{2}{n(n+1)}} \right)^{\frac{1}{q}} \right\} \right\},$$

and

$$\prod_{i=1}^{n} \prod_{j=i}^{n} (sd_i + td_j)^{\frac{2}{n(n+1)}} = \cup_{\gamma_i \in h_i, \gamma_j \in h_j, \eta_i \in g_i, \eta_j \in g_j} \left\{ \left\{ \prod_{i=1}^{n} \prod_{j=i}^{n} \left(1 - \left(1 - \gamma_i^q\right)^s \left(1 - \gamma_j^q\right)^t \right)^{\frac{2}{n(n+1)}} \right\}, \left\{ \left(1 - \prod_{i=1}^{n} \prod_{j=i}^{n} \left(1 - \left(\eta_i^s \eta_j^t\right)^q\right)^{\frac{2}{n(n+1)}} \right)^{\frac{1}{q}} \right\} \right\}$$

Therefore,

$$\frac{1}{s+t} \left(\prod_{i=1}^{n} \prod_{j=i}^{n} (sd_i + td_j)^{\frac{2}{n(n+1)}} \right) = \cup_{\gamma_i \in h_i, \gamma_j \in h_j, \eta_i \in g_i, \eta_j \in g_j} \left\{ \left\{ \left(1 - \left(1 - \prod_{i=1}^{n} \prod_{j=i}^{n} \left(1 - \left(1 - \gamma_i^q\right)^s \left(1 - \gamma_j^q\right)^t \right)^{\frac{2}{n(n+1)}} \right)^{\frac{1}{s+t}} \right)^{\frac{1}{q}} \right\}, \left\{ \left(1 - \prod_{i=1}^{n} \prod_{j=i}^{n} \left(1 - \left(\eta_i^s \eta_j^t\right)^q\right)^{\frac{2}{n(n+1)}} \right)^{\frac{1}{q(s+t)}} \right\} \right\}.$$

In the following, we present some desirable properties of the q-RDHFGHM operator. □

Theorem 8. *(Monotonicity) Let $d_j = (d_1, d_2, ..., d_n)$ and $d'_j = (d'_1, d'_2, ..., d'_n)$ be two collections of q-RDHFEs. If $d_j \geq d'_j$ for all $j = 1, 2, ..., n$, then*

$$q - RDHFGHM^{s,t}(d_1, d_2, ..., d_n) \geq q - RDHFGHM^{s,t}(d'_1, d'_2, ..., d'_n). \tag{21}$$

The proof of the Theorem 8 is similar to that of Theorem 2, which is omitted here.

Theorem 9. *(Idempotency) Let $d_j = (h_j, g_j)$, $j = 1, 2, ..., n$ be a collection of q-RDHFEs. If all the q-RDHFEs are equal, i.e., $d_j = d = (h, g)$, then*

$$q - RDHFGHM^{s,t}(d_1, d_2, ..., d_n) = d. \tag{22}$$

Proof. Since $d_j = d$ for all i, we have

$$\begin{aligned} q - RDHFGHM^{s,t}(d_1, d_2, ..., d_n) &= \frac{1}{s+t} \left(\prod_{i=1}^{n} \prod_{j=i}^{n} (sd_i + td_j)^{\frac{2}{n(n+1)}} \right) = \frac{1}{s+t} \left(\prod_{i=1}^{n} \prod_{j=i}^{n} (sd + td)^{\frac{2}{n(n+1)}} \right) \\ &= \frac{1}{s+t} \left(\prod_{i=1}^{n} \prod_{j=i}^{n} ((s+t)d)^{\frac{2}{n(n+1)}} \right) = \frac{1}{s+t}((s+t)d) = d \end{aligned}$$

□

Theorem 10. *(Boundedness) Let* $d_j = (h_j, g_j), j = 1, 2, \ldots, n$ *be a collection of q-RDHFEs. If* $d^+ = \max_j d_j$ *and* $d^- = \min_j d_j$*, then*

$$d^+ \geq q - RDHFGHM^{s,t}(d_1, d_2, \ldots, d_n) \geq d^-. \tag{23}$$

Analogous to the q-RDHFHM operator, the proposed q-RDHFGHM operator also exhibits high generality and flexibility. In the following, we shall discuss some special cases of the q-RDHFGHM operator.

1. If $t \to 0$, then the q-RDHFGHM reduces to a q-rung dual hesitant fuzzy generalized geometric linear descending weighted mean operator, and we can obtain

$$\begin{aligned}
&q - RDHFGHM^{s,0}(d_1, d_2, \ldots, d_n) \\
&= \lim_{t \to 0}\left\{ \cup_{\gamma_i \in h_i, \gamma_j \in h_j, \eta_i \in g_i, \eta_j \in g_j} \left\{\left\{ \left(1 - \left(1 - \prod_{i=1}^{n}\prod_{j=i}^{n}\left(1 - \left(1 - \gamma_i^q\right)^s \left(1 - r_j^q\right)^t\right)^{\frac{2}{n(n+1)}}\right)^{\frac{1}{s+t}}\right)^{\frac{1}{q}} \right\},\right.\right. \\
&\qquad \left.\left.\left\{\left(1 - \prod_{i=1}^{n}\prod_{j=i}^{n}\left(1 - \left(\mu_i^s \eta_j^t\right)^q\right)^{\frac{2}{n(n+1)}}\right)^{\frac{1}{q(s+t)}}\right\}\right\}\right\} \\
&= \cup_{\gamma_i \in h_i, \gamma_j \in h_j, \eta_i \in g_i, \eta_j \in g_j} \left\{\left\{ \left(1 - \left(1 - \left(\prod_{i=1}^{n}\left(1 - \left(1 - \gamma_i^q\right)^s\right)^{n+1-i}\right)^{\frac{2}{n(n+1)}}\right)^{\frac{1}{s}}\right)^{\frac{1}{q}} \right\},\right. \\
&\qquad \left.\left\{\left(1 - \left(\prod_{i=1}^{n}\left(1 - \left(\eta_i^s\right)^q\right)^{n+1-i}\right)^{\frac{2}{n(n+1)}}\right)^{\frac{1}{qs}}\right\}\right\}
\end{aligned} \tag{24}$$

2. If $s \to 0$, the q-RDHFGHM reduces to a q-rung dual hesitant fuzzy generalized geometric liner ascending weighted mean operator, and we can obtain

$$\begin{aligned}
&q - RDHFGHM^{0,t}(d_1, d_2, \ldots, d_n) \\
&= \lim_{s \to 0}\left\{ \cup_{\gamma_i \in h_i, \gamma_j \in h_j, \eta_i \in g_i, \eta_j \in g_j} \left\{\left\{ \left(1 - \left(1 - \prod_{i=1}^{n}\prod_{j=i}^{n}\left(1 - \left(1 - \gamma_i^q\right)^s \left(1 - \gamma_j^q\right)^t\right)^{\frac{2}{n(n+1)}}\right)^{\frac{1}{s+t}}\right)^{\frac{1}{q}} \right\},\right.\right. \\
&\qquad \left.\left.\left\{\left(1 - \prod_{i=1}^{n}\prod_{j=i}^{n}\left(1 - \left(\eta_i^s \eta_j^t\right)^q\right)^{\frac{2}{n(n+1)}}\right)^{\frac{1}{q(s+t)}}\right\}\right\}\right\} \\
&= \cup_{\gamma_i \in h_i, \gamma_j \in h_j, \eta_i \in g_i, \eta_j \in g_j} \left\{\left\{ \left(1 - \left(1 - \left(\prod_{i=1}^{n}\left(1 - \left(1 - \gamma_j^q\right)^t\right)^{i}\right)^{\frac{2}{n(n+1)}}\right)^{\frac{1}{t}}\right)^{\frac{1}{q}} \right\},\right. \\
&\qquad \left.\left\{\left(1 - \left(\prod_{i=1}^{n}\left(1 - \left(\eta_j^t\right)^q\right)^{i}\right)^{\frac{2}{n(n+1)}}\right)^{\frac{1}{qt}}\right\}\right\}
\end{aligned} \tag{25}$$

3. If $s = t = \frac{1}{2}$, the q-RDHFGHM reduces to a q-rung dual hesitant fuzzy basic geometric Heronian mean operator, and we can obtain

$$\begin{aligned}
&q - RDHFGHM^{\frac{1}{2},\frac{1}{2}}(d_1, d_2, \ldots, d_n) \\
&= \cup_{\gamma_i \in h_i, \gamma_j \in h_j, \eta_i \in g_i, \eta_j \in g_j} \left\{\left\{\prod_{i=1}^{n}\prod_{j=i}^{n}\left(1 - \sqrt{\left(1 - \gamma_i^q\right)\left(1 - \gamma_j^q\right)}\right)^{\frac{2}{nq(n+1)}}\right\}, \left\{\left(1 - \prod_{i=1}^{n}\prod_{j=i}^{n}\left(1 - \left(\sqrt{\eta_i \eta_j}\right)^q\right)^{\frac{2}{n(n+1)}}\right)^{\frac{1}{q}}\right\}\right\}
\end{aligned} \tag{26}$$

4. If $s = t = 1$, the q-RDHFGHM reduces to a q-rung dual hesitant fuzzy line Heronian mean operator, and it follows that

$$q - RDHFGHM^{1,1}(d_1, d_2, \ldots, d_n) = \cup_{\gamma_i \in h_i, \gamma_j \in h_j, \eta_i \in g_i, \eta_j \in g_j} \left\{ \left\{ \left(1 - \left(1 - \prod_{i=1}^{n} \prod_{j=i}^{n} \left(1 - \left(1 - \gamma_i^q\right)\left(1 - \gamma_j^q\right) \right)^{\frac{2}{n(n+1)}} \right)^{\frac{1}{2}} \right)^{\frac{1}{q}} \right\}, \left\{ \left(1 - \prod_{i=1}^{n} \prod_{j=i}^{n} \left(1 - \left(\eta_i \eta_j\right)^q \right)^{\frac{2}{n(n+1)}} \right)^{\frac{1}{2q}} \right\} \right\} \tag{27}$$

5. If $q = 2$, then the q-RDHFGHM reduces to the dual hesitant Pythagorean fuzzy Heronian mean operator, and can we can obtain

$$q - RDHFGHM^{s,t}(d_1, d_2, \ldots, d_n) = \cup_{\gamma_i \in h_i, \gamma_j \in h_j, \eta_i \in g_i, \eta_j \in g_j} \left\{ \left\{ \left(1 - \left(1 - \prod_{i=1}^{n} \prod_{j=i}^{n} \left(1 - \left(1 - \gamma_i^2\right)^s \left(1 - \gamma_j^2\right)^t \right)^{\frac{2}{n(n+1)}} \right)^{\frac{1}{s+t}} \right)^{\frac{1}{2}} \right\}, \left\{ \left(1 - \prod_{i=1}^{n} \prod_{j=i}^{n} \left(1 - \left(\eta_i^s \eta_j^t\right)^2 \right)^{\frac{2}{n(n+1)}} \right)^{\frac{1}{2(s+t)}} \right\} \right\} \tag{28}$$

6. If $q = 1$, then the q-RDHFGHM reduces to the dual hesitant fuzzy Heronian mean operator proposed by Yu et al. [47], and it follows that

$$q - RDHFHM^{s,t}(d_1, d_2, \ldots, d_n) = \cup_{\gamma_i \in h_i, \gamma_j \in h_j, \eta_i \in g_i, \eta_j \in g_j} \left\{ \left\{ 1 - \left(1 - \prod_{i=1}^{n} \prod_{j=i}^{n} \left(1 - \left(1 - \gamma_i\right)^s \left(1 - \gamma_j\right)^t \right)^{\frac{2}{n(n+1)}} \right)^{\frac{1}{s+t}} \right\}, \left\{ \left(1 - \prod_{i=1}^{n} \prod_{j=i}^{n} \left(1 - \left(\eta_i^s \eta_j^t\right) \right)^{\frac{2}{n(n+1)}} \right)^{\frac{1}{s+t}} \right\} \right\} \tag{29}$$

Similarly, the q-RDHFGHM does not consider the importance of the input arguments, which means the weights of the aggregated q-RDHFGHM are not taken into consideration. However, in real decision-making problems, the weight vector of the aggregated values plays an important role in the final ranking orders. Therefore, we propose the q-rung dual hesitant fuzzy weighted geometric Heronian mean (q-RDHFWGHM) operator, which can take the weights of the aggregated q-RDHFEs into account.

3.4. The q-Rung Dual Hesitant Fuzzy Weighted Geometric Heronian Mean Operator

Definition 12. *Let $s, t \geq 0$ and $d_j = (h_j, g_j)$ $(j = 1, 2, \ldots, n)$ be a collection of q-RDHFEs:*

$$q - RDHFWGHM^{s,t}(d_1, d_2, \ldots, d_n) = \frac{1}{s+t} \left(\prod_{i=1}^{n} \prod_{j=i}^{n} \left(s(d_i)^{nw_i} + t(d_j)^{nw_j} \right)^{\frac{2}{n(n+1)}} \right), \tag{30}$$

where $w = (w_1, w_2, \ldots, w_n)^T$ is the weight vector of $(d_1, d_2, \ldots, d_n)$, satisfying $w_j \in [0, 1]$, $\sum_{j=1}^{n} w_j = 1$.

Based on the operational laws of q-RDHFEs, the following theorem can be obtained.

Theorem 11. *Let $s,t \geq 0$ and $d_j = (h_j, g_j)$ $(j = 1, 2, ..., n)$ be a collection of q-RDHFEs. The aggregated value by the q-RDHFWGHM is also a q-RDHFE and*

$$q - RDHFWGHM^{s,t}(d_1, d_2, ..., d_n) = \\ \cup_{\gamma_i \in h_i, \gamma_j \in h_j, \eta_i \in g_i, \eta_j \in g_j} \left\{ \left\{ \left(1 - \left(1 - \prod_{i=1}^{n} \prod_{j=i}^{n} \left(1 - \left(1 - \gamma_i^{qnw_i}\right)^s \left(1 - \gamma_j^{qnw_j}\right)^t \right)^{\frac{2}{n(n+1)}} \right)^{\frac{1}{s+t}} \right)^{\frac{1}{q}} \right\}, \right. \\ \left. \left\{ \left(1 - \left(1 - \prod_{i=1}^{n} \prod_{j=i}^{n} \left(1 - \left(1 - \gamma_i^{qnw_i}\right)^s \left(1 - \gamma_j^{qnw_j}\right)^t \right)^{\frac{2}{n(n+1)}} \right)^{\frac{1}{s+t}} \right)^{\frac{1}{q}} \right\} \right\} \quad (31)$$

The proof of Theorem 11 is similar to Theorem 5, which is omitted here.

Theorem 12. *Suppose $w = (1/n, 1/n, ..., 1/n)^T$. Then,*

$$q - RDHFWGHM^{s,t}(d_1, d_2, ..., d_n) = q - RDHFGHM^{s,t}(d_1, d_2, ..., d_n). \quad (32)$$

The proof of Theorem 12 is similar to Theorem 6, which is omitted here.

Similarly, it is easy to prove that the q-RDHFWGHM has the properties of monotonicity and boundedness.

4. A Novel Approach to MAGDM with q-Rung Dual Hesitant Fuzzy Information

4.1. Description of a Typical MAGDM Problem with q-Rung Dual Hesitant Fuzzy Information

A typical MAGDM problem with q-rung dual hesitant fuzzy information can be described as follows: Let $A = \{A_1, A_2, ..., A_m\}$ and the set of attributes and $G = \{G_1, G_2, ..., G_n\}$ be a set of attributes. Let $w = (w_1, w_2, ..., w_n)^T$ be the weight vector of attributes, where $w_j \geq 0, j = 1, 2, ..., n$ and $\sum_{j=1}^{n} w_j = 1$. Suppose that $D = (d_{ij})_{m \times n} = (h_{ij}, g_{ij})_{m \times n}$ is the q-rung dual hesitant fuzzy decision matrix, where h_{ij} and g_{ij} indicate, respectively, the positive and negative degrees assessed by the decision maker that the alternative A_i satisfies the attribute G_j.

4.2. An Algorithm for q-Rung Dual Hesitant Fuzzy MAGDM Problems

In the following subsection, we present a novel algorithm for MAGDM based on the proposed operators.

Step 1. Standardize the original decision matrix according the following equation:

$$d_{ij} = \begin{cases} (h_{ij}, g_{ij}) & G_j \in I_1 \\ (g_{ij}, h_{ij}) & G_j \in I_2 \end{cases}, \quad (33)$$

where I_1 represents benefit attributes and I_2 represents cost attributes.

Step 2. For alternative $A_i (i = 1, 2, ..., m)$, utilize the q-RDHFWHM operator

$$d_i = q - RDHFWHM^{s,t}(d_{i1}, d_{i2}, \cdots, d_{in}), \quad (34)$$

or the q-RDHFWGHM operator

$$d_i = q - RDHFWGHM^{s,t}(d_{i1}, d_{i2}, \cdots, d_{in}), \quad (35)$$

to aggregate all the attributes values.

Step 3. Compute the score functions of all the alternatives and rank them.

Step 4. Rank the corresponding alternatives according to the rank of overall values and select the best alternative.

5. Numerical Example

In this section, to demonstrate the validity of the proposed method, we provide a numerical example adopted from [48]. A company wants to select a supplier, and after primary evaluation, four possible suppliers (A_1, A_2, A_3, and A_4) remain on the candidates list. To select the best supplier, a set of experts are invited to assess the four suppliers regarding four attributes: (1) relationship closeness (G_1); (2) product quality (G_2); (3) price competitiveness (G_3); and (4) delivery performance (G_4). The weight vector of the attributes is $w = (0.17, 0.32, 0.38, 0.13)^T$. The DMs are required to utilize DHFEs to express their preference information. The dual hesitant fuzzy decision matrix is shown in Table 1.

Table 1. The dual hesitant fuzzy decision matrix.

	G_1	G_2	G_3	G_4
A_1	{{0.3, 0.4}, {0.6}}	{{0.7, 0.9}, {0.1}}	{{0.4}, {0.2,0.3}}	{{0.5, 0.6}, {0.2}}
A_2	{{0.2, 0.3}, {0.5}}	{{0.6, 0.7}, {0.2}}	{{0.7, 0.8}, {0.2}}	{{0.6}, {0.1, 0.2, 0.3}}
A_3	{{0.4}, {0.2,0.3}}	{{0.2,0.3,0.4}, {0.6}}	{{0.7,0.8}, {0.1}}	{{0.7}, {0.2,0.3}}
A_4	{{0.6,0.7}, {0.3}}	{{0.5}, {0.4}}	{{0.3,0.4}, {0.5}}	{{0.4, 0.6}, {0.1,0.2}}

5.1. The Decision-Making Process

Step 1. As all the attributes are of the benefit type, the original decision matrix does not need to be normalized.

Step 2. Utilize the q-RDHFWHM operator to aggregate attributes values, so that the overall assessments are obtained (assume $s = t = 1$ and $q = 3$). Due to the relatively large numbers, the overall assessments are omitted.

Step 3. Calculate the scores of the overall assessments of alternatives to obtain $s(d_1) = 0.2235$, $s(d_2) = 0.2631$, $s(d_3) = 0.2097$, and $s(d_4) = 0.0780$.

Step 4. Rank the overall assessments so that we can obtain $A_2 \succ A_1 \succ A_3 \succ A_4$. Therefore, the best alternative is A_2.

In Step 2, if we utilize the q-RDHFWGHM operator to aggregate decision makers' assessments, we can obtain $s(d_1) = 0.1187$, $s(d_2) = 0.1819$, $s(d_3) = 0.0862$, and $s(d_4) = 0.0566$. Therefore, the ranking order is $A_2 \succ A_1 \succ A_3 \succ A_4$ and the best alternative is also A_2.

5.2. The Influence of the Parameters on the Results

Evidently, it is noted that the parameters s, t, and q play very important roles in the results. In the following subsection, we investigate the effect of parameters on the score functions and ranking results. To better illustrate the effect of the parameters s and t on the ranking results, we investigate the effects from the following three aspects: (1) We assign several fixed values to s and t and calculate the scores of the overall assessments. Further, we derive the ranking results of the alternatives. (2) Let $s \in (0, 10]$ and $t \in (0, 10]$, we investigate the influence of s and t on the ranking results. (3) Let s or t be a fixed value and investigate the influence of another parameter on the ranking results. Details can be found in Tables 1 and 2 and Figures 2–13.

Table 2. Scores and ranking results by using the q-rung dual hesitant fuzzy weighted Heronian mean (q-RDHFWHM) operator (q = 3).

Parameters	Score Function $s(d_i)(i = 1, 2, 3, 4)$	Ranking Results
$s = t = 1/2$	$s(d_1) = 0.1617$ $s(d_2) = 0.2086$ $s(d_3) = 0.1532$ $s(d_4) = 0.0685$	$A_2 \succ A_1 \succ A_3 \succ A_4$
$s = t = 1$	$s(d_1) = 0.2235$ $s(d_2) = 0.2631$ $s(d_3) = 0.2093$ $s(d_4) = 0.0780$	$A_2 \succ A_1 \succ A_3 \succ A_4$
$s = t = 2$	$s(d_1) = 0.3235$ $s(d_2) = 0.3375$ $s(d_3) = 0.2989$ $s(d_4) = 0.0942$	$A_2 \succ A_1 \succ A_3 \succ A_4$
$s = t = 5$	$s(d_1) = 0.4494$ $s(d_2) = 0.4363$ $s(d_3) = 0.4174$ $s(d_4) = 0.1198$	$A_1 \succ A_2 \succ A_3 \succ A_4$
$s = 1, t = 2$	$s(d_1) = 0.2769$ $s(d_2) = 0.3124$ $s(d_3) = 0.2626$ $s(d_4) = 0.0836$	$A_2 \succ A_1 \succ A_3 \succ A_4$
$s = 2, t = 1$	$s(d_1) = 0.3001$ $s(d_2) = 0.3079$ $s(d_3) = 0.2677$ $s(d_4) = 0.0930$	$A_2 \succ A_1 \succ A_3 \succ A_4$
$s = 1, t = 5$	$s(d_1) = 0.4024$ $s(d_2) = 0.4118$ $s(d_3) = 0.3719$ $s(d_4) = 0.1042$	$A_2 \succ A_1 \succ A_3 \succ A_4$
$s = 5, t = 1$	$s(d_1) = 0.4463$ $s(d_2) = 0.3952$ $s(d_3) = 0.3825$ $s(d_4) = 0.1214$	$A_1 \succ A_2 \succ A_3 \succ A_4$

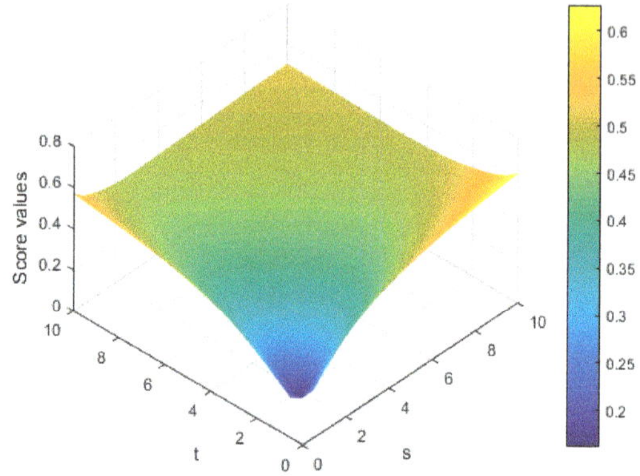

Figure 2. Scores of alternative A_1 when s, $t \in (0, 10)$ based on the q-RDHFWHM operator (q = 3).

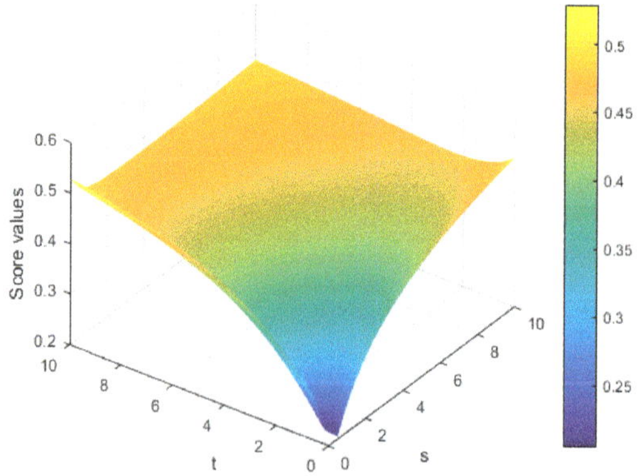

Figure 3. Scores of alternative A_2 when s, $t \in (0, 10)$ based on the q-RDHFWHM operator (q = 3).

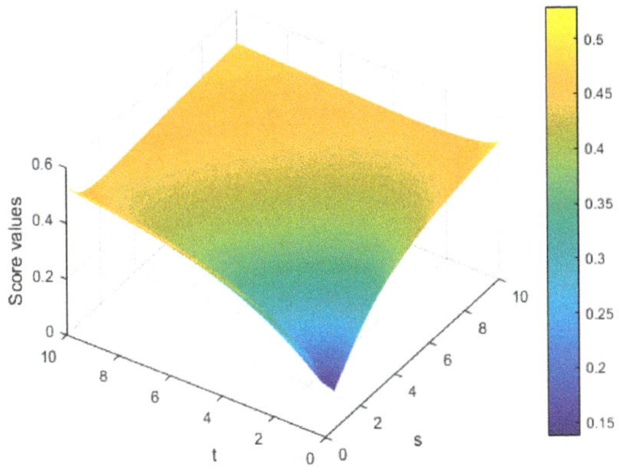

Figure 4. Scores of alternative A_3 when $s, t \in (0, 10)$ based on the q-RDHFWHM operator ($q = 3$).

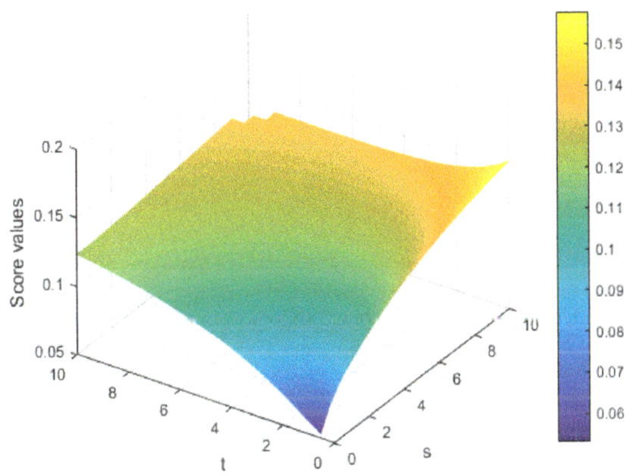

Figure 5. Scores of alternative A_4 when $s, t \in (0, 10)$ based on the q-RDHFWHM operator ($q = 3$).

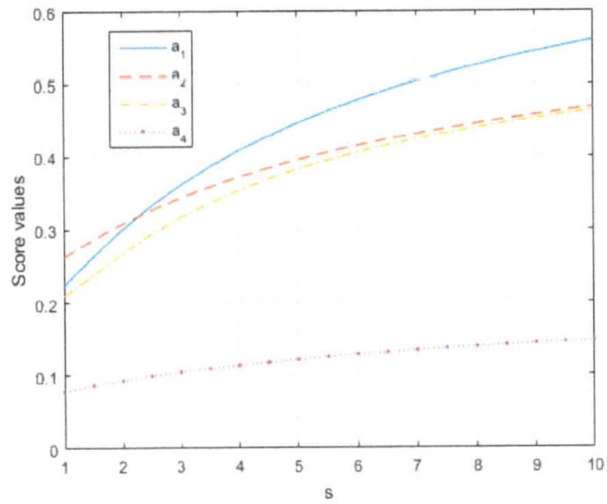

Figure 6. Scores of alternatives $A_i (i = 1, 2, 3, 4)$ when $t = 1$ and $s \in (1, 10)$ based on the q-RDHFWHM operator ($q = 3$).

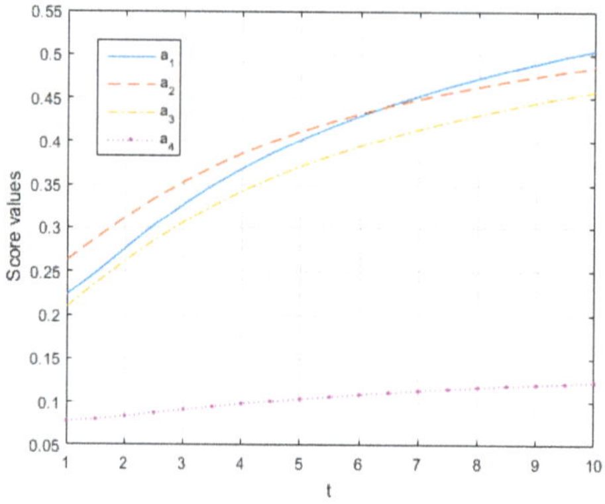

Figure 7. Scores of alternative $A_i(i = 1, 2, 3, 4)$ when $s = 1$ and $t \in (1, 10)$ based on the q-RDHFWHM operator ($q = 3$).

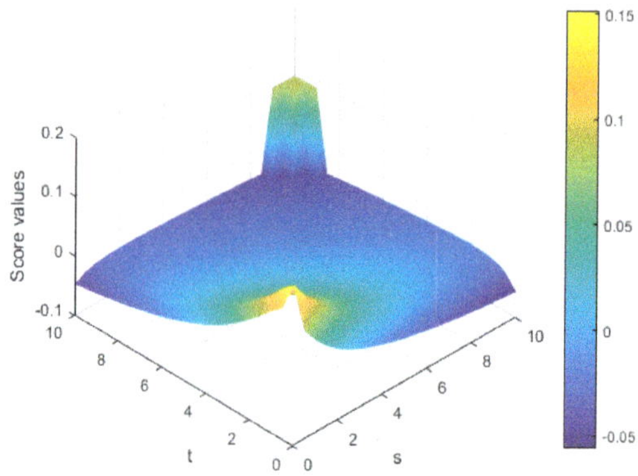

Figure 8. Scores of alternative A_1 when s, $t \in (0, 10)$ based on the q-rung dual hesitant fuzzy weighted geometric Heronian mean (q-RDHFWGHM) operator ($q = 3$).

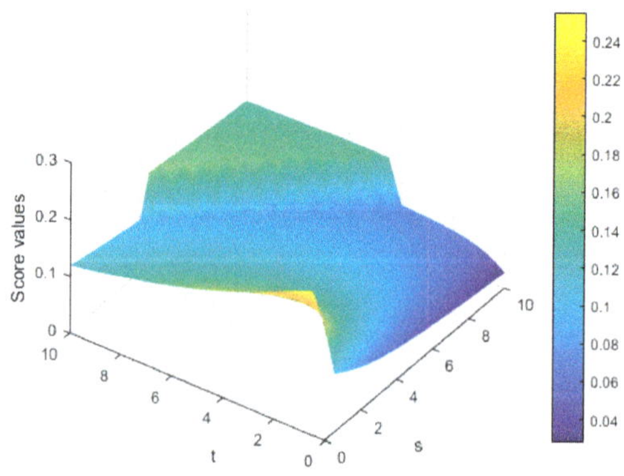

Figure 9. Scores of alternative A_2 when s, $t \in (0, 10)$ based on the q-RDHFWGHM operator ($q = 3$).

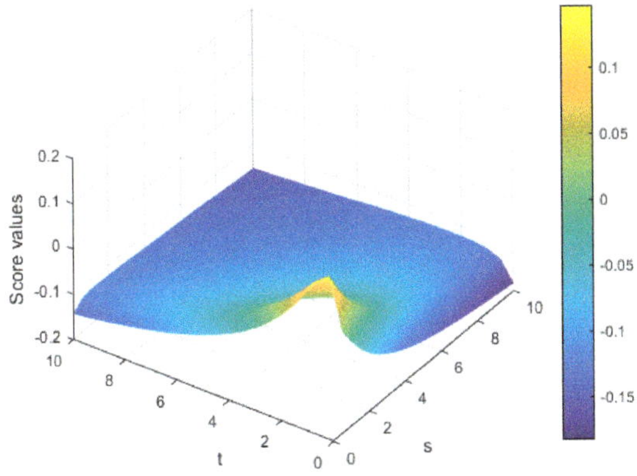

Figure 10. Scores of alternative A_3 when $s, t \in (0, 10)$ based on the q-RDHFWGHM operator ($q = 3$).

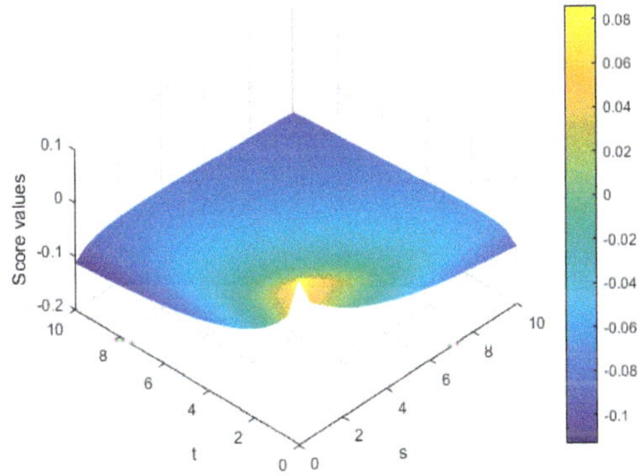

Figure 11. Scores of alternative A_4 when $s, t \in (0, 10)$ based on the q-RDHFWGHM operator ($q = 3$).

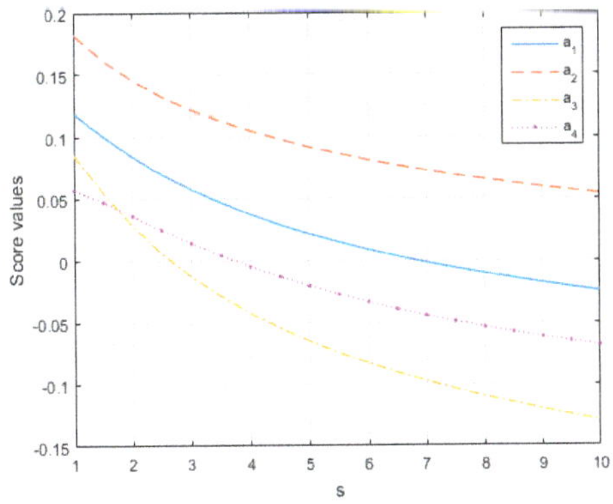

Figure 12. Scores of alternative $A_i (i = 1, 2, 3, 4)$ when $t = 1$ and $s \in (1, 10)$ based on the q-RDHFWGHM operator ($q = 3$).

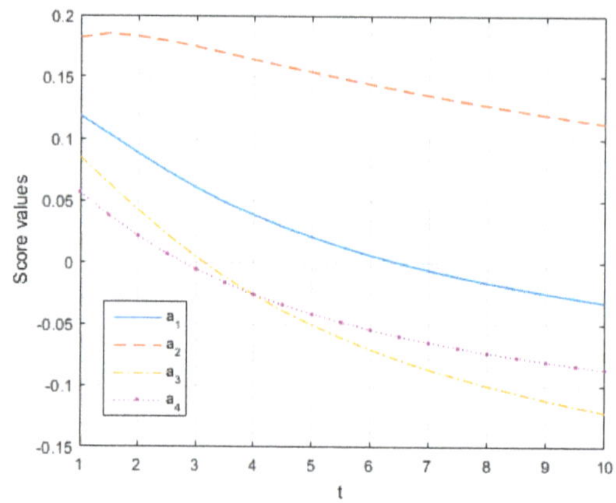

Figure 13. Scores of alternative $A_i(i = 1, 2, 3, 4)$ when $s = 1$ and $t \in (1, 10)$ based on the q-RDHFWGHM operator (q = 3).

From Table 2 and Figures 2–5, we can know that the scores and ranking results may be different for the different parameters s and t based on the q-RDHFWHM operator. However, the best alternative is A_2 or A_1. In addition, from Figures 6 and 7, we find that if we let t or s be a fixed value, then when s or t increases, the scores based on the q-RDHFWHM operator become greater and greater. Similarly, from Table 3 and Figures 8–11, we can obtain different scores and ranking results when s and t represent different values based on the q-RDHFWGHM operator. No matter what the values of s and t are, the best alternative is always A_2. However, what is opposite to the q-RDHFWHM operator is that if we let s or t be a fixed value, then when s or t increases, the scores based on the q-RDHFWGHM operator become smaller and smaller. The results shown in Tables 2 and 3 and Figures 2–13 demonstrate the flexibility of the aggregation processes by utilizing the q-RDHFWHM and q-RDHFWGHM operators. In real decision-making problems, DMs should choose the appropriate s and t according to their preference.

Table 3. Scores and ranking results by using the q-RDHFWGHM operator (q = 3).

Parameters	Score Function $s(d_i)(i = 1, 2, 3, 4)$	Ranking Results
$s = t = 1/2$	$s(d_1) = 0.1516$ $s(d_2) = 0.1972$ $s(d_3) = 0.1387$ $s(d_4) = 0.0861$	$A_2 \succ A_1 \succ A_3 \succ A_4$
$s = t = 1$	$s(d_1) = 0.1187$ $s(d_2) = 0.1819$ $s(d_3) = 0.0851$ $s(d_4) = 0.0566$	$A_2 \succ A_1 \succ A_3 \succ A_4$
$s = t = 2$	$s(d_1) = 0.0683$ $s(d_2) = 0.1544$ $s(d_3) = 0.0055$ $s(d_4) = 0.0132$	$A_2 \succ A_1 \succ A_4 \succ A_3$
$s = t = 5$	$s(d_1) = -0.0029$ $s(d_2) = 0.1054$ $s(d_3) = -0.0949$ $s(d_4) = -0.0521$	$A_2 \succ A_1 \succ A_4 \succ A_3$
$s = 1, t = 2$	$s(d_1) = 0.0884$ $s(d_2) = 0.1833$ $s(d_3) = 0.0423$ $s(d_4) = 0.0210$	$A_2 \succ A_1 \succ A_3 \succ A_4$
$s = 2, t = 1$	$s(d_1) = 0.0841$ $s(d_2) = 0.1456$ $s(d_3) = 0.0279$ $s(d_4) = 0.0358$	$A_2 \succ A_1 \succ A_4 \succ A_3$
$s = 1, t = 5$	$s(d_1) = 0.0204$ $s(d_2) = 0.1540$ $s(d_3) = -0.0507$ $s(d_4) = -0.0418$	$A_2 \succ A_1 \succ A_4 \succ A_3$
$s = 5, t = 1$	$s(d_1) = 0.0215$ $s(d_2) = 0.0917$ $s(d_3) = -0.0653$ $s(d_4) = -0.0206$	$A_2 \succ A_1 \succ A_4 \succ A_3$

In the following, we discuss the effects of the parameter q on the score function and ranking results based on q-RDHFWHM and q-RDHFWGHM operators. Details can be found in Figures 14 and 15.

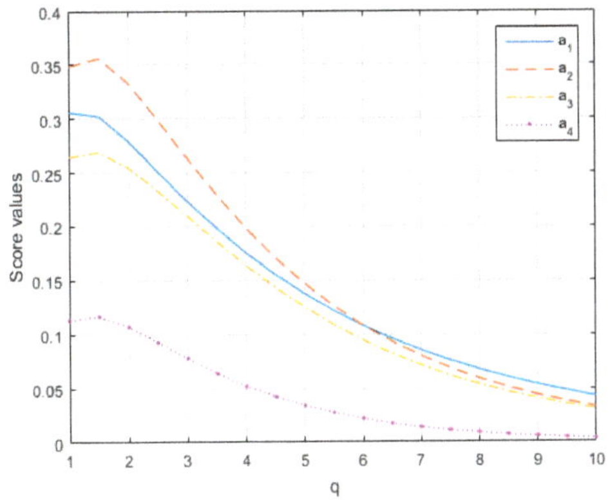

Figure 14. Scores of alternative $A_i (i = 1, 2, 3, 4)$ when s = t = 1 and $q \in (1, 10)$ based on the q-RDHFWHM operator.

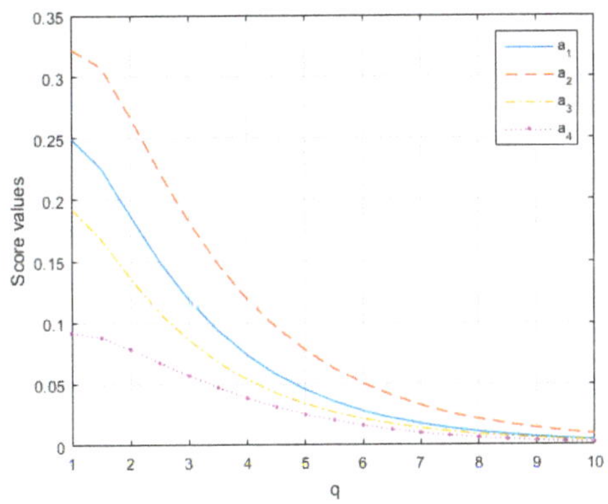

Figure 15. Scores of alternative $A_i (i = 1, 2, 3, 4)$ when s = t = 1 and $q \in (1, 10)$ based on the q-RDHFWGHM operator.

As seen in Figures 14 and 15, the scores and ranking results can be different for the different parameter q based on the q-RDHFWHM and q-RDHFWGHM operators. However, the best alternative is always A_2 or A_4 based on the q-RDHFWHM, whereas the best alternative is always A_2 based on the q-RDHFWGHM operators. In addition, when q increases, both the scores obtained by the q-RDHFWHM and q-RDHFWGHM operators have the tendency to decrease.

5.3. Compared with Exiting MAGDM Methods

To demonstrate the advantages and superiorities of the proposed method, we compared our method with that proposed by Wang et al. [48], which was based on the dual hesitant fuzzy weighted averaging (DHFWA) operator proposed by Yu et al. [47], which was based on the dual hesitant fuzzy weighted Heronian mean (DHFWHM) operator proposed by Tu et al. [49], which was based on the dual hesitant fuzzy weighted Bonferroni mean (DHFWBM) operator that proposed by Wei and Lu [42], which was based on the dual hesitant Pythagorean fuzzy Hamacher weighted averaging (DHPFHWA) operator. We utilized these methods to solve the above example, and the score functions and ranking methods can be found in Table 4.

Table 4. Score functions and ranking results by using different methods.

Methods	Score Function $s(d_i)(i = 1, 2, 3, 4)$	Ranking Results
Wang et al.' [48] method based on the DHFWA operator	$s(d_1) = 0.3915$ $s(d_2) = 0.4147$ $s(d_3) = 0.3573$ $s(d_4) = 0.1198$	$A_2 \succ A_1 \succ A_3 \succ A_4$
Yu et al.' s [47] method based on the DHFWHM operator ($s = t = 2$)	$s(d_1) = -0.3813$ $s(d_2) = -0.3916$ $s(d_3) = -0.3960$ $s(d_4) = -0.6147$	$A_1 \succ A_2 \succ A_3 \succ A_4$
Tu et al.'s [49] method based on the DHFWBM operator	$s(d_1) = 0.3152$ $s(d_2) = 0.3004$ $s(d_3) = 0.2978$ $s(d_4) = 0.0258$	$A_1 \succ A_2 \succ A_3 \succ A_4$
Wei and Lu's [42] method based on the DHPFHWA operator	$s(d_1) = 0.2369$ $s(d_2) = 0.2196$ $s(d_3) = 0.1284$ $s(d_4) = 0.0026$	$A_1 \succ A_2 \succ A_3 \succ A_4$
The proposed method in this paper	$s(d_1) = 0.2235$ $s(d_2) = 0.2631$ $s(d_3) = 0.2097$ $s(d_4) = 0.0780$	$A_2 \succ A_1 \succ A_3 \succ A_4$

First of all, Wang et al.'s [48], Yu et al.'s [47], and Tu et al.'s [49] methods are based on DHFSs. Wei and Lu's 429] method is based on DHPFSs. As mentioned above, DHFS and DHPFS are two special cases of q-RDHFS. When $q = 1$, then q-RDHFS is reduced to DHFS, and when $q = 2$, q-RDHFS is reduced to DHPFS. Evidently, q-RDHFS is more general and can describe a greater information range and process more information in the process of MAGDM. For instance, if an attribute value provided by DMs is $\{\{0.1, 0.2, 0.5, 0.8\}, \{0.1, 0.2, 0.7\}\}$, then obviously, the pair $\{\{0.1, 0.2, 0.5, 0.8\}, \{0.1, 0.2, 0.7\}\}$ is not valid for DHFSs and DHPFSs. Thus, our method is more general, powerful, and can process more information in MAGDM.

Wang et al.'s [48] and Wei and Lu's [42] methods are based on the simple weighted averaging operator. The drawback of the two methods is that they do not consider the interrelationship between arguments. In other words, they assume all attributes are independent, which is not correct to some extent. In the abovementioned example, when choosing the most appropriate supplier, we need to consider not only the attribute values of each supplier but also the correlation between these attributes. Thus, Wang et al.'s [48] and Wei and Lu's [42] methods are not suitable for dealing with this problem. As our method has the ability to capture variable correlations, it is more reasonable than Wang et al.'s [48] and Wei and Lu's [42] methods for addressing this problem.

Tu et al.'s [49] method is based on Bonferroni mean (BM), and Yu et al.'s [47] and our methods are based on HM. The prominent characteristic of BM and HM is that both can consider the interrelationship between arguments. Therefore, all the three can process the interrelationship among attribute values. However, Yu et al.'s [47] method and ours are better than Tu et al.'s [49] method. In addition, as Yu et al.'s [47] is a special case of our method (when $q = 1$), our method is more general, scientific, and applicable than Yu et al.'s [47] method.

In real decision-making problems, we may encounter situations in which DMs are hesitant between several possible values when determining the membership and nonmembership degrees. Additionally, the sum and square sum of membership and nonmembership degrees may be more than one. Moreover, as attributes are related, the interrelationship between attribute values should be considered. In this paper, we present a novel approach to MAGDM problems based on q-RDHFS, which is a powerful tool for expressing and denoting DMs' assessments. It can deal with DMs' hesitancy and its lax constraints give DMs more freedom to express their preference information. In addition, our method is based on HM so that the interrelationship between attributes can be processed. Therefore, our method has some advantages and superiorities compared with existing methods.

6. Conclusions

Supplier selection is very important for manufacturing companies. Choosing a suitable supplier can greatly enhance the competitiveness and vitality of the company. In modern society, the selection of an appropriate supplier often requires a comprehensive assessment of all suppliers from multiple perspectives. Thus, supplier selection is one of the most common types of MAGDM problems in daily life. The main contributions of this paper are threefold. Firstly, we proposed the concept of q-RDHFS by combining DHFS with q-ROFS. The q-RDHFS can not only deal with DMs' hesitancy when determining the membership and nonmembership degrees but also gives DMs' more freedom to express their assessments. Secondly, we proposed the q-RDHFHM, q-RDHFWHM, q-RDHFGHM, and q-RDHFWGHM operators to effectively aggregate q-RDHFEs. Thirdly, we developed a novel method for MAGDM with q-rung dual hesitant fuzzy information. Considering the supplier selection problem is essentially a MAGDM issue, we also applied the proposed method to a real MAGDM problem to show its performance. Additionally, through comparative analysis the superiorities and advantages of the newly proposed method over existing methods are illustrated. Compared with the existing methods, the proposed method is more general and powerful. In addition, it has three parameters—q, s, and t—making the process of information aggregation more flexible. In real decision-making problems, DMs can choose the appropriate values of the parameters according to their preference. It is worth pointing out that as the newly proposed method is based on the HM operator, it mainly focuses on the interrelationship between any two q-RDHFEs. In future works, we should investigate more aggregation operators for fusing q-RDHFEs, such as the q-rung dual hesitant fuzzy Maclaurin symmetric mean, the q-rung dual hesitant fuzzy Hamy mean, and the q-rung dual hesitant fuzzy Muirhead mean operators, which have the ability of capturing the interrelationship among multiple q-RDHFEs.

Author Contributions: Conceptualization, Y.X.; Formal Analysis, J.W.; All the authors have participated in writing the manuscript and have revised the final version. All authors read and approved the final manuscript.

Funding: This work was partially supported by the National Natural Science Foundation of China (Grant number 61702023), the Humanities and Social Science Foundation of the Ministry of Education of China (Grant number 17YJC870015), and the Fundamental Research Funds for the Central Universities (Grant number 2018JBM304).

Conflicts of Interest: The authors declare that there is no conflict of interest regarding the publication of this paper.

References

1. Atanassov, K.T. Intuitionistic fuzzy sets. *Fuzzy Sets Syst.* **1986**, *20*, 87–96. [CrossRef]
2. Zadeh, L.A. Fuzzy sets. *Inf. Control* **1965**, *8*, 338–356. [CrossRef]
3. Son, L.H.; Phong, P.H. On the performance evaluation of intuitionistic vector similarity measures for medical diagnosis 1. *J. Intell. Fuzzy Syst.* **2016**, *31*, 1597–1608. [CrossRef]
4. Own, C.M. Switching between type-2 fuzzy sets and intuitionistic fuzzy sets: An application in medical diagnosis. *Appl. Intell.* **2009**, *31*, 283–291. [CrossRef]
5. Chen, S.M.; Cheng, S.H.; Lan, T.C. A novel similarity measure between intuitionistic fuzzy sets based on the centroid points of transformed fuzzy numbers with applications to pattern recognition. *Inf. Sci.* **2016**, *343–344*, 15–40. [CrossRef]
6. Hwang, C.M.; Yang, M.S.; Hung, W.L.; Lee, M.G. A similarity measure of intuitionistic fuzzy sets based on the Sugeno integral with its application to pattern recognition. *Inf. Sci.* **2012**, *189*, 93–109. [CrossRef]
7. Wang, Z.; Xu, Z.S.; Liu, S.S.; Tang, J. A netting clustering analysis method under intuitionistic fuzzy environment. *Appl. Soft Comput.* **2011**, *11*, 5558–5564. [CrossRef]
8. Xu, D.W.; Xu, Z.S.; Liu, S.S.; Zhao, H. A spectral clustering algorithm based on intuitionistic fuzzy information. *Knowl. Based Syst.* **2013**, *53*, 20–26. [CrossRef]
9. Liu, P.D. Multiple attribute decision-making methods based on normal intuitionistic fuzzy interaction aggregation operators. *Symmetry* **2017**, *9*, 261. [CrossRef]
10. Wang, S.W.; Liu, J. Extension of the TODIM method to intuitionistic linguistic multiple attribute decision making. *Symmetry* **2017**, *9*, 95. [CrossRef]
11. Liu, P.D.; Mahmood, T.; Khan, Q. Multi-attribute decision-making based on prioritized aggregation operator under hesitant intuitionistic fuzzy linguistic environment. *Symmetry* **2017**, *9*, 270. [CrossRef]
12. Zhao, J.S.; You, X.Y.; Liu, H.C.; Wu, S.M. An extended VIKOR method using intuitionistic fuzzy sets and combination weights for supplier selection. *Symmetry* **2017**, *9*, 169. [CrossRef]

13. Yager, R.R. Pythagorean membership grades in multicriteria decision making. *IEEE Trans. Fuzzy Syst.* **2014**, *22*, 958–965. [CrossRef]
14. Zhang, X.; Xu, Z. Extension of TOPSIS to multiple criteria decision making with Pythagorean fuzzy sets. *Int. J. Intell. Syst.* **2014**, *29*, 1061–1078. [CrossRef]
15. Sajjad, A.K.M.; Abdullah, S.; Yousaf, A.M.; Hussain, I.; Farooq, M. Extension of TOPSIS method base on Choquet integral under interval-valued Pythagorean fuzzy environment. *J. Intell. Fuzzy Syst.* **2018**, *34*, 267–282. [CrossRef]
16. Ren, P.; Xu, Z.; Gou, X. Pythagorean fuzzy TODIM approach to multi-criteria decision making. *Appl. Soft Comput.* **2016**, *42*, 246–259. [CrossRef]
17. Ma, Z.M.; Xu, Z.S. Symmetric Pythagorean fuzzy weighted geometric/averaging operators and their application in multi-criteria decision making problems. *J. Intell. Fuzzy Syst.* **2016**, *31*, 1198–1219.
18. Xing, Y.P.; Zhang, R.T.; Wang, J.; Zhu, X.M. Some new Pythagorean fuzzy Choquet-Frank aggregation operators for multi-attribute decision making. *Int. J. Intell. Syst.* **2018**. [CrossRef]
19. Wei, G.W.; Lu, M. Pythagorean fuzzy Maclaurin symmetric mean operators in multiple attribute decision making. *Int. J. Intell. Syst.* **2017**, *33*, 1043–1070. [CrossRef]
20. Liang, D.C.; Xu, Z.S.; Darko, A.P. Projection model for fusing the information of Pythagorean fuzzy multicriteria group decision making based on geometric Bonferroni mean. *Int. J. Intell. Syst.* **2017**, *32*, 966–987. [CrossRef]
21. Zhang, R.T.; Wang, J.; Zhu, X.M.; Xia, M.M.; Yu, M. Some generalized Pythagorean fuzzy Bonferroni mean aggregation operators with their application to multi attribute group decision-making. *Complexity* **2017**, *2017*, 5937376. [CrossRef]
22. Li, L.; Zhang, R.T.; Wang, J.; Zhu, X.M.; Xing, Y.P. Pythagorean fuzzy power Muirhead mean operators with their application to multi-attribute decision making. *J. Intell. Fuzzy Syst.* **2018**, *35*, 2035–2050. [CrossRef]
23. Xu, Y.; Shang, X.P.; Wang, J. Pythagorean fuzzy interaction Muirhead means with their application to multi-attribute group decision making. *Information* **2018**, *9*, 7. [CrossRef]
24. Teng, F.; Liu, Z.; Liu, P. Some power Maclaurin symmetric mean aggregation operators based on Pythagorean fuzzy linguistic numbers and their application to group decision making. *Int. J. Intell. Syst.* **2018**, *33*, 1949–1985. [CrossRef]
25. Du, Y.; Hou, F.; Zafar, W.; Yu, Q.; Zhai, Y. A novel method for multiattribute decision making with interval-valued Pythagorean fuzzy linguistic information. *Int. J. Intell. Syst.* **2017**, *32*, 1085–1112. [CrossRef]
26. Xian, S.; Xiao, Y.; Yang, Z.; Li, Y.; Han, Z. A new trapezoidal Pythagorean fuzzy linguistic entropic combined ordered weighted averaging operator and its application for enterprise location. *Int. J. Intell. Syst.* **2018**, *33*, 1880–1899. [CrossRef]
27. Geng, Y.; Liu, P.; Teng, F.; Liu, Z. Pythagorean fuzzy uncertain linguistic TODIM method and their application to multiple criteria group decision making. *J. Intell. Fuzzy Syst.* **2017**, *33*, 3383–3395. [CrossRef]
28. Liu, C.; Tang, G.; Liu, P. An approach to multicriteria group decision-making with unknown weight information based on Pythagorean fuzzy uncertain linguistic aggregation operators. *Math. Probl. Eng.* **2017**. [CrossRef]
29. Liu, Z.; Liu, P.; Liu, W.; Pang, J. Pythagorean uncertain linguistic partitioned Bonferroni mean operators and their application in multi-attribute decision making. *J. Intell. Fuzzy Syst.* **2017**, *32*, 2779–2790. [CrossRef]
30. Wei, G.; Lu, M.; Alsaadi, F.E.; Hayat, T.; Alsaedi, A. Pythagorean 2-tuple linguistic aggregation operators in multiple attribute decision making. *J. Intell. Fuzzy Syst.* **2017**, *33*, 1129–1142. [CrossRef]
31. Huang, Y.H.; Wei, G.W. TODIM method for Pythagorean 2-tuple linguistic multiple attribute decision making. *J. Intell. Fuzzy Syst.* **2018**, *35*, 901–915. [CrossRef]
32. Tang, X.; Wei, G. Models for green supplier selection in green supply chain management with Pythagorean 2-tuple linguistic information. *IEEE Access* **2018**, *6*, 18042–18060. [CrossRef]
33. Yager, R.R. Generalized orthopair fuzzy sets. *IEEE Trans. Fuzzy Syst.* **2017**, *25*, 1222–1230. [CrossRef]
34. Liu, P.D.; Wang, P. Some *q*-rung orthopair fuzzy aggregation operators and their applications to multiple-attribute decision making. *Int. J. Intell. Syst.* **2018**, *33*, 259–280. [CrossRef]
35. Liu, P.D.; Liu, J.L. Some *q*-rung orthopair fuzzy Bonferroni mean operators and their application to multi-attribute group decision making. *Int. J. Intell. Syst.* **2018**, *33*, 315–347. [CrossRef]

36. Liu, Z.M.; Liu, P.D.; Liang, X. Multiple attribute decision-making method for dealing with heterogeneous relationship among attributes and unknown attribute weight information under q-rung orthopair fuzzy environment. *Int. J. Intell. Syst.* **2018**, *33*, 1900–1928. [CrossRef]
37. Bai, K.Y.; Zhu, X.M.; Wang, J.; Zhang, R.T. Some partitioned Maclaurin symmetric mean based on q-rung orthopair fuzzy information for dealing with multi-attribute group decision making. *Symmetry* **2018**, *10*, 383. [CrossRef]
38. Li, L.; Zhang, R.T.; Wang, J.; Shang, X.P. Some q-rung orthopair linguistic Heronian mean operators with their application to multi-attribute group decision making. *Arch. Control Sci.*. accepted.
39. Li, L.; Zhang, R.T.; Wang, J.; Shang, X.P.; Bai, K.Y. A novel approach to multi-attribute group decision-making with q-rung picture linguistic information. *Symmetry* **2018**, *10*, 172. [CrossRef]
40. Torra, V. Hesitant fuzzy sets. *Int. J. Intell. Syst.* **2010**, *25*, 529–539. [CrossRef]
41. Zhu, B.; Xu, Z.S.; Xia, M.M. Dual hesitant fuzzy sets. *J. Appl. Math.* **2012**, *2012*, 879629. [CrossRef]
42. Wei, G.; Lu, M. Dual hesitant Pythagorean fuzzy Hamacher aggregation operators in multiple attribute decision making. *Arch. Control Sci.* **2017**, *27*, 365–395. [CrossRef]
43. Khan, M.S.A.; Abdullah, S.; Ali, A.; Siddiqui, N.; Amin, F. Pythagorean hesitant fuzzy sets and their application to group decision making with incomplete weight information. *J. Intell. Fuzzy Syst.* **2017**, *33*, 3971–3985. [CrossRef]
44. Liang, D.C.; Xu, Z.S. The new extension of TOPSIS method for multiple criteria decision making with hesitant Pythagorean fuzzy sets. *Appl. Soft Comput.* **2017**, *60*, 167–179. [CrossRef]
45. Sykora, S. *Mathematical Means and Averages: Generalized Heronian Means*; Stan's Library: Castano Primo, Italy, 2009. Available online: http://www.ebyte.it/library/docs/math09/Means_Heronian.html (accessed on 9 October 2018).
46. Yu, D.J. Intuitionistic fuzzy geometric Heronian mean aggregation operators. *Appl. Soft Comput.* **2013**, *13*, 1235–1246. [CrossRef]
47. Yu, D.J.; Li, D.F.; Merigó, J.M. Dual hesitant fuzzy group decision making method and its application to supplier selection. *Int. J. Mach. Learn. Cybern.* **2016**, *7*, 819–831. [CrossRef]
48. Wang, H.J.; Zhao, X.F.; Wei, G.W. Dual hesitant fuzzy aggregation operators in multiple attribute decision making. *J. Intell. Fuzzy Syst.* **2014**, *26*, 2281–2290.
49. Tu, N.H.; Wang, C.Y.; Zhou, X.Q.; Tao, S.D. Dual hesitant fuzzy aggregation operators based on Bonferroni means and their applications to multiple attribute decision making. *Annl. Fuzzy Math. Inform.* **2017**, *14*, 265–278.

symmetry

MDPI

Article

New Similarity Measures of Single-Valued Neutrosophic Multisets Based on the Decomposition Theorem and Its Application in Medical Diagnosis

Qingqing Hu [1] and Xiaohong Zhang [1,2,*]

1 School of Arts and Sciences, Shaanxi University of Science & Technology, Xi'an 710021, China; huqingqing122@163.com
2 College of Arts and Sciences, Shanghai Maritime University, Shanghai 201306, China
* Correspondence: zhangxiaohong@sust.edu.cn or zhangxh@shmtu.edu.cn

Received: 16 September 2018; Accepted: 2 October 2018; Published: 9 October 2018

Abstract: Cut sets, decomposition theorem and representation theorem have a great influence on the realization of the transformation of fuzzy sets and classical sets, and the single-valued neutrosophic multisets (SVNMSs) as the generalization of fuzzy sets, which cut sets, decomposition theorem and representation theorem have the similar effects, so they need to be studied in depth. In this paper, the decomposition theorem, representation theorem and the application of a new similarity measures of SVNMSs are studied by using theoretical analysis and calculations. The following are the main results: (1) The notions, operation and operational properties of the cut sets and strong cut sets of SVNMSs are introduced and discussed; (2) The decomposition theorem and representation theorem of SVNMSs are established and rigorously proved. The decomposition theorem and the representation theorem of SVNMSs are the theoretical basis for the development of SVNMSs. The decomposition theorem provides a new idea for solving the problem of SVNMSs, and points out the direction for the principle of expansion of SVNMSs. (3) Based on the decomposition theorem and representation theorem of SVNMSs, a new notion of similarity measure of SVNMSs is proposed by applying triple integral. And this new similarity is applied to the practical problem of multicriteria decision-making, which explains the efficacy and practicability of this decision-making method. The new similarity is not only a way to solve the problem of multi-attribute decision-making, but also contains an important mathematical idea, that is, the idea of transformation.

Keywords: single-valued neutrosophic multiset (SVNMS); cut set; decomposition theorem; representation theorem; similarity measure; triple integral; multicriteria decision-making

1. Introduction

It is essential for medical experts to address incomplete and uncertain information included in actual medical diagnostic questions. In order to effectively use various uncertain diagnostic information, Smarandache [1] proposed neutrosophic set (NS), which is a generalization of fuzzy set (FS) and intuitionistic fuzzy set (IFS) [2]. NS is more flexible and applicable than FS and IFS. Nevertheless, it is hard to apply the NS to practical problems for the values of the functions with respect to truth, indeterminacy and falsity lie in $]0-,1+[$. Thus, Smarandache and Wang [3] introduced the notion of the single-valued neutrosophic set (SVNS), whose values belong to [0,1]. In the actual decision-making problems, scholars have obtained many inspiring research results according to the SVNS theories [4–9]. However, in the multicriteria decision-making problem, the application of SVNS has certain limitations. Fortunately, Yager [10] firstly discussed fuzzy multisets (FMSs), in which every element may appear more than once and may have the same or different membership values. Indeed, fuzzy multisets theories cannot cope with all types of uncertain and incomplete information. So, Ye [11] introduced the notion of the single-valued neutrosophic multisets (SVNMSs) by capitalizing on fuzzy multisets (FMs) [12,13]. So far, a large number of scholars have studied the similarity measures of SVNMSs from different angles and discuss its application in decision-making problems in [11,14–18], which is crucial for further in-depth analysis and research on SVNMSs in the future.

As we all know, the decomposition theorem, representation theorem and expansion theorem are three theoretical pillars of fuzzy mathematics. Decomposition theorem and representation theorem are the bond between fuzzy set theory and classical set theory, that is, any fuzzy set problem can be turned into a problem of classical set by taking a cut set and constructing a geometric set. The notion of λ-cut sets of FS, some basic properties of λ-cut sets, the decomposition theorem, the representation theorem of FSs had been proposed [1,19]. What is more, the definitions of cut sets, some basic properties of cut sets, the decomposition theorem and the representation theorem of IFS, interval intuitionistic fuzzy set (IIFS), interval value fuzzy set (IVFS) which as generations of FSs had been proposed [20–28]. After that D. Singh, A. J. Alkali and A. I. Isah introduced the definition of α-cuts for FMS, which is a generalization of λ-cut sets of FS, and proposed some properties of α-cuts, decomposition theorem for FMS [29]. However, the cut sets and its operational properties, decomposition theorem and representation theorem of the SVNMSs have not been studied yet. Thus, it is necessary to discuss the cut sets, decomposition theorem and representation theorem of SVNMSs. We have already been researching SVNMSs and proposed some new results in [30–32]. Moreover, this paper proposes a new similarity from the perspective of decomposition theorem which is different from [11–16]. This new method uses the decomposition theorem as the theoretical basis and the integral as the mathematical tool. The idea is simple, the calculation is convenient, and it contains important mathematical ideas, which is more practical [33–36].

The organization of this paper is as follows: In Section 2, some basic conceptions of FMS, IFM and SVNMS are reviewed. Section 3 discusses some new properties of SVNMS. Section 4 proposes the (α, β, γ)-cut sets for SVNMS, and investigates the decomposition theorem and the representation theorem of SVNMS. In Section 5, based on the established cut sets, a new method is proposed to calculate the similarity measure between SVNMSs. In Section 6, a practicable example is offered for medical diagnosis to illustrate the approach proposed in this paper. Section 7 presents final conclusions and further research.

2. Preliminaries

2.1. Some Basic Concepts of IFS, FMS

Definition 1 ([2]). *Let X be a nonempty set. An IFS M in X is given by*

$$M = \{\langle x, \mu_M(x), \nu_M(x)\rangle | x \in X\} \tag{1}$$

where $\mu_M : X \to [0,1]$ *and* $\nu_M : X \to [0,1]$ *with the condition* $0 \leq \mu_M + \nu_M \leq 1$ *for all* $x \in X$.

Here $\mu_M(x), \nu_M(x) \in [0,1]$ denote the membership and the non-membership functions of the fuzzy set M.

Definition 2 ([10]). *A fuzzy multiset M is a generation set of multisets over the universe X, which is denoted by pairs, where the first part of each pair is the element of X, and the second part is the membership of the element relative to M. Note that an element of X may occur more than once in the same or different membership values. For each $x \in X$, a membership sequence is defined to be the decreasing ordered sequence of the elements, that is,*

$$\left(\mu_M^1(x), \mu_M^2(x), \cdots, \mu_M^q(x)\right),$$

where $\mu_M^1(x) \geq \mu_M^2(x) \geq \cdots \geq \mu_M^q(x)$. *Hence, the FMS M is given by*

$$M = \left\{ \left(\mu_M^1(x), \mu_M^2(x), \cdots, \mu_M^q(x)\right) \middle| x \right\}, \text{ for all } x \in X. \tag{2}$$

2.2. Some Concepts of SVNMS

Definition 3 ([11]). *Let X be a nonempty set with a generic element in X denoted by x. A SVNMS M in X is characterized by three functions: count truth-membership of* CT_M*, count indeterminacy-membership of* CI_M*, and count falsity-membership of* CF_M*, such that* $CT_M(x) : X \to R$, $CI_M(x) : X \to R$, $CF_M(x) : X \to R$, *for every* $x \in X$*, where R is the set of all real number multisets in the real unit interval* $[0,1]$*. Then, a SVNMS M is given by*

$$M = \left\{ \left\langle x, \left(T_M^1(x), T_M^2(x), \cdots, T_M^k(x)\right), \left(I_M^1(x), I_M^2(x), \cdots, I_M^k(x)\right), \left(F_M^1(x), F_M^2(x), \cdots, F_M^k(x)\right) \right\rangle \middle| x \in X \right\},$$

where the truth-membership sequence $\left(T_M^1(x), T_M^2(x), \cdots, T_M^k(x)\right)$*, the indeterminacy-membership sequence* $\left(I_M^1(x), I_M^2(x), \cdots, I_M^k(x)\right)$*, and the falsity-membership sequence* $\left(F_M^1(x), F_M^2(x), \cdots, F_M^k(x)\right)$ *may be in decreasing order or not. Additionally, the* $T_M^j(x), I_M^j(x), F_M^j(x)$ *also satisfies the following condition*

$$0 \leq T_M^j(x) + I_M^j(x) + F_M^j(x) \leq 3, \text{ for all} x \in X,\ j = 1, 2, \cdots, k.$$

In order to express more concisely, a SVNMS M over X can be given by

$$M = \left\{ \left\langle x, T_M^j(x), I_M^j(x), F_M^j(x) \right\rangle \middle| x \in X, j = 1, 2, \cdots, k \right\} \tag{3}$$

Furthermore, we represent the set of all SVNMSs on X as SVNMS(X).

Definition 4 ([11]). *Let* $M \in SVNMS(X)$*, for every element x included in M, the length of x is defined as the cardinal number of* $CT_M(x)$ *or* $CI_M(x)$*, or* $CF_M(x)$*, and is expressed as* $l(x : M)$*. That is,* $l(x : M) = |CT_M(x)| = |CI_M(x)| = |CF_M(x)|$*. Suppose* $M, N \in SVNMS(X)$*, then,* $l(x : M, N) = \max\{l(x : M), l(x : N)\}$.

Definition 5 ([11]). *An absolute SVNMS* $\widetilde{M}$ *is a SVNMS, whose* $T_{\widetilde{M}}^j(x) = 1$, $I_{\widetilde{M}}^j(x) = 0$ *and* $F_{\widetilde{M}}^j(x) = 0$*, for all* $x \in X$ *and* $j = 1, 2, \cdots, l(x : \widetilde{M})$.

Definition 6 ([11]).*A null SVNMS* $\widetilde{\Phi}$ *is a SVNMS, whose* $T_{\widetilde{\Phi}}^j(x) = 0$, $I_{\widetilde{\Phi}}^j(x) = 1$ *and* $F_{\widetilde{\Phi}}^j(x) = 1$*, for all* $x \in X$ *and* $j = 1, 2, \cdots, l(x : \widetilde{\Phi})$.

Let $M, N \in SVNMS(X)$. In order to further study the operations between M and N, we must verify that $l(x : M) = l(x : N)$ is true for every $x \in X$, if not, we use a sufficient number of zeroes to fill the truth-membership values and a sufficient number of ones to fill the indeterminacy-membership values and falsity-membership values of the smaller-length sequences, respectively, so that the lengths of sequences are equal to facilitate computing.

Definition 7 ([11]). *Let* $M = \left\{ \left\langle x, T_M^j(x), I_M^j(x), F_M^j(x) \right\rangle \middle| x \in X, j = 1, 2, \cdots, l(x : M) \right\}$ *and* $N = \left\{ \left\langle x, T_N^j(x), I_N^j(x), F_N^j(x) \right\rangle \middle| x \in X, j = 1, 2, \cdots, l(x : N) \right\}$ *be two SVNMSs in X. Then, we have*

(1) Inclusion: $M \subseteq N$ *if and only if* $T_M^j(x) \leq T_N^j(x)$, $I_M^j(x) \geq I_N^j(x)$, $F_M^j(x) \geq F_N^j(x)$ *for* $j = 1, 2, \cdots, l(x : M, N)$*;*
(2) Equality: $M = N$ *if and only if* $M \subseteq N$ *and* $N \subseteq M$*;*

(3) *Complement:* $M^{\tilde{c}} = \left\{ \left\langle x, F_M^j(x), 1 - I_M^j(x), T_M^j(x) \right\rangle \middle| x \in X, j = 1, 2, \cdots, l(x : M) \right\}$;

(4) *Union:* $M \cup N = \left\{ \left\langle x, T_M^j(x) \vee T_N^j(x), I_M^j(x) \wedge I_N^j(x), F_M^j(x) \wedge F_N^j(x) \right\rangle \middle| x \in X, j = 1, 2, \cdots, l(x : M, N) \right\}$;

(5) *Intersection:* $M \cap N = \left\{ \left\langle x, T_M^j(x) \wedge T_N^j(x), I_M^j(x) \vee I_N^j(x), F_M^j(x) \vee F_N^j(x) \right\rangle \middle| x \in X, j = 1, 2, \cdots, l(x : M, N) \right\}$;

(6) *Addition:* $M \oplus N = \left\{ \left\langle x, T_M^j(x) + T_N^j(x) - T_M^j(x)T_N^j(x), I_M^j(x)I_N^j(x), F_M^j(x)F_N^j(x) \right\rangle \middle| x \in X, j = 1, 2, \cdots, l(x : M, N) \right\}$;

(7) *Multiplication:* $M \otimes N = \left\{ \left\langle x, T_M^j(x)T_N^j(x), I_M^j(x) + I_N^j(x) - I_M^j(x)I_N^j(x), F_M^j(x) + F_N^j(x) - F_M^j(x)F_N^j(x) \right\rangle \middle| x \in X, j = 1, 2, \cdots, l(M, N) \right\}$.

Definition 8 ([14]). *Let* $M = \left\{ \left\langle x_i, T_M^j(x_i), I_M^j(x_i), F_M^j(x_i) \right\rangle \middle| x_i \in X; i = 1, 2, \cdots, n; j = 1, 2, \cdots, l(x : M) \right\}$ *and* $N = \left\{ \left\langle x_i, T_N^j(x_i), I_N^j(x_i), F_N^j(x_i) \right\rangle \middle| x_i \in X; i = 1, 2, \cdots, n; j = 1, 2, \cdots, l(x : N) \right\}$ *be two SVNMSs in* $X = \{x_1, x_2, \cdots, x_n\}$. *Now, we propose the generalized distance measure between M and N as follows:*

$$D_P(M, N) = \left[\frac{1}{n} \sum_{i=1}^{n} \frac{1}{3l_i} \sum_{j=1}^{l_i} \left(\left| T_M^j(x_i) - T_N^j(x_i) \right|^P + \left| I_M^j(x_i) - I_N^j(x_i) \right|^P + \left| F_M^j(x_i) - F_N^j(x_i) \right|^P \right) \right]^{\frac{1}{P}}, \tag{4}$$

where $l_i = l(x_i : M, N) = \max\{l(x_i : M), l(x_i : N)\}$ *for* $i = 1, 2, \cdots, n$.

If $P = 1, 2$, it reduces to the Hamming distance and the Euclidean distance, which are usually applied to real science and engineering areas.

Based on the relationship between the distance measure and the similarity measure, we can introduce two distance-based similarity measures between M and N:

$$S_1(M, N) = 1 - D_P(M, N), \tag{5}$$

$$S_2(M, N) = \frac{1 - D_P(M, N)}{1 + D_P(M, N)}. \tag{6}$$

3. Some New Properties of SVNMS

The operation of SVNMS is discussed in depth and certain theoretical results are obtained. On this basis, this section generalizes the union and intersection operations of two SVNMSs to the general case, that is, for any indicator set. In addition, this section presents the arithmetic properties of SVNMSs.

Remark 1. *The union and intersection operations of the two SVNMSs can be extended to general case, that is, for any index set T, if* $M_t \in SVNMS(X)$, $\forall t \in T$, *we can define*

$$\cup_{t \in T} M_t = \left\{ \left\langle x, \vee_{t \in T} T_{M_t}^j(x), \wedge_{t \in T} I_{M_t}^j(x), \wedge_{t \in T} F_{M_t}^j(x) \right\rangle \middle| x \in X, j = 1, 2, \cdots, l_x \right\},$$

and

$$\cap_{t \in T} M_t = \left\{ \left\langle x, \wedge_{t \in T} T_{M_t}^j(x), \vee_{t \in T} I_{M_t}^j(x), \vee_{t \in T} F_{M_t}^j(x) \right\rangle \middle| x \in X, j = 1, 2, \cdots, l_x \right\},$$

where $l_x = \max\{ l(x : M_t) | t \in T\}$.

Proposition 1. *Let M, N and Q be three SVNMSs in X. We have the following operational properties:*

(1) *Commutation:* $M \cup N = N \cup M$, $M \cap N = N \cap M$;
(2) *Association:* $M \cup (N \cup Q) = (M \cup N) \cup Q$, $M \cap (N \cap Q) = (M \cap N) \cap Q$;
(3) *Idempotent:* $M \cup M = M$, $M \cap M = M$;
(4) *Absorption:* $M \cup (M \cap N) = M$, $M \cap (M \cup N) = M$;
(5) *Identity:* $M \cup \widetilde{M} = \widetilde{M}$; $M \cap \widetilde{M} = M$, $M \cup \widetilde{\Phi} = M$, $M \cap \widetilde{\Phi} = \widetilde{\Phi}$;
(6) *Distribution:* $M \cup (N \cap Q) = (M \cup N) \cap (M \cup Q)$, $M \cap (N \cup Q) = (M \cap N) \cup (M \cap Q)$;

(7) Involution: $\left(M^{\tilde{c}}\right)^{\tilde{c}} = M$, $\left(\widetilde{M}\right)^{\tilde{c}} = \widetilde{\Phi}$, $\left(\widetilde{\Phi}\right)^{\tilde{c}} = \widetilde{M}$;
(8) De Morgan: $(M \cap N)^{\tilde{c}} = M^{\tilde{c}} \cup N^{\tilde{c}}$, $(M \cup N)^{\tilde{c}} = M^{\tilde{c}} \cap N^{\tilde{c}}$.

Remark 2. *As we know, the complementation can be established in classical set, however, it is not true in SVNMS. For example, let* $X = \{x_1, x_2, x_3\}$, $M \in SVNMS(X)$ *as follows:*

$$M = \{\langle x_1, (0.5, 0.3), (0.1, 0.1), (0.7, 0.8)\rangle, \langle x_2, (0.7, 0.68, 0.62), (0.3, 0.45, 0.5), (0.34, 0.28, 0.49)\rangle, \langle x_3, (0.67, 0.5, 0.3), (0.2, 0.3, 0.4), (0.4, 0.5, 0.7)\rangle\}.$$

Obviously,

$$M \cup M^{\tilde{c}} = \{\langle x_1, (0.7, 0.8), (0.1, 0.1), (0.5, 0.3)\rangle, \langle x_2, (0.7, 0.68, 0.62), (0.3, 0.45, 0.5), (0.34, 0.28, 0.49)\rangle, \langle x_3, (0.67, 0.5, 0.7), (0.2, 0.3, 0.4), (0.4, 0.5, 0.3)\rangle\} \neq \widetilde{M};$$

$$M \cap M^{\tilde{c}} = \{\langle x_1, (0.5, 0.3), (0.9, 0.9), (0.7, 0.8)\rangle, \langle x_2, (0.34, 0.28, 0.49), (0.7, 0.55, 0.5), (0.7, 0.68, 0.62)\rangle, \langle x_3, (0.4, 0.5, 0.3), (0.8, 0.7, 0.6), (0.67, 0.5, 0.7)\rangle\} \neq \widetilde{\Phi}.$$

4. Decomposition Theorem and Representation Theorem of SVNMS

In this section, the notions of cut sets, strong cut sets of SVNMS are defined. Some properties of cut sets are proposed. We also investigate decomposition theorem and representation theorem of SVNMS based on cut sets.

4.1. Decomposition Theorem

Definition 9. *Let* $X = \{x_1, x_2, \cdots, x_n\}$, $A \in SVNMS(X)$ *and* $\alpha, \beta, \gamma \in [0, 1]$ *with* $0 \leq \alpha + \beta + \gamma \leq 3$. *The* α*-cut set of truth value function generated by* A *is defined as follows:*

$$A^{\alpha} = \left\{x_i \in X \middle| T_A^j(x_i) \geq \alpha; i = 1, 2, \cdots, n; j = 1, 2, \cdots, l(x_i : A)\right\}; \tag{7}$$

The strong α*-cut set of truth value function generated by* A *is defined as follows:*

$$A^{\alpha+} = \left\{x_i \in X \middle| T_A^j(x_i) > \alpha; i = 1, 2, \cdots, n; j = 1, 2, \cdots, l(x_i : A)\right\}; \tag{8}$$

The β*-cut set of indeterminacy value function generated by* A *is defined as follows:*

$$A\beta = \left\{x_i \in X \middle| I_A^j(x_i) \leq \beta; i = 1, 2, \cdots, n; j = 1, 2, \cdots, l(x_i : A)\right\}; \tag{9}$$

The strong β*-cut set of indeterminacy value function generated by* A *is defined as follows:*

$$A\beta+ = \left\{x_i \in X \middle| I_A^j(x_i) < \beta; i = 1, 2, \cdots, n; j = 1, 2, \cdots, l(x_i : A)\right\}; \tag{10}$$

The γ*-cut set of falsity value function generated by* A *is defined as follows:*

$$A_{\gamma} = \left\{x_i \in X \middle| F_A^j(x_i) \leq \gamma; i = 1, 2, \cdots, n; j = 1, 2, \cdots, l(x_i : A)\right\}; \tag{11}$$

The strong γ*-cut set of falsity value function generated by* A *is defined as follows:*

$$A_{\gamma+} = \left\{x_i \in X \middle| F_A^j(x_i) < \gamma; i = 1, 2, \cdots, n; j = 1, 2, \cdots, l(x_i : A)\right\}. \tag{12}$$

Next, we can define the (α,β,γ)-*cut* sets as follows:

$$A^{(\alpha,\,\beta,\,\gamma)} = \left\{x_i \in X \middle| T_A^j(x_i) \geq \alpha, I_A^j(x_i) \leq \beta, F_A^j(x_i) \leq \gamma; i = 1,2,\cdots,n; j = 1,2,\cdots,l(x_i : A)\right\}; \quad (13)$$

$$A^{(\alpha+,\,\beta,\,\gamma)} = \left\{x_i \in X \middle| T_A^j(x_i) > \alpha, I_A^j(x_i) \leq \beta, F_A^j(x_i) \leq \gamma; i = 1,2,\cdots,n; j = 1,2,\cdots,l(x_i : A)\right\}; \quad (14)$$

$$A^{(\alpha,\,\beta+,\,\gamma)} = \left\{x_i \in X \middle| T_A^j(x_i) \geq \alpha, I_A^j(x_i) < \beta, F_A^j(x_i) \leq \gamma; i = 1,2,\cdots,n; j = 1,2,\cdots,l(x_i : A)\right\}; \quad (15)$$

$$A^{(\alpha,\,\beta,\,\gamma+)} = \left\{x_i \in X \middle| T_A^j(x_i) \geq \alpha, I_A^j(x_i) \leq \beta, F_A^j(x_i) < \gamma; i = 1,2,\cdots,n; j = 1,2,\cdots,l(x_i : A)\right\}; \quad (16)$$

$$A^{(\alpha+,\,\beta+,\,\gamma)} = \left\{x_i \in X \middle| T_A^j(x_i) > \alpha, I_A^j(x_i) < \beta, F_A^j(x_i) \leq \gamma; i = 1,2,\cdots,n; j = 1,2,\cdots,l(x_i : A)\right\}; \quad (17)$$

$$A^{(\alpha+,\,\beta,\,\gamma+)} = \left\{x_i \in X \middle| T_A^j(x_i) > \alpha, I_A^j(x_i) \leq \beta, F_A^j(x_i) < \gamma; i = 1,2,\cdots,n; j = 1,2,\cdots,l(x_i : A)\right\}; \quad (18)$$

$$A^{(\alpha,\,\beta+,\,\gamma+)} = \left\{x_i \in X \middle| T_A^j(x_i) \geq \alpha, I_A^j(x_i) < \beta, F_A^j(x_i) < \gamma; i = 1,2,\cdots,n; j = 1,2,\cdots,l(x_i : A)\right\}; \quad (19)$$

$$A^{(\alpha+,\,\beta+,\,\gamma+)} = \left\{x_i \in X \middle| T_A^j(x_i) > \alpha, I_A^j(x_i) < \beta, F_A^j(x_i) < \gamma; i = 1,2,\cdots,n; j = 1,2,\cdots,l(x_i : A)\right\}. \quad (20)$$

The α-*cut* sets, β-*cut* sets, γ-*cut* sets of SVNMS satisfy the following properties:

Theorem 1. *Let $A, B \in SVNMS(X)$, $\alpha,\beta,\gamma \in [0,1]$ with $0 \leq \alpha+\beta+\gamma \leq 3$. Then,*

(1) $A \subseteq B \Rightarrow A^\alpha \subseteq B^\alpha$, $A\beta \subseteq B\beta$, $A_\gamma \subseteq B_\gamma$;

(2) $(A\cap B)^\alpha = A^\alpha \cap B^\alpha$, $(A\cap B)\beta = A\beta \cap B\beta$, $(A\cap B)_\gamma = A_\gamma \cap B_\gamma$;

(3) $(A\cup B)^\alpha = A^\alpha \cup B^\alpha$, $(A\cup B)\beta = A\beta \cup B\beta$, $(A\cup B)_\gamma = A_\gamma \cup B_\gamma$;

(4) $\left(\underset{t\in T}{\cap} A_t\right)^\alpha = \underset{t\in T}{\cap}(A_t)^\alpha$, $\left(\underset{t\in T}{\cap} A_t\right)\beta = \underset{t\in T}{\cap}(A_t)\beta$, $\left(\underset{t\in T}{\cap} A_t\right)_\gamma = \underset{t\in T}{\cap}(A_t)_\gamma$;

(5) $\left(\underset{t\in T}{\cup} A_t\right)^\alpha = \underset{t\in T}{\cup}(A_t)^\alpha$, $\left(\underset{t\in T}{\cup} A_t\right)\beta = \underset{t\in T}{\cup}(A_t)\beta$, $\left(\underset{t\in T}{\cup} A_t\right)_\gamma = \underset{t\in T}{\cup}(A_t)_\gamma$;

(6) $\alpha_1 \geq \alpha_2$, $\beta_1 \leq \beta_2$, $\gamma_1 \leq \gamma_2 \Rightarrow A^{\alpha_1} \subseteq A^{\alpha_2}$, $A\beta_1 \subseteq A\beta_2$, $A_{\gamma_1} \subseteq A_{\gamma_2}$.

Proof.

(1) Since $x \in A^\alpha$, we have $T_A^j(x) \geq \alpha$. From $A \subseteq B$, it follows that $T_A^j(x) \leq T_B^j(x)$. Thus, $T_B^j(x) \geq \alpha$. Thus, $x \in B^\alpha$. Therefore, $A^\alpha \subseteq B^\alpha$ for $j = 1,2,\cdots,l(x : A,B)$. Since $x \in A\beta$, we have $I_A^j(x) \leq \beta$. From $A \subseteq B$, it follows that $I_A^j(x) \geq I_B^j(x)$. Thus, $I_B^j(x) \leq \beta$. Thus, $x \in B\beta$. Hence, $A\beta \subseteq B\beta$ for $j = 1,2,\cdots,l(x : A,B)$.

(2) From $x \in (A\cap B)^\alpha$, we can obtain $T_{A\cap B}^j(x) \geq \alpha$. Then, $\min\left\{T_A^j(x), T_B^j(x)\right\} \geq \alpha$, that is, $T_A^j(x) \geq \alpha$, $T_B^j(x) \geq \alpha$. Thus, $x \in A^\alpha, x \in B^\alpha$. Hence, $x \in A^\alpha \cap B^\alpha$. On the other hand, since $x \in A^\alpha \cap B^\alpha$, we have $x \in A^\alpha, x \in B^\alpha$, that is, $T_A^j(x) \geq \alpha, T_B^j(x) \geq \alpha$. Then, $\min\left\{T_A^j(x), T_B^j(x)\right\} \geq \alpha$. Thus, $T_{A\cap B}^j(x) \geq \alpha$. Hence, $x \in (A\cap B)^\alpha$. Based on the above facts, we can check that $(A\cap B)^\alpha = A^\alpha \cap B^\alpha$ for $j = 1,2,\cdots,l(x : A,B)$.

Since $x \in (A\cap B)\beta$, we have $I_{A\cap B}^j(x) \leq \beta$. Then, $\max\left\{I_A^j(x), I_B^j(x)\right\} \leq \beta$, that is, $I_A^j(x) \leq \beta$, $I_B^j(x) \leq \beta$. Then, $x \in A\beta, x \in B\beta$. Hence, $x \in A\beta \cap B\beta$. On the other hand, from $x \in A\beta \cap B\beta$, we have $x \in A\beta, x \in B\beta$. Thus, $I_A^j(x) \leq \beta, I_B^j(x) \leq \beta$, that is, $\max\left\{I_A^j(x), I_B^j(x)\right\} \leq \beta$. Thus, $I_{A\cap B}^j(x) \leq \beta$. Thus, $x \in (A\cap B)\beta$. Therefore, we can check that $(A\cap B)\beta = A\beta \cap B\beta$ for $j = 1,2,\cdots,l(x : A,B)$.

(3) From $x \in (A\cup B)^\alpha$, we have $T_{A\cup B}^j(x) \geq \alpha$. Thus, $\max\left\{T_A^j(x), T_B^j(x)\right\} \geq \alpha$, that is, $T_A^j(x) \geq \alpha$ or $T_B^j(x) \geq \alpha$. Thus, $x \in A^\alpha$ or $x \in B^\alpha$. Hence, $x \in A^\alpha \cup B^\alpha$. On the other hand, since $x \in A^\alpha \cup B^\alpha$, we have $x \in A^\alpha$ or $x \in B^\alpha$. Thus, $T_A^j(x) \geq \alpha$ or $T_B^j(x) \geq \alpha$, that is, $\max\left\{T_A^j(x), T_B^j(x)\right\} \geq \alpha$. Thus,

$T^j_{A\cup B}(x) \geq \alpha$. Hence, $x \in (A \cup B)^\alpha$. Using the above facts, we can check that $(A \cup B)^\alpha = A^\alpha \cup B^\alpha$ for $j = 1, 2, \cdots, l(x : A, B)$.
Since $x \in (A \cup B)_\gamma$, we have $F^j_{A\cup B}(x) \leq \gamma$, that is, $\min\left\{F^j_A(x), F^j_B(x)\right\} \leq \gamma$. Thus, $F^j_A(x) \leq \gamma$ or $F^j_B(x) \leq \gamma$. Thus, $x \in A_\gamma$ or $x \in B_\gamma$. Hence, $x \in A_\gamma \cup B_\gamma$. On the other hand, from $x \in A_\gamma \cup B_\gamma$, we have $x \in A_\gamma$ or $x \in B_\gamma$. Thus, $F^j_A(x) \leq \gamma$ or $F^j_B(x) \leq \gamma$, that is, $\min\left\{F^j_A(x), F^j_B(x)\right\} \leq \gamma$. Thus, $F^j_{A\cup B}(x) \leq \gamma$. Hence, $x \in (A \cup B)_\gamma$. Therefore, we can check that $(A \cup B)_\gamma = A_\gamma \cup B_\gamma$ for $j = 1, 2, \cdots, l(x : A, B)$.
(4) From $x \in \left(\underset{t\in T}{\cap} A_t\right)^\alpha$, we have $T^j_{\underset{t\in T}{\cap} A_t}(x) \geq \alpha$, that is, $\underset{t\in T}{\inf}\left\{T^j_{A_t}(x)\right\} \geq \alpha$. Thus, $T^j_{A_t}(x) \geq \alpha$ for all $t \in T$, that is, $x \in (A_t)^\alpha$ for all $t \in T$. Hence, $x \in \underset{t\in T}{\cap}(A_t)^\alpha$. On the other hand, from $x \in \underset{t\in T}{\cap}(A_t)^\alpha$, it follows that $x \in (A_t)^\alpha$ for all $t \in T$. Then, $T^j_{A_t}(x) \geq \alpha$ for all $t \in T$, that is, $\underset{t\in T}{\inf}\left\{T^j_{A_t}(x)\right\} \geq \alpha$. Then, $T^j_{\underset{t\in T}{\cap} A_t}(x) \geq \alpha$. Thus, $x \in \left(\underset{t\in T}{\cap} A_t\right)^\alpha$. Based on the above facts, we can check that $\left(\underset{t\in T}{\cap} A_t\right)^\alpha = \underset{t\in T}{\cap}(A_t)^\alpha$ for $j = 1, 2, \cdots, l(l = \max\{l(x : A_t) | t \in T\})$.
Since $x \in \left(\underset{t\in T}{\cap} A_t\right)\beta$, we have $I^j_{\underset{t\in T}{\cap} A_t}(x) \leq \beta$, that is, $\underset{t\in T}{\sup}\left\{I^j_{A_t}(x)\right\} \leq \beta$. Thus, $I^j_{A_t}(x) \leq \beta$ for all $t \in T$, that is, $x \in (A_t)\beta$ for all $t \in T$. Thus, $x \in \underset{t\in T}{\cap}(A_t)\beta$. On the other hand, from $x \in \underset{t\in T}{\cap}(A_t)\beta$, we have $x \in (A_t)\beta$ for all $t \in T$. Thus, $I^j_{A_t}(x) \leq \beta$ for all $t \in T$, that is, $\underset{t\in T}{\sup}\left\{I^j_{A_t}(x)\right\} \leq \beta$. Thus, $I^j_{\underset{t\in T}{\cap} A_t}(x) \leq \beta$. Hence, $x \in \left(\underset{t\in T}{\cap} A_t\right)\beta$. Therefore, we can check that $\left(\underset{t\in T}{\cap} A_t\right)\beta = \underset{t\in T}{\cap}(A_t)\beta$ for $j = 1, 2, \cdots, l(l = \max\{l \ (x : A_t) | t \in T\})$.
(5) The proof of (5) is similar to Theorem 1 (4).
(6) The proof of (6) is obvious from Definition 9.

The (α, β, γ)-*cut* sets of SVNMS satisfy the following properties. □

Theorem 2. *Let $A, B \in SVNMS(X)$, $\alpha, \beta, \gamma \in [0, 1]$ with $0 \leq \alpha + \beta + \gamma \leq 3$. Then,*

(1) $A^{(\alpha+,\beta+,\gamma+)} \subseteq A^{(\alpha+,\beta+,\gamma)} \subseteq A^{(\alpha+,\beta,\gamma)} \subseteq A^{(\alpha,\beta,\gamma)}$, $A^{(\alpha+,\beta+,\gamma+)} \subseteq A^{(\alpha+,\beta+,\gamma)} \subseteq A^{(\alpha,\beta+,\gamma)} \subseteq A^{(\alpha,\beta,\gamma)}$, $A^{(\alpha+,\beta+,\gamma+)} \subseteq A^{(\alpha+,\beta,\gamma+)} \subseteq A^{(\alpha+,\beta,\gamma)} \subseteq A^{(\alpha,\beta,\gamma)}$, $A^{(\alpha+,\beta+,\gamma+)} \subseteq A^{(\alpha+,\beta,\gamma+)} \subseteq A^{(\alpha,\beta,\gamma+)} \subseteq A^{(\alpha,\beta,\gamma)}$, $A^{(\alpha+,\beta+,\gamma+)} \subseteq A^{(\alpha,\beta+,\gamma+)} \subseteq A^{(\alpha,\beta+,\gamma)} \subseteq A^{(\alpha,\beta,\gamma)}$, $A^{(\alpha+,\beta+,\gamma+)} \subseteq A^{(\alpha,\beta+,\gamma+)} \subseteq A^{(\alpha,\beta,\gamma+)} \subseteq A^{(\alpha,\beta,\gamma)}$;

(2) $A \subseteq B \Rightarrow A^{(\alpha,\beta,\gamma)} \subseteq B^{(\alpha,\beta,\gamma)}$, $A^{(\alpha+,\beta,\gamma)} \subseteq B^{(\alpha+,\beta,\gamma)}$, $A^{(\alpha,\beta+,\gamma)} \subseteq B^{(\alpha,\beta+,\gamma)}$, $A^{(\alpha,\beta,\gamma+)} \subseteq B^{(\alpha,\beta,\gamma+)}$, $A^{(\alpha+,\beta+,\gamma)} \subseteq B^{(\alpha+,\beta+,\gamma)}$, $A^{(\alpha+,\beta,\gamma+)} \subseteq B^{(\alpha+,\beta,\gamma+)}$, $A^{(\alpha,\beta+,\gamma+)} \subseteq B^{(\alpha,\beta+,\gamma+)}$, $A^{(\alpha+,\beta+,\gamma+)} \subseteq B^{(\alpha+,\beta+,\gamma+)}$;

(3) $\alpha_1 \leq \alpha_2, \beta_1 \geq \beta_2, \gamma_1 \geq \gamma_2 \Rightarrow A^{(\alpha_1,\beta_1,\gamma_1)} \supseteq A^{(\alpha_1+,\beta_1+,\gamma_1+)} \supseteq A^{(\alpha_2,\beta_2,\gamma_2)} \supseteq A^{(\alpha_2+,\beta_2+,\gamma_2+)}$;

(4) $A^{(\alpha,\beta,\gamma)} = A^\alpha \cap A\beta \cap A_\gamma$;

(5) $(A \cap B)^{(\alpha,\beta,\gamma)} = A^{(\alpha,\beta,\gamma)} \cap B^{(\alpha,\beta,\gamma)}$, $(A \cap B)^{(\alpha+,\beta,\gamma)} = A^{(\alpha+,\beta,\gamma)} \cap B^{(\alpha+,\beta,\gamma)}$, $(A \cap B)^{(\alpha,\beta+,\gamma)} = A^{(\alpha,\beta+,\gamma)} \cap B^{(\alpha,\beta+,\gamma)}$, $(A \cap B)^{(\alpha,\beta,\gamma+)} = A^{(\alpha,\beta,\gamma+)} \cap B^{(\alpha,\beta,\gamma+)}$, $(A \cap B)^{(\alpha+,\beta+,\gamma)} = A^{(\alpha+,\beta+,\gamma)} \cap B^{(\alpha+,\beta+,\gamma)}$, $(A \cap B)^{(\alpha+,\beta,\gamma+)} = A^{(\alpha+,\beta,\gamma+)} \cap B^{(\alpha+,\beta,\gamma+)}$, $(A \cap B)^{(\alpha,\beta+,\gamma+)} = A^{(\alpha,\beta+,\gamma+)} \cap B^{(\alpha,\beta+,\gamma+)}$, $(A \cap B)^{(\alpha+,\beta+,\gamma+)} = A^{(\alpha+,\beta+,\gamma+)} \cap B^{(\alpha+,\beta+,\gamma+)}$;

(6) $(A \cup B)^{(\alpha,\beta,\gamma)} \supseteq A^{(\alpha,\beta,\gamma)} \cup B^{(\alpha,\beta,\gamma)}$, $(A \cup B)^{(\alpha+,\beta,\gamma)} \supseteq A^{(\alpha+,\beta,\gamma)} \cup B^{(\alpha+,\beta,\gamma)}$, $(A \cup B)^{(\alpha,\beta+,\gamma)} \supseteq A^{(\alpha,\beta+,\gamma)} \cup B^{(\alpha,\beta+,\gamma)}$, $(A \cup B)^{(\alpha,\beta,\gamma+)} \supseteq A^{(\alpha,\beta,\gamma+)} \cup B^{(\alpha,\beta,\gamma+)}$, $(A \cup B)^{(\alpha+,\beta+,\gamma)} \supseteq A^{(\alpha+,\beta+,\gamma)} \cup B^{(\alpha+,\beta+,\gamma)}$, $(A \cup B)^{(\alpha+,\beta,\gamma+)} \supseteq A^{(\alpha+,\beta,\gamma+)} \cup B^{(\alpha+,\beta,\gamma+)}$, $(A \cup B)^{(\alpha,\beta+,\gamma+)} \supseteq A^{(\alpha,\beta+,\gamma+)} \cup B^{(\alpha,\beta+,\gamma+)}$, $(A \cup B)^{(\alpha+,\beta+,\gamma+)} \supseteq A^{(\alpha+,\beta+,\gamma+)} \cup B^{(\alpha+,\beta+,\gamma+)}$;

(7) $\left(\underset{t\in T}{\cap}A_t\right)^{(\alpha,\beta,\gamma)} = \underset{t\in T}{\cap}(A_t)^{(\alpha,\beta,\gamma)}$, $\left(\underset{t\in T}{\cap}A_t\right)^{(\alpha+,\beta,\gamma)} = \underset{t\in T}{\cap}(A_t)^{(\alpha+,\beta,\gamma)}$, $\left(\underset{t\in T}{\cap}A_t\right)^{(\alpha,\beta+,\gamma)} = \underset{t\in T}{\cap}(A_t)^{(\alpha,\beta+,\gamma)}$, $\left(\underset{t\in T}{\cap}A_t\right)^{(\alpha,\beta,\gamma+)} = \underset{t\in T}{\cap}(A_t)^{(\alpha,\beta,\gamma+)}$, $\left(\underset{t\in T}{\cap}A_t\right)^{(\alpha+,\beta+,\gamma)} = \underset{t\in T}{\cap}(A_t)^{(\alpha+,\beta+,\gamma)}$, $\left(\underset{t\in T}{\cap}A_t\right)^{(\alpha+,\beta,\gamma+)} = \underset{t\in T}{\cap}(A_t)^{(\alpha+,\beta,\gamma+)}$, $\left(\underset{t\in T}{\cap}A_t\right)^{(\alpha,\beta+,\gamma+)} = \underset{t\in T}{\cap}(A_t)^{(\alpha,\beta+,\gamma+)}$, $\left(\underset{t\in T}{\cap}A_t\right)^{(\alpha+,\beta+,\gamma+)} = \underset{t\in T}{\cap}(A_t)^{(\alpha+,\beta+,\gamma+)}$;

(8) $\left(\underset{t\in T}{\cup}A_t\right)^{(\alpha,\beta,\gamma)} \supseteq \underset{t\in T}{\cup}(A_t)^{(\alpha,\beta,\gamma)}$, $\left(\underset{t\in T}{\cup}A_t\right)^{(\alpha+,\beta,\gamma)} \supseteq \underset{t\in T}{\cup}(A_t)^{(\alpha+,\beta,\gamma)}$, $\left(\underset{t\in T}{\cup}A_t\right)^{(\alpha,\beta+,\gamma)} \supseteq \underset{t\in T}{\cup}(A_t)^{(\alpha,\beta+,\gamma)}$, $\left(\underset{t\in T}{\cup}A_t\right)^{(\alpha,\beta,\gamma+)} \supseteq \underset{t\in T}{\cup}(A_t)^{(\alpha,\beta,\gamma+)}$, $\left(\underset{t\in T}{\cup}A_t\right)^{(\alpha+,\beta+,\gamma)} \supseteq \underset{t\in T}{\cup}(A_t)^{(\alpha+,\beta+,\gamma)}$, $\left(\underset{t\in T}{\cup}A_t\right)^{(\alpha+,\beta,\gamma+)} \supseteq \underset{t\in T}{\cup}(A_t)^{(\alpha+,\beta,\gamma+)}$, $\left(\underset{t\in T}{\cup}A_t\right)^{(\alpha,\beta+,\gamma+)} \supseteq \underset{t\in T}{\cup}(A_t)^{(\alpha,\beta+,\gamma+)}$, $\left(\underset{t\in T}{\cup}A_t\right)^{(\alpha+,\beta+,\gamma+)} \supseteq \underset{t\in T}{\cup}(A_t)^{(\alpha+,\beta+,\gamma+)}$;

(9) $\underset{t\in T}{\cap}A^{(\alpha_t,\beta_t,\gamma_t)} = A^{(\alpha,\beta,\gamma)}$, $\underset{t\in T}{\cap}A^{(\alpha_t+,\beta_t,\gamma_t)} = A^{(\alpha+,\beta,\gamma)}$, $\underset{t\in T}{\cap}A^{(\alpha_t,\beta_t+,\gamma_t)} = A^{(\alpha,\beta+,\gamma)}$, $\underset{t\in T}{\cap}A^{(\alpha_t,\beta_t,\gamma_t+)} = A^{(\alpha,\beta,\gamma+)}$, $\underset{t\in T}{\cap}A^{(\alpha_t+,\beta_t+,\gamma_t)} = A^{(\alpha+,\beta+,\gamma)}$, $\underset{t\in T}{\cap}A^{(\alpha_t+,\beta_t,\gamma_t+)} = A^{(\alpha+,\beta,\gamma+)}$, $\underset{t\in T}{\cap}A^{(\alpha_t,\beta_t+,\gamma_t+)} = A^{(\alpha,\beta+,\gamma+)}$, $\underset{t\in T}{\cap}A^{(\alpha_t+,\beta_t+,\gamma_t+)} = A^{(\alpha+,\beta+,\gamma+)}$,

where e $\alpha = \vee_{t\in T}\alpha_t$, $\beta = \wedge_{t\in T}\beta_t$, $\gamma = \wedge_{t\in T}\gamma_t$.

Proof. The proofs of (1)~(4) are obtained directly from Definition 9. We denote,

$$A\cup B = \left\{\left\langle x, \max\left\{T_A^j(x), T_B^j(x)\right\}, \min\left\{I_A^j(x), I_B^j(x)\right\}, \min\left\{F_A^j(x), F_B^j(x)\right\}\right\rangle\right\},$$

$$A\cap B = \left\{\left\langle x, \min\left\{T_A^j(x), T_B^j(x)\right\}, \max\left\{I_A^j(x), I_B^j(x)\right\}, \max\left\{F_A^j(x), F_B^j(x)\right\}\right\rangle\right\},$$

where $j = 1, 2, \cdots, l(x : A, B)$.

$$\cup_{t\in T} A_t = \left\{\left\langle x, \underset{t\in T}{sup}\left\{T_{A_t}^j(x)\right\}, \underset{t\in T}{\inf}\left\{I_{A_t}^j(x)\right\}, \underset{t\in T}{\inf}\left\{F_{A_t}^j(x)\right\}\right\rangle\right\},$$

$$\cap_{t\in T} A_t = \left\{\left\langle x, \underset{t\in T}{\inf}\left\{T_{A_t}^j(x)\right\}, \underset{t\in T}{\sup}\left\{I_{A_t}^j(x)\right\}, \underset{t\in T}{\sup}\left\{F_{A_t}^j(x)\right\}\right\rangle\right\},$$

$j = 1, 2, \cdots, l$ where $l = \max\{l(x : A_t)|t \in T\}$.

(5) From $x \in (A\cap B)^{(\alpha,\beta,\gamma)}$, we have $\min\left\{T_A^j(x), T_B^j(x)\right\} \geq \alpha$, $\max\left\{I_A^j(x), I_B^j(x)\right\} \leq \beta$, $\max\left\{F_A^j(x), F_B^j(x)\right\} \leq \gamma$, that is, $T_A^j(x) \geq \alpha$ and $T_B^j(x) \geq \alpha, I_A^j(x) \leq \beta$ and $I_B^j(x) \leq \beta, F_A^j(x) \leq \gamma$ and $F_B^j(x) \leq \gamma$. Thus, $T_A^j(x) \geq \alpha$, $I_A^j(x) \leq \beta$, $F_A^j(x) \leq \gamma$ and $T_B^j(x) \geq \alpha$, $I_B^j(x) \leq \beta$, $F_B^j(x) \leq \gamma$, that is, $x \in A^{(\alpha,\beta,\gamma)}$, $x \in B^{(\alpha,\beta,\gamma)}$. Hence, $x \in A^{(\alpha,\beta,\gamma)} \cap B^{(\alpha,\beta,\gamma)}$. On the other hand, since $x \in A^{(\alpha,\beta,\gamma)} \cap B^{(\alpha,\beta,\gamma)}$, we have $x \in A^{(\alpha,\beta,\gamma)}$, $x \in B^{(\alpha,\beta,\gamma)}$, that is, $T_A^j(x) \geq \alpha$, $I_A^j(x) \leq \beta$, $F_A^j(x) \leq \gamma$ and $T_B^j(x) \geq \alpha$, $I_B^j(x) \leq \beta$, $F_B^j(x) \leq \gamma$. Thus, $T_A^j(x) \geq \alpha$ and $T_B^j(x) \geq \alpha$, $I_A^j(x) \leq \beta$ and $I_B^j(x) \leq \beta$, $F_A^j(x) \leq \gamma$ and $F_B^j(x) \leq \gamma$. Hence, $\min\left\{T_A^j(x), T_B^j(x)\right\} \geq \alpha, \max\left\{I_A^j(x), I_B^j(x)\right\} \leq \beta$, $\max\left\{F_A^j(x), F_B^j(x)\right\} \leq \gamma$. So, $x \in (A\cap B)^{(\alpha,\beta,\gamma)}$. Therefore, $(A\cap B)^{(\alpha,\beta,\gamma)} = A^{(\alpha,\beta,\gamma)} \cap B^{(\alpha,\beta,\gamma)}$ for $j = 1, 2, \cdots, l(x : A, B)$.

(6) Since $x \in A^{(\alpha,\beta,\gamma)} \cup B^{(\alpha,\beta,\gamma)}$, we have $x \in A^{(\alpha,\beta,\gamma)}$ or $x \in B^{(\alpha,\beta,\gamma)}$, that is, $T_A^j(x) \geq \alpha, I_A^j(x) \leq \beta, F_A^j(x) \leq \gamma$ or $T_B^j(x) \geq \alpha, I_B^j(x) \leq \beta, F_B^j(x) \leq \gamma$. Thus, $T_A^j(x) \geq \alpha$ or $T_B^j(x) \geq \alpha$, $I_A^j(x) \leq \beta$ or $I_B^j(x) \leq \beta$, $F_A^j(x) \leq \gamma$ or $F_B^j(x) \leq \gamma$, that is, $\max\left\{T_A^j(x), T_B^j(x)\right\} \geq \alpha$, $\min\left\{I_A^j(x), I_B^j(x)\right\} \leq$

β,$\min\left\{F_A^j(x), F_B^j(x)\right\} \leq \gamma$. Thus, $x \in (A \cup B)^{(\alpha,\beta,\gamma)}$. Therefore, $A^{(\alpha,\beta,\gamma)} \cup B^{(\alpha,\beta,\gamma)} \subseteq (A \cup B)^{(\alpha,\beta,\gamma)}$ for $j = 1, 2, \cdots, l(x : A, B)$.

(7) From $x \in \left(\underset{t\in T}{\cap} A_t\right)^{(\alpha,\beta,\gamma)}$, we have $\inf\limits_{t\in T}\left\{T_{A_t}^j(x)\right\} \geq \alpha, \sup\limits_{t\in T}\left\{I_{A_t}^j(x)\right\} \leq \beta, \sup\limits_{t\in T}\left\{F_{A_t}^j(x)\right\} \leq \gamma$, that is, $T_{A_t}^j(x) \geq \alpha, I_{A_t}^j(x) \leq \beta, F_{A_t}^j(x) \leq \gamma$ for all $t \in T$. Thus, $x \in (A_t)^{(\alpha,\beta,\gamma)}$ for all $t \in T$. Hence, $x \in \underset{t\in T}{\cap}(A_t)^{(\alpha,\beta,\gamma)}$. On the other hand, for any $x \in \underset{t\in T}{\cap}(A_t)^{(\alpha,\beta,\gamma)}$, we have $x \in (A_t)^{(\alpha,\beta,\gamma)}$ for all $t \in T$, that is, $T_{A_t}^j(x) \geq \alpha, I_{A_t}^j(x) \leq \beta, F_{A_t}^j(x) \leq \gamma$ for all $t \in T$. Thus,

$$\inf_{t\in T}\left\{F_{A_t}^j(x)\right\} \geq \alpha, \sup_{t\in T}\left\{I_{A_t}^j(x)\right\} \leq \beta, \sup_{t\in T}\left\{F_{A_t}^j(x)\right\} \leq \gamma.$$

Hence, $x \in \left(\underset{t\in T}{\cap} A_t\right)^{(\alpha,\beta,\gamma)}$. Therefore, $\left(\underset{t\in T}{\cap} A_t\right)^{(\alpha,\beta,\gamma)} = \underset{t\in T}{\cap}(A_t)^{(\alpha,\beta,\gamma)}$ for $j = 1, 2, \cdots, l(x : A, B)$.

(8) The proof of (8) is similar to that of (6).

(9) Since $x \in \underset{t\in T}{\cap} A^{(\alpha_t,\beta_t,\gamma_t)}$, we have $x \in A^{(\alpha_t,\beta_t,\gamma_t)}$ for all $t \in T$, that is,

$$T_A^j(x) \geq \alpha_t, I_A^j(x) \leq \beta_t, F_A^j(x) \leq \gamma_t \textit{ for all } t \in T.$$

Thus, $T_A^j(x) \geq \underset{t\in T}{\vee}\alpha_t, I_A^j(x) \leq \underset{t\in T}{\wedge}\beta_t, F_A^j(x) \leq \underset{t\in T}{\wedge}\gamma_t$, that is, $T_A^j(x) \geq \alpha, I_A^j(x) \leq \beta, F_A^j(x) \leq \gamma$. Thus, $x \in A^{(\alpha,\beta,\gamma)}$. On the other hand, from $x \in A^{(\alpha,\beta,\gamma)}$, we have $T_A^j(x) \geq \alpha, I_A^j(x) \leq \beta, F_A^j(x) \leq \gamma$, that is, $T_A^j(x) \geq \underset{t\in T}{\vee}\alpha_t, I_A^j(x) \leq \underset{t\in T}{\wedge}\beta_t, F_A^j(x) \leq \underset{t\in T}{\wedge}\gamma_t$ for all $t \in T$. Thus, $T_A^j(x) \geq \alpha_t, I_A^j(x) \leq \beta_t, F_A^j(x) \leq \gamma_t$ for all $t \in T$. Thus,$x \in A^{(\alpha_t,\beta_t,\gamma_t)}$ for all $t \in T$. Hence, $x \in \underset{t\in T}{\cap} A^{(\alpha_t,\beta_t,\gamma_t)}$. Therefore, $\underset{t\in T}{\cap} A^{(\alpha_t,\beta_t,\gamma_t)} = A^{(\alpha,\beta,\gamma)}$ for $j = 1, 2, \cdots, j = 1, 2, \cdots, l(l = \max\{l(x : A_t)|t \in T\})$. □

Remark 3. *In property (6)* $(A \cup B)^{(\alpha,\beta,\gamma)} \supseteq A^{(\alpha,\beta,\gamma)} \cup B^{(\alpha,\beta,\gamma)}$, *"$\supseteq$" cannot be strengthened as "=". For example, let* $X = \{x_1, x_2, x_3\}$, $A, B \in SVNMS(X)$ *as follows:*

$$A = \{\langle x_1, (0.5, 0.3), (0.1, 0.1), (0.7, 0.8)\rangle, \langle x_2, (0.7, 0.68, 0.62), (0.3, 0.45, 0.5), (0.34, 0.28, 0.49)\rangle,$$
$$\langle x_3, (0.67, 0.5, 0.3), (0.2, 0.3, 0.4), (0.4, 0.5, 0.7)\rangle\}$$

$$B = \{\langle x_1, 0.75, 0.2, 0.15\rangle, \langle x_2, (0.43, 0.37, 0.28), (0.5, 0.2, 0.3), (0.7, 0.8, 0.9)\rangle,$$
$$\langle x_3, (1.0, 0.86, 0.79), (0.01, 0.1, 0.2), (0.0, 0.3, 0.2)\rangle\}.$$

If we choose $\alpha = 0.4$, $\beta = 0.3$, $\gamma = 0.5$, *then,*

$$A^{(\alpha,\beta,\gamma)} = \{\langle x_1, (1,0), (1,1), (0,0)\rangle, \langle x_2, (1,1,1), (1,0,0), (1,1,1)\rangle, \langle x_3, (1,1,0), (1,1,0), (1,1,0)\rangle\},$$

$$B^{(\alpha,\beta,\gamma)} = \{\langle x_1, 1,1,1\rangle, \langle x_2, (1,0,0), (0,1,1), (0,0,0)\rangle, \langle x_3, (1,1,1), (1,1,1), (1,1,1)\rangle\},$$

$$(A \cup B)^{(\alpha,\beta,\gamma)} = \{\langle x_1, (1,0), (1,1), (1,0)\rangle, \langle x_2, (1,1,1), (1,1,1), (1,1,1)\rangle, \langle x_3, (1,1,1), (1,1,1), (1,1,1)\rangle\},$$

$$A^{(\alpha,\beta,\gamma)} \cup B^{(\alpha,\beta,\gamma)} = \{\langle x_1, (1,0), (1,1), (0,0)\rangle, \langle x_2, (1,1,1), (0,0,0), (0,0,0)\rangle, \langle x_3, (1,1,1), (1,1,0), (1,1,0)\rangle\}$$

Obviously, $(A \cup B)^{(\alpha,\beta,\gamma)} \neq A^{(\alpha,\beta,\gamma)} \cup B^{(\alpha,\beta,\gamma)}$.

In order to get the decomposition theorem of SVNMS, we also need to introduce the following important concepts.

Definition 10. *Let* $L = \{(\alpha,\beta,\gamma)|\alpha,\beta,\gamma \in [0,1], 0 \le \alpha+\beta+\gamma \le 3\}$, $(\alpha_1,\beta_1,\gamma_1) \le (\alpha_2,\beta_2,\gamma_2) \Leftrightarrow \alpha_1 \le \alpha_2, \beta_1 \ge \beta_2,\ \gamma_1 \ge \gamma_2$ *So, L is a complete lattice, and* $(1,0,0)$ *is the biggest element,* $(0,1,1)$ *is the smallest element.*

Definition 11. *Let* $(\alpha,\beta,\gamma) \in L$, $B \in 2^X$, $A = (\alpha,\beta,\gamma)B$. *And for any* $x \in X$,

$$T_A^j(x) = \begin{cases} \alpha, x \in B \\ 0, x \notin B \end{cases},\ I_A^j(x) = \begin{cases} \beta, x \in B \\ 1, x \notin B \end{cases},\ F_A^j(x) = \begin{cases} \gamma, x \in B \\ 1, x \notin B \end{cases}. \tag{21}$$

Then, $A = \left\{ \left\langle x, T_A^j(x), I_A^j(x), F_A^j(x) \right\rangle \middle| x \in X, j = 1,2,\cdots, l(x:A) \right\}$ is a SVNMS on the universe X, so we have the definition as follows:

Definition 12. *Suppose* $A \in SVNMS(X)$, $(\alpha,\beta,\gamma) \in L$, *the dot product (truncated product) of* (α,β,γ) *and* A *is defined as*

$$((\alpha,\beta,\gamma)A)(x) = \left\{ \left\langle x, \alpha \vee T_A^j(x), \beta \wedge I_A^j(x), , \gamma \wedge F_A^j(x) \right\rangle \middle| x \in X, j = 1,2,\cdots, l(x:A) \right\}. \tag{22}$$

That is $(\alpha,\beta,\gamma)A \in SVNMS(X)$.

Now, we can discuss the decomposition theorem of SVNMS based on the definitions and operational properties above.

Theorem 3. *Let* A *be a SVNMS. Then for any* $(\alpha,\beta,\gamma) \in L$, *we have*

$$\begin{aligned} A &= \underset{(\alpha,\beta,\gamma)\in L}{\cup}(\alpha,\beta,\gamma)A^{(\alpha,\beta,\gamma)} = \underset{(\alpha,\beta,\gamma)\in L}{\cup}(\alpha,\beta,\gamma)A^{(\alpha+,\beta,\gamma)} = \underset{(\alpha,\beta,\gamma)\in L}{\cup}(\alpha,\beta,\gamma)A^{(\alpha,\beta+,\gamma)} \\ &= \underset{(\alpha,\beta,\gamma)\in L}{\cup}(\alpha,\beta,\gamma)A^{(\alpha,\beta,\gamma+)} = \underset{(\alpha,\beta,\gamma)\in L}{\cup}(\alpha,\beta,\gamma)A^{(\alpha+,\beta+,\gamma)} = \underset{(\alpha,\beta,\gamma)\in L}{\cup}(\alpha,\beta,\gamma)A^{(\alpha+,\beta,\gamma+)} \\ &= \underset{(\alpha,\beta,\gamma)\in L}{\cup}(\alpha,\beta,\gamma)A^{(\alpha,\beta+,\gamma+)} = \underset{(\alpha,\beta,\gamma)\in L}{\cup}(\alpha,\beta,\gamma)A^{(\alpha+,\beta+,\gamma+)} \end{aligned} \tag{23}$$

Proof. With regard to $A = \underset{(\alpha,\beta,\gamma)\in L}{\cup}(\alpha,\beta,\gamma)A^{(\alpha,\beta,\gamma)}$, we just need to prove $A(x) = \left(\underset{(\alpha,\beta,\gamma)\in L}{\cup}(\alpha,\beta,\gamma)\ A^{(\alpha,\beta,\gamma)}\right)(x)$ for all $x \in X$. That is, $A(x) = \vee_{(\alpha,\beta,\gamma)\in L}\left((\alpha,\beta,\gamma)A^{(\alpha,\beta,\gamma)}\right)(x) = \left(\vee_{\alpha\in[0,1]}\left(\alpha \vee (A^\alpha)_j(x)\right), \wedge_{\beta\in[0,1]}\left(\beta \wedge (A\beta)_j(x)\right), \wedge_{\gamma\in[0,1]}\left(\gamma \wedge (A_\gamma)_j(x)\right)\right)$ for $x \in X$. Since $T_A^j(x) \in [0,1]$, we have $\vee_{\alpha\in[0,1]}(\alpha \vee (A^\alpha)_j(x)) = \left[\vee_{\alpha\in[0,T_A^j(x)]}\left(\alpha \vee (A^\alpha)_j(x)\right)\right] \vee \left[\vee_{\alpha\in[T_A^j(x),1]}\left(\alpha \vee (A^\alpha)_j(x)\right)\right]$. Indeed, taking $\alpha \le T_A^j(x)$, we have $(A^\alpha)_j(x) = 1$, otherwise, $(A^\alpha)_j(x) = 0$. Thus, $\vee_{\alpha\in[0,1]}\left(\alpha \vee (A^\alpha)_j(x)\right) = \vee_{\alpha\in[0,T_A^j(x)]}\left(\alpha \vee (A^\alpha)_j(x)\right) = \vee_{\alpha\in[0,T_A^j(x)]}\alpha = T_A^j(x)$. Similarly, $\wedge_{\beta\in[0,1]}\left(\beta \wedge (A\beta)_j(x)\right) = \wedge_{\beta\in[I_A^j(x),1]}\left(\beta \wedge (A\beta)_j(x)\right) = \wedge_{\beta\in[I_A^j(x),1]}\beta = I_A^j(x)$ and $\wedge_{\gamma\in[0,1]}\left(\gamma \wedge (A_\gamma)_j(x)\right) = \wedge_{\gamma\in[F_A^j(x),1]}\left(\gamma \wedge (A_\gamma)_j(x)\right) = \wedge_{\gamma\in[F_A^j(x),1]}\gamma = F_A^j(x)$.

Therefore, $\vee_{(\alpha,\beta,\gamma)\in L}\left((\alpha,\beta,\gamma)A^{(\alpha,\beta,\gamma)}\right)(x) = \left(T_A^j(x), I_A^j(x), F_A^j(x)\right) = A(x)$ for $j = 1,2,\cdots l(x:A)$. □

Next, we use an example to illustrate the idea of the decomposition theorem of SVNMS.

Example 1. *Let* $X = \{x_1, x_2, x_3\}$, $A \in SVNMS(X)$ *as follows:*

$$A = \{\langle x_1, (0.6, 0.4), (0.5, 0.3), (0.2, 0.3)\rangle, \langle x_2, 0.2, 0.4, 0.7\rangle, \langle x_3, (0.8, 0.6, 0.5), (0.2, 0.2, 0.3), (0.1, 0.3, 0.4)\rangle\}.$$

We show how A can be represented by180 special SVNMSs using (α, β, γ)*-cut sets. According to Definition 9, 11 and 12, we have:*

$$A^{(\alpha,\beta,\gamma)} = \{\langle x_1, (1,1), (0,0), (0,0)\rangle, \langle x_2, 1, 0, 0\rangle, \langle x_3, (1,1,1), (1,1,0), (1,0,0)\rangle\},$$

$$(\alpha, \beta, \gamma)A^{(\alpha,\beta,\gamma)} = \{\langle x_1, (0.2, 0.2), (1,1), (1,1)\rangle, \langle x_2, 0.2, 1, 1\rangle, \langle x_3, (0.2, 0.2, 0.2), (0.2, 0.2, 1), (0.1, 1, 1)\rangle\},$$

where $0 \le \alpha \le 0.2$, $0 \le \beta \le 0.2$, $0 \le \gamma \le 0.1$*;*

$$A^{(\alpha,\beta,\gamma)} = \{\langle x_1, (1,1), (0,1), (1,0)\rangle, \langle x_2, 0, 0, 0\rangle, \langle x_3, (1,1,1), (1,1,1), (1,0,0)\rangle\},$$

$$(\alpha, \beta, \gamma)A^{(\alpha,\beta,\gamma)} = \{\langle x_1, (0.4, 0.4), (1, 0.3), (0.2, 1)\rangle, \langle x_2, 1, 1, 1\rangle, \langle x_3, (0.4, 0.4, 0.4), (0.3, 0.3, 0.3), (0.2, 1, 1)\rangle\},$$

where $0.2 < \alpha \le 0.4$, $0.2 < \beta \le 0.3$, $0.1 < \gamma \le 0.2$*;*

$$A^{(\alpha,\beta,\gamma)} = \{\langle x_1, (1,0), (0,1), (1,1)\rangle, \ \langle x_2, 0, 1, 0\rangle, \langle x_3, (1,1,1), (1,1,1), (1,1,0)\rangle\},$$

$$(\alpha, \beta, \gamma)A^{(\alpha,\beta,\gamma)} = \{\langle x_1, (0.5, 0), (1, 0.4), (0.3, 0.3)\rangle, \langle x_2, 0, 0.4, 1\rangle, \langle x_3, (0.5, 0.5, 0.5), (0.4, 0.4, 0.4), (0.3, 0.3, 1)\rangle\},$$

where $0.4 < \alpha \le 0.5$, $0.3 < \beta \le 0.4$, $0.2 < \gamma \le 0.3$*;*

$$A^{(\alpha,\beta,\gamma)} = \{\langle x_1, (1,0), (1,1), (1,1)\rangle, \ \langle x_2, 0, 1, 0\rangle, \langle x_3, (1,1,0), (1,1,1), (1,1,1)\rangle\},$$

$$(\alpha, \beta, \gamma)A^{(\alpha,\beta,\gamma)} = \{\langle x_1, (0.6, 0), (0.5, 0.5), (0.4, 0.4)\rangle, \langle x_2, 0, 0.5, 1\rangle, \langle x_3, (0.6, 0.6, 0), (0.5, 0.5, 0.5), (0.4, 0.4, 0.4)\rangle\},$$

where $0.5 < \alpha \le 0.6$, $0.4 < \beta \le 0.5$, $0.3 < \gamma \le 0.4$*;*

$$A^{(\alpha,\beta,\gamma)} = \{\langle x_1, (0,0), (1,1), (1,1)\rangle, \langle x_2, 0, 1, 1\rangle, \langle x_3, (1,0,0), (1,1,1), (1,1,1)\rangle\},$$

$$(\alpha, \beta, \gamma)A^{(\alpha,\beta,\gamma)} = \{\langle x_1, (0,0), (1,1), (0.7, 0.7)\rangle, \langle x_2, 0, 1, 0.7\rangle, \langle x_3, (0.8, 0, 0), (1,1,1), (0.7, 0.7, 0.7)\rangle\},$$

where $0.6 < \alpha \le 0.8$, $0.5 < \beta \le 1$, $0.4 < \gamma \le 0.7$*;*

$$A^{(\alpha,\beta,\gamma)} = \{\langle x_1, (0,0), (1,1), (1,1)\rangle, \langle x_2, 0, 1, 1\rangle, \langle x_3, (0,0,0), (1,1,1), (1,1,1)\rangle\},$$

$$(\alpha, \beta, \gamma)A^{(\alpha,\beta,\gamma)} = \{\langle x_1, (0,0), (1,1), (1,1)\rangle, \langle x_2, 0, 1, 1\rangle, \langle x_3, (0,0,0), (1,1,1), (1,1,1)\rangle\},$$

where $0.8 < \alpha \le 1$, $0.5 < \beta \le 1$, $0.7 < \gamma \le 1$.

Similarly, we can get the rest of the results with special SVNMSs. It is obvious to see,

$$A = \bigcup_{(\alpha,\beta,\gamma)\in L} (\alpha, \beta, \gamma)A^{(\alpha,\beta,\gamma)}.$$

Definition 13. *Suppose* $H : L \to 2^X$, $(\lambda, \mu, \omega) \mapsto H(\lambda, \mu, \omega)$ *is a mapping, a neutrosophic nested set H can be defined in X if it satisfies the following conditions:*

(1) $(\lambda_1, \mu_1, \omega_1) \le (\lambda_2, \mu_2, \omega_2) \Rightarrow H(\lambda_1, \mu_1, \omega_1) \supseteq H(\lambda_2, \mu_2, \omega_2)$*;*

(2) $\cap_{t\in T} H(\lambda_t, \mu_t, \omega_t) \subseteq \{H(\lambda, \mu, \omega) | \lambda < \vee_{t\in T}\lambda_t, \mu > \wedge_{t\in T}\mu_t, \omega > \wedge_{t\in T}\omega_t\}$.

Remark 4. *Let* SVN_L *be a set which composed of all neutrosophic nested sets,* $A \in SVNMS(X)$*, then, all* (α, β, γ)*-cut sets of A are neutrosophic nested sets.*

Theorem 4. *Let* $A \in SVNMS(X)$, $H : L \to 2^X$, $(\alpha, \beta, \gamma) \mapsto H(\alpha, \beta, \gamma)$*, for any* $(\alpha, \beta, \gamma) \in L$ *satisfy* $A^{(\alpha+,\beta+,\gamma+)} \subseteq H(\alpha, \beta, \gamma) \subseteq A^{(\alpha,\beta,\gamma)}$*, then*

(1) $A = \underset{(\alpha,\beta,\gamma)\in L}{\cup}(\alpha,\beta,\gamma)H(\alpha,\beta,\gamma)$;

(2) $\alpha_1 < \alpha_2, \beta_1 > \beta_2, \gamma_1 > \gamma_2 \Rightarrow H(\alpha_1,\beta_1,\gamma_1) \supseteq H(\alpha_2,\beta_2,\gamma_2)$, *where* $\alpha_1,\alpha_2,\beta_1,\beta_2,\gamma_1,\gamma_2 \in [0,1]$, $0 \le \alpha_1+\beta_1+\gamma_1 \le 3, 0 \le \alpha_2+\beta_2+\gamma_2 \le 3$;

(3) (I) $A^{(\alpha,\beta,\gamma)} = \cap\{H(\lambda,\mu,\omega)|\lambda < \alpha, \mu > \beta, \omega > \gamma, 0 \le \lambda+\mu+\omega \le 3\}$, (II) $A^{(\alpha+,\beta+,\gamma+)} = \cup\{H(\lambda,\mu,\omega)|\lambda > \alpha, \mu < \beta, \omega < \gamma, 0 \le \lambda+\mu+\omega \le 3\}$;

(4) $\underset{t\in T}{\cap}H(\alpha_t,\beta_t,\gamma_t) \subseteq \cap\{H(\alpha,\beta,\gamma)|\alpha < \vee_{t\in T}\alpha_t, \beta > \wedge_{t\in T}\beta_t, \gamma > \wedge_{t\in T}\gamma_t\}$.

Proof.

(1) Since $A^{(\alpha+,\beta+,\gamma+)} \subseteq H(\alpha,\beta,\gamma) \subseteq A^{(\alpha,\beta,\gamma)}$ for all $(\alpha,\beta,\gamma) \in L$, we have

$$A = \underset{(\alpha,\beta,\gamma)\in L}{\cup}(\alpha,\beta,\gamma)A^{(\alpha+,\beta+,\gamma+)} \subseteq \underset{(\alpha,\beta,\gamma)\in L}{\cup}(\alpha,\beta,\gamma)H(\alpha,\beta,\gamma) \subseteq \underset{(\alpha,\beta,\gamma)\in L}{\cup}(\alpha,\beta,\gamma)A^{(\alpha,\beta,\gamma)} = A.$$

Thus, $A = \underset{(\alpha,\beta,\gamma)\in L}{\cup}(\alpha,\beta,\gamma)H(\alpha,\beta,\gamma)$.

(2) From $\alpha_1 < \alpha_2, \beta_1 > \beta_2, \gamma_1 > \gamma_2$, we can obtain

$$H(\alpha_1,\beta_1,\gamma_1) \supseteq A^{(\alpha_1+,\beta_1+,\gamma_1+)} \supseteq A^{(\alpha_2,\beta_2,\gamma_2)} \supseteq H(\alpha_2,\beta_2,\gamma_2).$$

(3) (*I*) Suppose $\sum = \{(\lambda,\mu,\omega)|\lambda < \alpha, \mu > \beta, \omega > \gamma, 0 \le \lambda+\mu+\omega \le 3\}$, then,

$$\vee_{(\lambda,\mu,\omega)\in\sum}(\lambda,\mu,\omega) = (\alpha,\beta,\gamma).$$

So, $\cap\{H(\lambda,\mu,\omega)|\lambda < \alpha, \mu > \beta, \omega > \gamma, 0 \le \lambda+\mu+\omega \le 3\} \subseteq \cap\left\{A^{(\lambda,\mu,\omega)}|\lambda < \alpha, \mu > \beta, \omega > \gamma, 0 \le \lambda+\mu+\omega \le 3\right\} = A^{(\alpha,\beta,\gamma)}$. On the other hand, since $x \in A^{(\alpha,\beta,\gamma)}$, we have $T_A^j(x) \ge \alpha, I_A^j(x) \le \beta, F_A^j(x) \le \gamma$. Thus, $T_A^j(x) \ge \alpha > \lambda, I_A^j(x) \le \beta < \mu, F_A^j(x) \le \gamma < \omega$. That is, $T_A^j(x) > \lambda, I_A^j(x) < \mu, F_A^j(x) < \omega$. Thus, $x \in A^{(\lambda+,\mu+,\omega+)}$. Thus, $x \in H(\lambda,\mu,\omega)$. Therefore, $x \in \cap\{H(\lambda,\mu,\omega)|\lambda < \alpha, \mu > \beta, \omega > \gamma, 0 \le \lambda+\mu+\omega \le 3\}$. Based on the above facts, we can obtain $A^{(\alpha,\beta,\gamma)} = \cap\{H(\lambda,\mu,\omega)|\lambda < \alpha, \mu > \beta, \omega > \gamma, 0 \le \lambda+\mu+\omega \le 3\}$.

(*II*) Since $A^{(\alpha+,\beta+,\gamma+)} \supseteq A^{(\lambda,\mu,\gamma)} \supseteq H(\lambda,\mu,\gamma)$ for any $\lambda > \alpha, \mu < \beta, \omega < \gamma, 0 \le \lambda+\mu+\omega \le 3$, we have

$A^{(\alpha+,\beta+,\gamma+)} \supseteq \cup\{H(\lambda,\mu,\omega)|\lambda > \alpha, \mu < \beta, \omega < \gamma, 0 \le \lambda+\mu+\omega \le 3\}$. On the other hand, from $x \in A^{(\alpha+,\beta+,\gamma+)}$ we have $T_A^j(x) > \alpha, I_A^j(x) < \beta, F_A^j(x) < \gamma$. It follows that there exists $\lambda > \alpha, \mu < \beta, \omega < \gamma, 0 \le \lambda+\mu+\omega \le 3$, such that $T_A^j(x) > \lambda > \alpha$, $I_A^j(x) < \mu < \beta$, $F_A^j(x) < \omega < \gamma$, that is, $x \in A^{(\lambda+,\mu+,\omega+)}$. Indeed, $A^{(\lambda+,\mu+,\omega+)} \subseteq H(\lambda,\mu,\omega)$, then, $x \in H(\lambda,\mu,\omega)$. Thus, $x \in \cup\{H(\lambda,\mu,\omega)|\lambda > \alpha, \mu < \beta, \omega < \gamma, 0 \le \lambda+\mu+\omega \le 3\}$. Thus,

$A^{(\alpha+,\beta+,\gamma+)} \subseteq \cup\{H(\lambda,\mu,\omega)|\lambda > \alpha, \mu < \beta, \omega < \gamma, 0 \le \lambda+\mu+\omega \le 3\}$. Therefore, we can obtain

$$A^{(\alpha+,\beta+,\gamma+)} = \cup\{H(\lambda,\mu,\omega)|\lambda > \alpha, \mu < \beta, \omega < \gamma, 0 \le \lambda+\mu+\omega \le 3\}.$$

(4) From $A^{(\alpha+,\beta+,\gamma+)} \subseteq H(\alpha,\beta,\gamma) \subseteq A^{(\alpha,\beta,\gamma)}$, we have $\underset{t\in T}{\cap}H(\alpha_t,\beta_t,\gamma_t) \subseteq \underset{t\in T}{\cap}A^{(\alpha_t,\beta_t,\gamma_t)} \subseteq A^{(\alpha',\beta',\gamma')}$ for $\alpha' = \vee_{t\in T}\alpha_t, \beta' = \wedge_{t\in T}\beta_t, \gamma' = \wedge_{t\in T}\gamma_t$. Applying (3) (I), we get

$$A^{(\alpha',\beta',\gamma')} = \cap\{H(\alpha,\beta,\gamma)|\alpha < \alpha', \beta > \beta', \gamma > \gamma', 0 \le \alpha+\beta+\gamma \le 3\}+\gamma \le 3\}.$$

Therefore,

$$\underset{t\in T}{\cap}H(\alpha_t,\beta_t,\gamma_t) \subseteq \cap\{H(\alpha,\beta,\gamma)|\alpha < \alpha', \beta > \beta', \gamma > \gamma', 0 \le \alpha+\beta+\gamma \le 3\} \square$$

Remark 5. *(1) The significance of Theorem 3 (Decomposition Theorem): A SVNMS can be composed of neutrosophic nested sets which consist of self-decomposed cut sets or strong cut sets. (2) The significance of*

Theorem 4 (Generalized Decomposition Theorem): A collection of family sandwiched between cut or strong cut sets of a SVNMS must be neutrosophic nested sets, and such nested sets can also compose the original SVNMS.

4.2. Representation Theorem of SVNMS

According to the relationship between the decomposition theorem and the representation theorem, we can obtain that each neutrosophic nested set can be combined into a single-valued neutrosophic multiset. Furthermore, its cut sets or strong cut sets can be constructed with the original neutrosophic nested set. In other words, it is theoretically explained: a family of special single-valued neutrosophic multisets can be used to completely depict and represent a single-valued neutrosophic multiset).

In this section, the representation theorem of SVNMS based on the decomposition theorem is proposed in this section.

Theorem 5. *Let $H \in SVN_L(X)$, $A \in SVNMS(X)$, and $\forall(\alpha,\beta,\gamma) \in L$. We have*

(I) $A^{(\alpha,\beta,\gamma)} = \cap\{H(\lambda,\mu,\omega)|\lambda < \alpha, \mu > \beta, \omega > \gamma, 0 \le \lambda+\mu+\omega \le 3\}$;
(II) $A^{(\alpha+,\beta+,\gamma+)} = \cup\{H(\lambda,\mu,\omega)|\lambda > \alpha, \mu < \beta, \omega < \gamma, 0 \le \lambda+\mu+\omega \le 3\}$.

Proof. Since $H(\alpha,\beta,\gamma) \in 2^X$ for all $(\alpha,\beta,\gamma) \in L$, and $(\alpha,\beta,\gamma)H(\alpha,\beta,\gamma) \in SVNMS(X)$, we have $\cup_{(\alpha,\beta,\gamma)\in L}(\alpha,\beta,\gamma)H(\alpha,\beta,\gamma) \in SVNMS(X)$, denoted by A. Applying Theorem 4, we only need to prove,

$$H : L \to 2^X \text{ satisfies } A^{(\alpha+,\beta+,\gamma+)} \subseteq H(\alpha,\beta,\gamma) \subseteq A^{(\alpha,\beta,\gamma)}.$$

Since $x \in A^{(\alpha+,\beta+,\gamma+)}$, we have $T_A^j(x) > \alpha, I_A^j(x) < \beta, F_A^j(x) < \gamma$. Thus, $\vee_{(\lambda,\mu,\omega)\in L}\left[\lambda \wedge (H(\lambda))_j(x)\right] > \alpha$,

$\wedge_{(\lambda,\mu,\omega)\in L}\left[\mu \vee (H(\mu))_j(x)\right] < \beta, \wedge_{(\lambda,\mu,\omega)\in L}\left[\omega \vee (H(\omega))_j(x)\right] < \gamma$. It follows that there exists $(\lambda_0,\mu_0,\omega_0) \in L$, such that $\lambda_0 \vee (H(\lambda_0))_j(x) > \alpha, \mu_0 \vee (H(\mu_0))_j(x) < \beta, \omega_0 \vee (H(\omega_0))_j(x) < \gamma$, that is, $\lambda_0 > \alpha, \mu_0 < \beta, \omega_0 < \gamma$. Taking $(H(\lambda_0,\mu_0,\omega_0))_j(x) = (1,1,1)$, we have $(\lambda_0,\mu_0,\omega_0) > (\alpha,\beta,\gamma)$. Thus, $x \in H(\lambda_0,\mu_0,\omega_0) \subseteq H(\alpha,\beta,\gamma)$. On the other hand, from $x \in H(\alpha,\beta,\gamma)$, we have $(H(\lambda,\mu,\omega))_j(x) = (1,1,1)$. Thus, $\vee_{(\lambda,\mu,\omega)\in L}\left[\lambda \vee (H(\lambda))_j(x)\right] \ge \alpha \wedge (H(\alpha))_j(x) = \alpha$, $\vee_{(\lambda,\mu,\omega)\in L}\left[\mu \vee (H(\mu))_j(x)\right] \le \beta \wedge (H(\beta))_j(x) = \beta$ and

$\vee_{(\lambda,\mu,\omega)\in L}\left[\omega \vee (H(\omega))_j(x)\right] \le \gamma \wedge (H(\gamma))_j(x) = \gamma$, that is, $T_A^j(x) \ge \alpha, I_A^j(x) \le \beta, F_A^j(x) \le \gamma$. Thus, $x \in A^{(\alpha,\beta,\gamma)}$. Therefore, $A^{(\alpha+,\beta+,\gamma+)} \subseteq H(\alpha,\beta,\gamma) \subseteq A^{(\alpha,\beta,\gamma)}$ for $j = 1,2,\cdots l(x : A,B)$. □

Theorem 5 (Representation Theorem) provides an effective method for constructing a SVNMS: Let $H \in SVN_L(X)$, we can construct a SVNMS with the following membership function:

$$A : X \to L,\ A(x) = \vee\{(\alpha,\beta,\gamma) \in L|x \in H(\alpha,\beta,\gamma)\},\ \forall x \in X$$

Example 2. *Suppose $X = \{x_1,x_2,x_3\}$. The neutrosophic nested sets on the given X is as follows:*

$$H(\alpha,\beta,\gamma) = \{\langle x_1,(1,1),(0,0),(0,0)\rangle, \langle x_2,1,0,0\rangle, \langle x_3,(1,1,1),(0,0,0),(0,0,0)\rangle\},$$

where $\alpha = \beta = \gamma = 0$;

$$H(\alpha,\beta,\gamma) = \{\langle x_1,(1,1),(0,0),(0,0)\rangle, \langle x_2,1,0,0\rangle, \langle x_3,(1,1,1),(1,1,0),(1,0,0)\rangle\},$$

where $0 < \alpha \le 0.2, 0 < \beta \le 0.2, 0 < \gamma \le 0.1$;

$$H(\alpha,\beta,\gamma) = \{\langle x_1,(1,1),(0,1),(1,0)\rangle, \langle x_2,0,0,0\rangle, \langle x_3,(1,1,1),(1,1,1),(1,0,0)\rangle\},$$

where $0.2 < \alpha \leq 0.4, 0.2 < \beta \leq 0.3, 0.1 < \gamma \leq 0.2$;

$$H(\alpha,\beta,\gamma) = \{\langle x_1,(1,0),(0,1),(1,1)\rangle,\ \langle x_2,0,1,0\rangle,\langle x_3,(1,1,1),(1,1,1),(1,1,0)\rangle\},$$

where $0.4 < \alpha \leq 0.5, 0.3 < \beta \leq 0.4, 0.2 < \gamma \leq 0.3$;

$$H(\alpha,\beta,\gamma) = \{\langle x_1,(1,0),(1,1),(1,1)\rangle,\ \langle x_2,0,1,0\rangle,\langle x_3,(1,1,0),(1,1,1),(1,1,1)\rangle\},$$

where $0.5 < \alpha \leq 0.6, 0.4 < \beta \leq 0.5, 0.3 < \gamma \leq 0.4$;

$$H(\alpha,\beta,\gamma) = \{\langle x_1,(0,0),(1,1),(1,1)\rangle,\langle x_2,0,1,1\rangle,\langle x_3,(1,0,0),(1,1,1),(1,1,1)\rangle\},$$

where $0.6 < \alpha \leq 0.8, 0.5 < \beta \leq 1, 0.4 < \gamma \leq 0.7$;

$$H(\alpha,\beta,\gamma) = \{\langle x_1,(0,0),(1,1),(1,1)\rangle,\langle x_2,0,1,1\rangle,\langle x_3,(0,0,0),(1,1,1),(1,1,1)\rangle\},$$

where $0.8 < \alpha \leq 1, 0.5 < \beta \leq 1, 0.7 < \gamma \leq 1$.

Similarly, we can give the remaining neutrosophic nested sets.

Then, the SVNMS A determined by H has the following membership function:

$$\begin{aligned}
&((T_A^1(x_1),T_A^2(x_1)),(I_A^1(x_1),I_A^2(x_1)),(F_A^1(x_1),F_A^2(x_1)))\\
&=(\vee\{\alpha\in[0,1]|x_1\in H(\alpha,\beta,\gamma)\},\wedge\{\beta\in[0,1]|x_1\in H(\alpha,\beta,\gamma)\},\wedge\{\gamma\in[0,1]|x_1\in H(\alpha,\beta,\gamma)\})\\
&=((0.6,0.4),(0.5,0.3),(0.2,0.3))\\
&(T_A^1(x_2),I_A^1(x_2),F_A^1(x_2))\\
&=(\vee\{\alpha\in[0,1]|x_2\in H(\alpha,\beta,\gamma)\},\wedge\{\beta\in[0,1]|x_2\in H(\alpha,\beta,\gamma)\},\wedge\{\gamma\in[0,1]|x_2\in H(\alpha,\beta,\gamma)\})\\
&=(0.2,0.4,0.7)\\
&((T_A^1(x_3),T_A^2(x_3),T_A^3(x_3)),(I_A^1(x_3),I_A^2(x_3),I_A^3(x_3)),(F_A^1(x_3),F_A^2(x_3),F_A^3(x_3)))\\
&=(\vee\{\alpha\in[0,1]|x_3\in H(\alpha,\beta,\gamma)\},\wedge\{\beta\in[0,1]|x_3\in H(\alpha,\beta,\gamma)\},\wedge\{\gamma\in[0,1]|x_3\in H(\alpha,\beta,\gamma)\})\\
&=((0.8,0.6,0.5),(0.2,0.2,0.3),(0.1,0.3,0.4))
\end{aligned}$$

Therefore,

$$A = \{\langle x_1,(0.6,0.4),(0.5,0.3),(0.2,0.3)\rangle,\langle x_2,0.2,0.4,0.7\rangle,\langle x_3,(0.8,0.6,0.5),(0.2,0.2,0.3),(0.1,0.3,0.4)\rangle\}.$$

5. New Similarity Measure between SVNMSs

On the basis of the decomposition theorem of SVNMS, this section presents a new similarity measure between SVNMSs. Then, we discuss the properties of this new similarity measure and give a concrete algorithm by example.

Definition 14. *Let* $M = \left\{\left\langle x,T_M^j(x),I_M^j(x),F_M^j(x)\right\rangle \middle| x\in X, j=1,2,\cdots,l(x:M)\right\}$ *and* $N = \left\{\left\langle x,T_N^j(x),I_N^j(x),F_N^j(x)\right\rangle \middle| x\in X, j=1,2,\cdots,l(x:N)\right\}$ *be two SVNMSs in X. Suppose* $V = [0,1]\times[0,1]\times[0,1]$. *Then, we define a new distance measure between M and N as follows:*

$$D_C(M,N) = \iiint_V f(\alpha,\beta,\gamma)dV$$

where $f(\alpha,\beta,\gamma) = D_P\left((\alpha,\beta,\gamma)M^{(\alpha,\beta,\gamma)},(\alpha,\beta,\gamma)N^{(\alpha,\beta,\gamma)}\right), \alpha\in[0,1], \beta\in[0,1], \gamma\in[0,1]$.

Proposition 2. *Let M, N be two SVNMSs in X. Then, the following properties hold (DC1-DC4):*

(DC1) $0 \leq D_C(M,N) \leq 1$;

(DC2) $D_C(M,N) = 0$ *if and only if* $M = N$;
(DC3) $D_C(M,N) = D_C(M,N)$;
(DC4) If Q *is a SVNMS in X and* $M \subseteq N \subseteq Q$, *then,* $D_C(M,Q) \le D_C(M,N) + D_C(N,Q)$ *for* $P > 0$.

According to the relationship between distance measure and similarity measures, we can introduce two distance-based similarity measures between M and N:

$$S_{C1}(M,N) = 1 - D_C(M,N) \tag{25}$$

$$S_{C2}(M,N) = \frac{1 - D_C(M,N)}{1 + D_C(M,N)} \tag{26}$$

Proposition 3. *Let* $M, N \in SVNMS(X)$. *The distance-based similarity measures* $S_{Cf}(M,N)$, $(f = 1,2)$ *hold the following properties (SC1-SC4):*

(SC1) $0 \le S_{Cf}(M,N) \le 1$;
(SC2) $S_{Cf}(M,N) = 1$ *if and only if* $M = N$;
(SC3) $S_{Cf}(M,N) = S_{Cf}(N,M)$;
(SC4) If Q *is a SVNMS in X and* $M \subseteq N \subseteq Q$, *then* $S_{Cf}(M,Q) \le S_{Cf}(M,N) + S_{Cf}(N,Q)$.

Proof. The proofs of proposition 2 and 3 are straightforward. □

This method is based on the cut sets, and uses the idea of the decomposition theorem to convert the similarity measure between the two SVNMSs into the similarity measure between the corresponding special SVNMSs. Now, let us use a concrete example to illustrate the specific algorithm.

Example 3. *Let* $X = \{x_1, x_2, x_3\}$, $M, N \in SVNMS(X)$. *That is,* $M = \{\langle x_1, (0.7,0.8), (0.1,0.2), (0.2,0.3)\rangle, \langle x_2, (0.5,0.6), (0.2,0.3), (0.4,0.5)\rangle\}$, $N = \{\langle x_1, (0.5,0.6), (0.1,0.2), (0.4,0.5)\rangle, \langle x_2, (0.6,0.7), (0.1,0.2), (0.7,0.8)\rangle\}$.

According to the values of $T_M^j(x_i)$, $T_N^j(x_i) (i = 1,2; j = 1,2)$, *we divide the interval* $[0,1]$ *of* α *into 5 subintervals:* $[0,0.5]$, $(0.5,0.6]$, $(0.6,0.7]$, $(0.7,0.8]$, $(0.8,1]$. *Similarly, we can obtain 4 subintervals of* β: $[0,0.1]$, $(0.1,0.2]$, $(0.2,0.3]$, $(0.3,1]$, *and 7 subintervals of* γ: $[0,0.2]$, $(0.2,0.3]$, $(0.3,0.4]$, $(0.4,0.5]$, $(0.5,0.7]$, $(0.7,0.8]$, $(0.8,1]$. *Thus, we have 140 interval combinations of* α, β, *and* γ, *take* $0 \le \alpha \le 0.5$, $0.2 < \beta \le 0.3$, $0.7 < \gamma \le 0.8$ for example. In this way, for each combination of interval, we can get the corresponding $M^{(\alpha,\beta,\gamma)}$, $N^{(\alpha,\beta,\gamma)}$ *and* $(\alpha,\beta,\gamma)M^{(\alpha,\beta,\gamma)}$, $(\alpha,\beta,\gamma)N^{(\alpha,\beta,\gamma)}$. Based on the above results, the process is as follows:

Step1, calculate $f(\alpha,\beta,\gamma) = D_P\left((\alpha,\beta,\gamma)M^{(\alpha,\beta,\gamma)}, (\alpha,\beta,\gamma)N^{(\alpha,\beta,\gamma)}\right)$ in every interval combination.

Step2, use Equation (24) to perform the integral operation on $f(\alpha,\beta,\gamma)$ *over* $V = [0,1] \times [0,1] \times [0,1]$, *and get* $D_C(M,N) = 0.2206$.

Step3, using Equation (25) and (26), we can get $S_{C1}(M,N) = 0.7794$ *and* $S_{C2}(M,N) = 0.6385$.

6. Application of New Similarity Measures in Multicriteria Decision-Making Problems

In this section, the new similarity measure is applied to a medical diagnosis problem. Next, we use the typical examples in [14] to verify the feasibility and effectiveness of the new similarity measure proposed in Section 5. Furthermore, we analyze the uniqueness of the new similarity measure by comparing the results with other similarity measures.

Assume that $I = \{I_1, I_2, I_3, I_4\}$ represents 4 patients, set $R = \{R_1, R_2, R_3, R_4\}$ = {viral fever, tuberculosis, typhoid, throat disease} indicates 4 diseases, and set $S = \{S_1, S_2, S_3, S_4\}$ = {temperature, cough, sore throat, headache, body pain} indicates 5 symptoms. In medical diagnosis, in order to obtain a more accurate diagnosis, the doctor collects symptom information for the same patient at different times of the day. Therefore, we use the following SVNMSs to indicate the affiliation between the patient and the symptom:

$$I_1 = \{\langle S_1, (0.8, 0.6, 05), (0.3, 0.2, 0.1), (0.4, 0.2, 0.1)\rangle, \langle S_2, (0.5, 0.4, 0.3), (0.4, 0.4, 0.3), (0.6, 0.3, 0.4)\rangle \langle S_3, (0.2, 0.1, 0.0), (0.3, 0.2, 0.2), (0.8, 0.7, 0.7)\rangle, \langle S_4, (0.7, 0.6, 0.5), (0.3, 0.2, 0.1), (0.4, 0.3, 0.2)\rangle \langle S_5, (0.4, 0.3, 0.2), (0.6, 0.5, 0.5), (0.6, 0.4, 0.4)\rangle;$$

$$I_2 = \{\langle S_1, (0.5, 0.4, 0.3), (0.3, 0.3, 0.2), (0.5, 0.4, 0.4)\rangle, \langle S_2, (0.9, 0.8, 0.7), (0.2, 0.1, 0.1), (0.2, 0.1, 0.0)\rangle \langle S_3, (0.6, 0.5, 0.4), (0.3, 0.2, 0.2), (0.4, 0.3, 0.3)\rangle, \langle S_4, (0.6, 0.4, 0.3), (0.3, 0.1, 0.1), (0.7, 0.7, 0.3)\rangle \langle S_5, (0.8, 0.7, 0.5), (0.4, 0.3, 0.1), (0.3, 0.2, 0.1)\rangle;$$

$$I_3 = \{\langle S_1, (0.2, 0.1, 0.1), (0.3, 0.2, 0.2), (0.8, 0.7, 0.6)\rangle, \langle S_2, (0.3, 0.2, 0.2), (0.4, 0.2, 0.2), (0.7, 0.6, 0.5)\rangle \langle S_3, (0.8, 0.8, 0.7), (0.2, 0.2, 0.2), (0.1, 0.1, 0.0)\rangle, \langle S_4, (0.3, 0.2, 0.2), (0.3, 0.3, 0.3), (0.7, 0.6, 0.6)\rangle \langle S_5, (0.4, 0.4, 0.3), (0.4, 0.3, 0.2), (0.7, 0.7, 0.5)\rangle;$$

$$I_4 = \{\langle S_1, (0.5, 0.5, 0.4), (0.3, 0.2, 0.2), (0.4, 0.4, 0.3)\rangle, \langle S_2, (0.4, 0.3, 0.1), (0.4, 0.3, 0.2), (0.7, 0.5, 0.3)\rangle \langle S_3, (0.7, 0.1, 0.0), (0.4, 0.3, 0.3), (0.7, 0.7, 0.6)\rangle, \langle S_4, (0.6, 0.5, 0.3), (0.6, 0.2, 0.1), (0.6, 0.4, 0.3)\rangle \langle S_5, (0.5, 0.1, 0.1), (0.3, 0.3, 0.2), (0.6, 0.5, 0.4)\rangle.$$

Then, the affiliation between the symptoms and the disease is represented by the following SVNMSs:

$$R_1 = \{\langle S_1, 0.8, 0.1, 0.1\rangle, \langle S_2, 0.2, 0.7, 0.1\rangle, \langle S_3, 0.3, 0.5, 0.2\rangle, \langle S_4, 0.5, 0.3, 0.2\rangle, \langle S_5, 0.5, 0.4, 0.1\rangle\};$$
$$R_2 = \{\langle S_1, 0.2, 0.7, 0.1\rangle, \langle S_2, 0.9, 0.0, 0.1\rangle, \langle S_3, 0.7, 0.2, 0.1\rangle, \langle S_4, 0.6, 0.3, 0.1\rangle, \langle S_4, 0.7, 0.2, 0.1\rangle\};$$
$$R_3 = \{\langle S_1, 0.5, 0.3, 0.2\rangle, \langle S_2, 0.3, 0.5, 0.2\rangle, \langle S_3, 0.2, 0.7, 0.1\rangle, \langle S_4, 0.2, 0.6, 0.2\rangle, \langle S_5, 0.4, 0.4, 0.2\rangle\};$$
$$R_4 = \{\langle S_1, 0.1, 0.7, 0.2\rangle, \langle S_2, 0.3, 0.6, 0.1\rangle, \langle S_3, 0.8, 0.1, 0.1\rangle, \langle S_4, 0.1, 0.8, 0.1\rangle, \langle S_5, 0.1, 0.8, 0.1\rangle\}.$$

Then, by Definition 14, we use Equations (24) and (25) to get the similarity $S_{C1}(I_i, R_j)$ between each patient $I_i (i = 1, 2, 3, 4)$ and disease $R_j (j = 1, 2, 3, 4)$, which are shown in Table 1. Similarly, we use Equations (24) and (26) to get the similarity $S_{C2}(I_i, R_j)$ between each patient $I_i (i = 1, 2, 3, 4)$ and disease $R_j (j = 1, 2, 3, 4)$, which are shown in Table 2.

Table 1. Similarity values of $S_{C1}\left(I_i, R_j\right)$.

	R_1 *(Viral fever)*	R_2 *(Tuberculosis)*	R_3 *(Typhoid)*	R_4 *(Troat Disease)*
I_1	0.6927	0.6616	**0.6934**	0.6694
I_2	0.6417	**0.6632**	0.6458	0.6414
I_3	0.6896	0.6820	0.6881	**0.7011**
I_4	0.6966	0.6850	**0.7156**	0.6923

Table 2. Similarity values of $S_{C2}\left(I_i, R_j\right)$.

	R_1 *(Viral fever)*	R_2 *(Tuberculosis)*	R_3 *(Typhoid)*	R_4 *(Troat Disease)*
I_1	0.5299	0.4943	**0.5307**	0.5031
I_2	0.4724	**0.4961**	0.4769	0.4721
I_3	0.5263	0.5175	0.5245	**0.5398**
I_4	0.5344	0.5209	**0.5571**	0.5294

It is well known that the closeness of the relationship between two SVNMSs can be described by the similarity between the two, that is, the greater the similarity, the closer the relationship is. As can be seen from Tables 1 and 2, for these four diseases, by comparison, we can determine the most similar disease to each patient and get the get the most realistic diagnosis: patient I_1 suffers from typhoid, patient I_2 suffers from tuberculosis, patient I_3 suffers from throat disease, and patient I_4 also suffers from typhoid.

The dice similarity measures proposed in [11] are applied to the decision-making example, and the diagnosis is that patient I_1 suffers from typhoid, patient I_2 suffers from viral fever, patient I_3 suffers from typhoid, and patient I_4 suffers from tuberculosis. The distance-based similarity measures proposed in [14] also are applied in this decision-making example, and the diagnosis is that patient I_1 suffers from viral fever, patient I_2 suffers from tuberculosis, patient I_3 suffers from typhoid, and patient I_4 suffers from typhoid.

By analyzing and comparing the diagnostic results obtained by the three methods, we found that when using the new similarity to calculate, the diagnosis of disease in patient I_1 is consistent with [11] and the diagnosis of patients I_2 and I_4 was consistent with [14], indicating that this method is more effective, because the results are closer to the actual situation.

According to the above comparative analysis, the method proposed in this paper has the following advantages: (1) The new similarity measure under the SVNMSs environment can deal with the indeterminacy and inconsistent information which exists in decision-making problems, that is, it can be effectively used in many practical applications. (2) The new similarity measure is based on the cut sets, with the decomposition theorem and the representation theorem as the main ideas, and the integral as the main mathematical tool. Therefore, it has a solid mathematical theoretical basis. (3) This method can make full use of all the information of SVNMSs, and use the idea of splitting and summing to simplify complex problem, provide a simple and effective method for solving practical problems.

7. Conclusions

This paper first systematically discussed 8 properties of the union, intersection and complement of the single valued neutrosophic multisets (SVNMSs), and showed that the complementation is no longer true in SVNMS by the counterexample. Secondly, this paper proposed the notions of cut sets and strong cut sets of SVNMSs and presented the related properties. On the basis of cut set sand strong cut sets, the decomposition theorem and representation theorem of SVNMSs were established and proved. The decomposition theorem realizes the transformation of SVNMSs and special SVNMSs. Thirdly, based on the decomposition theorem, we transformed the similarity between SVNMSs into the similarity between special SVNMSs. Therefore, we used the integral to give a new method to calculate the similarity between SVNMSs. The conceptions of new similarity measures were introduced, and its feasibility and effectiveness in multi-attribute decision making were verified accordinng to a typical example. Further, the uniqueness of the new similarity measure was analyzed by comparing the results with other similarity measures. The results obtained have a significant meaning for further theoretical research of SVNMSs. As the next research topic, we will explore the fuzzy measure and fuzzy integral of SVNMSs. In the future, we will discuss the integration of the related topics, such as neutrosophic set (multiset), fuzzy set (multiset), rough set, soft set and algebra systems (see [30–32,37–39]).

Author Contributions: All authors have contributed equally to this paper. The original idea of the study was proposed by X.Z., he also completed the preparation of the paper; X.Z. and Q.H conceived and designed the experiment; Q.H analyzed the experimental data and wrote the paper; The revision and submission of the paper was completed by X.Z. and Q.H.

Funding: This research was funded by National Natural Science Foundation of China grant number 61573240.

Conflicts of Interest: The authors declare no conflict of interest.

References

1. Smarandache, F. *Neutrosophy: Neutrosophic Probability, Set, and Logic: Analytic Synthesis and Synthetic Analysis*; American Research Press: Santa Fe, NM, USA, 1998.
2. Atanassov, K. Intuitionistic fuzzy sets. *Fuzzy Sets Syst.* **1986**, *20*, 87–96. [CrossRef]
3. Wang, H.; Smarandache, F.; Sunderraman, R.; Zhang, Y.Q. Interval neutrosophic sets and logic: Theory and applications in computing. *Comput. Sci.* **2012**, *65*, 87.

4. Pramanik, S.; Pramanik, S.; Giri, B.C. TOPSIS method for multi-attribute group decision-making under single-valued neutrosophic environment. *Neural Comput. Appl.* **2016**, *27*, 727–737.
5. Liu, P.; Wang, Y. Multiple attribute decision-making method based on single-valued neutrosophic normalized weighted Bonferroni mean. *Neural Comput. Appl.* **2014**, *25*, 2001–2010. [CrossRef]
6. Ye, J. Multicriteria decision-making method using the correlation coefficient under single-valued neutrosophic environment. *Int. J. Gen. Syst.* **2013**, *42*, 386–394. [CrossRef]
7. Ye, J. Single valued neutrosophic cross-entropy for multicriteria decision making problems. *Appl. Math. Model.* **2014**, *38*, 1170–1175. [CrossRef]
8. Pramanik, S.; Dalapati, S.; Alam, S.; Roy, T.K. NS-Cross entropy-based MAGDM under single-valued neutrosophic set environment. *Inf. Polity* **2018**, *9*, 37.
9. Ye, J. Single-valued neutrosophic minimum spanning tree and its clustering method. *J. Intell. Syst.* **2014**, *23*, 311–324. [CrossRef]
10. Yager, R. On the theory of bags. *Int. J. Gen. Syst.* **1986**, *13*, 23–27. [CrossRef]
11. Ye, S.; Ye, J. Dice similarity measure between single valued neutrosophic multisets and its application in medical diagnosis. *Neutrosophic Sets Syst.* **2014**, *6*, 48–53.
12. Miyamoto, S. Fuzzy multisets and their generalizations. *Workshop Multisets Process* **2000**, *353*, 225–236.
13. Miyamoto, S. Multisets and fuzzy multisets as a framework of information systems. *Model. Decis. Artif. Intell.* **2004**, *3131*, 27–40.
14. Ye, S.; Fu, J.; Ye, J. Medical diagnosis using distance-based similarity measures of single valued neutrosophic multisets. *Neutrosophic Sets Syst.* **2015**, *7*, 47–52.
15. Pramanik, S.; Dey, P.P.; Giri, B.C. Hybrid vector similarity measure of single valued refined neutrosophic sets to multi-attribute decision making problems. *Neural Comput. Appl.* **2015**, *28*, 1–14.
16. Fan, C.; Ye, J. The cosine measure of refined-single valued neutrosophic sets and refined-interval neutrosophic sets for multiple attribute decision-making. *J. Intell. Fuzzy Syst.* **2017**, *33*, 2281–2289. [CrossRef]
17. Garg, H.; Nancy. Some new biparametric distance measures on single-valued neutrosophic sets with applications to pattern recognition and medical diagnosis. *Information* **2017**, *8*, 162. [CrossRef]
18. Ye, J. Single-valued neutrosophic similarity measures based on cotangent function and their application in the fault diagnosis of steam turbine. *Soft Comput.* **2017**, *21*, 817–825. [CrossRef]
19. Zhang, X.; Pei, D.; Dai, J. Fuzzy mathematics and rough set theory. *Beijing Tsinghua Univ. Process.* **2013**, *20–59*, 100–204.
20. Szmidt, E.; Kacprzyk, J. Distances between intuitionistic fuzzy sets. *Fuzzy Sets Syst.* **2000**, *114*, 505–518. [CrossRef]
21. Li, M. Cut sets of intuitionistic fuzzy sets. *J. Liaoning Norm. Univ.* **2002**, *1*, 28–33.
22. Yuan, X.; Li, H.; Sun, K. The cut sets, decomposition theorems and representation theorems on intuitionistic fuzzy sets and interval valued fuzzy sets. *Sci. China (Inf. Sci.)* **2011**, *54*, 91–110. [CrossRef]
23. Zhang, C.; Guo, J.; Zhao, Z.; Pei, B. Decomposition theorem of interval-valued fuzzy sets and calculation of similarity measure. *J. Liaoning Tech. Univ. (Nat. Sci.)* **2015**, *34*, 1312–1315.
24. Yuan, X.; Li, H.; Lee, E.S. Three new cut sets of fuzzy sets and new theories of fuzzy sets. *Comput. Math. Appl.* **2009**, *57*, 691–701. [CrossRef]
25. Luka, G.; Vadim, K.; Makarov, K.A.; Kreå¡Imir, V. Representation theorems for indefinite quadratic forms revisited. *Mathematika* **2013**, *59*, 169–189.
26. Alcantud, J.C.R.; Torra, V. Decomposition theorems and extension principles for hesitant fuzzy sets. *Inf. Fusion* **2018**, *41*, 48–56. [CrossRef]
27. Li, J.; Li, H. The cut sets, decomposition theorems and representation theorems on R^--fuzzy sets. *Int. J. Inf. Syst. Sci.* **2010**, *6*, 61–71.
28. Wang, F.; Zhang, C.; Xia, X. Equivalence of the cut sets-based decomposition theorems and representation theorems on intuitionistic fuzzy sets and interval-valued fuzzy sets. *Math. Comput. Model.* **2013**, *57*, 1364–1370. [CrossRef]
29. Singh, D.; Alkali, A.; Isah, A. Some application of α-cuts in fuzzy multisets theory. *J. Emerg. Trends Comput. Inf. Sci.* **2014**, *5*, 328–335.
30. Zhang, X.H.; Bo, C.X.; Smarandache, F.; Dai, J.H. New inclusion relation of neutrosophic sets with applications and related lattice structure. *Int. J. Mach. Learn. Cybern.* **2018**, *9*, 1753–1763. [CrossRef]

31. Zhang, X.H.; Bo, C.X.; Smarandache, F.; Park, C. New operations of totally dependent-neutrosophic sets and totally dependent-neutrosophic soft sets. *Symmetry* **2018**, *10*, 187. [CrossRef]
32. Zhang, X.H.; Liang, X.; Smarandache, F. Neutrosophic duplet semi-group and cancellable neutrosophic triplet groups. *Symmetry* **2017**, *9*. [CrossRef]
33. Medina, J.; Ojeda-Aciego, M. Multi-adjoint t-concept lattices. *Inf. Sci.* **2010**, *180*, 712–725. [CrossRef]
34. Pozna, C.; Minculete, N.; Precup, R.E.; Kóczy, L.T.; Ballagi, A. Signatures: Definitions, operators and applications to fuzzy modelling. *Fuzzy Sets Syst.* **2012**, *201*, 86–104. [CrossRef]
35. Nowaková, J.; Prílepok, M.; Snášel, V. Correction to: Medical image retrieval using vector quantization and fuzzy s-tree. *J. Med. Syst.* **2017**, *41*, 18. [CrossRef] [PubMed]
36. Kumar, A.; Kumar, D.; Jarial, S.K. A hybrid clustering method based on improved artificial bee colony and fuzzy C-means algorithm. *Int. J. Artif. Intell.* **2017**, *15*, 40–60.
37. Zhang, X.H. Fuzzy anti-grouped filters and fuzzy normal filters in pseudo-BCI algebras. *J. Intel. Fuzzy Syst.* **2017**, *33*, 1767–1774. [CrossRef]
38. Zhang, X.H.; Park, C.; Wu, S.P. Soft set theoretical approach to pseudo-BCI algebras. *J. Intell. Fuzzy Syst.* **2018**, *34*, 559–568. [CrossRef]
39. Ma, X.L.; Zhan, J.M.; Ali, M.I. A survey of decision making methods based on two classes of hybrid soft set models. *Artif. Intell. Rev.* **2018**, *49*, 511–529. [CrossRef]

symmetry

MDPI

Article

Two Types of Intuitionistic Fuzzy Covering Rough Sets and an Application to Multiple Criteria Group Decision Making

Jingqian Wang [1] and Xiaohong Zhang [2,*]

1 College of Electrical and Information Engineering, Shaanxi University of Science and Technology, Xi'an 710021, China; 81157@sust.edu.cn or wangjingqianw@163.com
2 School of Arts and Sciences, Shaanxi University of Science and Technology, Xi'an 710021, China
* Correspondence: zhangxiaohong@sust.edu.cn or zxhonghz@263.net

Received: 7 September 2018; Accepted: 29 September 2018; Published: 4 October 2018

Abstract: Intuitionistic fuzzy rough sets are constructed by combining intuitionistic fuzzy sets with rough sets. Recently, Huang et al. proposed the definition of an intuitionistic fuzzy (IF) β-covering and an IF covering rough set model. In this paper, some properties of IF β-covering approximation spaces and the IF covering rough set model are investigated further. Moreover, we present a novel methodology to the problem of multiple criteria group decision making. Firstly, some new notions and properties of IF β-covering approximation spaces are proposed. Secondly, we study the characterizations of Huang et al.'s IF covering rough set model and present a new IF covering rough set model for crisp sets in an IF environment. The relationships between these two IF covering rough set models and some other rough set models are investigated. Finally, based on the IF covering rough set model, Huang et al. also defined an optimistic multi-granulation IF rough set model. We present a novel method to multiple criteria group decision making problems under the optimistic multi-granulation IF rough set model.

Keywords: intuitionistic fuzzy; covering; neighborhood system; decision making

1. Introduction

Rough set theory was proposed by Pawlak [1,2] in 1982 as a tool to conceptualize, organize and analyze various types of data in data mining. There are other generalizations of his original concepts, for example by general binary relations [3], multi-granulations [4] and coverings [5]. Aiming at covering-based rough sets [6,7], they have been applied to decision rule synthesis [8,9], knowledge reduction [10,11] and other fields [12,13]. In theory, covering-based rough set theory has been connected with other theories. For example, it has been connected with lattice theory [14,15], matroid theory [16,17] and fuzzy set theory [18–23].

Zadeh's fuzzy set theory [24] addresses the problem of how to understand and manipulate imperfect knowledge. Recent investigations have shown that rough set and fuzzy set theories can be combined into various models, which are used for incomplete information in information systems. Dübois and Prade [25] first presented a fuzzy rough set model. Based on their work, some extended models and corresponding applications have been investigated in [26,27]. As far as fuzzy covering rough sets, D'eer et al. [28] discussed some fuzzy covering-based rough set models. Ma [18] proposed two new types of fuzzy covering rough set models. Inspired by Ma's work, Yang and Hu [29] investigated some types of fuzzy covering-based rough sets. Then, Yang and Hu studied other problems in fuzzy covering-based rough sets [30].

Intuitionistic fuzzy set (IFS) theory, as a straightforward extension of fuzzy set theory, was proposed by Atanassov [31]. The combination of IFS and rough set theories has attracted more interesting studies. For example, Zhang et al. [32] studied a general frame of IF rough sets. Huang et al. [33] presented an IF rough set model by combining β-neighborhoods induced by an IF β-covering. There are also other new notions and properties in Huang et al.'s IF rough set model, so it is necessary to investigate the IF rough set model further in this paper.

Recently, many researchers have studied decision making problems by rough set models [19,34], especially multiple criteria group decision making [20,23]. Multiple criteria group decision making (MCGDM) involves ranking from all feasible alternatives in conflicting and interactive criteria. After presenting the IF covering rough set model, Huang et al. also defined an optimistic multi-granulation IF rough set model. By investigation, multi-granulation IF rough set models have not been used for MCGDM problems. According to the characterizations of MCGDM problems, we construct the multi-granulation IF decision information systems and present a novel approach to MCGDM problems based on the optimistic multi-granulation IF rough set model in this paper.

The rest of this paper is organized as follows. Section 2 recalls some notions about covering-based rough sets and intuitionistic fuzzy sets. In Section 3, some properties of IF β-covering approximation space are investigated further. In Section 4, we investigate two IF covering rough set models. Among these two models, one is presented by Huang et al. [33], which concerns the IF sets; and the other is proposed by us, which concerns the crisp sets in an IF environment. In Section 5, we present a novel approach to MCGDM problems based on Huang et al.'s optimistic multi-granulation IF rough set model. This paper is concluded and further work is indicated in Section 6.

2. Basic Definitions

This section reviews some fundamental notions related to covering-based rough sets and intuitionistic fuzzy sets. Suppose U is a nonempty and finite set called a universe.

Definition 1. *(Covering [35,36]) Let U be a universe and $\mathbf{C}$ a family of subsets of U. If none of the subsets in $\mathbf{C}$ are empty and $\bigcup \mathbf{C} = U$, then $\mathbf{C}$ is called a covering of U.*

Definition 2. *(Neighborhood [35]) Let $\mathbf{C}$ be a covering of U. For any $x \in U$,*

$$N_{\mathbf{C}}(x) = \bigcap \{K \in \mathbf{C} : x \in K\}$$

is called the neighborhood of x with respect to $\mathbf{C}$. When the covering is clear, we omit the lowercase $\mathbf{C}$ in the neighborhood.

Definition 3. *(Approximation operators [37]) Let $\mathbf{C}$ be a covering of U. For any $X \subseteq U$,*

$$SH_{\mathbf{C}}(X) = \{x \in U : N(x) \cap X \neq \emptyset\},$$
$$SL_{\mathbf{C}}(X) = \{x \in U : N(x) \subseteq X\}.$$

$SH_{\mathbf{C}}(X)$ and $SL_{\mathbf{C}}(X)$ are called the upper and lower approximation operators with respect to $\mathbf{C}$.

Definition 4. *(Intuitionistic fuzzy set [31]) Let U be a fixed set. An intuitionistic fuzzy set (IFS) A in U is defined as:*

$$A = \{\langle x, \mu_A(x), \nu_A(x)\rangle : x \in U\},$$

where $\mu_A : U \to [0,1]$ is called the degree of membership of the element $x \in U$ to A and $\nu_A : U \to [0,1]$ is called the degree of non-membership. They satisfy $\mu_A(x) + \nu_A(x) \leq 1$ for all $x \in U$. The family of all intuitionistic fuzzy sets in U is denoted by $IF(U)$.

We call $\langle a, b\rangle$ with $0 \leq a, b \leq 1$ and $a + b \leq 1$ an IF value. As is well known, for two IF values $\alpha = \langle a, b\rangle$ and $\beta = \langle c, d\rangle$, $\alpha \leq \beta \Leftrightarrow a \leq c$ and $b \geq d$.

For any family $\gamma_i \in [0,1], i \in I, I \subseteq \mathbb{N}^+$ ($\mathbb{N}^+$ is the set of all positive integers), we write $\vee_{i\in I}\gamma_i$ for the supremum of $\{\gamma_i : i \in I\}$ and $\wedge_{i\in I}\gamma_i$ for the infimum of $\{\gamma_i : i \in I\}$. Some basic operations on $IF(U)$ are shown as follows [31]: for any $A, B \in IF(U)$,

1. $A \subseteq B$ iff $\mu_A(x) \leq \mu_B(x)$ and $\nu_B(x) \leq \nu_A(x)$ for all $x \in U$;
2. $A = B$ iff $A \subseteq B$ and $B \subseteq A$;
3. $A \cup B = \{\langle x, \mu_A(x) \vee \mu_B(x), \nu_A(x) \wedge \nu_B(x)\rangle : x \in U\}$;
4. $A \cap B = \{\langle x, \mu_A(x) \wedge \mu_B(x), \nu_A(x) \vee \nu_B(x)\rangle : x \in U\}$;
5. $A' = \{\langle x, \nu_A(x), \mu_A(x)\rangle : x \in U\}$.

3. Some Properties of IF β-Covering Approximation Space

In this section, we introduce the notions of intuitionistic fuzzy (IF) β-covering approximation space. There are two concepts presented by Huang et al. in [33], which are IF β-covering and IF β-neighborhood in this approximation space. We mainly investigate some of their properties, and some new notions are presented.

3.1. IF β-Neighborhood and IF β-Neighborhood System

Definition 5. *([33]) Let U be a universe and $\beta = \langle a, b\rangle$ be an IF value. Then, we call $\widehat{C} = \{C_1, C_2, ..., C_m\}$, with $C_i \in IF(U)(i = 1, 2, ..., m)$, an IF β-covering of U, if for any $x \in U$, there exists $C_i \in \widehat{C}$, such that $C_i(x) \geq \beta$. We also call $(U, \widehat{C})$ an IF β-covering approximation space.*

Let $\Gamma_\beta = \{\widehat{C_1}, \widehat{C_2}, \cdots, \widehat{C_n}\}$. If any $\widehat{C}_i$ $(i = 1, 2, ..., n)$ is an IF β-covering of U, then we call (U, Γ_β) a n-IF β-covering approximation space.

Definition 6. *([33]) Let $\widehat{C}$ be an IF β-covering of U and $\widehat{C} = \{C_1, C_2, \ldots, C_m\}$. For each $x \in U$, the IF β-neighborhood $\widetilde{N}_x^\beta$ of x induced by $\widehat{C}$ can be defined as:*

$$\widetilde{N}_x^\beta = \cap\{C_i \in \widehat{C} : C_i(x) \geq \beta\}.$$

Note that $C_i(x)$ is an IF value $\langle \mu_{C_i}(x), \nu_{C_i}(x)\rangle$ in Definitions 5 and 6. Hence, $C_i(x) \geq \beta$ means $\mu_{C_i}(x) \geq a$ and $\nu_{C_i}(x) \leq b$ where IF value $\beta = \langle a, b\rangle$.

Remark 1. *Let $\widehat{C}$ be an IF β-covering of U, $\beta = \langle a, b\rangle$ and $\widehat{C} = \{C_1, C_2, \ldots, C_m\}$. For each $x \in U$:*

$$\widetilde{N}_x^\beta = \cap\{C_i \in \widehat{C} : \mu_{C_i}(x) \geq a, \nu_{C_i}(x) \leq b\}.$$

Example 1. *Let $U = \{x_1, x_2, x_3, x_4, x_5\}$ and $\widehat{C} = \{C_1, C_2, C_3, C_4\}$, where:*

$$C_1 = \frac{(0.7,0.2)}{x_1} + \frac{(0.5,0.3)}{x_2} + \frac{(0.4,0.5)}{x_3} + \frac{(0.6,0.1)}{x_4} + \frac{(0.3,0.2)}{x_5},$$
$$C_2 = \frac{(0.6,0.2)}{x_1} + \frac{(0.3,0.2)}{x_2} + \frac{(0.2,0.3)}{x_3} + \frac{(0.4,0.5)}{x_4} + \frac{(0.7,0.3)}{x_5},$$
$$C_3 = \frac{(0.4,0.1)}{x_1} + \frac{(0.4,0.5)}{x_2} + \frac{(0.5,0.2)}{x_3} + \frac{(0.3,0.6)}{x_4} + \frac{(0.6,0.3)}{x_5},$$
$$C_4 = \frac{(0.1,0.5)}{x_1} + \frac{(0.6,0.1)}{x_2} + \frac{(0.6,0.3)}{x_3} + \frac{(0.5,0.3)}{x_4} + \frac{(0.8,0.1)}{x_5}.$$

According to Definition 5, we know $\widehat{C}$ is an IF β-covering of U ($\beta = \langle a, b\rangle$ with $0 \leq a \leq 0.6, 0.3 \leq b \leq 1$). Let $\beta = \langle 0.5, 0.3\rangle$. Then, the IF β-neighborhoods are shown as follows:

$$\begin{aligned}
\widetilde{N}_{x_1}^{\beta} &= C_1 \cap C_2 = \frac{(0.6,0.2)}{x_1} + \frac{(0.3,0.3)}{x_2} + \frac{(0.2,0.5)}{x_3} + \frac{(0.4,0.5)}{x_4} + \frac{(0.3,0.3)}{x_5},\\
\widetilde{N}_{x_2}^{\beta} &= C_1 \cap C_4 = \frac{(0.1,0.5)}{x_1} + \frac{(0.5,0.3)}{x_2} + \frac{(0.4,0.5)}{x_3} + \frac{(0.5,0.3)}{x_4} + \frac{(0.3,0.2)}{x_5},\\
\widetilde{N}_{x_3}^{\beta} &= C_3 \cap C_4 = \frac{(0.1,0.5)}{x_1} + \frac{(0.4,0.5)}{x_2} + \frac{(0.5,0.3)}{x_3} + \frac{(0.3,0.6)}{x_4} + \frac{(0.6,0.3)}{x_5},\\
\widetilde{N}_{x_4}^{\beta} &= C_1 \cap C_4 = \frac{(0.1,0.5)}{x_1} + \frac{(0.5,0.3)}{x_2} + \frac{(0.4,0.5)}{x_3} + \frac{(0.5,0.3)}{x_4} + \frac{(0.3,0.2)}{x_5},\\
\widetilde{N}_{x_5}^{\beta} &= C_2 \cap C_3 \cap C_4 = \frac{(0.1,0.5)}{x_1} + \frac{(0.3,0.5)}{x_2} + \frac{(0.2,0.3)}{x_3} + \frac{(0.3,0.6)}{x_4} + \frac{(0.6,0.3)}{x_5}.
\end{aligned}$$

Theorem 1. *([33]) Let $\widehat{C}$ be an intuitionistic fuzzy β-covering of U and $\widehat{C} = \{C_1, C_2, \ldots, C_m\}$. Then, the following statements hold:*

1. *$\widetilde{N}_x^{\beta}(x) \geq \beta$ for any $x \in U$;*
2. *For $x, y, z \in U$, if $\widetilde{N}_x^{\beta}(y) \geq \beta$, $\widetilde{N}_y^{\beta}(z) \geq \beta$, then $\widetilde{N}_x^{\beta}(z) \geq \beta$;*
3. *For two IF values β_1, β_2, if $\beta_1 \leq \beta_2 \leq \beta$, then $\widetilde{N}_x^{\beta_1} \subseteq \widetilde{N}_x^{\beta_2}$ for all $x \in U$.*

For two different IF β-neighborhoods, a relationship between them is presented.

Proposition 1. *Let $\widehat{C}$ be an IF β-covering of U. For any $x, y \in U$, $\widetilde{N}_x^{\beta}(y) \geq \beta$ if and only if $\widetilde{N}_y^{\beta} \subseteq \widetilde{N}_x^{\beta}$.*

Proof. Suppose the IF value $\beta = \langle a, b\rangle$.

($\Rightarrow$): Since $\widetilde{N}_x^{\beta}(y) \geq \beta$, so

$$\mu_{\widetilde{N}_x^{\beta}}(y) = \mu_{\underset{\substack{\mu_{C_i}(x)\geq a\\ \nu_{C_i}(x)\leq b}}{\cap} C_i}(y) = \bigwedge_{\substack{\mu_{C_i}(x)\geq a\\ \nu_{C_i}(x)\leq b}} \mu_{C_i}(y) \geq a, \nu_{\widetilde{N}_x^{\beta}}(y) = \nu_{\underset{\substack{\mu_{C_i}(x)\geq a\\ \nu_{C_i}(x)\leq b}}{\cap} C_i}(y) = \bigvee_{\substack{\mu_{C_i}(x)\geq a\\ \nu_{C_i}(x)\leq b}} \nu_{C_i}(y) \leq b.$$

Hence,

$$\{C_i \in \widehat{C} : \mu_{C_i}(x) \geq a, \nu_{C_i}(x) \leq b\} \subseteq \{C_i \in \widehat{C} : \mu_{C_i}(y) \geq a, \nu_{C_i}(y) \leq b\}.$$

Therefore, for each $z \in U$,

$$\begin{aligned}
\mu_{\widetilde{N}_x^{\beta}}(z) &= \bigwedge_{\substack{\mu_{C_i}(x)\geq a\\ \nu_{C_i}(x)\leq b}} \mu_{C_i}(z) \geq \bigwedge_{\substack{\mu_{C_i}(y)\geq a\\ \nu_{C_i}(y)\leq b}} \mu_{C_i}(z) = \mu_{\widetilde{N}_y^{\beta}}(z),\\
\nu_{\widetilde{N}_x^{\beta}}(z) &= \bigvee_{\substack{\mu_{C_i}(x)\geq a\\ \nu_{C_i}(x)\leq b}} \nu_{C_i}(z) \leq \bigvee_{\substack{\mu_{C_i}(y)\geq a\\ \nu_{C_i}(y)\leq b}} \nu_{C_i}(z) = \nu_{\widetilde{N}_y^{\beta}}(z).
\end{aligned}$$

Hence, $\widetilde{N}_y^{\beta} \subseteq \widetilde{N}_x^{\beta}$.

($\Leftarrow$): For any $x, y \in U$, since $\widetilde{N}_y^{\beta} \subseteq \widetilde{N}_x^{\beta}$, so $\mu_{\widetilde{N}_x^{\beta}}(y) \geq \mu_{\widetilde{N}_y^{\beta}}(y) \geq a$, $\nu_{\widetilde{N}_x^{\beta}}(y) \leq \nu_{\widetilde{N}_y^{\beta}}(y) \leq b$. Therefore $\widetilde{N}_x^{\beta}(y) \geq \beta$. □

By the notion of IF β-neighborhood, we propose the following definition of the IF β-neighborhood system of $x \in U$.

Definition 7. *Let $\widehat{C}$ be an IF β-covering of U and $\widehat{C} = \{C_1, C_2, \ldots, C_m\}$. For each $x \in U$, the IF β-neighborhood system $\widetilde{\mathcal{N}}_{\widehat{C}}^{\beta}(x)$ of x induced by $\widehat{C}$ is defined as:*

$$\widetilde{\mathcal{N}}_{\widehat{C}}^{\beta}(x) = \{C_i \in \widehat{C} : C_i(x) \geq \beta\}.$$

According to Definition 6, we know $\widetilde{N}_x^{\beta} = \cap \widetilde{\mathcal{N}}_{\widehat{\mathbf{C}}}^{\beta}(x)$ for each $x \in U$. Let $\widehat{\mathbf{C}_1}$, $\widehat{\mathbf{C}_2}$ be two IF β-coverings of U. The statement does not hold: if $\widetilde{\mathcal{N}}_{\widehat{\mathbf{C}_1}}^{\beta}(x) = \widetilde{\mathcal{N}}_{\widehat{\mathbf{C}_2}}^{\beta}(x)$ for each $x \in U$, then $\widehat{\mathbf{C}_1} = \widehat{\mathbf{C}_2}$. The following example can illustrate it.

Example 2. *(Continued from Example 1) Let $\beta = \langle 0.5, 0.3 \rangle$, $\widehat{\mathbf{C}_1} = \widehat{\mathbf{C}} \cup \{C_5\}$, where $C_5 = \frac{(0.2,0.5)}{x_1} + \frac{(0.5,0.4)}{x_2} + \frac{(0.4,0.6)}{x_3} + \frac{(0.3,0.4)}{x_4} + \frac{(0.3,0.5)}{x_5}$. Then, $\widetilde{\mathcal{N}}_{\widehat{\mathbf{C}}}^{\beta}(x_i)$ and $\widetilde{\mathcal{N}}_{\widehat{\mathbf{C}_1}}^{\beta}(x_i)$ $(i = 1,2,3,4,5)$ are listed in Table 1.*

Table 1. $\widetilde{\mathcal{N}}_{\widehat{\mathbf{C}}}^{\beta}(x_i)$ and $\widetilde{\mathcal{N}}_{\widehat{\mathbf{C}_1}}^{\beta}(x_i)$ $(i = 1,2,3,4,5)$.

U	x_1	x_2	x_3	x_4	x_5
$\widetilde{\mathcal{N}}_{\widehat{\mathbf{C}}}^{\beta}$	$\{C_1, C_2\}$	$\{C_1, C_4\}$	$\{C_3, C_4\}$	$\{C_1, C_4\}$	$\{C_2, C_3, C_4\}$
$\widetilde{\mathcal{N}}_{\widehat{\mathbf{C}_1}}^{\beta}$	$\{C_1, C_2\}$	$\{C_1, C_4\}$	$\{C_3, C_4\}$	$\{C_1, C_4\}$	$\{C_2, C_3, C_4\}$

Hence, $\widetilde{\mathcal{N}}_{\widehat{\mathbf{C}}}^{\beta}(x_i) = \widetilde{\mathcal{N}}_{\widehat{\mathbf{C}_1}}^{\beta}(x_i)$ for any $i = 1,2,3,4,5$, but $\widehat{\mathbf{C}} \neq \widehat{\mathbf{C}_1}$.

Inspired by this statement, we consider in which conditions two IF β-coverings generate the same β-neighborhood system for any element of the universe. In order to find the conditions, we introduce two new concepts in IF β-covering approximation space firstly.

Definition 8. *Let $\widehat{\mathbf{C}}$ be an IF β-covering of U and $C \in \widehat{\mathbf{C}}$. If there exists $x \in U$ such that $C(x) \geq \beta$, then C is called an IF β-dependent element of $\widehat{\mathbf{C}}$; otherwise, C is called an IF β-independent element of $\widehat{\mathbf{C}}$. If every element in $\widehat{\mathbf{C}}$ is an IF β-dependent element, then $\widehat{\mathbf{C}}$ is IF β-dependent; otherwise, $\widehat{\mathbf{C}}$ is IF β-independent.*

Example 3. *(Continued from Example 2) Let $\beta = \langle 0.5, 0.3 \rangle$. Then, C_5 is an IF β-independent element of $\widehat{\mathbf{C}_1}$ and C_i $(i = 1,2,3,4)$ are IF β-dependent elements of $\widehat{\mathbf{C}_1}$. Hence, $\widehat{\mathbf{C}_1}$ is IF β-independent and $\widehat{\mathbf{C}}$ is IF β-dependent.*

Proposition 2. *Let $\widehat{\mathbf{C}}$ be an IF β-covering of U and $C \in \widehat{\mathbf{C}}$. If C is an IF β-independent element of $\widehat{\mathbf{C}}$, then $\widehat{\mathbf{C}} - \{C\}$ is still an IF β-covering of U.*

Proof. According to Definitions 5 and 8, it is straightforward. □

Proposition 3. *Let $\widehat{\mathbf{C}}$ be an IF β-covering of U, C be an IF β-independent element of $\widehat{\mathbf{C}}$ and $C_1 \in \widehat{\mathbf{C}} - \{C\}$. Then, C_1 is an IF β-independent element of $\widehat{\mathbf{C}}$ iff C_1 is an IF β-independent element of $\widehat{\mathbf{C}} - \{C\}$.*

Proof. According to Definitions 5 and 8, it is straightforward. □

According to Propositions 2 and 3, it is still an IF β-covering after deleting all IF β-independent elements of an IF β-covering.

Definition 9. *Let $\widehat{\mathbf{C}}$ be an IF β-covering of U and $\widehat{\mathbf{B}} \subseteq \widehat{\mathbf{C}}$. If $\widehat{\mathbf{B}}$ is the set of all IF β-dependent elements of $\widehat{\mathbf{C}}$, then $\widehat{\mathbf{B}}$ is called the IF β-base of $\widehat{\mathbf{C}}$ and is denoted as $\Delta^{\beta}(\widehat{\mathbf{C}})$.*

Proposition 4. *Let $\widehat{\mathbf{C}}$ be an IF β-covering of U. For any $x \in U$,*

$$\widetilde{\mathcal{N}}_{\widehat{\mathbf{C}}}^{\beta}(x) = \widetilde{\mathcal{N}}_{\Delta^{\beta}(\widehat{\mathbf{C}})}^{\beta}(x).$$

Proof. According to Definitions 8, 9 and Proposition 2, it is straightforward. □

Theorem 2. *Let $\widehat{\mathbf{C}_1}$, $\widehat{\mathbf{C}_2}$ be two IF β-coverings of U. For any $x \in U$,*

$$\widetilde{\mathcal{N}}^{\beta}_{\widehat{C_1}}(x) = \widetilde{\mathcal{N}}^{\beta}_{\widehat{C_2}}(x) \text{ iff } \Delta^{\beta}(\widehat{C_1}) = \Delta^{\beta}(\widehat{C_2}).$$

Proof. By Proposition 4 and Definition 7, it is straightforward. □

Corollary 1. *Let $\widehat{C_1}, \widehat{C_2}$ be two IF β-coverings of U. For any $x \in U$, if $\Delta^{\beta}(\widehat{C_1}) = \Delta^{\beta}(\widehat{C_2})$, then $\widetilde{N}^{\beta}_x = \widetilde{N'}^{\beta}_x$, where $\widetilde{N}^{\beta}_x$ and $\widetilde{N'}^{\beta}_x$ are the β-neighborhoods induced by $\widehat{C_1}$ and $\widehat{C_2}$, respectively.*

Proof. According to Theorem 2, $\Delta^{\beta}(\widehat{C_1}) = \Delta^{\beta}(\widehat{C_2}) \Rightarrow \widetilde{\mathcal{N}}^{\beta}_{\widehat{C_1}}(x) = \widetilde{\mathcal{N}}^{\beta}_{\widehat{C_2}}(x) \Rightarrow \cap\widetilde{\mathcal{N}}^{\beta}_{\widehat{C_1}}(x) = \cap\widetilde{\mathcal{N}}^{\beta}_{\widehat{C_2}}(x) \Rightarrow \widetilde{N}^{\beta}_x = \widetilde{N'}^{\beta}_x$. □

3.2. β-Neighborhood

In this subsection, the definition of β-neighborhood is presented by the IF β-neighborhood.

Definition 10. *Let $(U, \widehat{C})$ be an IF β-covering approximation space and $\widehat{C} = \{C_1, C_2, \ldots, C_m\}$. For each $x \in U$, we define the β-neighborhood $\overline{N}^{\beta}_x$ of x as:*

$$\overline{N}^{\beta}_x = \{y \in U : \widetilde{N}^{\beta}_x(y) \geq \beta\}.$$

Note that $\widetilde{N}^{\beta}_x(y)$ is an IF value $\langle \mu_{\widetilde{N}^{\beta}_x}(y), \nu_{\widetilde{N}^{\beta}_x}(y)\rangle$ in Definition 10. Based on this, the β-neighborhood can be represented by the following remark.

Remark 2. *Let $\widehat{C}$ be an IF β-covering of U, $\beta = \langle a, b\rangle$ and $\widehat{C} = \{C_1, C_2, \ldots, C_m\}$. For each $x \in U$,*

$$\overline{N}^{\beta}_x = \{y \in U : \mu_{\widetilde{N}^{\beta}_x}(y) \geq a, \nu_{\widetilde{N}^{\beta}_x}(y) \leq b\}.$$

Example 4. *(Continued from Example 1) Let $\beta = \langle 0.5, 0.3\rangle$, then we have:*

$$\overline{N}^{\beta}_{x_1} = \{x_1\}, \ \overline{N}^{\beta}_{x_2} = \{x_2, x_4\}, \ \overline{N}^{\beta}_{x_3} = \{x_3, x_5\},$$
$$\overline{N}^{\beta}_{x_4} = \{x_2, x_4\}, \ \overline{N}^{\beta}_{x_5} = \{x_5\}.$$

The following theorem shows the basic properties of β-neighborhoods.

Theorem 3. *Let $\widehat{C}$ be an IF β-covering of U and $\widehat{C} = \{C_1, C_2, \ldots, C_m\}$. Then:*

1. *$x \in \overline{N}^{\beta}_x$ for any $x \in U$;*
2. *For any $x, y, z \in U$, if $x \in \overline{N}^{\beta}_y$, $y \in \overline{N}^{\beta}_z$, then $x \in \overline{N}^{\beta}_z$.*

Proof. (1) According to the first statement in Theorem 1, we know $\widetilde{N}^{\beta}_x(x) \geq \beta$ for each $x \in U$. Hence, $x \in \{y \in U : \widetilde{N}^{\beta}_x(y) \geq \beta\} = \overline{N}^{\beta}_x$ for each $x \in U$.
(2) For any $x, y, z \in U$, $x \in \overline{N}^{\beta}_y \Leftrightarrow \widetilde{N}^{\beta}_y(x) \geq \beta \Leftrightarrow \widetilde{N}^{\beta}_x \subseteq \widetilde{N}^{\beta}_y$, and $y \in \overline{N}^{\beta}_z \Leftrightarrow \widetilde{N}^{\beta}_z(y) \geq \beta \Leftrightarrow \widetilde{N}^{\beta}_y \subseteq \widetilde{N}^{\beta}_z$. Hence, $\widetilde{N}^{\beta}_x \subseteq \widetilde{N}^{\beta}_z$. By Proposition 1, we have $\widetilde{N}^{\beta}_z(x) \geq \beta$, i.e., $x \in \overline{N}^{\beta}_z$. □

The following proposition shows a relationship between $\overline{N}^{\beta}_x$ and $\overline{N}^{\beta}_y$.

Proposition 5. *Let $\widehat{C}$ be an IF β-covering of U and $\widehat{C} = \{C_1, C_2, \ldots, C_m\}$. Then, for any $x \in U$, $x \in \overline{N}^{\beta}_y$ iff $\overline{N}^{\beta}_x \subseteq \overline{N}^{\beta}_y$.*

Proof. (⇒): For any $z \in \overline{N}_x^\beta$, we know $\widetilde{N}_x^\beta(z) \geq \beta$. Since $x \in \overline{N}_y^\beta$, so $\widetilde{N}_y^\beta(x) \geq \beta$. According to (2) in Theorem 1, we have $\widetilde{N}_y^\beta(z) \geq \beta$. Hence, $z \in \overline{N}_y^\beta$. Therefore, $\overline{N}_x^\beta \subseteq \overline{N}_y^\beta$.
(⇐): According to (1) in Theorem 3, $x \in \overline{N}_x^\beta$ for all $x \in U$. Since $\overline{N}_x^\beta \subseteq \overline{N}_y^\beta$, so $x \in \overline{N}_y^\beta$. □

A relationship between IF β-neighborhoods and β-neighborhoods is proposed in the following proposition.

Proposition 6. *Let $\widehat{C}$ be an IF β-covering of U. For any $x, y \in U$, $\widetilde{N}_x^\beta \subseteq \widetilde{N}_y^\beta$ iff $\overline{N}_x^\beta \subseteq \overline{N}_y^\beta$.*

Proof. For any $x, y \in U$, $\widetilde{N}_x^\beta \subseteq \widetilde{N}_y^\beta \Leftrightarrow \widetilde{N}_y^\beta(x) \geq \beta \Leftrightarrow x \in \overline{N}_y^\beta \Leftrightarrow \overline{N}_x^\beta \subseteq \overline{N}_y^\beta$. □

4. Two Intuitionistic Fuzzy Covering Rough Set Models

In this section, we investigate two IF covering rough set models on the basis of the IF β-neighborhoods and the β-neighborhoods, respectively. Firstly, one model is presented by Huang et al., and we study the properties of the IF lower and upper approximations of each IF set further. Secondly, we propose another new model, which concerns crisp sets in an IF environment. Moreover, some properties of the approximations are investigated. Finally, the relationships between these two IF covering rough set models with other rough set models are revealed.

4.1. Characterizations of Huang et al.'s Intuitionistic Fuzzy Covering Rough Set Model

Huang et al. [33] presented an IF covering rough set model.

Definition 11. *([33]) Let $(U, \widehat{C})$ be an IF β-covering approximation space. For each $A \in IF(U)$ where $A = \{\langle x, \mu_A(x), \nu_A(x)\rangle : x \in U\}$, we define the intuitionistic fuzzy (IF) covering upper approximation $\widetilde{C}(A)$ and lower approximation $\underset{\sim}{C}(A)$ of A as:*

$$\widetilde{C}(A) = \{\langle x, \vee_{y\in U}[\mu_{\widetilde{N}_x^\beta}(y) \wedge \mu_A(y)], \wedge_{y\in U}[\nu_{\widetilde{N}_x^\beta}(y) \vee \nu_A(y)]\rangle : x \in U\},$$
$$\underset{\sim}{C}(A) = \{\langle x, \wedge_{y\in U}[\nu_{\widetilde{N}_x^\beta}(y) \vee \mu_A(y)], \vee_{y\in U}[\mu_{\widetilde{N}_x^\beta}(y) \wedge \nu_A(y)]\rangle : x \in U\}.$$

If $\widetilde{C}(A) \neq \underset{\sim}{C}(A)$, then A is called the first type of IF covering rough set.

Example 5. *(Continued from Example 1) Let $\beta = \langle 0.5, 0.3\rangle$, $A = \frac{(0.6,0.3)}{x_1} + \frac{(0.4,0.5)}{x_2} + \frac{(0.3,0.2)}{x_3} + \frac{(0.5,0.3)}{x_4} + \frac{(0.7,0.2)}{x_5}$.*

$$\widetilde{C}(A) = \{\langle x_1, 0.6, 0.3\rangle, \langle x_2, 0.5, 0.2\rangle, \langle x_3, 0.6, 0.3\rangle, \langle x_4, 0.5, 0.2\rangle, \langle x_5, 0.6, 0.3\rangle\},$$
$$\underset{\sim}{C}(A) = \{\langle x_1, 0.4, 0.3\rangle, \langle x_2, 0.4, 0.5\rangle, \langle x_3, 0.3, 0.4\rangle, \langle x_4, 0.4, 0.5\rangle, \langle x_5, 0.3, 0.3\rangle\}.$$

Some characterizations of Huang et al.'s IF covering rough set model are shown in the following proposition.

Proposition 7. *([33]) Let $(U, \widehat{C})$ be an IF β-covering approximation space. Then, for all $A, B \in IF(U)$,*

1. $\underset{\sim}{C}(U) = U$, $\widetilde{C}(\emptyset) = \emptyset$;
2. $\widetilde{C}(A') = (\underset{\sim}{C}(A))'$, $\underset{\sim}{C}(A') = (\widetilde{C}(A))'$;
3. *If $A \subseteq B$, then* $\underset{\sim}{C}(A) \subseteq \underset{\sim}{C}(B)$, $\widetilde{C}(A) \subseteq \widetilde{C}(B)$;
4. $\underset{\sim}{C}(A \cap B) = \underset{\sim}{C}(A) \cap \underset{\sim}{C}(B)$, $\widetilde{C}(A \cup B) = \widetilde{C}(A) \cup \widetilde{C}(B)$;

5. $\underset{\sim}{C}(A \bigcup B) \supseteq \underset{\sim}{C}(A) \bigcup \underset{\sim}{C}(B)$, $\widetilde{C}(A \bigcap B) \subseteq \widetilde{C}(A) \bigcap \widetilde{C}(B)$.

Besides these characterizations shown in Proposition 7, there are other characterizations that should be investigated.

Example 6. *(Continued from Example 1) Let* $\beta = \langle 0.5, 0.3 \rangle$.

$$\widetilde{C}(U) = \{\langle x_1, 0.6, 0.2\rangle, \langle x_2, 0.5, 0.2\rangle, \langle x_3, 0.6, 0.3\rangle, \langle x_4, 0.5, 0.2\rangle, \langle x_5, 0.6, 0.3\rangle\},$$
$$\underset{\sim}{C}(\emptyset) = \{\langle x_1, 0.2, 0.6\rangle, \langle x_2, 0.2, 0.5\rangle, \langle x_3, 0.3, 0.6\rangle, \langle x_4, 0.2, 0.5\rangle, \langle x_5, 0.3, 0.6\rangle\}.$$

According to Example 6, we know $\widetilde{\mathbf{C}}(U) \neq U$ and $\underset{\sim}{\mathbf{C}}(\emptyset) \neq \emptyset$ in an IF β-covering approximation space. However, there are $\widetilde{\mathbf{C}}(U) = U$ and $\underset{\sim}{\mathbf{C}}(\emptyset) = \emptyset$ based on some conditions.

Proposition 8. *Let* $(U, \widehat{C})$ *be an IF* β*-covering approximation space with* $\beta = \langle a, b\rangle$ *and* $\widehat{C} = \{C_1, C_2, \ldots, C_m\}$. $\underset{\sim}{C}(\emptyset) = \emptyset$ *iff* $\vee_{y\in U}\mu_{\widetilde{N}_x^\beta}(y) = 1, \wedge_{y\in U}\nu_{\widetilde{N}_x^\beta}(y) = 0$ *for any* $x \in U$.

Proof.

$$\begin{aligned} \underset{\sim}{\mathbf{C}}(\emptyset) = \emptyset \quad &\Leftrightarrow \widetilde{\mathbf{C}}(U) = U \\ &\Leftrightarrow \vee_{y\in U}[\mu_{\widetilde{N}_x^\beta}(y) \wedge \mu_U(y)] = 1, \wedge_{y\in U}[\nu_{\widetilde{N}_x^\beta}(y) \vee \nu_U(y)] = 0, (\forall x \in U) \\ &\Leftrightarrow \vee_{y\in U}\mu_{\widetilde{N}_x^\beta}(y) = 1, \wedge_{y\in U}\nu_{\widetilde{N}_x^\beta}(y) = 0, (\forall x \in U). \end{aligned}$$

□

Remark 3. *Let* $(U, \widehat{C})$ *be an IF* β*-covering approximation space. The IF covering approximation operators* $\underset{\sim}{C}$ *and* $\widetilde{C}$ *in Definition 11 do not satisfy the following statements: for all* $A, B \in IF(U)$,

1. $\underset{\sim}{C}(\underset{\sim}{C}(A)) = \underset{\sim}{C}(A)$, $\widetilde{C}(\widetilde{C}(A)) = \widetilde{C}(A)$;
2. $\underset{\sim}{C}((\underset{\sim}{C}(A))') = (\underset{\sim}{C}(A))'$, $\widetilde{C}((\widetilde{C}(A))') = (\widetilde{C}(A))'$;
3. *For any* $C \in \widehat{C}$, $\underset{\sim}{C}(C) = C$ *and* $\widetilde{C}(C) = C$;

In order to illustrate this remark, the following example is introduced.

Example 7. *(Continued from Example 1) Let* $\beta = \langle 0.5, 0.3\rangle$, *and:*

$$A = \frac{\langle 0.6,0.3\rangle}{x_1} + \frac{\langle 0.4,0.5\rangle}{x_2} + \frac{\langle 0.3,0.2\rangle}{x_3} + \frac{\langle 0.5,0.3\rangle}{x_4} + \frac{\langle 0.7,0.2\rangle}{x_5},$$
$$C_1 = \frac{\langle 0.7,0.2\rangle}{x_1} + \frac{\langle 0.5,0.3\rangle}{x_2} + \frac{\langle 0.4,0.5\rangle}{x_3} + \frac{\langle 0.6,0.1\rangle}{x_4} + \frac{\langle 0.3,0.2\rangle}{x_5},$$

then:

$$
\begin{aligned}
&\widetilde{\mathrm{C}}(A)=\{\langle x_1,0.6,0.3\rangle,\langle x_2,0.5,0.2\rangle,\langle x_3,0.6,0.3\rangle,\langle x_4,0.5,0.2\rangle,\langle x_5,0.6,0.3\rangle\},\\
&\underset{\sim}{\mathrm{C}}(A)=\{\langle x_1,0.4,0.3\rangle,\langle x_2,0.4,0.5\rangle,\langle x_3,0.3,0.4\rangle,\langle x_4,0.4,0.5\rangle,\langle x_5,0.3,0.3\rangle\},\\
&\widetilde{\mathrm{C}}(\widetilde{\mathrm{C}}(A))=\{\langle x_1,0.6,0.3\rangle,\langle x_2,0.5,0.3\rangle,\langle x_3,0.6,0.3\rangle,\langle x_4,0.5,0.3\rangle,\langle x_5,0.6,0.3\rangle\},\\
&\underset{\sim}{\mathrm{C}}(\underset{\sim}{\mathrm{C}}(A))=\{\langle x_1,0.3,0.4\rangle,\langle x_2,0.3,0.5\rangle,\langle x_3,0.3,0.4\rangle,\langle x_4,0.3,0.5\rangle,\langle x_5,0.3,0.3\rangle\},\\
&\widetilde{\mathrm{C}}(C_1)=\{\langle x_1,0.6,0.2\rangle,\langle x_2,0.5,0.2\rangle,\langle x_3,0.4,0.3\rangle,\langle x_4,0.5,0.2\rangle,\langle x_5,0.3,0.3\rangle\},\\
&\underset{\sim}{\mathrm{C}}(C_1)=\{\langle x_1,0.3,0.3\rangle,\langle x_2,0.3,0.4\rangle,\langle x_3,0.3,0.5\rangle,\langle x_4,0.3,0.4\rangle,\langle x_5,0.3,0.3\rangle\},\\
&(\widetilde{\mathrm{C}}(A))'=\{\langle x_1,0.3,0.6\rangle,\langle x_2,0.2,0.5\rangle,\langle x_3,0.3,0.6\rangle,\langle x_4,0.2,0.5\rangle,\langle x_5,0.3,0.6\rangle\},\\
&(\underset{\sim}{\mathrm{C}}(A))'=\{\langle x_1,0.3,0.4\rangle,\langle x_2,0.5,0.4\rangle,\langle x_3,0.4,0.3\rangle,\langle x_4,0.5,0.4\rangle,\langle x_5,0.3,0.3\rangle\},\\
&\widetilde{\mathrm{C}}((\widetilde{\mathrm{C}}(A))')=\{\langle x_1,0.3,0.5\rangle,\langle x_2,0.3,0.5\rangle,\langle x_3,0.3,0.5\rangle,\langle x_4,0.3,0.5\rangle,\langle x_5,0.3,0.5\rangle\},\\
&\underset{\sim}{\mathrm{C}}((\underset{\sim}{\mathrm{C}}(A))')=\{\langle x_1,0.3,0.4\rangle,\langle x_2,0.3,0.4\rangle,\langle x_3,0.3,0.4\rangle,\langle x_4,0.3,0.4\rangle,\langle x_5,0.3,0.3\rangle\}.
\end{aligned}
$$

Hence, $\underset{\sim}{\mathrm{C}}(\underset{\sim}{\mathrm{C}}(A)) \neq \underset{\sim}{\mathrm{C}}(A)$*,* $\widetilde{\mathrm{C}}(\widetilde{\mathrm{C}}(A)) \neq \widetilde{\mathrm{C}}(A)$*,* $\underset{\sim}{\mathrm{C}}((\underset{\sim}{\mathrm{C}}(A))') \neq (\underset{\sim}{\mathrm{C}}(A))'$*,* $\widetilde{\mathrm{C}}((\widetilde{\mathrm{C}}(A))') \neq (\widetilde{\mathrm{C}}(A))'$*,* $\underset{\sim}{\mathrm{C}}(C_1) \neq C_1$ *and* $\widetilde{\mathrm{C}}(C_1) \neq C_1$*.*

A condition for $\underset{\sim}{\mathrm{C}}(A) \subseteq A \subseteq \widetilde{\mathrm{C}}(A)$ is proposed in the following proposition.

Proposition 9. *Let* $(U, \widehat{\mathrm{C}})$ *be an IF* β*-covering approximation space with* $\beta = \langle a, b\rangle$ *and* $\widehat{\mathrm{C}} = \{C_1, C_2, \ldots, C_m\}$*. Then, the following statements are equivalent:*

1. $\mu_{C_i}(x) \geq a, \nu_{C_i}(x) \leq b \Rightarrow \mu_{C_i}(x) = 1, \nu_{C_i}(x) = 0$ *for any* $x \in U, i \in \{1, 2, \ldots, m\}$*;*
2. $\underset{\sim}{\mathrm{C}}(A) \subseteq A$ *for any* $A \in IF(U)$*;*
3. $A \subseteq \widetilde{\mathrm{C}}(A)$ *for any* $A \in IF(U)$*.*

Proof. (1) ⇒ (2): According to (1), we know $\mu_{\widetilde{N}_x^\beta}(x) = 1$ and $\nu_{\widetilde{N}_x^\beta}(x) = 0$ for each $x \in U$. For each $A \in IF(U)$,

$$
\begin{aligned}
\mu_{\underset{\sim}{\mathrm{C}}(A)}(x) &= \wedge_{y\in U}[\nu_{\widetilde{N}_x^\beta}(y) \vee \mu_A(y)]\\
&= (\wedge_{y\in U-\{x\}}[\nu_{\widetilde{N}_x^\beta}(y) \vee \mu_A(y)]) \wedge [\nu_{\widetilde{N}_x^\beta}(x) \vee \mu_A(x)]\\
&\leq \nu_{\widetilde{N}_x^\beta}(x) \vee \mu_A(x)\\
&= \mu_A(x).
\end{aligned}
$$

$$
\begin{aligned}
\nu_{\underset{\sim}{\mathrm{C}}(A)}(x) &= \vee_{y\in U}[\mu_{\widetilde{N}_x^\beta}(y) \wedge \nu_A(y)]\\
&= (\vee_{y\in U-\{x\}}[\mu_{\widetilde{N}_x^\beta}(y) \wedge \nu_A(y)]) \vee [\mu_{\widetilde{N}_x^\beta}(x) \wedge \nu_A(x)]\\
&\geq \mu_{\widetilde{N}_x^\beta}(x) \wedge \nu_A(x)\\
&= \nu_A(x).
\end{aligned}
$$

Hence, $\underset{\sim}{\mathrm{C}}(A) \subseteq A$.

(2) ⇒ (3): According to (2) and the duality, it is immediate.

(3) ⇒ (1): For any $x \in U$, let $\mu_{1_x}(x) = 1$, $\nu_{1_x}(x) = 0$, elsewhere $\mu_{1_x}(y) = 0$, $\nu_{1_x}(y) = 1$ for all $y \in U - \{x\}$. Hence, $1_x \in IF(U)$. Since $A \subseteq \widetilde{\mathrm{C}}(A)$ for any $A \in IF(U)$, so:

$$
\mu_{\widetilde{N}_x^\beta}(x) = \mu_{\widetilde{\mathrm{C}}(1_x)}(x) \geq \mu_{1_x}(x) = 1, \nu_{\widetilde{N}_x^\beta}(x) = \nu_{\widetilde{\mathrm{C}}(1_x)}(x) \leq \nu_{1_x}(x) = 0.
$$

Therefore, $\mu_{C_i}(x) \geq a, \nu_{C_i}(x) \leq b \Rightarrow \mu_{C_i}(x) = 1, \nu_{C_i}(x) = 0$ for each $x \in U, i \in \{1, 2, \ldots, m\}$. □

4.2. An Intuitionistic Fuzzy Covering Rough Set Model for Crisp Subsets

In [18], Ma presented a fuzzy covering rough set model for crisp subsets. Inspired by his work, we propose an IF covering rough set model for crisp subsets.

Definition 12. *Let $(U, \widehat{C})$ be an IF β-covering approximation space. For each crisp subset $X \in P(U)$ ($P(U)$ is the power set of U), the IF covering upper approximation $\overline{C}(X)$ and lower approximation $\underline{C}(X)$ of X are defined as:*

$$\overline{C}(X) = \{x \in U : \overline{N}_x^\beta \cap X \neq \emptyset\},$$
$$\underline{C}(X) = \{x \in U : \overline{N}_x^\beta \subseteq X\}.$$

If $\overline{C}(X) \neq \underline{C}(X)$, then X is called the second type of IF covering rough set.

Example 8. *(Continued from Example 4) Let $\beta = \langle 0.5, 0.3\rangle$, $X = \{x_1, x_2\}$, $Y = \{x_2, x_4, x_5\}$. Then:*

$$\overline{C}(X) = \{x_1, x_2, x_4\}, \underline{C}(X) = \{x_1\},$$
$$\overline{C}(Y) = \{x_2, x_3, x_4, x_5\}, \underline{C}(Y) = \{x_2, x_4, x_5\},$$
$$\overline{C}(U) = U, \underline{C}(U) = U, \overline{C}(\emptyset) = \emptyset, \underline{C}(\emptyset) = \emptyset.$$

The basic characterizations of the IF covering rough set model for crisp subsets are investigated in the following proposition.

Proposition 10. *Let $(U, \widehat{C})$ be an IF β-covering approximation space. Then, for all $X, Y \in P(U)$,*

1. $\underline{C}(U) = U, \overline{C}(\emptyset) = \emptyset$;
2. $\underline{C}(\emptyset) = \emptyset, \overline{C}(U) = U$;
3. $\underline{C}(X') = (\overline{C}(X))', \overline{C}(X') = (\underline{C}(X))'$;
4. *If* $X \subseteq Y$, *then* $\underline{C}(X) \subseteq \underline{C}(Y), \overline{C}(X) \subseteq \overline{C}(Y)$;
5. $\underline{C}(X \bigcap Y) = \underline{C}(X) \bigcap \underline{C}(Y), \overline{C}(X \bigcup Y) = \overline{C}(X) \bigcup \overline{C}(Y)$;
6. $\underline{C}(X \bigcup Y) \supseteq \underline{C}(X) \bigcup \underline{C}(Y), \overline{C}(X \bigcap Y) \subseteq \overline{C}(X) \bigcap \overline{C}(Y)$;
7. $\underline{C}(X) \subseteq X \subseteq \overline{C}(X)$;
8. $\underline{C}(\underline{C}(X)) \subseteq \underline{C}(X), \overline{C}(\overline{C}(X)) \supseteq \overline{C}(X)$;
9. $X \subseteq Y$ *or* $Y \subseteq X \Leftrightarrow \underline{C}(X \cap Y) = \underline{C}(X) \cap \underline{C}(Y), \overline{C}(X \cup Y) = \overline{C}(X) \cup \overline{C}(Y)$.

Proof. According to Definitions 10 and 12, it is immediate. □

4.3. Relationships between These Two Models and Some Other Rough Set Models

These two types of IF covering rough set models introduced above can be viewed as a bridge linking intuitionistic fuzzy sets and covering-based rough sets. In these models, $\widetilde{C}$, $\underset{\sim}{C}$ are IF approximation operators, and $\overline{C}$, $\underline{C}$ are crisp approximation operators in the IF environment. Firstly, we present the relationship between the IF covering rough set model defined in Section 4.1 and the generalized IF rough set model proposed by Zhou et al. in [38].

Definition 13. *([38]) Let U be a universe of discourse and $R \in IFR(U \times U)$. For any $A \in IF(U)$, the upper approximation $\widetilde{R}(A)$ and lower approximation $\underset{\sim}{R}(A)$ of A are defined as:*

$$\widetilde{R}(A) = \{\langle x, \vee_{y \in U}[\mu_R(x,y) \wedge \mu_A(y)], \wedge_{y \in U}[\nu_R(x,y) \vee \nu_A(y)]\rangle : x \in U\},$$
$$\underset{\sim}{R}(A) = \{\langle x, \wedge_{y \in U}[\nu_R(x,y) \vee \mu_A(y)], \vee_{y \in U}[\mu_R(x,y) \wedge \nu_A(y)]\rangle : x \in U\}.$$

For an IF β-covering $\widehat{\mathbf{C}}$ of U, one can define an IF relation R on the universe U as:

$$\mu_R(x,y) = \mu_{\widetilde{N}_x^\beta(y)}, \nu_R(x,y) = \nu_{\widetilde{N}_x^\beta(y)} \text{ for any } x,y \in U.$$

The induced IF relation R is related to all $C \in \widehat{\mathbf{C}}$. Hence, the IF covering rough set model defined in Section 4.1 can be viewed as a generalized IF rough set model presented by Zhou and Wu in Definition 13: for each $A \in IF(U)$,

$$\widetilde{\mathbf{C}}(A) = \{\langle x, \vee_{y\in U}[\mu_R(x,y) \wedge \mu_A(y)], \wedge_{y\in U}[\nu_R(x,y) \vee \nu_A(y)]\rangle : x \in U\} = \widetilde{R}(A),$$
$$\underset{\sim}{\mathbf{C}}(A) = \{\langle x, \wedge_{y\in U}[\nu_R(x,y) \vee \mu_A(y)], \vee_{y\in U}[\mu_R(x,y) \wedge \nu_A(y)]\rangle : x \in U\} = \underset{\sim}{R}(A).$$

Then, we present the relationship between the IF covering rough set model defined in Section 4.2 and a covering-based rough set model in Definition 3 proposed by Samanta and Chakraborty in [37].

Proposition 11. *Let $(U, \widehat{\mathbf{C}})$ be an IF β-covering approximation space and $\widehat{\mathbf{C}} = \{C_1, C_2, \ldots, C_m\}$.*

1. *If $K_i = \{x \in U : C_i(x) \geq \beta\}$ $(i = 1,2,\ldots,m)$, then $\mathbf{C} = \{K_1, K_2, \ldots, K_m\}$ is a covering of U;*
2. *If (1) holds, then $N_{\mathbf{C}}(x) = \overline{N}_x^\beta$.*

Proof. (1) Since $\widehat{\mathbf{C}} = \{C_1, C_2, \ldots, C_m\}$ is an IF β-covering on U, so for any $x \in U$, there exists $C_i \in \widehat{\mathbf{C}}$ such that $C_i(x) \geq \beta$. Thus, $x \in K_i$. Hence, $\mathbf{C} = \{K_1, K_2, \ldots, K_m\}$ is a covering of U.

(2) Since $\mathbf{C} = \{K_1, K_2, \ldots, K_m\}$ is a covering of U with $K_i = \{x \in U : C_i(x) \geq \beta\}$, so:

$$\begin{aligned}
\overline{N}_x^\beta &= \{y \in U : \widetilde{N}_x^\beta(y) \geq \beta\} \\
&= \{y \in U : (\bigcap_{C_i(x)\geq\beta} C_i)(y) \geq \beta\} \\
&= \{y \in U : (\bigcap_{x\in K_i} C_i)(y) \geq \beta\} \\
&= \{y \in U : x \in K_i \Rightarrow y \in K_i, i = 1,2,\ldots,m\} \\
&= \cap\{K_i \in \mathbf{C} : x \in K_i\} \\
&= N_{\mathbf{C}}(x).
\end{aligned}$$

□

According to Proposition 11, for any fixed $\beta = \langle a,b\rangle$ $(0 \leq a,b \leq 1$ and $a+b \leq 1)$, an IF β-covering $\widehat{\mathbf{C}}$ of U induces a covering $\mathbf{C}$ of U. Then, the second type of intuitionistic fuzzy covering rough set model defined in Subsection 4.2 can be viewed as a covering-based rough set model in Definition 3: for each $X \in P(U)$,

$$\overline{\mathbf{C}}(X) = \{x \in U : \overline{N}_x^\beta \cap X \neq \emptyset\} = \{x \in U : N_{\mathbf{C}}(x) \cap X \neq \emptyset\} = SH_{\mathbf{C}}(X),$$
$$\underline{\mathbf{C}}(X) = \{x \in U : \overline{N}_x^\beta \subseteq X\} = \{x \in U : N_{\mathbf{C}}(x) \subseteq X\} = SL_{\mathbf{C}}(X).$$

5. An Application to Multiple Criteria Group Decision Making

In [33], Huang et al. also defined an optimistic multi-granulation IF rough set model. In this section, we present a novel approach to MCGDM based on the optimistic multi-granulation IF rough set model. We investigate the basic description of an MCGDM problem under the framework of multi-granulation spaces. Then, we put forth a general decision making methodology for MCGDM problems by means of the optimistic multi-granulation IF rough set model in the case of patient ranking.

5.1. An Optimistic Multi-Granulation IF Rough Set Model

Let $\widehat{\mathbf{C}}$ be an IF β-covering of U. For each $x \in U$,

$$\widetilde{N}_x^\beta = \cap\{C_i \in \widehat{\mathbf{C}} : C_i(x) \geq \beta\}.$$
$$\widetilde{N}_{\widehat{\mathbf{C}}}^\beta = \{\widetilde{N}_x^\beta : x \in U\}, \text{ where } \mu_{\widetilde{N}_{\widehat{\mathbf{C}}}^\beta}(x,y) = \mu_{\widetilde{N}_x^\beta}(y) \text{ and } \nu_{\widetilde{N}_{\widehat{\mathbf{C}}}^\beta}(x,y) = \nu_{\widetilde{N}_x^\beta}(y) \text{ for any } y \in U.$$

Definition 14. *([33]) Let $U = \{x_1, x_2, \cdots, x_m\}$ be a universe, (U, Γ_β) be a n-IF β-coverings approximation space with $\Gamma_\beta = \{\widehat{C_1}, \widehat{C_2}, \cdots, \widehat{C_n}\}$ and $(U, \mathfrak{N}_{\Gamma_\beta})$ be the IF β-neighborhood (graded neighborhood) approximation space induced by the n-IF β-covering approximation space (U, Γ_β). For each $A \in IF(U)$ where $A = \{\langle x_j, \mu_A(x_j), \nu_A(x_j)\rangle : 1 \leq j \leq m\}$, the optimistic upper approximation $\overline{\mathfrak{N}_{\Gamma_\beta}}^{(o)}(A)$ and lower approximation $\underline{\mathfrak{N}_{\Gamma_\beta}}^{(o)}(A)$ of A are defined as:*

$$\overline{\mathfrak{N}_{\Gamma_\beta}}^{(o)}(A) = \{\langle x_i, \mu_{\overline{\mathfrak{N}_{\Gamma_\beta}}^{(o)}(A)}(x_i), \nu_{\overline{\mathfrak{N}_{\Gamma_\beta}}^{(o)}(A)}(x_i)\rangle : 1 \leq i \leq m\},$$
$$\underline{\mathfrak{N}_{\Gamma_\beta}}^{(o)}(A) = \{\langle x_i, \mu_{\underline{\mathfrak{N}_{\Gamma_\beta}}^{(o)}(A)}(x_i), \nu_{\underline{\mathfrak{N}_{\Gamma_\beta}}^{(o)}(A)}(x_i)\rangle : 1 \leq i \leq m\},$$

where:

$$\mu_{\overline{\mathfrak{N}_{\Gamma_\beta}}^{(o)}(A)}(x_i) = \bigwedge_{k=1}^{n} \bigvee_{j=1}^{m} [\mu_{\widetilde{N}_{\widehat{C_k}}^\beta}(x_i, x_j) \wedge \mu_A(x_j)](1 \leq i \leq m),$$
$$\nu_{\overline{\mathfrak{N}_{\Gamma_\beta}}^{(o)}(A)}(x_i) = \bigvee_{k=1}^{n} \bigwedge_{j=1}^{m} [\nu_{\widetilde{N}_{\widehat{C_k}}^\beta}(x_i, x_j) \vee \nu_A(x_j)](1 \leq i \leq m),$$
$$\mu_{\underline{\mathfrak{N}_{\Gamma_\beta}}^{(o)}(A)}(x_i) = \bigvee_{k=1}^{n} \bigwedge_{j=1}^{m} [\nu_{\widetilde{N}_{\widehat{C_k}}^\beta}(x_i, x_j) \vee \mu_A(x_j)](1 \leq i \leq m),$$
$$\nu_{\underline{\mathfrak{N}_{\Gamma_\beta}}^{(o)}(A)}(x_i) = \bigwedge_{k=1}^{n} \bigvee_{j=1}^{m} [\mu_{\widetilde{N}_{\widehat{C_k}}^\beta}(x_i, x_j) \wedge \nu_A(x_j)](1 \leq i \leq m).$$

5.2. The Problem of Multiple Criteria Group Decision Making

Let $U = \{x_k : k = 1, 2, \cdots, l\}$ be the set of patients and $V = \{y_j | j = 1, 2, \cdots, m\}$ be the m main symptoms (for example, fever, cough, and so on) for a disease B. Assume that the duty doctor X invites n experts R_i $(i = 1, 2, \cdots, n)$ to evaluate every patient x_k $(k = 1, 2, \cdots, l)$.

Assume that every expert R_i $(i = 1, 2, \cdots, n)$ believes each patient $x_k \in U$ $(k = 1, 2, \cdots, l)$ has a symptom value C_{ij} $(j = 1, 2, \cdots, m)$, denoted by $C_{ij}(x_k) = \langle \mu_{C_{ij}}(x_k), \nu_{C_{ij}}(x_k)\rangle$, where $\mu_{C_{ij}}(x_k) \in [0, 1]$ is the degree that expert R_i confirms patient x_k has symptom y_j, $\nu_{C_{ij}}(x_k) \in [0, 1]$ is the degree that expert R_i confirms patient x_k does not have symptom y_j and $\mu_{C_{ij}}(x_k) + \nu_{C_{ij}}(x_k) \leq 1$.

Let $\beta = \langle a, b\rangle$ be the critical value. If any patient $x_k \in U$, there is at least one symptom $y_j \in V$ such that the symptom value C_{ij} for the patient x_k, which is diagnosed by the expert R_i, is not less than β, respectively, then $\Gamma_\beta = \{\widehat{C_1}, \widehat{C_2}, \cdots, \widehat{C_n}\}$, where $\widehat{\mathbf{C}}_i = \{C_{i1}, C_{i2}, \cdots, C_{im}\}$, for all $1 \leq i \leq n$, is a n-IF β-coverings of U for some IF value β.

The IF β-neighborhood $\widetilde{N}_x^\beta$ of x induced by $\widehat{\mathbf{C}}_i$ $(1 \leq i \leq n)$ is an IFS:

$$\widetilde{N}_x^\beta = \cap\{C_{ij} \in \widehat{\mathbf{C}}_i : C_{ij}(x) \geq \beta, j = 1, 2, \cdots, m\}.$$
$$\widetilde{N}_{\widehat{\mathbf{C}}_i}^\beta = \{\widetilde{N}_x^\beta : x \in U\}, \text{ where } \mu_{\widetilde{N}_{\widehat{\mathbf{C}}_i}^\beta}(x, x_t) = \mu_{\widetilde{N}_x^\beta}(x_t) \text{ and } \nu_{\widetilde{N}_{\widehat{\mathbf{C}}_i}^\beta}(x, x_t) = \nu_{\widetilde{N}_x^\beta}(x_t) \text{ for any } t = 1, 2, \cdots, l.$$

$\mu_{\widetilde{N}_{\widehat{\mathbf{C}}_i}^\beta}(x, x_t)$ denotes the minimum value among the degree of sickness of every patient x_t $(t = 1, 2, \cdots, l)$ according to the diagnoses of the expert R_i $(i = 1, 2, \cdots, n)$, respectively. $\nu_{\widetilde{N}_{\widehat{\mathbf{C}}_i}^\beta}(x, x_t)$ denotes the maximum value among the degree of non-sickness of every patient x_t $(t = 1, 2, \cdots, l)$ according to the diagnoses of the expert R_i $(i = 1, 2, \cdots, n)$, respectively.

If c is a possible degree and d is an impossible degree of the disease B of every patient $x_k \in U$ that is diagnosed by the duty doctor X, denoted by $A(x_k) = \langle c, d\rangle$, then the decision maker (the duty doctor X) for the MCGDM problem needs to know how to evaluate whether or not the patients $x_k \in U$ have the disease B.

5.3. Decision Making Methodology and Process

In this subsection, we give an approach to decision making for the problem of MCGDM with the above characterizations by means of the optimistic multi-granulation IF rough set model. According to the characterizations of the MCGDM problem in Subsection 5.2, we construct the multi-granulation intuitionistic fuzzy decision information systems and present the process of decision making under the framework of optimistic multi-granulation IF rough set model.

- **Input**: Multi-granulation fuzzy decision information systems $(U, \beta, \Gamma_\beta, A)$.
- **Output**: The score ordering for all alternatives.
- **Step 1**: Computing the IF β-neighborhood $\widetilde{N}_x^\beta$ of x induced by $\widehat{C}_i \in \Gamma_\beta$, for all $x \in U$ and $i = 1, 2, \cdots, n$.
- **Step 2**: Computing the optimistic upper approximation $\overline{\mathfrak{N}_{\Gamma_\beta}}^{(o)}(A)$ and the optimistic lower approximation $\underline{\mathfrak{N}_{\Gamma_\beta}}^{(o)}(A)$.
- **Step 3**: Giving the right weight value of ζ, where $\zeta \in [0, 1]$.
- **Step 4**: Computing:

$$\sum_{i=1}^{n} \widetilde{R}_i(A) = \zeta \underline{\mathfrak{N}_{\Gamma_\beta}}^{(o)}(A) + (1-\zeta)\overline{\mathfrak{N}_{\Gamma_\beta}}^{(o)}(A).$$

- **Step 5**: Computing:

$$s(x) = \mu_{\sum_{i=1}^{n} \widetilde{R}_i(A)}(x) - \nu_{\sum_{i=1}^{n} \widetilde{R}_i(A)}(x) \text{ for any } x \in U.$$

- **Step 6**: Obtain the ranking for all $s(x)$ by using the principle of numerical size.

According to the above process, we can get the decision making according to the ranking. In Step 3, ζ reflects the preference of the decision maker for the risk of decision making problems. The decision maker can adjust ζ according to the goal in the real world. In Step 4, $\sum_{i=1}^{n} \widetilde{R}_i(A)$ can be regarded as the compromise rule with the right weight ζ from the view of risk decision making with uncertainty. We define $\sum_{i=1}^{n} \widetilde{R}_i(A) = \{\langle x_k, \zeta\mu_{\underline{\mathfrak{N}_{\Gamma_\beta}}^{(o)}(A)}(x_k) + (1-\zeta)\mu_{\overline{\mathfrak{N}_{\Gamma_\beta}}^{(o)}(A)}(x_k), \zeta\nu_{\underline{\mathfrak{N}_{\Gamma_\beta}}^{(o)}(A)}(x_k) + (1-\zeta)\nu_{\overline{\mathfrak{N}_{\Gamma_\beta}}^{(o)}(A)}(x_k)\rangle : 1 \leq k \leq l\}$.

5.4. An Applied Example

Assume that $U = \{x_1, x_2, x_3, x_4, x_5\}$ is a set of patients. According to the patients' symptoms, we write $V = \{y_1, y_2, y_3, y_4\}$ to be four main symptoms (fever, cough, sore and headache) for a disease B. Assume that the duty doctor X invites two experts R_i $(i = 1, 2)$ to evaluate every patient x_k $(k = 1, 2, \cdots, 5)$ as in Tables 2 and 3.

Table 2. Symptom values of expert R_1 for every patient x_k $(k = 1, 2, \cdots, 5)$.

U	C_{11}	C_{12}	C_{13}	C_{14}
x_1	$\langle 0.7, 0.2\rangle$	$\langle 0.6, 0.2\rangle$	$\langle 0.4, 0.1\rangle$	$\langle 0.1, 0.5\rangle$
x_2	$\langle 0.5, 0.3\rangle$	$\langle 0.3, 0.2\rangle$	$\langle 0.4, 0.5\rangle$	$\langle 0.6, 0.1\rangle$
x_3	$\langle 0.4, 0.5\rangle$	$\langle 0.2, 0.3\rangle$	$\langle 0.5, 0.2\rangle$	$\langle 0.6, 0.3\rangle$
x_4	$\langle 0.6, 0.1\rangle$	$\langle 0.4, 0.5\rangle$	$\langle 0.3, 0.6\rangle$	$\langle 0.5, 0.3\rangle$
x_5	$\langle 0.3, 0.2\rangle$	$\langle 0.7, 0.3\rangle$	$\langle 0.6, 0.3\rangle$	$\langle 0.8, 0.1\rangle$

Table 3. Symptom values of expert R_2 for every patient x_k $(k = 1, 2, \cdots, 5)$.

U	C_{21}	C_{22}	C_{23}	C_{24}
x_1	$\langle 0.6, 0.2\rangle$	$\langle 0.7, 0.2\rangle$	$\langle 0.4, 0.2\rangle$	$\langle 0.2, 0.6\rangle$
x_2	$\langle 0.4, 0.2\rangle$	$\langle 0.5, 0.3\rangle$	$\langle 0.5, 0.4\rangle$	$\langle 0.6, 0.3\rangle$
x_3	$\langle 0.5, 0.4\rangle$	$\langle 0.3, 0.4\rangle$	$\langle 0.5, 0.2\rangle$	$\langle 0.7, 0.2\rangle$
x_4	$\langle 0.6, 0.1\rangle$	$\langle 0.4, 0.5\rangle$	$\langle 0.4, 0.6\rangle$	$\langle 0.5, 0.4\rangle$
x_5	$\langle 0.4, 0.2\rangle$	$\langle 0.6, 0.3\rangle$	$\langle 0.5, 0.3\rangle$	$\langle 0.7, 0.2\rangle$

Let $\beta = \langle 0.5, 0.3\rangle$ be the critical value. Then, $\Gamma_\beta = \{\widehat{C_1}, \widehat{C_2}\}$, where $\widehat{C}_i = \{C_{i1}, C_{i2}, C_{i3}, C_{i4}\}$, for all $i = 1, 2$, is a 2-IF β-coverings of U. $\widetilde{N}^\beta_{\widehat{C}_i}$ $(i = 1, 2)$ are shown in Tables 4 and 5, respectively.

Table 4. $\widetilde{N}^\beta_{\widehat{C}_1}$.

$\widetilde{N}^\beta_{\widehat{C}_1}$	x_1	x_2	x_3	x_4	x_5
$\widetilde{N}^\beta_{x_1}$	$\langle 0.6, 0.2\rangle$	$\langle 0.3, 0.3\rangle$	$\langle 0.2, 0.5\rangle$	$\langle 0.4, 0.5\rangle$	$\langle 0.3, 0.3\rangle$
$\widetilde{N}^\beta_{x_2}$	$\langle 0.1, 0.5\rangle$	$\langle 0.5, 0.3\rangle$	$\langle 0.4, 0.5\rangle$	$\langle 0.5, 0.3\rangle$	$\langle 0.3, 0.2\rangle$
$\widetilde{N}^\beta_{x_3}$	$\langle 0.1, 0.5\rangle$	$\langle 0.4, 0.5\rangle$	$\langle 0.5, 0.3\rangle$	$\langle 0.3, 0.6\rangle$	$\langle 0.6, 0.3\rangle$
$\widetilde{N}^\beta_{x_4}$	$\langle 0.1, 0.5\rangle$	$\langle 0.5, 0.3\rangle$	$\langle 0.4, 0.5\rangle$	$\langle 0.5, 0.3\rangle$	$\langle 0.3, 0.2\rangle$
$\widetilde{N}^\beta_{x_5}$	$\langle 0.1, 0.5\rangle$	$\langle 0.3, 0.5\rangle$	$\langle 0.2, 0.3\rangle$	$\langle 0.3, 0.6\rangle$	$\langle 0.6, 0.3\rangle$

Table 5. $\widetilde{N}^\beta_{\widehat{C}_2}$.

$\widetilde{N}^\beta_{\widehat{C}_2}$	x_1	x_2	x_3	x_4	x_5
$\widetilde{N}^\beta_{x_1}$	$\langle 0.6, 0.2\rangle$	$\langle 0.4, 0.3\rangle$	$\langle 0.3, 0.4\rangle$	$\langle 0.4, 0.5\rangle$	$\langle 0.4, 0.3\rangle$
$\widetilde{N}^\beta_{x_2}$	$\langle 0.2, 0.6\rangle$	$\langle 0.5, 0.3\rangle$	$\langle 0.3, 0.4\rangle$	$\langle 0.4, 0.5\rangle$	$\langle 0.6, 0.3\rangle$
$\widetilde{N}^\beta_{x_3}$	$\langle 0.2, 0.6\rangle$	$\langle 0.5, 0.4\rangle$	$\langle 0.5, 0.2\rangle$	$\langle 0.4, 0.6\rangle$	$\langle 0.5, 0.3\rangle$
$\widetilde{N}^\beta_{x_4}$	$\langle 0.6, 0.2\rangle$	$\langle 0.4, 0.2\rangle$	$\langle 0.5, 0.4\rangle$	$\langle 0.6, 0.1\rangle$	$\langle 0.4, 0.2\rangle$
$\widetilde{N}^\beta_{x_5}$	$\langle 0.2, 0.6\rangle$	$\langle 0.5, 0.4\rangle$	$\langle 0.3, 0.4\rangle$	$\langle 0.4, 0.6\rangle$	$\langle 0.5, 0.3\rangle$

Assume that the duty doctor X diagnosed the value $A = \frac{(0.6,0.3)}{x_1} + \frac{(0.4,0.5)}{x_2} + \frac{(0.3,0.2)}{x_3} + \frac{(0.5,0.3)}{x_4} + \frac{(0.7,0.2)}{x_5}$ of the disease B of every patient. Then:

$$\overline{\mathfrak{N}_{\Gamma_\beta}}^{(o)}(A) = \frac{(0.6,0.3)}{x_1} + \frac{(0.5,0.3)}{x_2} + \frac{(0.5,0.3)}{x_3} + \frac{(0.5,0.2)}{x_4} + \frac{(0.5,0.3)}{x_5},$$
$$\underline{\mathfrak{N}_{\Gamma_\beta}}^{(o)}(A) = \frac{(0.4,0.3)}{x_1} + \frac{(0.4,0.5)}{x_2} + \frac{(0.3,0.4)}{x_3} + \frac{(0.4,0.4)}{x_4} + \frac{(0.4,0.3)}{x_5}.$$

Let $\zeta = 0.7$. Then:

$$\begin{aligned}\sum_{i=1}^{2} \widetilde{R}_i(A) &= 0.7\underline{\mathfrak{N}_{\Gamma_\beta}}^{(o)}(A) + (1-0.7)\overline{\mathfrak{N}_{\Gamma_\beta}}^{(o)}(A)\\ &= \frac{(0.46,0.3)}{x_1} + \frac{(0.43,0.44)}{x_2} + \frac{(0.36,0.37)}{x_3} + \frac{(0.43,0.34)}{x_4} + \frac{(0.43,0.3)}{x_5}.\end{aligned}$$

Hence, we can obtain $s(x_k)$ $(k = 1, 2, \cdots, 5)$ in Table 6.

Table 6. $s(x_k)$ $(k = 1, 2, \cdots, 5)$.

U	x_1	x_2	x_3	x_4	x_5
$s(x_k)$	0.16	−0.01	−0.01	0.09	0.13

According to the principle of numerical size, we have:

$$s(x_2) = s(x_3) < s(x_4) < s(x_5) < s(x_1).$$

Therefore, the doctor X diagnoses the patient x_1 as more likely to be sick with the disease B.

6. Conclusions

Covering rough set models are important research topics, which investigate data mining in a more general manner. Huang et al. [33] presented an IF rough set model and an optimistic multi-granulation IF rough set model. By investigation, we have found that no one has applied multi-granulation IF rough set models to MCGDM problems. In this paper, by showing some new notions and properties of IF β-covering approximation spaces, we mainly study Huang et al.'s models and propose a novel approach to MCGDM problems. The main conclusions in this paper and the further work are listed as follows.

1. Some new notions and properties of IF β-covering approximation spaces are proposed. Aiming at the new notion of β-neighborhood systems, we present a necessary and sufficient condition for two IF β-coverings to induce the same IF β-neighborhood systems.
2. By introducing Huang et al.'s IF rough set model, some new characterizations of it are investigated. We present a new IF covering rough set model for crisp subsets, and the relationships between these two IF covering rough set models and some other rough set models are investigated. Neutrosophic sets and related algebraic structures [39–43] will be connected with the research content of this paper in further research.
3. We construct the multi-granulation intuitionistic fuzzy decision information systems and present a novel approach to MCGDM problems based on the optimistic multi-granulation IF rough set model. There are many MCGDM technologies by rough set models [20,23]. However, among these models, the multi-granulation IF rough set models are not used. We first use the optimistic multi-granulation IF rough set model to solve MCGDM problems.

Author Contributions: All authors have contributed equally to this paper. The individual responsibilities and contribution of all authors can be described as follows: the idea of this whole thesis was put forward by X.Z.; he also completed the preparatory work of the paper; J.W. analyzed the existing work of rough sets and IF sets, and wrote the paper.

Funding: This work is supported by the National Natural Science Foundation of China under Grant Nos. 61573240 and 61473239.

Conflicts of Interest: The authors declare no conflict of interest.

References

1. Pawlak, Z. Rough sets. *Int. J. Comput. Inf. Sci.* **1982**, *11*, 341–356. [CrossRef]
2. Pawlak, Z. *Rough Sets: Theoretical Aspects of Reasoning about Data*; Kluwer Academic Publishers: Boston, MA, USA, 1991.
3. Kondo, M. On the structure of generalized rough sets. *Inf. Sci.* **2006**, *176*, 589–600. [CrossRef]
4. Zhang, X.; Miao, D.; Liu, C.; Le, M. Constructive methods of rough approximation operators and multigranulation rough sets. *Knowl.-Based Syst.* **2016**, *91*, 114–125. [CrossRef]
5. Zhu, W. Relationship among basic concepts in covering-based rough sets. *Inf. Sci.* **2009**, *179*, 2478–2486. [CrossRef]
6. Bartol, W.; Miro, J.; Pioro, K.; Rossello, F. On the coverings by tolerance classes. *Inf. Sci.* **2004**, *166*, 193–211. [CrossRef]
7. Zhu, W.; Wang, F. Reduction and axiomatization of covering generalized rough sets. *Inf. Sci.* **2003**, *152*, 217–230. [CrossRef]
8. Wu, W. Attribute reduction based on evidence theory in incomplete decision systems. *Inf. Sci.* **2008**, *178*, 1355–1371. [CrossRef]

9. Cai, Z.; Zhu, W. Multi-label feature selection via feature manifold learning and sparsity regularization. *Int. J. Mach. Learn. Cybern.* **2018**, *9*, 1321–1334. [CrossRef]
10. Yao, Y.; Zhao, Y. Attribute reduction in decision-theoretic rough set models. *Inf. Sci.* **2008**, *178*, 3356–3373. [CrossRef]
11. Yang, T.; Li, Q.; Zhou, B. A new method for attribute reduction of covering information systems. *Inf. Sci.* **2013**, *228*, 175–191. [CrossRef]
12. Dai, J.; Wang, W.; Xu, Q.; Tian, H. Uncertainty measurement for interval-valued decision systems based on extended conditional entropy. *Knowl.-Based Syst.* **2012**, *27*, 443–450. [CrossRef]
13. Wang, J.; Zhu, W. Applications of bipartite graphs and their adjacency matrices to covering-based rough sets. *Fundam. Inform.* **2017**, *156*, 237–254. [CrossRef]
14. Tan, A.; Li, J.; Lin, G. Connections between covering-based rough sets and concept lattices. *Int. J. Approx. Reason.* **2015**, *56*, 43–58. [CrossRef]
15. Zhang, X.; Dai, J.; Yu, Y. On the union and intersection operations of rough sets based on various approximation spaces. *Inf. Sci.* **2015**, *292*, 214–229. [CrossRef]
16. Wang, J.; Zhang, X. Four operators of rough sets generalized to matroids and a matroidal method for attribute reduction. *Symmetry* **2018**, *10*, 418. [CrossRef]
17. Wang, J.; Zhu, W.; Wang, F.; Liu, G. Conditions for coverings to induce matroids. *Int. J. Mach. Learn. Cybern.* **2014**, *5*, 947–954. [CrossRef]
18. Ma, L. Two fuzzy covering rough set models and their generalizations over fuzzy lattices. *Fuzzy Sets Syst.* **2016**, *294*, 1–17. [CrossRef]
19. Zhan, J.; Xu, W. Two types of coverings based multigranulation rough fuzzy sets and applications to decision making. *Artif. Intell. Rev.* **2018**, 1–32. [CrossRef]
20. Zhang, L.; Zhan, J.; Alcantud, J.C.R. Novel classes of fuzzy soft β-coverings-based fuzzy rough sets with applications to multi-criteria fuzzy group decision making. *Soft Comput.* **2018**. [CrossRef]
21. Zhang, L.; Zhan, J. Fuzzy soft β-covering based fuzzy rough sets and corresponding decision-making applications. *Int. J. Mach. Learn. Cybern.* **2018**, 1–16. [CrossRef]
22. Zhan, J.; Wang, Q. Certain types of soft coverings based rough sets with applications. *Int. J. Mach. Learn. Cybern.* **2018**, 1–12. [CrossRef]
23. Zhan, J.; Alcantud, J.C.R. A novel type of soft rough covering and its application to multicriteria group decision making. *Artif. Intell. Rev.* **2018**, *4*, 1–30. [CrossRef]
24. Zadeh, L.A. Fuzzy sets. *Inf. Control* **1965**, *8*, 338–353. [CrossRef]
25. Dübois, D.; Prade, H. Rough fuzzy sets and fuzzy rough sets. *Int. J. Gener. Syst.* **1990**, *17*, 191–208. [CrossRef]
26. Mordeson, J. Rough set theory applied to (fuzzy) ideal theory. *Fuzzy Sets Syst.* **2001**, *121*, 315–324. [CrossRef]
27. Qian, Y.; Wang, Q.; Cheng, H.; Liang, J.; Dang, C. Fuzzy-rough feature selection accelerator. *Fuzzy Sets Syst.* **2015**, *258*, 61–78. [CrossRef]
28. D'eer, L.; Cornelis, C. A comprehensive study of fuzzy covering-based rough set models: Definitions, properties and interrelationships. *Fuzzy Sets Syst.* **2018**, *336*, 1–26. [CrossRef]
29. Yang, B.; Hu, B. On some types of fuzzy covering-based rough sets. *Fuzzy Sets Syst.* **2017**, *312*, 36–65. [CrossRef]
30. Yang, B.; Hu, B. Fuzzy neighborhood operators and derived fuzzy coverings. *Fuzzy Sets Syst.* **2018**. [CrossRef]
31. Atanassov, K. Intuitionistic fuzzy sets. *Fuzzy Sets Syst.* **1986**, *20*, 87–96. [CrossRef]
32. Zhang, X.; Zhou, B.; Li, P. A general frame for intuitionistic fuzzy rough sets. *Inf. Sci.* **2012**, *216*, 34–49. [CrossRef]
33. Huang, B.; Guo, C.; Li, H.; Feng, G.; Zhou, X. An intuitionistic fuzzy graded covering rough set. *Knowl.-Based Syst.* **2016**, *107*, 155–178. [CrossRef]
34. Zhang, Z. An approach to decision making based on intuitionistic fuzzy rough sets over two universes. *J. Oper. Res. Soc.* **2013**, *64*, 1079–1089.
35. Bonikowski, Z.; Bryniarski, E.; Wybraniec-Skardowska, U. Extensions and intentions in the rough set theory. *Inf. Sci.* **1998**, *107*, 149–167. [CrossRef]
36. Pomykala, J.A. Approximation operations in approximation space. *Bull. Pol. Acad. Sci.* **1987**, *35*, 653–662.
37. Samanta, P.; Chakraborty, M. Covering based approaches to rough sets and implication lattices. In Proceedings of the International Workshop on Rough Sets, Fuzzy Sets, Data Mining, and Granular-Soft Computing (RSFDGrC 2009), Delhi, India, 15–18 December 2009; pp. 127–134.

38. Zhou, L.; Wu, W. On generalized intuitionistic fuzzy rough approximation operators. *Inf. Sci.* **2008**, *178*, 2448–2465. [CrossRef]
39. Zhang, X.; Smarandache, F.; Liang, X. Neutrosophic duplet semi-group and cancellable neutrosophic triplet groups. *Symmetry* **2017**, *9*, 275. [CrossRef]
40. Zhang, X.; Bo, C.; Smarandache, F.; Park, C. New operations of totally dependent-neutrosophic sets and totally dependent-neutrosophic soft sets. *Symmetry* **2018**, *10*, 187. [CrossRef]
41. Zhang, X.; Bo, C.; Smarandache, F.; Dai, J. New inclusion relation of neutrosophic sets with applications and related lattice structure. *Int. J. Mach. Learn. Cybern.* **2018**, *9*, 1753–1763. [CrossRef]
42. Zhang, X. Fuzzy anti-grouped filters and fuzzy normal filters in pseudo-BCI algebras. *J. Intell. Fuzzy Syst.* **2017**, *33*, 1767–1774. [CrossRef]
43. Zhang, X.; Park, C.; Wu, S. Soft set theoretical approach to pseudo-BCI algebras. *J. Intell. Fuzzy Syst.* **2018**, *34*, 559–568. [CrossRef]

symmetry

MDPI

Article

A Novel Similarity Measure for Interval-Valued Intuitionistic Fuzzy Sets and Its Applications

Minxia Luo * and Jingjing Liang

Department of Information and Computing Science, China Jiliang University, Hangzhou 310018, China; s1608070103@cjlu.edu.cn
* Correspondence: mxluo@cjlu.edu.cn; Tel.: +86-571-8687-5634

Received: 29 August 2018; Accepted: 25 September 2018; Published: 27 September 2018

Abstract: In this paper, a novel similarity measure for interval-valued intuitionistic fuzzy sets is introduced, which is based on the transformed interval-valued intuitionistic triangle fuzzy numbers. Its superiority is shown by comparing the proposed similarity measure with some existing similarity measures by some numerical examples. Furthermore, the proposed similarity measure is applied to deal with pattern recognition and medical diagnosis problems.

Keywords: interval-valued intuitionistic fuzzy set; similarity measure; pattern recognition

1. Introduction

As a generalization concept of fuzzy set (FS) introduced by Zadeh [1], the definition of intuitionistic fuzzy set (IFS) was initiated by Atanassov [2] for dealing with vague and uncertain information, which elaborately describe uncertain information by membership degree, non-membership degree and hesitancy degree. In [3], Gau and Buehrer presented the definition of vague set. In [4], Bustince and Burillo have showed that the notion of IFSs and vague sets coincide with each other. In order to deal with indeterminate and inconsistent information, Smarandache [5] proposed a neutrosophic set (NS). In the NS, indeterminacy-membership $I_A(x)$ is independent, thus making the NS more flexible and the most suitable for solving some decision-making problems related to the use of incomplete and imprecise information, uncertainties, predictions and so on. Zhang [6,7] studied algebraic and lattice structure for neutrosophic sets.

The conception of similarity measure for IFSs is one of the most important subjects for degree of similarity between objects in IFS theory. Chen [8] proposed the similarity measure based on a vague set for the first time. Hong [9] introduced a new similarity measure based on vague set and overcame some drawbacks of Chen's similarity measure. Szmidt and Kacprzyk [10] extend Hamming distance and Euclidean distance to construct intuitionistic fuzzy similarity measure. However, Wang and Xin [11] implied that Szmidt and Kacprzyk's distance measure [10] were ineffective in some situations. Grzegorzewski [12] extended some novel similarity measures for IFSs based on Hausdorff distance. Chen [13] pointed out some defects of Grzegorzewski's similarity measure and show some counter examples. On the other hand, some studies defined new similarity measures for IFSs, rather than extending the well-known distance measures. Li and Cheng [14] presented a new similarity measure between IFSs and applied it to pattern recognition. Mitchell [15] indicated that similarity measure of Li and Cheng [14] had some counter-intuitive cases and modified that similarity measure based on a statistical perspective. Furthermore, Liang and Shi [16] presented some counter instances to indicate that the similarity measure of Li and Cheng [14] was not suitable for some situations, and proposed several new similarity measures for IFS. Ye [17] conducted a similarity comparative study of existing similarity measures for IFSs and proposed a cosine similarity measure and weighted cosine similarity measure. Xu [18] acquainted a sequence of similarity measures for IFSs and applied to solve multiple attribute decision-making problems. Boran et al. [19] proposed a new general type of similarity measures for IFSs with two parameters, expressing L_p-norm and give its relation with existing similarity measures. Zhang and Yu [20] presented a new distance measure based on interval comparison, where the IFSs were respectively transformed into the symmetric triangular fuzzy numbers. Comparison with the widely used methods indicated that the proposed method contained more information, with much less loss of information. Luo and Zhao [21] proposed a new distance measure for IFSs, which is based on a matrix norm and a strictly increasing (or decreasing) binary function, and applied it to solve pattern recognition problems.

As the development of IFSs, Atanassov introduced interval-valued intuitionistic fuzzy set (IVIFS) [22], which the membership degree, non-membership degree and hesitancy degree are represented by subinterval of $[0,1]$. It therefore can represent the dynamic character of features accurately. Due to the advantages of IVIFSs in practical application, various similarity measures based on IVIFSs were studied extensively by many researchers from different angles and applied to many areas such as medical diagnosis, pattern recognition problem and so on. Liu [23] proposed a set of axiomatic definitions for entropy measures between IVIFSs, which extends Szmidt and Kacprzyk's axioms formulated for entropy between IFSs. Xu [24] generalized some formulas of similarity measures of IFSs to IVIFSs. Wei [25] proposed an new similarity measure for IVIFSs, and also applied to solve problems on pattern recognitions, multi-criteria fuzzy decision-making and medical diagnosis. Singh [26] introduced a new cosine similarity measure for IVIFSs and applied to pattern recognition. Khalaf [27] advanced a new approach for medical diagnosis by IVIFSs, which is generalized by the application of IFS theory. Dhivya [28] presented a new similarity measure for IVIFSs based on the mid points of transformed triangular fuzzy numbers.

However, there are some drawbacks in some existing similarity measures for IVIFSs, most of which get counterintuitive results in some situations and they cannot get correct classification results for dealing with the pattern recognition problems and medical diagnosis problems. For example, letting $A =< [0.20, 0.30], [0.40, 0.60] >$, $B_1 =< [0.30, 0.40], [0.40, 0.60] >$ and $B_2 =< [0.30, 0.40], [0.30, 0.50] >$ be IVIFSs, we can compute the similarity measures between A and B_i $(i = 1, 2)$ by Formulas (1), (2) and (4) (see Section 3). Obviously, we have the result $B_1 \neq B_2$ because the membership degree of B_1 is identical to that of B_2, and the non-membership degree of B_1 is not identical to that of B_2. Therefore, we should obtain $S_i(A, B_1) \neq S_i(A, B_2) (i = 1, 2)$. However, we can obtain that $S_1(A, B_1) = S_1(A, B_2) = S_2(A, B_1) = S_2(A, B_2) = 0.9$ by the Formulas (1) and (2) (for $p = 1$), which is not reasonable. Meanwhile, we can get $S_D(A, B_1) = 1$ by Formula (4), which does not satisfy the second axiom of the definition for similarity measure. Therefore, we need to develop a new similarity measure to overcome these drawbacks.

The rest of the paper is organized as follows: Section 2 reviews some necessary definitions related to IVIFS. In Section 3, some existing similarity measures are reviewed. In Section 4, a novel similarity measure is introduced. The geometric interpretation of the new similarity measure and the explanation of parameters are briefly given in Section 5. Applications in pattern recognition and medical diagnosis are presented in Section 6. The conclusions for this paper are given in the last section.

2. Preliminary

In this section, we review the basic concepts related to IVIFSs that will be used in this paper.

Definition 1 ([1]). *A fuzzy set A in the unverse of discourse $X = \{x_1, x_2, \ldots, x_n\}$ is defined as follows:*

$$A = \{< x, \mu_A(x) > | x \in X\},$$

where $\mu_A(x) : X \to [0,1]$ is the membership degree.

Definition 2 ([2]). *An intuitionistic fuzzy set A in a universe of discourse* $X = \{x_1, x_2, \ldots, x_n\}$ *is defined as follows:*

$$A = \{< x, \mu_A(x), \nu_A(x) > | x \in X\},$$

where $\mu_A(x) : X \to [0,1]$ *and* $\nu_A(x) : X \to [0,1]$ *are membership and non-membership degree, respectively, such that:* $0 \leq \mu_A(x) + \nu_A(x) \leq 1$.

The third parameter of intuitionistic fuzzy set A is: $\pi_A(x) = 1 - \mu_A(x) - \nu_A(x)$, *which is known as the intuitionistic fuzzy index or the hesitation degree of whether x belongs to A or not. It is obviously seen that* $0 \leq \pi_A(x) \leq 1$. *If* $\pi_A(x)$ *is small; then, knowledge about x is more certain; if* $\pi_A(x)$ *is great, then knowledge about x is more uncertain.*

Definition 3 ([22]). *An interval-valued intuitionistic fuzzy set A in a universe of discourse* $X = \{x_1, x_2, \ldots, x_n\}$ *is defined as follows:*

$$A = \{< x, \mu_A(x), \nu_A(x) > | x \in X\} = \{< x, [\mu_A^-(x), \mu_A^+(x)], [\nu_A^-(x), \nu_A^+(x)] > | x \in X\},$$

where $\mu_A(x) \subseteq [0,1]$, $\nu_A(x) \subseteq [0,1]$, *which satisfies* $0 \leq \mu_A^+(x) + \nu_A^+(x) \leq 1$.

The intervals $\mu_A(x)$ *and* $\nu_A(x)$ *denote the membership degree and non-membership degree, respectively. Furthermore, for each* $x \in X$, *we can compute the hesitance degree* $\pi_A(x) = [\pi_A^-(x_i), \pi_A^+(x_i)] = [1 - \mu_A^+(x) - \nu_A^+(x), 1 - \mu_A^-(x) - \nu_A^-(x)]$.

Definition 4 ([29]). *For every two IVIFSs A and B in the universe of discourse X, we have the following relations:*

(1): $A \subseteq B$ *iff* $(\forall x \in X)\mu_A^-(x) \leq \mu_B^-(x)$ *and* $\mu_A^+(x) \leq \mu_B^+(x)$ *and* $\nu_A^-(x) \geq \nu_B^-(x)$ *and* $\nu_A^+(x) \geq \nu_B^+(x)$.
(2): $A \cup B - \langle x, [\max(\mu_A^-(x), \mu_B^-(x)), \max(\mu_A^+(x), \mu_B^+(x))], [\min(\nu_A^-(x), \nu_B^-(x)), \min(\nu_A^+(x), \nu_B^+(x))] \rangle$.
(3): $A \cap B = \langle x, [\min(\mu_A^-(x), \mu_B^-(x)), \min(\mu_A^+(x), \mu_B^+(x))], [\max(\nu_A^-(x), \nu_B^-(x)), \max(\nu_A^+(x), \nu_B^+(x))] \rangle$.
(4): $A = B$ *iff* $(\forall x \in X)\mu_A^-(x) = \mu_B^-(x)$ *and* $\mu_A^+(x) = \mu_B^+(x)$ *and* $\nu_A^-(x) = \nu_B^-(x)$ *and* $\nu_A^+(x) = \nu_B^+(x)$.
(5): $A^c = \langle x, [\nu_A^-(x), \nu_A^+(x)], [\mu_A^-(x), \mu_A^+(x)] \rangle$

Definition 5 ([18]). *Let A and B be interval-valued intuitionistic fuzzy sets in the unverse of discourse* $X = \{x_1, x_2, \ldots, x_n\}$, *a mapping* $S : IVIFS(X) \times IVIFS(X) \to [0,1]$, $S(A,B)$ *is called to be a similarity measure between A and B, if* $S(A,B)$ *satisfies the following properties:*

(S1): $0 \leq S(A,B) \leq 1$,
(S2): $S(A,B) = 1$ *if and only if* $A = B$,
(S3): $S(A,B) = S(B,A)$,
(S4): *If* $A \subseteq B \subseteq C$, *then* $S(A,C) \leq S(A,B)$, *and* $S(A,C) \leq S(B,C)$.

3. Some Existing Similarity Measures

In this section, we review some existing similarity measures.

Let $A = \{< x_i, [\mu_A^-(x_i), \mu_A^+(x_i)], [\nu_A^-(x_i), \nu_A^+(x_i)] > | x_i \in X\}$, $B = \{< x_i, [\mu_B^-(x_i), \mu_B^+(x_i)], [\nu_B^-(x_i), \nu_B^+(x_i)] > | x_i \in X\}$ be IVIFSs defined on a universe of discourse $X = \{x_1, x_2, \ldots, x_n\}$. The following Formulas (1)–(4) are similarity measures based on IVIFSs:

Xu's similarity measure([24]):

$$S_1(A,B) = 1 - \sqrt[p]{\frac{1}{4n}\sum_{i=1}^{n}(|\mu_A^-(x_i) - \mu_B^-(x_i)|^p + |\mu_A^+(x_i) - \mu_B^+(x_i)|^p + |\nu_A^-(x_i) - \nu_B^-(x_i)|^p + |\nu_A^+(x_i) - \nu_B^+(x_i)|^p)}, \quad (1)$$

$$S_2(A,B) = 1 - \sqrt[p]{\frac{1}{n}\sum_{i=1}^{n}\max(|\mu_A^-(x_i) - \mu_B^-(x_i)|^p, |\mu_A^+(x_i) - \mu_B^+(x_i)|^p, |\nu_A^-(x_i) - \nu_B^-(x_i)|^p, |\nu_A^+(x_i) - \nu_B^+(x_i)|^p)}. \quad (2)$$

Wei's similarity measure ([25]):

$$S_W(A,B) = \frac{1}{n}\sum_{i=1}^{n}\frac{2-\min(\mu_i^-,\nu_i^-)-\min(\mu_i^+,\nu_i^+)}{2+\max(\mu_i^-,\nu_i^-)+\max(\mu_i^+,\nu_i^+)}, \tag{3}$$

where

$$\mu_i^- = \left|\mu_A^-(x_i)-\mu_B^-(x_i)\right|,\ \mu_i^+ = \left|\mu_A^+(x_i)-\mu_B^+(x_i)\right|,$$
$$\nu_i^- = \left|\nu_A^-(x_i)-\nu_B^-(x_i)\right|,\ \nu_i^+ = \left|\nu_A^+(x_i)-\nu_B^+(x_i)\right|.$$

Dhivya's similarity measure ([28]):

$$\begin{aligned} S_D(A,B) \ = \ & 1-\tfrac{1}{n}\textstyle\sum_{i=1}^{n}(\tfrac{1}{2}(|\psi_A^-(x_i)-\psi_B^-(x_i)|+|\psi_A^+(x_i)-\psi_B^+(x_i)|)\cdot(1-\tfrac{\sigma_A(x_i)+\sigma_B(x_i)}{2})+ \\ & |\sigma_A(x_i)-\sigma_B(x_i)|\cdot(\tfrac{\sigma_A(x_i)+\sigma_B(x_i)}{2})), \end{aligned} \tag{4}$$

where

$$\psi_A^- = \frac{\mu_A^-(x_i)+1-\nu_A^-(x_i)}{2},\ \psi_A^+ = \frac{\mu_A^+(x_i)+1-\nu_A^+(x_i)}{2},$$
$$\psi_B^- = \frac{\mu_B^-(x_i)+1-\nu_B^-(x_i)}{2},\ \psi_B^+ = \frac{\mu_B^+ v+1-\nu_B^+(x_i)}{2},$$
$$\sigma_A(x_i) = 1-\frac{1}{2}(\mu_A^-(x_i)+\mu_A^+(x_i)+\nu_A^-(x_i)+\nu_A^+(x_i)),$$
$$\sigma_B(x_i) = 1-\frac{1}{2}(\mu_B^-(x_i)+\mu_B^+(x_i)+\nu_B^-(x_i)+\nu_B^+(x_i)).$$

4. A New Similarity Measure between Interval-Valued Intuitionistic Fuzzy Sets

Definition 6. *Let A, B be IVIFSs defined in universe of discourse* $X = \{x_1, x_2, \ldots, x_n\}$, *and* $A = \{< x_i, [\mu_A^-(x_i), \mu_A^+(x_i)], [\nu_A^-(x_i), \nu_A^+(x_i)] > | x_i \in X\}$, $B = \{< x_i, [\mu_B^-(x_i), \mu_B^+(x_i)], [\nu_B^-(x_i), \nu_B^+(x_i)] > | x_i \in X\}$. *We call*

$$S^p(A,B) = 1-\left\{\begin{array}{l} \frac{1}{2n}\sum\limits_{i=1}^{n}\left|\frac{t_1[(\mu_A^-(x_i)-\mu_B^-(x_i))+(\mu_A^+(x_i)-\mu_B^+(x_i))]-[(\nu_A^-(x_i)-\nu_B^-(x_i))+(\nu_A^+(x_i)-\nu_B^+(x_i))]}{2(t_1+1)}\right|^p \\ +\left|\frac{t_2[(\nu_A^-(x_i)-\nu_B^-(x_i))+(\nu_A^+(x_i)-\nu_B^+(x_i))]-[(\mu_A^-(x_i)-\mu_B^-(x_i))+(\mu_A^+(x_i)-\mu_B^+(x_i))]}{2(t_2+1)}\right|^p \end{array}\right\}^{\frac{1}{p}} \tag{5}$$

a similarity measure between A and B. $t_1, t_2, p \in [1, +\infty)$. *Here, three parameters: p is the* L_p*-norm and* t_1, t_2 *identifies the level of uncertainty.*

Theorem 1. $S^p(A,B)$ *is a similarity measure between IVIFSs A and B.*

Proof. Let A, B, C be IVIFSs defined on a universe of discourse $X = \{x_1, x_2, \ldots, x_n\}$, and $A = \{< x_i, [\mu_A^-(x_i), \mu_A^+(x_i)], [\nu_A^-(x_i), \nu_A^+(x_i)] > | x_i \in X\}$, $B = \{< x_i, [\mu_B^-(x_i), \mu_B^+(x_i)], [\nu_B^-(x_i), \nu_B^+(x_i)] > | x_i \in X\}$, and $C = \{< x_i, [\mu_C^-(x_i), \mu_C^+(x_i)], [\nu_C^-(x_i), \nu_C^+(x_i)] > | x_i \in X\}$.

(1) Firstly, we know that, for arbitrary $x_i \in X$:

$$\begin{aligned} & t_1[(\mu_A^-(x_i)-\mu_B^-(x_i))+(\mu_A^+(x_i)-\mu_B^+(x_i))]-[(\nu_A^-(x_i)-\nu_B^-(x_i))+(\nu_A^+(x_i)-\nu_B^+(x_i))] \\ = \ & [t_1(\mu_A^-(x_i)-\mu_B^-(x_i))-(\nu_A^-(x_i)-\nu_B^-(x_i))]+[t_1(\mu_A^+(x_i)-\mu_B^+(x_i))-(\nu_A^+(x_i)-\nu_B^+(x_i))]. \end{aligned}$$

For $\mu_A^-(x_i), \mu_B^-(x_i), \nu_A^-(x_i), \nu_B^-(x_i) \in [0,1]$, then we have $-t_1 \leq t_1(\mu_A^-(x_i) - \mu_B^-(x_i)) \leq t_1$, $-1 \leq \nu_A^-(x_i) - \nu_B^-(x_i) \leq 1$. Thus, we obtain that

$$-(t_1+1) \leq t_1(\mu_A^-(x_i) - \mu_B^-(x_i)) - (\nu_A^-(x_i) - \nu_B^-(x_i)) \leq t_1 + 1.$$

Similarly,

$$-(t_1+1) \leq t_1(\mu_A^+(x_i) - \mu_B^+(x_i)) - (\nu_A^+(x_i) - \nu_B^+(x_i)) \leq t_1 + 1.$$

Thus,

$$0 \leq \left| \frac{t_1[(\mu_A^-(x_i) - \mu_B^-(x_i)) + (\mu_A^+(x_i) - \mu_B^+(x_i))] - [(\nu_A^-(x_i) - \nu_B^-(x_i)) + (\nu_A^+(x_i) - \nu_B^+(x_i))]}{2(t_1+1)} \right|^p \leq 1.$$

By the same way, we have

$$0 \leq \left| \frac{t_2[(\nu_A^-(x_i) - \nu_B^-(x_i)) + (\nu_A^+(x_i) - \nu_B^+(x_i))] - [(\mu_A^-(x_i) - \mu_B^-(x_i)) + (\mu_A^+(x_i) - \mu_B^+(x_i))]}{2(t_2+1)} \right|^p \leq 1.$$

Therefore,

$$0 \leq \left\{ \begin{array}{l} \frac{1}{2n} \sum\limits_{i=1}^{n} \left| \frac{t_1[(\mu_A^-(x_i)-\mu_B^-(x_i))+(\mu_A^+(x_i)-\mu_B^+(x_i))]-[(\nu_A^-(x_i)-\nu_B^-(x_i))+(\nu_A^+(x_i)-\nu_B^+(x_i))]}{2(t_1+1)} \right|^p \\ + \left| \frac{t_2[(\nu_A^-(x_i)-\nu_B^-(x_i))+(\nu_A^+(x_i)-\nu_B^+(x_i))]-[(\mu_A^-(x_i)-\mu_B^-(x_i))+(\mu_A^+(x_i)-\mu_B^+(x_i))]}{2(t_2+1)} \right|^p \end{array} \right\} \leq 1.$$

That is, $0 \leq S^p(A,B) \leq 1$.

(2) $A = B$, if and only if for arbitrary $x_i \subset X$, we have $\mu_A^-(x_i) = \mu_B^-(x_i)$, $\mu_A^+(x_i) = \mu_B^+(x_i)$, $\nu_A^-(x_i) = \nu_B^-(x_i)$, $\nu_A^+(x_i) = \nu_B^+(x_i)$. It is obvious that $S^p(A,B) = 1$.

(3) For $S^p(A,B)$, we have

$$\begin{aligned} & \left| t_1[(\mu_A^-(x_i) - \mu_B^-(x_i)) + (\mu_A^+(x_i) - \mu_B^+(x_i))] - [(\nu_A^-(x_i) - \nu_D^-(x_i)) + (\nu_A^+(x_i) - \nu_B^+(x_i))] \right|^p \\ = & \left| -t_1[(\mu_A^-(x_i) - \mu_B^-(x_i)) + (\mu_A^+(x_i) - \mu_B^+(x_i))] + [(\nu_A^-(x_i) - \nu_B^-(x_i)) + (\nu_A^+(x_i) - \nu_B^+ v)] \right|^p \\ = & \left| t_1[(\mu_B^-(x_i) - \mu_A^-(x_i)) + (\mu_B^+(x_i) - \mu_A^+(x_i))] - [(\nu_B^-(x_i) - \nu_A^-(x_i)) - (\nu_B^+(x_i) - \nu_A^+(x_i))] \right|^p. \end{aligned}$$

Similarly,

$$\begin{aligned} & \left| t_2[(\nu_A^-(x_i) - \nu_B^-(x_i)) + (\nu_A^+(x_i) - \nu_B^+(x_i))] - [(\mu_A^-(x_i) - \mu_B^-(x_i)) + (\mu_A^+(x_i) - \mu_B^+(x_i))] \right|^p \\ = & \left| -t_2[(\nu_A^-(x_i) - \nu_B^-(x_i)) + (\nu_A^+(x_i) - \nu_B^+(x_i))] + [(\mu_A^-(x_i) - \mu_B^-(x_i)) + (\mu_A^+(x_i) - \mu_B^+(x_i))] \right|^p \\ = & \left| t_2[(\nu_B^-(x_i) - \nu_A^-(x_i)) - (\nu_B^+(x_i) - \nu_A^+(x_i))] - [(\mu_B^-(x_i) - \mu_A^-(x_i)) - (\mu_B^+(x_i) - \mu_A^+(x_i))] \right|^p. \end{aligned}$$

Thus, $S^p(A,B) = S^p(B,A)$.

(4) For A, B, C be IVIFSs, the similarity measure A and B, and A and C are the following:

$$S^p(A,B) = 1 - \left\{ \begin{array}{l} \frac{1}{2n} \sum\limits_{i=1}^{n} \left| \frac{t_1[(\mu_A^-(x_i)-\mu_B^-(x_i))+(\mu_A^+(x_i)-\mu_B^+(x_i))]-[(\nu_A^-(x_i)-\nu_B^-(x_i))+(\nu_A^+(x_i)-\nu_B^+(x_i))]}{2(t_1+1)} \right|^p \\ + \left| \frac{t_2[(\nu_A^-(x_i)-\nu_B^-(x_i))+(\nu_A^+(x_i)-\nu_B^+(x_i))]-[(\mu_A^-(x_i)-\mu_B^-(x_i))+(\mu_A^+(x_i)-\mu_B^+(x_i))]}{2(t_2+1)} \right|^p \end{array} \right\}^{\frac{1}{p}},$$

$$S^p(A,C) = 1 - \left\{ \begin{array}{l} \frac{1}{2n} \sum\limits_{i=1}^{n} \left| \frac{t_1[(\mu_A^-(x_i)-\mu_C^-(x_i))+(\mu_A^+(x_i)-\mu_C^+(x_i))]-[(\nu_A^-(x_i)-\nu_C^-(x_i))+(\nu_A^+(x_i)-\nu_C^+(x_i))]}{2(t_1+1)} \right|^p \\ + \left| \frac{t_2[(\nu_A^-(x_i)-\nu_C^-(x_i))+(\nu_A^+(x_i)-\nu_C^+(x_i))]-[(\mu_A^-(x_i)-\mu_C^-(x_i))+(\mu_A^+(x_i)-\mu_C^+(x_i))]}{2(t_2+1)} \right|^p \end{array} \right\}^{\frac{1}{p}}.$$

If $A \subseteq B \subseteq C$, then $\mu_A^-(x_i) \le \mu_B^-(x_i) \le \mu_C^-(x_i)$, $\mu_A^+(x_i) \le \mu_B^+(x_i) \le \mu_C^+(x_i)$, $\nu_C^-(x_i) \le \nu_B^-(x_i) \le \nu_A^-(x_i)$, and $\nu_C^+(x_i) \le \nu_B^+(x_i) \le \nu_A^+(x_i)$. Then, we have

$$\begin{aligned} & \left|t_1[(\mu_A^-(x_i) - \mu_B^-(x_i)) + (\mu_A^+(x_i) - \mu_B^+(x_i))] - [(\nu_A^-(x_i) - \nu_B^-(x_i)) + (\nu_A^+(x_i) - \nu_B^+(x_i))]\right| \\ = \; & t_1[(\mu_B^-(x_i) - \mu_A^-(x_i)) + (\mu_B^+(x_i) - \mu_A^+(x_i))] + [(\nu_A^-(x_i) - \nu_B^-(x_i)) + (\nu_A^+(x_i) - \nu_B^+(x_i))] \\ \le \; & t_1[(\mu_C^-(x_i) - \mu_A^-(x_i)) + (\mu_C^+(x_i) - \mu_A^+(x_i))] + [(\nu_A^-(x_i) - \nu_C^-(x_i)) + (\nu_A^+(x_i) - \nu_C^+(x_i))] \\ = \; & \left|t_1[(\mu_A^-(x_i) - \mu_C^-(x_i)) + (\mu_A^+(x_i) - \mu_C^+(x_i))] - [(\nu_A^-(x_i) - \nu_C^-(x_i)) + (\nu_A^+(x_i) - \nu_C^+(x_i))]\right|. \end{aligned}$$

By the same reason, we have

$$\begin{aligned} & \left|t_2[(\nu_A^-(x_i) - \nu_B^-(x_i)) + (\nu_A^+(x_i) - \nu_B^+(x_i))] - [(\mu_A^-(x_i) - \mu_B^-(x_i)) + (\mu_A^+(x_i) - \mu_B^+(x_i))]\right| \\ \le \; & \left|t_2[(\nu_A^-(x_i) - \nu_C^-(x_i)) + (\nu_A^+(x_i) - \nu_C^+(x_i))] - [(\mu_A^-(x_i) - \mu_C^-(x_i)) + (\mu_A^+(x_i) - \mu_C^+(x_i))]\right|. \end{aligned}$$

Therefore, $S^p(A,B) \ge S^p(A,C)$, and $S^p(B,C) \ge S^p(A,C)$.
In conclusion, $S^p(A,B)$ is a similarity measure between IVIFSs A and B. □

Remark 1. *If interval-valued intuitionistic fuzzy sets A and B degenerates to intuitionistic fuzzy set, i.e.,* $\mu_A^- = \mu_A^+, \nu_A^- = \nu_A^+$, *and* $\mu_B^- = \mu_B^+, \nu_B^- = \nu_B^+$, *then*

$$S^p(A,B) = 1 - \left\{ \frac{1}{n} \sum_{i=1}^{n} \left| \frac{t_1(\mu_A - \mu_B) - (\nu_A - \nu_B)}{2(t_1+1)} \right|^p + \left| \frac{t_2(\nu_A - \nu_B) - (\mu_A - \mu_B)}{2(t_2+1)} \right|^p \right\}^{\frac{1}{p}} \quad (6)$$

is a new similarity measure between intuitionistic fuzzy sets A and B.

Remark 2. *In the environment of IFSs, and when* $t_1 = t_2 = t$, *the proposed similarity measure*

$$S^p(A,B) = 1 - \left\{ \frac{1}{2n(t+1)^p} \sum_{i=1}^{n} \left(|t(\mu_A - \mu_B) - (\nu_A - \nu_B)|^p + |t(\nu_A - \nu_B) - (\mu_A - \mu_B)|^p \right) \right\}^{\frac{1}{p}} \quad (7)$$

is the similarity measure between intuitionistic fuzzy sets A and B in the literature ([19]).

Example 1. *Supposing that* A_i *and* B_i *are two IVIFSs, we can compute the similarity measures between* A_i *and* B_i *by different similarity measures listed in Table 1.*

Table 1. Comparison of similarity measures in the environment of IVIFSs (interval-valued intuitionistic fuzzy set) (counter-intuitive cases are in bold type; $p = 1$ in S_1 and S_2; $p = 1, t_1 = 2, t_2 = 3$ in S^p).

	1	2	3	4
A_i	$< [0.20, 0.30], [0.40, 0.60] >$	$< [0.20, 0.30], [0.40, 0.60] >$	$< [0.20, 0.30], [0.30, 0.50] >$	$< [0.20, 0.30], [0.30, 0.50] >$
B_i	$< [0.30, 0.40], [0.40, 0.60] >$	$< [0.30, 0.40], [0.30, 0.50] >$	$< [0.30, 0.40], [0.40, 0.60] >$	$< [0.30, 0.40], [0.30, 0.50] >$
S_1 [24]	**0.90**	**0.90**	0.90	0.95
S_2 [24]	**0.90**	**0.90**	**0.90**	**0.90**
S_D [28]	**1.00**	0.98	0.95	0.94
S^p	0.95	0.90	0.80	0.94

In Table 1, by comparing the first column and the second column, we can find that $S_i(A_1, B_1) = S_i(A_2, B_2) (i = 1, 2)$ *when* $A_1 = A_2$, $B_1 \ne B_2$. *Similarly, by comparing the third column and the fourth column, we can find* $S_2(A_3, B_3) = S_2(A_4, B_4)$ *when* $A_3 = A_4$, $B_3 \ne B_4$. *Therefore, we can determine that the similarity measure* S_1 *and* S_2 *is not reasonable. Meanwhile, we find that* $S_D(A_1, B_1) = 1$ *when* $A_1 \ne B_1$, *which is not satisfy the second axiom of the definition for similarity measure. Most importantly, we can observe that the proposed similarity measure* S^p *can overcome these drawbacks. Therefore, our novel similarity measure for IVIFSs is more reasonable than others.*

5. Geometric Interpretation of the Novel Similarity Measure

In this section, we briefly interpret the proposed similarity measure and explain the functionality of parameters t_1, t_2 and p defined in the proposed similarity measure.

Let $A =< [\mu_A^-, \mu_A^+], [\nu_A^-, \nu_A^+] >$, $B =< [\mu_B^-, \mu_B^+], [\nu_B^-, \nu_B^+] >$ be interval-valued intuitionistic fuzzy numbers. We can split A into two intuitionistic fuzzy numbers, i.e., $A^- =< \mu_A^-, \nu_A^- >$ and $A^+ =< \mu_A^+, \nu_A^+ >$. For intuitionistic fuzzy set A^-, μ_A^- can be equal to any value in $[\mu_A^-, \mu_A^- + \pi_A^+]$ and ν_A^- can be equal to any value in $[\nu_A^-, \nu_A^- + \pi_A^+]$, where $\pi_A^+ = 1 - \mu_A^- - \nu_A^-$. Similarly, μ_A^+ can be equal to any value in $[\mu_A^+, \mu_A^+ + \pi_A^-]$ and ν_A^+ can be equal to any value in $[\nu_A^+, \nu_A^+ + \pi_A^-]$ for intuitionistic fuzzy set A^+, where $\pi_A^- = 1 - \mu_A^+ - \nu_A^+$. Then, the possible values for A^- and A^+ illustrated in Figure 1 as the two triangles. As the center of gravity, D^- and D^+ are the most informative points in the triangle A^- and A^+, respectively.

However, $< \mu_A^- + \frac{\pi_A^+}{t_1+1}, \nu_A^- + \frac{\pi_A^+}{t_2+1} > (t_1, t_2 \in [1, +\infty))$ can represent any point in the triangle A^-. Especially when $t_1 = t_2 = t$, $< \mu_A^- + \frac{\pi_A^+}{t+1}, \nu_A^- + \frac{\pi_A^+}{t+1} >$ can denote the point of middle line of the triangle bevel. In the same way, $< \mu_A^+ + \frac{\pi_A^-}{t_1+1}, \nu_A^+ + \frac{\pi_A^-}{t_2+1} > (t_1, t_2 \in [1, +\infty))$ represents any point in the triangle A^+.

The following is the calculation process:

Firstly, $A'^- = \left\langle \mu_A^- + \frac{\pi_A^+}{t_1+1}, \nu_A^- + \frac{\pi_A^+}{t_2+1} \right\rangle$ denotes possible points of triangle A^-. By the same token, $A'^+ = \left\langle \mu_A^+ + \frac{\pi_A^-}{t_1+1}, \nu_A^+ + \frac{\pi_A^-}{t_2+1} \right\rangle$ denotes possible points of triangle A^+. Similarly, we can obtain that $B'^- = \left\langle \mu_B^- + \frac{\pi_B^+}{t_1+1}, \nu_B^- + \frac{\pi_B^+}{t_1+1} \right\rangle$ and $B'^+ = \left\langle \mu_B^+ + \frac{\pi_B^-}{t_1+1}, \nu_B^+ + \frac{\pi_B^-}{t_1+1} \right\rangle$ denote any points in triangles B^- and B^+, respectively.

Secondly, the average of A'^- and A'^+ can be computed as follows:

$$A'' =< \mu_A'', \nu_A'' >= \left\langle \frac{2 + t_1(\mu_A^- + \mu_A^+) - (\nu_A^- + \nu_A^+)}{2(t_1+1)}, \frac{2 + t_2(\mu_A^- + \mu_A^+) - (\nu_A^- + \nu_A^+)}{2(t_2+1)} \right\rangle.$$

We can also get the mean value of B'^- and B'^+:

$$B'' =< \mu_B'', \nu_B'' >= \left\langle \frac{2 + t_1(\mu_B^- + \mu_B^+) - (\nu_B^- + \nu_B^+)}{2(t_1+1)}, \frac{2 + t_2(\mu_B^- + \mu_B^+) - (\nu_B^- + \nu_B^+)}{2(t_2+1)} \right\rangle.$$

The absolute difference between A'' and B'' is calculated as follows:

$$\left|\mu_A'' - \mu_B''\right| = \left|\frac{t_1[(\mu_A^- - \mu_B^-) + (\mu_A^+ - \mu_B^+)] - [(\nu_A^- - \nu_B^-) + (\nu_A^+ - \nu_B^+)]}{2(t_1+1)}\right|,$$

$$\left|\nu_A'' - \nu_B''\right| = \left|\frac{t_2[(\nu_A^- - \nu_B^-) + (\nu_A^+ - \nu_B^+)] - [(\mu_A^- - \mu_B^-) + (\mu_A^+ - \mu_B^+)]}{2(t_2+1)}\right|.$$

$\left|\mu_A'' - \mu_B''\right|$ and $\left|\nu_A'' - \nu_B''\right|$ to the power of p is equal to the following:

$$\left|\mu_A'' - \mu_B''\right|^p = \frac{\left|t_1[(\mu_A^- - \mu_B^-) + (\mu_A^+ - \mu_B^+)] - [(\nu_A^- - \nu_B^-) + (\nu_A^+ - \nu_B^+)]\right|^p}{2^p(t_1+1)^p},$$

$$\left|\nu_A'' - \nu_B''\right|^p = \frac{\left|t_2[(\nu_A^- - \nu_B^-) + (\nu_A^+ - \nu_B^+)] - [(\mu_A^- - \mu_B^-) + (\mu_A^+ - \mu_B^+)]\right|^p}{2^p(t_2+1)^p}.$$

The average value of $\left|\mu_A^{''}-\mu_B^{''}\right|^p$ and $\left|v_A^{''}-v_A^{''}\right|^p$ is calculated as follows:

$$\begin{aligned}&\frac{1}{2}\left(\left|\mu_A^{''}-\mu_B^{''}\right|^p+\left|v_A^{''}-v_A^{''}\right|^p\right)\\=&\frac{1}{2}\left|\frac{t_1[(\mu_A^- -\mu_B^-)+(\mu_A^+ -\mu_B^+)]-[(v_A^- -v_B^-)+(v_A^+ -v_B^+)]}{2(t_1+1)}\right|^p\\&+\frac{1}{2}\left|\frac{t_2[(v_A^- -v_B^-)+(v_A^+ -v_B^+)]-[(\mu_A^- -\mu_B^-)+(\mu_A^+ -\mu_B^+)]}{2(t_2+1)}\right|^p.\end{aligned}$$

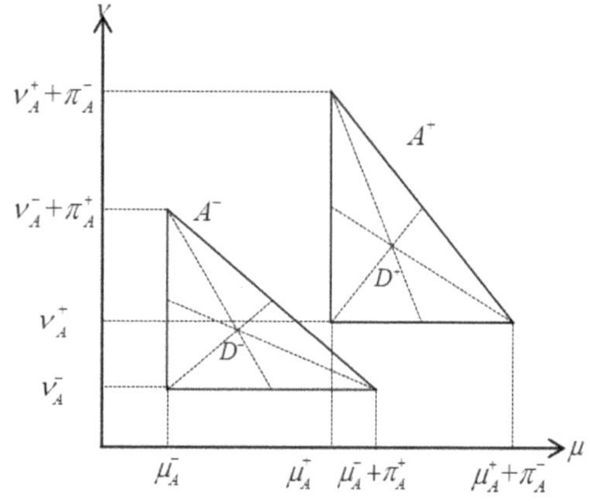

Figure 1. Possible value for A^- and A^+.

The p root of the average value of $\left|\mu_A^{''}-\mu_B^{''}\right|^p$ and $\left|v_A^{''}-v_A^{''}\right|^p$ is calculated as:

$$\left\{\frac{1}{2}\left(\left|\mu_A^{''}-\mu_B^{''}\right|^p+\left|v_A^{''}-v_B^{''}\right|^p\right)\right\}^{\frac{1}{p}}=\left\{\begin{array}{l}\left|\frac{t_1[(\mu_A^- -\mu_C^-)+(\mu_A^+ -\mu_C^+)]-[(v_A^- -v_C^-)+(v_A^+ -v_C^+)]}{2(t_1+1)}\right|^p\\+\left|\frac{t_2[(v_A^- -v_C^-)+(v_A^+ -v_C^+)]-[(\mu_A^- -\mu_C^-)+(\mu_A^+ -\mu_C^+)]}{2(t_2+1)}\right|^p\end{array}\right\}^{\frac{1}{p}}.$$

For an interval-valued intuitionistic fuzzy set instead of interval-valued intuitionistic fuzzy number, i.e., there is more than one feature in the discourse of universe, such as $X=\{x_1,x_2,\ldots,x_n\}$:

$$S^p(A,B)=1-\left\{\begin{array}{l}\frac{1}{2n}\sum\limits_{i=1}^{n}\left|\frac{t_1[(\mu_A^-(x_i)-\mu_B^-(x_i))+(\mu_A^+(x_i)-\mu_B^+(x_i))]-[(v_A^-(x_i)-v_B^-(x_i))+(v_A^+(x_i)-v_B^+(x_i))]}{2(t_1+1)}\right|^p\\+\left|\frac{t_2[(v_A^-(x_i)-v_B^-(x_i))+(v_A^+(x_i)-v_B^+(x_i))]-[(\mu_A^-(x_i)-\mu_B^-(x_i))+(\mu_A^+(x_i)-\mu_B^+(x_i))]}{2(t_2+1)}\right|^p\end{array}\right\}^{\frac{1}{p}}.$$

In particular, $A'^- = D^- = \left\langle \mu_A^- + \frac{1-\mu_A^- - v_A^-}{3}, v_A^- + \frac{1-\mu_A^- - v_A^-}{3}\right\rangle$ and $A'^+ = D^+ = \left\langle \mu_A^+ + \frac{1-\mu_A^+ - v_A^+}{3}, v_A^+ + \frac{1-\mu_A^+ - v_A^+}{3}\right\rangle$ when $t_1 = t_2 = 2$. Without a doubt, D^- and D^+ are the most concentrated points of information in triangle A^- and A^+, respectively; therefore, they are also the most significant points in all possible meaningful points.

6. Applications

In this section, the proposed similarity measure is used to solve the real life problems under the IVIFSs environment and obtained results have been compared with some existing similarity measures.

6.1. Pattern Recognition

6.1.1. Algorithms for Pattern Recognition

Letting $X = \{x_1, x_2, \ldots, x_n\}$ be a finite universe of discourse, there exists m patterns which are denoted by IVIFSs $A_j = \{< x_1, [\mu^-_{A_j}(x_1), \mu^+_{A_j}(x_1)], [\nu^-_{A_j}(x_1), \nu^+_{A_j}(x_1)] >, \ldots, < x_1, [\mu^-_{A_j}(x_n), \mu^+_{A_j}(x_n)], [\nu^-_{A_j}(x_n), \nu^+_{A_j}(x_n)] > | x_1, \ldots, x_n \in X\}$ $(j = 1, 2, \ldots, m)$ and there is a test sample to be classified which is denoted by an IVIFS $B = \{< x_1, [\mu^-_B(x_1), \mu^+_B(x_1)], [\nu^-_B(x_1), \nu^+_B(x_1)] >, \ldots, < x_1, [\mu^-_B(x_n), \mu^+_B(x_n)], [\nu^-_B(x_n), \nu^+_B(x_n)] > | x_1, \ldots, x_n \in X\}$. The recognition process is as follows:

Step 1. Calculate the similarity measure $S(B, A_j)$ between B and $A_j (j = 1, \ldots, m)$.

Step 2. Choose the maximum one $S(B, A_{j0})$ from $S(B, A_j)$ $(j = 1, 2, \ldots, m)$, i.e., $S(B, A_{j0}) = \max_{1 \le j \le m} S(B, A_j)$. Then, the test sample B is classified the pattern A_{j0}.

6.1.2. Applications for Pattern Recognition

Example 2. *Assume that there are four classes of ores $A_i (i = 1, 2, 3, 4)$ in the area developed by a coal mine company, for which the related feature information are expressed by IVIFSs, and $A_i = \{< x_1, \left[\mu^-_{A_i}(x_1), \mu^+_{A_i}(x_1)\right], \left[\nu^-_{A_i}(x_1), \nu^+_{A_i}(x_1)\right] >, \ldots, < x_4, \left[\mu^-_{A_i}(x_4), \mu^+_{A_i}(x_4)\right], \left[\nu^-_{A_i}(x_4), \nu^+_{A_i}(x_4)\right] > | x_1, x_2, x_3, x_4 \in X\}$, which are presented in Table 2. Now, there is an unknown ore B and our aim is to classify B into the four kinds of ores above.*

Table 2. Feature matrix of A_1, A_2, A_3, A_4 and B.

	Feature1	Feature2	Feature3	Feature4
A_1	$< [0.10, 0.50], [0.20, 0.30] >$	$< [0.10, 0.30], [0.00, 0.20] >$	$< [0.30, 0.50], [0.20, 0.40] >$	$< [0.20, 0.50], [0.10, 0.30] >$
A_2	$< [0.20, 0.40], [0.15, 0.35] >$	$< [0.20, 0.20], [0.05, 0.15] >$	$< [0.20, 0.60], [0.30, 0.30] >$	$< [0.30, 0.40], [0.15, 0.25] >$
A_3	$< [0.15, 0.30], [0.30, 0.40] >$	$< [0.20, 0.40], [0.50, 0.60] >$	$< [0.50, 0.60], [0.15, 0.35] >$	$< [0.25, 0.45], [0.30, 0.40] >$
A_4	$< [0.20, 0.35], [0.10, 0.65] >$	$< [0.35, 0.60], [0.05, 0.30] >$	$< [0.15, 0.30], [0.40, 0.55] >$	$< [0.15, 0.25], [0.45, 0.55] >$
B	$< [0.30, 0.40], [0.10, 0.50] >$	$< [0.10, 0.40], [0.25, 0.40] >$	$< [0.20, 0.30], [0.10, 0.35] >$	$< [0.15, 0.40], [0.20, 0.50] >$

Compute the similarity measures $S(A_i, B)$ between B and A_i. By analyzing the computed results in Table 3, we can easily see that, if S_1 is used for pattern recognition, we can obtain that $S_1(A_1, B) = S_1(A_2, B) = S_1(A_4, B) > S_1(A_3, B)$. In this way, we can not classify the sample B into a certain pattern accurately. If S_W is used for pattern recognition, we can obtain that $S_W(A_2, B) = S_W(A_4, B) > S_W(A_1, B) = S_W(A_3, B)$. In this way, we can not make sure if the sample B belongs to one of A_2 and A_4. If we use S_D for pattern recognition, we can get $S(A_3, B) = S(A_4, B) > S(A_2, B) > S(A_1, B)$. In this way, we can not classify the sample B into one of A_3 and A_4. If we use S^p for pattern recognition, we can get $S(A_1, B) > S(A_2, B) > S(A_3, B) > S(A_4, B)$. According to the principle of recognition, S_2 and S^p can get the same recognition result, i.e., the sample B can be classified into the pattern A_3. However, we can not distinguish which one is bigger between A_2 and A_4 when using S_2 to calculate the similarity measure. Therefore, we can assign the sample B to the pattern A_3.

Table 3. Pattern recognition result under different similarity measures (counter-intuitive cases are in bold type; $p = 1$ in S_1 and S_2; $p = 1, t_1 = 2, t_2 = 3$ in S^p; **N.A.** means method is not applicable).

	$S(A_1, B)$	$S(A_2, B)$	$S(A_3, B)$	$S(A_4, B)$	Classification Results
S_1 [24]	**0.87**	**0.87**	0.86	**0.87**	**N.A.**
S_2 [24]	0.75	**0.76**	0.79	**0.76**	A_3
S_W [25]	0.78	**0.79**	0.78	**0.79**	**N.A.**
S_D [28]	0.82	0.86	**0.88**	**0.88**	**N.A.**
S^p	0.82	0.81	0.88	0.75	A_3

Example 3 ([30]). *In this example, a pattern recognition example about classification of building materials is used to illustrate the proposed similarity measure. Suppose that there are four classes of building material, which are denoted by the IVIFSs* $A_j = \{< x_1, [\mu^-_{A_j}(x_1), \mu^+_{A_j}(x_1)], [\nu^-_{A_j}(x_1), \nu^+_{A_j}(x_1)] >, \ldots, < x_{12}, [\mu^-_{A_j}(x_{12}), \mu^+_{A_j}(x_{12})], [\nu^-_{A_j}(x_{12}), \nu^+_{A_j}(x_{12})] > | x_1, \ldots, x_{12} \in X\}$ $(j = 1, \ldots, 4)$ *in the feature space* $X = \{x_1, x_2, \ldots, x_{12}\}$, *and there is an unknown pattern B:*

$$\begin{aligned} A_1 = \{ &< x_1, [0.1, 0.2], [0.5, 0.6] >, < x_2, [0.1, 0.2], [0.7, 0.8] >, < x_3, [0.5, 0.6], [0.3, 0.4] >, \\ &< x_4, [0.8, 0.9], [0.0, 0.1] >, < x_5, [0.4, 0.5], [0.3, 0.4] >, < x_6, [0.0, 0.1], [0.8, 0.9] >, \\ &< x_7, [0.3, 0.4], [0.5, 0.6] >, < x_8, [1.0, 1.0], [0.0, 0.0] >, < x_9, [0.2, 0.3], [0.6, 0.7] >, \\ &< x_{10}, [0.4, 0.5], [0.4, 0.5] >, < x_{11}, [0.7, 0.8], [0.1, 0.2] >, < x_{12}, [0.4, 0.5], [0.4, 0.5] > \}, \end{aligned}$$

$$\begin{aligned} A_2 = \{ &< x_1, [0.5, 0.6], [0.3, 0.4] >, < x_2, [0.6, 0.7], [0.1, 0.2] >, < x_3, [1.0, 1.0], [0.0, 0.0] >, \\ &< x_4, [0.1, 0.2], [0.6, 0.7] >, < x_5, [0.0, 0.1], [0.8, 0.9] >, < x_6, [0.7, 0.8], [0.1, 0.2] >, \\ &< x_7, [0.5, 0.6], [0.3, 0.4] >, < x_8, [0.6, 0.7], [0.2, 0.3] >, < x_9, [1.0, 1.0], [0.0, 0.0] >, \\ &< x_{10}, [0.1, 0.2], [0.7, 0.8] >, < x_{11}, [0.0, 0.1], [0.8, 0.9] >, < x_{12}, [0.7, 0.8], [0.1, 0.2] > \}, \end{aligned}$$

$$\begin{aligned} A_3 = \{ &< x_1, [0.4, 0.5], [0.3, 0.4] >, < x_2, [0.6, 0.7], [0.2, 0.3] >, < x_3, [0.9, 1.0], [0.0, 0.0] >, \\ &< x_4, [0.0, 0.1], [0.8, 0.9] >, < x_5, [0.0, 0.1], [0.8, 0.9] >, < x_6, [0.6, 0.7], [0.2, 0.3] >, \\ &< x_7, [0.1, 0.2], [0.7, 0.8] >, < x_8, [0.2, 0.3], [0.6, 0.7] >, < x_9, [0.5, 0.6], [0.2, 0.4] >, \\ &< x_{10}, [1.0, 1.0], [0.0, 0.0] >, < x_{11}, [0.3, 0.4], [0.4, 0.5] >, < x_{12}, [0.0, 0.1], [0.8, 0.9] > \}, \end{aligned}$$

$$\begin{aligned} A_4 = \{ &< x_1, [1.0, 1.0], [0.0, 0.0] >, < x_2, [1.0, 1.0], [0.0, 0.0] >, < x_3, [0.8, 0.9], [0.0, 0.1] >, \\ &< x_4, [0.7, 0.8], [0.1, 0.2] >, < x_5, [0.0, 0.1], [0.7, 0.9] >, < x_6, [0.0, 0.1], [0.8, 0.9] >, \\ &< x_7, [0.1, 0.2], [0.7, 0.8] >, < x_8, [0.1, 0.2], [0.7, 0.8] >, < x_9, [0.4, 0.5], [0.3, 0.4] >, \\ &< x_{10}, [1.0, 1.0], [0.0, 0.0] >, < x_{11}, [0.3, 0.4], [0.4, 0.5] >, < x_{12}, [0.0, 0.1], [0.8, 0.9] > \}, \end{aligned}$$

$$\begin{aligned} B = \{ &< x_1, [0.9, 1.0], [0.0, 0.0] >, < x_2, [0.9, 1.0], [0.0, 0.0] >, < x_3, [0.7, 0.8], [0.1, 0.2] >, \\ &< x_4, [0.6, 0.7], [0.1, 0.2] >, < x_5, [0.0, 0.1], [0.8, 0.9] >, < x_6, [0.1, 0.2], [0.7, 0.8] >, \\ &< x_7, [0.1, 0.2], [0.7, 0.8] >, < x_8, [0.1, 0.2], [0.7, 0.8] >, < x_9, [0.4, 0.5], [0.3, 0.4] >, \\ &< x_{10}, [1.0, 1.0], [0.0, 0.0] >, < x_{11}, [0.3, 0.4], [0.4, 0.5] >, < x_{12}, [0.0, 0.1], [0.7, 0.9] > \}. \end{aligned}$$

Calculate the similarity measure $S(A_j, B)$ *between IVIFSs* A_j $(j = 1, 2, 3, 4)$ *and B by use of Formulas (1)–(5). It is obvious that the similarity measure in the literature ([30]) is the special case of* S_1 *and* S_2, *and the computed result is the same as ([30]). According to Table 4 and the recognition principle, the unknown pattern can be classified properly in* A_4 *by the computation of similarity measure. This conclusion coincides with that in [30].*

Table 4. Pattern recognition results under different similarity measures (counter-intuitive cases are in bold type; $p = 1$ in S_1 and S_2, $p = 1, t_1 = 2, t_2 = 3$ in S^p).

	$S(A_1, B)$	$S(A_2, B)$	$S(A_3, B)$	$S(A_4, B)$	Recognition Results
S_1 [24]	0.59	0.58	0.81	0.97	A_4
S_2 [24]	**0.53**	**0.53**	0.79	0.94	A_4
S_W [25]	0.48	0.47	0.74	0.94	A_4
S_D [28]	0.64	0.56	0.83	0.98	A_4
S^p	0.60	0.58	0.85	0.97	A_4

6.2. Applications for Medical Diagnosis

Researchers proposed a lot of methods from different points of view to deal with problems of medical diagnosis. Refs. [27,31–33] presented several ways to deal with the problems of medical diagnosis. In this section, the methods of pattern recognition are used for solving medical diagnosis problems, i.e., patients are unknown test samples, diseases are several patterns, and the symptom set is the set universe of discourse. Our aim is to classify patients in one of the illnesses, respectively.

Example 4. *Let* $A = \{A_1$ *(Viral fever),* A_2 *(Typhoid),* A_3 *(Pneumonia),* A_4 *(Stomach problem)*$\}$ *be a set of diagnoses and* $X = \{x_1$ *(Temperature),* x_2 *(Cough),* x_3 *(Headache),* x_4 *(Stomach pain)*$\}$ *be a set of symptoms. The disease–symptom matrix that is represented by IVIFSs is listed in Table 5.*

Table 5. Disease–symptom matrix.

	x_1 (Temperature)	x_2 (Cough)	x_3 (Headache)	x_4 (Stomach Pain)
A_1 (Viral fever)	$< [0.8, 0.9], [0.0, 0.1] >$	$< [0.7, 0.8], [0.1, 0.2] >$	$< [0.5, 0.6], [0.2, 0.3] >$	$< [0.6, 0.8], [0.1, 0.2] >$
A_2 (Typhoid)	$< [0.5, 0.6], [0.1, 0.3] >$	$< [0.8, 0.9], [0.0, 0.1] >$	$< [0.6, 0.8], [0.1, 0.2] >$	$< [0.4, 0.6], [0.1, 0.2] >$
A_3 (Pneumonia)	$< [0.7, 0.8], [0.1, 0.2] >$	$< [0.7, 0.9], [0.0, 0.1] >$	$< [0.4, 0.6], [0.2, 0.4] >$	$< [0.3, 0.5], [0.2, 0.4] >$
A_4 (Stomach problem)	$< [0.8, 0.9], [0.0, 0.1] >$	$< [0.7, 0.8], [0.1, 0.2] >$	$< [0.7, 0.9], [0.0, 0.1] >$	$< [0.8, 0.9], [0.0, 0.1] >$

Suppose the patient B can be represented as:

$B = \{< x_1, [0.4, 0.5], [0.1, 0.2] >, < x_2, [0.7, 0.8], [0.1, 0.2] >, < x_3, [0.9, 0.9], [0.0, 0.1] >, < x_4, [0.3, 0.5], [0.2, 0.4] >\}$.

Our aim is to classify the patient B in one of the illnesses A_1*,* A_2*,* A_3 *and* A_4*. Then, we can have the following results in the environment of IVIFSs, which are listed in Table 6.*

Table 6. Computed results under different similarity measures (counter-intuitive cases are in bold type; $p = 1$ in S_1 and S_2; $p = 1, t_1 = 2, t_2 = 3$ in S^p).

	$S(A_1, B)$	$S(A_2, B)$	$S(A_3, B)$	$S(A_4, B)$	Recognition Result
S_1 [24]	0.81	0.89	0.86	0.84	A_2
S_2 [24]	**0.73**	0.80	0.78	**0.73**	A_2
S_W [25]	0.82	0.80	0.79	0.77	A_2
S_D [28]	0.82	0.91	0.86	0.84	A_2
S^p	0.83	0.89	0.87	0.85	A_2

Considering the recognition principle of the maximum similarity degree for the IVIFSs, we can obtain the consequence that the similarity measure between A_2 *and B is the largest one. However, the similarity measures* S_2 *could not distinguish which one is bigger between* A_1 *and* A_4*. Thus, we can classify the patient B to illness* A_2 *due to the recognition principle. Therefore, we can diagnose that the patient's disease is typhoid.*

7. Conclusions

In this paper, a novel similarity measure for IVIFSs is proposed, which is obtained by splitting an IVIFS into two IFSs and computing the average value of the p power of any points in two triangles composed of the two intuitionistic fuzzy sets. Its superiority is presented by comparing the developed similarity measure with some existing similarity measures. Thus, we can use the similarity measure to deal with the problems with vagueness and uncertainty. For example, pattern recognition, medical diagnosis, game theory and so on.

In fact, we can choose different values of the three parameters (t_1, t_2 and p in Formula (5)) when facing different problems. However, there are some difficulties when choosing the value of parameters. This is also a problem to be solved in the future.

Author Contributions: M.L. initiated the research and provide the framework of this paper. J.L. wrote and complete this paper with M.L.'s validity and helpful suggeations.

Acknowledgments: The authors would like to thank the reviewers for their valuable comments and suggestions. This research was supported by the National Natural Science Foundation of China (Grant No. 61773019).

Conflicts of Interest: The authors declare no conflict of interest.

References

1. Zadeh, L.A. Fuzzy sets. *Inf. Control* **1965**, *8*, 338–353. [CrossRef]
2. Atanassov, K.T.; Rangasamy, P. Intuitionistic Fuzzy Sets. *Fuzzy Sets Syst.* **1986**, *20*, 87–96. [CrossRef]
3. Gau, W.L.; Buehrer, D.J. Vague sets. *IEEE Trans. Syst. Man Cybern.* **1993**, *23*, 610–614. [CrossRef]
4. Bustince, H.; Burillo, P. Vague sets are intuitionistic fuzzy sets. *Fuzzy Sets Syst.* **1996**, *79*, 403–405. [CrossRef]
5. Smarandache, F. *A Unifiying Field in Logics: Neutrosophic Logic, Neutrosophy, Neutrosophic Set, Neutrosophic Probability*; American Research Press: Rehoboth, DE, USA, 1999.
6. Zhang, X.H.; Smarandache, F.; Liang, X.L. Neutrosophic duplet semi-group and cancellable neutrosophic triplet groups. *Symmetry* **2017**, *9*, 275. [CrossRef]
7. Zhang, X.H.; Bo, C.X.; Smarandache, F.; Park, C. New operations of totally dependent- neutrosophic sets and totally dependent-neutrosophic soft sets. *Symmetry* **2018**, *10*, 187. [CrossRef]
8. Chen, S.M. Similarity measures between vague sets and between elements. *IEEE Trans. Syst. Man Cybern. Part B Cybern.* **1997**, *27*, 153-158. [CrossRef] [PubMed]
9. Hong, D.H.; Kim, C. A note on similarity measures between vague sets and between elements. *Inf. Sci.* **1999**, *115*, 83–96. [CrossRef]
10. Szmidt, E.; Kacprzyk, J. Distances between intuitionistic fuzzy sets. *Fuzzy Sets Syst.* **2000**, *114*, 505–518. [CrossRef]
11. Wang, W.Q.; Xin, X.L. Distance measure between intuitionistic fuzzy sets. *Pattern Recognit. Lett.* **2005**, *26*, 2063–2069. [CrossRef]
12. Grzegorzewski, P. Distances between intuitionistic fuzzy sets and/or interval-valued fuzzy sets based on the Hausdorff metric. *Fuzzy Sets Syst.* **2004**, *148*, 319–328. [CrossRef]
13. Chen, T.Y. A note on distances between intuitionistic fuzzy sets and/or interval-valued fuzzy sets based on the Hausdorff metric. *Fuzzy Sets Syst.* **2007**, *158*, 2523–2525. [CrossRef]
14. Li, D.F.; Cheng, C. New similarity measures of intuitionistic fuzzy sets and application to pattern recognitions. *Pattern Recognit. Lett.* **2002**, *23*, 221–225. [CrossRef]
15. Mitchell, H.B. On the Dengfeng-Chuntian similarity measure and its application to pattern recognition. *Pattern Recognit. Lett.* **2003**, *24*, 3101–3104. [CrossRef]
16. Liang, Z.Z.; Shi, P. Similarity measures on intuitionistic fuzzy sets. *Pattern Recognit. Lett.* **2003**, *24*, 2687–2693. [CrossRef]
17. Ye, J. Cosine similarity measures for intuitionistic fuzzy sets and their applications. *Math. Comput. Model.* **2011**, *53*, 91–97. [CrossRef]
18. Xu, Z.S. Some similarity measures of intuitionistic fuzzy sets and their applications to multiple attribute decision making. *Fuzzy Optim. Decis. Mak.* **2007**, *6*, 109–121. [CrossRef]
19. Boran, F.E.; Akay, D. A biparametric similarity measure on intuitionistic fuzzy sets with applications to pattern recognition. *Inf. Sci.* **2014**, *255*, 45–57. [CrossRef]
20. Zhang, H.M.; Yu, L.Y. New distance measures between intuitionistic fuzzy sets and interval-valued fuzzy sets. *Inf. Sci.* **2013**, *245*, 181–196. [CrossRef]
21. Luo, M.X.; Zhao, R.R. A distance measure between intuitionistic fuzzy sets and its application in medical diagnosis. *Artif. Intell. Med.* **2018**, *89*, 34–39. [CrossRef] [PubMed]

22. Atanassov, K.T.; Gargov, G. Interval-valued intuitionistic fuzzy sets. *Fuzzy Sets Syst.* **1989**, *31*, 343–349. [CrossRef]
23. Liu, X.D.; Zheng, S.H.; Xiong, F.L. Entropy and subsethood for general interval-valued intuitionistic fuzzy sets. *Int. Conf. Fuzzy Syst. Knowl. Discov.* **2005**, 42–52. [CrossRef]
24. Xu, Z.S.; Chen, J. An overview of distance and simiarity measures of intuitionistic fuzzy sets. *Int. J. Uncertain. Fuzziness Knowl.-Based Syst.* **2008**, *16*, 529–555. [CrossRef]
25. Wei, C.P.; Wang, P.; Zhang, Y.Z. Entropy, similarity measure of interval-valued intuitionistic fuzzy sets and their applications. *Inf. Sci.* **2011**, *181*, 4273-4286. [CrossRef]
26. Singh, P. A New Method on Measure of Similarity between Interval-Valued Intuitionistic Fuzzy Sets for Pattern Recognition. *J. Appl. Comput. Math.* **2012**, *1*. [CrossRef]
27. Khalaf, M. Medical diagnosis via interval valued intuitionistic fuzzy sets. *Ann. Fuzzy Math. Inf.* **2013**, *6*, 245–249.
28. Dhivya, J.; Sridevi, B. Similarity measure between interval-valued intuitionistic fzzy sets and their applications to medical diagnosis and pattern recognition. *Int. J. Math. Arch.* **2018**, *9*, 58–65.
29. Meng, F.Y.; Chen, X. Entropy and similarity measure for Atannasov's interval-valued intuitionistic fuzzy sets and their application. *Fuzzy Optim. Decis. Mak.* **2015**, *15*, 75–101. [CrossRef]
30. Xu, Z.S. On similarity measures of interval-valued intuitionistic fuzzy sets and their application to pattern recognitions. *J. Southeast Univ.* **2007**, *23*, 027.
31. Mathew, T.J.; Sherly, E.; Alcantud, J.C.R. A multimodal adaptive approach on soft set based diagnostic risk prediction system. *J. Intell. Fuzzy Syst.* **2018**, *34*, 1609–1618. [CrossRef]
32. Celik, Y.; Yamak, S. Fuzzy soft set theory applied to medical diagnosis using fuzzy arithmetic operations. *J. Inequal. Appl.* **2013** , *1*, 82. [CrossRef]
33. De, S.K.; Biswas, R.; Roy, A.R. An application of intuitionistic fuzzy sets in medical diagnosis. *Fuzzy Sets Syst.* **2001**, *117*, 209–213. [CrossRef]

symmetry

MDPI

Article

A New Method to Decision-Making with Fuzzy Competition Hypergraphs

Musavarah Sarwar [1], Muhammad Akram [2,*] and Noura Omair Alshehri [3]

1 Department of Mathematics, GC Women University, Sialkot 51310, Pakistan; musavarah656@gmail.com
2 Department of Mathematics, University of the Punjab, New Campus, Lahore 54590, Pakistan
3 Department of Mathematics, Faculty of Science, University of Jeddah, P.O. Box 80327, Jeddah 21589, Saudi Arabia; noal-shehri@uj.edu.sa
* Correspondence: m.akram@pucit.edu.pk

Received: 23 July 2018; Accepted: 12 September 2018; Published: 16 September 2018

Abstract: Hypergraph theory is the most developed tool for demonstrating various practical problems in different domains of science and technology. Sometimes, information in a network model is uncertain and vague in nature. In this paper, our main focus is to apply the powerful methodology of fuzziness to generalize the notion of competition hypergraphs and fuzzy competition graphs. We introduce various new concepts, including fuzzy column hypergraphs, fuzzy row hypergraphs, fuzzy competition hypergraphs, fuzzy *k*-competition hypergraphs and fuzzy neighbourhood hypergraphs, strong hyperedges, *k*th strength of competition and symmetric properties. We design certain algorithms for constructing different types of fuzzy competition hypergraphs. We also present applications of fuzzy competition hypergraphs in decision support systems, including predator–prey relations in ecological niche, social networks and business marketing.

Keywords: fuzzy competition hypergraph; fuzzy *k*-competition hypergraph; fuzzy open neighbourhood; fuzzy closed neighbourhood; ecological niches

MSC: 05C65; 05C85; 68R10; 03E72

1. Introduction

In mathematical modeling, competition graphs are sufficient to specify well defined behaviors of objects and specifically predator–prey relations. In 1968, while studying applications of graph theory in ecology, Cohen introduced the notion of a competition graph. Competition graphs have been applied to various fields of biological sciences and technology. After the strong motivation of energy and food competition in food webs between species, competition graphs were a part of active research in recent years. In 2004, Sonntag and Teichert [1] introduced the notion of competition hypergraphs. These representations are crisp hypergraphs that do not describe all the competitions of real-world problems. These models contain uncertainty and fuzzy in nature for problems that are more relevant to everyday life, including critical writing style of a writer, predator–prey relationship, trading relationship among different communities, honesty leadership quality of a politician and, signal strength of wireless devices. Motivating from this idea, we have applied the notion of fuzzy sets to competition hypergraphs to study the problems having nonlinear uncertainties.

In 1965, Zadeh [2] introduced the strong mathematical notion of fuzzy set in order to discuss the phenomena of vagueness and uncertainty in various real-life problems. Using the concept of fuzzy relations introduced by Zadeh [3], the idea of fuzzy graph was given by Kaufmann [4]. The fuzzy relations in fuzzy sets were studied by Rosenfeld [5] and he introduced the structure of fuzzy graphs, obtaining analysis of various graph theoretical concepts. Lee-kwang and Lee [6] redefined and extended the notion of fuzzy hypergraphs whose idea was first discussed by Kaufmann [4]. Later, the idea of fuzzy hypergraph was studied by Goetschel in [7,8]. The concept of interval-valued fuzzy hypergraphs was initiated by Chen [9] and Parvathi et al. [10] generalized the idea of hypergraphs to intuitionistic fuzzy hypergraphs. Moreover, Akram and Dudek [11], Akram and Luqman [12–14], and Akram and Shahzadi [15] have discussed certain extensions of fuzzy hypergraphs with applications.

Samanta and Pal [16] studied fuzzy k-competition graphs and p-competition graphs. Later, Samanta et al. [17] introduced the concept of m-step fuzzy competition graphs. Applying the idea of bipolar fuzzy sets to competition graphs, Alshehri and Akram [18] introduced the notion of bipolar fuzzy competition graphs and applied this idea to economic systems. Furthermore, the study of bipolar fuzzy competition graphs was discussed by Sarwar and Akram in [19]. Certain competition graphs based on neutrosophic environment were described in [20,21]. In this research paper, we introduce the concept of fuzzy competition hypergraphs as a generalized case of fuzzy competition graphs. We study various new concepts, including fuzzy column hypergraphs, fuzzy row hypergraphs, fuzzy competition hypergraphs, fuzzy k-competition hypergraphs and fuzzy neighbourhood hypergraphs and investigate some of their interesting properties. We design certain algorithms for the construction of different types of fuzzy competition hypergraphs. We also present applications of fuzzy competition hypergraphs in decision support systems, including food webs, social networks and business marketing.

We have used basic notions and terminologies in this research paper. For other terminologies, notations and definitions not given in the paper, the readers are referred to [2,3,5,9,10,17,19,22–36].

Definition 1. *A fuzzy hypergraph on a non-empty set X is a pair $H = (\mu, \rho)$ where $\mu = \{\mu_1, \mu_2, \dots, \mu_r\}$, $\mu_i : X \to [0,1]$ are fuzzy subsets on X such that $\bigcup_i supp(\mu_i) = X$, for all $\mu_i \in \mu$. ρ is a fuzzy relation on the fuzzy subsets μ_i such that*

$$\rho(E_i) \leq \min\{\mu_i(x_1), \mu_i(x_2), \dots, \mu_i(x_s)\}, \quad E_i = \{x_1, x_2, \dots, x_s\}, \textit{ for all } x_1, x_2, \dots, x_s \in X.$$

2. Fuzzy Competition Hypergraphs

In this section, we discuss various types of fuzzy competition hypergraphs with certain properties and algorithms.

Definition 2. *Let $A = [x_{ij}]_{n\times n}$ be the adjacency matrix of a fuzzy digraph $\vec{G} = (\mu, \vec{\lambda})$ on a non-empty set X. The fuzzy row hypergraph of $\vec{G}$, denoted by $\mathcal{R} \circ \mathcal{H}(\vec{G}) = (\mu, \lambda_r)$, having the same set of vertices as $\vec{G}$ and the set of hyperedges is defined as*

$$\Big\{\{x_1, x_2, \dots, x_r\} | A(x_{ij}) > 0, r \geq 2, \textit{ for each } 1 \leq i \leq r, x_i \in X, \textit{ for some } 1 \leq j \leq n\Big\}.$$

The degree of membership of hyperedges is defined as

$$\lambda_r(\{x_1, x_2, \dots, x_r\}) = [\mu(x_1) \wedge \mu(x_2) \wedge \dots \wedge \mu(x_s)] \times \max_j\{\vec{\lambda}(x_1x_j) \wedge \vec{\lambda}(x_2x_j) \wedge \dots \wedge \vec{\lambda}(x_rx_j)\}.$$

Definition 3. *The fuzzy column hypergraph of $\vec{G}$, denoted by $\mathcal{C} \circ \mathcal{H}(\vec{G}) = (\mu, \lambda_{cl})$, having the same set of vertices as $\vec{G}$ and the set of hyperedges is defined as*

$$\Big\{\{x_1, x_2, \dots, x_s\} | A(x_{ji}) > 0, s \geq 2, \textit{ for each } 1 \leq i \leq s, x_i \in X, \textit{ for some } 1 \leq j \leq n\Big\}.$$

The degree of membership of hyperedges is defined as

$$\lambda_{cl}(\{x_1, x_2, \dots, x_s\}) = [\mu(x_1) \wedge \mu(x_2) \wedge \dots \wedge \mu(x_s)] \times \max_j\{\vec{\lambda}(x_jx_1) \wedge \vec{\lambda}(x_jx_2) \wedge \dots \wedge \vec{\lambda}(x_jx_s)\}.$$

The methods for computing fuzzy row hypergraph and fuzzy column hypergraph are given in Algorithms A1 and A2, respectively.

Example 1. *Consider the universe* $X = \{x_1, x_2, x_3, x_4, x_5, x_6\}$, μ *a fuzzy set on X and* $\vec{\lambda}$ *a fuzzy relation in X as defined in Tables 1 and 2, respectively. The fuzzy digraph* $\vec{G} = (\mu, \vec{\lambda})$ *is shown in Figure 1. The adjacency matrix of* $\vec{G}$ *is given in Table 3.*

Using Algorithm A1 and Table 3, there are three hyperedges $E_2 = \{x_1, x_5, x_6\}$, $E_3 = \{x_2, x_5\}$ *and* $E_4 = \{x_3, x_5\}$, *corresponding to the columns* x_2, x_3 *and* x_4 *of adjacency matrix, in fuzzy row hypergraph of* $\vec{G}$. *The membership degree of the hyperedges is calculated as*

$\lambda_r(E_2) = [\mu(x_1) \wedge \mu(x_5) \wedge \mu(x_6)] \times [x_{12} \wedge x_{52} \wedge x_{62}] = 0.3 \times 0.3 = 0.09,$

$\lambda_r(E_3) = [\mu(x_2) \wedge \mu(x_5)] \times [x_{23} \wedge x_{53}] = 0.4 \times 0.1 = 0.04,$

$\lambda_r(E_4) = [\mu(x_3) \wedge \mu(x_5)] \times [x_{34} \wedge x_{54}] = 0.4 \times 0.4 = 0.16.$

The fuzzy row hypergraph is shown in Figure 2. Using Algorithm A2 and Table 3, the hyperedges in fuzzy column hypergraph of $\vec{G}$ *are* $E_1 = \{x_2, x_6\}$, $E_5 = \{x_2, x_3, x_4\}$ *and* $E_6 = \{x_2, x_5\}$, *corresponding to the rows* x_2, x_5 *and* x_6 *of the adjacency matrix. The membership degree of the hyperedges is calculated as*

$\lambda_{cl}(E_5) = [\mu(x_2) \wedge \mu(x_3) \wedge \mu(x_4)] \times [x_{52} \wedge x_{53} \wedge x_{54}] = 0.4 \times 0.3 = 0.12,$

$\lambda_{cl}(E_1) = [\mu(x_2) \wedge \mu(x_6)] \times [x_{12} \wedge x_{16}] = 0.3 \times 0.2 = 0.06,$

$\lambda_{cl}(E_6) = [\mu(x_2) \wedge \mu(x_5)] \times [x_{62} \wedge x_{65}] = 0.4 \times 0.1 = 0.04.$

The fuzzy column hypergraph is given in Figure 3.

Table 1. Fuzzy vertex set μ.

x	$\mu(x)$	x	$\mu(x)$
x_1	0.5	x_2	0.4
x_3	0.7	x_4	0.6
x_5	0.4	x_6	0.3

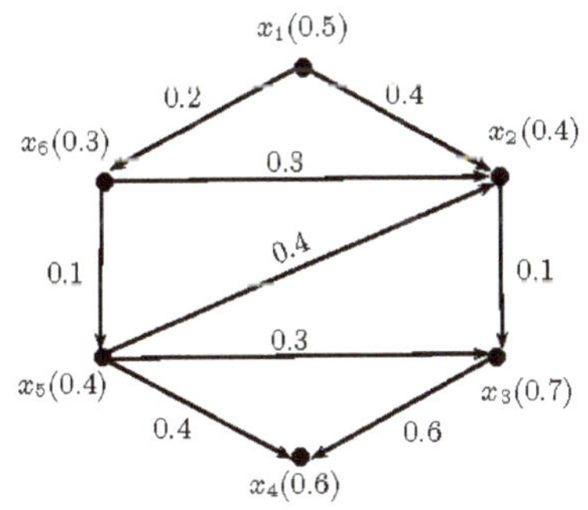

Figure 1. Fuzzy digraph $\vec{G}$.

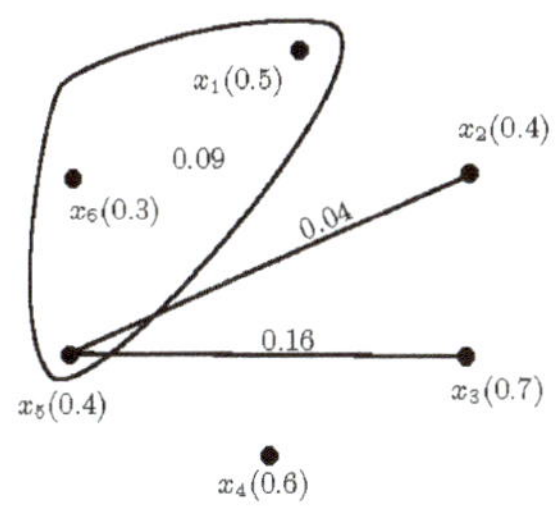

Figure 2. $\mathcal{R} \circ \mathcal{H}(\vec{G})$.

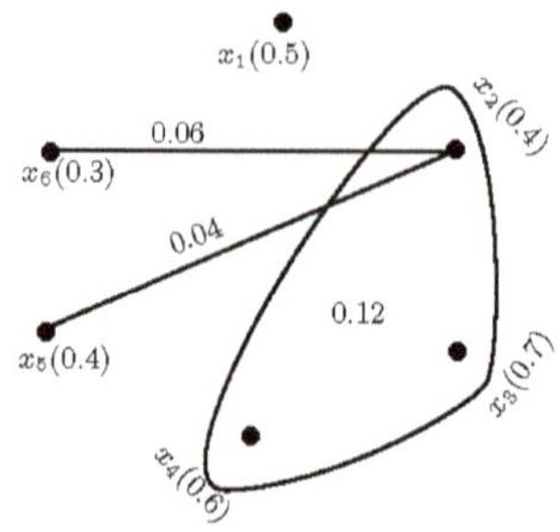

Figure 3. $\mathcal{C} \circ \mathcal{H}(\vec{G})$.

Table 2. Fuzzy relation $\vec{\lambda}$.

x	$\vec{\lambda}(x)$	x	$\vec{\lambda}(x)$
x_1x_2	0.4	x_6x_5	0.1
x_2x_3	0.1	x_1x_6	0.2
x_3x_4	0.6	x_6x_2	0.3
x_5x_4	0.4	x_5x_2	0.4
x_5x_3	0.3		

Table 3. Adjacency matrix.

A	x_1	x_2	x_3	x_4	x_5	x_6
x_1	0	0.4	0	0	0	0.2
x_2	0	0	0.1	0	0	0
x_3	0	0	0	0.6	0	0
x_4	0	0	0	0	0	0
x_5	0	0.4	0.3	0.4	0	0
x_6	0	0.3	0	0	0.1	0

Definition 4. *[25] A fuzzy digraph on a non-empty set X is a pair $\vec{G} = (\mu, \vec{\lambda})$ of functions $\mu : X \to [0,1]$ and $\vec{\lambda} : X \times X \to [0,1]$, such that for all $x, y \in X$, $\vec{\lambda}(xy) \leq \min\{\mu(x), \mu(y)\}$.*

Definition 5. *[16] A fuzzy out neighbourhood of a vertex x of a fuzzy digraph $\vec{G} = (\mu, \vec{\lambda})$ is a fuzzy set $\mathcal{N}^+(x) = (X_x^+, \mu_x^+)$, where $X_x^+ = \{y | \vec{\lambda}(xy) > 0\}$ and $\mu_x^+ : X_x^+ \to [0,1]$ is defined by $\mu_x^+(y) = \vec{\lambda}(xy)$.*

Definition 6. *[16] The fuzzy in neighbourhood of vertex x of a fuzzy digraph is a fuzzy set $\mathcal{N}^-(x) = (X_x^-, \mu_x^-)$, where $X_x^- = \{y | \vec{\lambda}(yx) > 0\}$ and $\mu_x^- : X_x^- \to [0,1]$ is defined by $\mu_x^-(y) = \vec{\lambda}(yx)$.*

Definition 7. *Let $\vec{G} = (\mu, \vec{\lambda})$ be a fuzzy digraph on a non-empty set X. The fuzzy competition hypergraph $\mathcal{CH}(\vec{G}) = (\mu, \lambda_c)$ on X having the same vertex set as $\vec{G}$ and there is a hyperedge consisting of vertices $x_1, x_2, \ldots, x_s$ if $\mathcal{N}^+(x_1) \cap \mathcal{N}^+(x_2) \cap \ldots \cap \mathcal{N}^+(x_s) \neq \emptyset$. The degree of membership of hyperedge $E = \{x_1, x_2, \ldots, x_s\}$ is defined as*

$$\lambda_c(E) = [\mu(x_1) \wedge \mu(x_2) \wedge \ldots \wedge \mu(x_s)] \times h(\mathcal{N}^+(x_1) \cap \mathcal{N}^+(x_2) \cap \ldots \cap \mathcal{N}^+(x_s)),$$

where $h(\mathcal{N}^+(x_1) \cap \mathcal{N}^+(x_2) \cap \ldots \cap \mathcal{N}^+(x_s))$ denotes the height of fuzzy set $\mathcal{N}^+(x_1) \cap \mathcal{N}^+(x_2) \cap \ldots \cap \mathcal{N}^+(x_s)$.

The method for constructing fuzzy competition hypergraph of a fuzzy digraph is given in Algorithm A3.

Lemma 1. *The fuzzy competition hypergraph of a fuzzy digraph $\vec{G}$ is a fuzzy row hypergraph of $\vec{G}$.*

Proof. Let $\vec{G} = (\mu, \vec{\lambda})$ be a fuzzy digraph; then, for any hyperedge $E = \{x_1, x_2, \ldots, x_s\}$ of $\mathcal{CH}(\vec{G})$,

$$\begin{aligned}\lambda_c(E) &= [\mu(x_1) \wedge \mu(x_2) \wedge \ldots \wedge \mu(x_s)] \times h(\mathcal{N}^+(x_1) \cap \mathcal{N}^+(x_2) \cap \ldots \cap \mathcal{N}^+(x_s)) \\ &= [\mu(x_1) \wedge \mu(x_2) \wedge \ldots \wedge \mu(x_s)] \times \max_j\{\mathcal{N}^+(x_1) \cap \mathcal{N}^+(x_2) \cap \ldots \cap \mathcal{N}^+(x_s)\} \\ &= [\mu(x_1) \wedge \mu(x_2) \wedge \ldots \wedge \mu(x_s)] \times \max_j\{\vec{\lambda}(x_1x_j) \wedge \vec{\lambda}(x_2x_j) \wedge \ldots \wedge \vec{\lambda}(x_nx_j)\} = \lambda_r(E).\end{aligned}$$

It follows that E is a hyperedge of fuzzy row hypergraph. □

Example 2. *Consider the fuzzy digraph given in Figure 1. The fuzzy out neighbourhood and fuzzy in neighbourhood of all the vertices are given in Table 4.*

Using Algorithm A3, the relation $f : X \to X$ of $\vec{G}$ is given in Figure 4. The construction of fuzzy competition hypergraph from $\vec{G}$ is given as follows:

1. *Since $f^{-1}(x_2) = E_2 = \{x_1, x_5, x_6\}$, $f^{-1}(x_3) = E_3 = \{x_2, x_5\}$ and $f^{-1}(x_4) = E_4 = \{x_3, x_5\}$, $\{x_1, x_5, x_6\}$, $\{x_2, x_5\}$ and $\{x_3, x_5\}$ are hyperedges in $\mathcal{CH}(\vec{G})$.*
2. *For hyperedge E_2: $\mathcal{N}^+(x_1) \cap \mathcal{N}^+(x_5) \cap \mathcal{N}^+(x_6) = \{(x_2, 0.3)\}$, $\lambda_c(E_2) = [\mu(x_1) \wedge \mu(x_5) \wedge \mu(x_6)] \times h(\mathcal{N}^+(x_1) \cap \mathcal{N}^+(x_5) \cap \mathcal{N}^+(x_6)) = 0.3 \times 0.3 = 0.09$.*
3. *Similarly, $\lambda_c(E_3) = [\mu(x_2) \wedge \mu(x_5)] \times h(\mathcal{N}^+(x_2) \cap \mathcal{N}^+(x_5)) = 0.04$ and $\lambda_c(E_4) = [\mu(x_3) \wedge \mu(x_5)] \times h(\mathcal{N}^+(x_3) \cap \mathcal{N}^+(x_5)) = 0.16$.*

The fuzzy competition hypergraph is given in Figure 5. From Figures 2 and 5, it is clear that fuzzy competition hypergraph is a fuzzy row hypergraph.

Table 4. Fuzzy out neighbourhood and fuzzy in neighbouhood of vertices in $\vec{G}$.

$x \in X$	$\mathcal{N}^+(x)$	$\mathcal{N}^-(x)$
x_1	$\{(x_2, 0.4), (x_6, 0.2)\}$	$\emptyset$
x_2	$\{(x_3, 0.1)\}$	$\{(x_1, 0.4), (x_5, 0.4), (x_6, 0.3)\}$
x_3	$\{(x_4, 0.6)\}$	$\{(x_2, 0.1), (x_5, 0.3)\}$
x_4	$\emptyset$	$\{(x_3, 0.6), (x_5, 0.4)\}$
x_5	$\{(x_2, 0.4), (x_3, 0.3), (x_4, 0.4)\}$	$\{(x_6, 0.1)\}$
x_6	$\{(x_2, 0.3), (x_5, 0.1)\}$	$\{(x_1, 0.2)\}$

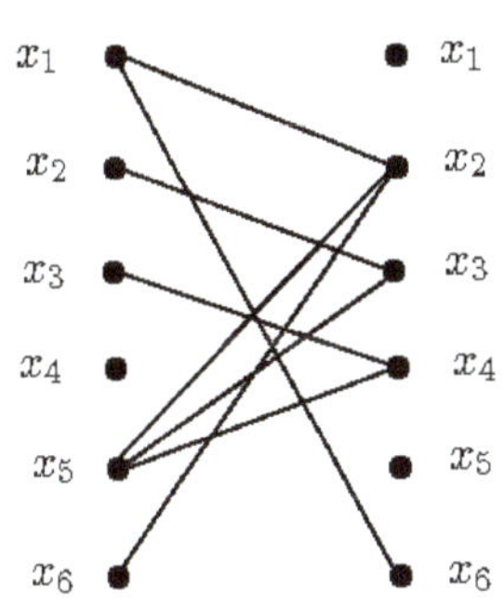

Figure 4. Representation of fuzzy relation in $\vec{G}$.

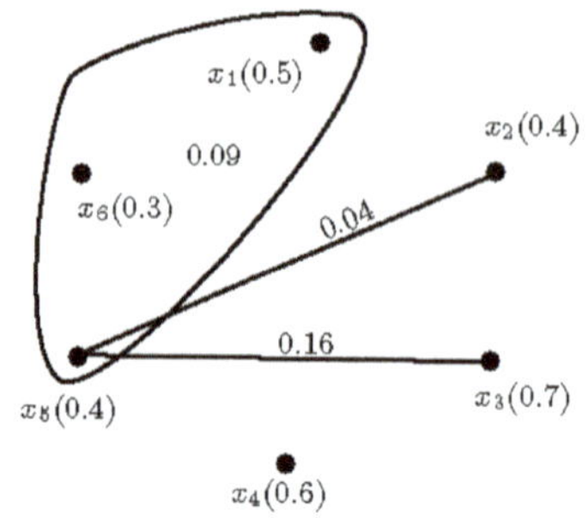

Figure 5. Fuzzy competition hypergraph $\mathcal{CH}(\vec{G})$.

Definition 8. *The fuzzy double competition hypergraph $\mathcal{DCH}(\vec{G}) = (\mu, \lambda_d)$ having same vertex set as $\vec{G}$ and there is a hyperedge consisting of vertices $x_1, x_2, \ldots, x_s$ if $\mathcal{N}^+(x_1) \cap \mathcal{N}^+(x_2) \cap \ldots \cap \mathcal{N}^+(x_s) \neq \emptyset$ and $\mathcal{N}^-(x_1) \cap \mathcal{N}^-(x_2) \cap \ldots \cap \mathcal{N}^-(x_s) \neq \emptyset$. The degree of membership of hyperedge $E = \{x_1, x_2, \ldots, x_s\}$ is defined as*

$$\begin{aligned}\lambda_d(E) =&[\mu(x_1) \wedge \mu(x_2) \wedge \ldots \wedge \mu(x_s)] \times \\ &[h(\mathcal{N}^+(x_1) \cap \mathcal{N}^+(x_2) \cap \ldots \cap \mathcal{N}^+(x_s)) \wedge h(\mathcal{N}^-(x_1) \cap \mathcal{N}^-(x_2) \cap \ldots \cap \mathcal{N}^-(x_s))].\end{aligned}$$

The method for the construction of fuzzy double competition hypergraph is given in Algorithm A4.

Lemma 2. *The fuzzy double competition hypergraph is the intersection of fuzzy row hypergraph and fuzzy column hypergraph.*

Proof. Let $\vec{G} = (\mu, \vec{\lambda})$ be a fuzzy digraph; then, for any hyperedge $E = \{x_1, x_2, \ldots, x_s\}$ of $\mathcal{CH}(\vec{G})$,

$$\begin{aligned}\lambda_d(E) =&[\mu(x_1) \wedge \mu(x_2) \wedge \ldots \wedge \mu(x_s)] \times \\ &[h(\mathcal{N}^+(x_1) \cap \mathcal{N}^+(x_2) \cap \ldots \cap \mathcal{N}^+(x_s)) \wedge h(\mathcal{N}^+(x_1) \cap \mathcal{N}^+(x_2) \cap \ldots \cap \mathcal{N}^+(x_s))]. \\ =&[\mu(x_1) \wedge \mu(x_2) \wedge \ldots \wedge \mu(x_s)] \times \\ &[\max_j\{\vec{\lambda}(x_1x_j) \wedge \vec{\lambda}(x_2x_j) \wedge \ldots \wedge \vec{\lambda}(x_nx_j)\} \wedge \max_k\{\vec{\lambda}(x_kx_1) \wedge \vec{\lambda}(x_kx_2) \wedge \ldots \wedge \vec{\lambda}(x_kx_n)\}]. \\ =&[\{\mu(x_1) \wedge \mu(x_2) \wedge \ldots \wedge \mu(x_s)\} \times \max_j\{\vec{\lambda}(x_1x_j) \wedge \vec{\lambda}(x_2x_j) \wedge \ldots \wedge \vec{\lambda}(x_nx_j)\}] \times \\ &[\{\mu(x_1) \wedge \mu(x_2) \wedge \ldots \wedge \mu(x_s)\} \times \max_k\{\vec{\lambda}(x_kx_1) \wedge \vec{\lambda}(x_kx_2) \wedge \ldots \wedge \vec{\lambda}(x_kx_n)\}] \\ =&\lambda_r(E) \wedge \lambda_{cl}(E).\end{aligned}$$

It follows that the fuzzy double competition hypergraph is the intersection of a fuzzy row hypergraph and fuzzy column hypergraph. □

Example 3. *Consider the example of fuzzy digraph shown in Figure 1. From Example 2, the fuzzy double competition hypergraph of Figure 1 is given in Figure 6. In addition, Figures 2, 3 and 6 show that the fuzzy double competition hypergraph is the intersection of fuzzy row hypergraph and fuzzy column hypergraph.*

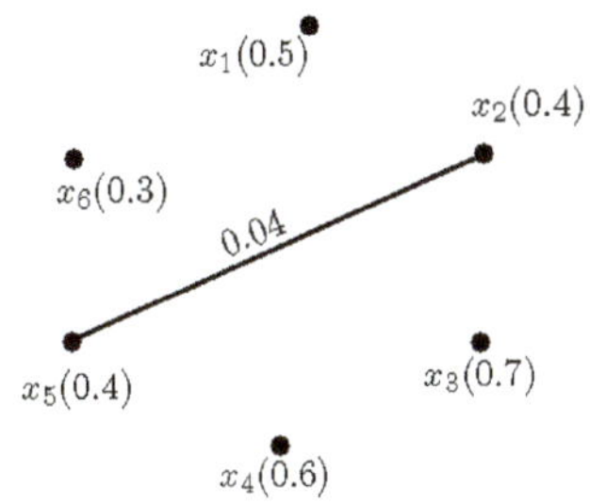

Figure 6. $\mathcal{DCH}(\vec{G})$.

Definition 9. *Let $\vec{G} = (\mu, \vec{\lambda})$ be a fuzzy digraph on a non-empty set X. The fuzzy niche hypergraph $\mathcal{NH}(\vec{G}) = (\mu, \lambda_n)$ has the same vertex set as $\vec{G}$ and there is hyperedge consisting of vertices $x_1, x_2, \ldots, x_s$ if either $\mathcal{N}^+(x_1) \cap \mathcal{N}^+(x_2) \cap \ldots \cap \mathcal{N}^+(x_s) \neq \emptyset$ or $\mathcal{N}^-(x_1) \cap \mathcal{N}^-(x_2) \cap \ldots \cap \mathcal{N}^-(x_s) \neq \emptyset$. The degree of membership of hyperedge $E = \{x_1, x_2, \ldots, x_s\}$ is defined as*

$$\begin{aligned}\lambda_n(E) = &[\mu(x_1) \wedge \mu(x_2) \wedge \ldots \wedge \mu(x_s)] \times \\ &\left[h\left(\mathcal{N}^+(x_1) \cap \mathcal{N}^+(x_2) \cap \ldots \cap \mathcal{N}^+(x_s)\right) \vee h\left(\mathcal{N}^-(x_1) \cap \mathcal{N}^-(x_2) \cap \ldots \cap \mathcal{N}^-(x_s)\right)\right].\end{aligned}$$

Lemma 3. *The fuzzy niche hypergraph is the union of fuzzy row hypergraph and fuzzy column hypergraph.*

Example 4. *The fuzzy niche hypergraph of Figure 1 is shown in Figure 7, which is the union of Figures 2 and 3.*

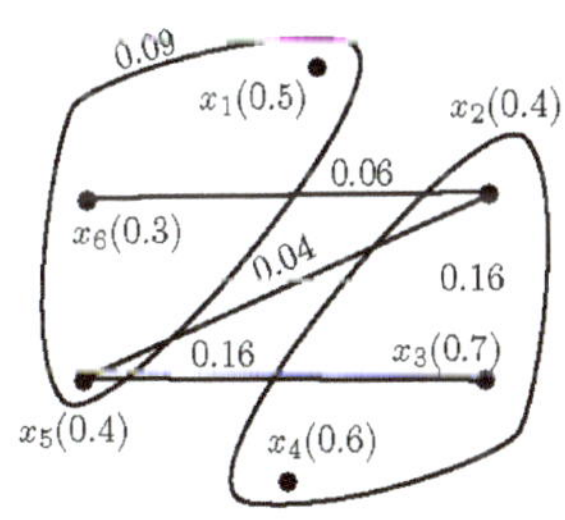

Figure 7. $\mathcal{NH}(\vec{G})$.

Definition 10. *Let H be a fuzzy hypergraph and t be the smallest non-negative number such that $H \cup I_t$ is a fuzzy niche hypergraph of some fuzzy digraph $\vec{G}$, where I_t is a fuzzy set on t isolated vertices X_t; then, t is called fuzzy niche number of H denoted by $\mathfrak{n}(H)$.*

Lemma 4. *Let H be a fuzzy hypergraph on a non-empty set X with $\mathfrak{n}(H) = t < \infty$ and $H \cup I_t$ is a fuzzy niche hypergraph of an acyclic digraph $\vec{G}$ then for all, $x \in X \cup X_t$,*

$$\mathcal{N}^+(y) \cap I_t \neq \emptyset \Rightarrow \exists\, z \in supp(I_t) \text{ such that } supp(\mathcal{N}^+(y)) = z,$$
$$\mathcal{N}^-(y) \cap I_t \neq \emptyset \Rightarrow \exists\, z \in supp(I_t) \text{ such that } supp(\mathcal{N}^-(y)) = z.$$

Proof. On the contrary, assume that, for some $y \in X$, either $supp(\mathcal{N}^+(y)) = \{z\} \cup X'$ or $supp(\mathcal{N}^-(y)) = \{z\} \cup X''$, where $\emptyset \neq X' \subseteq X \cup X_t \setminus \{z\}$. Then, by definition of a fuzzy niche hypergraph, z is adjacent to all vertices X' in $H \cup I_t$—a contradiction to the fact that $z \in X_t$. □

Lemma 5. *Let H be a fuzzy hypergraph with $\mathfrak{n}(H) = t < \infty$ and $H \cup I_t$ is a fuzzy niche hypergraph of an acyclic fuzzy digraph $\vec{G}$ then for all $z \in X_t$, $\mathcal{N}^+(z) = \emptyset$ and $\mathcal{N}^-(z) = \emptyset$.*

Proof. On the contrary, assume that $X_z^+ = \{y_1, y_2, \ldots, y_s\}$ and $X_z^- = \{y_1', y_2', \ldots, y_r'\}$. Clearly, $\mathcal{N}^+(z) \cap \mathcal{N}^-(z) = \emptyset$ because $\vec{G}$ is acyclic. According to Lemma 4, $\mathcal{N}^+(y_i) = \mathcal{N}^+(y_i')$.
Consider another fuzzy digraph $\vec{G}'$ such that $X_{\vec{G}'} = X_{\vec{G}} \setminus \{z\}$ and $E_{\vec{G}'} = (E_{\vec{G}} \setminus \{E_1\}) \cup E_2$, where

$$E_1 = \{\overrightarrow{zy_i} : 1 \leq i \leq s\} \cup \{\overrightarrow{y_i'z} : 1 \leq i \leq r\},$$
$$E_2 = \{\overrightarrow{y_1'y_i} : 1 \leq i \leq s\} \cup \{\overrightarrow{y_i'y_1} : 1 \leq i \leq r\}.$$

Clearly, $\mathcal{N}^+(z) = \mathcal{N}^+(y_1)$ and $\mathcal{N}^-(z) = \mathcal{N}^-(y_1')$. Thus, $\mathcal{NH}(\vec{G}') = H \cup I_{t-1}$ which contradicts the fact that $\mathfrak{n}(H) = t$. Hence, for all $z \in X_t$, $\mathcal{N}^+(z) = \emptyset$ and $\mathcal{N}^-(z) = \emptyset$. □

Definition 11. *Let $H = (\mu, \rho)$ be a fuzzy hypegraph on a non-empty set X. A hyperedge $E_i = \{x_1, x_2, \ldots, x_r\} \subseteq X$ is called strong if $\rho(E_i) \geq \frac{1}{2} \bigwedge_{k=1}^{r} \mu_i(x_k)$.*

Theorem 1. *Let $\vec{G} = (\mu, \vec{\lambda})$ be a fuzzy digraph. If $\mathcal{N}^+(x_1) \cap \mathcal{N}^+(x_2) \cap \ldots \cap \mathcal{N}^+(x_r)$ contains exactly one vertex, then the hyperedge $\{x_1, x_2, \ldots, x_r\}$ of $\mathcal{C}(\vec{G})$ is strong if and only if $|\mathcal{N}^+(x_1) \cap \mathcal{N}^+(x_2) \cap \ldots \cap \mathcal{N}^+(x_r)| > \frac{1}{2}$.*

Proof. Assume that $\mathcal{N}^+(x_1) \cap \mathcal{N}^+(x_2) \cap \ldots \cap \mathcal{N}^+(x_r) = \{(u, l)\}$, where l is degree of membership of u. As $|\mathcal{N}^+(x_1) \cap \mathcal{N}^+(x_2) \cap \ldots \cap \mathcal{N}^+(x_r)| = l = h(\mathcal{N}^+(x_1) \cap \mathcal{N}^+(x_2) \cap \ldots \cap \mathcal{N}^+(x_r))$; therefore, $\lambda_c(\{x_1, x_2, \ldots, x_r\}) = (\mu(x_1) \wedge \mu(x_2) \wedge \ldots \wedge \mu(x_r)) \times h(\mathcal{N}^+(x_1) \cap \mathcal{N}^+(x_2) \cap \ldots \cap \mathcal{N}^+(x_r)) = l \times (\mu(x_1) \wedge \mu(x_2) \wedge \ldots \wedge \mu(x_r)\})$. Thus, the hyperedge $\{x_1, x_2, \ldots, x_r\}$ in $\mathcal{C}(\vec{G})$ would be strong if $l > \frac{1}{2}$ by Definition 11. □

Definition 12. *Let k be a non-negative real number number; then, the fuzzy k-competition hypergraph of a fuzzy digraph $\vec{G} = (\mu, \vec{\lambda})$ is fuzzy hypergraph $\mathcal{C}_k(\vec{G}) = (\mu, \lambda_{kc})$, which has the same fuzzy vertex set as in $\vec{G}$ and there is a hyperedge $E = \{x_1, x_2, \ldots, x_r\}$ in $\mathcal{C}_k(\vec{G})$ if $|\mathcal{N}^+(x_1) \cap \mathcal{N}^+(x_2) \cap \ldots \cap \mathcal{N}^+(x_r)| > k$. The membership degree of the hyperedge E is defined as*

$$\lambda_{kc}(E) = \frac{l-k}{l}(\mu(x_1) \wedge \mu(x_2) \wedge \ldots \wedge \mu(x_r)) \times h(\mathcal{N}^+(x_1) \cap \mathcal{N}^+(x_2) \cap \ldots \cap \mathcal{N}^+(x_r)),$$

where $|\mathcal{N}^+(x_1) \cap \mathcal{N}^+(x_2) \cap \ldots \cap \mathcal{N}^+(x_r)| = l$.

Example 5. *The fuzzy $0.2-$competition hypergraph of Figure 1 is given in Figure 8.*

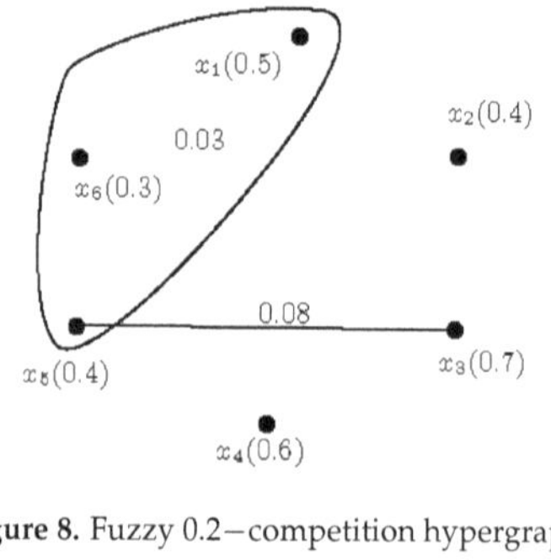

Figure 8. Fuzzy $0.2-$competition hypergraph.

Remark 1. *For $k = 0$, a fuzzy k-competition hypergraph is simply a fuzzy competition hypergraph.*

Theorem 2. *Let $\vec{G} = (\mu, \vec{\lambda})$ be a fuzzy digraph. If $h(\mathcal{N}^+(x_1) \cap \mathcal{N}^+(x_2) \cap \ldots \cap \mathcal{N}^+(x_r)) = 1$ and $|\mathcal{N}^+(x_1) \cap \mathcal{N}^+(x_2) \cap \ldots \cap \mathcal{N}^+(x_r)| > 2k$ for some $x_1, x_2, \ldots, x_r \in X$, then the hyperedge $\{x_1, x_2, \ldots, x_r\}$ is strong in $\mathcal{C}_k(\vec{G})$.*

Proof. Let $\mathcal{C}_k(\vec{G}) = (\mu, \lambda_{kc})$ be a fuzzy k-competition hypergraph of fuzzy digraph $\vec{G} = (\mu, \vec{\lambda})$. Suppose for $E = \{x_1, x_2, \ldots, x_r\} \subseteq X$, $|\mathcal{N}^+(x_1) \cap \mathcal{N}^+(x_2) \cap \ldots \cap \mathcal{N}^+(x_r)| = l$. Now,

$$\lambda_{kc}(E) = \frac{l-k}{l}(\mu(x_1) \wedge \mu(x_2) \wedge \ldots \wedge \mu(x_r)) \times h(\mathcal{N}^+(x_1) \cap \mathcal{N}^+(x_2) \cap \ldots \cap \mathcal{N}^+(x_r)),$$

$$\lambda_{kc}(E) = \frac{l-k}{l}(\mu(x_1) \wedge \mu(x_2) \wedge \ldots \wedge \mu(x_r)), \quad \because h(\mathcal{N}^+(x_1) \cap \mathcal{N}^+(x_2) \cap \ldots \cap \mathcal{N}^+(x_r)) = 1,$$

$$\Longrightarrow \frac{\lambda_{kc}(E)}{\mu(x_1) \wedge \mu(x_2) \wedge \ldots \wedge \mu(x_r)} > \frac{1}{2}, \qquad \because l > 2k.$$

Thus, the hyperedge E is strong in $\mathcal{C}_k(\vec{G})$. □

Fuzzy Neighbourhood Hypergraphs

The concepts of fuzzy open neighbourhood and fuzzy closed neighbourhood are given in Definition 13.

Definition 13. *[16] The fuzzy open neighbourhood of a vertex y in a fuzzy graph $G = (\mu, \lambda)$ is a fuzzy set $\mathcal{N}(y) = (X_y, \mu_y)$, where $X_y = \{w | \lambda(yw) > 0\}$ and $\mu_y : X_y \to [0, 1]$ a membership function defined by $\mu_y(w) = \lambda(yw)$.*

Definition 14. *[16] The fuzzy closed neighbourhood $\mathcal{N}[y]$ of a vertex y in a fuzzy graph $G = (\mu, \lambda)$ is defined as $\mathcal{N}[y] = \mathcal{N}(y) \cup \{(y, \mu(y))\}$.*

Definition 15. *The fuzzy open neighbourhood hypergraph of a fuzzy graph $G = (\mu, \lambda)$ is a fuzzy hypergraph $\mathcal{N}(G) = (\mu, \lambda')$ whose fuzzy vertex set is the same as G and there is a hyperedge $E = \{x_1, x_2, \ldots, x_r\}$ in $\mathcal{N}(G)$ if $\mathcal{N}(x_1) \cap \mathcal{N}(x_2) \cap \ldots \cap \mathcal{N}(x_r) \neq \emptyset$. The membership function $\lambda' : X \times X \to [0, 1]$ is defined as*

$$\lambda'(E) = (\mu(x_1) \wedge \mu(x_2) \wedge \ldots \wedge \mu(x_r)) \times h(\mathcal{N}(x_1) \cap \mathcal{N}(x_2) \cap \ldots \cap \mathcal{N}(x_r)).$$

The fuzzy closed neighbourhood hypergraph is defined on the same lines in the following definition.

Definition 16. *The fuzzy closed neighbourhood hypergraph of $G = (\mu, \lambda)$ is a fuzzy hypergraph $\mathcal{N}[G] = (\mu, \lambda^*)$ whose fuzzy set of vertices is same as G and there is a hyperedge $E = \{x_1, x_2, \ldots, x_r\}$ in $\mathcal{N}[G]$ if $\mathcal{N}[x_1] \cap \mathcal{N}[x_2] \cap X \ldots \cap \mathcal{N}[x_r] \neq \emptyset$. The membership function $\lambda^* : X \times X \to [0, 1]$ is defined as*

$$\lambda^*(E) = (\mu(x_1) \wedge \mu(x_2) \wedge \ldots \wedge \mu(x_r)) \times h(\mathcal{N}[x_1] \cap \mathcal{N}[x_2] \cap \ldots \cap \mathcal{N}[x_r]).$$

Example 6. *Consider the fuzzy graph $G = (\mu, \lambda)$ on set $Y = \{y_1, y_2, y_3, y_4\}$ as shown in Figure 9. The fuzzy open neighbourhoods are given in Table 5.*

Define a relation $f : X \to X$ by $f(y_i) = y_j$ if $y_j \in supp(\mathcal{N}(y_i))$ as shown in Figure 10. If, for $y_i \in X$, $|f^{-1}(y_i)| > 1$, then $f^{-1}(y_i)$ is a hyperedge of $\mathcal{N}[G]$. Since, from Figure 10, $f^{-1}(y_1) = \{y_2, y_3, y_4\} = E_1$, $f^{-1}(y_2) = \{y_1, y_4\} = E_2$ and $f^{-1}(y_4) = \{y_1, y_2\}_3$, therefore, E_1, E_2, E_3 are hyperedges of $\mathcal{N}(G)$. The degree of membership of each hyperedge can be computed using Definition 15 as follows.

For $f^{-1}(y_1) = E_1 = \{y_2, y_3, y_4\}$, $\lambda'(E_1) = (\mu(y_2) \wedge \mu(y_3) \wedge \mu(y_4)) \times h(\mathcal{N}(y_2) \cap \mathcal{N}(y_3) \cap \mathcal{N}(y_4)) = 0.4 \times 0.4 = 0.16$. Similarly, $\lambda'(\{y_1, y_4\}) = 0.4 \times 0.3 = 0.12$ and $\lambda'(\{y_1, y_2\}) = 0.5 \times 0.3 = 0.15$. The fuzzy open neighbourhood hypergraph constructed using Definition 13 from $\vec{G}$ is given in Figure 10.

Table 5. Fuzzy open neighbourhood of vertices.

y	$\mathcal{N}(y)$
y_1	$\{(y_2, 0.4), (y_3, 0.5), (y_4, 0.5)\}$
y_2	$\{(y_1, 0.4), (y_4, 0.3)\}$
y_3	$\{(y_1, 0.5)\}$
y_4	$\{(y_1, 0.4), (y_2, 0.3)\}$

The fuzzy closed neighbourhoods of all the vertices in G are given in Table 6. Since $\mathcal{N}[y_1] \cap \mathcal{N}[y_2] \cap \mathcal{N}[y_3] \cap \mathcal{N}[y_4] = \{(y_1, 0.4)\}$, therefore, $E = \{y_1, y_2, y_3, y_4\}$ is a hyperedge of $\mathcal{N}[G]$ and $\lambda^*(E) = 0.4 \times 0.4 = 0.16$. The fuzzy closed neighbourhood hypergraph is given in Figure 11.

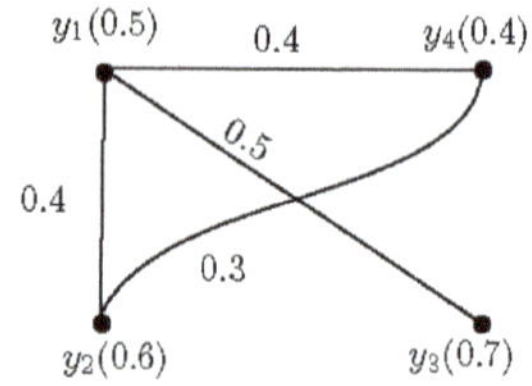

Figure 9. Fuzzy graph G.

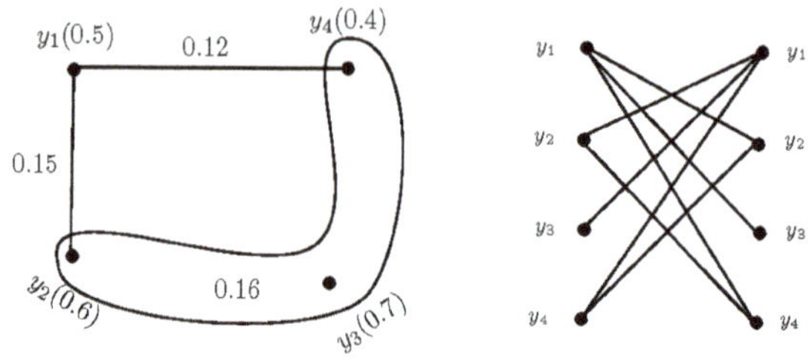

Figure 10. Fuzzy open neighbourhood hypergraph of G.

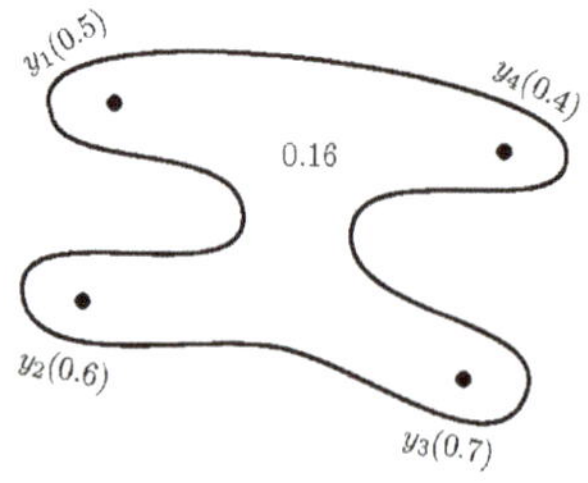

Figure 11. Fuzzy closed neighbourhood hypergraph.

Table 6. Fuzzy closed neighbourhood of vertices.

y	$\mathcal{N}[y]$
y_1	$\{(y_1, 0.5), (y_2, 0.4), (y_3, 0.5), (y_4, 0.5)\}$
y_2	$\{(y_2, 0.6), (y_1, 0.4), (y_4, 0.3)\}$
y_3	$\{(y_3, 0.7), (y_1, 0.5)\}$
y_4	$\{(y_4, 0.4), (y_1, 0.4), (y_2, 0.3)\}$

Using different types of fuzzy neighbourhood of the vertices, some other types of fuzzy hypergraphs are defined here.

Definition 17. *Let k be a non-negative real number; then, the fuzzy (k)-competition hypergraph of a fuzzy graph $G = (\mu, \lambda)$ is a fuzzy hypergraph $\mathcal{N}_k(G) = (\mu, \lambda'_{kc})$ having the same fuzzy set of vertices as G and there is a hyperedge $E = \{x_1, x_2, \ldots, x_r\}$ in $\mathcal{N}_k(G)$ if $|\mathcal{N}(x_1) \cap \mathcal{N}(x_2) \cap \ldots \cap \mathcal{N}(x_r)| > k$. The membership value of E is defined as*

$$\lambda'_{kc}(E) = \frac{l-k}{l}(\mu(x_1) \wedge \mu(x_2) \wedge \ldots \wedge \mu(x_r)) \times h(\mathcal{N}(x_1) \cap \mathcal{N}(x_2) \cap \ldots \cap \mathcal{N}(x_r)),$$

where $|\mathcal{N}(x_1) \cap \mathcal{N}(x_2) \cap \ldots \cap \mathcal{N}(x_r)| = l$.

Definition 18. *The fuzzy $[k]$-competition hypegraph of G is denoted by $\mathcal{N}_k[G] = (\mu, \lambda^*_{kc})$ and there is a hyperedge E in $\mathcal{N}_k[G]$ if $|\mathcal{N}[x_1] \cap \mathcal{N}[x_2] \cap \ldots \cap \mathcal{N}[x_r]| > k$. The membership value of E is defined as*

$$\lambda^*_{kc}(E) = \frac{p-k}{p}(\mu(x_1) \wedge \mu(x_2) \wedge \ldots \wedge \mu(x_r)) \times h(\mathcal{N}[x_1] \cap \mathcal{N}[x_2] \cap \ldots \cap \mathcal{N}[x_r]),$$

where $|\mathcal{N}[x_1] \cap \mathcal{N}[x_2] \cap \ldots \cap \mathcal{N}[x_r]| = p$.

Definition 19. *[16] Let $\vec{G} = (\mu, \vec{\lambda})$ be a fuzzy digraph. The underlying fuzzy graph of $\vec{G}$ is a fuzzy graph $\mathcal{U}(\vec{G}) = (\mu, \lambda)$ such that*

$$\lambda(xw) = \begin{cases} \vec{\lambda}(xw), & \text{if } \overrightarrow{wx} \notin \vec{E}, \\ \vec{\lambda}(wx), & \text{if } \overrightarrow{xw} \notin \vec{E}, \\ \vec{\lambda}(xw) \wedge \vec{\lambda}(wx), & \text{if } \overrightarrow{wx}, \overrightarrow{xw} \subset \vec{E}, \end{cases}$$

where $\vec{E} = supp(\vec{\lambda})$. The relations between fuzzy neighbourhood hypergraphs and fuzzy competition hypergraphs are given in the following theorems.

Theorem 3. *Let $\vec{G} = (\mu, \vec{\lambda})$ be a symmetric fuzzy digraph without any loops; then, $\mathcal{C}_k(\vec{G}) = \mathcal{N}_k(\mathcal{U}(\vec{G}))$, where $\mathcal{U}(\vec{G})$ is the underlying fuzzy graph of $\vec{G}$.*

Proof. Let $\mathcal{U}(\vec{G}) = (\mu, \lambda)$ correspond to the fuzzy graph $\vec{G} = (\mu, \vec{\lambda})$. In addition, let $\mathcal{N}_k(\mathcal{U}(\vec{G})) = (\mu, \lambda'_{kc})$ and $\mathcal{C}_k(\vec{G}) = (\mu, \lambda_{kc})$. Clearly, the fuzzy k-competition hypergraph $\mathcal{C}_k(\vec{G})$ and the underlying fuzzy graph have the same fuzzy set of vertices as $\vec{G}$. Hence, $\mathcal{N}_k(\mathcal{U}(\vec{G}))$ has the same vertex set as $\vec{G}$. It remains only to show that $\lambda_{kc}(xw) = \lambda'_{kc}(xw)$ for every $x, w \in X$. Thus, there are two cases.

Case 1: If, for each $x_1, x_2, \ldots, x_r \in X$, $\lambda_{kc}(\{x_1, x_2, \ldots, x_r\}) = 0$ in $\mathcal{C}_k(\vec{G})$, then $|\mathcal{N}^+(x_1) \cap \mathcal{N}^+(x_2) \cap \ldots \mathcal{N}^+(x_r)| \leq k$. Since $\vec{G}$ is symmetric, $|\mathcal{N}(x_1) \cap \mathcal{N}(x_2) \cap \ldots \mathcal{N}(x_r)| \leq k$ in $\mathcal{U}(\vec{G})$. Thus, $\lambda'_{kc}(\{x_1, x_2, \ldots, x_r\}) = 0$ and $\lambda_{kc}(E) = \lambda'_{kc}(E)$ for all $x_1, x_2, \ldots, x_r \in X$.

Case 2: If, for some $x_1, x_2, \ldots, x_r \in X$, $\lambda_{kc}(E) > 0$ in $\mathcal{C}_k(\vec{G})$, then $|\mathcal{N}^+(x_1) \cap \mathcal{N}^+(x_2) \cap \ldots \mathcal{N}^+(x_r)| > k$. Thus,

$$\lambda_{kc}(E) = \frac{l-k}{l}[\mu(x_1) \wedge \mu(x_2) \wedge \ldots \wedge \mu(x_r)] h(\mathcal{N}^+(x_1) \cap \mathcal{N}^+(x_2) \cap \ldots \cap \mathcal{N}^+(x_r)),$$

where $l = |\mathcal{N}^+(x_1) \cap \mathcal{N}^+(x_2) \cap \ldots \cap \mathcal{N}^+(x_r)|$. Since $\vec{G}$ is a symmetric fuzzy digraph, $|\mathcal{N}(x_1) \cap \mathcal{N}(x_2) \cap \ldots \mathcal{N}(x_r)| > k$. Hence, $\lambda_{kc}(E) = \lambda'_{kc}(E)$. Since $x_1, x_2, \ldots, x_r$ were taken to be arbitrary, the result holds for all hyperedges E of $\mathcal{C}_k(\vec{G})$. □

Theorem 4. *Let $\vec{G} = (C, \vec{D})$ be a symmetric fuzzy digraph having loops at every vertex; then, $\mathcal{C}_k(\vec{G}) = \mathcal{N}_k[\mathcal{U}(\vec{G})]$, where $\mathcal{U}(\vec{G})$ is the underlying fuzzy graph of $\vec{G}$.*

Proof. Let $\mathcal{U}(\vec{G}) = (\mu, \lambda)$ be an underlying fuzzy graph corresponding to fuzzy digraph $\vec{G} = (\mu, \vec{\lambda})$. Let $\mathcal{N}_k[\mathcal{U}(\vec{G})] = (\mu, \lambda'_{kc})$ and $\mathcal{C}_k(\vec{G}) = (\mu, \lambda_{kc})$. The fuzzy k-competition graph $\mathcal{C}_k(\vec{G})$ as well as the underlying fuzzy graph have the same vertex set as $\vec{G}$. It follows that $\mathcal{N}_k[\mathcal{U}(\vec{G})]$ has the same fuzzy vertex set as $\vec{G}$. It remains only to show that $\lambda_{kc}(\{x_1, x_2, \ldots, x_r\}) = \lambda'_{kc}(\{x_1, x_2, \ldots, x_r\})$ for every $x_1, x_2, \ldots, x_r \in X$. As the fuzzy digraph has a loop at every vertex, the fuzzy out neighbourhood contains the vertex itself. There are two cases.

Case 1: If, for all $x_1, x_2, \ldots, x_r \in X$, $\lambda_{kc}(E) = 0$ in $\mathcal{C}_k(\vec{G})$, then, $|\mathcal{N}^+(x_1) \cap \mathcal{N}^+(x_2) \cap \ldots \mathcal{N}^+(x_r)| \leq k$. As $\vec{G}$ is symmetric therefore, $|\mathcal{N}([x_1] \cap \mathcal{N}[x_2] \cap \ldots \mathcal{N}[x_r]| \leq k$ in $\mathcal{U}(\vec{G})$. Hence, $\lambda'_{kc}(E) = 0$ and so $\lambda_{kc}(E) = \lambda'_{kc}(E)$ for all $x_1, x_2, \ldots, x_r \in X$.

Case 2: If for some $x_1, x_2, \ldots, x_r \in X$, $\lambda_{kc}(E) > 0$ in $\mathcal{C}_k(\vec{G})$, then $|\mathcal{N}^+(x_1) \cap \mathcal{N}^+(x_2) \cap \ldots \mathcal{N}^+(x_r)| > k$. As $\vec{G}$ is symmetric fuzzy digraph and has loops at every vertex; therefore, $|\mathcal{N}([x_1] \cap \mathcal{N}[x_2] \cap \ldots \mathcal{N}[x_r]| > k$. Hence, $\lambda_{kc}(xy) = \lambda'_{kc}(xy)$. As $x_1, x_2, \ldots, x_r$ were taken to be arbitrary, the result holds for all hyperedges $E = \{x_1, x_2, \ldots, x_r\}$ of $\mathcal{C}_k(\vec{G})$. □

3. Applications of Fuzzy Competition Hypergraphs

In this section, we present several applications of fuzzy competition hypergraphs in food webs, business marketing and social networks.

3.1. *Identifying Predator–Prey Relations in Ecosystems*

We now present application of fuzzy competition hypergraphs in order to describe the interconnection of food chains between species, flow of energy and predator–prey relationship in ecosystems. The strength of competition between species represents the competition for food and common preys of species. We will discuss a method to give a description of species relationship, danger to the population growth rate of certain species, powerful animals in ecological niches and lack of food for weak animals.

Competition graphs arose in connection with an application in food webs. However, in some cases, competition hypergraphs provide a detailed description of predator–prey relations than competition graphs. In a competition hypergraph, it is assumed that vertices are defined clearly but in real-world problems, vertices are not defined precisely. As an example, species may be of different type like vegetarian, non-vegetarian, weak or strong.

Fuzzy food webs can be used to describe the combination of food chains that are interconnected by a fuzzy network of food relationship. There are many interesting variations of the notion of fuzzy competition hypergraph in ecological interpretation. For instance, two species may have a common prey *(fuzzy competition hypergraph)*, a common enemy (*fuzzy common enemy hypergraph*), both common prey and common enemy (*fuzzy competition common enemy hypergraph*), and either a common prey or a common enemy (*fuzzy niche hypergraph*). We now discuss a type of fuzzy competition hypergraph in which species have common enemies known as *fuzzy common enemy hypergraph*.

Let $\vec{G} = (\mu, \vec{\lambda})$ be a fuzzy food web. The fuzzy common enemy hypergraph $\mathcal{CH}(\vec{G}) = (\mu, \lambda_c)$ has the same vertex set as $\vec{G}$ and there is a hyperedge consisting of vertices $x_1, x_2, \ldots, x_s$ if $\mathcal{N}^+(x_1) \cap \mathcal{N}^+(x_2) \cap \ldots \cap \mathcal{N}^+(x_s) \neq \emptyset$. The degree of membership of hyperedge $E = \{x_1, x_2, \ldots, x_s\}$ is defined as

$$\lambda_c(E) = [\mu(x_1) \wedge \mu(x_2) \wedge \ldots \wedge \mu(x_s)] \times h(\mathcal{N}^+(x_1) \cap \mathcal{N}^+(x_2) \cap \ldots \cap \mathcal{N}^+(x_s)).$$

The strength of common enemies between species can be calculated using Algorithm A3. Consider the example of a fuzzy food web of 13 species giraffe, lion, vulture, rhinoceros, African skunk, fiscal shrike, grasshopper, baboon, leopard, snake, caracal, mouse and impala. The degree of membership of each species represents the species' ability of resource defence. The degree of membership of each directed edge represents the strength to which the prey is harmful for the predator. The fuzzy food web is shown in Figure 12. The directed edge between the giraffe and the lion shows that the giraffe is eaten by the lion and similarly.

The degree of membership of the lion is 0.9, which shows that the lion has 90% ability of resource defence, i.e., it can defend itself against other animals as well as survive many days if the lion does not find any food. The directed edge between giraffe and lion has degree of membership 0.25, which represents that the giraffe is 25% harmful for the lion because a giraffe can kill a lion with its long legs. This is an acyclic fuzzy digraph. The fuzzy out neighbourhoods are given in Table 7.

The fuzzy common-enemy hypergraph is shown in Figure 13. The hyperedges in Figure 13 show that there are common enemies between giraffe and rhinoceros, rhinoceros, African skunk and leopard, grasshopper and snake, mouse and impala, and baboon and impala. The membership value of each hyperedge represents the degree of common enemies among the species.

The hyperedge {impala, baboon} has a maximum degree of membership, which shows that the impala and the baboon have the largest number of common enemies, whereas the mouse and the impala have the least number of common enemies.

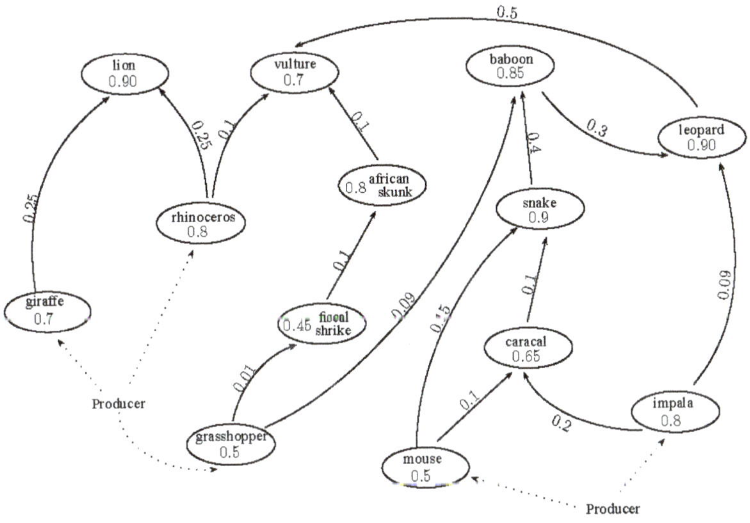

Figure 12. Fuzzy food web.

Table 7. Fuzzy out neighbourhoods of vertices.

Species	$\mathcal{N}^+(u)$: u is a specie
giraffe	$\{(\text{lion}, 0.25)\}$
lion	$\emptyset$
rhinoceros	$\{(\text{lion}, 0.25), (\text{vulture}, 0.1)\}$
vulture	$\emptyset$
African skunk	$\{(\text{vulture}, 0.1)\}$
fiscal shrike	$\{(\text{African skunk}, 0.1)\}$
grasshopper	$\{(\text{fiscal shrike}, 0.01), (\text{baboon}, 0.09)\}$
baboon	$\{(\text{leopard}, 0.3)\}$
leopard	$\{(\text{vulture}, 0.5)\}$
snake	$\{(\text{baboon}, 0.4)\}$
caracal	$\{(\text{snake}, 0.1)\}$
mouse	$\{(\text{caracal}, 0.1), (\text{snake}, 0.15)\}$
impala	$\{(\text{caracal}, 0.2), (\text{leopard}, 0.09)\}$

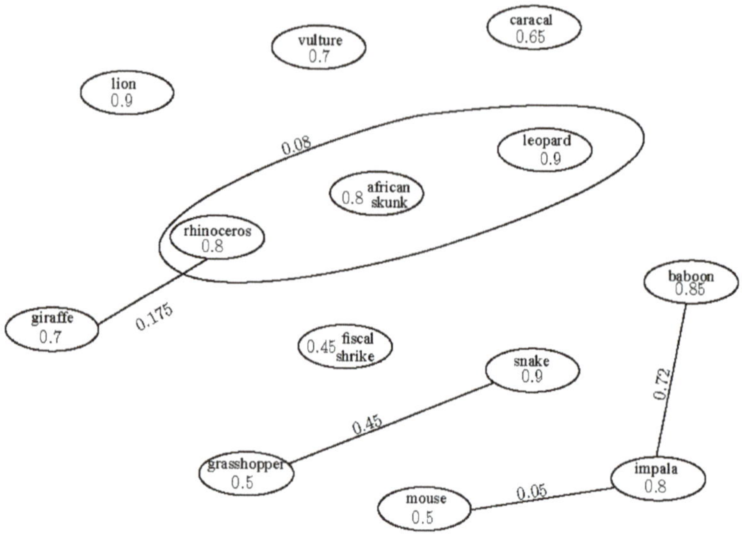

Figure 13. Fuzzy common enemy hypergraph.

3.2. *Identifying Competitors in the Business Market*

Fuzzy competition hypergraphs are a key approach to studying the competition, profit and loss, market power and rivalry among buyers and sellers using fuzziness in hypergraphical structures. We now discuss a method to study the business competition for power and profit, success and business failure, and demanding products in market.

In the business market, there are competitive rivalries among companies that are endeavoring to increase the demand and profit of their product. More than one company in the market sells identical products. Since various companies regularly market identical products, every company wants to attract a consumer's attention to its product. There is always a competitive situation in the business market. Hypergraph theory is a key approach to studying the competitive behavior of buyers and sellers using structures of hypergraphs. In some cases, these structures do not study the level of competition, profit and loss between the companies. As an example, companies may have different reputations in the market according to market power and rivalry. These are fuzzy concepts and motivates the necessity of fuzzy competition hypergraphs. The competition among companies can be studied using a fuzzy competition hypergraph known as *fuzzy enmity hypergraph*.

We present a method for calculating the strength of competition of companies in the following Algorithm 1.

Algorithm 1: Business competition hypegraph.

1. Input the adjacency matrix $[x_{ij}]_{n\times n}$ of bipolar fuzzy digraph $\vec{G} = (C, \vec{D})$ of n companies $x_1, x_2, \ldots, x_n$.
2. Construct the table of fuzzy out neighbourhoods of all the companies.
3. Construct fuzzy competition hypergraph using Algorithm A3.
4. **do** i from $1 \to n$
5. Calculate the degree of each vertex as, $S(x_i) = \sum_{x_i \in E} \lambda_c(E)$ where E is a hyperedge in fuzzy enmity hypergraph.
6. **end do**
7. $S(x_i)$ denotes the strength of competition of each company $x_i, 1 \leq i \leq n$.

Consider the example of a marketing competition between seven companies DEL, CB, HW, AK, LR, RP, SONY, RA, LR, three retailers, one retailer outlet and one multinational brand as shown in Figure 14.

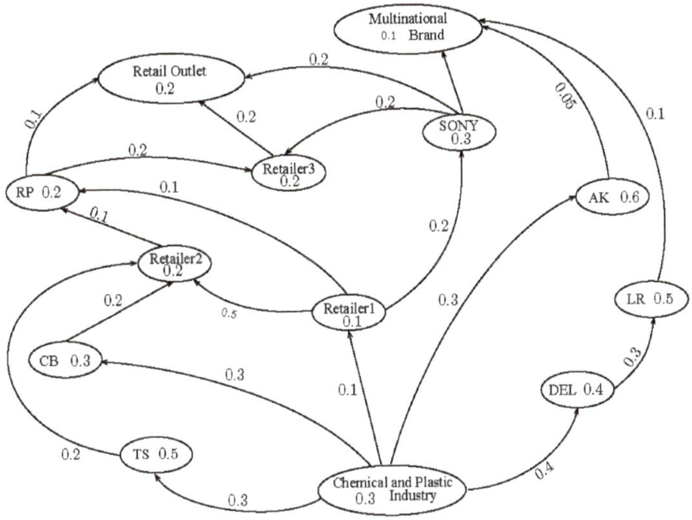

Figure 14. Fuzzy marketing digraph.

The vertices represent companies, retailers, outlets and brands. The degree of membership of each vertex represents the strength of rivalry (aggression) of each company in the market. The degree of membership of each directed edge $\overrightarrow{xy}$ represents the degree of rejectability of company's x product by company y. The strength of competition of each company can be discussed using fuzzy competition hypergraph known as *fuzzy enmity hypergraph*. The fuzzy out neighbouhoods are calculated in Table 8.

Table 8. Fuzzy out neighbourhoods of companies.

Company	$\mathcal{N}^+(u)$: u Is a Company
chemical and plastic industries	$\{(\text{DEL}, 0.4), (\text{AK}, 0.3), (\text{Retailer1}, 0.1), (\text{CB}, 0.3), (\text{TS}, 0.3)\}$
DEL	$\{(\text{LR}, 0.3)\}$
AK	$\{(\text{Multinational Brand}, 0.05)\}$
LR	$\{(\text{Multinational Brand}, 0.1)\}$
Retailer1	$\{(\text{SONY}, 0.2), (\text{RP}, 0.1), (\text{Retailer2}, 0.5)\}$
CB	$\{(\text{Retailer2}, 0.2)\}$
TS	$\{(\text{Retailer2}, 0.2)\}$
Retailer2	$\{(\text{RP}, 0.1)\}$
SONY	$\{(\text{Retailer3}, 0.2), (\text{R. Outlet}, 0.2), (\text{M. Brand}, 0.1)\}$
Retailer3	$\{(\text{R.Outlet}, 0.2)\}$
RP	$\{(\text{Retailer3}, 0.2), (\text{R. Outlet}, 0.1)\}$
M. Brand	$\emptyset$
R. Outlet	$\emptyset$

The fuzzy enmity hypergraph of Figure 14 is shown in Figure 15. The degree of membership of each hyperedge shows the strength of rivalry between the companies.

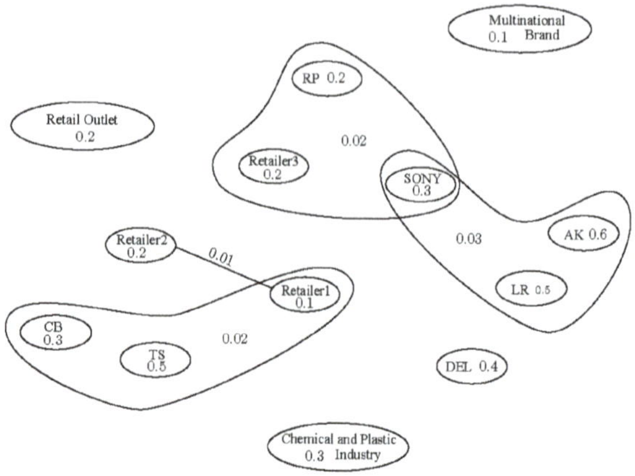

Figure 15. Fuzzy competition hypergraph.

The strength of rivalry of each company is calculated in Table 9, which shows its enmity value within the business market. Table 9 shows that SONY is the biggest rival company among other companies.

Table 9. Strength of rivalry between companies.

Company	Strength of Rivalry
LR	0.03
AK	0.03
SONY	0.05
Retailer3	0.02
RP	0.02
Retailer2	0.01
Retailer1	0.03
CB	0.02
TS	0.02

3.3. *Finding Influential Communities in a Social Network*

Fuzzy competition hypergraphs have a wide range of applications in decision-making problems and decision support systems based on social networking. To elaborate on the necessity of the idea discussed in this paper, we apply the notion of fuzzy competition hypergraphs to study the influence, centrality, socialism and proactiveness of human beings in any social network.

Social competition is a widespread mechanism to figure out a best-suited group economically, politically or educationally. Social competition occurs when individual's opinions, decisions and behaviors are influenced by others. Graph theory is a conceptual framework to study and analyze the units that are intensely or frequently connected in a network. Fuzzy hypergraphs can be used to study the influence and competition between objects more precisely. The social influence and conflict between different communities can be studied using a fuzzy competition hypergraph known as *fuzzy influence hypergraph*.

The fuzzy influence hypergraph $G = (\mu, \lambda_c)$ has the same set of vertices as $\vec{G}$ and there is a hyperedge consisting of vertices $x_1, x_2, \ldots, x_r$ if $\mathcal{N}^-(x_1) \cap \mathcal{N}^-(x_2) \cap \ldots \cap \mathcal{N}^-(x_r) \neq \emptyset$. The degree of membership of hyperedge $E = \{x_1, x_2, \ldots, x_r\}$ is defined as

$$\lambda_c(E) = [\mu(x_1) \wedge \mu(x_2) \wedge \ldots \wedge \mu(x_r)] \times h(\mathcal{N}^+(x_1) \cap \mathcal{N}^+(x_2) \cap \ldots \cap \mathcal{N}^+(x_r)).$$

The strength of influence between different objects in a fuzzy influence hypergraph can be calculated by the method presented in Algorithm 2. The complexity of algorithm is $O(n^2)$.

Algorithm 2: Fuzzy influence hypergraph.

1. Input the adjacency matrix $[x_{ij}]_{n\times n}$ of fuzzy digraph $\vec{G} = (C, \vec{D})$ of n families $x_1, x_2, \ldots, x_n$.
2. Using fuzzy in neighbourhoods, construct the fuzzy influence hypergraph following Algorithm A3.
3. **do** i from $1 \to n$
4. If x_i belongs to the hyperedge E in fuzzy influence hypergraph then calculate the degree of each vertex x_i as,
$$deg(x_i) = \sum_{x_i \in E} \lambda_c(E) \text{ and } A_i = \sum_{x_i \in E} (|E| - 1).$$
5. **end do**
6. **do** i from $1 \to n$
7. If $A_i > 1$ then calculate the degree of influence of each vertex x_i as,
$$S(x_i) = \frac{\deg(x_i)}{A_i}.$$
8. **end do**

Consider a fuzzy social digraph of Florientine trading families Peruzzi, Lambertes, Bischeri, Strozzi, Guadagni, Tornabuon, Castellan, Ridolfi, Albizzi, Barbadori, Medici, Acciaiuol, Salviati, Ginori and Pazzi. The vertices in a fuzzy network represent the name of trading families. The degree of membership of each family represents the strength of centrality in that network. The directed edge $\vec{xy}$ indicates that the family x is influenced by y. The degree of membership of each directed edge indicates to what extent the opinions and suggestions of one family influence the other. The degree of membership of Medici is 0.9, which shows that Medici has a 90% central position in a trading network. The degree of membership between Redolfi and Medici is 0.6, which indicates that Redolfi follows 60% of the suggestions of Medici. The fuzzy social digraph is shown in Figure 16.

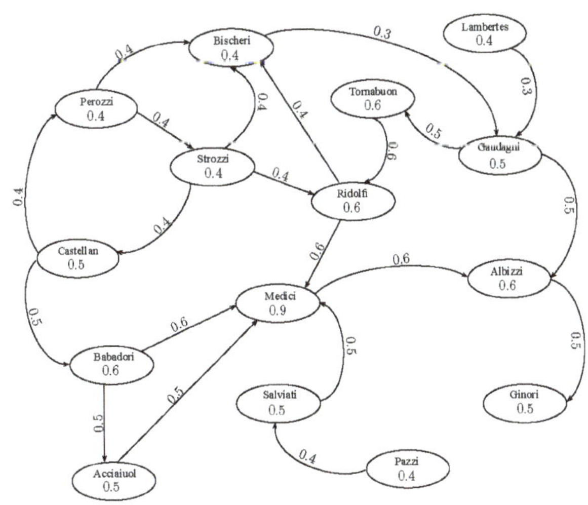

Figure 16. Fuzzy social digraph.

To find the most influential family in this fuzzy network, we construct its fuzzy influence hypergraph. The fuzzy in neighbourhoods are given in Table 10.

Table 10. Fuzzy in neighbourhoods of all vertices in social networks.

Family	$\mathcal{N}^-$(Family)	Family	$\mathcal{N}^-$(Family)
Acciaiuol	$\{(\text{Babadori}, 0.5,)\}$	Pazzi	$\emptyset$
Ginori	$\{(\text{Albizzi}, 0.5)\}$	Salviati	$\{(\text{Pazzi}, 0.4)\}$
Babadori	$\{(\text{Castellan}, 0.5)\}$	Castellan	$\{(\text{Strozzi}, 0.4)\}$
Tornabuon	$\{(\text{Gaudagni}, 0.5)\}$	Perozzi	$\{(\text{Castellan}, 0.5)\}$
Lambertes	$\emptyset$	Strozzi	$\{(\text{Perozzi}, 0.4)\}$
Medici	$\{(\text{Babadori}, 0.6), (\text{Acciaiuol}, 0.5), (\text{Salviati}, 0.5), (\text{Ridolfi}, 0.6)\}$		
Bischeri	$\{(\text{Perozzi}, 0.4), (\text{Strozzi}, 0.4), (\text{Redolfi}, 0.4)\}$		
Albizzi	$\{(\text{Medici}, 0.6), (\text{Gaudagni}, 0.5)\}$		
Redolfi	$\{(\text{Strozzi}, 0.4), (\text{Tornabuon}, 0.6)\}$		
Gaudgani	$\{(\text{Bischeri}, 0.3), (\text{Lambertes}, 0.3)\}$		

The fuzzy influence hypergraph is shown in Figure 17. The degree of membership of each hyperedge shows the strength of social competition between families to influence the other trading families. The strength of competition of vertices using Algorithm 2 is calculated in Table 11, where $S(x)$ represents the strength to which each trading family influences the other families. Table 11 shows that Acciaiuol and Medici are most influential families in the network.

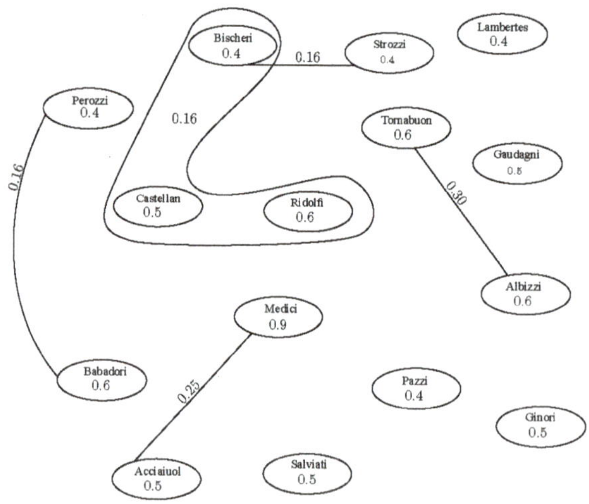

Figure 17. Fuzzy influence hypergraph.

Table 11. Degree of influence of vertices.

x	$deg(x)$	$S(x)$	x	$deg(x)$	$S(x)$
Acciaiuol	0.25	0.25	Medici	0.25	0.25
Babadori	0.16	0.16	Perozzi	0.16	0.16
Castellan	0.16	0.08	Redolfii	0.16	0.08
Strozzi	0.16	0.16	Besceri	0.32	0.12

A View of Fuzzy Competition Hypergraphs in Comparison with Fuzzy Competition Graphs

The concept of fuzzy competition graphs presented in [16,17] can be utilized successfully in different domains of applications. In the existing methods, we usually consider fuzziness in pairwise competition and conflicts between objects. However, in these representations, we miss some information about whether there is a conflict or a relation among three or more objects. For example, Figure 15 shows the strong competition for profit among SONY, LR and AK. However, if we draw the fuzzy competition graph of Figure 14, we cannot discuss the group-wise conflict among companies. Sometimes, we are not only interested in pair-wise relations but also in group-wise conflicts, influence and relations. The novel notion of fuzzy competition hypergraphs are a mathematical tool to overcome this difficulty. We have presented different methods for solving decision-making problems. These methods not only generalize the existing ones but also give better results regarding uncertainty.

4. Conclusions

In this research paper, we have applied the powerful technique of fuzziness to generalize the notion of competition hypergraphs and fuzzy competition graphs. Fuzzy models give more precision, flexibility and compatibility to the system as compared to the crisp models. We have mainly discussed the construction methods of various types of fuzzy hypergraphs using open and closed neighbourhoods, strong hyperedges, kth strength of competition and symmetric properties. We have also established strong relations among fuzzy k-competition hypergraphs and underlying fuzzy graphs along with fuzzy digraphs having loops at vertices. We have applied fuzzy competition hypergraphs to real-world problems for representation of fuzziness in different domains including identification of predator–prey relations, competitions in the business market and social networks which motivate the idea introduced in this research paper. We have designed certain algorithms to solve these decision-making problems.

Author Contributions: M.S., M.A. and N.O.A. conceived and designed the experiments; M.S. and N.O.A. wrote the paper.

Conflicts of Interest: The authors declare no conflict of interest.

Appendix

Algorithm A1: Method for construction of fuzzy row hypergraph

Begin
  Input the fuzzy set μ on set of vertices $X = \{x_1, x_2, \ldots, x_n\}$.
  Input the adjacency matrix $A = [x_{ij}]_{n\times n}$ of fuzzy digraph $\vec{G} = (\mu, \vec{\lambda})$ such that $\vec{\lambda}(x_i x_j) = x_{ij}$ as shown in Table A1.
do j from $1 \to n$
  Take a vertex x_j from first jth column.
  value1 $= \infty$, value2 $= \infty$, num $= 0$
  do i from $1 \to n$
    if $(x_{ij} > 0)$ **then**
      x_i belongs to the hyperedge E_j.
      num $=$ num $+ 1$
      value1 $=$ value1 $\wedge \mu(x_i)$
      value2 $=$ value2 $\wedge x_{ij}$
    end if
  end do
  if (num > 1) **then**
    $\lambda_r(E_j) =$ value1 $\times$ value2, where E_j is a hyperedge.
  end if
end do
If for some j, $supp(E_j) = supp(E_k), k \in \{j+1, j+2, \ldots, n\}$ then,

$$\lambda_r(E_j) = \max\{\lambda_r(E_j), \lambda_r(E_k), \ldots\}.$$

Table A1. Adjacency matrix.

A	x_1	x_2	$\ldots$	x_n
x_1	x_{11}	x_{12}	$\ldots$	x_{1n}
x_2	x_{21}	x_{22}	$\ldots$	x_{2n}
$\vdots$	$\vdots$	$\vdots$	$\ldots$	$\vdots$
x_n	x_{n1}	x_{n2}	$\ldots$	x_{nn}

Algorithm A2: Method for construction of fuzzy column hypergraph

1. Begin
2. Follow steps 2 and 3 of Algorithm A1.
3. **do** i from $1 \to n$
4. Take a vertex x_i from first ith row.
5. value1 $= \infty$, value2 $= \infty$, num $= 0$
6. **do** j from $1 \to n$
7. **if** $(x_{ij} > 0)$ **then**
8. x_j belongs to the hyperedge E_i.
9. num $=$ num $+ 1$
10. value1 $=$ value1 $\wedge \mu(x_j)$
11. value2 $=$ value2 $\wedge x_{ij}$
12. **end if**
13. **end do**
14. **if** (num > 1) **then**
15. $\lambda_{cl}(E_i) =$ value1 $\times$ value2, where E_i is a hyperedge.
16. **end if**
17. **end do**
18. If for some i, $supp(E_i) = supp(E_k), k \in \{j+1, j+2, \ldots, n\}$ then,

$$\lambda_{cl}(E_i) = \max\{\lambda_{cl}(E_j), \lambda_{cl}(E_k), \ldots\}.$$

Algorithm A3: Construction of fuzzy competition hypergraph

1. Begin
2. Input the adjacency matrix $A = [x_{ij}]_{n \times n}$ of a fuzzy digraph $\vec{G}$.
3. Define a relation $f : X \to X$ by $f(x_i) = x_j$, if $x_{ij} > 0$.
4. **do** i from $1 \to n$
5. **do** j from $1 \to n$
6. If $x_{ij} > 0$ then (x_j, x_{ij}) belongs to the fuzzy out neighbourhood $\mathcal{N}^+(x_i)$.
7. **end do**
8. **end do**
9. Compute the family of sets $\mathcal{S} = \{E_i = f^{-1}(x_i) : |f^{-1}(x_i)| \geq 2, x_i \in X\}$ where $E_i = \{x_{i_1}, x_{i_2}, \ldots, x_{i_r}\}$ is a hyperedge of $\mathcal{CH}(\vec{G})$.
10. For each hyperedge $E_i \in \mathcal{S}$, calculate the degree of membership of E_i as,

$$\lambda_c(E_i) = [\mu(x_{i_1}) \wedge \mu(x_{i_2}) \wedge \ldots \wedge \mu(x_{i_r})] \times h\left(\mathcal{N}^+(x_{i_1}) \cap \mathcal{N}^+(x_{i_2}) \cap \ldots \cap \mathcal{N}^+(x_{i_r})\right).$$

Algorithm A4: Construction of fuzzy double competition hypergraph

1. Input the adjacency matrix $A = [x_{ij}]_{n\times n}$ of a fuzzy digraph $\vec{G}$.
2. Define a relation $f : X \to X$ by $f(x_i) = x_j$, if $x_{ij} > 0$.
3. Compute the family of sets $\mathcal{S} = \{E_i = f^{-1}(x_i) : |f^{-1}(x_i)| \geq 2, x_i \in X\}$ where $E_i = \{x_{i_1}, x_{i_2}, \ldots, x_{i_r}\}$.
4. If $\mathcal{N}^+(x_{i_1}) \cap \mathcal{N}^+(x_{i_2}) \cap \ldots \cap \mathcal{N}^+(x_{i_r})$ and $\mathcal{N}^-(x_{i_1}) \cap \mathcal{N}^-(x_{i_2}) \cap \ldots \cap \mathcal{N}^-(x_{i_r})$ are non-empty then $E_i = \{x_{i_1}, x_{i_2}, \ldots, x_{i_r}\}$ is a hyperedge of $\mathcal{DCH}(\vec{G})$.
5. For each hyperedge $E_i \in \mathcal{S}$, calculate the degree of membership of hyperedge E_i,
$\lambda_d(E_i) = [\mu(x_{i_1}) \wedge \mu(x_{i_2}) \wedge \ldots \wedge \mu(x_{i_r})] \times h\left(\mathcal{N}^+(x_{i_1}) \cap \mathcal{N}^+(x_{i_2}) \cap \ldots \cap \mathcal{N}^+(x_{i_r})\right) \wedge$
$h\left(\mathcal{N}^-(x_{i_1}) \cap \mathcal{N}^-(x_{i_2}) \cap \ldots \cap \mathcal{N}^-(x_{i_r})\right).$

References

1. Sonntag, M.; Teichert, H.M. Competition hypergraphs of digraphs with certain properties. II. Hamiltonicity. *Discuss. Math. Graph Theory* **2008**, *28*, 23–34. [CrossRef]
2. Zadeh, L.A. Fuzzy sets. *Inf. Control* **1965**, *8*, 338–353. [CrossRef]
3. Zadeh, L.A. Similarity relations and fuzzy orderings. *Inf. Sci.* **1971**, *3*, 177–200. [CrossRef]
4. Kaufmann, A. *Introduction la Thorie des Sous-Ensembles Flous Lusage des Ingnieurs (Fuzzy Sets Theory)*; Masson: Paris, France, 1975.
5. Rosenfeld, A. *Fuzzy Graphs*; Fuzzy Sets and Their Applications; Academic Press: New York, NY, USA, 1975; pp. 77–95.
6. Lee-kwang, H.; Lee, K.-M. Fuzzy hypergraph and fuzzy partition. *IEEE Trans. Syst. Man Cybern.* **1995**, *25*, 96–102. [CrossRef]
7. Goetschel, R.H.; Craine, W.L.; Voxman, W. Fuzzy transversals of fuzzy hypergraphs. *Fuzzy Sets Syst.* **1996**, *84*, 235–254. [CrossRef]
8. Goetschel, R.H. Fuzzy colorings of fuzzy hypergraphs. *Fuzzy Sets Syst.* **1998**, *94*, 185–204. [CrossRef]
9. Chen, S.M. Interval-valued fuzzy hypergraph and fuzzy partition. *IEEE Trans. Syst. Man Cybern.* **1997**, *27*, 725–733. [CrossRef] [PubMed]
10. Parvathi, R.; Thilagavathi, S.; Karunambigai, M.G. Intuitionistic fuzzy hypergraphs. *Cybern. Inf. Technol.* **2009**, *9*, 46–48.
11. Akram, M.; Dudek, W.A. Intuitionistic fuzzy hypergraphs with applications. *Inf. Sci.* **2013**, *218*, 182–193. [CrossRef]
12. Akram, M.; Luqman, A. Certain concepts of bipolar fuzzy directed hypergraphs. *Mathematics* **2017**, *5*, 17. [CrossRef]
13. Akram, M.; Luqman, A. Intuitionistic single-valued neutrosophic hypergraphs. *OPSEARCH* **2017**, *54*, 799–815. [CrossRef]
14. Akram, M.; Luqman, A. Bipolar neutrosophic hypergraphs with applications. *J. Intell. Fuzzy Syst.* **2017**, *33*, 1699–1713. [CrossRef]
15. Akram, M.; Shahzadi, G. Hypergraphs in m-polar fuzzy environment. *Mathematics* **2018**, *6*, 28. math6020028. [CrossRef]
16. Samanta, S.; Pal, M. Fuzzy k-competition and p-competition graphs. *Fuzzy Inf. Eng.* **2013**, *2*, 191–204. [CrossRef]
17. Samanta, S., Akram, M.; Pal, M. m-step fuzzy competition graphs. *J. Appl. Math. Comput.* **2015**, *47*, 461–472.
18. Al-shehri, N.O.; Akram, M. Bipolar fuzzy competition graphs. *Ars Comb.* **2015**, *121*, 385–402.
19. Sarwar, M.; Akram, M. Novel concepts of bipolar fuzzy competition graphs. *J. Appl. Math. Comput.* **2017**, *54*, 511–547. [CrossRef]
20. Akram, M.; Nasir, M. Certain competition graphs based on intuitionistic neutrosophic environment. *Information* **2017**, *8*, 132. [CrossRef]
21. Akram, M.; Siddique, S. Neutrosophic competition graphs with applications. *J. Intell. Fuzzy Syst.* **2017**, *33*, 921–935. [CrossRef]
22. Akram, M.; Sarwar, M. New applications of m-polar fuzzy competition graphs. *New Math. Nat. Comput.* **2018**, *14*, 249–276. [CrossRef]

23. Berge, C. *Graphs and Hypergraphs*; North-Holland Publishing Company: Amsterdam, The Nertherland, 1973; Volume 7.
24. Harary, F. *Graph Theory*; Addison-Wesley: Reading, MA, USA, 1972.
25. Mordeson, J.N.; Nair, P.S. Successor and source of (fuzzy)finite state machines and (fuzzy)directed graphs. *Inf. Sci.* **1996**, *95*, 113–124.
26. Samanta, S.; Pal, M. Bipolar fuzzy hypergraphs. *Int. J. Fuzzy Log. Syst.* **2012**, *2*, 17–28. [CrossRef]
27. Sarwar, M.; Akram, M. New applications of m-polar fuzzy matroids. *Symmetry* **2018**, *9*, 319. sym9120319. [CrossRef]
28. Zhan, J.; Malik, H.M.; Akram, M. Novel decision-making algorithms based on intuitionistic fuzzy rough environment. *Int. J. Mach. Learn. Cybern.* **2018**. [CrossRef]
29. Radhamani, C.; Radhika, C. Isomorphism on fuzzy hypergraphs. *IOSR J. Math.* **2012**, *2*, 24–31.
30. Akram, M.; Sarwar, M. Novel applications of m-polar fuzzy competition graphs in decision support system. *Neural Comput. Appl.* **2018**. [CrossRef]
31. Bhattacharya, P. Some remarks on fuzzy graphs. *Pattern Recognit. Lett.* **1987**, 6, 297–302. [CrossRef]
32. Chen, S.M. A new approach to handling fuzzy decision-making problems. *IEEE Trans. Syst. Man Cybern.* **1988**, *18*, 1012–1016. [CrossRef]
33. Cohen, J.E. *Interval Graphs and Food Webs: A Finding and a Problem, Document 17696-PR*; RAND Coporation: Santa Monica, CA, USA, 1968.
34. Klir, G.; Yuan, B. *Fuzzy Sets and Fuzzy Logic*; Prentice Hall: Upper Saddle River, NJ, USA, 1995; Volume 4.
35. Mordeson, J.N.; Nair, P.S. *Fuzzy Graphs and Fuzzy Hypergraphs*, 2nd ed.; Springer: Heidelberg/Berlin, Germany, 2001.
36. Sonntag, M.; Teichert, H.M. Competition hypergraphs. *Discret. Appl. Math.* **2004**, *143*, 324–329. [CrossRef]

symmetry

MDPI

Article

Symmetric Triangular Interval Type-2 Intuitionistic Fuzzy Sets with Their Applications in Multi Criteria Decision Making

Sukhveer Singh and Harish Garg *

School of Mathematics, Thapar Institute of Engineering & Technology (Deemed University), Patiala 147004, India; sukhveer.singh@thapar.edu

* Correspondence: harish.garg@thapar.edu; Tel.: +91-86990-31147

Received: 26 August 2018; Accepted: 11 September 2018; Published: 14 September 2018

Abstract: Type-2 intuitionistic fuzzy set (T2IFS) is a powerful and important extension of the classical fuzzy set, intuitionistic fuzzy set to measure the vagueness and uncertainty. In a practical decision-making process, there always occurs an inter-relationship among the multi-input arguments. To deal with this point, the motivation of the present paper is to develop some new interval type-2 (IT2) intuitionistic fuzzy aggregation operators which can consider the multi interaction between the input argument. To achieve it, we define a symmetric triangular interval T2IFS (TIT2IFS), its operations, Hamy mean (HM) operator to aggregate the preference of the symmetric TIT2IFS and then shows its applicability through a multi-criteria decision making (MCDM). Several enviable properties and particular cases together with following different parameter values of this operator are calculated in detail. At last a numerical illustration is to given to exemplify the practicability of the proposed technique and a comparative analysis is analyzed in detail.

Keywords: type-2 fuzzy set; multi criteria decision-making; triangular interval type-2 intuitionistic fuzzy set; Hamy mean; aggregation operator

1. Introduction

Multiple criteria decision making (MCDM) is a hot research topic in the modern decision-making process to find the most suitable alternative(s) from the available ones. In this process, all the alternatives are to be evaluated under several attributes by both qualitatively and quantitatively [1,2]. Traditionally, the researchers offer his/her preference information towards the alternatives by using the crisp real numbers only. However, due to lack of knowledge, a time pressure, and other unavoidable factors, it is very difficult if not impossible to express the information precisely. Therefore, to handle the incomplete or incorrect information, the theory of fuzzy set (FS) also called as a type-1 fuzzy set (T1FS) [3] and its extensions as an intuitionistic FS (IFS) [4], type-2 FS (T2FS) [5] are widely used. Under these environments, authors have put forth the different techniques to solve the MCDM problems. For instance, geometric aggregation operators (AOs) for different intuitionistic fuzzy numbers (IFNs) are developed by Xu and Yager [6]. Garg [7,8] presented some Einstein norm based AOs for IFNs. Zhao et al. [9] presented some generalized AOs. Kaur and Garg [10] presented some generalized AOs using t-norm operations for cubic IFS information. However, apart from these, a comprehensive overview of the different approaches for solving the decision making (DM) problems by using aggregation operator (AOs) [11–21], information measures (IMs) [22–24] are summarized in these papers and their references.

In these existing works, authors have investigated the problem by taking quantitative environment to access the alternatives. However, not all the alternatives are accessed in terms of quantitative.

For this, there exists the concept of qualitative assessment in terms of linguistic variables/terms (LVs/LTs) [25,26]. By taking the advantages of LTs, Zhang [27] presented the linguistic IF (LIF) AOs to aggregate the LIF numbers. Chen et al. [28] presented an approach to solving the MCDM problem under LIFS environment. Garg and Kumar [29] presented AOs for LIF numbers (LIFNs) by using set pair analysis theory. Garg and Kumar [30] presented new possibility degree measure for LIFNs and an AO to aggregate the different LIFNs to solve MCDM problems. In many practical problems, it is not easy for any decision maker (DM) to discover an exact membership function of an FS corresponding to its element. To overthrow this limitation, type-2 fuzzy set (T2FS), an extension of T1FS, is applied to the model and is characterized by two functions: primary membership functions (PMF) and secondary membership function (SMF). Unfortunately, T2FSs are highly complex, it is troublesome for the DMs to implement it in the real situation; hence, their use is not yet widespread. To reduce the computational complexity, Interval type-2 fuzzy (IT2F) sets (IT2FSs) [31] is the most widely used in T2FSs. In past decades, many methods have been developed to extend the theory of MCDM under IT2FS environment. Chen et al. [32] built up an expanded QUALIFLEX strategy for taking care of DM issues in view of IT2FSs and gave a contextual analysis of medicinal basic leadership. Chen [33] built up an ELECTRE-base outranking strategy for decision-making problems using IT2FSs. Wu and Mendel [34] proposed a linguistic weighted average AOs to deal with analytical hierarchical process (AHP) process under IT2F environment. Qin and Liu [35] investigated a family of type-2 fuzzy AOs in light of Frank triangular norm and built up another way to deal with MCDM problems under the IT2FSs setting. Gong et al. [36] extended the generalized Bonferroni mean (GBM) operator to the trapezoidal IT2F environment. Apart from these, some other studies under T2FS environment are conducted which are summarized in [35–48].

In all these above AOs, researchers have described the information by considering the independent of argument assumptions during the aggregation. However, the interaction between the multi-input parameters have commonly occurred and thus, it is necessary to add their features into the process. In that direction, Bonferroni mean (BM) and generalized BM (GBM)-based operators are proposed by the researchers [49,50]. But from them, it has been observed that they have considered only two or three multi-parameter at a single time. However, they are unable to analyze the effect of the multi-input argument into one analysis. Furthermore, in BM and GBM, there is a need for two and three parameters from the irrational set during the process which increases the computational complexity. An alternative to BM operators, Hamy mean (HM) [51] or Maclaurin symmetric mean (MSM) or Muirhead mean (MM) operator has advantages of capturing the inter-relationship among the multiple input arguments. Qin [46] make a correlation between the HM and the MSM and conclude that the MSM is an instance of HM [16,17]. Garg and Nancy [52] develop MCDM method by prioritized MM aggregation operators. Additionally, the HM operator involves the parameter, which can provide more flexibility and robustness during the aggregation operator. The existing - arithmetic and geometric mean- operators can be easily deduced from the HM by setting a particular value to its parameter. Be that as it may, the HM just accomplished a couple of research results on the hypothesis and application of inequality [53,54]. Therefore, it is a means to study the AOs using the HM operator.

It is noted from the above studies that T2FS or IT2FS are examined by considering only the membership degree (MD) of an element. But in practical problems, it is sometimes not possible for a DM to give their preferences in terms of MD only as there may be some amount of hesitation also. For discussing this, a type-2 IFS (T2IFS) [39] has been introduced which simultaneously considers the MDs, non-membership degrees (NMDs) and the footprint of uncertainties (FOU) between them. Later on, due to the high complexity of T2IFS, Garg and Singh [55] introduced the concept of triangular interval T2IFS (TIT2IFS) has introduced by considering the MDs and NMDs as a triangular fuzzy number.

Based on the above analysis, we can know that the decision-making problems have become more tedious these days. So in order to make a better decision in terms of selecting the best alternative(s) for the MCDM problems, it is necessary to consider the various factors such as MDs, NMDs, FOU between the alternatives. By keeping the advantages of both the AOs and the TIT2IFS, it is necessary to extend the Hamy mean AOs to process the TIT2IFNs by using linguistic features of MDs and NMDs and hence to develop some MCDM methods. Until now, we have not seen any work based on the AOs used to aggregate the TIT2IFS information. Thus, keeping in mind the advantages of T2IFS and the multiple input interaction between the argument of HM operator, this paper has presented the concept of the symmetric TIT2IFS and their desired properties. These considerations have led us to consider the main objectives of this paper:

1. to propose the concept of the symmetric TIT2IFS (STIT2IFSs);
2. to propose some new AOs for STIT2IFSs under the linguistic intuitionistic features;
3. to develop an algorithm to solve the decision-making problems based on proposed operators;
4. to present some example to validate and compare the results.

To achieve the objective (1), we combine the T2IFSs and the symmetric triangular number to build a concept of the STIT2IFSs and studied their desired properties. To complete the objective (2), we presented the averaging AOs by using HM operations and named as symmetric triangular IT2IF HM averaging (STIT2IFHM) and weighted symmetric triangular IT2IF Hamy mean averaging (WSTIT2IFHM) operator for decision-making problems by keeping in mind the advantages of T2IFS and the multiple input interaction between the argument of HM operator. Several enviable properties and particular cases together with following different parameter values of this operator are calculated in detail. To cover the objective (3), we establish an MCDM method based on these proposed operators under the STIT2IFS environment where preferences related to each alternative is expressed in terms of linguistic STIT2IFNs. A numerical illustration is to given to exemplify the practicability of the proposed technique and a comparative analysis is analyzed in detail for fulfilling the Objective 4. Finally, the advantages of the proposed method in the state of the art are highlighted and discussed in detail.

The rest of the paper is organized as follows. In Section 2, some basic concepts on T2FS, IT2FS, T2IFS, and HM are reviewed briefly. In Section 3, we present the concept of the symmetric TIT2IF set and their desirable properties. Section 4 deals with new AOs based on HM operator to accommodate the STIT2IFN information and its special cases. In Section 5, we present an approach based on the WSTIT2IFHM operator to solve the MCDM problem. A practical example is discussed in Section 6 and some concluding remarks are summarized in Section 7.

2. Basic Concepts

In this section, we overview some basic definition of T2FSs, IT2FS and T2IFSs defined over the universal set X.

Definition 1 ([42]). *A type-2 fuzzy set (T2FS) $A \subseteq X$, defined as*

$$A = \{((x, u_A), \mu_A(x, u_A)) \mid x \in X, u_A \in j_x \subseteq [0,1]\} \tag{1}$$

where u_A denotes the primary membership function (PMF) of A, $\mu_A \in [0,1]$ is called as secondary membership function (SMF) $j_x \subseteq [0,1]$ is PMF of x.

Another equivalent expression for T2FS A is given as

$$A = \int_{x \in X} \frac{\mu_A(x)}{x} = \int_{x \in X} \left[\int_{u_A \in j_x} \frac{(f_x(u_A))}{u_A} \right] / x \tag{2}$$

Definition 2 ([20]). *The collection of all PMFs of T2FS is named as "footprint of uncertainty" (FOU), i.e., $FOU(A) = \bigcup_{x \in X} j_x$.*

However, because of high computational burden of T2FSs, researchers prefer using interval type-2 (IT2) fuzzy set (IT2FS) for real-world problems.

Definition 3 ([44])**.** *A T2FS transform into interval type-2 FS when the grades of all SMFs is equal to 1. Mathematically, an IT2FS A, with a membership function $\mu_A(x, u_A)$, may be expressed either as Equation (3) or as Equation (4) :*

$$A = \{(x, u_A), \mu_A(x, u_A) = 1 \mid \forall x \in X, \forall u_A \in j_x \subseteq [0,1]\} \quad (3)$$

$$A = \int_{x \in X} \int_{u_A \in j_x} 1/(x, u_A), j_x \subseteq [0,1] \quad (4)$$

Definition 4 ([44])**.** *An IT2 FS is normally described by a zone called as FOU, which is limited by two membership functions (MFs), known as lower MF (LMF) $\underline{\mu}_A(x, u_A)$ and the upper MF (UMF) $\overline{\mu}_A(x, u_A)$. That is FOU=[$\underline{\mu}_A(x, u_A), \overline{\mu}_A(x, u_A)$]. Figure 1 shows the graphical representation of IT2 fuzzy number (IT2 FN) with triangular MF shape.*

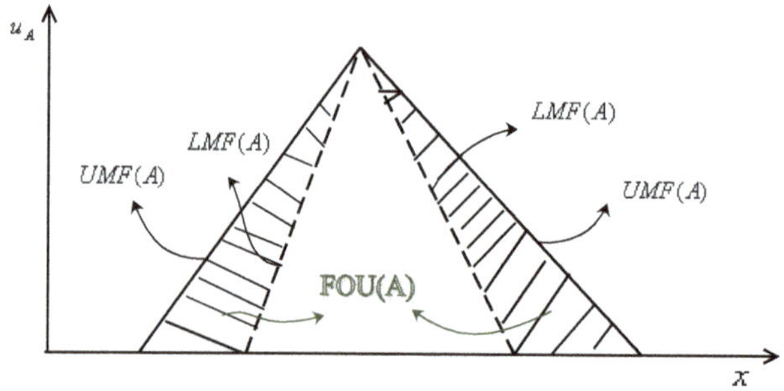

Figure 1. LMF (dashed), UMF (solid), FOU (shaded) for IT2FS *A*.

Definition 5 ([38,39])**.** *A T2IFS is a set of ordered pairs consisting of PMFs and SMFs of the element defined as*

$$A = \left\{ \langle (x, u_A, v_A), \mu_A(x, u_A), \nu_A(x, v_A) \rangle \mid x \in X, u_A \in j_x^1, v_A \in j_x^2 \right\} \quad (5)$$

where $u_A(v_A)$ represents the primary membership (non-membership) of A denoted by PMF(PNMF), $\mu_A(\nu_A)$ is secondary membership (non-membership) function of A, denoted by SMF (SNMF) and j_x^1, $j_x^2 \subseteq [0,1]$ are PMF and PNMF of x, respectively. When the SMFs $\mu_A(x, u_A) = 1$, and SNMF $\nu_A(x, v_A) = 0$, a T2IFS translates to an IT2 IFS.

Definition 6 ([55])**.** *An IT2 IFS, A, is described by a bounding functions of lower and upper membership and non-membership functions denoted by LMF, UMF, LNMF and UNMF defined as $\overline{\mu}_A$, $\underline{\mu}_A$ and $\overline{\nu}_A$, $\underline{\nu}_A$ with conditions: $0 \le \overline{\mu}_A + \underline{\nu}_A \le 1$ and $0 \le \underline{\mu}_A + \overline{\nu}_A \le 1$. The FOUs of an IT2IFS is illustrated in Figure 2 with triangular shape and defined mathematically as*

$$FOU(A) = \bigcup_{x \in X} \left[\underline{\mu}_A(x), \overline{\mu}_A(x), \underline{\nu}_A(x), \overline{\nu}_A(x) \right]$$

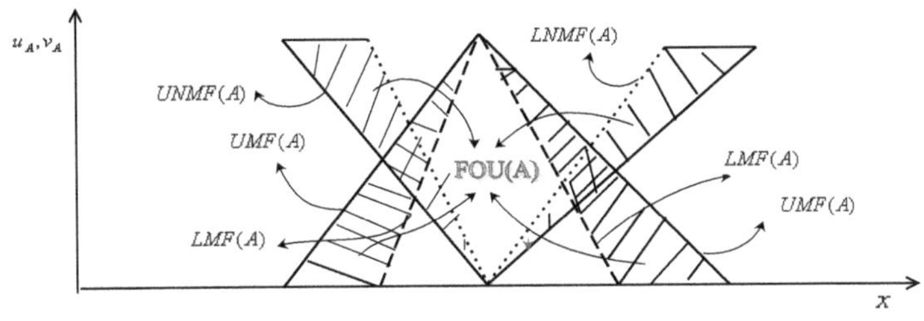

Figure 2. LMF (dashed), UMF (solid), LNMF (doted), UNMF (solid), FOU (shaded) for IT2IFS A.

Definition 7 ([51]). *For non-negative real numbers $x_i (i = 1, 2, \ldots, n)$, the Hamy mean (HM) is given as*

$$HM^{(k)}(x_1, x_2, \cdots, x_n) = \frac{\sum\limits_{\substack{1 \le i_1 < \\ \ldots < i_k \le n}} \left(\prod\limits_{j=1}^{k} x_{i_j} \right)^{\frac{1}{k}}}{\binom{n}{k}} \tag{6}$$

where k is the parameter, $\binom{n}{k} = \frac{n!}{k!(n-k)!}$ *and* $(i_1, i_2, \ldots, i_k)$ *crosses all the* $k-$*tuple mix of* $(1, 2, \ldots, n)$.

3. Proposed Symmetric Triangular Interval T2IFS

In this section, we present a symmetric triangular IT2IFS and characterize their fundamental operational laws.

Definition 8. *Let X be the universal set. A symmetric triangular interval T2 IFS (TIT2IFS) can be represented as follows:*

$$\alpha = \{(\zeta_\alpha(x), \varrho_\alpha(x), \varphi_\alpha(x), \varphi_\alpha^*(x), \vartheta_\alpha(x), \vartheta_\alpha^*(x)) \mid x \in X\} \tag{7}$$

where $\zeta_\alpha(x), \varrho_\alpha(x), \varphi_\alpha(x), \varphi_\alpha^*(x), \vartheta_\alpha(x), \vartheta_\alpha^*(x)$ *are the real numbers satisfying the inequalities,* $\zeta_\alpha(x) \ge \varrho_\alpha(x)$, $0 \le \varphi_\alpha(x) \le \varphi_\alpha^*(x) \le 1$, $0 \le \vartheta_\alpha^*(x) \le \vartheta_\alpha(x) \le 1$ *such that* $\varphi_\alpha(x) + \vartheta_\alpha(x) \le 1$ *and* $\varphi_\alpha^*(x) + \vartheta_\alpha^*(x) \le 1$.

For convenience, we represent this pair as $\alpha = (\zeta_\alpha, \varrho_\alpha, \varphi_\alpha, \varphi_\alpha^*, \vartheta_\alpha, \vartheta_\alpha^*)$ and called as symmetric triangular IT2 intuitionistic fuzzy (IT2IF) number (STIT2IFN) where $\zeta_\alpha \ge \varrho_\alpha$, $\varphi_\alpha + \vartheta_\alpha \le 1$, $\varphi_\alpha^* + \vartheta_\alpha^* \le 1$ and $\varphi_\alpha \le \varphi_\alpha^*$, $\vartheta_\alpha \ge \vartheta_\alpha^*$. The graphical representation of STIT2IFN is given in Figure 3.

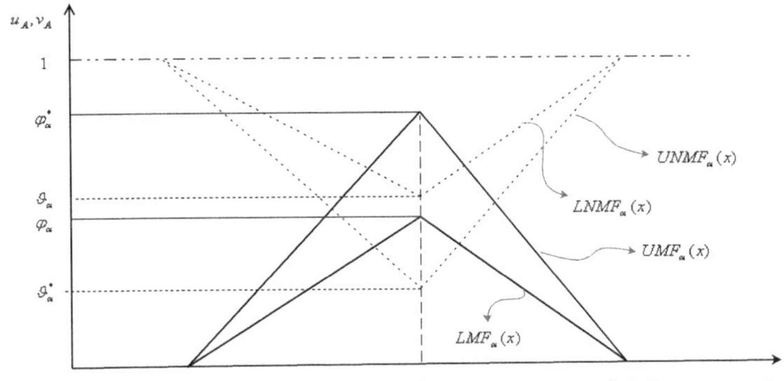

Figure 3. Representation of STIT2IFN α.

Definition 9. *For a STIT2IFN* $\alpha = (\zeta_\alpha, \varrho_\alpha, \varphi_\alpha, \varphi^*_\alpha, \vartheta_\alpha, \vartheta^*_\alpha)$, *the lower and upper membership and non-membership functions denoted by LMF, UMF, LNMF and UNMF are defined as*

$$UMF_\alpha(x) = \begin{cases} \frac{\varphi^*_\alpha}{\varrho_\alpha}(x - \zeta_\alpha + \varrho_\alpha), & \zeta_\alpha - \varrho_\alpha \le x < \zeta_\alpha \\ \varphi^*_\alpha, & x = \zeta_\alpha \\ \frac{\varphi^*_\alpha}{\varrho_\alpha}(\zeta_\alpha + \varrho_\alpha - x), & \zeta_\alpha < x \le \varrho_\alpha + \zeta_\alpha \end{cases}; \quad UNMF_\alpha(x) = \begin{cases} \frac{(\vartheta^*_\alpha - 1)(x - \zeta_\alpha + \varrho_\alpha) + \varrho_\alpha}{\varrho_\alpha}; & \zeta_\alpha - \varrho_\alpha \le x < \zeta_\alpha \\ \vartheta^*_\alpha; & x = \zeta_\alpha \\ \frac{(1 - \vartheta^*_\alpha)(x - \zeta_\alpha) + \vartheta^*_\alpha \varrho_\alpha}{\varrho_\alpha}; & \zeta_\alpha < x \le \varrho_\alpha + \zeta_\alpha \end{cases} \tag{8}$$

$$LMF_\alpha(x) = \begin{cases} \frac{\varphi_\alpha}{\varrho_\alpha}(x - \zeta_\alpha + \varrho_\alpha); & \zeta_\alpha - \varrho_\alpha \le x < \zeta_\alpha \\ \varphi_\alpha; & x = \zeta_\alpha \\ \frac{\varphi_\alpha}{\varrho_\alpha}(\zeta_\alpha + \varrho_\alpha - x); & \zeta_\alpha < x \le \varrho_\alpha + \zeta_\alpha \end{cases}; \quad LNMF_\alpha(x) = \begin{cases} \frac{(\vartheta_\alpha - 1)(x - \zeta_\alpha + \varrho_\alpha) + \varrho_\alpha}{\varrho_\alpha}; & \zeta_\alpha - \varrho_\alpha \le x < \zeta_\alpha \\ \vartheta_\alpha; & x = \zeta_\alpha \\ \frac{(1 - \vartheta_\alpha)(x - \zeta_\alpha) + \vartheta_\alpha \varrho_\alpha}{\varrho_\alpha}; & \zeta_\alpha < x \le \varrho_\alpha + \zeta_\alpha \end{cases} \tag{9}$$

Definition 10. *The score function of STIT2IFN* $\alpha = (\zeta_\alpha, \varrho_\alpha, \varphi_\alpha, \varphi^*_\alpha, \vartheta_\alpha, \vartheta^*_\alpha)$ *is defined as*

$$\begin{aligned} s(\alpha) &= (s_x(\alpha), s_y(\alpha)) \\ &= \left(\zeta_\alpha \frac{2\varphi_\alpha \varphi^*_\alpha}{\varphi_\alpha + \varphi^*_\alpha} - \zeta_\alpha \frac{2\vartheta_\alpha \vartheta^*_\alpha}{\vartheta_\alpha + \vartheta^*_\alpha}, \frac{\vartheta_\alpha + \varphi^*_\alpha}{2} - \frac{\varphi_\alpha + \vartheta^*_\alpha}{2}\right) \end{aligned} \tag{10}$$

Definition 11. *For two STIT2IFNs* α *and* β, *an order relation "*$(>)$*" to compare them is defined as*

1. *If* $s_x(\alpha) > s_x(\beta)$, *then* $\alpha > \beta$;
2. *If* $s_x(\alpha) = s_x(\beta)$, *then* $\begin{cases} s_y(\alpha) > s_y(\beta) & \Rightarrow \alpha > \beta; \\ s_y(\alpha) = s_y(\beta) & \Rightarrow \alpha = \beta; \end{cases}$

Definition 12. *For two STIT2IFNs* $\alpha = (\zeta_\alpha, \varrho_\alpha, \varphi_\alpha, \varphi^*_\alpha, \vartheta_\alpha, \vartheta^*_\alpha)$ *and* $\beta = \left(\zeta_\beta, \varrho_\beta, \varphi_\beta, \varphi^*_\beta, \vartheta_\beta, \vartheta^*_\beta\right)$, $\lambda > 0$, *then the operational laws of it are shown as follows:*

1. $\alpha \oplus \beta = \left(\zeta_\alpha + \zeta_\beta, \varrho_\alpha + \varrho_\beta, \varphi_\alpha \varphi_\beta, \varphi^*_\alpha + \varphi^*_\beta - \varphi^*_\alpha \varphi^*_\beta, \vartheta_\alpha + \vartheta_\beta - \vartheta_\alpha \vartheta_\beta, \vartheta^*_\alpha \vartheta^*_\beta\right)$;
2. $\alpha \otimes \beta = \left(\zeta_\alpha \zeta_\beta, \varrho_\alpha \varrho_\beta, \varphi_\alpha + \varphi_\beta - \varphi_\alpha \varphi_\beta, \varphi^*_\alpha \varphi^*_\beta, \vartheta_\alpha \vartheta_\beta, \vartheta^*_\alpha + \vartheta^*_\beta - \vartheta^*_\alpha \vartheta^*_\beta\right)$;
3. $\lambda\alpha = \left(\lambda\zeta_\alpha, \lambda\varrho_\alpha, (\varphi_\alpha)^\lambda, 1 - (1 - \varphi^*_\alpha)^\lambda, 1 - (1 - \vartheta_\alpha)^\lambda, (\vartheta^*_\alpha)^\lambda\right)$;
4. $\alpha^\lambda = \left(\zeta^\lambda_\alpha, \varrho^\lambda_\alpha, 1 - (1 - \varphi_\alpha)^\lambda, (\varphi^*_\alpha)^\lambda, (\vartheta_\alpha)^\lambda, 1 - (1 - \vartheta^*_\alpha)^\lambda\right)$

Theorem 1. *For STIT2IFNs* α *and* β, *the operations defined in Definition 12 are again STIT2IFNs.*

Proof. Consider two STIT2IFNs $\alpha = (\zeta_\alpha, \varrho_\alpha, \varphi_\alpha, \varphi^*_\alpha, \vartheta_\alpha, \vartheta^*_\alpha)$ and $\beta = \left(\zeta_\beta, \varrho_\beta, \varphi_\beta, \varphi^*_\beta, \vartheta_\beta, \vartheta^*_\beta\right)$. So by Definition 8, we have $\zeta_\alpha \ge \varrho_\alpha$, $\varphi_\alpha \le \varphi^*_\alpha$, $\vartheta_\alpha \ge \vartheta^*_\alpha$, $\varphi_\alpha + \vartheta_\alpha \le 1$, $\varphi^*_\alpha + \vartheta^*_\alpha \le 1$, $\zeta_\beta \ge \varrho_\beta$, $\varphi_\beta \le \varphi^*_\beta$, $\vartheta_\beta \ge \vartheta^*_\beta$ $\varphi_\beta + \vartheta_\beta \le 1$, $\varphi^*_\beta + \vartheta^*_\beta \le 1$.

Let $\alpha \oplus \beta = \gamma = \left(\zeta_\gamma, \varrho_\gamma, \varphi_\gamma, \varphi^*_\gamma, \vartheta_\gamma, \vartheta^*_\gamma\right)$ and thus by Definition 12, we get $\zeta_\gamma = \zeta_\alpha + \zeta_\beta$, $\varrho_\gamma = \varrho_\alpha + \varrho_\beta$, $\varphi_\gamma = \varphi_\alpha \varphi_\beta$, $\varphi^*_\gamma = \varphi^*_\alpha +^*_\beta - \varphi^*_\alpha \varphi^*_\beta$, $\vartheta_\gamma = \vartheta_\alpha + \vartheta_\beta - \vartheta_\alpha \vartheta_\beta$, $\vartheta^*_\gamma = \vartheta^*_\alpha \vartheta^*_\beta$. Now, to show $\alpha \oplus \beta$ is again an STIT2IFN, we need to prove that $\zeta_\gamma \ge \varrho_\gamma$, $\varphi_\gamma \le \varphi^*_\gamma$, $\vartheta_\gamma \ge \vartheta^*_\gamma$, $\varphi_\gamma + \vartheta_\gamma \le 1$, $\varphi^*_\gamma + \vartheta^*_\gamma \le 1$.

As $\zeta_\alpha \ge \varrho_\alpha$ and $\zeta_\beta \ge \varrho_\beta$ which implies that $\zeta_\gamma \ge \varrho_\gamma$. Further $\varphi_\alpha \le \varphi^*_\alpha$, $\varphi_\beta \le \varphi^*_\beta$, $\vartheta_\alpha \ge \vartheta^*_\alpha$, $\vartheta_\beta \ge \vartheta^*_\beta$, $\varphi_\alpha + \vartheta_\alpha \le 1$, $\varphi^*_\alpha + \vartheta^*_\alpha \le 1$ which gives that

$$\begin{aligned} \varphi_\gamma + \vartheta_\gamma &= \varphi_\alpha \varphi_\beta + \left(\vartheta_\alpha + \vartheta_\beta - \vartheta_\alpha \vartheta_\beta\right) \\ &= \varphi_\alpha \varphi_\beta + 1 - (1 - \vartheta_\alpha)\left(1 - \vartheta_\beta\right) \\ &\le \varphi_\alpha \varphi_\beta + 1 - \varphi_\alpha \varphi_\beta \\ &\le 1 \end{aligned}$$

and

$$\begin{aligned} \varphi_\gamma^* + \vartheta_\gamma^* &= \varphi_\alpha^*\varphi_\beta^* - \varphi_\alpha^*\varphi_\beta^* + \vartheta_\alpha^*\vartheta_\beta^* \\ &= 1 - (1-\varphi_\alpha^*)\left(1-\varphi_\beta^*\right) + \vartheta_\alpha^*\vartheta_\beta^* \\ &\leq 1 - \vartheta_\alpha^*\vartheta_\beta^* + \vartheta_\alpha^*\vartheta_\beta^* \\ &\leq 1 \end{aligned}$$

Finally, $\varphi_\gamma = \varphi_\alpha\varphi_\beta \leq \varphi_\alpha^*\varphi_\beta^* = \varphi_\gamma^*$ and $\vartheta_\gamma = \vartheta_\alpha + \vartheta_\beta - \vartheta_\alpha\vartheta_\beta = 1-(1-\vartheta_\alpha)(1-\vartheta_\beta) \geq 1-(1-\vartheta_\alpha^*)(1-\vartheta_\beta^*) = \vartheta_\gamma^*$.

Therefore, we conclude that $\alpha \oplus \beta$ becomes STIT2IFN. Similarly, we can prove that $\alpha \otimes \beta$, α^λ and $\lambda\alpha$ are also STIT2IFNs. □

4. TIT2IF Hamy Mean Aggregation Operators

Let Ω be the gathering of all non-empty STIT2IFNs $\alpha_i = \left(\zeta_i, \varrho_i, \varphi_i, \varphi_i^*, \vartheta_i, \vartheta_i^*\right), (i = 1(1)n)$. Here, we present HM-based AOs for STIT2IFNs.

4.1. STIT2IFHM Operator

Definition 13. *A STIT2IFHM is a mapping STIT2IFHM* : $\Omega^n \to \Omega$ *defined as*

$$STIT2IFHM^{(k)}(\alpha_1, \alpha_2, \ldots, \alpha_n) = \frac{\bigoplus\limits_{\substack{1 \leq i_1 < \\ \ldots < i_k \leq n}} \left(\bigotimes\limits_{j=1}^{k} \alpha_{i_j}\right)^{\frac{1}{k}}}{\binom{n}{k}} \tag{11}$$

then STIT2IHM$^{(k)}$ *is called the symmetric triangular IT2IF Hamy mean operator, where* $k = 1, 2, \ldots, n$ *is the parameter and* $\binom{n}{k} = \frac{n!}{k!(n-k)!}$ *represent the binomial coefficient.*

Theorem 2. *The aggregated value for n STIT2IFNs* $\alpha_i = \left(\zeta_i, \varrho_i, \varphi_i, \varphi_i^*, \vartheta_i, \vartheta_i^*\right)$ *by using Definition 13 is again STIT2IFN which is given as*

$$\begin{aligned} &STIT2IFHM^{(k)}(\alpha_1, \alpha_2, \ldots, \alpha_n) \\ &= \left(\begin{array}{l} \frac{\sum\limits_{\substack{1 \leq i_1 < \\ \ldots < i_k \leq n}} \left(\prod\limits_{j=1}^{k} \zeta_{\alpha_{i_j}}\right)^{\frac{1}{k}}}{\binom{n}{k}}, \frac{\sum\limits_{\substack{1 \leq i_1 < \\ \ldots < i_k \leq n}} \left(\prod\limits_{j=1}^{k} \varrho_{\alpha_{i_j}}\right)^{\frac{1}{k}}}{\binom{n}{k}}, \left(\prod\limits_{\substack{1 \leq i_1 < \\ \ldots < i_k \leq n}} \left(1 - \left(\prod\limits_{j=1}^{k}(1-\varphi_{\alpha_{i_j}})\right)^{\frac{1}{k}}\right)\right)^{\frac{1}{\binom{n}{k}}}, \\ 1 - \left(\prod\limits_{\substack{1 \leq i_1 < \\ \ldots < i_k \leq n}} \left(1 - \left(\prod\limits_{j=1}^{k}\varphi_{\alpha_{i_j}}^*\right)^{\frac{1}{k}}\right)\right)^{\frac{1}{\binom{n}{k}}}, 1 - \left(\prod\limits_{\substack{1 \leq i_1 < \\ \ldots < i_k \leq n}} \left(1 - \left(\prod\limits_{j=1}^{k}\vartheta_{\alpha_{i_j}}\right)^{\frac{1}{k}}\right)\right)^{\frac{1}{\binom{n}{k}}}, \\ \left(\prod\limits_{\substack{1 \leq i_1 < \\ \ldots < i_k \leq n}} \left(1 - \left(\prod\limits_{j=1}^{k}(1-\vartheta_{\alpha_{i_j}}^*)\right)^{\frac{1}{k}}\right)\right)^{\frac{1}{\binom{n}{k}}} \end{array} \right) \end{aligned} \tag{12}$$

Proof. The first part of the result can be easily obtained from Theorem 1. So, there is a need to prove only that Equation (12) is kept.

According to the operational laws of STIT2IFNs, we get

$$\bigotimes_{j=1}^{k}\alpha_{i_j} = \left(\prod_{j=1}^{k}\zeta_{\alpha_{i_j}}, \prod_{j=1}^{k}\varrho_{\alpha_{i_j}}, 1-\prod_{j=1}^{k}(1-\varphi_{\alpha_{i_j}}), \prod_{j=1}^{k}\varphi^*_{\alpha_{i_j}}, \prod_{j=1}^{k}\vartheta_{\alpha_{i_j}}, 1-\prod_{j=1}^{k}(1-\vartheta^*_{\alpha_{i_j}})\right)$$

and

$$\left(\bigotimes_{j=1}^{n}\alpha_{i_j}\right)^{\frac{1}{k}} = \begin{pmatrix} \left(\prod_{j=1}^{k}\zeta_{\alpha_{i_j}}\right)^{\frac{1}{k}}, \left(\prod_{j=1}^{k}\varrho_{\alpha_{i_j}}\right)^{\frac{1}{k}}, 1-\left(\prod_{j=1}^{k}(1-\varphi_{\alpha_{i_j}})\right)^{\frac{1}{k}}, \\ \left(\prod_{j=1}^{k}\varphi^*_{\alpha_{i_j}}\right)^{\frac{1}{k}}, \left(\prod_{j=1}^{k}\vartheta_{\alpha_{i_j}}\right)^{\frac{1}{k}}, 1-\left(\prod_{j=1}^{k}(1-\vartheta^*_{\alpha_{i_j}})\right)^{\frac{1}{k}} \end{pmatrix}$$

Therefore,

$$\bigoplus_{\substack{1\leq i_1< \\ \ldots<i_k\leq n}}\left(\bigotimes_{j=1}^{n}\alpha_{i_j}\right)^{\frac{1}{k}} = \begin{pmatrix} \sum\limits_{\substack{1\leq i_1< \\ \ldots<i_k\leq n}}\left(\prod_{j=1}^{k}\zeta_{\alpha_{i_j}}\right)^{\frac{1}{k}}, \sum\limits_{\substack{1\leq i_1< \\ \ldots<i_k\leq n}}\left(\prod_{j=1}^{k}\varrho_{\alpha_{i_j}}\right)^{\frac{1}{k}}, \prod\limits_{\substack{1\leq i_1< \\ \ldots<i_k\leq n}}\left(1-\left(\prod_{j=1}^{k}(1-\varphi_{\alpha_{i_j}})\right)^{\frac{1}{k}}\right), \\ 1-\prod\limits_{\substack{1\leq i_1< \\ \ldots<i_k\leq n}}\left(1-\left(\prod_{j=1}^{k}\varphi^*_{\alpha_{i_j}}\right)^{\frac{1}{k}}\right), 1-\prod\limits_{\substack{1\leq i_1< \\ \ldots<i_k\leq n}}\left(1-\left(\prod_{j=1}^{k}\vartheta_{\alpha_{i_j}}\right)^{\frac{1}{k}}\right), \\ \prod\limits_{\substack{1\leq i_1< \\ \ldots<i_k\leq n}}\left(1-\left(\prod_{j=1}^{k}(1-\vartheta^*_{\alpha_{i_j}})\right)^{\frac{1}{k}}\right) \end{pmatrix}$$

Subsequently, we have

$$\begin{aligned} &\mathrm{STIT2IFHM}^{(k)}(\alpha_1,\alpha_2,\ldots,\alpha_n) \\ &= \frac{\bigoplus\limits_{\substack{1\leq i_1< \\ \ldots<i_k\leq n}}\left(\bigotimes_{j=1}^{k}\alpha_{i_j}\right)^{\frac{1}{k}}}{\binom{n}{k}} \\ &= \begin{pmatrix} \frac{\sum\limits_{\substack{1\leq i_1< \\ \ldots<i_k\leq n}}\left(\prod_{j=1}^{k}\zeta_{\alpha_{i_j}}\right)^{\frac{1}{k}}}{\binom{n}{k}}, \frac{\sum\limits_{\substack{1\leq i_1< \\ \ldots<i_k\leq n}}\left(\prod_{j=1}^{k}\varrho_{\alpha_{i_j}}\right)^{\frac{1}{k}}}{\binom{n}{k}}, \left(\prod\limits_{\substack{1\leq i_1< \\ \ldots<i_k\leq n}}\left(1-\left(\prod_{j=1}^{k}(1-\varphi_{\alpha_{i_j}})\right)^{\frac{1}{k}}\right)\right)^{\frac{1}{\binom{n}{k}}}, \\ 1-\left(\prod\limits_{\substack{1\leq i_1< \\ \ldots<i_k\leq n}}\left(1-\left(\prod_{j=1}^{k}\varphi^*_{\alpha_{i_j}}\right)^{\frac{1}{k}}\right)\right)^{\frac{1}{\binom{n}{k}}}, 1-\left(\prod\limits_{\substack{1\leq i_1< \\ \ldots<i_k\leq n}}\left(1-\left(\prod_{j=1}^{k}\vartheta_{\alpha_{i_j}}\right)^{\frac{1}{k}}\right)\right)^{\frac{1}{\binom{n}{k}}}, \\ \left(\prod\limits_{\substack{1\leq i_1< \\ \ldots<i_k\leq n}}\left(1-\left(\prod_{j=1}^{k}(1-\vartheta^*_{\alpha_{i_j}})\right)^{\frac{1}{k}}\right)\right)^{\frac{1}{\binom{n}{k}}} \end{pmatrix} \end{aligned}$$

□

In what follows, we investigate the certain property of STIT2IFHM operator.

Theorem 3. *(Idempotency) If $\alpha_i = \alpha = (\zeta_\alpha, \varrho_\alpha, \varphi_\alpha, \varphi^*_\alpha, \vartheta_\alpha, \vartheta^*_\alpha)$ for all i, then*

$$STIT2IFHM^{(k)}(\alpha_1, \alpha_2, \ldots, \alpha_n) = \alpha.$$

Proof. Since $\alpha_i = \alpha = (\zeta_\alpha, \varrho_\alpha, \varphi_\alpha, \varphi^*_\alpha, \vartheta_\alpha, \vartheta^*_\alpha)$ for all i then based on Theorem 2, we have

$$\begin{aligned}
&\mathrm{STIT2IFHM}^{(k)}(\alpha, \alpha, \ldots, \alpha) \\
&= \left(\begin{array}{l}
\frac{\sum\limits_{\substack{1 \le i_1 < \\ \ldots < i_k \le n}} \left(\prod\limits_{j=1}^{k} \zeta_\alpha \right)^{\frac{1}{k}}}{\binom{n}{k}}, \frac{\sum\limits_{\substack{1 \le i_1 < \\ \ldots < i_k \le n}} \left(\prod\limits_{j=1}^{k} \varrho_\alpha \right)^{\frac{1}{k}}}{\binom{n}{k}}, \left(\prod\limits_{\substack{1 \le i_1 < \\ \ldots < i_k \le n}} \left(1 - \left(\prod\limits_{j=1}^{k} (1 - \varphi_\alpha) \right)^{\frac{1}{k}} \right) \right)^{\frac{1}{\binom{n}{k}}}, \\
1 - \left(\prod\limits_{\substack{1 \le i_1 < \\ \ldots < i_k \le n}} \left(1 - \left(\prod\limits_{j=1}^{k} \varphi^*_\alpha \right)^{\frac{1}{k}} \right) \right)^{\frac{1}{\binom{n}{k}}}, 1 - \left(\prod\limits_{\substack{1 \le i_1 < \\ \ldots < i_k \le n}} \left(1 - \left(\prod\limits_{j=1}^{k} \vartheta_\alpha \right)^{\frac{1}{k}} \right) \right)^{\frac{1}{\binom{n}{k}}}, \\
\left(\prod\limits_{\substack{1 \le i_1 < \\ \ldots < i_k \le n}} \left(1 - \left(\prod\limits_{j=1}^{k} (1 - \vartheta^*_\alpha) \right)^{\frac{1}{k}} \right) \right)^{\frac{1}{\binom{n}{k}}}
\end{array} \right) \\
&= \left(\begin{array}{l}
\frac{\sum\limits_{\substack{1 \le i_1 < \\ \ldots < i_k \le n}} \left(\zeta^k_\alpha \right)^{\frac{1}{k}}}{\binom{n}{k}}, \frac{\sum\limits_{\substack{1 \le i_1 < \\ \ldots < i_k \le n}} \left(\varrho^k_\alpha \right)}{\binom{n}{k}}, \left(\prod\limits_{\substack{1 \le i_1 < \\ \ldots < i_k \le n}} (1 - (1 - \varphi_\alpha)) \right)^{\frac{1}{\binom{n}{k}}}, 1 - \left(\prod\limits_{\substack{1 \le i_1 < \\ \ldots < i_k \le n}} (1 - \varphi^*_\alpha) \right)^{\frac{1}{\binom{n}{k}}}, \\
1 - \left(\prod\limits_{\substack{1 \le i_1 < \\ \ldots < i_k \le n}} (1 - \vartheta_\alpha) \right)^{\frac{1}{\binom{n}{k}}}, \left(\prod\limits_{\substack{1 \le i_1 < \\ \ldots < i_k \le n}} (1 - (1 - \vartheta^*_\alpha)) \right)^{\frac{1}{\binom{n}{k}}}
\end{array} \right) \\
&= \left(\frac{\binom{n}{k} (\zeta^k_\alpha)^{\frac{1}{k}}}{\binom{n}{k}}, \frac{\binom{n}{k} (\varrho^k_\alpha)^{\frac{1}{k}}}{\binom{n}{k}}, 1 - (1 - \varphi_\alpha)^{\frac{\binom{n}{k}}{\binom{n}{k}}}, 1 - (1 - \varphi^*_\alpha)^{\frac{\binom{n}{k}}{\binom{n}{k}}}, 1 - (1 - \vartheta_\alpha)^{\frac{\binom{n}{k}}{\binom{n}{k}}}, (1 - (1 - \vartheta^*_\alpha))^{\frac{\binom{n}{k}}{\binom{n}{k}}} \right) \\
&= (\zeta_\alpha, \varrho_\alpha, \varphi_\alpha, \varphi^*_\alpha, \vartheta_\alpha, \vartheta^*_\alpha) \\
&= \alpha
\end{aligned}$$

□

Theorem 4. *(Commutativity) Let $\alpha_i (i = 1, 2, \ldots, n)$ be a collection of STIT2IFNs, and $\overline{\alpha}_i$ be any permutation of α_i. Then*

$$STIT2IFHM^{(k)}(\overline{\alpha}_1, \overline{\alpha}_2, \ldots, \overline{\alpha}_n) = STIT2IFHM^{(k)}(\alpha_1, \alpha_2, \ldots, \alpha_n)$$

Proof. Based on the Definition 13, we have

$$\begin{aligned}
\mathrm{STIT2IFHM}^{(k)}(\overline{\alpha}_1, \overline{\alpha}_2, \ldots, \overline{\alpha}_n) &= \frac{\bigoplus\limits_{\substack{1 \le i_1 < \\ \ldots < i_k \le n}} \left(\bigotimes\limits_{j=1}^{k} \tilde{\alpha}_{i_j} \right)^{\frac{1}{k}}}{\binom{n}{k}} \\
&= \frac{\bigoplus\limits_{\substack{1 \le i_1 < \\ \ldots < i_k \le n}} \left(\bigotimes\limits_{j=1}^{k} \alpha_{i_j} \right)^{\frac{1}{k}}}{\binom{n}{k}} \\
&= \mathrm{STIT2IFHM}^{(k)}(\alpha_1, \alpha_2, \ldots, \alpha_n)
\end{aligned}$$

□

Theorem 5. *(Monotonicity) For two different STIT2IFNs* $\alpha_i = \left(\zeta_{\alpha_i}, \varrho_{\alpha_i}, \varphi_{\alpha_i}, \varphi^*_{\alpha_i}, \vartheta_{\alpha_i}, \vartheta^*_{\alpha_i}\right)$, *and* $\beta_i = \left(\zeta_{\beta_i}, \varrho_{\beta_i}, \varphi_{\beta_i}, \varphi^*_{\beta_i}, \vartheta_{\beta_i}, \vartheta^*_{\beta_i}\right)$, $(i = 1, 2, \ldots, n)$. *If* $\zeta_{\alpha_i} \leq \zeta_{\beta_i}$, $\varrho_{\alpha_i} \geq \varrho_{\beta_i}$, $\varphi_{\alpha_i} \geq \varphi_{\beta_i}$, $\varphi^*_{\alpha_i} \leq \varphi^*_{\beta_i}$, $\vartheta_{\alpha_i} \leq \vartheta_{\beta_i}$ *and* $\vartheta^*_{\alpha_i} \geq \vartheta^*_{\beta_i}$ *for all i, then*

$$STIT2IFHM^{(k)}(\alpha_1, \alpha_2, \ldots, \alpha_n) \leq STIT2IFHM^{(k)}(\beta_1, \beta_2, \ldots, \beta_n). \tag{13}$$

Proof. Let $A = \text{STIT2IFHM}^{(k)}(\alpha_1, \alpha_2, \ldots, \alpha_n)$ and $B = \text{STIT2IFHM}^{(k)}(\beta_1, \beta_2, \ldots, \beta_n)$. Then according to Theorem 2, we get

$$\begin{aligned} A &= \text{STIT2IFHM}^{(k)}(\alpha_1, \alpha_2, \ldots, \alpha_n) \\ &= \left(\begin{array}{c} \dfrac{\sum\limits_{\substack{1 \leq i_1 < \\ \ldots < i_k \leq n}} \left(\prod\limits_{j=1}^{k} \zeta_{\alpha_{i_j}} \right)^{\frac{1}{k}}}{\binom{n}{k}}, \dfrac{\sum\limits_{\substack{1 \leq i_1 < \\ \ldots < i_k \leq n}} \left(\prod\limits_{j=1}^{k} \varrho_{\alpha_{i_j}} \right)^{\frac{1}{k}}}{\binom{n}{k}}, \left(\prod\limits_{\substack{1 \leq i_1 < \\ \ldots < i_k \leq n}} \left(1 - \left(\prod\limits_{j=1}^{k} (1 - \varphi_{\alpha_{i_j}}) \right)^{\frac{1}{k}} \right) \right)^{\frac{1}{\binom{n}{k}}}, \\ 1 - \left(\prod\limits_{\substack{1 \leq i_1 < \\ \ldots < i_k \leq n}} \left(1 - \left(\prod\limits_{j=1}^{k} \varphi^*_{\alpha_{i_j}} \right)^{\frac{1}{k}} \right) \right)^{\frac{1}{\binom{n}{k}}}, 1 - \left(\prod\limits_{\substack{1 \leq i_1 < \\ \ldots < i_k \leq n}} \left(1 - \left(\prod\limits_{j=1}^{k} \vartheta_{\alpha_{i_j}} \right)^{\frac{1}{k}} \right) \right)^{\frac{1}{\binom{n}{k}}}, \\ \left(\prod\limits_{\substack{1 \leq i_1 < \\ \ldots < i_k \leq n}} \left(1 - \left(\prod\limits_{j=1}^{k} (1 - \vartheta^*_{\alpha_{i_j}}) \right)^{\frac{1}{k}} \right) \right)^{\frac{1}{\binom{n}{k}}} \end{array} \right) \end{aligned}$$

and

$$\begin{aligned} B &= \text{STIT2IFHM}^{(k)}(\beta_1, \beta_2, \ldots, \beta_n) \\ &= \left(\begin{array}{c} \dfrac{\sum\limits_{\substack{1 \leq i_1 < \\ \ldots < i_k \leq n}} \left(\prod\limits_{j=1}^{k} \zeta_{\beta_{i_j}} \right)^{\frac{1}{k}}}{\binom{n}{k}}, \dfrac{\sum\limits_{\substack{1 \leq i_1 < \\ \ldots < i_k \leq n}} \left(\prod\limits_{j=1}^{k} \varrho_{\beta_{i_j}} \right)^{\frac{1}{k}}}{\binom{n}{k}}, \left(\prod\limits_{\substack{1 \leq i_1 < \\ \ldots < i_k \leq n}} \left(1 - \left(\prod\limits_{j=1}^{k} (1 - \varphi_{\beta_{i_j}}) \right)^{\frac{1}{k}} \right) \right)^{\frac{1}{\binom{n}{k}}}, \\ 1 - \left(\prod\limits_{\substack{1 \leq i_1 < \\ \ldots < i_k \leq n}} \left(1 - \left(\prod\limits_{j=1}^{k} \varphi^*_{\beta_{i_j}} \right)^{\frac{1}{k}} \right) \right)^{\frac{1}{\binom{n}{k}}}, 1 - \left(\prod\limits_{\substack{1 \leq i_1 < \\ \ldots < i_k \leq n}} \left(1 - \left(\prod\limits_{j=1}^{k} \vartheta_{\beta_{i_j}} \right)^{\frac{1}{k}} \right) \right)^{\frac{1}{\binom{n}{k}}}, \\ \left(\prod\limits_{\substack{1 \leq i_1 < \\ \ldots < i_k \leq n}} \left(1 - \left(\prod\limits_{j=1}^{k} (1 - \vartheta^*_{\beta_{i_j}}) \right)^{\frac{1}{k}} \right) \right)^{\frac{1}{\binom{n}{k}}} \end{array} \right) \end{aligned}$$

Since $\zeta_{\alpha_i} \leq \zeta_{\beta_i}$ which implies that

$$\frac{\sum\limits_{\substack{1 \leq i_1 < \\ \ldots < i_k \leq n}} \left(\prod\limits_{j=1}^{k} \zeta_{\alpha_{i_j}} \right)^{\frac{1}{k}}}{\binom{n}{k}} \leq \frac{\sum\limits_{\substack{1 \leq i_1 < \\ \ldots < i_k \leq n}} \left(\prod\limits_{j=1}^{k} \zeta_{\beta_{i_j}} \right)^{\frac{1}{k}}}{\binom{n}{k}}$$

Also, $\varphi_{\alpha_i} \geq \varphi_{\beta_i}$ implies that

$$\prod_{j=1}^{k}(1-\varphi_{\alpha_{i_j}}) \leq \prod_{j=1}^{k}(1-\varphi_{\beta_{i_j}})$$
$$\Rightarrow \left(\prod_{j=1}^{k}(1-\varphi_{\alpha_{i_j}})\right)^{\frac{1}{k}} \leq \left(\prod_{j=1}^{k}(1-\varphi_{\beta_{i_j}})\right)^{\frac{1}{k}}$$
$$\Rightarrow \left(\prod_{\substack{1\leq i_1< \\ \ldots<i_k\leq n}}\left(1-\left(\prod_{j=1}^{k}(1-\varphi_{\alpha_{i_j}})\right)^{\frac{1}{k}}\right)\right)^{\frac{1}{\binom{n}{k}}} \geq \left(\prod_{\substack{1\leq i_1< \\ \ldots<i_k\leq n}}\left(1-\left(\prod_{j=1}^{k}(1-\varphi_{\beta_{i_j}})\right)^{\frac{1}{k}}\right)\right)^{\frac{1}{\binom{n}{k}}}$$

Similarly for $\varphi^*_{\alpha_i} \leq \varphi^*_{\beta_i}$, $\vartheta_{\alpha_i} \leq \vartheta_{\alpha_i}$ and $\vartheta^*_{\alpha_i} \geq \vartheta^*_{\beta_i}$ for all i, we have

$$1-\left(\prod_{\substack{1\leq i_1< \\ \ldots<i_k\leq n}}\left(1-\left(\prod_{j=1}^{k}\vartheta_{\alpha_{i_j}}\right)^{\frac{1}{k}}\right)\right)^{\frac{1}{\binom{n}{k}}} \leq 1-\left(\prod_{\substack{1\leq i_1< \\ \ldots<i_k\leq n}}\left(1-\left(\prod_{j=1}^{k}\vartheta_{\beta_{i_j}}\right)^{\frac{1}{k}}\right)\right)^{\frac{1}{\binom{n}{k}}};$$
$$\left(\prod_{\substack{1\leq i_1< \\ \ldots<i_k\leq n}}\left(1-\left(\prod_{j=1}^{k}(1-\vartheta^*_{\alpha_{i_j}})\right)^{\frac{1}{k}}\right)\right)^{\frac{1}{\binom{n}{k}}} \geq \left(\prod_{\substack{1\leq i_1< \\ \ldots<i_k\leq n}}\left(1-\left(\prod_{j=1}^{k}(1-\vartheta^*_{\beta_{i_j}})\right)^{\frac{1}{k}}\right)\right)^{\frac{1}{\binom{n}{k}}};$$

and

$$1-\left(\prod_{\substack{1\leq i_1< \\ \ldots<i_k\leq n}}\left(1-\left(\prod_{j=1}^{k}\varphi^*_{\alpha_{i_j}}\right)^{\frac{1}{k}}\right)\right)^{\frac{1}{\binom{n}{k}}} \leq 1-\left(\prod_{\substack{1\leq i_1< \\ \ldots<i_k\leq n}}\left(1-\left(\prod_{j=1}^{k}\varphi^*_{\beta_{i_j}}\right)^{\frac{1}{k}}\right)\right)^{\frac{1}{\binom{n}{k}}}.$$

Therefore, by using these inequalities and Definition 11, we get

$$\text{STIT2IFHM}^{(k)}(\alpha_1,\alpha_2,\ldots,\alpha_n) \leq \text{STIT2IFHM}^{(k)}(\beta_1,\beta_2,\ldots,\beta_n)$$

□

Theorem 6. *(Boundedness) For n STIT2IFNs α_i, $\alpha^- = \left(\min_i\{\zeta_i\}, \max_i\{\varrho_i\}, \min_i\{\varphi_i\}, \max_i\{\varphi^*_i\}, \max_i\{\vartheta_i\}, \min_i\{\vartheta^*_i\}\right)$, and $\alpha^+ = \left(\max_i\{\zeta_i\}, \min_i\{\varrho_i\}, \max_i\{\varphi_i\}, \min_i\{\varphi^*_i\}, \min_i\{\vartheta_i\}, \max_i\{\vartheta^*_i\}\right)$, we have*

$$\alpha^- \leq STIT2IFHM^{(k)}(\alpha_1,\alpha_2,\ldots,\alpha_n) \leq \alpha^+ \tag{14}$$

Proof. Clearly, we get $\alpha^- \leq \alpha_i \leq \alpha^+$. Thus, based on Theorems 4 and 5, we have

$$\text{STIT2IFHM}^{(k)}(\alpha_1,\alpha_2,\ldots,\alpha_n) \geq \text{STIT2IFHM}^{(k)}(\alpha^-,\alpha^-,\ldots,\alpha^-) = \alpha^-$$
$$\text{STIT2IFHM}^{(k)}(\alpha_1,\alpha_2,\ldots,\alpha_n) \leq \text{STIT2IFHM}^{(k)}(\alpha^+,\alpha^+,\ldots,\alpha^+) = \alpha^+$$

□

Lemma 1 ([51]). *For n non-negative real numbers x_i, we have*

$$HM^{(1)}(x_1, x_2, \ldots, x_n) \geq HM^{(2)}(x_1, x_2, \ldots, x_n) \geq \ldots \geq HM^{(n)}(x_1, x_2, \ldots, x_n) \tag{15}$$

with equality holding iff $x_1 = x_2 = \ldots = x_n$.

Lemma 2 ([54]). *Let $x_i, y_i > 0$ and $\sum\limits_{i=1}^{n} y_i = 1$. Then*

$$\prod_{i=1}^{n} x_i^{y_i} \leq \sum_{i=1}^{n} x_i y_i \tag{16}$$

Theorem 7. *For given STIT2IFNs α_i, the operator STIT2IFHM is monotonically decreasing with parameter k.*

Proof. For STIT2IFNs α_i and $k = 1, 2, \ldots, n$, we denote

$$C(k) = \frac{\sum\limits_{\substack{1 \leq i_1 < \\ \ldots < i_k \leq n}} \left(\prod\limits_{j=1}^{k} \zeta_{\alpha_{i_j}} \right)^{\frac{1}{k}}}{\binom{n}{k}}, \quad \Delta(k) = \frac{\sum\limits_{\substack{1 \leq i_1 < \\ \ldots < i_k \leq n}} \left(\prod\limits_{j=1}^{k} \varrho_{\alpha_{i_j}} \right)^{\frac{1}{k}}}{\binom{n}{k}},$$

$$T(k) = \left(\prod_{\substack{1 \leq i_1 < \\ \ldots < i_k \leq n}} \left(1 - \left(\prod_{j=1}^{k} (1 - \varphi_{\alpha_{i_j}}) \right)^{\frac{1}{k}} \right) \right)^{\frac{1}{\binom{n}{k}}}, \quad S(k) = 1 - \left(\prod_{\substack{1 \leq i_1 < \\ \ldots < i_k \leq n}} \left(1 - \left(\prod_{j=1}^{k} \varphi^*_{\alpha_{i_j}} \right)^{\frac{1}{k}} \right) \right)^{\frac{1}{\binom{n}{k}}},$$

$$T^*(k) = 1 - \left(\prod_{\substack{1 \leq i_1 < \\ \ldots < i_k \leq n}} \left(1 - \left(\prod_{j=1}^{k} \vartheta_{\alpha_{i_j}} \right)^{\frac{1}{k}} \right) \right)^{\frac{1}{\binom{n}{k}}}, \quad S^*(k) = \left(\prod_{\substack{1 \leq i_1 < \\ \ldots < i_k \leq n}} \left(1 - \left(\prod_{j=1}^{k} (1 - \vartheta^*_{\alpha_{i_j}}) \right)^{\frac{1}{k}} \right) \right)^{\frac{1}{\binom{n}{k}}}$$

Based on Theorem 2, we have

$$\begin{aligned} \text{STIT2IFHM}^{(k)}(\alpha_1, \alpha_2, \ldots, \alpha_n) &= (C(k), \Delta(k), T(k), S(k), T^*(k), S^*(k)) \\ \text{and } \text{STIT2IFHM}^{(k+1)}(\alpha_1, \alpha_2, \ldots, \alpha_n) &= (C(k+1), \Delta(k+1), T(k+1), S(k+1), T^*(k+1), S^*(k+1)) \end{aligned}$$

Following Definition 10 and Lemma 1, we obtained

$$\begin{aligned} s_x(\text{STIT2IFHM}^{(k)}(\alpha_1, \alpha_2, \ldots, \alpha_n)) &\geq \frac{\sum\limits_{\substack{1 \leq i_1 < \\ \ldots < i_k \leq n}} \left(\prod\limits_{j=1}^{k} \zeta_{\alpha_{i_j}} \right)^{\frac{1}{k}}}{\binom{n}{k}} \\ &\geq \frac{\sum\limits_{\substack{1 \leq i_1 < \\ \ldots < i_{k+1} \leq n}} \left(\prod\limits_{j=1}^{k+1} \zeta_{\alpha_{i_j}} \right)^{\frac{1}{k+1}}}{\binom{n}{k+1}} \\ &\geq s_x(\text{STIT2IFHM}^{(k+1)}(\alpha_1, \alpha_2, \ldots, \alpha_n)) \end{aligned}$$

Then, two cases are arisen:

Case 1 If $s_x\left(\text{STIT2IFHM}^{(k)}(\alpha_1, \alpha_2, \ldots, \alpha_n)\right) > s_x\left(\text{STIT2IFHM}^{(k+1)}(\alpha_1, \alpha_2, \ldots, \alpha_n)\right)$, following the Definition 11 we get

$$\text{STIT2IFHM}^{(k)}(\alpha_1, \alpha_2, \ldots, \alpha_n) > \text{STIT2IFHM}^{(k+1)}(\alpha_1, \alpha_2, \ldots, \alpha_n)$$

Case 2 If $s_x\left(\mathrm{STIT2IFHM}^{(k)}(\alpha_1,\alpha_2,\ldots,\alpha_n)\right) = s_x\left(\mathrm{STIT2IFHM}^{(k+1)}(\alpha_1,\alpha_2,\ldots,\alpha_n)\right)$. Then, by Lemmas 1 and 2, we get

$$S(k) = 1 - \left(\prod_{\substack{1\le i_1<\\ \ldots<i_k\le n}}\left(1-\left(\prod_{j=1}^{k}\varphi^*_{\alpha_{i_j}}\right)^{\frac{1}{k}}\right)\right)^{\frac{1}{\binom{n}{k}}} \geq 1 - \frac{\sum\limits_{\substack{1\le i_1<\\ \ldots<i_k\le n}}\left(1-\left(\prod\limits_{j=1}^{k}\varphi^*_{\alpha_{i_j}}\right)^{\frac{1}{k}}\right)}{\binom{n}{k}} = \sum_{\substack{1\le i_1<\\ \ldots<i_k\le n}}\frac{\left(\prod\limits_{j=1}^{k}\varphi^*_{\alpha_{i_j}}\right)^{\frac{1}{k}}}{\binom{n}{k}}$$

To check the monotonic behavior of $S(k)$, we assume that it is increasing with k, i.e.,

$$S(n) > S(n-1) > \ldots > S(1) \tag{17}$$

Also since

$$S(1) \geq 1 - \sum_{1\le i_1\le n}\frac{\prod\limits_{j=1}^{1}\left(1-\varphi^*_{\alpha_{i_j}}\right)}{\binom{n}{1}} = 1 - \frac{n-\sum\limits_{i=1}^{n}(\varphi^*_{\alpha_i})}{n} = \frac{\sum\limits_{i=1}^{n}\varphi^*_{\alpha_i}}{n} \tag{18}$$

which implies that

$$\begin{aligned} S(n) > S(1) &= \frac{\sum_{i=1}^{n}\varphi^*_{\alpha_i}}{n} \\ \Rightarrow \left(\prod_{i=1}^{n}\varphi^*_{\alpha_i}\right)^{\frac{1}{n}} &> \frac{\sum_{i=1}^{n}\varphi^*_{\alpha_i}}{n} \end{aligned}$$

which contradict the Lemma 2. Hence with parameter k, $S(k)$ is monotonically decreasing. Similarly, we can get $T^*(k)$ is also monotonically decreasing with parameter k. Also, the functions $T(k)$ and $S^*(k)$ are monotonically increasing with parameter k.

Therefore,

$$\begin{aligned} s_y\left(\mathrm{STIT2IFHM}^{(k)}(\alpha_1,\alpha_2,\ldots,\alpha_n)\right) &= \frac{S(k)+T^*(k)}{2} - \frac{T(k)+S^*(k)}{2} \\ &> \frac{S(k+1)+T^*(k+1)}{2} - \frac{T(k+1)+S^*(k+1)}{2} \\ &= s_y\left(\mathrm{STIT2IFHM}^{(k+1)}(\alpha_1,\alpha_2,\ldots,\alpha_n)\right) \end{aligned}$$

Thus, by both the cases, we get $\mathrm{STIT2IFHM}^{(k)}(\alpha_1,\alpha_2,\ldots,\alpha_n) \geq \mathrm{STIT2IFHM}^{(k+1)}(\alpha_1,\alpha_2,\ldots,\alpha_n)$. □

Furthermore, we will talk about a few special cases of the STIT2IFHM operator concerning the parameter the k.

1. When $k = 1$, Equation (12) reduces to the triangular IT2IF averaging operator.

$$\begin{aligned}
&\text{STIT2IFHM}^{(1)}(\alpha_1, \alpha_2, \cdots, \alpha_m) \\
&= \left(\begin{array}{l}
\frac{\sum\limits_{1 \le i_1 \le n} \left(\prod\limits_{j=1}^{1} \zeta_{\alpha_{i_j}} \right)^{\frac{1}{1}}}{\binom{n}{1}}, \frac{\sum\limits_{1 \le i_1 \le n} \left(\prod\limits_{j=1}^{1} \varrho_{\alpha_{i_j}} \right)^{\frac{1}{1}}}{\binom{m}{1}}, \left(\prod\limits_{1 \le i_1 \le n} \left(1 - \left(\prod\limits_{j=1}^{1} \left(1 - \varphi_{\alpha_{i_j}} \right) \right)^{\frac{1}{1}} \right) \right)^{\frac{1}{\binom{n}{1}}}, \\
1 - \left(\prod\limits_{1 \le i_1 \le n} \left(1 - \left(\prod\limits_{j=1}^{1} \varphi^*_{\alpha_{i_j}} \right)^{\frac{1}{1}} \right) \right)^{\frac{1}{\binom{n}{1}}}, 1 - \left(\prod\limits_{1 \le i_1 \le n} \left(1 - \left(\prod\limits_{j=1}^{1} \vartheta_{\alpha_{i_j}} \right)^{\frac{1}{1}} \right) \right)^{\frac{1}{\binom{n}{1}}}, \\
\left(\prod\limits_{1 \le i_1 \le n} \left(1 - \left(\prod\limits_{j=1}^{1} \left(1 - \vartheta^*_{\alpha_{i_j}} \right) \right)^{\frac{1}{1}} \right) \right)^{\frac{1}{\binom{n}{1}}}
\end{array} \right) \\
&= \left(\begin{array}{l}
\frac{\sum\limits_{i=1}^{n} \zeta_{\alpha_i}}{n}, \frac{\sum\limits_{i=1}^{n} \varrho_{\alpha_i}}{n}, \left(\prod\limits_{i=1}^{n} \left(1 - \left(1 - \varphi_{\alpha_{i_j}} \right) \right) \right)^{\frac{1}{n}}, 1 - \left(\prod\limits_{i=1}^{n} \left(1 - \varphi^*_{\alpha_{i_j}} \right) \right)^{\frac{1}{n}}, \\
1 - \left(\prod\limits_{i=1}^{n} \left(1 - \vartheta_{\alpha_{i_j}} \right) \right)^{\frac{1}{n}}, \left(\prod\limits_{i=1}^{n} \left(1 - \left(1 - \vartheta^*_{\alpha_{i_j}} \right) \right) \right)^{\frac{1}{n}}
\end{array} \right) \\
&= \left(\frac{\sum\limits_{i=1}^{r} \zeta_{\alpha_i}}{n}, \frac{\sum\limits_{i=1}^{n} \varrho_{\alpha_i}}{n}, \left(\prod\limits_{i=1}^{n} \varphi_{\alpha_{i_j}} \right)^{\frac{1}{n}}, 1 - \left(\prod\limits_{i=1}^{n} \left(1 - \varphi^*_{\alpha_{i_j}} \right) \right)^{\frac{1}{n}}, 1 - \left(\prod\limits_{i=1}^{n} \left(1 - \vartheta_{\alpha_{i_j}} \right) \right)^{\frac{1}{n}}, \left(\prod\limits_{i=1}^{n} \vartheta^*_{\alpha_{i_j}} \right)^{\frac{1}{n}} \right)
\end{aligned}$$

2. When $k = n$, Equation (12) will reduce to triangular IT2IF geometric operator.

$$\begin{aligned}
&\text{STIT2IFHM}^{(m)}(\alpha_1, \alpha_2, \cdots, \alpha_n) \\
&= \left(\begin{array}{l}
\frac{\sum\limits_{\substack{1 \le i_1 < \\ \ldots < i_k \le n}} \left(\prod\limits_{j=1}^{k} \zeta_{\alpha_{i_j}} \right)^{\frac{1}{n}}}{\binom{n}{n}}, \frac{\sum\limits_{\substack{1 \le i_1 < \\ \ldots < i_k \le n}} \left(\prod\limits_{j=1}^{k} \varrho_{\gamma_{i_j}} \right)^{\frac{1}{n}}}{\binom{n}{n}}, \left(\prod\limits_{\substack{1 \le i_1 < \\ \ldots < i_k \le n}} \left(1 - \left(\prod\limits_{j=1}^{k} \left(1 - \varphi_{\gamma_{i_j}} \right) \right)^{\frac{1}{n}} \right) \right)^{\frac{1}{\binom{n}{n}}}, \\
1 - \left(\prod\limits_{\substack{1 \le i_1 < \\ \ldots < i_k \le n}} \left(1 - \left(\prod\limits_{j=1}^{k} \varphi^*_{\alpha_{i_j}} \right)^{\frac{1}{n}} \right) \right)^{\frac{1}{\binom{n}{n}}}, 1 - \left(\prod\limits_{\substack{1 \le i_1 < \\ \ldots < i_k \le n}} \left(1 - \left(\prod\limits_{j=1}^{k} \vartheta_{\alpha_{i_j}} \right)^{\frac{1}{n}} \right) \right)^{\frac{1}{\binom{n}{n}}}, \\
\left(\prod\limits_{\substack{1 \le i_1 < \\ \ldots < i_k \le n}} \left(1 - \left(\prod\limits_{j=1}^{k} \left(1 - \vartheta^*_{\alpha_{i_j}} \right)^{\frac{1}{n}} \right) \right) \right)^{\frac{1}{\binom{n}{n}}}
\end{array} \right) \\
&= \left(\begin{array}{l}
\left(\prod\limits_{j=1}^{k} \zeta_{\alpha_{i_j}} \right)^{\frac{1}{n}}, \left(\prod\limits_{j=1}^{k} \varrho_{\alpha_{i_j}} \right)^{\frac{1}{n}}, \left(1 - \left(\prod\limits_{j=1}^{k} \left(1 - \varphi_{\alpha_{i_j}} \right) \right)^{\frac{1}{n}} \right), 1 - \left(1 - \left(\prod\limits_{j=1}^{k} \varphi^*_{\gamma_{i_j}} \right)^{\frac{1}{n}} \right), \\
1 - \left(1 - \left(\prod\limits_{j=1}^{k} \vartheta_{\alpha_{i_j}} \right)^{\frac{1}{n}} \right), \left(1 - \left(\prod\limits_{j=1}^{k} \left(1 - \vartheta^*_{\alpha_{i_j}} \right) \right)^{\frac{1}{n}} \right)
\end{array} \right) \\
&= \left(\begin{array}{l}
\left(\prod\limits_{j=1}^{k} \zeta_{\alpha_{i_j}} \right)^{\frac{1}{n}}, \left(\prod\limits_{j=1}^{k} \varrho_{\alpha_{i_j}} \right)^{\frac{1}{n}}, \left(1 - \left(\prod\limits_{j=1}^{k} \left(1 - \varphi_{\alpha_{i_j}} \right) \right)^{\frac{1}{n}} \right), \\
\left(\prod\limits_{j=1}^{k} \varphi^*_{\alpha_{i_j}} \right)^{\frac{1}{n}}, \left(\prod\limits_{j=1}^{k} \vartheta_{\alpha_{i_j}} \right)^{\frac{1}{n}}, \left(1 - \left(\prod\limits_{j=1}^{k} \left(1 - \vartheta^*_{\alpha_{i_j}} \right)^{\frac{1}{n}} \right) \right)
\end{array} \right)
\end{aligned}$$

4.2. WSTIT2IFHM Operator

Definition 14. *For a collection of n STIT2IFNs, α_i, $w = (w_1, w_2, \cdots, w_n)^T$ is weight vector of α_i, where $w_i > 0$ and $\sum_{i=1}^{n} w_i = 1$, we define WSTIT2IFHM operator as*

$$WSTIT2IFHM_w^{(k)}(\alpha_1, \alpha_2, \cdots, \alpha_m) = \begin{cases} \dfrac{\bigoplus\limits_{\substack{1 \le i_1 < \\ \ldots < i_k \le n}} \left(1 - \sum\limits_{j=1}^{k} w_{i_j}\right)\left(\bigotimes\limits_{j=1}^{k} \alpha_{i_j}\right)^{\frac{1}{k}}}{\binom{n-1}{k}} & ; 1 \le k < n \\ \bigotimes\limits_{j=1}^{k} \alpha_j^{\frac{1-w_j}{n-1}} & ; k = n \end{cases} \tag{19}$$

then $WSTIT2IFHM_w^{(k)}$ is stated as weighted symmetric triangular IT2IF Hamy mean operator.

Theorem 8. *For n STIT2IFNs $\alpha_i = (\zeta_i, \varrho_i, \varphi_i, \varphi_i^*, \vartheta_i, \vartheta_i^*)$ $(i = 1, 2, \ldots, n)$, the value obtained through Equation (19) is also STIT2IFN, and is given as*

$$WSTIT2IFHM_w^{(k)}(\alpha_1, \alpha_2, \cdots, \alpha_m)$$

$$= \left(\begin{array}{c} \dfrac{\sum\limits_{\substack{1 \le i_1 < \\ \ldots < i_k \le n}} \left(1 - \sum\limits_{j=1}^{k} w_{i_j}\right)\left(\prod\limits_{j=1}^{k} \zeta_{\alpha_{i_j}}\right)^{\frac{1}{k}}}{\binom{n-1}{k}}, \dfrac{\sum\limits_{\substack{1 \le i_1 < \\ \ldots < i_k \le n}} \left(1 - \sum\limits_{j=1}^{k} w_{i_j}\right)\left(\prod\limits_{j=1}^{k} \varrho_{\alpha_{i_j}}\right)^{\frac{1}{k}}}{\binom{n-1}{k}}, \\ \left(\prod\limits_{\substack{1 \le i_1 < \\ \ldots < i_k \le n}} \left(1 - \left(\prod\limits_{j=1}^{k}\left(1 - \varphi_{\alpha_{i_j}}\right)\right)^{\frac{1}{k}}\right)^{\left(1 - \sum\limits_{j=1}^{k} w_{i_j}\right)} \right)^{\frac{1}{\binom{n-1}{k}}}, 1 - \left(\prod\limits_{\substack{1 \le i_1 < \\ \ldots < i_k \le n}} \left(1 - \left(\prod\limits_{j=1}^{k} \varphi^*_{\alpha_{i_j}}\right)^{\frac{1}{k}}\right)^{\left(1 - \sum\limits_{j=1}^{k} w_{i_j}\right)} \right)^{\frac{1}{\binom{n-1}{k}}}, \\ 1 - \left(\prod\limits_{\substack{1 \le i_1 < \\ \ldots < i_k \le n}} \left(1 - \left(\prod\limits_{j=1}^{k} \vartheta_{\alpha_{i_j}}\right)^{\frac{1}{k}}\right)^{\left(1 - \sum\limits_{j=1}^{k} w_{i_j}\right)} \right)^{\frac{1}{\binom{n-1}{k}}}, \left(\prod\limits_{\substack{1 \le i_1 < \\ \ldots < i_k \le n}} \left(1 - \left(\prod\limits_{j=1}^{k}\left(1 - \vartheta^*_{\alpha_{i_j}}\right)\right)^{\frac{1}{k}}\right)^{\left(1 - \sum\limits_{j=1}^{k} w_{i_j}\right)} \right)^{\frac{1}{\binom{n-1}{k}}} \end{array} \right) ; \text{if } 1 \le k < n$$

and

$$WSTIT2IFHM_w^{(k)}(\alpha_1, \alpha_2, \cdots, \alpha_n)$$

$$= \left(\begin{array}{c} \prod\limits_{j=1}^{k} \zeta_{\alpha_j}^{\frac{1-w_j}{n-1}}, \prod\limits_{j=1}^{k} \varrho_{\alpha_j}^{\frac{1-w_j}{n-1}}, 1 - \prod\limits_{j=1}^{k}\left(1 - \varphi_{\alpha_j}\right)^{\frac{1-w_j}{n-1}}, \\ \prod\limits_{j=1}^{k}\left(\varphi^*_{\alpha_j}\right)^{\frac{1-w_j}{n-1}}, \prod\limits_{j=1}^{k}\left(\vartheta_{\alpha_j}\right)^{\frac{1-w_j}{n-1}}, 1 - \prod\limits_{j=1}^{k}\left(1 - \vartheta^*_{\alpha_j}\right)^{\frac{1-w_j}{n-1}} \end{array} \right) ; \text{if } k = n$$

Proof. Similar to the proof of Theorem 2. □

Theorem 9. *The operator STIT2IFHM is a special case of the WSTIT2IFHM operator.*

Proof. Assume that $w = \left(\frac{1}{n}, \frac{1}{n}, \cdots, \frac{1}{n}\right)^T$, then by Theorem 5, we have

1. if $1 \le k < n$, we have

$$
\begin{aligned}
&\mathrm{WSTIT2IFHM}_{w}^{(k)}(\alpha_1,\alpha_2,\cdots,\alpha_n)\\
=&\left(\begin{array}{l}
\dfrac{\sum\limits_{\substack{1\leq i_1<\\ \ldots<i_k\leq n}}\left(1-\frac{k}{n}\right)\left(\prod\limits_{j=1}^{k}\zeta_{\alpha_{i_j}}\right)^{\frac{1}{k}}}{\binom{n-1}{k}},\dfrac{\sum\limits_{\substack{1\leq i_1<\\ \ldots<i_k\leq n}}\left(1-\frac{k}{m}\right)\left(\prod\limits_{j=1}^{k}\varrho_{\alpha_{i_j}}\right)^{\frac{1}{k}}}{\binom{n-1}{k}},\\
\left(\prod\limits_{\substack{1\leq i_1<\\ \ldots<i_k\leq n}}\left(1-\left(\prod\limits_{j=1}^{k}\left(1-\varphi_{\alpha_{i_j}}\right)\right)^{\frac{1}{k}}\right)^{\left(1-\frac{k}{m}\right)}\right)^{\frac{1}{\binom{n-1}{k}}},1-\left(\prod\limits_{\substack{1\leq i_1<\\ \ldots<i_k\leq n}}\left(1-\left(\prod\limits_{j=1}^{k}\varphi^{*}_{\alpha_{i_j}}\right)^{\frac{1}{k}}\right)^{\left(1-\frac{k}{m}\right)}\right)^{\frac{1}{\binom{n-1}{k}}},\\
1-\left(\prod\limits_{\substack{1\leq i_1<\\ \ldots<i_k\leq n}}\left(1-\left(\prod\limits_{j=1}^{k}\vartheta_{\alpha_{i_j}}\right)^{\frac{1}{k}}\right)^{\left(1-\frac{k}{m}\right)}\right)^{\frac{1}{\binom{n-1}{k}}},\left(\prod\limits_{\substack{1\leq i_1<\\ \ldots<i_k\leq n}}\left(1-\left(\prod\limits_{j=1}^{k}\left(1-\vartheta^{*}_{\alpha_{i_j}}\right)\right)^{\frac{1}{k}}\right)^{\left(1-\frac{k}{m}\right)}\right)^{\frac{1}{\binom{n-1}{k}}}
\end{array}\right)\\
=&\left(\begin{array}{l}
\dfrac{\sum\limits_{\substack{1\leq i_1<\\ \cdots<i_k\leq n}}\left(1-\frac{k}{n}\right)\left(\prod\limits_{j=1}^{k}\zeta_{\alpha_{i_j}}\right)^{\frac{1}{k}}}{\binom{n}{k}\frac{n-k}{n}},\dfrac{\sum\limits_{\substack{1\leq i_1<\\ \cdots<i_k\leq n}}\left(1-\frac{k}{n}\right)\left(\prod\limits_{j=1}^{k}\varrho_{\alpha_{i_j}}\right)^{\frac{1}{k}}}{\binom{n}{k}\frac{n-k}{n}},\\
\left(\prod\limits_{\substack{1\leq i_1<\\ \cdots<i_k\leq n}}\left(1-\left(\prod\limits_{j=1}^{k}\left(1-\varphi_{\alpha_{i_j}}\right)\right)^{\frac{1}{k}}\right)^{\left(1-\frac{k}{n}\right)}\right)^{\frac{1}{\binom{n}{k}\frac{n-k}{n}}},1-\left(\prod\limits_{\substack{1\leq i_1<\\ \cdots<i_k\leq n}}\left(1-\left(\prod\limits_{j=1}^{k}\varphi^{*}_{\alpha_{i_j}}\right)^{\frac{1}{k}}\right)^{\left(1-\frac{k}{n}\right)}\right)^{\frac{1}{\binom{n}{k}\frac{n-k}{n}}},\\
1-\left(\prod\limits_{\substack{1\leq i_1<\\ \cdots<i_k\leq n}}\left(1-\left(\prod\limits_{j=1}^{k}\vartheta_{\alpha_{i_j}}\right)^{\frac{1}{k}}\right)^{\left(1-\frac{k}{n}\right)}\right)^{\frac{1}{\binom{n}{k}\frac{n-k}{n}}},\left(\prod\limits_{\substack{1\leq i_1<\\ \cdots<i_k\leq n}}\left(1-\left(\prod\limits_{j=1}^{k}\left(1-\vartheta^{*}_{\alpha_{i_j}}\right)\right)^{\frac{1}{k}}\right)^{1-\frac{k}{n}}\right)^{\frac{1}{\binom{n}{k}\frac{n-k}{n}}}
\end{array}\right)\\
=&\left(\begin{array}{l}
\dfrac{\sum\limits_{\substack{1\leq i_1<\\ \cdots<i_k\leq n}}\left(\prod\limits_{j=1}^{k}\zeta_{\alpha_{i_j}}\right)^{\frac{1}{k}}}{\binom{n}{k}},\dfrac{\sum\limits_{\substack{1\leq i_1<\\ \cdots<i_k\leq n}}\left(\prod\limits_{j=1}^{k}\varrho_{\alpha_{i_j}}\right)^{\frac{1}{k}}}{\binom{n}{k}},\left(\prod\limits_{\substack{1\leq i_1<\\ \cdots<i_k\leq n}}\left(1-\left(\prod\limits_{j=1}^{k}\left(1-\varphi_{\alpha_{i_j}}\right)\right)^{\frac{1}{k}}\right)\right)^{\frac{1}{\binom{n}{k}}},\\
1-\left(\prod\limits_{\substack{1\leq i_1<\\ \cdots<i_k\leq n}}\left(1-\left(\prod\limits_{j=1}^{k}\varphi^{*}_{\alpha_{i_j}}\right)^{\frac{1}{k}}\right)\right)^{\frac{1}{\binom{n}{k}}},1-\left(\prod\limits_{\substack{1\leq i_1<\\ \cdots<i_k\leq n}}\left(1-\left(\prod\limits_{j=1}^{k}\vartheta_{\alpha_{i_j}}\right)^{\frac{1}{k}}\right)\right)^{\frac{1}{\binom{n}{k}}},\\
\left(\prod\limits_{\substack{1\leq i_1<\\ \cdots<i_k\leq n}}\left(1-\left(\prod\limits_{j=1}^{k}\left(1-\vartheta^{*}_{\alpha_{i_j}}\right)\right)^{\frac{1}{k}}\right)\right)^{\frac{1}{\binom{n}{k}}}
\end{array}\right)\\
=&\ \mathrm{STIT2IFHM}^{(k)}(\alpha_1,\alpha_2,\cdots,\alpha_n)
\end{aligned}
$$

2. If $k=n$, we have

$$
\begin{aligned}
&\mathrm{WSTIT2IFHM}_{w}^{(k)}(\alpha_1,\alpha_2,\cdots,\alpha_n)\\
=&\left(\begin{array}{l}
\prod\limits_{j=1}^{k}\zeta_{\alpha_j}^{\frac{1-\frac{1}{n}}{n-1}},\prod\limits_{j=1}^{k}\varrho_{\alpha_j}^{\frac{1-\frac{1}{n}}{n-1}},1-\prod\limits_{j=1}^{k}\left(1-\varphi_{\alpha_j}\right)^{\frac{1-\frac{1}{n}}{n-1}},\\
\prod\limits_{j=1}^{k}\left(\varphi^{*}_{\alpha_j}\right)^{\frac{1-\frac{1}{n}}{n-1}},\prod\limits_{j=1}^{k}\left(\vartheta_{\alpha_j}\right)^{\frac{1-\frac{1}{n}}{n-1}},1-\prod\limits_{j=1}^{k}\left(1-\vartheta^{*}_{\alpha_j}\right)^{\frac{1-\frac{1}{n}}{n-1}}
\end{array}\right)\\
=&\left(\begin{array}{l}
\prod\limits_{j=1}^{k}\zeta_{\alpha_j}^{\frac{1}{n}},\prod\limits_{j=1}^{k}\varrho_{\alpha_j}^{\frac{1}{n}},1-\prod\limits_{j=1}^{k}(1-\varphi_{\alpha_j})^{\frac{1}{n}},\\
\prod\limits_{j=1}^{k}(\varphi^{*}_{\alpha_j})^{\frac{1}{n}},\prod\limits_{j=1}^{k}(\vartheta_{\alpha_j})^{\frac{1}{n}},1-\prod\limits_{j=1}^{k}(1-\vartheta^{*}_{\alpha_j})^{\frac{1}{n}}
\end{array}\right)\\
=&\ \mathrm{STIT2IFHM}^{(k)}(\alpha_1,\alpha_2,\cdots,\alpha_n)
\end{aligned}
$$

□

5. An Approach to MCDM Based on the Proposed WSTIT2IFHM Operator

In this section, an MCDM approach is developed under the triangular IT2IF (TIT2IF) environment. The description of the problem, as well as the procedure steps, are explained as below.

Assume an MCDM problem which consists of 'n' different alternatives $A_1, A_2, \ldots, A_n$ and a set of 'm' attributes $C_1, C_2, \ldots, C_m$ whose weight vector is $w = (w_1, w_2, \cdots, w_m)^T$, satisfying $w_j > 0$ and $\sum_{j=1}^{m} w_j = 1$. An expert has evaluated these given alternatives and rate them under TIT2IF environment denoted by $l_{pj}(p = 1, 2, \ldots, n; j = 1, 2, \ldots, m)$ where l_{pj} represent the linguistic information about the alternatives. Furthermore, the importance of the attributes plays a dominant role during the decision-making process. During handling the MCDM problems, if the sum of the relative coefficient w.r.t. each criterion is small, it relates that such criteria demonstrate a major impact on the overall values of the alternative. Similarly, if the relative coefficient sum is large then it shows such criterion play a less significant role. Hence, the relative coefficient of the alternative under the certain criteria is inversely proportional to the corresponding weights of criteria. Therefore, the weight of the criteria is determined by using the Spearman method [56] which main steps are summarized in Algorithm 1.

Algorithm 1 Weight determination using Spearman coefficient method.

1: Take two criteria C_k and C_j and then compute their relative coefficients as

$$\Delta_{kj} = 1 - \frac{6\sum_{p=1}^{n}(l_{pk} - l_{pj})^2}{m(m-1)} \tag{20}$$

and hence construct the matrix $\Delta_{m\times m} = (\Delta_{kj})_{m\times m}$ as

$$\Delta_{m\times m} = \begin{pmatrix} \Delta_{11} & \Delta_{12} & \cdots & \Delta_{1m} \\ \Delta_{21} & \Delta_{22} & \cdots & \Delta_{2m} \\ \ldots & \ldots & \ddots & \ldots \\ \Delta_{m1} & \Delta_{m2} & \cdots & \Delta_{mm} \end{pmatrix} \tag{21}$$

2: Compute the relative coefficient sum of each criteria by using Equation (22).

$$\Delta_j = \sum_{\substack{k=1 \\ k\neq j}}^{m} \Delta_{jk} \tag{22}$$

3: Compute the weight of each criteria as

$$w_j = \frac{\sigma_j}{\sum_{j=1}^{m}\sigma_j} \tag{23}$$

where $\sigma_j = \dfrac{1}{\Delta_j}$ represent the contribution index of the criteria.

By using this weight vector, we summarized the following steps based on the proposed AO to rank the alternatives under TIT2IFS environment.

Step 1: Arrange the information of each alternative in decision matrix $\overline{L}$ as

$$\overline{L} = \begin{array}{c} \\ A_1 \\ A_2 \\ \vdots \\ A_m \end{array} \begin{array}{c} \begin{array}{cccc} C_1 & C_2 & \ldots & C_n \end{array} \\ \begin{pmatrix} \overline{l}_{11} & \overline{l}_{12} & \ldots & \overline{l}_{1n} \\ \overline{l}_{21} & \overline{l}_{22} & \ldots & \overline{l}_{2n} \\ \vdots & \vdots & \ddots & \vdots \\ \overline{l}_{m1} & \overline{l}_{m2} & \ldots & \overline{l}_{mn} \end{pmatrix} \end{array} \tag{24}$$

where $\overline{l}_{pj} = \left(\overline{\zeta}_{pj}, \overline{\varrho}_{pj}, \overline{\varphi}_{pj}, \overline{\varphi}^*_{pj}, \overline{\vartheta}_{pj}, \overline{\vartheta}^*_{pj}\right)$ be the STIT2IFNs provided by an expert.

Step 2: Compute the normalized decision matrix L from $\overline{L}$ by using the normalized formula

$$l_{pj} = \begin{cases} \left(\overline{\zeta}_{pj}, \overline{\varrho}_{pj}, \overline{\varphi}_{pj}, \overline{\varphi}^*_{pj}, \overline{\vartheta}_{pj}, \overline{\vartheta}^*_{pj}\right) & ; \text{for the benefit type criteria} \\ \left(\overline{\zeta}_{pj}, \overline{\varrho}_{pj}, \overline{\vartheta}_{pj}, \overline{\vartheta}^*_{pj}, \overline{\varphi}_{pj}, \overline{\varphi}^*_{pj}\right) & ; \text{for the cost type criteria} \end{cases} \tag{25}$$

Step 3: Compute the weight vector to each criteria by using Algorithm 1.

Step 4: Combine the different values of STIT2IFNs $l_{pj}(j = 1, 2, \ldots, m)$ into the single one l_p of each alternative $A_p(p = 1, 2, \ldots, n)$ by using WSTIT2IFHM operator as follows:

$$\begin{aligned} l_p &= \text{WSTIT2IFHM}_w^{(k)}(l_{p1}, l_{p2}, \cdots, l_{pn}) \\ &= \left(\begin{array}{l} \dfrac{\sum\limits_{\substack{1 \le p_1 < \\ \ldots < p_k \le n}} \left(1 - \sum\limits_{j=1}^{k} w_{p_j}\right) \left(\prod\limits_{j=1}^{k} \zeta_{p_j}\right)^{\frac{1}{k}}}{\binom{n-1}{k}}, \dfrac{\sum\limits_{\substack{1 \le p_1 < \\ \ldots < p_k \le n}} \left(1 - \sum\limits_{j=1}^{k} w_{p_j}\right) \left(\prod\limits_{j=1}^{k} \varrho_{p_j}\right)^{\frac{1}{k}}}{\binom{n-1}{k}}, \\ \left(\prod\limits_{\substack{1 \le p_1 < \\ \ldots < p_k \le n}} \left(1 - \left(\prod\limits_{j=1}^{k} \left(1 - \varphi_{p_j}\right)\right)^{\frac{1}{k}}\right)^{\left(1 - \sum\limits_{j=1}^{k} w_{p_j}\right)} \right)^{\frac{1}{\binom{n-1}{k}}}, 1 - \left(\prod\limits_{\substack{1 \le p_1 < \\ \ldots < p_k \le n}} \left(1 - \left(\prod\limits_{j=1}^{k} \varphi^*_{p_j}\right)^{\frac{1}{k}}\right)^{\left(1 - \sum\limits_{j=1}^{k} w_{p_j}\right)} \right)^{\frac{1}{\binom{n-1}{k}}}, \\ 1 - \left(\prod\limits_{\substack{1 \le p_1 < \\ \ldots < p_k \le n}} \left(1 - \left(\prod\limits_{j=1}^{k} \vartheta_{p_j}\right)^{\frac{1}{k}}\right)^{\left(1 - \sum\limits_{j=1}^{k} w_{p_j}\right)} \right)^{\frac{1}{\binom{n-1}{k}}}, \left(\prod\limits_{\substack{1 \le p_1 < \\ \ldots < p_k \le n}} \left(1 - \left(\prod\limits_{j=1}^{k} \left(1 - \vartheta^*_{p_j}\right)\right)^{\frac{1}{k}}\right)^{\left(1 - \sum\limits_{j=1}^{k} w_{p_j}\right)} \right)^{\frac{1}{\binom{n-1}{k}}} \end{array} \right) \end{aligned}$$

Step 5: Compute the score value of the l_p by using Equation (10).

Step 6: Rank all the alternatives by using an order relation defined in Definition 11 and hence select the most feasible alternative(s).

6. Illustrative Example

The above mentioned approach has been illustrate with a numerical example which is stated as below.

6.1. A Case Study

Jharkhand is the eastern state of the India, which has the 40 percent mineral resources of the country and second leading state of the mineral wealth after Chhattisgarh state. It is also known for its vast forest resources. Jamshedpur, Bokaro and Dhanbad cities of the Jharkhand are famous for industries in all over the world. After that, it is the widespread poverty state of the India because it is the primarily a rural state as 76 percent of the population live in the villages which depend on the agriculture and wages. Only 30 percent villages are connected by roads while only 55 percent villages have accessed to electricity and other facilities. But in the today's life, everyone is changing fast to himself for a better life, therefore, everyone moves to the urban cities for a better job. To stop this emigration, Jharkhand government wants to set up the industries based on the agriculture in the rural areas. For this, the government has been organized "MOMENTUM JHARKHAND" global investor submit 2017 in Ranchi to invite the companies for investment in the rural areas. Government announced the various facilities for setup the five food processing plants in the rural areas and consider the six attributes required for company selection to setup them, namely, project cost (G_1), completion time (G_2), technical capability (G_3), financial status (G_4), company background (G_5), reference from previous project (G_6) and assign the weights of relative importance of each attributes. The six companies taken as in the form of the alternatives, namely, Surya Food and Agro Pvt. Ltd. (A_1), Mother Dairy Fruit and Vegetable Pvt. Ltd. (A_2), Parle Products Ltd. (A_3), Heritage Food Ltd. (A_4), Verka Pvt. Ltd. (A_5) and Reliance Pvt. Ltd. (A_6) interested for these projects. Then the main object of the government is to choose the best company among them for the task. In order to find the best feasible alternative(s) for the required task, the authority called an expert to evaluate these alternatives and rate their preferences in terms of linguistic terms (LTs). The standardized LTs such as "Very High" (VH), "High"(H), "Medium"(M), "Medium Low"(ML), "Low"(L), "Very Low"(VL) are defined in terms of STIT2IFNs given in Table 1. Furthermore, the complementary relation corresponding to LTs is presented in Table 2.

Table 1. Linguistic grade and coressponding values.

LTs	Triangular IT2IFNs
VL	(0.20,0.10,0.60,0.65,0.35,0.30)
L	(0.30,0.10,0.65,0.70,0.30,0.25)
ML	(0.40,0.20,0.70,0.75,0.20,0.18)
M	(0.50,0.20,0.75,0.80,0.16,0.15)
MH	(0.60,0.30,0.80,0.85,0.13,0.12)
H	(0.70,0.30,0.85,0.90,0.10,0.08)
VH	(0.80,0.40,0.90,0.95,0.07,0.03)

Table 2. Linguistic grades and compliments.

LT	VL	L	ML	M	MH	H	VH
Complemented LT	VH	H	MH	M	ML	L	VL

The above mentioned steps are executed to locate the best alternative(s).

Step 1. An expert has evaluated each alternative and present their rating values in terms of LTs which are summarized as

$$\overline{L} = \begin{array}{c} \\ A_1 \\ A_2 \\ A_3 \\ A_4 \\ A_5 \\ A_6 \end{array} \begin{array}{c} \begin{array}{ccccccc} C_1 & C_2 & C_3 & C_4 & C_5 & C_6 & C_7 \end{array} \\ \begin{pmatrix} VH & H & M & MH & H & VH & H \\ M & ML & H & VH & H & VH & VH \\ H & VH & VH & M & MH & L & VL \\ MH & VL & MH & H & VL & MH & H \\ VH & H & VL & H & M & VL & L \\ ML & VL & VH & M & VL & L & H \end{pmatrix} \end{array} \tag{26}$$

Step 2: As the criteria C_1 and C_2 are the cost type, so we normalize their rating values by using Table 2 and Equation (25), we get

$$L = \begin{matrix} & C_1 & C_2 & C_3 & C_4 & C_5 & C_6 & C_7 \\ A_1 \\ A_2 \\ A_3 \\ A_4 \\ A_5 \\ A_6 \end{matrix} \begin{pmatrix} VL & L & M & MH & H & VH & H \\ M & MH & H & VH & H & VH & VH \\ L & VL & VH & M & MH & L & VL \\ ML & VH & MH & H & VL & MH & H \\ VL & L & VL & H & M & VL & L \\ MH & VH & VH & M & VL & L & H \end{pmatrix} \quad (27)$$

Step 3: Apply the Algorithm 1 to compute the weight vector to each criteria. For it, we follows the steps of the algorithm and summarized as below

(a) By using Equation (20), construct the relative coefficient matrix Δ for each criteria as

$$\Delta = \begin{matrix} & C_1 & C_2 & C_3 & C_4 & C_5 & C_6 & C_7 \\ C_1 & 1 & 0.9666 & 0.9344 & 0.9094 & 0.9044 & 0.9174 & 0.9344 \\ C_2 & 0.9666 & 1 & 0.9344 & 0.9311 & 0.8444 & 0.9144 & 0.9694 \\ C_3 & 0.9344 & 0.9344 & 1 & 0.9344 & 0.9014 & 0.9144 & 0.9374 \\ C_4 & 0.9094 & 0.9314 & 0.9344 & 1 & 0.9414 & 0.9464 & 0.9574 \\ C_5 & 0.9044 & 0.8444 & 0.9014 & 0.9414 & 1 & 0.9504 & 0.9004 \\ C_6 & 0.9174 & 0.9144 & 0.9144 & 0.9464 & 0.9504 & 1 & 0.9714 \\ C_7 & 0.9344 & 0.9694 & 0.9374 & 0.9574 & 0.9004 & 0.9714 & 1 \end{matrix}$$

(b) The relative coefficient sum of each criteria is computed by using Equation (22) and get

$$\Delta_1 = 5.564,\ \Delta_2 = 5.558,\ \Delta_3 = 5.554,\ \Delta_4 = 5.618,$$
$$\Delta_5 = 5.440,\ \Delta_6 = 5.612,\ \Delta_7 = 5.668.$$

(c) By using Equation (23), the weight vector of each criteria is obtained as

$$w_1 = 0.1431,\ w_2 = 0.1432,\ w_3 = 0.1433,\ w_4 = 0.1417,$$
$$w_5 = 0.1463,\ w_6 = 0.1419,\ w_7 = 0.1405.$$

Step 4: Aggregate all the values by using WSTIT2IFHM operator into a collective one $l_p(p = 1, 2, \ldots, 6)$. Here, without loss of generality, we take $k = 2$ and the obtained results are

$$\begin{aligned} l_1 &= \text{WSTIT2IFHM}_w^{(2)}(l_{11}, l_{12}, \cdots, l_{17}) \\ &= \left(\begin{array}{l} \dfrac{\sum\limits_{1 \le p_1 < p_2 \le 7} \left(1 - \sum\limits_{j=1}^{2} w_{1_j}\right) \left(\prod\limits_{j=1}^{2} \zeta_{1_j}\right)^{\frac{1}{2}}}{\binom{6}{2}}, \dfrac{\sum\limits_{1 \le p_1 < p_2 \le 7} \left(1 - \sum\limits_{j=1}^{2} w_{p_j}\right) \left(\prod\limits_{j=1}^{2} \varrho_{p_j}\right)^{\frac{1}{2}}}{\binom{6}{2}}, \\ \left(\prod\limits_{1 \le p_1 < p_2 \le 7} \left(1 - \left(\prod\limits_{j=1}^{2} \left(1 - \varphi_{p_j}\right)\right)^{\frac{1}{2}}\right)^{\left(1 - \sum\limits_{j=1}^{2} w_{p_j}\right)} \right)^{\frac{1}{\binom{6}{2}}}, 1 - \left(\prod\limits_{1 \le p_1 < p_2 \le 7} \left(1 - \left(\prod\limits_{j=1}^{2} \varphi^*_{p_j}\right)^{\frac{1}{2}}\right)^{\left(1 - \sum\limits_{j=1}^{2} w_{p_j}\right)} \right)^{\frac{1}{\binom{6}{2}}}, \\ 1 - \left(\prod\limits_{1 \le p_1 < p_2 \le 7} \left(1 - \left(\prod\limits_{j=1}^{2} \vartheta_{p_j}\right)^{\frac{1}{2}}\right)^{\left(1 - \sum\limits_{j=1}^{2} w_{p_j}\right)} \right)^{\frac{1}{\binom{6}{2}}}, \left(\prod\limits_{1 \le p_1 < p_2 \le 7} \left(1 - \left(\prod\limits_{j=1}^{2} \left(1 - \vartheta^*_{p_j}\right)\right)^{\frac{1}{2}}\right)^{\left(1 - \sum\limits_{j=1}^{2} w_{p_j}\right)} \right)^{\frac{1}{\binom{6}{2}}} \end{array} \right) \\ &= (0.5154, 0.2276, 0.7820, 0.8314, 0.1596, 0.1339) \end{aligned}$$

Similarly, we have

$$
\begin{aligned}
l_2 &= (0.6950, 0.3239, 0.8546, 0.9053, 0.0974, 0.0687);\\
l_3 &= (0.3846, 0.1681, 0.7166, 0.7633, 0.2243, 0.1927);\\
l_4 &= (0.5481, 0.2612, 0.7952, 0.8449, 0.1436, 0.1210);\\
l_5 &= (0.3201, 0.1342, 0.6769, 0.7244, 0.2642, 0.2292);\\
l_6 &= (0.5272, 0.2390, 0.7914, 0.8414, 0.1536, 0.1232)
\end{aligned}
$$

Step 5: The score values of $l_p (p = 1, 2, \ldots, 6)$ are computed by Equation (10) and get

$$
\begin{aligned}
s(l_1) &= (0.3404, 0.0375); s(l_2) = (0.5550, 0.0396); s(l_3) = (0.2046, 0.0392)\\
s(l_4) &= (0.3770, 0.0362); s(l_5) = (0.1455, 0.0412); s(l_6) = (0.3579, 0.0402)
\end{aligned}
$$

Step 6: Since $s_x(l_2) > s_x(l_4) > s_x(l_6) > s_x(l_1) > s_x(l_3) > s_x(l_5)$ and thus by Definition 11, we get the ranking order of the alternatives as $A_2 \succ A_4 \succ A_6 \succ A_1 \succ A_3 \succ A_5$. Here "$\succ$" means "preferred to". Therefore, A_2 is the best alternative.

6.2. Influence of k on Alternatives

Keeping in mind the end goal to investigate the impact of the parameter k on to the final positioning order of the alternatives, we use an alternate estimation of k in our test. Here n is 7 in our case, so we shift k from 1 to 7 and their outcomes relating to the proposed technique have been outlined in Table 3. From this table, it is seen that with the expansion of the interaction of the multi-input options, the general score estimations of it diminishes which recommend that the proposed operator reflect the risk preferences to the decision makers. This examination will propose the distinctive decisions to the analyst as indicated by his/her decision. For example, in the event that he will cover the risk parameters during the aggregation then they will allocate a little incentive to the parameter k with the goal that score esteems increments while, if the analyst is pessimistic in nature towards the choice then the bigger estimation of k can be allocated during the procedure.

Table 3. Effect of k on to ranking of alternatives.

Value of k	Score Values (s_x, s_y) of the Alternatives						Ranking Order
	A_1	A_2	A_3	A_4	A_5	A_6	
1	(0.3615, 0.0762)	(0.5627, 0.0523)	(0.2268, 0.0872)	(0.3953, 0.0677)	(0.1577, 0.0702)	(0.3836, 0.0856)	$A_2 \succ A_4 \succ A_6 \succ A_1 \succ A_3 \succ A_5$
2	(0.3404, 0.0375)	(0.5550, 0.0396)	(0.2046, 0.0392)	(0.3770, 0.0362)	(0.1455, 0.0412)	(0.3579, 0.0402)	$A_2 \succ A_4 \succ A_6 \succ A_1 \succ A_3 \succ A_5$
3	(0.3324, 0.0241)	(0.5526, 0.0840)	(0.1997, 0.0250)	(0.3702, 0.0268)	(0.1427, 0.0321)	(0.3484, 0.0240)	$A_2 \succ A_4 \succ A_6 \succ A_1 \succ A_3 \succ A_5$
4	(0.3285, 0.0177)	(0.5507, 0.0329)	(0.1976, 0.0181)	(0.3656, 0.0203)	(0.1415, 0.0275)	(0.3437, 0.0161)	$A_2 \succ A_4 \succ A_6 \succ A_1 \succ A_3 \succ A_5$
5	(0.3260, 0.0138)	(0.5498, 0.0314)	(0.1964, 0.0141)	(0.3631, 0.0170)	(0.1409, 0.0247)	(0.3408, 0.0115)	$A_2 \succ A_4 \succ A_6 \succ A_1 \succ A_3 \succ A_5$
6	(0.3244, 0.0113)	(0.5492, 0.0304)	(0.1957, 0.0114)	(0.3613, 0.0148)	(0.1405, 0.0228)	(0.3389, 0.0086)	$A_2 \succ A_4 \succ A_6 \succ A_1 \succ A_3 \succ A_5$
7	(0.3232, 0.0095)	(0.5488, 0.0298)	(0.1952, 0.0094)	(0.3601, 0.0131)	(0.1402, 0.0215)	(0.3376, 0.0064)	$A_2 \succ A_4 \succ A_6 \succ A_1 \succ A_3 \succ A_5$

Furthermore, in some other existing Bonferroni mean (BM) and generalized Bonferroni mean (GBM) operators, the information takes only two or three arguments during an aggregation. Also, in BM operator there is need of two additional parameters (p, q) while the three parameters (p, q, r) for GBM from an infinite rational set. Thus, the computational complexity is too high in such cases. On the other hand, in the proposed operator, there is only one parameter k from a finite integer set and hence the computational complexity is low and easier to understand. Finally, the several operators such as averaging, BM and geometric for the T2IFNs can be deduced from the proposed ones by setting $k = 1$, $k = 2$ and $k = n$ respectively. Subsequently, our proposed operator and the strategy are more summed up and adaptable to tackle the decision-making problems.

6.3. Comparative Study

In this section, we perform some comparative analysis of the proposed method result with some of the existing approaches results in [36,46–48] under the uncertain environment. The results computed from them on to the considered problem are summarized as below:

1. In [36], authors proposed the weighted geometric Bonferroni mean operator under the type-2 fuzzy environment, denoted by IT2FWGBM, which is defined as

$$\begin{aligned} d_k &= \text{IT2FWGBM}_w^{p,q}(A_1, A_2, \ldots, A_m) \\ &= \frac{1}{p+q}\left(\bigotimes_{\substack{i,j=1 \\ i\neq j}}^{m} \left(p(A_i)^{w_i} \oplus q(A_j)^{w_j}\right)\right)^{1/m(m-1)} \end{aligned} \tag{28}$$

   By applying Equation (28) on to the considered data, we get the aggregated value corresponding to each alternative as

$$\begin{aligned} d_1 &= \text{IT2FWGBM}_w^{1,1}(A_{11}, A_{12}, A_{13}, A_{14}, A_{15}, A_{16}, A_{17}) \\ &= (0.8321, 0.9050, 0.9050, 0.9534, 0.6065) \\ d_2 &= \text{IT2FWGBM}_w^{1,1}(A_{21}, A_{22}, A_{23}, A_{24}, A_{25}, A_{26}, A_{27}) \\ &= (0.8671, 0.9486, 0.9486, 1.0000, 0.7500) \\ d_3 &= \text{IT2FWGBM}_w^{1,1}(A_{31}, A_{32}, A_{33}, A_{34}, A_{35}, A_{36}, A_{37}) \\ &= (0.7980, 0.8676, 0.8676, 0.9137, 0.6015) \\ d_4 &= \text{IT2FWGBM}_w^{1,1}(A_{41}, A_{42}, A_{43}, A_{44}, A_{45}, A_{46}, A_{47}) \\ &= (0.8317, 0.9131, 0.9131, 0.9656, 0.6080) \\ d_5 &= \text{IT2FWGBM}_w^{1,1}(A_{51}, A_{52}, A_{53}, A_{54}, A_{55}, A_{56}, A_{57}) \\ &= (0.7802, 0.8456, 0.8456, 0.8895, 0.6085) \\ d_6 &= \text{IT2FWGBM}_w^{1,1}(A_{61}, A_{62}, A_{63}, A_{64}, A_{65}, A_{66}, A_{67}) \\ &= (0.8318, 0.9073, 0.9073, 0.9569, 0.6000) \end{aligned}$$

   Therefore, the score values of these aggregated numbers are $s(d_1) = 0.5405$, $s(d_2) = 0.7079$, $s(d_3) = 0.5182$, $s(l_4) = 0.5450$, $s(l_5) = 0.5052$, and $s(l_6) = 0.5418$ and hence the final ranking of all alternatives $A_k (k = 1, 2, \ldots, 6)$ is found as

$$A_2 \succ A_4 \succ A_6 \succ A_1 \succ A_3 \succ A_5$$

2. If we use the existing WSTIT2FHM operator as proposed by Qin [46] under the T2FS environment

$$
\begin{aligned}
l_p &= \text{WSTIT2FHM}^{(k)}(A_1, A_2, \ldots, A_n) \\
&= \left(\begin{array}{l}
\dfrac{\sum\limits_{\substack{1 \le i_1 < \\ \ldots < i_k \le n}} \left(1 - \sum\limits_{j=1}^{k} w_{i_j}\right) \left(\prod\limits_{j=1}^{k} \zeta_{\alpha_{i_j}}\right)^{\frac{1}{k}}}{\binom{n-1}{k}}, \dfrac{\sum\limits_{\substack{1 \le i_1 < \\ \ldots < i_k \le n}} \left(1 - \sum\limits_{j=1}^{k} w_{i_j}\right) \left(\prod\limits_{j=1}^{k} \varrho_{\alpha_{i_j}}\right)^{\frac{1}{k}}}{\binom{n-1}{k}}, \\
\left(\prod\limits_{\substack{1 \le i_1 < \\ \ldots < i_k \le n}} \left(1 - \left(\prod\limits_{j=1}^{k} \left(1 - \varphi_{\alpha_{i_j}}\right)\right)^{\frac{1}{k}}\right)^{\left(1 - \sum\limits_{j=1}^{k} w_{i_j}\right)} \right)^{\frac{1}{\binom{n-1}{k}}}, \\
1 - \left(\prod\limits_{\substack{1 \le i_1 < \\ \ldots < i_k \le n}} \left(1 - \left(\prod\limits_{j=1}^{k} \varphi^{*}_{\alpha_{i_j}}\right)^{\frac{1}{k}}\right)^{\left(1 - \sum\limits_{j=1}^{k} w_{i_j}\right)} \right)^{\frac{1}{\binom{n-1}{k}}}
\end{array} \right)
\end{aligned} \tag{29}
$$

then, the aggregated values corresponding to each alternative (by taking $k = 2$) are obtained as

$$
\begin{aligned}
l_1 &= (0.5154, 0.2276, 0.7820, 0.8314); l_2 = (0.6950, 0.3239, 0.8546, 0.9054) \\
l_3 &= (0.3846, 0.1681, 0.7166, 0.7633); l_4 = (0.5481, 0.2612, 0.7951, 0.8449) \\
l_5 &= (0.3201, 0.1342, 0.6769, 0.7244); l_6 = (0.5272, 0.2390, 0.7914, 0.8414)
\end{aligned}
$$

Thus, the score values are

$$
\begin{aligned}
s(l_1) &= (0.2077, 0.8067); s(l_2) = (0.3055, 0.8799); s(l_3) = (0.1422, 0.7400) \\
s(l_4) &= (0.2245, 0.8200); s(l_5) = (0.1120, 0.7006); s(l_6) = (0.2150, 0.8164)
\end{aligned}
$$

and hence ordering is

$$A_2 \succ A_4 \succ A_6 \succ A_1 \succ A_3 \succ A_5$$

From the above examinations, it is revealed that the ranking order of the alternatives stays same yet the computational procedure is altogether unique. For instance, in [36,46] authors have introduced AOs under TIT2FNs by considering just the degree of membership during an examination. But it is quite recognizable that the level of non-membership likewise assumes a predominant part during the aggregation process. Thus, the outcomes processes by these methodologies [36,46] might be unreasonable under some specific constraints where the degree of non-membership pays a more significance than the degree of agreement.

However, apart from these, we give some characteristics comparison of our proposed method and the aforementioned methods, which are listed in Table 4.

Table 4. The characteristic comparisons of different methods.

Methods	Whether Captures Interrelationship of Two Aggregated Arguments	Whether Captures Interrelationship of Multiple Aggregated Arguments	Whether It Makes the Method Flexible by the Parameter Vector	Whether Criteria Weights Are Depends on the Collective Information	Whether Describe Information Using Linguistic Features	Whether Flexible to Express a Wider Range of Information
Gong et al. [36]	✓	×	×	×	×	×
Liu and Wang [47]	×	×	×	×	×	×
Pedrycz and Song [48]	×	×	×	✓	✓	×
Qin [46]	✓	✓	✓	×	✓	×
The proposed method	✓	✓	✓	✓	✓	✓

In [47], authors presented an analytical method for solving the problems by using the fuzzy weighted average. In [36], the authors have presented the BM by considering simultaneously the values of UMF and LMF to aggregate IT2FS information. On the other hand, the present study is based on the HM operator which is more adaptable and robustness in process of information fusion than others such as BM, GBM. The outstanding characteristic of the HM operator is to catch the inter-relationship between more than two input arguments with a parameter k from the finite integer set. Furthermore, in [46], the author developed HM operator by taking into account the membership degree only but in practical problems, it is sometimes not possible for DM to give their preferences in terms of acceptance degree only. Therefore, the non-membership degree is required for handling the problems in which rejection degree is not equal to one minus acceptance degree. Also by comparing with the AHP-based method [48], the proposed method does not require any software package to compute the results while the technique proposed in [48] requires it. Thus, the computation complexity of the proposed technique is comparatively easy. Furthermore, the AHP-based technique is usually dependent on various parameters and thus the final ranking may some time suffers from inconsistency, in the case of inappropriate parameter selection. On the other hand, the proposed method draws up a more authentic ranking result as it can terminate the difference, draws up for the flaws of already existing aggregation methods that do not capture experts utility or decision preference and achieves more stationary and commendable interrelationships result with less information loss. The proposed method takes into consideration the uniformity of the alternatives as well as highlights the significance and interactions in association with any solutions of alternatives. On the other hand, the AHP-based technique is good at calculating only the optimal ranking values of the alternatives beyond inter-relationships.

7. Conclusions

In this paper, an endeavor has been made to exhibit the some new AOs to accommodate the IT2IF conditions. IT2IFS is one of the augmentations of the conventional FS, IFS by considering grades of the primary membership functions also. On the other hand, in practical application problems, the criteria interrelationship phenomenon occurs frequently. To address it, Hamy means (HM) operator is a standout among the most critical operators that catches the inter-relationship together with the multi-input arguments. Furthermore, to diminish the computational complexity of the IT2IFS, we introduce symmetric IT2IFS and characterize some operation laws. Then, keeping the advantages of STIT2IFS and HM operators, we exhibit the symmetric TIT2 intuitionistic fuzzy HM (STIT2IFHM) operator and weighted symmetric TIT2 intuitionistic fuzzy HM (WSTIT2IFHM) operator under a provision of type-2 intuitionistic uncertain situation. Various beneficial characteristics of these operators have endorsed. Furthermore, in light of these operators, a decision-making approach is introduced to solve the MCDM problems. The presented approach has been tried and clarified with a numerical illustration and registered that it can efficiently deal with the available information by eliminating more amount of fuzziness as compared to the existing approaches. The major advantages of the proposed operator with respect to the existing ones are that it need only one parameter k from a finite integer set while other needs more than one from an infinite rational set such as BM and GBM etc., and hence the computational complexity is low and easier to understand. Additionally, a portion of the existing studies can be effectively concluded from the proposed operators by setting $k = 1$, $k = 2$ and $k = n$. Thus, it expresses a better technique for taking care of the decision-making problems with additional benefits.

In future research, we shall extend the present study to some more generalized environment and applied it to many other fields such as graph theory, transportation evaluation, resource management using different uncertain environments [57–62].

Author Contributions: Conceptualization, H.G. and S.S.; Methodology, H.G.; Validation and Investigation, S.S. and H.G.; Writing-Original Draft Preparation, H.G.; Writing-Review and Editing, H.G.

Funding: This research received no external funding.

Conflicts of Interest: The authors declare no conflict of interest.

References

1. Garg, H.; Kumar, K. An advanced study on the similarity measures of intuitionistic fuzzy sets based on the set pair analysis theory and their application in decision making. *Soft Comput.* **2018**, *22*, 4959–4970. [CrossRef]
2. Garg, H.; Arora, R. Dual hesitant fuzzy soft aggregation operators and their application in decision making. *Cogn. Comput.* **2018**, 1–21. [CrossRef]
3. Zadeh, L.A. Fuzzy sets. *Inf. Control* **1965**, *8*, 338–353. [CrossRef]
4. Atanassov, K.T. Intuitionistic fuzzy sets. *Fuzzy Sets Syst.* **1986**, *20*, 87–96. [CrossRef]
5. Zadeh, L.A. The concept of a linguistic variable and its application to approximate reasoning: Part-1. *Inf. Sci.* **1975**, *8*, 199–251. [CrossRef]
6. Xu, Z.S.; Yager, R.R. Some geometric aggregation operators based on intuitionistic fuzzy sets. *Int. J. Gen. Syst.* **2006**, *35*, 417–433. [CrossRef]
7. Garg, H. Novel intuitionistic fuzzy decision making method based on an improved operation laws and its application. *Eng. Appl. Artif. Intell.* **2017**, *60*, 164–174. [CrossRef]
8. Garg, H. Generalized intuitionistic fuzzy interactive geometric interaction operators using Einstein t-norm and t-conorm and their application to decision making. *Comput. Ind. Eng.* **2016**, *101*, 53–69. [CrossRef]
9. Zhao, H.; Xu, Z.; Ni, M.; Liu, S. Generalized aggregation operators for intuitionistic fuzzy sets. *Int. J. Intell. Syst.* **2010**, *25*, 1–30. [CrossRef]
10. Kaur, G.; Garg, H. Generalized cubic intuitionistic fuzzy aggregation operators using t-norm operations and their applications to group decision-making process. *Arab. J. Sci. Eng.* **2018**. [CrossRef]
11. Garg, H. Generalized interaction aggregation operators in intuitionistic fuzzy multiplicative preference environment and their application to multicriteria decision making. *Appl. Intell.* **2018**, *48*, 2120–2136. [CrossRef]
12. Xu, Z.S.; Yager, R.R. Intuitionistic fuzzy bonferroni means. *IEEE Trans. Syst. Man Cybern.* **2011**, *41*, 568–578.
13. Xia, M.; Xu, Z.; Zhu, B. Geometric bonferroni means with their application in multi criteria decision making. *Knowl.-Based. Syst.* **2013**, *40*, 88–100. [CrossRef]
14. Garg, H. New exponential operational laws and their aggregation operators for interval-valued Pythagorean fuzzy multicriteria decision making. *Int. J. Intell. Syst.* **2018**, *33*, 653–683. [CrossRef]
15. Garg, H. Some robust improved geometric aggregation operators under interval-valued intuitionistic fuzzy environment for multi-criteria decision-making process. *J. Ind. Manag. Optim.* **2018**, *14*, 283–308. [CrossRef]
16. Qin, J.; Liu, X. An approach to intuitionistic fuzzy multiple attribute decision making based on Maclaurin symmetric mean operators. *J. Intell. Fuzzy Syst.* **2014**, *27*, 2177–2190.
17. Qin, J.; Liu, X. Approaches to uncertain linguistic multiple attribute decision making based on dual Maclaurin symmetric mean. *J. Intell. Fuzzy Syst.* **2015**, *29*, 171–186. [CrossRef]
18. Garg, H.; Arora, R. Novel scaled prioritized intuitionistic fuzzy soft interaction averaging aggregation operators and their application to multi criteria decision making. *Eng. Appl. Artif. Intell.* **2018**, *71*, 100–112. [CrossRef]
19. Garg, H. Generalized Pythagorean fuzzy geometric interactive aggregation operators using Einstein operations and their application to decision making. *J. Exp. Theor. Artif. Intell.* **2018**, 1–32. [CrossRef]
20. Mendel, J.M. *Uncertain Rule-Based fuzzy Logic System: Introduction and New Directions*; Prentice-Hall: Englewood Cliffs, NJ, USA, 2001.
21. Xu, Z.S. Intuitionistic fuzzy aggregation operators. *IEEE Trans. Fuzzy Syst.* **2007**, *15*, 1179–1187.
22. Kumar, K.; Garg, H. Connection number of set pair analysis based TOPSIS method on intuitionistic fuzzy sets and their application to decision making. *Appl. Intell.* **2018**, *48*, 2112–2119. [CrossRef]
23. Peng, X.; Dai, J.; Garg, H. Exponential operation and aggregation operator for q-rung orthopair fuzzy set and their decision-making method with a new score function. *Int. J. Intell. Syst.* **2018**, 1–28. [CrossRef]
24. Kumar, K.; Garg, H. TOPSIS method based on the connection number of set pair analysis under interval-valued intuitionistic fuzzy set environment. *Comput. Appl. Math.* **2018**, *37*, 1319–1329. [CrossRef]

25. Herrera, F.; Martínez, L. A model based on linguistic 2-tuples for dealing with multigranular hierarchical linguistic contexts in multi-expert decision-making. *IEEE Trans. Syst. Man Cybern. Part B (Cybernetics)* **2001**, *31*, 227–234. [CrossRef] [PubMed]
26. Xu, Z. An approach based on similarity measure to multiple attribute decision making with trapezoid fuzzy linguistic variables. In Proceedings of the 2nd International Conference on Fuzzy Systems and Knowledge Discovery, Changsha, China, 27–29 August 2005; pp. 110–117.
27. Zhang, H. Linguistic Intuitionistic fuzzy sets and application in MAGDM. *J. Appl. Math.* **2014**, *2014*, 432092. [CrossRef]
28. Chen, Z.; Liu, P.; Pei, Z. An approach to multiple attribute group decision making based on linguistic intuitionistic fuzzy numbers. *J. Comput. Intell. Syst.* **2015**, *8*, 747–760. [CrossRef]
29. Garg, H.; Kumar, K. Some aggregation operators for linguistic intuitionistic fuzzy set and its application to group decision-making process using the set pair analysis. *Arab. J. Sci. Eng.* **2018**, *43*, 3213–3227. [CrossRef]
30. Garg, H.; Kumar, K. Group decision making approach based on possibility degree measures and the linguistic intuitionistic fuzzy aggregation operators using einstein norm operations. *J. Mult.-Valued. Log. Soft Comput.* **2018**, *31*, 175–209.
31. Mendel, J.M.; John, R.I.; LIu, F. Interval type-2 fuzzy logic systems made simple. *IEEE Trans. Fuzzy Syst.* **2006**, *14*, 808–821. [CrossRef]
32. Chen, T.Y.; Chang, C.H.; Lu, J.F.R. The extended QUALIFLEX method for multiple criteria decision analysis based on interval type-2 fuzzy sets and applications to medical decision making. *Eur. J. Oper. Res.* **2013**, *226*, 615–625. [CrossRef]
33. Chen, T.Y. An ELECTRE-based outranking method for multiple criteria group decision making using interval type-2 fuzzy sets. *Inf. Sci.* **2014**, *263*, 1–21. [CrossRef]
34. Wu, D.; Mendel, J.M. Aggregation using the linguistic weighted average and interval type-2 fuzzy sets. *IEEE Trans. Fuzzy Syst.* **2007**, *15*, 1145–1161. [CrossRef]
35. Qin, J.; Liu, X. Frank aggregation operators for triangular interval type-2 fuzzy set and its application in multiple attribute group decision making. *J. Appl. Math.* **2014**, *2014*, 923213. [CrossRef]
36. Gong, Y.; Hu, N.; Zhang, J.; Liu, G.; Deng, J. Multi-attribute group decision making method based on geometric Bonferroni mean operator of trapezoidal interval type-2 fuzzy numbers. *Comput. Ind. Eng.* **2015**, *81*, 167–176. [CrossRef]
37. Lee, L.W.; Chen, S.M. Fuzzy multiple attributes group decision-making based on the extension of TOPSIS method and interval type-2 fuzzy sets. In Proceedings of the 2008 International Conference on Machine Learning and Cybernetics, Kunming, China, 2–15 July 2008; pp. 3260–3265.
38. Zhao, T.; Xia, J. Type-2 intuitionistic fuzzy sets. *Control Theory Appl.* **2012**, *29*, 1215–1222.
39. Singh, S.; Garg, H. Distance measures between type-2 intuitionistic fuzzy sets and their application to multicriteria decision-making process. *Appl. Intell.* **2017**, *46*, 788–799. [CrossRef]
40. Mendel, J.M.; Wu, H. Type-2 fuzzistics for symmetric interval type-2 fuzzy sets: Part 1, forward problems. *IEEE Trans. Fuzzy Syst.* **2006**, *14*, 781–792. [CrossRef]
41. Jana, D.K. Novel arithmetic operations on type-2 intuitionistic fuzzy and its applications to transportation problem. *Pac. Sci. Rev. A Nat. Sci. Eng.* **2016**, *18*, 178–189. [CrossRef]
42. Mendel, J.M.; Mouzouris, G.C. Type-2 fuzzy logic systems. *IEEE Trans. Fuzzy Syst.* **1999**, *7*, 643–658.
43. Mendel, J.M.; Wu, H. Type-2 fuzzistics for symmetric interval type-2 fuzzy sets: Part 2, inverse problems. *IEEE Trans. Fuzzy Syst.* **2007**, *15*, 301–308. [CrossRef]
44. Chiao, K.P. Multiple criteria group decision making with triangular interval type-2 fuzzy sets. In Proceedings of the 2011 IEEE International Conference on Fuzzy Systems, Taipei, Taiwan, 27–30 June 2011; pp. 2572–2582.
45. Wang, W.; Liu, X.; Qin, Y. Multi-attribute group decision making models under interval type-2 fuzzy environment. *Knowl.-Based. Syst.* **2012**, *30*, 121–128. [CrossRef]
46. Qin, J. Interval type-2 fuzzy Hamy mean operators and their application in multiple criteria decision making. *Granul. Comput.* **2017**, *2*, 249–269. [CrossRef]
47. Liu, X.; Wang, Y.M. An analytical solution method for the generalized fuzzy weighted average problem. *Int. J. Uncertain. Fuzz. Knowl.-Based Syst.* **2013**, *21*, 455–480. [CrossRef]
48. Pedrycz, W.; Song, M. A granulation of linguistic information in AHP decision-making problems. *Inf. Fusion* **2014**, *17*, 93–101. [CrossRef]

49. Garg, H.; Arora, R. Bonferroni mean aggregation operators under intuitionistic fuzzy soft set environment and their applications to decision-making. *J. Oper. Res. Soc.* **2018**, 1–14. [CrossRef]
50. Kaur, G.; Garg, H. Multi-Attribute decision-making based on bonferroni mean operators under cubic intuitionistic fuzzy set environment. *Entropy* **2018**, *20*, 65. [CrossRef]
51. Hara, T.; Uchiyama, M.; Takahasi, S.E. A refinement of various mean inequalities. *J. Inequal. Appl.* **1998**, *1998*, 932025. [CrossRef]
52. Garg, H.; Nancy. Multi-criteria decision-making method based on prioritized Muirhead mean aggregation operator under Neutrosophic set environment. *Symmetry* **2018**, *10*, 280. [CrossRef]
53. Guan, K. The Hamy symmetric function and its generalization. *Math. Inequal. Appl.* **2006**, *9*, 797–805. [CrossRef]
54. Jiang, W.D. Some properties of dual form of the Hamy's symmetric function. *J. Math. Inequal.* **2007**, *1*, 117–125. [CrossRef]
55. Garg, H.; Singh, S. A novel triangular interval type-2 intuitionistic fuzzy sets and their aggregation operators. *Iran. J. Fuzzy Syst.* **2018**. [CrossRef]
56. Spearman, C. The proof and measurement of association between two things. *Am. J. Psychol.* **1904**, *15*, 72–101. [CrossRef]
57. Garg, H.; Arora, R. A nonlinear-programming methodology for multi-attribute decision-making problem with interval-valued intuitionistic fuzzy soft sets information. *Appl. Intell.* **2018**, *48*, 2031–2046. [CrossRef]
58. Peng, X.D.; Garg, H. Algorithms for interval-valued fuzzy soft sets in emergency decision making based on WDBA and CODAS with new information measure. *Comput. Ind. Eng.* **2018**, *119*, 439–452. [CrossRef]
59. Garg, H.; Rani, D. Some generalized complex intuitionistic fuzzy aggregation operators and their application to multicriteria decision-making process. *Arab. J. Sci. Eng.* **2018**, 1–20. [CrossRef]
60. Garg, H. Linguistic Pythagorean fuzzy sets and its applications in multiattribute decision-making process. *Int. J. Intell. Syst.* **2018**, *33*, 1234–1263. [CrossRef]
61. Selvachandran, G.; Garg, H.; Alaroud, M.H.S.; Salleh, A.R. Similarity measure of complex vague soft sets and its application to pattern recognition. *Int. J. Fuzzy Syst.* **2018**, *20*, 1901–1914. [CrossRef]
62. Selvachandran, G.; Garg, H.; Quek, S.G. Vague entropy measure for complex vague soft sets. *Entropy* **2018**, *20*, 403. [CrossRef]

symmetry

MDPI

Article

Probabilistic Hesitant Intuitionistic Linguistic Term Sets in Multi-Attribute Group Decision Making

M. G. Abbas Malik [1,2], Zia Bashir [3,*], Tabasam Rashid [4] and Jawad Ali [3]

1 Department of IT, AGI Education, Auckland 1051, New Zealand; abbas@agi.ac.nz or Muhammad.Malik@op.ac.nz
2 Department of Information Technology, Otago Polytechnic, Auckland International Campus, Auckland 1141, New Zealand
3 Department of Mathematics, Quaid-i-Azam University, Islamabad 45320, Pakistan; jawadsabirr@gmail.com
4 Department of Mathematics, University of Management and Technology, Lahore 54770, Pakistan; tabasam.rashid@umt.edu.pk
* Correspondence: zbashir@qau.edu.pk; Tel.: +92-331-482-3743

Received: 22 July 2018; Accepted: 03 September 2018; Published: 10 September 2018

Abstract: Decision making is the key component of people's daily life, from choosing a mobile phone to engaging in a war. To model the real world more accurately, probabilistic linguistic term sets (PLTSs) were proposed to manage a situation in which several possible linguistic terms along their corresponding probabilities are considered at the same time. Previously, in linguistic term sets, the probabilities of all linguistic term sets are considered to be equal which is unrealistic. In the process of decision making, due to the vagueness and complexity of real life, an expert usually hesitates and unable to express its opinion in a single term, thus making it difficult to reach a final agreement. To handle real life scenarios of a more complex nature, only membership linguistic decision making is unfruitful; thus, some mechanism is needed to express non-membership linguistic term set to deal with imprecise and uncertain information in more efficient manner. In this article, a novel notion called probabilistic hesitant intuitionistic linguistic term set (PHILTS) is designed, which is composed of membership PLTSs and non-membership PLTSs describing the opinions of decision makers (DMs). In the theme of PHILTS, the probabilities of membership linguistic terms and non membership linguistic terms are considered to be independent. Then, basic operations, some governing operational laws, the aggregation operators, normalization process and comparison method are studied for PHILTSs. Thereafter, two practical decision making models: aggregation based model and the extended TOPSIS model for PHILTS are designed to classify the alternatives from the best to worst, as an application of PHILTS to multi-attribute group decision making. In the end, a practical problem of real life about the selection of the best alternative is solved to illustrate the applicability and effectiveness of our proposed set and models.

Keywords: hesitant intuitionistic fuzzy linguistic term set; probabilistic hesitant intuitionistic linguistic term set; multi-attribute group decision making; aggregation operators; TOPSIS

---

1. Introduction

The choices we make today determine our future, therefore, to choose the best alternative subject to certain attributes is an important problem. Multi-attribute group decision making (MAGDM) has established its importance by providing the optimal solution considering different attributes in many real life problems. For this purpose, many sets and models have been designed to express and comprehend the opinions of DMs. The classical set theory is too restrictive to express one's opinion, as some real life scenarios are too complicated and the vague data are often involved, therefore the DMs are unable to form a definite opinion. Fuzzy set theory is proposed as a remedy for such kind of real life problems. Fuzzy set approaches are suitable to use when the modelling of human knowledge is necessary and when human evaluations are required. However, the usual fuzzy set theory is limited to the modelling in which the diversity of variants occurs at the same time.

To overcome such situation, different extensions of fuzzy set have been proposed to better model the real world, such as intuitionistic fuzzy set [1], hesitant fuzzy set [2], hesitant probabilistic fuzzy set [3], hesitant probabilistic multiplicative set [4], and necessary and possible hesitant fuzzy set [5]. Zadeh [6] suggested the concept of a linguistic variable that is more natural for humans to express there will in situations where data are imprecise. Thus far, linguistic environment has been extensively used to cope with the problems of decision making within [7]. Mergió et al. [8] used the Dempster–Shafer theory of evidence to construct an improved linguistic representation model for the sake of decision making process. Next, they introduced several linguistic aggregation operators. Zhu et al. [9] proposed a two-dimensional linguistic lattice implication algebra to determine implicitly and further the compilation of two-dimensional linguistic information decision in MAGDM dilemmas. Meng and Tang [10] generalized the 2-tuple linguistic aggregation operators and then used them in MAGDM dilemmas. Li and Dong [11] gave an introduction to the proportional 2-tuple linguistic form to make easy the solving of MAGDM dilemmas. Xu [12] introduced a dynamic linguistic weighted geometric operator to cumulate the linguistic information and then solved the problem of MAGDM when the judgment in different periods to change the linguistic information. Li [13] applied the concept of extended linguistic variables to construct an advanced way to cope with MAGDM dilemmas under linguistic environments. Agell et al. [14] used qualitative thinking approaches to perform and incorporate linguistic decision information and then applied it to MAGDM dilemmas.

Because of the uncertainty, vagueness and complexity of real world problems, it is troublesome for experts to grant linguistic judgment using a single linguistic term. Torra [2] managed the situation where several membership values of a fuzzy set are possible by defining hesitant fuzzy set (HFS). Experts may hesitate among several possible linguistic terms. For this purpose, Rodriguez et al. [15] introduced the concept of hesitant fuzzy linguistic term sets (HFLTS) to improve the flexibility of linguistic information within hesitant situation. Zhu and Li [16] designed hesitant fuzzy linguistic aggregation operators based on the Hamacher t-norm and t-conorm. Cui and Ye [17] proposed multiple-attribute decision-making method using similarity measures of hesitant linguistic neutrosophic numbers regarding least common multiple cardinality. Liu et al. [18] defined new kind of similarity and distance measures based on a linguistic scale function. However, in some cases, the probabilities of these possible terms are not equal. Given this reality, Peng et al. [19] proposed the more generalized concept, called probabilistic linguistic term sets (PLTSs). PLTSs allow DMs to state more than one linguistic term, as an assessment for linguistic variable. This increases the flexibility and the fruitfulness of the expression of linguistic information and it is more reasonable for DMs to state their preference in terms of PLTSs because the PLTSs can reflect different probabilities for each possible assessment of a given object. Therefore, the research on the PLTSs is necessary. Thus, they used PLTSs in multi-attribute group decision making problem and construct an extended TOPSIS method as well as an aggregation-based method for MAGDM. Recently, in 2017, Lin et al. [20] extended the PLTSs to probabilistic uncertain linguistic term set, which is designed as some possible uncertain linguistic terms coupled with the corresponding probabilities, and developed an extended approach for preference to rank the alternatives.

Atanassov [1,21] presented the concept of the intuitionistic fuzzy set (IFS) which has three main parts, membership function, non-membership function and hesitancy function, and is better suited to handling uncertainty than the usual fuzzy set. Many researchers have been applying IFS for multi-attribute decision making under various different fuzzy environments. Up to now, the intuitionistic fuzzy set has been applied extensively to decision making problems [22–27]. Beg and Rashid [28] generalized the concept of HFLTS by hesitant intuitionistic fuzzy linguistic term set (HIFLTS) which is characterized by a membership and non-membership function that is more applicable for dealing with uncertainty than the HFLTS. HIFLTS collects possible membership and non-membership linguistic values provided by the DMs. This approach is useful to model more complex real life scenarios.

In this article, we introduce the concept of PHILTS. The main idea is to facilitate DMs to provide their opinions about membership and non-membership linguistic terms more freely to cope with the vagueness and uncertainties of real life. To make meaningful decision making, the basic framework of PHILTS is developed. In this regard, normalization process for the purpose to equalize the length of PHILTSs, basic operations and their governing laws are presented. Furthermore, to deal with different scenarios, range of aggregation operators, i.e., probabilistic hesitant intuitionistic linguistic averaging operator, probabilistic hesitant intuitionistic linguistic weighted averaging operator, probabilistic hesitant intuitionistic linguistic geometric operator and probabilistic hesitant intuitionistic linguistic weighted geometric operator are proposed. The DM can choose the aggregation operator according to his preference. Lastly, for practical use of PHILTS in decision making, an extended TOPSIS method is derived, in which the DMs provide their opinions in PHILTSs which are further aggregated and processed according to the proposed mechanism of extended TOPSIS to find the best alternative.

This paper is organized as follows. In Section 2, we review some basic knowledge needed to understand our proposal. In Section 3, the concept of PHILTSs is firstly proposed and then some concepts concerning PHILTS, i.e., normalization process, deviation degree, score function, operations and comparison between probabilistic hesitant intuitionistic linguistic term elements (PHILTEs), are also discussed. In Section 4, aggregation operators, deviation degree between two PHILTEs and weight vector are derived. In Section 5, we propose an extended TOPSIS method and aggregation based method designed for MAGDM with probabilistic hesitant intuitionistic linguistic information. An example is provided in Section 6 to illustrate the usefulness and practicality of our methodology by ranking of alternatives. Section 7 is dedicated to highlighting the advantages of the proposed set and comparing proposed models with existing theory. Finally, some concluding remarks are given in Section 8.

2. Preliminaries

In this section, we give some concepts and operations related to HFLTSs, HIFLTSs and PLTSs that will be used in coming sections.

2.1. Hesitant Fuzzy Linguistic Term Set

The DMs may face such a problem where they hesitate with certain possible values. For this purpose, Rodriguez et al. [15] introduced the following concept of hesitant fuzzy linguistic term set (HFLTS).

Definition 1 ([15]). *Let $S = \{s_\alpha; \alpha = 0, 1, 2, \ldots, g\}$ be a linguistic term set; then, HFLTS, H_S, is a finite and ordered subset of the consecutive linguistic terms of S.*

Example 1. *Let* $S = \left\{ \begin{array}{c} s_0 = \textit{extremely poor}, s_1 = \textit{very poor}, s_2 = \textit{poor}, s_3 = \textit{medium}, s_4 = \textit{good}, s_5 = \textit{very good}, \\ s_6 = \textit{extremely good} \end{array} \right\}$
be a linguistic term set. Then, two different HFLTSs may be defined as:
$H_S(x) = \{s_1 = \textit{very poor}, s_2 = \textit{poor}, s_3 = \textit{medium}, s_4 = \textit{good}\}$ *and* $H_S(y) = \{s_3 = \textit{medium}, s_4 = \textit{good}, s_5 = \textit{very good}\}$.

Definition 2 ([15]). *Let $S = \{s_\alpha; \alpha = 0, 1, 2, \ldots, g\}$ be an ordered finite set of linguistic terms and E be an ordered finite subset of the consecutive linguistic terms of S. Then, the operators "max" and "min" on E can be defined as follows:*

(i) $\max(E) = \max(s_l) = s_m\,; s_l \in E \text{ and } s_l \leq s_m\ \forall l$
(ii) $\min(E) = \min(s_l) = s_n\,; s_l \in E \text{ and } s_l \geq s_n\ \forall l.$

2.2. Hesitant Intuitionistic Fuzzy Linguistic Term Set

In 2014, Beg and Rashid [28] introduced the concept of hesitant intuitionistic fuzzy linguistic term set (HIFLTS). This concept is actually based on HFLTS and intuitionistic fuzzy set.

Definition 3 ([28]). *Let X be a universe of discourse, and $S = \{s_\alpha; \alpha = 0, 1, 2, \ldots, g\}$ be a linguistic term set, then HIFLTS on X are two functions h and h' that when applied to an element of X return finite and ordered subsets of consecutive linguistic terms of S, this can be presented mathematically as:*

$$A = \left\{ \left\langle x, h(x), h'(x) \right\rangle | x \in X \right\},$$

where $h(x)$ and $h'(x)$ denote the possible membership and non-membership degree in terms of consecutive linguistic terms of the element $x \in X$ to the set A such that the following conditions are satisfied:

(i) $\max(h(x)) + \min\left(h'(x)\right) \le s_g$;

(ii) $\min(h(x)) + \max\left(h'(x)\right) \le s_g$.

2.3. Probabilistic Linguistic Term Sets

Recently, in 2016, Pang et al. [19] introduced the concept of PLTSs by attaching probabilities with each linguistic term, which is basically the generalization of HFLTS, and thus they opened a new dimension of research in decision theory.

Definition 4 ([19]). *Let $S = \{s_\alpha; \alpha = 0, 1, 2, \ldots, g\}$ be a linguistic term set, then a PLTS can be presented as follows:*

$$L(p) = \left\{ L^{(i)}\left(p^{(i)}\right) \mid L^{(i)} \in S,\ p^{(i)} \ge 0\ i = 1, 2, \ldots, \#L(p),\ \sum_{i=1}^{\#L(p)} p^{(i)} \le 1 \right\}. \quad (1)$$

where $L^{(i)}\left(p^{(i)}\right)$ is the ith linguistic term $L^{(i)}$ associated with the probability$p^{(i)}$, and $\#L(p)$ denotes the number of linguistic terms in $L(p)$.

Definition 5 ([19]). *Let $L(p) = \left\{L^{(i)}\left(p^{(i)}\right); i = 1, 2, \ldots, \#L(p)\right\}$, $r^{(i)}$ be the lower index of linguistic term $L^{(i)}$, $L(p)$ is called an ordered PLTS, if all the elements $L^{(i)}\left(p^{(i)}\right)$ in $L(p)$ are ranked according to the values of $r^{(i)} \times p^{(i)}$ in descending order.*

However, in a PLTS, it is possible for two or more linguistic terms with equal values of $r^{(i)} \times p^{(i)}$. Taking a PLTS $L(p) = \{s_1(0.4), s_2(0.2), s_3(0.4)\}$, here $r^{(1)} \times p^{(1)} = r^{(2)} \times p^{(2)} = 0.4$

According to the above rule, these two values cannot be arranged. To handle such type of problem, Zhang et al. [29] defined the following ranking rule.

Definition 6 ([29]). *Let $L(P) = \left\{L^{(i)}\left(p^{(i)}\right); i = 1, 2, \ldots, \#L(p)\right\}$, $r^{(i)}$ be the lower index of linguistic term $L^{(i)}$.*

(1) If the values of $r^{(i)}\left(p^{(i)}\right)$ are different for all elements in PLTS, then arrange all the elements according to the values of $r^{(i)}\left(p^{(i)}\right)$ directly.

(2) If all the values of $r^{(i)}\left(p^{(i)}\right)$ become equal for two or more elements, then

   (a) When the lower indices $r^{(i)}$ $(i = 1, 2, \ldots, \#L(p))$ are unequal, arrange $r^{(i)}\left(p^{(i)}\right)$ $(i = 1, 2, \ldots, \#L(p))$ according to the values of $r^{(i)}$ $(i = 1, 2, \ldots, \#L(p))$ in descending order.

(b) *When the lower indices* $r^{(i)}\ (i = 1, 2, \ldots, \#L(p))$ *are incomparable, arrange* $r^{(i)}\left(p^{(i)}\right)\ (i = 1, 2, \ldots, \#L(p))$ *according to the values of* $p^{(i)}\ (i = 1, 2, \ldots, \#L(p))$ *in descending order.*

Definition 7 ([19]). *Let* $L(p)$ *be a PLTS such that* $\sum_{i=1}^{\#L(p)} p^{(i)} < 1$, *then the associated PLTS is denoted and defined as*

$$L^{\cdot}(p) = \left\{ L^{(i)}\left(p^{\cdot(i)}\right);\ i = 1, 2, \ldots, \#L(p) \right\} \tag{2}$$

where $p^{\cdot(i)} = \frac{p^{(i)}}{\sum_{i=1}^{\#L(p)} p^{(i)}}, \forall i = 1, 2, \ldots, \#L(p)$.

Definition 8 ([19]). *Let* $L_1(p) = \left\{ L_1^{(i)}\left(p_1^{(i)}\right);\ i = 1, 2, \ldots, \#L_1(p) \right\}$ *and* $L_2(p) = \left\{ L_2^{(i)}\left(p_2^{(i)}\right);\ i = 1, 2, \ldots, \#L_2(p) \right\}$ *be two PLTSs, where* $\#L_1(p)$ *and* $\#L_2(p)$ *denote the number of linguistic terms in* $L_1(p)$ *and* $L_2(p)$, *respectively. If* $\#L_1(p) > \#L_2(p)$, *then* $\#L_1(p) - \#L_2(p)$ *linguistic terms will be added to* $L_2(p)$ *so that the number of elements in* $L_1(p)$ *and* $L_2(p)$ *becomes equal. The added linguistic terms are the smallest one's in* $L_2(p)$ *and the probabilities of all the linguistic terms are zero.*

Let $L_1(p) = \left\{ L_1^{(i)}\left(p_1^{(i)}\right);\ i = 1, 2, \ldots, \#L_1(p) \right\}$ and $L_2(p) = \left\{ L_2^{(i)}\left(p_2^{(i)}\right);\ i = 1, 2, \ldots, \#L_2(p) \right\}$, then the Normalized PLTSs denoted by $\widetilde{L_1}(p) = \left\{ \widetilde{L_1^{(i)}}\left(p_1^{(i)}\right);\ i = 1, 2, \ldots, \#L_1(p) \right\}$ and $\widetilde{L_2}(p) = \left\{ \widetilde{L_2^{(i)}}\left(p_2^{(i)}\right);\ i = 1, 2, \ldots, \#L_2(p) \right\}$ can be obtained according to the following two steps:

(1) If $\sum_{i=1}^{\#L_k(p)} p_k^{(i)} < 1$, then $L_k^{\cdot}(p), k = 1, 2$ is calculated according to Definition 7.
(2) If $\#L_1(p) \neq \#L_2(p)$, then according to Definition 8, add some linguistic terms to the one with the smaller number of elements.

The deviation degree between PLTSs, which is analogous to the Euclidean distance between hesitant fuzzy sets [30] can be defined as:

Definition 9 ([19]). *Let* $L_1(p) = \left\{ L_1^{(i)}\left(p_1^{(i)}\right);\ i = 1, 2, \ldots, \#L_1(p) \right\}$ *and* $L_2(p) = \left\{ L_2^{(i)}\left(p_2^{(i)}\right);\ i = 1, 2, \ldots, \#L_2(p) \right\}$ *be two PLTSs, where* $\#L_1(p)$ *and* $\#L_2(p)$ *denote the number of linguistic terms in* $L_1(p)$ *and* $L_2(p)$, *respectively, with* $\#L_1(p) = \#L_2(p)$. *Then, the deviation degree between these two PLTSs can be defined as*

$$d(L_1(p), L_2(p)) = \sqrt{\frac{1}{\#L_1(p)} \sum_{i=1}^{L_1(p)} \left(p_1^{(i)} r_1^{(i)} - p_2^{(i)} r_2^{(i)}\right)^2} \tag{3}$$

where $r_1^{(i)}$ *and* $r_2^{(i)}$ *denote the lower indices of linguistic terms* $L_1^{(i)}$ *and* $L_2^{(i)}$, *respectively.*

For further detail of PLTS, one can see Ref. [19].

3. Probabilistic Hesitant Intuitionistic Linguistic Term Set

Although HIFLTS allow the DM to state his assessments by using several linguistic terms, it cannot reflect the probabilities of the assessments of DM.

To overcome this issue, in this section, the concept of probabilistic hesitant intuitionistic linguistic term set (PHILTS) which is based on the concept of HIFLTS and PLTS is proposed. Furthermore, some basic operations for PHILTS are also designed.

Definition 10. *Let X be a universe of discourse, and $S = \{s_\alpha; \alpha = 0,1,2,\ldots,g\}$ be a linguistic term set, then a PHILTS on X are two functions l and l' that when applied to an element of X return finite and ordered subsets of the consecutive linguistic terms of S along with their occurrence probabilities, which can be mathematically expressed as*

$$A(p) = \left\{ \begin{array}{c} \left\langle x, l(x)(p(x)) = \left\{ l^{(i)}(x)\left(p^{(i)}(x)\right) \right\}, l'(x)\left(p'(x)\right) = \left\{ l'^{(j)}(x)\left(p'^{(j)}(x)\right) \right\} \right\rangle \\ |p^{(i)}(x) \geq 0, i = 1,2,\ldots,\#l(x)(p(x)), \sum_{i=1}^{\#l(x)(p(x))} p^i(x) \leq 1 \; \& \\ p'^{(j)}(x) \geq 0, j = 1,2,\ldots,\#l'(x)\left(p'(x)\right), \sum_{j=1}^{\#l'(x)\left(p'(x)\right)} p'^{(j)}(x) \leq 1 \end{array} \right\} \tag{4}$$

where $l(x)(p(x))$ and $l'(x)\,p'(x)$ are the PLTSs, denoting the membership and non-membership degree of the element $x \in X$ to the set $A(p)$ such that the following two conditions are satisfied:

(i) $\max(l(x)) + \min\left(l'(x)\right) \leq s_g$;

(ii) $\min(l(x)) + \max\left(l'(x)\right) \leq s_g$.

For the sake of simplicity and convenience, we call the pair $A(x)(p(x)) = \left\langle l(x)(p(x)), l'(x)\left(p'(x)\right)\right\rangle$ as the intuitionistic probabilistic linguistic term element (PHILTE), denoted by $A(p) = \left\langle l(p), l'\left(p'\right)\right\rangle$ for short.

Remark 1. *Particularly, if the probabilities of all linguistic terms in membership part and non-membership part become equal, then PHILTE reduces to HIFLTE.*

Example 2. *Let* $S = \left\{ \begin{array}{c} s_0 = \text{extremely poor}, s_1 = \text{very poor}, s_2 = \text{poor}, s_3 = \text{medium}, s_4 = \text{good}, s_5 = \text{very good}, \\ s_6 = \text{extremely good} \end{array} \right\}$
be a linguistic term set. A PHILTS is
$A(p) = \{\langle x_1, \{s_1(0.4), s_2(0.1), s_3(0.35)\}, \{s_3(0.3), s_4(0.4)\}\rangle, \langle x_2, \{s_4(0.33), s_5(0.5)\}, \{s_1(0.2), s_2(0.45)\}\rangle\}$

One can easily check the conditions of PHILTS for $A(p)$.

To illustrate the PHILTS more straightforwardly, in the following, a practical life example is given to depict the difference between the PHILTS and HIFLTS:

Example 3. *Take the evaluation of a vehicle on the comfortable degree attribute/criteria as an example. Let S be a linguistic term set used in the above example. An expert provides an HIFLTE $\langle\{s_1, s_2, s_3\}, \{s_3, s_4\}\rangle$ on the comfortable degree due to his/her hesitation for this evaluation. However, he/she is more confident in the linguistic term s_2 for the membership degree set and the linguistic term s_4 for the non-membership degree set. The HIFLTS fails to express his/her confidence. Therefore, we utilize the PHILTS to present his/her evaluations. In this case, his/her evaluations can be expressed as $A(p) = \langle\{s_1(0.2), s_2(0.6), s_3(0.2)\}, \{s_3(0.2), s_4(0.8)\}\rangle$.*

In the following, the ordered PHILTE is defined to make sure that the operational results among PHILTEs can be determined easily.

Definition 11. *A PHILTE $A(p) = \left\langle l(p), l'\left(p'\right)\right\rangle$ is known to be an ordered PHILTE, if $l(p)$ and $l'\left(p'\right)$ are ordered PLTSs.*

Example 4. *Consider a PHILTE $A(p) = \langle\{s_1(0.4), s_2(0.1), s_3(0.35)\}, \{s_3(0.3), s_4(0.4)\}\rangle$ used in the Example 2. Then, according to Definition 11 the ordered PHILTE is $A(p) = \langle\{s_3(0.35), s_1(0.4), s_2(0.1)\}, \{s_4(0.4), s_3(0.3)\}\rangle$*

3.1. The Normalization of PHILTEs

Ideally, the sum of the probabilities is one, but in PHILTE if either of the membership probabilities or non-membership probabilities have sum less than one than this issue is resolved as follows.

Definition 12. *Consider a PHILTE* $A(p) = \left\langle l(p), l'\left(p'\right)\right\rangle$*, the associated PHILTE* $A^{\cdot}(p) = \left\langle l(p^{\cdot}), l'\left(p'^{\cdot}\right)\right\rangle$ *is defined, where*

$$l(p^{\cdot}) = \left\{l^{(i)}\left(p^{\cdot(i)}\right) | i = 1, 2, \ldots, \#l(p)\right\}; p^{\cdot(i)} = \frac{p^{(i)}}{\sum_{i=1}^{\#l(p)} p^{(i)}}, \forall i = 1, 2, \ldots, \#l(p) \tag{5}$$

and

$$l'\left(p'^{\cdot}\right) = \left\{l'^{(j)}\left(\left(p'^{\cdot(j)}\right)\right) | j = 1, 2, \ldots, l'\left(p'\right)\right\}; p'^{\cdot(j)} = \frac{p'^{(j)}}{\sum_{j=1}^{l'(p')} p'^{(j)}}, \forall j = 1, 2, \ldots, l'\left(p'\right). \tag{6}$$

Example 5. *Consider a PHILTE* $A(p) = \langle\{s_1(0.4), s_2(0.1), s_3(0.35)\}, \{s_3(0.3), s_4(0.4)\}\rangle$*. Here, we see that* $\sum_{i=1}^{\#l(p)} p^{(i)} = 0.85 < 1$ *also* $\sum_{j=1}^{\#l'(p')} p'^{(j)} = 0.7 < 1$ *so the associated PHILTE* $A^{\cdot}(p) = \left\langle l(p^{\cdot}), l'\left(p'^{\cdot}\right)\right\rangle = \left\langle\left\{s_1\left(\frac{0.4}{0.85}\right), s_2\left(\frac{0.1}{0.85}\right), s_3\left(\frac{0.35}{0.85}\right)\right\}, \left\{s_3\left(\frac{0.3}{0.7}\right), s_4\left(\frac{0.4}{0.7}\right)\right\}\right\rangle$.

In decision making process, experts usually face such problems in which the length of PHILTEs is different. Let $A(p) = \left\langle l(p), l'\left(p'\right)\right\rangle$ and $A_1(p_1) = \left\langle l_1(p_1), l_1'\left(p_1'\right)\right\rangle$ be two PHILTEs of different lengths. Then, the following three cases are possible $(I)\ \#l(p) \neq \#l_1(p_1)$, $(II)\ \#l'\left(p'\right) \neq \#l_1'\left(p_1'\right)$, $(III)\ \#l(p) \neq \#l_1(p_1)$ and $\#l'\left(p'\right) \neq \#l_1'\left(p_1'\right)$. In such situation, they need to equalize their lengths by increasing the number of probabilistic linguistic terms in that PLTS in which the number of probabilistic linguistic terms are relatively small because PHILTEs of different lengths create great problems in operations, aggregation operators and finding the deviation degree between two PHILTEs.

Definition 13. *Given any two PHILTEs* $A(p) = \left\langle l(p), l'\left(p'\right)\right\rangle$ *and* $A_1(p_1) = \left\langle l_1(p_1), l_1'\left(p_1'\right)\right\rangle$ *if* $\#l(p) > \#l_1(p_1)$ *then* $\#l(p) - \#l_1(p_1)$ *linguistic terms should be added to* $l_1(p_1)$ *to make their cardinalities identical. The added linguistic terms are the smallest one(s) in* $l_1(p_1)$*, and the probabilities of all the linguistic terms are zero.*

The remaining cases are analogous to Case (I).

Let $A_1(p_1) = \left\langle l_1(p_1), l_1'\left(p_1'\right)\right\rangle$ and $A_2(p_2) = \left\langle l_2(p_2), l_2'\left(p_2'\right)\right\rangle$ be two PHILTEs. Then, the following two simple steps are involved in normalization process.

Step 1: If $\sum_{i=1}^{\#l_j(p_j)} p_j^{(i)} < 1$ or $\sum_{i=1}^{\#l_j'(p_j')} p_j'^{(i)} < 1$; $j = 1, 2$, then we calculate $l_j\left(p_j^{\cdot}\right), l_j'\left(p_j'^{\cdot}\right)$; $j = 1, 2$ using Equations (5) and (6).

Step 2: If $\#l_1(p_1) \neq \#l_2(p_2)$ or $\#l_1'\left(p_1'\right) \neq \#l_2'\left(p_2'\right)$, then we add some elements according to Definition 13 to the one with small number of elements.

The resultant PHILTEs are called the normalized PHILTEs which are denoted as $\widetilde{A}(p)$ and $\widetilde{A_1}(p_1)$.

Note, for the convenience of presentation, we denote the normalized PHILTEs by $A(p)$ and $A_1(p_1)$ as well.

Example 6. *Let* $A(p) = \langle\{s_2(0.3), s_3(0.7)\}, \{s_0(0.2), s_1(0.4), s_2(0.3)\}\rangle$ *and* $A_1(p_1) = \langle\{s_3(0.4), s_4(0.3), s_5(0.3)\}, \{s_1(0.4), s_2(0.6)\}\rangle$ *then*

Step 1: According to Equation (6) $l'\left(p'^{\cdot}\right) = \left\{s_0\left(\frac{0.2}{0.9}\right), s_1\left(\frac{0.4}{0.9}\right), s_2\left(\frac{0.3}{0.9}\right)\right\}$

Step 2: Since $\#l(p) < \#l_1(p_1)$, *so we add the linguistic term* s_2 *to* $l(p)$ *so that the number of linguistic terms in* $l(p)$ *and* $l_1(p_1)$ *becomes equal, thus* $l(p) = \{s_2(0.3), s_3(0.7), s_2(0)\}$. *In addition,* $\#l_1'\left(p_1'\right) < \#l'\left(p'\right)$ *so we add the linguistic term* s_1 *to* $l_1'\left(p_1'\right)$, $l_1'\left(p_1'\right) = \{s_1(0.4), s_2(0.6), s_1(0)\}$. *Therefore, after normalization, we have*

$A(p) = \langle\{s_2(0.3), s_3(0.7), s_2(0)\}, \{s_0(0.2), s_1(0.4), s_2(0.3)\}\rangle$ *and*

$A_1(p_1) = \langle\{s_3(0.4), s_4(0.3), s_5(0.3)\}, \{s_1(0.4), s_2(0.6), s_1(0)\}\rangle$.

3.2. The Comparison between PHILTEs

In this section, the comparison between two PHILTEs is presented. For this purpose, the score function and the deviation degree of the PHILTE are defined.

Definition 14. *Let* $A(p) = \left\langle l(p), l'\left(p'\right)\right\rangle = \left\langle l^{(i)}\left(p^{(i)}\right), l'^{(j)}\left(p'^{(j)}\right)\right\rangle; i = 1,2,\ldots,\#l(p), j = 1,2,\ldots,l'\left(p'\right)$ *be a PHILTE with a linguistic term set* $S = \{s_\alpha; \alpha = 0,1,2,\ldots,g\}$ *such that* $r^{(i)}$ *and* $r'^{(j)}$ *denote, respectively, the lower indices of linguistic terms* $l^{(i)}$ *and* $l'^{(j)}$, *then the score of* $A(p)$ *is denoted and defined as follows:*

$$E(A(p)) = s_{\overline{\gamma}} \tag{7}$$

where $\overline{\gamma} = \dfrac{g+\alpha-\beta}{2}$; $\alpha = \dfrac{\sum_{i=1}^{\#l(p)} r^{(i)} p^{(i)}}{\sum_{i=1}^{\#l(p)} p^{(i)}}$ *and* $\beta = \dfrac{\sum_{j=1}^{\#l'(p')} r'^{(j)} p'^{(j)}}{\sum_{j=1}^{\#l'(p')} p'^{(j)}}$.

It is easy to see that $0 \le \frac{g+\alpha-\beta}{2} \le g$ *which means* $s_{\overline{\gamma}} \in \overline{S} = \{s_\alpha | \alpha \in [0,g]\}$.

Apparently, the score function represents the averaging linguistic term of PHILTE.

For two PHILTEs $A(p)$ and $A_1(p_1)$, if $E(A(p)) > E(A_1(p_1))$, then $A(p)$ is superior to $A_1(p_1)$, denoted as $A(p) > A_1(p_1)$; if $E(A(p)) < E(A_1(p_1))$, then $E(A(p))$ is inferior to $A_1(p_1)$, denoted as $A(p) < A_1(p_1)$; and, if $E(A(p)) = E(A_1(p_1))$, then we cannot distinguish between them. Thus, in this case, we define another indicator, named as the deviation degree as follows:

Definition 15. *Let* $A(p) = \left\langle l(p), l'\left(p'\right)\right\rangle = \left\langle l^{(i)}\left(p^{(i)}\right), l'^{(j)}\left(p'^{(j)}\right)\right\rangle; i = 1,2,\ldots\#l(p), j = 1,2,\ldots,l'\left(p'\right)$ *be a PHILTE such that* $r^{(i)}$ *and* $r'^{(j)}$ *denote, respectively, the lower indices of linguistic terms* $l^{(i)}$ *and* $l'^{(j)}$, *then the deviation degree of* $A(p)$ *is denoted and defined as follows:*

$$\sigma(A(p)) = \left(\frac{\sum_{i=1}^{\#l(p)} \left(p^{(i)}\left(r^{(i)} - \overline{\gamma}\right)\right)^2}{\sum_{i=1}^{\#l(p)} p^{(i)}} + \frac{\sum_{j=1}^{\#l'(p')} \left(p'^{(j)}\left(r'^{(j)} - \overline{\gamma}\right)\right)^2}{\sum_{j=1}^{\#l'(p')} p'^{(j)}} \right)^{\frac{1}{2}} \tag{8}$$

The deviation degree shows the distance from the average value in the PHILTE. The greater value of σ implies lower consistency while the lesser value of σ indicates higher consistency.

Thus, $A(p)$ and $A_1(p_1)$ can be ranked by the following procedure:

(1) if $E(A(p)) > E(A_1(p_1))$, then $A(p) > A_1(p_1)$;
(2) if $E(A(p)) = E(A_1(p_1))$ and

(a) $\sigma(A(p)) > \sigma(A_1(p_1))$, then $A(p) < A_1(p_1)$;
(b) $\sigma(A(p)) < \sigma(A_1(p_1))$, then $A(p) > A_1(p_1)$;
(c) $\sigma(A(p)) = \sigma(A_1(p_1))$, then $A(p)$ is indifferent to $A_1(p_1)$ and is denoted as $A(p) \sim A_1(p_1)$.

Example 7. *Let* $A(p) = \left\langle l(p), l'\left(p'\right)\right\rangle = \langle\{s_1(0.12), s_2(0.26), s_3(0.62)\}, \{s_2(0.1), s_3(0.3), s_4(0.6)\}\rangle$ *,* $A_1(p_1) = \left\langle l_1(p_1), l_1'\left(p_1'\right)\right\rangle = \langle\{s_2(0.3), s_3(0.3)\}, \{s_3(0.35), s_4(0.35)\}\rangle$ *and S be the linguistic term set used in Example 2 then*

$$\alpha = \frac{1\times 0.12 + 2\times 0.26 + 3\times 0.62}{0.12+0.26+0.62} = 2.5,\ \beta = \frac{2\times 0.1 + 3\times 0.3 + 4\times 0.6}{0.6+0.3+0.1} = 3.5,$$

$$\overline{\gamma} = \frac{6+2.5-3.5}{2} = 2.5,\ E(A(p)) = s_{2.5}$$

$$\alpha_1 = \frac{2\times 0.3 + 3\times 0.3}{0.3+0.3} = 2.5,\ \beta_1 = \frac{0.35\times 3 + 0.35\times 4}{0.35+0.35} = 3.5,$$

$$\overline{\gamma}_1 = \frac{6+2.5-3.5}{2} = 2.5,\ E(A_1(P_1)) = s_{2.5}$$

Since $E(A(p)) = E(A_1(p_1))$*, we have to calculate the deviation degree of* $A(p)$ *and* $A_1(p_1)$.

$$\sigma(A(p)) = \sqrt{\frac{\left((0.12(1-2.5))^2+(0.26(2-2.5))^2+(0.62(3-2.5))^2\right)}{0.12+0.26+0.62} + \frac{\left((0.6(4-3.5))^2+(0.3(3-3.5))^2+(0.1(2-3.5))^2\right)}{0.6+0.3+0.1}} = 0.529,$$

$$\sigma(A_1(p_1)) = \sqrt{\frac{\left((0.3(2-2.5))^2+(0.3(3-2.5))^2\right)}{0.3+0.3} + \frac{\left((0.35(3-3.5))^2+(0.35(4-3.5))^2\right)}{0.35+0.35}} = 0.37$$

Thus, $\sigma(A(p)) > \sigma(A_1(p_1))$ *so* $A(p)$ *is inferior to* $A_1(p_1)$.

In the following, we present a theorem which shows that the association does not affect the score and deviation degree of PHILTE.

Theorem 1. *Let* $A(p) = \left\langle l(p), l'\left(p'\right)\right\rangle$ *be a PHILTE and* $A^{\bullet}(p) = \left\langle l(p^{\bullet}), l'\left(p''\right)\right\rangle$ *be the associated PHILTE then* $E(A(p)) = E(A^{\bullet}(p))$ *and* $\sigma(A(p)) = \sigma(A^{\bullet}(p))$.

Proof. $E(A^{\bullet}(p)) = s_{\dot{\overline{\gamma}}}$ where $\dot{\overline{\gamma}} = \frac{g+\dot{\alpha}-\dot{\beta}}{2}$ and $\dot{\alpha} = \frac{\sum_{i=1}^{\#l(p^{\bullet})} r^{(i)} p^{\bullet(i)}}{\sum_{i=1}^{\#l(p^{\bullet})} p^{\bullet(i)}}$. Since $\sum_{i=1}^{\#l(p^{\bullet})} p^{\bullet(i)} = 1$ and $p^{\bullet(i)} = \frac{p^{(i)}}{\sum_{i=1}^{\#l(p)} p^{(i)}}$, which implies that $\dot{\alpha} = \frac{\sum_{i=1}^{\#l(p)} r^{(i)} p^{(i)}}{\sum_{i=1}^{\#l(p)} p^{(i)}} = \alpha$ and $\dot{\beta} = \frac{\sum_{j=1}^{\#l'(p'^{\bullet})} r^{(j)} p'^{\bullet(j)}}{\sum_{j=1}^{\#l'(p'^{\bullet})} p'^{\bullet(j)}}$. Since $\sum_{j=1}^{\#l'(p'^{\bullet})} p'^{\bullet(j)} = 1$ and $p'^{\bullet(j)} = \frac{p'^{(j)}}{\sum_{j=1}^{\#l'(p')} p'^{(j)}}$ which further implies that $\dot{\beta} = \frac{\sum_{j=1}^{\#l'(p')} r^{(i)} p'^{(i)}}{\sum_{i=1}^{\#l'(p')} p'^{(i)}} = \beta$. Hence, $E(A^{\bullet}(p)) = E(A(p))$.

$$\text{Next, } \sigma\left(A^{\bullet}(p)\right)=\left(\frac{\sum_{i=1}^{\#l(p^{\bullet})}\left(p^{\bullet(i)}\left(r^{(i)}-\dot{\overline{\gamma}}\right)\right)^{2}}{\sum_{i=1}^{\#l(p^{\bullet})}p^{\bullet(i)}}+\frac{\sum_{j=1}^{\#l'(p'^{\bullet})}\left(p'^{\bullet(j)}\left(r'^{(j)}-\dot{\overline{\gamma}}\right)\right)^{2}}{\sum_{j=1}^{\#l'(p'^{\bullet})}p'^{\bullet(j)}}\right)^{\frac{1}{2}}$$

$$\text{Since } \sum_{i=1}^{\#l(p^{\bullet})}p^{\bullet(i)}=1,\ p^{\bullet(i)}=\frac{p^{(i)}}{\sum_{i=1}^{\#l(p)}p^{(i)}},\ \sum_{j=1}^{\#l'(p'^{\bullet})}p'^{\bullet(j)}=1,\ p'^{\bullet(j)}=\frac{p'^{(j)}}{\sum_{j=1}^{\#l'(p')}p'^{(j)}} \text{ and } \dot{\overline{\gamma}}=\overline{\gamma}.$$

$$\text{It yields that } \sigma\left(A^{\bullet}(p)\right)=\left(\frac{\sum_{i=1}^{\#l(p)}\left(p^{(i)}\left(r^{(i)}-\overline{\gamma}\right)\right)^{2}}{\sum_{i=1}^{\#l(p)}p^{(i)}}+\frac{\sum_{j=1}^{\#l'(p')}\left(p'^{(j)}\left(r'^{(j)}-\overline{\gamma}\right)\right)^{2}}{\sum_{j=1}^{\#l'(p')}p'^{(j)}}\right)^{\frac{1}{2}}=\sigma\left(A(p)\right). \quad \square$$

The following theorem shows that order of comparison between two PHILTEs remains unaltered after normalization.

Theorem 2. *Let* $A(p) = \left\langle l(p), l'\left(p'\right)\right\rangle$ *and* $A_1(p_1) = \left\langle l_1(p_1), l_1'\left(p_1'\right)\right\rangle$ *be any two PHILTEs,* $\widetilde{A}(p) = \left\langle \widetilde{l}(p), \widetilde{l}'\left(p'\right)\right\rangle$ *and* $\widetilde{A}_1(p_1) = \left\langle \widetilde{l}_1(p_1), \widetilde{l}'_1\left(p_1'\right)\right\rangle$ *be the corresponding normalized PHILTEs respectively, then* $A(p) < A_1(p_1) \iff \widetilde{A}(p) < \widetilde{A}_1(p_1)$.

Proof. The proof is quite clear because, according to Theorem 1, $E(A(p)) = E(A^{\bullet}(p))$ and $\sigma(A(p)) = \sigma(A^{\bullet}(p))$, so order of comparison in Step (1) of normalization process is preserved and so for Step (2) is concerned in that step we add some elements to PHILTEs though it does not change the order as we attach zero probabilities with the corresponding added elements so this means $E\left(\widetilde{A}(p)\right) = E\left(\widetilde{A}_1(p_1)\right)$ and $\sigma\left(\widetilde{A}(p)\right) = \sigma\left(\widetilde{A}_1(p_1)\right)$. Hence, the result holds true. □

In the following definition, we summarize the fact that comparison of any two PHILTEs can be done by their corresponding normalized PHILTEs.

Definition 16. *Let* $A(p) = \left\langle l(p), l'\left(p'\right)\right\rangle$ *and* $A_1(p_1) = \left\langle l_1(p_1), l_1'\left(p_1'\right)\right\rangle$ *be any two PHILTEs,* $\widetilde{A}(p) = \left\langle \widetilde{l}(p), \widetilde{l}'\left(p'\right)\right\rangle$ *and* $\widetilde{A}_1(p_1) = \left\langle \widetilde{l}_1(p_1), \widetilde{l}'_1\left(p_1'\right)\right\rangle$ *be the corresponding normalized PHILTEs, respectively, then*

(I) If $E\left(\widetilde{A}(p)\right) > E\left(\widetilde{A_1}(p_1)\right)$ then $A(p) > A_1(p_1)$.
(II) If $E\left(\widetilde{A}(p)\right) < E\left(\widetilde{A_1}(p_1)\right)$ then $A(p) < A_1(p_1)$.
(III) If $E\left(\widetilde{A}(p)\right) = E\left(\widetilde{A_1}(p_1)\right)$ then in this case we are unable to decide which one is superior. Thus, in this case, we do the comparison of PHILTEs on the bases of the deviation degree of normalized PHILTEs as follows.

(1) If $\delta\left(\widetilde{A}(p)\right) > \delta\left(\widetilde{A_1}(p_1)\right)$ then $A(p) < A_1(p_1)$.
(2) If $\delta\left(\widetilde{A}(p)\right) < \delta\left(\widetilde{A_1}(p_1)\right)$ then $A(p) > A_1(p_1)$.
(3) If $\delta\left(\widetilde{A}(p)\right) = \delta\left(\widetilde{A_1}(p_1)\right)$ in such case we say that $A(p)$ is indifferent to $A_1(p_1)$ and is denoted by $A(p) \sim A_1(p_1)$.

Example 8. *Let* S *be the linguistic term set used in Example 2,* $A(p) = \left\langle l(p), l'\left(p'\right)\right\rangle = \langle\{s_1(0.12), s_2(0.26), s_3(0.62)\}, \{s_2(0.1), s_3(0.3), s_4(0.5)\}\rangle$ *and* $A_1(p_1) = \left\langle l_1(p_1), l_1'\left(p_1'\right)\right\rangle =$

$\langle \{s_2(0.3), s_3(0.3)\}, \{s_3(0.35), s_4(0.35)\} \rangle$ *then the corresponding normalized PHILTEs are* $\widetilde{A}(p) = \left\langle \widetilde{l}(p), \widetilde{l'}\left(p'\right) \right\rangle = \langle \{s_1(0.12), s_2(0.26), s_3(0.62)\}, \{s_3(0.375), s_4(0.625), s_3(0)\} \rangle$ *and* $\widetilde{A}_1(p_1) = \left\langle \widetilde{l}_1(p_1), \widetilde{l'}_1\left(p'_1\right) \right\rangle = \langle \{s_2(.5), s_3(0.5), s_2(0)\}, \{s_3(0.5), s_4(0.5), s_3(0)\} \rangle$.

We calculate the score of these normalized PHILTEs

$$\alpha = \frac{1 \times 0.12 + 2 \times 0.26 + 3 \times 0.62}{0.12 + 0.26 + 0.62} = 2.5,\ \beta = \frac{3 \times 0.375 + 4 \times 0.625 + 3 \times 0}{0.375 + 0.625 + 0} = 3.625,$$

$$\overline{\gamma} = \frac{6 + 2.5 - 3.625}{2} = 2.437,\ E\left(\widetilde{A}(P)\right) = s_{2.437}$$

$$\alpha_1 = \frac{2 \times 0.5 + 3 \times 0.5 + 0 \times 2}{0.5 + 0.5} = 2.5,\ \beta_1 = \frac{0.5 \times 3 + 0.5 \times 4 + 0 \times 3}{0.5 + 0.5} = 3.5,$$

$$\overline{\gamma}_1 = \frac{6 + 2.5 - 3.5}{2} = 2.5,\ E\left(\widetilde{A_1}(p_1)\right) = s_{2.5}$$

Since $E\left(\widetilde{A}(p)\right) < E\left(\widetilde{A_1}(p_1)\right)$ *so* $A(p) < A_1(p_1)$.

3.3. Basic Operations of PHILTEs

Based on the operational laws of the PLTSs [19], we develop some basic operational framework of PHILTEs and investigate their properties in preparation for applications to the practical real life problems. Hereafter, it is assumed that all PHILTEs are normalized.

Definition 17. *Let* $A(p) = \left\langle l(p), l'\left(p'\right) \right\rangle = \left\langle l^{(i)}\left(p^{(i)}\right), l'^{(j)}\left(p'^{(j)}\right) \right\rangle; i = 1, 2, \ldots, \#l(p), j = 1, 2, \ldots, \#l'\left(p'\right)$ *and* $A_1(p_1) = \left\langle l_1(p_1), l'_1\left(p'_1\right) \right\rangle = \left\langle l_1^{(i)}\left(p_1^{(i)}\right), l_1'^{(j)}\left(p_1'^{(j)}\right) \right\rangle; i = 1, 2, \ldots, \#l_1(p_1), j = 1, 2, \ldots, \#l'_1\left(p'_1\right)$ *be two normalized and ordered PHILTEs, then*

Addition:

$$\begin{aligned} A(p) \oplus A_1(p_1) &= \left\langle l(p) \oplus l_1(p_1), l'\left(p'\right) \oplus l'_1\left(p'_1\right) \right\rangle \\ &= \left\langle \cup_{l^{(i)} \in l(p), l_1^{(i)} \in l_1(p_1)} \left\{ p^{(i)} l^{(i)} \oplus p_1^{(i)} l_1^{(i)} \right\}, \cup_{l'^{(j)} \in l'(p'), l_1'^{(j)} \in l_1'^{(j)}\left(p_1'^{(j)}\right)} \left\{ p'^{(j)} l'^{(j)} \oplus p_1'^{(j)} l_1'^{(j)} \right\} \right\rangle \end{aligned} \tag{9}$$

Multiplication:

$$\begin{aligned} A(p) \otimes A_1(p_1) &= \left\langle l(p) \otimes l_1(p_1), l'\left(p'\right) \otimes l'_1\left(p'_1\right) \right\rangle \\ &= \left\langle \cup_{l^{(i)} \in l(p), l_1^{(i)} \in l_1(p_1)} \left\{ \left(l^{(i)}\right)^{p^{(i)}} \otimes \left(l_1^{(i)}\right)^{p_1^{(i)}} \right\}, \cup_{l'^{(j)} \in l'(p'), l_1'^{(j)} \in l'_1(p_1)} \left\{ \left(l'^{(j)}\right)^{p'^{(j)}} \otimes \left(l_1'^{(j)}\right)^{p_1'^{(j)}} \right\} \right\rangle \end{aligned} \tag{10}$$

Scalar multiplication:

$$\gamma(A(p)) = \left\langle \gamma l(p), \gamma l'\left(p'\right) \right\rangle = \left\langle \cup_{l^{(i)} \in l(p)} \gamma p^{(i)} l^{(i)}, \cup_{l'^{(j)} \in l'(p')} \gamma p'^{(j)} l'^{(j)} \right\rangle \tag{11}$$

Scalar power:

$$(A(p))^{\gamma} = \left\langle (l(p))^{\gamma}, \left(l'\left(p'\right)^{\gamma}\right) \right\rangle = \left\langle \cup_{l^{(i)} \in l(p)} \left(l^{(i)}\right)^{\gamma p^{(i)}}, \cup_{l'^{(j)} \in l'(p')} \left(l'^{(j)}\right)^{\gamma p'^{(j)}} \right\rangle \tag{12}$$

where $l^{(i)}$ *and* $l_1^{(i)}$ *are the ith linguistic terms in* $l(p)$ *and* $l_1(p_1)$*, respectively;* $l'^{(j)}$ *and* $l_1'^{(j)}$ *are the jth linguistic terms in* $l'(p')$ *and* $l_1'(p_1')$*, respectively;* $p^{(i)}$ *and* $p_1^{(i)}$ *are the probabilities of the ith linguistic terms in* $l(p)$ *and* $l_1(p_1)$*, respectively;* $p'^{(j)}$ *and* $p_1'^{(j)}$ *are the probabilities of the jth linguistic terms in* $l'(p')$ *and* $l_1'(p_1')$*, respectively; and* γ *denote a nonnegative scalar.*

Theorem 3. *Let* $A(p) = \langle l(p), l'(p') \rangle$, $A_1(p_1) = \langle l_1(p_1), l_1'(p_1') \rangle$, $A_2(p_2) = \langle l_2(p_2), l_2'(p_2') \rangle$ *be any three ordered and normalized PHILTEs,* $\gamma_1, \gamma_2, \gamma_3 \geq 0$*, then*

(1) $A(p) \oplus A_1(p_1) = A_1(p_1) \oplus A(p)$;
(2) $A(p) \oplus (A_1(p_1) \oplus A_2(p_2)) = (A(p) \oplus A_1(p_1)) \oplus A_2(p_2)$;
(3) $\gamma(A(p) \oplus A_1(p_1)) = \gamma A(p) \oplus \gamma A_1(p_1)$;
(4) $(\gamma_1 + \gamma_2) A(p) = \gamma_1 A(p) \oplus \gamma_2 A(p)$;
(5) $A(p) \otimes A_1(p_1) = A_1(p_1) \otimes A(p)$;
(6) $A(p) \otimes (A_1(p_1) \otimes A_2(p_2)) = (A(p) \otimes A_1(p_1)) \otimes A_2(p_2)$;
(7) $(A(p) \otimes A_1(p_1))^{\gamma} = (A(p))^{\gamma} \otimes (A_1(p_1))^{\gamma}$;
(8) $(A(p))^{\gamma_1 + \gamma_2} = (A(p))^{\gamma_1} \otimes (A(p))^{\gamma_2}$.

Proof. (1) $A(p) \oplus A_1(p_1) = \langle l(p), l'(p') \rangle \oplus \langle l_1(p_1), l_1'(p_1') \rangle = \langle l(p) \oplus l_1(p_1), l'(p') \oplus l_1'(p_1') \rangle$

$$= \left\langle \cup_{l^{(i)} \in l(p), l_1^{(i)} \in l_1(p_1)} \left\{ p^{(i)} l^{(i)} \oplus p_1^{(i)} l_1^{(i)} \right\}, \cup_{l'^{(j)} \in l'(p'), l_1'^{(j)} \in l_1'(p_1)} \left\{ p'^{(j)} l'^{(j)} \oplus p_1'^{(j)} l_1'^{(j)} \right\} \right\rangle$$
$$= \left\langle \cup_{l^{(i)} \in l(p), l_1^{(i)} \in l_1(p_1)} \left\{ p_1^{(i)} l_1^{(i)} \oplus p^{(i)} l^{(i)} \right\}, \cup_{l'^{(j)} \in l'(p'), l_1'^{(j)} \in l_1'(p_1)} \left\{ p_1'^{(j)} l_1'^{(j)} \oplus p'^{(j)} l'^{(j)} \right\} \right\rangle$$
$$= \langle l_1(p_1) \oplus l(p), l_1'(p_1') \oplus l'(p') \rangle = \langle l_1(p_1), l_1'(p_1') \rangle \oplus \langle l(p), l'(p') \rangle$$
$$= A_1(P_1) \oplus A(p)$$

(2) $A(p) \oplus (A_1(p_1) \oplus A_2(p_2)) = \langle l(p), l'(p') \rangle \oplus (\langle l_1(p_1), l_1'(p_1') \rangle \oplus \langle l_2(p_2), l_2'(p_2') \rangle)$

$$= \langle l(p) \oplus (l_1(p_1) \oplus l_2(p_2)), l'(p') \oplus (l_1'(p_1') \oplus l_2'(p_2')) \rangle$$
$$= \left\langle \begin{array}{l} \cup_{l^{(i)} \in l(p), l_1^{(i)} \in l_1(p_1), l^{(i)}(z) \in l_2(p_2)} \left\{ p^{(i)} l^{(i)} \oplus \left(p_1^{(i)} l_1^{(i)} \oplus p_2^{(i)} l_2^{(i)} \right) \right\}, \\ \cup_{l'^{(j)} \in l'(p'), l_1'^{(j)} \in l_1'(p_1), l'^{(j)}(z) \in l_2'(p_2)} \left\{ p'^{(j)} l'^{(j)} \oplus \left(p_1'^{(j)} l_1'^{(j)} \oplus p_2'^{(j)} l_2'^{(j)} \right) \right\} \end{array} \right\rangle$$
$$= \left\langle \begin{array}{l} \cup_{l^{(i)} \in l(p), l_1^{(i)} \in l_1(p_1), l^{(i)}(z) \in l_2(p_2)} \left\{ \left(p^{(i)} l^{(i)} \oplus p_1^{(i)} l_1^{(i)} \right) \oplus p_2^{(i)} l_2^{(i)} \right\}, \\ \cup_{l'^{(j)} \in l'(p'), l_1'^{(j)} \in l_1'(p_1), l'^{(j)}(z) \in l_2'(p_2)} \left\{ \left(p'^{(j)} l'^{(j)} \oplus p_1'^{(j)} l_1'^{(j)} \right) \oplus p_2'^{(j)} l_2'^{(j)} \right\} \end{array} \right\rangle$$
$$= \langle (l(p) \oplus l_1(p_1)) \oplus l_2(p_2), (l'(p') \oplus l_1'(p_1')) \oplus l_2'(p_2') \rangle$$
$$= (\langle l(p), l'(p') \rangle \oplus \langle l_1(p_1), l_1'(p_1') \rangle) \oplus \langle l_2(p_2), l_2'(p_2') \rangle$$
$$= (A(p) \oplus A_1(p_1)) \oplus A_2(p_2)$$

(3) $\gamma(A(p) \oplus A_1(p_1)) = \gamma \langle l(p), l'(p') \rangle \oplus \langle l_1(p_1), l_1'(p_1') \rangle = \gamma \langle l(p) \oplus l_1(p_1), l'(p') \oplus l_1'(p_1') \rangle$

$$= \gamma \left\langle \cup_{l^{(i)} \in l(p), l_1^{(i)} \in l_1(p_1)} \left\{ p^{(i)} l^{(i)} \oplus p_1^{(i)} l_1^{(i)} \right\}, \cup_{l'^{(j)} \in l'(p'), l_1'^{(j)} \in l_1'(p_1)} \left\{ p'^{(j)} l'^{(j)} \oplus p_1'^{(j)} l_1'^{(j)} \right\} \right\rangle$$
$$= \left\langle \cup_{l^{(i)} \in l(p), l_1^{(i)} \in l_1(p_1)} \left\{ \gamma p^{(i)} l^{(i)} \oplus \gamma p_1^{(i)} l_1^{(i)} \right\}, \cup_{l'^{(j)} \in l'(p'), l_1'^{(j)} \in l_1'(p_1)} \left\{ \gamma p'^{(j)} l'^{(j)} \oplus \gamma p_1'^{(j)} l_1'^{(j)} \right\} \right\rangle$$
$$= \langle \gamma l(p) \oplus \gamma l_1(p_1), \gamma l'(p') \oplus \gamma l_1'(p_1') \rangle = \langle \gamma l(p), \gamma l'(p') \rangle \oplus \langle \gamma l_1(p_1), \gamma l_1'(p_1') \rangle$$
$$= \gamma A(p) \oplus \gamma A_1(p_1)$$

$$(4)\ (\gamma_1+\gamma_2)\,A(p)=(\gamma_1+\gamma_2)\left\langle l(p),l'\left(p'\right)\right\rangle$$
$$=\gamma\left\langle \cup_{l^{(i)}\in l(p)}\left\{(\gamma_1+\gamma_2)\,p^{(i)}l^{(i)}\right\},\cup_{l'^{(j)}\in l'(p')}\left\{(\gamma_1+\gamma_2)\,p'^{(j)}l'^{(j)}\right\}\right\rangle$$
$$=\left\langle \cup_{l^{(i)}\in l(p)}\left\{\gamma_1p^{(i)}l^{(i)}\oplus\gamma_2p^{(i)}l^{(i)}\right\},\cup_{l'^{(j)}\in l'(p')}\left\{\gamma_1p'^{(j)}l'^{(j)}\oplus\gamma_2p'^{(j)}l'^{(j)}\right\}\right\rangle$$
$$=\left\langle \cup_{l^{(i)}\in l(p)}\left\{\gamma_1p^{(i)}l^{(i)}\right\}\oplus\cup_{l^{(i)}\in l(p)}\left\{\gamma_2p^{(i)}l^{(i)}\right\},\cup_{l'^{(j)}\in l'(p')}\left\{\gamma_1p'^{(j)}l'^{(j)}\right\}\oplus\cup_{l'^{(j)}\in l'(p')}\left\{\gamma_2p'^{(j)}l'^{(j)}\right\}\right\rangle$$
$$=\left\langle \gamma_1l(p)\oplus\gamma_2l(p),\gamma_1l'\left(p'\right)\oplus\gamma_2l'\left(p'\right)\right\rangle=\left\langle \gamma_1l(p),\gamma_1l'\left(p'\right)\right\rangle\oplus\left\langle \gamma_2l(p),\gamma_2l'\left(p'\right)\right\rangle$$
$$=\gamma_1A(p)\oplus\gamma_2A(P)$$
$$(5)\ A(p)\otimes A_1(p_1)=\left\langle l(p),l'\left(p'\right)\right\rangle\otimes\left\langle l_1(p_1),l_1'\left(p_1'\right)\right\rangle=\left\langle l(p)\otimes l_1(p_1),l'\left(p'\right)\otimes l_1'\left(p_1'\right)\right\rangle$$
$$=\left\langle \cup_{l^{(i)}\in l(p),l_1^{(i)}\in l_1(p_1)}\left\{\left(l^{(i)}\right)^{p^{(i)}}\otimes\left(l_1^{(i)}\right)^{p_1^{(i)}}\right\},\cup_{l'^{(j)}\in l'(p'),l_1'^{(j)}\in l_1'(p_1)}\left\{\left(l'^{(j)}\right)^{p'^{(j)}}\otimes\left(l_1'^{(j)}\right)^{p_1'^{(j)}}\right\}\right\rangle$$
$$=\left\langle \cup_{l^{(i)}\in l(p),l_1^{(i)}\in l_1(p_1)}\left\{\left(l_1^{(i)}\right)^{p_1^{(i)}}\otimes\left(l^{(i)}\right)^{p^{(i)}}\right\},\cup_{l'^{(j)}\in l'(p'),l_1'^{(j)}\in l_1'(p_1)}\left\{\left(l_1'^{(j)}\right)^{p_1'^{(j)}}\otimes\left(l'^{(j)}\right)^{p'^{(j)}}\right\}\right\rangle$$
$$=\left\langle l(y)\,(p(y))\otimes l(p),l'(y)\left(p'(y)\right)\otimes l'\left(p'\right)\right\rangle=\left\langle l_1(p_1),l_1'\left(p_1'\right)\right\rangle\otimes\left\langle l(p),l'\left(p'\right)\right\rangle$$
$$=A_1(P_1)\otimes A(p)$$
$$(6)\ A(p)\otimes(A_1(p_1)\otimes A_2(p_2))=\left\langle l(p),l'\left(p'\right)\right\rangle\otimes\left(\left\langle l_1(p_1),l_1'\left(p_1'\right)\right\rangle\otimes\left\langle l_2(p_2),l_2'\left(p_2'\right)\right\rangle\right)$$
$$=\left\langle l(p)\otimes(l_1(p_1)\otimes l_2(p_2)),l'\left(p'\right)\otimes\left(l_1'\left(p_1'\right)\otimes l_2'\left(p_2'\right)\right)\right\rangle$$
$$=\left\langle \begin{array}{c}\cup_{l^{(i)}\in l(p),l_1^{(i)}\in l_1(p_1),l_2^{(i)}\in l_2(p_2)}\left\{\left(l^{(i)}\right)^{p^{(i)}}\otimes\left(l_1^{(i)}\right)^{p_1^{(i)}}\otimes\left(l_2^{(i)}\right)^{p_2^{(i)}}\right\},\\ \cup_{l'^{(j)}\in l'(p'),l_1'^{(j)}\in l_1'(p_1),l_2'^{(j)}\in l_2'(p_2)}\left\{\left(l'^{(j)}\right)^{p'^{(j)}}\otimes\left(\left(l_1'^{(j)}\right)^{p_1'^{(j)}}\otimes\left(l_2'^{(j)}\right)^{p_2'^{(j)}}\right)\right\}\end{array}\right\rangle$$
$$=\left\langle \begin{array}{c}\cup_{l^{(i)}\in l(p),l_1^{(i)}\in l_1(p_1),l_2^{(i)}\in l_2(p_2)}\left\{\left(\left(l^{(i)}\right)^{p^{(i)}}\otimes\left(l_1^{(i)}\right)^{p_1^{(i)}}\right)\otimes\left(l_2^{(i)}\right)^{p_2^{(i)}}\right\},\\ \cup_{l'^{(j)}\in l'(p'),l_1'^{(j)}\in l_1'(p_1),l_2'^{(j)}\in l_2'(p_2)}\left\{\left(\left(l'^{(j)}\right)^{p'^{(j)}}\otimes\left(l_1'^{(j)}\right)^{p_1'^{(j)}}\right)\otimes\left(l_2'^{(j)}\right)^{p_2'^{(j)}}\right\}\end{array}\right\rangle$$
$$=\left\langle (l(p)\otimes l_1(p_1))\otimes l_2(p_2),\left(l'\left(p'\right)\otimes l_1'\left(p_1'\right)\right)\otimes l_2'\left(p_2'\right)\right\rangle$$
$$=\left(\left\langle l(p),l'\left(p'\right)\right\rangle\otimes\left\langle l_1(p_1),l_1'\left(p_1'\right)\right\rangle\right)\otimes\left\langle l_2(p_2),l_2'\left(p_2'\right)\right\rangle$$
$$=(A(p)\otimes A_1(p_1))\otimes A_2(p_2)$$
$$(7)\ (A(p)\otimes A_1(p_1))^{\gamma}=\left(\left\langle l(p),l'\left(p'\right)\right\rangle\otimes\left\langle l_1(p_1),l_1'\left(p_1'\right)\right\rangle\right)^{\gamma}$$
$$=\left\langle (l(p)\otimes l_1(p_1))^{\gamma},\left(l'\left(p'\right)\otimes l_1'\left(p_1'\right)\right)^{\gamma}\right\rangle$$
$$=\left\langle \left(\cup_{l^{(i)}\in l(p),l_1^{(i)}\in l_1(p_1)}\left\{p^{(i)}l^{(i)}\otimes p_1^{(i)}l_1^{(i)}\right\}\right)^{\gamma},\left(\cup_{l'^{(j)}\in l'(p'),l_1'^{(j)}\in l_1'(p_1)}\left\{p'^{(j)}l'^{(j)}\otimes p_1'^{(j)}l_1'^{(j)}\right\}\right)^{\gamma}\right\rangle$$
$$=\left\langle \cup_{l^{(i)}\in l(p),l_1^{(i)}\in l_1(p_1)}\left\{\left(l^{(i)}\right)^{\gamma p^{(i)}}\otimes\left(l_1^{(i)}\right)^{\gamma p_1^{(i)}}\right\},\cup_{l'^{(j)}\in l'(p'),l'^{(j)}(y)\in l_1'(p_1)}\left\{\left(l'^{(j)}\right)^{\gamma p'^{(j)}}\otimes\left(l_1'^{(j)}\right)^{\gamma p_1'^{(j)}}\right\}\right\rangle$$
$$=\left\langle (l(p))^{\gamma}\otimes(l_1(p_1))^{\gamma},\left(l'\left(p'\right)\right)^{\gamma}\otimes\left(l_1'\left(p_1'\right)\right)^{\gamma}\right\rangle$$
$$=\left\langle (l(p))^{\gamma},\left(l'\left(p'\right)\right)^{\gamma}\right\rangle\otimes\left\langle (l_1(p_1))^{\gamma},\left(l_1'\left(p_1'\right)\right)^{\gamma}\right\rangle=(A(p))^{\gamma}\otimes(A_1(p_1))^{\gamma}$$
$$(8)\ (A(p))^{\gamma_1+\gamma_2}=\left(\left\langle l(p),l'\left(p'\right)\right\rangle\right)^{\gamma_1+\gamma_2}=\left\langle (l(p))^{\gamma_1+\gamma_2},\left(l'\left(p'\right)\right)^{\gamma_1+\gamma_2}\right\rangle$$

$$
\begin{aligned}
&= \left\langle \cup_{l^{(i)} \in l(p)} \left\{ \left(l^{(i)}\right)^{(\gamma_1+\gamma_2)p^{(i)}} \right\}, \cup_{l'^{(j)} \in l'(p')} \left\{ \left(l'^{(j)}(x)\right)^{(\gamma_1+\gamma_2)p'^{(j)}} \right\} \right\rangle \\
&= \left\langle \cup_{l^{(i)} \in l(p)} \left\{ \left(l^{(i)}\right)^{\gamma_1 p^{(i)}} \otimes \left(l^{(i)}\right)^{\gamma_2 p^{(i)}} \right\}, \cup_{l'^{(j)} \in l'(p')} \left\{ \left(l'^{(j)}\right)^{\gamma_1 p'^{(j)}} \otimes \left(l'^{(j)}\right)^{\gamma_2 p'^{(j)}} \right\} \right\rangle \\
&= \left\langle \begin{array}{c} \cup_{l^{(i)} \in l(p)} \left\{ \left(l^{(i)}\right)^{\gamma_1 p^{(i)}} \right\} \otimes \cup_{l^{(i)}(x) \in l(p)} \left\{ \left(l^{(i)}\right)^{\gamma_2 p^{(i)}} \right\}, \\ \cup_{l'^{(j)} \in l'(p')} \left\{ \left(l'^{(j)}\right)^{\gamma_1 p'^{(j)}} \right\} \otimes \cup_{l'^{(j)} \in l'(p')} \left\{ \left(l'^{(j)}\right)^{\gamma_2 p'^{(j)}} \right\} \end{array} \right\rangle \\
&= \left\langle (l(p))^{\gamma_1} \otimes (l(p))^{\gamma_2}, \left(l'\left(p'\right)\right)^{\gamma_1} \otimes \left(l'\left(p'\right)\right)^{\gamma_2} \right\rangle \\
&= \left\langle (l(p))^{\gamma_1}, \left(l'\left(p'\right)\right)^{\gamma_1} \right\rangle \otimes \left\langle (l(p))^{\gamma_2}, \left(l'\left(p'\right)\right)^{\gamma_2} \right\rangle = (A(p))^{\gamma_1} \otimes (A(p))^{\gamma_2}. \quad \square
\end{aligned}
$$

4. Aggregation Operators and Attribute Weights

This section is dedicated to discussion on some basic aggregation operators of PHILTS. Deviation degree between two PHILTEs is also defined in this section. Finally, we calculate the attribute weights in the light of PHILTEs.

4.1. The Aggregation Operators for PHILTEs

The aggregation operators are powerful tools to deal with linguistic information. To make a better usage of PHILTEs in real world problems, in the following, aggregation operators for PHILTEs have been developed.

Definition 18. *Let* $A_k(p_k) = \left\langle l_k(p_k), l'_k\left(p'_k\right)\right\rangle$ $(k = 1, 2, \ldots, n)$ *be* n *ordered and normalized PHILTEs. Then*

$$
\begin{aligned}
&PHILA\left(A_1(p_1), A_2(p_2), \ldots, A_n(p_n)\right) \\
&= \frac{1}{n}\left(\left\langle l_1(p_1), l'_1\left(p'_1\right)\right\rangle \oplus \left\langle l_2(p_2), l'_2\left(p'_2\right)\right\rangle \oplus \ldots \oplus \left\langle l_n(p_n), l'_n\left(p'_n\right)\right\rangle\right) \\
&= \frac{1}{n}\left\langle l_1(p_1) \oplus l_2(p_2) \oplus \ldots \oplus l_n(p_n), l'_1\left(p'_1\right) \oplus l'_2\left(p'_2\right) \oplus \ldots \oplus l'_n\left(p'_n\right)\right\rangle \\
&= \frac{1}{n}\left\langle \begin{array}{c} \cup_{l_1^{(i)} \in l_1(p_1), l_2^{(i)} \in l_2(p_2), \ldots, l_n^{(i)} \in l_n(p_n)} \left\{ p_1^{(i)} l_1^{(i)} \oplus p_2^{(i)} l_2^{(i)} \oplus \ldots \oplus p_n^{(i)} l_n^{(i)} \right\}, \\ \cup_{l_1'^{(j)} \in l'_1\left(p'_1\right), l_2'^{(j)} \in l'_2\left(p'_2\right), \ldots, l_n'^{(j)} \in l'_n\left(p'_n\right)} \left\{ p_1'^{(j)} l_1'^{(j)} \oplus p_2'^{(j)} l_2'^{(j)} \oplus \ldots \oplus p_n'^{(j)} l_n'^{(j)} \right\} \end{array} \right\rangle
\end{aligned}
\tag{13}
$$

is called the probabilistic hesitant intuitionistic linguistic averaging (PHILA) operator.

Definition 19. *Let* $A_k(p_k) = \left\langle l_k(p_k), l'_k\left(p'_k\right)\right\rangle$ $(k = 1, 2, \ldots, n)$ *be* n *ordered and normalized PHILTEs. Then*

$$
\begin{aligned}
&PHILWA\left(A_1(p_1), A_2(p_2), \ldots, A_n(p_n)\right) \\
&= w_1\left\langle l_1(p_1), l'_1\left(p'_1\right)\right\rangle \oplus w_2\left\langle l_2(p_2), l'_2\left(p'_2\right)\right\rangle \oplus \ldots \oplus w_n\left\langle l_n(p_n), l'_n\left(p'_n\right)\right\rangle \\
&= \left\langle w_1 l_1(p_1) \oplus w_2 l_2(p_2) \oplus \ldots \oplus w_n l_n(p_n), w_1 l'_1\left(p'_1\right) \oplus w_2 l'_2\left(p'_2\right) \oplus \ldots \oplus w_n l'_n\left(p'_n\right)\right\rangle \\
&= \left\langle \begin{array}{c} \cup_{l_1^{(i)} \in l_1(p_1)} \left\{ w_1 p_1^{(i)} l_1^{(i)} \right\} \oplus \cup_{l_2^{(i)} \in l_2(p_2)} \left\{ w_2 p_2^{(i)} l_2^{(i)} \right\} \oplus \ldots \oplus \cup_{l_n^{(i)} \in l_n(p_n)} \left\{ w_n p_n^{(i)} l_n^{(i)} \right\}, \\ \cup_{l_1'^{(j)} \in l'_1\left(p'_1\right)} \left\{ w_1 p_1'^{(j)} l_1'^{(j)} \right\} \oplus \cup_{l_2'^{(j)} \in l'_2\left(p'_2\right)} \left\{ w_2 p_2'^{(j)} l_2'^{(j)} \right\} \oplus \ldots \oplus \cup_{l_n'^{(j)} \in l'_n\left(p'_n\right)} \left\{ w_n p_n'^{(j)} l_n'^{(j)} \right\} \end{array} \right\rangle
\end{aligned}
\tag{14}
$$

is called the probabilistic hesitant intuitionistic linguistic weighted averaging (PHILWA) operator, where $w = (w_1, w_2, \ldots, w_n)^t$ *is the weight vector of* $A_k(p_k)$ $(k = 1, 2, \ldots, n)$, $w_k \geq 0, k = 1, 2, \ldots, n$, *and* $\sum_{k=1}^{n} w_k = 1$.

Particularly, if we take $w = \left(\frac{1}{n}, \frac{1}{n}, \ldots, \frac{1}{n}\right)^t$, then the PHILWA operator reduces to the PHILA operator.

Definition 20. *Let* $A_k(p_k) = \left\langle l_k(p_k), l'_k\left(p'_k\right)\right\rangle$ $(k = 1, 2, \ldots, n)$ *be* n *ordered and normalized PHILTEs. Then,*

$$
\begin{aligned}
&PHILG\left(A_1(p_1), A_2(p_2), \ldots, A_n(p_n)\right) \\
&= \left(\left\langle l_1(p_1), l'_1\left(p'_1\right)\right\rangle \otimes \left\langle l_2(p_2), l'_2\left(p'_2\right)\right\rangle \otimes \ldots \otimes \left\langle l_n(p_n), l'_n\left(p'_n\right)\right\rangle\right)^{\frac{1}{n}} \\
&= \left(\left\langle l_1(p_1) \otimes l_2(p_2) \otimes \ldots \otimes l_n(p_n), l'_1\left(p'_1\right) \otimes l'_2\left(p'_2\right) \otimes \ldots \otimes l'_n\left(p'_n\right)\right\rangle\right)^{\frac{1}{n}} \\
&= \left(\left\langle \begin{array}{c} \cup_{l_1^{(i)} \in l_1(p_1), l_2^{(i)} \in l_2(p_2), \ldots, l_n^{(i)} \in l_n(p_n)} \left\{\left(l_1^{(i)}\right)^{p_1^{(i)}} \otimes \left(l_2^{(i)}\right)^{p_2^{(i)}} \otimes \ldots \otimes \left(l_n^{(i)}\right)^{p_n^{(i)}}\right\}, \\ \cup_{l_1'^{(j)} \in l'_1\left(p'_1\right), l_2'^{(j)} \in l'_2\left(p'_2\right), \ldots, l_n'^{(j)} \in l'_n\left(p'_n\right)} \left\{\left(l_1'^{(j)}\right)^{p_1'^{(j)}} \otimes \left(l_2'^{(j)}\right)^{p_2'^{(j)}} \otimes \ldots \otimes \left(l_n'^{(j)}\right)^{p_n'^{(j)}}\right\} \end{array} \right\rangle\right)^{\frac{1}{n}}
\end{aligned}
\tag{15}
$$

is called the probabilistic hesitant intuitionistic linguistic geometric (PHILG) operator.

Definition 21. *Let* $A_k(p_k) = \left\langle l_k(p_k), l'_k\left(p'_k\right)\right\rangle$ $(k = 1, 2, \ldots, n)$ *be* n *ordered and normalized PHILTEs. Then*

$$
\begin{aligned}
&PHILWG\left(A_1(p_1), A_2(p_2), \ldots, A_n(p_n)\right) \\
&= \left\langle l_1(p_1), l'_1\left(p'_1\right)\right\rangle^{w_1} \otimes \left\langle l_2(p_2), l'_2\left(p'_2\right)\right\rangle^{w_2} \otimes \ldots \otimes \left\langle l_n(p_n), l'_n\left(p'_n\right)\right\rangle^{w_n} \\
&= \left\langle (l_1(p_1))^{w_1} \otimes (l_2(p_2))^{w_2} \otimes \ldots \otimes (l_n(p_n))^{w_n}, \left(l'_1\left(p'_1\right)\right)^{w_1} \otimes \left(l'_2\left(p'_2\right)\right)^{w_2} \otimes \ldots \otimes \left(l'_n\left(p'_n\right)\right)^{w_n} \right\rangle \\
&= \left\langle \begin{array}{c} \cup_{l_1^{(i)} \in l_1(p_1)} \left\{\left(l_1^{(i)}\right)^{w_1 p_1^{(i)}}\right\} \otimes \cup_{l_2^{(i)} \in l_2(p_2)} \left\{\left(l_2^{(i)}\right)^{w_2 p_2^{(i)}}\right\} \otimes \ldots \otimes \cup_{l_n^{(i)} \in l_n(p_n)} \left\{\left(l_n^{(i)}\right)^{w_n p_n^{(i)}}\right\}, \\ \cup_{l_1'^{(j)} \in l'_1\left(p'_1\right)} \left\{\left(l_1'^{(j)}\right)^{w_1 p_1'^{(j)}}\right\} \otimes \cup_{l_2'^{(j)} \in l'_2\left(p'_2\right)} \left\{\left(l_2'^{(j)}\right)^{w_2 p_2'^{(j)}}\right\} \otimes \ldots \otimes \cup_{l_n'^{(j)} \in l'_n\left(p'_n\right)} \left\{\left(l_n'^{(j)}\right)^{w_n p_n'^{(j)}}\right\} \end{array} \right\rangle
\end{aligned}
\tag{16}
$$

is called the probabilistic hesitant intuitionistic linguistic weighted geometric (PHILWG) operator, where $w = (w_1, w_2, \ldots, w_n)^t$ *is the weight vector of* $A_k(p_k)$ $(k = 1, 2, \ldots, n)$, $w_k \geq 0, k = 1, 2, \ldots, n$, *and* $\sum_{k=1}^{n} w_k = 1$.

Particularly, if we take $w = \left(\frac{1}{n}, \frac{1}{n}, \ldots, \frac{1}{n}\right)^t$, then the PHILWG operator reduces to the PHILG operator.

4.2. Maximizing Deviation Method for Calculating the Attribute Weights

The choice of weights directly affects the performance of weighted aggregation operators. For this purpose, in this subsection, the affective maximizing deviation method is adopted to calculate weight in MAGDM when weights are unknown or partly known. Based on Definition 9, the deviation degree between two PHILTEs is defined as follows:

Definition 22. *Let* $A(p)$ *and* $A_1(p_1)$ *be any two PHILTEs of equal length. Then, the deviation degree* D *between* $A(p)$ *and* $A_1(p_1)$ *is given by*

$$D\left(A(p), A_1(p_1)\right) = d\left(l(p), l_1(p_1)\right) + d\left(l'\left(p'\right), l'_1\left(p'_1\right)\right) \tag{17}$$

where

$$d\left(l\left(p\right),l_1\left(p_1\right)\right)=\sqrt{\frac{\sum_{i=1}^{\#l(p)}\left(p^{(i)}r^{(i)}-p_1^{(i)}r_1^{(i)}\right)}{\#l\left(p\right)}}, \tag{18}$$

$$d\left(l'\left(p'\right),l_1'\left(p_1'\right)\right)=\sqrt{\frac{\sum_{j=1}^{\#l'(p')}\left(p'^{(j)}r'^{(j)}-p_1'^{(j)}r_1'^{(j)}\right)}{\#l'\left(p'\right)}} \tag{19}$$

$r^{(i)}$ denote the lower index of the ith linguistic term of $l\left(p\right)$ and $r'^{(j)}$ denote the lower index of the jth linguistic term of $l'\left(p'\right)$.

Based on the above definition, in the following, we derive attribute weight vector because working on the probabilistic linguistic data to deal with the MAGDM problems, in which the weight information of attribute values is completely unknown or partly known, we must find the attribute weights in advance.

Given the set of alternatives $x=\{x_1,x_2,\ldots,x_m\}$ and the set of "n" attributes $c=\{c_1,c_2,\ldots,c_n\}$, respectively, then, by using Equation (17), the deviation measure between the alternative "x_i" and all other alternatives with respect to the attribute "c_j" can be given as:

$$D_{ij}\left(w\right)=\sum_{q=1,q\neq i}w_jD\left(h_{ij},h_{qj}\right),i=1,2,\ldots,m,j=1,2,\ldots,n \tag{20}$$

In accordance with the theme of the maximizing deviation method, if the deviation degree among alternatives is smaller for an attribute, then the attribute should give a smaller weight. This one shows that the alternatives are homologous to the attribute. Contrarily, it should give a larger weight. Let

$$\begin{aligned}D_j\left(w\right)&=\sum_{i=1}^{m}D_{ij}\left(w\right)=\sum_{i=1}^{m}\sum_{q\neq i}^{m}w_jD\left(h_{ij},h_{qj}\right)\\&=\sum_{i=1}^{m}\sum_{q\neq i}^{m}w_j\left(d\left(l_{ij}\left(p_{ij}\right),l_{qj}\left(p_{qj}\right)\right)+d\left(l_{ij}'\left(p_{ij}'\right),l_{qj}'\left(p_{qj}'\right)\right)\right)\end{aligned} \tag{21}$$

show the deviation degree of one alternative and others with respect to the attribute "c_j" and let

$$\begin{aligned}D\left(w\right)&=\sum_{j=1}^{n}D_j\left(w\right)=\sum_{j=1}^{n}\sum_{i=1}^{m}D_{ij}\left(w\right)=\sum_{j=1}^{n}\sum_{i=1}^{m}\sum_{q\neq i}^{m}w_jD\left(h_{ij},h_{qj}\right)\\&=\sum_{j=1}^{n}\sum_{i=1}^{m}\sum_{q\neq i}^{m}w_j\left(d\left(l_{ij}\left(p_{ij}\right),l_{qj}\left(p_{qj}\right)\right)+d\left(l_{ij}'\left(p_{ij}'\right),l_{qj}'\left(p_{qj}'\right)\right)\right)\\&=\sum_{j=1}^{n}\sum_{i=1}^{m}\sum_{q\neq i}^{m}w_j\left(\begin{array}{c}\sqrt{\frac{1}{\#l_{ij}\left(p_{ij}\right)}\sum_{k_1=1}^{\#l_{ij}\left(p_{ij}\right)}\left(p_{ij}^{(k_1)}r_{ij}^{(k_1)}-p_{qj}^{(k_1)}r_{qj}^{(k_1)}\right)^2}+\\\sqrt{\frac{1}{\#l_{ij}'\left(p_{ij}'\right)}\sum_{k_2=1}^{\#l_{ij}'\left(p_{ij}'\right)}\left(p_{ij}'^{(k_2)}r_{ij}'^{(k_2)}-p_{qj}'^{(k_2)}r_{qj}'^{(k_2)}\right)^2}\end{array}\right)\end{aligned} \tag{22}$$

express the sum of the deviation degrees among all attributes.

To obtain the attribute weights vector $w=\left(w_1,w_2,\ldots,w_n\right)^t$, we build the following single objective optimization model (named as M_1) to drive the deviation degree $d\left(w\right)$ as large as possible.

$$M_1=\begin{cases}\max D\left(w\right)=\sum_{j=1}^{n}\sum_{i=1}^{m}\sum_{q\neq i}^{m}w_jD\left(h_{ij},h_{qj}\right)\\w_j\geq 0,j=1,2,\ldots,n,\sum_{j=1}^{n}w_j^2=1\end{cases}$$

To solve the above model M_1, we use the Lagrange multiplier function:

$$L(w,\eta) = \sum_{j=1}^{n}\sum_{i=1}^{m}\sum_{q\neq i}^{m} w_j D\left(h_{ij}, h_{qj}\right) + \frac{\eta}{2}\left(\sum_{j=1}^{n} w_j^2 - 1\right) \tag{23}$$

where η is the Lagrange parameter.

Then, we compute the partial derivatives of Lagrange function with respect to w_j and η and let them be zero:

$$\begin{cases} \frac{\partial L(w,\eta)}{\partial w_j} = \sum_{i=1}^{m}\sum_{q\neq i}^{m} w_j D\left(h_{ij}, h_{qj}\right) + \eta w_j = 0, j = 1,2,\ldots,n. \\ \frac{\delta L(w,\eta)}{\partial \eta} = \sum_{j=1}^{n} w_j^2 - 1 = 0 \end{cases} \tag{24}$$

By solving Equation (24), one can obtain the optimal weight $w = (w_1, w_2, \ldots, w_n)^t$.

$$w_j = \frac{\sum_{i=1}^{m}\sum_{q\neq i}^{m} D\left(h_{ij}, h_{qj}\right)}{\sqrt{\sum_{j=1}^{n}\left(\sum_{i=1}^{m}\sum_{q\neq i} D\left(h_{ij}, h_{qj}\right)\right)^2}} = \frac{\sum_{i=1}^{m}\sum_{q\neq i}^{m}\left(d\left(l_{ij}(p_{ij}), l_{qj}(p_{qj})\right) + d\left(l'_{ij}\left(p'_{ij}\right), l'_{qj}\left(p'_{qj}\right)\right)\right)}{\sqrt{\sum_{j=1}^{n}\left(\sum_{i=1}^{m}\sum_{q\neq i}\left(d\left(l_{ij}(p_{ij}), l_{qj}(p_{qj})\right) + d\left(l'_{ij}\left(p'_{ij}\right), l'_{qj}\left(p'_{qj}\right)\right)\right)\right)^2}}$$

$$w_j = \frac{\sum_{i=1}^{m}\sum_{q\neq i}^{m}\left(\sqrt{\frac{1}{\#l_{ij}(p_{ij})}\sum_{k_1=1}^{\#l_{ij}(p_{ij})}\left(p_{ij}^{(k_1)} r_{ij}^{(k_1)} - p_{qj}^{(k_1)} r_{qj}^{(k_1)}\right)^2} + \sqrt{\frac{1}{\#l'_{ij}\left(p'_{ij}\right)}\sum_{k_2=1}^{\#l'_{ij}\left(p'_{ij}\right)}\left(p_{ij}^{\prime(k_2)} r_{ij}^{\prime(k_2)} - p_{qj}^{\prime(k_2)} r_{qj}^{\prime(k_2)}\right)^2}\right)}{\sqrt{\sum_{j=1}^{n}\left(\sum_{i=1}^{m}\sum_{q\neq i}\left(\sqrt{\frac{1}{\#l_{ij}(p_{ij})}\sum_{k_1=1}^{\#l_{ij}(p_{ij})}\left(p_{ij}^{(k_1)} r_{ij}^{(k_1)} - p_{qj}^{(k_1)} r_{qj}^{(k_1)}\right)^2} + \sqrt{\frac{1}{\#l'_{ij}\left(p'_{ij}\right)}\sum_{k_2=1}^{\#l'_{ij}\left(p'_{ij}\right)}\left(p_{ij}^{\prime(k_2)} r_{ij}^{\prime(k_2)} - p_{qj}^{\prime(k_2)} r_{qj}^{\prime(k_2)}\right)^2}\right)\right)^2}} \tag{25}$$

where $j = 1,2,\ldots,n$.

Obviously, $w_j > 0\ \forall\ j$. By normalizing Equation (25), we get:

$$w_j = \frac{\sum_{i=1}^{m}\sum_{q\neq i}^{m} D\left(h_{ij}, h_{qj}\right)}{\sum_{j=1}^{n}\sum_{i=1}^{m}\sum_{q\neq i} D\left(h_{ij}, h_{qj}\right)}$$

$$w_j = \frac{\sum_{i=1}^{m}\sum_{q\neq i}^{m}\left(\sqrt{\frac{1}{\#l_{ij}(p_{ij})}\sum_{k_1=1}^{\#l_{ij}(p_{ij})}\left(p_{ij}^{(k_1)} r_{ij}^{(k_1)} - p_{qj}^{(k_1)} r_{qj}^{(k_1)}\right)^2} + \sqrt{\frac{1}{\#l'_{ij}\left(p'_{ij}\right)}\sum_{k_2=1}^{\#l'_{ij}\left(p'_{ij}\right)}\left(p_{ij}^{\prime(k_2)} r_{ij}^{\prime(k_2)} - p_{qj}^{\prime(k_2)} r_{qj}^{\prime(k_2)}\right)^2}\right)}{\sum_{j=1}^{n}\sum_{i=1}^{m}\sum_{q\neq i}\left(\sqrt{\frac{1}{\#l_{ij}(p_{ij})}\sum_{k_1=1}^{\#l_{ij}(p_{ij})}\left(p_{ij}^{(k_1)} r_{ij}^{(k_1)} - p_{qj}^{(k_1)} r_{qj}^{(k_1)}\right)^2} + \sqrt{\frac{1}{\#l'_{ij}\left(p'_{ij}\right)}\sum_{k_2=1}^{\#l'_{ij}\left(p'_{ij}\right)}\left(p_{ij}^{\prime(k_2)} r_{ij}^{\prime(k_2)} - p_{qj}^{\prime(k_2)} r_{qj}^{\prime(k_2)}\right)^2}\right)} \tag{26}$$

where $j = 1,2,\ldots,n$.

The above end result can be applied to the situations where the information of attribute weights is completely unknown. However, in real life decision making problems, the weight information is usually partly known. In such cases, let H be a set of the known weight information, which can be given in the following forms based on the literature [31–34].

Form 1. A weak ranking: $\{w_i \geq w_j\}$ $(i \neq j)$.
Form 2. A strict ranking: $\{w_i - w_j \geq \beta_i\}$ $(i \neq j)$.
Form 3. A ranking of differences: $\{w_i - w_j \geq w_k - w_l\}$ $(j \neq k \neq l)$.
Form 4. A ranking with multiples: $\{w_i \geq \beta_i w_j\}$ $(i \neq j)$.
Form 5. An interval form: $\{\beta_i \leq w_j \leq \beta_i + \epsilon_i\}$ $(i \neq j)$.
β_i and ϵ_i denote the non-negative numbers.
With the set H, we can build the following model:

$$M_2 = \begin{cases} \max D(w) = \sum_{j=1}^{n} \sum_{i=1}^{m} \sum_{q \neq i}^{m} w_j D\left(h_{ij}, h_{qj}\right) \\ w_j \in H, w_j \geq 0, j = 1, 2, \ldots, n, \sum_{j=1}^{n} w_j^2 = 1 \end{cases}$$

from which the optimal weight vector $w = (w_1, w_2, \ldots, w_n)^t$ obtained.

5. MAGDM with Probabilistic Hesitant Intuitionistic Linguistic Information

In this section, two practical methods, i.e., an extended TOPSIS method and an aggregation based method, for MAGDM problems are proposed, where the opinions of DMs take the form of PHILTSs.

5.1. Extended TOPSIS Method for MAGDM with Probabilistic Hesitant Intuitionistic Linguistic Information

Of the numerous MAGDM methods, TOPSIS (Technique for Order of Preference by Similarity to Ideal Solution) is one of the effective methods for ranking and selecting a number of possible alternatives by measuring Euclidean distances. It has been successfully applied to solve evaluation problems with a finite number of alternatives and criteria [19,24,28] because it is easy to understand and implement, and can measure the relative performance for each alternative.

In the following, we discuss the complete construction of extended TOPSIS method in PHILTS regard. This methodology involves the following steps.

Step 1: Analyze the given MAGDM problem; since the problem is group decision making, so let there be "l" decision makers or experts $M = \{m_1, m_2, \ldots, m_l\}$ involved in the given problem. The set of alternatives is $x = \{x_1, x_2, \ldots, x_m\}$ and the set of attributes is $c = \{c_1, c_2, \ldots, c_n\}$. The experts provide their linguistic evaluation values for membership and non-membership by using linguistic term set $S = \{s_0, s_1, \ldots, s_g\}$ over the alternative x_i $(i = 1, 2, \ldots, m)$ with respect to the attribute c_j $(j = 1, 2, \ldots, n)$.

The DM m_k $(k = 1, 2, \ldots, l)$ states his membership and non-membership linguistic evaluation values keeping in mind all the alternatives and attributes in the form of PHILTEs. Thus, intuitionistic probabilistic linguistic decision matrix $H^k = \left[\left\langle l_{ij}^k(p_{ij}), l_{ij}^{'(k)}\left(p_{ij}^{'}\right)\right\rangle\right]_{m \times n}$ is constructed. It should be noted that preference of alternative "x_i" with respect to decision maker "m_k" and attribute "c_j" is denoted as PHILTE $A_{ij}^k(p_{ij})$ in a group decision making problem with "l" experts.

Step 2: Calculate the one probabilistic hesitant intuitionistic linguistic decision matrix H by aggregating the opinions of DMs$\left(H^{(1)}, H^{(2)}, \ldots, H^{(l)}\right)$; $H = \left[h_{ij}\right]$, where

$h_{ij} = \left\langle \left\{ s_{m_{ij}}(p_{ij}), s_{n_{ij}}(q_{ij}) \right\}, \left\{ s_{m_{ij}}^{'}\left(p_{ij}^{'}\right), s_{n_{ij}}^{'}\left(q_{ij}^{'}\right) \right\} \right\rangle$ where

$$s_{m_{ij}}(p_{ij}) = \min\left\{ \min_{k=1}^{l}\left(\max l_{ij}^k(p_{ij})\right), \max_{k=1}^{l}\left(\min l_{ij}^k(p_{ij})\right) \right\},$$

$$s_{n_{ij}}(q_{ij}) = \max\left\{ \min_{k=1}^{l}\left(\max l_{ij}^k(q_{ij})\right), \max_{k=1}^{l}\left(\min l_{ij}^k(q_{ij})\right) \right\},$$

$$s_{m_{ij}^{'}}\left(p_{ij}^{'}\right) = \min\left\{ \min_{k=1}^{l}\left(\max l_{ij}^{'k}\left(p_{ij}^{'}\right)\right), \max_{k=1}^{l}\left(\min l_{ij}^{'k}\left(p_{ij}^{'}\right)\right) \right\},$$

$$s_{n_{ij}^{'}}\left(q_{ij}^{'}\right) = \max\left\{ \min_{k=1}^{l}\left(\max l_{ij}^{'k}\left(q_{ij}^{'}\right)\right), \max_{k=1}^{l}\left(\min l_{ij}^{'k}\left(q_{ij}^{'}\right)\right) \right\},$$

Here, $\max l_{ij}^k(p_{ij})$ and $\min l_{ij}^k(p_{ij})$ are taken according to the maximum and minimum value of $p_{ij} \times r_{ij}^l, l = 1, 2, \ldots, \# l_{ij}^k(p_{ij})$, respectively, where r_{ij}^l denotes the lower index of the lth linguistic term and p_{ij} is its corresponding probability.

In this aggregated matrix H, the preference of alternative a_i with respect to attribute c_j is denoted as h_{ij}.

Each term of the aggregated matrix H i.e., h_{ij} is also an PHILTE; for this, we have to prove that $s_{m_{ij}}(p_{ij}) + s'_{n_{ij}}\left(q'_{ij}\right) \leq s_g$ and $s_{n_{ij}}(q_{ij}) + s'_{m_{ij}}\left(p'_{ij}\right) \leq s_g$. Since we know that $\left[l^k_{ij}(p_{ij}), l'_{ij}\left(p'_{ij}\right)\right]$ is a PHILTS for every k^{th} expert, i^{th} alternative and j^{th} attribute, a PHILTS it must satisfy the conditions $\min\left(l^{(k)}_{ij}\right) + \max\left(l'^{(k)}_{ij}\right) \leq s_g$, $\max\left(l^{(k)}_{ij}\right) + \min\left(l'^{(k)}_{ij}\right) \leq s_g$.

Thus, the above simple construction of $s_{m_{ij}}(p_{ij}), s_{n_{ij}}(q_{ij}), s'_{m_{ij}}\left(p'_{ij}\right)$, and $s_{n'_{ij}}\left(q'_{ij}\right)$ guarantees that the h_{ij} is a PHILTE.

Step 3: Normalize the probabilistic hesitant intuitionistic linguistic decision matrix $H = [h_{ij}]$ according to the method in Section 3.1.

Step 4: Obtain the weight vector $w = (w_1, w_2, \ldots, w_n)^t$ of the attributes c_j $(j = 1, 2, \ldots, n)$. $w_j = \dfrac{\sum_{i=1}^{m}\sum_{q\neq i} D(h_{ij}, h_{qj})}{\sum_{j=1}^{n}\sum_{i=1}^{m}\sum_{q\neq i} D(h_{ij}, h_{qj})} = \dfrac{\sum_{i=1}^{m}\sum_{q\neq i} d(l_{ij}(p_{ij}), l_{qj}(p_{qj})) + d\left(l'_{ij}\left(p'_{ij}\right), l'_{qj}\left(p'_{qj}\right)\right)}{\sum_{j=1}^{n}\sum_{i=1}^{m}\sum_{q\neq i} d(l_{ij}(p_{ij}), l_{qj}(p_{qj})) + d\left(l'_{ij}\left(p'_{ij}\right), l'_{qj}\left(p'_{qj}\right)\right)}, j = 1, 2, \ldots, n$

Step 5: The PHILTS positive ideal solution (PHILTS-PIS) of alternatives, denoted by $A^+ = \left\langle l^+(p), l'^+(p)\right\rangle$, is defined as follows:

$$A^+ = \left\langle l^+(p) = \left(l^+_1(p), l^+_2(p), \ldots, l^+_n(p)\right), l'^+(p) = \left(l'^+_1(p), l'^+_2(p), \ldots, l'^+_n(p)\right)\right\rangle \tag{27}$$

where $l^+_j(p) = \left\{\left(l^{(k_1)}_j\right)^+ | k_1 = 1, 2, \ldots, \#l_{ij}(p)\right\}$ and $\left(l^{(k_1)}_j\right)^+ = s_{\max_i}\left\{p^{(k_1)}_{ij} r^{(k_1)}_{ij}\right\}, k_1 = 1, 2, \ldots, \#l_{ij}(p), j = 1, 2, \ldots, n$ and $r^{(k_1)}_{ij}$ is lower index of the linguistic term $l^{(k_1)}_{ij}$ while $l'^+_j(p) = \left\{\left(l'^{(k_2)}_j\right)^+ | k_2 = 1, 2, \ldots, \#l'_{ij}(p)\right\}$ and $\left(l'^{(k_2)}_j\right)^+ = s_{\min_i}\left\{p'^{(k_2)}_{ij} r'^{(k_2)}_{ij}\right\}, k_2 = 1, 2, \ldots, \#l'_{ij}(p), j = 1, 2, \ldots, n$ and $r'^{(k_2)}_{ij}$ is lower index of the linguistic term $l'^{(k_2)}_{ij}$. Similarly, the PHILTS negative ideal solution (PHILTS-NIS) of alternatives, denoted by $A^- = \left\langle l^-(p), l'^-(p)\right\rangle$, is defined as follows:

$$A^- = \left\langle l^-(p) = \left(l^-_1(p), l^-_2(p), \ldots, l^-_n(p)\right), l'^-(p) = \left(l'^-_1(p), l'^-_2(p), \ldots, l'^-_n(p)\right)\right\rangle \tag{28}$$

where $l^-_j(p) = \left\{\left(l^{(k_1)}_j\right)^- | k_1 = 1, 2, \ldots, \#l_{ij}(p)\right\}$ and $\left(l^{(k_1)}_j\right)^- = s_{\min_i}\left\{p^{(k_1)}_{ij} r^{(k_1)}_{ij}\right\}, k_1 = 1, 2, \ldots, \#l_{ij}(p), j = 1, 2, \ldots, n$ and $r^{(k_1)}_{ij}$ is lower index of the linguistic term $l^{(k_1)}_{ij}$ while $l'^-_j(p) = \left\{\left(l'^{(k_2)}_j\right)^- | k_2 = 1, 2, \ldots, \#l'_{ij}(p)\right\}$ and $\left(l'^{(k_2)}_j\right)^+ = s_{\max_i}\left\{p'^{(k_2)}_{ij} r'^{(k_2)}_{ij}\right\}, k_2 = 1, 2, \ldots, \#l'_{ij}(p); j = 1, 2, \ldots, n$ and $r'^{(k_2)}_{ij}$ is lower index of the linguistic term $l'^{(k_2)}_{ij}$.

Step 6: Compute the deviation degree between each alternative x_i PHILTS-PIS A^+ as follows:

$$D(x_i, A^+) = \sum_{j=1}^{n} w_j D(h_{ij}, A^+) = \sum_{j=1}^{n} w_j \left(d\left(l_{ij}(p), l^+_j(p)\right) + d\left(l'_{ij}(p), l'^+_j(p)\right)\right)$$
$$= \sum_{j=1}^{n} w_j \left(\begin{array}{c} \sqrt{\dfrac{1}{\#l_{ij}(p)} \sum_{k_1=1}^{\#l_{ij}(p)} \left(p^{(k_1)}_{ij} r^{(k_1)}_{ij} - \left(p^{(k_1)}_j r^{(k_1)}_j\right)^+\right)^2} + \\ \sqrt{\dfrac{1}{\#l'_{ij}(p)} \sum_{k_2=1}^{\#l'_{ij}(p)} \left(p'^{(k_1)}_{ij} r'^{(k_2)}_{ij} - \left(p'^{(k_2)}_j r'^{(k_2)}_j\right)^+\right)^2} \end{array}\right) \tag{29}$$

The smaller is the deviation degree $D(x_i, A^+)$, the better is alternative x_i.

Similarly, compute the deviation degree between each alternative x_i PHILTS-NIS A^- as follows:

$$D\left(x_i, A^-\right) = \sum_{j=1}^{n} w_j D\left(h_{ij}, A^-\right) = \sum_{j=1}^{n} w_j \left(d\left(l_{ij}(p), l_j^-(p)\right) + d\left(l'_{ij}(p), l_j'^-(p)\right)\right)$$

$$= \sum_{j=1}^{n} w_j \left(\begin{array}{c} \sqrt{\frac{1}{\#l_{ij}(p)} \sum\limits_{k_1=1}^{\#l_{ij}(p)} \left(p_{ij}^{(k_1)} r_{ij}^{(k_1)} - \left(p_j^{(k_1)} r_j^{(k_1)}\right)^-\right)^2} + \\ \sqrt{\frac{1}{\#l'_{ij}(p)} \sum\limits_{k_2=1}^{\#l'_{ij}(p)} \left(p_{ij}^{'(k_1)} r_{ij}^{'(k_2)} - \left(p_j^{'(k_2)} r_j^{'(k_2)}\right)^-\right)^2} \end{array} \right) \tag{30}$$

The larger is the deviation degree $D\left(x_i, A^-\right)$, the better is alternative x_i.

Step 7: Determine $D_{\min}\left(x_i, A^+\right)$ and $D_{\max}\left(x_i, A^-\right)$, where

$$D_{\min}\left(x_i, A^+\right) = \min_{1 \le i \le m} D\left(x_i, A^+\right) \tag{31}$$

and

$$D_{\max}\left(x_i, A^-\right) = \max_{1 \le i \le m} D\left(x_i, A^-\right) \tag{32}$$

Step 8: Determine the closeness coefficient Cl of each alternative x_i to rank the alternatives.

$$Cl\left(x_i\right) = \frac{D\left(x_i, A^-\right)}{D_{\max}\left(x_i, A^-\right)} - \frac{D\left(x_i, A^+\right)}{D_{\min}\left(x_i, A^+\right)} \tag{33}$$

Step 9: Pick the best alternative x_i on the basis of the closeness coefficient Cl, where the larger is the closeness coefficient $Cl\left(x_i\right)$, the better is alternative x_i. Thus, the best alternative

$$x^b = \left\{ x_i \middle| \max_{1 \le i \le m} Cl\left(x_i\right) \right\} \tag{34}$$

5.2. The Aggregation-Based Method for MAGDM with Probabilistic Hesitant Intuitionistic Linguistic Information

In this subsection, the aggregation-based method for MAGDM is presented, where the preference opinions of DMs are represented by PHILTS. In Section 4, we have developed some aggregation operators, i.e., PHILA, PHILWA, PHILG and PHILWG. In this algorithm, we use PHILWA operator to aggregate the attribute values of each alternative x_i, into the overall attribute values. The following steps are involved in this algorithm. The first four Steps are similar to the extended TOPSIS method. Therefore, we go to Step 5.

Step 5: Determine the overall attribute values $\widetilde{Z}_i(w)\ (i = 1, 2, \ldots, m)$, where $w = (w_1, w_2, \ldots, w_n)^T$ is the weight vector of attributes, using PHILWA operator, this can be expressed as follows:

$$\begin{array}{l} \widetilde{Z}_i(w) = w_1 \left\langle l_{i1}(p), l'_{i1}\left(p'\right)\right\rangle \oplus w_2 \left\langle l_{i2}(p), l'_{i2}\left(p'\right)\right\rangle \oplus \ldots \oplus w_n \left\langle l_{in}(p), l'_{in}\left(p'\right)\right\rangle \\ = \left\langle w_1 l_{i1}(p) \oplus w_2 l_{i2}(p) \oplus \ldots \oplus w_n l_{in}(p), w_1 l'_{i1}\left(p'\right) \oplus w_2 l'_{i2}\left(p'\right) \oplus \ldots \oplus w_n l'_{in}\left(p'\right)\right\rangle \\ = \left\langle \begin{array}{c} \cup_{l_{i1}^{(k_1)} \in l_{i1}(p)} \left\{w_1 p_{i1}^{(k_1)} l_{i1}^{(k_1)}\right\} \oplus \cup_{l_{i2}^{(k_1)} \in l_{i2}(p)} \left\{w_2 p_{i2}^{(k_1)} l_{i2}^{(k_1)}\right\} \oplus \ldots \oplus \cup_{l_{in}^{(k_1)} \in l_{in}(p)} \left\{w_n p_{in}^{(k_1)} l_{in}^{(k_1)}\right\}, \\ \cup_{l_{i1}^{'(k_2)} \in l'_{i1}(p')} \left\{w_1 p_{i1}^{'(k_2)} l_{i1}^{'(k_2)}\right\} \oplus \cup_{l_{i2}^{'(k_2)} \in l'_{i2}(p')} \left\{w_2 p_{i2}^{'(k_2)} l_{i2}^{'(k_2)}\right\} \oplus \ldots \oplus \cup_{l_{in}^{'(k_2)} \in l'_{in}(p')} \left\{w_n p_{in}^{'(k_2)} l_{in}^{'(k_2)}\right\} \end{array} \right\rangle \end{array} \tag{35}$$

where $i = 1, 2, \ldots, m$.

Step 6: Compare the overall attribute values $\widetilde{Z}_i(w)\ (i = 1, 2, \ldots, m)$ mutually, based on their score function and deviation degree whose detail is given in Section 3.2.

Step 7: Rank the alternatives $x_i\ (i = 1, 2, \ldots, m)$ according to the order of $\widetilde{Z}_i(w)\ (i = 1, 2, \ldots, m)$ and pick the best alternative.

The flow chart of the proposed models is presented in Figure 1.

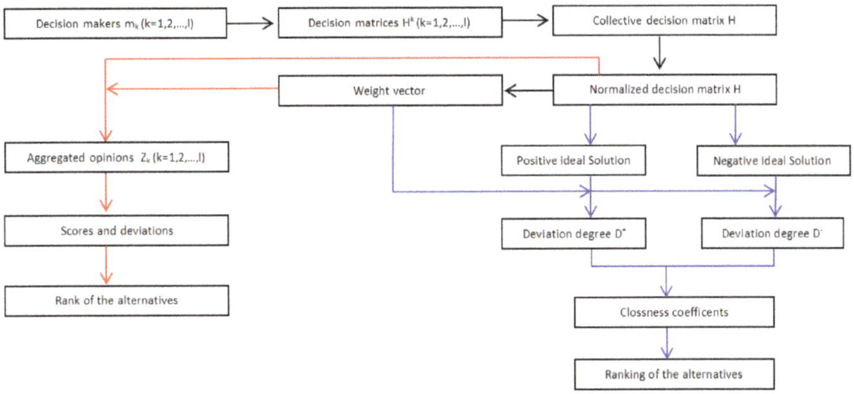

Figure 1. Extended TOPSIS and Aggregation-based models.

6. A Case Study

To validate the proposed theory and decision making models, in this section, a practical example taken from [28] is solved. A group of seven peoples m_l $(l = 1, 2, 3, \ldots, 7)$ need to invest their savings in a most profitable way. They considered five possibilities: x_1 is real estate, x_2 is stock market, x_3 is T-bills, x_4 is national saving scheme, and x_5 is insurance company. To determine best option, the following attributes are taken into account: c_1 is the risk factor, c_2 is the growth, c_3 is quick refund, and c_4 is complicated documents requirement. Base upon their knowledge and experience, they provide their opinion in terms of following HIFLTSs.

6.1. The Extended TOPSIS Method for the Considered Case

We handle the above problem by applying the extended TOPSIS method.

Step 1: The probabilistic hesitant intuitionistic linguistic decision matrices derived from Tables 1–3 are shown in Tables 4–6, respectively.

Table 1. Decision matrix provided by the DMs 1, 2, 3 (m_1, m_2, m_3).

	c_1	c_2	c_3	c_4
x_1	$\langle\{s_3,s_4,s_5\},\{s_1,s_2\}\rangle$	$\langle\{s_4,s_5\},\{s_0,s_1\}\rangle$	$\langle\{s_1,s_2\},\{s_3,s_4\}\rangle$	$\langle\{s_1,s_2\},\{s_3,s_4\}\rangle$
x_2	$\langle\{s_1,s_2\},\{s_3,s_4\}\rangle$	$\langle\{s_3,s_4,s_5\},\{s_1,s_2\}\rangle$	$\langle\{s_3,s_4\},\{s_0,s_1\}\rangle$	$\langle\{s_4,s_5\},\{s_1,s_2\}\rangle$
x_3	$\langle\{s_4,s_5)\},\{s_0,s_1,s_2\}\rangle$	$\langle\{s_3,s_4\},\{s_1,s_2\}\rangle$	$\langle\{s_5,s_6\},\{s_0\}\rangle$	$\langle\{s_1,s_2\},\{s_2,s_3,s_4\}\rangle$
x_4	$\langle\{s_5,s_6\},\{s_0,s_1\}\rangle$	$\langle\{s_1,s_2\},\{s_3,s_4\}\rangle$	$\langle\{s_1,s_2\},\{s_3,s_4\}\rangle$	$\langle\{s_3,s_4,s_5\},\{s_1,s_2\}\rangle$
x_5	$\langle\{s_6\},\{s_0\}\rangle$	$\langle\{s_1,s_2\},\{s_3,s_4,s_5\}\rangle$	$\langle\{s_0,s_1\},\{s_2,s_3\}\rangle$	$\langle\{s_4,s_5\},\{s_1,s_2\}\rangle$

Table 2. Decision matrix provided by the DMs 4, 5 (m_4, m_5).

	c_1	c_2	c_3	c_4
x_1	$\langle\{s_1,s_2\},\{s_3,s_4\}\rangle$	$\langle\{s_5,s_6\},\{s_0,s_1\}\rangle$	$\langle\{s_0,s_1\},\{s_3,s_4\}\rangle$	$\langle\{s_3,s_4\},\{s_1,s_2\}\rangle$
x_2	$\langle\{s_0,s_1\},\{s_2,s_3\}\rangle$	$\langle\{s_1,s_2\},\{s_2,s_3,s_4\}\rangle$	$\langle\{s_4,s_5\},\{s_0,s_1\}\rangle$	$\langle\{s_5,s_6\},\{s_0\}\rangle$
x_3	$\langle\{s_3,s_4\},\{s_0,s_1\}\rangle$	$\langle\{s_1,s_2\},\{s_3,s_4\}\rangle$	$\langle\{s_4,s_5\},\{s_1,s_2)\}\rangle$	$\langle\{s_0,s_1\},\{s_2,s_3\}\rangle$
x_4	$\langle\{s_5,s_6\},\{s_0\}\rangle$	$\langle\{s_3,s_4\},\{s_0,s_1,s_2\}\rangle$	$\langle\{s_1,s_2\},\{s_2,s_3,s_4\}\rangle$	$\langle\{s_4,s_5\},\{s_0\}\rangle$
x_5	$\langle\{s_4,s_5\},\{s_1,s_2\}\rangle$	$\langle\{s_3,s_4\},\{s_1,s_2,s_3\}\rangle$	$\langle\{s_1,s_2\},\{s_3,s_4\}\rangle$	$\langle\{s_5,s_6\},\{s_0\}\rangle$

Table 3. Decision matrix provided by the DMs 6, 7 (m_6, m_7).

	c_1	c_2	c_3	c_4
x_1	$\langle\{s_4, s_5\}, \{s_0, s_1\}\rangle$	$\langle\{s_5, s_6\}, \{s_0\}\rangle$	$\langle\{s_3, s_4\}, \{s_1, s_2\}\rangle$	$\langle\{s_0, s_1\}, \{s_3, s_4\}\rangle$
x_2	$\langle\{s_3, s_4\}, \{s_1, s_2, s_3\}\rangle$	$\langle\{s_1, s_2\}, \{s_3, s_4\}\rangle$	$\langle\{s_5, s_6\}, \{s_0\}\rangle$	$\langle\{s_3, s_4\}, \{s_1, s_2\}\rangle$
x_3	$\langle\{s_1, s_2\}, \{s_2, s_3, s_4\}\rangle$	$\langle\{s_5, s_6\}, \{s_0\}\rangle$	$\langle\{s_4, s_5\}, \{s_0, s_1\}\rangle$	$\langle\{s_0, s_1\}, \{s_3, s_4\}\rangle$
x_4	$\langle\{s_4, s_5\}, \{s_1, s_2\}\rangle$	$\langle\{s_4, s_5\}, \{s_0, s_1\}\rangle$	$\langle\{s_0, s_1, s_2\}, \{s_2, s_3\}\rangle$	$\langle\{s_3, s_4, s_5\}, \{s_1, s_2\}\rangle$
x_5	$\langle\{s_3, s_4\}, \{s_0, s_1, s_2\}\rangle$	$\langle\{s_1, s_2\}, \{s_2, s_3, s_4\}\rangle$	$\langle\{s_2, s_3\}, \{s_3, s_4\}\rangle$	$\langle\{s_6\}, \{s_0\}\rangle$

Table 4. Probabilistic hesitant intuitionistic linguistic decision matrix H_1 with respect to DMs 1, 2, 3 (m_1, m_2, m_3).

	c_1	c_2
x_1	$\langle\{(s_3(0.14), s_4(0.28), s_5(0.28))\}, \{s_1(0.28), s_2(0.14)\}\rangle$	$\langle\{s_4(0.14), s_5(0.42)\}, \{s_0(0.42), s_1(0.28)\}\rangle$
x_2	$\langle\{s_1(0.28), s_2(0.14)\}, \{s_3(0.42), s_4(0.14)\}\rangle$	$\langle\{s_3(0.14), s_4(.14), s_5(0.14)\}, \{s_1(0.14), s_2(0.28)\}\rangle$
x_3	$\langle\{s_4(0.28), s_5(0.14)\}, \{s_0(0.28), s_1(0.28), s_2(0.28)\}\rangle$	$\langle\{s_3(0.14), s_4(0.28)\}, \{s_1(0.14), s_2(0.14)\}\rangle$
x_4	$\langle\{s_5(0.42), s_6(0.28)\}, \{s_0(0.28), s_1(0.28)\}\rangle$	$\langle\{s_1(0.14), s_2(0.14)\}, \{s_3(0.14), s_4(0.14)\}\rangle$
x_5	$\langle\{s_6(0.14)\}, \{s_0(0.28)\}\rangle$	$\langle\{s_1(0.28), s_2(0.28)\}, \{s_3(0.42), s_4(0.28), s_5(0.14)\}\rangle$
	c_3	c_4
x_1	$\langle\{s_1(0.28), s_2(0.14)\}, \{s_3(0.28), s_4(0.28)\}\rangle$	$\langle\{s_1(0.28), s_2(0.14)\}, \{s_3(0.28), s_4(0.28)\}\rangle$
x_2	$\langle\{s_3(0.14), s_4(0.28)\}, \{s_0(0.42), s_1(0.28)\}\rangle$	$\langle\{s_4(0.14), s_5(0.28)\}, \{s_1(0.28), s_2(0.28)\}\rangle$
x_3	$\langle\{s_5(0.42), s_6(0.14)\}, \{s_0(0.28)\}\rangle$	$\langle\{s_1(0.42), s_2(0.14)\}, \{s_2(0.28), s_3(0.42), s_4(0.28)\}\rangle$
x_4	$\langle\{s_1(0.42), s_2(.42)\}, \{s_3(0.42), s_4(0.28)\}\rangle$	$\langle\{s_3(0.28), s_4(0.42), s_5(0.42)\}, \{s_1(0.28), s_2(0.28)\}\rangle$
x_5	$\langle\{s_0(0.14), s_1(0.28)\}, \{s_2(0.28), s_3(0.42)\}\rangle$	$\langle\{s_4(0.14), s_5(0.28)\}, \{s_1(0.14), s_2(0.14)\}\rangle$

Table 5. Probabilistic hesitant intuitionistic linguistic decision matrix H_2 with respect to DMs 4, 5 (m_4, m_5).

	c_1	c_2
x_1	$\langle\{s_1(0.14), s_2(0.14)\}, \{s_3(0.14), s_4(0.14)\}\rangle$	$\langle\{s_5(0.42), s_6(0.28)\}, \{s_0(0.42), s_1(0.28)\}\rangle$
x_2	$\langle\{s_0(0.14), s_1(0.28)\}, \{s_2(0.28), s_3(0.42)\}\rangle$	$\langle\{s_1(0.28), s_2(0.28)\}, \{s_2(0.28), s_3(0.28), s_4(0.28)\}\rangle$
x_3	$\langle\{s_3(0.14), s_4(.28)\}, \{s_0(0.28), s_1(0.28)\}\rangle$	$\langle\{s_1(0.14), s_2(0.14)\}, \{s_3(0.14), s_4(0.14)\}\rangle$
x_4	$\langle\{s_5(0.42), s_6(0.28)\}, \{s_0(0.28)\}\rangle$	$\langle\{s_3(0.14), s_4(0.28)\}, \{s_0(0.28), s_1(0.28), s_2(0.14)\}\rangle$
x_5	$\langle\{s_4(0.28), s_5(0.14)\}, \{s_1(0.28), s_2(0.28)\}\rangle$	$\langle\{s_3(0.14), s_4(0.14)\}, \{s_1(0.14), s_2(0.28), s_3(0.42)\}\rangle$
	c_3	c_4
x_1	$\langle\{s_{0.}(0.14), s_1(0.28)\}, \{s_3(0.28), s_4(0.28)\}\rangle$	$\langle\{s_3(0.14), s_4(0.14)\}, \{s_1(0.14), s_2(0.14)\}\rangle$
x_2	$\langle\{s_4(0.28), s_5(0.28)\}, \{s_0(0.42), s_1(0.28)\}\rangle$	$\langle\{s_5(0.28), s_6(0.14)\}, \{s_0(0.14)\}\rangle$
x_3	$\langle\{s_4(0.28), s_5(0.42)\}, \{s_1(0.28), s_2(0.14)\}\rangle$	$\langle\{s_0(0.28), s_1(0.42)\}, \{s_2(0.28), s_3(0.42)\}\rangle$
x_4	$\langle\{s_1(0.42), s_2(0.42)\}, \{s_2(0.28), s_3(0.42), s_4(0.28)\}\rangle$	$\langle\{s_4(0.42), s_5(0.42)\}, \{s_0(0.14)\}\rangle$
x_5	$\langle\{s_1(0.28), s_2(0.14)\}, \{s_3(0.42), s_4(0.28)\}\rangle$	$\langle\{s_5(0.28), s_6(0.28)\}, \{s_0(0.28)\}\rangle$

Table 6. Probabilistic hesitant intuitionistic linguistic decision matrix H_3 with respect to DMs 6, 7 (m_6, m_7).

	c_1	c_2
x_1	$\langle\{s_4(0.28), s_5(0.28)\}, \{s_0(0.14), s_1(0.28)\}\rangle$	$\langle\{s_5(0.42), s_6(0.28)\}, \{s_0(0.42)\}\rangle$
x_2	$\langle\{s_3(0.14), s_4(0.14)\}, \{s_1(0.14), s_2(0.28), s_3(0.42)\}\rangle$	$\langle\{s_1(0.28), s_2(0.28)\}, \{s_3(0.28), s_4(0.28)\}\rangle$
x_3	$\langle\{s_1(0.14), s_2(0.14)\}, \{s_2(0.28), s_3(0.14), s_4(0.14)\}\rangle$	$\langle\{s_5(0.28), s_6(0.14)\}, \{s_0(0.14)\}\rangle$
x_4	$\langle\{s_4(0.14), s_5(0.42)\}, \{s_1(0.28), s_2(0.14)\}\rangle$	$\langle\{s_4(0.28), s_5(0.14)\}, \{s_0(0.28), s_1(0.28)\}\rangle$
x_5	$\langle\{s_3(0.14), s_4(0.28)\}, \{s_0(0.28), s_1(0.28), s_2(0.28)\}\rangle$	$\langle\{s_1(0.28), s_2(0.28)\}, \{s_2(0.28), s_3(0.42), s_4(0.28)\}\rangle$
	c_3	c_4
x_1	$\langle\{s_3(0.14), s_4(0.14)\}, \{s_1(0.14), s_2(0.14)\}\rangle$	$\langle\{s_0(0.14), s_1(0.28)\}, \{s_3(0.28), s_4(0.28)\}\rangle$
x_2	$\langle\{s_5(0.28), s_6(0.14)\}, \{s_0(.42)\}\rangle$	$\langle\{s_3(0.14), s_4(0.28)\}, \{s_1(0.28), s_2(0.28)\}\rangle$
x_3	$\langle\{s_4(0.28), s_5(0.42)\}, \{s_0(0.28), s_1(0.28)\}\rangle$	$\langle\{s_0(0.28), s_1(0.42)\}, \{s_3(0.42), s_4(0.28)\}\rangle$
x_4	$\langle\{s_0(0.14), s_1(0.42), s_2(0.42)\}, \{s_2(0.28), s_3(0.42)\}\rangle$	$\langle\{s_3(0.28), s_4(0.42), s_5(0.42)\}, \{s_1(0.28), s_2(0.28)\}\rangle$
x_5	$\langle\{s_2(0.14), s_3(0.14)\}, \{s_3(0.28), s_4(0.28)\}\rangle$	$\langle\{s_6(0.28)\}, \{s_0(0.28)\}\rangle$

Step 2: The decision matrix H in Table 7 is constructed by utilizing Tables 4–6.

Table 7. Decision matrix (H).

	c_1	c_2
x_1	$\langle\{s_2(0.14), s_4(0.28)\}, \{s_1(0.28), s_3(0.14)\}\rangle$	$\langle\{s_6(0.28), s_5(0.42)\}, \{s_0(0.42), s_0(0.42)\}\rangle$
x_2	$\langle\{s_1(0.28), s_3(0.14)\}, \{s_4(0.14), s_3(0.42)\}\rangle$	$\langle\{s_2(0.28), s_3(0.14)\}, \{s_2(0.28), s_3(0.28)\}\rangle$
x_3	$\langle\{s_2(0.14), s_0(0.14)\}, \{s_1(0.28), s_3(0.14)\}\rangle$	$\langle\{s_2(0.14), s_6(0.14)\}, \{s_0(0.14), s_3(0.14)\}\rangle$
x_4	$\langle\{s_6(0.28), s_5(0.42)\}, \{s_0(0.28), s_1(0.28)\}\rangle$	$\langle\{s_2(0.14), s_5(0.14)\}, \{s_1(0.28), s_3(0.14)\}\rangle$
x_5	$\langle\{s_6(0.14), s_6(0.14)\}, \{s_0(0.28), s_1(0.28)\}\rangle$	$\langle\{s_3(0.14), s_2(0.28)\}, \{s_5(0.14), s_3(0.42)\}\rangle$
	c_3	c_4
x_1	$\langle\{s_1(0.28), s_3(0.14)\}, \{s_2(0.14), s_3(0.28)\}\rangle$	$\langle\{s_1(0.28), s_3(0.14)\}, \{s_2(0.14), s_3(0.28)\}\rangle$
x_2	$\langle\{s_4(0.28), s_4(0.14)\}, \{s_0(0.42), s_0(0.42)\}\rangle$	$\langle\{s_1(0.28), s_3(0.14)\}, \{s_0(0.14), s_3(0.28)\}\rangle$
x_3	$\langle\{s_4(0.28), s_5(0.42)\}, \{s_0(0.28), s_1(0.28)\}\rangle$	$\langle\{s_1(0.14), s_2(0.42)\}, \{s_4(0.28), s_3(0.42)\}\rangle$
x_4	$\langle\{s_1(0.42), s_2(0.42)\}, \{s_{4.}(0.28), s_3(0.42)\}\rangle$	$\langle\{s_4(0.42), s_5(0.42)\}, \{s_0(0.14), s_2(0.28)\}\rangle$
x_5	$\langle\{s_1(0.28), s_2(0.14)\}, \{s_{4.}(0.28), s_3(0.42)\}\rangle$	$\langle\{s_5(0.28), s_6(0.28)\}, \{s_0(0.28), s_1(0.14)\}\rangle$

Step 3: The normalized probabilistic hesitant intuitionistic linguistic decision matrix of the group is shown in Table 8.

Table 8. The normalized probabilistic hesitant intuitionistic linguistic decision matrix.

	c_1
x_1	$\langle\{s_4(0.6666667), s_2(0.3333333)\}, \{s_3(0.3333333), s_1(0.6666667)\}\rangle$
x_2	$\langle\{s_3(0.3333333), s_1(0.6666667)\}, \{s_3(0.75), s_4(0.25)\}\rangle$
x_3	$\langle\{s_0(0.5), s_2(0.5)\}, \{s_3(0.3333333), s_1(0.6666667)\}\rangle$
x_4	$\langle\{s_5(0.6), s_6(0.4)\}, \{s_1(0.5), s_0(0.5)\}\rangle$
x_5	$\langle\{s_6(0.5), s_6(0.5)\}, \{s_0(0.5), s_1(0.5)\}\rangle$
	c_2
x_1	$\langle\{s_5(0.6), s_6(0.4)\}, \{s_0(0.5), s_0(0.5)\}\rangle$
x_2	$\langle\{s_3(0.3333333), s_2(0.6666667)\}, \{s_3(0.5), s_2(0.5)\}\rangle$
x_3	$\langle\{s_6(0.5), s_2(0.5)\}, \{s_3(0.5), s_0(0.5)\}\rangle$
x_4	$\langle\{s_5(0.5), s_2(0.5)\}, \{s_3(0.3333333), s_1(0.6666667)\}\rangle$
x_5	$\langle\{s_2(0.6666667), s_3(0.3333333)\}, \{s_3(0.75), s_5(0.25)\}\rangle$
	c_3
x_1	$\langle\{s_3(0.3333333), s_1(0.6666667)\}, \{s_3(0.6666667), s_2(0.3333333)\}\rangle$
x_2	$\langle\{s_4(0.6666667), s_4(0.3333333)\}, \{s_0(0.5), s_0(0.5)\}\rangle$
x_3	$\langle\{s_5(0.6), s_4(0.4)\}, \{s_5(0.6), s_4(0.4)\}\rangle$
x_4	$\langle\{s_1(0.5), s_2(0.5)\}, \{s_3(0.6), s_{4.}(0.4)\}\rangle$
x_5	$\langle\{s_1(0.6666667), s_2(0.3333333)\}, \{s_3(0.6), s_4(0.4)\}\rangle$
	c_4
x_1	$\langle\{s_3(0.3333333), s_1(0.6666667)\}, \{s_3(0.6666667), s_2(0.3333333)\}\rangle$
x_2	$\langle\{s_3(0.3333333), s_1(0.6666667)\}, \{s_3(0.6666667), s_0(0.3333333)\}\rangle$
x_3	$\langle\{s_2(0.75), s_1(0.25)\}, \{s_3(0.6), s_4(0.4)\}\rangle$
x_4	$\langle\{s_5(0.5), s_4(0.5)\}, \{s_0(0.3333333), s_2(0.6666667)\}\rangle$
x_5	$\langle\{s_6(0.5), s_5(0.5)\}, \{s_1(0.3333333), s_0(0.6666667)\}\rangle$

Step 4: The weight vector is derived from Equation (26) as follows:
$w = (0.2715, 0.2219, 0.2445, 0.2621)^t$

Step 5: The PHILTS-PIS "A^+" and the PHILTS-NIS "A^-" of each alternative are derived using Equations (27) and (28) as follows:

$A^+ = (\langle\{3,3\},\{0,0\}\rangle, \langle\{3,2.4\},\{0,0\}\rangle, \langle\{3,1.6\},\{0,0\}\rangle, \langle\{3,2.5\},\{0,0\}\rangle)$

$A^- = (\langle\{0,0.661\},\{2.25,1\}\rangle, \langle\{1,1\},\{2.25,1.25\}\rangle, \langle\{.5,0.66\},\{2,1.6\}\rangle, \langle\{1,0.2\},\{2,1.6\}\rangle)$

$D(x_1, A^+) = 2.1211$, $D(x_2, A^+) = 2.5516$, $D(x_3, A^+) = 2.9129$, $D(x_4, A^+) = 1.7999$, $D(x_5, A^+) = 1.6494$

$D(x_1, A^-) = 2.0142$, $D(x_2, A^-) = 1.5861$, $D(x_3, A^-) = 1.6204$, $D(x_4, A^-) = 2.4056$, $D(x_5, A^-) = 2.2812$

Step 7: Calculate $D_{\min}(x_i, A^+)$ and $D_{\max}(x_i, A^-)$ by Equations (31) and (32) :

$D_{\min}(x_i, A^+) = 1.6494$, $D_{\max}(x_i, A^-) = 2.4050$

Step 8: Determine the closeness coefficient of each alternative x_i by Equation (33) :

$Cl(x_1) = -0.4486$, $Cl(x_2) = -0.8876$, $Cl(x_3) = -1.0924$, $Cl(x_4) = -0.0912$, $Cl(x_5) = -0.0519$

Step 9: Rank the alternatives according to the ranking of $Cl(x_i)$ $(i = 1, 2, \ldots, 5)$: $x_5 > x_4 > x_1 > x_2 > x_3$, and thus, x_5(insurance company) is the best alternative.

6.2. The Aggregation-Based Method for the Considered Case

We can also apply the aggregation-based method to attain the ranking of alternatives for the case study.

Step 1: Construct the probabilistic hesitant intuitionistic fuzzy decision matrices of the group as listed in Tables 4–6, and then aggregated and normalized as shown in Tables 7 and 8.

Step 2: Utilize Equation (26) to obtain the weight vector

$w = (0.2715, 0.2219, 0.2445, 0.2621)^t$.

Step 3: Derive the overall attribute value of each alternative x_i $(i = 1, 2, 3, 4, 5)$ by using Equation (35) :

$\widetilde{Z_1}(w) = \langle\{s_{1.8962}, s_{0.5187}\}, \{s_{1.2847}, s_{0.5187}\}\rangle$,

$\widetilde{Z_2}(w) = \langle\{s_{1.4074}, s_{0.9776}\}, \{s_{1.4679}, s_{0.4934}\}\rangle$,

$\widetilde{Z_3}(w) = \{s_{1.7923}, s_{1.1256}\}, \{s_{1.8096}, s_{0.9915}\}$,

$\widetilde{Z_4}(w) = \langle\{s_{2.1467}, s_{1.642}\}, \{s_{0.7977}, s_{0.8886}\}\rangle$,

$\widetilde{Z_5}(w) = \langle\{s_{2.0596}, s_{1.8546}\}, \{s_{1.0267}, s_{0.8043}\}\rangle$.

Step 4: Compute the score of each attribute value $\widetilde{Z_i}(w)$ by Definition 14:

$E\left(\widetilde{Z_1}(w)\right) = s_{3.1528}$, $E\left(\widetilde{Z_2}(w)\right) = s_{3.1059}$, $E\left(\widetilde{Z_3}(w)\right) = s_{3.0584}$, $E\left(\widetilde{Z_4}(w)\right) = s_{4.0512}$, $E\left(\widetilde{Z_5}(w)\right) = s_{5.8726}$

Step 5: Compare the overall attribute values of alternatives according to the values of the score function. It is obvious, that $x_5 > x_4 > x_1 > x_2 > x_3$. Thus, again, we get the best alternative x_5.

7. Discussions and Comparison

For the purpose of comparison, in this subsection, the case study is again solved by applying the TOPSIS method with traditional HIFLTSs.

Step 1: The decision matrix X in Table 9 is constructed by utilizing Tables 1–3 as follows:

Table 9. Decision matrix (X).

	c_1	c_2	c_3	c_4
x_1	$([s_2, s_4], [s_1, s_3])$	$([s_5, s_5], [s_0, s_0])$	$([s_1, s_3], [s_2, s_3])$	$([s_1, s_3], [s_2, s_3])$
x_2	$([s_1, s_3], [s_3, s_3])$	$([s_2, s_3], [s_2, s_3])$	$([s_4, s_5], [s_0, s_0])$	$([s_4, s_5], [s_0, s_1])$
x_3	$([s_2, s_4], [s_1, s_2])$	$([s_3, s_5], [s_0, s_3])$	$([s_5, s_5], [s_0, s_1])$	$([s_1, s_1], [s_3, s_3])$
x_4	$([s_5, s_5], [s_0, s_1])$	$([s_2, s_4], [s_1, s_3])$	$([s_1, s_2], [s_3, s_3])$	$([s_4, s_5], [s_1, s_2])$
x_5	$([s_4, s_6], [s_0, s_1])$	$([s_2, s_3], [s_3, s_3])$	$([s_1, s_2], [s_3, s_3])$	$([s_5, s_6], [s_0, s_1])$

Step 2: Determine the HIFLTS-PIS "P^+" and the HIFLTS-NIS "P^-" for cost criteria c_1,c_4 and benefit criteria c_2,c_3 as follows:

$P^+ = [([s_0, s_1], [s_3, s_4]), ([s_5, s_6], [s_0, s_0]), ([s_5, s_6], [s_0, s_0]), ([s_0, s_1], [s_3, s_4])]$

$P^- = [([s_6, s_6], [s_0, s_0]), ([s_1, s_2], [s_3, s_5]), ([s_0, s_1], [s_3, s_4]), ([s_6, s_6], [s_0, s_0])]$

Note: One can see the detail of HIFLTS-PIS "P^+" and the HIFLTS-NIS "P^-" in [28].

Step 3: Calculate the positive ideal matrix D^+ and the negative ideal matrix D^- as follows:

$$D^{+} = \begin{bmatrix} 8+1+12+5 \\ 4+11+2+14 \\ 9+7+2+2 \\ 15+9+14+12 \\ 15+12+14+16 \end{bmatrix} = \begin{bmatrix} 26 \\ 31 \\ 20 \\ 50 \\ 57 \end{bmatrix}$$

$D^{+}_{11} = d\left(x_{11}, v_1^{+}\right) + d\left(x_{12}, v_2^{+}\right) + d\left(x_{13}, v_3^{+}\right) + d\left(x_{14}, v_4^{+}\right)$ in which $d\left(x_{11}, v_1^{+}\right) = d\left(\left([s_2, s_4], [s_1, s_3]\right), \left([s_0, s_1], [s_3, s_4]\right)\right) = |2-0| + |4-1| + |1-3| + |3-4| = 8$

Other entries can be found by similar calculation.

$$D^{-} = \begin{bmatrix} 10+15+5+13 \\ 14+5+15+4 \\ 9+9+15+16 \\ 3+7+3+6 \\ 3+4+3+2 \end{bmatrix} = \begin{bmatrix} 43 \\ 38 \\ 49 \\ 19 \\ 12 \end{bmatrix}$$

Step 4: The relative closeness(RC) of each alternative to the ideal solution can be obtained as follows:

$RC(x_1) = 43/\left(26+43\right) = 0.6232$

$RC(x_2) = 38/\left(31+38\right) = 0.5507$

The RC of other alternatives can be find by similar calculations.

$RC(x_3) = 0.7101$, $RC(x_4) = 0.2754$, $RC(x_5) = 0.1739$.

Step 5: The ranking of alternatives of alternatives x_i $(i = 1, 2, \ldots, 5)$ according to the closeness coefficient $RC(x_i)$ is:

$x_3 > x_1 > x_2 > x_4 > x_5$.

- In Table 9, the disadvantages of HIFLTS are apparent because in HIFLTS the probabilities of the linguistic terms is not considered which means that all possible linguistic terms in HIFLTS have same occurrence possibility which is unrealistic, whereas the inspection of Table 7 shows that PHILTS not only contains the linguistic terms, but also considers the probabilities of linguistic terms, and, thus, PHILTS constitutes an extension of HIFLTS.
- The inspection of Table 10 reveals that the extended TOPSIS method and the aggregation-based method give the same best alternative x_5. The TOPSIS method with the traditional HIFLTSs gives x_3 as the best alternative.
- This difference of best alternative in Table 10 is due to the effect of probabilities of membership and non-membership linguistic terms, which highlight the critical role of probabilities. Thus, our methods are more rational to get the ranking of alternatives and further to find the best alternative.
- Extended TOPSIS method and aggregation-based method for MAGDM with PLTS information explained in [19] are more promising and better than extended TOPSIS method and aggregation-based method for MAGDM with HFLTS information. However, a clear superiority of PHILTS is that it assigns to each element the degree of belongingness and also the degree of non-belongingness along with probability. PLTS only assigns to each element a belongingness degree along with probability. Using PLTSs, various frameworks have been developed by DMs [19,29] but they are still intolerant, since there is no mean of attributing reliability or confidence information to the degree of belongingness.

Table 10. Comparison of Results.

TOPSIS [28]	$x_3 > x_1 > x_2 > x_4 > x_5$
Proposed extend TOPSIS	$x_5 > x_4 > x_1 > x_2 > x_3$
Proposed aggregation model	$x_5 > x_4 > x_1 > x_2 > x_3$

The comparisons and other aspects are summarized in Table 11.

Table 11. The advantages and limitations of the proposed methods.

Advantages	Limitations
1. PHILTS generalize the existing PLTS models since PHILTS take more information from the DMs into account. 2. PHILTS is not affected by partial vagueness. 3. PHILTS is more in line with people's language, leading to much more fruitful decisions. 4. The attribute weights are calculated with objectivity (without favor).	1. It is essential to take membership as well as non-membership probabilistic data. 2. Its computational index is high.

8. Conclusions

Because of the blurring of human thinking, sometimes it becomes difficult for experts to accurately measure the opinions in the area of the usual fuzzy set theory, even in the HIFLTSs and PLTSs. For this purpose, in this article, a new concept called PHILTS was introduced to extend the current HIFLTS and PLTS. To facilitate the calculation of the PHILTSs, a normalization process, basic operations and aggregation operators for PHILTSs are also designed. An extended TOPSIS method and aggregation based method have been proposed to solve decision ranking problems of the group with the multiple conflict criteria in PHILTS. The proposed models are compared with existing model of TOPSIS. The PLTS and HIFLTS are special cases of PHILTS, it grants the freedom to DMs to express their opinions in more dynamic way. Furthermore, the occurrence probabilities of membership and non-membership linguistic term sets greatly affects the decision making, validating the importance of designed theory and models in this manuscript. The probability is one of the best tool to handle uncertainty of future, thus our proposed models are more suitable of decision making related to the possible future scenarios. However, its arithmetic complexity is high.

In the future, all the work which has been done thus far PLTSs and HIFLTSs can be studied for PHILTS and then applied to decision making.

Author Contributions: Conceptualization, Z.B., T.R. and J.A.; Formal analysis, Z.B., M.G.A.M., T.R. and J.A.; Investigation, Z.B. and J.A.; Methodology, M.G.A.M.; Resources, T.R.; Software, M.G.A.M.

Funding: This work is partially supported by AGI Education, Auckland, New Zealand.

Acknowledgments: The authors would like to thank the editors and the anonymous reviewers, whose insightful comments and constructive suggestions helped us to significantly improve the quality of this paper. This research work is partially supported by AGI Education, Auckland, New Zealand.

Conflicts of Interest: The authors declare no conflict of interest.

References

1. Atanassov, K. Intuitionistic fuzzy sets. *Fuzzy Sets Syst.* **1986**, *20*, 87–96. [CrossRef]
2. Torra, V. Hesitant fuzzy sets. *Int. J. Intell. Syst.* **2010**, *25*, 529–539. [CrossRef]
3. Xu, Z.; Zhou, W. Consensus building with a group of decision makers under the hesitant probabilistic fuzzy environment. *Fuzzy Optim. Decis. Mak.* **2017**, *16*, 481–503. [CrossRef]
4. Bashir, Z.; Rashid, T.; Wątróbski, J.; Sałabun, W.; Malik, A. Hesitant Probabilistic Multiplicative Preference Relations in Group Decision Making. *Appl. Sci.* **2018**, *8*, 398. [CrossRef]
5. Alcantud, J.C.R.; Giarlotta, A. Necessary and possible hesitant fuzzy sets: A novel model for group decision making. *Inf. Fusion* **2019**, *46*, 63–76. [CrossRef]
6. Zadeh, L.A. The concept of a linguistic variable and its applications to approximate reasoning. *Inf. Sci. Part I II III* **1975**, *8–9*, 43–80, 199–249, 301–357.
7. Ju, Y.B.; Yang, S.H. Approaches for multi-attribute group decision making based on intuitionistic trapezoid fuzzy linguistic power aggregation operators. *J. Intell. Fuzzy Syst.* **2014**, *27*, 987–1000.

8. Merigó, J.M.; Casanovas, M.; Martínez, L. Linguistic aggregation operators for linguistic decision making based on the Dempster–Shafer theory of evidence. *Int. J. Uncertain. Fuzziness Knowl. Based Syst.* **2010**, *18*, 287–304.
9. Zhu, H.; Zhao, J.B.; Xu, Y. 2-Dimension linguistic computational model with 2-tuples for multi-attribute group decision making. *Knowl. Based Syst.* **2016**, *103*, 132–142. [CrossRef]
10. Meng, F.Y.; Tang, J. Extended 2-tuple linguistic hybrid aggregation operators and their application to multi-attribute group decision making. *Int. J. Comput. Intell. Syst.* **2014**, *7*, 771–784. [CrossRef]
11. Li, C.C.; Dong, Y. Multi-attribute group decision making methods with proportional 2-tuple linguistic assessments and weights. *Int. J. Comput. Intell. Syst.* **2014**, *7*, 758–770. [CrossRef]
12. Xu, Z.S. Multi-period multi-attribute group decision-making under linguistic assessments. *Int. J. Gen. Syst.* **2009**, *38*, 823–850. [CrossRef]
13. Li, D.F. Multiattribute group decision making method using extended linguistic variables. *Int. J. Uncertain. Fuzziness Knowl. Based Syst.* **2009**, *17*, 793–806. [CrossRef]
14. Agell, N.; Sánchez, M.; Prats, F.; Roselló, L. Ranking multi-attribute alternatives on the basis of linguistic labels in group decisions. *Inf. Sci.* **2012**, *209*, 49–60. [CrossRef]
15. Rodríguez, R.M.; Martínez, L.; Herrera, F. Hesitant fuzzy linguistic term sets for decision making. *IEEE Trans. Fuzzy Syst.* **2012**, *20*, 109–119.
16. Zhu, J.; Li, Y. Hesitant Fuzzy Linguistic Aggregation Operators Based on the Hamacher t-norm and t-conorm. *Symmetry* **2018**, *10*, 189. [CrossRef]
17. Cui, W.; Ye, J. Multiple-Attribute Decision-Making Method Using Similarity Measures of Hesitant Linguistic Neutrosophic Numbers Regarding Least Common Multiple Cardinality. *Symmetry* **2018**, *10*, 330. [CrossRef]
18. Liu, D.; Liu, Y.; Chen, X. The New Similarity Measure and Distance Measure of a Hesitant Fuzzy Linguistic Term Set Based on a Linguistic Scale Function. *Symmetry* **2018**, *10*, 367. [CrossRef]
19. Pang, Q.; Wang. H.; Xu, Z. Probabilistic linguistic term sets in multi-attribute group decision making. *Inf. Sci.* **2016**, *369*, 128–143. [CrossRef]
20. Lin, M.; Xu, Z.; Zhai, Y.; Zhai, T. Multi-attribute group decision-making under probabilistic uncertain linguistic environment. *J. Oper. Res. Soc.* **2017**. [CrossRef]
21. Atanassov, K. *Intuitionistic Fuzzy Sets*; Springer: Heidelberg, Germany, 1999.
22. Bashir, Z.; Rashid, T.; Wątróbski, J.; Sałabun, W.; Ali, J. Intuitionistic-fuzzy goals in zero-sum multi criteria matrix games. *Symmetry* **2017**, *9*, 158. [CrossRef]
23. Beg, I.; Rashid, T. Group Decision making Using Intuitionistic Hesitant Fuzzy Sets. *Int. J. Fuzzy Logic Intell. Syst.* **2014**, *14*, 181–187. [CrossRef]
24. Boran, F.E.; Gen, S.; Kurt, M.; Akay, D. A multi-criteria intuitionistic fuzzy group decision making for supplier selection with TOPSIS method. *Expert Syst. Appl.* **2009**, *36*, 11363–11368. [CrossRef]
25. De, S.K.; Biswas, R.; Roy, A.R. An application of intuitionistic fuzzy sets in medical diagnosis. *Fuzzy Sets Syst.* **2001**, *117*, 209–213. [CrossRef]
26. Li, D.F. Multiattribute decision making models and methods using intuitionistic fuzzy sets. *J. Comput. Syst. Sci.* **2005**, *70*, 73–85. [CrossRef]
27. Liu, P.; Mahmood, T.; Khan, Q. Multi-Attribute Decision-Making Based on Prioritized Aggregation Operator under Hesitant Intuitionistic Fuzzy Linguistic Environment. *Symmetry* **2017**, *9*, 270. [CrossRef]
28. Beg, I.; Rashid, T. Hesitant intuitionistic fuzzy linguistic term sets. *Notes Intuit. Fuzzy Sets* **2014**, *20*, 53–64.
29. Zhang, Y.; Xu, Z.; Wang, H.; Liao, H. Consistency-based risk assessment with probablistic linguistic prefrence relation. *Appl. Soft Comput.* **2016**, *49*, 817–833. [CrossRef]
30. Xu, Z.S.; Xia, M.M. On distance and correlation measures of hesitant fuzzy information. *Int. J. Intell. Syst.* **2011**, *26*, 410–425. [CrossRef]
31. Kim, S.H.; Ahn, B.S. Interactive group decision making procedure under incomplete information. *Eur. J. Oper. Res.* **1999**, *116*, 498–507. [CrossRef]
32. Kim, S.H.; Choi, S.H.; Kim, J.K. An interactive procedure for multiple attribute group decision making with incomplete information: Range-based approach. *Eur. J. Oper. Res.* **1999**, *118*, 139–152. [CrossRef]

33. Park, K.S. Mathematical programming models for characterizing dominance and potential optimality when multicriteria alternative values and weights are simultaneously incomplete. *IEEE Trans. Syst. Man Cybern.* **2004**, *34*, 601–614. [CrossRef]
34. Xu, Z.S. An interactive procedure for linguistic multiple attribute decision making with incomplete weight information. *Fuzzy Optim. Decis. Mak.* **2007**, *6*, 17–27. [CrossRef]

symmetry

MDPI

Article

Intertemporal Choice of Fuzzy Soft Sets

José Carlos R. Alcantud [1,*] and María José Muñoz Torrecillas [2]

1 BORDA Research Unit and Instituto Multidisciplinar de Empresa, University of Salamanca, E37007 Salamanca, Spain
2 Department of Economics and Business, University of Almería, E04120 Almería, Spain; mjmtorre@ual.es
* Correspondence: jcr@usal.es; Tel.: +34-923-294666

Received: 6 August 2018 ; Accepted: 27 August 2018 ; Published: 1 September 2018

Abstract: This paper first merges two noteworthy aspects of choice. On the one hand, soft sets and fuzzy soft sets are popular models that have been largely applied to decision making problems, such as real estate valuation, medical diagnosis (glaucoma, prostate cancer, etc.), data mining, or international trade. They provide crisp or fuzzy parameterized descriptions of the universe of alternatives. On the other hand, in many decisions, costs and benefits occur at different points in time. This brings about intertemporal choices, which may involve an indefinitely large number of periods. However, the literature does not provide a model, let alone a solution, to the intertemporal problem when the alternatives are described by (fuzzy) parameterizations. In this paper, we propose a novel soft set inspired model that applies to the intertemporal framework, hence it fills an important gap in the development of fuzzy soft set theory. An algorithm allows the selection of the optimal option in intertemporal choice problems with an infinite time horizon. We illustrate its application with a numerical example involving alternative portfolios of projects that a public administration may undertake. This allows us to establish a pioneering intertemporal model of choice in the framework of extended fuzzy set theories.

Keywords: fuzzy soft set; intertemporal choice; comparison table; decision making

1. Introduction

The scientific contribution of this paper is setting up a novel framework for making decisions that stems from the first cross-fertilization of two features: (a) intertemporal aspects of choice; and (b) extended fuzzy set models. We also give a novel adjustable algorithm that prioritizes alternatives with the aforementioned features.

Decisions whose consequences extend across multiple time periods are called intertemporal choices. The entry "Intertemporal choice" in the Palgrave Dictionary of Economics states: "Most choices require decision-makers to trade-off costs and benefits at different points in time. Decisions with consequences in multiple time periods are referred to as intertemporal choices. Decisions about savings, work effort, education, nutrition, exercise, and health care are all intertemporal choices" [1]. Although its analysis is preeminent in the standard crisp literature, to the best of our knowledge, the problem of intertemporal choice has never been modeled when the data are imprecise, uncertain or subjective in the sense of the extended fuzzy set theories. In this paper, we first put forward a model that fills this important gap. To prove that it can be used to make decisions, we propose a flexible mechanism that provides a ranking of the alternatives that are characterized by these characteristics. We also present some examples that illustrate the application of our decision making procedure.

To achieve our goals, we have selected the successful setting of fuzzy soft sets. (Other models of imprecise knowledge would require an ad hoc analysis, which we postpone for subsequent investigations to avoid confusions.) The new model that arises (from the amalgamation of the intertemporal setting of choice and data in the form of fuzzy soft sets) is called intertemporal fuzzy soft sets. We put forward various equivalent and complementary definitions of this concept. Some are more convenient for the purpose of algebraic manipulations and intuitions. Some are better suited to describe the computational machinery that produces the results from which the decisions are achieved.

In relation with the latter issue, the gist of standard intertemporal problems is that the consequences of a decision span along an infinite number of periods. However, to decide among various alternatives, their consequences across time are summarized by an amount called their respective Net Present Values. In this fashion, the infinite expansion that characterizes an alternative is summarized by a unique number, for example through a discounted sum. By inspiration of this widely accepted position, we propose to condense the information of intertemporal fuzzy soft sets into fuzzy soft sets in order to make optimal decisions. The tool that we introduce to achieve this target is called a reduction mechanism. Reduction mechanisms can both indicate symmetry in the valuation of a reward irrespective of the period when it is obtained, or a preference for earlier rewards (i.e., violation of the symmetric treatment of the periods). We provide several noteworthy examples of both behaviors. Once this reduction to a fuzzy soft set has been performed, our decision can rely on widely accepted solutions stated for that setting.

Actually, the main reason for choosing fuzzy soft sets in our pioneering approach is that there is a fully-developed theory for fuzzy soft set based decision making. To further assess the importance of this setting, in the next section, we review some general background about soft computing models with a more explicit explanation about soft set based modelizations and their decision making. We also dwell on the fundamentals of the intertemporal problem of choice and its applications.

This paper is organized as follows. In Section 2, we give some general background about the main notions that are used in our research. A fully developed real example helps to clarify the application of the discounted utility aggregation model. Section 3 recalls some terminology and definitions. In Section 4, we define our new model of intertemporal fuzzy soft sets and we also offer alternative formalizations. In Section 5, we define the notion of a reduction mechanism, which we use to state the decision algorithm that prioritizes alternatives in the framework of intertemporal fuzzy soft sets. We also illustrate the model with a numerical application to the selection of alternative portfolios of public projects. Finally, we conclude in Section 6.

2. Background

In this section, we give some background about various pertinent topics. Firstly, we provide some basic knowledge about the theory and practice of intertemporal choices, inclusive of a fully developed real example. Secondly, we give a general overview of fuzzy sets and other related models of uncertain information. Finally, we focus on the specific characteristics of the framework where we develop our contribution, namely, fuzzy soft sets and their decision making.

2.1. Intertemporal Choice: Theory and Practice

Intertemporal choices are decisions whose consequences (costs and benefits) are distributed over time [2]. Decisions about investments, spending and savings are standard examples of monetary intertemporal choices. However, there are also non-monetary intertemporal choices such as decisions related with sustainability (environmental issues such as forestry [3], climate policy [4] or the use of energy-using durables [5]), health (diet, exercise, and addictions [6,7]), job search [8], or work effort [9].

Discounted utility theory is the normative theory for intertemporal choice, or choices between outcomes accruing at different points in time; usually between immediate and delayed outcomes [10]. Since its introduction by Samuelson [11] in 1937, the discounted utility (DU) model has dominated the economic analysis of intertemporal choice (e.g., the aforementioned [4,5,9]). DU model was completed by Koopmans [12] who clarified its logic and main assumptions. This model presumes that people evaluate the pleasures and pains resulting from a decision in a similar way that financial markets evaluate gains and losses spread out over time. Anyone prefers to get 1000 dollars now rather than 1000 dollars in a year. However, people behave differently if they have to choose between receiving 1000 dollars now or 1100 dollars in a year. To compare choices made in different moments of time, under DU, it is assumed that agents exponentially discount these costs and benefits according to how delayed they are in time [13]. Although there is experimental evidence showing that this is not always the case [2,14,15], a fact that prompted the appearance of other explanatory models such as hyperbolic discounting [16–18] or q-exponential discounting [19–21], the DU model is nevertheless used as the common tool for public policy in the evaluation of public projects. The model can be calibrated with suitable discount factors, for example, using a decreasing sequence of discount rates for projects with very long-term impacts to account for intergenerational equity [22]. The governments of the United Kingdom and France, in line with the proposals of several authors for long-term valuations [23], recommend the use of decreasing discount rates in public projects with long time horizons [24,25].

In real practice, and provided that the DU model is adopted for the evaluation of intertemporal projects, the expected cash-flows of a project are always discounted to obtain its Net Present Value (NPV) at instant 0 (time of evaluation), see for example [26]. To that purpose, a well-known formula is applied which requires using an appropriate discount rate. For private projects, the weighted average cost of capital or a required profitability are usually used as a discount rate. For public projects, the social time preference rate is employed to calculate the discount rate, which is then called *social discount rate* (SDR). The social time preference rate is a rate used for discounting future benefits and costs, and it is based on comparisons of utility across different points in time or different generations [24]. The SDR "is used by society to give relative weight to social consumption or income accruing at different points in time" [3].

Since we are applying our intertemporal model of choice to concrete examples, we need to fix an appropriate discount rate. Our choice is not arbitrary. On the contrary, we take advantage of the fact that, for EU funded projects, the European Commission [26] recommends the use of the exponential discounting model and a constant 5% European social discount rate for the Cohesion Fund eligible countries and 3% for the others (countries non-eligible for the Cohesion Fund). Therefore, 5% is our reference rate unless otherwise stated.

According to the previous discussion, discounted utility computations made for choices in the present and at various moments in the future $(x_0, x_1, \ldots, x_T)$ adopt the form

$$\sum_{t=0}^{T} \beta^t u(x_t) \tag{1}$$

Here, x_t is the choice made at moment t, whose utility at that moment is $u(x_t)$; β^t is the discount factor for a time period of t periods, usually years (for example, $\beta = \frac{1}{1+0.05}$ under our assumption for the reference rate); choices are made along periods $t = 0$ (the present), $t = 1, 2, \ldots, T$; and T may be $+\infty$. Put otherwise, if we want to assess the temporal sequence $(x_0, x_1, \ldots, x_T)$ and the utility u gives us the degree of satisfaction of each choice x_t, which is of course $u(x_t)$, then $\sum_{t=0}^{T} \beta^t u(x_t)$ gives the discounted utility of this temporal sequence. The β parameter accounts for the fact that people prefer to enjoy utility as soon as possible.

Let us now give a real example that illustrates the application of our reference model for intertemporal choice.

Example 1. *On 7 November 2017, the Spanish infrastructure operator Ferrovial published a press release (see https://www.ferrovial.com/en/press-room/press_releases/500-million-euro-2-124-perpetual-hybrid-bond/, retrieved 18 August 2018.) It stated: "Taking advantage of a favorable market environment with low interest rates, Ferrovial today successfully priced a 500 million euro perpetual hybrid bond. The issue pays a* 2.124% *annual coupon until 14 May 2023. Subsequently, it will pay a fixed coupon equal to the applicable swap rate plus a spread of* 2.127% *until 14 May 2043 and of* 2.877% *thereafter. The swap rate will be updated every five years."*

Therefore, Ferrovial perpetual bonds, with a face value of 100 *euros, will pay a* 2.124% *annual coupon during the first six years,* 2.127% *(supposing a swap rate of* 0%*) the following* 20 *years, and a* 2.877% *thereafter. Table 1 expresses this intertemporal situation and gives the computations that produce the NPV at time 0 of such a bond when we assume a discount rate of* 3%*, hence* $\beta = (1 + 0.03)^{-1} = 0.97087$.

Table 1. Detailed computations of the NPV, assuming a discount rate of 3%, in the real Example 1. It is the sum of the present values at the right column of the table.

Period	Interest Rate	Cash-Flows Annual Coupon	Discount Factor	Present Value
1	2.124%	2.124 = 100 × 0.02124	$0.97087 = (1+0.03)^{-1} = \beta^1$	2.0621 = 2.124 × 0.97087
2	2.124%	2.124	$0.94260 = (1+0.03)^{-2} = \beta^2$	2.0021 = 2.124 × 0.94260
3	2.124%	2.124	$0.91514 = \beta^3$	1.9438
4	2.124%	2.124	$0.88849 = \beta^4$	1.8871
5	2.124%	2.124	$0.86261 = \beta^5$	1.8322
6	2.124%	2.124	$0.83748 = \beta^6$	1.7788
7	2.127%	2.127 = 100 × 0.02127	$0.81309 = \beta^7$	1.7294
8	2.127%	2.127	$0.78941 = \beta^8$	1.6791
⋮	⋮	⋮	⋮	⋮
26	2.127%	2.127	$0.46369 = \beta^{26}$	0.9863
27 and onwards	2.877%	2.877 = 100 × 0.02877		$44.4683 = 2.877\frac{\beta^{26}}{0.03}$
NPV				82.4761

2.2. A Concise Presentation of Fuzzy Sets and Related Notions

Since Zadeh [27] laid the foundations of fuzzy set theory, whose main feature is the introduction of partial membership degrees, many authors produced a large amount of literature on their advantages and potential applications in decision making. Mardani et al. [28] gave a summary of articles about fuzzy multi-criteria decision making from the period 1994–2014. Other classical references for the fundamentals of decision making in fuzzy set theory include Tanino [29] and Fodor and Roubens [30].

When imprecise individual or collective knowledge cannot be faithfully represented by fuzzy sets, extensions of this concept and multiple variations offer more suitable models. Atanassov [31,32] presented the idea of intuitionistic fuzzy sets. Afterwards, Chen et al. [33] or Wei [34] produced intuitionistic fuzzy multi-attribute group decision making methods, and De Miguel et al. [35] applied interval-valued Atanassov intuitionistic fuzzy sets in multi-expert decision making. Pythagorean fuzzy sets are surveyed in Peng and Selvachandran [36], and interval-valued Pythagorean fuzzy sets were studied by Peng and Yang [37] and Peng [38]. Hesitancy was first merged with fuzzy sets by Torra [39] (for more information, a good source is Rodríguez et al. [40]; see also Alcantud and Torra [41] for the first decomposition theorems and extension principles in the framework of hesitant fuzzy sets).

From a different position, rough set theory was established by Pawlak [42] and, in his first formulation, an equivalence binary relation is the source of granulation of the set of alternatives.

It is at this junction that a different, parameterized description of the alternatives made its appearance. The idea produced soft sets, extensions and hybrid models. Since they are a benchmark in our paper, we proceed to describe them succinctly in the next subsection.

2.3. Soft Sets, Extensions and Hybrid Models

The theory of soft sets originates with the seminal paper [43]. Feng and Zhou ([44], Section 1) cleverly described soft set theory in the following terms: it "is considered as a new mathematical tool for dealing with uncertainties which is free from the inadequacy of parameter tools. In soft set theory, the problem of setting the membership function simply does not arise as in fuzzy set theory, which makes the theory convenient and easy to use in practice." Its relevancy to decision making in various fields was already pointed out in [43], which also explained that the models by fuzzy sets and soft sets are linked to each other. Further relationships are proven in [45–48]. The early works of Maji et al. [49] and Aktaş and Çağman [50] among others expanded the basic theory of soft sets. Khameneh and Kılıçman [51] systematically reviewed multi-attribute decision-making based on soft set theory. Zhan and Alcantud [52] recently summarized parameter reduction of soft sets, a thriving area that allows for approaches in extended models.

Indeed, suitable extensions of soft sets come up from the incorporation of ideas such as the aforementioned fuzziness and hesitancy. Fuzzy soft sets were designed by Maji, Biswas and Roy [53]. Parameter reduction in the context of fuzzy soft sets was developed, e.g., by Khameneh and Kılıçman [54]. Wang, Li and Chen [55] produced hesitant fuzzy soft sets by adding up hesitancy to the latter concept. Because data collection and measurement often produce errors or are restricted, some studies (e.g., [56–58]) are concerned with another extended form of soft sets called incomplete soft sets, while other [59,60] are concerned with the natural extension called incomplete fuzzy soft sets.

Let us now consider fuzzy soft set based decision making. The performance of the pioneering analysis by Roy and Maji [61] has been improved by [62,63]. Feng et al. [64] put forward a flexible method based on level soft sets. They explained that [65] challenges the position in [61] by claiming that the criterion for making a decision should use scores instead of fuzzy choice values, a point of view that found little support among other scholars. Liu et al. [66] gave another methodology for fuzzy soft set based decision-making based on an ideal solution. Feng and Guo [67] intended to effectively resolve the natural group decision-making problem in the context of fuzzy soft sets. To achieve their goal, they first design another adjustable method for solving fuzzy soft set based decision-making problems.

Hybrid models combine the spirit of soft sets with other methodologies. Peng, Dai and Yuan ([68] and the references therein) contributed to interval-valued fuzzy soft decision making. Park, Kwun and Son [69] gave an approach to decision making problems based on generalized intuitionistic fuzzy soft sets (see also [70] for the operations in that framework). Feng et al. [48] used soft approximation spaces instead of binary relations in rough set theory. Zhan and Wang [71] built five new different types of soft coverings based rough sets and investigated relationships between soft rough sets and soft covering based rough sets. Zhan and Alcantud [72] designed a soft rough covering by means of soft neighborhoods, which they utilized to improve decision making in a multi-criteria group environment. Ma et al. [73] presented an updated summary of decision making methodologies based on two classes of hybrid soft set models. Fatimah et al. [74] studied the decision-making implications of probabilistic and dual probabilistic soft sets.

Now, we proceed to state the formal definitions that we shall use in the remaining of this paper.

3. Definitions: Soft Sets and Fuzzy Soft Sets

In soft set theory and their extensions, we start with U, which is a set of alternatives (or a universe of objects), and E, which is a universal set of attributes, parameters or characteristics. We let $\mathcal{P}(U)$ denote the set of parts of U, i.e., the set formed by all the subsets of U.

Definition 1 (Molodtsov [43]). *A soft set over U is a pair (F, A) with $A \subseteq E$ and $F : A \longrightarrow \mathcal{P}(U)$.*

A soft set over U is a parameterized family of subsets of U, and the set A contains the relevant parameters. For every $e \in A$, the subset $F(e)$ is the set of e-approximate elements, or also, the subset of U approximated by e. For example, if $U = \{s_1, s_2, s_3\}$ is a universe of shirts and A contains the parameter e that describes "blue color" and the parameter e' that describes "silk fabric" then $F(e) = \{s_1\}$ means that the only shirt with blue color is s_1 and $F(e') = \{s_1, s_3\}$ means that exactly s_1 and s_3 have silk fabric.

To model more general situations, the following notion is subsequently proposed and investigated in [53]:

Definition 2 (Maji, Biswas and Roy [53]). *Let $\mathbf{FS}(U)$ denote the set of all fuzzy sets on U. The pair (F, A) is a fuzzy soft set (FSS) over U when $A \subseteq E$ and $F : A \longrightarrow \mathbf{FS}(U)$.*

Needless to say, soft sets are an example of fuzzy soft sets. However, fuzzy soft sets are better suited to model subjectively perceived properties, since partial memberships are designed to express such subjectivity.

In the standard instance with finite U and A, both soft and fuzzy soft sets can be displayed in the form of a table, where rows correspond to the alternatives in U, columns correspond to the attributes in A, and there is a number from $[0,1]$ in each cell. Of course, the cells of these matrices contain either 0 or 1 when the fuzzy soft set is a soft set.

To illustrate these ideas and motivate the subsequent decisional analysis, let us put forward an example in terms of an object recognition problem:

Example 2. *A collection of objects* $U = \{o_1, ..., o_6\}$ *is characterized in terms of a space of attributes which is denoted as* $A = \{p_1, \ldots, p_7\}$*. Here, the attributes represent the relevant combinations of characteristics. The fuzzy soft set that describes the objects is* (F, A)*, which is given by the tabular representation in Table 2. For illustration, number* 0.650 *at the junction of Row* o_1 *and Column* p_1 *means that the degree of membership of* o_1 *to the objects that verify characteristic* p_1 *is* 0.650.

Table 2. Tabular representation of the fuzzy soft set (F, A) in Example 2.

	p_1	p_2	p_3	p_4	p_5	p_6	p_7
o_1	0.650	0.150	0.064	0.216	0.048	0.054	0.405
o_2	0.144	0.720	0.360	0.045	0.036	0.020	0.175
o_3	0.120	0.084	0.180	0.350	0.096	0.021	0.294
o_4	0.504	0.192	0.108	0.090	0.048	0.620	0.280
o_5	0.084	0.245	0.036	0.096	0.270	0.200	0.320
o_6	0.216	0.315	0.042	0.108	0.224	0.126	0.410

Fuzzy Soft Set Based Decision Making

Soft set based decision making relies on [44,75,76]. However, since the appearance of the seminal [61], there have been many remarkable approaches to decision making in the framework of fuzzy soft sets. The most successful contributions include [61–67]. We do not describe them all in detail here. For our purposes, it should suffice to know their general features and relative advantages, which are summarized in Table 3. It compares various noteworthy criteria with respect to their main characteristics.

Table 3. A critical summary of the main fuzzy soft set based decision making procedures.

Ref.	Aggregation	Methodology	Solution	Other Issues
[61]	Min operator	Scores from a comparison matrix	Unique	Many ties Loss of information [63] reformulates algorithm
[65]	Not discussed	Fuzzy choice values	Unique	It is too controversial
[64]	Not discussed	Choice value of level soft set	Not unique	Ties multiply Richness at the cost of indeterminacy Additional inputs needed (e.g., threshold fuzzy set)
[62]	Product operator	Scores from alternative comparison matrix	Unique	Improved power of discrimination
[66]	Not discussed	Similarity measure and substitutable	Unique	Application of subjective weights Low time complexity
[67]	Not discussed	Distance measure for Group Decision Making	Not unique	Two methods for obtaining appropriate experts' weights

As explained in Section 1, ultimately we need to make choices from a fuzzy soft set to solve the intertemporal choice problem for fuzzy soft sets that we present in Section 4. Therefore, it is convenient to be familiar with the machinery of at least one such procedure. For the sake of clarity, to apply fuzzy soft sets in decision making practice, we focus on the proposal by Alcantud [62], who stated a feasible algorithmic solution to solve problems in the format of Example 2.

To make this paper self-contained, we recall that the application of Alcantud's algorithm proceeds as follows (afterwards Example 3 illustrates the application of Algorithm 1 below to a concrete situation):

Algorithm 1 (Alcantud [62])

Input: a fuzzy soft set (F, A), which we place in the form of a table. Its cell (i, j) is represented by t_{ij}

1: For every attribute j, let M_j denote the maximum membership value of the alternatives, i.e., $M_j = \max_{i=1,...,k} t_{ij}$ for each $j = 1, ..., q$.
Produce a $k \times k$ comparison matrix $A = (a_{ij})_{k\times k}$ as follows: for every i, j, a_{ij} is the sum of the non-negative values in the following finite sequence:

$$\frac{t_{i1} - t_{j1}}{M_1}, \frac{t_{i2} - t_{j2}}{M_2}, ..., \frac{t_{iq} - t_{jq}}{M_q}.$$

We can display this matrix as a comparison table.

2: For each $i = 1, ..., k$, calculate R_i as the sum of the elements in row i of A, and T_i as the sum of the elements in column i of A. For every $i = 1, ..., k$, calculate the score $S_i = R_i - T_i$ of object i.

3: The result of the decision is any object o_k such that $S_k = \max_{i=1,...,k} S_i$.

Example 3. *Let us assume that we have the input data of Example 2. Its comparison table is given in Table 4, as computed by Algorithm 1 above. Then, Table 5 shows its associated scores.*

As a result, one concludes that o_4 should be selected when we consider the input data of Example 2 and we adhere to Algorithm 1.

Table 4. Comparison table of the fuzzy soft set (F, A) in Example 2 using Algorithm 1.

	o_1	o_2	o_3	o_4	o_5	o_6
o_1	0	1.93	1.23	0.89	1.5	1.04
o_2	1.61	0	1.42	1.43	1.65	1.45
o_3	0.88	1.39	0	1.15	1.18	1.07
o_4	1.09	1.95	1.71	0	1.52	1.42
o_5	1.19	1.66	1.22	1.01	0	0.29
o_6	1.01	1.73	1.39	1.19	0.57	0

Table 5. Score table of the fuzzy soft set (F, A), derived from its Comparison table by Algorithm 1.

	Row-Sum (R_i)	Column-Sum (T_i)	Score (S_i)
o_1	6.58	5.79	0.79
o_2	7.57	8.65	−1.08
o_3	5.68	6.97	−1.29
o_4	7.7	5.68	2.02
o_5	5.37	6.43	−1.06
o_6	5.9	5.27	0.63

4. A New Model: Intertemporal Fuzzy Soft Sets

Thus far, the literature has dealt with alternatives with a very simple structure: for each characteristic, we know the degree of membership of the alternative to the set of elements that verify the characteristic. However, it is not difficult to find examples where the performance of the options is far more complex. Particularly, in this paper, we are concerned with an intertemporal setting. Indeed, in the general framework of project selection (e.g., solar energy projects [77], environmental impact assessment [78], financial portfolio [79], etc.), each alternative has a performance along an indefinite number of periods, typically years. Therefore, if we want to decide which of a list of projects should be selected, we face an infinite number of fuzzy soft sets (one that represents each possible period). Neither of the existing approaches to fuzzy soft set based decision making can deal with this potentially infinite structure.

To tackle this new problem, now we proceed to formalize our model. It accounts for the intertemporal setting that we have motivated. Afterwards, we interpret the formal statement of the model in terms of tables when the number of alternatives and attributes is finite, which facilitates their computational manipulation. In the next section, we give a procedure for making decisions in this novel framework. In addition, in that section, an example illustrates the decision making algorithm in a situation motivated by public projects evaluation. For comparison, recall that Example 1 shows a recent, real situation in a financial environment with crisp data.

4.1. The Structure of Intertemporal Fuzzy Soft Sets

Molodtsov's notion of parameterized descriptions of the universe has already been combined with features that do not pertain to the original formulation of soft sets. Here, we propose a model where for each of a possibly infinite number of periods, each attribute produces a possibly different fuzzy parameterization of the universe.

The statement of the model is simple but powerful:

Definition 3. *An intertemporal fuzzy soft set (ItFSS) over U is a sequence $\mathbb{F} = \{(F_i, A)\}_{i \in \mathbb{N}}$ of fuzzy soft sets over the common universe U.*

In Section 4.2, we specify two alternative formulations of this definition, that are amenable for calculations with computers.

The basic idea is that the fuzzy parameterization of the universe is allowed to vary with time. According to this model, a fixed set of attributes is given, and then for each period i a parameterized description of the common universe produces a fuzzy soft set (F_i, A) which accounts for the situation at that period. To emphasize differences, on occasions, we refer to these standard FSSs as *static* FSSs.

Remark 1. *When the sequence in Definition 3 is constant (in other words, we use $\{(F_i, A)\}_{i \in \mathbb{N}}$ with $F_i = F$ for each i), then we identify the ItFSS with a standard FSS. Therefore, our model of ItFSSs is a natural extension of FSSs in Definition 2.*

Henceforth, we assume that the common attributes in A are all positive (otherwise, we can proceed as in [80] to convert the input into this format). Therefore, the higher is the membership degree, the better. The decision problem that arises consists of determining an optimal alternative from the list U. We propose a flexible solution in Section 5.

The next motivating example illustrates the structure of our model above. Observe that in this simplified statement it is possible to display the information pertaining to an intertemporal fuzzy soft set in one table, even though it concerns an infinite number of periods:

Example 4. *A civil project (e.g., building a bridge or a dam) will have a long-term impact on the population of a certain geographical area. There are two targets that should be achieved: environmental effects (e_1) and economic development (e_2). Let $A = \{e_1, e_2\}$. There are two possible projects, namely, p_1 and p_2. Their respective yearly effects are captured by the ItFSS $\mathbb{G} = \{(G_i, A)\}_{i \in \mathbb{N}}$ over U shown in Table 6, where $U = \{p_1, p_2\}$.*

It is not difficult to check that, in every period i, the positive effects of p_1 exceed the effects of p_2 at any of the two target attributes:

$$0.3 + \frac{1}{i+1} > 0.2 + \frac{1}{i+2} \text{ for each } i$$

$$0.4 + \frac{1}{i+1} > 0.3 + \frac{1}{i+2} \text{ for each } i$$

Therefore, project p_1 should be selected.

Table 6. A tabular representation of the intertemporal fuzzy soft set $\mathbb{G} = \{(G_i, A)\}_{i\in\mathbb{N}}$ in Example 4.

(G_i, A)	e_1	e_2
p_1	$0.3 + \frac{1}{i+1}$	$0.4 + \frac{1}{i+1}$
p_2	$0.2 + \frac{1}{i+2}$	$0.3 + \frac{1}{i+2}$

The streamlined Example above is very simple for two reasons. Firstly, even though there are an infinite number of different FSSs for the infinite periods, the ItFSS can be represented by one parametric table, which is often not the case. Secondly, we do not need to use the theoretical contribution of Section 5 because the decision about which project should be selected is trivial: there is a sort of domination of the first project over the second in all attributes and for all periods that makes the decision obvious. Now, we proceed to give a different example where the modelling power of ItFSSs is more apparent. We can also infer the need for a formal analysis of decision making in that context, hence the next example motivates Section 5.

Example 5. *A new regulation will have long-term effects on the population of a country. There are three groups that are potentially affected in terms of satisfaction: students* (e_1)*, working class* (e_2) *and retired people* (e_3)*. There are two possibilities, namely, passing the law* (p_1) *and retaining the current regulation* (p_2)*. Their respective yearly effects on the satisfaction across groups are captured by Table 7, which embodies an ItFSS* $\mathbb{F} = \{(F_i, A)\}_{i\in\mathbb{N}}$ *over* $U = \{p_1, p_2\}$ *where* $A = \{e_1, e_2, e_3\}$.

We read, for example, that in year 1 *the students' degree of membership to being satisfied with the new regulation is* 0.3*, and it is* 0.2 *under the current law. For the working class, the respective degrees of satisfaction are* 0.4 *and* 0.6*. For retired people, the respective degrees of satisfaction are* 0.7 *and* 0.8.

From the third year onwards, the students' degree of membership to being satisfied with the new regulation is 0.6*, and it is* 0.3 *under the current law. For the working class, the respective degrees of satisfaction are* 0.6 *and* 0.4*. For retired people, the respective degrees of satisfaction are* 0.8 *and* 0.6.

Table 7. Tabular representation of the intertemporal fuzzy soft set in Example 5.

(F_1, A)	e_1	e_2	e_3
p_1	0.3	0.4	0.7
p_2	0.2	0.6	0.8
(F_2, A)	e_1	e_2	e_3
p_1	0.4	0.4	0.7
p_2	0.3	0.5	0.7
(F_i, A)	e_1	e_2	e_3
p_1	0.6	0.6	0.8
p_2	0.3	0.4	0.6
$\forall i > 2$			

Example 5 shows the intrinsic difficulty of dealing with these problems. The snapshots at different moments can vary substantially from each other, and of course they are not always as obvious as the situation of Example 4. Additionally, the "attributes" can have different weights, for example because they represent characteristics of groups with different proportions in the society.

For the purpose of favoring implementability, now we proceed to state an equivalent formulation of our intertemporal model. It allows us to work with a tabular format that is amenable for computations.

4.2. An Alternative Representation of Intertemporal Fuzzy Soft Sets

In practical terms, when both $U = \{o_1, \ldots, o_m\}$ and $A = \{e_1, \ldots, e_n\}$ are finite, we can represent the information that describes an ItFSS in a table where the cells are either finite or infinite sequences of membership degrees. Table 8 gives the general form of such a representation.

Table 8. The tabular representation of our novel intertemporal model for fuzzy soft sets.

	e_1	e_2	$\ldots$	e_n
o_1	$(u^1_{11}, u^2_{11}, \ldots, u^t_{11}, \ldots)$	$(u^1_{12}, u^2_{12}, \ldots, u^t_{12}, \ldots)$	$\ldots$	$(u^1_{1n}, u^2_{1n}, \ldots, u^t_{1n}, \ldots)$
$\vdots$				
o_m	$(u^1_{m1}, u^2_{m1}, \ldots, u^t_{m1}, \ldots)$	$(u^1_{m2}, u^2_{m2}, \ldots, u^t_{m2}, \ldots)$	$\ldots$	$(u^1_{mn}, u^2_{mn}, \ldots, u^t_{mn}, \ldots)$

Let us analyze this alternative description. To that purpose, the set of infinite sequences of numbers from $[0,1]$ (or infinite utility streams [81–83]) is denoted by $\mathcal{S}$. Our intertemporal model of fuzzy soft sets over U can also be defined by $\bar{F} : A \longrightarrow \mathbf{S}(U)$ where $\mathbf{S}(U)$ represents the mappings $U \longrightarrow \mathcal{S}$. Consequently, for each attribute, we capture the degree of membership of any alternative in each moment of time.

Indeed, in Table 8, we can define $\bar{F}(e_j)(o_i) = (u^1_{ij}, u^2_{ij}, \ldots, u^t_{ij}, \ldots) \in \mathcal{S}$, hence u^t_{ij} means the degree of membership of alternative o_i to the fuzzy set of elements that verify attribute e_j in period t. Conversely, every $\bar{F} : A \longrightarrow \mathbf{S}(U)$ produces a table with the structure of Table 8 under the finiteness restriction for both A and U.

We can also swap between the tabular form and the notation of Definition 3.

From the aforementioned tabular representation of $\bar{F} : A \longrightarrow \mathbf{S}(U)$, we can define the corresponding ItFSS over U as the sequence $\mathbb{F} = \{(F_i, A)\}_{i \in \mathbb{N}}$ where the tabular form of each FSS (F_i, A) is described in Table 9.

Table 9. The tabular representation of the fuzzy soft set (F_i, A) corresponding to moment i in the ItFSS $\mathbb{F} = \{(F_i, A)\}_{i \in \mathbb{N}}$ represented by Table 8.

	e_1	e_2	$\ldots$	e_n
o_1	u^i_{11}	u^i_{12}	$\ldots$	u^i_{1n}
$\vdots$				
o_m	u^i_{m1}	u^i_{m2}	$\ldots$	u^i_{mn}

Conversely, an ItFSS given by the construction in Section 4.1 can be trivially transformed into the tabular form presented in this subsection. We do this in Table 10 for the case of $\mathbb{G}$ in Example 4.

Table 10. The tabular representation of $\mathbb{G} = \{(G_i, A)\}_{i \in \mathbb{N}}$ in Example 4.

$\mathbb{G}$	e_1	e_2
p_1	$(0.3 + \frac{1}{2}, 0.3 + \frac{1}{3}, 0.3 + \frac{1}{4}, \ldots)$	$(0.4 + \frac{1}{2}, 0.4 + \frac{1}{3}, 0.4 + \frac{1}{4}, \ldots)$
p_2	$(0.2 + \frac{1}{3}, 0.2 + \frac{1}{4}, 0.2 + \frac{1}{5}, \ldots)$	$(0.3 + \frac{1}{3}, 0.3 + \frac{1}{4}, 0.3 + \frac{1}{5}, \ldots)$

5. Choices from Intertemporal Fuzzy Soft Sets

When it comes to prioritizing alternatives in the framework of intertemporal fuzzy soft sets, the most natural course of action is to associate a FSS with our original ItFSS and then apply Algorithm 1 (or any other existing proposal, see Table 3) to the latter fuzzy soft set.

To implement this solution, now we proceed to describe some procedures that from each ItFSS produce a static or standard fuzzy soft set (cf., Section 5.1). Afterwards, we show how we can integrate these procedures with decision making based on FSSs to design an intertemporal fuzzy soft set based decision making procedure (cf., Section 5.2).

5.1. Static FSSs Associated with an Intertemporal FSS

In this subsection, we formalize some methods that associate a static FSS (Definition 2) with any intertemporal FSS (Definition 3). We refer to these methods as *reduction mechanisms*.

The simplest reduction mechanisms act cell-by-cell on the tabular representation. For instance, one can pinpoint the evaluation at a distinguished moment (e.g., the first period); or, in the case of finite time horizon, their lowest or highest evaluation, their (either arithmetic or geometric) average, etc. Under a genuine infinity of periods, we can use the natural modifications by infimum, supremum or discounted sums for the same purpose.

Let us now formalize a few explicit reduction mechanisms. We fix $\mathbb{F} = \{(F_i, A)\}_{i\in\mathbb{N}}$, an intertemporal fuzzy soft set over U. We can obtain a *reduced* FSS associated with $\mathbb{F}$ by the application of one of the following expressions:

1. The pessimistic FSS associated with $\mathbb{F}$ is (F^p, A) such that the fuzzy parameterization $F^p : A \longrightarrow \mathbf{FS}(U)$ verifies that for each $a \in A$, $F^p(A) = \inf\{F_i(a) : i \in \mathbb{N}\}$.
2. The optimistic FSS associated with $\mathbb{F}$ is (F^o, A) such that the fuzzy parameterization $F^o : A \longrightarrow \mathbf{FS}(U)$ verifies that for each $a \in A$, $F^o(A) = \sup\{F_i(a) : i \in \mathbb{N}\}$.
3. Let $\delta \in [0, 1)$ be a factor. The δ-discounted FSS associated with $\mathbb{F}$ is (F^δ, A), where the fuzzy parameterization $F^\delta : A \longrightarrow \mathbf{FS}(U)$ verifies that for each $a \in A$,

$$F^\delta(A) = \frac{1-\delta}{\delta} \sum_{i=1}^{\infty} \delta^i F_i(a). \tag{2}$$

Observe that these definitions are correct: the only non-trivial case is justified in the following auxiliary result.

Lemma 1. *The δ-discounted FSS is well-defined, i.e., it is a FSS.*

Proof. We just need to observe that, because each (F_i, A) is a FSS, $\sum_{i=1}^{\infty} \delta^i F_i(a)$ is bounded above by $\sum_{i=1}^{\infty} \delta^i = \frac{\delta}{1-\delta}$. Hence, $\sum_{i=1}^{\infty} \delta^i F_i(a)$ converges for each $a \in A$, and $\frac{1-\delta}{\delta} \sum_{i=1}^{\infty} \delta^i F_i(a) \leqslant 1$. Obviously, $\frac{1-\delta}{\delta} \sum_{i=1}^{\infty} \delta^i F_i(a) \geqslant 0$. □

The first two reduction mechanisms are symmetric, in the standard sense: when $\mathbb{F}_\sigma = \{(F_{\sigma(i)}, A)\}_{i\in\mathbb{N}}$ is the ItFSS derived from a permutation of the periods $\sigma : \mathbb{N} \longrightarrow \mathbb{N}$, one has $F^p_\sigma = F^p$ and $F^o_\sigma = F^o$. However, the δ-discounted reduction mechanism violates the symmetric treatment of the periods, i.e., the statement $F^\delta_\sigma = F^\delta$ is in general false in the aforementioned conditions.

However, the δ-discounted reduction mechanism has an important advantage over the pessimistic and optimistic reduction mechanisms. The pessimistic and optimistic reduction mechanisms produce a considerable loss of information because they discard the data about the degrees of membership that are not minimal and maximal respectively, whereas the δ-discounted reduction mechanism uses all the information available to produce the reduced FSS.

Our next remark insists on the importance of this mechanism in decision making.

Remark 2. *The deduction of the δ-discounted FSS is motivated by a successful solution to the problem of aggregating intergenerational utilities [81–83]. In Section 2.1, we explained that according to the popular DU model [11,12,84], decision-makers evaluate the alternatives on the basis of the weighted addition of utilities, these weights being discount factors based on temporal delays. In our benchmark case, these delays can extend to infinity.*

Due to the aforementioned advantages, henceforth we adopt the δ-discounted reduction mechanism as the standard mechanism for transforming an ItFSS into a FSS in practical problems.

5.2. Decision Making in Intertemporal FSSs

We are ready to put forward a procedure for ranking a finite list of alternatives when the decision-making problem is characterized by an intertemporal FSS. It consists of three basic steps.

Put shortly, the algorithm suggests to reduce the ItFSS to a FSS (Step 1) and then order the alternatives according to standard decision-making in this framework (Step 2). The ordering in the reduced FSS carries forward in the ItFSS for which it is a natural representation (Step 3), which solves our problem. The next subsection illustrates how this can be put into practice in a concrete example of project appraisal.

5.3. An Example of Decision Making in the Framework of ItFSSs

In this section, we develop an example that serves two purposes. Firstly, we describe the structure of the problem in a practical fashion that is different from the description in Section 4.2. Secondly, it illustrates the application of Algorithm 2.

Algorithm 2 Algorithm for decision making

Inputs: An intertemporal table of fuzzy soft sets (in the notation of Table 8 or, otherwise, see Section 5.3). A reduction mechanism (e.g., from Section 5.1). A fuzzy soft set decision making procedure (e.g., from Table 3).

1: Apply the selected reduction mechanism to the ItFSS in order to obtain a (reduced) FSS.
2: Rank the alternatives in this FSS by the decision making procedure that we have singled out.
3: Any object that is at the top of the ranking in the previous step is an optimal choice of the intertemporal statement of the problem.

Example 6. *For the convenience of presentation, we are going to evaluate two alternative portfolios of projects that a public administration may undertake. Portfolios 1 and 2 are parameterized in terms of four attributes along an infinite number of periods, and each of these characteristics can also be regarded as a project on its own (e.g. bike lanes, urban parks, sports facilities, and sewage treatment plants). The objective of this evaluation is to choose the best alternative.*

For each project, a value for its social suitability in each period is assigned. To simplify, we consider projects whose utilities follow the following patterns: increasing and then constant; decreasing and then constant; decreasing, then increasing and finally constant; and constant.

Tables 11 and 12, respectively, describe these two portfolios P_1 and P_2. At their bottoms they contain additional values whose meaning we explain below.

These tables jointly define an ItFSS that we have displayed in Table 13, albeit in incomplete form due to obvious restrictions: we cannot display the infinite digits that appear at each cell. According to Step 1 of Algorithm 2, with this element we associate a standard FSS that we denote as (S, P), by the application of the DU reduction mechanism with a 0.05 discount rate, hence $\delta = \frac{1}{1.05} \approx 0.952381$. The suitability of this rate has been argued in Section 2.1. Let us denote $\mathbf{T} = \frac{1-\delta}{\delta} = 0.05$. The reduced fuzzy soft set (S, P) is displayed in Table 14. The results of the computations that produce it already appear at the bottom of Tables 11 and 12, and now we explain how we have obtained them (see Equation (2), and Remark 2 for inspiration). In the case of Portfolio 1,

$$\mathbf{A}_1 = 13.085321 = 0.05\delta + 0.10\delta^2 + 0.15\delta^3 + 0.20\delta^4 + \ldots$$

$$\mathbf{A}_2 = 10.60642701 = 1\delta + 0.95\delta^2 + 0.90\delta^3 + 0.85\delta^4 + \ldots$$

$$\mathbf{A}_3 = 14 = 0.70\delta + 0.70\delta^2 + 0.70\delta^3 + 0.70\delta^4 + \ldots$$

$$\mathbf{A}_4 = 18.3258941 = 0.05\delta + 0.10\delta^2 + 0.15\delta^3 + 0.20\delta^4 + \ldots + 0.70\delta^9 + 0.75\delta^{10} + \ldots$$

Therefore, in the reduced FSS (S, P)*, Equation* (2) *states that the degrees of membership for Portfolio 1 are* $\mathbf{T} * \mathbf{A}_i$*. We proceed similarly to obtain the degrees of membership for Portfolio 2.*

We now apply Algorithm 1 to the FSS (S, P)*. Table 15 gives the necessary computations.*

From the last computation in Table 15, we conclude that the first portfolio should be selected.

Table 11. Portfolio 1.

Period	Attribute 1	Attribute 2	Attribute 3	Attribute 4
1	0.05	1.00	0.70	1.00
2	0.10	0.95	0.70	0.95
3	0.15	0.90	0.70	0.90
4	0.20	0.85	0.70	0.85
5	0.25	0.80	0.70	0.80
6	0.30	0.75	0.70	0.75
7	0.35	0.70	0.70	0.70
8	0.40	0.65	0.70	0.65
9	0.45	0.60	0.70	0.70
10	0.50	0.55	0.70	0.75
11	0.55	0.50	0.70	0.80
12	0.60	0.45	0.70	0.85
13	0.65	0.40	0.70	0.90
14	0.70	0.35	0.70	0.95
15	0.75	0.35	0.70	1.00
16	0.80	0.35	0.70	1.00
17	0.85	0.35	0.70	1.00
18	0.90	0.35	0.70	1.00
19	0.95	0.35	0.70	1.00
20	1.00	0.35	0.70	1.00
21	1.00	0.35	0.70	1.00
22	1.00	0.35	0.70	1.00
23 and onwards	1.00	0.35	0.70	1.00
$\mathbf{A}_i$	13.085321	10.60642701	14	18.3258941
$\mathbf{A}_i$*T	0.65427	0.530321351	0.7	0.91629471

Table 12. Portfolio 2.

Period	Attribute 1	Attribute 2	Attribute 3	Attribute 4
1	0.00	1.00	0.60	1.00
2	0.10	0.94	0.60	0.90
3	0.20	0.88	0.60	0.80
4	0.30	0.82	0.60	0.70
5	0.40	0.76	0.60	0.60
6	0.50	0.70	0.60	0.50
7	0.60	0.64	0.60	0.40
8	0.70	0.58	0.60	0.30
9	0.80	0.52	0.60	0.40
10	0.90	0.52	0.60	0.50
11	1.00	0.52	0.60	0.60
12	1.00	0.52	0.60	0.70
13	1.00	0.52	0.60	0.80
14	1.00	0.52	0.60	0.90
15	1.00	0.52	0.60	1.00
16	1.00	0.52	0.60	1.00
17	1.00	0.52	0.60	1.00
18	1.00	0.52	0.60	1.00
19	1.00	0.52	0.60	1.00
20	1.00	0.52	0.60	1.00
21	1.00	0.52	0.60	1.00
22	1.00	0.52	0.60	1.00
23 and onwards	1.00	0.52	0.60	1.00
$\mathbf{A}_i$	15.44	12.24	12.00	16.65
$\mathbf{A}_i$***T**	0.772173493	0.612207234	0.6	0.832589415

Table 13. A (necessarily incomplete) tabular representation of the intertemporal fuzzy soft set in Example 6. Each cell actually contains an infinite sequence.

	Attribute 1	Attribute 2	Attribute 3	Attribute 4
P_1	$(0.05, 0.10, 0.15, 0.20, 0.25, \ldots)$	$(1.00, 0.95, 0.90, 0.85, 0.80, \ldots)$	$(0.70, 0.70, 0.70, 0.70, \ldots)$	$(1.00, 0.95, 0.90, 0.85, 0.80, \ldots)$
P_2	$(0.00, 0.10, 0.20, 0.30, 0.40, \ldots)$	$(1.00, 0.94, 0.88, 0.82, 0, 76, \ldots)$	$(0.60, 0.60, 0.60, 0.60, \ldots)$	$(1.00, 0.90, 0.80, 0.70, 0.60, \ldots)$

Table 14. Tabular representation of the reduced fuzzy soft set (S, P) in Example 6.

	Attribute 1	Attribute 2	Attribute 3	Attribute 4
P_1	0.65	0.53	0.70	0.92
P_2	0.77	0.61	0.60	0.83

Table 15. Computing the Comparison table and scores of the reduced fuzzy soft set (S, P) through Algorithm 1.

	Attribute 1	Attribute 2	Attribute 3	Attribute 4
Diff. memberships P_1 vs. P_2	−0.12	−0.08	0.10	0.08
Diff. memberships P_2 vs. P_1	0.12	0.08	−0.10	−0.08
M_j	0.77	0.61	0.70	0.92

	P_1	P_2
P_1	0	0.29
P_2	0.23	0

	Row-Sum (R_i)	Column-Sum (T_i)	Score (S_i)
P_1	0.29	0.23	0.06
P_2	0.23	0.29	−0.06

6. Discussion and Concluding Remarks

We have designed a pioneering framework for making choices in soft computing models. For the first time in this broad area, we have considered the situation where the consequences of a decision extend along an unlimited number of periods, such as a financial investment or a social project. Existing models universally refer to a finite framework, hence they are incapable of dealing with these practical issues. We have set the grounds for a correct extension to this critical aspect of decision making.

In this paper, our reference model for uncertainty has been fuzzy soft sets, which allows for fuzzy parameterized description of the alternatives in terms of a list of attributes. We have opted for working with this environment because fuzzy soft sets are especially amenable for decision making, with plenty of interesting approaches in the literature. Future research should expand the scope of the intertemporal analysis that we have founded to other frameworks such as incomplete fuzzy soft sets, rough sets, hesitant fuzzy sets, or hesitant fuzzy soft sets among many others. Whatever the selected format for the input data, when choices extend along an infinite number of periods, the fundamental roadmap for making decisions has been established in this paper.

Obviously, it may also be possible to approach the exact problem that we have described in this paper by alternative methodologies to improve the performance of our proposal, or make it more faithfully adapted to the circumstances of the problem under inspection.

Overall, we believe that the intertemporal modelization may become a thriving area of research in the extended theories of fuzziness, vagueness and uncertainty.

Author Contributions: Conceptualization of the model, J.C.R.A.; Methodology, Validation and Formal Analysis, J.C.R.A. and M.J.M.T.; Writing—Original Draft Preparation, Review and Editing, J.C.R.A. and M.J.M.T.; and Funding Acquisition, M.J.M.T.

Funding: The research of Muñoz Torrecillas was funded by the Spanish Ministry of Economy and Competitiveness and the European Regional Development Fund ERDF/FEDER-UE (National R&D Project ECO2015-66504 and National R&D Project DER2016-76053-R).

Acknowledgments: We are grateful to three anonymous referees and the Academic Editor for their detailed comments on the original draft of this paper.

Conflicts of Interest: The authors declare no conflict of interest.

Abbreviations

The following abbreviations are used in this manuscript:

DU	Discounted Utility
FSS	Fuzzy soft set
NPV	Net Present Value
ItFSS	Intertemporal fuzzy soft set

References

1. Chabris, C.; Laibson, D.; Schuldt, J. Intertemporal Choice. In *Palgrave Dictionary of Economics*; Palgrave Macmillan: London, UK, 2008.
2. Loewenstein, G.; Thaler, R.H. Anomalies: Intertemporal Choice. *J. Econ. Perspect.* **1989**, *3*, 181–193. [CrossRef]
3. Price, C. Investment, reinvestment, and the social discount rate for forestry. *For. Ecol. Manag.* **1988**, *24*, 293–310. [CrossRef]
4. Dietz, S.; Asheim, G.B. Climate Policy Under Sustainable Discounted Utilitarianism. *J. Environ. Econ. Manag.* **2012**, *63*, 321–335. [CrossRef]
5. Hausman, J.A. Individual Discount Rates and the Purchase and Utilization of Energy-Using Durables. *Bell J. Econ.* **1979**, *10*, 33–54. [CrossRef]
6. Rambaud, S.C.; Torrecillas, M.J.M.; Takahashi, T. Observed and Normative Discount Functions in Addiction and other Diseases. *Front. Pharmacol.* **2017**, *8*, 416. [CrossRef] [PubMed]

7. Field, M.; Santarcangelo, M.; Sumnall, H.; Goudie, A.; Cole, J. Delay discounting and the behavioural economics of cigarette purchases in smokers: The effects of nicotine deprivation. *Psychopharmacology* **2006**, *186*, 255–263. [CrossRef] [PubMed]
8. Paserman, M. Job Search and Hyperbolic Discounting: Structural Estimation and Policy Evaluation. *Econ. J.* **2008**, *118*, 1418–1452. [CrossRef]
9. O'Donoghue, T.; Rabin, M. Doing it now or later. *Am. Econ. Rev.* **1999**, *89*, 103–124. [CrossRef]
10. Chapman, G.B. Temporal discounting and utility for health and money. *J. Exp. Psychol. Learn. Mem. Cogn.* **1996**, *22*, 771–791. [CrossRef] [PubMed]
11. Samuelson, P.A. A Note on Measurement of Utility. *Rev. Econ. Stud.* **1937**, *4*, 155–161. [CrossRef]
12. Koopmans, T.C. Stationary Ordinal Utility and Impatience. *Econometrica* **1960**, *28*, 287–309. [CrossRef]
13. Berns, G.S.; Laibson, D.; Loewenstein, G. Intertemporal choice: toward an integrative framework. *Trends Cogn. Sci.* **2007**, *11*, 482–488. [CrossRef] [PubMed]
14. Rambaud, S.C.; Torrecillas, M.J.M. An Analysis of the Anomalies in Traditional Discounting Models. *Int. J. Psychol. Psychol. Ther.* **2004**, *4*, 105–128.
15. Thaler, R. Some empirical evidence on dynamic inconsistency. *Econ. Lett.* **1981**, *8*, 201–207. [CrossRef]
16. Ainslie, G. Specious reward: A behavioral theory of impulsiveness and impulse control. *Econ. Lett.* **1975**, *82*, 463–496. [CrossRef]
17. Commons, M.L.; Mazur, J.E.; Nevin, J.A.E.; Rachlin, H.E. *Quantitative Analyses of Behavior: Vol. 5. The Effect of Delay and of Intervening Events on Reinforcement Value*; Chapter An Adjusting Procedure for Studying Delayed Reinforcement; Lawrence Erlbaum Associates, Inc.: Hillsdale, NJ, USA, 1987; pp. 55–73.
18. Rachlin, H. *Judgment, Decision, and Choice: A Cognitive/Behavioral Synthesis*; W. H. Freeman & Co.: New York, NY, USA, 1989.
19. Cajueiro, D.O. A note on the relevance of the q-exponential function in the context of intertemporal choices. *Phys. A Stat. Mech. Its Appl.* **2006**, *364*, 385–388. [CrossRef]
20. Rambaud, S.C.; Torrecillas, M.J.M. A generalization of the q-exponential discounting function. *Phys. A Stat. Mech. Its Appl.* **2013**, *392*, 3045–3050. [CrossRef]
21. Takahashi, T. A comparison of intertemporal choices for oneself versus someone else based on Tsallis' statistics. *Phys. A Stat. Mech. Its Appl.* **2007**, *385*, 637–644. [CrossRef]
22. Chichilnisky, G.; Heal, G. Social choice with infinite populations: construction of a rule and impossibility results. *Soc. Choice Welf.* **1997**, *14*, 303–318. [CrossRef]
23. Arrow, K.; Cropper, M.; Gollier, C.; Groom, B.; Heal, G.; Newell, R.; Nordhaus, W.; Pindyck, R.; Pizer, W.; Portney, P.; et al. Determining Benefits and Costs for Future Generations. *Science* **2013**, *341*, 349–350. [CrossRef] [PubMed]
24. Her Majesty's Treasury. *The Green Book: Appraisal and Evaluation in Central Government*; Her Majesty's Stationery Office: London, UK, 2003.
25. Baumstark, K.; Hirtzman, P.; Lebegue, D. *Revision du taux D'actualisation des Investissements Publics*; Commissariat general au Plan, Office International de l'Eau: Paris, France, 2005.
26. European Commission. Guide to Cost Benefit Analysis of Investment Projects: Economic Appraisal Tool for Cohesion Policy 2014–2020. Available online: http://ec.europa.eu/regional_policy/sources/docgener/studies/pdf/cba_guide.pdf (accessed on 18 August 2018).
27. Zadeh, L. Fuzzy sets. *Inf. Control* **1965**, *8*, 338–353. [CrossRef]
28. Mardani, A.; Jusoh, A.; Zavadskas, E.K. Fuzzy multiple criteria decision-making techniques and applications—Two decades review from 1994 to 2014. *Expert Syst. Appl.* **2015**, *42*, 4126–4148. [CrossRef]
29. Tanino, T. Fuzzy preference orderings in group decision making. *Fuzzy Sets Syst.* **1984**, *12*, 117–131. [CrossRef]
30. Fodor, J.; Roubens, M. *Fuzzy Preference Modelling and Multicriteria Decision Support*; Kluwer Academic Publishers: Alphen aan den Rijn, The Netherlands, 1994.
31. Atanassov, K. Intuitionistic fuzzy sets. *Fuzzy Sets Syst.* **1986**, *20*, 87–96. [CrossRef]
32. Atanassov, K.T. More on intuitionistic fuzzy sets. *Fuzzy Sets Syst.* **1989**, *33*, 37–45. [CrossRef]
33. Chen, S.M.; Cheng, S.H.; Chiou, C.H. Fuzzy multiattribute group decision making based on intuitionistic fuzzy sets and evidential reasoning methodology. *Inf. Fusion* **2016**, *27*, 215–227. [CrossRef]
34. Wei, G. Some induced geometric aggregation operators with intuitionistic fuzzy information and their application to group decision making. *Appl. Soft Comput.* **2010**, *10*, 423–431. [CrossRef]

35. De Miguel, L.; Bustince, H.; Fernández, J.; Induráin, E.; Kolesárová, A.; Mesiar, R. Construction of admissible linear orders for interval-valued Atanassov intuitionistic fuzzy sets with an application to decision making. *Inf. Fusion* **2016**, *27*, 189–197. [CrossRef]
36. Peng, X.; Selvachandran, G. Pythagorean fuzzy set: State of the art and future directions. *Artif. Intell. Rev.* **2017**. [CrossRef]
37. Peng, X.; Yang, Y. Fundamental Properties of Interval-Valued Pythagorean Fuzzy Aggregation Operators. *Int. J. Intell. Syst.* **2016**, *31*, 444–487. [CrossRef]
38. Peng, X. New Operations for Interval-valued Pythagorean Fuzzy Set. *Sci. Iran.* **2018**. [CrossRef]
39. Torra, V. Hesitant fuzzy sets. *Int. J. Intell. Syst.* **2010**, *25*, 529–539. [CrossRef]
40. Rodríguez, R.; Martínez, L.; Torra, V.; Xu, Z.; Herrera, F. Hesitant fuzzy sets: state of the art and future directions. *Int. J. Intell. Syst.* **2014**, *29*, 495–524. [CrossRef]
41. Alcantud, J.C.R.; Torra, V. Decomposition theorems and extension principles for hesitant fuzzy sets. *Inf. Fusion* **2018**, *41*, 48–56. [CrossRef]
42. Pawlak, Z. Rough sets. *Int. J. Comput. Inf. Sci.* **1982**, *15*, 341–356. [CrossRef]
43. Molodtsov, D. Soft set theory—First results. *Comput. Math. Appl.* **1999**, *37*, 19–31. [CrossRef]
44. Feng, Q.; Zhou, Y. Soft discernibility matrix and its applications in decision making. *Appl. Soft Comput.* **2014**, *24*, 749–756. [CrossRef]
45. Alcantud, J. Some formal relationships among soft sets, fuzzy sets, and their extensions. *Int. J. Approx. Reason.* **2016**, *68*, 45–53. [CrossRef]
46. Ali, M. A note on soft sets, rough soft sets and fuzzy soft sets. *Appl. Soft Comput.* **2011**, *11*, 3329–3332.
47. Feng, F.; Li, C.; Davvaz, B.; Ali, M. Soft sets combined with fuzzy sets and rough sets: A tentative approach. *Soft Comput.* **2010**, *14*, 899–911. [CrossRef]
48. Feng, F.; Liu, X.; Leoreanu-Fotea, V.; Jun, Y.B. Soft sets and soft rough sets. *Inf. Sci.* **2011**, *181*, 1125–1137. [CrossRef]
49. Maji, P.; Biswas, R.; Roy, A. Soft set theory. *Comput. Math. Appl.* **2003**, *45*, 555–562. [CrossRef]
50. Aktaş, H.; Çağman, N. Soft sets and soft groups. *Inf. Sci.* **2007**, *177*, 2726–2735. [CrossRef]
51. Khameneh, A.Z.; Kılıçman, A. Multi-attribute decision-making based on soft set theory: A systematic review. *Soft Comput.* **2018**. [CrossRef]
52. Zhan, J.; Alcantud, J.C.R. A survey of parameter reduction of soft sets and corresponding algorithms. *Artif. Intell. Rev.* **2017**. [CrossRef]
53. Maji, P.; Biswas, R.; Roy, A. Fuzzy soft sets. *J. Fuzzy Math.* **2001**, *9*, 589–602.
54. Khameneh, A.Z.; Kılıçman, A. Parameter Reduction of Fuzzy Soft Sets: An Adjustable Approach Based on the Three-Way Decision. *Int. J. Fuzzy Syst.* **2018**, *20*, 928–942. [CrossRef]
55. Wang, F.; Li, X.; Chen, X. Hesitant Fuzzy Soft Set and Its Applications in Multicriteria Decision Making. *J. Appl. Math.* **2014**, *2014*, 643785. [CrossRef]
56. Alcantud, J.C.R.; Santos-García, G. A new criterion for soft set based decision making problems under incomplete information. *Int. J. Comput. Intell. Syst.* **2017**, *10*, 394–404. [CrossRef]
57. Han, B.H.; Li, Y.; Liu, J.; Geng, S.; Li, H. Elicitation criterions for restricted intersection of two incomplete soft sets. *Knowl. Based. Syst.* **2014**, *59*, 121–131. [CrossRef]
58. Zou, Y.; Xiao, Z. Data analysis approaches of soft sets under incomplete information. *Knowl. Based. Syst.* **2008**, *21*, 941–945. [CrossRef]
59. Deng, T.; Wang, X. An object-parameter approach to predicting unknown data in incomplete fuzzy soft sets. *Appl. Math. Model.* **2013**, *37*, 4139–4146. [CrossRef]
60. Liu, Y.; Qin, K.; Rao, C.; Mahamadu, M.A. Object-parameter Approaches to Predicting Unknown Data in an Incomplete Fuzzy Soft Set. *Int. J. Appl. Math. Comput. Sci.* **2017**, *27*, 157–167. [CrossRef]
61. Roy, A.; Maji, P. A fuzzy soft set theoretic approach to decision making problems. *J. Comput. Appl. Math.* **2007**, *203*, 412–418. [CrossRef]
62. Alcantud, J.C.R. A novel algorithm for fuzzy soft set based decision making from multiobserver input parameter data set. *Inf. Fusion* **2016**, *29*, 142–148. [CrossRef]
63. Liu, Y.; Qin, K.; Martínez, L. Improving decision making approaches based on fuzzy soft sets and rough soft sets. *Appl. Soft Comput.* **2018**, *65*, 320–332. [CrossRef]
64. Feng, F.; Jun, Y.; Liu, X.; Li, L. An adjustable approach to fuzzy soft set based decision making. *J. Comput. Appl. Math.* **2010**, *234*, 10–20. [CrossRef]

65. Kong, Z.; Gao, L.; Wang, L. Comment on "A fuzzy soft set theoretic approach to decision making problems". *J. Comput. Appl. Math.* **2009**, *223*, 540–542. [CrossRef]
66. Liu, Z.; Qin, K.; Pei, Z. A Method for Fuzzy Soft Sets in Decision-Making Based on an Ideal Solution. *Symmetry* **2017**, *9*, 246. [CrossRef]
67. Feng, Q.; Guo, X. A Novel Approach to Fuzzy Soft Set-Based Group Decision-Making. *Complexity* **2018**, *2018*, 2501489. [CrossRef]
68. Peng, X.; Dai, J.; Yuan, H. Interval-valued Fuzzy Soft Decision Making Methods Based on MABAC, Similarity Measure and EDAS. *Fundam. Inform.* **2017**, *152*, 373–396. [CrossRef]
69. Park, J.H.; Kwun, Y.C.; Son, M.J. A generalized intuitionistic fuzzy soft set theoretic approach to decision making problems. *Int. J. Fuzzy Log. Intell. Syst.* **2011**, *11*, 71–76. [CrossRef]
70. Park, J. Operations on generalized intuitionistic fuzzy soft sets. *Int. J. Fuzzy Log. Intell. Syst.* **2011**, *11*, 184–189. [CrossRef]
71. Zhan, J.; Wang, Q. Certain types of soft coverings based rough sets with applications. *Int. J. Mach. Learn. Cybern.* **2018**. [CrossRef]
72. Zhan, J.; Alcantud, J.C.R. A novel type of soft rough covering and its application to multicriteria group decision making. *Artif. Intell. Rev.* **2018**. [CrossRef]
73. Ma, X.; Zhan, J.; Ali, M.I.; Mehmood, N. A survey of decision making methods based on two classes of hybrid soft set models. *Artif. Intell. Rev.* **2018**, *49*, 511–529. [CrossRef]
74. Fatimah, F.; Rosadi, D.; Hakim, R.F.; Alcantud, J.C.R. Probabilistic soft sets and dual probabilistic soft sets in decision-making. *Neural Comput. Appl.* **2017**. [CrossRef]
75. Maji, P.; Biswas, R.; Roy, A. An application of soft sets in a decision making problem. *Comput. Math. Appl.* **2002**, *44*, 1077–1083. [CrossRef]
76. Çağman, N.; Enginoğlu, S. Soft set theory and uni-int decision making. *Eur. J. Oper. Res.* **2010**, *207*, 848–855. [CrossRef]
77. Golabi, K.; Kirkwood, C.W.; Sicherman, A. Selecting a Portfolio of Solar Energy Projects Using Multiattribute Preference Theory. *Manag. Sci.* **1981**, *27*, 174–189. [CrossRef]
78. Liu, K.F.R.; Yu, C.W. Integrating case-based and fuzzy reasoning to qualitatively predict risk in an environmental impact assessment review. *Environ. Model. Softw.* **2009**, *24*, 1241–1251. [CrossRef]
79. Christoffersen, P.F. *Elements of Financial Risk Management*; Academic Press: Cambridge, MA, USA, 2012.
80. Alcantud, J.C.R.; Mathew, T.J. Separable fuzzy soft sets and decision making with positive and negative attributes. *Appl. Soft Comput.* **2017**, *59*, 586–595. [CrossRef]
81. Alcantud, J.C.R. Inequality averse criteria for evaluating infinite utility streams: The impossibility of Weak Pareto. *J. Econ. Theory* **2012**, *147*, 353–363. [CrossRef]
82. Alcantud, J.C.R.; García-Sanz, M.D. Paretian evaluation of infinite utility streams: An egalitarian criterion. *Econ. Lett.* **2010**, *106*, 209–211. [CrossRef]
83. Basu, K.; Mitra, T. Aggregating Infinite Utility Streams with Intergenerational Equity: The Impossibility of Being Paretian. *Econometrica* **2003**, *71*, 1557–1563. [CrossRef]
84. Ramsey, F.P. A Mathematical Theory of Saving. *Econ. J.* **1928**, *38*, 543–559. [CrossRef]

symmetry

MDPI

Article

A Three-Dimensional Constrained Ordered Weighted Averaging Aggregation Problem with Lower Bounded Variables

Yi-Fang Chen and Hui-Chin Tang *

Department of Industrial Engineering and Management, National Kaohsiung University of Science and Technology, Kaohsiung 80778, Taiwan; 1100403115@gm.kuas.edu.tw

* Correspondence: tang@nkust.edu.tw; Tel.: +886-9-12768867

Received: 25 July 2018; Accepted: 9 August 2018; Published: 13 August 2018

Abstract: We consider the constrained ordered weighted averaging (OWA) aggregation problem with a single constraint and lower bounded variables. For the three-dimensional constrained OWA aggregation problem with lower bounded variables, we present four types of solution depending on the number of zero elements. According to the computerized experiment we perform, the lower bounds can affect the solution types, thereby affecting the optimal solution of the three-dimensional constrained OWA aggregation problem with lower bounded variables.

Keywords: ordered weighted averaging (OWA) operators; constrained OWA aggregation problem; lower bounded variables

1. Introduction

An ordered weighted averaging (OWA) operator, proposed by Yager [1], is a general class of parametric aggregation operators that appears in many applications such as control, decision making, expert systems, fuzzy system, neural networks, regression analysis and risk analysis [2–6]. A citation-based survey of the literature in all types of optimization problems associated to OWA operators can be found in [7]. In 1996, Yager [8] investigated the constrained OWA aggregation problem [8–15] which is concerned with an optimization problem with an OWA operator. In particular, for the constrained OWA aggregation problem with a single constraint on the sum of all variables, Yager [8] presented the optimal solutions for the three-dimensional case. Furthermore, Carlsson, Fullér and Majlender [9] proposed a simple algorithm for obtaining the optimal solutions for any dimensions. Recently, Coroianu and Fullér [10] presented the optimal solution for the constrained OWA aggregation problem with a single constraint and any coefficients. However, in most practical problems the variables are usually bounded. This paper considers the three-dimensional constrained OWA aggregation problem with lower bounded variables.

The organization of this paper is as follows. Section 2 briefly reviews the constrained OWA aggregation problem. Section 3 discusses the constrained OWA aggregation problem with the same lower bounds. Section 4 presents the solution behaviors of three-dimensional constrained OWA aggregation problems with lower bounded variables. Section 5 outlines the design of the experiment and evaluates the optimal solution behaviors of the three-dimensional constrained OWA aggregation problems with the lower bounded variables. Finally, some concluding remarks are presented.

2. Constrained Ordered Weighted Averaging (OWA) Aggregation Problem

An OWA operator of dimension n is a mapping $F : \mathcal{R}^n \rightarrow \mathcal{R}$ that associates a weighting vector $W = (w_1, w_2, \ldots, w_n)$ satisfying:

$$w_1 + w_2 + \ldots + w_n = 1,\ 0 \le w_i \le 1,\ i = 1, 2, \ldots, n$$

and such that:

$$F(x_1, x_2, \ldots, x_n) = \sum_{i=1}^{n} w_i y_i, \tag{1}$$

with y_i being the ith largest of $\{x_1, x_2, \ldots, x_n\}$.

Consider the following constrained OWA aggregation problem:

$$\begin{array}{c} \text{Max } W^T Y \\ \text{s.t.} \boldsymbol{AX} \leq \boldsymbol{b} \\ X \geq \boldsymbol{0} \end{array} \tag{2}$$

where the column vectors $\boldsymbol{X}$, $\boldsymbol{Y}$, $\boldsymbol{W}$ and $\boldsymbol{b}$, and the $m \times n$ matrix $\boldsymbol{A}$ are:

$$\boldsymbol{X} = \begin{bmatrix} x_1 \\ x_2 \\ \vdots \\ x_n \end{bmatrix}, \boldsymbol{Y} = \begin{bmatrix} y_1 \\ y_2 \\ \vdots \\ y_n \end{bmatrix}, \boldsymbol{W} = \begin{bmatrix} w_1 \\ w_2 \\ \vdots \\ w_n \end{bmatrix}, \boldsymbol{b} = \begin{bmatrix} b_1 \\ b_2 \\ \vdots \\ b_m \end{bmatrix}, \boldsymbol{A} = \begin{bmatrix} a_{11} & a_{12} & \cdots & a_{1n} \\ a_{21} & a_{22} & & a_{2n} \\ & \vdots & \ddots & \vdots \\ a_{m1} & a_{m2} & \cdots & a_{mn} \end{bmatrix}.$$

By introducing the $(n-1) \times n$ matrix:

$$\mathbf{G} = \begin{bmatrix} -1 & 1 & 0 & 0 & \cdots & 0 & 0 \\ 0 & -1 & 1 & 0 & & 0 & 0 \\ & \vdots & & & \ddots & \vdots & \\ 0 & 0 & 0 & 0 & \cdots & -1 & 1 \end{bmatrix}$$

and the column binary vectors $Z_i \in \{0,1\}^n, i = 1, 2, \ldots, n$, Yager [8] transformed the above non-linear programing problem to the following mixed integer linear programming (MIP) problem:

$$\begin{array}{l} \text{Max } W^T Y \\ \text{s.t.} \boldsymbol{AX} \leq \boldsymbol{b} \\ \quad \mathbf{GY} \leq 0 \\ \quad y_i \mathcal{I} - X - MZ_i \leq 0,\ i = 1, 2, \ldots, n-1 \\ \quad y_n \mathcal{I} - X \leq 0 \\ \quad \mathcal{I}^T Z_i \leq n - i,\ i = 1, 2, \ldots, n-1 \\ \quad Z_i \in \{0,1\}^n,\ i = 1, 2, \ldots, n-1 \\ \quad X \geq \boldsymbol{0} \end{array} \tag{3}$$

where M is a huge positive number and $\mathcal{I}$ is the column vector with all elements equal 1.

For the MIP (3), the number of constraints is:

$$m + n - 1 + n^2 + n - 1 = m + n^2 + 2n - 2,$$

and the number of variables is:

$$n + n + (n-1)n = n^2 + n.$$

In the literature, the constrained OWA aggregation problem with a single constraint on the sum of all variables is as follows:

$$\begin{aligned} &\text{Max } W^T Y \\ &\text{s.t.} \mathcal{I}^T \mathbf{X} \leq 1 \\ &\quad \mathbf{GY} \leq 0 \\ &\quad y_i \mathcal{I} - X - MZ_i \leq 0,\ i = 1, 2, \ldots, n-1 \\ &\quad y_n \mathcal{I} - X \leq 0 \\ &\quad \mathcal{I}^T Z_i \leq n - i,\ i = 1, 2, \ldots, n-1 \\ &\quad Z_i \in \{0,1\}^n,\ i = 1, 2, \ldots, n-1 \\ &\quad X \geq \mathbf{0}. \end{aligned} \tag{4}$$

If:

$$\mathbf{X}^* = \begin{bmatrix} x_1^* \\ x_2^* \\ \vdots \\ x_n^* \end{bmatrix}$$

is an optimal solution of (4), then,

$$\begin{bmatrix} x_{\sigma 1}^* \\ x_{\sigma 2}^* \\ \vdots \\ x_{\sigma n}^* \end{bmatrix}$$

is also the optimal solution, for some $\sigma \in S_n$, where S_n is the set of all permutations of the set $\{1, 2, \ldots, n\}$. To reduce the multiple solutions of the MIP (4), we introduce the following constraints:

$$Z_{i+1} \leq Z_i,\ i = 1,\ 2,\ \ldots,\ n-2.$$

by inspecting the jth element of the constraint $y_i \mathcal{I} - X - MZ_i \leq 0$,

$$y_i - x_j - MZ_{ij} \leq 0,$$

if $Z_{ij} = 0$, then:

$$y_i \leq x_j.$$

From the optimal solution:

$$\mathcal{I}^T Z_i = n - i \text{ and } \mathcal{I}^T Z_{i+1} = n - i - 1$$

it follows that:

$$Z_{i+1,j} = 0$$

so,

$$y_{i+1} \leq x_j.$$

If $Z_{ij} = 1$, then no restriction is imposed on y_i, it implies that:

$$y_{i+1} \leq x_j \text{ and } y_{i+1} > x_j$$

so $Z_{i+1,j} = 0$ or 1.

Therefore, the more efficient MIP is as follows:

$$\begin{aligned}
&\text{Max } W^T Y \\
&\text{s.t.} \mathcal{I}^T \mathbf{X} \leq 1 \\
&\quad \mathbf{GY} \leq 0 \\
&\quad y_i \mathcal{I} - X - MZ_i \leq 0,\ i = 1,2,\ldots,n-1 \\
&\quad y_n \mathcal{I} - X \leq 0 \\
&\quad \mathcal{I}^T Z_i \leq n - i,\ i = 1,2,\ldots,n-1 \\
&\quad Z_{i+1} \leq Z_i,\ i = 1,2,\ldots,n-2 \\
&\quad Z_i \in \{0,1\}^n,\ i = 1,2,\ldots,n-1 \\
&\quad X \geq \mathbf{0}.
\end{aligned} \tag{5}$$

3. Constrained OWA Aggregation Problem with the Same Lower Bounds

In most practical problems the variables are usually bounded. A typical variable x_i is bounded from below by l_i and from above by u_i, where $l_i < u_i$ and $i = 1, 2, \ldots, n$. If we let $u_i = \infty$, we get the following constrained OWA aggregation problem with lower bounded variables:

$$\begin{aligned}
&\text{Max } W^T Y \\
&\text{s.t.} \mathcal{I}^T \mathbf{X} \leq 1 \\
&\quad \mathbf{GY} \leq 0 \\
&\quad y_i \mathcal{I} - X - MZ_i \leq 0,\ i = 1,2,\ldots,n-1 \\
&\quad y_n \mathcal{I} - X \leq 0 \\
&\quad \mathcal{I}^T Z_i \leq n - i,\ i = 1,2,\ldots,n-1 \\
&\quad Z_{i+1} \leq Z_i,\ i = 1,2,\ldots,n-2 \\
&\quad Z_i \in \{0,1\}^n,\ i = 1,2,\ldots,n-1 \\
&\quad X \geq \mathbf{L}
\end{aligned} \tag{6}$$

where the column vector:

$$\mathbf{L} = \begin{bmatrix} l_1 \\ l_2 \\ \vdots \\ l_n \end{bmatrix}.$$

By using the change of variable:

$$X' = X - \mathbf{L}$$

the lower bound vector can be transformed into the zero vector. The constrained OWA aggregation problem with lower bounded variables is:

$$\begin{aligned}
&\text{Max } W^T Y \\
&\text{s.t.} \mathcal{I}^T \mathbf{X}' \leq 1 - \mathcal{I}^T \mathbf{L} \\
&\quad \mathbf{GY} \leq 0 \\
&\quad y_i \mathcal{I} - X' - MZ_i \leq \mathbf{L},\ i = 1,2,\ldots,n-1 \\
&\quad y_n \mathcal{I} - X \leq 0 \\
&\quad \mathcal{I}^T Z_i \leq n - i,\ i = 1,2,\ldots,n-1 \\
&\quad Z_{i+1} \leq Z_i,\ i = 1,2,\ldots,n-2 \\
&\quad Z_i \in \{0,1\}^n,\ i = 1,2,\ldots,n-1 \\
&\quad X' \geq \mathbf{0}
\end{aligned} \tag{7}$$

If $1-\mathcal{I}^T\mathbf{L}<0$, the constrained OWA aggregation problem has no feasible solution. If $1-\mathcal{I}^T\mathbf{L}=0$, the unique optimal solution is $X'^{*}=\mathbf{0}$, so:

$$X^{*}=\mathbf{L}.$$

It remains to discuss the case that $1-\mathcal{I}^T\mathbf{L}>0$. More precisely, the three dimensional constrained OWA aggregation problem with lower bounded variables is as follows:

$$\begin{aligned}
&\text{Max } F=w_1y_1+w_2y_2+w_3y_3\\
&\text{s.t. } x_1'+x_2'+x_3'\le 1-l_1-l_2-l_3\\
&\quad y_2-y_1\le 0\\
&\quad y_3-y_2\le 0\\
&\quad y_3-x_1'\le l_1\\
&\quad y_3-x_2'\le l_2\\
&\quad y_3-x_3'\le l_3\\
&\quad y_2-x_1'-MZ_{21}\le l_1\\
&\quad y_2-x_2'-MZ_{22}\le l_2\\
&\quad y_2-x_3'-MZ_{23}\le l_3\\
&\quad Z_{21}+Z_{22}+Z_{23}\le 1\\
&\quad y_1-x_1'-MZ_{11}\le l_1\\
&\quad y_1-x_2'-MZ_{12}\le l_2\\
&\quad y_1-x_3'-MZ_{13}\le l_3\\
&\quad Z_{11}+Z_{12}+Z_{13}\le 2\\
&\quad Z_{21}\le Z_{11}\\
&\quad Z_{22}\le Z_{12}\\
&\quad Z_{23}\le Z_{13}\\
&\quad x_1',x_2',x_3'\ge 0,\ Z_{21},Z_{22},Z_{23},\ Z_{11},Z_{12},Z_{13}\in\{0,1\}.
\end{aligned}\tag{8}$$

For the special case that the same lower bounds $l_i=l, i=1,\,2,\ldots,n$, by the observing that the ith largest $(x_{\sigma i})$ of $\{x_1,x_2,\ldots,x_n\}$ is the same variable of the ith largest $(x'_{\sigma i})$ of $\{x_1',x_2',\ldots,x_n'\}$, let:

$$x_i''=\frac{x_i'}{1-nl}$$

it follows that the optimal solution is the same as that of the constrained OWA aggregation problem [8]. We establish the main results described as follows:

Theorem 1. *Consider the three-dimensional constrained OWA aggregation problem (8).*

(a) *If* $w_1=\max\limits_{i=1,2,3} w_i$, *then the optimal solutions are* $X''^{*}=\begin{bmatrix}1\\0\\0\end{bmatrix},\begin{bmatrix}0\\1\\0\end{bmatrix}$ *or* $\begin{bmatrix}0\\0\\1\end{bmatrix}$, $X^{*}=\begin{bmatrix}1-2l\\l\\l\end{bmatrix}$, $\begin{bmatrix}l\\1-2l\\l\end{bmatrix}$ *or* $\begin{bmatrix}l\\l\\1-2l\end{bmatrix}$, $Y^{*}=\begin{bmatrix}1-2l\\l\\l\end{bmatrix}$ *and* $F=w_1+l-3w_1l$.

(b) *If* $w_2=\max\limits_{i=1,2,3} w_i$, *then the optimal solutions are* $X''^{*}=\begin{bmatrix}1/2\\1/2\\0\end{bmatrix},\begin{bmatrix}1/2\\0\\1/2\end{bmatrix}$ *or* $\begin{bmatrix}0\\1/2\\1/2\end{bmatrix}$, $X^{*}=\begin{bmatrix}(1-l)/2\\(1-l)/2\\l\end{bmatrix},\begin{bmatrix}(1-l)/2\\l\\(1-l)/2\end{bmatrix}$ *or* $\begin{bmatrix}l\\(1-l)/2\\(1-l)/2\end{bmatrix}$, $Y^{*}=\begin{bmatrix}(1-l)/2\\(1-l)/2\\l\end{bmatrix}$ *and* $F=$

$(1-w_3-l+3w_3l)/2$ *for* $w_1+w_2 \geq 2w_3$, *and* $X''^* = \begin{bmatrix} 1/3 \\ 1/3 \\ 1/3 \end{bmatrix}$, $X^* = \begin{bmatrix} 1/3 \\ 1/3 \\ 1/3 \end{bmatrix}$, $Y^* = \begin{bmatrix} 1/3 \\ 1/3 \\ 1/3 \end{bmatrix}$, *and* $F = 1/3$ *for* $w_1 + w_2 \leq 2w_3$.

(c) *If* $w_3 = \max\limits_{i=1,2,3} w_i$, *then the optimal solutions are* $X''^* = \begin{bmatrix} 1 \\ 0 \\ 0 \end{bmatrix}, \begin{bmatrix} 0 \\ 1 \\ 0 \end{bmatrix}$ *or* $\begin{bmatrix} 0 \\ 0 \\ 1 \end{bmatrix}$, $X^* = \begin{bmatrix} 1-2l \\ l \\ l \end{bmatrix}$, $\begin{bmatrix} l \\ 1-2l \\ l \end{bmatrix}$ *or* $\begin{bmatrix} l \\ l \\ 1-2l \end{bmatrix}$, $Y^* = \begin{bmatrix} 1-2l \\ l \\ l \end{bmatrix}$ *and* $F = w_1 + l - 3w_1 l$ *for* $w_2 + w_3 \leq 2w_1$, *and* $X''^* = \begin{bmatrix} 1/3 \\ 1/3 \\ 1/3 \end{bmatrix}$, $X^* = \begin{bmatrix} 1/3 \\ 1/3 \\ 1/3 \end{bmatrix}$, $Y^* = \begin{bmatrix} 1/3 \\ 1/3 \\ 1/3 \end{bmatrix}$ *and* $F = 1/3$ *for* $w_2 + w_3 \geq 2w_1$.

Proof. For the three-dimensional constrained OWA aggregation problem, three cases are considered. Firstly, if:

$$w_1 = \max_{i=1,2,3} w_i$$

the optimal solutions are:

$$X''^* = \begin{bmatrix} 1 \\ 0 \\ 0 \end{bmatrix}, \begin{bmatrix} 0 \\ 1 \\ 0 \end{bmatrix} \text{ or } \begin{bmatrix} 0 \\ 0 \\ 1 \end{bmatrix}$$

So

$$X^* = \begin{bmatrix} 1-2l \\ l \\ l \end{bmatrix}, \begin{bmatrix} l \\ 1-2l \\ l \end{bmatrix} \text{ or } \begin{bmatrix} l \\ l \\ 1-2l \end{bmatrix}, Y^* = \begin{bmatrix} 1-2l \\ l \\ l \end{bmatrix}$$

and the most favorable value is:

$$F = w_1 + l - 3w_1 l.$$

Secondly, if:

$$w_2 = \max_{i=1,2,3} w_i$$

two subcases are considered. If:

$$w_1 + w_2 \geq 2w_3,$$

then the optimal solutions are:

$$X''^* = \begin{bmatrix} 1/2 \\ 1/2 \\ 0 \end{bmatrix}, \begin{bmatrix} 1/2 \\ 0 \\ 1/2 \end{bmatrix} \text{ or } \begin{bmatrix} 0 \\ 1/2 \\ 1/2 \end{bmatrix}$$

so:

$$X^* = \begin{bmatrix} (1-l)/2 \\ (1-l)/2 \\ l \end{bmatrix}, \begin{bmatrix} (1-l)/2 \\ l \\ (1-l)/2 \end{bmatrix} \text{ or } \begin{bmatrix} l \\ (1-l)/2 \\ (1-l)/2 \end{bmatrix}, Y^* = \begin{bmatrix} (1-l)/2 \\ (1-l)/2 \\ l \end{bmatrix}$$

and the largest objective function value is:

$$F = (1 - w_3 - l + 3w_3 l)/2.$$

If:

$$w_1 + w_2 \le 2w_3,$$

then the optimal solutions are:

$$X''^{*} = \begin{bmatrix} 1/3 \\ 1/3 \\ 1/3 \end{bmatrix}$$

so:

$$X^{*} = \begin{bmatrix} 1/3 \\ 1/3 \\ 1/3 \end{bmatrix}, Y^{*} = \begin{bmatrix} 1/3 \\ 1/3 \\ 1/3 \end{bmatrix} \text{ and } F = 1/3$$

Finally, if:

$$w_3 = \max_{i=1,2,3} w_i$$

two subcases are considered. If:

$$w_2 + w_3 \le 2w_1,$$

then the optimal solutions are:

$$X''^{*} = \begin{bmatrix} 1 \\ 0 \\ 0 \end{bmatrix}, \begin{bmatrix} 0 \\ 1 \\ 0 \end{bmatrix} \text{ or } \begin{bmatrix} 0 \\ 0 \\ 1 \end{bmatrix}$$

so:

$$X^{*} = \begin{bmatrix} 1-2l \\ l \\ l \end{bmatrix}, \begin{bmatrix} l \\ 1-2l \\ l \end{bmatrix} \text{ or } \begin{bmatrix} l \\ l \\ 1-2l \end{bmatrix}, Y^{*} = \begin{bmatrix} 1-2l \\ l \\ l \end{bmatrix},$$

and:

$$F = w_1 + l - 3w_1 l.$$

If:

$$w_2 + w_3 \ge 2w_1,$$

then the optimal solutions are:

$$X''^{*} = \begin{bmatrix} 1/3 \\ 1/3 \\ 1/3 \end{bmatrix}$$

so:

$$X^{*} = \begin{bmatrix} 1/3 \\ 1/3 \\ 1/3 \end{bmatrix}, Y^{*} = \begin{bmatrix} 1/3 \\ 1/3 \\ 1/3 \end{bmatrix} \text{ and } F = 1/3. \square$$

4. Constrained OWA Aggregation Problem with Lower Bounded Variables

For simplicity, we consider the three-dimensional constrained OWA aggregation problem with lower bounded variables. From the optimal solution of the first constraint of the model (8):

$$x'_1 + x'_2 + x'_3 = 1 - l_1 - l_2 - l_3 \tag{9}$$

there are four types (I, II, III and IV) of (x'_1, x'_2, x'_3) depending on the number of zero elements. The number of zero elements is two for type I, one for types II and III, and zero for type III. The solutions of $\mathbf{X}' = (x'_1, x'_2, x'_3)$ and $Y = (y_1, y_2, y_3)$ for the three-dimensional constrained OWA aggregation problem with lower bounded variables (8) are described as follows:

Theorem 2. *Consider the three-dimensional constrained OWA aggregation problem with lower bounded variables (8). For type I solution, there are three forms* $(1-l_1-l_2-l_3,0,0)$, $(0,1-l_1-l_2-l_3,0)$, $(0,0,1-l_1-l_2-l_3)$ *for* $\mathbf{X}'$ *and six forms* $(1-l_2-l_3,l_2,l_3)$, $(1-l_2-l_3,l_3,l_2)$, $(1-l_1-l_3,l_1,l_3)$, $(1-l_1-l_3,l_3,l_1)$, $(1-l_1-l_2,l_1,l_2)$, $(1-l_1-l_2,l_2,l_1)$ *for* $\mathbf{Y}$*. For type II, there are three forms* $\left(\frac{1-2l_1-l_3}{2},\frac{1-2l_2-l_3}{2},0\right)$, $\left(\frac{1-2l_1-l_2}{2},0,\frac{1-l_2-2l_3}{2}\right)$, $\left(0,\frac{1-l_1-2l_2}{2},\frac{1-l_1-2l_3}{2}\right)$ *for* $\mathbf{X}'$ *and six forms* $\left(\frac{1-l_3}{2},\frac{1-l_3}{2},l_3\right)$, $\left(l_3,\frac{1-l_3}{2},\frac{1-l_3}{2}\right)$, $\left(\frac{1-l_2}{2},\frac{1-l_2}{2},l_2\right)$, $\left(l_2,\frac{1-l_2}{2},\frac{1-l_2}{2}\right)$, $\left(\frac{1-l_1}{2},\frac{1-l_1}{2},l_1\right)$, $\left(l_1,\frac{1-l_1}{2},\frac{1-l_1}{2}\right)$ *for* $\mathbf{Y}$*. For type III, there are six forms* $(l_3-l_1,1-l_2-2l_3,0)$, $(1-l_1-2l_3,l_3-l_2,0)$, $(l_2-l_1,0,1-2l_2-l_3)$, $(1-l_1-2l_2,0,l_2-l_3)$, $(0,l_1-l_2,1-2l_1-l_3)$, $(0,1-2l_1-l_2,l_1-l_3)$ *for* $\mathbf{X}'$ *and six forms* $(l_3,l_3,1-2l_3)$, $(1-2l_3,l_3,l_3)$, $(l_2,l_2,1-2l_2)$, $(1-2l_2,l_2,l_2)$, $(l_1,l_1,1-2l_1)$, $(1-2l_1,l_1,l_1)$ *for* $\mathbf{Y}$*. For type IV, there are only one form* $(1/3-l_1,1/3-l_2,1/3-l_3)$ *for* $\mathbf{X}'$ *and one form* $(1/3,1/3,1/3)$ *for* $\mathbf{Y}$*.*

Proof. For type I, the possible values of (x'_1, x'_2, x'_3) are:

$$(1-l_1-l_2-l_3,0,0),\ (0,1-l_1-l_2-l_3,0) \text{ and } (0,0,1-l_1-l_2-l_3)$$

For the case of $(x'_1,x'_2,x'_3) = (1-l_1-l_2-l_3,0,0)$, we have:

$$(x'_1,x'_2,x'_3) = (1-l_2-l_3,l_2,l_3)(y_1,y_2,y_3) = (1-l_2-l_3,l_2,l_3) \text{ or } (1-l_2-l_3,l_3,l_2).$$

For the case of $(y_1,y_2,y_3) = (1-l_2-l_3,l_2,l_3)$, if:

$$l_1+l_2+l_3 \le 1,\ l_2 \ge l_3 \text{ and } 2l_2+l_3 \le 1,$$

then:

$$(x_1,x_2,x_3) = (1-l_2-l_3,l_2,l_3) \text{ and } (y_1,y_2,y_3) = (1-l_2-l_3,l_2,l_3)$$

is solution of MIP (8) and the objective value is:

$$F = w_1 + l_2(-w_1+w_2) + l_3(-w_1+w_3).$$

Since $w_1+w_2+w_3 = 1$, we can express the objective value F in only two weights. Then the other three formats of F are:

$$F = w_1 + l_2(-w_1+w_2) + l_3(1-2w_1-w_2),$$

$$F = w_1 + l_2(1-2w_1-w_3) + l_3(-w_1+w_3)$$

and:

$$F = 1 - w_2 - w_3 + l_2(-1+2w_1+w_3) + l_3(-1+w_2+2w_3).$$

Among these four formats, the explicit format adopted is $F = w_1 + l_2(-w_1+w_2) + l_3(-w_1+w_3)$ which is the most compact one.

If:

$$l_1+l_2+l_3 \le 1,\ l_2 \le l_3 \text{ and } l_2+2l_3 \le 1,$$

then:

$$(x_1,x_2,x_3) = (1-l_2-l_3,l_2,l_3) \text{ and } (y_1,y_2,y_3) = (1-l_2-l_3,l_3,l_2)$$

is the solution of MIP (8) and the objective value is:

$$F = w_1 + l_2(-w_1+w_3) + l_3(-w_1+w_2).$$

In Table 1, we display the possible solutions (x'_1, x'_2, x'_3), (x_1, x_2, x_3), (y_1, y_2, y_3), F and the conditions for the different choices of the type I.

We now consider that the number of zero elements is one. The possible values of (x'_1, x'_2, x'_3) are:

$$(x'_1, x'_2, 0),\ (x'_1, 0, x'_3) \text{ and } (0, x'_2, x'_3)$$

For the case of $(x'_1, x'_2, 0)$, we have:

$$(x_1, x_2, x_3) = (x'_1 + l_1, x'_2 + l_2, l_3).$$

At optimal, the possible choices of (x'_1, x'_2, x'_3) are:

$$x'_1 + l_1 = x'_2 + l_2,\ x'_1 + l_1 = l_3 \text{ or } x'_2 + l_2 = l_3.$$

We choose $x'_1 + l_1 = x'_2 + l_2$ for type II, and $x'_1 + l_1 = l_3$ or $x'_2 + l_2 = l_3$ for type III. For $x'_1 + l_1 = x'_2 + l_2$, from (9), it follows that:

$$(x'_1, x'_2, x'_3) = (\frac{1 - 2l_1 - l_3}{2}, \frac{1 - 2l_2 - l_3}{2}, 0).$$

so:

$$(x_1, x_2, x_3) = (\frac{1 - l_3}{2}, \frac{1 - l_3}{2}, l_3) \text{ and } (y_1, y_2, y_3) = (\frac{1 - l_3}{2}, \frac{1 - l_3}{2}, l_3) \text{ or } (l_3, \frac{1 - l_3}{2}, \frac{1 - l_3}{2})$$

More precisely, if:

$$l_3 \leq 1/3,\ 2l_2 + l_3 \leq 1 \text{ and } 2l_1 + l_3 \leq 1,$$

then:

$$(x_1, x_2, x_3) = (\frac{1 - l_3}{2}, \frac{1 - l_3}{2}, l_3) \text{ and } (y_1, y_2, y_3) = (\frac{1 - l_3}{2}, \frac{1 - l_3}{2}, l_3)$$

is the solution of MIP (8) and the objective value is:

$$F = \frac{1 - w_3 - l_3 + 3l_3 w_3}{2}.$$

If:

$$l_3 \geq 1/3,\ 2l_2 + l_3 \leq 1 \text{ and } 2l_1 + l_3 \leq 1,$$

then:

$$(x_1, x_2, x_3) = (\frac{1 - l_3}{2}, \frac{1 - l_3}{2}, l_3) \text{ and } (y_1, y_2, y_3) = (l_3, \frac{1 - l_3}{2}, \frac{1 - l_3}{2})$$

is the solution of MIP (8) and the objective value is:

$$F = \frac{1 - w_1 - l_3 + 3l_3 w_1}{2}.$$

For other cases of $(x'_1, 0, x'_3)$, $(0, x'_2, x'_3)$, the solutions and conditions are displayed in Table 2.

For type III, we have two possible $x'_1 + l_1 = l_3$ or $x'_2 + l_2 = l_3$. For $x'_1 + l_1 = l_3$, from (9), it follows that if:

$$l_3 \geq l_1,\ l_3 \geq 1/3 \text{ and } l_2 + 2l_3 \leq 1$$

then:

$$(x'_1, x'_2, x'_3) = (l_3 - l_1, 1 - l_2 - 2l_3, 0),\ (x_1, x_2, x_3) = (l_3, 1 - 2l_3, l_3),\ (y_1, y_2, y_3) = (l_3, l_3, 1 - 2l_3) \text{ and}$$

$$F = w_3 + l_3 - 3l_3 w_3.$$

If:

$$l_3 \geq l_1,\ l_3 \leq 1/3 \text{ and } l_2 + 2l_3 \leq 1$$

then:

$$(x_1', x_2', x_3') = (l_3 - l_1, 1 - l_2 - 2l_3, 0), (x_1, x_2, x_3) = (l_3, 1 - 2l_3, l_3), (y_1, y_2, y_3) = (1 - 2l_3, l_3, l_3) \text{ and}$$

$$F = w_1 + l_3 - 3l_3 w_1.$$

For different choices of type III, detailed results are presented in Table 3.

For type IV, from (9), it follows that:

$$(x_1', x_2', x_3') = (1/3 - l_1, 1/3 - l_2, 1/3 - l_3).$$

So, if:

$$l_1 \leq 1/3,\ l_2 \leq 1/3 \text{ and } l_3 \leq 1/3$$

then the solution of MIP (8) is:

$$(x_1, x_2, x_3) = (1/3, 1/3, 1/3) \text{ and } (y_1, y_2, y_3) = (1/3, 1/3, 1/3) \text{ and } F = 1/3. \ \square$$

Table 1. The values of (x_1', x_2', x_3'), (x_1, x_2, x_3), (y_1, y_2, y_3), F and the conditions for type I.

Type	(x_1', x_2', x_3')	(x_1, x_2, x_3)	(y_1, y_2, y_3)	F	Conditions
I1	$(1 - l_1 - l_2 - l_3, 0, 0)$	$(1 - l_2 - l_3, l_2, l_3)$	$(1 - l_2 - l_3, l_2, l_3)$	$w_1 + l_2(-w_1 + w_2) + l_3(-w_1 + w_3)$	$l_1 + l_2 + l_3 \leq 1$, $l_2 \geq l_3$, $2l_2 + l_3 \leq 1$
I2	$(1 - l_1 - l_2 - l_3, 0, 0)$	$(1 - l_2 - l_3, l_2, l_3)$	$(1 - l_2 - l_3, l_3, l_2)$	$w_1 + l_2(-w_1 + w_3) + l_3(-w_1 + w_2)$	$l_1 + l_2 + l_3 \leq 1$, $l_2 \leq l_3$, $l_2 + 2l_3 \leq 1$
I3	$(0, 1 - l_1 - l_2 - l_3, 0)$	$(l_1, 1 - l_1 - l_3, l_3)$	$(1 - l_1 - l_3, l_1, l_3)$	$w_1 + l_1(-w_1 + w_2) + l_3(-w_1 + w_3)$	$l_1 + l_2 + l_3 \leq 1$, $l_1 \geq l_3$, $2l_1 + l_3 \leq 1$
I4	$(0, 1 - l_1 - l_2 - l_3, 0)$	$(l_1, 1 - l_1 - l_3, l_3)$	$(1 - l_1 - l_3, l_3, l_1)$	$w_1 + l_1(-w_1 + w_3) + l_3(-w_1 + w_2)$	$l_1 + l_2 + l_3 \leq 1$, $l_1 \leq l_3$, $l_1 + 2l_3 \leq 1$
I5	$(0, 0, 1 - l_1 - l_2 - l_3)$	$(l_1, l_2, 1 - l_1 - l_2)$	$(1 - l_1 - l_2, l_1, l_2)$	$w_1 + l_1(-w_1 + w_2) + l_2(-w_1 + w_3)$	$l_1 + l_2 + l_3 \leq 1$, $l_1 \geq l_2$, $2l_1 + l_2 \leq 1$
I6	$(0, 0, 1 - l_1 - l_2 - l_3)$	$(l_1, l_2, 1 - l_1 - l_2)$	$(1 - l_1 - l_2, l_2, l_1)$	$w_1 + l_1(-w_1 + w_3) + l_2(-w_1 + w_2)$	$l_1 + l_2 + l_3 \leq 1$, $l_1 \leq l_2$, $l_1 + 2l_2 \leq 1$

Table 2. The values of (x_1', x_2', x_3'), (x_1, x_2, x_3), (y_1, y_2, y_3), F and the conditions for type II.

Type	(x_1', x_2', x_3')	(x_1, x_2, x_3)	(y_1, y_2, y_3)	F	Conditions
II1	$(\frac{1-2l_1-l_3}{2}, \frac{1-2l_2-l_3}{2}, 0)$	$(\frac{1-l_3}{2}, \frac{1-l_3}{2}, l_3)$	$(\frac{1-l_3}{2}, \frac{1-l_3}{2}, l_3)$	$\frac{1 - w_3 - l_3 + 3l_3 w_3}{2}$	$l_3 \leq 1/3, 2l_2 + l_3 \leq 1$, $2l_1 + l_3 \leq 1$
II2	$(\frac{1-2l_1-l_3}{2}, \frac{1-2l_2-l_3}{2}, 0)$	$(\frac{1-l_3}{2}, \frac{1-l_3}{2}, l_3)$	$(l_3, \frac{1-l_3}{2}, \frac{1-l_3}{2})$	$\frac{1 - w_1 - l_3 + 3l_3 w_1}{2}$	$l_3 \geq 1/3, 2l_2 + l_3 \leq 1$, $2l_1 + l_3 \leq 1$
II3	$(\frac{1-2l_1-l_2}{2}, 0, \frac{1-l_2-2l_3}{2})$	$(\frac{1-l_2}{2}, l_2, \frac{1-l_2}{2})$	$(\frac{1-l_2}{2}, \frac{1-l_2}{2}, l_2)$	$\frac{1 - w_3 - l_2 + 3l_2 w_3}{2}$	$l_2 \leq 1/3, 2l_1 + l_2 \leq 1$, $l_2 + 2l_3 \leq 1$
II4	$(\frac{1-2l_1-l_2}{2}, 0, \frac{1-l_2-2l_3}{2})$	$(\frac{1-l_2}{2}, l_2, \frac{1-l_2}{2})$	$(l_2, \frac{1-l_2}{2}, \frac{1-l_2}{2})$	$\frac{1 - w_1 - l_2 + 3l_2 w_1}{2}$	$l_2 \geq 1/3, 2l_1 + l_2 \leq 1$, $l_2 + 2l_3 \leq 1$
II5	$(0, \frac{1-l_1-2l_2}{2}, \frac{1-l_1-2l_3}{2})$	$(l_1, \frac{1-l_1}{2}, \frac{1-l_1}{2})$	$(\frac{1-l_1}{2}, \frac{1-l_1}{2}, l_1)$	$\frac{1 - w_3 - l_1 + 3l_1 w_3}{2}$	$l_1 \leq 1/3, l_1 + 2l_2 \leq 1$, $l_1 + 2l_3 \leq 1$
II6	$(0, \frac{1-l_1-2l_2}{2}, \frac{1-l_1-2l_3}{2})$	$(l_1, \frac{1-l_1}{2}, \frac{1-l_1}{2})$	$(l_1, \frac{1-l_1}{2}, \frac{1-l_1}{2})$	$\frac{1 - w_1 - l_1 + 3l_1 w_1}{2}$	$l_1 \geq 1/3, l_1 + 2l_2 \leq 1$, $l_1 + 2l_3 \leq 1$

Table 3. The values of (x'_1, x'_2, x'_3), (x_1, x_2, x_3), (y_1, y_2, y_3), F and the conditions for type III.

Type	(x'_1, x'_2, x'_3)	(x_1, x_2, x_3)	(y_1, y_2, y_3)	F	Conditions
III1	$(l_3 - l_1, 1 - l_2 - 2l_3, 0)$	$(l_3, 1 - 2l_3, l_3)$	$(l_3, l_3, 1 - 2l_3)$	$w_3 + l_3 - 3l_3 w_3$	$l_3 \geq l_1, l_3 \geq 1/3, l_2 + 2l_3 \leq 1$
III2	$(l_3 - l_1, 1 - l_2 - 2l_3, 0)$	$(l_3, 1 - 2l_3, l_3)$	$(1 - 2l_3, l_3, l_3)$	$w_1 + l_3 - 3l_3 w_1$	$l_3 \geq l_1, l_3 \leq 1/3, l_2 + 2l_3 \leq 1$
III3	$(1 - l_1 - 2l_3, l_3 - l_2, 0)$	$(1 - 2l_3, l_3, l_3)$	$(l_3, l_3, 1 - 2l_3)$	$w_3 + l_3 - 3l_3 w_3$	$l_3 \geq l_2, l_3 \geq 1/3, l_1 + 2l_3 \leq 1$
III4	$(1 - l_1 - 2l_3, l_3 - l_2, 0)$	$(1 - 2l_3, l_3, l_3)$	$(1 - 2l_3, l_3, l_3)$	$w_1 + l_3 - 3l_3 w_1$	$l_3 \geq l_2, l_3 \leq 1/3, l_1 + 2l_3 \leq 1$
III5	$(l_2 - l_1, 0, 1 - 2l_2 - l_3)$	$(l_2, l_2, 1 - 2l_2)$	$(l_2, l_2, 1 - 2l_2)$	$w_3 + l_2 - 3l_2 w_3$	$l_2 \geq l_1, l_2 \geq 1/3, 2l_2 + l_3 \leq 1$
III6	$(l_2 - l_1, 0, 1 - 2l_2 - l_3)$	$(l_2, l_2, 1 - 2l_2)$	$(1 - 2l_2, l_2, l_2)$	$w_1 + l_2 - 3l_2 w_1$	$l_2 \geq l_1, l_2 \leq 1/3, 2l_2 + l_3 \leq 1$
III7	$(1 - l_1 - 2l_2, 0, l_2 - l_3)$	$(1 - 2l_2, l_2, l_2)$	$(l_2, l_2, 1 - 2l_2)$	$w_3 + l_2 - 3l_2 w_3$	$l_2 \geq l_3, l_2 \geq 1/3, l_1 + 2l_2 \leq 1$
III8	$(1 - l_1 - 2l_2, 0, l_2 - l_3)$	$(1 - 2l_2, l_2, l_2)$	$(1 - 2l_2, l_2, l_2)$	$w_1 + l_2 - 3l_2 w_1$	$l_2 \geq l_3, l_2 \leq 1/3, l_1 + 2l_2 \leq 1$
III9	$(0, l_1 - l_2, 1 - 2l_1 - l_3)$	$(l_1, l_1, 1 - 2l_1)$	$(l_1, l_1, 1 - 2l_1)$	$w_3 + l_1 - 3l_1 w_3$	$l_2 \leq l_1, l_1 \geq 1/3, 2l_1 + l_3 \leq 1$
III10	$(0, l_1 - l_2, 1 - 2l_1 - l_3)$	$(l_1, l_1, 1 - 2l_1)$	$(1 - 2l_1, l_1, l_1)$	$w_1 + l_1 - 3l_1 w_1$	$l_2 \leq l_1, l_1 \leq 1/3, 2l_1 + l_3 \leq 1$
III11	$(0, 1 - 2l_1 - l_2, l_1 - l_3)$	$(l_1, 1 - 2l_1, l_1)$	$(l_1, l_1, 1 - 2l_1)$	$w_3 + l_1 - 3l_1 w_3$	$l_3 \leq l_1, l_1 \geq 1/3, 2l_1 + l_2 \leq 1$
III12	$(0, 1 - 2l_1 - l_2, l_1 - l_3)$	$(l_1, 1 - 2l_1, l_1)$	$(1 - 2l_1, l_1, l_1)$	$w_1 + l_1 - 3l_1 w_1$	$l_3 \leq l_1, l_1 \leq 1/3, 2l_1 + l_2 \leq 1$

For the three-dimensional constrained OWA aggregation problem with lower bounded variables (8), there are three forms for $\mathbf{X}'$ and six forms for $\mathbf{Y}$ for Type I solution. For type II, there are three forms for $\mathbf{X}'$ and six forms for $\mathbf{Y}$. For type III, there are six forms for $\mathbf{X}'$ and six forms for $\mathbf{Y}$. Type IV is that the number of zero elements of solution is zero, there are only one form for $\mathbf{X}'$ and one form for $\mathbf{Y}$.

We illustrate some concrete examples with various (l_1, l_2, l_3) and (w_1, w_2, w_3).

Example 1. *For the case of* $w_1 > \max_{i=2,3} w_i$*, we perform an exhaustive search for* $l_i \in \{-1, -0.9, -0.8, \ldots, 1\}$ *and* $w_i \in \{0,\ 0.1,\ 0.2, \ldots, 1\}$, $i = 1, 2, 3$*. The first type I is* $(l_1, l_2, l_3) = (-1, -1, -1)$ *and* $(w_1, w_2, w_3) = (0.9, 0, 0.1)$*. The optimal solution is* $(y_1, y_2, y_3) = (3, -1, -1)$, $(x'_1, x'_2, x'_3) = (4,\ 0,\ 0)$, $(x_1, x_2, x_3) = (3, -1, -1)$ *and* $F = 2.6$.

Example 2. *Consider the case of* $w_2 > \max_{i=1,3} w_i$*. Applying an exhaustive search for* $l_i \in \{-1, -0.9, -0.8, \ldots, 1\}$ *and* $w_i \in \{0,\ 0.1,\ 0.2, \ldots,\ 1\}$, $i = 1, 2, 3$*, the value of* $(l_1, l_2, l_3) = (-1, -1, 1)$ *and* $(w_1, w_2, w_3) = (0, 0.9, 0.1)$ *is the first one satisfies type I. The optimal solution is* $(y_1, y_2, y_3) = (1, 1, -1)$, $(x'_1, x'_2, x'_3) = (0,\ 2,\ 0)$, $(x_1, x_2, x_3) = (-1, 1, 1)$ *and* $F = 0.8$*. For* $(l_1, l_2, l_3) = (-1, -1, -1)$ *and* $(w_1, w_2, w_3) = (0,\ 0.9, 0.1)$*, the type II solution is* $(y_1, y_2, y_3) = (1, 1, -1)$, $(x'_1, x'_2, x'_3) = (0,\ 2,\ 2)$, $(x_1, x_2, x_3) = (-1, 1, 1)$ *and* $F = 0.8$*. For* $(l_1, l_2, l_3) = (-1, -1, 0.4)$ *and* $(w_1, w_2, w_3) = (0, 0.6, 0.4)$*, the type III solution is* $(y_1, y_2, y_3) = (0.4, 0.4, 0.2)$, $(x'_1, x'_2, x'_3) = (1.2, 1.4,\ 0)$, $(x_1, x_2, x_3) = (0.2, 0.4, 0.4)$ *and* $F = 0.32$*. For* $(l_1, l_2, l_3) = (-1, -1, -1)$ *and* $(w_1, w_2, w_3) = (0,\ 0.6, 0.4)$*, the type IV solution is* $(y_1, y_2, y_3) = \left(\frac{1}{3}, \frac{1}{3}, \frac{1}{3}\right)$, $(x'_1, x'_2, x'_3) = (4/3,\ 4/3,\ 4/3)$, $(x_1, x_2, x_3) = \left(\frac{1}{3}, \frac{1}{3}, \frac{1}{3}\right)$ *and* $F = 1/3$.

Example 3. *Consider the case of* $w_3 > \max_{i=2,3} w_i$*. For* $l_i \in \{-1, -0.9, -0.8, \ldots, 1\}$ *and* $w_i \in \{0,\ 0.1,\ 0.2, \ldots, 1\}$, $i = 1,\ 2,\ 3$*, the value of* $(l_1, l_2, l_3) = (-1, -1, -1)$ *and* $(w_1, w_2, w_3) = (0.4, 0, 0.6)$ *is the first one satisfies type I. The optimal solution is* $(y_1, y_2, y_3) = (3, -1, -1)$, $(x'_1, x'_2, x'_3) = (4,\ 0,\ 0)$, $(x_1, x_2, x_3) = (3, -1, -1)$ *and* $F = 0.6$*. For* $(l_1, l_2, l_3) = (-1, -1, 0.4)$ *and* $(w_1, w_2, w_3) = (0,\ 0.1,\ 0.9)$*, the type II solution is* $(y_1, y_2, y_3) = (0.4, 0.3, 0.3)$, $(x'_1, x'_2, x'_3) = (1.3,\ 1.3,\ 0)$, $(x_1, x_2, x_3) = (0.3, 0.3, 0.4)$ *and* $F = 0.3$*. For* $(l_1, l_2, l_3) = (-1, -0.9, -0.8)$ *and* $(w_1, w_2, w_3) = (0.4,\ 0, 0.6)$*, the type III solution is* $(y_1, y_2, y_3) = (2.8, -0.9, -0.9)$, $(x'_1, x'_2, x'_3) = (0.1,\ 0, 3.6)$, $(x_1, x_2, x_3) = (-0.9, -0.9, -2.8)$ *and* $F = 0.58$*. For* $(l_1, l_2, l_3) = (-1, -1, -1)$ *and* $(w_1, w_2, w_3) = (0,\ 0.1, 0.9)$*, the type IV solution is* $(y_1, y_2, y_3) = \left(\frac{1}{3}, \frac{1}{3}, \frac{1}{3}\right)$, $(x'_1, x'_2, x'_3) = (4/3,\ 4/3,\ 4/3)$, $(x_1, x_2, x_3) = \left(\frac{1}{3}, \frac{1}{3}, \frac{1}{3}\right)$ *and* $F = 1/3$.

Minimizing the objective function of the constrained OWA aggregation problem with bounded variables is also important. One interesting model is the constrained OWA aggregation problem with upper bounded variables described as follows:

$$\begin{aligned}
&\text{Min } W^T Y \\
&\text{s.t.} \mathcal{I}^T \mathbf{X} \leq 1 \\
&\quad \mathbf{GY} \leq 0 \\
&\quad y_i \mathcal{I} - X - MZ_i \leq 0,\ i = 1,2,\ldots,n-1 \\
&\quad y_n \mathcal{I} - X \leq 0 \\
&\quad \mathcal{I}^T Z_i \leq n - i,\ i = 1,2,\ldots,n-1 \\
&\quad Z_{i+1} \leq Z_i,\ i = 1,2,\ldots,n-2 \\
&\quad Z_i \in \{0,1\}^n,\ i = 1,2,\ldots,n-1 \\
&\quad X \leq \boldsymbol{U}
\end{aligned} \tag{10}$$

where the column vector:

$$\boldsymbol{U} = \begin{bmatrix} u_1 \\ u_2 \\ \vdots \\ u_n \end{bmatrix}.$$

By using the change of variable:

$$X' = U - \mathbf{X},\ y_i = -y'_{n+1-i} \text{ and } \boldsymbol{Y}' = \begin{bmatrix} y'_n \\ y'_{n-1} \\ \vdots \\ y'_1 \end{bmatrix}$$

minimizing the objective function of the constrained OWA aggregation problem with upper bounded variables is:

$$\begin{aligned}
&\text{Max } W^T Y' \\
&\text{s.t.} \mathcal{I}^T \geq \mathcal{I}^T \boldsymbol{U} - 1 \\
&\quad \mathbf{GY}' \leq 0 \\
&\quad y'_1 \mathcal{I} - X' \geq -\boldsymbol{U} \\
&\quad y'_i \mathcal{I} - X' + MZ_i \geq -\boldsymbol{U},\ i = 2,3,\ldots,n \\
&\quad \mathcal{I}^T Z_i \leq n - i,\ i = 1,2,\ldots,n-1 \\
&\quad Z_{i+1} \leq Z_i,\ i = 1,2,\ldots,n-2 \\
&\quad Z_i \in \{0,1\}^n,\ i = 1,2,\ldots,n-1 \\
&\quad X' \geq \mathbf{0}
\end{aligned} \tag{11}$$

If $\mathcal{I}^T \boldsymbol{U} - 1 < 0$, the constrained OWA aggregation problem has unbounded solution. If $\mathcal{I}^T \boldsymbol{U} - 1 = 0$, the unique optimal solution is $X'^* = \mathbf{0}$, so:

$$X^* = \boldsymbol{U}.$$

For the case of $1 - \mathcal{I}^T \boldsymbol{L} > 0$, the similar results as Theorem 2 can be derived.

5. Numerical Results

To evaluate the optimal solution behaviors of the three-dimensional constrained OWA aggregation problem with lower bounded variables, we present some numerical experiments.

In Table 4, we display the number of solution type I, II, III and IV for different choices of the weights and the lower bounds. To this end, we consider four types of solution forms I, II, III and IV and six types of weights:

$$w_1 = \max_{i=1,2,3} w_i,\ w_2 = \max_{i=1,2,3} w_i,\ w_3 = \max_{i=1,2,3} w_i,\ w_1 > \max_{i=2,3} w_i,\ w_2 > \max_{i=1,3} w_i,\ w_3 > \max_{i=2,3} w_i.$$

Each cell is associated to a pair (W, S) and gives the number of different instances of $(l_1, l_2, l_3, w_1, w_2, w_3)$ satisfying weight (W) and solution (S) conditions. We restrict our attention to:

$$W \in \left\{ w_1 = \max_{i=1,2,3} w_i,\ w_2 = \max_{i=1,2,3} w_i, w_3 = \max_{i=1,2,3} w_i,\ w_1 > \max_{i=2,3} w_i, w_2 > \max_{i=1,3} w_i,\ w_3 > \max_{i=2,3} w_i \right\},$$
$$S \in \{\text{I, II, III, IV}\},\ l_i \in \{-1, -0.9, -0.8, \ldots, 1\},\ w_i \in \{0, 0.1, 0.2, \ldots, 1\},\ i = 1, 2, 3.$$

For each cell, the instances $(l_1, l_2, l_3, w_1, w_2, w_3)$ of the test problem are 179,760 for $w_1 = \max_{i=1,2,3} w_i$, $w_2 = \max_{i=1,2,3} w_i$, $w_3 = \max_{i=1,2,3} w_i$ and 119,840 for $w_1 > \max_{i=2,3} w_i$, $w_2 > \max_{i=1,3} w_i$, $w_3 > \max_{i=2,3} w_i$. The total instances of the test problem are 898,800. An examination of the table reveals that the type IV is not optimal solution for $w_1 = \max_{i=1,2,3} w_i$. In particular, for $w_1 > \max_{i=2,3} w_i$, the optimal solution type is always type I solution. If the lower bounds $(l_1, l_2, l_3) = (0, 0, 0)$, then the optimal solution is types II, III and IV for $w_2 = \max_{i=1,2,3} w_i$ and $w_2 > \max_{i=1,3} w_i$, and types I and IV for $w_3 = \max_{i=1,2,3} w_i$ and $w_3 > \max_{i=2,3} w_i$. However, from Table 4, the possible optimal solutions are all the types I, II, III and IV for $w_2 = \max_{i=1,2,3} w_i$, $w_3 = \max_{i=1,2,3} w_i$, $w_2 > \max_{i=1,3} w_i$ and $w_3 > \max_{i=2,3} w_i$. Among a set of four optimal solution types, the largest number of instances of the test problem is the solution type II. Therefore, the optimal solution type is I for $w_1 = \max_{i=1,2,3} w_i$ and $w_1 > \max_{i=2,3} w_i$, and types I, II, III and IV for $w_2 = \max_{i=1,2,3} w_i$, $w_3 = \max_{i=1,2,3} w_i$, $w_2 > \max_{i=1,3} w_i$ and $w_3 > \max_{i=2,3} w_i$.

Table 4. The number of different instances satisfying weight (W) and solution type (S).

W	I	II	III	IV
$w_1 \geq w_2,\ w_1 \geq w_3$	168,101	6826	4833	0
$w_2 \geq w_1,\ w_2 \geq w_3$	47,133	114,240	7411	10,976
$w_3 \geq w_1,\ w_3 \geq w_2$	51,302	56,618	16,960	54,880
$w_1 > w_2,\ w_1 > w_3, w_2 \neq w_3$	119,840	0	0	0
$w_2 > w_1,\ w_2 > w_3, w_1 \neq w_3$	28,856	80,164	5332	5488
$w_3 > w_1,\ w_3 > w_2, w_1 \neq w_2$	30,720	40,656	10,048	38,416

For the three-dimensional constrained OWA aggregation problem with lower bounded variables, from the numerical experiments the solution type I is the same as that of the constrained OWA aggregation problem without lower bounded variables for $w_1 > \max_{i=2,3} w_i$. However, for $w_2 > \max_{i=1,3} w_i$ and $w_3 > \max_{i=2,3} w_i$, there are all solution types. For the constrained OWA aggregation problem without lower bounded variables, the solution are types II, III, IV and types I, IV, for $w_2 > \max_{i=1,3} w_i$ and $w_3 > \max_{i=2,3} w_i$, respectively. The four solution types may be too simple for the three-dimensional constrained OWA aggregation problem with lower bounded variables. From this result, we anticipate more complication in the higher dimensions of the constrained OWA aggregation problem with lower bounded variables.

6. Conclusions

For the constrained OWA aggregation problem with one constraint on the sum of all variables, this paper introduces some constraints to reduce the multiple solution problem. For the three-dimensional constrained OWA aggregation problem with the same lower bounds, by using the change of variables, the optimal solution is the same as that of the constrained OWA aggregation problem without lower bounded variables. For the three-dimensional constrained OWA aggregation problem with lower bounded variables, this paper presents four types (I, II, III and IV) of solutions depending on the number of zero elements. When the number of zero elements of solution is two (type I), there are three closed-form expressions of $\boldsymbol{X'}$ and six closed-form expressions of $\boldsymbol{Y}$. When the number of zero elements of the solution is one (types II and III), there are three closed-form expressions of $\boldsymbol{X'}$ and six closed-form expressions of $\boldsymbol{Y}$ for type II, and six closed-form expressions of $\boldsymbol{X'}$ and six closed-form expressions of $\boldsymbol{Y}$ for type III. When the number of zero elements of the solution is zero (type IV), there is only one closed-form expression of $\boldsymbol{X'}$ and one closed-form expression of $\boldsymbol{Y}$. According to the computerized experiment we perform for the three-dimensional constrained OWA aggregation problem with lower bounded variables, the optimal solution type is I for $w_1 = \max\limits_{i=1,2,3} w_i$ and $w_1 > \max\limits_{i=2,3} w_i$, and types I, II, III and IV for $w_2 = \max\limits_{i=1,2,3} w_i$, $w_3 = \max\limits_{i=1,2,3} w_i$, $w_2 > \max\limits_{i=1,3} w_i$ and $w_3 > \max\limits_{i=2,3} w_i$.

Worthy of future research is that the analysis is extended to the lower and upper bounded variables for the constrained OWA aggregation problem, especially for the three-dimensional constrained OWA aggregation problem with upper bounded variables. Thus, the analysis of the constrained OWA aggregation problem with bounded variables is a subject of considerable ongoing research.

Author Contributions: Y.-F.C. conceived the study and performed the experiments. H.-C.T. analyzed the method and wrote the paper.

Funding: This research received no external funding.

Conflicts of Interest: The authors declare that they have no competing interests.

References

1. Yager, R.R. On ordered weighted averaging aggregation operators in multicriteria decision making. *IEEE Trans. Syst. Man Cybern.* **1988**, *18*, 183–190. [CrossRef]
2. Merigó, J.M.; Gil-Lafuente, A.M.; Zhou, L.G.; Chen, H.Y. Induced and linguistic generalized aggregation operators and their application in linguistic group decision making. *Group Decis. Negot.* **2012**, *21*, 531–549. [CrossRef]
3. Ogryczak, W.; Śliwiński, T. On solving linear programs with the ordered weighted averaging objective. *Eur. J. Oper. Res.* **2003**, *148*, 80–91. [CrossRef]
4. Yager, R.R. On the analytic representation of the Leximin ordering and its application to flexible constraint propagation. *Eur. J. Oper. Res.* **1997**, *102*, 176–192. [CrossRef]
5. Yager, R.R.; Beliakov, G. OWA operators in regression problems. *IEEE Trans. Fuzzy Syst.* **2010**, *18*, 106–113. [CrossRef]
6. Zhou, L.; Chen, H.; Liu, J. Generalized multiple averaging operators and their applications to group decision making. *Group Decis. Negot.* **2013**, *22*, 331–358. [CrossRef]
7. Emrouznejad, A.; Marra, M. Ordered weighted averaging operators 1988–2014: A citation-based literature survey. *Int. J. Intell. Syst.* **2014**, *29*, 994–1014. [CrossRef]
8. Yager, R.R. Constrained OWA aggregation. *Fuzzy Sets Syst.* **1996**, *81*, 89–101. [CrossRef]
9. Carlsson, C.; Fullér, R.; Majlender, P. A note on constrained OWA aggregation. *Fuzzy Sets Syst.* **2003**, *139*, 543–546. [CrossRef]
10. Coroianu, L.; Fullér, R. On the constrained OWA aggregation problem with single constraint. *Fuzzy Sets Syst.* **2018**, *332*, 37–43. [CrossRef]
11. Ahn, B.S. Parameterized OWA operator weights: An extreme point approach. *Int. J. Approx. Reason.* **2010**, *51*, 820–831. [CrossRef]

12. Ahn, B.S. A new approach to solve the constrained OWA aggregation problem. *IEEE Trans. Fuzzy Syst.* **2017**, *25*, 1231–1238. [CrossRef]
13. Fullér, R.; Majlender, P. On obtaining minimal variability OWA operator weights. *Fuzzy Sets Syst.* **2003**, *136*, 203–215. [CrossRef]
14. Ogryczak, W.; Olender, P. Ordered median problem with demand distribution weights. *Optim. Lett.* **2016**, *10*, 1071–1086. [CrossRef]
15. Ogryczak, W.; Śliwiński, T. On efficient WOWA optimization for decision support under risk. *Int. J. Approx. Reason.* **2009**, *50*, 915–928. [CrossRef]

symmetry

MDPI

Article

A Multi-Level Privacy-Preserving Approach to Hierarchical Data Based on Fuzzy Set Theory

Jinyan Wang [1,2], Guoqing Cai [2], Chen Liu [2], Jingli Wu [1,2] and Xianxian Li [1,2,*]

1 Guangxi Key Lab of Multi-Source Information Mining & Security, Guangxi Normal University, Guilin 541004, China; wangjy612@gxnu.edu.cn (J.W.); wjlhappy@gxnu.edu.cn (J.W.)

2 School of Computer Science and Information Technology, Guangxi Normal University, Guilin 541004, China; caigq520@foxmail.com (G.C.); liuch1027@163.com (C.L.)

* Correspondence: lixx@gxnu.edu.cn; Tel.: +86-135-5723-8790

Received: 28 July 2018; Accepted: 9 August 2018; Published: 10 August 2018

Abstract: Nowadays, more and more applications are dependent on storage and management of semi-structured information. For scientific research and knowledge-based decision-making, such data often needs to be published, e.g., medical data is released to implement a computer-assisted clinical decision support system. Since this data contains individuals' privacy, they must be appropriately anonymized before to be released. However, the existing anonymization method based on *l*-diversity for hierarchical data may cause serious similarity attacks, and cannot protect data privacy very well. In this paper, we utilize fuzzy sets to divide levels for sensitive numerical and categorical attribute values uniformly (a categorical attribute value can be converted into a numerical attribute value according to its frequency of occurrences), and then transform the value levels to sensitivity levels. The privacy model (α^h_{lev}, k)-anonymity for hierarchical data with multi-level sensitivity is proposed. Furthermore, we design a privacy-preserving approach to achieve this privacy model. Experiment results demonstrate that our approach is obviously superior to existing anonymous approach in hierarchical data in terms of utility and security.

Keywords: fuzzy set theory; decision-making; hierarchical data; privacy model; anonymous approach; similarity attack

1. Introduction

Hospitals and other organizations often need to publish data, e.g., medical data or census data, for the purposes of scientific research and knowledge-based decision-making [1–10]. To avoid the leakage of individual privacy, explicit identifying information is removed when data is released. However, individual privacy still could be leaked by linking other public data [11]. Privacy-preserving data publishing provides methods and tools for publishing useful information while preserving individual privacy [12]. In recent years, the problem of privacy-preserving data publishing has been studied extensively. The existing privacy protection methods mainly focus on relational data, and many mature privacy models are proposed, such as k-anonymity [11], l-diversity [13], (α, k)-anonymity [14] and t-closeness [15]. However, data often has a complicated structure in the real world. With the advent of document-oriented databases (e.g., MongoDB) and the wide use of markup languages (e.g., XML), hierarchical data has become ubiquitous [16]. To avoid the leakage of individual privacy, the hierarchical data must be properly anonymized before it is released. At present, there are few researches on privacy protection for hierarchical data. Ozalp et al. [16] proposed l-diversity anonymous methods for hierarchical data. An example for hierarchical data is given in Figure 1. The schema for education data is obtained from Sabanci University [16] and the examples appearing in this paper are related to the schema. Figure 1a represents a student's record, which fits the education schema shown in Figure 1b. The student is born in 1990 and majors in Computer Science. He took two courses,

CS201 and CS305. For CS201, his evaluations are submitted for two instructors. For CS305, he submitted an evaluation and showed he bought a database book. The labels of vertices are all quasi-identifiers (QIs) of the student and the corresponding sensitive information is remarked in the side of every vertex. Quasi-identifier is a set of attributes that can potentially identify an individual [11]. Assume that an attacker knows some QIs of a victim, and his goal is to reason the sensitive information of the victim. In [16], they used suppression and generalization [11] to make the anonymous hierarchical dataset satisfy l-diversity, which ensures the frequency of every sensitive value for the union-compatible vertices (belonging to the same vertex in schema) in an equivalence class is not more than $1/l$. The constraint also can guarantee that every equivalence class contains at least l hierarchical data records. An equivalence class in an anonymous hierarchical dataset is a set of records with the same values for the QIs. However, the method does not consider the sensitivity of different sensitive attribute values, which lead to similarity attacks [15]. For example, an equivalence class contains three hierarchical data records and its class representative is shown in Figure 2, which satisfies 3-diversity. The sensitive values of their cumulative GPAs are 0.31, 0.15 and 0.09, respectively, where GPA is the grade point average. An attacker knows a victim in the equivalence class by linking with some QIs of the victim. Although the attacker does not infer the victim's specific sensitive value, he can know that the victim's academic performance is low with 100% probability and the victim's privacy is leaked. Similarly, the attacker can confirm that the grade of the victim in the course CS201 is very low according to the value $\{D, D+, D-\}$. Also, the attacker can infer that the victim is very dissatisfied with the DB Prof. by the value $\{0, 1/10, 2/10\}$. To avoid similarity attack, we propose a multi-level privacy-preserving approach in hierarchical data based on fuzzy sets.

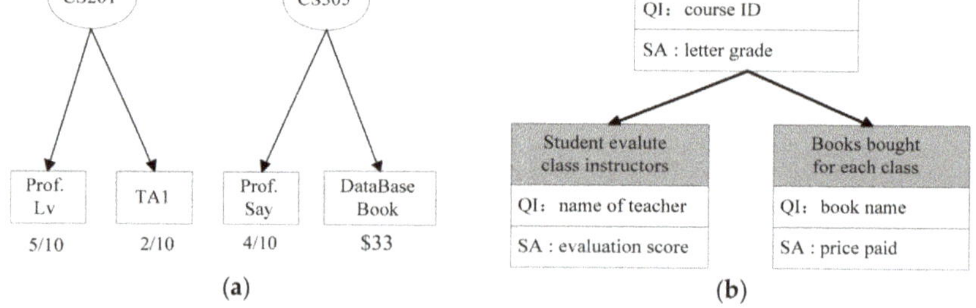

Figure 1. An example for hierarchical data: (**a**) A student's record; (**b**) Schema for education data.

The contributions of this paper are summarized as follows:

- We utilize the fuzzy set theory to obtain the sensitivity levels for sensitive numerical and categorical attribute values, and present the privacy model (α^h_{lev}, k)-anonymity for hierarchical data with multi-level sensitivity. This model can solve the similarity attack, and provide reasonable privacy protection for sensitive value in different sensitivity level.
- We improve the privacy-preserving approach in hierarchical data to obtain the anonymous data that satisfies (α^h_{lev}, k)-anonymity.

- We do experiments to compare our approach with the existing anonymous method ClusTree proposed in [16]. Experiment results demonstrate that our approach is superior to *ClusTree* in terms of utility and security.

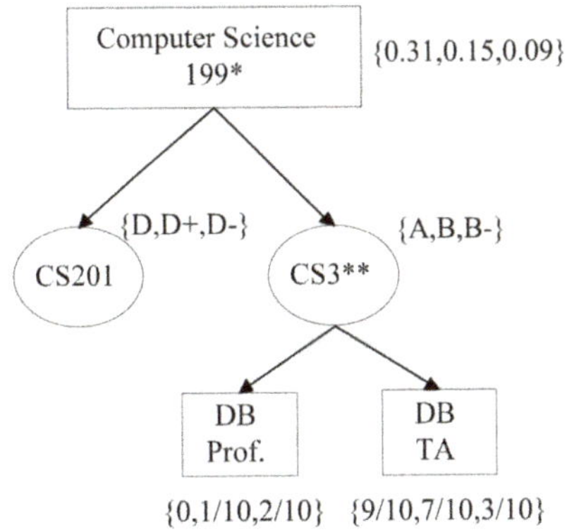

Figure 2. A class representative satisfying 3-diviersity.

2. Related Work

In this section, we review the related work about privacy preserving data publishing for relational data and hierarchical data.

2.1. Preserving Privacy for Publishing Relational Data

The first privacy model, proposed by Samarati and Sweeney [11] in 1998, is k-anonymity for relational data, which requires that every record in a table is indistinguishable from at least k-1 other records with respect to QI. There exist many anonymization methods to implement k-anonymity, such as bottom-up generalization, top-down specialization and anonymity by clustering technique [17–19]. k-anonymity can protect against identity disclosure, but cannot prevent attribute disclosure. Therefore, l-diversity has been proposed [13]. It requires that every equivalence class contains at least l different sensitive values. There are numerous methods for achieving l-diversity [20,21]. Furthermore, Wong et al. [14] extended k-anonymity to (α, k)-anonymity to limit the confidence of the implications from the QI to a sensitive value to within α in order to protect the sensitive information from being inferred by strong implications, and proposed a bottom-up generalization algorithm to achieve (α, k)-anonymity. Li et al. [15] pointed out that l-diversity does not prevent skewness attack and similarity attack, so they introduced t-closeness model, which requires that the distribution of a sensitive attribute in any equivalence class is close to the distribution of the attribute in the overall table. They also revised the Incognito algorithm [17], which is a top-down generalization method proposed for k-anonymity, to achieve t-closeness. However, t-closeness still does not prevent similarity attacks. Han et al. [22] considered the difference of sensitivity for sensitive values, and proposed multi-level l-diversity model for numerical sensitive attribute. Furthermore, Jin et al. [23] presented the (α_i, k)-anonymity privacy preservation based on sensibility grading. However, the levels are artificially assigned. Some researches proposed fuzzy based methods for privacy preserving [24,25]. They used fuzzy sets to transform sensitive values to semantic values and published the data with fuzzy sensitive information, which decreases the utility of sensitive information and still does not resist similarity attacks.

2.2. Preserving Privacy for Publishing Hierarchical Data

There are several studies about preserving privacy for publishing hierarchical or tree-structured data. Yang and Li [26] found that the dependencies between nodes in the XML data information may result in privacy leakage. They formally defined these dependencies as XML constraints, and designed an algorithm to sanitize XML documents by considering these constraints such that no privacy is leaked. However, their attack model is too weak. Our adversarial model assumes that the attacker has some information about the victim. Landberg et al. [27] proposed δ-dependency and extended the anatomy method in relational data to hierarchical data. But the dissection method will damage the original semantic structure of hierarchical data, and the generalization in sensitive attributes will affect the effectiveness of hierarchical data. Nergiz et al. [28] extended k-anonymity methods to a multi-relational database, and proposed multi-relational k-anonymity. Firstly, hierarchical data will be converted to multiple relational data tables, which related to each other by primary key or foreign key, then performed k-anonymity separately on each relational data. However, converting hierarchical data into relational data is not a simple matter, and will produce large amounts of data redundancy, which made the executive efficiency of algorithm extremely low. It will also lose a lot of structural information. Gkountouna and Terrovitistis [29] proposed the $k^{(m,\,n)}$-anonymity for tree-structured data. By using generalization and structure decomposition methods, they ensured that the number of matching records not less than k when the attacker knows up to m nodes in a tree and to n structural relations between these nodes. But the method cannot resist the attack with stronger background knowledge. In addition, they used structural decomposition that destroys the structural information of the hierarchical data. Ozalp et al. [16] extended l-diversity to hierarchical data. They utilized generalization and suppression to anonymize the hierarchical data, and make the hierarchical records in an equivalence class to be indistinguishable in terms of the QIs and structure and the sensitive values for the union-compatible vertices in an equivalence class satisfies the requirements of l-diversity. This method is very scalable for the general anonymous method of hierarchical data. However, this method does not consider the different sensitivity of sensitive attribute values in anonymous hierarchical data, so the anonymous hierarchical data still does not resist similarity attack. In this paper, we use fuzzy set theory to partition rank for sensitive values of union-compatible vertices, and propose a multi-level privacy-preserving approach in hierarchical data to solve similarity attacks.

3. Problem Descriptions

In this section, we describe the attack model, give some fundamental definitions, and introduce our privacy protection model.

3.1. Attack Model

We assume that an attacker knows a victim's QI information, which contains any combination of QI values in the same or different vertices of the victim's record. Also, the attacker can obtain some structural links. For example, the victim took two courses, and purchased only a book for course CS201. In addition, the attacker has some negative knowledge, e.g., the victim did not take CS305. Our anonymization approach can ensure that an attacker, who has this background knowledge about a victim, does not infer any sensitive value of the victim is in some level with the probability, which is greater than a given threshold.

3.2. Basic Definitions in Hierarchical Data

In this subsection, we give some basic definitions for hierarchical data [16]. Let T be a graph with n vertices. We say that T is a rooted tree if and only if (1) T is a directed acyclic graph with n-1 edges; (2) for every vertex (except root vertex), there is a single path from the root vertex to it in T; (3) there exists an edge $v \rightarrow c_i$ if $c_i \in children(v)$, where $children(v)$ is the children of vertex v. Such tree is denoted by $T(V, E)$, where V and E are the sets of vertices and edges in the tree, respectively.

A hierarchical data record satisfies the following conditions: (1) it follows a rooted tree structure; (2) each vertex v has two j-tuples ($j \geq 0$), v_{QIt} and v_{QI}, which contains the names of QI attributes and the values of corresponding QIs, respectively; (3) each vertex v also has two m-tuples ($0 \leq m \leq 1$), v_{SAt} and v_{SA}, which contains the name of sensitive attribute and the value of corresponding sensitive attribute, respectively; (4) assume that $|v_{QI}| + |v_{SA}| \geq 1$ to eliminate empty vertices. For a vertex v of a hierarchical data record, v_{QI} is the label of v and v_{SA} is next to v. For Figure 1, v_{QIt} = {*major program*, *year of birth*}, v_{Sat} = {*GPA*}, v_{QI} = {*Computer Science*, 1990}, and v_{SA} = {3.75}.

Definition 1 (Union-Compatibility) [16]. *Two vertices v and v' are union-compatible if and only if $v_{QIt} = v'_{QIt}$ and $v_{SAt} = v'_{SAt}$.*

Definition 2 (QI-isomorphism) [16]. *Let $T_1(V_1, E_1)$ and $T_2(V_2, E_2)$ are two hierarchical data records. $T_1(V_1, E_1)$ is isomorphic to $T_2(V_2, E_2)$ if and only if there exists a bijection f: $V_1 \rightarrow V_2$, such that:*

(1) *For $x, y \in V_1$, there exists an edge $e_i \in E_2$ from $f(x)$ to $f(y)$ if and only if there exists an edge $e_j \in E_1$ from x to y.*
(2) *$f(r_1) = r_2$, where $r_1 \in V_1$ and $r_2 \in V_2$ be the roots of $T_1(V_1, E_1)$ and $T_2(V_2, E_2)$, respectively.*
(3) *For all pairs (x, x'), where $x \in V_1$ and $x' = f(x)$, x and x' are union-compatible and $x_{QI} = x'_{QI}$.*

Definition 3 (Equivalence Class of Hierarchical Records) [16]. *Let $Q = \{T_1, T_2, ..., T_k\}$ is a collection of k hierarchical data records. We say Q is an equivalence class, if for $\forall i, j \in \{1, \ldots, k\}$, T_i and T_j are QI-isomorphic.*

Definition 4 (Class Representative) [16]. *Let $Q = \{T_1, T_2, ..., T_k\}$ be an equivalence class in hierarchical data, and f_i ($1 \leq i \leq k-1$) be a bijection that maps T_1's vertices to T_{i+1}'s vertices as in QI-isomorphism. $\hat{T}$ is the class representative for Q if $\hat{T}$ is QI-isomorphic to T_1 with a bijection function f and $\forall v \in \hat{T}$, $v_{SA} = \{f(v)_{SA}, f_1(f(v))_{SA}, ..., f_{k-1}(f(v))_{SA}\}$.*

Let $X = \{x_1, x_2, ..., x_o\}$ be a multiset of values from the domain of a sensitive attribute A. X satisfies l-diversity if $\forall x_i \in X$, $p(x_i) \leq 1/l$, where $p(x_i)$ is the frequency of s_i in X. For an equivalence class Q in hierarchical data, $\hat{T}$ is the class representative for Q. If for $\forall v \in \hat{T}$, v_{SA} satisfies l-diversity, then $\hat{T}$ satisfies l-diversity. Given a hierarchical data D, an anonymous hierarchical data D^* satisfies l-diversity, if the class representative of any equivalence class in D^* satisfies l-diversity. The l-diversity hierarchical data does not prevent similarity attack, since it does not consider the different sensitivity of sensitive attribute values.

3.3. Privacy Model

For every sensitive attribute, including numerical and categorical attributes, we partition sensitive values to five levels: *low*, *very low*, *middle*, *very high* and *high* (for some sensitive attributes, e.g., a student's grade in a course, the levels have been divided, and we do not need to handle it), and transform these value levels to corresponding sensitivity levels.

Let U be a universe of discourse. A mapping μ_A: $U \rightarrow [0, 1]$ is called a membership function on U, where the set A, which consists of $\mu_A(u)$ ($u \in U$), is a fuzzy set on U, and $\mu_A(u)$ is the membership degree of u to A [30–32]. The trapezoidal distribution [33] is used to give the membership functions for fuzzy sets *low*, *very low*, *middle*, *very high* and *high*, denoted by A_1, A_2, A_3, A_4, and A_5, respectively. Let U be the domain of a numerical attribute (for categorical attribute, a numerical attribute can be obtained according to the frequency of every value), and *min* and *max* be the minimum and maximum values in U, respectively. The five fuzzy sets have values in the range $[min, a_2]$, $[a_1, a_3]$, $[a_2, a_4]$, $[a_3, a_5]$ and $[a_4, max]$, respectively, where $a_3 = (min + max)/2$, $a_1 = min + (a_3\text{-}min)/3$, $a_2 = min + 2(a_3\text{-}min)/3$,

$a_4 = a_3 + (max\text{-}a_3)/3$, $a_5 = a_3 + 2(max\text{-}a_3)/3$. That is, a_1, a_2, a_3, a_4 and a_5 uniformly divide the interval $[min, max]$. The membership functions for A_i (i = 1, 2, ..., 5) are shown as follows.

$$\mu_{A_1}(u) = \begin{cases} 1 & u \le min \\ \frac{a_2 - u}{a_2 - min} & min < u < a_2 \\ 0 & u \ge a_2 \end{cases} \tag{1}$$

$$\mu_{A_i}(u) = \begin{cases} 0 & u \le a_{i-1} \\ \frac{u - a_{i-1}}{a_i - a_{i-1}} & a_{i-1} < u < a_i \\ 1 & u = a_i \qquad i = 1,\ 2,\ 3 \\ \frac{a_{i+1} - u}{a_{i+1} - a_i} & a_i < u < a_{i+1} \\ 0 & u \ge a_{i+1} \end{cases} \tag{2}$$

$$\mu_{A_5}(u) = \begin{cases} 0 & u \le min \\ \frac{u - a_4}{max - a_4} & a_4 < u < max \\ 1 & u \ge max \end{cases} \tag{3}$$

For any $u \in U$, argmax$\{u_{Ai}(u) \mid i \in \{1, 2, 3, 4, 5\}\}$ is the level which u belongs to. We transform the value level to sensitivity level. For some sensitive attributes, the higher the value level is, the larger the sensitivity level is, e.g., income; but it is reversed for other sensitive attributes, e.g., student's cumulative GPA. For a numerical attribute, we divide the five levels from 1 to 5 for sensitivity. Level 5 is the highest and level 1 is the lowest. The higher sensitivity level is, the stronger privacy protection will be given.

For example, for an equivalence class Q in a hierarchical data, we assume that the sensitive attribute of the root vertex in the class representative of Q is the cumulative GPA, whose value is {0.8, 1.6, 2.3, 2.7, 3.5, 3.9}, where the domain of the cumulative GPA is [0, 4]. We can obtain the min = 0, max = 4, a_3 = 2, a_1 = 2/3, a_2 = 4/3, a_4 = 8/3 and a_5 = 10/3. The membership degree of u_i to A_j are shown in Table 1, where $u_i \in$ {0.8, 1.6, 2.3, 2.7, 3.5, 3.9} and $A_j \in$ {*low*, *very low*, *middle*, *very high*, *high*}. We can know that 0.8, 1.6, 2.3, 2.7, 3.5 and 3.9 are belong to *low*, *very low*, *middle*, *very high*, *high* and *high*, respectively. Their sensitivity levels are 5, 4, 3, 2, 1 and 1, respectively.

Table 1. The membership degree of u_i to A_j.

GPA / Value Level	0.8	1.6	2.3	2.7	3.5	3.9
Low	0.40	0	0	0	0	0
Very low	0.20	0.60	0	0	0	0
Middle	0	0.40	0.55	0	0	0
Very high	0	0	0.45	0.95	0	0
High	0	0	0	0.025	0.625	0.925

In fact, for every sensitive value a numerical attribute A, we can confirm quickly its value level by using the membership functions. As shown in Figure 3, the $[min, max]$ is the domain of A, a_1, a_2, a_3, a_4 and a_5 equally divide the $[min, max]$. p_1, p_2, p_3 and p_4 are the points of intersection of membership functions μ_{A_1} and μ_{A_2}, μ_{A_2} and μ_{A_3}, μ_{A_3} and μ_{A_4}, and μ_{A_4} and μ_{A_5}, respectively. The ranges of *low*, *very low*, *middle*, *very high* and *high* are $[min, p_1]$, $[p_1, p_2]$, $[p_2, p_3]$, $[p_3, p_4]$ and $[p_4, max]$, respectively.

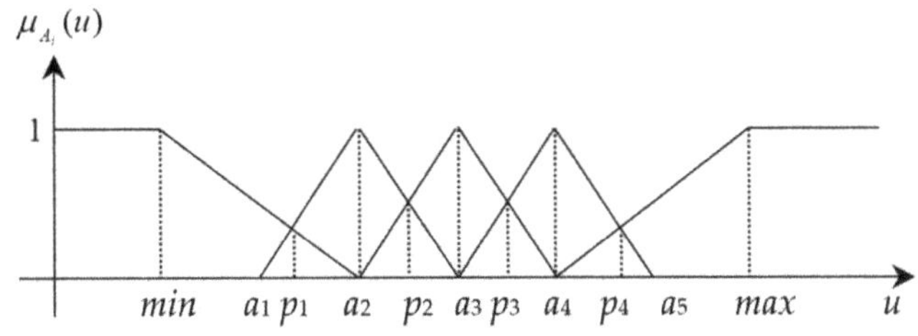

Figure 3. The membership functions for five value levels.

For example, for the cumulative GPA and evaluation score for a teacher, the domains are [0, 4] and [0, 1], respectively. Their value levels and sensitivity levels are shown in Table 2. The letter grade of a course has been divided five levels.

Table 2. The value levels and sensitivity levels for sensitive attributes.

Value Level	GPA	Letter Grade	Evaluation Score	Sensitivity Level	α^h_{lev}
Low	[0, 0.89)	E	[0, 0.25)	5	0.1
Very low	[0.89, 1.67)	D−, D, D+	[0.25, 0.42)	4	0.2
Middle	[1.67, 2.33)	C−, C, C+	[0.42, 0.58)	3	0.4
Very high	[2.33, 3.11)	B−, B, B+	[0.58, 0.78)	2	0.6
High	[3.11, 4]	A−, A, A+	[0.78, 1]	1	0.8

For a categorical attribute, e.g., *disease*, according to the frequency of every value, we obtain an attribute *Frequency*. The values of *Frequency* can be divided into 5 levels including *low*, *very low*, *middle*, *very high* and *high*. For the disease *HIV*, it is more sensitive than *flu*, and the frequency of *HIV* is less than one of *flu*. Therefore, we divide the values of *disease* into 5 sensitivity levels according to the value levels of *Frequency*. The lower the value level is, the larger the sensitivity level is.

Definition 5 ((α^h_{lev}, k)-anonymity in Hierarchical Data). *Given a hierarchical data H, a published anonymous hierarchical data H′ satisfies (α^h_{lev}, k)-anonymity if every equivalence class Q in H′ satisfies (α^h_{lev}, k)-anonymity. That is, Q contains at least k hierarchical data records, and for every vertex v in the class representative of Q, the frequency of the values in v_{SA} which belong to the sensitivity level i is less than or equal to $\alpha^h_{lev}[i]$, where $\alpha^h_{lev} = \{0.8, 0.6, 0.4, 0.2, 0.1\}$.*

4. The Anonymization Method

In this section, we introduce our anonymous method, which is divided into two parts. The first step is to realize the anonymization of two hierarchical data records or class representatives, and the second step is to anonymize the entire hierarchical data by using a clustering method.

The anonymization for two hierarchical data records is shown in Algorithm 1. The input is arbitrary two hierarchical data records T_1 and T_2. Without loss of generality, we assume that T_1 has fewer subtrees than T_2. The output is the information loss of anonymizing the two records.

We first check the root nodes of T_1 and T_2, stored in variables *a* and *b*, respectively, whether satisfy the anonymous constraint *check_cons*(*a*, *b*), shown as follows:

$$check\_cons(a,b) = \begin{cases} 1 & \text{if } a \text{ and } b \text{ are union-compatibility and } a_{SA} \cup \\ & b_{SA} \text{ is identical to } (\alpha^h_{lev}, k) - anonymity; \\ 0 & \text{Otherwise,} \end{cases} \quad (4)$$

where $a_{SA} \cup b_{SA}$ is identical to (α_{lev}^{h}, k)-anonymity, i.e., for any an vertex v in the class representative, the number of the values in v_{SA}, which lie in sensitivity level i, is less than or equal to $k^{*}\alpha_{lev}^{h}[i]$. If *check_cons(a, b)* is 0, *tree(a)* and *tree(b)* are suppressed, where *tree*(a_i) ($a_i \in \{a, b\}$) denotes the subtree rooted a_i; otherwise, the values in QI of a and b are generalized. Let *subtrees(a)* and *subtrees(b)* represent the set of subtrees under a and b, respectively. There are three cases: (1) *subtrees(a)* = $\emptyset$ and *subtrees(b)* = $\emptyset$, which indicates that a and b are leaves of hierarchical records, i.e., no vertex need to be processed, and algorithm returns the total cost in *tree(a)* and *tree(b)*; (2) *subtrees(a)* = $\emptyset$ and *subtrees(b)* $\neq \emptyset$, and we suppress all vertices under b to keep the structural consistency, and return the total cost; (3) *subtrees(a)* $\neq \emptyset$ and *subtrees(b)* $\neq \emptyset$, the subtrees under a and b need to be further processed. To minimize the information loss caused by anonymization, the subtrees under the a and b need to be optimally matched. Let *subtrees(a)* = $\{U_1, U_2, \ldots, U_m\}$ and *subtrees(b)* = $\{V_1, V_2, \ldots, V_n\}$ For every subtrees U_i of a, we find the subtrees V_j of b with minimum *MLevAnonytree*(U_i, V_j), as shown in lines 12–23. For every pair (i, j) in *pairs*, we call *MLevAnonytree*(U_i, V_j) to generalize them. In lines 26 and 27, we suppress the unpaired subtrees of b if they exist.

Algorithm 1. *MLevAnonytree*(T_1, T_2)

Input: Two hierarchical data records T_1 and T_2
Output: Anonymous information loss
1 $a \leftarrow root(T_1)$; $b \leftarrow root(T_2)$;
2 **if** *check_condition(a, b)* **then**
3 suppress *tree(a)* and *tree(b)*;
4 **return** *cost(tree(a))* + *cost(tree(b))*;
5 **for** $i = 1$ to $|a_{QI}|$ **do**
6 replace $a_{QI}[i]$ and $b_{QI}[i]$ with their generalized value;
7 **if** *subtrees(a)* = $\emptyset$ and *subtrees(b)* = $\emptyset$ **then**
8 **return** *cost(tree(a))* + *cost(tree(b))*;
9 **if** *subtrees(a)* = $\emptyset$ and *subtrees(b)* $\neq \emptyset$ **then**
10 suppress all vertices under b;
11 **return** *cost(tree(a))* + *cost(tree(b))*;
12 *pairs* $\leftarrow \emptyset$;
13 **for** $i = 1$ to m **do**
14 *min_cost* $\leftarrow \infty$;
15 *paired_index* $\leftarrow \emptyset$;
16 **for** $j = 1$ to n **do**
17 **if** $j \in$ pairs **then**
18 **continue**;
19 $x \leftarrow U_i$; $y \leftarrow V_j$;
20 *loss* $\leftarrow$ *MLevAnonytree(x, y)*;
21 **if** *loss* < *min_cost* **then**
22 *min_cost* $\leftarrow$ *loss*; *paired_index* $\leftarrow j$;
23 *pairs*.append(*i*, *paired_index*);
24 **for** $(i, j) \in$ *pairs* **do**
25 *MLevAnonytree* (U_i, V_j);
26 **if** there are unpaired subtrees in b **then**
suppress them;
27 **return** *cost(tree(a))* + *cost(tree(b))*;

An anonymous example of two hierarchical data records is shown in Figure 4, where Figure 4a–c are two raw hierarchical data records, with their anonymous results identical to $(\alpha_{lev}^{h}, 2)$-anonymity, and their class representative, respectively.

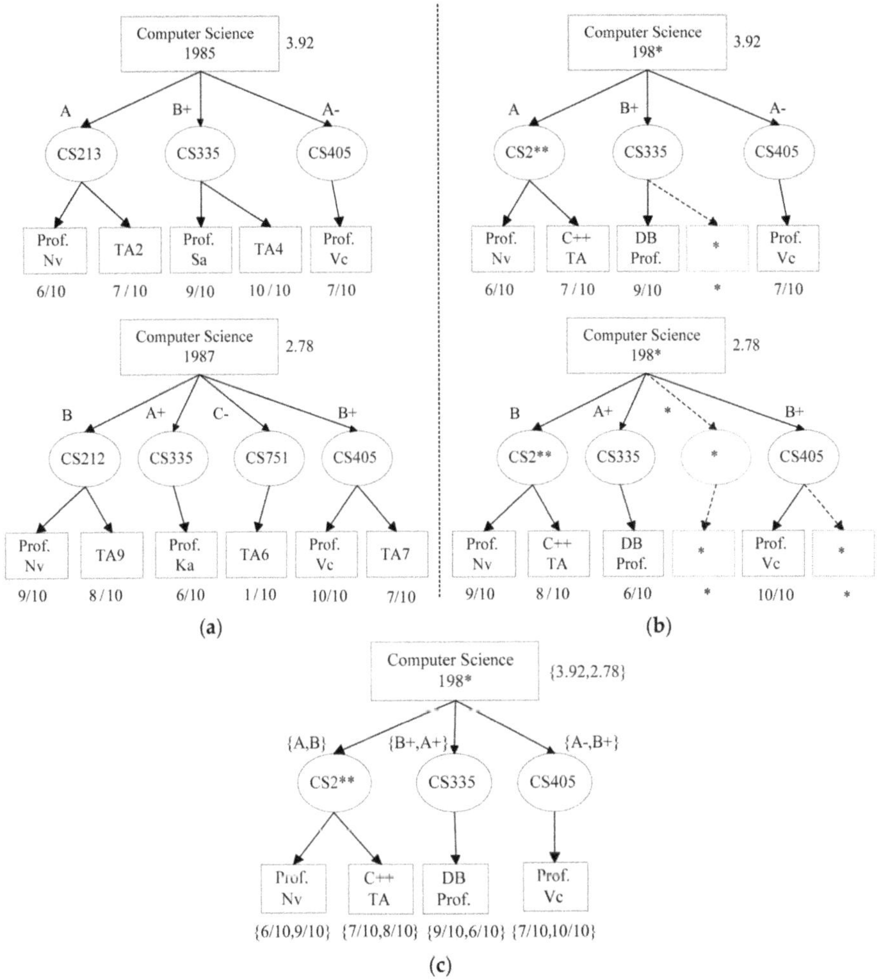

Figure 4. An anonymous example: (**a**) Two raw hierarchical data records; (**b**) The anonymous results; (**c**) Class representative of results.

Now, we give the clustering algorithm for anonymizing the entire hierarchical data, as shown in Algorithm 2. The input is a hierarchical data H and privacy parameters α^h_{lev} and k. The output is the anonymous data H' satisfies (α^h_{lev}, k)-anonymity. In lines 2–16, when the number of records in H is equal or larger than k, the algorithm creates an equivalence class from H. The first record is randomly picked in an equivalence class Q. For any residual record T_i in H, we compute the information loss by adding T_i to Q, and then sort H in ascending order according to the information loss. We select other k-1 records from the first 50 records to decrease the runtime of algorithm. In lines 17 and 18, when the number of records in H is less than k, the algorithm suppresses the all records in H.

Algorithm 2. *MLevCluTree*(H, α^h_{lev}, k)

Input: A hierarchical data $H = \{T_1, T_2, \ldots, T_n\}$, and privacy parameters α^h_{lev}, k;
Output: anonymous dataset H' which satisfies (α^h_{lev}, k)-anonymity

$H' \leftarrow \varnothing$;
while $H \geq k$ **do**
  pick randomly a record x from H; $H \leftarrow H\text{-}x$;
  initialize Q with x and $C_{rep} \leftarrow x$;
  $Q\_cost \leftarrow \varnothing$;
  for i = 1 to $|H|$ **do**
    $loss \leftarrow$ *MLevAnonytree*(copy(x), copy(T_i));
    Q_cost.append(*loss*);
  use *Q_cost* to sort H in ascending order;
  cand_set $\leftarrow$ H[1:50];
  for j = 2 to k **do**
    $y' \leftarrow \text{argmin}_{y \in cand\_set}$(*MLevAnonytree*(copy($C_{rep}$), copy($y$)));
  $H \leftarrow H\text{-}\ y'$; $cand\_set \leftarrow cand\_set\text{-}\ y'$; $Q \leftarrow Q \cup y'$;
  update C_{rep};
  $H' \leftarrow H' \cup Q$;
if $H \neq \varnothing$ **then**
  suppress all records in H;
return H';

5. Experimental Results

The objective of these experiments is to evaluate the performance of the proposed algorithm with respect to data utility, security and efficiency by comparing with existing anonymous approach *Clutree* [16] in hierarchical data which achieves l-diversity. The algorithms are implemented in Python, and ran on a computer with a four-core 3.4 GHz CPU and 8 GB RAM running Windows 7. We experimented on two synthetic datasets, which are obtained by the authors in [16]. They were modeled synthetically based on the real information of graduates from Sabanci University in Turkey. The synthetic dataset A has two levels (h = 2), in the order of (*major program*, *year of birth*) $\rightarrow$ *courses*, which contains 1000 students and nearly 20 courses per student. The synthetic data set B has three levels (h = 3), in the order of (*major program*, *year of birth*) $\rightarrow$ *courses* $\rightarrow$ *teachers*, in which there are 1000 students, every student studies nearly 20 courses, and every course has one to two teachers.

5.1. Evaluation Metrics

We evaluate data utility, security and efficiency of our method by using *LM* cost [16,28], dissimilarity degree of the equivalence class [22] and the execution time, respectively.

For a hierarchical data record T, the cost of T is computed as follows:

$$\cos t(T) = \sum_{v \in \Omega} \sum_{q \in v_{QI}} LM'(q) + \sum_{\omega \in \Psi} |\omega_{QI}| \quad (5)$$

where Ω and Ψ are the sets of vertices which are not suppressed and suppressed, respectively, $|\omega_{QI}|$ is the number of QI attributes in ω, and $LM'(q) = (|u_q| - 1)/(|u| - 1)$ is the information loss of generalizing q to u_q. The larger information loss is, the lower utility is. *LM* cost is an important index to evaluate the utility of the anonymous method.

The equivalence class dissimilarity is proposed in [22] for relational data, and we extend it to hierarchical data. Let Q be an equivalence class and its class representative be C_{rep}. v is a vertex in C_{rep},

m is the number of sensitive values in v, and z is the number of sensitivity levels. The dissimilarity degree of v is defined as:

$$DSimDegree(v) = \frac{\sum_{i=1}^{m-1} \sum_{j=i+1}^{m} m_{ij}}{\sum_{i=1}^{z-1} \sum_{j=i+1}^{z} z_{ij}} \quad (6)$$

where m_{ij} is the distance between the sensitivity levels of the ith and jth sensitive values, and z_{ij} is the distance between the ith and jth sensitivity levels. The dissimilarity degree of Q is

$$DSimDegree(Q) = \frac{\sum_{i=1}^{N} Degree(v_i)}{N} \quad (7)$$

where N is the number of vertices of C_{rep}. The larger *Degree(Q)* is, the larger the difference between the sensitive values is, the stronger the ability to resist attacks is and the higher the security is.

5.2. Experimental Analysis

We compare our algorithm *MLevCluTree* with *Clustree* in [16] with respect to data utility, security and efficiency. Because *l*-diversity can ensure there are at least *l* hierarchical data records in an equivalence class, we set $k = l$. k is varied from 2 to 6. The value of each point is the mean value on 10 experiments.

The average information loss of a hierarchical data record for algorithms *MLevClusTree* and *Clustree* is shown in Figure 5. From the two figures, we can see that the information loss increases when k increases. Because k increases, an equivalence class contains more hierarchical data records, and the possibility of providing more general values for every QI attributes increases. Therefore, the information loss increases. For the dataset B with $h = 3$, because more vertices for a hierarchical data record are needed to generalize, the information loss is higher than that of the dataset B with $h = 2$. Although *MLevClusTree* considers that multiple sensitive values lie in the same level, different sensitivity levels are evaluated with different constraints. So the information loss for our *MLevClusTree* is less than one for *Clustree*, i.e., the utility of *MLevClusTree* is better than that of *Clustree*.

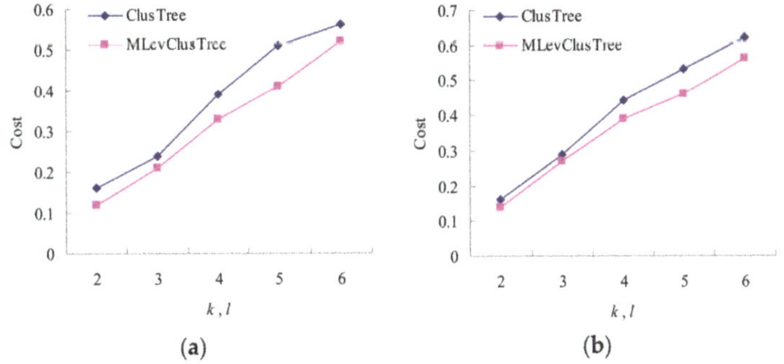

Figure 5. Information loss on two datasets: (**a**) Dataset A with $h = 2$; (**b**) Dataset B with $h = 3$.

The security of our *MLevClusTree* and *Clustree* is evaluated by the dissimilarity degree of equivalence class, and the results are shown in Figure 6. The ordinate denotes the average dissimilarity degree of an equivalence class. For an equivalence class, we can use Equation (7) to obtain its dissimilarity degree. Therefore, the results of dataset A with $h = 2$ and dataset B with $h = 3$ are not significantly different. As k increases, there are more sensitive values in different sensitivity levels, and the dissimilarity degree of a vertex in the class representative of an equivalence class increases. So the average dissimilarity degree of an equivalence class increases. From Figure 6, we can see that the average dissimilarity degree of an equivalence class for our *MLevClusTree* is higher than that for *Clustree*, since our approach restricts the proportion of sensitive values in different sensitivity levels. Therefore, our approach enhances the ability to resist similarity attacks and improves the data security.

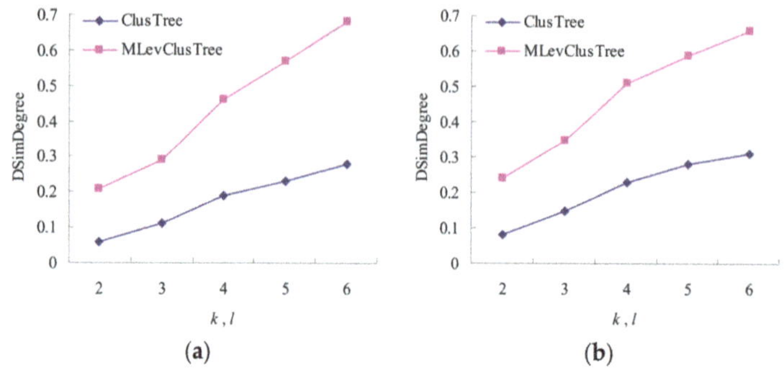

Figure 6. Dissimilarity degree of equivalence class on two datasets: (**a**) Dataset *A* with $h = 2$; (**b**) Dataset *B* with $h = 3$.

Finally, we evaluate the efficiency of our algorithm by the execution time. The experimental results are shown in Figure 7. We can see that the execution time of two algorithms increases with the increment of *k*. For every equivalence class *Q* in hierarchical data, the first hierarchical data record is randomly selected and we do not need to compute. For every other record in the equivalence class, we need to scan partial hierarchical data to find the record whose distance to current *Q* is approximately minimum. When *k* increases, the size of an equivalence class increases. Thus, the runtime increases. Also, we can see that the time for dataset *B* is more than that for dataset *A*, because the hierarchical data with more levels needs more time to find the record whose distance to current *Q* is approximately minimum. From Figure 7, we know that our *MLevClusTree* is slightly higher than that of *ClusTree* when *k* increases, since for every equivalence class *MLevClusTree* needs to decide whether the number of sensitive values in every sensitivity level exceeds the given threshold.

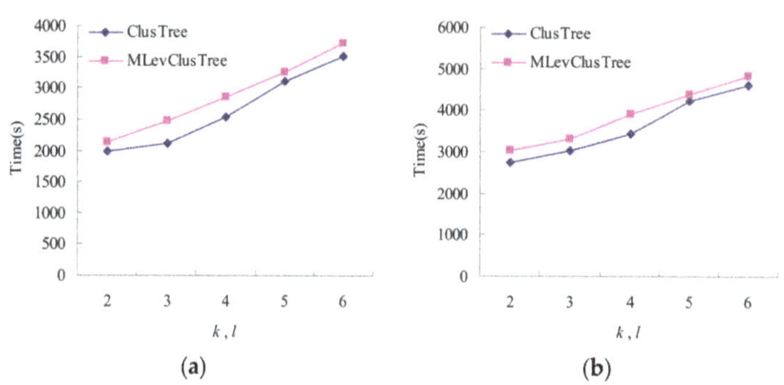

Figure 7. Execution time on two synthetic datasets: (**a**) Dataset *A* with $h = 2$; (**b**) Dataset *B* with $h = 3$.

From these experimental results, we can see that our *MlevClusTree* provides stronger privacy protection and has lower information loss, although it takes more time. It is acceptable because the anonymized process is offline.

6. Conclusions

Hierarchical data has become ubiquitous with the advent of document-oriented databases and the wide use of markup languages. However, this data contains privacy information, and so must be appropriately anonymized before it is to be published for scientific research and decision-making. To prevent similarity attacks in hierarchical data, in this paper, we use fuzzy set theory to partition sensitive values for a sensitive numerical or categorical attribute uniformly into five levels by converting the categorical attribute values into the numerical attribute values, and then map the five value levels to five sensitivity levels. According to these sensitivity levels, we propose privacy model (α_{lev}^{h}, k)-anonymity for hierarchical data with multi-level sensitivity and design a privacy-preserving approach to achieve (α_{lev}^{h}, k)-anonymity. Experimental results show that the average dissimilarity degree of these equivalence classes in anonymized hierarchical data obtained by our approach is higher than that for existing anonymous approaches in hierarchical data. Thus, our approach can effectively resist similarity attacks. Also, our approach causes less information loss and so improves the utility of anonymized hierarchical data.

Author Contributions: J.W. (Jinyan Wang), G.C. and X.L. put forward privacy model and the anonymization method, G.C. implemented the anonymization method with Python, J.W. (Jinyan Wang) and X.L. wrote the original manuscript and C.L. and J.W. (Jingli Wu) improved the writing.

Funding: This paper was supported by the National Natural Science Foundation of China (Nos. 61502111, 61763003, 61672176, 61762015, 61562007, 61662008), Guangxi Natural Science Foundation (Nos. 2016GXNSFAA380192, 2015GXNSFBA139246), Guangxi "Bagui Scholar" Teams for Innovation and Research Project, Guangxi Special Project of Science and Technology Base and Talents (AD16380008), and Guangxi Collaborative Innovation Center of Multisource Information Integration and Intelligent Processing.

Conflicts of Interest: The authors declare no conflict of interest.

References

1. Zavadskas, E.K.; Mardani, A.; Turskis, Z.; Jusoh, A.; Nor, K. Development of TOPSIS method to solve complicated decision-making problems—An overview on developments from 2000 to 2015. *Int. J. Inf. Technol. Decis. Mak.* **2016**, *15*, 645–682. [CrossRef]
2. Zavadskas, E.K.; Turskis, Z.; Kildienė, S. State of art surveys of overviews on MCDM/MADM methods. *Technol. Econ. Dev. Econ.* **2014**, *20*, 165–179. [CrossRef]
3. Li, Z.; Sun, D.; Zeng, H. Intuitionistic Fuzzy Multiple Attribute Decision-Making Model Based on Weighted Induced Distance Measure and Its Application to Investment Selection. *Symmetry* **2018**, *10*, 261. [CrossRef]
4. Li, D.; He, J.; Cheng, P.; Wang, J.; Zhang, H. A Novel Selection Model of Surgical Treatments for Early Gastric Cancer Patients Based on Heterogeneous Multicriteria Group Decision-Making. *Symmetry* **2018**, *10*, 223. [CrossRef]
5. Ma, X.; Zhan, J.; Ali, M.I.; Mehmood, N. A survey of decision making methods based on two classes of hybrid soft set models. *Artif. Intell. Rev.* **2018**, *49*, 511–529. [CrossRef]
6. Zhan, J.; Xu, W. Two types of coverings based multigranulation rough fuzzy sets and applications to decision making. *Artif. Intell. Rev.* **2018**. [CrossRef]
7. Zhan, J.; Liu, Q.; Herawan, T. A novel soft rough set: Soft rough hemirings and its multicriteria group decision making. *Appl. Soft Comput.* **2017**, *54*, 393–402. [CrossRef]
8. Hu, C.K.; Liu, F.B.; Hu, C.F. A Hybrid Fuzzy DEA/AHP Methodology for Ranking Units in a Fuzzy Environment. *Symmetry* **2017**, *9*, 273. [CrossRef]
9. Kang, J.; Han, J.; Park, J.H. Design of IP Camera Access Control Protocol by Utilizing Hierarchical Group Key. *Symmetry* **2015**, *7*, 1567–1586. [CrossRef]
10. Lee, G. Hierarchical Clustering Using One-Class Support Vector Machines. *Symmetry* **2015**, *7*, 1164–1175. [CrossRef]
11. Samarati, P.; Sweeney, L. Generalizing data to provide anonymity when disclosing information. In Proceedings of the ACM Symposium on Principles of Database Systems, Seattle, WA, USA, 1–3 June 1998; p. 188.

12. Fung, B.C.M.; Wang, K.; Chen, R.; Yu, P.S. Privacy-preserving data publishing: A survey of recent development. *ACM Comput. Surv.* **2010**, *42*, 14. [CrossRef]
13. Machanavajjhala, A.; Kifer, D.; Gehrke, J.; Venkitasubramaniam, M. *l*-diversity: Privacy beyond *k*-anonymity. *ACM Trans. Knowl. Discov Data* **2007**, *1*, 3. [CrossRef]
14. Wong, C.R.; Li, J.; Fu, A.; Wang, K. (α, k)-anonymity: An enhanced *k*-anonymity model for privacy preserving data publishing. In Proceedings of the 12th ACM SIGKDD International Conference on Knowledge Discovery and Data Mining, Philadelphia, PA, USA, 20–23 August 2006; pp. 754–759.
15. Li, N.; Li, T.; Venkatasubramanian, S. *t*-closeness: Privacy beyond *k*-anonymity and *l*-diversity. In Proceedings of the IEEE 23rd International Conference on Data Engineering, Istanbul, Turkey, 15–20 April 2007; pp. 106–115.
16. Ozalp, I.; Gursoy, M.E.; Nergiz, M.E.; Saygin, Y. Privacy-preserving publishing of hierarchical data. *ACM Trans. Priv. Secur.* **2016**, *19*, 7. [CrossRef]
17. Lefevre, K.; Dewitt, D.J.; Ramakrishnan, R. Incognito: Efficient full-domain *k*-anonymity. In Proceedings of the 2005 ACM SIGMOD International Conference on Management of Data, Baltimore, MD, USA, 14–16 June 2005; pp. 49–60.
18. Fung, B.C.M.; Wang, K.; Yu, P.S. Top-down specialization for information and privacy preservation. In Proceedings of the 21st International Conference on Data Engineering, Tokoyo, Japan, 5–8 April 2005; pp. 205–216.
19. Aggarwal, G.; Feder, T.; Kenthapadi, K.; Khuller, S.; Panigrahy, R.; Thomas, D.; Zhu, A. Achieving anonymity via clustering. *ACM Trans. Algorithms* **2010**, *6*, 49. [CrossRef]
20. Ghinita, G.; Karras, P.; Kalnis, P.; Mamoulis, N. A framework for efficient data anonymization under privacy and accuracy constraints. *ACM Trans. Database Syst.* **2009**, *32*, 9. [CrossRef]
21. Wang, J.; Du, K.; Luo, X.; Li, X. Two privacy-preserving approaches for data publishing with identity reservation. *Knowl. Inf. Syst.* **2018**. [CrossRef]
22. Han, J.; Yu, J.; Yu, H.; Jia, J. A multi-level *l*-diversity model for numerical sensitive attributes. *J. Comput. Res. Dev.* **2011**, *48*, 147–158.
23. Jin, H.; Zhang, Z.; Liu, S.; Ju, S. (α_i, k)-anonymity Privacy Preservation Based on Sensitivity Grading. *Comput. Eng.* **2011**, *37*, 12–17.
24. Wang, Q.; Yang, C.; Liu, H. Fuzzy based methods for privacy preserving. *Appl. Res. Comput.* **2013**, *30*, 518–520.
25. Kumari, V.V.; Rao, S.S.; Raju, K.; Ramana, K.V.; Avadhani, B.V.S. Fuzzy based approach for privacy preserving publication of data. *Int. J. Comput. Sci. Netw. Secur.* **2008**, *8*, 115–121.
26. Yang, X.; Li, C. Secure XML publishing without information leakage in the presence of data inference. In Proceedings of the Thirtieth International Conference on Very Large Data Bases, Toronto, ON, Canada, 31 August–3 September 2004; pp. 96–107.
27. Landberg, A.H.; Nguyen, K.; Pardede, E.; Rahayu, J.W. δ-dependency for privacy-preserving XML data publishing. *J. Biomed. Inform.* **2014**, *50*, 77–94. [CrossRef] [PubMed]
28. Nergiz, M.E.; Clifton, C.; Nergiz, A.E. Multirelational *k*-anonymity. *IEEE Trans. Knowl. Data Eng.* **2009**, *21*, 1104–1117. [CrossRef]
29. Gkountouna, O.; Terrovitis, M. Anonymizing collections of tree-structured data. *IEEE Trans. Knowl. Data Eng.* **2015**, *27*, 2034–2048. [CrossRef]
30. Zadeh, L.A. Fuzzy sets. *Inf. Control* **1965**, *8*, 338–353. [CrossRef]
31. Jorba, L.; Adillon, R. Interval Fuzzy Segments. *Symmetry* **2018**, *10*, 309. [CrossRef]
32. Bi, L.; Dai, S.; Hu, B. Complex Fuzzy Geometric Aggregation Operators. *Symmetry* **2018**, *10*, 251. [CrossRef]
33. Klir, G.J.; Clair, U.S.; Yuan, B. *Fuzzy Set Theory: Foundations and Applications*; Prentice-Hall, Inc.: Upper Saddle River, NJ, USA, 1997.

symmetry

MDPI

Article

Multiple-Attribute Decision-Making Method Using Similarity Measures of Hesitant Linguistic Neutrosophic Numbers Regarding Least Common Multiple Cardinality

Wenhua Cui and Jun Ye *

Department of Electrical Engineering and Automation, Shaoxing University, 508 Huancheng West Road, Shaoxing 312000, Zhejiang, China; wenhuacui@usx.edu.cn

* Correspondence: yejun@usx.edu.cn; Tel.: +86-575-8832-7323

Received: 10 July 2018; Accepted: 7 August 2018; Published: 9 August 2018

Abstract: Linguistic neutrosophic numbers (LNNs) are a powerful tool for describing fuzzy information with three independent linguistic variables (LVs), which express the degrees of truth, uncertainty, and falsity, respectively. However, existing LNNs cannot depict the hesitancy of the decision-maker (DM). To solve this issue, this paper first defines a hesitant linguistic neutrosophic number (HLNN), which consists of a few LNNs regarding an evaluated object due to DMs' hesitancy to represent their hesitant and uncertain information in the decision-making process. Then, based on the least common multiple cardinality (LCMC), we present generalized distance and similarity measures of HLNNs, and then develop a similarity measure-based multiple-attribute decision-making (MADM) method to handle the MADM problem in the HLNN setting. Finally, the feasibility of the proposed approach is verified by an investment decision case.

Keywords: hesitant linguistic neutrosophic number (HLNN); decision-making; similarity measure; distance measure; least common multiple cardinality (LCMC)

1. Introduction

In the real world, the linguistic expression is well-suited for the thinking and expressing patterns of human beings. Due to the vagueness of languages and the complexity of decision-making environments, the linguistic fuzzy theory has been well developed in the past decades and shows irreplaceable advantages in the fuzzy decision-making field. Linguistic variables (LVs) were defined for fuzzy reasoning and decision-making [1–4]. Linguistic uncertain variables [5,6] (interval-valued linguistic variables) were then defined to depict uncertain linguistic information in decision-making problems [7,8]. After that, a linguistic intuitionistic fuzzy number (LIFN) [9], which contains two independent LVs to describe the degrees of truth and falsity, respectively, was presented to handle the uncertainty and incompleteness in linguistic decision-making environments [10]. Furthermore, with the wide application of the neutrosophic theory in decision-making [11–13], Fang and Ye [14] proposed a linguistic neutrosophic number (LNN) by adding a new LV to the LIFN for representing the indeterminacy degree to do with the indeterminate and inconsistent linguistic information [15]. Although there exist some research works on LNNs [14,15], existing LNNs cannot depict the hesitancy of decision-makers (DMs) in the linguistic assessment of alternatives.

Concerning the handling of the human hesitant cognition in decision-making environments, many works have been published so far. Torra and Narukawa [16] and Torra [17] originally introduced hesitant fuzzy sets (HFSs) to express the hesitancy by allowing the membership to contain several possible values. Then, for linguistic decision-making problems, the expression of a hesitant fuzzy linguistic set (HFLS) [18] was obtained based on combining a linguistic term (LT) set with a HFS so as to satisfy the hesitant linguistic evaluation requirements [19,20] of DMs. After that, an interval-valued HFLS [21] was presented as an extension form by combining an interval-valued LT set with a HFS. Recently, Ye [22] proposed the hesitant neutrosophic linguistic number (HNLN) to carry out hesitant decision-making problems with the neutrosophic linguistic number that contains partial determinacy and partial indeterminacy. However, there is no definition or decision-making method for the hesitant sets of LNNs in the existing literature. Additionally, in the hesitant linguistic expressions of DMs, the components between two hesitant sets generally have difference in their length sizes, and thus it is difficult to directly perform measure calculations between hesitant sets. Thus, several researchers have proposed some extension methods to extend the shorter items in the two hesitant sets by adding the minimum values, maximum values, or any values [23,24] to reach their identical length. However, these extension methods depend too much on the subjective preferences and interests of the DMs. To solve this problem, we have already introduced the least common multiple cardinality (LCMC) to extend the hesitant fuzzy elements in our previous research works [22,25], which become more objective for the decision-making calculation of HFSs.

As aforementioned, there is a gap of hesitant LNNs in existing studies. For instance, suppose that we hesitate between two single-valued LNNs, $<h_7, h_3, h_4>$ and $<h_5, h_3, h_1>$, from the given LT set $H = \{h_s \mid s \in [0, 8]\}$ regarding an evaluated object. However, it is difficult to express the hesitation information and the LNN information of the DMs simultaneously by a unique LNN or a unique HFS. Therefore, for the purposes of satisfying the demand of hesitant decision-making with LNNs and ensuring the objectivity of the measure calculation, this paper aims to (i) define the concept of HLNNs by combining HFSs with LNNs, (ii) present the LCMC-based generalized distance and similarity measures of HLNNs for more objective measure calculation of HLNN information, and (iii) to propose a novel multiple-attribute decision-making (MADM) method based on the proposed LCMC-based similarity measure in the HLNN setting.

In order to do so, Section 2 briefly reviews LNNs. Section 3 defines a HLNN and a HLNN set. Then, in Section 4, the LCMC-based generalized distance and similarity measures of HLNNs are presented. In Section 5, a new MADM method was developed by using the proposed similarity measure of HLNNs. In Section 6, the feasibility of the proposed approach is demonstrated by an investment case. The conclusions and future research of HLNNs are discussed in the last section.

2. Linguistic Neutrosophic Numbers (LNNs)

Fang and Ye [14] originally presented the following definition of the LNN:

Definition 1 ([14]). *Let $H = \{h_0, h_1, \ldots, h_\tau\}$ be a LT set, where $\tau + 1$ is an odd cardinality. A LNN can be defined as $\vartheta = < h_T, h_U, h_F>$ for $h_T, h_U, h_F \in H$ and $T, U, F \in [0, \tau]$, where h_T, h_U, h_F represent the degrees of truth, indeterminacy, and falsity, respectively.*

For the comparison of LNNs, the score and accuracy functions of LNNs are defined as follows [14]:

Definition 2 ([14]). *Let $\vartheta = <h_T, h_U, h_F>$ be a LNN in H. Then its score function can be given by:*

$$S(\vartheta) = (2\tau + T - U - F)/3\tau \text{ for } S(\vartheta) \in [0,1], \tag{1}$$

and its accuracy function can be expressed as

$$V(\vartheta) = (T - F)/\tau \text{ for } V(\vartheta) \in [-1,1]. \tag{2}$$

Definition 3 ([14]). *Let $\vartheta_\alpha =< h_{T_\alpha}, h_{U_\alpha}, h_{F_\alpha} >$ and $\vartheta_\beta =< h_{T_\beta}, h_{U_\beta}, h_{F_\beta} >$ be two LNNs in H. There exist the following relations:*

(1) *If* $S(\vartheta_\alpha) < S(\vartheta_\beta)$*, then* $\vartheta_\alpha < \vartheta_\beta$*;*
(2) *If* $S(\vartheta_\alpha) > S(\vartheta_\beta)$*, then* $\vartheta_\alpha > \vartheta_\beta$*;*
(3) *If* $S(\vartheta_\alpha) = S(\vartheta_\beta)$ *and* $V(\vartheta_\alpha) < V(\vartheta_\beta)$*, then* $\vartheta_\alpha < \vartheta_\beta$*;*
(4) *If* $S(\vartheta_\alpha) = S(\vartheta_\beta)$ *and* $V(\vartheta_\alpha) > V(\vartheta_\beta)$*, then* $\vartheta_\alpha > \vartheta_\beta$*;*
(5) *If* $S(\vartheta_\alpha) = S(\vartheta_\beta)$ *and* $V(\vartheta_\alpha) = V(\vartheta_\beta)$*, then* $\vartheta_\alpha = \vartheta_\beta$*.*

3. Hesitant Linguistic Neutrosophic Numbers (HLNNs) and HLNN Set

Torra and Narukawa [16] and Torra [17] first defined the HFS as follows:

Definition 4 ([16,17]). *Assume S is a universe set, then a HFS N on S can be given by*

$$N = \{< s, E(s) > | s \in S\},$$

where E(s) is a hesitant component of N containing a set of some values in [0, 1], which represents all possible membership degrees of s.

By integrating HFS with LNN, we define a HLNN set as follows:

Definition 5. *Set a universe of discourse* $S = \{s_1, s_2, \ldots, s_q\}$ *and a finite LT set* $H = \{h_0, h_1, \ldots, h_\tau\}$*, and then a HLNN set* N_l *on S can be expressed as*

$$N_l = \{< s_j, E_l(s_j) > | s_j \in S, j = 1, 2, \cdots, q\}$$

where $E_l(s_j)$ *is a set of* m_j *LNNs, denoted by a HLNN* $E_l(s_j) = \{< h_{T_j^k}, h_{U_j^k}, h_{F_j^k} > h_{T_j^k} \in H, h_{U_j^k} \in H, h_{F_j^k} \in H, k = 1, 2, \cdots, m_j\}$ *for* $s_j \in S$*.*

4. LCMC-Based Distance and Similarity Measures of HLNNs

In most situations, the cardinal numbers (the number of LNNs) of HLNNs evaluated for the same object are usually different. Thus, it is necessary to make the cardinal numbers of the two HLNNs the same to satisfy the distance and similarity measures between them.

We assume that p HLNNs on $S = \{s_1, s_2, \ldots, s_q\}$ are $E_{l_1}(s_j), E_{l_2}(s_j), \cdots, E_{l_p}(s_j)$ for $s_j \in S$ (j = 1, 2, ..., q). Then, the HLNNs $E_{l_i}(s_j)$ for i = 1, 2, ... , p can be given by

$$\begin{aligned}
E_{l_1}(s_j) &= \{< h_{T_{1j}^1}, h_{U_{1j}^1}, h_{F_{1j}^1} >, < h_{T_{1j}^2}, h_{U_{1j}^2}, h_{F_{1j}^2} >, \cdots, < h_{T_{1j}^{m_{1j}}}, h_{U_{1j}^{m_{1j}}}, h_{F_{1j}^{m_{1j}}} >\}, \\
E_{l_2}(s_j) &= \{< h_{T_{2j}^1}, h_{U_{2j}^1}, h_{F_{2j}^1} >, < h_{T_{2j}^2}, h_{U_{2j}^2}, h_{F_{2j}^2} >, \cdots, < h_{T_{2j}^{m_{2j}}}, h_{U_{2j}^{m_{2j}}}, h_{F_{2j}^{m_{2j}}} >\}, \\
&\cdots, \\
E_{l_p}(s_j) &= \{< h_{T_{pj}^1}, h_{U_{pj}^1}, h_{F_{pj}^1} >, < h_{T_{pj}^2}, h_{U_{pj}^2}, h_{F_{pj}^2} >, \cdots, < h_{T_{pj}^{m_{pj}}}, h_{U_{pj}^{m_{pj}}}, h_{F_{pj}^{m_{pj}}} >\},
\end{aligned}$$

where m_{ij} is the cardinal number of $E_{l_i}(s_j)$ (i = 1, 2, ... , p and j = 1, 2, ... , q).

Provided that the LCMC of m_{ij} (i = 1, 2, ..., p and j = 1, 2, ..., q) is c_j (j = 1, 2, ... , q), by increasing the number of LNNs $< h_{T_{ij}^k}, h_{U_{ij}^k}, h_{F_{ij}^k} >$ (k = 1, 2, ..., m_{ij}) in $E_{l_i}(s_j)$ depending on c_j (j = 1, 2, ... , q), the extended HLNN $E_{l_i}^o(s_j)$ (i = 1, 2, ... , p and j = 1, 2, ... , q) will be obtained by the extension forms:

$$E_{l_1}^o(s_j) = \left\{ \overbrace{\underbrace{< h_{T_{1j}^1}, h_{U_{1j}^1}, h_{F_{1j}^1} >, \cdots}_{R_{1j}}, \underbrace{< h_{T_{1j}^2}, h_{U_{1j}^2}, h_{F_{1j}^2} >, \cdots}_{R_{1j}}, \cdots, \underbrace{< h_{T_{1j}^{m_{1j}}}, h_{U_{1j}^{m_{1j}}}, h_{F_{1j}^{m_{1j}}} >, \cdots}_{R_{1j}}}^{c_j} \right\},$$

$$E_{l_2}^o(s_j) = \left\{ \overbrace{\underbrace{< h_{T_{2j}^1}, h_{U_{2j}^1}, h_{F_{2j}^1} >, \cdots}_{R_{2j}}, \underbrace{< h_{T_{2j}^2}, h_{U_{2j}^2}, h_{F_{2j}^2} >, \cdots}_{R_{2j}}, \cdots, \underbrace{< h_{T_{2j}^{m_{2j}}}, h_{U_{2j}^{m_{2j}}}, h_{F_{2j}^{m_{2j}}} >, \cdots}_{R_{2j}}}^{c_j} \right\},$$

$$\cdots,$$

$$E_{l_p}^o(x_j) = \left\{ \overbrace{\underbrace{< h_{T_{pj}^1}, h_{U_{pj}^1}, h_{F_{pj}^1} >, \cdots}_{R_{pj}}, \underbrace{< h_{T_{pj}^2}, h_{U_{pj}^2}, h_{F_{pj}^2} >, \cdots}_{R_{pj}}, \cdots, \underbrace{< h_{T_{pj}^{m_{pj}}}, h_{U_{pj}^{m_{pj}}}, h_{F_{pj}^{m_{pj}}} >, \cdots}_{R_{pj}}}^{c_j} \right\},$$

where R_{ij} is the number of LNNs $< h_{T_{ij}^k}, h_{U_{ij}^k}, h_{F_{ij}^k} >$ (k = 1, 2, ..., m_{ij}) in $E_{l_i}^o(x_j)$ (i = 1, 2, ... , p and j = 1, 2, ... , q), calculated by:

$$R_{ij} = \frac{c_j}{m_{ij}}. \tag{3}$$

Additionally, the elements $\vartheta_{ij}^{\sigma(k)} = < h_{T_{ij}^{\sigma(k)}}, h_{U_{ij}^{\sigma(k)}}, h_{F_{ij}^{\sigma(k)}} >$ (k = 1, 2, ... , c_j) in $E_{l_i}^o(x_j)$ are arranged in an ascending order, denoted as $E_{l_i}^o(x_j) = \{\vartheta_{ij}^{\sigma(1)}, \vartheta_{ij}^{\sigma(2)}, \cdots, \vartheta_{ij}^{\sigma(c_j)}\}$ (i = 1, 2, ... , p and j = 1, 2, ... , q), where $\sigma : (1, 2, \ldots, c_j) \to (1, 2, \ldots, c_j)$ is a permutation satisfying $\vartheta_{ij}^{\sigma(k)} \le \vartheta_{ij}^{\sigma(k+1)}$ (k = 1, 2, ... , c_j).

Definition 6. *Let $N_{l_1} = \{E_{l_1}(s_1), E_{l_1}(s_2), \cdots, E_{l_1}(s_q)\}$ and $N_{l_2} = \{E_{l_2}(s_1), E_{l_2}(s_2), \cdots, E_{l_2}(s_q)\}$ be two HLNN sets on S = {s_1, s_2, ..., s_q}, where $E_{l_1}(s_j)$ and $E_{l_2}(s_j)$ (j = 1, 2, ... , q) are HLNNs in a LT set H = {h_0, h_1, ..., h_τ} for $h_j \in H$. Let $f(h_j) = j/\tau$ be a linguistic scale function. Then, the normalized generalized distance between N_{l_1} and N_{l_2} can be represented as:*

$$\begin{aligned} d(N_{l_1}, N_{l_2}) &= \left\{ \frac{1}{q}\sum_{j=1}^{q}\left[\frac{1}{3c_j}\sum_{k=1}^{c_j}\left(|f(h_{T_{1j}^{\sigma(k)}}) - f(h_{T_{2j}^{\sigma(k)}})|^\rho + |f(h_{U_{1j}^{\sigma(k)}}) - f(h_{U_{2j}^{\sigma(k)}})|^\rho + |f(h_{F_{1j}^{\sigma(k)}}) - f(h_{F_{2j}^{\sigma(k)}})|^\rho\right)\right]\right\}^{1/\rho} \\ &= \left\{ \frac{1}{q}\sum_{j=1}^{q}\left[\frac{1}{3c_j\tau^\rho}\sum_{k=1}^{c_j}\left(|T_{1j}^{\sigma(k)} - T_{2j}^{\sigma(k)}|^\rho + |U_{1j}^{\sigma(k)} - U_{2j}^{\sigma(k)}|^\rho + |F_{1j}^{\sigma(k)} - F_{2j}^{\sigma(k)}|^\rho\right)\right]\right\}^{1/\rho} \quad \text{for } \rho > 0. \end{aligned} \tag{4}$$

Obviously, $d(N_{l_1}, N_{l_2})$ degenerates to the normalized generalized distance of Hamming for ρ = 1 and to the normalized generalized distance of Euclidean for ρ = 2.

For the generalized distance $d(N_{l_1}, N_{l_2})$, there is a proposition as follows:

Proposition 1. *For any two HLNN sets $N_{l_1} = \{E_{l_1}(s_1), E_{l_1}(s_2), \cdots, E_{l_1}(s_q)\}$ and $N_{l_2} = \{E_{l_2}(s_1), E_{l_2}(s_2), \cdots, E_{l_2}(s_q)\}$, the generalized distance $d(N_{l_1}, N_{l_2})$ between N_{l_1} and N_{l_2} for ρ > 0 contains the following properties:*

(HP1) $0 \le d(N_{l_1}, N_{l_2}) \le 1$;
(HP2) $d(N_{l_1}, N_{l_2}) = 0$ *if and only if* $N_{l_1} = N_{l_2}$;
(HP3) $d(N_{l_1}, N_{l_2}) = d(N_{l_2}, N_{l_1})$;

(HP4) *Let* $N_{l_3} = \{E_{l_3}(s_1), E_{l_3}(s_2), \cdots, E_{l_3}(s_q)\}$ *be a HLNN set, then* $d(N_{l_1}, N_{l_2}) \le d(N_{l_1}, N_{l_3})$ *and* $d(N_{l_2}, N_{l_3}) \le d(N_{l_1}, N_{l_3})$ *if* $N_{l_1} \subseteq N_{l_2} \subseteq N_{l_3}$.

Proof. It is obvious that the properties (HP1)–(HP3) are satisfied for $d(N_{l_1}, N_{l_2})$. Thus, we only need to prove the property (HP4).

Since there is $N_{l_1} \subseteq N_{l_2} \subseteq N_{l_3}$, there exists $E^0_{l_1}(s_j) \le E^0_{l_2}(s_j) \le E^0_{l_3}(s_j)$ for $s_j \in S$ ($j = 1, 2, \ldots, q$), which implies $T_{3j}^{\sigma(k)} \ge T_{2j}^{\sigma(k)} \ge T_{1j}^{\sigma(k)}$, $U_{3j}^{\sigma(k)} \le U_{2j}^{\sigma(k)} \le U_{1j}^{\sigma(k)}$, $F_{3j}^{\sigma(k)} \le F_{2j}^{\sigma(k)} \le F_{1j}^{\sigma(k)}$ for $k = 1, 2, \ldots, c_j$. It follows that

$$\left|T_{1j}^{\sigma(k)} - T_{2j}^{\sigma(k)}\right|_\rho \le \left|T_{1j}^{\sigma(k)} - T_{3j}^{\sigma(k)}\right|^\rho, \left|T_{2j}^{\sigma(k)} - T_{3j}^{\sigma(k)}\right|_\rho \le \left|T_{1j}^{\sigma(k)} - T_{3j}^{\sigma(k)}\right|_\rho,$$
$$\left|U_{1j}^{\sigma(k)} - U_{2j}^{\sigma(k)}\right|_\rho \le \left|U_{1j}^{\sigma(k)} - U_{3j}^{\sigma(k)}\right|^\rho, \left|U_{2j}^{\sigma(k)} - U_{3j}^{\sigma(k)}\right|_\rho \le \left|U_{1j}^{\sigma(k)} - U_{3j}^{\sigma(k)}\right|_\rho,$$
$$\left|F_{1j}^{\sigma(k)} - F_{2j}^{\sigma(k)}\right|_\rho \le \left|F_{1j}^{\sigma(k)} - F_{3j}^{\sigma(k)}\right|^\rho, \left|F_{2j}^{\sigma(k)} - F_{3j}^{\sigma(k)}\right|_\rho \le \left|F_{1j}^{\sigma(k)} - F_{3j}^{\sigma(k)}\right|_\rho.$$

Then there are the following inequalities:

$$\left|T_{1j}^{\sigma(k)} - T_{2j}^{\sigma(k)}\right|_\rho + \left|U_{1j}^{\sigma(k)} - U_{2j}^{\sigma(k)}\right|_\rho + \left|F_{1j}^{\sigma(k)} - F_{2j}^{\sigma(k)}\right|_\rho \le \left|T_{1j}^{\sigma(k)} - T_{3j}^{\sigma(k)}\right|_\rho + \left|U_{1j}^{\sigma(k)} - U_{3j}^{\sigma(k)}\right|_\rho + \left|F_{1j}^{\sigma(k)} - F_{3j}^{\sigma(k)}\right|^\rho,$$
$$\left|T_{2j}^{\sigma(k)} - T_{3j}^{\sigma(k)}\right|_\rho + \left|U_{2j}^{\sigma(k)} - U_{3j}^{\sigma(k)}\right|_\rho + \left|F_{2j}^{\sigma(k)} - F_{3j}^{\sigma(k)}\right|_\rho \le \left|T_{1j}^{\sigma(k)} - T_{3j}^{\sigma(k)}\right|_\rho + \left|U_{1j}^{\sigma(k)} - U_{3j}^{\sigma(k)}\right|_\rho + \left|F_{1j}^{\sigma(k)} - F_{3j}^{\sigma(k)}\right|_\rho.$$

Thus, the following relations can be further obtained:

$$\frac{1}{3c_j\tau^\rho}\left[\sum_{k=1}^{c_j}\left(|(T_{1j}^{\sigma(k)}) - (T_{2j}^{\sigma(k)})|^\rho + |(U_{1j}^{\sigma(k)}) - (U_{2j}^{\sigma(k)})|^\rho + |(F_{1j}^{\sigma(k)}) - (F_{2j}^{\sigma(k)})|^\rho\right)\right]$$
$$\le \frac{1}{3c_j\tau^\rho}\left[\sum_{k=1}^{c_j}\left(|(T_{1j}^{\sigma(k)}) - (T_{3j}^{\sigma(k)})|^\rho + |(U_{1j}^{\sigma(k)}) - (U_{3j}^{\sigma(k)})|^\rho + |(F_{1j}^{\sigma(k)}) - (F_{3j}^{\sigma(k)})|^\rho\right)\right],$$
$$\frac{1}{3c_j\tau^\rho}\left[\sum_{k=1}^{c_j}\left(|(T_{2j}^{\sigma(k)}) - (T_{3j}^{\sigma(k)})|^\rho + |(U_{2j}^{\sigma(k)}) - (U_{3j}^{\sigma(k)})|^\rho + |(F_{2j}^{\sigma(k)}) - (F_{3j}^{\sigma(k)})|^\rho\right)\right]$$
$$\le \frac{1}{3c_j\tau^\rho}\left[\sum_{k=1}^{c_j}\left(|(T_{1j}^{\sigma(k)}) - (T_{3j}^{\sigma(k)})|^\rho + |(U_{1j}^{\sigma(k)}) - (U_{3j}^{\sigma(k)})|^\rho + |(F_{1j}^{\sigma(k)}) - (F_{3j}^{\sigma(k)})|^\rho\right)\right].$$

By Equation (4), there are $d(N_{l_1}, N_{l_2}) \le d(N_{l_1}, N_{l_3})$ and $d(N_{l_2}, N_{l_3}) \le d(N_{l_1}, N_{l_3})$ for $\rho > 0$. Therefore, the property (HP4) can hold. □

If we consider the weight w_j of an element $s_j \in S$ with $w_j \in [0, 1]$ and $\sum_{j=1}^{q} w_j = 1$, the generalized weighted distance between N_{l_1} and N_{l_2} is

$$\begin{aligned} d_w(N_{l_1}, N_{l_2}) &= \left\{\sum_{j=1}^{q} w_j\left[\frac{1}{3c_j}\sum_{k=1}^{c_j}\left(|f(h_{T_{1j}^{\sigma(k)}}) - f(h_{T_{2j}^{\sigma(k)}})|^\rho + |f(h_{U_{1j}^{\sigma(k)}}) - f(h_{U_{2j}^{\sigma(k)}})|^\rho + |f(h_{F_{1j}^{\sigma(k)}}) - f(h_{F_{2j}^{\sigma(k)}})|^\rho\right)\right]\right\}^{1/\rho} \\ &= \left\{\sum_{j=1}^{q} w_j\left[\frac{1}{3c_j\tau^\rho}\sum_{k=1}^{c_j}\left(|T_{1j}^{\sigma(k)} - T_{2j}^{\sigma(k)}|^\rho + |U_{1j}^{\sigma(k)} - U_{2j}^{\sigma(k)}|^\rho + |F_{1j}^{\sigma(k)} - F_{2j}^{\sigma(k)}|^\rho\right)\right]\right\}^{1/\rho} \quad for\ \rho > 0. \end{aligned} \tag{5}$$

Since the measures of similarity and distance are complementary with each other, the weighted measure of similarity between N_{l_1} and N_{l_2} can be represented by

$$\begin{aligned} S_w(N_{l_1}, N_{l_2}) &= 1 - d_w(N_{l_1}, N_{l_2}) \\ &= 1 - \left\{\sum_{j=1}^{q} w_j\left[\frac{1}{3c_j\tau^\rho}\sum_{k=1}^{c_j}\left(\left|T_{1j}^{\sigma(k)} - T_{2j}^{\sigma(k)}\right|_\rho + \left|U_{1j}^{\sigma(k)} - U_{2j}^{\sigma(k)}\right|_\rho + \left|F_{1j}^{\sigma(k)} - F_{2j}^{\sigma(k)}\right|^\rho\right)\right]\right\}^{1/\rho} \quad for\ \rho > 0. \end{aligned} \tag{6}$$

Similar to the properties (HP1)–(HP4) satisfied by the generalized distance measure in Proposition 1, the similarity measure $S_w(N_{l_1}, N_{l_2})$ also has the proposition as follows:

Proposition 2. *The similarity measure $S_w(N_{l_1}, N_{l_2})$ for $\rho > 0$ contains the following properties:*

(HP1) $0 \le S_w(N_{l_1}, N_{l_2}) \le 1$;

(HP2) $S_w(N_{l_1}, N_{l_2}) = 1$ *if and only if* $N_{l_1} = N_{l_2}$;

(HP3) $S_w(N_{l_1}, N_{l_2}) = S_w(N_{l_2}, N_{l_1})$;

(HP4) *Let* N_{l_3} *be a HLNN set, then there are* $S_w(N_{l_1}, N_{l_2}) \ge S_w(N_{l_1}, N_{l_3})$ *and* $S_w(N_{l_2}, N_{l_3}) \ge S_w(N_{l_1}, N_{l_3})$ *if* $N_{l_1} \subseteq N_{l_2} \subseteq N_{l_3}$.

Proof. It is clear that $S_w(N_{l_1}, N_{l_2})$ satisfies the properties (SP1)–(SP3). Thus, we only prove the property (SP4) here.

According to the proved property (HP4) in Proposition 1, if $N_{l_1} \subseteq N_{l_2} \subseteq N_{l_3}$, there exists the relations of $d_w(N_{l_1}, N_{l_2}) \le d_w(N_{l_1}, N_{l_3})$ and $d_w(N_{l_2}, N_{l_3}) \le d_w(N_{l_1}, N_{l_3})$ for $\rho > 0$. Since the similarity measure is the complement of the distance measure, both $S_w(N_{l_1}, N_{l_2}) \ge S_w(N_{l_1}, N_{l_3})$ and $S_w(N_{l_2}, N_{l_3}) \ge S_w(N_{l_1}, N_{l_3})$ can be easily obtained. Therefore, the property (SP4) can hold. □

5. MADM Method Using the Similarity Measure of HLNNs

For a MADM problem in the HLNN setting, some DMs need to evaluate p alternatives (denoted by $G = \{g_1, g_2, \dots, g_p\}$) over q attributes (denoted by $S = \{s_1, s_2, \dots, s_q\}$) from the LT set $H = \{h_0, h_1, \dots, h_\tau\}$. Then, a weight vector $W = (\omega_1, \omega_2, \dots, \omega_q)$, which is on the conditions of $0 \le \omega_j \le 1$ ($j = 1, 2, \dots, q$) and $\sum_{j=1}^{q} \omega_j = 1$, represents the importance of the attributes in S. Thus, the HLNN decision matrix M can be expressed as:

$$M = \left(E_{l_i}(s_j)\right)_{p\times q} = \begin{matrix} g_1 \\ g_2 \\ \vdots \\ g_p \end{matrix} \begin{bmatrix} E_{l_1}(s_1) & E_{l_1}(s_2) & \cdots & E_{l_1}(x_q) \\ E_{l_2}(s_1) & E_{l_2}(s_2) & \cdots & E_{l_2}(s_q) \\ \vdots & \vdots & \ddots & \vdots \\ E_{l_p}(s_1) & E_{l_p}(s_2) & \cdots & E_{l_p}(s_q) \end{bmatrix}.$$

where $E_{l_i}(s_j) = \{< h_{T_{ij}^1}, h_{U_{ij}^1}, h_{F_{ij}^1} >, < h_{T_{ij}^2}, h_{U_{ij}^2}, h_{F_{ij}^2} >, \cdots, < h_{T_{ij}^{m_{ij}}}, h_{U_{ij}^{m_{ij}}}, h_{F_{ij}^{m_{ij}}} >\}$ is a HLNN for $s_j \in S$, and m_{ij} is the number of LNNs in $E_{l_i}(s_j)$ ($i = 1, 2, \dots, p$ and $j = 1, 2, \dots, q$).

On the basis of the proposed similarity measure, a novel MADM method of HLNN is presented by the following steps:

Step 1: For any HLNN $E_{l_i}(s_j)$ ($j = 1, 2, \dots, q$) in M, rank all elements $\vartheta_{ij}^{\sigma(k)}$ ($k = 1, 2, \dots, m_{ij}$) in each HLNN $E_{l_i}(s_j)$ ($j = 1, 2, \dots, q$) in an ascending order according to their score and accuracy functions, then yield the corresponding extended HLNN $E_{l_i}^o(s_j)$ based on the LCMC c_j and the occurrence number R_{ij} of every LNN in $E_{l_i}(s_j)$ obtained by Equation (3). Hence, the extended decision matrix M° is

$$M^o = \left(E_{l_i}^o(s_j)\right)_{p\times q} = \begin{matrix} g_1 \\ g_2 \\ \vdots \\ g_p \end{matrix} \begin{bmatrix} E_{l_1}^o(s_1) & E_{l_1}^o(s_2) & \cdots & E_{l_1}^o(s_q) \\ E_{l_2}^o(s_1) & E_{l_2}^o(s_2) & \cdots & E_{l_2}^o(s_q) \\ \vdots & \vdots & \ddots & \vdots \\ E_{l_p}^o(s_1) & E_{l_p}^o(s_2) & \cdots & E_{l_p}^o(s_q) \end{bmatrix},$$

where $E_{l_i}^o(s_j) = \{\vartheta_{ij}^{\sigma(1)}, \vartheta_{ij}^{\sigma(2)}, \cdots, \vartheta_{ij}^{\sigma(c_j)}\}$ ($i = 1, 2, \dots, p$ and $j = 1, 2, \dots, q$) satisfies $\vartheta_{ij}^{\sigma(k)} \le \vartheta_{ij}^{\sigma(k+1)}$ ($k = 1, 2, \dots, c_j$).

Step 2: Specify an ideal HLNN set as $g^* = \{E_l^o(s_1), E_l^o(s_2), \dots, E_l^o(s_q)\} = \{\{\vartheta_1^{\sigma(1)}, \vartheta_1^{\sigma(2)}, \dots, \vartheta_1^{\sigma(c_1)}\}, \{\vartheta_2^{\sigma(1)}, \vartheta_2^{\sigma(2)}, \dots, \vartheta_2^{\sigma(c_2)}\}, \dots, \{\vartheta_q^{\sigma(1)}, \vartheta_q^{\sigma(2)}, \dots, \vartheta_q^{\sigma(c_q)}\}\}$ for all $\vartheta_j^{\sigma(k)} =< h_\tau, h_0, h_0 >$ ($k = 1, 2, \dots, c_j$ and $j = 1, 2, \dots, q$).

Hence, the similarity measure between g_i ($i = 1, 2, \ldots, p$) and g^* can be calculated by

$$\begin{aligned} S_w(g_i, g^*) &= 1 - d_w(g_i, g^*) \\ &= 1 - \left\{ \sum_{j=1}^{q} w_j \left[\frac{1}{3c_j} \sum_{k=1}^{c_j} \left(|f(h_{T_{a_{ij}}^{\sigma(k)}}) - f(h_\tau)|^\rho + |f(h_{U_{a_{ij}}^{\sigma(k)}}) - f(h_0)|^\rho + |f(h_{F_{a_{ij}}^{\sigma(k)}}) - f(h_0)|^\rho \right) \right] \right\}^{1/\rho} \\ &= 1 - \left\{ \sum_{j=1}^{q} w_j \left[\frac{1}{3c_j\tau^\rho} \sum_{k=1}^{c_j} \left(|T_{a_{ij}}^{\sigma(k)} - \tau|^\rho + |U_{a_{ij}}^{\sigma(k)}|^\rho + |F_{a_{ij}}^{\sigma(k)}|^\rho \right) \right] \right\}^{1/\rho} \quad for\ \rho > 0. \end{aligned} \tag{7}$$

Step 3: According to the similarity measure results, rank the alternatives in $G = \{g_1, g_2, \ldots, g_m\}$ in a descending order and choose the best one.

Step 4: End.

HLNN is a hybrid form of a LNN and HFS, which inherits the advantages of both the LNN and HFS, and expresses the decision-making information with a hesitant set of LNNs. The proposed LCMC-based distance and similarity measures can deal with not only the HLNN information, but also the LNN information, because the LNN is only a special case of the HLNN when the DMs have no hesitation; while all existing aggregation operators of LNNs [14] cannot aggregate HLNN information for the reason that the HLNN is a LNN set of any length. Furthermore, existing MADM methods cannot deal with decision-making problems in the HLNN setting.

Moreover, to ensure the objectivity of the measure calculational results, the proposed LCMC-based distance and similarity measures are based on the LCMC extension method in HLNNs rather than by simply adding special components, such as the maximum or the minimum or the average values, which heavily depend on the personal interests and preferences of DMs [23,24] so as to easily result in subjective decision-making results. Thus, the novel MADM method of HLNN provides a more general and objective decision-making process for decision-makers.

6. Actual Example

In this section, to verify whether the novel MADM approach with HLNNs is feasible and reasonable in practical applications, an investment decision-making case adapted from [14] is illustrated under a HLNN environment. In this case, the investment company makes an optimal selection in a set of four possible manufacturers, $G = \{g_1, g_2, g_3, g_4\}$, for producing computers (g_1), cars (g_2), food (g_3), and clothing (g_4), respectively. The four alternatives must satisfy a set of three attributes, $S = \{s_1, s_2, s_3\}$, including the risk (s_1), the growth (s_2), and the environmental impact (s_3), with the importance given by the weight vector $W = (0.35, 0.25, 0.4)$. Now, some DMs are assigned to assess the alternatives over the attributes by HLNN expressions from the given LT set $H = \{h_0$: *none*, h_1: *lowest*, h_2: *lower*, h_3: *low*, h_4: *moderate*, h_5: *high*, h_6: *higher*, h_7: *highest*, h_8: *perfect*$\}$. Then, the assessment results regarding the four alternatives g_1, g_2, g_3, and g_4 on the three attributes s_1, s_2, and s_3 can be constructed as

$$M = \begin{matrix} g_1 \\ g_2 \\ g_3 \\ g_4 \end{matrix} \begin{bmatrix} \{<h_6,h_1,h_2>,<h_6,h_1,h_2>,<h_7,h_3,h_4>\} & \{<h_7,h_2,h_1>,<h_6,h_1,h_1>,<h_7,h_3,h_3>\} & \{<h_6,h_2,h_2>,<h_4,h_2,h_3>\} \\ \{<h_7,h_1,h_1>,<h_7,h_2,h_3>,<h_6,h_3,h_4>\} & \{<h_7,h_3,h_2>,<h_6,h_1,h_1>\} & \{<h_7,h_3,h_2>,<h_6,h_1,h_1>\} \\ \{<h_6,h_2,h_2>,<h_5,h_1,h_2>\} & \{<h_7,h_1,h_1>,<h_5,h_1,h_2>\} & \{<h_6,h_2,h_2>,<h_5,h_4,h_2>\} \\ \{<h_7,h_1,h_2>,<h_6,h_1,h_1>,<h_7,h_2,h_3>\} & \{<h_7,h_2,h_3>,<h_5,h_1,h_1>\} & \{<h_7,h_2,h_1>,<h_5,h_2,h_3>\} \end{bmatrix}.$$

Thus, there are the following decision steps:

Step 1: According to the score and accuracy functions obtained by Equations (1) and (2), rank the LNNs $\vartheta_{ij}^{\sigma(k)}$ ($k = 1, 2, \ldots, m_{ij}$) in each HLNN $E_{l_i}(s_j)$ ($i = 1, 2, 3, 4$ and $j = 1, 2, 3$) in an ascending order, and obtain the following matrix:

$$M' = \begin{matrix} g_1 \\ g_2 \\ g_3 \\ g_4 \end{matrix} \begin{bmatrix} \{<h_7,h_3,h_4>,<h_6,h_1,h_2>,<h_6,h_1,h_2>\} & \{<h_7,h_3,h_3>,<h_6,h_1,h_1>,<h_7,h_2,h_1>\} & \{<h_4,h_2,h_3>,<h_6,h_2,h_2>\} \\ \{<h_6,h_3,h_4>,<h_7,h_2,h_3>,<h_7,h_1,h_1>\} & \{<h_7,h_3,h_2>,<h_6,h_1,h_1>\} & \{<h_4,h_2,h_3>,<h_6,h_2,h_3>,<h_7,h_2,h_1>\} \\ \{<h_5,h_1,h_2>,<h_6,h_2,h_2>\} & \{<h_5,h_1,h_2>,<h_7,h_1,h_1>\} & \{<h_5,h_4,h_2>,<h_6,h_2,h_2>\} \\ \{<h_7,h_2,h_3>,<h_7,h_1,h_2>,<h_6,h_1,h_1>\} & \{<h_7,h_2,h_3>,<h_5,h_1,h_1>\} & \{<h_5,h_2,h_3>,<h_7,h_2,h_1>\} \end{bmatrix}$$

Then, according to the LCMC $c_j = 6$ ($j = 1, 2, 3$) and the number of occurrences of LNNs R_{ij} of $E_{l_i}(s_j)$ ($i = 1, 2, 3, 4$ and $j = 1, 2, 3$) obtained by Equation (3), yield the following extended decision matrix M°:

$$M^o = \begin{matrix} g_1 \\ g_2 \\ g_3 \\ g_4 \end{matrix} \left[\begin{matrix} \{< h_7, h_3, h_4 >, < h_7, h_3, h_4 >, < h_6, h_1, h_2 >, < h_6, h_1, h_2 >, < h_6, h_1, h_2 >, < h_6, h_1, h_2 >\} \\ \{< h_6, h_3, h_4 >, < h_6, h_3, h_4 >, < h_7, h_2, h_3 >, < h_7, h_2, h_3 >, < h_7, h_1, h_1 >, < h_7, h_1, h_1 >\} \\ \{< h_5, h_1, h_2 >, < h_5, h_1, h_2 >, < h_5, h_1, h_2 >, < h_6, h_2, h_2 >, < h_6, h_2, h_2 >, < h_6, h_2, h_2 >\} \\ \{< h_7, h_2, h_3 >, < h_7, h_2, h_3 >, < h_7, h_1, h_2 >, < h_7, h_1, h_2 >, < h_6, h_1, h_1 >, < h_6, h_1, h_1 >\} \end{matrix} \right.$$
$$\begin{matrix} \{< h_7, h_3, h_3 >, < h_7, h_3, h_3 >, < h_6, h_1, h_1 >, < h_6, h_1, h_1 >, < h_7, h_2, h_1 >, < h_7, h_2, h_1 >\} \\ \{< h_7, h_3, h_2 >, < h_7, h_3, h_2 >, < h_7, h_3, h_2 >, < h_6, h_1, h_1 >, < h_6, h_1, h_1 >, < h_6, h_1, h_1 >\} \\ \{< h_5, h_1, h_2 >, < h_5, h_1, h_2 >, < h_5, h_1, h_2 >, < h_7, h_1, h_1 >, < h_7, h_1, h_1 >, < h_7, h_1, h_1 >\} \\ \{< h_7, h_2, h_3 >, < h_7, h_2, h_3 >, < h_7, h_2, h_3 >, < h_5, h_1, h_1 >, < h_5, h_1, h_1 >, < h_5, h_1, h_1 >\} \end{matrix}$$
$$\left. \begin{matrix} \{< h_4, h_2, h_3 >, < h_4, h_2, h_3 >, < h_4, h_2, h_3 >, < h_6, h_2, h_2 >, < h_6, h_2, h_2 >, < h_6, h_2, h_2 >\} \\ \{< h_4, h_2, h_3 >, < h_4, h_2, h_3 >, < h_6, h_2, h_3 >, < h_6, h_2, h_3 >, < h_7, h_2, h_1 >, < h_7, h_2, h_1 >\} \\ \{< h_5, h_4, h_2 >, < h_5, h_4, h_2 >, < h_5, h_4, h_2 >, < h_6, h_2, h_2 >, < h_6, h_2, h_2 >, < h_6, h_2, h_2 >\} \\ \{< h_5, h_2, h_3 >, < h_5, h_2, h_3 >, < h_5, h_2, h_3 >, < h_7, h_2, h_1 >, < h_7, h_2, h_1 >, < h_7, h_2, h_1 >\} \end{matrix} \right].$$

Step 2: Obtain the similarity measures between the alternatives g_1, g_2, g_3, and g_4 and the ideal solution g^* = {{<8,0,0>, <8,0,0>, <8,0,0>, <8,0,0>, <8,0,0>, <8,0,0>}, {<8,0,0>, <8,0,0>, <8,0,0>, <8,0,0>, <8,0,0>, <8,0,0>}, {<8,0,0>, <8,0,0>, <8,0,0>, <8,0,0>, <8,0,0>, <8,0,0>}} by Equation (7) for $\rho = 1$ and 2:

$$S_w(g_1, g^*) = 0.7354, S_w(g_2, g^*) = 0.7493, S_w(g_3, g^*) = 0.7406, S_w(g_4, g^*) = 0.7747 \text{ for } \rho = 1.$$
$$S_w(g_1, g^*) = 0.7121, S_w(g_2, g^*) = 0.7224, S_w(g_3, g^*) = 0.7217, S_w(g_4, g^*) = 0.7525 \text{ for } \rho = 2.$$

Step 3: Due to $S_w(g_4, g^*) > S_w(g_2, g^*) > S_w(g_3, g^*) > S_w(g_1, g^*)$ for $\rho = 1$ and 2, the ranking of the four alternatives is $g_4 > g_2 > g_3 > g_1$; thus, the best choice is g_4.

By following the above steps, the MADM calculations of $\rho \in [3, 100]$ are further performed for this example. The relative decision results, including the similarity measure, ranking order, average value (AV), standard deviation (SD), and the best alternative, are shown in Table 1. Obviously, the ranking order is $g_4 > g_2 > g_3 > g_1$ for $\rho = 1$ and 2, and then it becomes $g_4 > g_3 > g_2 > g_1$ for $\rho > 2$; while the best alternative is always g_4.

Table 1. Decision results of the proposed multiple-attribute decision-making (MADM) method for $\rho \in [1, 100]$ and $W = (0.35, 0.25, 0.4)$.

ρ [1]	$S_w(g_1, g^*)$, $S_w(g_2, g^*)$, $S_w(g_3, g^*)$, $S_w(g_4, g^*)$ [2]	Ranking Order	AV [3]	SD [4]	Best Alternative
1	0.7354, 0.7493, 0.7406, 0.7747	$g_4 > g_2 > g_3 > g_1$	0.7500	0.0151	g_4
2	0.7121, 0.7224, 0.7217, 0.7525	$g_4 > g_2 > g_3 > g_1$	0.7272	0.0152	g_4
3	0.6905, 0.6985, 0.7037, 0.7335	$g_4 > g_3 > g_2 > g_1$	0.7066	0.0163	g_4
4	0.6710, 0.6781, 0.6867, 0.7182	$g_4 > g_3 > g_2 > g_1$	0.6885	0.018	g_4
5	0.6539, 0.6608, 0.6710, 0.7061	$g_4 > g_3 > g_2 > g_1$	0.6730	0.0201	g_4
10	0.5972, 0.6047, 0.6133, 0.6722	$g_4 > g_3 > g_2 > g_1$	0.6219	0.0296	g_4
15	0.5690, 0.5754, 0.5817, 0.6575	$g_4 > g_3 > g_2 > g_1$	0.5959	0.0358	g_4
20	0.5531, 0.5582, 0.5631, 0.6497	$g_4 > g_3 > g_2 > g_1$	0.5810	0.0398	g_4
30	0.5361, 0.5397, 0.5432, 0.6417	$g_4 > g_3 > g_2 > g_1$	0.5652	0.0443	g_4
40	0.5273, 0.5301, 0.5327, 0.6376	$g_4 > g_3 > g_2 > g_1$	0.5569	0.0466	g_4
50	0.5220, 0.5242, 0.5264, 0.6351	$g_4 > g_3 > g_2 > g_1$	0.5519	0.048	g_4
100	0.5111, 0.5123, 0.5134, 0.6301	$g_4 > g_3 > g_2 > g_1$	0.5417	0.051	g_4

Notes: [1] ρ: parameter; [2] $S_w(g_i, g^*)$: the similarity measures between the alternatives g_i(i = 1, 2, 3, 4) and the ideal solution g^* = {{<8,0,0>, <8,0,0>, <8,0,0>, <8,0,0>, <8,0,0>, <8,0,0>}, {<8,0,0>, <8,0,0>, <8,0,0>, <8,0,0>, <8,0,0>, <8,0,0>}, {<8,0,0>, <8,0,0>, <8,0,0>, <8,0,0>, <8,0,0>, <8,0,0>}}; [3] AV: average value; [4] SD: standard deviation.

7. Discussion and Analysis

In this section, further discussion and analysis are carried out for the resolution and the sensitivity of the novel MADM method of HLNNs.

7.1. Resolution Analysis

According to Table 1, Figure 1 illustrates the SDs of the similarity measures for $\rho \in [1, 100]$. Clearly, the SD increases with increasing the value of ρ. Then, it reaches 0.051 for $\rho = 100$. Since the SD can reflect the resolution/discrimination level of the MADM method, it is obvious that the resolution/discrimination level will be enhanced with increasing the value of ρ so as to provide effective decision information for decision-makers in the MADM process. However, considering that the computational complexity of MADM increases with increasing the value of ρ, we recommend selecting the MADM method with some suitable value of ρ under the condition that the resolution degree meets some actual requirement and the DMs' preference.

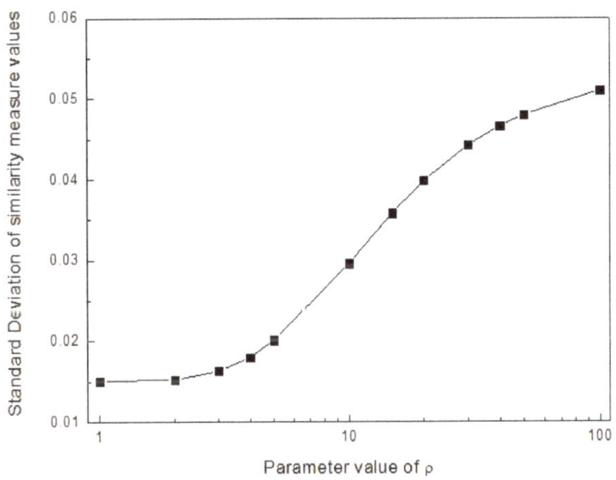

Figure 1. SD of similarity measure values for $\rho \in [1, 100]$ and $W = (0.35, 0.25, 0.4)$.

7.2. Sensitivity Analysis of Weights

The average weight vector of $W = (1/3, 1/3, 1/3)$ is applied to the actual example as a comparison with $W = (0.35, 0.25, 0.4)$ to illustrate the weight sensitivity of the MADM method. The decision results with $W = (1/3, 1/3, 1/3)$ are shown in Table 2. Then, the similarity measure values for $W = (0.35, 0.25, 0.4)$ and $W = (1/3, 1/3, 1/3)$ are further illustrated in Figure 2a,b.

From Figure 2, obviously, the similarity measure curves with $W = (0.35, 0.25, 0.4)$ are very similar to those with $W = (1/3, 1/3, 1/3)$. By carefully comparing Tables 1 and 2, we find that the ranking orders are identical except that of $\rho = 2$. For $\rho = 2$, the ranking orders of $g_4 > g_2 > g_3 > g_1$ for $W = (0.35, 0.25, 0.4)$ and $g_4 > g_3 > g_2 > g_1$ for $W = (1/3, 1/3, 1/3)$ indicate a little difference. Then, the best alternatives are the same within the entire range of ρ. Hence, the ranking orders in this example imply a little sensitivity to the attribute weights.

Table 2. Decision results of the proposed MADM method for $\rho \in [1, 100]$ and $W = (1/3, 1/3, 1/3)$.

ρ	$S_w(g_1, g^*)$, $S_w(g_2, g^*)$, $S_w(g_3, g^*)$, $S_w(g_4, g^*)$	Ranking Order	AV	SD	Best Alternative
1	0.7431, 0.7546, 0.7500, 0.7755	$g_4 > g_2 > g_3 > g_1$	0.7558	0.0121	g_4
2	0.7189, 0.7278, 0.7300, 0.7529	$g_4 > g_3 > g_2 > g_1$	0.7324	0.0125	g_4
3	0.6968, 0.7039, 0.7112, 0.7336	$g_4 > g_3 > g_2 > g_1$	0.7114	0.0138	g_4
4	0.6769, 0.6833, 0.6938, 0.7180	$g_4 > g_3 > g_2 > g_1$	0.693	0.0156	g_4
5	0.6596, 0.6659, 0.6779, 0.7057	$g_4 > g_3 > g_2 > g_1$	0.6773	0.0177	g_4
10	0.6019, 0.6088, 0.6193, 0.6717	$g_4 > g_3 > g_2 > g_1$	0.6254	0.0274	g_4
15	0.5727, 0.5786, 0.5865, 0.6572	$g_4 > g_3 > g_2 > g_1$	0.5988	0.0341	g_4
20	0.556, 0.5608, 0.5671, 0.6495	$g_4 > g_3 > g_2 > g_1$	0.5834	0.0384	g_4
30	0.5381, 0.5415, 0.5459, 0.6415	$g_4 > g_3 > g_2 > g_1$	0.5668	0.0432	g_4
40	0.5289, 0.5315, 0.5349, 0.6374	$g_4 > g_3 > g_2 > g_1$	0.5582	0.0458	g_4
50	0.5232, 0.5254, 0.5281, 0.6350	$g_4 > g_3 > g_2 > g_1$	0.5529	0.0474	g_4
100	0.5118, 0.5128, 0.5142, 0.6300	$g_4 > g_3 > g_2 > g_1$	0.5422	0.0507	g_4

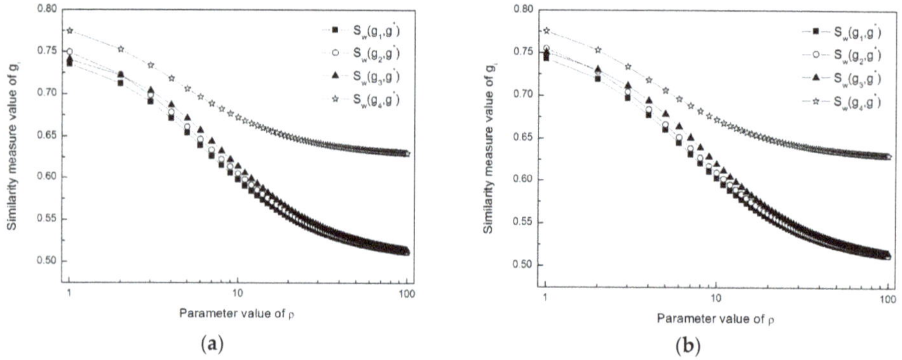

Figure 2. Similarity measure values of four alternatives for $\rho \in [1, 100]$. (**a**) $W = (0.35, 0.25, 0.4)$ and (**b**) $W = (1/3, 1/3, 1/3)$.

8. Conclusions

This paper firstly defined the concept of HLNNs by integrating a HFS with a LNN. Then, the normalized generalized distance and similarity measures of HLNNs were presented based on the LCMC method. Next, a novel MADM method based on the proposed similarity measure was presented under the HLNN environment. Finally, a MADM example of an investment problem was illustrated to demonstrate that the developed method is feasible and applicable. Since the HLNN combines the merits of the HFS and LNN, containing more information than the LNN, the MADM method of HLNNs based on the LCMC method is more objective and more suitable for the practical applications with HLNN information.

However, some advantages of the proposed HLNNs and MADM method based on the LCMC method are listed as follows:

(1) The proposed HLNN provides a new effective way to express more decision information than existing LNNs by considering the hesitancy of DMs.
(2) The proposed MADM method of HLNNs solves the MADM problems with HLNN information for the first time, as well as the gap of existing linguistic decision-making methods.
(3) The proposed distance and similarity measures of HLNNs based on the LCMC extension method for HLNNs are more objective and more reasonable than those reported in [23,24].

Future research on HLNNs will focus on the development of new aggregation operators and correlation coefficients of HLNNs, and their applications in fault diagnosis, medical diagnosis, decision-making, and so on in the HLNN setting.

Author Contributions: J.Y. originally proposed HLNNs and their operations, and W.C. presented the MADM method of HLNNs and the calculation and comparative analysis of an actual example. Both coauthors wrote the paper together.

Funding: This paper was supported by the National Natural Science Foundation of China (No. 61703280).

Conflicts of Interest: The authors declare no conflict of interest.

References

1. Zadeh, L.A. The concept of a linguistic variable and its application to approximate reasoning Part I. *Inf. Sci.* **1975**, *8*, 199–249. [CrossRef]
2. Herrera, F.; Herrera-Viedma, E.; Verdegay, J.L. A model of consensus in group decision making under linguistic assessments. *Fuzzy Set Syst.* **1996**, *78*, 73–87. [CrossRef]
3. Herrera, F.; Herrera-Viedma, E. Linguistic decision analysis: Steps for solving decision problems under linguistic information. *Fuzzy Set Syst.* **2000**, *115*, 67–82. [CrossRef]
4. Xu, Z.S. A method based on linguistic aggregation operators for group decision making with linguistic preference relations. *Inf. Sci.* **2004**, *166*, 19–30. [CrossRef]
5. Xu, Z.S. Uncertain linguistic aggregation operators based approach to multiple attribute group decision making under uncertain linguistic environment. *Inf. Sci.* **2004**, *168*, 171–184. [CrossRef]
6. Xu, Z.S. Induced uncertain linguistic OWA operators applied to group decision making. *Inf. Sci.* **2006**, *7*, 231–238. [CrossRef]
7. Wei, G.W. Uncertain linguistic hybrid geometric mean operator and its application to group decision making under uncertain linguistic environment. *Int. J. Uncertain. Fuzziness* **2009**, *17*, 251–267. [CrossRef]
8. Peng, B.; Ye, C.; Zeng, S. Uncertain pure linguistic hybrid harmonic averaging operator and generalized interval aggregation operator based approach to group decision making. *Knowl.-Based Syst.* **2012**, *36*, 175–181. [CrossRef]
9. Chen, Z.C.; Liu, P.H.; Pei, Z. An approach to multiple attribute group decision making based on linguistic intuitionistic fuzzy numbers. *Int. J. Comput. Intell. Syst.* **2015**, *8*, 747–760. [CrossRef]
10. Liu, P.D.; Liu, J.L.; Merigó, J.M. Partitioned Heronian means based on linguistic intuitionistic fuzzy numbers for dealing with multi-attribute group decision making. *Appl. Soft Comput.* **2018**, *62*, 395–422. [CrossRef]
11. Peng, X.; Liu, C. Algorithms for neutrosophic soft decision making based on EDAS, new similarity measure and level soft set. *J. Intell. Fuzzy Syst.* **2017**, *32*, 955–968. [CrossRef]
12. Zavadskas, E.K.; Bausys, R.; Juodagalviene, B.; Garnyte-Sapranaviciene, I. Model for residential house element and material selection by neutrosophic MULTIMOORA method. *Eng. Appl. Artif. Intell.* **2017**, *64*, 315–324. [CrossRef]
13. Bausys, R.; Juodagalviene, B. Garage location selection for residential house by WASPAS-SVNS method. *J. Civ. Eng. Manag.* **2017**, *23*, 421–429. [CrossRef]
14. Fang, Z.B.; Ye, J. Multiple attribute group decision-making method based on linguistic neutrosophic numbers. *Symmetry* **2017**, *9*, 111. [CrossRef]
15. Liu, P.D.; Mahmood, T.; Khan, Q. Group decision making based on power Heronian aggregation operators under linguistic neutrosophic environment. *Int. J. Fuzzy Syst.* **2018**, *20*, 970–985. [CrossRef]
16. Torra, V.; Narukawa, Y. On hesitant fuzzy sets and decision. In Proceedings of the 18th IEEE International Conference on Fuzzy Systems, Jeju Island, Korea, 20–24 August 2009; pp. 1378–1382. [CrossRef]
17. Torra, V. Hesitant fuzzy sets. *Int. J. Intell. Syst.* **2010**, *25*, 529–539. [CrossRef]
18. Rodríguez, R.M.; Martínez, L.; Herrera, F. Hesitant Fuzzy Linguistic Term Sets for Decision Making. *IEEE Trans. Fuzzy Syst.* **2012**, *20*, 109–119. [CrossRef]
19. Rodríguez, R.M.; MartíNez, L.; Herrera, F. A group decision making model dealing with comparative linguistic expressions based on hesitant fuzzy linguistic term sets. *Inf. Sci.* **2013**, *241*, 28–42. [CrossRef]
20. Lin, R.; Zhao, X.F.; Wei, G.W. Models for selecting an ERP system with hesitant fuzzy linguistic information. *J. Intell. Fuzzy Syst.* **2014**, *26*, 2155–2165. [CrossRef]
21. Wang, J.Q.; Wu, J.T.; Wang, J.; Zhang, H.; Chen, X. Interval-valued hesitant fuzzy linguistic sets and their applications in multi-criteria decision-making problems. *Inf. Sci.* **2014**, *288*, 55–72. [CrossRef]
22. Ye, J. Multiple Attribute Decision-Making Methods Based on the Expected Value and the Similarity Measure of Hesitant Neutrosophic Linguistic Numbers. *Cogn. Comput.* **2018**, *10*, 454–463. [CrossRef]

23. Zhu, B.; Xu, Z. Consistency Measures for Hesitant Fuzzy Linguistic Preference Relations. *IEEE Trans. Fuzzy Syst.* **2014**, *22*, 35–45. [CrossRef]
24. Liao, H.; Xu, Z.; Zeng, X.J.; Merigó, J.M. Qualitative decision making with correlation coefficients of hesitant fuzzy linguistic term sets. *Knowl.-Based Syst.* **2015**, *76*, 127–138. [CrossRef]
25. Ye, J. Multiple-attribute Decision-Making Method under a Single-Valued Neutrosophic Hesitant Fuzzy Environment. *J. Intell. Syst.* **2014**, *24*, 23–36. [CrossRef]

symmetry

MDPI

Article

Complex Fuzzy Geometric Aggregation Operators

Lvqing Bi [1], Songsong Dai [2,*] and Bo Hu [2,3]

[1] School of Electronics and Communication Engineering, Guangxi Colleges and Universities Key Laboratory of Complex System Optimization and Big Data Processing, Yulin Normal University, Yulin 537000, China; bilvqing108@163.com
[2] School of Information Science and Engineering, Xiamen University, Xiamen 361005, China; hb.gznu@gmail.com
[3] School of Mechanical and Electrical Engineering, Guizhou Normal University, Guiyang 550025, China
* Correspondence: ssdai@stu.xmu.edu.cn; Tel.: +86-592-258-0135

Received: 15 June 2018; Accepted: 28 June 2018; Published: 2 July 2018

Abstract: A complex fuzzy set is an extension of the traditional fuzzy set, where traditional [0,1]-valued membership grade is extended to the complex unit disk. The aggregation operator plays an important role in many fields, and this paper presents several complex fuzzy geometric aggregation operators. We show that these operators possess the properties of rotational invariance and reflectional invariance. These operators are also closed on the upper-right quadrant of the complex unit disk. Based on the relationship between Pythagorean membership grades and complex numbers, these operators can be applied to the Pythagorean fuzzy environment.

Keywords: complex fuzzy sets; aggregation operator; complex fuzzy geometric operators; rotational invariance; reflectional invariance

1. Introduction

Ramot et al. [1] introduced the innovative concept of complex fuzzy sets (CFSs), which is an extension of the traditional fuzzy sets [2] where traditional unit interval [0,1]-valued membership degrees are extended to the complex unit disk. CFSs are completely distinct from the fuzzy complex numbers discussed by Buckley [3–5]. The complex-valued membership grade has an amplitude term with the addition of a phase term. The phase term of complex valued membership grade is the key feature which essentially distinguishes complex fuzzy sets from other extensions of fuzzy sets. Ramot et al. [1,6] then introduced several operators of CFSs and a novel framework for complex fuzzy reasoning. Hu et al. [7] introduced the orthogonality relation for CFSs. Bi et al. [8] proposed the parallelity of CFSs and the parallelity-preserving operators. Zhang et al. [9] proposed the δ-equalities for CFSs. Alkouri and Salleh [10] and Hu et al. [11] defined several distances between CFSs. Tamir et al. [12] proposed a new interpretation of complex membership degree. They [13] then proposed complex fuzzy propositional and first-order logics. Dick [14] proposed the concept of rotational invariance for complex fuzzy operators. Recently, several scholars have developed extensions of CFSs. Greenfield et.al [15,16] introduced interval-valued complex fuzzy sets. Alkouri and Saleh [17] proposed complex intuitionistic fuzzy sets. Ali and Smarandache [18] introduced complex neutrosophic sets. Recently, CFSs and their extensions have been successfully applied in many fields, such as time series prediction [19–22], decision making [23], signal processing [1,7,9], and image restoration [24].

Yager and Abbsocv [25] discussed the relationship between CFSs and Pythagorean fuzzy sets (PFSs), which was developed by Yager [26,27] as an extension of Atanssov's intuitionistic fuzzy sets [28]. They showed that Pythagorean fuzzy membership grades can be viewed as complex numbers on the upper-right quadrant of the complex unit disk, named $\Pi - i$ numbers.

Dick, Yager, and Yazdanbahksh [29] then discussed several lattice-theoretic properties of PFSs and CFSs. Quantum information processing also allows for meaningful aggregation using complex numbers. Since qubits can be represented by unit vectors in the two-dimensional complex Hilbert space, geometric information or vector aggregation are used for meaningful clustering [30,31].

The information aggregation operator plays an important role in many fields of decision making. In the past several decades, many aggregation techniques for decision making have been developed. The ordered weighted averaging (OWA) operator introduced by Yager [32] is one of the well-known aggregation operators. Many different aggregation techniques have been applied in many different fuzzy environments, such as intuitionistic [33–35], Pythagorean [36–38], neutrosophic [39–41], interval-valued intuitionistic [42–45], and hesitant fuzzy environments [46–48].

As mentioned in [19], CFSs are suitable to represent information with uncertainty and periodicity, and thus this information aggregation procedure needs to simultaneously process the uncertainty and periodicity in the data. However, comparatively few aggregation techniques have been made in the complex fuzzy environment. Ramot et al. [6] defined the complex fuzzy aggregation operations as vectors aggregation. In particular, the complex weights are used in their definition. Ma et al. [19] developed a product-sum aggregation operator for multiple periodic factor prediction problems. They proved the continuity of this operator. However, they did not focus on techniques for complex fuzzy information aggregation in these two articles.

In this paper, we study aggregation operators in the complex fuzzy environment. Dick's [14] concept of rotational invariance is an intuitive and desirable feature for complex fuzzy operators. This feature is examined for complex fuzzy aggregation operators. This paper proposes a novel feature for complex fuzzy aggregation operators called *reflectional invariance*. Moreover, we study the aggregation operators of complex numbers in the upper-right quadrant of the complex unit disk.

The main contributions of the study include: (1) A concept of reflectional invariance for complex fuzzy operators. (2) Several complex fuzzy weighted geometric operators; we also show that these operators can also be used in a Pythagorean fuzzy environment. (3) A target location method which involves the complex fuzzy aggregation operators.

This paper is organized as follows. In Section 2, we review some basic and fundamental concepts of CFSs, rotational invariance, reflectional invariance, and Ramot et al.'s [6] complex fuzzy aggregation operators. In Section 3 we study the complex fuzzy weighted geometric (CFWG) operator on CFSs and its properties. In Section 4, we develop the complex fuzzy ordered weighted geometric (CFOWG) operator based on the traditional partial ordering by the modulus of a complex number, and study its properties. In Section 5, we study these operators in the domain of $\Pi - i$ numbers which belong to the upper-right quadrant of the complex unit disk. In Section 6, we present an application example in a target location. Conclusions are made in Section 7.

2. Preliminaries

In this section, we present some basic material, including the concepts of CFSs [1], rotational invariance [14], reflectional invariance, and complex fuzzy aggregation operators [1].

2.1. Complex Fuzzy Sets

Ramot et al. [1] defined the concept of CFSs as follows.

Definition 1 ([1]). *Let X be a universe, D be the set of complex numbers whose modulus is less than or equal to 1, i.e.,*

$$D = \{a \in C \big| |a| \leq 1\},$$

a complex fuzzy set A defined on X is a mapping: X → D, which can be denoted as below:

$$A = \{ < x, t_A(x) \cdot e^{j\nu_A(x)} > |x \in X\}, \tag{1}$$

where $j = \sqrt{-1}$, *the amplitude term* $t_A(x)$ *and the phase term* $v_A(x)$ *are both real-valued, and* $t_A(x) \in [0,1]$.

For convenience, we only consider the complex numbers on D, called complex fuzzy values (CFVs). Let $a = t_a \cdot e^{jv_a}$ be a CFV, then the amplitude of a is denoted by t_a and the phase of a is denoted by v_a. They are both real-valued, $t_a \in [0,1]$. The modulus of a is t_a, denoted by $|a|$.

Let $a = t_a \cdot e^{jv_a}$ and $b = t_b \cdot e^{jv_b}$ be two CFVs, then we have the following two commonly used binary operations.

(i) Algebraic product:

$$a \cdot b = t_a \cdot t_b \cdot e^{j(v_a+v_b)}. \tag{2}$$

(ii) Average:

$$\frac{a+b}{2} = \frac{t_a \cos v_a + t_b \cos v_b}{2} + j\frac{t_a \sin v_a + t_b \sin v_b}{2}. \tag{3}$$

The partial ordering of CFVs is the traditional partial ordering by the modulus of a complex number, that is, $a \leq b$ if and only if $|a| \leq |b|$, equivalently, $t_a \leq t_b$.

2.2. Rotational Invariance and Reflectional Invariance

Let $a = t_a \cdot e^{jv_a}$ be a CFV, then we have the following two commonly used unary operations:

(i) the rotation of a by θ radians, denoted $Rot_\theta(a)$, is defined as

$$Rot_\theta(a) = t_a \cdot e^{j(v_a+\theta)}; \tag{4}$$

(ii) the reflection of a, denoted $Ref(a)$, is defined as

$$Ref(a) = t_a \cdot e^{j-v_a}. \tag{5}$$

Then, based on the rotation operation, Dick [14] introduced the concept of rotational invariance for complex fuzzy operators.

Definition 2 ([14]). *A function* $f : D^2 \to D$ *is rotationally invariant if and only if*

$$f(Rot_\theta(a), Rot_\theta(b)) = Rot_\theta(f(a,b)), \tag{6}$$

for any θ.

We extend the above concept to multivariate operators.

Definition 3. *Let* $f : D^n \to D$ *be an n-order function.* f *is rotationally invariant if and only if*

$$f(Rot_\theta(a_1), Rot_\theta(a_2), ..., Rot_\theta(a_n)) = Rot_\theta(f(a_1, a_2, ..., a_n)), \tag{7}$$

for any θ.

In particular, since the periodicity of complex-valued membership grade, that is, $Rot_{2\pi}(x) = x$ for any $x \in D$, we have $f(Rot_{2\pi}(a_1), Rot_{2\pi}(a_2), ..., Rot_{2\pi}(a_n)) = f(a_1, a_2, ..., a_n) = Rot_{2\pi}(f(a_1, a_2, ..., a_n))$. This is a special case of rotational invariance.

Similar to the above definition, we introduce the concept of reflectional invariance for complex fuzzy operators based on the reflection operation.

Definition 4. *Let* $f : D^n \to D$ *be an n-order function.* f *is reflectionally invariant if and only if*

$$f(Ref(a_1), Ref(a_2), ..., Ref(a_n)) = Ref(f(a_1, a_2, ..., a_n)). \tag{8}$$

Rotational invariance and reflectional invariance are intuitive properties for complex fuzzy operators. It makes a great deal of sense that a operator is invariant under a rotation or a reflection. If we rotate two vectors by a common value, rotational invariance states that an aggregation of those vectors will be rotated by the same value, as shown in Figure 1a. If we reflect two vectors, reflectional invariance states that an aggregation of those vectors will be reflected as well, as shown in Figure 1b.

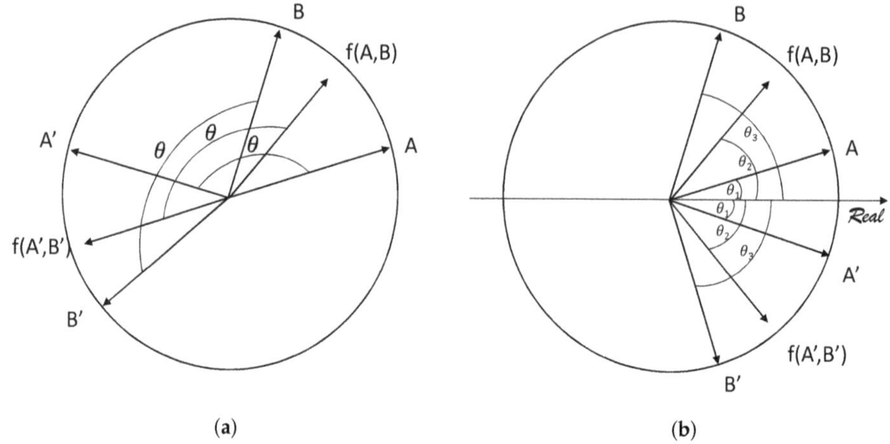

Figure 1. (**a**) Rotational invariance and (**b**) reflectional invariance.

Reflectional invariance and rotational invariance are two properties which are only concerned with the phase of CFVs.

These two properties of the algebraic product and average operators were examined, and the results are given as follows.

Theorem 1 ([14]). *The algebraic product is not rotationally invariant.*

Theorem 2. *The algebraic product is reflectionally invariant.*

Proof. For any $a, b \in D$, we have

$$Ref(a) \cdot Ref(b) = t_a \cdot e^{j-\nu_a} \cdot t_b \cdot e^{j-\nu_b} = t_a \cdot t_b \cdot e^{j(-\nu_a-\nu_b)},$$

$$Ref(a \cdot b) = Ref\left(t_a \cdot t_b \cdot e^{j(\nu_a+\nu_b)}\right) = t_a \cdot t_b \cdot e^{j(-\nu_a-\nu_b)},$$

then $Ref(a) \cdot Ref(b) = Ref(a \cdot b)$. □

Theorem 3. *The average operator is reflectionally invariant and rotationally invariant.*

Proof. (i) Let $a = r_a + j\omega_a, b = r_b + j\omega_b \in D$. We have

$$\frac{Ref(a) + Ref(b)}{2} = \frac{r_a + r_b}{2} + j\frac{-\omega_a - \omega_b}{2} = \frac{r_a + r_b}{2} - j\frac{\omega_a + \omega_b}{2} = Ref(\frac{a+b}{2}).$$

Then, the average operator is reflectionally invariant.

(ii) For any $a, b \in D$, we have

$$\frac{a \cdot e^{j\theta} + b \cdot e^{j\theta}}{2} = \frac{(a+b) \cdot e^{j\theta}}{2} = \frac{(a+b)}{2} \cdot e^{j\theta}.$$

Then, the average operator is rotationally invariant. □

2.3. Complex Fuzzy Aggregation Operators

Ramot et al. [6] defined the aggregation operation on CFSs as vector aggregation:

Definition 5 ([6]). *Let $A_1, A_2, ..., A_n$ be CFSs defined on X. Vector aggregation on $A_1, A_2, ..., A_n$ is defined by a function v.*

$$v: \ D^n \to D. \tag{9}$$

The function v produces an aggregate CFS A by operating on the membership grades of $A_1, A_2, ..., A_n$ for each $x \in X$. For all $x \in X$, v is given by

$$\begin{aligned} \mu_A(x) &= v\left(t_{A_1} \cdot e^{jv_{A_1}}, t_{A_2} \cdot e^{jv_{A_2}}, \cdots, t_{A_n} \cdot e^{jv_{A_n}},\right) \\ &= \sum_{i=1}^{n} \left(w_i \cdot t_{A_i} \cdot e^{jv_{A_i}}\right), \end{aligned} \tag{10}$$

where $w_i \in D$ for all i, and $\sum_{i=1}^{n} |w_i| = 1$.

The complex weights are used in Ramot et al.'s definition for the purpose of maintaining a definition that is as general as possible. In this paper, we only discuss the complex fuzzy aggregation operations with real-valued weights.

We notice that the above definition of Ramot et al's [6] aggregation operator is a complex fuzzy weighted arithmetic (CFWA) operator. For convenience, let $a_1, a_2, ..., a_n$ be CFVs. The CFWA operator is given as

$$CFWA(a_1, a_2, ..., a_n) = \sum_{i=1}^{n} (w_i \cdot a_i), \tag{11}$$

where $w_i \in [0,1]$ for all i, and $\sum_{i=1}^{n} w_i = 1$.

When $a_1, a_2, ..., a_n \in [0,1]$, the CFWA operator can reduce to a traditional fuzzy weighted arithmetic operator.

When $w_i = 1/n$ for all i, then the CFWA operator is the arithmetic average of $(a_1, a_2, ..., a_n)$, denoted by the complex fuzzy arithmetic average (CFAA) operator. That is,

$$CFAA(a_1, a_2, ..., a_n) = \frac{1}{n} \cdot \sum_{i=1}^{n} a_i. \tag{12}$$

When $a_1, a_2, ..., a_n \in [0,1]$ and $w_i = 1/n$ for all i, the CFAA operator is the arithmetic mean of numbers on [0,1].

As a special case of the CFWA operator, note that the average operator on CFVs is reflectionally invariant and rotationally invariant (see Theorem 3). We show that the CFWA operator also possesses these two properties.

Theorem 4. *The CFWA operator is reflectionally invariant and rotationally invariant.*

Proof. (i) Let $a_1 = r_{a_1} + j\omega_{a_1}, a_2 = r_{a_2} + j\omega_{a_2}, \cdots, a_n = r_{a_n} + j\omega_{a_n}$ be CFVs. We have

$$\begin{aligned} Ref(CFWA(a_1, a_2, ..., a_n)) &= Ref\Big(\sum_{i=1}^{n}(w_i \cdot r_{a_i}) + j\sum_{i=1}^{n}(w_i \cdot \omega_{a_i})\Big) \\ &= \sum_{i=1}^{n}(w_i \cdot r_{a_i}) - j\sum_{i=1}^{n}(w_i \cdot \omega_{a_i}) \\ &= \sum_{i=1}^{n}(w_i \cdot (r_{a_i} - \omega_{a_i})) \\ &= \sum_{i=1}^{n}(w_i \cdot Ref(a_i)). \end{aligned}$$

Then, the CFWA operator is reflectionally invariant.
(ii) For any CFVs $a_1, a_2, ..., a_n$, we have

$$\begin{aligned} CFWA(a_1 \cdot e^{j\theta}, a_2 \cdot e^{j\theta}, ..., a_n \cdot e^{j\theta}) &= w_1 \cdot a_1 \cdot e^{j\theta} + w_2 \cdot a_2 \cdot e^{j\theta} + ... + w_n \cdot a_n \cdot e^{j\theta} \\ &= (w_1 \cdot a_1 + w_2 \cdot a_2 + ... + w_n \cdot a_n) \cdot e^{j\theta} \\ &= CFWA(a_1, a_2, ..., a_n) \cdot e^{j\theta}. \end{aligned}$$

Then, the CFWA operator is rotationally invariant. □

3. Complex Fuzzy Weighted Geometric Operators

In this section, we introduce the weighted geometric aggregation operators in a complex fuzzy environment and discuss their fundamental characteristics.

Definition 6. *Let $a_1, a_2, ..., a_n$ be CFVs, a complex fuzzy weighted geometric (CFWG) operator is defined as:*

$$CFWG(a_1, a_2, ..., a_n) = \prod_{i=1}^{n} a_i^{w_i}, \tag{13}$$

where $w_i \in [0,1]$ for all i, and $\sum_{i=1}^{n} w_i = 1$.

Denoting $CFWG(a_1, a_2, ..., a_n) = t \cdot e^{jv}$, we have a weighted geometric aggregation (WGA) operator on [0,1], that is, $t = \prod_{i=1}^{n} t_{a_i}^{w_i}$ and a weighted arithmetic aggregation (WAA) operator on $\mathbb{R}$, that is, $v = \sum_{i=1}^{n} w_i \cdot v_{a_i}$.

When $a_1, a_2, ..., a_n \in [0,1]$, the CFWG operator can reduce to a traditional fuzzy weighted geometric operator.

When $w_i = 1/n$ for all i, then $t = \sqrt[n]{t_{a_1} \cdot t_{a_2} \cdots t_{a_n}}$ is the geometric mean of real numbers on unit interval [0,1], $v = \frac{1}{n} \cdot \sum_{i=1}^{n} v_{a_i}$ is the arithmetic mean of real numbers on $\mathbb{R}$.

When $a_1, a_2, ..., a_n \in [0,1]$ and $w_i = 1/n$ for all i, the CFWG operator is the geometric mean of real numbers on unit interval [0,1].

Theorem 5. *Let $a_1, a_2, ..., a_n$ be CFVs, then the aggregated value $CFWG(a_1, a_2, ..., a_n)$ is also a complex fuzzy value.*

Proof. Since $|CFWG(a_1, a_2, ..., a_n)| = \prod_{i=1}^{n} t_{a_i}^{w_i}$, which is a weighted arithmetic aggregation operator on unit interval [0,1], we have $|CFWG(a_1, a_2, ..., a_n)| \leq 1$. □

Similar to Theorem 4, the CFWA operator is reflectionally invariant and rotationally invariant. We show that the CFWG operator also possesses these two properties.

Theorem 6. *The CFWG operator is reflectionally invariant and rotationally invariant.*

Proof. (i) For any CFVs $a_1, a_2, ..., a_n$, we have

$$\begin{aligned} Ref(CFWG(a_1, a_2, ..., a_n)) &= Ref\Big(\prod_{i=1}^{n} t_{a_i}^{w_i} \cdot e^{j\sum_{i=1}^{n} w_i \cdot v_{a_i}}\Big) \\ &= \prod_{i=1}^{n} t_{a_i}^{w_i} \cdot e^{j - \sum_{i=1}^{n} w_i \cdot v_{a_i}} \\ &= \prod_{i=1}^{n} t_{a_i}^{w_i} \cdot e^{j\sum_{i=1}^{n} w_i \cdot (-v_{a_i})} \\ &= \prod_{i=1}^{n} Ref(a_i)^{w_i} \\ &= CFWG(Ref(a_1), Ref(a_2), ..., Ref(a_n)); \end{aligned}$$

(ii) and

$$\begin{aligned} CFWG(a_1 \cdot e^{j\theta}, a_2 \cdot e^{j\theta}, ..., a_n \cdot e^{j\theta}) &= a_1^{w_1} \cdot e^{jw_1\theta} \cdot a_2^{w_1} \cdot e^{jw_2\theta} \cdot ... \cdot a_n^{w_n} \cdot e^{jw_n\theta} \\ &= \Big(\prod_{i=1}^{n} a_i^{w_i}\Big) \cdot e^{j(w_1 \cdot \theta + w_2 \cdot \theta + ... + w_n \cdot \theta)} \\ &= CFWG(a_1, a_2, ..., a_n) \cdot e^{j\theta}. \end{aligned}$$

□

Idempotency, boundedness, and monotonicity are three important properties of aggregation operators. The CFWG operator satisfies the following properties.

Theorem 7. *Let $a_1, a_2, ..., a_n, b_1, b_2, ..., b_n$ be CFVs, CFWG weights are real values, that is, $w_i \in [0, 1]$ for all i, and $\sum_{i=1}^{n} w_i = 1$. Then, we have the following:*

(1) (Idempotency): If $a_1 = a_2 = ... = a_n$ then

$$CFWG(a_1, a_2, ..., a_n) = a_1.$$

(2) (Amplitude boundedness):

$$|CFWG(a_1, a_2, ..., a_n)| \leq a,$$

where $a = \max_i |a_i|$.

(3) (Amplitude monotonicity): If $|a_i| \leq |b_i|$ $i = 1, 2, ..., n$, then

$$|CFWAA(a_1, a_2, ..., a_n)| \leq |CFWAA(b_1, b_2, ..., b_n)|.$$

Proof. (1) Trivial form the facts that both WAA operator on [0,1] and WGA operator on $\mathbb{R}$ satisfy the property of idempotency.

(2) Trivial form the facts that $|CFWG(a_1, a_2, ..., a_n)| = \prod_{i=1}^{n} t_{a_i}^{w_i}$ and WGA operator on $\mathbb{R}$ satisfy the property of boundedness.

(3) Trivial form the facts that $|CFWG(a_1, a_2, ..., a_n)| = \prod_{i=1}^{n} t_{a_i}^{w_i}$ and WGA operator on $\mathbb{R}$ satisfy the property of monotonicity. □

In this paper, for complex fuzzy aggregation operators, boundedness and monotonicity are restricted exclusively to the amplitude of CFVs. They are two properties which are only concerned with the amplitude of CFVs. Idempotency is a property that is concerned with both the phase and amplitude of CFVs.

It is easy to prove that the CFWA operator satisfies idempotency and amplitude boundedness, but it does not satisfy the property of amplitude monotonicity.

Example 1. *Let* $a_1 = 0.4$, $a_2 = 0.4 \cdot e^{j2\pi/3}$, $b_1 = b_2 = 0.3$ *and weights be* $w_1 = w_2 = 0.5$. *Then,*

$$\begin{aligned} CFWA(a_1, a_2) &= 0.5 \cdot 0.4 + 0.5 \cdot 0.4 \cdot e^{j2\pi/3} \\ &= 0.2 \cdot e^{j\pi/3} \end{aligned}$$

and $CFWA(b_1, b_2) = 0.3$. *Then,* $|a_1| \geq |b_1|$, $|a_2| \geq |b_2|$, *but* $|CFWA(a_1, a_2)| \leq |CFWA(b_1, b_2)|$.

4. Complex Fuzzy Ordered Weighted Geometric Operators

Based on the partial ordering of complex numbers, we propose a complex fuzzy ordered weighted geometric (CFOWG) operator as follows:

Definition 7. *Let* $a_1, a_2, ..., a_n$ *be CFVs, a CFOWG operator is defined as*

$$CFOWG(a_1, a_2, ..., a_n) = \prod_{i=1}^{n} a_{\sigma(i)}^{w_i}, \tag{14}$$

where $w_i \in [0, 1]$ *for all i, and* $\sum_{i=1}^{n} w_i = 1$, $(\sigma(1), \sigma(2), ..., \sigma(n))$ *is a permutation of* $(1, 2, ..., n)$ *such that* $|a_{\sigma(i-1)}| \geq |a_{\sigma(i)}|$ *for all i.*

Especially when $w_i = 1/n$ for all i, then the CFOWG operator is reduced to the CFWG operator. Similar to Theorem 5, we have the following.

Theorem 8. *Let* $a_1, a_2, ..., a_n$ *be CFVs, then the aggregated value* $CFOWG(a_1, a_2, ..., a_n)$ *is also a complex fuzzy value.*

Similar to Theorem 6, the CFWG operator is reflectionally invariant and rotationally invariant. The CFOWG operator also possesses these two properties.

Theorem 9. *The CFOWG operator is reflectionally invariant and rotationally invariant.*

Similar to Theorem 7, the CFOWG operator satisfies idempotency, amplitude boundedness, and amplitude monotonicity.

Theorem 10. *Let* $a_1, a_2, ..., a_n$, $b_1, b_2, ..., b_n$ *be CFVs, CFOWAA weights are real values, that is,* $w_i \in [0, 1]$ *for all i, and* $\sum_{i=1}^{n} w_i = 1$. *Then, we have the following:*

(1) *(Idempotency): If* $a_1 = a_2 = ... = a_n$, *then*

$$CFOWG(a_1, a_2, ..., a_n) = a_1.$$

(2) *(Boundedness):*

$$\left|CFOWG(a_1, a_2, ..., a_n)\right| \leq a,$$

where $a = \max_i |a_i|$.

(3) *(Monotonicity): If* $|a_i| \leq |b_i|$ $i = 1, 2, ..., n$, *then*

$$|CFOWG(a_1, a_2, ..., a_n)| \leq |CFWAA(b_1, b_2, ..., b_n)|.$$

Besides the above properties, the CFOWG operator has the following.

Theorem 11. *Let* $a_1, a_2, ..., a_n$ *be CFVs, CFOWG weights are real values, that is,* $w_i \in [0, 1]$ *for all i, and* $\sum_{i=1}^{n} w_i = 1$. *Then, we have the following:*

(1) *If* $w = (1, 0, ..., 0)$, *then*

$$|CFOWG(a_1, a_2, ..., a_n)| = \max_i |a_i|;$$

(2) *If* $w = (0, 0, ..., 1)$, *then*

$$|CFOWG(a_1, a_2, ..., a_n)| = \min_i |a_i|;$$

(3) *If* $w_i = 1, w_k = 0, k \neq i$, *then*

$$|CFOWG(a_1, a_2, ..., a_n)| = |a_{\sigma(i)}|,$$

where $a_{\sigma(i)}$ *is the i-th (modulus-based) largest of* $a_1, a_2, ..., a_n$.

Now we give a brief summary of the properties of the CFWG and CFOWG operators with real-valued weights. The results are summarized in Table 1, in which $\surd$ and $\times$ represent that the corresponding property holds and does not hold, respectively.

Table 1. Properties of the complex fuzzy aggregation operators. $\surd$ and $\times$ represent that the corresponding property holds and does not hold, respectively.

	Idempotency	Amplitude Boundedness	Amplitude Monotonicity	Reflectional Invariance	Rotational Invariance
CFAA	$\surd$	$\surd$	$\times$	$\surd$	$\surd$
CFWA	$\surd$	$\surd$	$\times$	$\surd$	$\surd$
CFWG	$\surd$	$\surd$	$\surd$	$\surd$	$\surd$
CFOWG	$\surd$	$\surd$	$\surd$	$\surd$	$\surd$

5. Complex Fuzzy Values and Pythagorean Fuzzy Numbers

Yager and Abbasov [25] showed that Pythagorean membership grades can be expressed using complex numbers, called $\Pi - i$ numbers, which belong to the upper-right quadrant of the complex unit disk. Essentially, the CFWAA and CFOWAA operators are used to deal with special complex numbers, which belong to the complex unit disk.

In this section, we consider the CFWG and CFOWG operators in the domain of $\Pi - i$ numbers. We first recall the concepts of Pythagorean fuzzy sets (PFSs) and $\Pi - i$ numbers.

Definition 8 ([25]). *Let X be a universe. A PFS A is defined by*

$$A = \{ < x, p_A(x), \nu_A(x), > |x \in X\}, \tag{15}$$

where $p_A(x) \in [0,1]$ *and* $\nu_A(x) \in [0,1]$ *respectively represent the membership grade and nonmembership grade of the element x to set A, such that*

$$0 \leq (p_A(x))^2 + (\nu_A(x))^2 \leq 1$$

for all $x \in X$. *The degree of indeterminacy of the element x to set A is* $\pi_A(x)$, *defined by*

$$\pi_A(x) = \sqrt{1 - (p_A(x))^2 - (\nu_A(x))^2}.$$

For convenience, Zhang and Xu [49] referred to $(p_A(x), \nu_A(x))$ as a Pythagorean fuzzy number (PFN) simply denoted by $a = (p_a, \nu_a)$, where $p_a \in [0,1], \nu_a \in [0,1]$ and $(p_a)^2 + (\nu_a)^2 \leq 1$.

Yager and Abbasov [25] discussed the relationship between Pythagorean membership grades and complex numbers. They showed that the complex numbers of the form $z = r \cdot e^{j\theta}$ with conditions

$r \in [0,1]$ and $\theta \in [0, \pi/2]$ can be interpretable as PFNs $(r\cos\theta, r\sin\theta)$. They referred to these complex numbers as "$\Pi - i$ numbers", which are complex numbers in the upper-right quadrant of the complex unit disk.

As explained in [25], we should consider which aggregation operation is closed under $\Pi - i$ numbers.

Let us consider the closeness of $\Pi - i$ numbers under the CFWG and CFOWG operations. For the CFWG operator, we have the following result.

Theorem 12. *Let $z_1, z_2, ..., z_n$ be $\Pi - i$ numbers, and the CFWG weights are real values, that is, $w_i \in [0,1]$ for all i, and $\sum_{i=1}^{n} w_i = 1$. Then, the aggregated value $CFWG(z_1, z_2, ..., z_n)$ is also a $\Pi - i$ number.*

Proof. Denoting $CFWG(z_1, z_2, ..., z_n) = t \cdot e^{j\nu} = \prod_{i=1}^{n} t_{z_i}^{w_i} \cdot e^{j\sum_{i=1}^{n} w_i \cdot \nu_{z_i}}$, from Theorem 2, we have $t = |CFWG(z_1, z_2, ..., z_n)| \leq 1$.

Since $\nu = \sum_{i=1}^{n} w_i \cdot \nu_{z_i}$ is a weighted geometric aggregation (WGA) operator of real numbers on $[0, \pi/2]$, then we have $\nu \in [0, \pi/2]$. Thus, $CFWG(z_1, z_2, ..., z_n)$ is also a $\Pi - i$ number. □

Similar to the above Theorem, we have the following.

Theorem 13. *Let $z_1, z_2, ..., z_n$ be $\Pi - i$ numbers, and the CFOWG weights are real values, that is, $w_i \in [0,1]$ for all i, and $\sum_{i=1}^{n} w_i = 1$. Then, the aggregated value $CFOWG(z_1, z_2, ..., z_n)$ is also a $\Pi - i$ number.*

The above theorems show us that the CFWG and the CFOWG operators are closed under $\Pi - i$ numbers. When PFNs are interpreted as $\Pi - i$ numbers, then we can aggregate these PFNs to a PFN by using the CFWG or CFOWG operator.

From the above theorems, the CFWG and the CFOWG operators are closed on the upper-right quadrant of complex unit disk.

Consider other quadrants of the complex unit disk. Let

$$D_k = \left\{ z = r \cdot e^{j\theta} | r \in [0,1], \theta \in [\frac{(k-1)\pi}{2}, \frac{k\pi}{2}] \right\}$$

for $k = 1$ to 4. D_1 is the set of $\Pi - i$ numbers.

Now, we discuss the closeness of the CFWG and the CFOWG operators on other quadrants of the complex unit disk. Plainly, we have the following.

Theorem 14. *For any $k \in \{1,2,3,4\}$, if $z_1, z_2, ..., z_n \in D_k$, and the weights are real values, that is, $w_i \in [0,1]$ for all i, and $\sum_{i=1}^{n} w_i = 1$. Then, we have*

$$CFWG(z_1, z_2, ..., z_n) \in D_k,$$

$$CFOWG(z_1, z_2, ..., z_n) \in D_k.$$

Proof. Similar to Theorem 13. □

Theorem 15. *For any $k \in \{1,2,3,4\}$, if $z_1, z_2, ..., z_n, y_1, y_2, ..., y_n \in D_k$, and the weights are real values, that is, $w_i \in [0,1]$ for all i, and $\sum_{i=1}^{n} w_i = 1$. Then, we have the following:*

(1) (Idempotency): If $z_1 = z_2 = ... = z_n$, then

$$CFWG(z_1, z_2, ..., z_n) = z_1,$$

$$CFOWG(z_1, z_2, ..., z_n) = z_1.$$

(2) *(Amplitude boundedness):*

$$\left|CFWG(z_1, z_2, ..., z_n)\right| \leq z,$$

$$\left|CFOWG(z_1, z_2, ..., z_n)\right| \leq z,$$

where $z = \max_i |z_i|$.

(3) *(Amplitude monotonicity): If* $|z_i| \leq |y_i|$ $i = 1, 2, ..., n$, *then*

$$|CFWG(a_1, a_2, ..., a_n)| \leq |CFWG(y_1, y_2, ..., y_n)|,$$

$$|CFOWG(a_1, a_2, ..., a_n)| \leq |CFWG(y_1, y_2, ..., y_n)|.$$

Proof. Similar to Theorem 7. □

6. Example Application

In this section, we consider a target location application of the complex fuzzy aggregation operator. We do not intend to show the potential advantages of using complex fuzzy aggregation methods in comparison with existing alternative aggregation approaches in this section.

Assume the observer position is fixed. Using a position sensor and an angular sensor, the observer measures the distance and angle of the fixed target. To improve the target location accuracy, the observer repeatedly measures the same target. Then, the target position is estimated according to aggregation theory.

Assume n measurements $((d_1, \theta_1), (d_2, \theta_2), ..., (d_n, \theta_n))$ have been measured by the observer. The target position is estimated in the following five stages, as illustrated in Figure 2.

Step 1 Complexification of the measured results; each measurement is represented as $c_i = d_i \cdot e^{\theta_i}$.
Step 2 (Fuzzification) Normalize the amplitudes of all measurements. Let $d = \max_i d_i$, for each c_i, the normalized result is $a_i = c_i/d$, where $t_{a_i} = d_i/d$.
Step 3 (Aggregation) Produce an aggregate result. For simplicity, using the CFWG operator with weights $(1/n, 1/n, ..., 1/n)$. We obtain

$$a = \prod_{i=1}^{n} a_i^{1/n}, \tag{16}$$

where $t_a = \sqrt[n]{t_{a_1} \cdot t_{a_2} \cdots t_{a_n}}$ and $v_a = \frac{1}{n} \cdot \sum_{i=1}^{n} \theta_i$.
Step 4 (Defuzzification) Calculate $c = a \cdot d$, where $t_c = t_a \cdot d$.
Step 5 Decomplexification (or sometimes "realification") of c. We get the target position (p, v), where $p = t_c, v = v_a$.

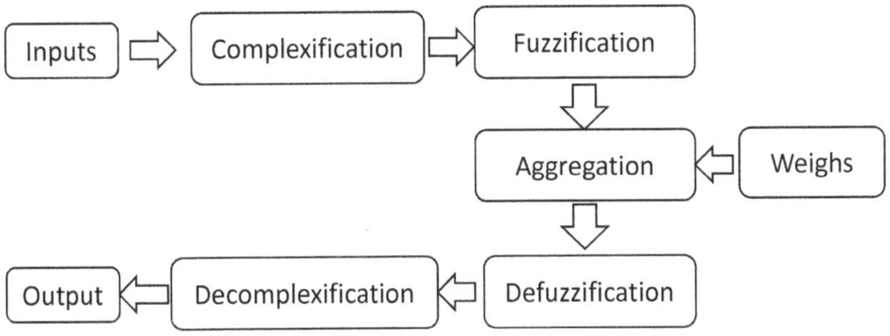

Figure 2. A simple method of target location based on complex fuzzy aggregation.

Numerical Example:Assume that the observer obtains five measurements as follows:

$$((865, 24°), (867, 25°), (871, 24°), (866, 25°), (869, 23°)),$$

where (d, θ) means that the target lies on the θ degrees east of south of the observer and d metres from the observer. Then,

Step 1 Complexifications of the measured results are calculated as

$$((865 \cdot e^{j2\pi 336/360}), (867 \cdot e^{j2\pi 335/360}), (871 \cdot e^{j2\pi 336/360}), (866 \cdot e^{j2\pi 335/360}), (869 \cdot e^{j2\pi 337/360})).$$

Step 2 Normalizations of the amplitudes of all measurements are calculated as

$$(0.9931 \cdot e^{j2\pi 336/360}, 0.9954 \cdot e^{j2\pi 335/360}, 1 \cdot e^{j2\pi 336/360}, 0.9943 \cdot e^{j2\pi 335/360}, 0.9977 \cdot e^{j2\pi 337/360}).$$

Step 3 Aggregation of CFVs is calculated as

$$0.9961 \cdot e^{j2\pi 335.8/360},$$

where the weights are $(1/5, 1/5, ..., 1/5)$.

Step 4 Defuzzification of the aggregate result is calculated as $867.6 \cdot e^{j2\pi 335.8/360}$.

Step 5 Decomplexification of the above result is calculated as $(867.6, 24.2)$.

Then, the target position is estimated at $(867.6, 24.2)$. That is, it lies 24.2 degrees east of south of the observer and 867.6 m from the observer.

Note that we do not discuss how to choose complex fuzzy aggregation functions, nor their weights.

7. Conclusions

In this paper, we propose two complex fuzzy aggregation operators: the CFWG and CFOWG operators. Their main properties are summarized in Table 1. In particular, both the CFWG and the CFOWG operators are reflectionally invariant and rotationally invariant. We also showed that the CFWG and the CFOWG operators are closed under $\Pi - i$ numbers.

As we know, complex data are frequently encountered in many different applications, such as engineering, management, finance, and medicine. CFSs are suitable to represent information with uncertainty and periodicity simultaneously. Complex fuzzy information aggregation techniques may be useful in these applications.

Author Contributions: Conceptualization, S.D.; Project administration, L.B. and B.H.; Writing-original draft, L.B., S.D. and B.H.; Writing-review and editing, L.B., S.D. and B.H.

Acknowledgments: The authors would like to thank the anonymous referees and academic editor for their very valuable comments.

Funding: This research was funded by the Opening Foundation of Guangxi Colleges and Universities Key Laboratory of Complex System Optimization and Big Data Processing (Grant No. 2017CSOBDP0103), the Science and Technology Foundation of Guizhou Province, China (LKS [2013] 35 and LKS (2012) 34), and the Guangxi University Science and Technology Research Project (Grant No. 2013YB193).

Conflicts of Interest: The author declares no conflict of interest.

References

1. Ramot, D.; Milo, R.; Friedman, M.; Kandel, A. Complex fuzzy sets. *IEEE Trans. Fuzzy Syst.* **2002**, *10*, 171–186. [CrossRef]
2. Zadeh, L.A. Fuzzy sets. *Inf. Control* **1965**, *8*, 338–353. [CrossRef]
3. Buckley, J.J. Fuzzy complex numbers. *Fuzzy Sets Syst.* **1989**, *33*, 333–345. [CrossRef]
4. Buckley, J.J. Fuzzy complex analysis I: Definition. *Fuzzy Sets Syst.* **1991**, *41*, 269–284. [CrossRef]

5. Buckley, J.J. Fuzzy complex analysis II: Integration. *Fuzzy Sets Syst.* **1992**, *49*, 171–179. [CrossRef]
6. Ramot, D.; Friedman, M.; Langholz, G.; Kandel, A. Complex fuzzy logic. *IEEE Trans. Fuzzy Syst.* **2003**, *11*, 450–461. [CrossRef]
7. Hu, B.; Bi, L.; Dai, S. The Orthogonality between Complex Fuzzy Sets and Its Application to Signal Detection. *Symmetry* **2017**, *9*, 175. [CrossRef]
8. Bi, L.; Hu, B.; Li, S.; Dai, S. The Parallelity of Complex Fuzzy sets and Parallelity Preserving Operators. *J. Intell. Fuzzy Syst.* **2018**, *34*. [CrossRef]
9. Zhang, G.; Dillon, T.S.; Cai, K.-Y.; Ma, J.; Lu, J. Operation properties and delta-equalities of complex fuzzy sets. *Int. J. Approx. Reason.* **2009**, *50*, 1227–1249. [CrossRef]
10. Alkouri, A.U.M.; Salleh, A.R. Linguistic variables, hedges and several distances on complex fuzzy sets. *J. Intell. Fuzzy Syst.* **2014**, *26*, 2527–2535.
11. Hu, B.; Bi, L.; Dai, S.; Li, S. Distances of Complex Fuzzy Sets and Continuity of Complex Fuzzy Operations. *J. Intell. Fuzzy Syst.* **2018**. [CrossRef]
12. Tamir, D.E.; Jin, L.; Kandel, A. A new interpretation of complex membership grade. *Int. J. Intell. Syst.* **2011**, *26*, 285–312. [CrossRef]
13. Tamir, D.E.; Kandel, A. Axiomatic theory of complex fuzzy logic and complex fuzzy classes. *Int. J. Comput. Commun. Control* **2011**, *4*, 562–576. [CrossRef]
14. Dick, S. Towards Complex Fuzzy Logic. *IEEE Trans. Fuzzy Syst.* **2005**, *13*, 405–414. [CrossRef]
15. Greenfield, S.; Chiclana, F.; Dick, S. Interval-valued complex fuzzy logic. In Proceedings of the 2016 IEEE International Conference on Fuzzy Systems (FUZZ-IEEE), Vancouver, BC, Canada, 24–29 July 2016.
16. Greenfield, S.; Chiclana, F.; Dick, S. Join and meet operations for interval-valued complex fuzzy logic. In Proceedings of the 2016 Annual Conference of the North American Fuzzy Information Processing Society (NAFIPS), El Paso, TX, USA, 31 October–4 November 2016; pp. 1–5.
17. Alkouri, A.; Salleh, A. Complex intuitionistic fuzzy sets. In Proceedings of the AIP Conference International Conference on Fundamental and Applied Sciences, Kuala Lumpur, Malaysia, 12–14 June 2012; pp. 464–470.
18. Ali, M.; Smarandache, F. Complex neutrosophic set. *Neural Comput. Appl.* **2017**, *28*, 1–18. [CrossRef]
19. Ma, J.; Zhang, G.; Lu, J. A method for multiple periodic factor prediction problems using complex fuzzy sets. *IEEE Trans. Fuzzy Syst.* **2012**, *20*, 32–45.
20. Chen, Z.; Aghakhani, S.; Man, J.; Dick, S. ANCFIS: A Neuro-Fuzzy Architecture Employing Complex Fuzzy Sets. *IEEE Trans. Fuzzy Syst.* **2011**, *19*, 305–322. [CrossRef]
21. Li, C. Complex Neuro-Fuzzy ARIMA Forecasting. A New Approach Using Complex Fuzzy Sets. *IEEE Trans. Fuzzy Syst.* **2013**, *21*, 567–584. [CrossRef]
22. Li, C.; Chiang, T.-W.; Yeh, L.-C. A novel self-organizing complex neuro-fuzzy approach to the problem of time series forecasting. *Neurocomputing* **2013**, *99*, 467–476. [CrossRef]
23. Al-Quran, A.; Hassan, N. The Complex Neutrosophic Soft Expert Relation and Its Multiple Attribute Decision-Making Method. *Entropy* **2018**, *20*, 101. [CrossRef]
24. Li, C.; Wu, T.; Chan, F.-T. Self-learning complex neuro-fuzzy system with complex fuzzy sets and its application to adaptive image noise canceling. *Neurocomputing* **2012**, *94*, 121–139. [CrossRef]
25. Yager, R.R.; Abbasov, A.M. Pythagorean membership grades, complex numbers and decision-making. *Int. J. Intell. Syst.* **2013**, *28*, 436–452. [CrossRef]
26. Yager, R.R. Pythagorean membership grades in multi-criteria decision making. *IEEE Trans Fuzzy Syst.* **2014**, *22*, 958–965. [CrossRef]
27. Yager, R.R. Pythagorean fuzzy subsets. In Proceedings of the Joint IFSA Congress and NAFIPS Meeting, Edmonton, AB, Canada, 24–28 June 2013; pp. 57–61.
28. Atanassov, K.T. Intuitionistic fuzzy sets. *IEEE Trans. Fuzzy Syst.* **1986**, *20*, 87–96. [CrossRef]
29. Dick, S.; Yager, R.R.; Yazdanbahksh, O. On Pythagorean and Complex Fuzzy Set Operations. *IEEE Trans. Fuzzy Syst.* **2016**, *24*, 1009–1021. [CrossRef]
30. Mayburov, S.N. Fuzzy Geometry of Commutative Spaces and Quantum Dynamics. *EPJ Web Conf.* **2016**, *126*, 05009. [CrossRef]
31. Bagarello, F.; Cinà, M.; Gargano, F. Projector operators in clustering. *Appl. Sci.* **2017**, *40*, 49–59. [CrossRef]
32. Yager, R.R. On ordered weighted averaging aggregation operators in multi-criteria decision making. *IEEE Trans. Syst. Man. Cybern.* **1988**, *18*, 183–190. [CrossRef]
33. Xu, Z.S. Intuitionistic fuzzy aggregation operators. *IEEE Trans. Fuzzy Syst.* **2007**, *15*, 1179–1187.

34. Xu, Z.S.; Yager, R.R. Some geometric aggregation operators based on intuitionistic fuzzy sets. *Int. J. Gen. Syst.* **2006**, *35*, 417–433. [CrossRef]
35. Xu, Z.S. *Intuitionistic Fuzzy Aggregation and Clustering*; Springer: Heidelberg/Berlin, Germany, 2013.
36. Zeng, S.; Mu, Z.; Baležentis, T. A novel aggregation method for Pythagorean fuzzy multiple attribute group decision making. *Int. J. Intell. Syst.* **2018**, *33*, 573–585. [CrossRef]
37. Zhang, X.L. A novel approach based on similarity measure for Pythagorean fuzzy multiple criteria group decision making. *Int. J. Intell. Syst.* **2016**, *31*, 593–611. [CrossRef]
38. Garg, H. A new generalized pythagorean fuzzy information aggregation using einstein operations and its application to decision making. *Int. J. Intell. Syst.* **2016**, *31*, 886–920. [CrossRef]
39. Ye, J. A multicriteria decision-making method using aggregation operators for simplified neutrosophic sets. *J. Intell. Fuzzy Syst.* **2014**, *26*, 2459–2466.
40. Tu, A.; Ye, J.; Wang, B. Multiple Attribute Decision-Making Method Using Similarity Measures of Neutrosophic Cubic Sets. *Symmetry* **2018**, *10*, 215. [CrossRef]
41. Tu, A.; Ye, J.; Wang, B. Symmetry Measures of Simplified Neutrosophic Sets for Multiple Attribute Decision-Making Problems. *Symmetry* **2018**, *10*, 144. [CrossRef]
42. Liu, P. Some Hamacher aggregation operators based on the interval-valued intuitionistic fuzzy numbers and their application to group decision making. *IEEE Trans. Fuzzy Syst.* **2014**, *22*, 83–97. [CrossRef]
43. Liu, P. Multiple Attribute Decision-Making Methods Based on Normal Intuitionistic Fuzzy Interaction Aggregation Operators. *Symmetry* **2017**, *9*, 261. [CrossRef]
44. Yu, D.; Wu, Y.; Lu, T. Interval-valued intuitionistic fuzzy prioritized operators and their application in group decision making. *Knowl. Based Syst.* **2012**, *30*, 57–66. [CrossRef]
45. Hussain, S.A.I.; Mandal, U.K.; Mondal, S.P. Decision Maker Priority Index and Degree of Vagueness Coupled Decision Making Method: A Synergistic Approach. *Int. J. Fuzzy Syst.* **2018**, *20*, 1551–1566. [CrossRef]
46. Xia, M.; Xu, Z. Hesitant fuzzy information aggregation in decision making. *Int. J. Approx. Reason.* **2011**, *52*, 395–407. [CrossRef]
47. Zhu, B.; Xu, Z.; Xia, M. Hesitant fuzzy geometric Bonferroni means. *Inf. Sci.* **2012**, *205*, 72–85. [CrossRef]
48. Song, C.; Xu, Z.; Zhao, H. A Novel Comparison of Probabilistic Hesitant Fuzzy Elements in Multi-Criteria Decision Making. *Symmetry* **2018**, *10*, 177. [CrossRef]
49. Zhang, X.; Xu, Z. Extension of TOPSIS to multiple criteria decision making with Pythagorean fuzzy sets. *Int. J. Intell. Syst.* **2014**, *29*, 1061–1078. [CrossRef]

symmetry

MDPI

Article

Another View of Aggregation Operators on Group-Based Generalized Intuitionistic Fuzzy Soft Sets: Multi-Attribute Decision Making Methods

Khizar Hayat [1], Muhammad Irfan Ali [2,*], Bing-Yuan Cao [1,3,*], Faruk Karaaslan [4] and Xiao-Peng Yang [5]

1 School of Mathematics and Information Sciences, Guangzhou University, Guangzhou 510000, China; khizar233@gmail.com or khizar233@e.gzhu.edu.cn

2 Department of Mathematics, Islamabad Model College for Girls F-6/2, Islamabad 44000, Pakistan

3 Department of Mathematics and Big Data Foshan University, Foshan 528000, and Guangzhou Vocational College of Science and Technology, Guangzhou 510550, China

4 Department of Mathematics, Faculty of Sciences, Çankırı Karatekin University, 18100 Çankırı, Turkey; fkaraaslan@karatekin.edu.tr

5 School of Mathematics and Statistics, Hanshan Normal University, Chaozhou 521041, China; happyyangxp@163.com

* Correspondence: mirfanali13@yahoo.com (M.I.A.); caobingy@163.com (B.-Y.C.)

Received: 5 November 2018; Accepted: 26 November 2018; Published: 14 December 2018

Abstract: In this paper, the existing definition of the group-based generalized intuitionistic fuzzy soft set is clarified and redefined by merging intuitionistic fuzzy soft set over the set of alternatives and a group of intuitionistic fuzzy sets on parameters. In this prospect, two new subsets of the group-based generalized intuitionistic fuzzy soft set are proposed and several operations are contemplated. The two new aggregation operators called generalized group-based weighted averaging and generalized group-based weighted geometric operator are introduced. The related properties of proposed operators are discussed. The recent research is emerging on multi-attribute decision making methods based on soft sets, intuitionistic fuzzy soft sets, and generalized intuitionistic fuzzy soft sets. An algorithm is structured and two case studies of multi-attribute decision makings are considered using proposed operators. Further, we provide the comparison and advantages of the proposed method, which give superiorities over recent major existing methods.

Keywords: decision-making; soft sets; intuitionistic fuzzy soft sets; group-based generalized intuitionistic fuzzy soft sets; aggregation operators

---

1. Introduction

The concept of fuzzy soft sets was popularized by Maji et al. [1], in the combination of fuzzy sets (Zadeh [2]) and soft sets (Molodtsov [3], Maji [4] and Ali [5]). To analyze the real-life problems, different types of uncertainties have been evaluated with fuzzy soft sets [6] and it has wide range of applications to deal with parameterizations and granularity. By virtue of robustness of fuzzy soft set theory in dealing with uncertain data, many researchers serve to integrate it with inductive learning techniques for better results. In recent past, fuzzy sets, soft sets and fuzzy soft sets are applied to evaluate vagueness in decision makings [7–18], algebraic structures [19,20], medical diagnosis [21] and differential equations [22]. Some hybrid models of fuzzy soft sets have been introduced and applied in several fields [23,24].

In 1986, Atanassov proposed the intuitionistic fuzzy set (IFS) [25], which appears as an inclusion of non-compatibility value with fuzzy set [2]. Every value of IFS is referred to a compatibility, a non-compatibility and a hesitancy, which assign it more dynamics in dealing with imprecise information. The initial aggregation instruments [26,27] on IFSs were introduced by Atanassov and Xu, and then applied in a various fields. The geometric [28], and arithmetic aggregation operators [27] have been studied in diverse fields and especially in multi-attribute decision making (MADM) problems in financial management, medical diagnosis, business and engineering designs [29–32].

IFSs with soft sets, that is, intuitionistic fuzzy soft sets (IFSSs) [33], are very instrumental and more realistic tools for uncertainty than fuzzy soft sets. As the dual-memberships structure of IFS allow marking hesitancy factors, the use of IFSSs in inductive learning techniques accounts for the degree of imprecision by assigning grades of compatibility and non-compatibility. In an inclusive way, several decision-making problems have been considered using IFSSs; some attempts are hybrid with intervals [34], multi-attributes [35] and nonlinear-programming [36]. Garg et al. [37,38] popularized aggregation operators on IFSSs and considered related decision-making methods. Several strategies are used to overcome the challenges of granularity and vagueness. Despite there being the applicability of IFSSs in diverse fields, an opinion of an expert who implicitly exercises his assessments on parameters of an IFSS is needed. On this motivation, Agarwal et al [39], who popularized generalized intuitionistic fuzzy soft set (GIFSS) by including assessment of a moderator on parameters, thus validating and supporting the information. Thus, the accumulation of generalized parameter can reduce possibility of errors which are occur due to imprecise data.

Although the definition of GIFSS in [39] is useful to tackle imprecise data, some difficulties appear in several notions [40,41]. Altogether, the assertions in [39] have been pointed out and a novel definition of GIFSS was established by Feng et al. [42]. They presented several operations and developed related multi-attribute decision-making methods by introducing operators on GIFSSs. On this prospect, a practical application of GIFSS for design concept evaluation was proposed by Hayat et al. [43]. Even though GIFSSs are applicable in diverse fields, sometimes assessments of more than one prospectors are needed in various problems. Thus, we consider the problem of validation of the notion of group-based GIFSS (GGIFSS) [44,45], and introduce a novel definition of GGIFSS, which is the generalization of the notion of GIFSS in [42]. Further, some basic properties are validated and aggregation instruments are proposed to determine the industrial applicability of GGIFSSs. Usually, an accurate aggregation process recommends the nature of MCDM model, which aggregates interdependent information and behaves in a linear manner. The prospect of proposing group-based generalized weighted averaging and geometric operators (hereafter, GBGWA and GBGWG) is to contemplate the information together with the influence of mathematical operations on GGIFSSs. The advantages of the given framework are to contemplate the prospector's demands or experts' judgments in an incorporated way such that establishing more operators constitutes the design concept of the evaluation mechanism of GGIFSSs.The results presented in this paper can be studied in several fields, such as electrical engineering, industrial designs, and construction engineering, as estimation of risk factors in risk management is a complex tasks.

The paper is organized as follows. Section 2 introduce basic concepts and notations. Section 3 clarify and redefine the notion of GGIFSS. Section 4 give operations on GGIFSSs, and introduce GBGWA(GBGWG) operators and related properties. Section 5 put forward the aggregation instruments of GGIFSSs into algorithm and discuss two different case studies. We present the comparison and benefits of method in Section 6. Advantages and superiorities are given in Section 7. Section 8 provide the conclusions of the paper.

2. Preliminaries

In this section, we present the basic definitions of fuzzy sets, IFSs, soft sets and GIFSS which would be useful for subsequent discussions. Throughout the paper, X is the universe.

A fuzzy set $\widetilde{t}$ in X is usually identified as its membership function $\widetilde{t} : X \longrightarrow [0,1]$ [2], each $x \in X$, where the membership grade $\widetilde{t}(x)$ indicates the degree to which the element x belongs to the fuzzy set $\widetilde{t}$. Here, we denote by $\mathcal{F}(X)$ the collection of all fuzzy sets in X. The subsets intersection, union and complement of fuzzy sets follow from Zadeh [2].

In 1999, Molodtsov [3] introduced the parameterization concept soft set theory, which is different from many traditional tools for dealing with uncertainties, such as fuzzy set theory [2], rough set theory [46], IFSs [25], and hesitant fuzzy sets. The main advantage of soft set theory is that it can be freely applied to characterized parameters, sentence, words and numbers. The natural manner of parameterization of this theory was augmented by the works of Maji et al. [4] and Ali et al. [5], among others.

Definition 1. *[3] Let E be the set of parameters, $\mathcal{A} \subseteq E$. A pair $(\mathcal{S}, \mathcal{A})$ is called a soft set over X, where $\mathcal{S}$ is a mapping given by $\mathcal{S} : \mathcal{A} \longrightarrow P(X)$. $P(X)$ is the set of all power sets of X.*

The set of all soft sets over X, with respect to subsets of E, is denoted by $\mathcal{SASS}^E(X)$.

2.1. Intuitionistic Fuzzy Sets

In the fuzzy set, only one compatibility degree exists, whereas intellectual insight in many cases suggests that non-compatibility degrees should be paired with compatibility degree. Atanassov [25] introduced the concept of IFS, which is an intellectual intuition to judge the uncertainty over the objects. Atanassov gave the definition of the IFS as follows:

Definition 2. *[25] An intuitionistic fuzzy set (IFS) in a universe X is defined as*

$$\mathcal{A} = \{\langle x, \widetilde{t}_{\mathcal{A}}(x), \widetilde{f}_{\mathcal{A}}(x)\rangle \mid x \in X\},$$

where the functions $\widetilde{t}, \widetilde{f} : X \longrightarrow [0,1]$ define, respectively, a membership function and a non-membership function of the element $x \in X$ to the set $\mathcal{A}$. Moreover, it is required that

$$0 \leq \widetilde{t}_{\mathcal{A}}(x) + \widetilde{f}_{\mathcal{A}}(x) \leq 1.$$

The function $\pi_{\mathcal{A}} = 1 - (\widetilde{t}_{\mathcal{A}}(x) + \widetilde{f}_{\mathcal{A}}(x))$ is called the degree of hesitancy of x to $\mathcal{A}$. The collection of all IFSs in X is denoted by $IFS(X)$.

Let $\mathcal{A}, \mathcal{B} \in IFS(X)$. Then,

$$\begin{aligned} \mathcal{A} \sqcup \mathcal{B} &= \{\langle x, \max\{\widetilde{t}_{\mathcal{A}}(x), \widetilde{t}_{\mathcal{B}}(x)\}, \min\{\widetilde{f}_{\mathcal{A}}(x), \widetilde{f}_{\mathcal{B}}(x)\}\rangle \mid x \in X\}, \\ \mathcal{A} \sqcap \mathcal{B} &= \{\langle x, \min\{\widetilde{t}_{\mathcal{A}}(x), \widetilde{t}_{\mathcal{B}}(x)\}, \max\{\widetilde{f}_{\mathcal{A}}(x), \widetilde{f}_{\mathcal{B}}(x)\}\rangle \mid x \in X\}, \\ \mathcal{A} \sqsubseteq \mathcal{B} &\iff \widetilde{t}_{\mathcal{A}}(x) \leq \widetilde{t}_{\mathcal{B}}(x) \text{ and } \widetilde{f}_{\mathcal{A}}(x) \geq \widetilde{f}_{\mathcal{B}}(x) \forall x \in X. \end{aligned}$$

Deschrijver and Kerre [47] defined that IFSs can be considered as L-fuzzy sets with respect to the complete lattice $(V^*, \leqslant_{V^*})$, where $V^* = \{\langle \mu_1, \mu_2\rangle \in [0,1]^2 \mid \mu_1 + \mu_2 \leq 1\}$, and the corresponding partial order $\leqslant_{V^*}$ is defined as $\langle \mu_1, \mu_2\rangle \leqslant_{V^*} \langle \nu_1, \nu_2\rangle \iff (\mu_1 \leq \nu_1) \wedge (\mu_2 \leq \nu_2)$ for all $\langle \mu_1, \mu_2\rangle, \langle \nu_1, \nu_2\rangle \in V^*$. Any ordered pair $\langle \mu_1, \mu_2\rangle \in V^*$ is called an intuitionistic fuzzy value (IFV) or intuitionistic fuzzy number (IFN).

Let V^* be the set of IFVs of IFS $\mathcal{A}$, such that $\langle \widetilde{t}_{\mathcal{A}}, \widetilde{f}_{\mathcal{A}}\rangle \in V^*$. Chen and Tan [48] presented score function, which was updated by Feng et al. [42] as follows:

Definition 3. *[42] Let $\langle \widetilde{t}_{\mathcal{A}}, \widetilde{f}_{\mathcal{A}}\rangle \in V^*$ be an IFV in a universe X. Then, expectation score function is a mapping $\delta : V^* \rightarrow [0,1]$, defined as follows:*

$$\delta_{\mathcal{A}} = \frac{\widetilde{t}_{\mathcal{A}} - \widetilde{f}_{\mathcal{A}} + 1}{2} \tag{1}$$

where $\delta_{\mathcal{A}}$ is called the decision value of $\langle \widetilde{t}_{\mathcal{A}}, \widetilde{f}_{\mathcal{A}}\rangle$ in $\mathcal{A}$. In addition, fuzzy set $\delta_{\mathcal{A}}$ is called the utility fuzzy set derived from the IFS $\mathcal{A}$.

Definition 4. [28] *Let $V_1 = \langle \widetilde{t}_A, \widetilde{f}_A \rangle, V_2 = \langle \widetilde{t}'_A, \widetilde{f}'_A \rangle \in V^*$ be two IFVs in a universe X. Then we have,*

(i) $V_1 \oplus V_2 = \langle \widetilde{t}_A + \widetilde{t}'_A - \widetilde{t}_A \widetilde{t}'_A, \widetilde{f}_A \widetilde{f}'_A \rangle$.
(ii) $V_1 \otimes V_2 = \langle \widetilde{t}_A \widetilde{t}'_A, \widetilde{f}_A + \widetilde{f}'_A - \widetilde{f}_A \widetilde{f}'_A \rangle$.
(iii) $\epsilon V_1 = \langle 1 - (1 - \widetilde{t}_A)^{\epsilon}, (\widetilde{f}_A)^{\epsilon} \rangle$, *where ϵ is a positive real number.*

More operations and properties of IFVs (or IF numbers (IFNs)) can be seen in [27,28,42]. Let $c_1, c_2, ..., c_m$ be the IFVs and $\phi = (\phi_1, \phi_2, ..., \phi_m)$ be the correlated weighted normalized vector, then, from Yager [28] and Xu [27], we denote and symbolize the following operators:

$$\text{IFWA}(c_1, c_2, ..., c_m) = \phi_1 c_1 \otimes \phi_2 c_3 \otimes, ..., \otimes \phi_m c_m = \langle 1 - \prod_{i}^{m} (1 - \widetilde{t}_{c_i})^{\phi_i}, \prod_{i}^{m} \widetilde{f}_{c_i}^{\phi_i} \rangle, \quad (2)$$

$$\text{IFWG}(c_1, c_2, ..., c_m) = c_1^{\phi_1} \otimes c_3^{\phi_3} \otimes, ..., \otimes c_m^{\phi_m} = \langle \prod_{i}^{m} \widetilde{t}_{c_i}^{\phi_i}, 1 - \prod_{i}^{m} (1 - \widetilde{f}_{c_i})^{\phi_i} \rangle. \quad (3)$$

IFWA and IFWG are the IF weighted averaging and geometric operators, respectively.

2.2. Intuitionistic Fuzzy Soft Sets and Generalized Intuitionistic Fuzzy Soft Sets

In this section, we present some basic notions in the theory of IFSS and GIFSS. The notion of IFSS is given as follows:

Definition 5. [33] *Let (X, E) be a soft universe and $\mathcal{A} \subseteq E$. A pair $\mathscr{F} = (\widetilde{\mathcal{S}}, \mathcal{A})$ is called intuitionistic fuzzy soft set (IFSS) over X, where $\widetilde{\mathcal{S}}$ is a mapping defined by $\widetilde{\mathcal{S}} : \mathcal{A} \longrightarrow IFS(X)$.*

Formally, $\widetilde{\mathcal{S}} : \mathcal{A} \longrightarrow IFS(X)$ is referred to as the approximate function of the IFSS $(\widetilde{\mathcal{S}}, \mathcal{A})$. It is easy to see that IFSSs extend both Atanassov's IFSs and Molodtsov's soft sets. The set of all IFSSs over X, with respect to subsets of E, is denoted by $\mathcal{IFSS}^E(X)$. Next, the two new subsets of a IFSS are presented as follows:

Definition 6. [42] *Let $\mathscr{F} = (\widetilde{\mathcal{S}}, \mathcal{A})$ and $\mathscr{G} = (\widetilde{\mathcal{T}}, \mathcal{B})$ be IFSSs over X and $\mathcal{A}, \mathcal{B} \subset E$. Then, $\mathscr{G}$ is anintuitionistic fuzzy soft F-subsetof $\mathscr{F}$, denoted by $\mathscr{G} \widetilde{\subseteq}_F \mathscr{F}$, if*

(i) $\mathcal{B} \subseteq \mathcal{A}$.
(ii) $\widetilde{\mathcal{T}}(a) \subseteq \widetilde{\mathcal{S}}(a) \; \forall a \in \mathcal{B}$.

Definition 7. [42] *Let $\mathscr{F} = (\widetilde{\mathcal{S}}, \mathcal{A})$ and $\mathscr{G} = (\widetilde{\mathcal{T}}, \mathcal{B})$ be IFSSs over X and $\mathcal{A}, \mathcal{B} \subset E$. Then, $\mathscr{G}$ is an intuitionistic fuzzy soft M-subsetof $\mathscr{F}$, denoted by $\mathscr{G} \widetilde{\subseteq}_M \mathscr{F}$, if*

(i) $\mathcal{B} \subseteq \mathcal{A}$.
(ii) $\widetilde{\mathcal{T}}(a) = \widetilde{\mathcal{S}}(a) \; \forall a \in \mathcal{B}$.

The related whole IFSS is denoted as $\widetilde{\mathcal{X}}^{\mathcal{A}^{(1,0)}}$, where all IFVs are $(1,0)$, and related null IFSS is denoted as $\widetilde{\mathcal{I}}^{\mathcal{A}^{(0,1)}}$, where all IFVs are $(0,1)$. The other definitions of union, intersection and complements of IFSSs follow [33,42]. It is required in many cases that an extra input of moderator with IFSS could be useful. The definition of GIFSS was given by Agarwal et al. [39] as follows:

Definition 8. [39] *Let (X, E) be a soft universe and $\mathcal{A} \subseteq E$. A generalized intuitionistic fuzzy soft set (GIFSS), $\mathscr{F}_\alpha$ over the soft universe (X, E) is defined as a mapping $\mathscr{F}_\alpha : \mathcal{A} \longrightarrow IFSS(X) \times IF$, $IFSS(X)$ the collection of all intuitionistic fuzzy subsets of X and the generalization parameter, $\alpha : \mathcal{A} \longrightarrow IF = (t_\alpha, f_\alpha)$, where IF is an IFS. The GIFSS is of the form $\mathscr{F}_\alpha(e_i) = (\widetilde{\mathcal{S}}(e_i), \alpha(e_i))$.*

The model of GIFSS is very fruitful in decision making, especially the input of an extra opinion of an expert works incentively. However, in Definition 8, $IFSS(X) \times IF$ is not a meaningful Cartesian product and generalized parameter $\widetilde{\alpha}$ is not well-defined. A more well-defined and flexible form of GIFSS was defined by Feng et al. [42]. They pointed out several assertions in [39], certified several notions and discussed GIFFSs theoretically. The definition of GIFSS is given as follows:

Definition 9. *[42] Let (X, E) be a soft universe and $\mathcal{A} \subseteq E$. A triple $\widetilde{\mathscr{F}} = (\widetilde{\mathcal{S}}, \mathcal{A}, \widetilde{\alpha})$ is called generalized intuitionistic fuzzy soft set (GIFSS) over X if $(\widetilde{\mathcal{S}}, \mathcal{A})$ is an IFSS over X and $\widetilde{\alpha}$ is an IFS in $\mathcal{A}$.*

This representation of GIFSS can be more significant to handle problems in which uncertain and unclear information are prevalent, and it enhances the accuracy and flexibility of results with opinions of experts as an IFS on the set of parameters. The two different types of subsets of GIFSS and several operations on GIFSSs are specified and categorized in [42]. Hayat et al. [43] presentes another form of GIFSSs and related notions.

3. Group-Based Generalized Intuitionistic Fuzzy Soft Sets

In this section, we clarifiy and reformulate the definition of GGIFSS presented in [44]. First, we recall the definition of GGIFSS that is given in [44];

Definition 10. *[44] A group-based generalized intuitionistic fuzzy soft sets (GGIFSS), $\mathcal{F}_G$, over the soft universe (X, E) is defined as $\mathcal{F}_G : E \to IFS(X) \times IF$ for all $v \in E$; we have $\mathcal{F}_G(v) = (\mathcal{F}(v), G_\varrho(v))$, where $\mathcal{F}(v) \in IFS(X)$ and $G_\varrho(v) \in IF$. Here, $G = (\varrho_1, \varrho_2, ..., \varrho_p)$ are intuitionistic fuzzy subset of set of the parameter E and $G_\varrho(v)$ denotes the opinion of experts on the elements of X in $\mathcal{F}(v)$.*

Remark 1. *The above definition of GGIFSS is very effective in many cases, due to its constructive scenario for decision making. However, this definition has some difficulties and dissensions on group of extra input of moderators, as well as on the mapping. Specifically, we identify the following:*

(i) *On the point that $G_\varrho(v)$ is an IFS, stated in Definition 10, but $G_\varrho(v)$ is a group of IFSs. $IFS(X) \times IF$ is not a meaningful product. In this way, mapping $\mathcal{F}_G : E \to IFS(X) \times IF$ is not well-defined.*
(ii) *As the Definition 10 is stated on group of extra opinions, which is an intuitionistic fuzzy subset of set of the parameter E, therefore $G_\varrho(v)$ is not defined in a precise way.*
(iii) *The extra inputs can be seen as another IFVs based data of alternatives.*

To clarify the problems mentioned in Remark 1, we reformulate the notion of GGIFSS as follows:

Definition 11. *Let (X, E) be a soft universe and $\mathcal{A} \subseteq E$. A triple $\widetilde{\mathscr{F}}_{\widetilde{g}} = (\widetilde{\mathcal{S}}, \mathcal{A}, \widetilde{g})$ is called group-based generalized intuitionistic fuzzy soft set (GGIFSS) over X, if $(\widetilde{\mathcal{S}}, \mathcal{A})$ is an elementary IFSS (EIFSS) over X and $\widetilde{g} = \{\widetilde{\alpha}_1, \widetilde{\alpha}_2, ..., \widetilde{\alpha}_p\}$ where $\widetilde{\alpha}_1, \widetilde{\alpha}_2, ..., \widetilde{\alpha}_p$ are the parameterized IFSs (PIFSs) of $\mathcal{A}$.*
In other words, $\widetilde{g}$ is a group of PIFSs considered by "p" number of experts/moderators.

Keeping the prospects of decision making in the mind, $(\widetilde{\mathcal{S}}, \mathcal{A})$ is basic IFS and $\widetilde{g}$ is a group of parameterized intuitionistic fuzzy sets (GPIFSs). The set of all GGIFSS over X obtained on E is denoted by $\mathcal{GGIFSS}_E(X)$. Further, the set of all GGIFSS over X obtained on subset $\mathcal{A} \subset E$ is denoted by $\mathcal{GGIFSS}_{\mathcal{A}}(X)$.

Example 1. *Let $X = \{\kappa_1, \kappa_2, ..., \kappa_6\}$ be the universe set, consisting six cellphones, under consideration and $E = \{v_1, v_2, v_3, v_4\}$ where $v_\imath (\imath = 1, 2, 3, 4)$, respectively, stand for "high battery timing", "low operating cost", "high quality of voice call" and "stylish look". Consider a set of attributes $\mathcal{B} = \{v_1, v_3, v_4\} \subset E$ chosen by an observer $\mathcal{M}$, which are anticipated to be most fruitful for judgment of cellphones. For $\mathcal{M}$, the evaluation of alternatives with rating values corresponding each parameters can be defined as EIFSS,*

$$\widetilde{\mathcal{S}}(v_1) = \{\frac{\kappa_1}{\langle 0.6,0.4\rangle}, \frac{\kappa_2}{\langle 0,0.3\rangle}, \frac{\kappa_3}{\langle 0.2,0.2\rangle}, \frac{\kappa_4}{\langle 0,1\rangle}, \frac{\kappa_5}{\langle 0.1,0.6\rangle}, \frac{\kappa_6}{\langle 0.6,0.2\rangle}\}$$
$$\widetilde{\mathcal{S}}(v_3) = \{\frac{\kappa_1}{\langle 0.6,0.2\rangle}, \frac{\kappa_2}{\langle 0.7,0.3\rangle}, \frac{\kappa_3}{\langle 0.3,0.6\rangle}, \frac{\kappa_4}{\langle 0.4,0.1\rangle}, \frac{\kappa_5}{\langle 0.4,0.2\rangle}, \frac{\kappa_6}{\langle 0.3,0.3\rangle}\}$$
$$\widetilde{\mathcal{S}}(v_4) = \{\frac{\kappa_1}{\langle 0.9,0\rangle}, \frac{\kappa_2}{\langle 0.5,0.1\rangle}, \frac{\kappa_3}{\langle 0.5,0.1\rangle}, \frac{\kappa_4}{\langle 0.2,0.5\rangle}, \frac{\kappa_5}{\langle 0.5,0.2\rangle}, \frac{\kappa_6}{\langle 0.6,0.1\rangle}\}$$

Consider three moderators d_1, d_2, d_3 for assessment of rating value, such that the opinion of each moderator on each parameter of $\mathcal{M}$ is analyzed and based on opinions, PIFSs $\widetilde{\alpha}_{d_1}, \widetilde{\alpha}_{d_2}$ and $\widetilde{\alpha}_{d_3}$ are defined on $\mathcal{A}$ as

$$\widetilde{g} = \begin{cases} \widetilde{\alpha}_{d_1} = \{(v_1, \langle 0.6, 0.2\rangle), (v_3, \langle 0.3, 0.4\rangle), (v_4, \langle 0.2, 0.2\rangle)\}, \\ \widetilde{\alpha}_{d_2} = \{(v_1, \langle 0.3, 0.4\rangle), (v_3, \langle 0.2, 0.4\rangle), (v_4, \langle 0.3, 0.5\rangle)\}, \\ \widetilde{\alpha}_{d_3} = \{(v_1, \langle 0.3, 0.4\rangle), (v_3, \langle 0.5, 0.4\rangle), (v_4, \langle 0.4, 0.1\rangle)\}. \end{cases}$$

Then, the GGIFSS is represented in Table 1.

Table 1. Tabular representation of the GGIFSS, $\widetilde{\mathscr{F}_{\widetilde{g}}} = (\widetilde{\mathcal{S}}, \mathcal{A}, \widetilde{g})$.

$X \setminus \mathcal{A}$	v_1	v_3	v_4
κ_1	$\langle 0.6, 0.4\rangle$	$\langle 0.6, 0.2\rangle$	$\langle 0.9, 0\rangle$
κ_2	$\langle 0, 0.3\rangle$	$\langle 0.7, 0.3\rangle$	$\langle 0.5, 0.1\rangle$
κ_3	$\langle 0.2, 0.2\rangle$	$\langle 0.3, 0.6\rangle$	$\langle 0.5, 0.1\rangle$
κ_4	$\langle 0, 1\rangle$	$\langle 0.4, 0.1\rangle$	$\langle 0.2, 0.5\rangle$
κ_5	$\langle 0.1, 0.6\rangle$	$\langle 0.4, 0.2\rangle$	$\langle 0.5, 0.2\rangle$
κ_6	$\langle 0.6, 0.2\rangle$	$\langle 0.3, 0.3\rangle$	$\langle 0.6, 0.1\rangle$
$\widetilde{\alpha}_{d_1}$	$\langle 0.6, 0.2\rangle$	$\langle 0.3, 0.4\rangle$	$\langle 0.2, 0.2\rangle$
$\widetilde{\alpha}_{d_2}$	$\langle 0.3, 0.4\rangle$	$\langle 0.2, 0.4\rangle$	$\langle 0.3, 0.5\rangle$
$\widetilde{\alpha}_{d_3}$	$\langle 0.3, 0.4\rangle$	$\langle 0.5, 0.4\rangle$	$\langle 0.4, 0.1\rangle$

If $p = 1$, then $\widetilde{\mathscr{F}_{\widetilde{g}}} = (\widetilde{\mathcal{S}}, \mathcal{A}, \widetilde{g})$ operates as GIFSS. In general, $\widetilde{\mathscr{F}_{\widetilde{g}}}$ can be sighted as a common form of generalized parameters with the information on IFSSs.

4. Operations on GGIFSSs and Aggregation Operators

In this section, several new operations on GGIFSSs and their examples are presented. As in Remark 1, it is pointed out that the group of extra assessments of experts in [36] is not defined in a precise way. In this scenario, we define two different subsets of a GGIFSS; for this purpose, a notion on group of generalized parameters of two GGIFSSs is defined as follows:

Definition 12. *Let (X, E) be a soft universe and $\mathcal{A}, \mathcal{B} \subseteq E$. Suppose that $\widetilde{\mathscr{F}_{\widetilde{g}}} = (\widetilde{\mathcal{S}}, \mathcal{A}, \widetilde{g}_1)$ and $\widetilde{\mathscr{G}_{\widetilde{g}}} = (\widetilde{\mathcal{T}}, \mathcal{B}, \widetilde{g}_2)$ are two GGIFSSs over X, where $\mathcal{A} \subseteq \mathcal{B}$, $\widetilde{g}_1 = \{\widetilde{\alpha}_{d_1}, \widetilde{\alpha}_{d_2}, ..., \widetilde{\alpha}_{d_p}\}$ and $\widetilde{g}_2 = \{\widetilde{\beta}_{d_1}, \widetilde{\beta}_{d_2}, ..., \widetilde{\beta}_{d_p}\}$ and $d_1, d_2, ..., d_p$ are "p" number of senior experts/members. If $\widetilde{g}_1$ is the group intuitionistic fuzzy subset of $\widetilde{g}_2$, then it is denoted and defined by $\widetilde{g}_1 \lll \widetilde{g}_2$ if and only if $t_{\widetilde{\alpha}_{d_1}}(v_i) \leq t_{\widetilde{\beta}_{d_1}}(v_i), f_{\widetilde{\alpha}_{d_1}}(v_i) \geq f_{\widetilde{\beta}_{d_1}}(v_i), t_{\widetilde{\alpha}_{d_2}}(v_i) \leq t_{\widetilde{\beta}_{d_2}}(v_i), f_{\widetilde{\alpha}_{d_2}}(v_i) \geq f_{\widetilde{\beta}_{d_2}}(v_i), \dots, t_{\widetilde{\alpha}_{d_p}}(v_i) \leq t_{\widetilde{\beta}_{d_p}}(v_i), f_{\widetilde{\alpha}_{d_p}}(v_i) \geq f_{\widetilde{\beta}_{d_p}}(v_i)$, $\forall\, i = 1, 2, ..., m$ and $v_i \in \mathcal{A}$.*

Based on Definition 12, the following two different kinds of group-based generalized intuitionistic fuzzy soft subsets can be presented.

Definition 13. *Let $\widetilde{\mathscr{F}_{\widetilde{g}}} = (\widetilde{\mathcal{S}}, \mathcal{A}, \widetilde{g}_1)$ and $\widetilde{\mathscr{G}_{\widetilde{g}}} = (\widetilde{\mathcal{T}}, \mathcal{B}, \widetilde{g}_2)$ be two GGIFSSs over X and $\mathcal{A}, \mathcal{B} \subseteq E$. Then, $\widetilde{\mathscr{F}_{\widetilde{g}}}$ is a group-based generalized intuitionistic fuzzy soft F-subsetof $\widetilde{\mathscr{G}_{\widetilde{g}}}$, denoted by $\widetilde{\mathscr{F}_{\widetilde{g}}} \widetilde{\sqsubseteq}_F \widetilde{\mathscr{G}_{\widetilde{g}}}$, if*

(i) $(\widetilde{\mathcal{S}}, \mathcal{A}) \widetilde{\subseteq}_F (\widetilde{\mathcal{T}}, \mathcal{B})$.
(ii) $\widetilde{g}_1 \lll \widetilde{g}_2$.

Now, an example is given to clarify group-based generalized intuitionistic fuzzy soft F-subset of a GGIFSS.

Example 2. *Let* $X = \{\kappa_1, \kappa_2, ..., \kappa_6\}$ *be the universe set, consisting six robots under consideration and* $E = \{v_1, v_2, v_3, v_4\}$ *where* v_i*, respectively, stand for "high capacity", "low degree of freedom", "high memory capacity" and "high repeatability". Consider two sets of parameters* $\mathcal{A} = \{v_1, v_4\} \subset E$*, and* $\mathcal{B} = \{v_1, v_3, v_4\} \subset E$ *chosen by observers* $\mathcal{M}_1$ *and* $\mathcal{M}_2$*, respectively, which are anticipated to be most fruitful for evaluation of robots. For* $\mathcal{M}_1$ *and* $\mathcal{M}_2$*, evaluation of alternatives with their ratting values corresponding each parameters can be defined, respectively, as EIFSSs,*

$$\widetilde{S}(v_1) = \{\tfrac{\kappa_1}{\langle 0.6,0.3\rangle}, \tfrac{\kappa_2}{\langle 0,0.3\rangle}, \tfrac{\kappa_3}{\langle 0.2,0.2\rangle}, \tfrac{\kappa_4}{\langle 0,1\rangle}, \tfrac{\kappa_5}{\langle 0.1,0.6\rangle}, \tfrac{\kappa_6}{\langle 0.6,0.2\rangle}\}$$
$$\widetilde{S}(v_4) = \{\tfrac{\kappa_1}{\langle 0.9,0\rangle}, \tfrac{\kappa_2}{\langle 0.5,0.1\rangle}, \tfrac{\kappa_3}{\langle 0.5,0.1\rangle}, \tfrac{\kappa_4}{\langle 0.3,0.5\rangle}, \tfrac{\kappa_5}{\langle 0.5,0.2\rangle}, \tfrac{\kappa_6}{\langle 0.6,0.1\rangle}\}$$

and

$$\widetilde{T}(v_1) = \{\tfrac{\kappa_1}{\langle 0.6,0.4\rangle}, \tfrac{\kappa_2}{\langle 0.1,0.5\rangle}, \tfrac{\kappa_3}{\langle 0.3,0.2\rangle}, \tfrac{\kappa_4}{\langle 0,1\rangle}, \tfrac{\kappa_5}{\langle 0.2,0.7\rangle}, \tfrac{\kappa_6}{\langle 0.7,0.3\rangle}\}$$
$$\widetilde{T}(v_3) = \{\tfrac{\kappa_1}{\langle 0.6,0.2\rangle}, \tfrac{\kappa_2}{\langle 0.7,0.3\rangle}, \tfrac{\kappa_3}{\langle 0.3,0.6\rangle}, \tfrac{\kappa_4}{\langle 0.4,0.1\rangle}, \tfrac{\kappa_5}{\langle 0.4,0.2\rangle}, \tfrac{\kappa_6}{\langle 0.3,0.3\rangle}\}$$
$$\widetilde{T}(v_4) = \{\tfrac{\kappa_1}{\langle 0.9,0.1\rangle}, \tfrac{\kappa_2}{\langle 0.6,0.2\rangle}, \tfrac{\kappa_3}{\langle 0.5,0.2\rangle}, \tfrac{\kappa_4}{\langle 0.3,0.5\rangle}, \tfrac{\kappa_5}{\langle 0.6,0.3\rangle}, \tfrac{\kappa_6}{\langle 0.6,0.1\rangle}\}$$

Consider three moderator, d_1 *from engineering department,* d_2 *from production department and* d_3 *from quality inspection department; their additional opinions for assessments of each observer are analyzed and, based on their opinions, PIFSs on* $\mathcal{M}_1$: $\widetilde{\alpha}_{d_1}, \widetilde{\alpha}_{d_2}, \widetilde{\alpha}_{d_3}$ *and IFSs of* $\mathcal{M}_2$: $\widetilde{\beta}_{d_1}, \widetilde{\beta}_{d_2}, \widetilde{\beta}_{d_3}$ *are defined.*

$$\widetilde{g}_1 = \begin{cases} \widetilde{\alpha}_{d_1} = \{(v_1, \langle 0.3, 0.2\rangle), (v_4, \langle 0.3, 0.4\rangle)\}, \\ \widetilde{\alpha}_{d_2} = \{(v_1, \langle 0.3, 0.4\rangle), (v_4, \langle 0.2, 0.4\rangle)\}, \\ \widetilde{\alpha}_{d_3} = \{(v_1, \langle 0.3, 0.4\rangle), (v_4, \langle 0.5, 0.4\rangle)\}, \end{cases}$$

$$\widetilde{g}_2 = \begin{cases} \widetilde{\beta}_{d_1} = \{(v_1, \langle 0.6, 0.2\rangle), (v_3, \langle 0.4, 0.5\rangle), (v_4, \langle 0.4, 0.4\rangle)\}, \\ \widetilde{\beta}_{d_2} = \{(v_1, \langle 0.3, 0.2\rangle), (v_3, \langle 0.4, 0.2\rangle), (v_4, \langle 0.4, 0.3\rangle)\}, \\ \widetilde{\beta}_{d_3} = \{(v_1, \langle 0.4, 0.2\rangle), (v_3, \langle 0.4, 0.4\rangle), (v_4, \langle 0.4, 0.2\rangle)\}. \end{cases}$$

Then, the GGIFSSs $\widetilde{\mathscr{F}}_{\widetilde{g}}$ *and* $\widetilde{\mathscr{G}}_{\widetilde{g}}$ *are tabulated in Tables 2, and 3, respectively.*

Table 2. Tabular representation of the GGIFSS $\widetilde{\mathscr{F}}_{\widetilde{g}} = (\widetilde{S}, \mathcal{A}, \widetilde{g}_1)$.

$X \setminus \mathcal{A}$	v_1	v_4
κ_1	$\langle 0.6, 0.3\rangle$	$\langle 0.9, 0\rangle$
κ_2	$\langle 0, 0.3\rangle$	$\langle 0.5, 0.1\rangle$
κ_3	$\langle 0.2, 0.2\rangle$	$\langle 0.5, 0.1\rangle$
κ_4	$\langle 0, 1\rangle$	$\langle 0.3, 0.5\rangle$
κ_5	$\langle 0.1, 0.6\rangle$	$\langle 0.5, 0.2\rangle$
κ_6	$\langle 0.6, 0.2\rangle$	$\langle 0.6, 0.1\rangle$
$\widetilde{\alpha}_{d_1}$	$\langle 0.3, 0.2\rangle$	$\langle 0.3, 0.4\rangle$
$\widetilde{\alpha}_{d_2}$	$\langle 0.3, 0.4\rangle$	$\langle 0.2, 0.4\rangle$
$\widetilde{\alpha}_{d_3}$	$\langle 0.3, 0.4\rangle$	$\langle 0.5, 0.4\rangle$

Table 3. Tabular representation of the GGIFSS, $\widetilde{\mathscr{G}_{\tilde{g}}} = (\widetilde{\mathcal{T}}, \mathcal{B}, \tilde{g}_2)$.

$X \setminus \mathcal{B}$	v_1	v_3	v_4
κ_1	$\langle 0.6, 0.2 \rangle$	$\langle 0.6, 0.2 \rangle$	$\langle 0.9, 0 \rangle$
κ_2	$\langle 0.1, 0.1 \rangle$	$\langle 0.7, 0.3 \rangle$	$\langle 0.6, 0.1 \rangle$
κ_3	$\langle 0.3, 0.2 \rangle$	$\langle 0.3, 0.6 \rangle$	$\langle 0.5, 0 \rangle$
κ_4	$\langle 0, 1 \rangle$	$\langle 0.4, 0.1 \rangle$	$\langle 0.4, 0.5 \rangle$
κ_5	$\langle 0.2, 0.3 \rangle$	$\langle 0.4, 0.2 \rangle$	$\langle 0.6, 0.1 \rangle$
κ_6	$\langle 0.7, 0.2 \rangle$	$\langle 0.3, 0.3 \rangle$	$\langle 0.8, 0.1 \rangle$
$\widetilde{\beta}_{d_1}$	$\langle 0.6, 0.2 \rangle$	$\langle 0.4, 0.5 \rangle$	$\langle 0.3, 0.4 \rangle$
$\widetilde{\beta}_{d_2}$	$\langle 0.3, 0.2 \rangle$	$\langle 0.4, 0.2 \rangle$	$\langle 0.3, 0.3 \rangle$
$\widetilde{\beta}_{d_3}$	$\langle 0.4, 0.2 \rangle$	$\langle 0.4, 0.4 \rangle$	$\langle 0.5, 0.2 \rangle$

One can easily check that $(\widetilde{\mathcal{S}}, \mathcal{A}) \widetilde{\subseteq}_F (\widetilde{\mathcal{T}}, \mathcal{B})$ and $\tilde{g}_1 \lll \tilde{g}_2$. Thus, $\widetilde{\mathscr{F}_{\tilde{g}}} = (\widetilde{\mathcal{S}}, \mathcal{A}, \tilde{g}_1)$ is group-based generalized intuitionistic fuzzy soft F-subset of $\widetilde{\mathscr{G}_{\tilde{g}}} = (\widetilde{\mathcal{T}}, \mathcal{B}, \tilde{g}_2)$.

Definition 14. *Let $\widetilde{\mathscr{F}_{\tilde{g}}} = (\widetilde{\mathcal{S}}, \mathcal{A}, \tilde{g}_1)$ and $\widetilde{\mathscr{G}_{\tilde{g}}} = (\widetilde{\mathcal{T}}, \mathcal{B}, \tilde{g}_2)$ be two GGIFSSs over X and $\mathcal{A}, \mathcal{B} \subset E$. Then, $\widetilde{\mathscr{F}_{\tilde{g}}}$ is a group-based generalized intuitionistic fuzzy soft M-subsetof $\widetilde{\mathscr{G}_{\tilde{g}}}$, denoted by $\widetilde{\mathscr{F}_{\tilde{g}}} \widetilde{\sqsubseteq}_M \widetilde{\mathscr{G}_{\tilde{g}}}$, if*

(i) $(\widetilde{\mathcal{S}}, \mathcal{A}) \widetilde{\subseteq}_M (\widetilde{\mathcal{T}}, \mathcal{B})$.
(ii) $\tilde{g}_1 \lll \tilde{g}_2$.

The complement of a GGIFSS is given as follows:

Definition 15. *Let $\widetilde{\mathscr{G}_{\tilde{g}}} = (\widetilde{\mathcal{T}}, \mathcal{A}, \tilde{g})$ be GGIFSS over X. The complement of $\widetilde{\mathscr{G}_{\tilde{g}}}$ is defined as the GGIFSS $\widetilde{\mathscr{G}^c_{\tilde{g}}} = (\widetilde{\mathcal{T}}^c, \mathcal{A}, \tilde{g}^c)$ where $(\widetilde{\mathcal{G}}^c, \mathcal{A})$ is the complement of the EIFSS $(\widetilde{\mathcal{G}}, \mathcal{A})$ and $\tilde{g}^c = \{\tilde{\alpha}^c{}_{d_1}, \tilde{\alpha}^c_{d_2}, ..., \tilde{\alpha}^c_{d_p}\}$ is the complement of $\tilde{g} = \{\tilde{\alpha}_{d_1}, \tilde{\alpha}_{d_2}, ..., \tilde{\alpha}_{d_p}\}$.*

Now, an example is given to clarify complement of a GGIFSS.

Example 3. *Consider GGIFSS $\widetilde{\mathscr{G}_{\tilde{g}}} = (\widetilde{\mathcal{T}}, \mathcal{A}, \tilde{g})$ defined in Example 1. Then, complement of $\widetilde{\mathscr{G}_{\tilde{g}}}$ is defined in Table 4.*

Table 4. Tabular representation of the GGIFSS $\widetilde{\mathscr{G}^c_{\tilde{g}}} = (\widetilde{\mathcal{T}}^c, \mathcal{A}, \tilde{g}^c)$.

$X \setminus \mathcal{A}$	v_1	v_3	v_4
κ_1	$\langle 0.4, 0.6 \rangle$	$\langle 0.2, 0.6 \rangle$	$\langle 0, 0.9 \rangle$
κ_2	$\langle 0.3, 0 \rangle$	$\langle 0.3, 0.7 \rangle$	$\langle 0.1, 0.5 \rangle$
κ_3	$\langle 0.2, 0.2 \rangle$	$\langle 0.6, 0.3 \rangle$	$\langle 0.1, 0.5 \rangle$
κ_4	$\langle 1, 0 \rangle$	$\langle 0.1, 0.4 \rangle$	$\langle 0.5, 0.2 \rangle$
κ_5	$\langle 0.6, 0.1 \rangle$	$\langle 0.2, 0.4 \rangle$	$\langle 0.2, 0.5 \rangle$
κ_6	$\langle 0.2, 0.6 \rangle$	$\langle 0.3, 0.3 \rangle$	$\langle 0.1, 0.6 \rangle$
$\tilde{\alpha}^c_{d_1}$	$\langle 0.2, 0.6 \rangle$	$\langle 0.4, 0.3 \rangle$	$\langle 0.2, 0.2 \rangle$
$\tilde{\alpha}^c_{d_2}$	$\langle 0.4, 0.3 \rangle$	$\langle 0.4, 0.2 \rangle$	$\langle 0.5, 0.3 \rangle$
$\tilde{\alpha}^c_{d_3}$	$\langle 0.4, 0.3 \rangle$	$\langle 0.4, 0.5 \rangle$	$\langle 0.1, 0.4 \rangle$

Next, the definitions of *extended union, extended intersection, restricted union* and*restricted intersection* are provided below.

Definition 16. *Let $\widetilde{\mathscr{F}_{\tilde{g}}} = (\widetilde{\mathcal{S}}, \mathcal{A}, \tilde{g}_1)$ and $\widetilde{\mathscr{G}_{\tilde{g}}} = (\widetilde{\mathcal{T}}, \mathcal{B}, \tilde{g}_2)$ be two GGIFSSs over X, $\mathcal{A}, \mathcal{B} \subseteq E$, $\mathcal{C} = \mathcal{A} \cup \mathcal{B}$ and $\tilde{g}_1 = \{\tilde{\alpha}_{d_1}, \tilde{\alpha}_{d_2}, ..., \tilde{\alpha}_{d_p}\}$, $\tilde{g}_2 = \{\widetilde{\beta}_{d_1}, \widetilde{\beta}_{d_2}, ..., \widetilde{\beta}_{d_p}\}$. The extended union of $\widetilde{\mathscr{F}_{\tilde{g}}}$ and $\widetilde{\mathscr{G}_{\tilde{g}}}$ is defined as the GGIFSS*

$$(\widetilde{\mathcal{H}}, \mathcal{C}, \tilde{g}) = (\widetilde{\mathcal{S}}, \mathcal{A}, \tilde{g}_1) \, \widetilde{\sqcup}_{\mathcal{E}} \, (\widetilde{\mathcal{T}}, \mathcal{B}, \tilde{g}_2)$$

such that

(i) $(\widetilde{\mathcal{H}},\mathcal{C}) = (\widetilde{\mathcal{S}},\mathcal{A}) \cup_{\mathcal{E}} (\widetilde{\mathcal{T}},\mathcal{B})$.
(ii) *For each moderator* d_k, $\widetilde{\gamma}_{d_k}$ $(k = 1,2,...,p)$ *can be defined* $\forall v \in \mathcal{C}$,

$$\widetilde{t}_{\widetilde{\gamma}_{d_k}}(v) = \begin{cases} \widetilde{t}_{\widetilde{\alpha}_{d_k}}(v), & \text{if } v \in \mathcal{A}\backslash\mathcal{B}, \\ \widetilde{t}_{\widetilde{\beta}_{d_k}}(v), & \text{if } v \in \mathcal{B}\backslash\mathcal{A}, \\ \max\{\widetilde{t}_{\widetilde{\alpha}_{d_k}}(v), \widetilde{t}_{\widetilde{\beta}_{d_k}}(v)\}, & \text{if } v \in \mathcal{A}\cap\mathcal{B}; \end{cases}$$

and

$$\widetilde{f}_{\widetilde{\gamma}_{d_k}}(v) = \begin{cases} \widetilde{f}_{\widetilde{\alpha}_{d_k}}(v), & \text{if } v \in \mathcal{A}\backslash\mathcal{B}, \\ \widetilde{f}_{\widetilde{\beta}_{d_k}}(v), & \text{if } v \in \mathcal{B}\backslash\mathcal{A}, \\ \min\{\widetilde{f}_{\widetilde{\alpha}_{d_k}}(v), \widetilde{f}_{\widetilde{\beta}_{d_k}}(v)\}, & \text{if } v \in \mathcal{A}\cap\mathcal{B}. \end{cases}$$

Definition 17. *Let* $\widetilde{\mathscr{F}}_{\widetilde{g}} = (\widetilde{\mathcal{S}},\mathcal{A},\widetilde{g}_1)$ *and* $\widetilde{\mathscr{G}}_{\widetilde{g}} = (\widetilde{\mathcal{T}},\mathcal{B},\widetilde{g}_2)$ *be two GGIFSSs over X, where* $\mathcal{A},\mathcal{B} \subseteq E$, $\mathcal{C} = \mathcal{A}\cap\mathcal{B}$ *and* $\widetilde{g}_1 = \{\widetilde{\alpha}_{d_1},\widetilde{\alpha}_{d_2},...,\widetilde{\alpha}_{d_p}\}$, $\widetilde{g}_2 = \{\widetilde{\beta}_{d_1},\widetilde{\beta}_{d_2},...,\widetilde{\beta}_{d_p}\}$. *The extended intersection of* $\widetilde{\mathscr{F}}_{\widetilde{g}}$ *and* $\widetilde{\mathscr{G}}_{\widetilde{g}}$ *is defined as the GGIFSS*

$$(\widetilde{\mathcal{R}},\mathcal{C},\widetilde{g}) = (\widetilde{\mathcal{S}},\mathcal{A},\widetilde{g}_1)\,\widetilde{\sqcap}_{\mathcal{E}}(\widetilde{\mathcal{T}},\mathcal{B},\widetilde{g}_2)$$

such that

(i) $(\widetilde{\mathcal{R}},\mathcal{C}) = (\widetilde{\mathcal{S}},\mathcal{A}) \cap_{\mathcal{E}} (\widetilde{\mathcal{T}},\mathcal{B})$.
(ii) *For each moderator* d_k, $\widetilde{\gamma}_{d_k}$ $(k = 1,2,...,p)$ *can be defined* $\forall v \in \mathcal{A}\cup\mathcal{B}$,

$$\widetilde{t}_{\widetilde{\gamma}_{d_k}}(v) = \begin{cases} \widetilde{t}_{\widetilde{\alpha}_{d_k}}(v), & \text{if } v \in \mathcal{A}\backslash\mathcal{B}, \\ \widetilde{t}_{\widetilde{\beta}_{d_k}}(v), & \text{if } v \in \mathcal{B}\backslash\mathcal{A}, \\ \min\{\widetilde{t}_{\widetilde{\alpha}_{d_k}}(v), \widetilde{t}_{\widetilde{\beta}_{d_k}}(v)\}, & \text{if } v \in \mathcal{A}\cap\mathcal{B}; \end{cases}$$

and

$$\widetilde{f}_{\widetilde{\gamma}_{d_k}}(v) = \begin{cases} \widetilde{f}_{\widetilde{\alpha}_{d_k}}(v), & \text{if } v \in \mathcal{A}\backslash\mathcal{B}, \\ \widetilde{f}_{\widetilde{\beta}_{d_k}}(v), & \text{if } v \in \mathcal{B}\backslash\mathcal{A}, \\ \max\{\widetilde{f}_{\widetilde{\alpha}_{d_k}}(v), \widetilde{f}_{\widetilde{\beta}_{d_k}}(v)\}, & \text{if } v \in \mathcal{A}\cap\mathcal{B}. \end{cases}$$

Definition 18. *Let* $\widetilde{\mathscr{F}}_{\widetilde{g}} = (\widetilde{\mathcal{S}},\mathcal{A},\widetilde{g}_1)$ *and* $\widetilde{\mathscr{G}}_{\widetilde{g}} = (\widetilde{\mathcal{T}},\mathcal{B},\widetilde{g}_2)$ *be two GGIFSSs over X, where* $\mathcal{A},\mathcal{B} \subseteq E$, $\mathcal{C} = \mathcal{A}\cap\mathcal{B}$ *and* $\widetilde{g}_1 = \{\widetilde{\alpha}_{d_1},\widetilde{\alpha}_{d_2},...,\widetilde{\alpha}_{d_p}\}$, $\widetilde{g}_2 = \{\widetilde{\beta}_{d_1},\widetilde{\beta}_{d_2},...,\widetilde{\beta}_{d_p}\}$. *The restricted union of* $\widetilde{\mathscr{F}}_{\widetilde{g}}$ *and* $\widetilde{\mathscr{G}}_{\widetilde{g}}$ *is defined as the GGIFSS*

$$(\widetilde{\mathcal{R}},\mathcal{C},g) = (\widetilde{\mathcal{S}},\mathcal{A},\widetilde{g}_1)\,\widetilde{\sqcup}_r(\widetilde{\mathcal{T}},\mathcal{B},\widetilde{g}_2)$$

such that

(i) $(\widetilde{\mathcal{R}},\mathcal{C}) = (\widetilde{\mathcal{S}},\mathcal{A}) \cup_r (\widetilde{\mathcal{T}},\mathcal{B})$;
(ii) *For each moderator* d_k, $\widetilde{\gamma}_{d_k}$ $(k = 1,2,...,p)$ *can be defined* $\forall v \in \mathcal{C}$,

$$\widetilde{t}_{\widetilde{\gamma}_{d_k}}(v) = \max\{\widetilde{t}_{\widetilde{\alpha}_{d_k}}(v), \widetilde{t}_{\widetilde{\beta}_{d_k}}(v)\}, \text{ for all } v \in \mathcal{A}\cap\mathcal{B};$$

and

$$\widetilde{f}_{\widetilde{\gamma}_{d_k}}(v) = \min\{\widetilde{f}_{\widetilde{\alpha}_{d_k}}(v), \widetilde{f}_{\widetilde{\beta}_{d_k}}(v)\}, \text{ for all } v \in \mathcal{A}\cap\mathcal{B}.$$

Definition 19. *Let* $\widetilde{\mathscr{F}}_{\widetilde{g}} = (\widetilde{\mathcal{S}}, \mathcal{A}, \widetilde{g}_1)$ *and* $\widetilde{\mathscr{G}}_{\widetilde{g}} = (\widetilde{\mathcal{T}}, \mathcal{B}, \widetilde{g}_2)$ *be two GGIFSSs over X, where* $\mathcal{A}, \mathcal{B} \subseteq E$, $\mathcal{C} = \mathcal{A} \cap \mathcal{B}$ *and* $\widetilde{g}_1 = \{\widetilde{\alpha}_{d_1}, \widetilde{\alpha}_{d_2}, ..., \widetilde{\alpha}_{d_p}\}$, $\widetilde{g}_2 = \{\widetilde{\beta}_{d_1}, \widetilde{\beta}_{d_2}, ..., \widetilde{\beta}_{d_p}\}$. *The restricted intersection of* $\widetilde{\mathscr{F}}_{\widetilde{g}}$ *and* $\widetilde{\mathscr{G}}_{\widetilde{g}}$ *is defined as the GGIFSS*

$$(\widetilde{\mathcal{R}}, \mathcal{C}, g) = (\widetilde{\mathcal{S}}, \mathcal{A}, \widetilde{g}_1) \widetilde{\sqcap}_r (\widetilde{\mathcal{T}}, \mathcal{B}, \widetilde{g}_2)$$

such that

(i) $(\widetilde{\mathcal{R}}, \mathcal{C}) = (\widetilde{\mathcal{S}}, \mathcal{A}) \cap_r (\widetilde{\mathcal{T}}, \mathcal{B})$.

(ii) *For each moderator* d_k, $\widetilde{\gamma}_{d_k}$ $(k = 1, 2, ..., p)$ *can be defined* $\forall v \in \mathcal{C}$,

$$\widetilde{t}_{\widetilde{\gamma}_{d_k}}(v) = \min\{\widetilde{t}_{\widetilde{\alpha}_{d_k}}(v), \widetilde{t}_{\widetilde{\beta}_{d_k}}(v)\}, \text{ for all } v \in \mathcal{A} \cap \mathcal{B};$$

and

$$\widetilde{f}_{\widetilde{\gamma}_{d_k}}(v) = \max\{\widetilde{f}_{\widetilde{\alpha}_{d_k}}(v), \widetilde{f}_{\widetilde{\beta}_{d_k}}(v)\}, \text{ for all } v \in \mathcal{A} \cap \mathcal{B}.$$

The definition of null GGIFSS and whole GGIFSS are specified below.

Definition 20. *Let* $\widetilde{\mathscr{G}}_{\widetilde{g}} = (\widetilde{\mathcal{S}}, \mathcal{A}, \widetilde{g})$ *be a GGIFSS over X, where* $\mathcal{A} \subseteq E$, *and* $\widetilde{g} = \{\widetilde{\alpha}_{d_1}, \widetilde{\alpha}_{d_2}, ..., \widetilde{\alpha}_{d_p}\}$. *Then,* $\widetilde{\mathscr{G}}_{\widetilde{g}}$ *is called the group-based generalized relative null intuitionistic fuzzy soft set, denoted by* $\widetilde{\mathscr{N}}_{\widetilde{g}}^{\mathcal{A}}$, *if*

(1) $(\widetilde{\mathcal{S}}, \mathcal{A}) = \widetilde{\mathcal{I}}^{\mathcal{A}^{(0,1)}}$.

(2) *For each moderator* d_k, $\widetilde{t}_{\widetilde{\alpha}_{d_k}}(v) = 0$ *and* $\widetilde{f}_{\widetilde{\alpha}_{d_k}}(v) = 1$ *for all* $v \in \mathcal{A}$.

Definition 21. *Let* $\widetilde{\mathscr{G}}_{\widetilde{g}} = (\widetilde{\mathcal{S}}, \mathcal{A}, \widetilde{g})$ *be a GGIFSS over X, where* $\mathcal{A} \subseteq E$ *and* $\widetilde{g} = \{\widetilde{\alpha}_{d_1}, \widetilde{\alpha}_{d_2}, ..., \widetilde{\alpha}_{d_p}\}$. *Then,* $\widetilde{\mathscr{G}}_{\widetilde{g}}$ *is called the group-based generalized relative whole intuitionistic fuzzy soft set, denoted by* $\widetilde{\mathscr{W}}_{\widetilde{g}}^{\mathcal{A}}$, *if*

(1) $(\widetilde{\mathcal{S}}, \mathcal{A}) = \widetilde{\mathcal{X}}^{\mathcal{A}^{(1,0)}}$.

(2) *For each moderator* d_k, $\widetilde{t}_{\widetilde{\alpha}_{d_k}}(v) = 1$ *and* $\widetilde{f}_{\widetilde{\alpha}_{d_k}}(v) = 0$ *for all* $v \in \mathcal{A}$.

Proposition 1. *Let* $\widetilde{\mathscr{G}}_{\widetilde{g}} = (\widetilde{\mathcal{S}}, \mathcal{A}, \widetilde{g})$ *be a GGIFSS over X, where* $\mathcal{A} \subseteq E$ *and* $\widetilde{g} = \{\widetilde{\alpha}_{d_1}, \widetilde{\alpha}_{d_2}, ..., \widetilde{\alpha}_{d_p}\}$. *Then,*

(i) $\widetilde{\mathscr{G}}_{\widetilde{g}} \widetilde{\sqcup}_{\mathcal{E}} \widetilde{\mathscr{G}}_{\widetilde{g}} = \widetilde{\mathscr{G}}_{\widetilde{g}} \widetilde{\sqcup}_r \widetilde{\mathscr{G}}_{\widetilde{g}} = \widetilde{\mathscr{G}}_{\widetilde{g}}$.

(ii) $\widetilde{\mathscr{G}}_{\widetilde{g}} \widetilde{\sqcap}_{\mathcal{E}} \widetilde{\mathscr{G}}_{\widetilde{g}} = \widetilde{\mathscr{G}}_{\widetilde{g}} \widetilde{\sqcap}_r \widetilde{\mathscr{G}}_{\widetilde{g}} = \widetilde{\mathscr{G}}_{\widetilde{g}}$.

(iii) $\widetilde{\mathscr{G}}_{\widetilde{g}} \widetilde{\sqcup}_{\mathcal{E}} \widetilde{\mathscr{N}}_{\widetilde{g}}^{\mathcal{A}} = \widetilde{\mathscr{G}}_{\widetilde{g}} \widetilde{\sqcup}_r \widetilde{\mathscr{N}}_{\widetilde{g}}^{\mathcal{A}} = \widetilde{\mathscr{G}}_{\widetilde{g}}$.

(iv) $\widetilde{\mathscr{G}}_{\widetilde{g}} \widetilde{\sqcap}_{\mathcal{E}} \widetilde{\mathscr{N}}_{\widetilde{g}}^{\mathcal{A}} = \widetilde{\mathscr{G}}_{\widetilde{g}} \widetilde{\sqcap}_r \widetilde{\mathscr{N}}_{\widetilde{g}}^{\mathcal{A}} = \widetilde{\mathscr{N}}_{\widetilde{g}}^{\mathcal{A}}$.

(v) $\widetilde{\mathscr{G}}_{\widetilde{g}} \widetilde{\sqcup}_{\mathcal{E}} \widetilde{\mathscr{W}}_{\widetilde{g}}^{\mathcal{A}} = \widetilde{\mathscr{G}} \widetilde{\sqcup}_r \widetilde{\mathscr{W}}_{\mathcal{A}} = \widetilde{\mathscr{W}}_{\mathcal{A}}$.

(vi) $\widetilde{\mathscr{G}}_{\widetilde{g}} \widetilde{\sqcap}_{\mathcal{E}} \widetilde{\mathscr{W}}_{\widetilde{g}}^{\mathcal{A}} = \widetilde{\mathscr{G}}_{\widetilde{g}} \widetilde{\sqcap}_r \widetilde{\mathscr{W}}_{\widetilde{g}}^{\mathcal{A}} = \widetilde{\mathscr{G}}_{\widetilde{g}}$.

Now, we introduce group-based generalized weighted averaging (GBGWA) and group-based generalized weighted geometric (GBGWG) operators on GGIFSSs. On these operators, we contemplate and discussed some properties as well. The definition of GBGWA operator is specified below.

Definition 22. GBGWA; *Let* $\widetilde{\mathscr{F}}_{\widetilde{g}} = (\widetilde{\mathcal{S}}, \mathcal{A}, \widetilde{g})$ *be a GGIFSS over X, where* $\widetilde{g} = \{\widetilde{\alpha}_{d_1}, \widetilde{\alpha}_{d_2}, ..., \widetilde{\alpha}_{d_p}\}$ *be the group of PIFSs. Assume that* $w = (w_1, w_2, ..., w_m)^T$ *is the normalized weight vector for* $\mathcal{A}$, *such that* $w_i > 0$ *and* $\sum_{i=1}^{m} w_i = 1$. *Let* $\widetilde{IFV}(\kappa_j) = \{c_{j1}, c_{j2}, ..., c_{jm}\}$ $(j = 1$ *to* $n)$ *be the set of IFVs in EIFSS* $(\widetilde{\mathcal{S}}, \mathcal{A})$ *for all* $\kappa_j \in X$. *For each senior moderator/prospector,* $\widetilde{\alpha}_{d_k}(v) = \{\langle \widetilde{t}_{\widetilde{\alpha}_{d_k}}(v), \widetilde{f}_{\widetilde{\alpha}_{d_k}}(v)\rangle \mid v \in \mathcal{A}\}$ $(k = 1$ *to* $p)$ *be the PIFS, it can be represented as* $\mathcal{IF}_k = \{a_{k1}, a_{k2}, ..., a_{km}\}$ $(k = 1$ *to* $p)$ *and* $\varpi = (\varpi_1, \varpi_2, ..., \varpi_p)^T$ *is the set of* *weights for moderators, such that* $\varpi_k > 0$ *and* $\sum_{k=1}^{p} \varpi_k = 1$*. Define* GBGWA: $\Gamma_s^m \longrightarrow \Gamma_s$, IFWA: $\Gamma^m \longrightarrow \Gamma$, *where*

$$\text{GBGWA}(c_{j1}, c_{j2}, ..., c_{jm}) = \text{IFWA}_k \begin{pmatrix} (\text{IFWA}_i(a_{11}, a_{12}, ..., a_{1m}) \otimes \text{IFWA}_i(c_{j1}, c_{j2}, ..., c_{jm})), \\ (\text{IFWA}_i(a_{21}, a_{22}, ..., a_{2m}) \otimes \text{IFWA}_i(c_{j1}, c_{j2}, ..., c_{jm})), ..., \\ (\text{IFWA}_i(a_{p1}, a_{p2}, ..., a_{pm}) \otimes \text{IFWA}_i(c_{j1}, c_{j2}, ..., c_{jm})) \end{pmatrix} \quad (4)$$

where GBGWA *is known as GGIFSS weighted averaging operator, then the set of all* GBGWAs *is denoted* $L = \{\ell'_1, \ell'_2, ..., \ell'_n\}$*. In addition,* IFWA_k *and* IFWA_i *are* IFWA *operators on set of moderators/prospectors and set of parameters, respectively. Note that* Γ_s^m *and* Γ *are families of GGIFSS and IFSs, respectively.*

Lemma 1. *Let* $\widetilde{\mathscr{F}}_{\widetilde{g}} = (\widetilde{\mathcal{S}}, \mathcal{A}, \widetilde{g})$ *be a GGIFSS over X, where* $\widetilde{g} = \{\widetilde{\alpha}_{d_1}, \widetilde{\alpha}_{d_2}, ..., \widetilde{\alpha}_{d_p}\}$ *be the group of IFSs. If* $p = 1$*, then* $\widetilde{\mathscr{F}}_{\widetilde{g}}$ *is a GIFSS and* GBGWA *is given as follows:*

$$\text{GBGWA}(c_{j1}, c_{j2}, ..., c_{jm}) = \text{IFWA}_i(a_{11}, a_{12}, ..., a_{1m}) \otimes \text{IFWA}_i(c_{j1}, c_{j2}, ..., c_{jm}). \quad (5)$$

Theorem 1. *If* $c_{ji} = \langle \widetilde{t}_{ji}, \widetilde{f}_{ji} \rangle$ *and* $a_{ki} = \langle \widetilde{t}_{ki}, \widetilde{f}_{ki} \rangle$ $(i = 1, 2, ..., m, j = 1, 2, ..., n, k = 1, 2, ..., p)$*, be the IFVs, then the accumulated value by* GBGWA *operator is given by*
$\text{GBGWA}(c_{j1}, c_{j2}, ..., c_{jm}) = \langle 1 - \prod_{k=1}^{p}(1 - (1 - \prod_{i=1}^{m}(1 - \widetilde{t}_{a_{ki}})) \cdot (1 - \prod_{i=1}^{m}(1 - \widetilde{t}_{c_{ji}}))), \prod_{k=1}^{p}(1 - (1 - \prod_{i=1}^{m} \widetilde{f}_{a_{ki}})(1 - \prod_{i=1}^{m} \widetilde{f}_{c_{ji}})) \rangle$.

Proof. Let $p = 1$ and $m = 2$. Firstly, we apply mathematical induction on m, we have $\text{GBGWA}(c_{j1}, c_{j2}) = \text{IFWA}_k(\text{IFWA}(a_{11}, a_{12}) \otimes \text{IFWA}(c_{j1}, c_{j2})) = (\text{IFWA}(a_{11}, a_{12}) \otimes \text{IFWA}(c_{j1}, c_{j2})) = \langle 1 - (1 - \widetilde{t}_{a_{11}})^{w_1} \cdot (1 - \widetilde{t}_{a_{12}})^{w_2}, \widetilde{f}_{a_{11}}^{w_1} \cdot \widetilde{f}_{a_{12}}^{w_2} \rangle \otimes \langle 1 - (1 - \widetilde{t}_{c_{j1}})^{w_1} \cdot (1 - \widetilde{t}_{c_{j2}})^{w_2}, \widetilde{f}_{c_{j1}}^{w_1} \cdot \widetilde{f}_{c_{j2}}^{w_2} \rangle$
$- \langle (1 - (1 - \widetilde{t}_{a_{11}})^{w_1} \cdot (1 - \widetilde{t}_{a_{12}})^{w_2}) \cdot (1 - (1 - \widetilde{t}_{c_{j1}})^{w_1} \cdot (1 - \widetilde{t}_{c_{j2}})^{w_2}), \widetilde{f}_{a_{11}}^{w_1} \cdot \widetilde{f}_{a_{12}}^{w_2} + \widetilde{f}_{c_{j1}}^{w_1} . \widetilde{f}_{c_{j2}}^{w_2} - \widetilde{f}_{a_{11}}^{w_1} . \widetilde{f}_{a_{12}}^{w_2} . \widetilde{f}_{c_{j1}}^{w_1} \cdot \widetilde{f}_{c_{j2}}^{w_2} \rangle = \langle (1 - \prod_{i=1}^{2}(1 - \widetilde{t}_{a_{1i}})^{w_i}) \cdot (1 - \prod_{i=1}^{2}(1 - \widetilde{t}_{c_{ji}})^{w_i}), \prod_{i=1}^{2} \widetilde{f}_{a_{1i}}^{w_i} + \prod_{i=1}^{2} \widetilde{f}_{c_{ji}}^{w_i} - \prod_{i=1}^{2} \widetilde{f}_{a_{1i}}^{w_i} . \prod_{i=1}^{2} \widetilde{f}_{c_{ji}}^{w_i} \rangle$.

Thus, theorem is true for $m = 2$; assuming that the result is true for $m = s'$, that is, $\text{GBGWA}(c_{j1}, c_{j2}, ..., c_{js'}) = \langle (1 - \prod_{i=1}^{s'}(1 - \widetilde{t}_{a_{1i}}) \cdot (1 - \prod_{i=1}^{s'}(1 - \widetilde{t}_{c_{ji}}))), 1 - ((1 - \prod_{i=1}^{s'} \widetilde{f}_{a_{1i}})(1 - \prod_{i=1}^{s'} \widetilde{f}_{c_{ji}})) \rangle$. then, for $m = s' + 1$, $\text{GBGWA}(c_{j1}, c_{j2}, ..., c_{j(s'+1)}) = \langle (1 - \prod_{i=1}^{s'+1}(1 - \widetilde{t}_{a_{1i}}) \cdot (1 - \prod_{i=1}^{s'+1}(1 - \widetilde{t}_{c_{ji}}))), 1 - ((1 - \prod_{i=1}^{s'+1} \widetilde{f}_{a_{1i}})(1 - \prod_{i=1}^{s'+1} \widetilde{f}_{c_{ji}})) \rangle$. Thus, by mathematical induction, Theorem 1 holds for all positive integer m. Similarly, we can prove this theorem for $k = 2, 3, ..., p$. □

Example 4. *Consider Example 1, where*
$\widetilde{IFV}(\kappa_1) = \{c_{11}, c_{12}, c_{13}\} = \{\langle 0.6, 0.4 \rangle, \langle 0.6, 0.2 \rangle, \langle 0.9, 0 \rangle\}$ *is a family of IFVs in second row of Table 1. The three IFSs of moderator's assessments are*

$$\begin{aligned} \mathcal{IF}_1 &= \{a_{11}, a_{12}, a_{13}\} = \{\langle 0.6, 0.2 \rangle, \langle 0.3, 0.4 \rangle, \langle 0.2, 0.2 \rangle\} \\ \mathcal{IF}_2 &= \{a_{21}, a_{22}, a_{23}\} = \{\langle 0.3, 0.4 \rangle, \langle 0.2, 0.4 \rangle, \langle 0.3, 0.5 \rangle\} \\ \mathcal{IF}_3 &= \{a_{31}, a_{32}, a_{33}\} = \{\langle 0.3, 0.4 \rangle, \langle 0.5, 0.4 \rangle, \langle 0.4, 0.1 \rangle\}, \end{aligned}$$

respectively. Let $w = \{w_1/0.29, w_2/0.35, w_3/0.36\}$ *be the weighted vector over E and* $\varpi = \{\varpi_1/0.25, \varpi_2/0.40, \varpi_3/0.35\}$ *be the weighted vector for three senior experts. Now, the GBGWA is given below.*

$$\ell'_1 = \text{GBGWA}(c_{11}, c_{12}, c_{13}) = \text{IFWA}_k \begin{pmatrix} (\text{IFWA}_i(a_{11}, a_{12}, a_{13}) \otimes \text{IFWA}_i(c_{11}, c_{12}, c_{13})), \\ (\text{IFWA}_i(a_{21}, a_{22}, a_{23}) \otimes \text{IFWA}_i(c_{11}, c_{12}, c_{13})), \\ (\text{IFWA}_i(a_{31}, a_{32}, a_{33}) \otimes \text{IFWA}_i(c_{11}, c_{12}, c_{13})) \end{pmatrix}$$

Next, calculate, $\text{IFWA}_i(c_{11}, c_{12}, c_{13}) = \text{IFWA}_i(\langle 0.6, 0.4 \rangle, \langle 0.6, 0.2 \rangle, \langle 0.9, 0.0 \rangle) = \langle 0.7572, 0.0000 \rangle$,
$\text{IFWA}_i(a_{11}, a_{12}, a_{13}) = \text{IFWA}_i(\langle 0.6, 0.2 \rangle, \langle 0.3, 0.4 \rangle, \langle 0.2, 0.2 \rangle) = \langle 0.3755, 0.2549 \rangle$,
$\text{IFWA}_i(a_{21}, a_{22}, a_{23}) = \text{IFWA}_i(\langle 0.3, 0.4 \rangle, \langle 0.2, 0.4 \rangle, \langle 0.3, 0.5 \rangle) = \langle 0.2665, 0.4334 \rangle$,

$$\mathrm{IFWA}_i(a_{31}, a_{32}, a_{33}) = \mathrm{IFWA}_i(\langle 0.3, 0.4\rangle, \langle 0.5, 0.4\rangle, \langle 0.4, 0.1\rangle) = \langle 0.4113, 0.2428\rangle.$$

Then,

$$\ell_1' = \mathrm{GBGWA}(\langle 0.6, 0.4\rangle, \langle 0.6, 0.2\rangle, \langle 0.9, 0\rangle) = \mathrm{IFWA}_k \begin{pmatrix} (\langle 0.2843, 0.2549\rangle, \\ \langle 0.2018, 0.4334\rangle, \\ \langle 0.3114, 0.2428\rangle) \end{pmatrix} = \langle 0.7441, 0.3099\rangle.$$

Similarly, we can calculate $\ell_2', \ell_3', \ell_4', \ell_5'$ and ℓ_6'.

Property 23. Idempotency; *If $c_{ji} = c_j$ and $a_{ki} = a_k = a$ for all i, then $GBGWA(c_{j1}, c_{j2}, ..., c_{jm}) = (a \otimes c_j)$.*

Proof. Since $c_{ji} = c_j \; \forall j$, that is, $\widetilde{t}_{c_{ji}} = \widetilde{t}_{c_j}$ and $\widetilde{f}_{c_{ji}} = \widetilde{f}_{c_j}$. Therefore, for $p = 2$, $a_{1i} = a$ and $a_{2i} = a$, this implies that $\widetilde{t}_{a_{1i}} = \widetilde{t}_a, \widetilde{f}_{a_{1i}} = \widetilde{f}_a$ and $\widetilde{t}_{a_{2i}} = \widetilde{t}_a, \widetilde{f}_{a_{2i}} = \widetilde{f}_a$. Then,

$$\mathrm{GBGWA}(c_{j1}, c_{j2}, ..., c_{jm}) = \mathrm{IFWA}_k \begin{pmatrix} \langle (1 - \prod_{i=1}^m (1 - \widetilde{t}_{a_{1i}})^{w_i}) \cdot (1 - \prod_{i=1}^m (1 - \widetilde{t}_{c_{ji}})^{w_i}), \\ \prod_{i=1}^m \widetilde{f}_{a_{1i}}^{w_i} + \prod_{i=1}^m \widetilde{f}_{c_{ji}}^{w_i} - \prod_{i=1}^m \widetilde{f}_{a_{1i}}^{w_i} . \prod_{i=1}^m \widetilde{f}_{c_{ji}}^{w_i} \rangle \\ , \langle (1 - \prod_{i=1}^m (1 - \widetilde{t}_{a_{2i}})^{w_i}) \cdot (1 - \prod_{i=1}^m (1 - \widetilde{t}_{c_{ji}})^{w_i}), \\ \prod_{i=1}^m \widetilde{f}_{a_{2i}}^{w_i} + \prod_{i=1}^m \widetilde{f}_{c_{ji}}^{w_i} - \prod_{i=1}^m \widetilde{f}_{a_{2i}}^{w_i} . \prod_{i=1}^m \widetilde{f}_{c_{ji}}^{w_i} \rangle \end{pmatrix}$$

$$= \mathrm{IFWA}_k \begin{pmatrix} \langle (1 - (1 - \widetilde{t}_a)^{\sum_{i=1}^m w_i}) \cdot (1 - (1 - \widetilde{t}_{c_j})^{\sum_{i=1}^m w_i}), \\ \widetilde{f}_a^{\sum_{i=1}^m w_i} + \widetilde{f}_{c_j}^{\sum_{i=1}^m w_i} - \widetilde{f}_a^{\sum_{i=1}^m w_i} . \widetilde{f}_{c_j}^{\sum_{i=1}^m w_i} \rangle \\ , \langle (1 - (1 - \widetilde{t}_a)^{\sum_{i=1}^m w_i}) \cdot (1 - (1 - \widetilde{t}_{c_j})^{\sum_{i=1}^m w_i}), \\ \widetilde{f}_a^{\sum_{i=1}^m w_i} + \widetilde{f}_{c_j}^{\sum_{i=1}^m w_i} - \widetilde{f}_a^{\sum_{i=1}^m w_i} . \widetilde{f}_{c_j}^{\sum_{i=1}^m w_i} \rangle \end{pmatrix}$$

$$= \mathrm{IFWA}_k \begin{pmatrix} \langle (1 - (1 - \widetilde{t}_a)) \cdot (1 - (1 - \widetilde{t}_{c_j})), \widetilde{f}_a + \widetilde{f}_{c_j} - \widetilde{f}_a . \widetilde{f}_{c_j} \rangle \\ , \langle (1 - (1 - \widetilde{t}_a)) \cdot (1 - (1 - \widetilde{t}_{c_j})), \widetilde{f}_a + \widetilde{f}_{c_j} - \widetilde{f}_a . \widetilde{f}_{c_j} \rangle \end{pmatrix}$$

$$= \mathrm{IFWA}_k \begin{pmatrix} \langle \widetilde{t}_a . \widetilde{t}_{c_j}, \widetilde{f}_a + \widetilde{f}_{c_j} - \widetilde{f}_a . \widetilde{f}_{c_j} \rangle \\ , \langle \widetilde{t}_a . \widetilde{t}_{c_j}, \widetilde{f}_a + \widetilde{f}_{c_j} - \widetilde{f}_a . \widetilde{f}_{c_j} \rangle \end{pmatrix}$$

$$= \begin{pmatrix} \langle 1 - (1 - \widetilde{t}_a . \widetilde{t}_{c_j})^{\varpi_1} \cdot (1 - \widetilde{t}_a . \widetilde{t}_{c_j})^{\varpi_2}, \\ (\widetilde{f}_a + \widetilde{f}_{c_j} - \widetilde{f}_a . \widetilde{f}_{c_j})^{\varpi_1} . (\widetilde{f}_a + \widetilde{f}_{c_j} - \widetilde{f}_a . \widetilde{f}_{c_j})^{\varpi_2} \rangle \end{pmatrix}$$

$$= \begin{pmatrix} \langle 1 - (1 - \widetilde{t}_a . \widetilde{t}_{c_j})^{\sum_{k=1}^2 \varpi_k}, \\ (\widetilde{f}_a + \widetilde{f}_{c_j} - \widetilde{f}_a . \widetilde{f}_{c_j})^{\sum_{k=1}^2 \varpi_k} \rangle \end{pmatrix}$$

$$= \left(\langle 1 - (1 - \widetilde{t}_a . \widetilde{t}_{c_j}), (\widetilde{f}_a + \widetilde{f}_{c_j} - \widetilde{f}_a . \widetilde{f}_{c_j}) \rangle \right)$$

$$= a \otimes c_j$$

Now, using operation laws between IFVs, assume that results hold for $p = p'$, that is,

$$\mathrm{GBGWA}(c_{j1}, c_{j2}, ..., c_{jm}) = (a \otimes c_j).$$

Then, for $p = p' + 1$,

$$\mathrm{GBGWA}(c_{j1}, c_{j2}, ..., c_{jm}) = (a \otimes c_j).$$

Thus, by mathematical induction, Theorem 23 holds for all positive integer p. □

Property 24. Boundedness; *If $c_j^+ = \langle \widetilde{t}^{\max}_{(a_{ki} \otimes c_{ji})}, \widetilde{f}^{\min}_{(a_{ki} \otimes c_{ji})} \rangle$ and $c_j^- = \langle \widetilde{t}^{\min}_{(a_{ki} \otimes c_{ji})}, \widetilde{f}^{\max}_{(a_{ki} \otimes c_{ji})} \rangle$, then $c_j^- \leq \mathrm{GBGWA}(c_{ji}, c_{j2}, ..., c_{jm}) \leq c_j^+$.*

Proof. Let $c_{ji} = \langle \widetilde{t}_{c_{ji}}, \widetilde{f}_{c_{ji}} \rangle$ and $a_{ki} = \langle \widetilde{t}_{a_{ki}}, \widetilde{f}_{a_{ki}} \rangle$ be IFVs, for all i, j, k. Then, $a_{ki} \otimes c_{ji} = \langle \widetilde{t}_{a_{ki}} \cdot \widetilde{t}_{c_{ji}}, 1 - (1 - \widetilde{f}_{a_{ki}})(1 - \widetilde{f}_{c_{ji}}) \rangle$ and denote $\widetilde{t}^{\max}_{(a_{ki} \otimes c_{ji})} = \widetilde{t}^{\max}_{a_{ki}} \cdot \widetilde{t}^{\max}_{c_{ji}}, \widetilde{t}^{\min}_{(a_{ki} \otimes c_{ji})} = \widetilde{t}^{\min}_{a_{ki}} \cdot \widetilde{t}^{\min}_{c_{ji}}, \widetilde{f}^{\max}_{(a_{ki} \otimes c_{ji})} = 1 - (1 - \widetilde{f}^{\max}_{a_{ki}}) \cdot (1 - \widetilde{f}^{\max}_{c_{ji}}), \widetilde{f}^{\min}_{(a_{ki} \otimes c_{ji})} = 1 - (1 - \widetilde{f}^{\min}_{a_{ki}}) \cdot (1 - \widetilde{f}^{\min}_{c_{ji}})$.

Now, $\widetilde{t}^{\min}_{c_{ji}} \leq \widetilde{t}_{c_{ji}} \leq \widetilde{t}^{\max}_{c_{ji}} \iff (1 - \widetilde{t}^{\max}_{c_{ji}}) \leq (1 - \widetilde{t}_{c_{ji}}) \leq (1 - \widetilde{t}^{\min}_{c_{ji}}) \iff (1 - \widetilde{t}^{\max}_{c_{ji}})^{\sum_{i=1}^{m} w_i} \leq \prod_{i=1}^{m}(1 - \widetilde{t}_{c_{ji}}) \leq (1 - \widetilde{t}^{\min}_{c_{ji}})^{\sum_{i=1}^{m} w_i} \iff 1 - (1 - \widetilde{t}^{\min}_{c_{ji}}) \leq 1 - \prod_{i=1}^{m}(1 - \widetilde{t}_{c_{ji}}) \leq 1 - (1 - \widetilde{t}^{\max}_{c_{ji}}) \iff \widetilde{t}^{\min}_{c_{ji}} \leq 1 - \prod_{i=1}^{m}(1 - \widetilde{t}_{c_{ji}}) \leq \widetilde{t}^{\max}_{c_{ji}}$. Similarly, we obtain $\widetilde{t}^{\min}_{a_{ki}} \leq 1 - \prod_{i=1}^{m}(1 - \widetilde{t}_{a_{ki}}) \leq \widetilde{t}^{\max}_{a_{ki}}$.

Therefore, $\widetilde{t}^{\min}_{a_{ki}} \cdot \widetilde{t}^{\min}_{c_{ji}} \leq (1 - \prod_{i=1}^{m}(1 - \widetilde{t}_{a_{ki}}))(1 - \prod_{i=1}^{m}(1 - \widetilde{t}_{c_{ji}})) \leq \widetilde{t}^{\max}_{a_{ki}} \cdot \widetilde{t}^{\max}_{c_{ji}} \iff \widetilde{t}^{\min}_{(a_{ki} \otimes c_{ji})} \leq (1 - \prod_{i=1}^{m}(1 - \widetilde{t}_{a_{ki}}))(1 - \prod_{i=1}^{m}(1 - \widetilde{t}_{c_{ji}})) \leq \widetilde{t}^{\max}_{(a_{ki} \otimes c_{ji})} \iff 1 - \widetilde{t}^{\max}_{(a_{ki} \otimes c_{ji})} \leq 1 - (1 - \prod_{i=1}^{m}(1 - \widetilde{t}_{a_{ki}})) \cdot (1 - \prod_{i=1}^{m}(1 - \widetilde{t}_{c_{ji}})) \leq 1 - \widetilde{t}^{\min}_{(a_{ki} \otimes c_{ji})} \iff (1 - \widetilde{t}^{\max}_{(a_{ki} \otimes c_{ji})})^{\sum_{k=1}^{p} \varpi_k} \leq \prod_{k=1}^{p}(1 - (1 - \prod_{i=1}^{m}(1 - \widetilde{t}_{a_{ki}})) \cdot (1 - \prod_{i=1}^{m}(1 - \widetilde{t}_{c_{ji}}))) \leq (1 - \widetilde{t}^{\min}_{(a_{ki} \otimes c_{ji})})^{\sum_{k=1}^{p} \varpi_k} \iff 1 - (1 - \widetilde{t}^{\min}_{(a_{ki} \otimes c_{ji})}) \leq 1 - \prod_{k=1}^{p}(1 - (1 - \prod_{i=1}^{m}(1 - \widetilde{t}_{a_{ki}})) \cdot (1 - \prod_{i=1}^{m}(1 - \widetilde{t}_{c_{ji}}))) \leq 1 - (1 - \widetilde{t}^{\max}_{(a_{ki} \otimes c_{ji})}) \iff$

$$\widetilde{t}^{\min}_{(a_{ki} \otimes c_{ji})} \leq 1 - \prod_{k=1}^{p}(1 - ((1 - \prod_{i=1}^{m}(1 - \widetilde{t}_{a_{ki}})) \cdot (1 - \prod_{i=1}^{m}(1 - \widetilde{t}_{c_{ji}})))) \leq \widetilde{t}^{\max}_{(a_{ki} \otimes c_{ji})} \tag{6}$$

In addition, $\widetilde{f}^{\min}_{c_{ji}} \leq \widetilde{f}_{c_{ji}} \leq \widetilde{f}^{\max}_{c_{ji}} \iff \widetilde{f}^{\min}_{c_{ji}} \leq \prod_{i=1}^{m} \widetilde{f}_{c_{ji}} \leq \widetilde{f}^{\max}_{c_{ji}} \iff 1 - \widetilde{f}^{\max}_{c_{ji}} \leq 1 - \prod_{i=1}^{m} \widetilde{f}_{c_{ji}} \leq 1 - \widetilde{f}^{\min}_{c_{ji}}$. Similarly, we obtain $1 - \widetilde{f}^{\max}_{a_{ki}} \leq 1 - \prod_{i=1}^{m} \widetilde{f}_{a_{ki}} \leq 1 - \widetilde{f}^{\min}_{a_{ki}}$.

Therefore, $1 - (1 - \widetilde{f}^{\min}_{a_{ki}})(1 - \widetilde{f}^{\min}_{c_{ji}}) \leq 1 - (1 - \prod_{i=1}^{m} \widetilde{f}_{a_{ki}})(1 - \prod_{i=1}^{m} \widetilde{f}_{c_{ji}}) \leq 1 - (1 - \widetilde{f}^{\max}_{a_{ki}})(1 - \widetilde{f}^{\max}_{c_{ji}}) \iff \widetilde{f}^{\min}_{(a_{ki} \otimes c_{ji})} \leq 1 - (1 - \prod_{i=1}^{m} \widetilde{f}_{a_{ki}})(1 - \prod_{i=1}^{m} \widetilde{f}_{c_{ji}}) \leq \widetilde{f}^{\max}_{(a_{ki} \otimes c_{ji})} \iff (\widetilde{f}^{\min}_{(a_{ki} \otimes c_{ji})})^{\sum_{k=1}^{p} \varpi_k} \leq \prod_{k=1}^{p}(1 - (1 - \prod_{i=1}^{m} \widetilde{f}_{a_{ki}})(1 - \prod_{i=1}^{m} \widetilde{f}_{c_{ji}})) \leq (\widetilde{f}^{\max}_{(a_{ki} \otimes c_{ji})})^{\sum_{k=1}^{p} \varpi_k} \iff$

$$\widetilde{f}^{\min}_{(a_{ki} \otimes c_{ji})} \leq \prod_{k=1}^{p}(1 - (1 - \prod_{i=1}^{m} \widetilde{f}_{a_{ki}})(1 - \prod_{i=1}^{m} \widetilde{f}_{c_{ji}})) \leq \widetilde{f}^{\max}_{(a_{ki} \otimes c_{ji})}. \tag{7}$$

If $\mathrm{GBGWA}(c_{j1}, c_{j2}, \ldots, c_{jm}) = \langle \widetilde{t}_\rho, \widetilde{f}_\rho \rangle$, therefore from Equations (6) and (7), we have $\widetilde{t}^{\min}_{(a_{ki} \otimes c_{ji})} \leq \widetilde{t}_\rho \leq \widetilde{t}^{\max}_{(a_{ki} \otimes c_{ji})}$ and $\widetilde{f}^{\min}_{(a_{ki} \otimes c_{ji})} \leq \widetilde{f}_\rho \leq \widetilde{f}^{\max}_{(a_{ki} \otimes c_{ji})}$. Further using score function $\delta(\mathrm{GBGWA}(c_{j1}, c_{j2}, \ldots, c_{jm})) = \widetilde{t}_\rho - \widetilde{f}_\rho \leq \widetilde{t}^{\max}_{(a_{ki} \otimes c_{ji})} - \widetilde{t}^{\min}_{(a_{ki} \otimes c_{ji})} = \delta(c_j^+)$, $\delta(\mathrm{GBGWA}(c_{j1}, c_{j2}, \ldots, c_{jm}) = \widetilde{t}_\rho - \widetilde{f}_\rho \geq \widetilde{t}^{\min}_{(a_{ki} \otimes c_{ji})} - \widetilde{t}^{\max}_{(a_{ki} \otimes c_{ji})} = \delta(c_j^-)$. Hence, by order relation $c_j^- \leq \mathrm{GBGWA}(c_{j1}, c_{j2}, \ldots, c_{jm}) \leq c_j^+$. □

Property 25. Monotonicity; *If c'_{ji} and c_{ji} are two IFVs such that $c'_{ji} \leq c_{ji}$, then* $\mathrm{GBGWA}(c'_{j1}, c'_{j2}, \ldots, c'_{jm}) \leq \mathrm{GBGWA}(c_{j1}, c_{j2}, \ldots, c_{jm})$.

Proof. It follows from Theorem 24, thus it is omitted from here. □

Proposition 2. *Let $\mathscr{G}_{\widetilde{g}} = (\widetilde{\mathcal{S}}, \mathcal{A}, \widetilde{g})$ be a GGIFSS over X. Then,*

(i) *If the assessments of each moderator/prospector on $\mathcal{A}$, are $\mathcal{IF}_k = \{\langle 1,0 \rangle, \langle 1,0 \rangle, \ldots, \langle 1,0 \rangle\}$, $k = 1, 2, \ldots, p$, then* $\mathrm{GBGWA}(c_{j1}, c_{j2}, \ldots, c_{jm}) = \mathrm{IFWA}_i(c_{j1}, c_{j2}, \ldots, c_{jm})$.

(ii) *If the assessments of each moderator/prospector on $\mathcal{A}$, are $\mathcal{IF}_k = \{\langle 0,1 \rangle, \langle 0,1 \rangle, \ldots, \langle 0,1 \rangle\}$, $k = 1, 2, \ldots, p$, then* $\mathrm{GBGWA}(c_{j1}, c_{j2}, \ldots, c_{jm}) = \langle 0, 1 \rangle$,

(iii) *If $(\widetilde{\mathcal{S}}, \mathcal{A}) = \widetilde{\mathcal{X}}^{\mathcal{A}^{(1,0)}}$, then*

$$\mathrm{GBGWA}(c_{j1}, c_{j2}, \ldots, c_{jm}) = \mathrm{IFWA}_k \left(\begin{array}{c} \mathrm{IFWA}_i(a_{11}, a_{12}, \ldots, a_{1m}), \\ \mathrm{IFWA}_i(a_{21}, a_{22}, \ldots, a_{2m}), \ldots, \mathrm{IFWA}_i(a_{p1}, a_{p2}, \ldots, a_{pm}) \end{array} \right).$$

(iv) *If $(\widetilde{\mathcal{S}}, \mathcal{A}) = \widetilde{\mathcal{I}}^{\mathcal{A}^{(0,1)}}$, then* $\mathrm{GBGWA}(c_{j1}, c_{j2}, \ldots, c_{jm}) = \langle 0, 1 \rangle$.

Proof. It is straightforward, thus it is omitted from here. □

Now, the definition of GBGWG operator is specified as follows:

Definition 26. GBGWG; *Let $\widetilde{\mathscr{F}}_{\widetilde{g}} = (\widetilde{\mathcal{S}}, \mathcal{A}, \widetilde{g})$ be a GGIFSS over X, where $\widetilde{g} = \{\widetilde{\alpha}_{d_1}, \widetilde{\alpha}_{d_2}, ..., \widetilde{\alpha}_{d_p}\}$ be the group of PIFSs. Assume that $w = (w_1, w_2, ..., w_m)^T$ be the normalized weight vector for $\mathcal{A}$, such that $w_i > 0$ and $\sum_{i=1}^{m} w_i = 1$. Let $\widetilde{IFV}(\kappa_j) = \{c_{j1}, c_{j2}, ..., c_{jm}\}$ ($j = 1$ to n) be the set of IFVs in EIFSS $(\widetilde{\mathcal{S}}, \mathcal{A})$ for all $\kappa_j \in X$. For each senior moderator/prospector, $\widetilde{\alpha}_{d_k}(v) = \{\langle \widetilde{t}_{\widetilde{\alpha}_{d_k}}(v), \widetilde{f}_{\widetilde{\alpha}_{d_k}}(v)\rangle \mid v \in \mathcal{A}\}$ ($k = 1$ to p) be the PIFS, it can be represented as $\mathcal{IF}_k = \{a_{k1}, a_{k2}, ..., a_{km}\}$ ($k = 1$ to p) and $\varpi = (\varpi_1, \varpi_2, ..., \varpi_p)^T$ is the set of weights for moderators, such that $\varpi_k > 0$ and $\sum_{k=1}^{p} \varpi_k = 1$. Define* GBGWG: $\Gamma_s^m \longrightarrow \Gamma_s$, IFWG: $\Gamma^m \longrightarrow \Gamma$, *where*

$$\mathrm{GBGWG}(c_{j1}, c_{j2}, ..., c_{jm}) = \mathrm{IFWG}_k \begin{pmatrix} (\mathrm{IFWG}_i(a_{11}, a_{12}, ..., a_{1m}) \otimes \mathrm{IFWG}_i(c_{j1}, c_{j2}, ..., c_{jm})), \\ (\mathrm{IFWG}_i(a_{21}, a_{22}, ..., a_{2m}) \otimes \mathrm{IFWG}_i(c_{j1}, c_{j2}, ..., c_{jm})), ..., \\ (\mathrm{IFWG}_i(a_{p1}, a_{p2}, ..., a_{pm}) \otimes \mathrm{IFWG}_i(c_{j1}, c_{j2}, ..., c_{jm})) \end{pmatrix} \tag{8}$$

where GBGWG *is known as GGIFSS weighted geometric operator, then the set of all* GBGWGs *is denoted $L = \{\ell_1'', \ell_2'', ..., \ell_n''\}$. In addition,* IFWA_k *and* IFWA_i *are* IFWG *operators on set of senior moderators/prospectors and set of parameters, respectively. Note that Γ_s^m and Γ are families of GGIFSS and IFSs, respectively.*

Lemma 2. *Let $\widetilde{\mathscr{F}}_{\widetilde{g}} = (\widetilde{\mathcal{S}}, \mathcal{A}, \widetilde{g})$ be a GGIFSS over X, where $\widetilde{g} = \{\widetilde{\alpha}_{d_1}, \widetilde{\alpha}_{d_2}, ..., \widetilde{\alpha}_{d_p}\}$ be the group of IFSs. If $p = 1$, then $\widetilde{\mathscr{F}}_{\widetilde{g}}$ is a GIFSS and* GBGWG *is specified below.*

$$\mathrm{GBGWG}(c_{j1}, c_{j2}, ..., c_{jm}) = \mathrm{IFWG}_i(a_{11}, a_{12}, ..., a_{1m}) \otimes \mathrm{IFWG}_i(c_{j1}, c_{j2}, ..., c_{jm})). \tag{9}$$

Theorem 2. *If $c_{ij} = (\widetilde{t}_{ji}, \widetilde{f}_{ji})$ and $a_{ki} = (\widetilde{t}_{ki}, \widetilde{f}_{ki})$ ($i = 1, 2, ..., m, j = 1, 2, ..., n, k = 1, 2, ..., p$), be an IFV, then the aggregated value by GBGWG operator is given by*
$\mathrm{GBGWG}(c_{j1}, c_{j2}, ..., c_{jm}) = \langle \prod_{k=1}^{p}(1 - (1 - \prod_{i=1}^{m} \widetilde{t}_{a_{ki}})(1 - \prod_{i=1}^{m} \widetilde{t}_{c_{ji}})), 1 - \prod_{k=1}^{p}(1 - (1 - \prod_{i=1}^{m}(1 - \widetilde{f}_{a_{ki}})) \cdot (1 - \prod_{i=1}^{m}(1 - \widetilde{f}_{c_{ji}})))\rangle$.

Proof. It follows from Theorem 1, thus it is omitted from here. □

In addition, the properties of idempotent, bounding and monotonicity for GBGWGs can be stated and proved in a similar manner as for GBGWAs.

Proposition 3. *Let $\widetilde{\mathscr{G}}_{\widetilde{g}} = (\widetilde{\mathcal{S}}, \mathcal{A}, \widetilde{g})$ be a GGIFSS over X. Then,*

(i) *If the assessments of each moderator on $\mathcal{A}$, are $\mathcal{IF}_k = \{\langle 1,0\rangle, \langle 1,0\rangle, ..., \langle 1,0\rangle\}$, $k = 1, 2, ..., p$, then* $\mathrm{GBGWG}(c_{j1}, c_{j2}, ..., c_{jm}) = \mathrm{IFWG}_i(c_{j1}, c_{j2}, ..., c_{jm})$.
(ii) *If the assessments of each moderator on $\mathcal{A}$, are $\mathcal{IF}_k = \{\langle 0,1\rangle, \langle 0,1\rangle, ..., \langle 0,1\rangle\}$, $k = 1, 2, ..., p$, then* $\mathrm{GBGWG}(c_{j1}, c_{j2}, ..., c_{jm}) = \langle 0, 1\rangle$.
(iii) *If $(\widetilde{\mathcal{S}}, \mathcal{A}) = \widetilde{\mathcal{X}}^{\mathcal{A}^{(1,0)}}$, then*
$$\mathrm{GBGWG}(c_{j1}, c_{j2}, ..., c_{jm}) = \mathrm{IFWG}_k \begin{pmatrix} \mathrm{IFWG}_i(a_{11}, a_{12}, ..., a_{1m}), \\ \mathrm{IFWG}_i(a_{21}, a_{22}, ..., a_{2m}), ..., \mathrm{IFWG}_i(a_{p1}, a_{p2}, ..., a_{pm}) \end{pmatrix}.$$
(iv) *If $(\widetilde{\mathcal{S}}, \mathcal{A}) = \widetilde{\mathcal{I}}^{\mathcal{A}^{(0,1)}}$, then* $\mathrm{GBGWG}(c_{j1}, c_{j2}, ..., c_{jm}) = \langle 0, 1\rangle$.

Proof. It is straightforward, thus is omitted here. □

As aggregation operators are used to create MCDM frameworks, based on proposed GBGWA or GBGWG operators, some multi-criteria decision making methods are discussed in next section.

5. Multi-Attribute Decision Making under GGIFSSs Environment

In this section, firstly we present our approach comprising of an algorithm by virtue of GGIFSSs, and GBGWA or GBGWG operators. Then, we conduct two illustrations on proposed method as in the case studies: (1) candidates evaluation for an insurance company; and (2) cinema selection for the customers.

5.1. Proposed Method

As stated above, the properties of boundedness and monotonicity are valid for proposed operators. Therefore, a comparison can be made among two or more GBGWA(GBGWG) operators. Let ς be the number of committees established comprising specialists, which are intended to classify each alternative $\kappa_j(j = 1, 2, ..., n)$, while making in account with the imperative attributes $v_i(i = 1, 2, ..., m)$, and by provision of their respective grades in terms of IFSSs. Consider $d_1, d_2, ..., d_p$ be the members/experts(directors or officers), who are in-charge of constituted committees. Thereafter, the subjective information (in the form of IFSSs) from committees is collected. The senior experts will examine it and give their judgements as a group of IFSs. Then, the information of each committee comprised the GGIFSS, and there will be ς number of GGIFSSs. The extended union on GGIFSSs is computed, denoted as $\widetilde{\mathscr{G}}_{\widetilde{g}}$ and expressed in a table. Here, two types of criteria occur in the $\widetilde{\mathscr{G}}_{\widetilde{g}}$, namely, benefit and cost criteria. To consolidate the criteria, the $\widetilde{\mathscr{G}}_{\widetilde{g}}$ must be normalized through the following equation:

$$r_{ji} = \begin{cases} \langle \widetilde{t}_{\mathcal{A}}(v_i), \widetilde{f}_{\mathcal{A}}(v_i) \rangle, & \text{if } v_i \text{ is a benefit criterion,} \\ \langle \widetilde{f}_{\mathcal{A}}(v_i), \widetilde{t}_{\mathcal{A}}(v_i) \rangle, & \text{if } v_i \text{ is a cost criterion,} \end{cases} \tag{10}$$

such that the normalized GGIFSS is denoted by $\widetilde{\mathscr{G}'}_{\widetilde{g}} = (\widetilde{\mathcal{T}}', E, \widetilde{g}')$, where $(\widetilde{\mathcal{T}}', E)$ is the normalization of $(\widetilde{\mathcal{T}}, E)$ and $\widetilde{g}'$ is the normalization of $\widetilde{g}$. Finally, GBGWA or GBGWG can be used to aggregate the data from $\widetilde{\mathscr{G}'}_{\widetilde{g}}$ and each ℓ'_j or ℓ''_j can be correlated through score function. Therefore, we propose our methodology as an algorithm as follows.

Algorithm 1 Multi-attribute decision making on GGIFSSs

Input: A set of alternatives
Output: The felicitous alternative for a problem

1: Let $X = \{\kappa_1, \kappa_2, ..., \kappa_n\}$ be the set of alternatives and $E = \mathcal{A}_1 \cup \mathcal{A}_2 \cup, ..., \cup \mathcal{A}_\varsigma = \{v_1, v_2, ..., v_m\}$ be the set of attributes. Constitute a mechanism of the specialists' judgements on attributes in the form of IFVs and establish IFSSs on each committee of specialists.
2: Obtain ς number of GGIFSSs, $\widetilde{\mathscr{F}}_{\widetilde{g}1} = (\widetilde{\mathcal{S}}_1, \mathcal{A}_1, \widetilde{g}_1), \widetilde{\mathscr{F}}_{\widetilde{g}2} = (\widetilde{\mathcal{S}}_2, \mathcal{A}_2, \widetilde{g}_2), ..., \widetilde{\mathscr{F}}_{\widetilde{g}\varsigma} = (\widetilde{\mathcal{S}}_\varsigma, \mathcal{A}_\varsigma, \widetilde{g}_\varsigma)$ over X, which are handled by ς number of committees of experts and specialists. Each group $\widetilde{g}_1, \widetilde{g}_2, ..., \widetilde{g}_\varsigma$ of IFSs is constituted by p number of senior members/moderators for available information on each $\mathcal{A}_{i'}(i' = 1, 2, ..., \varsigma)$, respectively.
3: Compute extended union $\widetilde{\mathscr{G}}_{\widetilde{g}} = (\widetilde{\sqcup}_{\mathcal{E}})_{i'=1}^{\varsigma} \widetilde{\mathscr{F}}_{\widetilde{g}i'}$ of GGIFSSs. Represent $\widetilde{\mathscr{G}}_{\widetilde{g}}$ in a table.
4: Normalized the data in $\widetilde{\mathscr{G}}_{\widetilde{g}}$ using Equation (3), and represent $\widetilde{\mathscr{G}'}_{\widetilde{g}}$ in a table.
5: Calculate GBGWA $\ell'_j(j = 1, 2, ..., n)$ or GBGWG $\ell''_j(j = 1, 2, ..., n)$ operators on GGIFFS $\widetilde{\mathscr{G}'}_{\widetilde{g}}$. There will be n operators.
6: Obtain score function on each operator ℓ'_j or ℓ''_j using Definition 3.
7: Rank the alternatives on score function; the best choice is obtained on a maximum score.

This algorithm is depicted as a flowchart in Figure 1. The Algorithm 1 can be formulated to select the best product or alternative for p number of customers. In this way, the extra inputs incorporate as the demands of customers in GGIFSS, and the Algorithm 1 will conduct on a GGIFSS, $\widetilde{\mathscr{F}}_{\widetilde{g}} = (\widetilde{\mathcal{S}}, E, \widetilde{g})$, from Step 4. To operate above methodology, we establish two case studies as below.

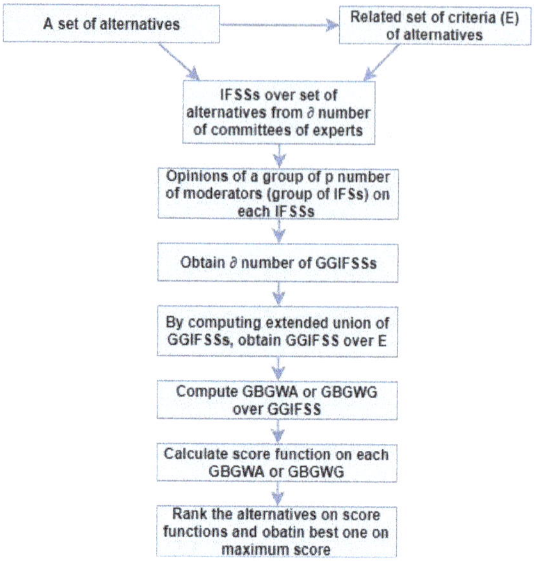

Figure 1. A flowchart for our algorithm.

5.2. Case Study: Candidate Selection Problem

In this case study, an example for evaluation of candidates is used to illustrate the applicability of the proposed method. An insurance company HG in Guangzhou, China is engaged for insurance of products, charging insurance premium, consultation on insurance, financial and other services for individuals and enterprises. Every year, this company recruits new staff for the post of insurance sales agents and consultants. To maintain the excellence and high admire reputation, the company consults with experts for their assessments and opinions to recruit the candidates. Furthermore, the insurance business department and human resources department are actively engaged in recruitment process.

Let $X = \{\kappa_1, \kappa_2, \kappa_3, \kappa_4, \kappa_5\}$ be the set of five candidates whom can be placed for the position of insurance sales consultant. A group of three senior members (directors, officers, etc.) d_1, d_2 and d_3 setup a committee of specialists and experts to appoint a felicitous candidate for this position. The set of criteria for committee to select the candidate is $E = \{v_1, v_2, v_3, v_4, v_5, v_6, v_7\}$, where

v_1 : english level;
v_2 : relevant problem solving skills;
v_3 : relevant working experience;
v_4 : communication skills;
v_5 : finance and insurance professional;
v_6 : score obtained in a college degree; and
v_7 : interpersonal skills.

On the set parameters, the weight vector is given and denoted by $w = (w_1/0.12, w_2/0.13, w_3/0.15, w_4/0.15, w_5/0.17, w_6/0.11, w_7/0.17)^T$ such that $\sum_{i=1}^{7} w_i = 1$. The three senior members arrange specialists into two groups; the first group consists of the specialists of insurance business management and the second group consists of the specialists of human resource management. The set of parameters $\mathcal{A} = \{v_2, v_3, v_5\}$ is assigned for first group and the set of parameters $\mathcal{B} = \{v_1, v_4, v_6, v_7\}$ is assigned for second group. These two groups give their judgments as IFSSs $(\widetilde{S}_1, \mathcal{A})$ and $(\widetilde{S}_2, \mathcal{B})$, respectively. Then, the group of senior members examine the data of IFSSs and then provide the two groups of IFSs, $g_1 = \{\widetilde{\alpha}_{d_1}, \widetilde{\alpha}_{d_2}, \widetilde{\alpha}_{d_3}\}$ and $g_2 = \{\widetilde{\beta}_{d_1}, \widetilde{\beta}_{d_2}, \widetilde{\beta}_{d_3}\}$ to complete the GGIFSSs, $\widetilde{\mathscr{F}}_{\widetilde{g}1} = (\widetilde{S}_1, \mathcal{A}, \widetilde{g}_1)$ and $\widetilde{\mathscr{F}}_{\widetilde{g}2} = (\widetilde{S}_2, \mathcal{B}, \widetilde{g}_2)$, as shown in Tables 5 and 6, respectively.

Table 5. Tabular representation of the GGIFSS $\widetilde{\mathscr{F}}_{\tilde{g}1} = (\tilde{S}_1, \mathcal{A}, \tilde{g}_1)$

$X \setminus \mathcal{A}$	v_2	v_3	v_5
κ_1	$\langle 0.7, 0.1 \rangle$	$\langle 0.7, 0.1 \rangle$	$\langle 0.6, 0.1 \rangle$
κ_2	$\langle 0.5, 0.3 \rangle$	$\langle 0.6, 0.3 \rangle$	$\langle 0.6, 0.4 \rangle$
κ_3	$\langle 0.5, 0.4 \rangle$	$\langle 0.5, 0.4 \rangle$	$\langle 0.9, 0.1 \rangle$
κ_4	$\langle 0.3, 0.4 \rangle$	$\langle 0.7, 0.2 \rangle$	$\langle 0.7, 0.2 \rangle$
κ_5	$\langle 0.6, 0.4 \rangle$	$\langle 0.6, 0.4 \rangle$	$\langle 0.8, 0.1 \rangle$
$\tilde{\alpha}_{d_1}$	$\langle 0.5, 0.2 \rangle$	$\langle 0.5, 0.4 \rangle$	$\langle 0.7, 0.2 \rangle$
$\tilde{\alpha}_{d_2}$	$\langle 0.4, 0.4 \rangle$	$\langle 0.4, 0.6 \rangle$	$\langle 0.3, 0.5 \rangle$
$\tilde{\alpha}_{d_3}$	$\langle 0.6, 0.4 \rangle$	$\langle 0.3, 0.5 \rangle$	$\langle 0.5, 0.5 \rangle$

Table 6. Tabular representation of the GGIFSS $\widetilde{\mathscr{F}}_{\tilde{g}2} = (\tilde{S}_2, \mathcal{B}, \tilde{g}_2)$

$X \setminus \mathcal{B}$	v_1	v_4	v_6	v_7
κ_1	$\langle 0.5, 0.5 \rangle$	$\langle 0.6, 0.1 \rangle$	$\langle 0.6, 0.4 \rangle$	$\langle 0.7, 0.1 \rangle$
κ_2	$\langle 0.5, 0.4 \rangle$	$\langle 0.3, 0.4 \rangle$	$\langle 0.5, 0.4 \rangle$	$\langle 0.8, 0.2 \rangle$
κ_3	$\langle 0.4, 0.4 \rangle$	$\langle 0.7, 0.2 \rangle$	$\langle 0.6, 0.3 \rangle$	$\langle 0.4, 0.4 \rangle$
κ_4	$\langle 0.5, 0.2 \rangle$	$\langle 0.7, 0.3 \rangle$	$\langle 0.7, 0.2 \rangle$	$\langle 0.5, 0.3 \rangle$
κ_5	$\langle 0.7, 0.2 \rangle$	$\langle 0.8, 0.1 \rangle$	$\langle 0.6, 0.3 \rangle$	$\langle 0.6, 0.1 \rangle$
$\tilde{\beta}_{d_1}$	$\langle 0.4, 0.3 \rangle$	$\langle 0.7, 0.3 \rangle$	$\langle 0.6, 0.2 \rangle$	$\langle 0.7, 0.2 \rangle$
$\tilde{\beta}_{d_2}$	$\langle 0.6, 0.4 \rangle$	$\langle 0.3, 0.4 \rangle$	$\langle 0.3, 0.6 \rangle$	$\langle 0.5, 0.3 \rangle$
$\tilde{\beta}_{d_3}$	$\langle 0.5, 0.5 \rangle$	$\langle 0.4, 0.5 \rangle$	$\langle 0.4, 0.4 \rangle$	$\langle 0.7, 0.2 \rangle$

To evaluate most felicitous candidate on provided information in Tables 5 and 6, the extended intersection of $\widetilde{\mathscr{F}}_{\tilde{g}1}$ and $\widetilde{\mathscr{F}}_{\tilde{g}2}$ is contemplated as follows:

$$\widetilde{\mathscr{G}}_{\tilde{g}} = \widetilde{\mathscr{F}}_{\tilde{g}1} \tilde{\sqcup}_{\mathcal{E}} \widetilde{\mathscr{F}}_{\tilde{g}2} = (\tilde{\mathcal{T}}, E, \tilde{g}) = (\tilde{S}_1, \mathcal{A}, \tilde{g}_1) \tilde{\sqcup}_{\mathcal{E}} (\tilde{S}_2, \mathcal{B}, \tilde{g}_2)$$

and shown in Table 7.

Table 7. Tabular representation of the GGIFSS $\widetilde{\mathscr{F}}_{\tilde{g}1} \tilde{\sqcup}_{\mathcal{E}} \widetilde{\mathscr{F}}_{\tilde{g}2}$.

$X \setminus E$	v_1	v_2	v_3	v_4	v_5	v_6	v_7
κ_1	$\langle 0.5, 0.5 \rangle$	$\langle 0.7, 0.1 \rangle$	$\langle 0.7, 0.1 \rangle$	$\langle 0.6, 0.1 \rangle$	$\langle 0.6, 0.1 \rangle$	$\langle 0.6, 0.4 \rangle$	$\langle 0.7, 0.1 \rangle$
κ_2	$\langle 0.5, 0.4 \rangle$	$\langle 0.5, 0.3 \rangle$	$\langle 0.6, 0.3 \rangle$	$\langle 0.3, 0.4 \rangle$	$\langle 0.6, 0.4 \rangle$	$\langle 0.5, 0.4 \rangle$	$\langle 0.8, 0.2 \rangle$
κ_3	$\langle 0.4, 0.4 \rangle$	$\langle 0.5, 0.4 \rangle$	$\langle 0.5, 0.4 \rangle$	$\langle 0.7, 0.2 \rangle$	$\langle 0.9, 0.1 \rangle$	$\langle 0.6, 0.3 \rangle$	$\langle 0.4, 0.4 \rangle$
κ_4	$\langle 0.5, 0.2 \rangle$	$\langle 0.3, 0.4 \rangle$	$\langle 0.7, 0.2 \rangle$	$\langle 0.7, 0.3 \rangle$	$\langle 0.7, 0.2 \rangle$	$\langle 0.7, 0.2 \rangle$	$\langle 0.5, 0.3 \rangle$
κ_5	$\langle 0.7, 0.2 \rangle$	$\langle 0.6, 0.4 \rangle$	$\langle 0.6, 0.4 \rangle$	$\langle 0.8, 0.1 \rangle$	$\langle 0.8, 0.1 \rangle$	$\langle 0.6, 0.3 \rangle$	$\langle 0.6, 0.1 \rangle$
$\tilde{\gamma}_{d_1}$	$\langle 0.4, 0.3 \rangle$	$\langle 0.5, 0.2 \rangle$	$\langle 0.5, 0.4 \rangle$	$\langle 0.7, 0.3 \rangle$	$\langle 0.7, 0.2 \rangle$	$\langle 0.6, 0.2 \rangle$	$\langle 0.7, 0.2 \rangle$
$\tilde{\gamma}_{d_2}$	$\langle 0.6, 0.4 \rangle$	$\langle 0.4, 0.4 \rangle$	$\langle 0.4, 0.6 \rangle$	$\langle 0.3, 0.4 \rangle$	$\langle 0.3, 0.5 \rangle$	$\langle 0.3, 0.6 \rangle$	$\langle 0.5, 0.3 \rangle$
$\tilde{\gamma}_{d_3}$	$\langle 0.5, 0.5 \rangle$	$\langle 0.6, 0.4 \rangle$	$\langle 0.3, 0.5 \rangle$	$\langle 0.4, 0.5 \rangle$	$\langle 0.5, 0.5 \rangle$	$\langle 0.4, 0.4 \rangle$	$\langle 0.7, 0.2 \rangle$

As the all criterion are benefit type, the normalized GGIFSS, $\widetilde{\mathscr{G}}_{\tilde{g}}$, is not needed. The weight vector for three senior members is given by $\varpi = (\varpi_1/0.33, \varpi_2/0.34, \varpi_3/0.33)^T$ such that $\sum_{k=1}^{3} \varpi_k = 1$. The GBGWA operator is used on integrated data in Table 7, and given as follows:

$\ell_1' = \text{GGWA}(c_{11}, c_{12}, ..., c_{17}) =$
$\text{IFWA}_k(\langle 0.390932, 0.353894 \rangle, \langle 0.260704, 0.518795 \rangle, \langle 0.325209, 0.489552 \rangle) = \langle 0.327079, 0.448601 \rangle$

$\ell_2' = \text{GGWA}(c_{21}, c_{22}, ..., c_{27}) =$
$\text{IFWA}_k(\langle 0.355421, 0.494395 \rangle, \langle 0.237022, 0.623437 \rangle, \langle 0.295668, 0.600553 \rangle) = \langle 0.297119, 0.570421 \rangle$

$\ell_3' = \text{GGWA}(c_{31}, c_{32}, ..., c_{37}) =$
$\text{IFWA}_k(\langle 0.390064, 0.455207 \rangle, \langle 0.260125, 0.594251 \rangle, \langle 0.324487, 0.569593 \rangle) = \langle 0.326346, 0.536655 \rangle$

$\ell_4' = \text{GGWA}(c_{41}, c_{42}, ..., c_{47}) =$

$\mathrm{IFWA}_k(\langle 0.374141, 0.435074\rangle, \langle 0.249507, 0.579256\rangle, \langle 0.311241, 0.553687\rangle) = \langle 0.312904, 0.519253\rangle$
$\ell'_5 = \mathrm{GGWA}(c_{51}, c_{52}, ..., c_{57}) =$
$\mathrm{IFWA}_k(\langle 0.422383, 0.383619\rangle, \langle 0.281678, 0.540934\rangle, \langle 0.351372, 0.513036\rangle) = \langle 0.353679, 0.474575\rangle$

Now, the score functions are calculated on above five operators and given as in the following: $\delta(\ell'_1) = 0.439239$, $\delta(\ell'_2) = 0.363349$, $\delta(\ell'_3) = 0.394845$, $\delta(\ell'_4) = 0.396825$, and $\delta(\ell'_5) = 0.439552$. The descending order is acquired as $\kappa_5 > \kappa_1 > \kappa_4 > \kappa_3 > \kappa_2$; thus, κ_5 is the felicitous candidate for the position because $\delta(\ell'_5) = 0.439552$ is the maximum score.

Next, a case study in a different scenario is given as follows.

5.3. Case Study: Alternative Evaluation on Customer Demands

Nowadays, the markets possess immense competition for the quality of service, besides the demands of customers are increased and widened in the different prospects. The service industries are booming and upgrading by entertainment, catering, tourism, and auction. Indeed, there is a fierce competition among the service industries, but currently film industry is in the most competitive position as customers always classify and compare cinemas on different parameters, such as convenience, environment, quality of service, upcoming movies, and expenses.

Let $X = \{\kappa_1, \kappa_2, \kappa_3, \kappa_4\}$ be the set of four cinemas. The set of attributes $E = \{v_1, v_2, v_3, v_4, v_5\}$, where

v_1 : quality of service;
v_2 : quality of expected films;
v_3 : environment in cinema;
v_4 : price reasonability; and
v_5 : convenience and luxuriousness.

A committee of experts and specialists from a cinema management organization give the judgment for cinemas on provided attributes as an IFSSs $(\widetilde{S}, E)$ (Table 8).

Table 8. Tabular representation of the IFSS, $(\widetilde{S}, E)$.

$X \setminus E$	v_1	v_2	v_3	v_4	v_5
κ_1	$\langle 0.4, 0.5\rangle$	$\langle 0.5, 0.5\rangle$	$\langle 0.6, 0.4\rangle$	$\langle 0.3, 0.4\rangle$	$\langle 0.3, 0.3\rangle$
κ_2	$\langle 0.6, 0.4\rangle$	$\langle 0.6, 0.3\rangle$	$\langle 0.5, 0.4\rangle$	$\langle 0.4, 0.4\rangle$	$\langle 0.6, 0.3\rangle$
κ_3	$\langle 0.3, 0.4\rangle$	$\langle 0.4, 0.4\rangle$	$\langle 0.5, 0.4\rangle$	$\langle 0.6, 0.4\rangle$	$\langle 0.5, 0.5\rangle$
κ_4	$\langle 0.5, 0.3\rangle$	$\langle 0.3, 0.3\rangle$	$\langle 0.3, 0.4\rangle$	$\langle 0.4, 0.6\rangle$	$\langle 0.5, 0.3\rangle$
κ_5	$\langle 0.6, 0.4\rangle$	$\langle 0.5, 0.3\rangle$	$\langle 0.5, 0.5\rangle$	$\langle 0.5, 0.4\rangle$	$\langle 0.3, 0.4\rangle$

Now, the two customers d_1 and d_2 desire to choose a most suitable cinema to watch movies; their demands comprise IFSs,

$$\widetilde{\alpha}_{d_1} = \{v_1/\langle 0.4, 0.6\rangle, v_2/\langle 0.5, 0.3\rangle, v_3/\langle 0.5, 0.4\rangle, v_4/\langle 0.4, 0.6\rangle, v_5/\langle 0.4, 0.4\rangle\},$$
$$\widetilde{\alpha}_{d_2} = \{v_1/\langle 0.5, 0.4\rangle, v_2/\langle 0.4, 0.5\rangle, v_3/\langle 0.5, 0.3\rangle, v_4/\langle 0.5, 0.4\rangle, v_5/\langle 0.4, 0.2\rangle\}.$$

The attribute v_4 belongs to cost criteria, therefore corresponding IFVs can be normalized using Equation (10). Let $\widetilde{g} = \{\widetilde{\alpha}_{d_1}, \widetilde{\alpha}_{d_2}\}$ and the normalization of $\widetilde{g}$ is expressed as $\widetilde{g}' = \{\widetilde{\alpha}'_{d_1}, \widetilde{\alpha}'_{d_2}\}$. Thereafter, Table 8 can be normalized. Then, the information can be extended into GGIFSS $\widetilde{\mathscr{F}}'_{\widetilde{g}} = (\widetilde{S}', E, \widetilde{g}')$, and specified in Table 9.

Table 9. Tabular representation of the GGIFSS, $(\widetilde{S'}, E, \widetilde{g'})$.

X\E	v_1	v_2	v_3	v_4	v_5
κ_1	$\langle 0.4, 0.5\rangle$	$\langle 0.5, 0.5\rangle$	$\langle 0.6, 0.4\rangle$	$\langle 0.4, 0.3\rangle$	$\langle 0.3, 0.3\rangle$
κ_2	$\langle 0.6, 0.4\rangle$	$\langle 0.6, 0.3\rangle$	$\langle 0.5, 0.4\rangle$	$\langle 0.4, 0.4\rangle$	$\langle 0.6, 0.3\rangle$
κ_3	$\langle 0.3, 0.4\rangle$	$\langle 0.4, 0.4\rangle$	$\langle 0.5, 0.4\rangle$	$\langle 0.4, 0.6\rangle$	$\langle 0.5, 0.5\rangle$
κ_4	$\langle 0.5, 0.3\rangle$	$\langle 0.3, 0.3\rangle$	$\langle 0.3, 0.4\rangle$	$\langle 0.6, 0.4\rangle$	$\langle 0.5, 0.3\rangle$
κ_5	$\langle 0.6, 0.4\rangle$	$\langle 0.5, 0.3\rangle$	$\langle 0.5, 0.5\rangle$	$\langle 0.4, 0.5\rangle$	$\langle 0.3, 0.4\rangle$
$\widetilde{\alpha}_{d_1}$	$\langle 0.4, 0.6\rangle$	$\langle 0.5, 0.3\rangle$	$\langle 0.5, 0.4\rangle$	$\langle 0.6, 0.4\rangle$	$\langle 0.4, 0.4\rangle$
$\widetilde{\alpha}_{d_2}$	$\langle 0.5, 0.4\rangle$	$\langle 0.4, 0.5\rangle$	$\langle 0.5, 0.3\rangle$	$\langle 0.4, 0.5\rangle$	$\langle 0.4, 0.2\rangle$

Let $w = (w_1/0.18, w_2/0.19, w_3/0.21, w_4/0.22, w_5/0.2)^T$ be the weight vector for attributes and $\varpi = (\varpi_1/0.52, \varpi_2/0.48)^T$ be the weight vector for customers. The GBGWA operator is used on integrated data in Table 9, and given as follows:
$\ell'_1 = \text{GGWA}(c_{11}, c_{21}, ..., c_{15}) =$
$\text{IFWA}_k(\langle 0.220934, 0.635540\rangle, \langle 0.199004, 0.605904\rangle) = \langle 0.210483, 0.621138\rangle$
$\ell'_2 = \text{GGWA}(c_{21}, c_{22}, ..., c_{52}) =$
$\text{IFWA}_k(\langle 0.265324, 0.619281\rangle, \langle 0.238988, 0.588324\rangle) = \langle 0.252798, 0.604223\rangle$
$\ell'_3 = \text{GGWA}(c_{31}, c_{32}, ..., c_{35}) =$
$\text{IFWA}_k(\langle 0.209414, 0.678382\rangle, \langle 0.188627, 0.652231\rangle) = \langle 0.199503, 0.665701\rangle$
$\ell'_4 = \text{GGWA}(c_{41}, c_{42}, ..., c_{45}) =$
$\text{IFWA}_k(\langle 0.223041, 0.608588\rangle, \langle 0.200901, 0.576761\rangle) = \langle 0.212491, 0.593098\rangle$
$\ell'_5 = \text{GGWA}(c_{51}, c_{52}, ..., c_{55}) =$
$\text{IFWA}_k(\langle 0.227867, 0.654431\rangle, \langle 0.205248, 0.626332\rangle) = \langle 0.217091, 0.640789\rangle$

Now, the score functions are calculated on above five operators and given as in the following: $\delta(\ell'_1) = 0.294672$, $\delta(\ell'_2) = 0.324287$, $\delta(\ell'_3) = 0.266901$, $\delta(\ell'_4) = 0.309696$, and $\delta(\ell'_5) = 0.288151$. One can check that κ_2 is the suitable cinema for both customers as $\delta(\ell'_2) = 0.324287$ is the maximum score.

Now, based on our results, comparisons with other methods are given in next section.

6. Comparisons and Discussions

In this section, we compare our framework and results with existing methodologies. At first, we make a comparison of our method with the framework presented in [44]. Then, we discuss the advantages of proposed technique.

6.1. Comparisons with the Method of Garg

Garg et al. [44] defined geometric and averaging operators on the GGIFSSs and then provided an algorithm for decision making methodology. Let $X = \{\kappa_1, \kappa_2, ..., \kappa_{n'}\}$ be the set of alternatives and $E = \{v_1, v_2, ..., v_{m'}\}$ be the set of criteria. To evaluate $\kappa_j (j = 1, ..., n')$ as a optimal choice, IFSS on E are given and assessments of moderators are given as an IFSs $G_\varrho(e)$, where $G = (\varrho_1, \varrho_2, ..., \varrho_p)$ and $G_\varrho(e)$ denotes the opinion of experts on the elements of X by virtue of IFSS on E. We recall the algorithm contemplated in [44] and given as follows:

Algorithm 2 Grag's Algorithm for faculty appointment

1: Make a framework of the specialists' judgment related to each possible choice (alternatives) in the form IFVs and then construct their corresponding decision matrix $[h]^{n' \times m'}$.
2: Get the generalization matrix $[G_\varrho]^{n' \times 1}$ by using the perceptions of the senior members/experts' committee on each κ_j $(j = 1, 2, ..., n')$.
3: Construct the new matrix $[h_G]^{n' \times (m'+1)}$ by placing $[G_\varrho]^{n' \times 1}$ in $[h]^{n' \times m'}$ with respect to κ_j $(j = 1, 2, ..., n')$.
4: Apply operators with respect to κ_j, and the results are denoted by $r'_j, j = 1, 2, ..., n'$.
5: Rank the $\kappa_j (j = 1, 2, ..., n')$ in descending order on their score values r'_j $(j = 1, 2, ..., n')$.

Under the approach established in [44], we provide some key points and compare Algorithm 1, with Algorithm 2 :

(i) In Algorithm 2, the generalized parameter matrix is obtained by incorporating preferences of experts on alternatives. In other words, information from moderator on alternatives is given; nevertheless, extra input can be sighted as another information over IFSS, and both information types (IFSS and extra inputs) deal with alternatives. Conversely, in Algorithm 1, clear and well-defined GGIFSSs are taken into account by incorporating IFSS and IFSs.
(ii) Operation of extended union is used in Algorithm 1 on two GGIFSSs, while, in Algorithm 2, there are some difficulties in defining an operation of extended union on two or more GGIFSSs.
(iii) It seems that GGWA or GGWG operators in Algorithm 2 are applied contrarily on two different information types, however, in Algorithm 1, an integrated manner is adopted to compile results through GBGWA or GBGWG.
(iv) In Algorithm 1, the generalized parameters can be applied as the demands of customers, and thus an integrated framework can be employed in industries. However, Algorithm 2 lacks creating such frameworks.

6.2. Comparisons with the Results of GIFSSs

The obtained results on the case studies in Sections 5.2 and 5.3 are compared with the outcomes that are achieved on GIFSSs as given below.

(i) As discussed earlier, GGIFSS with only single generalized parameter is known as GIFSS. Then, Algorithm 1 can be separated for each senior moderator/customer in the case studies in Sections 5.2 and 5.3.

1. Using the Lemma 1 and Algorithm 1, we obtained the results separately for each senior experts/members in the case study in Section 5.2. If only Senior Member 1 is taken into account during selection process, then $\delta(\ell'_1) = 0.5185$, $\delta(\ell'_2) = 0.4305$, $\delta(\ell'_3) = 0.4674$, $\delta(\ell'_4) = 0.4695$, and $\delta(\ell'_5) = 0.5194$. The descending order is acquired as $\kappa_5 > \kappa_1 > \kappa_4 > \kappa_3 > \kappa_2$; thus, κ_5 is the felicitous candidate for the position.

If only Senior Member 2 is taken into account during selection process in the case study in Section 5.2, then $\delta(\ell'_1) = 0.3709$, $\delta(\ell'_2) = 0.3068$, $\delta(\ell'_3) = 0.3329$, $\delta(\ell'_4) = 0.3351$, and $\delta(\ell'_5) = 0.3704$. The descending order is acquired as $\kappa_1 > \kappa_5 > \kappa_3 > \kappa_4 > \kappa_2$; thus, κ_1 is the felicitous candidate for the position.

If only Senior Member 3 is taken into account during selection process in the case study in Section 5.2, then $\delta(\ell'_1) = 0.4178$, $\delta(\ell'_2) = 0.3475$, $\delta(\ell'_3) = 0.3774$, $\delta(\ell'_4) = 0.3788$, and $\delta(\ell'_5) = 0.4192$. The descending order is acquired as $\kappa_5 > \kappa_1 > \kappa_4 > \kappa_2 > \kappa_3$; thus, κ_1 is the felicitous candidate for the position.

It can be observed from above discussion that κ_5 is the most suitable candidate as per individual opinions of Senior Members 1 and 3. Similarly, κ_1 is the most felicitous candidate on individual opinion of Senior Member 2, while κ_5 is on second place in descending order. Thus, in general, κ_5 is the most suitable candidate.

2. Using the Lemma 1 and Algorithm 1, we obtained the results separately for each customer for the case study in Section 5.3. If it is required to select cinema only for Customer 1, then $\delta(\ell_1') = 0.2927$, $\delta(\ell_2') = 0.3230$, $\delta(\ell_3') = 0.2655$, $\delta(\ell_4') = 0.3072$, and $\delta(\ell_5') = 0.2867$. The order is acquired as $\kappa_2 > \kappa_4 > \kappa_1 > \kappa_5 > \kappa_3$; thus, κ_2 is the best cinema for Customer 1.
If it is required to select cinema only for the Customer 2, then $\delta(\ell_1') = 0.2965$, $\delta(\ell_2') = 0.3253$, $\delta(\ell_3') = 0.2682$, $\delta(\ell_4') = 0.3121$, and $\delta(\ell_5') = 0.2894$. The order is acquired as $\kappa_2 > \kappa_4 > \kappa_1 > \kappa_5 > \kappa_3$; thus, κ_2 is the best cinema for Customer 2.
Thus, in general, κ_2 is the most suitable for both customers.

(ii) Feng et al. [42] introduced a framework of decision makings on GIFSSs. We correlate proposed results with their method as below. We acquired the results separately for each customer for the case study in Section 5.3. If it is required to select cinema only for Customer 1, then $\delta(Z_J(\kappa_1)) = 0.5344$, $\delta(Z_J(\kappa_2)) = 0.5928$, $\delta(Z_J(\kappa_3)) = 0.4857$, $\delta(Z_J(\kappa_1)) = 0.5549$, and $\delta(Z_J(\kappa_1)) = 0.5251$. The descending order acquired as $\kappa_2 > \kappa_4 > \kappa_1 > \kappa_5 > \kappa_3$; thus, κ_2 is the best cinema for Customer 1. If it is require to select cinema only for Customer 2, then $\delta(Z_J(\kappa_1)) = 0.5311$, $\delta(Z_J(\kappa_2)) = 0.5961$, $\delta(Z_J(\kappa_3)) = 0.4894$, $\delta(Z_J(\kappa_1)) = 0.5568$, and $\delta(Z_J(\kappa_1)) = 0.5262$. The descending order acquired as $\kappa_2 > \kappa_4 > \kappa_1 > \kappa_5 > \kappa_3$; thus, κ_2 is the best cinema for Customer 2.

(iii) A framework for the best concept selection in design process has been computed in [43], where GIFSSs are utilized to acquire integrated information on customers demands and design concepts. To meet their objectives, they introduced an algorithm, which we updated as follows:

Algorithm 3 Updated form of Algorithm in [43]

1: The demands of p number of customers are represented as IFSs $\widetilde{\alpha}_1, \widetilde{\alpha}_2, ..., \widetilde{\alpha}_p$.
2: Represent IFSS $(\widetilde{S}, E)$ over set of all possible choices X.
3: Represent GIFSSs for each customer $\widetilde{\mathscr{F}}_k = (\widetilde{S}, E, \widetilde{\alpha}_k)(k = 1, ..., p)$.
4: Compute $int - AND - product$ operation on $\widetilde{\mathscr{F}}_1, \widetilde{\mathscr{F}}_2, ..., \widetilde{\mathscr{F}}_p$, obtain GIFSS $\widetilde{\mathscr{F}}$ and show it in tabular form.
5: Derive the utility fuzzy set $\Delta_{\widetilde{\mathscr{F}}}$ from the GIFSS $\widetilde{\mathscr{F}}$.
6: Output $\kappa_{j'}$ as the optimal decision if $\Delta_{\widetilde{\mathscr{F}}}(\kappa_{j'}) = \max\{\Delta_{\widetilde{\mathscr{F}}}(\kappa_j) \mid \kappa_j \in X\}$.
7: If j' has more than one values then any one of κ_j may be chosen.

The case study in Section 5.3 can be contemplated through Algorithm 3. On this prospect, GGIFSS given in Table 9 can be separated into two GIFSSs. After adopting all steps of Algorithm 3, $\Delta(\kappa_1) = 0.2252$, $\Delta(\kappa_2) = 0.2678$, $\Delta(\kappa_3) = 0.2112$, $\Delta(\kappa_4) = 0.2419$, and $\Delta(\kappa_5) = 0.2261$. The descending order is acquired as $\kappa_2 > \kappa_4 > \kappa_5 > \kappa_1 > \kappa_3$; thus, κ_2 is the best cinema for both customers.

The superiorities and advantages of our method are given in next section.

7. Superiority of Proposed Method

In this section, we give some counter-examples to show the superiority of proposed method over recent approaches [42–44].

Example 5. *Assume a decision making problem by letting the two alternatives κ_1 and κ_2, which have to be evaluated by the committee of specialists over set of parameters $E = \{v_1, v_2, v_3\}$. The committee of specialists provide the judgments in the form of IFSS, given in Table 10;*

Table 10. Tabular representation of the IFSS, $(\widetilde{S}, E)$.

$X \diagdown E$	v_1	v_2	v_3
κ_1	$\langle 0.5, 0.5\rangle$	$\langle 0.4, 0.6\rangle$	$\langle 0.6, 0.4\rangle$
κ_2	$\langle 0.3, 0.7\rangle$	$\langle 0.5, 0.4\rangle$	$\langle 0.5, 0.3\rangle$

Here, we apply the approach provided by Feng et al. [42], by letting an extra input $\widetilde{\beta} = \{\langle 0.4, 0.2\rangle, \langle 0.5, 0.3\rangle, \langle 0.6, 0.4\rangle\}$ *of a moderator. Then, the GIFSS is consolidated as in Table 11;*

Table 11. Tabular representation of the IFSS, $(\widetilde{S}, E, \widetilde{\beta})$.

$X \diagdown E$	v_1	v_2	v_3
κ_1	$\langle 0.5, 0.5\rangle$	$\langle 0.4, 0.6\rangle$	$\langle 0.6, 0.4\rangle$
κ_2	$\langle 0.3, 0.7\rangle$	$\langle 0.5, 0.4\rangle$	$\langle 0.5, 0.3\rangle$
$\widetilde{\beta}$	$\langle 0.4, 0.2\rangle$	$\langle 0.6, 0.4\rangle$	$\langle 0.5, 0.3\rangle$

The score function on IFVs in $\widetilde{\beta}$ *are* 0.6, 0.6, *and* 0.6 *and the weights are* 0.33, 0.33, *and* 0.33, *respectively. It can be seen that, when we convert extra input into weights in initial stages of decision making, the importance of membership and non-membership diminish. Using the method of Feng et al. [42], we get* $\delta(Z_J(\kappa_1)) = 0.503 > \delta(Z_J(\kappa_2)) = 0.498$ *such that* $\kappa_1 > \kappa_2$. *Let* $w = \{w_1/0.32, w_2/0.33, w_3/0.35\}$ *be the weighted vector over E. Then, by proposed method,* $\delta(\ell_1') = 0.310 < \delta(\ell_2') = 0.314$ *such that* $\kappa_1 < \kappa_2$. *Therefore, the conversation of extra input into weighted vector in early process of decision making diminish the importance of membership and non-membership. Thus, proposed approach is better then the method of Feng et al. [42].*

Example 6. *Assume that* κ_1,κ_2 *and* κ_3 *are three products and* $E = \{v_1, v_2, v_3\}$ *is the set of parameters. The dependencies of products on criteria are provided in IFSS* $(\widetilde{S}, E)$ *and given in Table 12.*

Table 12. Tabular representation of the IFSS, $(\widetilde{S}, E)$.

$X \diagdown E$	v_1	v_2	v_3
κ_1	$\langle 0.3, 0.5\rangle$	$\langle 0.5, 0.3\rangle$	$\langle 0.6, 0.2\rangle$
κ_2	$\langle 0.2, 0.6\rangle$	$\langle 0.3, 0.4\rangle$	$\langle 0.6, 0.4\rangle$
κ_3	$\langle 0.4, 0.4\rangle$	$\langle 0.4, 0.3\rangle$	$\langle 0.5, 0.3\rangle$

Here, we consider the methodology of Hayat et al. [43]. To select best product for two customers d_1, d_2, *their demands are investigated as* $\widetilde{\beta}_{d_1} = \{\langle 0.3, 0.5\rangle, \langle 0.4, 0.4\rangle, \langle 0.6, 0.2\rangle\}$, $\widetilde{\beta}_{d_2} = \{\langle 0.3, 0.6\rangle, \langle 0.3, 0.4\rangle, \langle 0.5, 0.4\rangle\}$, *respectively. Then, the GIFSSs for* d_1 *and* d_2 *are given in Tables 13 and 14, respectively.*

Table 13. GIFSS $(\widetilde{S}, E, \widetilde{\beta}_{d_1})$.

$X \diagdown E$	v_1	v_2	v_3
κ_1	$\langle 0.3, 0.5\rangle$	$\langle 0.5, 0.3\rangle$	$\langle 0.6, 0.2\rangle$
κ_2	$\langle 0.2, 0.6\rangle$	$\langle 0.3, 0.4\rangle$	$\langle 0.6, 0.4\rangle$
κ_3	$\langle 0.4, 0.4\rangle$	$\langle 0.4, 0.3\rangle$	$\langle 0.5, 0.3\rangle$
$\widetilde{\beta}_{d_1}$	$\langle 0.3, 0.5\rangle$	$\langle 0.4, 0.4\rangle$	$\langle 0.6, 0.2\rangle$

Table 14. GIFSS $(\widetilde{S}, E, \widetilde{\beta}_{d_2})$.

$X \diagdown E$	v_1	v_2	v_3
κ_1	$\langle 0.3, 0.5 \rangle$	$\langle 0.5, 0.3 \rangle$	$\langle 0.6, 0.2 \rangle$
κ_2	$\langle 0.2, 0.6 \rangle$	$\langle 0.3, 0.4 \rangle$	$\langle 0.6, 0.4 \rangle$
κ_3	$\langle 0.4, 0.4 \rangle$	$\langle 0.4, 0.3 \rangle$	$\langle 0.5, 0.3 \rangle$
$\widetilde{\beta}_{d_2}$	$\langle 0.3, 0.6 \rangle$	$\langle 0.3, 0.4 \rangle$	$\langle 0.5, 0.4 \rangle$

In [43], AND operation is computed on two GIFSSs for product for two customers. One can check that

$$\widetilde{S}(v_1) \wedge \widetilde{S}(v_2) = \widetilde{S}(v_1) \;\; with\, \widetilde{t}_{\widetilde{\beta}_{d_1}}(v_1) \wedge \widetilde{t}_{\widetilde{\beta}_{d_2}}(v_2) = \widetilde{t}_{\widetilde{\beta}_{d_1}}(v_1), \widetilde{f}_{\widetilde{\beta}_{d_1}}(v_1) \vee \widetilde{f}_{\widetilde{\beta}_{d_2}}(v_2) = \widetilde{f}_{\widetilde{\beta}_{d_1}}(v_1) \tag{11}$$

$$\widetilde{S}(v_1) \wedge \widetilde{S}(v_3) = \widetilde{S}(v_1) \;\; with\, \widetilde{t}_{\widetilde{\beta}_{d_1}}(v_1) \wedge \widetilde{t}_{\widetilde{\beta}_{d_2}}(v_3) = \widetilde{t}_{\widetilde{\beta}_{d_1}}(v_1), \widetilde{f}_{\widetilde{\beta}_{d_1}}(v_1) \vee \widetilde{f}_{\widetilde{\beta}_{d_2}}(v_3) = \widetilde{f}_{\widetilde{\beta}_{d_1}}(v_1) \tag{12}$$

$$\widetilde{S}(v_2) \wedge \widetilde{S}(v_3) = \widetilde{S}(v_2) \;\; with\, \widetilde{t}_{\widetilde{\beta}_{d_1}}(v_2) \wedge \widetilde{t}_{\widetilde{\beta}_{d_2}}(v_3) = \widetilde{t}_{\widetilde{\beta}_{d_1}}(v_2), \widetilde{f}_{\widetilde{\beta}_{d_1}}(v_2) \vee \widetilde{f}_{\widetilde{\beta}_{d_2}}(v_3) = \widetilde{f}_{\widetilde{\beta}_{d_1}}(v_2). \tag{13}$$

It can be seen that, using AND operation, the importance of IFVs for v_2 and v_3 are diminished in Equations (11) and (12). The importance of IFVs for v_3 are diminished in Equation (13). Thus, such an approach is not valid in the initial stages of decision making; therefore, for this prospect, the proposed approach is better then Hayat et al. [43].

In [44], GWA is computed on two information; one from a committee of experts (in form of IFSS) and other from group of senior persons. The extra inputs can be seen as a so-called IFSS of a group of senior persons over alternatives. Consider Example 5, where IFSS from a committee of experts is given in Table 10. The extra input is given in Table 15, and can be seen as a so-called IFSS on alternatives. In the prospect of Garg et al. [44], the extra opinions of the two senior experts d_1, d_2 can be merged with IFSS in Table 10.

Table 15. Opinions of experts on alternatives.

$X \diagdown$ *Experts*	d_1	d_2
κ_1	$\langle 0.4, 0.5 \rangle$	$\langle 0.4, 0.4 \rangle$
κ_2	$\langle 0.3, 0.3 \rangle$	$\langle 0.2, 0.4 \rangle$

Clearly, the combination of two data ((i) IFSS and (ii) IFVs of experts) based matrix is analyzed as GGIFSS in [44]. In another way, if it might be recognized that the group of extra inputs of senior experts is a summarization of the data (IFSS obtained from a committee of specialists), then the results can be obtained from Table 15, thus why would we contemplate two data based matrix over alternatives? Nevertheless, there exist some serious difficulties in [44]. Noteworthily, the proposed results are superior in certain aspects and a well-defined manner is considered.

Advantages of Proposed Method

Based on correlative and comparative research, the following benefits of present framework are acquired and emphasized:

(i) The case study indicated in [44] is implemented on two IFVs based matrix but not for the GGIFSSs. The extra inputs are located as in IFVs type weights on alternatives and operators presumed to be collected on the two different IFVs based data. In this prospect, the proposed approach is based on well-defined GGIFSSs.
(ii) In [42], an extra input turns into the weighted vector in initial stages of decision making after calculation of score functions but it is not integrated with the information of experts to achieve

better results. In the proposed method, extra inputs are taken into account in an accurate way using GBGWA or GBGWG.

(iii) In [43], the *AND* operation is used on several GIFSSs. In many cases, *AND* or *OR* operations on IFVs provide instantaneous results but do not give comprehensively aggregated results.

(iv) The judgements/demands of senior prospectors/customers in GGIFSSs as managed with proposed operators are useful to rank the alternatives. The proposed framework can be correlated with the shortening of any number of existing senior prospectors/customers.

8. Conclusions

It has been noticed that the definition of GGIFSS, given by Garg et al. [44], did not provide supplementary information in a precise manner. Under this prospect, we have reformulated the existing definition of GGIFSS by establishing a novel notion of GGIFSS and related operation are also refined. We have aggregated GBGWA and GBGWG operators on GGIFSSs, which are employed to aggregate our techniques. Then, we formulated the framework of decision makings in an algorithm and two case studies have been handled by virtue of proposed methodology. We have given the advantages and comparison with existing techniques and correlated the results which are achieved on GIFSSs. The advantages of given framework are to contemplate the prospector's demands or expert's judgments in an incorporated way such that establishing more operators can be constituted the design concept evaluation mechanism on GGIFSSs. In this way, the results presented in this paper can be studied in several fields, such as electrical engineering, industrial designs, construction engineering, as estimation of risk factors in risk management is a complex tasks thus such problem can be considered.

Author Contributions: K.H. develops the idea and writes the paper. K.H.; methodology, K.H.; writing—original draft preparation. M.I.A., B.-Y.C., F.K. X.P.Y. review and edit the paper.

Funding: This paper is supported by the High Level Construction Fund of Guangzhou University China. Also this work was supported by the Natural Science Foundation (61877014), Natural Science Foundation of Guangdong Province (2017A030307020, 2016A030307037, 2016A030313552), the Guangdong Provincial Government to Guangdong International Student Scholarship (yuejiao [2014] 187), Guangzhou Vocational College of Science and Technology (no. 2016TD03) and the Foundation of Hanshan Normal University (QD20171001, LQ201702).

Acknowledgments: The authors would like to thank all the reviewers for their valuable comments.

Conflicts of Interest: The authors declare no conflict of interest.

Abbreviations

The following abbreviations are used in this manuscript:

GGIFSSs	Group-based generalized intuitionistic fuzzy soft sets
GIFSSs	Generalized intuitionistic fuzzy soft sets
GBGWA	Group-based generalized weighted averaging operators
GBGWG	Group-based generalized weighted geometric operators
IFSs	Intuitionistic fuzzy sets

References

1. Maji, P.K.; Biswas, R.; Roy, A.R. Fuzzy soft sets. *J. Fuzzy Math.* **2001**, *9*, 589–602.
2. Zadeh, L.A. Fuzzy Sets. *Inf. Control* **1965**, *8*, 338–353. [CrossRef]
3. Molodtsov, D. Soft set theory-first results. *Comput. Math. Appl.* **1999**, *37*, 19–31. [CrossRef]
4. Maji, P.K.; Biswas, R.; Roy, A.R. Soft set theory. *Comput. Math. Appl.* **2003**, *45*, 555–562. [CrossRef]
5. Ali, M.I.; Feng, F.; Liu, X.; Min, W.K.; Shabir, M. On some new operations in soft set theory. *Comput. Math. Appl.* **2009**, *57*, 1547–1553. [CrossRef]
6. Roy, A.R.; Maji, P.K. A fuzzy soft set theoretic approach to decision making problems. *J. Comput. Appl. Math.* **2007**, *203*, 412–418. [CrossRef]
7. Alcantud, J.C.R.; Mathew, T.J. Separable fuzzy soft sets and decision making with positive and negative attributes. *Appl. Soft Comput.* **2017**, *59*, 586–595. [CrossRef]

8. Alcantud, J.C.R.; Torra, V. Decomposition theorems and extension principles for hesitant fuzzy sets. *Inf. Fusion* **2018**, *41*, 48–56. [CrossRef]
9. Khameneh, A.Z.; Kılıçman, A. Parameter reduction of fuzzy soft sets: An adjustable approach based on the three-way decision. *Int. J. Fuzzy Syst.* **2018**, *20*, 928–942. [CrossRef]
10. Hassan, N.; Sayed, O.R.; Khalil, A.M.; Ghany, M.A. Fuzzy soft expert system in prediction of coronary artery disease. *Int. J. Fuzzy Syst.* **2017**, *19*, 1546–1559. [CrossRef]
11. Hayat, K.; Ali, M.I.; Cao, B.Y.; Karaaslan, F.; Qin, Z. Characterizations of certain types of type-2 soft graphs. *Discrete Dyn. Nat. Soc.* **2018**, *2018*, 8535703. [CrossRef]
12. Hayat, K.; Ali, M.I.; Cao, B.; Karaaslan, F. New results on type-2 soft sets. *J. Math. Stat.* **2018**, *484*, 47. [CrossRef]
13. Hayat, K.; Ali, M.I.; Cao, B.-Y.; Yang, X.-P. A new type-2 soft set: Type-2 soft graphs and their applications. *Adv. Fuzzy Sys.* **2017**. [CrossRef]
14. Kumar, A.; Kumar, D.; Jarial, S.K. A hybrid clustering method based on improved artificial bee colony and fuzzy C-means algorithm. *Int. J. Artif. Intell.* **2017**, *15*, 24–44.
15. Liu, Y.; Qin, K.; Martınez, L. Improving decision making approaches based on fuzzy soft sets and rough soft sets. *Appl. Soft Comput.* **2018**, *65*, 320–332. [CrossRef]
16. Medina, J.; Ojeda-Aciego, M. Multi-adjoint t-concept lattices. *Inf. Sci.* **2010**, *180*, 712–725. [CrossRef]
17. Pozna, C.; Minculete, N.; Precup, R.E.; Koczy, L.T.; Ballagi, A. Signatures: Definitions, operators and applications to fuzzy modelling. *Fuzzy Sets Syst.* **2012**, *201*, 86–104. [CrossRef]
18. Qinrong, F.; Fenfen, W. A discernibility matrix approach to fuzzy soft sets based decision making problems. *J. Intell. Fuzzy Syst.* **2018**, *22*, 59–74.
19. Vildan, C.; Halis, A. A topological view on L-fuzzy soft sets: Connectedness degree. *J. Intell. Fuzzy Syst.* **2018**, *34*, 1975–1983.
20. Vimala, J.; Arockia, J.R.; Anusuya, I.V.S. A study on fuzzy soft cardinality in lattice ordered fuzzy soft group and its application in decision making problems. *J. Intell. Fuzzy. Syst.* **2018**, *34*, 1535–1542. [CrossRef]
21. Xiao, F. A hybrid fuzzy soft sets decision making method in medical diagnosis. *IEEE Acess* **2018**, *6*, 25300–25312. [CrossRef]
22. Beg, I.; Rashid, T.; Jamil, R.N. Human attitude analysis based on fuzzy soft differential equations with Bonferroni mean. *Comput. Appl. Math.* **2018**, *37*, 2632–2647. [CrossRef]
23. Ma, X.; Zhan, J.; Ali, M.I. Applications of a kind of novel Z-soft fuzzy rough ideals to hemirings. *J. Intell. Fuzzy Syst.* **2017**, *32*, 2071–2080. [CrossRef]
24. Sun, B.; Ma, W. Soft fuzzy rough sets and its application in decision making. *Artif. Intell. Rev.* **2014**, *41*, 67–80. [CrossRef]
25. Atanassov, K. Intuitionistic fuzzy sets. *Fuzzy Sets Syst.* **1986**, *20*, 87–96. [CrossRef]
26. Atanassov, K. *Intuitionistic Fuzzy Sets, Theory, and Applications*; Series in Fuzziness and Soft Computing; Phisica-Verlag: Heidelberg, Germany, 1999.
27. Xu, Z. Intuitionistic fuzzy aggregation operators. *IEEE Trans. Fuzzy Syst.* **2007**, *15*, 1179–1187.
28. Xu, Z.; Yager, R.R. Some geometric aggregation operators based on intuitionistic fuzzy sets. *Int. J. Gen. Syst.* **2006**, *35*, 417–433. [CrossRef]
29. Jemal, H.; Kechaou, Z.; Ayed, M.B. Enhanced decision support systems in intensive care unit based on intuitionistic fuzzy sets. *Adv. Fuzzy Syst.* **2017**, *2017*, 7371634. [CrossRef]
30. Mukherjee, S. Selection of alternative fuels for sustainable urban transportation under multi-criteria intuitionistic fuzzy environment. *Fuzzy Inf. Eng.* **2017**, *9*, 117–135. [CrossRef]
31. Ren, H.P.; Chen, H.H.; Fei, W.; Li, D.F. A MAGDM method considering the amount and reliability information of interval-valued intuitionistic fuzzy sets. *Int. J. Fuzzy Syst.* **2017**, *19*, 15–25. [CrossRef]
32. Yun, S.M.; Lee, S.J. Intuitionistic fuzzy topologies induced by intuitionistic fuzzy approximation spaces. *Int. J. Fuzzy Syst.* **2017**, *9*, 285–291. [CrossRef]
33. Maji, P.K.; Biswas, R.; Roy, A.R. Intuitionistic fuzzy soft sets. *J. Fuzzy Math.* **2001**, *9*, 677–692.
34. Deli, I.; Karataş, S. Interval valued intuitionistic fuzzy parameterized soft set theory and its decision making. *J. Intell. Fuzzy Syst.* **2016**, *30*, 2073–2082. [CrossRef]
35. Akram, M.; Shahzadi, S. Novel intuitionistic fuzzy soft multiple-attribute decision-making methods. *Neural Comput. Appl.* **2018**, *29*, 435–447. [CrossRef]

36. Garg, H.; Arora, R. A nonlinear-programming methodology for multi-attribute decision-making problem with interval-valued intuitionistic fuzzy soft sets information. *Appl. Intell.* **2018**, *48*, 2031–2046. [CrossRef]
37. Garg, H.; Arora, R. Bonferroni mean aggregation operators under intuitionistic fuzzy soft set environment and their applications to decision-making. *J. Oper. Res. Soc.* **2018**, 1–4. [CrossRef]
38. Garg, H.; Arora, R. Novel scaled prioritized intuitionistic fuzzy soft interaction averaging aggregation operators and their application to multi criteria decision making. *Eng. Appl. Artif. Intell.* **2018**, *71*, 100–112. [CrossRef]
39. Agarwal, M.; Biswas, K.K.; Hanmandlu, M. Generalized intuitionistic fuzzy soft sets with applications in decision-making. *Appl. Soft Comput.* **2013**, *13*, 3552–3566. [CrossRef]
40. Khalil, A.M. Commentary on Generalized intuitionistic fuzzy soft sets with applications in decision-making. *Appl. Soft Comput.* **2015**, *37*, 519–520. [CrossRef]
41. Yang, Y.; Wang, Y.; Zhang, Y.; Zhang, D. Commentary on generalized intuitionistic fuzzy soft sets with applications in decision-making [Appl. Soft Comput. 37 (2015) 519–520]. *Appl. Soft Comput.* **2016**, *40*, 427–428. [CrossRef]
42. Feng, F.; Fujita, H.; Ali, M.I.; Yager, R.R.; Liu, X. Another View on generalized Intuitionistic Fuzzy Soft Sets and Related Multiattribute Decision Making Methods. *IEEE Trans. Fuzzy Syst.* **2018**. [CrossRef]
43. Hayat, K.; Ali, M.I.; Alcantud, J.C.R.; Cao, B.Y.; Tariq, K.U. Best concept selection in design process: An application of generalized intuitionistic fuzzy soft sets. *J. Intell. Fuzzy Syst.* **2018**. [CrossRef]
44. Garg, H.; Arora, R. Generalized and group-based generalized intuitionistic fuzzy soft sets with applications in decision-making. *Appl. Intell.* **2018**, *48*, 343–356. [CrossRef]
45. Selvachandran, G.; Maji, P.K.; Faisal, R.Q.; Salleh, A.R. Distance and distance induced intuitionistic entropy of generalized intuitionistic fuzzy soft sets. *Appl. Intell.* **2017**, *47*, 132–147. [CrossRef]
46. Pawlak, Z. Rough sets. *Int. J. Comput. Inf. Sci.* **1982**, *11*, 341–356. [CrossRef]
47. Deschrijver, G.; Kerre, E. On the relationship between some extensions of fuzzy set theory. *Fuzzy Sets Syst.* **2003**, *133*, 227–235. [CrossRef]
48. Chen, S.M.; Tan, J.M. Handling multicriteria fuzzy decision-making problems based on vague set theory. *Fuzzy Sets Syst.* **1994**, *67*, 163–172. [CrossRef]

symmetry

MDPI

Article

The New Similarity Measure and Distance Measure of a Hesitant Fuzzy Linguistic Term Set Based on a Linguistic Scale Function

Donghai Liu [1,*], Yuanyuan Liu [1] and Xiaohong Chen [2]

1 Department of Statistics, Hunan University of Science and Technology, Xiangtan 411201, Hunan, China; liuyuanyuanlyy111@163.com
2 Mobile E-business Collaborative Innovation Center of Hunan Province, Hunan University of Commerce, Changsha 410205, Hunan, China; csu_cxh@163.com
* Correspondence: donghailiu@126.com; Tel.: +86-731-5829-1382

Received: 4 August 2018; Accepted: 24 August 2018; Published: 29 August 2018

Abstract: The existing cosine similarity measure for hesitant fuzzy linguistic term sets (HFLTSs) has an impediment as it does not satisfy the axiom of similarity measure. Due to this disadvantage, a new similarity measure combining the existing cosine similarity measure and the Euclidean distance measure of HFLTSs is proposed, which is constructed based on a linguistic scale function; the related properties are also given. According to the relationship between the distance measure and the similarity measure, a corresponding distance measure between HFLTSs is obtained. Furthermore, we generalize the technique for order preference by similarity to an ideal solution (TOPSIS) method to the obtained distance measure of the HFLTSs. The principal advantages of the proposed method are that it cannot only effectively transform linguistic information in different semantic environments, but it can also avoid the shortcomings of existing the cosine similarity measure. Finally, a case study is conducted to illustrate the feasibility and effectiveness of the proposed method, which is compared to the existing methods.

Keywords: hesitant fuzzy linguistic term set; similarity measure; linguistic scale function; distance measure; TOPSIS method

1. Introduction

In many multi-criteria decision making (MCDM) problems, because of the incomplete information and the complexity of the decision-making environment, crisp numbers cannot describe the relevant decision information. Thus Zadeh [1] proposed the fuzzy set (FS) $A = \{ (x_j, \mu_A(x_j)) | x_j \in X \}$ $(0 \le \mu_A(x_j) \le 1)$ on $X = \{x_1, x_2, \cdots, x_n\}$, where $\mu_A(x_j)$ is the membership degree of $x_j \in X$. Since it was put forward, many scholars have generalized it. For example, Atanassov [2,3] introduced the concepts of the intuitionistic fuzzy set (IFS) and the interval-valued intuitionistic fuzzy set (IVIFS), and Torra [4] proposed the hesitant fuzzy set (HFS). In the past few years, the FS and its extensions have been applied in many fields, such as supplier selection, pattern recognition, and medical diagnosis. As the FS and its extensions mentioned above use crisp numbers to express decision information, they cannot express qualitative evaluation information. For instance, when one expert evaluates the performance of a company, he/she thinks that the performance of the company is very good. Because the evaluation expression is consistent with the human's cognitive process, it is suitable to express this in a linguistic term set (LTS). To describe the relevant information, Zadeh [5–7] proposed the LTS to express the relevant information. A general discrete LTS of seven terms can be represented as $S = \{s_0: \textit{very poor}, s_1: \textit{poor}, s_2: \textit{slightly poor}, s_3: \textit{fair}, s_4: \textit{slightly good}, s_5: \textit{good}, s_6: \textit{very good}\}$. Then, the expert's evaluation about the performance of the company can be represented as $\{s_6\}$.

However, due to the uncertainty of the problem in the decision making process, the decision makers cannot express their preferences using only one membership degree of a LTS. In order to express the decision makers' hesitation about the decision problem, Rodríguez et al. [8] proposed the hesitant fuzzy linguistic term set (HFLTS), which is based on LTS and HFS. The HFLTS makes the representation of the decision information more flexible. Since the HFLTS was proposed, a number of relevant studies and their applications [9–22] have been conducted. For example, Liu et al. [16] presented the fuzzy envelope for HFLTSs and applied it to choose the best alternative. Xu et al. [17] presented the hesitant fuzzy linguistic ordered weighted distance operator, and applied it to plan the selection of enterprise's large projects. Liao et al. [18] maked a survey on HFLTSs and reviewed the decision making process with hesitant fuzzy linguistic preference (HFLP) relations. Liao et al. [19] proposed the hesitant fuzzy linguistic preference utility set (HFLPUS) and applied the HFLP utility TOPSIS approach to choose the best fire rescue alternative. One thing that they have in common is that they use the subscript of linguistic terms directly in the process of operations, which may cause a loss of information. In order to overcome this problem, the linguistic scale function was introduced by Wang et al. [23], which can assign different numerical values to the linguistic terms set under different circumstances. The linguistic scale function can reflect the preferences of the decision makers in different environments. Since it was put forward, many scholars have studied this subject. For example, Wang et al. [24] presented the Hausdorff distance between the hesitant fuzzy linguistic numbers (HFLNs), based on the linguistic scale function, and developed the TOPSIS and TODIM approaches to it. Liu et al. [25] proposed a distance measure of HFLTSs, which also included the linguistic scale function. Furthermore, Liu et al. [26] proposed the intuitionistic fuzzy linguistic cosine similarity measure and the interval-valued intuitionistic fuzzy linguistic cosine similarity measure, they all contain the linguistic scale function. The research on this field has developed rapidly.

From another perspective, the similarity measure is also an important aspect in MCDM problems, which can measure the similarity degree between two different alternatives. It has been widely studied in the past few years. For example, Song et al. [27] considered the similarity measure between IFSs, and proposed the corresponding distance measure between IF belief functions. Liao et al. [9] presented some similarity measures and distance measures between HFLTSs; Lee et al. [10] proposed a similarity measure based on likelihood relations. For the other studies about the similarity measure, we can refer to [11–13]. The cosine similarity measure is also a significant similarity measure; it can be expressed as the inner product of two vectors divided by the product of their lengths [28]. Some scholars have studied the cosine similarity measure [29–31]. For instance, Ye [29] introduced a weighted cosine similarity measure between IFSs and they applied it to rank the alternative. Furthermore, Ye [30] presented the cosine similarity measure between interval-valued fuzzy sets (IVFSs) with risk preference, and altered its decision making method depending on decision makers' preferences. Liao et al. [31] defined the cosine similarity measure between HFLTSs and extended the TOPSIS approach and VIKOR approach to the cosine distance measure. It is already known that the cosine similarity measure proposed by Liao et al. [31] is not a regular similarity measure (because it is not satisfied with the axiom of the similarity measure; the example can be seen in Section 3). If it is applied in MCDM problems, it may cause the decision information to be distorted. Furthermore, the cosine similarity measure defined by Liao et al. [31] used the subscript of linguistic terms directly in process of operations; they did not consider the semantic decision environment, which may cause a loss of information in the decision making process.

Therefore, this paper introduces a new method to construct a similarity measure between HFLTSs; the main motivations and contributions of the paper are given as follows:

(1) In order to overcome the disadvantage of the similarity measure proposed by Liao et al. [31], a new similarity measure combining the existing cosine similarity measure [31] and the Euclidean distance measure of HFLTSs is proposed in this paper, which can improve the accuracy of the calculation to some extent.

(2) On the basis of the linguistic scale function, the paper proposes a new similarity measure between two HFLTSs; it is already known that the linguistic scale function can improve the flexibility of the transformation of the linguistic decision information in different semantic environments. The proposed method is capable of expressing the fuzzy linguistic information more flexibly and improving the adaptability of HFLTSs in practice.

(3) According to the relationship between the similarity measure and the distance measure, this paper proposes a new distance measure of HFLTSs and extends the TOPSIS method to it; it focuses on the differences between different alternatives, which can improve the effectiveness of solving MCDM problems.

The reminder of the paper is given as follows: the background on the MCDM problems, some concepts of LTS and HFLTS, the existing similarity measures of HFLTSs, and the linguistic scale function are reviewed in Section 2. In Section 3, a new score function of HFLTS based on the linguistic scale function, and a new approach to construct the similarity measure of HFLTSs, are presented. The corresponding distance measure is also constructed based on the relationship between the distance measure and the regular similarity measure. In Section 4, we extend the TOPSIS method to the proposed distance measure. In Section 5, a numerical example is given to illustrate the feasibility of the proposed method, and the same numerical example is examined to compare with other methods. Some conclusions and future research are proposed in Section 6.

2. Preliminaries

In this section, we will explain how the MCDM method works, and we review some basic knowledge, including LTS, HFLTS, the score function of HFLTS, and the linguistic scale function. Some existing distance measures and similarity measures of HFLTSs are also introduced. In this paper, we denote $X = \{x_1, x_2, \cdots, x_n\}$ as the discourse set.

2.1. MCDM

Multi-criteria decision making is an important branch of the decision-making field. There are some common aspects (alternatives and criteria) in MCDM; the typical MCDM problem can be described as follows:

Let $X = \{x_1, x_{72}, \cdots, x_n\}$ be a set of alternatives; let $C = \{c_1, c_2, \cdots, c_m\}$ be a set of criteria values. The decision matrix D is an $n \times m$ matrix, in which element d_{ij} indicates the performance of the alternative x_i when it is evaluated according to the decision criterion c_j ($i = 1, 2, \cdots n; j = 1, 2, \cdots m.$); the decision element d_{ij} is provided by the expert. It is also assumed that the expert has determined the weight of the criteria (denoted as $\omega_j, j = 1, 2, \cdots m$). There are three steps in utilizing the decision-making technique to rank the alternatives [32]: (1) Provide the relevant criteria and alternatives; (2) Collective information calculation; (3) Rank the alternative according to the collective information.

2.2. LTS

LTS is suitable for qualitative description of the decision-making problems, which can be defined as follows:

Definition 1. *Let $S = \{s_i | i = 0, 1, \cdots, 2t\}$ be a finite and totally ordered discrete linguistic term set, where s_i is a possible value for a linguistic variable, and t is a positive integer [33].*

(1) The LTS S satisfies the following properties:

(2) The set S is ordered: $s_i \le s_j$ if $i \le j$; $max(s_i, s_j) = s_i$ if $s_i \ge s_j$; $min(s_i, s_j) = s_i$ if $s_i \le s_j$;

(3) The negation operator is defined: $neg(s_i) = s_j$. satisfying with $i + j = 2t$.

In order to make the description of the given information more accurate, Xu [34] generalized the discrete linguistic term set S to the continuous linguistic term set $\overline{S} = \{s_i | i \in [0, \tau]\}$ $(\tau > 2t)$, where $s_i \leq s_j$ if $i \leq j$, and τ is a sufficiently large positive integer.

2.3. HFLTS

The HFLTS permits the membership of an element to be a set of several possible linguistic variable values. In the following, the concept of HFLTS and some related operations of HFLTS are reviewed.

Definition 2. *Given a fixed set X, let $S = \{s_i | i = 0, 1, \cdots, 2t\}$ be a LTS, then a HFLTS H_S on X is expressed by [8]:*

$$H_S = \{(x_j, h_S(x_j,)) | x_j \in X\}$$

where $h_S(x_j)$ is a subset of linguistic terms in S, it represents the membership degrees of the element x_j belongs to X. For convenience, the element of $h_S(x_j)$ is called the hesitant fuzzy linguistic element (HFLE).

Example 1. *Let $S = \{s_0 : very\ poor,\ s_1 : poor, s_2 : slightly\ poor,\ s_3 : fair, s_4\ :\ slightly\ good, s_5\ :\ good,\ s_6 : very\ good\}$ be a LTS. Two experts evaluate the performance of a company; one thinks the performance of a company is not less than good, the other thinks it is between fair and good. According to Definition 2, the above evaluation information can be represented as $H_S^1 = \{s_5,\ s_6\}$ and $H_S^2 = \{s_3, s_4, s_5\}$, respectively. The numbers of linguistic terms in H_S^1 and H_S^2 are not equal, which is not convenient for computing the similarity measure between H_S^1 and H_S^2.*

In order to solve this problem, for any two HFLTSs $H_S^1 = \{(x_j, h_S^1(x_j)) | x_j \in X\}$ and $H_S^2 = \{(x_j, h_S^2(x_j)) | x_j \in X\} (j = 1, 2, \cdots, n)$, where $h_S^k(x_j) = \{s_{\delta_l^k}(x_j) | s_{\delta_l^k}(x_j) \in S, l^k = 1, 2, \cdots, L_j^k\}$, if the numbers of $h_S^k(x_j)$ are not equal, we can let $L_j = max\{L_j^1, L_j^2\}$. Zhu et al. [35] proposed the rules of regulation: for the optimists, they extend the set with fewer numbers of elements by adding the maximum value $s_{\delta_l^k}^+(x_j) = \underset{l = L_j^1\ or\ l = L_j^2}{max} \{s_{\delta_l^k}(x_j)\}$ until the two sets have the same number of elements; while for the pessimists, they add the minimum value $s_{\delta_l^k}^-(x_j) = \underset{l = L_j^1\ or\ l = L_j^2}{min} \{s_{\delta_l^k}(x_j)\}$ to the set with fewer numbers of elements. In this paper, we assume that the largest element is added to the set with fewer elements until they have the same number.

The existing score function of HFLTSs is defined as follows:

Definition 3. *Let $S = \{s_i | i = 0, 1, \cdots, 2t\}$ be a LTS; $H_S = \{s_{\delta_l^k}(x_j) \big| s_{\delta_l^k}(x_j) \in S\}$ $(l = 1, 2, \cdots, L_j, j = 1, 2, \cdots,$ be a HFLTS on X, then the score function of H_S is [36]:*

$$F(H_S) = \frac{1}{n}\sum_{j=1}^{n} \overline{\delta}(x_j) - \frac{\sum_{j=1}^{n}\left(\frac{1}{L_j}\sum_{l=1}^{L_j}(\delta_l(x_j) - \overline{\delta}(x_j))^2\right)}{Var(2t)}$$

where $\overline{\delta} = \frac{1}{L_j}\sum_{l=1}^{L_j}\delta_l(x_j)$, $Var(2t) = \frac{\sum_{i=0}^{2t}(i-t)^2}{2t+1}$.

Lemma 1. *For two HFLTSs H_S^1 and H_S^2, the comparison rules between them are defined as follows [36]:*

(1) *$H_S^1 > H_S^2$ if and only if $F(H_S^1) > F(H_S^2)$;*
(2) *$H_S^1 = H_S^2$ if and only if $F(H_S^1) = F(H_S^2)$.*

2.4. Existing Distance and Similarity Measures Between HFLTSs

The Distance and similarity measure are effective tools for describing the deviation and closeness between different alternatives in MCDM problems; the definitions about the existing distance and similarity measures between HFLTSs are given as follows:

Definition 4. *Given a fixed set X, suppose that* $S = \{s_i | i = 0, 1, \cdots, 2t\}$ *be a LTS,* $h_S^k(x_j) = \{s_{\delta_l^k}(x_j) | s_{\delta_l^k}(x_j) \in S, l = 1, 2, \cdots, L_j\}$, *where* $L_j = max\{l(h_S^1(x_j)), l(h_S^2(x_j))\}$, $l(h_S^k(x_j))$ *represents the number of elements in* $h_S^k(x_j)$ $(k = 1, 2)$. *For any two HFLTSs* $H_S^1 = \{(x_j, h_S^1(x_j)) | x_j \in X\}$ *and* $H_S^2 = \{(x_j, h_S^2(x_j)) | x_j \in X\}$ $(j = 1, 2, \cdots, n)$, *assume the weight of different element* x_j *is* ω_j $(j = 1, 2, \cdots, n)$, *then the weighted Euclidean distance measure between* H_S^1 *and* H_S^2 *can be defined as follows [9]:*

$$D_{\omega HFL}\left(H_S^1, H_S^2\right) = \left(\sum_{j=1}^{n} \frac{\omega_j}{L_j} \sum_{l=1}^{L_j} \left(\frac{|\delta_l^1(x_j) - \delta_l^2(x_j)|}{2t+1}\right)^2\right)^{\frac{1}{2}}. \quad (1)$$

Remark 1. *For all* $j = 1, 2, \cdots, n$, *if the weight* $\omega_j = \frac{1}{n}$, *then the weighted Euclidean distance measure* $D_{\omega HFL}(H_S^1, H_S^2)$ *is reduced to the Euclidean distance measure* $D_{HFL}(H_S^1, H_S^2)$:

$$D_{HFL}\left(H_S^1, H_S^2\right) = \left(\frac{1}{n}\left(\sum_{j=1}^{n} \frac{1}{L_j} \sum_{l=1}^{L_j} \left(\frac{|\delta_l^1(x_j) - \delta_l^2(x_j)|}{2t+1}\right)^2\right)\right)^{\frac{1}{2}}$$

Liao et al. [31] defined a cosine similarity measure between HFLTSs as follows:

Definition 5. *Given a fixed set X, suppose that* $S = \{s_i | i = 0, 1, \cdots, 2t\}$ *is a LTS,* $h_S^k(x_j) = \{s_{\delta_l^k}(x_j) | s_{\delta_l^k}(x_j) \in S, l = 1, 2, \cdots, L_j\}$, *where* $L_j = max\{l(h_S^1(x_j)), l(h_S^2(x_j))\}$, $l(h_S^k(x_j))$ *represents the number of elements in* $h_S^k(x_j)$ $(k = 1, 2)$. *For any two HFLTSs* $H_S^1 = \{(x_j, h_S^1(x_j)) | x_j \in X\}$ *and* $H_S^2 = \{(x_j, h_S^2(x_j)) | x_j \in X\}$ $(j = 1, 2, \cdots, n)$, *if the weight of different element* x_j *is* ω_j $(j = 1, 2, \cdots, n)$, *then the weighted cosine similarity measure can be defined as [31]:*

$$Cos_{\omega HFL.}\left(H_S^1, H_S^2\right) = \frac{\sum_{j=1}^{n}\left(\frac{\omega_j}{L_j}\sum_{l=1}^{L_j}\left(\frac{\delta_l^1(x_j)}{2t+1}\cdot\frac{\delta_l^2(x_j)}{2t+1}\right)\right)}{\left(\sum_{j=1}^{n}\left(\frac{\omega_j}{L_j}\sum_{l=1}^{L_j}\left(\frac{\delta_l^1(x_j)}{2t+1}\right)^2\right)\cdot\sum_{j=1}^{n}\left(\frac{\omega_j}{L_j}\sum_{l=1}^{L_j}\left(\frac{\delta_l^2(x_j)}{2t+1}\right)^2\right)\right)^{\frac{1}{2}}} \quad (2)$$

Remark 2. *For all* $j = 1, 2, \cdots, n$, *if the weight* $\omega_j = \frac{1}{n}$, *then the weighted cosine similarity measure* $Cos_{\omega HFL}(H_S^1, H_S^2)$ *is reduced to the cosine similarity measure* $Cos_{HFL}(H_{S'}^1, H_{S'}^2)$:

$$Cos_{HFL.}\left(H_S^1, H_S^2\right) = \frac{\sum_{j=1}^{n}\left(\frac{1}{L_j}\sum_{l=1}^{L_j}\left(\frac{\delta_l^1(x_j)}{2t+1}\cdot\frac{\delta_l^2(x_j)}{2t+1}\right)\right)}{\left(\sum_{j=1}^{n}\left(\frac{1}{L_j}\sum_{l=1}^{L_j}\left(\frac{\delta_l^1(x_j)}{2t+1}\right)^2\right)\cdot\sum_{j=1}^{n}\left(\frac{1}{L_j}\sum_{l=1}^{L_j}\left(\frac{\delta_l^2(x_j)}{2t+1}\right)^2\right)\right)^{\frac{1}{2}}} \quad (3)$$

2.5. Linguistic Scale Function

In different semantic decision-making environments, linguistic terms have some differences in expressing alternatives. Bao et al. [37] thought that the information may be distorted when using the subscript of the linguistic term set directly in the process of operations. To solve this problem, Wang et al. [23] put forward the linguistic scale function to calculate the linguistic information.

According to the decision-making environment, the decision makers choose a different linguistic scale function, which can express the linguistic information more flexibly in different semantic situations.

Definition 6. *Let $S = \{s_i | i = 0, 1, \cdots, 2t\}$ be a LTS; if $\theta_i \in R^+ (R^+ = \{r | r \geq 0, r \in R\})$ is a real value, then the linguistic scale function f can be defined as follows [22]:*

$$f : s_i \to \theta_i \ (i = 0, 1, \cdots, 2t,)$$

where $0 \leq \theta_0 \leq \theta_1 \leq \cdots \leq \theta_{2t} \leq 1$. The linguistic scale function f is a strictly monotonically increasing function on the subscript of s_i. Actually, the function value θ_i represents the semantics of the linguistic terms.

Next we introduce three common linguistic scale functions as follows:

$$(1).\quad f_1(s_i) = \theta_i = \frac{i}{2t}\ (i = 0, 1, \cdots, 2t)_1).$$

$$(2).\quad f_2(s_i) = \theta_i = \begin{cases} \frac{a^t - a^{t-i}}{2a^t - 2}, & (i = 0, 1, \cdots, t); \\ \frac{a^t + a^{i-t} - 2}{2a^t - 2}, & (i = t+1, t+2, \cdots, 2t). \end{cases}$$

If the LTS is a set of seven terms, then $a \in [1.36, 1.4]$ [38]. In this paper, we assume that $a = 1.4$.

$$(3).\quad f_3(s_i) = \theta_i = \begin{cases} \frac{t^\alpha - (t-i)^\alpha}{2t^\alpha}, & (i = 0, 1, \cdots, t); \\ \frac{t^\beta - (t-i)^\beta}{2t^\beta}, & (i = t+1, t+2, \cdots, 2t), \end{cases}$$

where $\alpha, \beta \in (0, 1]$. If the LTS is a set of seven terms, then $\alpha = \beta = 0.8$ [39].

Example 2. *Assume that $S = \{s_i | i = 0, 1, \cdots, 2t\}$ be a LTS. When $t = 3$, the corresponding linguistic scale functions are $f_1(s_i)$, $f_2(s_i)$ $(a = 1.4)$, $f_3(s_i)$ $(\alpha = \beta = 0.8)$ respectively, and the characteristics of the three functions are shown in Figure 1.*

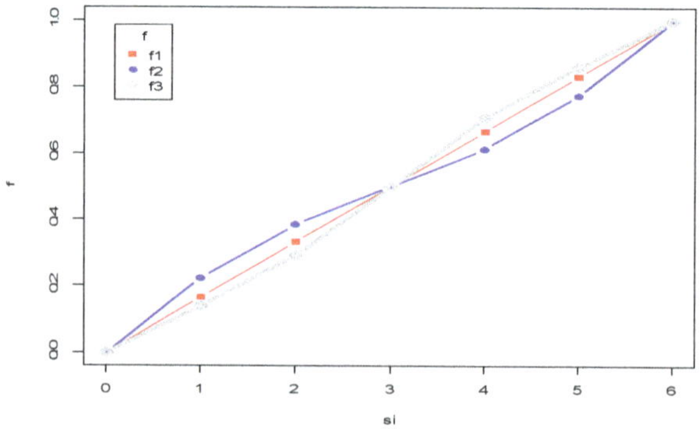

Figure 1. The change of the three linguistic scale functions.

Remark 3. *The linguistic scale function $f_1(s_i)$ can be explained as the decision maker's neutral attitude towards risk; the linguistic scale function $f_2(s_i)$ indicates that the decision maker's attitude towards risk is changing from aversion to appetite; the linguistic scale function $f_3(s_i)$ indicates that the decision maker's attitude towards risk is changing from appetite to aversion.*

3. The Score Function, Similarity Measure, and Distance Measure Between HFLTSs Based on a Linguistic Scale Function

In this section, we first propose the definition of a new score function of HFLTSs based on a linguistic scale function, then the new similarity measure and its properties are given. Furthermore, we construct a corresponding distance measure based on the relationship between the similarity measure and the distance measure.

3.1. The Score Function Between HFLTSs Based on the Linguistic Scale Function

Definition 7. *Let* $S = \{s_i | i = 0, 1, \cdots, 2t\}$ *be a LTS,* $H_S = \{s_{\delta_l^k}(x_j) | s_{\delta_l^k}(x_j) \in S\}$ $(l = 1, 2, \cdots, L_j, j = 1, 2, \cdots, n)$ *be the HFLTS on X, and f be a linguistic scale function, then the score function of* H_S *is defined as:*

$$F^*(H_S) = \frac{1}{n}\sum_{j=1}^{n}\overline{f}(s_{\delta_l}(x_j)) - \frac{\sum_{j=1}^{n}\left(\frac{1}{L_j}\sum_{l=1}^{L_j}\left(f(s_{\delta_l}(x_j)) - \overline{f}s_{\delta_l}(x_j)\right)^2\right)}{Var^*(2t)},$$

where $\overline{f}(s_{\delta_l}(x_j)) = \frac{1}{L_j}\sum_{l=1}^{L_j} f(s_{\delta_l}(x_j))$, $Var^*(2t) = \sum_{i=0}^{2t}(f(s_i) - f(s_t))^2$.

Theorem 1. *For any two HFLTSs* H_S^1 *and* H_S^2*, the comparison rules between them are defined as follows:*

(1) *If* $F^*(H_S^1) > F^*(H_S^2)$*, then* $H_S^1 > H_S^2$*;*
(2) *If* $F^*(H_S^1) = F^*(H_S^2)$*, then* $H_S^1 = H_S^2$*.*

Example 3. *Let* $S = \{s_0 : very\ poor,\ s_1 : poor,\ s_2 : slightly\ poor, s_3 : fair, s_4 : slightly\ good, s_5 : good,\ s_6 : very\ good\}$ *be a LTS, three HFLTSs are given as follows:* $H_S^1 = \{s_0, s_1, s_2\}$, $H_S^2 = \{s_2, s_3, s_4\}$ *and* $H_S^3 = \{s_0, s_2\}$*. By Definition 7, if the linguistic scale function* $f = f_1(s_i) = \frac{i}{2t}$ $(t = 3)$*, we obtain* $F^*(H_S^1) = 0.1429, F^*(H_S^2) = 0.4048, F^*(H_S^3) = 0.1310$*, then the ranking of the HFLTSs is* $H_S^2 > H_S^1 > H_S^3$*. By Definition 3, we can obtain* $F(H_S^1) = 0.8323, F(H_S^2) = 2.8333, F(H_S^3) = 0.75$*, according to Lemma 1, and it is clearly seen that* $H_S^2 > H_S^1 > H_S^3$*, which is same as the proposed score function in Theorem 1.*

3.2. The Similarity Measure Between HFLTSs Based on the Linguistic Scale Function

It is already known the regular similarity measure satisfies the following Lemma 2:

Lemma 2. *Let* $S = \{s_i | i = 0, 1, \cdots, 2t\}$ *be a LTS,* H_S^1 *and* H_S^2 *be any two HFLTSs; if the similarity measure* $S(H_S^1, H_S^2)$ *satisfies the following properties [9]:*

(1) $0 \le S(H_S^1, H_S^2) \le 1$,
(2) $S(H_S^1, H_S^2) = 1$ *if and only if* $H_S^1 = H_S^2$,
(3) $S(H_S^1, H_S^2) = S(H_S^2, H_S^1)$.

then the similarity measure $S(H_S^1, H_S^2)$ *is a regular similarity measure, and the corresponding distance measure* $D(H_S^1, H_S^2) = 1 - S(H_S^1, H_S^2)$.

The cosine similarity measure proposed by Liao et al. [31] is sometimes different from human intuition in practical decision-making problems, and we can determine this from the following Example 4.

Example 4. *When two experts evaluate the performance of a company, they provide their opinions with hesitant fuzzy linguistic information; for the given LTS,* $= \{s_0 : very\ poor,\ s_1 : poor, s_2 : slightly\ poor, s_3 :$

fair, s_4 : *slightly good*, s_5 : *good*, s_6 : *very good*$\}$, *and two experts' evaluations are represented as HFLTSs* $H_S^1 = \{s_1, s_2\}$ *and* $H_S^2 = \{s_2, s_4\}$, *respectively.*

It is already known $H_S^1 \neq H_S^2$, but from using Formula (3) to calculate the similarity measure between H_S^1 and H_S^2, we have $Cos_{HFL}(H_S^1, H_S^2) = 1$. That is to say, the property (2) in Lemma 2 is not satisfied. So, the similarity measure Cos_{HFL} proposed by Liao et al. [31] is not a regular similarity measure. On the other hand, the similarity measure Cos_{HFL} as defined by Liao et al. [31] used the subscript of linguistic terms directly in the process of operations; they did not consider the semantic environment, which may cause the loss of information in the decision process. In order to overcome its disadvantages, next we will construct a new similarity measure and derive a corresponding distance measure. A scheme of this process is shown in Figure 2.

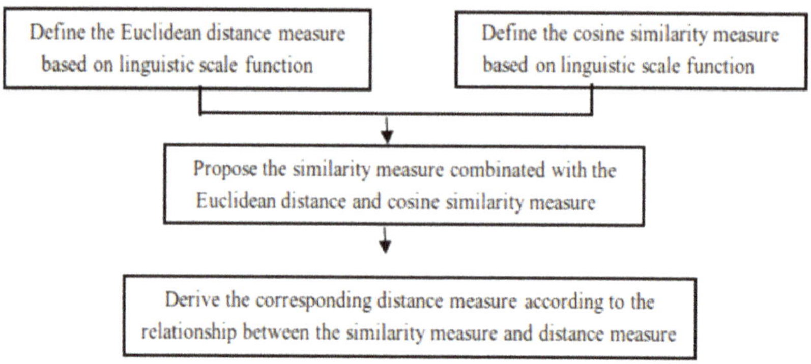

Figure 2. The scheme of the construction of the similarity measure.

At first, we improve the existing distance measure (1) and similarity measure (2) based on a linguistic scale function, which can be defined as follows:

Definition 8. *Given a fixed set X, let* $S = \{s_i | i = 0, 1, \cdots, 2t\}$ *be a LTS, and let f be a linguistic scale function,* $h_S^k(x_j) = \{s_{\delta_l^k}(x_j) \,|\, s_{\delta_l^k}(x_j) \in S, l = 1, 2, \cdots, L_j\}$, *where* $L_j = max\{l(h_S^1(x_j)), l(h_S^2(x_j))\}$, $l\left(h_S^k(x_j)\right)$ *represents the number of elements in* $h_S^k(x_j)$ $(k = 1, 2)$. *For any two HFLTSs* $H_S^1 = \{(x_j, h_S^1(x_j)) | x_j \in X\}$ *and* $H_S^2 = \{(x_j, h_S^2(x_j)) | x_j \in X\}$ $(j = 1, 2, \cdots, n)$, *if the weight of the different element* x_j *is* ω_j $(j = 1, 2, \cdots, n)$, *then the improved weighted distance measure between HFLTSs* H_S^1 *and* H_S^2 *can be defined as follows:*

$$D'_{\omega HFL}\left(H_S^1, H_S^2\right) = \left(\sum_{j=1}^{n} \frac{\omega_j}{L_j} \sum_{l=1}^{L_j} \left(\left|f\left(s_{\delta_l^1}(x_j)\right) - f\left(s_{\delta_l^2}(x_j)\right)\right|\right)^2\right)^{\frac{1}{2}}.$$

Theorem 2. *Let* H_S^1 *and* H_S^2 *be any two HFLTSs, and let f be a linguistic scale function; the distance measure* $D'_{\omega HFL}$ *between HFLTSs satisfies the following properties:*

(1) $0 \leq D'_{\omega HFL}(H_S^1, H_S^2) \leq 1$;
(2) $D'_{\omega HFL}(H_S^1, H_S^2) = 0$ *if and only if* $H_S^1 = H_S^2$;
(3) $D'_{\omega HFL}(H_S^1, H_S^2) = D'_{\omega HFL}(H_S^2, H_S^1)$.

Proof. Properties (1), (2), and (3) are obvious, and we omit the proof here. □

Remark 4. *For all $j = 1, 2, \cdots, n$, if the weight $\omega_j = \frac{1}{n}$, then the improved weighted distance measure $D'_{\omega HFL}(H_S^1, H_S^2)$ is reduced to the improved Euclidean distance measure $D'_{HFL}(H_S^1, H_S^2)$:*

$$D'_{HFL}\left(H_S^1, H_S^2\right) = \left(\frac{1}{n}\left(\sum_{j=1}^{n}\frac{1}{L_j}\sum_{l=1}^{L_j}\left(\left|f\left(s_{\delta_l^1}(x_j)\right) - f\left(s_{\delta_l^2}(x_j)\right)\right|\right)^2\right)\right)^{\frac{1}{2}}.$$

Definition 9. *Given a fixed set X, let $S = \{s_i | i = 0, 1, \cdots, 2t\}$ be a LTS, and let f be a linguistic scale function, $h_S^k(x_j) = \{s_{\delta_l^k}(x_j) | s_{\delta_l^k}(x_j) \in S, l = 1, 2, \cdots, L_j\}$, where $L_j = max\{l(h_S^1(x_j)), l(h_S^2(x_j))\}$, $l\left(h_S^k(x_j)\right)$ represents the number of elements in $h_S^k(x_j)$ $(k = 1, 2)$. For any two HFLTSs $H_S^1 = \{(x_j, h_S^1(x_j)) | x_j \in X\}$ and $H_S^2 = \{(x_j, h_S^2(x_j)) | x_j \in X\}$ $(j = 1, 2, \cdots, n)$, if the weight of different element x_j is ω_j $(j = 1, 2, \cdots, n)$, then the improved weighted cosine similarity measure between H_S^1 and H_S^2 can be defined as:*

$$Cos'_{\omega HFL}\left(H_S^1, H_S^2\right) = \frac{\sum_{j=1}^{n}\left(\frac{\omega_j}{L_j}\sum_{l=1}^{L_j} f\left(s_{\delta_l^1}(x_j)\right)\cdot f\left(s_{\delta_l^2}(x_j)\right)\right)}{\left(\sum_{j=1}^{n}\left(\frac{\omega_j}{L_j}\sum_{l=1}^{L_j}\left(f\left(s_{\delta_l^1}(x_j)\right)\right)^2\right)\cdot \sum_{j=1}^{n}\left(\frac{\omega_j}{L_j}\sum_{l=1}^{L_j}\left(f\left(s_{\delta_l^2}(x_j)\right)\right)^2\right)\right)^{\frac{1}{2}}}.$$

Remark 5. *For all $j = 1, 2, \cdots, n$, if the weight $\omega_j = \frac{1}{n}$, then the improved weighted cosine similarity measure $Cos'_{\omega HFL}(H_S^1, H_S^2)$ is reduced to the similarity measure $Cos'_{HFL}(H_S^1, H_S^2)$:*

$$Cos'_{HFL}\left(H_S^1, H_S^2\right) = \frac{\sum_{j=1}^{n}\left(\frac{1}{L_j}\sum_{l=1}^{L_j} f\left(s_{\delta_l^1}(x_j)\right)\cdot f\left(s_{\delta_l^2}(x_j)\right)\right)}{\left(\sum_{j=1}^{n}\left(\frac{1}{L_j}\sum_{l=1}^{L_j}\left(f\left(s_{\delta_l^1}(x_j)\right)\right)^2\right)\cdot \sum_{j=1}^{n}\left(\frac{1}{L_j}\sum_{l=1}^{L_j}\left(f\left(s_{\delta_l^2}(x_j)\right)\right)^2\right)\right)^{\frac{1}{2}}}.$$

In the following, we go on to propose a similarity measure between the HFLTSs, which combine the distance measure D'_{HFL} and the cosine similarity measure Cos'_{HFL}.

Definition 10. *Given a fixed set X, let $S = \{s_i | i = 0, 1, \cdots, 2t\}$ be a LTS, and let f be a linguistic scale function, $h_S^k(x_j) = \{s_{\delta_l^k}(x_j) | s_{\delta_l^k}(x_j) \in S, l = 1, 2, \cdots, L_j\}$, where $L_j = max\{l(h_S^1(x_j)), l(h_S^2(x_j))\}$, $l\left(h_S^k(x_j)\right)$ represents the number of elements in $h_S^k(x_j)$ $(k = 1, 2)$. For any two HFLTSs $H_S^1 = \{(x_j, h_S^1(x_j)) | x_j \in X\}$ and $H_S^2 = \{(x_j, h_S^2(x_j)) | x_j \in X\}$ $(j = 1, 2, \cdots, n)$, then the new similarity measure $S^*_{HFL}(H_S^1, H_S^2)$ can be defined as follows:*

$$S^*_{HFL,}\left(H_S^1, H_S^2\right) = \frac{1}{2}\left(Cos'_{HFL}\left(H_S^1, H_S^2\right) + 1 - D'_{HFL}\left(H_S^1, H_S^2\right)\right)$$

where $Cos'_{HFL}(H_S^1, H_S^2) = \frac{\sum_{j=1}^{n}\left(\frac{1}{L_j}\sum_{l=1}^{L_j} f\left(s_{\delta_l^1}(x_j)\right)\cdot f\left(s_{\delta_l^2}(x_j)\right)\right)}{\left(\sum_{j=1}^{n}\left(\frac{1}{L_j}\sum_{l=1}^{L_j}\left(f\left(s_{\delta_l^1}(x_j)\right)\right)^2\right)\cdot \sum_{j=1}^{n}\left(\frac{1}{L_j}\sum_{l=1}^{L_j}\left(f\left(s_{\delta_l^2}(x_j)\right)\right)^2\right)\right)^{\frac{1}{2}}}$, $D'_{HFL}(H_S^1, H_S^2) = \left(\frac{1}{n}\left(\sum_{j=1}^{n}\frac{1}{L_j}\sum_{l=1}^{L_j}\left(\left|f\left(s_{\delta_l^1}(x_j)\right) - f\left(s_{\delta_l^2}(x_j)\right)\right|\right)^2\right)\right)^{\frac{1}{2}}$.

Theorem 3. *The similarity measure $S^*(H_S^1, H_S^2)$ is a regular similarity measure.*

Proof. According to Lemma 2, we will prove it by three steps as follows:

(1) Since $0 \leq f \leq 1, Cos'_{HFL}(H_S^1, H_S^2)$ can be considered as the extension of cosine function, then $0 \leq Cos'_{HFL}(H_S^1, H_S^2) \leq 1$. According to Theorem 2, we know that $D'_{HFL}(H_S^1, H_S^2)$ is a distance measure, then $0 \leq 1 - D'_{HFL}(H_S^1, H_S^2) \leq 1$. Thus, we get $0 \leq Cos'_{HFL}(H_S^1, H_S^2) + 1 - D'_{HFL}(H_S^1, H_S^2) \leq 2$, so $0 \leq S^*_{HFL}(H_S^1, H_S^2) \leq 1$ is obvious.
(2) If $H_S^1 = H_S^2$, we have $f(s_{\delta_l^1}(x_j)) = f(s_{\delta_l^2}(x_j))$, $Cos'_{HFL}(H_S^1, H_S^2) = 1$, $D'_{HFL}(H_S^1, H_S^2) = 0$, then $S^*_{HFL}(H_S^1, H_S^2) = 1$. On the other hand, when $S^*_{HFL}(H_S^1, H_S^2) = 1$, we have $Cos'_{HFL}(H_S^1, H_S^2) + 1 - D'_{HFL}(H_S^1, H_S^2) = 2$; that is, $Cos'_{HFL}(H_S^1, H_S^2) = 1 + D'_{HFL}(H_S^1, H_S^2)$. Because $0 \leq Cos'_{HFL}(H_S^1, H_S^2) \leq 1$, $D'_{HFL}(H_S^1, H_S^2) \geq 0$ hold simultaneously, then we have $D'_{HFL}(H_S^1, H_S^2) = 0$, $Cos'_{HFL}(H_S^1, H_S^2) = 1$. When $Cos'_{HFL}(H_S^1, H_S^2) = 1$, we know that $H_S^1 = kH_S^2$ and k is a constant; while $D'_{HFL}(H_S^1, H_S^2) = 0$, we know that $H_S^1 = H_S^2$. That is to say, when $S^*_{HFL}(H_S^1, H_S^2) = 1$, $H_S^1 = H_S^2$. Thus, $S^*_{HFL}(H_S^1, H_S^2) = 1$ if and only if $H_S^1 = H_S^2$.
(3) According to Remark 5, $Cos'_{HFL}(H_S^1, H_S^2) = Cos'_{HFL}(H_S^2, H_S^1)$ is obvious. From Theorem 2, it is already known that when $D'_{HFL}(H_S^1, H_S^2) = D'_{HFL}(H_S^2, H_S^1)$, then $S^*_{HFL}(H_S^1, H_S^2) = S^*_{HFL}(H_S^2, H_S^1)$ are proven. □

From Theorem 3, we know that the proposed similarity measure S^*_{HFL} is a regular similarity measure, which overcomes the disadvantages of the similarity measure as defined by Liao et al. [31].

Remark 6. *According to the relation between the distance measure and the regular similarity measure, we can obtain a new distance measure $D^*_{HFL}(H_S^1, H_S^2)$, which is based on the proposed similarity measure $S^*_{HFL}(H_S^1, H_S^2)$:*

$$D^*_{HFL}(H_S^1, H_S^2) = 1 - S^*_{HFL}(H_S^1, H_S^2) = \frac{1}{2}(1 - Cos'_{HFL}(H_S^1, H_S^2) + D'_{HFL}(H_S^1, H_S^2))$$

$$= \frac{1}{2}\left(1 - \frac{\sum_{j=1}^{n}\left(\frac{1}{L_j}\sum_{l=1}^{L_j} f\left(s_{\delta_l^1}(x_j)\right)\cdot f\left(s_{\delta_l^2}(x_j)\right)\right)}{\left(\sum_{j=1}^{n}\left(\frac{1}{L_j}\sum_{l=1}^{L_j}\left(f\left(s_{\delta_l^1}(x_j)\right)\right)^2\right)\cdot\sum_{j=1}^{n}\left(\frac{1}{L_j}\sum_{l=1}^{L_j}\left(f\left(s_{\delta_l^2}(x_j)\right)\right)^2\right)\right)^{\frac{1}{2}}} + \left(\frac{1}{n}\left(\sum_{j=1}^{n}\frac{1}{L_j}\sum_{l=1}^{L_j}\left(\left|f\left(s_{\delta_l^1}(x_j)\right) - f(s_{\delta_l^2}(x_j))\right|\right)^2\right)\right)^{\frac{1}{2}}\right).$$

Theorem 4. *The new distance measure $D^*_{HFL}(H_S^1, H_S^2)$ satisfies the following properties:*

(1) $0 \leq D^*_{HFL}(H_S^1, H_S^2) \leq 1$;
(2) $D^*_{HFL}(H_S^1, H_S^2) = 0$ *if and only if* $H_S^1 = H_S^2$;
(3) $D^*_{HFL}(H_S^1, H_S^2) = D^*_{HFL}(H_S^2, H_S^1)$.

Proof. Properties (1) and (3) are obvious, here we only present the proof of property (2).
If $H_S^1 = H_S^2$, we have $S^*_{HFL}(H_S^1, H_S^2) = 1$, then $D^*_{HFL}(H_S^1, H_S^2) = 1 - S^*_{HFL}(H_S^1, H_S^2) = 0$. On the other hand, when $D^*_{HFL}(H_S^1, H_S^2) = 0$, we have $S^*_{HFL}(H_S^1, H_S^2) = 1 - D^*_{HFL}(H_S^1, H_S^2) = 1$. Because S^*_{HFL} is a regular similarity measure, according to Lemma 2, we have $H_S^1 = H_S^2$.
Thus, we obtain $D^*_{HFL}(H_S^1, H_S^2) = 0$ if and only if $H_S^1 = H_S^2$. □

Definition 11. *Given a fixed set X, let $S = \{s_i | i = 0, 1, \cdots, 2t\}$ be a LTS, and let f be a linguistic scale function. $h_S^k(x_j) = \{s_{\delta_l^k}(x_j) \big| s_{\delta_l^k}(x_j) \in S, l = 1, 2, \cdots, L_j\}$, where $L_j = max\{l(h_S^1(x_j)), l(h_S^2(x_j))\}$, $l\left(h_S^k(x_j)\right)$ represents the number of elements in $h_S^k(x_j)$ $(k = 1, 2)$. For any two HFLTSs $H_S^1 = \{(x_j, h_S^1(x_j)) | x_j \in X\}$ and $H_S^2 = \{(x_j, h_S^2(x_j)) | x_j \in X\}$ $(j = 1, 2, \cdots, n)$, the associated weighting vector $\omega = (\omega_1, \omega_2, \cdots, \omega_j)$ satisfying with $\sum_{j=1}^{n}\omega_j = 1$ $(0 \leq \omega_j \leq 1)$, then the weighted similarity measure between H_S^1 and H_S^2 can be defined as:*

$$S^*_{\omega HFL}\left(H_S^1, H_S^2\right) = \frac{1}{2}\left(Cos'_{\omega HFL}\left(H_S^1, H_S^2\right) + 1 - D'_{\omega HFL}\left(H_S^1, H_S^2\right)\right)$$

Theorem 5. *The weighted similarity measure $S^*_{\omega HFL}(H_S^1, H_S^2)$ is also a regular similarity measure.*

Proof. The proof is similar to Theorem 3; we omit it here. □

Remark 7. *If the weight of the different element* x_j *is* ω_j $(j = 1, 2, \cdots, n)$, *satisfying* $\sum_{j=1}^{n} \omega_j = 1$ $(0 \le \omega_j \le 1)$, *then the weighted distance measure between* H_S^1 *and* H_S^2 *can be obtained by:*

$$D^*_{\omega HFL}\left(H_S^1, H_S^2\right) = 1 - S^*_{\omega HFL}\left(H_S^1, H_S^2\right)$$

Remark 8. *If we take the weight* $\omega_j = \frac{1}{n}$ $(j = 1, 2, \cdots, n)$ *in* $S^*_{\omega HFL}(H_S^1, H_S^2)$ *and* $D^*_{\omega HFL}(H_S^1, H_S^2)$, *then* $S^*_{\omega HFL}(H_S^1, H_S^2)$ *and* $D^*_{\omega HFL}(H_S^1, H_S^2)$ *are reduced to* $S^*_{HFL}(H_S^1, H_S^2)$ *and* $D^*_{HFL}(H_S^1, H_S^2)$, *respectively.*

Next, we utilize the medical diagnosis example to illustrate the application of the proposed similarity measure.

Example 5. *In traditional Chinese medical diagnosis, doctors diagnose patients by watching, smelling, asking and touching, so the doctor always get some imprecise information about patients' symptoms. Let us consider a set of diagnoses* $G = \{\ Viral\ fever,\ Typhoid,\ Pneumonia, Stomach\ problem\}$ *and a set of symptoms* $X = \{\ temperature,\ headache,\ cough,\ stomach\ pain\}$. *Assume that a patient, with respect to all symptoms, can be depicted as the following LTS, respectively:* $S_1 = \{\ s_0 : very\ low,\ s_1 : low,\ s_2 : slightly\ low,\ s_3 : normal, s_4 : slightly\ high,\ s_5 : high, s_6 : very\ high\}, S_j = \{s_0 : none, s_1 : very\ slight, s_2 : slight,\ s_3 : normal,\ s_4 : slightly\ terrible, s_5 : terrible, s_6 : very\ terrible\}$ $(j = 2, 3, 4)$. *Furthermore, let* $\omega_j = (0.25, 0.25, 0.25, 0.25)$ $(j = 1, 2, 3, 4)$ *be the weight vector of symptoms.*

Suppose that the patient $P = \{Richard, Catherine, Nicle, Kevin\}$ has all of the symptoms, which are represented by a HFLTS and are given in Table 1.

Table 1. Symptoms characteristic for the patients.

	Viral Fever	Typhoid	Pneumonia	Stomach Problem
Richard	$\{s_5\}$	$\{s_5\}$	$\{s_4,s_5\}$	$\{s_0\}$
Catherine	$\{s_3\}$	$\{s_0\}$	$\{s_0\}$	$\{s_4,s_5\}$
Nicle	$\{s_6\}$	$\{s_4\}$	$\{s_5\}$	$\{s_0\}$
Kevin	$\{s_4\}$	$\{s_2,s_3\}$	$\{s_5\}$	$\{s_0\}$

According to experience, each patient's symptoms diagnosis can be viewed as a HFLTS, and these are shown in Table 2.

Table 2. Symptoms characteristic for the diagnosis.

	Viral Fever	Typhoid	Pneumonia	Stomach Problem
Richard	$\{s_4,s_5,s_6\}$	$\{s_3,s_4,s_5\}$	$\{s_4,s_5,s_6\}$	$\{s_0\}$
Catherine	$\{s_5,s_6\}$	$\{s_1,s_2,s_3\}$	$\{s_4,s_5,s_6\}$	$\{s_0,s_1\}$
Nicle	$\{s_3,s_4\}$	$\{s_2,s_3\}$	$\{s_5,s_6\}$	$\{s_0\}$
Kevin	$\{s_3\}$	$\{s_0\}$	$\{s_0\}$	$\{s_4,s_5,s_6\}$

In order to diagnose what kind of symptoms that the patients belong to, we can calculate the similarity measure between each patient's symptoms and the diagnosis. If the linguistic scale function $f = f_1(s_i) = \frac{i}{2t}$ $(t = 3)$, we apply the proposed similarity measure $S^*_{\omega HFL}$ to calculate the degree of similarity between each patient's symptoms and the diagnosis; the results are shown in Table 3.

Table 3. Hesitant fuzzy linguistic similarity measure.

	Viral Fever	Typhoid	Pneumonia	Stomach Problem
Richard	0.9574	0.9702	0.8973	0.8606
Catherine	0.7865	0.7856	0.8323	0.9954
Nicle	0.9558	0.9517	0.8846	0.7741
Kevin	0.9279	0.8728	0.9814	0.8370

It is already known that the larger value of similarity measure, the higher the possibility of diagnosis for the patient. From the above results of Table 3, the symptoms of Richard, Catherine, Nicole, and Kevin indicate that they are suffering from typhoid, stomach problems, viral fever, and pneumonia, respectively.

4. The TOPSIS Method with the Proposed Distance Measure $D^*_{\omega HFL}$

In Section 4, we will present the TOPSIS method [40] to the proposed distance measure $D^*_{\omega HFL}$ for hesitant fuzzy linguistic multi-criteria decision-making problems.

Suppose that a panel of decision makers are invited to evaluate the alternatives $H = \{H_1, H_2, \cdots, H_m\}$ with respect to the criteria $C = \{C_1, C_2, \cdots, C_n\}$. Let $S = \{s_i | i = 0, 1, \cdots, 2t\}$ be a LTS, let ω_j $(j = 1, 2, \cdots, n)$ be the weight of criteria C_j, where $0 \le \omega_j \le 1$ $(j = 1, 2, \cdots, n)$ and $\sum_{j=1}^{n} \omega_j = 1$; the hesitant fuzzy linguistic information decision matrix H are given as follows:

$$H = \begin{pmatrix} H_S^{11} & H_S^{12} & \cdots & H_S^{1n} \\ H_S^{21} & H_S^{22} & \cdots & H_S^{2n} \\ \vdots & \vdots & \ddots & \vdots \\ H_S^{m1} & H_S^{m2} & \cdots & H_S^{mn} \end{pmatrix},$$

where $H_S^{ij} = \{s_{\delta_l}^{ij} \big| l = 1, 2, \cdots, L_j\} (i = 1, 2, \cdots, m; j = 1, 2, \cdots, n)$ are HFLTSs, representing the evaluation about alternative H_i with respect to the criterion C_j.

Next, we present the TOPSIS method with the distance measure $D^*_{\omega HFL}$ for MCDM problems. In general, it includes the following steps:

Step 1. Normalize the hesitant fuzzy linguistic decision matrix H.

If the criteria belong to the benefit-type, we need not do anything; if the criteria belong to the cost-type, we should use $neg(s_i) = s_j (i + j = 2t)$ to normalize the decision matrix.

Step 2. For $i = 1, 2, \cdots, m, j = 1, 2, \cdots, n$, the hesitant fuzzy linguistic positive ideal solution (HFLPIS) $H^+ = \{H_S^{1+}, H_S^{2+}, \cdots, H_S^{n+}\}$ and hesitant fuzzy linguistic negative ideal solution (HFLNIS) $\mathrm{H}^- = \{\mathrm{H}_S^{1-}, \mathrm{H}_S^{2-}, \cdots, \mathrm{H}_S^{n-}\}$ are given in the following:

$$H_S^{j+} = H_S^{ij}, \; H_S^{j-} = H_S^{ij}.$$

For criteria $C_j (j = 1, 2, \cdots, n)$, by the score function proposed in Definition 7, we can get the value of $F^*(H_S^{ij})$ $(i = 1, 2, \cdots, m)$. According to Theorem 1, the order relationship for HFLTSs can be given as: if $F^*(H_S^1) > F^*(H_S^2)$, then $H_S^1 > H_S^2$, so that H_S^{j+} and H_S^{j-} can be obtained.

Step 3. Use the distance measure to calculate the separation of each alternative between the HFLPIS $H^+ = \{H_S^{1+}, H_S^{2+}, \cdots, H_S^{n+}\}$ and HFLNIS $H^- = \{H_S^{1-}, H_S^{2-}, \cdots, H_S^{n-}\}$, respectively.

The distance measure between $H_i (i = 1, 2, \cdots, m)$ and H^+ can be given as: $D_i^+ = \sum_{j=1}^{n} D^*_{\omega HFL}(H_S^{ij}, H^+)$. Similarly to the distance measure D_i^+, the distance measure between the alternative $H_i (i = 1, 2, \cdots, m)$ and H^- is obtained as: $D_i^- = \sum_{j=1}^{n} D^*_{\omega HFL}(H_S^{ij}, H^-)$.

For the given H_i $(i = 1, 2, \cdots, m)$,

$$D_i^+ = \sum_{j=1}^n D^*_{\omega HFL}(H_S^{ij}, H^+) = \sum_{j=1}^n (1 - \tfrac{1}{2}(Cos'_{\omega HFL}(H_S^{ij}, H^+) + 1 - D'_{\omega HFL}(H_S^{ij}, H^+))) = \sum_{j=1}^n \tfrac{1}{2}(1-$$
$$Cos'_{\omega HFL}\left(H_S^{ij}, H^+\right) + D'_{\omega HFL}\left(H_S^{ij}, H^+\right)\,) = \sum_{j=1}^n \tfrac{1}{2}\left(1 - \frac{\sum_{j=1}^n \left(\frac{1}{L_j}\sum_{l=1}^{L_j} f\left(s_{\delta_l^{ij}}(x_j)\right)\cdot f\left(s_{\delta_l^+}(x_j)\right)\right)}{\left(\sum_{j=1}^n\left(\frac{1}{L_j}\sum_{l=1}^{L_j}\left(f\left(s_{\delta_l^{ij}}(x_j)\right)\right)^2\right)\cdot\sum_{j=1}^n\left(\frac{1}{L_j}\sum_{l=1}^{L_j}\left(f\left(s_{\delta_l^+}(x_j)\right)\right)^2\right)\right)^{\frac{1}{2}}} + \right.$$
$$\left.\left(\sum_{j=1}^n \frac{\omega_j}{L_j}\sum_{l=1}^{L_j}\left(\left|f\left(s_{\delta_l^{ij}}(x_j)\right) - f\left(s_{\delta_l^+}(x_j)\right)\right|\right)^2\right)^{\frac{1}{2}}\right);$$

$$D_i^- = \sum_{j=1}^n D^*_{\omega HFL}\left(H_S^{ij}, H^-\right) = \sum_{j=1}^n\left(1 - \tfrac{1}{2}\left(Cos'_{\omega HFL}\left(H_S^{ij}, H^-\right) + 1 - D'_{\omega HFL}\left(H_S^{ij}, H^-\right)\right)\right) = \sum_{j=1}^n \tfrac{1}{2}(1-$$
$$Cos'_{\omega HFL}\left(H_S^{ij}, H^-\right) + D'_{\omega HFL}\left(H_S^{ij}, H^-\right)\,) = \sum_{j=1}^n \tfrac{1}{2}\left(1 - \frac{\sum_{j=1}^n \left(\frac{1}{L_j}\sum_{l=1}^{L_j} f\left(s_{\delta_l^{ij}}(x_j)\right)\cdot f\left(s_{\delta_l^-}(x_j)\right)\right)}{\left(\sum_{j=1}^n\left(\frac{1}{L_j}\sum_{l=1}^{L_j}\left(f\left(s_{\delta_l^{ij}}(x_j)\right)\right)^2\right)\cdot\sum_{j=1}^n\left(\frac{1}{L_j}\sum_{l=1}^{L_j}\left(f\left(s_{\delta_l^-}(x_j)\right)\right)^2\right)\right)^{\frac{1}{2}}} + \right.$$
$$\left.\left(\sum_{j=1}^n \frac{\omega_j}{L_j}\sum_{l=1}^{L_j}\left(\left|f\left(s_{\delta_l^{ij}}(x_j)\right) - f\left(s_{\delta_l^-}(x_j)\right)\right|\right)^2\right)^{\frac{1}{2}}\right).$$

Step 4. Calculate the closeness coefficient Φ_i of each alternative $H_i (i = 1, 2, \cdots, m)$:

$$\Phi_i = \frac{D_i^-}{D_i^+ + D_i^-}.$$

Step 5. Rank the alternatives by decreasing order of the closeness coefficient Φ_i; the greater value Φ_i is, the better alternative H_i will be.

5. Numerical Example

In this section, we give a numerical example that concerns logistics outsourcing (adapted from Wang et al. [38]) to illustrate the feasibility of the TOPSIS method with the proposed distance measure $D^*_{\omega HFL}$.

5.1. Background

The ABC Limited Company is a passenger car manufacturer in China. To improve the competitiveness of products and reduce production costs, ABC decides to choose a third-party logistics service provider for logistics outsourcing. Through preliminary selection, five possible logistics providers $H = \{H_1, H_2, H_3, H_4, H_5\}$ are provided for further evaluation with respect to the following four criteria: service (C_1), relationship (C_2), quality (C_3), and equipment systems (C_4). Furthermore, assume that the weight vector of criteria C_j (j =1, 2, 3, 4) is $\omega = (0.4, 0.3, 0.2, 0.1)$. Three experts with different backgrounds are invited by the company to evaluate the TPLSP. Since these criteria are all qualitative, it is suitable for the experts to express their views in linguistic term sets. The ABC Company uses a LTS of seven terms to evaluate the TPLSP, which can be expressed by $S = \{s_0 : \textit{very poor}, s_1 : \textit{poor},\ s_2 : \textit{slightly poor}, s_3 : \textit{fair}, s_4 : \textit{slightly good},\ s_5 : \textit{good}, s_6 : \textit{very good}\}$. The final judgment of the five providers with the hesitant fuzzy linguistic decision matrix $H = (H_S^{ij})_{5\times 4}$ are given in Table 4.

To verify the feasibility and effectiveness of the decision method proposed in Section 4, at first, we assume $f = f_1(s_i) = \frac{i}{2t}$ $(t = 3)$.

Table 4. The hesitant fuzzy linguistic decision matrix provided by experts.

	C_1	C_2	C_3	C_4
H_1	$\{s_2,s_3,s_4\}$	$\{s_3,s_6\}$	$\{s_4,s_6\}$	$\{s_0,s_1,s_2\}$
H_2	$\{s_3,s_4\}$	$\{s_4,s_6\}$	$\{s_0,s_1\}$	$\{s_1,s_4\}$
H_3	$\{s_0,s_1\}$	$\{s_4\}$	$\{s_0,s_1,s_3\}$	$\{s_2\}$
H_4	$\{s_5\}$	$\{s_1,s_3\}$	$\{s_4,s_6\}$	$\{s_0,s_1,s_4\}$
H_5	$\{s_4,s_5\}$	$\{s_2,s_3\}$	$\{s_1,s_3,s_4\}$	$\{s_0,s_2\}$

Step 1. Normalize the hesitant fuzzy linguistic decision matrix.

It is already known the criteria C_1, C_2, C_3, C_4 are all benefit-type criteria, and thus we do not need to do anything.

Step 2. According to the score function in Theorem 1, we can calculate the HFLPIS H^+ and the HFLNIS H^-, which are given as follows:

$$H^+ = \{\{s_5\},\{s_4, s_6\},\{s_4, s_6\},\{s_1, s_4\}\}$$

$$H^- = \{\{s_0, s_1\},\{s_1, s_3\},\{s_0, s_1, s_3\},\{s_0, s_2\}\}$$

Step 3. Calculate the distance measure $D^*_{\omega HFL}(H_S^{ij}, H^+)$ and $D^*_{\omega HFL}(H_S^{ij}, H^-)$ for different alternative $H_i(i = 1,2,3,4,5)$ respectively, which are given in Table 5.

Table 5. The distance measure of each alternative.

	D_i^+	D_i^-
H_1	0.1524	0.2907
H_2	0.2395	0.3169
H_3	0.4201	0.1694
H_4	0.1753	0.4624
H_5	0.1884	0.3740

Step 4. Calculate the closeness coefficient Φ_i of each alternative H_i; they are obtained in Table 6.

Table 6. The closeness coefficient of each alternative.

	H_1	H_2	H_3	H_4	H_5
Φ_i	0.6561	0.5695	0.2873	0.7251	0.6650

Step 5. Rank the alternatives H_i and utilize Φ_i $(i = 1, 2, 3, 4, 5)$.

It is already known that $H_4 \succ H_5 \succ H_1 \succ H_2 \succ H_3$, which means that the best choice is H_4.

In order to illustrate the impact of the linguistic scale function f on MCDM, we use the different linguistic scale functions $f = f_2(s_i)$ $(a = 1.4, t = 3)$ and $f = f_3(s_i)$ $(\alpha = \beta = 0.8)$ to calculate the distance measure between HFLTSs, the results are given in Table 7.

Table 7. Results obtained by different linguistic scale functions.

	Ranking
$f = f_1(s_i)$	$H_4 \succ H_5 \succ H_1 \succ H_2 \succ H_3$
$f = f_2(s_i)$	$H_4 \succ H_1 \succ H_5 \succ H_2 \succ H_3$
$f = f_3(s_i)$	$H_4 \succ H_5 \succ H_1 \succ H_2 \succ H_3$

The results between the different linguistic scale functions are shown in Figure 3.

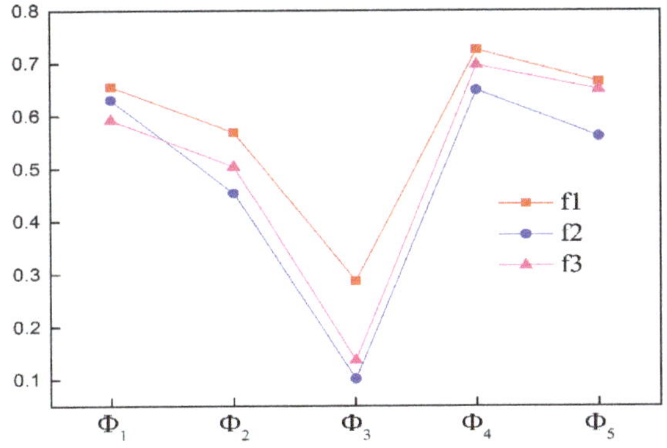

Figure 3. The differences between the different linguistic scale functions.

5.2. Comparison Analysis

To illustrate the feasibility and effectiveness of the proposed method, different approaches are used to compare with the same numerical example. The comparison is displayed in Table 8.

From Table 8, we know that the optimal alternative obtained by the proposed method is H_4; it is same as Liao et al. [31], Wang et al. [38], and Zhang et al. [41], which illustrates the feasibility and effectiveness of the proposed decision method.

Table 8. Comparison of different methods.

	Ranking
Approach from Liao et al. [9]	$H_2 \succ H_1 \succ H_4 \succ H_5 \succ H_3$
Approach from Liao et al. [31]	$H_4 \succ H_5 \succ H_1 \succ H_2 \succ H_3$
Approach from Wang et al. [38]	$H_4 \succ H_1 \succ H_2 \succ H_5 \succ H_3$
Approach from Zhang et al. [41]	$H_4 \succ H_1 \succ H_2 \succ H_5 \succ H_3$
Proposed approach based on $D^*_{\omega HFL}$	$H_4 \succ H_5 \succ H_1 \succ H_2 \succ H_3$

In Liao et al. [9], we can see the best alternative is different from other methods. The reason is that the approach from Liao et al. [9] only considers the algebraic relations of two HFLTSs, and they use the subscript of the linguistic terms directly in the process of operations, which may cause the loss of decision information. The method proposed in this paper is superior to the method in Liao et al. [9] for considering the distance measure, not only from the point of view of algebra, but also from the point of view of geometry.

Furthermore, in the MCDM method proposed by Liao et al. [31], the cosine similarity measure defined by them is not a regular similarity measure, as it cannot precisely deal with the hesitant fuzzy linguistic information that the subscripts of two linguistic terms have in the linear relationship, so that the result obtained in Liao et al. [31] seems unreliable. The proposed similarity measure combining the existing cosine similarity measure and the Euclidean distance measure overcomes this disadvantage; it can improve the accuracy of calculations to some extent, and it appears that the similarity measure that is proposed in this paper outperforms the existing similarity measure of HFLTSs.

In Wang et al. [38], the ranking results are a little different from the proposed method. Because the TODIM method in Wang et al. [38] has complicated parameters, the parameters selected by the expert will affect the ranking results. The proposed approach in this paper is capable of expressing the fuzzy linguistic information more flexibly; it can improve the adaptability of HFLTSs in practice.

In the method proposed by Zhang et al. [41], the evaluation values of each provider were aggregated independently. Because the best evaluation information under one criterion were usually offset by the worst evaluation information under another criterion in the process of aggregation, this may cause the decision information to be distorted. Compared with the method in Zhang et al. [41], the proposed method takes notice of the differences between different alternatives, and it is more meaningful in representing practical examples.

According to the results of comparative analysis, the benefits and advantages of this approach can be given in the following:

(1) The distance measure $D^*_{\omega HFL}$ is derived from the cosine function and the Euclidean distance measure; it considers the distance measure not only from the point of view of algebra, but also from the point of view of geometry. It shows a better performance when the subscripts of the linguistic term sets in the two HFLTS have the linear relationship.
(2) The similarity measure $S^*_{\omega HFL}$ and distance measure $D^*_{\omega HFL}$ based on the linguistic scale function can express information better under different circumstances, and the decision makers can select the appropriate linguistic scale function f on the basis of their preferences. It also can be applied more widely in the decision-making field than the existing distance measure and cosine similarity measure.
(3) The proposed method focuses on the differences of each alternative, which can improve the effectiveness of solving MCDM problems.

6. Conclusions

The similarity measure and distance measure are widely used in MCMD problems. Considering that the cosine similarity measure proposed by Liao et al. [31] is not a regular similarity measure, a new similarity measure combining the existing cosine similarity measure [31] and the Euclidean distance measure of HFLTSs is proposed, which is constructed based on the linguistic scale function. The proposed similarity measure in this paper considers the distance measure, not only from the point of view of algebra, but also from the point of view of geometry, and it also satisfies the axiom of the similarity measure. As far as we know, the new similarity measure between HFLTSs can express the fuzzy linguistic information more flexibly, which can improve the adaptability of HFLTSs in practice. Furthermore, the TOPSIS method with the corresponding distance measure is developed, and it focuses on the differences for each alternative, which can improve the effectiveness of solving MCDM problems. Finally, a numerical example is given to demonstrate the feasibility and the effectiveness of the proposed method, which is compared to the existing methods. In future research, efforts are continued to find other applications of the proposed similarity measure in the fields of supplier selection, pattern recognition, and so on.

Author Contributions: The research is conceptualized and performed by D.L. and Y.L. The paper is written by D.L. and Y.L., and finally checked and revised by X.C. The paper is supervised by X.C. All authors read and approved the final manuscript.

Funding: This research was funded by Natural Science Foundation of Hunan (No. 2017JJ2096, 2018JJ3137), the National Natural Science Foundation of China (No. 11501191), the National Social Science Foundation of China (No. 15BTJ028), the outstanding youth project of the Hunan Education Department (No. 1713092) and by Major projects of the National Social Science Foundation of China (17ZDA046).

Conflicts of Interest: The authors declare no conflict of interest regarding the publication for the paper.

References

1. Zadeh, L.A. Fuzzy sets. *Inf. Control.* **1965**, *8*, 338–353. [CrossRef]
2. Atanassov, K.T. Intuitionistic fuzzy sets. *Fuzzy Sets Syst.* **1986**, *20*, 87–96. [CrossRef]
3. Atanassov, K.T.; George, G. Interval valued intuitionistic fuzzy sets. *Fuzzy Sets Syst.* **1989**, *31*, 343–349. [CrossRef]
4. Torra, V. Hesitant fuzzy sets. *Int. J. Intell. Syst.* **2010**, *25*, 529–539. [CrossRef]
5. Zadeh, L.A. The concept of a linguistic variable and its application to approximate reasoning—I. *Inf. Sci.* **1975**, *8*, 199–249. [CrossRef]

6. Zadeh, L.A. The concept of a linguistic variable and its application to approximate reasoning—II. *Inf. Sci.* **1975**, *8*, 301–357. [CrossRef]
7. Zadeh, L.A. The concept of a linguistic variable and its application to approximate reasoning—III. *Inf. Sci.* **1975**, *9*, 43–80. [CrossRef]
8. Rodriguez, R.M.; Luis, M.; Francisco, H. Hesitant Fuzzy Linguistic Term Sets for Decision Making. *IEEE Trans. Fuzzy Syst.* **2012**, *20*, 109–119. [CrossRef]
9. Liao, H.C.; Xu, Z.S.; Zeng, X.J. Distance and similarity measures for hesitant fuzzy linguistic term sets and their application in multi-criteria decision making. *Inf. Sci.* **2014**, *271*, 125–142. [CrossRef]
10. Lee, L.W.; Chen, S.M. Fuzzy decision making based on likelihood-based comparison relations of hesitant fuzzy linguistic term sets and hesitant fuzzy linguistic operators. *Inf. Sci.* **2015**, *294*, 513–529. [CrossRef]
11. Hesamian, G.; Shams, M. Measuring Similarity and Ordering based on Hesitant Fuzzy Linguistic Term Sets. *J. Intell. Fuzzy Syst.* **2015**, *28*, 983–990.
12. Song, Y.M.; Hu, J.; Schwenker, F. Vector similarity measures of hesitant fuzzy linguistic term sets and their applications. *PLoS ONE* **2017**, *12*, 1–13. [CrossRef] [PubMed]
13. Wei, C.P.; Zhao, N.; Tang, X.J. Operators and Comparisons of Hesitant Fuzzy Linguistic Term Sets. *IEEE Trans. Fuzzy Syst.* **2014**, *22*, 575–585. [CrossRef]
14. Liao, H.C.; Xu, Z.S.; Zeng, X.J.; Merigó, J.M. Qualitative decision making with correlation coefficients of hesitant fuzzy linguistic term sets. *Knowl. Based Syst.* **2015**, *76*, 127–138. [CrossRef]
15. Liu, D.H.; Liu, Y.Y.; Chen, X.H. The new similarity measure and distance measure between hesitant fuzzy linguistic term set and their application in multi-criteria decision making. *J. Intell. Fuzzy. Syst.* **2018**, submitted.
16. Liu, H.B.; Rosa, M.R. A fuzzy envelope for hesitant fuzzy linguistic term set and its application to multicriteria decision making. *Inf. Sci.* **2014**, *258*, 220–238. [CrossRef]
17. Xu, Y.J.; Xu, A.W.; Merigo, J.M.; Wang, H.M. Hesitant fuzzy linguistic ordered weighted distance operators for group decision making. *J. Appl. Math. Comput.* **2015**, *49*, 285–308. [CrossRef]
18. Liao, H.C.; Xu, Z.S.; Herrera, V.E.; Francisco, H. Hesitant Fuzzy Linguistic Term Set and Its Application in Decision Making: A State of the Art Survey. *Int. J. Fuzzy Syst.* **2017**, *12*, 1–27. [CrossRef]
19. Liao, H.C.; Si, G.; Xu, Z.S.; Fujita, H. Hesitant Fuzzy Linguistic Preference Utility Set and Its Application in Selection of Fire Rescue Plans. *Int. J. Environ. Res. Public Health* **2018**, *15*, 664. [CrossRef] [PubMed]
20. Boulid, Y.; Souhar, A.; Elkettani, M.E. Multi-agent Systems for Arabic Handwriting Recognition. *Int. J. Int. Multimed. Artif. Intell.* **2017**, *4*, 31–32. [CrossRef]
21. Idrus, A.; Mahmoud, M.A.; Ahmad, M.S.; Yahya, A.; Husen, H. A Solution Generator Algorithm for Decision Making based Automated Negotiation in the Construction Domain. *Int. J. Int. Multimed. Artif. Intell.* **2017**, *4*, 18–23. [CrossRef]
22. Zhu, J.H.; Li, Y.L. Hesitant fuzzy linguistic aggregation operators based on the Hamacher t-norm and t-conorm. *Symmetry* **2018**, *10*, 189. [CrossRef]
23. Wang, J.Q.; Wu, J.T.; Wang, J.; Chen, X.H. Interval-valued hesitant fuzzy linguistic sets and their applications in multi-criteria decision-making problems. *Inf. Sci.* **2014**, *288*, 55–72. [CrossRef]
24. Wang, J.Q.; Wu, J.T.; Wang, J.; Zhang, H.Y.; Chen, X.H. Multi-criteria decision-making methods based on the Hausdorff distance of hesitant fuzzy linguistic numbers. *Soft Comput.* **2016**, *20*, 1621–1633. [CrossRef]
25. Liu, D.H.; Chen, X.H.; Peng, D. Distance Measures for Hesitant Fuzzy Linguistic Sets and Their Applications in Multiple Criteria Decision Making. *Int. J. Fuzzy Syst.* **2018**, 1–11. [CrossRef]
26. Liu, D.H.; Chen, X.H.; Peng, D. The Intuitionistic Fuzzy Linguistic Cosine Similarity Measure and Its Application in Pattern Recognition. *Complexity* **2018**, *2018*, 1–11. [CrossRef]
27. Song, Y.F.; Wang, X.D.; Zhang, H.L. A distance measure between intuitionistic fuzzy belief functions. *Knowl. Based Syst.* **2015**, *86*, 288–298. [CrossRef]
28. Bhattacharyya, A. On a Measure of Divergence between Two Multinomial Populations. *Sankhyā Indian J. Stat.* **1946**, *7*, 401–406.
29. Ye, J. Cosine similarity measures for intuitionistic fuzzy sets and their applications. *Math. Comput. Model.* **2011**, *53*, 91–97. [CrossRef]
30. Ye, J. Multicriteria decision-making method based on cosine similarity measures between interval-Valued Fuzzy Sets with Risk Preference. *Econ. Comput. Econ. Cybern. Stud. Res.* **2016**, *50*, 205–215.

31. Liao, H.C.; Xu, Z.H. Approaches to manage hesitant fuzzy linguistic information based on the cosine distance and similarity measures for HFLTSs and their application in qualitative decision making. *Expert Syst. Appl.* **2015**, *42*, 5328–5336. [CrossRef]
32. Morente-Molinera, A.; Kou, G.; González-Crespo, R.; Corchado, J.M. Solving multi-criteria group decision making problems under environments with a high number of alternatives using fuzzy ontologies and multi-granular linguistic modelling methods. *Knowl. Based Syst.* **2017**, *13*, 54–64. [CrossRef]
33. Herrera, F.; Herrera-Viedma, E. Linguistic decision analysis: Steps for solving decision problems under linguistic information. *Fuzzy Sets Syst.* **2000**, *115*, 67–82. [CrossRef]
34. Xu, Z.H. A method based on linguistic aggregation operators for group decision making with linguistic preference relations. *Inf. Sci.* **2004**, *166*, 19–30. [CrossRef]
35. Zhu, B.; Xu, Z.S. Consistency Measures for Hesitant Fuzzy Linguistic Preference Relations. *IEEE Trans. Fuzzy Syst.* **2014**, *22*, 35–45. [CrossRef]
36. Wei, C.P.; Ren, Z.L.; Rosa, M.R. A Hesitant Fuzzy Linguistic TODIM Method Based on a Score Function. *Int. J. Comput. Intell. Syst.* **2015**, *8*, 701–712. [CrossRef]
37. Bao, G.Y.; Lian, X.L.; He, M. Improved two-tuple linguistic representation model based on new linguistic evaluation scale. *Control Decis.* **2010**, *25*, 780–784. (In Chinese)
38. Wang, J.Q.; Wang, J.; Zhang, H.Y. A likelihood-based TODIM approach based on multi-hesitant fuzzy linguistic information for evaluation in logistics outsourcing. *Comput. Ind. Eng.* **2016**, *99*, 287–299. [CrossRef]
39. Kahneman, D.; Amos, T. Prospect Theory: An Analysis of Decision under Risk. *Econometrica* **1979**, *47*, 263–291. [CrossRef]
40. Beg, I.; Tabasam, R. TOPSIS for Hesitant Fuzzy Linguistic Term Sets. *Int. J. Intell. Syst.* **2013**, *28*, 1162–1171. [CrossRef]
41. Zhang, Z.M.; Wu, C. Hesitant fuzzy linguistic aggregation operators and their applications to multiple attribute group decision making. *J. Intell. Fuzzy Syst.* **2014**, *26*, 2185–2202.

symmetry

MDPI

Article

Some Interval Neutrosophic Linguistic Maclaurin Symmetric Mean Operators and Their Application in Multiple Attribute Decision Making

Yushui Geng [1,*], Xingang Wang [2], Xuemei Li [2], Kun Yu [2] and Peide Liu [3,*]

1 Graduate School, Qilu University of Technology, Jinan 250353, China
2 School of Information, Qilu University of Technology, Jinan 250353, China; Wxg@qlu.edu.cn (X.W.); leexm0911@gmail.com (X.L.); ytyukun@gmail.com (K.Y.)
3 School of Management Science and Engineering, Shandong University of Finance and Economics, Jinan 250353, China
* Correspondence: gys@qlu.edu.cn (Y.G.); liupd@sdufe.edu.cn or peide.liu@gmail.com (P.L.)

Received: 10 April 2018; Accepted: 17 April 2018; Published: 22 April 2018

Abstract: There are many practical decision-making problems in people's lives, but the information given by decision makers (DMs) is often unclear and how to describe this information is of critical importance. Therefore, we introduce interval neutrosophic linguistic numbers (INLNs) to represent the less clear and uncertain information and give their operational rules and comparison methods. In addition, since the Maclaurin symmetric mean (*MSM*) operator has the special characteristic of capturing the interrelationships among multi-input arguments, we further propose an *MSM* operator for INLNs (*INLMSM*). Furthermore, considering the weights of attributes are the important parameters and they can influence the decision results, we also propose a weighted *INLMSM* (*WINLMSM*) operator. Based on the *WINLMSM* operator, we develop a multiple attribute decision making (MADM) method with INLNs and some examples are used to show the procedure and effectiveness of the proposed method. Compared with the existing methods, the proposed method is more convenient to express the complex and unclear information. At the same time, it is more scientific and flexible in solving the MADM problems by considering the interrelationships among multi-attributes.

Keywords: multiple attribute decision making (MADM); neutrosophic number; Maclaurin symmetric mean; linguistic variables

1. Introduction

The unclear set (FS) theory was put forward by Zadeh [1] in 1965. In this theory, the membership degree (MD) $T(x)$ is used to describe fuzzy information and it has also been widely used in practice. However, the inadequacies of FS are evident. For example, it is difficult to express the non-membership degree (NMD) $F(x)$. In order to fix this problem, Intuitionistic FS (IFS) was proposed by Atanassov [2] in 1986. It is made up of two parts: MD and NMD. IFS is an extension and development of Zadeh' FS and Zadeh' FS is a special case of IFS [3]. IFS needs to meet two conditions: (1) $T(x), F(x) \in [0,1]$; (2) $0 \leq T(x) + F(x) \leq 1$ [2]. Subsequently, the IFS theory was further extended such as Zadeh [4] proposed interval IFS (IIFS). Zwick et al. [5] put forward the triangular IFS while Zeng and Li [6] defined trapezoidal IFS. However, under some circumstances due to the limited cognitive ability of the DMs, they may hesitate in the two choices for accuracy and uncertainty. Since they choose both of them at the same time, this can produce an imprecise or contradictory evaluation result. Therefore, Smarandache [7,8] introduced a concept called neutrosophic set (NS), which included MD, NMD, and indeterminacy membership degree (IMD) in a non-standard unit interval [9]. Clearly, the NS is the generalization of FS and IFS. Furthermore, Wang [10] proposed the definition of interval NS (INS) which uses the standard interval to express the function of MD, IMD, and NMD. Broumi and Smarandache [11] presented the correlation coefficient of INS.

When dealing with the MADM problems with qualitative information, it is difficult for DMs to describe their own ideas with precise values. Generally, DMs ordinarily uses some linguistic terms (LTs) like "excellent", "good", "bad", "very bad", or "general" to indicate their evaluations. For example, when we look at a person's height, we usually describe him as "high" or "very high" by visual inspection, but we will not give the exact value. In order to easily process the qualitative information, Herrera and Herrera-Viedma [12] proposed the LTs to deal with this kind of information instead of numerical computation. However, because LT such as "high" is not with MD, or we can think its MD is 1, which means LTs cannot describe the MD and NMD. Therefore, in order to facilitate DMs to describe the MD and NMD for one LT, Liu and Chen [13] defined the linguistic intuitionistic fuzzy number (LIFN), which combined the advantages of intuitionistic fuzzy numbers (IFNs) and linguistic variables (LVs). Therefore, LIFN can fully express the complex fuzzy information and there is a good prospect in MADM. After that, Ye [14] came up with the single-valued neutrosophic linguistic number (SVNLN). The most striking feature of the SVNLN is that it used LTs to describe the MD, IMD, and NMD. Sometimes, the three degrees are not expressed in a single real number, but is expressed in intervals [15]. And then, Ye [16] defined an interval neutrosophic linguistic set (INLS) and INLNs. INLNs is used to represent three values of MD, IMD, and NMD in the form of intervals. Clearly, INLS is a generalization of FS, IFS, NS, INS, LIFN, and SVNLN. It is general and beneficial for describing practical problems.

The aggregation operators (AOs) are an efficient way to handle MADM problems [17,18]. Many AOs are proposed for achieving some special functions. Yager [19] employed the ordered weighted average (OWA) operator for MADM. Bonferroni [20] proposed the Bonferroni mean (BM) operator, which can capture the correlation between input variables very well. Then BM operators have been extended to process different uncertain information such as IFS [21,22], interval-valued IFS [23], q-Rung Orthopai Fuzzy set [24], and Multi-valued Ns [25]. In addition, Beliakov [26] presented the Heronian mean (HM) operators, which have the same feature as the BM (i.e., they can capture the interrelationship between input parameters). Some HM operators have been proposed [27–30]. Furthermore, Yu [31] gave the comparison of BM with HM. However, since the BM operator and the HM operator can only reflect the relationship between any two parameters, they cannot process the MADM problems, which require the relationship for multiple inputs. In order to solve this shortcoming, Maclaurin [32] proposed the *MSM* operator, which has prominent features of capturing the relationship among multiple input parameters. Afterward, Qin and Liu [33] developed some *MSM* operators for uncertain LVs. Liu and Qin [34] developed some *MSM* for LIFNs. Liu and Zhang [35] proposed some *MSM* operators for single valued trapezoidal neutrosophic numbers.

Since the INLNs are superior to other ways of expressing complex uncertain information [16] and the *MSM* has good flexibility and adaptability, it can capture the relationship among multiple input parameters. However, now the *MSM* cannot deal with INLNs. Therefore, the objectives of this paper are to extend the *MSM* and weighted *MSM* (*WMSM*) operators to INLNs and to propose the *INLMSM* operator and the *WINLMSM* operator, to prove some properties of them and discuss some special cases, to propose a MADM approach with INLNs, and show the advantages of the proposed approach by comparing with other studies.

In Section 2 of this paper, we introduce some basic concepts about NS, INS, INLS, and *MSM*. In Section 3, we introduce the INLN and its operations including a new scoring function and a comparison method of INLN. In Section 4, we introduce an operator of *INLMSM*. Additionally, in order to improve flexibility, we propose the *INLGMSM* operator based on the *GMSM* operator. Furthermore, we develop the *WINLMSM* operator and the *WINLGMSM* operator to compare with operators that lack weight. Afterwards, we use examples to prove our theories. In Section 5, we give a MADM method for INLNs. In Section 6, we provide an example to demonstrate the effectiveness of the proposed method. Lastly, we provide the conclusions.

2. Preliminaries

In this section, we will introduce some existing definitions and basic concepts in order to understand this study.

2.1. The NS and INS

Definition 1 [7–9]. *Let X be a space of points (objects) with a generic element in X denoted by x. A NS A in X is expressed by a MD $T_A(x)$, an IMD $I(x)$, and a NMD $F_A(x)$.*

Then a NS A is denoted below.

$$A = \{\langle x, T_A(x), I_A(x), F_A(x)\rangle | x \in X\} \tag{1}$$

$T_A(x)$, $I(x)$, and $F_A(x)$ are real standard or non-standard subsets of $]^-0, 1^+[$. That is

$$T_A : X \to]^-0, 1^+[;\ I_A : X \to]^-0, 1^+[;\ F_A : X \to]^-0, 1^+[$$

With the condition $^-0 \le T_A(x) + I_A(x) + F_A(x) \le 3^+$.

Definition 2 [10,11]. *Let X be a space of points (objects) with a generic element in X denoted by x. For convenience, the lower and upper ends of T, I, F are expressed as $T_A^L(x)$, $T_A^U(x)$, $I_A^L(x)$, $I_A^U(x)$, $F_A^L(x)$, and $F_A^U(x)$. An INS A in X is defined below.*

$$A = \left\{x, \left\langle \left[T_A^L(x), T_A^U(x)\right], \left[I_A^L(x), I_A^U(x)\right], \left[F_A^L(x), F_A^U(x)\right]\right\rangle \middle| x \in X\right\} \tag{2}$$

For each point x in X, we have that $[T_A^L(x), T_A^U(x)] \subseteq [0,1]$, $[I_A^L(x), I_A^U(x)] \subseteq [0,1]$, $[F_A^L(x), F_A^U(x)] \subseteq [0,1]$, and $0 \le T_A^U(x) + I_A^U(x) + F_A^U(x) \le 3$.

Definition 3 [10,11]. *An INS A is contained in the INS B, $A \subseteq B$, if and only if $T_A^L(x) \le T_B^L(x)$, $T_A^U(x) \le T_B^U(x)$, $I_A^L(x) \ge I_B^L(x)$, $I_A^U(x) \ge I_B^U(x)$, $F_A^L(x) \ge F_B^L(x)$, and $F_A^U(x) \ge F_B^U(x)$. If $A \subseteq B$ and $A \supseteq B$, then $A = B$.*

2.2. LVs

Definition 4 [36,37]. *Let $S = \{s_i | i = 0, 1, \ldots, l, l \in N^*\}$ be a LT set (LTS) where N^* is a set of positive integers and s_i represents LV.*

Because the LTS is convenient and efficient, it is widely used by DMs in decision making. For instance, when we evaluate the production quality, we can set l = 9, then S is given below.

$$S = \{s_0 = extremely\ bad, s_1 = very\ bad, s_2 = bad, s_3 = slightly\ bad, s_4 = fair, s_5 = slightly\ good,$$
$$s_6 = good, s_7 = very\ good, s_8 = extremely\ good\}$$

To relieve the loss of linguistic information in operations, Xu [38,39] extended LTS S to continuous LTS $\overline{S} = \{s_\theta | 0 \le \theta \le l\}$. About the characteristics of LTS, please refer to References [38–40].

Definition 5 [13]. *Let s_α and s_β be any two LVs in $\overline{S}$. The related operations can be defined below.*

$$s_\alpha \oplus s_\beta = s_{\alpha+\beta-\frac{\alpha\cdot\beta}{l}} \tag{3}$$

$$\lambda s_\alpha = s_{l-l\cdot(1-\frac{\alpha}{l})^\lambda}, \lambda > 0 \tag{4}$$

$$s_\alpha \otimes s_\beta = s_{\frac{\alpha \cdot \beta}{l}} \tag{5}$$

$$(s_\alpha)^\lambda = s_{l \cdot (\frac{\alpha}{l})^\lambda}, \lambda > 0 \tag{6}$$

2.3. MSM Operator

Definition 6 [15,32]. *Let $x_i (i = 1, 2, \ldots, n)$ be the set of the non-negative real number. An MSM operator of dimension n is a mapping $MSM^{(m)} : (R^+)^n \to R^+$ and it can be defined below.*

$$MSM^{(m)}(x_1, \ldots, x_n) = \left(\frac{\sum_{1 \le i_1 < \ldots < i_m \le n} \prod_{j=1}^{m} x_{i_j}}{C_n^m} \right)^{\frac{1}{m}} \tag{7}$$

where $(i_1, i_2, \ldots, i_m)$ traverses all the m-tuple combination of $(1, 2, \ldots, n)$ and $C_n^m = \frac{n!}{m!(n-m)!}$ is the binomial coefficient. In addition, x_{i_j} refers to i_jth element in a particular arrangement.

There are some properties of the $MSM^{(m)}$ operator, which are defined below.

(1) Idempotency. If $x_i = x$ for each i, and then $MSM^{(m)}(x, x, \ldots, x) = x$;
(2) Monotonicity. If $x_i <= y_i$ for all i, $MSM^{(m)}(x_1, x_2, \ldots, x_n) \le MSM^{(m)}(y_1, y_2, \ldots, y_n)$;
(3) Boundedness. $\min\{x_1, x_2, \ldots x_n\} \le MSM^{(m)}\{x_1, x_2, \ldots x_n\} \le \max\{x_1, x_2, \ldots x_n\}$.

Furthermore, the $MSM^{(m)}$ operator would degrade some particular forms when m takes some special values, which are shown as follows.

1. When $m = 1$, the $MSM^{(m)}$ operator would become the average operator.

$$MSM^{(1)}(x_1, x_2, \ldots x_n) = \left(\frac{\sum_{1 \le i_1 \le n} x i_1}{C_n^1} \right) = \frac{\sum\limits_{i=1}^{n} xi}{n} \tag{8}$$

2. When $m = 2$, the $MSM^{(m)}$ operator would become the following BM operator ($p = q = 1$).

$$\begin{aligned} MSM^{(2)}(x_1, \ldots, x_n) &= \left(\frac{\sum_{1 \le i_1 < i_2 \le n} \prod_{j=1}^{2} x_{i_j}}{C_n^2} \right)^{\frac{1}{2}} = \left(\frac{2 \sum_{1 \le i_1 < i_2 \le n} x i_1 x i_2}{n(n-1)} \right)^{\frac{1}{2}} \\ &= \left(\frac{\sum_{i,j=1, i \ne j}^{n} x i x j}{n(n-1)} \right)^{\frac{1}{2}} = BM^{1,1}(x_1, \ldots, x_n) \end{aligned} \tag{9}$$

3. When $m = n$, the $MSM^{(m)}$ operator would become the geometric mean.

$$MSM^{(n)}(x_1, \ldots, x_n) = \left(\prod_{j=1}^{n} x_j \right)^{\frac{1}{n}} \tag{10}$$

Definition 7 [15]. *Let $x_i (i = 1, 2, \ldots, n)$ be the set of non-negative real numbers and $p_1, p_2, \ldots, p_m \ge 0$. A generalized MSM operator of dimension n is a mapping $GMSM^{(m, p_1, p_2, \ldots, p_m)} : (R^+)^n \to R^+$ and it is defined below.*

$$GMSM^{(m, p_1, p_2, \ldots, p_m)}(x_1, \ldots, x_n) = \left(\frac{\sum_{1 \le i_1 < \ldots < i_m \le n} \prod_{j=1}^{m} x_{i_j}^{p_j}}{C_n^m} \right)^{\frac{1}{p_1 + p_2 + \ldots p_m}} \tag{11}$$

where $(i_1, i_2, \ldots, i_m)$ *traverses all the m-tuple combination of* $(1, 2, \ldots, n)$ *and* $C_n^m = \frac{n!}{m!(n-m)!}$ *is the binomial coefficient.*

There are some properties of the $GMSM^{(m,P_1,P_2,\ldots,P_m)}$ operator below.

(1) Idempotency. If $x_i = x$ for each i, and then $GMSM^{(m,P_1,P_2,\ldots,P_m)}(x, x, \ldots, x) = x$;
(2) Monotonicity. If $x_i \leq y_i$ for all i, $GMSM^{(m,p_1,p_2,\ldots,p_m)}(x_1, x_2, \ldots, x_n) \leq GMSM^{(m,P_1,P_2,\ldots,P_m)}(y_1, y_2, \ldots, y_n)$;
(3) Boundedness. $\min\{x_1, x_2, \ldots x_n\} \leq GMSM^{(m,p_1,p_2,\ldots,p_m)}\{x_1, x_2, \ldots x_n\} \leq \max\{x_1, x_2, \ldots x_n\}$.

In addition, the $GMSM^{(m,P_1,P_2,\ldots,P_m)}$ operator would degrade to some particular forms when m takes some special values, which are shown below.

1. When $m = 1$, we have the formula below.

$$GMSM^{(1,P_1)}(x_1, x_2, \ldots x_n) = \left(\frac{\sum_{1 \leq i_1 \leq n} {x_{i_1}}^{p_1}}{C_n^1}\right)^{\frac{1}{p_1}} = \left(\frac{\sum_{i=1}^{n} {x_i}^{p_1}}{n}\right)^{\frac{1}{p_1}} \tag{12}$$

2. When $m = 2$, the $GMSM^{(m,P_1,P_2,\ldots,P_m)}$ operator would become the following BM operator.

$$\begin{aligned} GMSM^{(2,p_1,p_2)}(x_1, \ldots, x_n) &= \left(\frac{\sum_{1 \leq i_1 < i_2 \leq n} x_{i_1}^{p_1} x_{i_2}^{p_2}}{C_n^2}\right)^{\frac{1}{p_1+p_2}} = \left(\frac{2\sum_{1 \leq i < j \leq n} x_i^{p_1} x_j^{p_2}}{n(n-1)}\right)^{\frac{1}{p_1+p_2}} \\ &= \left(\frac{\sum_{i,j=1, i \neq j}^{n} x_i^{p_1} x_j^{p_2}}{n(n-1)}\right)^{\frac{1}{p_1+p_2}} = BM^{p_1,p_2} \end{aligned} \tag{13}$$

3. When $m = n$, the $MSM^{(m)}$ operator would become the following formula.

$$GMSM^{(n,p_1,p_2,\ldots,p_n)}(x_1, \ldots, x_n) = \left(\prod_{j=1}^{n} x_j^{p_j}\right)^{\frac{1}{p_1+p_2+\ldots p_n}} \tag{14}$$

4. When $p_1 = p_2 = \ldots = p_m = 1$, the $GMSM^{(m,P_1,P_2,\ldots,P_m)}$ operator would degenerate to the MSM operator and the parameter is m below.

$$GMSM^{(m,1,1,\ldots,1)}(x_1, \ldots, x_n) = \left(\frac{\sum_{1 \leq i_1 < \ldots < i_m \leq n} \prod_{j=1}^{m} x_{ij}^{1}}{C_n^m}\right)^{\frac{1}{m}} = MSM^{(m)}(x_1, \ldots, x_n). \tag{15}$$

3. INLNs and Operations

Definition 8 [16,41]. *Let X be a finite universal set. An INLS in X is defined by the equation below.*

$$A = \left\{x, \left\langle s_{\theta(x)}, [T_A(x), I_A(x), F_A(x)]\right\rangle | x \in X\right\} \tag{16}$$

where $s_{\theta(x)} \in \overline{S}$, $T_A(x) = [T_A^L(x), T_A^U(x)] \subseteq [0,1]$, $I_A(x) = [I_A^L(x), I_A^U(x)] \subseteq [0,1]$, $F_A(x) = [F_A^L(x), F_A^U(x)] \subseteq [0,1]$ *represent the MD, the IMD, and the NMD of the element x in X to the LV* $s_{\theta(x)}$, *respectively, with the condition* $0 \leq T_A^U(x) + I_A^U(x) + F_A^U(x) \leq 3$ *for any* $x \in X$.

Then the seven tuple $\left\langle s_{\theta(x)}, ([T_A^L(x), T_A^U(x)], [I_A^L(x), I_A^U(x)], [F_A^L(x), F_A^U(x)])\right\rangle$ in A is called an INLN. For convenience, an INLN can be represented as $a = \left\langle s_{\theta(a)}, ([T^L(a), T^U(a)], [I^L(a), I^U(a)], [F^L(a), F^U(a)])\right\rangle$.

Then we introduced the operational rules of operators of INLNs.

Definition 9 [16,37,42]. *Let* $a_1 = \left\langle s_{\theta(a_1)}, \left(\left[T^L(a_1), T^U(a_1)\right], \left[I^L(a_1), I^U(a_1)\right], \left[F^L(a_1), F^U(a_1)\right]\right)\right\rangle$ *and* $a_2 = \left\langle s_{\theta(a_2)}, \left(\left[T^L(a_2), T^U(a_2)\right], \left[I^L(a_2), I^U(a_2)\right], \left[F^L(a_2), F^U(a_2)\right]\right)\right\rangle$ *be two INLNs and* $\lambda \geq 0$*. Then the operation of the INLNs can be expressed by the equation below.*

$$a_1 \oplus a_2 = \left\langle s_{\theta(a_1)+\theta(a_2)}, \left(\left[T^L(a_1) + T^L(a_2) - T^L(a_1) \times T^L(a_2), T^U(a_1) + T^U(a_2) - T^U(a_1) \times T^U(a_2)\right],\right.\right. \\ \left.\left. \left[I^L(a_1) \times I^L(a_2), I^U(a_1) \times I^U(a_2)\right], \left[F^L(a_1) \times F^L(a_2), F^U(a_1) \times F^U(a_2)\right]\right)\right\rangle \tag{17}$$

$$I^U(a_1) + I^U(a_2) - I^U(a_1) \times I^U(a_2)], [F^L(a_1) + F^L(a_2) - F^L(a_1) \times F^L(a_2), \\ a_1 \otimes a_2 = \left\langle s_{\theta(a_1)\times\theta(a_2)}, \left(\left[T^L(a_1) \times T^L(a_2), T^U(a_1) \times T^U(a_2)\right], \left[I^L(a_1) + I^L(a_2) - I^L(a_1) \times I^L(a_2),\right.\right.\right. \\ \left.\left.\left. F^U(a_1) + F^U(a_2) - F^U(a_1) \times F^U(a_2)\right]\right)\right\rangle \tag{18}$$

$$\lambda a_1 = \left\langle s_{\lambda\times\theta(a_1)}, \left(\left[1 - (1 - T^L(a_1))^\lambda, 1 - (1 - T^U(a_1))^\lambda\right], \left[(I^L(a_1))^\lambda, (I^U(a_1))^\lambda\right],\right.\right. \\ \left.\left. \left[(F^L(a_1))^\lambda, (F^U(a_1))^\lambda\right]\right)\right\rangle (\lambda > 0) \tag{19}$$

$$a_1^\lambda = s_{\theta^\lambda(a_1)}, \left(\left[(T^L(a_1))^\lambda, (T^U(a_1))^\lambda\right], \left[1 - (1 - I^L(a_1))^\lambda, 1 - (1 - I^U(a_1))^\lambda\right],\right. \\ \left. \left[1 - (1 - F^L(a_1))^\lambda, 1 - (1 - F^U(a_1))^\lambda\right]\right)\rangle, (\lambda > 0) \tag{20}$$

Example 1. *Let* $a_1 = \langle s_3, ([0.1, 0.2], [0.2, 0.3], [0.4, 0.5])\rangle$ *and* $a_2 = \langle s_4, ([0.3, 0.5], [0.3, 0.4], [0.5, 0.6])\rangle$ *be two INLNs and* $S = \{s_0 = \textit{very bad}, s_1 = \textit{bad}, s_2 = \textit{slightly bad}, s_3 = \textit{fair}, s_4 = \textit{slightly good}, s_5 = \textit{good}, s_6 = \textit{very good}\}$*, then we have the equations below.*

$$a_1 \oplus a_2 = \langle s_{3+4}, ([0.1 + 0.3 - 0.1 \times 0.3, 0.2 + 0.5 - 0.2 \times 0.5], [0.2 \times 0.3, 0.3 \times 0.4], [0.4 \times 0.5, 0.5 \times 0.6]\rangle \\ = \langle s_7, ([0.37, 0.6], [0.06, 0.12], [0.2, 0.3]\rangle$$

$$a_1 \otimes a_2 = \langle s_{3\times4}, ([0.1 \times 0.3, 0.2 \times 0.5], [0.2 + 0.3 - 0.2 \times 0.3, 0.3 + 0.4 - 0.3 \times 0.4], \\ [0.4 + 0.5 - 0.4 \times 0.5, 0.5 + 0.6 - 0.5 \times 0.6])\rangle \\ = \langle s_{12}, ([0.03, 0.1], [0.44, 0.58], [0.7, 0.8])\rangle$$

As seen from the above examples, these results are not reasonable because they exceed the range of LTS. In order to overcome these limitations, we will improve these operations by Definition 10.

Definition 10. *Let* $a_1 = \left\langle s_{\theta(a_1)}, \left(\left[T^L(a_1), T^U(a_1)\right], \left[I^L(a_1), I^U(a_1)\right], \left[F^L(a_1), F^U(a_1)\right]\right)\right\rangle$ *and* $a_2 = \left\langle s_{\theta(a_2)}, \left(\left[T^L(a_2), T^U(a_2)\right], \left[I^L(a_2), I^U(a_2)\right], \left[F^L(a_2), F^U(a_2)\right]\right)\right\rangle$ *be two INLNs and* $\lambda \geq 0$*. Then the operations of the INLNs can be defined by the equations below.*

$$a_1 \oplus a_2 = \left\langle s_{\theta(a_1)+\theta(a_2)-\frac{\theta(a_1)\cdot\theta(a_2)}{l}}, \left(\left[T^L(a_1) + T^L(a_2) - T^L(a_1) \times T^L(a_2), T^U(a_1) + T^U(a_2) - T^U(a_1) \times T^U(a_2)\right],\right.\right. \\ \left.\left. \left[I^L(a_1) \times I^L(a_2), I^U(a_1) \times I^U(a_2)\right], \left[F^L(a_1) \times F^L(a_2), F^U(a_1) \times F^U(a_2)\right]\right)\right\rangle \tag{21}$$

$$a_1 \otimes a_2 = \left\langle s_{\frac{\theta(a_1)\times\theta(a_2)}{l}}, \left(\left[T^L(a_1) \times T^L(a_2), T^U(a_1) \times T^U(a_2)\right], \left[I^L(a_1) + I^L(a_2) - I^L(a_1) \times I^L(a_2),\right.\right.\right. \\ I^U(a_1) + I^U(a_2) - I^U(a_1) \times I^U(a_2)], [F^L(a_1) + F^L(a_2) - F^L(a_1) \times F^L(a_2), \\ \left.\left.\left. F^U(a_1) + F^U(a_2) - F^U(a_1) \times F^U(a_2)\right]\right)\right\rangle \tag{22}$$

$$\lambda a_1 = \left\langle s_{l - l\cdot(1-\frac{\theta(a_1)}{l})^\lambda}, \left(\left[1 - (1 - T^L(a_1))^\lambda, 1 - (1 - T^U(a_1))^\lambda\right],\right.\right. \\ \left.\left. \left[(I^L(a_1))^\lambda, (I^U(a_1))^\lambda\right], \left[(F^L(a_1))^\lambda, (F^U(a_1))^\lambda\right]\right)\right\rangle, (\lambda > 0) \tag{23}$$

$$a_1^\lambda = s_{l\cdot(\frac{\theta(a_1)}{l})^\lambda}, \left(\left[(T^L(a_1))^\lambda, (T^U(a_1))^\lambda\right], \left[1 - (1 - I^L(a_1))^\lambda, 1 - (1 - I^U(a_1))^\lambda\right],\right. \\ \left. \left[1 - (1 - F^L(a_1))^\lambda, 1 - (1 - F^U(a_1))^\lambda\right]\right)\rangle, (\lambda > 0). \tag{24}$$

Based on the operational rules above, the above example is recalculated as follow.

Example 2. *Let* $a_1 = \langle s_3, ([0.1, 0.2], [0.2, 0.3], [0.4, 0.5])\rangle$ *and* $a_2 = \langle s_4, ([0.3, 0.5], [0.3, 0.4], [0.5, 0.6])\rangle$ *be two INLNs and* $S = \{s_0 = very\ bad, s_1 = bad, s_2 = slightly\ bad, s_3 = fair, s_4 = slightly\ good, s_5 = good, s_6 = very\ good\}$*, then we have the equations below.*

$$a_1 \oplus a_2 = \left\langle s_{3+4-\frac{3\times4}{6}}, ([0.1 + 0.3 - 0.1 \times 0.3, 0.2 + 0.5 - 0.2 \times 0.5], [0.2 \times 0.3, 0.3 \times 0.4], [0.4 \times 0.5, 0.5 \times 0.6]\right\rangle$$
$$= \langle s_5, ([0.37, 0.6], [0.06, 0.12], [0.2, 0.3]\rangle$$

$$a_1 \otimes a_2 = \left\langle s_{\frac{3\times4}{6}}, ([0.1 \times 0.3, 0.2 \times 0.5], [0.2 + 0.3 - 0.2 \times 0.3, 0.3 + 0.4 - 0.3 \times 0.4], \right.$$
$$[0.4 + 0.5 - 0.4 \times 0.5, 0.5 + 0.6 - 0.5 \times 0.6]))$$
$$= \langle s_2, ([0.03, 0.1], [0.44, 0.58], [0.7, 0.8])\rangle$$

From the above example, the results are more reasonable than the previous ones.

In the following definitions, a new scoring function and a comparison method of INLN are described.

Definition 11. [37]. *Let* $a = \left\langle s_{\theta(a)}, \left([T^L(a), T^U(a)], [I^L(a), I^U(a)], [F^L(a), F^U(a)]\right)\right\rangle$ *be an INLN. Then the score function of a can be expressed by the equation below.*

$$S(a) = \alpha \cdot \frac{\theta(a)}{6}\left[0.5(T^U(a) + 1 - F^L(a)) + \alpha I^U(a)\right] + (1-\alpha) \cdot \frac{\theta(a)}{6}\left[0.5(T^L(a) + 1 - F^U(a)) + \alpha I^L(a)\right] \quad (25)$$

where the values of $\alpha \in [0, 1]$ *reflect the attitudes of the decision makers.*

Definition 12. [37]. *Let a and b be two INLNs. Then the INLN comparison method can be expressed by the statements below.*

$$\text{If } S(a) > S(b), \text{ then } a \succ b; \quad (26)$$

$$\text{If } S(a) = S(b), \text{ then } a \sim b; \quad (27)$$

$$\text{If } S(a) < (b), \text{ then } a \prec b; \quad (28)$$

4. Some Interval Neutrosophic Linguistic MSM Operators

In this section, we will propose *INLMSM* operators and *INLGMSM* operators.

4.1. The INLMSM Operators

Definition 13. *Let* $a_i = \langle s_{\theta_i}, ([T^L(a_i), T^U(a_i)], [I^L(a_i), I^U(a_i)], [F^L(a_i), F^U(a_i)])\rangle$ $(i = 1, 2, \ldots, n)$ *be a set of INLNs. Then the INLMSM operator:* $\Omega^n \to \Omega$ *is shown below.*

$$INLMSM^{(m)}(a_1, \ldots, a_n) = \left(\frac{\underset{1\le i_1<\ldots<i_m\le n}{\oplus}\left(\overset{m}{\underset{j=1}{\otimes}} a_{i_j}\right)}{C_n^m}\right)^{\frac{1}{m}} \quad (29)$$

Ω *is a set of INLNs and* $m = 1, 2, \ldots, n$.

According to the operational laws of INLNs in Definition 10, we can get the expression of the INLMSM operator shown below.

Theorem 1. *Let* $a_i = \langle s_{\theta_i}, ([T^L(a_i), T^U(a_i)], [I^L(a_i), I^U(a_i)], [F^L(a_i), F^U(a_i)]) \rangle$ $(i = 1, 2, ..., n)$ *be a set of INLNs and* $m = 1, 2, ..., n$. *Then the value aggregated from Definition 13 is still an INLN.*

$$\begin{gathered} INLMSM^{(m)}(a_1, \ldots, a_n) = \\ \left\langle s_{l\cdot(1-\prod_{k=1}^{C_n^m}(1-\prod_{j=1}^m(\frac{\theta_{i_j(k)}}{l}))^{\frac{1}{C_n^m}})^{\frac{1}{m}}}, \left(\left[\left(1-\prod_{k=1}^{C_n^m}\left(1-\prod_{j=1}^m T^L{}_{i_j(k)}\right)^{\frac{1}{C_n^m}}\right)^{\frac{1}{m}}, \left(1-\prod_{k=1}^{C_n^m}\left(1-\prod_{j=1}^m T^U{}_{i_j(k)}\right)^{\frac{1}{C_n^m}}\right)^{\frac{1}{m}}\right],\right.\right. \\ \left[1-\left(1-\prod_{k=1}^{C_n^m}\left(1-\prod_{j=1}^m\left(1-I^L{}_{i_j(k)}\right)\right)^{\frac{1}{C_n^m}}\right)^{\frac{1}{m}}, 1-\left(1-\prod_{k=1}^{C_n^m}\left(1-\prod_{j=1}^m\left(1-I^U{}_{i_j(k)}\right)\right)^{\frac{1}{C_n^m}}\right)^{\frac{1}{m}}\right], \\ \left.\left.\left[1-\left(1-\prod_{k=1}^{C_n^m}\left(1-\prod_{j=1}^m\left(1-F^L{}_{i_j(k)}\right)\right)^{\frac{1}{C_n^m}}\right)^{\frac{1}{m}}, 1-\left(1-\prod_{k=1}^{C_n^m}\left(1-\prod_{j=1}^m\left(1-F^U{}_{i_j(k)}\right)\right)^{\frac{1}{C_n^m}}\right)^{\frac{1}{m}}\right]\right)\right\rangle \end{gathered} \tag{30}$$

where $k = 1, 2, \ldots C_n^m$, $a_{i_j(k)}$ *is the* i_j*th element of* k *th permutation.*

Proof.

Because

$$a_{i_j(k)} = \left\langle s_{\theta i_j(k)}, ((T^L i_j(k), T^U i_j(k)), (I^L i_j(k), I^U i_j(k)), (F^L i_j(k), F^U i_j(k)))\right\rangle (j = 1, 2, ..., m)$$

$$\Rightarrow \overset{m}{\underset{j=1}{\otimes}} a_{i_j(k)} = \left\langle s_{l\cdot\prod_{j=1}^m(\frac{\theta i_j(k)}{l})}, \left(\left[\prod_{j=1}^m T^L i_j(k), \prod_{j=1}^m T^U i_j(k)\right],\right.\right.$$

$$\left.\left.\left[1-\prod_{j=1}^m\left(1-I^L{}_{i_j(k)}\right), 1-\prod_{j=1}^m\left(1-I^U{}_{i_j(k)}\right)\right], \left[1-\prod_{j=1}^m\left(1-F^L{}_{i_j(k)}\right), 1-\prod_{j=1}^m\left(1-F^U{}_{i_j(k)}\right)\right]\right)\right\rangle$$

$$\Rightarrow \underset{1\le i_1<\ldots<i_m\le n}{\oplus}\left(\overset{m}{\underset{j=1}{\otimes}} ai_j\right) = \left\langle s_{l-l\cdot\prod_{k=1}^{C_n^m}(1-\prod_{j=1}^m(\frac{\theta i_j(k)}{l}))},\right.$$

$$\left(\left[1-\prod_{k=1}^{C_n^m}\left(1-\prod_{j=1}^m T^L{}_{i_j(k)}\right), 1-\prod_{k=1}^{C_n^m}\left(1-\prod_{j=1}^m T^U{}_{i_j(k)}\right)\right],\right.$$

$$\left[\prod_{k=1}^{C_n^m}\left(1-\prod_{j=1}^m\left(1-I^L{}_{i_j(k)}\right)\right), \prod_{k=1}^{C_n^m}\left(1-\prod_{j=1}^m\left(1-I^U{}_{i_j(k)}\right)\right)\right],$$

$$\left.\left.\left[\prod_{k=1}^{C_n^m}\left(1-\prod_{j=1}^m\left(1-F^L{}_{i_j(k)}\right)\right), \prod_{k=1}^{C_n^m}\left(1-\prod_{j=1}^m\left(1-F^U{}_{i_j(k)}\right)\right)\right]\right)\right\rangle$$

$$\Rightarrow \left(\frac{\underset{1\le i_1<\ldots<i_m\le n}{\oplus}\left(\overset{m}{\underset{j=1}{\otimes}} a_{i_j}\right)}{C_n^m}\right)^{\frac{1}{m}} = \left\langle s_{l\cdot(1-\prod_{k=1}^{C_n^m}(1-\prod_{j=1}^m(\frac{\theta i_j(k)}{l}))^{\frac{1}{C_n^m}})^{\frac{1}{m}}},\right.$$

$$\left(\left[\left(1-\prod_{k=1}^{C_n^m}\left(1-\prod_{j=1}^m T^L{}_{i_j(k)}\right)^{\frac{1}{C_n^m}}\right)^{\frac{1}{m}}, \left(1-\prod_{k=1}^{C_n^m}\left(1-\prod_{j=1}^m T^U{}_{i_j(k)}\right)^{\frac{1}{C_n^m}}\right)^{\frac{1}{m}}\right],\right.$$

$$\left[1-\left(1-\prod_{k=1}^{C_n^m}\left(1-\prod_{j=1}^m\left(1-I^L{}_{i_j(k)}\right)\right)^{\frac{1}{C_n^m}}\right)^{\frac{1}{m}}, 1-\left(1-\prod_{k=1}^{C_n^m}\left(1-\prod_{j=1}^m\left(1-I^U{}_{i_j(k)}\right)\right)^{\frac{1}{C_n^m}}\right)^{\frac{1}{m}}\right],$$

$$\left[1-\left(1-\prod_{k=1}^{C_n^m}\left(1-\prod_{j=1}^{m}\left(1-F^L{}_{i_j(k)}\right)\right)^{\frac{1}{C_n^m}}\right)^{\frac{1}{m}}, 1-\left(1-\prod_{k=1}^{C_n^m}\left(1-\prod_{j=1}^{m}\left(1-F^U{}_{i_j(k)}\right)\right)^{\frac{1}{C_n^m}}\right)^{\frac{1}{m}}\right]\right)\right\rangle$$

Therefore, Theorem 1 is kept. □

Property 1. *Let* $x_i = \langle s_{\alpha_i}, ([T^L(x_i), T^U(x_i)], [I^L(x_i), I^U(x_i)], [F^L(x_i), F^U(x_i)])\rangle$ $(i = 1, 2, ..., n)$ *and* $y_i = \langle s_{\beta_i}, ([T^L(y_i), T^U(y_i)], [I^L(y_i), I^U(y_i)], [F^L(y_i), F^U(y_i)])\rangle$ $(i = 1, 2, ..., n)$ *be sets of INLNs. There are four properties of* $INLMSM^{(m)}$ *operator, which is shown below.*

(1) *Idempotency. If the INLNs* $x_i = x = \langle s_{\theta_x}, ([T^L{}_x, T^U{}_x], [I^L{}_x, I^U{}_x], [F^L{}_x, F^U{}_x])\rangle$ *for each* $i(i = 1, 2, ..., n)$ *and then* $INLMSM^{(m)} = x = \langle s_{\theta_x}, (T_x, I_x, F_x)\rangle$.

(2) *Commutativity. If* x_i *is a permutation of* y_i *for all* i $(i = 1, 2, ..., n)$ *and then* $INLMSM^{(m)}(x_1, x_2, ..., x_n) = INLMSM^{(m)}(y_1, y_2, ..., y_n)$.

(3) *Monotonicity. If* $\alpha_i \le \beta_i$, $T^L(x_i) \le T^L(y_i)$, $T^U(x_i) \le T^U(y_i)$, $I^L(x_i) \ge I^L(y_i)$, $I^U(x_i) \ge I^U(y_i)$, $F^L(x_i) \ge F^L(y_i)$ *and* $F^U(x_i) \ge F^U(y_i)$ *for all* i $(i = 1, 2, ..., n)$, *then* $x_i \le y_i$ *and* $INLMSM^{(m)}(x_1, x_2, ..., x_n) \le INLMSM^{(m)}(y_1, y_2, ..., y_n)$.

(4) *Boundedness.* $\min\{x_1, x_2, ...x_n\} \le INLMSM^{(m)}\{x_1, x_2, ...x_n\} \le \max\{x_1, x_2, ...x_n\}$..

Proof.

1 If each $a_i = x$, then we get the equation below.

$$INLMSM^{(m)}(x, x, ..., x) =$$

$$\left\langle s_{l\cdot\left(1-\prod_{k=1}^{C_n^m}\left(1-\prod_{j=1}^{m}\left(\frac{\theta_x}{l}\right)\right)^{\frac{1}{C_n^m}}\right)^{\frac{1}{m}}}, \left(\left[\left(1-\prod_{k=1}^{C_n^m}\left(1-\prod_{j=1}^{m}T^L{}_x\right)^{\frac{1}{C_n^m}}\right)^{\frac{1}{m}}, \left(1-\prod_{k=1}^{C_n^m}\left(1-\prod_{j=1}^{m}T^U{}_x\right)^{\frac{1}{C_n^m}}\right)^{\frac{1}{m}}\right],\right.\right.$$

$$\left[1-\left(1-\prod_{k=1}^{C_n^m}\left(1-\prod_{j=1}^{m}(1-I^L{}_x)\right)^{\frac{1}{C_n^m}}\right)^{\frac{1}{m}}, 1-\left(1-\prod_{k=1}^{C_n^m}\left(1-\prod_{j=1}^{m}(1-I^U{}_x)\right)^{\frac{1}{C_n^m}}\right)^{\frac{1}{m}}\right],$$

$$\left.\left.\left[1-\left(1-\prod_{k=1}^{C_n^m}\left(1-\prod_{j=1}^{m}(1-F^L{}_x)\right)^{\frac{1}{C_n^m}}\right)^{\frac{1}{m}}, 1-\left(1-\prod_{k=1}^{C_n^m}\left(1-\prod_{j=1}^{m}(1-F^U{}_x)\right)^{\frac{1}{C_n^m}}\right)^{\frac{1}{m}}\right]\right)\right\rangle$$

$$= \langle s_{\theta_x}, (Tx, Ix, Fx)\rangle = x.$$

2 This property is clear and it is now omitted.

3 If $\alpha_i \le \beta_i$, $T^L(x_i) \le T^L(y_i)$, $T^U(x_i) \le T^U(y_i)$, $I^L(x_i) \ge I^L(y_i)$, $I^U(x_i) \ge I^U(y_i)$, $F^L(x_i) \ge F^L(y_i)$ and $F^U(x_i) \ge F^U(y_i)$ for all i, according to Theorem 1. Since

$$\prod_{j=1}^{m}\alpha_i \le \prod_{j=1}^{m}\beta_i, \prod_{j=1}^{m}T^L(x_i) \le \prod_{j=1}^{m}T^L(y_i), \prod_{j=1}^{m}T^U(x_i) \le \prod_{j=1}^{m}T^U(y_i), \prod_{j=1}^{m}I^L(x_i) \ge \prod_{j=1}^{m}I^L(y_i),$$
$$\prod_{j=1}^{m}I^U(x_i) \ge \prod_{j=1}^{m}I^U(y_i), \prod_{j=1}^{m}F^L(x_i) \ge \prod_{j=1}^{m}F^L(y_i), \prod_{j=1}^{m}F^U(x_i) \ge \prod_{j=1}^{m}F^U(y_i)$$

$$\text{then } l\cdot\left(1-\prod_{k=1}^{C_n^m}\left(1-\prod_{j=1}^{m}\frac{\alpha_i}{l}\right)^{\frac{1}{C_n^m}}\right)^{\frac{1}{m}} \le l\cdot\left(1-\prod_{k=1}^{C_n^m}\left(1-\prod_{j=1}^{m}\frac{\beta_i}{l}\right)^{\frac{1}{C_n^m}}\right)^{\frac{1}{m}},$$

$$\left(1-\prod_{k=1}^{C_n^m}\left(1-\prod_{j=1}^{m}T^L(x_i)\right)^{\frac{1}{C_n^m}}\right)^{\frac{1}{m}} \le \left(1-\prod_{k=1}^{C_n^m}\left(1-\prod_{j=1}^{m}T^L(y_i)\right)^{\frac{1}{C_n^m}}\right)^{\frac{1}{m}},$$

$$\left(1-\prod_{k=1}^{C_n^m}\left(1-\prod_{j=1}^{m}T^U(x_i)\right)^{\frac{1}{C_n^m}}\right)^{\frac{1}{m}} \leq \left(1-\prod_{k=1}^{C_n^m}\left(1-\prod_{j=1}^{m}T^U(y_i)\right)^{\frac{1}{C_n^m}}\right)^{\frac{1}{m}},$$

$$1-\left(1-\prod_{k=1}^{C_n^m}\left(1-\prod_{j=1}^{m}\left(1-I^L(x_i)\right)\right)^{\frac{1}{C_n^m}}\right)^{\frac{1}{m}} \geq 1-\left(1-\prod_{k=1}^{C_n^m}\left(1-\prod_{j=1}^{m}\left(1-I^L(y_i)\right)\right)^{\frac{1}{C_n^m}}\right)^{\frac{1}{m}},$$

$$1-\left(1-\prod_{k=1}^{C_n^m}\left(1-\prod_{j=1}^{m}\left(1-I^U(x_i)\right)\right)^{\frac{1}{C_n^m}}\right)^{\frac{1}{m}} \geq 1-\left(1-\prod_{k=1}^{C_n^m}\left(1-\prod_{j=1}^{m}\left(1-I^U(y_i)\right)\right)^{\frac{1}{C_n^m}}\right)^{\frac{1}{m}},$$

$$1-\left(1-\prod_{k=1}^{C_n^m}\left(1-\prod_{j=1}^{m}\left(1-F^L(x_i)\right)\right)^{\frac{1}{C_n^m}}\right)^{\frac{1}{m}} \geq 1-\left(1-\prod_{k=1}^{C_n^m}\left(1-\prod_{j=1}^{m}\left(1-F^L(y_i)\right)\right)^{\frac{1}{C_n^m}}\right)^{\frac{1}{m}},$$

$$1-\left(1-\prod_{k=1}^{C_n^m}\left(1-\prod_{j=1}^{m}\left(1-F^U(x_i)\right)\right)^{\frac{1}{C_n^m}}\right)^{\frac{1}{m}} \geq 1-\left(1-\prod_{k=1}^{C_n^m}\left(1-\prod_{j=1}^{m}\left(1-F^U(y_i)\right)\right)^{\frac{1}{C_n^m}}\right)^{\frac{1}{m}}.$$

Therefore, we can get the following conclusion.

$$INLMSM^{(m)}(x_1, x_2, ..., x_n) \leq INLMSM^{(m)}(y_1, y_2, ..., y_n)$$

4 According to the idempotency, let $\min\{x_1, x_2, ...x_n\} = x_a = INLMSM^{(m)}(x_a, x_a, ..., x_a)$ and $\max\{x_1, x_2, ...x_n\} = x_b = INLMSM^{(m)}(x_b, x_b, ..., x_b)$. According to the monotonicity, if $x_a \leq x_i$ and $x_b \geq x_i$ for all i, then we have $x_a = INLMSM^{(m)}(x_a, x_a, ..., x_a) \leq INLMSM^{(m)}(x_1, x_2, ..., x_n)$ and

$$INLMSM^{(m)}(x_1, x_2, ..., x_n) \leq x_b = INLMSM^{(m)}(x_b, x_b, ..., x_b).$$

Therefore, we can get the conclusion below.

$$\min\{x_1, x_2, ...x_n\} \leq INLMSM^{(m)}\{x_1, x_2, ...x_n\} \leq \max\{x_1, x_2, ...x_n\}.$$

Furthermore, the $INLMSM^{(m)}$ operator would degrade to some particular forms when m takes some special values.

(1) When $m = 1$, we have the formula below.

$$\begin{gathered} INLMSM^{(1)}(x_1, x_2, ...x_n) = \left(\frac{\oplus_{i=1}^n x_i}{C_n^1}\right) = \\ \left\langle s_{l\cdot(1-\prod_{k=1}^n(1-\frac{k}{l})^{\frac{1}{n}})}, \left(\left[1-\prod_{k=1}^n\left(1-T^L{}_k\right)^{\frac{1}{n}}, 1-\prod_{k=1}^n\left(1-T^U{}_k\right)^{\frac{1}{n}}\right],\right.\right. \\ \left.\left.\left[\prod_{k=1}^n\left(I^L{}_k\right)^{\frac{1}{n}}, \prod_{k=1}^n\left(I^U{}_k\right)^{\frac{1}{n}}\right], \left[\prod_{k=1}^n\left(F^L{}_k\right)^{\frac{1}{n}}, \prod_{k=1}^n\left(F^U{}_k\right)^{\frac{1}{n}}\right]\right)\right\rangle \end{gathered} \tag{31}$$

(2) When $m = 2$, we have the formula below.

$$\begin{gathered}INLMSM^{(2)}(x_1, x_2, \ldots x_n) = \\ \left\langle s_{l\cdot\left(1-\prod_{k=1}^{C_n^2}\left(1-\left(\frac{\theta i_1(k)}{l}\right)\cdot\left(\frac{\theta i_2(k)}{l}\right)\right)^{\frac{1}{C_n^2}}\right)^{\frac{1}{2}}}, \left(\left[\left(1-\prod_{k=1}^{C_n^2}\left(1-T^L{}_{i_1(k)}\cdot T^L{}_{i_2(k)}\right)^{\frac{1}{C_n^2}}\right)^{\frac{1}{2}}, \left(1-\prod_{k=1}^{C_n^2}\left(1-T^U{}_{i_1(k)}\cdot T^U{}_{i_2(k)}\right)^{\frac{1}{C_n^2}}\right)^{\frac{1}{2}}\right],\right.\right. \\ \left[1-\left(1-\prod_{k=1}^{C_n^2}(1-(1-I^L i_1(k))\cdot(1-I^L i_2(k)))^{\frac{1}{C_n^2}}\right)^{\frac{1}{2}}, 1-\left(1-\prod_{k=1}^{C_n^2}(1-(1-I^U i_1(k))\cdot(1-I^U i_2(k)))^{\frac{1}{C_n^2}}\right)^{\frac{1}{2}}\right], \\ \left.\left.\left[1-\left(1-\prod_{k=1}^{C_n^2}(1-(1-F^L i_1(k))\cdot(1-F^L i_2(k)))^{\frac{1}{C_n^2}}\right)^{\frac{1}{2}}, 1-\left(1-\prod_{k=1}^{C_n^2}(1-(1-F^U i_1(k))\cdot(1-F^U i_2(k)))^{\frac{1}{C_n^2}}\right)^{\frac{1}{2}}\right]\right)\right\rangle\end{gathered} \tag{32}$$

(3) When $m = n$, the $INLMSM^{(m)}$ operator would reduce to the following form.

$$\begin{gathered}INLMSM^{(n)}(x_1, \ldots, x_n) = \\ \left\langle s_{l\cdot\left(\prod_{j=1}^{n}\left(\frac{\theta_j}{l}\right)\right)^{\frac{1}{n}}}, \left(\left[\left(\prod_{j=1}^{n} T^L{}_j\right)^{\frac{1}{n}}, \left(\prod_{j=1}^{n} T^U{}_j\right)^{\frac{1}{n}}\right], \left[1-\left(\prod_{j=1}^{n}(1-I^L{}_j)\right)^{\frac{1}{n}}, 1-\left(\prod_{j=1}^{n}(1-I^U{}_j)\right)^{\frac{1}{n}}\right],\right.\right. \\ \left.\left.\left[1-\left(\prod_{j=1}^{n}(1-F^L{}_j)\right)^{\frac{1}{n}}, 1-\left(\prod_{j=1}^{n}(1-F^U{}_j)\right)^{\frac{1}{n}}\right]\right)\right\rangle\end{gathered} \tag{33}$$

□

Definition 14. *Let* $a_i = \langle s_{\theta_i}, ([T^L(a_i), T^U(a_i)], [I^L(a_i), I^U(a_i)], [F^L(a_i), F^U(a_i)])\rangle$ $(i = 1, 2, \ldots, n)$ *be a set of INLNs. Then the INLGMSM operator:* $\Omega^n \to \Omega$ *is shown below.*

$$INLGMSM^{(m,p_1,p_2,\ldots,p_m)}(a_1, \ldots, a_n) = \left(\frac{\underset{1\le i_1<\ldots<i_m<n}{\oplus}\left(\overset{m}{\underset{j=1}{\otimes}} a_{i_j}^{p_j}\right)}{C_n^m}\right)^{\frac{1}{p_1+p_2+\ldots+p_m}}, \tag{34}$$

Ω is a set of INLNs and $m = 1, 2, \ldots, n$.

According to the operational laws of INLNs in Definition 10, we can get the expression of the INLMSM operator shown below.

Theorem 2. *Let* $a_i - \langle s_{\theta_i}, ([T^L(a_i), T^U(a_i)], [I^L(a_i), I^U(a_i)], [F^L(a_i), F^U(a_i)])\rangle$ $(i = 1, 2, \ldots, n)$ *be a set of INLNs and* $m = 1, 2, \ldots, n$. *Then the value aggregated from Definition 14 is still an INLN.*

$$\begin{gathered}INLGMSM^{(m,p_1,p_2,\ldots,p_m)}(a_1, \ldots, a_n) = \left\langle s_{l\cdot\left(1-\prod_{k=1}^{C_n^m}\left(1-\prod_{j=1}^{m}\left(\frac{\theta i_j(k)}{l}\right)^{p_j}\right)^{\frac{1}{C_n^m}}\right)^{\frac{1}{p_1+p_2+\ldots+p_m}}},\right. \\ \left(\left[\left(1-\prod_{k=1}^{C_n^m}\left(1-\prod_{j=1}^{m}\left(T^L{}_{i_j(k)}\right)^{p_j}\right)^{\frac{1}{C_n^m}}\right)^{\frac{1}{p_1+p_2+\ldots+p_m}},\right.\right. \\ \left.\left(1-\prod_{k=1}^{C_n^m}\left(1-\prod_{j=1}^{m}\left(T^U{}_{i_j(k)}\right)^{p_j}\right)^{\frac{1}{C_n^m}}\right)^{\frac{1}{p_1+p_2+\ldots+p_m}}\right], \\ \left[1-\left(1-\prod_{k=1}^{C_n^m}\left(1-\prod_{j=1}^{m}\left(1-I^L{}_{i_j(k)}\right)^{p_j}\right)^{\frac{1}{C_n^m}}\right)^{\frac{1}{p_1+p_2+\ldots+p_m}},\right. \\ \left.1-\left(1-\prod_{k=1}^{C_n^m}\left(1-\prod_{j=1}^{m}\left(1-I^U{}_{i_j(k)}\right)^{p_j}\right)^{\frac{1}{C_n^m}}\right)^{\frac{1}{p_1+p_2+\ldots+p_m}}\right], \\ \left.\left.\left[1-\left(1-\prod_{k=1}^{C_n^m}\left(1-\prod_{j=1}^{m}\left(1-F^L{}_{i_j(k)}\right)^{p_j}\right)^{\frac{1}{C_n^m}}\right)^{\frac{1}{p_1+p_2+\ldots+p_m}}, 1-\left(1-\prod_{k=1}^{C_n^m}\left(1-\prod_{j=1}^{m}\left(1-F^U{}_{i_j(k)}\right)^{p_j}\right)^{\frac{1}{C_n^m}}\right)^{\frac{1}{p_1+p_2+\ldots+p_m}}\right]\right)\right\rangle\end{gathered} \tag{35}$$

where $k = 1,2,...C_n^m$, $a_{i_j(k)}$ *is the* i_j*th element of* k_j*th permutation. Therefore, Theorem 2 is kept. The process of proof is similar to Theorem 1 and is now omitted.*

Property 2. *Let* $x_i = \langle s_{\alpha_i}, ([T^L(x_i), T^U(x_i)], [I^L(x_i), I^U(x_i)], [F^L(x_i), F^U(x_i)]) \rangle$ $(i = 1,2,...,n)$ *and* $y_i = \langle s_{\beta_i}, ([T^L(y_i), T^U(y_i)], [I^L(y_i), I^U(y_i)], [F^L(y_i), F^U(y_i)]) \rangle$ $(i = 1,2,...,n)$ *be two sets of INLNs. There are four properties of* $INLGMSM^{(m,p_1,p_2,...,p_m)}$ *operator shown as follows.*

1. Idempotency. If the INLNs $x_i = x = \langle s_{\theta_x}, ([T^L{}_x, T^U{}_x], [I^L{}_x, I^U{}_x], [F^L{}_x, F^U{}_x]) \rangle$ for each $i (i = 1,2,...,n)$ and then $INLGMSM^{(m,p_1,p_2,...,p_m)} = x = \langle s_{\theta_x}, (T_x, I_x, F_x) \rangle$.
2. Commutativity. If x_i is a permutation of y_i for all I $(i = 1,2,...,n)$, and then $INLGMSM^{(m,p_1,p_2,...,p_m)}(x_1, x_2, ..., x_n) = INLGMSM^{(m,p_1,p_2,...,p_m)}(y_1, y_2, ..., y_n)$.
3. Monotonicity. If $\alpha_i \le \beta_i$, $T^L(x_i) \le T^L(y_i)$, $T^U(x_i) \le T^U(y_i)$, $I^L(x_i) \ge I^L(y_i)$, $I^U(x_i) \ge I^U(y_i)$, $F^L(x_i) \ge F^L(y_i)$ and $F^U(x_i) \ge F^U(y_i)$ for all i $(i = 1,2,...,n)$, then $x_i \le y_i$ and $INLGMSM^{(m,p_1,p_2,...,p_m)}(x_1, x_2, ..., x_n) \le INLGMSM^{(m,p_1,p_2,...,p_m)}(y_1, y_2, ..., y_n)$.
4. Boundedness. $\min\{x_1, x_2, ...x_n\} \le INLGMSM^{(m,p_1,p_2,...,p_m)}\{x_1, x_2, ...x_n\} \le \max\{x_1, x_2, ...x_n\}$.

The proofs are similar to Property 1, which are now omitted.

Furthermore, the $INLGMSM^{(m,p_1,p_2,...,p_m)}$ operator would degrade to some particular forms when m takes some special values.

(1) When $m = 1$, we have the following formula.

$$INLGMSM^{(1)}(x_1, x_2, ..., x_n) = \left(\frac{\oplus_{i=1}^n x i^{p_1}}{C_n^1}\right)^{\frac{1}{p_1}} = \\ \left\langle s_{l \cdot (1 - \prod_{k=1}^n (1 - (\frac{k}{l})^{p_1})^{\frac{1}{n}})^{\frac{1}{p_1}}}, \left(\left[\left(1 - \prod_{k=1}^n \left(1 - (T^L{}_k)^{p_1}\right)^{\frac{1}{n}}\right)^{\frac{1}{p_1}}, \left(1 - \prod_{k=1}^n \left(1 - (T^U{}_k)^{p_1}\right)^{\frac{1}{n}}\right)^{\frac{1}{p_1}} \right], \right. \right. \\ \left[1 - \left(1 - \prod_{k=1}^n \left(1 - (1 - I^L{}_{i_1}(k))^{p_1}\right)^{\frac{1}{n}}\right)^{\frac{1}{p_1}}, 1 - \left(1 - \prod_{k=1}^n \left(1 - (1 - I^U{}_{i_1}(k))^{p_1}\right)^{\frac{1}{n}}\right)^{\frac{1}{p_1}} \right], \\ \left. \left. \left[1 - \left(1 - \prod_{k=1}^n \left(1 - (1 - F^L{}_{i_1}(k))^{p_1}\right)^{\frac{1}{n}}\right)^{\frac{1}{p_1}}, 1 - \left(1 - \prod_{k=1}^n \left(1 - (1 - F^U{}_{i_1}(k))^{p_1}\right)^{\frac{1}{n}}\right)^{\frac{1}{p_1}} \right] \right) \right\rangle \tag{36}$$

(2) When $m = 2$, we have the following formula.

$$INLMSM^{(2)}(x_1, x_2, ..., x_n) = \\ \left\langle s_{l \cdot (1 - \prod_{k=1}^{C_n^2} (1 - (\frac{\theta i_1(k)}{l})^{p_1} \cdot (\frac{\theta i_2(k)}{l})^{p_2})^{\frac{1}{C_n^2}})^{\frac{1}{p_1+p_2}}}, \right. \\ \left(\left[\left(1 - \prod_{k=1}^{C_n^2} \left(1 - \left(T^L{}_{i_1(k)}\right)^{p_1} \cdot \left(T^L{}_{i_2(k)}\right)^{p_2}\right)^{\frac{1}{C_n^2}}\right)^{\frac{1}{p_1+p_2}}, \left(1 - \prod_{k=1}^{C_n^2} \left(1 - \left(T^U{}_{i_1(k)}\right)^{p_1} \cdot \left(T^U{}_{i_2(k)}\right)^{p_2}\right)^{\frac{1}{C_n^2}}\right)^{\frac{1}{p_1+p_2}} \right] \right. \\ \left[1 - \left(1 - \prod_{k=1}^{C_n^2} \left(1 - (1 - I^L{}_{i_1}(k))^{p_1} \cdot (1 - I^L{}_{i_2}(k))^{p_2}\right)^{\frac{1}{C_n^2}}\right)^{\frac{1}{p_1+p_2}}, \right. \\ \left. 1 - \left(1 - \prod_{k=1}^{C_n^2} \left(1 - (1 - I^U{}_{i_1}(k))^{p_1} \cdot (1 - I^U{}_{i_2}(k))^{p_2}\right)^{\frac{1}{C_n^2}}\right)^{\frac{1}{p_1+p_2}} \right], \\ \left[1 - \left(1 - \prod_{k=1}^{C_n^2} \left(1 - (1 - F^L{}_{i_1}(k))^{p_1} \cdot (1 - F^L{}_{i_2}(k))^{p_2}\right)^{\frac{1}{C_n^2}}\right)^{\frac{1}{p_1+p_2}}, \right. \\ \left. \left. \left. 1 - \left(1 - \prod_{k=1}^{C_n^2} \left(1 - (1 - F^U{}_{i_1}(k))^{p_1} \cdot (1 - F^U{}_{i_2}(k))^{p_2}\right)^{\frac{1}{C_n^2}}\right)^{\frac{1}{p_1+p_2}} \right] \right) \right\rangle \tag{37}$$

When $m = 2$, the $INLGMSM^{(m,p_1,p_2,...,p_m)}$ operator would reduce to the BM for INLNs (INLGBM) operator.

(3) When $m = n$, the $INLMSM^{(m)}$ operator would reduce to the form below.

$$INLGMSM^{(n,p_1,p_2,\ldots,p_m)}(a_1,\ldots,a_n) = \left\langle s_{l\cdot\left(\prod_{j=1}^{n}\left(\frac{\theta_{i_j(k)}}{l}\right)^{p_j}\right)^{\frac{1}{p_1+p_2+\ldots+p_n}}}, \left(\left[\left(\prod_{j=1}^{n}\left(T^L{}_{i_j(k)}\right)^{p_j}\right)^{\frac{1}{p_1+p_2+\ldots+p_n}}, \left(\prod_{j=1}^{n}\left(T^U{}_{i_j(k)}\right)^{p_j}\right)^{\frac{1}{p_1+p_2+\ldots+p_n}}\right], \left[1-\left(\prod_{j=1}^{m}\left(1-I^L{}_{i_j(k)}\right)^{p_j}\right)^{\frac{1}{p_1+p_2+\ldots+p_n}}, 1-\left(\prod_{j=1}^{m}\left(1-I^U{}_{i_j(k)}\right)^{p_j}\right)^{\frac{1}{p_1+p_2+\ldots+p_m}}\right], \left[1-\left(\prod_{j=1}^{m}\left(1-F^L{}_{i_j(k)}\right)^{p_j}\right)^{\frac{1}{p_1+p_2+\ldots+p_n}}, 1-\left(\prod_{j=1}^{m}\left(1-F^U{}_{i_j(k)}\right)^{p_j}\right)^{\frac{1}{p_1+p_2+\ldots+p_n}}\right]\right)\right\rangle \quad (38)$$

4.2. Some Weighted INLMSM Operators

We will introduce two operators, which are the weighted forms of the *INLMSM* operator and *INLGMSM* operator.

Definition 15. *Let* $a_i = \langle s_{\theta_i}, ([T^L(a_i), T^U(a_i)], [I^L(a_i), I^U(a_i)], [F^L(a_i), F^U(a_i)])\rangle$ i $(i = 1,2,\ldots,n)$ *be a set of INLNs. Let* $\omega = (\omega_1, \omega_2, \ldots, \omega_n)$ T *is the weight vector and satisfies* $\sum_{i=1}^{n}\omega_i = 1$ *with* $\omega_i > 0$ $(i = 1,2,\ldots,n)$*. Each* ω_i *represents the importance of* a_i*. Then the WINLMSM operator:* $\Omega^n \to \Omega$ *is defined below.*

$$WINLMSM^{(m)}(a_1,\ldots,a_n) = \left(\frac{\underset{1\le i_1<\ldots<i_m\le n}{\oplus}\left(\overset{m}{\underset{j=1}{\otimes}}\left(n\omega_{i_j}\right)a_{i_j}\right)}{C_n^m}\right)^{\frac{1}{m}}, \quad (39)$$

Ω is a set of INLNs and $m = 1,2,\ldots,n$.

According to the operational laws of INLNs in Definition 10, we can get the expression of the WINLMSM operator, which is shown below.

Theorem 3. *Let* $a_i = \langle s_{\theta_i}, ([T^L(a_i), T^U(a_i)], [I^L(a_i), I^U(a_i)], [F^L(a_i), F^U(a_i)])\rangle$ i $(i = 1,2,\ldots,n)$ *be a set of INLNs and* $m = 1,2,\ldots,n$*, then the value aggregated from Definition 15 is still a WINLMSM operator.*

$$WINLMSM^{(m)}(a_1,\ldots,a_n) = \left\langle s_{l\cdot\left(1-\prod_{k=1}^{C_n^m}\left(1-\prod_{j=1}^{m}\left(1-\left(1-\frac{\theta_{i_j(k)}}{l}\right)^{n\cdot\omega_{i_j}}\right)\right)^{\frac{1}{C_n^m}}\right)^{\frac{1}{m}}}, \left(\left[\left(1-\prod_{k=1}^{C_n^m}\left(1-\prod_{j=1}^{m}\left(1-\left(1-T^L{}_{i_j(k)}\right)^{n\cdot\omega_{ij}}\right)\right)^{\frac{1}{C_n^m}}\right)^{\frac{1}{m}}, \left(1-\prod_{k=1}^{C_n^m}\left(1-\prod_{j=1}^{m}\left(1-\left(1-T^U{}_{i_j(k)}\right)^{n\cdot\omega_{ij}}\right)\right)^{\frac{1}{C_n^m}}\right)^{\frac{1}{m}}\right], \left[1-\left(1-\prod_{k=1}^{C_n^m}\left(1-\prod_{j=1}^{m}\left(1-\left(I^L{}_{i_j(k)}\right)^{n\cdot\omega_{ij}}\right)\right)^{\frac{1}{C_n^m}}\right)^{\frac{1}{m}}, 1-\left(1-\prod_{k=1}^{C_n^m}\left(1-\prod_{j=1}^{m}\left(1-\left(I^U{}_{i_j(k)}\right)^{n\cdot\omega_{ij}}\right)\right)^{\frac{1}{C_n^m}}\right)^{\frac{1}{m}}\right], \left[1-\left(1-\prod_{k=1}^{C_n^m}\left(1-\prod_{j=1}^{m}\left(1-\left(F^L{}_{i_j(k)}\right)^{n\cdot\omega_{ij}}\right)\right)^{\frac{1}{C_n^m}}\right)^{\frac{1}{m}}, 1-\left(1-\prod_{k=1}^{C_n^m}\left(1-\prod_{j=1}^{m}\left(1-\left(F^U{}_{i_j(k)}\right)^{n\cdot\omega_{ij}}\right)\right)^{\frac{1}{C_n^m}}\right)^{\frac{1}{m}}\right]\right)\right\rangle \quad (40)$$

where $k = 1,2,\ldots,C_n^m$, $a_{i_j(k)}$ *is the* i_j*th element of* k*th permutation. The process of proof is similar to Theorem 1. Now it is omitted.*

Property 3. *Let* $x_i = \langle s_{\alpha_i}, ([T^L(x_i), T^U(x_i)], [I^L(x_i), I^U(x_i)], [F^L(x_i), F^U(x_i)]) \rangle$ $(i = 1, 2, ..., n)$ *and* $y_i = \langle s_{\beta_i}, ([T^L(y_i), T^U(y_i)], [I^L(y_i), I^U(y_i)], [F^L(y_i), F^U(y_i)]) \rangle$ $(i = 1, 2, ..., n)$ *be sets of INLNs. There are some properties of the* $WINLMSM^{(m)}$ *operator as shown below.*

1 *Reducibility. When* $\omega = (\frac{1}{n}, \frac{1}{n}, ..., \frac{1}{n})$, *then* $WINLMSM^{(m)}(a_1, a_2, ..., a_n) = INLMSM^{(m)}(a_1, a_2, ..., a_n)$.

2 *Monotonicity. If* $\alpha_i \leq \beta_i$, $T^L(x_i) \leq T^L(y_i)$, $T^U(x_i) \leq T^U(y_i)$, $I^L(x_i) \geq I^L(y_i)$, $I^U(x_i) \geq I^U(y_i)$, $F^L(x_i) \geq F^L(y_i)$ *and* $F^U(x_i) \geq F^U(y_i)$ *for all* i $(i = 1, 2, ..., n)$, *then* $x_i \leq y_i$ *and* $WINLMSM^{(m)}(x_1, x_2, ..., x_n) \leq WINLMSM^{(m)}(y_1, y_2, ..., y_n)$.

3 *Boundedness.* $\min\{x_1, x_2, ... x_n\} \leq WINLMSM^{(m)}\{x_1, x_2, ... x_n\} \leq \max\{x_1, x_2, ... x_n\}$.

Proof.

1 If $\omega = (\frac{1}{n}, \frac{1}{n}, ..., \frac{1}{n})$, then $WINLMSM^{(m)}(a_1, a_2, ..., a_n) =$

$$\left\langle s_{l \cdot \left(1 - \prod_{k=1}^{C_n^m} \left(1 - \prod_{j=1}^{m} \left(1 - \left(1 - \frac{\theta_{i_j(k)}}{l}\right)^{n \cdot \frac{1}{n}}\right)\right)^{\frac{1}{C_n^m}}\right)^{\frac{1}{m}}}, \left(\left[\left(1 - \prod_{k=1}^{C_n^m}\left(1 - \prod_{j=1}^{m}\left(1 - \left(1 - T^L{}_{i_j(k)}\right)\right)\right)^{\frac{1}{C_n^m}}\right)^{\frac{1}{m}}, \left(1 - \prod_{k=1}^{C_n^m}\left(1 - \prod_{j=1}^{m}\left(1 - \left(1 - T^U{}_{i_j(k)}\right)\right)\right)^{\frac{1}{C_n^m}}\right)^{\frac{1}{m}}\right], \left[1 - \left(1 - \prod_{k=1}^{C_n^m}\left(1 - \prod_{j=1}^{m}\left(1 - \left(I^L{}_{i_j(k)}\right)\right)\right)^{\frac{1}{C_n^m}}\right)^{\frac{1}{m}}, 1 - \left(1 - \prod_{k=1}^{C_n^m}\left(1 - \prod_{j=1}^{m}\left(1 - \left(I^U{}_{i_j(k)}\right)\right)\right)^{\frac{1}{C_n^m}}\right)^{\frac{1}{m}}\right] \left[1 - \left(1 - \prod_{k=1}^{C_n^m}\left(1 - \prod_{j=1}^{m}\left(1 - \left(F^L{}_{i_j(k)}\right)\right)\right)^{\frac{1}{C_n^m}}\right)^{\frac{1}{m}}, 1 - \left(1 - \prod_{k=1}^{C_n^m}\left(1 - \prod_{j=1}^{m}\left(1 - \left(F^U{}_{i_j(k)}\right)\right)\right)^{\frac{1}{C_n^m}}\right)^{\frac{1}{m}}\right]\right)\right\rangle = INLMSM^{(m)}(a_1, a_2, ..., a_n).$$

2 The proofs of Monotonicity and Boundedness are similar to Property 1, which are now omitted.

Furthermore, the $WINLMSM^{(m)}$ operator would degrade a particular form when m takes some special values.

(1) When $m = 1$, we have the formula below.

$$WINLMSM^{(1)}(a_1, \ldots, a_n) = \left\langle s_{l \cdot \left(1 - \prod_{i=1}^{n}\left(1 - \frac{\theta_i}{l}\right)^{\omega_i}\right)}, \left(\left[\left(1 - \prod_{i=1}^{n}\left(1 - T_i^L\right)^{\omega_i}\right), \left(1 - \prod_{i=1}^{n}\left(1 - T_i^U\right)^{\omega_i}\right)\right], \left[\prod_{i=1}^{n}\left(I_i^L\right)^{\omega_i}, \prod_{i=1}^{n}\left(I_i^U\right)^{\omega_i}\right], \left[\prod_{i=1}^{n}\left(F_i^L\right)^{\omega_i}, \prod_{i=1}^{n}\left(F_i^U\right)^{\omega_i}\right]\right)\right\rangle \tag{41}$$

(2) When $m = 2$, we have the formula below.

$$WINLMSM^{(2)}(a_1, \ldots, a_n) = \left\langle s_{l \cdot \left(1 - \prod_{k=1}^{C_n^2}\left(1 - \left(1 - \left(1 - \frac{\theta_{i_1(k)}}{l}\right)^{n \cdot \omega_{i_1}}\right) \cdot \left(1 - \left(1 - \frac{\theta_{i_2(k)}}{l}\right)^{n \cdot \omega_{i_2}}\right)\right)^{\frac{1}{C_n^2}}\right)^{\frac{1}{2}}}, \left(\left[\left(1 - \prod_{k=1}^{C_n^2}\left(1 - \left(1 - \left(1 - T^L_{i_1(k)}\right)^{n \cdot \omega_{i_1}}\right) \cdot \left(1 - \left(1 - T^L_{i_2(k)}\right)^{n \cdot \omega_{i_2}}\right)\right)^{\frac{1}{C_n^2}}\right)^{\frac{1}{2}}, \left(1 - \prod_{k=1}^{C_n^2}\left(1 - \left(1 - \left(1 - T^U_{i_1(k)}\right)^{n \cdot \omega_{i_1}}\right) \cdot \left(1 - \left(1 - T^U_{i_2(k)}\right)^{n \cdot \omega_{i_2}}\right)\right)^{\frac{1}{C_n^2}}\right)^{\frac{1}{2}}\right], \left[1 - \left(1 - \prod_{k=1}^{C_n^2}\left(1 - \left(1 - \left(I^L_{i_1(k)}\right)^{n \cdot \omega_{i_1}}\right) \cdot \left(1 - \left(I^L_{i_2(k)}\right)^{n \cdot \omega_{i_2}}\right)\right)^{\frac{1}{C_n^2}}\right)^{\frac{1}{2}}, 1 - \left(1 - \prod_{k=1}^{C_n^2}\left(1 - \left(1 - \left(I^U_{i_1(k)}\right)^{n \cdot \omega_{i_1}}\right) \cdot \left(1 - \left(I^U_{i_2(k)}\right)^{n \cdot \omega_{i_2}}\right)\right)^{\frac{1}{C_n^2}}\right)^{\frac{1}{2}}\right], \left[1 - \left(1 - \prod_{k=1}^{C_n^2}\left(1 - \left(1 - \left(F^L_{i_1(k)}\right)^{n \cdot \omega_{i_1}}\right) \cdot \left(1 - \left(F^L_{i_2(k)}\right)^{n \cdot \omega_{i_2}}\right)\right)^{\frac{1}{C_n^2}}\right)^{\frac{1}{2}}, 1 - \left(1 - \prod_{k=1}^{C_n^2}\left(1 - \left(1 - \left(F^U_{i_1(k)}\right)^{n \cdot \omega_{i_1}}\right) \cdot \left(1 - \left(F^U_{i_2(k)}\right)^{n \cdot \omega_{i_2}}\right)\right)^{\frac{1}{C_n^2}}\right)^{\frac{1}{2}}\right]\right)\right\rangle \tag{42}$$

(3) When $m = n$, we have the formula below.

$$\begin{aligned} WINLMSM^{(n)}(a_1,\ldots,a_n) = \Bigg\langle & s_{l\cdot\left(\prod_{j=1}^{n}\left(1-\left(1-\frac{\theta_j}{l}\right)^{n\cdot\omega_j}\right)\right)^{\frac{1}{n}}}, \\ & \Bigg(\Bigg[\left(\prod_{j=1}^{n}\left(1-\left(1-T_j^L\right)^{n\cdot\omega_j}\right)\right)^{\frac{1}{n}}, \left(\prod_{j=1}^{n}\left(1-\left(1-T_j^U\right)^{n\cdot\omega_j}\right)\right)^{\frac{1}{n}}\Bigg], \\ & \Bigg[1-\left(\prod_{j=1}^{n}\left(1-\left(I_j^L\right)^{n\cdot\omega_j}\right)\right)^{\frac{1}{n}}, 1-\left(\prod_{j=1}^{n}\left(1-\left(I_j^U\right)^{n\cdot\omega_j}\right)\right)^{\frac{1}{n}}\Bigg], \\ & \Bigg[1-\left(\prod_{j=1}^{n}\left(1-\left(F_j^L\right)^{n\cdot\omega_j}\right)\right)^{\frac{1}{n}}, 1-\left(\prod_{j=1}^{n}\left(1-\left(F_j^U\right)^{n\cdot\omega_j}\right)\right)^{\frac{1}{n}}\Bigg]\Bigg)\Bigg\rangle. \end{aligned} \tag{43}$$

□

Definition 16. *Let* $a_i = \langle s_{\theta_i}, ([T^L(a_i), T^U(a_i)], [I^L(a_i), I^U(a_i)], [F^L(a_i), F^U(a_i)])\rangle$ i $(i = 1, 2, \ldots, n)$ *be a set of INLNs. Let* $\omega = (\omega_1, \omega_2, \ldots, \omega_n)$ T *is the weight vector and it satisfies* $\sum_{i=1}^{n}\omega_i = 1$ *with* $\omega_i > 0$ $(i = 1, 2, \ldots, n)$*. Each* ω_i *represents the importance of* a_i*. Then the WINLGMSM operator:* $\Omega^n \to \Omega$ *is defined below.*

$$WINLGMSM^{(m,p_1,p_2,\ldots p_m)}(a_1,\ldots,a_n) = \left(\frac{\bigoplus_{1\le i_1<\ldots<i_m\le n}\left(\bigotimes_{j=1}^{m}\left(n\omega_{i_j}\cdot a_{i_j}\right)^{p_j}\right)}{C_n^m}\right)^{\frac{1}{p_1+p_2+\ldots+p_m}} \tag{44}$$

Ω is a set of INLNs and $m = 1, 2, \ldots, n$.

According to the operational laws of INLNs in Definition 10, we can get the expression of WINLMSM operator shown below.

Theorem 4. *Let* $a_i = \langle s_{\theta_i}, ([T^L(a_i), T^U(a_i)], [I^L(a_i), I^U(a_i)], [F^L(a_i), F^U(a_i)])\rangle$ i $(i = 1, 2, \ldots, n)$ *be a set of INLNs and* $m = 1, 2, \ldots, n$*. Then the value aggregated from Definition 16 is still an WINLGMSM.*

$$\begin{aligned} WINLGMSM^{(m,p_1,p_2,\ldots,p_m)}(a_1,\ldots,a_n) = \Bigg\langle & s_{l\cdot\left(1-\prod_{k=1}^{C_n^m}\left(1-\prod_{j=1}^{m}\left(1-\left(1-\frac{\theta_{i_j}(k)}{l}\right)^{n\cdot\omega_{i_j}}\right)^{p_j}\right)^{\frac{1}{C_n^m}}\right)^{\frac{1}{p_1+p_2+\ldots+p_m}}}, \\ & \Bigg(\Bigg[\left(1-\prod_{k=1}^{C_n^m}\left(1-\prod_{j=1}^{m}\left(1-\left(1-T^L i_j(k)\right)^{n\cdot\omega_{i_j}}\right)^{p_j}\right)^{\frac{1}{C_n^m}}\right)^{\frac{1}{p_1+p_2+\ldots+p_m}}, \\ & \left(1-\prod_{k=1}^{C_n^m}\left(1-\prod_{j=1}^{m}\left(1-\left(1-T^U i_j(k)\right)^{n\cdot\omega_{i_j}}\right)^{p_j}\right)^{\frac{1}{C_n^m}}\right)^{\frac{1}{p_1+p_2+\ldots+p_m}}\Bigg], \\ & \Bigg[1-\left(1-\prod_{k=1}^{C_n^m}\left(1-\prod_{j=1}^{m}\left(1-\left(I^L i_j(k)\right)^{n\cdot\omega_{i_j}}\right)^{p_j}\right)^{\frac{1}{C_n^m}}\right)^{\frac{1}{p_1+p_2+\ldots+p_m}}, \\ & 1-\left(1-\prod_{k=1}^{C_n^m}\left(1-\prod_{j=1}^{m}\left(1-\left(I^U i_j(k)\right)^{n\cdot\omega_{i_j}}\right)^{p_j}\right)^{\frac{1}{C_n^m}}\right)^{\frac{1}{p_1+p_2+\ldots+p_m}}\Bigg], \\ & \Bigg[1-\left(1-\prod_{k=1}^{C_n^m}\left(1-\prod_{j=1}^{m}\left(1-\left(F^L i_j(k)\right)^{n\cdot\omega_{i_j}}\right)^{p_j}\right)^{\frac{1}{C_n^m}}\right)^{\frac{1}{p_1+p_2+\ldots+p_m}}, \\ & 1-\left(1-\prod_{k=1}^{C_n^m}\left(1-\prod_{j=1}^{m}\left(1-\left(F^U i_j(k)\right)^{n\cdot\omega_{i_j}}\right)^{p_j}\right)^{\frac{1}{C_n^m}}\right)^{\frac{1}{p_1+p_2+\ldots+p_m}}\Bigg]\Bigg)\Bigg\rangle \end{aligned} \tag{45}$$

where $k = 1, 2, \ldots, C_n^m$, $a_{i_j(k)}$ *is the* i_j*th element of* k*th permutation. The process of proof is similar to Theorem 1. It is now omitted.*

Property 4. *Let* $x_i = \langle s_{\alpha_i}, ([T^L(x_i), T^U(x_i)], [I^L(x_i), I^U(x_i)], [F^L(x_i), F^U(x_i)]) \rangle$ $(i = 1, 2, ..., n)$ *and* $y_i = \langle s_{\beta_i}, ([T^L(y_i), T^U(y_i)], [I^L(y_i), I^U(y_i)], [F^L(y_i), F^U(y_i)]) \rangle$ $(i = 1, 2, ..., n)$ *be two sets of INLNs. There are some properties of the* $WINLGMSM^{(m,p_1,p_2,...,p_m)}$ *operator shown below.*

1 Reducibility. When $\omega = \left(\frac{1}{n}, \frac{1}{n}, ..., \frac{1}{n}\right)$. Additionally, $WINLGMSM^{(m,p_1,p_2,...,p_m)}(a_1, a_2, ..., a_n) = INLGMSM^{(m,p_1,p_2,...,p_m)}(a_1, a_2, ..., a_n)$.

2 Monotonicity. If $\alpha_i \leq \beta_i$, $T^L(x_i) \leq T^L(y_i)$, $T^U(x_i) \leq T^U(y_i)$, $I^L(x_i) \geq I^L(y_i)$, $I^U(x_i) \geq I^U(y_i)$, $F^L(x_i) \geq F^L(y_i)$ and $F^U(x_i) \geq F^U(y_i)$ for all i $(i = 1, 2, ..., n)$, then $x_i \leq y_i$ and $WINLGMSM^{(m,p_1,p_2,...,p_m)}(x_1, x_2, ..., x_n) \leq WINLGMSM^{(m,p_1,p_2,...,p_m)}(y_1, y_2, ..., y_n)$.

3 Boundedness. $\min\{x_1, x_2, ...x_n\} \leq WINLGMSMSM^{(m,p_1,p_2,...,p_m)}\{x_1, x_2, ...x_n\} \leq \max\{x_1, x_2, ...x_n\}$.

The process of proof is similar to Property 3 and is now omitted.

Furthermore, the $WINLGMSM^{(m,p_1,p_2,...,p_m)}$ operator would degrade some particular forms when m takes some special values.

(1) When $m = 1$, we have the following formula.

$$\begin{aligned}
&WINLGMSM^{(1,p_1)}(a_1, \ldots, a_n) = \Bigg\langle s_{l \cdot \left(1 - \prod_{k=1}^{n}\left(1-\left(1-\left(1-\frac{\theta_1(k)}{l}\right)^{n\cdot\omega_1}\right)^{p_1}\right)^{\frac{1}{n}}\right)^{\frac{1}{p_1}}}, \\
&\Bigg(\Bigg[\left(1 - \prod_{k=1}^{n}\left(1 - \left(1 - \left(1 - T^L_{i_j(k)}\right)^{n\cdot\omega_1}\right)^{p_1}\right)^{\frac{1}{n}}\right)^{\frac{1}{p_1}}, \left(1 - \prod_{k=1}^{n}\left(1 - \left(1 - \left(1 - T^U_{i_j(k)}\right)^{n\cdot\omega_1}\right)^{p_1}\right)^{\frac{1}{n}}\right)^{\frac{1}{p_1}}\Bigg], \\
&\Bigg[1 - \left(1 - \prod_{k=1}^{n}\left(1 - \left(1 - \left(I^L_{i_j(k)}\right)^{n\cdot\omega_1}\right)^{p_1}\right)^{\frac{1}{n}}\right)^{\frac{1}{p_1}}, 1 - \left(1 - \prod_{k=1}^{n}\left(1 - \left(1 - \left(I^U_{i_j(k)}\right)^{n\cdot\omega_1}\right)^{p_1}\right)^{\frac{1}{n}}\right)^{\frac{1}{p_1}}\Bigg], \\
&\Bigg[1 - \left(1 - \prod_{k=1}^{n}\left(1 - \left(1 - \left(F^L_{i_j(k)}\right)^{n\cdot\omega_1}\right)^{p_1}\right)^{\frac{1}{n}}\right)^{\frac{1}{p_1}}, 1 - \left(1 - \prod_{k=1}^{n}\left(1 - \left(1 - \left(F^U_{i_j(k)}\right)^{n\cdot\omega_1}\right)^{p_1}\right)^{\frac{1}{n}}\right)^{\frac{1}{p_1}}\Bigg]\Bigg)\Bigg\rangle
\end{aligned} \quad (46)$$

(2) When $m = 2$, we have the formula below.

$$\begin{aligned}
&WINLGMSM^{(2,p_1,p_2)}(a_1, \ldots, a_n) = \Bigg\langle s_{l \cdot \left(1 - \prod_{k=1}^{C_n^2}\left(1-\left(1-\left(1-\frac{\theta i_1(k)}{l}\right)^{n\cdot\omega_{i_1}}\right)^{p_1} \cdot \left(1-\left(1-\frac{\theta i_2(k)}{l}\right)^{n\cdot\omega_{i_2}}\right)^{p_2}\right)^{\frac{1}{C_n^2}}\right)^{\frac{1}{p_1+p_2}}}, \\
&\Bigg(\Bigg[\left(1 - \prod_{k=1}^{C_n^2}\left(1 - \left(1 - \left(1 - T^L_{i_1(k)}\right)^{n\cdot\omega_{i_1}}\right)^{p_1} \cdot \left(1 - \left(1 - T^L_{i_2(k)}\right)^{n\cdot\omega_{i_2}}\right)^{p_2}\right)^{\frac{1}{C_n^2}}\right)^{\frac{1}{p_1+p_2}}, \\
&\left(1 - \prod_{k=1}^{C_n^2}\left(1 - \left(1 - \left(1 - T^U_{i_1(k)}\right)^{n\cdot\omega_{i_1}}\right)^{p_1} \cdot \left(1 - \left(1 - T^U_{i_2(k)}\right)^{n\cdot\omega_{i_2}}\right)^{p_2}\right)^{\frac{1}{C_n^2}}\right)^{\frac{1}{p_1+p_2}}\Bigg], \\
&\Bigg[1 - \left(1 - \prod_{k=1}^{C_n^2}\left(1 - \left(1 - \left(I^L_{i_1(k)}\right)^{n\cdot\omega_{i_1}}\right)^{p_1} \cdot \left(1 - \left(I^L_{i_2(k)}\right)^{n\cdot\omega_{i_2}}\right)^{p_2}\right)^{\frac{1}{C_n^2}}\right)^{\frac{1}{p_1+p_2}}, \\
&1 - \left(1 - \prod_{k=1}^{C_n^2}\left(1 - \left(1 - \left(I^U_{i_1(k)}\right)^{n\cdot\omega_{i_1}}\right)^{p_1} \cdot \left(1 - \left(I^U_{i_2(k)}\right)^{n\cdot\omega_{i_2}}\right)^{p_2}\right)^{\frac{1}{C_n^2}}\right)^{\frac{1}{p_1+p_2}}\Bigg], \\
&\Bigg[1 - \left(1 - \prod_{k=1}^{C_n^2}\left(1 - \left(1 - \left(F^L_{i_1(k)}\right)^{n\cdot\omega_{i_1}}\right)^{p_1} \cdot \left(1 - \left(F^L_{i_2(k)}\right)^{n\cdot\omega_{i_2}}\right)^{p_2}\right)^{\frac{1}{C_n^2}}\right)^{\frac{1}{p_1+p_2}}, \\
&1 - \left(1 - \prod_{k=1}^{C_n^2}\left(1 - \left(1 - \left(F^U_{i_1(k)}\right)^{n\cdot\omega_{i_1}}\right)^{p_1} \cdot \left(1 - \left(F^U_{i_2(k)}\right)^{n\cdot\omega_{i_2}}\right)^{p_2}\right)^{\frac{1}{C_n^2}}\right)^{\frac{1}{p_1+p_2}}\Bigg]\Bigg)\Bigg\rangle
\end{aligned} \quad (47)$$

(3) When $m = n$, we have the formula below.

$$WINLGMSM^{(n,p_1,p_2,\ldots,p_n)}(a_1,\ldots,a_n) = \left\langle s_{l\cdot\left(\prod_{j=1}^{n}\left(1-\left(1-\frac{\theta_j}{l}\right)^{n\cdot\omega_j}\right)^{p_j}\right)^{\frac{1}{p_1+p_2+\ldots+p_n}}}, \right.$$
$$\left(\left[\left(\prod_{j=1}^{n}\left(1-\left(1-T_j^L\right)^{n\cdot\omega_j}\right)^{p_j}\right)^{\frac{1}{p_1+p_2+\ldots+p_n}}, \left(\prod_{j=1}^{n}\left(1-\left(1-T_j^U\right)^{n\cdot\omega_j}\right)^{p_j}\right)^{\frac{1}{p_1+p_2+\ldots+p_n}}\right],\right.$$
$$\left[1-\left(\prod_{j=1}^{n}\left(1-\left(I_j^L\right)^{n\cdot\omega_j}\right)^{p_j}\right)^{\frac{1}{p_1+p_2+\ldots+p_n}}, 1-\left(\prod_{j=1}^{n}\left(1-\left(I_j^U\right)^{n\cdot\omega_j}\right)^{p_j}\right)^{\frac{1}{p_1+p_2+\ldots+p_n}}\right], \tag{48}$$
$$\left.\left.\left[1-\left(\prod_{j=1}^{n}\left(1-\left(F_j^L\right)^{n\cdot\omega_j}\right)^{p_j}\right)^{\frac{1}{p_1+p_2+\ldots+p_n}}, 1-\left(\prod_{j=1}^{n}\left(1-\left(F_j^U\right)^{n\cdot\omega_j}\right)^{p_j}\right)^{\frac{1}{p_1+p_2+\ldots+p_n}}\right]\right)\right\rangle$$

5. MADM Method Based on INLMSM Operator

In this section, we introduce the MADM method based on the *WINLMSM* and *WINLGMSM* operators. Let $d = \{d_1, d_2, \ldots, d_m\}$ be a collection of alternatives and $c = \{c_1, c_2, \ldots, c_n\}$ is a collection of n criteria. The weight vector is $\omega = (\omega_1, \omega_2, \ldots, \omega_n)^T$ with satisfying $\sum_{i=1}^{n}\omega_i = 1 (\omega_i \geq 0,\ i = 1, 2, \ldots, n)$, and each ω_i represents the importance of c_j. The performance of alternative d_j in criteria c_j is surveyed by INLNs and the decision matrix is $A = (a_{ij})_{m\times n}$, where $a_{ij} = \langle s_{\theta_{ij}}, ([T^L(r_{ij}), T^U(r_{ij})], [I^L(r_{ij}), I^U(r_{ij})], [F^L(r_{ij}), F^U(r_{ij})])\rangle$. The objective is to rank the alternatives.

The detailed steps are shown below.

Step 1 Normalize the decision matrix.

We should normalize the decision-making information in the matrix. The benefit (the bigger the better) and the cost (the smaller the better) are the two possible types. In order to keep the consistency of the types, it is necessary to convert the decision matrix A into a standardized matrix $R = (r_{ij})m \times n$.

If c_j is cost type, then $r_{ij} = \langle s_{\theta_{ij}}, ([F^L(r_{ij}), F^U(r_{ij})], [1 - I^U(r_{ij}), 1 - I^L(r_{ij})][T^L(r_{ij}), T^U(r_{ij})])\rangle$

else $r_{ij} = \langle s_{\theta_{ij}}, ([T^L(r_{ij}), T^U(r_{ij})], [I^L(r_{ij}), I^U(r_{ij})], [F^L(r_{ij}), F^U(r_{ij})])\rangle$.

Step 2 Aggregate the criterion values of each alternative. We would use Definition 15 and Definition 16 to aggregate $r_{ij}(j = 1, 2, \ldots, n)$ of the ith alternative and get the overall value r_i.

Step 3 Calculate the score values of $r_i (i = 1, 2, \ldots, m)$ according to Definition 11. If two score values are equal, then calculate the accuracy values and certainty values.

Step 4 According to Step 3 and Definition 12, rank the alternatives.

6. Illustrative Example

There are many decision-making problems to be solved in the current society, which requires some decision-making methods.

In this section, we investigate an example (adapted from Ref [43]) about the MADM. In a MADM problem, there are four possible alternatives for an investment company including a car company (A_1), a food company (A_2), a computer company (A_3), and an arms company (A_4). The following three attributes can be used to evaluate alternatives by the investment company: the risk (C_1), the growth (C_2), and the environmental impact (C_3) where C_1 and C_2 are benefit types and C_3 is cost type. Then the evaluation values of alternatives are shown in Table 1 where the LTS is $S = \{s_0 = extremely\ poor(EP),\ s_1 = very\ poor(VP),\ s_2 = poor(P),\ s_3 = medium(M),\ s_4 = good(G),\ s_5 = very\ good(VG),\ s_6 = extremely\ good(EG)\}$, and the weight vector of criteria is $\omega = (0.35, 0.25, 0.4)^T$. Now we will use the method proposed in this paper, according to the above LTs and three criteria. Then we evaluate and sort the four options in Table 1.

Table 1. Evaluation values of alternatives.

	C_1	C_2	C_3
A_1	$\langle s_5, ([0.4, 0.5], [0.2, 0.3], [0.3, 0.4])\rangle$	$\langle s_6, ([0.4, 0.6], [0.1, 0.2], [0.2, 0.4])\rangle$	$\langle s_5, ([0.2, 0.3], [0.1, 0.2], [0.5, 0.6])\rangle$
A_2	$\langle s_6, ([0.5, 0.7], [0.1, 0.2], [0.2, 0.3])\rangle$	$\langle s_5, ([0.6, 0.7], [0.1, 0.2], [0.2, 0.3])\rangle$	$\langle s_5, ([0.5, 0.7], [0.2, 0.2], [0.1, 0.2])\rangle$
A_3	$\langle s_6, ([0.3, 0.5], [0.1, 0.2], [0.3, 0.4])\rangle$	$\langle s_5, ([0.5, 0.6], [0.1, 0.3], [0.3, 0.4])\rangle$	$\langle s_4([0.5, 0.6], [0.1, 0.3], [0.1, 0.3])\rangle$
A_4	$\langle s_4, ([0.7, 0.8], [0.0, 0.1], [0.1, 0.2])\rangle$	$\langle s_4, ([0.5, 0.7], [0.1, 0.2], [0.2, 0.3])\rangle$	$\langle s_6([0.3, 0.4], [0.1, 0.2], [0.1, 0.2])\rangle$

6.1. The Method Based on the WINLMSM Operator

Generally, we can give $m = \frac{n}{2}$, so $m = 1$ and $m = 2$. Then, according to Section 5, we have the statements below.

(1) When $m = 1$, the steps are shown below.

Step 1 Normalize the decision matrix.

From the example, the risk (C_1) and the growth (C_2) are benefit types while the environmental impact (C_3) is cost type. We set up the decision matrix as shown below.

$$R = \begin{bmatrix} \langle s_5, ([0.4, 0.5], [0.2, 0.3], [0.3, 0.4])\rangle & \langle s_6, ([0.4, 0.6], [0.1, 0.2], [0.2, 0.4])\rangle & \langle s_5, ([0.2, 0.3], [0.1, 0.2], [0.5, 0.6])\rangle \\ \langle s_6, ([0.5, 0.7], [0.1, 0.2], [0.2, 0.3])\rangle & \langle s_5, ([0.6, 0.7], [0.1, 0.2], [0.2, 0.3])\rangle & \langle s_5, ([0.5, 0.7], [0.2, 0.2], [0.1, 0.2])\rangle \\ \langle s_6, ([0.3, 0.5], [0.1, 0.2], [0.3, 0.4])\rangle & \langle s_5, ([0.5, 0.6], [0.1, 0.3], [0.3, 0.4])\rangle & \langle s_4([0.5, 0.6], [0.1, 0.3], [0.1, 0.3])\rangle \\ \langle s_4, ([0.7, 0.8], [0.0, 0.1], [0.1, 0.2])\rangle & \langle s_4, ([0.5, 0.7], [0.1, 0.2], [0.2, 0.3])\rangle & \langle s_6([0.3, 0.4], [0.1, 0.2], [0.1, 0.2])\rangle \end{bmatrix}$$

Step 2 Aggregate all attribute values of each alternative and get the overall value of each alternative a_i denoted as $r_i (i = 1, 2, 3, 4)$.

$$\begin{aligned} r_1 &= \langle s_6, ([0.3268, 0.4590], [0.1275, 0.2305], [0.3325, 0.4704])\rangle, \\ r_2 &= \langle s_6, ([0.5271, 0.7000], [0.1320, 0.2000], [0.1516, 0.2551])\rangle, \\ r_3 &= \langle s_6, ([0.4375, 0.5675], [0.1000, 0.2603], [0.1933, 0.3565])\rangle, \\ r_4 &= \langle s_6, ([0.5216, 0.6565], [0.0000, 0.1569], [0.1189, 0.2213])\rangle \end{aligned}$$

Step 3 According to Definition 11, we assume $\alpha = 0.7$ and calculate the score values of $r_i (i = 1, 2, 3, 4)$ below.

$$S_{(r_1)} = s_{0.6228}, S_{(r_2)} = s_{0.8306}, S_{(r_3)} = s_{0.7462}, S_{(r_4)} = s_{0.7778}$$

Step 4 According to Step 3 and Definition 12, we would get the ranking of the alternatives, which are $A_2 \succ A_4 \succ A_3 \succ A_1$.

(2) When $m = 2$, the steps are shown below.

Step 1 Normalize the decision matrix.

From the example, the risk (C_1) and the growth (C_2) are benefit types while the environmental impact (C_3) is cost type. We set up the decision matrix as shown below.

$$R = \begin{bmatrix} \langle s_5, ([0.4, 0.5], [0.2, 0.3], [0.3, 0.4])\rangle & \langle s_6, ([0.4, 0.6], [0.1, 0.2], [0.2, 0.4])\rangle & \langle s_5, ([0.2, 0.3], [0.1, 0.2], [0.5, 0.6])\rangle \\ \langle s_6, ([0.5, 0.7], [0.1, 0.2], [0.2, 0.3])\rangle & \langle s_5, ([0.6, 0.7], [0.1, 0.2], [0.2, 0.3])\rangle & \langle s_5, ([0.5, 0.7], [0.2, 0.2], [0.1, 0.2])\rangle \\ \langle s_6, ([0.3, 0.5], [0.1, 0.2], [0.3, 0.4])\rangle & \langle s_5, ([0.5, 0.6], [0.1, 0.3], [0.3, 0.4])\rangle & \langle s_4([0.5, 0.6], [0.1, 0.3], [0.1, 0.3])\rangle \\ \langle s_4, ([0.7, 0.8], [0.0, 0.1], [0.1, 0.2])\rangle & \langle s_4, ([0.5, 0.7], [0.1, 0.2], [0.2, 0.3])\rangle & \langle s_6([0.3, 0.4], [0.1, 0.2], [0.1, 0.2])\rangle \end{bmatrix}$$

Step 2 Aggregate all attribute values of each alternative and get the overall value of each alternative a_i denoted as $r_i(i = 1,2,3,4)$.

$$\begin{aligned} r_1 &= \langle s_{5.4841}, ([0.3190, 0.4520], [0.1406, 0.2420], [0.3391, 0.4771])\rangle, \\ r_2 &= \langle s_{5.3016}, ([0.5260, 0.6922], [0.1366, 0.2083], [0.1791, 0.2772])\rangle, \\ r_3 &= \langle s_{4.9567}, ([0.4224, 0.5587], [0.1077, 0.2741], [0.2494, 0.3754])\rangle, \\ r_4 &= \langle s_{4.4896}, ([0.4794, 0.6190], [0.0711, 0.1739], [0.1415, 0.2416])\rangle \end{aligned}$$

Step 3 According to Definition 11, we assume $\alpha = 0.7$ and calculate the score values of $r_i(i = 1,2,3,4)$. We get the values below.

$$S_{(r_1)} = s_{0.5695}, S_{(r_2)} = s_{0.7170}, S_{(r_3)} = s_{0.6004}, S_{(r_4)} = s_{0.5765}.$$

Step 4 According to Step 3 and Definition 12, we get the ranking of the alternatives below.

$$A_2 \succ A_3 \succ A_4 \succ A_1$$

6.2. The Method Based on the WINLGMSM Operator

When $m = 1, p = 1$, the $WINLGMSM^{(1)}$ operator is the same as the $WINLMSM^{(1)}$ operator. The steps are omitted here. When $m = 2$, the steps are below.

Step 1 Normalize the decision matrix.

From the example, the risk (C_1), the growth (C_2) are benefit types and the environmental impact (C_3) is cost type, so we set up the matrix as step 1 of Section 6.1.

Step 2 Aggregate all attribute values of each alternative by the $WINLMSM^{(2)}$ operator and get the overall value of each alternative a_i denoted as $r_i(i = 1,2,3,4)$

$$\begin{aligned} r_1 &= \langle s_{5.4988}, ([0.3221, 0.4549], [0.1387, 0.2401], [0.3374, 0.4752])\rangle, \\ r_2 &= \langle s_{5.3735}, ([0.5264, 0.6938], [0.1358, 0.2070], [0.1772, 0.2745])\rangle, \\ r_3 &= \langle s_{5.0083}, ([0.4296, 0.5610], [0.1069, 0.2702], [0.2449, 0.3721])\rangle, \\ r_4 &= \langle s_{4.5371}, ([0.4892, 0.6244], [0.0634, 0.1698], [0.1390, 0.2381])\rangle \end{aligned}$$

Step 3 According to Definition 11, we assume $\alpha = 0.7$, calculate the score values of $r_i(i = 1,2,3,4)$, and get the values shown below.

$$S_{(r_1)} = s_{0.5722}, S_{(r_2)} = s_{0.7276}, S_{(r_3)} = s_{0.6087}, S_{(r_4)} = s_{0.5839}$$

Step 4 According to Step 3 and Definition 12, we get the rankings of the alternatives, which are shown below.

$$A_2 \succ A_3 \succ A_4 \succ A_1$$

6.3. Comparative Analysis and Discussion

(1) Based on the results in Sections 6.1 and 6.2, we can show them by using Table 2. From Table 2, we know that there are the same ranking results in two methods when $m = 1$ or $m = 2$. However, the result when $m = 1$ is different from the one when $m = 2$. It can be explained that, when $m = 1$, the interrelationship between the attributes doesn't need to be considered when $m = 2$. We can consider the interrelationship between two attributes.

(2) Furthermore, we get the comparisons for different values of P_1 and P_2 when $m = 2$, which are shown in Table 3. From Table 3, we know when $m = 2$ and P_1 and P_2 are not equal to zero, we can get the same ranking results, i.e., $A_2 \succ A_4 \succ A_3 \succ A_1$. However, when $P_1 = 0$ or $P_2 = 0$, the ranking results are different from the ones when P_1 and P_2 are not equal to zero. When $P_1 = 0$

or $P_2 = 0$, the interrelationship between the attributes doesn't need to be considered, so it can get the same ranking results as the ones when $m = 1$.

Table 2. Comparison of different operator.

Operator	m	P_1	P_2	Ranking
$WINLMSM^{(m)}$	1	-	-	$A_2 \succ A_4 \succ A_3 \succ A_1$
	2	-	-	$A_2 \succ A_3 \succ A_4 \succ A_1$
$WINLGMSM^{(m,p_1,p_2)}$	1	1	-	$A_2 \succ A_4 \succ A_3 \succ A_1$
	2	1	2	$A_2 \succ A_3 \succ A_4 \succ A_1$

Table 3. Comparisons of different values of P_1 and P_2 when $m = 2$.

Operator	P_1	P_2	$S_{(r_i)}(i = 1,2,3,4)$	Ranking
$WINLGMSM^{(m,p_1,p_2)}$	0	1	$S_{(r_1)} = s_{0.6228}$ $S_{(r_2)} = s_{0.8306}$ $S_{(r_3)} = s_{0.7462}$ $S_{(r_4)} = s_{0.7778}$	$A_2 \succ A_4 \succ A_3 \succ A_1$
	1	0	$S_{(r_1)} = s_{0.6228}$ $S_{(r_2)} = s_{0.8306}$ $S_{(r_3)} = s_{0.7462}$ $S_{(r_4)} = s_{0.7778}$	$A_2 \succ A_4 \succ A_3 \succ A_1$
	1	1	$S_{(r_1)} = s_{0.5695}$ $S_{(r_2)} = s_{0.7170}$ $S_{(r_3)} = s_{0.6004}$ $S_{(r_4)} = s_{0.5765}$	$A_2 \succ A_3 \succ A_4 \succ A_1$
	1	2	$S_{(r_1)} = s_{0.5722}$ $S_{(r_2)} = s_{0.7276}$ $S_{(r_3)} = s_{0.6087}$ $S_{(r_4)} = s_{0.5839}$	$A_2 \succ A_3 \succ A_4 \succ A_1$
	1	3	$S_{(r_1)} = s_{0.5769}$ $S_{(r_2)} = s_{0.7387}$ $S_{(r_3)} = s_{0.6227}$ $S_{(r_4)} = s_{0.6027}$	$A_2 \succ A_3 \succ A_4 \succ A_1$
	2	1	$S_{(r_1)} = s_{0.5745}$ $S_{(r_2)} = s_{0.7199}$ $S_{(r_3)} = s_{0.6116}$ $S_{(r_4)} = s_{0.6004}$	$A_2 \succ A_3 \succ A_4 \succ A_1$
	2	2	$S_{(r_1)} = s_{0.5717}$ $S_{(r_2)} = s_{0.7196}$ $S_{(r_3)} = s_{0.6037}$ $S_{(r_4)} = s_{0.5837}$	$A_2 \succ A_3 \succ A_4 \succ A_1$
	2	3	$S_{(r_1)} = s_{0.5733}$ $S_{(r_2)} = s_{0.7256}$ $S_{(r_3)} = s_{0.6079}$ $S_{(r_4)} = s_{0.5859}$	$A_2 \succ A_3 \succ A_4 \succ A_1$
	3	1	$S_{(r_1)} = s_{0.5806}$ $S_{(r_2)} = s_{0.7276}$ $S_{(r_3)} = s_{0.6269}$ $S_{(r_4)} = s_{0.6280}$	$A_2 \succ A_4 \succ A_3 \succ A_1$
	3	2	$S_{(r_1)} = s_{0.5751}$ $S_{(r_2)} = s_{0.7206}$ $S_{(r_3)} = s_{0.6101}$ $S_{(r_4)} = s_{0.5997}$	$A_2 \succ A_3 \succ A_4 \succ A_1$
	3	3	$S_{(r_1)} = s_{0.5741}$ $S_{(r_2)} = s_{0.7223}$ $S_{(r_3)} = s_{0.6071}$ $S_{(r_4)} = s_{0.5909}$	$A_2 \succ A_3 \succ A_4 \succ A_1$

Furthermore, in order to verify the validity of the methods proposed in this paper, we can compare them with methods from Ye [16] and the ranking results are shown in Table 4.

From Table 4, we know that the best choice is A_2 for all methods, which is the same as the results produced above. However, the ranking results are different. Compared with the approach proposed by Ye [16], when $m = 1$, our ranking results have the same values as that of Ye [16], but when $m = 2$, our ranking results are different from the Ye method [16]. When $m = 1$, all methods don't consider the interrelationship. They produce the same results, however, when $m = 2$. Our methods in this paper can take into account the interrelationship while the method by Ye [16] doesn't consider the interrelationship. Therefore, there are different ranking results. Therefore, our methods are more suitable for the different applications.

Table 4. Comparison of different methods.

Methods	Operator	Ranking
Methods in this paper	$WINLMSM^{(m)} m = 1$	$A_2 \succ A_4 \succ A_3 \succ A_1$
	$WINLMSM^{(m)} m = 2$	$A_2 \succ A_3 \succ A_4 \succ A_1$
	$WINLGMSM^{(m,p_1,p_2)}\ m = 1$	$A_2 \succ A_4 \succ A_3 \succ A_1$
	$WINLGMSM^{(m,p_1,p_2)}\ m = 2$	$A_2 \succ A_3 \succ A_4 \succ A_1$
Method in [16]	*INLWAA*	$A_2 \succ A_4 \succ A_3 \succ A_1$
	INLWGA	$A_2 \succ A_4 \succ A_3 \succ A_1$

From the above comparison results, we can obtain that the methods proposed by this paper are feasible and adaptable for the MADM problems. Additionally, they have better reliability and wider application space than other existing methods.

7. Conclusions

In this study, we propose the concept of *INLMSM*, which can not only adapt to the cognitive situation of decision maker, but also provide convenience for decision making. We introduce the basic concept of *INLMSM* and its generalized form, give some operators based on *INLMSM*, and introduce the theory of weight to investigate *WINLMSM* and *WINLGMSM*. Afterwards, we put forward the *INLMSM* operator, the *INLGMSM* operator, the *WINLMSM* operator, and the *WINLGMSM* operator. In addition, we proved these operators. In addition, we introduce the MADM methods with *INLMSM* in detail and illustrate their usefulness and effectiveness by showing examples. Finally, we compare other methods to demonstrate our approach. From this paper, we can see that *WINLGMSM* is more practical and flexible in application and *INLMSM* can express fuzzy information more conveniently. In further study, we can use the *INLMSM* operator to solve practical problems and pattern recognition. We should develop other aggregation operators for future research.

Acknowledgments: This paper is supported by the National Natural Science Foundation of China (Nos. 71771140 and 71471172), the Special Funds of Taishan Scholars Project of Shandong Province (No. ts201511045), Shandong Provincial Social Science Planning Project (Nos. 17BGLJ04,16CGLJ31 and 16CKJJ27), the Natural Science Foundation of Shandong Province (Nos. ZR2013FM017, ZR2017MG007), and the Teaching Reform Research Project of Undergraduate Colleges and Universities in Shandong Province (2015Z057).

Author Contributions: Yushui Geng proposed extended MSM operators for INLNs; Xingang Wang designed the experiments; Xuemei Li and Kun Yu analyzed the data, and wrote the paper; Peide Liu proposed the ideas of this paper.

Conflicts of Interest: The authors declare no conflict of interest.

Compliance with Ethical Standards: (1) Disclosure of potential conflicts of interest. We declare that we have no commercial or associative interests that represent a conflict of interest in connection with this manuscript. There are no professional or other personal interests that can inappropriately influence our submitted work. (2) Research involving human participants and/or animals. This article does not contain any studies with human participants or animals performed by any of the authors.

References

1. Zadeh, L.A. Fuzzy sets. *Inf. Control* **1965**, *8*, 338–356. [CrossRef]
2. Atanassov, T.K. Intuitionistic fuzzy sets. *Fuzzy Sets Syst.* **1986**, *20*, 87–96. [CrossRef]
3. Abbas, S.E. On intuitionistic fuzzy compactness. *Inf. Sci.* **2005**, *173*, 75–91. [CrossRef]
4. Zadeh, L.A. Similarity relations and fuzzy ordering. *Inf. Sci.* **1971**, *3*, 177–200. [CrossRef]
5. Zwick, R.; Carlstein, E.; Budescu, D.V. Measures of similarity among fuzzy concepts: A comparative analysis. *Int. J. Approx. Reason.* **1987**, *1*, 221–242. [CrossRef]
6. Zeng, W.; Li, H. Relationship between similarity measure and entropy of interval-valued fuzzy sets. *Fuzzy Sets Syst.* **2006**, *157*, 1477–1484. [CrossRef]
7. Smarandache, F. *Neutrosophy: Neutrosophic Probability, Set, and Logic: Analytic Synthesis and Synthetic Analysis*; ProQuest Information & Learning: Ann Harbor, MI, USA, 1998; p. 105.
8. Smarandache, F. A unifying field in logic: Neutrsophic logic. Neutrosophy, Neutrosophic Set, Neutrosophic Probability. *J. Mol. Biol.* **2001**, *332*, 489–503.
9. Smarandache, F. A geometric interpretation of the neutrosophic set—A generalization of the intuitionistic fuzzy set. *IEEE Int. Conf. Granul. Comput.* **2003**, *28*, 1496.
10. Wang, H.; Smarandache, F.; Zhang, Y.Q.; Sunderraman, R. *Interval Neutrosophic Sets and Logic: Theory and Applications in Computing*; ProQuest Information & Learning: Ann Harbor, MI, USA, 2005.
11. Broumi, S.; Smarandache, F. Correlation Coefficient of Interval Neutrosophic Set. *Appl. Mech. Mater.* **2013**, *436*, 511–517. [CrossRef]
12. Herrera, F.; Herrera-Viedma, E. Aggregation operators for linguistic weighted information. *IEEE Trans. Syst. Man Cybern. Part A Syst. Hum.* **1997**, *27*, 646–656. [CrossRef]
13. Chen, Z.C.; Liu, P.H. An approach to multiple attribute group decision making based on linguistic intuitionistic fuzzy numbers. *Int. J. Comput. Intell. Syst.* **2015**, *8*, 747–760. [CrossRef]
14. Ye, J. An extended TOPSIS method for multiple attribute group decision making based on single valued neutrosophic linguistic numbers. *J. Intell. Fuzzy Syst.* **2015**, *28*, 247–255.
15. Wang, J.Q.; Yang, Y.; Li, L. Multi-criteria decision-making method based on single-valued neutrosophic linguistic Maclaurin symmetric mean operators. *Neural Comput. Appl.* **2016**, *2016*, 1–19. [CrossRef]
16. Ye, J. Some aggregation operators of interval neutrosophic linguistic numbers for multiple attribute decision making. *J. Intell. Fuzzy Syst.* **2014**, *27*, 2231–2241.
17. Liu, P.; Chen, S.M. Multiattribute Group Decision Making Based on Intuitionistic 2-Tuple Linguistic Information. *Inf. Sci.* **2018**, *430–431*, 599–619. [CrossRef]
18. Liu, P.; Wang, P. Some q-Rung Orthopair Fuzzy Aggregation Operators and Their Applications to Multiple-Attribute Decision Making. *Int. J. Intell. Syst.* **2018**, *33*, 259–280. [CrossRef]
19. Yager, R.R. On ordered weighted averaging aggregation operators in multicriteria decisionmaking. *IEEE Trans. Syst. Man Cybern.* **1988**, *18*, 183–190. [CrossRef]
20. Bonferroni, C. Sulle medie multiple di potenze. *Boll. Mater. Ital.* **1950**, *5*, 267–270.
21. Liu, P.; Liu, J.; Chen, S.M. Some Intuitionistic Fuzzy Dombi Bonferroni Mean Operators and Their Application to Multi-Attribute Group Decision Making. *J. Oper. Res. Soc.* **2018**, *69*, 1–24. [CrossRef]
22. Liu, P.; Chen, S.M.; Liu, J. Some intuitionistic fuzzy interaction partitioned Bonferroni mean operators and their application to multi-attribute group decision making. *Inf. Sci.* **2017**, *411*, 98–121. [CrossRef]
23. Liu, P.; Li, H. Interval-valued intuitionistic fuzzy power Bonferroni aggregation operators and their application to group decision making. *Cognit. Comput.* **2017**, *9*, 494–512. [CrossRef]
24. Liu, P.; Liu, J. Some q-Rung Orthopai Fuzzy Bonferroni Mean Operators and Their Application to Multi-Attribute Group Decision Making. *Int. J. Intell. Syst.* **2018**, *33*, 315–347. [CrossRef]
25. Liu, P.; Zhang, L.; Liu, X.; Wang, P. Multi-valued Neutrosophic number Bonferroni mean operators and their application in multiple attribute group decision making. *Int. J. Inf. Technol. Decis. Mak.* **2016**, *15*, 1181–1210. [CrossRef]
26. Beliakov, G.; Pradera, A.; Calvo, T. Aggregation functions: A guide for practitioners. *Stud. Fuzziness Soft Comput.* **2010**, *2010*, 139–141.
27. Liu, P.; Liu, J.; Merigó, J.M. Partitioned Heronian means based on linguistic intuitionistic fuzzy numbers for dealing with multi-attribute group decision making. *Appl. Soft Comput.* **2018**, *62*, 395–422. [CrossRef]

28. Liu, P. Multiple attribute group decision making method based on interval-valued intuitionistic fuzzy power Heronian aggregation operators. *Comput. Ind. Eng.* **2017**, *108*, 199–212. [CrossRef]
29. Liu, P.; Chen, S.M. Group decision making based on Heronian aggregation operators of intuitionistic fuzzy numbers. *IEEE Trans. Cybern.* **2017**, *47*, 2514–2530. [CrossRef] [PubMed]
30. Liu, P.; Shi, L. Some Neutrosophic uncertain linguistic number Heronian mean operators and their application to multi-attribute group decision making. *Neural Comput. Appl.* **2017**, *28*, 1079–1093. [CrossRef]
31. Yu, D. Interval-valued intuitionistic fuzzy Heronian mean operators and their application in multi-criteria decision making. *Afr. J. Bus. Manag.* **2012**, *6*, 4158–4168. [CrossRef]
32. Maclaurin, C. A second letter to Martin Folkes, Esq., Concerning the roots of equations, with demonstration of other rules of algebra. *Philos. Trans. R. Soc. Lond. Ser. A* **1972**, *36*, 59–96.
33. Qin, J.; Liu, X. Approaches to uncertain linguistic multiple attribute decision making based on dual Maclaurin symmetric mean. *J. Intell. Fuzzy Syst.* **2015**, *29*, 171–186. [CrossRef]
34. Liu, P.; Qin, X. Maclaurin symmetric mean operators of linguistic intuitionistic fuzzy numbers and their application to multiple-attribute decision-making. *J. Exp. Theor. Artif. Intell.* **2017**, *29*, 1173–1202. [CrossRef]
35. Liu, P.; Zhang, X. Some Maclaurin symmetric mean operators for single valued trapezoidal neutrosophic numbers and their applications to group decision making. *Int. J. Fuzzy Syst.* **2017**. [CrossRef]
36. Delgado, M.; Verdegay, J.L.; Vila, M.A. Linguistic decision-making models. *Int. J. Intell. Syst.* **2010**, *7*, 479–492. [CrossRef]
37. Ma, Y.X.; Wang, J.Q.; Wang, J.; Wu, X.H. An interval neutrosophic linguistic multi-criteria group decision-making method and its application in selecting medical treatment options. *Neural Comput. Appl.* **2017**, *28*, 2745–2765. [CrossRef]
38. Xu, Z.S. A note on linguistic hybrid arithmetic averagingoperator in multiple attribute group decision making with linguistic information. *Group Decis. Negot.* **2006**, *15*, 593–604. [CrossRef]
39. Xu, Z.S. Fuzzy harmonic mean operators. *Int. J. Intell. Syst.* **2009**, *24*, 152–172. [CrossRef]
40. Xu, Z.S. Group decision making based on multiple types of linguistic preference relations. *Inf. Sci.* **2008**, *178*, 452–467. [CrossRef]
41. Broumi, S.; Ye, J.; Smarandache, F. An Extended TOPSIS Method for Multiple Attribute Decision Making based on Interval Neutrosophic Uncertain Linguistic Variables. *Neutrosophic Sets Syst.* **2015**, *8*, 22–31.
42. Wang, J.Q.; Wu, J.T.; Wang, J.; Zhang, H.Y.; Chen, X.H. Interval-valued hesitant fuzzy linguistic sets and their applications in multi-criteria decision-making problems. *Inf. Sci.* **2014**, *288*, 55–72. [CrossRef]
43. Ye, J. Similarity measures between interval neutrosophic sets and their applications in multicriteria decision-making. *J. Intell. Fuzzy Syst.* **2014**, *26*, 165–172.

symmetry

Article

Picture Hesitant Fuzzy Set and Its Application to Multiple Criteria Decision-Making

Rui Wang [1] **and Yanlai Li** [1,2,*]

[1] School of Transportation and Logistics, Southwest Jiaotong University, Chengdu 610031, China; wryuedi@my.swjtu.edu.cn

[2] National Lab of Railway Transportation, Southwest Jiaotong University, Chengdu 610031, China

* Correspondence: yanlaili@home.swjtu.edu.cn; Tel.: +86-156-8007-2505

Received: 21 June 2018; Accepted: 18 July 2018; Published: 20 July 2018

Abstract: To address the complex multiple criteria decision-making (MCDM) problems in practice, this article proposes the picture hesitant fuzzy set (PHFS) theory based on the picture fuzzy set and the hesitant fuzzy set. First, the concept of PHFS is put forward, and its operations are presented, simultaneously. Second, the generalized picture hesitant fuzzy weighted aggregation operators are developed, and some theorems and reduced operators of them are discussed. Third, the generalized picture hesitant fuzzy prioritized weighted aggregation operators are put forward to solve the MCDM problems that the related criteria are at different priorities. Fourth, two novel MCDM methods combined with the proposed operators are constructed to determine the best alternative in real life. Finally, two numerical examples and an application of web service selection are investigated to illustrate the effectiveness of the proposed methods. The sensitivity analysis shows that the different values of the parameter λ affect the ranking of alternatives, and the proposed operators are compared with several existing MCDM methods to illustrate their advantages.

Keywords: multiple criteria decision-making; picture hesitant fuzzy set; generalized picture hesitant fuzzy weighted aggregation operator; generalized picture hesitant fuzzy prioritized weighted aggregation operator

1. Introduction

Multiple criteria decision-making (MCDM) problems occur in numerous practical fields [1–3]. For a specific purpose, several possible plans may be presented as the alternatives; then, decision makers assess the alternatives concerning the related criteria to determine the best one. Traditionally, crisp numbers are utilized to express the evaluation information. However, in real life, the data are inevitably incomplete and complex, and decision makers may be uncertain when evaluating the alternatives. To deal with the fuzziness of evaluation information, the fuzzy set (FS) [4] was proposed to improve the information form. During the past decades, many scholars devoted themselves to the study of the fuzzy MCDM problems [1]. Furthermore, in recent years, along with the complexity of the MCDM problems, how to improve the FS theory to deal with different specific situations has been a hot topic.

Although FS is a valid form to express the uncertain evaluation information, it cannot solve several complex situations in real life. For more effective expression of the evaluation information, many generalized forms of FS were proposed [5–10]. The purpose of this paper is to propose a new information form; the picture hesitant fuzzy set (PHFS) theory is put forward combined with the concepts of picture fuzzy set (PFS) [7] and hesitant fuzzy set (HFS) [8]. As a generalized form of FS, intuitionistic fuzzy set (IFS) [5], PFS, and HFS, PHFS can express the uncertainty and complexity of human opinions in practice; furthermore, the positive, neutral, negative, and refusal membership degrees are represented by several possible values that are given by decision makers.

In practice, the uncertain and complex evaluation information will be inevitably given by decision makers. For example, ten business managers discuss an investment project; five suggest agreement, two present disagreement, and the other business managers choose to abstain. Obviously, FS can only indicate the membership degree of evaluation information; thus, the opinions of the 10 business managers cannot be represented by FS. For overcoming the limitation of FS, Atanassov [5] put forward the non-membership function and developed the IFS. Then, the evaluation information in the aforementioned example can be expressed by IFS accurately. Later, the interval numbers were used to substitute the crisp numbers in IFS; then, the interval-valued intuitionistic fuzzy set (IVIFS) was developed [6]. To convey the indeterminate information of decision makers more effectively, Ye [9] and Liu and Yuan [11] extended the FS to triangular and trapezoidal intuitionistic fuzzy set, respectively. However, in some particular situations, it is not convincing to represent the evaluation information combined with IFS or IVIFS. For instance, there is a vote for a specific matter, the voting opinions of voters can be divided into four types, namely, vote for, abstain, vote against, and a refusal of the voting [12]. Therefore, Cuong [7,13] put forward the PFS, which is composed by the positive, neutral, negative, and refusal membership functions; thus, PFS can express the opinions of decision makers accurately in the example above. Subsequently, the correlation coefficient, distance measure, and cross-entropy measure of PFS were investigated in detail [14–16].

On the other hand, sometimes the accurate membership degree of evaluation information is difficult to be determined, which is also another shortcoming of FS. Therefore, the HFS was developed [17], in which the membership degrees are represented by several possible crisp numbers. Next, the interval numbers were introduced to extend the membership function of HFS and the interval-valued hesitant fuzzy set (IVHFS) theory was proposed [18]. According to the IFS and HFS, several potential membership and non-membership functions were expressed to put forward the dual hesitant fuzzy set (DHFS) [10]. Later, Farhadinia [19] constructed the dual interval-valued hesitant fuzzy set (DIVHFS) combined with DHFS. Nevertheless, HFS in the existing research cannot express all types of human opinions in the aforementioned example.

According to the evaluation information of the individual decision makers, the collective evaluation information of each alternative is obtained through the information fusion. Due to the important role of aggregation tools in MCDM problems, many scholars have investigated the aggregation operators of different fuzzy information. For example, Xu and Yager [20] developed the operations of intuitionistic fuzzy numbers (IFNs) and proposed the intuitionistic fuzzy geometric aggregation operators. Later, Xu [21] put forward the intuitionistic fuzzy weighted averaging aggregation operators to aggregate the IFNs. Next, several interval-valued intuitionistic fuzzy aggregation operators were constructed to deal with the MCDM [22–24]. With respect to the picture fuzzy (PF) evaluation information, Wei [25] defined the operations of picture fuzzy numbers (PFNs) according to the study of [21] and proposed the picture fuzzy weighted aggregation operators. In addition, several PF aggregation operators according to different operations were put forward [12,26]. Besides, a great time of hesitant fuzzy aggregation operators and their generalized forms were constructed [27], and several aggregation operators under dual hesitant fuzzy and dual interval-valued hesitant fuzzy environment were developed [28–30].

In some practical MCDM problems, the related criteria may be at different priority levels. For instance, a young couple wants to choose a toy for their child, the criteria of the toy they will consider are safety and price; obviously, the criteria safety has a higher priority than price. However, the aforementioned aggregation operators cannot fuse the aggregated arguments that are in different priority levels. In response to these situations, Yager [31] proposed the prioritized averaging (PA) operator. Inspired by Yager [31], Yu et al. [32,33] constructed the intuitionistic fuzzy prioritized fuzzy and interval-valued intuitionistic fuzzy prioritized fuzzy aggregation operators. Besides, the hesitant fuzzy prioritized aggregation operators were proposed to aggregate the evaluation information that is at different priorities [34]. Nevertheless, to our best knowledge, few researches have extended the PA operator to solve the MCDM problems under PF environment.

In summary, this paper defines the PHFS based on the PFS and HFS and develops the operations laws of picture hesitant fuzzy elements (PHFEs) according to the operations of IFNs [21]. Then, the generalized picture hesitant fuzzy aggregation operators and generalized picture hesitant fuzzy prioritized aggregation operators are put forward, and the properties and reduced operators of them are investigated. Furthermore, the proposed operators are utilized to solve diverse situations during MCDM processes under picture hesitant fuzzy (PHF) environment.

The rest of this paper is structured as follows. Definitions of the PFS, HFS, and PA operator are presented in Section 2. The concept of PHFS is defined, and the comparison method and operations of PHFEs are proposed in Section 3. Section 4 constructs the generalized picture hesitant fuzzy weighted averaging (GPHFWA), generalized picture hesitant fuzzy weighted geometric (GPHFWG), generalized picture hesitant fuzzy prioritized weighted averaging (GPHFPWA), and generalized picture hesitant fuzzy prioritized weighted geometric (GPHFPWG) operators. In Section 5, two MCDM methods are constructed according to the proposed operators. Section 6 applies the proposed methods into two numerical examples and an application of web service selection to show the effectiveness and advantages of the proposed methods. Finally, some conclusions are summarized in Section 7.

2. Preliminaries

To make this paper as self-contained as possible, we recall the definitions of the PFS, HFS, and PA operator, which will be utilized in the subsequent research.

2.1. PFS

Atanassov [5] applied the non-membership degree to extend FS; however, expressing the evaluation information depend on IFS is unreasonable in practice, at times. Therefore, Cuong [13] proposed the PFS theory based on FS and IFS, which can represent more information of decision makers, including yes, abstain, no, and refusal.

Definition 1. *Let X be a non-empty and finite set, a PFS P on X is defined by*

$$P = \{\langle x, \mu_P(x), \eta_P(x), v_P(x)\rangle | x \in X\}, \tag{1}$$

where $\mu_P(x)$, $\eta_P(x)$, and $v_P(x)$ are the positive, neutral, and negative membership functions that are belonging to $[0,1]$, respectively, and they meet the condition of $0 \leq \mu_P(x) + \eta_P(x) + v_P(x) \leq 1$. Furthermore, $\pi_P(x) = 1 - \mu_P(x) - \eta_P(x) - v_P(x)$ is the refusal membership function.

Definition 2. *A picture fuzzy number (PFN) is represented by $a = (\mu_a, \eta_a, v_a)$, where $\mu_a \in [0,1]$, $\eta_a \in [0,1]$, $v_a \in [0,1]$, and $\mu_a + \eta_a + v_a \leq 1$ [25].*

Wei [25] proposed the operations of PFNs based on the operations of IFNs in [21].

Definition 3. *Let $a_1 = (\mu_1, \eta_1, v_1)$, $a_2 = (\mu_2, \eta_2, v_2)$, and $a = (\mu, \eta, v)$ be three PFNs, $\lambda > 0$, and a^c is the complementary set of a, then [25]*

$$a^c = (v, \eta, \mu); \tag{2}$$

$$a_1 \oplus a_2 = (\mu_1 + \mu_2 - \mu_1\mu_2, \eta_1\eta_2, v_1v_2); \tag{3}$$

$$a_1 \otimes a_2 = (\mu_1\mu_2, \eta_1 + \eta_2 - \eta_1\eta_2, v_1 + v_2 - v_1v_2); \tag{4}$$

$$\lambda a = \left(1 - (1-\mu)^\lambda, \eta^\lambda, v^\lambda\right); \tag{5}$$

$$a^\lambda = \left(\mu^\lambda, 1 - (1-\eta)^\lambda, 1 - (1-v)^\lambda\right). \tag{6}$$

2.2. HFS

Due to the complexity of the evaluated object in practice, decision makers may have difficulty determining an accurate value of the membership level. To deal with this situation, Torra [8] developed the HFS theory in which the membership degree is expressed by several possible values.

Definition 4. *Let $\wp([0,1])$ be the set of all subsets of the unitary interval and X be a non- empty set. Let $h_A : X \to \wp([0,1])$, then an HFS A on X is defined by*

$$A = \{\langle x, h_A(x)\rangle | x \in X\}. \tag{7}$$

Definition 5. *A hesitant fuzzy element (HFE) is a non-empty and finite subset of $[0,1]$ [27].*

Although a HFE can be given by any subset of $[0,1]$, in practice, HFS is commonly restricted to finite set in the MCDM problems [27]. Therefore, Bedregal at al. [35] proposed the typical hesitant fuzzy set (THFS), which is the finite and non-empty HFS. Later, Alcantud and Torra [36] defined the uniformly typical hesitant fuzzy set (UTHFS) that can simplify many theoretical and practical arguments, which is a generalized form of THFS. In this paper, the evaluation information of decision makers is expressed by UTHFS during the MCDM processes under hesitant fuzzy environment.

Definition 6. *Let $\mathrm{H} \subseteq \wp([0,1])$ be the set of all finite and non-empty subsets of $[0,1]$, and let X be a non-empty set. Then, a THFS A on X is defined by Equation (7), where $h_A : \mathrm{X} \to \mathrm{H}$. Each $h \in \mathrm{H}$ is called a typical hesitant fuzzy element (THFE) [35].*

Definition 7. *Let A be a THFS on X, if there is N such that the cardinality of the THFS $l_A(x) \leq N$ for each $x \in X$. Then, the THFS A is an UTHFS. Each $h \in \mathrm{H}$ is called an uniformly typical hesitant fuzzy element (UTHFE) [36].*

To aggregate the hesitant fuzzy evaluation information, Xia and Xu [27] investigated the operations of HFEs, which is also valid for fusing UTHFEs.

Definition 8. *Let h, h_1, and h_2 be three UTHFEs, $\lambda > 0$, then*

$$h_1 \oplus h_2 = \underset{\gamma_1 \in h_1, \gamma_2 \in h_2}{\cup} \{\gamma_1 + \gamma_2 - \gamma_1\gamma_2\}; \tag{8}$$

$$h_1 \otimes h_2 = \underset{\gamma_1 \in h_1, \gamma_2 \in h_2}{\cup} \{\gamma_1\gamma_2\}; \tag{9}$$

$$\lambda h = \underset{\gamma \in h}{\cup} \left\{1 - (1-\gamma)^\lambda\right\}; \tag{10}$$

$$h^\lambda = \underset{\gamma \in h}{\cup} \left\{\gamma^\lambda\right\}. \tag{11}$$

2.3. The PA Operator

Aggregation operator plays a crucial role in the process of information fusion. Sometimes, the criteria have different priorities according to their important degree; thus, Yager [31] constructed the PA operator to address these situations.

Definition 9. *Let $C = \{C_1, C_2, \ldots, C_n\}$ be a set of criteria, which are divided into several priority levels, i.e., the priority of C_p is higher than C_q when $p < q$. The $C_j(x) \in [0,1]$ is the evaluation value of the alternative x concerning the criteria C_j. Thus, the PA operator is expressed by*

$$PA(C_1(x), C_2(x), \ldots, C_n(x)) = \sum_{j=1}^{n} w_j C_j(x). \tag{12}$$

where $w_j = T_j / \sum_{j=1}^{n} T_j$, $T_j = \prod_{k=1}^{j-1} C_k(x)$, and $T_1 = 1$.

3. PHFS

According to the PFS and UTHFS, we can define the PHFS that is composed by four membership functions, namely, positive, neutral, negative, and refusal membership functions. The four membership degrees are denoted by several values belonging to $[0,1]$, respectively, which can convey the hesitancy of decision makers.

Definition 10. *Let X be a non-empty and finite set, a PHFS N on X is defined by*

$$N = \{\langle x, \widetilde{\mu}(x), \widetilde{\eta}(x), \widetilde{v}(x)\rangle | x \in X\}, \tag{13}$$

where $\widetilde{\mu}(x) = \{\alpha | \alpha \in \widetilde{\mu}(x)\}$, $\widetilde{\eta}(x) = \{\beta | \beta \in \widetilde{\eta}(x)\}$, and $\widetilde{v}(x) = \{\gamma | \gamma \in \widetilde{v}(x)\}$ are three sets of several values in $[0,1]$, representing the potential positive, neutral, and negative membership degrees. The degrees above satisfy the condition of $0 \le \alpha^+ + \beta^+ + \gamma^+ \le 1$, where $\alpha^+ = \cup_{\alpha \in \widetilde{\mu}(x)} \max\{\alpha\}$, $\beta^+ = \cup_{\beta \in \widetilde{\eta}(x)} \max\{\beta\}$, and $\gamma^+ = \cup_{\gamma \in \widetilde{v}(x)} \max\{\gamma\}$. For convenience, we call $\widetilde{n} = \{\widetilde{\mu}(x), \widetilde{\eta}(x), \widetilde{v}(x)\}$ is a PHFE, denoted by $\widetilde{n} = \{\widetilde{\mu}, \widetilde{\eta}, \widetilde{v}\}$.

During the process of applying the PHFEs to the practical MCDM problems, it is necessary to rank the PHFEs; thus, we develop the score and accuracy functions of PHFEs.

Definition 11. *Let $\widetilde{n} = \{\widetilde{\mu}, \widetilde{\eta}, \widetilde{v}\}$ be a PHFE, the numbers of values in $\widetilde{\mu}, \widetilde{\eta}, \widetilde{v}$ are l, p, q, respectively. Thus, the score function is defined as*

$$s(\widetilde{n}) = \left(1 + \frac{1}{l}\sum_{i=1}^{l} \alpha_i - \frac{1}{p}\sum_{i=1}^{p} \beta_i - \frac{1}{q}\sum_{i=1}^{q} \gamma_i\right)/2, s(\widetilde{n}) \in [0,1]. \tag{14}$$

the accuracy function is expressed as

$$h(\widetilde{n}) = \frac{1}{l}\sum_{i=1}^{l} \alpha_i + \frac{1}{p}\sum_{i=1}^{p} \beta_i + \frac{1}{q}\sum_{i=1}^{q} \gamma_i, h(\widetilde{n}) \in [0,1]. \tag{15}$$

Based on the score and accuracy values of PHFEs, we can determine the order relations between two PHFEs as in the following.

Definition 12. *Let $\widetilde{n}_1 = \{\widetilde{\mu}_1, \widetilde{\eta}_1, \widetilde{v}_1\}$ and $\widetilde{n}_2 = \{\widetilde{\mu}_2, \widetilde{\eta}_2, \widetilde{v}_2\}$ be two PHFEs, then*

(1) If $s(\widetilde{n}_1) > s(\widetilde{n}_2)$, then $\widetilde{n}_1 > \widetilde{n}_2$;
(2) If $s(\widetilde{n}_1) = s(\widetilde{n}_2)$, then

a. If $h(\widetilde{n}_1) > h(\widetilde{n}_2)$, then $\widetilde{n}_1 > \widetilde{n}_2$;
b. If $h(\widetilde{n}_1) = h(\widetilde{n}_2)$, then $\widetilde{n}_1 = \widetilde{n}_2$;
c. If $h(\widetilde{n}_1) < h(\widetilde{n}_2)$, then $\widetilde{n}_1 < \widetilde{n}_2$

For example, let $\tilde{n}_1 = \{\{0.3, 0.4\}, \{0.2\}, \{0.2, 0.3\}\}\pm$ and $\tilde{n}_2 = \{\{0.3\}, \{0.2, 0.3\}, \{0.1, 0.2\}\}$ be two PHFEs, according to the Definition 11, we have $s(\tilde{n}_1) = s(\tilde{n}_2) = 0.45$, $h(\tilde{n}_1) = 0.4$, and $h(\tilde{n}_2) = 0.35$, then $\tilde{n}_1 > \tilde{n}_2$.

Inspired by the operational laws of PFNs and UTHFEs, i.e., the Definition 3 and 8, we propose the operational laws of PHFEs as follows.

Definition 13. *Let $\tilde{n} = \{\tilde{\mu}, \tilde{\eta}, \tilde{v}\}$, $\tilde{n}_1 = \{\tilde{\mu}_1, \tilde{\eta}_1, \tilde{v}_1\}$, and $\tilde{n}_2 = \{\tilde{\mu}_2, \tilde{\eta}_2, \tilde{v}_2\}$ be three PHFEs, $\lambda > 0$, and $\tilde{n}^c$ is the complementary set of $\tilde{n}$, and the operations of PHFEs are represented as*

$$\tilde{n}^c = \bigcup_{\alpha \in \tilde{\mu}, \beta \in \tilde{\eta}, \gamma \in \tilde{v}} \{\{\gamma\}, \{\beta\}, \{\alpha\}\}; \tag{16}$$

$$\tilde{n}_1 \oplus \tilde{n}_2 = \{\tilde{\mu}_1 \oplus \tilde{\mu}_2, \tilde{\eta}_1 \otimes \tilde{\eta}_2, \tilde{v}_1 \otimes \tilde{v}_2\} = \bigcup_{\alpha_1 \in \tilde{\mu}_1, \beta_1 \in \tilde{\eta}_1, \gamma_1 \in \tilde{v}_1, \alpha_2 \in \tilde{\mu}_2, \beta_2 \in \tilde{\eta}_2, \gamma_2 \in \tilde{v}_2} \{\{\alpha_1 + \alpha_2 - \alpha_1\alpha_2\}, \{\beta_1\beta_2\}, \{\gamma_1\gamma_2\}\}; \tag{17}$$

$$\tilde{n}_1 \otimes \tilde{n}_2 = \{\tilde{\mu}_1 \otimes \tilde{\mu}_2, \tilde{\eta}_1 \oplus \tilde{\eta}_2, \tilde{v}_1 \oplus \tilde{v}_2\} = \bigcup_{\alpha_1 \in \tilde{\mu}_1, \beta_1 \in \tilde{\eta}_1, \gamma_1 \in \tilde{v}_1, \alpha_2 \in \tilde{\mu}_2, \beta_2 \in \tilde{\eta}_2, \gamma_2 \in \tilde{v}_2} \{\{\alpha_1\alpha_2\}, \{\beta_1 + \beta_2 - \beta_1\beta_2\}, \{\gamma_1 + \gamma_2 - \gamma_1\gamma_2\}\}; \tag{18}$$

$$\lambda\tilde{n} = \bigcup_{\alpha \in \tilde{\mu}, \beta \in \tilde{\eta}, \gamma \in \tilde{v}} \left\{\left\{1 - (1-\alpha)^\lambda\right\}, \left\{\beta^\lambda\right\}, \left\{\gamma^\lambda\right\}\right\}; \tag{19}$$

$$\tilde{n}^\lambda = \bigcup_{\alpha \in \tilde{\mu}, \beta \in \tilde{\eta}, \gamma \in \tilde{v}} \left\{\left\{\alpha^\lambda\right\}, \left\{1 - (1-\beta)^\lambda\right\}, \left\{1 - (1-\gamma)^\lambda\right\}\right\}. \tag{20}$$

For example, let $\tilde{n}_1 = \{\{0.3, 0.4\}, \{0.2\}, \{0.2, 0.3\}\}$ and $\tilde{n}_2 = \{\{0.3\}, \{0.2, 0.3\}, \{0.1, 0.2\}\}$ be two PHFEs, $\lambda = 2$, then

(1) $\tilde{n}_1{}^c = \{\{0.2, 0.3\}, \{0.2\}, \{0.3, 0.4\}\}$, $\tilde{n}_2{}^c = \{\{0.1, 0.2\}, \{0.2, 0.3\}, \{0.3\}\}$;
(2) $\tilde{n}_1 \oplus \tilde{n}_2 = \{\{0.51, 0.58\}, \{0.04, 0.06\}, \{0.02, 0.03, 0.04, 0.06\}\}$;
(3) $\tilde{n}_1 \otimes \tilde{n}_2 = \{\{0.09, 0.12\}, \{0.36, 0.44\}, \{0.28, 0.36, 0.37, 0.44\}\}$;
(4) $\lambda\tilde{n}_1 = \{\{0.51, 0.64\}, \{0.04\}, \{0.04, 0.09\}\}$, $\lambda\tilde{n}_2 = \{\{0.51\}, \{0.04, 0.09\}, \{0.01, 0.04\}\}$;
(5) $\tilde{n}_1{}^\lambda = \{\{0.09, 0.16\}, \{0.36\}, \{0.36, 0.51\}\}$, $\tilde{n}_2{}^\lambda = \{\{0.09\}, \{0.36, 0.51\}, \{0.19, 0.36\}\}$.

Obviously, the following theorem can be obtained based on the Definition 13.

Theorem 1. *Let $\tilde{n} = \{\tilde{\mu}, \tilde{\eta}, \tilde{v}\}$, $\tilde{n}_1 = \{\tilde{\mu}_1, \tilde{\eta}_1, \tilde{v}_1\}$, and $\tilde{n}_2 = \{\tilde{\mu}_2, \tilde{\eta}_2, \tilde{v}_2\}$ be three PHFEs, $\lambda, \lambda_1, \lambda_2 > 0$, then*

(1) $\tilde{n}_1 \oplus \tilde{n}_2 = \tilde{n}_2 \oplus \tilde{n}_1$;
(2) $\tilde{n}_1 \otimes \tilde{n}_2 = \tilde{n}_2 \otimes \tilde{n}_1$;
(3) $\lambda(\tilde{n}_1 \oplus \tilde{n}_2) = \lambda\tilde{n}_1 \oplus \lambda\tilde{n}_2$;
(4) $(\tilde{n}_1 \otimes \tilde{n}_2)^\lambda = \tilde{n}_1{}^\lambda \otimes \tilde{n}_2{}^\lambda$;
(5) $\lambda_1\tilde{n} \oplus \lambda_2\tilde{n} = (\lambda_1 + \lambda_2)\tilde{n}$;
(6) $\tilde{n}^{\lambda_1} \otimes \tilde{n}^{\lambda_2} = \tilde{n}^{(\lambda_1 + \lambda_2)}$;
(7) $\left(\tilde{n}^{\lambda_1}\right)^{\lambda_2} = \tilde{n}^{\lambda_1\lambda_2}$.

4. Generalized Picture Hesitant Fuzzy Aggregation Operators

Combined with the concept and operations of PHFS, the GPHFWA, GPHFWG, GPHFPWA, and GPHFPWG operators are developed. Then, several properties of them are discussed, and some other aggregation operators under PHF environment that reduced by the proposed operators are presented.

4.1. The GPHFWA Operator

Definition 14. *Let $\tilde{n}_j (j = 1, 2, \ldots, n)$ be a collection of PHFEs, the GPHFWA operator is a mapping $\Omega^n \to \Omega$ as*

$$GPHFWA_\lambda(\tilde{n}_1, \tilde{n}_2, \ldots, \tilde{n}_n) = \left(w_1 \tilde{n}_1{}^\lambda \oplus w_2 \tilde{n}_2{}^\lambda \oplus \cdots \oplus w_n \tilde{n}_n{}^\lambda\right)^{1/\lambda} = \mathop{\oplus}_{j=1}^{n} \left(w_j \tilde{n}_j{}^\lambda\right)^{1/\lambda}. \tag{21}$$

where $w = (w_1, w_2, \ldots, w_n)$ is the weight vector of PHFEs $\tilde{n}_j$, and satisfies the conditions of $w_j > 0$ and $\sum_{j=1}^{n} w_j = 1$.

Based on the Definition 13, we can obtain the theorems as follows.

Theorem 2. *Let $\tilde{n}_j (j = 1, 2, \ldots, n)$ be a collection of PHFEs, then their aggregated value by using the GPHFWA operator is also a PHFE, and*

$$\begin{gathered} GPHFWA_\lambda(\tilde{n}_1, \tilde{n}_2, \ldots, \tilde{n}_n) = \bigcup_{\alpha_1 \in \tilde{\mu}_1, \alpha_2 \in \tilde{\mu}_2, \ldots, \alpha_n \in \tilde{\mu}_n, \beta_1 \in \tilde{\eta}_1, \beta_2 \in \tilde{\eta}_2, \ldots, \beta_n \in \tilde{\eta}_n, \gamma_1 \in \tilde{v}_1, \gamma_2 \in \tilde{v}_2, \ldots, \gamma_n \in \tilde{v}_n} \left\{ \left\{ \left(1 - \prod_{j=1}^{n} (1 - \alpha_j{}^\lambda)^{w_j}\right)^{1/\lambda} \right\}, \right. \\ \left. \left\{ 1 - \left(1 - \prod_{j=1}^{n} \left(1 - (1 - \beta_j)^\lambda\right)^{w_j}\right)^{1/\lambda} \right\}, \left\{ 1 - \left(1 - \prod_{j=1}^{n} \left(1 - (1 - \gamma_j)^\lambda\right)^{w_j}\right)^{1/\lambda} \right\} \right\}. \end{gathered} \tag{22}$$

Proof. See Appendix A. □

Theorem 3. *(Idempotency) Let $\tilde{n}_j (j = 1, 2, \ldots, n)$ be a collection of PHFEs, if all the PHFEs are equal, i.e., $\tilde{n}_j = \tilde{n} = \{\tilde{\mu}, \tilde{\eta}, \tilde{v}\}$, $\tilde{\mu} = \alpha, \tilde{\eta} = \beta, \tilde{v} = \gamma$, then*

$$GPHFWA_\lambda(\tilde{n}_1, \tilde{n}_2, \ldots, \tilde{n}_n) = \tilde{n} = \{\tilde{\mu}, \tilde{\eta}, \tilde{v}\}. \tag{23}$$

Proof. See Appendix B. □

Theorem 4. *(Boundedness) Let $\tilde{n}_j (j = 1, 2, \ldots, n)$ be a collection of PHFEs, if $\tilde{n}^- = \{\{\alpha^-\}, \{\beta^+\}, \{\gamma^+\}\}$ and $\tilde{n}^+ = \{\{\alpha^+\}, \{\beta^-\}, \{\gamma^-\}\}$, where $\alpha^- = \cup_{\alpha_j \in \tilde{\mu}_j} \min\{\alpha_j\}$, $\beta^- = \cup_{\beta_j \in \tilde{\eta}_j} \min\{\beta_j\}$, $\gamma^- = \cup_{\gamma_j \in \tilde{v}_j} \min\{\gamma_j\}$, $\alpha^+ = \cup_{\alpha_j \in \tilde{\mu}_j} \max\{\alpha_j\}$, $\beta^+ = \cup_{\beta_j \in \tilde{\eta}_j} \max\{\beta_j\}$, and $\gamma^- = \cup_{\gamma_j \in \tilde{v}_j} \min\{\gamma_j\}$, thus*

$$\tilde{n}^- \leq GPHFWA_\lambda(\tilde{n}_1, \tilde{n}_2, \ldots, \tilde{n}_n) \leq \tilde{n}^+. \tag{24}$$

Proof. See Appendix C. □

Theorem 5. *(Monotonicity) Let $\tilde{n}_j (j = 1, 2, \ldots, n)$ and $\tilde{n}_j{}^* (j = 1, 2, \ldots, n)$ be two collections of PHFEs, if $\tilde{n}_j \leq \tilde{n}_j{}^*$, then*

$$GPHFWA_\lambda(\tilde{n}_1, \tilde{n}_2, \ldots, \tilde{n}_n) \leq GPHFWA_\lambda(\tilde{n}_1{}^*, \tilde{n}_2{}^*, \ldots, \tilde{n}_n{}^*). \tag{25}$$

Proof. Theorem 5 can be obtained by the Theorem 4. □

Under some specific situations, we can obtain the reduced operators of the GPHFWA operator.

Case 1. If $\lambda = 1$, then the GPHFWA operator is reduced to the picture hesitant fuzzy weighted averaging (PHFWA) operator

$$PHFWA(\tilde{n}_1, \tilde{n}_2, \ldots, \tilde{n}_n) = (w_1 \tilde{n}_1 \oplus w_2 \tilde{n}_2 \oplus \cdots \oplus w_n \tilde{n}_n) = \mathop{\oplus}_{j=1}^{n} (w_j \tilde{n}_j). \tag{26}$$

Case 2. If $\lambda = 1$ and $w = (1/n, 1/n, \ldots, 1/n)$, then the GPHFWA operator is reduced to the picture hesitant fuzzy arithmetic averaging (PHFAA) operator

$$PHFAA(\widetilde{n}_1, \widetilde{n}_2, \ldots, \widetilde{n}_n) = \left(\frac{1}{n}\widetilde{n}_1 \oplus \frac{1}{n}\widetilde{n}_2 \oplus \cdots \oplus \frac{1}{n}\widetilde{n}_n\right). \tag{27}$$

4.2. The GPHFWG Operator

Similarly, the GPHFWG operator can be defined as in the following.

Definition 15. *Let $\widetilde{n}_j (j = 1, 2, \ldots, n)$ be a collection of PHFEs, the GPHFWG operator is a mapping $\Omega^n \to \Omega$ as*

$$GPHFWG_\lambda(\widetilde{n}_1, \widetilde{n}_2, \ldots, \widetilde{n}_n) = \frac{1}{\lambda}(\lambda \widetilde{n}_1{}^{w_1} \otimes \lambda \widetilde{n}_2{}^{w_2} \otimes \cdots \otimes \lambda \widetilde{n}_n{}^{w_n}) = \frac{1}{\lambda} \underset{j=1}{\overset{n}{\otimes}} (\lambda \widetilde{n}_j{}^{w_j}). \tag{28}$$

where $w = (w_1, w_2, \ldots, w_n)$ is the weight vector of PHFEs $\widetilde{n}_j$, and satisfies the conditions of $w_j > 0$ and $\sum_{j=1}^n w_j = 1$.

According to the operational laws of PHFEs, the theorem can be obtained as follows.

Theorem 6. *Let $\widetilde{n}_j (j = 1, 2, \ldots, n)$ be a collection of PHFEs, then their aggregated value by using the GPHFWG operator is also a PHFE, and*

$$\begin{gathered} GPHFWG_\lambda(\widetilde{n}_1, \widetilde{n}_2, \ldots, \widetilde{n}_n) = \\ \bigcup_{\alpha_1 \in \widetilde{\mu}_1, \alpha_2 \in \widetilde{\mu}_2, \ldots, \alpha_n \in \widetilde{\mu}_n, \beta_1 \in \widetilde{\eta}_1, \beta_2 \in \widetilde{\eta}_2, \ldots, \beta_n \in \widetilde{\eta}_n, \gamma_1 \in \widetilde{v}_1, \gamma_2 \in \widetilde{v}_2, \ldots, \gamma_n \in \widetilde{v}_n} \left\{ \left\{ 1 - \left(1 - \prod_{j=1}^n \left(1 - (1-\alpha_j)^\lambda\right)^{w_j}\right)^{1/\lambda} \right\}, \left\{ \left(1 - \prod_{j=1}^n (1 - \beta_j{}^\lambda)^{w_j}\right)^{1/\lambda} \right\}, \right. \\ \left. \left\{ \left(1 - \prod_{j=1}^n (1 - \gamma_j{}^\lambda)^{w_j}\right)^{1/\lambda} \right\} \right. \end{gathered} \tag{29}$$

It can be proven by the same process as Theorem 3–5 that the GPHFWG operator also has several properties.

Theorem 7. *(Idempotency) Let $\widetilde{n}_j (j = 1, 2, \ldots, n)$ be a collection of PHFEs, if all the PHFEs are equal, i.e., $\widetilde{n}_j = \widetilde{n} = \{\widetilde{\mu}, \widetilde{\eta}, \widetilde{v}\}$, $\widetilde{\mu} = \alpha$, $\widetilde{\eta} = \beta$, $\widetilde{v} = \gamma$, then*

$$GPHFWG_\lambda(\widetilde{n}_1, \widetilde{n}_2, \ldots, \widetilde{n}_n) = \widetilde{n} = \{\widetilde{\mu}, \widetilde{\eta}, \widetilde{v}\}. \tag{30}$$

Theorem 8. *(Boundedness) Let $\widetilde{n}_j (j = 1, 2, \ldots, n)$ be a collection of PHFEs, if $\widetilde{n}^- = \{\{\alpha^-\}, \{\beta^+\}, \{\gamma^+\}\}$ and $\widetilde{n}^+ = \{\{\alpha^+\}, \{\beta^-\}, \{\gamma^-\}\}$, where $\alpha^- = \cup_{\alpha_j \in \widetilde{\mu}_j} \min\{\alpha_j\}$, $\beta^- = \cup_{\beta_j \in \widetilde{\eta}_j} \min\{\beta_j\}$, $\gamma^- = \cup_{\gamma_j \in \widetilde{v}_j} \min\{\gamma_j\}$, $\alpha^+ = \cup_{\alpha_j \in \widetilde{\mu}_j} \max\{\alpha_j\}$, $\beta^+ = \cup_{\beta_j \in \widetilde{\eta}_j} \max\{\beta_j\}$, and $\gamma^- = \cup_{\gamma_j \in \widetilde{v}_j} \min\{\gamma_j\}$, thus*

$$\widetilde{n}^- \le GPHFWG_\lambda(\widetilde{n}_1, \widetilde{n}_2, \ldots, \widetilde{n}_n) \le \widetilde{n}^+. \tag{31}$$

Theorem 9. *(Monotonicity) Let $\widetilde{n}_j (j = 1, 2, \ldots, n)$ and $\widetilde{n}_j{}^* (j = 1, 2, \ldots, n)$ be two collections of PHFEs, if $\widetilde{n}_j \le \widetilde{n}_j{}^*$, then*

$$GPHFWG_\lambda(\widetilde{n}_1, \widetilde{n}_2, \ldots, \widetilde{n}_n) \le GPHFWG_\lambda(\widetilde{n}_1{}^*, \widetilde{n}_2{}^*, \ldots, \widetilde{n}_n{}^*). \tag{32}$$

Several reduced operators of the GPHFWG operator are presented as:

Case 3. If $\lambda = 1$, then the GPHFWG operator is reduced to the picture hesitant fuzzy weighted geometric (PHFWG) operator

$$PHFWG(\widetilde{n}_1, \widetilde{n}_2, \ldots, \widetilde{n}_n) = (\widetilde{n}_1{}^{w_1} \otimes \widetilde{n}_2{}^{w_2} \otimes \cdots \otimes \widetilde{n}_n{}^{w_n}) = \bigotimes_{j=1}^{n} (\widetilde{n}_j{}^{w_j}). \tag{33}$$

Case 4. If $\lambda = 1$ and $w = (1/n, 1/n, \ldots, 1/n)$, then the GPHFWG operator is reduced to the picture hesitant fuzzy geometric averaging (PHFGA) operator

$$PHFGA(\widetilde{n}_1, \widetilde{n}_2, \ldots, \widetilde{n}_n) = (\widetilde{n}_1 \otimes \widetilde{n}_2 \otimes \cdots \otimes \widetilde{n}_n)^{1/n}. \tag{34}$$

4.3. The GPHFPWA Operator

In real life, the criteria sometimes have different priority levels. For example, safety has a higher priority than price when a couple chooses a toy for their child. Obviously, the GPHFWA and GPHFWG operators cannot deal with this situation; then, the GPHFPWA and GPHFPWG operators are developed according to the PA operator proposed by Yager [31].

Definition 16. *Let $\widetilde{n}_j (j = 1, 2, \ldots, n)$ be a collection of PHFEs, the GPHFPWA operator is a mapping $\Omega^n \to \Omega$ as*

$$GPHFPWA_\lambda(\widetilde{n}_1, \widetilde{n}_2, \ldots, \widetilde{n}_n) = \left(\frac{T_1}{\sum_{j=1}^{n} T_j} \widetilde{n}_1{}^\lambda \oplus \frac{T_2}{\sum_{j=1}^{n} T_j} \widetilde{n}_2{}^\lambda \oplus \cdots \oplus \frac{T_n}{\sum_{j=1}^{n} T_j} \widetilde{n}_n{}^\lambda \right)^{1/\lambda}. \tag{35}$$

where $T_j = \prod_{k=1}^{j-1} s(\widetilde{n}_k) (j = 2, \ldots, n)$, $T_1 = 1$, and $s(\widetilde{n}_k)$ is the score value of PHFE $\widetilde{n}_k$.

Similarly, the following theorem can be put forward.

Theorem 10. *Let $\widetilde{n}_j (j = 1, 2, \ldots, n)$ be a collection of PHFEs, then their aggregated value by using the GPHFPWA operator is also a PHFE, and*

$$\begin{aligned} &GPHFPWA_\lambda(\widetilde{n}_1, \widetilde{n}_2, \ldots, \widetilde{n}_n) = \\ &\bigcup_{\alpha_1 \in \widetilde{\mu}_1, \alpha_2 \in \widetilde{\mu}_2, \ldots, \alpha_n \in \widetilde{\mu}_n, \beta_1 \in \widetilde{\eta}_1, \beta_2 \in \widetilde{\eta}_2, \ldots, \beta_n \in \widetilde{\eta}_n, \gamma_1 \in \widetilde{v}_1, \gamma_2 \in \widetilde{v}_2, \ldots, \gamma_n \in \widetilde{v}_n} \left\{ \left\{ \left(1 - \prod_{j=1}^{n} (1 - \alpha_j{}^\lambda)^{\frac{T_j}{\sum_{j=1}^{n} T_j}} \right)^{1/\lambda} \right\}, \left\{ 1 - \left(1 - \prod_{j=1}^{n} \left(1 - (1 - \beta_j)^\lambda\right)^{\frac{T_j}{\sum_{j=1}^{n} T_j}} \right)^{1/\lambda} \right\}, \right. \\ &\left. \left\{ 1 - \left(1 - \prod_{j=1}^{n} \left(1 - (1 - \gamma_j)^\lambda\right)^{\frac{T_j}{\sum_{j=1}^{n} T_j}} \right)^{1/\lambda} \right\} \right\}. \end{aligned} \tag{36}$$

The GPHFPWA operator also has the properties as follows.

Theorem 12. *(Idempotency) Let $\widetilde{n}_j (j = 1, 2, \ldots, n)$ be a collection of PHFEs, if all the PHFEs are equal, i.e., $\widetilde{n}_j = \widetilde{n} = \{\widetilde{\mu}, \widetilde{\eta}, \widetilde{v}\}$, $\widetilde{\mu} = \alpha, \widetilde{\eta} = \beta, \widetilde{v} = \gamma$, then*

$$GPHFPWA_\lambda(\widetilde{n}_1, \widetilde{n}_2, \ldots, \widetilde{n}_n) = \widetilde{n} = \{\widetilde{\mu}, \widetilde{\eta}, \widetilde{v}\}. \tag{37}$$

Theorem 13. *(Boundedness) Let $\widetilde{n}_j (j = 1, 2, \ldots, n)$ be a collection of PHFEs, $\widetilde{n}^- = \{\{\alpha^-\}, \{\beta^+\}, \{\gamma^+\}\}$ and $\widetilde{n}^+ = \{\{\alpha^+\}, \{\beta^-\}, \{\gamma^-\}\}$, where $\alpha^- = \cup_{\alpha_j \in \widetilde{\mu}_j} \min\{\alpha_j\}$, $\beta^- = \cup_{\beta_j \in \widetilde{\eta}_j} \min\{\beta_j\}$, $\gamma^- = \cup_{\gamma_j \in \widetilde{v}_j} \min\{\gamma_j\}$, $\alpha^+ = \cup_{\alpha_j \in \widetilde{\mu}_j} \max\{\alpha_j\}$, $\beta^+ = \cup_{\beta_j \in \widetilde{\eta}_j} \max\{\beta_j\}$, and $\gamma^- = \cup_{\gamma_j \in \widetilde{v}_j} \min\{\gamma_j\}$, thus*

$$\tilde{n}^- \leq GPHFPWA_\lambda(\tilde{n}_1, \tilde{n}_2, \ldots, \tilde{n}_n) \leq \tilde{n}^+. \tag{38}$$

Theorem 14. *(Monotonicity) Let $\tilde{n}_j (j = 1, 2, \ldots, n)$ and $\tilde{n}_j{}^* (j = 1, 2, \ldots, n)$ be two collections of PHFEs, if $\tilde{n}_j \leq \tilde{n}_j{}^*$, then*

$$GPHFPWA_\lambda(\tilde{n}_1, \tilde{n}_2, \ldots, \tilde{n}_n) \leq GPHFPWA_\lambda(\tilde{n}_1{}^*, \tilde{n}_2{}^*, \ldots, \tilde{n}_n{}^*). \tag{39}$$

Then, the reduced operators of the GPHFPWA operator can be obtained.

Case 5. If $\lambda = 1$, then the GPHFPWA operator is reduced to the picture hesitant fuzzy prioritized weighted averaging (PHFPWA) operator

$$PHFPWA(\tilde{n}_1, \tilde{n}_2, \ldots, \tilde{n}_n) = \left(\frac{T_1}{\sum_{j=1}^n T_j} \tilde{n}_1 \oplus \frac{T_2}{\sum_{j=1}^n T_j} \tilde{n}_2 \oplus \cdots \oplus \frac{T_n}{\sum_{j=1}^n T_j} \tilde{n}_n \right). \tag{40}$$

Case 6. If $\lambda = 1$ and the criteria are at the same priority, then the GPHFPWA operator is reduced to the PHFWA operator

$$PHFWA(\tilde{n}_1, \tilde{n}_2, \ldots, \tilde{n}_n) = (w_1 \tilde{n}_1 \oplus w_2 \tilde{n}_2 \oplus \cdots \oplus w_n \tilde{n}_n) = \bigoplus_{j=1}^{n} (w_j \tilde{n}_j). \tag{41}$$

Case 7. If $\lambda = 1$, $w = (1/n, 1/n, \ldots, 1/n)$, and the criteria are at the same priority, then the GPHFPWA operator is reduced to the PHFAA operator

$$PHFAA(\tilde{n}_1, \tilde{n}_2, \ldots, \tilde{n}_n) = \left(\frac{1}{n} \tilde{n}_1 \oplus \frac{1}{n} \tilde{n}_2 \oplus \cdots \oplus \frac{1}{n} \tilde{n}_n \right). \tag{42}$$

4.4. The GPHFPWG Operator

Similarly, the GPHFPWG operator is constructed as below.

Definition 17. *Let $\tilde{n}_j (j = 1, 2, \ldots, n)$ be a collection of PHFEs, the GPHFPWG operator is a mapping $\Omega^n \to \Omega$ as*

$$GPHFPWG_\lambda(\tilde{n}_1, \tilde{n}_2, \ldots, \tilde{n}_n) = \frac{1}{\lambda} \left((\lambda \tilde{n}_1)^{\frac{T_1}{\sum_{j=1}^n T_j}} \otimes (\lambda \tilde{n}_2)^{\frac{T_2}{\sum_{j=1}^n T_j}} \otimes \cdots \otimes (\lambda \tilde{n}_n)^{\frac{T_n}{\sum_{j=1}^n T_j}} \right). \tag{43}$$

where $T_j = \prod_{k=1}^{j-1} s(\tilde{n}_k) (j = 2, \ldots, n)$, $T_1 = 1$, and $s(\tilde{n}_k)$ is the score value of PHFE $\tilde{n}_k$.

Combined with the operations of PHFEs, the following theorems are obtained.

Theorem 15. *Let $\widetilde{n}_j (j = 1, 2, \ldots, n)$ be a collection of PHFEs, then their aggregated value by using the GPHFPWG operator is also a PHFE, and*

$$GPHFPWG_\lambda(\widetilde{n}_1, \widetilde{n}_2, \ldots, \widetilde{n}_n) = \bigcup_{\alpha_1 \in \widetilde{\mu}_1, \alpha_2 \in \widetilde{\mu}_2, \ldots, \alpha_n \in \widetilde{\mu}_n, \beta_1 \in \widetilde{\eta}_1, \beta_2 \in \widetilde{\eta}_2, \ldots, \beta_n \in \widetilde{\eta}_n, \gamma_1 \in \widetilde{v}_1, \gamma_2 \in \widetilde{v}_2, \ldots, \gamma_n \in \widetilde{v}_n} \left\{ \left\{ 1 - \left(1 - \prod_{j=1}^{n} \left(1 - (1-\alpha_j)^\lambda \right)^{\frac{T_j}{\sum_{j=1}^{n} T_j}} \right)^{1/\lambda} \right\}, \left\{ \left(1 - \prod_{j=1}^{n} (1 - \beta_j{}^\lambda)^{\frac{T_j}{\sum_{j=1}^{n} T_j}} \right)^{1/\lambda} \right\}, \left\{ \left(1 - \prod_{j=1}^{n} (1 - \gamma_j{}^\lambda)^{\frac{T_j}{\sum_{j=1}^{n} T_j}} \right)^{1/\lambda} \right\} \right\}. \tag{44}$$

Theorem 16. *(Idempotency) Let $\widetilde{n}_j (j = 1, 2, \ldots, n)$ be a collection of PHFEs, if all the PHFEs are equal, i.e., $\widetilde{n}_j = \widetilde{n} = \{\widetilde{\mu}, \widetilde{\eta}, \widetilde{v}\}$, $\widetilde{\mu} = \alpha, \widetilde{\eta} = \beta, \widetilde{v} = \gamma$, then*

$$GPHFPWG_\lambda(\widetilde{n}_1, \widetilde{n}_2, \ldots, \widetilde{n}_n) = \widetilde{n} = \{\widetilde{\mu}, \widetilde{\eta}, \widetilde{v}\}. \tag{45}$$

Theorem 17. *(Boundedness) Let $\widetilde{n}_j (j = 1, 2, \ldots, n)$ be a collection of PHFEs, if $\widetilde{n}^- = \{\{\alpha^-\}, \{\beta^+\}, \{\gamma^+\}\}$ and $\widetilde{n}^+ = \{\{\alpha^+\}, \{\beta^-\}, \{\gamma^-\}\}$, where $\alpha^- = \cup_{\alpha_j \in \widetilde{\mu}_j} \min\{\alpha_j\}$, $\beta^- = \cup_{\beta_j \in \widetilde{\eta}_j} \min\{\beta_j\}$, $\gamma^- = \cup_{\gamma_j \in \widetilde{v}_j} \min\{\gamma_j\}$, $\alpha^+ = \cup_{\alpha_j \in \widetilde{\mu}_j} \max\{\alpha_j\}$, $\beta^+ = \cup_{\beta_j \in \widetilde{\eta}_j} \max\{\beta_j\}$, and $\gamma^- = \cup_{\gamma_j \in \widetilde{v}_j} \min\{\gamma_j\}$, thus*

$$\widetilde{n}^- \leq GPHFPWG_\lambda(\widetilde{n}_1, \widetilde{n}_2, \ldots, \widetilde{n}_n) \leq \widetilde{n}^+. \tag{46}$$

Theorem 18. *(Monotonicity) Let $\widetilde{n}_j (j = 1, 2, \ldots, n)$ and $\widetilde{n}_j{}^* (j = 1, 2, \ldots, n)$ be two collections of PHFEs, if $\widetilde{n}_j \leq \widetilde{n}_j{}^*$, then*

$$GPHFPWG_\lambda(\widetilde{n}_1, \widetilde{n}_2, \ldots, \widetilde{n}_n) \leq GPHFPWG_\lambda(\widetilde{n}_1{}^*, \widetilde{n}_2{}^*, \ldots, \widetilde{n}_n{}^*). \tag{47}$$

Several reduced operators of the GPHFPWG operator are presented as below:

Case 8. If $\lambda = 1$, then the GPHFPWG operator is reduced to the picture hesitant fuzzy prioritized weighted geometric (PHFPWG) operator

$$PHFPWG(\widetilde{n}_1, \widetilde{n}_2, \ldots, \widetilde{n}_n) = \left((\widetilde{n}_1)^{\frac{T_1}{\sum_{j=1}^{n} T_j}} \otimes (\widetilde{n}_2)^{\frac{T_2}{\sum_{j=1}^{n} T_j}} \otimes \cdots \otimes (\widetilde{n}_n)^{\frac{T_n}{\sum_{j=1}^{n} T_j}} \right). \tag{48}$$

Case 9. If $\lambda = 1$ and the criteria are at the same priority, then the GPHFPWG operator is reduced to the PHFWG operator

$$PHFWG(\widetilde{n}_1, \widetilde{n}_2, \ldots, \widetilde{n}_n) = (\widetilde{n}_1{}^{w_1} \otimes \widetilde{n}_2{}^{w_2} \otimes \cdots \otimes \widetilde{n}_n{}^{w_n}) = \bigotimes_{j=1}^{n} (\widetilde{n}_j{}^{w_j}). \tag{49}$$

Case 10. If $\lambda = 1, w = (1/n, 1/n, \ldots, 1/n)$ and the criteria are at the same priority, then the GPHFPWG operator is reduced to the PHFGA operator

$$PHFGA(\widetilde{n}_1, \widetilde{n}_2, \ldots, \widetilde{n}_n) = (\widetilde{n}_1 \otimes \widetilde{n}_2 \otimes \cdots \otimes \widetilde{n}_n)^{1/n}. \tag{50}$$

5. MCDM Methods under PHF Environment

We utilize the proposed operators to deal the different MCDM problems under PHF environment in this section. Let $A = \{A_1, A_2, \ldots, A_m\}$ be a collection of alternatives and $C = \{C_1, C_2, \ldots, C_n\}$ be a set of criteria; decision maker evaluates the m alternatives concerning the n criteria by using the PHFEs. Thus, suppose that $N = (\tilde{n}_{ij})(i = 1, 2, \ldots, m; j = 1, 2, \ldots, n)$ is the PHF evaluation matrix, and $\tilde{n}_{ij} = \{\tilde{\mu}_{ij}, \tilde{\eta}_{ij}, \tilde{v}_{ij}\}$ is the evaluation information when the alternative A_i is evaluated concerning the criteria C_j. In general, the criteria can be divided into two types in practice, namely, the cost criteria and benefit criteria; therefore, the evaluation information concerning the cost criteria should be transformed into the evaluation information concerning the benefit criteria to obtain the standardized PHF evaluation matrix $\overline{N} = (\overline{n}_{ij})$ as

$$\overline{n}_{ij} = \begin{cases} \tilde{n}_{ij}, & \text{for the benefit criteria;} \\ (\tilde{n}_{ij})^c, & \text{for the cost criteria.} \end{cases} \tag{51}$$

According to the aforementioned assumptions, when the criteria of a specific MCDM problem are in same priority level, and let $w = (w_1, w_2, \ldots, w_n)$ be the weight vector of the criteria. We can construct a novel approach, i.e., Algorithm 1 to solve it based on the GPHFWA or the GPHFWG operator. The flow diagram of the Algorithm 1 is presented in Figure 1, and the ranking result can be obtained by the following steps.

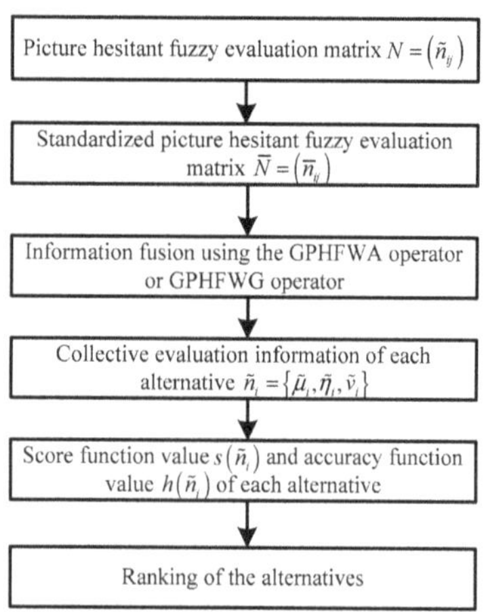

Figure 1. Flow diagram of the Algorithm 1.

Algorithm 1. MCDM method based on the GPHFWA or the GPHFWG operator.

1: Normalize the PHF evaluation matrix N to obtain the standardized PHF evaluation matrix $\overline{N}$ combined with Equation (51).

2: Utilize the GPHFWA operator

$$\begin{gathered} GPHFWA_\lambda(\overline{n}_{i1}, \overline{n}_{i2}, \ldots, \overline{n}_{in}) = \left(w_1 \overline{n}_{i1}{}^\lambda \oplus w_2 \overline{n}_{i2}{}^\lambda \oplus \cdots \oplus w_n \overline{n}_{in}{}^\lambda\right)^{1/\lambda} = \overset{n}{\underset{j=1}{\oplus}} \left(w_j \overline{n}_{ij}{}^\lambda\right)^{1/\lambda} = \\ \bigcup_{\alpha_{i1} \in \tilde{\mu}_{i1}, \alpha_{i2} \in \tilde{\mu}_{i2}, \ldots, \alpha_{in} \in \tilde{\mu}_{in}, \beta_{i1} \in \tilde{\eta}_{i1}, \beta_{i2} \in \tilde{\eta}_{i2}, \ldots, \beta_{in} \in \tilde{\eta}_{in}, \gamma_{i1} \in \tilde{v}_{i1}, \gamma_{i2} \in \tilde{v}_{i2}, \ldots, \gamma_{in} \in \tilde{v}_{in}} \left\{ \left\{ \left(1 - \prod_{j=1}^{n} \left(1 - \overline{\alpha}_{ij}{}^\lambda\right)^{w_j}\right)^{1/\lambda} \right\}, \left\{ 1 - \left(1 - \prod_{j=1}^{n} \left(1 - \left(1 - \overline{\beta}_{ij}\right)^\lambda\right)^{w_j}\right)^{1/\lambda} \right\}, \right. \\ \left. \left\{ 1 - \left(1 - \prod_{j=1}^{n} \left(1 - \left(1 - \overline{\gamma}_{ij}\right)^\lambda\right)^{w_j}\right)^{1/\lambda} \right\} \right\} \end{gathered} \tag{52}$$

or the GPHFWG operator

$$GPHFWG_{\lambda}(\overline{n}_{i1},\overline{n}_{i2},\ldots,\overline{n}_{in}) = \frac{1}{\lambda}(\lambda\overline{n}_{i1}{}^{w_1} \otimes \lambda\overline{n}_{i2}{}^{w_2} \otimes \cdots \otimes \lambda\overline{n}_{in}{}^{w_n}) = \frac{1}{\lambda}\overset{n}{\underset{j=1}{\otimes}}\left(\lambda\overline{n}_{ij}{}^{w_j}\right) = \bigcup_{\alpha_{i1}\in\widetilde{\mu}_{i1},\alpha_{i2}\in\widetilde{\mu}_{i2},\ldots,\alpha_{in}\in\widetilde{\mu}_{in},\beta_{i1}\in\widetilde{\eta}_{i1},\beta_{i2}\in\widetilde{\eta}_{i2},\ldots,\beta_{in}\in\widetilde{\eta}_{in},\gamma_{i1}\in\widetilde{v}_{i1},\gamma_{i2}\in\widetilde{v}_{i2},\ldots,\gamma_{in}\in\widetilde{v}_{in}}\left\{\left\{1-\left(1-\prod_{j=1}^{n}\left(1-\left(1-\overline{\alpha}_{ij}\right)^{\lambda}\right)^{w_j}\right)^{1/\lambda}\right\},\left\{\left(1-\prod_{j=1}^{n}\left(1-\overline{\beta}_{ij}{}^{\lambda}\right)^{w_j}\right)^{1/\lambda}\right\},\left\{\left(1-\prod_{j=1}^{n}\left(1-\overline{\gamma}_{ij}{}^{\lambda}\right)^{w_j}\right)^{1/\lambda}\right\}\right\} \quad (53)$$

to aggregate the standardized PHF evaluation matrix $\overline{N}$ to obtain the collective evaluation information of each alternative, i.e., $\widetilde{n}_i = \{\widetilde{\mu}_i, \widetilde{\eta}_i, \widetilde{v}_i\}$.
3: Compute the score and accuracy values of each alternative using Equation (14) and (15).
4: Based on the comparison method of PHFEs, rank the alternatives.

When the criteria are in different priorities, we can solve the MCDM problem combined with the Algorithm 2 based on the GPHFPWA or the GPHFPWG operator. The flow diagram of Algorithm 2 is presented in Figure 2, and the ranking result can be obtained by the following steps.

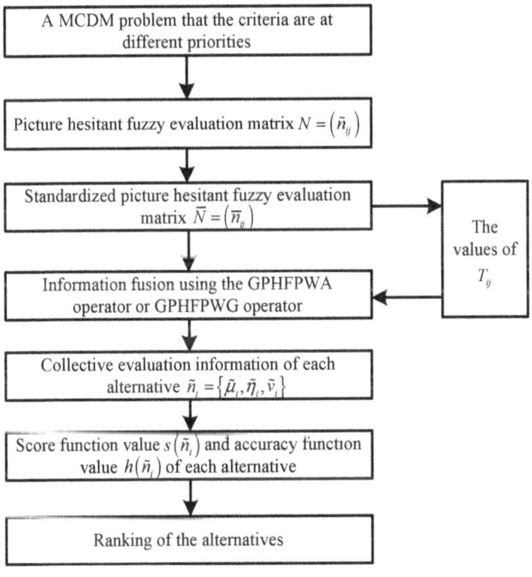

Figure 2. Flow diagram of the Algorithm 2.

Algorithm 2. MCDM method based on the GPHFPWA or the GPHFPWG operator.

1: Normalize the PHF evaluation matrix N to obtain the standardized PHF evaluation matrix combined with Equation (51).
2: Compute the values of T_{ij} using the equations as

$$T_{ij} = \prod_{k=1}^{j-1} s(\widetilde{n}_{ik}), T_{i1} = 1. \quad (54)$$

3: Utilize the GPHFPWA operator

$$GPHFPWA_{\lambda}(\tilde{n}_{i1},\tilde{n}_{i2},\ldots,\tilde{n}_{in}) = \left(\frac{T_{i1}}{\sum_{j=1}^{n}T_{ij}}\tilde{n}_{i1}^{\lambda} \oplus \frac{T_{i2}}{\sum_{j=1}^{n}T_{ij}}\tilde{n}_{i2}^{\lambda} \oplus \cdots \oplus \frac{T_{in}}{\sum_{j=1}^{n}T_{ij}}\tilde{n}_{in}^{\lambda}\right)^{1/\lambda}$$
$$= \bigcup_{\alpha_{i1}\in\tilde{\mu}_{i1},\alpha_{i2}\in\tilde{\mu}_{i2},\ldots,\alpha_{in}\in\tilde{\mu}_{in},\beta_{i1}\in\tilde{\eta}_{i1},\beta_{i2}\in\tilde{\eta}_{i2},\ldots,\beta_{in}\in\tilde{\eta}_{in},\gamma_{i1}\in\tilde{v}_{i1},\gamma_{i2}\in\tilde{v}_{i2},\ldots,\gamma_{in}\in\tilde{v}_{in}} \left\{\left\{\left(1-\prod_{j=1}^{n}\left(1-\tilde{\alpha}_{ij}{}^{\lambda}\right)^{\frac{T_{ij}}{\sum_{j=1}^{n}T_{ij}}}\right)^{1/\lambda}\right\},\left\{1-\left(1-\prod_{j=1}^{n}\left(1-\left(1-\tilde{\beta}_{ij}\right)^{\lambda}\right)^{\frac{T_{ij}}{\sum_{j=1}^{n}T_{ij}}}\right)^{1/\lambda}\right\}, \left\{1-\left(1-\prod_{j=1}^{n}\left(1-\left(1-\tilde{\gamma}_{ij}\right)^{\lambda}\right)^{\frac{T_{ij}}{\sum_{j=1}^{n}T_{ij}}}\right)^{1/\lambda}\right\}\right\} \quad (55)$$

or the GPHFPWG operator

$$GPHFPWG_{\lambda}(\tilde{n}_{i1},\tilde{n}_{i2},\ldots,\tilde{n}_{in}) = \frac{1}{\lambda}\left((\lambda\tilde{n}_{i1})^{\frac{T_{i1}}{\sum_{j=1}^{n}T_{ij}}} \otimes (\lambda\tilde{n}_{i2})^{\frac{T_{i2}}{\sum_{j=1}^{n}T_{ij}}} \otimes \cdots \otimes (\lambda\tilde{n}_{in})^{\frac{T_{in}}{\sum_{j=1}^{n}T_{ij}}}\right) =$$
$$\bigcup_{\alpha_{i1}\in\tilde{\mu}_{i1},\alpha_{i2}\in\tilde{\mu}_{i2},\ldots,\alpha_{in}\in\tilde{\mu}_{in},\beta_{i1}\in\tilde{\eta}_{i1},\beta_{i2}\in\tilde{\eta}_{i2},\ldots,\beta_{in}\in\tilde{\eta}_{in},\gamma_{i1}\in\tilde{v}_{i1},\gamma_{i2}\in\tilde{v}_{i2},\ldots,\gamma_{in}\in\tilde{v}_{in}} \left\{\left\{1-\left(1-\prod_{j=1}^{n}\left(1-\left(1-\tilde{\alpha}_{ij}\right)^{\lambda}\right)^{\frac{T_{ij}}{\sum_{j=1}^{n}T_{ij}}}\right)^{1/\lambda}\right\},\left\{\left(1-\prod_{j=1}^{n}\left(1-\tilde{\beta}_{ij}{}^{\lambda}\right)^{\frac{T_{ij}}{\sum_{j=1}^{n}T_{ij}}}\right)^{1/\lambda}\right\}, \left\{\left(1-\prod_{j=1}^{n}\left(1-\tilde{\gamma}_{ij}{}^{\lambda}\right)^{\frac{T_{ij}}{\sum_{j=1}^{n}T_{ij}}}\right)^{1/\lambda}\right\}\right\} \quad (56)$$

to aggregate the standardized PHF evaluation matrix $\overline{N}$ to obtain the collective evaluation information of each alternative, i.e., $\tilde{n}_i = \{\tilde{\mu}_i, \tilde{\eta}_i, \tilde{v}_i\}$.
4: Compute the score and accuracy values of each alternative using Equation (14) and (15).
5: Based on the comparison method of PHFEs, rank the alternatives.

6. Numerical Examples

We adopt two numerical examples of MCDM problems from the study of [25] and [34] and an application of web service selection [37] to show the feasibility and advantages of the proposed methods.

6.1. Implementation

Example 1. *Suppose that an organization wants to construct the enterprise resource planning (ERP) system [25]. After investigating the existing vendors of ERP systems on the market, five potential ERP systems are primary determined to be chosen from, i.e.,* A_i $(i = 1, 2, 3, 4, 5)$. *Decision makers utilize the PHFEs to evaluate the five alternatives with respect to four criteria, namely, function and technology* (C_1), *strategic fitness* (C_2), *ability of vendor* (C_3), *and reputation of vendor* (C_4), *and the weight vector of the criteria is* $w = (0.2, 0.1, 0.3, 0.4)$. *Subsequently, the PHF evaluation matrix* $N = (\tilde{n}_{ij})$ *is obtained as shown in Table 1.*

Table 1. PHF evaluation matrix of Example 1.

Alternatives	C_1	C_2	C_3	C_4
A_1	{{0.43,0.53},{0.33},{0.06,0.09}}	{{0.76,0.89},{0.05,0.08},{0.03}}	{{0.42},{0.35},{0.12,0.18}}	{{0.08},{0.75,0.89},{0.02}}
A_2	{{0.53,0.65,0.73},{0.10,0.12},{0.08}}	{{0.13},{0.53,0.64},{0.12,0.21}}	{{0.03},{0.77,0.82},{0.10,0.13}}	{{0.58,0.73},{0.15},{0.08}}
A_3	{{0.72,0.86,0.91},{0.03},{0.02}}	{{0.07},{0.05,0.09},{0.05}}	{{0.04},{0.65,0.72,0.85},{0.05,0.10}}	{{0.45,0.68},{0.18,0.26},{0.06}}
A_4	{{0.77,0.85},{0.09},{0.05}}	{{0.65,0.74},{0.10,0.16},{0.10}}	{{0.02},{0.78,0.89},{0.05}}	{{0.08},{0.65,0.84},{0.06}}
A_5	{{0.70,0.81,0.90},{0.05},{0.02}}	{{0.68},{0.08},{0.13,0.21}}	{{0.05},{0.77,0.87},{0.06}}	{{0.13},{0.65,0.75},{0.09}}

Then, we can determine the ranking of the five potential ERP systems using the Algorithm 1, which are presented as below.

Step 1: Because of all the criteria are the benefit type, the standardized PHF evaluation matrix $\overline{N}$ is as same as the PHF evaluation matrix N.

Step 2: Use the GPHFWA ($\lambda = 1$) operator to aggregate the standardized PHF evaluation matrix $\overline{N}$, and the collective evaluation information of each alternative is obtained as

$$\tilde{n}_1 = \{\{0.3636, 0.3877, 0.4113, 0.4336\}, \{0.3862, 0.4048, 0.4136, 0.4335\}, \{0.0444, 0.0482, 0.0502, 0.0544\}\};$$
$$\tilde{n}_2 = \{\{0.4061, 0.4401, 0.4684, 0.5023, 0.5308, 0.5545\}, \{0.2563, 0.2612, 0.2612, 0.2659, 0.2662, 0.2709, 0.2709, 0.2761\}, \{0.0891, 0.0942, 0.0964, 0.1019\}\};$$
$$\tilde{n}_3 = \{\{0.4014, 0.4789, 0.5180, 0.5230, 0.5804, 0.6159\}, \{0.1627, 0.1677, 0.1725, 0.1763, 0.1779, 0.1870, 0.1884, 0.1943, 0.1999, 0.2042, 0.2061, 0.2166\}, \{0.0448, 0.0551\}\};$$
$$\tilde{n}_4 = \{\{0.3549, 0.3738, 0.4077, 0.4251\}, \{0.3834, 0.3989, 0.4018, 0.4181, 0.4248, 0.4420, 0.4452, 0.5632\}, \{0.0576\}\};$$
$$\tilde{n}_5 = \{\{0.3468, 0.4038, 0.4756\}, \{0.3321, 0.3444, 0.3516, 0.3647\}, \{0.0612, 0.0642\}\}.$$

Step 3: Compute the score values of each alternative combined with Equation (14):

$$s(\tilde{n}_1) = 0.4701, s(\tilde{n}_2) = 0.5611, s(\tilde{n}_3) = 0.6409, s(\tilde{n}_4) = 0.4553, s(\tilde{n}_5) = 0.4989.$$

Step 4: According to the score values, the ranking result of the five ERP systems is determined as $A_3 \succ A_2 \succ A_5 \succ A_1 \succ A_4$.

If the GPHFWG operator is utilized in the steps above to complete the information fusion, the ranking procedures are presented as follows.

Step 1′: See **Step 1**.
Step 2′: Use the GPHFWG ($\lambda = 1$) operator to aggregate the standardized PHF evaluation matrix $\overline{N}$, and the collective evaluation information of each alternative is obtained as

$$\tilde{n}_1 = \{\{0.2307, 0.2343, 0.2405, 0.2443\}, \{0.5365, 0.6663, 0.5380, 0.6673\}, \{0.0600, 0.0797, 0.0661, 0.0856\}\};$$
$$\tilde{n}_2 = \{\{0.2017, 0.2101, 0.2151, 0.2212, 0.2304, 0.2358\}, \{0.4526, 0.4550, 0.4670, 0.4693, 0.4914, 0.4936, 0.5047, 0.5070\}, \{0.0901, 0.0993, 0.0999, 0.1090\}\};$$
$$\tilde{n}_3 = \{\{0.1986, 0.2057, 0.2081, 0.2342, 0.2427, 0.2454\}, \{0.3334, 0.3363, 0.3602, 0.3630, 0.3766, 0.3792, 0.4016, 0.4042, 0.4830, 0.4852, 0.5038, 0.5059\}, \{0.0481, 0.0634\}\};$$
$$\tilde{n}_4 = \{\{0.1024, 0.1037, 0.1044, 0.1058\}, \{0.5949, 0.5977, 0.6709, 0.6732, 0.7038, 0.7058, 0.7594, 0.7611\}, \{0.0591\}\};$$
$$\tilde{n}_5 = \{\{0.1613, 0.1660, 0.1696\}, \{0.5850, 0.6372, 0.6503, 0.6943\}, \{0.0716, 0.0805\}\}$$

Step 3′: Compute the score values of each alternative combined with Equation (14):

$$s(\tilde{n}_1) = 0.2813, s(\tilde{n}_2) = 0.3197, s(\tilde{n}_3) = 0.3778, s(\tilde{n}_4) = 0.1808, s(\tilde{n}_5) = 0.2240.$$

Step 4′: According to the score values, the ranking result of the five ERP systems is determined as $A_3 \succ A_2 \succ A_1 \succ A_5 \succ A_4$.

Example 2. *Suppose a university wants to introduce excellent foreign professors to improve the level of teaching and scientific research [34]. There are five foreign professors who are selected by the University's human resources department. Based on the priority level, the criteria of investigation is successively morality (C_1), research ability (C_2), teaching capacity (C_3), and educational experience (C_4); a priority relationship $C_1 \succ C_2 \succ C_3 \succ C_4$ exists between the criteria. Then, the PHF evaluation matrix $N = (\tilde{n}_{ij})$ is presented in Table 2.*

Table 2. PHF evaluation matrix in Example 2.

Alternatives	C_1	C_2	C_3	C_4
A_1	{{0.40,0.50,0.70},{0.05},{0.10,0.20}}	{{0.65},{0.05,0.08}, {0.15}}	{{0.40,0.50,0.60},{0.03},{0.10,0.20}}	{{0.55},{0.10,0.15}, {0.15}}
A_2	{{0.65,0.75},{0.02,0.04},{0.15}}	{{0.60},{0.05,0.10}, {0.10,0.20}}	{{0.75,0.80},{0.06}, {0.05,0.08}}	{{0.40,0.50},{0.20}, {0.15,0.25}}
A_3	{{0.70},{0.06,0.10}, {0.10,0.15}}	{{0.20,0.30,0.50},{0.04},{0.30,0.40}}	{{0.50},{0.03,0.06}, {0.30,0.35}}	{{0.50,0.70},{0.10}, {0.10}}
A_4	{{0.50,0.60,0.70},{0.08},{0.10}}	{{0.40,0.50},{0.20}, {0.10,0.20}}	{{0.85},{0.03,0.07}, {0.05}}	{{0.45},{0.10,0.20}, {0.15,0.30}}
A_5	{{0.65},{0.05,0.10}, {0.15,0.20}}	{{0.50,0.70},{0.08}, {0.20}}	{{0.70,0.80},{0.04}, {0.10}}	{{0.35},{0.10,0.20},{0.30,0.40}}

Subsequently, we can determine the ranking of the five foreign professors using the Algorithm 2, which are presented as follows.

Step 1: Because of all the criteria are the benefit type, the standardized PHF evaluation matrix $\overline{N}$ is as same as the PHF evaluation matrix N.
Step 2: Compute the values of T_{ij} using the Equation (54)

$$T_{ij} = \begin{bmatrix} 1.000 & 0.6667 & 0.4783 & 0.3157 \\ 1.000 & 0.7600 & 0.5225 & 0.4311 \\ 1.000 & 0.7475 & 0.3526 & 0.1992 \\ 1.000 & 0.7100 & 0.3905 & 0.3417 \\ 1.000 & 0.7000 & 0.4620 & 0.3719 \end{bmatrix}.$$

Step 3: Use the GPHFPWA ($\lambda = 1$) operator to aggregate the standardized PHF evaluation matrix $\overline{N}$, and the collective evaluation information of each alternative is obtained as

$$\widetilde{n}_1 = \{\{0.5003, 0.5177, 0.5360, 0.5382, 0.5522, 0.6230, 0.6361, 0.6516, 0.6516\}, \{0.0495, 0.0521, 0.0562, 0.0592\}, \{0.1176, 0.1345, 0.1558, 0.1783\}\};$$

$$\widetilde{n}_2 = \{\{0.6290, 0.6396, 0.6446, 0.6548, 0.6723, 0.6816, 0.6861, 0.6950\}, \{0.0460, 0.0559, 0.0594, 0.0722\}, \{0.1084, 0.1175, 0.1186, 0.1287, 0.1316, 0.1427, 0.1441, 0.1562\}\};$$

$$\widetilde{n}_3 = \{\{0.5335, 0.5533, 0.5537, 0.5727, 0.5996, 0.6169\}, \{0.0494, 0.0550, 0.0617, 0.0686\}, \{0.1692, 0.1732, 0.1857, 0.1902, 0.2018, 0.2066, 0.2216, 0.2269\}\};$$

$$\widetilde{n}_4 = \{\{0.5593, 0.5820, 0.5978, 0.6185, 0.6425, 0.6609\}, \{0.0921, 0.1015, 0.1055, 0.1162\}, \{0.0947, 0.1044, 0.1159, 0.1277\}\};$$

$$\widetilde{n}_5 = \{\{0.5888, 0.6181, 0.6429, 0.6683\}, \{0.0605, 0.0670, 0.0796, 0.0881\}, \{0.1670, 0.1742, 0.1871, 0.1951\}\}.$$

Step 4: Compute the score values of each alternative combined with Equation (14)

$$s(\widetilde{n}_1) = 0.6888, s(\widetilde{n}_2) = 0.7368, s(\widetilde{n}_3) = 0.6580, s(\widetilde{n}_4) = 0.6978, s(\widetilde{n}_5) = 0.6874.$$

Step 5: According to the score values, the ranking result of the five foreign professors is determined as $A_2 \succ A_4 \succ A_1 \succ A_5 \succ A_3$.

If the GPHFPWG operator is utilized in the steps above to complete the information fusion, the ranking procedures are presented as follows.

Step 1′: See **Step 1**.
Step 2′: See **Step 2**.

Step 3′: Use the GPHFPWG ($\lambda = 1$) operator to aggregate standardized the PHF evaluation matrix $\overline{N}$, and the collective evaluation information of each alternative is obtained as

$$\widetilde{n}_1 = \{\{0.4753, 0.4963, 0.5142, 0.5204, 0.5434, 0.5966, 0.6231, 0.6455, 0.6455\}, \{0.0527, 0.0597, 0.0609, 0.0678\}, \{0.1203, 0.1402, 0.1614, 0.1804\}\};$$
$$\widetilde{n}_2 = \{\{0.6049, 0.6124, 0.6267, 0.6345, 0.6376, 0.6456, 0.6606, 0.6689\}, \{0.0668, 0.0739, 0.0809, 0.0878\}, \{0.1176, 0.1230, 0.1350, 0.1403, 0.1462, 0.1515, 0.1630, 0.1682\}\};$$
$$\widetilde{n}_3 = \{\{0.4297, 0.4424, 0.4902, 0.5047, 0.5788, 0.5959\}, \{0.0526, 0.0571, 0.0703, 0.0748\}, \{0.2020, 0.2110, 0.2216, 0.2304, 0.2410, 0.2495, 0.2596, 0.2680\}\};$$
$$\widetilde{n}_4 = \{\{0.5026, 0.5363, 0.5416, 0.5769, 0.5779, 0.6155\}, \{0.1119, 0.1178, 0.1264, 0.1322\}, \{0.0994, 0.1236, 0.1297, 0.1531\}\};$$
$$\widetilde{n}_5 = \{\{0.5596, 0.5733, 0.6141, 0.6292\}, \{0.0640, 0.0801, 0.0838, 0.0995\}, \{0.1791, 0.1975, 0.1985, 0.2164\}\}.$$

Step 4′: Compute the score values of each alternative combined with Equation (14)

$$s(\widetilde{n}_1) = 0.6757, s(\widetilde{n}_2) = 0.7080, s(\widetilde{n}_3) = 0.6040, s(\widetilde{n}_4) = 0.6550, s(\widetilde{n}_5) = 0.6572.$$

Step 5′: According to the score values, the ranking result of the five foreign professors is obtained as $A_2 \succ A_1 \succ A_5 \succ A_4 \succ A_3$.

6.2. Sensitivity Analysis

To explore the impact of the parameter λ on the ranking results, different possible values of λ are used in the algorithms of two aforementioned numerical examples, such as 0.001, 0.5, 1, 2, 3, 5, 10, 20, and 50. Then, combined with the proposed methods, the different rankings of alternatives are presented in Tables 3–6. From Tables 3 and 4, we can find that the best potential ERP system in Example 1 is always A_3 using both the GPHFWA operator and GPHFWG operator; however, some differences exist between the ranking results concerning different values of λ. Tables 5 and 6 show that when we utilize the GPHFPWA operator to complete the information fusion, the best foreign professor is A_2 for $0.001 \le \lambda \le 10$, but the best alternative is A_4 for $20 \le \lambda \le 50$. In addition, when the GPHFPWG operator is used in Algorithm 2, the best foreign professor is A_2 for $0.001 \le \lambda \le 3$, but the best alternative is A_1 for $5 \le \lambda \le 50$. On the other hand, the score values of all the alternatives vary with different values of λ; the reason is that the aggregation processes of the proposed operators have changed. For instance, when $\lambda = 2$, the GPHFWA operator can be reduced to the picture hesitant fuzzy weighted quadratic averaging (PHFWQA) operator as

$$PHFWQA(\overline{n}_{i1}, \overline{n}_{i2}, \ldots, \overline{n}_{in}) = \left(w_1\overline{n}_{i1}{}^2 \oplus w_2\overline{n}_{i2}{}^2 \oplus \cdots \oplus w_n\overline{n}_{in}{}^2\right)^{1/2}.$$

when $\lambda = 3$, the GPHFWA operator can be reduced to the picture hesitant fuzzy weighted cubic averaging (PHFWCA) operator as

$$PHFWCA(\overline{n}_{i1}, \overline{n}_{i2}, \ldots, \overline{n}_{in}) = \left(w_1\overline{n}_{i1}{}^3 \oplus w_2\overline{n}_{i2}{}^3 \oplus \cdots \oplus w_n\overline{n}_{in}{}^3\right)^{1/3}.$$

Besides, the following results can be obtained from Tables 3–6:

(1) In Example 1, the score values of each alternative obtained by the GPHFWA operator are bigger than those obtained by the GPHFWG operator, and the difference between them increases along with the increasing of λ. It means that the GPHFWA operator is more suitable to aggregate the PHFEs of optimistic decision makers, while the GPHFWG operator can reflect the opinion of pessimistic decision makers. Furthermore, the level of optimism and pessimism are greater with the bigger value of λ.

(2) In Example 2, the score values of each alternative obtained by the GPHFPWA and GPHFPWG operators are relatively stable when the different values of λ are used; the parameter λ cannot

reflect the attitude of decision makers. In addition, the best alternative varies when the value of λ is relatively high, while the best alternative is always the same in Example 1. It means that the rankings obtained by the GPHFPWA and GPHFPWG operators are more affected by the parameter λ than those obtained by the GPHFWA and GPHFWG operators.

The aforementioned sensitivity analysis results show that the value of λ plays a very important role in MCDM problems, especially when the value of λ is relatively high. The value of λ can be determined based on the personal preference of decision makers to obtain different ranking results; thus, the proposed methods are highly flexible to deal with different situations in practice.

Table 3. Sensitivity analysis results obtained by the GPHFWA operator.

Values	$s(\tilde{n}_1)$	$s(\tilde{n}_2)$	$s(\tilde{n}_3)$	$s(\tilde{n}_4)$	$s(\tilde{n}_5)$	Ranking
$\lambda = 0.001$	0.4117	0.4988	0.5787	0.3388	0.4058	$A_3 \succ A_2 \succ A_1 \succ A_5 \succ A_4$
$\lambda = 0.5$	0.4412	0.5328	0.6124	0.3977	0.4526	$A_3 \succ A_2 \succ A_5 \succ A_1 \succ A_4$
$\lambda = 1$	0.4701	0.5611	0.6409	0.4553	0.4989	$A_3 \succ A_2 \succ A_5 \succ A_1 \succ A_4$
$\lambda = 2$	0.5197	0.6001	0.6812	0.5426	0.5743	$A_3 \succ A_2 \succ A_5 \succ A_4 \succ A_1$
$\lambda = 3$	0.5578	0.6242	0.7074	0.5981	0.6255	$A_3 \succ A_5 \succ A_2 \succ A_4 \succ A_1$
$\lambda = 5$	0.6133	0.6522	0.7406	0.6613	0.6852	$A_3 \succ A_5 \succ A_4 \succ A_2 \succ A_1$
$\lambda = 10$	0.6965	0.6831	0.7837	0.7274	0.7481	$A_3 \succ A_5 \succ A_4 \succ A_1 \succ A_2$
$\lambda = 20$	0.7679	0.7062	0.8197	0.7722	0.7916	$A_3 \succ A_5 \succ A_4 \succ A_1 \succ A_2$
$\lambda = 50$	0.8243	0.7274	0.8527	0.8063	0.8280	$A_3 \succ A_5 \succ A_1 \succ A_4 \succ A_2$

Table 4. Sensitivity analysis results obtained by the GPHFWG operator.

Values	$s(\tilde{n}_1)$	$s(\tilde{n}_2)$	$s(\tilde{n}_3)$	$s(\tilde{n}_4)$	$s(\tilde{n}_5)$	Ranking
$\lambda = 0.001$	0.3227	0.3791	0.4486	0.2142	0.2720	$A_3 \succ A_2 \succ A_1 \succ A_5 \succ A_4$
$\lambda = 0.5$	0.3019	0.3492	0.4120	0.1960	0.2457	$A_3 \succ A_2 \succ A_1 \succ A_5 \succ A_4$
$\lambda = 1$	0.2813	0.3197	0.3778	0.1808	0.2240	$A_3 \succ A_2 \succ A_1 \succ A_5 \succ A_4$
$\lambda = 2$	0.2451	0.2690	0.3237	0.1587	0.1926	$A_3 \succ A_2 \succ A_1 \succ A_5 \succ A_4$
$\lambda = 3$	0.2168	0.2314	0.2858	0.1439	0.1716	$A_3 \succ A_2 \succ A_1 \succ A_5 \succ A_4$
$\lambda = 5$	0.1780	0.1829	0.2389	0.1249	0.1441	$A_3 \succ A_2 \succ A_1 \succ A_5 \succ A_4$
$\lambda = 10$	0.1301	0.1265	0.1878	0.0991	0.1093	$A_3 \succ A_1 \succ A_2 \succ A_5 \succ A_4$
$\lambda = 20$	0.1252	0.1158	0.1923	0.1216	0.1095	$A_3 \succ A_1 \succ A_4 \succ A_2 \succ A_5$
$\lambda = 50$	0.1458	0.1377	0.1692	0.1041	0.1351	$A_3 \succ A_1 \succ A_2 \succ A_5 \succ A_4$

Table 5. Sensitivity analysis results obtained by the GPHFPWA operator.

Values	$s(\tilde{n}_1)$	$s(\tilde{n}_2)$	$s(\tilde{n}_3)$	$s(\tilde{n}_4)$	$s(\tilde{n}_5)$	Ranking
$\lambda = 0.001$	0.6866	0.7337	0.6477	0.6911	0.6829	$A_2 \succ A_4 \succ A_1 \succ A_5 \succ A_3$
$\lambda = 0.5$	0.6877	0.7352	0.6529	0.6944	0.6852	$A_2 \succ A_4 \succ A_1 \succ A_5 \succ A_3$
$\lambda = 1$	0.6888	0.7368	0.6580	0.6978	0.6874	$A_2 \succ A_4 \succ A_1 \succ A_5 \succ A_3$
$\lambda = 2$	0.6913	0.7399	0.6680	0.7054	0.6920	$A_2 \succ A_4 \succ A_5 \succ A_1 \succ A_3$
$\lambda = 3$	0.6940	0.7432	0.6769	0.7135	0.6964	$A_2 \succ A_4 \succ A_5 \succ A_1 \succ A_3$
$\lambda = 5$	0.6994	0.7495	0.6915	0.7298	0.7045	$A_2 \succ A_4 \succ A_5 \succ A_1 \succ A_3$
$\lambda = 10$	0.7108	0.7634	0.7151	0.7628	0.7203	$A_2 \succ A_4 \succ A_5 \succ A_3 \succ A_1$
$\lambda = 20$	0.7243	0.7831	0.7369	0.7976	0.7407	$A_4 \succ A_2 \succ A_5 \succ A_3 \succ A_1$
$\lambda = 50$	0.7408	0.8092	0.7570	0.8332	0.7693	$A_4 \succ A_2 \succ A_5 \succ A_3 \succ A_1$

Table 6. Sensitivity analysis results obtained by the GPHFPWG operator.

Values	$s(\tilde{n}_1)$	$s(\tilde{n}_2)$	$s(\tilde{n}_3)$	$s(\tilde{n}_4)$	$s(\tilde{n}_5)$	Ranking
$\lambda = 0.001$	0.6814	0.7230	0.6242	0.6733	0.6706	$A_2 \succ A_1 \succ A_4 \succ A_5 \succ A_3$
$\lambda = 0.5$	0.6789	0.7163	0.6147	0.6647	0.6646	$A_2 \succ A_1 \succ A_4 \succ A_5 \succ A_3$
$\lambda = 1$	0.6757	0.7080	0.6040	0.6550	0.6572	$A_2 \succ A_1 \succ A_5 \succ A_4 \succ A_3$
$\lambda = 2$	0.6680	0.6885	0.5815	0.6344	0.6393	$A_2 \succ A_1 \succ A_5 \succ A_4 \succ A_3$
$\lambda = 3$	0.6597	0.6692	0.5611	0.6155	0.6194	$A_2 \succ A_1 \succ A_5 \succ A_4 \succ A_3$
$\lambda = 5$	0.6442	0.6378	0.5306	0.5873	0.5815	$A_1 \succ A_2 \succ A_4 \succ A_5 \succ A_3$
$\lambda = 10$	0.6190	0.5931	0.4921	0.5524	0.5213	$A_1 \succ A_2 \succ A_4 \succ A_5 \succ A_3$
$\lambda = 20$	0.6727	0.5774	0.5122	0.5451	0.5002	$A_1 \succ A_2 \succ A_4 \succ A_3 \succ A_5$
$\lambda = 50$	0.7825	0.7355	0.6732	0.7214	0.6872	$A_1 \succ A_2 \succ A_4 \succ A_5 \succ A_3$

6.3. Comparative Analysis

To prove the feasibility of the proposed MCDM methods, the rankings of Example 1 in this paper are compared with the rankings obtained by the existing MCDM methods as presented in Table 7; including the PFWA and PFWG operators [25], and the picture fuzzy cross-entropy method [16]. Similarly, a comparison of Example 2 between the GPHFPWA and GPHFPWG operators and the HFPWA and HFPWG operators [34] is presented in Table 8.

Table 7. Comparison result of Example 1.

MCDM method	Ranking
The GPHFWA operator ($\lambda = 1$)	$A_3 \succ A_2 \succ A_5 \succ A_1 \succ A_4$
The GPHFWG operator ($\lambda = 1$)	$A_3 \succ A_2 \succ A_1 \succ A_5 \succ A_4$
The PFWA operator	$A_3 \succ A_2 \succ A_1 \succ A_5 \succ A_4$
The PFWG operator	$A_3 \succ A_1 \succ A_2 \succ A_5 \succ A_4$
The picture fuzzy cross-entropy method	$A_3 \succ A_1 \succ A_2 \succ A_5 \succ A_4$

Table 8. Comparison result of Example 2.

MCDM method	Ranking
The GPHFPWA operator ($\lambda = 1$)	$A_2 \succ A_4 \succ A_1 \succ A_5 \succ A_3$
The GPHFPWG operator ($\lambda = 1$)	$A_2 \succ A_1 \succ A_5 \succ A_4 \succ A_3$
The HFPWA operator	$A_5 \succ A_2 \succ A_1 \succ A_4 \succ A_3$
The HFPWG operator	$A_2 \succ A_5 \succ A_1 \succ A_4 \succ A_3$

Table 7 shows that the best alternative of Example 2 obtained by the MCDM methods based on the GPHFWA and GPHFWG operators is always A_3, which is consistent with the existing methods; the results can demonstrate the feasibility of the proposed method. Compared with the PFS that is used in the study of [25] and [16], PHFS proposed in this paper can convey the human opinions more effectively, including yes, abstain, no, and refusal. For instance, the evaluation information of the alternative A_1 concerning the criteria C_1 that are given by decision maker is expressed as a PFN (0.53,0.33,0.09) [16,25]. In practice, decision maker may feel doubtful to determine an exact value of each membership level. Obviously, PFS cannot deal with this situation; however, we can use PHFS to represent the evaluation information as a PHFE {{0.43,0.53}, {0.33}, {0.06,0.09}} as shown in Table 1. Consequently, the proposed method can solve the MCDM problems when decision makers feel difficulty to determine the accurate value of each membership level. On the other hand, when the numbers of the criteria are relatively large, the aggregation process of the proposed operators will be more complicated than the existing methods and the data size will be relatively large; it is the limitation of the proposed method. Table 8 shows that the best alternative of Example 2 obtained by the HFPWA operator is A_5, but the result of other MCDM methods is A_2. The main reason of the difference is that the MCDM methods combined with the HFPWA and HFPWG operators ignore some complex evaluation information of decision makers in practice. UTHFS allows the decision makers to give several values of positive membership level, for instance, the evaluation information of the alternative A_1 concerning the criteria C_1 that are given by decision maker is expressed as a UTHFE (0.4,0.5,0.7) [34]. Nevertheless, in some particular situations, it is not convincing to express the evaluation information that only considers the positive membership level of decision makers; many scholars have focused on this problem and made some improvements to UTHFS [10,19]. Thus, we can overcome the limitation of UTHFS combined with the proposed method. It is worth noting that the GPHFPWA and GPHFPWG operators also have the same disadvantage as the GPHFWA and GPHFWG operators.

According to the aforementioned comparison results, we can summarize the advantages and disadvantages of the different MCDM methods (see Table 9), as well as their respective fields of application (see Table 10). In addition, the benefits of the aggregation process by using the proposed operators are presented as in the following.

(1) The Expansion of the Evaluation Information

The GPHFWA, GPHFWG, GPHFPWA, and GPHFPWG operators can solve the MCDM problems under PHF environment. PHFS proposed in this paper can express the different human opinions in real life and allow the decision makers to give several possible values of the different membership levels; thus, it can simultaneously depict the uncertainty and hesitancy of decision makers' evaluation information, which cannot be achieved by PFS and UTHFS. Therefore, when decision makers are not fully aware of the evaluation target and feel doubtful about each membership level, it is reasonable to deal with these MCDM problems combined with the proposed methods. Furthermore, as a generalized form of FS, IFS, PFS, and UTHFS, we can transform the proposed methods into the existing MCDM methods if necessary.

(2) The Flexibility of Information Aggregation with Different Values of λ

Recall the sensitivity analysis in Section 6.2, the proposed operators can be reduced to other specific PHF aggregation operators by varying the value of λ; thus, the proposed methods are highly flexible to deal with different situations. Furthermore, the parameter λ can also be regarded as a measure of the optimism and pessimism level of decision makers in the information fusion of the GPHFWA and GPHFWG operators; and the value of λ can be determined by decision makers according to their preferences in practice.

(3) The Simplicity of Dealing with Different Types of Criteria

During the MCDM process, the weight values of criteria play an important role and will affect the final ranking results. The criteria can be divided into two categories: one is in the same priority, the other is in different priorities. On the one hand, when the criteria have the same priority level, we can utilize the proposed method based on the GPHFWA and GPHFWG operators combined with the weight vector of criteria to solve the MCDM problem. On the other hand, when the criteria have different priority levels, the GPHFPWA and GPHFPWG operators can be introduced to determine the ranking of alternatives. In practice, we can use different aggregation operators in this paper to deal with different situations.

6.4. Application of Web Service Selection

To investigate the applications of the proposed methods in a more realistic scenario, we use the proposed methods to solve the Quality of Service (QoS) based web service selection problem [37]. According to the study of [37], the evaluation information of QoS is measured by a crisp number scale of 1–9, and the related criteria are availability (C_1), throughput (C_2), successability (C_3), reliability (C_4), compliance (C_5), best practices (C_6), documentation (C_7), latency (C_8), and response time (C_9).

Due to the criteria latency and response time are the cost type criteria, the closer the evaluation values concerning these two criteria are to 1, the better the alternative.

Table 9. Comparison of each MCDM methods.

Methods	Advantages	Disadvantages
The GPHFWA/ GPHFWG operator	• The human opinions including yes, abstain, no, and refusal can be expressed, and each membership functions can be represented by several possible values. • The operators can be reduced to other forms by varying the value of λ. • The PHFS can be transformed into its special cases, i.e., FS, IFS, PFS, and UTHFS.	• The calculating process is complex when the numbers of criteria are relatively large. • The size of data is relatively large.
The PFWA/ PFWG operator	• The human opinions including yes, abstain, no, and refusal can be expressed. • The PFS can be transformed into IFS and FS.	• It cannot express the evaluation information when decision makers have difficulty determining an accurate value of each membership level.
Picture fuzzy cross-entropy	• The human opinions including yes, abstain, no, and refusal can be expressed. • The PFS can be transformed into IFS and FS. • The ranking is obtained without aggregating the evaluation information; it can avoid the loss of information. • The step of normalizing the evaluation information can be omitted.	• It cannot express the evaluation information when decision makers have difficulty determining an accurate value of each membership level. • It cannot solve the multiple criteria group decision-making problems.
The GPHFPWA/ GPHFPWG operator	• The human opinions including yes, abstain, no, and refusal can be expressed, and each membership functions can be represented by several possible values. • The operators can be reduced to other forms by varying the value of the λ. • The PHFS can be transformed into its special cases, i.e., FS, IFS, PFS, and UTHFS. • It can solve the MCDM problem that the criteria are in different priorities.	• The calculating process is complex when the numbers of criteria are relatively large. • The size of data is relatively large.
The HFPWA/ HFPWG operator	• The positive membership function can be expressed by several possible values. • It can solve the MCDM problem that the criteria are in different priorities.	• It cannot express the human opinions including abstain, no, and refusal.

Table 10. MCDM application fields of each MCDM method.

Methods	MCDM Application Fields
The GPHFWA/GPHFWG operator	• The evaluation information of decision makers is diverse. • Decision makers feel doubtful to determine the accurate value of each membership level. • The numbers of the criteria are relatively small.
The PFWA/PFWG operator	• The evaluation information of decision makers is diverse.
Picture fuzzy cross entropy	• The evaluation information of decision maker is diverse. • The alternatives are evaluated by an individual decision maker.
The GPHFPWA/GPHFPWG operator	• The evaluation information of decision makers is diverse. • Decision makers feel doubtful to determine the accurate value of each membership level. • The numbers of the criteria are relatively small. • The criteria are in different priorities.
The HFPWA/HFPWG operator	• Decision makers feel doubtful to determine the accurate value of positive membership level. • The criteria are in different priorities.

Suppose there are 20 web services to be evaluated concerning the aforementioned nine criteria, i.e., $WS_i (i = 1,2,\ldots,20)$; the evaluation information of each web service is presented in Table 11. As each evaluation value in [37] is expressed by an exact crisp number, the PHFS can be reduced to the PFS to represent the evaluation information of each web service. Based on the relationship between the linguistic variables and IFNs [38], we develop the transformation relationship between the linguistic variables and PFNs as presented in Table 12. Then, the evaluation information in Table 11 can be transformed into a PF evaluation matrix $A = (a_{ij}) (i = 1,2,\ldots,20;\ j = 1,2,\ldots,9)$, and the ranking of the 20 web services can be obtained by the Algorithm 1 in this paper. Subsequently, the ranking result will be compared with the rankings determined by AHP, TOPSIS, COPRAS, VIKOR, and SAW methods in [37]. It is worth noting that, in order to compare different MCDM methods, more effectively we suppose each criteria is considered equally important, i.e., $w_j = 1/9 (j = 1,2,\ldots,9)$. Then, the ranking of the 20 web services can be determined by the following steps.

Step 1: According to the Definition 3, normalize the PF evaluation matrix $A = (a_{ij})$ to the standardized PF evaluation matrix $\overline{A} = (\overline{a}_{ij})$ as

$$\overline{a}_{ij} = \begin{cases} a_{ij}, & \text{for the benefit criteria;} \\ (a_{ij})^c, & \text{for the cost criteria.} \end{cases} \tag{57}$$

Step 2: Utilize the GPFWA ($\lambda = 1$) operator

$$GPFWA_{\lambda=1}(\overline{a}_{i1}, \overline{a}_{i2}, \ldots, \overline{a}_{i9}) = a_i = \left(1 - \prod_{j=1}^{9}\left(1 - \overline{\mu}_{ij}\right)^{w_j}, \prod_{j=1}^{9}\left(\overline{\eta}_{ij}\right)^{w_j}, \prod_{j=1}^{9}(\overline{v}_{ij})^{w_j}\right)$$

to aggregated the PF evaluation matrix $\overline{A} = (\overline{a}_{ij})$, and the collective PFNs of each web service are obtained.

Step 3: Compute the score values of each web service using the equation

$$s(a_i) = (1 + \mu_i - \eta_i - v_i)/2. \tag{58}$$

Table 11. Evaluation information of each web service.

Alternatives	C_1	C_2	C_3	C_4	C_5	C_6	C_7	C_8	C_9
WS_1	3	4	2	5	6	7	6	2	8
WS_2	4	5	2	3	4	6	8	3	4
WS_3	3	5	6	8	3	2	2	4	5
WS_4	4	4	5	5	6	2	6	7	8
WS_5	5	6	2	4	7	7	8	8	3
WS_6	4	4	3	7	6	8	6	8	6
WS_7	4	4	5	7	8	7	7	4	3
WS_8	6	7	6	6	5	7	8	6	3
WS_9	5	3	2	2	6	5	2	4	7
WS_{10}	8	8	7	6	2	2	3	4	3
WS_{11}	5	8	4	4	5	7	4	5	8
WS_{12}	4	5	5	5	6	8	7	6	6
WS_{13}	6	4	7	6	4	5	4	4	5
WS_{14}	4	6	5	4	5	4	6	7	4
WS_{15}	7	5	6	2	7	6	5	5	2
WS_{16}	3	6	4	7	2	3	8	2	4
WS_{17}	4	4	8	4	4	5	2	4	8
WS_{18}	4	3	5	3	5	4	4	6	6
WS_{19}	5	4	4	6	6	7	6	7	2
WS_{20}	6	2	3	5	4	6	5	5	5

Table 12. Transformation between linguistic variables and PFNs.

Crisp numbers	Linguistic variables	PFNs
1	Extremely low (EL)	(0.05,0.05,0.85)
2	Very low (VL)	(0.15,0.05,0.75)
3	Low (L)	(0.25,0.05,0.65)
4	Medium low (ML)	(0.35,0.05,0.55)
5	Medium (M)	(0.45,0.05,0.45)
6	Medium high (MH)	(0.55,0.05,0.35)
7	High (H)	(0.65,0.05,0.25)
8	Very high (VH)	(0.75,0.05,0.15)
9	Extremely high (EH)	(0.85,0.05,0.05)

Then, the ranking of the 20 web services can be determined; the lager the score value, the better the web service. The related data of the ranking are presented in Table 13.

Table 13. Ranking results obtained by the proposed method.

Alternatives	Collective Evaluation Information	Score Values	Ranking
WS_1	(0.4677,0.0500,0.4210)	0.4983	12
WS_2	(0.4833,0.0500,0.4068)	0.5133	9
WS_3	(0.4313,0.0500,0.4579)	0.4617	15
WS_4	(0.3776,0.0500,0.5192)	0.4042	18
WS_5	(0.5244,0.0500,0.3647)	0.5548	5
WS_6	(0.4736,0.0500,0.4159)	0.5038	11
WS_7	(0.5817,0.0500,0.3119)	0.6099	2
WS_8	(0.5871,0.0500,0.3079)	0.6146	1
WS_9	(0.3485,0.0500,0.5475)	0.3755	20
WS_{10}	(0.5447,0.0500,0.3414)	0.5767	3
WS_{11}	(0.4683,0.0500,0.4222)	0.4980	13
WS_{12}	(0.5045,0.0500,0.3879)	0.5333	8
WS_{13}	(0.4828,0.0500,0.4145)	0.5091	10
WS_{14}	(0.4371,0.0500,0.4609)	0.4631	14
WS_{15}	(0.5426,0.0500,0.3498)	0.5714	4
WS_{16}	(0.5191,0.0500,0.3668)	0.5511	6
WS_{17}	(0.4154,0.0500,0.4744)	0.4455	16
WS_{18}	(0.3535,0.0500,0.5459)	0.3788	19
WS_{19}	(0.5186,0.0500,0.3737)	0.5474	7
WS_{20}	(0.4179,0.0500,0.4798)	0.4440	17

To verify the accuracy of the ranking obtained by the proposed method, we use AHP, TOPSIS, COPRAS, VIKOR, and SAW methods to solve the web service selection problem combined with the evaluation information in Table 11. Subsequently, the ranking results of different MCDM methods are presented in Table 14. The Spearman's rank correlation coefficient is a powerful tool for measuring the similarity between two MCDM methods [39]. Then, we can calculate the Spearman's rank correlation coefficients between the proposed method and the other five MCDM methods as shown in Table 15. Table 15 shows that the Spearman's rank correlation coefficients between the proposed method and AHP and TOPSIS are 0.9722 and 0.9549, respectively, which demonstrate that the proposed method is highly correlated with these two methods. AHP and TOPSIS methods have been approved to be the most suitable two methods to solve web service selection problems [39]; thus, the comparison results above illustrate the feasibility of the proposed method.

Table 14. Ranking of web services of different MCDM methods.

Alternatives	The Proposed Method	AHP	TOPSIS	COPRAS	VIKOR	SAW
WS_1	12	13	14	12	15	14
WS_2	9	10	11	6	12	12
WS_3	15	16	15	11	17	19
WS_4	18	18	18	20	19	18
WS_5	5	6	8	9	9	9
WS_6	11	11	12	16	14	13
WS_7	2	2	2	1	3	2
WS_8	1	1	1	5	1	1
WS_9	20	20	20	19	20	20
WS_{10}	3	4	4	4	8	10
WS_{11}	13	11	10	15	13	8
WS_{12}	8	6	7	10	4	6
WS_{13}	10	8	6	8	2	5
WS_{14}	14	13	13	13	6	7
WS_{15}	4	3	3	3	7	3
WS_{16}	6	8	9	2	11	11
WS_{17}	16	17	16	17	18	16
WS_{18}	19	19	19	18	10	17
WS_{19}	7	4	5	7	5	4
WS_{20}	17	15	17	14	16	15

Table 15. Spearman's rank correlation coefficients between the proposed method and the other MCDM methods.

Existing MCDM Methods	Spearman's Rank Correlation Coefficients
AHP	0.9722
TOPSIS	0.9549
COPRAS	0.9023
VIKOR	0.7429
SAW	0.8165

From the information aggregation of the proposed method, we can find that the calculating procedure of the proposed method is more complicated than AHP and TOPSIS methods. In addition, TOPSIS method does not require the transformation of the evaluation information concerning cost and benefit type criteria. However, when decision makers are not sure if it is 3 or 4 about the evaluation information of the web service WS_1 concerning the criteria C_1, AHP and TOPSIS methods cannot deal with this situation in practice; we can use PHFS to express the evaluation information above, i.e., $\{\{0.25,0.35\},\{0.05\},\{0.55,0.65\}\}$. On the other hand, when the criteria are in different priorities, the GPHFPWA and GPHFPWG operators can be used to aggregation the evaluation information. Thus, the AHP, TOPSIS, and proposed methods have their own advantages and disadvantages; in real life, decision makers can determine to utilize which MCDM methods to solve problems according to the actual situations.

7. Conclusions

Combined with the picture fuzzy set and uniformly typical hesitant fuzzy set, this paper develops the picture hesitant fuzzy set, in which the positive, neutral, negative, and refusal membership degrees are expressed by several possible values. Then, the operations and comparison method of picture hesitant fuzzy elements are developed. To solve the multiple criteria decision-making problems under picture hesitant fuzzy environment, the generalized picture hesitant fuzzy weighted averaging and generalized picture hesitant fuzzy weighted geometric operators are put forward to aggregate the picture hesitant elements given by decision maker. Furthermore, considering the different priorities between the related criteria in practice, the generalized picture hesitant fuzzy prioritized weighted averaging and generalized picture hesitant fuzzy prioritized weighted geometric operators are proposed. Meanwhile, some desirable properties and the reduced operators of them are investigated in detail. Finally, two kinds of multiple criteria decision-making methods combined with the proposed operators are constructed to solve the multiple criteria decision-making problems in different situations. Subsequently, two numerical examples and an application of web service selection are provided to indicate the applications and advantages of the proposed methods.

In future research, we will investigate other operations of picture hesitant fuzzy elements and develop different aggregation operators to aggregate picture hesitant fuzzy elements. In addition, we will propose the consensus model to improve the proposed methods; then, the non-consensus evaluation information of decision makers will be revised to obtain a more accurate ranking result.

Author Contributions: R.W. put forward the picture hesitant fuzzy set, and explored the operational laws and comparison method of the picture hesitant fuzzy set. R.W. and Y.L. developed the aggregation operators under picture hesitant fuzzy environment. Then, R.W. wrote the original manuscript, and Y.L. improved the writing.

Funding: This study was supported by the National Natural Science Foundation of China (no. 71371156) and the Doctoral Innovation Fund Program of Southwest Jiaotong University (D-CX201727).

Acknowledgments: Thank the editor and reviewers for the positive comments.

Conflicts of Interest: The authors declare no conflict of interest.

Appendix A

Proof.

a. For $n = 1$, according to Theorem 1, since

$$GPHFWA_\lambda(\widetilde{n}_1) = \left(w_1\widetilde{n}_1{}^\lambda\right)^{1/\lambda} = \left(\widetilde{n}_1{}^\lambda\right)^{1/\lambda} = \widetilde{n}_1.$$

Obviously, Equation (22) holds for $n = 1$.

b. For $n = 2$, since

$$\widetilde{n}_1 = \bigcup_{\alpha_1\in\widetilde{\mu}_1,\beta_1\in\widetilde{\eta}_1,\gamma_1\in\widetilde{v}_1}\left\{\left\{\alpha_1{}^\lambda\right\},\left\{1-(1-\beta_1)^\lambda\right\},\left\{1-(1-\gamma_1)^\lambda\right\}\right\},$$
$$\widetilde{n}_2 = \bigcup_{\alpha_2\in\widetilde{\mu}_2,\beta_2\in\widetilde{\eta}_2,\gamma_2\in\widetilde{v}_2}\left\{\left\{\alpha_2{}^\lambda\right\},\left\{1-(1-\beta_2)^\lambda\right\},\left\{1-(1-\gamma_2)^\lambda\right\}\right\}.$$

we have

$$w_1\widetilde{n}_1{}^\lambda = \bigcup_{\alpha_1\in\widetilde{\mu}_1,\beta_1\in\widetilde{\eta}_1,\gamma_1\in\widetilde{v}_1}\left\{\left\{1-\left(1-\alpha_1{}^\lambda\right)^{w_1}\right\},\left\{\left(1-(1-\beta_1)^\lambda\right)^{w_1}\right\},\left\{\left(1-(1-\gamma_1)^\lambda\right)^{w_1}\right\}\right\},$$
$$w_2\widetilde{n}_2{}^\lambda = \bigcup_{\alpha_2\in\widetilde{\mu}_2,\beta_2\in\widetilde{\eta}_2,\gamma_2\in\widetilde{v}_2}\left\{\left\{1-\left(1-\alpha_2{}^\lambda\right)^{w_2}\right\},\left\{\left(1-(1-\beta_2)^\lambda\right)^{w_2}\right\},\left\{\left(1-(1-\gamma_2)^\lambda\right)^{w_2}\right\}\right\}.$$

then,

$$w_1\widetilde{n}_1{}^\lambda \oplus w_2\widetilde{n}_2{}^\lambda = \bigcup_{\alpha_1\in\widetilde{\mu}_1,\beta_1\in\widetilde{\eta}_1,\gamma_1\in\widetilde{v}_1,\alpha_2\in\widetilde{\mu}_2,\beta_2\in\widetilde{\eta}_2,\gamma_2\in\widetilde{v}_2}\left\{\left\{1-\left(1-\alpha_1{}^\lambda\right)^{w_1}+1-\left(1-\alpha_2{}^\lambda\right)^{w_2}-\left(1-\left(1-\alpha_1{}^\lambda\right)^{w_1}\right)\left(1-\left(1-\alpha_2{}^\lambda\right)^{w_2}\right)\right\},\right.$$
$$\left.\left\{\left(1-(1-\beta_1)^\lambda\right)^{w_1}\left(1-(1-\beta_2)^\lambda\right)^{w_2}\right\},\left\{\left(1-(1-\gamma_1)^\lambda\right)^{w_1}\left(1-(1-\gamma_2)^\lambda\right)^{w_2}\right\}\right\}.$$
$$w_1\widetilde{n}_1{}^\lambda \oplus w_2\widetilde{n}_2{}^\lambda = \bigcup_{\alpha_1\in\widetilde{\mu}_1,\beta_1\in\widetilde{\eta}_1,\gamma_1\in\widetilde{v}_1,\alpha_2\in\widetilde{\mu}_2,\beta_2\in\widetilde{\eta}_2,\gamma_2\in\widetilde{v}_2}\left\{\left\{1-\left(1-\alpha_1{}^\lambda\right)^{w_1}\left(1-\alpha_2{}^\lambda\right)^{w_2}\right\},\right.$$
$$\left.\left\{\left(1-(1-\beta_1)^\lambda\right)^{w_1}\left(1-(1-\beta_2)^\lambda\right)^{w_2}\right\},\left\{\left(1-(1-\gamma_1)^\lambda\right)^{w_1}\left(1-(1-\gamma_2)^\lambda\right)^{w_2}\right\}\right\}.$$

and

$$GPHFWA_\lambda(\tilde{n}_1,\tilde{n}_2) = \left(w_1\tilde{n}_1{}^\lambda \oplus w_2\tilde{n}_2{}^\lambda\right)^{1/\lambda} = \bigcup_{\alpha_1\in\tilde{\mu}_1,\beta_1\in\tilde{\eta}_1,\gamma_1\in\tilde{v}_1,\alpha_2\in\tilde{\mu}_2,\beta_2\in\tilde{\eta}_2,\gamma_2\in\tilde{v}_2}\left\{\left\{\left(1-(1-\alpha_1{}^\lambda)^{w_1}(1-\alpha_2{}^\lambda)^{w_2}\right)^{1/\lambda}\right\},\left\{1-\left(1-\left(1-(1-\beta_1)^\lambda\right)^{w_1}\left(1-(1-\beta_2)^\lambda\right)^{w_2}\right)^{1/\lambda}\right\},\left\{1-\left(1-\left(1-(1-\gamma_1)^\lambda\right)^{w_1}\left(1-(1-\gamma_2)^\lambda\right)^{w_2}\right)^{1/\lambda}\right\}\right\}.$$

i.e., Equation (22) holds for $n = 2$.

c. If Equation (22) holds for $n = k$, we have

$$GPHFWA_\lambda(\tilde{n}_1,\tilde{n}_2,\ldots,\tilde{n}_k) = \bigcup_{\alpha_1\in\tilde{\mu}_1,\alpha_2\in\tilde{\mu}_2,\ldots,\alpha_k\in\tilde{\mu}_k,\beta_1\in\tilde{\eta}_1,\beta_2\in\tilde{\eta}_2,\ldots,\beta_k\in\tilde{\eta}_k,\gamma_1\in\tilde{v}_1,\gamma_2\in\tilde{v}_2,\ldots,\gamma_k\in\tilde{v}_k}\left\{\left\{\left(1-\prod_{j=1}^{k}(1-\alpha_j{}^\lambda)^{w_j}\right)^{1/\lambda}\right\},\left\{1-\left(1-\prod_{j=1}^{k}\left(1-(1-\beta_j)^\lambda\right)^{w_j}\right)^{1/\lambda}\right\},\left\{1-\left(1-\prod_{j=1}^{k}\left(1-(1-\gamma_j)^\lambda\right)^{w_j}\right)^{1/\lambda}\right\}\right\},$$

when $n = k+1$, according to the operations of PHFEs, we have

$$w_1\tilde{n}_1{}^\lambda \oplus w_2\tilde{n}_2{}^\lambda \oplus\cdots\oplus w_k\tilde{n}_k{}^\lambda \oplus w_{k+1}\tilde{n}_{k+1}{}^\lambda = \bigcup_{\alpha_1\in\tilde{\mu}_1,\alpha_2\in\tilde{\mu}_2,\ldots,\alpha_k\in\tilde{\mu}_k,\beta_1\in\tilde{\eta}_1,\beta_2\in\tilde{\eta}_2,\ldots,\beta_k\in\tilde{\eta}_k,\gamma_1\in\tilde{v}_1,\gamma_2\in\tilde{v}_2,\ldots,\gamma_k\in\tilde{v}_k}\left\{\left\{1-\prod_{j=1}^{k}(1-\alpha_j{}^\lambda)^{w_j}\right\},\left\{\prod_{j=1}^{k}\left(1-(1-\beta_j)^\lambda\right)^{w_j}\right\},\left\{\prod_{j=1}^{k}\left(1-(1-\gamma_j)^\lambda\right)^{w_j}\right\}\right\}\oplus\bigcup_{\alpha_{k+1}\in\tilde{\mu}_{k+1},\beta_{k+1}\in\tilde{\eta}_{k+1},\gamma_{k+1}\in\tilde{v}_{k+1}}\left\{\left\{1-(1-\alpha_{k+1}{}^\lambda)^{w_{k+1}}\right\},\left\{\left(1-(1-\beta_{k+1})^\lambda\right)^{w_{k+1}}\right\},\left\{\left(1-(1-\gamma_{k+1})^\lambda\right)^{w_{k+1}}\right\}\right\}$$

$$= \bigcup_{\alpha_1\in\tilde{\mu}_1,\alpha_2\in\tilde{\mu}_2,\ldots,\alpha_{k+1}\in\tilde{\mu}_{k+1},\beta_1\in\tilde{\eta}_1,\beta_2\in\tilde{\eta}_2,\ldots,\beta_{k+1}\in\tilde{\eta}_{k+1},\gamma_1\in\tilde{v}_1,\gamma_2\in\tilde{v}_2,\ldots,\gamma_{k+1}\in\tilde{v}_{k+1}}\left\{\left\{1-\prod_{j=1}^{k+1}(1-\alpha_j{}^\lambda)^{w_j}\right\},\left\{\prod_{j=1}^{k+1}\left(1-(1-\beta_j)^\lambda\right)^{w_j}\right\},\left\{\prod_{j=1}^{k+1}\left(1-(1-\gamma_j)^\lambda\right)^{w_j}\right\}\right\}.$$

then,

$$GPHFWA_\lambda(\tilde{n}_1,\tilde{n}_2,\ldots,\tilde{n}_k) = \left(w_1\tilde{n}_1{}^\lambda \oplus w_2\tilde{n}_2{}^\lambda \oplus\cdots\oplus w_k\tilde{n}_k{}^\lambda \oplus w_{k+1}\tilde{n}_{k+1}{}^\lambda\right)^{1/\lambda} = \bigcup_{\alpha_1\in\tilde{\mu}_1,\alpha_2\in\tilde{\mu}_2,\ldots,\alpha_{k+1}\in\tilde{\mu}_{k+1},\beta_1\in\tilde{\eta}_1,\beta_2\in\tilde{\eta}_2,\ldots,\beta_{k+1}\in\tilde{\eta}_{k+1},\gamma_1\in\tilde{v}_1,\gamma_2\in\tilde{v}_2,\ldots,\gamma_{k+1}\in\tilde{v}_{k+1}}\left\{\left\{\left(1-\prod_{j=1}^{k+1}(1-\alpha_j{}^\lambda)^{w_j}\right)^{1/\lambda}\right\},\left\{1-\left(1-\prod_{j=1}^{k+1}\left(1-(1-\beta_j)^\lambda\right)^{w_j}\right)^{1/\lambda}\right\},\left\{1-\left(1-\prod_{j=1}^{k+1}\left(1-(1-\gamma_j)^\lambda\right)^{w_j}\right)^{1/\lambda}\right\}\right\}.$$

i.e., Equation (22) holds for $n = k+1$; we can demonstrate that Equation (22) holds for all values of n.

□

Appendix B

Proof. According to Theorem 2, since $\tilde{n}_j = \tilde{n} = \{\tilde{\mu},\tilde{\eta},\tilde{v}\}$, we have

$$GPHFWA_\lambda(\tilde{n}_1,\tilde{n}_2,\ldots,\tilde{n}_n) = \left(w_1\tilde{n}^\lambda \oplus w_2\tilde{n}^\lambda \oplus\cdots\oplus w_n\tilde{n}^\lambda\right)^{1/\lambda} = \bigcup_{\alpha\in\tilde{\mu},\beta\in\tilde{\eta},\gamma\in\tilde{v}}\left\{\left\{\left(1-\prod_{j=1}^{n}(1-\alpha^\lambda)^{w_j}\right)^{1/\lambda}\right\},\left\{1-\left(1-\prod_{j=1}^{n}\left(1-(1-\beta)^\lambda\right)^{w_j}\right)^{1/\lambda}\right\},\left\{1-\left(1-\prod_{j=1}^{n}\left(1-(1-\gamma)^\lambda\right)^{w_j}\right)^{1/\lambda}\right\}\right\}$$

$$= \bigcup_{\alpha\in\tilde{\mu},\beta\in\tilde{\eta},\gamma\in\tilde{v}}\left\{\left\{\left(1-(1-\alpha^\lambda)\right)^{1/\lambda}\right\},\left\{1-\left(1-\left(1-(1-\beta)^\lambda\right)\right)^{1/\lambda}\right\},\left\{1-\left(1-\left(1-(1-\gamma)^\lambda\right)\right)^{1/\lambda}\right\}\right\}$$

$$= \bigcup_{\alpha\in\tilde{\mu},\beta\in\tilde{\eta},\gamma\in\tilde{v}}\left\{\left\{(\alpha^\lambda)^{1/\lambda}\right\},\left\{1-\left((1-\beta)^\lambda\right)^{1/\lambda}\right\},\left\{1-\left((1-\gamma)^\lambda\right)^{1/\lambda}\right\}\right\} = \bigcup_{\alpha\in\tilde{\mu},\beta\in\tilde{\eta},\gamma\in\tilde{v}}\{\{\alpha\},\{\beta\},\{\gamma\}\} = \tilde{n}.$$

□

Appendix C

Proof. For $\lambda \in (0,\infty)$, since $\alpha^- \le \alpha_j \le \alpha^+$, then

$$\alpha_j{}^\lambda \ge (\alpha^-)^\lambda, 1-\alpha_j{}^\lambda \le 1-(\alpha^-)^\lambda, (1-\alpha_j{}^\lambda)^{w_j} \le \left(1-(\alpha^-)^\lambda\right)^{w_j},$$
$$\prod_{j=1}^{n}(1-\alpha_j{}^\lambda)^{w_j} \le \prod_{j=1}^{n}\left(1-(\alpha^-)^\lambda\right)^{w_j},$$
$$1-\prod_{j=1}^{n}(1-\alpha_j{}^\lambda)^{w_j} \ge 1-\prod_{j=1}^{n}\left(1-(\alpha^-)^\lambda\right)^{w_j},$$
$$\left(1-\prod_{j=1}^{n}(1-\alpha_j{}^\lambda)^{w_j}\right)^{1/\lambda} \ge \left(1-\prod_{j=1}^{n}\left(1-(\alpha^-)^\lambda\right)^{w_j}\right)^{1/\lambda} = \alpha^-.$$

similarly, we have

$$\left(1-\prod_{j=1}^{n}\left(1-\alpha_j{}^\lambda\right)^{w_j}\right)^{1/\lambda} \le \left(1-\prod_{j=1}^{n}\left(1-(\alpha^+)^\lambda\right)^{w_j}\right)^{1/\lambda} = \alpha^+.$$

As $\beta^- \le \beta_j \le \beta^+$, then

$$1-\beta_j \le 1-\beta^-, (1-\beta_j)^\lambda \le (1-\beta^-)^\lambda, 1-(1-\beta_j)^\lambda \ge 1-(1-\beta^-)^\lambda,$$
$$\left(1-(1-\beta_j)^\lambda\right)^{w_j} \ge \left(1-(1-\beta^-)^\lambda\right)^{w_j},$$
$$\prod_{j=1}^{n}\left(1-(1-\beta_j)^\lambda\right)^{w_j} \ge \prod_{j=1}^{n}\left(1-(1-\beta^-)^\lambda\right)^{w_j},$$
$$1-\prod_{j=1}^{n}\left(1-(1-\beta_j)^\lambda\right)^{w_j} \le 1-\prod_{j=1}^{n}\left(1-(1-\beta^-)^\lambda\right)^{w_j},$$
$$\left(1-\prod_{j=1}^{n}\left(1-(1-\beta_j)^\lambda\right)^{w_j}\right)^{1/\lambda} \le \left(1-\prod_{j=1}^{n}\left(1-(1-\beta^-)^\lambda\right)^{w_j}\right)^{1/\lambda},$$
$$1-\left(1-\prod_{j=1}^{n}\left(1-(1-\beta_j)^\lambda\right)^{w_j}\right)^{1/\lambda} \ge 1-\left(1-\prod_{j=1}^{n}\left(1-(1-\beta^-)^\lambda\right)^{w_j}\right)^{1/\lambda} = \beta^-.$$

similarly, we have

$$1-\left(1-\prod_{j=1}^{n}\left(1-(1-\beta_j)^\lambda\right)^{w_j}\right)^{1/\lambda} \le 1-\left(1-\prod_{j=1}^{n}\left(1-(1-\beta^+)^\lambda\right)^{w_j}\right)^{1/\lambda} = \beta^+.$$

and, as $\gamma^- \le \gamma_j \le \gamma^+$, we have

$$\gamma^- \le 1-\left(1-\prod_{j=1}^{n}\left(1-(1-\gamma_j)^\lambda\right)^{w_j}\right)^{1/\lambda} \le \gamma^+.$$

let $GPHFWA_\lambda(\tilde{n}_1, \tilde{n}_2, \ldots, \tilde{n}_n) = \tilde{n} = \{\{\alpha\}, \{\beta\}, \{\gamma\}\}$, then

$$s(\tilde{n}) = \frac{1+\frac{1}{l}\sum_{i=1}^{l}\alpha_i - \frac{1}{p}\sum_{i=1}^{p}\beta_i - \frac{1}{q}\sum_{i=1}^{q}\gamma_i}{2} \ge \frac{1+\frac{1}{l^-}\sum_{i=1}^{l^-}\alpha^- - \frac{1}{p^-}\sum_{i=1}^{p^-}\beta^+ - \frac{1}{q^-}\sum_{i=1}^{q^-}\gamma^+}{2} = s(\tilde{n}^-),$$

and

$$s(\tilde{n}) = \frac{1+\frac{1}{l}\sum_{i=1}^{l}\alpha_i - \frac{1}{p}\sum_{i=1}^{p}\beta_i - \frac{1}{q}\sum_{i=1}^{q}\gamma_i}{2} \le \frac{1+\frac{1}{l^+}\sum_{i=1}^{l^+}\alpha^+ - \frac{1}{p^+}\sum_{i=1}^{p^+}\beta^- - \frac{1}{q^+}\sum_{i=1}^{q^+}\gamma^-}{2} = s(\tilde{n}^+).$$

we obtain

$$\widetilde{n}^- \leq GPHFWA_\lambda(\widetilde{n}_1, \widetilde{n}_2, \ldots, \widetilde{n}_n) \leq \widetilde{n}^+, \lambda \in (0, \infty).$$

Similarly, we have

$$\widetilde{n}^- \leq GPHFWA_\lambda(\widetilde{n}_1, \widetilde{n}_2, \ldots, \widetilde{n}_n) \leq \widetilde{n}^+, \lambda \in (-\infty, 0).$$

□

References

1. Mardani, A.; Jusoh, A.; Zavadskas, E.K. Fuzzy multiple criteria decision-making techniques and applications–two decades review from 1994 to 2014. *Expert Syst. Appl.* **2015**, *42*, 4126–4148. [CrossRef]
2. Morente-Molinera, J.A.; Kou, G.; González-Crespo, R.; Corchado, J.M.; Herrera-Viedma, E. Solving multi-criteria group decision making problems under environments with a high number of alternatives using fuzzy ontologies and multi-granular linguistic modelling methods. *Knowl. Syst.* **2017**, *137*, 54–64. [CrossRef]
3. Hernández, F.L.; Ory, E.G.D.; Aguilar, S.R.; González-Crespo, R. Residue properties for the arithmetical estimation of the image quantization table. *Appl. Soft Comput.* **2018**, *67*, 309–321. [CrossRef]
4. Zadeh, L.A. Fuzzy sets. *Inf. Control* **1965**, *8*, 338–353. [CrossRef]
5. Atanassov, K.T. Intuitionistic fuzzy sets. *Fuzzy sets Syst.* **1986**, *20*, 87–96. [CrossRef]
6. Atanassov, K.T. Gargov, G. Interval valued intuitionistic fuzzy sets. *Fuzzy Sets Syst.* **1989**, *31*, 343–349. [CrossRef]
7. Cường, B.C. Picture fuzzy sets. *J. Comput. Sci.Cybern.* **2015**, *30*, 409–420. [CrossRef]
8. Torra, V. Hesitant fuzzy sets. *Int. J. Intell. Syst.* **2010**, *25*, 529–539. [CrossRef]
9. Ye, J. Trapezoidal neutrosophic set and its application to multiple attribute decision-making. *Neural Comput. Appl.* **2015**, *26*, 1157–1166. [CrossRef]
10. Zhu, B.; Xu, Z.S.; Xia, M.M. Dual hesitant fuzzy sets. *J. Appl. Math.* **2012**, *2012*, 2607–2645. [CrossRef]
11. Liu, F.; Yuan, X.H. Fuzzy number intuitionistic fuzzy set. *Fuzzy Syst. Math.* **2007**, *21*, 88–91.
12. Garg, H. Some picture fuzzy aggregation operators and their applications to multicriteria decision-making. *Arab. J. Sci. Eng.* **2017**, *42*, 1–16. [CrossRef]
13. Cuong, B. Picture fuzzy sets-first results. Part 1. In Proceedings of the Third World Congress on Information and Communication WICT'2013, Hanoi, Vietnam, 15–18 December 2013; pp. 1–6.
14. Singh, P. Correlation coefficients for picture fuzzy sets. *J. Intell. Fuzzy Syst.* **2015**, *28*, 1–12.
15. Son, L.H. Generalized picture distance measure and applications to picture fuzzy clustering. *Appl. Soft Comput.* **2016**, *46*, 284–295. [CrossRef]
16. Wei, G.W. Picture fuzzy cross-entropy for multiple attribute decision making problems. *J. Bus. Econ. Manag.* **2016**, *17*, 491–502. [CrossRef]
17. Torra, V.; Narukawa, Y. On hesitant fuzzy sets and decision. In Proceedings of the IEEE International Conference on Fuzzy Systems, Jeju Island, Korea, 20–24 August 2009; pp. 1378–1382.
18. Chen, N.; Xu, Z.S.; Xia, M.M. Interval-valued hesitant preference relations and their applications to group decision making. *Knowl. Syst.* **2013**, *37*, 528–540. [CrossRef]
19. Farhadinia, B. Correlation for dual hesitant fuzzy sets and dual interval-valued hesitant fuzzy sets. *Int. J. Intell. Syst.* **2013**, *29*, 184–205. [CrossRef]
20. Xu, Z.S.; Yager, R.R. Some geometric aggregation operators based on intuitionistic fuzzy sets. *Int. J. Intell. Syst.* **2006**, *35*, 417–433. [CrossRef]
21. Xu, Z.S. Intuitionistic fuzzy aggregation operators. *IEEE Trans. Fuzzy Syst.* **2007**, *15*, 1179–1187.
22. Xu, Z.S. Methods for aggregating interval-valued intuitionistic fuzzy information and their application to decision making. *Control Decis.* **2007**, *22*, 215–219.
23. Xu, Z.S.; Chen, J. Approach to group decision making based on interval-valued intuitionistic judgment matrices. *Syst. Eng. Theory Pract.* **2007**, *27*, 126–133. [CrossRef]
24. Xu, Z.S.; Chen, J. On geometric aggregation over interval-valued intuitionistic fuzzy information. In Proceedings of the International Conference on Fuzzy Systems and Knowledge Discovery, Haikou, China, 24–27 August 2007; pp. 466–471.

25. Wei, G.W. Picture fuzzy aggregation operators and their application to multiple attribute decision making. *J. Intell. Fuzzy Syst.* **2017**, *33*, 713–724. [CrossRef]
26. Wang, C.; Zhou, X.; Tu, H.; Tao, S. Some geometric aggregation operators based on picture fuzzy sets and their application in multiple attribute decision making. *Ital. J. Pure Appl. Math.* **2017**, *37*, 477–492.
27. Xia, M.M.; Xu, Z.S. Hesitant fuzzy information aggregation in decision making. *Int. J. Approx. Reason.* **2011**, *52*, 395–407. [CrossRef]
28. Yu, D.J.; Zhang, W.Y.; Huang, G. Dual hesitant fuzzy aggregation operators. *Technol. Econ. Develop. Econ.* **2016**, *22*, 194–209. [CrossRef]
29. Wang, H.J.; Zhao, X.F.; Wei, G.W. Dual hesitant fuzzy aggregation operators in multiple attribute decision making. *J. Intell. Fuzzy Syst. Appl. Eng. Technol.* **2014**, *26*, 2281–2290.
30. Zhang, W.K.; Li, X.; Ju, Y.B. Some aggregation operators based on Einstein operations under interval-valued dual hesitant fuzzy setting and their application. *Math. Probl. Eng.* **2014**. [CrossRef]
31. Yager, R.R. Prioritized aggregation operators. *Int. J. Approx. Reason.* **2008**, *48*, 263–274. [CrossRef]
32. Yu, D.J. Intuitionistic fuzzy prioritized operators and their application in multi-criteria group decision making. *Technol. Econ. Develop. Econ.* **2013**, *19*, 1–21. [CrossRef]
33. Yu, D.J.; Wu, Y.Y.; Lu, T. Interval-valued intuitionistic fuzzy prioritized operators and their application in group decision making. *Knowl. Syst.* **2012**, *30*, 57–66. [CrossRef]
34. Wei, G.W. Hesitant fuzzy prioritized operators and their application to multiple attribute decision making. *Knowl. Syst.* **2012**, *31*, 176–182. [CrossRef]
35. Bedregal, B.; Reiser, R.; Bustince, H.; Lopez-Molina, C.; Torra, V. Aggregation functions for typical hesitant fuzzy elements and the action of automorphisms. *Infor. Sci.* **2014**, *255*, 82–99. [CrossRef]
36. Alcantud, J.C.R.; Torra, V. Decomposition theorems and extension principles for hesitant fuzzysets. *Infor. Fusion.* **2018**, *41*, 48–56. [CrossRef]
37. Bagga, P.; Joshi, A.; Hans, R. QoS based web service selection and multi-criteria decision making methods. *Int. J. Int. Multimed. Artif. Intell.* **2017**. [CrossRef]
38. Zhao, H.; You, J.X.; Liu, H.C. Failure mode and effect analysis using MULTIMOORA method with continuous weighted entropy under interval-valued intuitionistic fuzzy environment. *Soft Comput.* **2017**, *12*, 5355–5367. [CrossRef]
39. Kou, G.; Lu, Y.Q.; Peng, Y.; Shi, Y. Evaluation of classification algorithms using MCDM and rank correlation. *Int. J. Inf. Technol. Decis. Mak.* **2012**, *11*, 197–225. [CrossRef]

symmetry MDPI

Article

Selecting the Optimal Mine Ventilation System via a Decision Making Framework under Hesitant Linguistic Environment

Weizhang Liang [1], Guoyan Zhao [1,*] and Suizhi Luo [2]

[1] School of Resources and Safety Engineering, Central South University, Changsha 410083, China; wzlian@csu.edu.cn

[2] School of Systems Engineering, National University of Defense Technology, Changsha 410073, China; szlluo@csu.edu.cn

* Correspondence: gyzhao@csu.edu.cn

Received: 15 June 2018; Accepted: 10 July 2018; Published: 13 July 2018

Abstract: Ventilation systems are amongst the most essential components of a mine. As the indicators of ventilation systems are in general of ambiguity or uncertainty, the selection of ventilation systems can therefore be regarded as a complex fuzzy decision making problem. In order to solve such problems, a decision making framework based on a new concept, the hesitant linguistic preference relation (HLPR), is constructed. The basic elements in the HLPR are hesitant fuzzy linguistic numbers (HFLNs). At first, new operational laws and aggregation operators of HFLNs are defined to overcome the limitations in existing literature. Subsequently, a novel comparison method based on likelihood is proposed to obtain the order relationship of two HFLNs. Then, a likelihood-based consistency index is introduced to represent the difference between two hesitant linguistic preference relations (HLPRs). It is a new way to express the consistency degree for the reason that the traditional consistency indices are almost exclusively based on distance measures. Meanwhile, a consistency-improving model is suggested to attain acceptable consistent HLPRs. In addition, a method to receive reasonable ranking results from HLPRs with acceptable consistency is presented. At last, this method is used to pick out the best mine ventilation system under uncertain linguistic decision conditions. A comparison and a discussion are conducted to demonstrate the validity of the presented approach. The results show that the proposed method is effective for selecting the optimal mine ventilation system, and provides references for the construction and management of mines.

Keywords: mine ventilation systems; hesitant linguistic environment; likelihood; preference relations

---

1. Introduction

The ventilation system is one of the most important technologies to ensure the safety of mines [1]. In the process of mining, it is necessary to provide enough fresh air and exclude harmful gases, heat and dust [2]. Then, a good working environment can be created to guarantee the health and safety of underground workers. Therefore, choosing an applicable mine ventilation system is essential and important for mines. Since there is much ambiguity and uncertainty in the evaluation process, the selection of mine ventilation systems can be deemed as a fuzzy decision making problem.

In the process of decision making, experts or decision makers (DMs) may prefer to do comparisons among each pair of systems or construct a preference matrix when expressing their opinions [3,4]. On the other hand, because of the complexity of alternatives and the fuzziness of human cognitions, many people may tend to give preference information in the form of language phrases, such as "good", "bad" and so on [5–7]. Thereafter, the decision making problems based on linguistic preference relations (LPRs) have attracted extensive attention [8–10]. However, there is a hypothesis that the membership degree of each element in LPRs is a certain number "one". It is unable to accurately depict the professionals' supporting or hesitant degrees of linguistic assessment information.

Accordingly, Rodriguez et al. [11] put forward the concept of hesitant fuzzy linguistic term set (HFLTS) to express the experts' hesitation or inconsistency. HFLTS is an orderly limited subset of linguistic terms. Different varieties of aggregation operators [12–14], measures [15,16] and decision making approaches [17–19] based on HFLTS and its extensions were proposed. For instance, Liu et al. [20] defined the distance measures for HFLTS to deal with hesitant fuzzy linguistic multi-criteria decision making problems; Adem et al. [21] proposed an integrated model using SWOT analysis and HFLTS for evaluating occupational safety risks in the life cycle of wind turbines. Besides, Zhu and Xu [22] came up with the concept of hesitant fuzzy linguistic preference relations (HFLPRs), where the basic elements are in the form of HFLTS.

Subsequently, numerous researchers had great interest in studying preference matrices under hesitant fuzzy linguistic conditions. Zhang and Wu [23] introduced the multiplicative consistency of HFLPRs based on distance measures. Wang and Xu [24] defined the additive and weak consistency of extended HFLPRs on the basis of graph theory. Wu and Xu [25] discussed the consistency and consensus of HFLPRs in a group decision environment. Gou et al. [26] proposed the compatibility measures and weights determination approach for HFLPRs, and then applied them in selecting a desirable aspect in the medical and health system reform process. Li et al. [27] recommended an approach of obtaining the interval consistency degree of HFLPRs. Xu et al. [28] constructed a group decision support model for HFLPRs to reach consistency and consensus.

Nevertheless, HFLTS cannot reflect the membership degree of an element that belongs to a specific concept [29], such as a certain ventilation system in this paper. It is only a collection of several linguistic evaluation values, and it has strong subjectivity and fuzziness [30]. In order to overcome the inherent defects of linguistic variables and HFLTS, hesitant fuzzy linguistic sets (HFLSs) were introduced by Lin et al. [31]. They can describe hesitant degrees of DMs with some membership degrees based on a given linguistic term. Compared with uncertain linguistic variables, they have the edge on describing the fuzziness [29]. HFLSs combine linguistic term sets with hesitant fuzzy sets (HFSs), which include both the quantitative and qualitative evaluation information [32]. Each element in the HFLSs can be called a hesitant fuzzy linguistic number (HFLN). For instance, half of the specialists in Group A think that vs_1 is a good ventilation system, and 80 percent in Group B think so. In this case, it can be expressed with a HFLN $< good, \{0.5, 0.8\} >$.

The motivations of this paper are mainly two-fold. (1) The mine ventilation systems selection context requires dealing with fuzzy evaluation information and building appropriate decision making models. Hesitant fuzzy linguistic numbers (HFLNs) have advantages in describing the fuzziness and hesitancy of experts [29]. Moreover, preference relations are among the most powerful tools to select the best system. (2) Currently, researches on HFLNs are relatively insufficient compared with other fuzzy sets. Wang et al. [29] developed a decision making method based on the Hausdorff distance of HFLNs. In addition, Wang et al. [30] put forward the concept of interval-valued HFLNs to deal with complex decision making issues. Yet, there are still some limitations with existent operational laws and comparison methods of HFLNs [29,31].

Taking the aforementioned motivations into account, this paper concentrates on selecting the optimal mine ventilation system under a hesitant linguistic environment.

The novelty and contributions of this paper are listed as follows.

(1) New operational laws and aggregation operators of HFLNs are presented. These new operations can reflect the relationship of the linguistic term and its corresponding membership degrees. Furthermore, a hesitant fuzzy linguistic likelihood is presented to compare two arbitrary HFLNs. It can effectively overcome the limitations of the existing comparison method based on score function and accuracy function.
(2) The concept of HLPRs is proposed to tackle decision making issues under hesitant fuzzy linguistic circumstances. A consistency index using likelihood is defined to check the consistency degree of

HLPRs and a consistency-improving model is introduced to get acceptable consistency. Besides, a likelihood-based method is adopted to obtain the final ranking result.

(3) The proposed method is applied in the engineering field of choosing appropriate mine ventilation systems. Thereafter, an in-depth comparison analysis is conducted to demonstrate the validity and merits of the presented method.

The remainder of this paper is arranged as follows: Introductory knowledge about HFLSs and preferences relations are briefly reviewed in Section 2. Section 3 proposes new operations and comparison method of HFLNs. A consistency index is put forward for checking the consistency level of HLPRs in Section 4. A consistency-improving process may be carried out when a HLPR's consistency is unacceptable. And a likelihood-based approach is presented to get the ranking results subsequently. Section 5 illustrates an example of ventilation systems selection and makes a comparative analysis to express the effectiveness of the proposed method. Necessary discussions and brief comments are informed in Section 6.

2. General Concepts

In this section, general concepts related to linguistic variables, HFSs, HFPRs and HFLSs are recalled.

2.1. Linguistic Variables

Assume lv_i stands for a possible linguistic value in a finite and entirely ordered separate term set $LV = \{lv_i | i = -t, \ldots, -1, 0, 1, \ldots, t\}$ [33,34]. It is usually required to meet the following conditions:

(1) There is an order: $lv_i > lv_j$, when $i > j$;
(2) A negation operator exists: $ne(lv_i) = lv_{-i}$.

When the aggregated information is used in the process of decision making, it usually does not go with the values in the predefined evaluation scope. To reserve all the obtained values, Xu [33] changed the preceding term set LV into a continuous one $\overline{LV} = \{lv_i | i \in [-p, p]\}$, where $p(p > t)$ is a adequately great positive integer.

Taking two linguistic terms $lv_i, lv_j \in \overline{LV}$ into account, some operations are proposed in the following:

(1) $lv_i \oplus_{Xu} lv_j = lv_{i+j}$;
(2) $lv_i \oplus_{Xu} lv_j = lv_j \oplus_{Xu} lv_i$;
(3) $\rho lv_i = lv_{\rho i}, \rho \in [0, 1]$.

2.2. Hesitant Fuzzy Sets

Since Zadeh [35] proposed fuzzy sets, it has been widely applied in various fields [36–40] and many extensions based on fuzzy set have been developed [41,42]. HFSs, as extensions of fuzzy sets, were firstly presented by Torra [32]. They are defined in coping with several numerical values permitted to indicate an element's membership degree [43–45]. The definition of HFSs is given as follows.

Definition 1 [32]. *If X is a fixed set, then a hesitant fuzzy set (HFS) on X is in relation to the function, which can go back a set of numbers between zero and one. It is described as the mathematical sign in the following:*

$$F = \{< x, h_F(x) > | x \in X\} \tag{1}$$

where $h_F(x)$ is a subset of several values between zero and one, which represents the probable membership degrees of an element $x \in X$ to a certain set F. Xia and Xu [46] believe that it is convenient to call $h_F(x)$ a hesitant fuzzy element (HFE).

Preference relations are impactful tools in respect to modeling the decision making process. On the basis of HFSs, Zhu [47] came up with the concept of hesitant fuzzy preference relations (HFPRs), which is given as follows.

Definition 2 [47]. *Let $X = \{x_1, x_2, \ldots, x_n\}$ be a reference set, then a HFPR G on X is denoted by a matrix $G = (g_{ij})_{n \times n} \subset X \times X$, where $g_{ij} = \{[q_{ij}^{\sigma(l)} | l = 1, \ldots, |l_{ij}|]\}$ is a HFE expressing whole possible preference degree(s) of the object x_i over x_j. Furthermore, g_{ij} $(i, j = 1, 2, \ldots, n;\ i < j)$ should meet the following requirements:*

$$q_{ij}^{\sigma(l)} + q_{ji}^{\sigma(l)} = 1,\ q_{ii}^{\sigma(l)} = 0.5,\ |l_{ij}| = |l_{ji}| \tag{2}$$

$$q_{ij}^{\sigma(l)} < q_{ij}^{\sigma(l+1)},\ q_{ji}^{\sigma(l+1)} < q_{ji}^{\sigma(l)} \tag{3}$$

where $q_{ij}^{\sigma(l)}$ is the l-th largest element in g_{ij}, and $|l_{ij}|$ is the number of elements in g_{ij}.

2.3. Hesitant Fuzzy Linguistic Sets

The concept, operational laws and comparison method of HFLNs are recalled in this section. Moreover, the limitations of them are discussed in the corresponding places.

Definition 3 [31]. *Let $X = \{x_1, x_2, \ldots, x_n\}$ be a fixed set, and $lv_{\theta(x)} \in \overline{LV}$. Then, the hesitant fuzzy linguistic set (HFLS) U in X can be described as the subsequent object:*

$$U = \left\{ < x, lv_{\theta(x)}, h_U(x) > | x \in X \right\} \tag{4}$$

where $h_U(x)$ is a set of finite numbers in [0,1] and signifies the possible degrees of membership that x belongs to $lv_{\theta(x)}$.

There are two special cases of HFLNs: (1) A hesitant fuzzy linguistic number (HFLN): There is only one element in the set $X = \{x_1, x_2, \ldots, x_n\}$, and HFLS U is reduced to $< lv_{\theta(x)}, h_U(x) >$; (2) A fuzzy linguistic number: There is only one element in $h_B(x)$, like $h_U(x) = \{u\}$, and HFLS U is reduced to $< lv_{\theta(x)}, u >$. For example, $< lv_3, 0.5 >$ shows that the membership degree of x belongs to lv_3 is 0.5.

The operational laws about HFLNs are introduced in literature [31] as follows. Based on them, many aggregation operators are also presented in this paper.

Definition 4 [31]. *Given two HFLNs $a =< lv_{\theta(a)}, h_a >$ and $b =< lv_{\theta(b)}, h_b >$ arbitrarily, and $\lambda \in [0, 1]$, then*

(1) $a \oplus_{Lin} b =< lv_{\theta(a)+\theta(b)}, \bigcup_{r_1 \in h_a, r_2 \in h_b} \{r_1 + r_2 - r_1 \cdot r_2\} >$;

(2) $\lambda a =< lv_{\lambda \cdot \theta(a)}, \bigcup_{r \in h_a} \{1 - (1 - r)^{\lambda}\} >$.

It is clear that the operations mentioned above are not very reasonable as the linguistic values and the membership degrees are operated separately. In fact, the membership degrees should be related to the homologous linguistic values in the operation process.

Definition 5 [29]. *If $a =< lv_{\theta(a)}, h_a >$ is a HFLN, then the score function $E(a)$ of a can be described as follows:*

$$E(a) = s(h_a) \times f^*(lv_{\theta(a)}) \tag{5}$$

where $s(h_a) = \frac{1}{\#h_a} \sum_{r \in h_a} r$, $s(h_a)$ is the score function of h_a, $\#h_a$ is the number of values in h_a, $f^*(lv_i) = \frac{1}{2} + \frac{i}{2t}$ is one of the three different expressions of the linguistic scale function defined by Wang et al. [29], and it can be replaced by another expressions under different semantics. For more details please refer to literature [29].

Definition 6 [29]. *Let $a = < lv_{\theta(a)}, h_a > = < lv_{\theta(a)}, \cup_{r \in h_a}\{r\} >$ be a HFLN, and the variance function is represented as $V^*(h_a) = \frac{1}{\#h_a}\sum_{r \in h_a}[r - s(h_a)]^2$. Hence, the accuracy function $V(a)$ of a can be shown as follows:*

$$V(a) = f^*(lv_{\theta(a)}) \cdot [1 - V^*(h_a)] \quad (6)$$

where $\#h_\alpha$ is the number of the values in h_a.

The accuracy function $V(\alpha)$ is analogous to the sample variance statistically and can display the fluctuation of assessment values of h_a. The greater the volatility is, the larger the hesitation will be. Then, the ranking order of HFLNs can be derived by using the score function and accuracy function as follows.

Definition 7 [29]. *If $a = < lv_{\theta(a)}, h_a >$ and $b = < lv_{\theta(b)}, h_b >$ are two arbitrary HFLNs, $r_a^{\sigma(l)}$ and $r_b^{\sigma(l)}$ are regarded as the lth number in h_a and h_b respectively, and all membership degrees are arranged in ascending order. Then the comparison method is*

(1) *If $lv_{\theta(a)} \leq lv_{\theta(b)}$, $r_a^{\sigma(l)} \leq_b^{\sigma(l)}$ and $r_a^{\sigma(\#h_a)} \leq r_b^{\sigma(\#h_b)}$, then $a < b$, where at least one of "<" exists, $r_a^{\sigma(l)} \in h_a$, $r_b^{\sigma(l)} \in h_b$, $l = 1, 2, \ldots, \min(\#h_a, \#h_b)$, $\#h_a$ and $\#h_b$ are the numbers of values in h_a and h_b respectively;*
(2) *If $E(a) < E(b)$ but $a < b$, then $a \prec b$;*
(3) *If $E(a) = E(b)$ and $V(a) < V(b)$, then $a \prec b$;*
(4) *If $E(a) = E(b)$ and $V(a) = V(b)$, then $a = b$.*

Example 1. *Suppose $a = < lv_0, \{0.1, 0.4\} >$, $b = < lv_{-3}, \{0.1, 0.4\} >$ and $c = < lv_0, \{0.2, 0.3\} >$ are three HFLNs. Let $f^*(lv_i) = \frac{1}{2} + \frac{i}{2t}$ and $t = 3$, then:*

(1) $lv_{\theta(b)} = lv_{-3} < lv_{\theta(a)} = lv_0$, $r_b^{\sigma(1)} = r_a^{\sigma(1)} = 0.1$, $r_b^{\sigma(2)} = r_a^{\sigma(2)} = 0.4$, *thus $b < a$;*
(2) $E(b) = 0$, $E(c) = 0.125$, *i.e.*, $E(b) < E(c)$, *thus $b \prec c$;*
(3) $E(a) = E(c) = 0.125$, $V(a) = 0.48875$, $V(c) = 0.49875$, *i.e.*, $V(a) < V(c)$, *thus $a \prec c$.*

There is no doubt that the amounts of calculations are increased when the score function or even the accuracy function needs to be calculated. Besides, according to this comparison method, if $E(a) = E(b)$ and $V(a) = V(b)$ are true simultaneously, a conclusion is that $a = b$. It is reasonable in most conditions. However, it is not well tenable when the linguistic scale function $f^*(lv_i) = 0$ and the possible memberships in a certain HFLN are not strictly superior to the memberships in another HFLN. For instance, assume $\alpha = < lv_{-3}, \{0.1, 0.7\} >$, $\beta = < lv_{-3}, \{0.1, 0.9\} >$ and $\eta = < lv_{-3}, \{0.5, 0.6\} >$ are three HFLNs. Let $f^*(lv_i) = \frac{1}{2} + \frac{i}{2t}$ and $t = 3$, then $E(\alpha) = E(\beta) = E(\eta) = 0$ and $V(\alpha) = V(\beta) = V(\eta) = 0$ are true, a decision is that $\alpha < \beta$ according to part (1) of the comparison method, and a decision is that $\alpha = \eta$ and $\beta = \eta$ according to part (3) of this method. It is clear that these conclusions are self-contradictory and counterintuitive.

3. New Operations and Comparison Method

As mentioned in Section 2.3, there are some weaknesses in the existent operational laws and comparison method with HFLNs. Thus, new operations and comparison methods are presented in this section.

3.1. New Operational Laws and Aggregation Operators

To overcome the limitations of operations proposed in Section 2.3, some new operational laws on the HFLNs are raised in this section. Afterwards, the hesitant fuzzy linguistic weighted average (HFLWA) operator and hesitant fuzzy linguistic average (HFLA) operator based on them are presented.

Definition 8. *If* $a = < lv_{\theta(a)}, h_a >$ *and* $b = < lv_{\theta(b)}, h_b >$ *are HFLNs, and* $\lambda \in [0,1]$, *then*

(1) $a \oplus b = < lv_{\theta(a)+\theta(b)}, \bigcup_{r_1 \in h_a, r_2 \in h_b} \left\{ \frac{(\theta(a)+t) \cdot r_1 + (\theta(b)+t) \cdot r_2}{(\theta(a)+t)+(\theta(b)+t)} \right\} >$;

(2) $\lambda a = < lv_{\lambda \cdot \theta(a)}, h_a >$.

It is easily verified that all operational results mentioned above are still HFLNs. Although there are no practical meanings with the operational results, the basic operations are necessary to be defined in practice. When these operations are used together, the actual significance can be reflected in reality.

In view of Definition 8, the equivalent relations can be further acquired as follows.

(1) Commutativity: $a \oplus b = b \oplus a$;
(2) Associativity: $(a \oplus b) \oplus c = a \oplus (b \oplus c)$;
(3) Distributivity: $\lambda(a \oplus b) = \lambda a \oplus \lambda b$, $\lambda \in [0,1]$;
(4) Distributivity: $\lambda_1 a \oplus \lambda_2 a = (\lambda_1 + \lambda_2) a$, $\lambda_1, \lambda_2 \in [0,1]$.

Definition 9. *Let* $a_i = < lv_{\theta(a_i)}, h_{a_i} >$ *be a group of HFLNs with* $i = 1, 2, \ldots, n$. *The HFLWA operator can be denoted as follows:*

$$HFLWA(a_1, a_2, \ldots, a_n) = \omega_1 a_1 \oplus \omega_2 a_2 \oplus \cdots \oplus \omega_n a_n \tag{7}$$

where $\omega = (\omega_1, \omega_2, \ldots, \omega_n)^T$ *is the weight vector of* a_i $(i = 1, 2, \ldots, n)$, $\omega_i \in [0,1]$ *and* $\sum_{i=1}^{n} \omega_i = 1$.

Particularly, if $\omega = \left(\frac{1}{n}, \frac{1}{n}, \ldots, \frac{1}{n}\right)^T$, then the HFLWA operator is degenerated to the HFLA operator as follows:

$$HFLA(a_1, a_2, \ldots, a_n) = \frac{1}{n}(a_1 \oplus a_2 \oplus \cdots \oplus a_n) \tag{8}$$

Theorem 1. *Assume* $a_i = < lv_{\theta(a_i)}, h_{a_i} >$ *are a set of HFLNs, and* $\omega = (\omega_1, \omega_2, \ldots, \omega_n)^T$ *is the weight vector of* a_i $(i = 1, 2, \ldots, n)$, $\omega_i \in [0,1]$ *and* $\sum_{i=1}^{n} \omega_i = 1$, *then the aggregated result through applying the HFLWA operator is still a HFLN, and*

$$HFLWA(a_1, a_2, \ldots, a_n) = < lv_{\sum_{i=1}^{n} C_i}, \bigcup_{r_1 \in h_{a_1}, r_2 \in h_{a_2}, \ldots, r_n \in h_{an}} \left\{ \frac{\sum_{i=1}^{n} D_i \cdot r_i}{\sum_{i=1}^{n} D_i} \right\} > \tag{9}$$

where $C_i = \omega_i \theta(a_i)$ *and* $D_i = \omega_i(\theta(a_i) + t)$ *for all* $i = 1, 2, \ldots, n$.

Proof. Clearly, by Definition 8, the aggregated data by exploiting the HFLWA operator remains a HFLN. Next, Equation (9) is proved through utilizing mathematical induction on n.

(1) When $n = 2$: we have $\omega_1 a_1 = < lv_{C_1}, h_{a_1} >$ and $\omega_2 a_2 = < lv_{C_2}, h_{a_2} >$, then $HFLWA(a_1, a_2) = \omega_1 a_1 \oplus \omega_2 a_2 = < lv_{C_1 + C_2}, \bigcup_{r_1 \in h_{a_1}, r_2 \in h_{a_2}} \left\{ \frac{D_1 \cdot r_1 + D_2 \cdot r_2}{D_1 + D_2} \right\} > = < lv_{\sum_{i=1}^{2} C_i}, \bigcup_{r_1 \in h_{a_1}, r_2 \in h_{a_2}} \left\{ \frac{\sum_{i=1}^{2} D_i \cdot r_i}{\sum_{i=1}^{2} D_i} \right\} >$.

(2) For $n = k$: If Equation (9) holds, then $HFLA(a_1, a_2, \ldots, a_k) = < lv_{\sum_{i=1}^{k} C_i}, \bigcup_{r_1 \in h_{a_1}, r_2 \in h_{a_2}, \ldots, r_k \in h_{ak}} \left\{ \frac{\sum_{i=1}^{k} D_i \cdot r_i}{\sum_{i=1}^{k} D_i} \right\} >$. Hence, for $n = k + 1$, from Definition 8, that is $HFLA(a_1, a_2, \ldots, a_{k+1}) = < lv_{\sum_{i=1}^{k} C_i}, \bigcup_{r_1 \in h_{a_1}, r_2 \in h_{a_2}, \ldots, r_k \in h_{ak}} \left\{ \frac{\sum_{i=1}^{k} D_i \cdot r_i}{\sum_{i=1}^{k} D_i} \right\} >$

$$\oplus(\omega_{k+1}\cdot a_{k+1}), = < lv_{\sum\limits_{i=1}^{k} C_i + C_{k+1}}, \bigcup_{r_1\in h_{a_1}, r_2\in h_{a_2},\ldots,r_{k+1}\in h_{a_{k+1}}} \left\{ \frac{\sum\limits_{i=1}^{k} D_i \cdot \frac{\sum\limits_{i=1}^{k} D_i \cdot r_i}{\sum\limits_{i=1}^{k} D_i} + D_{k+1}\cdot r_{k+1}}{\sum\limits_{i=1}^{k} D_i + D_{k+1}} \right\} > =$$

$$< lv_{\sum\limits_{i=1}^{k+1} C_i}, \bigcup_{r_1\in h_{a_1}, r_2\in h_{a_2},\ldots,r_{k+1}\in h_{a_{k+1}}} \left\{ \frac{\sum\limits_{i=1}^{k+1} D_i \cdot r_i}{\sum\limits_{i=1}^{k+1} D_i} \right\} >.$$

i.e., for $n = k+1$, Equation (9) follows.

Therefore, combined (1) with (2), Equation (9) follows for all $n \in N$, then the proof of Theorem 1 is completed. □

3.2. Likelihood of Hesitant Fuzzy Linguistic Numbers

The likelihood-based comparison method is an effective way to compare fuzzy numbers. Inspired by literature [48,49], a new method based on likelihood to compare HFLNs is proposed. From an example, it can be seen that the limitations of the comparison method mentioned in Section 2.3 have been overcome when the proposed likelihood-based comparison method is adopted.

The likelihood between two HFLNs is described in the following:

Definition 10. *If $a =< lv_{\theta(a)}, h_a >$ and $b =< lv_{\theta(b)}, h_b >$ are two optional HFLNs, then the likelihood between a and b can be demonstrated as follows:*

$$L(a \geq b) = \begin{cases} 1, & 0v_{\theta(a)} > l_{v_{\theta(b)}}, h_a^+ > h_b^- \\ \frac{1}{\#h_a \#h_b} \sum\limits_{i=1}^{\#h_a} \sum\limits_{j=1}^{\#h_b} \frac{r_a^{\sigma(\imath)}}{r_a^{\sigma(i)} + r_b^{\sigma(j)}}, & lv_{\theta(\alpha)} = lv_{\theta(\beta)} \\ \frac{1}{\#h_a \#h_b} \sum\limits_{i=1}^{\#h_a} \sum\limits_{j=1}^{\#h_b} \frac{f^*(lv_{\theta(a)})\cdot r_a^{\sigma(i)}}{f^*(lv_{\theta(a)})\cdot r_a^{\sigma(i)} + f^*(lv_{\theta(b)})\cdot r_b^{\sigma(j)}}, & lv_{\theta(a)} \neq lv_{\theta(b)} \\ 1, & lv_{\theta(a)} < lv_{\theta(a)}, h_a^- < h_b^+ \end{cases} \quad (10)$$

where $\gamma_a^{\sigma(i)}$ and $\gamma_b^{\sigma(j)}$ are the i-th and j-th largest value, $\#h_a$ and $\#h_b$ are the numbers of element in h_a and h_b respectively.

Property 1. *Suppose Ω is a set with all HFLNs, $\forall_{a,b,c} \in \Omega$, the likelihood satisfies the following properties:*

(1) $0 \leq L(a \geq b) \leq 1$;
(2) If $lv_{\theta(a)} \leq lv_{\theta(b)}$, $h_a^+ < h_b^-$, then $L(a \geq b) = 0$;
(3) If $lv_{\theta(a)} \geq lv_{\theta(b)}$, $h_a^- > h_b^+$, then $L(a \geq b) = 1$;
(4) $L(a \geq b) + L(b \geq a) = 1$;
(5) If $L(a \geq b) = L(b \geq a)$, then $L(a \geq b) = L(b \geq a) = 0.5$;
(6) If $L(a \geq c) \geq 0.5$, and $L(c \geq b) \geq 0.5$, then $L(a \geq b) \geq 0.5$.

Proof. We only prove (4) of Property 1 in the paper, as the other properties can be easily proven.

(1) If $lv_{\theta(a)} < lv_{\theta(b)}$, $h_a^+ < h_b^-$ or $lv_{\theta(a)} > lv_{\theta(b)}$, $h_a^- < h_b^+$, according to Definition 10, it is true that $L(a \geq b) + L(b \geq a) = 1$.

(2) If $lv_{\theta(a)} = lv_{\theta(b)}$, the following deduction can be derived: $L(a \geq b) = \frac{1}{\#h_a \#h_b} \sum_{i=1}^{\#h_a} \sum_{j=1}^{\#h_b} \frac{r_a^{\sigma(i)}}{r_a^{\sigma(i)} + r_b^{\sigma(j)}}$ and $L(a \leq b) = L(b \geq a) = \frac{1}{\#h_a \#h_b} \sum_{i=1}^{\#h_a} \sum_{j=1}^{\#h_b} \frac{r_a^{\sigma(i)}}{r_a^{\sigma(i)} + r_b^{\sigma(j)}}$, then $L(a \geq b) + L(a \leq b) = \frac{1}{\#h_a \#h_b} \sum_{i=1}^{\#h_a} \sum_{j=1}^{\#h_b} \frac{r_a^{\sigma(i)}}{r_a^{\sigma(i)} + r_b^{\sigma(j)}} + \frac{1}{\#h_a \#h_b} \sum_{j=1}^{\#h_b} \sum_{i=1}^{\#h_a} \frac{r_b^{\sigma(j)}}{r_a^{\sigma(i)} + r_b^{\sigma(j)}} = \frac{1}{\#h_a \#h_b} \sum_{j=1}^{\#h_b} \sum_{i=1}^{\#h_a} \frac{r_a^{\sigma(i)} + r_b^{\sigma(j)}}{r_a^{\sigma(i)} + r_b^{\sigma(j)}} = 1$.

(3) If $lv_{\theta(a)} \neq lv_{\theta(b)}$, similar to proof (2), we can obtain the following: $L(a \leq b) = L(b \geq a) = \frac{1}{\#h_a \#h_\beta} \sum_{j=1}^{\#h_b} \sum_{i=1}^{\#h_a} \frac{f^*(lv_{\theta(b)}) \cdot r_b^{\sigma(j)}}{f^*(lv_{\theta(a)}) \cdot r_a^{\sigma(i)} + f^*(lv_{\theta(b)}) \cdot r_b^{\sigma(j)}}$, $L(a \geq b) + L(a \leq b) = \frac{1}{\#h_a \#h_\beta} \sum_{i=1}^{\#h_a} \sum_{j=1}^{\#h_b} \frac{f^*(lv_{\theta(a)}) \cdot r_a^{\sigma(i)}}{f^*(lv_{\theta(a)}) \cdot r_a^{\sigma(i)} + f^*(lv_{\theta(b)}) \cdot r_b^{\sigma(j)}} + \frac{1}{\#h_a \#h_\beta} \sum_{j=1}^{\#h_b} \sum_{i=1}^{\#h_a} \frac{f^*(lv_{\theta(b)}) \cdot r_b^{\sigma(j)}}{f^*(lv_{\theta(a)}) \cdot r_a^{\sigma(i)} + f^*(lv_{\theta(b)}) \cdot r_b^{\sigma(j)}} = \frac{1}{\#h_a \#h_\beta} \sum_{j=1}^{\#h_b} \sum_{i=1}^{\#h_a} \frac{f^*(lv_{\theta(a)}) \cdot r_a^{\sigma(i)} + f^*(lv_{\theta(b)}) \cdot r_b^{\sigma(j)}}{f^*(lv_{\theta(a)}) \cdot r_a^{\sigma(i)} + f^*(lv_{\theta(b)}) \cdot r_b^{\sigma(j)}} = 1$.

Therefore, $L(a \geq b) + L(a \leq b) = 1$.
Now, the proof is completed. □

Definition 11. *If $a = < lv_{\theta(a)}, h_a >$ and $b = < lv_{\theta(b)}, h_b >$ are two HFLNs. The new comparison method for HFLNs can be defined as follows:*

(1) *If $L(a \geq b) > 0.5$, then a is superior to b, expressed by $a > b$;*
(2) *If $L(a \geq b) < 0.5$, then a is inferior to b, expressed by $a < b$;*
(3) *If $L(a \geq b) = 0.5$, then a is indifferent to b, expressed by $a = b$.*

Example 2. *Suppose that three HFLNs are the same as Example 1, the comparison results with new proposed comparison method are given as follows.*

(1) $L(a \geq b) = 1$, $L(b \geq a) = 0$, *then* $b < a$.
(2) $L(b \geq c) = 0$, $L(c \geq b) = 1$, *then* $b < c$.
(3) $L(a \geq c) = 0.455$, $L(c \geq a) = 0.5446$, *then* $a < c$.

It is true that the results in Examples 1 and 2 are the same, which verifies the validity of the presented comparison method. Moreover, assume $\alpha = < lv_{-3}, \{0.1, 0.7\} >$, $\beta = < lv_{-3}, \{0.1, 0.9\} >$ and $\eta = < lv_{-3}, \{0.5, 0.6\} >$ are three HFLNs, $f^*(lv_i) = \frac{1}{2} + \frac{i}{2t}$ and $t = 3$, then $L(\alpha \geq \beta) = 0.4781$, $L(\alpha \geq \eta) = 0.3578$ and $L(\beta \geq \eta) = 0.3881$. So we get a conclusion that $\alpha < \beta < \eta$, which is more reasonable than the results obtained by using the previous comparison method.

4. Decision Making Framework

In this section, a decision making framework is proposed to handle decision making problems under a hesitant linguistic environment. Original preference information is expressed by HLPRs and the consistency level is checked and improved. Then, a likelihood-based model is suggested to derive a ranking from HLPRs with acceptable consistency.

4.1. Original Preference Information

When making evaluations for some alternatives under a hesitant linguistic environment, DMs can provide original preference information with HLPRs. To facilitate the following discussions, the concepts of HLPRs and consistent HLPRs are defined as follows.

Definition 12. *If $X = \{x_1, x_2, \ldots, x_n\}$ is a set of alternatives, then the HLPR K on X can be described as a matrix $K = (k_{ij})_{n \times n} \subset X \times X$. Each element $k_{ij} = < lv_{ij}, r_{ij} >$ is a HFLN, where lv_{ij} and r_{ij} demonstrate* *respectively, the degree of x_i preferred to x_j and the possible membership degrees that x belongs to lv_{ij}. Then, for k_{ij} $(i, j = 1, 2, \ldots, n,\ i < j)$, the following requirements should be met:*

$$lv_{ij} \oplus lv_{ji} = lv_0,\ lv_{ii} = lv_0,\ r_{ij}^{\sigma(l)} = r_{ji}^{\sigma(l)},\ r_{ii}^{\sigma(l)} = 1,\ |k_{ij}| = |k_{ji}| \tag{11}$$

where $r_{ij}^{\sigma(l)}$ is the l-th element in r_{ij}, and $|k_{ij}|$ is the number of values in k_{ij}.

Definition 13. *Let $K = (k_{ij})_{n\times n}$ be a HLPR, if*

$$r_{ik}^{\sigma(l)} \cdot lv_{ik} \oplus r_{kj}^{\sigma(l)} \cdot lv_{kj} = r_{ij}^{\sigma(l)} \cdot lv_{ij}\ (i, j, k = 1, 2, \ldots n) \tag{12}$$

then K is a consistent HLPR.

Example 3. *Given a HLPR* $K_1 = \begin{bmatrix} < lv_0, \{1\} > & < lv_1, \{0.3, 0.9\} > & < lv_{-2}, \{0.1, 0.6\} > \\ < lv_{-1}, \{0.3, 0.9\} > & < lv_0, \{1\} > & < lv_2, \{0.4, 0.9\} > \\ < lv_2, \{0.1, 0.6\} > & < lv_{-2}, \{0.4, 0.9\} > & < lv_0, \{1\} > \end{bmatrix}$.
Since $r_{13}^{\sigma(1)} \cdot lv_{13} = lv_{-0.2}$, $r_{12}^{\sigma(1)} \cdot lv_{12} \oplus r_{23}^{\sigma(1)} \cdot lv_{23} = lv_{1.1}$, $r_{13}^{\sigma(1)} \cdot lv_{13} \neq r_{12}^{\sigma(1)} \cdot lv_{12} \oplus r_{23}^{\sigma(1)} \cdot lv_{23}$, then K_1 is not a consistent HLPR.

Theorem 2. *Assume a HLPR $K = (k_{ij})_{n\times n}$, if*

$$\max\left\{\oplus_{k=1}^{n}(r_{ik}^{\sigma(l)} \cdot lv_{ik} \oplus r_{kj}^{\sigma(l)} \cdot lv_{kj})\right\} < lv_0 \text{ or } \min\left\{\oplus_{k=1}^{n}(r_{ik}^{\sigma(l)} \cdot lv_{ik} \oplus r_{kj}^{\sigma(l)} \cdot lv_{kj})\right\} > lv_0 \quad (i, j, k = 1, 2, \ldots n) \tag{13}$$

then $K = (k_{ij})_{n\times n}$ has a corresponding consistent HLPR.

Proof. The proof is straightforward. According to Equation (11), if $\min\left\{\oplus_{k=1}^{n}(r_{ik}^{\sigma(l)} \cdot lv_{ik} \oplus r_{kj}^{\sigma(l)} \cdot lv_{kj})\right\} < lv_0$ and $\max\left\{\oplus_{k=1}^{n}(r_{ik}^{\sigma(l)} \cdot lv_{ik} \oplus r_{kj}^{\sigma(l)} \cdot lv_{kj})\right\} > lv_0$, some calculated membership degrees will be less than zero. Clearly, it is unreasonable. Therefore, when $\max\left\{\oplus_{k=1}^{n}(r_{ik}^{\sigma(l)} \cdot lv_{ik} \oplus r_{kj}^{\sigma(l)} \cdot lv_{kj})\right\} < lv_0$ or, $\min\left\{\oplus_{k=1}^{n}(r_{ik}^{\sigma(l)} \cdot lv_{ik} \oplus r_{kj}^{\sigma(l)} \cdot lv_{kj})\right\} > lv_0$, the corresponding consistent HLPR of $K = (k_{ij})_{n\times n}$ exists. □

Example 4. *Given a HLPR* $K_2 = \begin{bmatrix} < lv_0, \{1\} > & < lv_1, \{0.3, 0.5\} > & < lv_{-1}, \{0.1, 0.7\} > \\ < lv_{-1}, \{0.3, 0.5\} > & < lv_0, \{1\} > & < lv_1, \{0.4, 0.8\} > \\ < lv_1, \{0.1, 0.7\} > & < lv_{-1}, \{0.4, 0.8\} > & < lv_0, \{1\} > \end{bmatrix}$.
Since$\oplus_{k=1}^{3}(r_{1k}^{\sigma(1)} \cdot lv_{1k} \oplus r_{k3}^{\sigma(1)} \cdot lv_{k3}) = lv_{0.5} > lv_0$ and $\oplus_{k=1}^{n}(r_{1k}^{\sigma(2)} \cdot lv_{1k} \oplus r_{k3}^{\sigma(2)} \cdot lv_{k3}) = lv_{-0.1} < lv_0$, then K_2 does not have a consistent HLPR.

Note that: when a HLPR $K = (k_{ij})_{n\times n}$ does not have the corresponding consistent HLPR, it should be adjusted based on Equation (14) until a consistent HLPR exists.

Theorem 3. *Assume a HLPR $K = (k_{ij})_{n\times n}$ has the consistent HLPR, for all $i, j, k = 1, 2, \ldots n$, if*

$$r_{ij}^{*\sigma(l)} \cdot lv_{ij}^{*} = \frac{1}{n} \oplus_{k=1}^{n} (r_{ik}^{\sigma(l)} \cdot lv_{ik} \oplus_{Xu} r_{kj}^{\sigma(l)} \cdot lv_{kj}), \tag{14}$$

$$lv_{ij}^{*} = \max\left\{r_{1k}^{\sigma(1)} \cdot lv_{1k} \oplus r_{k3}^{\sigma(1)} \cdot lv_{k3}\right\}\ (if\ \oplus_{k=1}^{n}(r_{1k}^{\sigma(1)} \cdot lv_{1k} \oplus r_{k3}^{\sigma(1)} \cdot lv_{k3}) > lv_0) \tag{15}$$

$$lv_{ij}^{*} = \min\left\{r_{1k}^{\sigma(2)} \cdot lv_{1k} \oplus r_{k3}^{\sigma(2)} \cdot lv_{k3}\right\}\ (if\ \oplus_{k=1}^{n}(r_{1k}^{\sigma(2)} \cdot lv_{1k} \oplus r_{k3}^{\sigma(2)} \cdot lv_{k3}) < lv_0) \tag{16}$$

then $K^* = (k^*_{ij})_{n \times n} = (lv^*_{ij}, r^*_{ij})_{n \times n}$ *is a consistent HLPR.*

Proof. Since $r^{*\sigma(l)}_{ik} \cdot lv^*_{ik} \oplus r^{*\sigma(l)} \cdot lv^*_{kj} = \frac{1}{n}(\oplus^n_{e=1}(r^{\sigma(l)}_{ie} \cdot lv_{ie} \oplus r^{\sigma(l)}_{ek} \cdot lv_{ek})) \oplus \frac{1}{n}(\oplus^{\,n}_{e=1}(r^{\sigma(l)}_{ke} \cdot lv_{ke} \oplus r^{\sigma(l)}_{ej} \cdot lv_{ej})) = \frac{1}{n}(\oplus^{\,n}_{e=1}(r^{\sigma(l)}_{ie} \cdot lv_{ie} \oplus r^{\sigma(l)}_{ek} \cdot lv_{ek} \oplus r^{\sigma(l)}_{ke} \cdot lv_{ke} \oplus r^{\sigma(l)}_{ej} \cdot lv_{ej})) = \frac{1}{n}(\oplus^{\,n}_{e=1}(r^{\sigma(l)}_{ie} \cdot lv_{ie} \oplus r^{\sigma(l)}_{ej} \cdot lv_{ej} \oplus r^{\sigma(l)}_{ek} \cdot lv_0)) = \frac{1}{n}(\oplus^n_{e=1}(r^{\sigma(l)}_{ie} \cdot lv_{ie} \oplus r^{\sigma(l)}_{ej} \cdot lv_{ej})) = r^{\sigma(l)}_{ij} \cdot lv^*_{ij}$ based on Definition 13, $K^* = (k^*_{ij})_{n \times n} = (lv^*_{ij}, r^*_{ij})_{n \times n}$ is a consistent HLPR. □

Example 5. *Assume a HLPR is the same in Example 3. Based on Equation (14), the consistent HLPR* K^*_1 *is obtained as follows:* $K^*_1 = \begin{bmatrix} < lv_0, \{1\} > & < lv_{-1}, \{2/15, 7/15\} > & < lv_{1.1}, \{7/33, 1/11\} > \\ < lv_1, \{2/15, 7/15\} > & < lv_0, \{1\} > & < lv_{0.8}, \{11/24, 5/8\} > \\ < lv_{-1.1}, \{7/33, 1/11\} > & < lv_{-0.8}, \{11/24, 5/8\} > & < lv_0, \{1\} > \end{bmatrix}$.

4.2. Consistency Checking and Improving Models

When an initial preference matrix is constructed, checking and improving its consistency is necessary and vital [50–52]. The consistency of preference relations reflects the rationality of DMs' judgments, and inconsistent preference matrices may generate undesirable or improper conclusions. In this section, a likelihood-based consistency index is defined to test the consistency degree and a consistency-improving process is presented to modify the consistency level.

Definition 14. *Given two arbitrary HLPRs* $A = (a_{ij})_{n \times n}$ *and* $B = (b_{ij})_{n \times n}$*, then*

$$L(A \geq B) = \frac{2}{n(n-1)} \sum_{i<j}^{n} L(a_{ij} \geq b_{ij}) \tag{17}$$

is called the likelihood between two HLPRs.

The likelihood $L(A \geq B)$ satisfies Theorem 4 as follows.

Theorem 4. *Assume A and B are two HLPRs, the likelihood between them can be represented as* $L(A \geq B)$*, then*

(1) $0 \leq L(A \geq B) \leq 1$;
(2) $L(A \geq B) + L(B \geq A) = 1$;
(3) *If* $L(A \geq B) = L(B \geq A)$*, then* $L(A \geq B) = L(B \geq A) = 0.5$.

Definition 15. *Suppose a HLPR K and its corresponding consistent HLPR* K^**; a consistency index is used to calculate the deviation between K and* K^**, which is defined as*

$$CI(K) = \frac{1}{n(n-1)} \sum_{i \neq j}^{n} |L(k_{ij} \geq k^*_{ij}) - \frac{1}{2}| \tag{18}$$

It is true that $0 \leq CI(K) \leq \frac{1}{2}$. Based on Definition 15, a smaller value of $CI(K)$ means a more consistent HLPR K. As the DMs would be often influenced by many uncertainties when they make decisions, HLPRs provided by the DMs are not always perfectly consistent.

Definition 16. *Given a HLPR K and the corresponding threshold value CI, when the consistency index meets:*

$$CI(K) < CI \tag{19}$$

then K is regarded as a HLPR whose consistency is acceptable.

Note: There is an attractive subject about how to determine the value of CI. It may be confirmed in accordance with the DMs' knowledge, experience and other conditions.

In some circumstances, the HLPR K constructed by the DMs is always with unacceptable consistency due to the lack of knowledge or other reasons. Hence, a consistency-improving model is built to acquire a reasonable solution. Some critical steps in Algorithm 1 can be taken repeatedly until the predefined consistency threshold is satisfied.

The main steps of this consistency-improving process are shown as follows.

Algorithm 1. Consistency improving model of HLPRs

Input: The original HLPR $K = (k_{ij})_{n\times n}$, the threshold value $CI = CI_0$ and the maximum number of iterative times $s_{\max} \geq 1$.
Output: The adjusted HLPR K_a and its consistency index $CI(K_a)$.
Step 1: Let the iterative times $s = 0$, and the original HLPR $K = K^{(0)} = (k_{ij}^{(0)})_{n\times n}$.
Step 2: According to Equation (14), obtain the corresponding consistent HLPR $K^{*(s)} = (k_{ij}^{*(s)})_{n\times n} = (< lv_{ij}^{*(s)}, r_{ij}^{*(s)} >)_{n\times n}$ of HLPR $K^{(s)} = (k_{ij}^{(s)})_{n\times n}$.
Step 3: Based on Equation (10), calculate the likelihood $L(k_{ij}^{(s)} \geq k_{ij}^{*(s)})$ of the corresponding elements (e.g., $k_{ij}^{(s)}$ and $k_{ij}^{*(s)}$) in the HLPR $K^{(s)} = (k_{ij}^{(s)})_{n\times n}$ and its consistent HLPR $K^{*(s)} = (k_{ij}^{*(s)})_{n\times n}$. Then, construct the likelihood matrix $L^{(s)} = (l_{ij}^{(s)})_{n\times n} = (L(k_{ij}^{(s)} \geq k_{ij}^{*(s)}))_{n\times n}$ of HLPR $K^{(s)}$.
Step 4: Calculate the consistency index $CI(K^{(s)})$ of HLPR $K^{(s)}$ by Equation (18).
Step 5: If the consistency level of $K^{(s)}$ is acceptable, namely $CI(K^{(s)}) < CI_0$ or the iterative times is maximum, namely $s > s_{\max}$, then go to Step 7; or else, go to the next step.
Step 6: Find an element $l_{ij}^{(s)}$ in the likelihood matrix $L^{(s)} = (l_{ij}^{(s)})_{n\times n}$, which has the maximum deviation on the diagonal, namely $\max\left\{|l_{ij}^{(s)} - \frac{1}{2}| + |l_{ji}^{(s)} - \frac{1}{2}|\right\}$. If $l_{ij}^{(s)} + l_{ij}^{(s)} - 1 < 0$, then the DMs may increase their preference of $k_{ij}^{(s)}$; if $l_{ij}^{(s)} + l_{ij}^{(s)} - 1 > 0$, then the DMs can decrease their values of $k_{ij}^{(s)}$. And the modified HLPR is denoted as $K^{(s+1)} = (k_{ij}^{(s+1)})_{n\times n} = (< lv_{ij}^{(s+1)}, r_{ij}^{(s+1)} >)_{n\times n}$. Let $s = s + 1$, then return to Step 2.
Step 7: Let the final adjusted HLPR $K^{(s)} = K_a$, Output K_a and its consistency index $CI(K_a)$.

Theorem 5. *Given a HFPR K, which is unacceptably consistent. If $CI = CI_0$ is the consistency threshold, $\left\{K^{(s)}\right\}$ is a HFPR sequence, and $CI(K^{(s)})$ is the consistency index of $K^{(s)}$. Therefore, we can obtain that for any s: $CI(K^{(s+1)}) < CI(K^{(s)})$ and $\lim_{s\to\infty} CI(K^{(s)}) = 0$.*

The proof is straightforward. There is no less than one position where $|l_{ij_1}^{(s)} - \frac{1}{2}| < |l_{ij_1}^{(s+1)} - \frac{1}{2}|$ can be obtained. It follows that $CI(K^{(s+1)}) < CI(K^{(s)})$.

Theorem 5 guarantees that any HLPR with insupportable consistency can be converted into an acceptable HLPR. The speed and times may be influenced by the values of the adjusted elements, which are recommended by the DMs or specialists according to the practical situation. How to determine the value of adjusted elements more reasonably is also a controversial issue and deserves to be further investigated.

Example 6. *Given an original HLPR* $K = \begin{bmatrix} < lv_0, \{1\} > & < lv_3, \{0.6, 0.7\} > & < lv_{-2}, \{0.8, 0.9\} > \\ < lv_{-3}, \{0.6, 0.7\} > & < lv_0, \{1\} > & < lv_1, \{0.2, 0.3\} > \\ < lv_2, \{0.8, 0.9\} > & < lv_{-1}, \{0.2, 0.3\} > & < lv_0, \{1\} > \end{bmatrix}$.
Suppose the threshold $CI_0 = 0.25$ and the maximum number of iterative times $s_{\max} = 3$, check and improve its consistency. The detailed procedures are listed as follows.

Step 1: Let $s = 0$ and $K^{(0)} = K$.

Step 2: Based on Equation (14), obtain the consistent HLPR

$$K^{*(0)} = \begin{bmatrix} < lv_0, \{1\} > & < lv_{2.1}, \{2/7, 1/3\} > & < lv_{-1.8}, \{2/9, 2/9\} > \\ < lv_{-2.1}, \{2/7, 1/3\} > & < lv_0, \{1\} > & < lv_{-3.9}, \{10/39, 11/39\} > \\ < lv_{1.8}, \{2/9, 2/9\} > & < lv_{3.9}, \{10/39, 11/39\} > & < lv_0, \{1\} > \end{bmatrix}.$$

Step 3: Based on Equation (10), the likelihood matrix is $L^{(0)} = \begin{bmatrix} 0.5 & 1 & 0.7762 \\ 0.5249 & 0.5 & 0.9781 \\ 1 & 0.2590 & 0.5 \end{bmatrix}$.

Step 4: Based on Equation (18), calculate the consistency index $CI(K^{(0)}) \approx 0.2534$.

Step 5: Since $CI(K^{(0)}) > CI_0$*, then go to the next step.*

Step 6: Since $l_{13}^{(0)} = \max\left\{|l_{ij}^{(0)} - \frac{1}{2}| + |l_{ji}^{(0)} - \frac{1}{2}|\right\}$ *and* $l_{13}^{(0)} + l_{31}^{(0)} - 1 > 0$*, then the DMs decrease their preference. The modified HLPR is* $K^{(1)} = \begin{bmatrix} < lv_0, \{1\} > & < lv_3, \{0.6, 0.7\} > & < lv_{-2}, \{0.1, 0.2\} > \\ < lv_{-3}, \{0.6, 0.7\} > & < lv_0, \{1\} > & < lv_1, \{0.2, 0.3\} > \\ < lv_2, \{0.1, 0.2\} > & < lv_{-1}, \{0.2, 0.3\} > & < lv_0, \{1\} > \end{bmatrix}$ *and* $CI(K^{(1)}) \approx 0.2230 < CI_0$.

Step 7: Let $K_a = K^{(1)}$*, Output* $K_a = \begin{bmatrix} < lv_0, \{1\} > & < lv_3, \{0.6, 0.7\} > & < lv_{-2}, \{0.1, 0.2\} > \\ < lv_{-3}, \{0.6, 0.7\} > & < lv_0, \{1\} > & < lv_1, \{0.2, 0.3\} > \\ < lv_2, \{0.1, 0.2\} > & < lv_{-1}, \{0.2, 0.3\} > & < lv_0, \{1\} > \end{bmatrix}$ *and* $CI(K_a) \approx 0.2230$.

4.3. Likelihood-Based Ranking Method

As the likelihood between two HFLNs is a useful tool to make comparisons, a likelihood-based method is introduced to derive a ranking from the consistent HLPRs in this section.

Pondering over the decision making problem within the hesitant fuzzy linguistic context, assume that the DMs' plan to select the optimal alternative or get a ranking order from n objects. Let $X = \{x_1, x_2, \ldots, x_n\}$ be a discrete set of alternatives being chosen and $K = (k_{ij})_{n \times n}$ $(i, j = 1, 2, \ldots, n)$ is the preference matrix, where k_{ij} is the preference value in the form of HFLNs. The entire procedures of earning the ideal order of alternatives are shown in Algorithm 2.

Algorithm 2. Likelihood-based ranking method

Input: The initial HLPR $K = (k_{ij})_{n \times n}$.
Output: The optimal alternative x^*.
Step 1: Obtain the acceptable HLPR K_a by Algorithm 1.
Step 2: Utilize the HFLA operator based on Equation (8) to aggregate each row of the HLPR K_a, then determine the overall preference degree p_i of each alternative x_i $(i = 1, 2, \ldots, n)$.
Step 3: According to Equation (10), calculate the likelihood $l_{ij} = L(p_i \geq p_j)$ between p_i and p_j $(i = 1, 2, \ldots, n,$ $j = 1, 2, \ldots, n)$, then construct a likelihood matrix $L = (l_{ij})_{n \times n}$.
Step 4: Calculate the dominance degree $\varphi(x_i) = \frac{1}{n}\sum_{j=1}^{n} l_{ij}$ of alternative $x_i (i = 1, 2, \ldots, n)$, where $\varphi(x_i)$ represents the degree of x_i preferred to other alternatives. Obviously, the greater the value of $\varphi(x_i)$, the better the alternative x_i.
Step 5: Rank all the alternatives on the basis of the dominance degree $\varphi(x_i)$ of each alternative $x_i (i = 1, 2, \ldots, n)$. Then obtain the ranking results and the optimal alternative(s) is denoted as x^*.

5. Selection of Mine Ventilation Systems

In this section, an example of mine ventilation systems selection is afforded for voicing the application of the suggested method.

Sanshandao gold mine is the first subsea hard rock mine in China, which lies in Sanshandao Town, Laizhou City, Shandong Province, China [53]. As the mine is going into the stage of deep exploitation, the distance of ventilation becomes longer and the temperature also rises severely. Therefore, some problems are beginning to appear after using the traditional ventilation systems, for instance, the temperature is so high that laborers find it hard to work efficiently; exhaust gas emitted by diesel equipment pollutes underground air seriously; and the concentration of dust exceeds the national standard. Accordingly, a better ventilation system needs to be adopted.

After a thorough survey, four ventilation systems, i.e., $\{vs_1, vs_2, vs_3, vs_4\}$, are under consideration, and a group of professionals are invited to select the optimal ventilation system. The linguistic term set $lv = \{lv_{-4} = tremendously\ worse,\ \ lv_{-3} = a\ lot\ worse, lv_{-2} = worse, lv_{-1} = a\ little\ worse, v_0 = fair,$ $lv_1 = a\ little\ better, lv_2 = better, lv_3 = a\ lot\ better,\ lv_4 = tremendously\ better\}$ is used. The preference values are shown in the form of HFLNs. Suppose all DMs have a consensus on the selected linguistic term, and all teams provided their membership degrees (preference) in line with the researches of the above four systems and their preference simultaneously. Then, all of the probable membership degrees are gathered with the previous linguistic set. When a team does not give a membership degree, we consider it as 0.5. And when the same membership degrees about identical linguistic terms are given, we may regard them as different data in a HFLN.

Consequently, after a heated discussion, experts decided the threshold value $CI_0 = 0.18$ and the maximum number of iterative times $s_{\max} = 3$. Then, the preference information was given in Table 1.

Table 1. Original HLPR VS.

VS	vs_1	vs_2	vs_3	vs_4
vs_1	$< lv_0, \{1\} >$	$< lv_3, \{0.2, 0.3, 0.6\} >$	$< lv_1, \{0.4, 0.6, 0.8\} >$	$< lv_2, \{0.3, 0.4, 0.8\} >$
vs_2	$< lv_{-3}, \{0.2, 0.3, 0.6\} >$	$< lv_0, \{1\} >$	$< lv_{-2}, \{0.3, 0.4, 0.7\} >$	$< lv_3, \{0.2, 0.5, 0.6\} >$
vs_3	$< lv_{-1}, \{0.4, 0.6, 0.8\} >$	$< lv_2, \{0.3, 0.4, 0.7\} >$	$< lv_0, \{1\} >$	$< lv_{-1}, \{0.4, 0.5, 0.9\} >$
vs_4	$< lv_{-2}, \{0.3, 0.4, 0.8\} >$	$< lv_{-3}, \{0.2, 0.5, 0.6\} >$	$< lv_1, \{0.4, 0.5, 0.9\} >$	$< lv_0, \{1\} >$

5.1. Illustrative Example

Steps outlined in Section 4.3 are completed to get satisfied ventilation system(s) in this section.

Step 1: Obtain the acceptable HLPR VS_a by Algorithm 1.

Based on Equation (14), the consistent HLPR VS^* is shown in Table 2. And the likelihood matrix $L^{(0)}$ is calculated based on Equation (10), as shown in Table 3. Then, calculate the consistency index $CI(VS^{(0)}) \approx 0.1956 > 0.18$ by Equation (18). Since $l_{34}^{(0)} = \max\left\{|l_{ij}^{(0)} - \frac{1}{2}| + |l_{ji}^{(s)} - \frac{1}{2}|\right\}$ and $l_{34}^{(0)} + l_{43}^{(0)} - 1 > 0$, then the DMs decrease their preference, and the modified HLPR $VS^{(1)}$ is in Table 4. Since $CI(VS^{(1)}) \approx 0.1754 < 0.18$, let $VS_a = VS^{(1)}$.

Table 2. Consistent HLPR VS^*.

VS^*	vs_1	vs_2	vs_3	vs_4
vs_1	$< lv_0, \{1\} >$	$< lv_{2.2}, \left\{\frac{1}{4}, \frac{25}{88}, \frac{7}{11}, \frac{7}{11}\right\} >$	$< lv_{2.5}, \left\{\frac{9}{50}, \frac{13}{50}, \frac{9}{20}\right\} >$	$< lv_{3.6}, \left\{\frac{1}{6}, \frac{41}{159}, \frac{67}{159}\right\} >$
vs_2	$< lv_{-2.2}, \left\{\frac{1}{4}, \frac{25}{88}, \frac{7}{11}\right\} >$	$< lv_0, \{1\} >$	$< lv_{-1.4}, \left\{\frac{3}{14}, \frac{9}{56}, \frac{29}{56}\right\} >$	$< lv_{1.8}, \left\{\frac{1}{36}, \frac{2}{9}, \frac{11}{72}\right\} >$
vs_3	$< lv_{-2.5}, \left\{\frac{9}{50}, \frac{13}{50}, \frac{9}{20}\right\} >$	$< lv_{1.4}, \left\{\frac{3}{14}, \frac{9}{56}, \frac{29}{56}\right\} >$	$< lv_0, \{1\} >$	$< lv_{3.2}, \left\{\frac{3}{64}, \frac{15}{128}, \frac{11}{64}\right\} >$
vs_4	$< lv_{-3.6}, \left\{\frac{1}{6}, \frac{41}{159}, \frac{67}{159}\right\} >$	$< lv_{-1.8}, \left\{\frac{1}{36}, \frac{2}{9}, \frac{11}{72}\right\} >$	$< lv_{-3.2}, \left\{\frac{3}{64}, \frac{15}{128}, \frac{11}{64}\right\} >$	$< lv_0, \{1\} >$

Table 3. Likelihood matrix $L^{(0)}$.

$L^{(0)}$	vs_1	vs_2	vs_3	vs_4
vs_1	0.5000	0.5076	0.6078	0.5693
vs_2	0.3656	0.5000	0.5642	0.7844
vs_3	0.8069	0.6378	0.5000	0.6862
vs_4	0.9287	0.6195	1.0000	0.5000

Table 4. Modified HLPR $VS^{(1)}$.

$VS^{(1)}$	vs_1	vs_2	vs_3	vs_4
vs_1	$< lv_0, \{1\} >$	$< lv_3, \{0.2, 0.3, 0.6\} >$	$< lv_1, \{0.4, 0.6, 0.8\} >$	$< lv_2, \{0.3, 0.4, 0.8\} >$
vs_2	$< lv_{-3}, \{0.2, 0.3, 0.6\} >$	$< lv_0, \{1\} >$	$< lv_{-2}, \{0.3, 0.4, 0.7\} >$	$< lv_3, \{0.2, 0.5, 0.6\} >$
vs_3	$< lv_{-1}, \{0.4, 0.6, 0.8\} >$	$< lv_2, \{0.3, 0.4, 0.7\} >$	$< lv_0, \{1\} >$	$< lv_{-1}, \{0.1, 0.2, 0.3\} >$
vs_4	$< lv_{-2}, \{0.3, 0.4, 0.8\} >$	$< lv_{-3}, \{0.2, 0.5, 0.6\} >$	$< lv_1, \{0.1, 0.2, 0.3\} >$	$< lv_0, \{1\} >$

Step 2: Utilize the HFLA operator based on Equation (8) to aggregate each row of the HLPR VS_a, then the overall preference degree p_i of each alternative is acquired as follows:

$$p_1 = \begin{pmatrix} lv_{1.5}, 0.4182, 0.4455, 0.4500, 0.4636, 0.4773, 0.4909, 0.4955, 0.5091, , 0.5227, \\ 0.5364, 0.5409, 0.5455, 0.5545, 0.5682, 0.5727, 0.5864, 0.5909, 0.6000, \\ 0.6182, 0.6318, 0.6364, 0.6455, 0.6636, 0.6773, 0.6818, 0.7273, 0.7727 \end{pmatrix},$$

$$p_2 = \begin{pmatrix} lv_{-0.5}, 0.4429, 0.4500, 0.4571, 0.4643, 0.4714, 0.4857, 0.5000, 0.5071, 0.5286, \\ 0.5929, 0.6000, 0.6071, 0.6143, 0.6214, 0.6357, 0.6429, 0.6500, 0.6500, \\ 0.6571, 0.6571, 0.6643, 0.6714, 0.6786, 0.6857, 0.7000, 0.7071, 0.7286 \end{pmatrix},$$

$$p_3 = \begin{pmatrix} lv_0, 0.4563, 0.4750, 0.4938, 0.4938, 0.4938, 0.5125, 0.5125, 0.5313, 0.5313, \\ 0.5313, 0.5313, 0.5500, 0.5500, 0.5688, 0.5688, 0.5688, 0.5875, 0.6063, \\ 0.6063, 0.6250, 0.6438, 0.6438, 0.6625, 0.6813, 0.6813, 0.7000, 0.7188 \end{pmatrix},$$

and

$$p_4 = \begin{pmatrix} lv_{-1}, & 0.4417, 0.4583, 0.4667, 0.4750, 0.4833, 0.4833, 0.4917, 0.5000, 0.5083, \\ & 0.5167, 0.5250, 0.5250, 0.5250, 0.5333, 0.5417, 0.5500, 0.5500, 0.5583, \\ & 0.5583, 0.5667, 0.5667, 0.5750, 0.5917, 0.6000, 0.6083, 0.6333, 0.6417 \end{pmatrix}.$$

Step 3: According to Equation (10), calculate the likelihood between p_i and p_j ($i = 1, 2, 3, 4$, $j = 1, 2, 3, 4$), then the likelihood matrix $L = (l_{ij})_{4\times4}$ is constructed in Table 5.

Table 5. Likelihood matrix L.

L	p_1	p_2	p_3	p_4
p_1	0.5000	0.6001	0.5756	0.6586
p_2	0.3999	0.5000	0.4745	0.5620
p_3	0.4244	0.5255	0.5000	0.5872
p_4	0.3414	0.4380	0.4128	0.5000

Step 4: Calculate the dominance degree of each alternative with $\varphi(vs_i) = \frac{1}{4}\sum_{j=1}^{4} l_{ij} (i = 1, 2, 3, 4)$ as: $\varphi(vs_1) \approx 0.5836$, $\varphi(vs_2) \approx 0.4841$, $\varphi(vs_3) \approx 0.5093$, $\varphi(vs_4) \approx 0.4231$.

Step 5: Since $\varphi(vs_1) > \varphi(vs_3) > \varphi(vs_2) > \varphi(vs_4)$, then the ranking is $vs_1 \succ vs_3 \succ vs_2 \succ vs_4$ and the optimal system is $vs^* = vs_1$.

5.2. Comparative Analysis

Since the HLPR presented in this paper is a new type of preference relation, no related researches have been conducted so far. To testify the validity and advantages of the proposed method, several methods for HFLPRs [23–28] can be made for comparisons.

Note: The definitions of HLPRs and HFLPRs are not the same. The basic elements in HLPRs are HFLNs, whereas those in HFLPRs are HFLTS. As a result, each HFLN in the HLPR should be transformed into the corresponding HFLTS by using the linguistic term multiplies the corresponding membership degrees successively. For example, $< lv_2, \{0.3, 0.5\} >$ can be converted into $\{lv_{0.6}, lv_1\}$. Then, a same illustration is applied in these methods and detailed comparisons are provided in Table 6.

Table 6. Comparisons with different methods.

Methods	Consistency Checking	Consistency Improving	Ranking Approaches	Ranking Results
Zhang and Wu [23]	Distance measure	Iterative algorithm	Score functions	$vs_1 \succ vs_3 \succ vs_4 \succ vs_2$
Wang and Xu [24]	Graph theory	Not given	Not given	Unavailable
Wu and Xu [25]	Distance measure	Feedback mechanism	Score functions	Uncertain
Gou et al. [26]	Compatibility measure	Not given	Complementary matrix	$vs_1 \succ vs_2 \succ vs_3 \succ vs_4$
Li et al. [27]	Linear programing model	Not given	Not given	Unavailable
Xu et al. [28]	Distance measure	Iterative algorithm	Score functions	$vs_1 \succ vs_3 \succ vs_2 \succ vs_4$
The proposed method	likelihood	Feedback mechanism	Likelihood matrix	$vs_1 \succ vs_3 \succ vs_2 \succ vs_4$

(1) Comparison with literature [24,25,27]

In literature [24], Wang and Xu provided a visible interpretation of additive consistency and weak consistency of extended HFLPRs based on graph theory. In literature [27], Li et al. defined an interval consistency index of HFLPRs based on the linear programming model. However, the methods of improving consistency degrees and getting ranking orders are not mentioned in literature [24,27]. Thus, the rankings are unavailable in these cases. In literature [25], Wu and Xu discussed some issues of HFLPRs on consistency and consensus, and defined a consistency index based on distance measure. Nevertheless, dissimilar ranking results may occur with different adjust preference when the feedback mechanism was adopted to improve the consistency level in this literature [25]. Note: Feedback mechanisms are presented to improve the consistency level of preference relations in both literature [25] and this paper. Different from existing feedback approaches [25], people can directly adjust their preference with our method according to the values of elements in the likelihood matrix.

(2) Comparison with literature [23,26,28]

From Table 6, it is clear that the best alternative in different methods is always vs_1, which reveals the effectiveness of the proposed method. In literature [26], Gou et al. defined the consistency index on the basis of compatibility measure and then got the ranking result based on a complementary matrix; however, the approach of improving consistency level of HFLPRs was not given. Even though the rankings obtained in literature [28] and this paper are the same, there are still some differences between these two methods. First, in literature [23,28], a consistency index based on distance measure was defined to check consistency level of HFLPRs, while a likelihood-based index is suggested in this paper. Compared with compatibility or distance measure, the largest advantage of the likelihood is that not only the deviation degree, but also that the order relationship of two elements can be directly indicated. Second, compared with automatic iterative algorithms [23,28], the feedback mechanism proposed in this paper reduces the loss of original information, and DMs can understand their current status in each round. Besides, an approach of using aggregation operators and then calculating score functions was adopted to get the ranking order in both literature [23,28]. By contrast, a likelihood matrix is constructed in this paper to avoid the second calculations and information distortions.

The advantages of the proposed approach are summarized as follows:

(1) The HFLNs can closely depict the experts' preferences as the membership degrees of a certain linguistic value are given. And they can reserve the completeness of initial information in some extents, which is the guarantee for obtaining ideal results.
(2) Only one element which greatly affects the consistency needs to be adjusted by professionals. The revised alternatives may be diverse according to the reality. Specialists make a decision in the light of a recommended direction as they are acquainted with their current positions.
(3) The experts may change the linguistic scale function under different semantics on the basis of their preferences and reality. Then different ranking results may be achieved if another linguistic scale function is applied. The flexibility and practicability of the method can be reflected.

Overall, the proposed method brings up a new and useful way to resolve complex fuzzy decision making issues under a hesitant linguistic environment, especially when experts or decision makers (DMs) readily make comparisons among each pair of alternatives but hardly provide direct evaluation information.

6. Conclusions

Ventilation systems selection is an essential decision for a mining project. However, the influence characteristics of a ventilation system are complex and of strong fuzziness or uncertainty. Since preference relations play a significant role in the decision making process, HLPRs were proposed to deal with mine ventilation systems selection problems. HLPRs can be regarded as extensions of LPRs. They provide not only the priority intensity of alternatives, but also the possible membership degrees of this priority intensity. The likelihood-based index was defined to test the consistency of experts and the improving model was constructed to modify consistency level of HLPRs. Preference information in HFPRs is in the form of HFLNs. For the accuracy of HFLNs' computation, new operational laws and the comparison method were presented after reviewing the relevant literature. Furthermore, the decision making framework based on HLPRs was built to select a proper ventilation system for mines. Finally, an illustration and some comparisons with other methods were drawn to highlight the applicability and advantages of the developed approach. In the future, more engineering applications with this proposed method could be researched or more decision making methods can be developed to address complex decision making problems in mines.

Author Contributions: W.L., G.Z. and S.L. conceived and worked together to achieve this work, W.L. wrote the paper, and G.Z. made contribution to the case study.

Funding: This work was supported by National Natural Science Foundation of China (51774321).

Acknowledgments: We sincerely thank the anonymous reviewers for their helpful and constructive suggestions and the editors for their careful and patient work.

Conflicts of Interest: The authors declare that there is no conflict of interests regarding the publication of this paper.

References

1. Euler, D.S. Application of ventilation management programs for improved mine safety. *Int. J. Min. Sci. Technol.* **2017**, *27*, 647–650.
2. Geng, F.; Luo, G.; Zhou, F.B.; Zhao, P.T.; Ma, L.; Chai, H.L.; Zhang, T.T. Numerical investigation of dust dispersion in a coal roadway with hybrid ventilation system. *Powder Technol.* **2017**, *313*, 260–271. [CrossRef]
3. Xia, M.M.; Xu, Z.S.; Liao, H.C. Preference relations based on intuitionistic multiplicative information. *IEEE Trans. Fuzzy Syst.* **2013**, *21*, 113–133.
4. Luo, S.Z.; Cheng, P.F.; Wang, J.Q.; Huang, Y.J. Selecting project delivery systems based on simplified neutrosophic linguistic preference relations. *Symmetry* **2017**, *9*, 151. [CrossRef]
5. Zadeh, L.A. The concept of a linguistic variable and its application to approximate reasoning. *Inf. Sci.* **1974**, *8*, 199–249. [CrossRef]
6. Liang, W.Z.; Zhao, G.Y.; Hong, C.S. Selecting the optimal mining method with extended multi-objective optimization by ratio analysis plus the full multiplicative form (MULTIMOORA) approach. *Neural Comput. Appl.* **2018**, 1–16. [CrossRef]
7. Cabrerizo, F.J.; Morente-Molinera, J.A.; Pedrycz, W.; Taghavi, A.; Herrera-Viedma, E. Granulating linguistic information in decision making under consensus and consistency. *Expert Syst. Appl.* **2018**, *99*, 83–92. [CrossRef]
8. Wang, H.; Xu, Z.S. Interactive algorithms for improving incomplete linguistic preference relations based on consistency measures. *Appl. Soft Comput.* **2016**, *42*, 66–79. [CrossRef]
9. Massanet, S.; Riera, J.V.; Torrens, J.; Herrera-Viedma, E. A model based on subjective linguistic preference relations for group decision making problems. *Inf. Sci.* **2016**, *355*, 249–264. [CrossRef]
10. Xu, Y.J.; Wang, H. A group consensus decision support model for hesitant 2-tuple fuzzy linguistic preference relations with additive consistency. *J. Intell. Fuzzy Syst.* **2017**, *33*, 41–54. [CrossRef]

11. Rodriguez, R.M.; Martinez, L.; Herrera, F. Hesitant fuzzy linguistic term sets for decision making. *IEEE Trans. Fuzzy Syst.* **2012**, *20*, 109–119. [CrossRef]
12. Zhang, Z.M.; Wu, C. Hesitant fuzzy linguistic aggregation operators and their applications to multiple attribute group decision making. *J. Intell. Fuzzy Syst.* **2014**, *26*, 2185–2202.
13. Lee, L.W.; Chen, S.M. Fuzzy decision making based on likelihood-based comparison relations of hesitant fuzzy linguistic term sets and hesitant fuzzy linguistic operators. *Inf. Sci.* **2015**, *294*, 513–529. [CrossRef]
14. Liu, P.D.; Teng, F. Some Interval neutrosophic hesitant uncertain linguistic Bonferroni aggregation operators for multiple attribute decision-making. *Int. J. Uncertain. Quantif.* **2017**, *7*, 525–572. [CrossRef]
15. Liao, H.C.; Xu, Z.S.; Zeng, X.J. Distance and similarity measures for hesitant fuzzy linguistic term sets and their application in multi-criteria decision making. *Inf. Sci.* **2014**, *271*, 125–142. [CrossRef]
16. Lee, L.W.; Chen, S.M. Fuzzy decision making and fuzzy group decision making based on likelihood-based comparison relations of hesitant fuzzy linguistic term sets 1. *J. Intell. Fuzzy Syst.* **2015**, *29*, 1119–1137. [CrossRef]
17. Chen, S.M.; Hong, J.A. Multicriteria linguistic decision making based on hesitant fuzzy linguistic term sets and the aggregation of fuzzy sets. *Inf. Sci.* **2014**, *286*, 63–74. [CrossRef]
18. Adar, T.; Delice, E.K. Evaluating mental work load using multi-criteria hesitant fuzzy linguistic term set (HFLTS). *Int. J. Fuzzy Syst.* **2017**, *8*, 90–101.
19. Joshi, D.K.; Kumar, S. Trapezium cloud TOPSIS method with interval-valued intuitionistic hesitant fuzzy linguistic information. *Granul. Comput.* **2018**, *3*, 139–152. [CrossRef]
20. Liu, D.H.; Chen, X.H.; Peng, D. Distance measures for hesitant fuzzy linguistic sets and their applications in multiple criteria decision making. *Int. J. Fuzzy Syst.* **2018**. [CrossRef]
21. Adem, A.; Çolak, A.; Dağdeviren, M. An integrated model using SWOT analysis and hesitant fuzzy linguistic term set for evaluation occupational safety risks in life cycle of wind turbine. *Saf. Sci.* **2018**, *106*, 184–190. [CrossRef]
22. Zhu, B.; Xu, Z.S. Consistency measures for hesitant fuzzy linguistic preference relations. *IEEE Trans. Fuzzy Syst.* **2014**, *22*, 35–45. [CrossRef]
23. Zhang, Z.M.; Wu, C. On the use of multiplicative consistency in hesitant fuzzy linguistic preference relations. *Knowl.-Based Syst.* **2014**, *72*, 13–27. [CrossRef]
24. Wang, H.; Xu, Z.S. Some consistency measures of extended hesitant fuzzy linguistic preference relations. *Inf. Sci.* **2015**, *297*, 316–331. [CrossRef]
25. Wu, Z.B.; Xu, J.P. Managing consistency and consensus in group decision making with hesitant fuzzy linguistic preference relations. *Omega* **2016**, *65*, 28–40. [CrossRef]
26. Gou, X.J.; Xu, Z.S.; Liao, H.C. Group decision making with compatibility measures of hesitant fuzzy linguistic preference relations. *Soft Comput.* **2017**, 1–17. [CrossRef]
27. Li, C.C.; Rodríguez, R.M.; Martínez, L.; Dong, Y.C.; Herrera, F. Consistency of hesitant fuzzy linguistic preference relations: An interval consistency index. *Inf. Sci.* **2018**, *432*, 347–361. [CrossRef]
28. Xu, Y.J.; Wen, X.; Sun, H.; Wang, H.M. Consistency and consensus models with local adjustment strategy for hesitant fuzzy linguistic preference relations. *Int. J. Fuzzy Syst.* **2018**, 1–18. [CrossRef]
29. Wang, J.Q.; Wu, J.T.; Wang, J.; Zhang, H.Y.; Chen, X.H. Multi-criteria decision-making methods based on the Hausdorff distance of hesitant fuzzy linguistic numbers. *Soft Comput.* **2016**, *20*, 1621–1633. [CrossRef]
30. Wang, J.Q.; Wu, J.T.; Wang, J.; Zhang, H.Y.; Chen, X.H. Interval-valued hesitant fuzzy linguistic sets and their applications in multi-criteria decision-making problems. *Inf. Sci.* **2014**, *288*, 55–72. [CrossRef]
31. Lin, R.; Zhao, X.F.; Wei, G.W. Models for selecting an ERP system with hesitant fuzzy linguistic information. *J. Intell. Fuzzy Syst.* **2014**, *26*, 2155–2165.
32. Torra, V. Hesitant fuzzy sets. *Int. J. Intell. Syst.* **2010**, *25*, 529–539. [CrossRef]
33. Xu, Z.S. Deviation measures of linguistic preference relations in group decision making. *Omega* **2005**, *33*, 249–254. [CrossRef]
34. Liang, W.Z.; Zhao, G.Y.; Wu, H. Evaluating investment risks of metallic mines using an extended TOPSIS method with linguistic neutrosophic numbers. *Symmetry* **2017**, *9*, 149. [CrossRef]
35. Zadeh, L.A. Fuzzy sets. *Inf. Control* **1965**, *8*, 338–353. [CrossRef]
36. Wang, H.Y.; Chen, S.M. Evaluating students' answerscripts using fuzzy numbers associated with degrees of confidence. *IEEE Trans. Fuzzy Syst.* **2008**, *16*, 403–415. [CrossRef]

37. Chen, S.M.; Wang, N.Y.; Pan, J.S. Forecasting enrollments using automatic clustering techniques and fuzzy logical relationships. *Expert Syst. Appl.* **2009**, *36*, 11070–11076. [CrossRef]
38. Chen, S.M.; Chen, C.D. Handling forecasting problems based on high-order fuzzy logical relationships. *Expert Syst. Appl.* **2011**, *38*, 3857–3864. [CrossRef]
39. Chen, S.M.; Tanuwijaya, K. Fuzzy forecasting based on high-order fuzzy logical relationships and automatic clustering techniques. *Expert Syst. Appl.* **2011**, *38*, 15425–15437. [CrossRef]
40. Chen, S.M.; Munif, A.; Chen, G.S.; Liu, H.C.; Kuo, B.C. Fuzzy risk analysis based on ranking generalized fuzzy numbers with different left heights and right heights. *Expert Syst. Appl.* **2012**, *39*, 6320–6334. [CrossRef]
41. Liang, W.Z.; Zhao, G.Y.; Hong, C.S. Performance assessment of circular economy for phosphorus chemical firms based on VIKOR-QUALIFLEX method. *J. Clean. Prod.* **2018**, *196*, 1365–1378. [CrossRef]
42. Liang, W.Z.; Zhao, G.Y.; Wu, H.; Chen, Y. Assessing the risk degree of goafs by employing hybrid TODIM method under uncertainty. *Bull. Eng. Geol. Environ.* **2018**. [CrossRef]
43. Rodríguez, R.M.; Bedregal, B.; Bustince, H.; Dong, Y.C.; Farhadinia, B.; Kahraman, C.; Martinez, L.; Torra, V.; Xu, Y.J.; Xu, Z.S.; et al. A position and perspective analysis of hesitant fuzzy sets on information fusion in decision making. Towards high quality progress. *Inf. Fusion* **2016**, *29*, 89–97. [CrossRef]
44. Liu, N.; Meng, S.S. Approaches to the selection of cold chain logistics enterprises under hesitant fuzzy environment based on decision distance measures. *Granul. Comput.* **2018**, *3*, 27–38. [CrossRef]
45. Alcantud, J.C.R.; Torra, V. Decomposition theorems and extension principles for hesitant fuzzy sets. *Inf. Fusion* **2018**, *41*, 48–56. [CrossRef]
46. Xia, M.M.; Xu, Z.S. Hesitant fuzzy information aggregation in decision making. *Int. J. Approx. Reason.* **2011**, *52*, 395–407. [CrossRef]
47. Zhu, B. Studies on consistency measure of hesitant fuzzy preference relations. *Proc. Comput. Sci.* **2013**, *17*, 457–464. [CrossRef]
48. Tian, Z.P.; Wang, J.; Wang, J.Q.; Zhang, H.Y. A likelihood-based qualitative flexible approach with hesitant fuzzy linguistic information. *Cogn. Comput.* **2016**, *8*, 670–683. [CrossRef]
49. Zhang, Z.M. Multi-criteria decision-making using interval-valued hesitant fuzzy QUALIFLEX methods based on a likelihood-based comparison approach. *Neural Comput. Appl.* **2017**, *28*, 1835–1854. [CrossRef]
50. Yang, Y.; Hu, J.H.; An, Q.X.; Chen, X.H. Group decision making with multiplicative triangular hesitant fuzzy preference relations and cooperative games method. *Int. J. Uncertain. Quantif.* **2017**, *7*, 271–284. [CrossRef]
51. Dong, Y.C.; Li, C.C.; Chiclana, F.; Herrera-Viedma, E. Average-case consistency measurement and analysis of interval-valued reciprocal preference relations. *Knowl.-Based Syst.* **2016**, *114*, 108–117. [CrossRef]
52. Liu, F.; Pedrycz, W.; Wang, Z.X.; Zhang, W.G. An axiomatic approach to approximation-consistency of triangular fuzzy reciprocal preference relations. *Fuzzy Set. Syst.* **2017**, *322*, 1–18. [CrossRef]
53. Li, X.B.; Li, D.Y.; Liu, Z.X.; Zhao, G.Y.; Wang, W.H. Determination of the minimum thickness of crown pillar for safe exploitation of a subsea gold mine based on numerical modelling. *Int. J. Rock Mech. Min.* **2013**, *57*, 42–56. [CrossRef]

symmetry

Article

Intuitionistic Fuzzy Multiple Attribute Decision-Making Model Based on Weighted Induced Distance Measure and Its Application to Investment Selection

Zhengmao Li [1], **Dechao Sun** [2,*] **and Shouzhen Zeng** [3,*]

1 School of Finance, Zhejiang Gongshang University, Hangzhou 310018, China; lzm@ctsec.com
2 College of Computer and Information, Zhejiang Wanli University, Ningbo 315100, China
3 School of Business, Ningbo University, Ningbo 315211, China
* Correspondence: sundechao123@163.com (D.S.); zengshouzhen@nbu.edu.cn (S.Z.); Tel.: +86-574-86300396 (S.Z.)

Received: 11 June 2018; Accepted: 2 July 2018; Published: 4 July 2018

Abstract: This paper investigates an intuitionistic fuzzy multiple attribute decision-making method based on weighted induced distance and its application to investment selection. Specifically, an intuitionistic fuzzy weighted induced ordered weighted averaging operator is proposed to eliminate the drawbacks of existing methods by extending the functions of the order-induced variables. The main advantage of the proposed operator is its dual roles of the order-inducing variables that can simultaneously induce arguments and moderate associated weights. A further extension of the proposed operator is its adaptation towards measuring intuitionistic fuzzy information more effectively. In addition, a multiple attribute decision-making model based on the proposed distance operators is proposed. Finally, the practicability and validity of the proposed model are illustrated by using a numerical example related to investment selection.

Keywords: intuitionistic fuzzy set; multiple attribute decision-making; weighted induced distance; investment selection

---

1. Introduction

The multiple attribute decision-making (MADM) technique is a widely used method in solving real-world problems, in which a variety of attributes are involved to consider from finite feasible alternatives according to the evaluated attributes' evaluation or preference information given by multiple decision-makers. Clearly, fuzziness and vagueness are inevitably integrated into the MADM process due to the vagueness and uncertainty of evaluated objects and the ambiguous nature of human thinking. Intuitionistic fuzzy set (IFS), initially developed by Atanassov [1], has proven to be a powerful and useful tool for processing complex-type information in day-to-day life. The IFS is described by a membership degree ($0 \leq \mu \leq 1$) and a non-membership degree ($0 \leq v \leq 1$) that satisfies the condition $\mu^2 + v^2 \leq 1$. To date, various MADM methods related to IFS have appeared in well-known publications and conferences. Several authors have conducted valuable scientific investigations and literature reviews on the development of IFS from different viewpoints [2–4].

As one of the important aspects of fuzzy theory, the distance measured between IFSs has received continuous attention for decades in both the theory and application areas. Existing IFS distance measures are mostly investigated from the weighted averaging perspective [5–8]. Recently, Zeng and Su [9] proposed a new intuitionistic fuzzy distance measure from the ordered weighted viewpoint, namely the intuitionistic fuzzy ordered weighted distance (IFOWD) operator, whose prominent feature is that it can incorporate a decision-maker's attitudinal characters into the MADM process.

Later, by combining the weighted average and IFOWD methods, Zeng and Xiao [10] developed the intuitionistic ordered weighted averaging-weighted average distance (IFOWAWAD) operator and explored its usefulness in solving MADM problems. More recently, motivated by the induced ordered weighted averaging distance (IOWAD) measure [11], Zeng et al. [12] proposed the intuitionistic fuzzy induced ordered weighted averaging distance (IFIOWAD) measure that enables us to consider the complex attitude of decision-makers using order-induced variables. The essence of the IFIOWAD as well as the IOWAD operator is to enable the decision-makers to incorporate their complex attitude into the aggregation process, using the order-induced variables on the ordered arguments. Thus, the interests of the decision-makers are taken into account during the decision-making process. Although it is a relatively new MADM approach, the induced aggregation distance operator has been successfully applied in various fields of research. Recent literature contains a number of extensions and subsequent applications in MADM problems, as listed in Table 1.

Table 1. Induced aggregation distance methodology in multiple attribute decision-making (MADM) problems.

Author, Year	Induced Aggregation Distance Methodology
Merigó and Casanovas, 2011 [13], Casanovas et al., 2016 [14]	Induced Minkowski ordered weighted averaging distance (IMOWAD)
Merigó and Casanovas, 2011 [15]	Induced Euclidean ordered weighted averaging distance (IEOWAD)
Zeng et al., 2013 [16]	Uncertain IOWAD (UIOWAD) operator, Fuzzy IOWAD (FIOWAD)
Su et al., 2013 [17]	Uncertain IEOWAD (UIEOWAD)
Zeng et al., 2014 [18]	Induced heavy ordered weighted averaging distance (IHOWAD)
Li et al., 2014 [19]	2-tuple linguistic IOWAD (2TLIOWAD) operator
Xian and Sun, 2014 [20]	Fuzzy linguistic IOWAD (FLIOWAD)
Su et al., 2015 [21]	Uncertain IHOWAD (UIHOWAD)
Zeng et al., 2017 [12]	Intuitionistic fuzzy IOWA weighted averaging distance (IFIOWAWAD)
Xian et al., 2016 [22]	Fuzzy linguistic IMOWAD (FLIMOWAD)
Xian et al., 2017 [23]	Novel intuitionistic fuzzy IEOWAD distance (NIFIEOWAD)

It is clearly shown in previous reviews that the existing induced aggregation distance methods, such as the IFIOWAD operator, are popular techniques that have been applied successfully in many real-world problems. However, one can observe that the above-mentioned induced aggregated distances share a similar problem that must be solved: their order-inducing variables are not involved in the actual aggregation of results. As a consequence of this, the results obtained by these distance operators cannot account for the variation derived from a change of the order-inducing variables. The latter issue is especially important whenever variation degrees of property regarding alternative-attribution pairs, such as confidence, consistency or importance, are represented in terms of order-inducing variables and need to be considered. To circumvent this defect, this paper develops a revised induced aggregated distance measure between IFSs, termed as an intuitionistic fuzzy weighted induced ordered weighted averaging distance (IFWIOWAD) operator that takes into account the intrinsic variations in the order-inducing variables during the aggregation process. Further, to enrich the theory and application of the developed IFWIOWAD operator, we propose an intuitionistic weighted induced ordered weighted averaging weighted average distance (IFWIOWAWAD) operator that can integrate the weighted average approach with the IFWIOWAD measure. Therefore, it can address the complex attitude of experts and the importance of attributes in the decision-making framework.

The rest of this paper is structured as follows. In Section 2, some definitions of the IFS and induced aggregation distance operators are reviewed. Section 3 presents the IFWIOWAD operator and explores its main properties. Section 4 develops the IFWIOWAWAD operator, based on which a MADM model is represented in Section 5. An example concerning investment selection is presented in Section 6. In the final section, we summarize the paper's main results.

2. Preliminaries

This section reviews several basic concepts concerning the IFS and the induced aggregated distance methods.

Definition 1. *An IFS P in a set* $Z = \{z_1, z_2, \ldots, z_n\}$ *is defined as in (1) [1]:*

$$P = \{\langle z, (\mu_P(z), v_P(z))\rangle | z \in Z\} \tag{1}$$

where the function $0 \leq \mu_P(z) \leq 1$ *and* $0 \leq v_P(z) \leq 1$ *denote as the degree of membership and the non-membership, respectively, and satisfy* $0 \leq \mu_P(z) + v_P(z) \leq 1$. *For convenient calculation, the pair* $\alpha = (\mu_\alpha, v_\alpha)$ *is signed as an intuitionistic fuzzy number (IFN) [24], where* $\mu_\alpha, v_\alpha \in [0,1]$ *and* $\mu_\alpha + v_\alpha \leq 1$.

Definition 2. *The intuitionistic fuzzy distance (IFD) between IFNs* α_1 *and* α_2 *is given by the following formula [9]:*

$$d_{IFD}(\alpha_1, \alpha_2) = |\alpha_1 - \alpha_2| = \frac{1}{2}(|\mu_{\alpha_1} - \mu_{\alpha_2}| + |v_{\alpha_1} - v_{\alpha_2}|) \tag{2}$$

As one of the most widely used and effective extensions of the ordered weighted averaging (OWA) methods [25], the IOWA operator [26] aggregates information by its reordering rule, performed with the order-inducing variables to accommodate a more complicated attitude of decision-makers.

Definition 3. *An IOWA is defined as follows:*

$$\text{IOWA}(\langle u_1, a_1\rangle, \ldots, \langle u_n, a_n\rangle) = \sum_{j=1}^{n} w_j b_j \tag{3}$$

where $W = (w_1, w_2, \ldots, w_n)^T$ *is the weight vector satisfying* $w_1 + \ldots + w_n = 1$ *and* $w_j \in [0,1]$. b_j *is the reordered value of* a_i *in the argument* $\langle u_i, a_i\rangle$ *having the jth largest order-inducing variable* u_i.

Based on the work of IOWAD proposed by Merigó and Casanovas [11], Zeng and Su [12] introduced the IFIOWAD operator by combining the advantages of the induced aggregation and the IFD. For IFSs $A = (\alpha_1, \ldots, \alpha_n)$ and $B = (\beta_1, \ldots, \beta_n)$, it is formulated as follows:

Definition 4. *An IFIOWAD operator is defined by a weight vector W with* $0 \leq w_j \leq 1$ *and* $w_1 + \ldots + w_n = 1$; *and an order-inducing vector* $U = (u_1, \ldots, u_n)$, *such that:*

$$\text{IFIOWAD}(\langle u_1, \alpha_1, \beta_1\rangle, \ldots, \langle u_n, \alpha_n, \beta_n\rangle) = \sum_{j=1}^{n} w_j d_{IFD}(\alpha_{\sigma(j)}, \beta_{\sigma(j)}) \tag{4}$$

where $d_{IFD}(\alpha_{\sigma(j)}, \beta_{\sigma(j)})$ *is the reordering of* $d_{IFD}(\alpha_j, \beta_j)$ *induced by the decreasing the order of* u_j, *and* $d_{IFD}(\alpha_j, \beta_j)$ *is the IF distance between IFNs* α_j *and* β_j.

Although the IFIOWAD operator is considered a useful and powerful measure tool, its inherent defect often leads to the loss of information and biased results that can be observed from the following example.

Example 1. *Let* $A = \{(0.3, 0.5), (0.5, 0.2), (0.7, 0.1), (0.4, 0.5)\}$ *and* $B = \{(0.4, 0.6), (0.7, 0.4), (0.2, 0.7), (0.6, 0.2)\}$ *be two sections of IFNs, and let the order-inducing variables be* $U = (7, 8, 3, 5)$. *The main steps for the aggregation of the above arguments based on the IFIOWAD operator are shown as follows:*

1. Calculate the distances $d_{IFD}(\alpha_j, \beta_j)$ $(j = 1, 2, 3, 4)$ using Equation (2):

$$d_{IFD}(\alpha_1, \beta_1) = \frac{1}{2}(|0.3 - 0.4| + |0.5 - 0.6|) = 0.1,$$

Similarly, we have

$$d_{IFD}(\alpha_2, \beta_2) = 0.2, d_{IFD}(\alpha_3, \beta_3) = 0.55, d_{IFD}(\alpha_4, \beta_4) = 0.25.$$

2. Reordering the $d_{IFD}(\alpha_j, \beta_j)$ $(j = 1, 2, 3, 4)$ according to the decreasing values of the variable u_j yields:

$$d_{IFD}(\alpha_{\sigma(1)}, \beta_{\sigma(1)}) = d_{IFD}(\alpha_2, \beta_2) = 0.2, d_{IFD}(\alpha_{\sigma(2)}, \beta_{\sigma(2)}) = d_{IFD}(\alpha_1, \beta_1) = 0.1,$$

$$d_{IFD}(\alpha_{\sigma(3)}, \beta_{\sigma(3)}) = d_{IFD}(\alpha_4, \beta_4) = 0.25, d_{IFD}(\alpha_{\sigma(4)}, \beta_{\sigma(4)}) = d_{IFD}(\alpha_3, \beta_3) = 0.55.$$

3. Let the associated weighting vector be $W = (0.3, 0.4, 0.1, 0.2)^T$, then the aggregation result yields:

$$\text{IFIOWAD}(U, A, B) = 0.3 \times 0.2 + 0.4 \times 0.1 + 0.1 \times 0.25 + 0.2 \times 0.55 = 0.225.$$

If we adjust the values of the order-inducing variables to $U' = (8, 10, 1, 6)$, then the aggregation result would be:

$$\text{IFIOWAD}(U', A, B) = 0.3 \times 0.2 + 0.4 \times 0.1 + 0.1 \times 0.25 + 0.2 \times 0.55 = 0.225.$$

One can be observed that we get the same aggregated results for different values of the order-inducing variables. The reason is that the order-inducing variables in the IFIOWAD operator only play the induced role and are not integrated into the actual aggregation results, thus corresponding aggregation results cannot embody the variation caused by a change of order-inducing variables. In the next section we will develop a new method to overcome this drawback.

3. The IFWIOWAD Operator

To solve the feedback problem of the existing IFIOWAD operator, we propose an improved aggregation method, named the intuitionistic fuzzy weighted IOWA distance (IFWIOWAD) operator. It can be formulated as follows:

Definition 5. *Let $A = (\alpha_1, \ldots, \alpha_n)$ and $B = (\beta_1, \ldots, \beta_n)$ be two sets of IFNs. An IFWIOWAD operator is defined by W with $0 \leq w_j \leq 1$ and $w_1 + \ldots + w_n = 1$, and an order-inducing vector $U = (u_1, \ldots, u_n)$, such that:*

$$\text{IFWIOWAD}(\langle u_1, \alpha_1, \beta_1\rangle, \ldots, \langle u_n, \alpha_n, \beta_n\rangle) = \sum_{j=1}^{n} \varpi_j d_{IFD}(\alpha_{\sigma(j)}, \beta_{\sigma(j)}) \tag{5}$$

where ϖ_j $(j = 1, \ldots, n)$ is a moderated weight that relatively depends on weight $w_j \in W$ and order-inducing variable $u_j \in U$, defined as:

$$\varpi_j = \frac{w_j u_{\sigma(j)}}{\sum_{j=1}^{n} w_j u_{\sigma(j)}} \tag{6}$$

where $(\sigma(1), \ldots, \sigma(n))$ is any possible permutation of $(1, \ldots, n)$, and clearly satisfies $u_{\sigma(j-1)} \geq u_{\sigma(j)}$ for $j > 1$. The distance $d_{IFD}(\alpha_{\sigma(j)}, \beta_{\sigma(j)})$ $(j = 1, \ldots, n)$ is the reordering of $d_{IFD}(\alpha_j, \beta_j)$ induced by $u_{\sigma(j)}$.

Example 2. *Assume the same collections of IFNS and order-inducing variables as defined in Example 1. Then the aggregation process by the IFWIOWAD is illustrated as follows:*

1. Record the order-inducing variables:

$$u_{\sigma(1)} = u_3 = 8, u_{\sigma(2)} = u_1 = 7, u_{\sigma(3)} = u_4 = 5, u_{\sigma(4)} = u_2 = 3.$$

2. Calculate the moderated weight ϖ_j using Equation (6):

$$\varpi_1 = \frac{w_1 u_{\sigma(1)}}{\sum_{j=1}^{4} w_j u_{\sigma(j)}} = \frac{0.3 \times 8}{0.3 \times 8 + 0.4 \times 7 + 0.1 \times 5 + 0.2 \times 3} = 0.381.$$

Similarly,

$$\varpi_2 = 0.445, \varpi_3 = 0.079, \varpi_4 = 0.095.$$

3. Compute the distance between α_i and β_i using Equation (2) (note that we can get these distances directly from Example 1:

$$d_{IFD}(\alpha_1, \beta_1) = 0.1, d_{IFD}(\alpha_2, \beta_2) = 0.2, d_{IFD}(\alpha_3, \beta_3) = 0.55, d_{IFD}(\alpha_4, \beta_4) = 0.25.$$

4. Rank $d_{IFD}(\alpha_j, \beta_j)$ $(j = 1, 2, 3, 4)$ according to associated value of $u_{\sigma(j)}$:

$$d_{IFD}(\alpha_{\sigma(1)}, \beta_{\sigma(1)}) = d_{IFD}(\alpha_2, \beta_2) = 0.2, d_{IFD}(\alpha_{\sigma(2)}, \beta_{\sigma(2)}) = d_{IFD}(\alpha_1, \beta_1) = 0.1,$$

$$d_{IFD}(\alpha_{\sigma(3)}, \beta_{\sigma(3)}) = d_{IFD}(\alpha_4, \beta_4) = 0.25, d_{IFD}(\alpha_{\sigma(4)}, \beta_{\sigma(4)}) = d_{IFD}(\alpha_3, \beta_3) = 0.55.$$

Employ the IFWIOWAD operator defined in Equation (5) to obtain the aggregation result:

$$\text{IFWIOWAD}(U, A, B) = 0.381 \times 0.2 + 0.445 \times 0.1 + 0.079 \times 0.25 + 0.095 \times 0.55 = 0.1927.$$

It is easy to see that we get a different aggregation value compared to the IFIOWAD operator. In addition, the variables $u_j (j = 1, \ldots, n)$ in the IFWIOWAD operator play dual functions, one is to induce the collection of arguments while the other moderates the weights that can overcome the drawback of the IFIOWAD operator caused by the limited role of the order-inducing variables.

Moreover, if the values of the order-inducing variables are changed to $U' = (8, 10, 1, 6)$, then we can recalculate the moderated weights:

$$\varpi_1 = \frac{w_1 u_{\sigma(1)}}{\sum_{j=1}^{4} w_j u_{\sigma(j)}} = \frac{0.3 \times 10}{0.3 \times 10 + 0.4 \times 8 + 0.1 \times 6 + 0.2 \times 1} = 0.429.$$

Similarly,

$$\varpi_2 = 0.456, \varpi_3 = 0.086, \varpi_4 = 0.029.$$

Thus, the aggregation of the IFWIOWAD operator will yield the following result:

$$\text{IFWIOWAD}(U', A, B) = 0.429 \times 0.2 + 0.456 \times 0.1 + 0.086 \times 0.25 + 0.029 \times 0.55 = 0.16885.$$

As can be seen, in comparison to the IFIOWAD operator, the aggregation result of the IFWIOWAD is changed based on the adjustment of the values of $u_j (j = 1, \ldots, n)$, thus it can accommodate the variation caused by a change of order-inducing variables and yield better results.

Depending on the operational laws defined for the IFNs, one can drive some properties of the IFWIOWAD operator that are illustrated by the following theorems.

Theorem 1. *(Commutativity—distance measures). Let $\widetilde{F}$ be the IFWIOWAD operator, then*

$$\widetilde{F}(\langle u_1, \alpha_1, \beta_1 \rangle, \ldots, \langle u_n, \alpha_n, \beta_n \rangle) = \widetilde{F}(\langle u_1, \beta_1, \alpha_1 \rangle, \ldots, \langle u_n, \beta_n, \alpha_n \rangle) \tag{7}$$

Theorem 2. *(Commutativity—IOWA aggregation). Let $(\langle u_1, s_1, t_1 \rangle, \ldots, \langle u_n, s_n, t_n \rangle)$ is any possible permutation of argument vector $(\langle u_1, \alpha_1, \beta_1 \rangle, \ldots, \langle u_n, \alpha_n, \beta_n \rangle)$, then*

$$\widetilde{F}(\langle u_1, \alpha_1, \beta_1 \rangle, \ldots, \langle u_n, \alpha_n, \beta_n \rangle) = \widetilde{F}(\langle u_1, s_1, t_1 \rangle, \ldots, \langle u_n, s_n, t_n \rangle) \tag{8}$$

Theorem 3. *(Monotonicity). If $|\alpha_i - \beta_i| \le |\alpha_i' - \beta_i'|$ for all i, then*

$$\widetilde{F}(\langle u_1, \alpha_1, \beta_1 \rangle, \ldots, \langle u_n, \alpha_n, \beta_n \rangle) \le \widetilde{F}(\langle u_1, s_1, t_1 \rangle, \ldots, \langle u_n, s_n, t_n \rangle) \tag{9}$$

Theorem 4. *(Boundedness). Let $\min\limits_i(|\alpha_i - \beta_i|) = d$ and $\max\limits_i((|\alpha_i - \beta_i|)) = D$, then*

$$d \le \widetilde{F}(\langle u_1, \alpha_1, \beta_1 \rangle, \ldots, \langle u_n, \alpha_n, \beta_n \rangle) \le D \tag{10}$$

Theorem 5. *(Idempotency). If all $\widetilde{d}_i = |\alpha_i - \beta_i| = \widetilde{d}$ for all i, then*

$$\widetilde{F}(\langle u_1, \alpha_1, \beta_1 \rangle, \ldots, \langle u_n, \alpha_n, \beta_n \rangle) = \widetilde{d} \tag{11}$$

It is straightforward to prove these theorems and therefore omitted for sake of brevity. Moreover, some particular cases of the IFWIOWAD operator can be explored by analyzing the order-inducing values and the weight vector. For example,

- If $U = (u, 0, \cdots, 0)$ $(u \ne 0)$, then

$$\text{IFWIOWAD}(\langle u_1, \alpha_1, \beta_1 \rangle, \ldots, \langle u_n, \alpha_n, \beta_n \rangle) = d_{IFD}(\alpha_{\sigma(1)}, \beta_{\sigma(1)}) \tag{12}$$

- If $U = (0, \cdots, 0, u)$ $(u \ne 0)$, then

$$\text{IFWIOWAD}(\langle u_1, \alpha_1, \beta_1 \rangle, \ldots, \langle u_n, \alpha_n, \beta_n \rangle) = d_{IFD}(\alpha_{\sigma(n)}, \beta_{\sigma(n)}) \tag{13}$$

- If $w_j = 0$ and $w_k = 1$, for all $j \ne k$, then

$$\text{WIEOWD}(\langle u_1, p_1, q_1 \rangle, \ldots, \langle u_n, p_n, q_n \rangle) = d_{IFD}(\alpha_{\sigma(k)}, \beta_{\sigma(k)}) \tag{14}$$

Especially, if $D_k = \max\limits_i\{|\alpha_i - \beta_i|\}$, then we get the intuitionistic fuzzy maximum distance; if $D_k = \min\limits_i\{|\alpha_i - \beta_i|\}$, the intuitionistic fuzzy minimum distance.

Other a parameterized family of the IFWIOWAD operator can be described by similar methods, as applied in references [27–31].

4. The IFWIOWAWAD Operator

From the examples illustrated in the Section 3, we can see that the proposed IFWIOWAD operator can effectively eliminate the defects of the existing methods. However, further analysis indicates that the IFWIOWAD operator also has some shortcomings; i.e., it cannot integrate the weight of integrated arguments—and thus the importance of the integrated date cannot be reflected in the aggregation process. Recently, Merigó [32] presented a unification of the OWA and the IOWA operators, and termed it the induced ordered weighted averaging–weighted average (IOWAWA) operator. The prominent feature of the IOWAWA operator is that it unifies the IOWA operator and weighted average (WA) in the same formula, and allows each of the two concepts to be assigned a degree of importance in the aggregation. The IOWAWA operator has been receiving increasing attention to date. For example, Zeng et al. [33] explored the usefulness of the IOWAWA in the intuitionistic fuzzy situation. Merigó et al. [34] studied the application of the IOWAWA in entrepreneurial fuzzy group decision-making problems. Merigó et al. [35] presented some new IOWAWA–based methods to compute variance and covariance. Zeng et al. [36] proposed some aggregation operators based on the IOWAWA method in Pythagorean fuzzy environment. Motivated the idea of the IOWAWA operator, in this section we present the IFWIOWAWAD operator that comprises a unified model that employs the main advantages of IFWIOWAD operator and the weighted average (WA) methods. Thus, it can perform the importance of attributes and complex attitude of experts in the decision-making framework.

Definition 6. *Let $A = (\alpha_1, \ldots, \alpha_n)$ and $B = (\beta_1, \ldots, \beta_n)$ be two sets of IFNs defined in set $Z = \{z_1, z_2, \ldots, z_n\}$ and δ_i be the weight of the element $z_i (i = 1, \ldots, n)$, satisfying $\delta_1 + \ldots + \delta_n = 1$ and $\delta_i \in [0, 1]$. Then, the IFWIOWAWAD is termed intuitionistic fuzzy weighted IOWA weighted average distance operator and defined as*

$$\text{IFWIOWAWAD}(\langle u_1, \alpha_1, \beta_1 \rangle, \ldots, \langle u_n, \alpha_n, \beta_n \rangle) = \sum_{j=1}^{n} \widetilde{w}_j d_{IFD}(\alpha_{\sigma(j)}, \beta_{\sigma(j)}) \tag{15}$$

where $d_{IFD}(\alpha_{\sigma(j)}, \beta_{\sigma(j)})$ is the argument value of $d_{IFD}(\alpha_j, \beta_j)$ reordered by the order-inducing variable $u_{\sigma(j)}$ such that $u_{\sigma(j-1)} \geq u_{\sigma(j)}$ for $1 < j \leq n$. The combined weight of $\widetilde{w}$ is defined as follows:

$$\widetilde{w}_j = \lambda \varpi_j + (1 - \lambda) \delta_{\sigma(j)} \tag{16}$$

where $\lambda \in [0, 1]$, $W = (w_1, \ldots, w_n)^T$ is the associated weighting vector that simply satisfies the condition $0 \leq w_j \leq 1$ and $w_1 + \ldots + w_n = 1$. ϖ_j is defined by Equation (6), that is

$$\varpi_j = \frac{w_j u_{\sigma(j)}}{\sum_{j=1}^{n} w_j u_{\sigma(j)}} \tag{17}$$

The IFWIOWAWAD operator can also be explicitly illustrated in terms of the two underlying rules of aggregation (i.e., WA and IOWA). Thus, the IFWIOWAWAD can be separated into a linear combination of the IF weighted distance (IFWD) [15] and the IFWIOWAD:

$$\begin{gathered} \text{IFWIOWAWAD}(\langle u_1, \alpha_1, \beta_1 \rangle, \ldots, \langle u_n, \alpha_n, \beta_n \rangle) = \\ \lambda \sum_{j=1}^{n} \varpi_j d_{IFD}(\alpha_{\sigma(j)}, \beta_{\sigma(j)}) + (1 - \lambda) \sum_{i=1}^{n} \delta_i d_{IFD}(\alpha_i, \beta_i) \end{gathered} \tag{18}$$

Example 3. *(Continuing from Example 2). Let the weighting vector $\delta = (\delta_1, \delta_2, \delta_3, \delta_4)^T = (0.2, 0.3, 0.15, 0.35)^T$ and $\lambda = 0.6$, then with the help of Example 2, the rest steps using the IFWIOWAWAD operator are given as follows:*

1. Compute the combined weight $\widetilde{w}_j$ $(j = 1, 2, 3, 4)$ using Equation (16):

$$\begin{aligned} \widetilde{w}_1 &= \lambda \varpi_1 + (1 - \lambda) \delta_{\sigma(1)} = 0.6 \times 0.429 + (1 - 0.6) \times 0.3 = 0.3774, \\ \widetilde{w}_2 &= \lambda \varpi_2 + (1 - \lambda) \delta_{\sigma(2)} = 0.6 \times 0.456 + (1 - 0.6) \times 0.2 = 0.3536, \\ \widetilde{w}_3 &= \lambda \varpi_3 + (1 - \lambda) \delta_{\sigma(3)} = 0.6 \times 0.086 + (1 - 0.6) \times 0.35 = 0.1916, \\ \widetilde{w}_4 &= \lambda \varpi_4 + (1 - \lambda) \delta_{\sigma(4)} = 0.6 \times 0.029 + (1 - 0.6) \times 0.15 = 0.0774. \end{aligned}$$

2. Employ the IFWIOWAWAD operator defined in Equation (15) to perform the aggregation as follows:

$$\text{IFWIOWAWAD}(U, A, B) = 0.3774 \times 0.2 + 0.3536 \times 0.1 + 0.1916 \times 0.25 + 0.0774 \times 0.55 = 0.20131$$

The aggregation of IFWIOWAWAD can also be performed using Equation (19) as following:

$$\text{IFWIOWAWAD}(U, A, B) = 0.6 \times \text{IFWIOWAD} + 0.4 \times \text{IFWD}$$
$$= 0.6 \times 0.16885 + 0.4 \times (0.2 \times 0.1 + 0.3 \times 0.2 + 0.15 \times 0.55 + 0.35 \times 0.25) = 0.20131$$

Evidently, we get the same aggregate values for both methods. Moreover, we can see that, contrary to the IFWIOWAD operator, the IFWIOWAWAD operator cannot only consider the attitudinal character represented by the order induced variable, but also take into account the importance of the argument based on the weighted average method.

In the following results, we show some of the most important properties of the IFWIOWAWAD operator.

Proposition 1. *The IFWIOWAWAD is commutative if it follows (let φ be the IFWIOWAWAD operator for a simple notation):*

$$\varphi(\langle u_1, \alpha_1, \beta_1 \rangle, \ldots, \langle u_n, \alpha_n, \beta_n \rangle) = \varphi(\langle u_1, \beta_1, \alpha_1 \rangle, \ldots, \langle u_n, \beta_n, \alpha_n \rangle) \tag{19}$$

or

$$\varphi(\langle u_1, \alpha_1, \beta_1 \rangle, \ldots, \langle u_n, \alpha_n, \beta_n \rangle) = \varphi(\langle u_1, s_1, t_1 \rangle, \ldots, \langle u_n, s_n, t_n \rangle) \tag{20}$$

where $(\langle u_1, s_1, t_1 \rangle, \ldots, \langle u_n, s_n, t_n \rangle)$ is a possible permutation of the argument vector $(\langle u_1, \alpha_1, \beta_1 \rangle, \ldots, \langle u_n, \alpha_n, \beta_n \rangle)$.

Proposition 2. *If $|\alpha_i - \beta_i| \le |s_1 - t_1|$ for all i, it follows that:*

$$\varphi(\langle u_1, \alpha_1, \beta_1 \rangle, \ldots, \langle u_n, \alpha_n, \beta_n \rangle) \le \varphi(\langle u_1, s_1, t_1 \rangle, \ldots, \langle u_n, s_n, t_n \rangle) \tag{21}$$

Then the IFWIOWAWAD is monotonic.

Proposition 3. *The IFWIOWAWAD is bounded if it follows that:*

$$\min_i(|\alpha_i - \beta_i|) \le \varphi(\langle u_1, \alpha_1, \beta_1 \rangle, \ldots, \langle u_n, \alpha_n, \beta_n \rangle) \le \max_i(|\alpha_i - \beta_i|) \tag{22}$$

Proposition 4. *If all $\tilde{d}_i = |\alpha_i - \beta_i| = \tilde{d}$ for $i \in [1, n]$, it follows that:*

$$\tilde{F}(\langle u_1, \alpha_1, \beta_1 \rangle, \ldots, \langle u_n, \alpha_n, \beta_n \rangle) = \tilde{d} \tag{23}$$

Then the IFWIOWAWAD operator is idempotent.

By selecting different values for the weights and parameters in the IFWIOWAWAD operator, we can derive some special intuitionistic fuzzy distance operators. For example:

- When $\lambda = 1$, the IFWIOWAWAD reduces to the IFWIOWAD operator.
- When $\lambda = 0$, we get the IFWD operator.

Equivalently, many other special cases can be derived by analyzing the weighting vectors W, V and the order inducing variable vector U in a similar way (see [33–36]).

5. A MADM Model Based on the IFWIOWAWAD Operator

A framework of the MADM model based on the IFWIOWAWAD is presented in this section. The main process for the model is structured as follows:

Step 1. Each decision maker e_k provides their opinions and thus forms the individual decision matrix, constructed as in (24):

$$D_k = \begin{array}{c} \\ A_1 \\ \vdots \\ A_n \end{array} \begin{array}{c} \begin{array}{ccc} C_1 & \cdots & C_n \end{array} \\ \left(\begin{array}{ccc} \alpha_{11}^{(k)} & \cdots & \alpha_{1n}^{(k)} \\ \vdots & \ddots & \vdots \\ \alpha_{m1}^{(k)} & \cdots & \alpha_{mn}^{(k)} \end{array} \right) \end{array} \tag{24}$$

where A_i and C_j indicate the alternative $i(i = 1, \ldots, m)$ and the attribute $j(j = 1, \ldots, n)$, respectively. Meanwhile the IFNs $\alpha_{ij}^{(k)} = (\mu_{ij}^{(k)}, v_{ij}^{(k)})$ represents the preference for A_i with respect to the attribute C_j.

Step 2. Employ the IF weighted average (IFWA) operator [24] to convert individual opinions of each decision makers into a group decision matrix $D = (\alpha_{ij})_{m \times n}$, where

$$\alpha_{ij} = \text{IFWA}\left(\alpha_{ij}^{(1)}, \ldots, \alpha_{ij}^{(t)}\right) \; i = 1, \ldots, m, j = 1, \ldots, n. \tag{25}$$

Step 3. Construct the ideal alternative I and determine the order-inducing variables and weights used for the IFWIOWAWAD operator.

Step 4. Compute the weighted distance between the ideal alternative I and each $A_i(i = 1, \ldots, 5)$ using the IFWIOWAWAD operator.

Step 5. Establish a ranking for the alternative $A_i(i = 1, \ldots, 5)$ in accordance with the IFWIOWAWAD(I, A_i) obtained in step 4. The alternative with the smallest distance will be selected as the best.

6. An Example of Investment Selection

Decision-making related to the selection of a suitable investment from finite feasible alternatives constitutes one of the most common and important activities in various business fields. The complexity of the assessment and selection process for investment projects necessitates a complex method: i.e., the multiple attribute decision-making (MADM) technique provides an efficient tool for decision makers to solve problems based on an evaluation or preference information given by multiple experts. In the past, many authors have proposed different MADM approaches for solving the selection of investment problems [37–41]. Previous findings have shown that the applications of the induced aggregation distance operators are very heartening and widely used in the decision-making process. This paper presents the application of the proposed model in the process of selecting investments in which a group of decision makers (or experts) are invited for the selection of a suitable strategy (adapted from Ref. [32]). Based on the market research and preliminary screening, there are five companies (alternatives) to be considered as potential investment options, namely a chemical company (A_1), a food company (A_2), a car company (A_3), a furniture company (A_4) and a computer company (A_5). The main situations of company for investment are evaluated by the world economic growth rate: C_1 = High growth rate, C_2 = Medium growth rate, C_3 = Low growth rate, C_4 = Growth rate near 0 and C_5 = Negative growth rate. The assessment of the alternatives with respect to each attribute given by three decision makers, are given in Tables 2–4. For example, the decision maker e_1 called ten experts together to assess the situations (attributes) for these five companies. As for the C_1 of the company A_1, if six experts consider C_1 strong while three experts consider C_1 low and one expert do not judge whether C_1 is strong or not, then the evaluation of company A_1 relative to C_1 can be represented by IFN (0.6,0.3) by using the statistical approach.

Table 2. Decision matrix D_1 .

Alternatives	C_1	C_2	C_3	C_4	C_5
A_1	(0.2,0.6)	(0.5,0.3)	(0.4,0.4)	(0.5,0.4)	(0.3,0.5)
A_2	(0.6,0.2)	(0.7,0.3)	(0.6.0.2)	(0.7,0.2)	(0.4,0.5)
A_3	(0.6,0.2)	(0.6,0.4)	(0.5,0.3)	(0.6,0.3)	(0.4,0.4)
A_4	(0.4,0.2)	(0.7,0.2)	(0.5,0.2)	(0.4,0.4)	(0.6,0.3)
A_5	(0.7,0.3)	(0.4,0.3)	(0.6,0.3)	(0.5,0.4)	(0.6,0.2)

Table 3. Decision matrix D_2 .

Alternatives	C_1	C_2	C_3	C_4	C_5
A_1	(0.5,0.3)	(0.7,0.2)	(0.5,0.4)	(0.7,0.3)	(0.4,0.3)
A_2	(0.7,0.2)	(0.6,0.2)	(0.8,0.1)	(0.5,0.4)	(0.6,0.2)
A_3	(0.4,0.4)	(0.4,0.4)	(0.4,0.2)	(0.6,0.3)	(0.4,0.4)
A_4	(0.6,0.2)	(0.6,0.2)	(0.7,0.2)	(0.6,0.2)	(0.5,0.3)
A_5	(0.8,0.2)	(0.5,0.3)	(0.6,0.1)	(0.6,0.2)	(0.6,0.2)

Table 4. Decision matrix D_3 .

Alternatives	C_1	C_2	C_3	C_4	C_5
A_1	(0.6,0.2)	(0.8,0.2)	(0.7,0.2)	(0.6,0.3)	(0.5,0.4)
A_2	(0.7,0.3)	(0.6,0.2)	(0.3,0.4)	(0.7,0.1)	(0.8,0.2)
A_3	(0.8,0.1)	(0.7,0.2)	(0.7,0.1)	(0.3,0.4)	(0.6,0.3)
A_4	(0.5,0.5)	(0.3,0.4)	(0.6,0.2)	(0.4,0.5)	(0.5,0.2)
A_5	(0.6,0.3)	(0.8,0.2)	(0.6,0.2)	(0.5,0.3)	(0.7,0.2)

In this problem, the weighting vector of the three experts is assumed to $V = (0.3, 0.4, 0.3)^T$ while, the collective results performed by the IFWA operator are listed in Table 5.

Table 5. Collective decision matrix D .

Alternatives	C_1	C_2	C_3	C_4	C_5
A_1	(0.46,0.33)	(0.69,0.26)	(0.55,0.32)	(0.62,0.33)	(0.41,0.38)
A_2	(0.67,0.23)	(0.63,0.23)	(0.64,0.19)	(0.63,0.21)	(0.63,0.26)
A_3	(0.68,0.20)	(0.59,0.23)	(0.54,0.18)	(0.53,0.33)	(0.47,0.37)
A_4	(0.43,0.26)	(0.57,0.25)	(0.62,0.20)	(0.49,0.32)	(0.53,0.27)
A_5	(0.72,0.26)	(0.60,0.27)	(0.60,0.17)	(0.54,0.22)	(0.63,0.20)

The order-inducing variables and the ideal alternative determined by the group of experts are shown in Tables 6 and 7, respectively.

Table 6. Order-inducing variables.

Varaible	C_1	C_2	C_3	C_4	C_5
U	0.8	0.9	0.4	0.7	0.6

Table 7. Ideal alternative.

Ideal Alternative	C_1	C_2	C_3	C_4	C_5
I	(0.8,0.1)	(0.9,0.1)	(0.9,0)	(0.8,0.1)	(0.9,0.1)

The weights δ_i for the attributes are given as 0.1, 0.25, 0.2, 0.35, 0.1 while the ordered weights, w_j are assumed to be 0.15, 0.25, 0.2, 0.1, 0.3. Table 8 shows the aggregated results performed by the IFWIOWAWAD operator ($\lambda = 0.4$).

Table 8. Aggregate results and ranking rendered by the IFWIOWAWAD operator.

Results	A_1	A_2	A_3	A_4	A_5
IFWIOWAWAD(A_i, I)	0.25745	0.187427	0.231064	0.25484	0.195926
Ranking	5	1	3	4	2

Thus, A_2 appears to be the best choice as it is closest to the ideal alternative while, the ranking of the five alternatives is $A_2 \succ A_5 \succ A_3 \succ A_4 \succ A_1$.

To conduct a comparative analysis, we employ the IFIOWAWAD and IFOWAWAD operators in identical decision information to further explore the effectiveness of the order-inducing variables on the aggregation results. The results are shown in Table 9.

Thus, the rankings of the alternatives obtained by the IFIOWAWAD and IFOWAWAD operators are $A_2 \succ A_5 \succ A_3 \succ A_1 \succ A_4$ and $A_5 \succ A_2 \succ A_3 \succ A_4 \succ A_1$, respectively. From Tables 8 and 9, it is clear that the orderings of the alternatives may change if a different distance operator is used. It should be pointed out that the order-inducing variables in the IFIOWAWAD operator only perform a single induced function during the aggregation process. The IFOWAWAD operator integrates the importance of attributes and ordered weights into the formula to evaluate the IFS information, but fails to account for the attitudinal characters as it cannot infuse the order-inducing variables. However, the IFWIOWAWAD not only integrates both of the weights, but also captures the variation in the order-inducing variables, and thus achieves a more scientific and accurate result in comparison with other approaches.

Table 9. Aggregate results driven by the IFIOWAWAD and the IFOWAWAD operators.

Results	A_1	A_2	A_3	A_4	A_5
IFIOWAWAD(A_i, I)	0.265624	0.182068	0.213909	0.275726	0.185689
IFOWAWAD(A_i, I)	0.261	0.1975	0.2355	0.25085	0.19625

Moreover, it is possible to conduct a sensitive analysis to explore the robustness of the ranking of the alternative with regards to the parameter λ, $\lambda \in [0, 1]$. The computation results are illustrated in Table 10.

Table 10. Ranking rendered by the IFWIOWAWAD operator with different values of λ.

λ	**Ranking of Alternative**
$\lambda = 0$	$A_2 \succ A_5 \succ A_3 \succ A_1 \succ A_4$
$\lambda = 0.1$	$A_2 \succ A_5 \succ A_3 \succ A_1 \succ A_4$
$\lambda = 0.2$	$A_2 \succ A_5 \succ A_3 \succ A_4 \succ A_1$
$\lambda = 0.3$	$A_2 \succ A_5 \succ A_3 \succ A_4 \succ A_1$
$\lambda = 0.4$	$A_2 \succ A_5 \succ A_3 \succ A_4 \succ A_1$
$\lambda = 0.5$	$A_2 \succ A_5 \succ A_3 \succ A_4 \succ A_1$
$\lambda = 0.6$	$A_2 \succ A_5 \succ A_3 \succ A_4 \succ A_1$
$\lambda = 0.7$	$A_2 \succ A_5 \succ A_3 \succ A_4 \succ A_1$
$\lambda = 0.8$	$A_2 \succ A_5 \succ A_3 \succ A_4 \succ A_1$
$\lambda = 0.9$	$A_2 \succ A_5 \succ A_3 \succ A_4 \succ A_1$
$\lambda = 1$	$A_5 \succ A_2 \succ A_3 \succ A_4 \succ A_1$

As can be seen, the ranking of alternatives may be different based on the different values of λ. Thus, the decision maker can select suitable values of λ to meet their interests or actual needs at hand.

Therefore, this model is rather flexible as it provides more choices to decision makers for the selection of aggregation schemes by adjusting different values of the parameters.

7. Conclusions

To effectively deal with and process intuitionistic fuzzy information, in this study we have proposed the intuitionistic fuzzy weighted induced ordered weighted averaging distance operator, which improves the existing aggregation operators by extending the role of the order-inducing variables. In the proposed operator, the order-inducing variables induce the order of arguments and moderate the associated weights simultaneously. Thus, it enables us to capture the variations in the final aggregation results caused by the order-inducing variables. A generation of intuitionistic fuzzy weighted induced ordered weighted averaging distance operator has been further developed, based on which, a novel model for intuitionistic fuzzy multiple attribute decision making problems was developed. This model presents a useful and adaptable way to integrate subjective opinions and complex attitudinal characters in real situations. The comparative analysis illustrates that this model is expected to lead to more realistic and accurate results in intuitionistic fuzzy situations. Thus, this paper offers a significant contribution in regards to the development of MADM frameworks for investment selection problems.

In future research efforts, we will consider extending the approach with probabilities or other kinds of distance measures. We may also consider other situations based on the presented procedures and tools, such as the Pythagorean fuzzy set [36,42] and Neutrosophic set [43,44].

Author Contributions: Z.L. drafted the initial manuscript and conceived the MADM framework. D.S. provided the relevant literature review and the illustrated example. S.Z. revised the manuscript and analyzed the data.

Funding: This paper is supported by National Social Science Fund of China (No.18BTJ027).

Conflicts of Interest: The authors declare no conflict of interest.

References

1. Atanassov, K. Intuitionistic fuzzy sets. *Fuzzy Sets Syst.* **1986**, *20*, 87–96. [CrossRef]
2. Yu, D.J.; Liao, H.C. Visualization and quantitative research on intuitionistic fuzzy studies. *J. Intell. Fuzzy Syst.* **2016**, *30*, 3653–3663. [CrossRef]
3. Yu, D.J.; Shi, S.S. Researching the development of Atanassov intuitionistic fuzzy set: Using a citation network analysis. *Appl. Soft Comput.* **2015**, *32*, 189–198. [CrossRef]
4. Mehdi, K.G.; Maghsoud, A.; Edmundas, K.Z.; Jurgita, A. Supplier evaluation and selection in fuzzy environments: A review of MADM approaches. *Econ. Res. Ekon. Istraž.* **2017**, *30*, 1073–1118.
5. Shen, F.; Ma, X.S.; Li, Z.; Xu, Z.S.; Cai, D.L. An extended intuitionistic fuzzy TOPSIS method based on a new distance measure with an application to credit risk evaluation. *Inf. Sci.* **2018**, *428*, 105–119. [CrossRef]
6. Song, Y.F.; Wang, X.D.; Zhang, H.L. A distance measure between intuitionistic fuzzy belief functions. *Knowl.-Based Syst.* **2015**, *86*, 288–298. [CrossRef]
7. Zhang, H.M.; Yu, L.Y. New distance measures between intuitionistic fuzzy sets and interval-valued fuzzy sets. *Inf. Sci.* **2013**, *245*, 181–196. [CrossRef]
8. Hung, W.L.; Yang, M.S. Similarity measures of intuitionistic fuzzy sets based on Hausdorff distance. *Pattern Recognit. Lett.* **2004**, *25*, 1603–1611. [CrossRef]
9. Zeng, S.Z.; Su, W.H. Intuitionistic fuzzy ordered weighted distance operator. *Knowl.-Based Syst.* **2011**, *24*, 1224–1232. [CrossRef]
10. Zeng, S.Z.; Xiao, Y. TOPSIS method for intuitionistic fuzzy multiple-criteria decision making and its application to investment selection. *Kybernetes* **2016**, *45*, 282–296. [CrossRef]
11. Merigó, J.M.; Casanovas, M. Decision making with distance measures and induced aggregation operators. *Comput. Ind. Eng.* **2011**, *60*, 66–76. [CrossRef]
12. Zeng, S.Z.; Merigó, J.M.; Palacios-Marques, D.; Jin, H.H.; Gu, F.J. Intuitionistic fuzzy induced ordered weighted averaging distance operator and its application to decision making. *J. Intell. Fuzzy Syst.* **2017**, *32*, 11–22. [CrossRef]

13. Merigó, J.M.; Casanovas, M. A new Minkowski distance based on induced aggregation operators. *Int. J. Comput. Intell. Syst.* **2011**, *4*, 123–133. [CrossRef]
14. Casanovas, M.; Torres-Martínez, A.; Merigó, J.M. Decision making in reinsurance with induced OWA operators and Minkowski distances. *Cybern. Syst.* **2016**, *47*, 460–477. [CrossRef]
15. Merigó, J.M.; Casanovas, M. Induced aggregation operators in the Euclidean distance and its application in financial decision making. *Expert Syst. Appl.* **2011**, *38*, 7603–7608. [CrossRef]
16. Zeng, S.Z.; Li, W.; Merigó, J.M. Extended induced ordered weighted averaging distance operators and their application to group decision-making. *Int. J. Inf. Technol. Decis. Mak.* **2013**, *12*, 1973–6845. [CrossRef]
17. Su, W.H.; Zeng, S.Z.; Ye, X.J. Uncertain group decision-making with induced aggregation operators and Euclidean distance. *Technol. Econ. Dev. Econ.* **2013**, *19*, 431–447. [CrossRef]
18. Zeng, S.Z.; Su, W.H.; Chen, J. Fuzzy decision making with induced heavy aggregation operators and distance measures. *J. Intell. Fuzzy Syst.* **2014**, *26*, 127–135.
19. Li, C.G.; Zeng, S.Z.; Pan, T.J.; Zheng, L.N. A method based on induced aggregation operators and distance measures to multiple attribute decision making under 2-tuple linguistic environment. *J. Comput. Syst. Sci.* **2014**, *80*, 1339–1349. [CrossRef]
20. Xian, S.D.; Sun, W.J. Fuzzy linguistic induced euclidean OWA distance operator and its application in group linguistic decision making. *Int. J. Intell. Syst.* **2014**, *29*, 478–491. [CrossRef]
21. Su, W.; Zhang, C.; Zeng, S. Uncertain induced heavy aggregation distance operator and its application to decision making. *Cybern. Syst.* **2015**, *46*, 172–187. [CrossRef]
22. Xian, S.D.; Sun, W.J.; Xu, S.H.; Gao, Y.Y. Fuzzy linguistic induced OWA Minkowski distance operator and its application in group decision making. *Pattern Anal. Appl.* **2016**, *19*, 325–335. [CrossRef]
23. Xue, W.T.; Xian, S.D.; Dong, Y.F. A novel intuitionistic fuzzy induced ordered weighted Euclidean distance operator and its application for group decision making. *Int. J. Intell. Syst.* **2017**, *32*, 739–753. [CrossRef]
24. Xu, Z.S. Intuitionistic fuzzy aggregation operators. *IEEE Trans. Fuzzy Syst.* **2007**, *15*, 1179–1187.
25. Yager, R.R. On ordered weighted averaging aggregation operators in multi-criteria decision making. *IEEE Trans. Syst. Man Cybern. B* **1988**, *18*, 183–190. [CrossRef]
26. Yager, R.R.; Filev, D.P. Induced ordered weighted averaging operators. *IEEE Trans. Syst. Man Cybern. Part B* **1999**, *29*, 141–150. [CrossRef] [PubMed]
27. Aggarwal, M. A New Family of Induced OWA Operators. *Int. J. Intell. Syst.* **2015**, *30*, 170–205. [CrossRef]
28. Zeng, S.Z. Pythagorean fuzzy multiattribute group decision making with probabilistic information and OWA approach. *Int. J. Intell. Syst.* **2017**, *32*, 1136–1150. [CrossRef]
29. Merigó, J.M.; Palacios-Marqués, D.; Soto-Acosta, P. Distance measures, weighted averages, OWA operators and Bonferroni means. *Appl. Soft Comput.* **2017**, *50*, 356–366. [CrossRef]
30. Zeng, S.Z.; Xiao, Y. A method based on TOPSIS and distance measures for hesitant fuzzy multiple attribute decision making. *Technol. Econ. Dev. Econ.* **2018**, *24*, 969–983. [CrossRef]
31. Zeng, S.Z.; Wang, N.; Zhang, C.H.; Su, W.H. A novel method based on induced aggregation operator for classroom teaching quality evaluation with probabilistic and Pythagorean fuzzy information. *Eurasia J. Math. Sci. Technol. Educ.* **2018**, *14*, 3205–3212. [CrossRef]
32. Merigó, J.M. A unified model between the weighted average and the induced OWA operator. *Expert Syst. Appl.* **2011**, *38*, 11560–11572. [CrossRef]
33. Zeng, S.Z.; Wang, Q.; Merigó, J.M.; Pan, T. Induced intuitionistic fuzzy ordered weighted averaging—Weighted average operator and its application to business decision-making. *Comput. Sci. Inf. Syst.* **2014**, *11*, 839–857.
34. Merigó, J.M.; Peris-Ortiz, M.; Palacios-Marqués, D. Entrepreneurial fuzzy group decision-making under complex environments. *J. Intell. Fuzzy Syst.* **2014**, *27*, 901–912.
35. Merigó, J.M.; Guillén, M.; Sarabia, J.M. The ordered weighted average in the variance and the covariance. *Int. J. Intell. Syst.* **2015**, *30*, 985–1005. [CrossRef]
36. Zeng, S.Z.; Mu, Z.M.; Baležentis, T. A novel aggregation method for Pythagorean fuzzy multiple attribute group decision making. *Int. J. Intell. Syst.* **2018**, *33*, 573–585. [CrossRef]
37. Renna, P. A Decision Investment Model to Design Manufacturing Systems based on a genetic algorithm and Monte-Carlo simulation. *Int. J. Comput. Integr. Manuf.* **2017**, *30*, 590–605. [CrossRef]

38. Chatterjee, P.; Mondal, S.; Boral, S.; Banerjee, A.; Chakraborty, S. A novel hybrid method for non-traditional machining process selection using factor relationship and multi-attributive border approximation method. *Facta Univ. Ser. Mech. Eng.* **2017**, *15*, 439–456. [CrossRef]
39. Petkovic, D.; Madic, M.; Radovanovic, M.; Gecevska, V. Application of the performance selection index method for solving machining MCDM problems. *Facta Univ. Ser. Mech. Eng.* **2017**, *15*, 97–106. [CrossRef]
40. Roy, J.; Adhikary, K.; Kar, S.; Pamučar, D. A rough strength relational DEMATEL model for analysing the key success factors of hospital service quality. *Decis. Mak. Appl. Manag. Eng.* **2018**, *1*, 121–142. [CrossRef]
41. Vasiljević, M.; Fazlollahtabar, H.; Stević, Ž.; Vesković, S. A rough multicriteria approach for evaluation of the supplier criteria in automotive industry. *Decis. Mak. Appl. Manag. Eng.* **2018**, *1*, 82–96. [CrossRef]
42. Zhang, X.L.; Xu, Z.S. Extension of TOPSIS to multiple criteria decision making with Pythagorean fuzzy sets. *Int. J. Intell. Syst.* **2014**, *29*, 1061–1078. [CrossRef]
43. Ye, J. Multiple attribute decision-making method using correlation coefficients of normal Neutrosophic sets. *Symmetry* **2017**, *9*, 80. [CrossRef]
44. Wang, J.; Wei, G.W.; Wei, Y. Models for green supplier selection with some 2-tuple linguistic Neutrosophic number Bonferroni mean operators. *Symmetry* **2018**, *10*, 131. [CrossRef]

symmetry

MDPI

Article

A Hybrid Fuzzy Analytic Network Process (FANP) and Data Envelopment Analysis (DEA) Approach for Supplier Evaluation and Selection in the Rice Supply Chain

Chia-Nan Wang [1,2,*], Van Thanh Nguyen [1,3,*], Duy Hung Duong [3] and Hanh Tuong Do [3]

1 Department of Industrial Engineering and Management, National Kaohsiung University of Science and Technology, Kaohsiung 80778, Taiwan
2 Department of Industrial Engineering and Management, Fortune Institute of Technology, Kaohsiung 83160, Taiwan
3 Department of Industrial Systems Engineering, CanTho University of Technology, CanTho 900000, Vietnam; ddhung.htcn0114@student.ctuet.edu.vn (D.H.D.); hanhtuong58@gmail.com (H.T.D.)
* Correspondence: cn.wang@nkust.edu.tw (C.N.W.); jenny9121989@gmail.com (V.T.N.); Tel: +886-906-942-769 (V.T.N)

Received: 18 May 2018; Accepted: 12 June 2018; Published: 14 June 2018

Abstract: In the market economy, competition is typically due to the difficulty in selecting the most suitable supplier, one that is capable to help a business to develop a profit to the highest value threshold and capable to meet sustainable development features. In addition, this research discusses a wide range of consequences from choosing an effective supplier, including reducing production cost, improving product quality, delivering the product on time, and responding flexibly to customer requirements. Therefore, the activities noted above are able to increase an enterprise's competitiveness. It can be seen that selecting a supplier is complex in that decision-makers must have an understanding of the qualitative and quantitative features for assessing the symmetrical impact of the criteria to reach the most accurate result. In this research, the multi-criteria group decision-making (MCGDM) approach was proposed to solve supplier selection problems. The authors collected data from 25 potential suppliers, and the four main criteria within contain 15 sub-criteria to define the most effective supplier, which has viewed factors, including financial efficiency guarantee, quality of materials, ability to deliver on time, and the conditioned response to the environment to improve the efficiency of the industry supply chain. Initially, fuzzy analytic network process (ANP) is used to evaluate and rank these criteria, which are able to be utilized to clarify important criteria that directly affect the profitability of the business. Subsequently, data envelopment analysis (DEA) models, including the Charnes Cooper Rhodes model (CCR model), Banker Charnes Cooper model (BCC model), and slacks-based measure model (SBM model), were proposed to rank suppliers. The result of the model has proposed 7/25 suppliers, which have a condition response to the enterprises' supply requirements.

Keywords: fuzzy analytic network process (FANP); data envelopment analysis (DEA); supplier selection; multi-criteria group decision-making (MCGDM)

---

1. Introduction

The task of selecting suppliers becomes more important in today's competitive and global environment when it is impractical or virtually impossible to create high-quality, low-cost, successful products without a vendor. For businesses today, vendor selection is one of the most important and indispensable components of the supply chain function of Florez–Lopez [1]. The enterprises' expected goal of selecting a vendor is necessary to reduce the risk in buying, making an optimum decision, and establishing a sustainable alliance between buyers and suppliers [2]. Basically, choosing suppliers is a decision-making process because a business expects to obtain a supplier [3]. Additionally, it requires a powerful analytical approach, via utilizing decision-support tools, which is capable of addressing multiple criteria [4]. Incidentally, the supplier's price includes many qualitative and qualitative conflicts.

The author represents two techniques, i.e., DEA and the FANP, which are used to design a method for evaluating suppliers. In order to obtain the accurate result as the chosen supplier based on the frontier point of the DEA model from input and output decision-makers (DMUs) [5]. The drawback of DEA, related to this study, is the requirement of data for various inputs and outputs to be in a quantitative format. This DEA limitation is addressed by analyzing the qualitative factors/attributes associated with the supplier using FANP. FANP is a more general form of the decentralized process, which includes the feedback and interdependencies of decision attributes and alternatives. This additional feature provides a more accurate and robust approach when modeling a complex decision-making environment [6].

The decision-making process is designed to provide a holistic approach in which the relevant factors and criteria are integrated into the FANP's decentralized network. Different relationships are combined in these structures and then both judgment and logic are used to estimate the relative effect from which the overall response is derived [7]. The FANP model used here provides a unique quantitative value for vendor-specific qualitative factors and is based on buyers' preferences and perceptions. This quantitative value from FANP for each supplier is used as a qualitative benefit in the DEA model to obtain the ranking or performance of different suppliers.

This research proposed hybrid FANP and DEA approaches for supplier selection in the rice supply chain, which also considers green issues under uncertain environment conditions. The aim of this research is to provide a useful guideline for supplier selection based on qualitative and quantitative factors (including the main criteria, such as financial, delivery services, qualitative factors, and environmental management systems) to improve the efficiency of supplier selection in the rice supply chain and other industries.

In the remainder of this paper, this research provides the platform data to further support the need of the development of a decision approach. Then, the synthetic supplier evaluation approach was applied to a case study of a company, which could be used for the explanation of the findings. Finally, this paper ends with a summary, and conclusions are made.

2. Literature Review

2.1. Supplier Selection Methods

Aissaoui et al. [8] presented a literature review that covers the entire purchasing process, considered both parts and services outsourcing activities, and covers Internet-based procurement environments, such as electronic marketplace auctions. Govindan et al. [9] presented a literature review for multi-criteria decision-making approaches for green supplier evaluation and selection. Chai et al. [10] provided a systematic literature review on articles published from 2008 to 2012 on the application of DM techniques for supplier selection.

Wu and Blackhurst [11] proposed a methodology termed augmented DEA, which has enhanced discriminatory power over basic DEA models to rank suppliers. Amirteimoori and Khoshandam [12] developed a DEA model for evaluating the performance of suppliers and manufacturers in supply chain operations. Lin et al. [13] provided a MCDM model by combining the Delphi method and the ANP method for evaluating and selecting suppliers for the sustainable operation and development of enterprises in the aerospace industry. Galankashi et al. [14] proposed an integrated balanced scorecard (BSC) and fuzzy analytic hierarchical process (FAHP) model to select suppliers in the automotive industry. Kilincci and Onal [15] used a fuzzy AHP approach for supplier selection in a washing machine company.

Tyagi et al. [16] proposed fuzzy AHP and AHP methods to prioritize the alternatives of the supply chain performance system. Karsak and Dursun [17] proposed a fuzzy MCDM model including the quality function deployment (QFD), fusion of fuzzy information, and 2-tuple linguistic representation for supplier evaluation and selection. Chen et al. [18] proposed a hybrid AHP and TOPSIS for evaluating and ranking the potential suppliers. Guo et al. [19] used fuzzy MCDM approaches for green supplier selection in apparel manufacturing. Wu et al. [20] constructed a multiple criteria decision-making model for the selection of fishmeal suppliers. Hu et al. [21] proposed a hybrid fuzzy DEA/AHP methodology for ranking units in a fuzzy environment. He and Zhang [22] used a hybrid evaluation model based on factor analysis (FA), data envelopment analysis (DEA), with analytic hierarchy process (AHP) for a supplier selection from the perspective of a low-carbon supply chain.

Parkouhi et al. [23] used the fuzzy analytic network process and VlseKriterijuska Optimizacija I Komoromisno Resenje (VIKOR) techniques for supplier selection. Wan et al. [24] proposed a hybrid DEA and Grey Model (1,1) approach for partner selection in the supply chain of Vietnam's textile and apparel industry. Wu et al. [25] used the fuzzy Delphi method, ANP, and TOPSIS for supplier selection. Rezaeisaray et al. [26] proposed a hybrid DEMATLE, FANP, and DEA model for outsourcing supplier selection in pipe and fittings manufacturing. Rouyendegh and Erol [27] applied the DEA-fuzzy ANP for department ranking at Iran Amirkabir University. Fuzzy set theory formalized by Zadeh [28] is an effective tool, which has been widely used in the supplier selection decision process because it provides a suitable language to transform imprecise criteria to precise criteria.

Junior et al. [29] presented a comparison between fuzzy AHP and fuzzy TOPSIS methods to supplier selection. The linear programming of data envelopment analysis (DEA), which is proposed by Charnes et al. [30], and is able to produce the result of measured efficiency without having specific weights for inputs and outputs or specify the form of the production function, is a nonparametric technique used to measure the relative efficiency of peer decision-making units with multiple inputs and outputs [31,32]. In the supplier's evaluation and selection process, many researchers calculated the supplier's performance by using the ratio of weighted outputs to weighted inputs [32]. Thus, the integrated FANP and DEA method is used to determine supplier selection criteria and select supplier in this paper.

Talluri et al. [33] provided vendor evaluation models by presenting a chance-constrained data envelopment analysis (CCDEA) approach in the presence of multiple performance measures that are uncertain. Saen [34] applied a DEA model for ranking suppliers in the presence of nondiscretionary factors. Saen [35] also proposed a new AR-IDEA model for supplier selection. Saen and Zohrehbandian [36] proposed a DEA approach for supplier selection. Saen [37] proposed an innovative method, which is based on imprecise data envelopment analysis (IDEA) to select the best suppliers in the presence of both cardinal and ordinal data. Lo Storto [38] proposed a double DEA framework to support decision making in the choice of advanced manufacturing technologies. Adler et al. [39] reviewed of ranking method in the data envelopment analysis context. Lo Storto [40] presented a peeling DEA-game cross efficiency procedure to classify suppliers.

Kuo et al. [41] developed a supplier selection system through integrating fuzzy AHP and fuzzy DEA on an auto lighting System Company in Taiwan. Kuo and Lin [42] used ANP and DEA for supplier selection.

Taibi and Atmani [43] proposed a MCDM model combining fuzzy AHP with GIS and decision rules for industrial site selection. Molinera et al. [44] used fuzzy ontologies and multi-granular linguistic modelling methods for solving MCGDM problems under environments with a high number of alternatives. Adrian et al. [45] proposed a conceptual model development of big data analytics implementation assessment effect on decision making.

Staníčková and Melecký used DEA models to evaluate the performances of Visegrad Four (V4) countries and regions [46]. Schaar and Sherry have been shown to contribute to the overall performance efficiency of the air transportation network by used three DEA models (CCR, BCC and SBM) [47].

2.2. Criteria and Sub-Criteria for Supplier Selection

The initial criteria for the supplier set are developed based on a literature study.

Financial: The firm should require its suppliers to have a sound financial position. Financial strength can be a good indicator of the supplier's long-term stability. A solid financial position also helps ensure that performance standards can be maintained and that products and services will continue to be available [48].

Delivery and service: A firm can use service performance criteria to evaluate the benefits provided by supplier services. When considering services, a firm needs to clearly define its expectations since there are few uniform, established service standards to draw upon. Since any purchase involves some degree of service, such as order processing, delivery, and support, a firm should always include some service criteria in its evaluation. If the supplier provides a solution combining products and services, the firm should be sure to adequately represent its service needs in the selection criteria [48]. The suppliers have to follow the predefined delivery schedule for achieving on-time delivery. All the manufacturers want to work with the supplier who can manage the supply chain system on time and has the ability for following the exact delivery schedule table [49].

Qualitative: Qualitative criteria are developed to measure important aspects of the supplier's business: business experience and position among competitors, expert labor, technical capabilities and facilities, operational control, and quality [50].

Environmental management system: Due to increasing awareness about environmental degradation manufacturing companies and customers are both becoming alert of environmental protection [51]. This has led stakeholders of companies to ensure safe practices, like pollution control, reuse, recovery, etc. It includes criteria like pollution control: resource consumption of raw materials, use of environmentally friendly technology and materials, design capability for reduced consumption of materials/energy, reuse, and recycling of materials. To reduce the harm to the environment, organizations should also consider factors like permit requirements, compliance requirements, strategic considerations, climatic considerations, and government policy [52,53].

There are four main criteria and some sub-criteria, as shown in Table 1.

Table 1. Criteria for supplier selection.

Criteria	Sub-Criteria	Researcher
Financial	Capital and financial power of supplier company	Ho et al. [54], Dickson [55], Weber et al. [56]
	Proposed raw material price	Banaeian et al. [50], Dickson [55], Weber et al. [56], Ho et al. [54]
	Transportation cost to the geographical location	Dickson [55], Weber et al. [56]
Delivery and service	Communication system	Dickson [55], Weber et al. [56]
	Lead time	Handfield [57], Choi & Hartley [58], Verma & Pullman [59], Bharadwa [60], Kannan et al. [61], Chu & Varma [62], Tam & Tummala [63], Shahgholian et al. [64]
	Production capacity	Kannan [61], Dickson [55], Weber et al. [56]
	After sales service	Dzever et al. [65], Choi & Hartley [58], Bevilacqua & Petroni [66], Bharadwaj [60], Rezaei & Ortt [67], Roshandel et al. [68]
Qualitative	Business experience and position among competitors	Banaeian et al. [50], Dickson [55], Weber et al. [56]
	Expert labor, technical capabilities and facilities	Banaeian et al. [50], Dickson [55], Weber et al. [56]
	Operational control	Dickson [55], Weber et al. [56]
	Quality	Grover et al. [55], Dickson [55]
Environmental management system	Environmental friendly technology	Rajesri Govindaraju et al. [69], Grover et al. [53]
	Environmental planning	Banaeian et al. [50], Nielsen et al. [70]
	Environmentally friendly material	Grover et al. [53]
	Environmental prerequisite	Banaeian et al. [50]

3. Material and Methodology

3.1. Research Development

Figure 1 illustrates the selection process, which is sequentially presented in three steps. In the first step, the decision-maker examines the material, interviews the experts, and surveys managers to determine the criteria and sub-criteria affecting to decision making. In the second step data are then processed using the FANP method to rank the criteria. Results from the FANP method are used for the input and output of the DEA model. The DEA model is implemented in the final stage.

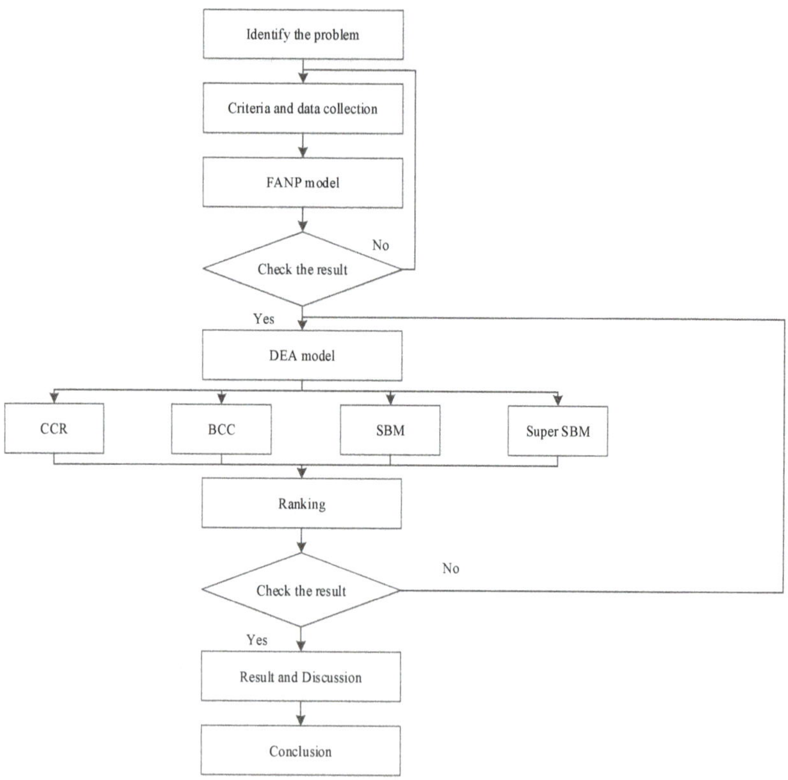

Figure 1. Research process.

Step 1: *Determining evaluation criteria and sub-criteria*

Determine the key criteria and sub-criteria for a comprehensive assessment of the potential supplier. At this stage, the identification of key criteria and sub-criteria is based on a review of the literature and scientific reports related to the content of the research to determine the necessary criteria for the topic [50]. After identifying the groups of criteria required, the decision-maker should select the potential supplier that matches the set criteria. Here, the criteria are defined as four main criteria and 15 sub-criteria, as shown in Figure 2.

Step 2: *Implementing the FANP technique*

Incorporating hybrid fuzzy set theory into the ANP model is the most effective tool for addressing complex problems of decision-making, which has a connection with various qualitative criteria [37].

As can be seen from the solution algorithm in this technique, as presented in Figure 3, at first, the decision-making hierarchical structure is determined to assist the selection [71].

Step 3: *Implementation of the DEA model*

In this study, the FANP and DEA techniques for efficiency measurement have advantages over other fuzzy ANP approaches. In this step, several DEA models, including the Charnes–Cooper–Rhodes model (CCR model), Banker–Charnes–Cooper model (BCC model), Slacks-Based Measure model (SBM model), and Super Slacks-Based Measure model (Super SBM model) are applied to rank suppliers and potential suppliers.

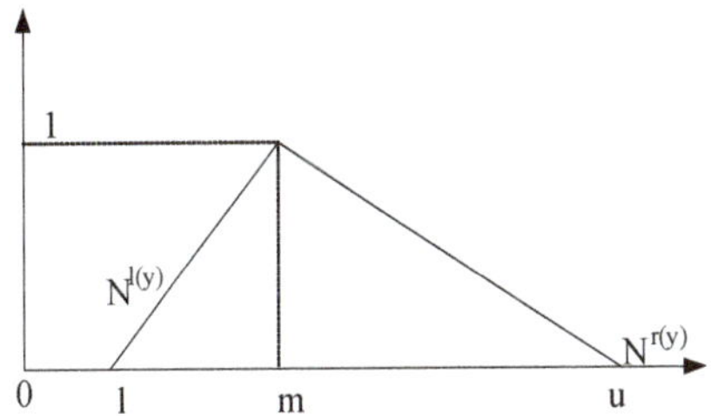

Figure 2. A triangular fuzzy number (TFN).

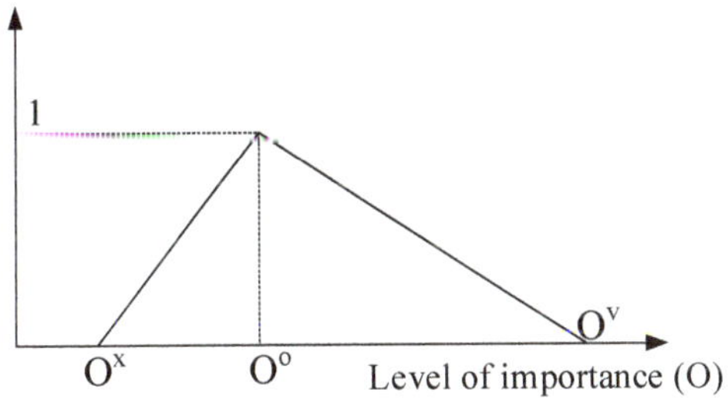

Figure 3. Triangular fuzzy number (TFN).

3.2. Methodology

3.2.1. Fuzzy Set Theory

Fuzzy set was proposed by Zadeh to solve problems existing in uncertain environments. Fuzzy sets are functions that show the dependence degree of one fuzzy number on a set number. A tilde (~) is placed above any symbol representing a fuzzy set number. If $\widetilde{A}$ is a TFN, each value of the membership function is between [0, 1] and can be explained, as shown in Equation (1):

$$\mu_{\widetilde{A}}(x) = \begin{cases} \frac{(x-l)}{(m-l)} & l \le x \le m \\ \frac{(u-x)}{(u-m)} & m \le x \le u \\ 0 & 0.W \end{cases} \quad (1)$$

Each degree of membership includes a left- and right-side representation of a TFN, as shown here:

$$\widetilde{N} = (N^{1(y)}, N^{r(y)}) = (1 - (m - l)y,\ u + (m - u)y), y \in [0,\ 1].$$

A TFN is shown in Figure 2.

3.2.2. Fuzzy Analytic Network Process

ANP does not require a strict hierarchical structure, such as AHP. It allows elements to control, and be controlled, by different levels or clusters of attributes. Several control elements are also present at the same level. Interdependence between factors and their level is defined as a systematic approach to feedback or interactions between elements.

During the ANP process, the elements will be compared pairwise using the expert rating scale, from which the weighting matrix is established. The weights are then adjusted by defining the product of the super matrix.

The AHP method provides a structured framework to set priorities for each level of the hierarchy by using pairwise comparisons quantitated with a priority scale of 1–9, as shown. In contrast, the ANP approach allows for more complex relationships between the elements and their ranks. The 1–9 scale for AHP is shown in Table 2.

Table 2. The 1–9 scale for AHP [6].

Importance Intensity	Definition
1	Equally importance
3	Moderate importance
5	Strongly more importance
7	Very strong more importance
9	Extremely importance
2, 4, 6, 8	Intermediate values

It is clear that the disadvantage of ANP in dealing with the impression and objectiveness in the pairwise comparison process has been improved in the fuzzy analytic network process. The FANP applies a range of values to incorporate the decision-makers' uncertainly [38], whereas the ANP model shows a crisp value. The author assigns the fuzzy conversion scale of this formula, which will be used in the Saaty [72] fuzzy prioritization approach, as shown in Table 2, where $O_{ab} = (O^{x}_{ab}, O^{o}_{ab}, O^{v}_{ab})$ is a triangular fuzzy number with the core O^{o}_{ab}, the support $[O^{x}_{ab}, O^{v}_{ab}]$, and the triangular fuzzy number, as shown in Figure 3.

The 1–9 fuzzy conversion scale is shown in Table 3:

Table 3. The 1–9 fuzzy conversion scale [72].

Importance Intensity	Triangular Fuzzy Scale
1	(1, 1, 1)
2	(1, 1, 2)
3	(1, 2, 3)
4	(2, 3, 4)
5	(3, 4, 5)
6	(4, 5, 6)
7	(5, 6, 7)
8	(7, 8, 9)
9	(9, 9, 9)

The reversed degree to O_{ab} expressing the non-preference is also expressed by a triangular fuzzy number: $(1/O^{v}_{ab}, 1/O^{o}_{ab}, 1/O^{x}_{ab}.)$. By the way, the weights of criteria from the fuzzy Saaty's matrix can be divided into four steps [73]:

1. Fuzzy synthetic extension calculation will transformed into TNT, called fuzzy synthetic extensions $K_a(k^{x}_{a}, k^{o}_{a}, k^{uv}_{a})$. using Equations (2)–(4) [74]:

$$K_a = \sum_{b=1}^{n} O_{ab} \otimes \left(\sum_{a=1}^{n} \sum_{b=1}^{n} O_{ab}\right)^{-1} \quad (2)$$

$$\sum_{j=1}^{n} O_{ab} = \left(\sum_{b=1}^{n} M_{ab}^{x}, \sum_{j=1}^{n} O_{ab}^{o}, \sum_{b=1}^{n} O_{ab}^{v} \right) \quad (3)$$

$$O_{ab}^{-1} = 1/O_{ab}^{v}, 1/O_{ab}^{o}, 1/O_{ab}^{x} \quad (4)$$

$$O \bigotimes N = (O_x \,.\, N_x,\ O_0.\, N_0,\ O_v.N_v) \quad (5)$$

Assign $a = 1, 2, \ldots, n$, in which a and b specifically are triangular fuzzy number (O_x, O_o, O_v) and (N_x, N_0, N_v).

2. Weights of criteria are addressed by using relations of the fuzzy-valued. In this step, fuzzy synthetic extensions are blurred by using the min fuzzy extension of the valued relation $\leq$ given by Equation (5), and weights W_i are calculated (for more detail, see [75]):

$$Q_a = min_b \left\{ \frac{k_b^b - k_a^v}{(k_a^o - k_a^v) - (k_b^o - k_b^x)} \right\} \quad (6)$$

For $a, b = 1, 2, \ldots, n$.

3. The standardization of the weights. If we expect to obtain the sum of weights within one matrix equal to 1, final weights w_i are solved using Equation (7):

$$q_i = Q_i / \sum_{a=1}^{n} Q_a \quad (7)$$

For $a, b = 1, 2, \ldots, n$.

4. An assessment of a Saaty's matrix consistency. In the line with [74], a consistency of the matrix is sufficient if inequality from Equation (8) holds:

$$RT = \frac{CT}{RR} = \frac{\overline{\lambda} - n}{(n-1).RR} \leq 0.1 \quad (8)$$

where $\overline{\lambda}$ is a symbol for the arithmetic mean of the maximum real eigenvalues of the matrices $(a_{ab}^{\xi})_{1 \leq a,b \leq n}, \xi \in \{x, o, v\}$ for $a, b = 1, 2, \ldots, n$ is the size of the Saaty's matrix, and RR represents a random index whose value depends on [74].

3.3. Data Envelopment Analysis

3.3.1. Charnes-Cooper-Rhodes Model (CCR Model)

Charnes, Cooper, and Rhodes (1978) [30] proposed a basic DEA model, called the CCR model:

$$\begin{aligned} &\max_{f.g} \gamma = \frac{f^V y_0}{g^V x_0} \\ \text{S.t.} \quad & \\ &f^V y_b - g^V x_b \leq 0,\ b = 1, 2, \ldots, n \\ &f \geq 0 \\ &g \geq 0 \end{aligned} \quad (9)$$

Due to constraints, the optimal value γ^* is a maximum of 1.

DMU_0 is efficient if $\gamma^* = 1$ and have at least one optimal $f^* > 0$ and $g^* > 0$. In addition, the fractional program can be presented as follows [76]:

$$\begin{array}{l} \min_{g \cdot f} \gamma = g^v y_0 \\ \text{St.} \\ g^v x_0 - 1 = 0 \\ f^v y_j - g^v x_j \leq 0,\ j = 1, 2, \cdots, n \\ g \geq 0 \\ f \geq 0 \end{array} \tag{10}$$

The Farrell [77] model of Equation (10) with variable γ and a nonnegative vector $\beta = \beta_1, \beta_2, \beta_3, \ldots, \beta_n$ is expressed as [76].

$$\begin{array}{l} max \sum_{i=1}^{m} d_i^- + \sum_{r=1}^{s} d_r^+ \\ \text{S.t} \\ \sum_{j=1}^{n} x_{ij}\beta_j + d_i^- = \gamma x_{i0},\ i = 1, 2, \ldots, p \\ \sum_{j=1}^{n} y_{rj}\beta_j - d_r^+ = y_{r0},\ r = 1, 2, \ldots, q \\ \beta_j \geq 0,\ j = 1, 2, \ldots, n \\ d_i^- \geq 0, i = 1, 2, \ldots, p \\ d_r^+ \geq 0, r = 1, 2, \ldots, q \end{array} \tag{11}$$

Equation (11) has a feasible solution, $\gamma^* = 1, \beta_0^* = 1, \beta_j^* = 0, (j \neq 0)$, which effects the optimal value γ^* not greater than 1. The process will be repeated for each DMUj, $j = 1, 2, \ldots, n$. DMUs are inefficient when $\gamma^* < 1$, while DMUs are boundary points if $\gamma^* = 1$. We avoid the weakly efficient frontier point by invoking a linear program as follows [76]:

$$\begin{array}{l} max \sum_{i=1}^{m} d_i^- + \sum_{r=1}^{d} d_r^+ \\ \text{S.t} \\ \sum_{j=1}^{n} x_{ij}\beta_j + d_i^- = \gamma x_{i0},\ i = 1, 2, \ldots, p \\ \sum_{j=1}^{n} y_{rj}\beta_j - d_r^+ = y_{r0},\ r = 1, 2, \ldots, q \\ \beta_j \geq 0,\ j = 1, 2, \ldots, n \\ d_i^- \geq 0, i = 1, 2, \ldots, p \\ d_r^+ \geq 0, r = 1, 2, \ldots, q \end{array} \tag{12}$$

In this case, note that the choices the d_i^- and d_r^+ do not affect the optimal γ^*. The performance of DMU_0 achieves 100% efficiency if, and only if, both (1) $\gamma^* = 1$ and (2) $d_i^{-*} = d_r^+ = 0$. The performance of DMU_0 is weakly efficient if, and only if, both (1) $\gamma^* = 1$ and (2) $d_i^{-*} \neq 0$ and $d_r^+ \neq 0$ for i or r in optimal alternatives. Thus, the preceding development amounts to solving the problem as follows [76]:

$$
\begin{aligned}
& min\theta - \alpha\left(\sum_{i=1}^{m} d_i^- + \sum_{r=1}^{d} d_r^+\right) \\
\text{S.t} \quad & \\
& \sum_{j=1}^{n} x_{ij}\beta_j + d_i^- = \gamma x_{i0},\ i = 1,\ 2, \ldots, p \\
& \sum_{j=1}^{n} y_{rj}\beta_j - d_r^+ = y_{r0},\ r = 1,\ 2, \ldots, q \\
& \beta_j \geq 0,\ j = 1,\ 2, \ldots, n \\
& d_i^- \geq 0, i = 1, 2, \ldots, p \\
& d_r^+ \geq 0, r = 1, 2, \ldots, q
\end{aligned}
\tag{13}
$$

In this case, d_i^- and d_r^+ variables will be used to convert the inequalities into equivalent equations. This is similar to solving Equation (11) in two stages by first minimizing γ and then fixing $\gamma = \gamma^*$ as in Equation (12). This would reset the objective from max to min, as in Equation (9), to obtain [76]:

$$
\begin{aligned}
& \max_{g \cdot f} \gamma = \frac{g^V x_0}{f^V y_j} \\
\text{S.t} \quad & \\
& g^V x_0 \leq g^V y_j,\ j = 1,\ 2, \ldots, n \\
& g \geq \varepsilon > 0 \\
& f \geq \varepsilon > 0
\end{aligned}
\tag{14}
$$

If the $\alpha > 0$ and the non-Archimedean element is defined, the input models are similar to Equations (10) and (13), as follows [76]:

$$
\begin{aligned}
& \max_{g \cdot f} \gamma = g^V x_0 \\
\text{S.t} \quad & \\
& f^V y_0 = 1 \\
& g^V x_o - f^V y_j \geq 0,\ j = 1,\ 2, \ldots, n \\
& g \geq \varepsilon > 0 \\
& f \geq \varepsilon > 0
\end{aligned}
\tag{15}
$$

and:

$$
\begin{aligned}
& max\phi - \varepsilon\left(\sum_{i=1}^{m} d_i^- + \sum_{r=1}^{d} d_r^+\right) \\
\text{S.t} \quad & \\
& \sum_{j=1}^{n} x_{ij}\beta_j + d_i^- = x_{i0},\ i = 1,\ 2, \ldots, p \\
& \textstyle\sum_{j=1}^{n} y_{rj}\beta_j - d_r^+ = \varnothing y_{r0},\ r = 1,\ 2, \ldots, q \\
& \beta_j \geq 0,\ j = 1,\ 2, \ldots, n \\
& d_i^- \geq 0, i = 1, 2, \ldots, p \\
& d_r^+ \geq 0, r = 1, 2, \ldots, q
\end{aligned}
\tag{16}
$$

The input-oriented CCR (CCR-I) has the dual multiplier model, expressed as [76]:

$$\begin{aligned} & \max z = \sum_{r=1}^{q} g_r y_{r0} \\ \text{S.t} & \\ & \sum_{r=1}^{q} g_r y_{rj} - \sum_{r=1}^{q} f_r y_{rj} \leq 0 \\ & \sum_{i=1}^{p} f_i x_{i0} = 1 \\ & g_r, f_i \geq \varepsilon > 0 \end{aligned} \tag{17}$$

The output-oriented CCR (CCR-O) has the dual multiplier model, expressed as [76]:

$$\begin{aligned} & \min q = \sum_{i=1}^{p} f_i x_{i0} \\ \text{S.t} & \\ & \sum_{i=1}^{p} f_i x_{ij} - \sum_{r=1}^{q} g_r y_{rj} \leq 0 \\ & \sum_{r=1}^{q} g_r y_{r0} = 1 \\ & g_r, f_i \geq \varepsilon > 0 \end{aligned} \tag{18}$$

3.3.2. Banker–Charnes–Cooper Model (BCC Model)

Banker et al. proposed the input-oriented BBC model (BCC-I) [30], which is able to assess the efficiency of DMU_0 by solving the following linear program [76]:

$$\begin{aligned} & \gamma_B = min\gamma \\ \text{S.t} & \\ & \sum_{j=1}^{n} x_{ij}\beta_j + d_i^- = \gamma x_{i0},\ i = 1,\ 2, \ldots, p \\ & \sum_{j=1}^{n} y_{rj}\beta_j - d_r^+ = y_{r0},\ r = 1,\ 2, \ldots, q \\ & \sum_{k=1}^{n} \beta_k = 1 \\ & \beta_k \geq 0, k = 1, 2, \ldots, n \end{aligned} \tag{19}$$

We avoid the weakly efficient frontier point by invoking the linear program as follows [76]:

$$\begin{aligned} & max \sum_{i=1}^{m} d_i^- + \sum_{r=1}^{d} d_r^+ \\ \text{S.t} & \\ & \sum_{j=1}^{n} x_{ij}\beta_j + d_i^- = \gamma x_{i0},\ i = 1,\ 2, \ldots, p \\ & \sum_{j=1}^{n} y_{rj}\beta_j - d_r^+ = y_{r0},\ r = 1,\ 2, \ldots, q \\ & \sum_{k=1}^{n} \beta_k = 1 \\ & \beta_k \geq 0,\ k = 1,\ 2, \ldots, n \\ & d_i^- \geq 0, i = 1, 2, \ldots, p \\ & d_r^+ \geq 0, r = 1, 2, \ldots, q \end{aligned} \tag{20}$$

Therefore, this is the first multiplier form to solve the problem as follows [76]:

$$
\begin{gathered}
min\gamma - \varepsilon\left(\sum_{i=1}^{m} d_i^- + \sum_{r=1}^{d} d_r^+\right) \\
\text{S.t} \\
\sum_{j=1}^{n} x_{ij}\beta_j + d_i^- = \gamma x_{i0},\ i = 1,\ 2, \ldots, p \\
\sum_{j=1}^{n} y_{rj}\beta_j - d_r^+ = y_{r0},\ r = 1,\ 2, \ldots, q \\
\sum_{k=1}^{n} \beta_k = 1 \\
\beta_k \geq 0,\ k = 1,\ 2, \ldots, n \\
d_i^- \geq 0, i = 1, 2, \ldots, p \\
d_r^+ \geq 0, r = 1, 2, \ldots, q
\end{gathered} \tag{21}
$$

The linear program in Equation (17) gives us the second multiplier form, which is expressed as [76]:

$$
\begin{gathered}
\max_{g.f, f_0} \gamma_B = f^V y_0 - f_0 \\
\text{S.t} \\
g^V x_0 = 1 \\
f^V y_j - g^V x_j - f_0 \leq 0,\ j = 1,\ 2, \ldots, n \\
g \geq 0 \\
f \geq 0
\end{gathered} \tag{22}
$$

If g and f, which are mentioned in Equation (22), are vectors, the scalar v_0 may be positive or negative (or zero). Thus, the equivalent BCC fractional program is obtained from the dual program in Equation (22) as [76]:

$$
\begin{gathered}
\max_{g.f} \gamma = \frac{f^V y_0 - f_0}{g^V x_0} \\
\text{S.t} \\
\frac{f^V y_j - f_0}{g^V x_j} \leq 1,\ j = 1,\ 2, \ldots, n \\
g \geq 0 \\
f \geq 0
\end{gathered} \tag{23}
$$

The DMU_0 can be called BCC-efficient if an optimal solution $(\gamma_B^*,\ d^{-*}, d^{+*})$ is claimed in this two-phase process for Equation (17) satisfies $\gamma_B^* = 1$ and has no slack $d^{-*} = d^{+*} = 0$, then. The improved activity $(\gamma^* x - d^{-*}, y + d^{+*})$ also can be illustrated as BCC-efficient [76].

The output-oriented BCC model (BCC–O) is:

$$
\begin{gathered}
\max \eta \\
\text{S.t} \\
\sum_{j=1}^{n} x_{ij}\beta_j + d_i^- = \gamma x_{i0},\ i = 1,\ 2, \ldots, p \\
\sum_{j=1}^{n} y_{rj}\beta_j - d_r^+ = \eta y_{r0},\ r = 1,\ 2, \ldots, q \\
\sum_{k=1}^{n} \beta_k = 1 \\
\beta_k \geq 0,\ k = 1,\ 2, \ldots, n
\end{gathered} \tag{24}
$$

From Equation (24), we have the associate multiplier form, which is expressed as [76]:

$$\begin{aligned} &\min_{g\cdot f,g_0} f^V y_0 - f_0 \\ \text{S.t} \\ &f^V y_0 = 1 \\ &g^V x_j - f^V y_j - f_0 \leq 0,\ j = 1,\ 2, \ldots, n \\ &g \geq 0 \\ &f \geq 0 \end{aligned} \tag{25}$$

f_0 is the scalar associated with $\sum_{k=1}^{n} \beta_k = 1$. In conclusion, the authors achieve the equivalent (BCC) fractional programming formulation for Equation (25) [76]:

$$\begin{aligned} &\min_{g\cdot f,g_0} \frac{g^V x_0 - f_0}{f^V y_0} \\ \text{S.t} \\ &\frac{f^V x_j - f_0}{f^V y_j} \leq 1,\ j = 1,\ 2, \ldots, n \\ &g \geq 0 \\ &f \geq 0 \end{aligned} \tag{26}$$

3.3.3. Slacks-Based Measure Model (SBM Model)

The SBM model was introduced by Tone [78] (see also Pastor et al. [79]).

Input-Oriented SBM (SBM-I-C)

The input-oriented SBM under a constant-returns-to-scale assumption [76] is described as follows:

$$\begin{aligned} &\rho_I^* = \min_{\beta,\, d^-, d^+} 1 - \frac{1}{m} \sum_{i=1}^{m} \frac{d_i^-}{x_{ih}} \\ \text{S.t} \\ &x_{ic} = \sum_{j=1}^{m} x_{ic}\beta_i + d_i^-\ ,\ i = 1,\ 2,\ \ldots m \\ &y_{rc} = \sum_{j=1}^{m} y_{rc}\beta_i - d_r^+\ ,\ i = 1,\ 2,\ \ldots d \\ &\beta_j \geq 0,\ k\ (\forall j),\ d_i^- \geq 0\ (\forall j), d_r^+ \geq 0\ (\forall j) \end{aligned} \tag{27}$$

The DMUs in the reference set R of $(x_c,\ y_c)$ are SBM-input-efficient. In addition, the SBM-input-efficiency score must is lower than the CCR efficiency score.

Output-Oriented SBM (SBM-O-C)

The output-oriented SBM efficiency ρ_O^* of DMU$_c$ = (x_c, y_c) is defined by [SBM-O-C] [76]:

$$\begin{aligned} &1/\rho_O^* = \max_{\lambda, s^-, s^+} 1 + \frac{1}{s} \sum_{r=1}^{s} \frac{s_r^+}{y_{rh}} \\ \text{S.t.} \\ &x_{ic} = \sum_{j=1}^{n} x_{ij}\beta_j + d_j^-\,(i = 1, ..m) \\ &y_{ic} = \sum_{j=1}^{n} y_{ij}\beta_j + d_i^+\,(i = 1, \ldots m) \\ &\beta_j \geq 0(\forall j),\ d_i^- \geq 0(\forall i),\ d_i^+ \geq 0\ (\forall r) \end{aligned} \tag{28}$$

The optimal solution of $[\mathrm{SBM} - \mathrm{O} - \mathrm{C}]\beta^*, d^{-*}, d^{+*})$.

3.3.4. Super-Slacks-Based Measure Model (Super SBM Model)

Tone's super SBM model [78] has proposed a slacks-based measure of efficiency (SBM model) that measures the efficiency of the units under evaluation using slack variables only. The super efficiency SBM model removes the evaluated unit DMUq from the set of units and looks for a DMU* with inputs x_i^*, $i = 1, \ldots, m$, and outputs y_k^*, $k = 1, \ldots, r$, being SBM (and CCR) efficient after this removal. The super SBM model is formulated as follow:

$$\text{minimize } \theta_q^{SBM} = \frac{\frac{1}{p}\sum_{i=1}^{m} x_i^*/x_{i0}}{\frac{1}{q}\sum_{k=1}^{r} y_k^*/y_{k0}} \quad (29)$$

S.t

$$\begin{aligned}
&\sum_{j=1}^{n} x_{ij}\beta_j + d_i^- = \gamma x_{i0},\ i = 1,\ 2, \ldots, p \\
&\sum_{j=1}^{n} y_{rj}\beta_j - d_r^+ = y_{r0},\ r = 1,\ 2, \ldots, q \\
&x_i^* \geq x_{i0},\ i = 1,\ 2, \ldots, n \\
&y_k^* \leq y_{k0},\ k = 1,\ 2, \ldots, n \\
&\beta_k \geq 0,\ k = 1,\ 2, \ldots, n \\
&d_i^- \geq 0, i = 1, 2, \ldots, p \\
&d_r^+ \geq 0, r = 1, 2, \ldots, q
\end{aligned} \quad (30)$$

The numerator in the ratio in Equation (29) can be explained as the distance of units DMUq and DMU* in input space and the average reduction rate of inputs of DMU* to inputs of DMUq.

4. Case Study

In this research, the authors collected 25 suppliers (DMU) in Vietnam. Information about the suppliers is shown in Table 4.

Table 4. Number of suppliers (DMU).

No	Company Name	Address	Turnover (USD)	Employees	Market Geographical Area	Symbol
1	An Gia Phu Food and Cereal Limited Liability Company	Vinh Long Province, Vietnam	616,894	25	Vietnam	DMU 1
2	VINA Fragrant Rice Limited Liability Company	Can Tho City, Vietnam	877,662	39	Vietnam	DMU 2
3	Thai Hung Cereal Co-operative Company	Can Tho City, Vietnam	616,309	31	Vietnam	DMU 3
4	Sang Mai Agricultural Production Limited Liability Company	Hai Phong Provice, Vietnam	686,350	39	Vietnam	DMU 4
5	FAS Vietnam Cereal Limited Liability Company	Vinh Long Province, Vietnam	729,349	24	Vietnam	DMU 5
6	S1000 Food Commercial and Service Limited Liability Company	Ho Chi Minh City, Vietnam	590,814	21	Vietnam	DMU 6
7	Khau Thien Thanh Phat Production and Commercial Export-Import Company	Ho Chi Minh City, Vietnam	3,180,926	121	Vietnam, Malaysia, Japan, Australia	DMU 7
8	Gia Son Phat Commercial and Service Limited Liability Company	Kien Giang, Vietnam	613,654	33	Vietnam	DMU 8
9	VILACONIC Cereal Joint Stock Company	Nghe An Province, Vietnam	717,780	31	Vietnam	DMU 9
10	Binh Minh Cereal Joint Stock Company	Can Tho City, Vietnam	658,272	26	Vietnam	DMU 10
11	Phu Thai Huong Joint Stock Company	Long An Province, Vietnam	1.347,621	57	Vietnam	DMU 11
12	Long Tra Agroforestry Food Production Limited Liability Company	Ho Chi Minh City, Vietnam	4,650,698	234	Vietnam, Asia	DMU 12
13	Huong Chien Rice Production Limited Liability Company	Long An Province, Vietnam	674,388	18	Vietnam	DMU 13
14	Loc Troi Joint Stock Incorporated Company	An Giang Province, Vietnam	3,077,786	179	Vietnam, Lao, Cambodia	DMU 14
15	Ngoc Oanh Rice Private Business	Ho Chi Minh City, Vietnam	502,448	23	Vietnam	DMU 15
16	Khanh Tam Rice Private Business	Ho Chi Minh City Vietnam	589,577	16	Vietnam	DMU 16
17	Thien Ngoc Cereal Limited Liability Company	Long An Province, Vietnam	1,094,880	31	Vietnam	DMU 17
18	Xuyen Giang Commercial and Service Limited Liability Company	Ho Chi Minh City, Vietnam	1,475,431	59	Vietnam	DMU 18
19	Viet Lam Commercial and Service Limited Liability Company	Vinh Long Province, Vietnam	1,502,043	42	Vietnam	DMU 19
20	Long An Export-Production Joint Stock Company	Ha Noi City, Vietnam	2,125,825	89	Vietnam, EU	DMU 20
21	Phat Tai Limited Liability Company	Dong Thap Province, Vietnam	1,054,156	29	Vietnam	DMU 21
22	Thai Binh Rice Joint Stock Company	Thai Binh Province, Vietnam	1,777,244	51	Vietnam	DMU 22
23	Angimex Kitoku Limited Liability Company	Tien Giang Province, Vietnam	1,098,978	38	Vietnam	DMU 23
24	Hoa Lua Rice Commercial Limited Liability Company	Ho Chi Minh City, Vietnam	1,029,622	59	Vietnam	DMU 24
25	Phuong Quan Production Limited Liability Company	Long An Province, Vietnam	1,733,256	61	Vietnam	DMU 25

The data collection of the FANP and hierarchical structure are introduced in Figure 4.

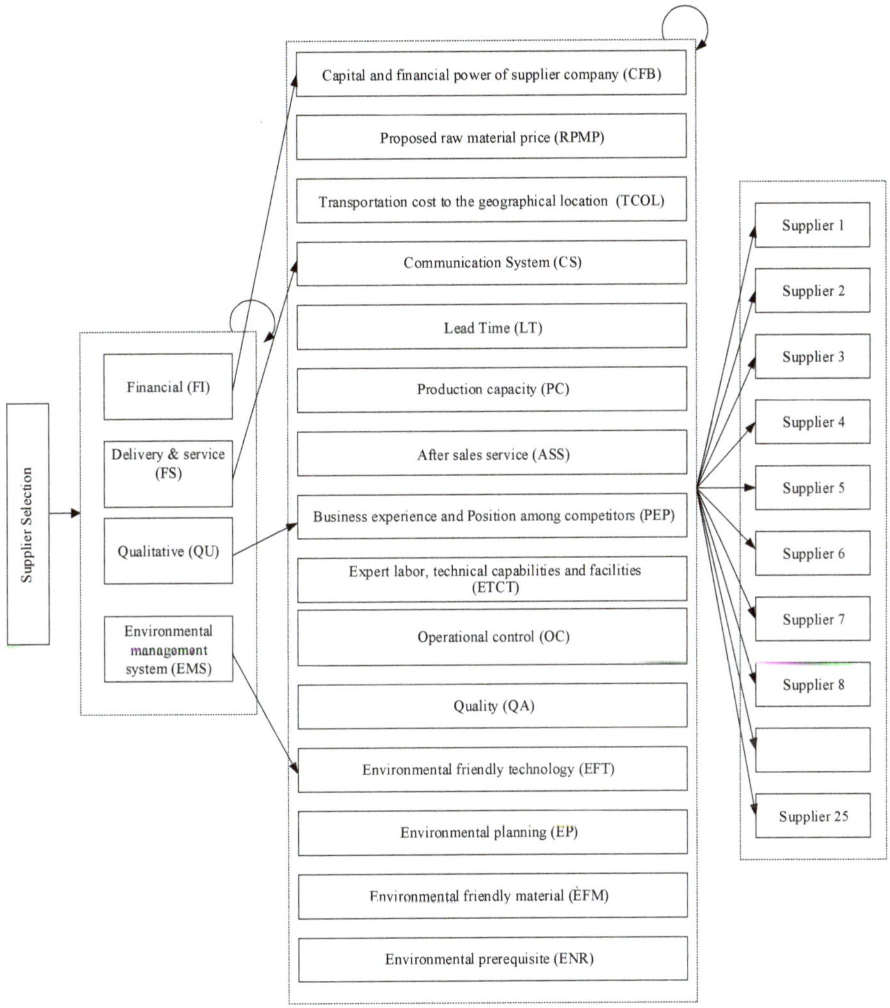

Figure 4. Hierarchical structure to select best suppliers.

A fuzzy comparison matrix for all criteria is shown in Table 5.

Table 5. Fuzzy comparison matrix for criteria.

Criteria	FS	EMS	FI	QU
FS	(1, 1, 1)	(1/8, 1/7, 1/6)	(1/9, 1/8, 1/7)	(1/3, 1/2, 1)
EMS	(6, 7, 8)	(1, 1, 1)	(1/6, 1/5, 1/4)	(1, 2, 3)
FI	(7, 8, 9)	(4, 5, 6)	(1, 1, 1)	(4, 5, 6)
QU	(1, 2, 3)	(1/3, 1/2, 1)	(1/6, 1/5, 1/4)	(1, 1, 1)

During the defuzzification, we obtain the coefficients $\alpha = 0.5$ and $\beta = 0.5$ (Tang and Beynon) [80]. In it, α represents the uncertain environment conditions, and β represents the attitude of the evaluator is fair.

$$g_{0.5,0.5}(\overline{a_{EMS,FS}}) = [(0.5 \times 6.5) + (1 - 0.5) \times 7.5] = 7$$
$$f_{0.5}(L_{EMS,FS}) = (7 - 6) \times 0.5 + 6 = 6.5$$
$$f_{0.5}(U_{EMS,FS}) = 8 - (8 - 7) \times 0.5 = 7.5$$
$$g_{0.5,0.5}(\overline{a_{EMS,FS}}) = 1/7$$

The remaining calculations are similar to the above, as well as the fuzzy number priority points. The real number priorities when comparing the main criteria pairs are presented in Table 6.

Table 6. Real number priority.

Criteria	FS	EMS	FI	QU
FS	1	1/7	1/8	1/2
EMS	7	1	1/6	2
FI	8	6	1	5
QU	2	1/2	1/5	1

We calculate the maximum individual values as follows:

$$GM1 = (1 \times 1/7 \times 1/8 \times 1/2)^{1/4} = 0.03073$$
$$GM2 = (7 \times 1 \times 1/6 \times 2)^{1/4} = 1.2359$$
$$GM3 = (8 \times 6 \times 1 \times 5)^{1/4} = 3.9359$$
$$GM4 = (2 \times 1/2 \times 1/5 \times 1)^{1/4} = 0.6687$$
$$\sum GM = GM1 + GM2 + GM3 + GM4 = 6.1478$$

$$\omega_1 = \frac{0.3073}{6.1478} = 0.0499$$
$$\omega_2 = \frac{1.2359}{6.1478} = 0.2010$$
$$\omega_3 = \frac{3.9359}{6.1478} = 0.6402$$
$$\omega_4 = \frac{0.6687}{6.1478} = 0.1087$$

$$\begin{bmatrix} 1 & 1/7 & 1/8 & 1/2 \\ 7 & 1 & 1/6 & 2 \\ 8 & 6 & 1 & 5 \\ 2 & 1/2 & 1/5 & 1 \end{bmatrix} \times \begin{bmatrix} 0.0499 \\ 0.2010 \\ 0.6402 \\ 0.1087 \end{bmatrix} = \begin{bmatrix} 0.2129 \\ 0.8744 \\ 2.7889 \\ 0.4370 \end{bmatrix} \begin{bmatrix} 0.2129 \\ 0.8744 \\ 2.7889 \\ 0.4370 \end{bmatrix} / \begin{bmatrix} 0.0499 \\ 0.2010 \\ 0.6402 \\ 0.1087 \end{bmatrix} = \begin{bmatrix} 4.2665 \\ 4.3502 \\ 4.3562 \\ 4.0202 \end{bmatrix}$$

with the number of criteria is 4, we obtain $n = 4$, and λ_{max} and CI are calculated as follows:

$$\lambda_{max} = \frac{4.2665 + 4.3502 + 4.3562 + 4.0202}{4} = 4.2482$$

$$CI = \frac{4.2482 - 4}{4 - 1} = 0.0827$$

For CR, with $n = 4$ we obtain $RI = 0.9$:

$$CR = \frac{0.0827}{1.12} = 0.0919$$

We have $CR = 0.0919 \leq 0.1$, so the pairwise comparison data is consistent and does not need to be re-evaluated. The results of the pair comparison between the main criteria are presented in Tables 7–11.

Table 7. Fuzzy comparison matrices for the criteria.

Criteria	FS	EMS	FI	QU	Weight
FS	(1, 1, 1)	(1/8, 1/7, 1/6)	(1/9, 1/8, 1/7)	(1/3, 1/2, 1)	0.04929
EMS	(6, 7, 8)	(1, 1, 1)	(1/7, 1/6, 1/5)	(1, 2, 3)	0.20144
FI	(7, 8, 9)	(5, 6, 7)	(1, 1, 1)	(4, 5, 6)	0.64816
QU	(1, 2, 3)	(1/3, 1/2, 1/1)	(1/6, 1/5, 1/4)	(1, 1, 1)	0.10111
Total					1
CR = 0.09480					

Table 8. Comparison matrix for the financial criteria.

Criteria	CFB	RPMP	TCOOL	Weight
CFB	(1, 1, 1)	(1/5, 1/4, 1/3)	(3, 4, 5)	0.2290
RPMP	(3, 4, 5)	(1, 1, 1)	(6, 7, 8)	0.6955
TCOOL	(1/5, 1/4, 1/3)	(1/8, 1/7, 1/6)	(1, 1, 1)	0.0754
Total				1
CR = 0.07348				

Table 9. Comparison matrix for the delivery and service criteria.

Criteria	CS	LT	PC	ASS	Weight
CS	(1, 1, 1)	(1/9, 1/8, 1/7)	(1/5, 1/4, 1/3)	(2, 3, 4)	0.0924
LT	(7, 8, 9)	(1, 1, 1)	(1/3, 1/2, 1)	(6, 7, 8)	0.3956
PC	(3, 4, 5)	(1, 2, 3)	(1, 1, 1)	(7, 8, 9)	0.4672
ASS	(1/4, 1/3, 1/2)	(1/8, 1/7, 1/6)	(1/9, 1/8, 1/7)	(1, 1, 1)	0.0448
Total					1
CR = 0.09456					

Table 10. Comparison matrix for the qualitative criteria.

Criteria	PEP	ETCT	OC	QA	Weight
PEP	(1, 1, 1)	(2, 3, 4)	(4, 5, 6)	(1/5, 1/4, 1/3)	0.2136
ETCT	(1/4, 1/3, 1/2)	(1, 1, 1)	(1/4, 1/3, 1/2)	(1, 1, 1)	0.0436
OC	(1/6, 1/5, 1/4)	(2, 3, 4)	(1, 1, 1)	(1/9, 1/8, 1/7)	0.0791
QA	(3, 4, 5)	(1, 1, 1)	(7, 8, 9)	(1, 1, 1)	0.6638
Total					1
CR = 0.09005					

Table 11. Comparison matrix for the environmental management systems criteria.

Criteria	EFT	EP	EFM	ENR	Weight
EFT	(1, 1, 1)	(1/9, 1/9, 1/9)	(1/6, 1/5, 1/4)	(1/6, 1/5, 1/4)	0.0445
EP	(9, 9, 9)	(1, 1, 1)	(1, 2, 3)	(5, 6, 7)	0.5345
EFM	(4, 5, 6)	(1/3, 1/2, 1)	(1, 1, 1)	(3, 4, 5)	0.3009
ENR	(4, 5, 6)	(1/7, 1/6, 1/5)	(1/5, 1/4, 1/3)	(1, 1, 1)	0.1201
Total					1
CR = 0.0838					

Based on how the hierarchical structure was built, the pairwise comparison matrix was built through completing a questionnaire. Then, the received data to calculate the weight of supplier's indices and to ensure the accuracy of judged inconsistency rate and other constraints are presented.

In summary, a graphic of the DEA model for analysis of DMUs (suppliers) along with three inputs and three outputs is shown in Figure 4. The results of the FANP model for the ranking of various suppliers on qualitative attributes are utilized in the output qualitative benefits of the DEA model [71,81]. In our situation, inputs are those factors that organizations would consider as an improvement if they were decreased in value (i.e., smaller values are better), whereas outputs are those factors that organizations would consider as improvements if they were increased in value (i.e., larger is better). This is a standard approach when seeking to use DEA as a discrete alternative multiple criteria decision-making tool [71]. There are three inputs and three outputs, as shown in Figure 5.

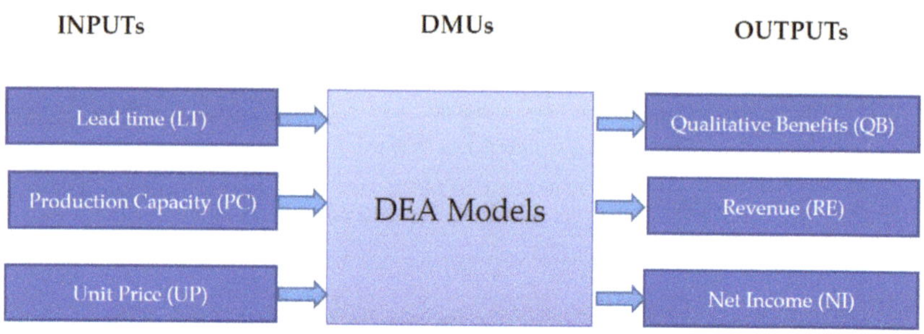

Figure 5. Data envelopment analysis model.

To aid in reducing scaling errors associated with the mathematical programming software packages, the dataset is mean normalized for each factor, i.e., each value in each column is divided by that column's mean score. This normalization procedure does not change the efficiency scores of the ratio-based DEA models. As previously mentioned, to help model the analysis as inputs and outputs, instead of the standard productivity efficiency measurement approach, assume that the inputs are those factors that improve as their values decrease and the outputs are those values that improve as their values increase [71]. Raw data are provided by the case organization, as shown in Table 12.

Table 12. Raw data provided by case organization used to assess the relative efficiency of various suppliers.

A Supplier (DMU)	Input			Output		
	LT (Days)	UP (USD)	PC (Tons)	QB (%)	NI (USD)	RE (USD)
DMU 1	3	347.3	50	3.7221	44.03	58.71
DMU 2	5	391.45	70	1.3459	25.20	33.60
DMU 3	4	332.4	50	0.8243	26.03	34.70
DMU 4	4	321.5	40	1.7611	22.95	30.60
DMU 5	4	213.5	50	1.0023	40.05	53.40
DMU 6	4	312.6	50	1.6047	30.45	40.60
DMU 7	5	345.3	40	2.5748	48.00	68.20
DMU 8	5	342.9	70	2.0095	44.03	58.71
DMU 9	3	343.6	50	3.2401	32.70	43.60
DMU 10	3	354.1	30	3.0687	44.29	59.05
DMU 11	5	320.10	30	4.0040	32.78	43.70
DMU 12	3	346.30	70	2.9141	44.02	58.70
DMU 13	4	340.60	50	4.0194	44.12	58.83
DMU 14	4	315.05	40	5.1484	34.88	46.50

Table 12. *Cont.*

A Supplier (DMU)	Input			Output		
	LT (Days)	UP (USD)	PC (Tons)	QB (%)	NI (USD)	RE (USD)
DMU 15	5	332.40	60	4.6604	43.02	57.36
DMU 16	4	350.90	40	5.4623	50.00	74.30
DMU 17	4	320.00	71	6.1238	44.01	58.68
DMU 18	5	344.60	50	4.7115	44.12	58.82
DMU 19	5	314.03	50	7.4178	44.15	58.86
DMU 20	4	342.30	40	4.7039	44.06	58.75
DMU 21	5	310.80	50	3.2497	44.15	58.86
DMU 22	4	312.40	50	6.8631	43.93	58.57
DMU 23	5	342.00	50	7.4577	43.92	58.56
DMU 24	5	337.60	70	6.5602	43.11	57.48
DMU 25	5	340.10	50	5.5501	43.02	57.36

4.1. Isotonicity Test

The variables of input and output for the correlation coefficient matrix should comply with the isotonicity premise. In other words, the increase of an input will not cause the decreasing output of another item. The results of the Pearson correlation coefficient test are shown in Table 13.

Table 13. The results of the Pearson correlation coefficient.

Inputs/Outputs	LT	UP	PC	QB	NI	RE
LT	1	0.02484	0.16149	0.24257	0.0776	0.07681
UP	0.02484	1	0.14105	0.09301	0.00725	0.03435
PC	0.16149	0.14105	1	0.01713	0.04728	0.00201
QB	0.24257	0.09301	0.01713	1	0.54664	0.51879
NI	0.0776	0.00725	0.04728	0.54664	1	0.98863
RE	0.07681	0.03435	0.00201	0.51879	0.98863	1

Based on the results of Pearson correlation test, the results of all correlation coefficients are positive, thus meeting a basic assumption of the DEA model. Hence, we do not to change the input and output.

4.2. Results and Discussion

Supplier evaluation and selection have been identified as important issues that could affect the efficiency of a supply chain. It can be seen that selecting a supplier is complicated in that decision-makers must understand qualitative and quantitative features for assessing the symmetrical impact of the criteria to reach the most accurate result.

For the performance in an empirical study, the authors collected data from 25 suppliers in Vietnam. A hierarchical structure to select the best suppliers is built with four main criteria (including 15 sub-criteria). Completion of a questionnaire for analyzing the FANP model is done by interviewing experts, and surveying the managers and company's databases. The ANP model is combined with a fuzzy set, to evaluate the supplier selection criteria and define the priorities of each supplier, which are able to be utilized to clarify important criteria that directly affect the profitability of the business. Then, several DEA models are proposed for ranking suppliers. As a result, DMU 1, DMU 5, DMU 10, DMU 16, DMU 19, DMU 22, and DMU 23 are identified as efficient in all nine models, as shown in Table 7 [78], which have a conditioned response to the enterprises' supply requirements. Whereas for other DMUs, there were differences in the results, so the next research should include an improvement or review of data inputs to produce appropriate outputs, so that suppliers remain efficient. This integration model supports a great deal of corporate decision-making because of the effectiveness and the complication of this model, for exactly choosing the most suitable supplier. The ranking list of 25 DMUs as shown in Table 14.

Table 14. Ranking list of suppliers by using nine DEA models (CCR, BCC, and SBM, Super SBM).

Supplier	CCR-I	CCR-O	BCC-I	BCC-O	SBM-I-C	SBM-O-C	Super SBM-I-C	Super SBM-AR-C	Super SBM-AR-V
DMU 1	**1**	**1**	**1**	**1**	**1**	**1**	**1**	**1**	**1**
DMU 2	25	25	25	24	25	24	25	24	24
DMU 3	23	23	22	23	23	25	23	25	25
DMU 4	24	24	15	25	24	23	24	23	21
DMU 5	**1**	**1**	**1**	**1**	**1**	**1**	**1**	**1**	**1**
DMU 6	22	22	20	22	22	22	22	22	23
DMU 7	9	9	12	12	9	20	9	19	20
DMU 8	21	21	24	20	20	21	20	21	22
DMU 9	20	20	1	11	21	18	21	20	11
DMU 10	**1**	**1**	**1**	**1**	**1**	**1**	**1**	**1**	**1**
DMU 11	11	11	1	1	18	11	18	16	1
DMU 12	1	1	1	1	8	1	8	1	1
DMU 13	13	13	16	18	13	14	13	14	16
DMU 14	16	16	1	1	12	15	12	12	1
DMU 15	19	19	23	20	19	17	19	17	18
DMU 16	**1**	**1**	**1**	**1**	**1**	**1**	**1**	**1**	**1**
DMU 17	10	10	13	13	11	9	11	10	13
DMU 18	18	18	21	19	17	16	17	15	17
DMU 19	**1**	**1**	**1**	**1**	**1**	**1**	**1**	**1**	**1**
DMU 20	12	12	14	16	10	12	10	9	12
DMU 21	14	14	18	15	15	19	15	18	19
DMU 22	**1**	**1**	**1**	**1**	**1**	**1**	**1**	**1**	**1**
DMU 23	**1**	**1**	**1**	**1**	**1**	**1**	**1**	**1**	**1**
DMU 24	15	15	19	14	16	10	16	13	15
DMU 25	17	17	17	17	14	13	14	11	14

The optimal weights and the slacks for each DMU using nine DEA models (CCR, BCC, and SBM, Super SBM) are shown from Tables A1–A18 in appendix section.

5. Conclusions

Many studies have applied the MCDM approach to various fields of science and engineering, and their numbers have been increasing over the past years. The fuzzy MCDM model has been applied to supplier selection problems. Although some studies have considered a review of applications of MCDM approaches in this field, little work has focused on this problem in a fuzzy environment. This is a reason why hybrid ANP with fuzzy logic and DEA is proposed in this study.

Initially, we proposed the ANP model combined with a fuzzy set, to evaluate supplier selection criteria and define a priority of each supplier, which are able to be utilized to clarify important criteria that directly affect the profitability of the business. The FANP can be used for ranking suppliers but the number of supplier selections is practically limited because of the number of pairwise comparisons that need to be made, and a disadvantage of the FANP approach is that input data, expressed in linguistic terms, depend on the experience of decision-makers and, thus, involves subjectivity. This is a reason why several DEA models are proposed for ranking suppliers in the final stage. The DEA model can handle hundreds of suppliers with multiple inputs and outputs for the best supplier rating. The FANP-DEA integration model supports a great deal of corporate decision-making because of the effectiveness and complication of this model, for exactly choosing the most suitable supplier. Finally, this research will provide a potential suppliers list, which has a conditioned response to the enterprises' supply requirements.

The main contribution of this research is to develop complete approaches for supplier evaluation and selection of the rice supply chain as a typical example. This is a useful proposed model on an academic and practical front. The FANP-DEA method not only provides reasonable results but also allows the decision-maker to visualize the impact of different criteria in the final result. Furthermore, this integrated model may offer valuable insights, as well as provide methods for other sectors to select and evaluate suppliers. This model can also be applied to many different industries for future research directions.

For improving these MCDM model, outlier detection and the curse of dimensionality of the DEA model will be considered in future research. Moreover, different methodologies, such as the preference ranking organization method for enrichment of evaluations (PROMETHEE), fuzzy data envelopment analysis (FDEA), etc., can also been combined for different scenarios.

Author Contributions: In this research, C.-N.W. contributed to generating the research ideas and designing the theoretical verifications, and reviewed the manuscript; V.T.N. contributed the research ideas, designed the framework, collected data, analyzed the data, and summarized and wrote the manuscript; D.H.D. collected data, analyzed the data, wrote the manuscript; and H.T.D. wrote and formatted the manuscript.

Conflicts of Interest: The authors declare no conflict of interest.

Appendix A

Table A1. The optimal weights for each DMU using the CCR-I model.

DMU	Score	Rank	V (1)	V (2)	V (3)	U (1)	U (2)	U (3)
DMU 1	1	1	0.312446	0	1.25×10^{-3}	4.57×10^{-2}	0	1.41×10^{-2}
DMU 2	0.4245	25	9.96×10^{-2}	1.25×10^{-3}	2.05×10^{-3}	0	1.68×10^{-2}	0
DMU 3	0.5329	23	0.121082	1.51×10^{-3}	2.49×10^{-4}	0	2.05×10^{-2}	0
DMU 4	0.4876	24	0	2.12×10^{-3}	7.97×10^{-3}	0	0.021246	0
DMU 5	1	1	7.31×10^{-3}	4.55×10^{-3}	0	5.74×10^{-2}	0	1.76×10^{-2}
DMU 6	0.6428	22	0.124062	1.57×10^{-3}	2.79×10^{-4}	6.24×10^{-4}	2.11×10^{-2}	0
DMU 7	0.9708	9	0	2.02×10^{-3}	7.59×10^{-3}	0	2.02×10^{-2}	0
DMU 8	0.79	21	0.105865	1.37×10^{-3}	0	2.51×10^{-3}	1.78×10^{-2}	0
DMU 9	0.7934	20	0.333333	0	0	0.10656	1.37×10^{-2}	0
DMU 10	1	1	0.303641	0	2.97×10^{-3}	0.097177	1.41×10^{-2}	1.34×10^{-3}
DMU 11	0.9529	11	0	0	3.33×10^{-2}	0.186293	6.31×10^{-3}	0
DMU 12	1	1	0.136388	1.71×10^{-3}	0	0	2.27×10^{-2}	0
DMU 13	0.8941	13	0.118819	1.54×10^{-3}	0	2.82×10^{-3}	2.00×10^{-2}	0
DMU 14	0.8845	16	8.30×10^{-2}	0	1.67×10^{-2}	0.139719	4.74×10^{-3}	0
DMU 15	0.8357	19	1.53×10^{-2}	2.29×10^{-3}	2.72×10^{-3}	2.77×10^{-2}	1.64×10^{-2}	0
DMU 16	1	1	0.112767	1.49×10^{-3}	6.21×10^{-4}	0	2.00×10^{-2}	0
DMU 17	0.9683	10	8.09×10^{-2}	2.11×10^{-3}	0	2.32×10^{-2}	1.88×10^{-2}	0
DMU 18	0.858	18	0	2.33×10^{-3}	3.91×10^{-3}	2.59×10^{-2}	1.67×10^{-2}	0
DMU 19	1	1	0	3.18×10^{-3}	0	0.134811	0	0
DMU 20	0.8967	12	0	2.44×10^{-3}	4.10×10^{-3}	2.71×10^{-2}	1.75×10^{-2}	0
DMU 21	0.8909	14	0	2.53×10^{-3}	4.25×10^{-3}	2.81×10^{-2}	0.018109	0
DMU 22	1	1	0.25	0	0	0.145707	0	0
DMU 23	1	1	0	1.20×10^{-4}	1.92×10^{-2}	0.10872	4.31×10^{-3}	0
DMU 24	0.8906	15	1.85×10^{-3}	2.69×10^{-3}	0	3.52×10^{-2}	4.82×10^{-3}	7.86×10^{-3}
DMU 25	0.8705	17	0	2.36×10^{-3}	3.96×10^{-3}	2.62×10^{-2}	1.69×10^{-2}	0

Table A2. The slacks for each DMU using the CCR-I model.

DMU	Score	Rank	LT	UP	PC	QB	NI	RE
DMU 1	1	1	0	0	0	0	0	0
DMU 2	0.4245	25	0	0	0	0.012	0	0.001
DMU 3	0.5329	23	0	0	0	0.558	0	0.006
DMU 4	0.4876	24	0.064	0	0	0.531	0	3.114
DMU 5	1	1	0	0	0	0	0	0
DMU 6	0.6428	22	0	0	0	0	0	0.275
DMU 7	0.9708	9	0.995	0	0	2.588	0	2.98
DMU 8	0.79	21	0	0	2.474	0	0	0.221
DMU 9	0.7934	20	0	19.826	2.518	0	0	0.001
DMU 10	1	1	0	0	0	0	0	0
DMU 11	0.9529	11	1.906	69.61	0	0	0	3.945
DMU 12	1	1	0	0	0	0	0	0
DMU 13	0.8941	13	0	0	3.512	0	0	4.416
DMU 14	0.8845	16	0	16.896	0	0	0	1.959
DMU 15	0.8357	19	0	0	0	0	0	0.231

Table A2. *Cont.*

DMU	Score	Rank	LT	UP	PC	QB	NI	RE
DMU 16	1	1	0	0	0	0	0	0
DMU 17	0.9683	10	0	0	22.934	0	0	1.983
DMU 18	0.858	18	0.222	0	0	0	0	3.722
DMU 19	1	1	0	0	0	0	0	0
DMU 20	0.8967	12	0.034	0	0	0	0	6.509
DMU 21	0.8909	14	0.489	0	0	0	0	3.554
DMU 22	1	1	0	0	0	0	0	0
DMU 23	1	1	0	0	0	0	0	0
DMU 24	0.8906	15	0	0	13.043	0	0	0
DMU 25	0.8705	17	0.116	0	0	0	0	2.629

Table A3. The optimal weights for each DMU using the CCR-O model.

DMU	Score	Rank	V (1)	V (2)	V (3)	U (1)	U (2)	U (3)
DMU 1	1	1	0.219376	9.79×10^{-4}	3.76×10^{-5}	0	2.27×10^{-2}	0
DMU 2	0.4245	25	0.234678	2.93×10^{-3}	4.82×10^{-4}	0	3.97×10^{-2}	0
DMU 3	0.5329	23	0.227195	2.84×10^{-3}	4.66×10^{-4}	0	3.84×10^{-2}	0
DMU 4	0.4876	24	0	4.35×10^{-3}	1.63×10^{-2}	0	4.36×10^{-2}	0
DMU 5	1	1	9.80×10^{-2}	2.85×10^{-3}	0	0	0	1.87×10^{-2}
DMU 6	0.6428	22	0.193011	2.44×10^{-3}	4.34×10^{-4}	9.71×10^{-4}	3.28×10^{-2}	0
DMU 7	0.9708	9	0	2.08×10^{-3}	7.82×10^{-3}	0	2.08×10^{-2}	0
DMU 8	0.79	21	0.134006	1.74×10^{-3}	0	3.18×10^{-3}	2.26×10^{-2}	0
DMU 9	0.7934	20	0.420146	0	0	0.134312	1.73×10^{-2}	0
DMU 10	1	1	0.333333	0	0	0	0	1.69×10^{-2}
DMU 11	0.9529	11	0	0	3.50×10^{-2}	0.195497	6.63×10^{-3}	0
DMU 12	1	1	0.136388	1.71×10^{-3}	0	0	2.27×10^{-2}	0
DMU 13	0.8941	13	0.132886	1.72×10^{-3}	0	3.16×10^{-3}	2.24×10^{-2}	0
DMU 14	0.8845	16	0	0	2.83×10^{-2}	0.157959	5.35×10^{-3}	0
DMU 15	0.8357	19	1.83×10^{-2}	2.74×10^{-3}	3.26×10^{-3}	0.033178	1.97×10^{-2}	0
DMU 16	1	1	0.116963	1.52×10^{-3}	0	2.78×10^{-3}	1.97×10^{-2}	0
DMU 17	0.9683	10	8.36×10^{-2}	2.18×10^{-3}	0	2.40×10^{-2}	1.94×10^{-2}	0
DMU 18	0.858	18	0	2.72×10^{-3}	4.56×10^{-3}	3.02×10^{-2}	1.94×10^{-2}	0
DMU 19	1	1	0	3.18×10^{-3}	0	0.134811	0	0
DMU 20	0.8967	12	0	2.72×10^{-3}	4.57×10^{-3}	3.02×10^{-2}	0.019468	0
DMU 21	0.8909	14	0	2.84×10^{-3}	4.77×10^{-3}	3.16×10^{-2}	2.03×10^{-2}	0
DMU 22	1	1	8.23×10^{-2}	2.15×10^{-3}	0	2.36×10^{-2}	1.91×10^{-2}	0
DMU 23	1	1	6.40×10^{-2}	1.37×10^{-4}	1.27×10^{-2}	0.114556	0	2.49×10^{-3}
DMU 24	0.8906	15	2.08×10^{-2}	3.02×10^{-3}	0	3.96×10^{-2}	5.41×10^{-3}	8.82×10^{-3}
DMU 25	0.8705	17	0	2.71×10^{-3}	4.54×10^{-3}	3.01×10^{-2}	1.94×10^{-2}	0

Table A4. The slacks for each DMU using the CCR-O model.

No.	DMU	Score	Rank	LT	UP	PC	QB	NI	RE
1	DMU 1	1	1	0	0	0	0	0	0
2	DMU 2	0.4245	25	0	0	0	0.029	0	0.002
3	DMU 3	0.5329	23	0	0	0	1.047	0	0.012
4	DMU 4	0.4876	24	0.131	0	0	1.09	0	6.387
5	DMU 5	1	1	0	0	0	0	0	0
6	DMU 6	0.6428	22	0	0	0	0	0	0.428
7	DMU 7	0.9708	9	1.025	0	0	2.665	0	3.07
8	DMU 8	0.79	21	0	0	3.132	0	0	0.279
9	DMU 9	0.7934	20	0	24.989	3.174	0	0	0.002
10	DMU 10	1	1	0	0	0	0	0	0
11	DMU 11	0.9529	11	2	73.049	0	0	0	4.14
12	DMU 12	1	1	0	0	0	0	0	0
13	DMU 13	0.8941	13	0	0	3.928	0	0	4.939
14	DMU 14	0.8845	16	0	19.102	0	0	0	2.214

Table A4. *Cont.*

No.	DMU	Score	Rank	LT	UP	PC	QB	NI	RE
15	DMU 15	0.8357	19	0	0	0	0	0	0.277
16	DMU 16	1	1	0	0	0	0	0	0
17	DMU 17	0.9683	10	0	0	23.686	0	0	2.048
18	DMU 18	0.858	18	0.259	0	0	0	0	4.338
19	DMU 19	1	1	0	0	0	0	0	0
20	DMU 20	0.8967	12	0.037	0	0	0	0	7.259
21	DMU 21	0.8909	14	0.549	0	0	0	0	3.989
22	DMU 22	1	1	0	0	0	0	0	0
23	DMU 23	1	1	0	0	0	0	0	0
24	DMU 24	0.8906	15	0	0	14.645	0	0	0
25	DMU 25	0.8705	17	0.133	0	0	0	0	3.019

Table A5. The optimal weights for each DMU using the BBC-I model.

DMU	Score	Rank	V (1)	V (2)	V (3)	U (0)	U (1)	U (2)	U (3)
DMU 1	1	1	0.333333	0	0	0	9.02×10^{-2}	0	1.13×10^{-2}
DMU 2	0.7047	25	0.120608	9.27×10^{-4}	4.87×10^{-4}	0.7047	0	0	0
DMU 3	0.8647	22	0.148001	1.14×10^{-3}	5.97×10^{-4}	0.8647	0	0	0
DMU 4	0.9274	15	2.67×10^{-2}	1.57×10^{-3}	9.71×10^{-3}	0.9274	0	0	0
DMU 5	1	1	0	4.68×10^{-3}	0	0	5.84×10^{-2}	0	1.76×10^{-2}
DMU 6	0.8847	20	0.151411	1.16×10^{-3}	6.11×10^{-4}	0.8847	0	0	0
DMU 7	0.9792	12	0	1.81×10^{-3}	9.41×10^{-3}	0.2364	0	1.55×10^{-2}	0
DMU 8	0.792	24	0.106413	1.36×10^{-3}	0	0.03772	0	1.88×10^{-2}	0
DMU 9	1	1	0.165486	1.47×10^{-3}	0	0.9636	1.12×10^{-2}	0	0
DMU 10	1	1	0.132633	1.68×10^{-3}	2.98×10^{-4}	0	6.68×10^{-4}	2.25×10^{-2}	0
DMU 11	1	1	0	0	3.33×10^{-2}	0.1448	0.179671	4.14×10^{-3}	0
DMU 12	1	1	0.244753	7.67×10^{-4}	0	0	0	3.06×10^{-3}	1.47×10^{-2}
DMU 13	0.9087	16	0.10788	1.53×10^{-3}	9.40×10^{-4}	0.297	1.77×10^{-2}	1.23×10^{-2}	0
DMU 14	1	1	4.61×10^{-7}	7.35×10^{-4}	1.46×10^{-2}	0.5676	8.40×10^{-2}	0	0
DMU 15	0.8389	23	0.017621	2.74×10^{-3}	0	0.34973	3.02×10^{-2}	2.44×10^{-4}	0
DMU 16	1	1	7.04×10^{-2}	2.05×10^{-3}	0	0	0	0	1.35×10^{-2}
DMU 17	0.9747	13	0.115365	1.68×10^{-3}	0	0.2037	1.85×10^{-2}	1.49×10^{-2}	0
DMU 18	0.8811	21	0	1.76×10^{-3}	7.86×10^{-3}	0.5174	2.40×10^{-2}	5.68×10^{-3}	0
DMU 19	1	1	0	3.18×10^{-3}	0	0	0.134811	0	0
DMU 20	0.9679	14	1.71×10^{-2}	1.79×10^{-3}	7.96×10^{-3}	0.6404	0.027225	4.53×10^{-3}	0
DMU 21	0.8998	18	0	1.87×10^{-3}	0.008356	0.5501	2.55×10^{-2}	6.04×10^{-3}	0
DMU 22	1	1	7.72×10^{-2}	2.21×10^{-3}	0	0	0.145707	0	0
DMU 23	1	1	0	0	2.00×10^{-7}	0	0.117199	0	2.15×10^{-3}
DMU 24	0.8989	19	2.05×10^{-2}	2.66×10^{-3}	0	0.6047	0.04485	0	0
DMU 25	0.9038	17	1.23×10^{-2}	1.68×10^{-3}	7.35×10^{-3}	0.5734	2.54×10^{-2}	4.40×10^{-3}	0

Table A6. The slacks for each DMU using the BCC-I model.

DMU	Score	Rank	LT	UP	PC	QB	NI	RE
DMU 1	1	1	0	0	0	0	0	0
DMU 2	0.7047	25	0	0	0	0.717	11.736	15.648
DMU 3	0.8647	22	0	0	0	1.331	13.962	18.622
DMU 4	0.9274	15	0	0	0	0.74	17.793	23.721
DMU 5	1	1	0	0	0	0	0	0
DMU 6	0.8847	20	0	0	0	0.381	9.551	12.733
DMU 7	0.9792	12	1.076	0	0	2.021	0	1.311
DMU 8	0.792	24	0	0	8.479	0.782	0	2.928
DMU 9	1	1	0	0	0	0	0.001	0.001
DMU 10	1	1	0	0	0	0	0	0
DMU 11	1	1	0	0.002	0	0	0	0
DMU 12	1	1	0	0	0	0	0	0
DMU 13	0.9087	16	0	0	0	0	0	1.281
DMU 14	1	1	0	0	0	0	0.001	0.001

Table A6. *Cont.*

DMU	Score	Rank	LT	UP	PC	QB	NI	RE
DMU 15	0.8389	23	0	0	1.157	0	0	0.626
DMU 16	1	1	0	0	0	0	0	0
DMU 17	0.9747	13	0	0	19.705	0	0	0.379
DMU 18	0.8811	21	0.033	0	0	0	0	2.897
DMU 19	1	1	0	0	0	0	0	0
DMU 20	0.9679	14	0	0	0	0	0	2.287
DMU 21	0.8998	18	0.424	0	0	0	0	3.248
DMU 22	1	1	0	0	0	0	0	0
DMU 23	1	1	0	0	0	0	0	0
DMU 24	0.8989	19	0	0	12.923	0	0.546	0.724
DMU 25	0.9038	17	0	0	0	0	0	1.467

Table A7. The optimal weights for each DMU using the BBC-O model.

DMU	Score	Rank	V (0)	V (1)	V (2)	V (3)	U (1)	U (2)	U (3)
DMU 1	1	1	0	0.333333	0	0	0.106578	2.16×10^{-3}	8.66×10^{-3}
DMU 2	0.504	24	1.98413	0	0	0	0	3.97×10^{-2}	0
DMU 3	0.5349	23	0.0769	0.216938	2.78×10^{-3}	0	0	3.84×10^{-2}	0
DMU 4	0.4928	25	-0.6658	0	5.08×10^{-3}	2.65×10^{-2}	0	4.36×10^{-2}	0
DMU 5	1	1	0	0	2.49×10^{-3}	9.37×10^{-3}	0	2.50×10^{-2}	0
DMU 6	0.6448	22	0.06573	0.185448	2.38×10^{-3}	0	0	3.28×10^{-2}	0
DMU 7	0.9727	12	-0.3183	0	2.43×10^{-3}	1.27×10^{-2}	0	2.08×10^{-2}	0
DMU 8	0.8909	20	0.55846	0	1.64×10^{-3}	0	0	2.27×10^{-2}	0
DMU 9	0.9997	11	-26.435	4.540095	4.02×10^{-2}	0	0.308632	0	0
DMU 10	1	1	0	0.133527	1.67×10^{-3}	2.74×10^{-4}	0	2.26×10^{-2}	0
DMU 11	1	1	-0.1694	0	0	3.90×10^{-2}	0.2101	4.84×10^{-3}	0
DMU 12	1	1	0	0.242562	7.86×10^{-4}	0	0	2.27×10^{-2}	0
DMU 13	0.8958	18	0.04537	0.127989	1.64×10^{-3}	0	0	2.27×10^{-2}	0
DMU 14	1	1	-1.3128	0.106691	1.70×10^{-3}	3.38×10^{-2}	0.194235	0	0
DMU 15	0.8909	20	0.38995	0	2.20×10^{-3}	0	2.11×10^{-2}	2.10×10^{-2}	0
DMU 16	1	1	0	0	2.81×10^{-3}	3.35×10^{-4}	3.48×10^{-2}	0	1.09×10^{-2}
DMU 17	0.9683	13	0	0.083582	2.18×10^{-3}	0	2.40×10^{-2}	1.94×10^{-2}	0
DMU 18	0.8911	19	0.38077	0	2.15×10^{-3}	0	2.06×10^{-2}	2.05×10^{-2}	0
DMU 19	1	1	0	0	1.92×10^{-4}	0.018792	0.134811	0	0
DMU 20	0.9106	16	-1.9554	5.23×10^{-2}	5.47×10^{-3}	0.02431	8.31×10^{-2}	1.38×10^{-2}	0
DMU 21	0.9374	15	0.55694	0	1.64×10^{-3}	0	0	2.27×10^{-2}	0
DMU 22	1	1	0	7.72×10^{-2}	2.21×10^{-3}	0	0.145707	0	0
DMU 23	1	1	0	6.11×10^{-2}	1.20×10^{-4}	1.31×10^{-2}	0.10872	4.31×10^{-3}	0
DMU 24	0.9456	14	1.05747	0	0	0	4.77×10^{-2}	1.59×10^{-2}	0
DMU 25	0.8987	17	1.11266	0	0	0	5.02×10^{-2}	1.68×10^{-2}	0

Table A8. The slacks for each DMU using the BCC-O model.

DMU	Score	Rank	LT	UP	PC	QB	NI	RE
DMU 1	1	1	0	0	0	0	0	0
DMU 2	0.504	24	1	40.546	30	2.792	0	7.633
DMU 3	0.5349	23	0	0	8.652	3.321	0	6.618
DMU 4	0.4928	25	0.219	0	0	0.387	0	4.289
DMU 5	1	1	0	0	0	0	0	0
DMU 6	0.6448	22	0	0	7.211	1.73	0	5.505
DMU 7	0.9727	12	1.042	0	0	2.529	0	2.679
DMU 8	0.8909	20	1	0	29.417	2.947	0	7.185
DMU 9	0.9997	11	0	0	0	0	0.014	0.018
DMU 10	1	1	0	0	0	0	0	0
DMU 11	1	1	0	0.003	0	0	0	0
DMU 12	1	1	0	0	0	0	0	0
DMU 13	0.8958	18	0	0	9.249	0.641	0	7.057
DMU 14	1	1	0	0	0	0	0.001	0.002

Table A8. *Cont.*

DMU	Score	Rank	LT	UP	PC	QB	NI	RE
DMU 15	0.8909	20	0.883	0	17.796	0	0	5.95
DMU 16	1	1	0	0	0	0	0	0
DMU 17	0.9683	13	0	0	23.686	0	0	2.048
DMU 18	0.8911	19	0.991	0	9.472	0	0	7.238
DMU 19	1	1	0	0	0	0	0	0
DMU 20	0.9106	16	0	0	0	0	0	6.369
DMU 21	0.9374	15	1	0	7.081	0.694	0	5.413
DMU 22	1	1	0	0	0	0	0	0
DMU 23	1	1	0	0	0	0	0	0
DMU 24	0.9456	14	0.246	14.506	22.457	0	0	1.871
DMU 25	0.8987	17	0.635	2.642	6.353	0	0	4.847

Table A9. The optimal weights for each DMU using the SBM-I-C model.

DMU	Score	Rank	V (1)	V (2)	V (3)	U (1)	U (2)	U (3)
DMU 1	1	1	15.13416	9.60×10^{-4}	6.67×10^{-3}	0.583152	0	0.747719
DMU 2	0.3666	25	6.67×10^{-2}	8.52×10^{-4}	4.76×10^{-3}	0	9.58×10^{-3}	3.73×10^{-3}
DMU 3	0.4732	23	8.33×10^{-2}	1.00×10^{-3}	6.67×10^{-3}	0	1.82×10^{-2}	0
DMU 4	0.4537	24	8.33×10^{-2}	1.04×10^{-3}	8.33×10^{-3}	2.58×10^{-2}	1.78×10^{-2}	0
DMU 5	1	1	8.33×10^{-2}	6.08×10^{-3}	6.67×10^{-3}	0	0	3.68×10^{-2}
DMU 6	0.569	22	8.33×10^{-2}	1.07×10^{-3}	6.67×10^{-3}	0	1.17×10^{-2}	5.22×10^{-3}
DMU 7	0.8934	9	6.67×10^{-2}	1.88×10^{-3}	8.33×10^{-3}	0	2.52×10^{-2}	0
DMU 8	0.6873	20	6.67×10^{-2}	1.43×10^{-3}	4.76×10^{-3}	0	1.92×10^{-2}	0
DMU 9	0.6775	21	0.111111	9.70×10^{-4}	6.67×10^{-3}	3.08×10^{-2}	1.77×10^{-2}	0
DMU 10	1	1	0.111111	9.41×10^{-4}	5.479046	10.57818	1.577778	1.061762
DMU 11	0.7471	18	6.67×10^{-2}	1.04×10^{-3}	1.11×10^{-2}	9.77×10^{-2}	1.09×10^{-2}	0
DMU 12	0.9036	8	17.93597	0.111646	4.76×10^{-3}	0	1.111111	0.747719
DMU 13	0.8148	13	8.33×10^{-2}	9.79×10^{-4}	6.67×10^{-3}	2.21×10^{-2}	1.65×10^{-2}	0
DMU 14	0.8334	12	8.33×10^{-2}	1.06×10^{-3}	8.33×10^{-3}	0.081691	1.18×10^{-2}	0
DMU 15	0.7229	19	6.67×10^{-2}	1.00×10^{-3}	5.56×10^{-3}	1.30×10^{-2}	1.54×10^{-2}	0
DMU 16	1	1	8.33×10^{-2}	9.50×10^{-4}	3.575827	7.933635	0.823944	0.796321
DMU 17	0.8534	11	8.33×10^{-2}	1.04×10^{-3}	4.69×10^{-3}	5.55×10^{-2}	0.011673	0
DMU 18	0.7683	17	6.67×10^{-2}	9.67×10^{-4}	6.67×10^{-3}	1.7×10^{-2}	1.56×10^{-2}	0
DMU 19	1	1	6.67×10^{-2}	0.280893	6.67×10^{-3}	6.346908	0.946667	0
DMU 20	0.8856	10	8.33×10^{-2}	9.74×10^{-4}	8.33×10^{-3}	2.73×10^{-2}	1.72×10^{-2}	0
DMU 21	0.8027	15	6.67×10^{-2}	1.67×10^{-3}	6.67×10^{-3}	0	2.24×10^{-2}	0
DMU 22	1	1	3.741093	1.07×10^{-3}	0.705214	6.346908	0	0.119497
DMU 23	1	1	6.67×10^{-2}	9.75×10^{-4}	8.86×10^{-2}	0.683238	0	0
DMU 24	0.789	16	6.67×10^{-2}	9.87×10^{-4}	4.76×10^{-3}	5.19×10^{-2}	0.0104	0
DMU 25	0.8106	14	6.67×10^{-2}	9.80×10^{-4}	6.67×10^{-3}	6.56×10^{-2}	1.04×10^{-2}	0

Table A10. The slacks for each DMU using the SBM-I-C model.

DMU	Score	Rank	LT	UP	PC	QB	NI	RE
DMU 1	1	1	0	0	0	0	0	0
DMU 2	0.3666	25	3.293	189.987	52.929	0.401	0	0
DMU 3	0.4732	23	2.237	124.289	32.368	0.979	0	0.005
DMU 4	0.4537	24	2.393	142.194	23.93	0	0	0.652
DMU 5	1	1	0	0	0	0	0	0
DMU 6	0.569	22	1.937	69.166	29.373	0.506	0	0
DMU 7	0.8934	9	1.266	0	2.659	2.324	0	1.809
DMU 8	0.6873	20	1.903	0	39.031	1.414	0	1.419
DMU 9	0.6775	21	0.486	105.988	24.86	0	0	3.722
DMU 10	1	1	0	0	0	0	0	0
DMU 11	0.7471	18	2.278	89.21	0.732	0	0	3.636
DMU 12	0.9036	8	0	0	20.238	0.812	0	0
DMU 13	0.8148	13	0.716	11.392	17.161	0	0	3.672
DMU 14	0.8334	12	0.895	67.605	2.466	0	0	0.982

Table A10. *Cont.*

DMU	Score	Rank	LT	UP	PC	QB	NI	RE
DMU 15	0.7229	19	1.57	29.523	25.705	0	0	6.417
DMU 16	1	1	0	0	0	0	0	0
DMU 17	0.8534	11	0.17	8.524	26.32	0	0	2.441
DMU 18	0.7683	17	1.504	32.315	15.037	0	0	6.328
DMU 19	1	1	0	0	0	0	0	0
DMU 20	0.8856	10	0.509	30.41	5.088	0	0	6.305
DMU 21	0.8027	15	1.48	0	14.8	1.534	0	6.598
DMU 22	1	1	0	0	0	0	0	0
DMU 23	1	1	0	0	0	0	0	0
DMU 24	0.789	16	1.116	31.378	22.191	0	0	0.565
DMU 25	0.8106	14	1.358	36.497	9.463	0	0	3.803

Table A11. The optimal weights for each DMU using the SBM-O-C model.

DMU	Score	Rank	V (1)	V (2)	V (3)	U (1)	U (2)	U (3)
DMU 1	1	1	272.1957	0	0	7.006445	7.57×10^{-3}	13.45895
DMU 2	0.2795	24	0.715473	0	0	0.247666	1.32×10^{-2}	9.92×10^{-3}
DMU 3	0.2564	25	0.975128	0	0	0.404384	1.28×10^{-2}	9.61×10^{-3}
DMU 4	0.4089	23	0	4.22×10^{-3}	2.72×10^{-2}	0.189276	1.45×10^{-2}	1.09×10^{-2}
DMU 5	1	1	0	2.67×10^{-2}	0	0.332568	8.32×10^{-3}	9.42×10^{-2}
DMU 6	0.4189	22	0.477617	0	9.54×10^{-3}	0.207723	1.09×10^{-2}	8.21×10^{-3}
DMU 7	0.7104	20	0	1.74×10^{-3}	2.01×10^{-2}	0.12946	6.94×10^{-3}	4.89×10^{-3}
DMU 8	0.4919	21	0.087746	4.65×10^{-3}	0	0.165879	7.57×10^{-3}	5.68×10^{-3}
DMU 9	0.7822	18	0.356897	6.04×10^{-4}	0	0.102877	1.02×10^{-2}	7.65×10^{-3}
DMU 10	1	1	0	0	69.73439	134.0896	20	13.45895
DMU 11	0.8709	11	0	0	3.94×10^{-2}	9.18×10^{-2}	1.02×10^{-2}	7.63×10^{-3}
DMU 12	1	1	405.5784	1.320418	0	1.220084	20	13.45895
DMU 13	0.8021	14	0.27006	4.89×10^{-4}	0	0.082931	7.56×10^{-3}	5.67×10^{-3}
DMU 14	0.7971	15	3.51×10^{-2}	3.06×10^{-3}	3.77×10^{-3}	6.47×10^{-2}	9.56×10^{-3}	7.17×10^{-3}
DMU 15	0.7845	17	3.77×10^{-2}	3.27×10^{-3}	0	7.15×10^{-2}	7.75×10^{-3}	5.81×10^{-3}
DMU 16	1	1	0	0	60.44946	134.0896	13.71082	13.45895
DMU 17	0.9518	9	0.139904	1.53×10^{-3}	0	0.054432	7.57×10^{-3}	5.68×10^{-3}
DMU 18	0.7918	16	3.88×10^{-2}	2.32×10^{-3}	5.36×10^{-3}	7.07×10^{-2}	7.56×10^{-3}	5.67×10^{-3}
DMU 19	1	1	0	5.980267	0	134.0896	20	5.66×10^{-3}
DMU 20	0.857	12	3.88×10^{-2}	2.33×10^{-3}	5.37×10^{-3}	7.09×10^{-2}	7.57×10^{-3}	5.67×10^{-3}
DMU 21	0.7152	19	0	9.07×10^{-3}	1.29×10^{-2}	0.102574	5.44×10^{-2}	5.66×10^{-3}
DMU 22	1	1	79.70503	0	14.9138	134.0896	7.59×10^{-3}	2.457001
DMU 23	1	1	0	5.24×10^{-4}	7.74×10^{-2}	0.453676	7.59×10^{-3}	5.69×10^{-3}
DMU 24	0.8857	10	2.68×10^{-2}	2.95×10^{-3}	0	5.08×10^{-2}	7.73×10^{-3}	5.80×10^{-3}
DMU 25	0.838	13	3.27×10^{-2}	2.44×10^{-3}	3.98×10^{-3}	6.01×10^{-2}	7.75×10^{-3}	5.81×10^{-3}

Table A12. The slacks for each DMU using the SBM-O-C model.

DMU	Score	Rank	LT	UP	PC	QB	NI	RE
DMU 1	1	1	0	0	0	0	0	0
DMU 2	0.2795	24	0	0.95	7.5	7.233	29.712	39.612
DMU 3	0.2564	25	0	20	0	6.039	17.9	23.87
DMU 4	0.4089	23	0	0	0	3.84	22.72	35.674
DMU 5	1	1	0	0	0	0	0	0
DMU 6	0.4189	22	0	0.2	0	5.258	13.48	17.97
DMU 7	0.7104	20	1	0	0	2.914	1.175	4.571
DMU 8	0.4919	21	0	0	15.281	5.847	4.183	5.569
DMU 9	0.7822	18	0	0	0.88	0.498	11.043	14.979
DMU 10	1	1	0	0	0	0	0	0
DMU 11	0.8709	11	2	48.724	0	0	5.332	12.325
DMU 12	1	1	0	0	0	0	0	0
DMU 13	0.8021	14	0	0	7.325	1.818	4.256	11.262
DMU 14	0.7971	15	0	0	0	0.484	9.84	18.013

Table A12. *Cont.*

DMU	Score	Rank	LT	UP	PC	QB	NI	RE
DMU 15	0.7845	17	0	0	6.997	3.036	3.715	4.948
DMU 16	1	1	0	0	0	0	0	0
DMU 17	0.9518	9	0	0	22.974	0.463	1.118	2.995
DMU 18	0.7918	16	0	0	0	2.562	4.532	8.386
DMU 19	1	1	0	0	0	0	0	0
DMU 20	0.857	12	0	0	0	0.799	4.673	13.202
DMU 21	0.7152	19	0.044	0	0	3.877	0	0.095
DMU 22	1	1	0	0	0	0	0	0
DMU 23	1	1	0	0	0	0	0	0
DMU 24	0.8857	10	0	0	16.147	1.215	4.357	5.804
DMU 25	0.838	13	0	0	0	1.744	4.97	8.617

Table A13. The optimal weights for each DMU using the Super SBM-I-C model.

No.	DMU	Score	Rank	V (1)	V (2)	V (3)	U (1)	U (2)	U (3)
1	DMU 1	1	1	15.13416	9.60×10^{-4}	6.67×10^{-3}	0.583152	0	0.747719
2	DMU 2	0.3666	25	6.67×10^{-2}	8.52×10^{-4}	4.76×10^{-3}	0	9.58×10^{-3}	3.73×10^{-3}
3	DMU 3	0.4732	23	8.33×10^{-2}	1.00×10^{-3}	6.67×10^{-3}	0	1.82×10^{-2}	0
4	DMU 4	0.4537	24	8.33×10^{-2}	1.04×10^{-3}	8.33×10^{-3}	2.58×10^{-2}	1.78×10^{-2}	0
5	DMU 5	1	1	8.33×10^{-2}	6.08×10^{-3}	6.67×10^{-3}	0	0	3.68×10^{-2}
6	DMU 6	0.569	22	8.33×10^{-2}	1.07×10^{-3}	6.67×10^{-3}	0	1.17×10^{-2}	5.22×10^{-3}
7	DMU 7	0.8934	9	6.67×10^{-2}	1.88×10^{-3}	8.33×10^{-3}	0	2.52×10^{-2}	0
8	DMU 8	0.6873	20	6.67×10^{-2}	1.43×10^{-3}	4.76×10^{-3}	0	1.92×10^{-2}	0
9	DMU 9	0.6775	21	0.111111	9.70×10^{-4}	6.67×10^{-3}	3.08×10^{-2}	1.77×10^{-2}	0
10	DMU 10	1	1	0.111111	9.41×10^{-4}	5.479046	10.57818	1.577778	1.061762
11	DMU 11	0.7471	18	6.67×10^{-2}	1.04×10^{-3}	1.11×10^{-2}	9.77×10^{-2}	1.09×10^{-2}	0
12	DMU 12	0.9036	8	17.93597	0.111646	4.76×10^{-3}	0	1.111111	0.747719
13	DMU 13	0.8148	13	8.33×10^{-2}	9.79×10^{-4}	6.67×10^{-3}	2.21×10^{-2}	1.65×10^{-2}	0
14	DMU 14	0.8334	12	8.33×10^{-2}	1.06×10^{-3}	8.33×10^{-3}	0.081691	1.18×10^{-2}	0
15	DMU 15	0.7229	19	6.67×10^{-2}	1.00×10^{-3}	5.56×10^{-3}	1.30×10^{-2}	1.54×10^{-2}	0
16	DMU 16	1	1	8.33×10^{-2}	9.50×10^{-4}	3.575827	7.933635	0.823944	0.796321
17	DMU 17	0.8534	11	8.33×10^{-2}	1.04×10^{-3}	4.69×10^{-3}	5.55×10^{-2}	0.011673	0
18	DMU 18	0.7683	17	6.67×10^{-2}	9.67×10^{-4}	6.67×10^{-3}	1.73×10^{-2}	1.56×10^{-2}	0
19	DMU 19	1	1	6.67×10^{-2}	0.280893	6.67×10^{-3}	6.346908	0.946667	0
20	DMU 20	0.8856	10	8.33×10^{-2}	9.74×10^{-4}	8.33×10^{-3}	2.73×10^{-2}	1.72×10^{-2}	0
21	DMU 21	0.8027	15	6.67×10^{-2}	1.67×10^{-3}	6.67×10^{-3}	0	2.24×10^{-2}	0
22	DMU 22	1	1	3.741093	1.07×10^{-3}	0.705214	6.346908	0	0.119497
23	DMU 23	1	1	6.67×10^{-2}	9.75×10^{-4}	8.86×10^{-2}	0.683238	0	0
24	DMU 24	0.789	16	6.67×10^{-2}	9.87×10^{-4}	4.76×10^{-3}	5.19×10^{-2}	0.0104	0
25	DMU 25	0.8106	14	6.67×10^{-2}	9.80×10^{-4}	6.67×10^{-3}	6.56×10^{-2}	1.04×10^{-2}	0

Table A14. The slacks for each DMU using the Super SBM-I-C model.

No.	DMU	Score	Rank	LT	UP	PC	QB	NI	RE
1	DMU 1	1	1	0	0	0	0	0	0
2	DMU 2	0.3666	25	3.293	189.987	52.929	0.401	0	0
3	DMU 3	0.4732	23	2.237	124.289	32.368	0.979	0	0.005
4	DMU 4	0.4537	24	2.393	142.194	23.93	0	0	0.652
5	DMU 5	1	1	0	0	0	0	0	0
6	DMU 6	0.569	22	1.937	69.166	29.373	0.506	0	0
7	DMU 7	0.8934	9	1.266	0	2.659	2.324	0	1.809
8	DMU 8	0.6873	20	1.903	0	39.031	1.414	0	1.419
9	DMU 9	0.6775	21	0.486	105.988	24.86	0	0	3.722
10	DMU 10	1	1	0	0	0	0	0	0
11	DMU 11	0.7471	18	2.278	89.21	0.732	0	0	3.636
12	DMU 12	0.9036	8	0	0	20.238	0.812	0	0
13	DMU 13	0.8148	13	0.716	11.392	17.161	0	0	3.672
14	DMU 14	0.8334	12	0.895	67.605	2.466	0	0	0.982

Table A14. *Cont.*

No.	DMU	Score	Rank	LT	UP	PC	QB	NI	RE
15	DMU 15	0.7229	19	1.57	29.523	25.705	0	0	6.417
16	DMU 16	1	1	0	0	0	0	0	0
17	DMU 17	0.8534	11	0.17	8.524	26.32	0	0	2.441
18	DMU 18	0.7683	17	1.504	32.315	15.037	0	0	6.328
19	DMU 19	1	1	0	0	0	0	0	0
20	DMU 20	0.8856	10	0.509	30.41	5.088	0	0	6.305
21	DMU 21	0.8027	15	1.48	0	14.8	1.534	0	6.598
22	DMU 22	1	1	0	0	0	0	0	0
23	DMU 23	1	1	0	0	0	0	0	0
24	DMU 24	0.789	16	1.116	31.378	22.191	0	0	0.565
25	DMU 25	0.8106	14	1.358	36.497	9.463	0	0	3.803

Table A15. The optimal weights for each DMU using the SBM-AR-C model.

No.	DMU	Score	V (1) LT	V (2) UP	V (3) PC	U (1) QB	U (2) NI	U (3) RE
1	DMU 1	1	0.7774838	9.60×10^{-4}	6.67×10^{-3}	0.2139225	4.25×10^{-2}	5.68×10^{-3}
2	DMU 2	0.269326	6.67×10^{-2}	8.52×10^{-4}	4.76×10^{-3}	6.67×10^{-2}	3.56×10^{-3}	2.67×10^{-3}
3	DMU 3	0.251235	8.33×10^{-2}	1.00×10^{-3}	6.67×10^{-3}	0.1015951	3.22×10^{-3}	2.41×10^{-3}
4	DMU 4	0.401049	8.33×10^{-2}	1.04×10^{-3}	8.33×10^{-3}	7.59×10^{-2}	5.82×10^{-3}	4.37×10^{-3}
5	DMU 5	1	8.33×10^{-2}	9.26×10^{-2}	6.67×10^{-3}	1.1563198	8.32×10^{-3}	0.3546177
6	DMU 6	0.418778	8.33×10^{-2}	1.07×10^{-3}	6.67×10^{-3}	8.70×10^{-2}	4.58×10^{-3}	3.44×10^{-3}
7	DMU 7	0.662241	6.67×10^{-2}	9.65×10^{-4}	8.33×10^{-3}	8.57×10^{-2}	4.60×10^{-3}	3.24×10^{-3}
8	DMU 8	0.448251	6.67×10^{-2}	9.72×10^{-4}	4.76×10^{-3}	7.44×10^{-2}	3.39×10^{-3}	2.55×10^{-3}
9	DMU 9	0.641302	0.1111111	9.70×10^{-4}	6.67×10^{-3}	6.60×10^{-2}	6.54×10^{-3}	4.90×10^{-3}
10	DMU 10	1	1.2031017	9.41×10^{-4}	1.11×10^{-2}	0.3587726	6.42×10^{-2}	5.64×10^{-3}
11	DMU 11	0.703643	6.67×10^{-2}	1.04×10^{-3}	1.11×10^{-2}	0.0585784	7.16×10^{-3}	5.37×10^{-3}
12	DMU 12	1	66.742332	0.2530347	4.76×10^{-3}	0.1143864	6.5315673	5.68×10^{-3}
13	DMU 13	0.753682	8.33×10^{-2}	9.79×10^{-4}	6.67×10^{-3}	6.25×10^{-2}	5.69×10^{-3}	4.27×10^{-3}
14	DMU 14	0.771137	8.33×10^{-2}	1.90×10^{-3}	8.33×10^{-3}	0.1013309	7.37×10^{-3}	5.53×10^{-3}
15	DMU 15	0.694923	6.67×10^{-2}	1.00×10^{-3}	5.56×10^{-3}	4.97×10^{-2}	5.38×10^{-3}	4.04×10^{-3}
16	DMU 16	1	8.33×10^{-2}	9.50×10^{-4}	8.33×10^{-3}	6.10×10^{-2}	6.67×10^{-3}	4.49×10^{-3}
17	DMU 17	0.837507	8.33×10^{-2}	1.55×10^{-3}	4.69×10^{-3}	7.19×10^{-2}	6.34×10^{-3}	4.76×10^{-3}
18	DMU 18	0.73634	6.67×10^{-2}	9.67×10^{-4}	6.67×10^{-3}	5.21×10^{-2}	5.56×10^{-3}	4.17×10^{-3}
19	DMU 19	1	6.67×10^{-2}	4.73×10^{-3}	6.67×10^{-3}	0.1230422	7.55×10^{-3}	1.54×10^{-2}
20	DMU 20	0.849581	8.33×10^{-2}	9.74×10^{-4}	8.33×10^{-3}	6.02×10^{-2}	6.43×10^{-3}	4.82×10^{-3}
21	DMU 21	0.669586	6.67×10^{-2}	1.07×10^{-3}	6.67×10^{-3}	6.87×10^{-2}	5.06×10^{-3}	3.79×10^{-3}
22	DMU 22	1	8.33×10^{-2}	1.98×10^{-3}	6.67×10^{-3}	9.00×10^{-2}	7.59×10^{-3}	0.0056912
23	DMU 23	1	0.428732	9.75E-04	7.86×10^{-2}	0.7697859	7.59×10^{-3}	5.69×10^{-3}
24	DMU 24	0.769097	6.67×10^{-2}	1.54×10^{-3}	4.76×10^{-3}	6.75×10^{-2}	5.95×10^{-3}	4.46×10^{-3}
25	DMU 25	0.772696	6.67×10^{-2}	1.55×10^{-3}	6.67×10^{-3}	8.13×10^{-2}	5.99×10^{-3}	4.49×10^{-3}

Table A16. The slacks for each DMU using the SBM-AR-C model.

No.	DMU	Score	Excess	Excess	Excess	Shortage	Shortage	Shortage
			LT	UP	PC	QB	NI	RE
			S−(1)	S−(2)	S−(3)	S+(1)	S+(2)	S+(3)
1	DMU 1	1	0	0	0	0	0	0
2	DMU 2	0.269326	0	0.95	7.5	7.232975	29.7125	39.6125
3	DMU 3	0.251235	0	20	0	6.0388	17.9	23.87
4	DMU 4	0.401049	0.3351382	0	3.351382	3.2435436	22.86077	37.47481
5	DMU 5	1	0	0	0	0	0	0
6	DMU 6	0.418778	0	0.2	0	5.2584	13.48	17.97
7	DMU 7	0.662241	1.16	8.436	1.6	2.669008	0	3.128
8	DMU 8	0.448251	0.609475	0	15.11844	5.523653	4.18894	5.578262
9	DMU 9	0.641302	0.384	114.1114	23.84	0.3322442	0	4.9922
10	DMU 10	1	0	0	0	0	0	0

Table A16. *Cont.*

No.	DMU	Score	Excess	Excess	Excess	Shortage	Shortage	Shortage
			LT	UP	PC	QB	NI	RE
			S−(1)	S−(2)	S−(3)	S+(1)	S+(2)	S+(3)
11	DMU 11	0.703643	2	56.925	0	9.27×10^{-2}	4.72	12.025
12	DMU 12	1	0	0	0	0	0	0
13	DMU 13	0.753682	0.4704	30.96584	14.704	0.8005335	0	6.73232
14	DMU 14	0.771137	0.3550889	0	2.330154	0	9.940403	19.28401
15	DMU 15	0.694923	1.2108863	0	22.10886	0.513919	4.343921	13.02279
16	DMU 16	1	0	0	0	0	0	0
17	DMU 17	0.837507	0.1015444	0	26.30323	0	1.253382	4.748504
18	DMU 18	0.73634	1.4704	34.96584	14.704	0.1084335	0	6.74232
19	DMU 19	1	0	0	0	0	0	0
20	DMU 20	0.849581	9.80E-02	0	0.980336	0.6245277	4.71458	13.72903
21	DMU 21	0.669586	1.4571103	0	14.5711	1.5883816	0.136121	6.949176
22	DMU 22	1	0	0	0	0	0	0
23	DMU 23	1	0	0	0	0	0	0
24	DMU 24	0.769097	0.8652768	0	22.12846	0	4.613784	9.059744
25	DMU 25	0.772696	1.0669574	0	9.390053	0	5.366362	13.68358

Table A17. The optimal weights for each DMU using the SBM-AR-V model.

No.	DMU	Score	V (1) LT	V (2) UP	V (3) PC	U (1) QB	U (2) NI	U (3) RE
1	DMU 1	1	2.2055847	2.40×10^{-2}	6.67×10^{-3}	8.96×10^{-2}	0.3318029	5.68E-03
2	DMU 2	0.269326	6.67×10^{-2}	8.52×10^{-4}	4.76×10^{-3}	6.67×10^{-2}	3.56×10^{-3}	2.67×10^{-3}
3	DMU 3	0.251235	8.33×10^{-2}	1.00×10^{-3}	6.67×10^{-3}	0.1015951	3.22×10^{-3}	2.41×10^{-3}
4	DMU 4	0.45679	8.33×10^{-2}	5.91×10^{-3}	2.56×10^{-2}	8.65×10^{-2}	6.63×10^{-3}	4.98×10^{-3}
5	DMU 5	1	0.1763621	2.50×10^{-2}	6.67×10^{-3}	0.3325684	7.64×10^{-2}	5.90×10^{-2}
6	DMU 6	0.418778	8.33×10^{-2}	1.07×10^{-3}	6.67×10^{-3}	8.70×10^{-2}	4.58×10^{-3}	3.44×10^{-3}
7	DMU 7	0.677494	6.67×10^{-2}	4.83×10^{-3}	2.28×10^{-2}	8.77×10^{-2}	4.70×10^{-3}	3.31×10^{-3}
8	DMU 8	0.448556	6.67×10^{-2}	9.72×10^{-4}	4.76×10^{-3}	7.44×10^{-2}	8.89×10^{-3}	2.55×10^{-3}
9	DMU 9	0.999453	9.7808186	7.58×10^{-2}	6.67×10^{-3}	0.1028212	1.02×10^{-2}	7.64×10^{-3}
10	DMU 10	1	0.1111111	9.41×10^{-4}	8.65×10^{-2}	0.1086236	7.53×10^{-3}	4.40×10^{-2}
11	DMU 11	1	6.67×10^{-2}	1.04×10^{-3}	9.61×10^{-2}	0.3550234	1.02×10^{-2}	7.63×10^{-3}
12	DMU 12	1	17.038047	0.2170056	4.76×10^{-3}	0.1143864	2.9286539	5.68×10^{-3}
13	DMU 13	0.762943	8.33×10^{-2}	2.10×10^{-3}	6.67×10^{-3}	6.33×10^{-2}	5.76×10^{-3}	4.32×10^{-3}
14	DMU 14	1	0.2244187	1.37×10^{-2}	5.83×10^{-2}	0.2180815	9.56×10^{-3}	7.17×10^{-3}
15	DMU 15	0.712474	6.67×10^{-2}	2.15×10^{-3}	5.56×10^{-3}	5.10×10^{-2}	5.52×10^{-3}	4.14×10^{-3}
16	DMU 16	1	8.33×10^{-2}	9.50×10^{-4}	8.33×10^{-3}	6.10×10^{-2}	6.67×10^{-3}	4.49×10^{-3}
17	DMU 17	0.849109	8.33×10^{-2}	2.52×10^{-3}	4.69×10^{-3}	4.62×10^{-2}	6.43×10^{-3}	4.82×10^{-3}
18	DMU 18	0.744086	6.67×10^{-2}	2.43×10^{-3}	6.67×10^{-3}	5.26×10^{-2}	5.62×10^{-3}	4.22×10^{-3}
19	DMU 19	1	6.67×10^{-2}	4.73×10^{-3}	6.67×10^{-3}	0.1230422	7.55×10^{-3}	1.54×10^{-2}
20	DMU 20	0.876299	8.33×10^{-2}	4.66×10^{-3}	1.94×10^{-2}	6.21×10^{-2}	6.63×10^{-3}	4.97×10^{-3}
21	DMU 21	0.693331	6.67×10^{-2}	5.64×10^{-3}	6.67×10^{-3}	0.0711174	3.10×10^{-2}	3.93×10^{-3}
22	DMU 22	1	8.33×10^{-2}	1.98×10^{-3}	6.67×10^{-3}	9.00×10^{-2}	7.59×10^{-3}	0.0056912
23	DMU 23	1	0.428732	9.75×10^{-4}	7.86×10^{-2}	0.7697859	7.59×10^{-3}	5.69×10^{-3}
24	DMU 24	0.778043	6.67×10^{-2}	9.87×10^{-4}	4.76×10^{-3}	0.0835919	6.02×10^{-3}	4.51×10^{-3}
25	DMU 25	0.78211	6.67×10^{-2}	2.83×10^{-3}	6.67×10^{-3}	4.70×10^{-2}	6.06×10^{-3}	4.55×10^{-3}

Table A18. The slacks for each DMU using the SBM-AR-V model.

No.	DMU	Score	Excess	Excess	Excess	Shortage	Shortage	Shortage
			LT	UP	PC	QB	NI	RE
			S−(1)	S−(2)	S−(3)	S+(1)	S+(2)	S+(3)
1	DMU 1	1	0	0	0	0	0	0
2	DMU 2	0.269326	1	79.05	20	5.5172	18.73	24.97
3	DMU 3	0.251235	0	20	0	6.0388	17.9	23.87
4	DMU 4	0.45679	0.2190489	0	0	2.1999675	23.619423	35.781249
5	DMU 5	1	0	0	0	0	0	0

Table A18. *Cont.*

No.	DMU	Score	Excess	Excess	Excess	Shortage	Shortage	Shortage
			LT	UP	PC	QB	NI	RE
			S−(1)	S−(2)	S−(3)	S+(1)	S+(2)	S+(3)
6	DMU 6	0.418778	0	0.2	0	5.2584	13.48	17.97
7	DMU 7	0.677494	0.9193048	0	0	2.8828591	0.1189338	2.3581649
8	DMU 8	0.448556	1	29.86573	20.16474	4.8305226	0	0.1191433
9	DMU 9	0.999453	0	0	0	8.57×10^{-4}	2.24×10^{-2}	2.98×10^{-2}
10	DMU 10	1	0	0	0	0	0	0
11	DMU 11	1	8.79×10^{-5}	0	0	0	0	5.42×10^{-4}
12	DMU 12	1	0	0	0	0	0	0
13	DMU 13	0.762943	4.00×10^{-5}	0	7.325987	1.8174772	4.2561312	11.262405
14	DMU 14	1	0	0	0	0	3.01×10^{-3}	4.01×10^{-3}
15	DMU 15	0.712474	1.00004	0	15.19612	1.4748294	4.0633	9.3821192
16	DMU 16	1	0	0	0	0	0	0
17	DMU 17	0.849109	0	0	22.97534	0.4625961	1.118287	2.9958335
18	DMU 18	0.744086	1.00004	0	8.364948	0.9798395	4.8867805	12.906691
19	DMU 19	1	0	0	0	0	0	0
20	DMU 20	0.876299	6.41×10^{-2}	0	0	0.3194245	4.9366059	13.23423
21	DMU 21	0.693331	1.00004	0	0.621362	3.2900805	0	0.4775992
22	DMU 22	1	0	0	0	0	0	0
23	DMU 23	1	0	0	0	0	0	0
24	DMU 24	0.778043	1	16.87501	22.16234	0	2.1325378	4.4913542
25	DMU 25	0.78211	1.00004	0	7.196117	0.3049694	5.2773	12.528119

References

1. Flórez-López, R. Strategic supplier selection in the added-value perspective: A CI approach. *Inf. Sci. Int. J.* **2007**, *177*, 1169–1179.
2. Chen, C.T.; Lin, C.T.; Huang, S.F. A fuzzy approach for supplier evaluation and selection in supply chain management. *Int. J. Prod. Econ.* **2006**, *102*, 289–301. [CrossRef]
3. Wu, C.; Barnes, D. A literature review of decision-making models and approaches for partner selection in agile supply chains. *J. Purch. Supply Manag.* **2011**, *17*, 256–274. [CrossRef]
4. Bhattacharya, A.; Geraghty, J.; Young, P. Supplier selection paradigm: An integrated hierarchical QFD methodology under multiple-criteria environment. *Appl. Soft Comput.* **2010**, *10*, 1013–1027. [CrossRef]
5. Liu, J.; Ding, F.-Y.; Lall, V. Using data envelopment analysis to compare suppliers for supplier selection and performance improvement. *Supply Chain Manag. Int. J.* **2000**, *5*, 143–150. [CrossRef]
6. Saaty, T.L. *Decision Making with Dependence and Feedback: The Analytic Network Process*; RWS Publications: Pittsburgh, PA, USA, 1996.
7. Agarwal, A.; Shankar, R. Analyzing alternatives for improvement in supply chain performance. *Work Study* **2002**, *51*, 32–37. [CrossRef]
8. Aissaoui, N.; Haouari, M.; Hassini, E. Supplier selection and order lot sizing modeling: A review. *Comp. Oper. Res.* **2007**, *34*, 3516–3540. [CrossRef]
9. Govindan, K.; Rajendran, S.; Sarkis, J.; Murugesan, P. Multi criteria decision making approaches for green supplier evaluation and selection: A literature review. *J. Clean. Prod.* **2015**, *98*, 66–83. [CrossRef]
10. Chai, J.; Liu, J.N.K.; Ngai, E. Application of decision-making techniques in supplier selection: A systematic review of literature. *Expert Syst. Appl.* **2013**, *40*, 3872–3885. [CrossRef]
11. Wu, T.; Blackhurst, J. Supplier evaluation and selection: An augmented DEA approach. *Int. J. Prod. Res.* **2009**, *47*, 4593–4608. [CrossRef]
12. Amirteimoori, A.; Khoshandam, L. A Data Envelopment Analysis Approach to Supply Chain Efficiency. *Adv. Decis. Sci.* **2011**, *2011*. [CrossRef]
13. Lin, C.-T.; Hung, K.-P.; Hu, S.-H. A Decision-Making Model for Evaluating and Selecting Suppliers for the Sustainable Operation and Development of Enterprises in the Aerospace Industry. *Sustainability* **2018**, *10*, 735.
14. Galankashi, M.R.; Helmi, S.A.; Hashemzahi, V. Supplier selection in automobile industry: A mixed balanced scorecard–fuzzy AHP approach. *Alex. Eng. J.* **2016**, *55*, 93–100. [CrossRef]

15. Kilincci, O.; Onal, S.A. Fuzzy AHP approach for supplier selection in a washing machine company. *Expert Syst. Appl.* **2011**, *38*, 9656–9664. [CrossRef]
16. Tyagi, M.; Kumar, P.; Kumar, D. Permutation of fuzzy AHP and AHP methods to prioritizing the alternatives of supply chain performance system. *Int. J. Ind. Eng. Theory Appl. Pract.* **2015**, *22*, 24.
17. Karsak, E.E.; Dursun, M. An integrated fuzzy MCDM approach for supplier evaluation and selection. *Comput. Ind. Eng.* **2015**, *82*, 82–93. [CrossRef]
18. Chen, H.M.W.; Chou, S.-Y.; Luu, Q.D.; Yu, T.H.-K. A Fuzzy MCDM Approach for Green Supplier Selection from the Economic and Environmental Aspects. *Math. Probl. Eng.* **2016**, *2016*. [CrossRef]
19. Guo, Z.G.; Liu, H.; Zhang, D.; Yang, J. Green Supplier Evaluation and Selection in Apparel Manufacturing Using a Fuzzy Multi-Criteria Decision-Making Approach. *Sustainability* **2017**, *9*, 650.
20. Wu, T.-H.; Chen, C.-H.C.; Mao, N.; Lu, S.-T. Fishmeal Supplier Evaluation and Selection for Aquaculture Enterprise Sustainability with a Fuzzy MCDM Approach. *Symmetry* **2017**, *9*, 286. [CrossRef]
21. Hu, C.-K.; Liu, F.-B.; Hu, C.-F. A Hybrid Fuzzy DEA/AHP Methodology for Ranking Units in a Fuzzy Environment. *Symmetry* **2017**, *9*, 273. [CrossRef]
22. He, X.; Zhang, J. Supplier Selection Study under the Respective of Low-Carbon Supply Chain: A Hybrid Evaluation Model Based on FA-DEA-AHP. *Sustainability* **2018**, *10*, 564. [CrossRef]
23. Parkouhi, S.V.; Ghadikolaei, A.S. A resilience approach for supplier selection: Using Fuzzy Analytic Network Process and grey VIKOR techniques. *J. Clean. Prod.* **2017**, *161*, 431–451. [CrossRef]
24. Wang, C.-N.; Nguyen, H.-K.; Liao, R.-Y. Partner Selection in Supply Chain of Vietnam's Textile and Apparel Industry: The Application of a Hybrid DEA and GM (1,1) Approach. *Math. Probl. Eng.* **2017**, *2017*. [CrossRef]
25. Wu, C.-M.; Hsieh, C.-L.; Chang, K.-L. A Hybrid Multiple Criteria Decision Making Model for Supplier Selection. *Math. Probl. Eng.* **2013**, *2013*. [CrossRef]
26. Rezaeisaray, M.; Ebrahimnejad, S.; Khalili-Damghani, K. A novel hybrid MCDM approach for outsourcing supplier selection: A case study in pipe and fittings manufacturing. *J. Model. Manag.* **2016**, *11*, 536–559. [CrossRef]
27. Rouyendegh, B.D.; Erol, S. The DEA—FUZZY ANP Department Ranking Model Applied in Iran Amirkabir University. *Acta Polytech. Hung.* **2010**, *7*, 103–114.
28. Zadeh, L. Fuzzy sets. *Inf. Contr.* **1965**, *8*, 338–353. [CrossRef]
29. Junior, F.R.L.; Osiro, L.; Carpinetti, L.C.R. A comparison between Fuzzy AHP and Fuzzy TOPSIS methods to supplier selection. *Appl. Soft Comp.* **2014**, *21*, 194–209. [CrossRef]
30. Charnes, A.; Cooper, W.; Rhodes, E. Measuring the efficiency of decision making units. *Eur. J. Oper. Res.* **1978**, *2*, 429–444. [CrossRef]
31. Rouyendegh, B.D.; Oztekin, A.O.; Ekong, J.; Dag, A. Measuring the efficiency of hospitals: A fully-ranking DEA–FAHP approach. *Ann. Oper. Res.* **2016**. [CrossRef]
32. Mohaghar, A.; Fathi, M.R.; Jafarzadeh, A.H. A Supplier Selection Method Using AR DEA and Fuzzy VIKOR. *Int. J. Ind. Eng. Theory Appl. Pract.* **2013**, *20*, 387–400.
33. Talluri, S.; Narasimhan, R.; Nair, A. Vendor performance with supply risk: A chance-constrained DEA approach. *Int. J. Prod. Econ.* **2006**, *100*, 212–222. [CrossRef]
34. Saen, R.F. Restricting Weights in Supplier Selection Decisions in the presence of Dual-Role factors. *Appl. Math. Model.* **2010**, *2*, 229–243. [CrossRef]
35. Saen, R.F. Supplier selection by the new AR-IDEA model. *Int. J. Adv. Manuf. Tech.* **2008**, *39*, 1061–1070. [CrossRef]
36. Farzipoor, R.; Zohrehbandian, M. A Data Envelopment Analysis Approach to Supplier Selection in Volume Discount Environments. *Int. J. Procure. Manag.* **2008**, *1*, 472–488.
37. Saen, R.F. Suppliers Selection in the Presence of Both Cardinal and Ordinal Data. *Eur. J. Oper. Res.* **2007**, *183*, 741–747. [CrossRef]
38. Storto, C.L. A double-DEA framework to support decision-making in the choice of advanced manufacturing technologies. *Manag. Decis.* **2018**, *56*, 488–507. [CrossRef]
39. Adler, N.; Friedman, L.; Sinuany-Stern, Z. Review of ranking methods in the data envelopment analysis context. *Eur. J. Oper. Res.* **2002**, *140*, 249–265. [CrossRef]
40. Storto, C.L. A peeling DEA-game cross efficiency procedure to classify suppliers. In Proceedings of the 21st Innovative Manufacturing Engineering and Energy International Conference (IManE 2017), Iasi, Romania, 25–26 May 2017.

41. Kuo, R.J.; Lee, L.Y.; Hu, T.-L. Developing a supplier selection system through integrating fuzzy AHP and fuzzy DEA: A case study on an auto lighting system company in Taiwan. *Prod. Plan. Control* **2010**, *21*, 468–484. [CrossRef]
42. Kuo, R.J.; Lin, Y.J. Supplier selection using analytic network process and data envelopment analysis. *Int. J. Prod. Res.* **2011**, *50*, 1–12. [CrossRef]
43. Taibi, A.; Atmani, B. Combining Fuzzy AHP with GIS and Decision Rules for Industrial Site Selection. *Int. Interact. Multimed. Artif. Intell.* **2017**, *4*, 60–69. [CrossRef]
44. Morente-Molinera, J.; Kou, G.; González-Crespo, R.; Corchado, J.; Herrera-Viedma, E. Solving multi-criteria group decision making problems under environments with a high number of alternatives using fuzzy ontologies and multi-granular linguistic modelling methods. *Knowl. Based Syst.* **2017**, *137*, 54–64. [CrossRef]
45. Adrian, C.; Abdullah, R.; Atan, R.; Jusoh, Y.Y. Conceptual Model Development of Big Data Analytics Implementation Assessment Effect on Decision-Making. *Int. J. Interact. Multimed. Artif. Intell.* **2018**, *5*, 101–106. [CrossRef]
46. Staníčková, M.; Melecký, L. Assessment of efficiency in Visegrad countries and regions using DEA models. *Ekonomická Revue Cent. Eur. Rev. Econ.* **2012**, *15*, 145–156. [CrossRef]
47. Schaar, D.; Sherry, L. Comparison of Data Envelopment Analysis Methods Used in Airport Benchmarking. In Proceedings of the Third International Conference on Research In Air Transportation, Fairfax, VA, USA, 1–4 June 2008.
48. Kahraman, C.; Cebeci, U.; Ulukan, Z. Multi-criteria supplier selection using fuzzy AHP. *Logist. Inf. Manag.* **2003**, *16*, 382–394. [CrossRef]
49. Hsu, C.-W.; Hu, A.H. Applying hazatdous substance management to supplier selection using analytic network process. *J. Clean. Prod.* **2009**, *17*, 255–264. [CrossRef]
50. Banaeian, N.; Mobli, H.; Nielsen, I.E.; Omid, M. Criteria definition and approaches in green supplier selection—A case study for raw material and packaging of food industry. *Prod. Manuf. Res.* **2015**, *3*, 149–168. [CrossRef]
51. Frosch, R. Industrial Ecology: Minimising the Impact of Industrial Waste. *Phys. Today* **1994**, *47*, 63–68. [CrossRef]
52. Mani, V.; Agarwal, R.; Sharma, V. Supplier selection using social sustainbility: AHP based approach in India. *Int. Strateg. Manag. Rev.* **2014**, *2*. [CrossRef]
53. Grover, R.; Grover, R.; Rao, V.B.; Kejriwal, K. Supplier Selection Using Sustainable Criteria in Sustainable Supply Chain Management. *Int. J. Econ. Manag. Eng.* **2016**, *10*, 1775–1780.
54. Ho, W.; Xu, X.; Dey, P.K. Multi-criteria decision making approaches for supplier evaluation and selection: A literature review. *Eur. J. Oper. Res.* **2010**, *202*, 16–24. [CrossRef]
55. Dickson, G. An analysis of vendor selection systems and decisions. *J. Purch.* **1966**, *2*, 5. [CrossRef]
56. Weber, C.C.J.; Weber, B.W. Vendor selection criteria and methods. *Eur. J. Oper. Res.* **1991**, *50*, 2–18. [CrossRef]
57. Handfield, R.B. US Global Sourcing: Patterns of Development. *Int. J. Oper. Prod. Manag.* **1994**, *14*, 40–51. [CrossRef]
58. Choi, T.Y.; Hartley, J.L. An exploration of supplier selection practices across the supply chain. *J. Oper. Manag.* **1996**, *14*, 333–343. [CrossRef]
59. Verma, R.; Pullman, M.E. An Analysis of the Supplier Selection Process. *Omega* **1998**, *26*, 739–750. [CrossRef]
60. Bharadwaj, N. Investigating the decision criteria used in electronic components procurement. *Ind. Mark. Manag.* **2004**, *33*, 317–323. [CrossRef]
61. Kannan, D.; Khodaverdi, R.; Olfat, L.; Jafarian, A.; Diabat, A. Integrated fuzzy multi criteria decision making method and multi-objective programming approach for supplier selection and order allocation in a green supply chain. *J. Clean. Prod.* **2013**, *47*, 355–367. [CrossRef]
62. Chu, T.-C.; Varma, R. Evaluating suppliers via a multiple levels multiple criteria decision making method under fuzzy environment. *Comput. Ind. Eng.* **2012**, *62*, 653–660. [CrossRef]
63. Tam, M.; Tummala, V.R. An Application of the AHP in Vendor Selection of a Telecommunications System. *Omega Int. J. Manag. Sci.* **2001**, *29*, 171–182. [CrossRef]
64. Shahgholian, K. A model for supplier selection based on fuzzy multi-criteria group decision making. *Afr. J. Bus. Manag.* **2012**, *6*. [CrossRef]
65. Dzever, S.; Merdji, M.; Saives, A. Purchase decision making and buyer-seller relationship development in the french food processing industry. *Supply Chain Manag. Int. J.* **2001**, *6*, 216–229. [CrossRef]

66. Bevilacqua, M.; Petroni, A. From traditional purchasing to supplier management: A fuzzy logic-based approach to supplier selection. *Int. J. Logist. Res. Appl.* **2002**, *5*, 235–255. [CrossRef]
67. Rezaei, J.; Ortt, R. Multi-criteria supplier segmentation using a fuzzy preference relations based ahp. *Eur. J. Oper. Res.* **2013**, *225*, 75–84. [CrossRef]
68. Roshandel, J.; Miri-Nargesi, S.S.; Hatami-Shirkouhi, L. Evaluating and selecting the supplier in detergent production industry using hierarchical fuzzy topsis. *Appl. Math. Model.* **2013**, *37*, 10170–10181. [CrossRef]
69. Govindaraju, R.; Akbar, M.I.; Gondodiwiryo, L.; Simatupang, T. The Application of a Decision-making Approach based on Fuzzy ANP and TOPSIS for Selecting a Strategic Supplier. *J. Eng. Technol. Sci.* **2015**, *47*, 406–425. [CrossRef]
70. Nielsen, I.E.; Banaeian, N.; Golińska, P.; Mobli, H.; Omid, M. Green supplier selection criteria: From a literature review to a flexible framework for determination of suitable criteria. *Logist. Oper. Supply Chain Manag. Sustain.* **2014**, 79–99. [CrossRef]
71. Sarkis, J. A methodological framework for evaluating environmentally conscious manufacturing programs. *Comput. Ind. Eng.* **1999**, *36*, 793–810. [CrossRef]
72. Saaty, T.L. Decision making with the analytic hierarchy process. *Int. J. Services Sci.* **2008**, *1*, 83–98. [CrossRef]
73. Zhu, K.J.; Jing, Y.; Chang, D. A discussion on extent analysis method and applications of fuzzy AHP. *Eur. J. Oper. Res.* **1999**, *116*, 450–456. [CrossRef]
74. Tang, Y.C.; Lin, T. Application of the fuzzy analytic hierarchy process to the lead-free equipment selection decision. *Int. J. Bus. Syst. Res.* **2011**, *5*, 35–56. [CrossRef]
75. Fiedler, M.; Nedoma, J.; Ramik, J.; Rohn, J.; Karel, Z. *Linear Optimization Problems with Inexact Data*, 1st ed.; Springer: New York, NY, USA, 2006.
76. Wen, M. Uncertain Data Envelopment Analysis. In *Uncertainty and Operations Research*; Springer: Berlin/Heidelberg, Germany, 2015.
77. Farrell, M.J. The Measurement of Productive Efficiency. *J. R. Stat. Soc.* **1957**, *120*, 253–281. [CrossRef]
78. Tone, K. A slacks-based measure of efficiency in data envelopment analysis. *Eur. J. Oper. Res.* **2001**, *130*, 498–509. [CrossRef]
79. Pastor, J.T.; Ruiz, J.L.; Sirvent, I. An enhanced DEA Russell graph efficiency measure. *Eur. J. Oper. Res.* **1999**, *115*, 596–607. [CrossRef]
80. Tang, Y.-C.; Beynon, M.J. Application and Development of a Fuzzy Analytic Hierarchy Process within a Capital Investment Study. *J. Econ. Manag.* **2005**, *1*, 207–230.
81. Shang, J.; Sueyoshi, T. A unified framework for the selection of a flexible manufacturing system. *Eur. J. Oper. Res.* **1995**, *85*, 297–315. [CrossRef]

symmetry

MDPI

Article

Hesitant Fuzzy Linguistic Aggregation Operators Based on the Hamacher t-norm and t-conorm

Jianghong Zhu [1,*] and Yanlai Li [1,2]

[1] School of Transportation and Logisitics, Southwest Jiaotong University, Chengdu 611756, China; lyl_2001@163.com
[2] National Laboratory of Railway Transportation, Southwest Jiaotong University, Chengdu 611756, China
* Correspondence: zhujianghong007@163.com; Tel.: +86-186-9696-3770

Received: 11 May 2018; Accepted: 30 May 2018; Published: 31 May 2018

Abstract: Hesitant fuzzy linguistic (HFL) term set, as a very flexible tool to represent the judgments of decision makers, has attracted the attention of many researchers. In recent years, some HFL aggregation operators have been developed to aggregate the HFL information. However, most of these operators are proposed based on the Algebraic product and Algebraic sum. In this paper, we presented some HFL aggregation operators to handle HFL information based on Hamacher triangle norms. We first define new operational laws on the HFL element according to Hamacher triangle norms. Then we present a family of HFL Hamacher aggregation operators, including the HFL Hamacher weighted averaging, HFL Hamacher weighted geometric, HFL Hamacher power weighted averaging and HFL Hamacher power weighted geometric operators and their generalized forms. We also investigate some special cases and properties of these operators in detail. Furthermore, we develop two approaches based on the proposed operators to deal with the multi criteria decision-making problem with HFL information. Finally, a numerical example with regard to choosing a suitable city to release sharing car is provided to illustrate the feasibility of the proposed method, and the advantages of the proposed methods are shown by conducting a sensitivity and comparative analysis.

Keywords: hesitant fuzzy linguistic term set; Hamacher t-norm and t-conorm; power aggregation operator; multi-criteria decision-making

---

1. Introduction

Multi-criteria decision-making (MCDM) problems with different kinds of fuzzy information is handled by utilizing Zadeh's fuzzy set [1] and their various extensions, including the interval-valued fuzzy set [2], intuitionistic fuzzy set [3,4], Pythagorean fuzzy set [5,6], Type-2 fuzzy set [7,8], fuzzy multi set [9], and hesitant fuzzy set (HFS) [10,11]. However, these fuzzy tools are only suitable to deal with quantitative situations rather than qualitative situations. The Fuzzy linguistic method (FLM) [2,12,13], which decision makers prefer to provide an evaluation for using a linguistic term, is a more suitable approach than the above fuzzy set to handle qualitative situations and has been extensively applied in various fields and applications [14–18]. In some cases, the modeling capacity of fuzzy linguistic is also quite limited because simple linguistic terms find it hard to express the hesitation of decision makers. For instance, a customer is invited to evaluate the satisfying degree of a service product with respect to a given criterion. Suppose $S = \{s_{-2} = \mathit{very\ low},\ s_{-1} = \mathit{low},\ s_0 = \mathit{medium},\ s_1 = \mathit{high},\ s_2 = \mathit{very\ high}\}$ is a linguistic term set (LTS). The customer regards s_0 or s_1 as the evaluation value of the satisfying degree for a service product, but he/she quietly finds it difficult to choose one of them as the final evaluation value. In this situation, an effective method is that the evaluation value of the satisfying degree provided by the customer should consist of the two possible values. To handle this situation, Rodríguez et al. [19] proposed the concept of hesitant fuzzy linguistic term set (HFLTS), which uses a linguistic term to replace the numerical elements of HFS. Subsequently, Liao et al. [20] gave the mathematical form of the HFLTS according to the concept of HFLTS and utilized the hesitant fuzzy linguistic element (HFLE) to represent the elements of HFLTS. For the above example, the customer's evaluation can be expressed by an HFLE $\{s_0, s_1\}$. The HFLTS, which is a combination of HFS and FLM, has the advantages of HFS and FLM at the same time. Therefore, it is a useful tool for a decision maker to express his/her judgment under the hesitation and fuzziness environment.

Recently, HFLTS has been used by more and more researchers to handle MCDM problems with uncertain information [21]. In this situation, the hesitant fuzzy linguistic (HFL) aggregation operator that is applied to aggregate the criteria's value into a comprehensive value of the alternative is one of the core issues. Therefore, the investigation of HFL aggregation operator is one of the hot topics. Various HFL aggregation operators have been developed from four respects as follows (1) Rodríguez et al. [19] defined the operational rules on HFLTS and proposed the min_upper and max_lower operators to select the worst of the superior values and the best of the inferior values, respectively; (2) based on the likelihood-based comparison relation between two HFLEs, Wei et al. [22] proposed the HFL weighted averaging (HFLWA) and HFL order weighted averaging (HFLOWA) operators, and Lee and Chen [23] presented the HFLWA, HFLOWA, and HFL weighted geometric (HFLWG), and HFL order weighted geometric operators; (3) according to the operational laws defined on HFLTS in [24,25], Zhang and Wu [24] proposed a family of operators for HFLEs, such as HFLWA, HFLWG, and generalized HFLWA operators. Wang [25] developed an extending HFLTS according to the definition of HFLTS, and defined the extending HFLWA, extending HFLWG and their ordered weighted forms. Shi and Xiao [26] presented the HFL reducible weighted Bonferroni mean, HFL generalized the reducible weighted Bonferroni mean, and HFL weighted power Bonferroni mean operators. Xu et al. [27] proposed an HFL order weighted distance operator and utilized to deal with multi-attribute group decision-making (MAGDM) problems. Liu et al. [28] developed the HFLWA, HFLWG, and HFL harmonic operators and their order weighted and hybrid weighted forms; (4) Based on the equivalent transformation function between HFLE and hesitant fuzzy element (HFE), Zhang and Qi [29] presented the HFLWA and HFLWG operators, and applied to solve a production strategy decision-making problem; Gou et al. [30] introduced the Bonferroni mean operator into the HFLTS environment and defined the HFL Bonferroni mean and HFL weighted Bonferroni mean operators.

It's worth noting that these existing HFL aggregation operators are constructed by the algebraic product and algebraic sum operational laws of HFLEs, which are a pair of special t-norm and t-conorm. A generalized intersection and union on HFLEs can be constructed by a generalized t-norm and t-conorm. For an intersection and union, a good alternative and approximation to the algebraic product and algebraic sum are the Einstein product and Einstein sum, respectively [31,32]. Recently, Wang and Liu [31,32] proposed the intuitionistic fuzzy Einstein weighted averaging and intuitionistic fuzzy Einstein weighted geometric operators. Further, Zhang [33] presented the intuitionistic fuzzy Einstein hybrid weighted averaging and intuitionistic fuzzy Einstein hybrid weighted geometric operators and their quasi-forms. Yu [34] introduced the Einstein operations into the HFS and developed the hesitant fuzzy Einstein weighted averaging and hesitant fuzzy Einstein weighted geometric operators and their ordered forms. Jin et al. [35] derived some interval-valued hesitant fuzzy Einstein prioritized operators and applied to solve MAGDM problems. On the other hand, Hamacher [36] presented a Hamacher t-norm and Hamacher t-conorm, which can be transformed into the algebraic and Einstein t-norms and t-conorms when the parameter $v = 1$ and $v = 2$ in Hamacher t-norm and t-conorm, respectively. Therefore, as general and flexible continuous triangular norms, Hamacher t-norm and t-conorm have been explored by many researchers in various fuzzy environments. Tan et al. [37] defined some hesitant fuzzy operational laws based on Hamacher operations and presented a family of hesitant fuzzy Hamacher aggregation operators, such as hesitant fuzzy Hamacher weighted averaging and hesitant fuzzy Hamacher weighted geometric operators. Ju et al. [38] proposed the dual hesitant fuzzy Hamacher weighted averaging and dual hesitant fuzzy Hamacher weighted geometric operators, and their order and hybrid forms. Liu et al. [39] proposed the improved interval-valued hesitant fuzzy Hamacher ordered weighted averaging and improved interval-valued hesitant fuzzy Hamacher ordered weighted geometric operators. Moreover, Hamacher operations are also introduced to other fuzzy environments, such as the intuitionistic fuzzy set [40], interval-valued intuitionistic fuzzy set [41], Pythagorean fuzzy set [42], and single-valued neutrosophic 2-tuple linguistic set [43]. From the above analysis, we can see that it is of important theoretical significance to explore the aggregation operators of HFLTS based on Hamacher operational laws and their application to MCDM problems, which is justly the first focus of this paper.

In practical MCDM process, it is extensively important to employ a suitable aggregation operator to drive the comprehensive preference value of each alternative. Various aggregation operators have been developed by many researchers to perform this process in MCDM problems. In these operators, the power average (PA) operator was originally presented by Yager [44], which allows the input data to support and strengthen one another, and the weight vectors in PA operator are associated with the input arguments. Inspired by the PA operator, Xu and Yager [45] presented a power geometric (PG) operator and a power ordered weighted geometric operator. The prominent characteristic of PA and PG operators is that they consider the relationships between the input arguments. Based on this advantages, many extending forms of PA and PG operators have been proposed, such as Xu [46] developing the intuitionistic fuzzy power weighted averaging and intuitionistic fuzzy power weighted geometric operators and their ordered forms. Further, Wei and Liu [47] introduced the PA and PG operators into a Pythagorean fuzzy environment and proposed a family of Pythagorean fuzzy power aggregation operators, including the Pythagorean fuzzy power averaging and Pythagorean fuzzy power geometric operators and their weighted, ordered weighted, and hybrid weighted forms. Zhang [48] presented a series of hesitant fuzzy power aggregation operators, such as hesitant fuzzy power averaging and hesitant fuzzy power geometric operators, and their ordered, weighted, and generalized forms. Furthermore, PA and PG operators have also been extended to other fuzzy environments to propose some new operators, such as intuitionistic fuzzy power aggregation operators based on entropy [49], linguistic hesitant fuzzy power aggregation operators [50], linguistic intuitionistic fuzzy power aggregation operators [51], dual hesitant fuzzy power aggregation operators based on Archimedean t-norm and t-conorm [52], and simplified neutrosophic power aggregation operators [53]. However, there is no one has explored the power aggregation operators on HFLTS, especially based on the Hamacher operations. Therefore, extending the power aggregation operators to HFLTS environments, especially based on Hamacher operational laws, is also very meaningful work and another focus of this paper.

According to the analysis above, this paper extends the Hamacher t-norm and t-conorm to an HFL environment and presents several new HFL aggregation operators to handle MCDM problem with HFL information. The main advantage of these operators is that they provide a good compensation to the existing HFL aggregation operators, and the HFL power aggregation operators can capture the relationships between the input arguments. The organization of this paper is arranged as follows. In Section 2, we briefly introduce the Hamacher t-norm and t-conorm and review some basic concepts of an HFL term set. We develop some HFL Hamacher aggregation operators and some HFL Hamacher power aggregation operators in Sections 3 and 4, respectively, and also discuss their special cases and investigate their basic properties. Section 5 utilizes these proposed operators to present two methods to handle MCDM problems with HFL information. We perform the developed methods on a numerical example and compare them with some existing HFL MCDM approaches in Section 6. Section 7 provides the conclusions of this paper.

2. Preliminaries

In this section, we briefly introduce the Hamacher t-norm and t-conorm and some basic concepts of HFLTS.

2.1. Hamacher Operations

There is an important concept in fuzzy set theory, that is, t-norm and t-conorm, which are utilized to define a generalized intersection and union of fuzzy sets [54]. A number of t-norm and t-conorm have been proposed, including Algebraic product T_A and Algebraic sum S_A [1], Einstein product T_E and Einstein sum S_E [55], and drastic product T_D and drastic sum S_D [56]. Further, Hamacher [36] developed a more generalized t-norm and t-conorm, that is, the Hamacher product (Hamacher t-norm) and Hamacher sum (Hamacher t-conorm), which are calculated as follows:

$$T_H^v(a,b) = a \otimes b = \frac{ab}{v + (1-v)(a+b-ab)},\ v > 0$$

$$S_H^v(a,b) = a \oplus b = \frac{a+b-ab-(1-v)ab}{1-(1-v)ab},\ v > 0$$

In particular, when $v = 1$, then the Hamacher t-norm and t-conorm are transformed into the Algebraic product T_A and Algebraic sum S_A [1].

$$T_A(a,b) = a \cdot b$$

$$S_A(a,b) = a + b - a \cdot b$$

When $v = 2$, then the Hamacher t-norm and t-conorm are transformed into the Einstein product T_E and Einstein sum S_E [55].

$$T_E(a,b) = a \otimes b = \frac{ab}{1+(1-a)(1-b)}$$

$$S_E(a,b) = a \oplus b = \frac{ab}{1+ab}$$

2.2. Hesitant Fuzzy Linguistic Term Set

Motivated by the HFS and fuzzy linguistic method, Rodríguez et al. [19] introduced the notion of HFLTS.

Definition 1. *[19]. Let $S = \{s_t | t = -\tau, \cdots, -1, 0, 1, \cdots, \tau\}$ be an LTS. An HFLTS, H_S, is constructed by a finite subset of the continuous linguistic terms of S.*

In order to help understand the concept of HFLTS, Liao et al. [20] gave the mathematical expression of HFLTS.

Definition 2. *[20]. Let $X = \{x_1, x_2, \cdots, x_n\}$ be a fixed set and $S = \{s_t | t = -\tau, \cdots, -1, 0, 1, \cdots, \tau\}$ be an LTS. An HFLST on X, H_S, is defined as the following*

$$H_S = \{< x, h_S(x_i) > | x_i \in X\}, i = 1, 2, \cdots, n. \quad (1)$$

where $h_S(x_i)$ is a collection of some linguistic terms in S and can be defined as $h_S(x_i) = \{s_t^i | s_t^i \in S, i = 1, 2, \cdots, L\}$ with L being the number of linguistic term in $h_S(x_i)$. For convenience, $h_S(x_i)$ is referred to as the HFLE.

To perform the equivalent conversion between HFLE and HFE, Gou [30] defined two equivalent conversion functions.

Definition 3. [30]. *Let* $S = \{s_t | t = -\tau, \cdots, -1, 0, 1, \cdots, \tau\}$ *be an LTS,* $h_S = \{s_t | t \in [-\tau, \tau]\}$ *be an HFLE, and* $h_\sigma = \{\sigma | \sigma \in [0,1]\}$ *be an HFE. The equivalent transformation from HFLE* h_S *to HFE* h_σ *is performed by the following function* g

$$g : [-\tau, \tau] \to [0,1], h_\sigma = g(h_S) = \{\sigma = g(s_t) = \frac{t}{2\tau} + \frac{1}{2}\}$$

Similarly, the equivalent transformation from HFE h_σ to HFLE h_S is performed by the following inverse function g^{-1}.

$$g^{-1} : [0,1] \to [-\tau, \tau],\ h_S = g^{-1}(h_\sigma) = \{s_t = g^{-1}(\sigma) = s_{(2\sigma-1)\tau}\}$$

Definition 4. [57]. *For any three HFLEs,* h_S, h_{S_1}, *and* h_{S_2}, g *and* g^{-1} *are the equivalent conversion functions between HFLE and HFE, and* $\lambda > 0$*; the operational rules on HFLEs are defined as follows:*

(1) $h_{S_1} \oplus h_{S_2} = \underset{\sigma_1 \in g(h_{S_1}), \sigma_2 \in g(h_{S_2})}{\cup} \{g^{-1}(\sigma_1 + \sigma_2 - \sigma_1\sigma_2)\}$;

(2) $h_{S_1} \otimes h_{S_2} = \underset{\sigma_1 \in g(h_{S_1}), \sigma_2 \in g(h_{S_2})}{\cup} \{g^{-1}(\sigma_1\sigma_2)\}$;

(3) $\lambda h_S = \underset{\sigma \in g(h_S)}{\cup} \{g^{-1}(1 - (1-\sigma)^\lambda)\}$;

(4) $(h_S)^\lambda = \underset{\sigma \in g(h_S)}{\cup} \{g^{-1}(\sigma^\lambda)\}$.

In the following, we introduce the Hamacher t-norm and t-conorm to the HFLTS environment and define some new operational rules on HFLEs.

Definition 5. *For any three HFLEs,* h_S, h_{S_1}, *and* h_{S_2}, g *and* g^{-1} *are the equivalent conversion functions between HFLE and HFE, and* $v > 0$*. According to the Hamacher t-norm and t-conorm, some operational rules on HFLEs are defined as follows:*

(1) $h_{S_1} \oplus_H h_{S_2} = \underset{\sigma_1 \in g(h_{S_1}), \sigma_2 \in g(h_{S_2})}{\cup} \left\{g^{-1}\left(\frac{\sigma_1+\sigma_2-\sigma_1\sigma_2-(1-v)\sigma_1\sigma_2}{1-(1-v)\sigma_1\sigma_2}\right)\right\}$;

(2) $h_{S_1} \otimes_H h_{S_2} = \underset{\sigma_1 \in g(h_{S_1}), \sigma_2 \in g(h_{S_2})}{\cup} \left\{g^{-1}\left(\frac{\sigma_1\sigma_2}{v+(1-v)(\sigma_1+\sigma_2-\sigma_1\sigma_2)}\right)\right\}$;

(3) $\lambda h_S = \underset{\sigma \in g(h_S)}{\cup} \left\{g^{-1}\left(\frac{(1+(v-1)\sigma)^\lambda-(1-\sigma)^\lambda}{(1+(v-1)\sigma)^\lambda+(v-1)(1-\sigma)^\lambda}\right)\right\}$, $\lambda > 0$;

(4) $(h_S)^\lambda = \underset{\sigma \in g(h_S)}{\cup} \left\{g^{-1}\left(\frac{v\sigma^\lambda}{(1+(v-1)(1-\sigma))^\lambda+(v-1)\sigma^\lambda}\right)\right\}$, $\lambda > 0$.

Remark 1. *When* $v = 1$*, we can see that these operations of HFLEs in Definition 5 are transformed into those in Definition 4. In other words, the operations in Definition 4 are a special case of Definition 5 by comparing Definition 4 with Definition 5.*

In addition, when $v = 2$*, these basic operations of HFLEs in Definition 5 are transformed into the Einstein operations on HFLEs.*

(1) $h_{S_1} \oplus_E h_{S_2} = \underset{\sigma_1 \in g(h_{S_1}), \sigma_2 \in g(h_{S_2})}{\cup} \left\{g^{-1}\left(\frac{\sigma_1+\sigma_2}{1+\sigma_1\sigma_2}\right)\right\}$;

(2) $h_{S_1} \otimes_E h_{S_2} = \underset{\sigma_1 \in g(h_{S_1}), \sigma_2 \in g(h_{S_2})}{\cup} \left\{g^{-1}\left(\frac{\sigma_1\sigma_2}{1-(1-\sigma_1)(1-\sigma_2)}\right)\right\}$;

(3) $\lambda h_S = \underset{\sigma \in g(h_S)}{\cup} \left\{g^{-1}\left(\frac{(1+\sigma)^\lambda-(1-\sigma)^\lambda}{(1+\sigma)^\lambda+(1-\sigma)^\lambda}\right)\right\}$, $\lambda > 0$;

(4) $(h_S)^\lambda = \underset{\sigma \in g(h_S)}{\cup} \left\{ g^{-1}\left(\frac{2\sigma^\lambda}{(2-\sigma)^\lambda + \sigma^\lambda} \right) \right\}$, $\lambda > 0$.

To compare the two HFLEs, Gou [30] defined the score function of HFLE as follows.

Definition 6. *[30]. Let $S = \{s_t | t = -\tau, \cdots, -1, 0, 1, \cdots, \tau\}$ be an LTS and $h_S = \{s_t | t \in [-\tau, \tau]\}$ be an HFLE, then the score function of h_S* is defined as the following

$$s(h_S) = \sum_{i=1}^{L} g(s_i)/L \tag{2}$$

where L is the number of the elements of h_S. Therefore, the comparative relation for two HFLEs is determined as follows:

(1) *If $s(h_{S_1}) > s(h_{S_2})$, then h_{S_1} is superior h_{S_2}, denoted by $h_{S_1} > h_{S_2}$;*
(2) *If $s(h_{S_1}) = s(h_{S_2})$, then h_{S_1} is equal to h_{S_2}, denoted by $h_{S_1} = h_{S_2}$.*

Definition 7. *[58]. Let $S = \{s_t | t = -\tau, \cdots, -1, 0, 1, \cdots, \tau\}$ be an LTS, and $h_{S_1} = \{s_{1t}^l | s_{1t}^l \in S, l = 1, 2, \cdots, L_1\}$, and $h_{S_2} = \{s_{2t}^l | s_{2t}^l \in S, l = 1, 2, \cdots, L_2\}$ be the two HFLEs. If $L_1 = L_2$ and $\lambda > 0$, then the generalized hesitant fuzzy linguistic distance between h_{S_1} and h_{S_2} is defined as follows*

$$d(h_{S_1}, h_{S_2}) = \left(\frac{1}{L} \sum_{i=1}^{L} \left(\left| g(s_{1t}^i) - g(s_{2t}^i) \right| \right)^\lambda \right)^{\frac{1}{\lambda}} \tag{3}$$

where g is the equivalent conversion function gave in Definition 3. When $\lambda = 2$, $d(h_{S_1}, h_{S_2})$ is called the HFL Euclidean distance between h_{S_1} and h_{S_2}.

When applying Equation (3), if $L_1 \neq L_2$, then the shorter one ($L_1 < L_2$) needs to be extended by adding the linguistic terms given as $s^1 = \left(s_{1t}^1 + s_{1t}^{L_1} \right)/2$, where s_{1t}^1 and $s_{1t}^{L_1}$ are the smallest and biggest linguistic terms in h_{S_1}, respectively.

3. Hesitant Fuzzy Linguistic Hamacher Aggregation Operators

In this part, we present a hesitant fuzzy linguistic Hamacher weighted averaging (HFLHWA) and a hesitant fuzzy linguistic Hamacher weighted geometric (HFLHWG), a generalized hesitant fuzzy linguistic Hamacher weighted averaging (GHFLHWA) and a generalized hesitant fuzzy linguistic Hamacher weighted geometric (GHFLHWG) operators. Furthermore, we also discuss some special cases of these operators and explore some properties of these operators.

3.1. HFLHWA and HFLHWG Operators

Definition 8. *Let $H_S = \{h_{S_1}, h_{S_2}, \cdots, h_{S_n}\}$ be a collection of HFLEs and $v > 0$. $w_i (i = 1, 2, \cdots, n)$ is the weight of $h_{S_i} (i = 1, 2, \cdots, n)$, satisfying $w_i \in [0, 1]$ and $\sum_{i=1}^{n} w_i = 1$. If*

$$HFLHWA_w^v(h_{S_1}, h_{S_2}, \cdots, h_{S_n}) = w_1 h_{S_1} \oplus_H w_2 h_{S_2} \oplus_H \cdots \oplus_H w_n h_{S_n} = \overset{n}{\underset{i=1}{\oplus_H}} (w_i h_{S_i}) \tag{4}$$

Then, $HFLHWA_w^v$ is designated as the HFL Hamacher weighted averaging (HFLHWA) operator.

Theorem 1. *Let $H_S = \{h_{S_1}, h_{S_2}, \cdots, h_{S_n}\}$ be a set of HFLEs and $v > 0$. $w_i (i = 1, 2, \cdots, n)$ is the weight of $h_{S_i} (i = 1, 2, \cdots, n)$, satisfying $w_i \in [0, 1]$ and $\sum_{i=1}^{n} w_i = 1$. g and g^{-1} are the equivalent transformation functions between HFLEs and HFEs. Then the aggregated value by the HFLHWA operator is also an HFLE, and*

$$HFLHWA_w^v(h_{S_1}, h_{S_2}, \cdots, h_{S_n}) = \bigcup_{\sigma_i \in g(h_{S_i})} \left\{ g^{-1}\left(\frac{\prod_{i=1}^{n}(1+(v-1)\sigma_i)^{w_i} - \prod_{i=1}^{n}(1-\sigma_i)^{w_i}}{\prod_{i=1}^{n}(1+(v-1)\sigma_i)^{w_i} + (v-1)\prod_{i=1}^{n}(1-\sigma_i)^{w_i}} \right) \right\} \tag{5}$$

Proof. According to mathematical induction method, Equation (5) can be proved as follows.

For $n = 1$, the result of Equation (5) clearly holds. Suppose Equation (5) hold for $n = k$, namely

$$HFLHWA_w^v(h_{S_1}, h_{S_2}, \cdots, h_{S_k}) = \bigcup_{\sigma_i \in g(h_{S_i})} \left\{ g^{-1}\left(\frac{\prod_{i=1}^{k}(1+(v-1)\sigma_i)^{w_i} - \prod_{i=1}^{k}(1-\sigma_i)^{w_i}}{\prod_{i=1}^{k}(1+(v-1)\sigma_i)^{w_i} + (v-1)\prod_{i=1}^{k}(1-\sigma_i)^{w_i}} \right) \right\}$$

Then, for $n = k + 1$, by Equation (4), we can get

$$HFLHWA_w^v(h_{S_1}, h_{S_2}, \cdots, h_{S_k}, h_{S_{k+1}}) = w_1 h_{S_1} \oplus_H w_2 h_{S_2} \oplus_H \cdots \oplus_H w_{k+1} h_{S_{k+1}} = \mathop{\oplus_H}_{i=1}^{k} (w_i h_{S_i}) \oplus_H w_{k+1} h_{S_{k+1}}$$
$$= \bigcup_{\sigma_i \in g(h_{S_i})} \left\{ g^{-1}\left(\frac{\prod_{i=1}^{k}(1+(v-1)\sigma_i)^{w_i} - \prod_{i=1}^{k}(1-\sigma_i)^{w_i}}{\prod_{i=1}^{k}(1+(v-1)\sigma_i)^{w_i} + (v-1)\prod_{i=1}^{k}(1-\sigma_i)^{w_i}} \right) \right\} \oplus_H \bigcup_{\sigma_{k+1} \in g(h_{S_{k+1}})} \left\{ g^{-1}\left(\frac{(1+(v-1)\sigma_{k+1})^{w_{k+1}} - (1-\sigma_{k+1})^{w_{k+1}}}{(1+(v-1)\sigma_{k+1})^{w_{k+1}} + (v-1)(1-\sigma_{k+1})^{w_{k+1}}} \right) \right\}$$

Let $\prod_{i=1}^{k}(1+(v-1)\sigma_i)^{w_i} = \alpha_1$, $\prod_{i=1}^{k}(1-\sigma_i)^{w_i} = \beta_1$, $(1+(v-1)\sigma_{k+1})^{w_{k+1}} = \alpha_2$, and $(1-\sigma_{k+1})^{w_{k+1}} = \beta_2$, then

$$\mathop{\oplus_H}_{i=1}^{k} (w_i h_{S_i}) = \bigcup_{\sigma_i \in g(h_{S_i})} \left\{ g^{-1}\left(\frac{\alpha_1 - \beta_1}{\alpha_1 + (v-1)\beta_1} \right) \right\} \text{ and } w_{k+1} h_{S_{k+1}} = \bigcup_{\sigma_{k+1} \in g(h_{S_{k+1}})} \left\{ g^{-1}\left(\frac{\alpha_2 - \beta_2}{\alpha_2 + (v-1)\beta_2} \right) \right\}$$

Further, the operational law (1) in Definition 5 yields

$$\mathop{\oplus_H}_{i=1}^{k} (w_i h_{S_i}) \oplus_H w_{k+1} h_{S_{k+1}} = \bigcup_{\sigma_1 \in g(h_{S_1}), \sigma_2 \in g(h_{S_2}), \cdots, \sigma_k \in g(h_{S_k})} \left\{ g^{-1}\left(\frac{\alpha_1 - \beta_1}{\alpha_1 + (v-1)\beta_1} \right) \right\} \oplus_H \bigcup_{\sigma_{k+1} \in g(h_{S_{k+1}})} \left\{ g^{-1}\left(\frac{\alpha_2 - \beta_2}{\alpha_2 + (v-1)\beta_2} \right) \right\}$$
$$= \bigcup_{\sigma_1 \in g(h_{S_1}), \sigma_2 \in g(h_{S_2}), \cdots, \sigma_{k+1} \in g(h_{S_{k+1}})} \left\{ g^{-1}\left(\frac{(\alpha_1-\beta_1)(\alpha_2+(v-1)\beta_2) + (\alpha_2-\beta_2)(\alpha_1+(v-1)\beta_1) - (2-v)(\alpha_1-\beta_1)(\alpha_2-\beta_2)}{(\alpha_1+(v-1)\beta_1)(\alpha_2+(v-1)\beta_2) - (1-v)(\alpha_1-\beta_1)(\alpha_2-\beta_2)} \right) \right\}$$
$$= \bigcup_{\sigma_1 \in g(h_{S_1}), \sigma_2 \in g(h_{S_2}), \cdots, \sigma_{k+1} \in g(h_{S_{k+1}})} \left\{ g^{-1}\left(\frac{\alpha_1\alpha_2 - \beta_1\beta_2}{\alpha_1\alpha_2 + (v-1)\beta_1\beta_2} \right) \right\}$$
$$= \bigcup_{\sigma_i \in g(h_{S_i})} \left\{ g^{-1}\left(\frac{\prod_{i=1}^{k+1}(1+(v-1)\sigma_i)^{w_i} - \prod_{i=1}^{k+1}(1-\sigma_i)^{w_i}}{\prod_{i=1}^{k+1}(1+(v-1)\sigma_i)^{w_i} + (v-1)\prod_{i=1}^{k+1}(1-\sigma_i)^{w_i}} \right) \right\}$$

That is, Equation (5) holds for $n = k + 1$. Therefore, Equation (5) holds for all n. □

Remark 2. *When $v = 1$, then the HFLHWA operator is transformed into the following:*

$$HFLWA_w(h_{S_1}, h_{S_2}, \cdots, h_{S_k}) = \mathop{\oplus_H}_{i=1}^{n} (w_i h_{S_i}) = \bigcup_{\sigma_i \in g(h_{S_i})} \left\{ g^{-1}\left(1 - \prod_{i=1}^{k}(1-\sigma_i)^{w_i} \right) \right\}$$

where $HFLWA_w$ is called the HFLWA operator by Zhang and Qi [29]. When $v = 2$, the HFLHWA operator is transformed into to the following:

$$HFLEWA_w(h_{S_1}, h_{S_2}, \cdots, h_{S_n}) = \bigcup_{\sigma_i \in g(h_{S_i})} \left\{ g^{-1}\left(\frac{\prod_{i=1}^{n}(1+\sigma_i)^{w_i} - \prod_{i=1}^{n}(1-\sigma_i)^{w_i}}{\prod_{i=1}^{n}(1+\sigma_i)^{w_i} + \prod_{i=1}^{n}(1-\sigma_i)^{w_i}} \right) \right\}$$

Here, $HFLEWA_w$ is called the HFLEWA operator. Especially when $w_i = 1/n$, then the HFLHWA operator is transformed into the hesitant fuzzy Hamacher averaging (HFLHA) operator.

$$HFLHA_w^v(h_{S_1}, h_{S_2}, \cdots, h_{S_n}) = \underset{\sigma_i \in g(h_{S_i})}{\cup} \left\{ g^{-1}\left(\frac{\prod_{i=1}^n (1+(v-1)\sigma_i)^{\frac{1}{n}} - \prod_{i=1}^n (1-\sigma_i)^{\frac{1}{n}}}{\prod_{i=1}^n (1+(v-1)\sigma_i)^{\frac{1}{n}} + (v-1)\prod_{i=1}^n (1-\sigma_i)^{\frac{1}{n}}} \right) \right\}$$

Example 1. *Let $S = \{s_t | t = -\tau, \cdots, -1, 0, 1, \cdots, \tau\}$ be an LTS and $\tau = 3$. $h_{S_1} = \{s_{-1}, s_1\}$ and $h_{S_2} = \{s_{-2}, s_0\}$ are two HFLEs; $w = (0.4, 0.6)$ are the weights of h_{S_1} and h_{S_2}, respectively. Then we can aggregate them by employing the HFLHWA ($v = 3$) operator.*

$$HFLHWA_w^3(h_{S_1}, h_{S_2}) = \underset{\sigma_1 \in g(h_{S_1}), \sigma_2 \in g(h_{S_2})}{\cup} \left\{ g^{-1}\left(\frac{\prod_{i=1}^2 (1+(3-1)\sigma_i)^{w_i} - \prod_{i=1}^2 (1-\sigma_i)^{w_i}}{\prod_{i=1}^2 (1+(3-1)\sigma_i)^{w_i} + (3-1)\prod_{i=1}^2 (1-\sigma_i)^{w_i}} \right) \right\}$$

$$= \underset{\sigma_1 \in g(h_{S_1}), \sigma_2 \in g(h_{S_2})}{\cup} \left\{ g^{-1}\left(\begin{array}{cc} \frac{(1+2\times\frac{1}{3})^{0.4}(1+2\times\frac{1}{6})^{0.6} - (\frac{2}{3})^{0.4}(\frac{5}{6})^{0.6}}{(1+2\times\frac{1}{3})^{0.4}(1+2\times\frac{1}{6})^{0.6} + 2\times(\frac{2}{3})^{0.4}(\frac{5}{6})^{0.6}}, & \frac{3\times(\frac{1}{3})^{0.4}\times(\frac{1}{2})^{0.6}}{(1+2\times\frac{2}{3})^{0.4}\times(1+2\times\frac{1}{2})^{0.6} + 2\times(\frac{1}{3})^{0.4}\times(\frac{1}{2})^{0.6}} \\ \frac{3\times(\frac{2}{3})^{0.4}\times(\frac{1}{6})^{0.6}}{(1+2\times\frac{1}{3})^{0.4}\times(1+2\times\frac{5}{6})^{0.6} + 2\times(\frac{2}{3})^{0.4}\times(\frac{1}{6})^{0.6}}, & \frac{3\times(\frac{2}{3})^{0.4}\times(\frac{1}{2})^{0.6}}{(1+2\times\frac{1}{3})^{0.4}\times(1+2\times\frac{1}{2})^{0.6} + 2\times(\frac{2}{3})^{0.4}\times(\frac{1}{2})^{0.6}} \end{array} \right) \right\}$$

$$= \{g^{-1}(0.2333,\ 0.3862,\ 0.4355,\ 0.5716)\}$$

$$= \{s_{-1.6004},\ s_{-0.6829},\ s_{-0.3870},\ s_{0.4299}\}$$

Idempotent 1. *Let $h_{S_i} (i = 1, 2, \cdots, n)$ be equal and each h_{S_i} which have only one value, namely, $h_{S_i} = h_S = \{s_t\}$ for any i, then*

$$HFLHWA_w^v(h_{S_1}, h_{S_2}, \cdots, h_{S_n}) = h_S \tag{6}$$

Proof. According to Definition 3, we have

$$g: [-\tau, \tau] \to [0, 1],\ g(s_t) = \left\{ \frac{t}{2\tau} + \frac{1}{2} = \sigma \middle| t \in [-\tau, \tau] \right\} = h_\sigma$$

Then, $HFLHWA_w^v(h_{S_1}, h_{S_2}, \cdots, h_{S_n}) = \underset{\sigma \in g(h_S)}{\cup} \left\{ g^{-1}\left(\frac{\prod_{i=1}^n (1+(v-1)\sigma)^{w_i} - \prod_{i=1}^n (1-\sigma)^{w_i}}{\prod_{i=1}^n (1+(v-1)\sigma)^{w_i} + (v-1)\prod_{i=1}^n (1-\sigma)^{w_i}} \right) \right\} =$ $\underset{\sigma \in g(h_S)}{\cup} \left\{ g^{-1}\left(\frac{(1+(v-1)\sigma)^{\sum_{i=1}^n w_i} - (1-\sigma)^{\sum_{i=1}^n w_i}}{(1+(v-1)\sigma)^{\sum_{i=1}^n w_i} + (v-1)(1-\sigma)^{\sum_{i=1}^n w_i}} \right) \right\} = \underset{\sigma \in g(h_S)}{\cup} \{g^{-1}(\sigma)\} = \{s_t\} = h_S.$

Therefore, we have $HFLHWA_w^v(h_{S_1}, h_{S_2}, \cdots, h_{S_n}) = h_S$. □

Remark 3. *Note that the HFLHWA operator is not idempotent in general; the following example is provided to demonstrate this case.*

Example 2. *Let $S = \{s_t | t = -\tau, \cdots, -1, 0, 1, \cdots, \tau\}$ be an LTS, $\tau = 3$, $h_{S_1} = h_{S_2} = h_S = \{s_{-1}, s_1\}$ and $w = (0.4, 0.6)^T$. Then $HFLHWA_w^3(h_{S_1}, h_{S_2}) = \{0.3333, 0.4804, 0.5480, 0.6667\}$, $s(HFLHWA_w^3(h_{S_1}, h_{S_2})) = 0.5071$ and $s(h_S) = 0.5$. Therefore, $HFLHWA_w^3(h_{S_1}, h_{S_2}) > h_S$.*

Monotonic 1. *Let $h_S^a = \{h_S^{a_1}, h_S^{a_2}, \cdots, h_S^{a_n}\}$ and $h_S^b = \{h_S^{b_1}, h_S^{b_2}, \cdots, h_S^{b_n}\}$ be two any collection of HFLEs. If for any $s_t^{a_i} \in h_S^{a_i}$ and $s_t^{b_i} \in h_S^{b_i}$, and $s_t^{a_i} \le s_t^{b_i}$ for any i, then*

$$HFLHWA_w^v(h_S^{a_1}, h_S^{a_2}, \cdots, h_S^{a_n}) \le HFLHWA_w^v(h_S^{b_1}, h_S^{b_2}, \cdots, h_S^{b_n}) \tag{7}$$

Proof. Let $f(x) = \frac{1+(v-1)x}{1-x}$, $x \in [0, 1)$ and $v > 0$. Since $f'(x) = \frac{v}{(1-x)^2} > 0$, $f(x)$ is an increasing function.

According to Definition 3, we have

$$g:[-\tau,\tau]\to[0,1],\ g\left(s_t^{\rho_i}\right)=\frac{t}{2\tau}+\frac{1}{2}=\sigma_{\rho_i},\ g\left(h_S^{\rho_i}\right)=\left\{\frac{t}{2\tau}+\frac{1}{2}=\sigma_{\rho_i}\middle|t\in[-\tau,\tau]\right\}=h_{\rho_i}$$

where $i=1,2,\cdots,n$ and $\rho=a$ or $\rho=b$. Then for any $s_t^{a_i}\le s_t^{b_i}$, we have $\sigma_{a_i}\le\sigma_{b_i}$, further, $f(\sigma_{a_i})\le f(\sigma_{b_i})$.

Suppose $w_i(i=1,2,\cdots,n)$ be the weight of h_{S_i}, satisfying $w_i\in[0,1]$ and $\sum_{i=1}^n w_i=1$. Based on the above condition, we can get

$$\begin{aligned}
&\bigcup_{\sigma_{a_i}\in g(h_S^{a_i})}\left\{g^{-1}\left(\prod_{i=1}^n\left(\tfrac{1+(v-1)\sigma_{a_i}}{1-\sigma_{a_i}}\right)^{w_i}\right)\right\}\le\bigcup_{\sigma_{b_i}\in g(h_S^{b_i})}\left\{g^{-1}\left(\prod_{i=1}^n\left(\tfrac{1+(v-1)\sigma_{b_i}}{1-\sigma_{b_i}}\right)^{w_i}\right)\right\}\\
\Rightarrow&\bigcup_{\sigma_{a_i}\in g(h_S^{a_i})}\left\{g^{-1}\left(\prod_{i=1}^n\left(\tfrac{1+(v-1)\sigma_{a_i}}{1-\sigma_{a_i}}\right)^{w_i}+(v-1)\right)\right\}\le\bigcup_{\sigma_{b_i}\in g(h_S^{b_i})}\left\{g^{-1}\left(\prod_{i=1}^n\left(\tfrac{1+(v-1)\sigma_{b_i}}{1-\sigma_{b_i}}\right)^{w_i}+(v-1)\right)\right\}\\
\Rightarrow&\bigcup_{\sigma_{a_i}\in g(h_S^{a_i})}\left\{g^{-1}\left(1-\tfrac{v\prod_{i=1}^n(1-\sigma_{a_i})^{w_i}}{\prod_{i=1}^n(1+(v-1)\sigma_{a_i})^{w_i}+(v-1)\prod_{i=1}^n(1-\sigma_{a_i})^{w_i}}\right)\right\}\\
\le&\bigcup_{\sigma_{b_i}\in g(h_S^{b_i})}\left\{g^{-1}\left(1-\tfrac{v\prod_{i=1}^n(1-\sigma_{b_i})^{w_i}}{\prod_{i=1}^n(1+(v-1)\sigma_{b_i})^{w_i}+(v-1)\prod_{i=1}^n(1-\sigma_{b_i})^{w_i}}\right)\right\}\\
\Rightarrow&\bigcup_{\sigma_{a_i}\in g(h_S^{a_i})}\left\{g^{-1}\left(\tfrac{\prod_{i=1}^n(1+(v-1)\sigma_{a_i})^{w_i}-\prod_{i=1}^n(1-\sigma_{a_i})^{w_i}}{\prod_{i=1}^n(1+(v-1)\sigma_{a_i})^{w_i}+(v-1)\prod_{i=1}^n(1-\sigma_{a_i})^{w_i}}\right)\right\}\\
\le&\bigcup_{\sigma_{b_i}\in g(h_S^{b_i})}\left\{g^{-1}\left(\tfrac{\prod_{i=1}^n(1+(v-1)\sigma_{b_i})^{w_i}-\prod_{i=1}^n(1-\sigma_{b_i})^{w_i}}{\prod_{i=1}^n(1+(v-1)\sigma_{b_i})^{w_i}+(v-1)\prod_{i=1}^n(1-\sigma_{b_i})^{w_i}}\right)\right\}
\end{aligned}$$

Therefore, based on Theorem 1, we have $HFLHWA_w^v(h_S^{a_1},h_S^{a_2},\cdots,h_S^{a_n})\le HFLHWA_w^v\left(h_S^{b_1},h_S^{b_2},\cdots,h_S^{b_n}\right)$. □

Bounded 1. *Let $h_{S_i}(i=1,2,\cdots,n)$ be a set of HFLEs, if $h_S^+=\{s^+\}=\max\left(\bigcup_{s_t^i\in h_{S_i}}\max\{s_t^i\}\right)$ and $h_S^-=\{s^-\}=\left(\bigcup_{s_t^i\in h_{S_i}}\min\{s_t^i\}\right)$, then*

$$h_S^-\le HFLHWA_w^v(h_{S_1},h_{S_2},\cdots,h_{S_n})\le h_S^+\tag{8}$$

Proof. According to Definition 3, we have

$$g:[-\tau,\tau]\to[0,1],\ g\left(s_t^i\right)=\frac{t}{2\tau}+\frac{1}{2}=\sigma_i,\ g(h_{S_i})=\left\{\frac{t}{2\tau}+\frac{1}{2}=\sigma_i\middle|t\in[-\tau,\tau]\right\}=h_i$$

where $i=1,2,\cdots,n$. Then, $s^-\le s_t^i\le s^+$ for any i, we have $\sigma^-\le\sigma_i\le\sigma^+$.

Suppose $w_i(i=1,2,\cdots,n)$ be the weight of h_{S_i}, satisfying $w_i\in[0,1]$ and $\sum_{i=1}^n w_i=1$. Based on the monotonic of HFLHWA, we can get

$$\begin{aligned}
HFLHWA_w^v(h_{S_1},h_{S_2},\cdots,h_{S_n})&=\bigcup_{\sigma_i\in g(h_{S_i})}\left\{g^{-1}\left(\tfrac{\prod_{i=1}^n(1+(v-1)\sigma_i)^{w_i}-\prod_{i=1}^n(1-\sigma_i)^{w_i}}{\prod_{i=1}^n(1+(v-1)\sigma_i)^{w_i}+(v-1)\prod_{i=1}^n(1-\sigma_i)^{w_i}}\right)\right\}\\
&\ge\bigcup_{\sigma^-\in g(h_{S_i})}\left\{g^{-1}\left(\tfrac{\prod_{i=1}^n(1+(v-1)\sigma^-)^{w_i}-\prod_{i=1}^n(1-\sigma^-)^{w_i}}{\prod_{i=1}^n(1+(v-1)\sigma^-)^{w_i}+(v-1)\prod_{i=1}^n(1-\sigma^-)^{w_i}}\right)\right\}\\
=\bigcup_{\sigma^-\in g(h_{S_i})}&\left\{g^{-1}\left(\tfrac{(1+(v-1)\sigma^-)^{\sum_{i=1}^n w_i}-(1-\sigma^-)^{\sum_{i=1}^n w_i}}{(1+(v-1)\sigma^-)^{\sum_{i=1}^n w_i}+(v-1)(1-\sigma^-)^{\sum_{i=1}^n w_i}}\right)\right\}=\bigcup_{\sigma^-\in g(h_{S_i})}\{g^{-1}(\sigma^-)\}=h_S^-
\end{aligned}$$

Similarly, $HFLHWA_w^v(h_{S_1}, h_{S_2}, \cdots, h_{S_n}) \leq h_S^+$. Therefore, $h_S^- \leq HFLHWA_w^v(h_{S_1}, h_{S_2}, \cdots, h_{S_n}) \leq h_S^+$. □

Commutative 1. *Let $h_{S_i}(i = 1, 2, \cdots, n)$ be a set of HFLEs, and $(\overline{h}_{S_1}, \overline{h}_{S_2}, \cdots, \overline{h}_{S_n})$ be any permutation of $(h_{S_1}, h_{S_2}, \cdots, h_{S_n})$, then*

$$HFLHWA_w^v(h_{S_1}, h_{S_2}, \cdots, h_{S_n}) = HFLHWA_w^v(\overline{h}_{S_1}, \overline{h}_{S_2}, \cdots, \overline{h}_{S_n}) \tag{9}$$

Proof. Equation (9) clearly holds and the proof is omitted here. □

Lemma 1. *[59]. Let $y_i > 0$ $(i = 1, 2, \cdots, n)$ and w_i be the weight of y_i, satisfying $w_i \in [0, 1]$ and $\sum_{i=1}^n w_i = 1$, then*

$$\prod_{i=1}^n (y_i)^{w_i} \leq \sum_{i=1}^n (w_i y_i) \tag{10}$$

with equality if and only if $y_1 = y_2 = \cdots = y_n$.

Theorem 2. *Let $H_S = \{h_{S_1}, h_{S_2}, \cdots, h_{S_n}\}$ be a set of HFLEs and $w_i(i = 1, 2, \cdots, n)$ be the weight of $h_{S_i}(i = 1, 2, \cdots, n)$, satisfying $w_i \in [0, 1]$ and $\sum_{i=1}^n w_i = 1$. g and g^{-1} are the equivalent conversion functions between HFLEs and HFEs, and $v > 0$. Then*

$$HFLHWA_w^v(h_{S_1}, h_{S_2}, \cdots, h_{S_n}) \leq HFLWA_w(h_{S_1}, h_{S_2}, \cdots, h_{S_n}) \tag{11}$$

Proof. For any $s_t^i \in h_{S_i}$, based on Definition 3, we have

$$g : [-\tau, \tau] \rightarrow [0, 1],\ g(h_{S_i}) = \left\{ \frac{t}{2\tau} + \frac{1}{2} = \sigma_i \middle| t \in [-\tau, \tau] \right\} = h_i$$

Further, according to Equation (10), we have

$$\prod_{i=1}^n (1 + (v-1)\sigma_i)^{w_i} + (v-1)\prod_{i=1}^n (1-\sigma_i)^{w_i} \leq \sum_{i=1}^n w_i(1 + (v-1)\sigma_i) + (v-1)\sum_{i=1}^n w_i(1-\sigma_i) = v$$

then,

$$\begin{aligned} HFLHWA_w^v(h_{S_1}, h_{S_2}, \cdots, h_{S_n}) &= \bigcup_{\sigma_i \in g(h_{S_i})} \left\{ g^{-1}\left(\frac{\prod_{i=1}^n (1+(v-1)\sigma_i)^{w_i} - \prod_{i=1}^n (1-\sigma_i)^{w_i}}{\prod_{i=1}^n (1+(v-1)\sigma_i)^{w_i} + (v-1)\prod_{i=1}^n (1-\sigma_i)^{w_i}} \right) \right\} \\ &= \bigcup_{\sigma_i \in g(h_{S_i})} \left\{ g^{-1}\left(1 - \frac{v\prod_{i=1}^n (1-\sigma_i)^{w_i}}{\prod_{i=1}^n (1+(v-1)\sigma_i)^{w_i} + (v-1)\prod_{i=1}^n (1-\sigma_i)^{w_i}} \right) \right\} \\ \leq \bigcup_{\sigma_i \in g(h_{S_i})} \left\{ g^{-1}\left(1 - \frac{v\prod_{i=1}^n (1-\sigma_i)^{w_i}}{v} \right) \right\} &= \bigcup_{\sigma_i \in g(h_{S_i})} \left\{ g^{-1}\left(1 - \prod_{i=1}^n (1-\sigma_i)^{w_i} \right) \right\} = HFLWA_w(h_{S_1}, h_{S_2}, \cdots, h_{S_n}) \end{aligned}$$

Therefore, Equation (11) holds. □

Definition 9. *Let $H_S = \{h_{S_1}, h_{S_2}, \cdots, h_{S_n}\}$ be a collection of HFLEs and $v > 0$. $w_i(i = 1, 2, \cdots, n)$ be the weight of $h_{S_i}(i = 1, 2, \cdots, n)$, satisfying $w_i \in [0, 1]$ and $\sum_{i=1}^n w_i = 1$. If*

$$HFLHWG_w^v(h_{S_1}, h_{S_2}, \cdots, h_{S_n}) = (h_{S_1})^{w_1} \otimes_H (h_{S_2})^{w_2} \otimes_H \cdots \otimes_H (h_{S_n})^{w_n} = \underset{i=1}{\overset{n}{\otimes}}_H (h_{S_i})^{w_i} \tag{12}$$

then $HFLHWG_w^v$ is designated as the HFL Hamacher weighted geometric (HFLHWG) operator.

Theorem 3. *Let $H_S = \{h_{S_1}, h_{S_2}, \cdots, h_{S_n}\}$ be a set of HFLEs and $w_i(i = 1, 2, \cdots, n)$ be the weight of $h_{S_i}(i = 1, 2, \cdots, n)$, satisfying $w_i \in [0, 1]$ and $\sum_{i=1}^n w_i = 1$. g and g^{-1} are the equivalent conversion* *functions between HFLEs and HFEs, and $v > 0$. Then the aggregated value by the HFLHWG operator is also an HFLE, and*

$$HFLHWG_w^v(h_{S_1}, h_{S_2}, \cdots, h_{S_n}) = \bigcup_{\sigma_i \in g(h_{S_i})} \left\{ g^{-1}\left(\frac{v\prod_{i=1}^n (\sigma_i)^{w_i}}{\prod_{i=1}^n (1+(v-1)(1-\sigma_i))^{w_i} + (v-1)\prod_{i=1}^n (\sigma_i)^{w_i}} \right) \right\} \tag{13}$$

Proof. According to the mathematical induction method, Equation (13) can be proved as follows.

For $n = 1$, the result of Equation (13) clearly holds. Suppose Equation (13) holds for $n = k$, namely

$$HFLHWG_w^v(h_{S_1}, h_{S_2}, \cdots, h_{S_k}) = \bigcup_{\sigma_i \in g(h_{S_i})} \left\{ g^{-1}\left(\frac{v\prod_{i=1}^k (\sigma_i)^{w_i}}{\prod_{i=1}^k (1+(v-1)(1-\sigma_i))^{w_i} + (v-1)\prod_{i=1}^k (\sigma_i)^{w_i}} \right) \right\}$$

Then, for $n = k + 1$, by Equation (12), we can get

$$HFLHWG_w^v(h_{S_1}, h_{S_2}, \cdots, h_{S_k}, h_{S_{k+1}}) = (h_{S_1})^{w_1} \otimes_H (h_{S_2})^{w_2} \otimes_H \cdots \otimes_H (h_{S_n})^{w_n} \otimes_H (h_{S_{k+1}})^{w_{k+1}} = \overset{n}{\underset{i=1}{\otimes_H}} (h_{S_i})^{w_i} \otimes_H (h_{S_{k+1}})^{w_{k+1}}$$
$$= \bigcup_{\sigma_i \in g(h_{S_i})} \left\{ g^{-1}\left(\frac{v\prod_{i=1}^k (\sigma_i)^{w_i}}{\prod_{i=1}^k (1+(v-1)(1-\sigma_i))^{w_i} + (v-1)\prod_{i=1}^k (\sigma_i)^{w_i}} \right) \right\} \otimes_H \bigcup_{\sigma_{k+1} \in g(h_{S_{k+1}})} \left\{ g^{-1}\left(\frac{v\sigma_{k+1}^{w_{k+1}}}{(1+(v-1)(1-\sigma_{k+1}))^{w_{k+1}} + (v-1)\sigma_{k+1}^{w_{k+1}}} \right) \right\}$$

Let $\prod_{i=1}^k (1+(v-1)(1-\sigma_i))^{w_i} = \alpha_1$, $\prod_{i=1}^k (\sigma_i)^{w_i} = \beta_1$, $(1+(v-1)(1-\sigma_{k+1}))^{w_{k+1}} = \alpha_2$ and $\sigma_{k+1}^{w_{k+1}} = \beta_2$, then

$$\overset{k}{\underset{i=1}{\otimes_H}} (h_{S_i})^{w_i} = \bigcup_{\sigma_1 \in g(h_{S_1}), \sigma_2 \in g(h_{S_2}), \cdots, \sigma_k \in g(h_{S_k})} \left\{ g^{-1}\left(\frac{v\beta_1}{\alpha_1 + (v-1)\beta_1} \right) \right\} \text{ and } (h_{S_{k+1}})^{w_{k+1}} = \bigcup_{\sigma_{k+1} \in g(h_{S_{k+1}})} \left\{ g^{-1}\left(\frac{v\beta_2}{\alpha_2 + (v-1)\beta_2} \right) \right\}$$

Further, the operational law (2) in Definition 5 yields

$$\overset{k}{\underset{i=1}{\otimes_H}} (h_{S_i})^{w_i} \otimes_H (h_{S_{k+1}})^{w_{k+1}}$$
$$= \bigcup_{\sigma_1 \in g(h_{S_1}), \sigma_2 \in g(h_{S_2}), \cdots, \sigma_{k+1} \in g(h_{S_{k+1}})} \left\{ g^{-1}\left(\frac{\frac{v^2\beta_1\beta_2}{(\alpha_1+(v-1)\beta_1)(\alpha_2+(v-1)\beta_2)}}{v+(1-v)\left(\frac{v\beta_1(\alpha_2+(v-1)\beta_2)+v\beta_2(\alpha_1+(v-1)\beta_1)-v^2\beta_1\beta_2}{(\alpha_1+(v-1)\beta_1)(\alpha_2+(v-1)\beta_2)} \right)} \right) \right\}$$
$$= \bigcup_{\sigma_1 \in g(h_{S_1}), \sigma_2 \in g(h_{S_2}), \cdots, \sigma_{k+1} \in g(h_{S_{k+1}})} \left\{ g^{-1}\left(\frac{v\beta_1\beta_2}{(\alpha_1+(v-1)\beta_1)(\alpha_2+(v-1)\beta_2)-(v-1)(\alpha_2\beta_1+\alpha_1\beta_2+(v-2)\beta_1\beta_2)} \right) \right\}$$
$$= \bigcup_{\sigma_1 \in g(h_{S_1}), \sigma_2 \in g(h_{S_2}), \cdots, \sigma_{k+1} \in g(h_{S_{k+1}})} \left\{ g^{-1}\left(\frac{v\beta_1\beta_2}{\alpha_1\alpha_2+(v-1)\beta_1\beta_2} \right) \right\}$$
$$= \bigcup_{\sigma_1 \in g(h_{S_1}), \sigma_2 \in g(h_{S_2}), \cdots, \sigma_{k+1} \in g(h_{S_{k+1}})} \left\{ g^{-1}\left(\frac{v\prod_{i=1}^{k+1} (\sigma_i)^{w_i}}{\prod_{i=1}^{k+1} (1+(v-1)(1-\sigma_i))^{w_i} + (v-1)\prod_{i=1}^{k+1} (\sigma_i)^{w_i}} \right) \right\}$$

That is, Equation (13) holds for $n = k + 1$. Therefore, Equation (13) holds for all n. □

Remark 4. *When $v = 1$, then the HFLHWG operator transforms into the following:*

$$HFLWG_w(h_{S_1}, h_{S_2}, \cdots, h_{S_n}) = \bigcup_{\sigma_i \in g(h_{S_i})} \left\{ g^{-1}\left(\prod_{i=1}^n (\sigma_i)^{w_i} \right) \right\}$$

where $HFLWG_w$ is called the HFLWG operator [29]. When $v = 2$, then the HFLHWG operator transforms into the following:

$$HFLEWG_w(h_{S_1}, h_{S_2}, \cdots, h_{S_n}) = \bigcup_{\sigma_i \in g(h_{S_i})} \left\{ g^{-1}\left(\frac{2\prod_{i=1}^n (\sigma_i)^{w_i}}{\prod_{i=1}^n (2-\sigma_i)^{w_i} + \prod_{i=1}^n (\sigma_i)^{w_i}} \right) \right\}$$

where $HFLEWG_w$ is called the HFL Einstein weighted geometric (HFLEWG) operator. Especially when $w_i = 1/n$, then the HFLHWG operator is transformed into the hesitant fuzzy Hamacher geometric (HFLHG) operator.

$$HFLHG_w^v(h_{S_1}, h_{S_2}, \cdots, h_{S_n}) = \bigcup_{\sigma_i \in g(h_{S_i})} \left\{ g^{-1}\left(\frac{v\prod_{i=1}^n (\sigma_i)^{\frac{1}{n}}}{\prod_{i=1}^n (1+(v-1)(1-\sigma_i))^{\frac{1}{n}} + (v-1)\prod_{i=1}^n (\sigma_i)^{\frac{1}{n}}} \right) \right\}$$

Example 3. *Let $S = \{s_t | t = -\tau, \cdots, -1, 0, 1, \cdots, \tau\}$ be an LTS and $\tau = 3$. $h_{S_1} = \{s_{-1}, s_1\}$ and $h_{S_2} = \{s_{-2}, s_0\}$ are two HFLEs, and $w = (0.4, 0.6)$ is the weight of h_{S_1} and h_{S_2}, respectively. Then we can aggregate them by employing the HFLHWG ($v = 3$) operator.*

$$HFLHWG_w^3(h_{S_1}, h_{S_2}) = \bigcup_{\sigma_1 \in h_{S_1}, \sigma_2 \in h_{S_2}} \left\{ g^{-1}\left(\frac{3\prod_{i=1}^2 (\sigma_i)^{w_i}}{\prod_{i=1}^2 (1+(3-1)(1-\sigma_i))^{w_i} + (3-1)\prod_{i=1}^2 (\sigma_i)^{w_i}} \right) \right\}$$

$$= \bigcup_{\sigma_1 \in g(h_{S_1}), \sigma_2 \in g(h_{S_2})} \left\{ g^{-1}\left(\begin{array}{c} \frac{3\times(\frac{1}{3})^{0.4}\times(\frac{1}{6})^{0.6}}{(1+2\times\frac{2}{3})^{0.4}\times(1+2\times\frac{5}{6})^{0.6}+2\times(\frac{1}{3})^{0.4}\times(\frac{1}{6})^{0.6}}, \frac{3\times(\frac{1}{3})^{0.4}\times(\frac{1}{2})^{0.6}}{(1+2\times\frac{2}{3})^{0.4}\times(1+2\times\frac{1}{2})^{0.6}+2\times(\frac{1}{3})^{0.4}\times(\frac{1}{2})^{0.6}} \\ \frac{3\times(\frac{2}{3})^{0.4}\times(\frac{1}{6})^{0.6}}{(1+2\times\frac{1}{3})^{0.4}\times(1+2\times\frac{5}{6})^{0.6}+2\times(\frac{2}{3})^{0.4}\times(\frac{1}{6})^{0.6}}, \frac{3\times(\frac{2}{3})^{0.4}\times(\frac{1}{2})^{0.6}}{(1+2\times\frac{1}{3})^{0.4}\times(1+2\times\frac{1}{2})^{0.6}+2\times(\frac{2}{3})^{0.4}\times(\frac{1}{2})^{0.6}} \end{array} \right) \right\}$$

$$= \{g^{-1}(0.2223, 0.3120, 0.4284, 0.5645)\}$$

$$= \{s_{-1.6662}, s_{-1.1279}, s_{-0.4299}, s_{0.3870}\}$$

Idempotent 2. *Let $h_{S_i}(i = 1, 2, \cdots, n)$ be equal with each h_{S_i} having only one value, namely, $h_{S_i} = h_S = \{s_t\}$ for any i, then*

$$HFLHWG_w^v(h_{S_1}, h_{S_2}, \cdots, h_{S_n}) = h_S \tag{14}$$

Proof. The proof of Equation (14) is similar to Equation (6) and is omitted here. □

Remark 5. *Note that the HFLHWG operator is not idempotent when h_{S_i} includes more than one value; the following example is provided to demonstrate this case.*

Example 4. *Let $S = \{s_t | t = -\tau, \cdots, -1, 0, 1, \cdots, \tau\}$ be an LTS, $\tau = 3$, $h_{S_1} = h_{S_1} = h_S = \{s_{-1}, s_1\}$ and $w = (0.4, 0.6)^T$. Then $HFLHWG_w^3(h_{S_1}, h_{S_2}) = \{0.3333, 0.4520, 0.5196, 0.6667\}$, $s(HFLHWG_w^3(h_{S_1}, h_{S_2})) = 0.4929$, and $s(h_S) = 0.5$. Therefore, $HFLHWG_w^3(h_{S_1}, h_{S_2}) < h_S$.*

Monotonic 2. *Let $h_S^a = \{h_S^{a_1}, h_S^{a_2}, \cdots, h_S^{a_n}\}$ and $h_S^b = \{h_S^{b_1}, h_S^{b_2}, \cdots, h_S^{b_n}\}$ be two of any collection of HFLEs. If for any $s_t^{a_i} \in h_S^{a_i}$ and $s_t^{b_i} \in h_S^{b_i}$, and $s_t^{a_i} \leq s_t^{b_i}$ for any i, then*

$$HFLHWG_w^v(h_S^{a_1}, h_S^{a_2}, \cdots, h_S^{a_n}) \leq HFLHWG_w^v(h_S^{b_1}, h_S^{b_2}, \cdots, h_S^{b_n}) \tag{15}$$

Proof. Let $f(x) = \frac{1+(v-1)(1-x)}{x}$, $x \in (0,1]$ and $v > 0$. Since $f'(x) = \frac{-v}{x^2} < 0$, hence $f(x)$ is a decreasing function.

According to Definition 3, we have

$$g: [-\tau, \tau] \to [0,1],\ g\left(s_t^{\rho_i}\right) = \frac{t}{2\tau} + \frac{1}{2} = \sigma_{\rho_i},\ g\left(h_S^{\rho_i}\right) = \{\frac{t}{2\tau} + \frac{1}{2} = \sigma_{\rho_i} \Big| t \in [-\tau, \tau]\} = h_{\rho_i}$$

where $i = 1, 2, \cdots, n$ and $\rho = a$ or $\rho = b$. Then for any $s_t^{a_i} \leq s_t^{b_i}$, we have $\sigma_{a_i} \leq \sigma_{b_i}$, further, $f(\sigma_{a_i}) \geq f(\sigma_{b_i})$.

Suppose $w_i (i = 1, 2, \cdots, n)$ is the weight of h_{S_i}, satisfying $w_i \in [0,1]$ and $\sum_{i=1}^{n} w_i = 1$. Based on the above condition, we have

$$\begin{aligned}
&\bigcup_{\sigma_{a_i} \in g(h_S^{a_i})} \left\{ g^{-1}\left(\prod_{i=1}^{n} \left(\frac{1+(v-1)(1-\sigma_{a_i})}{\sigma_{a_i}} \right)^{w_i} \right) \right\} \geq \bigcup_{\sigma_{b_i} \in g(h_S^{b_i})} \left\{ g^{-1}\left(\prod_{i=1}^{n} \left(\frac{1+(v-1)(1-\sigma_{b_i})}{\sigma_{b_i}} \right)^{w_i} \right) \right\} \\
\Rightarrow &\bigcup_{\sigma_{a_i} \in g(h_S^{a_i})} \left\{ g^{-1}\left(\prod_{i=1}^{n} \left(\frac{1+(v-1)(1-\sigma_{a_i})}{\sigma_{a_i}} \right)^{w_i} + (v-1) \right) \right\} \geq \bigcup_{\sigma_{b_i} \in g(h_S^{b_i})} \left\{ g^{-1}\left(\prod_{i=1}^{n} \left(\frac{1+(v-1)(1-\sigma_{a_i})}{\sigma_{a_i}} \right)^{w_i} + (v-1) \right) \right\} \\
\Rightarrow &\bigcup_{\sigma_{a_i} \in g(h_S^{a_i})} \left\{ g^{-1}\left(\frac{v\prod_{i=1}^{n} (\sigma_{a_i})^{w_i}}{\prod_{i=1}^{n} (1+(v-1)(1-\sigma_{a_i}))^{w_i} + (v-1)\prod_{i=1}^{n} (\sigma_{a_i})^{w_i}} \right) \right\} \\
\leq &\bigcup_{\sigma_{b_i} \in g(h_S^{b_i})} \left\{ g^{-1}\left(\frac{v\prod_{i=1}^{n} (\sigma_{b_i})^{w_i}}{\prod_{i=1}^{n} (1+(v-1)(1-\sigma_{b_i}))^{w_i} + (v-1)\prod_{i=1}^{n} (\sigma_{b_i})^{w_i}} \right) \right\}
\end{aligned}$$

Therefore, based on Theorem 3, we have $HFLHWG_w^v(h_S^{a_1}, h_S^{a_2}, \cdots, h_S^{a_n}) \leq HFLHWG_w^v\left(h_S^{b_1}, h_S^{b_2}, \cdots, h_S^{b_n}\right)$. □

Bounded 2. *Let $h_{S_i} (i = 1, 2, \cdots, n)$ be a set of HFLEs, if $h_S^+ = \{s^+\} = \max\left(\bigcup_{s_t^i \in h_{S_i}} \max\{s_t^i\} \right)$ and $h_S^- = \{s^-\} = \left(\bigcup_{s_t^i \in h_{S_i}} \min\{s_t^i\} \right)$, then*

$$h_S^- \leq HFLHWG_w^v(h_{S_1}, h_{S_2}, \cdots, h_{S_n}) \leq h_S^+ \tag{16}$$

Proof. The proof of Equation (16) is similar to Equation (8) and is omitted here. □

Commutative 2. *Let $h_{S_i} (i = 1, 2, \cdots, n)$ be a collection of HFLEs, and $(\overline{h}_{S_1}, \overline{h}_{S_2}, \cdots, \overline{h}_{S_n})$ be any permutation of $(h_{S_1}, h_{S_2}, \cdots, h_{S_n})$, then*

$$HFLHWG_w^v(h_{S_1}, h_{S_2}, \cdots, h_{S_n}) = HFLHWG_w^v(\overline{h}_{S_1}, \overline{h}_{S_2}, \cdots, \overline{h}_{S_n}) \tag{17}$$

Proof. Equation (17) clearly holds and the proof of Equation (17) is omitted here. □

Theorem 4. *Let $H_S = \{h_{S_1}, h_{S_2}, \cdots, h_{S_n}\}$ be a set of HFLEs and $w_i (i = 1, 2, \cdots, n)$ be the weight of $h_{S_i} (i = 1, 2, \cdots, n)$, satisfying $w_i \in [0,1]$ and $\sum_{i=1}^{n} w_i = 1$. g and g^{-1} are the equivalent conversion functions between HFLEs and HFEs, and $v > 0$. Then*

$$HFLHWG_w^v(h_{S_1}, h_{S_2}, \cdots, h_{S_n}) \geq HFLWG_w(h_{S_1}, h_{S_2}, \cdots, h_{S_n}) \tag{18}$$

Proof. For any $s_t^i \in h_{S_i}$, based on Definition 3, we have

$$g : [-\tau, \tau] \to [0,1],\ g(h_{S_i}) = \left\{ \frac{t}{2\tau} + \frac{1}{2} = \sigma_i \middle| t \in [-\tau, \tau] \right\} = h_i$$

Further, according to Equation (10), we have

$$\prod_{i=1}^{n} (1 + (v-1)(1-\sigma_i))^{w_i} + (v-1)\prod_{i=1}^{n} (\sigma_i)^{w_i} \leq \sum_{i=1}^{n} w_i(1 + (v-1)(1-\sigma_i)) + (v-1)\sum_{i=1}^{n} w_i(\sigma_i) = v$$

then

$$\begin{aligned}
HFLHWG_w^v(h_{S_1}, h_{S_2}, \cdots, h_{S_n}) &= \bigcup_{\sigma_i \in g(h_{S_i})} \left\{ g^{-1}\left(\frac{v\prod_{i=1}^{n} (\sigma_i)^{w_i}}{\prod_{i=1}^{n} (1+(v-1)(1-\sigma_i))^{w_i} + (v-1)\prod_{i=1}^{n} (\sigma_i)^{w_i}} \right) \right\} \\
\geq \bigcup_{\sigma_i \in g(h_{S_i})} \left\{ g^{-1}\left(\frac{v\prod_{i=1}^{n} (\sigma_i)^{w_i}}{v} \right) \right\} &= \bigcup_{\sigma_i \in g(h_{S_i})} \{ g^{-1}(\prod_{i=1}^{n} (\sigma_i)^{w_i}) \} = HFLWG_w(h_{S_1}, h_{S_2}, \cdots, h_{S_n})
\end{aligned}$$

Therefore, Equation (18) holds. □

3.2. GHFLHWA and GHFLHWG Operators

Definition 10. *Let $H_S = \{h_{S_1}, h_{S_2}, \cdots, h_{S_n}\}$ be a collection of HFLEs, $v > 0$ and $\lambda > 0$. $w_i(i = 1, 2, \cdots, n)$ is the weight of $h_{S_i}(i = 1, 2, \cdots, n)$, satisfying $w_i \in [0, 1]$ and $\sum_{i=1}^{n} w_i = 1$. If*

$$GHFLHWA_w^{v,\lambda}(h_{S_1}, h_{S_2}, \cdots, h_{S_n}) = w_1(h_{S_1}^{\lambda}) \oplus_H w_2(h_{S_2}^{\lambda}) \oplus_H \cdots \oplus_H w_n(h_{S_n}^{\lambda}) = \left(\overset{n}{\underset{i=1}{\oplus_H}}\left(w_i(h_{S_i}^{\lambda})\right)\right)^{\frac{1}{\lambda}} \tag{19}$$

then $GHFLHWA_w^{v,\lambda}$ is designated as the generalized HFL Hamacher weighted averaging (GHFLHWA) operator.

Theorem 5. *Let $H_S = \{h_{S_1}, h_{S_2}, \cdots, h_{S_n}\}$ be a set of HFLEs and $w_i(i = 1, 2, \cdots, n)$ be the weight of $h_{S_i}(i = 1, 2, \cdots, n)$, satisfying $w_i \in [0, 1]$ and $\sum_{i=1}^{n} w_i = 1$. g and g^{-1} are the equivalent conversion functions between HFLEs and HFEs, and $v > 0$. Then the aggregated value by the GHFLHWA operator is also an HFLE and*

$$GHFLHWA_w^{v,\lambda}(h_{S_1}, h_{S_2}, \cdots, h_{S_n})$$
$$= \underset{\sigma_i \in g(h_{S_i})}{\cup}\left\{g^{-1}\left(\left(\frac{\prod_{i=1}^{n}\left(1+\frac{v(v-1)\sigma_i^{\lambda}}{(1+(v-1)(1-\sigma_i))^{\lambda}+(v-1)\sigma_i^{\lambda}}\right)^{w_i}-\prod_{i=1}^{n}\left(1-\frac{v\sigma_i^{\lambda}}{(1+(v-1)(1-\sigma_i))^{\lambda}+(v-1)\sigma_i^{\lambda}}\right)^{w_i}}{\prod_{i=1}^{n}\left(1+\frac{v(v-1)\sigma_i^{\lambda}}{(1+(v-1)(1-\sigma_i))^{\lambda}+(v-1)\sigma_i^{\lambda}}\right)^{w_i}+(v-1)\prod_{i=1}^{n}\left(1-\frac{v\sigma_i^{\lambda}}{(1+(v-1)(1-\sigma_i))^{\lambda}+(v-1)\sigma_i^{\lambda}}\right)^{w_i}}\right)^{\frac{1}{\lambda}}\right)\right\} \tag{20}$$

Proof. According to the mathematical induction method, the proof of Equation (20) is similar to that of Theorem 1 and is omitted here. □

Remark 6. *When $\lambda = 1$, the GHFLHWA operator is reduced to the HFLHWA operator; when $\lambda \to 0$, GHFLHWA operator is reduced to the HFLHWG operator.*

When $v = 1$, the GHFLHWA operator is reduced to the following:

$$GHFLWA_w^{\lambda}(h_{S_1}, h_{S_2}, \cdots, h_{S_n}) = \underset{\sigma_i \in g(h_{S_i})}{\cup}\left\{g^{-1}\left(\left(1-\prod_{i=1}^{n}(1-\sigma_i^{\lambda})^{w_i}\right)^{\frac{1}{\lambda}}\right)\right\}$$

where $GHFLWA_w^{\lambda}$ is called the generalized HFL weighted averaging (GHFLWA) operator. Particularly, when $\lambda = 1$, the GHFLHWA operator is further transformed into the HFLWA operator; when $\lambda \to 0$, the GHFLHWA operator is further transformed into the HFLWG operator.

When $v = 2$, the GHFLHWA operator is transformed into the following:

$$GHFLEWA_w^{\lambda}(h_{S_1}, h_{S_2}, \cdots, h_{S_n}) = \underset{\sigma_i \in g(h_{S_i})}{\cup}\left\{g^{-1}\left(\left(\frac{\prod_{i=1}^{n}\left(1+\frac{2\sigma_i^{\lambda}}{(2-\sigma_i)^{\lambda}+\sigma_i^{\lambda}}\right)^{w_i}-\prod_{i=1}^{n}\left(1-\frac{2\sigma_i^{\lambda}}{(2-\sigma_i)^{\lambda}+\sigma_i^{\lambda}}\right)^{w_i}}{\prod_{i=1}^{n}\left(1+\frac{2\sigma_i^{\lambda}}{(2-\sigma_i)^{\lambda}+\sigma_i^{\lambda}}\right)^{w_i}+\prod_{i=1}^{n}\left(1-\frac{2\sigma_i^{\lambda}}{(2-\sigma_i)^{\lambda}+\sigma_i^{\lambda}}\right)^{w_i}}\right)^{\frac{1}{\lambda}}\right)\right\}$$

where $GHFLEWA_w^{\lambda}$ is designated as the generalized HFL Einstein weighted averaging (GHFLEWA) operator. Particularly, when $\lambda = 1$, the GHFLHWA operator is further transformed into the HFLEWA operator; when $\lambda \to 0$, GHFLHWA is further transformed into the HFLEWG operator.

Definition 11. *Let $H_S = \{h_{S_1}, h_{S_2}, \cdots, h_{S_n}\}$ be a collection of HFLEs and $v > 0$. $w_i(i = 1, 2, \cdots, n)$ is the weight of $h_{S_i}(i = 1, 2, \cdots, n)$, satisfying $w_i \in [0, 1]$ and $\sum_{i=1}^{n} w_i = 1$. If*

$$GHFLHWG_w^{v,\lambda}(h_{S_1}, h_{S_2}, \cdots, h_{S_n}) = \frac{1}{\lambda}(\lambda h_{S_1})^{w_1} \otimes_H (\lambda h_{S_2})^{w_2} \otimes_H \cdots \otimes_H (\lambda h_{S_n})^{w_n} = \frac{1}{\lambda}\left(\overset{n}{\underset{i=1}{\otimes_H}}(\lambda h_{S_i})^{w_i}\right) \tag{21}$$

then $GHFLHWG_w^{v,\lambda}$ *is designated as the generalized HFL Hamacher weighted geometric (GHFLHWG) operator.*

Theorem 6. *Let* $H_S = \{h_{S_1}, h_{S_2}, \cdots, h_{S_n}\}$ *be a set of HFLEs and* $w_i (i = 1, 2, \cdots, n)$ *be the weight of* $h_{S_i} (i = 1, 2, \cdots, n)$, *satisfying* $w_i \in [0, 1]$ *and* $\sum_{i=1}^n w_i = 1$. g *and* g^{-1} *are the equivalent conversion functions between HFLEs and HFEs, and* $v > 0$. *Then the aggregated value by the GHFLHWG operator is also an HFLE, and*

$$GHFLHWG_w^{v,\lambda}(h_{S_1}, h_{S_2}, \cdots, h_{S_n})$$
$$= \bigcup_{\sigma_i \in g(h_{S_i})} \left\{ g^{-1}\left(1 - \left(1 - \frac{v\prod_{i=1}^n \left(\frac{(1+(v-1)\sigma_i)^\lambda - (1-\sigma_i)^\lambda}{(1+(v-1)\sigma_i)^\lambda + (v-1)(1-\sigma_i)^\lambda}\right)^{w_i}}{\prod_{i=1}^n \left(1 + \frac{v(v-1)(1-\sigma_i)^\lambda}{(1+(v-1)\sigma_i)^\lambda + (v-1)(1-\sigma_i)^\lambda}\right)^{w_i} + (v-1)\prod_{i=1}^n \left(\frac{(1+(v-1)\sigma_i)^\lambda - (1-\sigma_i)^\lambda}{(1+(v-1)\sigma_i)^\lambda + (v-1)(1-\sigma_i)^\lambda}\right)^{w_i}}\right)^{\frac{1}{\lambda}}\right)\right\} \quad (22)$$

Proof. According to mathematical induction method, the proof of Equation (21) is similar to that of Theorem 3 and is omitted here. □

Remark 7. *When* $\lambda = 1$, *the GHFLHWG operator is transformed into the HFLHWG operator; when* $\lambda \to 0$; *the GHFLHWG operator is transformed into the HFLHWA operator.*

When $v = 1$, *GHFLHWG operator is transformed into the following:*

$$GHFLWG_w^\lambda = \bigcup_{\sigma_i \in g(h_{S_i})} \left\{ g^{-1}\left(1 - \left(1 - \prod_{i=1}^n (1 - (1-\sigma_i)^\lambda)^{w_i}\right)^{\frac{1}{\lambda}}\right)\right\}$$

where $GHFLWG_w^\lambda$ *is designated as the generalized HFL weighted geometric (GHFLWG) operator. Particularly, when* $\lambda = 1$, *the GHFLHWG operator is further transformed into the HFLWG operator; when* $\lambda \to 0$, *GHFLHWG operator is further transformed into the HFLWA operator.*

When $v = 2$, *the GHFLHWG operator is transformed into the following:*

$$GHFLEWG_w^\lambda = \bigcup_{\sigma_i \in g(h_{S_i})} \left\{ g^{-1}\left(1 - \left(1 - \frac{2\prod_{i=1}^n \left(\frac{(1+\sigma_i)^\lambda - (1-\sigma_i)^\lambda}{(1+\sigma_i)^\lambda + (1-\sigma_i)^\lambda}\right)^{w_i}}{\prod_{i=1}^n \left(2 - \frac{(1+\sigma_i)^\lambda - (1-\sigma_i)^\lambda}{(1+\sigma_i)^\lambda + (1-\sigma_i)^\lambda}\right)^{w_i} + \prod_{i=1}^n \left(\frac{(1+\sigma_i)^\lambda - (1-\sigma_i)^\lambda}{(1+\sigma_i)^\lambda + (1-\sigma_i)^\lambda}\right)^{w_i}}\right)^{\frac{1}{\lambda}}\right)\right\}$$

where $GHFLWG_w^\lambda$ *is designated as the generalized HFL Einstein weighted geometric (GHFLEWG) operator. Particularly, when* $\lambda = 1$, *the GHFLHWG operator is transformed into the HFLEWG operator; when* $\lambda \to 0$, *GHFLHWG operator is reduced to the HFLEWA operator.*

4. Hesitant Fuzzy Linguistic Hamacher Power Aggregation Operators

This section defines an HFL Hamacher power weighted averaging (HFLHPWA) operator, an HFL Hamacher power weighted geometric (HFLHPWG) operator, a generalized HFL Hamacher power weighted averaging (GHFLHPWA) operator, and a generalized HFL Hamacher power weighted geometric (GHFLHPWG) operator. In addition, we discuss some special cases withthese operators.

4.1. The HFLHPWA and HFLHPWG Operators

Definition 12. *Let* $H_S = \{h_{S_1}, h_{S_2}, \cdots, h_{S_n}\}$ *be a collection of HFLEs and* $w_i (i = 1, 2, \cdots, n)$ *be the weight of* $h_{S_i} (i = 1, 2, \cdots, n)$, *satisfying* $w_i \in [0, 1]$ *and* $\sum_{i=1}^n w_i = 1$. *Then the hesitant fuzzy linguistic Hamacher power weighted averaging (HFLHPWA) operator is defined as follows:*

$$HFLHPWA_w(h_{S_1}, h_{S_2}, \cdots, h_{S_n}) = \bigoplus_{i=1}^n \left(w_i(1 + T(h_{S_i}))h_{S_i} / \sum_{i=1}^n w_i(1 + T(h_{S_i}))\right) \quad (23)$$

where $T(h_{S_i}) = \sum_{i=1, j\neq i}^{n} Sup(h_{S_i}, h_{S_j})$ *and* $Sup(h_{S_i}, h_{S_j})$ *expresses the support degree for* h_{S_i} *from* h_{S_j}*, which satisfies the following three properties.*

(1) $0 \leq Sup(h_{S_i}, h_{S_j}) \leq 1$;
(2) $Sup(h_{S_i}, h_{S_j}) = Sup(h_{S_j}, h_{S_i})$;
(3) $Sup(h_{S_i}, h_{S_j}) \geq Sup(h_{S_x}, h_{S_y})$, *if* $d(h_{S_i}, h_{S_j}) \leq d(h_{S_x}, h_{S_y})$.

Theorem 7. *Let* $H_S = \{h_{S_1}, h_{S_2}, \cdots, h_{S_n}\}$ *be a set of HFLEs and* $v > 0$. $w_i (i = 1, 2, \cdots, n)$ *is the weight of* $h_{S_i} (i = 1, 2, \cdots, n)$*, satisfying* $w_i \in [0, 1]$ *and* $\sum_{i=1}^{n} w_i = 1$*. Then the aggregated value by the HFLHPWA operator is also an HFLE, and*

$$HFLHPWA_w^v(h_{S_1}, h_{S_2}, \cdots, h_{S_n}) = \bigcup_{\sigma_i \in g(h_{S_i})} \left\{ g^{-1}\left(\frac{\prod_{i=1}^{n}(1+(v-1)\sigma_i)^{p_i} - \prod_{i=1}^{n}(1-\sigma_i)^{p_i}}{\prod_{i=1}^{n}(1+(v-1)\sigma_i)^{p_i} + (v-1)\prod_{i=1}^{n}(1-\sigma_i)^{p_i}} \right) \right\} \tag{24}$$

where $p_i = w_i(1 + T(h_{S_i})) / \sum_{i=1}^{n} w_i(1 + T(h_{S_i}))$, $p_i \geq 0$ *and* $\sum_{i=1}^{n} p_i = 1$.

Proof. According to mathematical induction method, the proof of Equation (24) is similar to Theorem 1 and is omitted here. □

Remark 8. *If* $Sup(h_{S_i}, h_{S_j}) = c$*, for all* $i \neq j$*, then HFLHPWA operator is transformed into the following:*

$$HFLHA^v(h_{S_1}, h_{S_2}, \cdots, h_{S_n}) = \bigcup_{\sigma_i \in g(h_{S_i})} \left\{ g^{-1}\left(\frac{\prod_{i=1}^{n}(1+(v-1)\sigma_i)^{\frac{1}{n}} - \prod_{i=1}^{n}(1-\sigma_i)^{\frac{1}{n}}}{\prod_{i=1}^{n}(1+(v-1)\sigma_i)^{\frac{1}{n}} + (v-1)\prod_{i=1}^{n}(1-\sigma_i)^{\frac{1}{n}}} \right) \right\}$$

where $HFLHA^v$ *is called the HFLHA operator.*

When $v = 1$*, then the HFLHPWA operator is transformed into the following:*

$$HFLPWA_w(h_{S_1}, h_{S_2}, \cdots, h_{S_n}) = \bigcup_{\sigma_i \in g(h_{S_i})} \left\{ g^{-1}\left(1 - \prod_{i=1}^{n}(1-\sigma_i)^{p_i}\right) \right\}$$

where $HFLPWA_w$ *is called the HFL power weighted averaging (HFLPWA) operator.*

When $v = 2$*, then the HFLHPWA operator is transformed into the following:*

$$HFLEPWA_w(h_{S_1}, h_{S_2}, \cdots, h_{S_n}) = \bigcup_{\sigma_i \in g(h_{S_i})} \left\{ g^{-1}\left(\frac{\prod_{i=1}^{n}(1+\sigma_i)^{p_i} - \prod_{i=1}^{n}(1-\sigma_i)^{p_i}}{\prod_{i=1}^{n}(1+\sigma_i)^{p_i} + \prod_{i=1}^{n}(1-\sigma_i)^{p_i}} \right) \right\}$$

where $HFLEPWA_w$ *is designated as the HFL Einstein power weighted averaging (HFLEPWA) operator.*

Remark 9. *The HFLHPWA operator is neither idempotent, monotonic, bounded, nor commutative with regard to the input arguments, which are shown in Example 5.*

Example 5. *Let* $S = \{s_t | t = -\tau, \cdots, -1, 0, 1, \cdots, \tau\}$ *be an LTS,* $\tau = 3$, $h_{S_1} = \{s_1, s_2\}$, $h_{S_2} = \{s_0, s_3\}$, $h_{S_3} = \{s_0, s_2\}$*, and* $h_{S_4} = \{s_0, s_1\}$ *be four HFLEs. Let* $w = (0.3, 0.5, 0.2)^T$ *and* $v = 3$.

Based on Definition 3, according to Equation (2), we have $s(h_{S_1}) = s(h_{S_2}) = 0.75$, $s(h_{S_3}) = 0.6667$ and $s(h_{S_4}) = 0.5833$. Then, by employing HFLHPWA operator yields

$$s(HFLHPWA^3(h_{S_1}, h_{S_1}, h_{S_1})) = 0.7572$$

$$s(HFLHPWA^3(h_{S_1}, h_{S_3}, h_{S_4})) = 0.6903$$

$$s(HFLHPWA^3(h_{S_1}, h_{S_4}, h_{S_3})) = 0.6657$$

$$s(HFLHPWA^3(h_{S_1}, h_{S_1}, h_{S_3})) = 0.7452$$

$$s(HFLHPWA^3(h_{S_2}, h_{S_2}, h_{S_4})) = 0.8793$$

Since $s(HFLHPWA^3(h_{S_1}, h_{S_1}, h_{S_1})) \neq s(h_{S_1})$, the HFLHPWA operator is not idempotent.

It is obvious that $s(HFLHPWA^3(h_{S_2}, h_{S_2}, h_{S_4})) > s(HFLHPWA^3(h_{S_1}, h_{S_1}, h_{S_3}))$, therefore, the HFLHPWA operator is not monotonic. On the other hand, since $s(HFLHPWA^3(h_{S_2}, h_{S_2}, h_{S_4})) > s(h_{S_2}) > s(h_{S_4})$, the HFLHPWA operator is not bounded.

Furthermore, $s(HFLHPWA^3(h_{S_1}, h_{S_3}, h_{S_4})) \neq s(HFLHPWA^3(h_{S_1}, h_{S_4}, h_{S_3}))$, the HFLHPWA operator is not commutative.

Theorem 8. *Let $H_S = \{h_{S_1}, h_{S_2}, \cdots, h_{S_n}\}$ be a set of HFLEs and $w_i (i = 1, 2, \cdots, n)$ be the weight of $h_{S_i} (i = 1, 2, \cdots, n)$, satisfying $w_i \in [0, 1]$ and $\sum_{i=1}^{n} w_i = 1$. g and g^{-1} are the equivalent transformation functions between HFLEs and HFEs, and $v > 0$. Then*

$$HFLHPWA_w^v(h_{S_1}, h_{S_2}, \cdots, h_{S_n}) \leq HFLPWA_w(h_{S_1}, h_{S_2}, \cdots, h_{S_n}) \tag{25}$$

Proof. According to Equation (10), we have

$$\prod_{i=1}^{n} (1 + (v-1)\sigma_i)^{p_i} + (v-1)\prod_{i=1}^{n} (1 - \sigma_i)^{p_i} \leq \sum_{i=1}^{n} p_i(1 + (v-1)\sigma_i) + (v-1)\sum_{i=1}^{n} p_i(1 - \sigma_i) = v$$

$$HFLHPWA_w^v(h_{S_1}, h_{S_2}, \cdots, h_{S_n}) = \bigcup_{\sigma_i \in g(h_{S_i})} \left\{ g^{-1}\left(\frac{\prod_{i=1}^{n} (1+(v-1)\sigma_i)^{p_i} - \prod_{i=1}^{n} (1-\sigma_i)^{p_i}}{\prod_{i=1}^{n} (1+(v-1)\sigma_i)^{p_i} + (v-1)\prod_{i=1}^{n} (1-\sigma_i)^{p_i}} \right) \right\}$$

$$= \bigcup_{\sigma_i \in g(h_{S_i})} \left\{ g^{-1}\left(1 - \frac{v\prod_{i=1}^{n} (1-\sigma_i)^{p_i}}{\prod_{i=1}^{n} (1+(v-1)\sigma_i)^{p_i} + (v-1)\prod_{i=1}^{n} (1-\sigma_i)^{p_i}} \right) \right\}$$

$$\leq= \bigcup_{\sigma_i \in g(h_{S_i})} \left\{ g^{-1}\left(1 - \frac{v\prod_{i=1}^{n} (1-\sigma_i)^{p_i}}{v} \right) \right\} = \bigcup_{\sigma_i \in g(h_{S_i})} \{ g^{-1}(1 - \prod_{i=1}^{n} (1 - \sigma_i)^{p_i}) \} = HFLPWA_w(h_{S_1}, h_{S_2}, \cdots, h_{S_n})$$

Therefore, Equation (25) holds. □

Definition 13. *Let $H_S = \{h_{S_1}, h_{S_2}, \cdots, h_{S_n}\}$ be a set of HFLEs, and $w_i (i = 1, 2, \cdots, n)$ be the weight of $h_{S_i} (i = 1, 2, \cdots, n)$, satisfying $w_i \in [0, 1]$ and $\sum_{i=1}^{n} w_i = 1$. Then the HFL Hamacher power weighted geometric (HFLHPWG) operator is defined as follows:*

$$HFLHPWG_w(h_{S_1}, h_{S_2}, \cdots, h_{S_n}) = \bigotimes_{i-1}^{n} (h_{S_i})^{w_i(1+T(h_{S_i}))/\sum_{i=1}^{n} w_i(1+T(h_{S_i}))} \tag{26}$$

where $T(h_{S_i}) = \sum_{i=1, j \neq i}^{n} Sup(h_{S_i}, h_{S_j})$ and $Sup(h_{S_i}, h_{S_j})$ expresses the support degree for h_{S_i} from h_{S_j}, which is also satisfy the three properties in Definition 12.

Theorem 9. *Let $H_S = \{h_{S_1}, h_{S_2}, \cdots, h_{S_n}\}$ be a collection of HFLEs and $v > 0$. $w_i (i = 1, 2, \cdots, n)$ is the weight of $h_{S_i} (i = 1, 2, \cdots, n)$, satisfying $w_i \in [0, 1]$ and $\sum_{i=1}^{n} w_i = 1$. Then the aggregated value by the HFLHPWG operator is also an HFLE, and*

$$HFLHPWG_w^v(h_{S_1}, h_{S_2}, \cdots, h_{S_n}) = \bigcup_{\sigma_i \in h_{S_i}} \left\{ g^{-1}\left(\frac{v\prod_{i=1}^{n} (\sigma_i)^{p_i}}{\prod_{i=1}^{n} (1+(v-1)(1-\sigma_i))^{p_i} + (v-1)\prod_{i=1}^{n} (\sigma_i)^{p_i}} \right) \right\} \tag{27}$$

where $p_i = w_i(1 + T(h_{S_i}))/\sum_{i=1}^{n} w_i(1 + T(h_{S_i}))$, $p_i \geq 0$ and $\sum_{i=1}^{n} p_i = 1$.

Proof. According to mathematical induction method, the proof of Equation (27) is similar to Theorem 3 and is omitted here. □

Remark 10. *If* $Sup(h_{S_i}, h_{S_j}) = c$, *for all* $i \neq j$, *then the HFLHPWG operator is transformed into the following:*

$$HFLHG^{v}(h_{S_1}, h_{S_2}, \cdots, h_{S_n}) = \underset{\sigma_i \in g(h_{S_i})}{\cup} \left\{ g^{-1}\left(\frac{v\prod_{i=1}^{n}(\sigma_i)^{\frac{1}{n}}}{\prod_{i=1}^{n}(1+(v-1)(1-\sigma_i))^{\frac{1}{n}} + (v-1)\prod_{i=1}^{n}(\sigma_i)^{\frac{1}{n}}} \right) \right\}$$

where $HFLHG^{v}$ *is called the HFL Hamacher geometric (HFLHG) operator.*

When $v = 1$, *then the HFLHPWG operator is transformed into the following:*

$$HFLPWG_w(h_{S_1}, h_{S_2}, \cdots, h_{S_n}) = \underset{\sigma_i \in g(h_{S_i})}{\cup} \left\{ g^{-1}\left(\prod_{i=1}^{n}(\sigma_i)^{p_i} \right) \right\}$$

where $HFLPWG_w$ *is called the HFL power weighted geometric (HFLPWG) operator.*

When $v = 2$, *then the HFLHPWG operator is transformed into the following:*

$$HFLEPWG_w(h_{S_1}, h_{S_2}, \cdots, h_{S_n}) = \underset{\sigma_i \in g(h_{S_i})}{\cup} \left\{ g^{-1}\left(\frac{2\prod_{i=1}^{n}(\sigma_i)^{p_i}}{\prod_{i=1}^{n}(2-\sigma_i)^{p_i} + \prod_{i=1}^{n}(\sigma_i)^{p_i}} \right) \right\}$$

where $HFLEPWG_w$ *is designated as the HFL Einstein power geometric (HFLEPWG) operator.*

Remark 11. *Similar to the HFLHPWA operator, the HFLHPWG operator is neither idempotent, monotonic, bounded, nor commutative with regard to the input arguments.*

Theorem 10. *Let* $H_S = \{h_{S_1}, h_{S_2}, \cdots, h_{S_n}\}$ *be a collection of HFLEs and* $w_i(i = 1, 2, \cdots, n)$ *be the weight of* $h_{S_i}(i = 1, 2, \cdots, n)$, *satisfying* $w_i \in [0, 1]$ *and* $\sum_{i=1}^{n} w_i = 1$. g *and* g^{-1} *are the equivalent transformation functions between HFLEs and HFEs, and* $v > 0$. *Then*

$$HFLHPWG_w^{v}(h_{S_1}, h_{S_2}, \cdots, h_{S_n}) \geq HFLPWG_w(h_{S_1}, h_{S_2}, \cdots, h_{S_n}) \tag{28}$$

Proof. According to Equation (10), we have

$$\prod_{i=1}^{n}(1+(v-1)\sigma_i)^{p_i} + (v-1)\prod_{i=1}^{n}(1-\sigma_i)^{p_i} \leq \sum_{i=1}^{n} p_i(1+(v-1)\sigma_i) + (v-1)\sum_{i=1}^{n} p_i(1-\sigma_i) = v$$

$$HFLHPWG_w^{v}(h_{S_1}, h_{S_2}, \cdots, h_{S_n}) = \underset{\sigma_i \in g(h_{S_i})}{\cup} \left\{ g^{-1}\left(\frac{v\prod_{i=1}^{n}(\sigma_i)^{p_i}}{\prod_{i=1}^{n}(1+(v-1)(1-\sigma_i))^{p_i} + (v-1)\prod_{i=1}^{n}(\sigma_i)^{p_i}} \right) \right\}$$

$$\geq \underset{\sigma_i \in g(h_{S_i})}{\cup} \left\{ g^{-1}\left(\frac{v\prod_{i=1}^{n}(\sigma_i)^{p_i}}{v} \right) \right\} = \underset{\sigma_i \in g(h_{S_i})}{\cup} \left\{ g^{-1}\left(\prod_{i=1}^{n}(\sigma_i)^{p_i} \right) \right\} = HFLPWG_w(h_{S_1}, h_{S_2}, \cdots, h_{S_n})$$

Therefore, Equation (28) holds. □

4.2. The GHFLHPWA and GHFLHPWG Operators

Definition 14. *Let* $H_S = \{h_{S_1}, h_{S_2}, \cdots, h_{S_n}\}$ *be a collection of HFLEs and* $w_i(i = 1, 2, \cdots, n)$ *be the weight of* $h_{S_i}(i = 1, 2, \cdots, n)$, *satisfying* $w_i \in [0, 1]$ *and* $\sum_{i=1}^{n} w_i = 1$. *Then the generalized hesitant fuzzy linguistic Hamacher power weighted averaging (GHFLHPWA) operator is defined as follows:*

$$GHFLHPWA_w^{\lambda}(h_{S_1}, h_{S_2}, \cdots, h_{S_n}) = \left(\overset{n}{\underset{i=1}{\oplus}} \left(\left(w_i(1+T(h_{S_i}))(h_{S_i})^{\lambda} \right) / \sum_{i=1}^{n} w_i(1+T(h_{S_i})) \right) \right)^{\frac{1}{\lambda}} \tag{29}$$

where $T(h_{S_i}) = \sum_{i=1, j \neq i}^{n} Sup(h_{S_i}, h_{S_j})$ *and* $Sup(h_{S_i}, h_{S_j})$ *expresses the support degree for* h_{S_i} *from* h_{S_j}, *which satisfies the three properties in Definition 12.*

Theorem 11. *Let $H_S = \{h_{S_1}, h_{S_2}, \cdots, h_{S_n}\}$ be a collection of HFLEs and $v > 0$. $w_i(i = 1, 2, \cdots, n)$ is the weight of $h_{S_i}(i = 1, 2, \cdots, n)$, satisfying $w_i \in [0, 1]$ and $\sum_{i=1}^n w_i = 1$. Then the aggregated value by the GHFLHPWA operator is also an HFLE, and*

$$GHFLHPWA_w^{v,\lambda}(h_{S_1}, h_{S_2}, \cdots, h_{S_n}) = \bigcup_{\sigma_i \in g(h_{S_i})} \left\{ g^{-1}\left(\left(\frac{\prod_{i=1}^n \left(1+\frac{v(v-1)\sigma_i^\lambda}{(1+(v-1)(1-\sigma_i))^\lambda+(v-1)\sigma_i^\lambda}\right)^{p_i} - \prod_{i=1}^n \left(1-\frac{v\sigma_i^\lambda}{(1+(v-1)(1-\sigma_i))^\lambda+(v-1)\sigma_i^\lambda}\right)^{p_i}}{\prod_{i=1}^n \left(1+\frac{v(v-1)\sigma_i^\lambda}{(1+(v-1)(1-\sigma_i))^\lambda+(v-1)\sigma_i^\lambda}\right)^{p_i} + (v-1)\prod_{i=1}^n \left(1-\frac{v\sigma_i^\lambda}{(1+(v-1)(1-\sigma_i))^\lambda+(v-1)\sigma_i^\lambda}\right)^{p_i}}\right)^{\frac{1}{\lambda}}\right)\right\} \quad (30)$$

where $p_i = w_i(1 + T(h_{S_i}))/\sum_{i=1}^n w_i(1 + T(h_{S_i}))$, $p_i \geq 0$ and $\sum_{i=1}^n p_i = 1$.

Proof. According to the mathematical induction method, the proof of Equation (30) is similar to Theorem 1 and is omitted here. □

Remark 12. *$Sup(h_{S_i}, h_{S_j}) = c$, for all $i \neq j$, then GHFLHPWA operator is transformed into the following:*

$$GHFLHA^{v,\lambda}(h_{S_1}, h_{S_2}, \cdots, h_{S_n}) = \bigcup_{\sigma_i \in g(h_{S_i})} \left\{ g^{-1}\left(\left(\frac{\prod_{i=1}^n \left(1+\frac{v(v-1)\sigma_i^\lambda}{(1+(v-1)(1-\sigma_i))^\lambda+(v-1)\sigma_i^\lambda}\right)^{\frac{1}{n}} - \prod_{i=1}^n \left(1-\frac{v\sigma_i^\lambda}{(1+(v-1)(1-\sigma_i))^\lambda+(v-1)\sigma_i^\lambda}\right)^{\frac{1}{n}}}{\prod_{i=1}^n \left(1+\frac{v(v-1)\sigma_i^\lambda}{(1+(v-1)(1-\sigma_i))^\lambda+(v-1)\sigma_i^\lambda}\right)^{\frac{1}{n}} + (v-1)\prod_{i=1}^n \left(1-\frac{v\sigma_i^\lambda}{(1+(v-1)(1-\sigma_i))^\lambda+(v-1)\sigma_i^\lambda}\right)^{\frac{1}{n}}}\right)^{\frac{1}{\lambda}}\right)\right\}$$

where $GHFLHA^{v,\lambda}$ is designated as the generalized HFL Hamacher averaging (GHFLHA) operator.

When $v = 1$, then the GHFLHPWA operator is transformed into the following:

$$GHFLPWA_w^\lambda(h_{S_1}, h_{S_2}, \cdots, h_{S_n}) = \bigcup_{\sigma_i \in g(h_{S_i})} \left\{ g^{-1}\left(\left(1 - \prod_{i=1}^n (1-\sigma_i^\lambda)^{p_i}\right)^{\frac{1}{\lambda}}\right)\right\}$$

where $GHFLPWA_w^\lambda$ is designated as the generalized HFL power weighted averaging (GHFLPWA) operator. Particularly, when $\lambda = 1$, the GHFLHPWA operator is further transformed into the HFLPWA operator; when $\lambda \to 0$, GHFLHPWA operator is further transformed into the HFLPWG operator.

When $v = 2$, then GHFLHPWA operator is transformed to the following:

$$GHFLEPWA_w^\lambda(h_{S_1}, h_{S_2}, \cdots, h_{S_n}) = \bigcup_{\sigma_i \in g(h_{S_i})} \left\{ g^{-1}\left(\left(\frac{\prod_{i=1}^n \left(1+\frac{2\sigma_i^\lambda}{(2-\sigma_i)^\lambda+\sigma_i^\lambda}\right)^{p_i} - \prod_{i=1}^n \left(1-\frac{2\sigma_i^\lambda}{(2-\sigma_i)^\lambda+\sigma_i^\lambda}\right)^{p_i}}{\prod_{i=1}^n \left(1+\frac{2\sigma_i^\lambda}{(2-\sigma_i)^\lambda+\sigma_i^\lambda}\right)^{p_i} + \prod_{i=1}^n \left(1-\frac{2\sigma_i^\lambda}{(2-\sigma_i)^\lambda+\sigma_i^\lambda}\right)^{p_i}}\right)^{\frac{1}{\lambda}}\right)\right\}$$

where $GHFLEPWA_w^\lambda$ is designated as the generalized HFL Einstein power weighted averaging (GHFLEPWA) operator. Particularly, when $\lambda = 1$, the GHFLHPWA operator is further transformed into the HFLEPWA operator; when $\lambda \to 0$, GHFLHPWA operator is further transformed into the HFLEPWG operator.

Remark 13. Similar to the HFLHPWA operator, the GHFLHPWA operator is neither idempotent, monotonic, bounded, nor commutative with regard to the input arguments.

Definition 15. *Let $H_S = \{h_{S_1}, h_{S_2}, \cdots, h_{S_n}\}$ be a set of HFLEs and $w_i(i = 1, 2, \cdots, n)$ be the weight of $h_{S_i}(i = 1, 2, \cdots, n)$, satisfying $w_i \in [0, 1]$ and $\sum_{i=1}^n w_i = 1$. Then the generalized hesitant fuzzy linguistic Hamacher power weighted geometric (GHFLHPWG) operator is defined as follows:*

$$GHFLHPWG_w^\lambda(h_{S_1}, h_{S_2}, \cdots, h_{S_n}) = \frac{1}{\lambda}\left(\bigotimes_{i=1}^n (\lambda h_{S_i})^{w_i(1+T(h_{S_i}))/\sum_{i=1}^n w_i(1+T(h_{S_i}))}\right) \quad (31)$$

where $T(h_{S_i}) = \sum_{i=1, j \neq i}^{n} Sup(h_{S_i}, h_{S_j})$, *and* $Sup(h_{S_i}, h_{S_j})$ *expresses the support degree for* h_{S_i} *from* h_{S_j}, *which satisfies the three properties in Definition 12.*

Theorem 12. *Let* $H_S = \{h_{S_1}, h_{S_2}, \cdots, h_{S_n}\}$ *be a set of HFLEs and* $v > 0$. $w_i (i = 1, 2, \cdots, n)$ *is the weight of* $h_{S_i} (i = 1, 2, \cdots, n)$, *satisfying* $w_i \in [0, 1]$ *and* $\sum_{i=1}^{n} w_i = 1$. *Then the aggregated value by the GHFLHPWG operator is also an HFLE, and*

$$GHFLHPWG_w^{v,\lambda}(h_{S_1}, h_{S_2}, \cdots, h_{S_n})$$

$$= \bigcup_{\sigma_i \in g(h_{S_i})} \left\{ g^{-1}\left(1 - \left(1 - \frac{v\prod_{i=1}^{n}\left(\frac{(1+(v-1)\sigma_i)^\lambda - (1-\sigma_i)^\lambda}{(1+(v-1)\sigma_i)^\lambda + (v-1)(1-\sigma_i)^\lambda}\right)^{p_i}}{\prod_{i=1}^{n}\left(1 + \frac{v(v-1)(1-\sigma_i)^\lambda}{(1+(v-1)\sigma_i)^\lambda + (v-1)(1-\sigma_i)^\lambda}\right)^{p_i} + (v-1)\prod_{i=1}^{n}\left(\frac{(1+(v-1)\sigma_i)^\lambda - (1-\sigma_i)^\lambda}{(1+(v-1)\sigma_i)^\lambda + (v-1)(1-\sigma_i)^\lambda}\right)^{p_i}}\right)^{\frac{1}{\lambda}}\right)\right\} \quad (32)$$

where $p_i = w_i(1 + T(h_{S_i})) / \sum_{i=1}^{n} w_i(1 + T(h_{S_i}))$, $p_i \geq 0$ *and* $\sum_{i=1}^{n} p_i = 1$.

Proof. According to the mathematical induction method, the proof of Equation (32) is similar to Theorem 3 and is omitted here. □

Remark 14. $Sup(h_{S_i}, h_{S_j}) = c$, *for all* $i \neq j$, *then the GHFLHPWG operator is transformed into the following:*

$$GHFLHG^{v,\lambda}(h_{S_1}, h_{S_2}, \cdots, h_{S_n})$$

$$= \bigcup_{\sigma_i \in g(h_{S_i})} \left\{ g^{-1}\left(1 - \left(1 - \frac{v\prod_{i=1}^{n}\left(\frac{(1+(v-1)\sigma_i)^\lambda - (1-\sigma_i)^\lambda}{(1+(v-1)\sigma_i)^\lambda + (v-1)(1-\sigma_i)^\lambda}\right)^{\frac{1}{n}}}{\prod_{i=1}^{n}\left(1 + \frac{v(v-1)(1-\sigma_i)^\lambda}{(1+(v-1)\sigma_i)^\lambda + (v-1)(1-\sigma_i)^\lambda}\right)^{\frac{1}{n}} + (v-1)\prod_{i=1}^{n}\left(\frac{(1+(v-1)\sigma_i)^\lambda - (1-\sigma_i)^\lambda}{(1+(v-1)\sigma_i)^\lambda + (v-1)(1-\sigma_i)^\lambda}\right)^{\frac{1}{n}}}\right)^{\frac{1}{\lambda}}\right)\right\}$$

where $GHFLHG^{v,\lambda}$ *is designated as the generalized HFL Hamacher geometric (GHFLHG) operator.*

When $v = 1$, *then the GHFLHPWG operator is transformed into the following:*

$$GHFLPWG_w^{\lambda}(h_{S_1}, h_{S_2}, \cdots, h_{S_n}) = \bigcup_{\sigma_i \in g(h_{S_i})} \left\{ g^{-1}\left(1 - \left(1 - \prod_{i=1}^{n}(1 - (1 - \sigma_i)^\lambda)^{p_i}\right)^{\frac{1}{\lambda}}\right)\right\}$$

where $GHFLPWG_w^{\lambda}$ *is designated as the generalized HFL power weighted geometric (GHFLPWG) operator. Particularly, when* $\lambda = 1$, *the GHFLHPWG operator is further transformed into the HFLPWG operator; when* $\lambda \to 0$, *the GHFLHPWG operator is further transformed into the HFLPWA operator.*

When $v = 2$, *then the GHFLHPWG operator is transformed into the following:*

$$GHFLEPWG_w^{\lambda}(h_{S_1}, h_{S_2}, \cdots, h_{S_n})$$

$$= \bigcup_{\sigma_i \in g(h_{S_i})} \left\{ g^{-1}\left(1 - \left(1 - \frac{2\prod_{i=1}^{n}\left(\frac{(1+\sigma_i)^\lambda - (1-\sigma_i)^\lambda}{(1+\sigma_i)^\lambda + (1-\sigma_i)^\lambda}\right)^{p_i}}{\prod_{i=1}^{n}\left(2 - \frac{(1+\sigma_i)^\lambda - (1-\sigma_i)^\lambda}{(1+\sigma_i)^\lambda + (1-\sigma_i)^\lambda}\right)^{p_i} + (v-1)\prod_{i=1}^{n}\left(\frac{(1+\sigma_i)^\lambda - (1-\sigma_i)^\lambda}{(1+\sigma_i)^\lambda + (1-\sigma_i)^\lambda}\right)^{p_i}}\right)^{\frac{1}{\lambda}}\right)\right\}$$

where $GHFLEPWG_w^{\lambda}$ *is designated as the generalized HFL Einstein power weighted geometric (GHFLEPWG) operator. Particularly, when* $\lambda = 1$, *the GHFLHPWG operator is further transformed into the HFLEPWG operator; when* $\lambda \to 0$, *the GHFLHPWG operator is further transformed into the HFLPWA operator.*

Remark 15. *Similar to the HFLHPWA operator, the GHFLHPWG operator is neither idempotent, monotonic, bounded, nor commutative with regard to the input arguments.*

5. Methods for MCDM Based on the Hesitant Fuzzy Linguistic Hamacher Operators

In this part, we develop two methods based on the presented operators to handle an MCDM problem with hesitant fuzzy linguistic information.

A general MCDM problem under the hesitant fuzzy linguistic environment can be depicted as follows.

Let $A = \{A_1, A_2, \cdots, A_m\}$ be the set of m candidates alternatives, and $C = \{C_1, C_2, \cdots, C_n\}$ be the set of n evaluation criteria, which have the weight vector $w = (w_1, w_2, \cdots, w_n)^T$ satisfying $w_j \in [0,1]$ and $\sum_{j=1}^{n} w_j = 1$. Suppose that $\hat{H}_S = (\hat{h}_{S_{ij}})_{m \times n}$ be the hesitant fuzzy linguistic evaluation matrix, where $\hat{h}_{S_{ij}}$ is an HFLE and expresses the evaluation value of alternative A_i with respect to the criterion C_j.

Generally, there are two types of criteria, the benefit criterion and cost criterion, in an MCDM problem. When all the criteria are not of the same types, the values of the cost criterion need to be transformed into the values of the benefit criterion to construct a decision-making matrix $H_S = (h_{S_{ij}})_{m \times n}$ by employing Equation (33).

$$h_{S_{ij}} = \{ \begin{array}{l} \hat{h}_{S_{ij}}, \text{ for benefit criterion} \\ (\hat{h}_{S_{ij}})^C, \text{ for cost criterion} \end{array}, (i = 1, 2, \cdots, m; j = 1, 2, \cdots, n) \tag{33}$$

In order to yield the best alternative, the GHFLHWA operator or the GHFLHWG operator, which was developed based on the Hamacher operations, is utilized for the proposed MCDM approach under the hesitant fuzzy linguistic environment. The proposed method includes the following steps.

Method 1. (The flowchart of Method 1 is shown in Figure 1.)

Step 1. Determine the linguistic term set that is applied to evaluate each alternative with respect to each criterion; then the hesitant fuzzy linguistic evaluation matrix $\hat{H}_S = (\hat{h}_{S_{ij}})_{m \times n}$ is obtained.

Step 2. Normalized the evaluation matrix $\hat{H}_S = (\hat{h}_{S_{ij}})_{m \times n}$ according to Equation (33).

Step 3. Aggregate the criteria values by the GHFLHWA or GHFLHWG operator as follow:

$$h_{S_i} = GHFLHWA(h_{S_{i1}}, h_{S_{i2}}, \cdots, h_{S_{in}}) \text{ or } h_{S_i} = GHFLHWG(h_{S_{i1}}, h_{S_{i2}}, \cdots, h_{S_{in}}) \tag{34}$$

Step 4. Compute the score value of each alternative by Equation (2).

Step 5. Obtained the ranking order of alternatives by the decreasing of the score value.

To reflect the correlation between the input arguments in MCDM problem, we use the GHFLHPWA or GHFLHPWG operator for the proposed MCDM approach. The steps involved are depicted as follows.

Method 2. (The flowchart of Method 2 is shown in Figure 1.)

Step 1. Determine the linguistic term set that is applied to evaluate each alternative with respect to each criterion; then the hesitant fuzzy linguistic evaluation matrix $\hat{H}_S = (\hat{h}_{S_{ij}})_{m \times n}$ is obtained.

Step 2. Normalize the evaluation matrix $\hat{H}_S = (\hat{h}_{S_{ij}})_{m \times n}$ according to Equation (33).

Step 3. Calculate the support degree of h_{S_i} using the following formula.

$$T(h_{S_{ij}}) = \sum_{j=1, k \neq j}^{n} Sup(h_{S_{ij}}, h_{S_{ik}}) \tag{35}$$

$$Sup(h_{S_{ij}}, h_{S_{ik}}) = 1 - d(h_{S_{ij}}, h_{S_{ik}}) \tag{36}$$

Step 4. Obtained the power weight vector p by the following formula.

$$p_{ij} = w_j(1 + T(h_{S_{ij}})) / \sum_{j=1}^{n} w_j(1 + T(h_{S_{ij}})) \tag{37}$$

Step 5. Aggregate the criteria values by the GHFLHPWA or GHFLHPWG operators.

$$h_{S_i} = GHFLHPWA(h_{S_{i1}}, h_{S_{i2}}, \cdots, h_{S_{in}}) \text{ or } h_{S_i} = GHFLHPWG(h_{S_{i1}}, h_{S_{i2}}, \cdots, h_{S_{in}}) \quad (38)$$

Step 6. Compute the score value of each alternative by Equation (2).
Step 7. Determined the priority order of alternatives by the decreasing of score value.

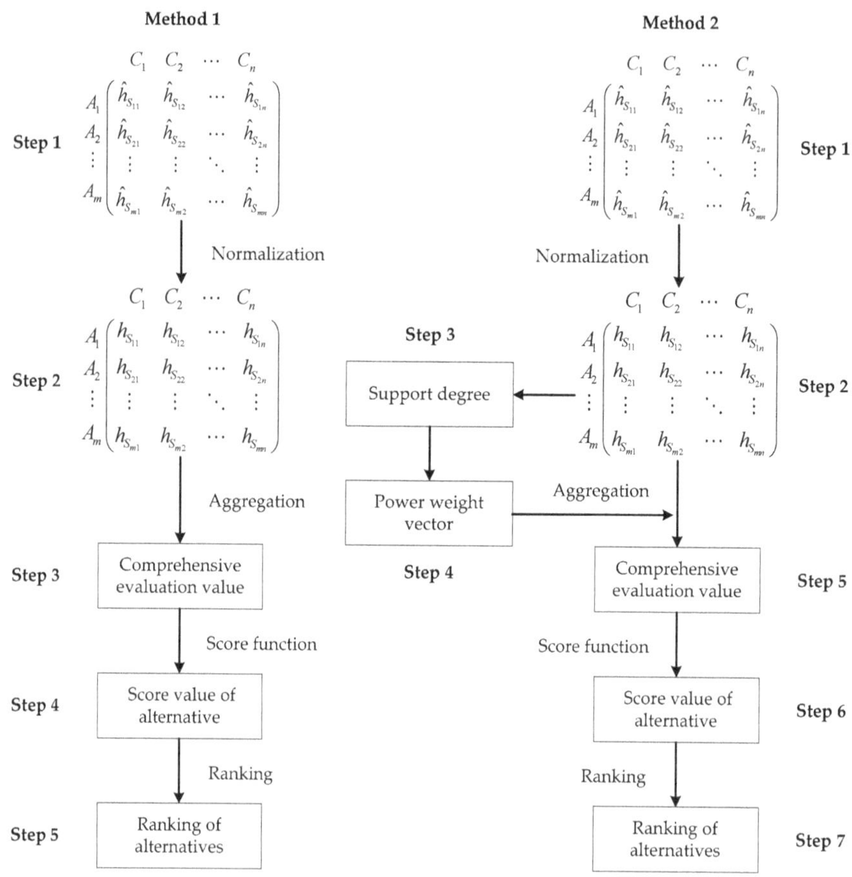

Figure 1. The flowcharts of the Method 1 and Method 2.

6. An Application of the Proposed Operators to MCDM

6.1. Numeric Example

A board of directors of a venture capital company is planning to choose a suitable city to invest in a project of sharing cars in the next three years. The venture capital company determined four alternative cities $A_i(i = 1,2,3,4)$ through preliminary market research. In order to evaluate and rank these cities, four criteria (all of them are benefit criteria) are identified by the board of directors including the economic development level (C_1), the public transportation development level (C_2), the number of public parking lots (C_3), and the urban road resources (C_4). Assume that the weight vector of these criteria is $w = (0.3, 0.1, 0.4, 0.2)^T$.

In what follows, we employ Method 1 to determine the most suitable city without considering the correlations of the input arguments.

Step 1. The board of directors constructs a nine-point linguistic term set to evaluate the ratings of cities, that is, $S = \{s_{-4} = worst,\ s_{-3} = very\ bad,\ s_{-2} = bad,\ s_{-1} = slightly\ bad,\ s_0 = medium,\ s_1 = slightly\ good,\ s_2 = good,\ s_3 = very\ good,\ s_4 = best\}$. Then the decision makers utilize the linguistic term to evaluate the ratings of the cities and the obtained hesitant fuzzy linguistic evaluation matrix $\hat{H}_S = (\hat{h}_{S_{ij}})_{m\times n}$ is presented in Table 1.

Step 2. Since these criteria are all benefit criterions, the evaluate matrix $\hat{H}_S = (\hat{h}_{S_{ij}})_{m\times n}$ is not necessary to be normalized.

Step 3. Let $\lambda = 2$ and $v = 3$, aggregate all of the criteria evaluation values according to the GHFLHWA operator into the total evaluation value $h_{S_i}(i = 1,2,3,4)$ of alternative $A_i(i = 1,2,3,4)$.

Step 4. Calculate the score values $s(h_{S_i})$ of h_{S_i} by Definition 6.The obtained results are as follows:

$$s(h_{S_1}) = 0.5080, s(h_{S_2}) = 0.6534,\ s(h_{S_3}) = 0.5685,\ s(h_{S_4}) = 0.7340$$

Step 5. Based on the decreasing order of score values, we have $h_{S_4} > h_{S_2} > h_{S_3} > h_{S_1}$. Therefore, the best city is A_4.

Table 1. The hesitant fuzzy linguistic evaluation matrix.

Cities	C_1	C_2	C_3	C_4
A_1	$\{s_0, s_1\}$	$\{s_0\}$	$\{s_1, s_2\}$	$\{s_{-2}, s_{-1}\}$
A_2	$\{s_1, s_2\}$	$\{s_2, s_3\}$	$\{s_2\}$	$\{s_{-1}, s_1\}$
A_3	$\{s_1\}$	$\{s_0, s_1\}$	$\{s_1, s_2\}$	$\{s_0, s_1\}$
A_4	$\{s_2, s_3\}$	$\{s_0, s_2\}$	$\{s_1, s_3\}$	$\{s_2\}$

The parameter v in the GHFLHWA operator indicates the experts' preference over the alternative with respect to each criterion. In order to explore how the different preference parameter v in the GHFLHWA operator influences the score values of the alternatives, we utilized different values of $v \in (0, 10]$, which are commonly determined by decision makers. The relative results are shown in Figure 2. It is easy to observe from Figure 2 that the score values of the alternatives become smaller with the increasing values of parameter v. In addition, for the GHFLHWA operator, we can also ascertain from Figure 2 that the final ranking of alternatives for the different parameter v values does not change. Therefore, the value of parameter v can be chosen by the decision maker according to their preference.

If we use the GHFLHWG operator instead of the GHFLHWA operator to aggregate the criteria values, the variation of score values of the alternatives is shown in Figure 3. From Figure 3, for the GHFLHWG operator, we can see that the score values of the alternatives become greater with the increase of parameter v, which is just the opposite of the GHFLHWA operator. Furthermore, the priority order of alternatives is also not influenced by the different values of parameter v.

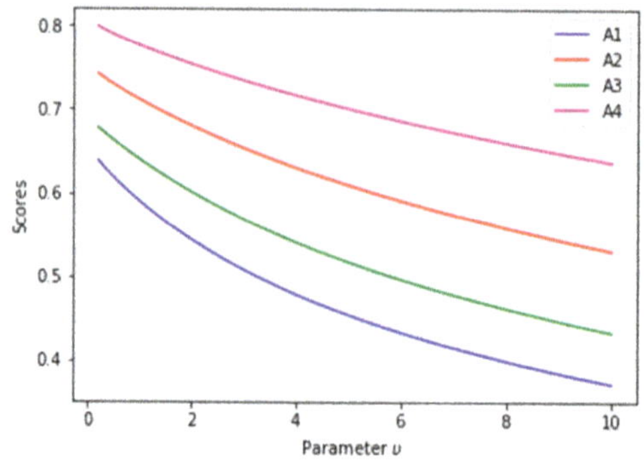

Figure 2. The variation of score values of alternatives with regard to v in the GHFLHWA operator.

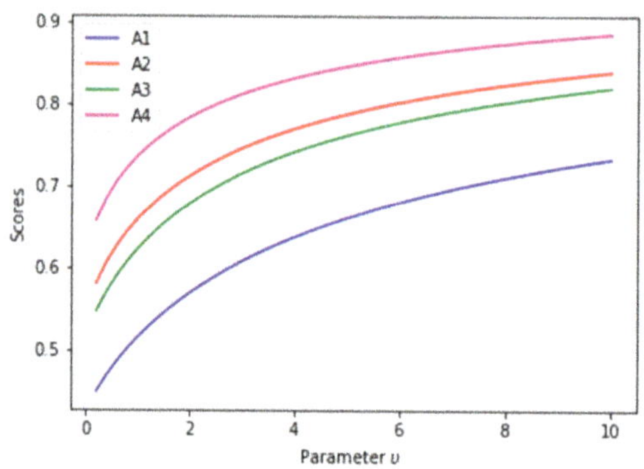

Figure 3. The variation of score values of alternatives with regard to v in the GHFLHWG operator.

When the relationships of the input data are taken into account, we apply Method 2 to resolve the above numerical example.

The first two steps are the same as Method 1.

Step 3. Compute the support degree $Sup(h_{S_j}, h_{S_k})(j = 1,2,3,4; j \neq k)$.

$$Sup_{1j} = \begin{bmatrix} 0 & 0.9116 & 0.8750 & 0.7500 \\ 0.9116 & 0 & 0.8024 & 0.8024 \\ 0.8750 & 0.8024 & 0 & 0.6250 \\ 0.7500 & 0.8024 & 0.6250 & 0 \end{bmatrix}, \quad Sup_{2j} = \begin{bmatrix} 0 & 0.8750 & 0.9116 & 0.8024 \\ 0.8750 & 0 & 0.9116 & 0.6813 \\ 0.9116 & 0.9116 & 0 & 0.7205 \\ 0.8024 & 0.6813 & 0.7205 & 0 \end{bmatrix}$$

$$Sup_{3j} = \begin{bmatrix} 0 & 0.9116 & 0.9116 & 0.9116 \\ 0.9116 & 0 & 0.8750 & 1 \\ 0.9116 & 0.8750 & 0 & 0.8750 \\ 0.9116 & 1 & 0.8750 & 0 \end{bmatrix}, \quad Sup_{4j} = \begin{bmatrix} 0 & 0.8024 & 0.9116 & 0.9116 \\ 0.8024 & 0 & 0.8750 & 0.8232 \\ 0.9116 & 0.8750 & 0 & 0.8750 \\ 0.9116 & 0.8232 & 0.8750 & 0 \end{bmatrix}$$

then

$$T = \begin{bmatrix} 2.5366 & 2.5163 & 2.3024 & 2.1774 \\ 2.5890 & 2.4679 & 2.5437 & 2.2042 \\ 2.7348 & 2.7866 & 2.6616 & 2.7866 \\ 2.6256 & 2.5006 & 2.6616 & 2.6908 \end{bmatrix}$$

Step 4. Calculate the power weight matrix.

$$P = \begin{bmatrix} 0.3149 & 0.1044 & 0.3921 & 0.1886 \\ 0.3092 & 0.0996 & 0.4071 & 0.1841 \\ 0.3011 & 0.1018 & 0.3936 & 0.2035 \\ 0.3001 & 0.0966 & 0.4041 & 0.1992 \end{bmatrix}$$

Step 5. Let $\lambda = 2$ and $v = 3$, aggregate all of the criteria values into the total evaluation value $h_{S_i}(i = 1,2,3,4)$ of alternative $A_i(i = 1,2,3,4)$ by the GHFLHPWA operator.

Step 6. Calculate the score values $s(h_{S_i})$ of h_{S_i} by Definition 6; the obtained results are as follows: $s(h_{S_1}) = 0.5085$, $s(h_{S_2}) = 0.6563$, $s(h_{S_3}) = 0.5677$, $s(h_{S_4}) = 0.7344$.

Step 7. Based on the decreasing order of score values, we have $h_{S_4} > h_{S_2} > h_{S_3} > h_{S_1}$. Therefore, the best city is A_4.

When $\lambda = 2$, let $v = 0.1, 0.7, 2, 5, 9$, respectively. From one hand, the score values and priority orders of all alternatives determined by the GHFLHPWA operator are shown in Table 2. When the value of parameter v becomes greater, we can obtain a smaller score value of the alternative. We can also see that the ranking order of alternatives is not affected by the different values of parameter v.

Table 2. The score values and rankings of alternatives obtained by the GHFLHPWA operator.

GHFLHPWA	A_1	A_2	A_3	A_4	Ranking
$GHFLHPWA_w^{0.1}$	0.6464	0.7480	0.6832	0.8034	$A_4 > A_2 > A_3 > A_1$
$GHFLHPWA_w^{0.7}$	0.6066	0.7231	0.6529	0.7843	$A_4 > A_2 > A_3 > A_1$
$GHFLHPWA_w^{2}$	0.5441	0.6816	0.6005	0.7542	$A_4 > A_2 > A_3 > A_1$
$GHFLHPWA_w^{5}$	0.4550	0.6118	0.5173	0.7016	$A_4 > A_2 > A_3 > A_1$
$GHFLHPWA_w^{9}$	0.3856	0.5467	0.4470	0.6493	$A_4 > A_2 > A_3 > A_1$

On the other hand, if the GHFLHPWG operator is employed to replace the GHFLHPWA operator in the above calculation, Table 3 gives the score values and the final ranking of the alternatives. In Table 3, we can observe that the score values of alternatives become greater when the value of parameter v increases. In addition, the priority order of alternatives does not change when the value of parameter v changes. Hence, the ranking order of alternatives is robust for the parameters $v = 0.1, 0.7, 2, 5, 9$ in this example.

Table 3. The score values and rankings of alternatives obtained by the GHFLHPWG operator.

GHFLHPWG	A_1	A_2	A_3	A_4	Ranking
$GHFLHPWG_w^{0.1}$	0.4409	0.5693	0.5328	0.6400	$A_4 > A_2 > A_3 > A_1$
$GHFLHPWG_w^{0.7}$	0.4969	0.6400	0.5985	0.7143	$A_4 > A_2 > A_3 > A_1$
$GHFLHPWG_w^{2}$	0.5727	0.7162	0.6779	0.7836	$A_4 > A_2 > A_3 > A_1$
$GHFLHPWG_w^{5}$	0.6638	0.7909	0.7608	0.8454	$A_4 > A_2 > A_3 > A_1$
$GHFLHPWG_w^{9}$	0.7253	0.8347	0.8107	0.8796	$A_4 > A_2 > A_3 > A_1$

Based on the above analysis, we can conclude that the priority order of alternatives obtained by the GHFLHWA and GHFLHWG operators are the same as that obtained by the GHFLHPWA and GHFLHPWG operators, that is, the ranking order of alternatives is $A_4 > A_2 > A_3 > A_1$. Further, the results also indicate that the correlations between the input arguments are not enough to affect the ranking order of alternatives in this example.

6.2. Comparison with Existing Methods of Hesitant Fuzzy Linguistic MCDM

In this section, we use the proposed methods comparison with the previously developed HFL MCDM approaches. The previous methods include the proposed approach with Zhang and Wu [24], where the HFL weighted averaging and HFL weighted geometric operators were employed to aggregate the HFL evaluation information, and the HFL TOPSIS method [22].

The linguistic term set in these two methods is subscript-asymmetric, however, the linguistic term set used in this paper is subscript-symmetric. Therefore, we need to transform the evaluation matrix into another form for the use of these two approaches. The transformed HFL evaluation matrix is shown in Table 4.

Table 4. The transformed hesitant fuzzy linguistic evaluation matrix.

Cities	C_1	C_2	C_3	C_4
A_1	$\{s_4, s_5\}$	$\{s_4\}$	$\{s_5, s_6\}$	$\{s_2, s_3\}$
A_2	$\{s_5, s_6\}$	$\{s_6, s_7\}$	$\{s_6\}$	$\{s_3, s_5\}$
A_3	$\{s_5\}$	$\{s_4, s_5\}$	$\{s_5, s_6\}$	$\{s_4, s_5\}$
A_4	$\{s_6, s_7\}$	$\{s_4, s_6\}$	$\{s_5, s_7\}$	$\{s_6\}$

In the following, we utilize the HFLWA operator [24] instead of the GHFLHWA operator in Method 1 based on the operational laws in Definition 4 to solve the numerical example. That is

$$hs_i = HFLWA(hs_{i1}, hs_{i2}, hs_{i3}, hs_{i4}) = \mathop{\oplus}_{j=1}^{4} (w_j hs_{ij}) = \bigcup_{\sigma_{ij} \in g(hs_{ij})} \left\{ g^{-1}\left(1 - \prod_{j=1}^{4} (1-\sigma_{ij})^{w_j}\right) \right\}$$

then, we can obtain the score values of the alternatives as follows:

$$s(hs_1) = 0.5790,\ s(hs_2) = 0.7060,\ s(hs_3) = 0.6376,\ s(hs_4) = 0.7731$$

In this situation, the priority order of alternatives is $A_4 > A_2 > A_3 > A_1$, and the best city is A_4. If we use the HFLWG operator [24] instead of the GHFLHWA operator in Method 1, we get

$$hs_i = HFLWG(hs_{i1}, hs_{i2}, hs_{i3}, hs_{i4}) = \mathop{\otimes}_{j=1}^{4} (hs_{ij})^{w_j} = \bigcup_{\sigma_{ij} \in g(hs_{ij})} \left\{ g^{-1}\left(\prod_{j=1}^{n} (\sigma_{ij})^{w_j}\right) \right\}$$

Then, we can obtain the score values of the alternatives as follows:

$$s(hs_1) = 0.5326,\ s(hs_2) = 0.6749,\ s(hs_3) = 0.6275,\ s(hs_4) = 0.7094$$

In this situation, the priority order of alternatives is $A_4 > A_2 > A_3 > A_1$, and the best city is A_4.

Based on the above analyses, we can see that the best city and the ranking order of alternatives obtained by the HFLWA and HFLWG operators are the same for Methods 1 and 2, which illustrate the validity of Methods 1 and 2. In addition, we should note that the GHFLHWA and GHFLHWG operators reduce to the HFLWA and HFLWG operator, respectively, when $\lambda = 1$ and $v = 1$. It indicates that the method based on the GHFLHWA or GHFLHWG operators is more general and flexible than the HFLWA or HFLWG operators.

In the following, we apply the HFL TOPSIS method [22] to solve the numerical example. First, we review the HFL TOPSIS approach as follows:

Step 1. For an MCDM problem with HFL information, let $X = \{x_1, x_2, \cdots, x_m\}$ be a collection of m alternatives and $C = \{c_1, c_2, \cdots, c_n\}$ be a collection of n criteria with weight vector $w = (w_1, w_2, \cdots, w_n)^T$ satisfying $w_j \in [0,1]$ and $\sum_{j=1}^{n} w_j = 1$. Suppose $R = (hs_{ij})_{m \times n}$ is an HFL evaluation matrix provided by the decision makers, where hs_{ij} is an HFLE.

Step 2. Based on the evaluation matrix R, an HFL positive ideal solution (HFLPIS) and an HFL negative ideal solution (HFLNIS) can be determined by

$$H_S^+ = (h_{S_1}^+, h_{S_2}^+, \cdots, h_{S_n}^+) \tag{39}$$

where $h_{S_j}^+ = h_{S_{1j}} \vee h_{S_{2j}} \vee \cdots \vee h_{S_{mj}}$ if c_j is a benefit criterion and $h_{S_j}^+ = h_{S_{1j}} \wedge h_{S_{2j}} \wedge \cdots \wedge h_{S_{mj}}$ if c_j is a cost criterion.

$$H_S^- = (h_{S_1}^-, h_{S_2}^-, \cdots, h_{S_n}^-) \tag{40}$$

where $h_{S_j}^- = h_{S_{1j}} \wedge h_{S_{2j}} \wedge \cdots \wedge h_{S_{mj}}$ if c_j is a benefit criterion and $h_{S_j}^- = h_{S_{1j}} \vee h_{S_{2j}} \vee \cdots \vee h_{S_{mj}}$ if c_j is a cost criterion. Where $\vee$ and $\wedge$ are defined by Definition 3 [22].

Step 3. The distance from each alternative to HFLPIS and HFLNIS are calculated as follows:

$$d_i^+ = \sum_{j=1}^n w_j d(h_{S_{ij}}, h_{S_j}^+) \tag{41}$$

$$d_i^- = \sum_{j=1}^n w_j d(h_{S_{ij}}, h_{S_j}^-) \tag{42}$$

where $d(h_{S_{ij}}, h_{S_j}^+)$ and $d(h_{S_{ij}}, h_{S_j}^-)$ are determined by Definition 7.

Step 4. The closeness coefficients d_i of alternatives x_i can be calculated by

$$cc_i = \frac{d_i^-}{d_i^+ + d_i^-} \tag{43}$$

Step 5. Determine the priority orders of all alternatives in the light of the decrease of the closeness coefficient d_i.

In what follows, we utilize the HFL TOPSIS approach to resolve the numerical example. The detailed steps are described as follows:

Step 1. The hesitant fuzzy linguistic evaluation matrix R is shown in Table 4.

Step 2. Based on the hesitant fuzzy linguistic evaluation matrix R, the HFLPIS and the HFLNIS are determined as

$$H_S^+ = (\{s_6, s_7\}, \{s_6, s_7\}, \{s_6, s_7\}, \{s_6\})$$
$$H_S^- = (\{s_4, s_5\}, \{s_4\}, \{s_5, s_6\}, \{s_2, s_3\})$$

Step 3. The distance from each alternative to HFLPIS and HFLNIS are obtained as

$$d_1^+ = 0.2453, d_2^+ = 0.1288, d_3^+ = 0.1738, d_4^+ = 0.0551$$

$$d_1^- = 0.0000, d_2^- = 0.1443, d_3^- = 0.0854, d_4^- = 0.2164$$

Step 4. Employ Equation (43) to compute the closeness coefficient of alternative x_i.

$$cc_1 = 0.0000, cc_2 = 0.5284, cc_3 = 0.3293, cc_4 = 0.7970$$

Step 5. The final priority order of all alternatives obtained as follows: $A_4 > A_2 > A_3 > A_1$.

Based on the above calculation, we can see that the best city is A_4.

From the obtained results above, we can ascertain that the results determined by the HFL TOPSIS are the same as that of the proposed methods, which also validates the effectiveness of the presented methods in this paper. Furthermore, the GHFLHPWA or GHFLHPWG operators in Method 2 consider the relationships between the input arguments through the weight vector determined by the support degree.

Compared with the HFLWA or HFLWG operators and the HFL TOPSIS method, the presented Methods in this paper have the following two advantages. First, decision makers can determine the parameter value v in the operators of Methods1 and 2 according to their subjective preferences, which increases the flexibility of the proposed methods to handle practical decision-making problems. Second, Method 2 reduces the influences of unreasonable input arguments by using the support measure assigning a lower weight to them and reflects the correlations between the input arguments by applying the weight vector allowing the input arguments to support and reinforce each other, both of which rendering the decision result more reasonable.

7. Conclusions

This paper investigates the information aggregation problem of MCDM problems in which the value of the criterion is expressed with HFLEs. Inspired by the idea of Hamacher t-norm and t-conorm, we defined some new basic operational laws on HFLEs based on the Hamacher t-norm and t-conorm. Then, based on these operational laws, we present several hesitant fuzzy linguistic Hamacher aggregation operators which are more general and flexible aggregation operators, including the HFLHWA, HFLHWG, GHFLHWA, GHFLHWG, HFLHPWA, HFLHPWG, GHFLHPWA, and GHFLHPWG operators. We also discuss some special cases of these operators and explore some of their desirable properties. Further, we propose two methods based on the GHFLHWA, GHFLHWG, GHFLHPWA, and GHFLHPWG operators to deal with the MCDM problem with HFLE information. Ultimately, a numerical example is provided to demonstrate the process of the developed methodology, and the influence of distinct parameters v on the score function of the alternative is discussed. In the future, we will extend the presented operators to other uncertain environments and apply these operators to other fields, such as supply chain management, risk management, and fuzzy cluster analysis.

Author Contributions: Conceptualization, J.Z. and Y.L; Methodology, J.Z.; Formal Analysis, J.Z.; Data Curation, J.Z.; Writing-Original Draft Preparation, J.Z.; Writing-Review & Editing, J.Z. and Y.L.; Visualization, J.Z.; Supervision, Y.L.

Funding: This research was funded by National Natural Science Foundation of China (No. 71371156) the Doctoral Innovation Fund Program of Southwest Jiaotong University (D-CX201729).

Conflicts of Interest: The authors declare no conflict of interest.

References

1. Zadeh, L.A. Fuzzy sets. *Inf. Control* **1965**, *8*, 338–353. [CrossRef]
2. Zadeh, L.A. The concept of a linguistic variable and its application to approximate reasoning—I. *Inf. Sci.* **1974**, *8*, 199–249. [CrossRef]
3. Atanassov, K.T.; Rangasamy, P. Intuitionistic fuzzy sets. *Fuzzy Sets Syst.* **1986**, *20*, 87–96. [CrossRef]
4. Atannasov, K. *Intuitionistic Fuzzy Sets: Theory and Applications*; Physica-Verlag: Heidelberg, Germany, 1999.
5. Yager, R.R. Pythagorean fuzzy subsets. In Proceedings of the IFSA World Congress and NAFIPS Annual Meeting, Edmonton, AB, Canada, 24–28 June 2013; pp. 57–61.
6. Yager, R.R. Pythagorean membership grades in multicriteria decision making. *IEEE Trans. Fuzzy Syst.* **2014**, *22*, 958–965. [CrossRef]
7. Dubois, D.; Prade, H. *Fuzzy Sets and Systems: Theory and Applications*; Academic Press: Cambridge, MA, USA, 1980; pp. 370–374.
8. Mizumoto, M.; Tanaka, K. Fuzzy sets and type 2 under algebraic product and algebraic sum. *Fuzzy Sets Syst.* **1981**, *5*, 277–290. [CrossRef]
9. Yager, R.R. On the theory of bags. *Int. J. Gen. Syst.* **1986**, *13*, 23–37. [CrossRef]
10. Torra, V.; Narukawa, Y. On hesitant fuzzy sets and decision. In Proceedings of the 18th IEEE International Conference on Fuzzy Systems, Jeju Island, Korea, 20–24 August 2009; pp. 1378–1382.
11. Torra, V. Hesitant fuzzy sets. *Int. J. Intell. Syst.* **2010**, *25*, 529–539. [CrossRef]

12. Zadeh, L.A. The concept of a linguistic variable and its application to approximate reasoning—II. *Inf. Sci.* **1975**, *8*, 301–357. [CrossRef]
13. Zadeh, L.A. The concept of a linguistic variable and its application to approximate reasoning—III. *Inf. Sci.* **1975**, *9*, 43–80. [CrossRef]
14. Ciasullo, M.V.; Fenza, G.; Loia, V.; Orciuoli, F.; Troisi, O.; Herrera-Viedma, E. Business process outsourcing enhanced by fuzzy linguistic consensus model. *Appl. Soft Comput.* **2018**, *64*, 436–444. [CrossRef]
15. Cui, J. Model for evaluating the security of wireless network with fuzzy linguistic information. *J. Intell. Fuzzy Syst.* **2017**, *32*, 2697–2704. [CrossRef]
16. Peiris, H.O.W.; Chakraverty, S.; Perera, S.S.N.; Ranwala, S.M.W. Novel fuzzy linguistic based mathematical model to assess risk of invasive alien plant species. *Appl. Soft Comput.* **2017**, *59*, 326–339. [CrossRef]
17. Wang, G.; Tian, X.; Hu, Y.; Evans, R.D.; Tian, M.; Wang, R. Manufacturing process innovation-oriented knowledge evaluation using mcdm and fuzzy linguistic computing in an open innovation environment. *Sustainability* **2017**, *9*, 1630. [CrossRef]
18. Pei, Z.; Liu, J.; Hao, F.; Zhou, B. Flm-topsis: The fuzzy linguistic multiset topsis method and its application in linguistic decision making. *Inf. Fusion* **2018**, *45*, 266–281. [CrossRef]
19. Rodriguez, R.M.; Martinez, L.; Herrera, F. Hesitant fuzzy linguistic term sets for decision making. *IEEE Trans. Fuzzy Syst.* **2012**, *20*, 109–119. [CrossRef]
20. Liao, H.; Xu, Z.; Zeng, X.J. Qualitative decision making with correlation coefficients of hesitant fuzzy linguistic term sets. *Knowl. Based Syst.* **2015**, *76*, 127–138. [CrossRef]
21. Zhang, B.; Liang, H.; Zhang, G. Reaching a consensus with minimum adjustment in magdm with hesitant fuzzy linguistic term sets. *Inf. Fusion* **2017**, *42*, 12–23. [CrossRef]
22. Wei, C.; Zhao, N.; Tang, X. Operators and comparisons of hesitant fuzzy linguistic term sets. *IEEE Trans. Fuzzy Syst.* **2014**, *22*, 575–585. [CrossRef]
23. Lee, L.W.; Chen, S.M. Fuzzy decision making based on likelihood-based comparison relations of hesitant fuzzy linguistic term sets and hesitant fuzzy linguistic operators. *Inf. Sci.* **2015**, *294*, 513–529. [CrossRef]
24. Zhang, Z.; Wu, C. Hesitant fuzzy linguistic aggregation operators and their applications to multiple attribute group decision making. *J. Intell. Fuzzy Syst.* **2014**, *26*, 2185–2202.
25. Wang, H. Extended hesitant fuzzy linguistic term sets and their aggregation in group decision making. *Int. J. Comput. Intell. Syst.* **2015**, *8*, 14–33. [CrossRef]
26. Shi, M.; Xiao, Q. Hesitant fuzzy linguistic aggregation operators based on global vision. *J. Intell. Fuzzy Syst.* **2017**, *33*, 193–206. [CrossRef]
27. Xu, Y.; Xu, A.; Merigó, J.M.; Wang, H. Hesitant fuzzy linguistic ordered weighted distance operators for group decision making. *J. Appl. Math. Comput.* **2015**, *49*, 285–308. [CrossRef]
28. Liu, X.; Zhu, J.; Zhang, S.; Liu, G. Multiple attribute group decision-making methods under hesitant fuzzy linguistic environment. *J. Intell. Syst.* **2017**, *26*, 387–406. [CrossRef]
29. Zhang, J.L.; Qi, X.W. In Research on multiple attribute decision making under hesitant fuzzy linguistic environment with application to production strategy decision making. In *Advanced Materials Research*; Trans Tech Publications: Zürich, Switzerland, 2013; Volume 753–755, pp. 2829–2836.
30. Gou, X.; Xu, Z.; Liao, H. Multiple criteria decision making based on bonferroni means with hesitant fuzzy linguistic information. *Soft Comput.* **2017**, *21*, 6515–6529. [CrossRef]
31. Wang, W.; Liu, X. Intuitionistic fuzzy information aggregation using Einstein operations. *IEEE Trans. Fuzzy Syst.* **2012**, *20*, 923–938. [CrossRef]
32. Wang, W.; Liu, X. Intuitionistic fuzzy geometric aggregation operators based on Einstein operations. *Int. J. Intell. Syst.* **2011**, *26*, 1049–1075. [CrossRef]
33. Zhang, Z. Multi-criteria group decision-making methods based on new intuitionistic fuzzy Einstein hybrid weighted aggregation operators. *Neural Comput. Appl.* **2017**, *28*, 3781–3800. [CrossRef]
34. Yu, D. Some hesitant fuzzy information aggregation operators based on Einstein operational laws. *Int. J. Intell. Syst.* **2014**, *29*, 320–340. [CrossRef]
35. Jin, F.; Ni, Z.; Chen, H. *Interval-Valued Hesitant Fuzzy Einstein Prioritized Aggregation Operators and Their Applications to Multi-Attribute Group Decision Making*; Springer-Verlag: Berlin, Germany, 2016; pp. 1863–1878.
36. Hamacher, H. Uber logische verknunpfungenn unssharfer aussagen undderen zugenhorige bewertungsfunktione. *Prog. Cybern. Syst. Res.* **1978**, *3*, 267–288.

37. Tan, C.; Yi, W.; Chen, X. Hesitant fuzzy hamacher aggregation operators for multicriteria decision making. *Appl. Soft Comput.* **2015**, *26*, 325–349. [CrossRef]
38. Ju, Y.; Zhang, W.; Yang, S. Some dual hesitant fuzzy hamacher aggregation operators and their applications to multiple attribute decision making. *J. Intell. Fuzzy Syst.* **2014**, *27*, 2481–2495.
39. Liu, J.; Zhou, N.; Zhuang, L.H.; Li, N.; Jin, F.F. Interval-valued hesitant fuzzy multiattribute group decision making based on improved hamacher aggregation operators and continuous entropy. *Math. Probl. Eng.* **2017**, *2017*, 2931482. [CrossRef]
40. Huang, J.Y. Intuitionistic fuzzy hamacher aggregation operators and their application to multiple attribute decision making. *J. Intell. Fuzzy Syst.* **2014**, *27*, 505–513.
41. Liu, P. Some hamacher aggregation operators based on the interval-valued intuitionistic fuzzy numbers and their application to group decision making. *IEEE Trans. Fuzzy Syst.* **2014**, *22*, 83–97. [CrossRef]
42. Wei, G.; Lu, M.; Tang, X.; Wei, Y. Pythagorean hesitant fuzzy hamacher aggregation operators and their application to multiple attribute decision making. *Int. J. Intell. Syst.* **2018**, *33*, 1197–1233. [CrossRef]
43. Wu, Q.; Wu, P.; Zhou, L.; Chen, H.; Guan, X. Some new Hamacher aggregation operators under single-valued neutrosophic 2-tuple linguistic environment and their applications to multi-attribute group decision making. *Comput. Ind. Eng.* **2018**, *116*, 144–162. [CrossRef]
44. Yager, R.R. The power average operator. *IEEE Trans. Syst. Man Cybern. Part A Syst. Hum.* **2001**, *31*, 724–731. [CrossRef]
45. Xu, Z.; Yager, R.R. Power-geometric operators and their use in group decision making. *IEEE Trans. Fuzzy Syst.* **2010**, *18*, 94–105.
46. Xu, Z. Approaches to multiple attribute group decision making based on intuitionistic fuzzy power aggregation operators. *Knowl. Based Syst.* **2011**, *24*, 749–760. [CrossRef]
47. Wei, G.; Lu, M. Pythagorean fuzzy power aggregation operators in multiple attribute decision making. *Int. J. Intell. Syst.* **2018**, *33*, 169–186. [CrossRef]
48. Zhang, Z. Hesitant fuzzy power aggregation operators and their application to multiple attribute group decision making. *Inf. Sci.* **2013**, *234*, 150–181. [CrossRef]
49. Jiang, W.; Wei, B.; Liu, X.; Li, X.; Zheng, H. Intuitionistic fuzzy power aggregation operator based on entropy and its application in decision making. *Int. J. Intell. Syst.* **2018**, *33*, 49–67. [CrossRef]
50. Zhu, C.; Zhu, L.; Zhang, X. Linguistic hesitant fuzzy power aggregation operators and their applications in multiple attribute decision-making. *Inf. Sci.* **2016**, *367*, 809–826. [CrossRef]
51. Liu, P.; Qin, X. Power average operators of linguistic intuitionistic fuzzy numbers and their application to multiple-attribute decision making. *J. Intell. Fuzzy Syst.* **2017**, *32*, 1029–1043. [CrossRef]
52. Wang, L.; Shen, Q.; Zhu, L. Dual hesitant fuzzy power aggregation operators based on archimedean t-conorm and t-norm and their application to multiple attribute group decision making. *Appl. Soft Comput.* **2016**, *38*, 23–50. [CrossRef]
53. Liu, C.; Luo, Y. Power aggregation operators of simplified neutrosophic sets and their use in multi-attribute group decision making. *IEEE/CAA J. Autom. Sin.* **2017**. [CrossRef]
54. Deschrijver, G.; Cornelis, C.; Kerre, E.E. On the representation of intuitionistic fuzzy t-norms and t-conorms. *IEEE Trans. Fuzzy Syst.* **2004**, *12*, 45–61. [CrossRef]
55. Klement, E.P.; Mesiar, R.; Pap, E. Triangular norms. Position paper I: Basic analytical and algebraic properties. *Fuzzy Sets Syst.* **2004**, *143*, 5–26. [CrossRef]
56. Jenei, S. A note on the ordinal sum theorem and its consequence for the construction of triangular norms. *Fuzzy Sets Syst.* **2002**, *126*, 199–205. [CrossRef]
57. Gou, X.; Xu, Z. Novel basic operational laws for linguistic terms, hesitant fuzzy linguistic term sets and probabilistic linguistic term sets. *Inf. Sci.* **2016**, *372*, 407–427. [CrossRef]
58. Gou, X.; Xu, Z.; Liao, H. Group decision making with compatibility measures of hesitant fuzzy linguistic preference relations. *Soft Comput.* **2017**. [CrossRef]
59. Torra, V.; Narukawa, Y. Modeling decisions—Information fusion and aggregation operators. *Cogn. Technol.* **2007**, *61*, 1090–1093.

symmetry

MDPI

Article

A Novel Approach to Multi-Attribute Group Decision-Making with *q*-Rung Picture Linguistic Information

Li Li [1], Runtong Zhang [1,*], Jun Wang [1], Xiaopu Shang [1] and Kaiyuan Bai [2]

1 School of Economics and Management, Beijing Jiaotong University, Beijing 100044, China; 16120608@bjtu.edu.cn (L.L.); 14113149@bjtu.edu.cn (J.W.); sxp@bjtu.edu.cn (X.S.)

2 School of Mechanical, Electronic and Control Engineering, Beijing Jiaotong University, Beijing 100044, China; 16116373@bjtu.edu.cn

* Correspondence: rtzhang@bjtu.edu.cn; Tel.: +86-010-5168-3854

Received: 20 April 2018; Accepted: 14 May 2018; Published: 18 May 2018

Abstract: The proposed *q*-rung orthopair fuzzy set (*q*-ROFS) and picture fuzzy set (PIFS) are two powerful tools for depicting fuzziness and uncertainty. This paper proposes a new tool, called *q*-rung picture linguistic set (*q*-RPLS) to deal with vagueness and impreciseness in multi-attribute group decision-making (MAGDM). The proposed *q*-RPLS takes full advantages of *q*-ROFS and PIFS and reflects decision-makers' quantitative and qualitative assessments. To effectively aggregate *q*-rung picture linguistic information, we extend the classic Heronian mean (HM) to *q*-RPLSs and propose a family of *q*-rung picture linguistic Heronian mean operators, such as the *q*-rung picture linguistic Heronian mean (*q*-RPLHM) operator, the *q*-rung picture linguistic weighted Heronian mean (*q*-RPLWHM) operator, the *q*-rung picture linguistic geometric Heronian mean (*q*-RPLGHM) operator, and the *q*-rung picture linguistic weighted geometric Heronian mean (*q*-RPLWGHM) operator. The prominent advantage of the proposed operators is that the interrelationship between *q*-rung picture linguistic numbers (*q*-RPLNs) can be considered. Further, we put forward a novel approach to MAGDM based on the proposed operators. We also provide a numerical example to demonstrate the validity and superiorities of the proposed method.

Keywords: *q*-rung picture linguistic set; Heronian mean; *q*-rung picture linguistic Heronian mean; multi-attribute group decision-making

1. Introduction

Decision-making is a common activity in daily life, aiming to select the best alternative from several candidates. As one of the most important branches of modern decision-making theory, multi-attribute group decision-making (MAGDM) has been widely investigated and successfully applied to economics and management due to its high capacity to model the fuzziness and uncertainty of information [1–15]. In actual decision-making problems, decision-makers usually rely on their intuition and prior expertise to make decisions. Owing to the complicacy of decision-making problems, the precondition is to represent the fuzzy and vague information appropriately in the process of MAGDM. Atanassov [16] originally proposed the concept of intuitionistic fuzzy set (IFS), characterized by a membership degree and a non-membership degree. Since its appearance, IFS has received substantial attention and has been studied by thousands of scientists worldwide in theoretical and practical aspects [17–20]. Thereafter, Yager [21] proposed the concept of Pythagorean fuzzy set (PYFS). The constraint of PYFS is that the square sum of membership and non-membership degrees is less than or equal to one, making PYFS more effective and powerful than IFS. Due to its merits and advantages, PYFS has been widely applied to decision-making [22–25].

PYFSs can effectively address some real MAGDM problems. However, there are quite a few cases that PYFSs cannot deal with. For instance, the membership and non-membership degrees provided by a decision-maker are 0.7 and 0.8 respectively. Evidently, the ordered pair (0.7, 0.8) cannot be represented by Pythagorean fuzzy numbers (PYFNs), as $0.7^2 + 0.8^2 = 1.13 > 1$. In other words, PYFSs do not work for some circumstances in which the square sum of membership and non-membership degrees is greater than one. To effectively deal with these cases, Yager [26] proposed the concept of q-ROFS, whose constraint is the sum of qth power of membership degree and qth power of the degree of non-membership is less than or equal to one. Thus, q-ROFSs relax the constraint of PYFSs and widen the information range. In other words, all intuitionistic fuzzy membership degrees and Pythagorean fuzzy membership degrees are a part of q-rung orthopair fuzzy membership degrees. This characteristic makes q-ROFSs more powerful and general than IFSs and PYFSs. Subsequently, Liu and Wang [27] developed some simple weighted averaging operators to aggregate q-rung orthopair fuzzy numbers (q-ROFNs) and applied these operators to MAGDM. Considering these operators cannot capture the interrelationship among aggregated q-ROFNs, Liu P.D. and Liu J.L. [28] proposed a family of q-rung orthopair fuzzy Bonferroni mean operators.

As real decision-making problems are too complicated, we may face the following issues. The first issue is that although IFSs, PYFSs and q-ROFSs have been successfully applied in decision-making, there are situations that cannot be addressed by IFSs, PYFSs and q-ROFSs. For example, human voters may be divided into groups of those who: vote for, abstain, refusal of in a voting. In other words, in a voting we have to deal with more answers of the type: yes, abstain, no, refusal. Evidently, IFSs, PYFSs and q-ROFSs do not work in this case. Recently, Cuong [29] proposed the concept of PIFS, characterized by a positive membership degree, a neutral membership degree, and a negative membership degree. Since its introduction, PIFSs have drawn much scholars' attention and have been widely investigated [30–37]. Therefore, motivated by the ideas of q-ROFS and PIFS, we propose the concept of q-rung picture fuzzy set (q-RPFS), which takes the advantages of both q-ROFS and PIFS. The proposed q-RPFS can not only express the degree of neutral membership, but also relax the constraint of PIFS that the sum of the three degrees must not exceed 1. The lax constraint of q-RPFS is that the sum of qth power of the positive membership, neutral membership and negative membership degrees is equal to or less than 1. In other words, the proposed q-RPFS enhances Yager's [26] q-ROFS by taking the neutral membership degree into consideration. For instance, if a decision-maker provides the degrees of positive membership, neutral membership and negative membership as 0.6, 0.3, and 0.5 respectively. Then the ordered pair (0.6, 0.3, 0.5) is not valid for q-ROFSs or PIFSs, whereas valid for the proposed q-RPFSs. This instance reveals that q-RPFS has a higher capacity to model fuzziness than q-ROFSs and PIFSs. The second issue is that, in some situations, decision-makers prefer to make qualitative decisions instead of quantitative decisions due to time shortage and a lack of prior expertise. Zadeh's [38] linguistic variables are powerful tools to model these circumstances. However, Wang and Li [39] pointed out that linguistic variables can only express decision-makers' qualitative preference but cannot consider the membership and non-membership degrees of an element to a particular concept. And subsequently, they proposed the concept of intuitionistic linguistic set. Other extensions are interval-valued Pythagorean fuzzy linguistic set proposed by Du et al. [40] and picture fuzzy linguistic set proposed by Liu and Zhang [41]. Therefore, this paper proposes the concept of q-RPLS by combining linguistic variables with q-RPFSs. The third issue is that in most real MAGDM problems, attributes are dependent, meaning that the interrelationship among aggregated values should be considered. The Bonferroni mean (BM) [42] and Heronian mean (HM) [43] are two effective aggregation technologies which can capture the interrelationship among fused arguments. However, Yu and Wu [44] pointed out that HM has some advantages over BM. Therefore, we utilize HM to aggregate q-rung picture linguistic information.

The main contribution of this paper is that a novel decision-making model is proposed. In the proposed model, attribute values take the form q-RPLNs, and weights of attributes take the form of crisp numbers. The motivations and aims of this paper are: (1) to provide the definition of q-RPLS and operations for q-RPLNs; (2) to develop a family of q-rung picture linguistic Heronian mean operators; (3) to put forward a novel approach to MAGDM with q-rung picture linguistic information on the basis of the proposed operators. In order to do this, the rest of this paper is organized as follows. Section 2 briefly recalls some basic concepts. In Section 3, we develop some q-rung picture linguistic aggregation operators. In addition, we present and discuss some desirable properties of the proposed operators. In Section 4, we introduce a novel method to MAGDM problems based on the proposed operators. In Section 5, a numerical instance is provided to show the validity and superiority of the proposed method. The conclusions are given in Section 6.

2. Preliminaries

In this section, we briefly review concepts about q-ROFS, PIFS, linguistic term sets and HM. Meanwhile, we provide the definitions of q-PRFS and q-RPLS.

2.1. q-Rung Orthopair Fuzzy Set (q-ROFS) and q-Rung Picture Fuzzy Set (q-RPFS)

Definition 1 [26]. *Let X be an ordinary fixed set, a q-ROFS A defined on X is given by*

$$A = \{\langle x, u_A(x), v_A(x)\rangle \,|x \in X\}, \tag{1}$$

where $u_A(x)$ and $v_A(x)$ represent the membership degree and non-membership degree respectively, satisfying $u_A(x) \in [0,1]$, $v_A(x) \in [0,1]$ and $0 \le u_A(x)^q + v_A(x)^q \le 1$, $(q \ge 1)$. The indeterminacy degree is defined as $\pi_A(x) = \left(u_A(x)^q + v_A(x)^q - u_A(x)^q v_A(x)^q\right)^{1/q}$. For convenience, $(u_A(x), v_A(x))$ is called a q-ROFN by Liu and Wang [27], *which can be denoted by $\widetilde{a} = (u, v)$.*

Liu and Wang [27] also proposed some operations for q-ROFNs.

Definition 2 [27]. *Let $\widetilde{a}_1 = (u_1, v_1)$, $\widetilde{a}_2 = (u_2, v_2)$ be two q-ROFNs, and λ be a positive real number, then*

1. $\widetilde{a}_1 \oplus \widetilde{a}_2 = \left(\left(u_1^q + u_2^q - u_1^q u_2^q\right)^{1/q}, v_1 v_2\right)$,
2. $\widetilde{a}_1 \otimes \widetilde{a}_2 = \left(u_1 u_2, \left(v_1^q + v_2^q - v_1^q v_2^q\right)^{1/q}\right)$,
3. $\lambda \widetilde{a}_1 = \left(\left(1 - \left(1 - u_1^q\right)^{\lambda}\right)^{1/q}, v_1^{\lambda}\right)$,
4. $\widetilde{a}_1^{\lambda} = \left(u_1^{\lambda}, \left(1 - \left(1 - v_1^q\right)^{\lambda}\right)^{1/q}\right)$.

To compare two q-ROFNs, Liu and Wang [27] proposed a comparison method for q-ROFNs.

Definition 3 [27]. *Let $\widetilde{a} = (u_a, v_a)$ be a q-ROFN, then the score of $\widetilde{a}$ is defined as $S(\widetilde{a}) = u_a^q - v_a^q$, the accuracy of $\widetilde{a}$ is defined as $H(\widetilde{a}) = u_a^q + v_a^q$. For any two q-ROFNs, $\widetilde{a}_1 = (u_1, v_1)$ and $\widetilde{a}_2 = (u_2, v_2)$. Then*

1. *If $S(\widetilde{a}_1) > S(\widetilde{a}_2)$, then $\widetilde{a}_1 > \widetilde{a}_2$;*
2. *If $S(\widetilde{a}_1) = S(\widetilde{a}_2)$, then*
   (1) *If $H(\widetilde{a}_1) > H(\widetilde{a}_2)$, then $\widetilde{a}_1 > \widetilde{a}_2$;*
   (2) *If $H(\widetilde{a}_1) = H(\widetilde{a}_2)$, then $\widetilde{a}_1 = \widetilde{a}_2$.*

The PIFS, constructed by a positive membership degree, a neutral membership degree as well as a negative membership degree, was originally proposed by Cuong [29].

Definition 4 [29]. *Let X be an ordinary fixed set, a picture fuzzy set (PIFS) B defined on X is given as follows*

$$B = \{\langle x, u_B(x), \eta_B(x), v_B(x)\rangle | x \in X\}, \tag{2}$$

where $u_B(x) \in [0,1]$ is called the degree of positive membership of B, $\eta_B(x) \in [0,1]$ is called the degree of neutral membership of B and $v_B(x) \in [0,1]$ is called the degree of negative membership of B, and $u_B(x), \eta_B(x), v_B(x)$ satisfy the following condition: $0 \le u_B(x) + \eta_B(x) + v_B(x) \le 1$, $\forall x \in X$. Then for $x \in X$, $\pi_B(x) = 1 - (u_B(x) + \eta_B(x) + v_B(x))$ is called the degree of refusal membership of x in B.

Motivated by the concepts of q-ROFS and PIFS, we give the definition of q-RPFS.

Definition 5. *Let X be an ordinary fixed set, a q-rung picture fuzzy set (q-RPFS) C defined on X is given as follows*

$$C = \{\langle x, u_C(x), \eta_C(x), v_C(x)\rangle | x \in X\}, \tag{3}$$

where $u_C(x), \eta_C(x)$ and $v_C(x)$ represent degree of positive membership, degree of neutral membership and degree of negative membership respectively, satisfying $u_C(x) \in [0,1]$, $\eta_C(x) \in [0,1]$, $v_C(x) \in [0,1]$ and $0 \le u_C(x)^q + \eta_C(x)^q + v_C(x)^q \le 1$ $(q \ge 1)$, $\forall x \in X$. Then for $x \in X$, $\pi_C(x) = \left(1 - \left(u_C(x)^q + \eta_C(x)^q + v_C(x)^q\right)\right)^{1/q}$ is called the degree of refusal membership of x in C.

2.2. Linguistic Term Sets and q-Rung Picture Linguistic Set (q-RPLS)

Let $S = \{s_i | i = 1, 2, ..., t\}$ be a linguistic term set with odd cardinality and t is the cardinality of S. The label s_i represents a possible value for a linguistic variable. For instance, a possible linguistic term set can be defined as follows:

$$S = (s_1, s_2, s_3, s_4, s_5, s_6, s_7) = \{\text{verypoor, poor, slightly poor, fair, slightly good, good, very good}\}.$$

Motivated by the concept of picture linguistic set [41], we shall define the concept of q-RPLS by combining the linguistic term set with q-RPFS.

Definition 6. *Let X be an ordinary fixed set, $\overline{S}$ be a continuous linguistic term set of $S = \{s_i | i = 1, 2, ..., t\}$, then a q-rung picture linguistic set (q-RPLS) D defined on X is given as follows*

$$D = \left\{\left\langle s_{\theta(x)}, u_D(x), \eta_D(x), v_D(x)\right\rangle | x \in X\right\}, \tag{4}$$

where $s_{\theta(x)} \in \overline{S}$,$u_D(x) \in [0,1]$ is called the degree of positive membership of D, $\eta_D(x) \in [0,1]$ is called the degree of neutral membership of D and $v_D(x) \in [0,1]$ is called the degree of negative membership of D, and $u_D(x), \eta_D(x), v_D(x)$ satisfy the following condition: $0 \le u_D(x)^q + \eta_D(x)^q + v_D(x)^q \le 1 (q \ge 1)$, $\forall x \in X$. Then $\left\langle s_{\theta(x)}, (u_D(x), \eta_D(x), v_D(x))\right\rangle$ is called a q-RPLN, which can be simply denoted by $\alpha = \langle s_\theta, (u, \eta, v)\rangle$. When $q = 1$, then D is reduced to the picture linguistic set (PFLS) proposed by Liu and Zhang [41].

In the following, we provide some operations for q-RPLNs.

Definition 7. *Let $\alpha = \langle s_\theta, (u, \eta, v)\rangle$, $\alpha_1 = \langle s_{\theta_1}, (u_1, \eta_1, v_1)\rangle$ and $\alpha_2 = \langle s_{\theta_2}, (u_2, \eta_2, v_2)\rangle$ be three q-RPLNs and λ be a positive real number, then*

1. $\alpha_1 \oplus \alpha_2 = \left\langle s_{\theta_1+\theta_2}, \left(\left(u_1^q + u_2^q - u_1^q u_2^q\right)^{1/q}, \eta_1\eta_2, v_1 v_2\right)\right\rangle,$
2. $\alpha_1 \otimes \alpha_2 = \left\langle s_{\theta_1\times\theta_2}, \left(u_1 u_2, \left(\eta_1^q + \eta_2^q - \eta_1^q\eta_2^q\right)^{1/q}, \left(v_1^q + v_2^q - v_1^q v_2^q\right)^{1/q}\right)\right\rangle,$
3. $\lambda\alpha = \left\langle s_{\lambda\times\theta}, \left(\left(1 - (1 - u^q)^\lambda\right)^{1/q}, \eta^\lambda, v^\lambda\right)\right\rangle,$

4. $\alpha^\lambda = \left\langle s_{\theta\lambda}, \left(u^\lambda, \left(1-(1-\eta^q)^\lambda\right)^{1/q}, \left(1-(1-v^q)^\lambda\right)^{1/q}\right)\right\rangle.$

To compare two q-RPLNs, we first propose the concepts of score function and accuracy function of a q-RPLN and based on which we propose a comparison law for q-RPLNs.

Definition 8. *Let $\alpha = \langle s_\theta, (u, \eta, v)\rangle$ be a q-RPLN, then the score function of α is defined as*

$$S(\alpha) = (u^q + 1 - v^q) \times \theta \tag{5}$$

Definition 9. *Let $\alpha = \langle s_\theta, (u, \eta, v)\rangle$ be a q-RPLN, then the accuracy function of α is defined as*

$$H(\alpha) = (u^q + \eta^q + v^q) \times \theta \tag{6}$$

Definition 10. *Let $\alpha_1 = \langle s_{\theta_1}, (u_1, \eta_1, v_1)\rangle$ and $\alpha_2 = \langle s_{\theta_2}, (u_2, \eta_2, v_2)\rangle$ be two q-RPLNs, $S(\alpha_1)$ and $S(\alpha_2)$ be score functions of α_1 and α_2 respectively, $H(\alpha_1)$ and $H(\alpha_2)$ be the accuracy functions of α_1 and α_2 respectively, then*

1. *if $S(\alpha_1) > S(\alpha_2)$, then $\alpha_1 > \alpha_2$;*
2. *if $S(\alpha_1) = S(\alpha_2)$, then*

    (1) *if $H(\alpha_1) > H(\alpha_2)$, then $\alpha_1 > \alpha_2$;*
    (2) *if $H(\alpha_1) > H(\alpha_2)$, then $\alpha_1 = \alpha_2$.*

2.3. Heronian Mean

Definition 11 [43,45]. *Let $a_i (i = 1, 2, ..., n)$ be a collection of crisp numbers, and $s, t \geq 0$, then the Heronian mean (HM) is defined as follows:*

$$HM^{s,t}(a_1, a_2, ..., a_n) = \left(\frac{2}{n(n+1)}\sum_{i=1}^{n}\sum_{j=i}^{n} a_i^s a_j^t\right)^{1/(s+t)} \tag{7}$$

Definition 12 [46]. *Let $a_i (i = 1, 2, ..., n)$ be a collection of crisp numbers, and $s, t \geq 0$, then the geometric Heronian mean (GHM) is defined as follows:*

$$GHM^{s,t}(a_1, a_2, ..., a_n) = \frac{1}{s+t}\prod_{i=1}^{n}\prod_{j=i}^{n}(sa_i + ta_j)^{\frac{1}{n(n+2)}} \tag{8}$$

3. The q-Rung Picture Linguistic Heronian Mean Operators

In this section, we extend the HM to q-rung picture linguistic environment and propose a family of q-rung picture linguistic Heronian mean operators. Moreover, some desirable properties of the proposed aggregation operators are presented and discussed.

3.1. The q-Rung Picture Linguistic Heronian Mean (q-RPLHM) Operator

Definition 13. *Let $\alpha_i = \langle s_{\theta_i}, (\mu_i, \eta_i, v_i)\rangle (i = 1, 2, ..., n)$ be a collection of q-RPLNs, and $s, t > 0$. If*

$$q-RPLHM^{s,t}(\alpha_1, \alpha_2, ..., \alpha_n) = \left(\frac{2}{n(n+1)}\sum_{i=1}^{n}\sum_{j=i}^{n} \alpha_i^s \alpha_j^t\right)^{1/(s+t)}, \tag{9}$$

then $q-RPLHM^{s,t}$ is called the q-rung picture linguistic Heronian mean (q-RPLHM) operator.

According to the operations for q-RPLNs, the following theorem can be obtained.

Theorem 1. *Let $\alpha_i = \langle s_{\theta_i}, (\mu_i, \eta_i, v_i)\rangle (i = 1, 2, ..., n)$ be a collection of q-RPLNs, then the aggregated value by using q-RPLHM operator is also a q-RPLN and*

$$q-RPLHM^{s,t}(\alpha_1, \alpha_2, ..., \alpha_n) = \left\langle s_{\left(\frac{2}{n(n+1)}\sum_{i=1}^{n}\sum_{j=i}^{n}\theta_i^s\theta_j^t\right)^{1/(s+t)}}, \right.$$
$$\left(\left(1-\prod_{i=1}^{n}\prod_{j=1}^{n}\left(1-u_i^{sq}u_j^{tq}\right)^{\frac{2}{n(n+1)}}\right)^{1/q(s+t)},\right.$$
$$\left(1-\left(1-\prod_{i=1}^{n}\prod_{j=1}^{n}\left(1-\left(1-\eta_i^q\right)^s\left(1-\eta_j^q\right)^t\right)^{\frac{2q}{n(n+1)}}\right)^{1/(s+t)}\right)^{1/q}, \tag{10}$$
$$\left.\left.\left(1-\left(1-\prod_{i=1}^{n}\prod_{j=1}^{n}\left(1-\left(1-v_i^q\right)^s\left(1-v_j^q\right)^t\right)^{\frac{2q}{n(n+1)}}\right)^{1/(s+t)}\right)^{1/q}\right)\right\rangle.$$

Proof. According to the operations for q-RPLNs, we can obtain the followings

$$\alpha_i^s = \left\langle s_{\theta_i^s}, \left(u_i^s, \left(1-\left(1-\eta_i^q\right)^s\right)^{1/q}, \left(1-\left(1-v_i^q\right)^s\right)^{1/q}\right)\right\rangle,$$
$$\alpha_j^t = \left\langle s_{\theta_j^t}, \left(u_j^t, \left(1-\left(1-\eta_j^q\right)^t\right)^{1/q}, \left(1-\left(1-v_j^q\right)^t\right)^{1/q}\right)\right\rangle.$$

Therefore,

$$\alpha_i^s\alpha_j^t = \left\langle s_{\theta_i^s\theta_j^t}, \left(u_i^s u_j^t, \left(1-\left(1-\eta_i^q\right)^s\left(1-\eta_j^q\right)^t\right), \left(1-\left(1-v_i^q\right)^s\left(1-v_j^q\right)^t\right)\right)\right\rangle.$$

Further,

$$\sum_{j=i}^{n} a_i^s a_j^t = \left\langle s_{\sum_{j=i}^{n}\theta_i^s\theta_j^t}, \left(\left(1-\prod_{j=i}^{n}\left(1-u_i^{sq}u_j^{tq}\right)\right)^{1/q},\right.\right.$$
$$\left.\left.\prod_{j=i}^{n}\left(1-\left(1-\eta_i^q\right)^s\left(1-\eta_j^q\right)^t\right), \prod_{j=i}^{n}\left(1-\left(1-v_i^q\right)^s\left(1-v_j^q\right)^t\right)\right)\right\rangle.$$

In addition,

$$\sum_{i=1}^{n}\sum_{j=i}^{n} a_i^s a_j^t = \left\langle s_{\sum\limits_{i=1}^{n}\sum\limits_{j=i}^{n}\theta_i^s\theta_j^t}, \left(\left(1-\prod_{i=1}^{n}\left(\prod_{j=i}^{n}\left(1-u_i^{sq}u_j^{tq}\right)\right)\right)^{1/q},\right.\right.$$

$$\left.\left.\prod_{i=1}^{n}\prod_{j=i}^{n}\left(1-\left(1-\eta_i^q\right)^s\left(1-\eta_j^q\right)^t\right), \prod_{i=1}^{n}\prod_{j=i}^{n}\left(1-\left(1-v_i^q\right)^s\left(1-v_j^q\right)^t\right)\right)\right\rangle.$$

Thus,

$$\frac{2}{n(n+1)}\sum_{i=1}^{n}\sum_{j=i}^{n} a_i^s a_j^t = \left\langle s_{\frac{2}{n(n+1)}\sum\limits_{i=1}^{n}\sum\limits_{j=i}^{n}\theta_i^s\theta_j^t}, \left(\left(1-\left(\prod_{i=1}^{n}\prod_{j=i}^{n}\left(1-u_i^{sq}u_j^{tq}\right)\right)^{\frac{2}{n(n+1)}}\right)^{1/q},\right.\right.$$

$$\left.\left.\left(\prod_{i=1}^{n}\prod_{j=i}^{n}\left(1-\left(1-\eta_i^q\right)^s\left(1-\eta_j^q\right)^t\right)\right)^{\frac{2}{n(n+1)}}, \left(\prod_{i=1}^{n}\prod_{j=i}^{n}\left(1-\left(1-v_i^q\right)^s\left(1-v_j^q\right)^t\right)\right)^{\frac{2}{n(n+1)}}\right)\right\rangle.$$

So,

$$q-RPLHM^{s,t}(\alpha_1,\alpha_2,...,\alpha_n) = \left(\frac{2}{n(n+1)}\sum_{i=1}^{n}\sum_{j=i}^{n} a_i^s a_j^t\right)^{1/(s+t)}$$

$$= \left\langle s_{\left(\frac{2}{n(n+1)}\sum\limits_{i=1}^{n}\sum\limits_{j=i}^{n}\theta_i^s\theta_j^t\right)^{1/(s+t)}}, \left(\left(1-\prod_{i=1}^{n}\prod_{j=1}^{n}\left(1-u_i^{sq}u_j^{tq}\right)^{\frac{2}{n(n+1)}}\right)^{1/q(s+t)},\right.\right.$$

$$\left(1-\left(1-\prod_{i=1}^{n}\prod_{j=1}^{n}\left(1-\left(1-\eta_i^q\right)^s\left(1-\eta_j^q\right)^t\right)^{\frac{2q}{n(n+1)}}\right)^{1/(s+t)}\right)^{1/q},$$

$$\left.\left.\left(1-\left(1-\prod_{i=1}^{n}\prod_{j=1}^{n}\left(1-\left(1-v_i^q\right)^s\left(1-v_j^q\right)^t\right)^{\frac{2q}{n(n+1)}}\right)^{1/(s+t)}\right)^{1/q}\right)\right\rangle.$$

□

In addition, the q-RPLHM operator has the following properties.

Theorem 2 (Monotonicity). *Let α_i and β_i $(i = 1,2,...,n)$ be two collections of q-RPLNs, if $\alpha_i \le \beta_i$ for all $i = 1,2,\cdots,n$, then*

$$q-RPLHM^{s,t}(\alpha_1,\alpha_2,...,\alpha_n) \le q-RPLHM^{s,t}(\beta_1,\beta_2,...,\beta_n). \tag{11}$$

Proof. As $\alpha_i = \alpha$ for all i, we can obtain

Since $\alpha_i \le \beta_i$ and $\alpha_j \le \beta_j$ for $i = 1,2,\cdots,n$ and $j = i, i+1,\cdots,n$, we have $\alpha_i^s\alpha_j^t \le \beta_i^s\beta_j^t$. Then

$$\frac{2}{n(n+1)}\sum_{i=1}^{n}\sum_{j=i}^{n}\alpha_i^s\alpha_j^t \le \frac{2}{n(n+1)}\sum_{i=1}^{n}\sum_{j=i}^{n}\beta_i^s\beta_j^t.$$

So,

$$\left(\frac{2}{n(n+1)}\sum_{i=1}^{n}\sum_{j=i}^{n}\alpha_i^s\alpha_j^t\right)^{1/(s+t)} \le \left(\frac{2}{n(n+1)}\sum_{i=1}^{n}\sum_{j=i}^{n}\beta_i^s\beta_j^t\right)^{1/(s+t)},$$

i.e.,

$$q-RPLHM^{s,t}(\alpha_1,\alpha_2,...,\alpha_n) \le q-RPLHM^{s,t}(\beta_1,\beta_2,...,\beta_n).$$

□

Theorem 3 (Idempotency). *Let $\alpha_i (i = 1, 2, ..., n)$ be a collection of q-RPLNs, if $\alpha_i = \alpha$, for all $i = 1, 2, ..., n$, then*

$$q - RPLHM^{s,t}(\alpha_1, \alpha_2, ..., \alpha_n) = \alpha. \tag{12}$$

Proof. Since $\alpha_i = \alpha$, for all i, we have

$$q - RPLHM^{s,t}(\alpha_1, \alpha_2, ..., \alpha_n) = \left(\frac{2}{n(n+1)} \sum_{i=1}^{n} \sum_{j=i}^{n} \alpha_i^s \alpha_j^t \right)^{1/(s+t)}$$

$$= \left(\frac{2}{n(n+1)} \sum_{i=1}^{n} \sum_{j=i}^{n} \alpha_i^s \alpha_j^t \right)^{1/(s+t)} = \left(\alpha^{s+t} \right)^{1/(s+t)} = \alpha.$$

□

Theorem 4 (Boundedness). *The q-RPLHM operator lies between the max and min operators*

$$\min(\alpha_1, \alpha_2, ..., \alpha_n) \le q - RPLHM^{s,t}(\alpha_1, \alpha_2, ..., \alpha_n) \le \max(\alpha_1, \alpha_2, ..., \alpha_n). \tag{13}$$

Proof. Let $a = \min(\alpha_1, \alpha_2, ..., \alpha_n)$, $b = \max(\alpha_1, \alpha_2, ..., \alpha_n)$ according to Theorem 2, we have

$$q - RPLHM^{s,t}(a, a, ..., a) \le q - RPLHM^{s,t}(\alpha_1, \alpha_2, ..., \alpha_n) \le q - RPLHM^{s,t}(b, b, ..., b).$$

Further, $q - RPLHM^{s,t}(a, a, ..., a) = a$ and $q - RPLHM^{s,t}(b, b, ..., b) = b$.
So,

$$a \le q - RPLHM^{s,t}(\alpha_1, \alpha_2, ..., \alpha_n) \le b,$$

i.e.,

$$\min(\alpha_1, \alpha_2, ..., \alpha_n) \le q - RPLHM^{s,t}(\alpha_1, \alpha_2, ..., \alpha_n) \le \max(\alpha_1, \alpha_2, ..., \alpha_n).$$

□

The parameters s and t play a very important role in the aggregated results. In the followings, we discuss some special cases of the q-RPLHM operator with respect to the parameters s and t.

Case 1: When $t \to 0$, then the q-RPLHM operator reduces to the followings,

$$\begin{aligned} q - RPLHM^{s,0}(\alpha_1, \alpha_2, ..., \alpha_n) = \lim_{t \to 0} \Bigg\langle & s_{\left(\frac{2}{n(n+1)} \sum_{i=1}^{n} \sum_{j=i}^{n} \theta_i^s \theta_j^t\right)^{1/(s+t)}}, \\ & \left(\left(1 - \left(\prod_{i=1}^{n} \prod_{j=i}^{n} \left(1 - u_i^{sq} u_j^{tq} \right) \right)^{\frac{2}{n(n+1)}} \right)^{1/q(s+t)} \right), \\ & \left(1 - \left(1 - \left(\prod_{i=1}^{n} \prod_{j=i}^{n} \left(1 - \left(1 - \eta_i^q\right)^s \left(1 - \eta_j^q\right)^t \right) \right)^{\frac{2q}{n(n+1)}} \right)^{1/(s+t)} \right)^{1/q}, \\ & \left(1 - \left(1 - \left(\prod_{i=1}^{n} \prod_{j=i}^{n} \left(1 - \left(1 - v_i^q\right)^s \left(1 - v_j^q\right)^t \right) \right)^{\frac{2q}{n(n+1)}} \right)^{1/(s+t)} \right)^{1/q} \Bigg) \Bigg\rangle \\ = \Bigg\langle & s_{\left(\frac{2}{n(n+1)} \sum_{i=1}^{n} (n+1-i) \theta_i^s\right)^{1/s}}, \left(\left(1 - \left(\prod_{i=1}^{n} \left(1 - u_i^{sq} \right)^{n+1-i} \right)^{\frac{2}{n(n+1)}} \right)^{1/qs}, \right. \\ & \left(1 - \left(1 - \left(\prod_{i=1}^{n} \left(1 - \left(1 - \eta_i^q\right)^s \right)^{n+1-i} \right)^{\frac{2q}{n(n+1)}} \right)^{1/s} \right)^{1/q}, \left(1 - \left(1 - \left(\prod_{i=1}^{n} \left(1 - \left(1 - v_i^q\right)^s \right)^{n+1-i} \right)^{\frac{2q}{n(n+1)}} \right)^{1/s} \right)^{1/q} \Bigg) \Bigg\rangle. \end{aligned} \tag{14}$$

which is a q-rung picture linguistic generalized linear descending weighted mean operator. Evidently, it is equivalent to weight the information $(\alpha_1^s, \alpha_2^s, \dots, \alpha_n^s)$ with $(n, n-1, \dots, 1)$.

Case 2: When $s \to 0$, then the q-RPLHM operator reduces to the followings,

$$
\begin{aligned}
&q-RPLHM^{0,t}(\alpha_1,\alpha_2,\dots,\alpha_n) = \lim_{s\to 0}\Bigg\langle s_{\left(\frac{2}{n(n+1)}\sum_{i=1}^{n}\sum_{j=i}^{n}\theta_i^s\theta_j^t\right)^{1/(s+t)}},\\
&\left(\left(1-\left(\prod_{i=1}^{n}\prod_{j=i}^{n}\left(1-u_i^{sq}u_j^{tq}\right)\right)^{\frac{2}{n(n+1)}}\right)^{1/q(s+t)}\right),\\
&\left(1-\left(1-\left(\prod_{i=1}^{n}\prod_{j=i}^{n}\left(1-\left(1-v_i^q\right)^s\left(1-v_j^q\right)^t\right)\right)^{\frac{2q}{n(n+1)}}\right)^{1/(s+t)}\right)^{1/q},\\
&\left(1-\left(1-\left(\prod_{i=1}^{n}\prod_{j=i}^{n}\left(1-\left(1-v_i^q\right)^s\left(1-v_j^q\right)^t\right)\right)^{\frac{2q}{n(n+1)}}\right)^{1/(s+t)}\right)^{1/q}\Bigg\rangle\\
&=\Bigg\langle s_{\left(\frac{2}{n(n+1)}\sum_{i=1}^{n}i\theta_i^t\right)^{1/t}}, \left(\left(1-\left(\prod_{i=1}^{n}\left(1-u_i^{tq}\right)^i\right)^{\frac{2}{n(n+1)}}\right)^{1/qt}\right),\\
&\left(1-\left(1-\left(\prod_{i=1}^{n}\left(1-\left(1-v_i^q\right)^t\right)^i\right)^{\frac{2q}{n(n+1)}}\right)^{1/t}\right)^{1/q}, \left(1-\left(1-\left(\prod_{i=1}^{n}\left(1-\left(1-v_i^q\right)^t\right)^i\right)^{\frac{2q}{n(n+1)}}\right)^{1/t}\right)^{1/q}\Bigg\rangle
\end{aligned}
\tag{15}
$$

which is a q-rung picture linguistic generalized linear ascending weighted mean operator. Obviously, it is equivalent to weight the information $(\alpha_1^t, \alpha_2^t, \dots, \alpha_n^t)$ with $(1, 2, \dots, n)$, i.e., when $t \to 0$ or $s \to 0$, the q-RPLHM operator has the linear weighted function for input data.

Case 3: When $s = t = 1$, then the q-RPLHM operator reduces to the followings,

$$
\begin{aligned}
&q-RPLHM^{1,1}(\alpha_1,\alpha_2,\dots,\alpha_n) = \Bigg\langle s_{\left(\frac{2}{n(n+1)}\sum_{i=1}^{n}\sum_{j=i}^{n}\theta_i\theta_j\right)^{1/(s+t)}},\\
&\left(\left(1-\left(\prod_{i=1}^{n}\prod_{j=1}^{n}\left(1-\left(u_iu_j\right)^q\right)\right)^{\frac{2}{n(n+1)}}\right)^{1/2q}\right),\\
&\left(1-\left(1-\left(\prod_{i=1}^{n}\prod_{j=1}^{n}\left(1-\left(1-\eta_i^q\right)\left(1-\eta_j^q\right)\right)\right)^{\frac{2q}{n(n+1)}}\right)^{1/2}\right)^{1/q},\\
&\left(1-\left(1-\left(\prod_{i=1}^{n}\prod_{j=1}^{n}\left(1-\left(1-v_i^q\right)\left(1-v_j^q\right)\right)\right)^{\frac{2q}{n(n+1)}}\right)^{1/2}\right)^{1/q}\Bigg\rangle.
\end{aligned}
\tag{16}
$$

which is a q-rung picture linguistic line Heronian mean operator.

Case 4: When $s = t = 1/2$, then the q-RPLHM operator reduces to the followings

$$
\begin{aligned}
&q-RPLHM^{\frac{1}{2},\frac{1}{2}}(\alpha_1,\alpha_2,\dots,\alpha_n) = \Bigg\langle s_{\frac{2}{n(n+1)}\sum_{i=1}^{n}\sum_{j=i}^{n}\sqrt{\theta_i\theta_j}}, \left(\left(1-\left(\prod_{i=1}^{n}\prod_{j=i}^{n}\left(1-\sqrt{u_i^q u_j^q}\right)\right)^{\frac{2}{n(n+1)}}\right)^{1/q}\right),\\
&\left(\prod_{i=1}^{n}\prod_{j=i}^{n}\left(1-\sqrt{\left(1-\eta_i^q\right)\left(1-\eta_j^q\right)}\right)\right)^{\frac{2}{n(n+1)}}, \left(\prod_{i=1}^{n}\prod_{j=i}^{n}\left(1-\sqrt{\left(1-v_i^q\right)\left(1-v_j^q\right)}\right)\right)^{\frac{2}{n(n+1)}}\Bigg\rangle,
\end{aligned}
\tag{17}
$$

which is a q-rung picture linguistic basic Heronian mean operator.

Case 5: When $q = 2$, then the q-RPLHM operator reduces to the followings,

$$\begin{aligned} q - RPLHM^{s,t}(\alpha_1, \alpha_2, \ldots, \alpha_n) = \Bigg\langle & s_{\left(\frac{2}{n(n+1)} \sum\limits_{i=1}^{n} \sum\limits_{j=1}^{n} \theta_i^s \theta_j^t\right)^{\frac{1}{s+t}}}, \\ & \left(\left(1 - \left(\prod_{i=1}^{n} \prod_{j=i}^{n} \left(1 - u_i^{2s} u_j^{2t} \right) \right)^{\frac{2}{n(n+1)}} \right)^{1/2(s+t)} \right), \\ & \left(1 - \left(1 - \left(\prod_{i=1}^{n} \prod_{j=i}^{n} \left(1 - (1-\eta_i^2)^s \left(1 - \eta_j^2 \right)^t \right) \right)^{\frac{4}{n(n+1)}} \right)^{1/(s+t)} \right)^{1/2}, \\ & \left(1 - \left(1 - \left(\prod_{i=1}^{n} \prod_{j=i}^{n} \left(1 - (1-v_i^2)^s \left(1 - v_j^2 \right)^t \right) \right)^{\frac{4}{n(n+1)}} \right)^{1/(s+t)} \right)^{1/2} \Bigg) \Bigg\rangle, \end{aligned} \tag{18}$$

which is the Pythagorean picture linguistic Heronian mean operator.

Case 6: When $q = 1$, then the q-ROLHM operator reduces to the followings,

$$\begin{aligned} q - RPLHM^{s,t}(\alpha_1, \alpha_2, \ldots, \alpha_n) = \Bigg\langle & s_{\left(\frac{2}{n(n+1)} \sum\limits_{i=1}^{n} \sum\limits_{j=1}^{n} \theta_i^s \theta_j^t\right)^{\frac{1}{s+t}}}, \\ & \left(\left(1 - \left(\prod_{i=1}^{n} \prod_{j=1}^{n} \left(1 - u_i^{s} u_j^{t} \right) \right)^{\frac{2}{n(n+1)}} \right)^{\frac{1}{s+t}}, \right. \\ & 1 - \left(1 - \left(\prod_{i=1}^{n} \prod_{j=1}^{n} \left(1 - (1-\eta_i)^s \left(1 - \eta_j \right)^t \right) \right)^{\frac{2}{n(n+1)}} \right)^{\frac{1}{s+t}}, \\ & \left. 1 - \left(1 - \left(\prod_{i=1}^{n} \prod_{j=1}^{n} \left(1 - (1-v_i)^s \left(1 - v_j \right)^t \right) \right)^{\frac{2}{n(n+1)}} \right)^{\frac{1}{s+t}} \right) \Bigg\rangle, \end{aligned} \tag{19}$$

which is the picture linguistic Heronian mean operator.

3.2. The q-Rung Picture Linguistic Weighted Heronian Mean (q-RPLWHM) Operator

It is noted that the proposed q-RPLHM operator does not consider the self-importance of the aggregated arguments. Therefore, we put forward the weighted Heronian mean for q-RPLNs, which also considers the weights of aggregated arguments.

Definition 14. *Let $\alpha_i (i = 1, 2, \ldots, n)$ be a collection of q-RPLNs, and $s, t > 0$, $w = (w_1, w_2, \ldots, w_n)^T$ be the weight vector, satisfying $w_i \in [0, 1]$ and $\sum_{i=1}^{n} w_i = 1$. If*

$$q - RPLWHM^{s,t}(\alpha_1, \alpha_2, \ldots, \alpha_n) = \left(\frac{2}{n(n+1)} \sum_{i=1}^{n} \sum_{j=i}^{n} (n w_i \alpha_i)^s \left(n w_j \alpha_j \right)^t \right)^{1/(s+t)}, \tag{20}$$

then $q - RPLWHM^{s,t}(\alpha_1, \alpha_2, \ldots, \alpha_n)$ is called the q-RPLWHM.

According to the operations for q-RPLNs, the following theorems can be obtained.

Theorem 5. *Let $\alpha_i (i = 1, 2, ..., n)$ be a collection of q-RPLNs, $w = (w_1, w_2, ..., w_n)^T$ be the weight vector, satisfying $w_i \in [0,1]$ and $\sum_{i=1}^{n} w_i = 1$, then the aggregated value by using q-RPLWHM is also a q-RPLN and*

$$q - RPLWHM^{s,t}(\alpha_1, \alpha_2, ..., \alpha_n) = \left\langle s_{\left(\frac{2}{n(n+1)}\sum_{i=1}^{n}\sum_{j=i}^{n}(nw_i\theta_i)^s \times (nw_j\theta_j)^t\right)^{1/(s+t)}}, \right.$$
$$\left(\left(1-\prod_{i=1}^{n}\prod_{j=i}^{n}\left(1-\left(1-\left(1-u_i^q\right)^{nw_i}\right)^{\frac{2s}{n(n+1)}}\left(1-\left(1-u_j^q\right)^{nw_j}\right)^{\frac{2t}{n(n+1)}}\right)\right)\right)^{1/(s+t)q},$$
$$\left(1-\left(1-\prod_{i=1}^{n}\prod_{j=i}^{n}\left(1-\left(1-\eta_i^{nw_iq}\right)^s\left(1-\eta_j^{nw_jq}\right)^t\right)^{\frac{2}{n(n+1)}}\right)^{1/(s+t)}\right)^{1/q}, \tag{21}$$
$$\left.\left(1-\left(1-\prod_{i=1}^{n}\prod_{j=i}^{n}\left(1-\left(1-v_i^{nw_iq}\right)^s\left(1-v_j^{nw_jq}\right)^t\right)^{\frac{2}{n(n+1)}}\right)^{1/(s+t)}\right)^{1/q}\right\rangle.$$

The proof of Theorem 5 is similar to that of Theorem 1, which is omitted here.

Similarly, q-RPLWHM has the following properties.

Theorem 6 (Monotonicity). *Let α_i and $\beta_i (i = 1, 2, ..., n)$ be two collections of q-RPLNs, if $\alpha_i \leq \beta_i$ for all i, then*

$$q - RPLWHM^{s,t}(\alpha_1, \alpha_2, ..., \alpha_n) \leq q - RPLWHM^{s,t}(\beta_1, \beta_2, ..., \beta_n). \tag{22}$$

Theorem 7 (Boundedness). *The q-RPLWHM operator lies between the max and min operators*

$$\min(\alpha_1, \alpha_2, ..., \alpha_n) \leq q - RPLWHM^{s,t}(\alpha_1, \alpha_2, ..., \alpha_n) \leq \max(\alpha_1, \alpha_2, ..., \alpha_n). \tag{23}$$

3.3. The q-Rung Picture Linguistic Geometric Heronian Mean (q-RPLGHM) Operator

Definition 15. *Let $\alpha_i (i = 1, 2, ..., n)$ be a collection of q-RPLNs, and $s, t > 0$. If*

$$q - RPLGHM^{s,t}(\alpha_1, \alpha_2, ..., \alpha_n) = \frac{1}{s+t}\prod_{i=1}^{n}\prod_{j=i}^{n}(s\alpha_i + t\alpha_j)^{\frac{2}{n(n+1)}}, \tag{24}$$

then $q - RPLGHM^{s,t}$ is called the q-rung picture linguistic geometric Heronian mean (q-RPLGHM) operator.

Similarly, the following theorem can be obtained according to Definition 7.

Theorem 8. *Let $\alpha_i (i = 1, 2, ..., n)$ be a collection of q-RPLNs, then the aggregated value by using q-RPLGHM is also a q-RPLN and*

$$q - RPLGHM^{s,t}(\alpha_1, \alpha_2, ..., \alpha_n) =$$
$$\left\langle s_{\frac{1}{s+t}\prod_{i=1}^{n}\prod_{j=i}^{n}(s\theta_i+t\theta_j)^{\frac{2}{n(n+1)}}}, \left(\left(1-\left(1-\prod_{i=1}^{n}\prod_{j=i}^{n}\left(1-\left(1-u_i^q\right)^s\left(1-u_j^q\right)^t\right)^{\frac{2}{n(n+1)}}\right)^{\frac{1}{s+t}}\right)\right)^{1/q}, \right. \tag{25}$$
$$\left.\left(1-\prod_{i=1}^{n}\prod_{j=i}^{n}\left(1-v_i^{sq}v_j^{tq}\right)^{\frac{2}{n(n+1)}}\right)^{\frac{1}{(s+t)q}}, \left(1-\prod_{i=1}^{n}\prod_{j=i}^{n}\left(1-v_i^{sq}v_j^{tq}\right)^{\frac{2}{n(n+1)}}\right)^{\frac{1}{(s+t)q}}\right\rangle.$$

The proof of Theorem 8 is similar to that of Theorem 1. In the following, we present some desirable properties of the q-RPLGHM operator.

Theorem 9 (Idempotency). *Let $\alpha_i = \langle s_{\theta_i}, (\mu_i, \eta_i, v_i) \rangle (i = 1, 2, ..., n)$ be a collection of q-RPLNs, if all the q-RPLNs are equal,* i.e., *$\alpha_i = \alpha$ for all i, then*

$$q - RPLGHM^{s,t}(\alpha_1, \alpha_2, ..., \alpha_n) = \alpha. \tag{26}$$

The proof of Theorem 9 is similar to that of Theorem 2.

Theorem 10 (Monotonicity). *Let α_i and $\beta_i (i = 1, 2, ..., n)$ be two collections of q-RPLNs, if $\alpha_i \leq \beta_i$ for all i, then*

$$q - RPLGHM^{s,t}(\alpha_1, \alpha_2, ..., \alpha_n) \leq q - RPLGHM^{s,t}(\beta_1, \beta_2, ..., \beta_n). \tag{27}$$

The proof of Theorem 10 is similar to that of Theorem 3.

Theorem 11 (Boundedness). *Let $\alpha_i = \langle s_{\theta_i}, (\mu_i, \eta_i, v_i) \rangle (i = 1, 2, ..., n)$ be a collection of q-RPLNs, then*

$$\min(\alpha_1, \alpha_2, ..., \alpha_n) \leq q - RPLGHM^{s,t}(\alpha_1, \alpha_2, ..., \alpha_n) \leq \max(\alpha_1, \alpha_2, ..., \alpha_n). \tag{28}$$

The proof of Theorem 11 is similar to that of Theorem 4. In the followings, we discuss some special cases of the q-RPLGHM operator.

Case 1: When $t \to 0$, then the q-RPLGHM operator reduces to the followings,

$$\begin{aligned}
q - RPLGHM^{s,0}(\alpha_1, \alpha_2, ..., \alpha_n) = \lim_{t \to 0} \Bigg\langle & s_{\frac{1}{s+t} \prod_{i=1}^{n} \prod_{j=i}^{n} (s\theta_i + t\theta_j)^{\frac{2}{n(n+1)}}}, \\
& \left(\left(1 - \left(1 - \prod_{i=1}^{n} \prod_{j=i}^{n} \left(1 - \left(1 - u_i^q\right)^s \left(1 - u_j^q\right)^t \right)^{\frac{2}{n(n+1)}} \right)^{\frac{1}{s+t}} \right)^{1/q}, \right. \\
& \left. \left(1 - \prod_{i=1}^{n} \prod_{j=i}^{n} \left(1 - \eta_i^{sq} \eta_j^{tq} \right)^{\frac{2}{n(n+1)}} \right)^{\frac{1}{(s+t)q}}, \left(1 - \prod_{i=1}^{n} \prod_{j=i}^{n} \left(1 - v_i^{sq} v_j^{tq} \right)^{\frac{2}{n(n+1)}} \right)^{\frac{1}{(s+t)q}} \right) \Bigg\rangle \\
= \Bigg\langle & s_{\frac{1}{s} \left(\prod_{i=1}^{n} (s\theta_i)^{n+1-i} \right)^{\frac{2}{n(n+1)}}}, \left(\left(1 - \left(1 - \left(\prod_{i=1}^{n} \left(1 - \left(1 - u_i^q\right)^s \right)^{n+1-i} \right)^{\frac{2}{n(n+1)}} \right)^{\frac{1}{s}} \right)^{1/q}, \right. \\
& \left. \left(1 - \left(\prod_{i=1}^{n} \left(1 - \eta_i^{sq} \right)^{n+1-i} \right)^{\frac{2}{n(n+1)}} \right)^{\frac{1}{sq}}, \left(1 - \left(\prod_{i=1}^{n} \left(1 - v_i^{sq} \right)^{n+1-i} \right)^{\frac{2}{n(n+1)}} \right)^{\frac{1}{sq}} \right) \Bigg\rangle,
\end{aligned} \tag{29}$$

which is a q-rung picture linguistic generalized geometric linear descending weighted mean operator.

Case 2: When $s \to 0$, then the q-RPLGHM operator reduces to the followings,

$$
\begin{aligned}
q-RPLGHM^{0,t}(\alpha_1,\alpha_2,\ldots,\alpha_n) &= \lim_{s\to 0}\Bigg\langle s_{\frac{1}{s+t}\prod\limits_{i=1}^{n}\prod\limits_{j=i}^{n}(s\theta_i+t\theta_j)^{\frac{2}{n(n+1)}}}, \\
&\left(\left(1-\left(1-\prod_{i=1}^{n}\prod_{j=i}^{n}\left(1-\left(1-u_i^q\right)^s\left(1-u_j^q\right)^t\right)\right)^{\frac{2}{n(n+1)}}\right)^{\frac{1}{s+t}}\right)^{1/q}, \\
&\left(1-\prod_{i=1}^{n}\prod_{j=i}^{n}\left(1-\eta_i^{sq}\eta_j^{tq}\right)^{\frac{2}{n(n+1)}}\right)^{\frac{1}{(s+t)q}}, \left(1-\prod_{i=1}^{n}\prod_{j=i}^{n}\left(1-v_i^{sq}v_j^{tq}\right)^{\frac{2}{n(n+1)}}\right)^{\frac{1}{(s+t)q}}\Bigg\rangle \\
&= \Bigg\langle s_{\frac{1}{t}\left(\prod\limits_{i=1}^{n}(t\theta_j)^i\right)^{\frac{2}{n(n+1)}}}, \left(\left(1-\left(1-\left(\prod_{i=1}^{n}\left(1-\left(1-u_j^q\right)^t\right)\right)^i\right)^{\frac{2}{n(n+1)}}\right)^{\frac{1}{t}}\right)^{1/q}, \\
&\left(1-\left(\prod_{i=1}^{n}\left(1-\eta_j^{tq}\right)^i\right)^{\frac{2}{n(n+1)}}\right)^{\frac{1}{tq}}, \left(1-\left(\prod_{i=1}^{n}\left(1-v_j^{tq}\right)^i\right)^{\frac{2}{n(n+1)}}\right)^{\frac{1}{tq}}\Bigg\rangle,
\end{aligned} \tag{30}
$$

which is a q-rung picture linguistic generalized geometric linear ascending weighted mean operator.

Case 3: When $s = t = 1$, then the q-RPLGHM operator reduces to the followings,

$$
\begin{aligned}
q-RPLGHM^{1,1}(\alpha_1,\alpha_2,\ldots,\alpha_n) &= \Bigg\langle s_{\frac{1}{2}\prod\limits_{i=1}^{n}\prod\limits_{j=i}^{n}(\theta_i+\theta_j)^{\frac{2}{n(n+1)}}}, \\
&\left(\left(1-\left(1-\prod_{i=1}^{n}\prod_{j=i}^{n}\left(1-\left(1-u_i^q\right)\left(1-u_j^q\right)\right)\right)^{\frac{2}{n(n+1)}}\right)^{\frac{1}{2}}\right)^{1/q}, \\
&\left(1-\left(\prod_{i=1}^{n}\prod_{j=i}^{n}\left(1-\eta_i^q\eta_j^q\right)\right)^{\frac{2}{n(n+1)}}\right)^{\frac{1}{2q}}, \left(1-\left(\prod_{i=1}^{n}\prod_{j=i}^{n}\left(1-v_i^q v_j^q\right)\right)^{\frac{2}{n(n+1)}}\right)^{\frac{1}{2q}}\Bigg\rangle,
\end{aligned} \tag{31}
$$

which is a q-rung picture linguistic geometric line Heronian mean operator.

Case 4: When $s = t = 1/2$, then the q-RPLGHM operator reduces to the followings,

$$
\begin{aligned}
q-RPLGHM^{\frac{1}{2},\frac{1}{2}}(\alpha_1,\alpha_2,\ldots,\alpha_n) &= \Bigg\langle s_{\left(\prod\limits_{i=1}^{n}\prod\limits_{j=i}^{n}\left(\frac{1}{2}\theta_i+\frac{1}{2}\theta_j\right)\right)^{\frac{2}{n(n+1)}}}, \left(\left(\prod_{i=1}^{n}\prod_{j=i}^{n}\left(1-\sqrt{\left(1-u_i^q\right)\left(1-u_j^q\right)}\right)\right)^{\frac{2}{n(n+1)}}\right)^{1/q}, \\
&\left(1-\prod_{i=1}^{n}\prod_{j=i}^{n}\left(1-\sqrt{\eta_i^q\eta_j^q}\right)^{\frac{2}{n(n+1)}}\right)^{\frac{1}{q}}, \left(1-\prod_{i=1}^{n}\prod_{j=i}^{n}\left(1-\sqrt{v_i^q v_j^q}\right)^{\frac{2}{n(n+1)}}\right)^{\frac{1}{q}}\Bigg\rangle,
\end{aligned} \tag{32}
$$

which is a q-rung picture linguistic basic geometric Heronian mean operator.

Case 5: When $q = 2$, then the q-RPLGHM operator reduces to the followings

$$q-RPLGHM^{s,t}(\alpha_1,\alpha_2,...,\alpha_n)=\left\langle s_{\frac{1}{s+t}(\prod_{i=1}^{n}\prod_{j=1}^{n}(s\theta_i+t\theta_j))^{\frac{2}{n(n+1)}}},\right.$$
$$\left(\left(1-\left(1-\prod_{i=1}^{n}\prod_{j=i}^{n}\left(1-(1-u_i^2)^s\left(1-u_j^2\right)^t\right)^{\frac{2}{n(n+1)}}\right)^{\frac{1}{s+t}}\right)^{1/2},\tag{33}$$
$$\left.\left(1-\prod_{i=1}^{n}\prod_{j=i}^{n}\left(1-\eta_i^{2s}\eta_j^{2t}\right)^{\frac{2}{n(n+1)}}\right)^{\frac{1}{2(s+t)}},\left(1-\prod_{i=1}^{n}\prod_{j=i}^{n}\left(1-v_i^{2s}v_j^{2t}\right)^{\frac{2}{n(n+1)}}\right)^{\frac{1}{2(s+t)}}\right)\right\rangle,$$

which is the Pythagorean picture linguistic geometric Heronian mean operator.

Case 6: When $q=1$, then the q-RPLGHM operator reduces to the followings

$$q-RPLGHM^{s,t}(\alpha_1,\alpha_2,...,\alpha_n)=\left\langle s_{\frac{1}{s+t}(\prod_{i=1}^{n}\prod_{j=1}^{n}(s\theta_i+t\theta_j))^{\frac{2}{n(n+1)}}},\right.$$
$$\left(1-\left(1-\left(\prod_{i=1}^{n}\prod_{j=i}^{n}\left(1-(1-u_i)^s(1-u_j)^t\right)^{\frac{2}{n(n+1)}}\right)^{\frac{1}{s+t}}\right),\right.\tag{34}$$
$$\left.\left(1-\left(\prod_{i=1}^{n}\prod_{j=i}^{n}\left(1-\eta_i^s\eta_j^t\right)\right)^{\frac{2}{n(n+1)}}\right)^{\frac{1}{s+t}},\left(1-\left(\prod_{i=1}^{n}\prod_{j=i}^{n}\left(1-v_i^s v_j^t\right)\right)^{\frac{2}{n(n+1)}}\right)^{\frac{1}{s+t}}\right)\right\rangle,$$

which is the picture linguistic geometric Heronian mean operator.

3.4. The q-Rung Picture Linguistic Weighted Geometric Heronian Mean (q-RPLWGHM) Operator

Definition 16. *Let $\alpha_i(i=1,2,...,n)$ be a collection of q-RPLNs, and $s,t>0$, $w=(w_1,w_2,...,w_n)^T$ be the weight vector, satisfying $w_i\in[0,1]$ and $\sum_{i=1}^{n}w_i=1$. If*

$$q-RPLWGHM^{s,t}(\alpha_1,\alpha_2,...,\alpha_n)=\frac{1}{s+t}\prod_{i=1}^{n}\prod_{j=i}^{n}\left(s a_i^{nw_i}+t a_j^{nw_j}\right)^{\frac{2}{n(n+1)}},\tag{35}$$

then $q-RPLWGHM^{s,t}$ is called the q-RPLWGHM.

Additionally, q- RPLWGHM has the following theorem.

Theorem 12. *Let $\alpha_i(i=1,2,...,n)$ be a collection of q-RPLNs, $w=(w_1,w_2,...,w_n)^T$ be the weight vector, satisfying $w_i\in[0,1]$ and $\sum_{i=1}^{n}w_i=1$, then the aggregated value by using q-RPLWGHM is also a q-RPLN and*

$$q-RPLWGHM^{s,t}(\alpha_1,\alpha_2,...,\alpha_n)=\left\langle s_{\frac{1}{s+t}\prod\limits_{i=1}^{n}\prod\limits_{j=i}^{n}(\theta_i^{nw_i}+t\theta_j^{nw_j})^{\frac{2}{n(n+1)}}},\right.$$
$$\left(\left(1-\left(1-\prod_{i=1}^{n}\prod_{j=i}^{n}\left(1-\left(1-u_i^{nw_iq}\right)^s\left(1-u_j^{nw_jq}\right)^t\right)^{\frac{2}{n(n+1)}}\right)^{\frac{1}{s+t}}\right)^{1/q},\right.$$
$$\left(1-\prod_{i=1}^{n}\prod_{i=i}^{n}\left(1-\left(1-\left(1-\eta_i^q\right)^{nw_i}\right)^s\left(1-\left(1-\eta_j^q\right)^{nw_j}\right)^t\right)^{\frac{2}{n(n+1)}}\right)^{\frac{1}{(s+t)q}},\tag{36}$$
$$\left.\left.\left(1-\prod_{i=1}^{n}\prod_{i=i}^{n}\left(1-\left(1-\left(1-v_i^q\right)^{nw_i}\right)^s\left(1-\left(1-v_j^q\right)^{nw_j}\right)^t\right)^{\frac{2}{n(n+1)}}\right)^{\frac{1}{(s+t)q}}\right)\right\rangle.$$

The proof of Theorem 12 is similar to that of Theorem 1, which is omitted here.

In addition, the q-RPLWGHM operator has the following properties.

Theorem 13 (Monotonicity). *Let α_i and $\beta_i(i=1,2,...,n)$ be two collections of q-RPLNs, if $\alpha_i\leq\beta_i$ for all i, then*

$$q-RPLWGHM^{s,t}(\alpha_1,\alpha_2,...,\alpha_n)\leq q-RPLWGHM^{s,t}(\beta_1,\beta_2,...,\beta_n).\tag{37}$$

Theorem 14 (Boundedness). *The q-RPLWGHM operator lies between the max and min operators*

$$\min(\alpha_1,\alpha_2,...,\alpha_n)\leq q-RPLWGHM^{s,t}(\alpha_1,\alpha_2,...,\alpha_n)\leq\max(\alpha_1,\alpha_2,...,\alpha_n).\tag{38}$$

4. A Novel Approach to MAGDM Based on the Proposed Operators

In this section, we shall apply the proposed aggregation operators to solving MAGDM problems in q-rung picture linguistic environment. Considering a MAGDM process in which the attribute value take the form of q-RPLNs.: let $A=\{A_1,A_2,...,A_m\}$ be a set of all alternatives, and $C=\{c_1,c_2,...,c_n\}$ be a set of attributes with the weight vector being $w=(w_1,w_2,...,w_n)^T$, satisfying $w_i\in[0,1]$ and $\sum_{i=1}^{n}w_i=1$. A set of decision-makers D_k are organized to make the assessment for every attribute $c_j(j=1,2,...,n)$ of all alternatives by q-RPLNs $\alpha_{ij}^k=\left\langle s_{\theta_{ij}}^k,\left(u_{ij}^k,\eta_{ij}^k,v_{ij}^k\right)\right\rangle$, and $\lambda=(\lambda_1,\lambda_2,\ldots,\lambda_k)$ is the weight vector of decision-makers $D_k(k=1,2,...,p)$. Therefore, the q-rung picture linguistic decision matrices can be denoted by $A^k=\left(\alpha_{ij}^k\right)_{m\times n}$. The main steps to solve MAGDM problems based on the proposed operators are given as follows.

Step 1. Standardize the original decision matrices. There are two types of attributes, benefit and cost attributes. Therefore, the original decision matrix should be normalized by

$$\alpha_{ij}^k=\begin{cases}\left\langle s_{\theta_{ij}}^k,\left(u_{ij}^k,\eta_{ij}^k,v_{ij}^k\right)\right\rangle & y_j\in I_1\\ \left\langle s_{\theta_{ij}}^k,\left(v_{ij}^k,\eta_{ij}^k,u_{ij}^k\right)\right\rangle & y_j\in I_2\end{cases},\tag{39}$$

where I_1 and I_2 represent the benefit attributes and cost attributes respectively.

Step 2. Utilize the q-RPLWHM operator

$$\alpha_{ij}=q-RPLWHM^{s,t}\left(\alpha_{ij}^1,\alpha_{ij}^2,...,\alpha_{ij}^p\right),\tag{40}$$

or the q-RPLWGHM operator

$$\alpha_{ij} = q - RPLWGHM^{s,t}\left(\alpha_{ij}^1, \alpha_{ij}^2, \ldots, \alpha_{ij}^p\right), \quad (41)$$

to aggregate all the decision matrices $A^k(k = 1, 2, \ldots p)$ into a collective decision matrix $A = \left(\alpha_{ij}\right)_{m \times n}$.

Step 3. Utilize the q-RPLWHM operator

$$\alpha_i = q - RPLWHM^{s,t}(\alpha_{i1}, \alpha_{i2}, \ldots, \alpha_{in}), \quad (42)$$

or the q-RPLWGHM operator

$$\alpha_i = q - RPLWGHM^{s,t}(\alpha_{i1}, \alpha_{i2}, \ldots, \alpha_{in}), \quad (43)$$

to aggregate the assessments $\alpha_{ij}(j = 1, 2, \ldots, n)$ for each A_i so that the overall preference values $\alpha_i(i = 1, 2, \ldots, m)$ of alternatives can be obtained.

Step 4. Calculate the score functions of the overall values $\alpha_i(i = 1, 2, \ldots, m)$.

Step 5. Rank all alternatives according to the score functions of the corresponding overall values and select the best one(s).

Step 6. End.

5. Numerical Instance

In this part, to validate the proposed method, we provide a numerical instance about choosing an Enterprise resource planning (ERP) system adopted from Liu and Zhang [41]. After primary evaluation, there are four possible systems provided by different companies remained on the candidates list and they are $\{A_1, A_2, A_3, A_4\}$. Four experts $D_k(k = 1, 2, 3, 4)$ are invited to evaluate the candidates under four attributes, they are (1) technology C_1; (2) strategic adaptability C_2; (3) supplier's ability C_3; (4) supplier's reputation C_4. Weight vector of the four attributes is $w = (0.25, 0.3, 0.25, 0.2)^T$. The decision-makers are required to use picture fuzzy linguistic numbers (PFLNs) on the basic of the linguistic term set $S = \{s_0$ = terrible, s_1 = bad, s_2 = poor, s_3 = neutral, s_4 = good, s_5 = well, s_6 = excellent$\}$ to express their preference information. Decision-makers' weight vector is $\lambda = (0.3, 0.2, 0.2, 0.3)^T$. After evaluation, the individual picture fuzzy linguistic decision matrix $A^k = \left(\alpha_{ij}^k\right)_{4\times 4}$ can be obtained, which are shown in Tables 1–4.

Table 1. Decision matrix A^1 provided by D_1.

	C_1	C_2	C_3	C_4
A_1	$\langle s_4, (0.53, 0.33, 0.09)\rangle$	$\langle s_2, (0.89, 0.08, 0.03)\rangle$	$\langle s_1, (0.42, 0.35, 0.18)\rangle$	$\langle s_3, (0.08, 0.89, 0.02)\rangle$
A_2	$\langle s_2, (0.73, 0.12, 0.08)\rangle$	$\langle s_4, (0.13, 0.64, 0.21)\rangle$	$\langle s_2, (0.03, 0.82, 0.13)\rangle$	$\langle s_4, (0.73, 0.15, 0.08)\rangle$
A_3	$\langle s_5, (0.91, 0.03, 0.02)\rangle$	$\langle s_1, (0.07, 0.79, 0.05)\rangle$	$\langle s_4, (0.04, 0.85, 0.10)\rangle$	$\langle s_2, (0.68, 0.26, 0.06)\rangle$
A_4	$\langle s_5, (0.85, 0.09, 0.05)\rangle$	$\langle s_3, (0.74, 0.16, 0.10)\rangle$	$\langle s_6, (0.02, 0.89, 0.05)\rangle$	$\langle s_1, (0.08, 0.84, 0.06)\rangle$

Table 2. Decision matrix A^2 provided by D_2.

	C_1	C_2	C_3	C_4
A_1	$\langle s_3, (0.53, 0.33, 0.09)\rangle$	$\langle s_3, (0.73, 0.12, 0.08)\rangle$	$\langle s_2, (0.91, 0.03, 0.02)\rangle$	$\langle s_4, (0.85, 0.09, 0.05)\rangle$
A_2	$\langle s_1, (0.89, 0.08, 0.03)\rangle$	$\langle s_3, (0.13, 0.64, 0.21)\rangle$	$\langle s_3, (0.77, 0.09, 0.05)\rangle$	$\langle s_4, (0.74, 0.16, 0.10)\rangle$
A_3	$\langle s_4, (0.42, 0.35, 0.18)\rangle$	$\langle s_2, (0.03, 0.82, 0.13)\rangle$	$\langle s_4, (0.04, 0.85, 0.10)\rangle$	$\langle s_3, (0.02, 0.89, 0.05)\rangle$
A_4	$\langle s_5, (0.33, 0.51, 0.12)\rangle$	$\langle s_2, (0.53, 0.31, 0.16)\rangle$	$\langle s_6, (0.68, 0.26, 0.06)\rangle$	$\langle s_3, (0.08, 0.84, 0.06)\rangle$

Table 3. Decision matrix A^3 provided by D_3.

	C_1	C_2	C_3	C_4
A_1	$\langle s_4, (0.33, 0.52, 0.12)\rangle$	$\langle s_2, (0.52, 0.31, 0.16)\rangle$	$\langle s_4, (0.31, 0.39, 0.25)\rangle$	$\langle s_5, (0.64, 0.16, 0.10)\rangle$
A_2	$\langle s_4, (0.17, 0.53, 0.13)\rangle$	$\langle s_3, (0.51, 0.24, 0.21)\rangle$	$\langle s_4, (0.31, 0.39, 0.25)\rangle$	$\langle s_5, (0.64, 0.16, 0.10)\rangle$
A_3	$\langle s_2, (0.90, 0.05, 0.02)\rangle$	$\langle s_1, (0.68, 0.08, 0.21)\rangle$	$\langle s_5, (0.05, 0.87, 0.06)\rangle$	$\langle s_3, (0.13, 0.75, 0.09)\rangle$
A_4	$\langle s_3, (0.15, 0.73, 0.08)\rangle$	$\langle s_3, (0.70, 0.20, 0.10)\rangle$	$\langle s_5, (0.91, 0.03, 0.05)\rangle$	$\langle s_3, (0.18, 0.64, 0.06)\rangle$

Table 4. Decision matrix A^4 provided by D_4.

	C_1	C_2	C_3	C_4
A_1	$\langle s_3, (0.90, 0.05, 0.02)\rangle$	$\langle s_1, (0.68, 0.08, 0.21)\rangle$	$\langle s_3, (0.05, 0.87, 0.06)\rangle$	$\langle s_1, (0.13, 0.75, 0.09)\rangle$
A_2	$\langle s_6, (0.77, 0.13, 0.10)\rangle$	$\langle s_2, (0.62, 0.24, 0.11)\rangle$	$\langle s_2, (0.10, 0.75, 0.10)\rangle$	$\langle s_4, (0.64, 0.16, 0.10)\rangle$
A_3	$\langle s_3, (0.80, 0.15, 0.02)\rangle$	$\langle s_4, (0.68, 0.18, 0.05)\rangle$	$\langle s_5, (0.05, 0.87, 0.06)\rangle$	$\langle s_1, (0.12, 0.65, 0.20)\rangle$
A_4	$\langle s_6, (0.15, 0.73, 0.08)\rangle$	$\langle s_3, (0.61, 0.25, 0.10)\rangle$	$\langle s_5, (0.91, 0.03, 0.05)\rangle$	$\langle s_4, (0.28, 0.44, 0.16)\rangle$

5.1. The Decision-Making Process

Step 1. As the four attributes are benefit types, the original decision matrices do not need normalization.

Step 2. Utilize Equation (40) to calculate the comprehensive value α_{ij} of each attribute for every alternative. The collective decision matrix $A = (\alpha_{ij})_{4\times 4}$ is shown in Table 5 (suppose $s = t = 1, q = 3$):

Table 5. Collective picture fuzzy linguistic decision matrix (by q-rung Picture Linguistic Weighted Geometric Heronian Mean (q-RPLWHM) operator).

	C_1	C_2	C_3	C_4
A_1	$\langle s_{3.52}, (0.71, 0.29, 0.18)\rangle$	$\langle s_{1.91}, (0.76, 0.15, 0.14)\rangle$	$\langle s_{2.44}, (0.63, 0.45, 0.39)\rangle$	$\langle s_{2.83}, (0.57, 0.61, 0.44)\rangle$
A_2	$\langle s_{3.56}, (0.75, 0.19, 0.15)\rangle$	$\langle s_{3.04}, (0.48, 0.44, 0.30)\rangle$	$\langle s_{2.60}, (0.51, 0.52, 0.38)\rangle$	$\langle s_{4.21}, (0.69, 0.17, 0.14)\rangle$
A_3	$\langle s_{3.67}, (0.83, 0.13, 0.11)\rangle$	$\langle s_{2.22}, (0.56, 0.42, 0.23)\rangle$	$\langle s_{4.52}, (0.05, 0.86, 0.74)\rangle$	$\langle s_{2.11}, (0.49, 0.62, 0.52)\rangle$
A_4	$\langle s_{4.97}, (0.63, 0.49, 0.43)\rangle$	$\langle s_{2.83}, (0.66, 0.24, 0.20)\rangle$	$\langle s_{5.53}, (0.80, 0.22, 0.09)\rangle$	$\langle s_{2.76}, (0.21, 0.69, 0.47)\rangle$

Step 3. Utilize Equation (42) to obtain the overall values of each alternative, we can get

$$\alpha_1 = \langle s_{3.86}, (0.75, 0.27, 0.24)\rangle \ \alpha_2 = \langle s_{2.53}, (0.63, 0.32, 0.29)\rangle$$
$$\alpha_3 = \langle s_{3.65}, (0.61, 0.51, 0.47)\rangle \ \alpha_4 = \langle s_{3.10}, (0.57, 0.51, 0.47)\rangle.$$

Step 4. Compute the score functions of the overall values, which are shown as follows:

$$S(\alpha_1) = 5.42,\ S(\alpha_2) = 3.11,\ S(\alpha_3) = 4.11,\ S(\alpha_4) = 3.35.$$

Step 5. Then the rank of the four alternatives is obtained

$$A_1 \succ A_3 \succ A_4 \succ A_2.$$

Therefore, the optimal alternative is A_1.

In step 2, if we utilize Equation (41) to aggregate the assessments, then we can derive the following collective decision matrix in Table 6 (suppose $s = t = 1, q = 3$).

Table 6. Collective picture fuzzy linguistic decision matrix (by q-RPLWGHM operator).

	C_1	C_2	C_3	C_4
A_1	$\langle s_{3.52}, (0.59, 0.37, 0.36)\rangle$	$\langle s_{1.80}, (0.72, 0.20, 0.20)\rangle$	$\langle s_{2.22}, (0.32, 0.70, 0.70)\rangle$	$\langle s_{2.55}, (0.24, 0.76, 0.70)\rangle$
A_2	$\langle s_{3.09}, (0.64, 0.34, 0.34)\rangle$	$\langle s_{2.96}, (0.34, 0.53, 0.49)\rangle$	$\langle s_{2.50}, (0.23, 0.69, 0.64)\rangle$	$\langle s_{4.23}, (0.69, 0.16, 0.15)\rangle$
A_3	$\langle s_{3.58}, (0.77, 0.23, 0.23)\rangle$	$\langle s_{1.89}, (0.28, 0.67, 0.62)\rangle$	$\langle s_{4.56}, (0.05, 0.86, 0.87)\rangle$	$\langle s_{1.96}, (0.20, 0.71, 0.71)\rangle$
A_4	$\langle s_{5.01}, (0.37, 0.62, 0.62)\rangle$	$\langle s_{2.82}, (0.65, 0.24, 0.24)\rangle$	$\langle s_{5.61}, (0.48, 0.67, 0.32)\rangle$	$\langle s_{2.51}, (0.16, 0.74, 0.70)\rangle$

Then we utilize Equation (43) to obtain the following overall values of alternatives:

$$\alpha_1 = \langle s_{3.65}, (0.60, 0.44, 0.44)\rangle \ \alpha_2 = \langle s_{2.36}, (0.49, 0.52, 0.52)\rangle$$
$$\alpha_3 = \langle s_{3.36}, (0.26, 0.75, 0.75)\rangle \ \alpha_4 = \langle s_{2.87}, (0.32, 0.67, 0.66)\rangle.$$

In addition, we calculate the score functions of the overall assessments and we can get

$$S(\alpha_1) = 4.13,\ S(\alpha_2) = 2.31,\ S(\alpha_3) = 2.01,\ S(\alpha_4) = 2.14.$$

Therefore, the rank of the four alternatives is $A_1 \succ A_2 \succ A_4 \succ A_3$ and the best alternative is A_1.

5.2. The Influence of the Parameters on the Results

The parameters q, s and t play significant roles in the final ranking results. In the following, we shall investigate the influence of the parameters on the overall assessments of alternatives and the final ranking results. First, we discuss the effects of the parameter q on the ranking results (suppose $s = t = 1$). Details are presented in Figures 1 and 2.

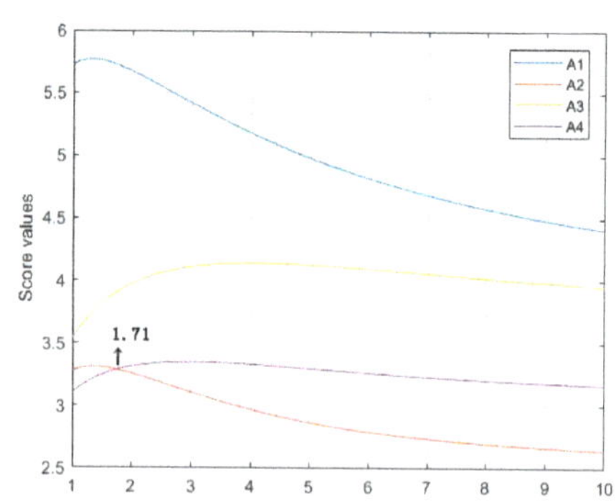

Figure 1. Score values of the alternatives when $q \in [1, 10], s = t = 1$ based on q-RPLWHM operator.

As seen in Figures 1 and 2, the score values of the overall assessments are different with different value of q, leading to different ranking results based the q-RFLWHM operator and the q-RFLWGHM operator. However, the best alternative is always A_1. The decision-makers can choose the appropriate parameter value q according to their preferences. From Figure 1, we can find that when $q \in [1, 1.71]$ the ranking order is $A_1 \succ A_3 \succ A_2 \succ A_4$ and when $q \in [1.71, 10]$ the ranking order is $A_1 \succ A_3 \succ A_4 \succ A_2$ by the q-RFLWHM operator. In addition, from Figure 2 we know when $q \in [1, 4.12]$ the ranking order is $A_1 \succ A_2 \succ A_4 \succ A_3$; when $q \in [4.12, 4.34]$ the ranking order is $A_1 \succ A_4 \succ A_2 \succ A_3$; when $q \in [4.34, 4.66]$ the ranking order is $A_1 \succ A_4 \succ A_3 \succ A_2$, and when $q \in [4.66, 10]$ the ranking order is $A_1 \succ A_3 \succ A_4 \succ A_2$ by the q-RFLWGHM operator.

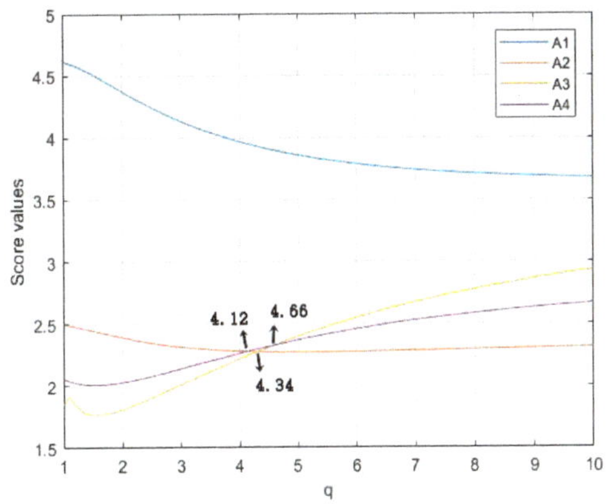

Figure 2. Score values of the alternatives when $q \in [1, 10], s = t = 1$ using q-RPLWGHM operator.

In the followings, we investigate influence of the parameters s and t on the score functions and ranking orders respectively (suppose q = 3). Details are presented in Tables 7 and 8.

Table 7. Ranking orders by utilizing different values of s and t in the q-RPLWHM operator.

s, t	Score Functions of $S(\alpha_i)(i = 1, 2, 3, 4)$	Ranking Results
$s \to 0,\ t = 1$	$S(\alpha_1) = 5.37, S(\alpha_2) = 3.03, S(\alpha_3) = 5.21, S(\alpha_4) = 2.95$	$A_1 \succ A_3 \succ A_2 \succ A_4$
$s = 1,\ t \to 0$	$S(\alpha_1) = 5.11, S(\alpha_2) = 3.42, S(\alpha_3) = 3.67, S(\alpha_4) = 4.24$	$A_1 \succ A_4 \succ A_3 \succ A_2$
$s = t = 1/2$	$S(\alpha_1) = 4.75, S(\alpha_2) = 2.74, S(\alpha_3) = 3.53, S(\alpha_4) = 2.85$	$A_1 \succ A_3 \succ A_4 \succ A_2$
$s = t = 1$	$S(\alpha_1) = 5.42, S(\alpha_2) = 3.11, S(\alpha_3) = 4.11, S(\alpha_4) = 3.35$	$A_1 \succ A_3 \succ A_4 \succ A_2$
$s = t = 2$	$S(\alpha_1) = 6.55, S(\alpha_2) = 3.81, S(\alpha_3) = 4.95, S(\alpha_4) = 4.15$	$A_1 \succ A_3 \succ A_4 \succ A_2$
$s = t = 5$	$S(\alpha_1) = 9.15, S(\alpha_2) = 5.68, S(\alpha_3) = 6.48, S(\alpha_4) = 5.65$	$A_1 \succ A_3 \succ A_2 \succ A_4$
$s = 1,\ t = 2$	$S(\alpha_1) = 6.13, S(\alpha_2) = 3.40, S(\alpha_3) = 4.83, S(\alpha_4) = 3.86$	$A_1 \succ A_3 \succ A_4 \succ A_2$
$s = 2,\ t = 1$	$S(\alpha_1) = 6.02, S(\alpha_2) = 3.67, S(\alpha_3) = 4.48, S(\alpha_4) = 4.13$	$A_1 \succ A_3 \succ A_4 \succ A_2$
$s = 1,\ t = 5$	$S(\alpha_1) = 8.02, S(\alpha_2) = 4.50, S(\alpha_3) = 6.18, S(\alpha_4) = 5.00$	$A_1 \succ A_3 \succ A_4 \succ A_2$
$s = 5,\ t = 1$	$S(\alpha_1) = 7.89, S(\alpha_2) = 5.36, S(\alpha_3) = 5.55, S(\alpha_4) = 5.42$	$A_1 \succ A_3 \succ A_4 \succ A_2$

Table 8. Ranking orders by utilizing different values of s and t in the q-RPLWGHM operator.

s, t	Score Functions of $S(\alpha_i)(i = 1, 2, 3, 4)$	Ranking Results
$s \to 0,\ t = 1$	$S(\alpha_1) = 3.43, S(\alpha_2) = 2.32, S(\alpha_3) = 2.40, S(\alpha_4) = 1.92$	$A_1 \succ A_3 \succ A_2 \succ A_4$
$s = 1,\ t \to 0$	$S(\alpha_1) = 3.72, S(\alpha_2) = 2.02, S(\alpha_3) = 1.71, S(\alpha_4) = 2.23$	$A_1 \succ A_4 \succ A_2 \succ A_3$
$s = t = 1/2$	$S(\alpha_1) = 4.35, S(\alpha_2) = 2.50, S(\alpha_3) = 2.44, S(\alpha_4) = 2.48$	$A_1 \succ A_2 \succ A_4 \succ A_3$
$s = t = 1$	$S(\alpha_1) = 4.13, S(\alpha_2) = 2.31, S(\alpha_3) = 2.07, S(\alpha_4) = 2.24$	$A_1 \succ A_2 \succ A_4 \succ A_3$
$s = t = 2$	$S(\alpha_1) = 3.79, S(\alpha_2) = 2.03, S(\alpha_3) = 1.74, S(\alpha_4) = 1.95$	$A_1 \succ A_2 \succ A_4 \succ A_3$
$s = t = 5$	$S(\alpha_1) = 3.24, S(\alpha_2) = 1.64, S(\alpha_3) = 1.41, S(\alpha_4) = 1.57$	$A_1 \succ A_2 \succ A_4 \succ A_3$
$s = 1,\ t = 2$	$S(\alpha_1) = 3.86, S(\alpha_2) = 2.13, S(\alpha_3) = 1.93, S(\alpha_4) = 2.03$	$A_1 \succ A_2 \succ A_4 \succ A_3$
$s = 2,\ t = 1$	$S(\alpha_1) = 3.91, S(\alpha_2) = 2.15, S(\alpha_3) = 1.76, S(\alpha_4) = 2.09$	$A_1 \succ A_2 \succ A_4 \succ A_3$
$s = 1,\ t = 5$	$S(\alpha_1) = 3.30, S(\alpha_2) = 1.78, S(\alpha_3) = 1.64, S(\alpha_4) = 1.70$	$A_1 \succ A_2 \succ A_4 \succ A_3$
$s = 5,\ t = 1$	$S(\alpha_1) = 3.38, S(\alpha_2) = 1.81, S(\alpha_3) = 1.35, S(\alpha_4) = 1.73$	$A_1 \succ A_2 \succ A_4 \succ A_3$

As seen in Tables 7 and 8, when different values are assigned to the parameters s and t, different scores and corresponding ranking results can be obtained. However, the best alternative is always A_1. Especially, in the q-RPLWHM operator, the increase of the parameters s and t leads to increase of the score functions, whereas a decrease of the score functions is witnessed using q-RPLWGHM operator. Furthermore, there is a difference in the ranking orders of A_2, A_3 and A_4 when $s \to 0, t = 1$ or $s = 1$, $t \to 0$ for the linear weighting by q-RPLWHM or q-RPLWGHM operator. Therefore, the parameters

s and t can be also viewed a decision-makers' optimistic or pessimistic attitude to their assessments. This demonstrates the flexibility in the aggregation processes using the proposed operators.

5.3. Comparative Analysis

To further demonstrate the merits and superiorities of the proposed methods, we conduct the following comparative analysis.

5.3.1. Compared with the Method Proposed by Liu and Zhang [41]

We utilize Liu and Zhang's [41] method to solve the above problem and results can be found in Table 9. From Table 9, we can find out that the results by using Liu and Zhang's [41] method and the proposed method in this paper are quite different. The reasons can be explained as follows: (1) Our method is based on the HM, which considers the interrelationship among attribute values, whereas the method based Archimedean picture fuzzy linguistic weighted arithmetic averaging (A-PFLWAA) operator proposed by Liu and Zhang [41] can only provide the arithmetic weighting function. In other words, Liu and Zhang's [41] method assumes that attributes are independent. In most real decision-making problems, attributes are correlated so that the interrelationship among attributes should be taken into consideration. Therefore, our proposed method is more reasonable than Liu and Zhang's [41] method. (2) Liu and Zhang's [41] method is based on PFLS, which is only a special case of q-RPLS (when q = 1). Therefore, our method is more general, flexible and reasonable than that proposed by Liu and Zhang [41].

Table 9. Score values and ranking results using our methods and the method in Liu and Zhang [41].

Method	Score Values	Ranking Result
Liu and Zhang's [41] method based on the A-PFLWAA operator	$S(\alpha_1) = 1.74, S(\alpha_2) = 1.86,$ $S(\alpha_3) = 1.79, S(\alpha_4) = 2.62$	$A_4 \succ A_2 \succ A_3 \succ A_1$
The proposed method based on q-RPLWHM operator in this paper	$S(\alpha_1) = 5.42, S(\alpha_2) = 3.11,$ $S(\alpha_3) = 4.11, S(\alpha_4) = 3.35$	$A_1 \succ A_3 \succ A_4 \succ A_2$
The proposed method based on q-RPLWGHM operator in this paper	$S(\alpha_1) = 4.13, S(\alpha_2) = 2.31,$ $S(\alpha_3) = 2.01, S(\alpha_4) = 2.14$	$A_1 \succ A_2 \succ A_4 \succ A_3$

5.3.2. Compared with the Methods Proposed by Wang et al. [47], Liu et al. [48], and Ju et al. [49]

To further demonstrate the effectiveness and validity of the proposed methods in this paper, we will deal with the problems in Wang et al. [47], Liu et al. [48], and Ju et al. [49] by using our methods respectively. Given that there is no method to aggregate q-rung picture linguistic information and there are various methods based on the intuitionistic linguistic numbers (ILNs), which are special cases of q-RPLNs when q equals to one and the neutral membership degree η equals to zero, we use the following three cases described by intuitionistic fuzzy numbers (ILNs) to verify our methods. For instance, when an ILN is $\langle s_1, (0.6, 0.4)\rangle$, it can be transformed into a q-RPLN $\langle s_1, (0.6, 0, 0.4)\rangle$. The score values and ranking results by different methods are shown in Tables 10–12.

Table 10. Score values and ranking results using our method and the method in Wang et al. [47].

Method	Score Values	Ranking Result
Wang et al.'s [47] method based on the ILHA operator	$S(\beta_1) = 2.35, S(\beta_2) = 2.72,$ $S(\beta_3) = 3.35, S(\beta_4) = 3.09$	$A_3 \succ A_4 \succ A_2 \succ A_1$
The proposed method based on q-RPLWHM operator in this paper	$S(\beta_1) = 6.98, S(\beta_2) = 6.61,$ $S(\beta_3) = 8.07, S(\beta_4) = 7.31$	$A_3 \succ A_4 \succ A_1 \succ A_2$
The proposed method based on q-RPLWGHM operator in this paper	$S(\beta_1) = 6.80, S(\beta_2) = 6.43,$ $S(\beta_3) = 7.88, S(\beta_4) = 7.19$	$A_3 \succ A_4 \succ A_1 \succ A_2$

Table 11. Score values and ranking results using our method and the method in Liu et al. [48].

Method	Score Values	Ranking Result
Liu et al.'s [48] method based on ILWBM operator	$S(\tilde{r}_1) = 0.483, S(\tilde{r}_2) = 0.412,$ $S(\tilde{r}_3) = 0.383, S(\tilde{r}_4) = 0.381$	$A_1 \succ A_2 \succ A_3 \succ A_4$
The proposed method based on q-RPLWHM operator in this paper	$S(\tilde{r}_1) = 5.42, S(\tilde{r}_2) = 4.75,$ $S(\tilde{r}_3) = 4.40, S(\tilde{r}_4) = 4.12$	$A_1 \succ A_2 \succ A_3 \succ A_4$
The proposed method based on q-RPLWGHM operator in this paper	$S(\tilde{r}_1) = 5.42, S(\tilde{r}_2) = 4.64,$ $S(\tilde{r}_3) = 4.04, S(\tilde{r}_4) = 4.05$	$A_1 \succ A_2 \succ A_4 \succ A_3$

Table 12. Score values and ranking results using our method and the method in Ju et al. [49].

Method	Score Values	Ranking Result
Ju et al.'s [49] method based on WILMSM operator	$S(r_1) = 0.18, S(r_2) = 0.20,$ $S(r_3) = 0.14, S(r_4) = 0.19$	$A_2 \succ A_4 \succ A_1 \succ A_3$
The proposed method based on q-RPLWHM operator in this paper	$S(r_1) = 3.98, S(r_2) = 4.05,$ $S(r_3) = 3.68, S(r_4) = 3.94$	$A_2 \succ A_1 \succ A_4 \succ A_3$
The proposed method based on q-RPLWGHM operator in this paper	$S(r_1) = 3.74, S(r_2) = 3.92,$ $S(r_3) = 3.52, S(r_4) = 3.82$	$A_2 \succ A_4 \succ A_1 \succ A_3$

From Tables 10–12, it is obvious to find that the ranking results produced by our method are little different to those produced by other methods. However, their optimal selections are the same, which can prove the effectiveness of our methods very well. As mentioned above, the q-RPLN contains more information than ILN, and it is a generalization of the ILN. Thus, our method based on q-RPLNs can be utilized in a wider range of environments.

Wang et al.'s [47] method is based on intuitionistic linguistic hybrid averaging (ILHA) operator, which cannot consider the interrelationship among attribute values. Because our proposed method can make up for this disadvantage, our method is more reasonable than Wang et al.'s [47] method.

Liu et al.'s [48] method is based on intuitionistic linguistic weighted Bonferroni mean (ILWBM) operator. It can cope with the interrelationship between augments, which is same as our method. However, as Yu and Wu [44] pointed out that HM has some advantages over BM, our method is better than Liu et al.'s [48] method.

Ju et al.'s [49] method is based on weighted intuitionistic linguistic Maclaurin symmetric mean (WILMSM) operator and when k = 2, the interrelationship between any two arguments can be considered, which is the same as our proposed method. However, our methods are based on the q-RPLWHM and q-RPLWGHM operators, which have two parameters (s and t). The prominent advantage of our methods is that we can control the degree of the interactions of attribute values that are emphasized. The increase of values of the parameters means the interactions of attribute values are more emphasized. Therefore, the decision-making committee can properly select the desirable alternative according to their interests and the actual needs by determining the values of parameters. Moreover, in the WILMSM operator proposed by Ju et al. [49], the balancing coefficient n is not considered, leading to some unreasonable results. In our proposed operators, the coefficient n is considered so that our methods are more reliable and reasonable.

From above analysis, we can find out that our proposed methods can be successfully applied to actual decision-making problems. Compared with other methods, our methods are more flexible and suitable for addressing MAGDM problems. The advantages and merits of the proposed methods can be concluded as followings. Firstly, the proposed methods are based on q-RPLSs. The prominent characteristic of q-RPLS is that it allows the sum and square sum of positive membership degree, neutral membership degree, and negative membership degree to be greater than one, providing more freedom for decision-makers to express their evaluations, and further leading to less information loss in the process of MAGDM. Secondly, considering the fact that decision-makers prefer to make qualitative decisions due to lack of time and expertise, the proposed q-RPLSs not only express decision-makers' qualitative assessments, but also reflect decision-makers' quantitative ideas. Therefore, q-RPLSs are suitable and sufficient for modeling decision-makers' evaluations on alternatives. Thirdly, in most real decision-making problems, attributes are correlated, so that the interrelationship between attribute values should be taken into account when fusing them. Our method to MAGDM is based on q-RPLWHM or q-RPLWGHM operators, which consider the interrelationship between arguments. Therefore, our method can effectively actual MAGDM problems. In a word, the proposed method not only provides a new tool for decision-makers to express their assessments, but also effectively model the process of real MAGDM problems. Therefore, our method is more general, powerful and flexible than other methods.

6. Conclusions

The main contribution of this paper is that a novel MAGDM model is proposed. In the proposed model, q-RPLNs are utilized to represent decision-makers' assessments of alternatives and weights of attributes and decision-makers take the form of crisp numbers. In addition, q-RPLHM, q-RPLWHM, q-RPLGHM, and q-RPLWGHM operators are proposed to aggregate attribute values to obtain overall assessments of alternatives. In order to do this, we proposed the q-RPFS and q-RPLS, which are powerful and effective tools for coping with uncertainty and vagueness. Subsequently, the operations and comparison law for q-RPLNs were introduced. We also proposed some aggregation operators for fusing q-rung picture linguistic information. The prominent characteristic of these operators is that they can capture the interrelationship between q-RPLNs. Moreover, we have studied some desirable properties and special cases of the proposed operators. Thereafter, we utilized the proposed operators to establish a novel method to solve MAGDM problems. To illustrate the validity of the proposed method, we used the proposed method to solve an ERP system selection problem. In addition, we conducted comparative analysis to demonstrate the effectiveness and superiorities of the proposed method. Due to the high ability of q-RPLSs for describing fuzziness and expressing decision-makers' assessments over alternatives, and the powerfulness of q-rung picture linguistic Heronian mean operators, the proposed method can be applied to solving real decision-making problems, such as supplier selection, low carbon supplier selection, hospital-based post-acute care, risk management, medical diagnosis, and resource evaluation, etc. In future works, considering the advantages of q-RPLSs, we should investigate more aggregation operators for fusing q-rung picture linguistic information such as the q-rung picture linguistic Bonferroni mean, q-rung picture linguistic Maclaurin symmetric mean, q-rung picture linguistic Hamy mean, and q-rung picture linguistic Muirhead mean. Additionally, we should investigate more methods of MAGDM with q-rung picture linguistic information.

Author Contributions: The idea of the whole thesis was put forward by L.L. She also wrote the paper. R.Z. analyzed the existing work and J.W. provided the numerical instance. The computation of the paper was conducted by X.S. and K.B.

Acknowledgments: This work was partially supported by National Natural Science Foundation of China (Grant number 71532002, 61702023), and the Fundamental Fund for Humanities and Social Sciences of Beijing Jiaotong University (Grant number 2016JBZD01).

Conflicts of Interest: The authors declare no conflict of interest.

References

1. Mardani, A.; Nilashi, M.; Zavadskas, E.K.; Awang, S.R.; Zare, H.; Jamal, N.M. Decision making methods based on fuzzy aggregation operators: Three decades review from 1986 to 2017. *Int. J. Inf. Technol. Decis. Mak.* **2018**, *17*, 391–466. [CrossRef]
2. Liu, P.D.; Liu, J.L.; Merigó, J.M. Partitioned Heronian means based on linguistic intuitionistic fuzzy numbers for dealing with multi-attribute group decision making. *Appl. Soft Comput.* **2018**, *62*, 395–422. [CrossRef]

3. Yu, D.J. Hesitant fuzzy multi-criteria decision making methods based on Heronian mean. *Technol. Econ. Dev. Econ.* **2017**, *23*, 296–315. [CrossRef]
4. Liu, P.D.; Teng, F. Some Muirhead mean operators for probabilistic linguistic term sets and their applications to multiple attribute decision-making. *Appl. Soft Comput.* **2018**, *68*, 396–431. [CrossRef]
5. Liang, X.; Jiang, Y.P.; Liu, P.D. Stochastic multiple-criteria decision making with 2-tuple aspirations: A method based on disappointment stochastic dominance. *Int. Trans. Oper. Res.* **2018**, *25*, 913–940. [CrossRef]
6. Li, Y.; Liu, P.D. Some Heronian mean operators with 2-tuple linguistic information and their application to multiple attribute group decision making. *Technol. Econ. Dev. Econ.* **2015**, *21*, 797–814.
7. Liu, W.H.; Liu, H.B.; Li, L.L. A multiple attribute group decision making method based on 2-D uncertain linguistic weighted Heronian mean aggregation operator. *Int. J. Comput. Commun.* **2017**, *12*, 254–264. [CrossRef]
8. Geng, Y.S.; Wang, X.G.; Li, X.M.; Yu, K.; Liu, P.D. Some interval neutrosophic linguistic Maclaurin symmetric mean operators and their application in multiple attribute decision making. *Symmetry* **2018**, *10*, 127. [CrossRef]
9. Garg, H. Some methods for strategic decision-making problems with immediate probabilities in Pythagorean fuzzy environment. *Int. J. Intell. Syst.* **2018**, *33*, 687–712. [CrossRef]
10. Liu, P.D.; Chen, S.M. Multi-attribute group decision making based on intuitionistic 2-tuple linguistic information. *Inf. Sci.* **2018**, *430*, 599–619. [CrossRef]
11. Liu, P.D.; Gao, H. Multicriteria decision making based on generalized Maclaurin symmetric means with multi-hesitant fuzzy linguistic information. *Symmetry* **2018**, *10*, 81. [CrossRef]
12. Xing, Y.P.; Zhang, R.T.; Xia, M.M.; Wang, J. Generalized point aggregation operators for dual hesitant fuzzy information. *J. Intell. Fuzzy Syst.* **2017**, *33*, 515–527. [CrossRef]
13. Liu, P.D.; Mahmood, T.; Khan, Q. Group decision making based on power Heronian aggregation operators under linguistic neutrosophic environment. *Int. J. Fuzzy Syst.* **2018**, *20*, 970–985. [CrossRef]
14. Mahmood, T.; Liu, P.D.; Ye, J.; Khan, Q. Several hybrid aggregation operators for triangular intuitionistic fuzzy set and their application in multi-criteria decision making. *Granul. Comput.* **2018**, *3*, 153–168. [CrossRef]
15. Liu, P.D.; Teng, F. Multiple attribute decision making method based on normal neutrosophic generalized weighted power averaging operator. *Int. J. Mach. Learn. Cybern.* **2018**, *9*, 281–293. [CrossRef]
16. Atanassov, K.T. Intuitionistic fuzzy sets. *Fuzzy Sets Syst.* **1986**, *20*, 87–96. [CrossRef]
17. Liu, P.D. Multiple attribute decision-making methods based on normal intuitionistic fuzzy interaction aggregation operators. *Symmetry* **2017**, *9*, 261. [CrossRef]
18. Liu, P.D.; Mahmood, T.; Khan, Q. Multi-attribute decision making based on prioritized aggregation operator under hesitant intuitionistic fuzzy linguistic environment. *Symmetry* **2017**, *9*, 270. [CrossRef]
19. Zhao, J.; You, X.Y.; Liu, H.C.; Wu, S.M. An extended VIKOR method using intuitionistic fuzzy sets and combination weights for supplier selection. *Symmetry* **2017**, *9*, 169. [CrossRef]
20. Wang, S.W.; Liu, J. Extension of the TODIM Method to Intuitionistic Linguistic Multiple Attribute Decision Making. *Symmetry* **2017**, *9*, 95. [CrossRef]
21. Yager, R.R. Pythagorean membership grades in multicriteria decision making. *IEEE Trans. Fuzzy Syst.* **2014**, *22*, 958–965. [CrossRef]
22. Wei, G.W.; Lu, M.; Tang, X.Y.; Wei, Y. Pythagorean hesitant fuzzy Hamacher aggregation operators and their application to multiple attribute decision making. *Int. J. Intell. Syst.* **2018**, *33*, 1197–1233. [CrossRef]
23. Liang, D.; Xu, Z.S. The new extension of TOPSIS method for multiple criteria decision making with hesitant Pythagorean fuzzy sets. *Appl. Soft Comput.* **2017**, *60*, 167–179. [CrossRef]
24. Zhang, R.T.; Wang, J.; Zhu, X.M.; Xia, M.M.; Yu, M. Some generalized Pythagorean fuzzy Bonferroni mean aggregation operators with their application to multi-attribute group decision-making. *Complexity* **2017**, *2017*, 5937376. [CrossRef]
25. Ren, P.J.; Xu, Z.S.; Gou, X.J. Pythagorean fuzzy TODIM approach to multi-criteria decision making. *Appl. Soft Comput.* **2016**, *42*, 246–259. [CrossRef]
26. Yager, R.R. Generalized orthopair fuzzy sets. *IEEE Trans. Fuzzy Syst.* **2017**, *25*, 1222–1230. [CrossRef]
27. Liu, P.D.; Wang, P. Some q-rung orthopair fuzzy aggregation operators and their applications to multiple-attribute decision making. *Int. J. Intell. Syst.* **2018**, *33*, 259–280. [CrossRef]
28. Liu, P.D.; Liu, J.L. Some q-rung orthopair fuzzy Bonferroni mean operators and their application to multi-attribute group decision making. *Int. J. Intell. Syst.* **2018**, *33*, 315–347. [CrossRef]

29. Cuong, B.C. Picture fuzzy sets-first results. part 1. *Semin. Neuro-Fuzzy Syst. Appl.* **2013**, *4*, 2013.
30. Wei, G.W. Picture uncertain linguistic Bonferroni mean operators and their application to multiple attribute decision making. *Kybernetes* **2017**, *46*, 1777–1800. [CrossRef]
31. Son, L.H. Measuring analogousness in picture fuzzy sets: From picture distance measures to picture association measures. *Fuzzy Optim. Decis. Mak.* **2017**, *16*, 359–378. [CrossRef]
32. Wei, G.W. Picture fuzzy aggregation operators and their application to multiple attribute decision making. *J. Intell. Fuzzy Syst.* **2017**, *33*, 713–724. [CrossRef]
33. Wei, G.W. Picture fuzzy cross-entropy for multiple attribute decision making problems. *J. Bus. Econ. Manag.* **2016**, *17*, 491–502. [CrossRef]
34. Garg, H. Some picture fuzzy aggregation operators and their applications to multicriteria decision-making. *Arab. J. Sci. Eng.* **2017**, *42*, 5275–5290. [CrossRef]
35. Wei, G.W. Picture fuzzy Hamacher aggregation operators and their application to multiple attribute decision making. *Fund. Inform.* **2018**, *157*, 271–320. [CrossRef]
36. Li, D.X.; Dong, H.; Jin, X. Model for evaluating the enterprise marketing capability with picture fuzzy information. *J. Intell. Fuzzy Syst.* **2017**, *33*, 3255–3263. [CrossRef]
37. Thong, P.H. Picture fuzzy clustering: A new computational intelligence method. *Soft Comput.* **2016**, *20*, 3549–3562. [CrossRef]
38. Zadeh, L.A. The concept of a linguistic variable and its application to approximate reasoning-Part II. *Inf. Sci.* **1965**, *8*, 301–357. [CrossRef]
39. Wang, J.Q.; Li, J.J. The multi-criteria group decision making method based on multi-granularity intuitionistic two semantics. *Sci. Technol. Inf.* **2009**, *33*, 8–9.
40. Du, Y.Q.; Hou, F.J.; Zafar, W.; Yu, Q.; Zhai, Y.B. A novel method for multi-attribute decision making with interval-valued Pythagorean fuzzy linguistic information. *Int. J. Intell. Syst.* **2017**, *32*, 1085–1112. [CrossRef]
41. Liu, P.D.; Zhang, X.H. A novel picture fuzzy linguistic aggregation operator and its application to group decision-making. *Cognit. Comput.* **2018**, *10*, 242–259. [CrossRef]
42. Bonferroni, C. Sulle medie multiple di potenze. *Boll. Unione Mat. Ital.* **1950**, *5*, 267–270.
43. Sykora, S. *Mathematical Means and Averages: Generalized Heronian Means*; Stan's Library: Italy, 2009. Available online: http://www.ebyte.it/library/docs/math09/Means_Heronian.html (accessed on 16 May 2018).
44. Yu, D.J.; Wu, Y.Y. Interval-valued intuitionistic fuzzy Heronian mean operators and their application in multi-criteria decision making. *Afr. J. Bus. Manag.* **2012**, *6*, 4158. [CrossRef]
45. Liu, P.D.; Shi, L.L. Some neutrosophic uncertain linguistic number Heronian mean operators and their application to multi-attribute group decision making. *Neural Comput. Appl.* **2017**, *28*, 1079–1093. [CrossRef]
46. Yu, D.J. Intuitionistic fuzzy geometric Heronian mean aggregation operators. *Appl. Soft Comput.* **2013**, *13*, 1235–1246. [CrossRef]
47. Wang, X.F.; Wang, J.Q.; Yang, W.E. Multi-criteria group decision making method based on intuitionistic linguistic aggregation operators. *J. Intell. Fuzzy Syst.* **2014**, *26*, 115–125.
48. Liu, P.D.; Rong, L.L.; Chu, Y.C.; Li, Y.W. Intuitionistic linguistic weighted Bonferroni mean operator and its application to multiple attribute decision making. *Sci. World J.* **2014**, *2014*, 545049. [CrossRef] [PubMed]
49. Ju, Y.B.; Liu, X.Y.; Ju, D.W. Some new intuitionistic linguistic aggregation operators based on Maclaurin symmetric mean and their applications to multiple attribute group decision making. *Soft Comput.* **2016**, *22*, 4521–4548. [CrossRef]

MDPI
St. Alban-Anlage 66
4052 Basel
Switzerland
Tel. +41 61 683 77 34
Fax +41 61 302 89 18
www.mdpi.com

Symmetry Editorial Office
E-mail: symmetry@mdpi.com
www.mdpi.com/journal/symmetry

www.ingramcontent.com/pod-product-compliance
Lightning Source LLC
Chambersburg PA
CBHW061444220926
51850CB00004B/13
9783039364152